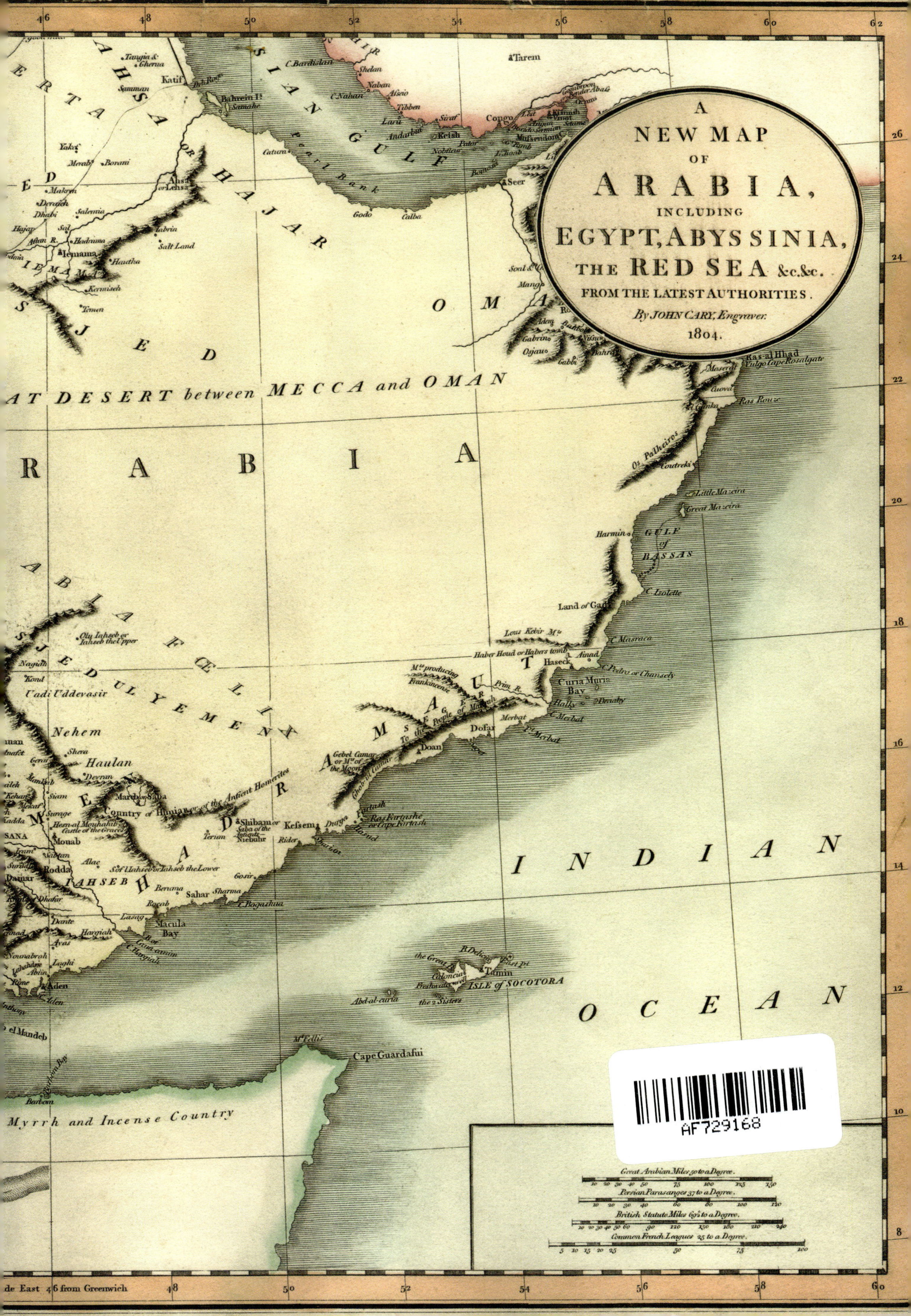
A NEW MAP
OF
ARABIA,
INCLUDING
EGYPT, ABYSSINIA,
THE RED SEA &c.&c.
FROM THE LATEST AUTHORITIES.
By JOHN CARY, Engraver.
1804.
PERSIAN GULF
HAJAR
OMAN
GREAT DESERT between MECCA and OMAN
ARABIA
ARABIA FELIX
NEDJED UL YEMEN
HADRAMAUT
INDIAN OCEAN
GULF of BASSAS
ISLE of SOCOTORA
Cape Guardafui
Myrrh and Incense Country
Ras al Hhad
Vulgo Cape Rosalgate
Ras Rouze
Os Patheiros
Little Mazeira
Great Mazeira
Land of Gad
C. Isolette
C. Masraca
Curia Muria Bay
C. Pedro or Chansely
Ras Fartashe or Cape Fartash
Macula Bay
Aden
SANA
Katif
Bahrein I.
Tarem
Seer
Great Arabian Miles 50 to a Degree.
Persian Parasanges 37 to a Degree.
British Statute Miles 69½ to a Degree.
Common French Leagues 25 to a Degree.
Longitude East 46 from Greenwich

ETHNOFLORA OF THE SOQOTRA ARCHIPELAGO

SOQOTRAN DRAGON'S BLOOD TREE (DRACAENA CINNABARI) DICCON ALEXANDER

ANTHONY G MILLER
MIRANDA MORRIS

DESIGNED BY DICCON ALEXANDER

EDITED BY RUTH ATKINSON

ETHNOFLORA OF THE SOQOTRA ARCHIPELAGO

ANTHONY G MILLER
MIRANDA MORRIS

DESIGNED BY DICCON ALEXANDER
EDITED BY RUTH ATKINSON

MARTIN PULLAN (BIOINFORMATICS)
ROGER HYAM (BIOINFORMATICS)
ALI MUQADDAM KOWTER (LOGISTICS)

ROYAL BOTANIC GARDEN EDINBURGH

الجمهورية اليمنية
مجلس حماية البيئة

GEF

DARWIN INITIATIVE

First published in the UK in 2004 by

The Royal Botanic Garden Edinburgh
20A Inverleith Row, Edinburgh EH3 5LR, UK
www.rbge.org.uk

A catalogue record for this book is available from the British Library.

ISBN 1 872291 59 7

Printed and bound by The Charlesworth Group, Huddersfield, UK
(tel: 01484 517077)

This book is dedicated to the people of the Soqotra Archipelago who have over many generations been the careful guardians of their islands' biodiversity.

That the value of their biodiversity has now been globally recognised is a testament to the success of their stewardship.

Soqotran dragon's blood tree (*Dracaena cinnabari*).
Tab. XCVI from Balfour's *Botany of Socotra* (1888), based on a sketch by Lieutenant C.J. Cockburn.

CONTENTS

Forewords viii

By Dr Abd al Karim al Eryani, Professor David Ingram and Professor Stephen Blackmore

Preface xi

Acknowledgements xii

List of new species and combinations xvi

INTRODUCTION 1

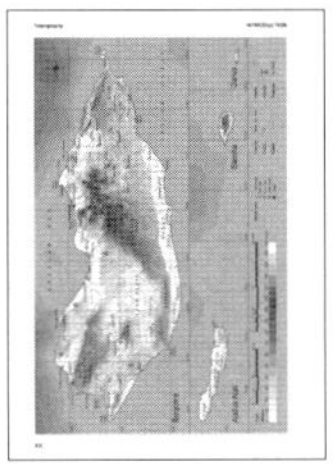

An introduction to all aspects of the Soqotran environment, including geography, floristics and vegetation, people's use of plants, and conservation issues

Introductory colour plates 45

ILLUSTRATED KEY 61

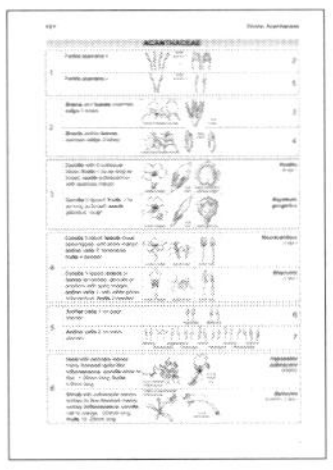

A user-friendly illustrated guide to identify all Soqotra's 825 higher plant species, cross-referenced to colour plant portraits and the plant encyclopaedia

PLANT PORTRAITS 233

Colour photographs of most endemic and some non-endemic plant species in the Archipelago, arranged alphabetically by family

ENCYCLOPAEDIA 283

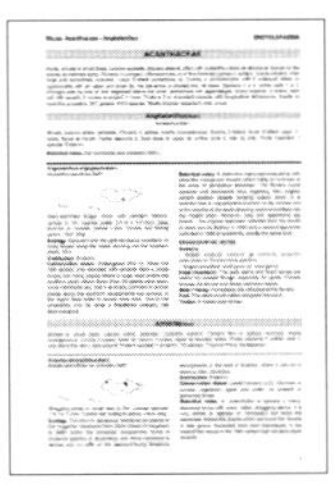

Full plant descriptions, including taxonomy, ecology, distribution and ethnobotany

Pteridophytes 285

Gymnosperm 296

Monocotyledons 297

Dicotyledons 400

APPENDICES 719

1 Notes on lichens 721

2 New species and combinations 723

References 727

Select bibliography 731

Index to plant names 733

Index to Soqotri names 749

FOREWORD

Soqotra has been famed from ancient times as a source of precious plant substances, such as bitter aloes juice and dragon's blood. More recently it has fascinated the scientific community as the home of such botanical oddities as the wild pomegranate, the cucumber tree and a great number of endemic species and genera. Now the Archipelago has attained international recognition as a major centre of biodiversity, and is recognised as the 'Galápagos of the Indian Ocean'.

However, Soqotra is not only home to a unique flora and fauna of global significance. It is also home to some 50 000 people who have co-existed with this rich natural environment and been its careful guardians. For millennia they have lived off the island's resources and have developed complex, traditional rules to manage the environment, which have played a vital role in maintaining the biodiversity we celebrate today. This makes Soqotra unique. In most parts of the world this kind of traditional expertise is either already lost or disappearing faster than the plants themselves. However, it is just such knowledge that provides the key to the survival of islands, and this book attempts to record, in a scientific framework, the vital and integral role that plants have played in the life of the islands and those who live there.

This book is a record of local and international expertise on the plants of the islands and as such can provide a baseline against which future change can be measured. However, it can also be used as a tool to help Soqotra face the challenges of the future, and to play a part in helping Soqotrans avoid the fate of other oceanic islands which have been decimated by unsustainable development and exploitation.

It is also an example of what can result from collaboration between countries and between many different disciplines and layers of expertise. The links between the Royal Botanic Garden Edinburgh (RBGE) and the Republic of Yemen go back to 1880, when the Edinburgh botanist Isaac Bayley Balfour first visited the islands and later recorded their fascinating plant life. More recently a team from the RBGE has been closely collaborating with Yemeni and international scientists in major biodiversity conservation programmes on the islands. Without this and the generous co-operation of the islanders it is true to say that the book could not have been written.

The many illustrations and colour photographs highlight the beauty and strangeness of the plants of Soqotra. But, vitally, the pictures also help to make the flora accessible to non-specialists who may be directly interested in decisions about managing the island or who may simply be interested in such a remarkable place.

This book demonstrates that 'conservation' does not only have to be about the survival or protection of rare species of plant or animal. By combining the expertise of the local people who depend on a healthy ecosystem with that of the scientific community who can provide a systematic framework for understanding that ecosystem, conservation becomes instead a practical, multidisciplinary process, the key to human as well as to environmental survival. We hope that Yemen will lead the world in this 'model approach' to conservation.

DR ABD AL KARIM AL ERYANI
Political Advisor to HE the President of the Republic of Yemen
Director of the Soqotra Conservation Fund
February 2004

FOREWORD

The Soqotra Archipelago has a wonderfully rich flora. That it has remained so whilst many other island floras have been damaged beyond repair demonstrates the vital role of the 50 000 islanders as custodians of their environment.

I first encountered the project to write an ethnoflora of the Archipelago almost ten years ago, when I was Regius Keeper of the Royal Botanic Garden Edinburgh. The Garden had a long history of research in the region, largely through the studies of Anthony Miller, and this farsighted botanist had already established a collaboration with the ethnologist Miranda Morris. They understood that by working together in the field and sharing their knowledge and expertise, it was possible to produce a flora that would break entirely new ground. The result of their collaboration is not only a sophisticated catalogue of the plants growing in the Soqotran islands, but also achieves a complete integration of our knowledge of the plants themselves with the accumulated wisdom of the people of the islands about how those plants may be utilised in a sustainable way. The term 'ethnoflora' is a reflection not so much of the content of the book, however, but rather an approach to flora writing in which it is recognised that people have lived together with the natural world for centuries and can continue to do so for ever, if we care enough.

The Royal Botanic Garden Edinburgh had its origins in the late 17th century as a physic garden for the cure of local people and the teaching of local student doctors. It is most fitting that in the 21st century it should, through publication of this *Ethnoflora*, continue the tradition of working for the benefit of both plants and people.

In addition to Anthony Miller and Miranda Morris, two other experts have played key roles in the production of this book: the designer Diccon Alexander and the editor Ruth Atkinson. Through the use of sophisticated design, IT and careful editing, the authors, designer and editor have striven to create a model of how complex botanical and ethnological information may be presented in a friendly and accessible way, truly reaching out to as wide an audience as possible. The combination of careful layout, simple language and high quality illustrations has enabled the team to more than fulfil their ambition. I found the *Ethnoflora* easy to use and a pleasure to read. Indeed, I was so fascinated by the descriptions of the rare and exotic plants that produce equally rare and exotic medicines, foods, spices and perfumes that I could hardly put the book down.

Finally, the *Ethnoflora of the Soqotra Archipelago* should be seen as a glorious celebration of the plants and people of the islands at the dawn of a new century. The wealth of 'information' that it contains provides not only a baseline of knowledge that will make it possible to monitor any changes that may occur in the future, but also serves as a timely reminder of the continuing need for vigilance in conserving the islands' rich but fragile natural and cultural heritage.

PROFESSOR DAVID INGRAM, OBE, ScD, VMH, FRSE
Master, St Catharine's College, Cambridge
Chairman, The Darwin Initiative for the Survival of Species
Former Regius Keeper of the Royal Botanic Garden Edinburgh
March 2004

FOREWORD

The twenty-first century is proving to be a time of global change during which both biodiversity itself and our vital cultural knowledge of that diversity are threatened by the pace of development and change. Since the Earth Summit in Rio de Janeiro in 1992, biodiversity has steadily emerged as a global priority for international programmes to conserve the diversity of life on our planet and to learn how to use and benefit from it in truly sustainable ways. However, whilst the Convention on Biological Diversity refers explicitly to traditional knowledge and practices, and the study of ethnobotany has grown in importance as a discipline, cultural knowledge continues to prove difficult to systematise and document. One of the most basic challenges is simply to connect the many traditional names that refer to an individual species or group of species with a precise scientific name. Given the number of languages that are falling out of use around the world it is scarcely surprising that we are losing the very tools by which civilisations add to and communicate their knowledge and wisdom. Such knowledge, built up over millennia and passed on from generation to generation, is of enormous intrinsic value not only as part of the rich cultural diversity of our own species but as a guide to the continuing and future uses of plants.

Happily, in the Soqotra Archipelago, one of the world's most important island groups with its high level of unique, endemic biodiversity, the plants and the rich cultural traditions surrounding them have survived and are being protected for the future. During the course of research for the *Ethnoflora of the Soqotra Archipelago* it became apparent that few, if any, of the plants encountered by the first botanists to visit Soqotra have become extinct, in marked contrast to the situation in most other biologically important islands. Furthermore, the researchers have continued to discover and describe many new plant species.

This remarkable new publication provides a bridge between the ethnobotanical knowledge of the people of Soqotra and the application of plant taxonomy to place the plants within the framework of a scientific classification, reflecting their patterns of evolutionary history. This is done in a deceptively simple but innovative way that provides for communication between cultures and between scientists and non-specialists. It is often said that a picture can convey a thousand words. When the words concerned are complex botanical terms the benefits of accessibility to all users of the *Ethnoflora* are abundantly apparent. The authors are to be congratulated for having produced a work of great beauty, matched to the beauty and diversity of the Soqotran flora, but also one that sets a new standard in presenting complex scientific information about biodiversity in a highly comprehensible manner.

PROFESSOR STEPHEN BLACKMORE FRSE
Regius Keeper, Royal Botanic Garden Edinburgh
March 2004

PREFACE

Anthony G. Miller and Miranda Morris have worked for many years on Soqotra prior to the inception of this project, and their botanical and ethnographic research formed the foundation for this book. Based on their previous collaboration on the plants of Dhofar (Oman), they conceived the project of doing the same for the plants of the Soqotra Archipelago, but as a more rigorous Flora. However, the book would not have been produced in its current form without the contributions of Diccon Alexander and Ruth Atkinson.

Anthony G. Miller conducted the fieldwork and much of the original botanical work on which the book is based. He also contributed many taxonomic accounts, and all notes on ecology, conservation status and local plant distributions.

Miranda Morris was responsible for all the ethnographic work in the book, including all original data collection on uses of plants and the plants' Soqotri names. She also collated all the information on comparative uses for species or their relatives outside of Soqotra.

Diccon Alexander provided the original design concept for the book and established the layout of the key. He drew the majority of illustrations for the key and set the high standard for these illustrations. He also prepared the colour plates and maps.

Ruth Atkinson acted as overall editor of the book by pulling all the material together and organising it into its current structure. She also contributed taxonomic accounts for a large number of families and some illustrations for the key.

ACKNOWLEDGEMENTS

The book was a complex undertaking requiring help and input from a wide range of experts, who without exception made available to us their time and expertise. We have attempted to acknowledge them all here – but we offer our apologies to those we may have overlooked.

There were many logistical and financial difficulties to overcome when working on Soqotra. Many people made our trips possible and gave us aid and friendship in the field. We would like especially to thank Ali Muqaddam Ahmad Kowter for all the complicated logistical arrangements he made to enable us to carry out our fieldwork: on foot, by camel, by car and by dhow. We would also like to thank the staff at the Global Environment Facility (GEF) project office on Soqotra, and in particular to recognise the help of Ahmad Said Sulaiman, Catherine Cheung, Salih Hakim, Noah Adham Ali, Ahmad Issa Ali, Abdul Raqib Shamsan Al-Selowe and Nadim Mohammed Abdullah Talib and Salim Ali Deheq. For help with managing funds on Soqotra, we would like to thank the manager of the Soqotra branch of the National Bank of Yemen, Muhammad Salim Hammood. We would also like to thank Edoardo Zandri and staff of the Soqotra Biodiversity Project Office in Sana'a for help with many things, including obtaining visas, visa extensions and Soqotra Permits and not least for arranging transport on and off Soqotra. Much-needed institutional support was also provided by staff of the Environmental Protection Agency and more recently the Environmental Protection Authority in Yemen and the GEF, UNDP and UNOPS.

Information technology was essential for organising the original data and considerable computing input was required. This was initially provided by Roger Hyam and then, more recently, by Martin Pullan at the Royal Botanic Garden Edinburgh (RBGE), whose technical skills and help were invaluable.

The final stages were greatly helped by Erica Schwarz, whose organisational and copyediting skills were greatly needed, and appreciated.

BOTANY – Anthony G. Miller

This book is largely based on information gathered in the field. This would not have been possible without the many islanders who shared their knowledge and expertise with us, and at all times showed us endless patience and the greatest hospitality. I welcome this opportunity to acknowledge their unique contribution; indeed it is true to say that without their input the book could not have been realised in its current form.

Several colleagues at the RBGE undertook fieldwork in the Archipelago, including Diccon Alexander, Ruth Atkinson, Mark Hughes, Roger Hyam, Jane Nyberg and Vanessa Plana. For their help in organising and participating in these trips I am grateful.

I would also like to thank the scientists with whom we have worked on Soqotra over the years, including: Abdul Habib Mahyoub Ali Al-Kadsi, Abdul Karim Nasher, Abdul Raqib Shamsahn Al-Selowe, Abdul-Wali Ahmad Al Khulaidi, Adel Al-Areefi, Ahmad Issa Ali, Ahmad Said Sulaiman, Ahmed Mohamed A-Sallam, Sue Christie, Roderic Dutton, Mark Hughes, Ismael Mohammed Al-Mosanif, Jamil Salim Mubarak, Jawad Abdul Rahman Qaid, Mohsin Bazara'a Mussa Al-Gumaili, Nabil Abdul-Aziz Othman Al-Garadi, Nabil Obadi, Nadim Mohammed Abdullah Talib, Nageeb Ali Saeed Al-Saghier, Noah Adham Ali, Omar Al-Saghier, Richard Porter, Sadiq Al Mahdi, Kay Van Damme, Wadie Abdul-Ghanie Saeed, Wolfgang Wranik, and last but not least Luigi Guarino of IPGRI with whom I made my first trip to Soqotra and discussed early stages of this book.

The taxonomic accounts were mostly prepared by Ruth Atkinson and myself but major contributions were made by the following people: Sultan Abedin (Sterculiaceae and Tiliaceae); Hannah Atkins (Gramineae and Scrophulariaceae); Peter Bruyns (Asclepiadaceae); Tom Cope (Gramineae), Paul Goetghebeur (Cyperaceae); Roger Hyam (Gramineae); Norbert Kilian (Compositae), Robert Mill (Boraginaceae, Convolvulaceae and providing the Latin diagnoses), Vanessa Plana (Rubiaceae), Tom Schils (marine angiosperms), Liesbeth Vandershaeve (Cyperaceae); and Mark Watson (Umbelliferae).

During the production of this book a number of MSc, PhD and summer students have undertaken research projects at the RBGE on the Soqotran flora. Diccon Alexander first became involved in the project through his MSc thesis on a user-friendly key for Acanthaceae, which later developed into the full illustrated key. Other contributions were also made by: Emma Barclay (Umbelliferae); James Bond (ethnobotany); Emma Cocklin (*Polycarpaea*); Helen Cortés-Burns (*Indigofera*, *Leucas* and particularly her notes on biogeography which formed the basis for this section in the Introduction); Mark Hughes (*Begonia*); Katherine Liston (vegetation classification); Reynaldo Palomino Linares (*Kickxia* and *Nanorrhinum*); Vanessa Plana (vegetation classification); Lillis Urban (*Echiochilon*); and Katherine Vint (*Boswellia*).

We are grateful to the herbaria which have sent material on loan or allowed us to work on their collections. In particular we are indebted to the directors of the following institutions: Royal Botanic Gardens, Kew; the Natural History Museum London; Naturhistorisches Museum, Wien; Botanisches Garten, Berlin; and Uppsala Universitet.

Lastly I would like to thank my wife for her support and forbearance over my long absences on Soqotra and my daughters Anna and Harriet for keeping things in proportion.

ETHNOGRAPHY – Miranda Morris

Innumerable Soqotrans helped write the ethnobotanical content of the book. I am extremely grateful to them for their sustained efforts over the years and their patience as teachers. Those who endured repeated and more detailed working sessions are listed below. Women made an important and valuable contribution; however, it was felt inappropriate by many Soqotrans to name them individually.

EASTERN SOQOTRA
Abdullah Salim Musa
Ahmad Eesa Qasim
Ali Eesa Ahmad
Ali Salim Ali Sowqer
Ahmad Sa'ad Ahmad
Ahmad Sa'ad Khamis
Ahmad Sa'id Ali
Ahmad Salim Ali Shermat
Amir Hamad Ali Sowqer
Amir Ahmad Eesa
Hamad Salim Bu Fehed
Hamoodi Hamidehan
Muhammad Salim Thabit
Muhammad 'Umar Ali Leybihi
Qowqehiyn Hasan
Qasim Sa'ad Sa'id
Sa'ad Muqaddam Ahmad
Sa'ad Sa'id Umar Shetehi
Sa'ad Sa'id Silihem
Sa'ad Hilal Sa'ad
Salih Tani
Silyif Sa'ad Khamis
Salim Hammud
Salim Ali Zeykoni Sowqer
Umr Sa'ad Khamis Tahki

WESTERN SOQOTRA
Abdullah Sa'id Abdullah Al Harsi
Abdullah Sa'ad Balhewreher
Ahmad Nuh Howrihin
Ahmad Abdullah Salim Gizelhi
De'ef Ahmad Ahmad Abdullah 'Agmihi
Fidenhan Ahmad Abdullah Belugehur
Hamad Salim Nowna Higihi
Muhammad Eesa Nuh Belugehur
Nuh Abdullah Dow'han Girided
Niyh Hazmi Belugehur
Dr Sa'ad Ahmad Al Qaddomi
Sheneyhan Sa'ad Sheneyhan Fa'arhi
Salim Dow'han Sikki
the late Sulaiman Hamad Hazmihi
Sa'ad Sa'id Sa'ad Iderhen
Sa'ad Salim Bit Dir'iy
Sa'ad Etan Sa'id Zeydihi
Sa'ad Salim Ahmad Rismi
Salim Sa'ad Salim Rismi
Sa'ad Sulaiman Ahmad Al Biseri
Sulaiman Salim Nuh
Suwdad Musallim Bit Dir'ay
Tanuf Salim Nuh Il Kishin
Sa'id Nuh
Eesa Salim Di Ba'a

CENTRAL SOQOTRA
Abdullah Ahmad Fa'amhi
Abdullah Muhammad Sa'id
Abdullah Fidan Ali
Abdullah Faisal Abdullah
Abdullah Sa'ad Bal Hewreher
Abdullah Itman Salim Ba Sowter
Abdullah Sa'ad Abdullah
Abdullah Ali Sa'ad
Ablas Abullah Sittit
Adeeb Abdullah
Ahmad Hammud Musa Bede'hen
Ahmad Sa'ad Khamis
Ahmad Sa'id Qiyhen Bin Molik
Ahmad Sa'ad Salim Siyyid
Ahmad Ali Sa'id
Ahmad Qiyhen
Ahmad Ali Sa'id Qirbeb
Ahmad Abdullah Salim
Ahmad Salim Ali
Ahmad Sa'ad Salim Siyyid
Ali Ahmad Amir
Ali Ahmad Seliman
Ali Salim Muhammad
Ali Abdullah Rigdihi

ACKNOWLEDGEMENTS

Ali Abdullah Ahmad Ba Sowter
Ali Eesa Ali Azerhi
Ali Mubarak Sa'ad Ali
Ali Sa'id Salim Sebeb
Amir Ahmad Abdullah Bin Habshi
Amir Ahmad bin Habshi
Amir Sa'ad Hammudi
Eesa Amir Ahmad Da'arhi
Eesa Lam Ali Hazmehi
Eesa Salim Ahmad Di Meroh
Eesa Ali Abdullah
Eesa Amir Di Asmi
Eesa Ablas Sittit
Fidehan Ahmad Abdullah
Hasan Abdullah Sa'id Rigdihi
Itman Muhammad Itman
Itman Jum'an Itman
Itman Qiyhen Salim Belbelhan
Khamis Ahmad Ali
Khatim Abdullah Ahmad
Mubarak Eesa Walad
Muhammad Amir Da'arhi
Muhammad Zetoon
Muhammad Mesded Salim Sha'abihi
Muhammad Abdullah Agmihi
Muhammad Ahmad Salim Sha'abihi
Muhammad Abdullah Sa'id Rigdihi
Muhammad Zeytun Tamer Bedebo
Muhammad Abdullah Bin Towfel
Muhammad Sa'ad Amir Remhi
Muhammad Salim Ahmad Di Meroh
Musa Salim Muhammad di Merk'ho
Niyh Ma'alhe Kedehe
Sa'ad Ahmad Salimin Bedebo
Sa'ad Ahmad
Sa'ad Abdullah Za'benote
Sa'ad Abdullah Qiyhen Za'benote
Sa'id Ali Ahmad
Sa'id Hasan Abdullah
Salim Ali Ahmad
Salim 'Aziz Sulaiman
Salim Amir Hazmehi
Salim Hammud Di Merk'ho
Salim Muhammad 'Itman Re'esh
Salim Ali Sa'id Za'benote
Salim Magid Sa'id bin Magid
Saleem Majid Sa'id Bin Majid
Sa'yuf Siydun Abdullah Sheqma'
Sulaiman Muhammad Salim Di Meroh

THE OUTER ISLANDS

ABD AL KURI
Ghalib Ahmad Sa'id
Muhammad Ahmad Sa'id
Salim Salih
Awadh Salim Ba Musa
Salih Ali Salim
Hamad Salim Eesa

SAMHA
Sa'id Mansur Ahmad Mahlil
Abdullah Salim Mahmus
Abood Salih
Sa'ad Hilal
Mubarak Hilal Sa'id Samhi

In addition, for help and advice with all health matters, the medical glossary, terminology and descriptions in general, I would like to thank my husband, Dr Hugh Morris. For advice on veterinary matters, I am indebted to Ian McLeish, and for guidance on zoology, Wolfgang Wranik. For first introducing me to Soqotra, I owe special thanks to Ahmad Sa'ad Khamis Tahki and his family.

ILLUSTRATED KEY AND DESIGN – Diccon Alexander

Many people have contributed to the *Ethnoflora* design since its inception as a MSc project at the RBGE in 1994. Initially I am indebted to Miller and Morris whose enthusiasm for the Soqotran flora and mutual interest in user-friendly design first inspired the illustrated key concept. Kerry Walter of the RBGE helped solve the technical challenges of early designs, Mary Mendum supervised my first forays into botanical illustration, and many other colleagues offered refinements to the concept including Sam Bridgewater, Quentin Cronk, Mark Newman and Toby Pennington.

In the field I am grateful to the many botanists and non-botanists alike who tested the key over the years and gave me invaluable comments and corrections. These include Abdul Karim Nasher, Abdul Raqib Shamsahn Al-Selowe, Abdul-Wali Ahmad Al Khulaidi, Ahmad Issa Ali, Ahmad Said Sulaiman, Mark Hughes, Nadim Mohammed Abdullah Talib, Nageeb Ali Saeed Al-Saghier, Vanessa Plana and Kay Van Damme.

The key features over 2000 original illustrations and it is a task I happily shared. I am particularly indebted to the following artists who, often working with inadequate material and to unrealistic timescales, produced such fine results: Ruth Atkinson, Liz Caddick, Linda Gurr, Jane Nyberg, Christina Oliver, Louise Olley and Susanna Stuart-Smith.

Photographs were mainly taken by Miller, Morris and myself, but I am grateful to the following for making their slides available to us: Sue Christie, Dylan Hannon, Neil Oakham and Mats Thulin. In addition Debbie White took many studio shots of plants cultivated at the RBGE.

I must thank Roderic Dutton of the University of Durham for his generosity and dedication in providing the raw data for the new maps and GIS views, and Omar Ali-Shah for allowing us to reproduce his copy of an 1804 map of the Archipelago which appears on the inner cover.

Finally I would like to thank my partner China Williams and daughter Freya for their patience and support over many years of fieldwork and late-night illustration.

EDITORIAL – Ruth Atkinson

Editing this book was a complex undertaking and several people were essential in helping to ensure that the volume stands as a coherent whole. Martin Pullan provided the computing skills to enable the ethnographic and taxonomic information to be brought together from many sources and combined in a single document. This allowed the encyclopaedia section to be generated automatically and greatly facilitated the editorial process. Erica Schwarz acted as copyeditor in the final stages, and her attention to detail was much appreciated. William Milliken and James Ratter both provided useful advice on writing style and editing. Editing the illustrated key required help from many people, especially Kerry Walter and Rob Cubey. Finally I would like to thank Mary Gibby, head of science at the RBGE, who has provided essential support in the latter stages of the project.

FUNDING AND INSTITUTIONAL SUPPORT

The project would not have been possible without the support of the RBGE. In particular we would like to acknowledge the support and guidance of Steve Blackmore, Mary Gibby, Richard Bateman, David Ingram, David Mann, Henry Noltie and Mark Watson. Thanks are also due to Jane Hutcheon and the staff in the Library, Steve Scott and Ursula Duncan who looked after the living collections and Jane Neville, Rose Colangeli and the late Malcolm Newport who made sense of the particularly complex finances of this project.

The research for this book was largely carried out under the auspices of two programmes: the GEF/UNDP/UNOPS programme 'Conservation and Sustainable Use of the Biodiversity of the Soqotra Archipelago' and the UK Government's Darwin Initiative project 'Biodiversity Inventory of the Soqotra Archipelago', which was a collaborative programme between the RBGE and Birdlife International. In particular we would like to thank Valerie Richardson and Sylvia Smith of the Darwin Initiative; Roger Aertgeerts, Nahid Choudhury and Andrew Menz of UNOPS; Edoardo Zandri of GEF; and Richard Porter of Birdlife International.

Finally, the *Ethnoflora* could not have been undertaken without the generous financial support from the UK Darwin Initiative for the Survival of Species, the GEF and the Gatsby Charitable Foundation.

NEW SPECIES DESCRIBED IN THIS PUBLICATION

Acanthaceae
Justicia alexandri R.Atkinson **sp. nov.**
Justicia takhinensis R.Atkinson **sp. nov.**

Aloaceae
Aloe jawiyon Christie, Hannon & Oakham **sp. nov.**

Boraginaceae
Echiochilon pulvinata A.G.Mill. & L.Urb. **sp. nov.**

Capparaceae
Cadaba insularis A.G.Mill. **sp. nov.**

Compositae
Pulicaria dioscorides R.Atkinson **sp. nov.**

Cruciferae
Farsetia inconspicua A.G.Mill. **sp. nov.**

Euphorbiaceae
Euphorbia hamaderoensis A.G.Mill. **sp. nov.**

Geraniaceae
Pelargonium insularis Gibby & A.G.Mill. **sp. nov.**

Hyacinthaceae
Dipcadi kuriensis A.G.Mill. **sp. nov.**
Ledebouria insularis A.G.Mill. **sp. nov.**

Labiatae
Leucas hagghierensis Gifri & Cortés-Burns **sp. nov.**
Leucas penduliflora Gifri & Cortés-Burns **sp. nov.**
Leucas samhaensis Cortés-Burns & A.G.Mi l. **sp. nov.**

Malvaceae
Hibiscus dioscorides A.G.Mill. **sp. nov.**
Hibiscus diriffan A.G.Mill. **sp. nov.**
Hibiscus noli-tangere A.G.Mill. **sp. nov.**
Hibiscus quattenensis A.G.Mill. & Thulin **sp. nov.**

Polygalaceae
Polygala kuriense A.G.Mill. **sp. nov.**

Portulacaeae
Portulaca samhaensis A.G.Mill. **sp. nov.**
Portulaca sedifolia A.G.Mill. **sp. nov.**

Resedaceae
Ochradenus socotranus A.G.Mill. **sp. nov.**

Verbenaceae
Coelocarpum haggierensis A.G.Mill. **sp. nov.**

NEW COMBINATIONS MADE

Anthericaceae
Chlorophytum graptophyllum (Baker) A.G.Mill.

Hyacinthaceae
Drimia porphyrostachys (Baker) A.G.Mill.

Umbelliferae
Rughidia cordata (Balf.f.) M.F.Watson & E.L.Barclay

INTRODUCTION

SOQOTRAN DESERT ROSE (ADENIUM OBESUM SUBSP. SOKOTRANUM) DICCON ALEXANDER

BACKGROUND

This project was inspired by a remarkable situation. When AM and MJM first visited Soqotra in 1989 they saw an unparalleled opportunity to investigate a group of islands rich in unexplored botanical diversity with a unique people and culture and which had not yet entered a period of modern development. Until very recent times the islands were, and to a large extent still are, managed according to age-old traditional practices based on an intimate knowledge of the local environment. The flora of the islands is the result of a complex interaction between the natural environment and the human population. This book was conceived as a record of this balance, reflecting the situation on the islands in the late 20th and early 21st centuries within a historical and geographical context. Just as importantly, perhaps more so, it is also designed for use as a practical manual, based on a comprehensive and accurate gathering of information. As a manual it can be used both as a guide and inspiration to the beautiful and often bizarre flora and, more practically, we hope it will be used to help planners and conservationists working on Soqotra make well-considered decisions.

Interest in Soqotra for the two authors grew out of their collaboration in writing *Plants of Dhofar* (Miller & Morris 1988). This was an ethnobotanical study of an area of mainland Arabia that was, in the late 1970s, in a similar situation to Soqotra now, i.e. an isolated and self-sufficient traditional pastoralist culture just entering a period of rapid modern development. As this development has progressed, much of the knowledge recorded in the *Plants of Dhofar* has now been lost from the oral culture of the region. The approach developed during fieldwork for *Plants of Dhofar* is particularly appropriate for a place like Soqotra where so much knowledge about plants, not only their uses but also their distribution and ecology, is still available.

Plants of Dhofar was a detailed treatment of a limited number of species and, although the book is lavishly illustrated, no keys were included. A more rigorous taxonomic approach was taken on Soqotra and the current work is a flora in the traditional sense of the word, i.e. it attempts to record all species found on the Archipelago. However, it differs from a traditional flora in several ways:

- It attempts to provide comprehensive ethnographic information for all plant species used by the islanders and their livestock.
- There is a fully illustrated key to families and species. Thus all species on the islands are illustrated with at least one, and usually several, line drawing(s) (plus colour plates of most endemic species).
- Conventional taxonomic descriptions are brief, technical terms are kept to a minimum and only the most important synonymy is given.
- All endemic species are given a rating, based on the IUCN categories of threat.

The term 'Ethnoflora' (first coined by Hew Prendergast) as it is used here indicates the approach taken whereby botanists (AM, RA, DA, Roger Hyam) worked closely in the field with an ethnographer/linguist (MJM). The addition of the ethnographic research and the illustrated key made this a far bigger project than a conventional flora. However, this additional work was considered necessary to make the *Ethnoflora* as 'user-friendly' and useful for as wide an audience as possible.

A traditional taxonomic approach provides a rigorous framework for handling information and this was particularly advantageous in this project, where the ethnographic data were based on accurate identifications from voucher specimens. It also enabled all the information gathered to be incorporated into a database specially developed for handling taxonomic information (PANDORA). This means that considerable flexibility could be introduced into the final stages of production of the *Ethnoflora*. Additional information not included in this printed version (e.g. cited specimens, scanned pictures of herbarium specimens, etc.) can still be included in non-printed versions on the Worldwide Web. Smaller-scale printed works can also be produced, aimed specifically at fieldworkers, e.g. manuals of rangeland plants, woody plants, etc.

SOQOTRI CONTRIBUTORS TO THE *ETHNOFLORA*

The true authors of the 'ethno' part of the *Ethnoflora* are of course the Soqotrans. What they describe is what they remember from their own lives, from the teachings of their elders, or from oral tradition. Many of the traditional uses described have undergone modification or have already been lost. One result of recent development on the islands is that certain traditions or procedures are now seen as unsuitable or 'backward' and at odds with the more conservative views of modern Islam. Other uses and customs are seen as representing a time of desperation and poverty which many would prefer to forget. In this way, the recent rapid changes affecting the islands threaten to obliterate expertise and knowledge that have passed down the generations over hundreds of years. Unless recorded, all this detailed information and the islanders' comprehensive understanding of their environment will be irrevocably lost.

The *Ethnoflora* is thus an attempt to begin this process of documenting the islanders' expertise. By now many of the uses discussed here may well have become redundant, especially in the richer areas of the Archipelago. Nevertheless, as Soqotrans themselves are well aware, it would be rash to assume that such skills and knowledge will never be needed again. The enthusiasm shown by so many of the islanders for helping to record their knowledge, and their perception of the fact that such information was at risk of being lost unless they helped to do so, was a great encouragement.

The Soqotran islands and their people cannot be viewed as a single, uniform entity. They must be seen instead as a mosaic of different traditions and customs, each with their own conventions and priorities. The rich diversity of flora and fauna is mirrored by an equally rich diversity of human behaviour and activity. Nor is this region any freer of bias and prejudice than other areas. Thus what is found to be true for one area of the Archipelago will not necessarily be so for another; a practice documented in one locality will be regarded with disbelief, even horror or scorn, elsewhere. Views on what is regarded as suitable food or good livestock fodder, or what is seen as useful medicinally, can be expected to differ from area to area.

METHODS USED FOR COLLECTING ETHNO-BOTANICAL INFORMATION

Ethnobotanical information was collected during:

- Fieldwork, working alongside botanists and island experts (with fresh plant material).
- Fieldwork, working with herders and their families (with fresh plant material).
- Personal and participant observation during periods spent living with herding families.
- Group interviews, both semi-structured and open-ended.
- Structured interviews with one participant at a time and with a checklist of questions.
- Detailed one-to-one work with individual island experts and specialists.

When working in the field at times when no botanist was present, voucher specimens were collected for future identification.

The research was carried out in Soqotri and Arabic. It was explained to all participants that their information was contributing to a book on the ethnobotany of the islands, and their permission was sought before beginning work. For detailed work, older participants were preferred, but younger participants of both sexes were also consulted.

It was not uncommon for ethnobotanical work with participants to lead to the discovery of new species. Their knowledge also helped to clarify where particular plants were likely to be found and when in flower or fruit. In the field the sharp eyes of expert herders often led to the finding of plants which might otherwise have been overlooked.

Ethnobotanical information was always checked with other people from the same area as the original informant, and with people outside that area. As a general rule, when information was independently confirmed by a minimum of three informants, it was included in the *Ethnoflora*.

[For Soqotri terms see 'Notes on the ethnology text', p. 44]

PLAN OF THE *ETHNOFLORA*

The *Ethnoflora* is divided into two main sections: an illustrated key to identify plants and an encyclopaedia of plant information. Identification is a practical requirement and can be facilitated by using the key alone. The encyclopaedia then provides confirmation of that identification. The encyclopaedia also stands alone as a record of the status of all plant taxa, by providing a written description, all basic synonymy, and information on distribution, ecology, conservation status and uses.

ILLUSTRATED KEY

From the inception of the project the authors wanted to reach as broad an audience as possible, including many users whose first language was not English or who were unfamiliar with the many technical terms found in conventional Floras. The concept of the illustrated key was born and early drafts of the design were soon being tested in the field by taxonomists and non-taxonomists alike. Guided by feedback from these potential users the design was gradually revised and refined. Today it stands as a dichotomous key where every choice on the way to identifying a species is accompanied by an illustration, backed up by the user-friendly encyclopaedia and colour plates. For many users this makes the process of identification much more attractive and intuitive, and avoids the need for lengthy glossaries of botanical terms.

For ease of use the main key is divided into seven major groups: ferns, freshwater monocots, marine monocots, terrestrial monocots, herbaceous dicots, woody dicots with opposite leaves, and woody dicots with alternate leaves. In the first three groups the keys are straight to individual species, while in the monocot and dicot groups the keys are first to families, then genera, and finally species. In the latter groups the families and genera are all arranged alphabetically. Monotypic families (i.e. families represented by only one species in the Archipelago) occur twice, once in each group key to families and once in a single descriptive lead in each family key to genera. Monotypic genera (i.e. genera represented by only one species) occur only once in each family key to genera.

Once a particular species is reached in the key, the user is given the scientific name and whether the species is endemic to the Archipelago. Then a choice of references to confirm the identification are offered – first one or more page number references to the species in the encyclopaedia, followed for most endemics by a page number reference (in bold) to a colour plate.

ENCYCLOPAEDIA

All families, genera and species recorded from the Soqotra Archipelago are described. Family circumscription generally follows Engler & Prantl (1964) apart from monocotyledonous families, which are arranged mostly according to Dahlgren *et al.* (1985). All ferns are treated together, followed by the single gymnosperm and a split between monocots and dicots. Within these very broad groups the families are arranged alphabetically, with genera and species in corresponding alphabetical order within them. This arrangement was used to maximise ease of use, especially for non-taxonomists. An index is given for all taxa but refers only to entries in the encyclopaedic part.

Descriptions are diagnostic and based on Soqotran material (unless otherwise stated) so they are not appropriate for plants outside Soqotra. The descriptions are backed by illustrations in the key and comprehensive colour plates, particularly of endemic species. Specimens and types have not been cited but have been recorded in the database. Global distributions of all taxa are given and the distribution of all species is shown on maps of the Archipelago. Notes are given on ecology, altitudinal range for all species and conservation status (for endemics), with additional notes on taxonomic problems where appropriate. The ethnographic information is organised in categories according to how the plant is used on Soqotra (discussed further in 'People and plants' below) followed by comparative uses, which refers to literature records of uses in the region but outside the Archipelago.

Maps

All records are shown on an outline map of the Archipelago. Some critical groups are under-recorded, notably grasses and Cyperaceae, as well as many herbs and weeds. However, the maps are complete enough to provide a reasonable overview of the distribution of most species on the islands.

HISTORY OF BOTANICAL EXPLORATION

The earliest botanical collections and floristic observations from Soqotra were made by Lieutenant J.R. Wellstead and Captain Haines from the East India Company's survey ship *Palinurus* in 1834. Wellstead (1835) made a detailed account of the island and included notes on the traditional methods of collecting the 'Socotrine aloes' (*Aloe perryi*), as well as providing the first description of the dragon's blood tree, *Dracaena cinnabari*.

The French collector L.-H. Boivin visited briefly in 1847 and Commander Wykeham Perry of the British Royal Navy visited the island in 1876 and made a small collection of plants. He was followed in 1877 by Dr J. Collins, a British economic botanist who stopped off on his way back from the botanic garden in Singapore and collected some non-flowering specimens of *Aloe perryi*.

But it was not until 1880 that the first detailed scientific survey of Soqotra was undertaken, by the British botanist Isaac Bayley Balfour, accompanied by Lieutenant C.J. Cockburn of the British Army and Alexander Scott, a gardener from the Royal Botanic Garden Edinburgh. Balfour reached the island on 11th February and departed on 30th March having spent 48 days on the island. During this time he made zoological, geological and botanical collections which included 500 to 600 specimens of flowering plants and resulted in the description of over 200 species and 20 genera new to science. These are described and meticulously illustrated in the monumental *Botany of Socotra* (Balfour 1888).

Balfour's visit to Soqotra was quickly followed by a second major expedition led by the German linguist Dr Riebeck in the spring of 1881. This expedition was primarily to study the people and language of the island but it was accompanied by Dr Georg August Schweinfurth, the famous German naturalist, scholar and traveller, who investigated the botany of the island. Schweinfurth learned of Balfour's recent visit only when he arrived at Aden and, on his return, he very generously sent his collection to Balfour so that the two sets of specimens could be investigated together. Schweinfurth's plants were collected at a slightly different season to Balfour's, in April and May rather than February to March, and so they frequently supplemented the specimens collected by Balfour in flowering and fruiting characters. They are a major contributing factor to the completeness and lasting importance of Balfour's *Botany of Socotra*.

At the close of the 19th century there was another burst of botanical interest in Soqotra and in 1896–97 Dr Theodore Bent, accompanied by his wife Mabel, made a collection of plants whilst studying the archaeology of the island. The joint British and Liverpool museums expedition to Soqotra and Abd al Kuri, led by Dr Henry Forbes and W.R. Ogilvie-Grant, reached Abd al Kuri on 3rd December 1897 and departed on 23rd February 1898 having spent six weeks on Soqotra. The primary objective of their expedition was zoological but they also made a collection of plants. This was sent to Balfour for identification and he contributed the botanical sections to Forbes's *Natural History of Socotra and Abd-el Kuri* (1903). Also in 1898 Dr Stefan Paulay and Professor Oskar Simony from the Vienna Academy of Science made a collection of plants on Soqotra, which were described by Dr Fritz Vierhapper (1907). They also visited Abd al Kuri and made the first collection of plants from the island of Samha.

After all these expeditions it was to be over 50 years until a botanist again visited the island. George Basil Popov, a British ecologist working for the Desert Locust Survey, undertook a general ecological survey of the island in 1953 and published the first systematic description of its vegetation (Popov 1957). Dr Michael Gwynne, a British tropical ecologist and member of the University of Oxford Expedition to Soqotra in 1956, made biological collections on the island including a collection of plants, which are deposited at the British Museum. Gwynne (1968) produced the first vegetation map of Soqotra based largely on Popov's vegetation survey.

During the next few years, the island was visited by several British Army and Political officers, many of whom made small collections of plants. This period ended with the Middle East Command Expedition to the Soqotra Archipelago in the spring of 1967, the same year as the British withdrawal from Aden. Two botanists accompanied the expedition: Alan Radcliffe-Smith from the Royal Botanic Gardens, Kew and John Lavranos. Following the 1967 expedition Soqotra remained effectively closed to botanical exploration until the mid-1980s. The geographical and biological departments of the University of Aden sent a scientific mission to the island in 1982 but apart from this there was no activity. Indeed, following the pessimistic reports by the 1967 expedition, it was generally believed that the vegetation had been decimated by the unrestrained cutting of trees and the vast, uncontrolled flocks of goats which were being allowed to roam freely over the island. This impression was questioned by Quentin Cronk, a botanist from Cambridge University, who made a short visit to the island in 1985. He thought that the vegetation had changed little, if at all, since Balfour's pioneering expedition in 1880. This impression was confirmed by MJM and AM in 1989, and since then there has been a major resurgence of interest in the biodiversity of Soqotra, with a large number of expeditions to the Archipelago. The most important of these have been a series of zoological expeditions led by the German zoologist Wolfgang Wranik, and a series of botanical expeditions by staff from the Royal Botanic Garden Edinburgh in collaboration with scientists from Aden University, Sana'a University and the Agricultural Research Extension Authority of the Yemen Ministry of Agriculture. Other important expeditions have been undertaken by the German botanists Bruno Mies and Norbert Kilian and the Swedish botanist Mats Thulin.

GEOGRAPHY OF THE ARCHIPELAGO

LOCATION AND TOPOGRAPHY

The Soqotra Archipelago consists of four islands: Soqotra, Abd al Kuri, Samha and Darsa and a number of small sea stacks. It is situated in the northern part of the Indian Ocean between 12°06'–12°42'N and 52°03'–54°32'E (see maps below).

The islands are separated from one another by relatively shallow seas but from the mainland of Africa by a narrow but deep trench of several hundred fathoms. Abd al Kuri lies c.80km due east of Cape Guardafui in Somalia, and Soqotra c.380km south of Ras Fartak on the mainland of Arabia.

Soqotra can be divided into three physiographic zones: the coastal plains, the limestone plateau and the igneous Haggeher mountains.

The coastal plains

These vary in width and length but are about 8km at their widest. The largest unbroken stretch is the Noged in the south, which is c.80km long and averages 6km in width. Along other parts of the coast, shorter stretches of plain are broken by rocky headlands and by the escarpment cliffs of the limestone plateau. The wadis are mostly choked by boulders and debris so no permanent streams cross the plains although they do flow to the sea when in flood. Many of the north-flowing wadis terminate in small fresh or brackish lakes separated from the sea by spits and bars. The soils of the plains are alluvial, consisting of compacted gravels, stones and coarse sands with an admixture of fine grey soil. In some areas fine soils carried down from the uplands have become shallowly deposited. These soils were important for local agriculture (Popov 1957). Shallow deposits of wind-blown sand occur along most plains, particularly near Qalansiyah, Hallah on the east coast and on the Ras Hawlaf promontory where they have become banked against the limestone to a height of about 100m. On the Noged plain white sands of shell and coral origin form a belt along the coast whilst further inland large accumulations of yellow sand form dunes up to 70m high in places.

The limestone plateau

This extends across a large part of the island, averaging 300–700m in elevation and reaching over 800m at its highest point in the west. The plateau drops in steep, often almost vertical escarpments either to the coastal plain or directly to the sea. Wadis flow from the central mountains, particularly in the south, cutting deep into the plateau. The summit is rolling or, in some areas, broken by well-vegetated cliffs and gullies. The surface exhibits typical karst features, with in some areas (e.g. Rewgid) large areas of bare pavement, gullies, swallow holes and cliffs.

In other areas (particularly in the east), where deeper soils have accumulated in broad, shallow depressions, there are extensive areas of flat grassland.

Few streams are visible on the surface and caves are common. In the west there is a large, dry and poorly vegetated drainage basin (the Zahr). Soils are poorly developed but weathering has produced a fertile red or grey, clay soil which accumulates in fissures and solution hollows in the limestone. In the valleys and some depressions a greater depth of reddish clay soils (terra rossa) has accumulated.

The Haggeher mountains

The Haggeher mountains are a high igneous plateau (generally above 750m) and dominate the eastern and central part of Soqotra island, being perhaps the most spectacular feature of the islands. They rise in a series of dramatic pinnacles radiating from the base of which are deep, moist gullies, steep boulder-strewn slopes and, on gentler gradients, cleared grassland.

There are several peaks over 1300m and Jebel Skand at over 1550m is the highest point in the Archipelago. The slopes of the Haggeher are well vegetated and there are several permanent streams. Decomposition of the granite has led to the formation of deep and fertile red soils in the valleys and on the gentler slopes whilst on the pinnacles and steeper slopes soils are thin or absent. The rocks of the Haggeher are predominantly red but often appear white from the lichens that cover them.

Abd al Kuri

Abd al Kuri is the second largest island in the Archipelago. It is approximately 33 x 5km and covers 133km^2. In aspect it is desertic with little or no vegetation visible from a distance. The coastal plains are composed primarily of sandy and gravelly soils with frequent rocky outcrops. The mountains are fairly steep and their slopes are covered with loose rocks and scree. The summit ridge of Jebel Saleh consists of a cap of limestone split into large blocks by cliffs and deep gullies providing a variety of relatively hospitable niches for plants. There are no permanent streams on the island and only a few freshwater seepages at higher elevations.

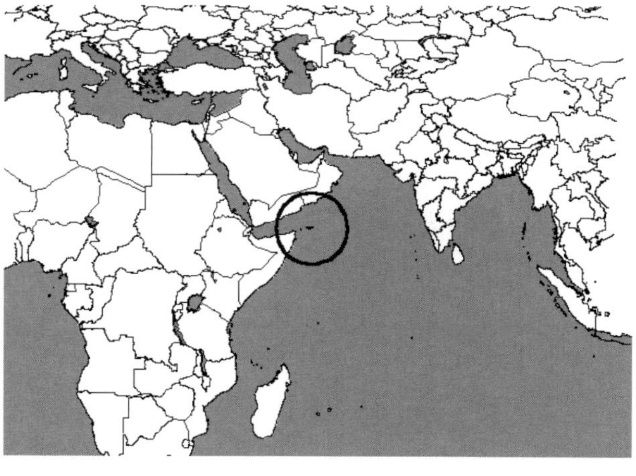

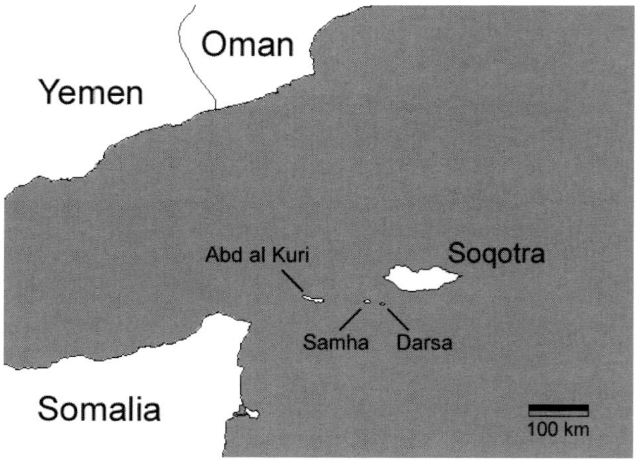

Samha and Darsa

Samha has a land surface of 41km^2 and is the third largest island in the Archipelago. It lies approximately 50km southwest of Soqotra and 60km east of Abd al Kuri. Its closest neighbour is the small uninhabited island of Darsa, which lies some 17km to the east. Samha and Darsa are collectively known as 'The Brothers'. Samha is largely desertic in aspect with little or no vegetation visible from a distance. The centre of the island is dominated by a flat-topped limestone plateau the summit of which slopes gently upwards from the east to the highest point at c.680m in the west. The plateau is surrounded on all sides by steep cliffs dropping to the boulder-strewn foothills that cover the rest of the island. Along most of the coastline the foothills drop directly to the sea and there is little or no coastal plain, except in the west where a low, sandy peninsula extends c.2km into the sea.

GEOLOGY

Soqotra

The geology of Soqotra has been described in detail by Beydoun & Bichan (1970). They summarise the main features as follows:

'Soqotra for the most part is covered by a thin veneer of Cretaceous and Tertiary limestones [with some marls and sandstones] reaching up to 700m in thickness; these form an undulating, irregular and in part, block faulted upland plateau ranging between 300 and 900m in elevation overlying a much older igneous and metamorphic basement [with gneisses, schists, gabbros and granites] which is particularly well exposed in three main structural uplift areas. These are the coastal/plain valley areas extending inland from Bandar Shu'ub and Qalansiyah on the west coast, and in the spectacular Haggeher range rising to peaks of up to 1500m and forming the backbone of Soqotra. Other more limited exposures of the basement complex occur elsewhere. In other areas where part of the limestone cover has been removed, recent deposits largely obscure the earlier rocks in particular along the northern and southern coastal plains and in some inland depressions.'

On the coastal plains, valleys and inland depressions Quaternary and recent deposits of Pleistocene and Holocene age overlie an eroded substratum of older rocks. These are heterogeneous and include marine, fluvial and continental deposits.

Samha and Darsa

No account of the geology of Darsa or Samha is available. However, they are similarly igneous and metamorphic basement complex rocks capped by limestone of Cretaceous age.

Abd al Kuri

The following short description is based on the account of the geology of Abd al Kuri by Professor J.W. Gregory (in Forbes 1903): the island is for the most part composed of igneous and metamorphic basement rocks including amphibolite schists and gneisses, which are cut by pegmatite dykes. The basement complex rocks are capped by limestones of Cretaceous age.

CLIMATE

General description

Few climatic data are available for Soqotra but some general comments can be made. The climate is monsoonal, being strongly influenced by both the SW (Indian or summer) monsoon and the NE (winter) monsoon. The SW monsoon blows from June until September bringing hot, dry winds (often averaging 70mph) from Africa. During these months the island is effectively cut off from the outside world and life is very harsh. On the exposed plateaus and coastal plains there is little precipitation and the extreme desiccating conditions brought by the dry winds have an important effect on the development of the vegetation. However, at higher elevations, and particularly in areas facing the southwest, the season brings drizzle and cloud.

Winds are predominantly northeasterly from November until January. These winds are lighter than those of the SW monsoon and most rain falls between November and January. As one might expect, the islanders have a detailed knowledge and descriptive terminology for the local climate; this is described under 'People and plants' below.

Rainfall

Rainfall is very sporadic and varies across the Archipelago with season, aspect and elevation. Although there are no rainfall figures for the interior or the smaller islands we can make an estimate by comparison with the rainfall of similar areas in Somalia. According to Hemming (1966) the evergreen woodland dominated by *Buxanthus pedicellatus*, which occurs on the escarpment mountains of northern Somalia, receives 500–650mm of rain. On Soqotra *Buxanthus* is frequent in sheltered places on the limestone plateau and is common above about 400m on the Haggeher mountains. It therefore seems probable that middle elevations on the Haggeher and favoured sites on the mountains in other areas receive rainfall of this order.

Measurements for the coastal plain (airport) are around 125–175mm per annum (Popov 1957) with in some years rain being as little as 7.5mm or failing altogether (Cronk 1986). The estimate of 500–650mm of rain per annum accords favourably with the estimate made by Gwynne (1968) of 380–624mm for the mountains. The highest peaks of the Haggeher (above 1000m) undoubtedly receive more rainfall than this, probably in the region of 1000mm. Cloud cover is constant over the Haggeher for long periods each year and, in some years, almost continual during the period of the NE monsoon. Heavy dews are common in summer in all high areas. Tropical storms and cyclones bring torrential rain to the Archipelago every few years. They occur mainly in the transitional periods between the NE and SW monsoons.

Temperature

Some temperature measurements are available for the coastal plain; these give a mean annual temperature of 17–26°C minimum and 27–37°C maximum (Griffiths 1972). No systematic measurements are available for the mountains although it is considerably cooler there. Popov (1957) reports a minimum of 13.5°C at night at 1000m in March. Frost, but not snow, is reported. The hottest months are in the transitional periods between the NE and SW monsoons from April to June and again in September and October.

Winds

The contrast between the winds of the SW and NE monsoons is a major feature of the climate of the Archipelago. During the NE monsoon the winds are relatively light, with the wind speed seldom exceeding 25–35km per hour. However, Gwynne (1968) cites observations during the SW monsoon of gale force winds and above making up over 50% of all records for Soqotra. There were, for instance, 23 gale days during July. Cronk (1986) mentions how, during the SW monsoon, the winds average 70mph.

FLORISTICS AND VEGETATION

CLASSIFICATION OF THE SOQOTRAN FLORA

Takhtajan (1986) includes Soqotra in the Eritreo-Arabian subregion, which includes southern Arabia, Somalia, Ethiopia, most of Kenya and large parts of Tanzania, and has two endemic families, Barbeyaceae (not present on Soqotra) and Dirachmaceae, and about 50 endemic genera, including *Poskea* in Soqotra, the Yemen mainland and Somalia. Several large genera are particularly diverse here including *Acacia*, *Aloe*, *Boswellia*, *Cadaba*, *Euphorbia*, *Farsetia*, *Indigofera*, *Maerua*, *Neuracanthus*, *Commicarpus*, *Commiphora* and *Jatropha*. According to White (1983, 1986) the Soqotra Archipelago falls within the Somalia-Masai Regional Centre of Endemism and enough species are confined to the islands to make them stand out as a local centre of endemism.

Soqotra has one subendemic family – the Dirachmaceae, which was known only from one Soqotran species until the recent discovery of a second species in Somalia. There are 15 endemic genera and 307 endemic species. Endemism is not evenly distributed across plant families. Grasses form the biggest family on the Archipelago, with 112 species, but have only 10 endemic species and no endemic genera. On the other hand Acanthaceae is represented by 34 species and 15 genera, of which 24 species and three genera are endemic.

Table 1 gives a breakdown of the flora by species, genera and families and shows the distribution of endemics.

Table 1 Breakdown of the Soqotran flora by species, genera and families, and the distribution of endemics

	Families	Genera	Species	Endemics
Ferns	10	18	26	1
Gymnosperms	1	1	1	
Dicots	82	321	625	276
Monocots	21	90	173	30
Total	114	430	825	307

Table 2 Endemism in oceanic islands

	Size (km^2)	Total plant spp.	Endemics	% endemics
Soqotra	**3799**	**825**	**307**	**37**
Azores	2235	600	55	9
Canary Is.	7273	2000	569	28
Madeira	769	760	760	17
Ascension	94	25	11	44
St Helena	121	320	49	15
Mauritius	1865	800–900	280	31–35
Rodriguez	104	145	48	33
Galápagos	7844	543	229	42
Norfolk Is.	13	174	48	28
Juan Fernandez	93	147	118	80

The figures for botanical diversity confirm the Soqotra Archipelago's global status as one of the world's most botanically important island groups, comparing favourably in floristic richness and endemism with such famous islands as Mauritius, the Galápagos and the Canary Islands (see Table 2).

LINKS WITH THE SOQOTRAN FLORA

Although the Soqotran flora has many unique features and a high proportion of endemics there are still clear indications of relationships with several geographical areas, some far distant from Soqotra. There are strong links between the Soqotran flora and its nearest geographical neighbours in southern Arabia and the Horn of Africa, but relatives of Soqotran taxa are also found in NW Africa and Macaronesia, SW Africa, S Asia, Madagascar and the New World. Large-scale factors such as climatic variation, fluctuating sea-levels and continental drift all help to explain how the patterns we see today have come about, but there is very little direct fossil evidence for past distributions of any components of the flora. Thus if we want to understand the origin of the Soqotran flora we must rely largely on evidence inferred from current distribution patterns in conjunction with an understanding of the ecology and evolution of the plants in question in relation to historical changes in climate and geography.

PAST CLIMATE IN THE REGION

The Archipelago is situated in an arid area that has experienced fluctuations in rainfall, particularly over the last 150 000 years, and this has caused widespread desertification in the region. The varied topography of the Archipelago, especially of Soqotra itself, and its position in the path of the Indian Ocean monsoons, allows the islands to trap moisture far more effectively than much of the neighbouring continents. As a result they have acted as a refuge for many plants that have otherwise been unable to survive the fluctuating aridity of the area. There are several areas in the region that also trap moisture from the monsoons, notably Dhofar and Mahra in western Oman and eastern Yemen and the coastal escarpments in Somalia. Both these areas have many species and genera in common with Soqotra. More distant disjunctions are the result of many factors, requiring an understanding of past climate, varying sea-levels, dispersal abilities and ecological tolerance of the plants concerned, and continental drift.

GEOLOGICAL HISTORY

The land surface of the islands is very old and represents a fragment of the ancient southern supercontinent of Gondwana. Soqotra occupied a central position in Gondwana, close to what are now Africa, Arabia, India and Madagascar. Its current position as a continental fragment arose 20 to 15 million years ago as Arabia split from the African continental plate, opening up the Gulf of Aden and leaving Soqotra on the same continental plate as Africa. Later, the Red Sea axis and the East African Rift Valley system started to split apart, causing further separation of Arabia and the Horn of Africa from the rest of Africa. Fluctuations in sea-levels may have allowed a land bridge to develop between the Archipelago to the Horn of Africa at various times, particularly during the glaciations of the last 150 000 years.

EXPLANATIONS FOR THE ORIGIN OF THE SOQOTRAN FLORA

The endemic flora of Soqotra appears to largely represent relict populations of formerly more widespread taxa. So where did the endemics, or their relatives, originally come from? There are two possible explanations: that the flora is an ancient remnant, representing relicts of the flora that originated on Gondwana; or that it has migrated into the islands from neighbouring continents. The theories are not mutually exclusive: different taxa have different histories and both theories are supported by the distributions of related taxa from Soqotra and a wide range of other geographical areas. Given the proximity to Africa and Arabia and the way the monsoon winds blow it seems likely that migration in and out of the islands has played an important role. However, the stable geology and relatively moist, oceanic climate must have allowed plants to survive and evolve here over much longer time periods than in the harsher environments of the neighbouring continents.

It has been suggested (Klackenberg 1985) that the current distribution of *Exacum* in Madagascar, Arabia, India, S Asia and Soqotra can be explained by its origin in Gondwana and subsequent movement with the continental fragments into their present positions, taking species into what are now widely disjunct areas. However, more recent evidence based on molecular data and the fossil record indicates that the Gentianaceae had not evolved until after the break up of Gondwana (Magallón *et al.* 1999). An unpublished molecular study of *Exacum* indicates that the genus evolved in Madagascar and its current distribution is the result of long-distance dispersal. However, the species on Soqotra are closely related to species on the Arabian mainland, and this relationship suggests either that there has been dispersal from Arabia to Soqotra or that divergence occurred as a result of Soqotra's separation from the Arabian mainland.

Evolutionary pressures are still acting and may be the cause of the taxonomic difficulties found in several species complexes in the Archipelago such as those in *Hibiscus* and *Heliotropium.* Poor differentiation between species may be explained by poorly defined barriers to gene flow between populations. Detailed population studies of these groups would be very rewarding in helping to clarify the extent of active speciation on Soqotra.

GEOGRAPHICAL AREAS WITH LINKS TO SOQOTRA

Links between NE Africa and Arabia and Soqotra are not surprising, considering the proximity of the regions, but some taxa particularly emphasise these links, notably:

NE Africa: *Buxanthus pedicellatus*, Dirachmaceae, *Neuracanthus argyrophyllus*, *Indigofera nugalensis*, *Wellstedia*, *Thamnosma*, *Chapmannia*, *Acacia*, *Campylanthus*.

Arabia: *Ochradenus*, *Adenium*, *Pulicaria*, *Boswellia* (also NE Africa), *Commicarpus* (also NE Africa), *Exacum*, *Lavandula*, *Campylanthus*, *Cocculus balfourii*, *Poskea socotrana*.

Other geographical areas with links to the Soqotran flora are perhaps less expected. NW Africa and the Canaries are now separated from Soqotra by a vast expanse of desert but wetter climates in the past help to explain alliances in the floras of the two regions. It is less easy to explain disjunctions between Soqotra and SW Africa and even more difficult to find conclusive explanations of disjunctions with the New World. Nevertheless such disjunctions do exist and hypotheses have been proposed to explain them.

NW Africa and Macaronesia: The closest relative of the Soqotran dragon's blood tree (*Dracaena cinnabari*) is *D. draco* (L.) L. from the Canaries and Morocco. *Dracaena* fossils have been found in southern France dating from the Tertiary (Eocene-Neogene) when much of what is now southern Europe was covered in subtropical forests, probably rather similar to the laurel forests still surviving in the Canaries. It has been suggested (Marrero *et al.* 1998) that the arborescent *Dracaena* species were dispersed around the Tethys, an ancient sea between Africa and Laurasia that has become the Mediterranean as Africa and Arabia have moved northwards, reducing the size of the sea and enclosing it in the east. The increased aridity in N Africa has forced the ancient Tethyan forests to fragment, and relicts, such as *Dracaena* in the Canaries and Soqotra, now survive only at the margins of this ancient distribution.

Nesocrambe is a genus discovered during fieldwork for this flora (Miller *et al.* 2002), whose closest relative is apparently *Hemicrambe. Hemicrambe* has two species, one from Soqotra and the other from Morocco. *Campylanthus* has 15 species with restricted distributions, found in the Canaries, NE Africa, Arabia and east to Pakistan. *Camptoloma* is found in the Canaries and SW Africa.

SW Africa: Another striking disjunction found in several elements of the Soqotran and NE African floras occurs with SW Africa. *Wellstedia* is a somewhat anomalous genus, usually placed in the Boraginaceae but sometimes separated into its own family, which has four species in Soqotra, Somalia and Ethiopia and one species in Namibia. This disjunction could be the result of long-distance dispersal or it could represent a relict of a more ancient link between the floras (Thulin & Johansson 1996). The fact that other Soqotran genera, such as *Thamnosma*, *Camptoloma* and *Chorisochora*, show similar links suggests that there was indeed a climatic corridor linking NE and SW Africa in the Tertiary.

New World: One of the most extreme disjunctions is found in *Chapmannia* (Leguminosae: Papilionoideae). The genus has recently been recircumscribed (Thulin 1999b) to include seven species, four on Soqotra, one in nearby Somalia and two in Central America (one of which reaches into northern S America) and the southern USA. There are a number of possible explanations for this curious distribution, including long-distance dispersal or migration over now extinct land bridges. Recent molecular evidence (Lavin *et al.* 2000) indicates that *Chapmannia* and its relatives originated in northern tropical areas (boreotropics), which covered large areas of the Eurasian and N American landmasses in Cretaceous and Tertiary times. These boreotropical areas were home to extensive areas of dry tropical woodland and barriers to dispersal were relatively weak within this climatic and ecological band, which stretched across the two continental landmasses. Several recent phylogenetic studies have suggested that many distribution patterns can be explained by a boreotropical origin (including N America) and subsequent migration southwards into Africa (Cortés-Burns 2000; Lavin *et al.* 2000). Continental drift northwards, glaciation in northern areas and increased aridity in northern Africa and the Middle East has led to many extinctions and barriers to continued dispersal across the region. Other papilionoid legumes from Soqotra have been included in molecular analyses and indicate a similar pattern, e.g. *Zygocarpum coeruleum* (Lavin *et al.* 2000) and Soqotran *Indigofera* (Cortés-Burns 2000). The recent discovery of a species of *Erythroxylon*, a largely Neotropical genus, also suggests a link with the New World. *Thamnosma* is another genus distributed in southern N America, southern Africa and Soqotra.

Madagascar: Several Soqotran genera are also found in Madagascar, notably *Exacum*, *Begonia*, *Coelocarpum* and *Distephanus*.

PATTERNS OF DISTRIBUTION AND ENDEMISM WITHIN THE ARCHIPELAGO

There are several marked distribution patterns across the Archipelago. These reflect variations in climate and topography combined with the effects of isolation, caused by sea or inhospitable habitats.

Widespread plants

The most widespread plants are found on the coastal plains, dry foothills and escarpments, a continuously distributed habitat across Soqotra. This in large part explains the wide distribution and abundance of species such as *Justicia rigida*, *Indigofera nephrocarpa*, *Dendrosicyos socotrana*, *Aizoon canariense*, *Pulicaria stephanocarpa*, *Ballochia* species, *Trichocalyx* species, *Commiphora ornifolia*, *Maerua angolensis*, *Croton socotranus* and *Jatropha unicostata* (the last two are absent from the outer islands).

Wet centres

Perhaps the most marked patterns of distribution are those exhibited by plants restricted to 'wet centres' – that is, areas, usually mountainous, which typically catch monsoon precipitation (see map opposite). These vary in size from just a few square kilometres, such as the north-facing limestone escarpment on Samha or the limestone plateau of Jebel Saleh on Abd al Kuri, to several hundred square kilometres, such as the granitic Central Highlands and the adjacent limestone plateaus on Soqotra.

Seven principal 'wet centres' have been identified:

1. Granitic Central Highlands and adjacent limestone plateaus (in north-central Soqotra).
2. Limestone plateaus and escarpments (principally Reqadihon and Hamaderoh) in NE Soqotra.
3. Ma'alah in NW Soqotra.
4. Hager, limestone ridge above Bandar Shu'ub in W Soqotra.
5. Qatariyah limestone plateau in SW Soqotra.
6. North-facing limestone escarpment on Samha.
7. Limestone plateau and cliffs of Jebel Saleh on Abd al Kuri.

These areas can all be considered as relatively wet 'islands' surrounded by inhospitable drier areas. During times of greater rainfall in the geological past they would have increased in extent, and at certain times become continuous with one another, allowing exchange of plants, and then, during times of decreased rainfall, they would have contracted in area. The age of their isolation from one another can be inferred by the occurrence of: disjunctions (i.e. a species found in two or more centres but not in intervening areas); vicariance (i.e. clearly related but distinct species occurring in different areas); and endemics (i.e. species restricted to a given area).

Many species are found in more than one 'wet centre' but a few are restricted to and found in all or nearly all of the wet centres, e.g. *Adiantum balfourii*, *Erucastrum rostratum* (except Abd al Kuri) and *Habenaria socotrana*. Some species are considered good indicators of 'wet centres' on Soqotra, e.g. *Rhus thyrsiflora*, *Cystostemon socotranus*, *Eurieandra balfourii*, *Euclea divinorum* and *Punica protopunica*.

Pairs of vicariant species occur between refugia on either Samha or Abd al Kuri and Soqotra, e.g. *Begonia samhaensis* (Samha) and *B. socotrana* (Soqotra); *Ledebouria samhaensis* (Samha) and *L. grandifolia* (Soqotra); and *Dipcadi kuriensis* (Abd al Kuri) and *D. balfourii* (Soqotra).

Some species, which are common and characteristic of the Central Highlands and limestone plateaus, are known from only one or two other refugia on Soqotra, e.g. *Anisotes diversifolius*, *Crinum balfourii*, *Kalanchoe rotundifolia*, *Dioscorea lanata*, *Dirachma socotrana* and *Heliotropium nigricans*.

All seven wet centres have their own endemics:

1. Central Highlands and adjacent limestone plateaus: c.80 endemics, e.g. *Begonia socotrana* and *Exacum caeruleum*.

2. Limestone plateaus and escarpments of NE Soqotra: c.5 endemics, e.g. *Dorstenia socotrana* and *Boswellia nana*.

3. Ma'alah: c.5 endemics, e.g. *Gaillonia thymoides* and *Prenanthes amabilis*.

4. Hager, limestone ridge above Bandar Shu'ub: c.4 endemics, e.g. *Nesocrambe socotrana* and *Pulicaria dioscorides*.

5. Qatariyah: 1 endemic, *Maytenus* sp. *A*.

6. Limestone escarpment on Samha: c.5 endemics, e.g. *Begonia samhaensis* and *Pelargonium insularis*.

7. Limestone plateau and cliffs of Jebel Saleh on Abd al Kuri: c.3 endemics, e.g. *Oldenlandia ocellata* and *Dipcadi kuriensis*.

Dry centres

There is a less marked trend for species with narrow or restricted distributions in drier areas, since dry conditions are more or less continuously distributed across the islands, but certain areas can be classed as 'dry centres' (see map opposite). On Soqotra these are either high dry limestone plateaus (e.g. Qatariyah) or isolated desertic (e.g. the Zahr basin) or coastal plains (e.g. Bandar Shu'ub and the plains of NE Soqotra). Abd al Kuri and Samha can also be considered as 'dry centres'. During times of reduced rainfall in the geological past these areas would have increased, particularly in their altitudinal range, and during times of increased rainfall they

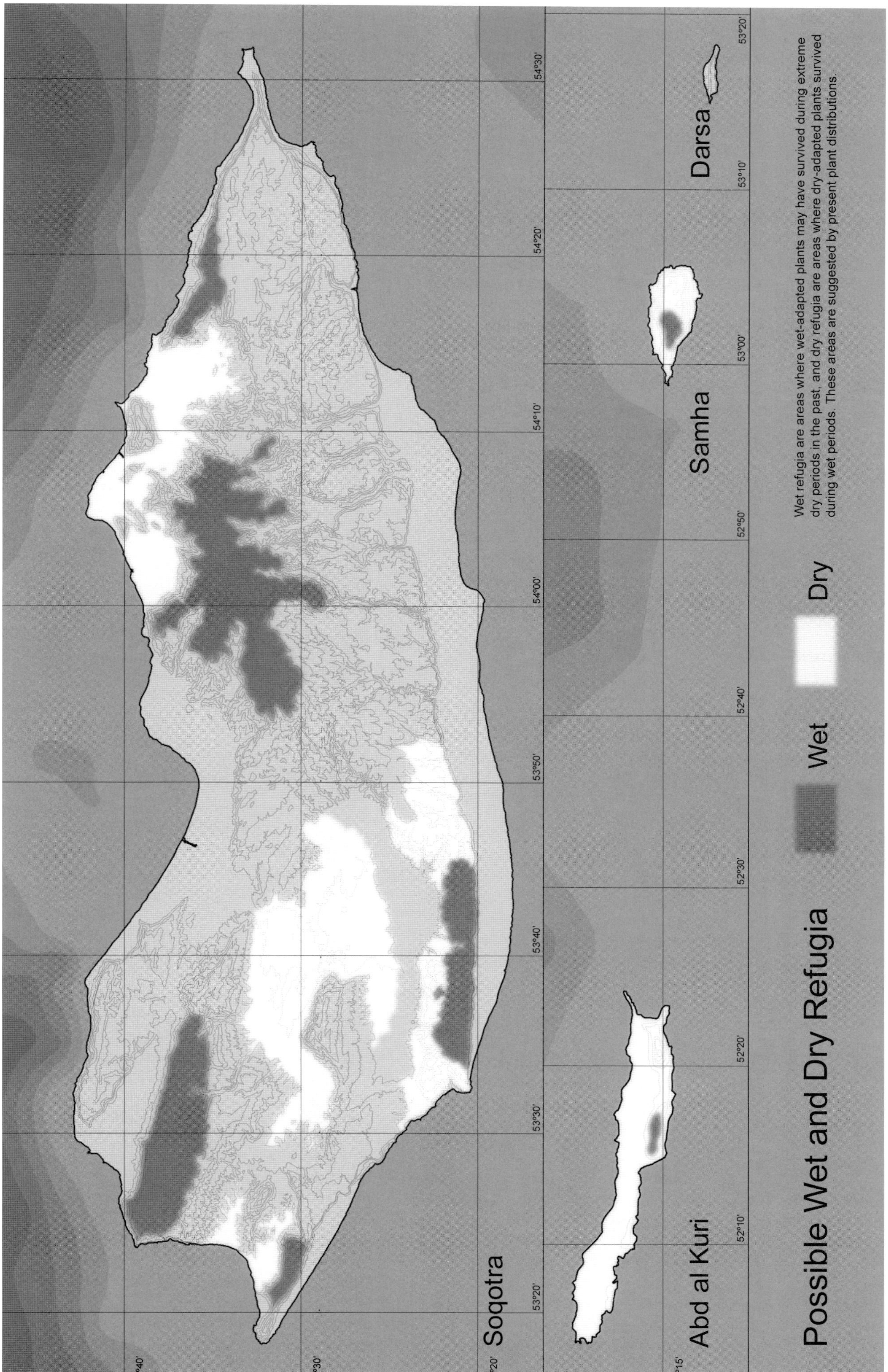
Possible Wet and Dry Refugia
Wet
Dry
Wet refugia are areas where wet-adapted plants may have survived during extreme dry periods in the past, and dry refugia are areas where dry-adapted plants survived during wet periods. These areas are suggested by present plant distributions.
Soqotra
Abd al Kuri
Samha
Darsa
12°40'
12°30'
12°20'
12°15'
53°20'
53°30'
53°40'
53°50'
54°00'
54°10'
54°20'
54°30'
52°10'
52°20'
52°30'
52°40'
52°50'
53°00'
53°10'
53°20'

would have contracted in area. As with the 'wet centres' the age of their isolation from one another can be inferred from plants with disjunct distributions, vicariant species and endemics. Six principal 'dry centres' can be identified.

1. NE coastal plains: These areas are particularly threatened by development. Endemics include *Blepharis spiculifolia* and *Neuracanthus aculeatus* as well as several species that have not been seen since they were first discovered in the 19th century, including *Barleria argentea* and *Neuracanthus capitatus*.

2. Dry western plateau: Much of the summit and the drier, more exposed, dip slopes of Qatariyah are covered with bare limestone pavement where there has been little or no soil accumulation. In these areas the vegetation is sparse and dominated by very open *Buxanthus pedicellatus* shrubland. One of the most interesting vegetation types of the more exposed areas is a community of low, mainly cushion plants found in loose rock debris overlying the limestone pavement. Characteristic species here are: *Echiochilon pulvinata*, *Distephanus qazmi*, *Lotus mollis*, *Launaea crepoides*, *Exacum gracilipes*, *Leucas flagellifolia*, *Convolvulus sarmentosus*, *Helichrysum balfourii*, *Campylanthus spinosus*, *Wellstedia socotrana* (in east) and *Oldenlandia pulvinata.*

3. Zahr basin: A large drainage basin in western Soqotra: plants restricted to the basin include *Metaporana obtusa*, *Acacia tortilis*, *Seddera glomerata* and *Socotrella dolicocnema.*

4. Bandar Shu'ub: Although not rich in endemic species the coastal plains and coastal strip of Bandar Shu'ub contain some of the most fragile and threatened vegetation types on Soqotra. They are also home to some of the most threatened species on the island. It is likely that the dwarf shrubland found along the coast at Bandar Shu'ub represents a vegetation type which was at one time more widespread in coastal plains in other parts of Soqotra before it was degraded by the development of fishing communities and overgrazing. Bandar Shu'ub is thus important as a refugium of vegetation types which were once more widespread. It is the only known locality for *Tamarix aphylla* in the Soqotra Archipelago, a common species in surrounding regions. Other species of note include *Taverniera sericophylla*, *Farsetia socotrana*, *Leucas flagellifolia* and excellent specimens of *Limonium paulayanum* and *Limonium socotranum* which here form shrubs up to 1.5m in height but which are elsewhere rarely above 50cm in height. The best examples of mangrove woodland on Soqotra, dominated by *Avicennia marina*, are found in Bandar Shu'ub, at Qow'har and further around the coast at Nit. This mangrove vegetation is not unique to Soqotra but it is nevertheless a vulnerable vegetation type that supports a rich endemic fauna and is also an important breeding ground for various animal species.

5. Samha: Five species are endemic to the dry areas of Samha including *Seddera samhaensis* and *Portulaca samhaensis*. Dry parts of Samha are also an important refuge for certain species endemic to the Archipelago: *Lachnocapsa spathulata*, *Leucas flagellifolia*, *Farsetia socotrana*, *Echiochilon pulvinata* and *Helichrysum* sp. *A*. The altitudinal range of some species is greatly increased on Samha. For example, *Aerva microphylla* and *Atriplex griffithii* are typically not found above about 50m on Soqotra whereas on Samha they are common from sea-level to over 500m. This is probably due to lack of competition.

6. Abd al Kuri: Abd al Kuri is distinct in both its floristic composition and vegetation from any other area in the Archipelago. About 10 species are endemic to the dry areas of Abd al Kuri, including *Seddera spinosa* and *Convolvulus grantii* and undoubtedly the most famous endemic on the island, *Euphorbia abdelkuri.* Interestingly, several species which are widespread outside the Archipelago have their only 'Soqotran' station on Abd al Kuri. These include *Cadaba farinosa*, *Sphaerocoma hookeri*, *Anabasis ehrendbergii*, *Arthrocnemum macrostachyum* and *Lagenanthera cycloptera*. Conversely, few plants are found on both Abd al Kuri and Soqotra, although one such is *Acacia sarcophylla*, a species of coastal areas with thick woody pods which are probably able to survive in the sea between the two islands.

Cliff refugia

Cliffs are important refugia on Soqotra and certain species are restricted to them, e.g. *Aloe squarrosa*, *Boswellia* species, *Dorstenia* species, *Exacum socotranus*, *Hemicrambe fruticosa*, *Hypericum tortuosum*, *Nesocrambe socotrana*, *Oreofraga morrisiana*, *Pulicaria aromatica*, *Rughidia milleri* and *Turraea socotrana*. See also discussion under *Boswellia* species and *Nesocrambe socotrana* (smooth rock refugia).

Adaptive radiation

Genera that are clearly adaptively radiating on Soqotra include *Campylanthus*, *Chapmannia*, *Cissus*, *Echidnopsis*, *Euphorbia*, *Exacum*, *Helichrysum*, *Heliotropium*, *Hibiscus*, *Polycarpaea*, *Seddera* and *Trichodesma*.

CLASSIFICATION OF VEGETATION

In general the islands of the Soqotra Archipelago are sparsely vegetated and dominated by xeromorphic lifeforms. The development of the vegetation is strongly influenced by the desiccating winds that sweep the islands in summer. In consequence it is only in sheltered valleys, escarpments and the high mountains that woodland develops. In general the exposed summits of the limestone plateau are covered with sparse shrubland or low woody-herb communities, with more luxuriant vegetation developing on cliffs and in gullies. The coastal plains are largely subdesertic with open shrubland or in some areas more or less devoid of vegetation.

There are several distinctive vegetation formations on Soqotra, particularly with respect to their physiognomy and structure. Perhaps the most famous, because of its often bizarre appearance, is the shrubland dominated by tree succulents such as *Dendrosicyos socotranus*, *Adenium obesum* subsp. *sokotranum* and *Euphorbia arbuscula*. This occurs mainly on the foothills of the mountains and the limestone escarpments and is the vegetation type for which Soqotra is most renowned. Another unique vegetation formation is the evergreen woodland dominated by the dragon's blood tree, *Dracaena cinnabari*, commented upon by White (1983, 1986) as the most 'singular community on Soqotra'. The finest examples of this are found on the higher slopes of the limestone mountains, particularly in the centre and east of the island.

The description of vegetation presented problems. There is some altitudinal zonation, but over most of the island there is a mosaic of related vegetation types, which although floristically related can be physiognomically very distinct. The system below is based on physiognomy but reflects floristic differences as far as possible. Photographic panoramas and releves were used to record these vegetation types.

We have recognised six principal vegetation types on Soqotra. The vegetation types of Abd al Kuri and Samha differ in composition and structure from those found on the island of Soqotra.

1. Coastal vegetation

a. Mangroves: Mangrove thicket [<5m]. Dominants: *Avicennia marina.*

b. Coastal plains: Mosaic of low succulent shrubs [<1.5m] and woody-based herbs [<0.5m] and patches of shrubland [<3m]. Dominants: various, including *Limonium* species, *Tamarix nilotica*, *Suaeda* species and *Atriplex* species.

2. Croton shrubland: Deciduous shrubland [<2.5m] with scattered emergents [<5m] and open dwarf shrub layer [<1m] below. Dominated by *Croton socotranus.* If heavily grazed the herb cover is dominated by *Cassia holosericea* and *Tephrosia apollinea.*

3. Succulent shrubland: Open deciduous succulent shrubland [<4m] with emergent trees [<6m] and a lower layer [<1m] of cushion vegetation and subshrubs. Dominants: *Jatropha unicostata* and *Croton socotranus* with *Euphorbia arbuscula*, *Adenium obesum* and *Tephrosia apollinea.*

4. Semi-evergreen woodland: Semi-deciduous thicket [<5m] with emergent trees [<8m]. Dominants: *Rhus thyrsiflora*, *Buxanthus pedicellatus*, *Carphalea obovata*, *Sterculia africana.*

5. Open and woody-based herb communities: Mosaic of woody-based herb communities [<0.5m], cleared grassland and dwarf shrubland [<2m]. Relict woodlands, typically with *Dracaena cinnabari* and *Boswellia elongata*, are also occasionally found in this zone. Dominants: *Jatropha unicostata*, *Lycium sokotranum*, *Gnidia socotrana*, *Buxanthus pedicellatus*, *Croton socotranus*, *Leucas virgata* and *Cissus hamaderohensis.*

6. Submontane shrubland: Semi-deciduous shrubland [<5m] with emergent *Dracaena* [<8m] and cleared grassland. Dominants: *Dracaena cinnabari*, *Rhus thyrsiflora*, *Euryops arabicus*, *Buxanthus pedicellatus*, *Gnidia socotrana* and *Cocculus balfourii.*

7. Montane mosaic of evergreen woodland, grassland, dwarf shrubland and cushion vegetation: Mosaic of dense evergreen woodland and thicket [<5m], patches of cleared grassland, dwarf shrubland [<1.5m] and cushion vegetation on exposed summits. Dominants: *Dracaena cinnabari*, *Rhus thyrsiflora*, *Hypericum* species, *Helichrysum* species and *Euryops arabicus.*

PEOPLE AND PLANTS

BACKGROUND

The island environment is challenging and often hostile. The majority of the islanders depended until very recently on their livestock for survival, and thus of necessity on the rangeland which nourished their animals. They coped with the precariousness of their environment in various ways. They reared different livestock species (goats, sheep, cattle and camels) to provide milk and meat, as well as their principal trading commodity, clarified butter. Each kind of livestock has distinct dietary needs which enables them to exploit different areas of the range at different seasons of the year. The circulation of livestock through mutual support networks also helped to offset risks. The islanders developed diverse food-procuring strategies, including fishing, small-scale rainfed cultivation, hunting, trapping, and gathering wild foods. They supplemented this by bartering any surplus for basic foodstuffs such as dates and cereals, and by producing goods for both internal and external trade such as sheep's wool cloth, clay pottery, basketry, leather goods, civet, aloes juice and various gums and resins.

Careful management of the vegetation was the key to survival on the islands. Plants act as a microclimate for other plants, for living creatures and for soils; they prevent wind and water erosion, and provide a means of increasing or maintaining soil fertility and productivity. Quite apart from their primary value of supplying livestock with the protein, minerals and vitamins necessary for surviving a dry season of five to eleven months annually, plants were the source of most other necessities for the islanders. Plants provided shade from the sun, protection from rain and cold, and shelter from the strong winds of the monsoon. They provided timber and fuel, their wood was used to make tools and utensils, and their fibres were worked into cordage, matting and basketwork. They provided food, medicaments, dyes, tannins, adhesives, pesticides, cleansing materials and fertilisers, as well as cosmetics and toys. In addition to being the source of material goods plants were also important in ritual and in dealing with the spirit world.

THE ISLANDERS

The people of Soqotra have been the guardians of their islands for generations, and it is due to their efforts and custodial care that they are as we see them today. Their complex traditional rules governing the management of livestock and pasture, access to water and dry season fodders, and the exploitation of the flora and fauna in general have been instrumental in maintaining the biodiversity of the islands. That the Soqotra Archipelago has been declared one of the world's most important island groups, with a flora and fauna of global significance, demonstrates the success of their stewardship.

The population of the Soqotra Archipelago is estimated to be around 50 000. At present, the bulk of the population lives in and around the capital area on the northern plain, and in the coastal fishing villages. However, this is a recent phenomenon. Before the 1970s, the vast majority of Soqotrans were seasonally transhumant pastoralists, rearing sheep and goats, and, where possible, cattle and camels. In some areas they also bred camels and donkeys as pack animals. Even now, most Soqotrans are still dependent on livestock rearing, though fishing has assumed a greater importance, and in some areas of Soqotra and on the outer islands of Samha and Abd al Kuri it has become the main source of livelihood.

The system of seasonal transhumance, which permits periodical resting of the pastures, has played a major role in the conservation of the vegetation. Different species of livestock feed on different plants and are rarely in competition. The traditional separation of livestock – sheep being herded on the upland plateau sheepwalks, goats browsing the steeper mountain slopes, and cattle grazing on the upland pastures or along the valleys – has also played its part in conserving the rangeland. Other factors combined to prevent livestock numbers spiralling out of control. For instance, since livestock were managed to maximise milk production, most male young (of all species except small numbers of camels and donkeys retained as pack animals) were slaughtered soon after birth, only a very small number being reared as breeding stock. Breeding itself was tightly controlled, mature males wearing a 'contraceptive apron' out of season, and being free to breed only when this would result in young being born during the rainy season. The absence of veterinary care meant that infectious diseases spread rapidly through the flocks and herds. Finally, the weather itself often acted as a harsh brake on livestock productivity. The long dry season and frequent severe droughts decimated livestock.

Population groupings

Apart from in some of the larger coastal villages, Soqotrans are generally divided into tribal groupings, a tribe commonly being made up of a number of clans. Clan and tribal membership is determined in principle by kinship, often in terms of ability to trace descent from a common ancestor. Individual clans and tribes, over time, gained control of a known area of land with rights of access to known sources of water. The people of the major coastal settlements of the northern plain are more diverse. They are a mix of the descendants of: traders and merchants of mainland origin; the former slave population; shipwrecked and deserting seamen; immigrants from the African seaboard; and fishermen of various origins. This cosmopolitan mix is most noticeable in the larger settlements of Qalansiyah, Qadheb and Hadiboh and its satellite villages.

LANGUAGE

Soqotrans speak their own Semitic language, Soqotri, which has a variety of local dialects. It is a pre-literate language with no written form and, with the rapid development of the Archipelago and its opening up to the wider Arab world, it is likely to become extinct. This has already happened to closely related, pre-literate languages of the south Arabian mainland, largely as a result of schooling, radio and television, all of which are in the Arabic language.

LOCAL DESCRIPTION OF CLIMATE

The climate of the Archipelago is monsoonal, heavily influenced by the major wind and water circulation movements within the Arabian Sea and Indian Ocean. During the SW summer monsoon which lasts from May to September, the wind speeds reach an average of Force 7 (Beaufort Scale), exceeding gale force for over 50% of the time, and shipping is unable to land. The winds can be magnified as they pass over the mountains, causing gusts on the northern side of the island of much greater wind speed. These hot, dry winds effectively close the island to sea traffic and fishing for the four to five months of the monsoon. The salt-laden air and salty deposits on the pastureland make livestock excessively thirsty, thereby imposing an extra burden on their herders who find it increasingly hard to find water as the water holes dry up. The winds of the NE monsoon, November to March, are much milder and bring the precious winter rains, but also in some years cyclonic storms with torrential rain. The spring transitional period between the two monsoons (end of March to early May) in most years brings the greatly valued gentle summer rains to some parts of the islands, though otherwise this is the hottest time of the year. The autumn transitional period (September to October) is the most trying time for the islanders. They have already had to survive a five-month dry season, longer if the summer rains failed in their particular area, water levels are at their lowest, and the rangeland is bare.

Seasons

The exact terminology used to describe the seasons differs from area to area according to the dialect of the Soqotri language. However, the following terms are broadly understood throughout the Archipelago. Briefly, the seasons are as follows:

***ṣerˈebhen* (*zerebhen*):** The transitional post-monsoon period between the two monsoons as the SW winds die down and before the SE winds start to blow. The first day is marked by a short period of dead calm with no wind at all from any direction. Then the winds start to shift around and clouds pile up and disperse repeatedly, the last of the southern winds alternating with the northern winds. It is a time of heavy dewfall in many areas and is also the worst time for flies. It is often very hot with periods of great stillness. Sometimes there are showers and rain in *zerebhen* bringing up vegetation that commonly causes scour in livestock.

***gyaḥś* (*gyahsh*):** The NE winds finally settle down and start to blow steadily (in mid-September). The same term is used to describe the beginning of both the NE and SW monsoon winds. Sometimes there are short heavy rainstorms during this period. If it rains now after a totally dry summer, the new grazing that comes up causes severe scour in livestock. The first rains bring further hardship. Livestock are weakened by hunger and chilled by the cold and damp, and the rains spoil any remaining standing hay before any useful new vegetation comes up. This is the period when most pastoralists begin to transhume to their winter quarters.

***ṣεrb* (*serb*):** The season of the NE monsoon that runs approximately from late September to late January or early February. This winter season falls into two halves with rains in the first half followed by a second half when it becomes steadily hotter as summer approaches. Thus the coldest period is usually the second half of November, December and early January, when most rains fall. In the early *serb* period it rains most in the high central plateaus and interior valleys followed by the Central Highlands. Towards the end of winter, it rains mostly in the lowland areas. The Central Highlands are covered in cloud and mists for much of the season and experience steady drizzle interspersed with rainfall.

ḳeyʌṭ **(*qeyat*):** The short summer season which follows straight on from *serb*, approximately from February to April. It is generally uncomfortably hot, especially in low-lying areas, but is still regarded as pleasant in comparison to the *horf* monsoon season that follows. Unless the winter rains were very poor, there is still some green on the trees and shrubs, and livestock can find their own forage.

The beginning of summer (late February, early March) is also called *nibithitin* (*nibit'hitin*) 'pollination', because it is now that the majority of the dates on the island are pollinated (along the northern plains, the Qalansiyah area, the eastern interior plains and much of the interior valleys). It is often very hazy at this time of year, especially in upland areas. Rain is rare, but any showers are called *šidmi* (*shidmi*). Even one day of rain, enough to leave puddles, revives the perennial grasses. The rains of *qeyat* are quite unpredictable, and can fall anywhere.

mən ḳeyʌṭ **(*minqeyat*):** This is a period of intense heat before the sky clouds over for the *doti* season. It is very still and extremely hot, even more so than *serebhen*, when there is often some late afternoon breeze. Towards the end of this short period, at the very end of the summer, either the sea becomes rough and the sky clouds over, or it is glassy with a hazy sky and then a fierce, short-lived storm erupts. These are seen as sure signs that the season of *doti* will now begin. This period is also called *ḥal idumhur* (*hal idumhur*) 'the time when the rain clouds begin to gather', in the hopes that rain will fall.

dɔ̄ti **(*doti*):** The transitional pre-monsoon season which runs approximately from the second half of April to the end of May or early June. Summer rains mostly fall during this period. They are unpredictable, but in areas where they occur the weather becomes cool and there are gentle breezes. In areas where it remains dry, the heat steadily increases until the monsoon breaks. It rarely rains in the lowland areas and there is high humidity in some coastal areas, where it is particularly stifling at this time.

The rains of this season are the preferred rains throughout the islands. They bring up the best and most nutritious grasses, and ensure a supply of graze and browse for much of the long, hot and windy monsoon to follow. Livestock come into season and sheep and goats are bred now to give birth in the winter. Indeed traditionally the best of all times on the island were when there had been good *doti* rains. Then livestock were fat, healthy and pregnant, dates had begun to appear on the palms so a food supply was assured for the next few months, and there was plenty of milk, dried cheese and butter. There is a lot of movement at this season, as tradition demands free transhumance of people and their animals from dry and pastureless areas into areas where good rains have fallen.

In years when the *doti* rains are light or fail all over the island, this presages drought. If the *doti* rains are good, they are said to 'protect' the islanders and foretell good winter rains. If they are poor or absent, though, it is said that the monsoon season will be very hard, and the winter rains likewise will be poor, or that the rain they bring will not cause the animals to thrive.

At the end of *doti* the sky begins to cloud over and the sea becomes restless, a sign that the *medeh* winds of the summer monsoon are about to blow.

ḥorf **(*horf*):** The season of the SW monsoon which runs approximately from June to October. The winds are hot and desiccating, and graze and browse become scarce, especially in areas where there were no summer rains. It is also the season when the majority of the date palms of the island are harvested. The season falls into two halves. In the first the winds are at their fiercest, and in the second they start to moderate. In the higher regions of the island there is constant drizzle and the mountains are shrouded in mist and cloud. In the middle of the *horf* season here the water flow in springs increases, and shrubs and trees put out new leaves. If rain falls during this season the vegetation that results is widely regarded as dangerous for livestock, as many *'uqabah*-causing plants come up, and although the rains bring up new greenery the animals grow thin and give less milk. A certain amount of cloud and mist is also experienced in the high limestone plateau areas, especially along the Noged escarpment and in the hinterland of the Central Highlands. The monsoon winds are felt most fiercely in the high plateaus overlooking the east and north coasts (at Shibereh in Momi and in Ma'alah above Qalansiyah). They are also very strong in the Central Highlands on the seaward-facing side. The winds are mildest in the interior valleys, along the southern plain, and in the lowland grazings of the interior generally.

The sea is closed to shipping and to fishing boats from *doti* to the beginning of *serebhen*, when the winds calm down and the sea once more becomes calm.

Winds

The four main winds of Soqotra are:

East wind *eʿeris* (*e'eris*): The east wind is dreaded because it is very hot and damages green vegetation, burning off young grass in particular. If it blows when there is dewfall it is especially harmful to the pastureland. Commonly it blows for three to six days, stops, and then starts up again. It can blow very strongly, especially in the winter, and, if it brings rain, this can reach storm proportions. At sea it is known for causing a strong swell and seasickness. It is most common in early summer and if it blows in the late summer, there is a risk that it will disperse the rain clouds. It is an unpopular wind on the island and to avoid the worst of its effects the entrance to living quarters, for people and livestock alike, normally faces northwards.

North wind *ṣɛrbihi* (*serbihi*): The north wind is the prevailing wind of the NE monsoon and brings winter rains. It occasionally whips up severe storms. It blows at other times of year but not when the winds of the SW monsoon are blowing. When it blows 'out of season', i.e. not in winter, it is a well-liked and cooling wind. If it blows in *doti* it sometimes brings rain.

South wind *mɛdɛh* (*medeh*): The south wind is the prevailing wind of the SW monsoon and blows for four and a half months or more. When it occurs 'out of season' it is much appreciated. When it blows in the short *qeyat* summer season especially, it brings cool and showers, and is welcomed for its effect on the rangeland, reviving both grasses and foliage. However, it can also make fishing impossible on the south coast.

West wind *maʿaribo* (*ma'aribo*): The west wind is feared, especially on the south coast where it stirs up waves, blows boats out to sea and makes fishing difficult or impossible. It can be stronger than the *e'eris* winds. It commonly blows in the short *qeyat* summer season and sometimes brings either *shidmi* or *doti* rains. It is a hot wind and harmful to tender green vegetation. When it blows in *serb* or *doti* it can bring welcome rain, usually to the central part of the island.

Any wind that blows strongly when the sky is clouding over is most feared since it may disperse the cloud or blow it out to sea.

Dewfall *ẓ́ˈerbə* (*dherbih*)

Dewfall is heaviest at times when the sea is still and calm, during *serebhen* at the end of the monsoon for instance, or at the end of the summer at *menqeyat*. It is more common in the upland areas, and especially in the Central Highlands and high plateau areas behind, and is often accompanied by morning mists. If the dry season is severe there is little or no dewfall. Dew is not seen as being of any practical use to livestock or the vegetation. It causes chills in both livestock and their owners and when the sun comes up it 'scorches' the vegetation, especially the younger, tender foliage. In the Central Highlands, dew commonly falls from summer through to *gyahsh* and is welcomed by cattle. However, goats everywhere are reluctant to browse damp foliage, and at times of heavy dew, lose condition, give less milk and seek out areas of rock to avoid lying on the damp ground. In low-lying areas, small stock tend to become sickly when the dew falls in summer and at *gyahsh*; they seek shelter and do not leave it to feed until the sun has burned off the dew.

Rain *mˈesse* (*messeh*)

As might be expected, the Soqotran vocabulary describing every sort of rain in great detail is extensive. For the islanders an ideal year of rains would be:

- *Qeyat* summer rains (*shidmi*) bringing livestock into milk. However, herders must be careful to control breeding in small stock at this time or they will give birth in the hungry months of *horf*.
- Post-summer *doti* rains bringing small stock into season; the breed males are freed to run with the females; all livestock put on weight and are healthy and well prepared for the forthcoming dry season.
- *Serebhen* or *gyahsh* rains bringing up grazing for small stock that became pregnant in *doti* and now give birth, and for cows bred in the winter of the previous year. There will now be plenty of grazing for the animals and plenty of milk and meat for their owners at a time which otherwise is extremely hard for both.
- *Serb* winter rains that are both plentiful and prolonged, bringing sufficient graze and browse to last through the summer.

Plants and rain

The islanders recognise that different plants respond differently to differing types of rainfall. For example, in the Central Highlands, the gentle *'ilhil* rains of the monsoon season bring up species of *Commelina*, *Pouzolzia*, *Pavonia*, *Solanum*, *Eclipta*, *Acalypha*, *Poskea* and *Tephrosia*. Many of these cause *'uqabah* and, as this is a period when the goats are very hungry, they tend to gorge on them and then fall ill. However, after a few weeks of these light rains, more useful plants and grasses come up. The same group of dangerous plants respond to the summer *shidmi* and *doti* rains, but at this time of year other grasses and herbs come up too and crowd out the less welcome plants.

RANGELAND

The rangeland is regarded as being shared between tribal and subtribal groups of pastoralists who have exclusive communal control over rights of residence, graze and browse, and access to water. They may allow nonmembers to make temporary use of their resources, but these incomers are there by permission only, and such permission, even if repeatedly given over a prolonged period, does not eventually confer membership of that group. Throughout the islands, there is reciprocity of access to grazing and water during periods of stress. Host groups, by agreeing to assist, expect reciprocal rights in the future. Plants which provide products other than graze or browse (such as edible parts, fibre, timber, fuel, dyes, tanning, gums and resins, medicaments and so on) also belong to the community in whose territory they grow, and are secured as though grazing were being accessed. The exception to this is products, such as aloes juice or dracaena resin, which are extracted by specialists with separately established rights, or trees that have specific property value, such as date palms.

Rangeland is broadly divided into five types:

- Mountains *fˈidəhān* (*fidihan*).
- High limestone plateaus *dimˈidi* (*dimidi*).
- Lowland grazings *śε̄təh* (*shetih*).
- Coastal grazings *ṭʌḥḥ* or *ḥāllah* (*tahh* or *hallah*).
- Watercourses *śˈaʿab* or *gˈεḥi* (*sha'ab* or *gehi*).

The pastoralists move their animals between the upland and lowland grazings according to season, weather, available graze, browse and water. The animals are entirely dependent on the naturally occurring graze and browse, cattle and sheep relying principally on grasses, with camels and goats consuming more browse. All pastoralists own goats and they say that a combination of browse and grasses in equal quantities produces the healthiest animals; a diet of grasses and herbs alone makes them lose condition.

Mountains are divided into two kinds: 'dry' mountains (especially in the east and west, such as Fidehan di Mtolo in the west or Bit Gobihir in the east) which are poorly vegetated and where water is absent or extremely rare; and 'wet' mountains (mainly of the centre, such as the Haggeher mountains) which are richly vegetated, and where lack of water is rarely a serious problem.

High limestone plateaus cover much of the island interior. They are cut by deep ravines in the west, and by broad, well-watered valleys in the centre. Such upland areas generally lack water, or where water is present, it is often brackish.

Lowland grazings are described as 'lowland' relative to 'upland' pasturelands. Thus they can be well above sea-level, such as the plains areas of Mshilhi in Momi (eastern Soqotra). To those who live in the Central Highlands, the limestone plateaus are regarded as 'lowland grazings', though to those who live below in the *shetih* areas, they are regarded as 'upland' grazings.

Coastal grazings are generally referred to as *tahh* when lying along the seashore, and as *hallah* when referring to the whole coastal plain. A coastal plain is regarded as running from the sea back to the mountain foothills (as on the northern coast) or to the base of the cliff escarpment (as on the eastern and southern coasts).

Watercourses or *wadis* criss-cross the island. Those that are broad enough for settlement are generally referred to as *śaʿab* (*sha'ab*) or *gɛḥi* (*gehi*), while the deeper ravines are called *śiṣehur* (*shizehur*).

Salt and rangeland

Salt and saline grazings are seen as vital to livestock health. A period once or twice a year grazing halophytic rangeland is considered essential, for the health of small stock and camels especially. Indeed, there is considerable evidence, for camels at least, that unless they spend at least one-third of their time browsing on saltbush species such as *Atriplex*, *Salsola* and *Suaeda* their intake of salt is inadequate for good health. The animals are taken to feed in suitable areas such as the coastal *tahh* grazings, which also provide excellent *'aazel* grazing (see below). Areas that suffer strong monsoon winds are also regarded as 'salty' due to the fine coating of salt deposited on the vegetation by the wind. However, these are not seen as being as beneficial to livestock as are the coastal grazings.

However, excessive salt has to be avoided. Grasses and herbs growing in *Croton socotranus* shrubland are little grazed during the monsoon or in early winter because the droplets of salty water dripping from the leaves and stems of the trees above make them too salty. (Indeed herders who had no salt for cooking were able to make emergency use of *Croton* foliage at such times.) Similarly, livestock herders who live along the coast always remove their animals from the coastal grazings and take them inland or to the upland grazings for part of the year to give them a rest from the salt, and herders who spend much of the year on the plateau overlooking the southern Noged plain move their animals away from this area as soon after the monsoon as possible. Towards the end of the monsoon the amount of salt in the air increases, and the salty rime that develops around the trunks of the trees at this time is a sign that the winter rains will soon come.

During the monsoon season itself, small stock ideally are not moved. This is because the increase in salt in their home area does not trouble them, but they suffer if moved to a different 'salt' area at this time. Even goats accustomed to one kind of 'salt' area suffer initially if moved to a different 'salt' area (for instance, from a 'salt' coastal plain to a 'salt' high plateau). In general small stock native to a 'salty' area cope well, but incoming goats develop diarrhoea, lose weight and often suffer from sores on the skin. However, this is seen as being beneficial and as 'removing impurities from the blood'. The animals eventually recover; they put on weight and begin to grow fat and healthy with glossy coats. Cattle especially do not thrive on salty grazings if left there for too long, though they can benefit from a short period. For instance, the cattle of the Central Highlands are taken down to the limestone plateau area below especially for this reason.

It has been suggested that some of the plants of the saline grazings have a vermifuge action, which could explain the animal's initial poor condition and subsequent recovery.

GENERAL FORAGE TERMINOLOGY AND CONCEPTS ABOUT PLANTS IN THE ISLANDS

Graze and browse

Grazing and browse plants are divided into two basic categories:

- *ruʿud* (*ru'ud*): generally annual plants, grasses, and small perennial plants;
- *širəhām* (*shiriham*): generally trees, shrubs, and woody perennial plants.

The principal defining characteristics are that '*ru'ud*' plants come up after rain and die back in the dry season. This includes all Gramineae and Cyperaceae, the more ephemeral herbs, small green ferns and the aerial parts of bulbs and corms. Another characteristic of '*ru'ud*' plants is that when emergent they can be classed as *ḳəṣeho* (*kiziho*) (see below). Immature *ru'ud* plants are called *raʿadihon* (*ra'adihon*), a diminutive form of the term *ru'ud*. Some herbaceous plants, for example species of *Ruellia*, *Pergularia* and *Aerva*, are somewhat on the borderline, sometimes being included in the *ru'ud* plants, sometimes being classed as 'small *shireham*' or *širemhən*, the diminutive form of the term *širəhām*.

Widely used categories to describe groups or types of plant

***ḳəṣəho* and *śaʿar* (*kiziho* and *sha'ar*):** *Ru'ud* plants develop from *kiziho*, the first green shoots of the plant emerging from the ground after rain. At the end of their life they become *sha'ar*, dried-out herbs and standing hay. *Kiziho* grazing produces the best milk.

***ritrit*, *rižəbeb* and *ʿažerib* (*ritrit*, *rijibeb* and *'ajirib*):** These terms are used by cowherds to describe *ru'ud* herbs that are beneficial to their animals, and by shepherds and goatherds to describe plants that can be harmful to their animals if they dominate the vegetation or if the animals are not accustomed to them. The terms are more common in the better-vegetated areas of the island, especially in cattle areas. They are also sometimes used to describe the same plants at different stages of development, *ritrit* being the lowest growing and *'ajirib* the tallest. These plants are rare after the lighter summer rains, but are widespread in upland areas in the winter, and in response to the monsoon rains and mists of the Central Highlands.

All these plants tend to be rather weedy and come up around settlements, caves and livestock quarters. They are also well-known weeds of cultivated areas. If shepherds and goatherds are discussing the state of the rangeland in an area, they will not transhume there if told that these plants had come up, even if their animals are hungry. This is because they are short-lived, and once the clouds disperse and the sun comes out quickly become limp and die back. Also, without other forage, they are not seen as valuable grazing for small stock, and can be harmful to sickly stock or to animals unaccustomed to them, causing scour or *'uqabah* (see below). Even cattle in poor condition can develop severe diarrhoea from eating too many of these plants if there is little other forage. Once the grasses and woody herbs have come up, this group of plants is no longer harmful to small stock.

***ritrit* (*ritrit*):** Immature herbs that grow in thick patches or spread to form mats, such as Commelinaceae, Convolvulaceae and Portulacaceae. They are much liked by cattle, but many cause *'uqabah* (see below) and scour in small stock if there is little other vegetation present.

***rižebeb* (*rijibeb*):** Annual, short-stemmed and leafy herbs which either do not grow tall or which have not yet reached full height. Examples are smaller species of

Compositae, Euphorbiaceae, Geraniaceae and Scrophulariaceae.

***ʿažerib* ('ajirib):** Annual, long-stemmed and leafy herbs, many of which when immature are called *rižəbeb* (examples are taller species of Compositae, Euphorbiaceae, Cruciferae, small species of Acanthaceae, Amaranthaceae, and Urticaceae).

***ʿašbub* ('ashbub, pl. 'ashebeb):** Many of the cattle herders of the Central Highlands distinguish this group, which they describe as growing in 'sticky black soil' or in the 'crumbly black soil' of cattle paddocks and around byres and cattle handling areas. A diet of these plants is seen to improve lactation in cows and to give the milk a recognisable and sweet flavour. However, as with the other plant groups described above, cattle in poor condition can develop severe diarrhoea from eating too many of the plants when other forage is scarce.

***kabs* and *təbəb di ḥa* (*kabs* and *tibib di ha*):** Forage of little or no intrinsic nutritional value but which provides a bare maintenance or survival feed for livestock. This is called *kabs* because it *təkōbis mer* 'packs the stomach' and gives hungry animals the feeling of 'having something in the belly'. *Tibib di ha*, 'causing salivation', is a small amount of feed given to a hungry animal to keep it going.

***yəʿōbərin* or *taʿabəhor* (*yi'owbirin* or *ta'abihor*):** Particularly shrubs and trees, which are either evergreen or which come into leaf at the end of the monsoon season, before the winter rains. As such, they can be an extremely valuable source of forage to livestock in the drier areas of the island. Examples are *Adenium obesum*, *Tamarindus indica*, *Ziziphus spina-christi*, *Ficus vasta*, *Hibiscus* species, *Dovyalis abyssinica*, *Carphalea* species, some *Boswellia* and *Commiphora* species, *Anisotes diversifolius*, *Jatropha unicostata*, *Dorstenia gigas*, *Ballochia* species, *Trichocalyx* species, *Euclea divinorum*, *Maerua angolensis*, *Capparis cartilaginea*, and *Acridocarpus socotranus*. In the better-vegetated highlands, others come into leaf at the very end of summer and in the early monsoon period. Examples are *Carphalea* species, *Ballochia* species, *Trichocalyx* species, *Anisotes diversifolius*, *Hibiscus* species, *Sideroxylon fimbriatum*, *Spiniluma discolor*, *Cryptolepis macrophylla*, *Pittosporum viridiflorum*, *Hypericum balfourii* and *H. scopulorum*, *Turraea socotrana* and *Tragia balfouriana*.

***šeber* (*sheber*):** Refers to a group of 'bitter' plants, usually with copious, often poisonous, milky exudate. Examples are the tree euphorbias, smaller euphorbias such as *E. schimperi* and the various forms of *E. schweinfurthii*, and *E. spiralis*. *Adenium obesum* and the various *Cissus* are also included in this group. An area with many such plants is referred to as '*shibiher*', and honey from these places has a recognisably bitter aftertaste.

***dūr* and *śaḥaz* (*duwr* and *shahaz*):** Any plant exudate is generally called *duwr* 'blood'. If the exudate is a gum or resin, hardens on exposure to the air, and can be chewed, it is called *shahaz*, for instance the gum resin of *Boswellia* and *Commiphora* species.

***ɛrsīmo* (*ersiymo*):** Refers to the widespread *Croton socotranus* shrubland at lower altitudes where grasses such as *Aristida adscensionis* and *Eragrostis* species dominate.

***šʌṭeher* (*shateher*):** A term used to describe areas of pure grassland in the Central Highlands. ('*Shateher*' otherwise describes the meat of a carcass freed from bones, offal and fat.) Such grassland areas can occur naturally, or are a result of clearing rocks and shrubs from the area at some unspecified time in the past.

***kor inābut* (*kor inabut*):** This means 'so that it may recuperate and come into flower' and refers to the reason underlying the traditional seasonal transhumance between upper and lower pastures, i.e. to allow the rangeland to recover.

***ṭʌmhum* or *ṭʌmhur* (*tamhum* or *tamhur*):** The dense shrubland of the high pinnacles of the Central Highland mountains composed of spiny plants growing so closely together that they present an impenetrable barrier.

***ʿūd əl kəsīr* ('*uwd il kisiyr*), sometimes *ʿabd əl ḳəsīr* ('*abd il kisiyr*):** A mythical plant or group of plants that possess magical qualities that influence the productivity of the animal eating them. They increase the production of fat in livestock and of butterfat in their milk. Even the smallest fragment of plant in the milk-skin will make prodigious amounts of superlative butter. A herder may suspect that his animal, commonly a goat, has eaten some of this plant because it will become exceptionally fat and its milk will change colour and become especially rich. Any herder noticing such changes in one of his animals will do his best to track it to see if he can find the plant. They are rare and found only in the most inaccessible places such as cliff ledges. Ravens are said to search out these plants to use as nest-building material. However, looking around under their nests for fallen plant fragments is useless, because if any part of the plant falls to the ground it promptly vanishes. There are many stories current in the islands about people coming across pieces of these plants and the good fortune that results.

fīlʌṭ and *ʿāṣəl* (*fiylet* and *'aazel*) grazing

Over much of Soqotra the rangeland is divided into *fiylet* and *'aazel* grazings (these are the most widely understood terms, but others exist). *Fiylet* grazing tends to be in the upland areas and *'aazel* in the lowland areas. *Fiylet* rangeland tends to include a lot of trees and shrubs, while *'aazel* grazings are mainly made up of grass and herbs.

Fiylet grazings are seen as being 'weak' and 'cold'; they are said to produce an animal with friable bones, one whose milk is low in butterfat, and on slaughter, a carcass with little fat and less palatable meat. *Fiylet* grazings are always associated with the greatest cold and damp.

'Aazel grazings are seen as being 'strong' and 'hot'; they are said to produce an animal with strong bones; a healthy animal whose milk is high in butterfat, and on slaughter, a carcass with plentiful and firm fat and meat. It is said that if a goat has been feeding on *'aazel* grazing, 'a big rock will be needed to break open its bones to get at the marrow'. The term *'aazel* comes from a root with the basic meaning 'of noble origin, good blood', 'genuine, authentic'. *'Aazel* grazings are always associated with the greatest heat and dryness.

The cold/hot duality is clearly illustrated by the difference in incidence of *'uqabah* (see below) in the winter rains grazing and that resulting from the lighter summer rains. The latter are said never to cause *'uqabah*, even in upland pastures.

'MALE' AND 'FEMALE' WOODY PLANTS

Trees and shrubs are often described as being 'male' or 'female'. This distinction is generally made on several criteria (though some plants have more specific 'sexual' characteristics).

'Male' plants are thin and tall with small and deeply wrinkled leaves and usually smaller flowers. They often do not fruit or the fruit has no or few seeds. If they have gum or latex they generally produce less exudate. In thorny species 'male' plants are extremely thorny and in all species where a male/female dichotomy is recognised the wood and bark are harder than in the 'female'.

'Female' plants are fat, short, have broader and smoother leaves and usually larger flowers. They always bear fruit, which has many seeds. Gum or latex is produced in copious quantities and thorny species have fewer and softer thorns than 'males'. The wood and bark are softer and easier to chop or break.

Some plants, such as *Maerua angolensis*, *Adenium obesum*, *Ficus salicifolium*, *Flueggea* species, *Euphorbia socotrana, Hibiscus* species, *Vernonia* species, *Euclea divinorum, Dendrosicyos socotranus*, *Spiniluma discolor*, *Cryptolepis* species, *Boswellia socotrana*, *Ballochia* species, *Trichocalyx* species, *Buxanthus pedicellatus*, *Jatropha unicostata*, *Lycium sokotranum*, *Paracalyx balfourii*, *Clerodendrum* species, *Angkalanthus oligophylla* and *Placopoda virgata*, are not generally regarded as possessing this male/female dichotomy.

PLANT COMMUNITIES

Like botanists, the islanders commonly group plants into 'families'. Members of the same family are likely to occur together (for instance, *Euphorbia schweinfurthii* species with *Ochradenus* species, or on the high limestone plateaus, *Punica protopunica*, *Jatropha unicostata*, *Buxanthus pedicellatus* and *Zygocarpum coeruleum*, or, in raised plains areas, *Asphodelus* with the grasses of the *di qertomo* and *'ibihir* groups). Other plants are recognised as being 'solitary', such as *Ocimum forskolei*, *Lavandula nimmoi* or the *'uqabah* plants.

Habit, altitude and soil

Islanders attribute many of the differences in appearance between plants otherwise regarded as being closely related to differences in habit, type and depth of soil, and altitude. The underlying rock also plays a part; thus plants growing on granite look different from those growing on limestone, which in turn little resemble those growing in gravel or sand. Differences in the age and 'sex' of a particular plant also contribute to disparities in appearance. This is discussed more fully under 'Soqotran plant names' below.

'UQABAH AND *'UQABAH* PLANTS

'Uqabah is a sickness of livestock, especially goats. It is believed by island herders to be caused by (a) dew and damp grazing in general and especially that of *fiylet* grazings; (b) certain plants. The scientific basis of this disease is not known, and no diagnosis has been made, but it is suggested (Ian McLeish pers. comm.) that it might be either *Pestes des petits ruminants* or *Mycoplasma* F38. It has also been suggested (A. Hunter & B. Hosie pers. comm.) that it may be a complex type syndrome, with possible contributory causes of *CCPP*, *Coccidiosis*, *Helminthiasis*, *Besnoitiosis*, *Paramphistomiasis* and perhaps also certain mineral deficiencies. Only laboratory serology analysis will determine the exact nature of the disease (or group of diseases).

'Uqabah and *'aazel* and *fiylet* grazings

Sheep are much less affected by *'uqabah* than goats. Unlike small stock, cattle thrive in both *fiylet* and *'aazel* grazings. Goats accustomed to *'aazel* grazings nearly always fall ill when moved to *fiylet* grazings, though the reverse is not true. Goats native to a *fiylet* area do not get *'uqabah*, only incoming goats. *'Aazel* areas can never cause *'uqabah* even when *'uqabah* plants are found in small numbers in such areas. The most superlative *'aazel* grazings of all are those of the coastal and interior plains. Whereas in the interior lambs and kids commonly develop a heavy worm burden, this is rare in *'uqabah* grazings. To the untutored eye, the best *'aazel* grazings look hot, hard, stony and unproductive, but it is in just such areas that goats do best. Here, the islanders say, plants grow only very slowly and are tough and resilient.

The *'uqabah* illness

Its main signs are general loss of condition, loss of weight, a distended stomach, and often also diarrhoea. The animal makes an unusual but characteristic grunting sound deep in the throat, and if the goat is pregnant, abortion is likely to occur. Milk yield drops sharply and the milk is rather thicker than usual.

'Uqabah is especially prevalent in periods of drought, and during the first rains of the winter season when goats are stressed by the long dry season. It is not immediately apparent that an animal is suffering from *'uqabah*, as the signs take some time to appear. Sometimes it only becomes apparent when a goat is hand fed in the dry season. If it is fed on lopped *Ziziphus spina-christi*, *Euphorbia arbuscula* or *Dichrostachys dehiscens*, instead of improving in health, a goat with *'uqabah* develops severe diarrhoea and this can be fatal. Indeed, if a goat being fed *Euphorbia arbuscula* sickens, this is taken as evidence that it has the disease, since goats normally grow fat and healthy on this diet.

Diagnosis is confirmed on slaughtering the sick animal, when sores and pus-filled blisters (*qamtat*) are seen in the stomachs (*bi mer*, *šeriś*, *di rīḳob*, *ərbʌź*, *dʌrz*) and intestines, especially the latter. The pustules exude a watery pus-like material on being squeezed. Usually worms are also present, and the stomach is commonly filled with fluid. It is said that if the blister-like sores are found only on the outer wall of the intestines, the disease is less advanced, but if they have penetrated the intestinal wall, then it is severe. Even kids of one to two months get *'uqabah*, and if they are slaughtered are seen to be full of worms.

Experts say that there are two kinds of *'uqabah*: the one described above which results in raw-looking sores and adhesive patches in the digestive tract, and another which is accompanied by a heavy burden of intestinal worms, both 'thread-like' and 'worm-like'. This worm infestation is sometimes also accompanied by blistered patches in the lungs, which are distended. This is a well-recognised condition in goats, producing hoarse night-time coughing with audible stridor. The meat of such an animal is usually inedible, and only the liver, tongue, heart, brain and kidneys are eaten. Untreated, *'uqabah* remains with the animal for two months or more before it finally disappears.

Cattle develop similar sores in the intestines, mainly in very cold and wet weather. These are said to be caused by grazing heavily on plants such as Cyperaceae of damp areas, and the spiny *sherishiroh* group of plants (such as *Medicago sativa*, *Tribulus terrestris* and *Acanthospermum hispidum*). Sometimes the cushion-forming plants of the *kibidinoh* group, such as *Kohautia subverticellata*, *Aizoon canariensis*, *Pulicaria diversifolia* and various *Oldenlandia* species, are also implicated. Unlike a goat with *'uqabah*, externally the cow is unaffected, her appetite remains good and her milk unchanged. The small *qamtat* found in the intestine are soft, odour free, friable and can be popped between the fingers. They are dark-coloured and contain a pale, pus-like substance inside.

'Uqabah and plants

There is a widespread belief that *'uqabah* is caused by grazing certain plants, which goatherds call 'the cattle plants'. These are more common in the higher, cooler pastures than in the drier, lower ones. It is said that the recent rise in cattle numbers, coupled with the decrease in restrictions on livestock movement, have led to the migration of such plants from the high grazings to the lower ones. It must be said, however, that some elderly and more reflective herders are sceptical about the claim that the signs and symptoms of this condition are caused principally by plants.

Some *'uqabah* plants are seen as more pernicious than others: the *di 'aardehan* and *di hebaalinoh* plants are blamed all over the island, while the *qowquh* and *ezeninoh* group are seen as especially dangerous in some areas but not others. If there are good, heavy rains, the *di 'aardehan* plants especially become a serious problem as they spread very rapidly under these conditions. Some plants, such as *'ibelit*, *'irsib*, *leza'han*, *qowquh* and *ezeninoh*, are seen as causing the problem only under certain conditions. In general *'uqabah* plants are safe when immature, becoming dangerous only when in flower or fruit. When rains grazing first comes up in the *fiylet* areas where the *'uqabah* plants occur, small stock eat all fresh greenery indiscriminately. They generally flourish on this green pasture, even when it contains young *'uqabah* plants. However, *'uqabah* plants commonly remain green longer than the grasses, and it is at this time, when the grasses and more palatable herbs begin to die back, that livestock turn to them.

'Uqabah plants have something in common: they are either spiny or prickly, or have stellate hairs, characteristics which might disturb a delicate digestive system. Or, they turn to a sticky slime when rubbed between the fingers, and herders say that it is this which damages the animal. Grazing on some *'uqabah* plants, such as *Commelina* species, when they are green and lush is said to cause intestinal worms, though once they have dried out, these species are a safe and nutritious forage.

Damp and cold; dew; unseasonal rains

A condition similar to *'uqabah*, and thought by some to be another kind of *'uqabah*, occurs when the pastures are damp from heavy dew, mist, low cloud or unseasonal rain. All herders agree that dew is harmful to goats. Traditionally small stock were kept penned with a smudge burning until any dew had dried off. Dewfall is most common in the hottest summer months and at the very end of the monsoon, a time when animals are already stressed. With the exception of the goats of the Central Highlands, which are accustomed to damp forage, goats will not feed on damp graze or browse, and generally eat less when the weather is damp or cold. If it rains during the monsoon, in upland pastures especially, the resultant grazing causes *'uqabah*, even more so if there were summer rains. The goats are hungry at this time, and lack their usual discrimination.

Scour and *'uqabah*

A variety of plants are known to cause diarrhoea in small stock. For example, *Dovyalis abyssinica* fruit cause shivering attacks accompanied by severe diarrhoea, and *Portulaca* species and *Orthosiphon ferrugineus* can cause chronic diarrhoea. Small stock develop scour when they feed on the first flush of new grazing brought up by winter rain. This is seen as normal, however, and not a sign of *'uqabah*. Unless in very poor condition, the animal recovers in time without human intervention, and on slaughter no intestinal or lung damage is seen.

Treatment of *'uqabah*

The first attack of *'uqabah* is the worst and the hardest to treat. Thereafter the animal in some way becomes acclimatised, and with care, the attacks are less severe and prolonged. The key to treatment is heat, as much and for as long as possible.

The most successful and widespread treatment is called *ṣɔ̄wnaḥ* (*zownah*), meaning 'to extract moisture'. In its simplest form *zownah* involves moving the affected goats from *'uqabah* grazings to a hotter and drier area: down to the *'aazel* grazings in the plains, or, in the hotter months, up to the hot, dry mountains. The affected goats are left there for a month or more. If there is adequate *'aazel* grazing, they will slowly recover. If the sick goats are not moved, and if they have sickened in the long dry season, they become progressively thinner and stop giving milk.

A more rigorous form of treatment involves herding the goats into a cave or covered pen made as draught-free as possible by blocking up any gaps. The smaller the area and the closer packed the animals, the more effective the treatment. A fire is then lit inside, using woods known to produce the greatest heat. It is kept burning until the animals begin to pant and their noses to run, a reliable sign that the cure is taking effect. Traditionally, many herders used to do this for a few hours every morning in winter before letting the goats out to graze. Careful goatherds would do this whenever the weather was cold or damp, or if there was dew, and the island-wide rise in the incidence of *'uqabah* is attributed to the collapse of this tradition. Some goatherds now herd their goats to the open goat-pen in the morning and leave them there in full sun before milking them in the hopes that this will have the same prophylactic effect, but goat experts say that this is never as successful.

Drinking salty water also helps to cure *'uqabah*. In general, goats and cattle are said to do better on brackish water. The saline water of some well-known coastal wells, such as the Bi'r Guwhah well in Nowgiher, beside Qadheb, used to be sold to goatherds for this reason. It is one of the reasons for the popularity of the Abd al Kuri goats on Soqotra since all the available water on that island is brackish.

SOQOTRAN PLANT NAMES

Collecting and interpreting vernacular plant names is always complex. On the Soqotra islands there are several major difficulties:

- It is common for the same plant to bear several different names and, conversely, for different plants to bear the same name.
- Some names refer to a group of taxonomically different species that are nevertheless locally seen as having something in common.
- Variant names for the same plant appear in the different localities and sometimes even in the same locality. Thus *Flueggea* species are variously called *ᶜudf*, *ᶜutf*, *ᶜufd* and *ᶜuˈft*; *Ochradenus* species appear as *gerśīˀo* and *gēśīˀo*; *Leucas virgata* as *nōḥʌl* and *nōmḥel*; and *Dendrosicyos socotranus* as *ḳˈʌmhīn* and *ḳˈʌmhīm*.
- Differences in pronunciation of a name can suggest that different plants are being referred to. However, many dissimilarities reflect the phonetic rules operating within the different dialects. Thus in western areas, *ḥˈimiher* for *Zygocarpum coeruleum* and *Chapmannia* species becomes *khˈaymǝher*; *mǝśḥˈermǝhim* for *Caralluma socotrana* and *Edithcolea grandis* becomes *mǝś'khˈayrim*; *ṣˈaᶜalhil* for *Metaporana obtusa* becomes *ṣˈaghalhil*; and *ᶜekḳǝhīn* for various asclepiad climbers becomes *ġˈekḳǝhīn*.

Nevertheless, allowing for regional dialectical differences and minor differences in pronunciation, the names for many of the 'key' plants are the same all over the islands. For instance, *ˈiśḥib* for *Maerua angolensis*, *sˈǝbǝra* for *Jatropha unicostata*, *mˈiṭǝhʌn* for *Buxanthus pedicellatus*, *ṭayf* for aloes, *aᶜarhīyib* for *Dracaena cinnabari*, *zūrif* for *Rhus thyrsiflora*, *zˈirkin* for *Allophylus rubifolius*, *żˈoˀod* for *Ziziphus spina-christi*, *śīhiṭ* for *Carphalea* species, *mˈitrer* for *Croton socotranus*, *ˈikšǝh* for *Commiphora ornifolia*, *ˈitˀib* for *Ficus cordata*, *ṭiḳ* for *Ficus vasta*, *kōlǝ* for *Euclea divinorum*, *lǝṣāfǝ* for *Capparis* species, *ᶜǝrǝhik* for *Salvadora persica*, *bōhīn* for *Sterculia africana*, *lˈekǝhem* for small-leaved *Commiphora*, *ˈimṭǝhe* for *Euphorbia arbuscula*, *ḥfˈoˀ* for *Vernonia* sensu lato species, *ṣˈubǝhur* for *Tamarindus indica*, *ᶜˈašbak* for *Euphorbia schweinfurthii* species complex and *ḥˈaymǝher* for *Zygocarpum coeruleum* and *Chapmannia* species. The same names are also used all over the island for some of the useful herbs, such as *ˈirǝhem* for *Pergularia tomentosa* and *fˈǝḳǝla* for *Glossonema revoili*.

Another problem is that local plant experts do not necessarily distinguish between superficially similar plants as botanists do; conversely they can occasionally differentiate between plants which botanists regard as identical. Both over-differentiation and under-differentiation occur, but the former is less common, and most typically encountered with culturally important plants. This can be due to differentiating between 'male' and 'female' plants according to a system which is foreign to taxonomists, for example *ˀˈeggǝhur* for the 'male' *Euphorbia arbuscula* and *ˈimṭǝhe* for the 'female'; or it can be a result of plants having alternative names at different stages of development according to whether they are more or less palatable to livestock. It is not that local experts do not appreciate the differences between certain plants as described by taxonomists, but they attribute these to differences in altitude, habitat and type or depth of soil, not to differences in plant species.

Examples of categories of plant name

A similar name with a female suffix is used to differentiate similar plants perceived to have different characteristics. Thus *ḥˈimiher* and *ḥamirmˈiro* are used for larger and smaller *Chapmannia* species and *Zygocarpum coeruleum*; *ḥˈarśīm* and *ḥariśˈemo* are used for larger and smaller *Grewia* species (with even smaller specimens bearing the diminutive form of the same name, *di ḥereśśiyun*); *ḥˈamḥam* is applied to various small, spiny 'white' plants (such as *Dicoma cana*, *Campylanthus spinosa*, etc.) and *ḥamaḥˈamo* refers to various small, spiny 'black' ones (such as *Neuracanthus aculeatus* and *Rhinacanthus scoparius*).

The same name is used for different plants with similar characteristics. Thus *ḥˈarḥeyr* is used for thorny acacias as well as for *Dichrostachys dehiscens* or the spiky *Commiphora habessinica*; *mǝdˈǝwhīn* is used for *Bidens* species, but also refers to the spiked seeds of some grasses and other herbaceous plants; *ᶜˈʌlǝhʌl* is the name for both *Anisotes diversifolius* and *Trichocalyx* species; *di ḥśāwǝ* refers to all herbs producing cucumber-like fruit; *ḳˈʌlḳǝhʌl* is used for a wide variety of halophytic plants; *ginsˈibiher* or *gesˈibiher* is the name for all large ferns in the Central Highlands; and *zˈeydiroh* is used for a variety of prickly or spiny, low, greyish herbs, such as *Indigofera marmorata*, *I. pseudointricata*, *Seddera spinosa* and *Campylanthus spinosus*, and also for stiff, perennial grasses such as *Dactyloctenium* species or *Panicum rigidum*.

The same name is used for different species within a family, even when they can appear dissimilar. Thus it is common to call all *Cissus* species *ᶜˈaṭerhe*, all *Hibiscus* shrubs, regardless of size, *di rˈʌffʌ̄n*; and all asclepiad climbers *ᶜˈekḳǝhīn*.

There is commonly a marked difference between the eastern and western names for the same plant. Thus *Leucas virgata* is called *ḥarˈiṭreṭ* in the east and *nōḥʌl* or *nōmḥʌl* in western Soqotra; *Gnidea socotrana* is *ḳˈiṭǝhǝr* in the east and *lǝˀēyǝf* in the west; *Adenium obesum* is *tǝrīmo* in the east and *ǝsfǝd* in the west; and *Punica protopunica* is *rihīni* in the east and *mˈiśǝṭ* in the west.

Sometimes the same name appears in different guises though based on the same root letters. Thus Iridaceae and Hyacinthaceae are variously called *bǝṭˈaḥ*, *di beṭˈaḥun* and *bǝṭˈaḥiyun*; and Portulacaceae appear as *ˈirsib*, *irsˈibihon* and *mǝrsūbǝhān*; and a variety of unrelated sticky or clinging plants are called *lǝsēyʌḳ*, *lūwsʌḳ* or *meltūwseḳ*.

Uncommon plants commonly have a wide variety of possible names, reflecting their relative rarity. Thus *Turraea socotrana* can be called *ḳˈerᶜʌš*, *bˈǝššǝ*, *ḥalˈebleb*, *mǝstˈǝwrǝher*, *kˈǝnǝhor* or just *śirǝhām ḥˈahǝr* ('dark-coloured tree'); and *Jasminum fluminense* has several names, *ḳˈerᶜʌš*, *rˈikǝbǝ*, *niśˈazǝ* and *bˈǝššǝ* again (this name is given to various plants with either dramatic flowers or which grow in cliff habitats). Interestingly, the same can be true of extremely common but unimportant plants. Thus the widespread plant *Convolvulus sarmentosus* is variously called *ḥalēbˈoh*, *ḳˈʌzḳʌz*, *tˈaᶜabǝhur*, *ḥalebˈeboh*, *mˈʌḥǝriś*, *sedˈʌḥan* and *ḳʌˀ* (or its plural *ḳōwnˈitin*).

The Soqotran language is distinguished by its widespread use of diminutive forms, and plant names are no exception. Thus as a group, the larger Cyperaceae are called *ḥˈǝwgeb*, and the smaller ones *di ḥǝgˈubgub*, the diminutive form; the tamarind with its edible sour fruit and foliage is called *ṣˈubǝhur*; many species of small, sour-leaved plants (such as *Begonia* and *Oxalis* species) which are chewed for their lemony flavour or used to curdle milk are called *ṣebirbəher*; the many unpalatable *Tephrosia* species are called *ṭifher*, and smaller unpalatable legume species *ṭefəririhun*. However, one can fall into the trap of taking it for granted that two related plant names bear this relationship. For example, the small plant *ṭaʿiyiḳ* is not just a possible diminutive form of *ṭiḳ*, *Ficus vasta*, but also refers to a group of herbs, such as *Pouzolzia auriculata* and *Parietaria debilis*, which flourish after the rains and cause severe diarrhoea; *ṭaʿiyiḳ* in fact describing the very sloppy droppings which result in animals that eat them. The preferred diminutive form for a young fig tree is *ṭɔ̄wḳʌḳ* and this appears in many island place names.

Some names are used to refer to more than one species of small herbs that are of little importance, either culturally or as livestock fodder. Examples of such names are *mʌhariś* for plants such as *Nanorrhinum hastatum*, *Ruellia* species, *Priva socotrana* and *Hypoestes pubescens*, and *ṣəfṣifo* for plants such as *Fagonia luntii*, *Teucrium sokotranum*, *Kohautia* species, and *Polycarpaea* species. Many of the smaller, inconspicuous plants which cling to the bare rock surface of drier areas, and which are greyish in colour, are loosely grouped together under such broadly descriptive names as *ḥʌlḥʌləhin* ('of a grey-brown colour'), *libənhan* ('pale, white'), *mərḥɔ̄mhəm* or *di raḥimimihon* ('grey, ash-coloured'). Similarly, small plants growing near water have such general names as *di-rihoytin* ('of the water') or *ṣefiririhon di gɛḥi* ('little flowers of the *wadi*'); and many varieties of small, clump-forming plants which become covered in tiny flowers after rain are simply called *di ṣʌfəririhon* ('many-flowered').

Some plants are grouped together in 'families' that bear little relation to taxonomic families. For example, the group containing plants such as *Caralluma socotrana*, *Edithcolea grandis*, *Euphorbia spiralis*, *Orobanche cernua* and *Echidnopsis* species is referred to in general as *məśḥermihim*.

Some plants are of such key importance that their name has entered the general vocabulary. Thus the important Amaranthaceae food plants are called *ʾibəlit*. There is also a verbal form that describes the activity of harvesting the plant, and another nominal form for the pot in which they are cooked. Similarly the thorny *Lycium sokotranum*, *suʿhur*, so important for erecting barriers, enclosing and fencing, has entered the vocabulary as a verb meaning 'to enclose or divide an area with thornbush material'.

Plant names are often informative

Names can be used to indicate many attributes of plants: where and when they occur; the appearance of the fruit; their taste; whether they are delicate and sweet-smelling, or whether they are unpleasant, poisonous or useless. The name can also refer to the general appearance of the plant and, more specifically, whether they can be differentiated by the size of their leaves. Names can also indicate what the plant can be used for; its cultural significance; its use as livestock fodder; which animal most likes to eat it; or a specific quality of the plant. An understanding of plant names can be of use in understanding place names and in clarifying taxonomic problems. Some examples are given below.

Where the plant occurs: *nɔ̄wḥel ḥegirhi*, *'nɔ̄wḥel* 'of the Central Highlands' distinguishes *Leucas penduliflora* from other *Leucas* species, as does *fəʾ ḥagərhīyo*, *Aerva lanata* from the more widespread *A. javanica*. *ʿašbak dimidi* ('of the limestone plateau') for *Kleinia scottii* distinguishes it from *ʿašbak* (*Euphorbia schimperi*). Various plants are called *di ʾitəhi*, *di tīyə* ('cliff-growing'), and *śḳaʿhinitin di məźɔ̄wźʌ* ('prickly plant of damp places') is used for *Leucas spiculifolia*.

When the plant occurs: The group of edible leafy herbs that come up after the winter rains (*ṣɛrb*) are called a *məṣirbihon*.

Taste: Sour-tasting, astringent plants are commonly called names such as *ḥʌmź*, *ḥɔ̄wmʌź*, *ḥəmʌźəhen* or *di ḥʌmźəhʌn*, reflecting their similarity in taste to *'ḥʌmź'*, soured milk; or they are called something like *ṣəberhon*, *ṣʌberbəher*, *ṣʌbəher* or *məṣirbihān*, 'sour'. Similarly salty plants such as *Dyerophytum pendulum* are given names such as *əmlaḥ* from *milḥo*, salt.

Delicate and sweet-smelling plants: Before sugar was available, dates with nectar and wild honey were the only sources of sweetness. So perhaps it is not surprising that 'date' (*'tāmər'*) appears as a name for certain appreciated species: plants called *timihor* or *tamerhen* tend to be delicate with sweet-smelling flowers, and are usually regarded as excellent grazing. Herbs with the name *fegəhun* or *nɔ̄i* also have perfumed flowers or leaves. On the other hand, a name can also warn of the opposite. Thus the common name for all the foul-smelling *Gaillonia* species is *źerṭhanitin* ('farts') or *di ṣīnem* ('with a foul smell').

Unpleasant, poisonous or useless plants: Such a plant is often named after something despised or feared. Thus, the raven (*aʿarib*), a troublesome predator of weak livestock, appears in the name, *di aʿarib* ('belonging to the raven'), for the showy, large *Crinum* species, which have unpalatable broad leaves and an inedible bulb. The feral or wild cat (*girbaḳ*) attacks young livestock, and any plant name with the epithet *'di girbaḳ'* is likely to be inedible and probably also have some other unpleasant characteristic. *Edithcolea grandis*, a prostrate succulent with hard spines and an evil-smelling flower, is called *məśḥerməhum di girbaḳ*, or *girbiḳhitin* ('wild cats'). Unpalatable ferns are commonly called after the feared giant centipede (*di ḥʌṣʌhʌn*). Other creatures also appear in names: the Egyptian vulture (*sūʿīdoh*) lends its name to *ʾīdʾəhʌntən di ṣūʿīdoh* ('vulture's ears') for all sorts of toadstools. Edible and inedible species can be distinguished in this way; for instance, men tend to have a low opinion of sheep (in contrast to women, who look after them), and differentiate between *tabaʿš*, the edible *Echidnopsis*, and *tabaʿš di tɛtin*, 'sheep's *Echidnopsis*', which are not edible.

Appearance: Examples are *ṣaʿələffən* ('with tiny leaves'), *gifiʾhin* ('stunted, twisted') and *śḳaʾḥinitin* or *śḳaʿinitin* ('thorny', 'spiny'). Small cushion-forming plants are commonly called *di kubəhur* ('small ball of spun sheep's wool'), and plants that sprawl across the ground *ḳʌʾ* ('thread'), *ḳūwnuh* or *ḳɔ̄wnitin* ('small thread or threads') or *ḳūwnu rxamma* or *ḳʌʾ di mərḥūmhəm* ('grey threads'). Vines and twining climbers are often called *ʿekəhīn* or *məzuwkan* ('twisting and twining around something else'), and plants such as Cyperaceae and *Chlorophytum* species are called *di ḥagubgub* ('miniature date palms').

Appearance of the fruit: *ḳewrēni* or *di ḳerəhān* ('horned') for species of *Digera*, *Cleome* and *Senna*; *śəriśiroh* ('with burr-like fruit like *Tribulus terrestris*') for some species of *Medicago*, *Euphorbia* and *Acanthospermum*; and *di gilʿamihun*, *giləʿamihin* or *gilʿəmhənitin* ('with little round fruit') for species of *Polygala*, *Crotalaria* and *Melhania*.

Size of the leaves: *ḳīḳəhe* refers to smaller-leaved varieties and *ṣʌʿəlhil* to larger-leaved ones. Thus the smaller-leaved *Kalanchoe farinacea* is called *bəgələhān ḳiḳəhe* and the larger-leaved *Kalanchoe robusta* is *bəgələhān ṣʌʿəlhʌl*; and the widespread, small-leaved *Croton socotranus*, called *mitrer* throughout the island, becomes *mitrer ḳīḳəhe* in the Central Highlands to differentiate it from the larger-leaved species of *Croton* which grow there and which are called *mitrer ṣʌʿəlhʌl*.

Uses: Trees and shrubs that produce good, hard wood are commonly called *ḥaher* ('black, dark-coloured'), while soft woods of no value for timber, and which are also usually unpalatable or toxic to livestock, are called *ahtəreś* or *gənigəno*. Plants regarded as especially rich fodder are often called *di məḥaḥōn* from the word '*mʌḥḥ*', for rich and fatty bone marrow. Painful plants such as *Tragia balfouriana* or *Urtica urens* are called *ṣafaḳa* from the verb 'to slap hard'. Plants used to cure skins and hides are called *ṭʌbṭʌb*, from the verb 'to cure leather'; woody plants with tiny leaves are called *gisso*, from the verb '(young weaned kids and lambs) to begin to nibble'.

Cultural significance: A common name for *Heliotropium odorata* is *ḳeṭəmʌmhiyon* ('small lumps of butter'). This small plant is used in tanning leather and especially for preparing the skin butter-churns, and has been given this name as a form of 'sympathetic magic' to encourage the formation of butter lumps.

Plants most liked by animals: *ṣafāḳa di iləhe* ('the *Tragia* of cattle') for *Scrophularia arguta* and *ḥʌźḳeḳ di ilhitin* ('*ḥʌźḳeḳ* of the cattle') for *Pulicaria diversifolia*; *eṣənīno di bəʿeyr* ('*eṣənīno* of the pack camel') for *Ipomoea cairica*, *d-ibʿar* ('of the camels') for *Heliotropium riebeckii* and *zēydiro di bəʿeyr* for *Indigofera marmorata*. We also find *zēydiro di ḥəmār* ('*Indigofera* of the donkey') for *Indigofera pseudointricata* and *ḳʌʾ di tētin* ('thread for sheep') for prostrate *Convolvulus* on the high plateau sheepwalks.

Livestock fodder: Many plant names based on the root '*lḥlb*' 'to milk', such as *di ḥalbeb*, its plural *di ḥalbibihon* and its various diminutive forms, *ḥalebleb* or *ḥalebeboh* ('one who milks quickly'). Plants with this name are often succulent and well liked by livestock, and are also believed to be good for milking animals. A name can also indicate that a plant is not good as livestock fodder. Thus some small and unpalatable ferns and Cyperaceae are called *ḥamžig*, 'milkless', 'mean, lacking in generosity'. Such plants are of little or no nutritive value and are only browsed by very hungry animals in the dry season.

Specific quality of a plant: The tough *Campylanthus spinosus* is called *mirigo* ('with, or causing, gaps in the teeth'), and the painful spines also give it the name *zimgilo* ('a swarm of bees'). Spiked plants are often called *ṣaʾiyud* or *məṣūʾidān* (from the verb 'to poke, prick or pierce'); crisp plants such as *Lachnocapsa spathulata* are called *źʌḥyīm* ('to chew noisily on something hard'); and *Launaea* species, with their milky exudate, are commonly called *dūwrīr* ('bleeding a little blood') or *šiberhen* ('with bitter latex').

Place names: It is of practical use to understand the meaning of such names as *di ḥamź* (a group of settlements on the northern plain), meaning 'area of saline grazings'; *di ṣebərhen*, meaning a settlement 'with tamarind trees'; *di bōwnino*, a pass 'with *Sterculia africana* trees'; or *ṭōwḳʌḳ*, a settlement 'with young *Ficus vasta* tree(s)'. Saline grazings are a seasonally important source of grazing, tamarinds and figs provide valuable human foods, while *Sterculia* foliage is a key dry season fodder.

Another example is the place name *di-rəfontən*, Direfonten, which has been claimed as demonstrating Portuguese influence on the island. However, this name is the plural form of *di rʌffān*, the name for those *Hibiscus* species which have leaves covered with extremely irritating and penetrating hairs. It is customary courtesy for those in front to warn those behind of any such shrubs encountered along the way. The hillside area of 'Direfonten' is indeed rich in these shrubs.

Taxonomy: Understanding plant names can also help resolve botanical puzzles. For example, there are eight *Boswellia* (frankincense) species on Soqotra which, as regards Soqotran nomenclature, fall into various groupings. However, over and above this grouping, many *Boswellia* species are given the name *źifḥa*. In other words, a single taxonomic species is sometimes given this name, and sometimes not. However, this is not because Soqotrans cannot tell one *Boswellia* from another, but because *źifḥa* comes from the verb *źāfaḥ*, 'to be squashed, squeezed, have insufficient room, to not stand free'. This term is therefore applied to any *Boswellia* which grows out of a rock crevice or on a cliff face.

PLANT USES

Individual species entries include information on their uses under the following headings:

Names: Soqotri and Soqotri simplified (see 'Notes on the ethnology text' on p. 44). Where a majority of those consulted said that a plant had no specific name, this is recorded as *nil*. Possible names suggested by individual informants follow.

Food (human): edibility or otherwise; which part is eaten and how it is prepared as a food.

Food (livestock): whether it is important as a livestock fodder, for which livestock and in which season.

Bees / honey: whether it is attractive to wild honeybees and what sort of honey it produces.

Flavouring / food additive: whether it affects the taste of milk or meat from livestock that graze on it; whether it can be used as a salt or sugar substitute; whether it can be used to flavour food in other ways.

Fuel: how it is valued as a fuel and what its main characteristics are.

Timber: how it is valued as a timber; how it is prepared for timber use; how it is used; whether it is resistant to termites.

Fibre / cordage: whether it is used to make fibres or cordage; how it is prepared; what the cordage is used for.

Leatherwork: whether it is used in preparing skins and hides; how it is prepared.

Wood: whether it is used for charcoal, firesticks, quicklime, or to make implements and utensils.

Medicinal / veterinary: whether it is used to treat sick people or animals; how it is prepared for use.

Bedding, stuffing: whether it is used for bedding, padding, or as stuffing.

Cleansing: whether it is used as a soap or shampoo substitute; to remove or disguise odours; as an aromatic or fumigant.

Cosmetic: whether it is used for the eyes, skin, teeth and gums, to dress the hair, or chewed to freshen the breath.

Commercial value (past, present): whether it has, or used to have, any commercial significance for the islanders; how it was harvested and traded.

Pest control: whether it is used as a pesticide to control household vermin, agricultural pests, livestock parasites, or as an insecticide.

Other: uses in addition to the above, such as ritual use and in dealings with the supernatural; whether or not it is toxic; as a water substitute or indicator species; as a fertiliser; as a toy and in games; as an adhesive or as ink or dyes.

Comparative uses: the significance and uses of the plant or closely related species in the surrounding region. Comparative uses are put under genus if there is no direct reference to species occurring on Soqotra but there is relevant information for other species in the genus.

Notes: any additional material.

Human food

Wild foods played a very important role in the diet of the islanders and contributed valuable minerals and vitamins to an otherwise limited diet. Wild foods were crucial in the long dry season, when livestock were unproductive, the dates of the previous year were exhausted and this year's dates were not yet ripe. In times of drought, or when the grain-bearing trading boats failed to turn up as expected, wild foods could be all that stood between the islander and death. Studies of some of the plants eaten in Soqotra have shown them to be very nutritious, for example the fruit of *Ziziphus spina-christi*, and *Amaranthus* species have been shown to be a good source of iron, calcium, vitamin C, folic acid and other minerals and micronutrients.

Perceived palatability is a matter of regional preference and not all roots or fruits described as edible are invariably eaten. The need for supplementing the diet with wild foods was more acute in the poorer and less well-vegetated areas of the island, often the same areas where cultivation of food plants such as finger millet or date palms was impossible. People from botanically richer areas of the island regard some of the plants eaten in the more arid areas as unimportant. Thus the situation not infrequently arises where a plant eaten in one part of the island is said never to be eaten in another: both statements being true in the places where they are expressed.

Some wild foods were carefully harvested (such as tamarind, ziziphus and *Rhus thyrsiflora* fruits, the aerial parts of Amaranthaceae, and the underground tubers and bulbs of Dioscoraceae and Iridaceae). The consumption of others was mainly opportunistic, sometimes gathered as a delicacy or eaten to break the monotony of the normal diet. Some edible plants do not occur in sufficient quantities to have ever been of real nutritional significance. Sour-tasting foods were especially liked by pregnant and breast-feeding women, and it is likely that such plants contain vitamins, minerals and trace elements needed by the female body and developing foetus at this time. Indeed a taste for such sour foods was regarded as one of the most reliable signs of pregnancy.

Livestock fodder

The primary use of wild plants in the eyes of the Soqotran herder is as livestock forage. Most Soqotrans are still dependent on their animals for part, or all, of their livelihood, and they differentiate carefully between the plants which support their animals. They recognise plants that help animals to put on weight, promote lactation, have recuperative properties for sick livestock, enable animals to go longer without water, plants that are suitable for recently weaned animals, affect the taste of the milk or meat and fodder plants which are harmful to their animals. They recognise that some fodders are nutritious, while others offer only subsistence for the long months of the dry season, providing bulk and stilling hunger. They know which plants are able to survive the longest drought and where to find them. They are expert in where specific plants are to be found after different kinds of rain and which plants respond to light showers, to mist and cloud, and which only to heavier or more persistent rain.

Graze and browse: Cattle and sheep rely in large part on grass, supplemented by herbs and browse in the dry season. Camels and goats rely mainly on browse, and donkeys, while essentially grazers, will also feed on small browse plants. Sheep and cattle cannot browse for long periods. They need a definite source of grazing, and, moreover, a larger supply of water than other livestock on Soqotra. Cattle in particular drink three times the quantity of water used by a sheep, goat or a camel, and cannot go without drinking for a long period or reduce their water requirement by eating green plants, as goats and sheep do.

Browse: Woody plants have a number of advantages in arid regions. For instance, woody vegetation varies less inter-seasonally in general, and woody litter remains for longer than grass litter. Apart from the early growing season, the crude protein content of most browse is usually higher than that for the grasses, and much woody vegetation breaks dormancy and begins to produce the new season's flush of leaves well before the winter rains begin. The development of new leaves is very rapid and does not constitute growth as such since it is independent of rainfall and redistributes stored carbohydrate and protein from the roots and woody tissues. For the hot months before the winter rains, when the quantity and quality of grasses is low, woody vegetation can often provide a good quality food source.

Dry season forage collected as supplementary feed: During the dry season, as the rangeland is progressively depleted, herders begin to gather supplementary forage to feed their livestock. This can only be done within a herder's own area, or, once his own area is exhausted, by permission in someone else's area. It is this necessary collection of supplementary feed that makes the dry season such an exhausting time of year, especially for those in the more poorly vegetated areas, and, in years when summer rains fail, all across the island. Men and women climb down cliffs or are lowered down on ropes, scale the most inaccessible peaks, penetrate ravines and edge along rock ledges to gather fodder. Plants with too many spines and prickles are usually avoided, as these can become lodged in the mouth or throat, and thereby cause ulceration and loss of appetite. Standing hay is also cut. Generally speaking, this is mainly gathered by women and fed to milking cows and goats. Fallen foliage is mainly gathered by women and fed to milking or pregnant cows and goats. Trees and shrubs are lopped and pruned by men and fed principally to pregnant and milking goats. Scarce fodder is also given to favourite or sickly animals.

Supplementary forage consists of:

- Dried grasses and herbs *śaˈʿar* (*sha'ar*).
- Green foliage lopped from trees and shrubs *ʿʌlf* (*'alf*), and leafless twigs and tips of succulent shrubs *ḳɔ̄wnem* (*qownem*).
- Dried and fallen foliage *maˈḥliṭ* (*mahlit*).
- Pulped branches and trunks of larger trees, and of entire smaller plants *kˈidihir* (*kidihir*), although this is only when everything else is exhausted.

Different plants are resorted to in different areas of the islands, and plants acceptable to livestock in one area may be of little interest in others: the palatability of the same species varies from one vegetation type to another and livestock become accustomed to different tastes. However, some general statements can be made.

Green foliage and twigs and tips of succulent shrubs are exploited differently at different times. Earlier in the dry season, the crown of the tree or shrub is beaten with a heavy club to break off leaves, stalks and smaller twigs, or foliage is brought within reach of the animals by partially cutting through a branch, pulling it down, and anchoring it in place with rocks or pegging it down with a V-shaped branch. Later on, the leafy stems and the tips of branches which do not yet contain heartwood are cut and fed to the waiting animals below or carried back to the livestock pens. Trees such as the smaller-leaved *Commiphora* species and ziziphus are pruned even when leafless, and the prunings fed to goats.

Dried and fallen foliage that is palatable and lies within reach of livestock is eaten as and when it is found. When this is exhausted, herders collect it in sacks or wool blankets.

Smaller plants are chopped up and fed straight away, while the branches of larger trees and shrubs are brought to the settlement and pieces cut off and pulped as and when needed. When there are no branches left, the trunk is usually carried back to the settlement and sections cut off and pulped one at a time.

Dried grasses and herbs are discussed under Gramineae.

Lagoon weed as fodder: In areas around the coast where there are inlets with mixed salt and sweet water, livestock herders make use of the weed that grows in them to feed their animals in the dry season. This weed is mainly of two sorts, one 'rough and scratchy' (probably *Najas graminea* or *N. marina*), the other 'smooth' (probably *Potamogeton pectinatus*), but both produce long fronds. These do not grow in sweet or only slightly brackish water, or in a lagoon recently washed out by flooding. During the summer *doti* season tides are high, often coming up and over the lagoons. If there is no rain at this season the water becomes more saline, and the weeds begin to grow. The weed is gathered by cutting it off above the roots at the bottom and then brought to the surface and any mud clinging to it washed off. Some dry it and feed it straight away, others store it for the long, dry monsoon months to come. All livestock of the coast eat it and some cows and donkeys can be observed standing up to their thighs in the water to feed on the living plants. It is seen as being a good and nutritious feed.

De-barking woody plants: Hungry goats and sheep both chew the bark of certain trees when driven by hunger. The most palatable of these, for goats in particular, are *Dendrosicyos socotrana*, *Sterculia africana*, smaller-leaved *Commiphora* and *Boswellia* species, *Euphorbia schweinfurthii* species complex, *Angkalanthus oligophylla*, *Cryptolepis intricata*, *Dirachma socotrana* and species of *Chorisochora.* If the goats are very hungry and there is little else to eat, they continue to feed on a tree in this way until they cause it severe damage or kill it. Apparently the *Commiphora* species survive such an onslaught better than the others.

Bees and honey

Wild honey has always been highly regarded as a medicine on the islands and is prescribed for all sorts of illnesses: two fingers dipped in honey first thing in the morning before breakfast is the usual prescription. Its great value, medicinally and as a luxury food, and the fact that it can only be obtained by robbing the nests of wild bees, means that the islanders are very observant about the behaviour of honeybees and are well aware of which plants they favour. Bees feeding on different plants at different seasons produce honey of differing viscosity, colour and flavour. Some of these honeys were considered by the Soqotrans to be better than others for treating certain conditions. Before the arrival of sugar on Soqotra honey was always the most prized amongst the limited sources of sweetness available.

Flavourings or food additives

Sweetness: Before the arrival of sugar, the only source of sweetness was from wild honey, ripe dates, or the fruits, nectar and barks of certain wild plants. In good years this was supplemented by a sweet and sticky substance that appears on certain grasses (and some herbs, such as *Boerhavia diffusa*) in the mid-morning in the early summer period before the summer rains, or, at some altitudes, in the winter rainy season. It persists for up to a month at a time, and is much liked by livestock and their owners alike. It is perceived as a sign of God's blessing and a good year.

Salt: Large quantities of salt were extracted from seawater to salt fish, both to preserve it for eating during periods when the sea was too rough to fish, and for selling to the trading boats which plied the East African coast and the southern coast of Arabia. Salt was always seen as essential for cereal dishes, and ghee was greatly improved both in flavour and in keeping qualities if salt was used in its preparation. Those furthest away from the sea, such as the people of the Central Highlands and their hinterland, were renowned for their sparse use of salt. Substitutes were occasionally found in the absence of sea salt. Near the coast seawater would be used if no other salt was available, and further inland, rock salt, though this last was mainly used medicinally. But sometimes plants were used in place of salt. In the monsoon months, the salt-encrusted foliage of *Croton socotranus* or *Pulicaria stephanocarpa* or other plants exposed to salty monsoon winds can be briefly dipped in the cooking water to flavour it. The effect is better if the plants are leafless, otherwise they tend to give a bitter flavour to the water.

Souring milk: The leaves and fruits of certain plants, or roots and small whole plants, are used to sour milk. They are crushed and added to a little milk in the butter-skin or a small pot of milk, and are left overnight to coagulate the milk. This thickened and soured milk is then added to fresh milk to ripen it before churning it to make butter.

Fuel

Until recently, firewood was the only source of heat, light, warmth and cooking fuel on the islands. The islanders were expert in recognising which plants made the best firewood for different purposes: which burned slowly and steadily; which made the least smoke for cooking and for warming living quarters; which burned even in damp weather; which tainted cooking food; which produced the greatest illumination; which made the best tinder and kindling; and which produced the most pungent smoke to use as an insecticide. All firewood gathered was dead wood, the only exception to this ruling being small pieces gathered for burning as an insecticide which were often put onto the fire green. However, plants for this purpose are usually unpalatable to livestock. A store of firewood, tinder and kindling was gathered throughout the year and stored at home ready for the cold and wet months. In cold weather it was important to keep the fire going for as long as possible, preferably overnight. In the days of widespread livestock theft, this was also important during the dry season, the time when the theft of animals was most common. The night fire gave good visibility and also demonstrated to any potential thief that people were on guard. The islanders recognised which woods were best for this purpose, and would keep a good, thick log of slow-burning wood ready for use.

Timber

After stone, wood was the main construction material. All sizes of cave living quarters were improved and modified for human and livestock occupation by using wood or brushwood. Wood provided the framework for livestock shelters and byres, for fencing, and with brushwood, even for entire houses. In the less well-vegetated areas of the island, people had to economise on scarce timber, and there was a greater use of brushwood. Climbing plants were encouraged to grow up and over the walls to form a matted layer which held together the rather unstable roofing and walling, and provided added protection against wind and rain.

The islanders knew which timbers were best for which use: which were weak and soft or hard and strong; which cracked under pressure and which split over time; which rotted if allowed to become damp; which had to be aged or fire-tempered before use to give them extra strength, and which were attractive to termites and other wood-boring insects and so had to be de-barked first.

When they moved out of their caves, islanders from the interior began to construct stone housing. The houses were built with low drystone walls, often with inner and outer faces infilled with smaller stones. Undressed trunks and branches provided the pillars supporting the roof, the kingposts, roof beams, rafters, cross beams and door posts. Layers of small sticks were laid across the roof beams and rafters; above these a layer of stones was placed to fill in any gaps or cracks, and then a layer of gravel and pebbles was placed on top. Finally fine, powdery, pale earth was mixed with water to a paste, and a layer of this was laid to form a thick coating over the whole roof. Once that had dried, a further layer of the dry earth was laid on top and stamped down. All that remained to complete the roofing after this was to lay the roof edging stones and gutters. Gutters were made from hollowed lengths of branches cut from soft-wooded trees. They were cut in half longitudinally and placed all around the roof edge except immediately overlying the front entrance. The main roof-supporting pillars or roof trees, on which the whole structure of the roof depended, livestock byres as well as human living quarters, were precious possessions, and if the family moved to new quarters, they would normally take these with them.

The sale of timber and firewood

The selling of wood as timber only began with the arrival of 'modern' government in the late 1960s, and was a result of a sharp rise in demand as new types of building were introduced and as Hadiboh and the northern coast settlements began to expand. Timber was brought down to the plain and then sold by the '*kurgeh*'-load, consisting of 20 pieces of timber. The timber came from species such as *Croton socotranus*, *Cephalocroton socotranus*, ziziphus, *Maerua angolensis*, *Zygocarpum coeruleum*, *Chapmannia* species, *Grewia bilocularis*, *Acridocarpus socotranus*, and *Ballochia*, *Trichocalyx*, *Carphalea* and *Flueggea* species. These were mainly cut in the seaward foothills of the Central Highlands, which for the heavily populated northern plain was the nearest source of good timber. The southern plains at that time were very lightly inhabited, mostly seasonally, except for the small settlements clustered around the few saline wells where date palms could be cultivated. Communities around date gardens made use of date palm material to build their living quarters. There was no demand for timber from eastern or western Soqotra, as people here primarily made use of caves modified with locally available wood. Most regions of the island, apart from the Central Highlands, are not rich in woodland, and traditionally there was a marked reluctance on the part of the inhabitants to permit the felling of trees and shrubs for timber when these played such an important role as livestock fodder, or yielded food or other products for the human population.

Fibre and cordage

The strongest cordage was prepared from dracaena and date palm fibre. Cordage was also made from *Hibiscus* species, *Jatropha unicostata*, *Ficus vasta*, *Gossypium barbadense* and various rushes and reeds. Non-plant materials were also used, such as goats' hair, sheep's wool, leather, and fibres teased from cotton cloth and sacking. The strongest ropes were used for letting men down over sea cliffs to hunt nesting seabirds and their young; to access cliff-growing plants in droughts as livestock fodder; to reach an otherwise inaccessible pool of water, or to climb the taller trees to harvest their products and gather drought fodder. Finer cordage was used to patch and repair leatherwork or for rough sewing, such as stitching sackcloth to make saddle bags, for instance.

It was also used for pulling teeth, to remove warts by tying fine threads around them, to make slingshots to protect planted areas from birds and other predators, and for temporary or emergency repairs to fishing tackle or pack equipment.

Date palm fibre

Most household items in everyday use were made from plaited date palm fibre, especially in date-growing areas. Strands of dried leaflets are plaited in narrow strips sewn together to make a variety of objects, such as:

ḥāṣir (*hazir*): A large oblong mat used as a floor covering.

mˈeraḥt (*meraht*): A circular mat on which food is served.

mˈosfi (*mosfi*): A very large container usually used to store cereals, and especially the locally grown finger millet. When full it takes at least four men to carry. The base was made like the circular food mat and then the sides were gradually built up with strips of plaited fibre. When filled with grain it was sewn up tightly, leaving a stretch of lacing which could be undone to allow access.

dˈizihen (*dizihen*): An oblong container with a cover that exactly fits over it. It has a carrying handle extending from one (long) side to the other. It was usually used by a woman to store her private possessions.

mˈillug (*millug*): A container similar to the *dizihen* but smaller. It was most commonly used to hold the Holy Qur'an, or sections of it.

dˈaḥanə (*dahanih*): A two-handled open basket, generally used for carrying fish.

gˈifir (*gifir*): A cone-shaped food cover used in the coastal villages to protect cooked food.

mṣˈʌllə (*mizallih*): An oblong mat used mainly in the coastal villages as a prayer mat.

ˈśitfəh (*shitfeh*): A small circular, cylindrical, unlidded basket for general use.

Date palm leaflets were also plaited and stitched to make equipment for pack animals.

Leatherwork

There is detailed knowledge throughout the Archipelago of which plants are suitable for curing leather. These plants contain tannins, usually concentrated in the bark and leaves. These are complex compounds that can be used to cure or tan leather, a process in which the proteins in skins and hides are broken down and rendered impervious to putrefaction. Tannins also have astringent and antiseptic properties, so it is not surprising that many of the tanning plants are also used medicinally. Different plants work best at different seasons and each has its particular properties as regards curing, dyeing and strengthening leather. The age and condition of the animal from which the hide has been removed has to be taken into account, as has the purpose of the item to be made. Throughout the Archipelago, goatskins were the most widely used. Any stitching required was done by piercing the leather with a heated, sharpened nail, or using an awl or a needle made from goats' horn. Fine thongs of leather or string worked from plant fibre were used as threads. Finer leather items were softened by rubbing repeatedly through fingers lubricated with butter and oil.

Preparation of skins and hides

Removing hair: After the skin has been removed from the carcass, it is rolled up, haired side on the inside, and put aside until there is time to deal with it. If freshly removed, the skin is spread out, skin side uppermost, and a paste of ash and water applied in a thick layer over it – the fresher the skin the sooner the hair can be plucked off. If there is not time to apply the ash dressing immediately, and the skin has dried out and hardened, it is soaked in water to soften it and then the ash is rubbed in. The skin is then folded over and rubbed back and forth on itself until thoroughly impregnated with the ash, and then put aside to allow the depilatory time to work. In all leatherwork, the use of ash is vital, not only to loosen the hair, but also to toughen the leather. The islanders know which woods produce the best ash for this. *Pergularia tomentosa* is used in place of ash as a depilatory when the leather needs to be exceptionally fine, usually for items of personal apparel, as belts and fine skin containers. If there is an immediate need for the leather, the ashed skin is left in the sun for the ash coat to dry out, when it is removed to the shade and left until the skin is nearly dry. The hair is then plucked off by hand, the last vestiges being removed by scraping the skin with the edge of a knife blade or a sharp stone. If the skin is not to be processed until later, it is wrapped loosely, still with its ash covering, in a cloth or some sacking. Once the hair has been removed, the legs are tied at their extremities to make a rough bag. It is inflated and hung up in the sun to dry.

Smoke-treating skins and hides: This is considered necessary to deter mites, leather beetles and other pests which attack leather. The skins are hung above the fireplace until thoroughly impregnated with smoke. They are then given a further dusting of ash, and are rolled up and wrapped in a piece of cloth or sacking. They are stored in the roof, poked amongst the roofing beams, or placed on a shelf made from sticks or date palm ribs suspended above the fireplace. A skin treated in this way will keep almost indefinitely, as long as it never gets damp.

Curing and dyeing leather with plant material: The chosen plant material is pounded to a coarse paste and for best results is applied to a fresh skin. It is first softened in water and then wrung out, and the paste smeared over the skin. It is rubbed into the outer side if the skin is to be used for liquid or food to prevent the plant material tainting the contents. The skin is then folded and refolded and rubbed back and forth on itself, rather like kneading dough. It is then rolled into a ball, weighed down with a rock and left for some hours. It is then kneaded again – the longer the kneading, the stronger the leather. The skin is then tied and inflated and hung up in the sun to dry. The plant debris is knocked off and the skin is now ready for use. Normally a skin treated in this way will last for up to two years, after which it has to be treated again. Some plants not only cure the leather and enhance its tensile strength, but also colour it a rich, mahogany red-brown. This is done when the finished leather is intended to be beautiful as well as functional.

Leather items that are normally treated with plant material are:

- Milk-skins, both those used to carry fresh milk back to the settlement, and those in which successive milkings are stored until enough has accumulated to churn to butter.

- Butter-churning skins.
- Intricately plaited belts worn by adult men.
- Satchels or leather bags in which men carry their personal possessions.
- Leather buckets used for fetching water and for watering.

The haired side of the skin is always on the inside of bags used for milk, water and dates.

Leather items that are not normally treated with plant material are:

- Water-skins, though some people are prepared to put up with the taste passed on to the water by the plant material in exchange for stronger leather.
- Skins for storing dates.
- Skins for storing buttermilk.
- Items made using entire skins and hides.

Whole goatskins are used to make bags for storing milk and buttermilk, for churning milk to make butter, for fetching and storing water and for storing dates. Islanders who owned a lot of sheep also used their skins to make butter- and milk-skins. To make a leather bag from a skin, the front legs and the back legs are tied together and the smaller neck opening and the wider anus opening are closed by tying. A carrying handle is made by extending a length of leather or fibre string from the front to the back of the bag. The wider opening is used for pouring milk or water, and the narrower end for pouring the buttermilk.

Leather bags for milk products: The skin used for churning milk constantly oozes moisture and suffers harsh handling during churning. As a result it soon becomes holed or torn. When grease stains begin to show through on the outside it has reached the end of its life. Although it can no longer be used to churn butter, it can be dried and re-ashed and used for other purposes, to store a small quantity of dates, or to transport water if the water source is not too distant, or as a milk-skin. The butter-churning skin generally lasts for two or three months, or, if made from a mature male goat or a fat and healthy mature female, for four or five months. It then has to be replaced. The skins used for storing milk and buttermilk have a longer life. The skins for storing fresh butter and ghee are usually made from the smaller skins of kids and lambs. Once enough ghee has been processed it is stored in clay or china pots.

Leather bags for water: The skin used to transport water lasts for two to four months, depending on how far away the source of water is and how rough the terrain. The huge weight of the water carried stretches the skin very thin, and as a result it tears easily. If it is used only to store water it lasts longer. Once a water-skin has become too thin it can be re-ashed to thicken and strengthen it and then be used to store a few dates.

Leather bags for dates: Skins for storing dates are nearly always made from a mature animal. Care is always taken in selecting the skin to use for protecting this vital food. Best of all is a skin that has been buried for about two years in the ash of a hard wood. Those preparing skins for dates generally process 10 or so at the same time, and store them until needed, placing ash in between and all around them. After the dates have been harvested the surplus ash is knocked off and the skins washed. They are packed with the dates while still damp. The date-filled skins are stored covered in ash or with fine red earth. Once the dates have been eaten, the skin can be washed, re-ashed and re-used until it wears out. The skins can last for up to three years if they are kept quite dry and away from cold or damp. When the skin becomes dotted with small perforations this is a sign that it is reaching the end of its life, and the dates must be transferred to another skin.

Leather bags for personal possessions: A bag made from the finer skin of an adult sheep or a young goat was carried by men to hold their personal possessions (such as tobacco and pipe, flint, firesticks, toothstick, a ball of dates or some dried fish). The hair was removed with *Pergularia tomentosa* and the skin then thoroughly washed to remove every last trace of the caustic latex. While still damp, the skin was then tanned and dyed. Women also had a leather bag in which to keep their possessions (comb, toothstick, a lump of dried aloes, cosmetics – antimony for the eyes, turmeric powder, *Euclea divinorum* chewsticks – small pieces of cotton fibre, needle and thread, for instance). Some also carried a spindle, balls of wool and other weaving materials, though many used a special bag for this. This large leather bag was made by spreading a skin out flat and then folding it to make an oblong pocket, with a flap covering the open edge, like a briefcase. Traditionally it was said that a reliable sign of a wife being really angry was that she took this bag with her when leaving home.

Floor coverings: Hides of bull calves, adult bulls and adult male goats were used entire as floor mats, mattresses and waterproof roofing material. A fresh hide is prepared by spreading it out, hair side down, on a patch of flat earth. It is pegged down with slivers of wood to stretch it flat and to prevent vultures pulling it around and crumpling it. It is left for two or three days and then, when quite dry, cured with plant material while still pegged down. It is then unpegged and rolled up, hair side inwards, and stored. This cannot be done when the weather is damp otherwise the skin will rot. For the same reason, during rains or when it is misty and cloudy, the mat must be stored deep in a dry cave and covered with a layer of dry earth.

The harness for climbing date palms: This is worn by those climbing date palms to allow them to prop themselves against the trunk with their lower back supported, leaving their hands free. It is made from a fresh cow or camel hide. The hair is not removed and the leather is not cured in any way. The damp hide is laboriously cut into as long strands as possible. These are then left to dry and are stored until needed. To make the harness, the strands are soaked in water to soften them and are then plaited to form a single piece, a thick band which goes around the body and plaited ropes that extend from each side of the central section. It is generally made by specialists for sale or barter.

Items made using pieces of leather

Small mats: A variety of small leather mats were made.

Mats for treading grain or dates: Men trod the freshly harvested finger-millet heads on leather mats to separate the grains. They also trod the new dates into a solid mass on leather mats before storing them in the goatskin bags. Sometimes these mats were made from date palm fibre, but leather was generally preferred as being more robust.

Mats for 'beating' sheep's wool: Women 'beat' the curls of shorn wool with the 'bow' on a small leather mat.

Mats used in the highlands when making butter in the cold season: A small mat usually made from a piece of haired calfskin was placed beneath the butter-churning skin to protect the contents from the ground. Cold and damp prevent the milk from ripening properly. The non-haired side of the mat was placed next to the ground, the bag resting on the warmer, haired side. Butter-skin and mat were then wrapped around with a sheep's wool rug and placed close to the fire

Knife sheaths: The sheath for the three-knife set carried by the adult Soqotran is generally made from the skin of the head of a calf or lamb, or from the skin of a goat's leg removed in one piece, in a tube. The hair is removed with *Pergularia tomentosa* and the leather tanned and dyed. The expert leather worker uses three pieces of leather as well as the tube of skin and has to work quickly while all the pieces are damp, or, better still, fresh. The three pieces of leather are folded over to make three long tubes, and these are pressed flat, one on top of the other. These three are then inserted into the larger tube of skin, one end of which is folded back on itself. The whole is then moulded, squeezed and trimmed to form a neat 'pocket' with three partitions, one for each knife. It is then left to dry out. If expertly and quickly made there is no need for stitching. Those who are less expert simply take the single damp piece of leather and wrap it tightly around the three knives and then sew or tie it firmly in place. It sets hard and shrinks to a closer fit as it dries. Smaller offcuts of leather are then used to make dividers between the knives, and a small front pocket is added to hold an awl if wanted. Sometimes the sheath is made even redder by crushing a piece of dracaena resin to powder with a lump of a certain kind of soft red stone. This is rubbed hard and long into the leather until the required depth of colour has been achieved. It is then oiled and polished.

Belts – boys: From the age of about four months most uncircumcised boys wore a waist thong, usually made from a single strip of lambskin. Sometimes calfskin was used, but never goatskin. More elaborate waist thongs were plaited from three strips cut from a larger skin, or were twisted from two leather strips. The leather for this was tanned and sometimes also dyed with plant materials. The belts were secured by inserting the knobble of the knot at one end through the loop at the other. In some parts of the island wool or twisted cotton fibre was used instead.

Belts – men: Two belts called *siyr* and *sibet* were first worn after circumcision. The *siyr* was made of more strands than the *sibet* and was generally considered to be more beautiful and desirable. Preferably it was made from the skin of a young kid, while the *sibet* was usually plaited from the skin of a young lamb. To make the belts, the hair was removed from the skin with *Pergularia tomentosa*, the paste being left on longer than usual to make the leather very thin and fine. It was then tanned and dyed with plant materials. Once the leather was dry, the horn handle of a knife covered with a folded strip of leather was scraped back and forth along the whole length of the skin as hard and as steadily as possible. This caused the colour to lighten as the leather was stretched. The skin was then cut into as fine strips as possible with the sharp little knife used for shaving and circumcising. The preferred way of doing this was to cut the skin into two and then, starting at the circumference, cut round and round in a circle to produce a single long strip from each half skin. The *sibet* was usually made from six of these leather thongs, though some were made from 20 or more strands. The six thongs were either plaited in two groups of three all along the belt, or were initially plaited in two groups of three but then all six plaited together for a distance, then in two groups of three again and so on, making a series of regular loops like the links of a chain. The belts were secured by inserting the knobble of the knot at one end through a loop at the other. This meant that each belt had to be made exactly to fit its wearer.

The *siyr* was made of a large number of tiny plaited strands. The best examples, and those that took the most skill to make, used the finest thongs to make the greatest number of very thin three-ply plaits. A well-made *siyr* belt had between 30 and 50 plaited strands. The very best belt had double this: two belts, each of 30 or more plaits, joined to make a single belt. They were highly prized and there was a great demand for them.

Bellows: Bellows used by the blacksmith were commonly made from a sheep- or goatskin. *Pergularia tomentosa* was used as the depilatory, and the skin was then inflated and hung up to dry, usually overnight so that the night dews would help to soften it. The next day it was tanned. To make the bellows, two straight lengths of wood were cut and whittled as flat as possible to make narrow blades. The neck end of the bag was tied and the blades laid one each side of the wide anus opening of the skin, the skin being rolled back over the blade and secured by stitching. Four loops of cloth were then made and one stitched at each end of the two blades. The bellows were worked with the ring and middle fingers in the loops of cloth on one blade and the thumb and index finger in the loops of the blade on the other side, leaving only the little finger free. A length of metal tubing was inserted into the neck end of the skin and bound firmly in position. To inflate and deflate the bellows, the fingers of the left hand were inserted into the loops as described, and the skin pumped up and down. The movement also involved folding the leather bag over towards the end with the metal tube while deflating or pressing down, and then inflating it by pulling up away from the tube end and straightening it out.

Drums and tambourines: Skins of sheep or calves were used to cover the Soqotri drums and tambourines.

Feeding bottle: The skin of the forearm or hind leg of a goat or sheep was used to make a feeding bottle for suckling kids, calves and lambs.

Tulchan: When a calf dies, or a male calf has been slaughtered to conserve the milk for human use, a tulchan made using the skin of the calf is presented to the bereaved cow to persuade her to let down her milk. Some cows can be milked with little or no trouble after the death of their calf but others are extremely stubborn. The tulchan is made from the calfskin and sweet-smelling plants. In its simplest form, a piece of the dried skin is wrapped around sweet-smelling herbs (such as *Lavandula nimmoi* or *Micromeria remota*) and given to the cow to sniff at milking time. Alternatively, the whole skin of the dead calf is used, the head removed and the haired skin cured with plant material. The skin is then shaped into a rough cylinder, stuffed with fragrant herbs and stitched securely. At milking time, the herder covers both ends of the cylinder with his hands and offers it to the cow to sniff, or tucks one end of the tulchan under his armpit and covers the other with a hand. Some of the cow's milk is sprayed over it, and she is then given it to sniff and nuzzle while the herder tries to milk her.

Genital 'contraceptive apron': Outside the breeding season all sexually mature male livestock had this 'apron' strapped on to prevent them mating and causing the females to give birth at an inappropriate time. The apron was made from pieces of leather (often from cow or calf hide, or for larger livestock, from camel hide) shaped into a roughly oblong shape. Leather or fibre straps were used to attach it to the animal, placed in such a way that the leather flap hung beneath the belly and in front of the genitals. This apparatus was also made from pieces of sackcloth or was woven from sheep's wool or plaited from date palm fibre.

Watering bucket: The bucket used for lifting water from wells and water pools, and for watering livestock, cultivated plots and young date palms, was usually made from sheepskin. It was treated with ash to remove the hair, and, by those anxious to prolong the life of the bucket, was also tanned. A single skin was cut in half around the middle to make two buckets. The longest edge was rolled over a hoop of wood (a supple wood such as the stems of *Secamone socotrana* or *Cynanchum gerardii*) and stitched in place. It was then curved around to make the circular opening of the bucket, and the rest stitched into a bucket shape. A carrying handle was made from date palm fibre or from a leather thong.

Wool-spinners' bag: Women span wool as they moved around shepherding their sheep. The necessary equipment for doing this and for keeping the small balls of spun thread was a small leather bag made from leather offcuts.

Male headband: Traditionally, younger men and those in the prime of life wore their hair long. For everyday work, it was roughly tied back with anything to hand, but for celebrations, such as circumcisions and weddings, it was combed out, oiled with butter and then tied back from the face with a headband. This was a plaited circlet made from fine leather thongs cut from the skin of a lamb or a kid, though some wore one plaited from sheep's wool or goats' hair. The circlet was pressed down over the head to hold the hair back and the hair below teased out to make it thick and bushy.

Leather as food

In times of great hunger, skins that had been used to hold milk or to churn or store butter were used as food, *mi̍kəlit* (*mikilit*). The skins were either cut into pieces and roasted at the fire before being eaten, or were cooked in water, the liquid drunk and the softened leather chewed.

Wood

Wood was the most widely available raw material for the islanders and they made full use of it. They distinguished carefully between woods in terms of flexibility, straightness and lack of knots, durability, liability to warp and split, hardness and softness, response to tempering by fire, suitability for carving into implements of various kinds, and resistance to termites. Individual, well-grown trees are also important in their own right. Many settlements were built around a large, shade-giving tree, which acted as an outdoor focal point for many of the activities of those working around the home. Many distinctive trees are also given individual names and act as road markers. If they are tall and offer good shade, they also act as a resting or meeting place for those travelling around the island.

Firesticks: Before the days of matches, fire was made by rubbing two sticks together to produce a spark, a vertical drilling stick *mošk* (*moshk*) and a horizontal stick *źo̍hiro* (*dhohiro*). The best wood for this was dead and completely dry, and the best results were obtained by using the same wood for both sticks. The most renowned wood for making fire was from *Commiphora ornifolia*, but other woods could be used, and, in the hands of a very energetic and persistent person, almost any dead and dry wood (other than latex-bearing ones such as *Euphorbia arbuscula* and *Ficus vasta*) could be worked to produce a spark.

Charcoal: Charcoal was not made commercially on Soqotra. It was only regularly used by blacksmiths, who usually made their own, or in some of the kitchens of the larger coastal settlements. However, herders would often make charcoal and use it to pay blacksmiths for work done for them, such as putting a keen edge on their knives. To make charcoal in any quantity, a site with a plentiful supply of sand or sandy soil was selected. This was heaped up in a circle around where the wood was to be fired. The wood was stacked in the middle, ends together, forming a triangular stack. Twigs and other inflammable material were gathered as firelighters and placed strategically here and there in the woodpile, which was then lit from the bottom. As soon as the flames had died down and the wood was blackened but glowing, and before it burned down to ash, it was removed from the fire and spread out in a single layer on the ground inside the circle. It was then quickly covered with a thick layer of the surrounding sandy material to exclude the oxygen, and was left to cool for a day or two. The charcoal was then picked out and bagged up.

Use of wood in fishing

Boats: The basic dugout canoe used for fishing was imported, but modified locally. Ribs were shaped to fit and were nailed inside to brace the sides and strengthen the canoe; a runner with holes bored along its length was nailed along the keel to give the boat added protection when dragged between sea and shore. Oars were made from straight, slim limbs of a hard wood and a circular blade shaped from a piece of driftwood planking. If a sail was wanted, a mast was made from a stout, straight piece of hard wood and a sail from the strong, unbleached calico imported for clothing. Most canoe owners also made a set of three rollers to protect the keel of the valuable boat when it was dragged right up to the beach area above the tide-line at the onset of the monsoon. Here it was thoroughly oiled with sharks-liver oil and turned upside down before being laid across the rollers to raise it from the ground and allow air to circulate. It was stored in this way, out of the sun if possible, for the five months of the monsoon wind when all sea fishing was impossible.

Rafts: Many fishermen had no access to a canoe and they either line-fished from cliffs and rocks, used cast-nets, or made do with rafts. These were made from date palm trunks or branches of local woods lashed together. If available, a flat surface was made by strapping on driftwood planking. Rafts did not venture beyond the shallow, inshore waters where they were used as fishing platforms or as a diving platform when diving for oysters or other shellfish. They could also be used like the canoes to unload larger boats anchored offshore and to transfer goods to the shore and back.

Fishing equipment: A bailer was made from a half coconut or from a knot of wood hollowed out to make a bowl. The canoe also carried a date palm fibre basket for bait, and a stout club for stunning larger fish. If the fishermen were after shark, they carried a harpoon and gaff with a shaft shaped from a local hard wood. Wooden spears were also made and used to spear fish and snakes and trapped wild and feral cats.

Fish traps: Woven fish traps were made by specialists using either fruiting stalks from date palms, or strips of various local woods whittled to the correct length and then softened in water before weaving. In rocky areas fish traps were used mainly during the monsoon season, when fishing from a boat was possible only in a few sheltered areas. Along sandy shores, they were used the year round, with the exception of the monsoon season when the crashing surf made this unfeasible.

Staffs, herding crooks and sticks

mˈərḳaḥ (*mirqah*): All Soqotran herders carry herding staffs. The best ones are prized possessions and there is great competition to own the smoothest, straightest and most highly polished stick. It varies in length and circumference according to the species of livestock owned. It is carried at all times and can be used variously as a walking stick, to guide and control livestock, to emphasise points in discussion, and to correct naughty children. It is the most carefully made of all. A suitably straight and slim branch is cut from a hardwood tree and then laid in a watertight cave for a week or so. If the wood is green, not of the best quality, or the branch is a little crooked, it is fire-treated. It is held over the fire and turned this way and that until thoroughly heated through and any sap expelled. Then the bark is removed, and, if the stick needs to be straightened, one end is wedged under a heavy rock, or in a rock crevice, and the stick twisted and turned using the full weight of the body to make it as straight as possible. It is then held in position with rocks and left to cool. The ends are then trimmed neatly and the bark removed. The stick is then hung over the fireplace or stuck in the rafters above for a period of two to four weeks until thoroughly impregnated with smoke. This strengthens it further and darkens the colour of the wood. It is then polished and rubbed until it gleams with hands greased with fat or ghee.

ḳˈariṣ̌hīyo (*qarij'hiyoh*): A thick and heavy stick, shorter than the *mirqah*. It is used when working with baggage donkeys or camels and to control a rutting camel or donkey.

rɔ̄ṭel (*rowtel*): A heavy club or short, stout stick. This is mainly used for beating date palm fibre, or, by fishermen, to stun shark and large fish. A thick staff is also used by livestock herders to beat the crown of trees and shrubs to knock foliage and fruit onto a cloth spread beneath.

śinɛ̄gə (*shinegih*): A fodder hook or crook made from a long stick of a hard wood with a naturally occurring V-shape at one end. One side of the V is left long and the other cut short to make the hook. If no such branch can be found, a shorter branch can be tied to the tip of a longer stick, but this is rarely as effective. This stick is used to hook branches down to within reach of the herder so he can lop foliage for livestock, or pick fruit from them.

Utensils and implements

Harder woods from trees such as the tamarind, ziziphus, *Grewia* species, *Zygocarpum coeruleum* and *Chapmannia* species were carved or whittled to make a variety of utensils and implements. These were sometimes polished to a fine gloss with ghee.

Grain and ghee measures: Bowls of different sizes were made for measuring out cereals and ghee. They were carefully worked to hold an exact measure and were used by both seller and purchaser, as well as within the home to measure out the cereals for the main meal in order to control consumption.

Wooden locks: Locks with comb-like teeth that meshed together were made from hard woods or from planks that drifted ashore from the sea. In the more permanent homesteads these were fitted to the door of the food store where the valuable cereals, dates and ghee were stored.

Spoons and stirrers: Rough spoons and stirrers were made from all sorts of wood, even – for temporary use – from soft woods such as from smaller *Commiphora* species or green *Jatropha unicostata* wood. Stirrers were used for stirring maize and finger millet as they cooked, or for buttermilk as it was processed to make feed for young calves or thickened over the fire to make cheese. Spoons were mainly used for serving food.

Combs: Combs with teeth along one side, or with different-sized teeth along both, were made from local hardwoods.

Hair dividers: These were broad at one end and narrowed to a point at the other. They were made from a slim twig of a hard wood and were used to dress the hair and to divide it into sections for plaiting.

Antimony containers and applicators: Small sections of a branch of soft woods such as from the smaller *Commiphora* species or green *Jatropha unicostata* wood were hollowed out with a knife or burned out to make a small container. This was used to store the rare imported antimony used along the rims of the eyes to improve vision and to beautify the eyes. The stopper was usually made from a small pad of material or a piece of wood whittled to fit the mouth.

Bowls: A knot from the wood of a tree such as *Commiphora ornifolia* was used to make a bowl in which cereals were served or prepared, or smaller bowls used for ritual ablutions or to hold water being passed round for drinking. Small bowls for serving milk or buttermilk were also made especially for the use of young children.

Tambourines: Hard woods such as tamarind and ziziphus were considered to be the best woods for making tambourines. A large branch was cut into lengths or slats of different sizes that were de-barked and then whittled until their sides were parallel. They were put to soak in water for a week to 10 days, then removed from the water and, while still soft, were curved around a rock or boulder and held in position with weights and ropes. They were then left to dry out. Holes were bored along one edge of the curved slats to take the wooden pegs. Then a previously prepared and dampened sheepskin was stretched over the edge along which the holes had been bored. The skin was stretched tight and the pegs hammered through the skin into the bored holes, leaving the outer side of the pegs flush and the inner part projecting. Leather thongs were then wound around the projecting stumps to hold them taut. Once the sheepskin had dried and shrunk taut, the metal rattles were added and fixed in place by cutting notches in the upper edge of the tambourine. The tambourine was 'tuned' by warming it over a fire.

Tool handles: Handles for agricultural and building tools were made from any hard wood. The wood was shaped so that the projecting part of the blade of the implement fitted as exactly as possible, and any gaps were packed with sticky *Euphorbia* or other latex, mixed with animal hair. The metal part was then hammered into the wood and the handle bound, usually with date palm cordage extending from the point where the metal meets the wood right to the other end of the handle.

Stakes: Fencing stakes are used to enclose cultivated plots, pens, date gardens and date-drying grounds. Stakes are also cut and wedged into cracks in the cave or hut walls to serve as hooks for hanging up household goods. Smaller stakes were hammered into the trunk of trees like the tall *Sterculia africana* to make a sort of makeshift ladder so that a herder could reach the crown of the tree to lop off foliage. Stakes were also used in this way to reach a beehive in a cliff face. The strongest and most durable woods were used for this purpose, such as *Buxanthus pedicellatus*, *Croton socotranus* and *Carphalea* species.

Weaning muzzle: This is a bar of wood to each end of which is tied a cord of goats' hair or some other soft material. The bar is placed across the tongue of the young animal and the cords passed under the chin, across the muzzle and around the back of the head where they are tied.

Digging sticks: Slim stems of hardwood are shaped to a point at one end and used as digging sticks: to turn the earth and prepare the ground for planting seed, to dig out clay, or to break up dry earth for roofing.

Crutches: Suitably shaped pieces of wood were also made into crutches and splints.

Camel saddles: Two kinds of camel saddle were made, the 'Arab' one and the 'Soqotri' one. Both are placed in front of the camel's hump.

The 'Arab' saddle was either imported or made in the capital area, Hadiboh, from driftwood planking. It consisted of two inverted Vs joined by a single length of wood. This sort of saddle was much simpler than the Soqotri saddle, needing only five lengths of wood to make it.

The 'Soqotri' saddle was made from local hardwoods and consisted of two A-shaped frames made from two sticks with a bracing stick across the middle. The two A-frames were joined at their apex by a longer stick, and braced by two others halfway down the frames. At the base of the two A-frames holes or loops were made through which the girth was fed. This sort of saddle required nine lengths of wood to make it, although fewer could be used if naturally V-shaped pieces of wood were found.

Weaving

All the woollen cloth produced in the Archipelago is woven from sheep's wool. The fibres are elastic, a good insulator, and can absorb a lot of moisture without feeling wet, especially if, as on Soqotra, the grease is not removed. Sheep were very numerous formerly, especially in the high limestone plateau sheepwalks. Their numbers have fallen drastically over the past decades. The shearing of sheep, the preparation of the wool and the weaving was carried out by women.

Wool-beating bow: The curls of newly shorn wool are 'beaten' with a 'bow' to separate out the individual strands, smooth the fibres and remove dirt and debris. The wool is laid across a leather mat and then 'beaten' with the 'bow' in a series of very swift plucking movements. The bow is made by women from a strong but pliable wood such as *Cryptolepis arbuscula*, or one of the climbing asclepiad vines. The 'string' of the 'bow' is made from a leather thong, or a thread twisted from calico cloth or cotton from the cotton bush.

Loom: Weaving is done on a horizontal ground loom. Strips some six inches wide and six to eight feet long are woven, six (usually) of these then being sewn together to make the wool rugs which have been exported from the island for over a thousand years. The simple looms are extremely suitable to a transhumant way of life, since they can be set up and dismantled as required, the unfinished work and the loom parts simply being rolled up, carried to the new site, and unrolled again, ready for the weaving to continue. The wooden parts of the loom frame are made from local hardwoods, held in place by four large rocks or four stakes hammered into the ground as warping posts.

Sword beater: The sword beater over which the threads are stretched is made from a shaped piece of wood from *Chapmannia* species or *Zygocarpum coeruleum*. This is the only part of the loom which needs to be carved, and is usually made by men.

Shuttle: Finer threads are wound on a shuttle, usually made from a date palm rib.

Heddle bar: This is made from a local hardwood, from trees such as ziziphus, *Chapmannia* species or *Zygocarpum coeruleum*.

Breast beam, warp beam and shed stick: These are usually made from date palm ribs.

Medicine for people and livestock

Plants have made a great contribution to medicine. Still today about a third of the pharmacopoeia have a plant component, and another third are derived from compounds using synthesised versions of plant chemicals. Biochemical analysis of traditional remedies based on plants has often verified their specific curative properties.

It is likely that plants on Soqotra have been used medicinally from the earliest times, their use determined empirically and the expertise passed on through the generations. However, despite the islanders' undoubted intimate knowledge of the properties of their plants, and their unique and rich vegetation, the role of herbal medicine seems to be limited. This is not to say that plants were not used to treat illness – they certainly were – but Soqotrans primarily regarded plants as a source of fodder for their animals, and then as a source of food, fuel and timber, rather than as a possible remedy for illness. Honey, seen as the essence of all that was beneficial in plants, was highly regarded as a remedy, and was prescribed for almost any illness. Nevertheless the plants themselves did not play as major a role in the Soqotri *Materia Medica* as might have been expected, or as they do in related cultures nearby, such as the Yemeni mainland or Dhofar in Oman.

The medicinal value of some groups of plants is universally recognized, while other plants are used in one locality only. Where the same plants are used in the same way in different areas, or for similar ailments in other parts of the world, this might lend authority to their use. In some cases the medicinal use of plants is based on the 'doctrine of signatures', an old belief which perceives a relationship between the shape, colour or odour of a plant and features of an illness; for example plants with red sap or resin being used for disorders of the blood or to control bleeding, such as is the case for dracaena on Soqotra.

Illness on Soqotra

The worst time of year for illness was the end of the monsoon and in the transitional period before the winter rains fell, when people were hungry, livestock unproductive, fish scarce and everyone exhausted by the long dry months of ceaseless wind. It is said that those who lived on the coast used to suffer more illness than those in the interior and that the healthiest people were those who lived in the cooler, higher areas.

The islanders agree that the most common illnesses of people formerly were:

Children: Diarrhoea, unexplained fevers, skin complaints, coughs and eye infections. Formerly babies were breastfed for three years and even longer. This helped to space out pregnancies but also gave a certain degree of protection from some of the commoner ailments.

Men: Unexplained fevers, headaches, chest complaints, skin complaints, eye problems and joint pains in the elderly.

Women: As for men, plus mastitis and problems in childbirth.

The winter rains generate a sudden explosion in the fly population and this leads to a similar increase in eye infections, diarrhoea and skin infections, especially in children. The most dangerous diseases are seen as the epidemics of measles and whooping cough that occasionally sweep the islands, killing many children and even the occasional adult. If a family member is suffering from one of these infectious illnesses, their movement is restricted because people of other areas are reluctant to risk catching the disease by allowing them into their area. Apart from these infectious diseases, the greatest killer of earlier years was said to be severe headaches with accompanying high fever, which struck adults and children alike and often proved fatal.

The perception of illness has changed. Now islanders see the main problems as being malaria (a recent diagnosis), and a vague debility which they largely ascribe to the change in diet to one whose basic ingredients are processed white sugar, flour and rice in place of the milk products, maize, dates and finger millet of earlier times. Babies are now usually breastfed for six months to a year, and some young mothers feed powdered milk from birth. Islanders feel that the incidence of severe and chronic diarrhoea in infants, of fevers and of childbirth-related problems has risen, but that the incidence of skin complaints and chest complaints has remained about the same.

Sorcery and illness

A breakdown in health, in humans or livestock, was traditionally seen as being the result of errors of commission or omission by the sufferer, or of malign influence. This was the case even where the agent was readily apparent. Thus falling out of a tree and breaking a leg, for instance, was not an accident, but rather the result of a deliberate wish on the part of someone that this should happen. Ultimately all disease could be laid at the door of some malicious individual, commonly someone known to the victim, and usually an old woman. Mishap and illness were also caused by the (usually malevolent) casting of the Evil Eye, though the Evil Eye can also be invoked inadvertently: in shock, or when innocent admiration is improperly expressed. Although an islander was generally powerless to prevent a witch's attack or to avoid the Evil Eye, he may well have been guilty of having laid himself open to this by inappropriate behaviour, or by the omission of some necessary prophylactic measure.

The most intractable illnesses, however, were considered to be the direct result of sorcery. Soqotran sorcery has long been famous (indeed is mentioned as a characteristic of the island by Marco Polo). Sorcerers were always female, nearly always old women. They had entered into an alliance with a *jinn* who was able to take many forms, usually animal, and could thus approach the victim unnoticed. They could be discovered, held in check or defeated only by a small number of *mεkóli* (*mekoli*). These were men who were specialists in sorcery, who also had powerful *jinn* allies on whose assistance they could call. The basic difference between the two was that whereas the witch always went about her business in the greatest secrecy, the *mekoli* worked in public and his labours were rewarded by those who had appealed to him for help.

Thus a cure for serious illness depended on the defeat of one contender, the *jinn* of the witch, by the other, that of the *mekoli*. This gladiatorial contest was usually carried out in an idiom foreign to all but the two locked in combat. If the *mekoli* lost, the only solution was to employ further *mekoli* until one powerful enough to name and thereby overcome the witch was found. A named and formally accused witch, if she denied the charge, had to undergo trial by ordeal. She was taken out to sea and thrown out, bound hand and foot and weighed down with a rock. A witch who did not sink to the bottom was thereby proven guilty and was banished from the island (often to Sur on the coast of Oman).

Diagnosis and treatment

There is a basic understanding of bodily structure and function in Soqotra, a result of the close contact Soqotrans have with their livestock, both living and when butchering a carcass. However, except in cases of simple trauma, such as a fracture, sprain or burn, there is little understanding of how and why the body 'goes wrong'. Diagnosis and treatment concentrate firstly on signs obvious to the observer (such as inflammation, swellings, the spots of measles); secondly on symptoms experienced by the sufferer (such as problems with vision or hearing, feelings of pain, dizziness or nausea), and thirdly on the history given by the patient and those close to him (such as a raised temperature, breathlessness, loss of appetite, coughing or vomiting). Successful treatment may well occur independent of diagnosis, but this is most unlikely for the majority of serious infections or malignancies, for instance, or for degenerative, metabolic and nutritional disorders. That some diseases were infectious was realised, and led to the appropriate avoidance of contact. However, illness was generally perceived as a conglomerate of disparate symptoms and signs, any or all of which might or might not respond to various treatments. Traditional healers in general aimed to treat such overt signs rather than any underlying disease, whose pathology remained unknown to them. For instance, 'stomach pain' could mean indigestion, an ulcer, or cancer of the stomach. Indeed the very concept of diagnosis, as understood in modern medicine, which presupposes the existence of an underlying distinct functional disorder – as opposed to theoretical causal phenomena – is usually absent. Thus remedies tend to progress from the pragmatic (cooling fevers, setting fractures, purging a painful stomach) to, if these were unsuccessful, the more specific and specialised treatments (such as cauterisation). Initially, help was sought from within the extended family and then from within the local community. If the condition persisted or worsened, help had to be sought from further away. This could be from healers with specialist skills, in bone-setting or cauterisation, for instance, or from those endowed with more supernatural skills, such as the *mekoli*.

The most widespread treatments for illness were:

- Cauterisation, mostly with heated cauterising irons, but also with caustic plant latex and hot ash or embers from the fire.
- Offering up of an animal in the name of the patient, entreating the Creator to send better health.
- Honey, taken first thing in the morning on an empty stomach, or applied to the painful or affected place.
- Giving alms to those in need.
- If none of these work the specialist healer and sorcery detector, the *mekoli*, was called in to discover who lay behind the illness, and, by disabling its author, to cure it.

The smoke produced by burning certain plant and other materials was widely used therapeutically, the smoke being inhaled or body parts being fumigated. Many remedies were taken for their laxative or emetic effect, to 'cleanse the system of impurities'. Bleeding and cupping were also practised with the intention of ridding the body of the 'bad' or 'excess' blood, believed to be the principal cause of the symptoms. Shock treatment and massage were also practised. Traditional plant remedies focused particularly on gastro-intestinal, inflammatory, dermatological, ocular, pulmonary and obstetric/gynaecological disorders and ailments. The many remedies for burns reflect not only the use of open fires but also widespread use of cautery as a remedial treatment for a variety of illnesses.

Other less common procedures for protecting oneself from illness and discovering the cause of the illness also existed:

- Asking the *muʿallim* (*mu'allim*) (traditionally anyone who could recite some Qur'anic verses or repeat pious formulae) for a protective amulet.
- Consulting one of the seers who could discover hidden truth by casting pebbles, or by 'reading' soot marks on the shoulder-blade of a carcass held over the flames, or by interpreting the stars.

Those accused of sorcery by the *mekoli* or others could choose to undergo trial by ordeal to establish their innocence. This was most commonly done by two tests: the *mūrʌt* (*muwrat*) test which involved laying a heated knife blade on the tongue, guilt or innocence being determined by whether the tongue was burned or not; and the "*sūrat yāsīn*" (*surat yasin*) test in which a booklet of Qur'anic verses was suspended from a thread, and guilt or innocence was determined by how it swung.

Procedures such as these are likely to have reached the island from overseas, since they are more prevalent in the larger coastal villages and involve much non-Soqotri terminology.

Cauterisation

In theory, anyone can do this, on people and on animals, but in practice it was recognised that some had more expertise in this area than others. These were people who were more informed about anatomy and the diagnosis and development of disease, so knew exactly where to place the heated iron, what size of iron to use, how deeply to burn and over how extensive an area. The general aim of cauterisation is to stop the disease spreading along the '*'aruq*', or 'nerve pathways' that link the different parts of the body. This is achieved by 'blocking' its progress by an application of a heated iron at one or more points along the 'pathway'. This method of treatment continues to be practised in several areas of the world, and is seen by many as being of genuine curative, or at least palliative, value. This may well appear to be the case when an illness is anyway self-limiting, or of psychosomatic origin, or in cases where the referred pain of the cauterisation successfully supersedes the pain of the original complaint.

Livestock illness and veterinary medicine

Apart from simple trauma, the greatest problems with livestock were those that affected milk production, such as mastitis. These conditions were of great economic consequence in a population so dependent on milk as a food source and as the basic material for their principal saleable commodity, ghee. The search for a reliable cure for this problem has occupied generations of local healers. The problem is exacerbated by the traditional reluctance to wash the hands between milking different animals, since this is seen as 'repudiating the blessing of the milk'. Other common livestock problems were diarrhoea, coughs and colds and eye infections. The main concern of livestock herders was to keep their animals adequately nourished in order to protect them from the many parasites, both internal and external. Healthy animals were seen as less likely to suffer from infections and their pregnancies less likely to end in abortion. Care had to be taken to avoid toxic plants, to move animals in time from *'uqabah* or salt grazings, and to give them adequate shelter and warmth at night. When, despite all care, their animals fell ill, by far the most common treatment was cauterisation.

Soqotri non-plant medicaments

Some non-plant therapeutics were widely used, such as breast milk and early morning saliva of people themselves; similarly the milk products of livestock, as well as their hair, horn, hoof and dung. Some popular medicaments – such as myrrh, antimony, turmeric, cloves, asafoetida, sesame oil, ginger, coffee beans, resin from various *Ferula* species and the seeds of *Trigonella foenum-graecum* – were imported and could only be obtained at the coastal markets. More recently, tomato paste, hair cream, sugar, garlic, onions and fizzy drinks have joined the Soqotran *Materia Medica*. Minerals and soils were also used medicinally, such as sea salt, rock salt, termite earth, various soft stones, quicklime, ash, charcoal, clay and fine red earth. Some remedies were available from the sea, such as sharks-liver oil and dried shark-meat, the fat of dolphins and turtles and the highly prized ambergris. Some were provided by the island fauna, such as the Egyptian vulture, the green pigeon, snails, beetles, the chameleon, the wild goat, even ticks and the humble head louse. An important medicament, and one that was sold commercially, was the thick, yellow-brown secretion squeezed from the anal pouch of the civet cat. This paste was used on the island and elsewhere to treat a variety of ailments, but especially headaches and fever. It is taken by sniffing it, by rubbing small amounts on the skin, or by adding some to food and taking internally.

The traditional healer

There was no class of 'traditional healers' as such on the islands. There were people known for their skill in specific fields – cauterisation, bone-setting, massage, or insight into the supernatural, for instance. Nevertheless, some individuals became renowned island wide for their expertise in healing, and sick people would travel across the island to consult them. They were highly regarded, and still have a role to play today.

Abayomi Sofowora (1982) discusses some of the advantages of traditional medical practitioners. They:

- Fill the vacuum in health care created by the shortage of manpower and the high cost of training modern health care workers.
- Have developed traditional skills in the dispensing of curative, preventative and rehabilitative care.
- Use an astute approach to human ecology and health in their treatment.
- Belong to the same culture as their patients, sharing common beliefs, values and symbols of communication.
- Are effective in some aspects of psychosomatic medicine and in the use of local herbs, roots and barks for symptomatic treatment.
- Have skill in interpersonal relations, including counselling with sympathy, identification and concern.
- Are unhampered by inadequate transportation in rural areas.

Much of this is relevant for Soqotra. Nevertheless, it is as well to remember that improved health for the islanders depends more on greater access to clean drinking water, better nutrition, vaccination and antenatal care, for example, than on developing herbal medicine or their traditional medical skills. In an ideal world the best of both traditional and modern systems would be combined.

Bedding and stuffing

Some plants are laid on the ground and covered with a wool rug or skin to make a mattress. Other plants are used as stuffing for bolsters, or for pack animal equipment. Sprigs of sweet-smelling plants are often laid amongst bedding or added to the stuffing for bolsters and pillows to give it a pleasant smell. Fragrant plants are particularly used in the bedding for sick people, as the fragrance is believed to relieve their discomfort and hasten recovery.

Cleansing

Plants are used as shampoos, soaps, and to remove or disguise other odours. Certain succulent plants are crushed and used to cleanse the skin, or to rub clean tainted clay pots and dishes. Strong-smelling and juicy plants are much used by people handling fish or dealing with carcasses to remove the smell and the blood. Sprigs of fragrant plants are tucked into clothing or rubbed on the skin to perfume it. The sniffing of sweet-smelling plants is held to protect against 'noxious air' which is suspected of causing disease, and to act as a safeguard against infection. All sweet-smelling plants are endowed to some extent with beneficial medicinal and prophylactic qualities. A paste of fragrant plants is rubbed over the back of a suckling animal to disguise its smell before presenting it to a milker not its mother to encourage her to accept it as her own and allow it to feed.

Cosmetics

Plants are used cosmetically to dye the lips and gums, to darken the eyes, to make the teeth gleam, to freshen the breath, dress the hair and to stain the skin.

Commercial value

Some Soqotran plants have historically been of considerable economic importance, and some still are today, though on a much-reduced scale. These include the dried juice of aloes, dracaena resin, *Boswellia* resins and, to a lesser extent, resins from some *Commiphora*. Nowadays there is also a market for the wild bitter oranges, the ziziphus fruit, toothsticks, chewsticks, and, increasingly, for timber.

Pest control

Latex as a pesticide

The caustic latex of some plants is used as a rodenticide or insecticide.

Smoke as an insecticide for living quarters

Fumigation with smoke is seen as the most useful means of pest control, effective against a wide range of vermin, as well as against disease generally. Before leaving a settlement when transhuming, and before moving into new quarters, it is customary to carry fire around in a dish and thoroughly fumigate human and livestock living quarters. This is repeated at regular intervals to keep vermin numbers down.

Smoke as an insecticide for livestock

The dense and acrid smoke produced by burning certain plants is used to protect livestock. As well as discouraging ticks, fleas and the biting flies that come with the winter rains, the heat of the smudge warms the animals and relaxes them for milking. Smudges are also seen as soothing and comforting for sick animals, and causing the noses of small stock in particular to run. This is believed to cleanse them of 'impurities' and to eliminate any botfly larvae from their nasal passages. Warmth is always important, since cold of any sort is seen as being very harmful to livestock. At times of the year when the animals are left out overnight, the smell of the burning wood and the sight of the rising smoke act as a signal to livestock to come to the milking yard. Smudges are lit in the milking and handling areas and inside the byres and other night-time quarters, every day during the cold and wet winter season and wherever the monsoon brings low cloud and drizzle. What is burned, when and where, differs according to season and type of livestock.

Smoke to drive out noxious creatures

Certain woods are burned to rid living quarters of creatures believed to be harmful, such as the giant centipede, biting ant, scorpion, certain spiders such as the wolf spider, and reptiles such as skinks and snakes. It is said that they cannot abide this type of smoke and are driven from their hiding places by it.

Smoke-treating timber

Timber is smoke-treated to deter termites and other wood-boring insects. Smoke treatment is also seen as making the wood harder.

Smoke and ash as an insecticide in cultivation

The ash of certain plants is scattered over growing plants as an insecticide. The very acrid smoke given off by certain woods or aged dung is used in the same way. Sometimes fertilising the soil and adding the insecticide are combined. When preparing a seedbed or working the soil of the plot before sowing seed, dried dung of cattle or sheep was placed in small heaps across the plot and the insecticide plants laid on top. The heaps were set alight, and once they had burned down, the ash was spread over the plot and dug in.

Other uses

Inks

The resin or crushed bast of certain plants was used to make inks, often with the addition of wood ash or soot.

Dyes

Plants were used to stain leather or to dye the coarse, unbleached calico that used to be imported for clothing. Cotton fibre, both locally produced and that imported as cotton twist, was dyed and used in some areas in the manufacture of sheep's wool cloth. Wool thread was sometimes dyed and used to weave coloured stripes in rugs. The colour was never as strong or as clear as that produced by imported dyes. Plants were used to dye and strengthen cotton fishing equipment to make it less visible when in use. The dyes were extracted from the plants in the simplest manner, by prolonged soaking in water. There was no use or understanding of mordants. This meant that the dye leached out with washing and rapidly became bleached on exposure to the sun.

Adhesives

The latex and sap of certain plants was used as a general all-purpose adhesive.

Plant latex as bird lime

The latex of certain trees, such as *Euphorbia socotrana*, *Ficus cordata* and *F. vasta*, was used as bird lime. *Euphorbia socotrana* latex has the unfortunate characteristic of becoming runny in direct sun, allowing the stronger birds to escape, while that of *F. cordata* withstands the midday heat better so is preferred for liming sites in full sun. The expertise of the trappers, usually boys, lay in knowing where to find the different species of bird at different times of the day and where they were likely to perch. The latex is removed about halfway along a well-developed branch, the bark being cut or nicked and then left to ooze. The latex is gathered by twisting a stick around in it. It is then put in a pot of water and left there until early the next morning when it is taken out and the latex peeled off. This is stretched out to make long strips, like a bandage. The best time to set the traps is early dawn, or just before midday around the water pools where birds come to drink. The strips of bird lime are laid on rocks in and around water pools, along the branches of date palms close to any ripening dates especially, in the date-drying areas, and along the stone wall or palm-rib fencing enclosing cultivation. They are pressed into place with wetted hands to prevent the lime from sticking to them. It is better if the trapper remains nearby to keep an eye on the trap so that he can immediately remove any bird that gets caught in the lime to stop its struggling putting off others. It is essential that at least one wing becomes glued to the lime; otherwise the larger birds, such as the starling or green pigeon, can extract their feet from the lime and fly away. If the birds are quickly removed as soon as they are caught, and the bird lime stuck down again neatly, it is possible on a good day to spend many hours in the same area, catching one bird after another. However, the most sought-after bird, the pigeon, is difficult to catch. It spends most of its time in date palm gardens and rarely lands on the ground. It can be trapped in bird lime if this is smeared on its accustomed perch high in the date palm. It is more easily taken when it has young. Once a nest has been discovered, the area around the nest is limed to trap the adults when they come to feed their young. Boys also climb into the palms to remove the nestlings. Otherwise it is caught by being chased from tree to tree by throwing stones at it until it is exhausted, or hit by a stone.

Plant latex to stun lagoon fish

This was usually done during the monsoon months, when livestock were unproductive, sea fishing was impossible and most people were down at the coast to work at the date harvest. Each lagoon was only worked once a year, when all those in the neighbourhood had to be informed so that they could come and get their share. Most of the smaller lagoons of the island were fished in this way: Khawr Hadiboh, Erhino, Shiq, Dibini, Tow'aq (above Qariyah: the lagoon at Qariyah is too big), Matyif, Di Kibshib (at Di Setmoh) and Di Nowtit (near Qalansiyah). *Euphorbia* latex was the most widely used, especially that of *E. arbuscula*, though *Adenium obesum* sap or crushed twigs could also be used. In the heavily populated northern plain, it was a very organised affair. *Euphorbia arbuscula* was used and a group of men set off early in the morning to gather the latex, having first prepared a thick liquid of dates mashed up in water and rubbed this over their forearms and face to protect themselves from the very caustic latex. They covered the rest of the body with oldest clothing, or with sackcloth, and took with them several clay pots and a goatskin of water. When they reached the *E. arbuscula* trees they would each take a pot, fill it about a third of the way up with water, and then bleed the trees by hitting the bark with a rock and collecting the latex in a cloth held below. The cloth was then squeezed out into the pot of water. When all the pots were full of latex, they were tipped into the emptied goatskin, and the men then went to the lagoon where by now many people were ready and waiting. One of them walked out into the lagoon towing the latex-filled goatskin behind him, the others following behind with clay pots filled with latex from the goatskin. They all then walked or swam around the lagoon where the water was deep, tipping out the contents of the bowls as they went. When the goatskin was light enough, its contents were tipped directly into the middle of the lagoon. The latex works quickly, bringing the fish gasping to the surface. The first pick of the fish went to the men who had gathered the latex and worked the lagoon as their reward. After that it was a very splashy free for all. The taste of the fish is not affected by the latex, nor does the water treated in this way become undrinkable, though some animals are said to be reluctant to drink more than a small amount for a few days afterwards.

Quicklime

Quicklime and lime mortar of good quality has long been produced on the islands. Its manufacture was a source of income to anyone who knew how to make it and had access to the necessary materials. The demand for quicklime remains high today, and although it initially slumped with the introduction of cement, has risen again over recent years. This is partly because imported cement is expensive, and partly because cement has proven to be less successful than the more flexible quicklime for waterproofing the flat-roofed houses. Quicklime is now mainly used for roofing and as a lime wash for the walls of buildings, especially mosques, cement having largely replaced it as a mortar. Some islanders are now successfully using a combination of cement and quicklime, covering the roof with a layer of cement and then with a thick coating of quicklime. Whereas cement often cracks after rain, quicklime is more elastic, contracting and expanding in heat and cold, and is thus less susceptible to cracking. The materials used are wood and coralline rock or limestone. Quicklime is more expensive now than formerly, as there is a shortage of suitable wood for making it. The best wood is considered to be dead palm-tree trunks, but these are in short supply nowadays, and other dead woods are also used, in the main those which are of little use to livestock or which are abundant and quick-growing. There are many new date palm gardens now but they are all much of an age, and therefore as yet have few old or infertile palms to sell for quicklime manufacture.

Quicklime manufacture: The length and width of the limekiln are not important, but the deeper the pit, the better. The walls are lined with stones of a hard rock that will not shatter when fired. The standard proportion of wood to rock is two to one. The wood is cut to a size that fits the pit, and then, starting from the bottom, the layers are built up. The first layer is of dry date palm leaflets; then a layer of wood interspersed with more dry date palm leaflets; then a layer of stone; then a layer of wood alone; then a final top layer of stone. The stone layer is laid with great care so that it does not touch the walls of the limekiln, while the wood layers are built up right against the walls. As the layers are built up, two holes running from the top of the pit to the bottom, or four holes, one at each corner of the pit, must be left. These are filled with bundles of whole dried date palm branches, which have been tied one above the other to make a single 'pole' tall enough to reach the bottom of the pit and still protrude well beyond the top layer of stone. The holes act as channels for the air and the tops of the 'poles' are set alight, and the flames left to work their way slowly down to the bottom of the pit.

The pit is then left to burn for at least two days. When still alight and red hot, cold water is poured all over it. Although some use the brackish water of lagoons, the quicklime is considered to be of a better quality if fresh water is used. The kiln is then left to cool down. The next day it is inspected to see if it is still alight. If there is smoke still rising from the pit more water is poured over. Once the last of the fire has definitely been put out, the pit is left for at least a week to allow it to 'ferment and settle'. If it rains while the quicklime is still in the pit, it must be left untouched until it has completely dried out. Finally, all the pieces of unburned stone and wood are removed, and the powdery mass remaining is scooped up and bagged. Sometimes the burned-down contents of the limekiln are sold and the job of extracting and bagging the lime becomes that of the purchaser. Unused quicklime turns hard in the sack, but can be broken down to a powder again by beating it with a heavy stick.

To prepare quicklime for use as a lime wash: Two or three sacks, depending on sack size, are poured into a drum of water, and the mixture stirred daily for about three weeks until it has formed a rather liquid, smooth, white paste. It is usually applied with a pad made from the fibrous leaf sheath of the date palm.

Uses of quicklime: After mixing with sand and water, it was formerly used as a lime mortar for walling in much the same way cement is used now. It was used to lime rock pools and cisterns (and, in the former Sultans' gardens at Ileheh, to line the water channels). It was used as a lime wash for both internal and external walling, and to render the flat roofs watertight, a layer of thick paste was spread all over the roofing area at least once a year. Cotton fishing nets were strengthened and bleached by soaking them in a weak quicklime solution.

Medicinal use of quicklime: It was ground to a fine powder and applied to infected, deep or extensive lacerations to dry them out, to fresh cuts, or the surgical wound of circumcision to dry them out and speed their healing. The best lime for medicinal use was considered to be that made from *Ficus cordata*, *Lannea transulta*, *Ziziphus spina-christi*, *Euphorbia socotrana* or any species of *Commiphora.*

Plants in pottery repair

Certain plants of a viscid nature were used to repair holes and cracks in clay pottery.

Plants as fertilisers

Although the main fertilisers were animal dung and bat guano, the dark earth dug from underneath and around certain plants, such as *Euclea divinorum*, *Dovyalis abyssinica*, *Ficus cordata*, *Ficus vasta*, *Citrus aurantium*, *Acridocarpus socotranus*, *Rhus thyrsiflora* or tamarind trees, was also used as a soil improver. The soils from around trees such as *Sterculia africana*, *Dracaena cinnabari*, *Boswellia* species and the larger *Commiphora* species, and from around smaller plants such as *Hypericum scopularum* and large clumps of *Cissus* species, were also used. Date palm foliage was crumbled when dry and added to soil to improve it.

TOXIC PLANTS

Some plants are recognised by the islanders as being toxic to livestock. The toxicity varies according to animal and time of year, the worst effects occurring when the animals are very hungry, when the toxic plants are dried in hay, or when animals eat the poisonous underground parts. Among the more important factors affecting prevalence of poisoning by plants are:

- **Season of year:** Some plants may grow well in one season and be dormant in another; they may be highly toxic at the early stage of new growth, when the flowers and seeds are forming, or at all seasons and at all stages of growth.
- **Climate:** Generally, high humidity and temperature favour growth of poisonous plants and development of the poisonous principle in them. This does not necessarily lead to a higher incidence of poisoning – similar conditions also favour the growth of palatable forage and so reduce the risk of animals ingesting the poisonous plant.
- **Accessibility.**
- **Palatability:** This is a relative term: animals on a satisfactory plane of nutrition will not eat poisonous plants when other more nutritious species are available, whereas when starving in the middle of a drought will often graze toxic plants to satisfy their hunger. Some highly toxic plants are never consumed because of their unpalatability which in itself may be due to bitterness, the presence of spines, the woodiness of the plant, or the ability to cause blistering or a rash around the mouthparts of an animal.
- **Species and age of animal:** Some animals are highly tolerant of some poisonous plants while others are susceptible – sheep and goats are usually more tolerant to most toxic plant foods. Young animals are usually more susceptible.
- **Familiarity:** Animals may become familiar with a poisonous plant through past association. Certainly animals moved from an area in which a poisonous plant does not grow to one where it does will consume it and sicken, whereas animals born and raised on such an area will be quite safe, a set of circumstances frequently occurring with transhumance.
- **Toxic part of the plant:** Many plants are toxic in all their parts, some only in one or two parts (e.g. seeds or flowers or roots).
- **Parasitic fungi:** These can invade and render toxic an otherwise harmless plant.
- **Variability of toxic principle:** Many plants that are perfectly good forage under certain conditions of climate, soil type, etc. become highly toxic. This even applies to certain well-known grasses. Some highly toxic plants when dried and treated in some way are rendered harmless. Sometimes plants will change from the toxic to the non-toxic or vice versa for no apparent reason.
- **Soil type:** Type of soil has a marked influence on the toxicity or otherwise of plants which grow on it (Hall 1977).

Some of the common poisonous plants on the islands of Soqotra, and their effects, are:

***Ricinus communis*:** Ingestion of the seeds can cause tetanic spasms, bloody diarrhoea and profuse sweating. Cattle may die in convulsions.

***Tribulus terrestris*:** In Australia it is reported to cause photosensitisation; in South America animals eating it were reported to show inability to eat and drink, and have fever, oedema of the limbs especially, blindness, jaundice, purulent dermatitis and in advanced stages asphyxiation resulting in death. In Asia it is not believed to be poisonous. In South Africa it is reported to cause photosensitisation and liver damage. In West Africa it is reported that it appears to be most toxic during humid changeable weather at the beginning of the rains.

***Argemone mexicana*:** This does not contain morphine but is rich in other toxic alkaloids and the seeds are particularly toxic. Sheep eating it during drought have been poisoned, the main effect being severe emphysema and oedema of the lungs. In all animals there is intense capillary dilatation leading to loss of fluid from the tissues.

Asclepiadaceae: The toxic principle is a mixture of glycosides. Animals ingesting the plant suffer from gastroenteritis and heart conditions: weakness, paraplegia, laboured respiration, convulsions and death from respiratory failure.

Liliaceae sensu lato: Most of the lilies are poisonous; they are also unpalatable, but many grow on pastureland and may be eaten accidentally; they can cause convulsions, vomiting and diarrhoea. Animals die from cardiac arrest.

Solanaceae: Symptoms vary according to the plant ingested, but generally there is dullness followed by increased pulse and respiration, nervousness, muscular tremors, followed by paralysis, a drop in temperature, slowed pulse and respiration, relaxation of sphincters, recumbency and death. Animals will usually avoid eating these plants, but in times of drought when they are starving they may eat them (Hall 1977).

SOWING AND TRANSPLANTING VALUED PLANTS

Certain plants were cultivated, either sown from seed or brought in from the rangeland and transplanted close to the family settlement. The best time for this was at the onset of the winter rains when there is a lot of cloud and the air is cool. More recently, attempts have been made to cultivate new species, such as dracaena, tamarind, *Sterculia africana* and *Maerua angolensis*.

Euphorbia arbuscula is principally cultivated as drought fodder for goats, and is grown from branch tips with a bunch of leaflets. These are placed in a hole dug in the ground – the best soil is fine red earth – and watered until established. It must be enclosed to protect it from livestock.

Alternatively a branch is cut from the tree and planted beside it in a hole filled with water, or a bunch of leaflets is removed from the tree and planted in the middle of a clump of *Cissus* to protect it from livestock. Cultivation is generally successful, and thickets of these plants have been the result. It is widely practised in the drier areas of the island.

Aloe perryi is cultivated for medicinal use, and new aloe colonies are started in areas where aloes juice is commercially harvested. It is grown from seed or vegetative offset from a young plant or whole mature plant. Offsets are transplanted to an area of similar habitat to that of the parent and watered lightly until established. When offsets are removed from the parent plant, care must be taken to ensure that the rooting system that joins them is cut cleanly, leaving adequate roots to support both. Seed is planted in a small hole filled with water and then covered over. In areas where goats are known to eat them, and when planted close to the settlement, new plants must be enclosed to protect them from livestock. Cultivation is generally successful, especially the transplanting of young plants and vegetative offsets.

Commiphora ornifolia is cultivated as livestock fodder, for its fruit and for its medicinal qualities and, in earlier times, also for firesticks. It is transplanted to provide a living support for the roof of a shelter for people or animals. It is grown from seed or cuttings and transplanted into soft ground with gravel and pebbles. It is always better if dung is crumbled up into the hole dug to receive it. It is watered and protected from livestock until established. Young seedlings or a number of seeds are sometimes planted in the middle of a clump of *Cissus* species for protection. Cuttings are taken from large branches and sunk in the ground into suitable earth with added fertiliser as above, and watered as often as possible until it roots and/or puts out new leaves. Or the entire tree is cut off above ground and transplanted as above. Roots usually form if this is done before the winter rains and it is most successful when the selected tree is in vigorous green leaf at the end of the dry season. The stump of the original tree usually survives and puts out new shoots in the rain.

Small-leaved species of *Commiphora* are cultivated as dry season livestock fodder and for their medicinal qualities. They are grown from seedlings or saplings or from a young tree cut off above the ground. They are best transplanted to an area whose habitat is similar to that of the parent tree. Otherwise they are treated as for *C. ornifolia*.

Boswellia ameero and other large-leaved *Boswellia* species are cultivated as dry season fodder. They are grown from saplings or from a young tree cut off near ground level. The sapling is transplanted and watered until established and must be enclosed to protect it from livestock. Otherwise the entire tree is cut off above ground and transplanted as for *C. ornifolia*.

Citrus aurantium is cultivated for its fruit which now have commercial value as well as being liked for their own sake. It is grown from seed, seedling or sapling and sown or transplanted into fine red soil to which powdered dung and dark earth collected from under a mature orange tree have been added as a fertiliser. It must be well watered at regular intervals until well established. It is enclosed to protect it from livestock until it is growing on strongly. Cultivation from a transplanted sapling is usually successful.

Ziziphus spina-christi is cultivated as a source of dry season fodder and for its fruit which now have commercial value as well as being liked for their own sake. It is grown from seedlings or saplings, which are transplanted to an area of good and preferably gravelly soil, not too far from a water source. Water is needed at regular intervals and small side branches are regularly pruned so that the sapling grows tall and straight with a single trunk. It must be enclosed to protect it from livestock. Transplanting is most successful when a straggly young sapling is found, its side branches cut off and then used to make an enclosure for the transplant. Cultivation of this plant is successful only in the better-watered areas of the island. Attempts elsewhere have failed and planting seeds has not been successful in the interior.

Gossypium barbadense was formerly widely cultivated for its fibre, which was used cosmetically, medicinally and in laying out the dead. It was also used in weaving woollen and cotton cloth and by fishing communities to make fishing lines. It is grown from seed, seedling or a young plant. Seeds or seedlings were planted in an area of soft soil, or in a corner of an enclosed date palm or finger-millet garden. It must be watered well and must be enclosed to protect it from livestock – goats in particular are very keen on *G. barbadense* foliage and flowers. Cultivation is usually successful.

Ocimum forskolei is cultivated for its fragrance and is grown from seed. It is planted in any soil. It needs regular and plentiful watering to produce most foliage, and it should be protected from livestock. Cultivation is usually successful.

Jatropha unicostata is cultivated for medicinal use. It is grown from seedlings or saplings, which are transplanted into a habitat similar to that of its original one and watered until established. This unpalatable plant does not need to be protected from livestock. Cultivation is usually successful but is little practised since these plants are readily available in most of the island.

Dracaena cinnabari Recent attempts have been made to cultivate this out of curiosity and in the hopes that if successful its resin could be harvested and its fruit used as a source of dry season feed. It is grown from seed, seedling or a young plant. It is planted beside a rock or a crack in the stone pavement and watered until established, and enclosed to protect it from livestock. Seeds will sprout and young plants survive transplanting. However, as yet none of these plants have grown to any height or begun to branch.

Sterculia africana Attempts have been made to cultivate this as dry season fodder. It is grown from a sapling, which is transplanted and watered until established. It must be enclosed to protect it from livestock. This is rarely successful and consequently is little practised.

Maerua angolensis Attempts have been made to cultivate this for its timber and as dry season fodder. It is grown from a seedling or sapling. It is transplanted and watered until established. It must be enclosed to protect it from livestock. This has not yet proven successful and consequently is little practised.

Tamarindus indica is cultivated for its fruit, which have medicinal and commercial value as well as being liked for their own sake. Full-grown tamarinds also provide excellent shade and are a popular meeting place. It is grown from seed, seedling or sapling. Seed of ripe fruit is planted in rich black soil gathered from around the roots of a mature tree. Once a seedling is transplanted to its final site it must be watered well and regularly until it is growing on well and must be enclosed to protect it from livestock. Attempts to cultivate tamarind are relatively recent; there are many seedlings and saplings already growing in different areas of the island.

CONSERVATION AND THREATS

The isolation of the islands and the strong traditional land-use management practices employed by the indigenous population have both served to protect the vegetation and biodiversity of the Soqotra Archipelago. However, with the building of the seaport, the development of the airport and building of roads on the island, Soqotra is fast losing its isolation. An inevitable consequence of this is that traditional land-use management practices will be modified to adapt to these developments, with the risk that they may break down altogether.

NOTES ON APPLICATION OF IUCN THREATENED LIST CATEGORIES

The conservation status of the 307 endemic species and seven endemic infraspecific taxa (varieties and subspecies) has been assessed based on the IUCN red list categories and criteria (Version 3.1; IUCN 2001). These are a widely understood method for classifying species under the threat of extinction and by so doing focusing attention on conservation measures designed to protect them. The assessments made here are provisional and still have to be approved by the relevant committees of the Species Survival Commission (a commission of the IUCN). Plants have been placed in seven categories. The results are summarised in Table 3.

Table 3 IUCN categories for Soqotran endemic plants

Extinct	4
Critically Endangered	3
Endangered	27
Vulnerable	115
Near Threatened	14
Least Concern	118
Data Deficient	33
Total	**314**

The categories are defined below. Critically Endangered, Endangered and Vulnerable are the 'Threatened' categories. These categories are further divided into a series of quantitative criteria and subcriteria by which species are evaluated.

1. Extinct (EX). *A taxon is Extinct when there is no reasonable doubt that the last individual has died. A taxon is presumed Extinct when exhaustive surveys in known and/or expected habitat, at appropriate times (diurnal, seasonal, annual), throughout its historic range have failed to record an individual. Surveys should be over a time frame appropriate to the taxon's life cycle and life form.*

Four plants (*Barleria argentea* Balf.f., *Pluchea glutinosa* Balf.f., *Psiadia schweinfurthii* Balf.f. and *Valerianella affinis* Balf.f.) are considered to be extinct. These species have not been seen for 100 years or more after repeated searches.

2. Critically Endangered (CR). *A taxon is Critically Endangered when the best available evidence indicates that it meets any of the criteria A to E for Critically Endangered, and it is therefore considered to be facing an extremely high risk of extinction in the wild.*

Three taxa (*Cadaba insularis* A.G.Mill., *Pelargonium insularis* M.Gibby & A.G.Mill. and *Duvaliandra dioscoridis* (Lavranos) M.G.Gilbert) have been placed in this category. These are the plants under the most serious threat of extinction on the islands.

3. Endangered (EN). *A taxon is Endangered when the best available evidence indicates that it meets any of the criteria A to E for Endangered, and it is therefore considered to be facing a very high risk of extinction in the wild.*

Twenty-seven taxa have been placed in this category. These are plants with a very restricted distribution and also under a threat, most commonly by a decline in the quality of habitat due to overgrazing (*Neuracanthus aculeatus*) or by the trend of lower rainfall in the region and would be particularly vulnerable to the effects of global warming (*Begonia samhaensis* M.Hughes & A.G.Mill., *Ledebouria insularis* A.G.Mill. and *Pelargonium insularis*).

4. Vulnerable (VU). *A taxon is Vulnerable when the best available evidence indicates that it meets any of the criteria A to E for Vulnerable, and it is therefore considered to be facing a high risk of extinction in the wild.*

A total of 115 taxa have been placed in this category. A number of different criteria have been applied. The commonest category of threatened status is VU D2 with 106 taxa. This is rather a 'catch-all' category, which includes those narrowly distributed endemics which have 'populations with a very restricted area of occupancy such that they are prone to the effects of human activities or stochastic events within a very short period in an uncertain future'.

Using the definition of VU D2 is needed to assess potential threat. It seems likely that in general the greater threat would apply to the 'rarer' – that is, more narrowly distributed – taxa. Therefore, it became clear to us that an objective approach was needed to assess rarity. We decided to use the system developed by Rabinowitz (1981) (see Table 4). She has constructed a system using three criteria for assessing the 'rarity' of a species: geographic range, habitat specificity and local population size. The categories most relevant to island endemics would be those for plants with a small geographic range.

Table 4 Forms of rarity exhibited by island endemics (after Rabinowitz 1981)

Locally abundant in several habitats	[Rarity 1]
Constantly sparse in several habitats	[Rarity 2]
Locally abundant in a specific habitat	[Rarity 3]
Constantly sparse in a specific habitat	[Rarity 4]

We applied these four 'forms of rarity' to each of the Soqotran taxa in categories VU D2 and Least Concern. In the particular situation on Soqotra, it was decided that only the most extreme form of rarity (Rarity 4) would warrant Vulnerable status. Furthermore, on Soqotra we decided that it was appropriate to interpret 'habitat' as equivalent to one of the vegetation types defined in the vegetation classification developed for Soqotra (Miller & Hyam 2001). These vegetation types each include a number of more specific habitats and niches. If a narrower interpretation of habitat is used then plants classified as having a broader form of rarity [Rarity 3] could also be considered Vulnerable.

5. Near Threatened (NT). *A taxon is Near Threatened when it has been evaluated against the criteria but does not qualify for Critically Endangered, Endangered or Vulnerable now, but is close to qualifying for or is likely to qualify for a threatened category in the near future.*

Fourteen taxa have been placed in this category. It includes several trees (e.g. *Maerua angolensis* var. *socotrana*) which are abundant at present but which are likely to suffer as a consequence of the development of the island and the almost inevitable breakdown of traditional practices. There is minimal regeneration of *Maerua angolensis* var. *socotrana* in the more inaccessible areas of the island, but in the absence of alternative building woods, this straight-trunked, hardwood species continues to be under threat for building purposes. Likewise *Acacia pennivenia* and *Commiphora ornifolia*, whilst abundant at present, are both lopped as livestock fodder in dry periods (as is the evergreen *Maerua angolensis*); *Commiphora ornifolia* also suffers from lack of regeneration for reasons perhaps connected with the progressive desiccation of the islands. If livestock numbers increase greatly, or a succession of drought years should occur, all three of these species would be greatly threatened.

6. Least Concern (LC). *A taxon is Least Concern when it has been evaluated against the criteria and does not qualify for Critically Endangered, Endangered, Vulnerable or Near Threatened. Widespread and abundant taxa are included in this category.*

This category includes a number of common, widely distributed endemic species, such as *Croton socotranus* and *Jatropha unicostata*, which are under no immediate or perceived threat. Least Concern here contains taxa whose form of rarity is classified as 1 to 3.

7. Data Deficient (DD). *A taxon is Data Deficient when there is inadequate information to make a direct, or indirect, assessment of its risk of extinction based on its distribution and/or population status. Listing of taxa in this category indicates that more information is required and acknowledges the possibility that future research will show that threatened classification is appropriate.*

This category has been applied when data are uncertain; we have restricted its use to two main categories of plants: where their taxonomy is unclear (e.g. *Ballochia*, *Trichocalyx*) or where they have not been recognised in the field (e.g. grass species). We have tried to keep taxa in this category to a minimum (33 taxa) following the instruction that 'the absence of high quality data should not deter attempts at applying the criteria'. The bulk of taxa placed here are taxonomically poorly defined species (such as *Ballochia* species, *Trichocalyx* species and certain *Heliotropium* species) where no reliance can be placed on field records. A smaller number of species are included when they cannot be identified with certainty because they are only rarely seen in flower and are similar in the vegetative state to other species (for instance *Crinum balfourii*).

THREATS TO THE FLORA

We have identified five types of threat. The first four categories can be considered human in origin, the fifth is climatic:

- Development of infrastructure, including uncontrolled and unmonitored sinking of wells, construction of rainwater catchments and piping of water from springs.
- Breakdown of traditional land management practices governing seasonal transhumance and resting of the rangeland, timber and fuel extraction, plant products, and livestock and water management.
- Introduction of exotic plants.
- Plant collecting.
- Climate change.

The following sections discuss the likely impacts of these threats on the vegetation of Soqotra.

Development of infrastructure

The most obvious direct consequences of development are the immediate destruction of habitats, for instance by the clearance of vegetation for building, road construction, new date palm plantations and other cultivation. There are more subtle (and difficult to predict) effects caused by changes in land use such as traditional areas of rangeland being denied to livestock and the widespread sinking of wells and installation of pumps to create new date gardens. The effect of this new water management on the availability of water elsewhere and its possible effects on the vegetation should be carefully monitored. At present the direct effects of development are most evident on the coastal plains, particularly near large settlements, but they are spreading to other areas of the coast as people from the interior start to spend increasing amounts of time engaged in fishing – currently the only profitable activity on the island. The adverse effects of development on vegetation are likely to be felt over a wider area and increasingly into the interior as new roads are constructed and older roads improved. In the Red Data Book (Lucas & Synge 1978) *Taverniera sericophylla* was categorised as extinct. However, in 1998 it was rediscovered on an isolated coastal plain, some 15km from the type locality (Balfour 1888), where it now occupies a narrow belt parallel with the beach. The failure to find it at its original site suggests that its area of occupancy has declined since its original discovery. The new site was inaccessible by road but this has changed with the construction of a new vehicle track. There are also plans for developing the plain that would severely threaten this delicate coastal habitat and place *T. sericophylla* under critical threat of extinction. However, the recent discovery of the plant on the dry adjoining mountains much reduces the perceived threat to this species.

Another potential threat is likely to be destruction of vegetation as a direct consequence of new experiments in agriculture, which often involve heavy use of water, insecticides and chemical fertilisers and may introduce alien species that could pose a threat to the native vegetation.

Finally, there is a particular risk to some very restricted endemics through the development of springs. For instance, *Scaevola socotraensis*, *Helichrysum* sp. *C* and *Erythroxylon socotranum* are restricted to seepages next to springs on the escarpment cliffs above the Noged plain. These springs are already being modified for use by local communities and further development could lead to the destruction of this specialised niche and the extinction of these species.

Breakdown of traditional land management practices

The traditional land-use management practices employed by the indigenous population play a vital role in protecting against the overexploitation of natural resources and the diminution of biodiversity. However, as people increasingly cluster together in hamlets, villages and towns rather than living in small and highly dispersed family settlements, many traditional land management practices are being modified or abandoned altogether. This, together with the higher concentration of population in the larger settlements, increases local pressure on timber, fuel, water and livestock grazing.

Grazing

Pastoralists have an intimate knowledge of the island's vegetation, and appreciate its crucial importance to their continued survival. Livestock are grazed within traditional tribal boundaries, alternating between lowland and upland grazings at different seasons and thus allowing exhausted pastures a period of recuperation. However, changes caused in part by changing expectations as regards standards of living, and in part by a decline in the economic importance of the livestock sector, are already happening. In some areas signs of overgrazing are apparent, especially around the larger coastal settlements but also in other areas of the coast, as more people from the interior spend increasing amounts of time engaged in fishing. This is evidenced by an increase in weedy species and indicators of overgrazing such as *Tephrosia apollinea* and *Solanum incanum*. There has been a sharp decrease in sheep numbers all over the island. This means that the graze and browse of the high sheepwalks everywhere (the limestone plateaus) is now principally available for goats.

In some areas of the island, and on the outer islands, as livestock decrease in importance, increasing numbers of sheep and goats are left to graze unattended the year round. This is leading to an increase in the numbers of feral sheep and goats. Until very recently goats and sheep were actively herded and were moved down from the limestone plateau during periods of drought. In Samha, both the sheep (now largely feral) and the goats (actively managed only when in milk) are now left to roam freely for most of the year and now spend much of their time grazing the limestone plateau. This could quickly lead to overgrazing and even the eradication of vulnerable species such as *Begonia samhaensis* from all but the most inaccessible ledges.

Timber and fuel extraction

Traditional rules, supported by government, exist about cutting the more valuable species of live trees for building and other purposes – genuine need must be proved to the local community, and permission is not always given. However, enforcement of such rules is patchy and lopped branches are in great demand for a variety of building purposes, and there is no doubt that this poses a serious threat to the vegetation. Imported wood is expensive and not always readily available, and local woods command a good price. There has been a building boom on the island since the 1970s and consequent pressure on timber resources. Disputes over land ownership and consequent breakdown of traditional controls is likely to place an increasing pressure on all trees.

Maerua angolensis var. *socotrana* is an important source of timber. It is widespread on the island and abundant in places. Nevertheless, in some areas all the large trees have been selectively removed and only small trees remain. Another striking example is the mangroves of the northern coast, which have recently declined drastically in extent and quality. Other trees such as *Ziziphus spina-christi* and *Grewia* species are also likely to suffer from indiscriminate cutting for either fodder or fuel wood and timber as a consequence of the changes and population movement on the islands.

Most people on the islands rely to a lesser or greater extent on wood for their heating and cooking requirements. Pastoralists in addition burn certain woods for the abundant smoke they produce as livestock smudges. However, the lack of other forms of fuel, the rapid growth of many coastal settlements and the rise of new ones poses an obvious threat to the vegetation. In some of the inland villages the inhabitants have to go further and further in their search for fuel wood.

The dead wood of *Dirachma socotrana* is collected for the aromatic smoke it produces when burned. Although small quantities are offered for sale this is not considered a problem at present. However, should the demand for this scented wood increase it is possible that live trees might be cut for sale and this would, of course, seriously threaten its survival.

Plant products

The commercial exploitation of the vegetation is limited to certain resins and gums for local sale, some fruits (e.g. ziziphus berries) and roots (e.g. *Euclea divinorum*), various lichens, dracaena products for export, products from *Boswellia* species, certain scented woods, aloes, and wood for building and fuel. With the possible exception of aloes and wood for building and fuel, we do not consider the current rate of extraction for commercial purposes to pose any threat to the vegetation. Certain other fruits, roots, etc. which are collected for local consumption and not for sale (and which are unlikely to be offered for sale in the future) also pose no threat to the vegetation.

Bitter aloes – the juice of *Aloe perryi* – has important pharmaceutical and medicinal uses. At present this species is widespread and abundant in places and is not threatened by the sustainable, small-scale traditional exploitation. However, there is a real threat of overexploitation leading to extinction if the market for it opened up further.

A potential threat to the long-term survival of *Dracaena cinnabari* is the overexploitation of 'dragon's blood' – a resin obtained from the bark of the tree. At present the demand is small but a major increase in demand could be very damaging. Another threat is from the use of the trunks to make beehives. There was recently a case when over 20 trees were felled to make beehives for export to the mainland. This incident was generally condemned on the island but illustrates how a breakdown in traditional practices poses a very real threat to species on the islands.

Livestock management

Livestock are currently carefully managed but if the traditional controls are relaxed or lost there is great risk to vegetation and environmental quality on the island. The main threats are associated with:

- Uncontrolled breeding.
- Rises in livestock numbers due to supplementary cereal feed and rearing male young rather than slaughtering them.
- Livestock running free rather than being closely herded.
- Human population movements to larger settlements, where the practice of keeping a household herd in addition to the main one increases the pressure on grazing around these settlements.

Introduction of exotic plants

At present introduced exotic plants pose no threat to the vegetation. However, with the opening up of the island, they could become a problem. Examples abound on other islands where introduced plants have invaded and outcompeted the native vegetation. The introduction of living plants (including seeds) should be closely monitored. The only exotic (apart from weeds of agriculture) which is presently widespread is *Argemone mexicana*; however, this is not a new introduction and at present is not a problem. A few newly introduced shrubs of *Prosopis juliflora* were seen near the airport. This legume is a major problem on the nearby mainland. A programme to eradicate the species on the island has been undertaken.

Plant collecting

The uncontrolled collection of plants by amateur and professional collectors for private collections is a potential threat to certain succulent species. *Euphorbia abdelkuri* is a remarkable succulent endemic to the island of Abd al Kuri, where between 250 and 300 mature individuals are found at a single location covering less than 10km^2. The latex of *E. abdelkuri* is toxic so it is not eaten by livestock and is under no immediately perceivable threat except perhaps from collection by succulent enthusiasts. It is not found in similar habitats at other locations on the island, suggesting that there has been a decline in its extent of occurrence. That it was once more abundant is confirmed by Forbes (1903) who comments: 'This *Euphorbia* grows all up the side of Jebel Saleh [where it still grows] from about the upper two-thirds (500–1500ft), sending up its leafless stems like a forest of green candles'. It was classed as endangered in the Red Data Book (Lucas & Synge 1978).

Climate change

Long- and medium-term climatic change in the region leading to a gradual drying of the Archipelago is an unproven but likely scenario. This process of desiccation is likely to be compounded and possibly speeded up by the more recent effects of global warming.

The evidence for increasing aridity is indirect. No direct palaeoclimatic data are available from Soqotra, but there is strong circumstantial evidence that the islands have been getting drier over the last few hundred years. Evidence from surrounding regions points to extremely arid conditions from about 12 000 to 7000 years BP since which time the climate has remained arid with minor fluctuations (Hare 1977). Hemming (1966) in a major survey of the vegetation of Somalia suggests that the most recent process of climatic change in the country has been one of decreasing rainfall and increasing temperature.

In some areas of Soqotra there are disused field systems with associated farmsteads, sophisticated irrigation networks and laboriously constructed thoroughfares linking the main regions of the island. Archaeological activity is still in its infancy on the island but these structures suggest that there used to be farms with rainfed agriculture in these areas, which would have required more rain than falls at present.

Evidence for drying also comes from the distribution of *Dracaena cinnabari*. Local experts recognise that the distribution of dracaena coincides with the highland areas affected by mists and constant drizzle during the monsoon months, a time when the rest of the Archipelago suffers strong and desiccating winds. Oral history and tradition reports that not only has the extent of this mist and drizzle diminished over the years, but also its duration and continuity: it used to last for a predictable six to seven months (four to five months of the summer *horf* monsoon followed by two or so months of the *serb* winter monsoon), resulting in the total evacuation of both people and livestock from the region down to the lower hills, but is now patchy and discontinuous.

Dracaena cinnabari is fairly widespread, occurring in several vegetation types in central and eastern parts of Soqotra at altitudes above 500m. It has a fragmented distribution with different subpopulations showing varying degrees of 'vigour'. The following variation can be identified in population structure:

- Good age structure and active regeneration in the Central Highlands (particularly in the centre and western parts). *D. cinnabari* is abundant at higher altitudes and regeneration is good, particularly in sheltered gullies and in the dense woodland amongst the granite pinnacles.
- Good age structure but with no sign of regeneration (e.g. Rokeb di Firmihin).
- No sign of regeneration except in places inaccessible to livestock. These populations consist almost entirely of mature trees (e.g. Hamaderoh).
- Small relict populations or isolated individuals (south-central plateau). It is difficult to say whether the decline in quality of habitat in these relict populations is an ancient or recent phenomenon since it is hard to assess the age of these trees.

It is likely that *D. cinnabari* was formerly widely distributed over large parts of Soqotra but has declined in recent times. What then could account for this decline? Information from local informants and the historic record suggests that there has been no change in traditional land management practices over the last hundred years. Other uses of the tree have declined: harvesting of resin has lessened over recent years as the market for it has declined, and the use of the growing leaflets for cordage has almost died out with the introduction of nylon rope. During severe droughts, leaflets are removed and cracked open as fodder for goats but this causes scour and can only be fed in very small quantities. The same goes for the fruit, which are gathered in the summer and fed to goats and especially to cattle, but large quantities make both cattle and goats ill, so they must be fed in very small quantities (typically five to eight berries for a goat, for instance). Thus it can be seen that dracaena on the whole is at little risk from livestock, and although a starving goat will indeed nibble at the softer, younger leaflets of a young plant if it can find absolutely nothing else to eat, this is not a common occurrence.

Regeneration in relict populations that lack a dense understorey of shrubs is problematic since this understorey provides a protective 'nursery' for young plants in healthy woodland. Herders who feed their cattle on dracaena fruit report observing seeds sprouting in areas where there appears to be no active regeneration but these young seedlings do not survive the subsequent dry season, presumably because they lack protection from desiccation and grazing provided by the understorey. It therefore seems that grazing is a factor preventing regeneration but only when populations are already under stress because of increasing aridity. Man and his livestock (including goats) have been on the islands for at least 2000 years and it is, after all, the Soqotrans' traditional practices that have played the larger part in preserving the vegetation and floristic diversity for which the Archipelago is famed. Yet the extent of woodland on Soqotra has declined in recent times. A change in the climate therefore seems to be the most likely cause of recent changes in the vegetation.

The impact of increasing aridity is lessened on Soqotra by orographic rainfall (caused by mountains standing in the path of moisture-laden air) and the buffering effect of the surrounding seas. Furthermore, many plants on the islands are adapted to xerophytic conditions and presumably survived the extremely arid conditions which affected the region between 12 000 and 7000 BP. Extinctions caused by drying may not be on a catastrophic scale but many species could experience a drastic reduction in the size and extent of their populations. Such a decline for species such as *D. cinnabari*, which characterise the vegetation, would have a major impact on the ecology of Soqotra.

The most direct and immediate effect of drying is likely to be an increase in the incidence and duration of times of drought and a consequent increase of pressure on the pastureland and on plants used as drought fodder. This has already led, in recent times, to the complete elimination of some species from certain areas. *Dendrosicyos socotranus* is the best example of the swollen-trunked 'bottle trees' for which Soqotra is famous. It is widely distributed in several vegetation types and, in some places, is abundant. At present, the main threat to this species is its use as a drought fodder. In times of severe drought trees are cut down, pulped and fed to livestock. We have evidence (from local informants) that this has almost eradicated the species from some areas; for instance, at Kilissan the few trees of this very arid area were cut down during recent severe droughts and now only a single tree remains.

Change of habitat is an important factor in preventing regeneration. Typically, on the plains, *D. socotranus* seedlings germinate and are protected from livestock by the cover of dense vegetation provided by spiny shrubs such as *Lycium sokotranum* or by colonies of the succulent shrub *Cissus*. So, for instance, where there is no *Cissus* or *L. sokotranum* there is little, if any, regeneration. Along the drier southern region of Soqotra *Cissus* species are used as drought fodder for livestock and the recovery of populations of *D. socotranus* after drought can be impeded by this. When other drought fodders are exhausted, either stems are lopped and part-roasted at a fire to make them palatable to goats, or a fire is lit under living clumps of *Cissus* so that the starving goats can feed. Informants report that after a particularly bad drought in the 1980s they became aware that they had lost numerous saplings of *D. socotranus* in this way.

NOTES ON THE ETHNOLOGY TEXT

USE OF PAST AND PRESENT TENSE

Many of the traditional uses of plants described here have already been lost or are undergoing change. This state of flux is reflected in the text, where both past and present tenses are used. The past tense is used in much of the text, partly for the sake of consistency, but partly because this reflects present day realities. Few now depend on plants alone to treat their own and their animals' ailments; nylon has replaced much of the cordage and basketwork from plants, and a greater access to imported food has made many wild foods less necessary. However, for the many islanders who still rely on livestock for their survival, the knowledge of which plants are best for which animals at what season, which are toxic and which lessen the need for frequent watering is as important as before. The expertise that distinguishes between which woods are good to burn and which make a good timber is still relevant, as indeed is an ability to select specific plants for treating skins and hides, and for recognising which flowers attract honeybees.

TRANSCRIPTION OF SOQOTRI

All Soqotri plant names have been presented in two forms: firstly transcribed phonetically based on the International Phonetic Alphabet and using the Times New Roman font, and secondly in a much simplified form using the Roman script in Arial font. In Soqotri, there are sounds which cannot be adequately represented in the Roman alphabet; also, stress and differentiation of vowel length influence meaning. These and much else are lost in the simplified transcription. However, for ease of reading by those unfamiliar with the more complex transliteration systems it was felt that a simplified version was desirable.

OTHER NOTES

Regional dialects mean that variant pronunciations of the same plant name are common. For reasons of space, these have not all been noted. However, where regional names differ sharply, the alternatives have been included.

The phrase 'Not seen, no information recorded' that appears under a handful of plants does not mean that no information on this plant exists on Soqotra; only that the plant has not been discussed with islanders. We hope that future research will fill these and other gaps in our text.

Computer-simulated aerial view of the main island Soqotra from the northwest showing the diverse topography of lowland coastal plains, limestone plateaux and high granite mountains (enhanced from GIS and remote sensing data with elevation exaggerated by a factor of 2.5).

Photo credits: DA - Diccon Alexander; RH - Roger Hyam; AM - Anthony Miller; MM - Miranda Morris

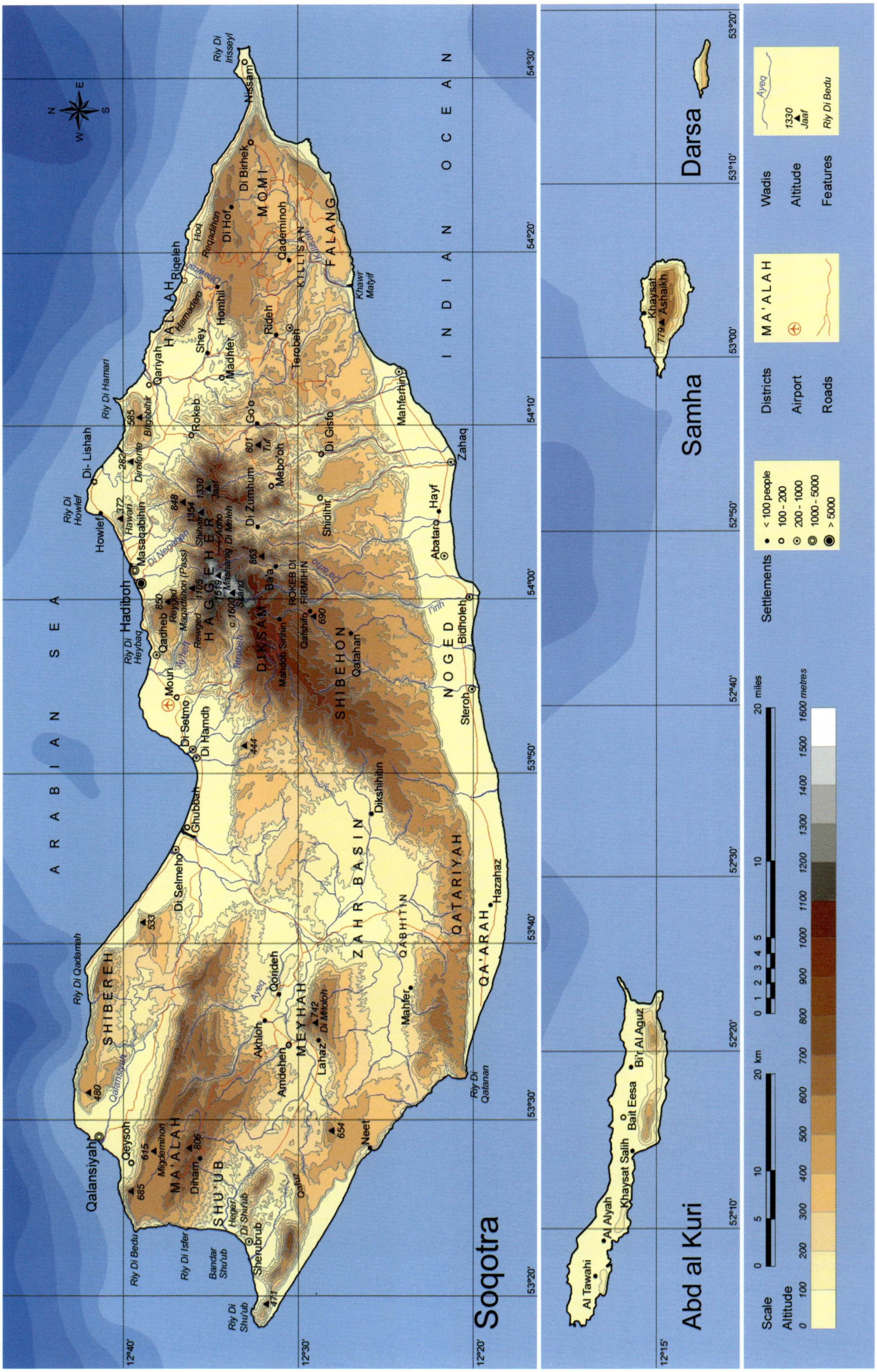
Soqotra
Abd al Kuri
Samha
Darsa
ARABIAN SEA
INDIAN OCEAN
HAGGEHER
DIKSAM
SHIBEHON
NOGED
ZAHR BASIN
QATARIYAH
MA'ALAH
SHIBEREH
SHU'UB
MEYHAH
MOMI
FALANG
Hadiboh
Qalansiyah
Scale
0 5 10 20 km
0 1 2 3 4 5 10 20 miles
Altitude
0 100 200 300 400 500 600 700 800 900 1000 1100 1200 1300 1400 1500 1600 metres
Settlements
< 100 people
100 - 200
200 - 1000
1000 - 5000
> 5000
Districts
Airport
Roads
Wadis
Altitude
Features
MA'ALAH
Ayeq
1330 Jaaf
Riy Di Bedu

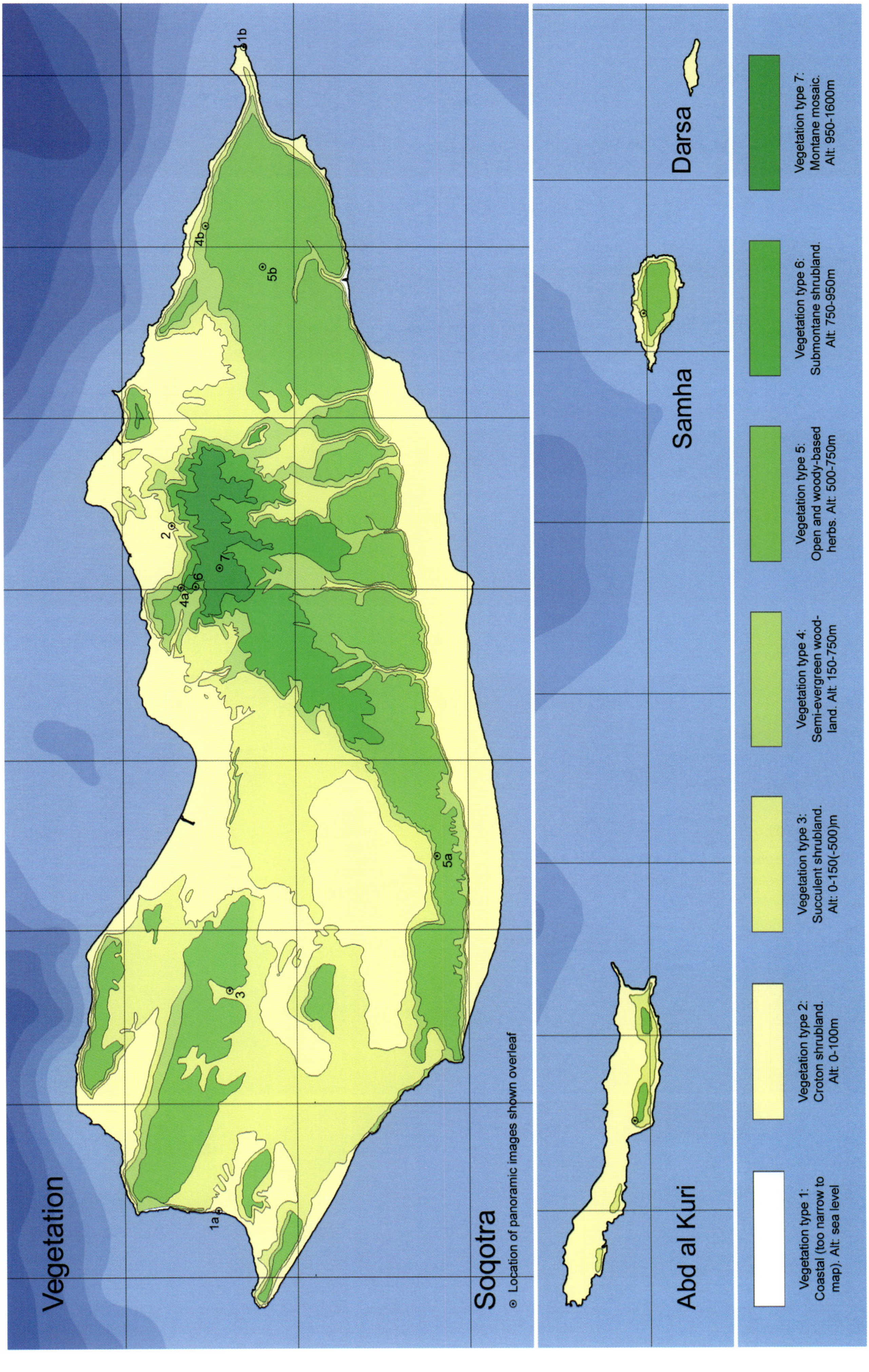
Vegetation
Soqotra
⊙ Location of panoramic images shown overleaf
Abd al Kuri
Samha
Darsa
1a
1b
2
3
4a
4b
5a
5b
6
7
Vegetation type 1: Coastal (too narrow to map). Alt: sea level
Vegetation type 2: Croton shrubland. Alt: 0-100m
Vegetation type 3: Succulent shrubland. Alt: 0-150(-500)m
Vegetation type 4: Semi-evergreen woodland. Alt: 150-750m
Vegetation type 5: Open and woody-based herbs. Alt: 500-750m
Vegetation type 6: Submontane shrubland. Alt: 750-950m
Vegetation type 7: Montane mosaic. Alt: 950-1600m

Vegetation type 1: Coastal **Top:** Mangrove woodland dominated by *Avicennia marina*. A rare vegetation type, scattered in small patches mainly in western Soqotra. **Above:** Mosaic of low succulent shrubs and woody-based herbs, here dominated by *Limonium* species with *Zygophyllum qatarense, Aeluropus lagopoides* and *Lachnocapsa spathulata.* (RH)

Vegetation type 2: Croton shrubland on the coastal plains dominated by *Croton socotranus,* here with low shrubs including *Barleria spiculifera, Indigofera pseudointricata, Justicia rigida* and *Placopoda virgata* and emergent trees *Euphorbia arbuscula, Zizyphus spina-christi, Dendrosicyos socotrana* and *Commiphora* spp. (RH)

Vegetation type 3: Succulent shrubland on limestone cliffs and slopes, with a lower layer of cushion plants and sub-shrubs, here dominated by *Jatropha unicostata* and *Croton socotranus* with emergent trees *Dendrosicyos socotrana, Euphorbia arbuscula* and *Adenium obesum* subsp. *sokotranum*. (DA)

Vegetation type 4: Semi-evergreen woodland on limestone escarpments and sheltered ravines. **Top:** Dominated by *Rhus thyrsifolia* and *Buxus pedicellatus* with *Punica protopunica, Commiphora socotrana* and *Boswellia* spp. **Above:** Dominated by *Jatropha unicostata* with *Sterculia africana, Carphalea obovata, Lannea transulta, and Acacia pennivenia*. (RH)

Vegetation type 5: Open and woody-based herb communities on the limestone plateau. **Top:** Dominated by *Buxus pedicellatus, Cissus hamaderohensis, Allophyllus rhoidiphyllus* and *Cryptolepis intricata.* **Above:** Low herbs with relict woodland dominated by *Dracaena cinnabari* with *Buxus pedicellatus, Lycium sokotranum* and *Adenium obesum.* (RH)

Vegetation type 6: Submontane shrubland on the slopes and cliffs of the granite mountains. Semi-deciduous shrubland with cleared grassland, here dominated by *Rhus thyrsiflora, Buxus pedicellatus* and *Gnidia socotrana*, with emergent *Dracaena cinnabari*. (RH)

Vegetation type 7: Montane mosaic of shrubland, woodland and grassland in the granite mountains. Dense evergreen dwarf shrubland and woodland with patches of cleared grassland, here dominated by *Dracaena cinnabari* with *Hypericum scopulorum, Rhus thyrsiflora, Euryops arabicus* and *Pteridium aquilinum.* (RH)

Vegetation of the outer islands: Top: Succulent shrubland on Abd al Kuri on the limestone slopes with *Euphorbia abdelkuri, E. oblanceolata* and *E. balsamifera*. **Above:** Low succulent shrubland on Samha on the slopes below the limestone escarpment, with *Aloe perryi, Euphorbia oblanceolata, Atriplex griffithii, Pulicaria stephanocarpa* and occasional trees of *E. arbuscula*. (RH)

The granite pinnacles of the Haggeher mountains, rising to over 1500m, set against the Arabian Sea. (DA)

DIVERSE TOPOGRAPHY OF PLATEAU, PLAIN AND MOUNTAIN

Date palm grove *(Phoenix dactylifera)* by Wadi Di Negehen, with the granite peaks of the Haggeher beyond. (DA)

The arid limestone plateau, at over 500m, where many uniquely adapted species survive the rocky terrain. (DA)

The broad expanse of the lowland coastal plain near Qalansiyah, dominated by *Croton socotranus* shrubland. (DA)

The peak of Mashanig (c.1519m) and a dragon's blood tree (*Dracaena cinnabari*) rising above the mists at sunrise. (DA)

A lone desert rose (*Adenium obesum* subsp. *sokotranum*) in the windswept, parched landscape of the plateau. (DA)

Dragon's blood trees, with the primitive branching pattern typical of early trees, and desert rose in eastern Soqotra. (DA)

Dragon's blood forest at sunset, at Rokeb di Firmihin in the foothills of the Haggeher mountains. (DA)

SOQOTRAN LANDSCAPES FROM DAWN TO DUSK

The extremely arid Zahr basin, where dry-adapted plants may have found refuge during past wetter climates. (DA)

Dry refugia xerophytic plants on the limestone plateau, including *Caralluma*, *Aloe*, *Kalanchoe*, *Adenium* and *Dracaena*. (AM)

Typical gorge in the plateau at Ma'alah, where wet-adapted plants may have survived during past dry periods. (AM)

The highest peak of the Haggeher, Skand at c.1600m, in a lush fertile area where almost all plant species are endemic. (DA)

WET AND DRY REFUGIA FOR PLANTS AND ANIMALS

Westward view of Abd al Kuri, the second largest island in the Archipelago with high levels of endemism. (DA)

OUTER ISLANDS OF ABD AL KURI, SAMHA AND DARSA

Darsa coastline with the table-top island of Samha beyond, the islands collectively known as "the Brothers". (DA)

Wet refugia below the summit plateau of Samha, where many of the island's endemic plants are found. (AM)

Lowland coastal plains of Abd al Kuri, one of the driest areas in the Archipelago. (AM)

Typical house interior with *Zizyphus* pillar, *Croton* and date palm roofing, date palm matting, *Zizyphus* saddle and lime wash. (DA)

PEOPLE AND PLANTS IN THE ARCHIPELAGO

Rolling hills and vallies of the interior in the east of Soqotra, with a village in the mid-ground. (MM)

Rural village with houses built of drystone walls and date palm roofing, and a grove of *Zizyphus* trees. (MM)

Local plant experts discussing plant uses, in the Semitic language of Soqotri spoken by the 40,000 or so islanders. (DA)

FOOD (HUMAN): collecting fruit from a *Zizyphus spina-christi* tree, a good source of zinc, iron and vitamin C. (MM) (DA)

FOOD (HUMAN): digging for tubers of *Dioscorea lanata*, one of many wild tuberous plants with nutritious value. (MM)

FOOD (HUMAN): a woman and child using a two-stone grindstone to grind finger millet. Finger millet (*Eleusine coracana*) and dates (*Phoenix dactylifera*) are the only two crops traditionally cultivated in the Archipelago. (MM)

FOOD (LIVESTOCK): collecting *Euphorbia arbuscula* as goat fodder. (MM)

FOOD (LIVESTOCK): a herd of goats being moved across the arid Zahr Basin in search of grazing plants and water. (MM)

FOOD (LIVESTOCK): Soqotra's unique breed of miniature cattle are ideally suited to the high grassland areas of the Haggeher mountains. (DA)

FUEL: Making fire using traditional fire-sticks. (DA)

FUEL: Stacked wood ready to be transported for sale. (MM)

FUEL: A large evening feast cooked on open fires. Until recently firewood was the islanders' only source of heat, light and fuel. (DA)

WOOD: Tools of cultivation, including a Soqotran knife with a goathorn handle. (MM)

WOOD: Fencing around a date palm plantation made from brushwood and palm ribs, bound by plaited strips of palm fibre (*Phoenix dactylifera*). (MM)

WOOD: A ground loom made from local woods, including *Zizyphus,* and wool coloured with natural plant dyes, used to woollen cloth. (MM)

TIMBER: House with palm rib door, timber door lintel and central stone pillar. (MM)

TIMBER: Date palm housing on the southern coastal plain at Noged. (MM)

TIMBER: The trunk of a *Zizyphus* tree destined for use as a central house pillar. Only after much local discussion can living trees be felled. (MM)

LEATHERWORK: The skin of a bullock stretched out ready for the application of ash to toughen the leather. (MM)

LEATHERWORK: A whole goatskin used to store dates. (DA)

LEATHERWORK: Preparation of frankincense bark (*Boswellia* sp.) to cure and tan leather. Knife scars can still be seen on the trunk. (MM)

LEATHERWORK: Leather bag for spinning carried by women while herding their sheep. (MM)

FIBRE/ CORDAGE: Bull wearing a woven contraceptive apron to control breeding. (MM)

FIBRE/ CORDAGE: Camels laden with date palm fibre bags (*Phoenix dactylifera*) and tied with dragon's blood and palm cord. (MM)

MEDICINAL/ CLEANSING: Resin of a frankincense tree (*Boswellia elongata*) burnt as incense to cleanse and cure the sick. (DA) (MM)

PEST CONTROL: Smudge lit to keep insects away from cattle during morning milking. (MM)

PEST CONTROL: Stripping the bark of a desert rose (*Adenium obesum* subsp. *sokotranum*) to be made into a goat collar to protect against ravens and wild cats. (MM)

OTHER USES: Tobacco pipe made from the hollowed-out branch of *Jatropha unicostata*. (DA)

OTHER USES: Wooden measuring bowl for grain, Soqotran pottery, and rugs woven from sheepswool. (MM)

OTHER USES: Quicklime manufacture. (MM)

OTHER USES: Pottery decorated with dragon's blood resin. (MM)

FUTURE CHALLENGES: The future will bring modern development as in the busy capital (top left), but also perennial dangers such as drought (top right). Strengthening traditional practices will be critical to the next generation (above). (MM) (DA)

ILLUSTRATED KEY

SOQOTRAN FIG (DORSTENIA GIGAS) DICCON ALEXANDER

INTRODUCTION TO ILLUSTRATED KEY

The Illustrated Key enables easy identification of all 825 higher plant species found in the Soqotra Archipelago. The key is basically dichotomous, that is with two alternative choices (leads) at each step. However, occasionally, when it helps clarify the key, there may be more than two choices. All choices are illustrated, which makes the key accessible to those who are not familiar with technical botanical language. It is important to study all of the leads carefully before deciding which to follow.

The first part of the key is divided into seven major groups. These are A: Pteridophytes (ferns), B: Freshwater monocots, C: Marine monocots, D: Terrestrial monocots, E: Herbaceous dicots, F: Woody dicots with opposite leaves, and G: Woody dicots with alternate leaves. Groups A to C are keys to individual species (since the groups are relatively small), whereas the larger groups D to G are keys to families. The remainder of the key consists of an alphabetic arrangement of individual terrestrial monocot and dicot family keys where genera and species can be identified.

Once a species identification has been reached, the species name is given, as well as its endemic status and further page references to confirm the identification – a colour photograph in the Plant Portraits section (for most endemics) and a full written description in the Encyclopaedia section. Page references are given in **bold** font for Plant Portraits and plain font for the Encyclopaedia.

KEY TO MAJOR GROUPS

1	**Plants** reproducing by spores	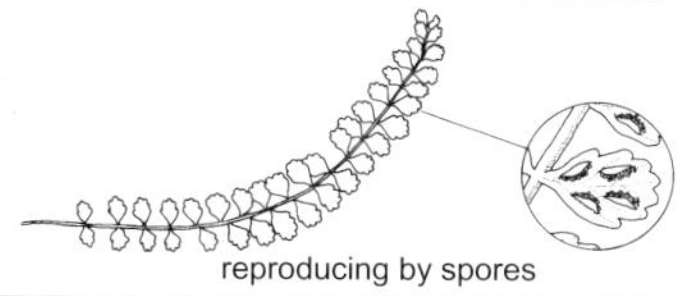	**Group A:** Pteridophytes
	Plants reproducing by cones	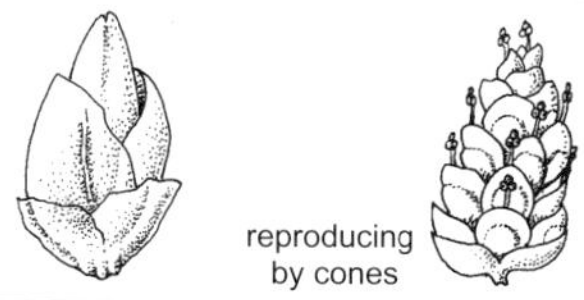	***Ephedra foliata*** (Gymnosperm) 296
	Plants reproducing by flowers	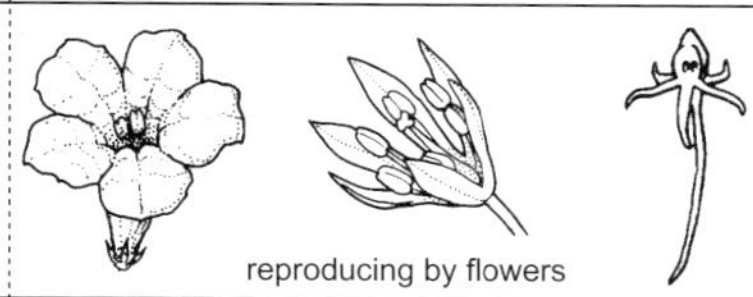	2
2	Submerged marine or freshwater **plants**	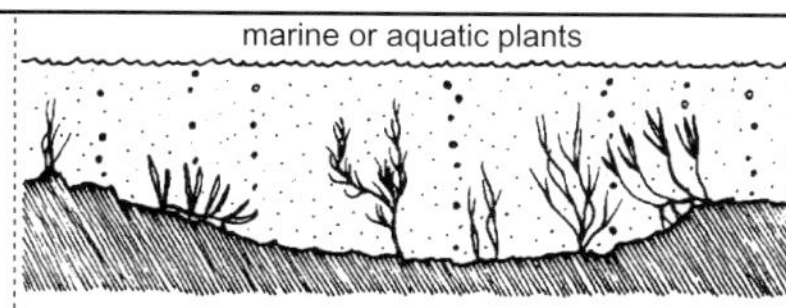	3
	Terrestrial or epiphytic **plants** (sometimes rooted in water but never submerged)	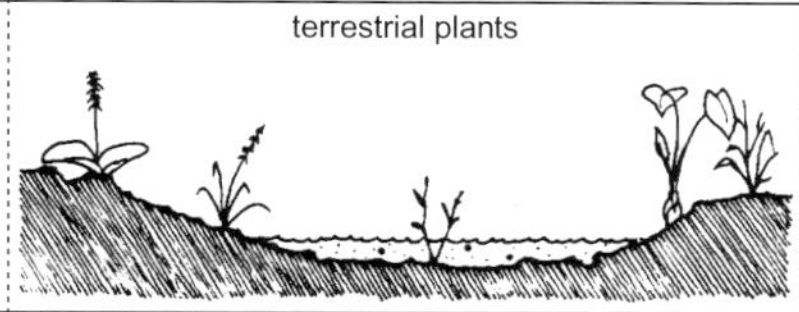	4
3	**Plants** submerged in fresh or brackish water		**Group B:** Fresh or brackish water monocots
	Plants submerged in the sea	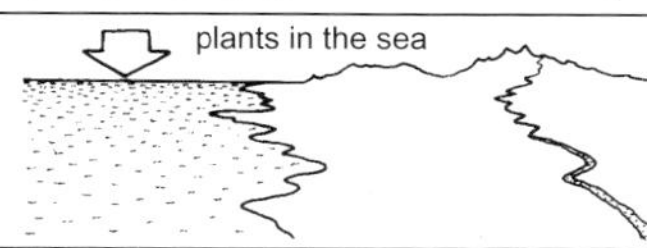	**Group C:** Marine monocots
4	**Leaves** with parallel veins; **flowers** with parts in 3s	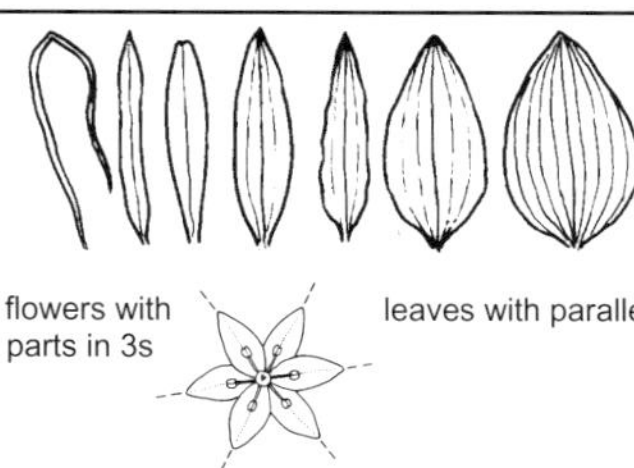	**Group D:** Terrestrial monocots
	Leaves with pinnate veins; **flowers** with parts in 2s, 4s or 5s, or parts many (dicots)		5
5	Annual or perennial **herbs**, sometimes woody at base	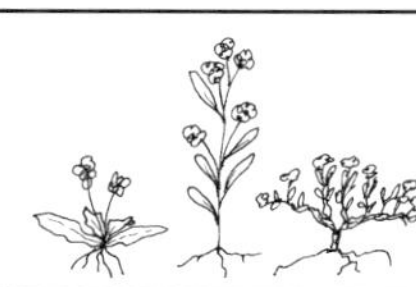	**Group E:** Herbaceous dicots
	Shrubs or **trees**		6
6	**Leaves** opposite	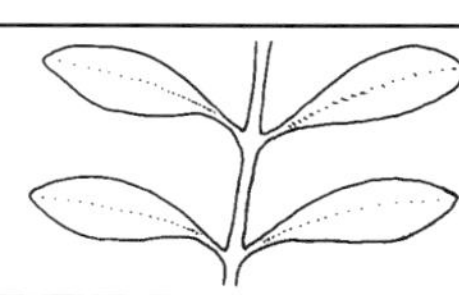	**Group F:** Woody dicots with opposite leaves
	Leaves alternate	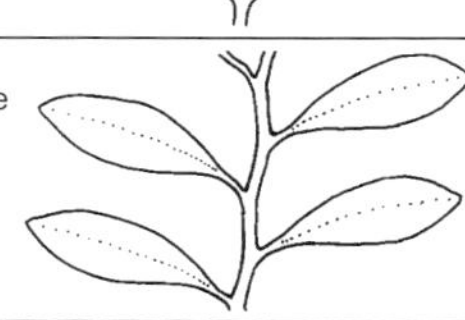	**Group G:** Woody dicots with alternate leaves

GROUP A: PTERIDOPHYTES

1	**Fronds** simple or divided into leaflets or dichotomously divided into linear segments	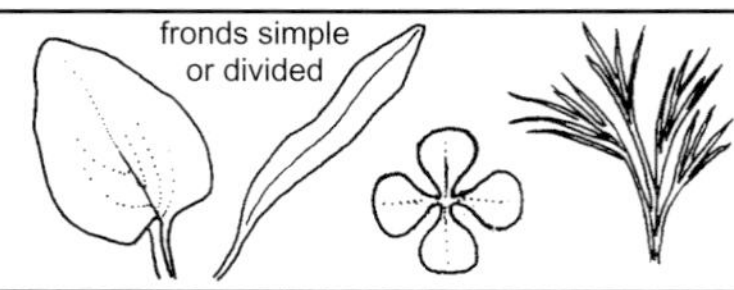	2
	Fronds 1-pinnate	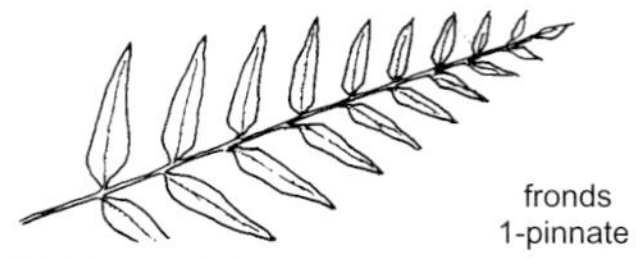	3
	Fronds >1-pinnate		4

2	**Fronds** ovate to elliptic with a cuneate base 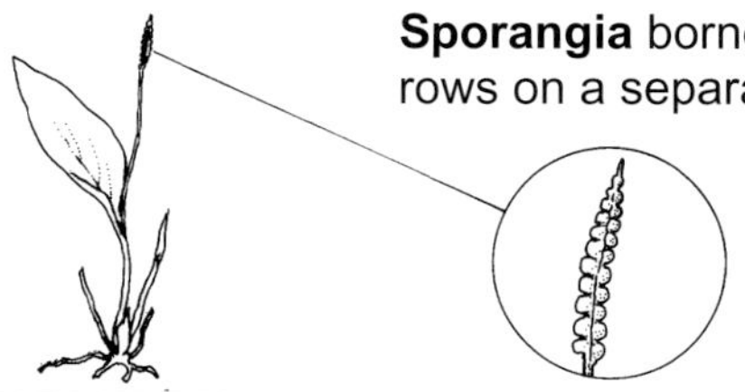	**Sporangia** borne in two rows on a separate spike	***Ophioglossum polyphyllum*** (Ophioglossaceae) 292
	Fronds simple, broadly ovate with a cordate or truncate base 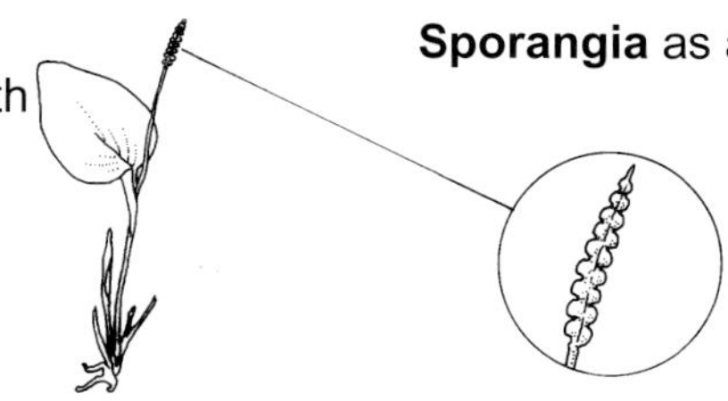	**Sporangia** as above	***Ophioglossum reticulatum*** (Ophioglossaceae) 292
	Fronds simple, 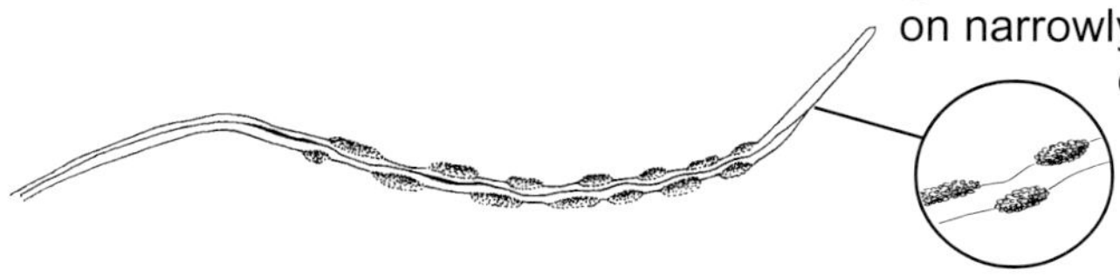	**Sporangia** borne leathery on narrowly elliptic, elongate sori	***Loxogramme abyssinica*** (Polypodiaceae) 293
	Fronds simple, narrowly elliptic 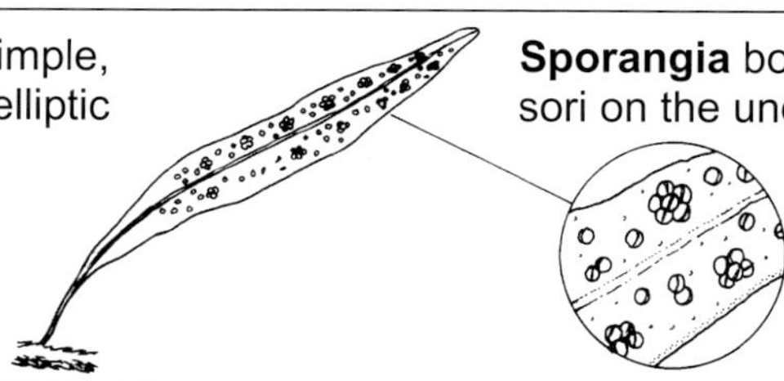	**Sporangia** borne in round sori on the underside of the frond	***Pleopeltis macrocarpa*** (Polypodiaceae) 293
	Fronds divided into 4 obovate leaflets 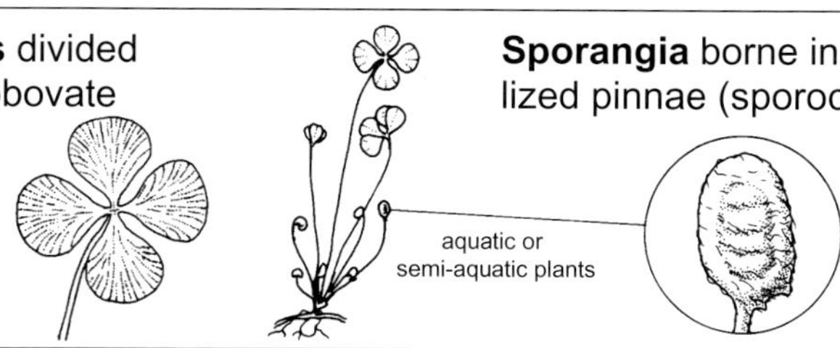	**Sporangia** borne in specialized pinnae (sporocarps)	***Marsilea coromandeliana*** (Marsileaceae) 291
	Fronds dichotomously divided into linear segments 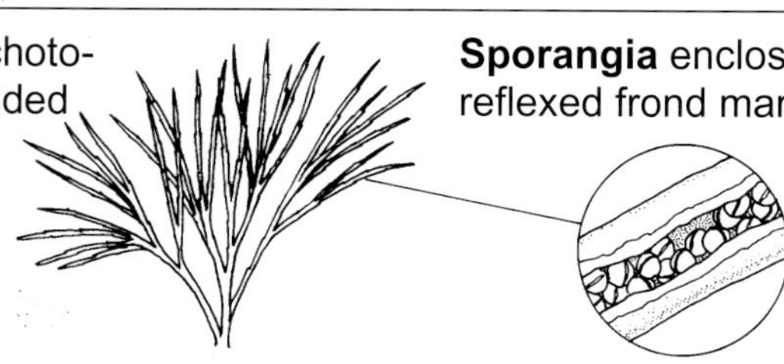	**Sporangia** enclosed in reflexed frond margins	***Actiniopteris semiflabellata*** (Actiniopteridaceae) **233**, 285

3	**Pinnae** obovate, crenate-dentate rounded at the tip, <1cm long rhachis reddish brown to black 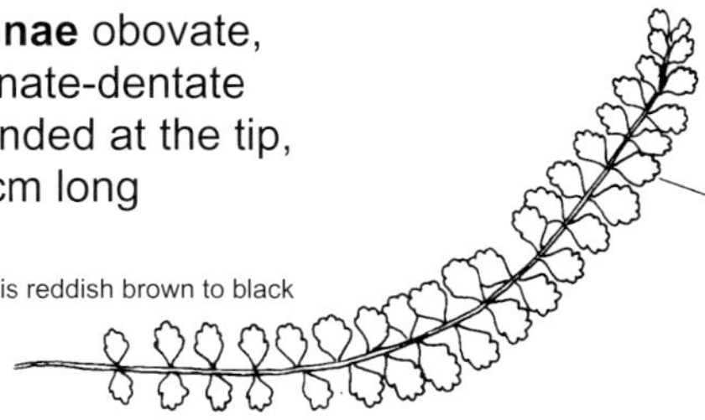	**Sori** elongate, borne on the underside 	***Asplenium trichomanes*** (Aspleniaceae) 289
	Pinnae narrowly oblong-triangular, with dense scales beneath, <1cm long 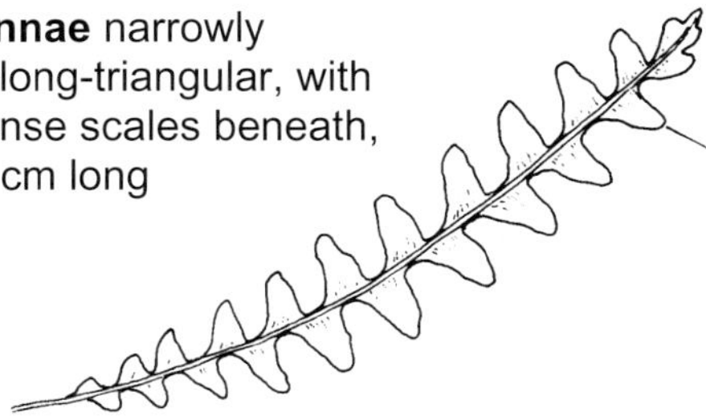	**Sori** round, borne on the underside, hidden by scales 	***Ceterach phillipsianum*** (Aspleniaceae) 290
	Pinnae fan-shaped, with flabellate venation, <2cm long 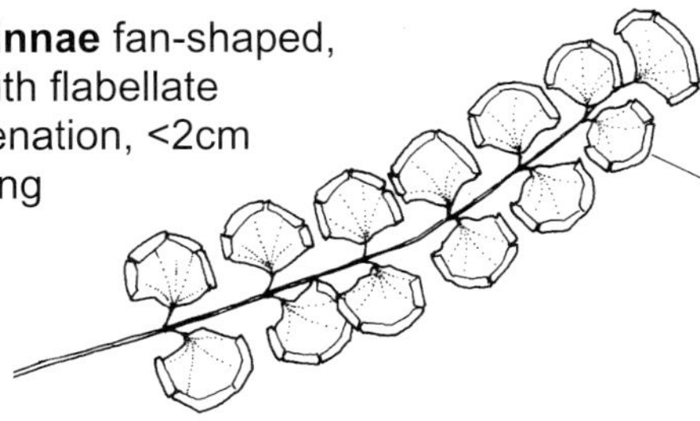	**Sori** borne on the inner surface of the reflexed frond margins 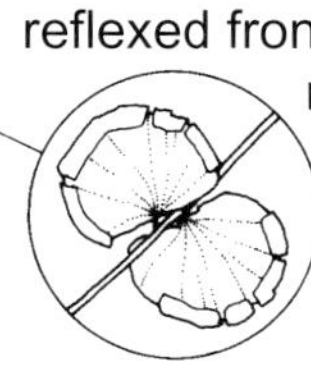	***Adiantum balfourii*** (Adiantaceae) **233**, 286
	Pinnae lanceolate, sessile, 1–4cm long 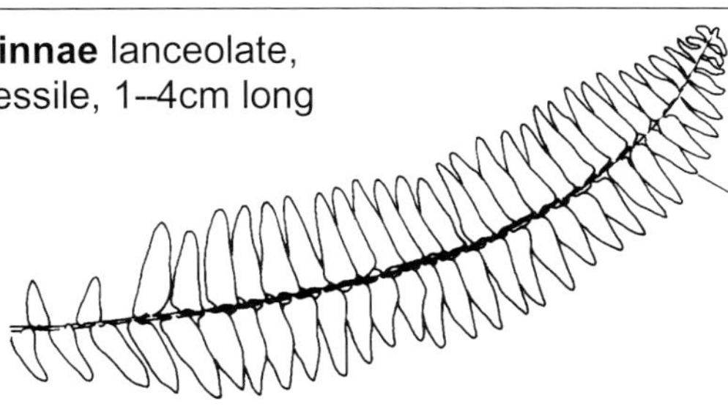	**Sori** round, borne on the underside 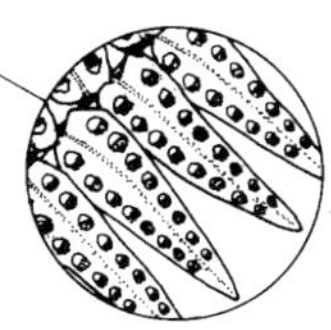	***Nephrolepis undulata*** (Oleandraceae) **233**, 291
	Pinnae lanceolate, petiolate, 1–6cm long rhachis green 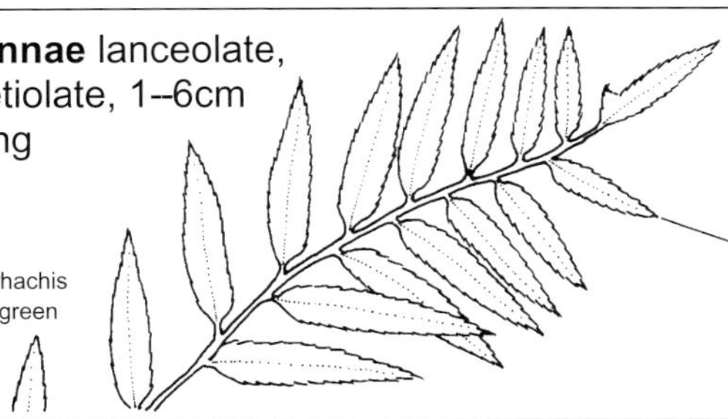	**Sori** elongate, dense, borne on the underside 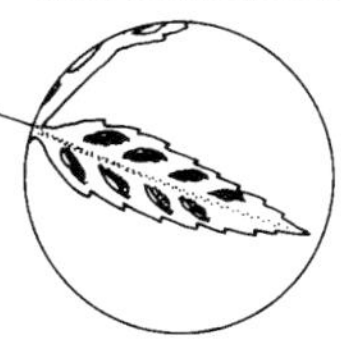	***Asplenium schweinfurthii*** (Aspleniaceae) Endemic **233**, 289
	Pinnae linear-elliptic, 2–6cm long, lowest pinnae 2-lobed 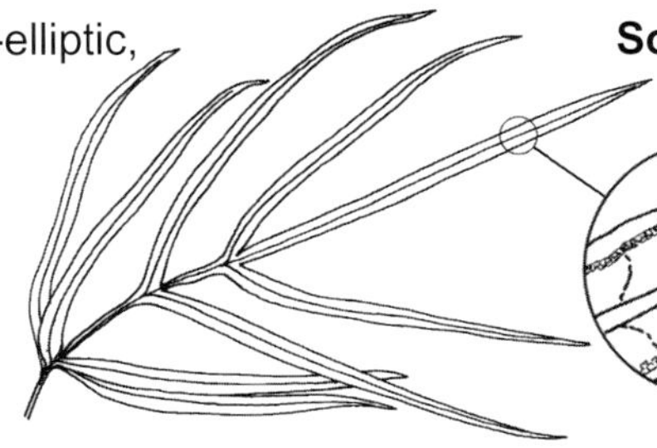	**Sori** elongate, restricted to margin 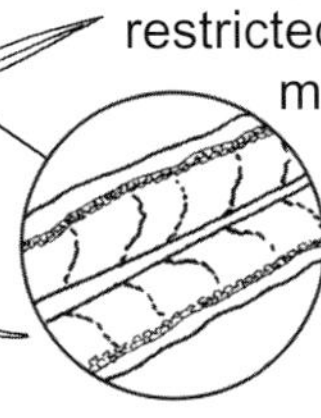	***Pteris cretica*** (Pteridaceae) 294
	Pinnae linear, sessile, 5–15cm long 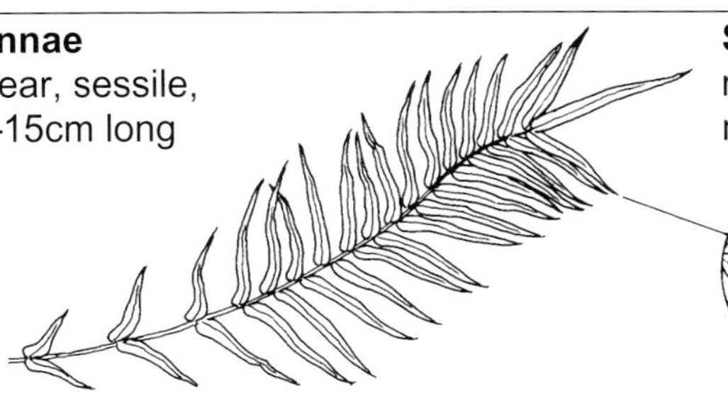	**Sori** elongate, restricted to the margin, except at the tip 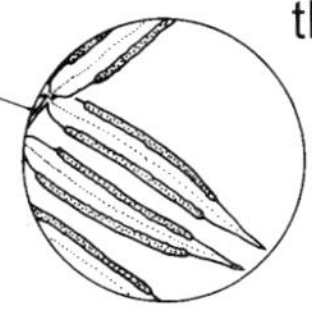	***Pteris vittata*** (Pteridaceae) **233**, 294

4	**Fronds** of 2 types present (sterile and fertile); **pinnules** various, with pinnate venation		5
	Fronds all similar (fertile); **pinnules** orbicular, fan- or wedge-shaped, with flabellate venation	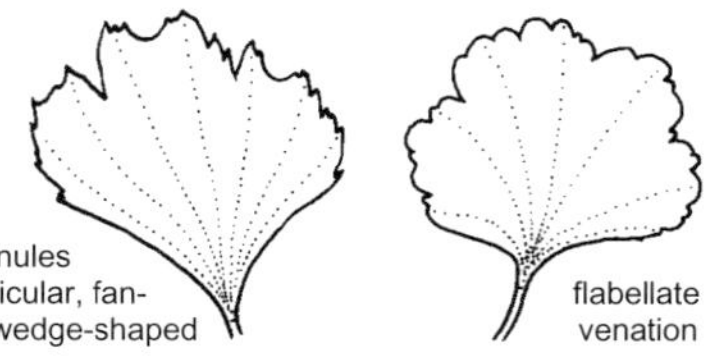	6
	Fronds all similar (fertile); **pinnules** various, with pinnate venation	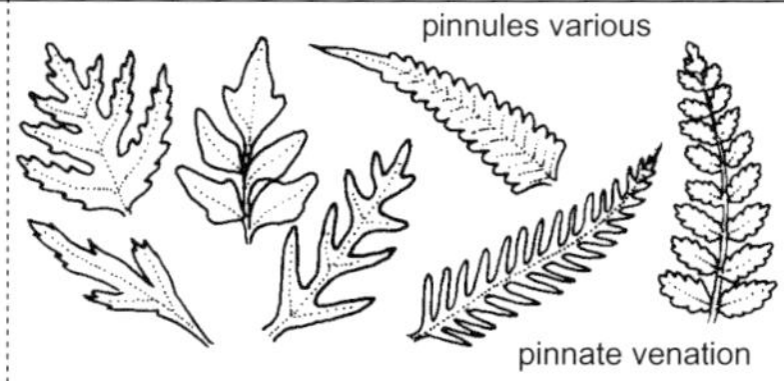	7
5	Large aquatic **fern**; **fertile pinnae** linear, **sterile pinnae** triangular or ovate	**Sori** enclosed in the reflexed frond margins 	***Ceratopteris cornuta*** (Parkeriaceae) 293
	Small terrestrial **fern**; **fertile pinnae** narrowly elliptic, **sterile pinnae** narrowly oblong	**Sori** enclosed in the reflexed frond margins 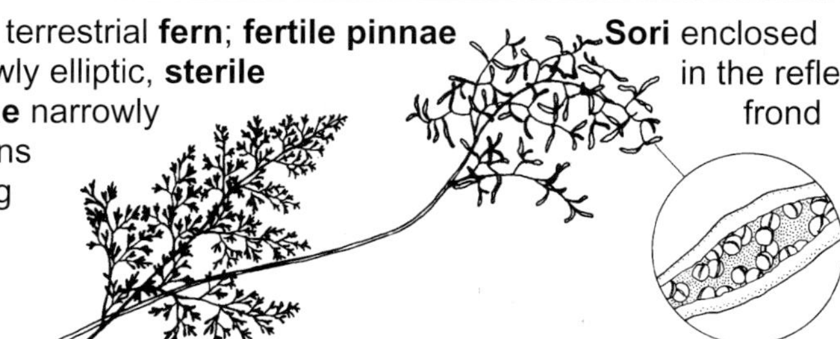	***Onychium divaricatum*** (Adiantaceae) 288
6	**Fronds** 1 to 2(to 3)-pinnate; **margin** crenate-dentate	**Sori** enclosed in the reflexed frond margins 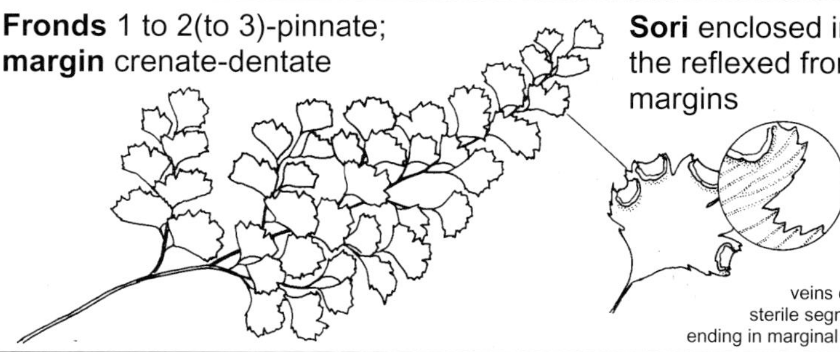	***Adiantum capillus-veneris*** (Adiantaceae) 286
	Fronds 3(or 4)-pinnate; **margin** crenate	**Sori** enclosed in the reflexed frond margins 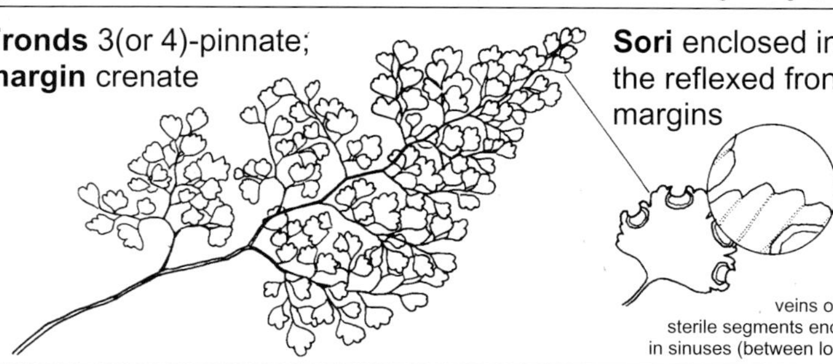	***Adiantum poiretii*** (Adiantaceae) 287

7	**Fronds** 2 or 3 pinnatifid, green above, powdery white beneath 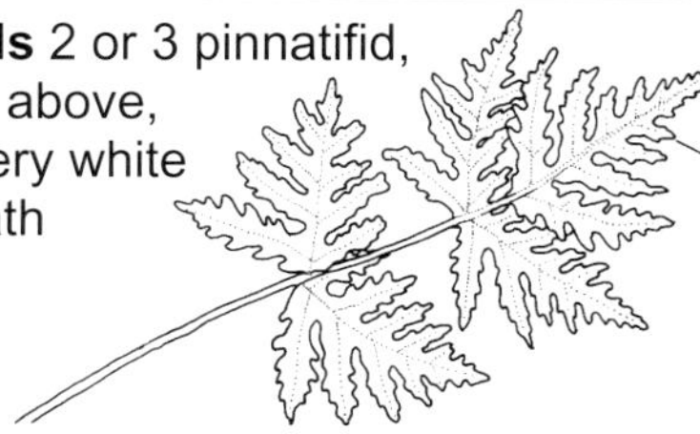	**Sori** round, restricted to the margin, shiny black 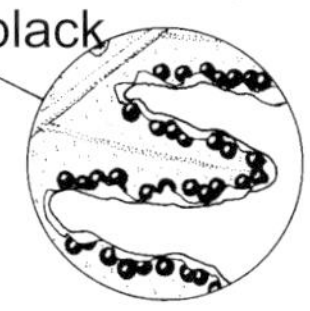	***Negripteris sciona*** (Adiantaceae) **233**, 287
	Fronds 2 to 4 pinnatifid, pentagonal in outline 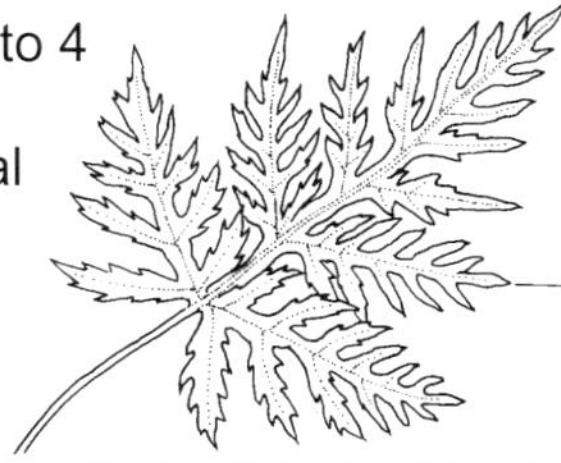	**Sori** round, restricted to the margin 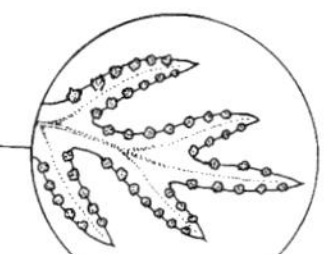	***Doryopteris concolor*** (Adiantaceae) 287
	Fronds 2 or 3 pinnatifid, with unevenly deeply lobed pinnae 	**Sori** elongate, evenly distributed over the underside 	***Asplenium aethiopicum*** (Aspleniaceae) 289
	Fronds 2 pinnatifid 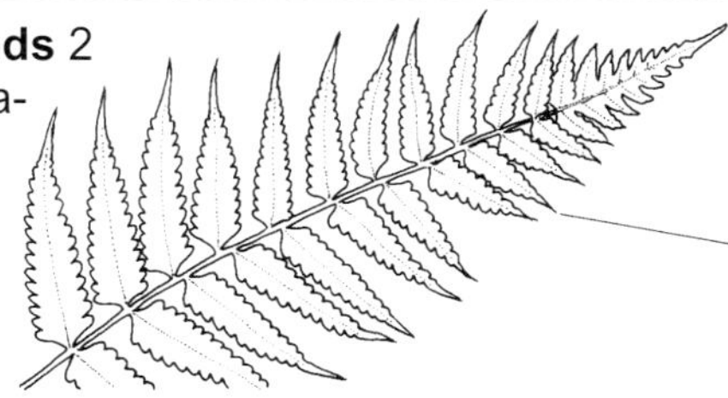	**Sori** round, evenly distributed over the underside 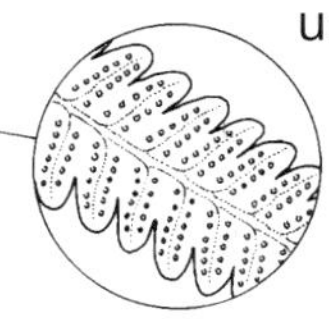	***Christella dentata*** (Thelypteridaceae) 295
	Fronds 2 to 4 pinnatifid  upper surface with minute spines on the midrib and veins	**Sori** elongate, restricted to the margin 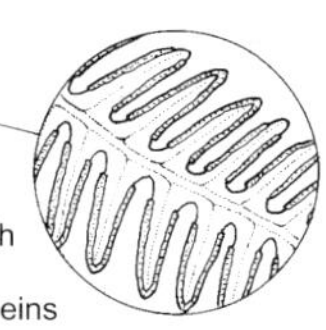	***Pteris quadriaurita*** (Pteridaceae) 294
	Fronds 3 or 4 pinnatifid  frond glabrescent above, tomentose below	**Sori** linear, restricted to and covered by reflexed margin 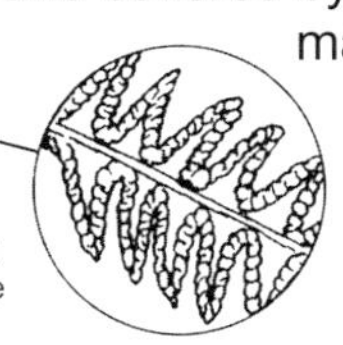	***Pteridium aquilinum*** (Dennstaedtiaceae) **233**, 290
	Fronds 2 to 4 pinnatifid 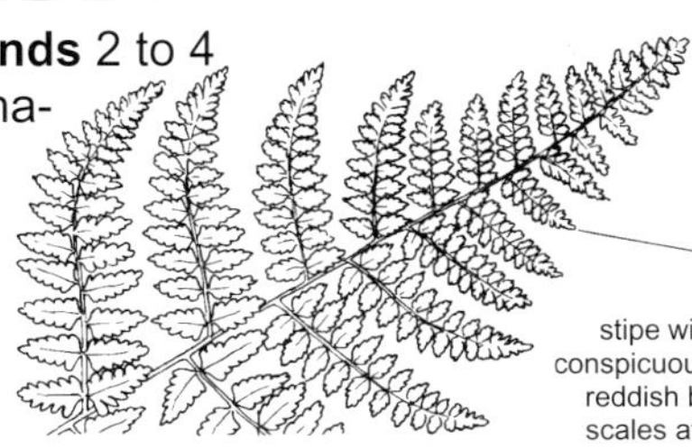 stipe with a conspicuous tuft of reddish brown scales at base	**Sori** round, evenly distributed over the underside 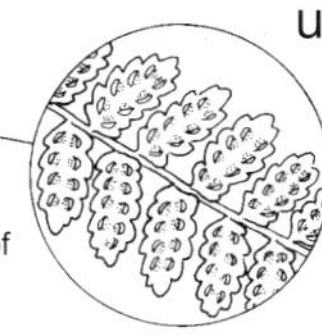	***Hypodematium crenatum*** (Dryopteridaceae) 291
	Fronds at least 2-pinnate 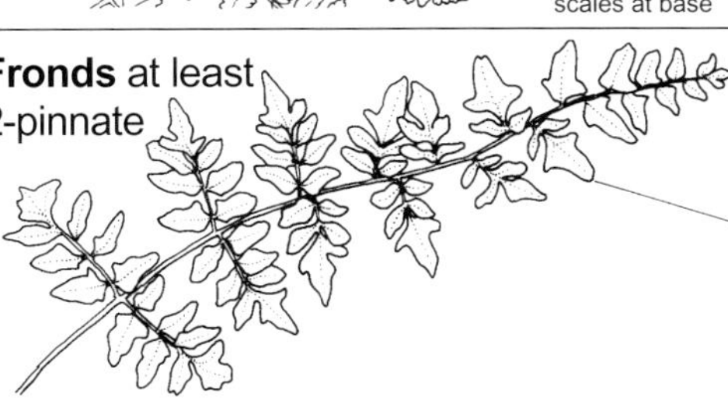	**Sori** elongate, restricted to the margin 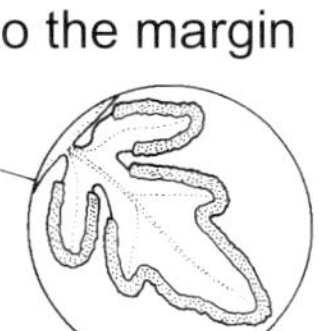	8

8	Mature **stipe** and **rhachis** glabrous or with a few scales at base of stipe; **rhizome-scales** minutely serrulate	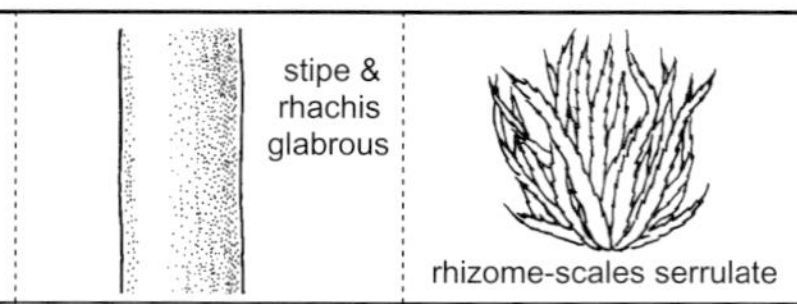	***Pellaea viridis*** (Adiantaceae) 288
	Mature **stipe** and **rhachis** clothed with scales; **rhizome-scales** entire	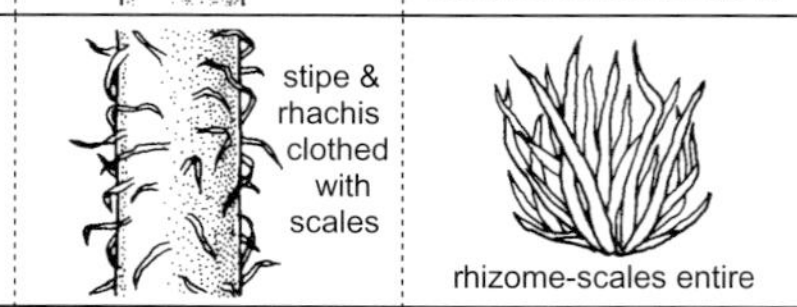	***Pellaea involuta*** (Adiantaceae) 288

GROUP B: FRESHWATER MONOCOTS

1	**Leaves** oblong-elliptic, petiolate; **leaf sheath** absent, translucent stipule present; **flowers** in dense spikes	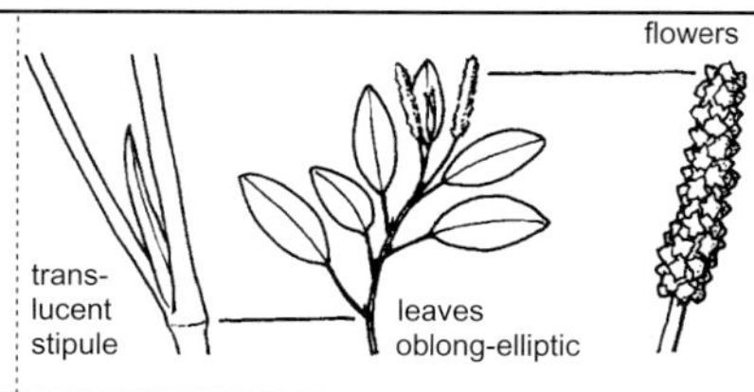	***Potamogeton nodosus*** (Potamogetonaceae) 398
	Leaves filiform, entire; **leaf sheath** tubular, fused, ligule present; **flowers** in interrupted spikes	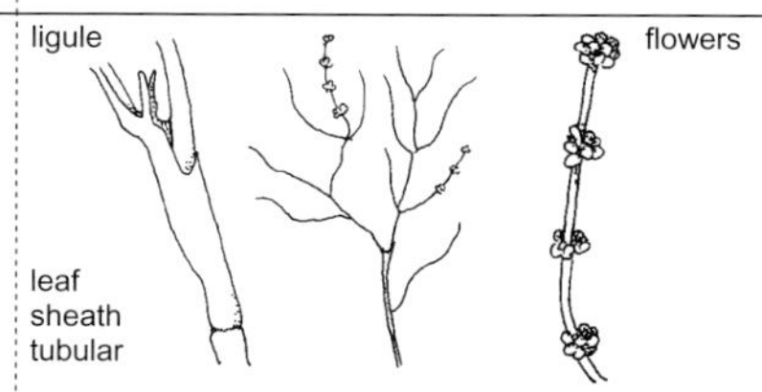	***Potamogeton filiformis*** (Potamogetonaceae) 398
	Leaves linear, entire; **leaf sheath** tubular, not fused, ligule present; **flowers** in interrupted spikes	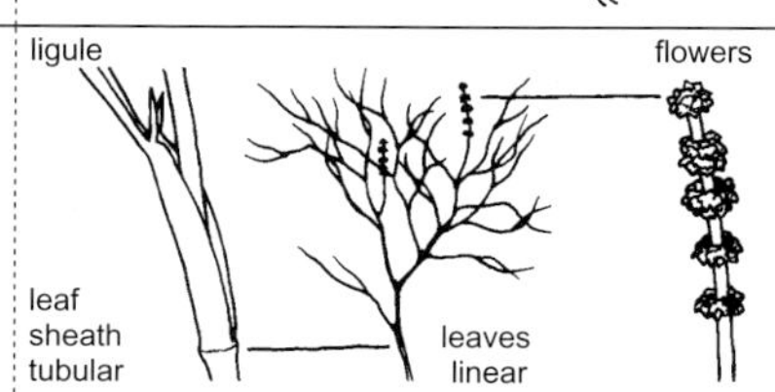	***Potamogeton pectinatus*** (Potamogetonaceae) 398
	Leaves linear, denticulate at apex; **leaf sheath** tubular; **flowers** in sub-umbellate racemes, long-pedicellate	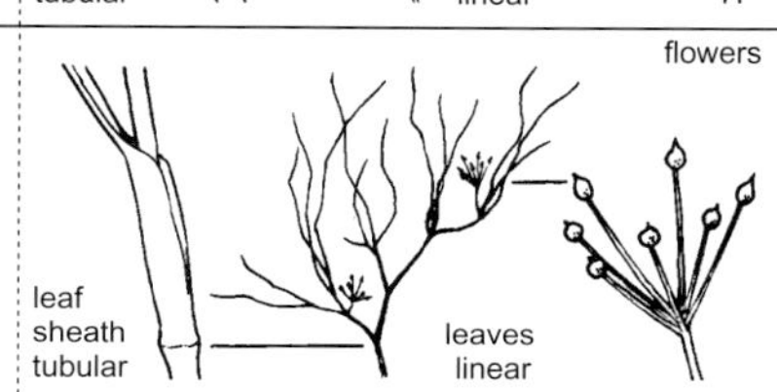	***Ruppia maritima*** (Potamogetonaceae) 399
	Leaves linear, minutely denticulate; **leaf sheath** open with appendages at tip; **flowers** 1 to 3 in leaf axils, inconspicuous	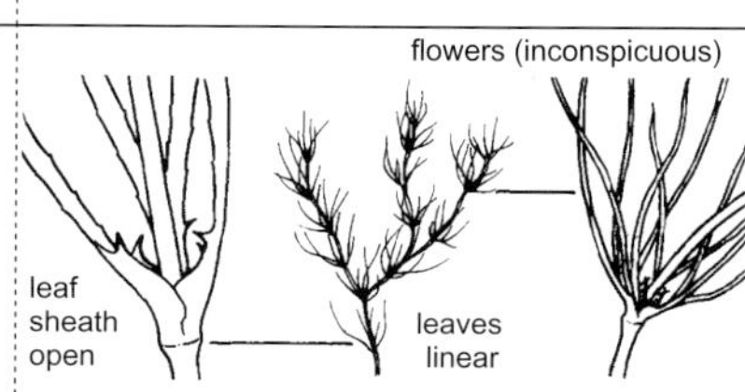	***Najas graminea*** (Najadaceae) 390
	Leaves linear-lanceolate, coarsely spinose-dentate; **leaf sheath** open; **flowers** solitary in leaf axils, inconspicuous	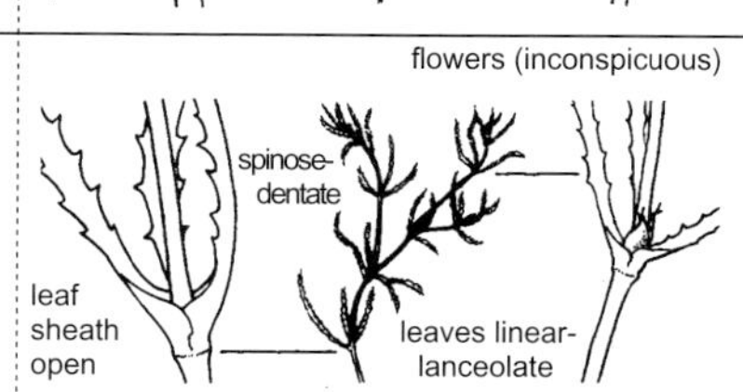	***Najas marina*** (Najadaceae) 391
	Leaves linear-lanceolate; **leaf sheath** open; **flowers** at first hidden then rising to the surface on filiform stalks	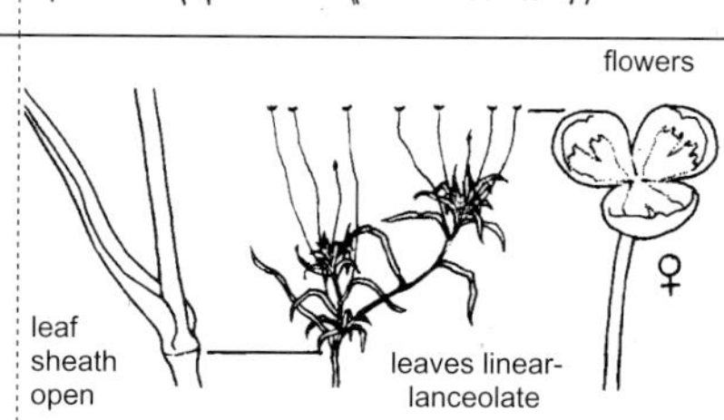	***Lagarosiphon roxburghii*** (Hydrocharitaceae) 387

GROUP C: MARINE MONOCOTS

1	**Leaves** oblong-elliptic, long-petiolate	***Halophila ovalis*** (Hydrocharitaceae) (not recorded but likely to occur)
	Leaves linear or strap-shaped, without distinct petioles	2

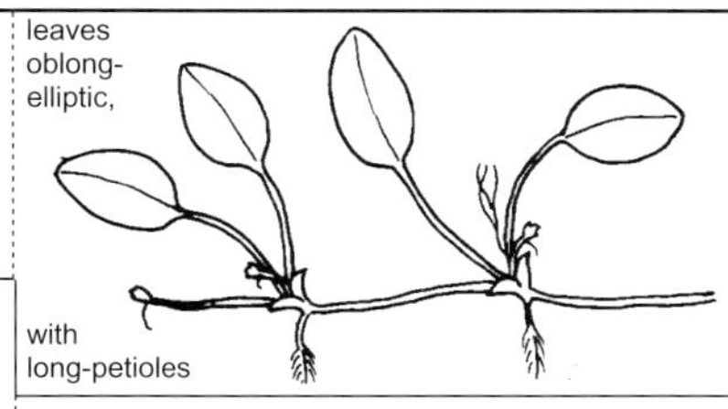

2	**Rhizome** covered in long, stiff bristles; **leaves** 30–100cm long x 1–2cm wide	***Enhalus acoroides*** (Hydrocharitaceae) (not recorded but likely to occur)
	Rhizome without bristles; **leaves** <30(–40)cm long x 1–10mm wide	3

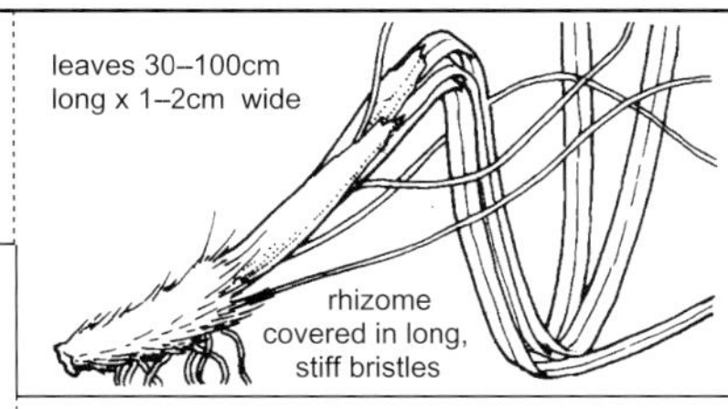

3	**Leaves** cylindrical; **flowers** in a branched inflorescence	***Syringodium isoetifolium*** (Cymodoceaceae) 312
	Leaves flat; **flowers** solitary or paired	4

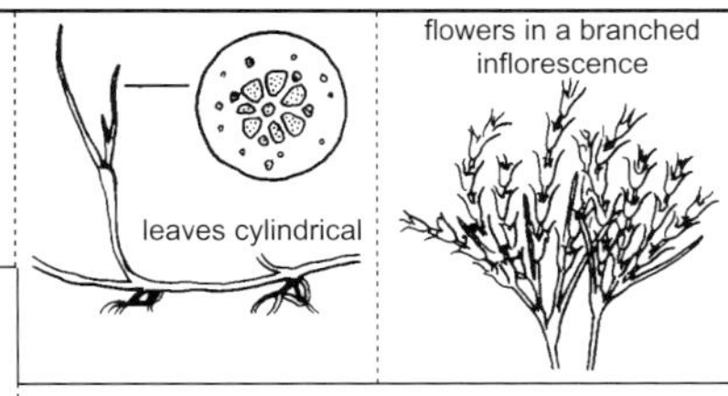

4	**Leaves** grasslike; **leaf tips** truncate, with 3 short teeth	***Halodule uninervis*** (Cymodoceaceae) 312
	Leaves strap-shaped or rarely grass-like; **leaf tips** rounded, entire or serrate	5

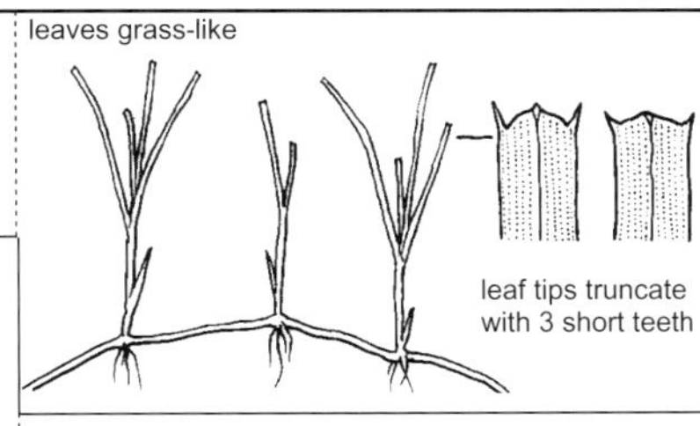

5	**Leaves** in pairs, enclosed at base by 2 membranous scales	***Halophila stipulacea*** (Hydrocharitaceae) (not recorded but likely to occur)
	Leaves 1 to 6, sometimes enclosed by persistent leaf bases but not by scales	6

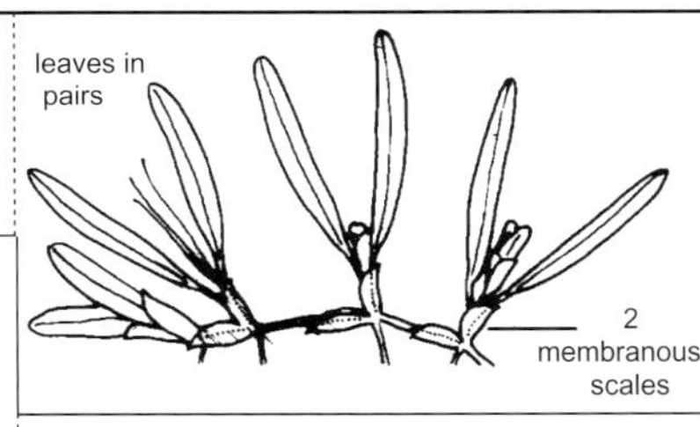

6	**Leaves** without ligules, tip entire; **base of shoots** enclosed by a shaggy sheath of old leaf bases	***Thalassia hemprichii*** (Hydrocharitaceae) (not recorded but likely to occur)
	Leaves with ligules, tip entire or serrate; **base of shoots** sometimes with old leaf bases but never a shaggy sheath	7

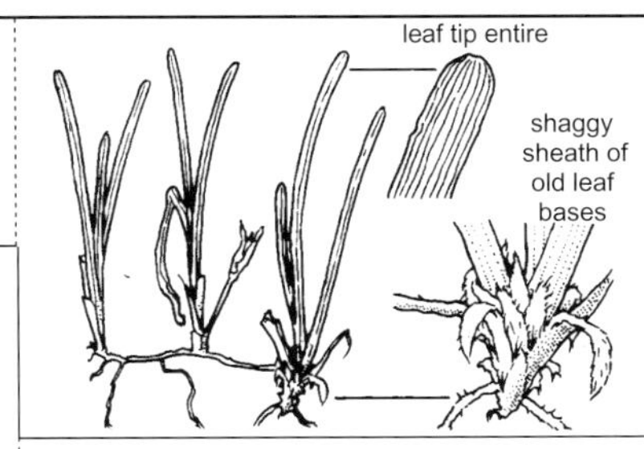

7	**Rhizome** bearing an elongated shoot at every fourth node, with conspicuous annular scars; **leaf blade** and **sheath** shed together	***Thalassodendron ciliatum*** (Cymodoceaceae) 313
	Rhizome bearing a contracted shoot at every node, without conspicuous scars; **leaf blade** shed before the **sheath**	8

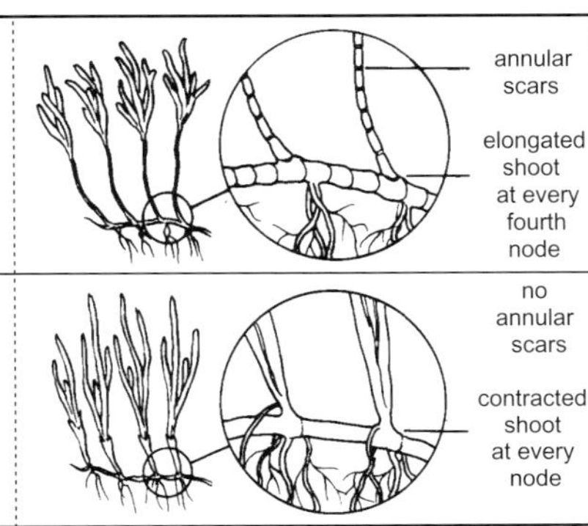

8	**Leaf blade** 2–4mm across; **leaf tip** entire; persistent **leaf sheath** closed	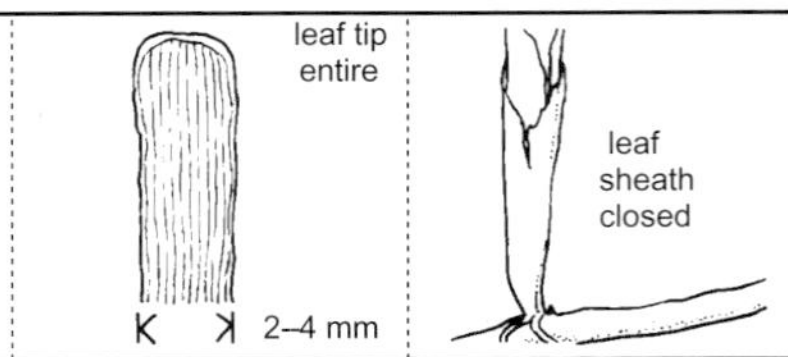	***Cymodocea rotundata*** (Cymodoceaceae) 312
	Leaf blade 4–10mm across; **leaf tip** toothed; persistent **leaf sheath** open	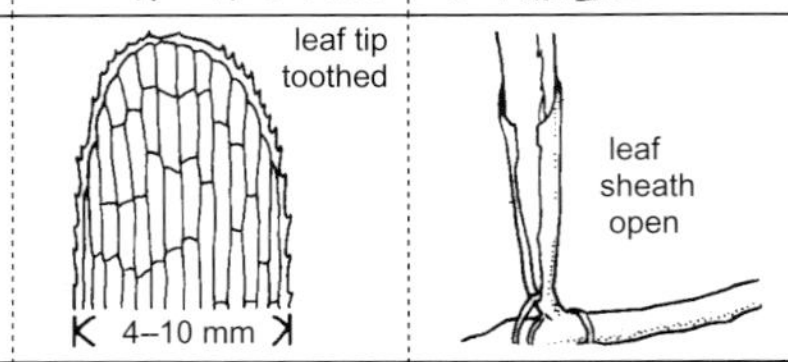	***Cymodocea serrulata*** (Cymodoceaceae) 312

GROUP D: TERRESTRIAL MONOCOTS

1	**Trees** with a single trunk		2
	Herbs, sometimes woody but never with a trunk	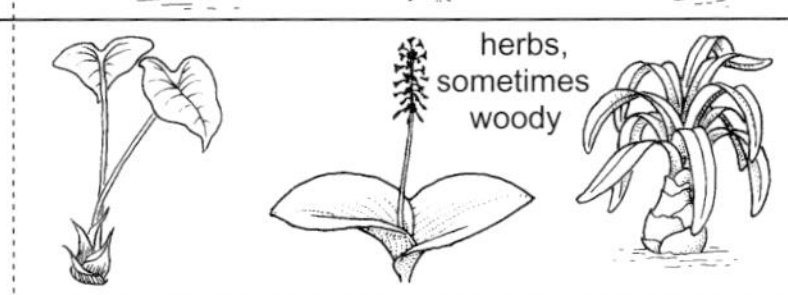	3
2	**Trunk** branched above; **leaves** linear	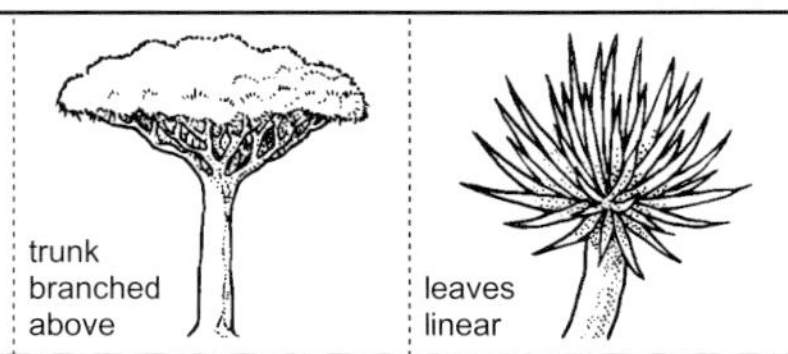	**Dracaenaceae** (1 sp.) (*Dracaena cinnabari*)
	Trunk not branched; **leaves** palmate (fan-shaped) or pinnate		**Palmae** (2 spp.)
3	**Leaves** large, succulent, with a toothed margin, arrranged in a basal rosette	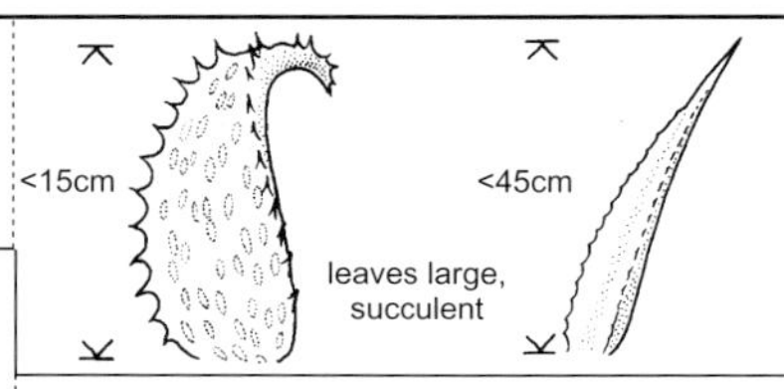	**Aloaceae** (3 spp.)
	Leaves not large or succulent, never with a toothed margin		4
4	**Leaves** heart-shaped	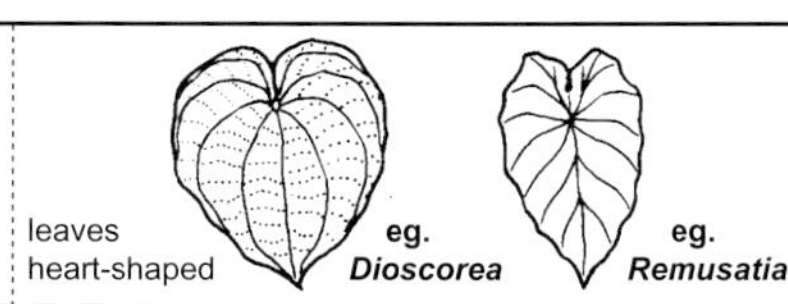	5
	Leaves linear or if broader then rounded at the base	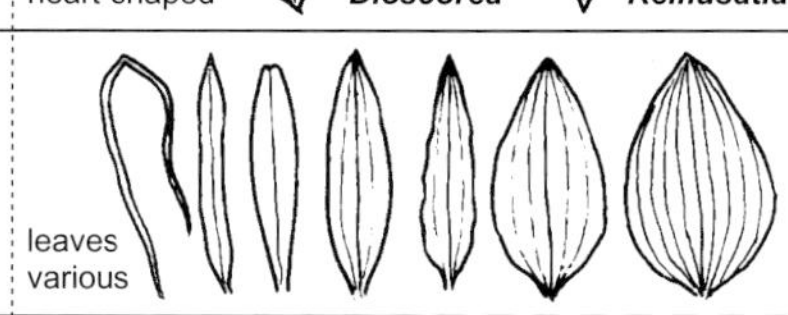	6
5	**Stems** twining; **leaves** alternate; **flowers** in racemes, never with hooked organs	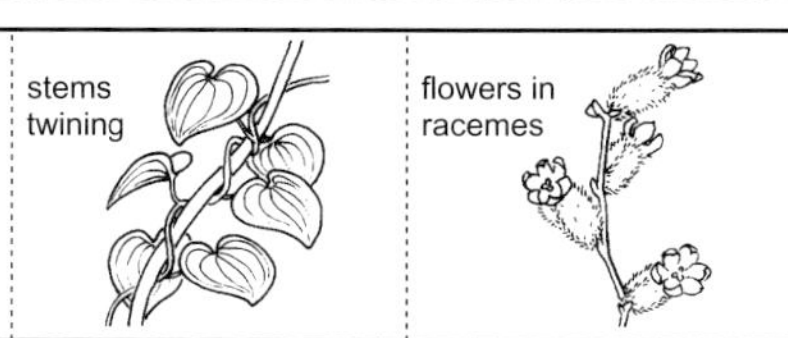	**Dioscoriaceae** (1 sp.) (*Dioscorea lanata*)
	Stems absent; **leaves** in basal rosette; **flowers** hidden by a spathe, asexually spread by hooked organs	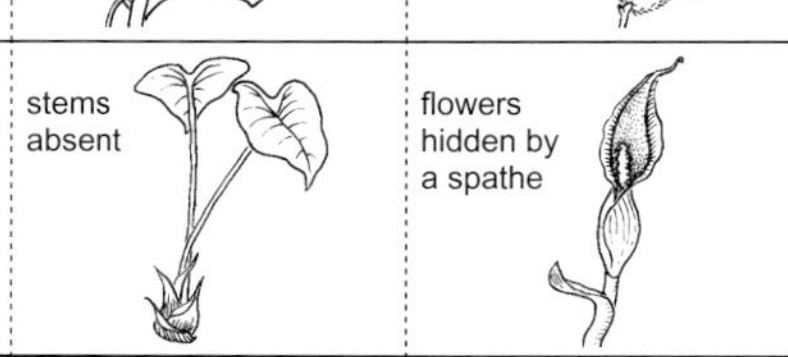	**Araceae** (2 spp.)

6	**Flowers** with petaloid perianth parts	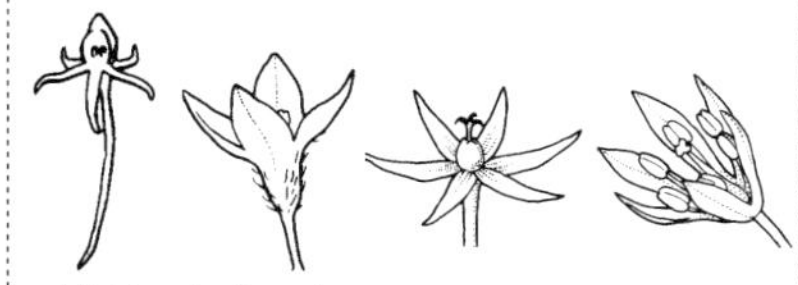	7
	Flowers with scarious perianth parts (grasses, sedges and rushes)	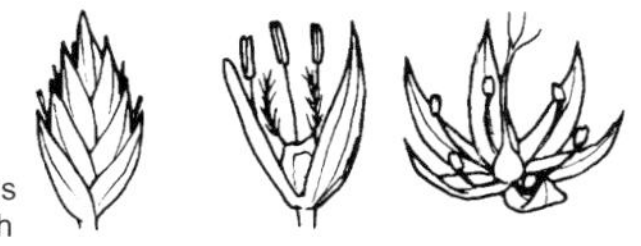	16
7	**Ovary** inferior		8
	Ovary superior	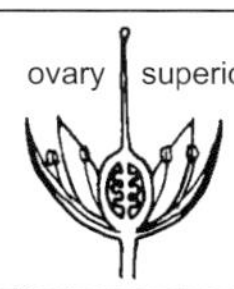	11
8	**Flowers** zygomorphic	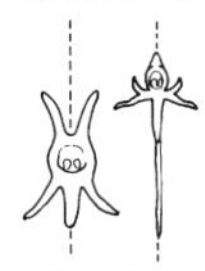	**Orchidaceae** (4 spp.)
	Flowers actinomorphic	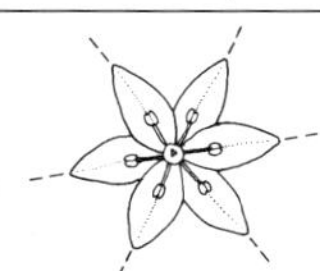	9
9	**Flowers** large, in an umbel-like inflorescence subtended by a pair of bracts; **filaments** fused into a cylindrical "corona"	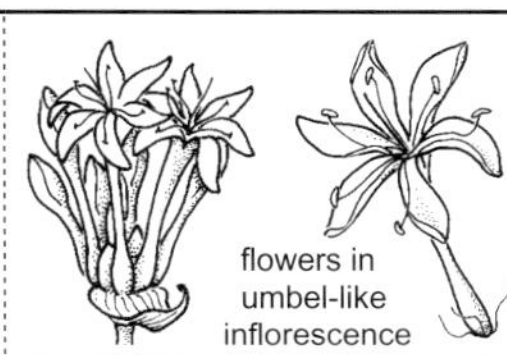 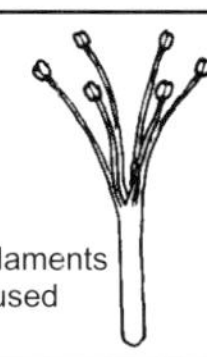	**Amaryllidaceae** (1 sp.) (*Crinum balfourii*)
	Flowers solitary or in spikes, not subtended by a pair of bracts; **filaments** free	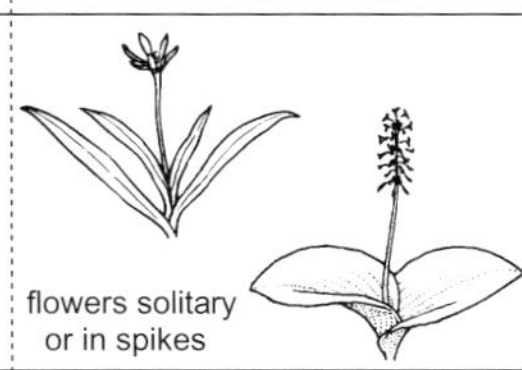 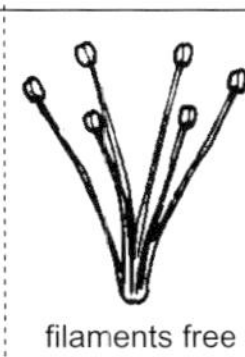	10
10	**Flowers** small, blue; **stamens** 3; **leaves** glabrous	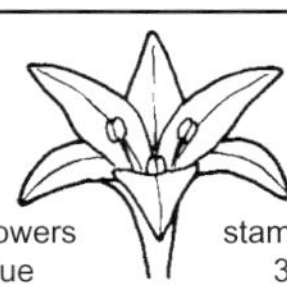 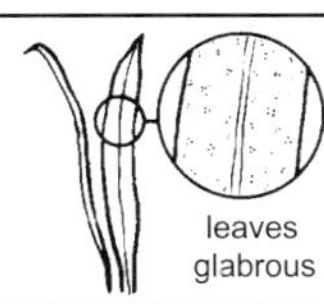	**Iridaceae** (2 spp.)
	Flowers yellow; **stamens** 4 (to 6); **leaves** hairy	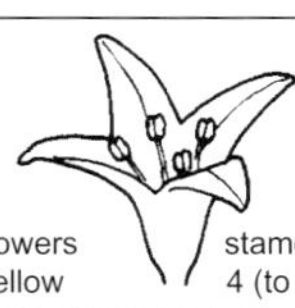 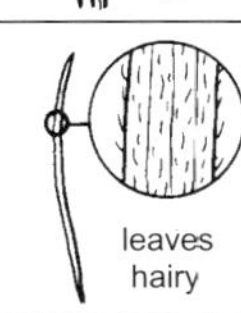	**Hypoxidaceae** (1 sp.) (*Hypoxis angustifolia*)
11	**Shrublet**; **leaves** reduced to spines or scales, leaflike branches (cladodes) present	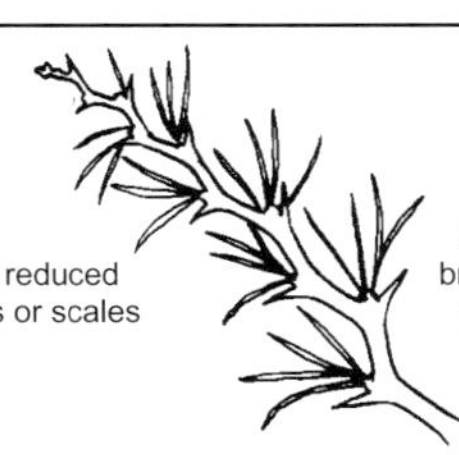	**Asparagaceae** (2 spp.)
	Herbs; leaves present, spines never present		12

12	**Flower** with perianth parts in two dissimilar whorls (sepals and petals); **stems** prostrate, jointed	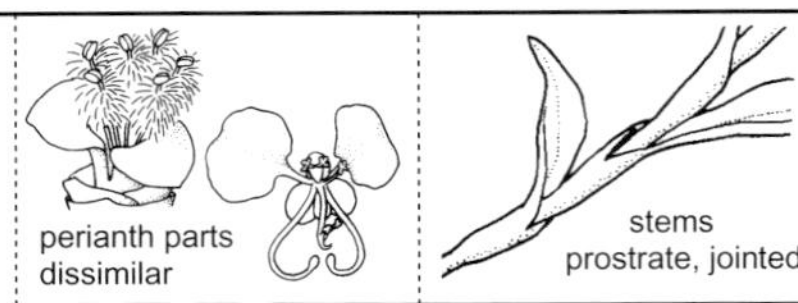	**Commelinaceae** (4 spp.)
	Flower with perianth parts all similar, all petaloid (tepals); **stems** erect or underground, not jointed	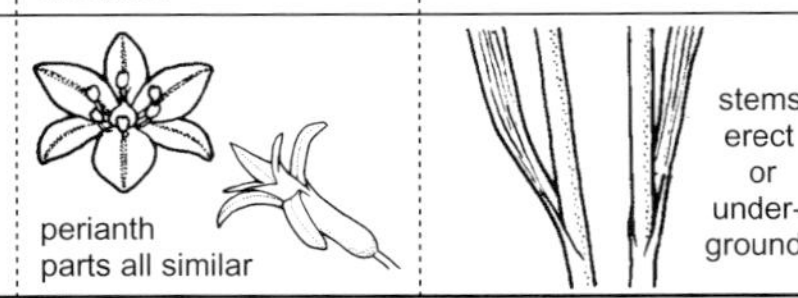	13
13	**Styles** 3; **stems** leafy; **flowers** in the axils of the leaves	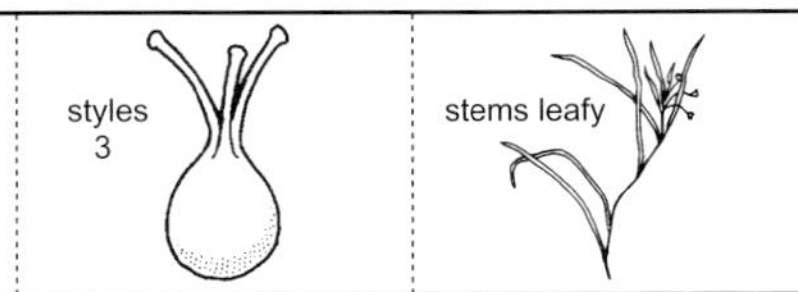	**Colchicaceae** (1 sp.) (*Iphigenia socotrana*)
	Style simple, the stigma sometimes 3-lobed; **leaves** all basal; **flowers** in leafless racemes	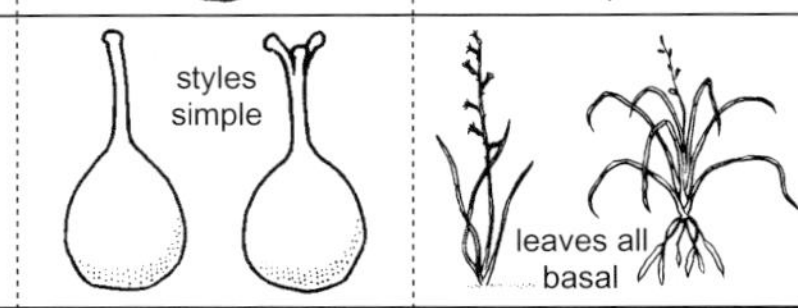	14
14	**Leaves** linear, terete (rounded in section), hollow; **plant** with a rhizome which is yellow when cut	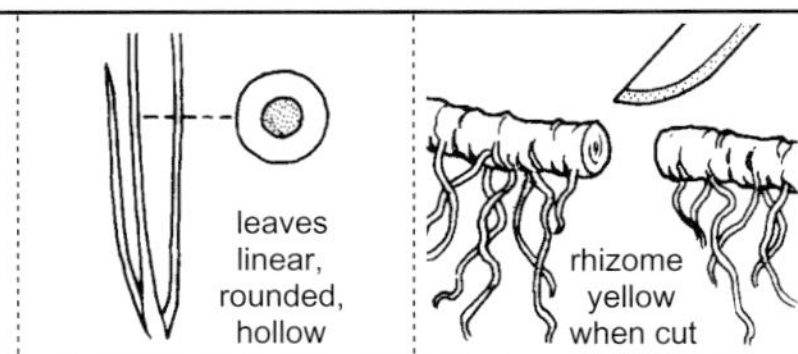	**Asphodelaceae** (1 sp.) (*Asphodelus fistulosus*)
	Leaves linear to broadly ovate, always flattened, never hollow; **plant** with a bulb or with a rhizome which is not yellow when cut	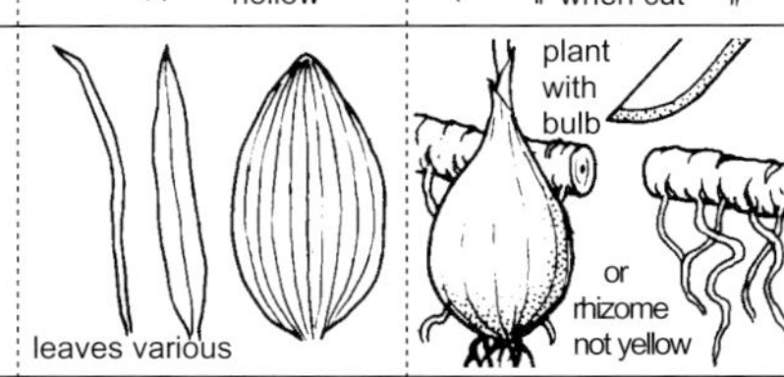	15
15	**Plant** with a bulb; **leaves** glabrous; **capsule** globose or ovoid, rounded or lobed	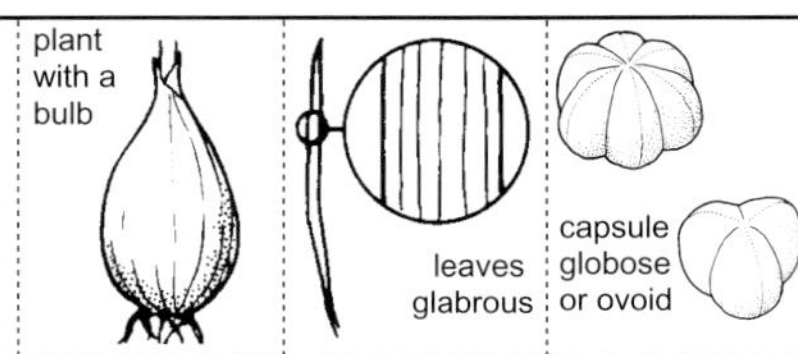	**Hyacinthaceae** (8 spp.)
	Plant with a rhizome; **leaves** hairy; **capsule** 3-angled	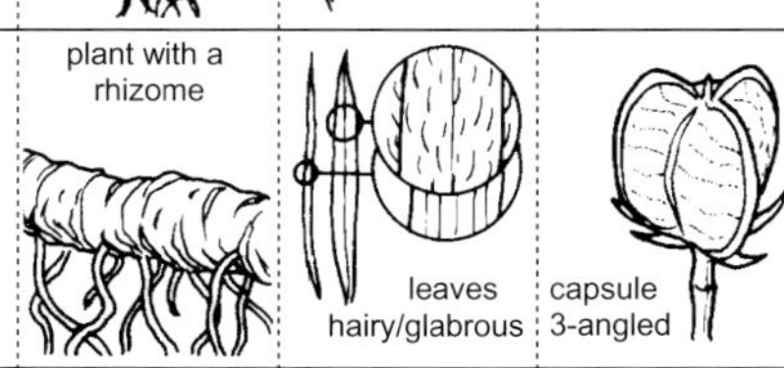	**Anthericaceae** (3 spp.)
16	**Perianth** a series of bracts; **flowers** arranged in spikelets; **stems** rounded in section; **leaf sheaths** open	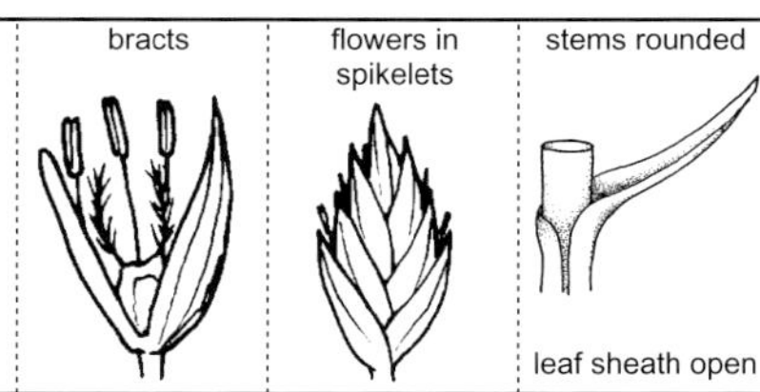	**Gramineae** (96 spp.)
	Perianth absent or of bristles; **flowers** arranged in spikes or heads; **stems** triangular in section; **leaf sheaths** closed	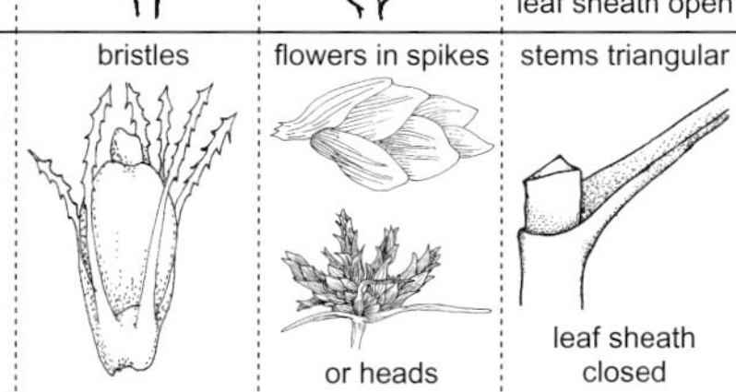	**Cyperaceae** (27 spp.)
	Perianth of 6 tepals; **flowers** arranged in panicles; **stems** rounded in section; **leaf sheaths** closed	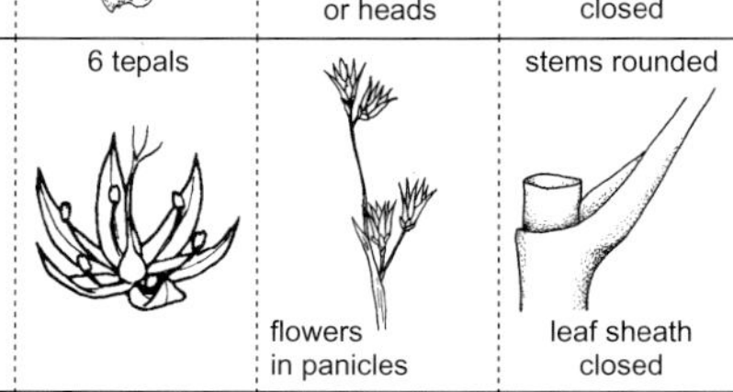	**Juncaceae** (3 spp.)

GROUP E: HERBACEOUS DICOTS

1	Parasitic **plants,** often without chlorophyll	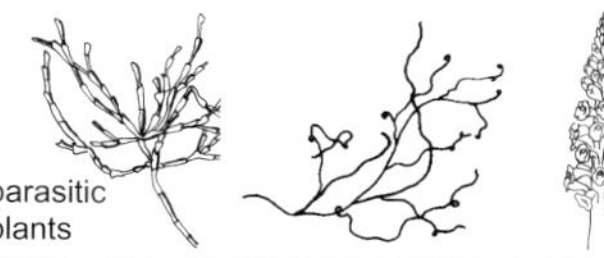 parasitic plants	2
	Free-living **plants,** always with chlorophyll	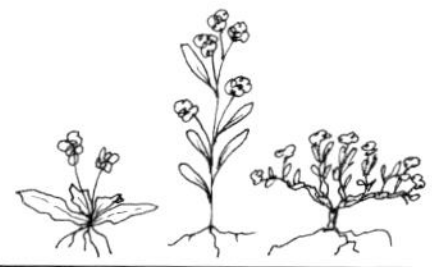 free-living plants	4

2	Leafless **epiphyte**	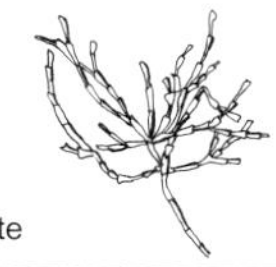 leafless epiphyte	**Viscaceae** (1 sp.) (*Korthalsella taenioides*)
	Leafless **twiner**	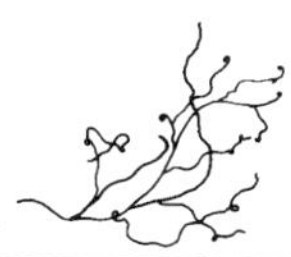 leafless twiner	**Cuscutaceae** (2 spp.)
	Erect **herb**, leafy	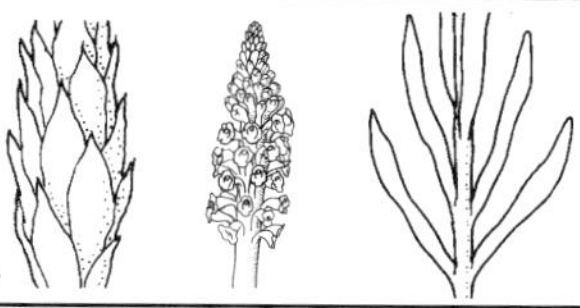 leafy herb	3

3	**Plants** green; **corolla tube** long, cylindrical, sharply bent: **capsule** without internal walls	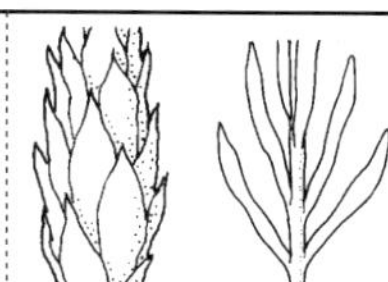 plants green	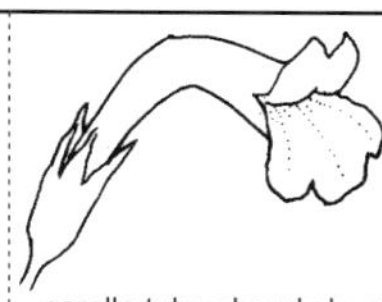 corolla tube sharply bent	**Scrophulariaceae** (19 spp.)
	Plants without chlorophyll; **corolla tube** not sharply bent; **capsule** with internal walls	 plants without chlorophyll	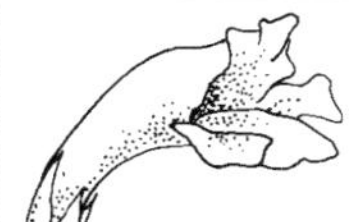 corolla tube not sharply bent	**Orobanchaceae** (4 spp.)

4	**Flowers** in spikes or heads, borne on a common receptacle, individual flowers small and parts often reduced	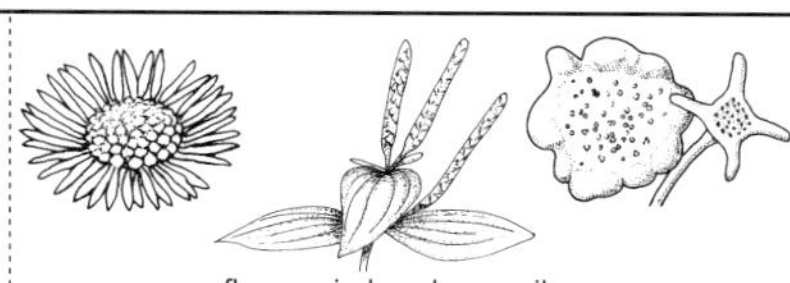 flowers in heads or spikes	5
	Flowers borne separately, not in heads or spikes and not borne on a common receptacle	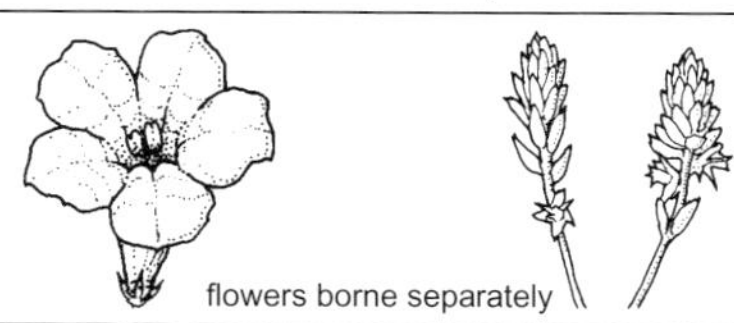 flowers borne separately	8

5	**Flowers** borne on a fleshy spike, difficult to separate from each other; **spike** not surrounded by bracts	flowers on a fleshy spike 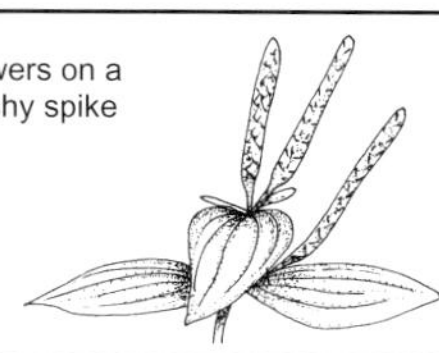	**Piperaceae** (2 spp.)
	Flowers borne on flattened receptacles**,** usually circular; **receptacle** surrounded by whorls, bracts or specialised appendages	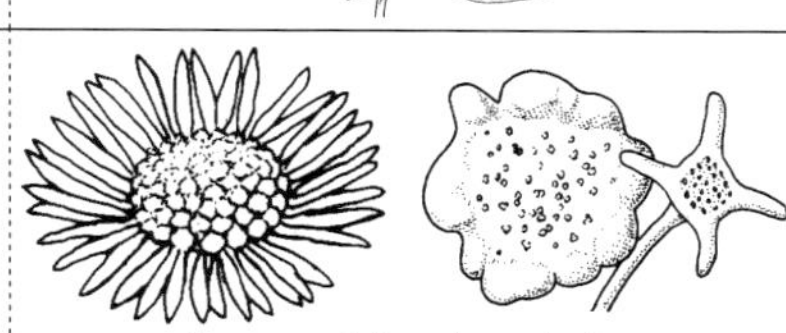 flowers on flattened receptacles	6

6	**Flowers** with coloured petals; **receptacle** surrounded by whorls of greenish or papery bracts	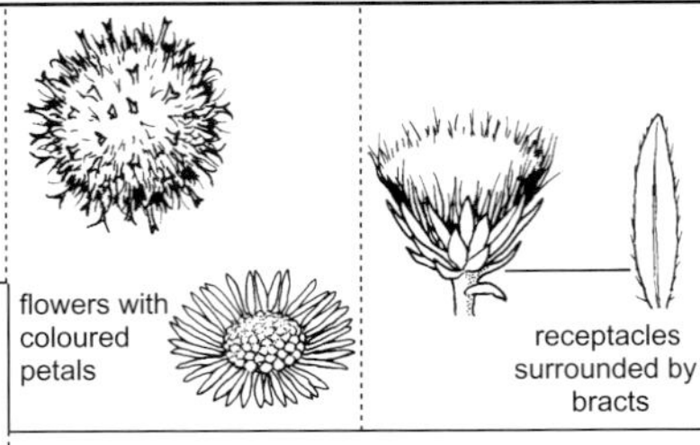	**Compositae** (55 spp.)
	Flowers inconspicuous; **receptacle** lacking bracts		7
7	**Flowers** on axillary receptacles; **receptacles** densely woolly within, surrounded by an involucre of 2 to 8 bracts; **herbs** with hooked hairs	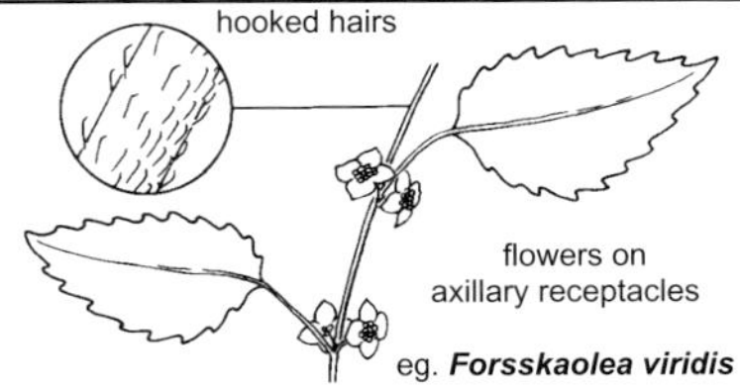	**Urticaceae** (5 spp.)
	Flowers on a pedunculate, circular receptacle; **receptacle** surrounded by linear appendages; stemless **herbs** from underground tubers	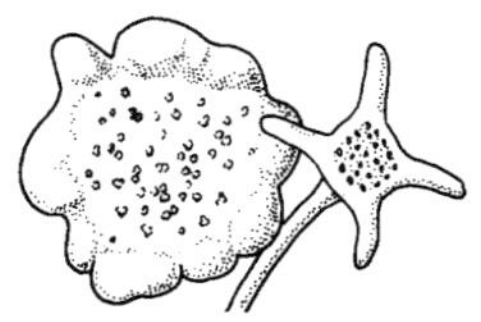 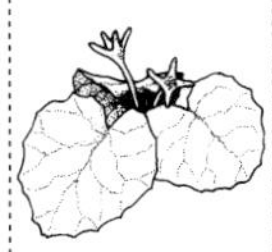	**Moraceae** (4 spp.)
	Flowers arranged on a circular receptacle; **receptacle** surrounded by 4 or 5 glands**,** alternating with entire or fringed lobes; **herbs** or **subshrubs** with milky latex	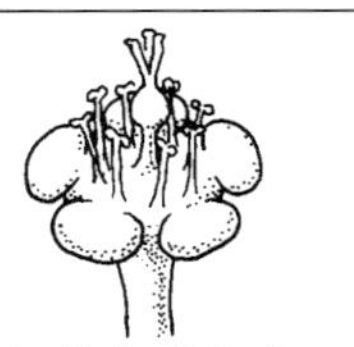 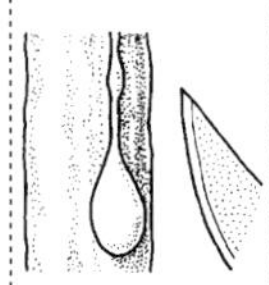	**Euphorbiaceae** (33 spp.)
8	**Leaves** opposite (or in a basal rosette)	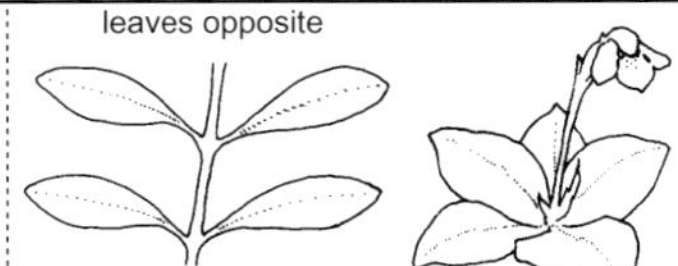	9
	Leaves alternate	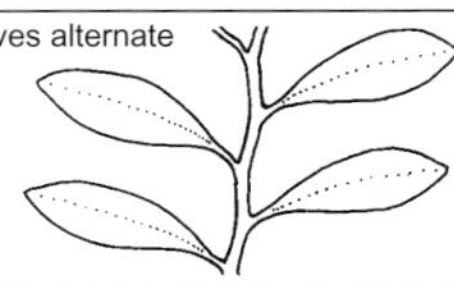	33
9	**Flowers** zygomorphic	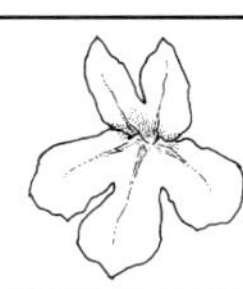	10
	Flowers actinomorphic		16
10	**Petals** free	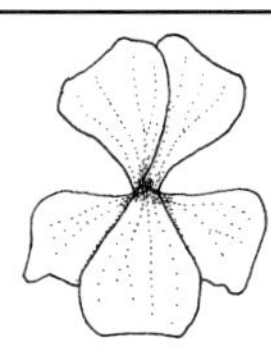	**Geraniaceae** (5 spp.)
	Petals fused	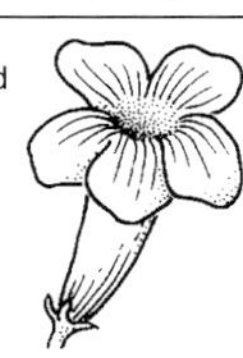	11

11

Fruit dry, 6-hooked

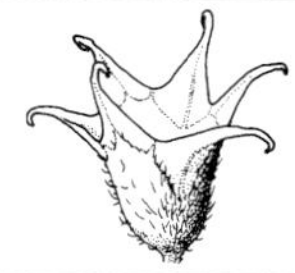

Valerianaceae
(1 sp.)
(*Valerianella affinis*)

Fruit a berry or dry and splitting into 1-seeded nutlets when ripe

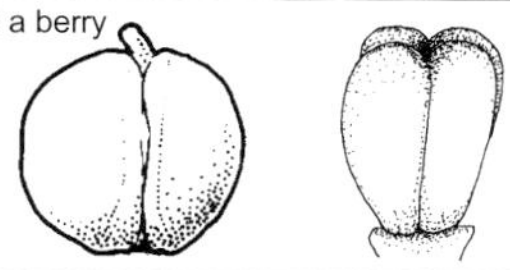

12

Fruit a capsule

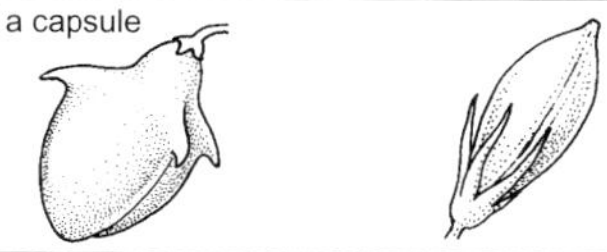

13

12

Flowers in spikes; **styles** attached to top of ovary

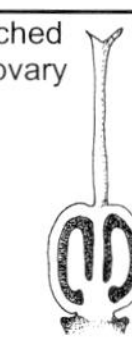

Verbenaceae
(6 spp.)

Flowers in axillary cymes; **styles** attached to base of ovary

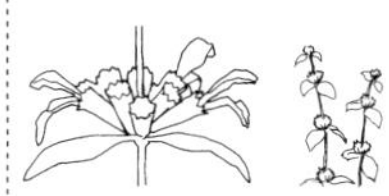

Labiatae
(19 spp.)

13

Flowers solitary; **fruit** spiny

Pedaliaceae
(1 sp.)
(*Pedalium murex*)

Flowers in groups; **fruit** unarmed

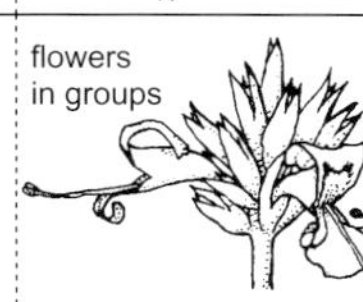

14

14

Fruit a woody, 2 or 4-seeded capsule; **seeds** on hook-like projections

Acanthaceae
(34 spp.)

Fruit a leathery, usually many-seeded capsule; **seeds** not on hook-like projections

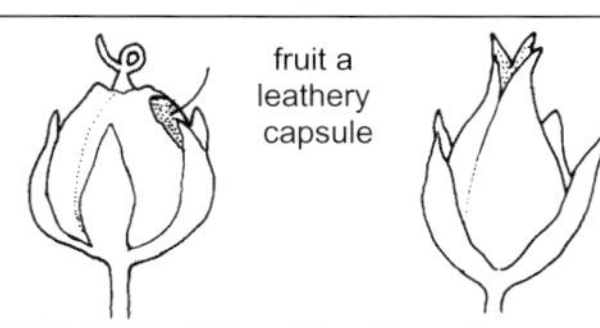

Scrophulariaceae
(19 spp.)

15

Petals absent (petaloid staminodes sometimes present in Aizoaceae)

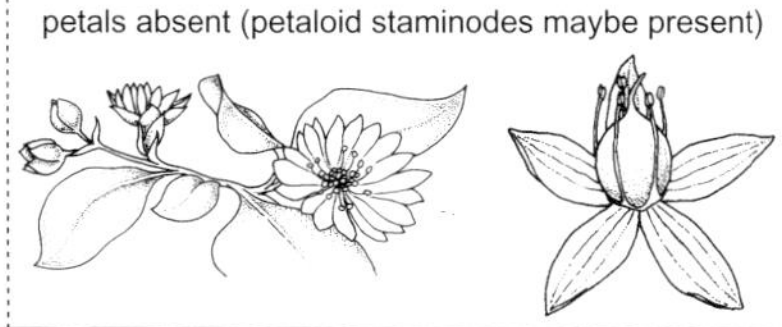

16

Petals present

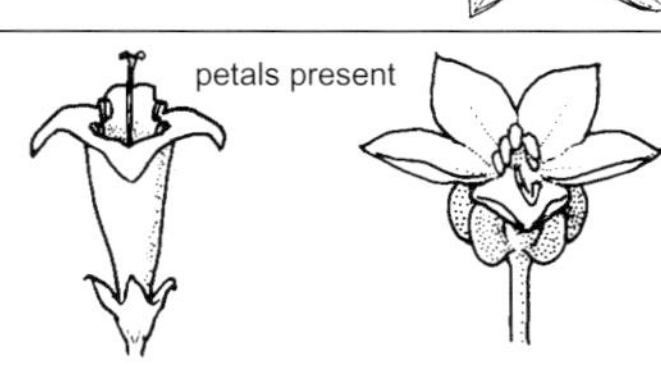

18

16	**Ovary** inferior; **flowers** solitary in leaf axils; **stems** and **petioles** winged	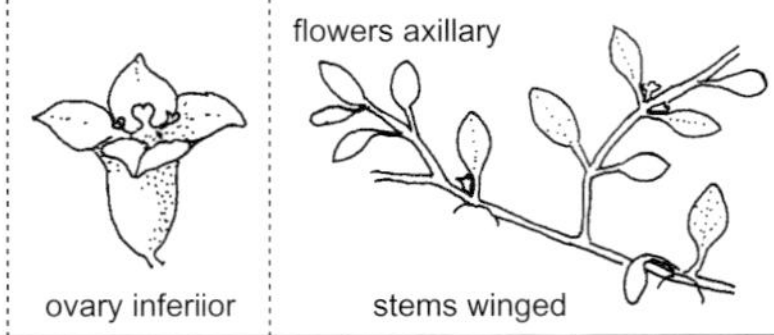	**Onagraceae** (1 sp.) (*Ludwigia palustris*)
	Ovary superior; **flowers** clustered or in spikes or panicles; **stems** and **petioles** not winged	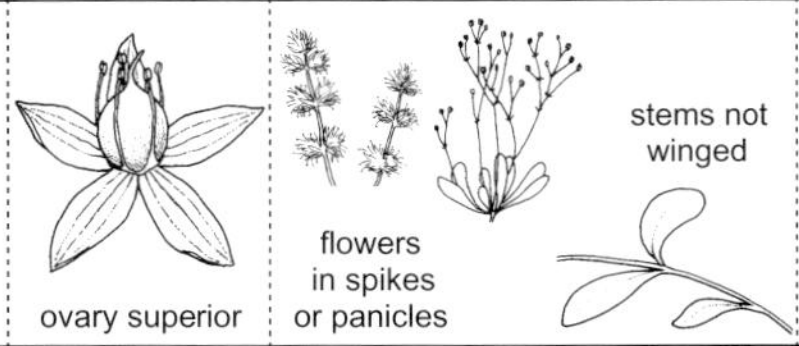	17
17	**Capsule** 2 to 5-celled or a cluster of 5 achenes	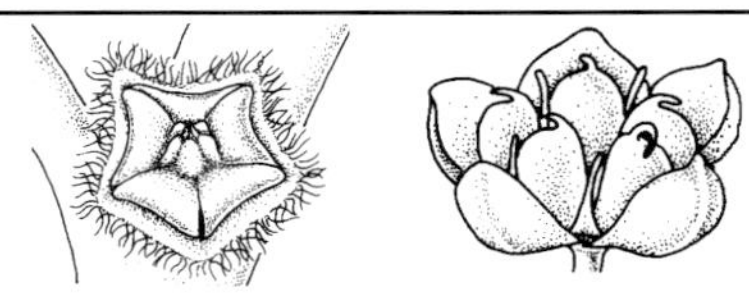	**Aizoaceae** (8 spp.)
	Capsule 1-celled	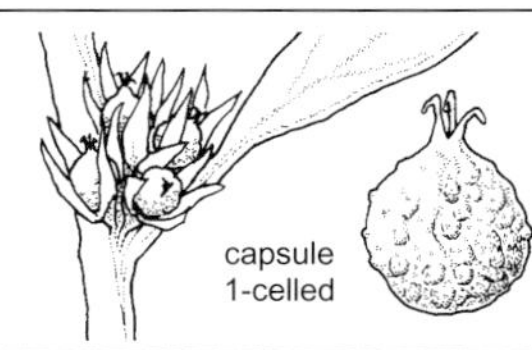	**Amaranthaceae** (13 spp.)
18	**Petals** fused	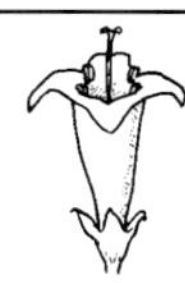	19
	Petals free		27
19	**Ovary** inferior		20
	Ovary superior	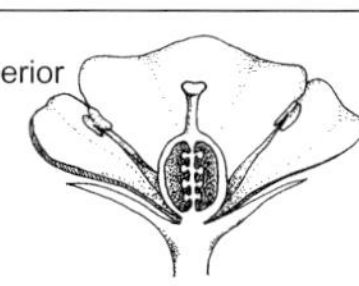	22
20	**Stipules** fused between opposite petioles, or **leaves** whorled	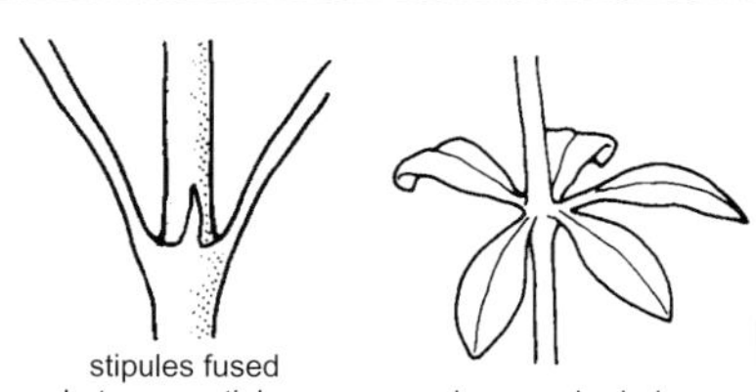	**Rubiaceae** (22 spp.)
	Stipules absent; **leaves** never whorled		21
21	**Fruits** with 6 hooks	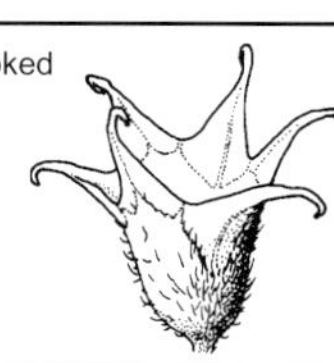	**Valerianaceae** (1sp.) (*Valerianella affinis*)
	Fruit 5 or 10-ribbed, glandular	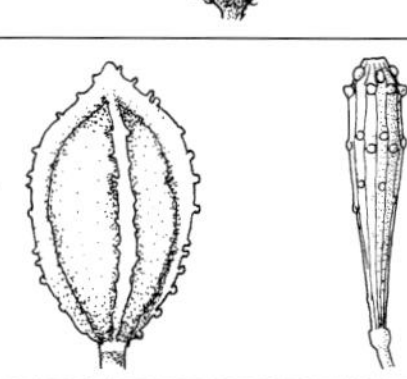	**Nyctaginaceae** (5 spp.)

22	**Stamens** fused to stigma to form a central column; **fruit** a pair of follicles; **seeds** with a tuft of feathery hairs	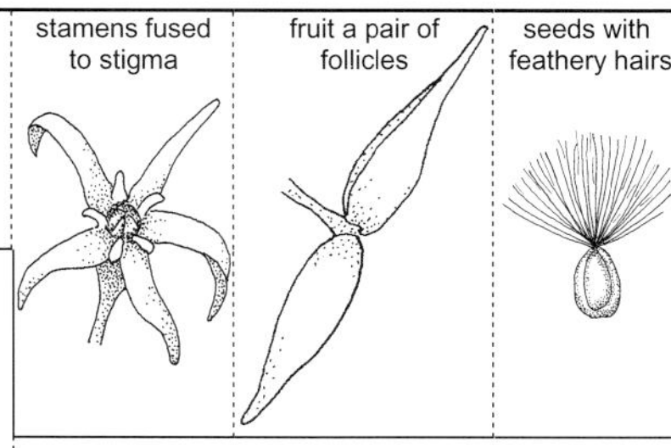	**Asclepiadaceae** (27 spp.)
	Stamens free from stigma; **fruit** not a pair of follicles; **seeds** without a tuft of hairs		23

23	**Fruit** a pair of smooth or spiny nutlets	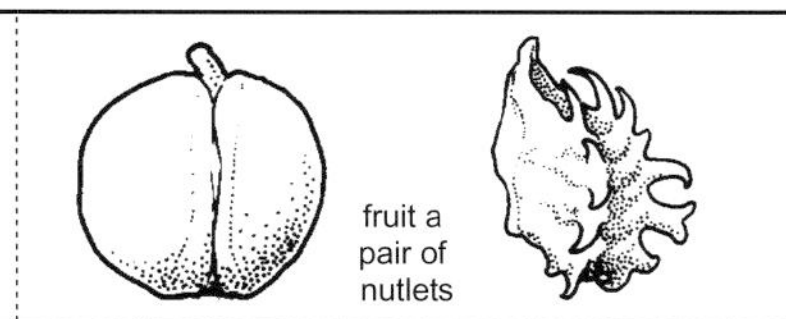	**Verbenaceae** (6 spp.)
	Fruit of 3 to 5 free follicles	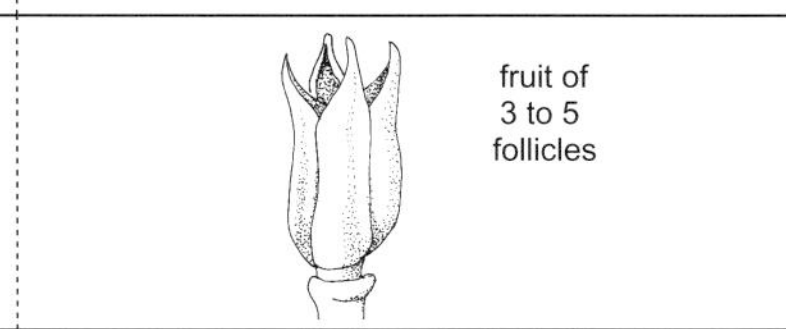	**Crassulaceae** (6 spp.)
	Fruit a capsule	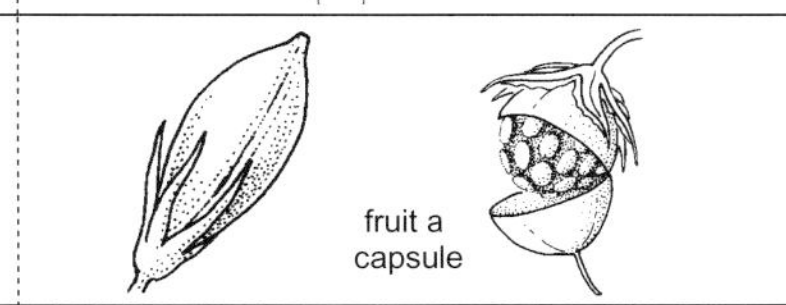	24

24	**Petals** small (or absent) and borne on the rim of the perianth tube	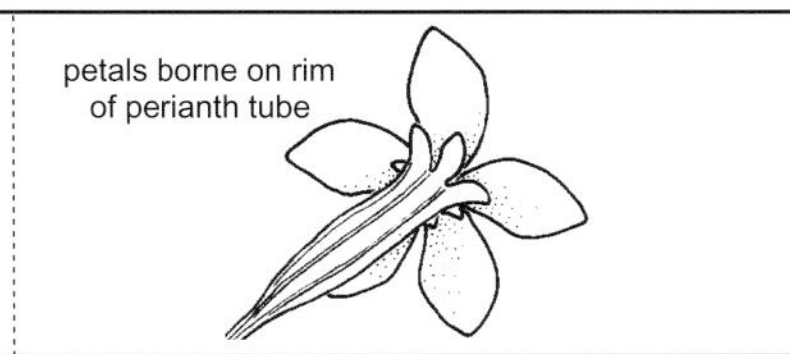	**Lythraceae** (3 spp.)
	Petals conspicuous, fused into a corolla tube (maybe very short)	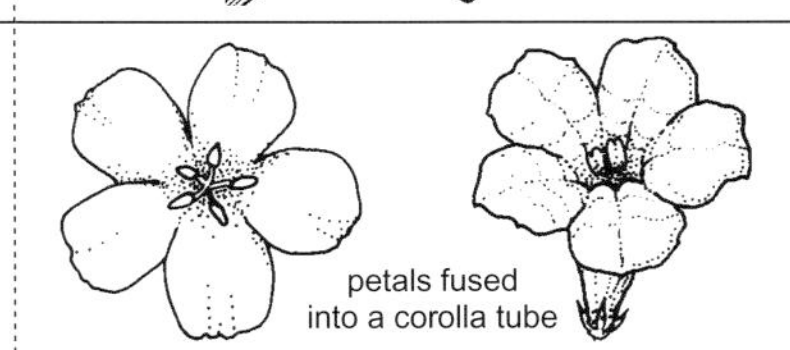	25

25	**Stamens** 2 or 4; **fruit** a woody, 2 or 4-seeded capsule; **seeds** on hook-like projections	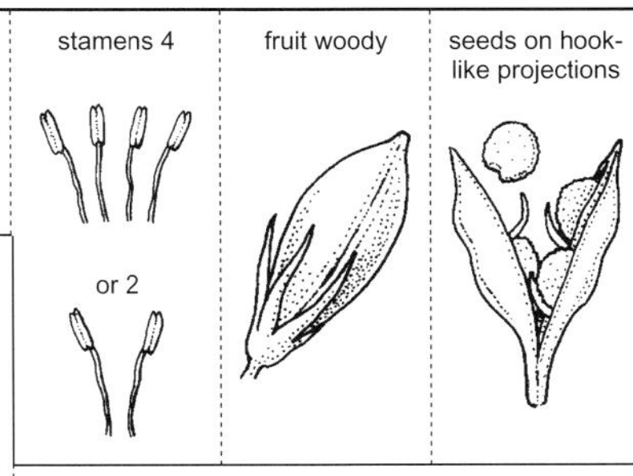	**Acanthaceae** (34 spp.)
	Stamens 5; **fruit** a many-seeded membranous capsule; **seeds** not on hook-like projections		26

26	**Stamens** 5, opposite corolla lobes; **ovary** 1-celled; **fruit** opening by a lid	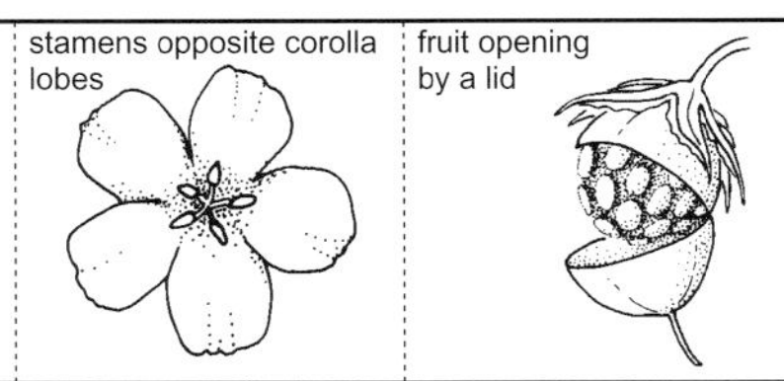	**Primulaceae** (1 sp.) (*Anagallis arvensis*)
	Stamens 5, alternating with corolla lobes; **ovary** 2-celled; **fruit** opening by valves or splits	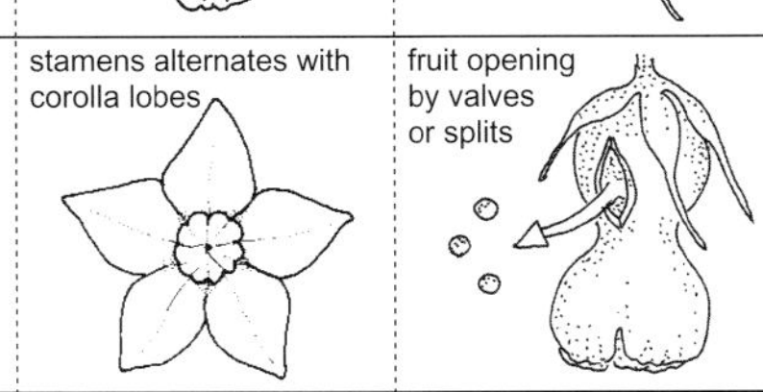	**Gentianaceae** (5 spp.)

27	**Leaves** pinnately or palmately lobed		28
	Leaves simple	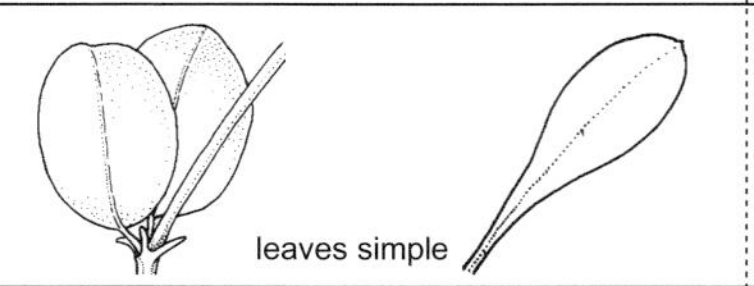	29
28	**Fruit** splitting into 5 dehiscent, unarmed segments (mericarps)	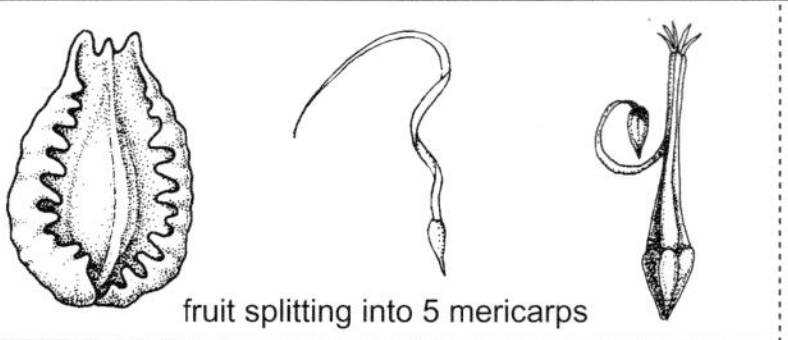	**Geraniaceae** (5 spp.)
	Fruit a capsule or splitting into 5 indehiscent, spiny segments	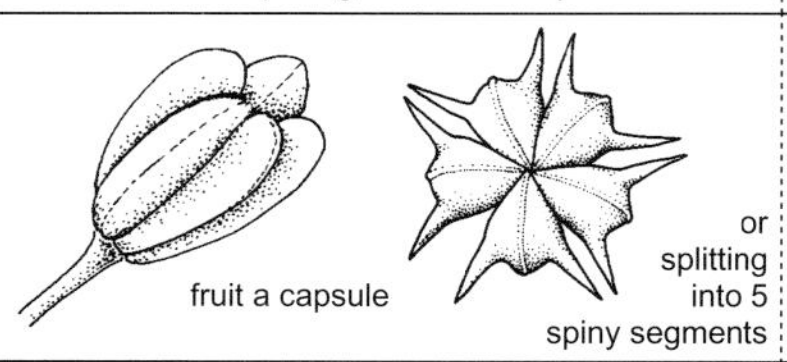	**Zygophyllaceae** (6 spp.)
29	**Sepals** 2	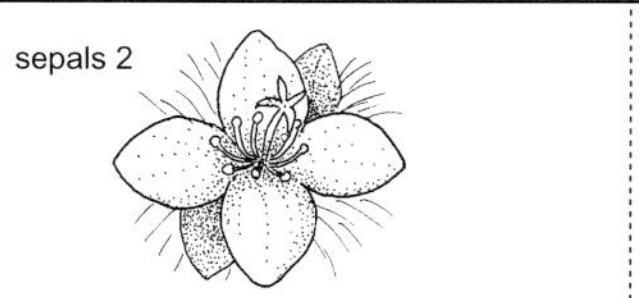	**Portulacaceae** (6 spp.)
	Sepals 4 or 5	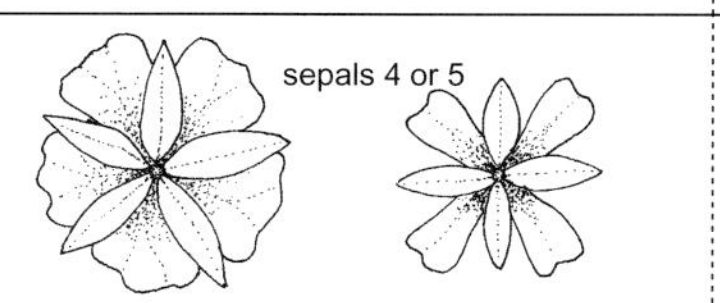	30
30	**Ovules** on free central or axile placentas	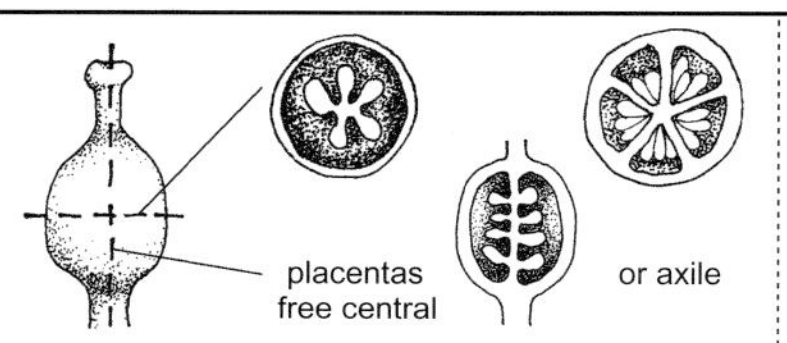	31
	Ovules on parietal placentas	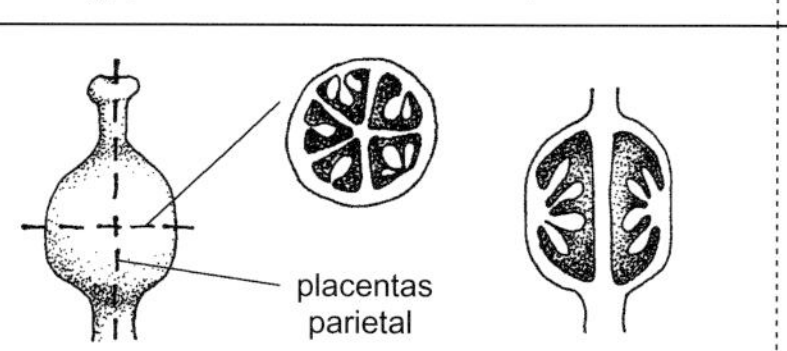	**Frankeniaceae** (1 sp.) (*Frankenia pulverulenta*)
31	**Ovules** on free central placentas; **styles** 2 or 3 or solitary with a 2 or 3-lobed stigma; **stipules** never spinescent, if absent then leaves not succulent	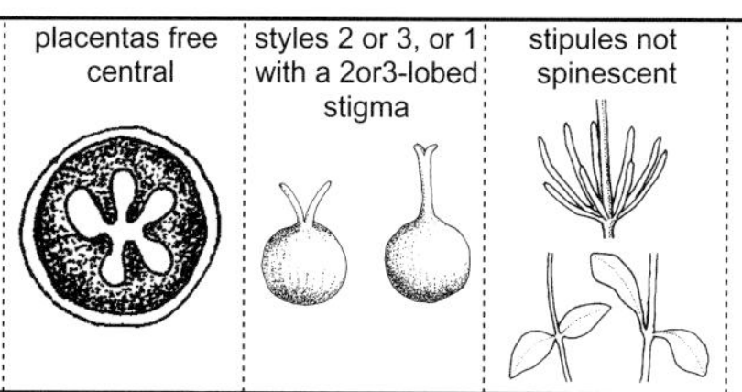	**Caryophyllaceae** (17 spp.)
	Ovules on axile placentas; **style** 1, stigma not lobed; **stipules** spinescent or if absent then leaves succulent	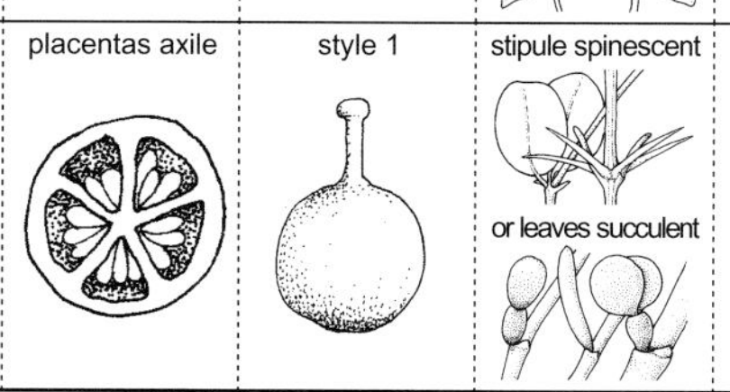	**Zygophyllaceae** (6 spp.)

Couplet	Character	Illustration	Go to / Family
32	**Flowers** zygomorphic	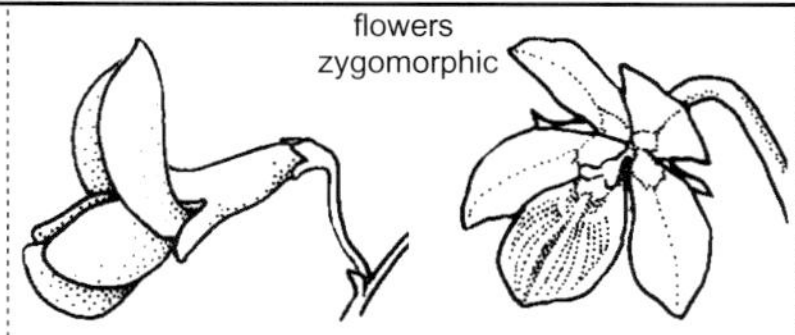	33
	Flowers actinomorphic	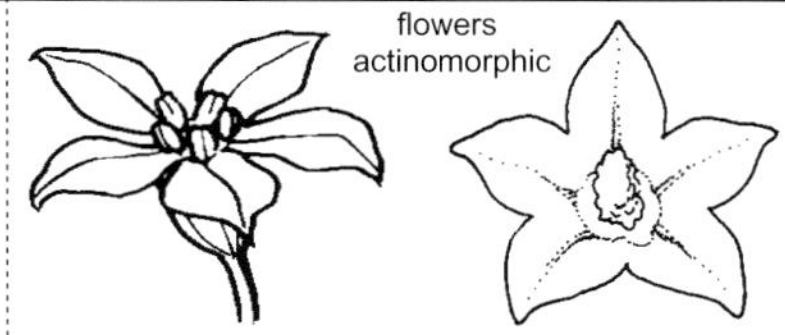	41
33	**Leaves** with stipules (sometimes spiny)	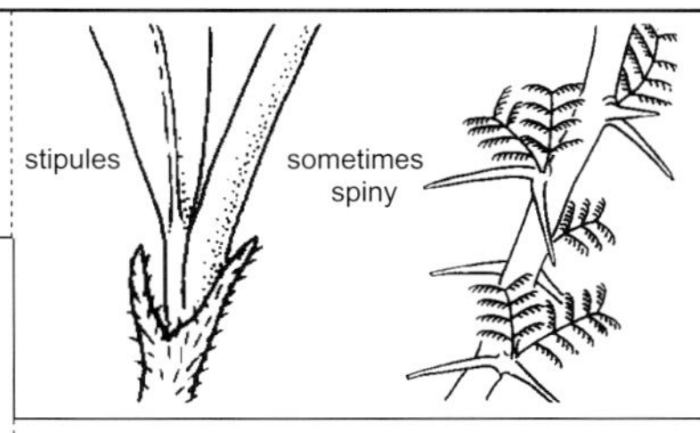	34
	Leaves without stipules		35
34	**Corolla** not spurred; **lower petals** forming a keel	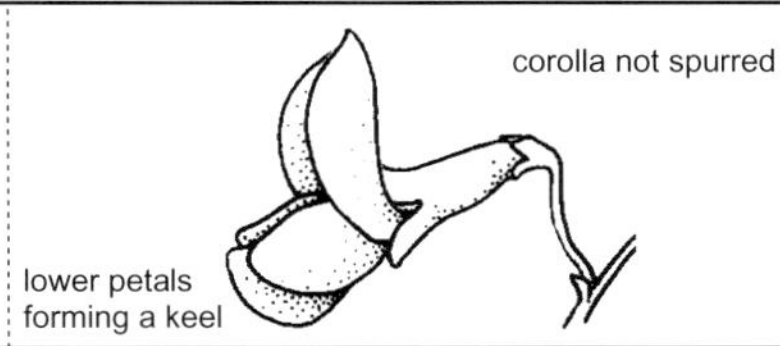	**Leguminosae (Papilionoideae)** (57 spp.)
	Corolla spurred; **lower petals** stalked	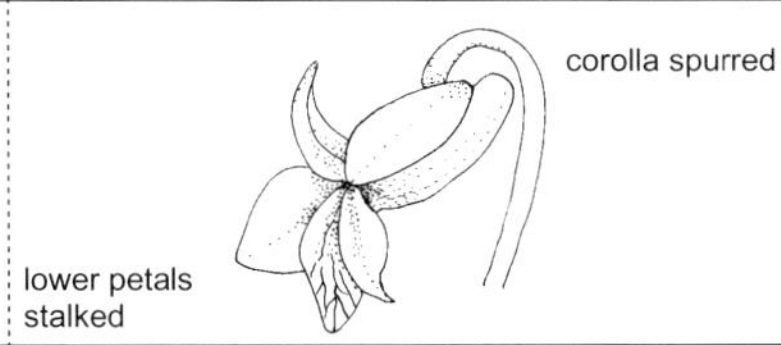	**Violaceae** (2 spp.)
35	**Petals** free (sometimes 2 cohering)	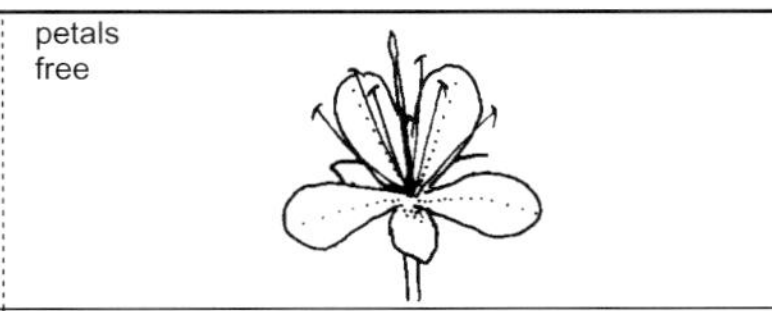	36
	Petals fused, usually into a tube	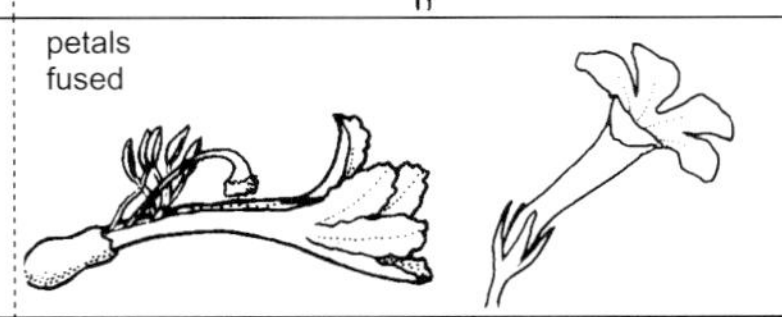	38
36	**Lateral sepals** enlarged and forming petaloid wings enclosing the flower; **lower petal** with a crest of threads, other petals obscure	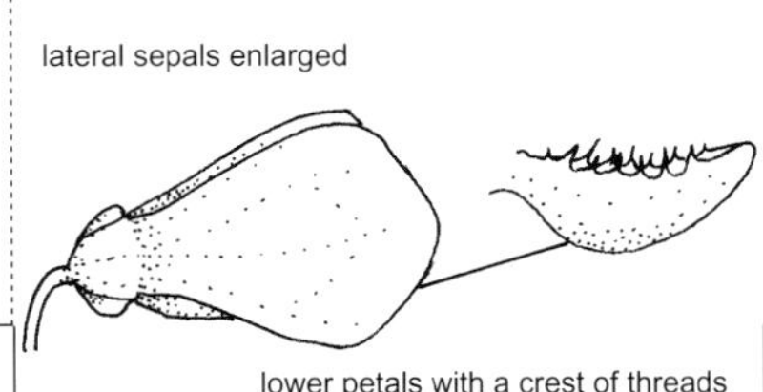	**Polygalaceae** (3 spp.)
	Sepals and **petals** similar, sometimes unequal in size		37
37	**Fruit** an elongated, rounded capsule, not open at tip; **petals** entire	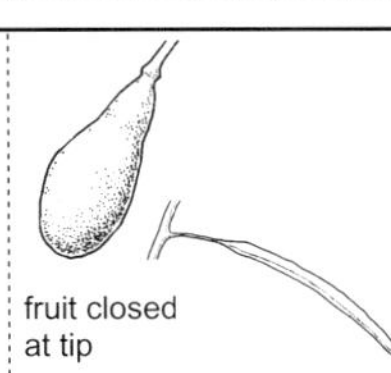 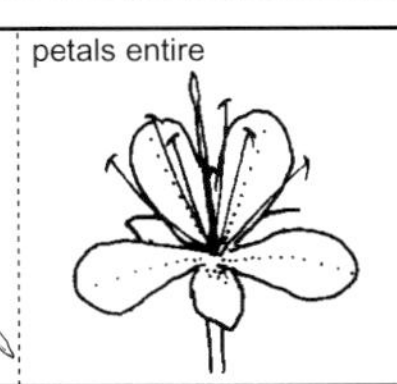	**Capparaceae** (14 spp.)
	Fruit an ellipsoid capsule, 3-angled, open at tip; **petals** laciniate	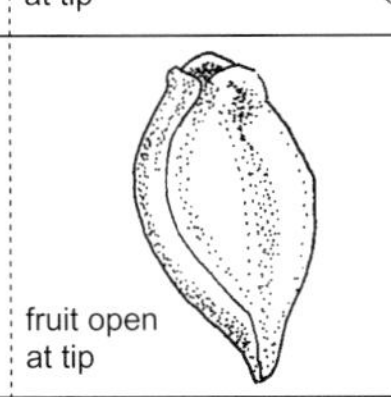 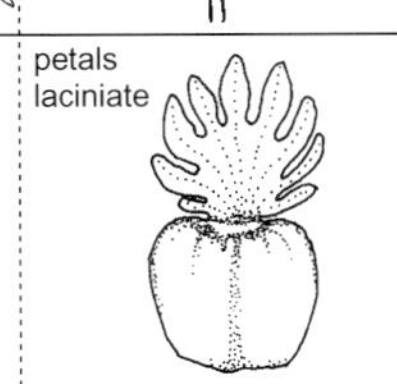	**Resedaceae** (3 spp.)

38	Tiny moss-like insectivorous plant bearing bladder-like **traps**	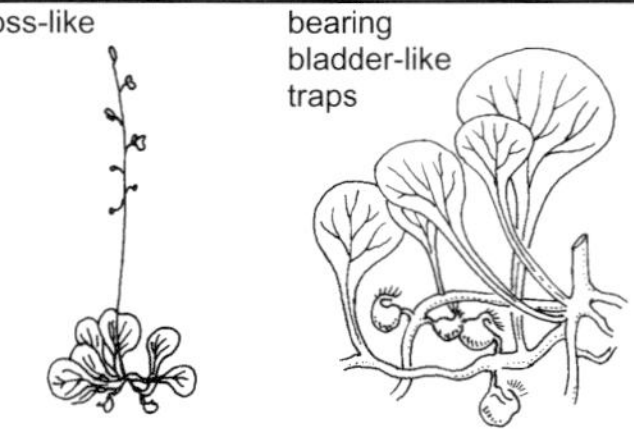	**Lentibulariaceae** (1 sp.) (*Utricularia striatula*)
	Erect herbs, not bearing **traps**		39

39	**Flowers** in a dense spike		**Globulariaceae** (1 sp.) (*Poskea socotrana*)
	Flowers never in dense spikes		40

40	**Corolla** 5-lobed, slit almost to base on one side; **ovary** inferior		**Goodeniaceae** (1 sp.) (*Scaevola socotraensis*)
	Corolla 2-lipped or 5-lobed, never split down one side; **ovary** superior	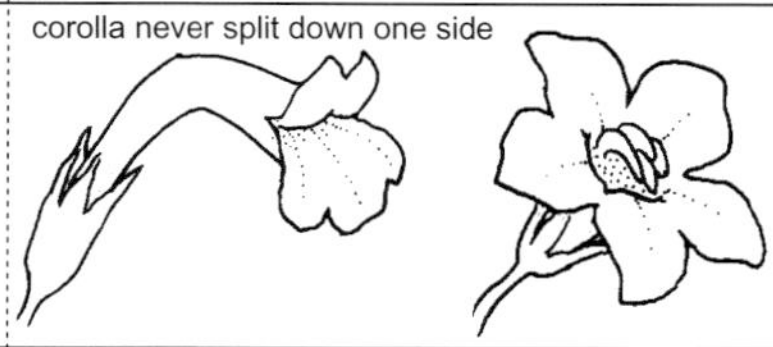	**Scrophulariaceae** (19 spp.)

41	**Petals** fused	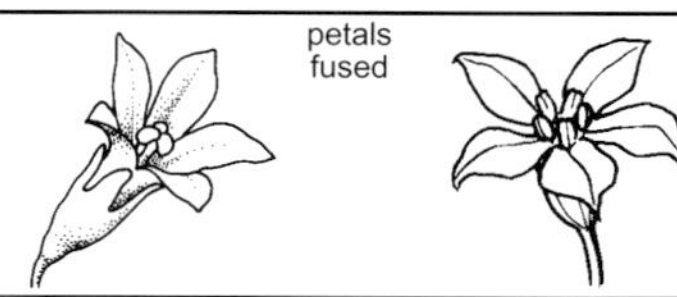	42
	Petals free	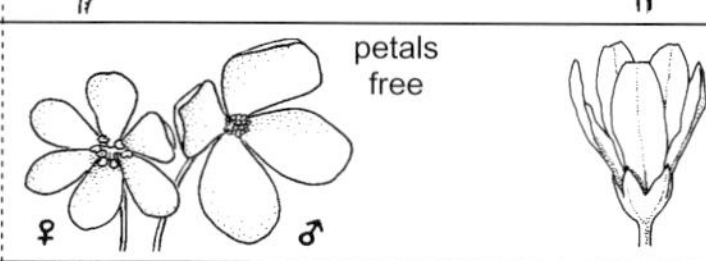	50
	Petals absent and replaced by one row of tepals or petaloid staminodes (Aizoaceae)	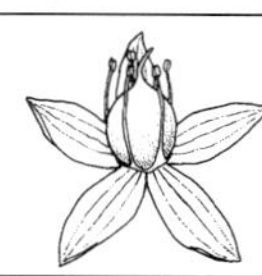	65

42	**Ovary** inferior	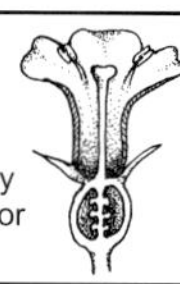	43
	Ovary superior	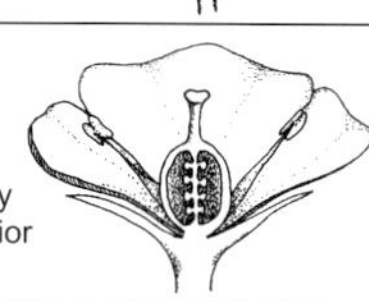	44

43	Erect annual **herb**; **tendrils** absent; **fruit** a dehiscent capsule	 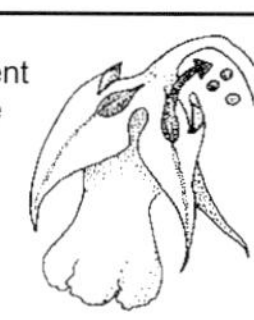	**Campanulaceae** (2 spp.)
	Climbing or trailing **herb**; **tendrils** present; **fruit** a fleshy indehiscent berry	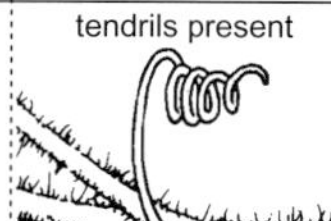 	**Cucurbitaceae** (10 spp.)

44	**Leaves** trifoliolate	**Oleaceae** (1 sp.) (*Jasminum fluminense*)
	Leaves simple	45

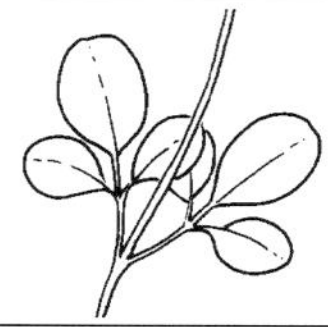

45	**Flowers** sessile in a spike; **leaves** all basal; **corolla** papery, translucent	**Plantaginaceae** (2 spp.)
	Flowers stalked; **leaves** not all basal; **corolla** not papery	46

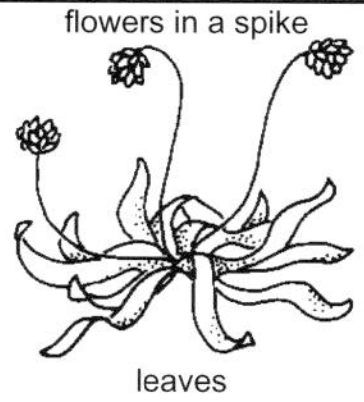

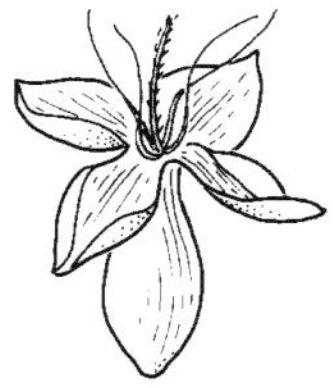

46	**Fruit** of 4 nutlets	**Boraginaceae** (29 spp.)
	Fruit a berry or capsule	47

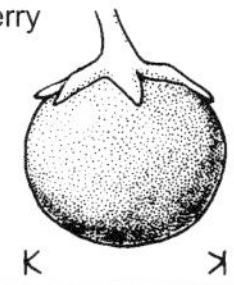

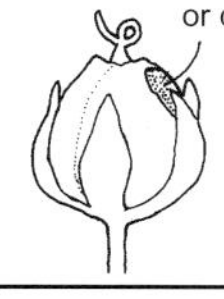

47	**Stamens** opposite the corolla lobes; **calyx** scarious	**Plumbaginaceae** (4 spp.)
	Stamens alternating with the corolla lobes; **calyx** not scarious	48

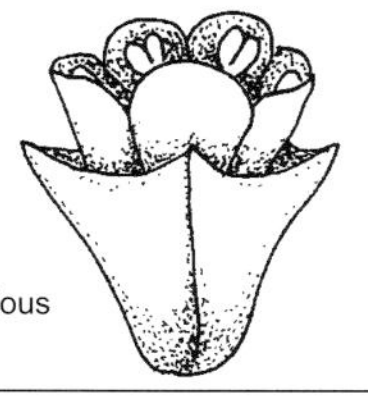

48	**Stamens** 2 (inserted in corolla); **branches** spinescent	**Scrophulariaceae** (19 spp.) (*Campylanthus spinosus*)
	Stamens 5; **branches** never spiny	49

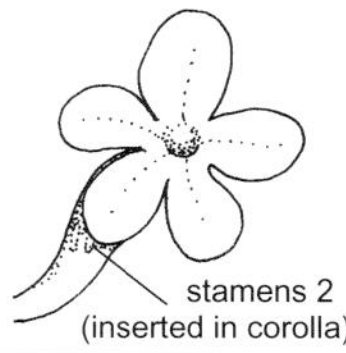

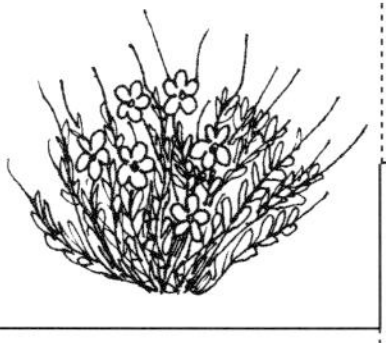

49	**Sepals** free; **fruit** a capsule	**Convolvulaceae** (22 spp.)
	Sepals fused; **fruit** a berry	**Solanaceae** (11 spp.)

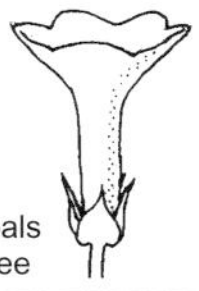

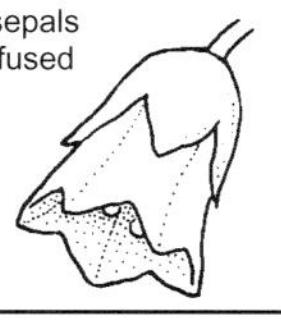

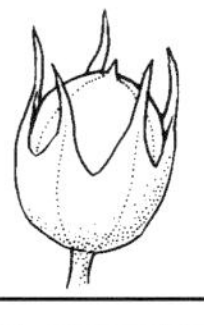

50	**Stipules** present, sometimes as tendrils	51
	Stipules absent	58

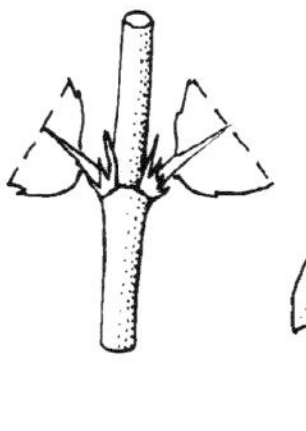

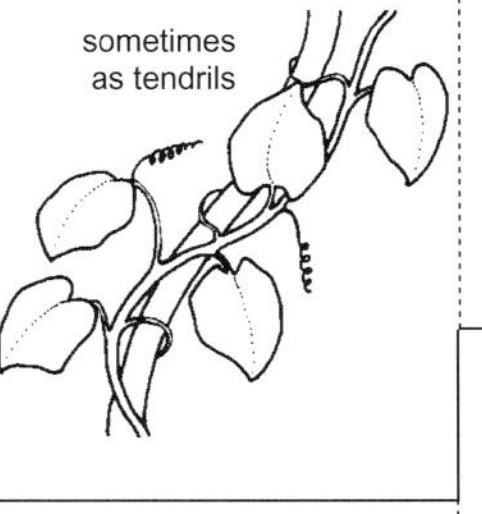

51	**Plant** climbing with tendrils	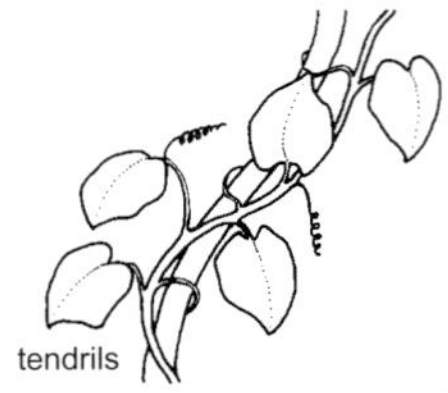	**Vitaceae** (3 spp.)
	Plant not climbing, or if climbing tendrils absent		52
52	**Leaves** peltate; **flowers** unisexual; **ovary** inferior	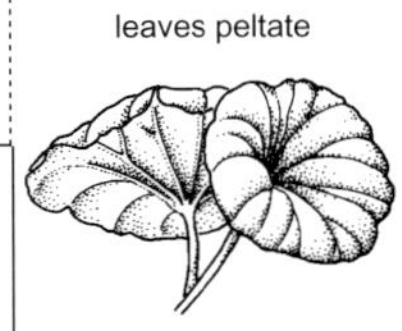 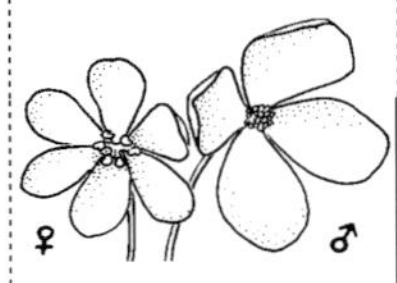	**Begoniaceae** (2 spp.)
	Leaves with petioles attached at base; **flowers** unisexual or bisexual; **ovary** superior		53
53	**Stamens** 5	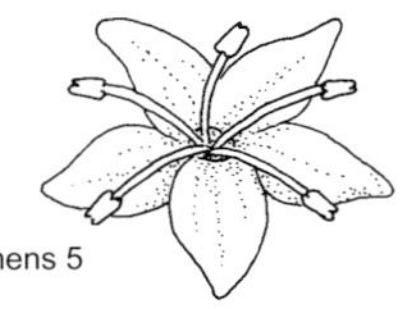	54
	Stamens 10 or more	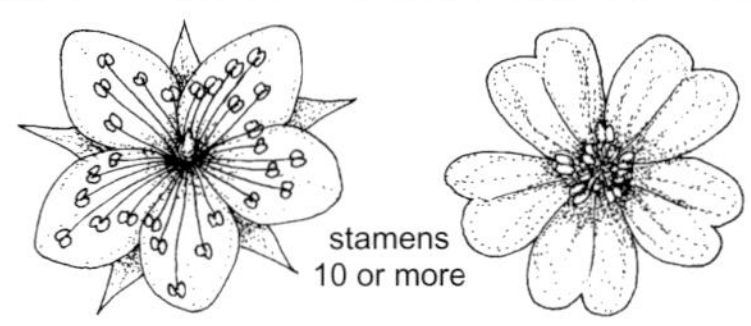	55
54	**Flowers** unisexual; **petals** minute, white; **fruits** 3-lobed, 3-valved	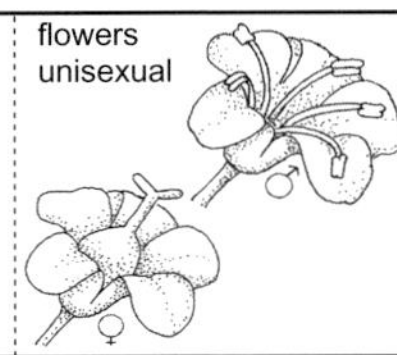 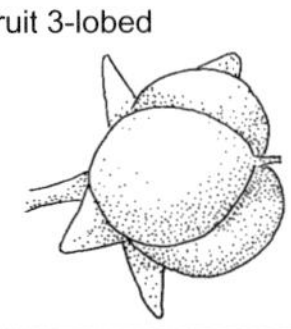	**Euphorbiaceae** (33 spp.) (*Andrachne* & *Chrozophora*)
	Flowers bisexual; **petals** obvious, yellow; **fruits** globose, 1 or 5-valved	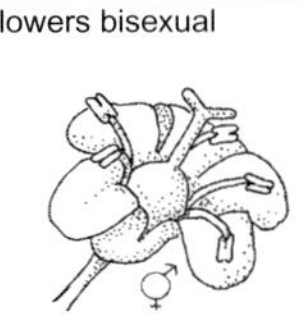 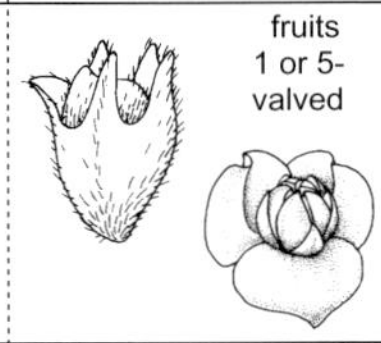	**Sterculiaceae** (4 spp.) (*Waltheria* & *Melhania*)
55	**Sepals** 2; **leaves** succulent	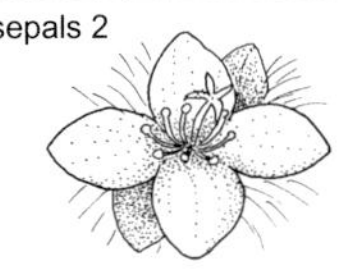 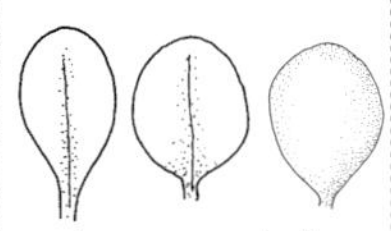	**Portulacaceae** (6 spp.)
	Sepals 5; **leaves** not succulent	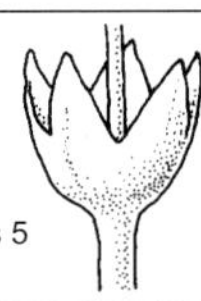 	56
56	**Fruit** an elongate capsule; **flowers** yellow	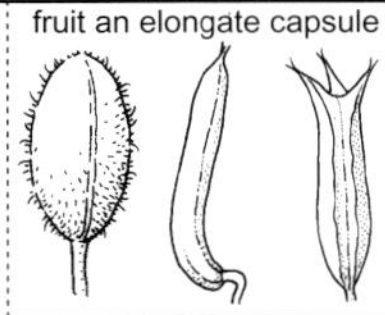 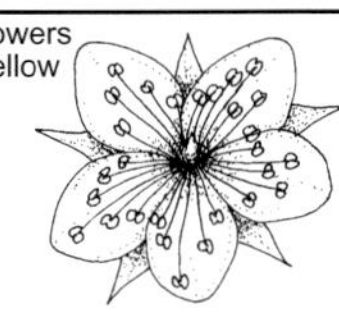	**Tiliaceae** (9 spp.) (*Corchorus*)
	Fruit a rounded capsule or schizocarp splitting into dehiscent segments; **flowers** pink or mauve	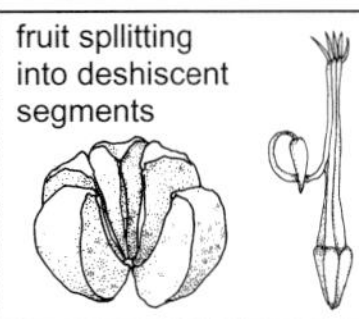 	57

57	**Fruits** with elongated beaks; **stamens** free; **flowers** in umbel-like groups	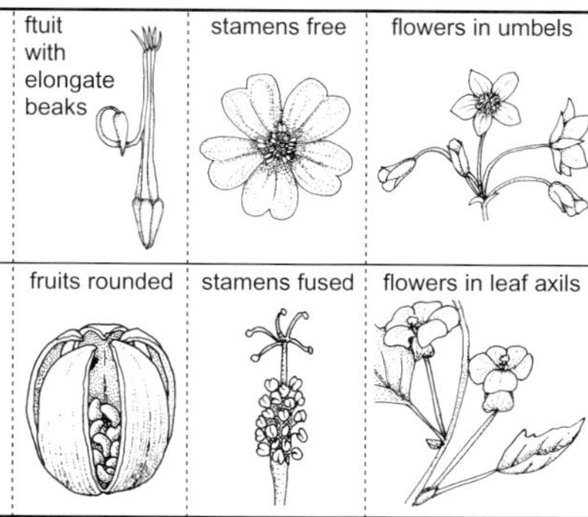	**Geraniaceae** (5 spp.)
	Fruits rounded or very shortly beaked; **stamens** fused into tube; **flowers** solitary or clustered in leaf axils		**Malvaceae** (25 spp.)
58	**Climber**; **flowers** unisexual; **fruit** fleshy, red	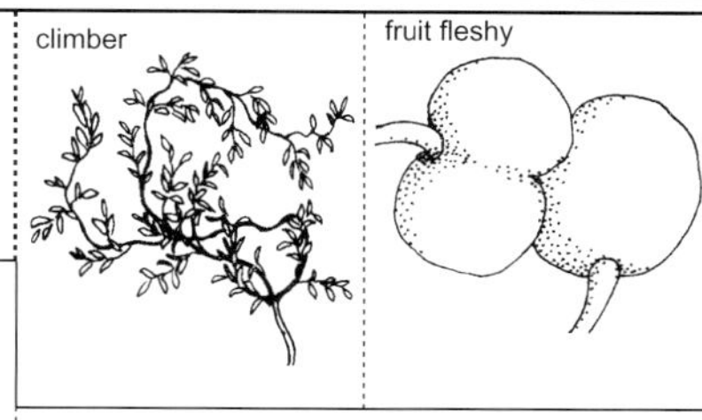	**Menispermaceae** (2 spp.)
	Not a **climber**; **flowers** bisexual; **fruit** a dry capsule		59
59	**Fruit** splitting into 2 single-seeded segments; **ovary** inferior; **flowers** in umbels	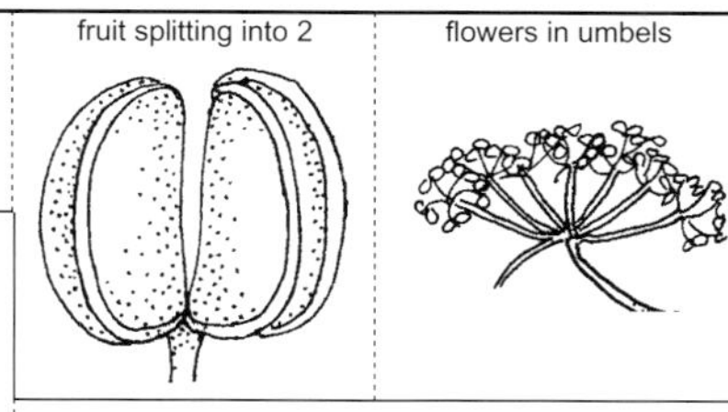	**Umbelliferae** (7 spp.)
	Fruit a capsule,not splitting into 2; **ovary** superior; **flowers** variously arranged		60
60	**Leaves** spiny; **flowers** showy, yellow	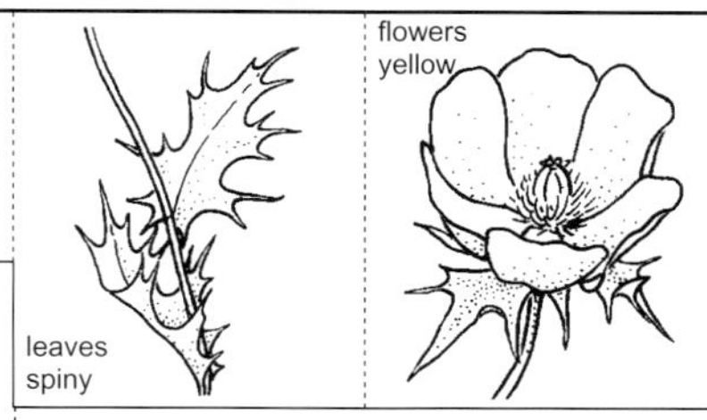	**Papaveraceae** (1 sp.) (*Argemone mexicana*)
	Leaves never spiny; **flowers** variously coloured		61
61	**Fruit** elongated, >3x as long as broad	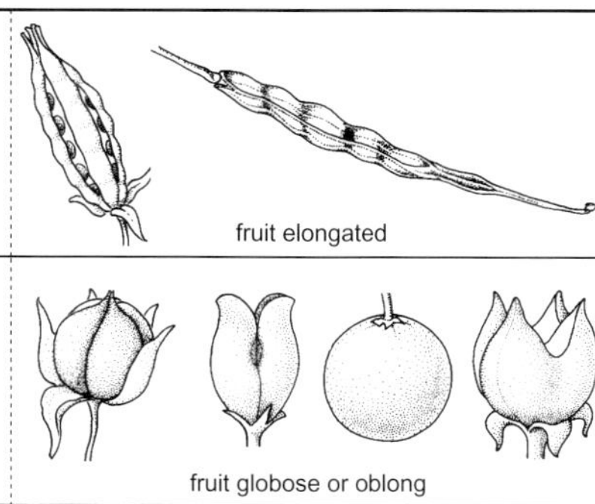	62
	Fruit globose or oblong, <3x as long as broad		64
62	**Fruit** a 5-celled capsule, tip with 5 styles	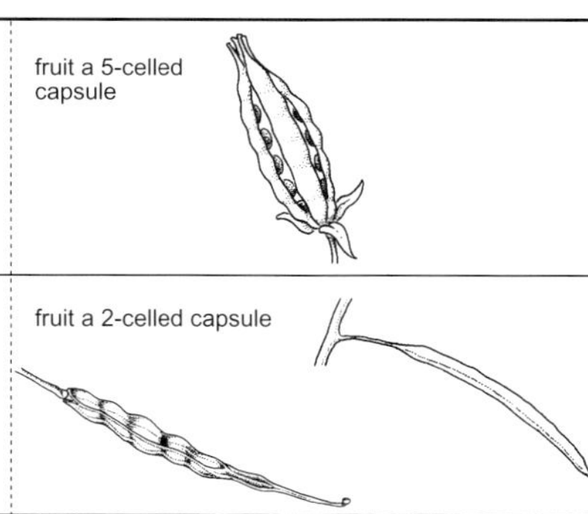	**Oxalidaceae** (1 sp.) (*Oxalis corniculata*)
	Fruit a 2-celled capsule (dehiscent or indehiscent), tip entire or bifid		63

63	**Leaves** 3 to 5-foliolate, or if simple then **fruits** glandular-hairy	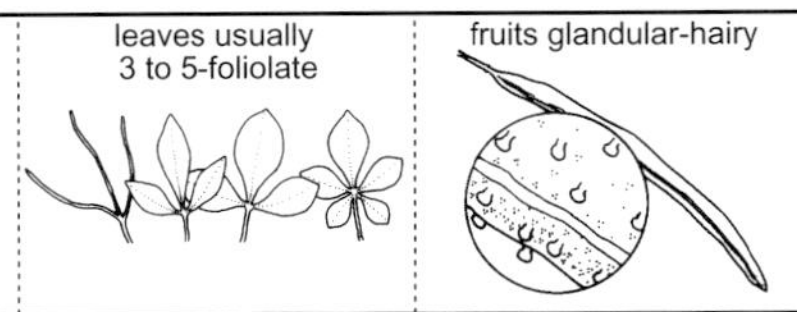	**Capparaceae** (14 spp.)
	Leaves simple or pinnately lobed or divided; **fruits** never glandular	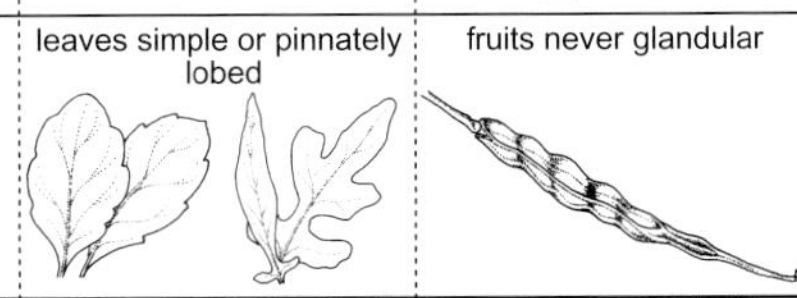	**Cruciferae** (11 spp.)
64	**Fruit** a globose 5-celled capsule; **plants** not strongly aromatic	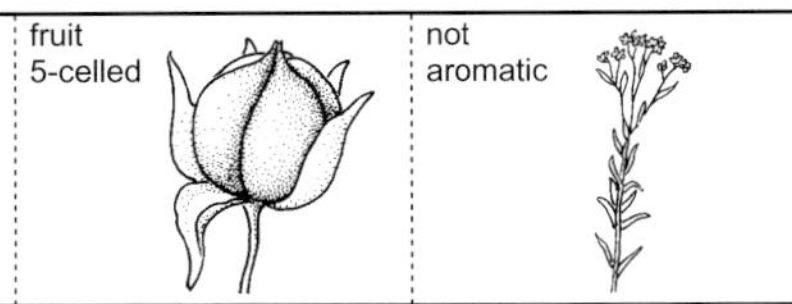	**Linaceae** (1 sp.) (*Linum trigynum*)
	Fruit a 2 or 4-lobed capsule; **plants** glandular and strongly aromatic	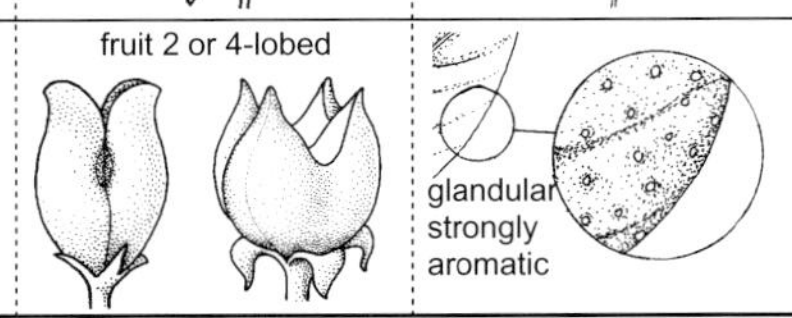	**Rutaceae** (4 spp.)
65	**Stipules** present	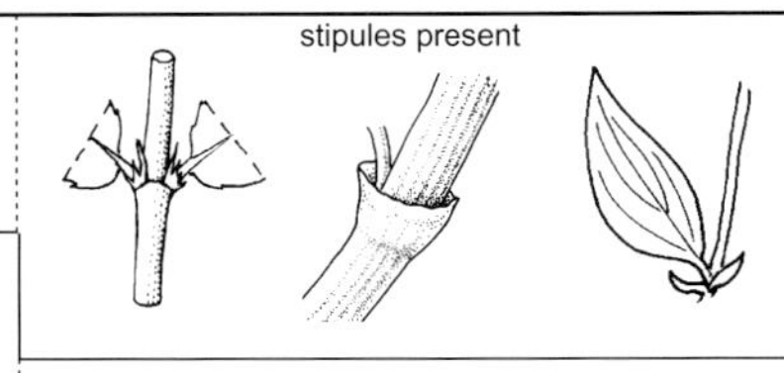	66
	Stipules absent		68
66	**Stipules** sheathing stems at leaf base	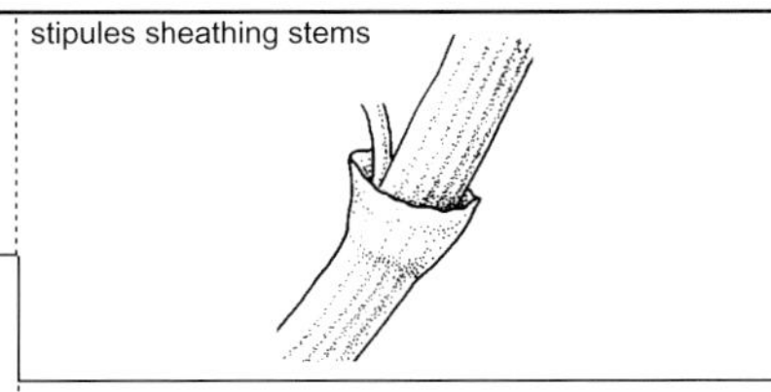	**Polygonaceae** (5 spp.)
	Stipules not sheathing		67
67	**Fruit** 3-lobed	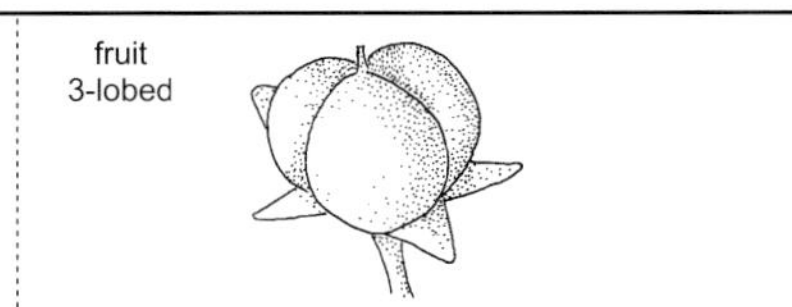	**Euphorbiaceae** (33 spp.)
	Fruit an achene (indehiscent, 1-seeded)	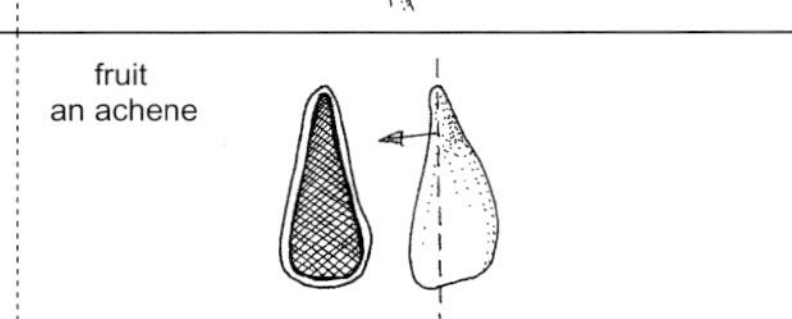	**Urticaceae** (5 spp.)
68	**Sepals** and **bracts** papery	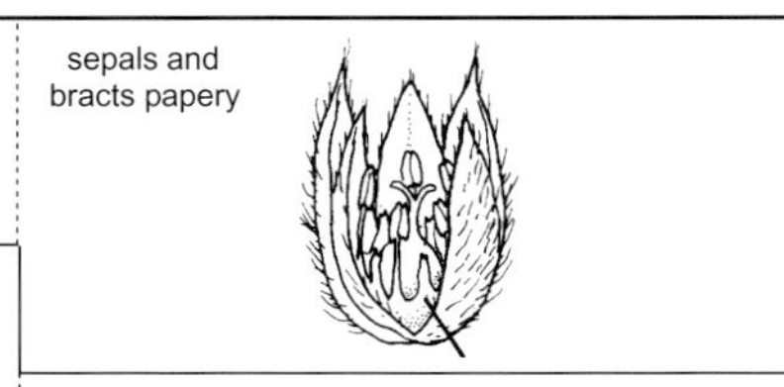	**Amaranthaceae** (13 spp.)
	Sepals and **bracts** not papery		69
69	**Ovary** superior, 1-celled with 1 ovule; **fruit** a nutlet, often enclosed by persistent perianth, sometimes winged	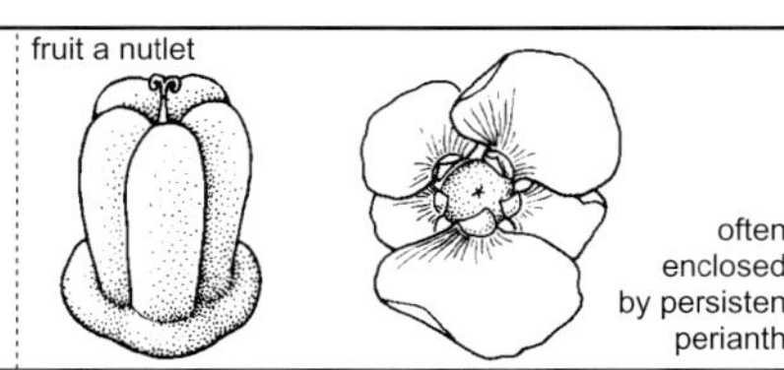	**Chenopodiaceae** (12 spp.)
	Ovary superior, 5-celled with many ovules; **fruit** a many-seeded, 5-valved capsule or indehiscent, or (*Tetragonia*) ovary inferior, 1-celled	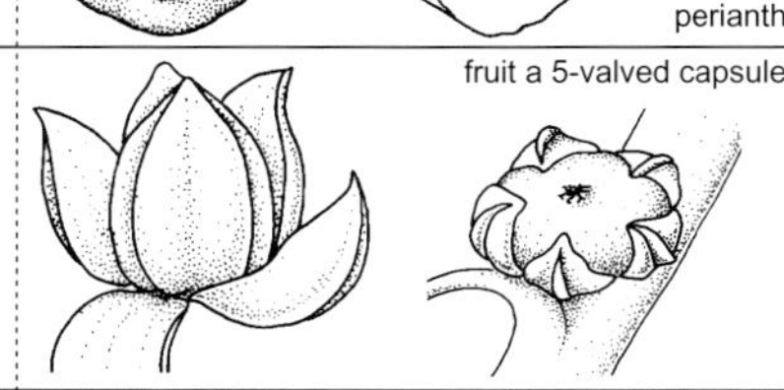	**Aizoaceae** (8 spp.)

GROUP F: WOODY DICOTS WITH OPPOSITE LEAVES

1	Parasitic **shrubs**	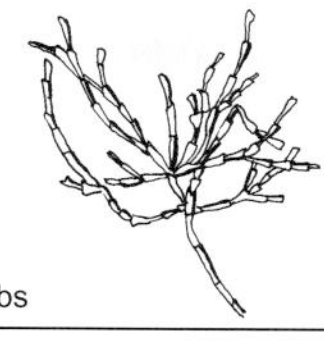	**Viscaceae** (1 spp.) (*Korthalsella taenioides*)
	Mangrove **tree,** on coastal mud with erect breathing roots		**Avicenniaceae** (1 spp.) (*Avicennia marina*)
	Free-living **shrubs** or **trees**, not in coastal mud	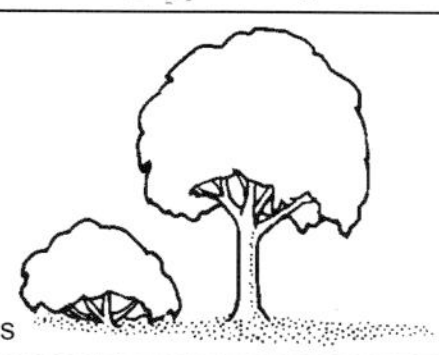	2
2	**Leaves** trifoliolate or rarely simple; climbing **plant**	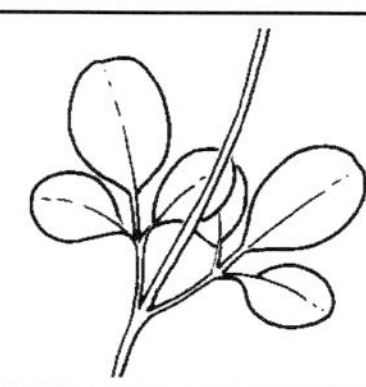	**Oleaceae** (1 spp.) (*Jasminum fluminense*)
	Leaves simple; **plant** not climbing		3
3	**Flowers** inconspicuous; **perianth** 1 whorl of 2 to 5 lobes		4
	Flowers conspicuous; **perianth** of 2 whorls of 4 to 8 petals and sepals	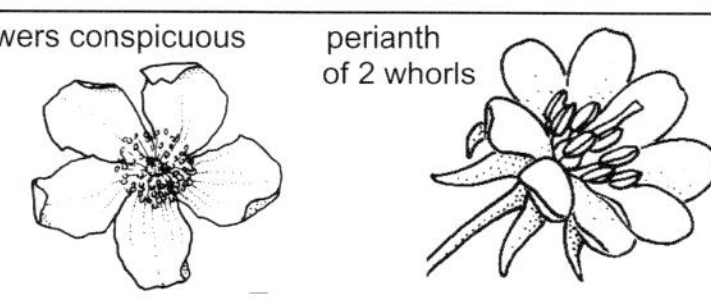	5
4	**Flowers** in dense, spike-like panicles, hidden by hairs	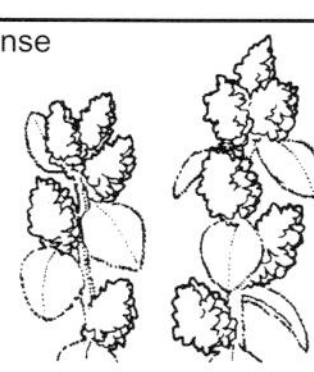	**Amaranthaceae** (13 spp.) (*Aerva*)
	Flowers axillary, not in spike-like panciles, not hidden in hairs		**Chenopodiaceae** (12 spp.)
5	**Flowers** unisexual	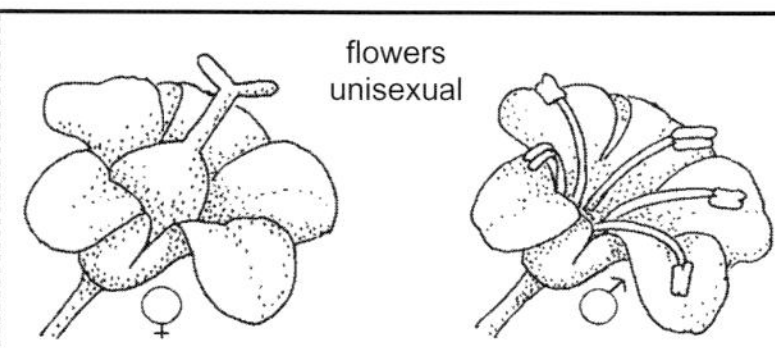	6
	Flowers bisexual	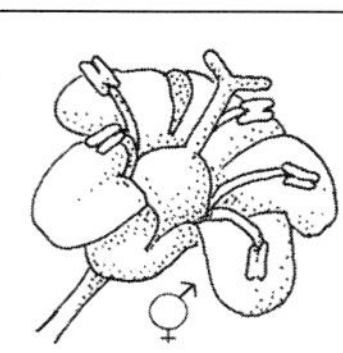	7

6

Plant monoecious; **perianth** of 4 sepals (corolla absent); **styles** 3; **stamens** 4 to 6; **fruit** a 3-horned capsule; **leaves** stiff, tough

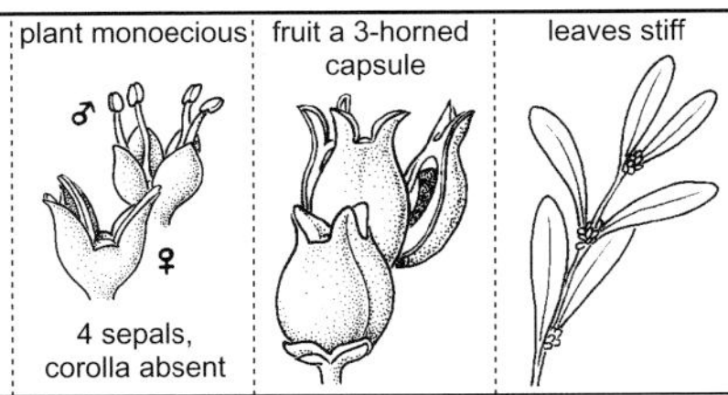

Buxaceae
(1 sp.)
(*Buxanthus pedicellatus*)

Plant dioecious; **perianth** of 4 or 5 fused petals and 4 or 5 fused sepals; **styles** 2; **stamens** 10 to 20; **fruit** subglobose; **leaves** leathery

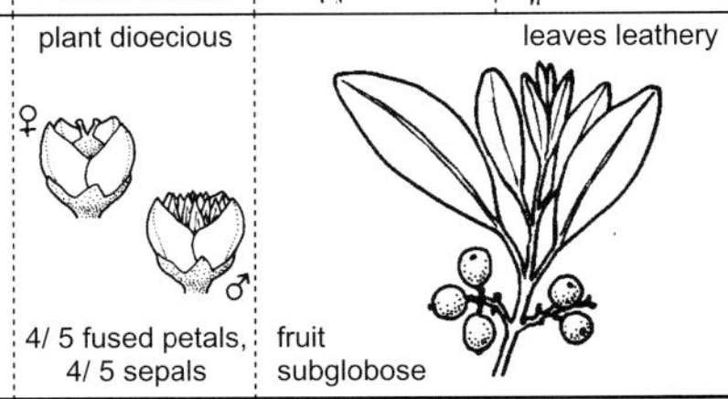

Ebenaceae
(1 sp.)
(*Euclea divinorum*)

7

Stamens numerous

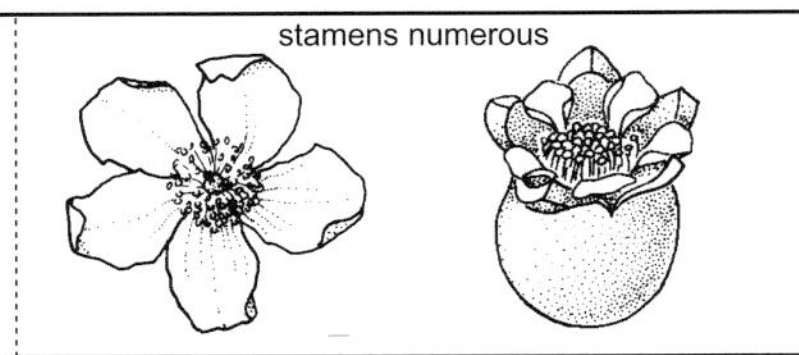

8

Stamens 10 or less

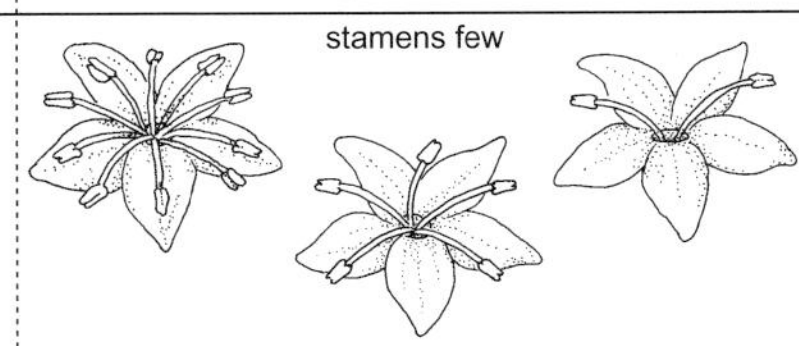

9

8

Flowers red; **fruit** a 'pomegranate'

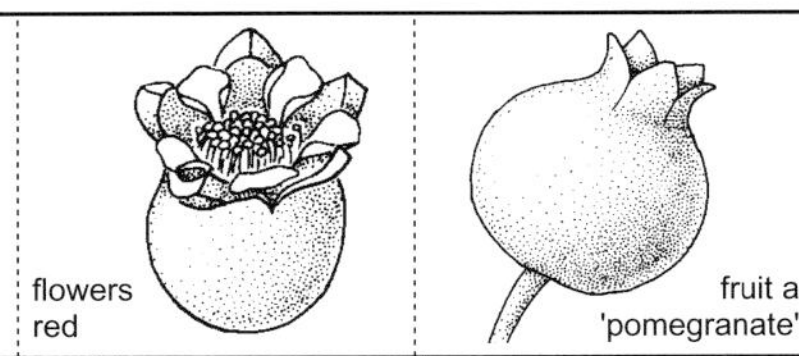

Punicaceae
(1 sp.)
(*Punica protopunica*)

Flowers yellow; **fruit** a capsule

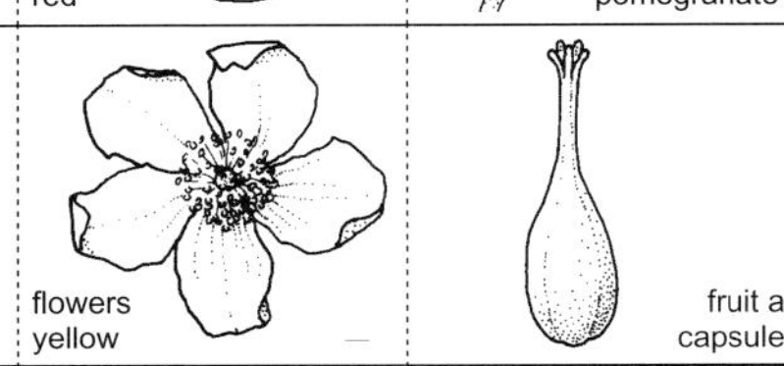

Guttiferae
(5 spp.)

9

Flowers zygomorphic

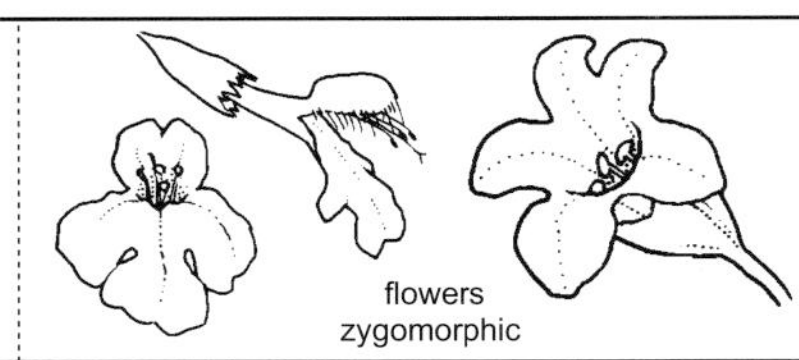

10

Flowers actinomorphic

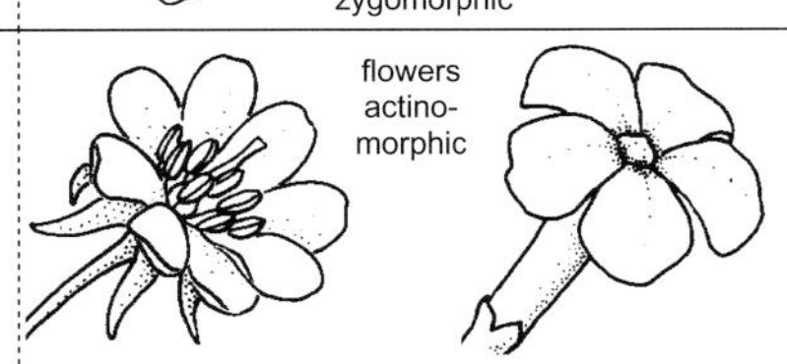

13

10

Leaves glandular, aromatic; **fruit** with 1 seed in each compartment

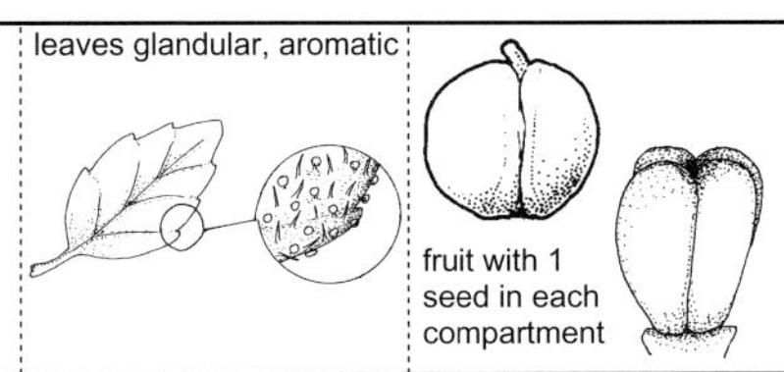

11

Leaves non-glandular, not aromatic; **fruit** many-seeded

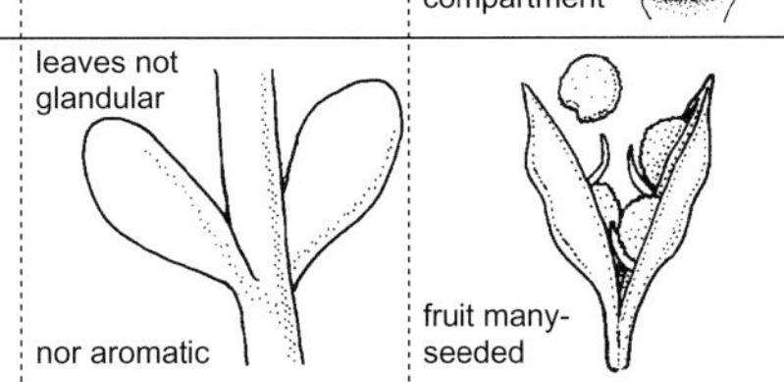

12

11

Ovary 4-lobed with style attached at base; **fruit** 4-lobed — **Labiatae** (19 spp.)

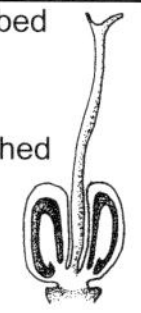

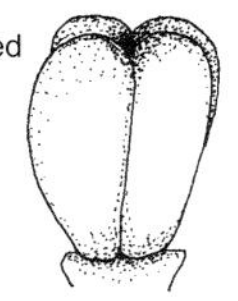

Ovary 2-lobed with style arising from top; **fruit** 2-lobed, dry or fleshy — **Verbenaceae** (6 spp.)

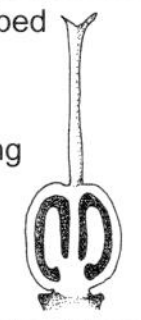

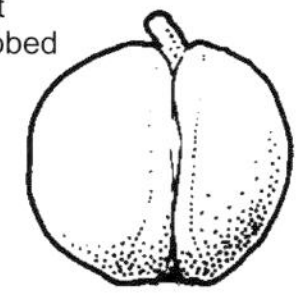

12

Seeds on hook-like projections — **Acanthaceae** (34 spp.)

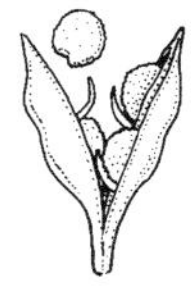

Seeds not on hook-like projections — **Scrophulariaceae** (19 spp.)

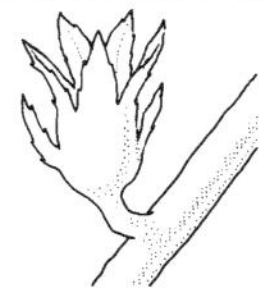

13

Petals free — 14

Petals fused — 17

14

Flowers 8-merous; **fruit** of 8 mericarps; **leaves** with crenate margins and linear stipules — **Dirachmaceae** (1 sp.) (*Dirachma socotrana*)

Flowers 4 to 6-merous; **fruit** a berry, drupe or capsule — 15

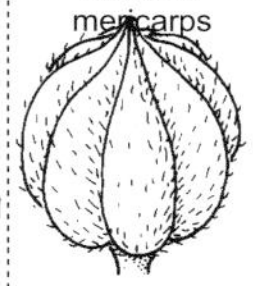

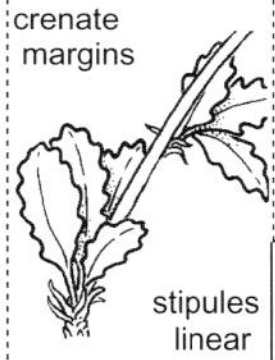

15

Flowers in panicles; **fruit** a berry or drupe; **leaves** astringent when chewed — **Salvadoraceae** (1 sp.) (*Salvadora persica*)

Flowers solitary; **fruit** a 5-lobed capsule; **leaves** not astringent — **Zygophyllaceae** (6 spp.)

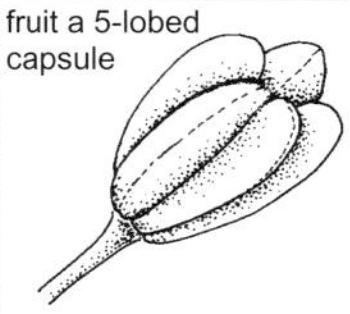

16

Ovary inferior; **interpetiolar stipules** often present; **anthers** free; **fruit** a berry — **Rubiaceae** (22 spp.)

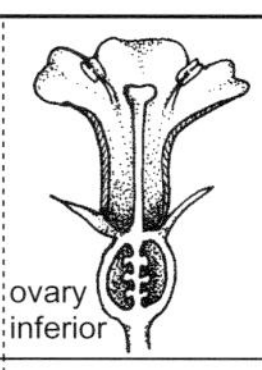

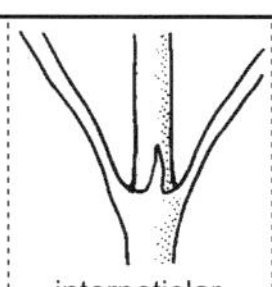

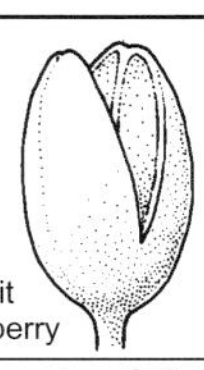

Ovary superior; **interpetiolar** stipules absent; **anthers** usually fused; **fruit** a pair of follicles or a 2-lobed berry — 17

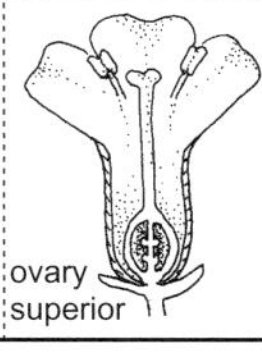

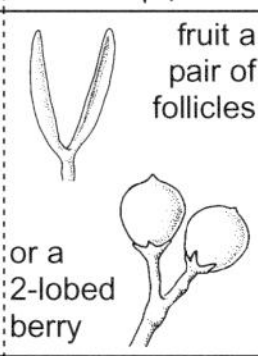

17

- **Climbers** or unarmed **trees** or **shrubs,** never swollen-stemmed trees — **Asclepiadaceae** (27 spp.)

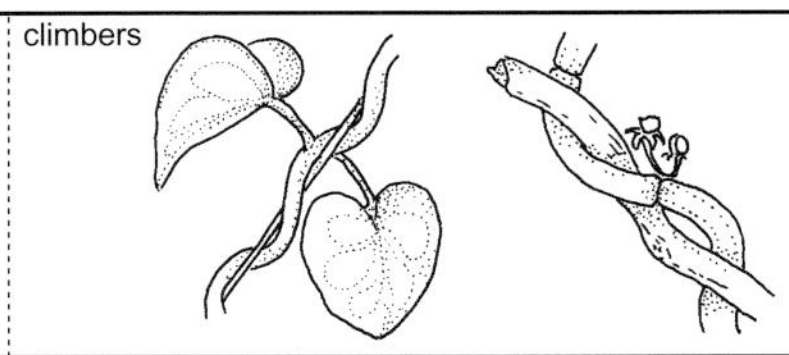

- Spiny **shrub** or swollen-stemmed **tree** — **Apocynaceae** (3 spp.)

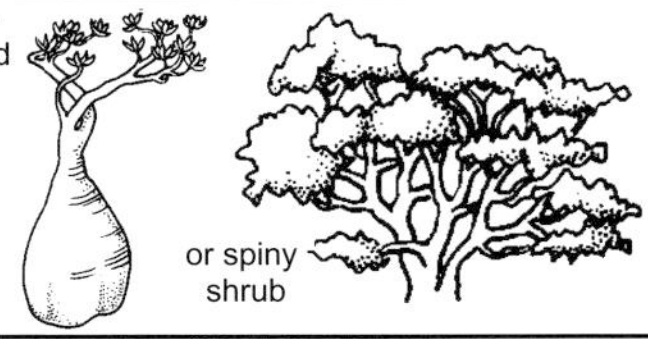

GROUP G: WOODY DICOTS WITH ALTERNATE LEAVES

1

- **Leaves** compound — 2

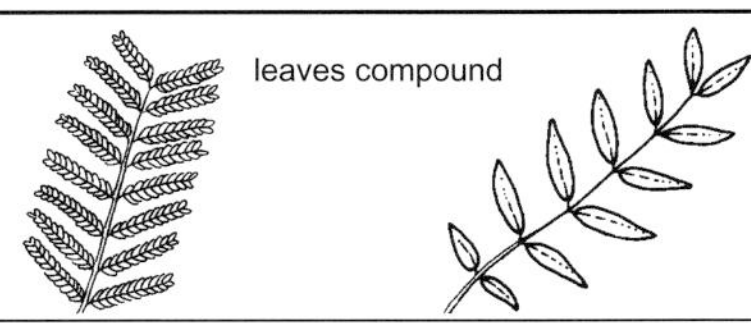

- **Leaves** simple — 9

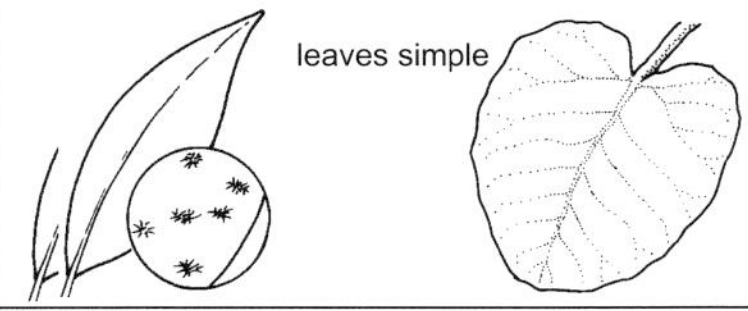

- **Leaves** palmatisect with glandular margin — **Euphorbiaceae** (33 spp.) (*Ricinus communis*)

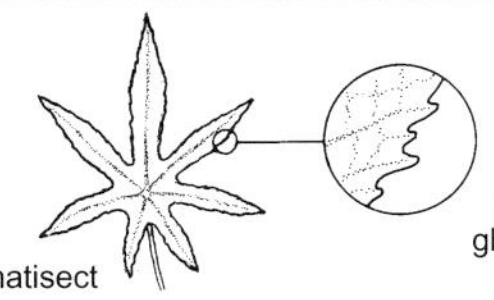

2

- **Flowers** in umbels — **Umbelliferae** (7 spp.)
- **Flowers** in other forms of inflorescence — 3

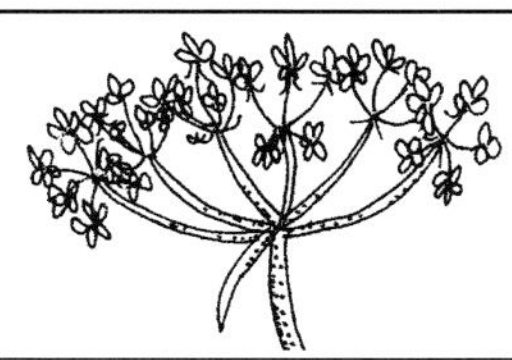

3

- **Stamens** many; **leaves** 2-pinnate — **Leguminosae** (Mimosoideae) (8 spp.)

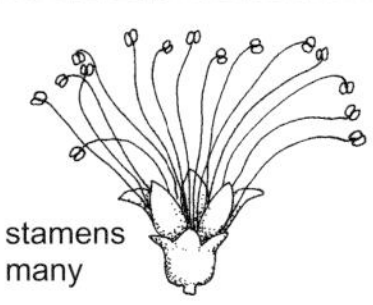

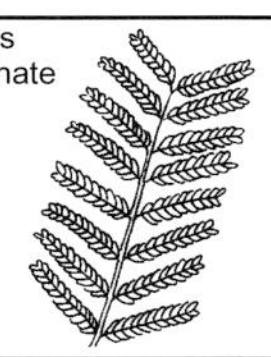

- **Stamens** </= 10; **leaves** 3-foliolate or 1-pinnate — 4

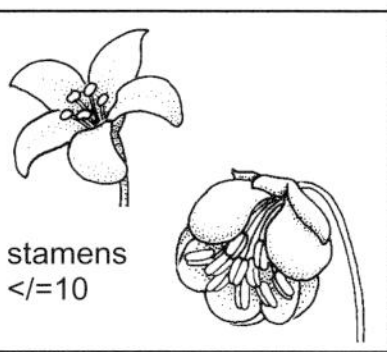

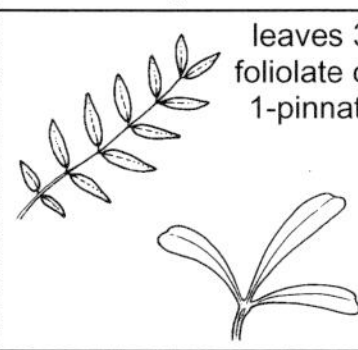

4

- **Fruit** a pod; **flowers** usually zygomorphic, if actinomorphic stamens 10 — **Leguminosae** (Papilionoideae & Caesalpinioideae) (57 & 6 spp.)

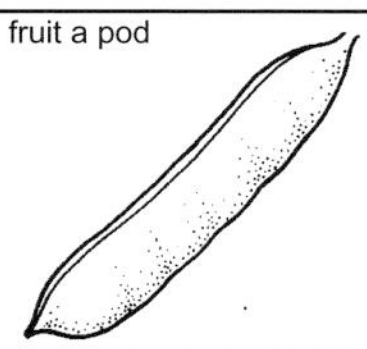

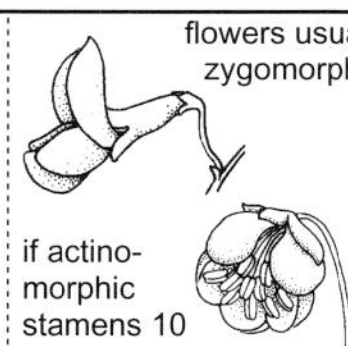

- **Fruit** a drupe or capsule; **flowers** actinomorphic with <10 stamens — 5

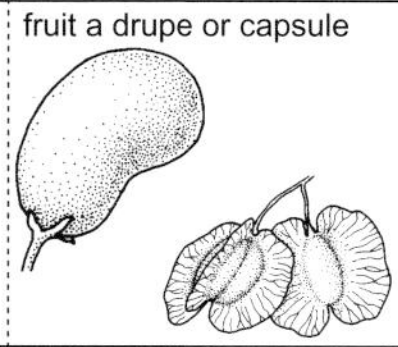

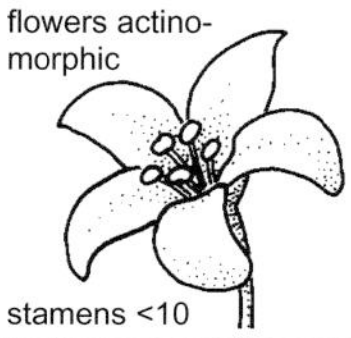

5	**Petals** fused; **fruit** a dehiscent drupe; **seeds** arillate	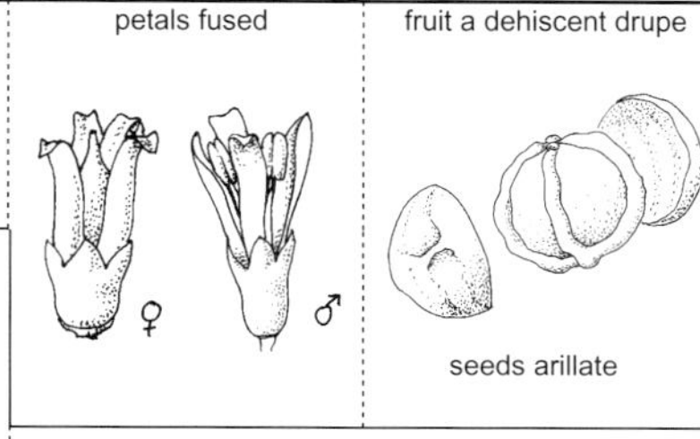	**Burseraceae** (13 spp.) (*Commiphora*)
	Petals free; **fruit** a capsule or indehiscent drupe; **seeds** without arils		6
6	**Fruit** a 3-angled capsule; **flowers** showy	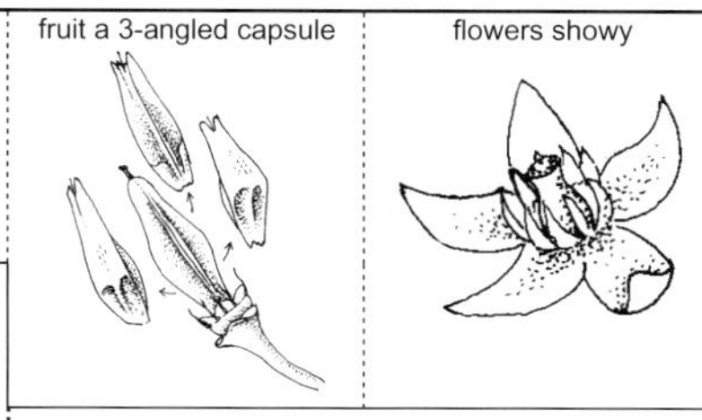	**Burseraceae** (13 spp.) (*Boswellia*)
	Fruit a drupe; **flowers** inconspicuous		7
7	**Leaves** not glandular, with simple or stellate hairs; **flowers** in panicles	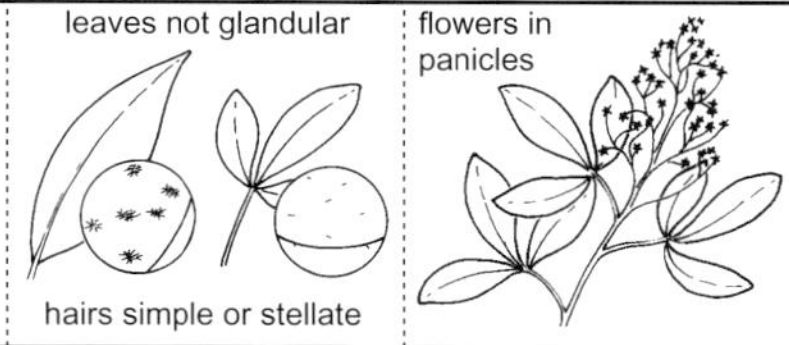	**Anacardiaceae** (3 spp.)
	Leaves glandular, sticky; **flowers** in racemes	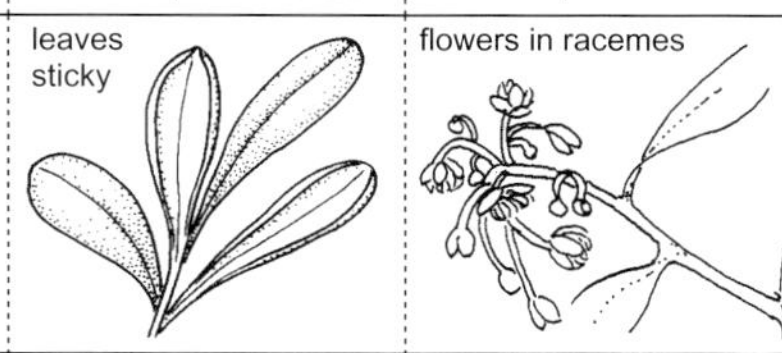	**Sapindaceae** (2 spp.) (*Dodonaea viscosa*)
8	**Flowers** borne on a common receptacle, individual flowers small and parts often reduced	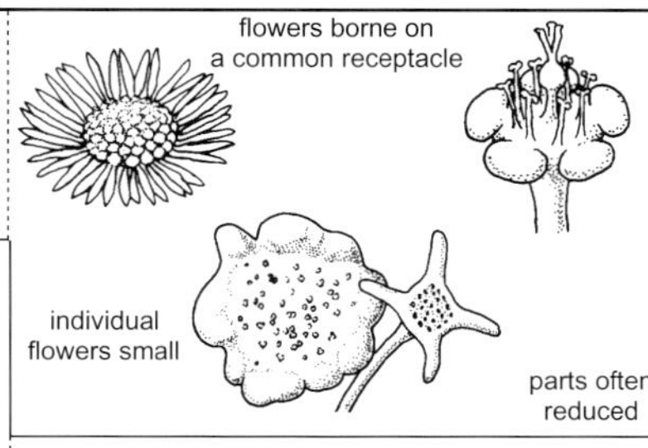	9
	Flowers not borne on a common receptacle		10
9	**Flowers** with coloured petals; receptacle surrounded by whorls of greenish or papery **bracts**	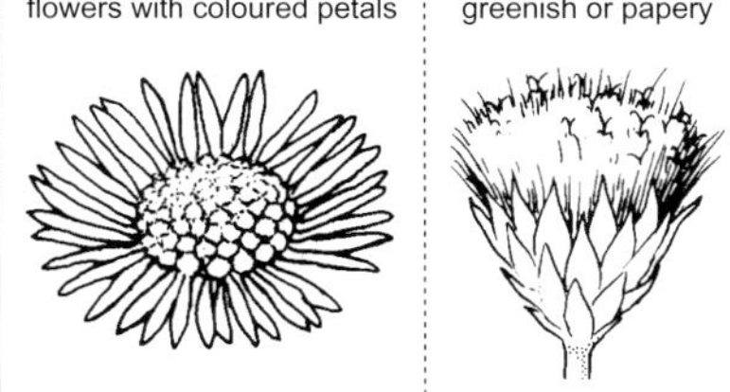	**Compositae** (55 spp.)
	Flowers inconspicuous, perianth absent or green or brownish; scarious or papery **bracts** absent		10
10	**Flowers** enclosed in a specialized organ (fig) or much reduced and on a fleshy, flattened receptacle; clear or white **sap** present	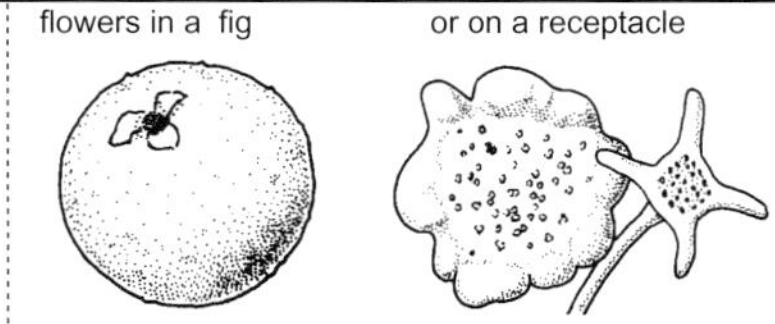	**Moraceae** (4 spp.)
	Flowers much reduced and on a receptacle (cyathia) surrounded by 5 glands; white **sap** present	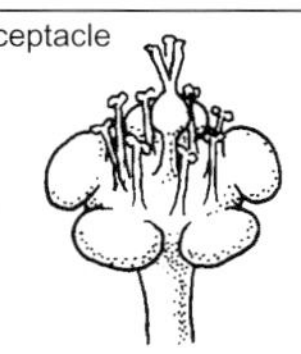	**Euphorbiaceae** (33 spp.) (*Euphorbia*)

11

- **Leaves** scale-like; **flowers** pink, in spikes; **fruit** a capsule, the seeds with a tuft of long hairs at the tip — **Tamaricaceae** (2 spp.)
- **Leaves** free; **flowers** and **fruits** various — 12

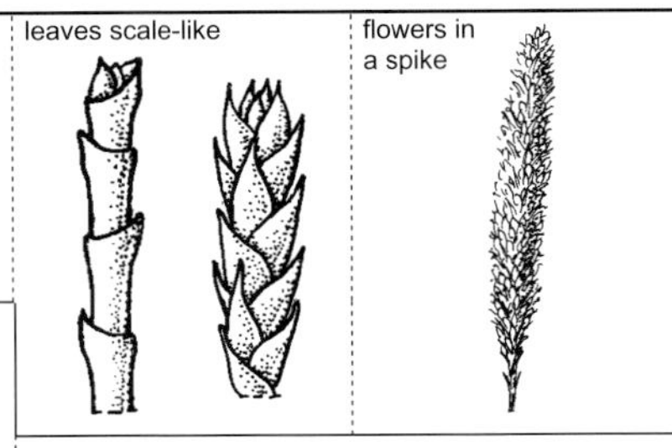

12

- **Plants** with tendrils (usually climbing or scrambling) — 13
- **Plants** lacking tendrils — 14

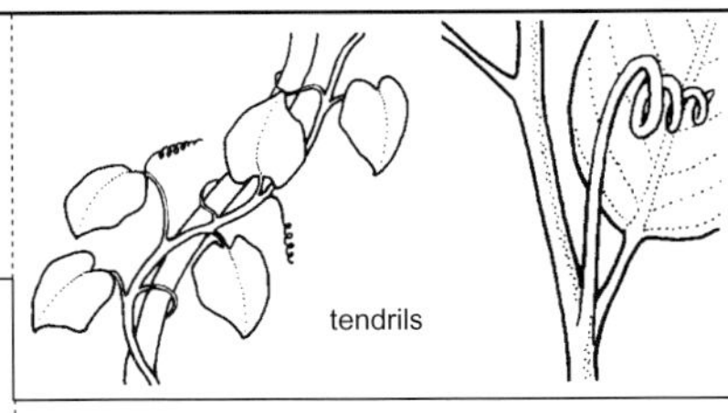

13

- **Plants** with rounded, succulent stems; **flowers** 4-merous — **Vitaceae** (3 spp.) (*Cissus*)

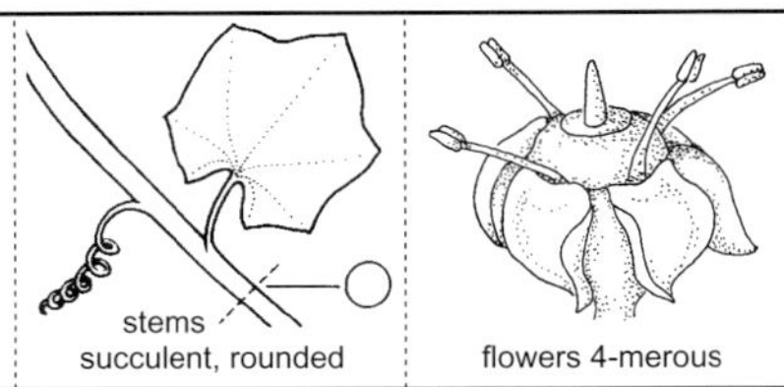

- **Plants** without succulent stems**; flowers** 5-merous — **Rhamnaceae** (2 spp.) (*Helinus integrifolius*)

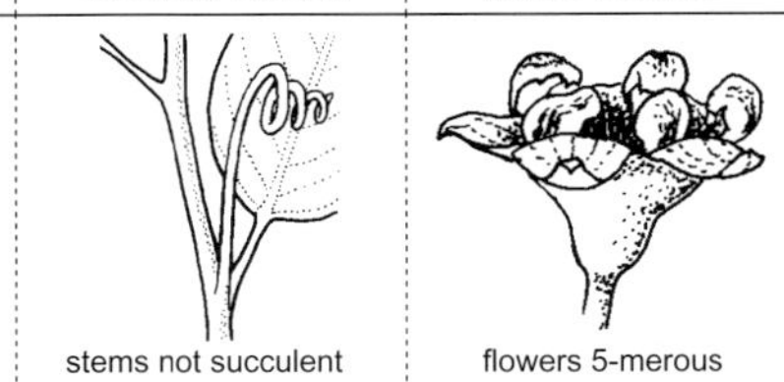

14

- **Perianth** of 1 whorl — 15

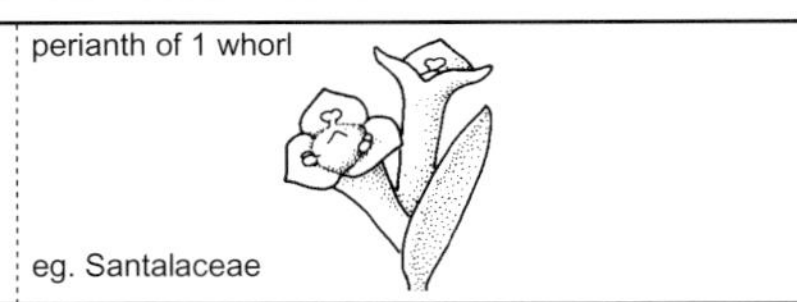

- **Perianth** in 2 whorls, both calyx and corolla — 21

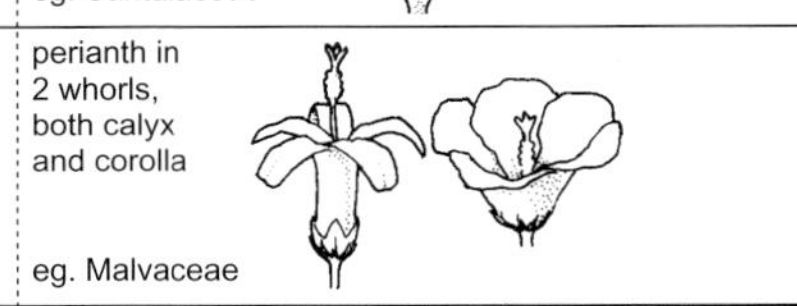

15

- **Leaves** lobed; **hairs** stellate or stinging or plant glabrous — 16

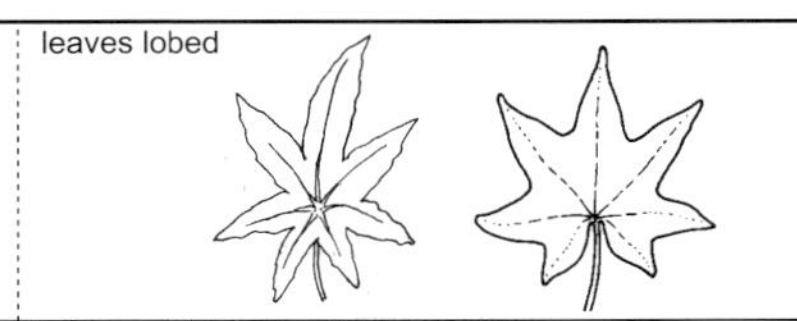

- **Leaves** simple, not lobed; **hairs** simple — 17

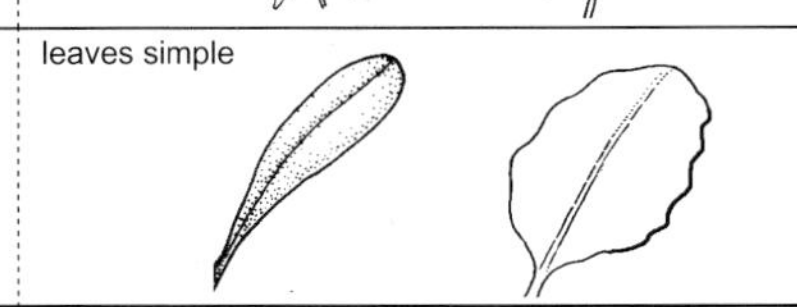

16

- Large **tree**; **fruit** star-shaped, of 5 follicles; **leaves** shallowly lobed — **Sterculiaceae** (4 spp.) (*Sterculia africana*)

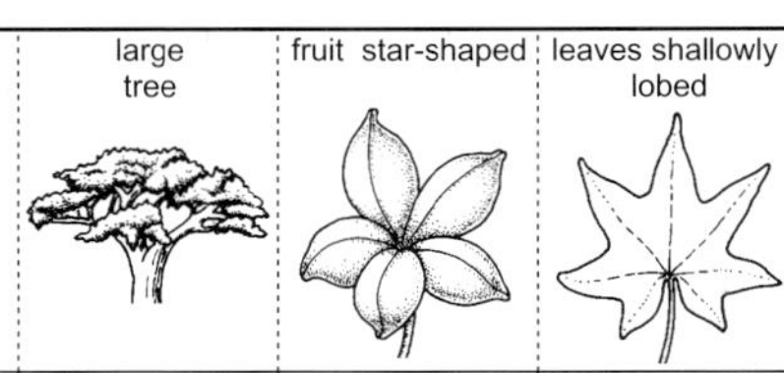

- Small to medium-sized **tree**; **fruit** a 3-lobed capsule; **leaves** deeply pinnatisect — **Euphorbiaceae** (33 spp.) (*Ricinus communis*)

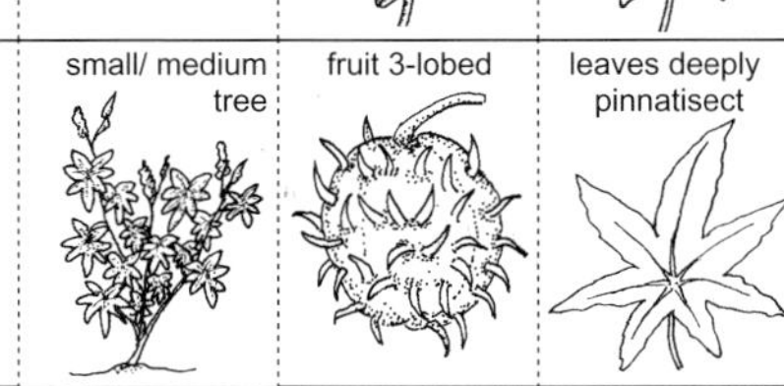

17

Leaves sticky; **fruit** a 2 or 3-winged capsule — **Sapindaceae** (2 spp.)

Leaves not sticky; **fruit** a drupe or berry or 5-winged — 18

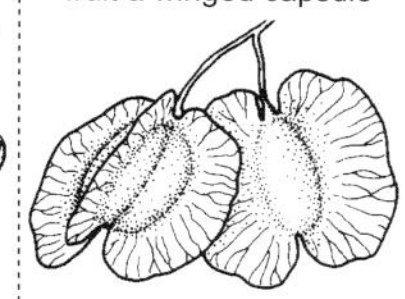

18

Leaves succulent or mealy; **fruit** a thin-walled utricle, sometimes 5-winged (*Salsola,* etc) or enclosed in 2 bracts (*Atriplex*) — **Chenopodiaceae** (12 spp.)

Leaves not succulent or mealy; **fruit** a drupe or berry — 19

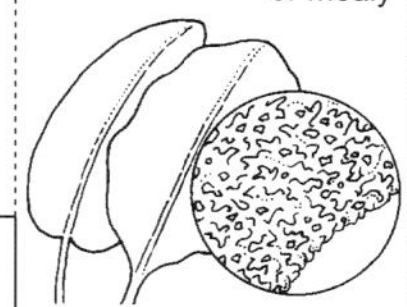

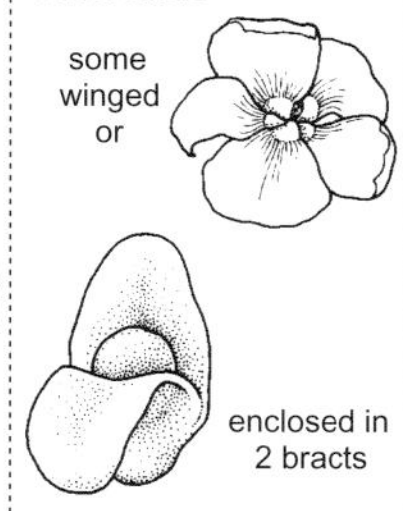

19

Flowers bisexual, bright yellow, arranged in a head surrounded by overlapping bracts; **calyx** tubular — **Thymelaeaceae** (1 spp.) (*Gnidia socotrana*)

Flowers unisexual, greenish or pale yellow, not in a head; **calyx** free or fused only at base — 20

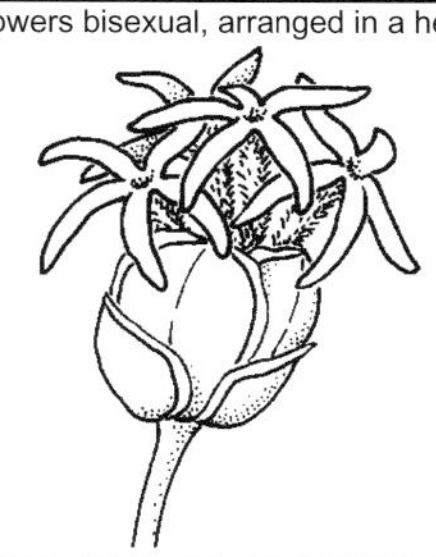

20

Stamens many; **perianth** 5 to 8-lobed; **ovary** superior; **fruit** a berry — **Flacourtiaceae** (1 sp.) (*Dovyalis abyssinica*)

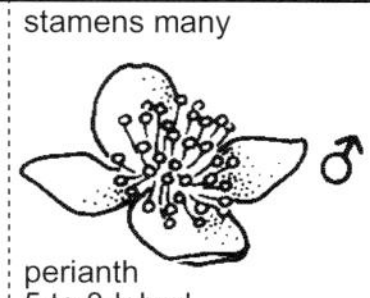

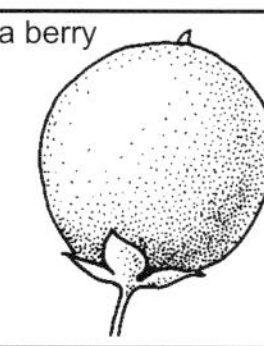

Stamens 3; **perianth** 3-lobed; **ovary** inferior; **fruit** a drupe — **Santalaceae** (1 sp.) (*Osyris quadripartita*)

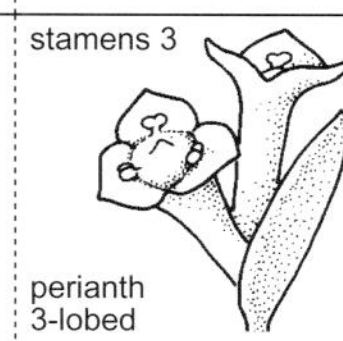

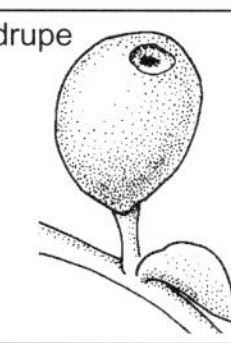

21

Flowers unisexual — 22

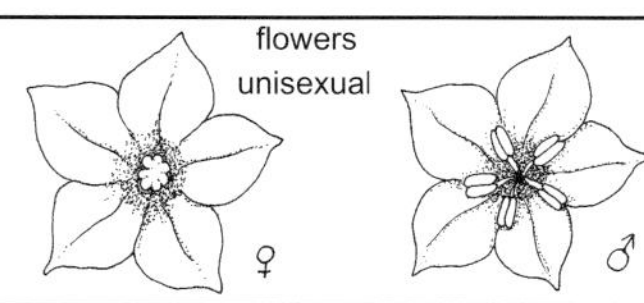

Flowers bisexual — 25

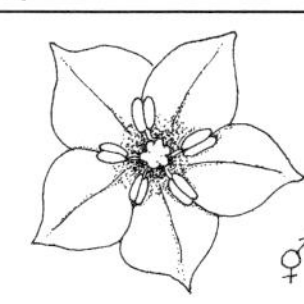

22

Tree with a swollen trunk; **fruit** a small cucumber; **anthers** 5; **leaves** hispid — **Cucurbitaceae** (10 spp.) (*Dendrosicyos socotrana*)

Tree without swollen trunk; **fruit** a drupe, berry or capsule; **anthers** 6 or more; **leaves** not hispid — 23

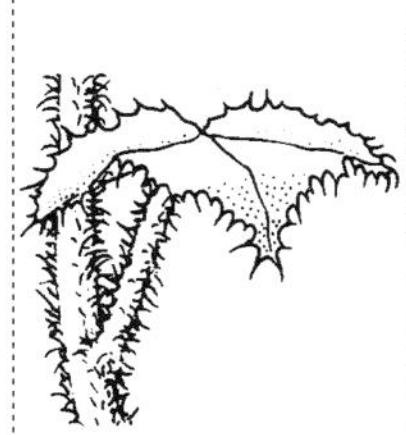

	Character	Illustration	Result
23	**Leaves** +/- absent, replaced by spiny cladodes; **flowers** minute, greenish	leaves replaced by spiny cladodes 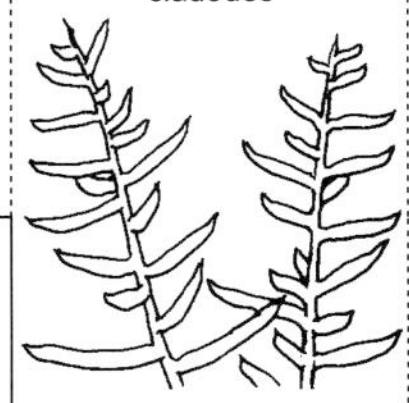 flowers minute, greenish 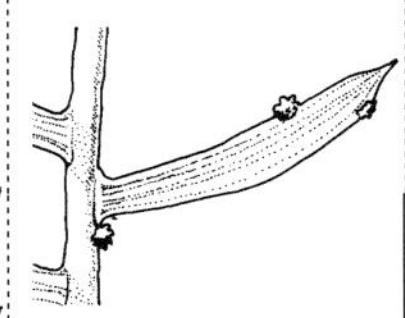	**Menispermaceae** (2 spp.) (*Cocculus balfourii*)
	Leaves conspicuous, cladodes absent; **flowers** large, or if small then red		24
24	Aromatic **shrub** with spiny stems and clear sap; **flowers** small, red; **fruit** a fleshy 2-lobed capsule; **plant** glabrous	flowers red 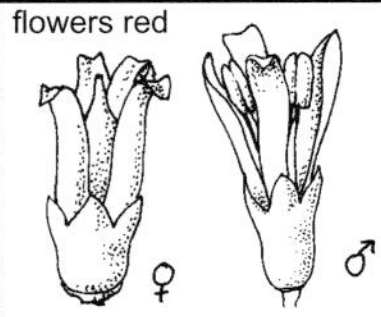 fruit 2-lobed 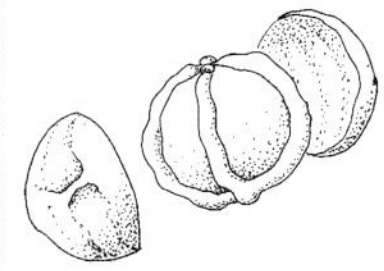	**Burseraceae** (13 spp.) (*Commiphora*)
	Trees or shrubs, often with milky latex; **flowers** yellowish or green; **fruit** a 3-lobed capsule or berry; **plant** often with stellate hairs	flowers yellowish or green  fruit 3-lobed 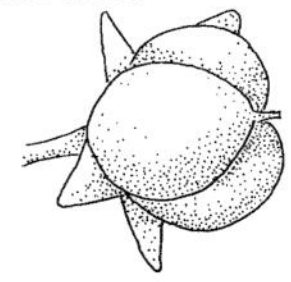	**Euphorbiaceae** (33 spp.)
25	**Petals** free	petals free 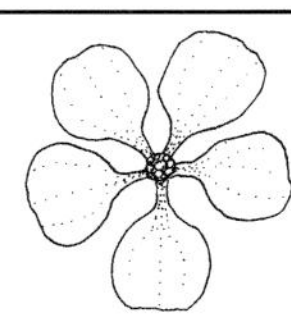	26
	Petals fused	petals fused 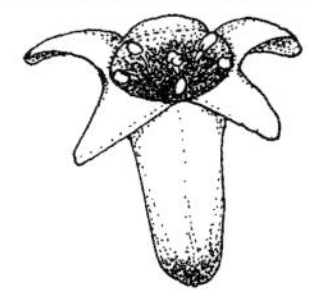	35
26	**Stamens** 2x as many as petals or fewer	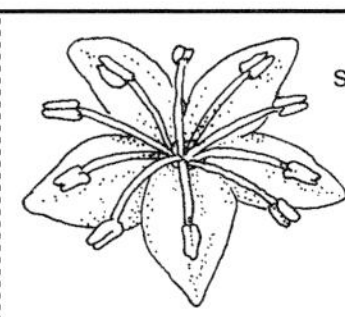 stamens <2x petals 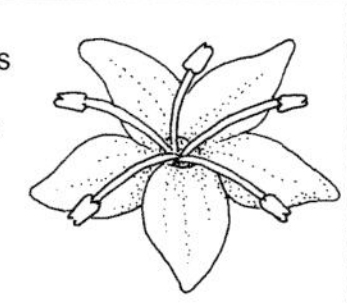	27
	Stamens >2x as many as petals	stamens >2x petals 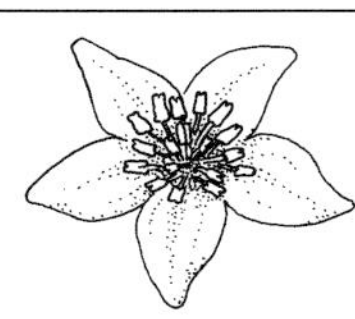	31
27	**Leaves** with pellucid glands; **fruit** a glandular, 2-lobed capsule	leaves with pellucid glands 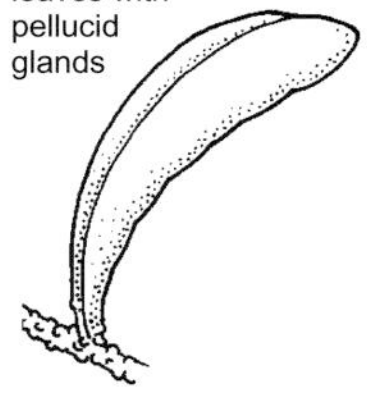 fruit a 2-lobed capsule 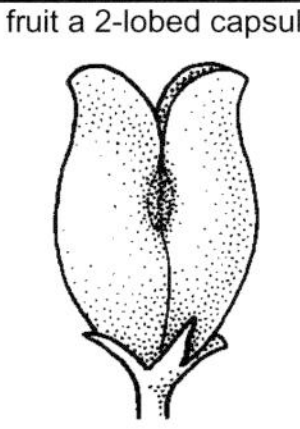	**Rutaceae** (4 spp.) (*Thamnosma*)
	Leaves without glands; **fruit** winged or an unlobed drupe or capsule		28
28	**Stamens** 10	stamens 10 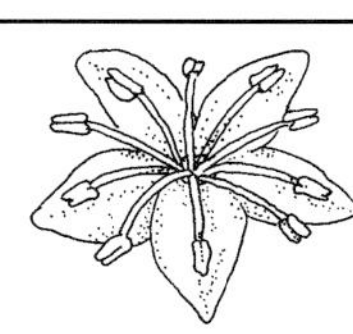	29
	Stamens 5	stamens 5 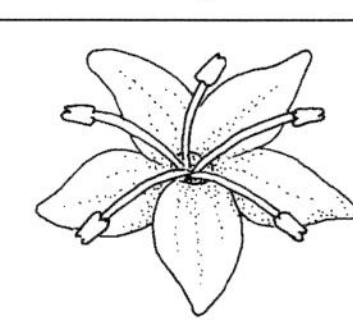	30

29	Small **tree**; **flowers** large (10–15mm across); **petals** bright yellow, fringed; **fruit** winged	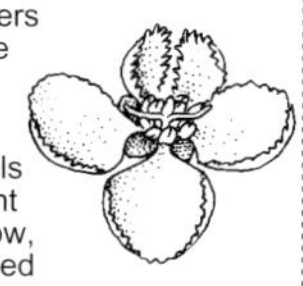	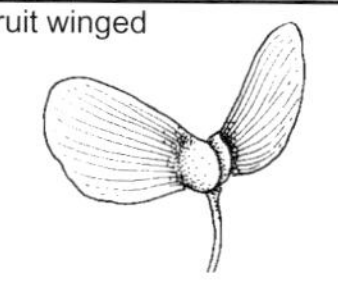	**Malpighiaceae** (1 sp.) (*Acridocarpus socotranus*)
	Shrub; **flowers** small (c.3mm diam.); **petals** cream, entire; **fruit** fleshy, ellipsoid	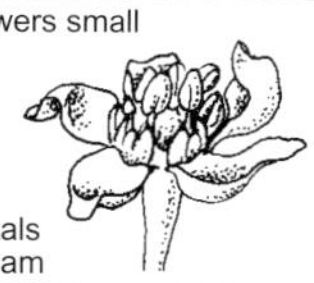		**Erythroxylaceae** (1 sp.) (*Erythroxylon socotranum*)
30	**Fruit** spherical; **seeds** without arils	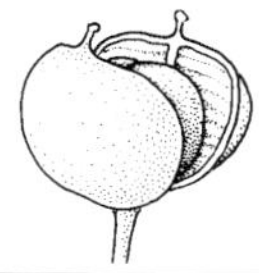		**Pittosporaceae** (1 sp.) (*Pittosporum viridiflorum*)
	Fruit obconic, 6-lobed; **seeds** arillate	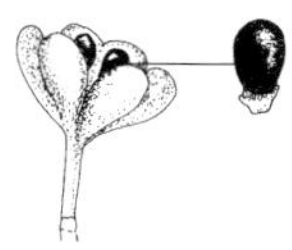		**Celastraceae** (1 sp.) (*Maytenus* sp. A)
31	**Anthers** fused into a tube below	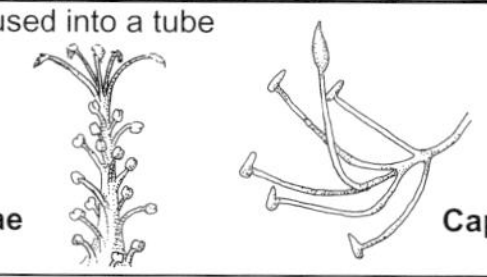		32
	Anthers free			33
32	**Plant** with stellate hairs; **flowers** bright yellow; **fruit** a capsule	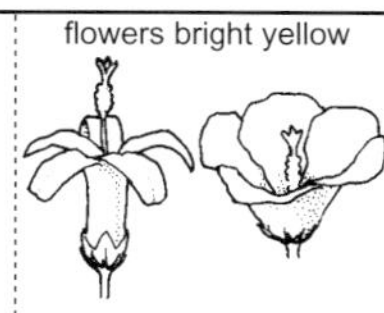		**Malvaceae** (25 spp.) (*Hibiscus*)
	Plant glabrous; **flowers** white; **fruit** a linear capsule	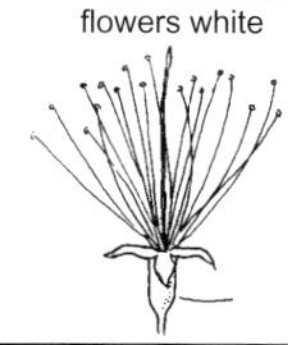	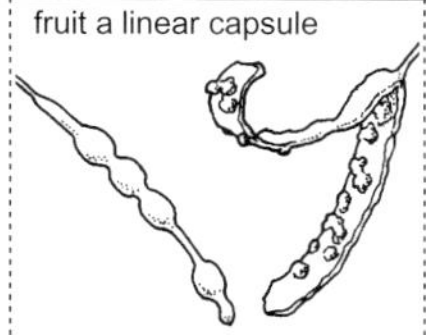	**Capparaceae** (14 sp.) (*Maerua angolensis*)
33	**Petiole** winged; **leaves** gland-dotted; **stipules** spiny; **fruit** a pyxidium or orange	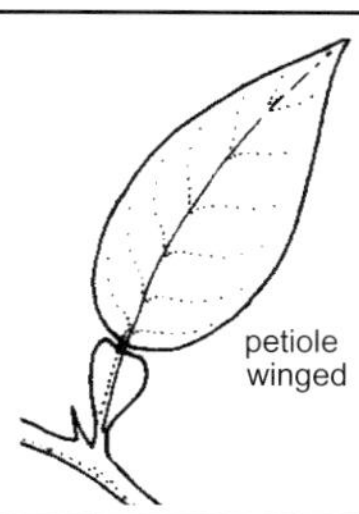		**Rutaceae** (4 spp.) (*Citrus aurantium*)
	Petiole not winged; **leaves** not gland-dotted; **stipules** not spiny; **fruit** a capsule or berry			34
34	**Petals** fringed; **fruit** a gaping capsule	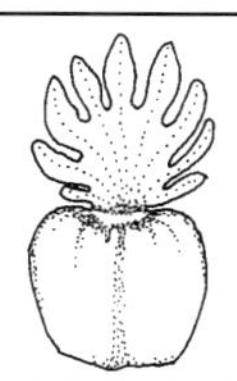		**Resedaceae** (3 spp.) (*Reseda*)
	Petals not fringed; **fruit** a 1 to 4-lobed berry	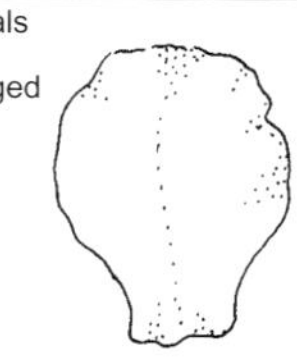	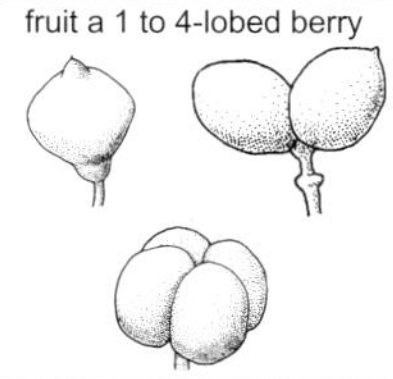	**Tiliaceae** (9 spp.) (*Grewia*)

35	**Flowers** zygomorphic, in **spikes**	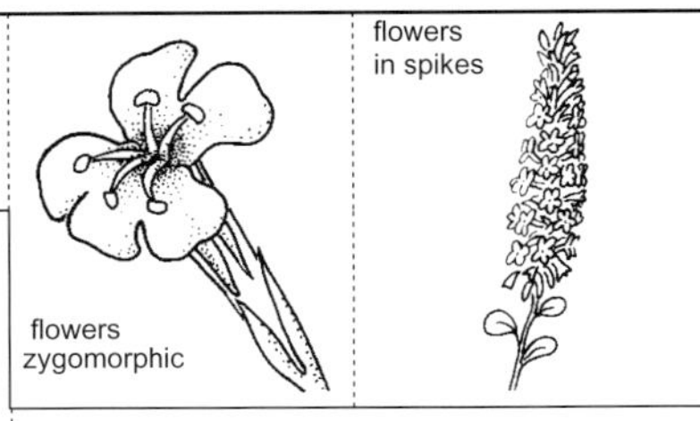	**Globulariaceae** (1 sp.) (*Poskea socotrana*)
	Flowers actinomorphic, arranged in various types of **open inflorescence**		36

36	**Anthers** united into a tube; **petals** reflexed; **fruit** a berry	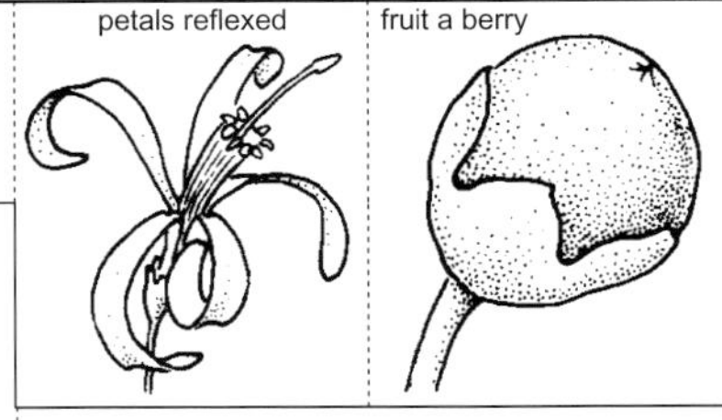	**Meliaceae** (1 sp.) (*Turraea socotrana*)
	Anthers free or attached to the corolla; **petals** various; **fruits** various		37

37	**Calyx** and **corolla** 4-merous; **corolla** long-tubular with spreading lobes; **flowers** bright red; **fruit** of 4 follicles	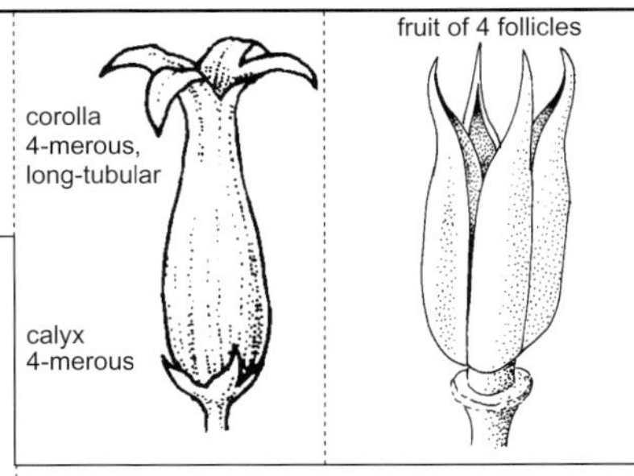	**Crassulaceae** (6 spp.) (*Kalanchoe*)
	Calyx and **corolla** 5-merous; **corolla** short tubular or tube absent; **flower** colour various; **fruit** various but not 4 follicles		38

38	**Anthers** opposite corolla lobes	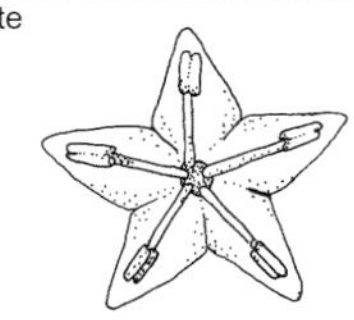	39
	Anthers alternate with corolla lobes	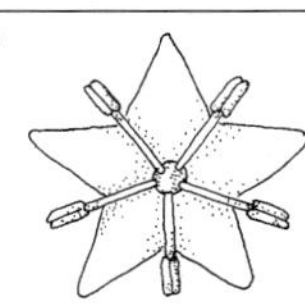	41

39	**Flowers** with papery bracts and calyx; **styles** 5; **fruit** a capsule	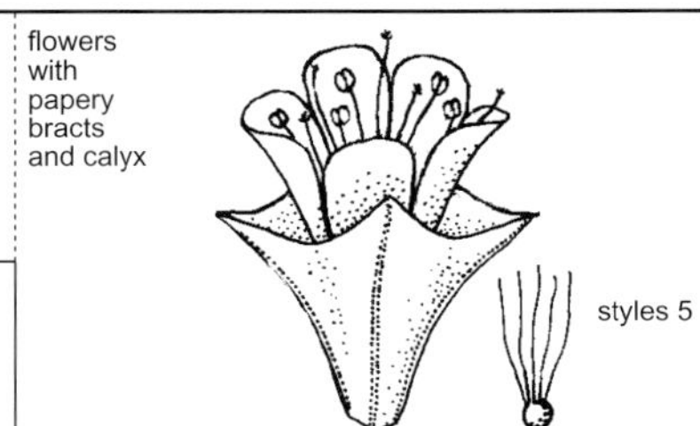	**Plumbaginaceae** (4 spp.)
	Flowers without papery bracts and calyx; **styles** 1; **fruit** a berry		40

40	**Staminodes** present; **fruit** a 1-seeded, fleshy berry	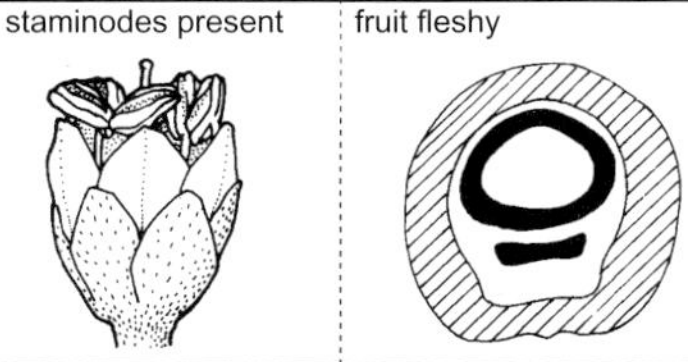	**Sapotaceae** (2 spp.)
	Staminodes absent; **fruit** a dry, apiculate berry	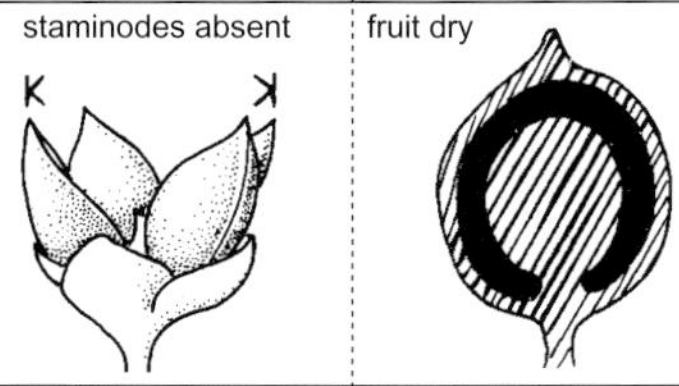	**Myrsinaceae** (1 sp.) (*Myrsine africana*)

41	**Shrub** spiny; **corolla** narrowly tubular; **fruit** a fleshy, red berry	corolla narrowly tubular  fruit a fleshy red berry 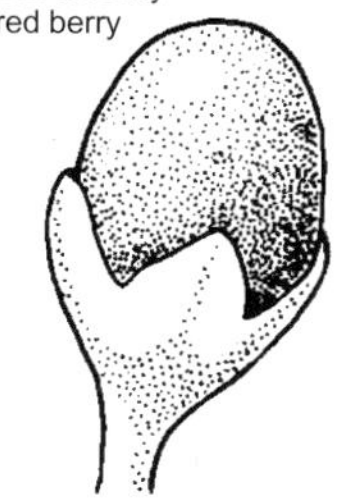	**Solanaceae** (11 spp.) (*Lycium sokotranum*)
	Shrub without spines; **corolla** lobes spreading, not narrowly tubular; **fruit** a capsule, or a drupe (sticky when crushed) or of 4 nutlets		42

42	**Corolla** folded in bud; **rough hairs** absent, silky hairs sometimes present; **fruit** a capsule	corolla folded in bud 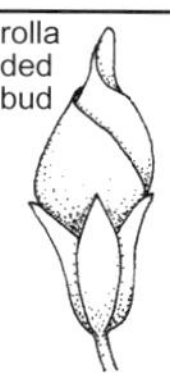 hairs absent or silky  fruit a capsule 	**Convolvulaceae** (22 spp.)
	Corolla lobes overlapping in bud; **rough hairs** often present; **fruit** of 4 nutlets (*Trichocalyx*) or a drupe (*Cordia* and *Ehretia*)	corolla lobes overlapping in bud 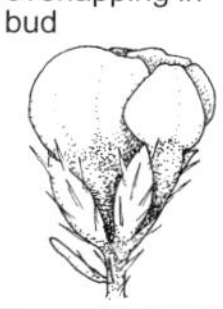 rough hairs  fruit of 4 nutlets 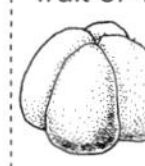 or a drupe 	**Boraginaceae** (29 spp.)

TERRESTRIAL MONOCOTYLEDON FAMILIES

ALOACEAE

1	**Leaves** <15cm, strongly recurved at tip, green with white speckles, along a contracted stem; **inflorescence** <20cm; **plant** of cliff faces	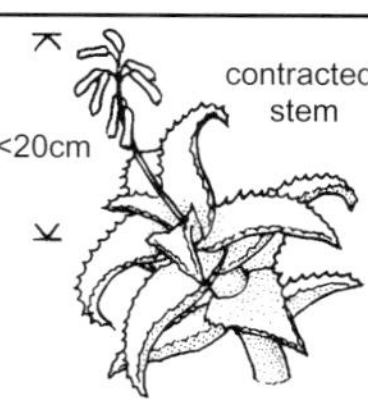 <20cm; contracted stem 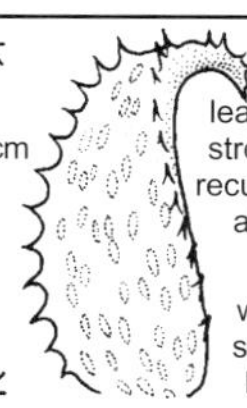 <15cm; leaves strongly recurved at tip, white spec-kles	***Aloe squarrosa*** Endemic **234**, 303
	Leaves 20–45cm, weakly recurved at tip, not speckled, in a basal rosette; **inflorescence** <1.5m, sometimes branched; **plant** of plains or rocky slopes	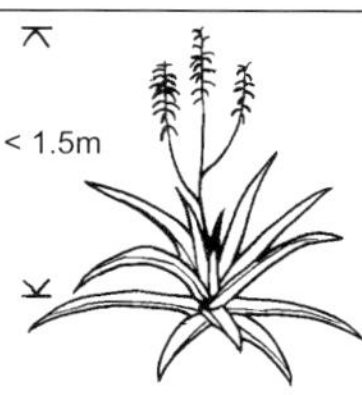 < 1.5m 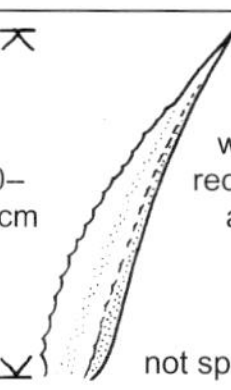 20–45cm; weakly recurved at tip; not speckled	***Aloe perryi*** Endemic **234**, 297
	Leaves 20–30cm, weakly recurved at tip, not speckled, in a basal rosette; **inflorescence** <20cm; **plant** of plains or rocky slopes	< 20cm 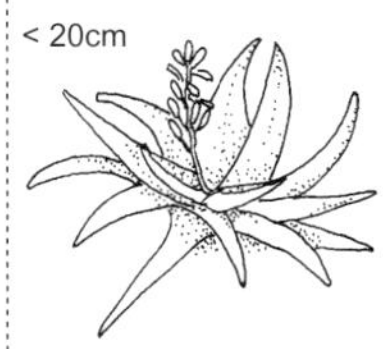 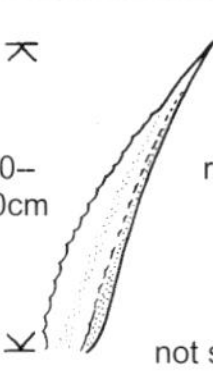 20–30cm; weakly recurved at tip; not speckled	***Aloe jawiyon*** Endemic **234**, 297

AMARYLLIDACEAE

1	**Bulb**; **inflorescence** umbel-like; **flowers** white with 6 stamens and a long tube	bulb   inflorescence umbel-like 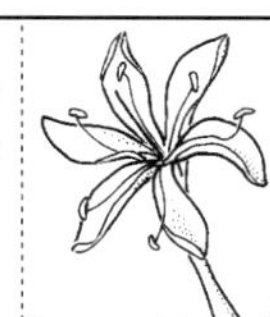 flowers with 6 stamens and a long tube	***Crinum balfourii*** Endemic **234**, 303

ANTHERICACEAE

1	**Leaves** linear, 3--15 x <0.2cm, pale green folded, margin minutely denticulate; **flowering spike** prostrate, lax,appearing with the leaves		***Chlorophytum filifolium*** **234**, 304
	Leaves lanceolate, 3--13 x1.5--2.2cm, green, margin minutely denticulate; **flowering spike** erect,rather densely flowered, appearing with the leaves	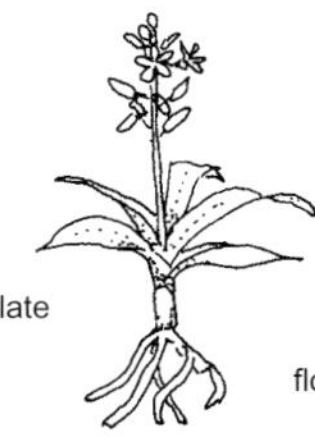	***Chlorophytum* sp. A** Endemic **234**, 305
	Leaves narrowly oblanceolate, 10--11.5 x 0.8--1.2cm, pale green with darker broken stripes, margins undulate with conspicuous hairs; **flowering spike** erect, open, appearing before the leaves	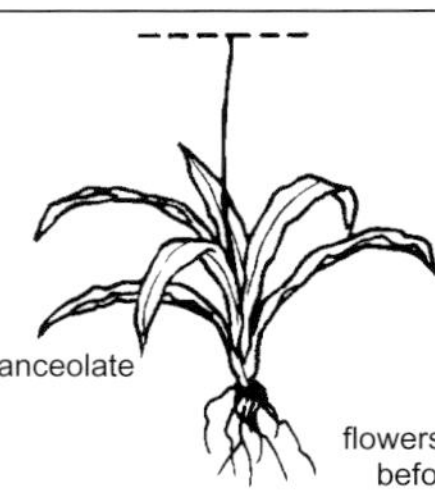	***Chlorophytum graptophyllum*** Endemic **234**, 305

ARACEAE

1	**Leaves** large, <30 cm; starchy **tuber** absent; **bulbil-bearing shoots** usually present, bulbils covered in scales with hooked tips	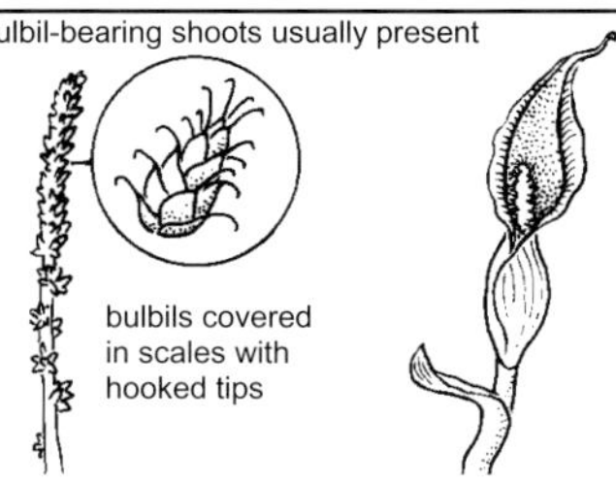	***Remusatia vivipara*** 306
	Leaves very large, <45cm; starchy **tuber** present; **bulbil-bearing shoots** absent	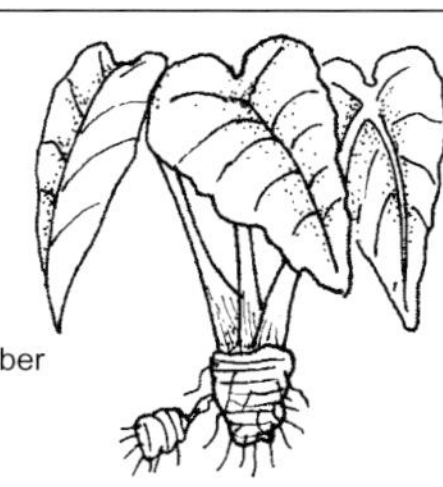	***Colocasia esculenta*** 306

ASPARAGACEAE

1	Low intricately branched **shrub** to 50cm; **branches** very spiny; **cladodes** stiff, needle-like	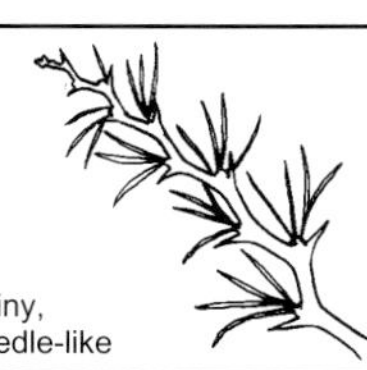	***Asparagus africanus*** Endemic var. **235**, 307
	Climbing **shrub** with long flexuous branches to 3m; **branches** weakly spiny; **cladodes** flexible, flattened	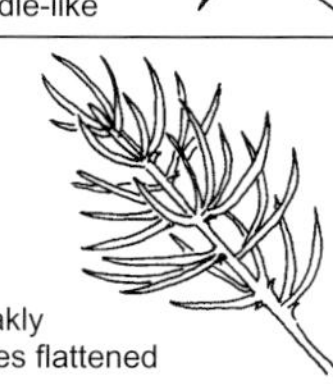	***Asparagus* sp. A** Endemic 308

ASPHODELACEAE

1 **Leaves** linear, rounded, hollow, succulent; **tepals** 5–12mm, white with a brownish midvein; **capsules** on short, erect pedicels

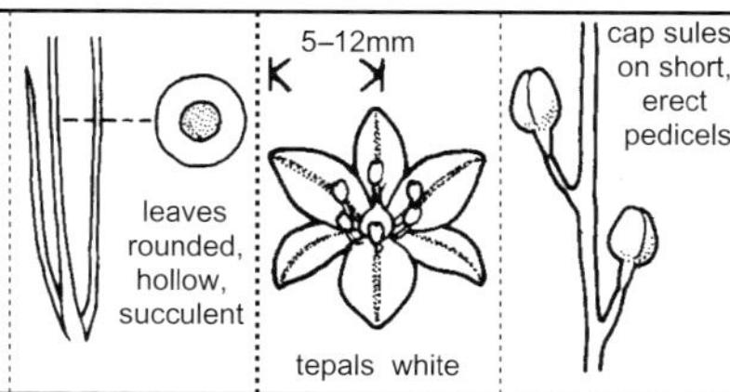

Asphodelus fistulosus 308

COLCHICACEAE

1 **Leaves** linear, <18cm long, arranged along stem; **inflorescence** umbel-like, 1 to 3-flowered; **flowers** purplish-brown, c.1mm across; **fruits** on recurved pedicels

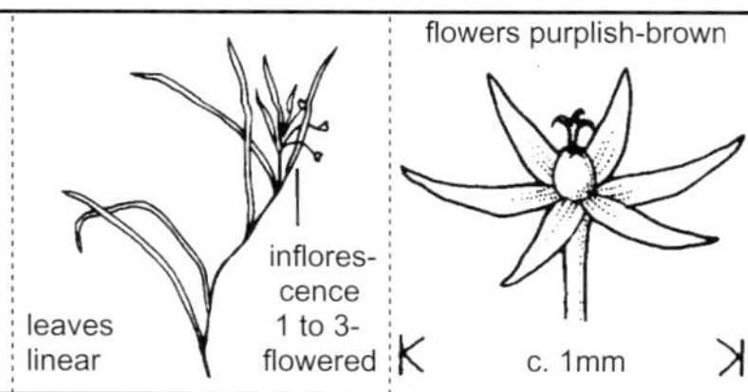

Iphigenia socotrana Endemic **235**, 309

COMMELINACEAE

1 **Flowers** actinomorphic; **petals** fused at base, equal in size; fertile **stamens** 6, all similar, the **filaments** hairy

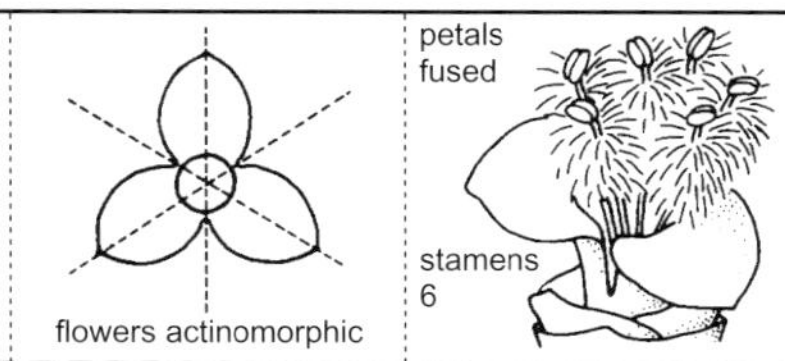

Cyanotis cristata 311

Flowers zygomorphic; **petals** free, 2 large and 1 small; fertile **stamens** 3, the middle one different in shape and size, the **filaments** glabrous

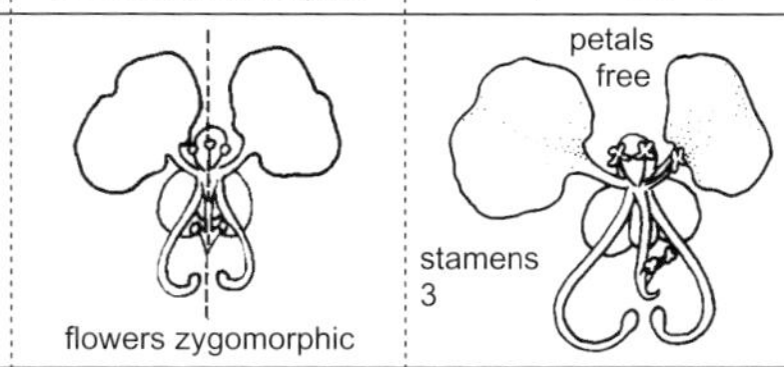

Commelina (3 spp.)

Commelina

1 **Leaves** ovate to lanceolate, not undulate; **leaf sheath** with long reddish hairs at apex; angle at base of **spathe** <90°, the margins ± straight; **flowers** blue

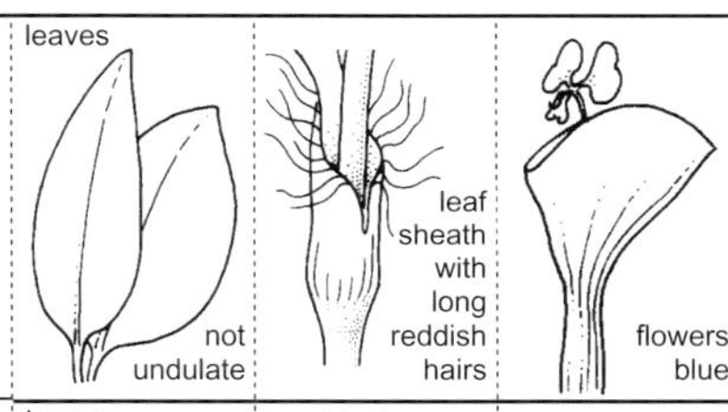

C. benghalensis 310

Leaves narrowly oblong-elliptic, often undulate; **leaf sheath** with short white hairs at apex; angle at base of **spathe** >90°, the margins ± straight; **flowers** blue

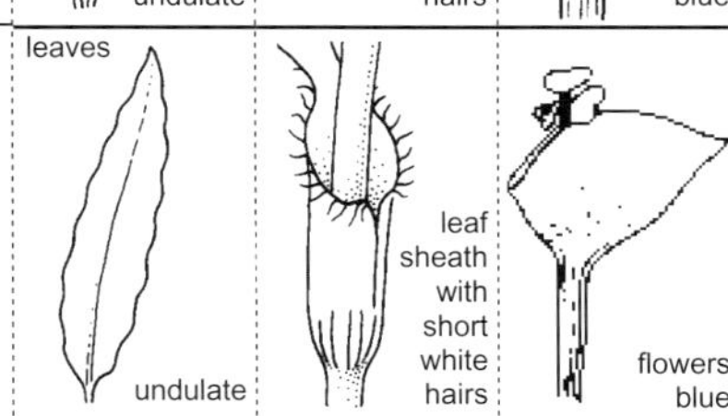

C. forskalaei 311

Leaves linear, not undulate; **leaf sheath** puberulous at apex; angle at base of **spathe** <90°, the margin strongly recurved; **flowers** mauve or pale blue

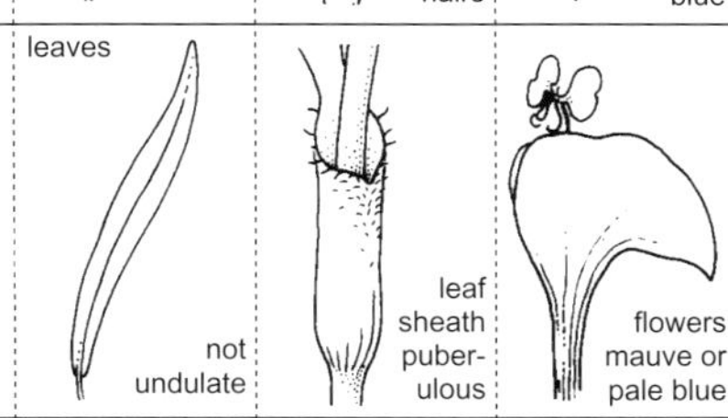

C. albescens 310

CYPERACEAE

1

Glumes spirally arranged, or inflorescence spirally arranged → 2

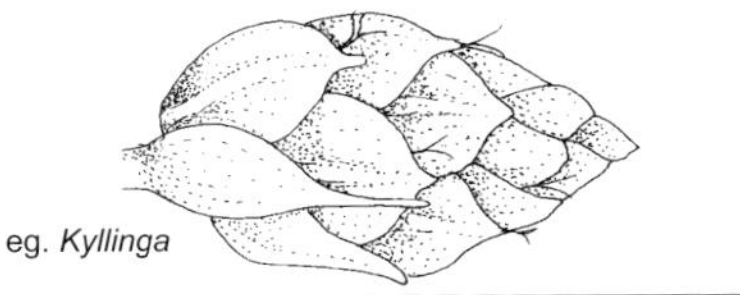

Glumes distichous (in 2 rows) → 7

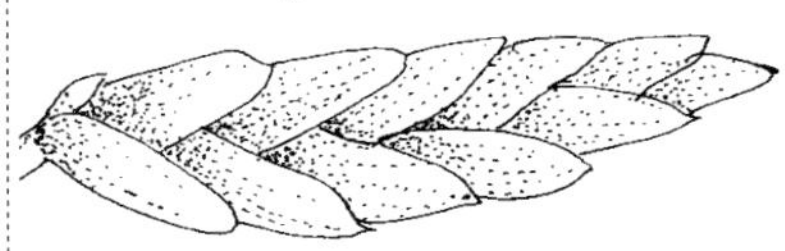

2

Leaf blade absent; **inflorescence** a solitary spike → ***Eleocharis*** (2 spp.)

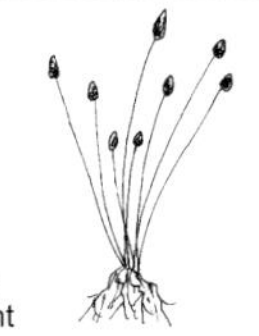

Leaf blade conspicuous; **inflorescence** of 2 to many spikes or spikelets → 3

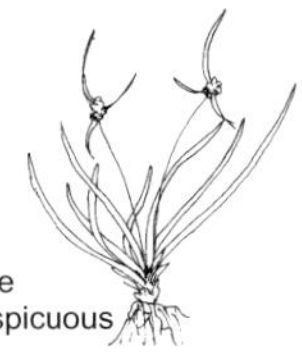

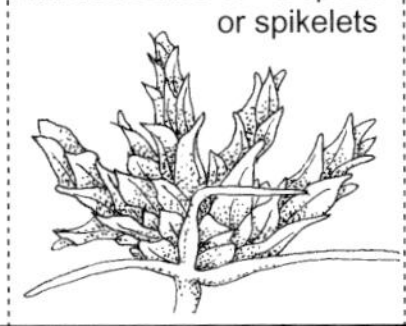

3

Leaves conspicuously arranged along stem → 4

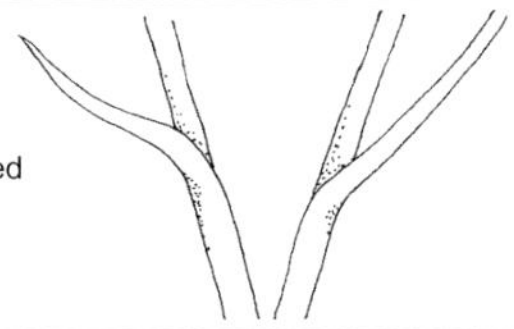

Leaves mostly basal → 5

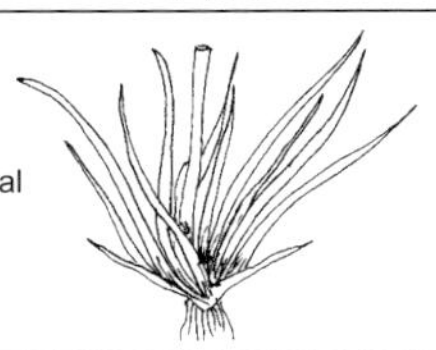

4

Annual herb <25cm; strongly **hairy** throughout; **leaf margin** entire → ***Fuirena ciliaris*** 321

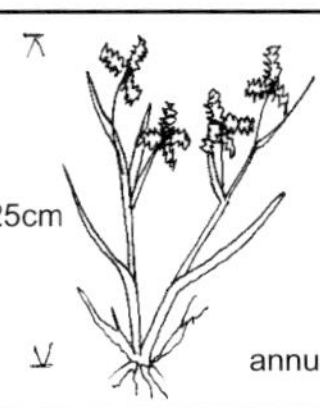

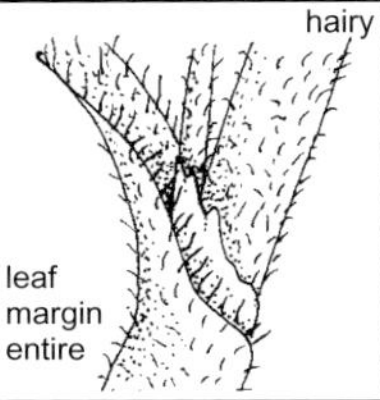

Robust **perennial** forming large clumps <2m; **glabrous** throughout; **leaf margin** toothed → ***Cladium mariscus*** 314

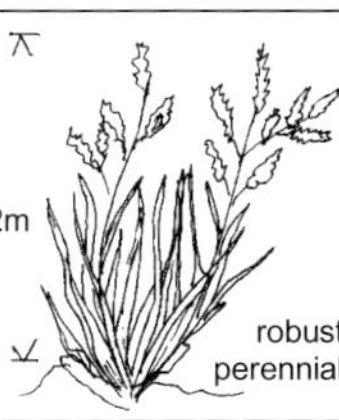

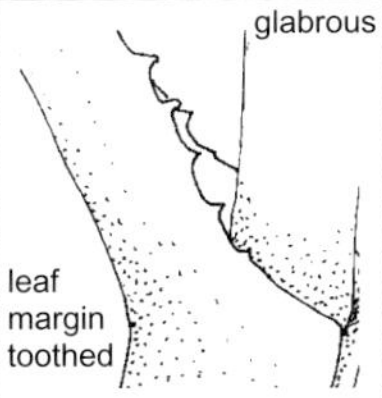

5

Leaf sheath with long flexuous hairs at opening → ***Bulbostylis*** (2 spp.)

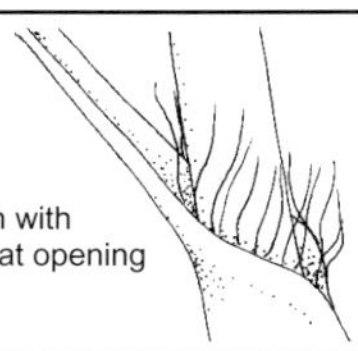

Leaf sheath glabrous or shortly hairy at opening → 6

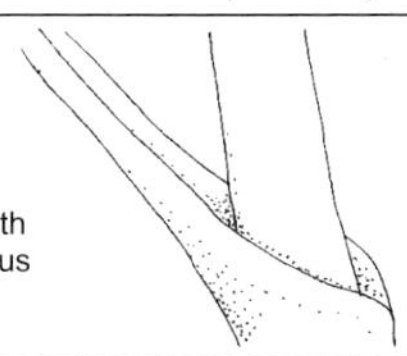

6

Inflorescence open, panicle-like, some spikes stalked; **inflorescence bracts** various, mostly only 1 overtopping the inflorescence — ***Fimbristylis*** (4 spp.)

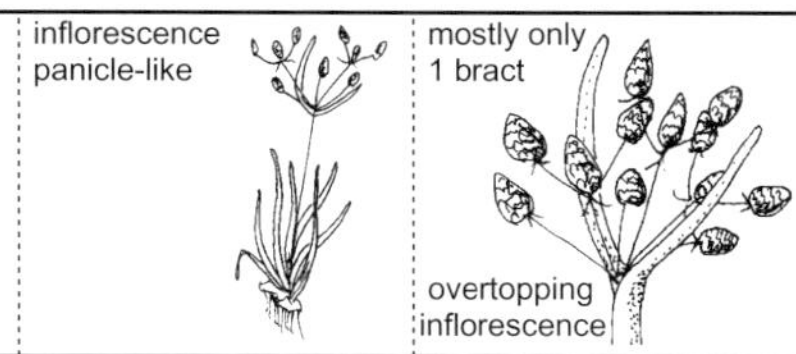

Inflorescence a compact head of 1 to 3 sessile spikes; **inflorescence bracts** usually 3, all overopping inflorescence — ***Kyllinga*** (2 spp.)

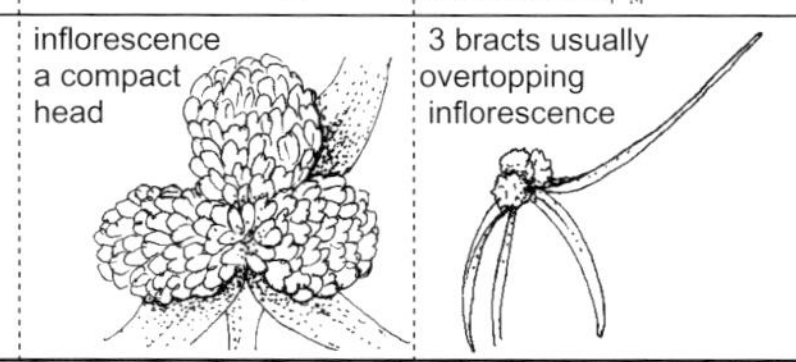

7

Tufted **annual**; **leaf blades** flat, grass-like; **glumes** c. 1mm long, mucronate — ***Pycreus pumilus*** 322

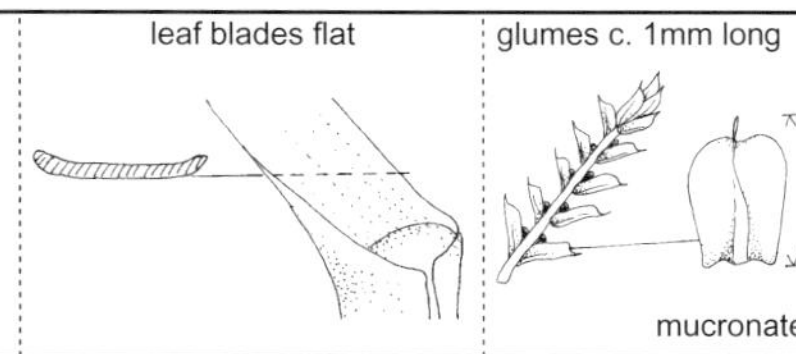

Annuals or **perennials**; **leaf blades** triangular; **glumes** various — ***Cyperus*** (14 spp.)

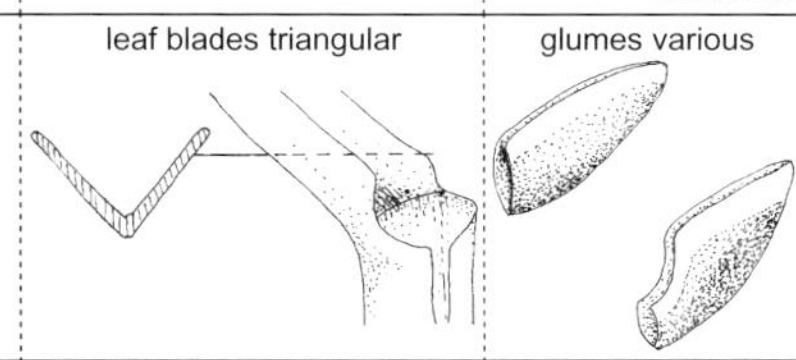

Bulbostylis

1

Inflorescence a solitary head of 1 to 4 sessile spikelets; **style base** persistent as a distinct knob on the nutlet — ***B. barbata*** 313

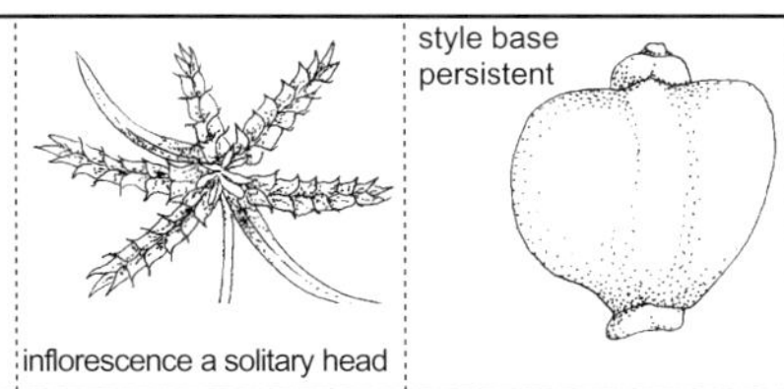

Inflorescence a reduced 'panicle' with 1 to 3 sessile spikelets and 1 to 2 stalked spikelets (the latter sometimes missing); **style base** dehiscent — ***B. hispidula*** 314

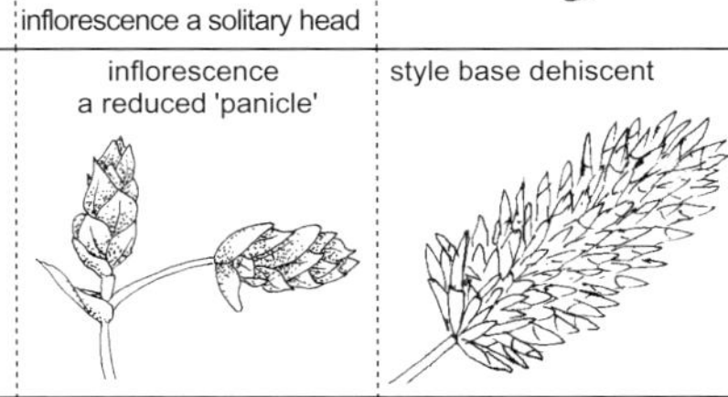

Cyperus

1

Stigmas 2; **nutlets** flattened in section; **leaf blades** absent or very short, if present inrolled and resembling the stem — ***C. laevigatus*** 317

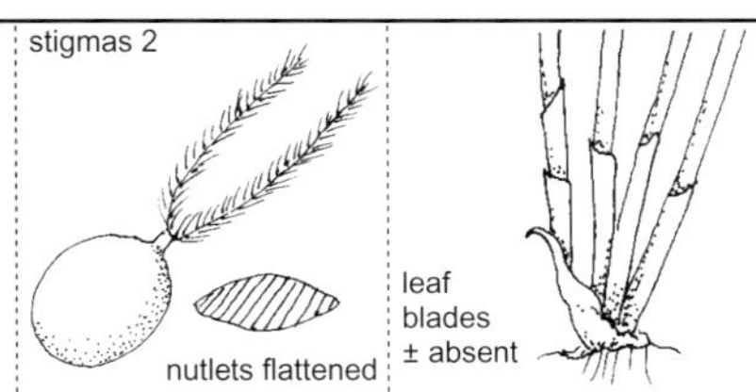

Stigmas 3; **nutlets** triangular in section; **leaf blades** well developed — 2

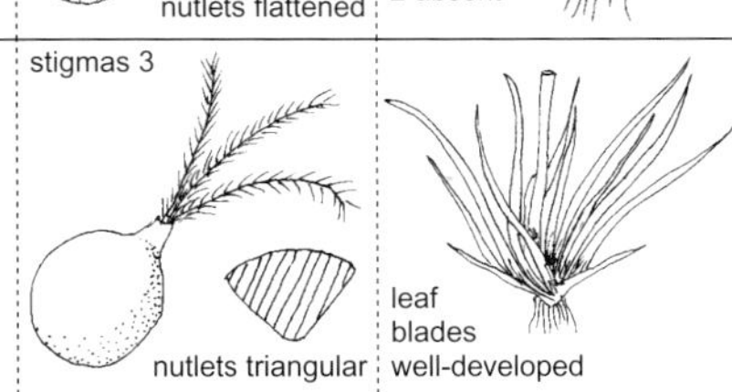

2

Glumes with long recurved tip; **dried leaves** smelling of curry — ***C. squarrosus*** 318

Glumes entire or with straight tip; **dried leaves** not smelling of curry — 3

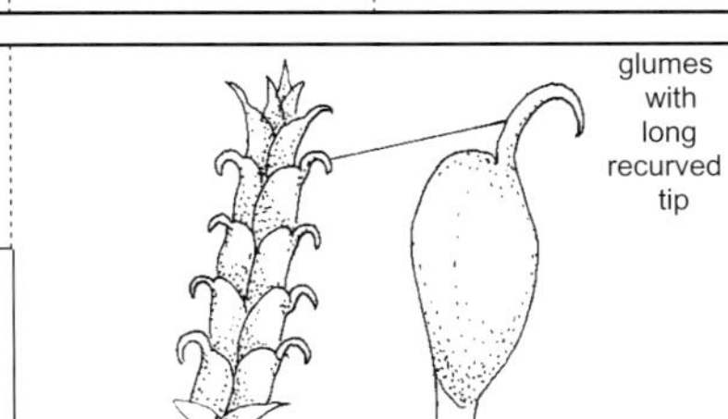

3

Spikelets in spikes

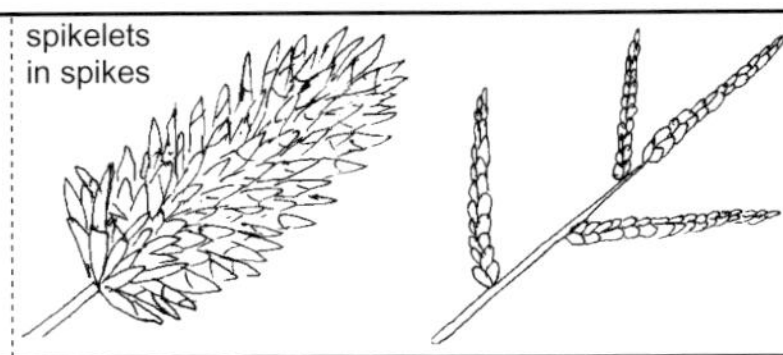

→ 4

Spikelets in digitate clusters

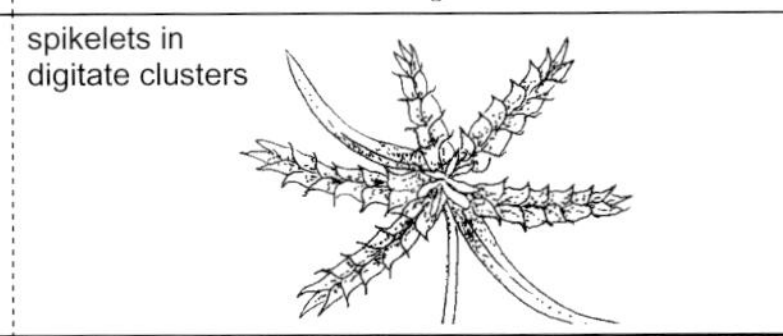

→ 8

4

Spikes cylindrical; **spikelets** crowded, 0.3--0.5cm long, 1 to 3-flowered

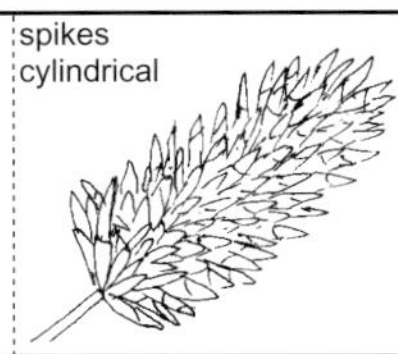

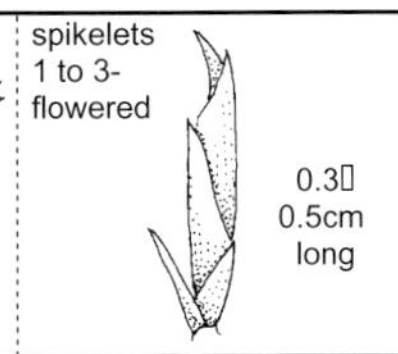

→ ***C. cyperoides*** 316

Spikes flattened; **spikelets** distant, spreading, 0.5--5cm long, 6 to many-flowered

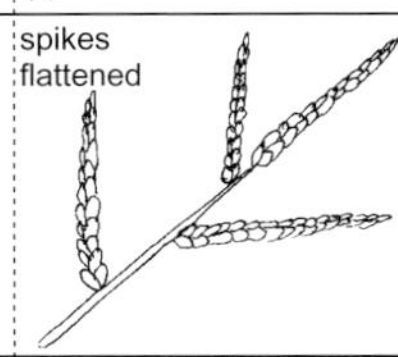

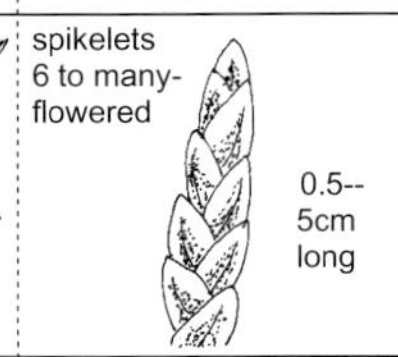

→ 5

5

Spikelet continuously growing, <5cm long

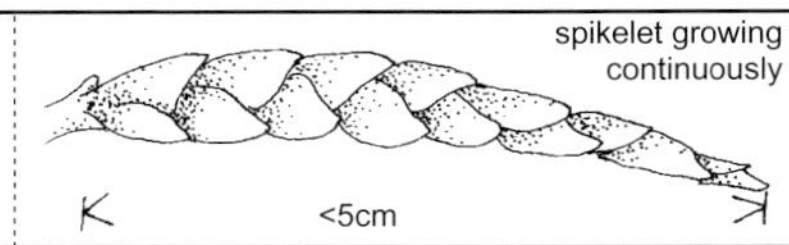

→ ***C. tenuiculmis*** 319

Spikelet not continuously growing, <2 cm long

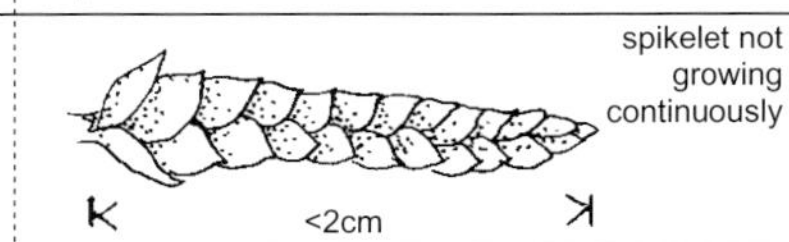

→ 6

6

Plants with stems arising from a bulb

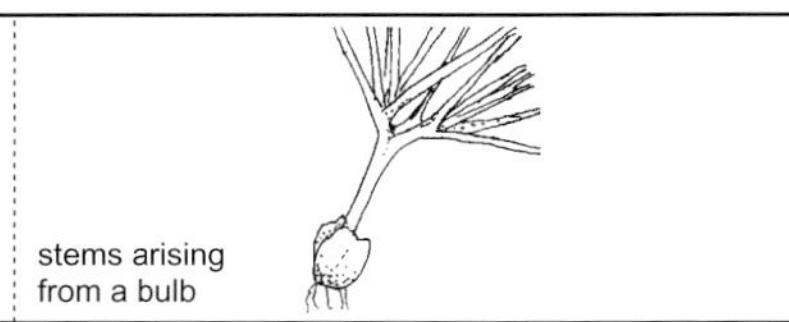

→ ***C. bulbosus*** 315

Plants without bulbs

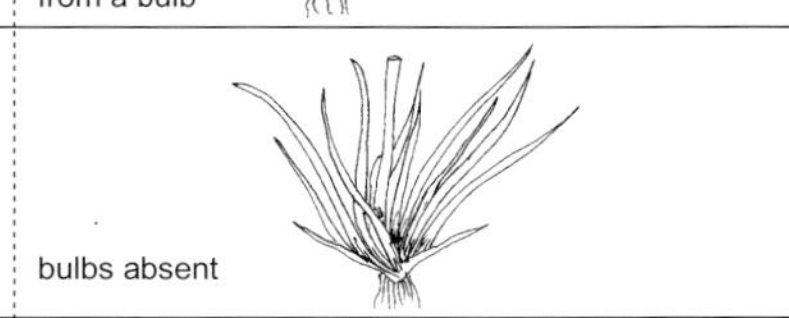

→ 7

7

Stem base not swollen; **stolon** stout, rhizome-like; **spikelets** with crowded closely imbricate glumes; **glumes** with a broad colourless margin

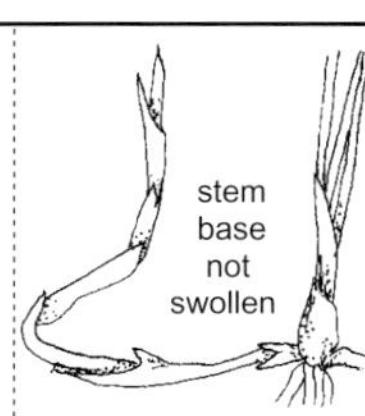

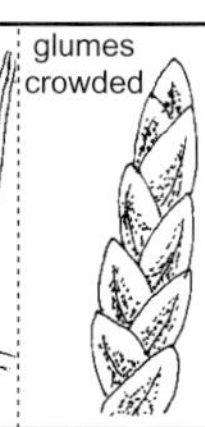

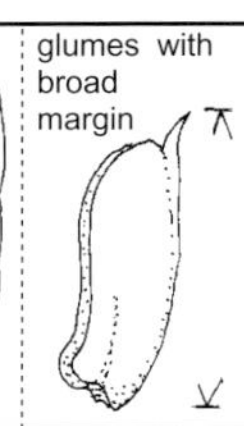

→ ***C. longus*** 317

Stem base swollen; **stolon** slender; **spikelets** with crowded closely imbricate glumes; **glumes** with a narrow colourless margin

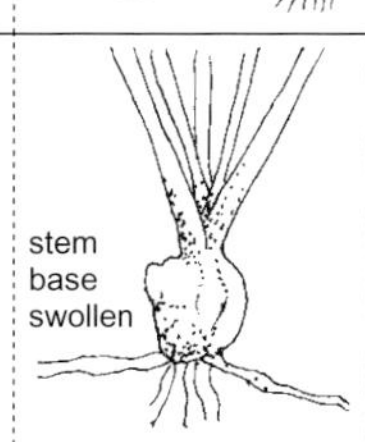

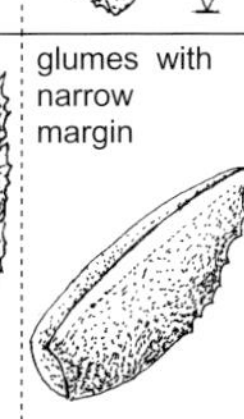

→ ***C. rotundus*** 318

Stem base swollen; **stolon** slender; **spikelets** with distant spreading glumes; **glumes** with a broad colourless margin

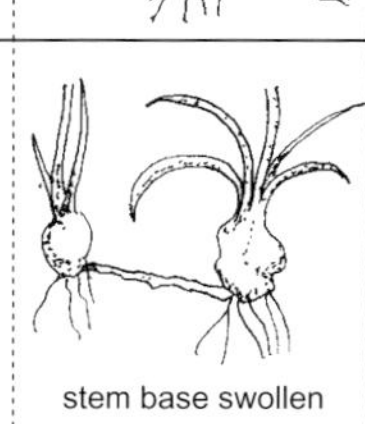

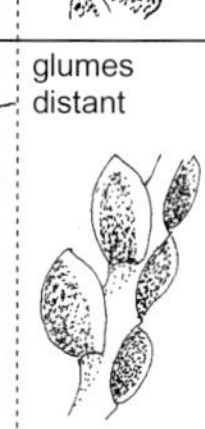

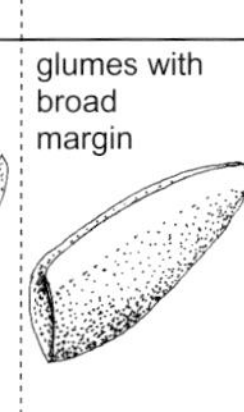

→ ***C. dilatatus*** 317

8	**Plant** annual, old leaf bases absent	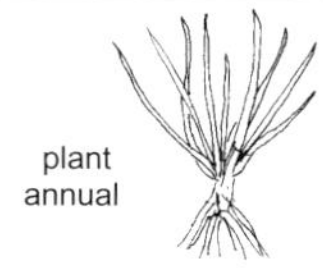	9
	Plant perennial, old leaf bases present	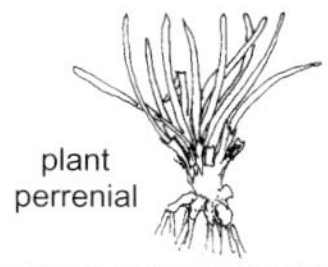	11
9	**Inflorescence** a head of digitate sessile spikes; **glumes** reddish-brown with a recurved tip	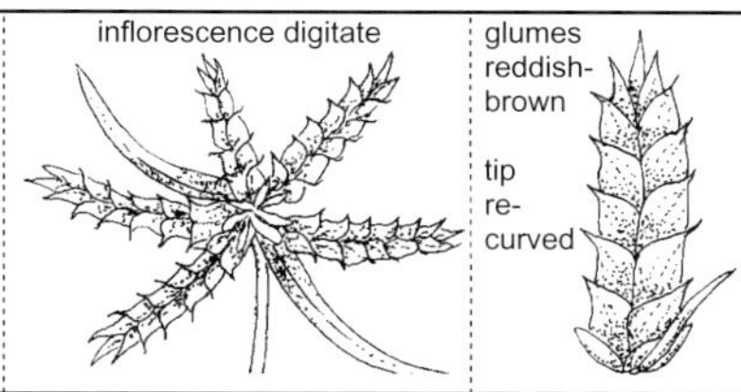	***C. rubicundus*** 318
	Inflorescence open, panicle-like, with at least some spikes stalked; **glumes** with a straight apex	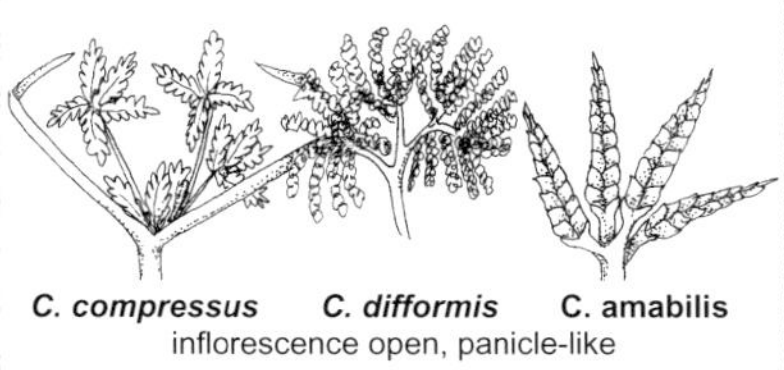	10
10	**Spikelets** strongly compressed; **glumes** 3.5--5mm long, mucronate; **partial inflorescence** of digitate spikelets; **stamens** 3	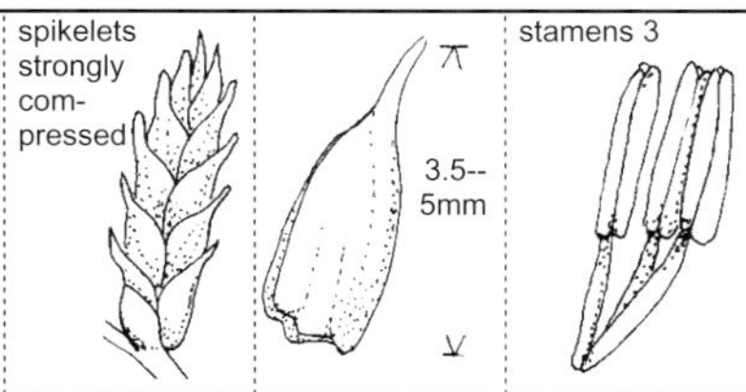	***C. compressus*** 316
	Spikelets strongly compressed; **glumes** 1--1.5mm long, mucronate; **partial inflorescences** of digitate spikelets; **stamen** 1	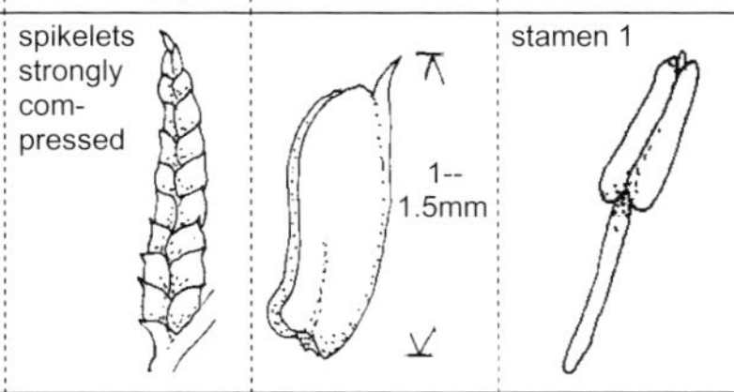	***C. amabilis*** 315
	Spikelets quadrangular; **glumes** 0.6--0.8mm long, rounded; **partial inflorescences** of spherical heads; **stamens** 2	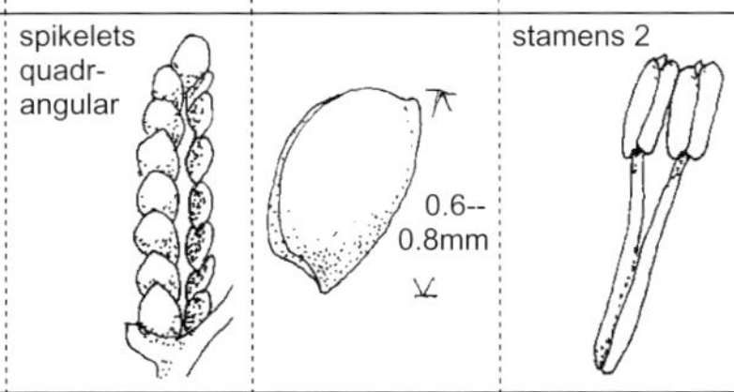	***C. difformis*** 317
11	Robust tussock-forming **perennial**; **leaves** thick often wiry, strongly curving; **roots** covered with dense grey hairs; **glumes** straw-coloured	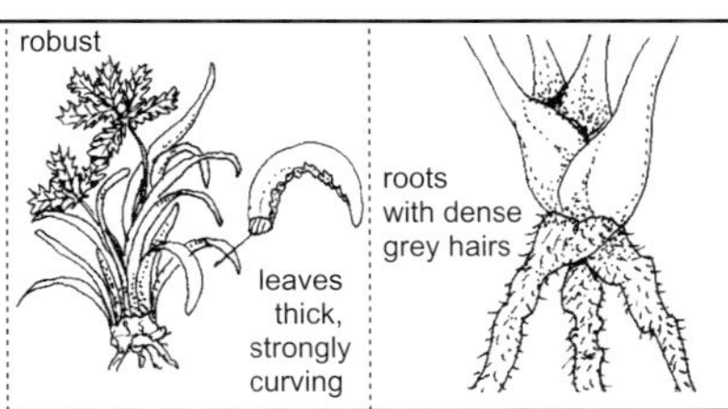	***C. conglomeratus*** 316
	Slender tussock-forming **perennial**; **leaves** filiform, straight; **roots** fibrous; **glumes** reddish-brown	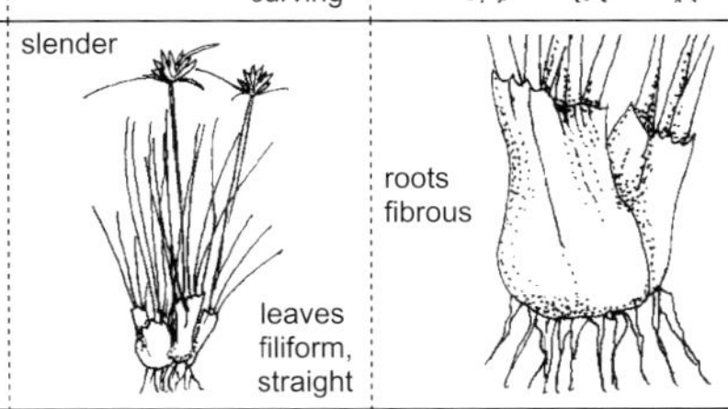	***C. stramineo-ferrugineus*** 318

Eleocharis

1	**Spikelet** many-flowered, globose with rounded tip, 2.5--3.5mm across; **perianth bristles** glabrous or shortly hairy	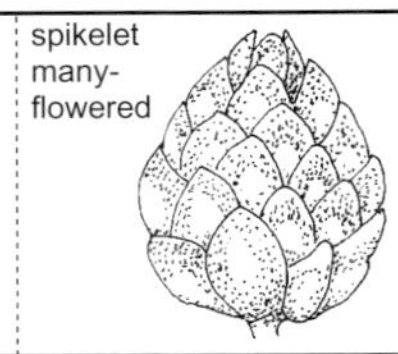	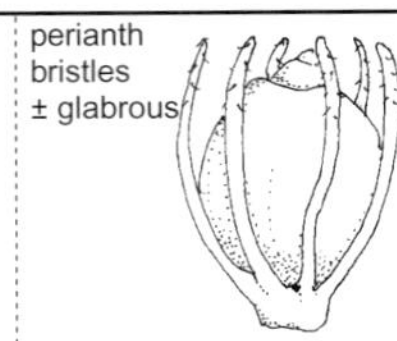	***E. geniculata*** 319
	Spikelet <10-flowered, ovoid with acute tip, 1--2mm across; **perianth bristles** with numerous recurved teeth	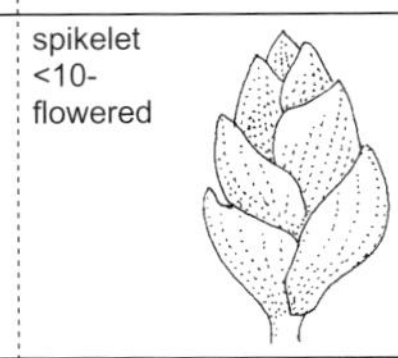	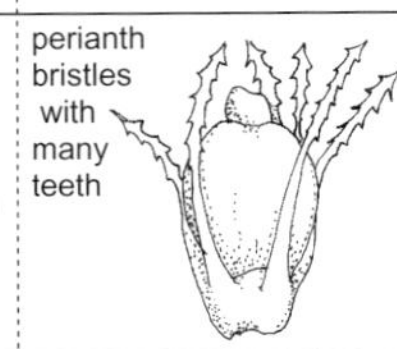	***E. caduca*** 319

Fimbristylis

1	**Style** branches 3; **nutlet** triangular; **stem** flattened	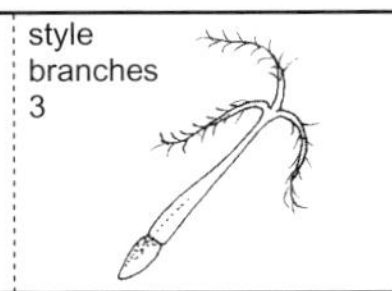	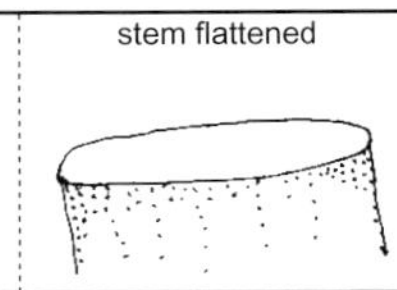	***F. complanata*** 320
	Style branches 2; **nutlet** flattened; **stem** rounded or angular	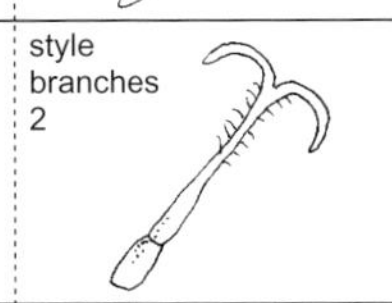	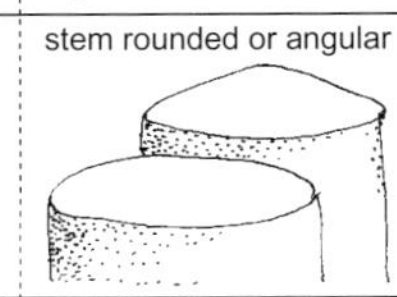	2
2	**Ligule** absent; **spikelets** in pedunculate clusters; **glumes** rounded	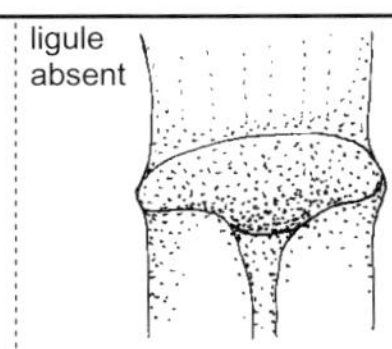	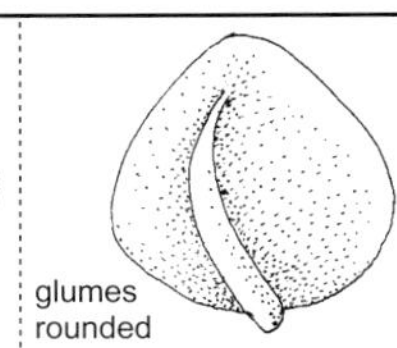	***F. cymosa*** 320
	Ligule a rim of hairs; **spikelets** not clustered; **glumes** shortly mucronate	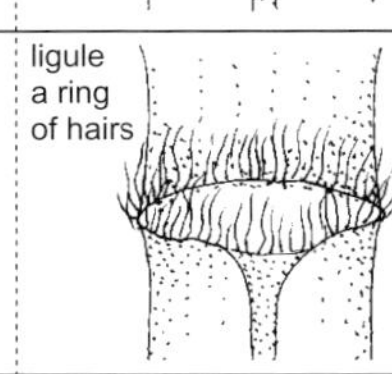	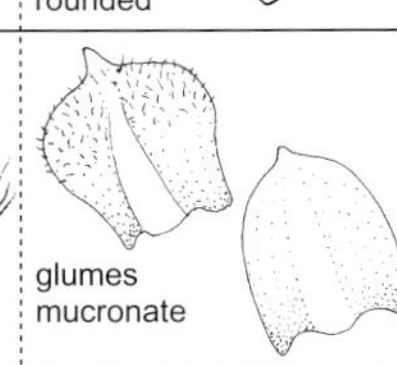	3
3	**Glumes** hairy in the upper half; **nutlets** smooth	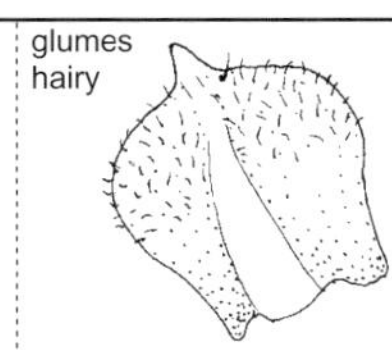	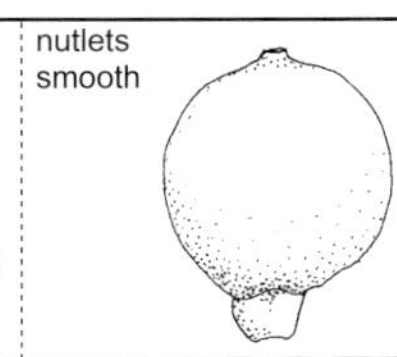	***F. ferruginea*** 321
	Glumes glabrous; **nutlets** longitudinally grooved with transverse ribs over the grooves	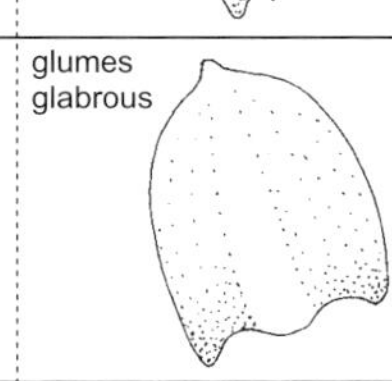	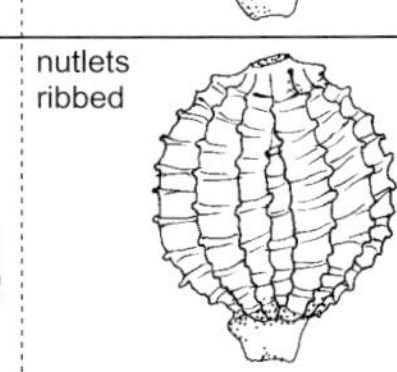	***F. dichotoma*** 320

Kyllinga

1	**Rhizomes** creeping, stems arising at intervals along it; **inflorescence** greenish; **glumes** 2--3mm long (lower <1mm)	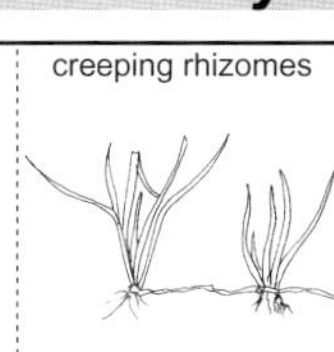	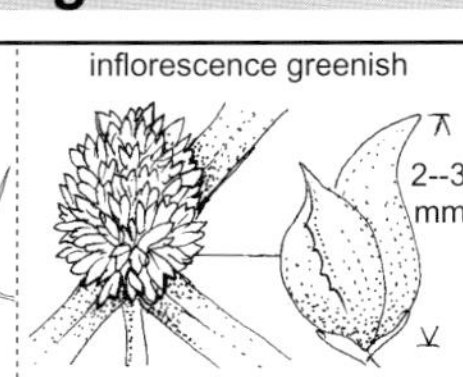	***K. brevifolia*** 321
	Rhizomes barely developed, stems tufted; **inflorescence** yellowish; **glumes** c. 0.8mm long	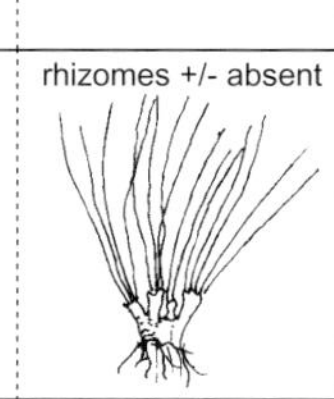	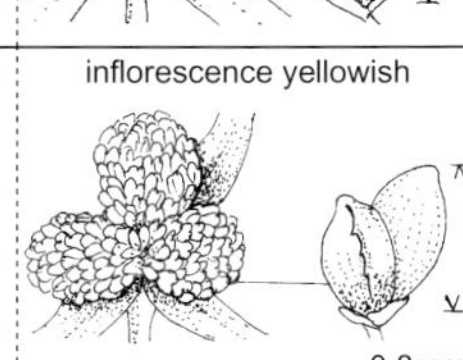	***K. microstyla*** 322

DIOSCOREACEAE

1	**Stems** twining, from an underground tuber; **leaves** heart-shaped, glossy; **flowers** cream, inconspicuous, 3-merous; **fruits** 3-lobed	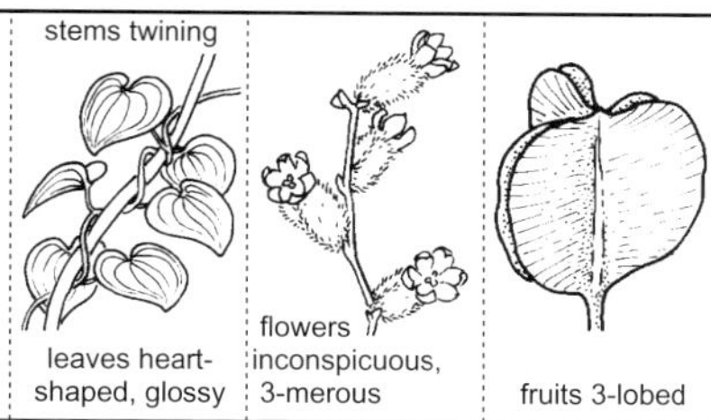	***Dioscorea lanata*** Endemic **235**, 322

DRACAENACEAE

1	**Tree** with rounded canopy; **leaves** in dense rosettes; **flowers** in a panicle, cream; **fruit** a berry, red, ripening black	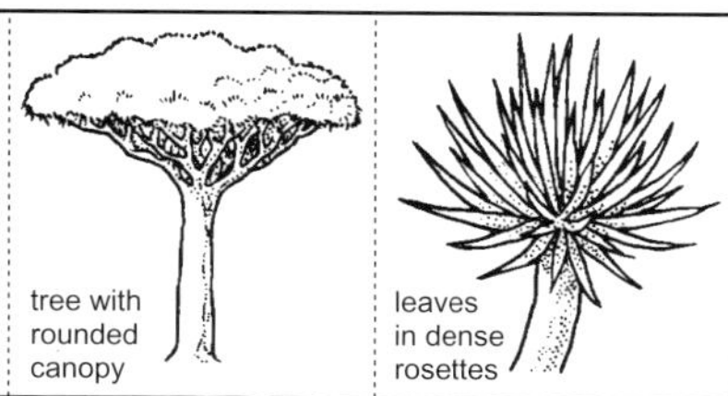	***Dracaena cinnabari*** Endemic **235**, 324

GRAMINEAE

		Awns absent or < ½ floret length	**Awns** or **bristles** conspicuous > ½ floret length
1	**Inflorescence** made up of one 'raceme'	 41	 31
	Inflorescence made up of a series of spikes arising from a single point 'fingers',	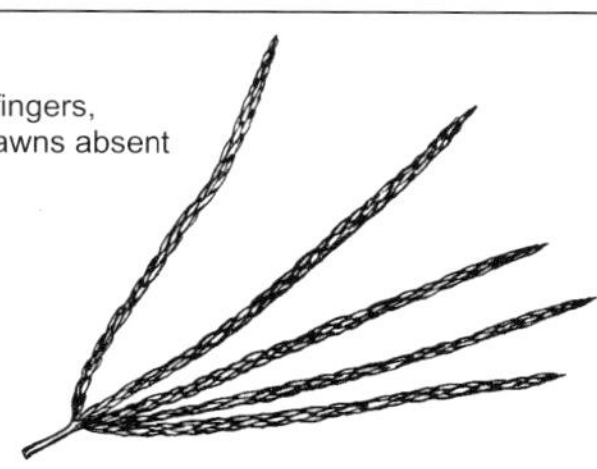 14	 7
	Inflorescence with a series of 'racemes' arising from main axis	 46	 12
	Inflorescence an open panicle	 2	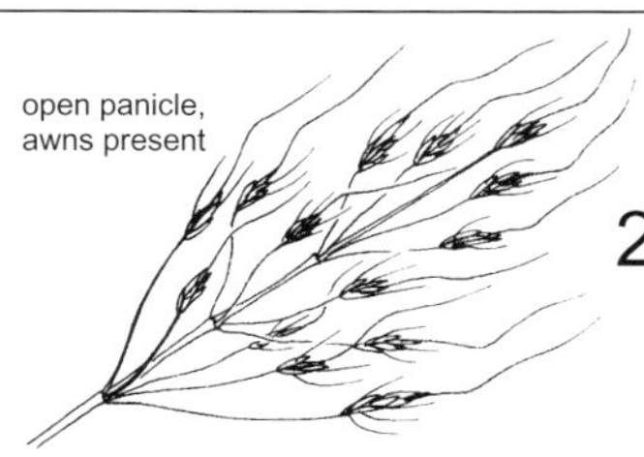 26
	Inflorescence a contracted panicle or complex head	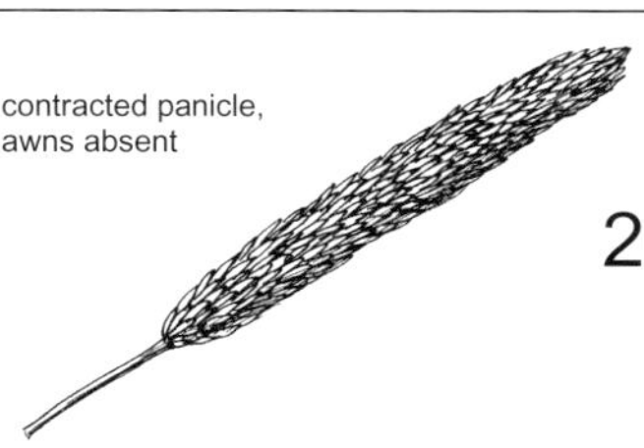 20	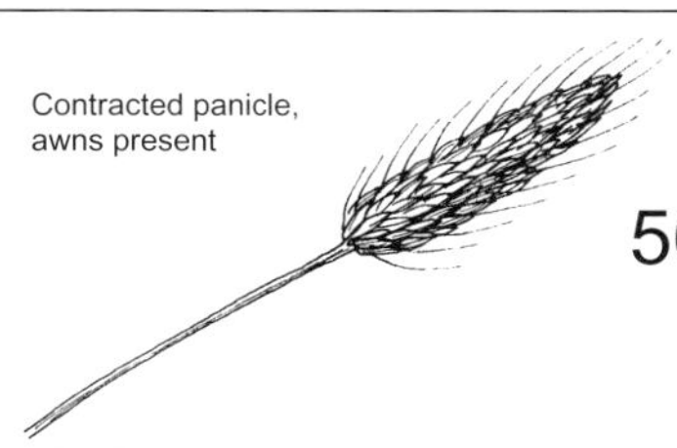 56

2	**Spikelets** 3 to many-flowered	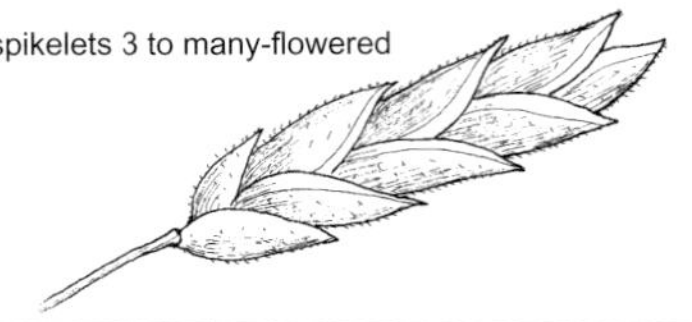	***Eragrostis*** (7 spp. excluding *E. ciliaris*)
	Spikelets 1 or 2-flowered	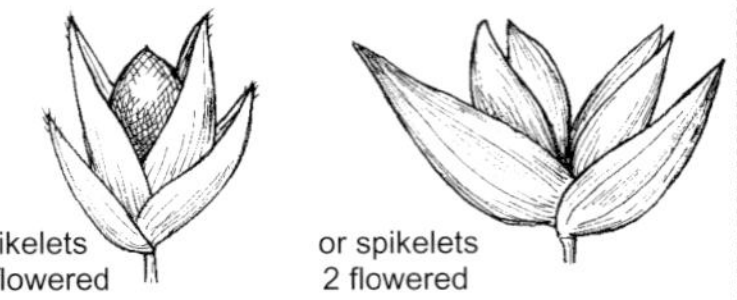	3
3	**Spikelets** silky-hairy		4
	Spikelets glabrous or shortly hairy	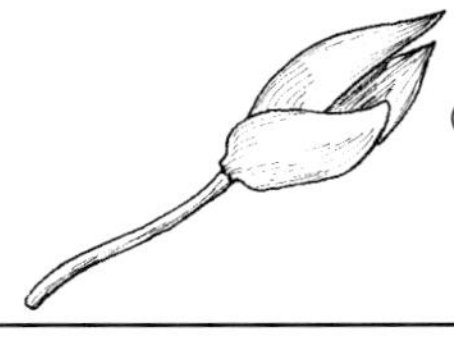	5
4	**Pedicels** glabrous; **upper floret** dorsally compressed	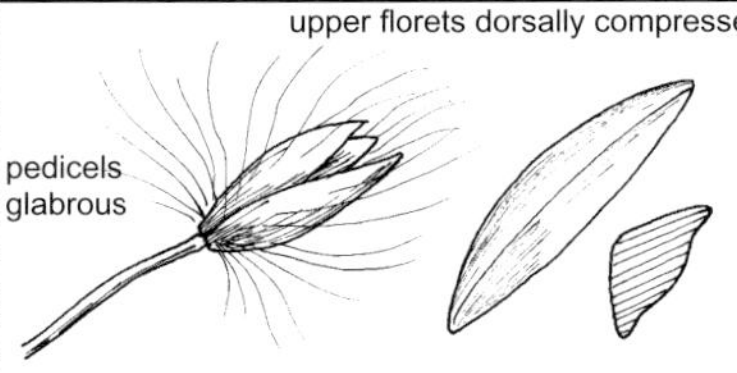	***Tricholaena*** (2 spp.)
	Pedicels with several long hairs; **upper florets** laterally compressed	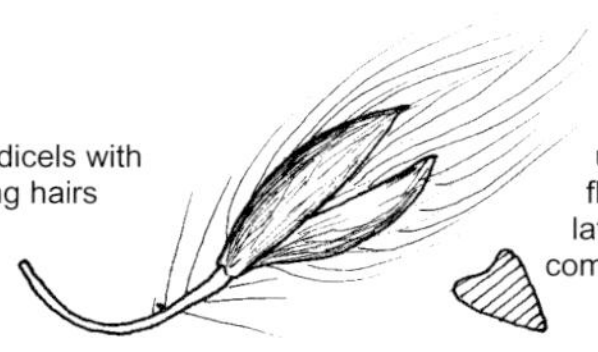	***Melinis repens*** 372
5	**Spikelets** globose and wrinkled		***Hackelochloa granularis*** 366
	Spikelets not globose and wrinkled	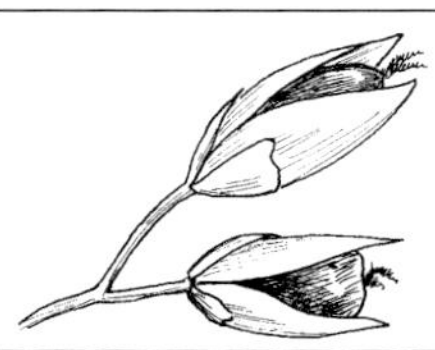	6
6	**Spikelets** 2-flowered; **spikelets** falling with glumes	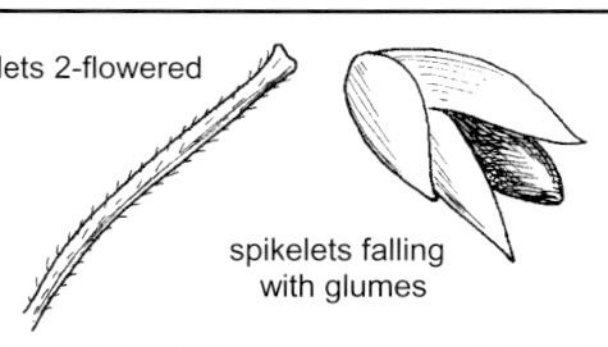	***Panicum*** (5 spp.)
	Spikelets 1-flowered; **spikelets** disarticulating above the glumes (glumes persistent on inflorescence, at least for a while)	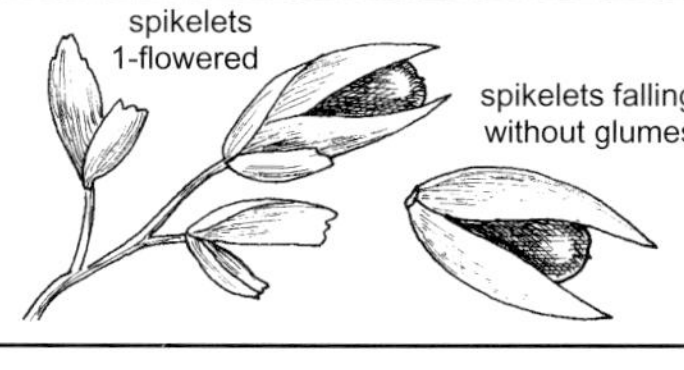	***Sporobolus*** (4 spp. excluding *S. spicatus*)

7	**Leaves** broad elliptic, 'herbaceous' - soft clasping		***Arthraxon*** (2 spp.)
	Leaves linear	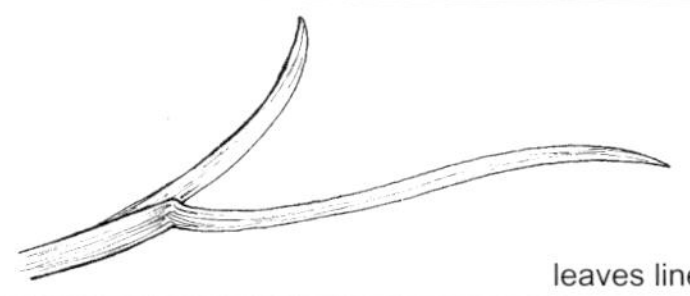	8
8	**Stem nodes** hairy; **lemmas** with or without a distinct central pit	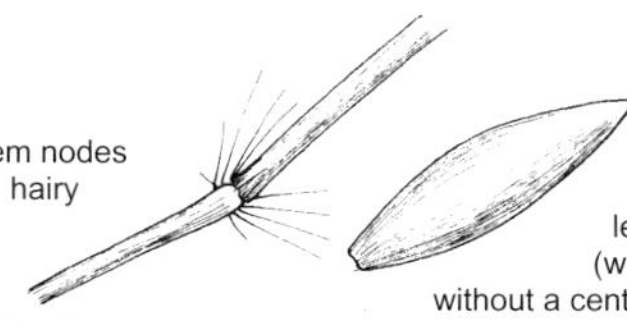	***Dichanthium*** (2 spp.)
	Stem nodes glabrous; **lemmas** with a distinct central pit	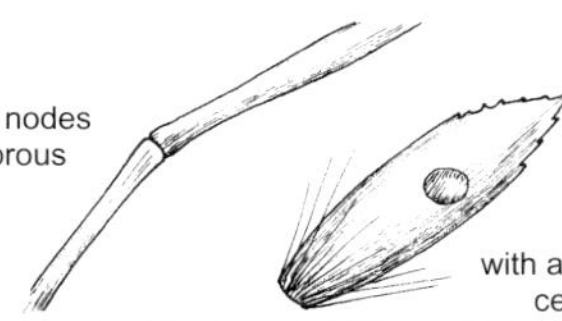	***Bothriochloa insculpta*** 350
	Stem nodes glabrous; **lemmas** without a distinct central pit	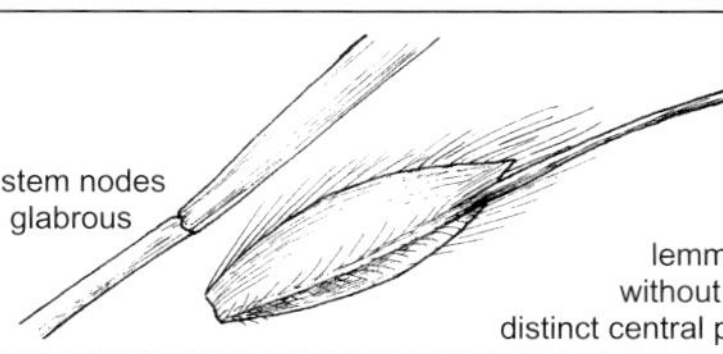	9
9	**'Racemes'** more than 2 or 3		***Chloris*** (2 spp.)
	'Racemes' typically 2 or 3	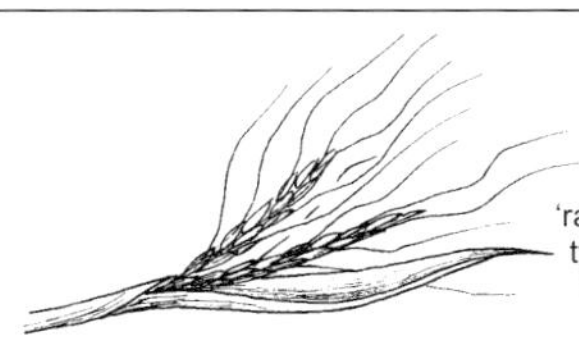	10
10	**Inflorescence** spatheate	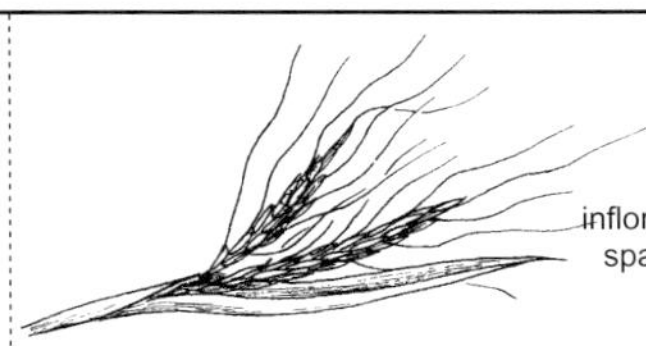	***Hyparrhenia hirta*** 368
	Inflorescence not spatheate		11
11	**Awns** straight; **leaves** glabrous	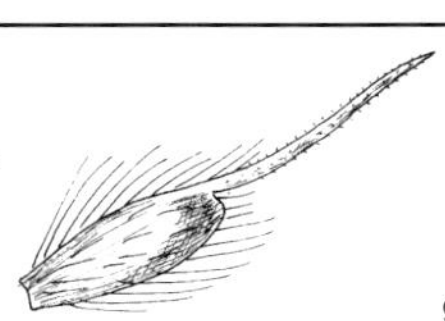	***Tetrapogon*** (2 spp.)
	Awns twisted; **leaves** hairy		***Ischaemum* sp. A** Endemic 369

12	**Spikelets** 4 to 9-flowered		***Schmidtia pappophoroides*** 377
	Spikelets 2-flowered		13
13	Tufted/ erect **annuals**; **spikelets** subtended by thick, dark bristles; **glumes** awnless	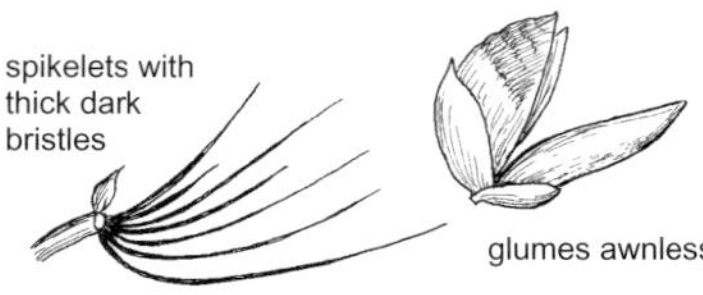	***Setaria*** (3 spp.)
	Trailing **plants**; **spikelets** subtended by fine, white hairs; **glumes** awned		***Oplismenus*** (2 spp.)
14	**Spikelets** 1 or 2-flowered	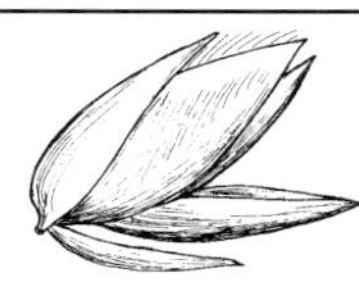	18
	Spikelets 3 to many-flowered		15
15	**Tip of 'raceme'** bare of spikelets		***Dactyloctenium*** (4 spp.)
	Tip of 'raceme' not bare of spikelets	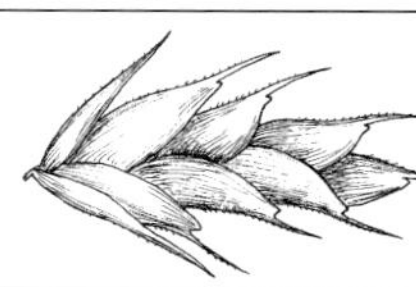	16
16	**'Racemes'** form a dense head	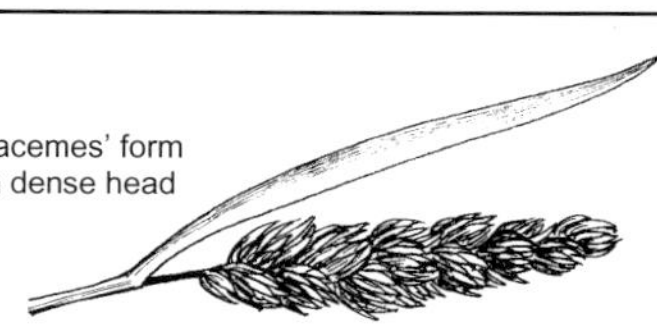	***Aeluropus lagopodioides*** 346
	'Racemes' do not form a dense head - 'racemes' clearly separated	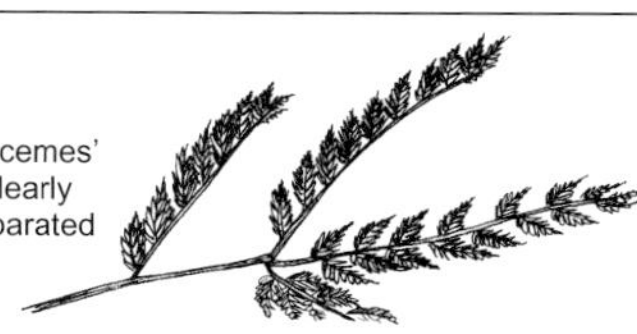	17
17	Tuft forming, robust **plant**; **lemma** awnless; **spikelets** disarticulating between florets	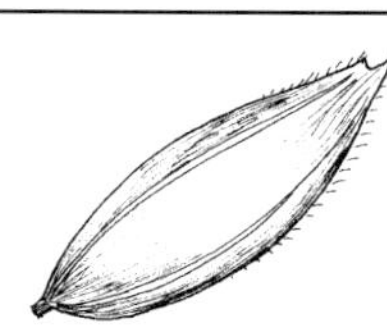	***Eleusine coracana*** 361
	Delicate **annual**, not tuft forming; **lemma** awned (giving 'raceme' a spikey appearance); **lemmas** disarticulating leaving paleas persistent on 'raceme'		***Acrachne racemosa*** 346

18	**'Racemes'** forming a distinct star; **glumes** subequal; **ligule** a fringed membrane	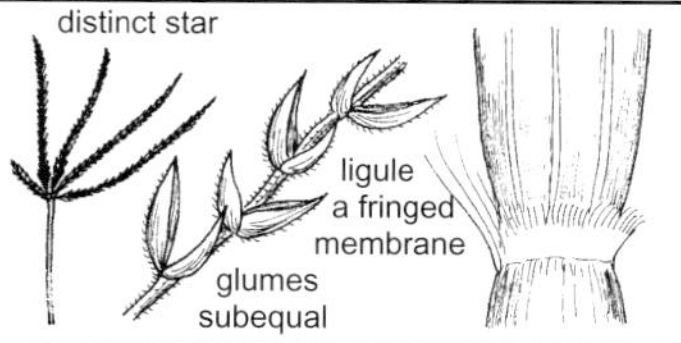	***Cynodon dactylon*** 356
	'Racemes' not forming a distinct star; **lower glume** very small or absent; **ligule** membranous	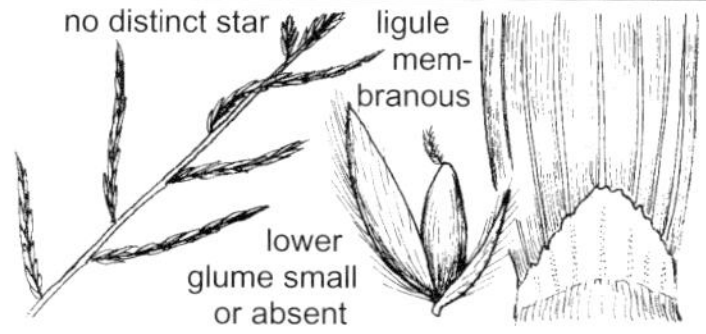	19
19	**Spikelets** sessile in 2 rows		***Paspalum*** (2 spp.)
	Spikelets sessile or shortly pedicellate, not in distinct rows		***Digitaria*** (3 spp.)
20	**Spikelets** 1 or 2-flowered	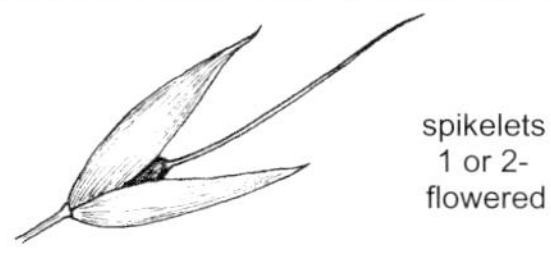	23
	Spikelets 3 to many-flowered	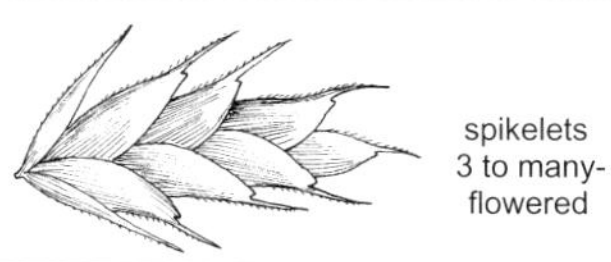	21
21	**Tip of 'raceme'** bare of spikelets	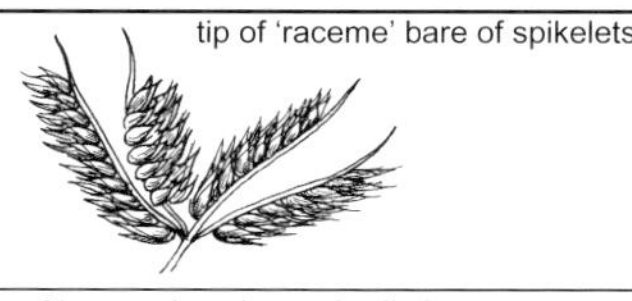	***Dactyloctenium*** (4 spp.)
	Tip of 'raceme' not bare of spikelets		22
22	**Glumes** glabrous but palea conspicuously fringed; **leaf blades** flat, not distichous or pungent	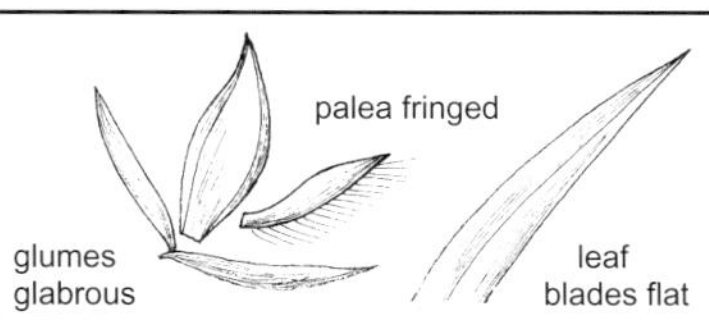	***Eragrostis ciliaris*** 365
	Glumes hairy, lemmas and paleas also hairy; **leaf blades** flat or inrolled, distichous, pungent		***Aeluropus lagopodioides*** 346
23	**Inflorescence** to 1m, large, covered in long white hairs		***Imperata cylindrica*** 368
	Inflorescence a more or less dense cylinder	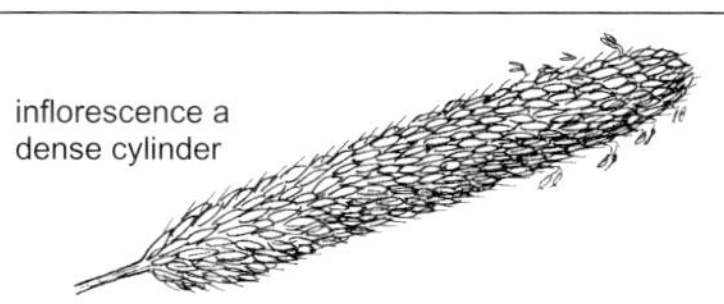	24

24	**Annual**; **ligule** membranous	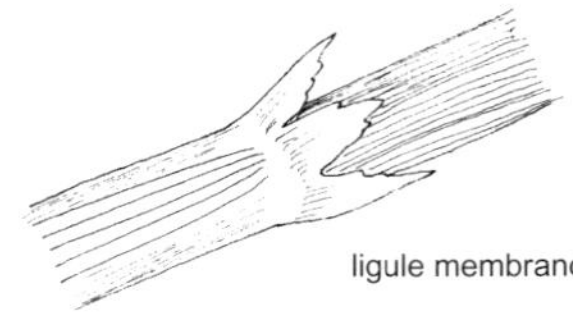	***Gastridium phleoides*** 366
	Tufted **perennials**; **ligule** a ring of hairs	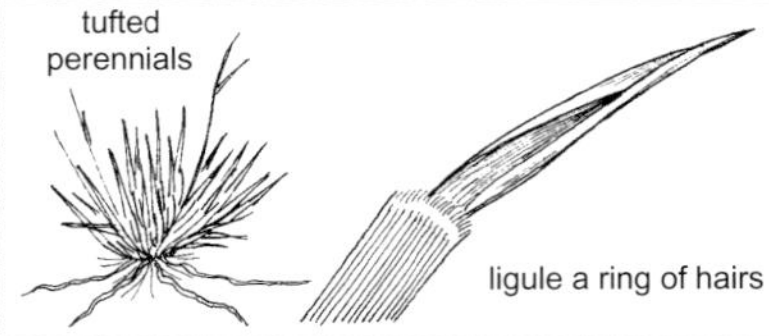	25
25	**Leaf blades** <6 cm long, forming a spiny ball; **plant** spreading by smooth, looping stolons; **spikelets** <2.5 mm long	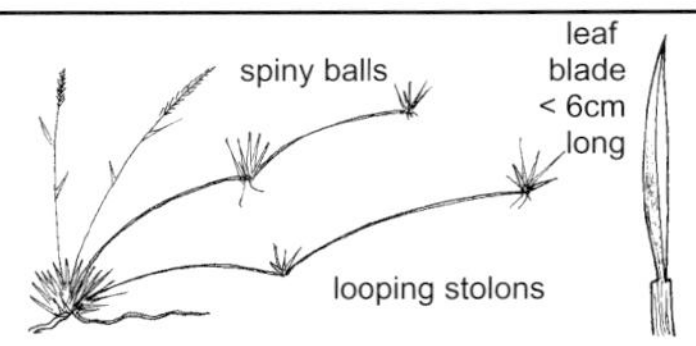	***Sporobolus spicatus*** 380
	Leaf blades >6 cm long, not forming a spiny ball; **plant** not spreading by looping stolons; **spikelets** >2.5 mm long	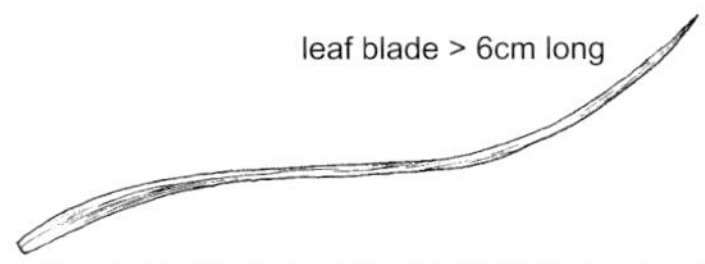	***Urochondra setulosa*** 383
26	**Spikelets** 4 to 9-flowered	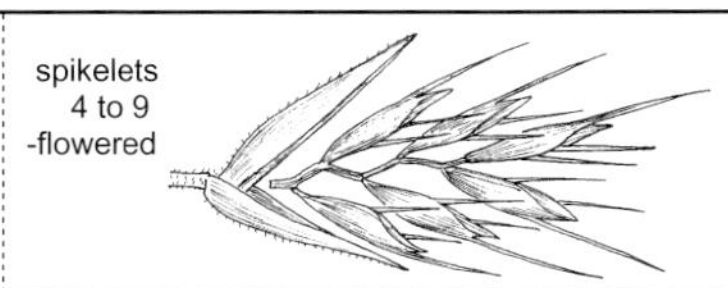	***Schmidtia pappophoroides*** 377
	Spikelets 1 or 2-flowered		27
27	**Awns** simple, glabrous or hairy	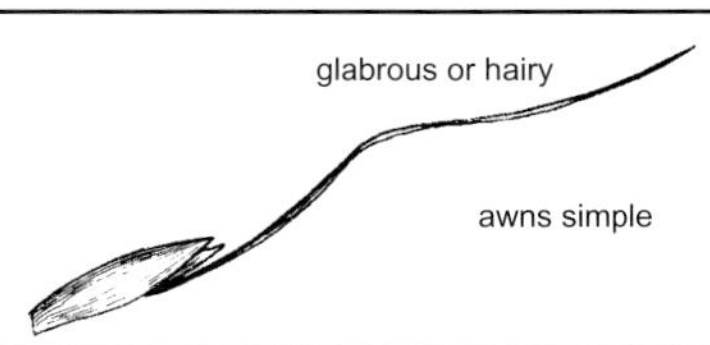	28
	Awns trifid, with 2 short and 1 long branch, long branch hairy or plumose towards tip	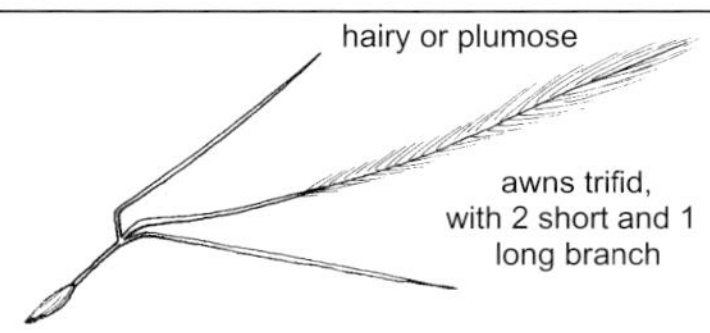	***Stipagrostis sokotrana*** 380
	Awns trifid, with equal branches, glabrous	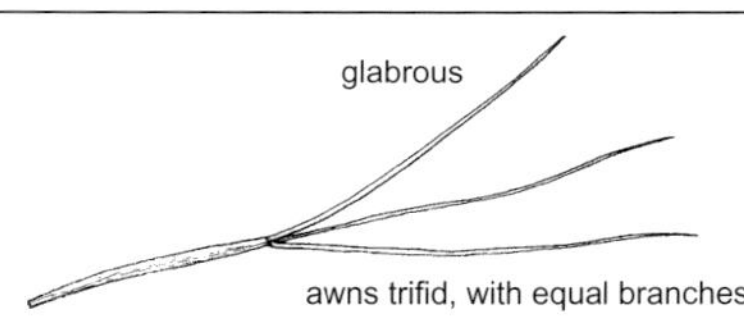	***Aristida*** (3 spp.)
28	**Inflorescence** not spatheate	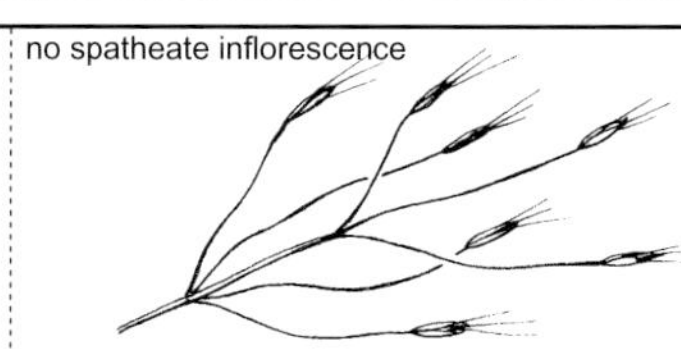	***Chrysopogon*** (3 spp.)
	Inflorescence spatheate	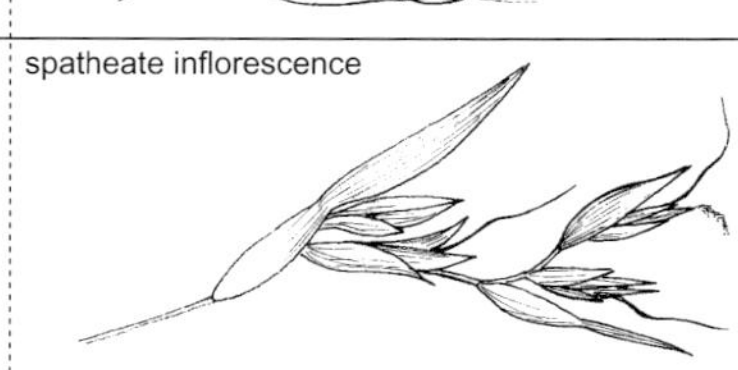	29

29	**Awns** glabrous, fine, <1 cm; **spikelets** in groups of 3, one sessile and 2 on short, flat pedicels; **inflorescence** an open panicle of 'racemes' each enclosed in a boat-shaped bract	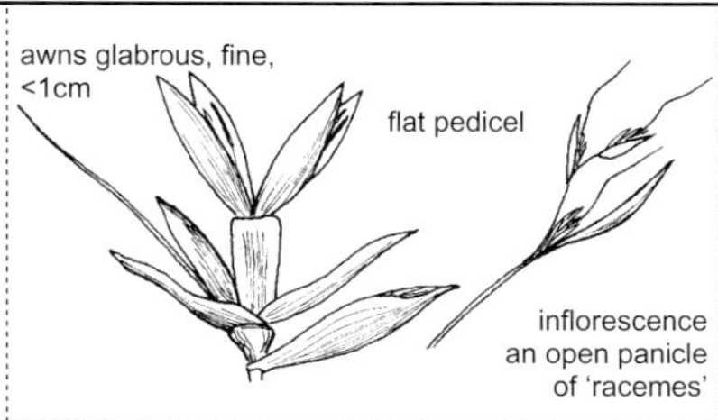	***Apluda mutica*** 347
	Awns hairy, >2 cm; **spikelets** in pairs, one sessile and one pedicelled; **inflorescence** of paired finger-like 'racemes' enclosed by a large bract		***Hyparrhenia hirta*** 368
	Awns hairy, >2cm; 'raceme' of 2 pairs of identical **spikelets** at the base and paired pedicelled and sessile spikelets above; **inflorescence** an open panicle of 'racemes' each enclosed in a boat-shaped bract		***Themeda*** (2 spp.)

31	**Awns** very long, 4 to 5x spikelet length, one per floret	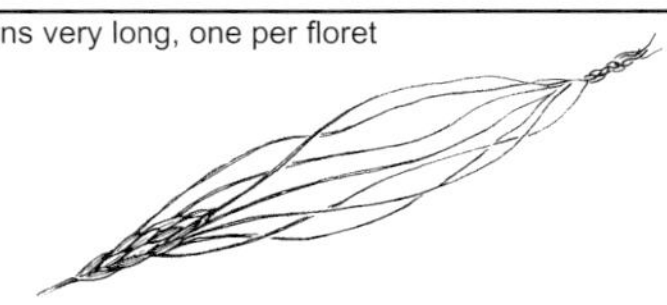	32
	Awns > 1/2x floret length but < 4x spikelet length, several per floret	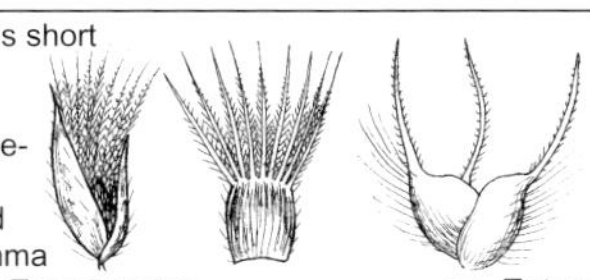	34

32	**Awns** hairy, 5--10 cm long, the tips of successive awns often twisted tightly together	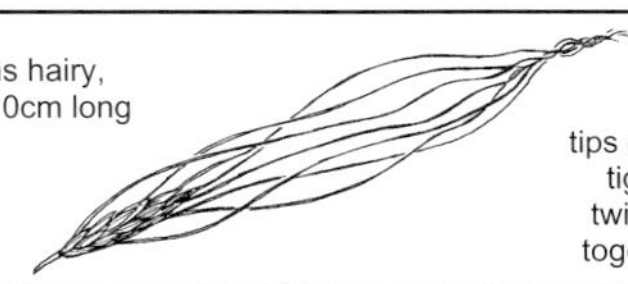	***Heteropogon contortus*** 367
	Awns glabrous or slightly hairy, <2.5 cm long, the tips of successive awns not twisted together	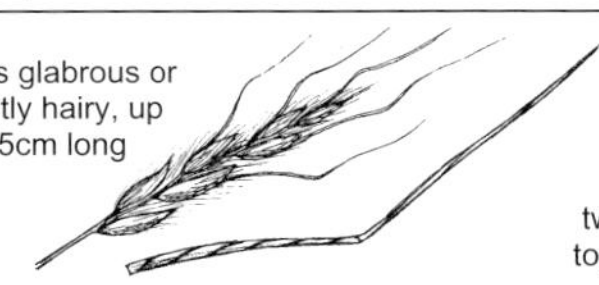	33

33	**Rhachis internodes** and **pedicels** stout and slightly inflated	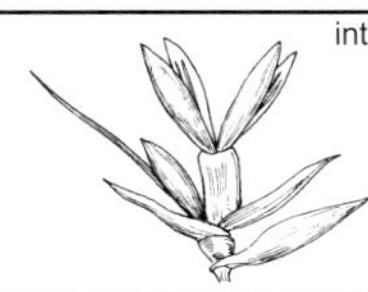	***Ischaemum* sp. A** Endemic 369
	Rhachis internodes and **pedicels** slender	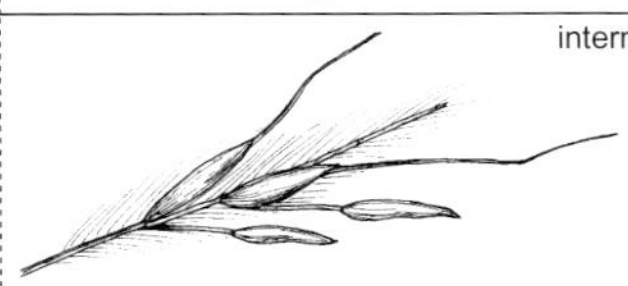	***Dichanthium*** (2 spp.)

34	**Inflorescence** flattened; **spikelets** clearly distichous and alternating	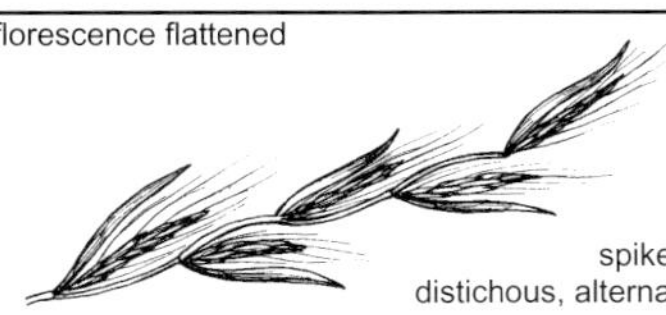	***Lolium*** (2 spp.)
	Inflorescence not flattened; **spikelets** not clearly distichous or alternating	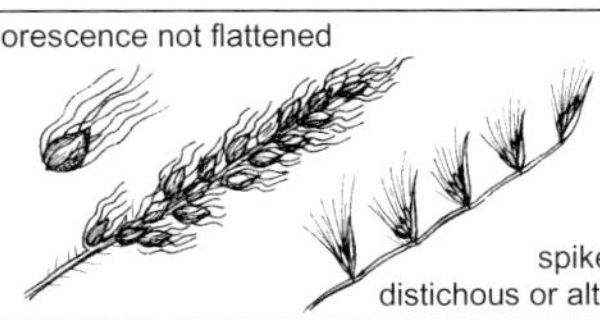	35

35	**Spikelets** subtended by a whorl of bristles or hairs	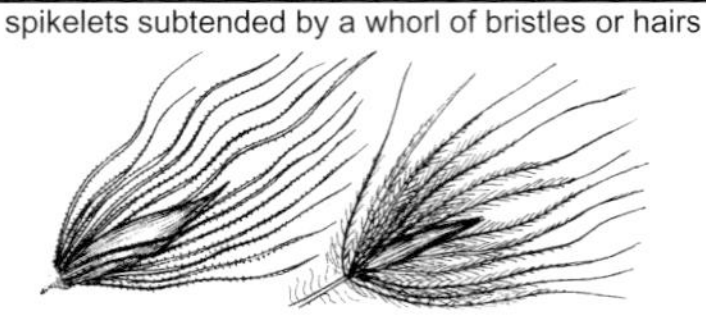	37
	Spikelets may be bristly-hairy but bristles not subtending spikelet	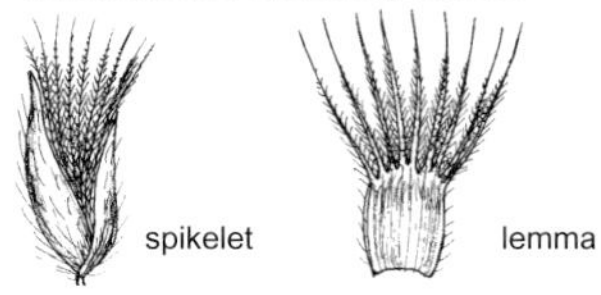	36

36	**Lemmas** bearing 1 or 2 awns; **leaves** usually flat	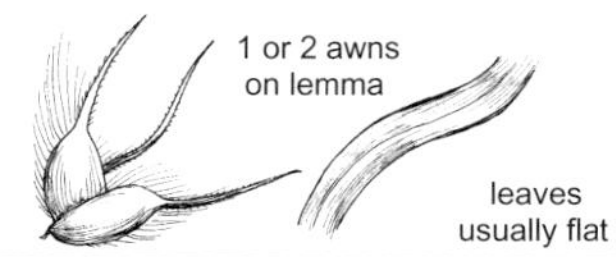	***Tetrapogon*** (2 spp.)
	Lemmas bearing 9 awns; **leaves** usually inrolled	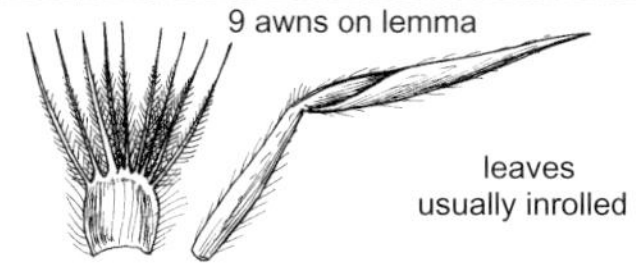	***Enneapogon*** (2 spp.)

37	**Spikelets** falling to leave subtending bristles or hairs on inflorescence axis	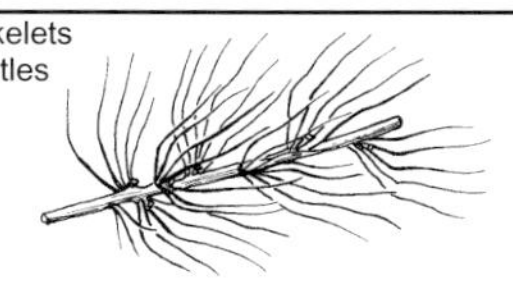	38
	Spikelets falling complete with subtending bristles or hairs	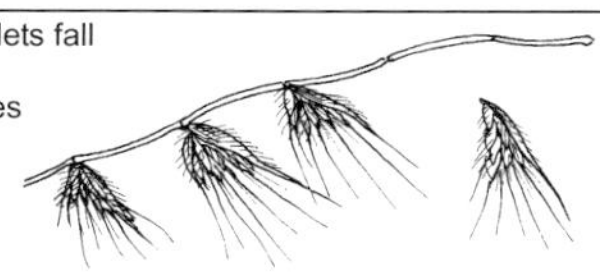	39

38	**Annuals**; **spikelets** subtended by thick, dark bristle-like hairs; **spikelets** not paired	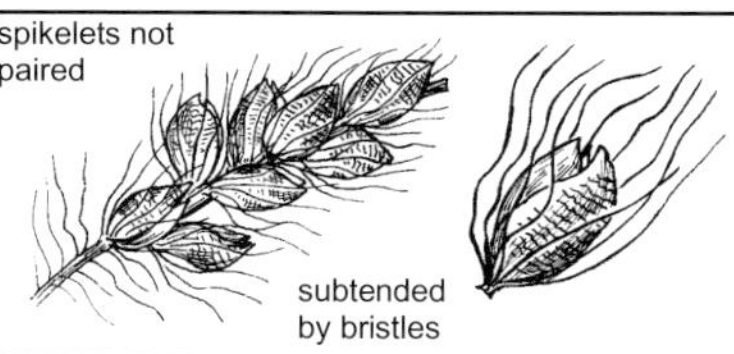	***Setaria*** (3 spp.)
	Shrubby, much-branched **perennial**; **spikelets** subtended by fine, white hairs; **spikelets** paired, one sessile and one pedicelled	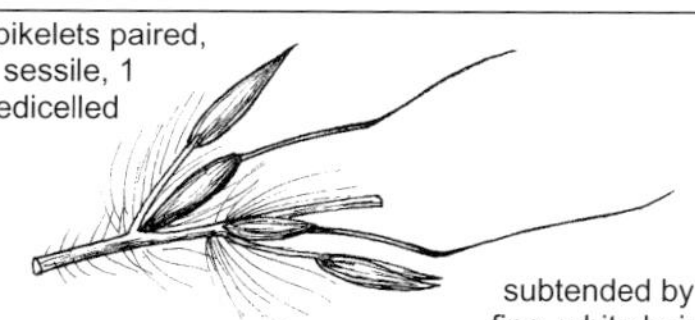	***Andropogon bentii*** Endemic **236**, 347

39	**Spikelets** secund (ie. borne on one side of the 'raceme')	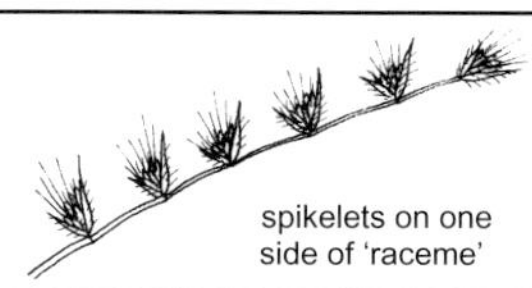	***Melanocenchris jacquemontii*** 371
	Spikelets not clearly secund (ie. not borne on one side of the 'raceme')	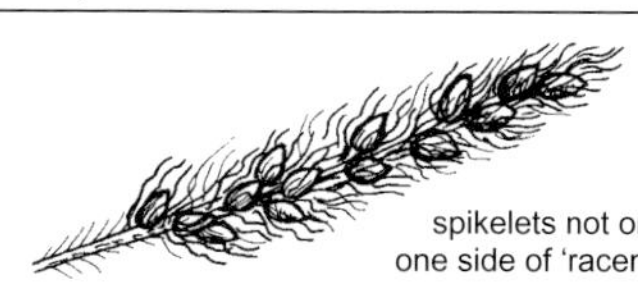	40

40	**Bristles** subtending spikelets united above base (not clear in *C. ciliaris*)	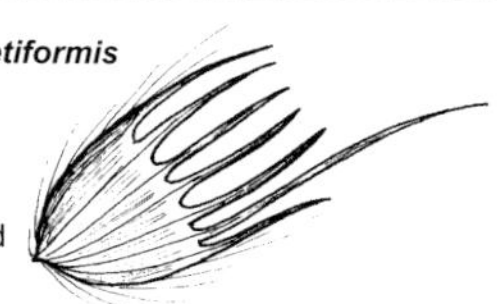	***Cenchrus*** (3 spp.)
	Bristles subtending spikelets free to the base	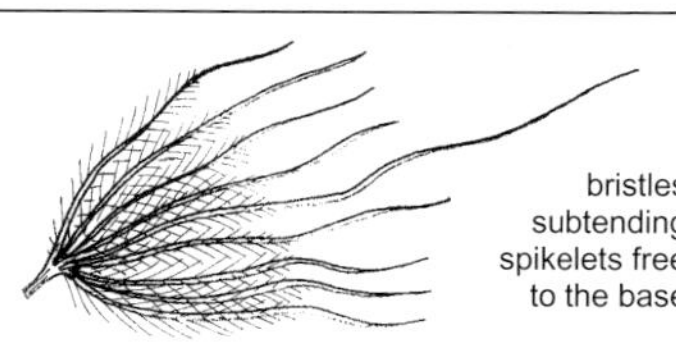	***Pennisetum*** (2 spp.)

41	**Glumes** not spiny or tuberculate	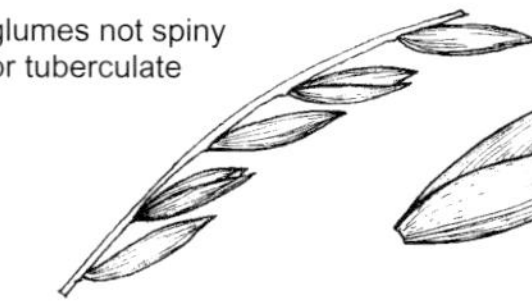	42
	Glumes spiny or tuberculate		45
42	**Spikelets** 1 or 2-flowered	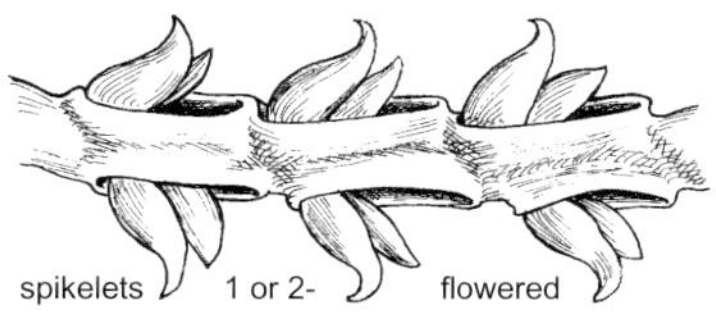	43
	Spikelets 3 to many-flowered		44
43	**'Raceme'** curved, one-sided.	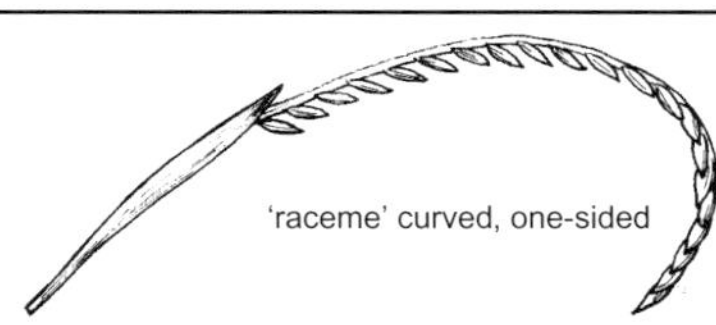	***Microchloa kunthii*** 372
	'Raceme' straight, spikelets on both sides of 'raceme'	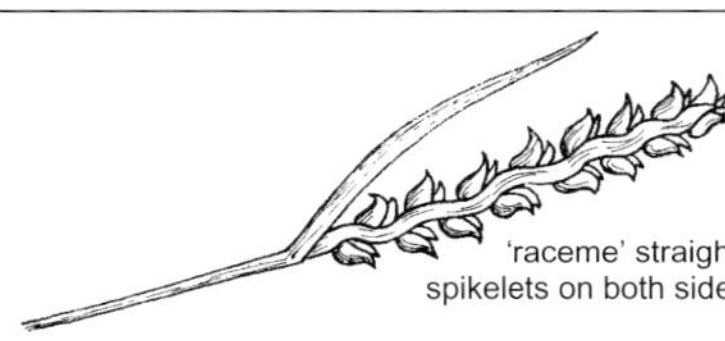	***Lepturus*** (3 spp.)
44	**Spikelets** large, 1--3 cm long, glabrous		***Halopyrum mucronatum*** 367
	Spikelets small, <0.5 cm long, hairy	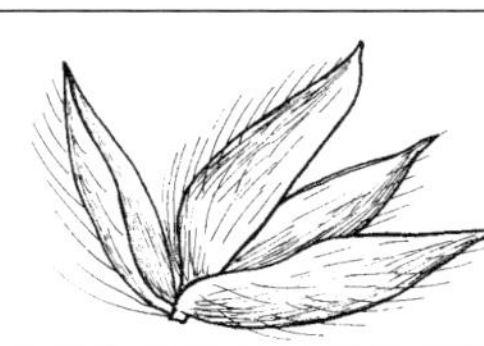	***Aeluropus lagopodioides*** 346
45	Tip of lower **glume** prominently recurved; **spikelets** on distinctive flattened, green peduncles	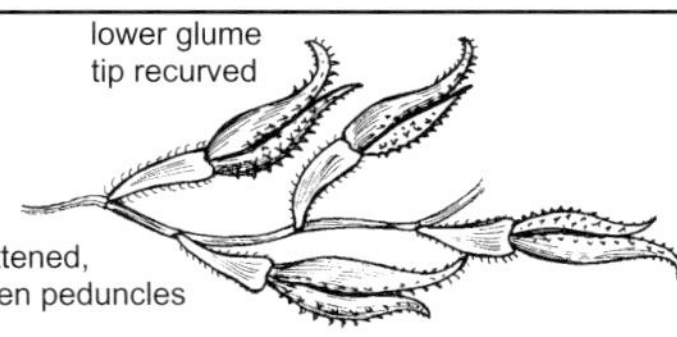	***Leptothrium senegalense*** 369
	Upper **glume** densely covered in hooked spines; **spikelets** in pairs	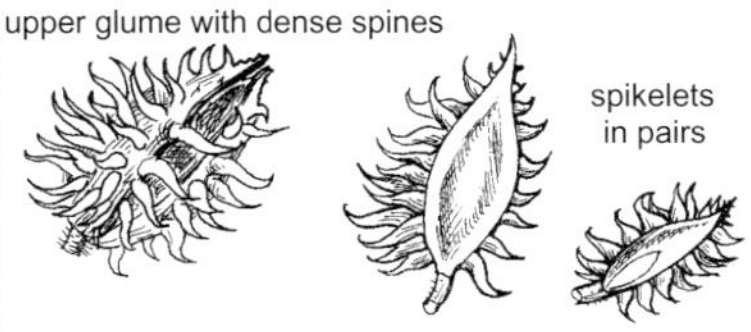	***Tragus berteronianus*** 382

46

Spikelets 1 or 2-flowered

49

Spikelets 3 to many-flowered

47

47

Perennial; **leaves** inrolled

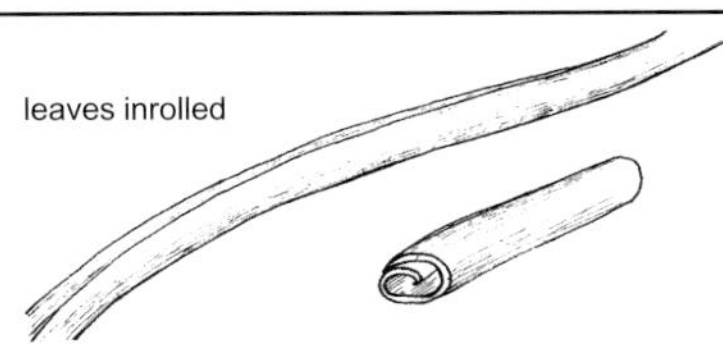

Desmostachya bipinnata
358

Annual; **leaves** flat

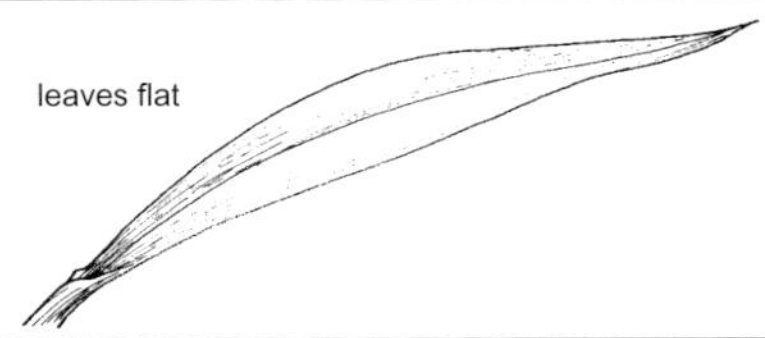

48

48

Leaves linear; **glumes** longer than lemmas; **lemmas** membranous, sparsely to densely hairy

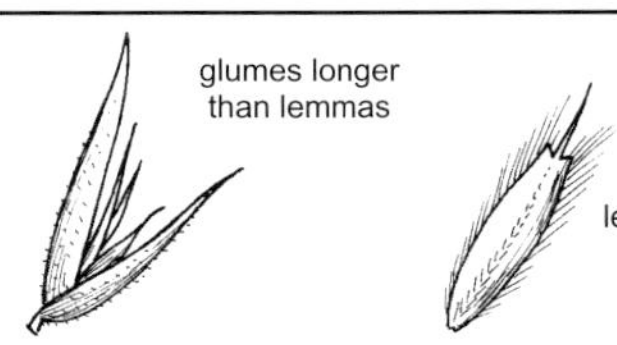

Trichoneura mollis
383

Leaves elliptic-ovate; **glumes** ± shorter than lemmas; **lemmas** firm, glabrous

Acrachne racemosa
346

49

Spikelets spaced out along 'raceme'

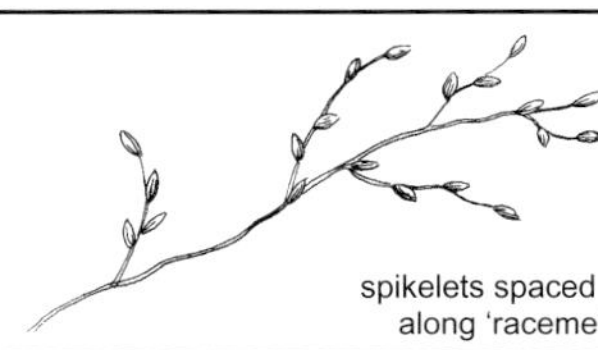

50

Spikelets densely packed on 'raceme'

51

50

Upper lemma leathery, hardened, with narrow inrolled margins clasping the edge of the palea

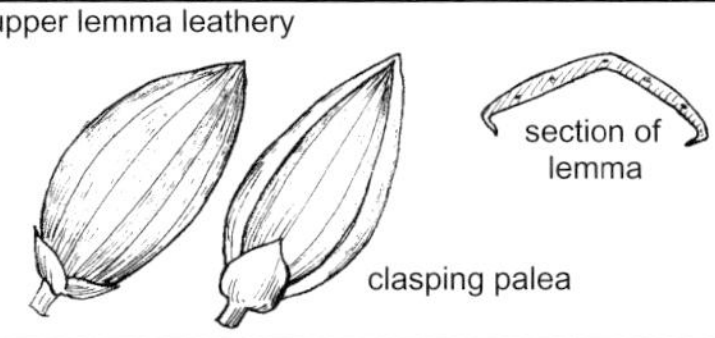

Brachiaria
(5 spp. excluding *B. lata*)

Upper lemma papery, not hardened, with thin, flat margins covering most of the palea

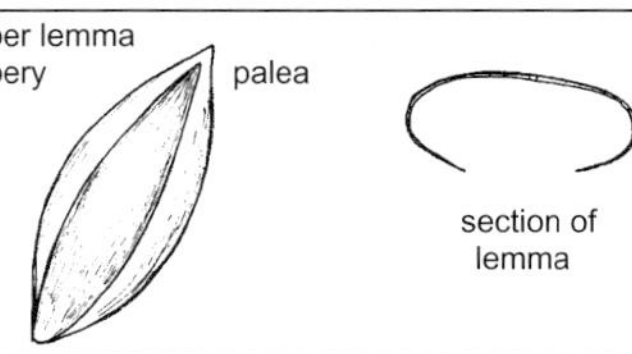

Digitaria
(3 spp.)

51

Spikelets globose and wrinkled

Hackelochloa granularis 366

Spikelets not globose and wrinkled

52

52

'Raceme' not ending in a spikelet

Paspalidium geminatum 375

'Raceme' ending in a spikelet

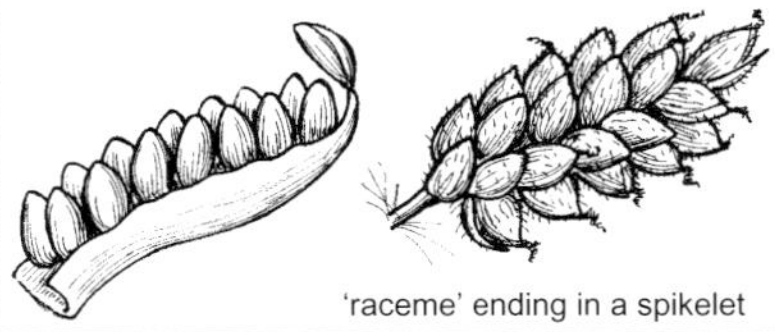

53

53

Lower glume very small (<1/3 spikelet length) or absent; **ligule** membranous

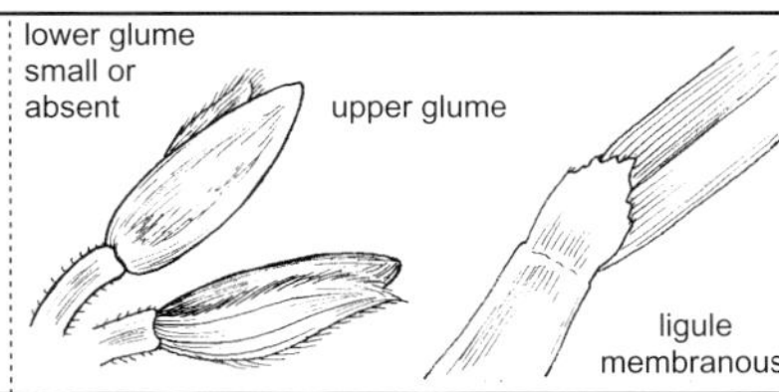

54

Lower glume 1/3 to 1/2 spikelet length; **ligule** absent or a fringed membrane

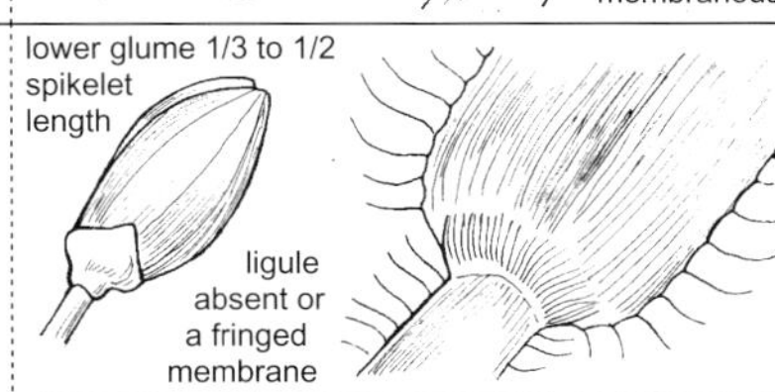

55

54

Spikelets sessile in 2 rows

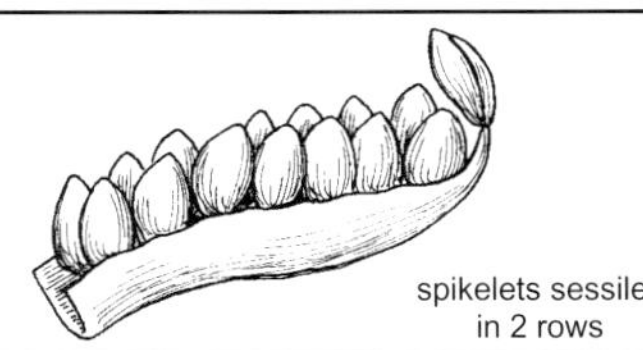

Paspalum (2 spp.)

Spikelets sessile or shortly pedicellate not in distinct rows

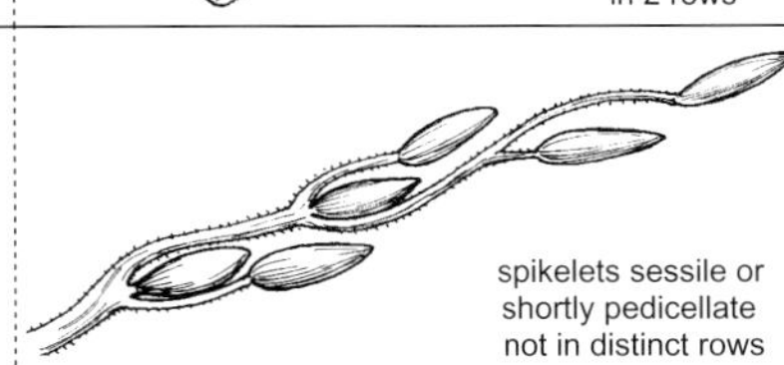

Digitaria (3 spp.)

55

Spikelets in 4 untidy rows; **ligule** absent

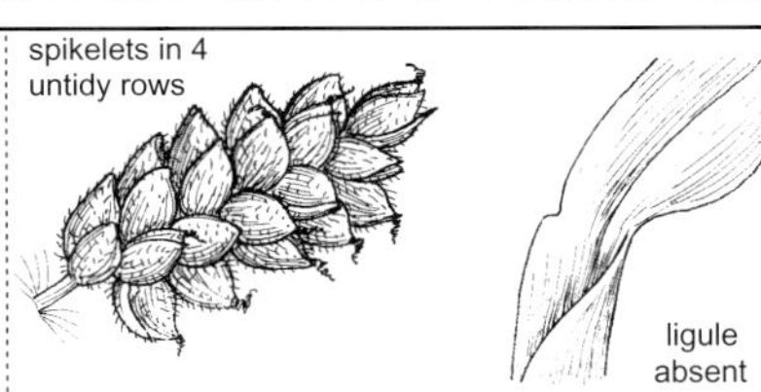

Echinochloa colona 361

Spikelets in 2 rows; **ligule** a fringed membrane

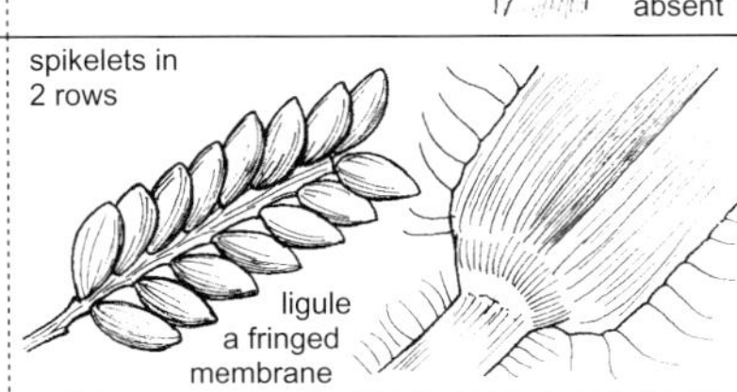

Brachiaria lata 351

56	**Spikelets** 1-flowered; **lemma** with a single awn	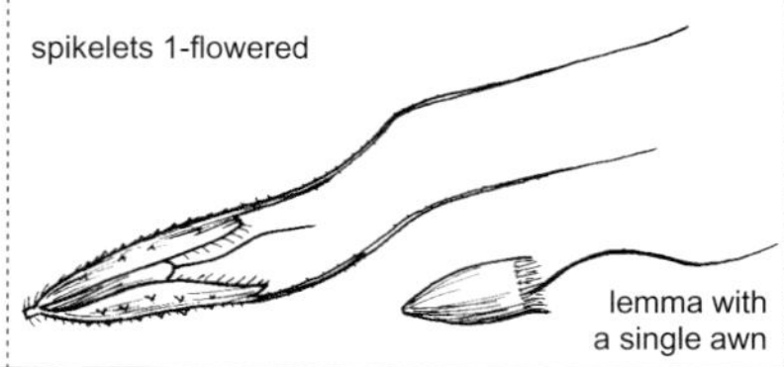	***Polypogon monspeliensis*** 377
	Spikelets 2-flowered; **lemmas** and **glumes** enveloped in long white silky hairs	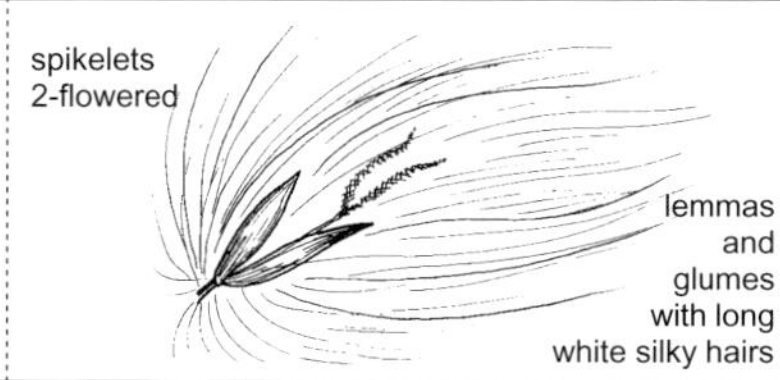	***Imperata cylindrica*** 368
	Spikelets 3 or 4-flowered; **lemmas** with 9 awns	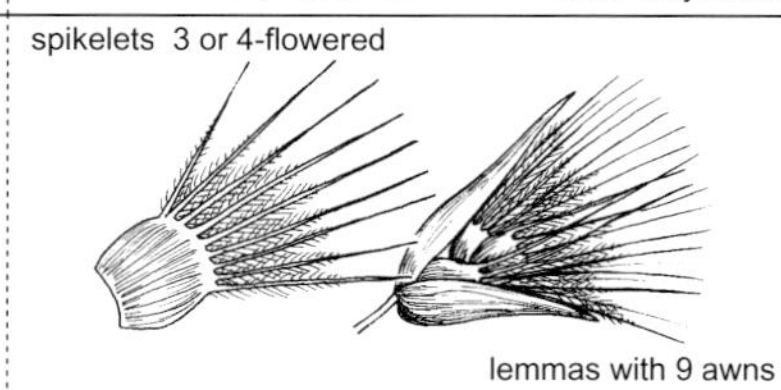	***Enneapogon*** (2 spp.)

Aristida

1	**Awn** strongly twisted, articulated at the base where it separates from the lemma; **lower glume** longer than the upper	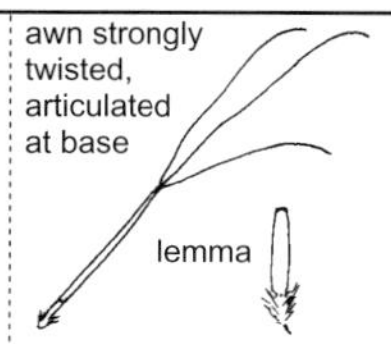	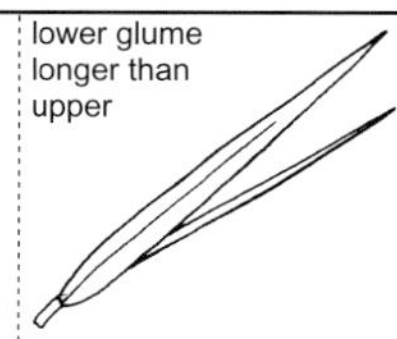	***A. funiculata*** 349
	Awn without column or articulation passing directly into the lemma; **lower glume** shorter than the upper	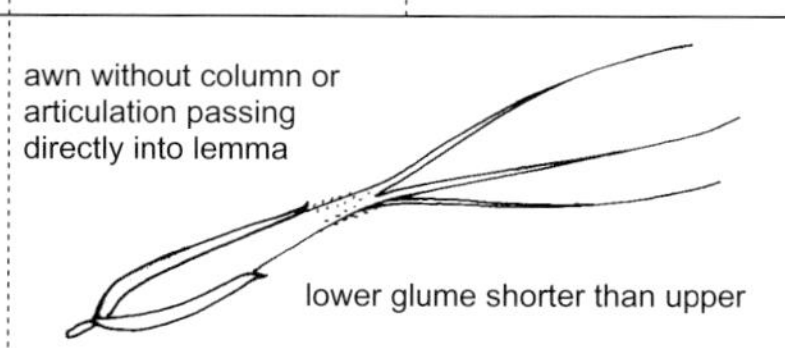		2
2	**Central awn** not reflexed; **panicle** linear <25 cm long, many branched and bearing many spikelets	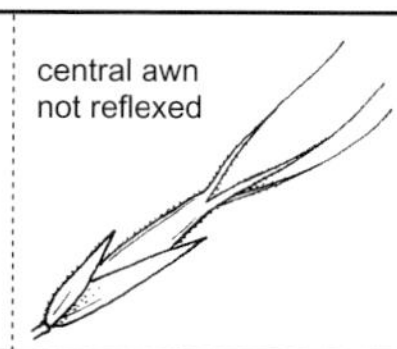		***A. adscensionis*** 348
	Central awn reflexed; **panicle** ovate, 5--7 cm long, sparingly branched and bearing few spikelets	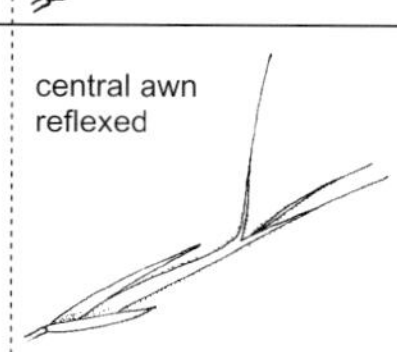	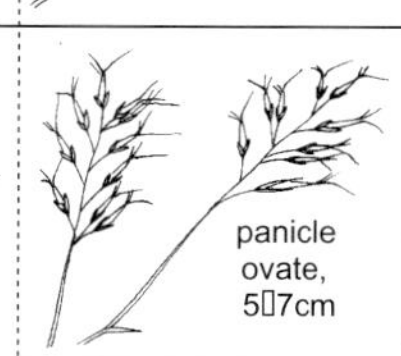	***A. anaclasta*** Endemic 348

Arthraxon

1	**Pedicelled spikelets** absent (arrangement not immediately clear)	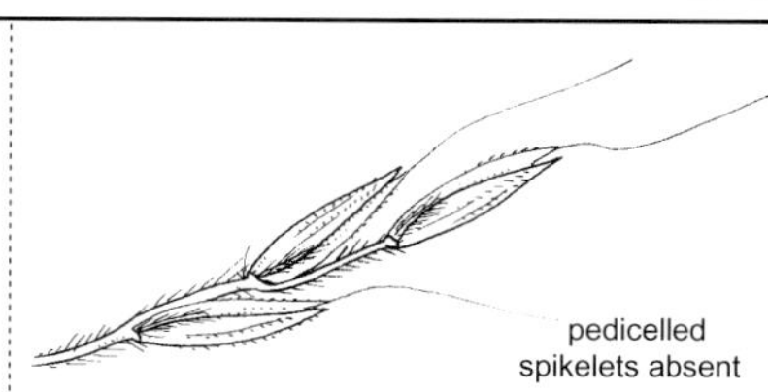	***A. micans*** 349
	Pedicelled spikelets well developed, at least towards the tip of the 'raceme'	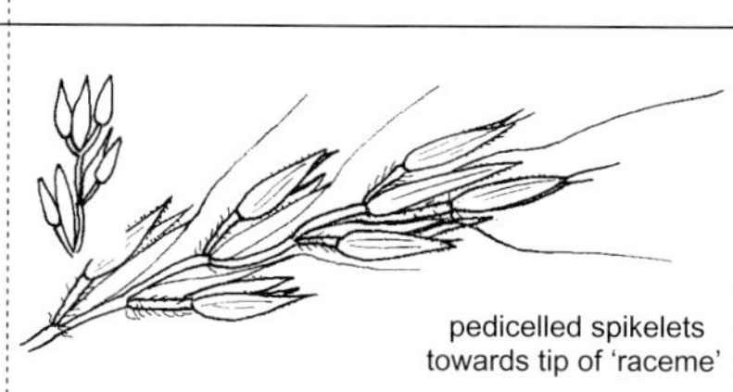	***A. lancifolius*** 349

Brachiaria

1

Upper lemma smooth, shiny; **upper floret** easily deciduous; **lower glume** tiny

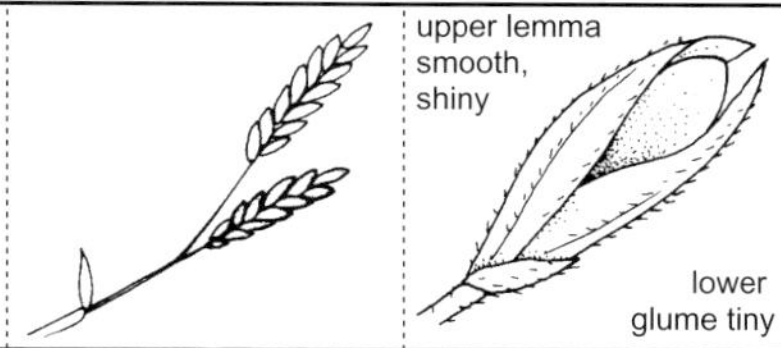

B. eruciformis 351

Upper lemma rugose; **upper floret** not easily deciduous; **lower glume** at least 1/4 spikelet length

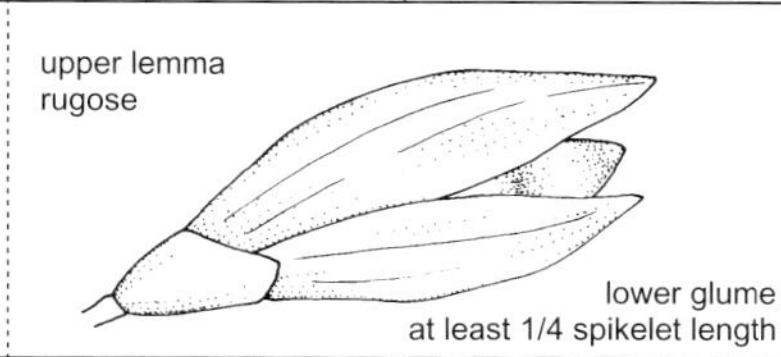

2

2

Leaves linear; **spikelets** glabrous, rhachis scaberulous, ± glabrous; **glumes** separated by a short internode

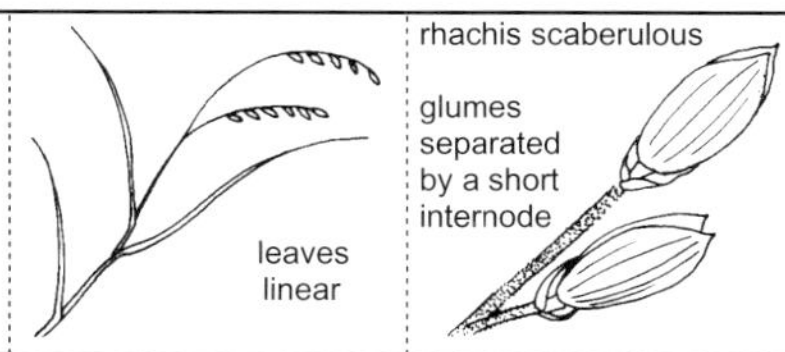

B. arida 350

Leaves lanceolate; **spikelets** glabrous or hairy, rhachis bristly or shortly hairy; **glumes** not separated by a short internode

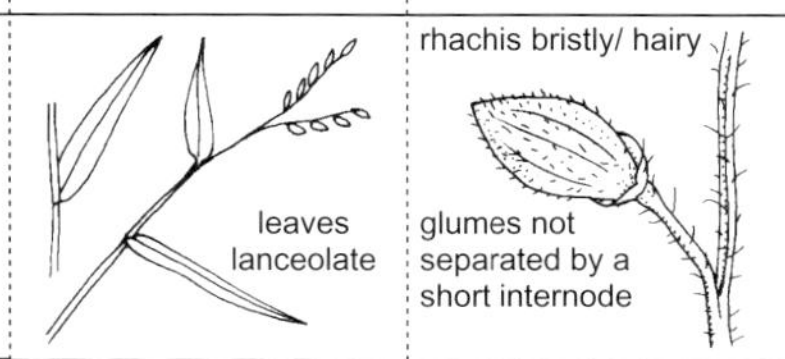

3

3

'Raceme' compound, some spikelets on short secondary branchlets

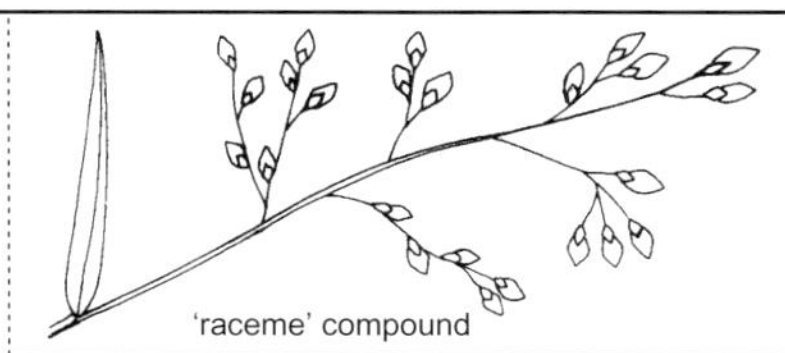

B. deflexa 351

'Raceme' simple, spikelets not (or very rarely) on short secondary branchlets and then only on the longest 'raceme'

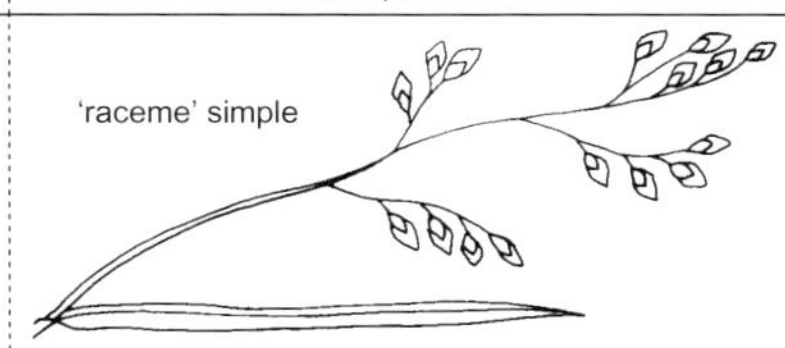

4

4

Spikelets cartilaginous, smooth and slightly shiny

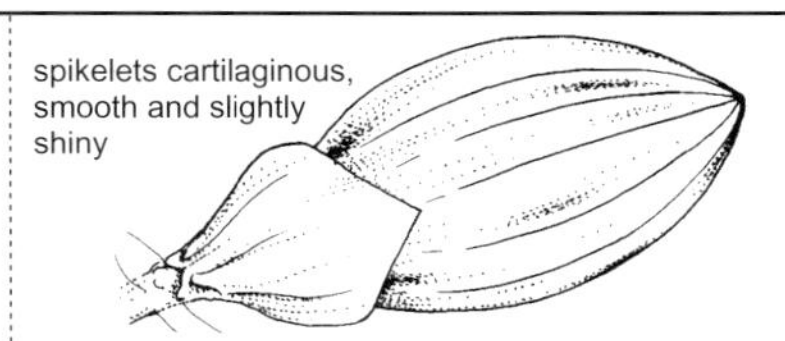

B. ovalis 351

Spikelets membranous, not shiny

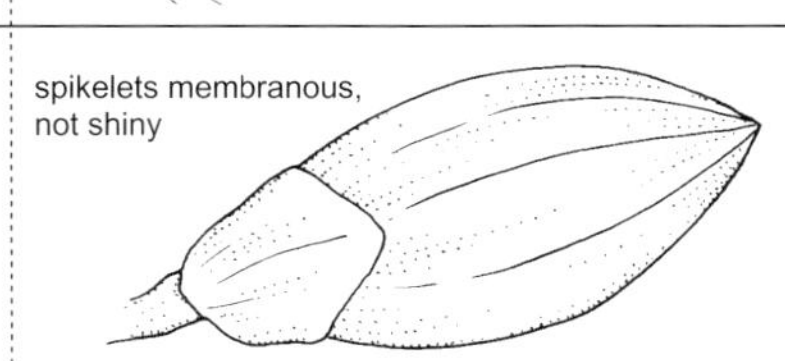

5

5

Spikelets not densely packed on 'raceme'; **spikelets** usually shortly hairy; **leaf blades** velvety, not fringed above the ligule

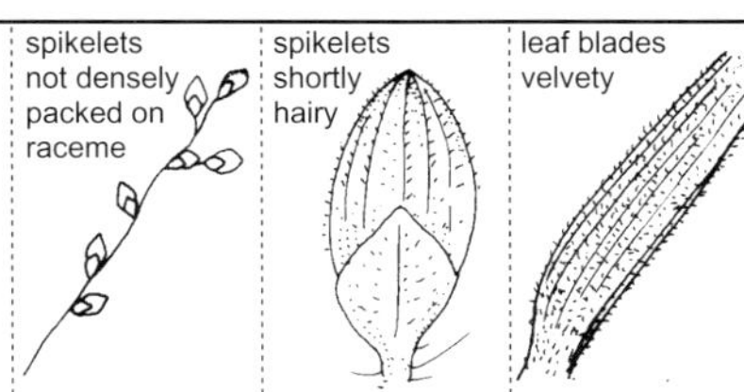

B. ramosa 352

Spikelets densely packed on 'raceme'; **spikelets** usually glabrous; **leaf blades** glabrous or bristly, fringed above the ligule

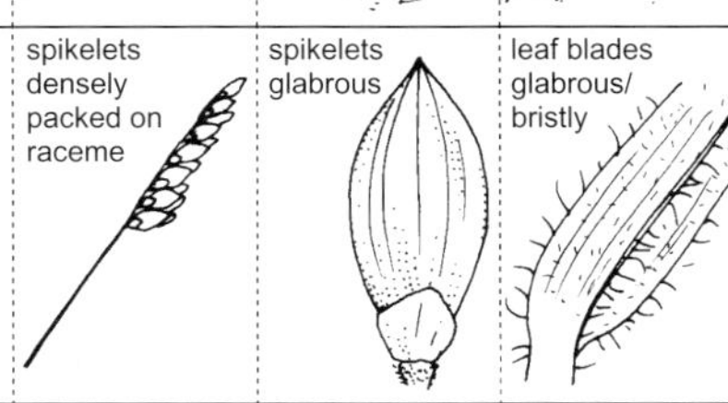

B. lata 351

Cenchrus

1	**Inner bristles** of involucres short, stiff and flattened like spines; **outer bristles**, when present, few, much shorter than inner bristles	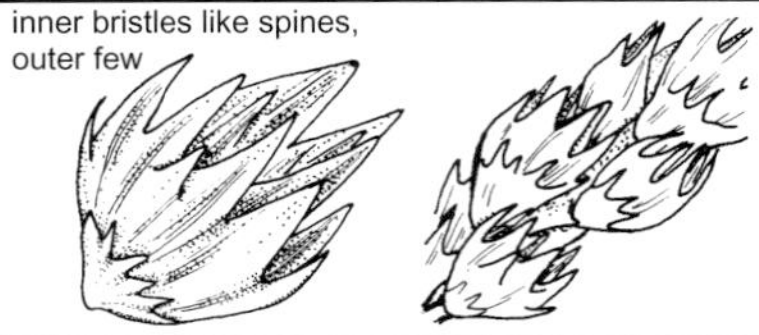	***C. setigerus*** 353
	Inner bristles of involucres long and flexuous, extended well beyond the spikelet; **outer bristles** numerous	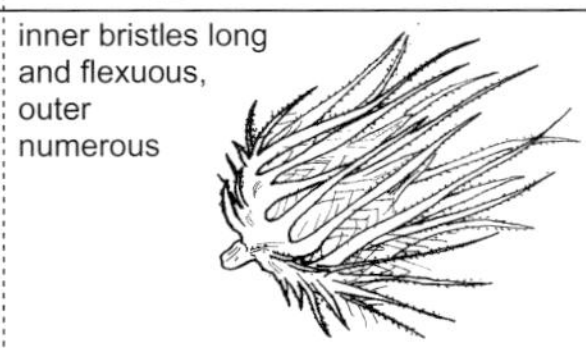	2

2	**Inner bristles** united only at the base to form a shallow disc only 0.5mm deep	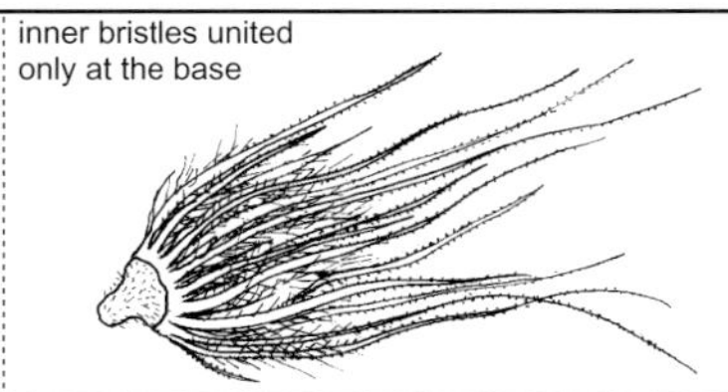	***C. ciliaris*** 352
	Inner bristles united to 2.5mm above the base	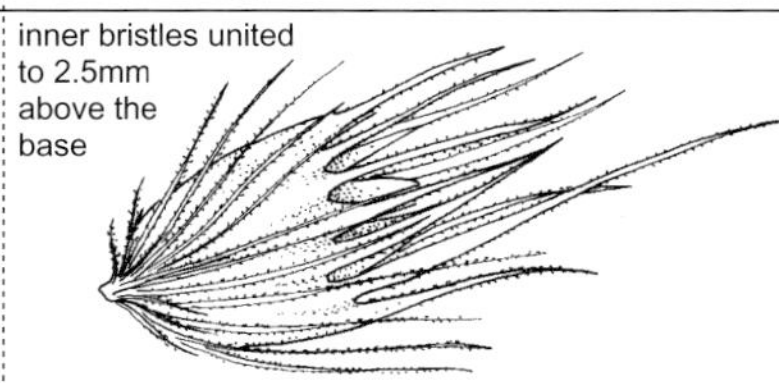	***C. pennisetiformis*** 353

Chloris

1	**Leaves** obtuse at apex	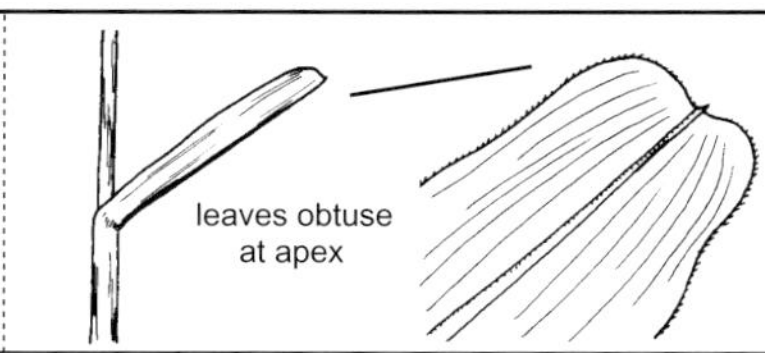	***C. pycnothrix*** 353
	Leaves acute or tapering at apex	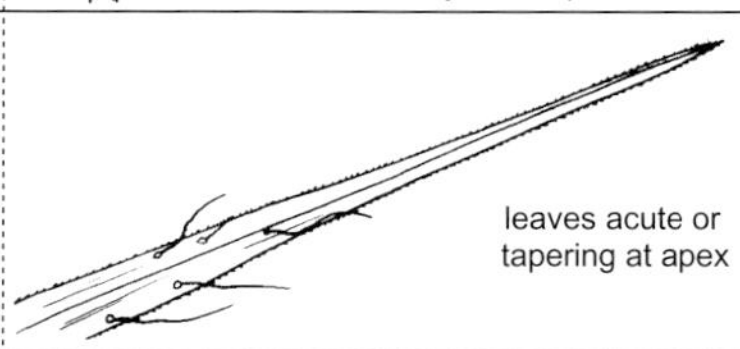	***C. virgata*** 354

Chrysopogon

1	Some **awns** plumose	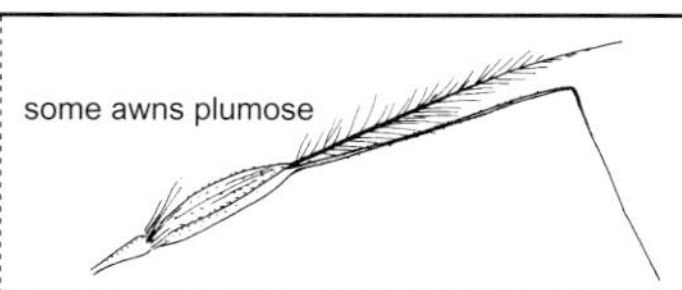	***C. plumulosus*** 354
	Awns hairy but not plumose	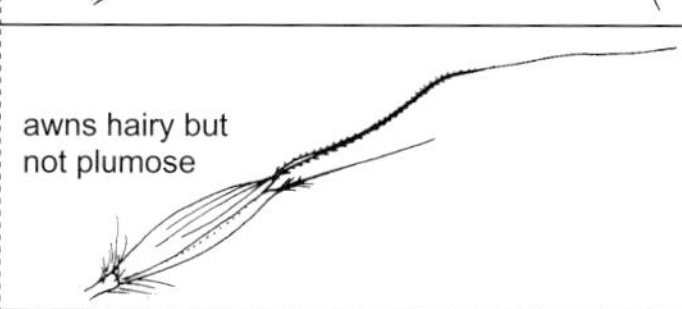	2

2	**Leaves** glabrous or sparsely hairy	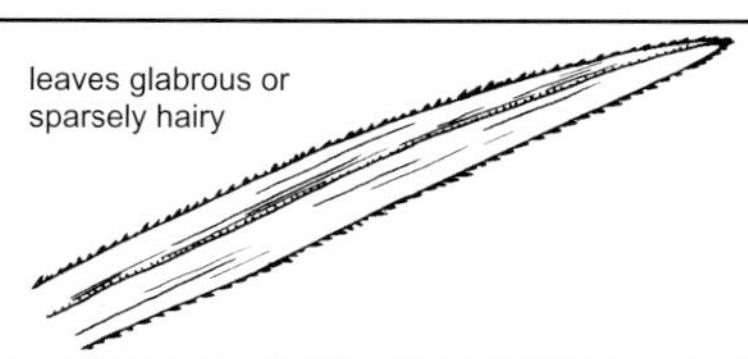	***C. serrulatus*** 355
	Leaves velvety-hairy	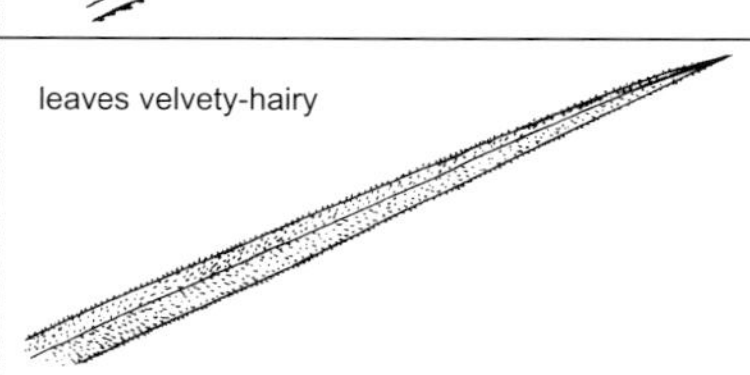	***C. aucheri*** 354

Dactyloctenium

1	**Leaves** short, rigid, distichous and pungent	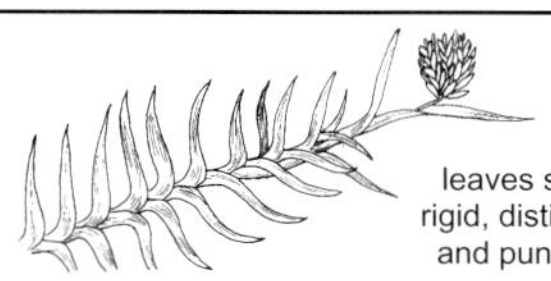	2
	Leaves flat, soft	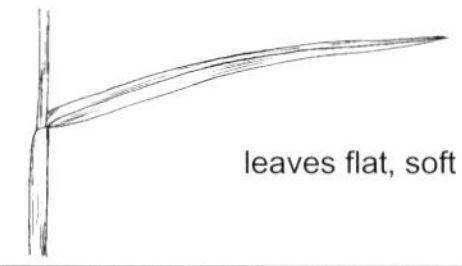	3
2	**Plant** with short woody stems arising from scaly, grey stolons; **leaf sheaths** glabrous		***D. robecchii*** 358
	Plant with short woody stems arising from smooth, pale stolons; **leaf sheaths** sparsely covered with long stiff hairs	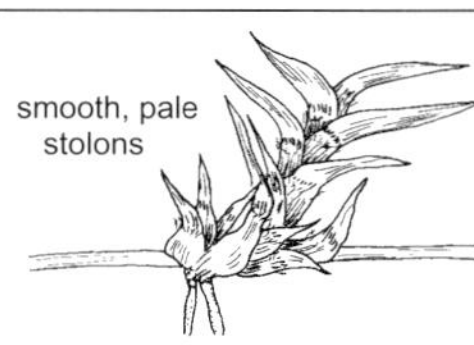 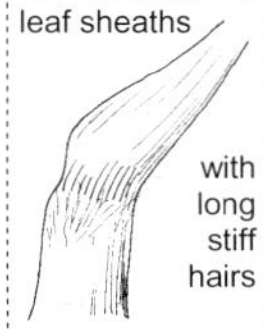	***D. hackelii*** Endemic 357
3	**Inflorescence** relatively open, 'racemes' spreading,1--5 cm long; **lemmas** not conspicuously pointed; **grain** transversely rugose	 	***D. aegyptium*** 357
	Inflorescence compact, 'racemes' in dense head, 1--2 cm long; **lemmas** conspicuously pointed; **grain** granular	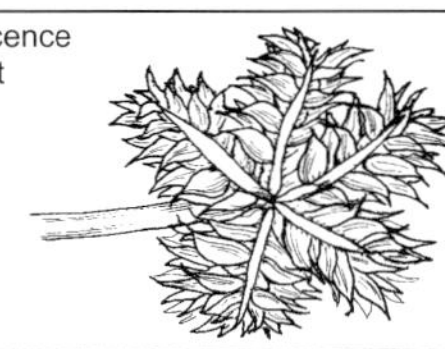	***D. aristatum*** 357

Dicanthium

1	**Inflorescence** a cluster of 2 to 9 finger-like 'racemes' at the tip of the stem; **lower glume** of sessile spikelet pilose, purple, rounded at the apex, not pitted		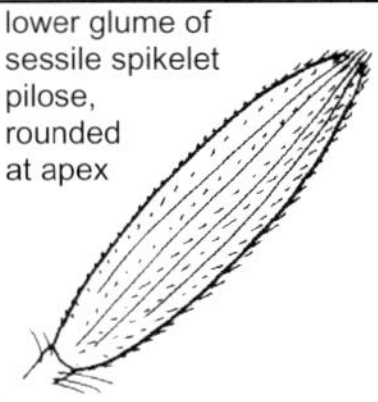	***D. annulatum*** 359
	Inflorescence a solitary 'raceme'; **lower glume** of sessile spikelet ± glabrous, with an acute, orange tip and usually with a single pit	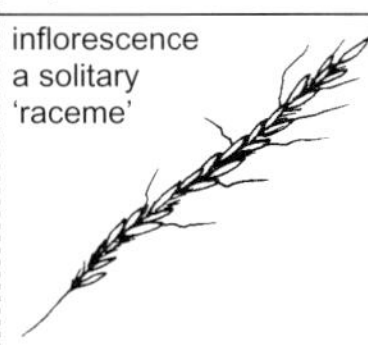	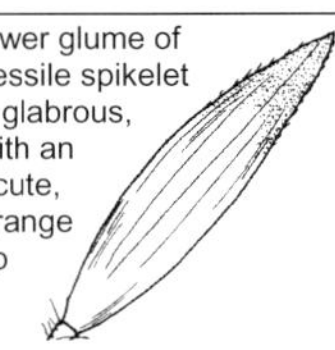	***D. foveolatum*** 359

Digitaria

1	**Spikelets** conspicuously silky-hairy, some hairs exceeding spikelet	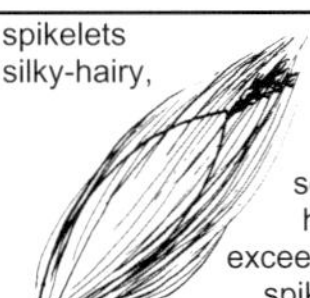		***D. rivae*** 360
	Spikelets glabrous or shortly hairy, no hairs exceeding spikelet	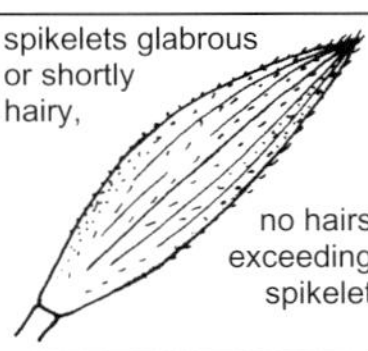	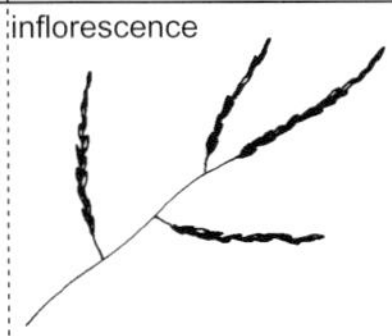	2

2

Spikelets 1.5--2.2 mm long, loosely imbricate (overlapping by ± half their length); **leaves** hairy

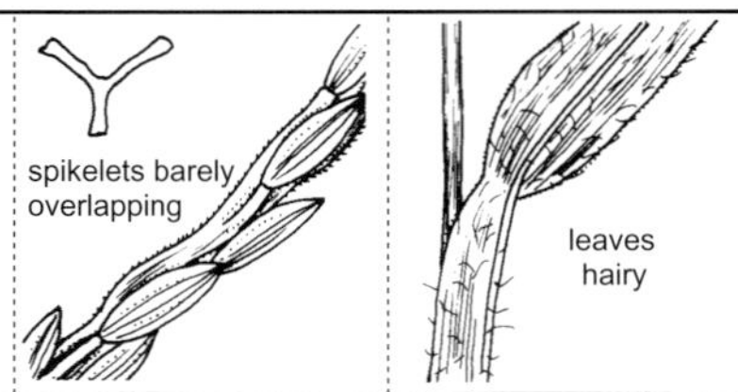

D. velutina 360

Spikelets 2.5--3.5 mm long, closely overlapping (by more than half their length); **leaves** glabrous or sparsely hairy

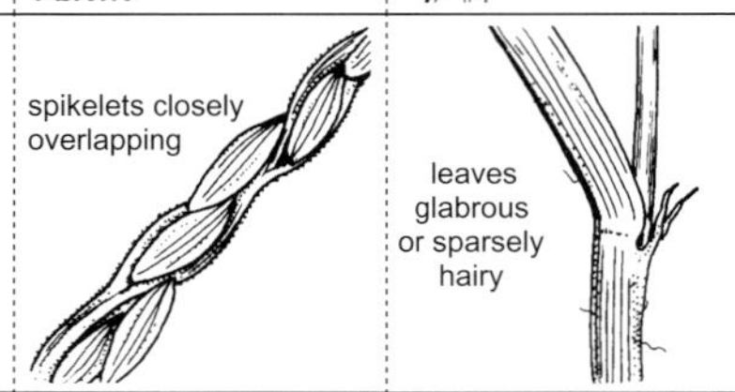

D. ciliaris 360

Enneapogon

1

Leaves often diverging at a wide angle from the stem; **spikelets** 3 or 4-flowered, third floret sterile but well developed, 3--5(--9) mm long, fourth floret, when present, vestigial

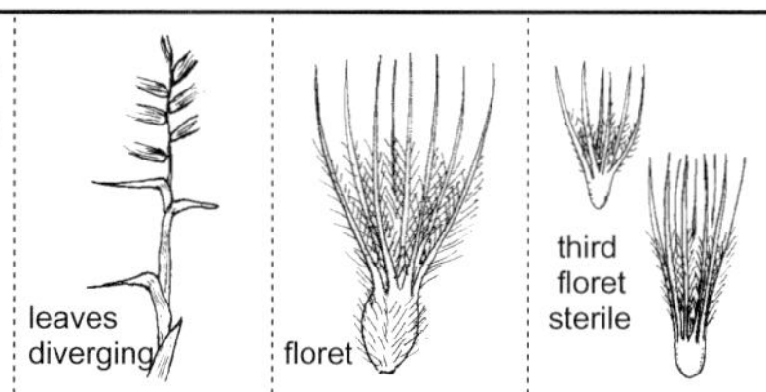

E. persicus 363

Leaves not diverging at a wide angle from the stem; **spikelets** 3-flowered, third floret reduced to a small tuft of awns, 0.5--2 mm long

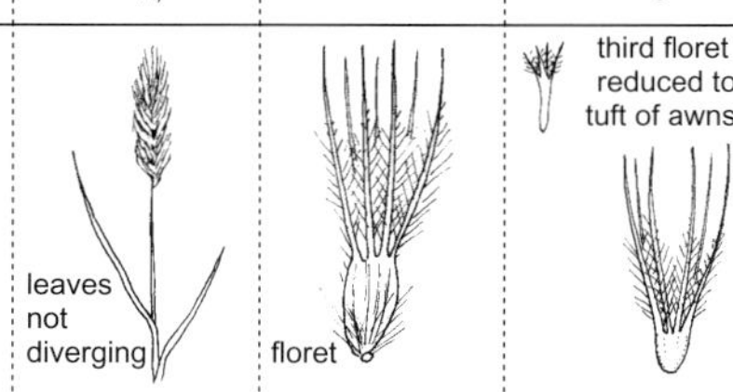

E. desvauxii 363

Eragrostis

1

Spikelets with a conspicuously fringed palea

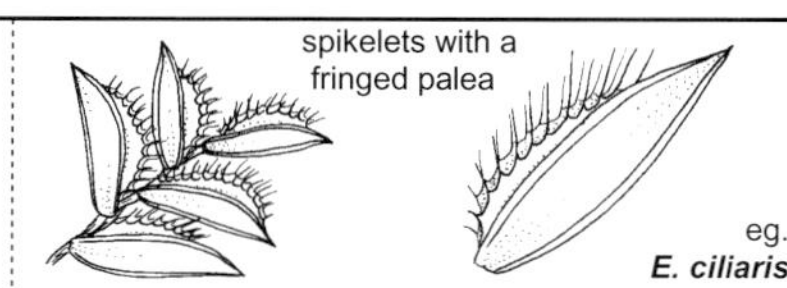

2

Spikelets glabrous

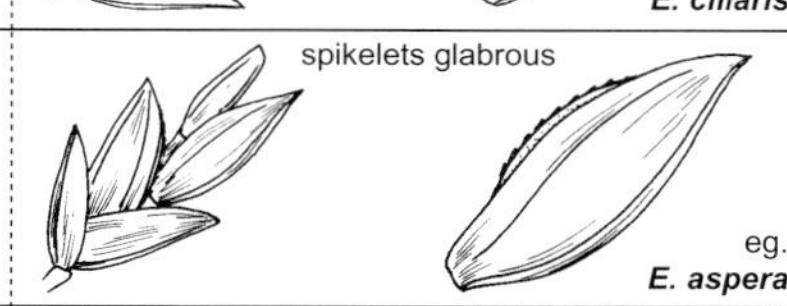

4

2

Inflorescence a contracted panicle (either spicate or a sub-globose head), **spikelets** densely packed, primary branches very short

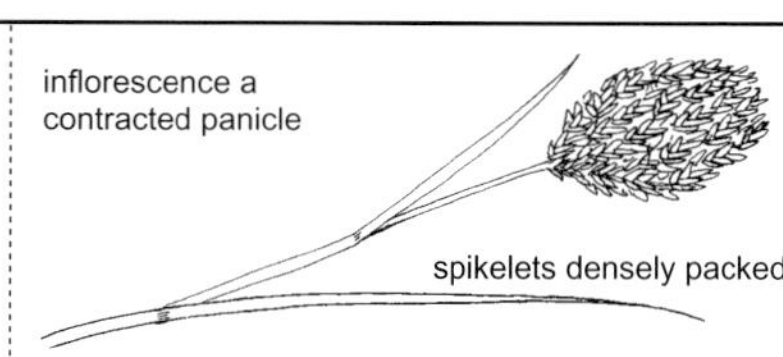

E. ciliaris 365

Inflorescence an open panicle, **spikelets** not densely packed, primary branches relatively long

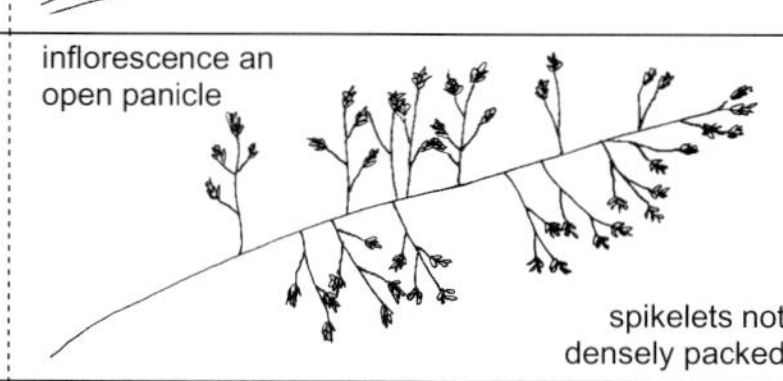

3

3

Stem and **pedicels** sticky (sand grains and debris often stuck to plant)

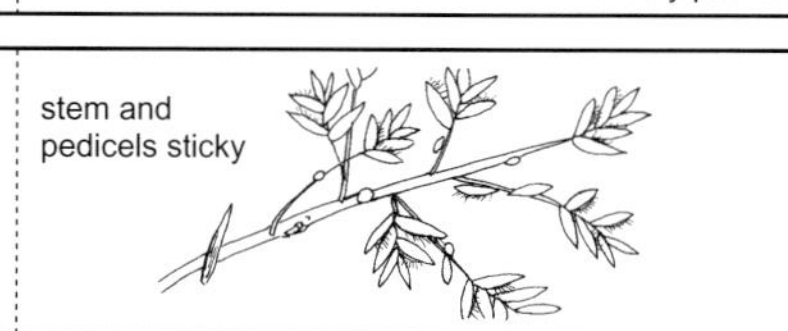

E. viscosa 366

Stem and **pedicels** not sticky

E. lepida 365

4	**Spikelets** disarticulating from the top, rhachis fragile between the florets; no persistent **paleas**		***E. aspera*** 364
	Spikelets disarticulating from the base; **paleas** persistent on the rhachis	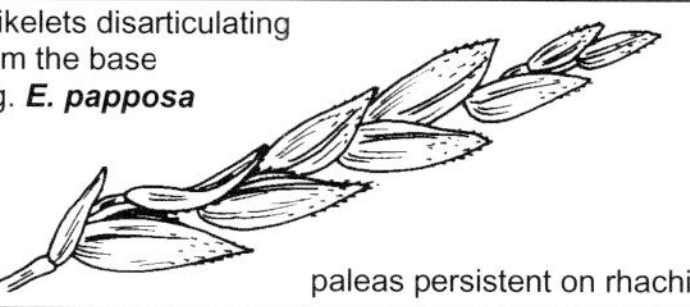	5

5	**Glumes** markedly unequal, the lower shorter than the upper		***E. papposa*** 365
	Glumes subequal or slightly unequal	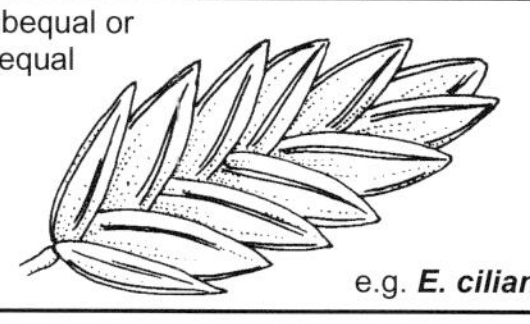	6

6	**Spikelets** narrowly oblong to ovate, 2.5--4 mm wide; **lemmas** 2--3.2 mm long		***E. cilianensis*** 364
	Spikelets linear, 1--2.2(--2.5) mm wide; **lemmas** 1.2--2.3 mm long	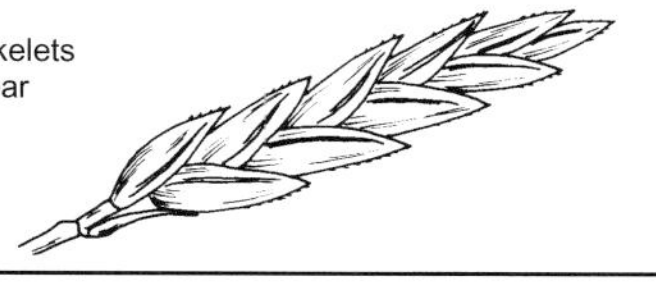	7

7	Wart-like **glands** on leaf margins	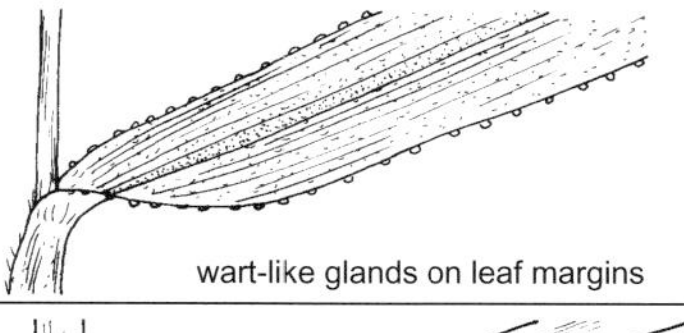	***E. minor*** 365
	No wart-like **glands** on leaf margins	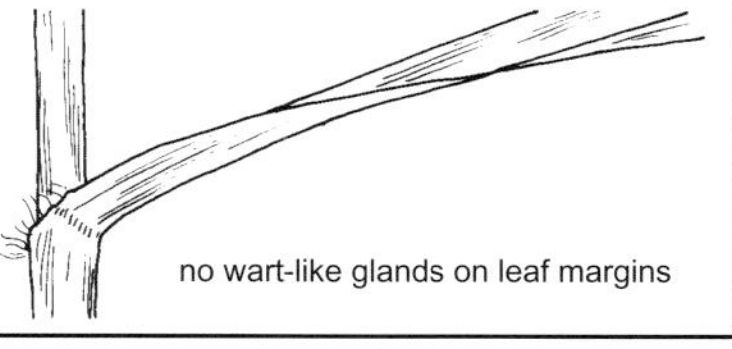	***E. barrelieri*** 364

Lepturus

1	**Upper glume** lanceolate, 5--5.5 mm long, convex, scarcely embraced by a narrow wing on each margin of the rhachis joint; **inflorescence** green	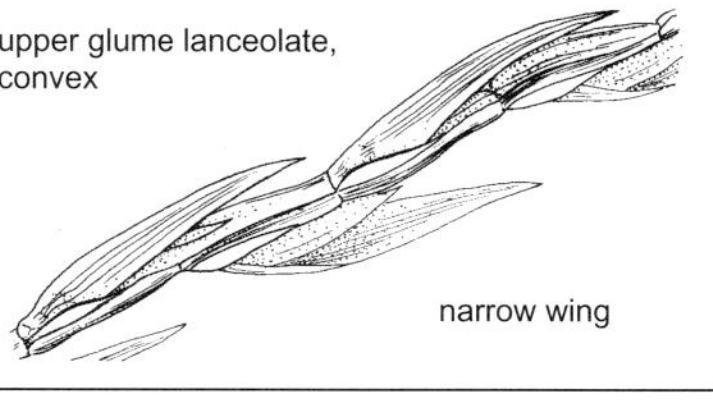	***L. tenuis*** Endemic 370
	Upper glume oblong and truncate, 1.5--2 mm long, concave on the back, embraced by a triangular wing on each margin of the rhachis-joint; **inflorescence** green	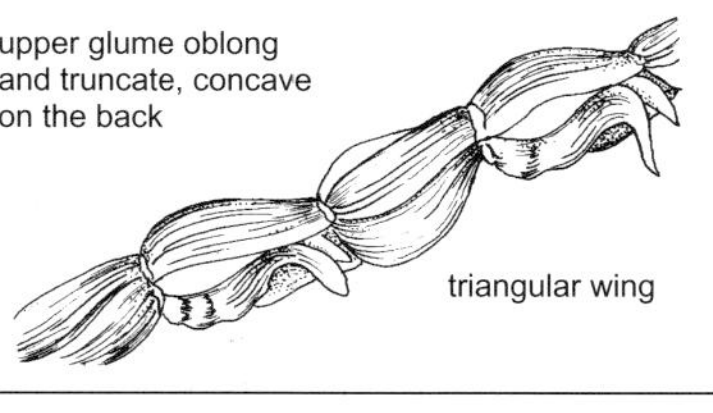	***L. pulchellus*** Endemic 370
	Upper glume broadly ovate, 1.6--1.8 mm long, "margins lie flush with those of the rhachis joint"; **inflorescence** grey-purple	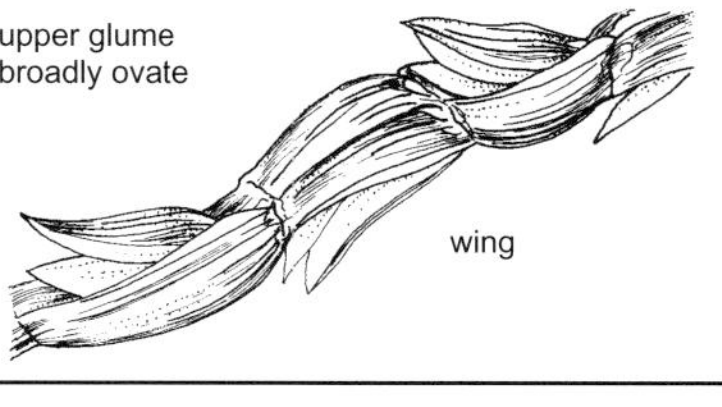	***L. calcareus*** Endemic 369

Lolium

1	**Spikelets** <30 mm long; **glume** toughened and usually as long as or longer than the spikelet; **lemma** turgid at maturity; **plant** sometimes tufted but often a single stem		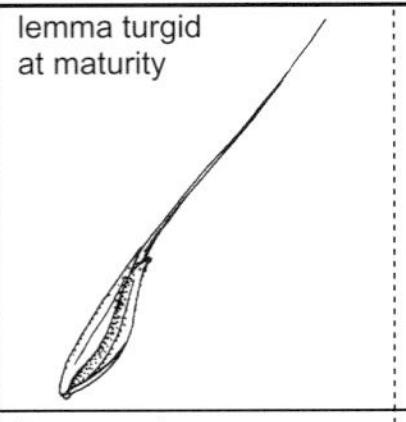	***L. temulentum*** 371
	Spikelets <22 mm long; **glume** not toughened and usually as long as or shorter than the spikelet; **lemma** not turgid at maturity; **plant** growing in tufts	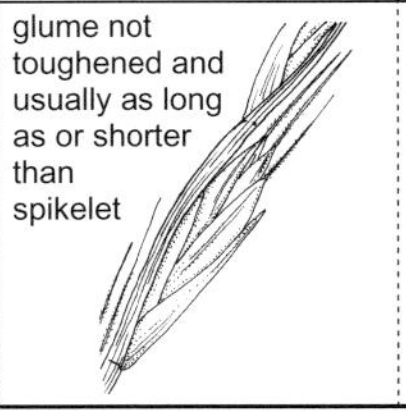	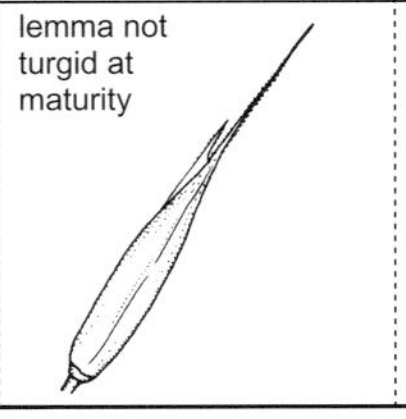	***L. persicum*** 370

Oplismenus

1	**Awns** stout but smooth; **lemma** without a fringe of hairs across it; **rhachis** and **pedicels** glabrous or shortly hairy; **perennial**	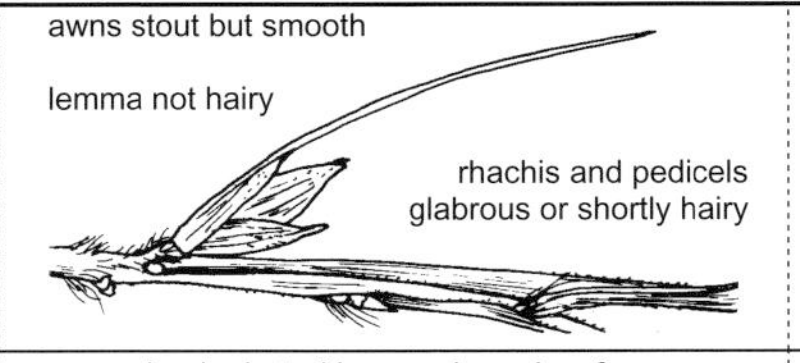	***O. compositus*** 373
	Awns slender but with a roughened surface; **lemma** with a band of hairs across the middle; **rhachis** and **pedicels** silky-hairy; **annual**	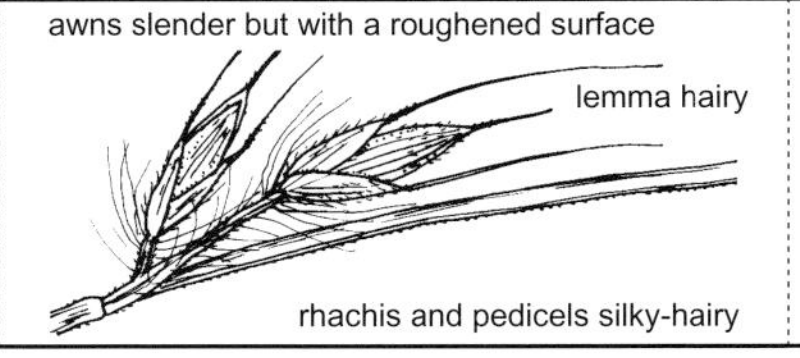	***O. burmannii*** 373

Panicum

1	**Leaves** hairy (tubercle based hairs)	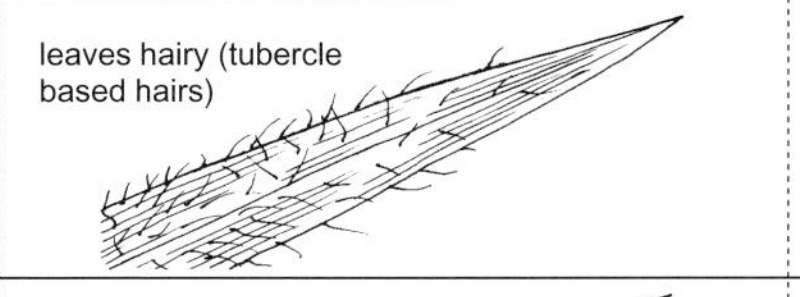	***P. atrosanguineum*** 373
	Leaves ± glabrous	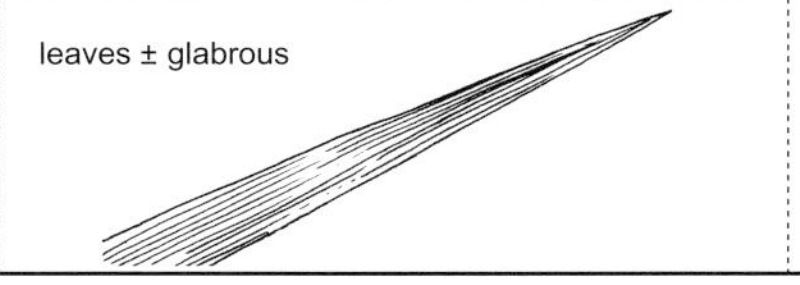	2
2	**Spikelets** >3mm long	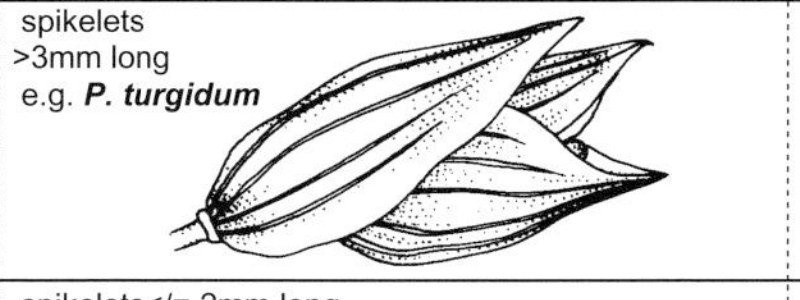	3
	Spikelets </= 2mm long	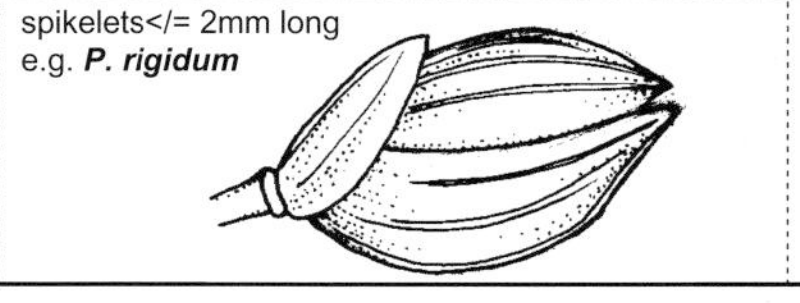	4
3	**Lower glume** nearly as long as spikelet; **upper lemma** smooth	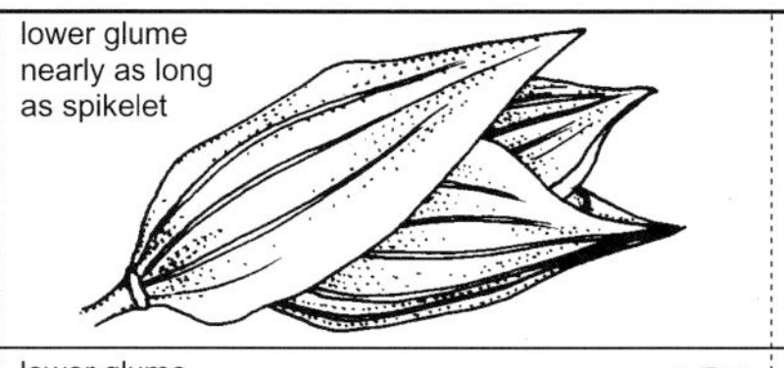	***P. turgidum*** 375
	Lower glume 1/4 to 1/3 spikelet length; **upper lemma** rugose	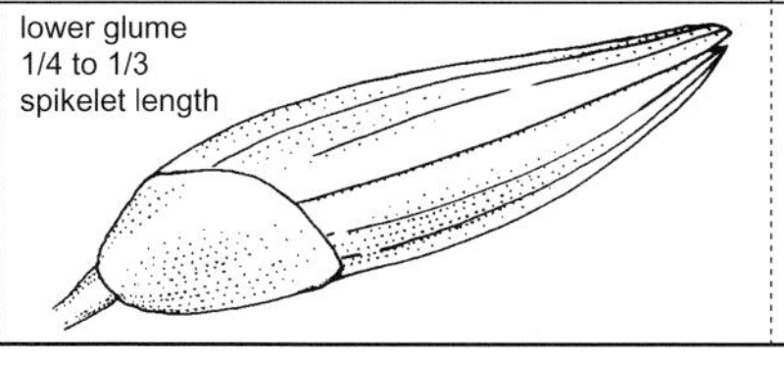	***P.maximum*** 374

4	Lowermost branches of **inflorescence** whorled, **inflorescence** 4.5--5.5 cm long	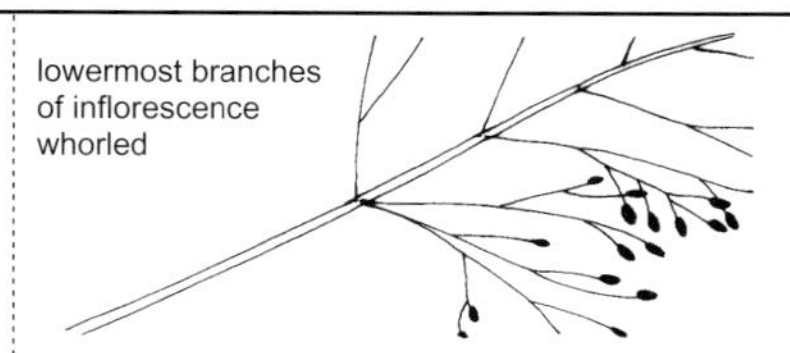	***P. socotranum*** Endemic 374
	Lowermost branches of **inflorescence** not whorled; **inflorescence** 1.5--4 cm long	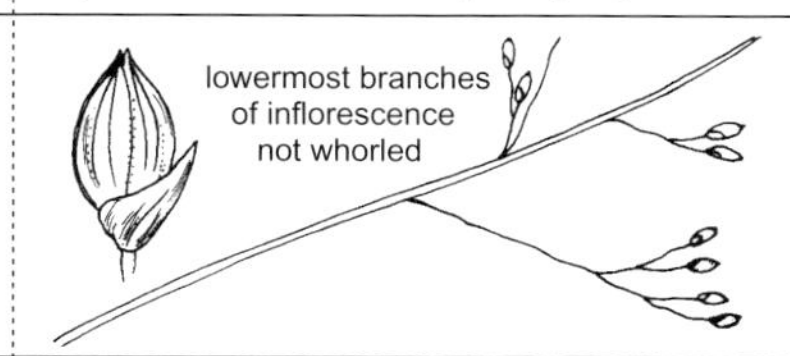	***P. rigidum*** Endemic 374

Paspalum

1	**Racemes** 3-- cm long, usually >2 (1 to 5); **spikelets** rounded; **upper lemma** striate, brown	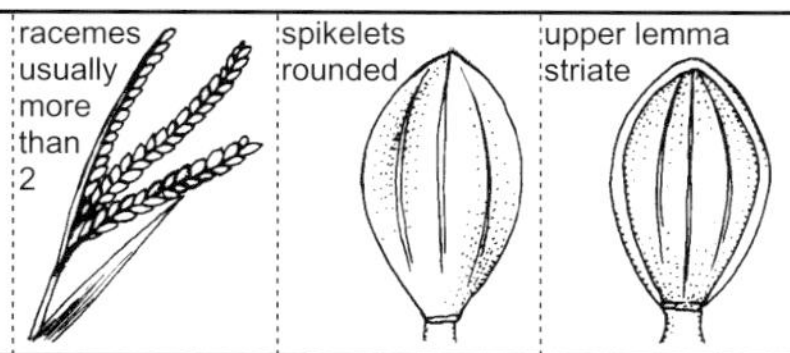	***P. scrobiculatum*** 376
	Racemes <2 cm, usually in pairs; **spikelets** elliptic; **upper lemma** smooth, pale	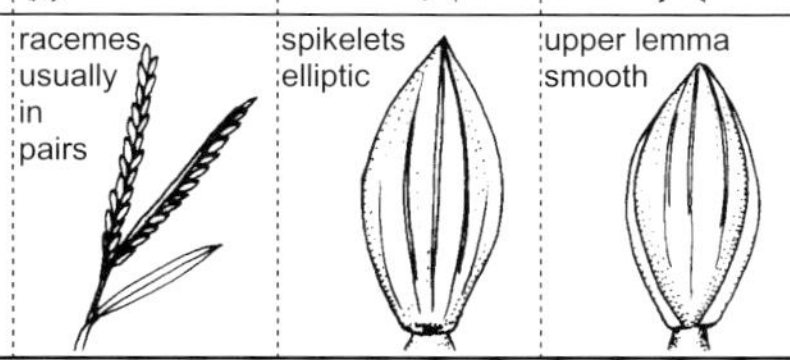	***P. vaginatum*** 376

Pennisetum

1	**Leaves** <3 mm wide; **leaves** rigid, inrolled, midrib noticeably thickened on upper surface	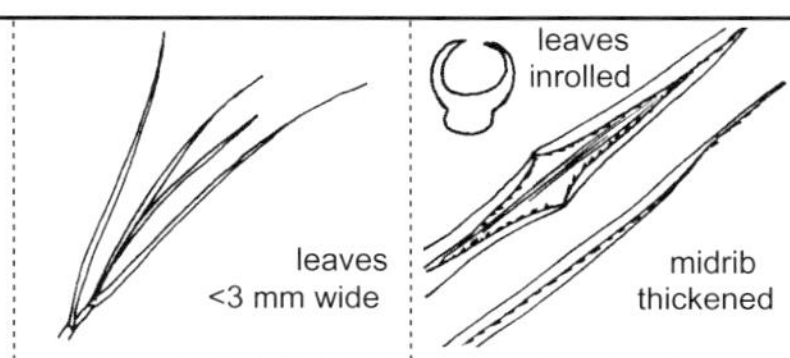	***P. setaceum*** 377
	Leaves <15 mm wide; **leaves** flat or loosely folded, midrib not noticeably thickened on upper surface (at least on fresh material)	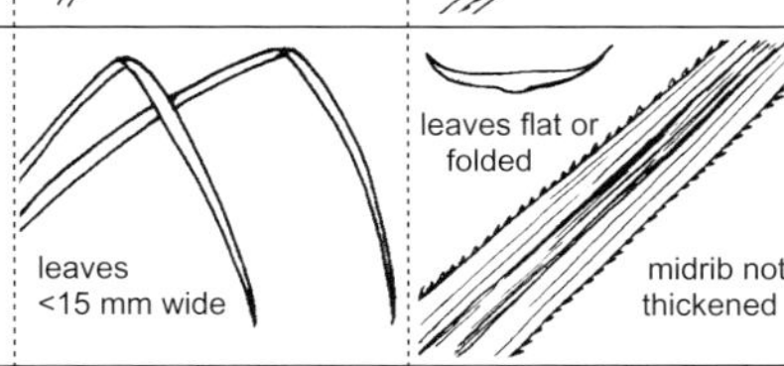	***P. orientale*** 376

Setaria

1	**Bristles** subtending spikelet retrorsely scabrid, clinging to clothing	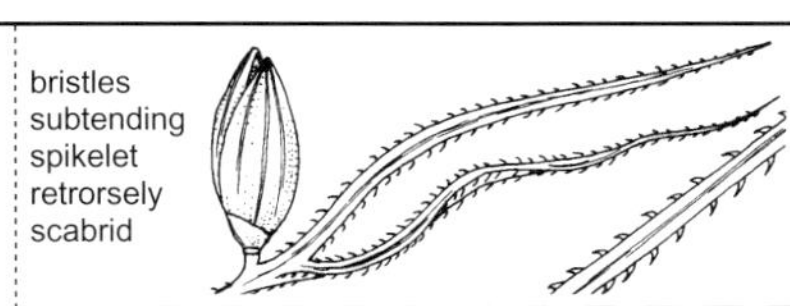	***S. verticillata*** 378
	Bristles subtending spikelet antrorsely scabrid	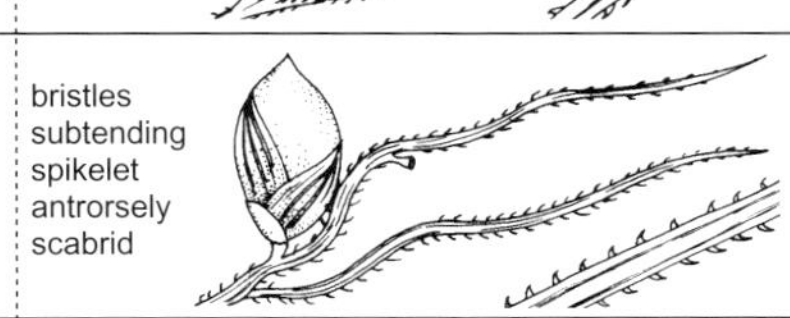	2

2	**Leaves** with a fairly dense covering of tubercle-based hairs along the main nerves; **bristles** pale	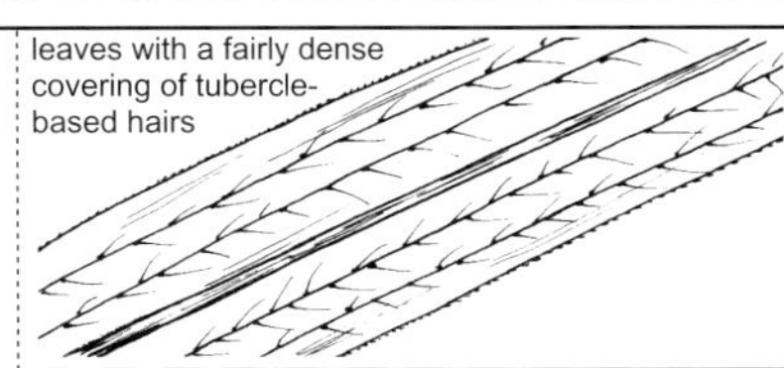	***S. intermedia*** 378
	Leaves more or less glabrous; **bristles** brownish-yellow	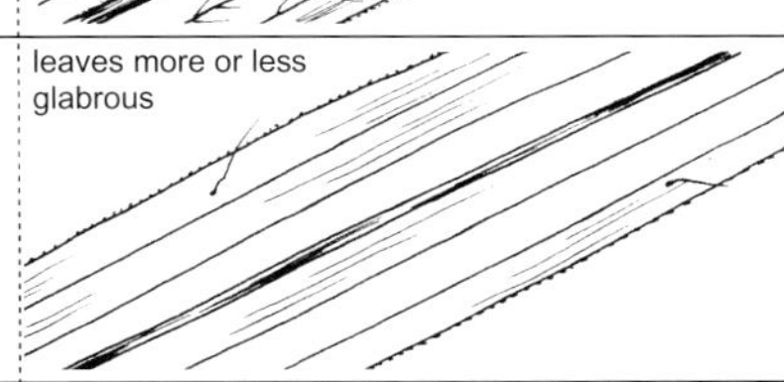	***S. pumila*** 378

Sporobolus

<table>
<tr><td rowspan="2">1</td><td>Panicle very dense and spike-like</td><td>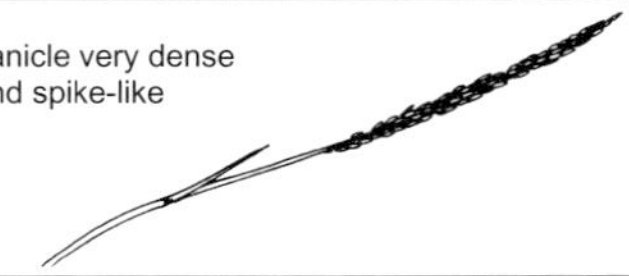</td><td>S. spicatus
380</td></tr>
<tr><td>Panicle open, not dense and spike-like</td><td></td><td>2</td></tr>
<tr><td rowspan="2">2</td><td>Tiny tufted annual <10cm in height; panicle 2--3.5 cm long</td><td>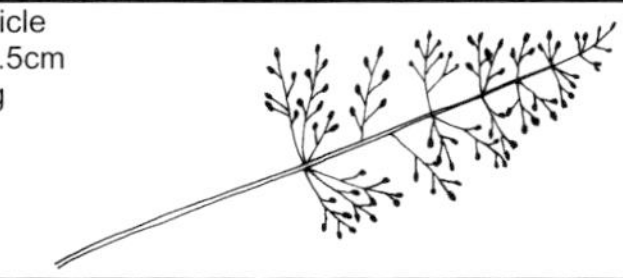</td><td>S. minutus
380</td></tr>
<tr><td>Tufted perennial <80cm in height; panicle 3--20cm long</td><td></td><td>3</td></tr>
<tr><td rowspan="2">3</td><td>Primary branches whorled</td><td></td><td>S. ioclados
379</td></tr>
<tr><td>Primary branches not whorled</td><td></td><td>4</td></tr>
<tr><td rowspan="2">4</td><td>Glumes scabrid; lower glume 1/2 spikelet length; upper glume 2/3 to 3/4 spikelet length</td><td>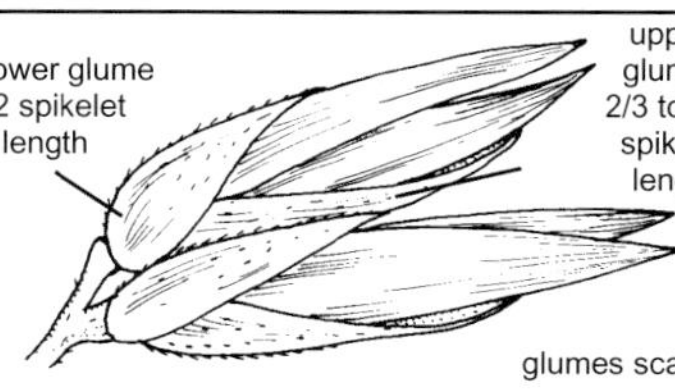</td><td>S. ruspolianus
380</td></tr>
<tr><td>Glumes smooth; lower glume 1/4 to 1/3 spikelet length; upper glume 1/2 spikelet length</td><td>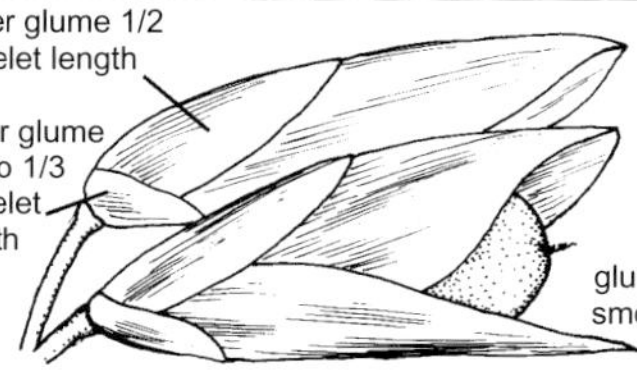</td><td>S. diander
379</td></tr>
</table>

Tetrapogon

<table>
<tr><td rowspan="2">1</td><td>Lowest, and most clearly visible, lemma glabrous or with short hairs on the nerves; 'racemes', if in pairs, clearly separated; upper glume apiculate</td><td>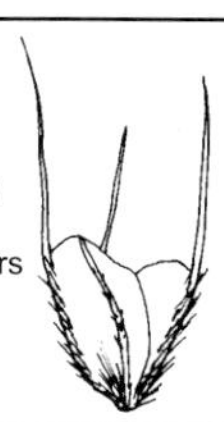 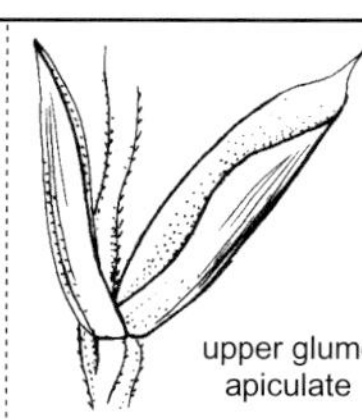</td><td>T. tenellus
381</td></tr>
<tr><td>Lowest, and most clearly visible, lemma conspicuously silky-hairy; 'racemes', if in pairs, stuck together; upper glume mucronate</td><td>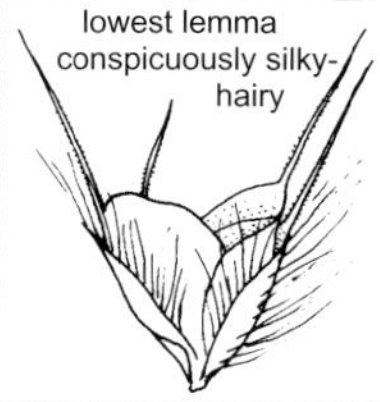 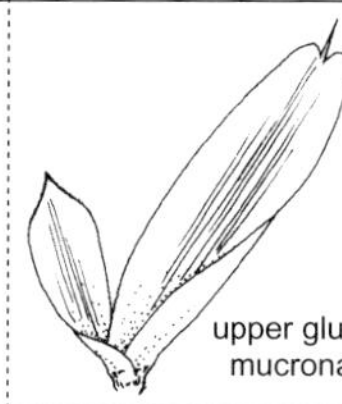</td><td>T. villosus
381</td></tr>
</table>

Themeda

1	**Annual**; **stems** tall and stout, often supported at base by stilt roots; **involucral spikelets** 8--14mm long, glabrous or with a few tubercle-based hairs		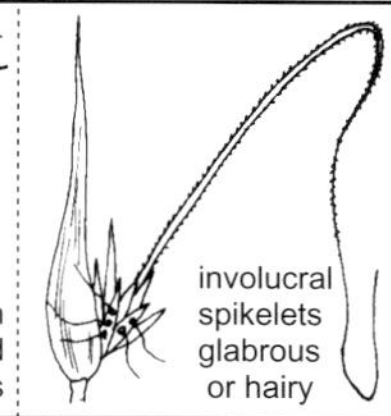	***T. quadrivalvis*** 382
	Perennial; flowering **stems** slender, **leaves** mostly confined to a basal cushion; **involucral spikelets** 5--8mm long, always with tubercle-based hairs	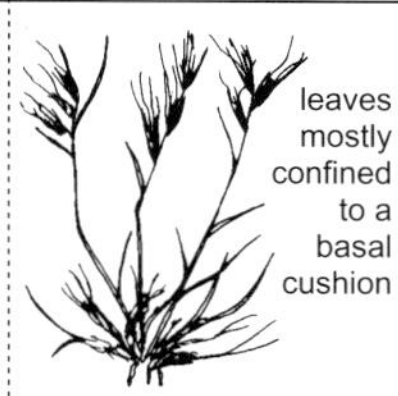	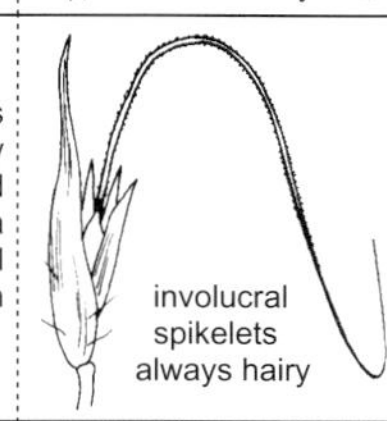	***T. triandra*** 382

Tricholaena

1	**Spikelets** long-hairy, hairs usually at least 2x length of spikelet; erect or ascending, tufted, **perennial** to 60cm, glabrous or loosely hairy	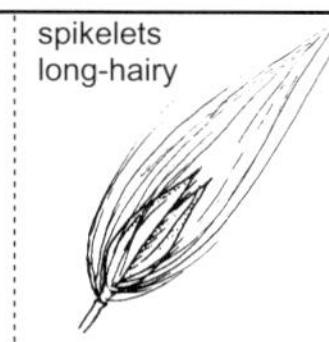	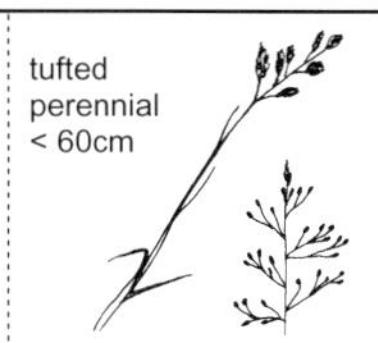	***T. teneriffae*** 383
	Spikelets short-hairy, hairs scarcely > tip of the spikelet; low-growing **perennial** to 15cm, shortly hairy		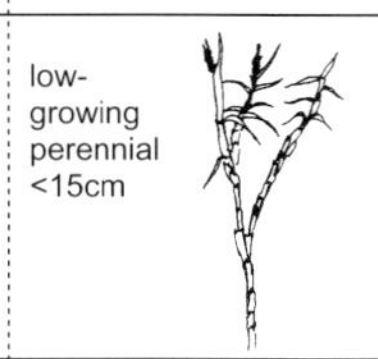	***T. vestita*** Endemic 383

HYACINTHACEAE

1	**Leaves** 1 to 3, ovate, usually >3cm wide; **flowers** pink; **fruit** 6-lobed	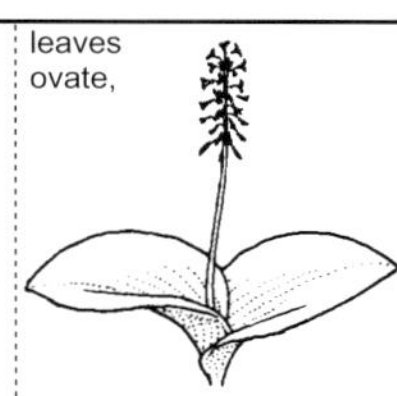	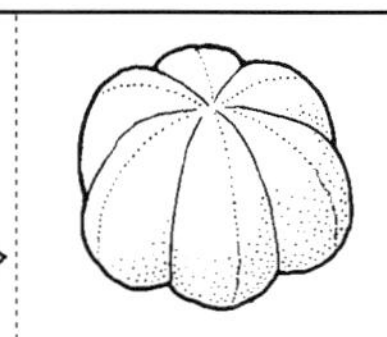	***Ledebouria*** (3 spp.)
	Leaves linear to linear-lanceolate, <3cm wide; **flowers** white, dull green or brown; **fruit** 3-lobed	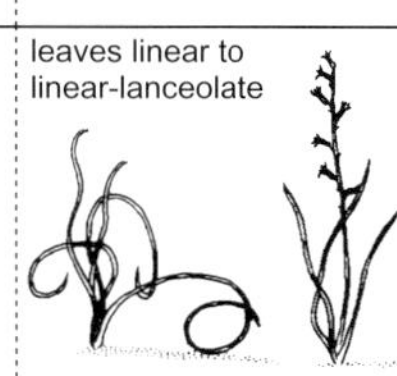	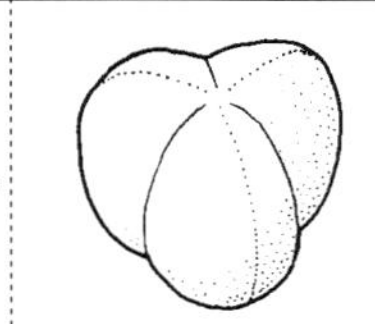	2
2	**Tepals** in 2 whorls, the inner usually erect, the outer spreading; **bracts** not spurred		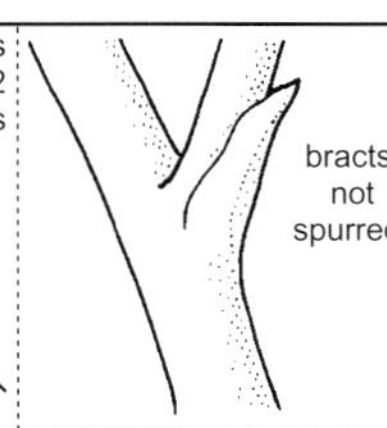	***Dipcadi*** (4 spp.)
	Tepals all spreading; **bracts** minutely spurred at the base	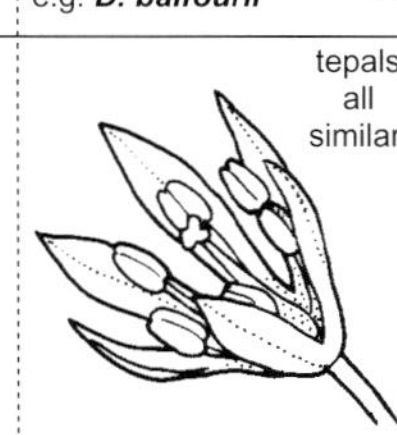	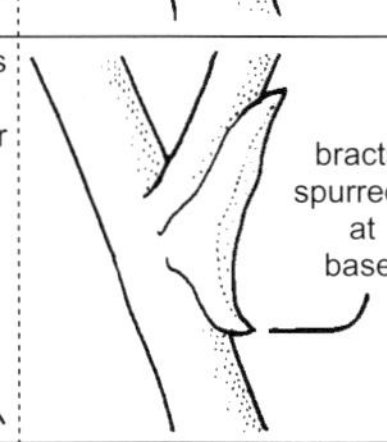	***Drimia porphyrostachys*** Endemic 385

Dipcadi

1	**Leaves** spirally coiled; **flowers** 2 or 3 on a short (1--5cm) peduncle	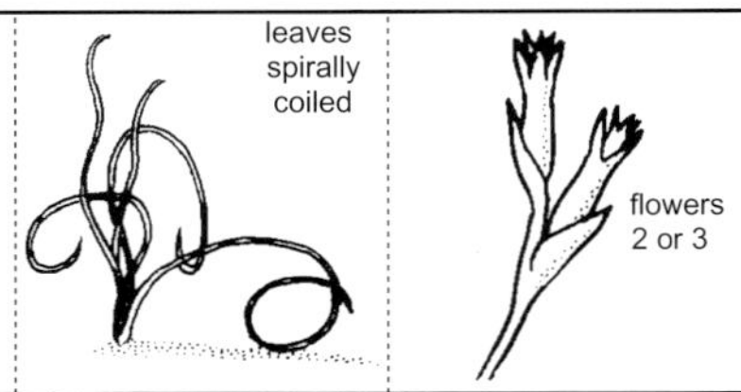	***D. biflorum*** **236**, 384
	Leaves straight; **flowers** several to many on a long (10--30cm) peduncle	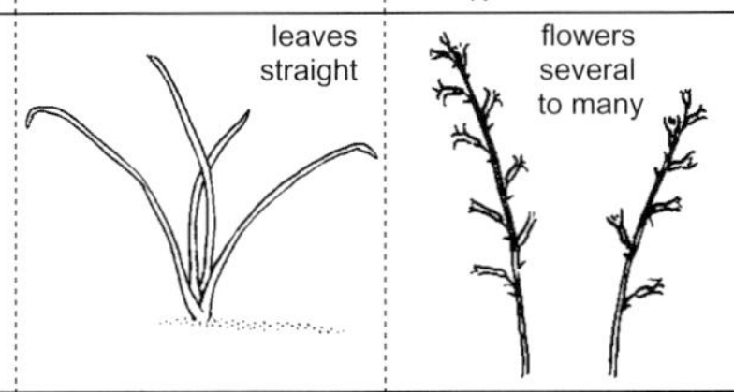	2

2	Robust **plant** <45cm; **leaves** dark green, tinged red, <0.5cm wide; **flowers** yellowish-green, c.15mm long	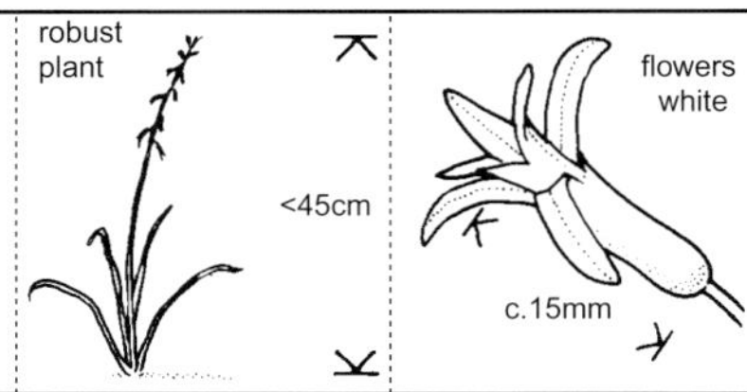	***D. balfourii*** Endemic **236**, 384
	Robust **plant** <45cm; **leaves** pale green, <0.4--1cm wide; **flowers** white, <15mm long	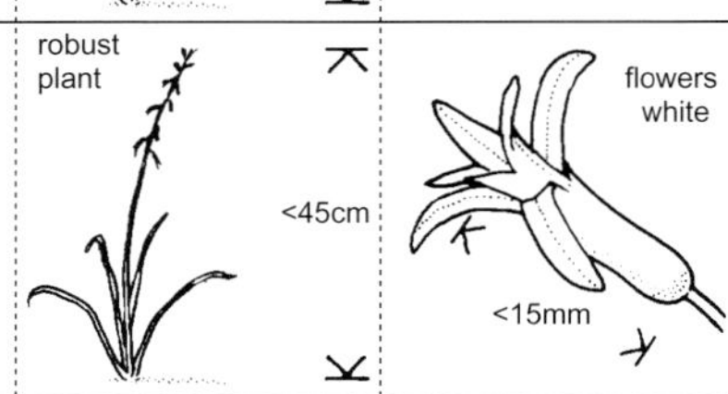	***D. kuriensis*** Endemic **236**, 385
	Delicate **plant** <15cm; **leaves** filiform; **flowers** greenish-brown, c.10mm long	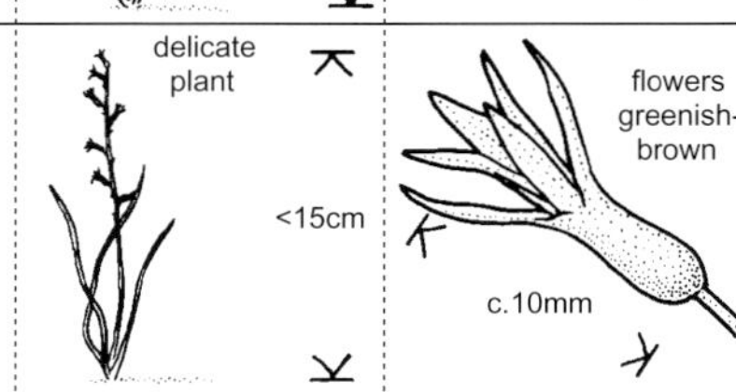	***D. guichardii*** Endemic **236**, 385

Ledebouria

1	**Leaves** 2, 10--25 x 8--15cm, not spotted; **inflorescence** 30 to 100-flowered, <35cm long; **flowers** 6--7mm long	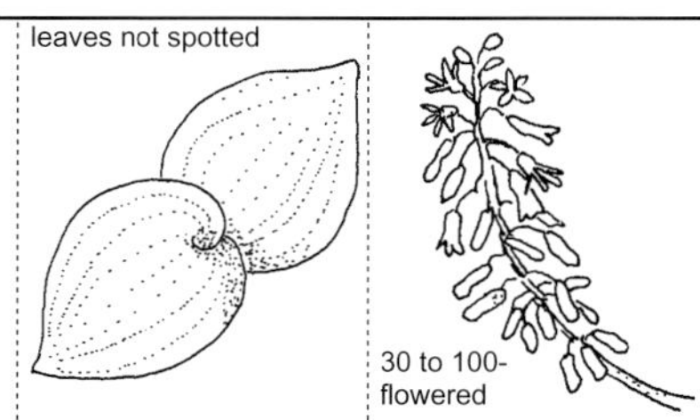	***L. grandifolia*** Endemic **236**, 386
	Leaves 2 or 3, 5 x 3.5cm, not spotted; **inflorescence** 15 to 25-flowered, <10cm long; **flowers** c. 4.5mm long	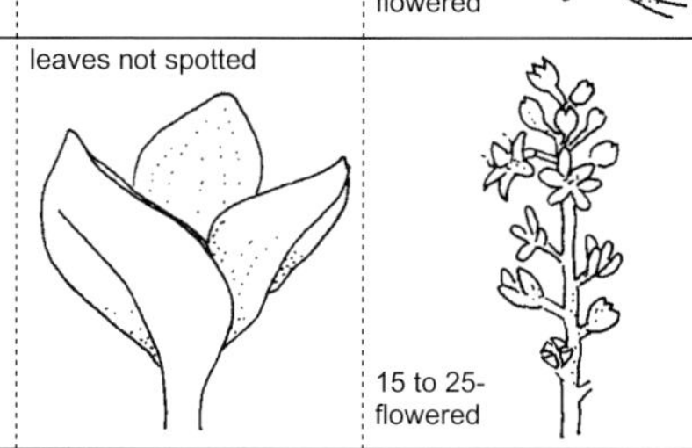	***L. insularis*** Endemic **236**, 386
	Leaves 1 or 2, 3--5 x 2--3cm, spotted purplish green; **inflorescence** many-flowered; **flowers** 3--7mm long	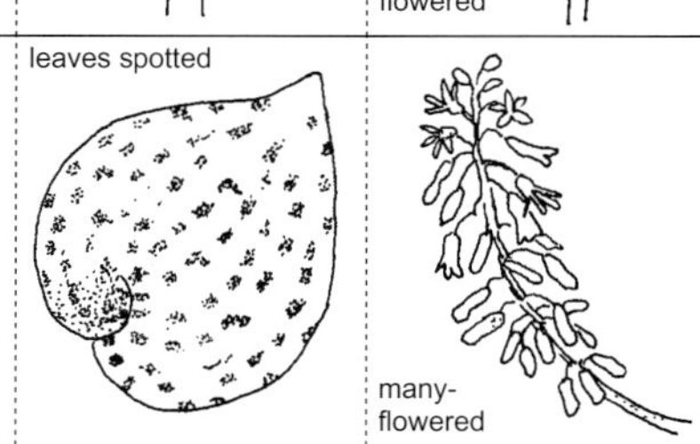	***L. aff. revoluta*** **236**, 386

HYPOXIDACEAE

	Character	Illustration	Taxon
1	**Leaves** grasslike, hairy; **flowers** 4(to 6)-merous, yellow; **capsule** dehiscing by slits, topped by remains of perianth	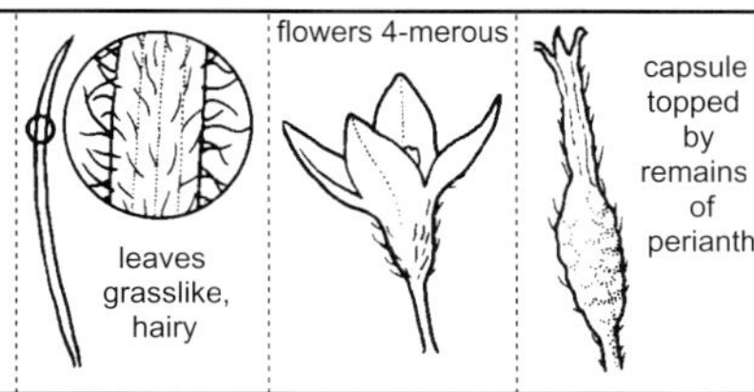	***Hypoxis angustifolia*** **237**, 387

IRIDACEAE

	Character	Illustration	Taxon
1	**Leaves** lanceolate, 2-ranked; **corm** covered with fibrous scales; **flowers** blue; **perianth tube** long (<5cm), narrow, arising from the leaves	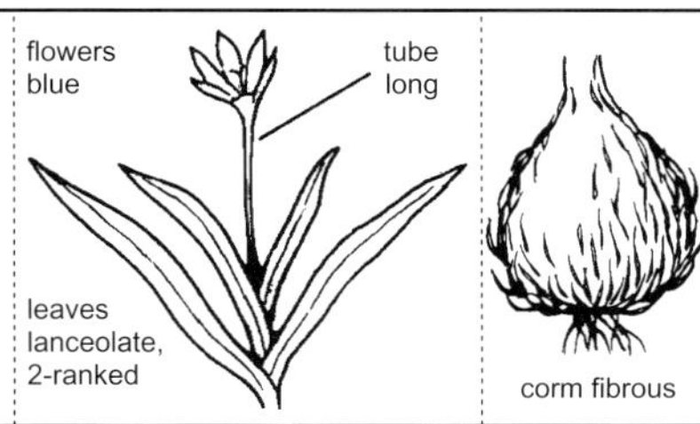	***Babiana socotrana*** Endemic **237**, 388
	Leaves linear, not 2-ranked; **corm** covered with membranous scales; **flowers** blue/violet with a yellow centre; **perianth tube** short (c.5mm), arising from a pair of bracts	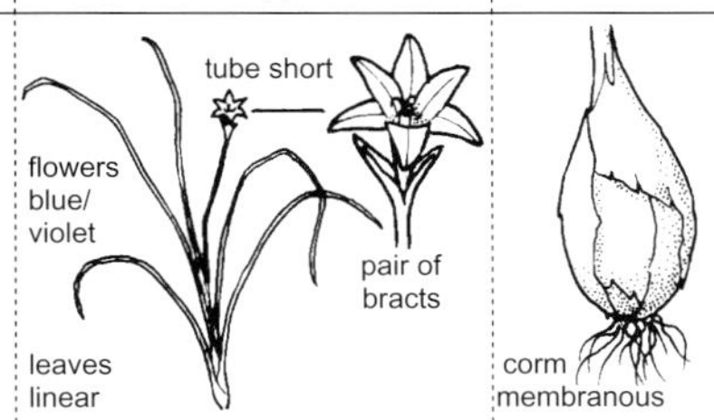	***Romulea fischeri*** **237**, 389

JUNCACEAE

	Character	Illustration	Taxon
1	Slender **annual** to 20cm; **leaves** soft, filiform; **inflorescence** an open, leafy panicle	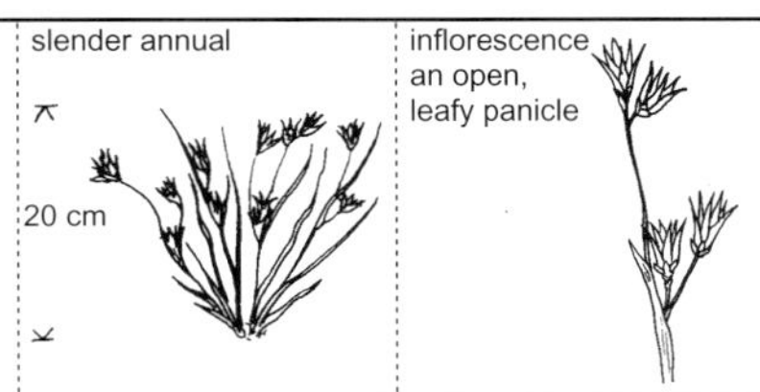	***Juncus bufonius*** 389
	Stout **perennial** to 150cm; **leaves** stiff, spinescent; **inflorescence** a many-flowered panicle, appearing lateral	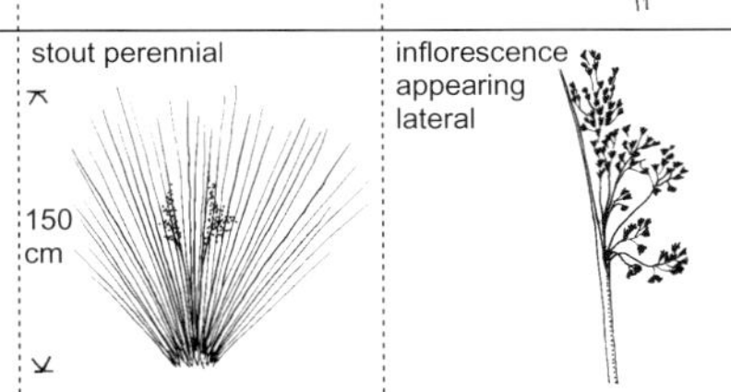	2

	Character	Illustration	Taxon
2	**Tepals** 3.5--5mm; **capsule** 3.5--5mm, narrowly ellipsoid, pale green	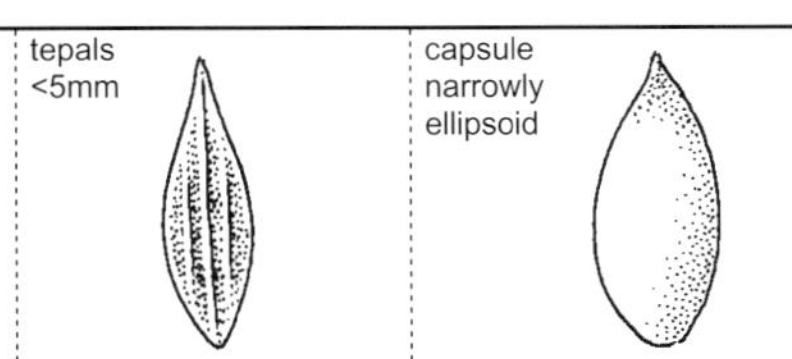	***Juncus rigidus*** 389
	Tepals 2--3mm; **capsule** 2.5--3mm, subglobose, glossy brown	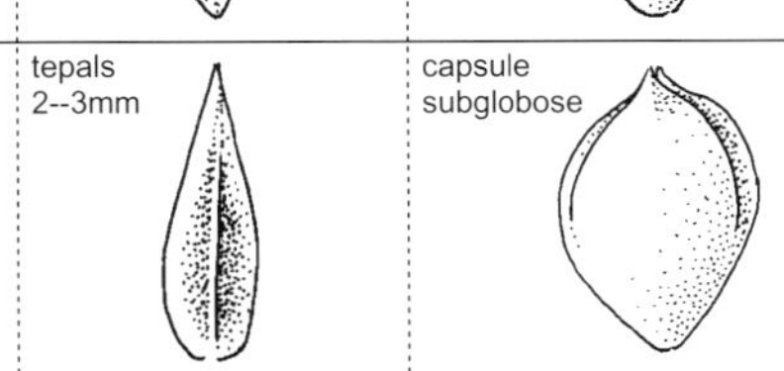	***Juncus socotranus*** **237**, 390

ORCHIDACEAE

1	**Plant** growing on trees (epiphytic) or rocks (lithophytic); **leaves** leathery arising from prostrate stem, rounded at tip; **flowers** in spikes <25cm	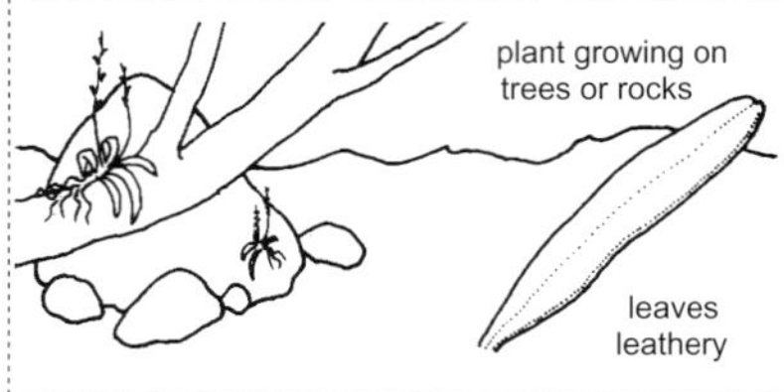	***Angraecum dives*** **237**, 391
	Plant growing on ground (terrestrial) or on cliffs; **leaves** leathery arising from cylindrical pseudobulbs, acute at tip; **flowers** in erect spikes, <1m	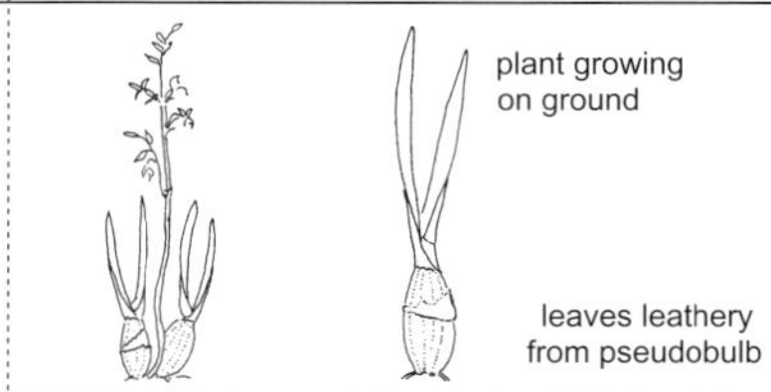	***Eulophia petersii*** **237**, 391
	Plant growing on ground (terrestrial); **leaves** thin-textured; **flowers** in spikes <20cm	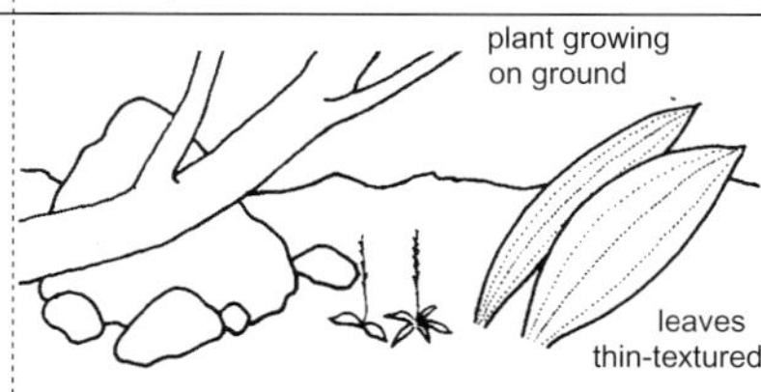	2
2	**Leaves** and **stem** glabrous; **leaves** 3 to 5; **spur** slender, c.8mm long	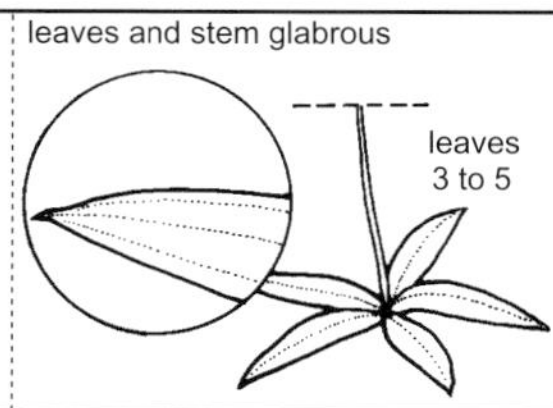 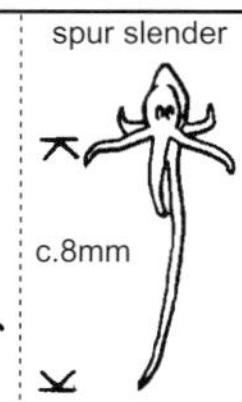	***Habenaria socotrana*** **237**, 392
	Leaves and **stem** hairy; **leaves** 2; **spur** short, c.2mm long	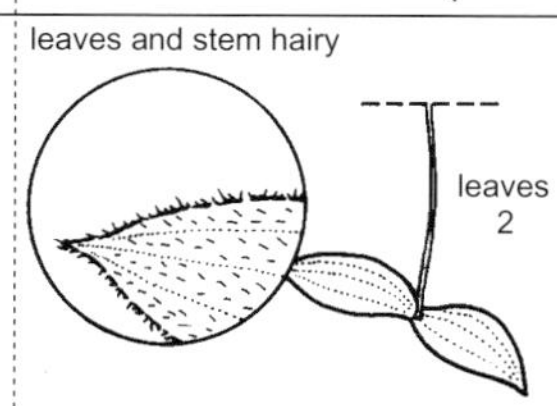 	***Holothrix socotrana*** Endemic **237**, 392

PALMAE

1	Tall **tree** with persistent leaf bases near the crown; **leaves** pinnate	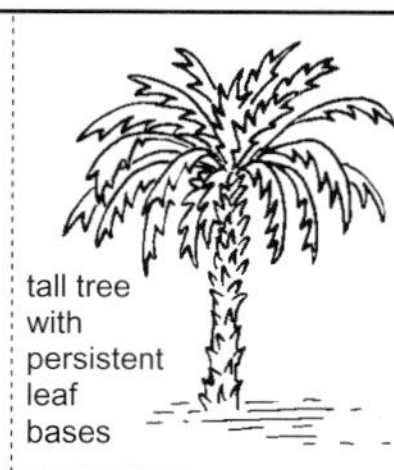 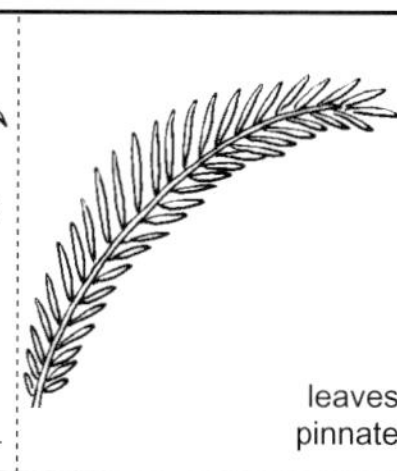	***Phoenix dactylifera*** **237**, 393
	Very tall **tree** without persistent leaf bases near the crown; **leaves** palmate	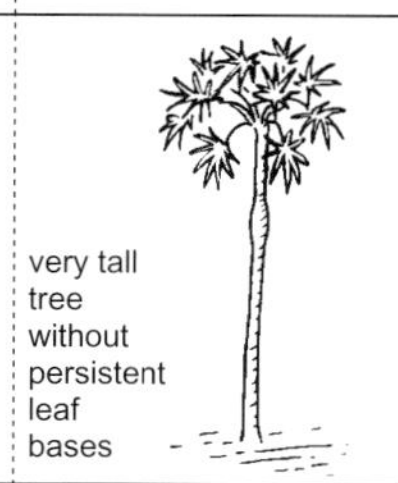 	***Borassus aethiopum*** **237**, 393

DICOTYLEDON FAMILIES

ACANTHACEAE

1 **Fertile stamens** 4

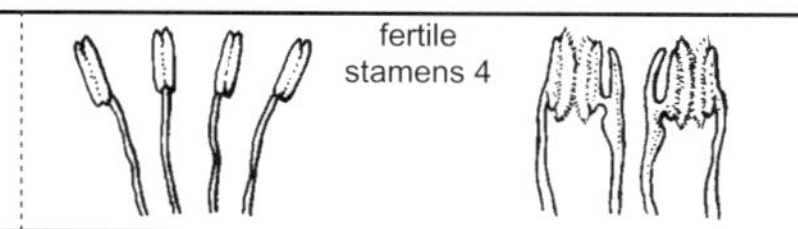

2

Fertile stamens 2

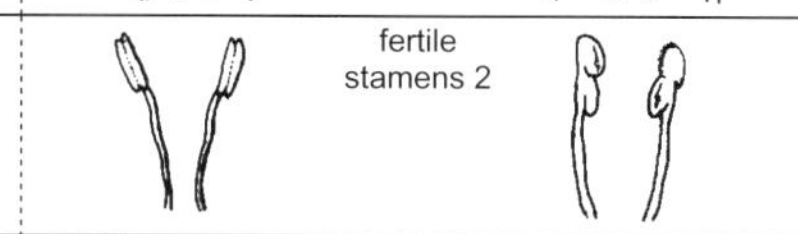

5

2 **Bracts** and **leaves** unarmed; **calyx** 5-lobed

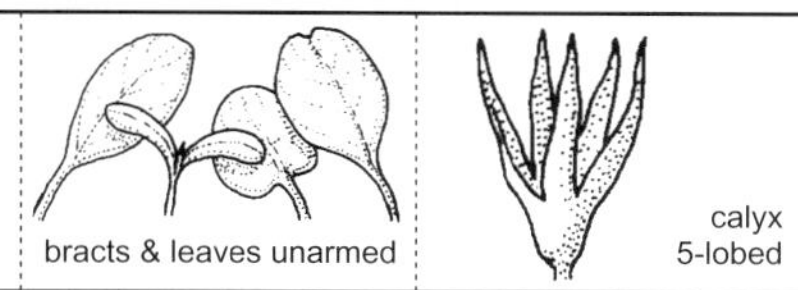

3

Bracts and/or **leaves** spinose; **calyx** 2-lobed

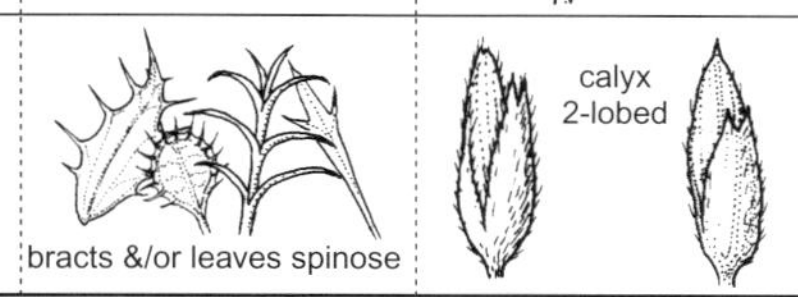

4

3 **Corolla** with 5 subequal lobes; **fruits** < 3x as long as broad; **seeds** pubescent or with scarious margin

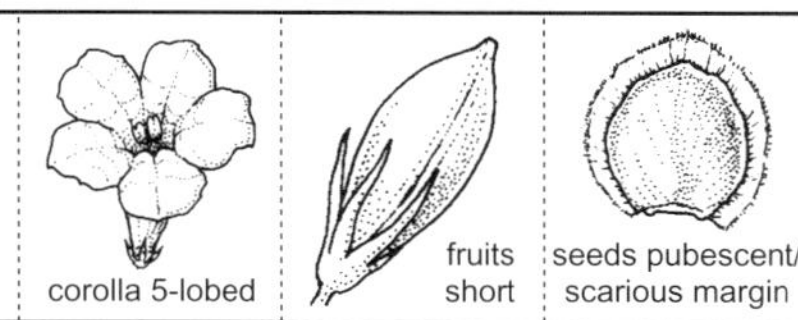

Ruellia
(6 spp.)

Corolla 2-lipped; **fruits** > 3x as long as broad; **seeds** glabrous, rough

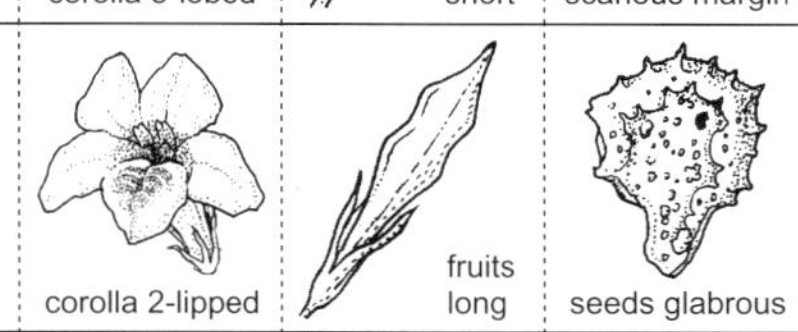

Asystasia gangetica
401

4 **Corolla** 5-lobed; **bracts** linear, spine-tipped, with entire margin; **anther cells** 2, tomentose; **fruits** 4-seeded

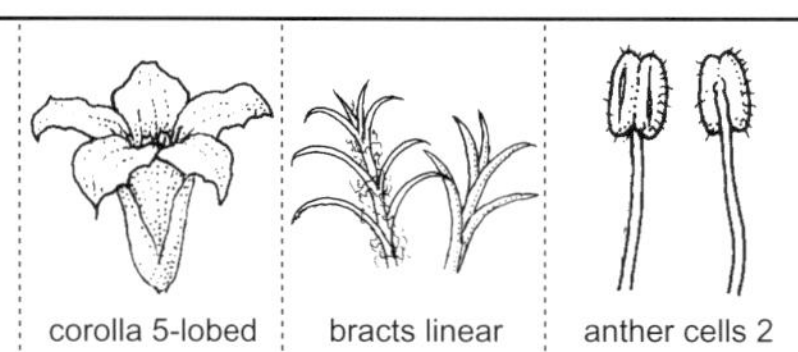

Neuracanthus
(3 spp.)

Corolla 1-lipped; **bracts** or **leaves** lanceolate, obovate or ensiform with spiny margin; **anther cells** 1, with white pilose indumentum; **fruits** 2-seeded

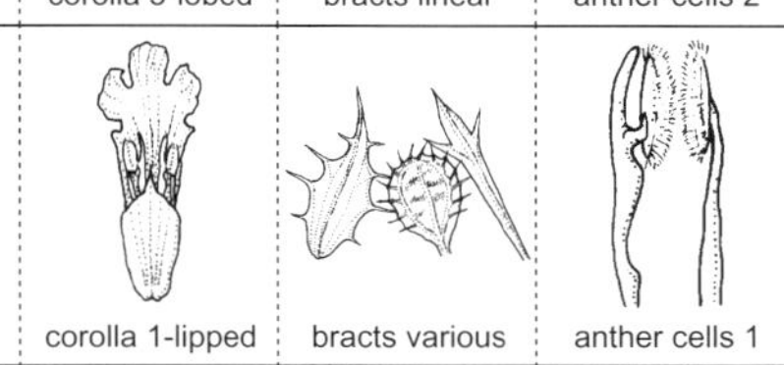

Blepharis
(3 spp.)

5 **Anther cells** 1 on each stamen

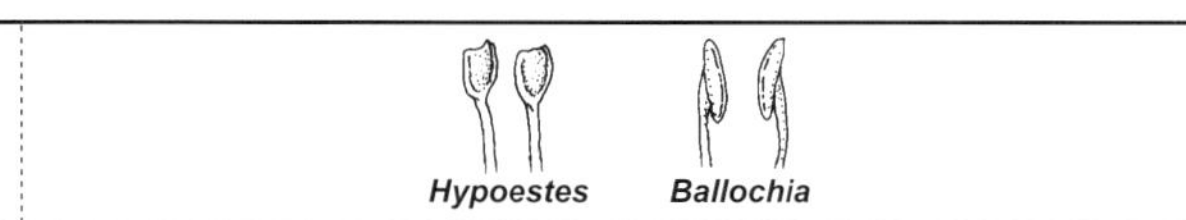

6

Anther cells 2 on each stamen

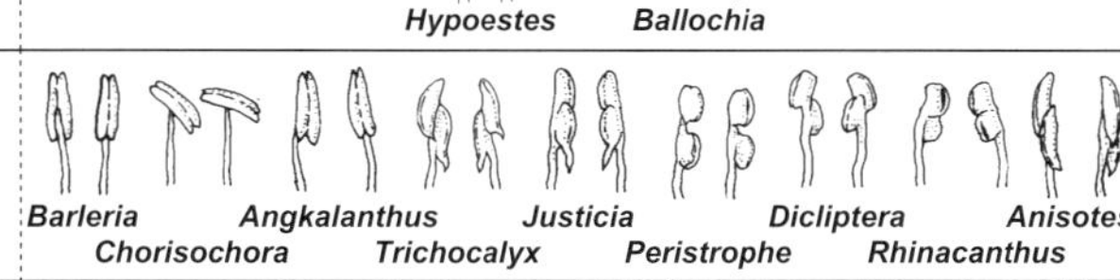

7

6 **Herb** with petiolate leaves; many-flowered spike-like **inflorescence**; **corolla** white to lilac, < 20mm long; **fruits** c.8mm long

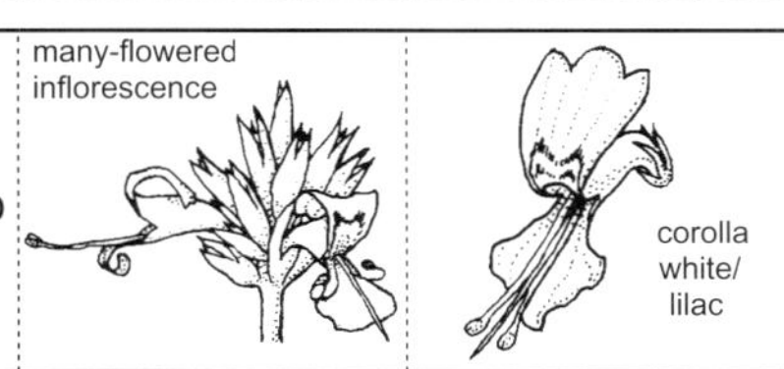

Hypoestes pubescens
Endemic
239, 407

Shrub with subsessile leaves; solitary to few-flowered mainly axillary **inflorescence**; **corolla** red to orange, >20mm long; **fruits** 15--25mm long

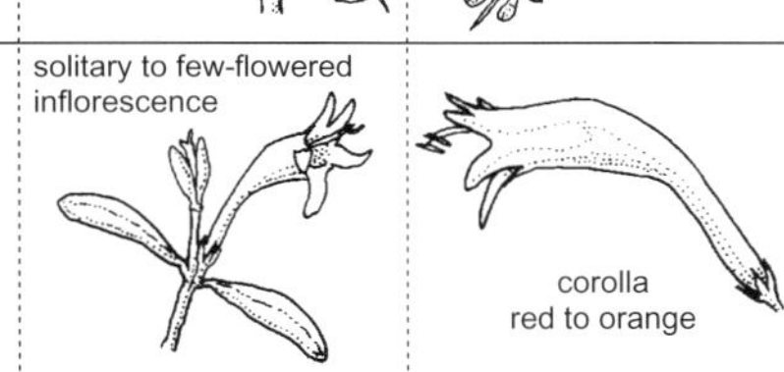

Ballochia
Endemic genus
(3 spp.)

7	**Anther cells** side by side on each stamen			8
	Anther cells one above the other on each stamen	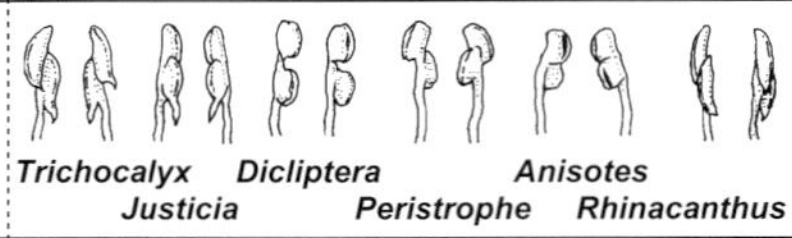		10
8	**Calyx** deeply 4-lobed, outer pair longer than inner; **leaves** linear or spinose, green, or if lanceolate then grey	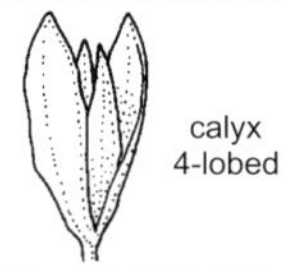	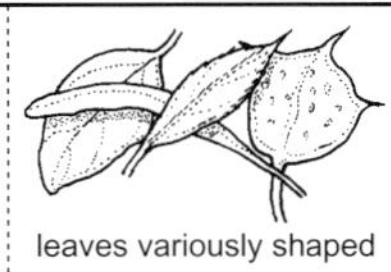	***Barleria*** (4 spp.)
	Calyx 5-lobed, lobes more or less equal; **leaves** lanceolate, unarmed, green	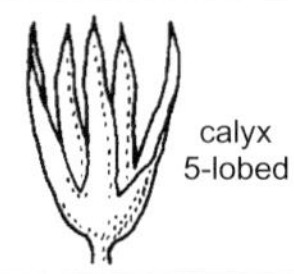	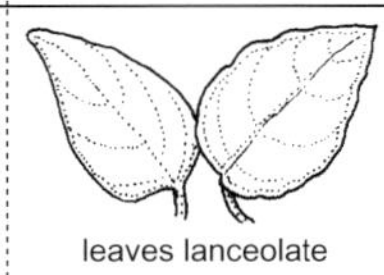	9
9	Perennial **herb** with erect or ascending branches; **corolla** white to pink, with 5 subequal lobes; **bracts** reddish-green, 12--15mm long	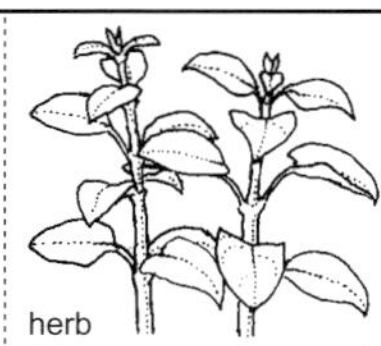	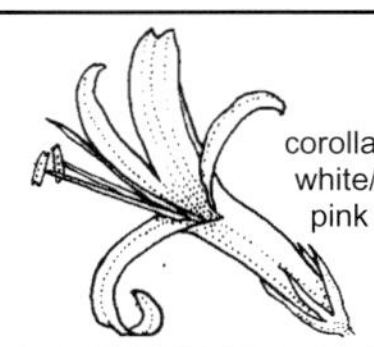	***Chorisochora*** (2 spp.)
	Thin-stemmed **shrub** with pendant branches; **corolla** red fading to yellow, 2-lipped; **bracts** green, c.1mm long	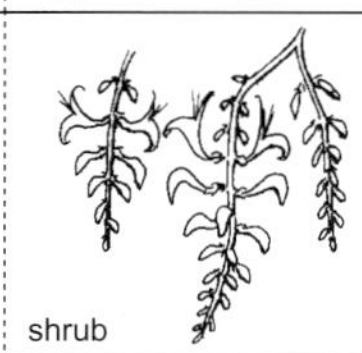	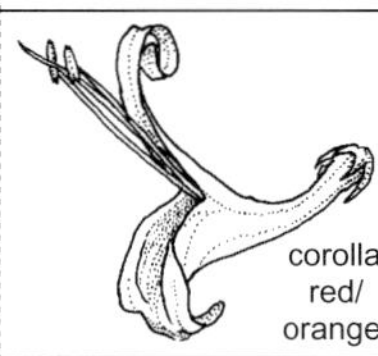	***Angkalanthus oligophylla*** Endemic Endemic genus **238**, 400
10	**Anthers** with tails	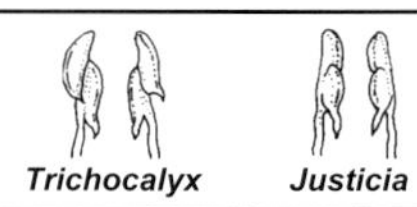		11
	Anthers without tails	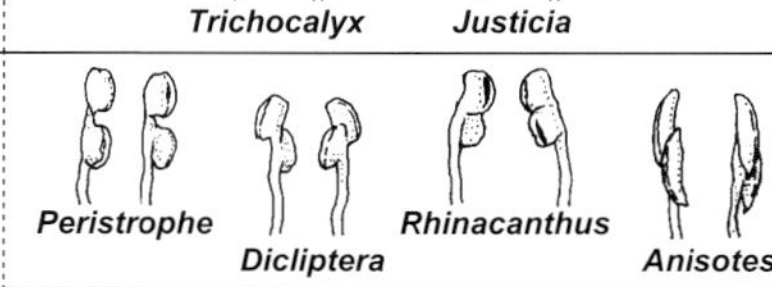		12
11	**Corolla** red fading to orange, 20--25mm long; **bracts** threadlike, densely clustered, 10--25mm long; **fruits** 13--20mm long	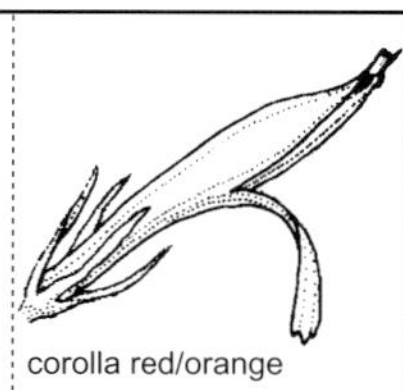	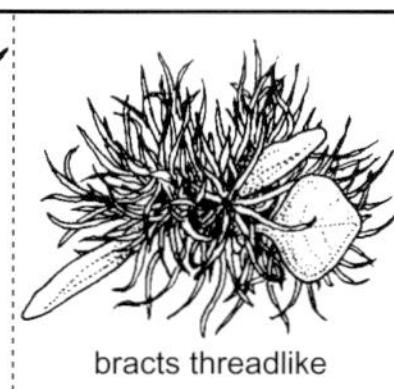	***Trichocalyx*** Endemic genus (2 spp.)
	Corolla white or lilac, 4--5mm long; **bracts** triangular, few, c.1mm long; **fruits** 3--6mm long	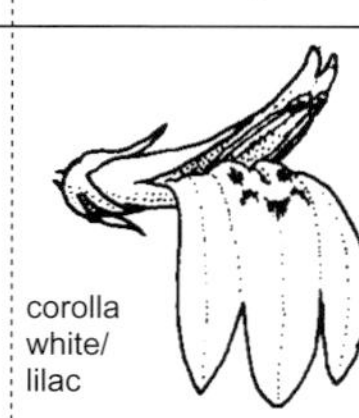	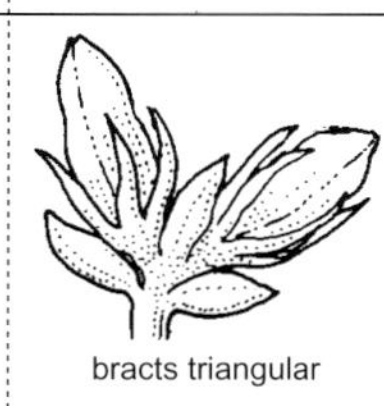	***Justicia*** (4 spp.)
12	**Corolla** upper lip 3-lobed, lower lip entire or 2-lobed; **leaves** ovate, lanceolate or elliptic	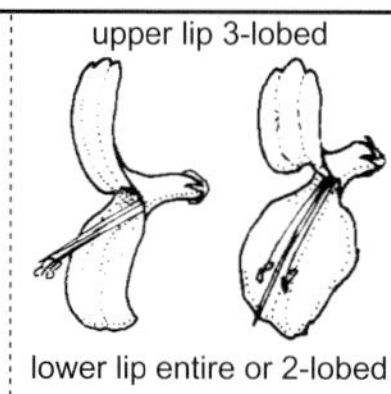	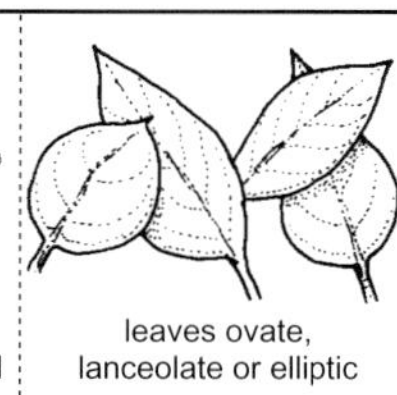	13
	Corolla upper lip entire or 2-lobed, lower lip 3-lobed; **leaves** linear; spatulate or obovate	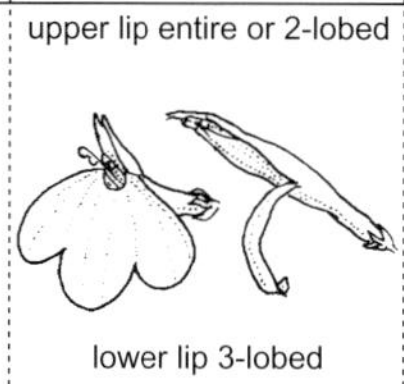	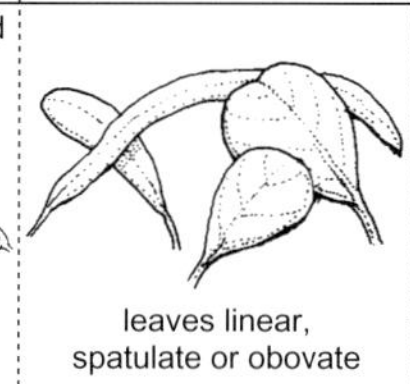	14

13

Shrub; lower lip of **corolla** narrow, 2-lobed; outer 2 **bracts**, unequal, fused at base, linear-lanceolate, unarmed; **fruits** c.9mm long — ***Peristrophe paniculata*** 410

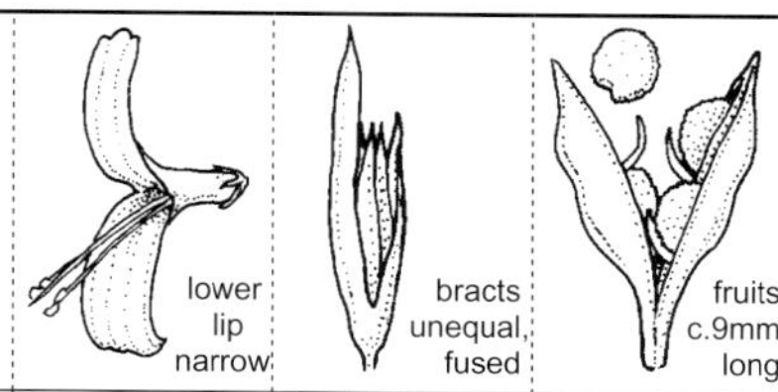

Herb; lower lip of **corolla** broad, entire; all **bracts**, +/- equal, free at base, elliptic, spine-tipped; **fruits** c.6mm long — ***Dicliptera effusa*** Endemic **239**, 407

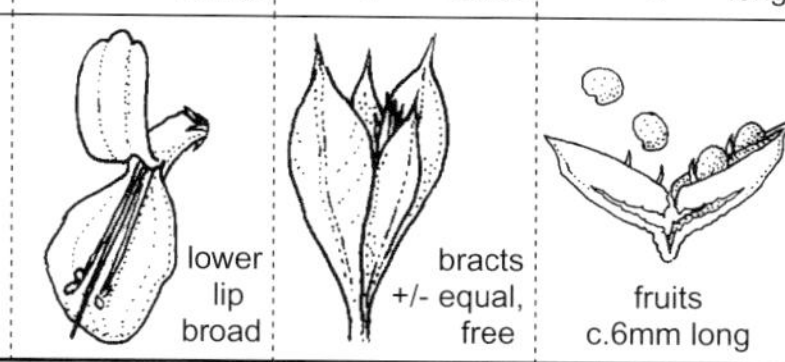

14

Perennial **herb**; **corolla** white, <20mm long, with broad lower lip; **leaves** linear-spatulate, acute to obtuse at tip — ***Rhinacanthus scoparius*** Endemic **240**, 410

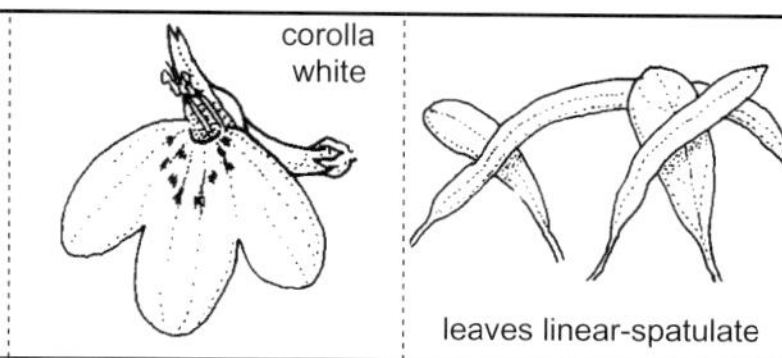

Shrub; **corolla** red fading to yellow, >20mm long, with narrow lower lip; **leaves** obovate, rounded to retuse at tip — ***Anisotes diversifolius*** Endemic **238**, 400

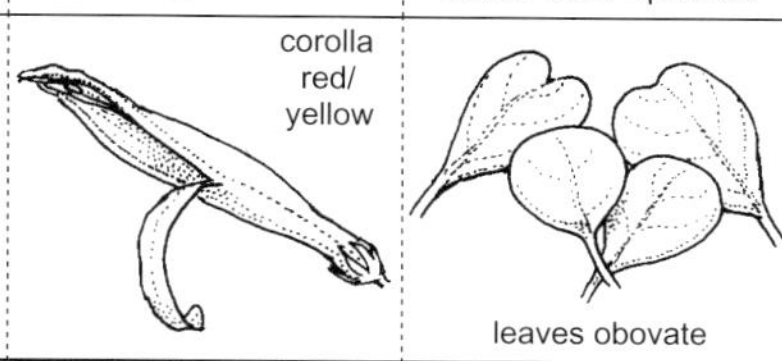

Ballochia

1

Leaves small, </= 3x as long as broad, linear, usually fasciculate — ***B. amoena*** Endemic **238**, 402

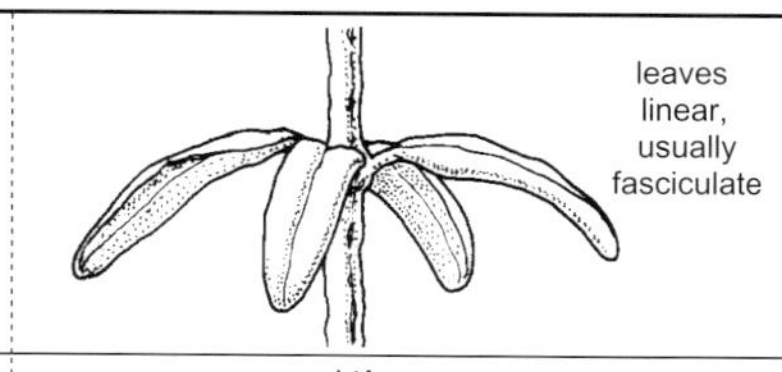

Leaves variable but always <2x as long as broad, oblong to orbicular — 2

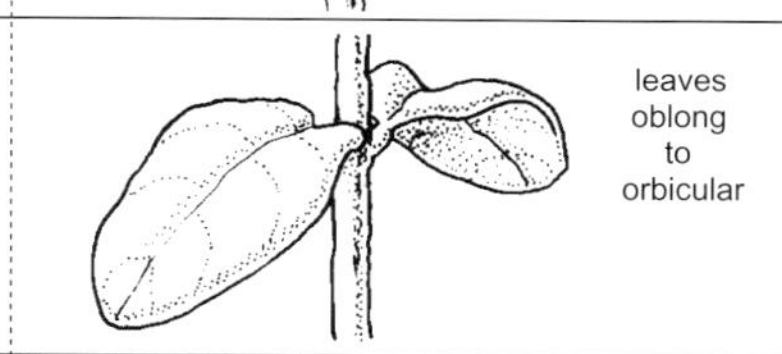

2

Low **shrub** to 50cm; **calyx** covered in glandular club-shaped hairs — ***B. rotundifolia*** Endemic 403

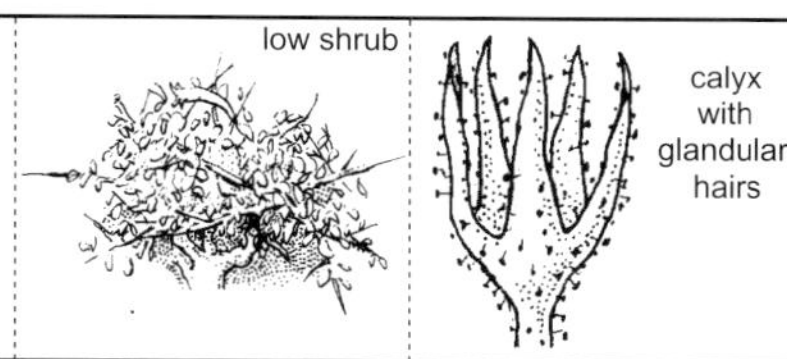

Shrub or small **tree** to 3m; **calyx** covered in non-glandular hairs — ***B. atro-virgata*** Endemic 403

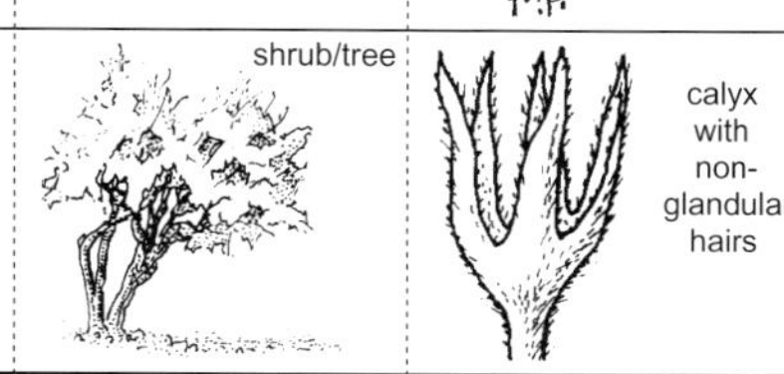

Barleria

1

Leaves spinose; **corolla** subequally 5-lobed — 2

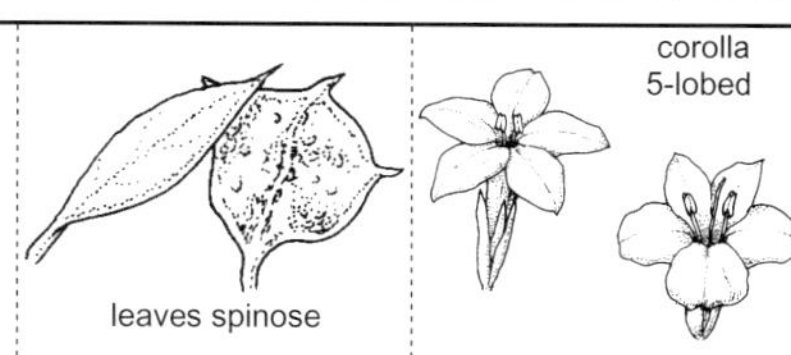

Leaves unarmed; **corolla** 2-lipped — 3

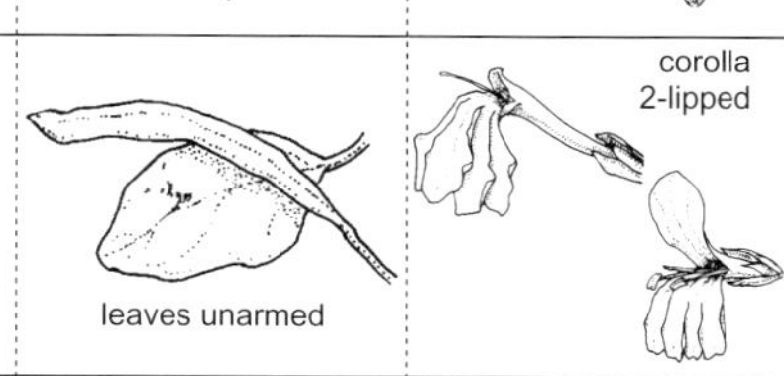

2	**Corolla** orange-yellow, with equal lobes; **leaves** elliptic, spine-tipped, with an entire margin; **spined branches** in the leaf axils; **fruit** 2-seeded	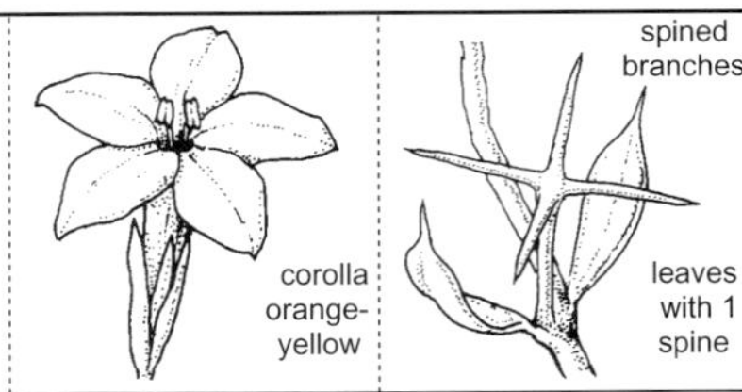	***B. tetracantha*** Endemic **238**, 404
	Corolla lilac, with subequal lobes; **leaves** ± orbicular with up to 3 spines on the margin; no **spined branches** in the leaf axils; **fruit** 4-seeded	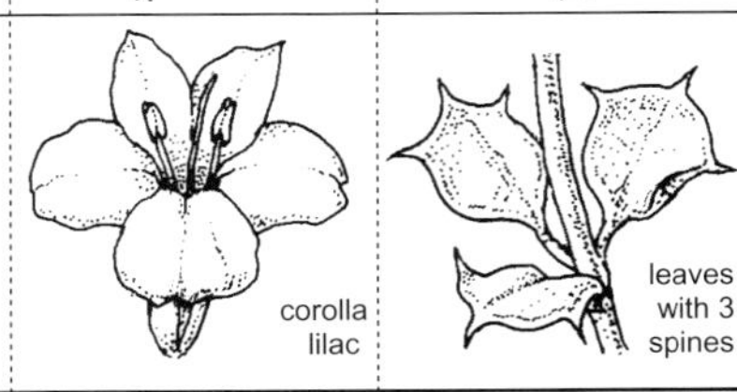	***B. aculeata*** Endemic **238**, 403

3	Low **undershrub**; **corolla** blue, c.15mm long; **leaves** linear to spatulate, green, with oblong cystoliths; **fruit** c.10mm long, beaked	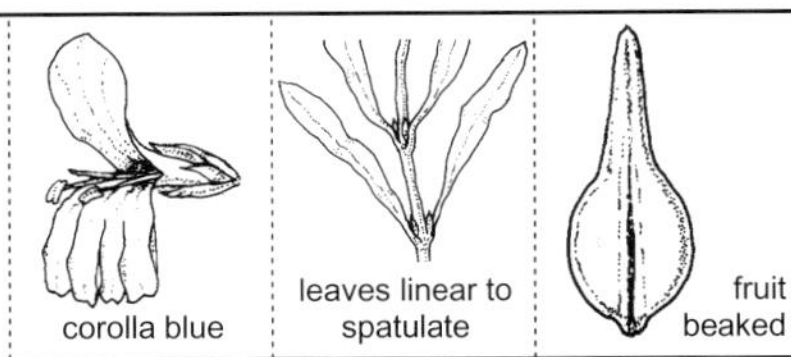	***B. argentea*** Endemic 404
	Small **tree**; **corolla** white, 45--55mm long; **leaves** ovate to lanceolate, grey-green, cystoliths absent; **fruit** c.22mm long, not beaked	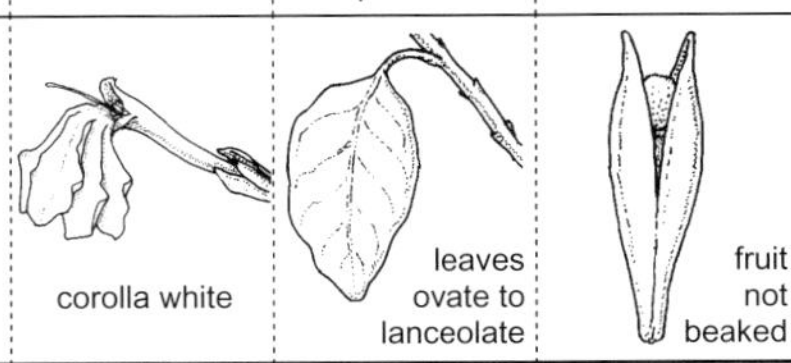	***B. popovii*** Endemic **238**, 404

Blepharis

1	Prostrate woody-based perennial **herb**; **corolla** lilac, 3-lobed; **bracts** 10--45mm long, ovate to lanceolate, spinose; **leaves** linear to spatulate, spinose	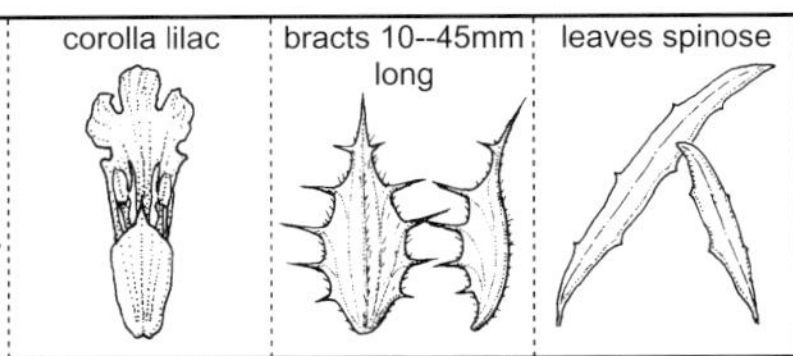	***B. kuriensis*** **239**, 405
	Erect **shrub**; **corolla** white to cream, 5-lobed; **bracts** c.6mm long, linear, spine-tipped; **leaves** oblanceolate, spinose	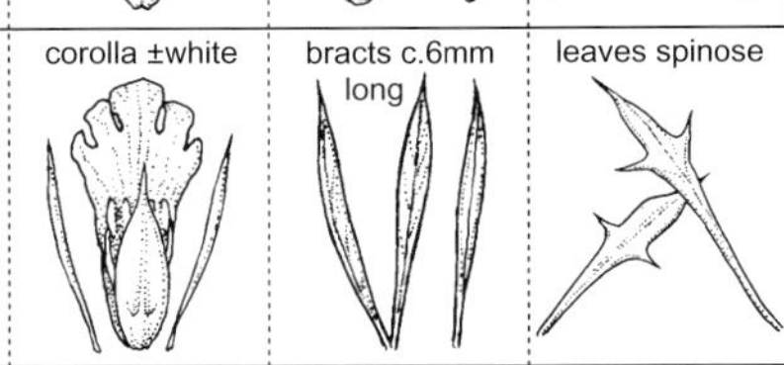	***B. spiculifolia*** Endemic **239**, 405
	Prostrate **shrub**; **corolla** white, 3-lobed; **bracts** 2--10mm long, obovate, spinose; **leaves** elliptic, unarmed	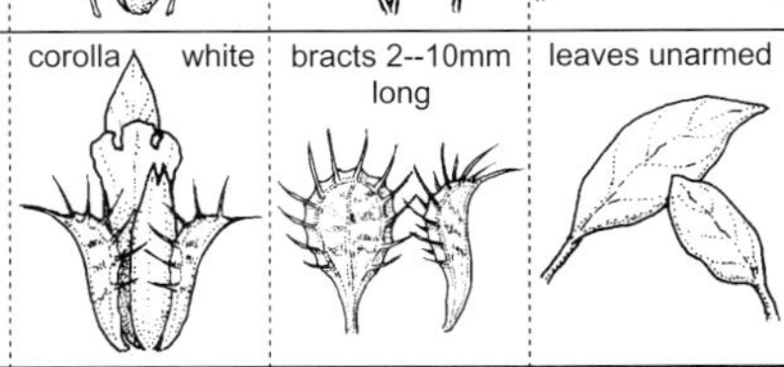	***B. maderaspatensis*** 405

Chorisochora

1	**Corolla** white, 15--20mm long; **bracts** not imbricate (overlapping), axes visible, lower bracts 3--7mm long; **fruit** 8--10mm long	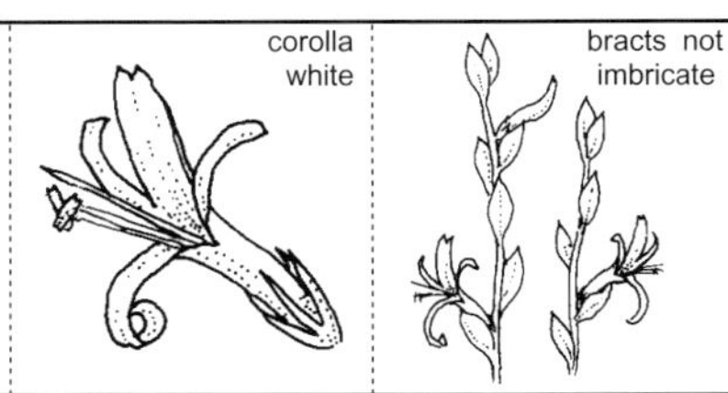	***C. minor*** Endemic 406
	Corolla pink to magenta, 25--35mm long; **bracts** imbricate (overlapping), axes hidden, lower bracts 12--20 mm long; **fruit** 9--13mm long	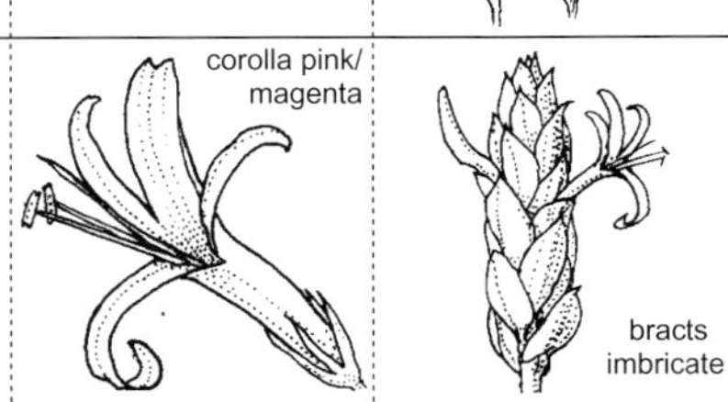	***C. striata*** Endemic **239**, 406

Justicia

1	Annual, thin-stemmed **herb**; **fruit** of 2 types, either elliptic and 4-seeded, or winged and hooked, 1-seeded	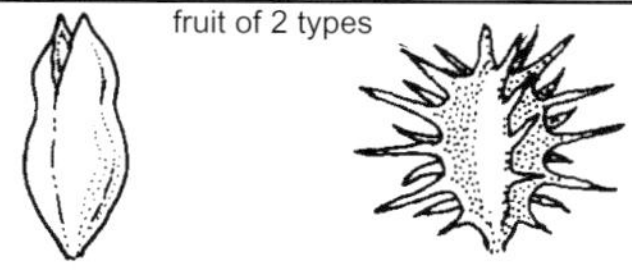	***J. heterocarpa*** 408
	Woody, intricately-branched **shrubs**; **fruit** all elliptic, 4-seeded	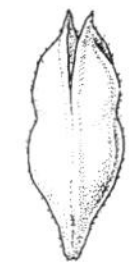	2

2	**Leaves** ovate, <3x1.5cm, not fasciculate, grey below, cystoliths not prominent; **corolla** magenta	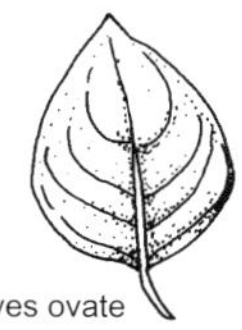	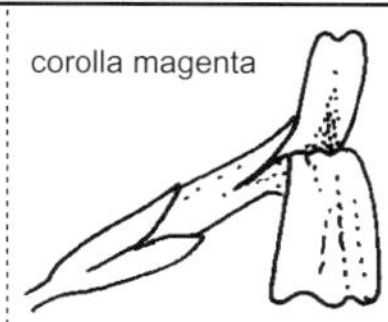	***J. alexandri*** Endemic **239**, 408
	Leaves obovate or round, <1x1cm, fasciculate, fleshy, cystoliths prominent; **corolla** magenta	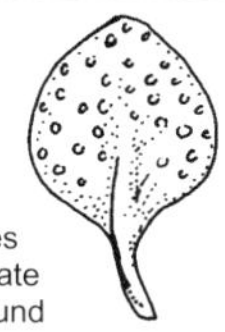	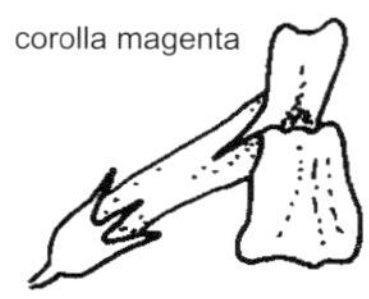	***J. takhinensis*** Endemic **239**, 409
	Leaves obovate to spatulate, grey-green, <1cm long, cystoliths not prominent; **corolla** white		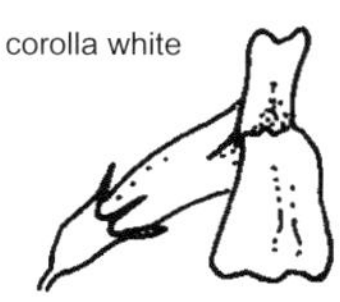	***J. rigida*** Endemic **239**, 408

Neuracanthus

1	Erect **herb**; **flower spikes** in dense clusters on erect close-set branches; **inflorescence bracts** grey pubescent		***N. aculeatus*** Endemic **240**, 409
	Cushion-forming **herb**; **flower spikes** in dense heads, partially hidden by leaves; **inflorescence bracts** white tomentose	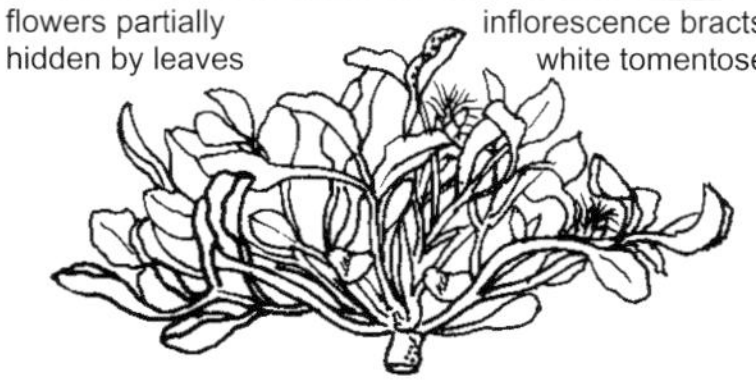	***N. argyrophyllus*** 409
	Prostrate **herb**; **flower spikes** in dense globose heads on prostrate branches; **inflorescence bracts** grey pubescent	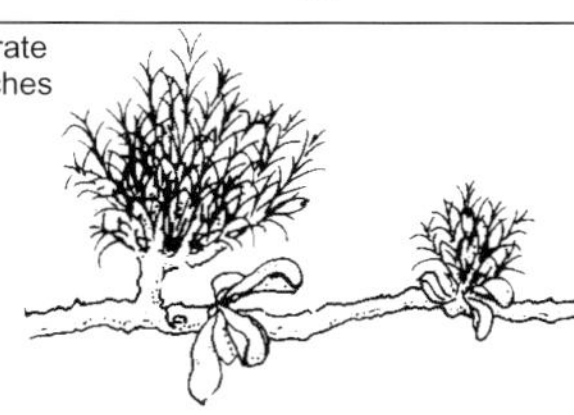	***N. capitatus*** 410

Ruellia

1	**Stem**, **leaves**, **bracts** and **calyces** densely hairy with whitish grey hairs	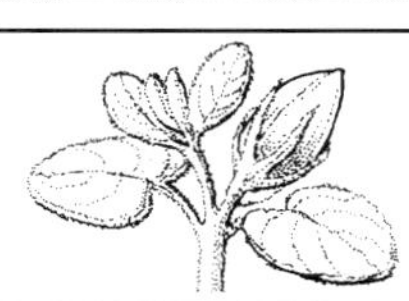	2
	Stem, **leaves**, **bracts** and **calyces** glabrous or tomentose with greenish hairs	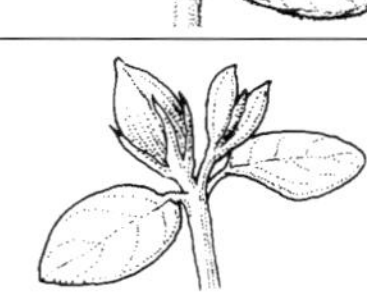	3

2	**Corolla** red; **leaves** cordate at base; **hairs** stellate or glandular; **fruit** tomentose, c.15mm long	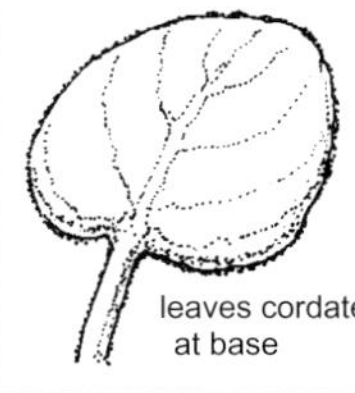 leaves cordate at base 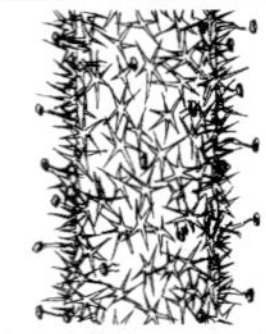 hairs stellate or glandular	***R. carnea*** **240**, 411
	Corolla lilac; **leaves** cuneate at base; **hairs** simple, strigose; **fruit** glabrous, c.10mm long	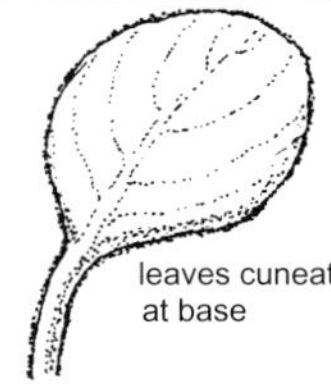 leaves cuneate at base 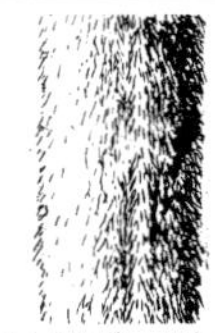 hairs simple, strigose	***R. paulayana*** Endemic **240**, 412

3	Erect **shrub** or small **tree**; **corolla** lilac, lobes strongly overlapping, 35--45mm long; **leaves** 12--35mm long; **fruit** c.16mm long	 shrub or small tree  corolla lilac 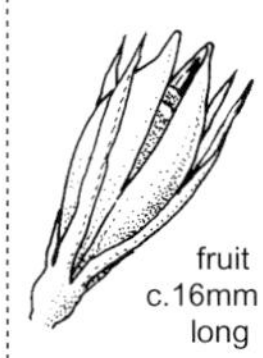 fruit c.16mm long	***R. insignis*** Endemic **240**, 411
	Low **shrub** or woody-based **herb**; **corolla** blue, lilac or white, lobes ± overlapping, 10--30mm long; **leaves** 5--12mm long; **fruit** 5--10mm long	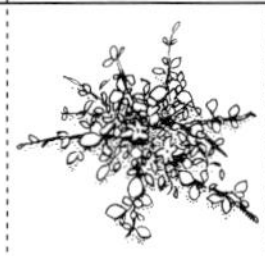 low shrub or woody-based herb 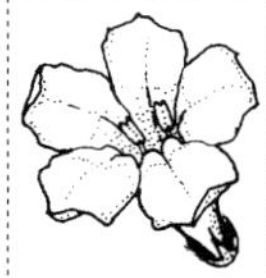 corolla blue, lilac or white  fruit 5--10mm long	4

4	**Subshrub** with corky branches; **leaves** fleshy; **corolla** lilac, <3cm long, tube broadened above; **anthers** included in tube	subshrub, branches corky 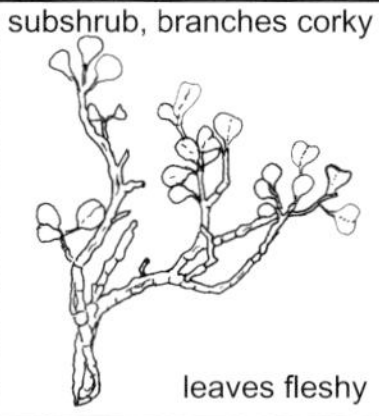 leaves fleshy; corolla lilac 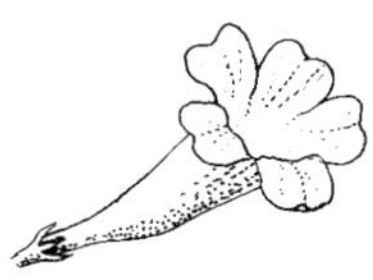	***R. kuriensis*** Endemic **240**, 412
	Prostrate **herb**, branches not corky; **leaves** not fleshy; **corolla** lilac to blue, rarely white, <2.5cm long, tube broadened above; **anthers** included in corolla tube	prostrate herb  leaves not fleshy; corolla lilac/blue, rarely white 	***R. patula*** 412
	Prostrate **herb**, branches not corky; **leaves** not fleshy; **corolla** white, rarely lilac, <1.5cm long, tube narrow; **anthers** level with top of corolla tube	prostrate herb  leaves not fleshy; corolla white, rarely lilac 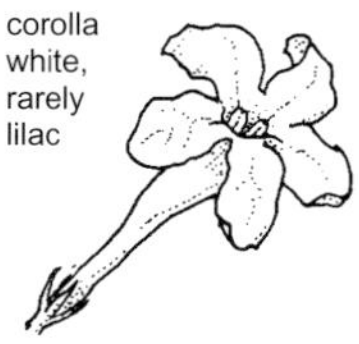	***R. dioscoridis*** Endemic **240**, 411

Trichocalyx

1	**Leaves** tending to be orbicular	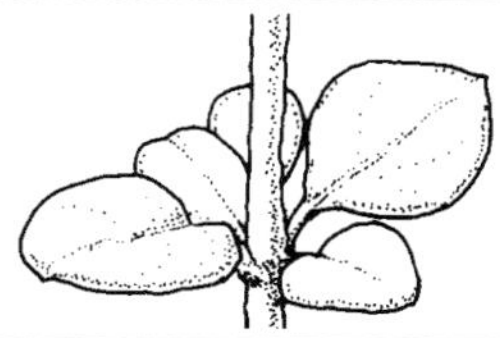 leaves orbicular	***T. orbiculatus*** Endemic 413
	Leaves tending to be obovate	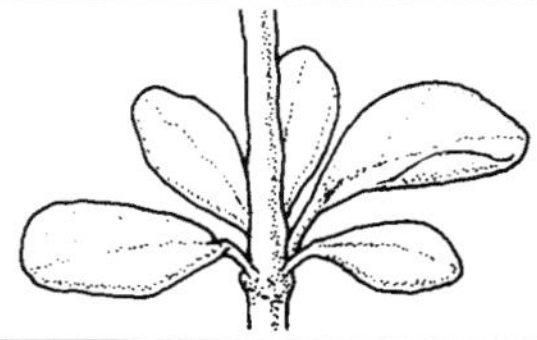 leaves obovate	***T. obovatus*** Endemic **241**, 413

AIZOACEAE

1 Erect annual **herbs**; **leaves** in a basal rosette; **stem leaves** absent or clustered in inflorescence axils → ***Mollugo*** (2 spp.)

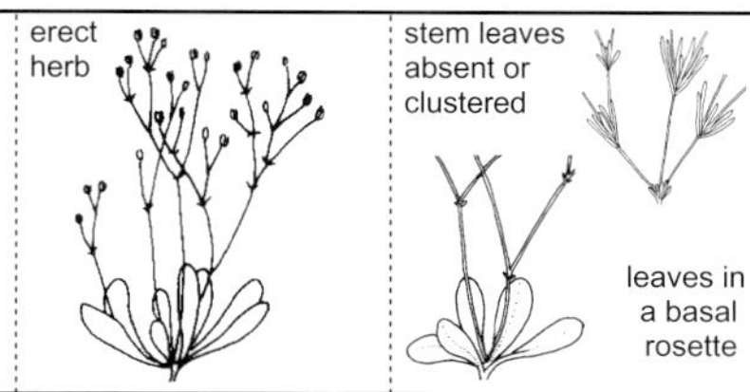

1 Prostrate or decumbent annual or perennial **herbs**; **leaves** various, never in a basal rosette; **stem leaves** present, not clustered in inflorescence axils → 2

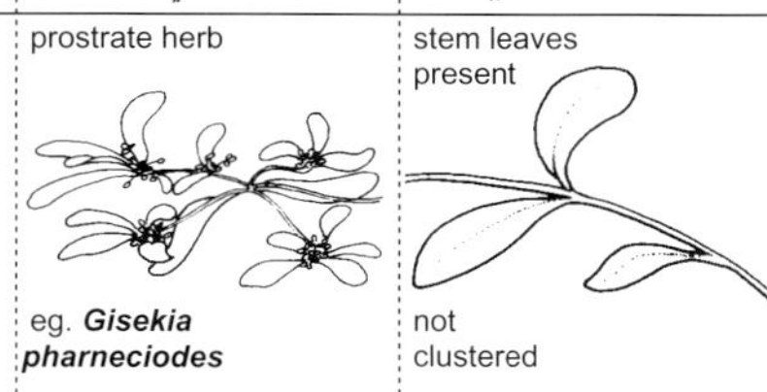

2 **Leaves** alternate → 3

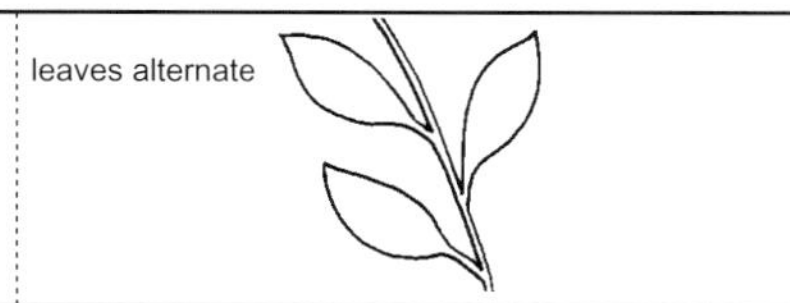

2 **Leaves** opposite → 5

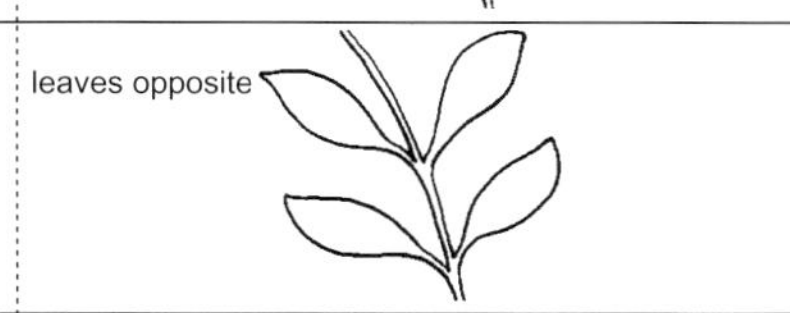

3 Many pink or magenta petal-like **staminodes** present → ***Corbichonia decumbens*** **241**, 414

3 Petal-like **staminodes** absent → 4

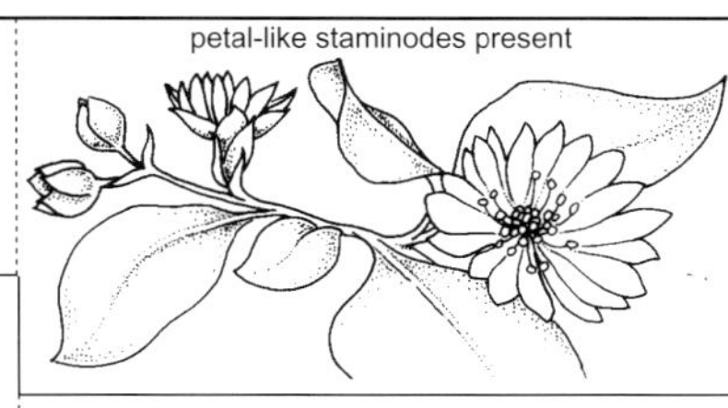

4 **Fruit** dry and indehiscent, rounded; **leaves** glabrous, broadly elliptic or ovate, contracted abruptly below into the petiole → ***Tetragonia pentandra*** 416

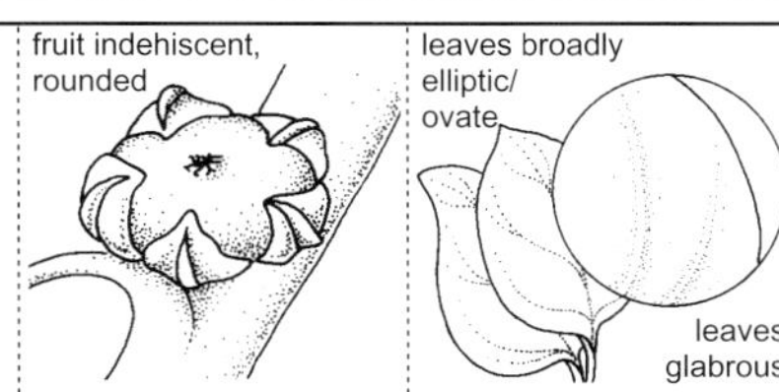

4 **Fruit** a many-seeded capsule dehiscing by 5 valves, 5-angled; **leaves** densely or thinly pilose, orbicular to obovate, attenuate below → ***Aizoon canariense*** **241**, 414

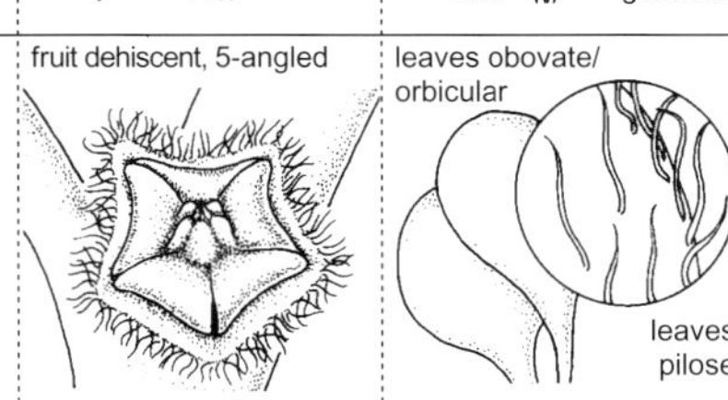

5 **Carpels** free; **fruit** a cluster of 5 achenes; **plant** marked throughout with numerous linear raphides → ***Gisekia pharnaceoides*** 415

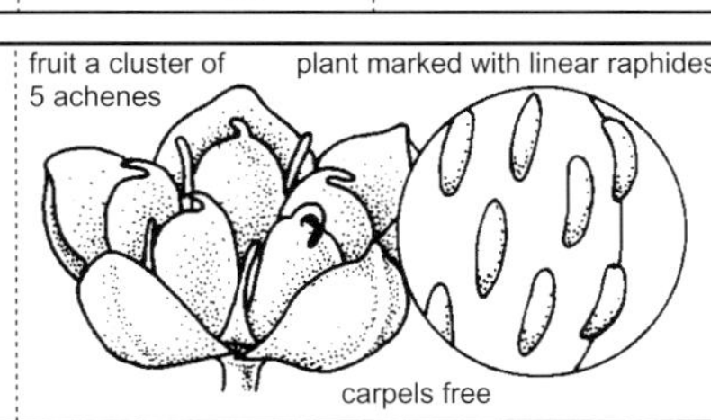

5 **Carpels** united; **fruit** a capsule; **plant** without linear raphides → 6

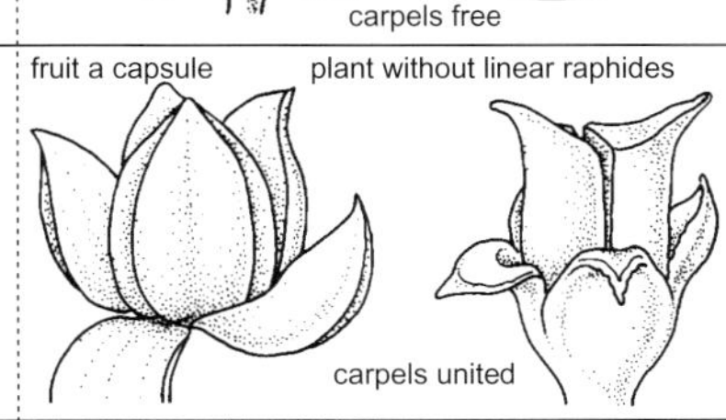

6

Plant softly stellate-tomentose; **sepals** free; **fruit** a many-seeded capsule, dehiscing by (3 to) 5 valves — ***Glinus lotoides*** 415

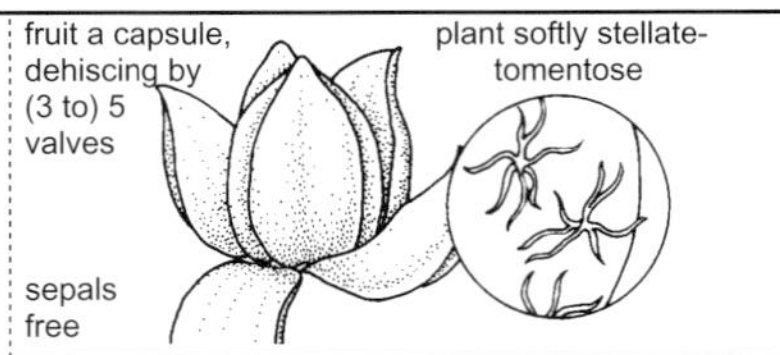

Plant glabrous or papillose; **sepals** united into a tube below; **fruit** a 4-seeded capsule, dehiscing by a 2-valved lid, the valves usually separating — ***Zaleya pentandra*** 417

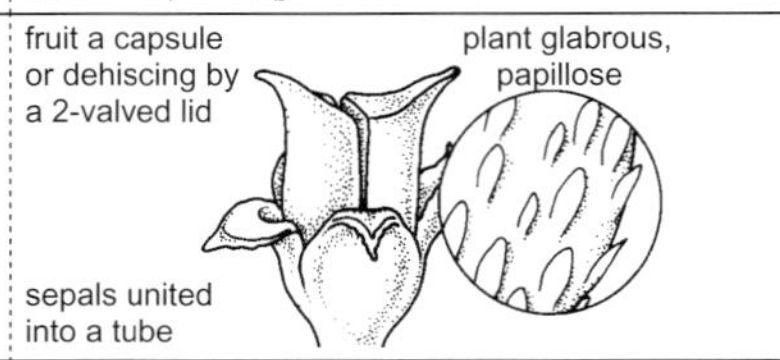

Mollugo

1

Delicately branched erect or ascending annual **herb**; **leaves** obovate to oblanceolate, confined to the basal rosette; **stem leaves** absent — ***Mollugo nudicaulis*** 416

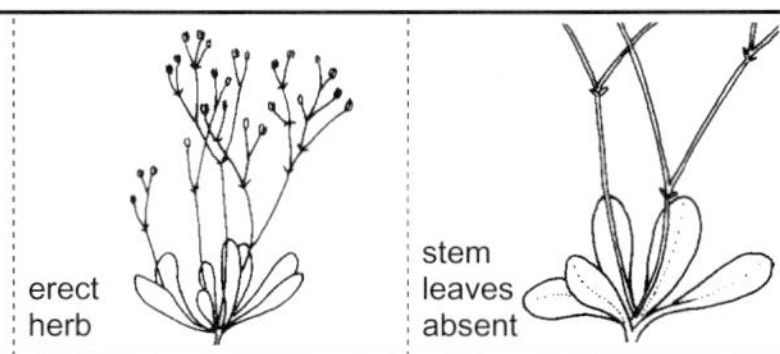

Delicate spreading annual **herb**; **leaves** linear-lanceolate; **stem leaves** present in inflorescence axils — ***Mollugo cerviana*** 416

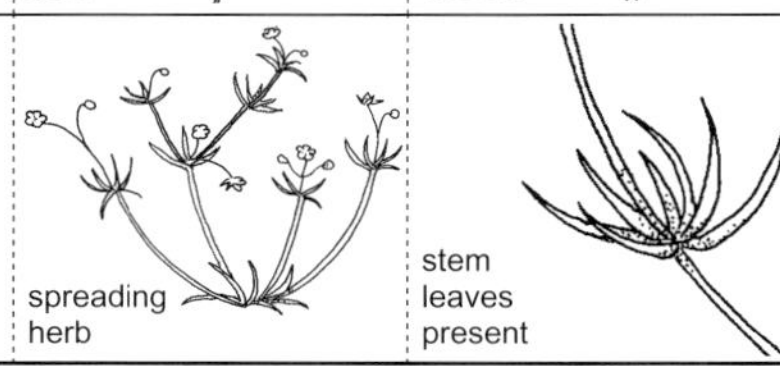

AMARANTHACEAE

1

Leaves opposite or absent — 2

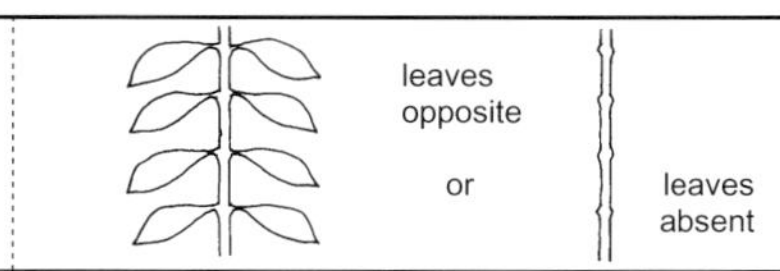

Leaves alternate — 3

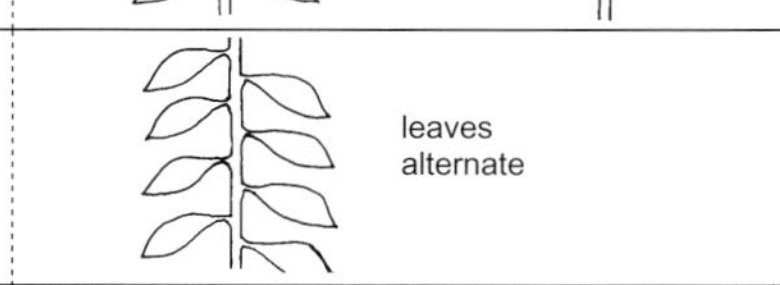

2

Flowers in open spikes, subtended by rigid, hooked spines; **fruit** a highly adhesive burr — ***Pupalia lappacea*** 423

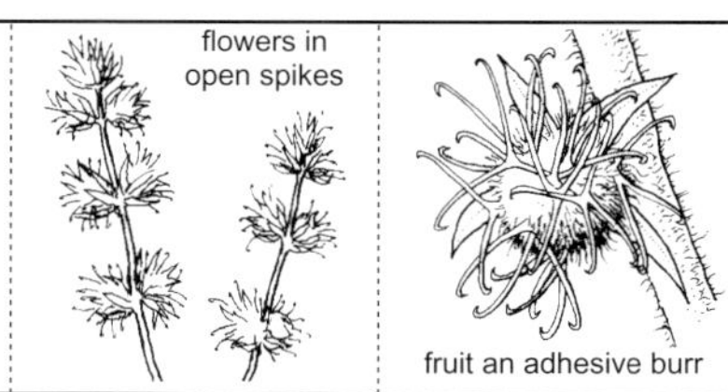

Flowers in globose heads, becoming cylindrical in fruit; **bracteoles** with a small, irregularly serrulate crest — ***Gomphrena celosioides*** 422

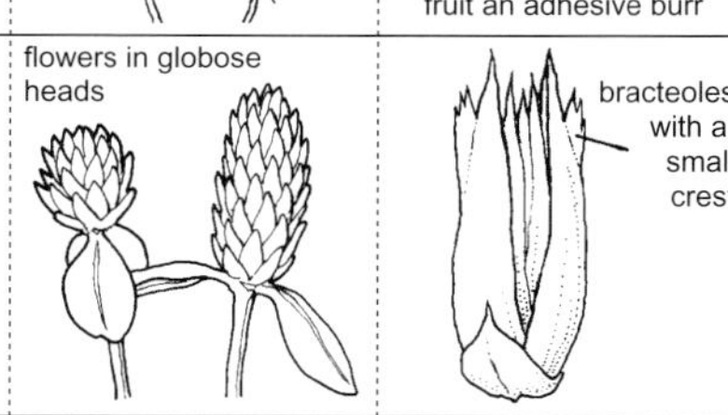

Flowers in elongated spikes, becoming deflexed and appressed to the axis; **pseudostaminodes** (alternating with the stamens) present — ***Achyranthes aspera*** 417

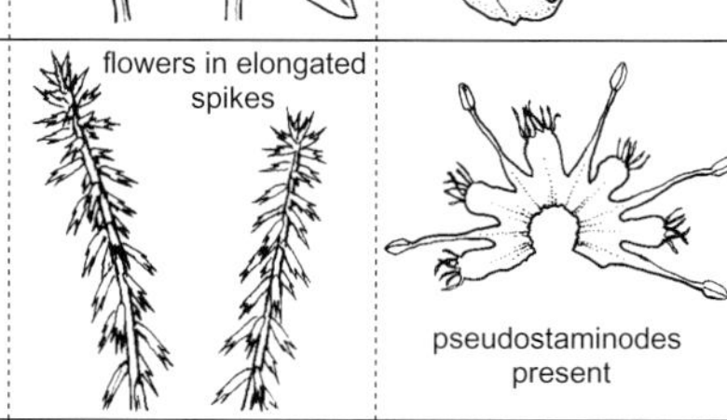

Flowers in lax spikes forming open panicles or in compact (< 6-flowered) spikes, never deflexed; **tepals** distinctly 3-nerved — ***Psilotrichum*** (2 spp.)

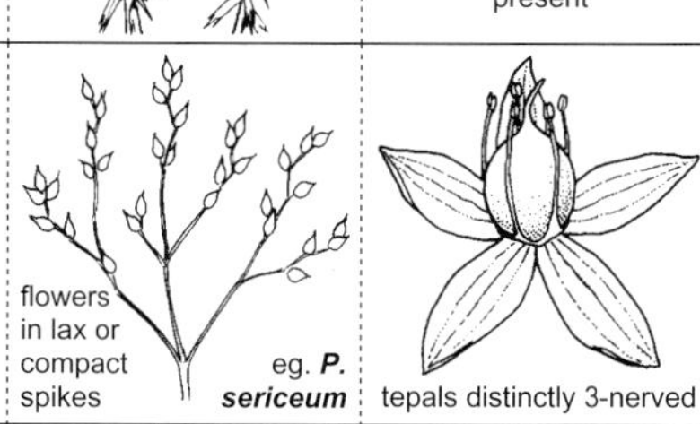

3	Annual **herbs**; **flowers** glabrous and subtended by 2 antler-shaped appendages; **pseudostaminodes** absent	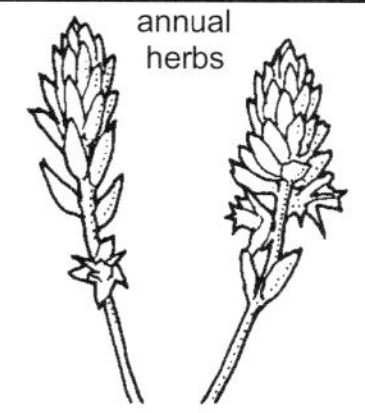	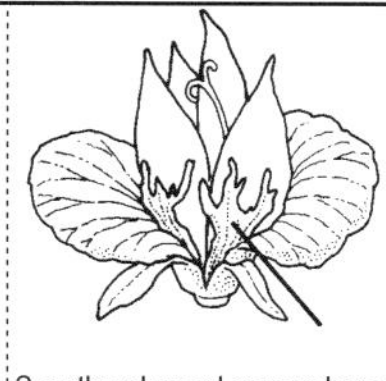	***Digera muricata*** 421
	Annual **herbs**; **flowers** glabrous, without appendages; **pseudostaminodes** absent	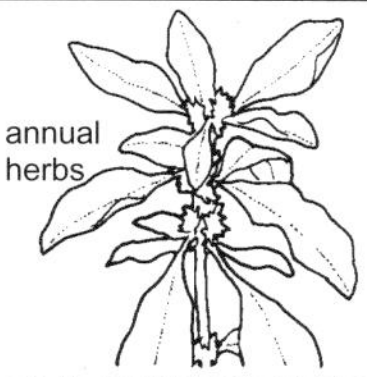	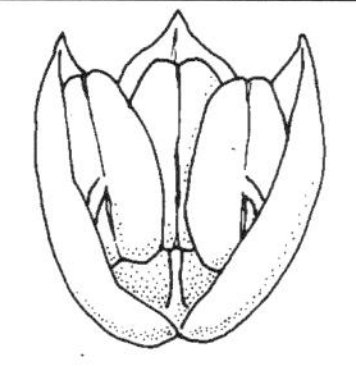	***Amaranthus*** (3 spp.)
	Shrubs or **herbs**; **flowers** lanate or tomentose, without appendages; **pseudostaminodes** (alternating with the stamens) present	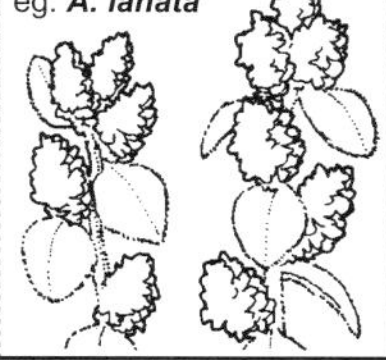		***Aerva*** (4 spp.)

Aerva

1	**Stems** and **leaves** glabrous except for tufts of hairs in the leaf axils	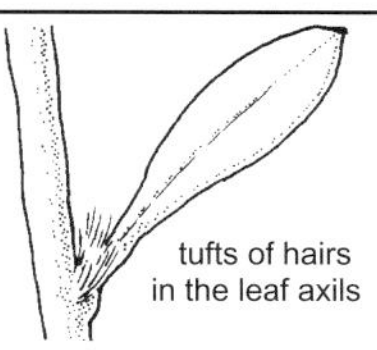	***A. microphylla*** Endemic **241**, 419
	Stems and **leaves** lanate or tomentose	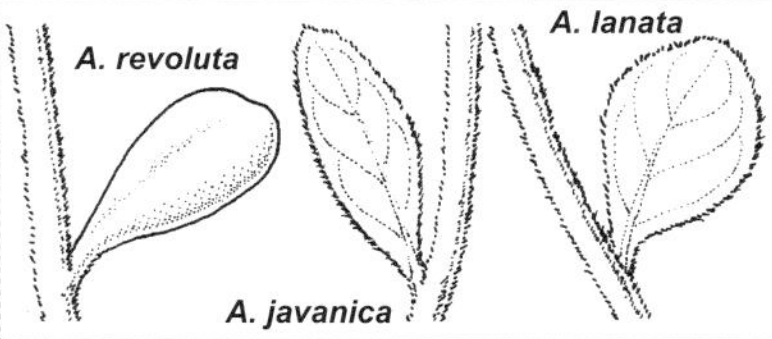	2

2	**Subshrub**; **leaves** glabrous above, shortly tomentose beneath, margins revolute	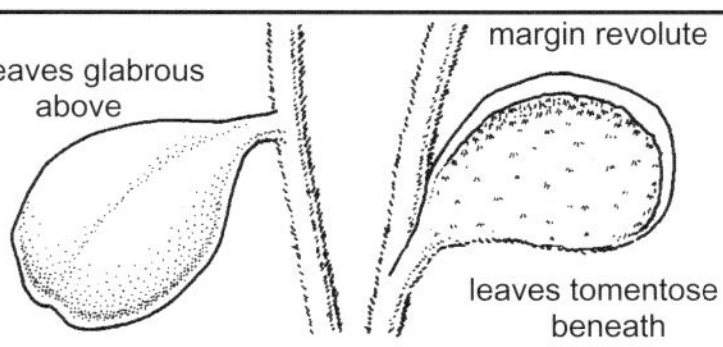	***A. revoluta*** Endemic **241**, 419
	Annual or perennial **herbs**; **leaves** lanate or tomentose above and beneath, margin not revolute	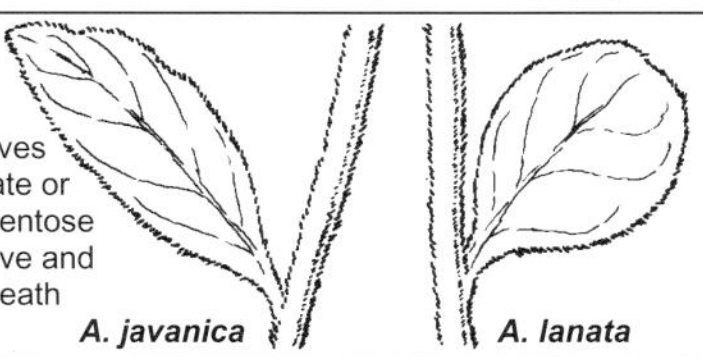	3

3	**Inflorescence** usually leafless; **spikes** 1--15cm long; **flowers** unisexual; outer 2 **tepals** acute or obtuse, inner 3 tepals with narrow midrib and broad membranous margin	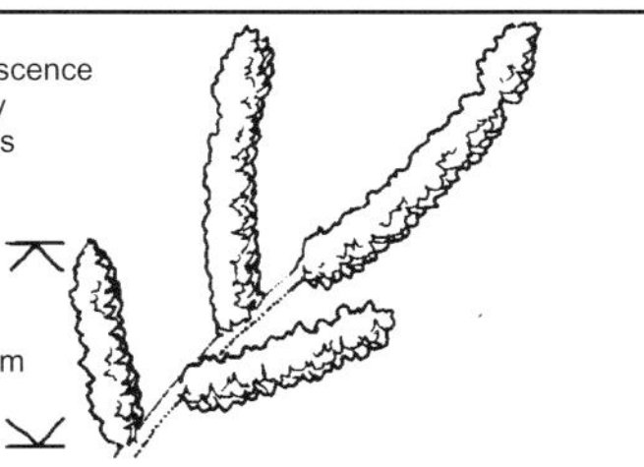	***A. javanica*** 418
	Inflorescence usually leafy; **spikes** 0.5--1(--2)cm long; **flowers** usually bisexual; outer 2 **tepals** mucronate, inner 3 with broad midrib and narrow membranous margin	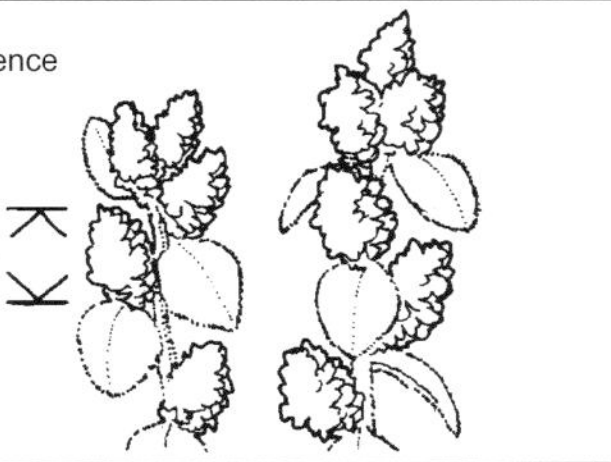	***A. lanata*** 419

Amaranthus

1	**Fruits** ± woody, longitudinally winged, arranged in stellately divergent clusters in the axils of the leaves	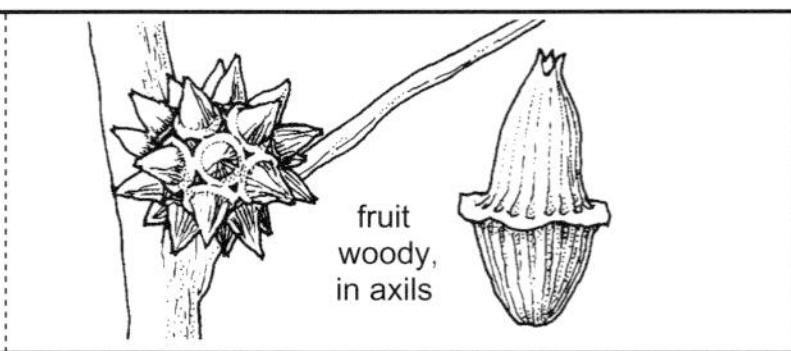	***A. sparganiocephalus*** 421
	Fruits thin-walled, strongly wrinkled, clustered in the axils of the leaves	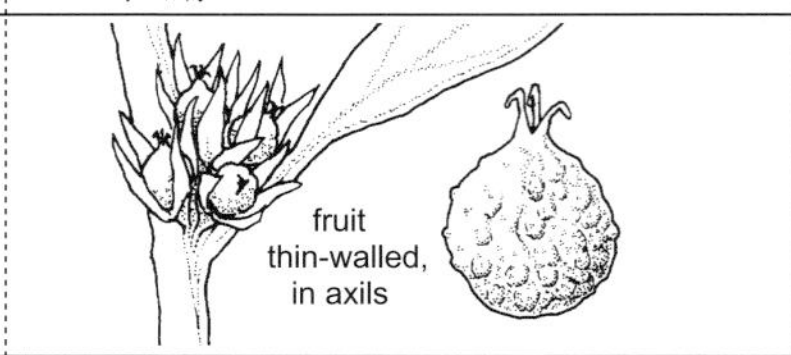	***A. graecizans*** 420
	Fruits with an elongated beak, wrinkled, arranged in terminal spikes or in the leaf axils	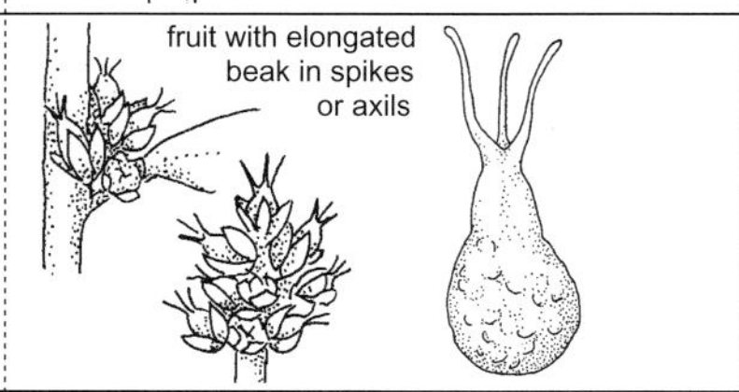	***A. dubius*** 420

Psilotrichum

1	**Plant** with ovate leaves**,** silvery-sericeous beneath; **flowers** 2--2.5mm long, in lax spikes forming large, open panicles; **nodes** with tufts of long fine hairs	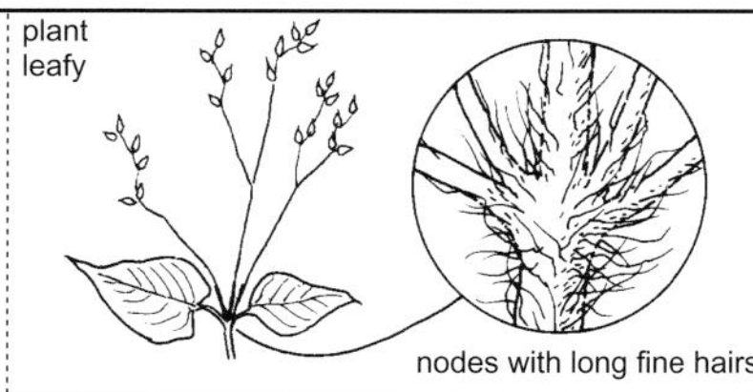	***P.sericeum*** 422
	Plant leafless; **flowers** c.5mm long, in short, compact, up to 6-flowered spikes; **nodes** without long hairs	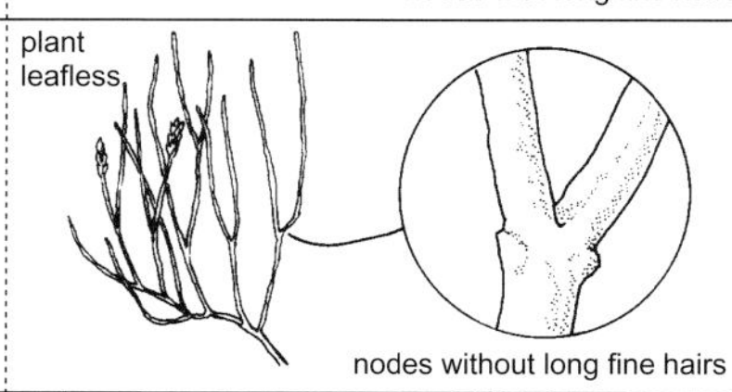	***P. aphyllum*** Endemic **241**, 422

ANACARDIACEAE

1	Large **tree**; **leaves** simple with stellate hairs; **flowers** with 4 free petals; **fruit** ovoid, with 4 prominent style bases	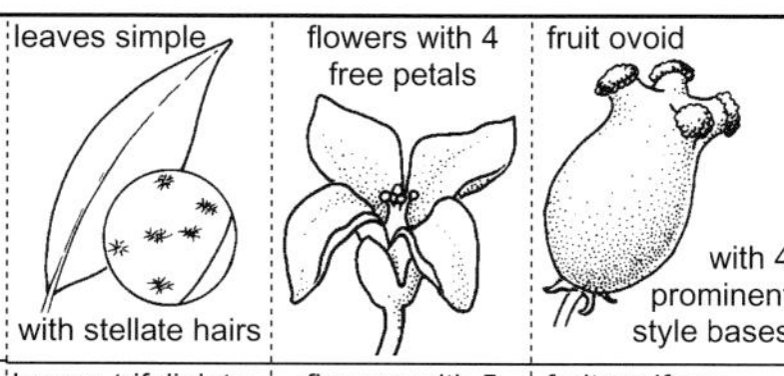	***Lannea transulta*** Endemic **241**, 424
	Small **tree**; **leaves** trifoliolate, glabrous or with simple leaves; **flowers** with 5 free petals; **fruit** reniform, without prominent style bases	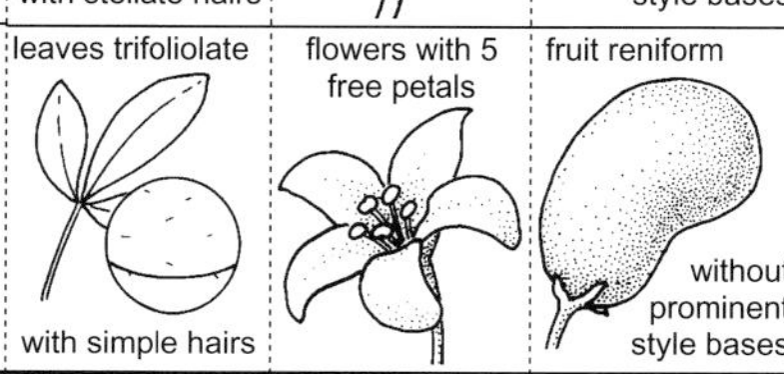	***Rhus*** (2 spp.)

Rhus

1	**Leaflets** <2cm long with recurved margins; **flowers** in little-branched panicles, shorter than leaves	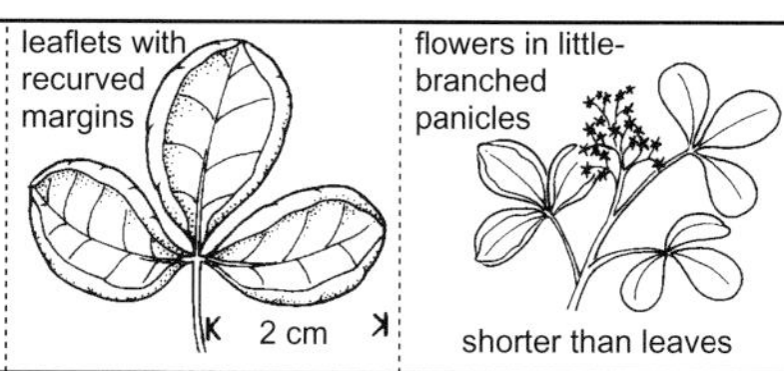	***Rhus* sp. A** Endemic 425
	Leaflets >3cm long, margins not recurved; **flowers** in open panicles, exceeding leaves	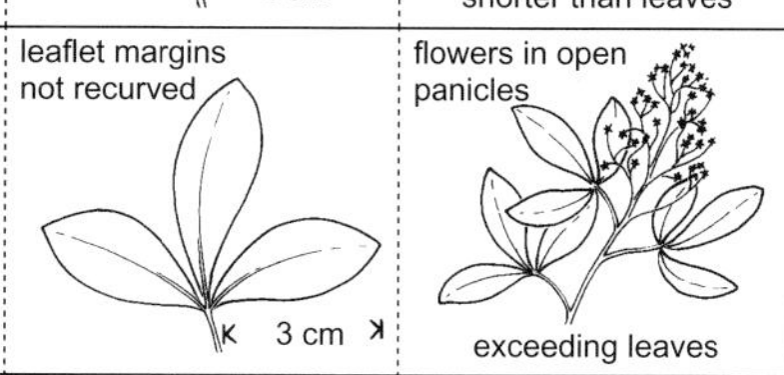	***Rhus thyrsiflora*** Endemic **242**, 425

APOCYNACEAE

1 **Shrub** with bottle-like, swollen trunks; **leaves** spirally arranged, obovate; **flowers** pink, 3--6cm long; **fruit** a pair of cylindrical follicles — ***Adenium obesum*** Endemic subsp. **242**, 427

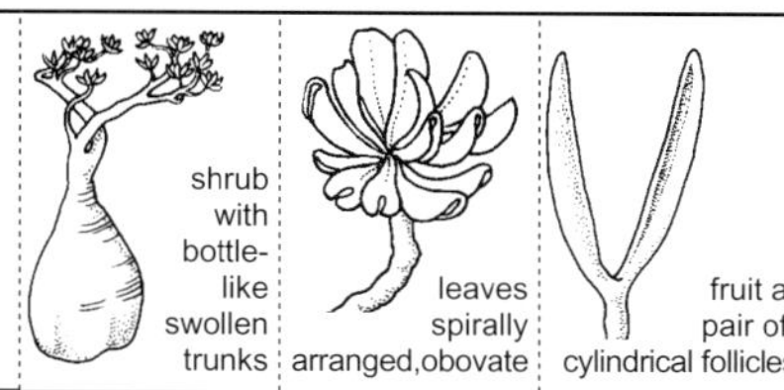

1 **Shrub** or small **tree**, trunks not swollen; **leaves** opposite; **flowers** white or pinkish, c.1cm long; **fruit** ellipsoid or ovoid berries — 2

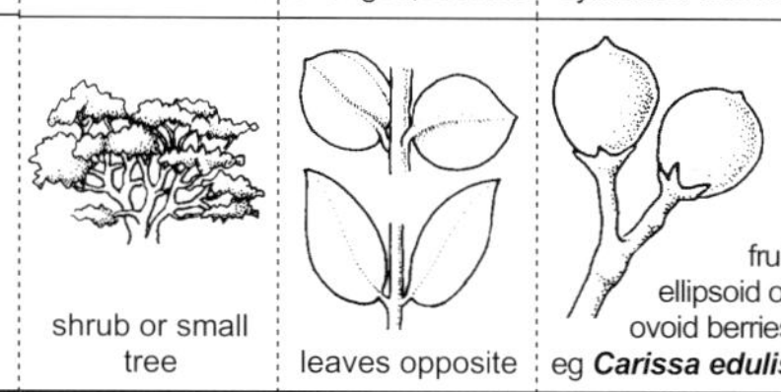

2 **Branches** spiny; **corolla lobes** lanceolate — ***Carissa spinarum*** 429

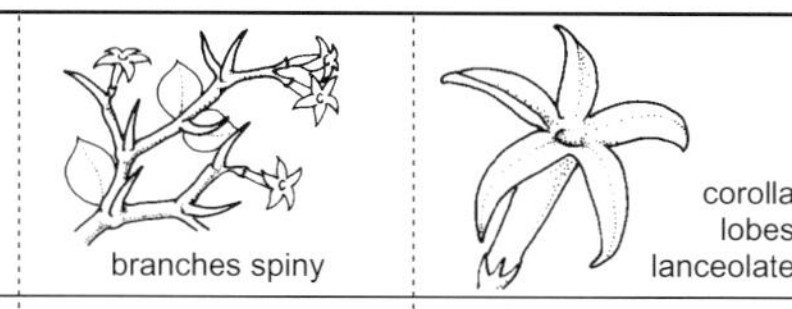

2 **Branches** unarmed; **corolla lobes** broadly ovate — ***Acokanthera schimperi*** 427

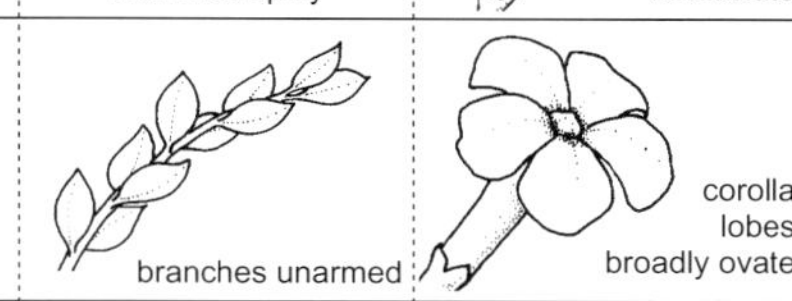

ASCLEPIADACEAE

1 **Stems** pencil-like and rigid or succulent and twining; **leaves** absent or scale-like — 2

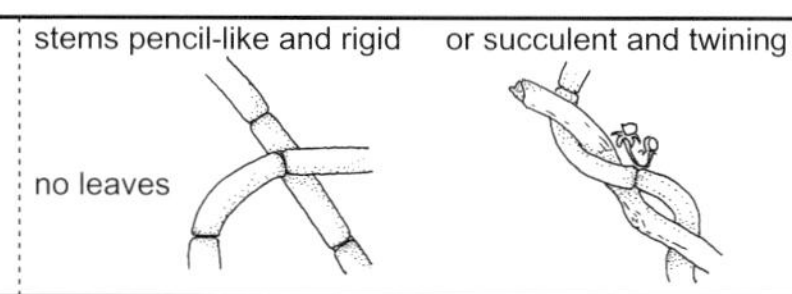

1 **Stems** not succulent or pencil-like; **leaves** well-developed — 3

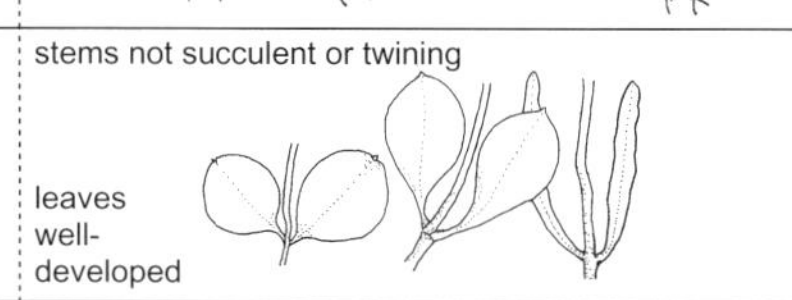

1 **Stems** succulent, with angles and tubercles; **leaves** absent or reduced to tiny scales — 10

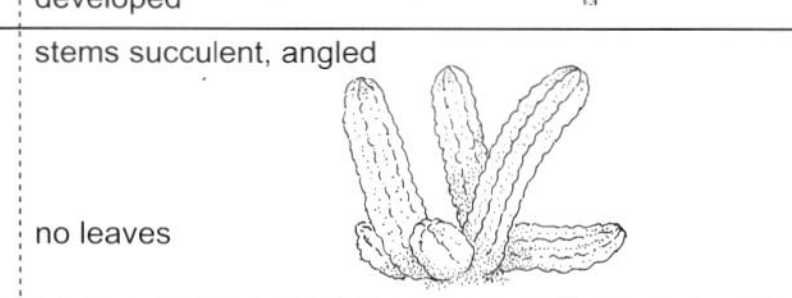

2 Densely-branched **shrub** with pencil-like stems; **corolla lobes** free at apex; **stamens** woolly — ***Periploca visciformis*** **245**, 440

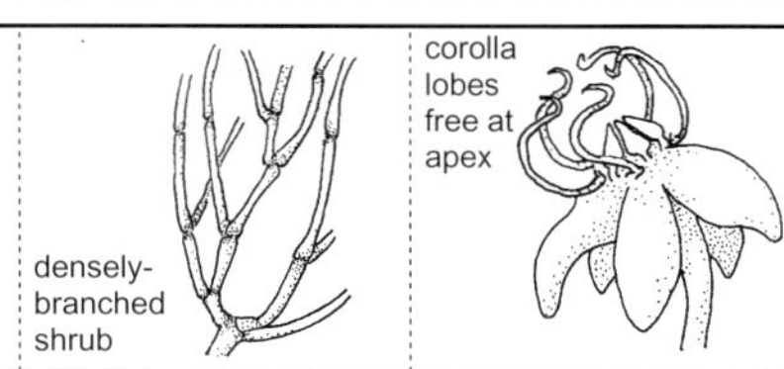

2 Densely-branched **shrub** with pencil-like stems; **corolla lobes** free at apex; **stamens** glabrous — ***Sarcostemma*** (2 spp.)

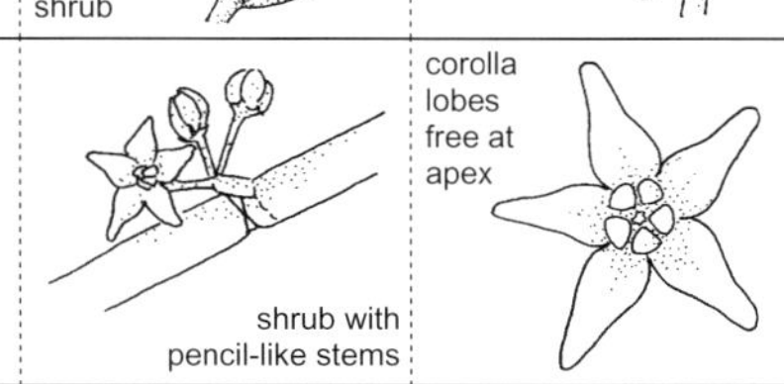

2 Succulent **climber**; **corolla lobes** free at apex; **stamens** glabrous — ***Cynanchum gerardii*** 435

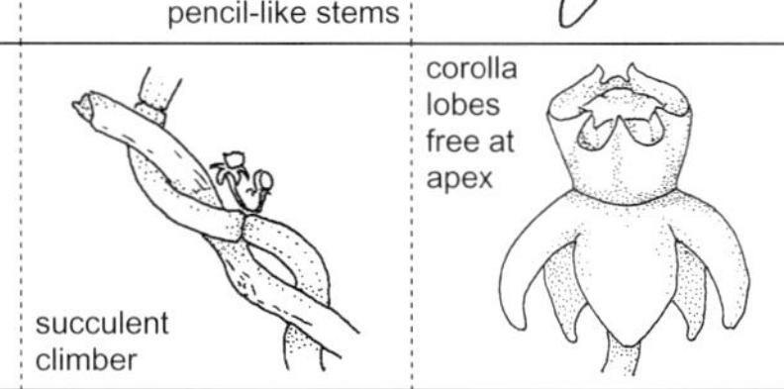

2 Succulent **climber**; **corolla lobes** fused at apex; **stamens** glabrous — ***Ceropegia*** (2 spp.)

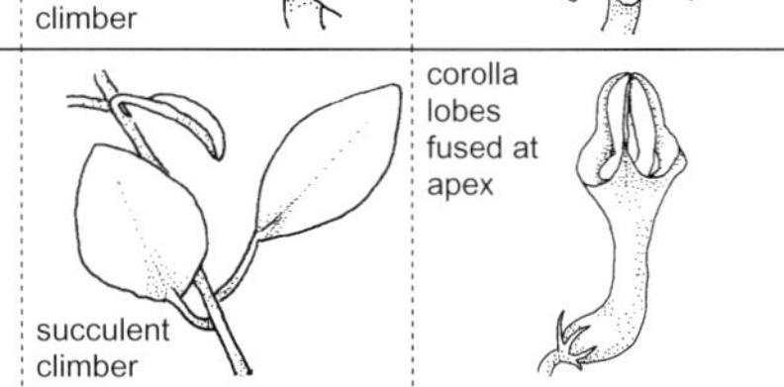

3

- **Sap** white → 4

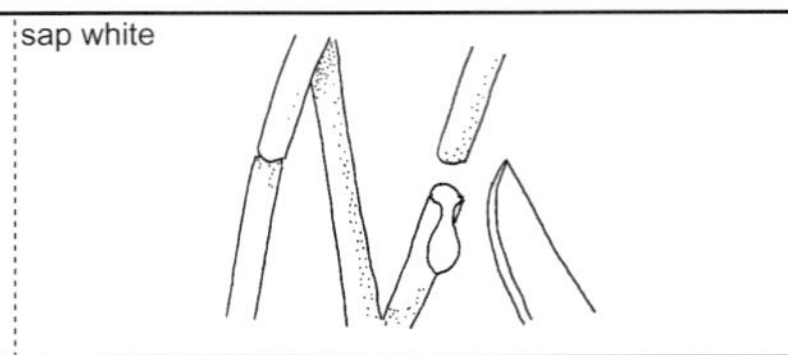

- **Sap** clear → 9

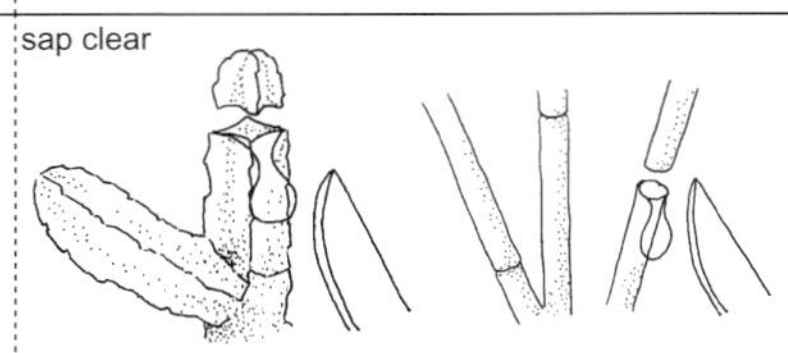

4

- Large **shrub** or **tree** with fissured bark; **leaves** 10--15cm long; **flowers** c.20mm diameter; **fruits** large (c. 10cm long), globose, much inflated → ***Calotropis procera*** **243**, 429
- **Habit** various; **leaves** <10cm long; **flowers** <10mm diameter; **fruits** not inflated → 5

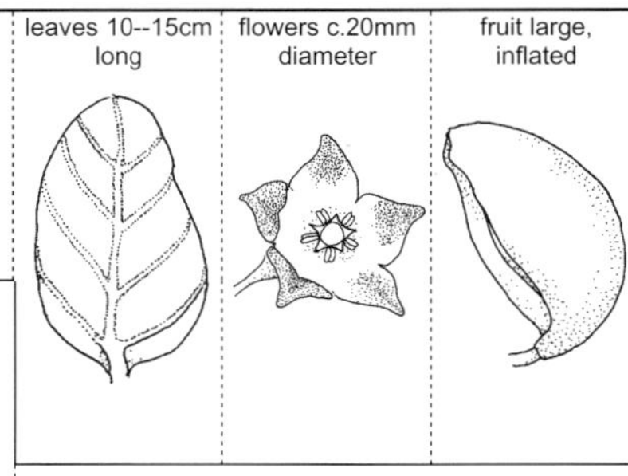

5

- Densely branched climbing **herb**; **leaves** linear; **flowers** reddish-brown, <2mm diameter → ***Vincetoxicum linifolium*** Endemic **245**, 442
- **Herbs** or **shrubs**, if climbing then not densely branched; **leaves** not linear; **flowers** larger → 6

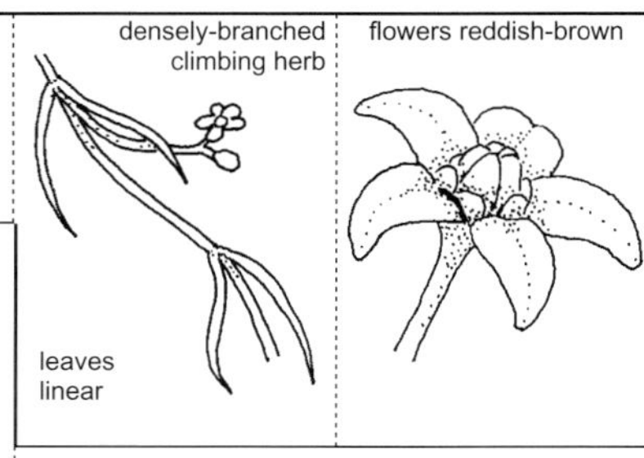

6

- **Herb** or herbaceous **climber**; **stems** and **leaves** densely-tomentose; **fruit** softly spiny → 7

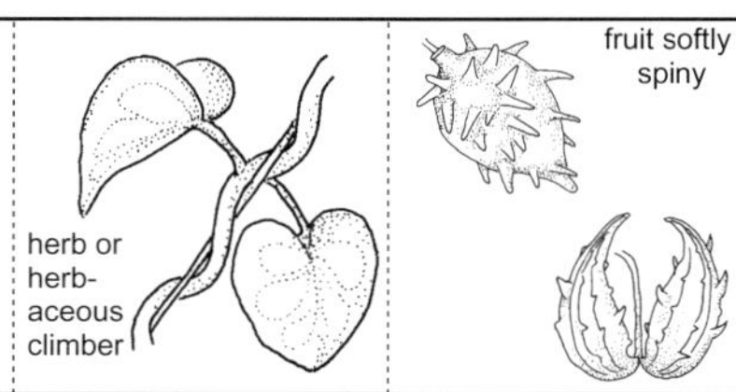

- **Shrub** or woody **climber**; **stems** and **leaves** glabrous; **fruit** smooth → 8

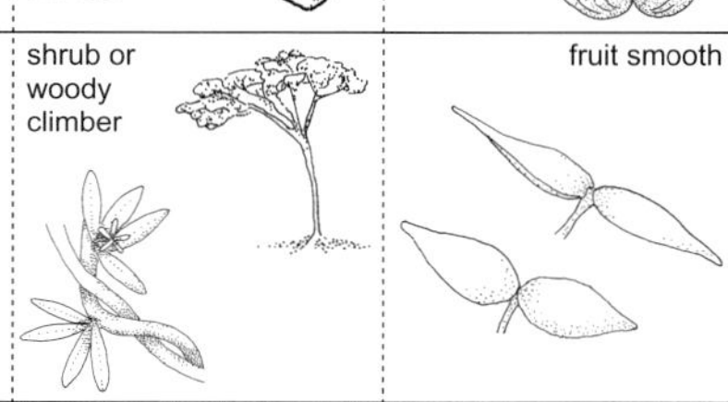

7

- Ascending **herb** <20cm; **leaves** oblong-elliptic; **Flowers** yellowish, glabrous within → ***Glossonema revoili*** 438

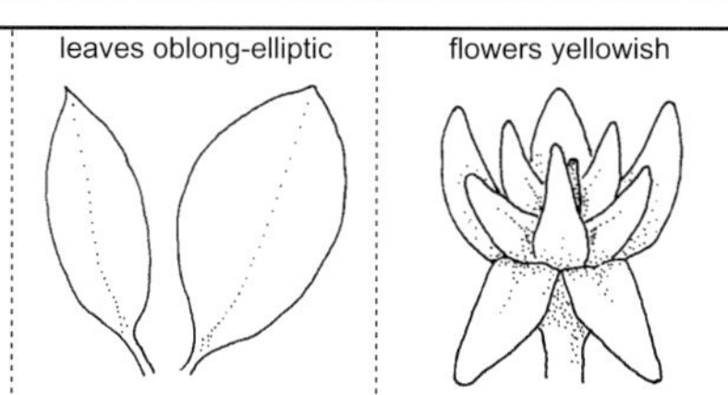

- Twining **herb** <1m; **leaves** cordate; **flowers** greenish with conspicuous hairs within → ***Pergularia tomentosa*** 439

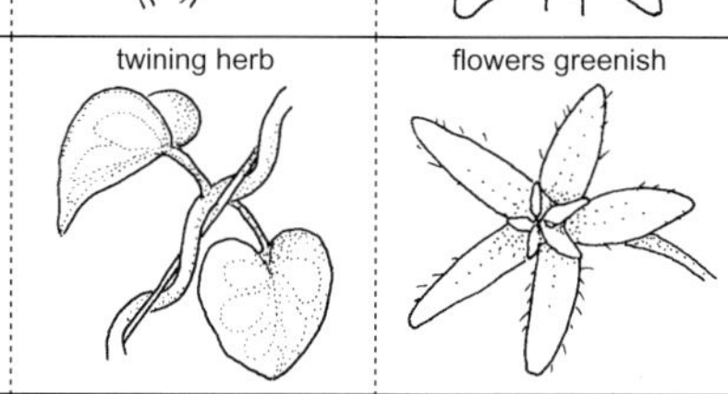

8	**Style head** not exserted; **coronal lobes** filiform or club-shaped, straight or bent	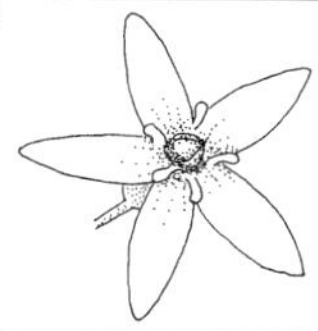	***Cryptolepis*** (5 spp.)
	Style head exserted from flower; **coronal lobes** inconspicuous	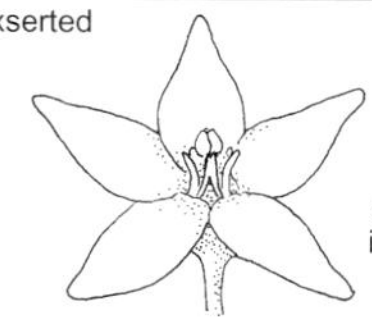	***Secamone*** (2 spp.)
9	**Shrub** or small **tree**; **leaves** ovate; **corolla lobes** spreading	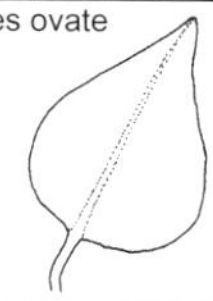 	***Marsdenia robusta*** Endemic 438
	Herb with twining stems; **leaves** oblong or scale-like; **corolla lobes** fused at tips	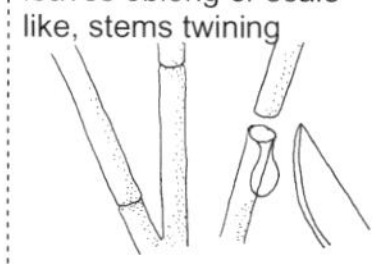 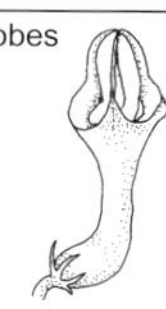	***Ceropegia*** (2 spp.)
10	**Stems** 4-angled with soft fleshy teeth	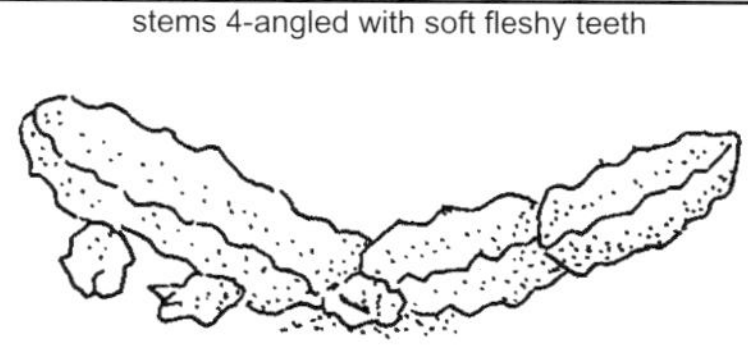	11
	Stems 5 to 8-angled, nearly cylindrical, with either a hard conicle tooth on each tubercle or tubercles soft	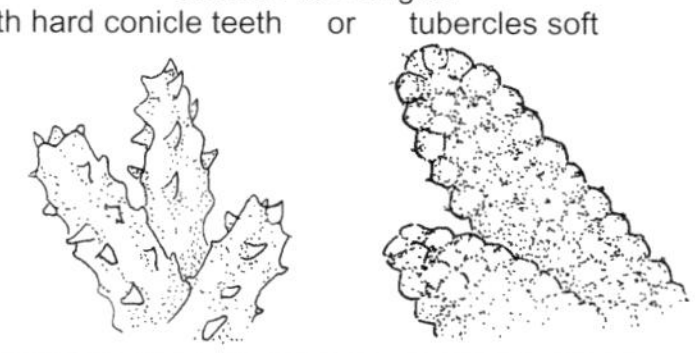	13
11	**Flowers** produced at or below middle of stems; **corolla** rotate, 35--55mm diameter	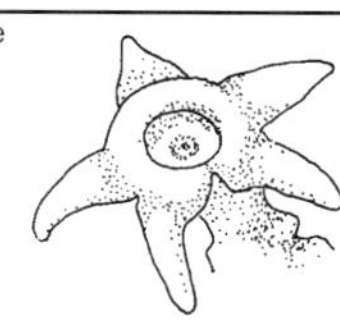	***Duvaliandra dioscoridis*** Endemic Endemic genus **244**, 435
	Flowers produced towards apex of stems; **corolla** cup-shaped, 16--22mm diameter	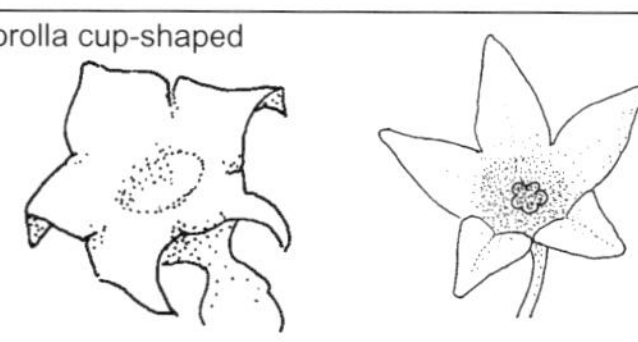	12
12	**Stems** erect and rooting from a a central stem only; **pedicels** 1--3mm long; **corolla** with cup-shaped tube 6--8mm deep, bright red or yellowish-tinged	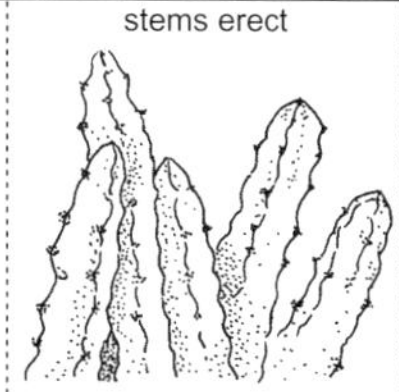 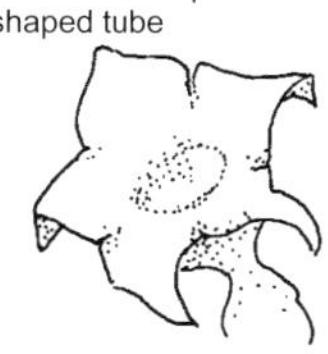	***Caralluma socotrana*** **243**, 430
	Stems ascending from horizontal and often rhizomatous base; **pedicels** 25--42mm long; **corolla** with shallow tube 2mm deep, bright yellow with red markings	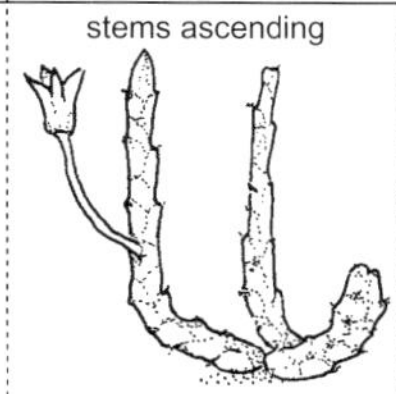 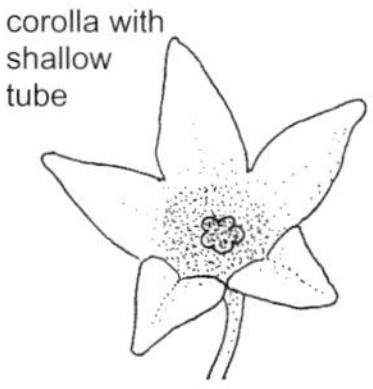	***Socotrella dolichocnema*** Endemic Endemic genus **245**, 442

13	**Stems** roughly 5-angled to cylindrical, tubercles each tipped with a hard conical tooth; **flowers** large, 4--10cm diameter	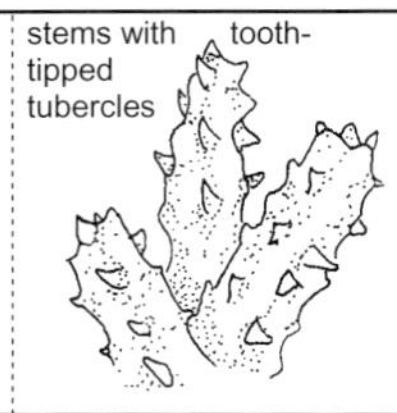	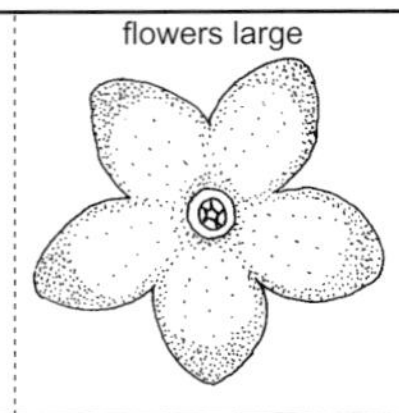	***Edithcolea grandis*** Endemic **245**, 437
	Stems 6 to 8-angled, covered with soft, obtuse tubercles; **flowers** small, 0.2--1.3cm diameter	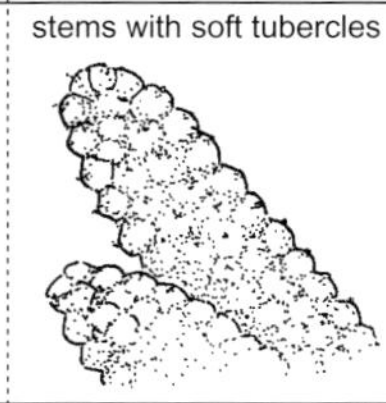	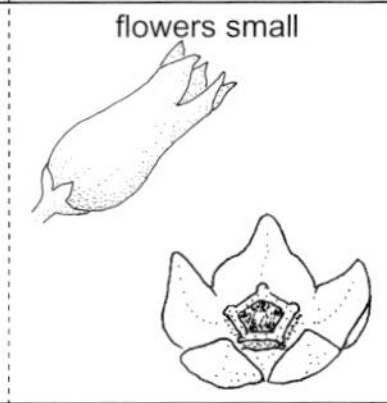	***Echidnopsis*** (5 spp.)

Ceropegia

	Character	Illustration	Result
1	**Stems** thin (<1.5mm diameter), <2m long; **leaves** narrowly oblong, usually >15mm long, persistent; lower part of **corolla tube** scarcely inflated, lobes nearly horizontal at apex	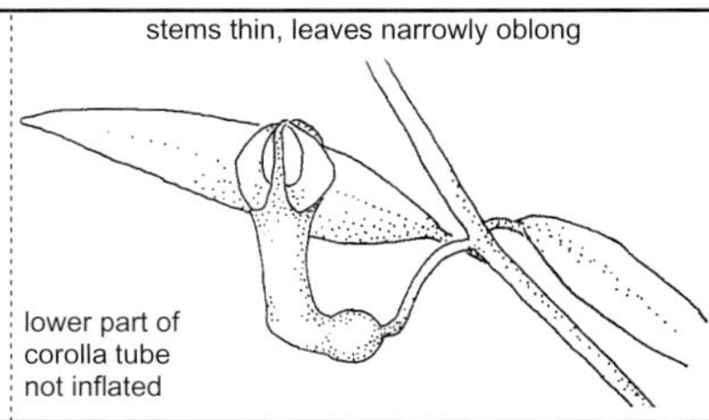	***C. affinis*** **243**, 431
	Stems thick (c. 3mm diameter), <0.5m; **leaves** scale-like, mostly 3--6mm long, soon dropping; lower part of **corolla tube** noticeably inflated, lobes ascending steeply towards tips	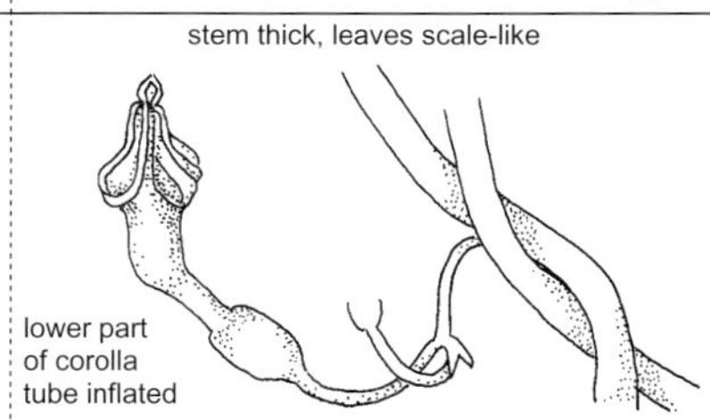	***C. tihamana*** **243**, 432

Cryptolepis

1	**Corona lobes** comparatively long (much longer than gynostegium), slender and narrowing towards tips	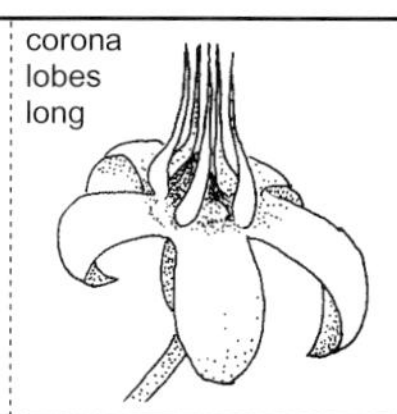	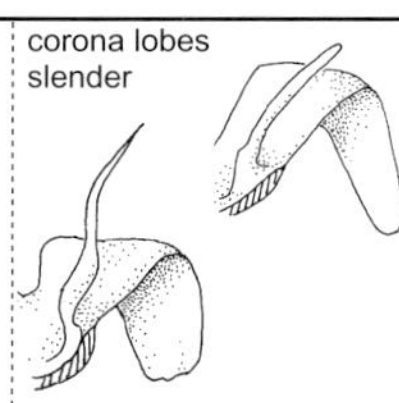	2
	Corona lobes short (shorter than or slightly longer than gynostegium), somewhat club-shaped or distinctly bent outwards near base	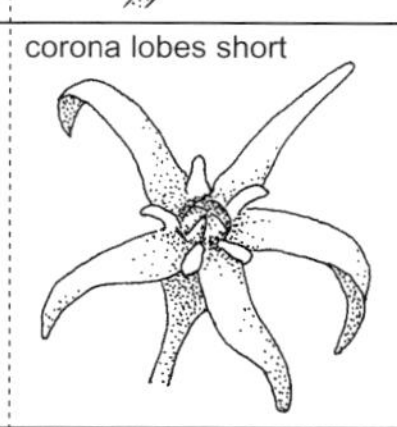	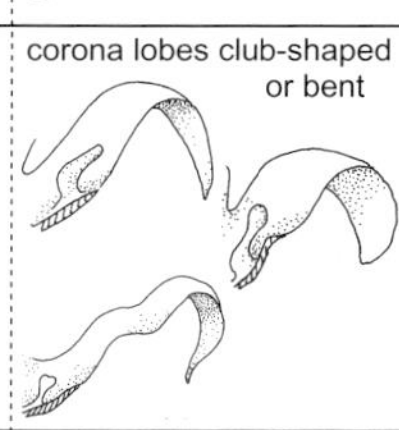	3
2	Intricately branched **shrub** <1m; **leaves** linear-oblong to linear-obovate, c.1cm x <3mm; **corolla lobes** c. 4mm long	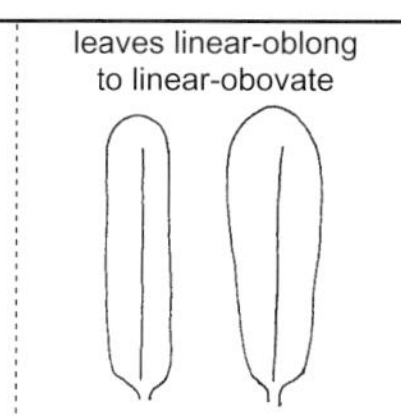	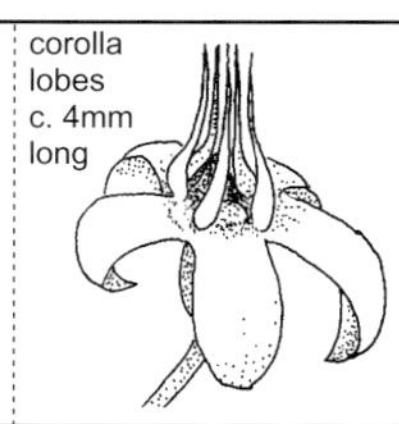	***C. intricata*** Endemic **243**, 433
	Shrub or small **tree** <2.5m; **leaves** ovate to obovate, 1--2cm x 4--6mm; **corolla lobes** 10□12 mm long	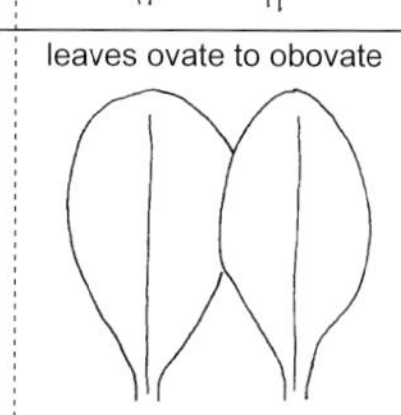	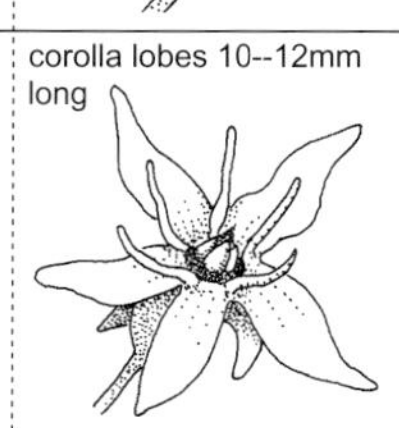	***C. arbuscula*** Endemic 432

3

Climbing **shrub**; **leaves** linear-elliptic to narrowly obovate or obtriangular, <1(--1.5)cm across; **flowers** in axillary clusters; **corolla lobes** c.4mm long

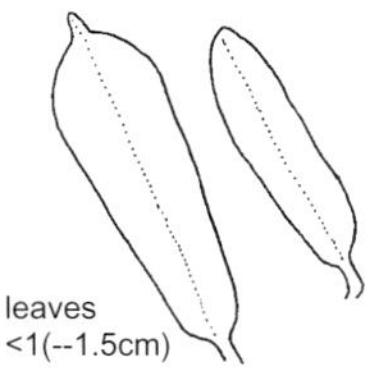

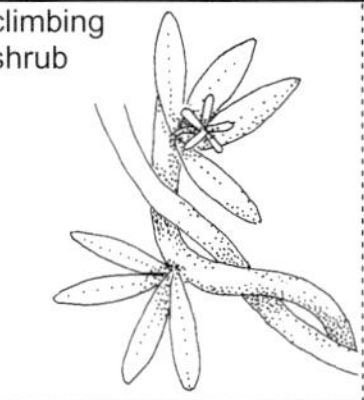

C. volubilis
434

Small **tree** or hanging **shrub**; **leaves** obovate or orbicular, >2cm across; **flowers** in terminal panicles; **corolla lobes** c.9mm long

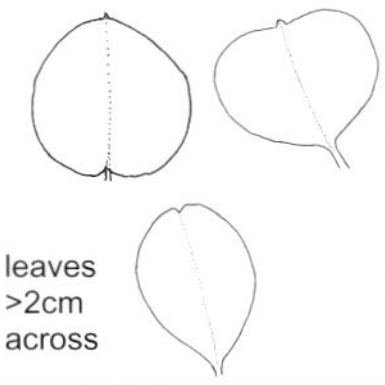

4

4

Hanging **shrub**; **leaves** sessile; **corolla lobes** twisted; **fruits** ovoid

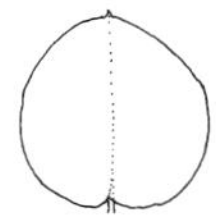

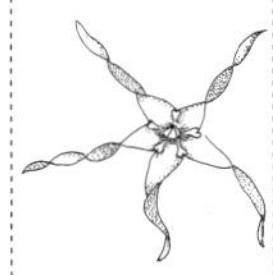

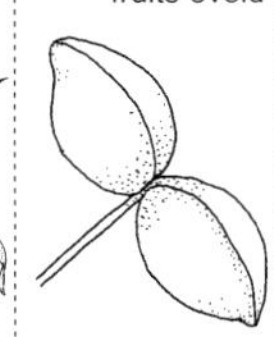

C. socotrana
Endemic
244, 434

Erect **tree**; **leaves** with petiole 0.5--1cm long; **corolla lobes** straight; **fruits** narrowly cylindrical

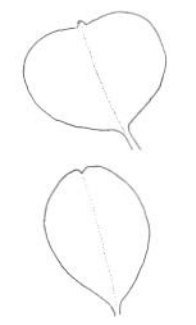

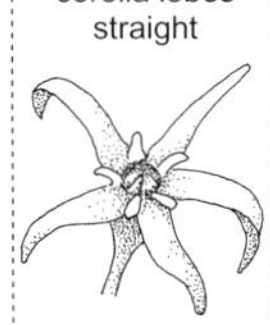

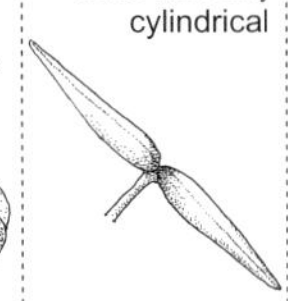

C. macrophylla
Endemic
243, 433

Echidnopsis

1

Corolla tube >6mm long, greenish yellow to pink

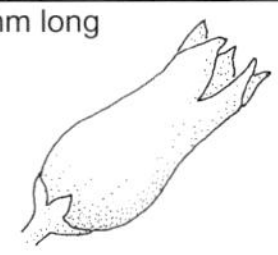

E. insularis
Endemic
244, 436

Corolla tube <2mm long, colour various

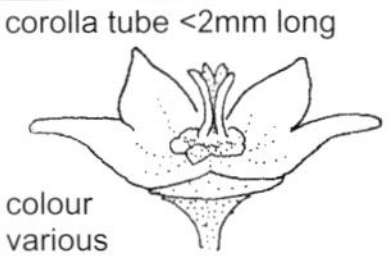
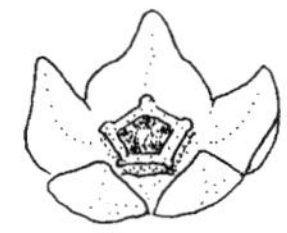

2

2

Inner corona lobes </=1.5mm long, rising up in small column in centre of flower; **corolla** often rotate (flattened)

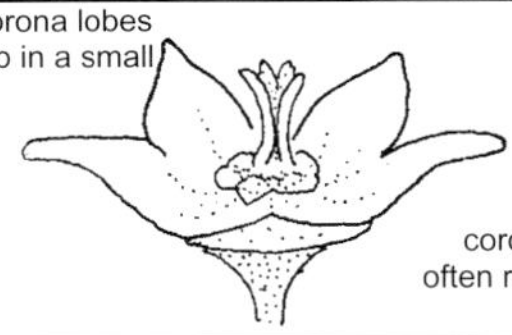

3

Inner corona lobes <0.5mm long, not rising up in centre of flower; **corolla** usually +/- campanulate

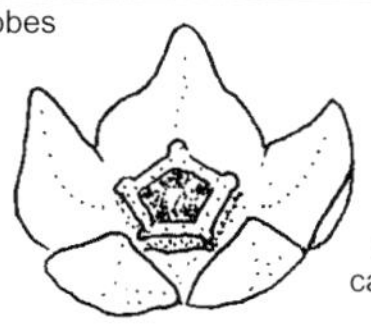

4

3

Corolla pale pink and smooth inside; free part of small **outer corona lobes** spreading on surface of corolla, clearly divided by deep groove opposite inner lobes

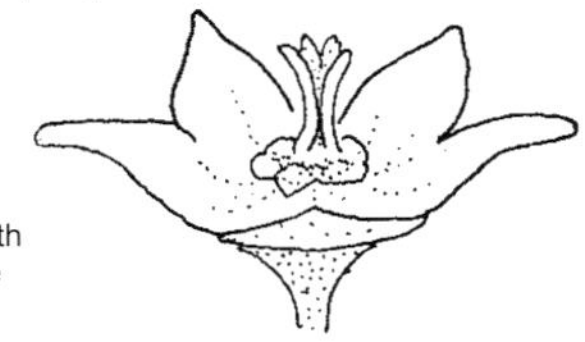

E. milleri
Endemic
244, 437

Corolla deep maroon and finely papillate inside, **outer corona lobes** without any clearly free part, not at all divided opposite inner lobes

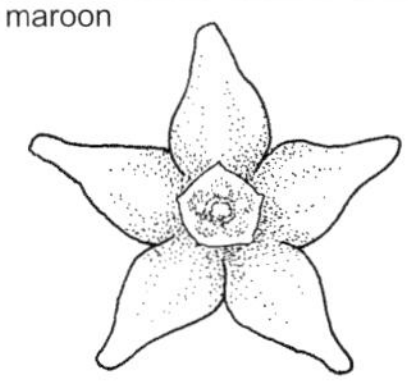

E. bentii
Endemic
244, 436

4	**Corolla** cream to pale yellow with dark maroon ring; **outer corona lobes** forming a continuous ring around top of column	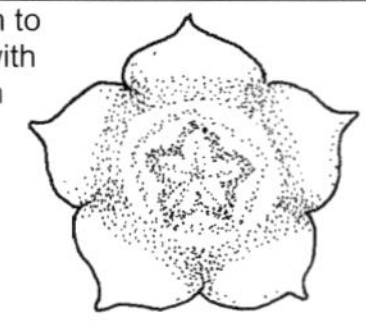	***E. inconspicua*** Endemic **244**, 436
	Corolla yellowish green within; **outer corona lobes** forming 5 discrete pockets between swollen lower parts of inner lobes	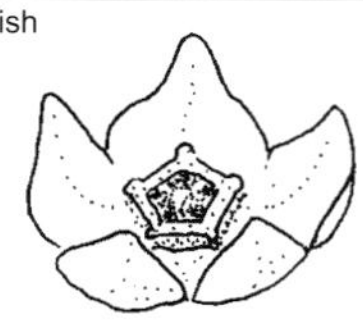	***E. socotrana*** Endemic **244**, 437

Sarcostemma

1	**Outer corona** forming low continuous ring around bases of swollen inner lobes; **corolla** >10mm diameter	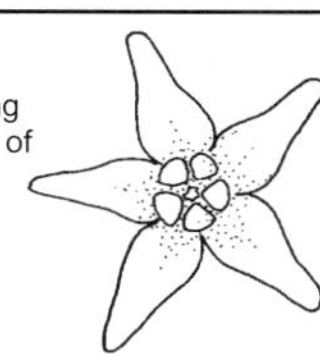	***S. viminale*** 441
	Outer corona forming 5 small ridges joining swollen inner lobes; **corolla** usually </=8mm diameter	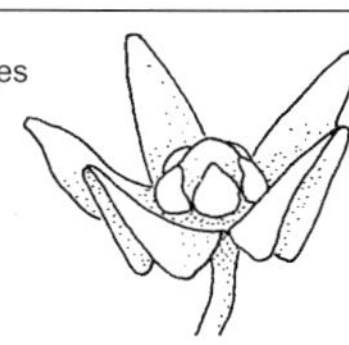	***S. socotranum*** Endemic 440

Secamone

1	**Leaves** linear with notched or truncate tip; **corolla lobes** <1.3mm long, staminal corona lobes c. 0.15mm long	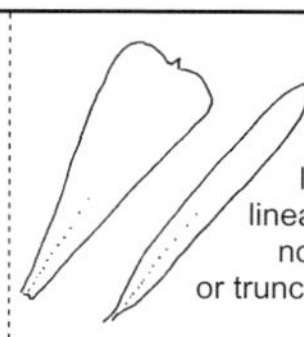	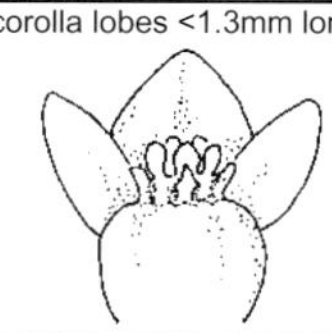	***S. cuneifolia*** Endemic **243**, 441
	Leaves obovate with rounded or acute tip; **corolla lobes** >/=2.5mm long, staminal corona lobes c. 1mm long	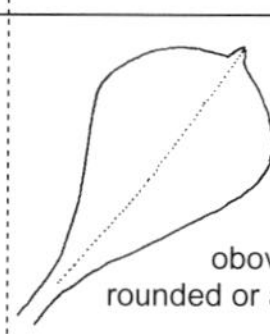	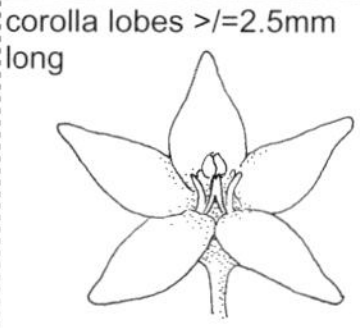	***S. socotrana*** Endemic **245**, 441

AVICENNIACEAE

1	Mangrove **tree** with erect breathing **roots**; **leaves** opposite, greyish below; **flowers** yellow, in globose heads, 4-lobed			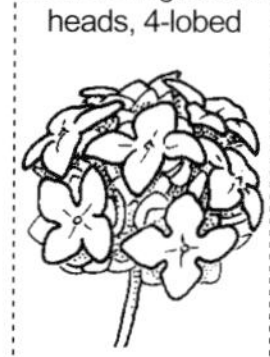	***Avicennia marina*** **245**, 443

BEGONIACEAE

1	**Leaves** round; **male tepals** ± equal	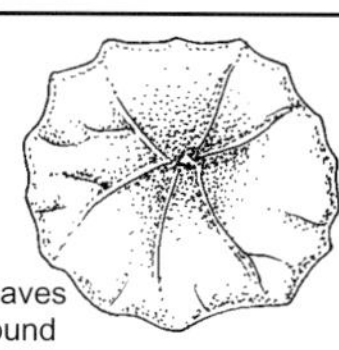		***Begonia socotrana*** Endemic **246**, 444
	Leaves broadly ovate, asymmetric with an acute; **male tepals** strongly unequal	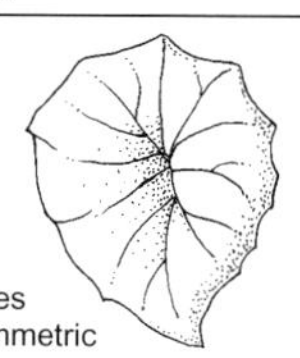	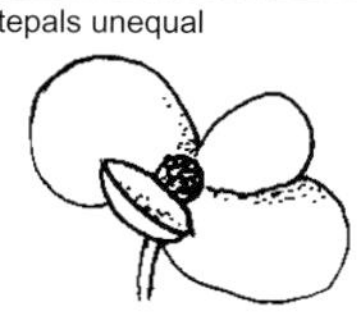	***Begonia samhaensis*** Endemic **246**, 444

BORAGINACEAE

1

Trees or **shrubs**; **fruit** a drupe — 2

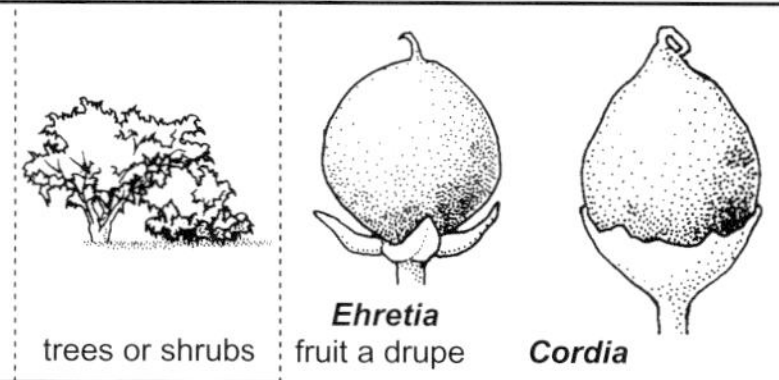

Herbs or low **subshrubs**, rarely shrubs (*Trichodesma scottii*); **fruit** of nutlets or a loculicidal capsule — 3

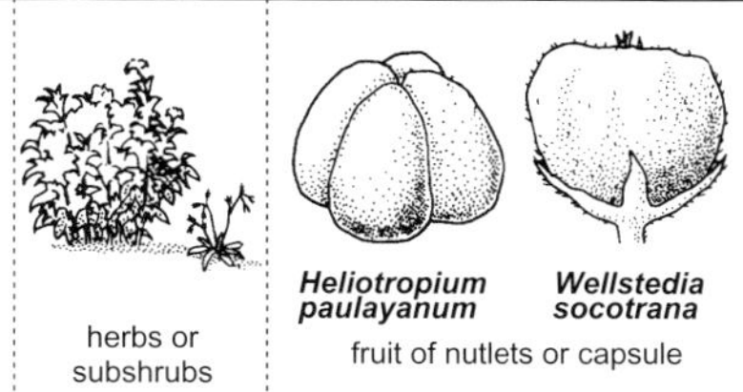

2

Style once divided with 2 stigmas; **calyx** not accrescent in fruit — ***Ehretia obtusifolia*** 448

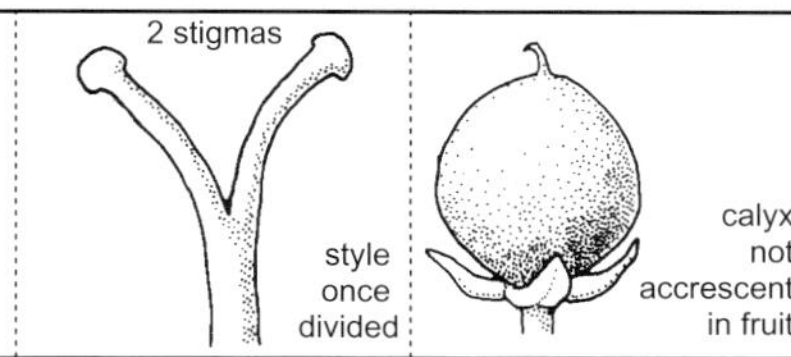

Style twice divided with 4 stigmas; **calyx** accrescent and cupular in fruit — ***Cordia*** (3 spp.)

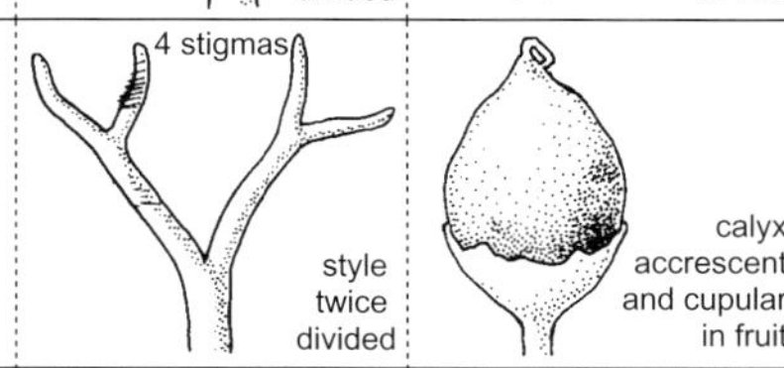

3

Flowers 4-merous; **fruit** a loculicidal capsule — ***Wellstedia socotrana*** Endemic **247**, 456

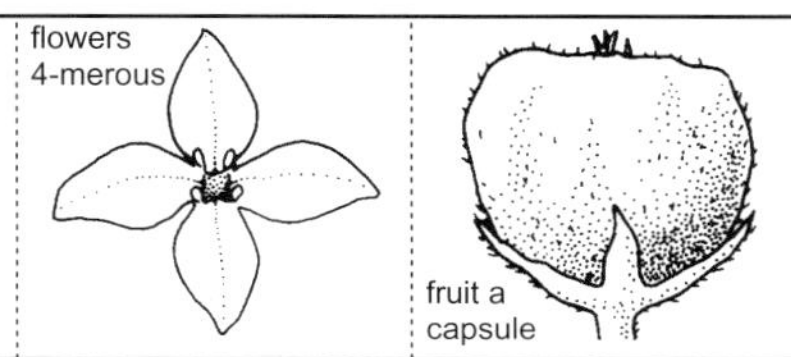

Flowers 5-merous; **fruit** of nutlets — 4

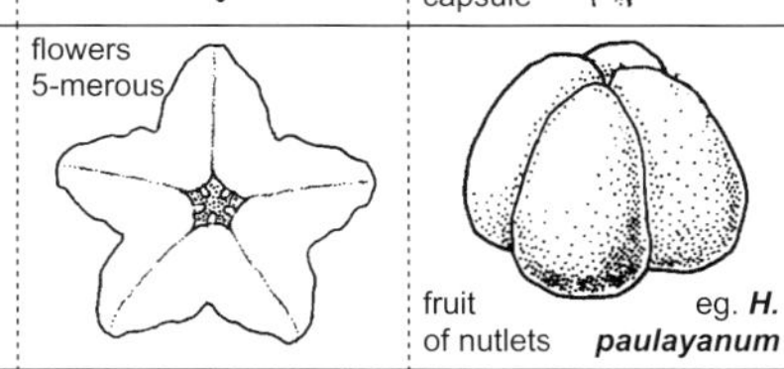

4

Stamens with sterile terminal appendages which form a cone-like appendage — 5

Stamens without terminal appendages — 6

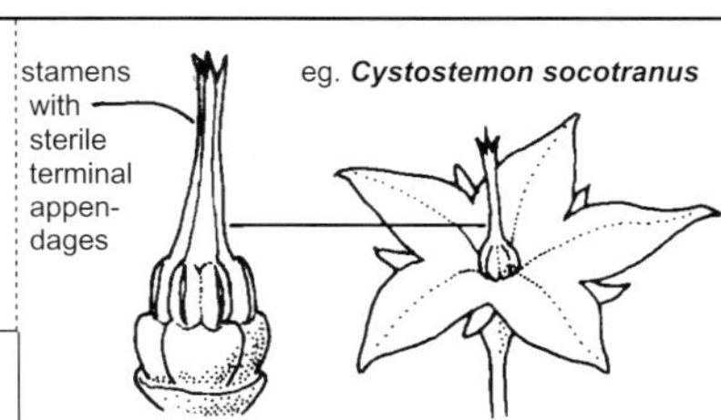

5

Filaments with basal appendages; **plant** coarsely hispid; **nutlets** triangular with an oblique beak and humped back — ***Cystostemon socotranus*** Endemic **246**, 447

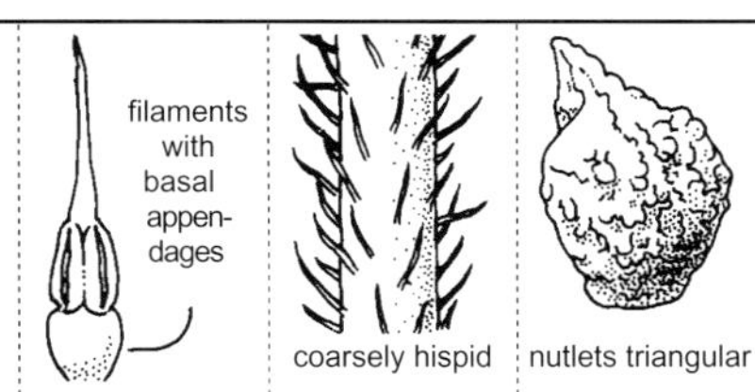

Filaments without basal appendages; **plant** strigose or weakly hispid; **nutlets** triangular-ovate with a flattened and winged or rounded back — ***Trichodesma*** (3 spp.)

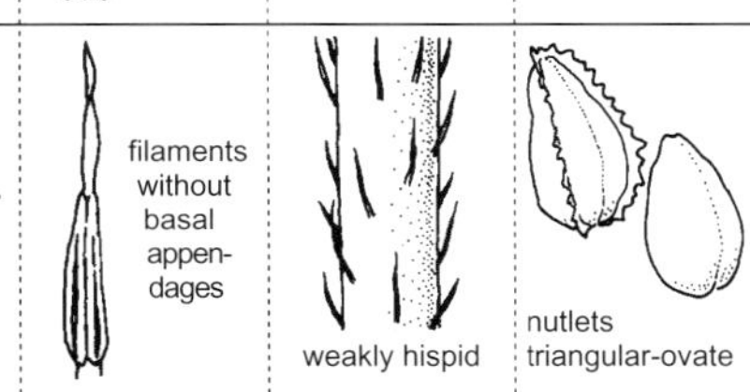

6	**Style** terminal on the ovary, with a ring below the stigma	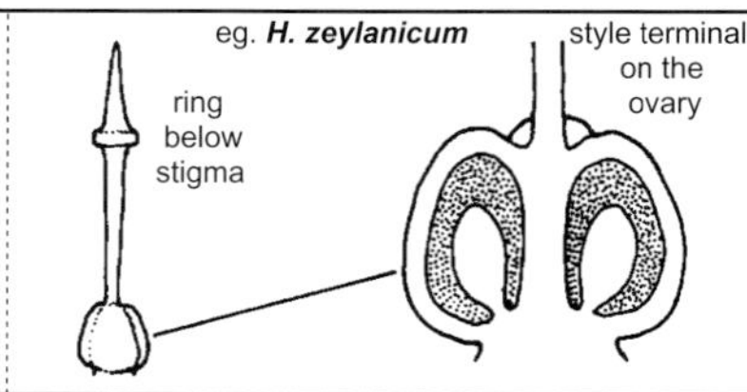	***Heliotropium*** (c. 15 spp.)
	Style inserted at the base of the ovary-lobes, lacking a stigmatic ring	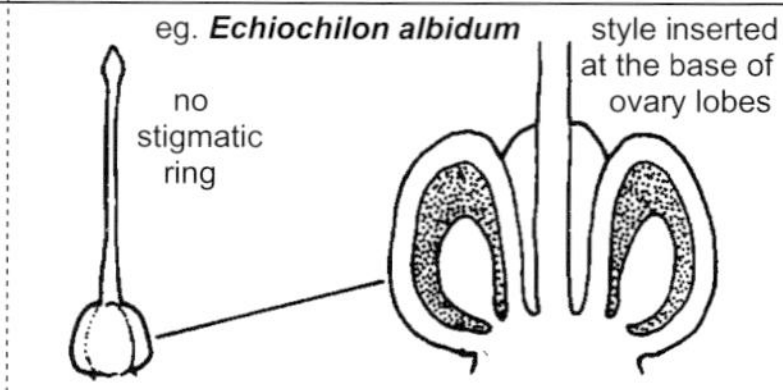	7

7	Cushion-forming **perennial** or **shrublet**; **flowers** white; **leaves** and **stems** silvery with appressed hairs	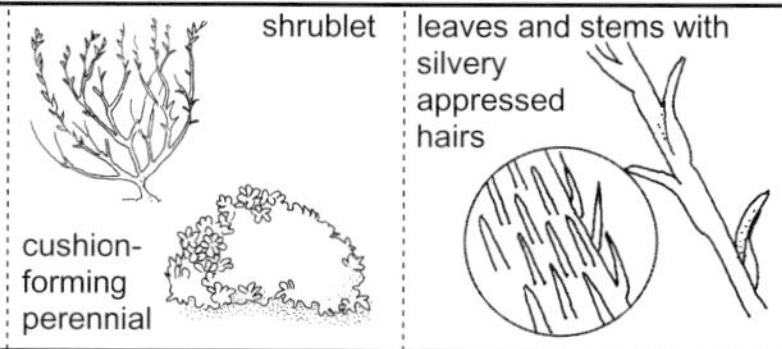	***Echiochilon*** (2 spp.)
	Erect **annual**; **flowers** yellow; **leaves** and **stems** green, with spreading hispid hairs	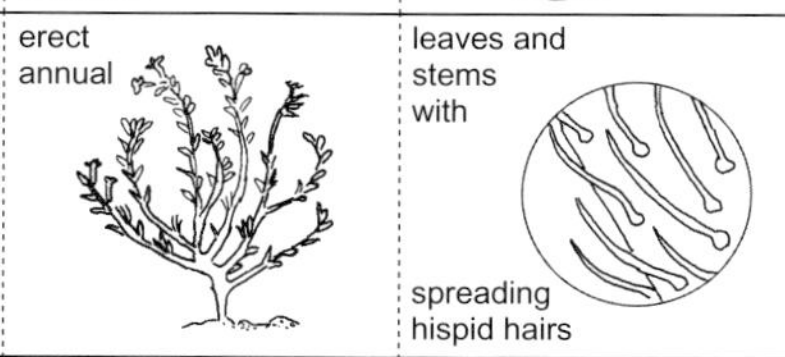	***Arnebia hispidissima*** 445

Cordia

1	**Leaves** opposite or subopposite, densely reticulate on both surfaces; **corolla tube** 3--4mm long	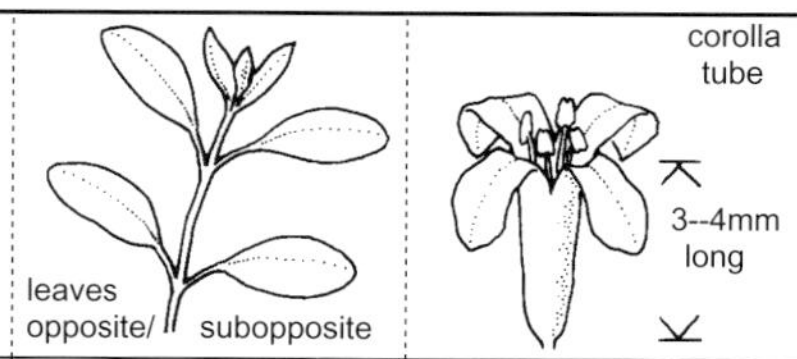	***C. quercifolia*** 447
	Leaves alternate, densely reticulate only below; **corolla tube** 5.5--7mm long	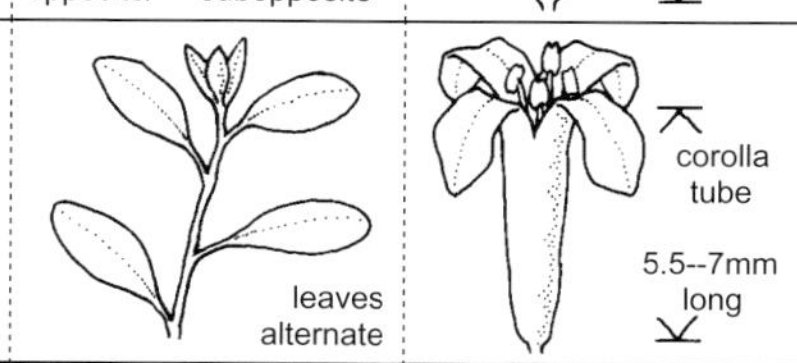	2

2	**Leaves** scabrid above; **calyx** densely tomentose	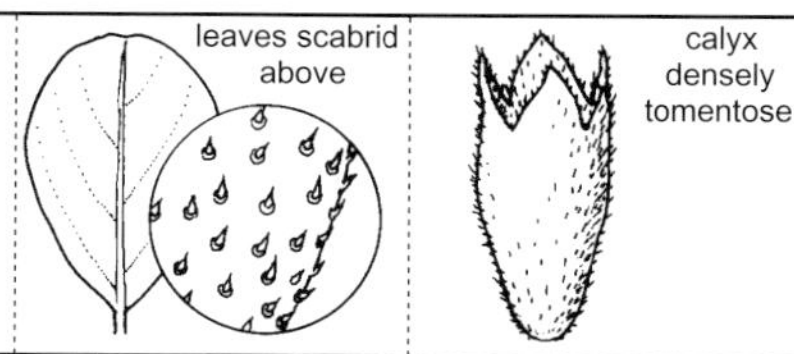	***C. obovata*** 446
	Leaves glabrous above; **calyx** glabrous or papillose with pilose tips to the lobes	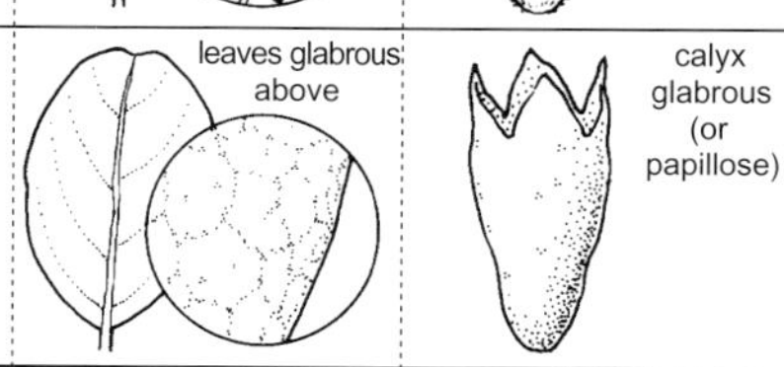	***C. obtusa*** **246**, 447

Echiochilon

1	Erect **shrublet**; **leaves** alternate with simple hairs	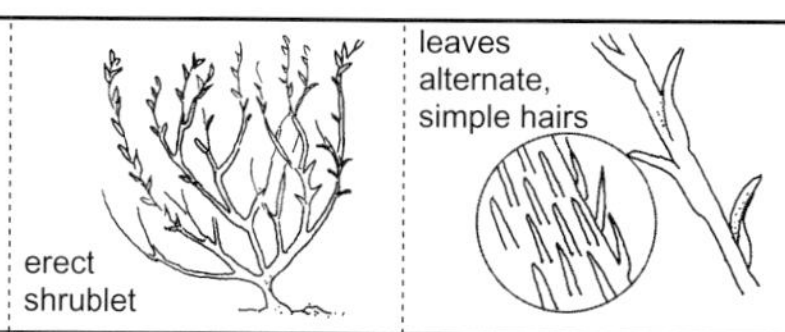	***E. persicum*** 448
	Cushion-forming **perennial**; **leaves** spirally arranged, congested with bulbous-based hairs	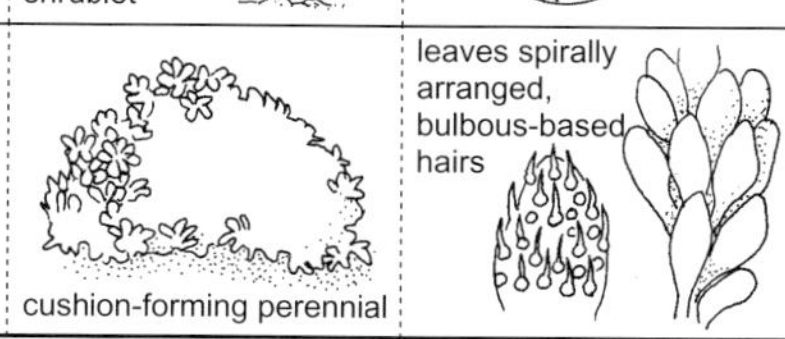	***E. pulvinata*** Endemic **246**, 448

Heliotropium

1

Corolla lobes subulate — ***H. zeylanicum*** 455

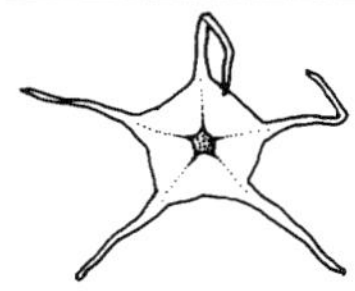

Corolla lobes rounded or acute — 2

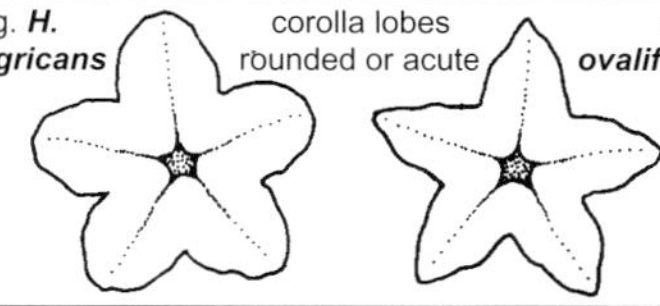

2

Leaves with an undulate-dentate margin,several x longer than broad; **inflorescence** few-flowered and open — ***H. dentatum*** Endemic **247**, 453

Leaves entire, rarely ± undulate; **inflorescence** open or congested — 3

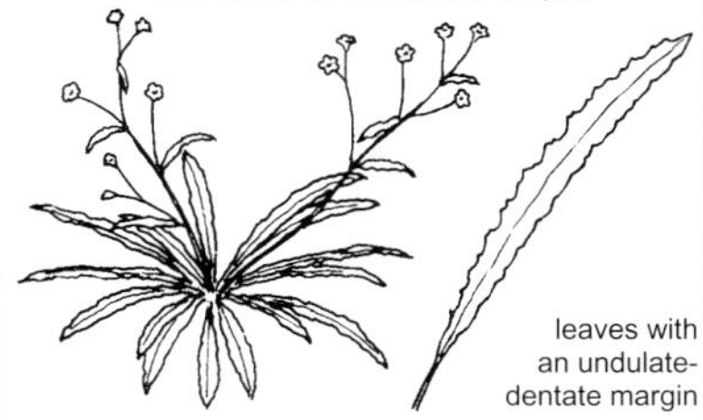

3

Calyx with long spreading hairs; **stigma lobes** long, filiform, exserted — ***H. ophioglossum*** 453

Calyx with appressed or short spreading hairs; **stigma lobes** short, never exserted — 4

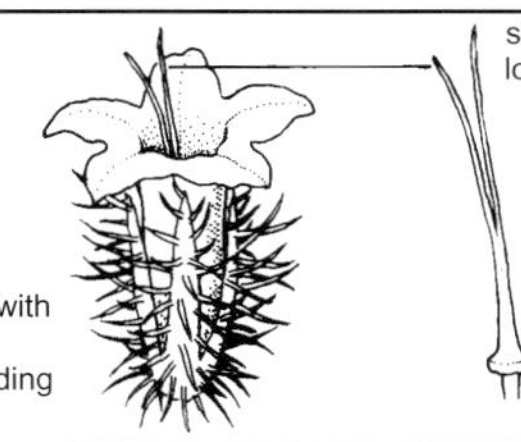

4

Leaves subopposite, bright green, glossy, suborbicular, thinly appressed hairy; **flowers** c.8mm across, long-pedicellate, solitary or in groups of 2 or 3 — ***H. nigricans*** Endemic **247**, 453

Leaves alternate, colour, shape and indumentum various; **flowers** <5mm diameter, in distinct or indistinct cymes — 5

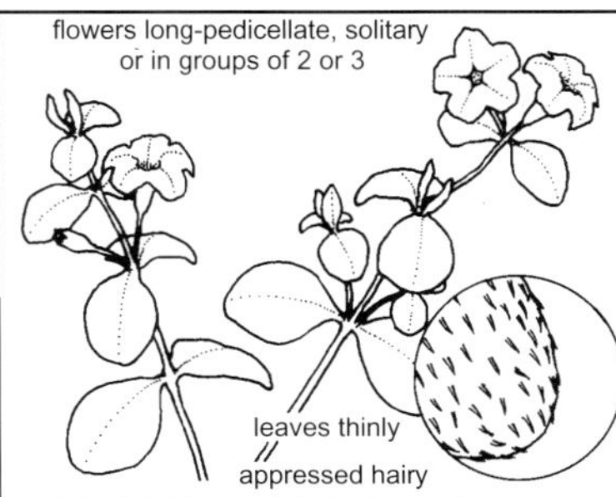

5

Leaves linear; **stigma** globose; **stems** with distinct white peeling bark — ***H. rariflorum*** 454

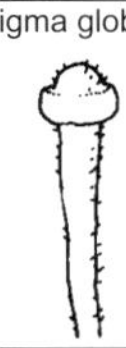

Leaves broader; **stigma** triangular or cylindrical; **stems** without white peeling bark — 6

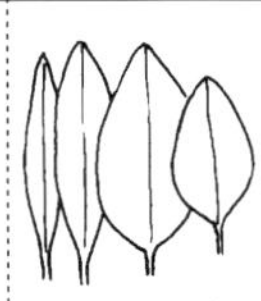

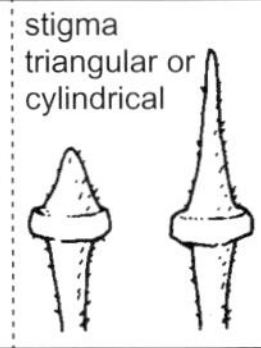

6

Erect annual **herb or shrublet**, usually >30cm; **leaves** ovate, distinctly petiolate, 3--6cm long; **style** long — 7

Prostrate or ascending perennial **herbs**, usually <20cm; **leaves** variable, if distinctly petiolate then <2cm long; **style** short — 8

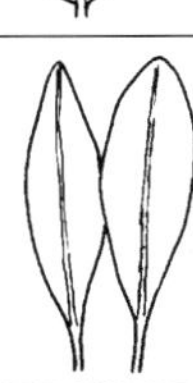

7	Annual **herb**; **leaves** softly grey-hairy throughout; **fruit** splitting into 2 fused pairs	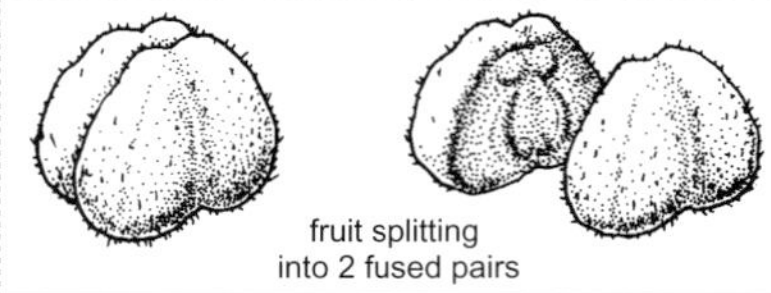	***H. aegyptiacum*** **246**, 449
	Shrublet; **leaves** thinly-appressed hairy, dark green; **fruit** splitting into 4 nutlets	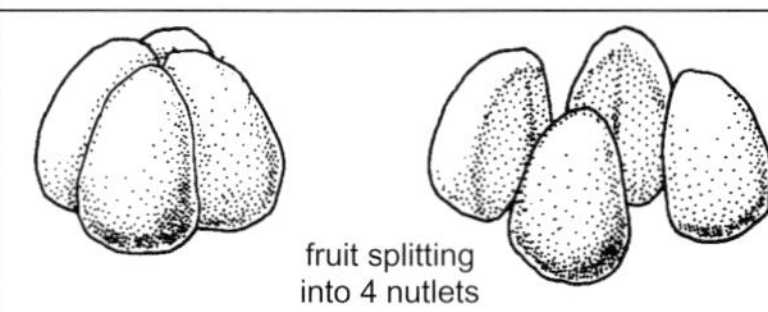	***H.balfourii*** Endemic **246**, 452
8	**Calyx** with sparse appressed hairs	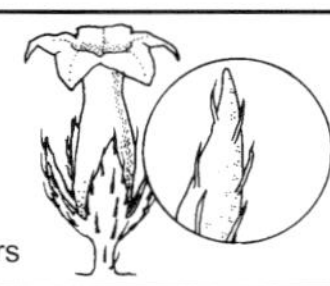	9
	Calyx densely hirsute with short hairs and long bulbous-based hairs	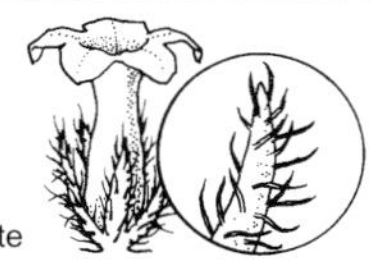	10
9	**Style** obsolete; **flowers** congested; **leaves** elliptic to obovate	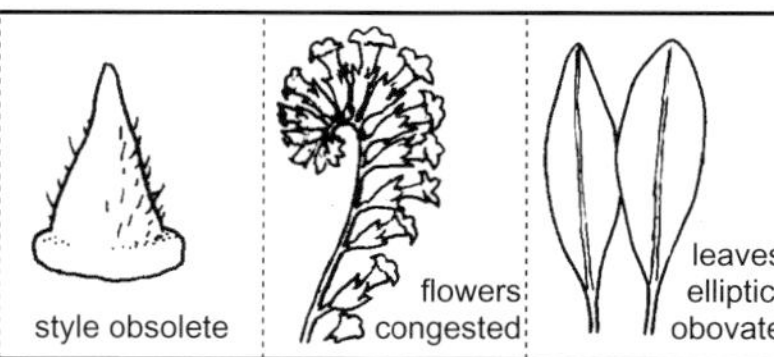	***H. ovalifolium*** 454
	Style distinct; **flowers** distant; **leaves** linear to linear-lanceolate	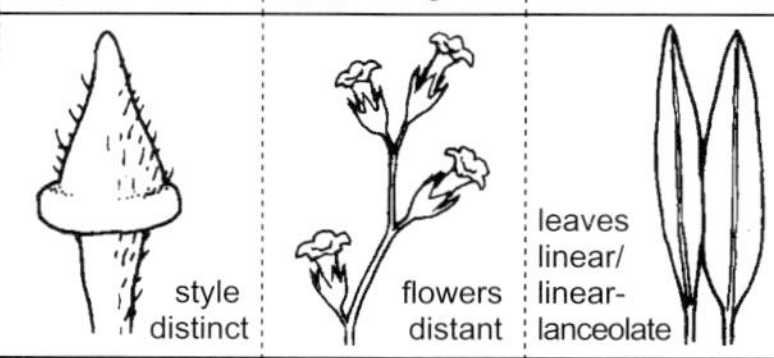	***H. strigosum*** 454
10	**Fruit** splitting into 2 fused pairs	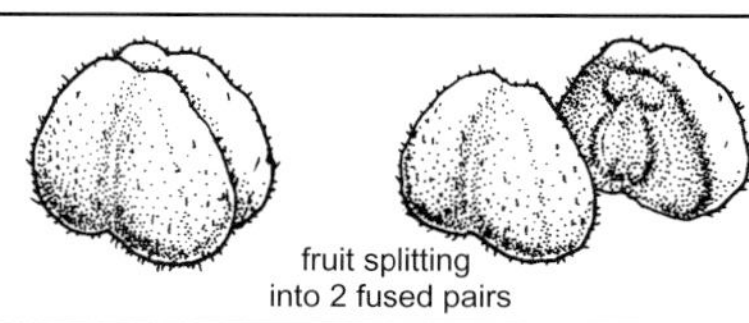	11
	Fruit splitting into 4 nutlets	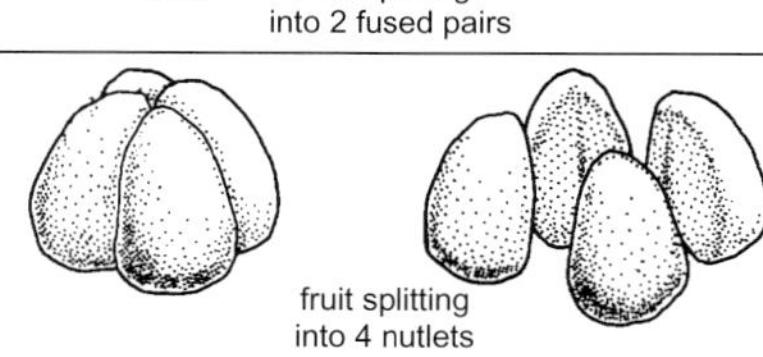	12
11	**Flowers** white; **stigma** long; **nutlets** glabrous, sometimes warty	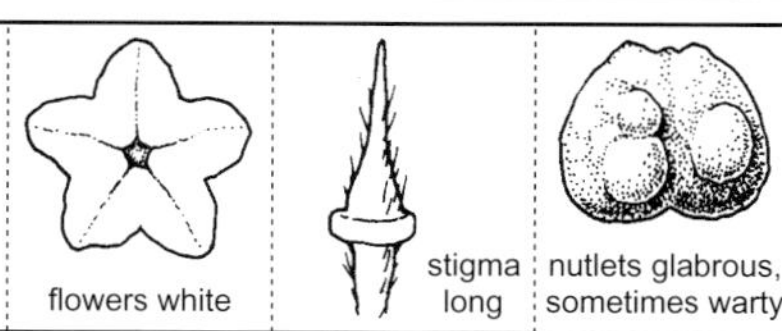	***H. riebeckii*** Endemic 451
	Flowers yellow; **stigma** long; **nutlets** glabrous	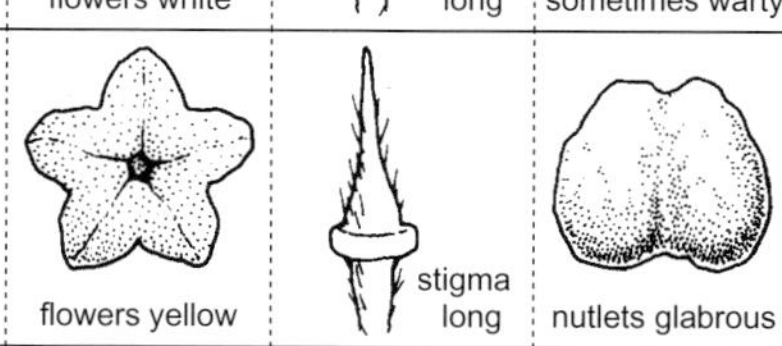	***H. wagneri*** Endemic 450
	Flowers yellow; **stigma** short; **nutlets** glabrous or hairy, warted	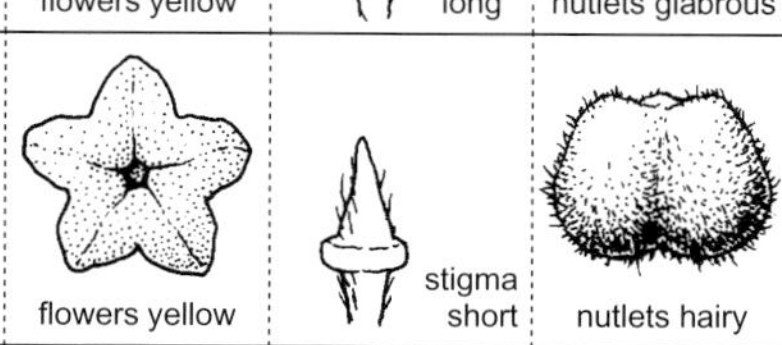	***H. shoabense*** Endemic 452
	Flowers yellow or white; **stigma** short or long; **nutlets** glabrous or hairy	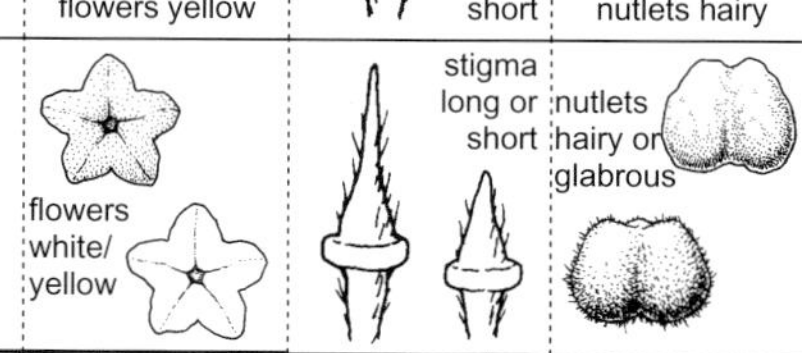	***H. socotranum* s.l.** Endemic **247**, 452

12	**Inflorescence** poorly-defined; **leaves** broadly ovate to orbicular, base truncate; some **petioles** equalling or longer than lamina	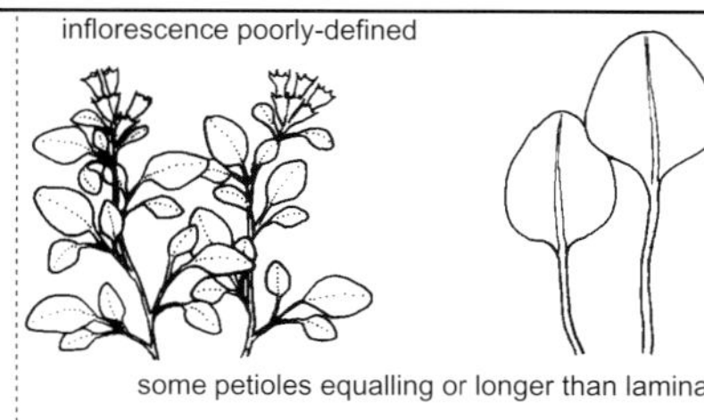	***H. paulayanum*** Endemic 450
	Inflorescence well-defined; **leaves** ovate with an acute base; some **petioles** shorter than lamina	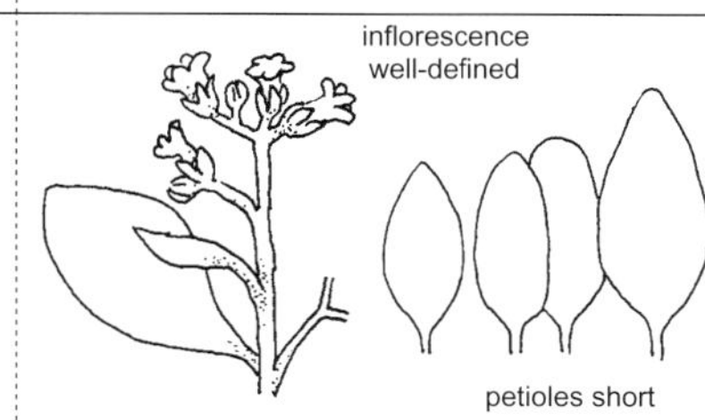	***H. kuriense*** Endemic 450
	Inflorescence congested or few-flowered and poorly defined; **leaves** narrowly oblong to elliptic; **petioles** shorter than lamina	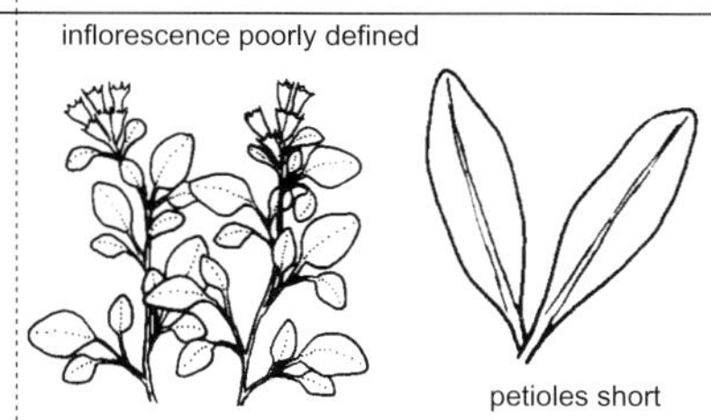	***H. socotranum* s.l.** Endemic **247**, 452

Trichodesma

1	**Shrub**; **flowers** subumbellate, pendulous, 1--3cm diameter, white; **nutlets** c.15 x 10mm		***T. scottii*** Endemic **247**, 456
	Herb; **flowers** in an open inflorescence, spreading or pendulous, up to 1cm across, white or blue; **nutlets** 3--5 x 2--4mm	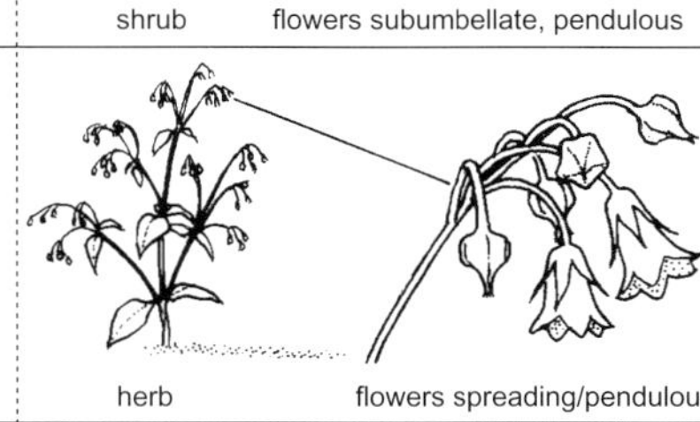	2
2	**Fruits** unwinged; **indumentum** dense and spreading	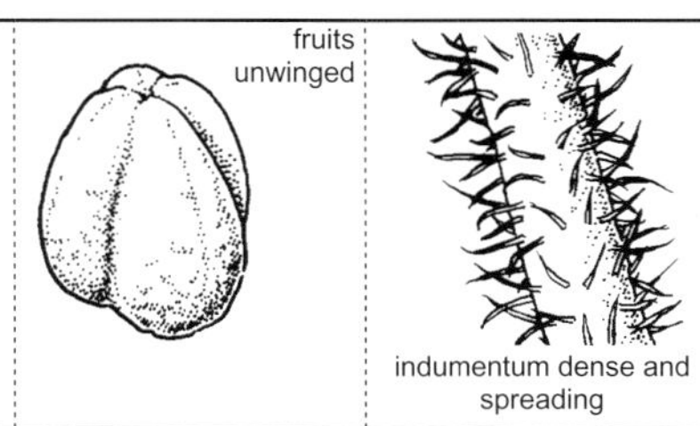	***T. microcalyx*** Endemic **247**, 455
	Fruits with toothed wings; **indumentum** strigose	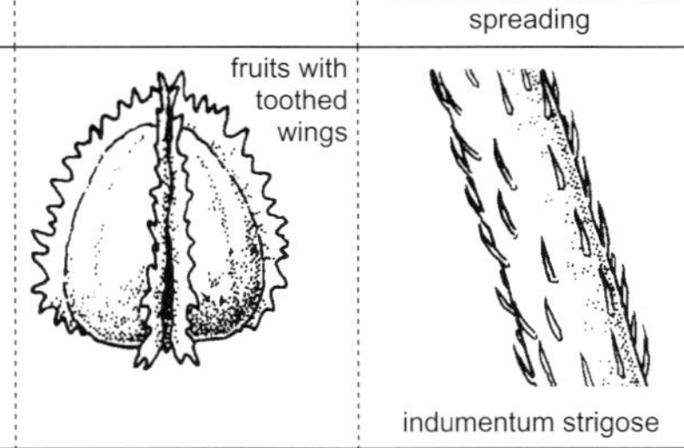	***T. laxiflorum*** Endemic **247**, 455

BURSERACEAE

1	Calyx lobes and petals 5, **stamens** 10; **flowers** conspicuous, bisexual; **fruit** rather woody, 3 to 5-valved with 3 to 5 stones	flowers conspicuous  bisexual	 fruit woody, 3 to 5-valved	***Boswellia*** (8 spp.)
	Calyx lobes and petals 5, **stamens** 4 (or 8); **flowers** inconspicuous, unisexual; **fruit** leathery or fleshy, 2-valved, with a single stone	flowers inconspicuous 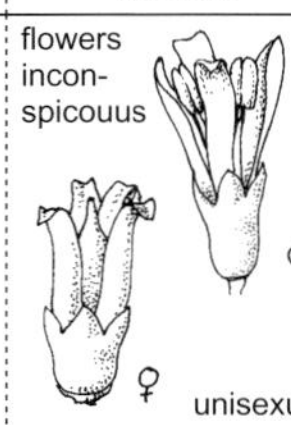 ♂ ♀ unisexual	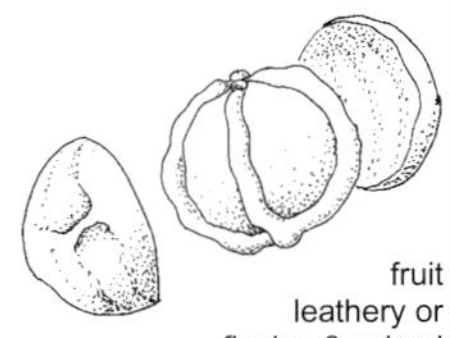 fruit leathery or fleshy, 2-valved	***Commiphora*** (5 spp.)

Boswellia

1	**Trees** rooted on cliffs and large rocks	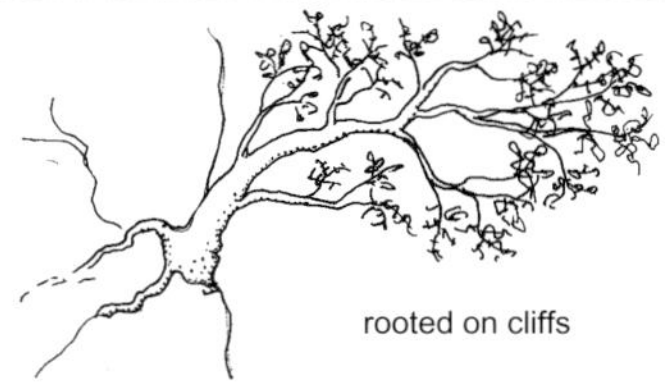 rooted on cliffs	2
	Trees rooted on the ground	 rooted on the ground	5
2	**Leaves** simple or sinuately lobed, sometimes pinnae also present	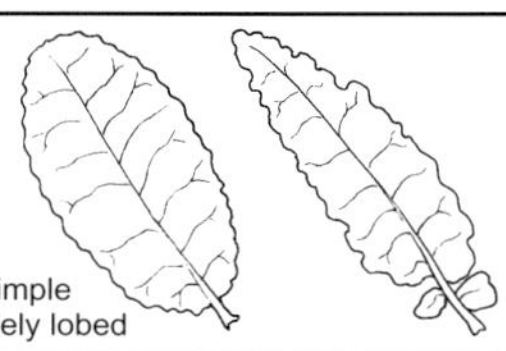 leaves simple or sinuately lobed	3
	Leaves all pinnate	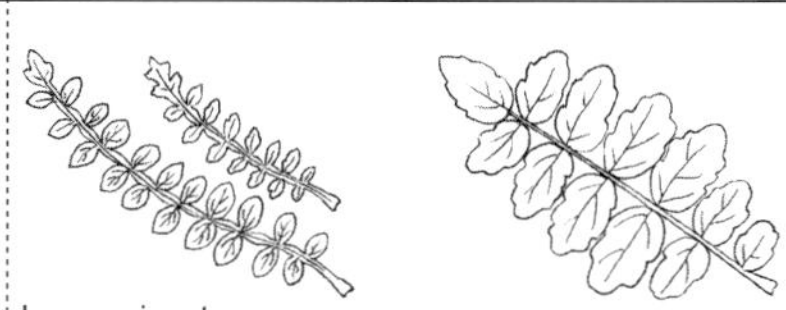 leaves pinnate	4

3	**Leaves** glabrous or minutely puberulous; **flowers** in short (<5cm long) few-flowered panicles	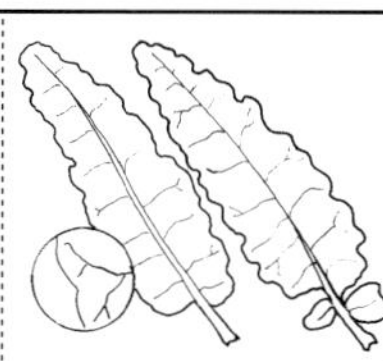 leaves +/- glabrous	few-flowered panicles 	***B. nana*** Endemic **249**, 460
	Leaves densely white-felted beneath; **flowers** in long (<20cm), many-flowered panicles	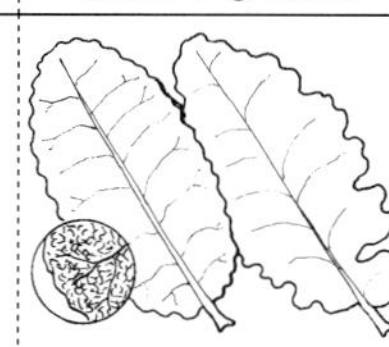 leaves white-felted beneath	many-flowered panicles 	***B. popoviana*** Endemic **249**, 461

4	**Leaflets** 11 to 17, glaucous, 2--4 x 2--3cm, obscurely crenate-serrate, glabrous; **petals** cream or white, sometimes tinged pink, glabrous	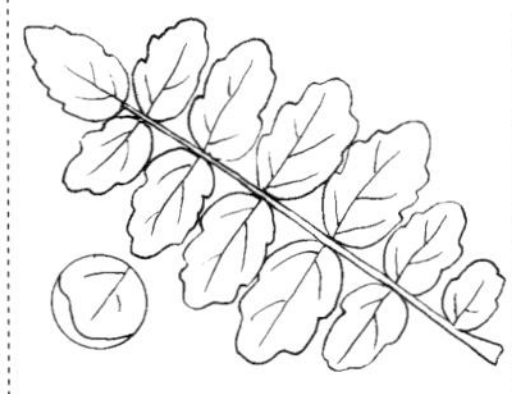 leaves glabrous	petals cream or white 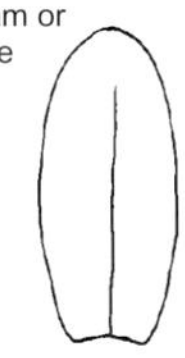 glabrous	***B. dioscorides*** Endemic **248**, 460
	Leaflets 17 to 19, dark green, 2--6 x 2--3cm, obscurely crenate, glabrous or with scattered hairs on veins above; **petals** pale pink, glabrous	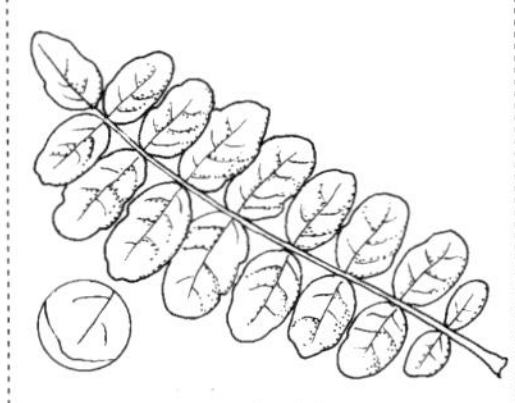 leaves +/- glabrous	petals pale pink 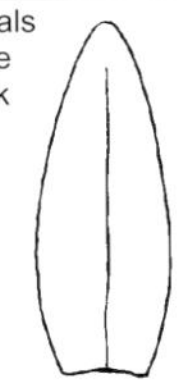 glabrous	***B.* sp. A** Endemic **249**, 464
	Leaflets 13 to 17, dark green, 1--4 x 0.5--1.5cm, deep, irregular crenate-serrate, puberulent above,white tomentose below; **petals** greenish yellow, puberulous outside	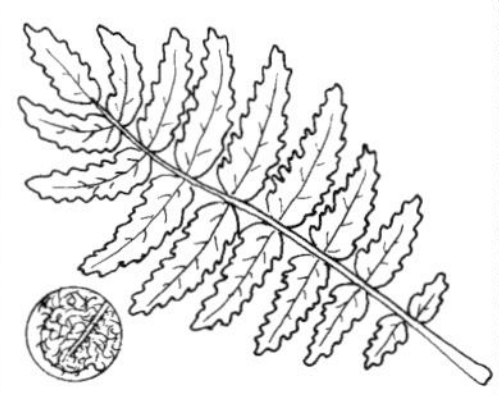 leaves white-felted below	petals greenish yellow 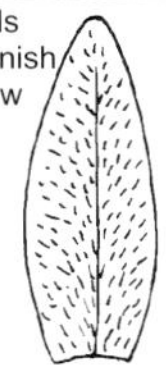 puberulous outside	***B. bullata*** Endemic **248**, 460

5	**Rhachis** winged; leaflets 0.2--1.5 x 0.1--0.7cm, entire to lobed, pale or greyish-green, glabrous or puberulent; **flowers** in short (c.5cm), sparse panicles	rhachis winged 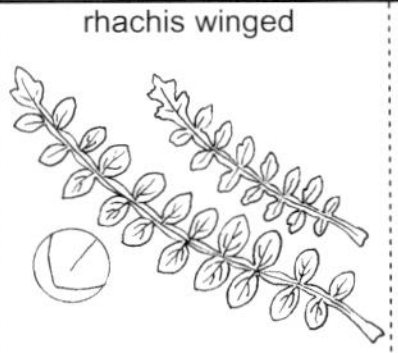 leaves glabrous or puberulent	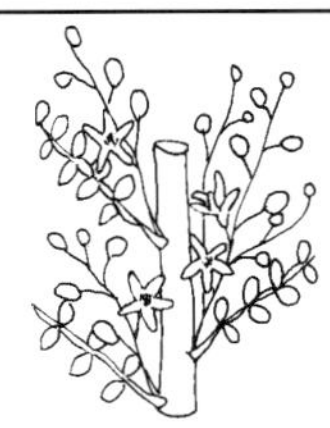 flowers in short panicles	***B. socotrana*** Endemic **250**, 461
	Rhachis unwinged; leaflets 3--10 x 1--3cm, shallowly crenate to deeply crenate-serrate, green and glabrescent above, densely white-tomentose below; **flowers** in long (<15cm) panicles or racemes	rhachis unwinged  leaves white-tomentose below	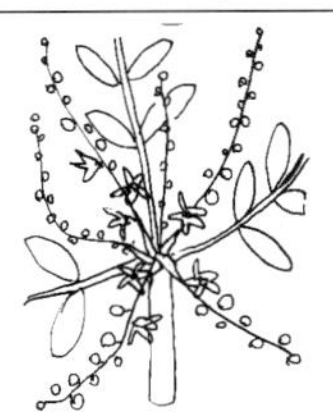 flowers in long panicles or racemes	***B. elongata*** Endemic **249**, 460
	Rhachis unwinged; leaflets 1.5--9 x (1.5--)3--6cm, entire to crenate-serrate, dark green, glabrescent above, thinly tomentose below; **flowers** in long (<15cm) dense, branched racemes	rhachis unwinged  leaves thinly tomentose below	 flowers in long, dense racemes	***B. ameero*** Endemic **248**, 458

Commiphora

1	**Leaves** simple or with < 5 leaflets	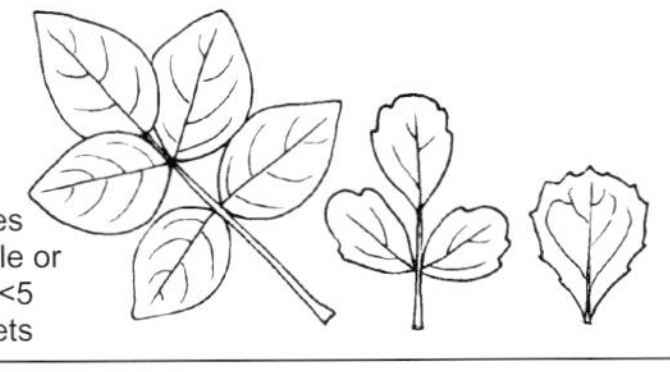 leaves simple or with <5 leaflets	2
	Leaves with > 7 leaflets	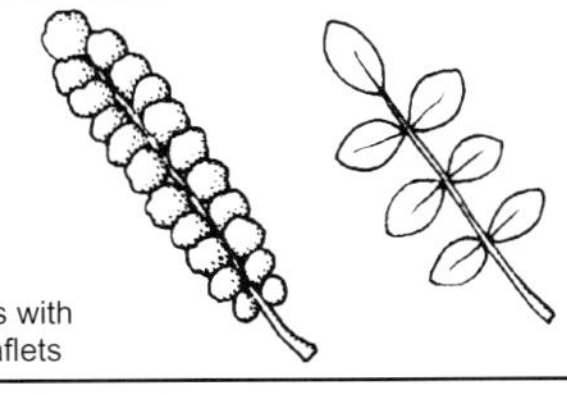 leaves with >7 leaflets	4

2	**Leaflets** usually 5, 2--10cm long; **fruits** globose, apex rounded, bright green	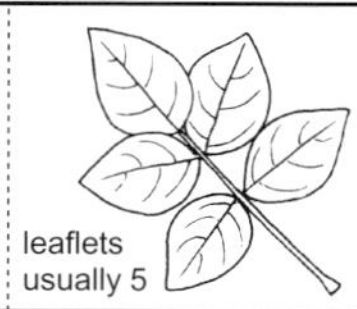 leaflets usually 5	 fruit apex rounded	***C. ornifolia*** Endemic **250**, 466
	Leaflets 1 to 5, <3cm long; **fruits** ellipsoid, apex apiculate, reddish-green	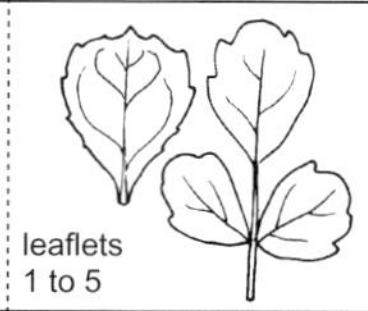 leaflets 1 to 5	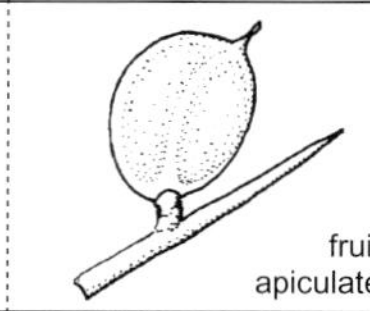 fruit apiculate	3
3	Low **spiny shrub**; **leaves** simple; **bark** strongly peeling	 spiny shrub	leaves simple 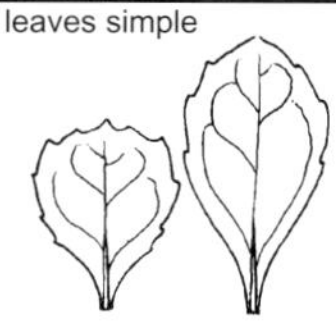	***C. kua*** **250**, 465
	Tree (or shrub), not spiny (branches sometimes spinescent); **leaves** simple or with 2 (rarely 4) leaflets; **bark** not peeling	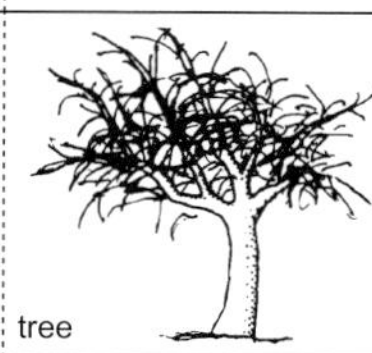 tree	leaves simple or with 2(4) leaflets 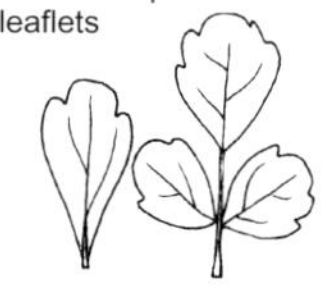	***C. socotrana*** Endemic 471
4	**Leaflets** 9 to 23, imbricate, margins strongly revolute, often appearing globular; **branches** horizontal	margins revolute 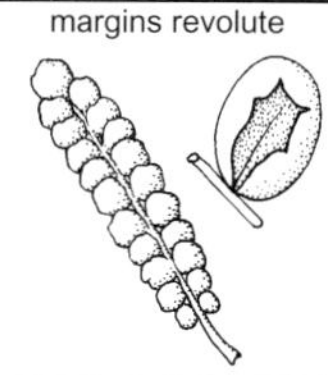	 branches horizontal	***C. planifrons*** Endemic **250**, 471
	Leaflets 7 to 11, spaced, margins not revolute; **branches** spreading	margins not revolute 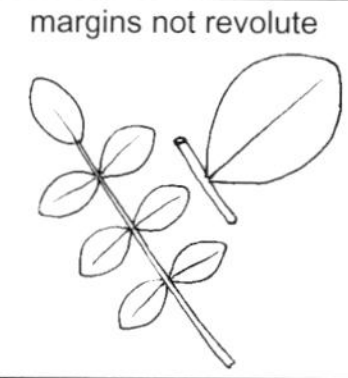	 branches spreading	***C. parvifolia*** Endemic **250**, 468

BUXACEAE

1	**Shrub**; **leaves** opposite, evergreen, leathery, stiffly erect; male **flowers** with 4 sepals, female with 3 sepals; **fruit** 3-horned	leaves opposite 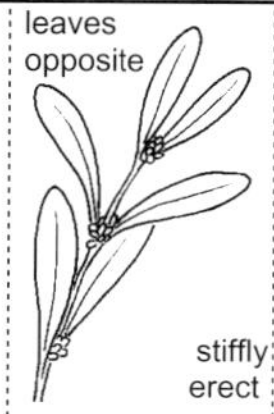 stiffly erect	male flowers with 4 sepals 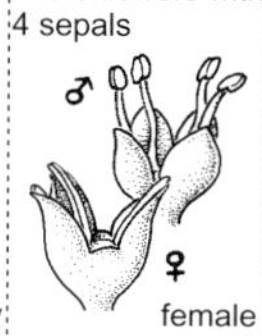 female with 3 sepals	fruit 3-horned 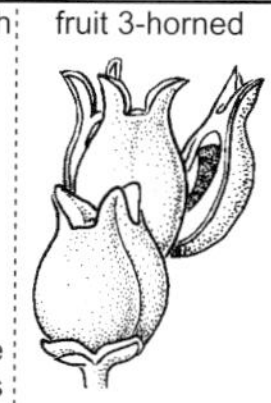	***Buxanthus pedicellatus*** **251**, 472

CAMPANULACEAE

1	**Sepals** with a deflexed margin between the lobes; **leaves** broadly oblong-elliptic to orbicular; **fruit** dehiscing by basal pores	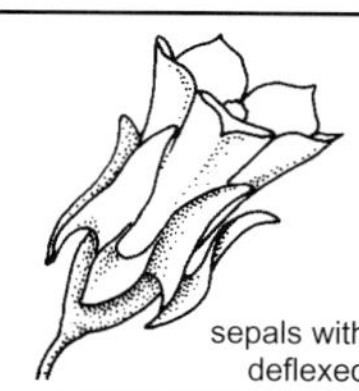 sepals with deflexed margin between lobes	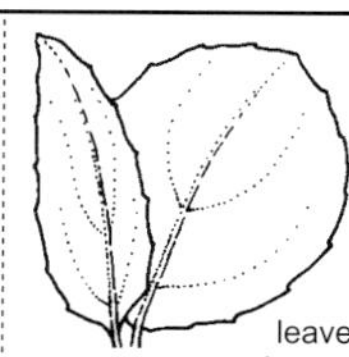 leaves broadly oblong-elliptic to orbicular	***Campanula balfourii*** Endemic **251**, 473
	Sepals without a deflexed margin; **leaves** linear to lanceolate; **fruit** dehiscing by apical valves	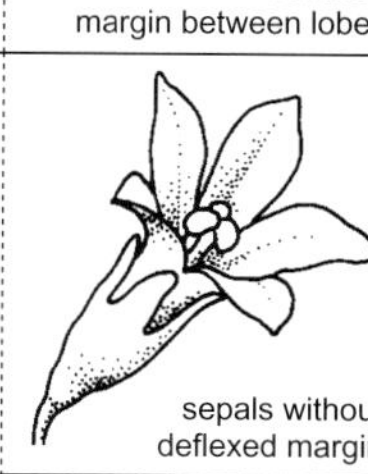 sepals without deflexed margin	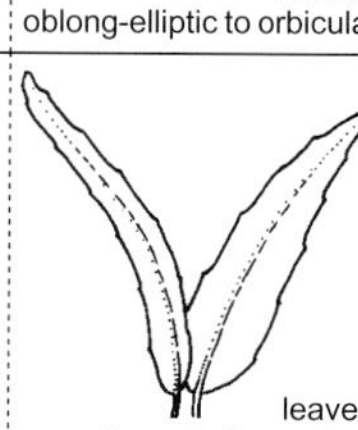 leaves linear to lanceolate	***Wahlenbergia lobelioides*** 474

CAPPARACEAE

1 **Herb**; **fruit** a dehiscent linear capsule, the seeds readily dispersed — ***Cleome*** (7 spp.)

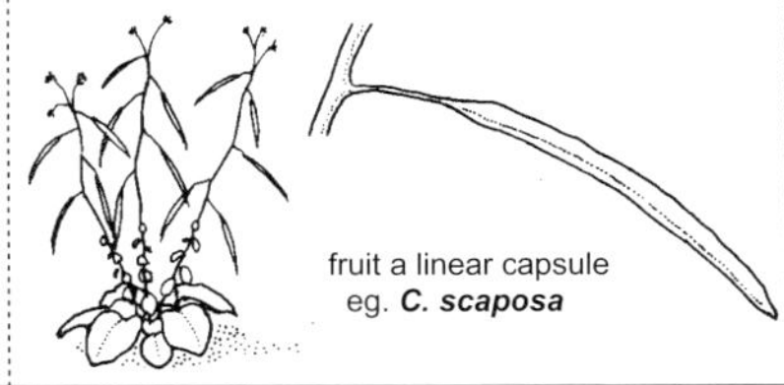

Trees or **shrubs**; **fruit** obovoid or linear to cylindrical, indehiscent or dehiscent with the seeds embedded in a fleshy pulp — 2

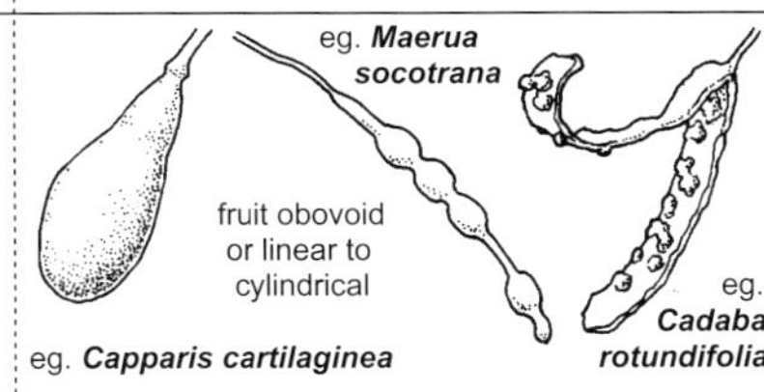

2 **Fruit** obovoid; **leaves** with paired stipular thorns at the base of the petiole; **sepals** dissimilar (lower sepal hooded) — ***Capparis*** (2 spp.)

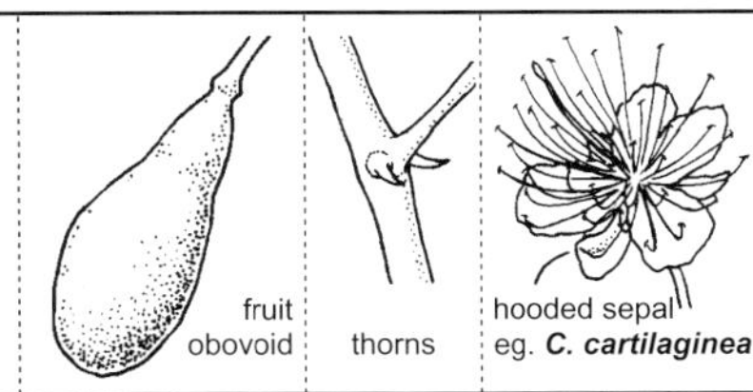

Fruit linear to cylindrical; **leaves** without paired thorns; **sepals** all similar — 3

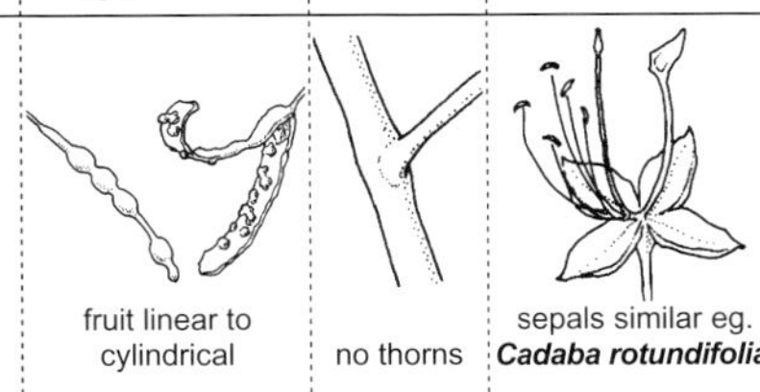

3 Small **tree** or **shrub** with pendulous branches; **receptacle** (the hypanthium) cylindrical or campanulate; **stamens** many — ***Maerua angolensis*** Endemic subsp. **251**, 479

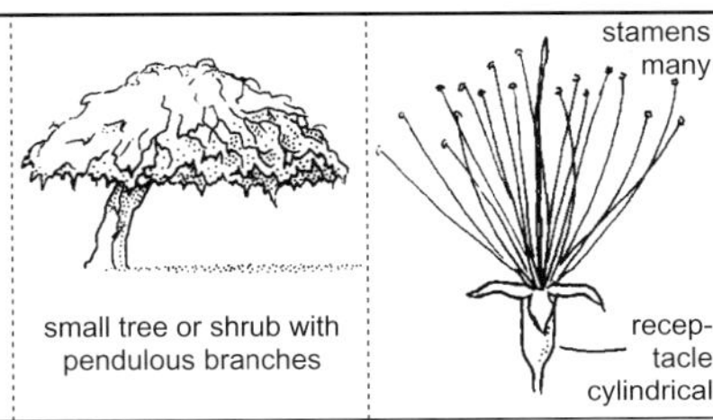

Shrub, branches never pendulous; **receptacle** disc-like; **stamens** 4 or 5 — ***Cadaba*** (4 spp.)

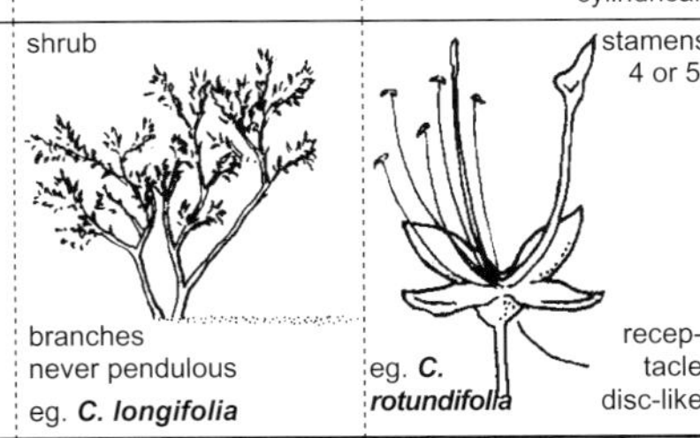

Cadaba

1 **Stamens** attached towards the base of the gynophore; only a single **petal** present — ***C. rotundifolia*** 475

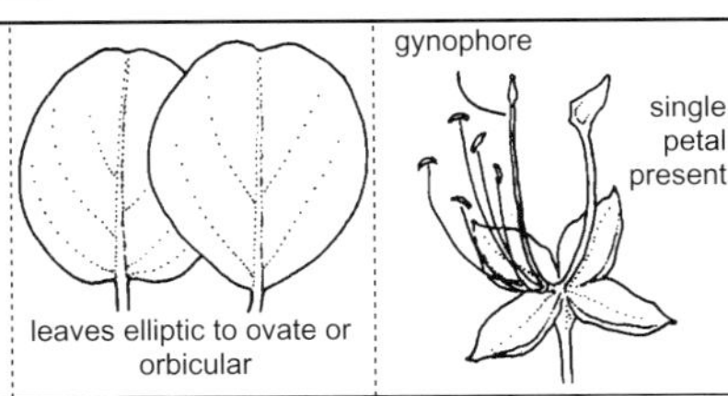

Stamens attached near the middle of the gynophore; 4 **petals** present — 2

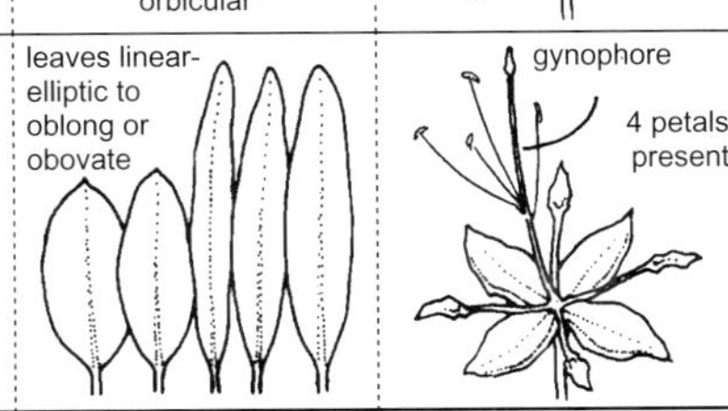

2	**Leaves** subsucculent, oblong-elliptic to suborbicular, <20x15mm, glandular beneath; **fruits** with stalked glands	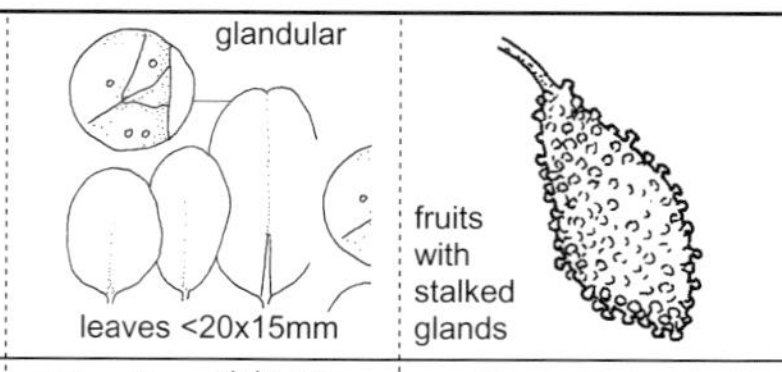	***C. insularis*** Endemic **251**, 475
	Leaves bright green, linear-elliptic, 30--70mm long, glabrous; **fruits** densely glandular	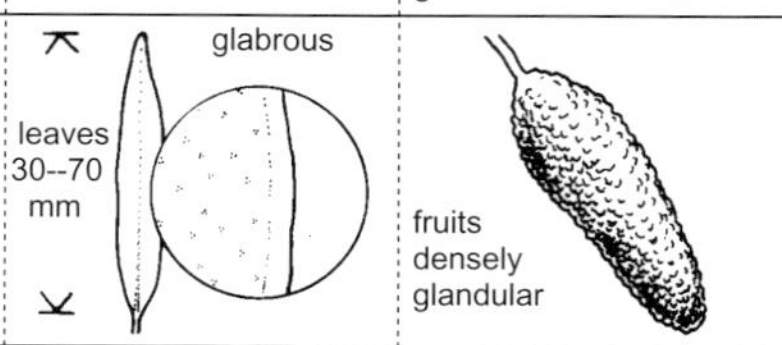	***C. longifolia*** 475
	Leaves grey-green, narrowly oblong or narrowly obovate, 5 --25mm long, farinose; **fruits** non-glandular	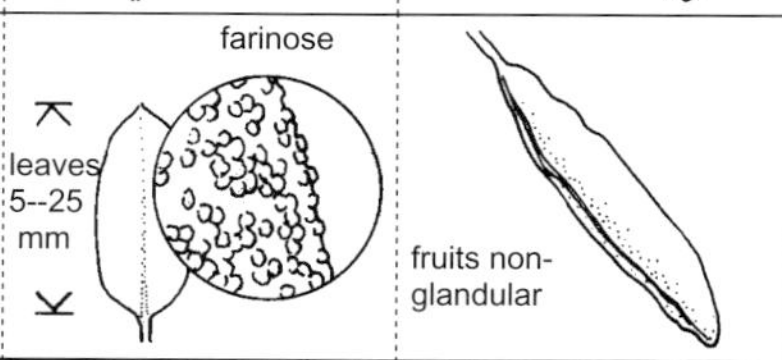	***C. farinosa*** 474

Capparis

1	**Leaves** soon falling, when present linear-oblong; **petals** red	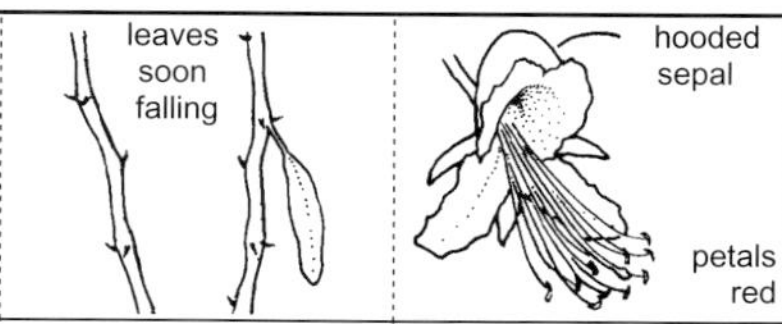	***C. decidua*** **251**, 477
	Leaves present, broadly ovate to orbicular; **petals** white or cream, often becoming pink with age	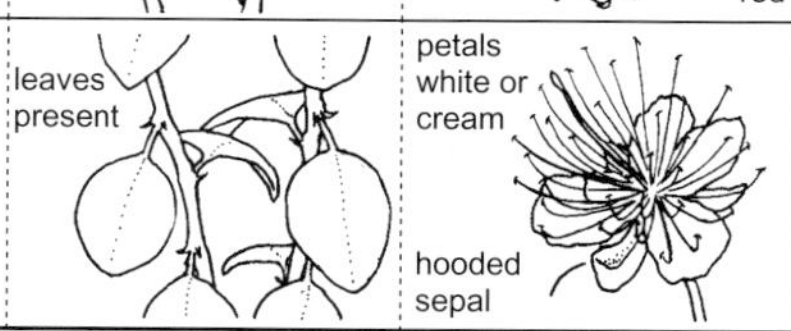	***C. cartilaginea*** **251**, 476

Cleome

1	**Leaves** 3 to 5-foliate, sometimes simple above	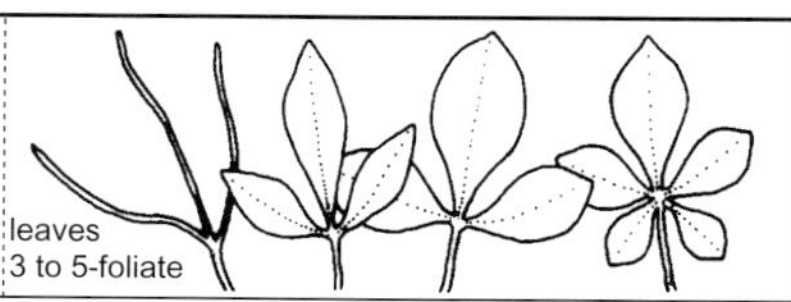	2
	Leaves all simple, broadly ovate to orbicular	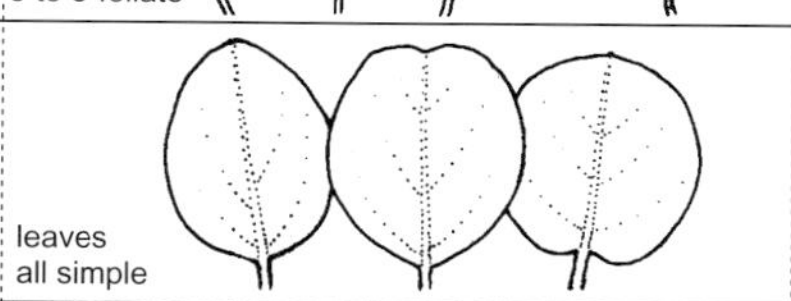	6

2	Delicate **annual**; **leaflets** filiform	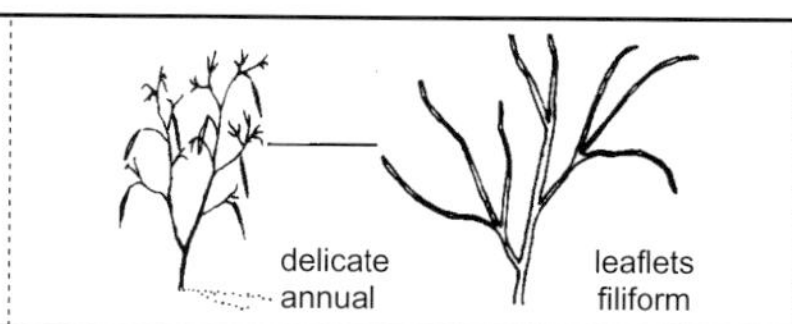	***C. tenella*** 479
	Robust **annual** or **perennial**; **leaflets** ovate or elliptic to broadly ovate	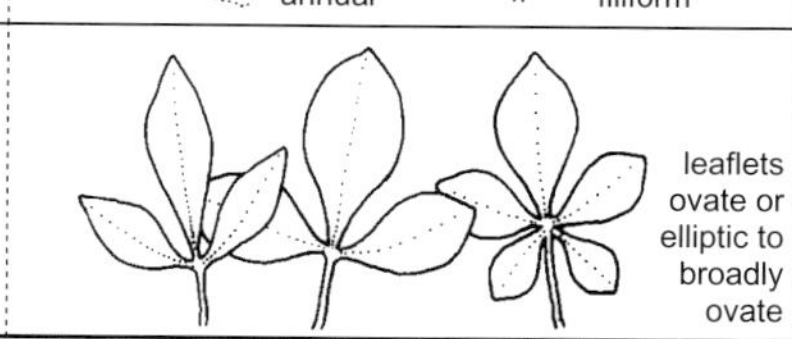	3

3	**Fruit** 4--12mm long, oblong with a rounded base and reticulate venation	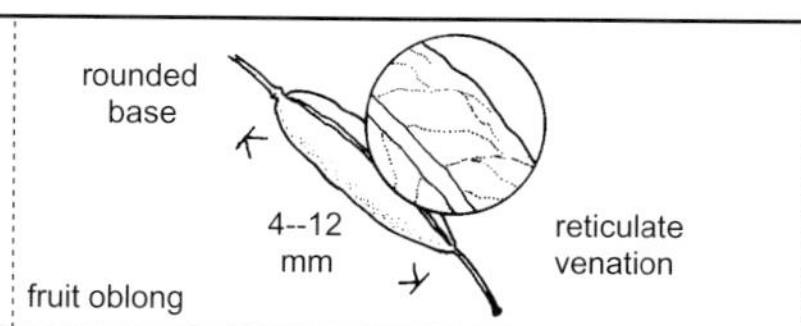	***C. brachycarpa*** 478
	Fruit 10--100mm long, narrowly elliptic to narrowly oblong, base acute, venation mostly parallel	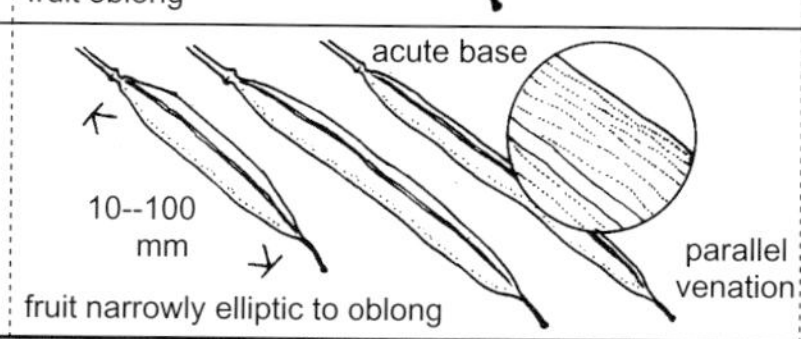	4

4	**Stamens** and **style** borne on an androgynophore (<1cm long in fruit)	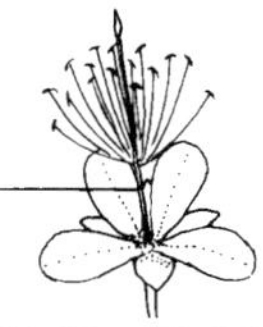	***C. gynandra*** 478
	Stamens and **style** not borne on an androgynophore		5

5	**Stems** sparsely clothed in short glandular hairs; **stamens** 6 to 8; **seeds** hairy	 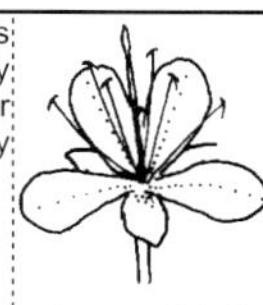 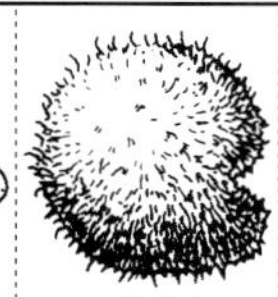	***C. socotrana*** Endemic 479
	Stems densely clothed in long and short glandular hairs; **stamens** 10 to 20; **seeds** glabrous	 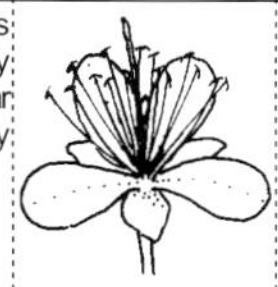 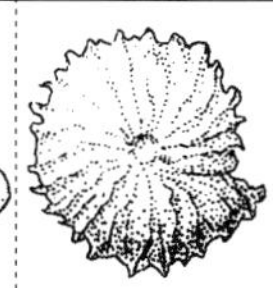	***C. viscosa*** 479

6	**Annual** (rarely **perennial**) herb; **stamens** 6; **fruit** linear, 0.5--1mm across, sparsely clothed in short glandular hairs	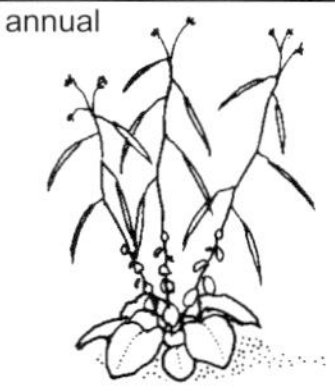 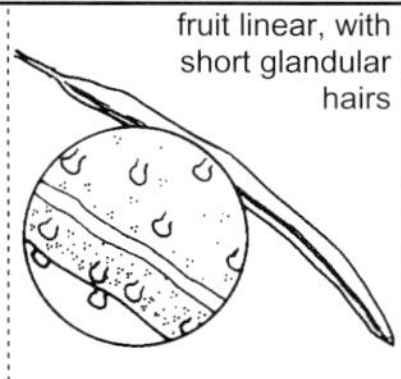	***C. scaposa*** 478
	Woody-based **perennial** herb; **stamens** 4; **fruit** oblong-ellipsoid, 3--4.5mm across, densely clothed in stipitate glands	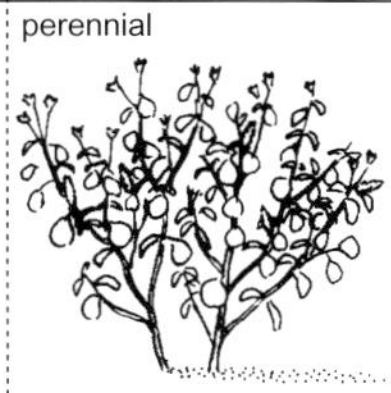 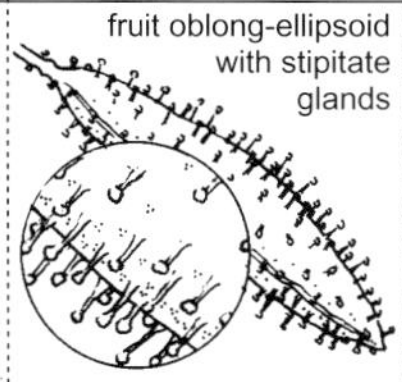	***C. austroarabica*** 477

CARYOPHYLLACEAE

1	**Stamens** 10; **styles** 2 or 3	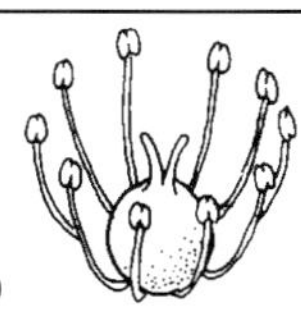	2
	Stamens 5; **style** solitary with a 2 or 3-lobed stigma	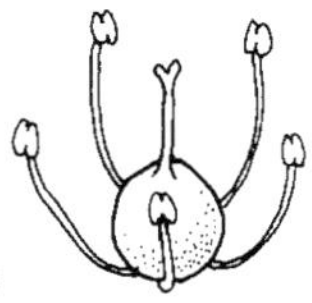	5

2	**Petals** bilobed almost to the base; **sepals** 8--10mm long; **ovary** stalked	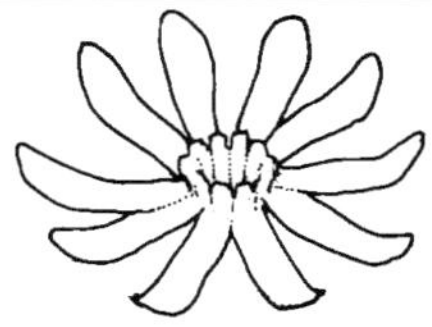 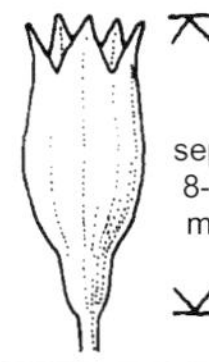	***Silene apetala*** 486
	Petals entire; **sepals** 2--5mm long; **ovary** sessile	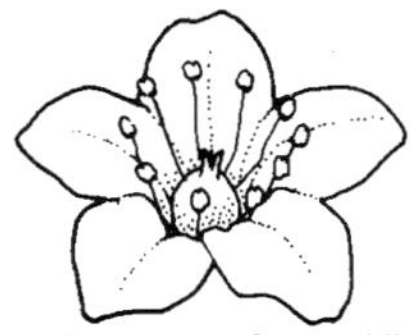 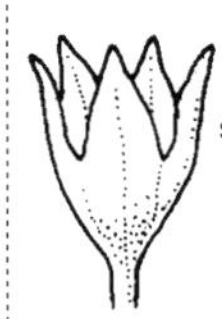	3

3

Leaves linear, whorled; **seeds** winged — ***Spergula fallax*** 487

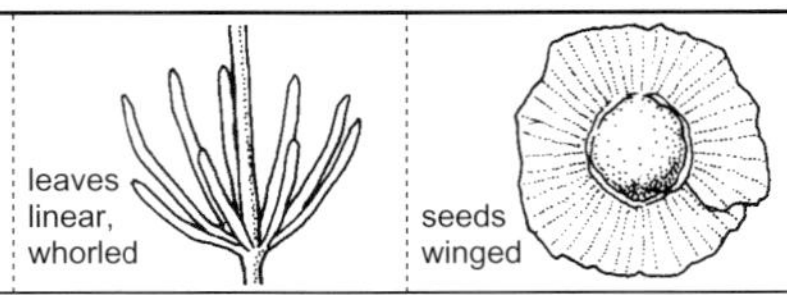

Leaves ovate to lanceolate or obovate, opposite; **seeds** not winged — 4

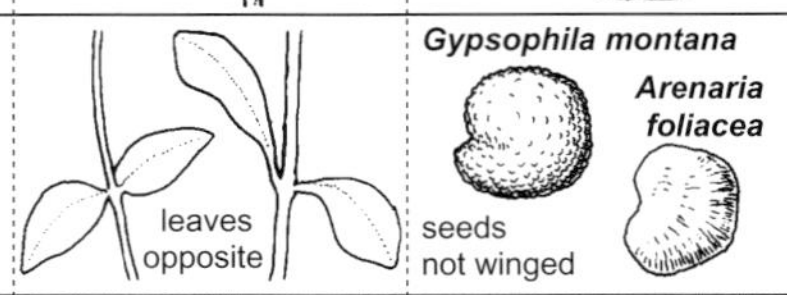

4

Delicate **annual** with prostrate stems; **leaves** ovate to lanceolate; **sepals** free — ***Arenaria foliacea*** 481

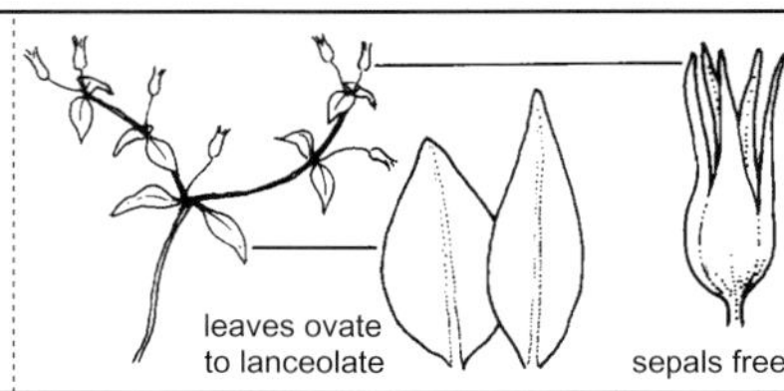

Robust woody-based **perennial** with erect stems; **leaves** obovate; **sepals** fused — ***Gypsophila montana*** 483

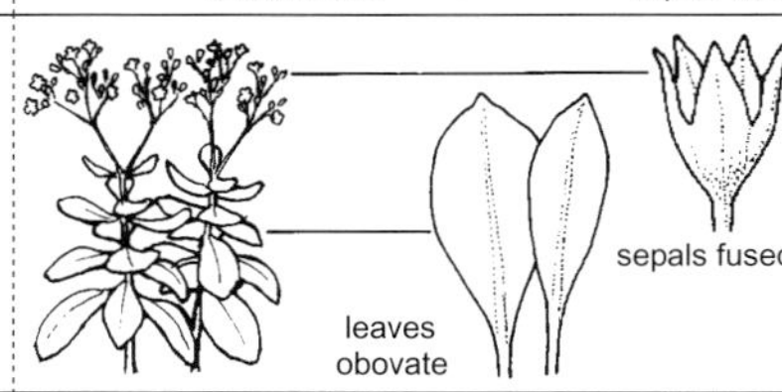

5

Flowers surrounded by feather-like bracts which conspicuously lengthen in fruit — ***Cometes abyssinica*** 481

Flowers without such bracts — 6

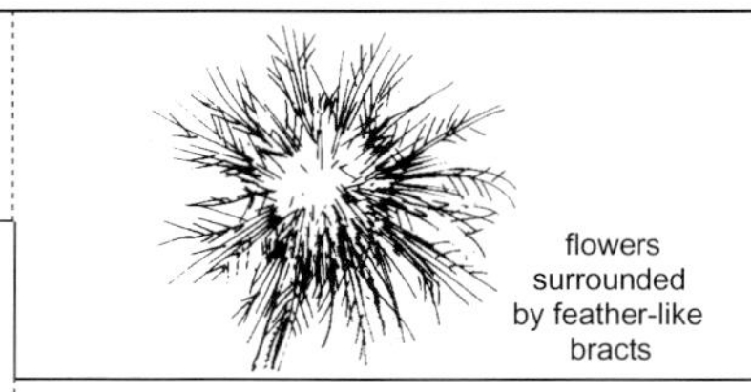

6

Capsule many-seeded — ***Polycarpaea*** (8 spp.)

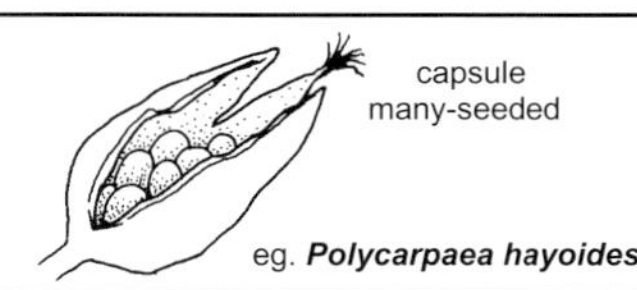

Capsule 1-seeded

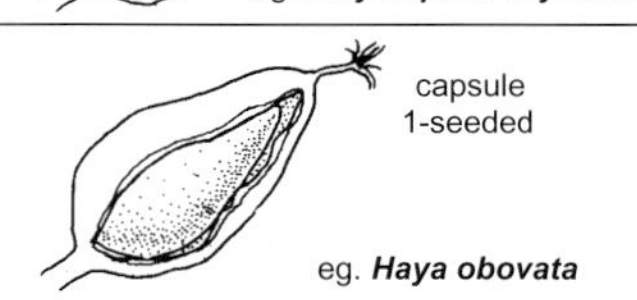

7

Prostrate **perennial**; **leaves** obovate to broadly oblanceolate, in whorls of 3 to 5 — ***Haya obovata*** Endemic Endemic genus **252**, 483

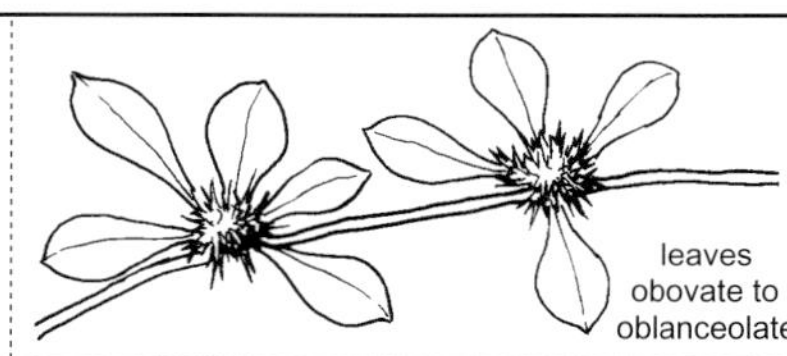

Subshrub; **leaves** linear to circular, fleshy, opposite or fascicled — 8

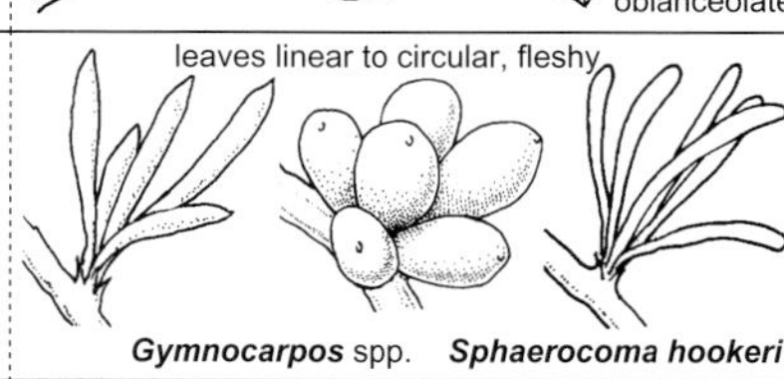

8

Sepals narrowly triangular to oblong-lanceolate with a membranous margin and a mucronate apex; **inflorescence** sessile — ***Gymnocarpos*** (2 spp.)

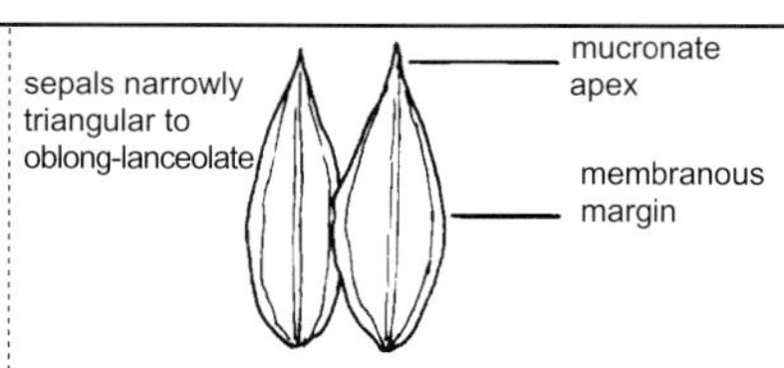

Sepals ovate to oblong-ovate with a fimbriate margin and a conspicuously deflexed apical awn; **inflorescence** pedunculate — ***Sphaerocoma hookeri*** **252**, 487

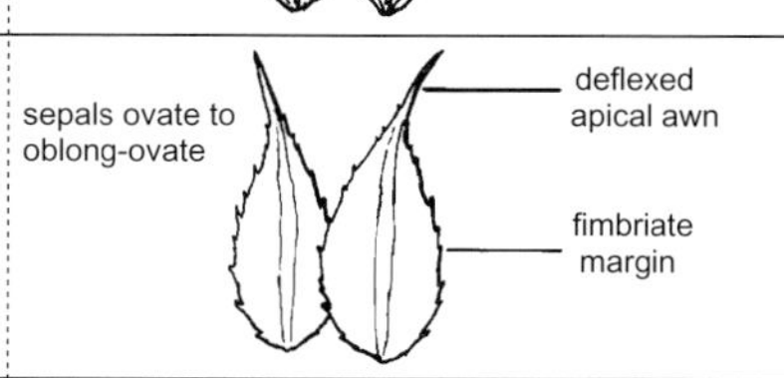

Gymnocarpos

1	**Leaves** green, linear-lanceolate; **bracts** brown or with a narrow white margin in the upper part		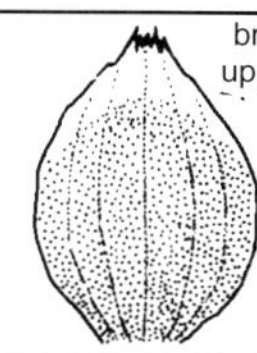	***G. bracteatus*** Endemic **252**, 482
	Leaves grey-green, narrowly oblanceolate to suborbicular; **bracts** with narrow to broad white margins all around		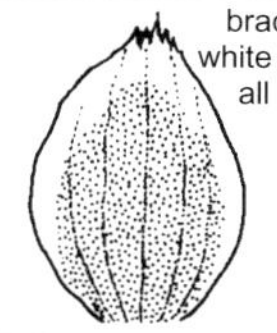	***G. kuriensis*** Endemic **252**, 482

Polycarpaea

1	**Plant** thinly woolly	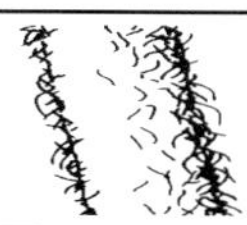		***P. corymbosa*** 484
	Plant glabrous			2
2	**Stem leaves** linear	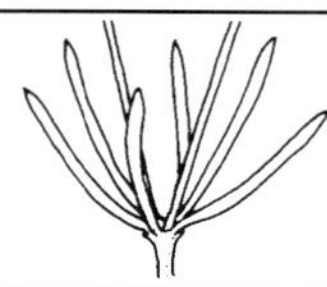		3
	Stem leaves narrowly spathulate to suborbicular	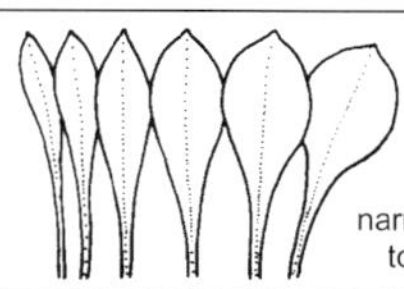		5
3	Woody-based perennial **herb** or **shrublet**; **leaves** all linear	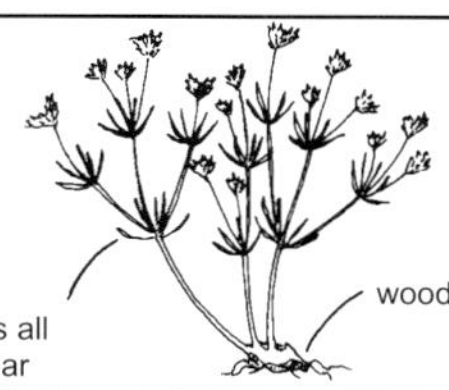		***P. caespitosa*** Endemic **252**, 484
	Annual **herb**; basal **leaves** linear, stem leaves spathulate			4
4	**Sepals** and **bracts** 2--2.5mm long, ovate; **stem leaves** more or less 1/3 as long as the internodes	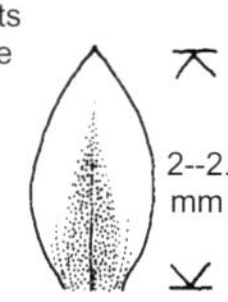	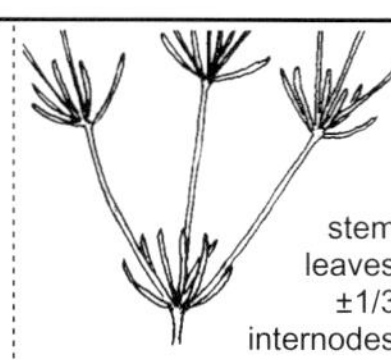	***P. spicata*** Endemic 486
	Sepals and **bracts** 3--4mm long, narrowly ovate; **stem leaves** >1/2 as long as the internodes	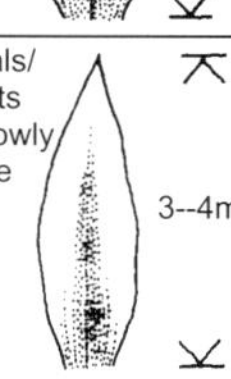	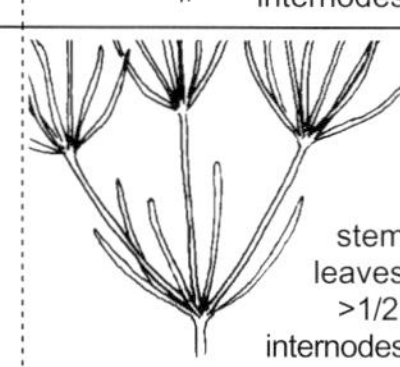	***P. balfourii*** Endemic 484
5	Annual **herb**	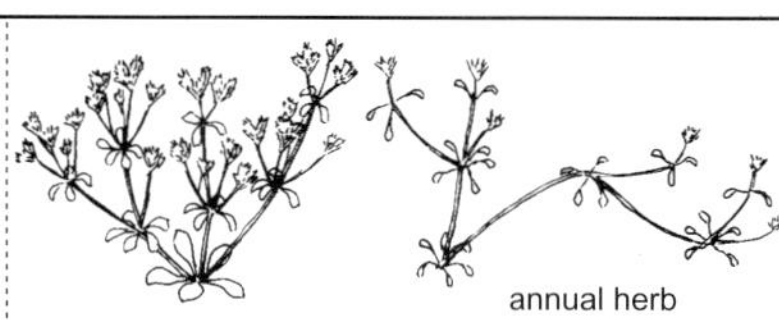		6
	Woody-based perennial **herb**			7

6

Stems prostrate, diffusely branched, <45cm long; **sepals** with a rounded tip

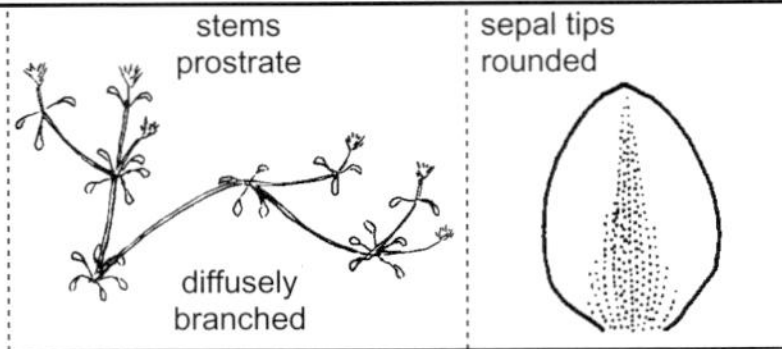

P. paulayana
Endemic
485

Stems erect, ascending or prostrate, repeatedly branched, <10 cm long; **sepals** with an acute or acuminate tip

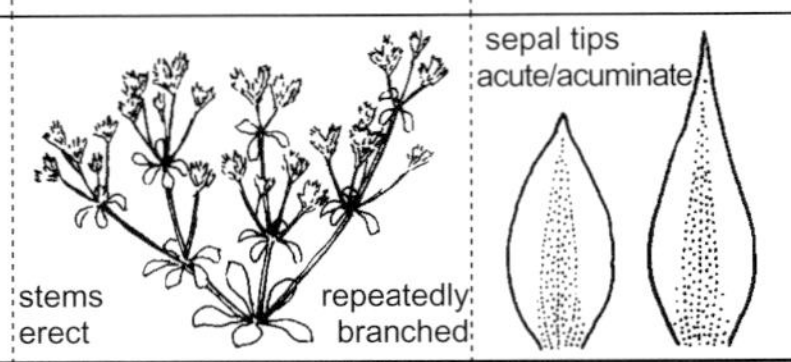

P. hayoides
Endemic
485

7

Stems prostrate; **sepals** and **bracts** narrowly ovate, 3--4 x 0.8--1mm; **peduncles** swollen above

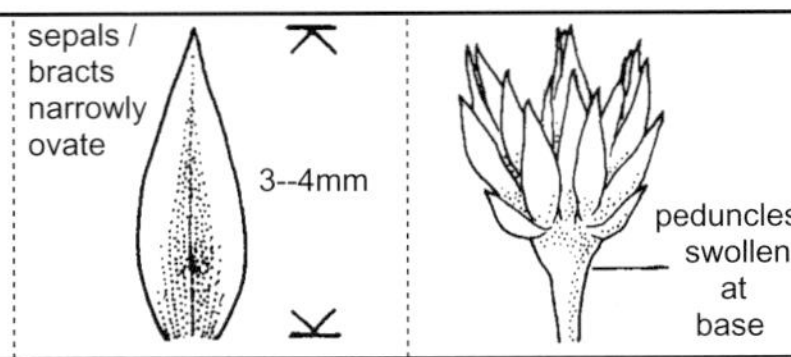

P. kuriensis
Endemic
252, 485

Stems erect; **sepals** and **bracts** ovate, 2.25--2.5 x 1--1.2mm; **peduncles** not swollen above

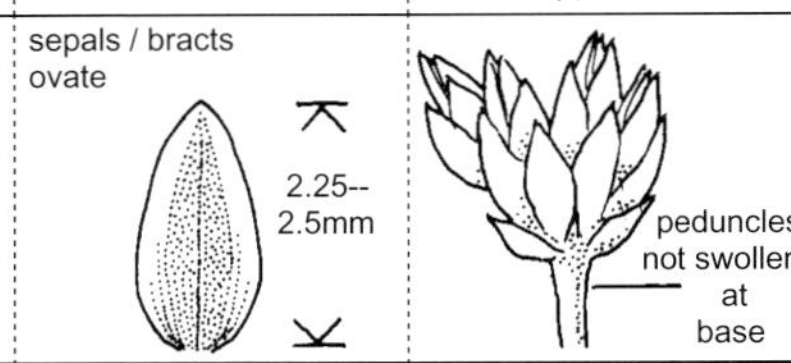

P. hassalensis
Endemic
485

CELASTRACEAE

1

Small spiny **tree** to 3m; **leaves** clustered, obovate to orbicular, <2 x1.5cm; **fruit** a 6-lobed, obconic capsule; **seeds** with a basal aril

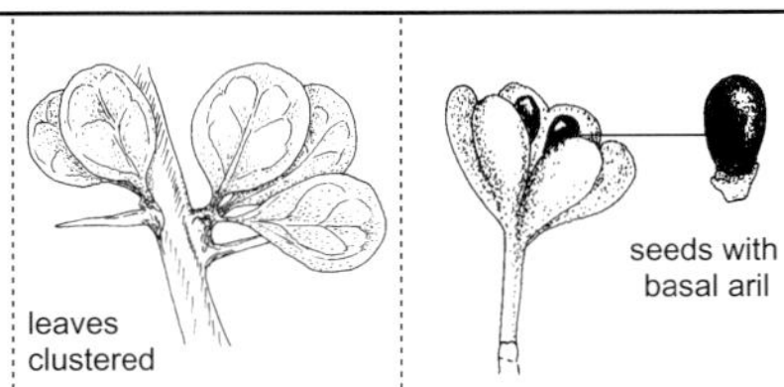

***Maytenus* sp. A**
Endemic
488

CHENOPODIACEAE

1

Stems jointed, succulent; **leaves** inconspicuous, reduced to short cupules at the tips of the stem-segments; **flowers** in elongated spikes

Arthrocnemum macrostachyum
252, 489

Stems not jointed, leafy; **leaves** usually conspicuous, sometimes scale-like; **flowers** in panicles, axillary clusters or spikes

2

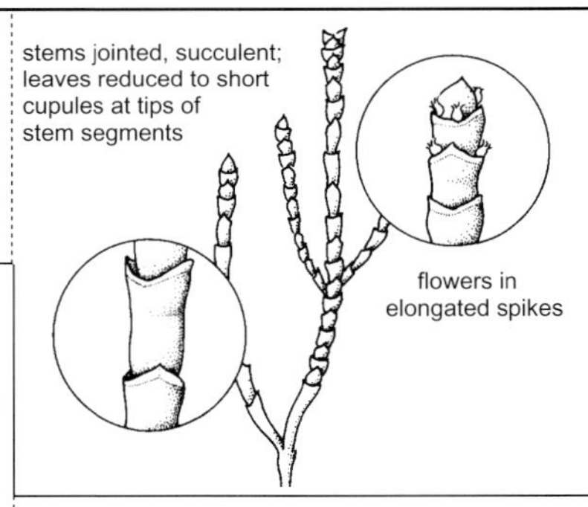

2

Leaves broad and flat

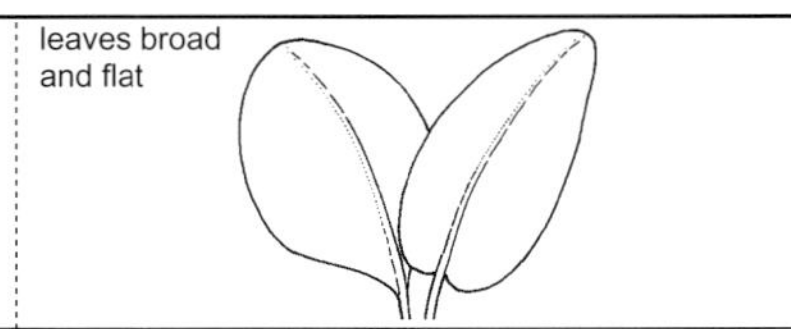

3

Leaves cylindrical, usually constricted at base into a short petiole

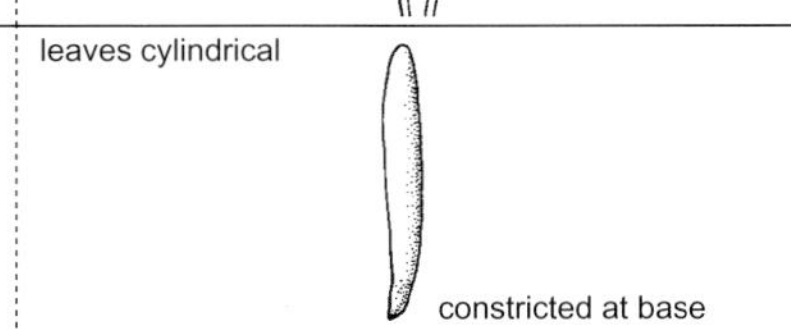

4

Leaves subglobose, ovoid or scale-like, clasping stem at base

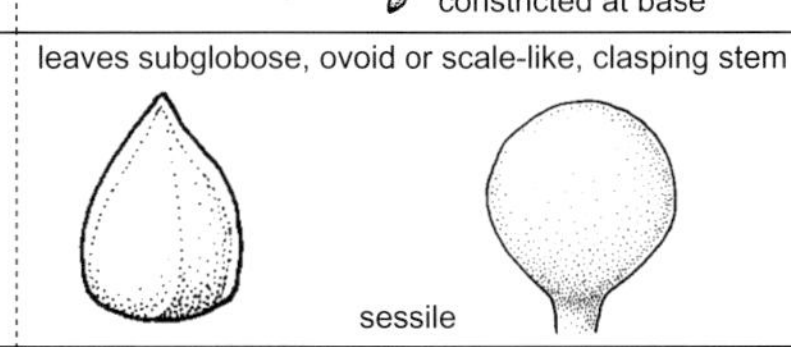

7

3	**Leaves** whitish grey, densely mealy, entire or undulate; **fruit** included in the enlarged **bracteoles**	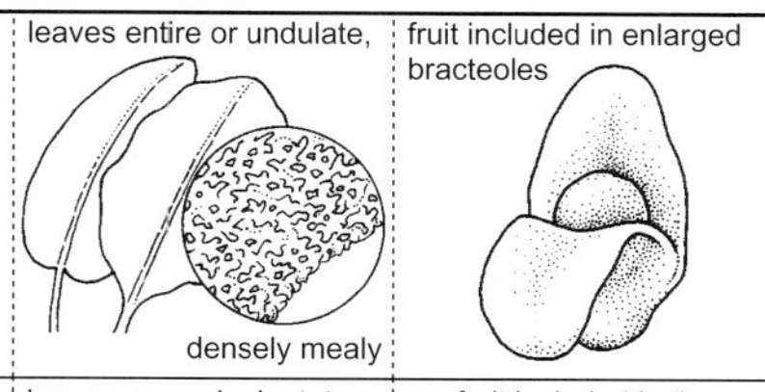	***Atriplex*** (2 spp.)
	Leaves dark green, thinly mealy (farinose) at first then glabrous, coarsely dentate; **fruit** included in the perianth, **bracteoles** absent	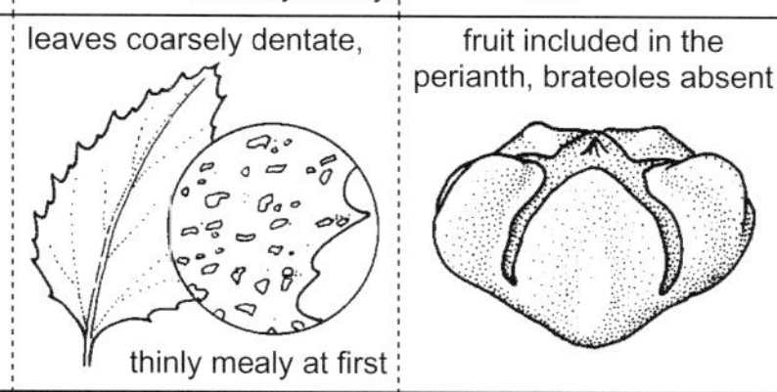	***Chenopodium murale*** 491
4	**Perianth segments** with conspicuous, membranous wings in fruit	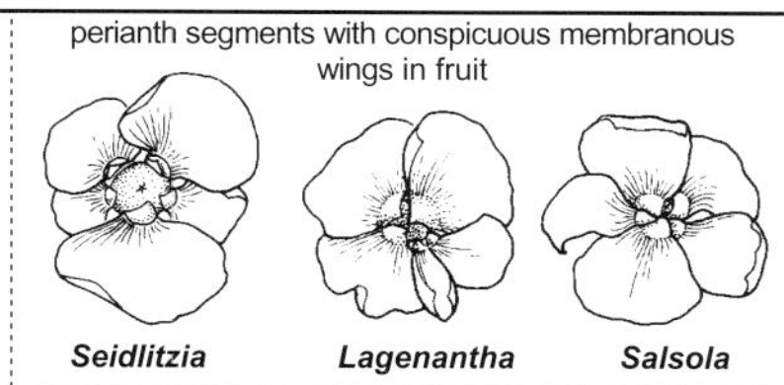	5
	Perianth segments without conspicuous, membranous wings in fruit	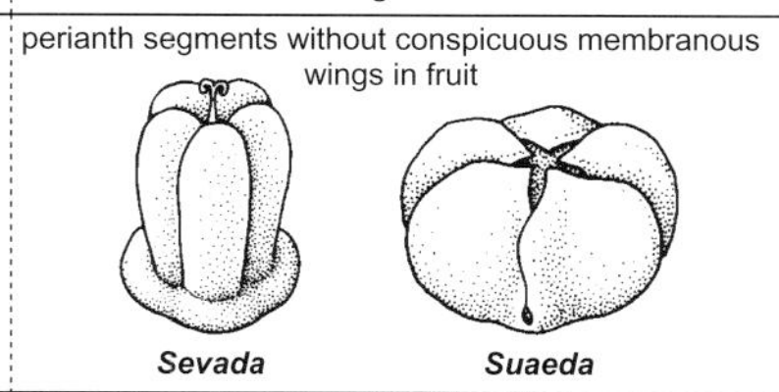	6
5	**Leaves** all opposite, cylindrical, 4--35mm long, glabrous; **leaf bases** not thickened; **fruiting perianth tube** inconspicuous, **wings** very unequal	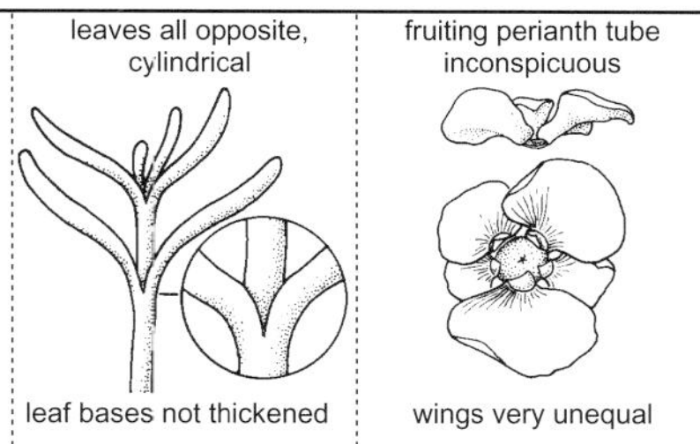	***Seidlitzia rosmarinus*** 493
	Leaves alternate or some opposite, subglobose to narrowly cylindrical, 3--20mm long, glabrous; **leaf bases** thickened; **fruiting perianth tube** well-developed with a conspicuous swollen base, **wings** subequal	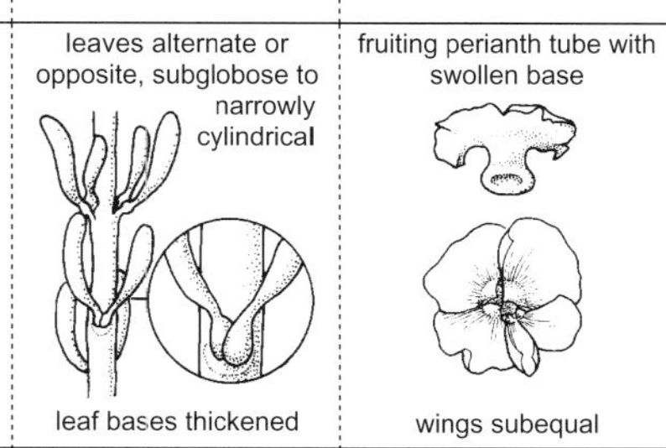	***Lagenantha cycloptera*** **252**, 492
6	**Leaf bases** thickened; **leaves** with a tuft of minute hairs in the axils; **fruiting perianth** well-developed, slightly constricted towards the expanded base	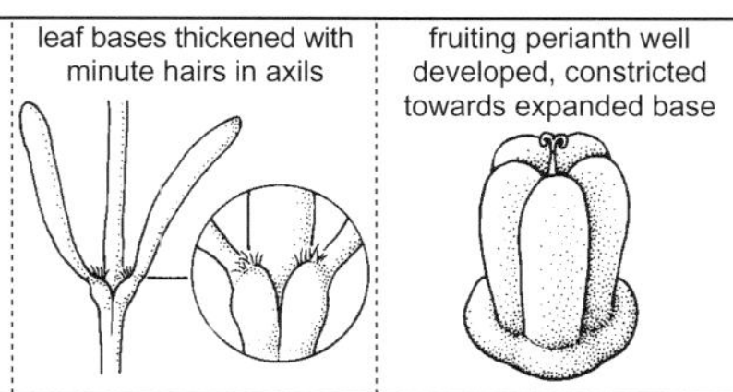	***Sevada schimperi*** 493
	Leaf bases not thickened; **leaves** without a tuft of hairs in the axils; **fruiting perianth** inconspicuous	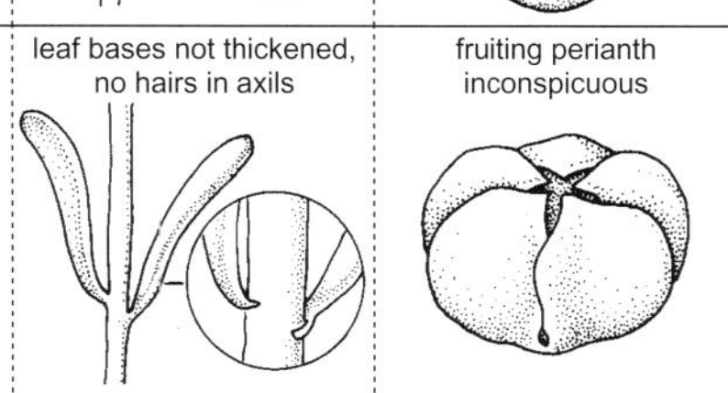	***Suaeda*** (2 spp.)

7	**Leaves** alternate, leathery, triangular, spine-tipped, c.3mmm long, glabrous; **leaf base** fused to and hiding stem	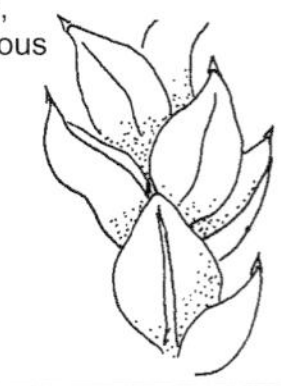		***Cornulaca ehrenbergii*** 491
	Leaves opposite, succulent, obovoid c.3mm long, densely overlapping on lateral branches, glabrous; **leaf bases** of opposite leaves fused	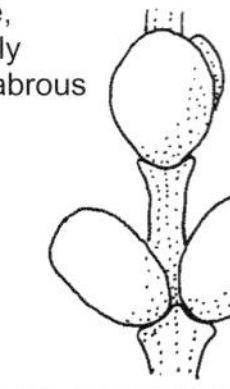		***Anabasis ehrendbergii*** 489
	Leaves alternate, succulent, ovate, 1.5--2mm long, densely covered with medifixed hairs when young; **leaf base** clasping the stem	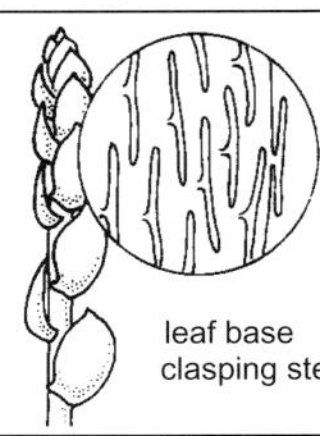		***Salsola spinescens*** 492

Atriplex

1	At least some **leaves** cordate at the base; **fruiting bracteoles** ovate, entire, connate at the base	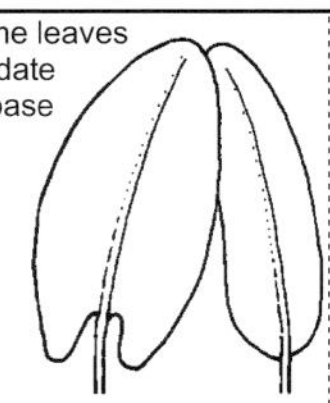	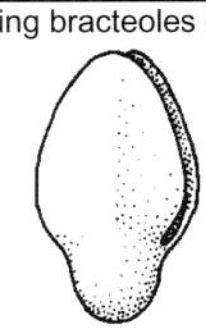	***A. farinosa*** 490
	Leaves not cordate; **fruiting bracteoles** broadly triangular-ovate to cordate, entire or with 2 small basal teeth	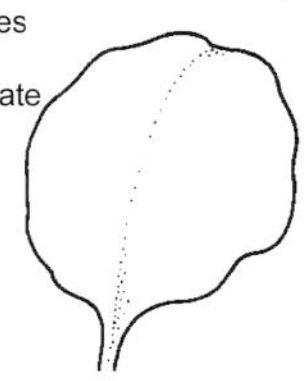	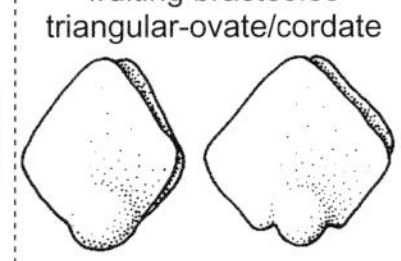	***A. griffithii*** 490

Suaeda

1	Small **tree** or **shrub** to 5m; **leaves** linear to linear-oblong, flattened on both sides; **flowers** unisexual with a large male flower surrounded by several, smaller female flowers	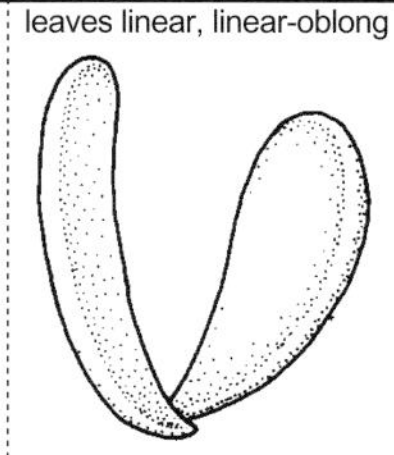	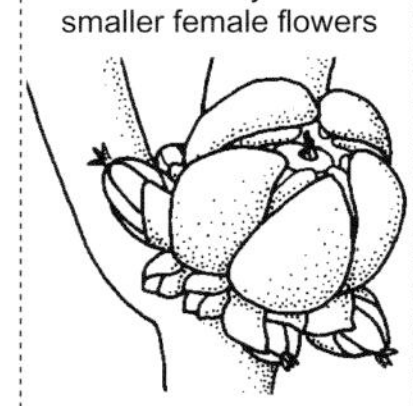	***S. monoica*** 493
	Shrub to 1m; **leaves** obovate-oblong to subglobose, terete or flattened on one side only; **flowers** bisexual, similar in size	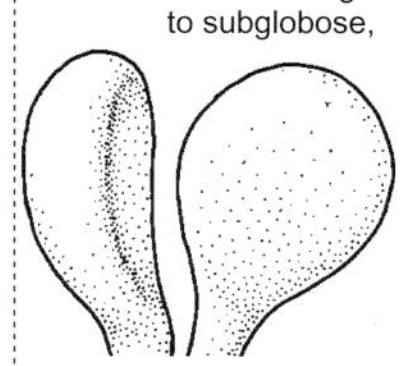	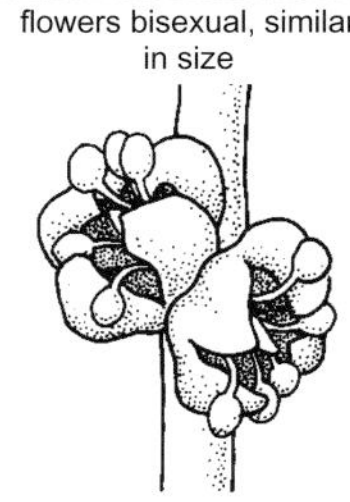	***S. vermiculata*** 494

COMPOSITAE

1	**Leaves** opposite	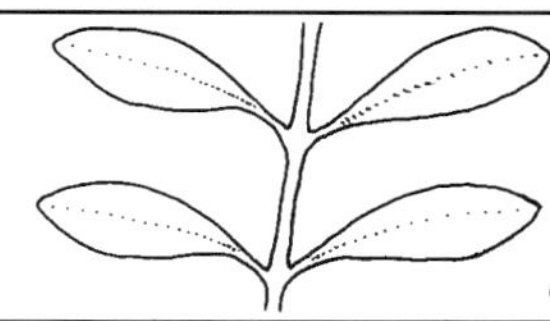	2
	Leaves alternate or in a basal rosette	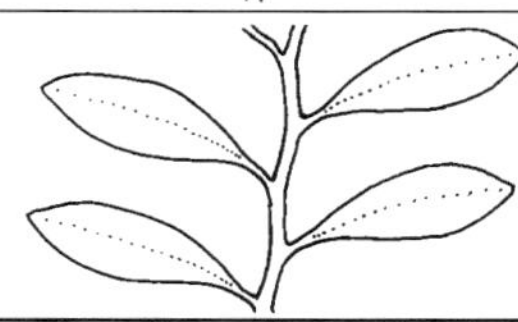	5

2	**Ray flowers** present		3
	Ray flowers absent		4

3	**Pappus** of 2 or more erect bristles bearing retrorse barbs which adhere to clothing; **flowers** white; **achenes** narrowed, black; **leaves** pinnate; **phyllaries** non-glandular	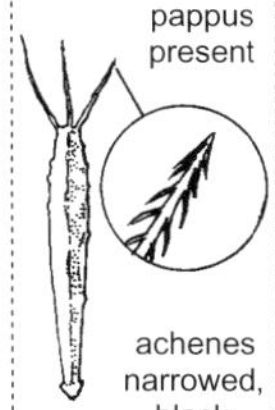 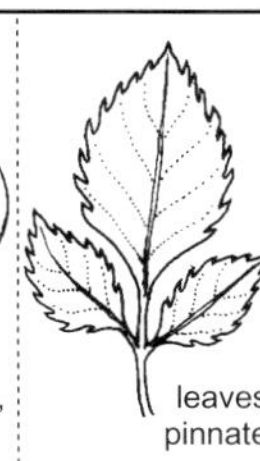 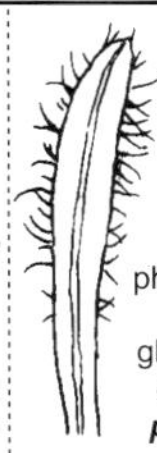	***Bidens*** (2 spp.)
	Pappus absent; **flowers** white; **achenes** columnar; **leaves** simple; **phyllaries** non-glandular (often found near water)	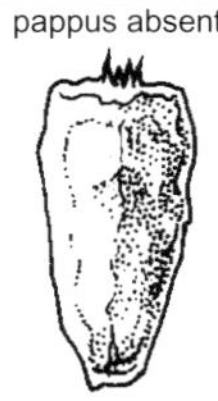 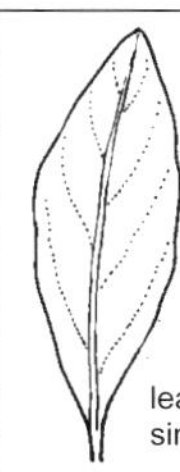 	***Eclipta prostrata*** 500
	Pappus absent; **flowers** yellow; **achenes** obconical; **leaves** simple; **phyllaries** glandular	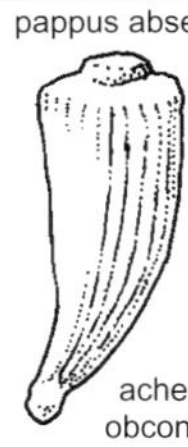 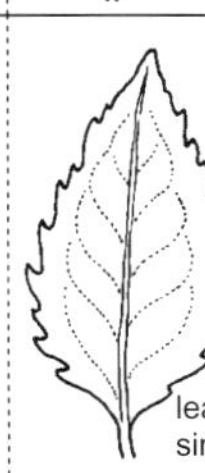 	***Sigesbeckia orientalis*** 514

4	**Flowers** yellow; **pappus** of 2 to 4 bristles; **achenes** columnar, 3 or 4-angled	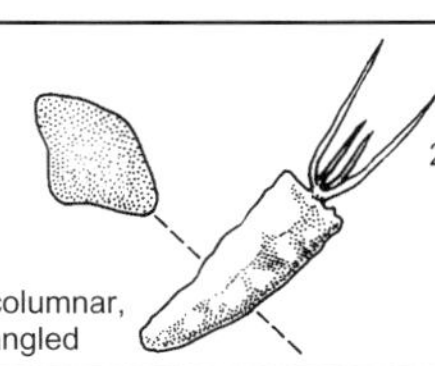	***Blainvillea gayana*** **253**, 497
	Flowers blue; **pappus** 5-awned; **achenes** columnar, 4 or 5-angled	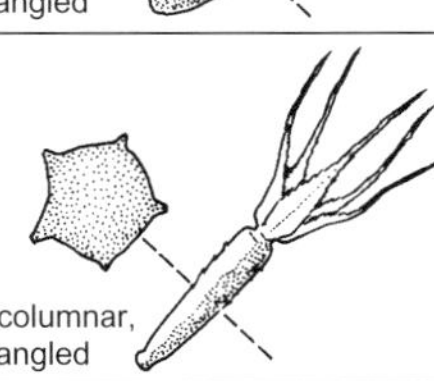	***Ageratum conyzoides*** **253**, 496
	Flowers yellow; **pappus** absent; **achenes** spiny	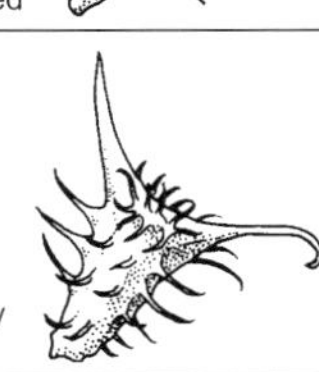	***Acanthospermum hispidum*** 495

5	**Phyllaries** spine-tipped; **plant** forming low spiny clumps	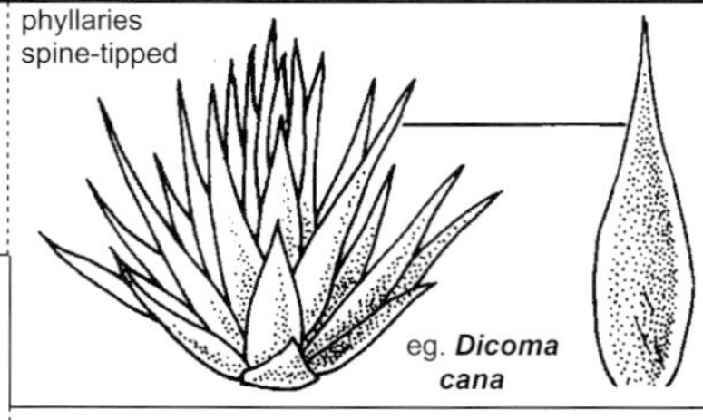	***Dicoma*** (2 spp.)
	Phyllaries not spine-tipped; **plant** habit various		6
6	**Achenes** winged	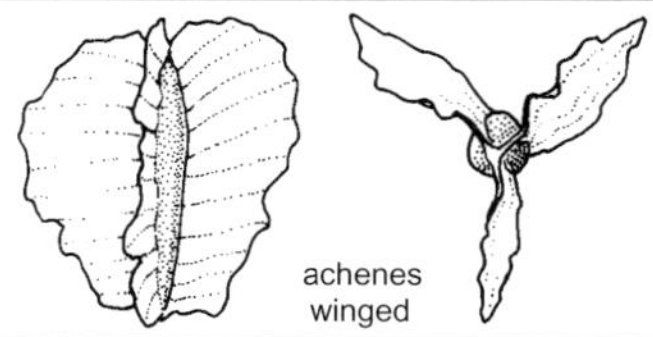	***Tripteris vaillantii*** **255**, 515
	Achenes not winged	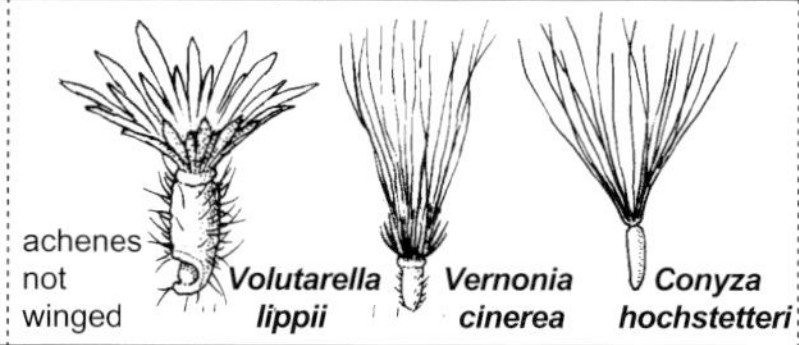	7
7	Succulent stemmed **shrub**; **leaves** reduced, soon falling; **flowers** pink	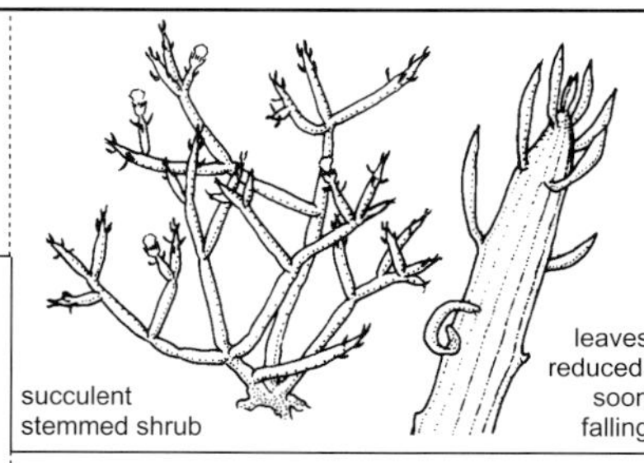	***Kleinia scottii*** Endemic **255**, 505
	Herbs or **shrubs**, stems never succulent (leaves sometimes fleshy); **leaves** present; **flowers** white, yellow, pinkish or purple		8
8	**Leaves** ending in 2 or 3 linear lobes; **shrub** with the stems marked with leaf scars	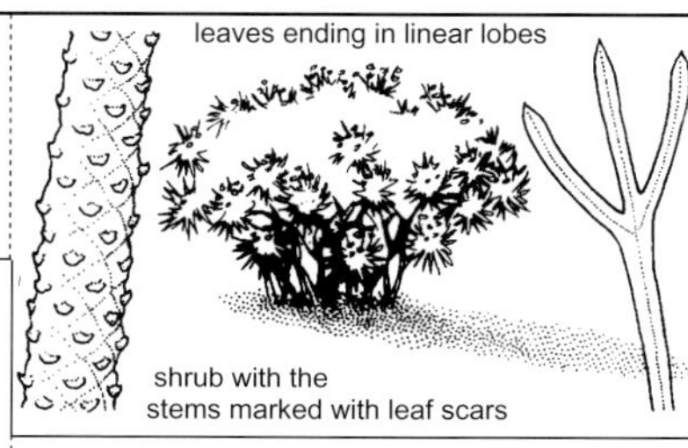	***Euryops arabicus*** **253**, 500
	Leaves never ending in linear lobes; **herbs** or **shrubs**, stems not prominently scarred		9
9	**Florets** all ray, strap-shaped: elongated into a 5-toothed, petal-like structure; **white latex** often present	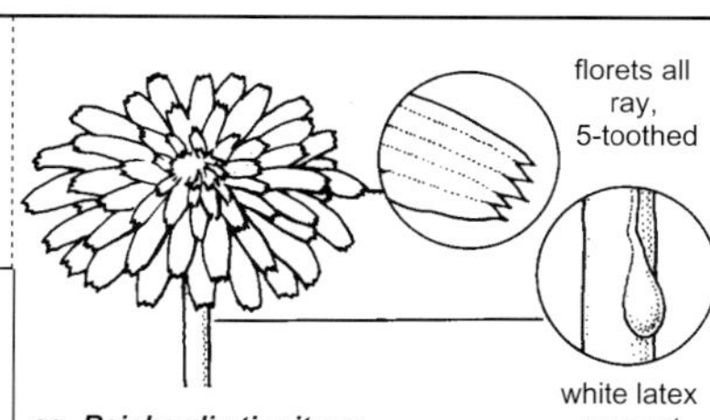	10
	Florets all tubular or a mixture of tubular and ray [strap-shaped], never all ray; **white latex** never present		13
10	**Flowers** pink or purple		***Prenanthes amabilis*** Endemic **256**, 510
	Flowers yellow	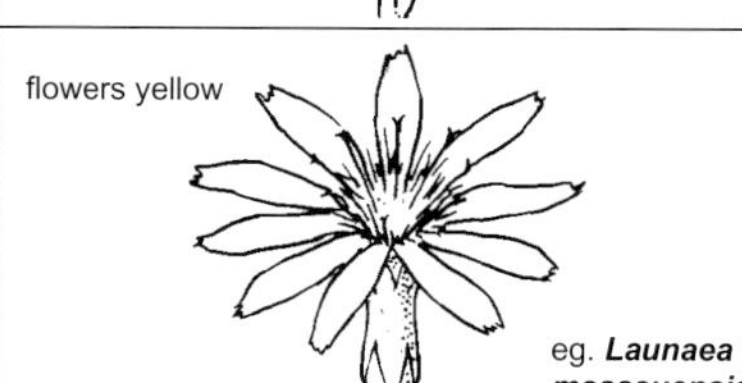	11

11 **Peduncles** of capitula conspicuously expanded above; **phyllaries** glabrous with broad white margins (amplexicaul) — ***Reichardia tingitana*** 513

Peduncles conspicuously expanded above; **phyllaries** glabrous or hairy, white margins narrow or absent — 12

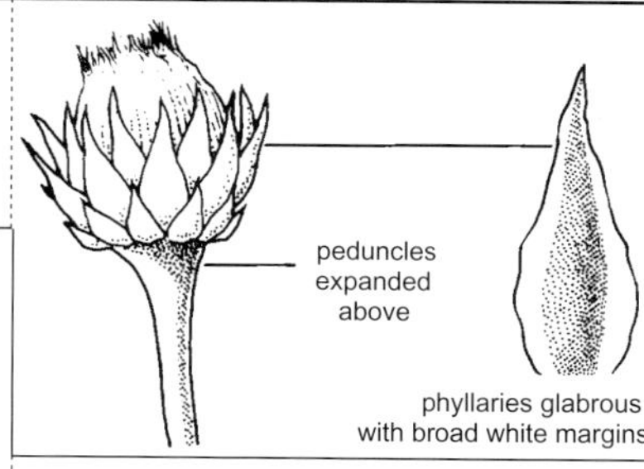

12 Robust **herb**; **leaf base** with lobes clasping stem; **achenes** not notched at base — ***Sonchus oleraceus*** 514

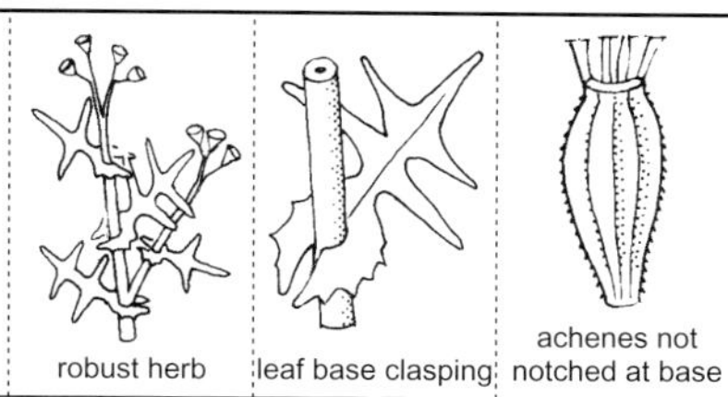

Delicate **herbs**; **leaf base** without clasping lobes; **achenes** notched at base — ***Launaea*** (7 spp.)

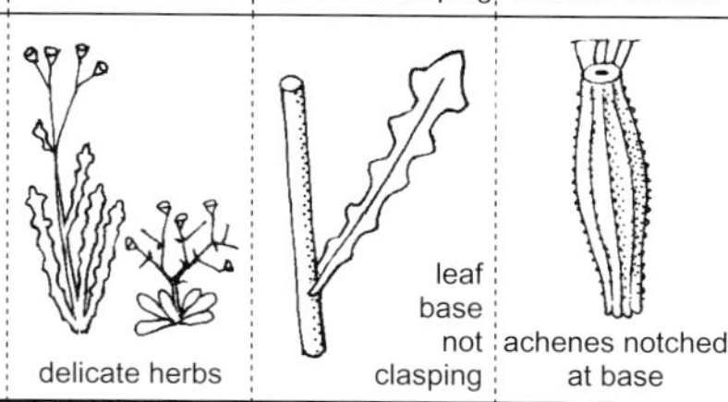

13 **Phyllaries** scarious or glossy — ***Helichrysum*** including *Achyrocline* (15 spp.)

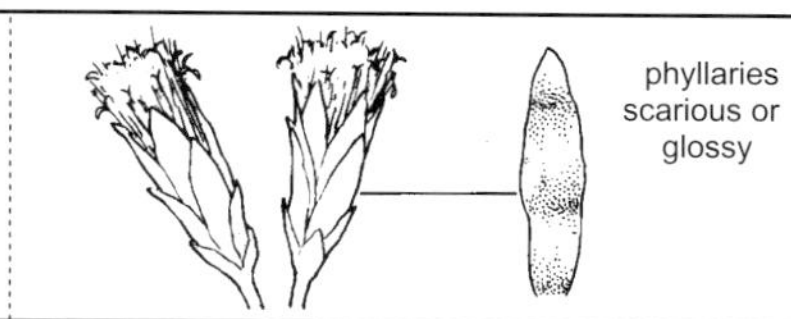

Phyllaries not scarious or glossy (herbaceous) — 14

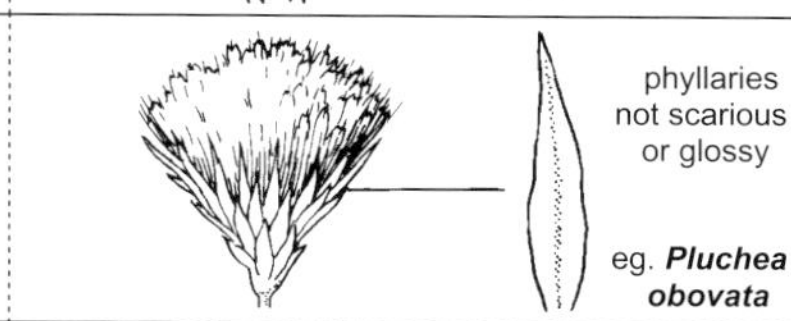

14 **Flowers** white or purple — 15

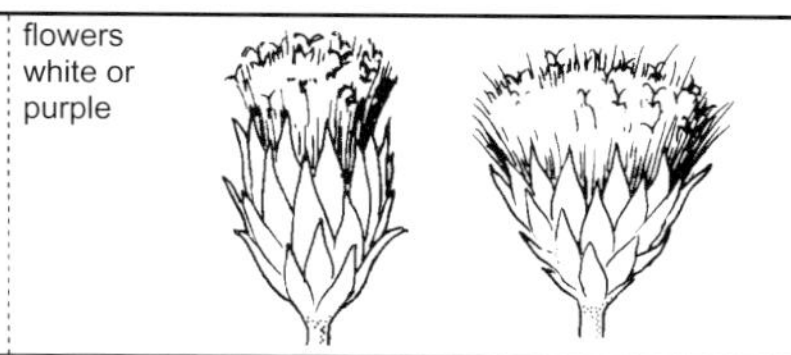

Flowers yellow — 16

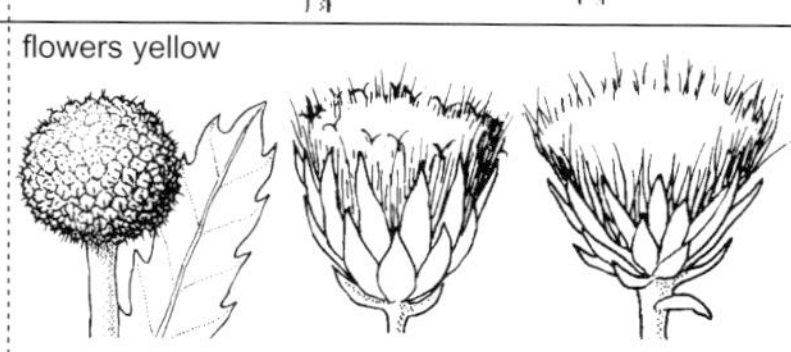

15 **Pappus** of scales; **receptacle** densely hairy; **phyllaries** with white margins and black tips — ***Volutaria lippii*** 516

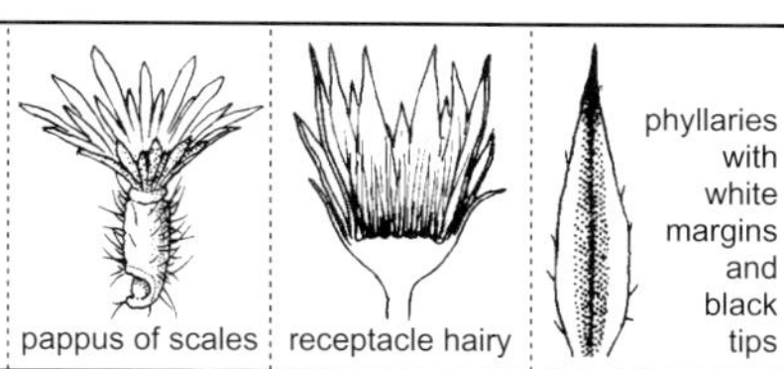

Pappus of two rows of bristles; **receptacle** naked; **phyllaries** neither with white margins nor black tips — ***Vernonia* s.l.** (4 spp.)

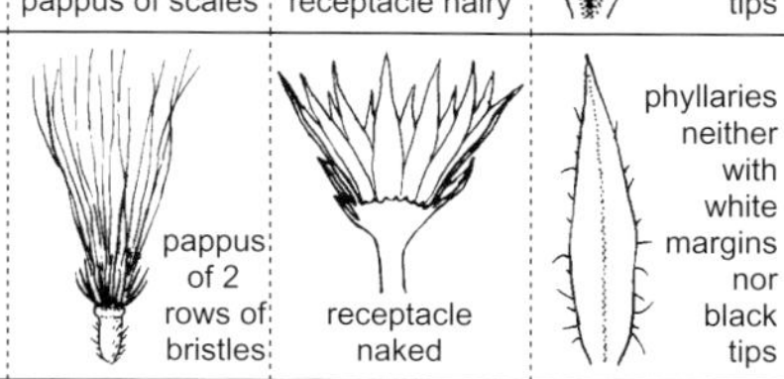

Pappus of one row of bristles; **receptacle** naked; **phyllaries** neither with white margins nor black tips — ***Pluchea*** (2 spp.)

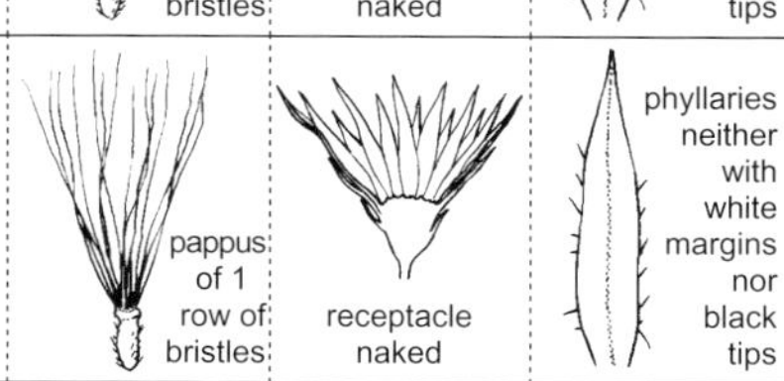

16	**Capitula** globose; **pappus** absent	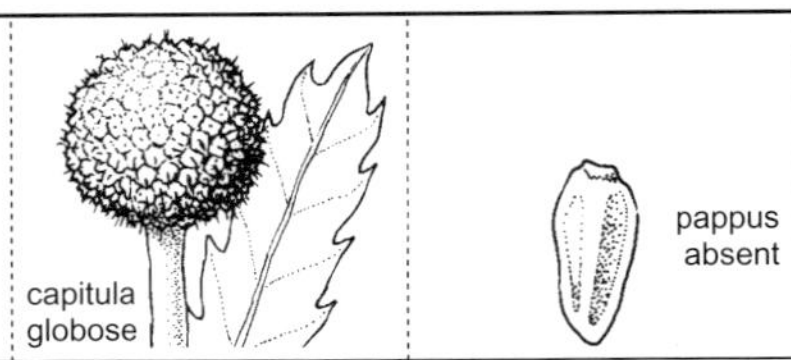	***Dichrocephala chrysanthemifolia*** 498
	Capitula conical or cylindrical; **pappus** present	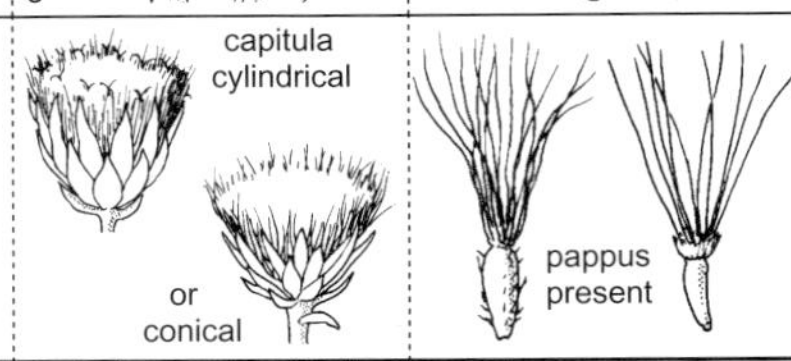	17

17	**Leaves** glabrous, sticky; **phyllaries** straw-coloured	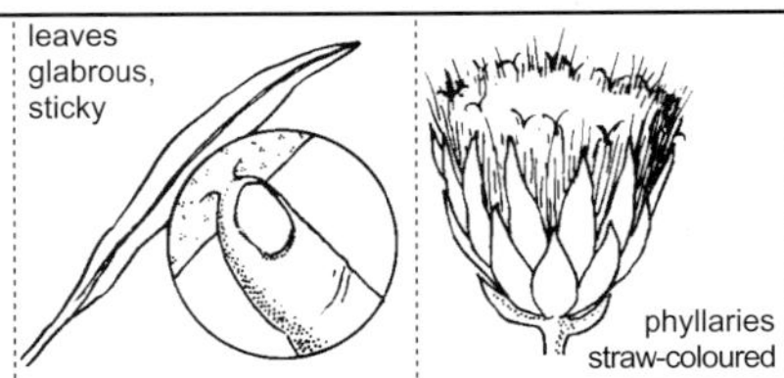	***Psiadia schweinfurthii*** Endemic 510
	Leaves glabrous or hairy, not sticky; **phyllaries** pale green or purplish-tinged	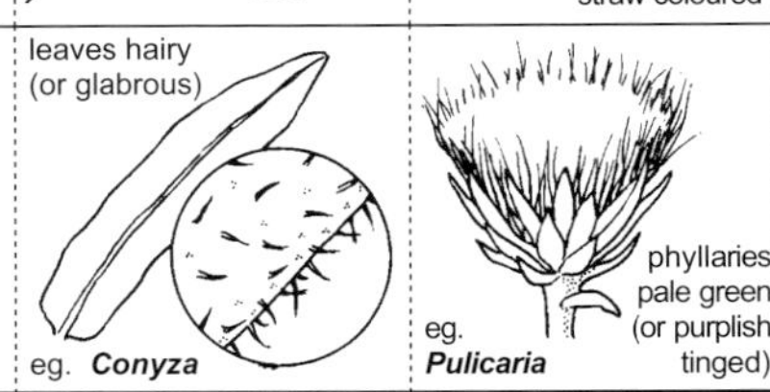	18

18	Tiny rosette-forming **herb** to 3cm with thick woody **taproot**	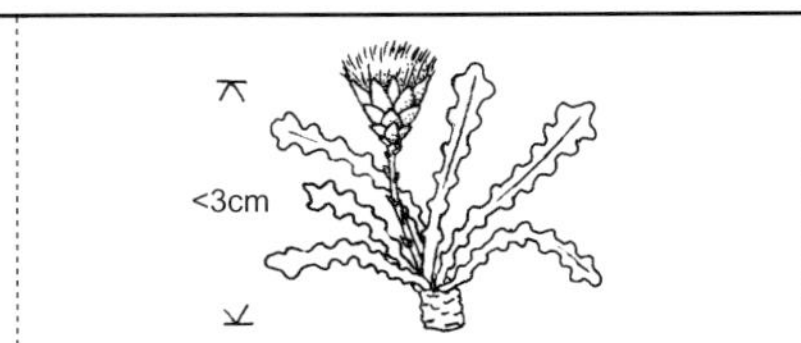	***Distephanus qazmi*** Endemic **253**, 499
	Herb or **subshrub** to 10cm, **taproot** absent	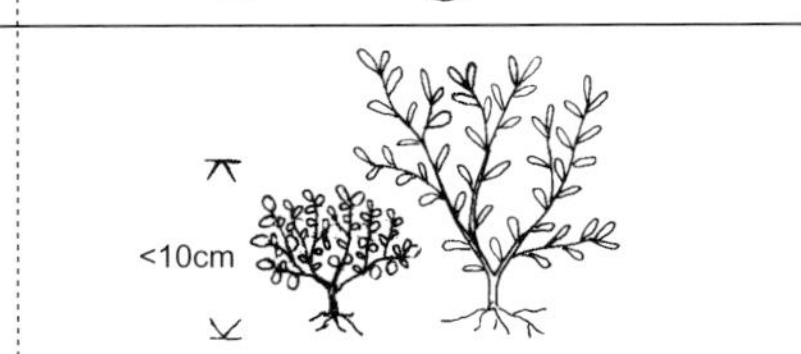	9

19	**Capitula** borne singly or in small groups; **pappus** in 2 whorls, inner of bristles and outer a ring of scales (sometimes reduced)	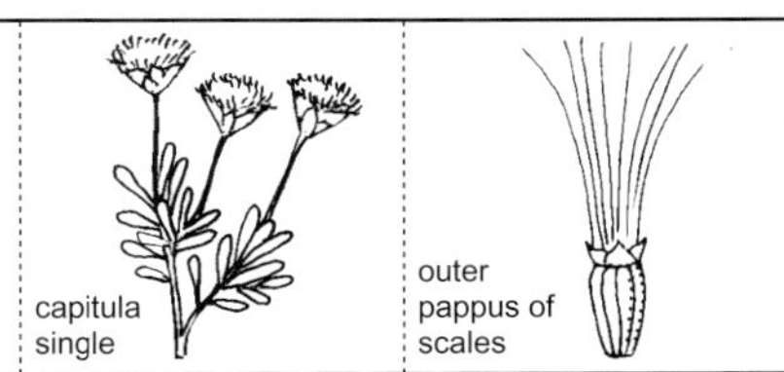	***Pulicaria*** (7 spp.)
	Capitula in dense groups; **pappus** entirely of hairs	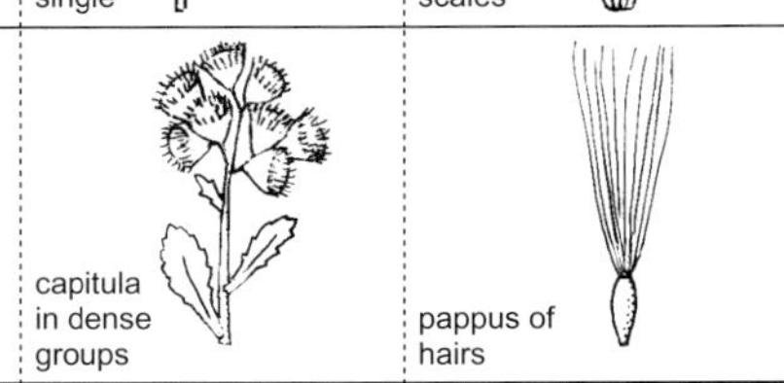	***Conyza hochstetteri*** 498

Bidens

1	Outer **phyllaries** linear or oblanceolate, longer than inner; inner **achenes** longer than outer; **ray florets** yellow	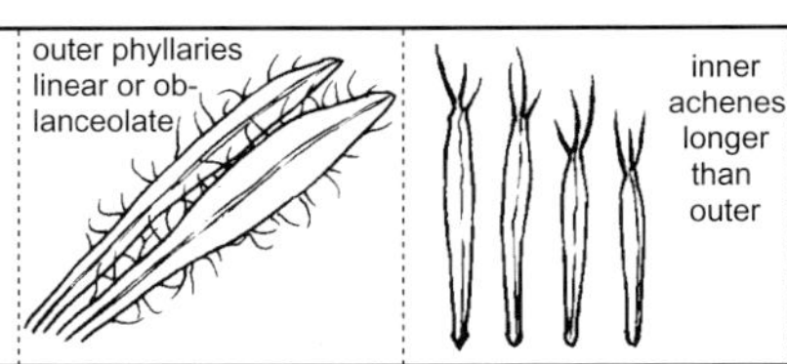	***B. biternata*** 496
	Outer **phyllaries** spathulate, not longer than inner; **achenes** uniform in size; **ray florets** white	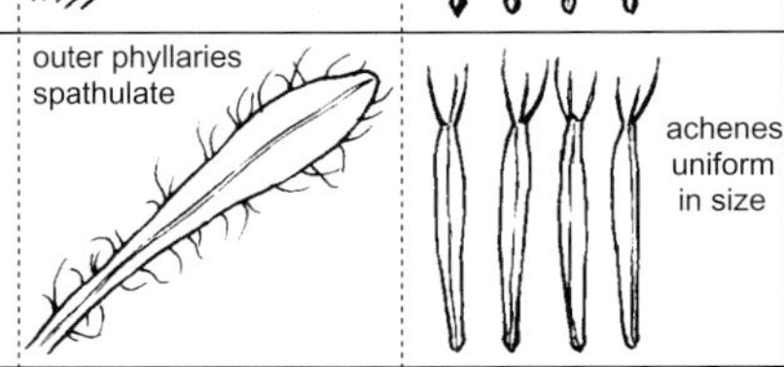	***B. pilosa*** 497

Dicoma

1	Clump-forming **perennial**; **leaves** linear, entire; **florets** red	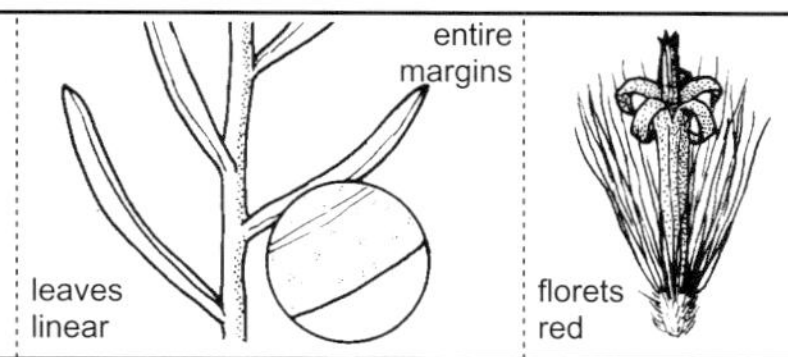	***D. cana*** Endemic **253**, 499
	Ascending **annual** or **perennial** with open branching; **leaves** oblanceolate with spinose margins; **florets** yellow	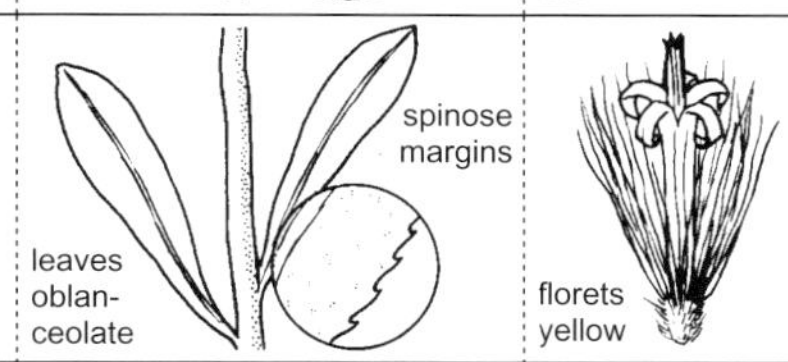	***D. tomentosa*** 499

Helichrysum (including *Achyrocline*)

1	Cushion-forming perennial **herbs**	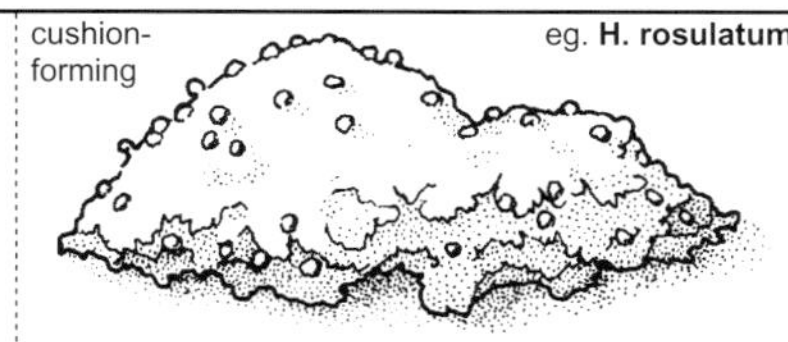	2
	Annual and non-cushion-forming perennial **herbs**	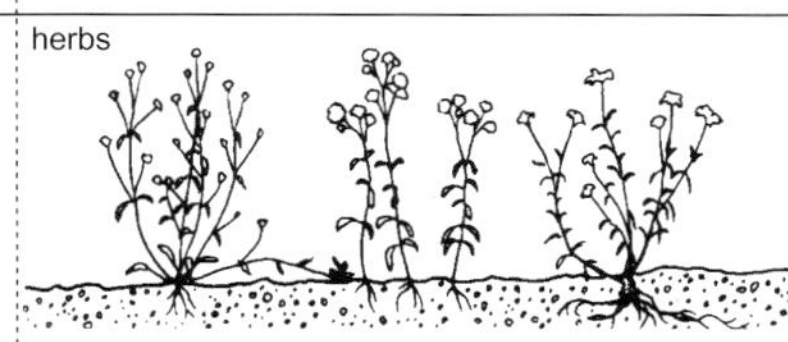	3
	Shrubs	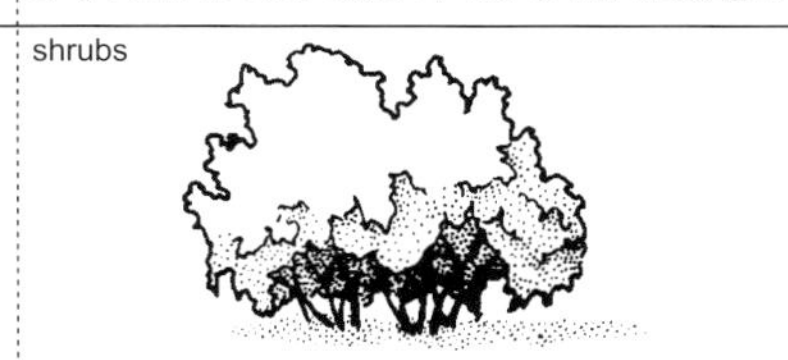	9

2	**Leaves** linear, needle-like; **capitula** on long peduncles		***H. aciculare*** Endemic **253**, 501
	Leaves narrowly elliptic, soft; **capitula** on short peduncles	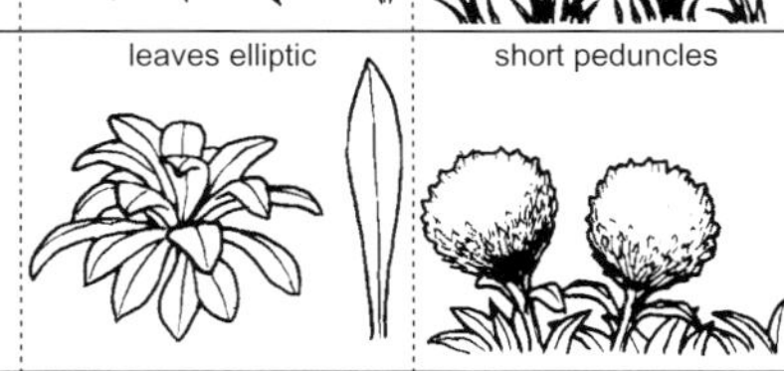	***H. rosulatum*** Endemic **254**, 503

3	**Capitula** solitary; **outer florets** zygomorphic	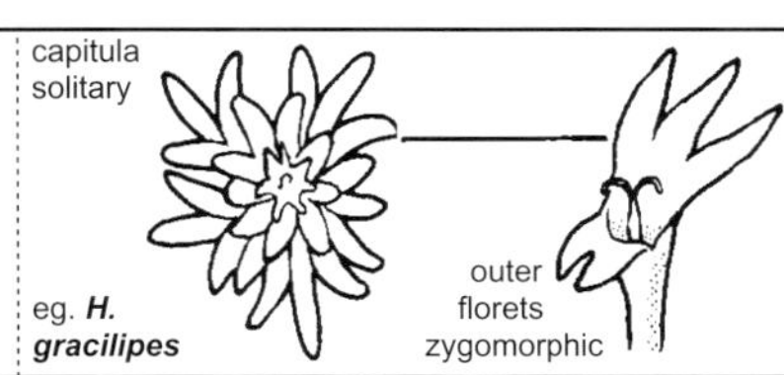	4
	Capitula in clusters; **outer florets** actinomorphic	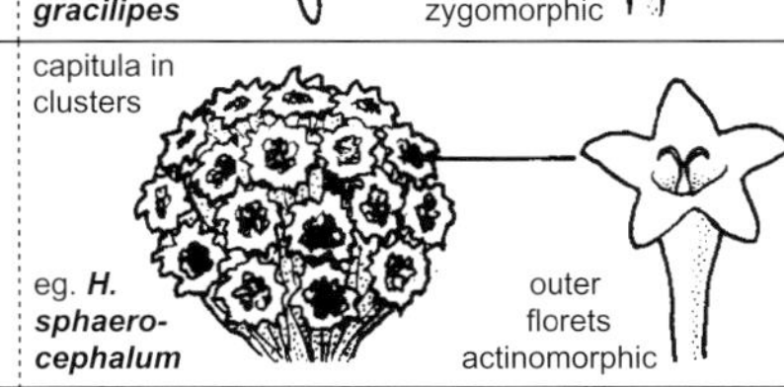	5

4	**Phyllaries** spathulate, dark below, straw-coloured above; **peduncles** curving with age; **capitula** 5--8mm across; **outer florets** weakly zygomorphic	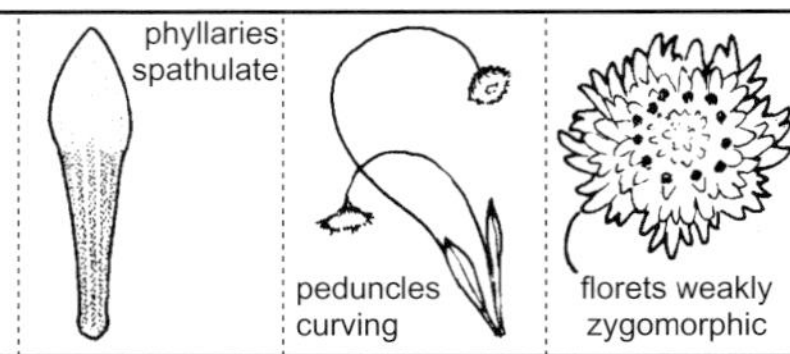	***H. balfourii*** Endemic **254**, 502
	Phyllaries narrowly ovate, straw-coloured with a dark midrib; **peduncles** straight; **capitula** 3--4mm across; **outer florets** weakly zygomorphic	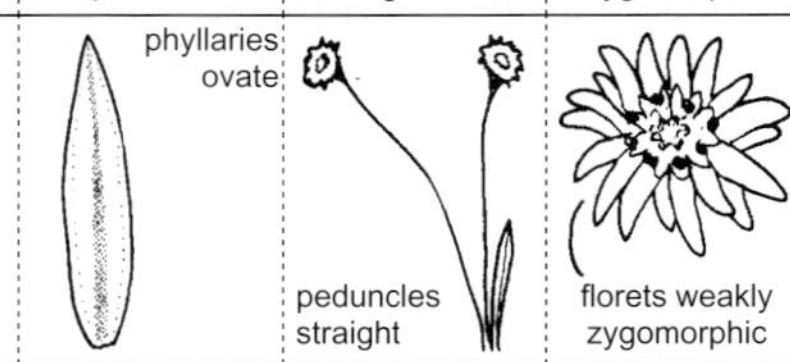	***H. paulayanum*** Endemic **254**, 503
	Phyllaries narrowly ovate, uniformly straw-coloured; **peduncles** straight; **capitula** 2 --4mm across; **outer florets** strongly zygomorphic	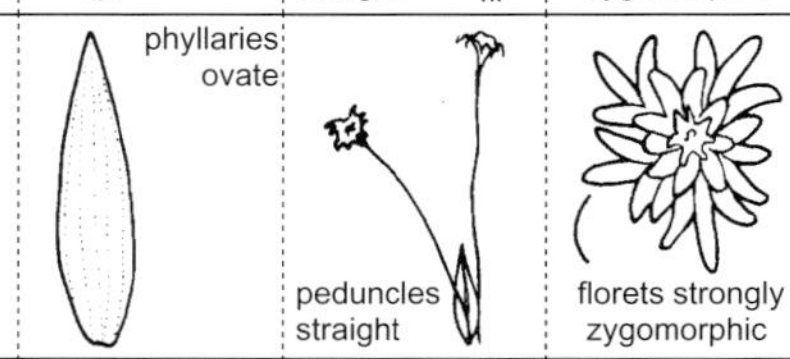	***H. gracilipes*** **254**, 502
5	**Inner phyllaries** rounded, petaloid	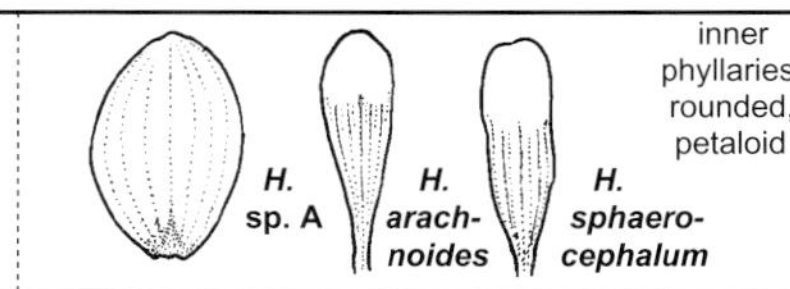	6
	Inner phyllaries acute, not petaloid (*Achyrocline*)	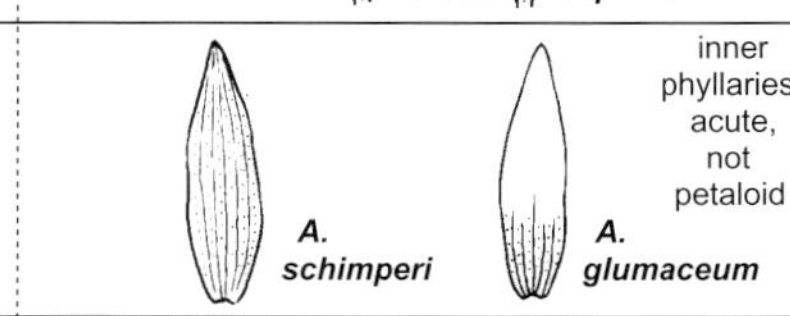	8
6	All **phyllaries** petaloid, **margins** entire; **leaves** fleshy, narrowly spathulate, overlapping	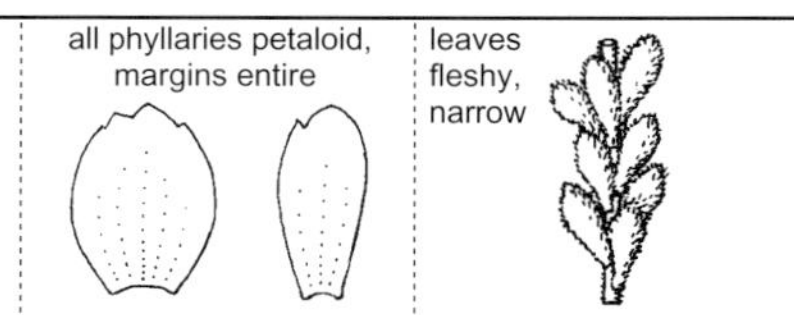	***H. sp. A*** Endemic **254**, 503
	All **phyllaries** petaloid, **margins** ragged; **leaves** fleshy, broadly spathulate, overlapping	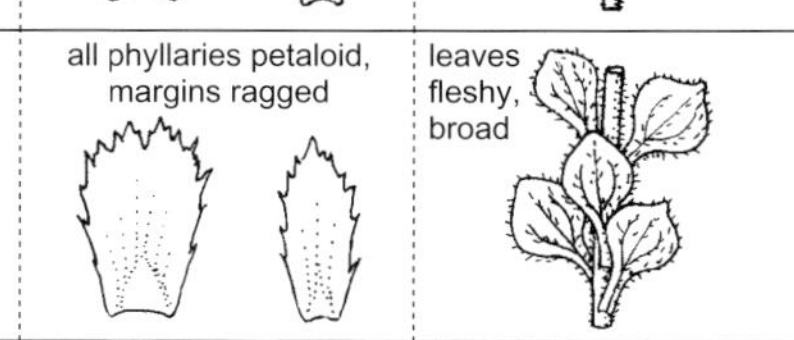	***H. sp. D*** Endemic **254**, 504
	Inner **phyllaries** petaloid, outer straw-coloured, **margins** entire; **leaves** not fleshy, not spathulate, distant	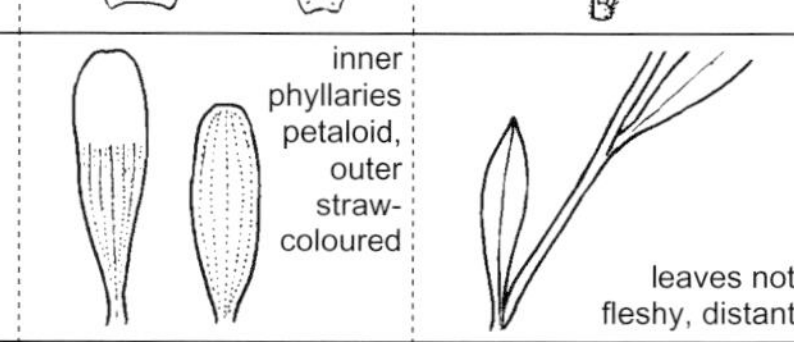	7
7	Delicate **annual**; **flowering heads** 0.5--1cm diameter	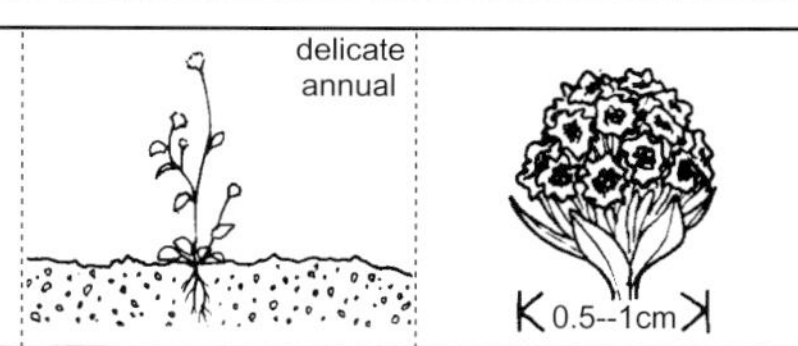	***H. arachnoides*** Endemic **253**, 501
	Somewhat robust **perennial**; **flowering heads** 1--2cm diameter	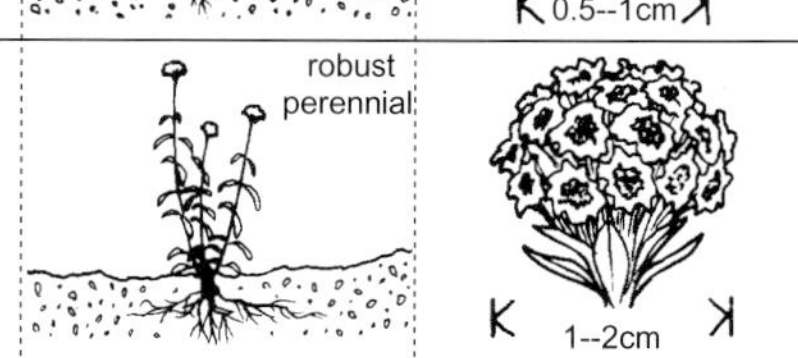	***H. sphaerocephalum*** Endemic **255**, 505
8	**Annual**; **leaves** ovate; **phyllaries** straw-coloured, translucent	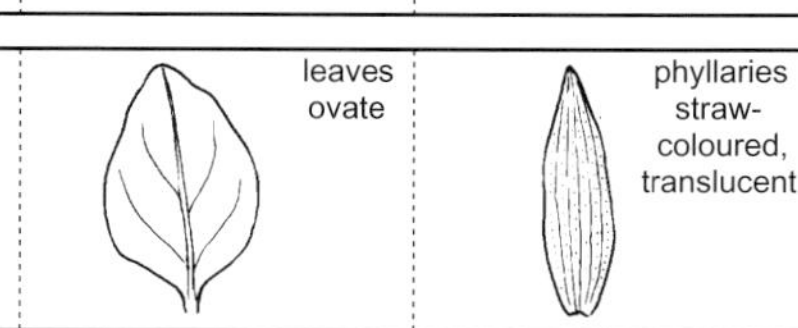	***Achyrocline schimperi*** 496
	Perennial; **leaves** linear to narrowly oblong; **phyllaries** opaque-white above	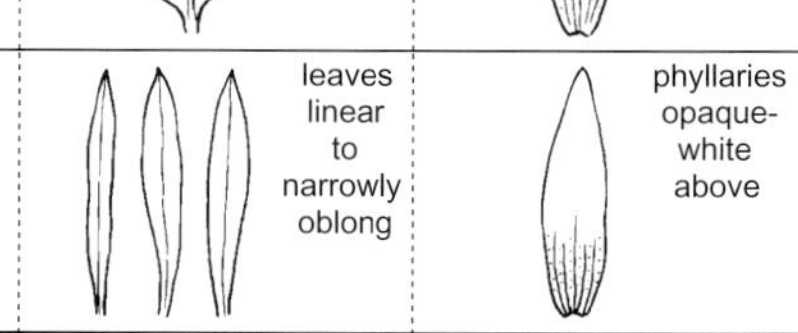	***Achyrocline luzuloides*** 495

9

Leaves white tomentose, venation obscure, **leaves** obovate, gradually narrowing below

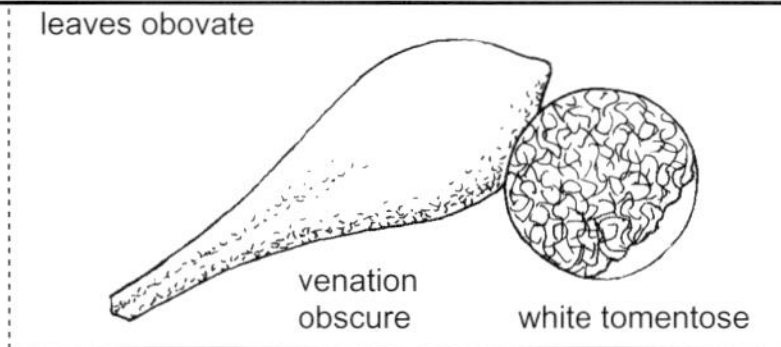

***H.* sp. C**
Endemic
504

Leaves white tomentose, nerves ± radiating from base; **leaves** spathulate, gradually narrowing below, entire

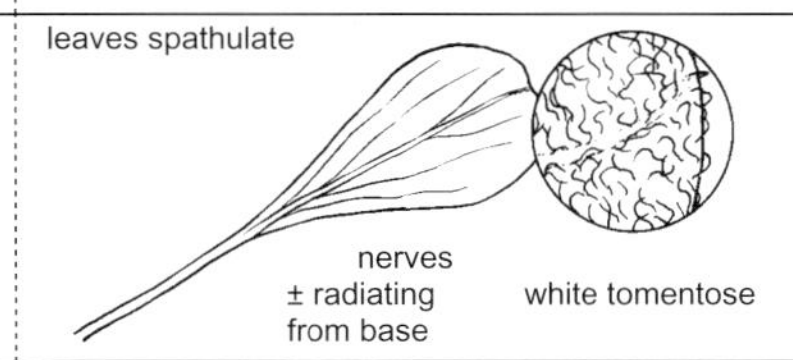

***H.* sp. B**
Endemic
254, 504

Leaves white tomentose below, becoming glabrous above, venation pinnate; **leaves** usually elliptic, gradually narrowing below, entire

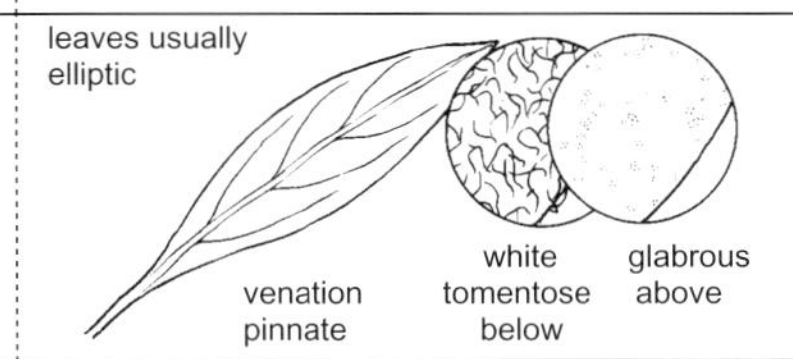

H. nimmoanum
Endemic
254, 502

Leaves white tomentose below, glandular above, venation pinnate; **leaves** ovate, abruptly narrowing below into a winged petiole, crenate

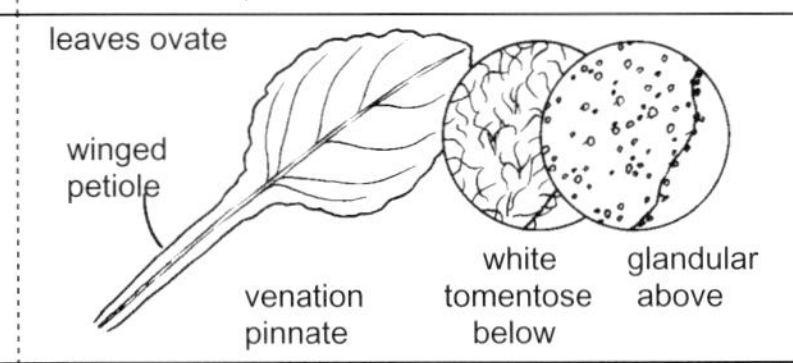

H. suffruticosum
Endemic
505

Launaea

1

Basal leaves entire (margins sometimes sinuate); **stem leaves** reduced, bract-like

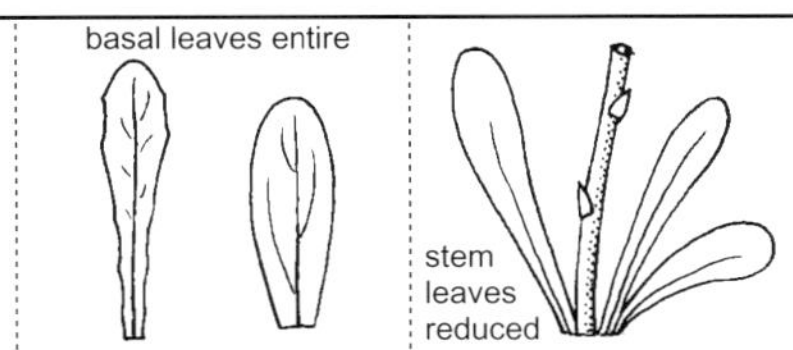

2

Basal leaves divided; **stem leaves** usually well-developed

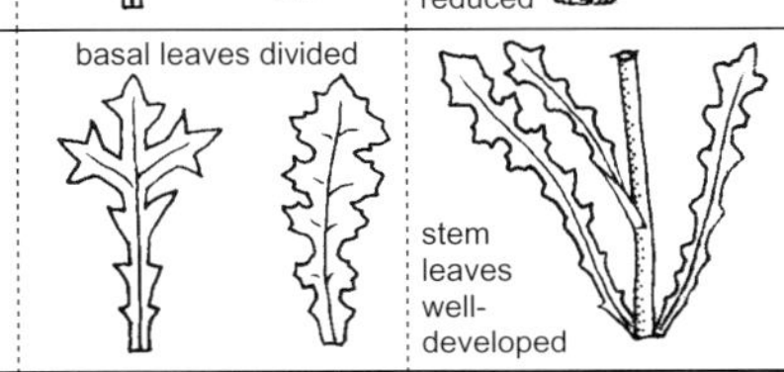

3

2

Basal leaves entire, glaucous; **stem leaves** auriculate at base; **rootstock** thick, woody

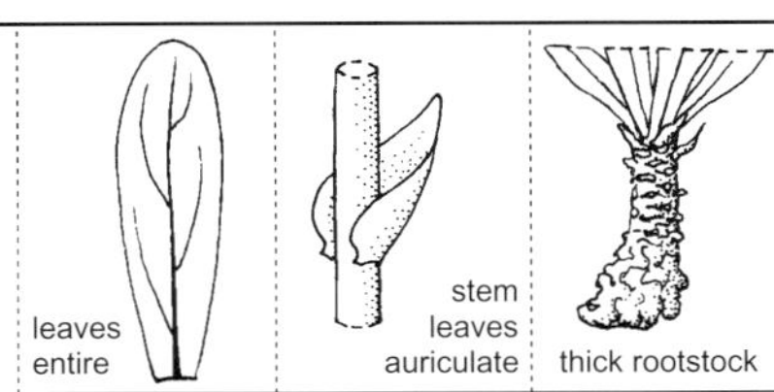

***L.* sp. A**
Endemic
508

Basal leaves somewhat sinuate, bright green; **stem leaves** narrowed at base; **rootstock** narrow, fleshy

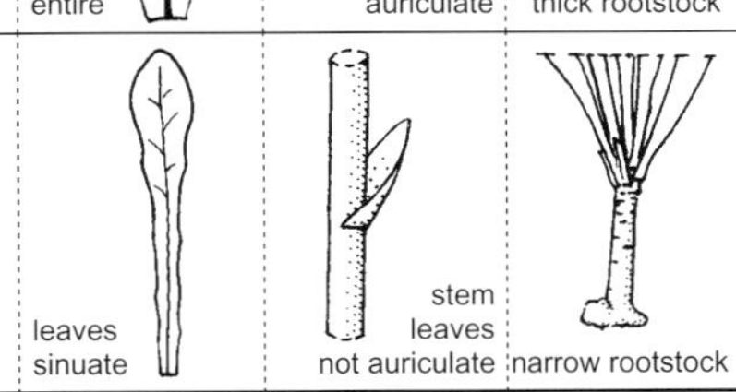

L. crepoides
Endemic
255, 506

3

Inflorescence widely branched, becoming spinescent; **outer achenes** long papillose

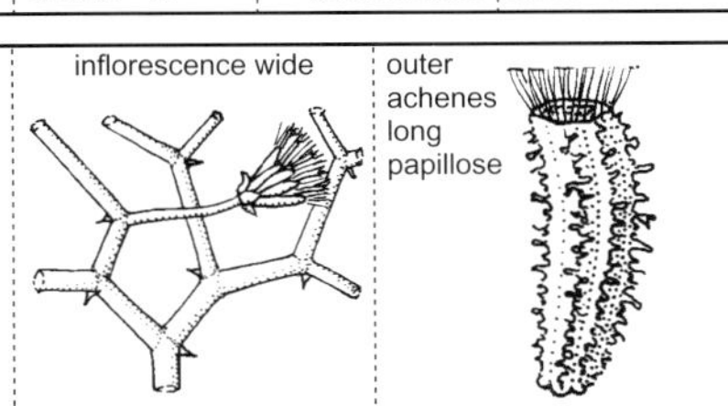

L. hafunensis
Endemic
507

Inflorescence narrowly branched, not spinescent; **outer achenes** glabrous or shortly papillose

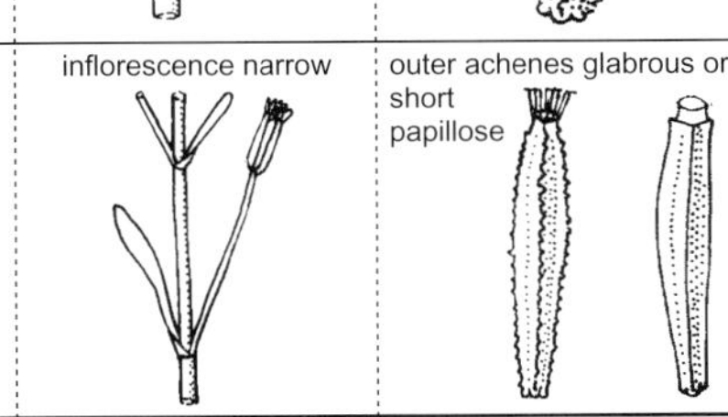

4

4

Capitula ovate, becoming swollen and hardened at the base in fruit; **phyllaries** with reflexed tips

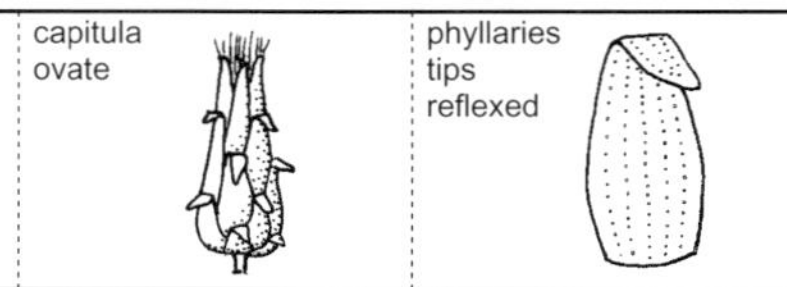

L. crassifolia 506

Capitula narrowly cylindrical, not swelling in fruit; **phyllaries** straight

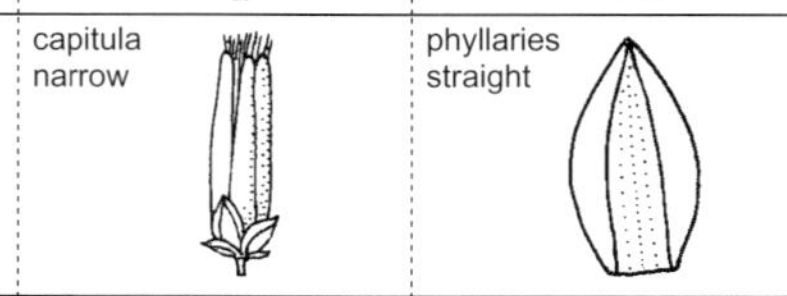

5

5

Achenes beaked; **leaves** with many oblong lateral lobes

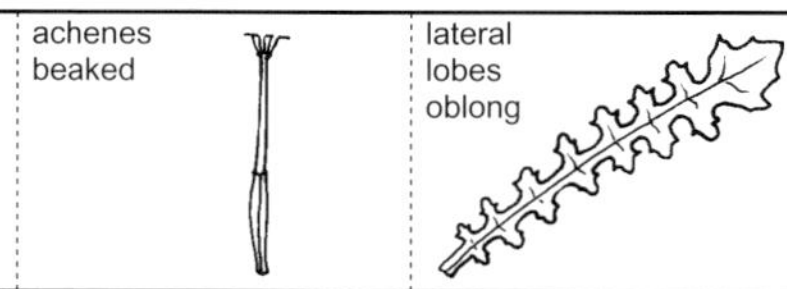

L. rhynchocarpa Endemic 507

Achenes not beaked; **leaves** with few triangular lobes

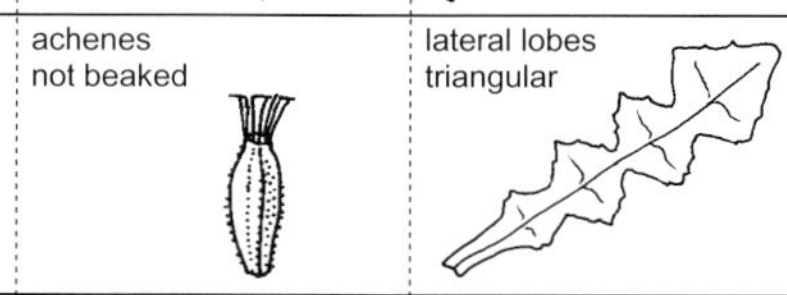

6

6

Outer **achenes** black, inner achenes brown; **annual**

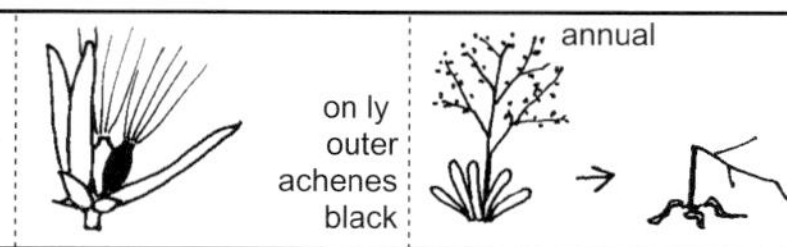

L. massauensis 507

Outer and inner **achenes** black; **perennial**

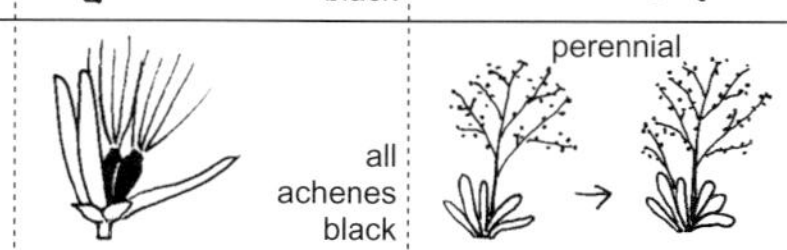

L. socotrana Endemic **255**, 507

Pluchea

1

Sprawling subshrub; **capitula** 7.5--11mm long, solitary or loosely clustered; **marginal florets** 4--5.5mm long, corolla clearly lobed

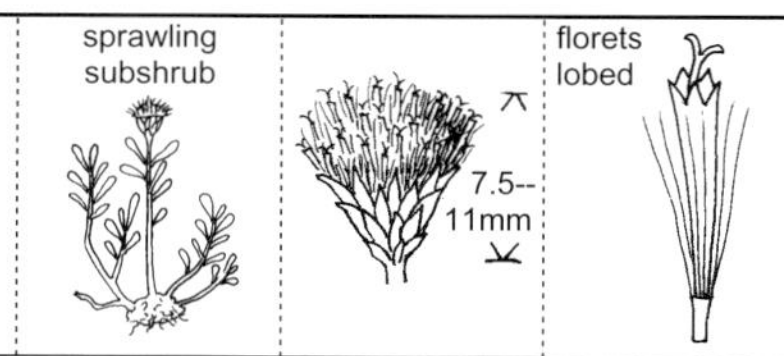

P. obovata Endemic **255**, 509

Erect subshrub; **capitula** 6--8mm long, in dense corymbose panicles; **marginal florets** 3.5--4.5mm long, corolla lobes minute

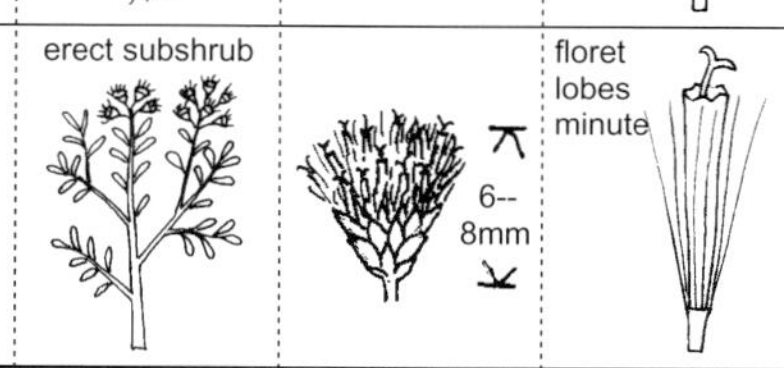

P. glutinosa Endemic 509

Pulicaria

1

Capitula with both ray and disc florets

2

Capitula without ray florets

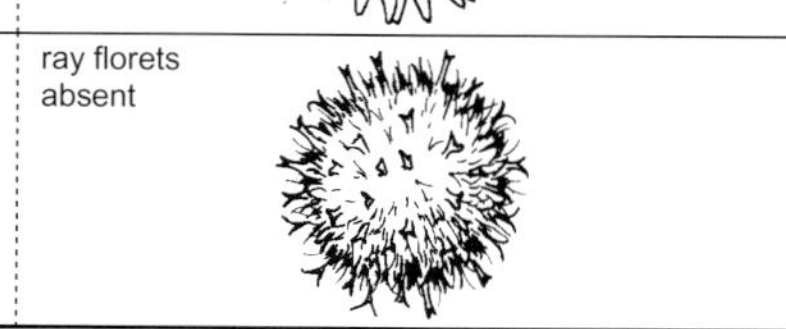

5

2

Erect **herb**; **branches** not woody or bearing leaf scars

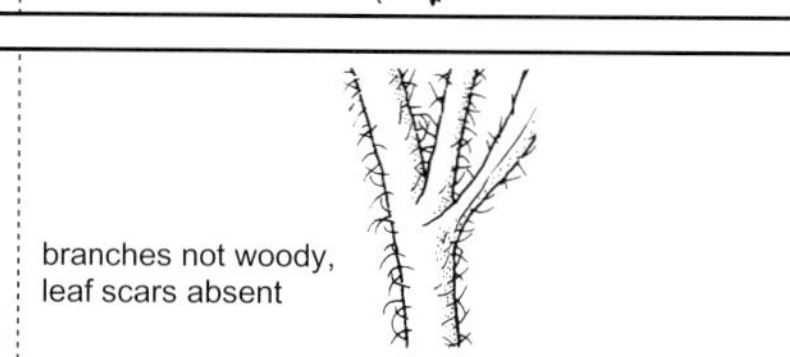

P. diversifolia Endemic **256**, 511

Prostrate, erect or clump-forming **herbs** or **subshrubs**; older **branches** woody, bearing prominent leaf scars

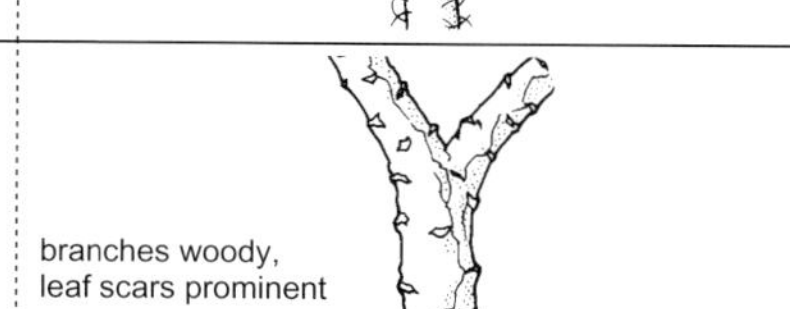

3

3	**Leaves** sessile, clasping stem at base; **leaves** with long woolly hairs	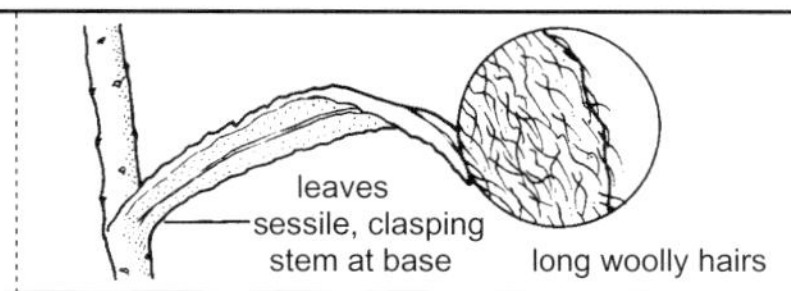	4
	Leaves petiolate, not clasping stem; **leaves** glandular or shortly hairy	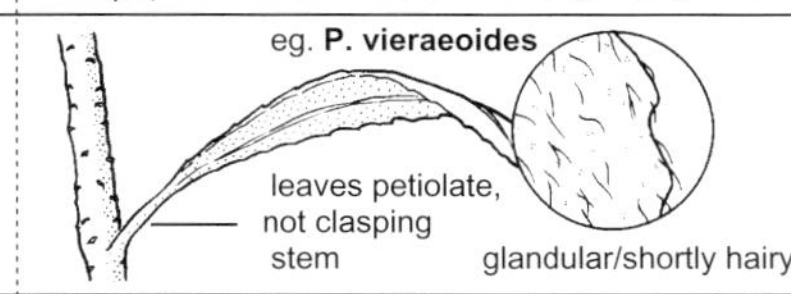	5

4	Small **shrub** or erect woody **herb**; **leaves** not crowded, <4 x 1cm	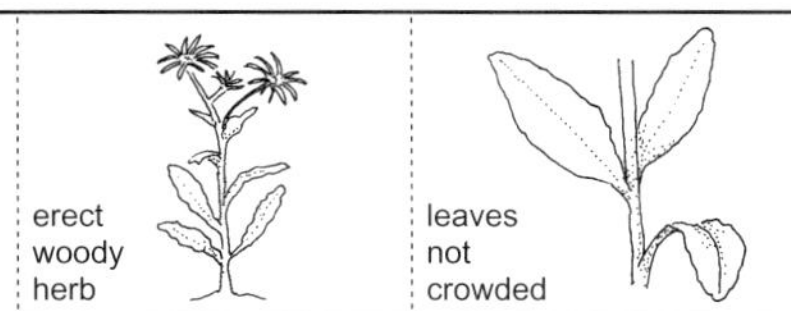	***P. lanata*** Endemic **256**, 512
	Cushion-forming woody **herb**; **leaves** crowded, <1.2 x 0.4cm	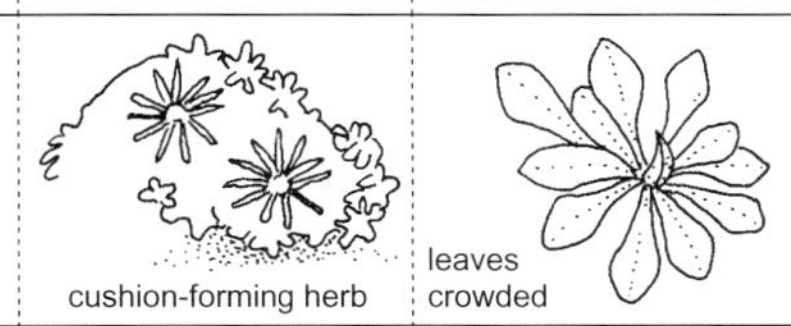	***P. dioscorides*** Endemic **256**, 511

5	**Branches** erect, leafy only at tips, bearing close-set leaf scars below; **leaves** shortly white-hairy; **capitula** c.2cm across	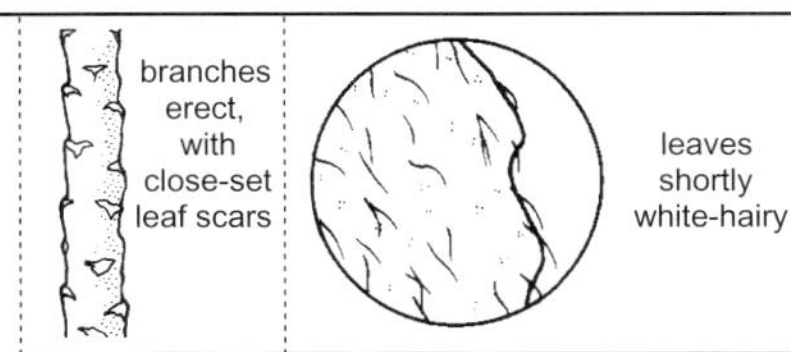	***P. vieraeoides*** Endemic **256**, 513
	Branches prostrate, mostly leafy, bearing distant leaf scars below; **leaves** shortly glandular; **capitula** c.1cm across	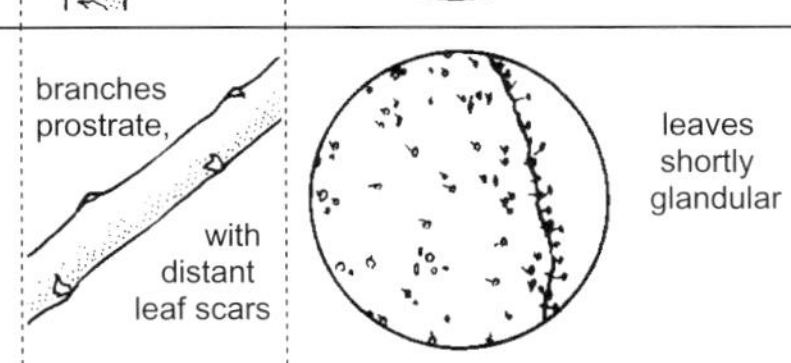	***P. elegans*** Endemic **256**, 512

6	**Leaves** ovate, not fleshy, distinctly petiolate; **outer pappus** reduced to a narrow rim on top of achene	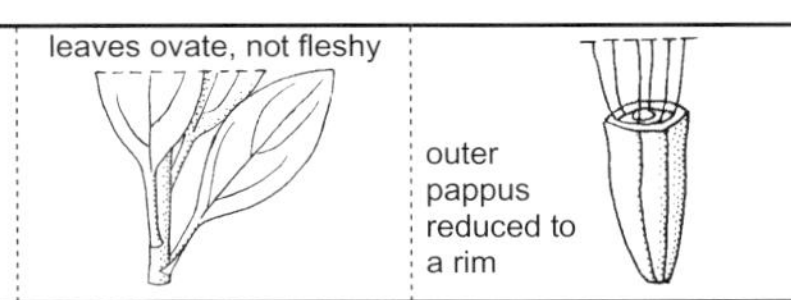	***P. aromatica*** Endemic **256**, 510
	Leaves spathulate, somewhat flesh, not distinctly petiolate; **outer pappus** a row of scales	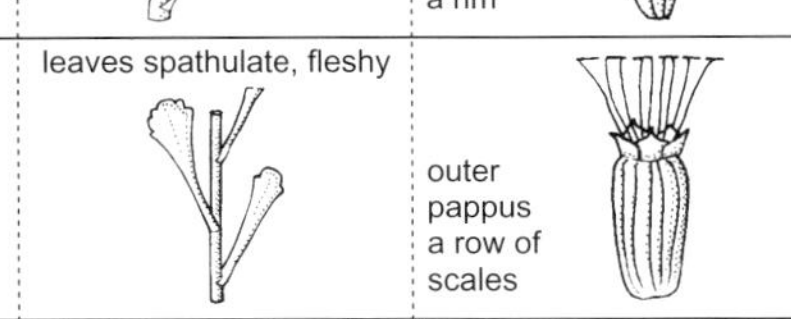	***P. stephanocarpa*** Endemic **256**, 513

Vernonia s. l.

1	Weak annual **herb**; **leaves** abruptly narrowing at base; **pappus** pure white	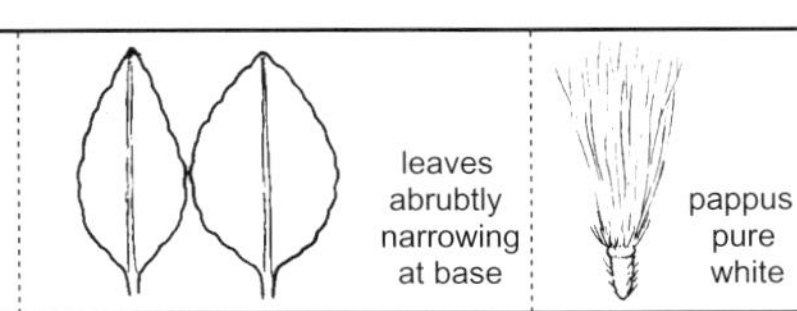	***Cyanthillium cinereum*** 498
	Woody-based **herbs**, **shrubs** or small **trees**; **leaves** gradually narrowing at base; **pappus** white or coloured	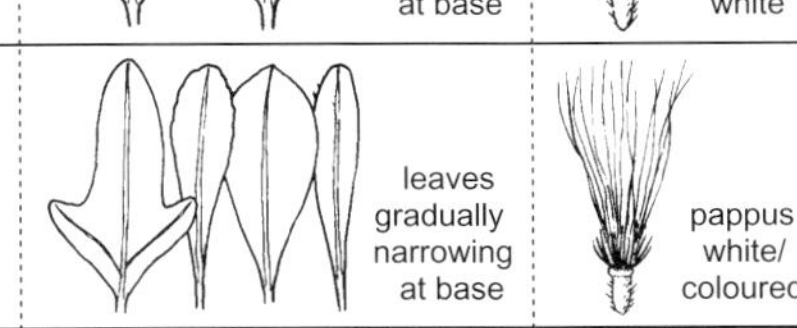	2

2	Perennial **herb** or small **shrub**; **pappus** white; **leaves** green both sides	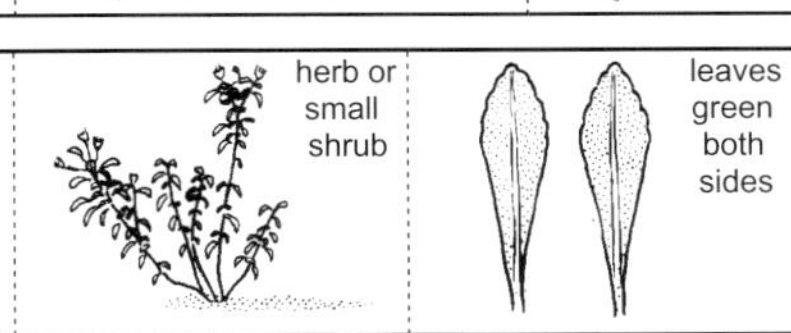	***Orbivestus cinerascens*** 508
	Shrubs or small **trees**; **pappus** brown; **leaves** green above, whitish below	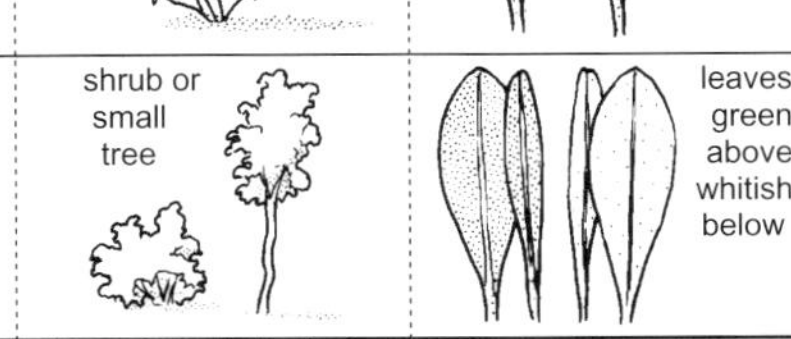	3

3

Achenes glabrous; **phyllaries** acute; **leaves** obovate

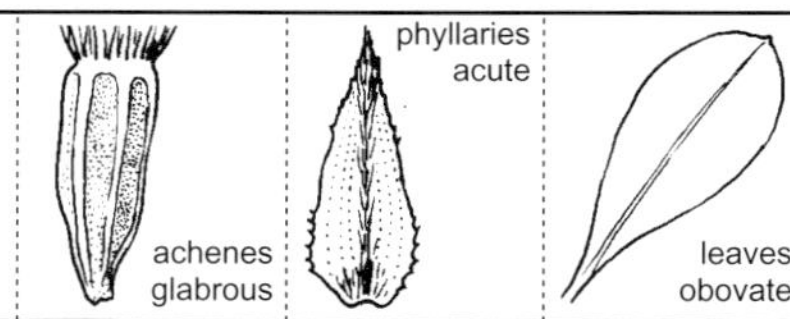

V. unicata
Endemic
516

Achenes glandular; **phyllaries** obtuse to subacute; **leaves** narrowly obovate

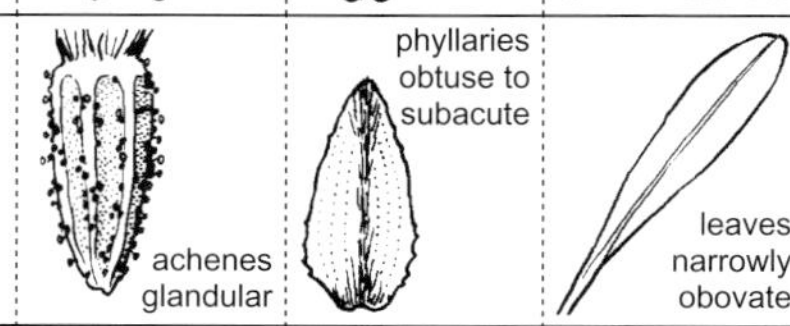

V. cockburniana
Endemic
257, 515

CONVOLVULACEAE

1

Annual or short-lived perennial **herb** of salty places; **stamens** and **styles** exserted from the corolla

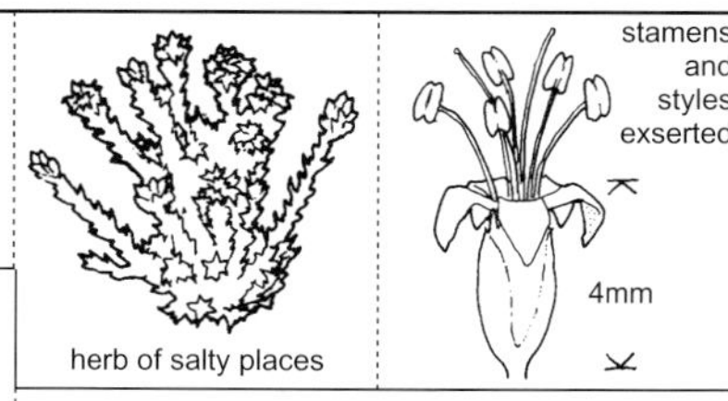

Cressa cretica
519

Combination of characters not as above → 2

2

Leaves reniform, cordate at base, arranged along one side of prostrate, mat-forming stems; **flowers** small, greenish or cream, axillary

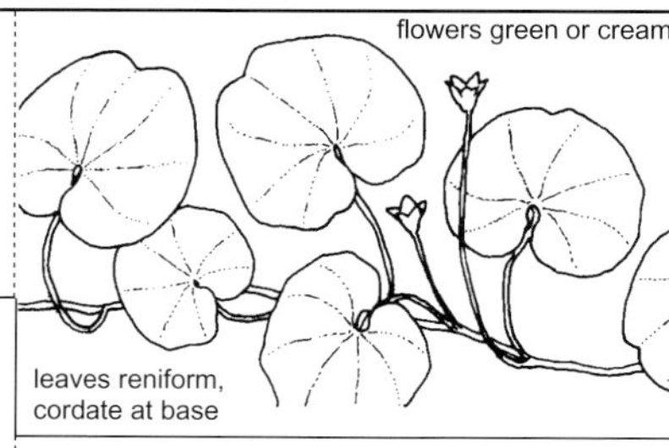

Dichondra repens
519

Combination of characters not as above → 3

3

Styles 2, obviously separate → 4

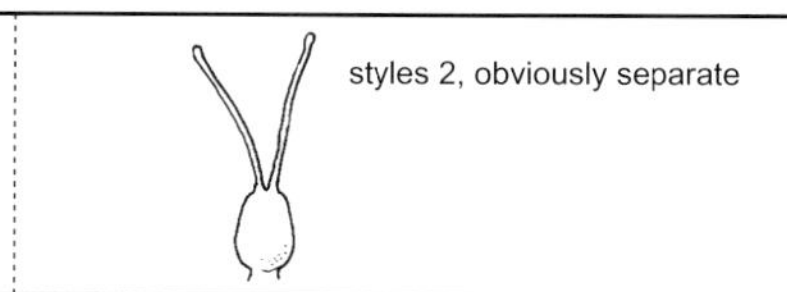

Style 1 (or apparently 1), with 2 stigmas → 5

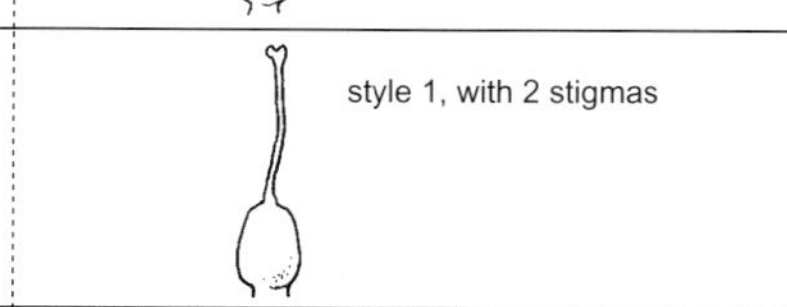

4

Delicate annual or non-woody perennial **herb**; **stems** with long silky hairs; **styles** bifid at tip

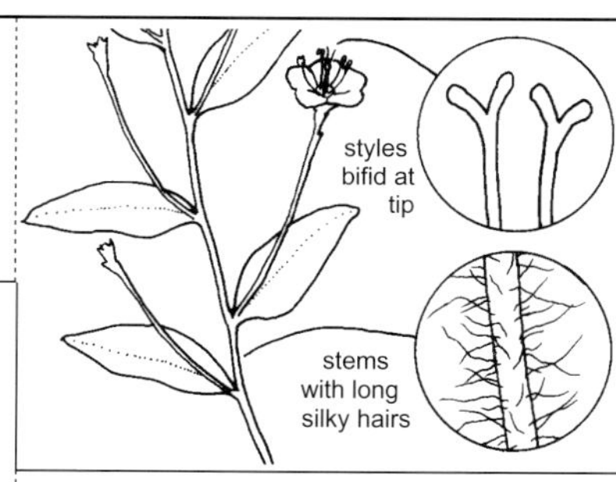

Evolvulus alsinoides
519

Shrubs or woody-based perennial **herbs**; **stem** hairs short; **styles** undivided at tip → 6

5

Low shrubs, never climbing; **leaf blade** 1--8mm wide, sessile or almost sessile; **calyx** not lengthening in fruit

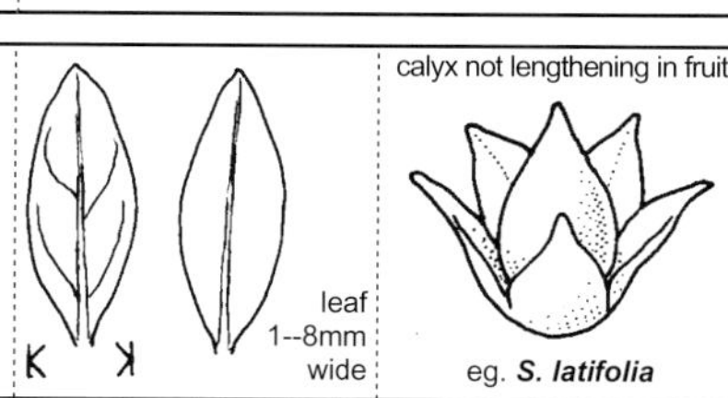

Seddera
(6 spp.)
(sect. ***Seddera***)

Small tree, sometimes climbing; **leaf blade** 10--50mm wide, petiolate; **calyx** lengthening in fruit

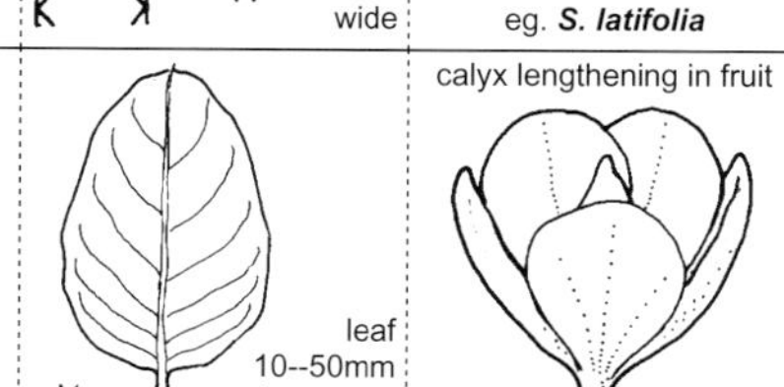

Metaporana obtusa
Endemic
257, 522

6	Spinescent **shrubs** forming rounded clumps; **stigmas** pyriform	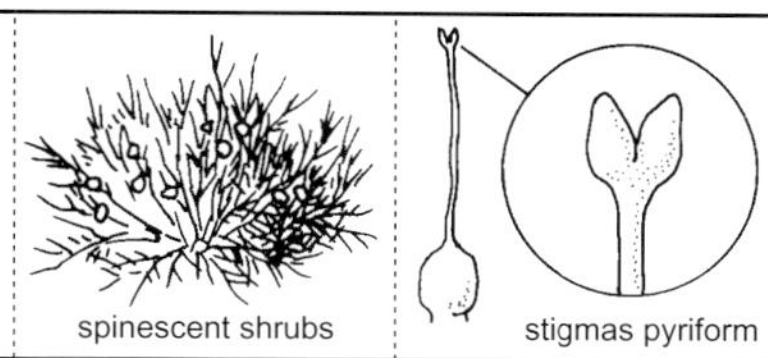	***Seddera*** (6 spp.) (sect. ***Socotroseddera***)
	Non-spinescent **herbs** or **shrubs**, not forming rounded clumps; **stigmas** globose or bi-globose or rarely 5-lobed (*I. kotschyana*)	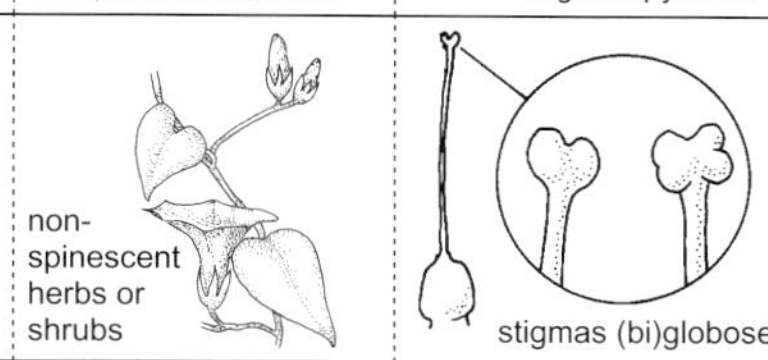	***Ipomoea*** (6 spp.)
	Non-spinescent **herbs** or **shrubs**, not forming rounded clumps; **stigmas** linear-cylindrical	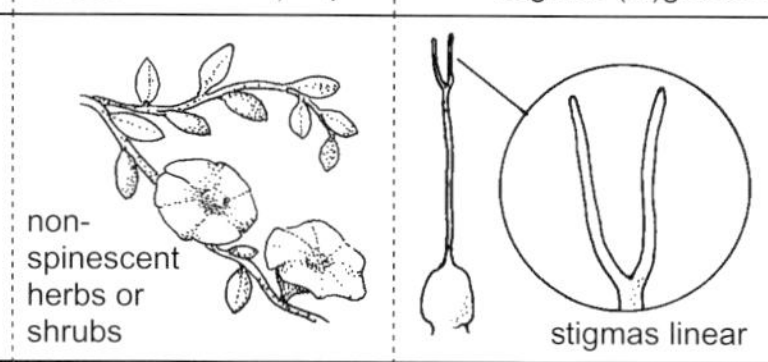	***Convolvulus*** (6 spp.)

Convolvulus

1	**Flowers** grouped in clusters of 3 to 8	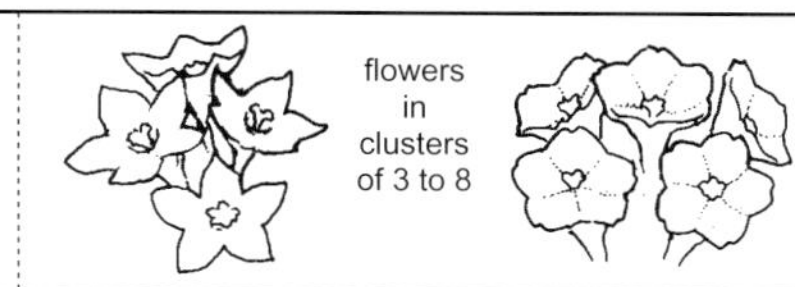	2
	Flowers solitary or in axillary 2 to 3-flowered cymes	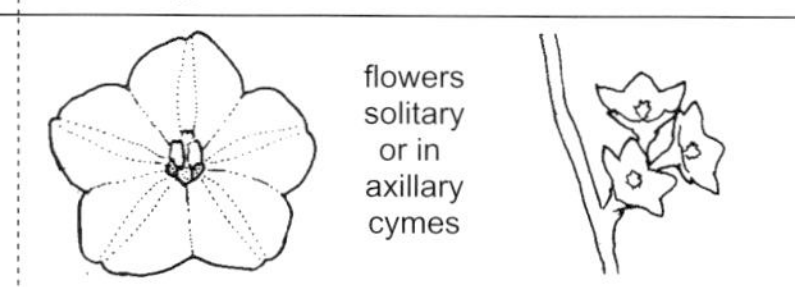	3

2	**Flower clusters** sessile or very shortly pedunculate; **leaf base** attenuate or cuneate; **corolla tube** shorter than the deeply lobed limb	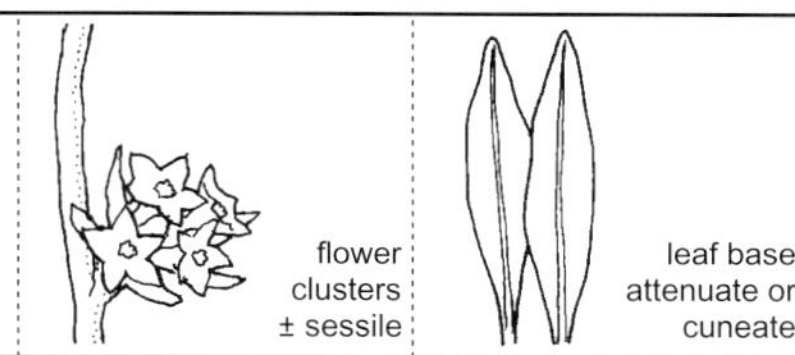	***C. rhyniospermus*** 518
	Flower clusters long-pedunculate; **leaf base** truncate or slightly hastate; **corolla tube** 2 to 4 x as long as ±unlobed limb	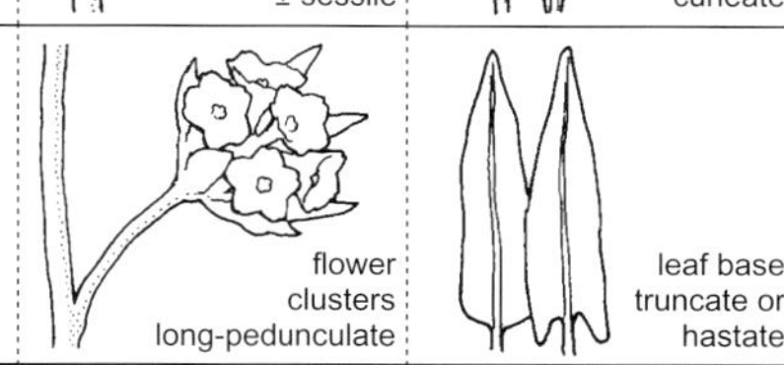	***C. arabicus*** 517

3	**Corolla** pale blue to pink, deeply lobed, the lobes hairy at their tips; **inner sepals** pale green below, green above	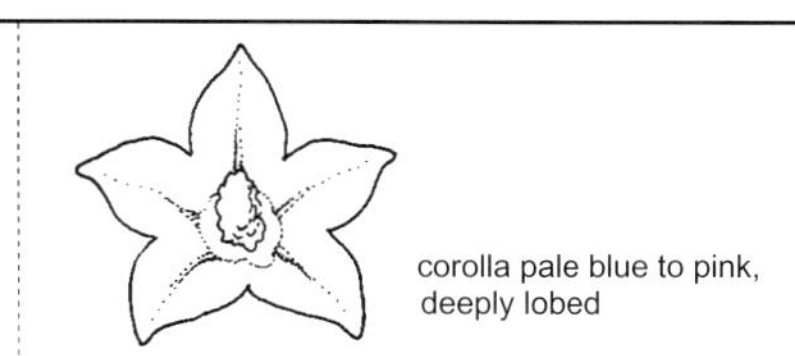	***C. siculus*** 518
	Corolla white, unlobed or shallowly lobed, hairy throughout; **inner sepals** green with paler membranous margins	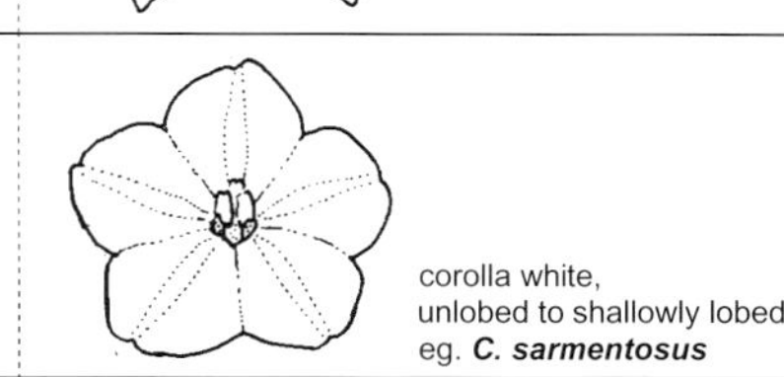	4

4	**Stem leaves** linear, sessile; woody-based **perennial** with many ascending, sometimes hanging branches; **corolla** sparsely hairy	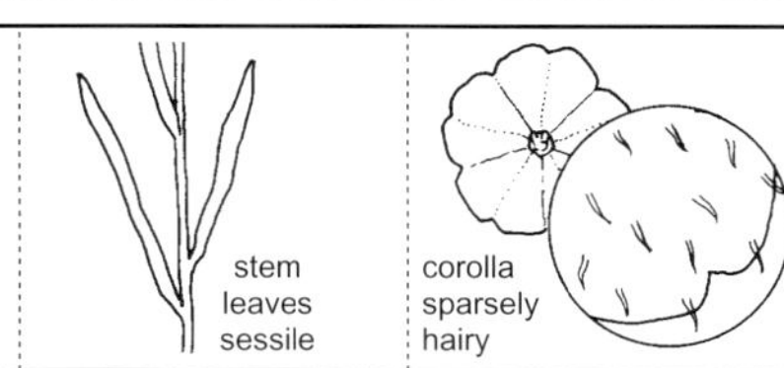	***C. hildebrandtii*** Endemic **257**, 517
	Stem leaves elliptic to oblanceolate, shortly petiolate; **annual** or **perennial** with prostrate stems; **corolla** sericeous	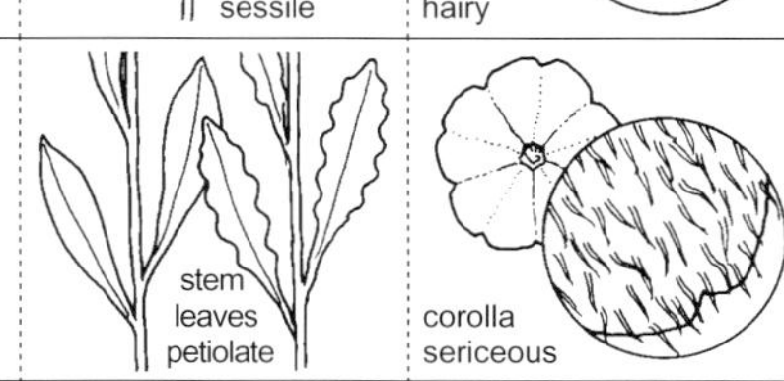	5

5

Rosette-forming **annual**; **stem leaves** undulate, sometimes ±entire; **stems** with short appressed hairs and long spreading hairs

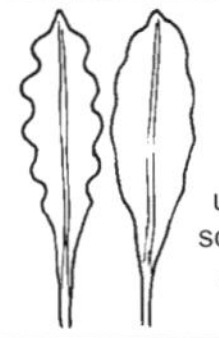

C. grantii
Endemic
257, 517

Woody-based **perennial**; **stem leaves** entire; **stems** with grey or white silky hairs (all appressed)

C. sarmentosus
Endemic
257, 518

Ipomoea

1

Corolla yellow, orange or apricot, with a dark red or blackish centre — ***I. obscura*** 521

Corolla blue,purple, mauve, lilac, pink or white, sometimes with a differently-coloured centre — 2

corolla yellow, orange or apricot, with a dark red or blackish centre

2

Leaves obreniform; **plant** forming mats on beaches; **corolla** pink/ mauve, 40--55mm — ***I. pes-caprae*** 521

Leaves never obreniform; **plants** usually climbing over shrubs; **corolla** not as above — 3

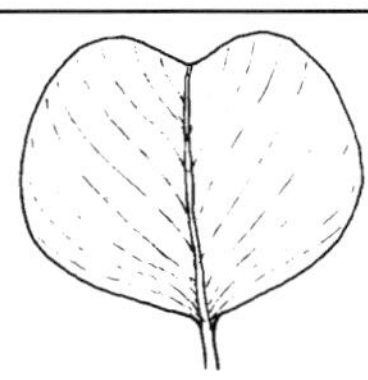

leaves obreniform

plants forming mats on beaches

3

Leaves pinnatifid or pinnatisect; **corolla** narrowly tubular (tube scarcely 1mm wide) — ***I. kotschyana*** 520

Leaves palmately lobed or simple; **corolla** funnel-shaped or tubular below and abruptly expanded above — 4

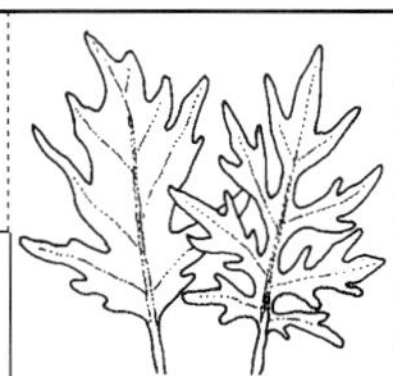

leaves pinnatifid or pinnatisect

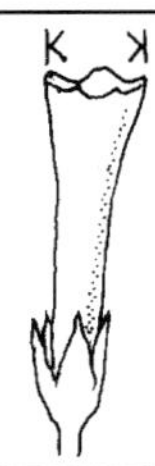

4

Leaves palmately 3 or 5-lobed almost to the base; **stems** glabrous, sometimes with projections — ***I. cairica*** 520

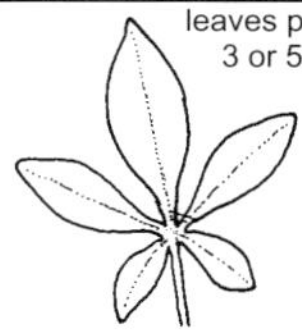

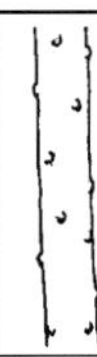

Leaves unlobed or 3-lobed but never deeply so; **stems** hairy, never with projections — 5

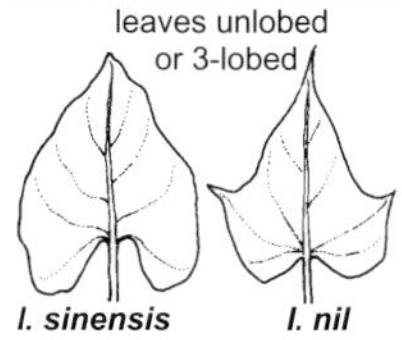

5

Corolla 7--15mm, tubular below and abruptly expanded above, pink with purple centre; **sepals** truncate or cordate at base, cuspidate at tip — ***I. sinensis*** 521

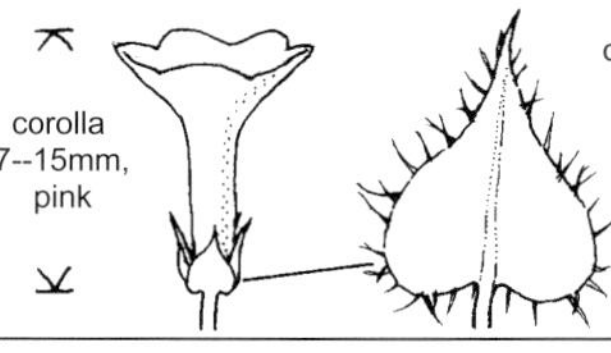

Corolla 30--50mm, funnel-shaped, blue fading to pink with white stripes; **sepals** not cordate at base, attenuate at tip — ***I. nil*** 521

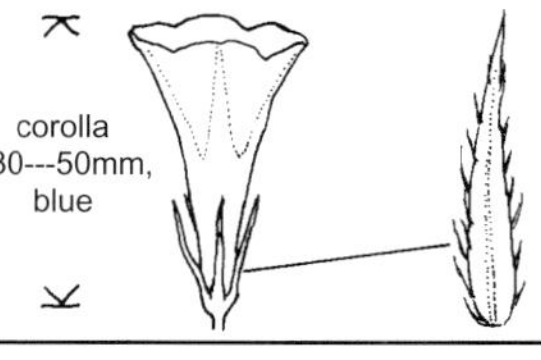

Seddera

1

Styles apparently fused into a single style; **flowers** solitary, axillary; **plants** large, dark green, spinescent mounds (sect. *Socotroseddera*) — **2**

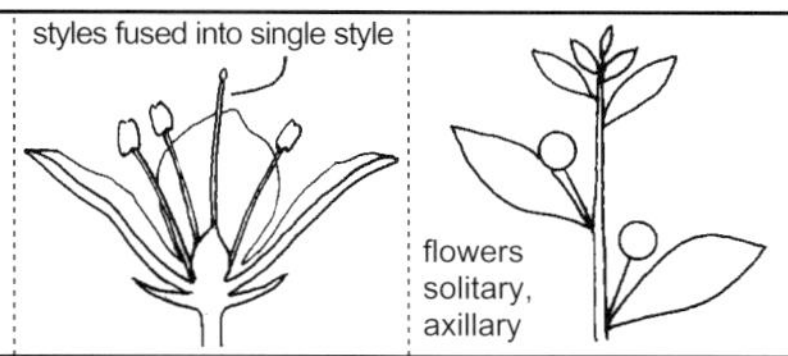

Styles 2, free to the base; **flowers** in terminal heads or racemes; **plants** not forming spinescent mounds — **4**

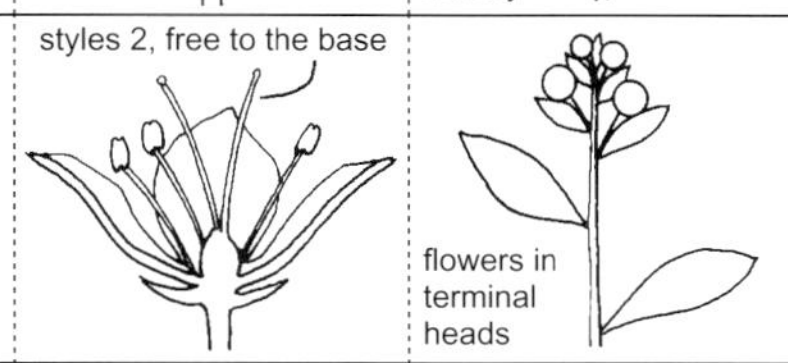

2

Corolla >2x the length of calyx — ***S. spinosa*** Endemic **257**, 524

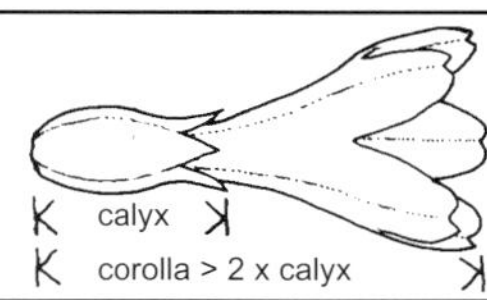

Corolla <2x the length of calyx — **3**

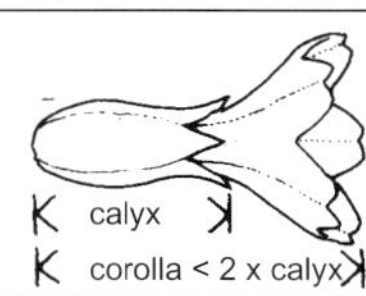

3

Corolla glabrous below; **leaf bases** shortly cuneate; **spreading tips of sepals** c.1mm long — ***S. fastigiata*** Endemic 523

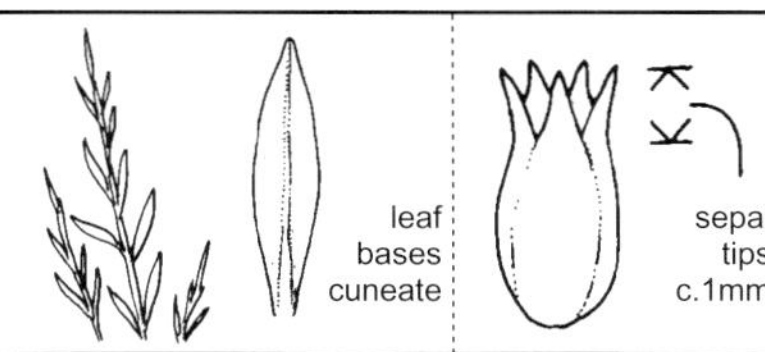

Corolla hairy below; **leaf bases** obtuse; **spreading tips of sepals** 1.5--3mm long — ***S. semhahensis*** Endemic **257**, 524

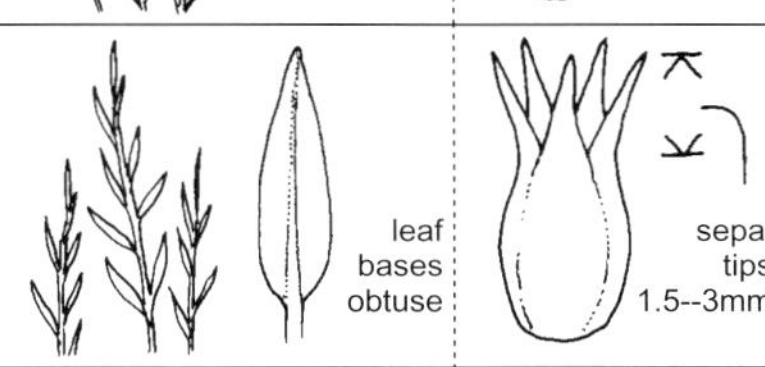

4

Flowers in dense, woolly, head-like inflorescences, the flowers partly hidden by bracteoles; **leaves** with c.2 pairs of distinct lateral veins — ***S. glomerata*** 523

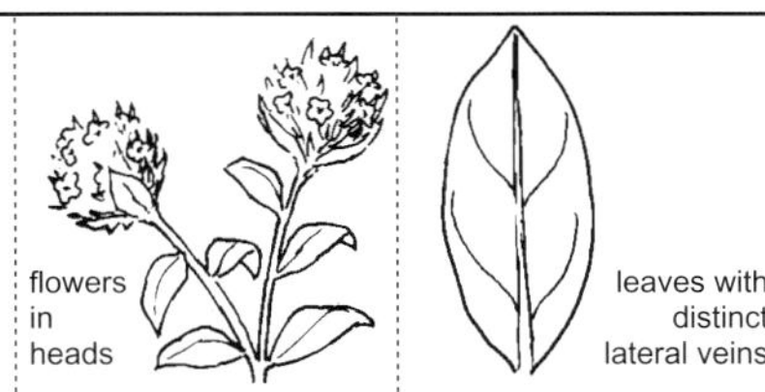

Flowers in long, terminal racemes, each flower clearly visible; **leaves** with only a midvein — **5**

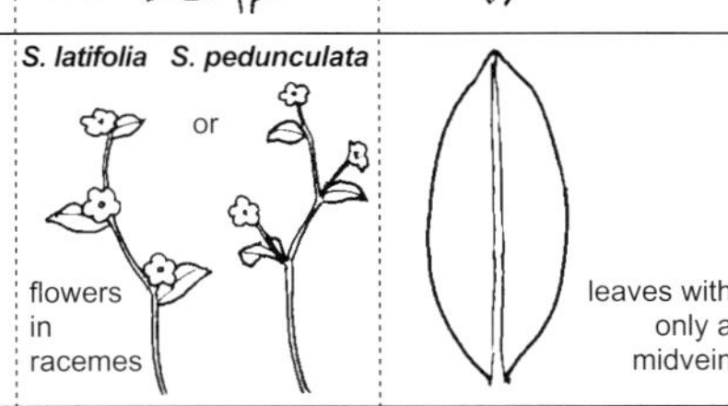

5

Flowers sessile; **bracts** longer than the flower — ***S. latifolia*** 523

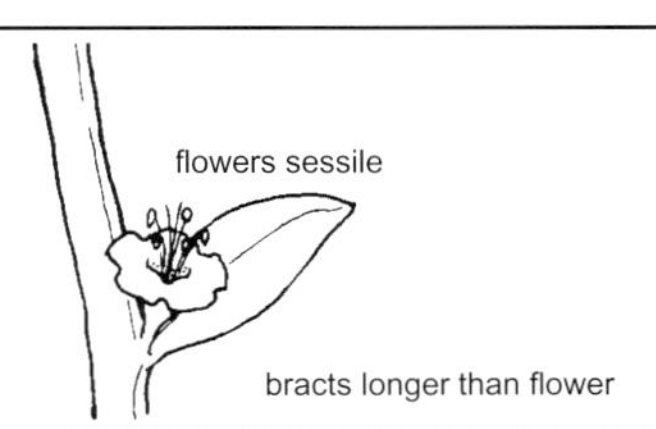

Flowers pedicellate; **bracts** shorter than the flower — ***S. pedunculata*** Endemic 524

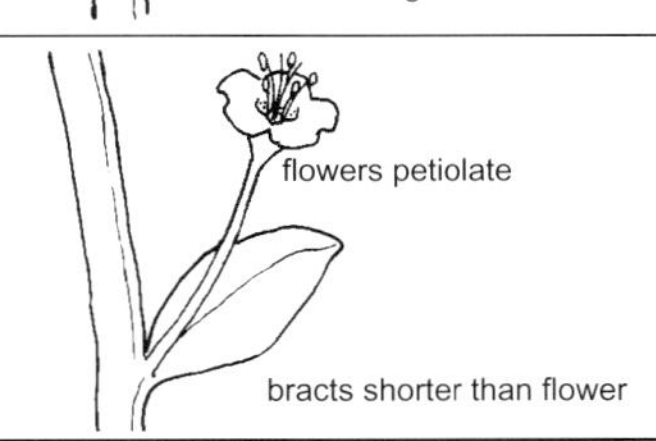

CRASSULACEAE

1	**Leaves** circular, peltate with long petioles; **flowers** in a terminal many-flowered raceme	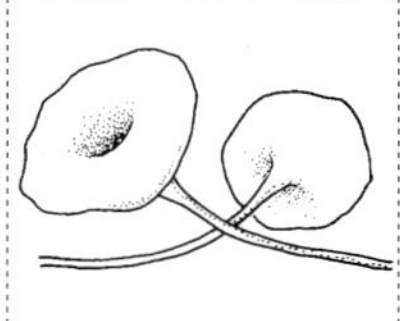 leaves circular, peltate	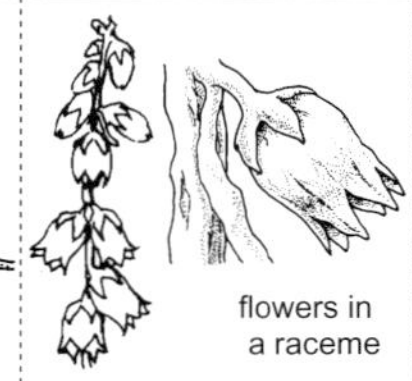 flowers in a raceme	***Umbilicus horizontalis*** 527
	Leaves various, narrowed at the base in to a short petiole or sessile, never peltate or circular; **flowers** in axillary clusters or a terminal many-flowered panicle	*Crassula* 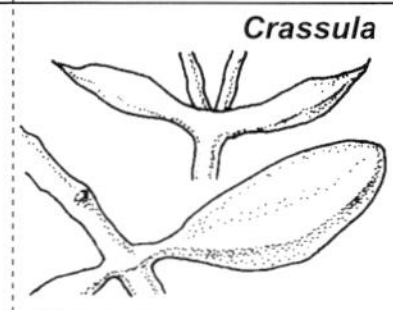 *Kalanchoe* leaves various	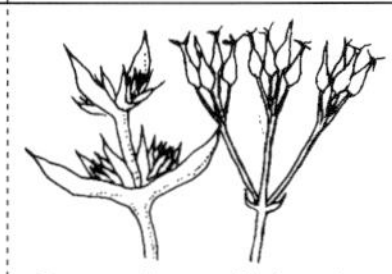 *Crassula* *Kalanchoe* flowers in axillary clusters or panicle	2

2	**Corolla** whitish, greenish or pinkish; **petals** free or fused only at the base; **stamens** same in number as petals; **leaves** fused at the base to form a sheath around the stem	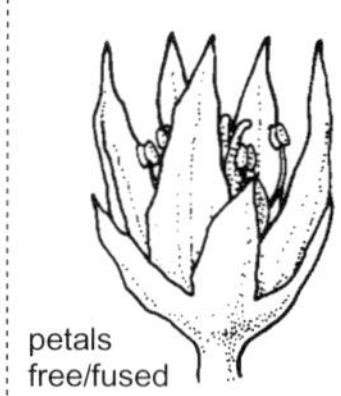 petals free/fused	 leaves forming sheath around stem	***Crassula*** (2 spp.)
	Corolla orange or red; **petals** fused at least to the middle; **stamens** twice as many as petals; **leaves** not forming a sheath around the stem	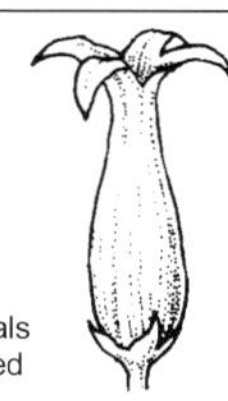 petals fused	 leaves not forming sheath around stem	***Kalanchoe*** (3 spp.)

Crassula

1	Perennial **herb**; **stems** usually with peeling brown bark at base, often rooting at the nodes; **seeds** smooth	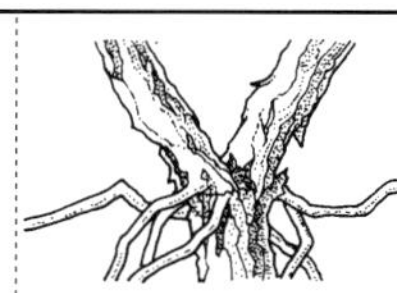 stems rooting at nodes	seeds smooth 	***C. schimperi*** **258**, 525
	Ephemeral **herb**; **stems** wiry, without peeling bark, not rooting at the nodes; **seeds** longitudinally ridged	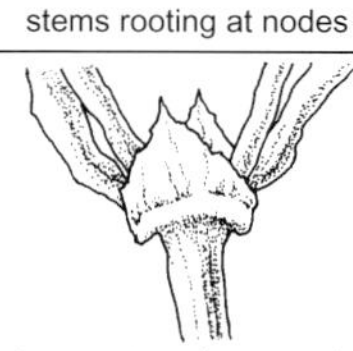 stems not rooting at nodes	seeds ridged 	***C. alata*** 525

Kalanchoe

1	**Corolla tube** 8--10mm long, actinomorphic; **flowers** held more or less erect; **leaves** not imbricate, the petiole not broadened at the base	flowers erect 	leaves not imbricate, petiole not broadened at base 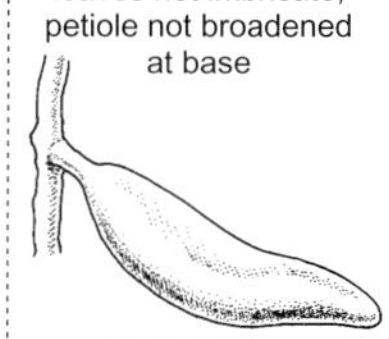	***K. rotundifolia*** **258**, 527
	Corolla tube 10--15mm long, slightly zygomorphic; **flowers** held horizontally; **leaves** not imbricate, the petiole slightly broadened at the base	flowers horizontal 	leaves not imbricate, petiole broadened at base 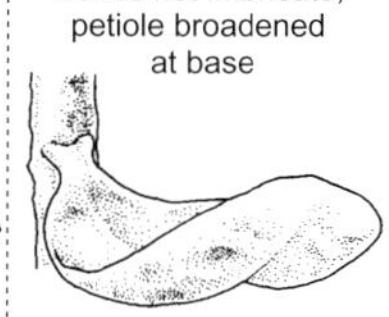	***K. robusta*** Endemic **258**, 526
	Corolla tube 20--30mm long, actinomorphic; **flowers** held more or less erect; **leaves** imbricate, the petiole not broadened at the base	flowers erect 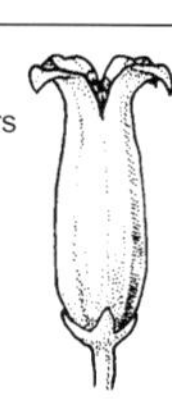	leaves imbricate, petiole broadened at base 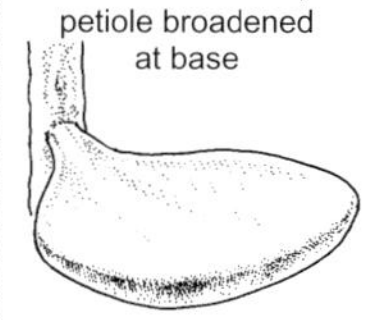	***K. farinacea*** Endemic **258**, 526

CRUCIFERAE

1	**Fruit** <3 x as long as broad(a silicula)	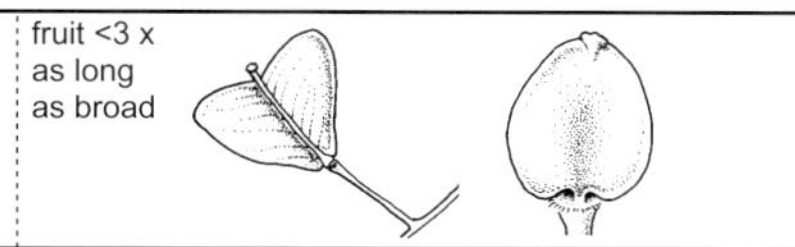	2
	Fruit >3 x as long as broad (a siliqua)	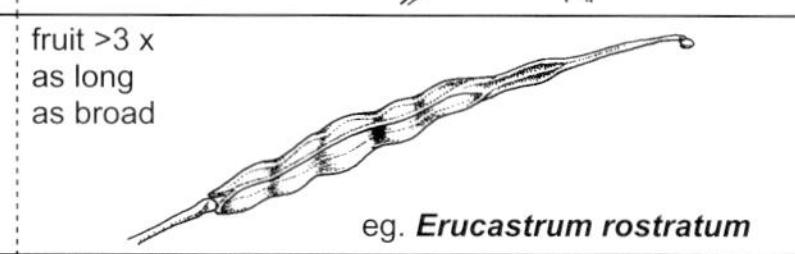	3
2	Annual **herb**; **fruit** obtriangular; **leaves** toothed or pinnatisect, sparsely covered with scattered simple and branched hairs	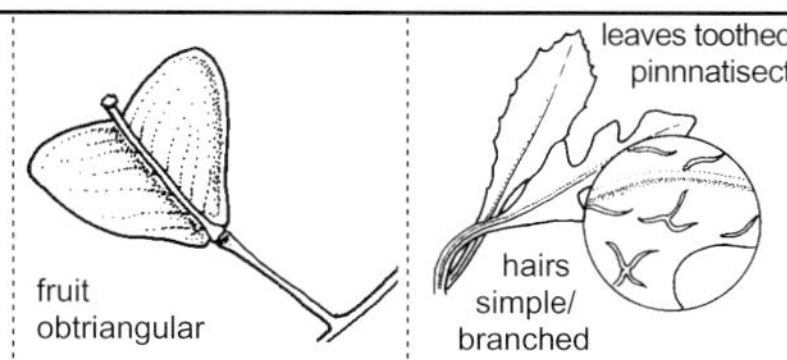	***Capsella bursa-pastoris*** 528
	Shrublet; **fruit** broadly elliptic; **leaves** succulent, entire, densely white-tomentose with branched and stellate hairs	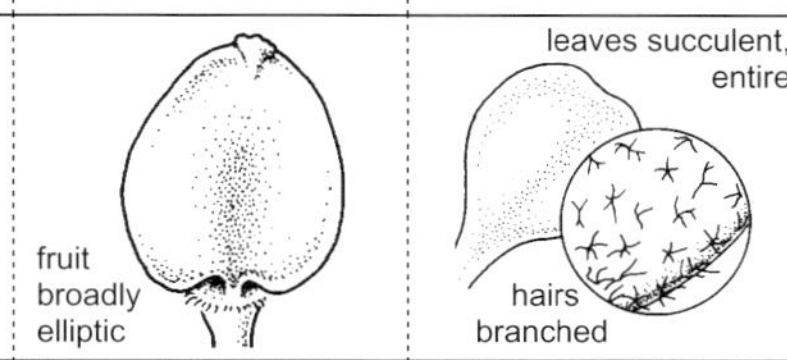	***Lachnocapsa spathulata*** Endemic Endemic genus **259**, 531
3	**Fruit** with a pair of apical horns; **leaves** densely white-tomentose with simple and branched hairs	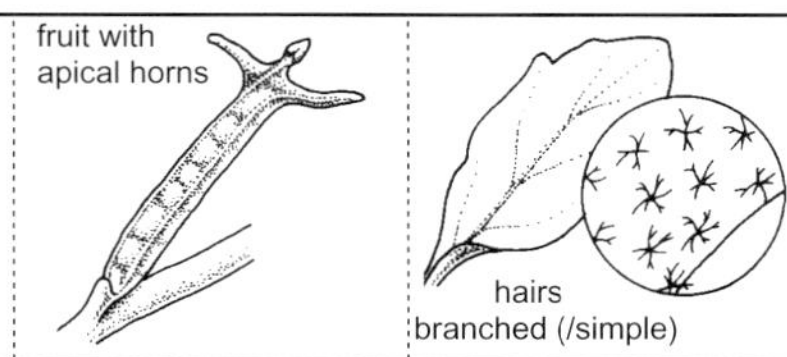	***Diceratella incana*** **258**, 528
	Fruit lacking apical horns; **leaves** green, glabrous or thinly hairy with simple or medifixed hairs	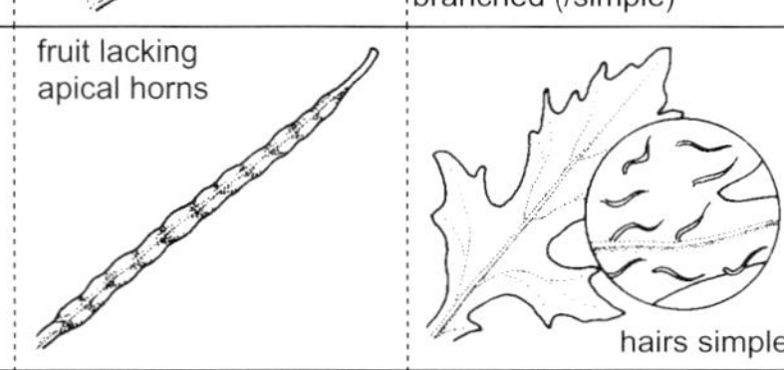	4
4	**Fruit** divided into 2 segments; **hairs** simple or absent	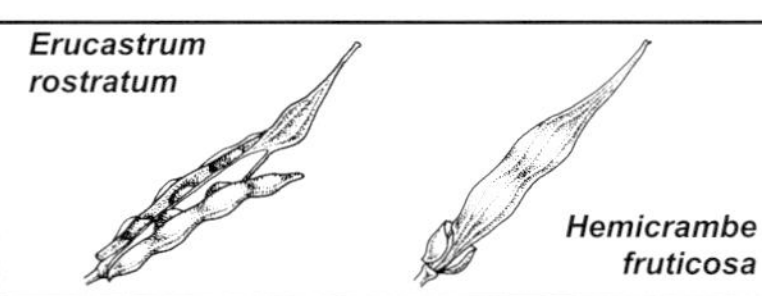	5
	Fruit not segmented; **hairs** simple or medifixed, rarely absent	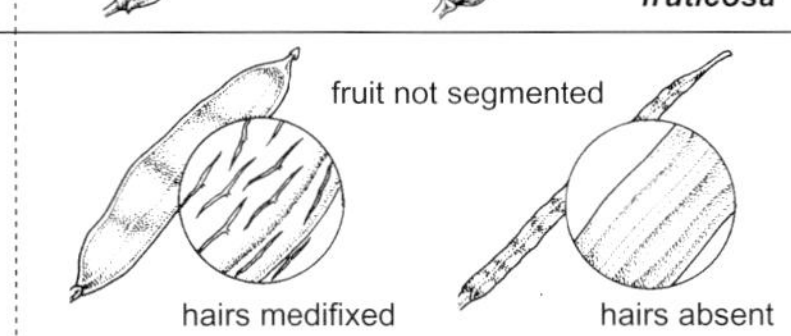	6
5	**Upper segment of fruit** 0 or 1-seeded, attenuate into an acute tip; **lower segment** dehiscent, 3 to 9-seeded, much longer than the upper; **leaves** lyrate-pinnatifid	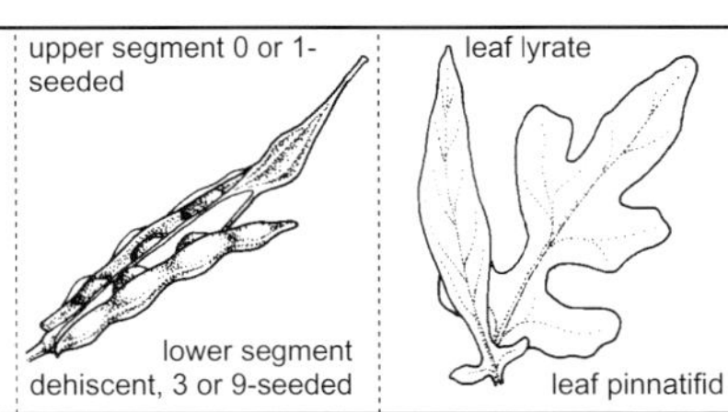	***Erucastrum rostratum*** Endemic **258**, 529
	Upper segment of fruit 1 or 2-seeded, beak-like; **lower segment** dehiscent, sterile, much shorter than the upper; **leaves** crenate-dentate	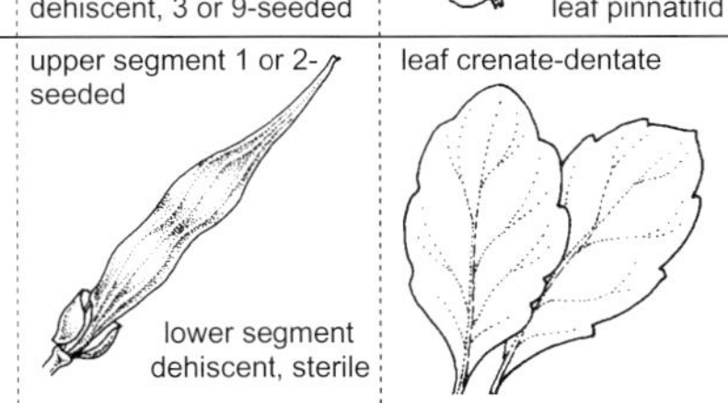	***Hemicrambe fruticosa*** Endemic 531
	Upper segment of fruit 1-seeded; **lower segment** indehiscent, 1-seeded, c. same length as upper; **leaves** simple or lyrate-pinnatifid	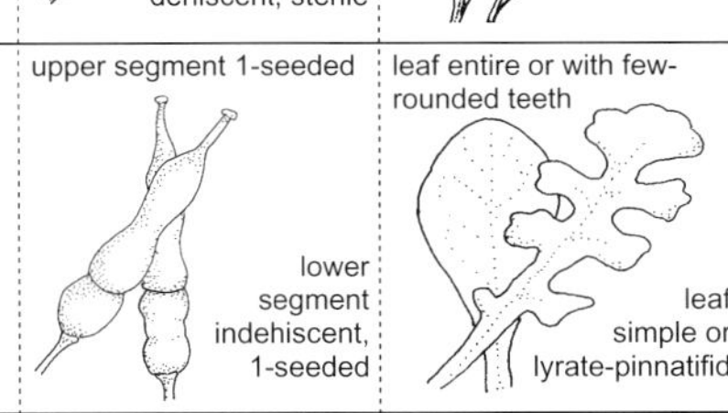	***Nesocrambe socotrana*** Endemic Endemic genus 532

6	**Leaves** sinuately lobed; **hairs** simple or absent	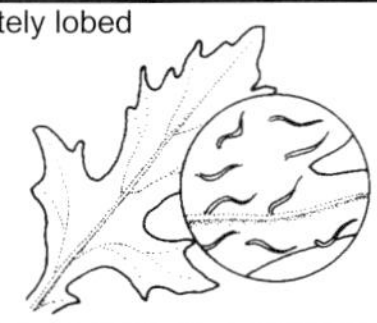	***Sisymbrium erysimoides*** 532
	Leaves simple; **hairs** parallel, medifixed	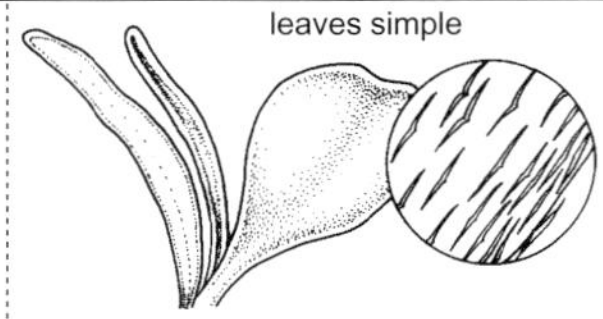	***Farsetia*** (4 spp.)

Farsetia

1	**Leaves** broadly obovate; **fruit** 10--30 x 4--6mm, the style c. 0.5mm with decurrent stigmatic lobes; **sepals** c. 6mm	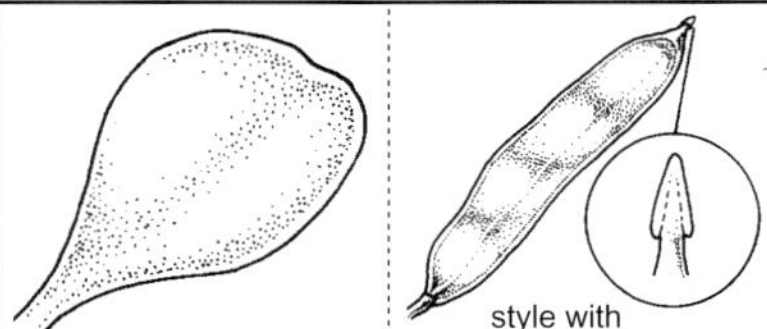	***F. socotrana*** Endemic **258**, 530
	Leaves linear to linear-elliptic; **fruit** 7--30 x2.5--4mm, the style <3.5mm with a capitate stigma; **sepals** c. 4mm		***F. stylosa*** 530
	Leaves linear; **fruit** 20--50 x 3--5mm, the style <3mm with decurrent stigmatic lobes; **sepals** 6--10mm	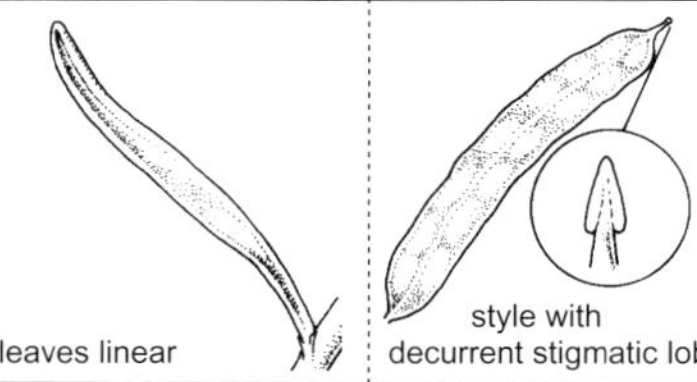	***F. longisiliqua*** 530
	Leaves linear-elliptic to narrowly oblanceolate; **fruit** <18 x 5mm, the style c. 0.5mm with decurrent stigmatic lobes; **sepals** c. 5mm	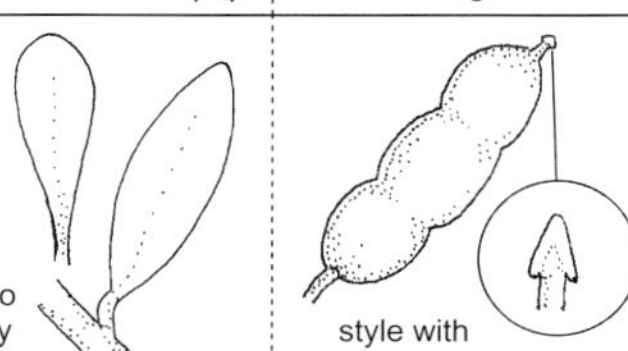	***F. inconspicua*** Endemic **259**, 529

CUCURBITACEAE

1	**Tree** to 6m tall with a swollen trunk <2m in diameter and pendant branches; **tendrils** absent	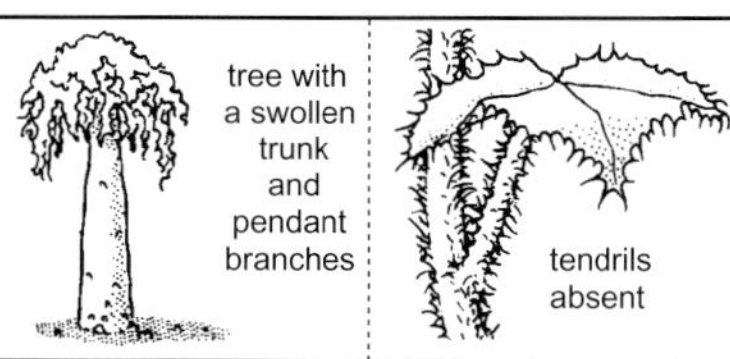	***Dendrosicyos socotrana*** Endemic Endemic genus **259**, 535
	Annual or perennial, climbing or trailing **herbs**; **tendrils** present	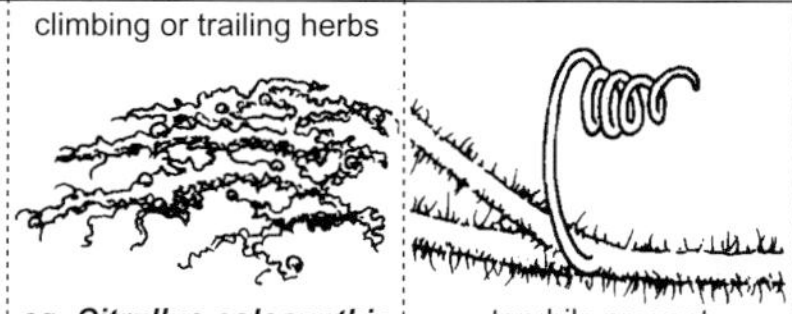	2

2	**Fruits** smooth	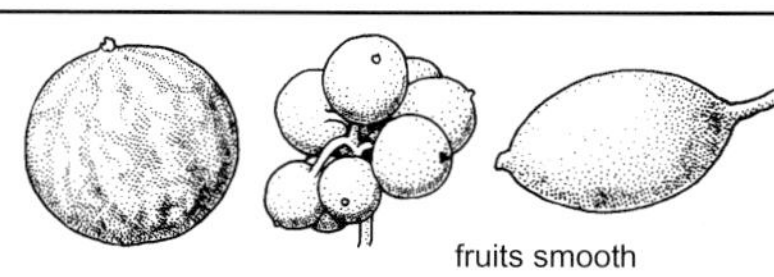	3
	Fruits warted or spiny	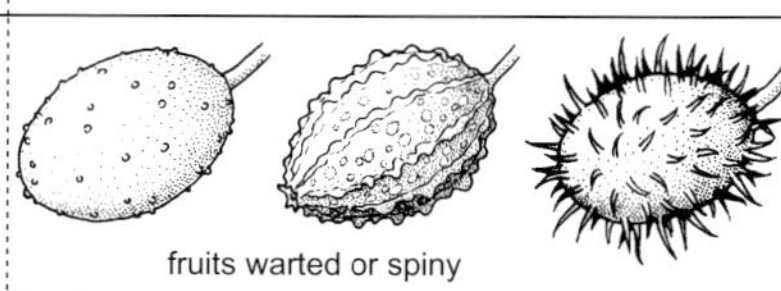	5

3	**Leaves** deeply pinnately lobed; **fruits** globose, <30cm across; ripening yellow	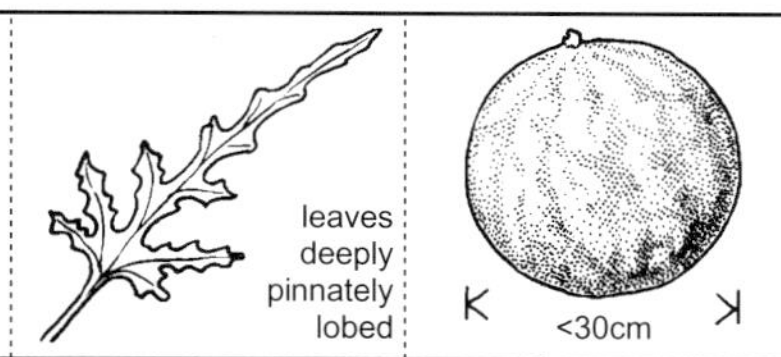	***Citrullus*** (2 spp.)
	Leaves palmately lobed; **fruits** globose or ellipsoid, <5cm across, ripening red	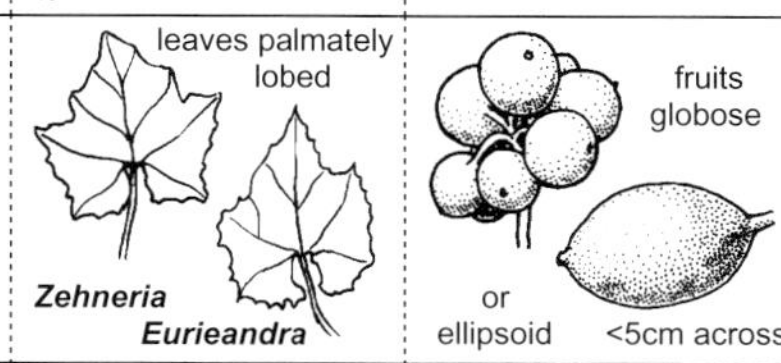	4

4	**Flowers** c.0.5cm across; **stamens** 3; **fruits** globose, rounded at apex, 0.5--1cm across, scarlet when mature;	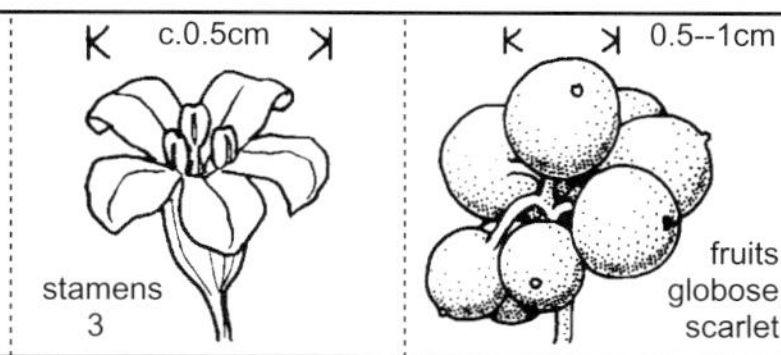	***Zehneria scabra*** 537
	Flowers c.3cm across; **stamens** 5; **fruits** ellipsoid, beaked, c.0.5cm x 2.5cm, orange-red when mature	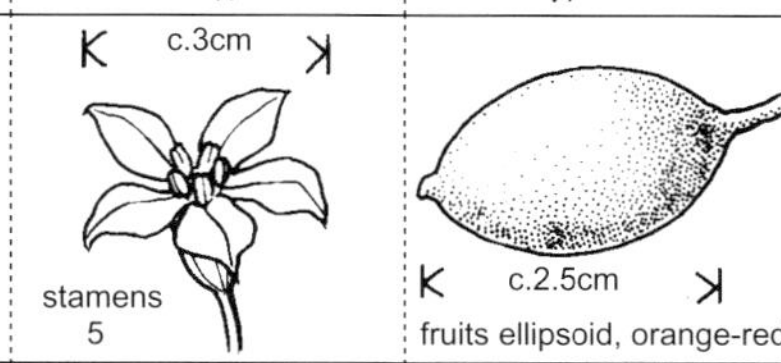	***Eurieandra balfourii*** Endemic **260**, 536

5	**Stems** scabrid or hispid (rough to the touch); **fruits** with bristle-tipped spines, ripening yellow; **seeds** smooth	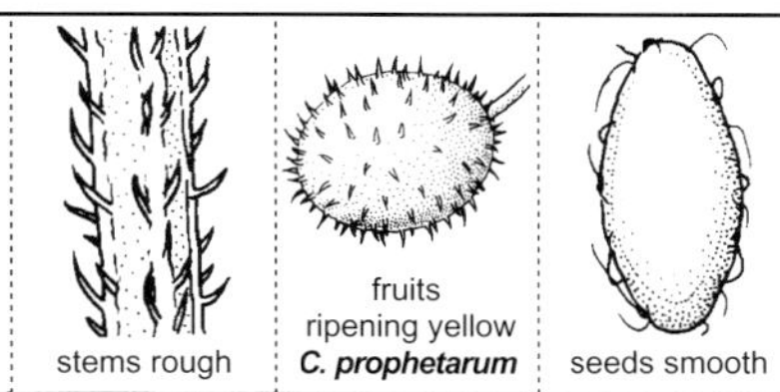	***Cucumis*** (3 spp.)
	Stems hairy or glabrous (smooth to the touch); **fruits** without bristles or spines, ripening orange or red; **seeds** sculptured	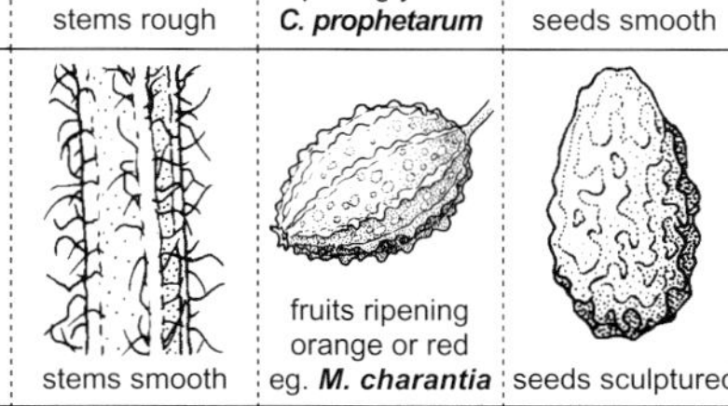	***Momordica*** (2 spp.)

Citrullus

1	**Perennial**; **leaves** narrowly pinnate; **tendrils** simple; **fruit** globose <10cm diameter	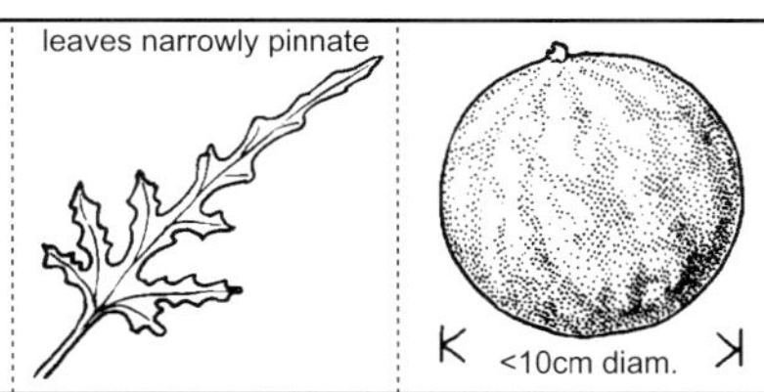	***C. colocynthis*** 533
	Annual; **leaves** broadly pinnate; **tendrils** bifid; **fruit** globose or ellipsoid <60 x 30cm	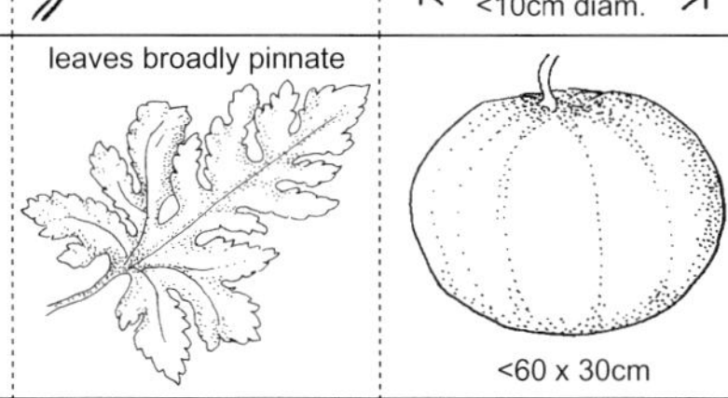	***C. lanatus*** 534

Cucumis

1	**Fruits** covered with wart-like protuberances	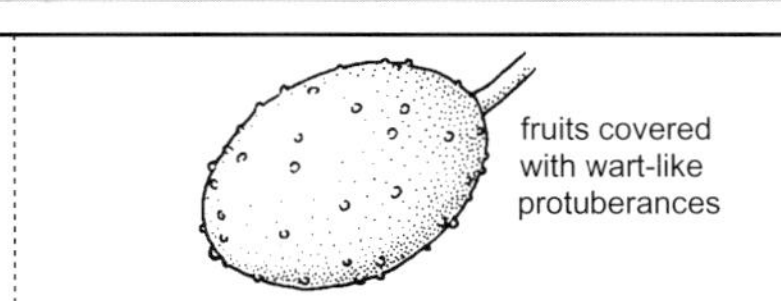	***C. ficifolius*** 534
	Fruits spiny	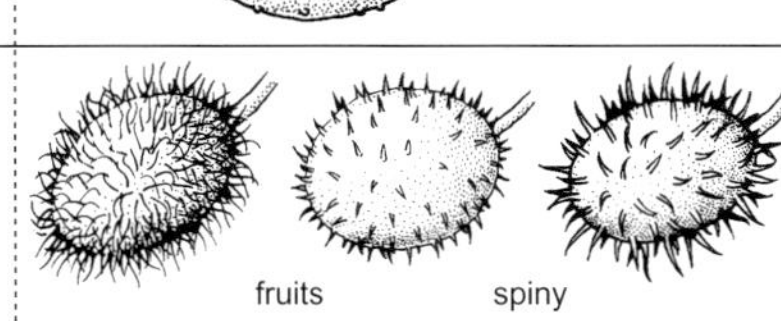	2

2

Stems markedly ridged with stiff spreading hairs; **fruits** with long, dense spines obscuring the surface

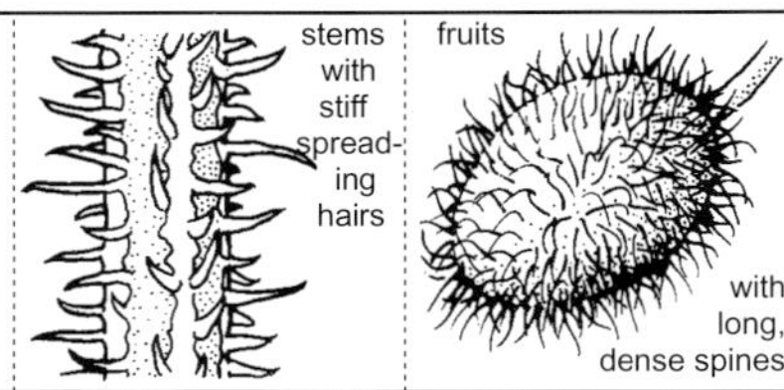

C. dispsaceus 534

Stems somewhat ridged with adpressed or spreading hairs; **fruits** with short scattered spines not obscuring the surface

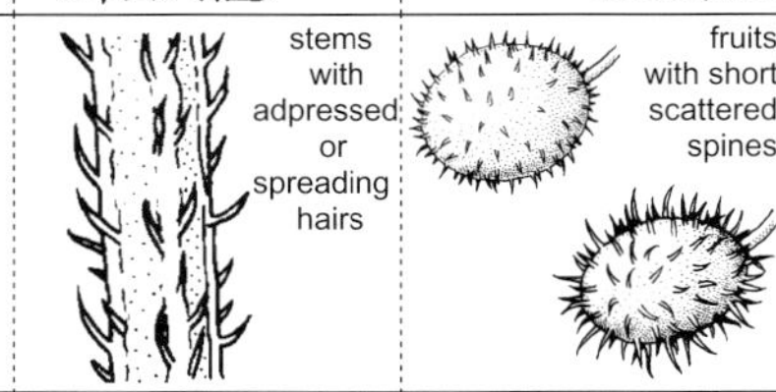

3

3

Plant grey-green; **fruits** ellipsoid to subglobose, with spines up to 4mm, pale green with dark green stripes, ripening clear yellow, 2--4cm diam.

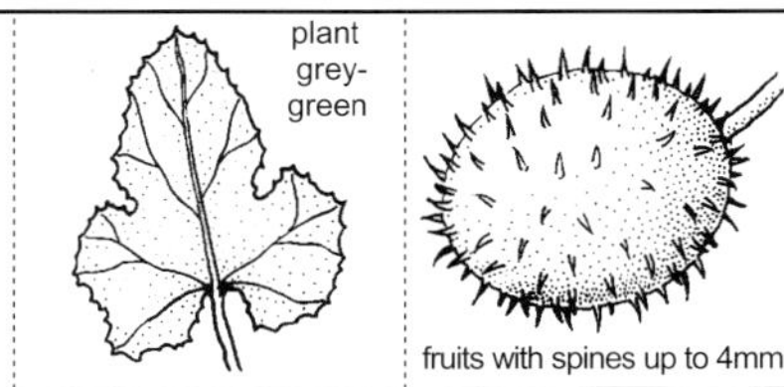

C. prophetarum subsp. ***prophetarum*** **259**, 535

Plant green; **fruits** ellipsoid with spines 4--6mm, pale green with dark green or purple-tinged stripes, ripening dull yellow, 1--2.5cm diam.

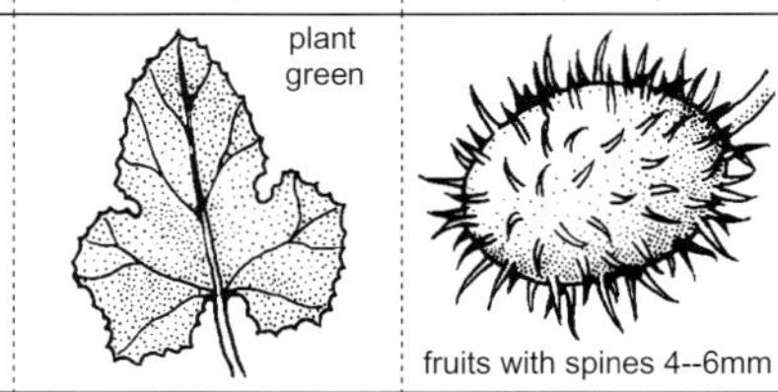

C. prophetarum subsp. ***dissectum*** 535

Momordica

1

Leaves sharply dentate or lobed; **fruits** with rows of small tubercles but otherwise smooth

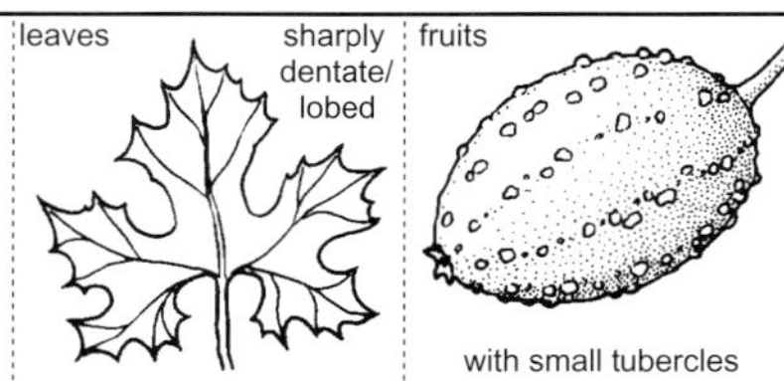

M. balsamina 537

Leaves sinuately toothed or lobed; **fruits** with rows of conspicuous fleshy tubercles and scattered smaller tubercles

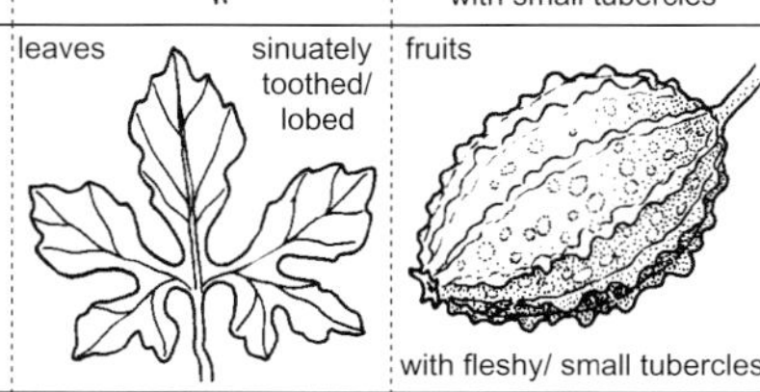

M. charantia 537

CUSCUTACEAE

1

Stigmas globose, smaller than styles; **styles** equalling or longer than the ovary

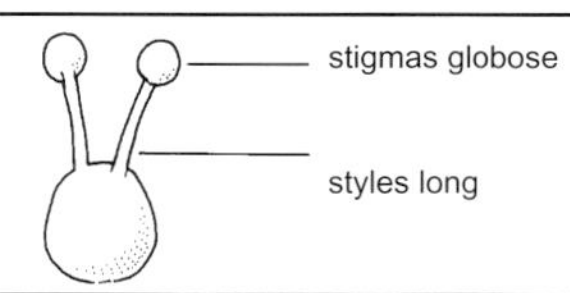

Cuscuta chinensis 538

Stigmas linear, longer than styles; **styles** much shorter than the ovary

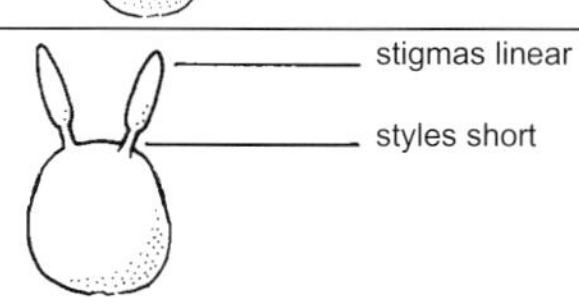

Cuscuta planiflora 538

DIRACHMACEAE

1

Small **tree; leaves** simple, alternate, with linear stipules; **flowers** white, 8-merous; **fruit** dehiscent, of 8 segments

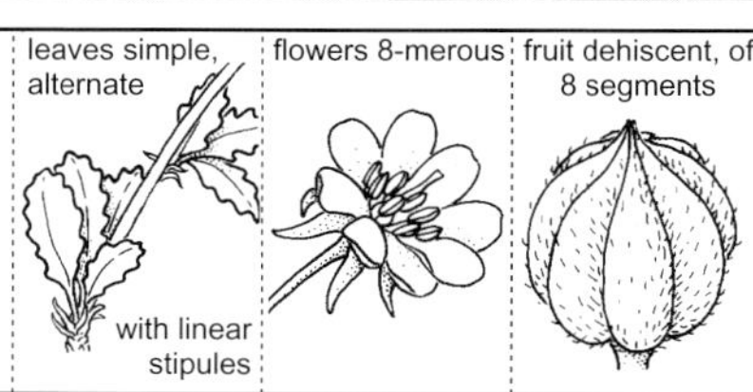

Dirachma socotrana Endemic **260**, 539

EBENACEAE

1 **Shrub** or small **tree**; **flowers** white, male with 10 to 20 stamens, female with 2 styles; **fruit** ± globose, ripening red — ***Euclea divinorum*** **260**, 540

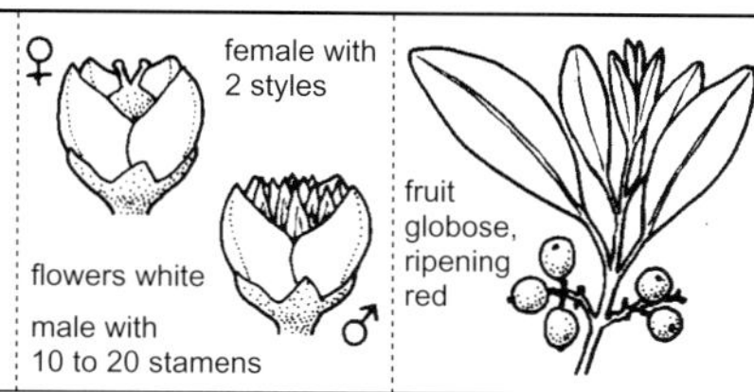

ERYTHROXYLACEAE

1 **Shrub**; **leaves** obovate, reticulate; **leaf scars** prominent; **flowers** cream, small, c. 3mm diameter; **fruit** fleshy, ellipsoid — ***Erythroxylon socotranum*** Endemic 541

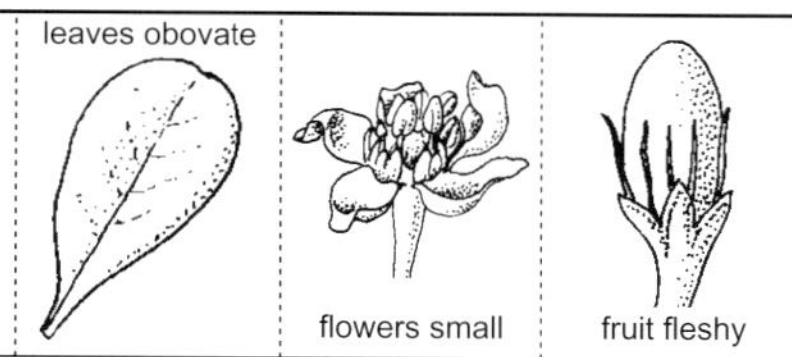

EUPHORBIACEAE

1 **Flowers** minute, in a specialised structure, the cyathium, resembling a single flower; **plants** with milky latex — ***Euphorbia*** (16 spp.)

Flowers not in a cyathium; **plants** without latex or if latex present then this clear, never milky — 2

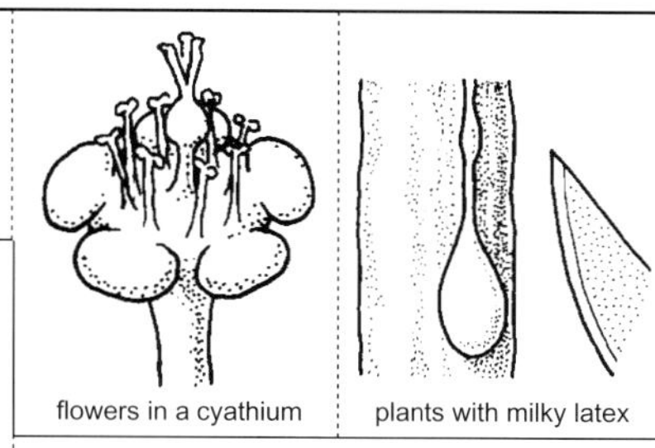

2 Scrambling **herb** with stinging **hairs** — ***Tragia balfouriana*** Endemic **263**, 567

Herb, shrub or tree, never scrambling, **hairs** not stinging — 3

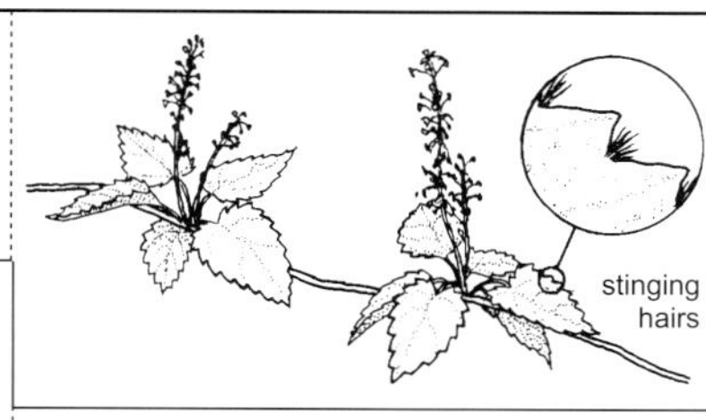

3 **Hairs** on leaf and stem stellate or lepidote (scale-like) — 10

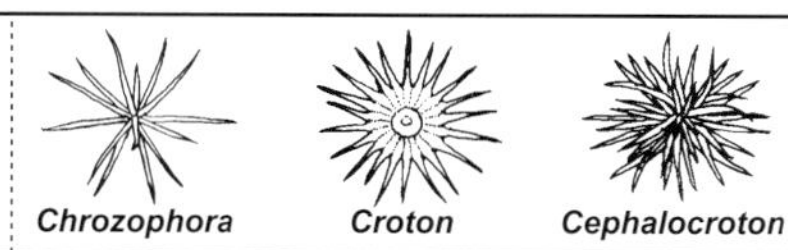

Hairs on leaf and stem unbranched or absent — 4

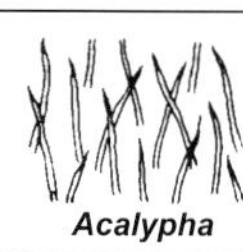

4 **Leaves** palmately lobed with glandular-serrate margins; **stamens** up to 1000 — ***Ricinus communis*** 566

Leaves simple, margin entire or if serrate then not glandular; **stamens** few — 5

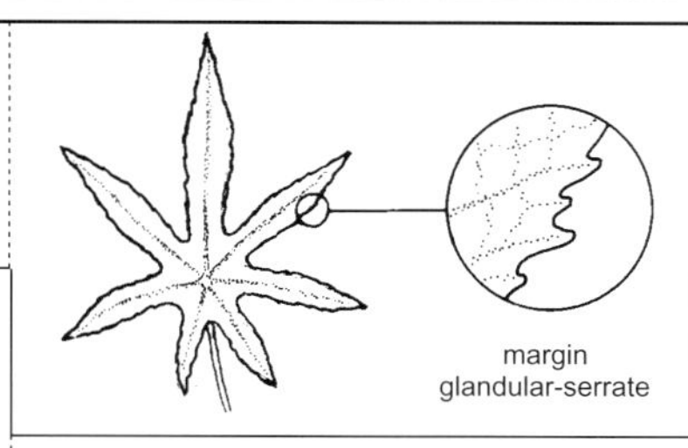

5 **Leaves** serrate; **hairs** present; **female flowers** in the axils of accrescent leaf-like bracts — ***Acalypha indica*** **260**, 542

Leaves entire; **hairs** absent; **female flowers** not in the axils of leaf-like bracts — 6

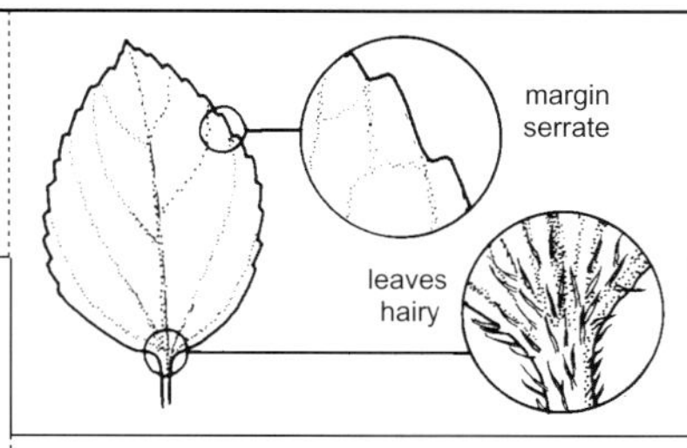

6	**Leaves** large, >5cm long; **flowers** in terminal cymes	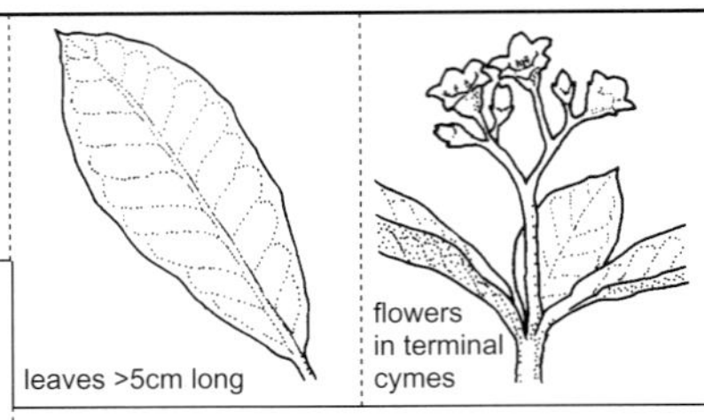	***Jatropha unicostata*** Endemic **263**, 561
	Leaves small, <5cm long; **flowers** in axillary clusters		7
7	**Tree**; **fruit** fleshy, indehiscent	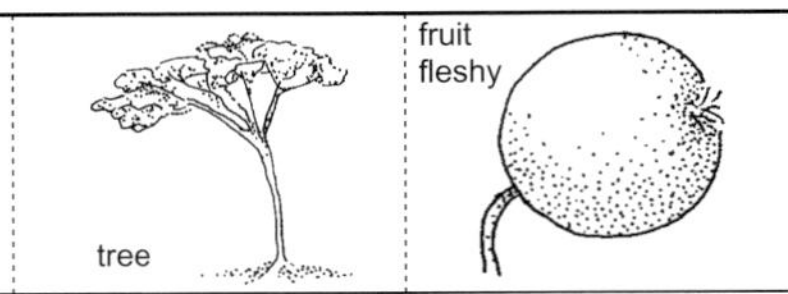	***Flueggea*** (2 spp.)
	Prostrate **herb**; **fruit** dry, dehiscent	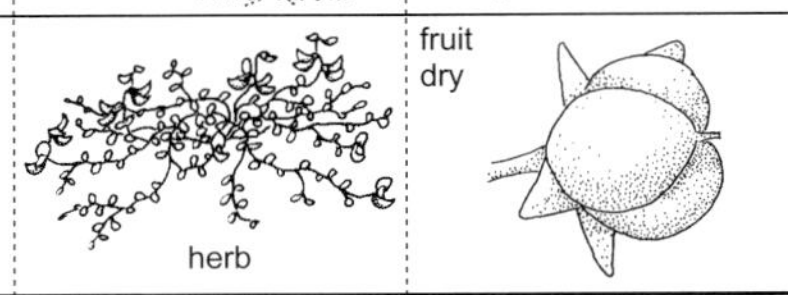	8
8	**Petals** present, at least in male flowers; **stipules** laciniate or lobed, minutely spurred	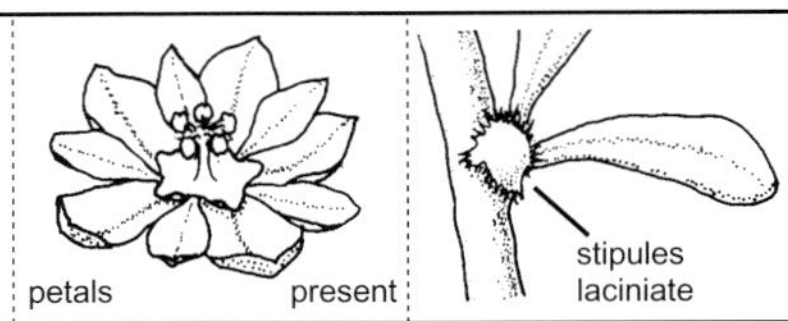	***Andrachne*** (2 spp.)
	Petals absent; **stipules** entire or dentate	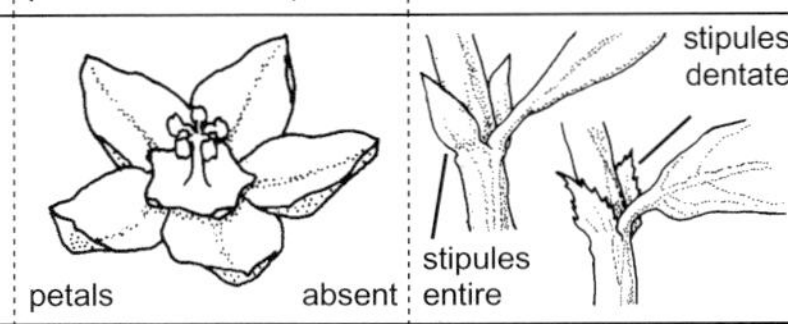	9
9	**Pedicels** short, usually <2mm long; **stipules** entire; **pistillode** absent in male flowers	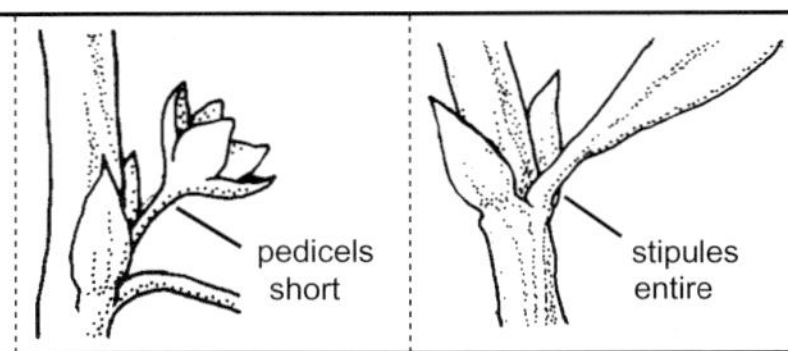	***Phyllanthus*** (2 spp.)
	Pedicels long, slender, usually >5mm; **stipules** dentate; **pistillode** present in male flowers	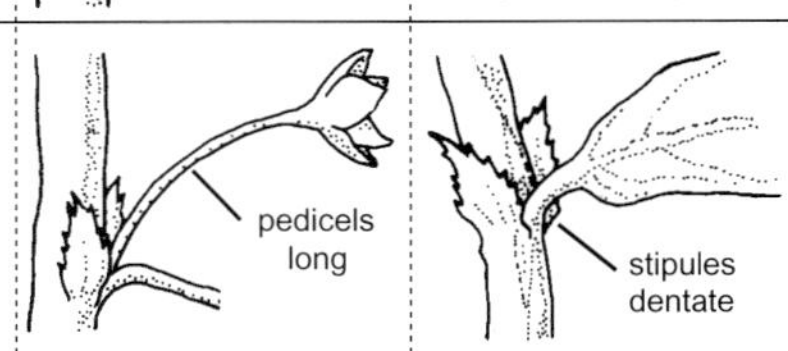	***Meineckia filipes*** Endemic 565
10	Erect **herb**; **styles** bifid	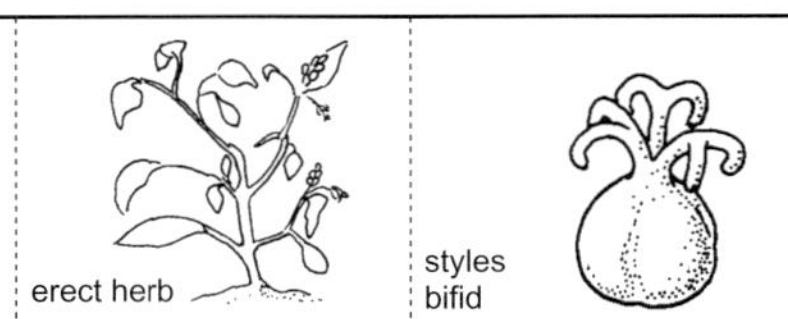	***Chrozophora oblongifolia*** 544
	Tree or **shrub**; **styles** more than twice-divided	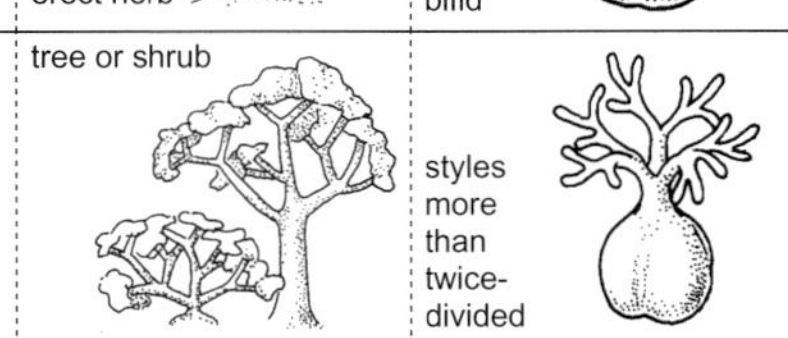	11
11	**Male flowers** in racemes or open clusters; **petals** present; **flowers** white or cream; **seeds** caruncular	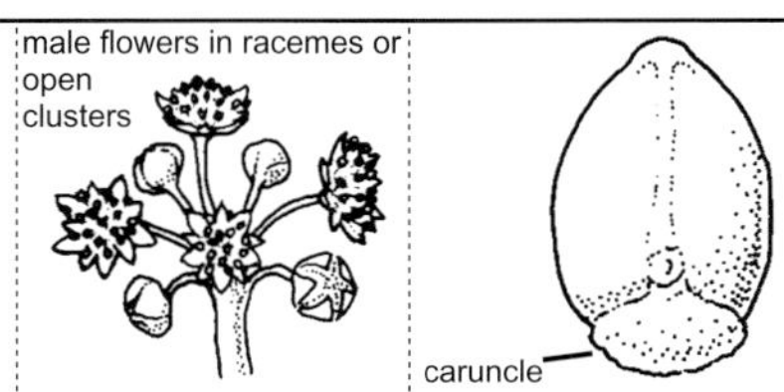	***Croton*** (3 spp.)
	Male flowers in dense, caruncular heads; **petals** absent; **flowers** yellow; **seeds** not caruncular	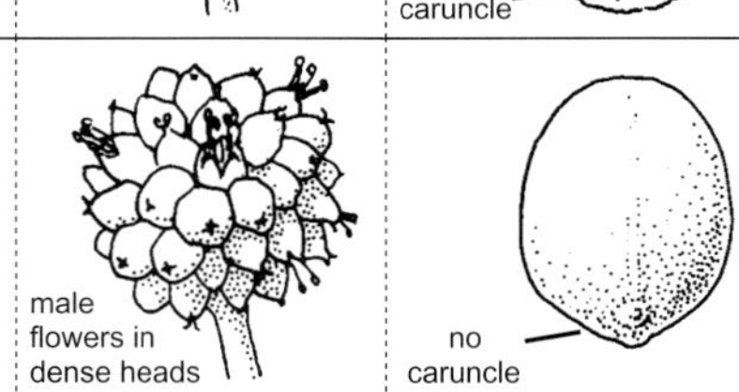	***Cephalocroton socotranus*** Endemic **260**, 543

Andrachne

1	**Leaves** suborbicular; **stipules** white with a brown centre, lobed	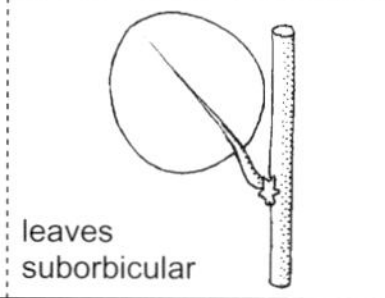	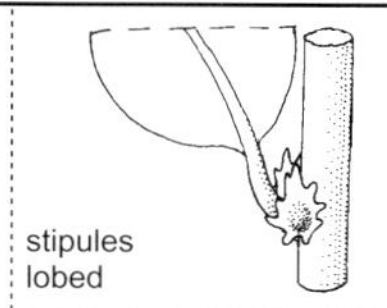	***A. telephioides*** 543
	Leaves oblanceoate; **stipules** reddish-brown, fimbriate	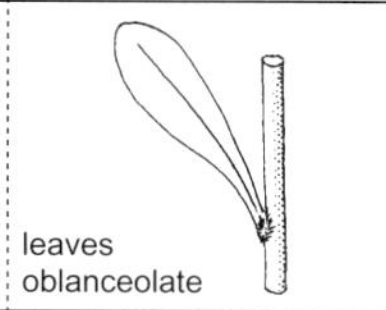	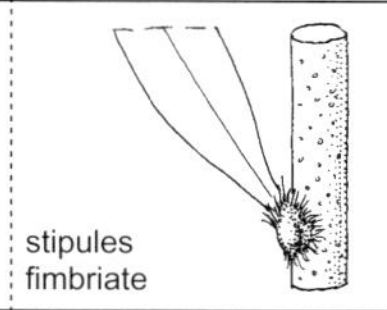	***A. schweinfurthii*** Endemic 542

Croton

1	**Fruits** with sessile stellate hairs; **flowers** in racemes; monoecious **trees**	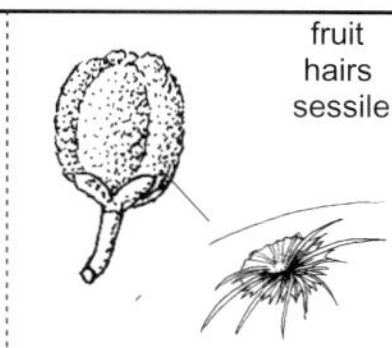	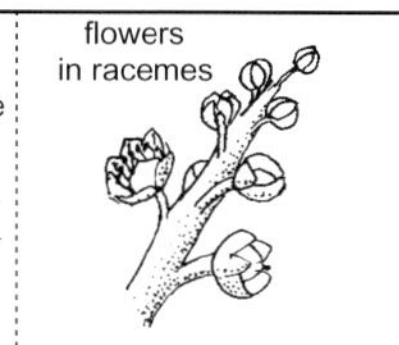	***C. sulcifructus*** Endemic **261**, 548
	Fruits with stalked stellate hairs; **flowers** in umbellate clusters; dioecious **trees** or **shrubs**	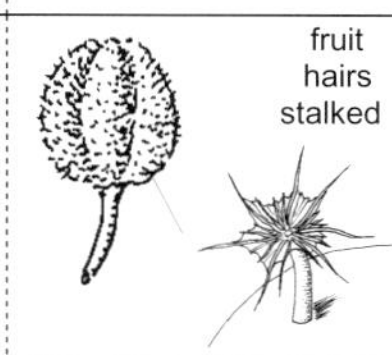	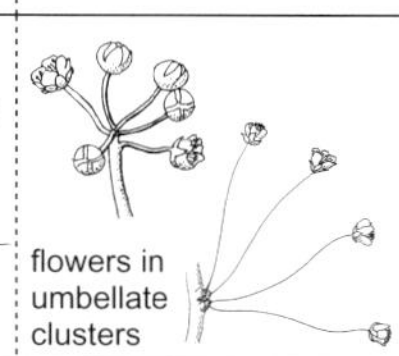	2

2	**Leaves** cordate, <4 x 2.5cm, thinly (rarely densely) hairy, not strongly discolorous, green; **fruit** <6 x 6mm; small **tree or shrub**	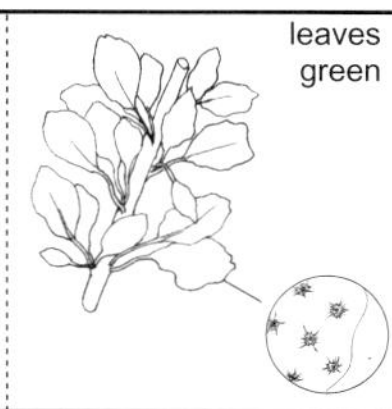	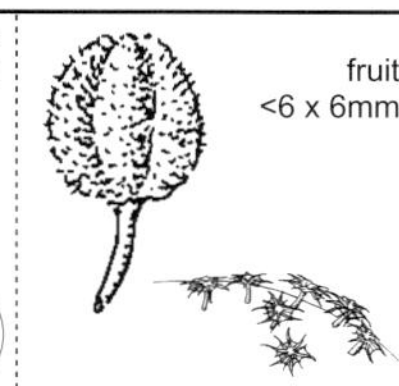	***C. socotranus*** Endemic **260**, 545
	Leaves narrowly-ovate to orbicular, <10 x 6cm, discolourous, green above, silvery below; **fruit** coarsely hairy, <12 x 10mm; medium-sized **tree**	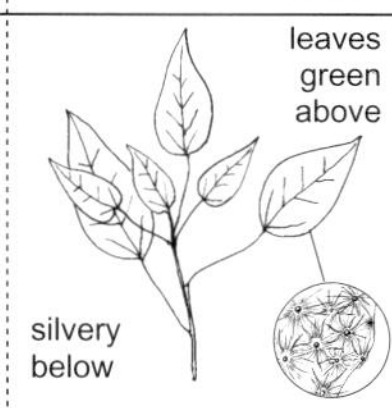	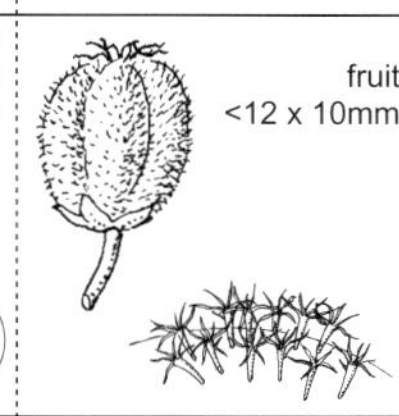	***C. sarcocarpus*** Endemic **260**, 545

Euphorbia

1	**Herbs, shrubs, trees** or **succulents**; at least **lower leaves** alternate or absent; **glands of cyathium** without petaloid appendages	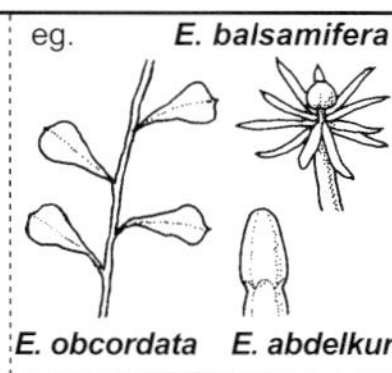	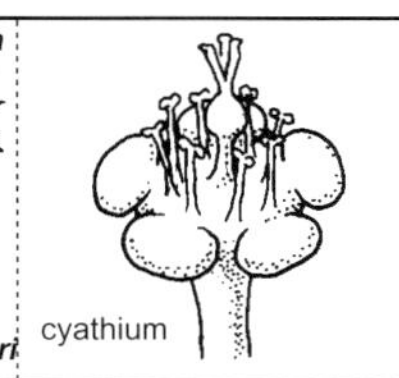	2
	Low, often prostrate annual or perennial **herbs**, sometimes woody-based; all **leaves** opposite; **glands of cyathium** with petaloid appendages	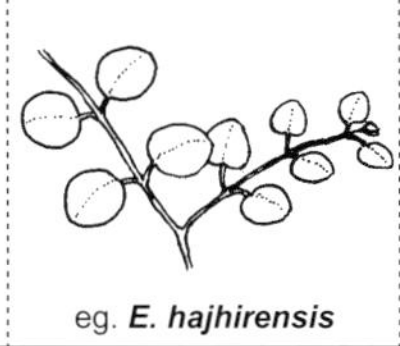		9

2	**Stems** succulent, spiny	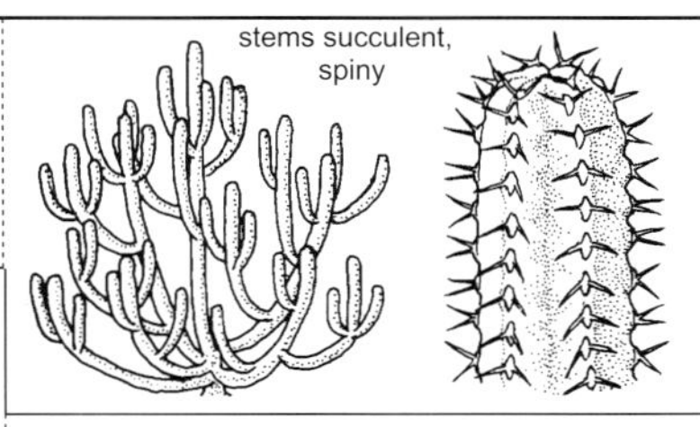		***E. spiralis*** Endemic **263**, 559
	Stems succulent, woody or herbaceous but never spiny			3

3	**Branches** succulent, cylindrical; **leaves** soon falling		4
	Branches herbaceous or woody, if succulent then not cylindrical; **leaves** persistent	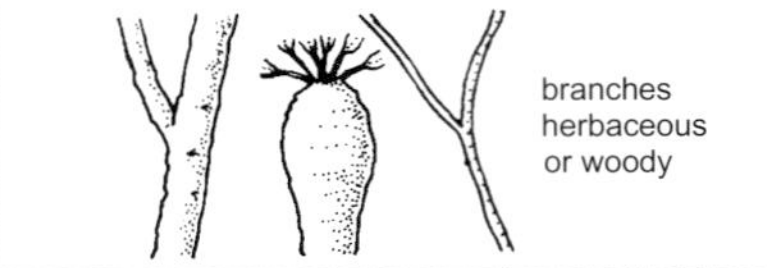	6
4	Large **shrub** with erect stems; **ultimate branches** >5cm in diameter, with latitudinal markings; **latex** yellowish green	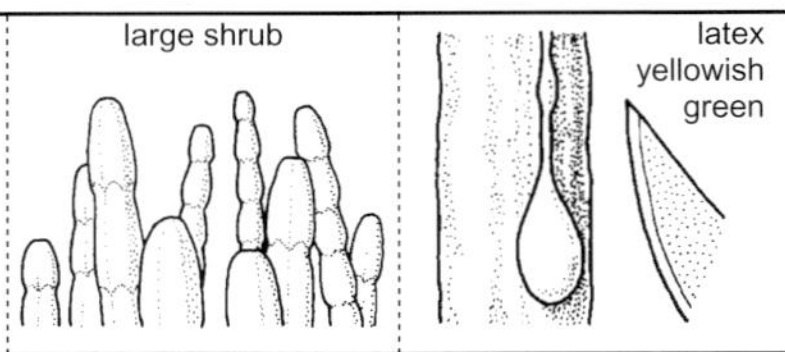	***E. abdelkuri*** Endemic **261**, 549
	Tree or **shrub**; **ultimate branches** <2cm in diameter, longitudinally striate; **latex** white	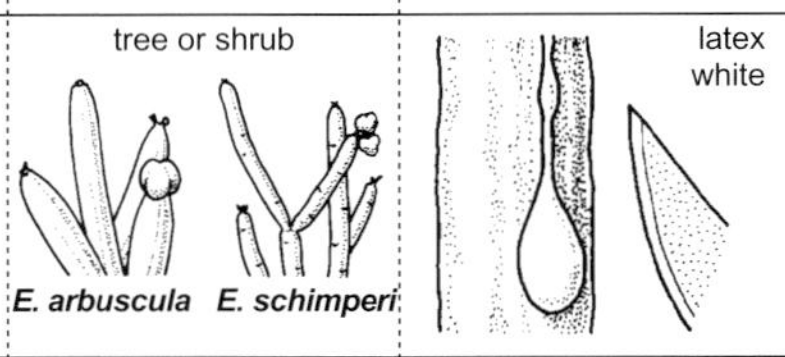	5
5	**Tree** with distinct trunk and much-branched crown; **fruits** densely pubescent	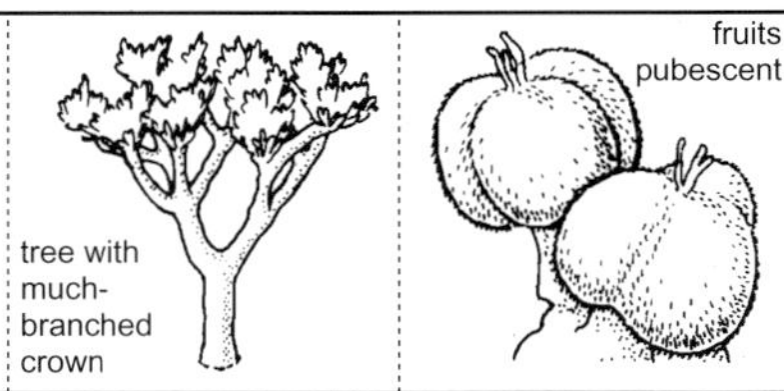	***E. arbuscula*** Endemic **261**, 549
	Virgately-branched **shrub** lacking a distinct trunk and little-branched above; **fruits** glabrous or glabrescent	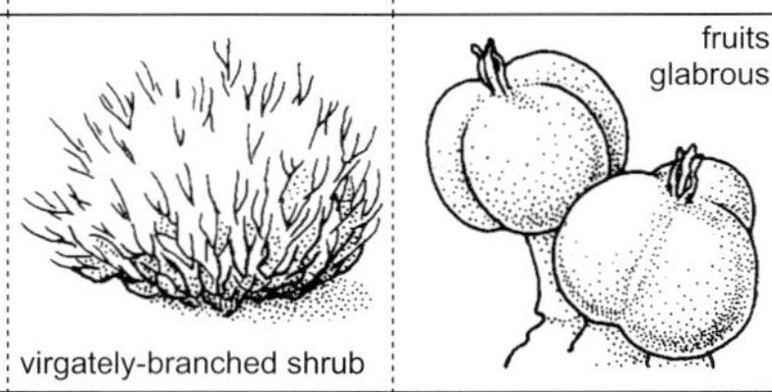	***E. schimperi*** **262**, 556
6	**Herb**; **stem** swollen, cigar-shaped, bearing numerous leaf scars, not or little branched; **leaves** undulate in an apical tuft	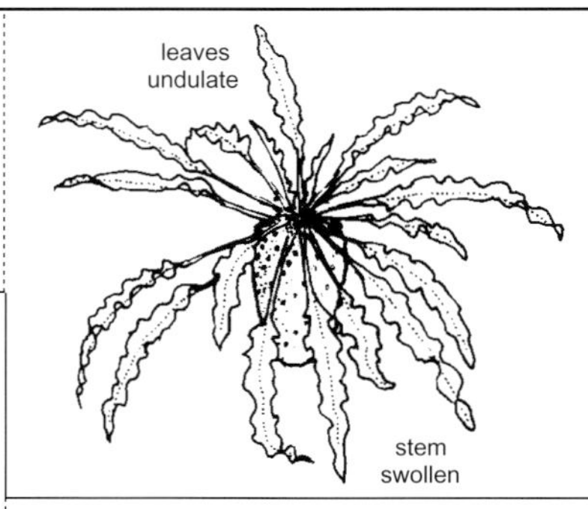	***E. hadramautica*** **261**, 553
	Shrubs or **trees**; **stem** woody or semi-succulent; **leaves** entire, not undulate		7
7	**Inflorescence** well-developed, branched, leafy; **ultimate branches** thin; **cyathia** subtended by a pair of cordate bracts	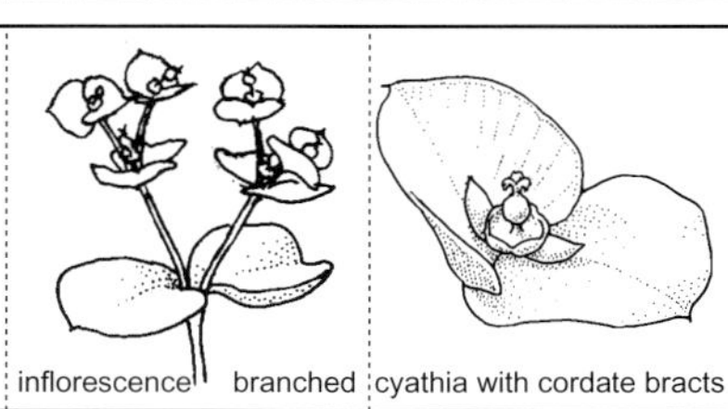	***E. obcordata*** Endemic **262**, 556
	Cyathia solitary or grouped at the ends of branches, not in branched **inflorescences; ultimate branches** thick; **cyathia** surrounded by several elliptic or obovate bracts	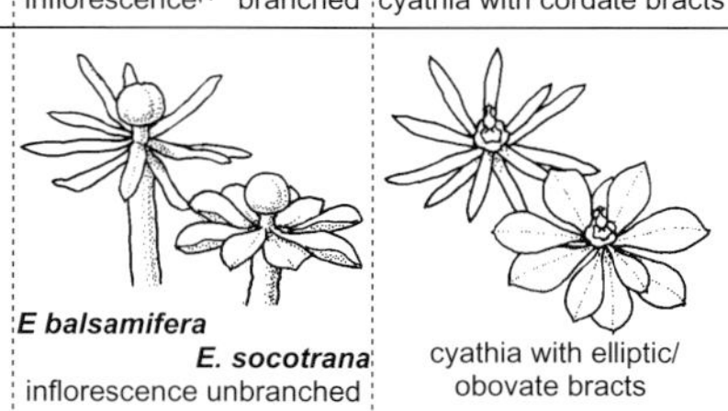	8

8	Rounded, regularly-branched **shrub**; **leaves** linear-lanceolate to narrowly obovate, <3cm long; **cyathia** surrounded by narrowly elliptic bracts	 regularly-branched shrub	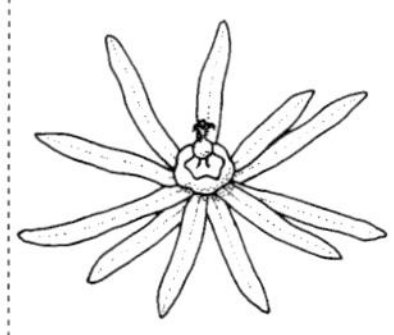 cordate bracts	***E. balsamifera*** **261**, 552
	Low, intricately branched **subshrub**; **leaves** broadly obovate to obtriangular, <1.5cm long; **cyathia** surrounded by leaf-like broadly obovate bracts	 intricately branched subshrub	 broadly obovate bracts	***E. hamaderoensis*** Endemic **262**, 554
	Tree rather randomly-branched; **leaves** broadly obovate to obtriangular, <1.5cm long; **cyathia** surrounded by broadly elliptic or obovate bracts	 randomly-branched tree	 elliptic/ obovate bracts	***E. socotrana*** Endemic **263**, 558
9	**Cyathia** clustered in cymes subtended by small bracts; erect annual herb; **leaves** toothed	Cyathia clustered in cymes 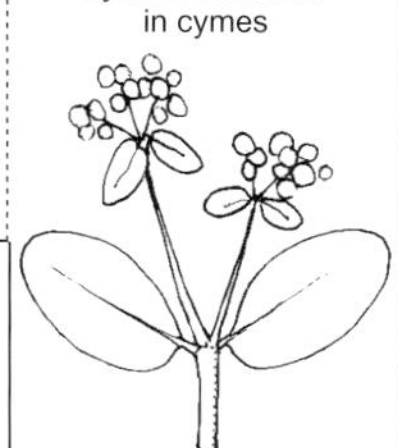 cymes with small bracts	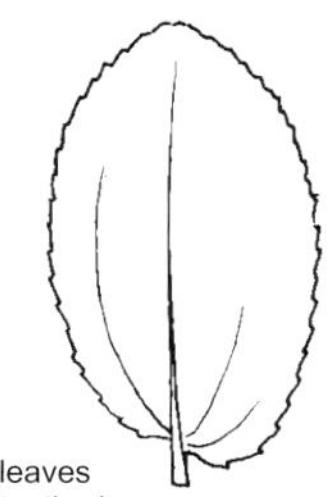 leaves toothed	***E. indica*** 555
	Cyathia solitary or in small groups; prostrate herbs or subshrubs; **leaves** entire (*E. inaequilatera* toothed)			10
10	**Leaves** suborbicular, 8--15mm diameter, green above, white below; **petaloid appendages** green	leaves suborbicular 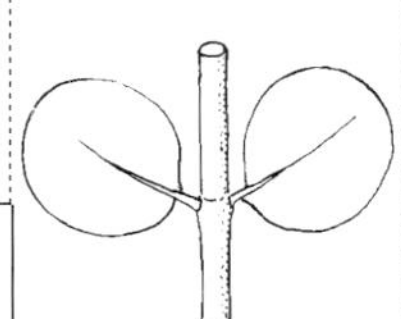	petaloid appendages green 	***E. hajhirensis*** Endemic 554
	Leaves variously shaped, <5mm diameter, variously coloured; **petaloid appendages** white or pink			11
11	Intricately branched **subshrub**; **leaves** glaucous, glabrous or minutely puberulous, minutely papillose beneath (lens needed)	intricately branched subshrub 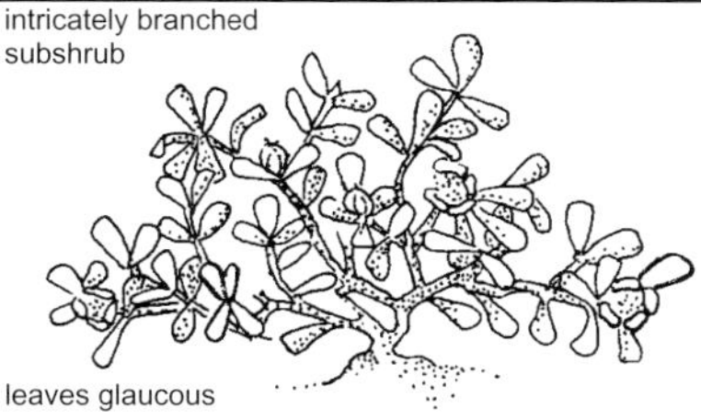 leaves glaucous		12
	Prostrate **herbs**, sometimes woody at base; **leaves** green or whitish-green, glabrous or variously hairy but never papillose	prostrate herb 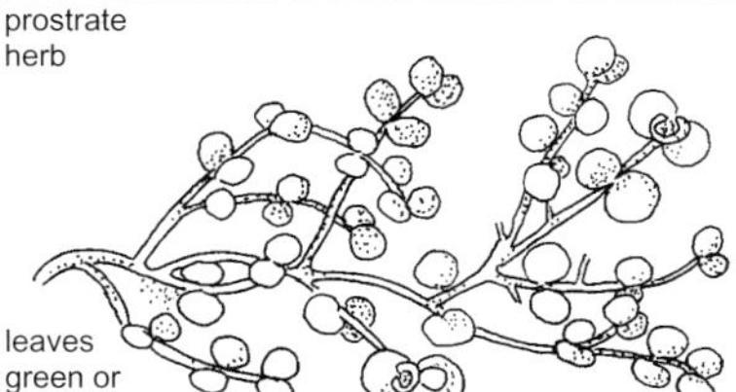 leaves green or whitish-green		13

12

Leaves linear-oblong, apex rounded or retuse; **petaloid appendages** entire; **stems** puberulous

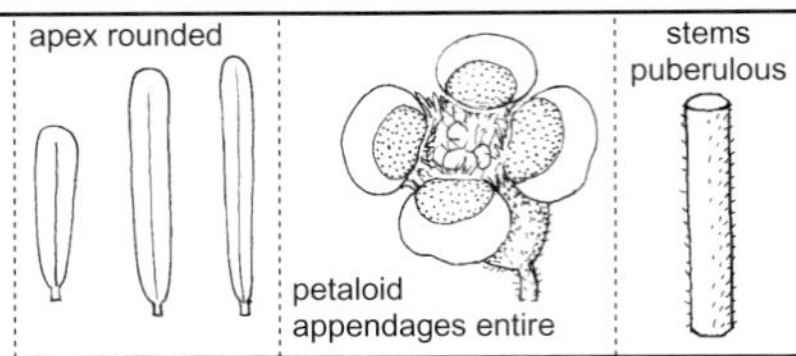

***E. schweinfurthii* form A**
Endemic
262, 558

Leaves obovate to oblong-obovate, apex rounded or retuse; **petaloid appendages** entire; **stems** glabrous

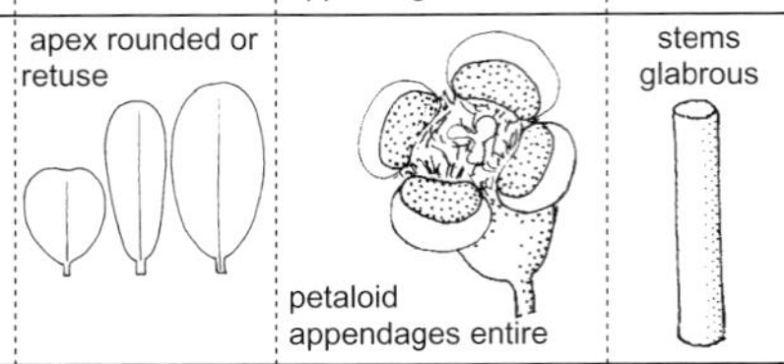

***E. schweinfurthii* form B**
Endemic
558

Leaves narrowly obovate to linear-oblong, apex notched; **petaloid appendages** toothed; **stems** puberulous

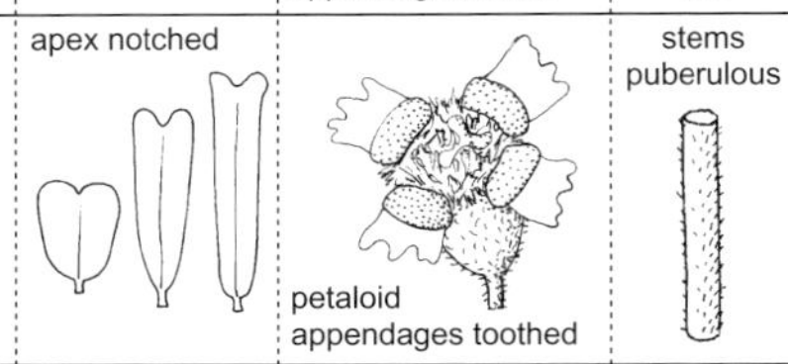

***E. schweinfurthii* form C**
Endemic
558

Leaves orbicular to broadly elloptic, apex rounded or retuse; **petaloid appendages** entire or undulate; **stems** puberulous

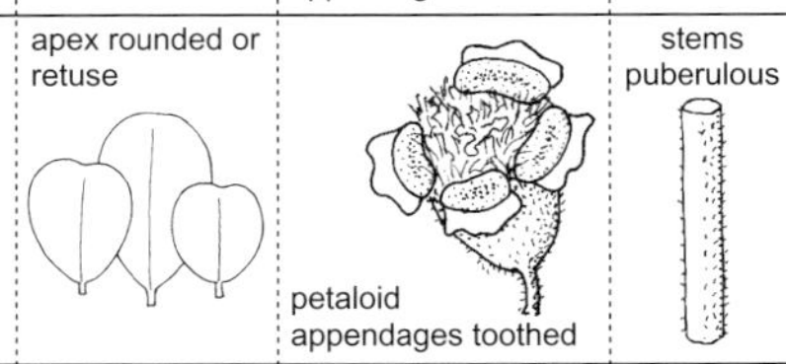

***E. schweinfurthii* form D**
Endemic
263, 558

13

Leaves orbicular to kidney-shaped; **petaloid appendages** with acute teeth

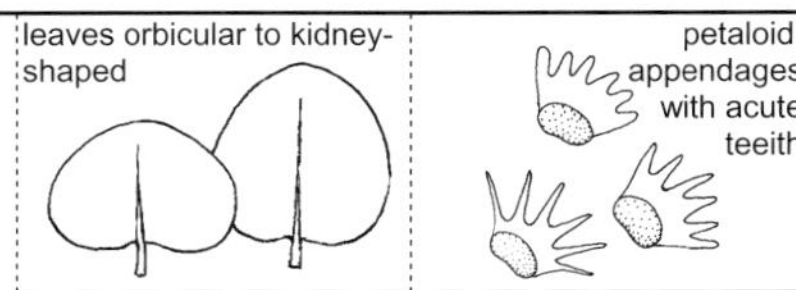

14

Leaves elliptic to oblong-obovate; **petaloid appendages** with rounded teeth or lobes

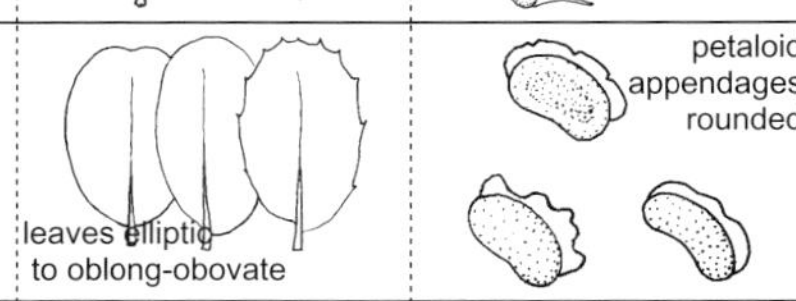

15

14

Leaves uniformly green; **stems** glabrous or puberulous

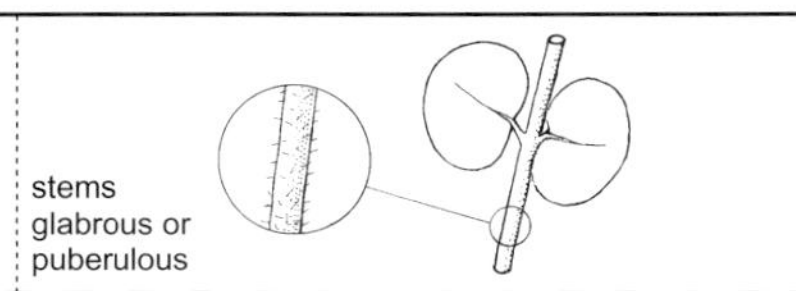

E. kischenensis
Endemic
262, 555

Leaves green above, whitish-green below; **stems** villous

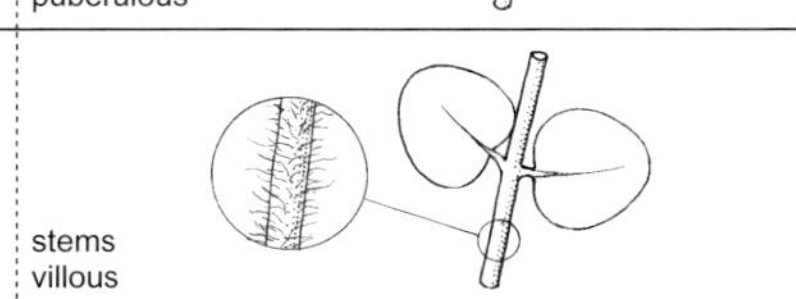

E. leptoclada
Endemic
556

15

Perennial with woody taproot; **leaves** entire, base symmetric, green; **petaloid appendages** white, lobes rounded; **stems** glabrous; **stipules** triangular, ciliate

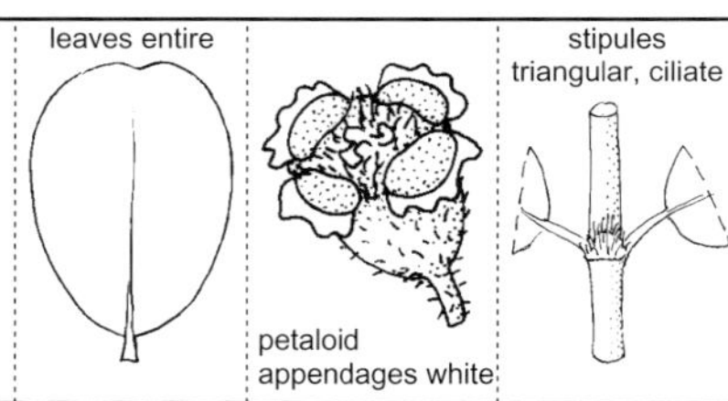

E. kuriensis
Endemic
262, 555

Annual; **leaves** entire, base asymmetric, grey-green; **petaloid appendages** pink, obscurely lobed; **stems** hairy; **stipules** linear, glabrous

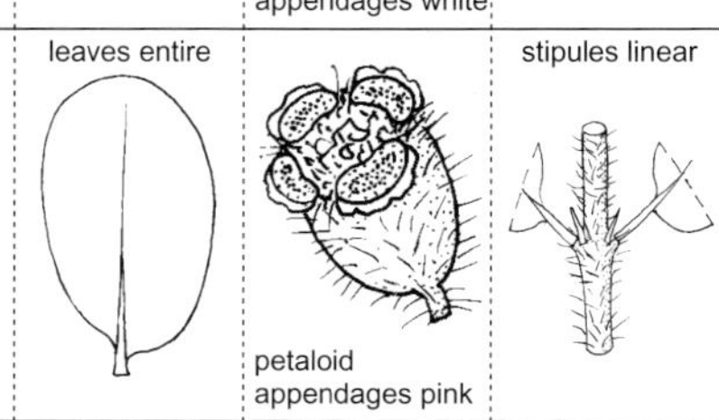

E. granulata
553

Annual; **leaves** serrate, base asymmetric (at least sometimes); **petaloid appendages** obscure, pink or white; **stems** hairy; **stipules** linear, divided, glabrous

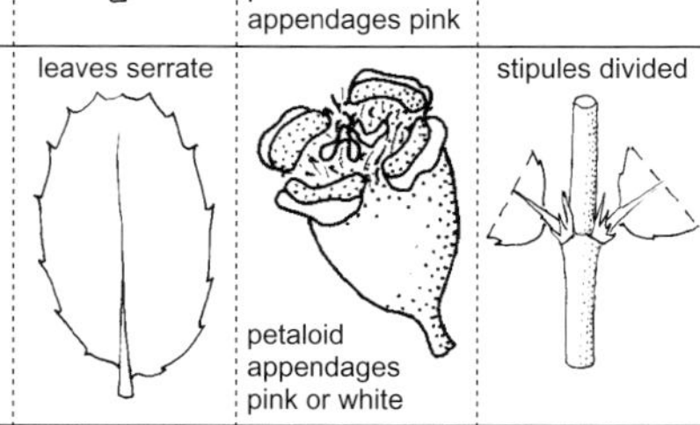

E. inaequilatera
554

Flueggea

1	**Leaves** <1.5(--2)cm long, obovate, with truncate or emarginate tips	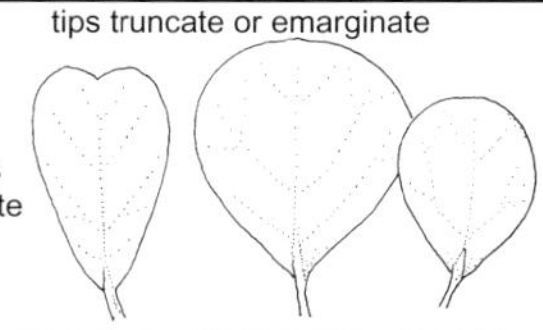		***F. leucopyrus*** 560
	Leaves <3.5(--6)cm long, elliptic to obovate, with rounded or shortly acuminate tips	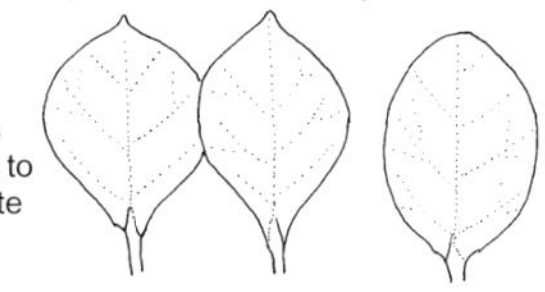		***F. virosa*** 561

Phyllanthus

1	**Leaves** lanceolate; **sepals** suborbicular	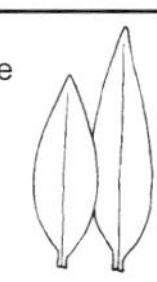	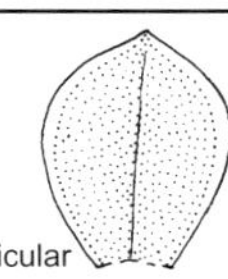	***P. maderaspatensis*** 565
	Leaves obovate to orbicular; **sepals** ovate to obovate	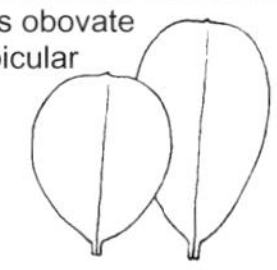	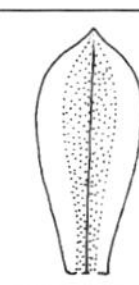	***P. rotundifolius*** **263**, 565

FLACOURTIACEAE

1	Dioecious **shrub** or **tree**; **leaves** alternate, entire; **flowers** cream, without petals, the male with many **stamens**, the female with 4 or 5 styles; **fruit** a globose berry	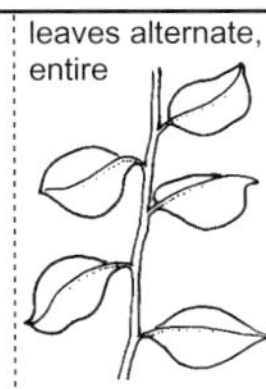		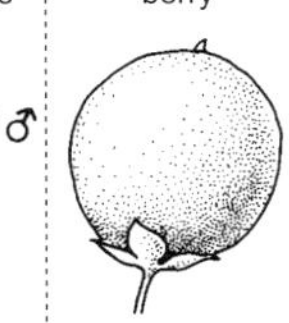	***Dovyalis abyssinica*** **263**, 568

FRANKENIACEAE

1	Prostrate annual **herb**; **leaves** opposite, 2--6mm, often covered with salt crystals; **flower** c.4mm across; **calyx** tubular; **petals** 5, pink	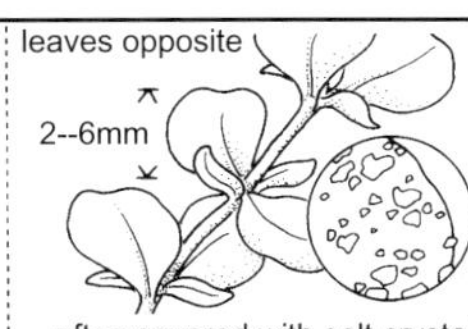	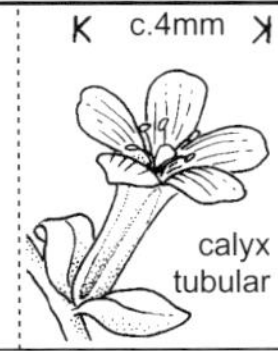	***Frankenia pulverulenta*** 569

GENTIANACEAE

1	**Corolla** pink, **tube** longer than corolla lobes; **calyx** unwinged; **leaves** petiolate, venation indistinct			***Centaurium tenuiflorum*** 569
	Corolla pale yellow; **calyx** unwinged; **leaves** sessile, venation indistinct			***Sebaea microphylla*** 571
	Corolla lilac to blue, **tube** shorter than corolla lobes; **calyx** winged; **leaves** sessile or petiolate, with 3 to 5 parallel veins	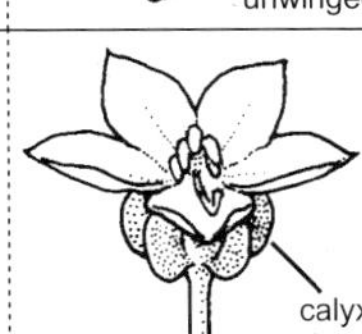	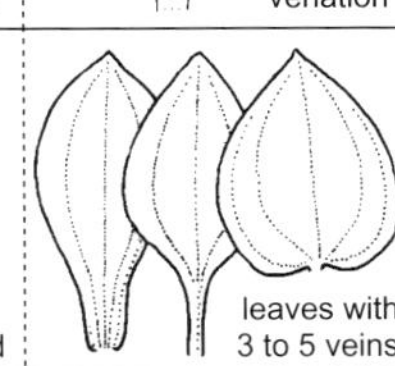	***Exacum*** (3 spp.)

Exacum

1	**Leaves** sessile, with a leathery texture	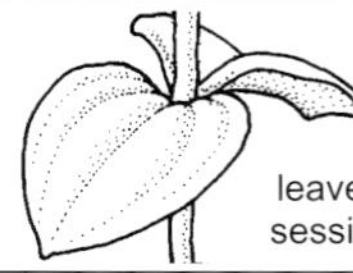	***E. caeruleum*** Endemic **264**, 570
	Leaves petiolate, with a papery texture	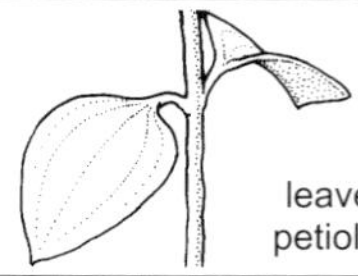	2
2	**Free tip of calyx lobes** indistinct, <1mm long; **upper leaves** elliptic to spathulate, with a rather short, broad petiole	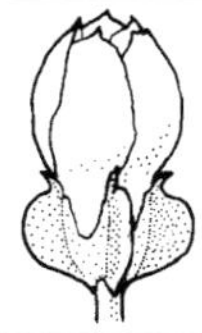 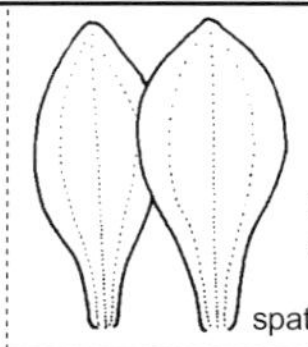	***E. socotranum*** Endemic 571
	Free tip of calyx lobes distinct, >2mm long; **upper leaves** ovate to elliptic, with a rather long, narrow petiole	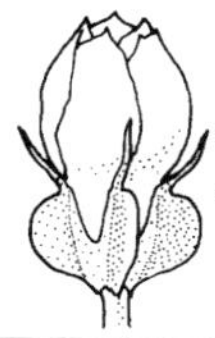 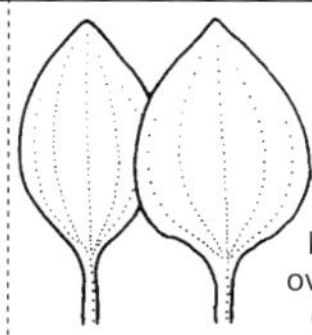	***E. affine*** Endemic **264**, 570

GERANIACEAE

1	**Annual herb**; **leaves** pinnately divided; **petals** pin, 8--10mm long; **mericarps** with long spirally twisted awns	 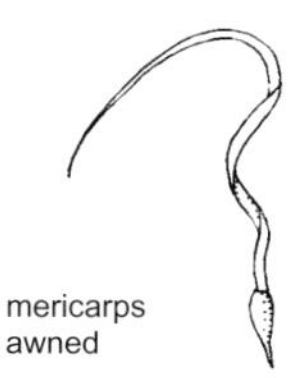	***Erodium cicutarium*** 571
	Robust **perennial herb; petals** pink, <25mm long, unequal; **leaves** shallowly lobed; **mericarps** with long spirally twisted awns	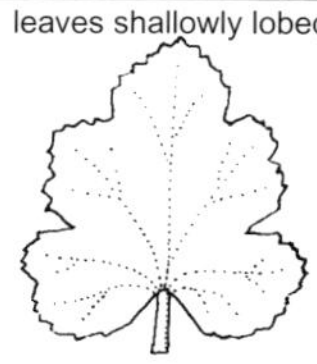 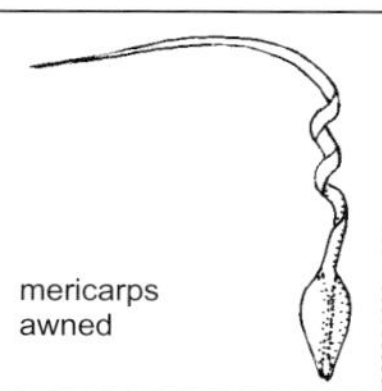	***Pelargonium insularis*** Endemic **264**, 573
	Annual herb; **petals** mauve with dark centre, 6--8mm long; **leaves** deeply lobed; **mericarps** without long awns	 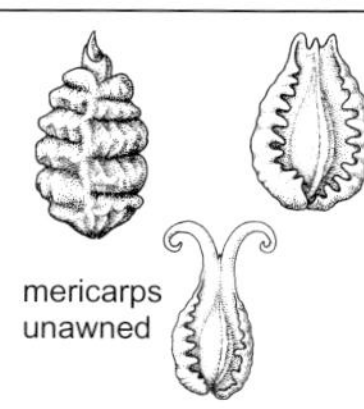	***Geranium*** (3 spp.)

Geranium

1	**Mericarps** 2--3.5mm long, deeply tranversely rugose, the ridges with short longitudinal crests		***G. mascatense*** 572
	Mericarps 4--6mm long, minutely ridged and with a revolute toothed margin	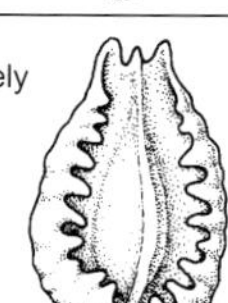	***G. trilophum*** 572
	Mericarps 3--5mm long, minutely ridged and with a revolute margin and 2 hooked apical horns	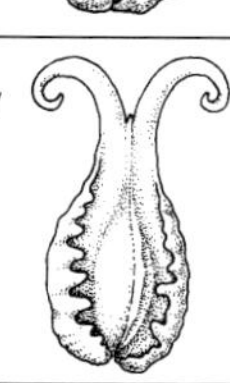	***G. biuncinatum*** **264**, 572

GLOBULARIACEAE

1	Woody-based **herb** or small **shrub**; **leaves** entire, alternate; **flowers** white, in a dense spike; **corolla** 2-lipped	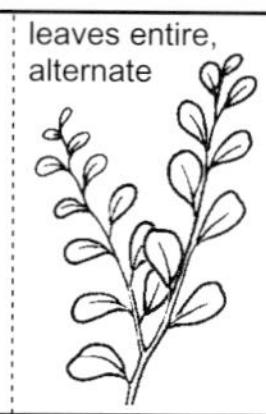	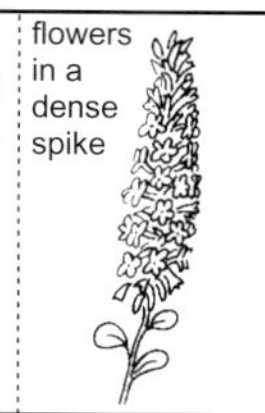	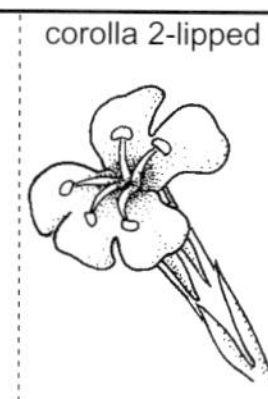	***Poskea socotrana*** **264**, 573

GOODENIACEAE

1	Pendulous **shrub**; **leaves** elliptic to obovate, slightly fleshy; **corolla** white, c.18mm long, tubular with 5 lobes, split down one side	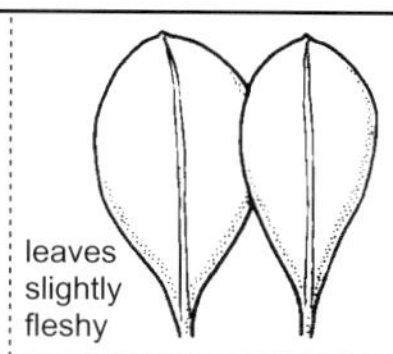	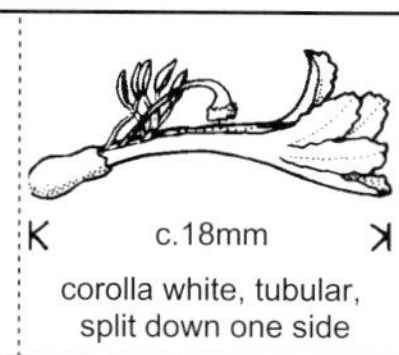	***Scaevola socotraensis*** Endemic **264**, 574

GUTTIFERAE

Hypericum

1	**Flowers** 1--1.5cm across; **styles** 3, free, outcurving or spreading	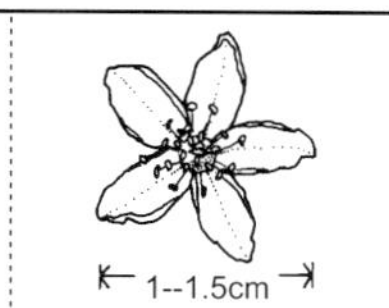	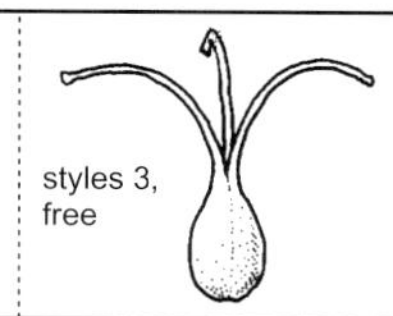	2
	Flowers (2.5--)4--7.5cm across; **styles** 5, wholly or partly fused	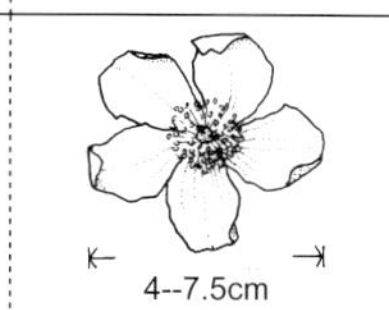	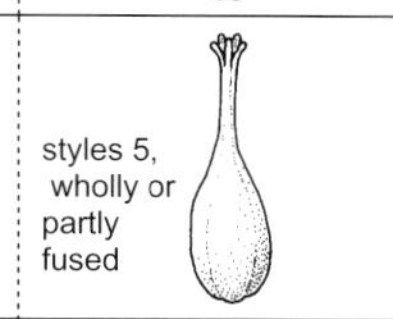	4
2	**Stems** and **leaves** pubescent or puberulous	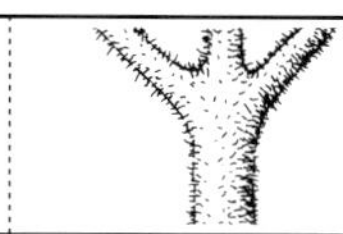		***H. fierense*** Endemic 575
	Stems and **leaves** glabrous	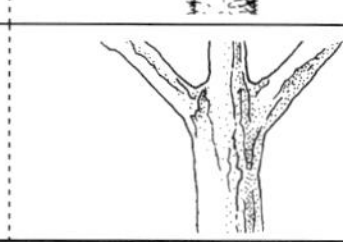		3
3	**Flowers** 1(or 2), terminal, and in the axils of older leaves; **habit** erect; **leaves** sessile	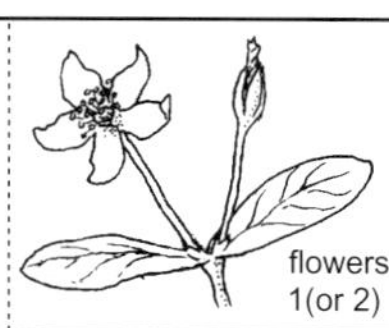	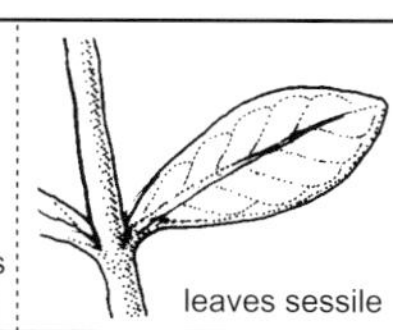	***H. scopulorum*** Endemic **265**, 575
	Flowers 7 to 13 in terminal subumbellate inflorescence; **habit** spreading or straggling; **leaves** petiolate	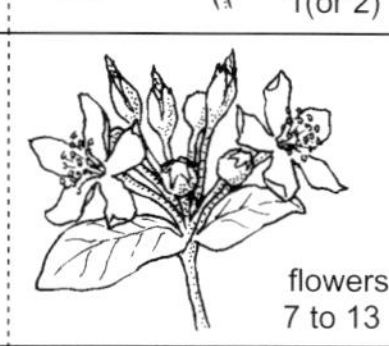	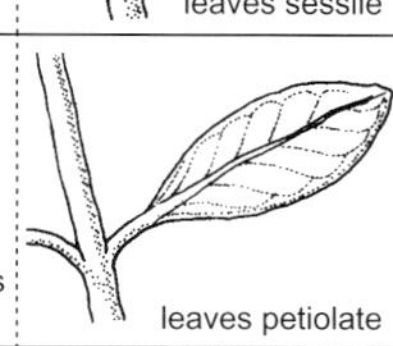	***H. tortuosum*** Endemic **265**, 576
4	**Leaves** narrowly elliptic or lanceolate, acute; **leaf venation** parallel with 3 to 7 main veins towards the base, becoming pinnate along the midrib towards the tip	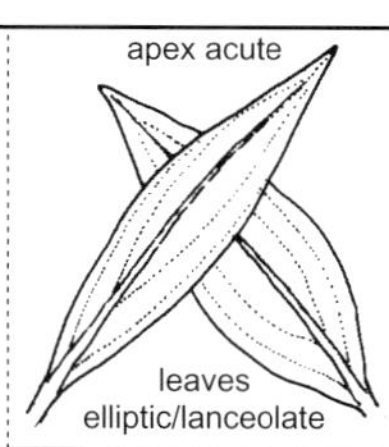	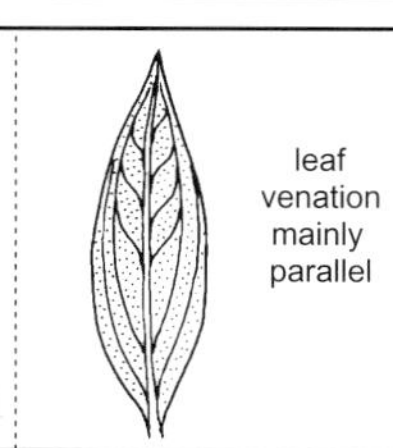	***H. balfourii*** Endemic **265**, 575
	Leaves elliptic to obovate or subcircular; **leaf venation** mainly pinnate with 1 or 2 pairs of lateral veins running around the margin to the tip	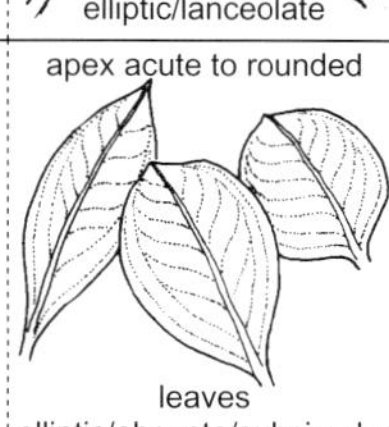	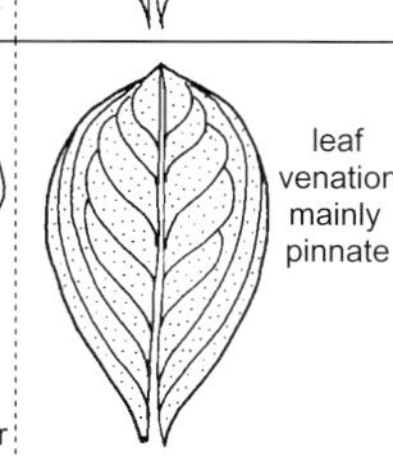	5

5

Leaves elliptic to oblanceolate, 5--10mm wide, acute to obtuse; **pedicels** 1--6mm long; **flowers** solitary

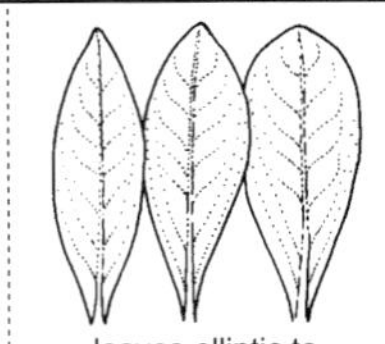

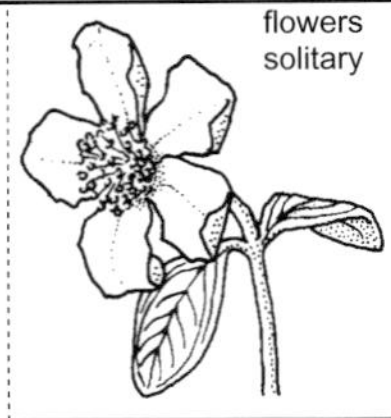

H. socotranum subsp. ***socotranum***
Endemic
576

Leaves obovate to subcircular, 8--22mm wide, apiculate to rounded; **pedicels** 5--25mm long; **flowers** 1 to 5

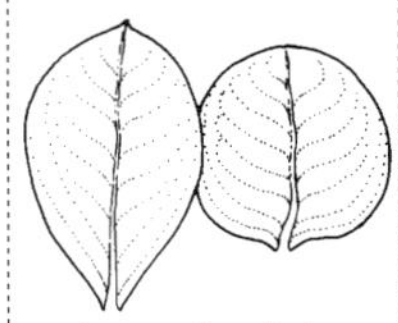

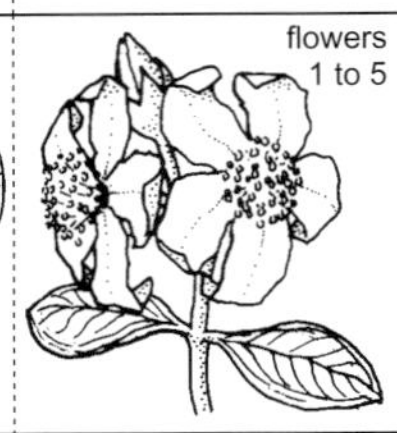

H. socotranum subsp. ***smithii***
Endemic
265, 576

LABIATAE

1

Calyx ± zygomorphic (2-lipped)

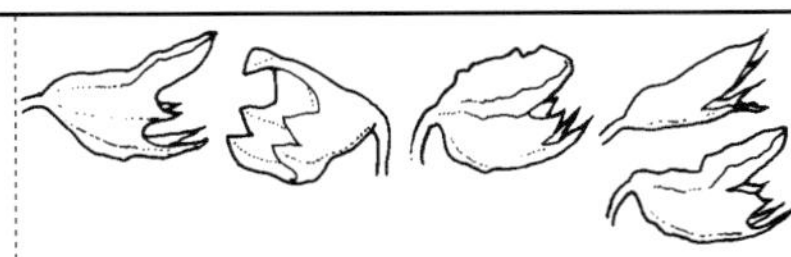

→ 2

Calyx ± actinomorphic

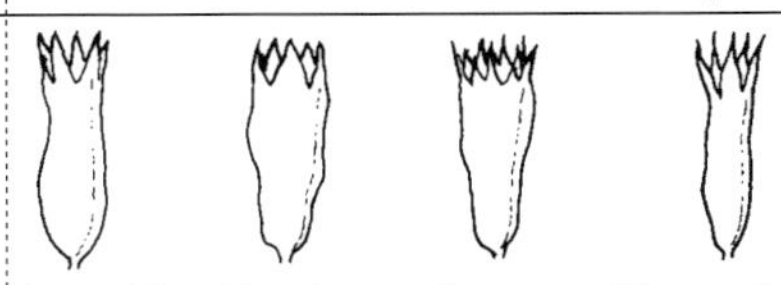

→ 5

2

Corolla with 4 or 5 subequal lobes, not distinctly 2-lipped; **stamens** included in the corolla tube; **calyx** with tuft of white hairs

→ ***Endostemon tenuiflorus***
265, 577

Corolla distinctly 2-lipped; **stamens** exserted from the corolla tube; **calyx** without tuft of white hairs

→ 3

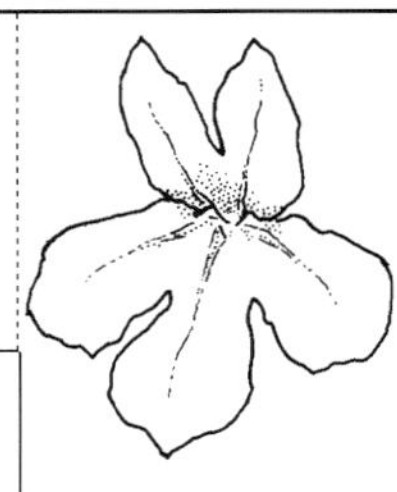

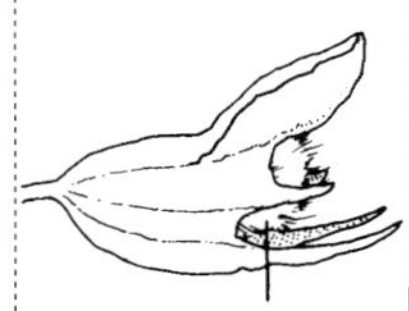

3

Succulent **herb** or **shrub**; **corolla** lilac to purple, tube curved or bent down at the mouth of the calyx

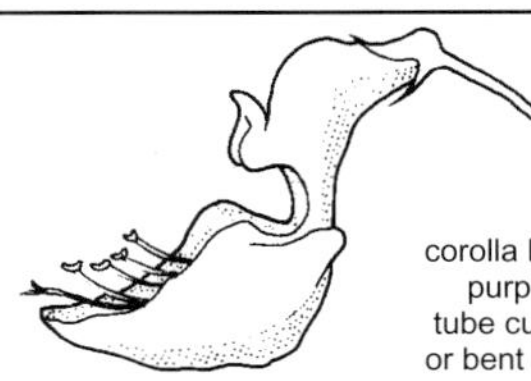

Plectranthus socotranus
Endemic
584

Non-succulent **herb** or **shrub**; **corolla** white to pink, tube straight

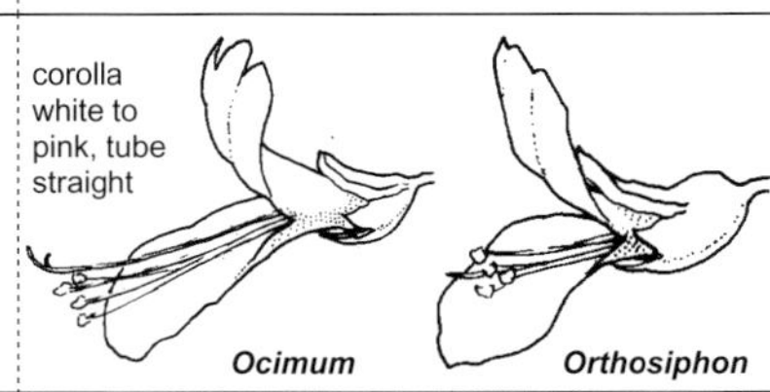

→ 4

4

Stamens exceeding the corolla lip; **style** 2-lobed at the tip

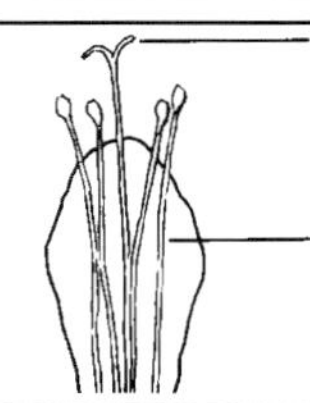

Ocimum forskolei
583

Stamens shorter than the corolla lip; **style** undivided at the tip

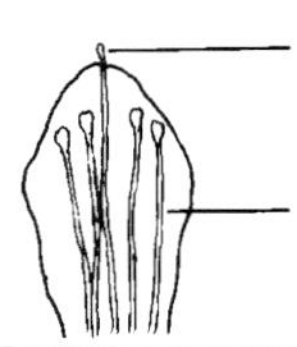

Orthosiphon
(2 spp.)

5

Corolla subequally 5-lobed; **leaves** deeply dissected — ***Lavandula nimmoi*** Endemic **265**, 577

Corolla 1 or 2-lipped; **leaves** simple — 6

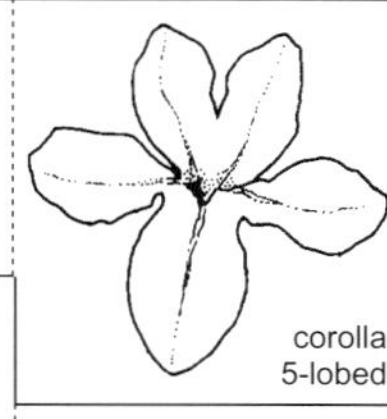

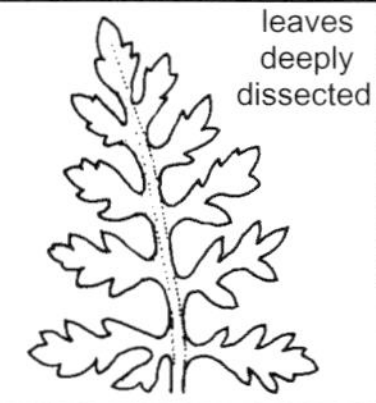

6

Corolla 1-lipped; **flowers** in terminal subcapitate inflorescences — ***Teucrium*** (2 spp.)

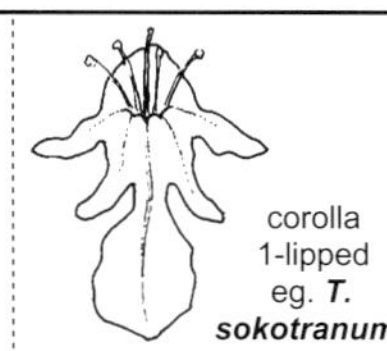

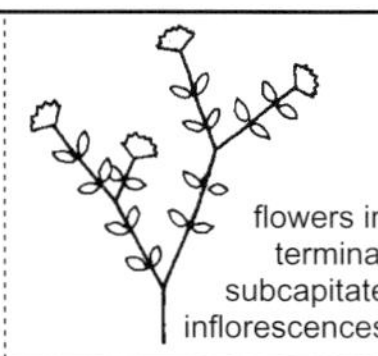

Corolla 2-lipped; **flowers** in distinct clusters along the stems — 7

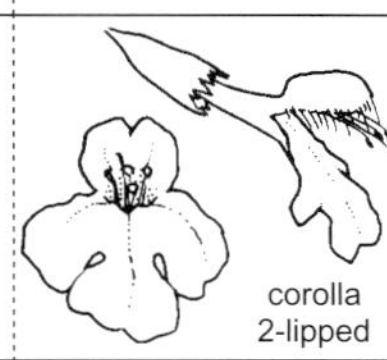

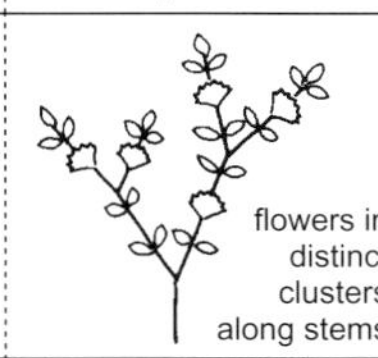

7

Upper corolla lip distinctly fringed with white hairs; **style** very unequally 2-lobed at tip; **flowers** large — ***Leucas*** (10 spp.)

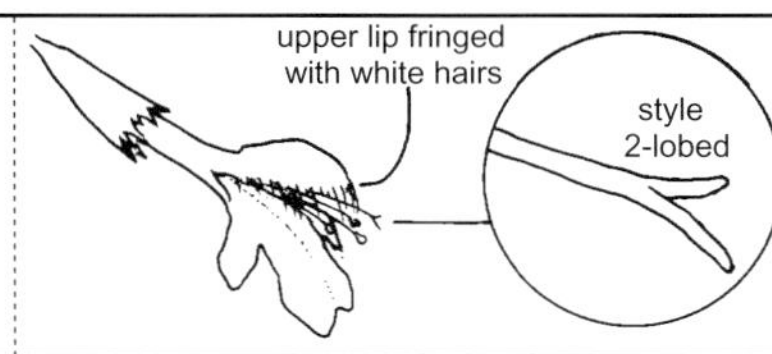

Upper corolla lip glabrous or glabrescent; **style** undivided at tip; **flowers** minute — ***Micromeria remota*** Endemic **266**, 582

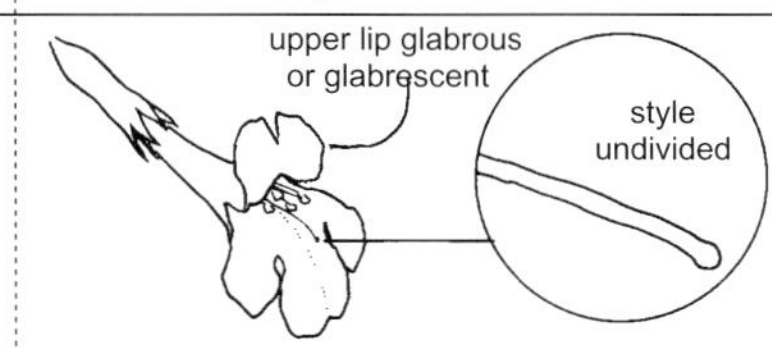

Leucas

1

Erect annual **herbs** with simple or little-branched stems; **flowers** many in axillary whorls (usually at least 10) — 2

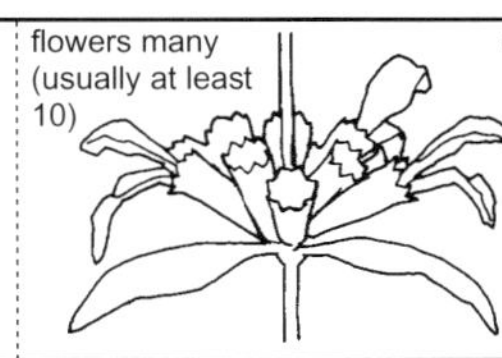

Prostrate or erect perennial **herbs** and **shrubs** with much-branched stems; **flowers** 1 to 5 in the axils of leaf-like bracts — 3

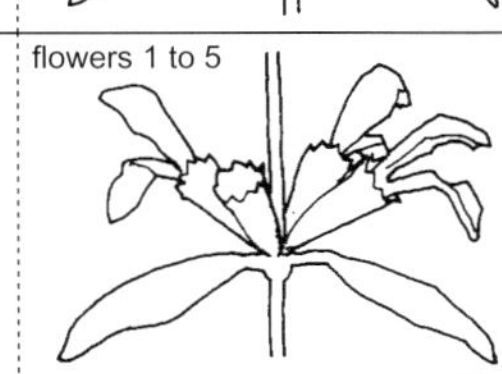

2

Leaves narrowly obovate to spatulate; **bracts** minute; **calyx** straight — ***L. neuflizeana*** 580

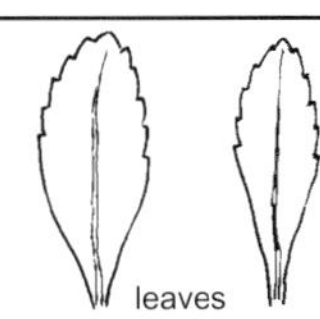

Leaves ovate to elliptic; **bracts** conspicuous, 1-veined, linear; **calyx** bent — ***L. martinicensis*** 579

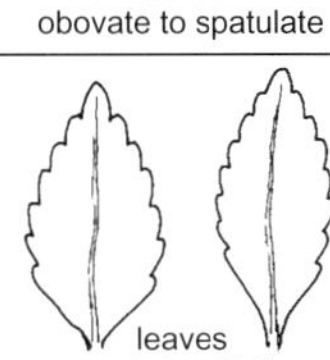

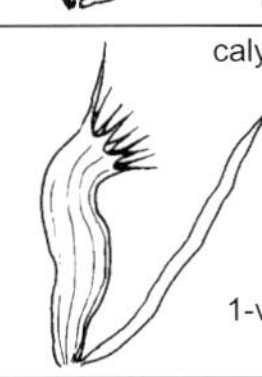

Leaves ovate to elliptic; **bracts** conspicuous, 3-veined, linear-lanceolate; **calyx** straight — ***L. urticaefolia*** 581

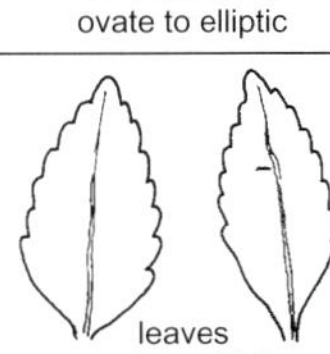

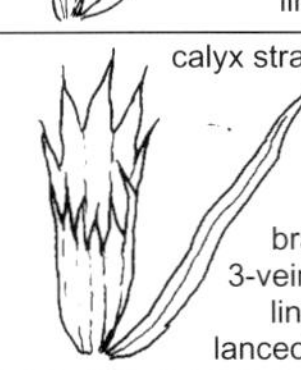

3	**Leaves** spinose	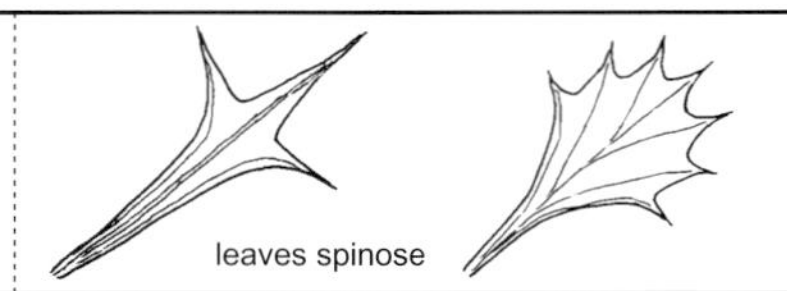	4
	Leaves entire or crenate	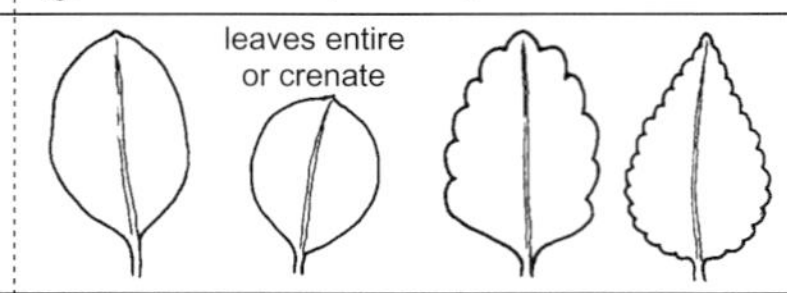	5
4	**Leaves** oblanceolate; **stems** glabrous	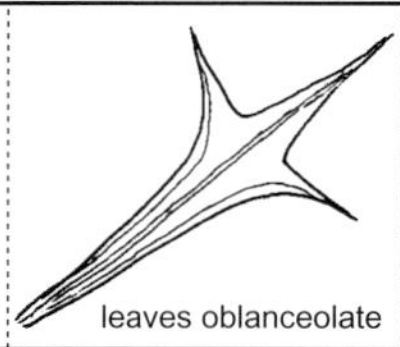 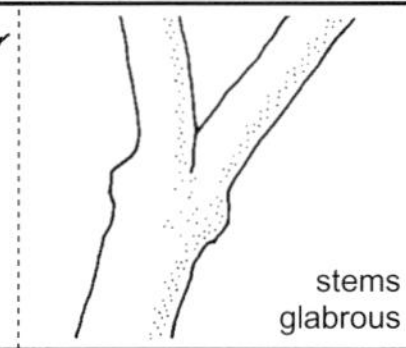	***L. spiculifera*** Endemic **266**, 580
	Leaves obovate; **stems** velutinous	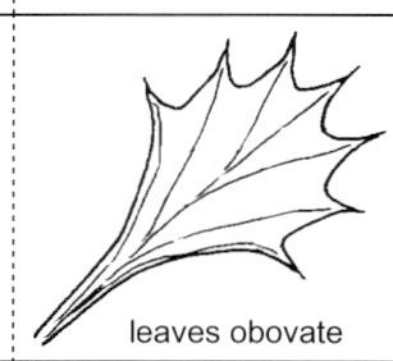 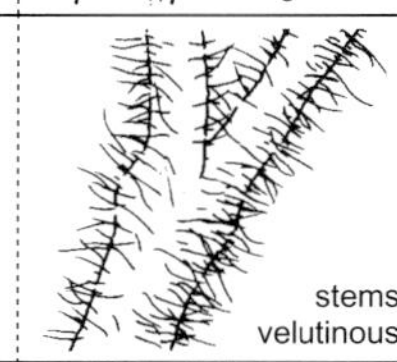	***L. flagellifolia*** Endemic **266**, 578
5	**Calyx** equally 10-toothed	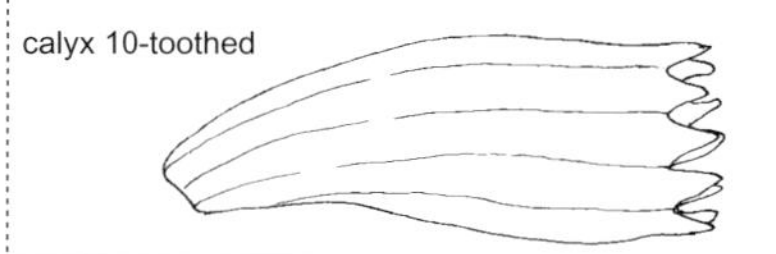	6
	Calyx with 5 large teeth alternating with 5 small teeth	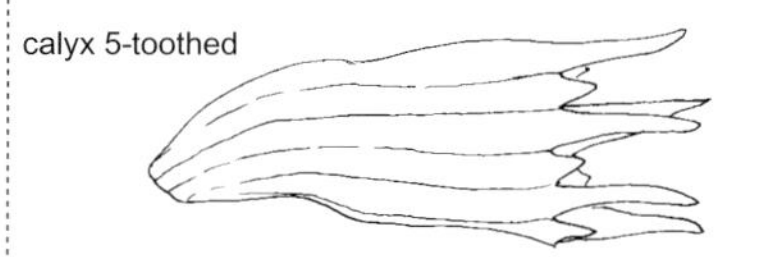	7
6	**Leaves** elliptic to orbicular, entire, <1.5cm long; **flowers** solitary or few in leaf axils	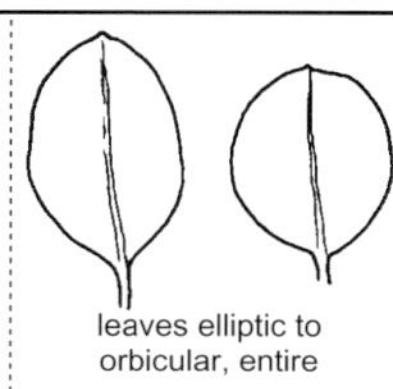 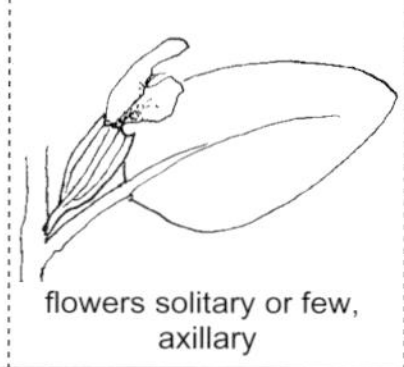	***L. virgata*** Endemic **266**, 581
	Leaves orbicular to kidney-shaped, entire, <1cm long; **flowers** 3 or 4 in subterminal clusters	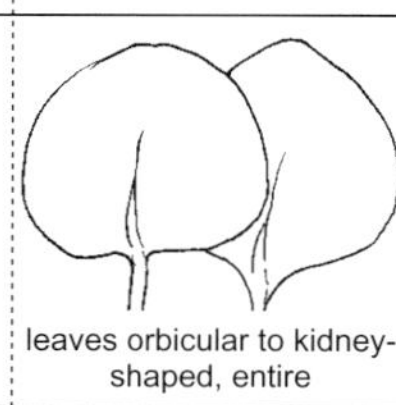 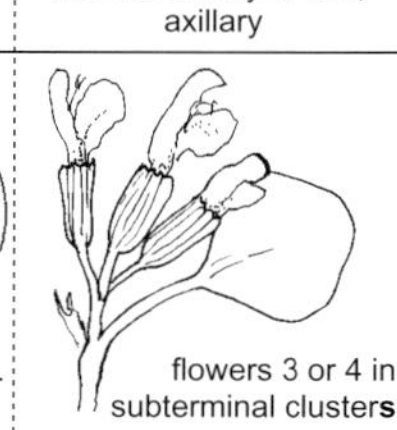	***L. samhaensis*** Endemic 580
	Leaves elliptic to orbicular, coarsely crenate, >2cm long; **flowers** few in leaf axils	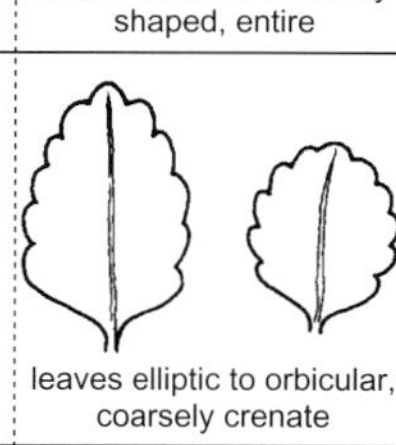 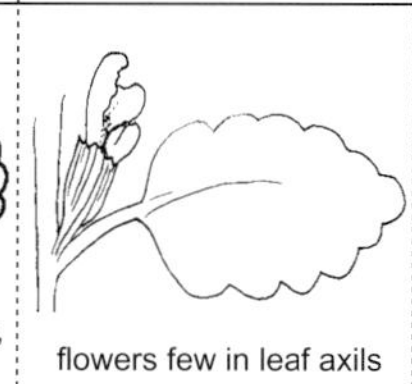	***L. kishenensis*** Endemic **266**, 579
7	**Leaves** elliptic to orbicular, entire, less than 1.5cm long	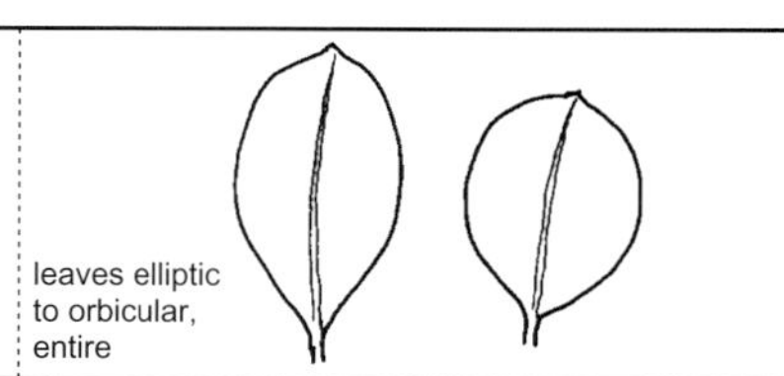	***L. penduliflora*** Endemic 580
	Leaves ovate, crenate, more than 2cm long	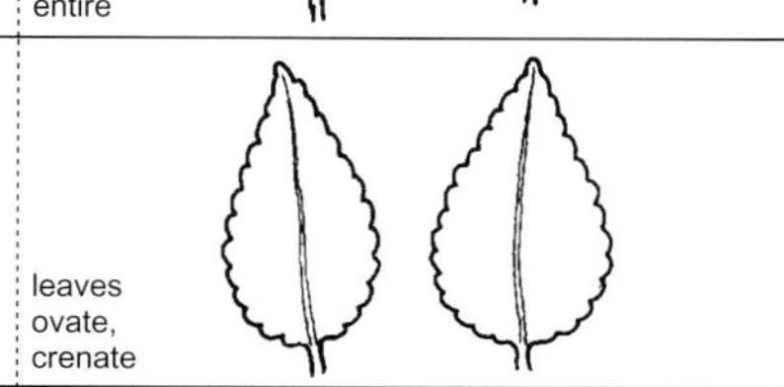	***L. hagghierensis*** Endemic **266**, 579

Orthosiphon

1	**Flowers** solitary or in short racemes; **calyx** weakly 2-lipped; **corolla** pink, c.10mm, tube long exserted	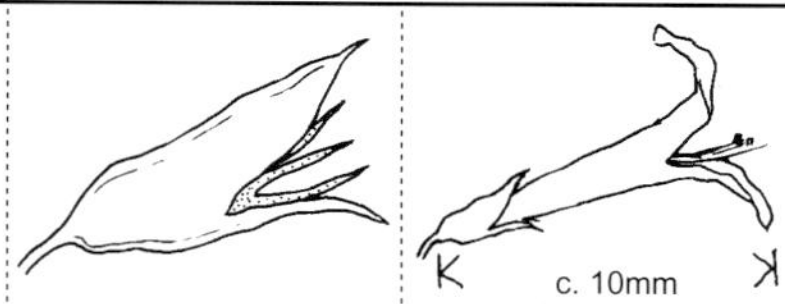	***O. ferrugineus*** Endemic **266**, 584
	Flowers in a terminal spike-like racemes; **calyx** strongly 2-lipped; **corolla** white, c.5mm, tube shortly exserted	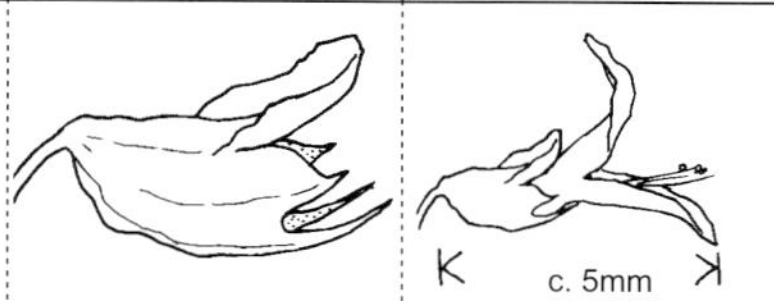	***O. pallidus*** 584

Teucrium

1	**Leaves** linear with revolute margins, <0.5cm long		***T. balfourii*** Endemic **266**, 585
	Leaves elliptic with crenate margins, <1.5cm long	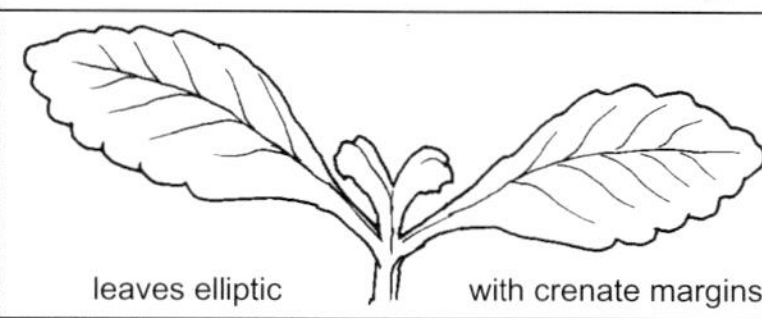	***T. sokotranum*** Endemic **267**, 586

LEGUMINOSAE

1	**Flowers** actinomorphic, very small, in spikes or heads; **petals** insignificant, **leaves** bi-pinnate	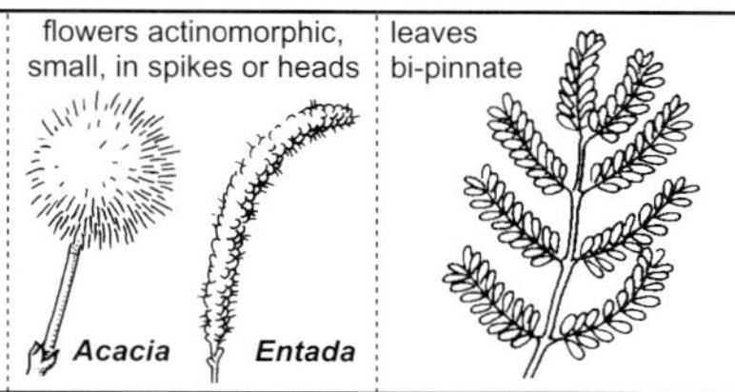	**MIMOSOIDEAE** (subfamily)
	Flowers zygomorphic, usually larger, rarely in spikes or heads, **petals** conspicuous, **leaves** simple, trifoliolate or pinnate	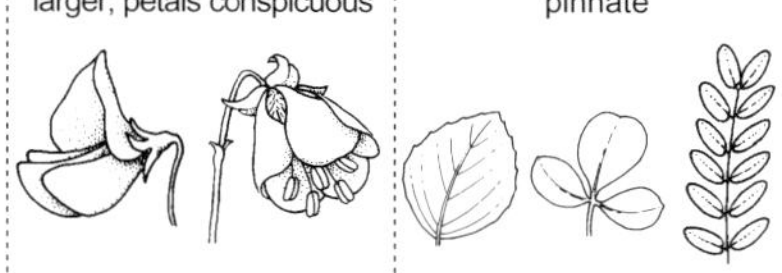	2

2	**Flowers** pea-shaped, consisting of a standard, keel and wings; **sepals** fused; **stamens** variously fused	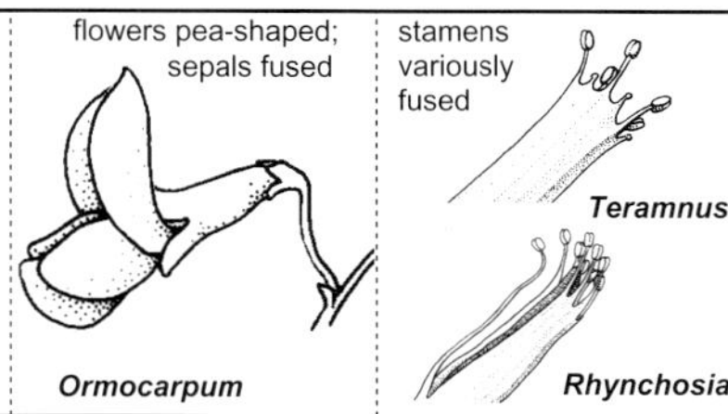	**PAPILIONOIDEAE** (subfamily)
	Flowers not pea-shaped; **sepals** free; **stamens** free	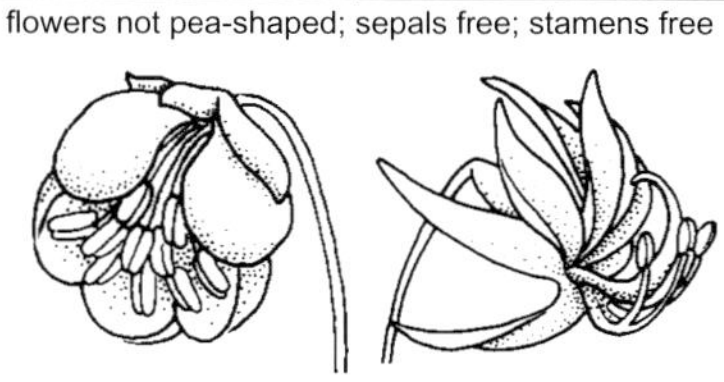	**CAESALPINIOIDEAE** (subfamily)

CAESALPINIOIDEAE

1	**Tree**; **petals** reddish brown, 3 large and 2 minute; fertile **stamens** 3; **pods** thick and pulpy	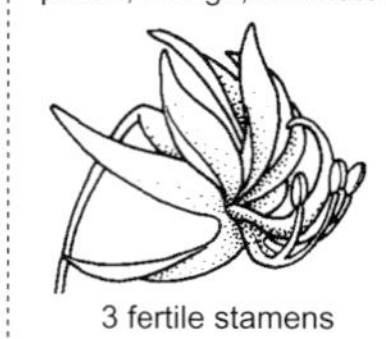	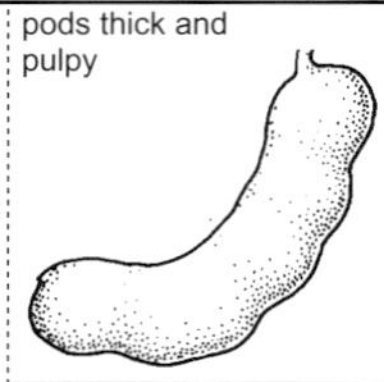	***Tamarindus indica*** 589
	Herb or **shrub; petals** yellow, 5; fertile **stamens** 5 or 10; **pods** terete or compressed, dehiscent or indhiscent	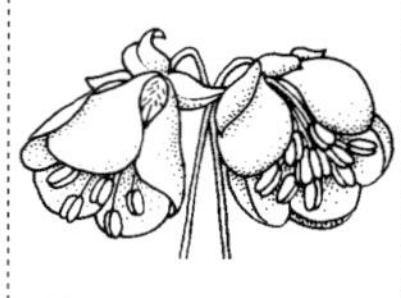	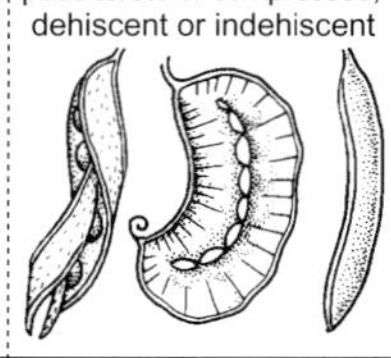	2
2	**Stamens** 10; **pedicels** without bracteoles; **pods** indehiscent or dehiscing with difficulty, never with twisted valves	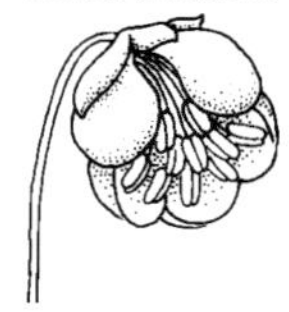	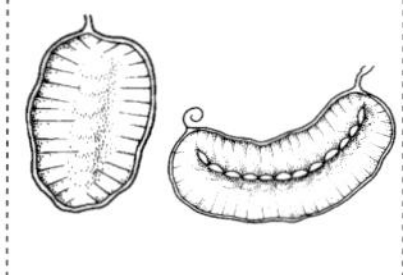	***Senna*** (4 spp.)
	Stamens 5; **pedicels** with 2 bracteoles; **pod** elastically dehiscent with twisting valves	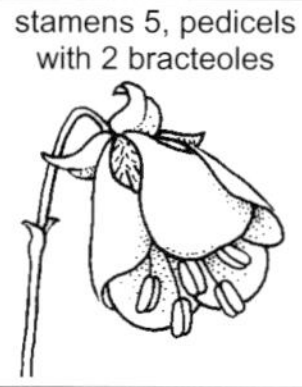	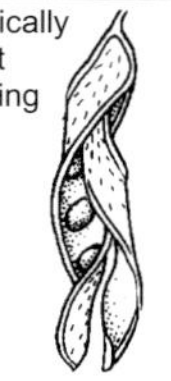	***Chamaecrista absus*** 586

Senna

1	**Pods** linear, ± round in section; **leaf petiole** with conspicuous glands	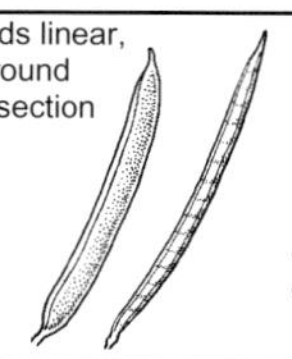	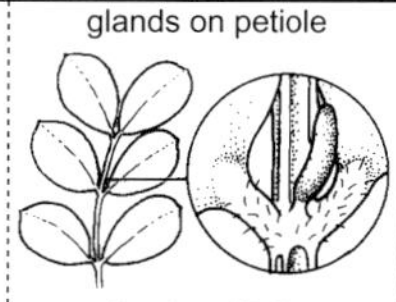	2
	Pods oblong, flattened; **leaves** without conspicuous glands	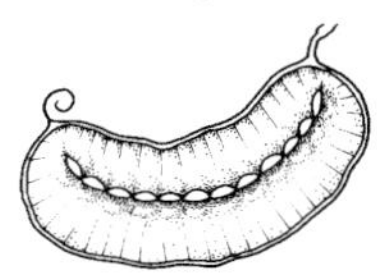	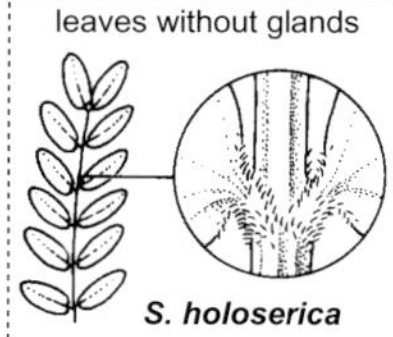	3
2	**Leaflets** obovate, obtuse; **glands** positioned on rhachis between some pairs of leaflets	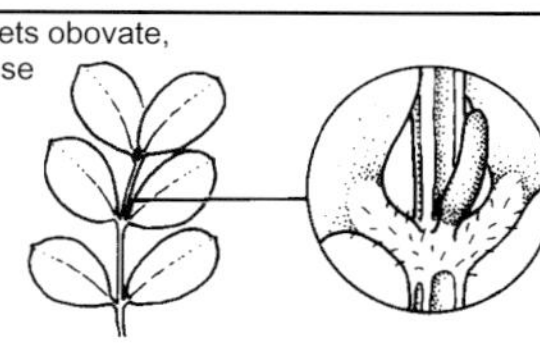		***S. obtusifolia*** 588
	Leaflets lanceolate, acute; **gland** positioned near base of petiole	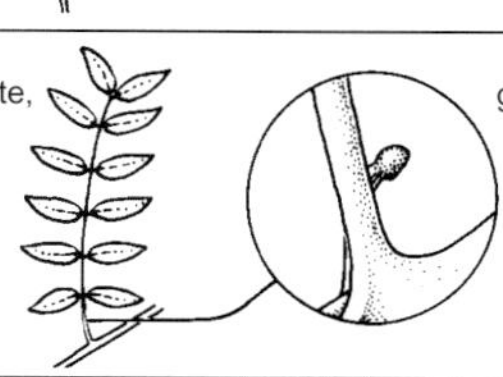		***S. socotrana*** Endemic **269**, 588
3	**Pods** with a crest running along middle	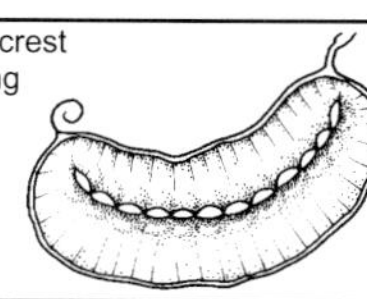		***S. italica*** 588
	Pods without crest	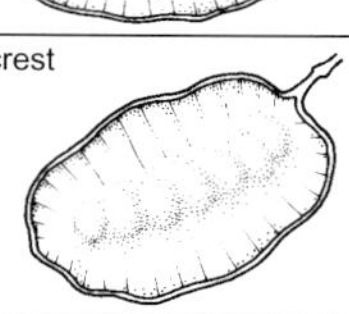		***S. holosericea*** **269**, 587

MIMOSOIDEAE

1	**Flowers** in heads, all similar; **branches** armed with prickles or stipular spines; **fruits** dehiscent or indehiscent	flowers in heads, all similar 	branches armed 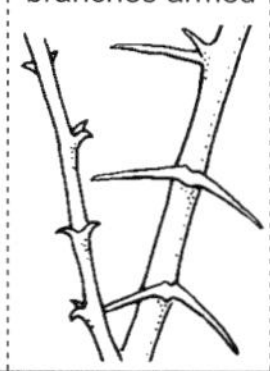	fruits dehiscent or indehiscent 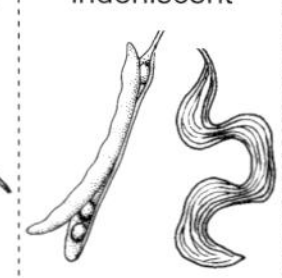	***Acacia*** (5 spp.)
	Flowers in spikes, the lower sterile, pink, the upper fertile, yellow, smaller; **branches** armed, the spines formed from short lateral twigs; **fruits** dehiscent	flowers in spikes, lower sterile, upper fertile 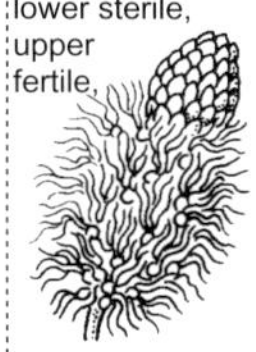	branches armed 	fruits dehiscent 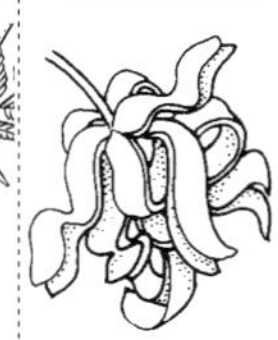	***Dichrostachys dehiscens*** Endemic **268**, 594
	Flowers in spikes, all similar; **branches** armed; **fruits** indehiscent	flowers in spikes, all similar 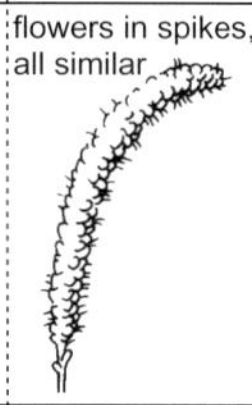	branches armed 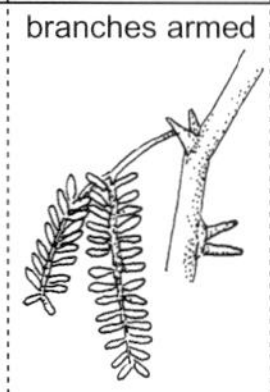	fruits indehiscent 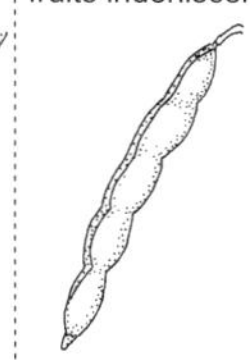	***Prosopis juliflora*** 595

Acacia

1	Spreading **shrub**; **stipular spines** straight, 1--6 cm; **pinnae** 4 to 12 pairs; **flowers** white; **pods** thick, woody ± straight, dehiscent	stipular spines straight 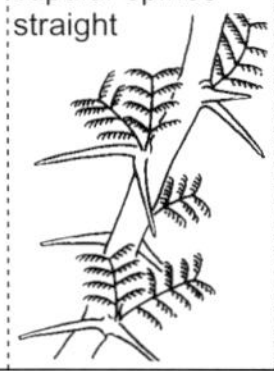	pinnae 4 to 12 pairs 	pod thick, woody, ± straight, dehiscent 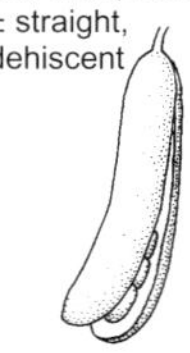	***A. edgeworthii*** **267**, 591
	Small **tree**; **stipular spines** very small or absent; **pinnae** 2 or 3 pairs; **flowers** white; **pods** not seen	stipular spines very small 	pinnae 2 or 3 pairs 		***A. pennivenia*** Endemic **267**, 592
	Shrub; **stipular spines** straight, <3.5cm; **pinnae** 1 or 2 pairs; **leaflets** fleshy; **flowers** cream; **pods** straight, constricted, dehiscent	stipular spines straight 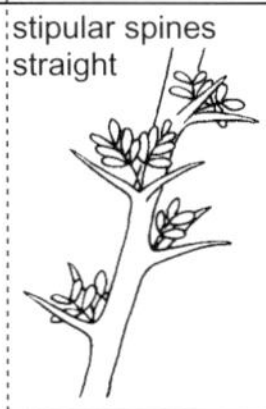	pinnae 1 or 2 pairs, leaflets fleshy 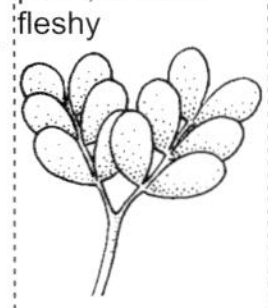	pods straight, constricted, dehiscent 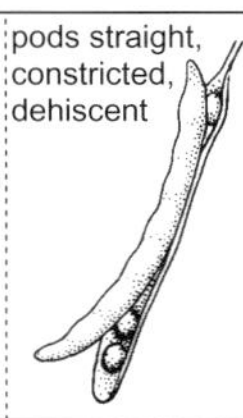	***A. sarcophylla*** **267**, 592
	Small **tree**; **stipular spines** short, recurved or long, straight; **pinnae** 2 to 10 pairs; **leaflets** very small; **flowers** cream; **pods** spirally contorted, slowly dehiscent or indehiscent	stipular spines recurved or straight 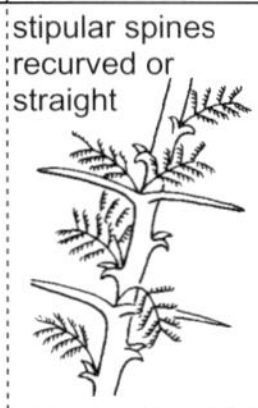	pinnae 2 to 10 pairs 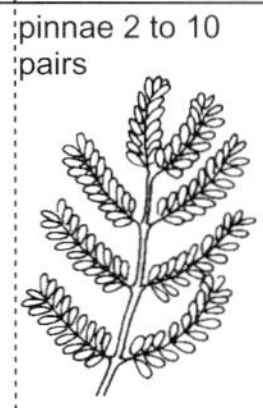	pods spirally contorted 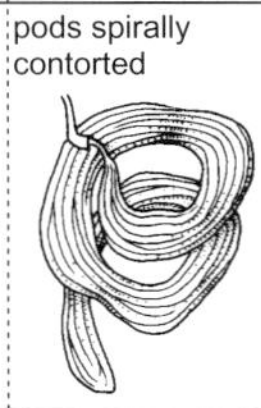	***A. aff. tortilis*** 593
	Large **tree**; **stipular spines** straight, up to 8cm long; **pinnae** 2 to 4 pairs; **flowers** yellow; **pods** indehiscent	stipular spines straight 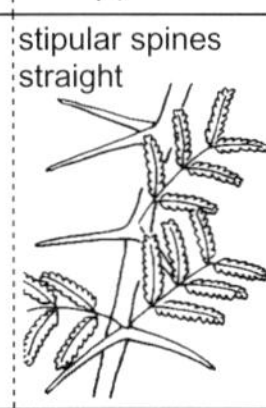	pinnae 2 to 4 pairs 		***A. nilotica*** 591
	Low **shrub**; **stipular spines** <0.4cm long; **pinnae** 2 or 3 pairs; **leaflets** very small with conspicuous midrib; **flowers** and **pods** not seen	stipular spines small 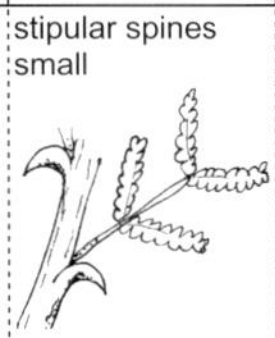	pinnae 2 or 3 pairs 		***A. sp. A*** 593

PAPILIONOIDEAE

1	**Leaves** simple (upper leaves on plant sometimes trifoliolate)	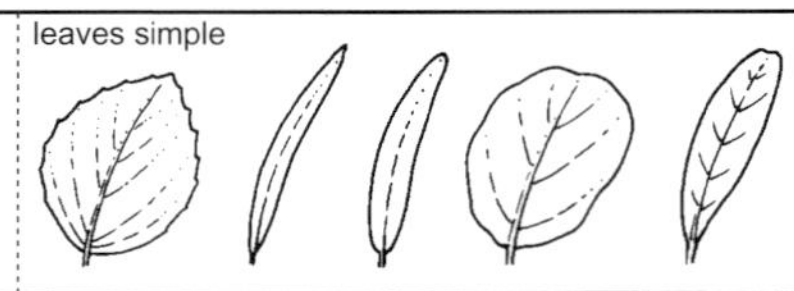	2
	Leaves with 2 leaflets; fruit constricted between seeds, hispid	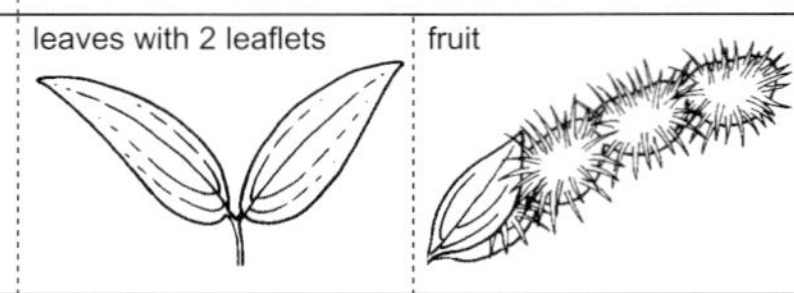	***Zornia glochidiata*** 616
	Leaves trifoliolate	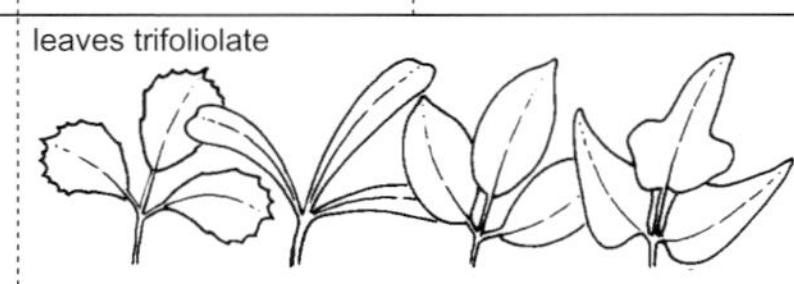	7
	Leaves pinnate with more than 3 leaflets		18
2	**Leaves** toothed; **fruit** 1-seeded, indehiscent	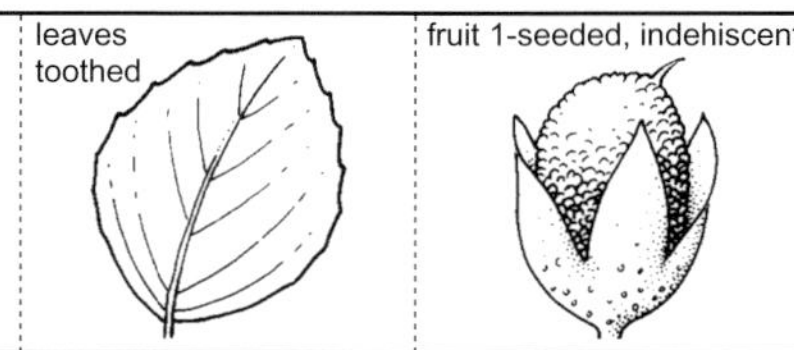	***Cullen corylifolia*** 601
	Leaves entire; **fruit** > 2-seeded, dehiscent or breaking transversely	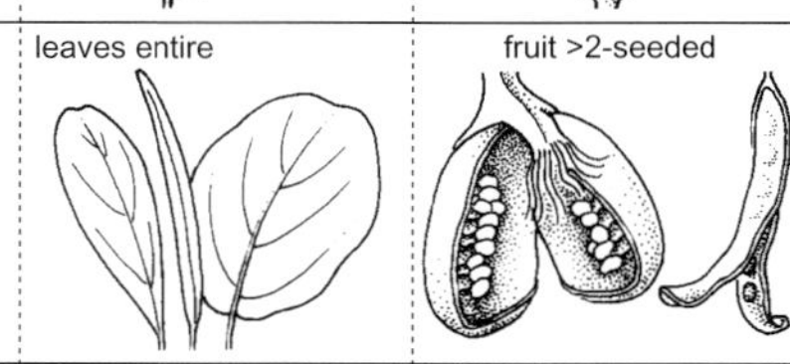	3
3	**Flowers** yellow; **fruits** inflated; **keel** strongly hooked	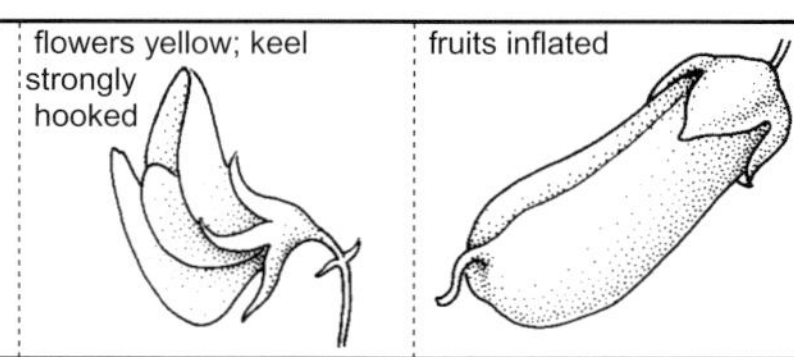	***Crotalaria*** (6 spp.)
	Flowers red or purple; **fruits** not inflated; **keel** rounded	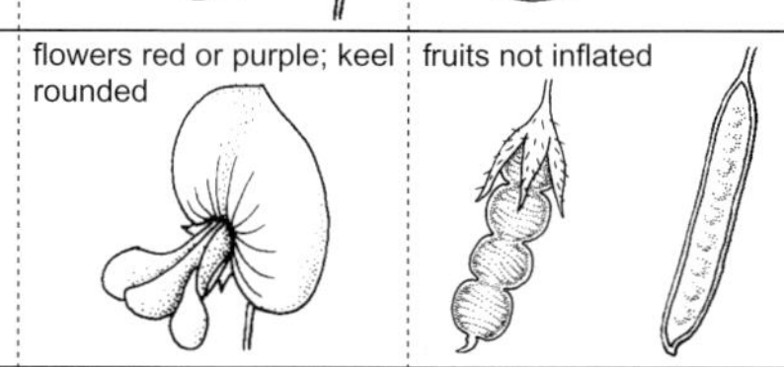	4
4	**Stipules** and **calyx** scarious; **pods** jointed	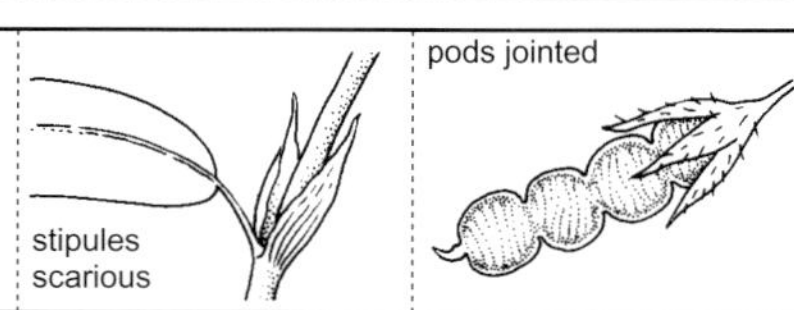	***Alysicarpus*** (2 spp.)
	Stipules and **calyx** not scarious; **pod** not jointed	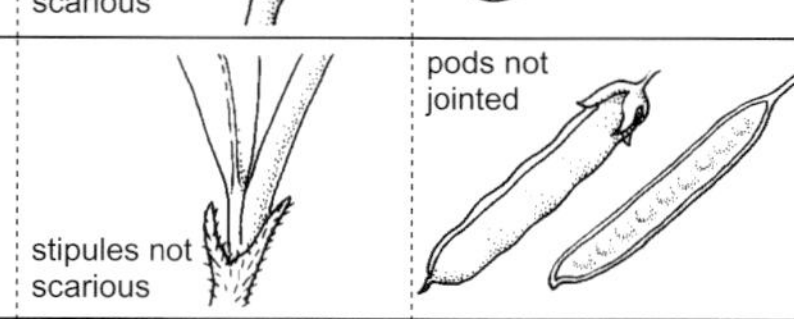	5
5	**Pods** rounded in section; **hairs** medifixed, parallel	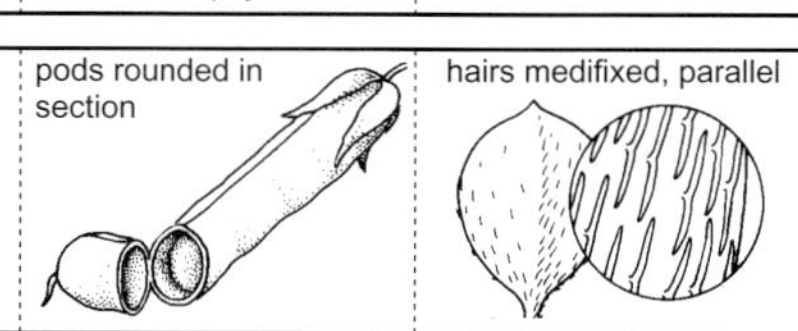	***Tephrosia*** (7 spp.)
	Pod flattened; **hairs** simple	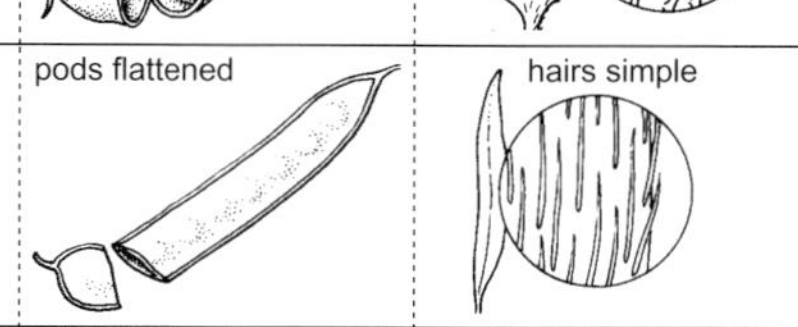	5

6	**Flowers** held on erect pedicels; **petals** hairy	flowers held on erect pedicels 	***Indigofera*** (15 spp.)
	Flowers held on bent pedicels; **petals** glabrous	flowers held on bent pedicels 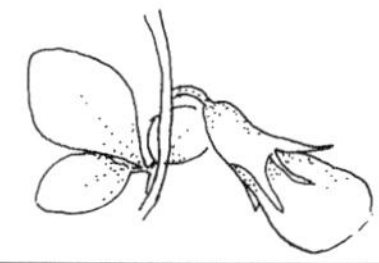	***Microcharis disjuncta*** 611
7	**Leaflets** toothed	leaflets toothed 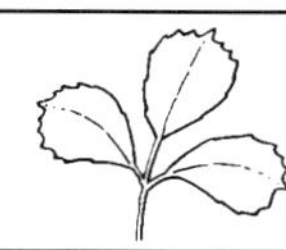	8
	Leaflets entire	leaflets entire 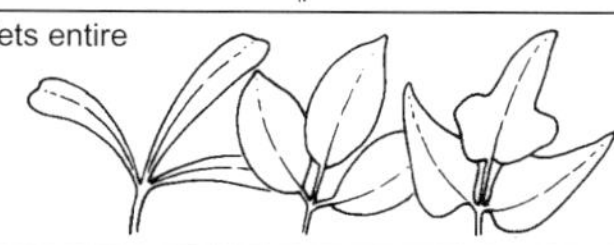	9
8	**Pod** spiralled	pod spiralled 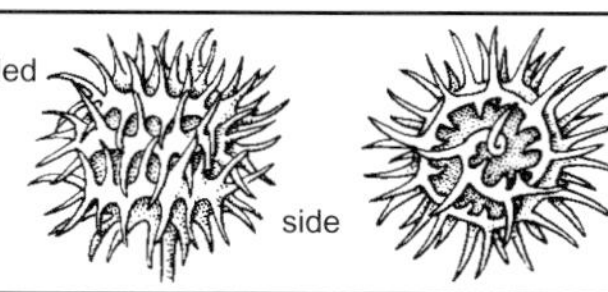	***Medicago*** (3 spp.)
	Pod ovoid	pod ovoid 	***Meliolotus indica*** 610
	Pod curved	pod curved 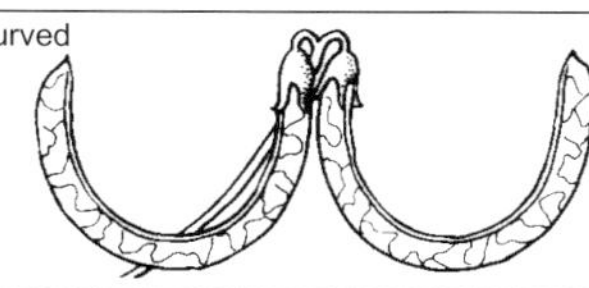	***Trigonella falcata*** Endemic 616
9	**Pods** jointed	pods jointed 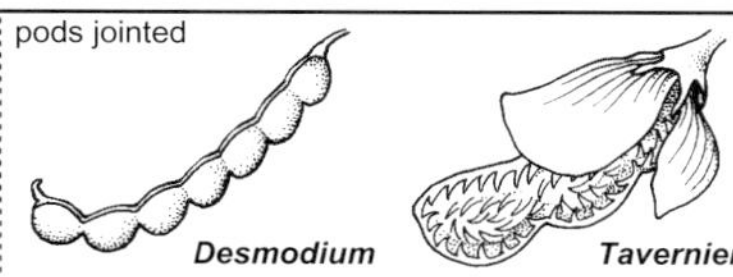	10
	Pods not jointed	pods not jointed 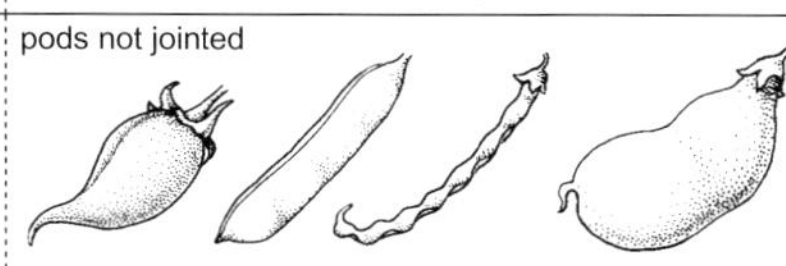	11
10	**Fruit** spiny, hidden by the calyx; **flowers** pink	fruit spiny, hidden by calyx 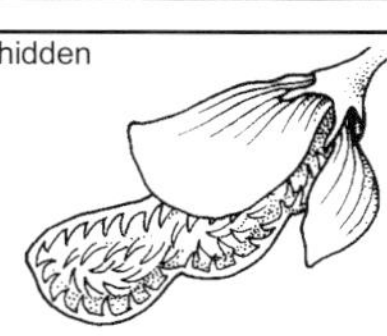	***Taverniera sericophylla*** Endemic **269**, 612
	Fruit not spiny or hidden; **flowers** blue	fruit not spiny, visible 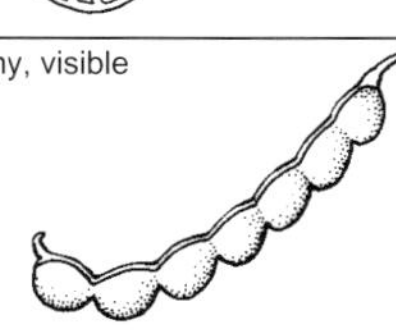	***Desmodium triflorum*** 601
11	**Tree,** the branches armed with curved prickles; **flowers** large (>4cm), red; **calyx** with ferruginous, stellate indumentum	tree; branches armed with prickles 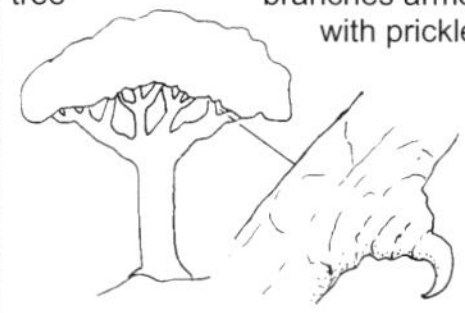 flowers large, red 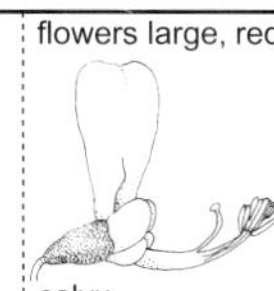 calyx ferruginous	***Erythrina melanacantha*** 602
	Herb, **shrub** or **climber**, unarmed; **flowers** usually smaller; **calyx** glabrous or with simple or 2-armed hairs	herb, shrub or climber 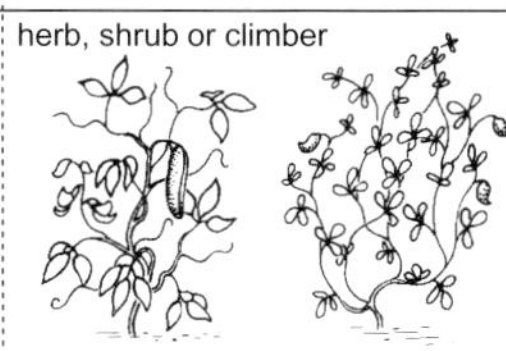 calyx glabrous or with simple hairs 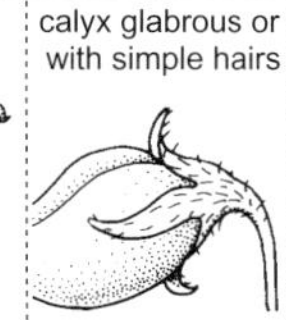	12

12

Plant with parallel, medifixed hairs; **anthers** apiculate; **flowers** usually bright red, sometimes pink

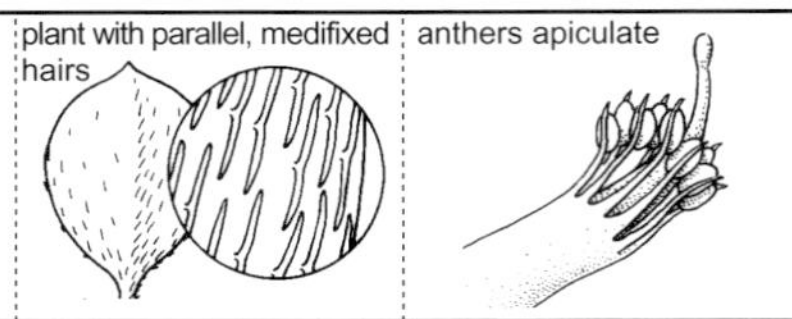

Indigofera (15 spp.)

Plant glabrous or with simple hairs only; **anthers** not apiculate; **flowers** yellow, mauve or someimes pink

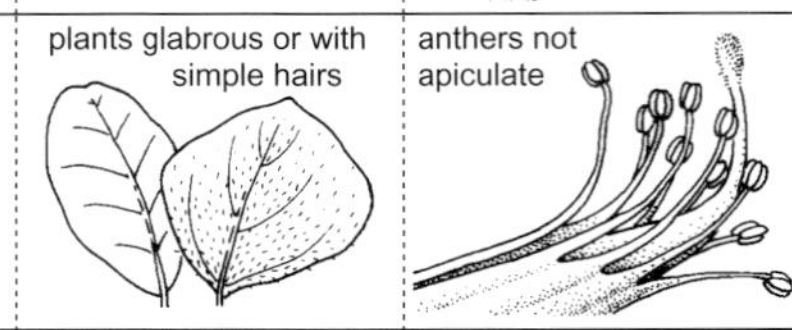

13

13

Herb, **shrub** or **tree**; **pods** inflated; **leaves** digitately 3-foliolate; **keel** beaked, strongly angled (hooked)

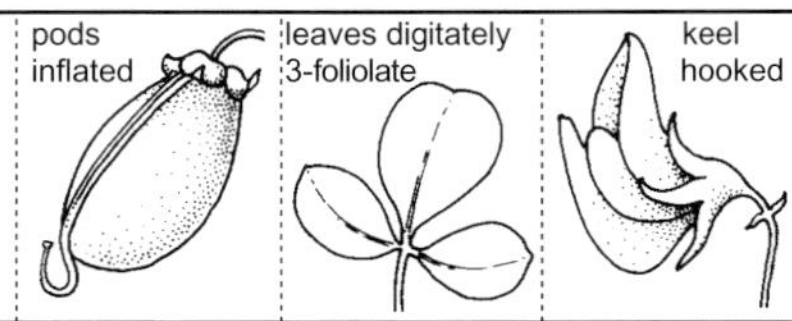

Crotalaria (6 spp.)

Twining **herb** or **shrub**; **pods** not usually inflated; **leaves** pinnately 3-foliolate; **keel** not hooked

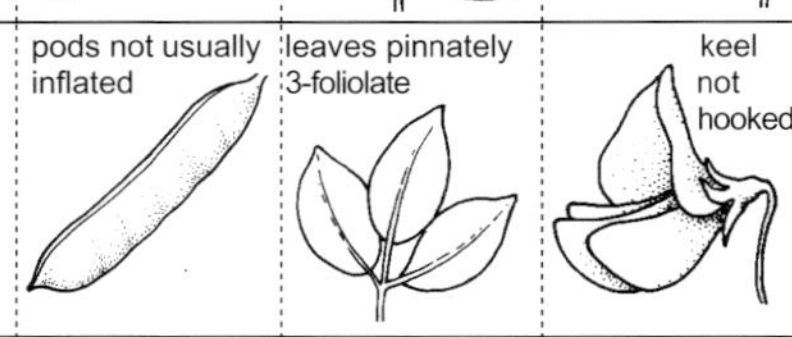

14

14

Vigorous **shrub** with twining or trailing branches; **calyx** scarious, strongly accrescent, the lower lobe boat-shaped; **fruit** short, enclosed within the calyx

Paracalyx balfourii Endemic **269**, 611

Twining **shrub** or **herb; calyx** not as above; **fruit** not enclosed in the calyx

15

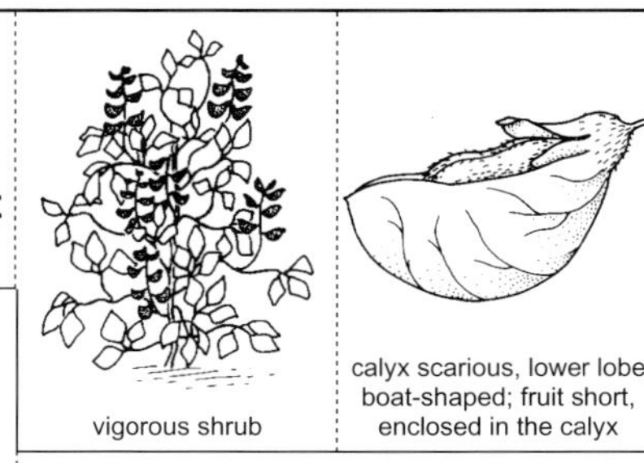

15

Robust **twiner**; **pods** large, >10cm long

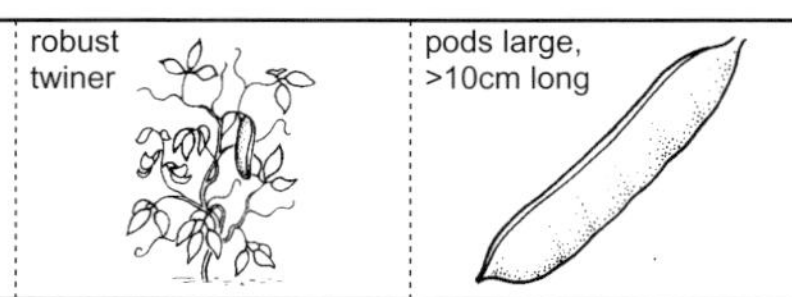

Canavalia africana 596

Herb; pods small <10cm long

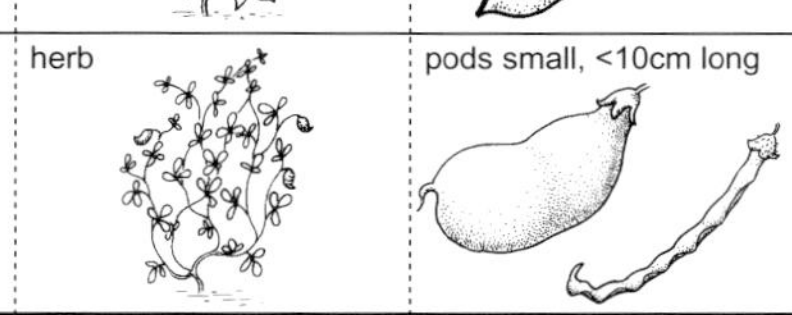

16

16

5 fertile **anthers** alternating with 5 sterile anthers; **pods** with an upturned beak

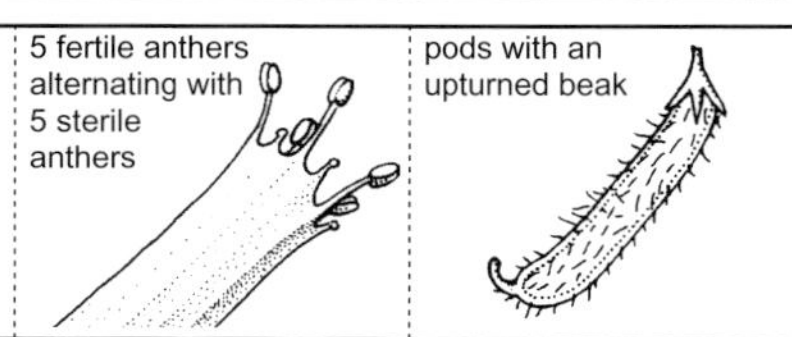

Teramnus repens 615

9 **stamens** fused together with 1 free; **pods** without an upturned beak

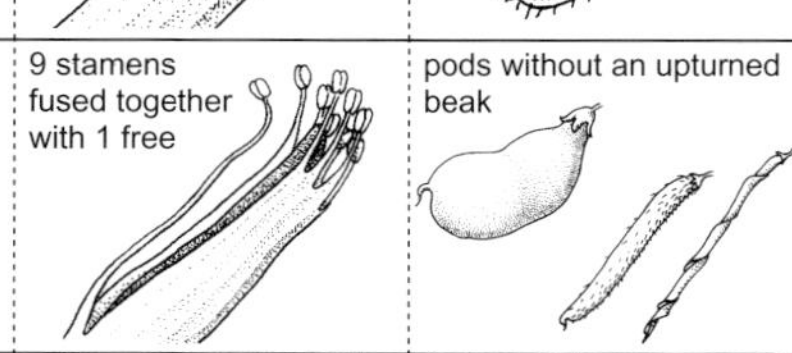

17

17

Leaves gland-dotted beneath; **pods** circular or shortly oblong, compressed

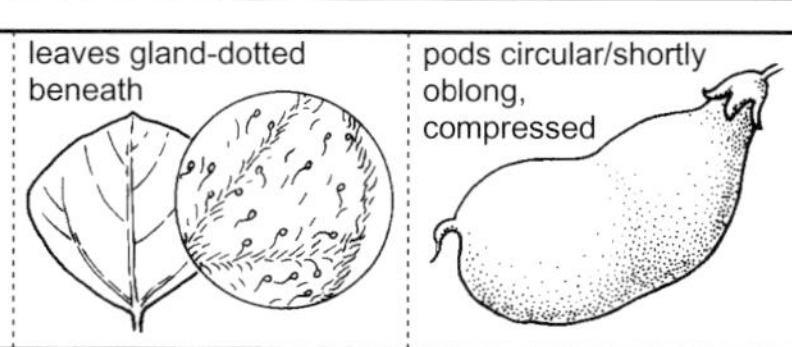

Rhynchosia (2 spp.)

Leaves not gland-dotted beneath; **pods** linear-cylindrical or flattened

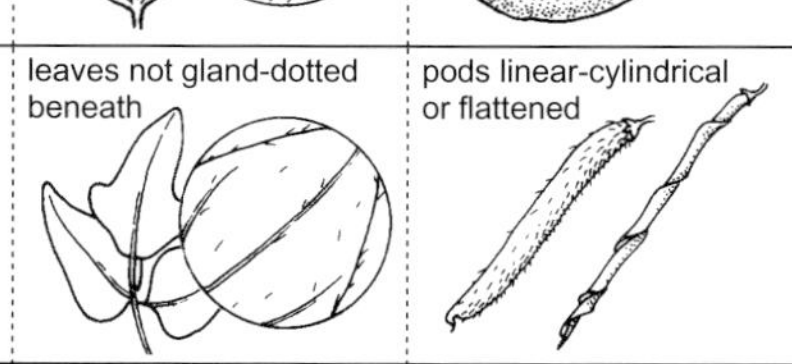

Vigna membranacea 616

18 **Plants** with parallel, medifixed hairs; **anthers** apiculate, **flowers** small, red — ***Indigofera*** (15 spp.)

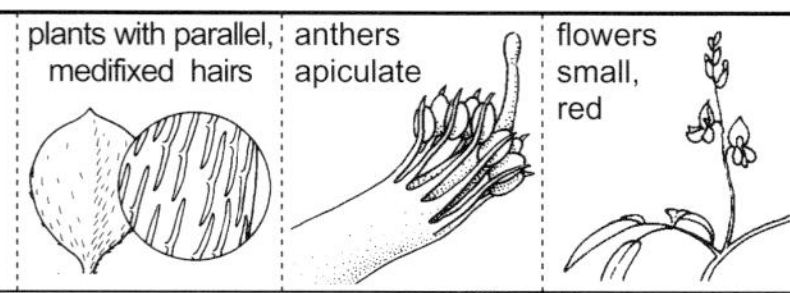

Plants glabrous or with simple hairs; **anthers** not apiculate, **flowers** variously coloured, usually large — 19

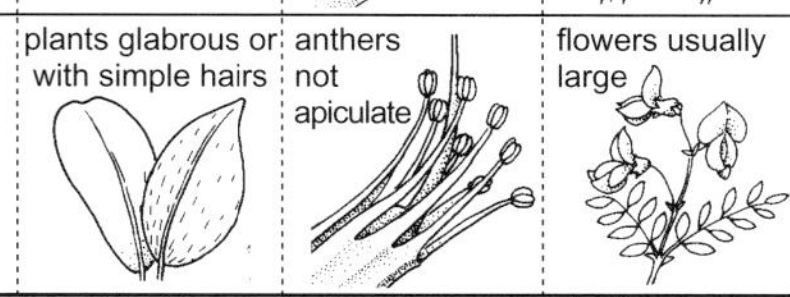

19 **Herb**; **fruits** dehiscent — 20

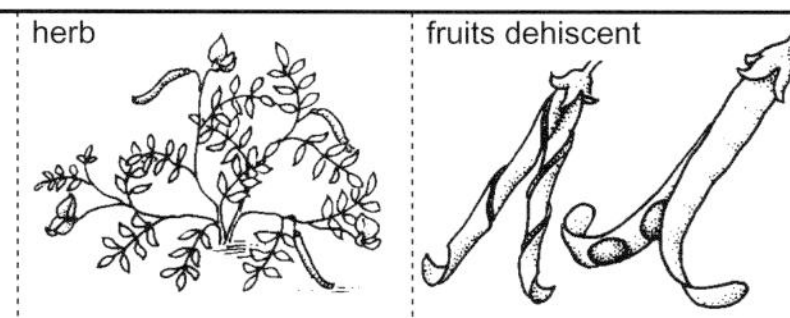

Shrub; **fruits** jointed, breaking transversely into 1-seeded segments — 21

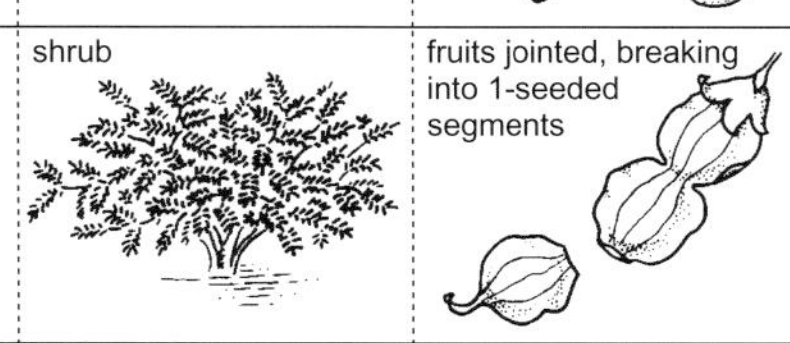

20 **Leaflets** 5 to 9, all similar; **pods** longitudinally grooved — ***Astragalus hauarensis*** 596

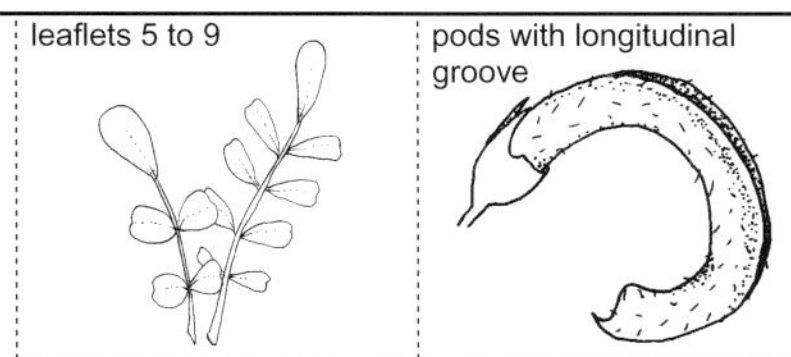

Leaflets 5, the lower 2 stipule-like, glabrous or hairy with spreading hairs; **pods** rounded — ***Lotus*** (4 spp.)

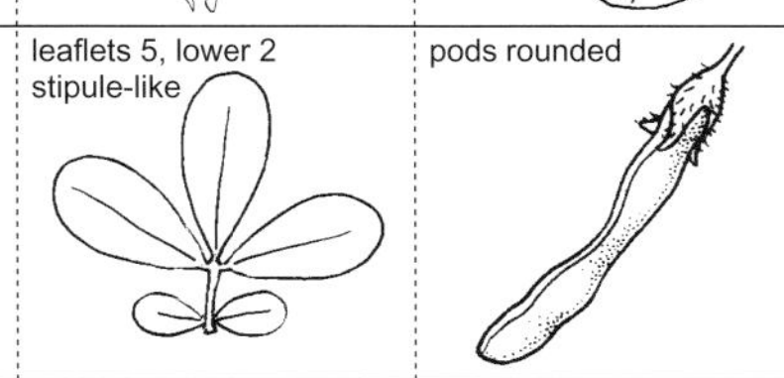

Leaflets 5 to many, all similar, silky hairy; **pods** narrowly oblong, flattened — ***Tephrosia*** (6 spp.)

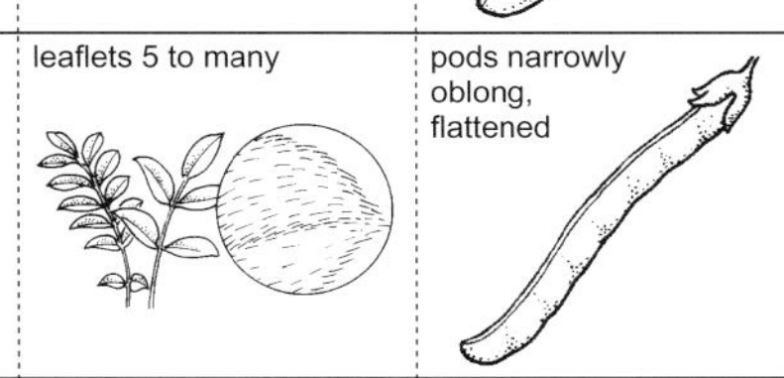

21 **Flowers** yellow, sessile; **leaflets** 3 to 5, obcordate-obovate — ***Chapmannia*** (4 spp.)

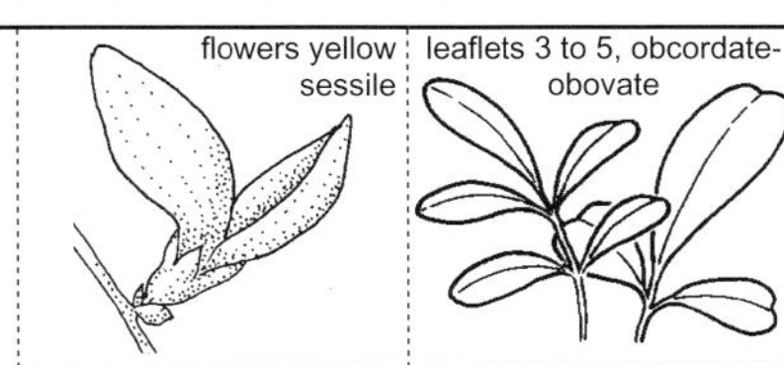

Flowers purple, pedicellate; **leaflets** 3 to 7, obovate-orbicular — ***Zygocarpum coeruleum*** Endemic **269**, 617

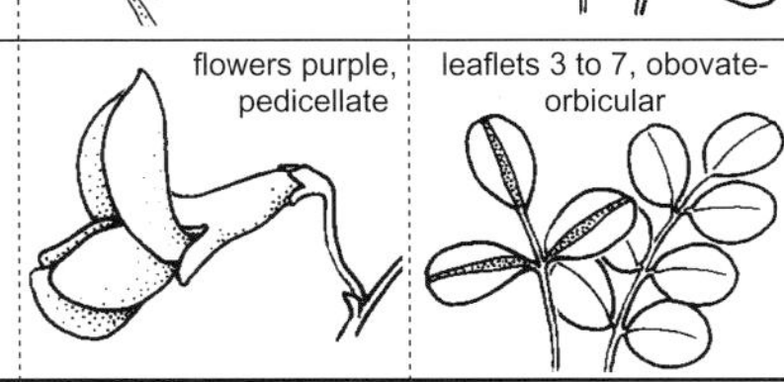

Alysicarpus

1 **Pods** not constricted between the segments; **stem** glabrescent — ***A. vaginalis*** 596

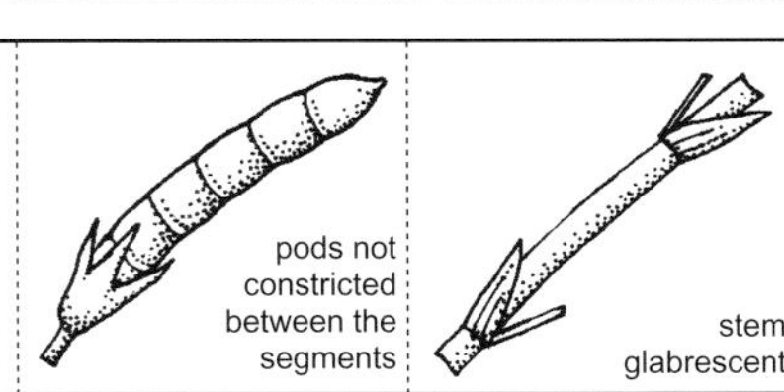

Pods constricted between the segments; **stem** appressed-pubescent — ***A. monilifer*** 595

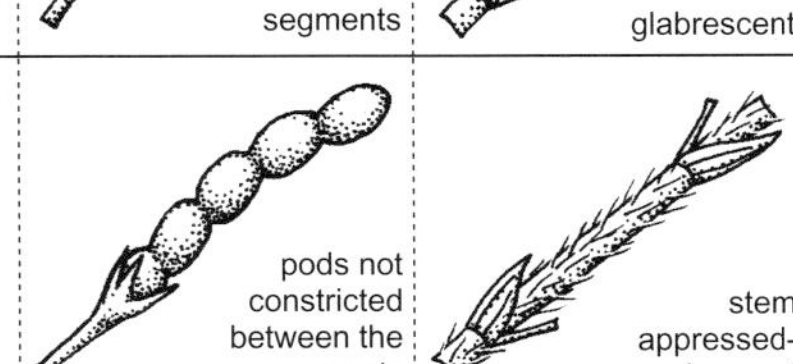

Chapmannia

1	**Leaves** silvery-grey, sericeous above and below; **standard petal** finely hairy at tip	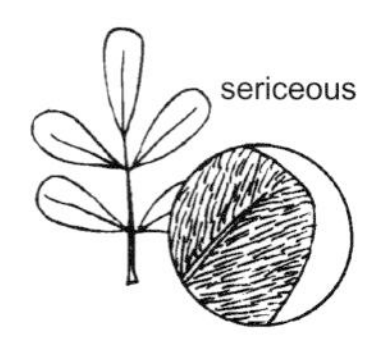	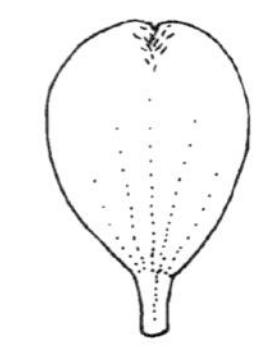	***C. sericea*** Endemic **267**, 598
	Leaves green, glabrous (sometimes pubescent below); **standard petal** glabrous	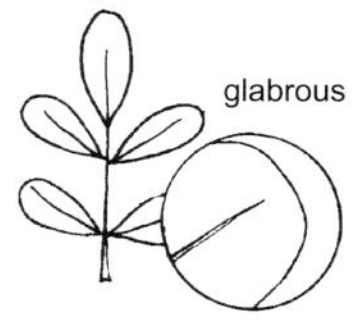	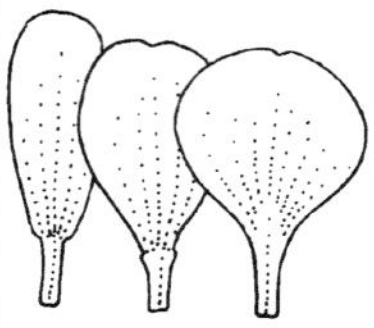	2

2	**Calyx** densely covered in tubercle-based hairs, upper lip 4-lobed	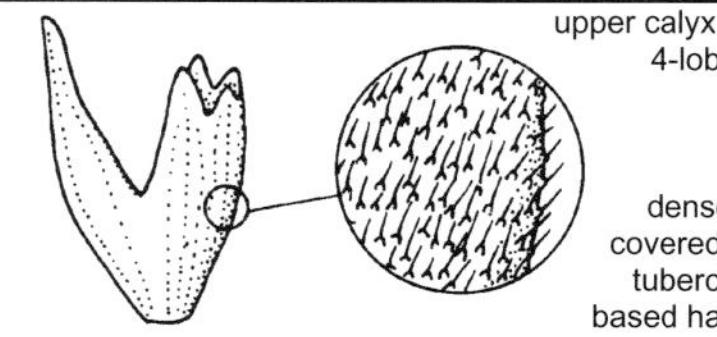	***C. tinireana*** Endemic 599
	Calyx with ordinary hairs (some tubercle-based hairs may be present), upper lip 3-lobed	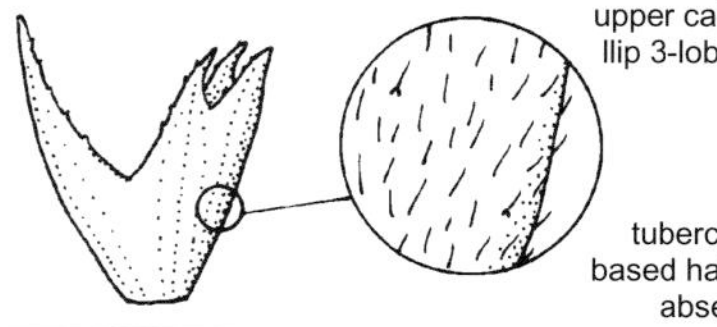	3

3	**Inflorescence** 1 or 2-flowered; **blade of standard** petal suborbicular, c. 8 x 10mm	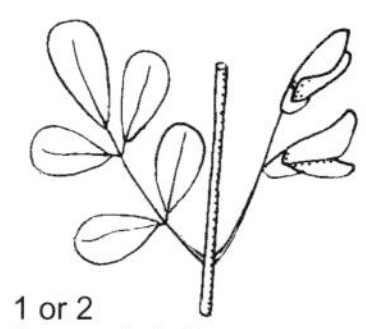	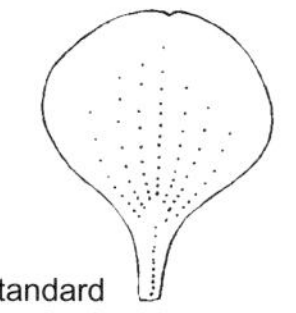	***C. gracilis*** Endemic 597
	Inflorescence 10 to 20-flowered; **blade of standard** petal ovate, c. 13 x 6.5mm	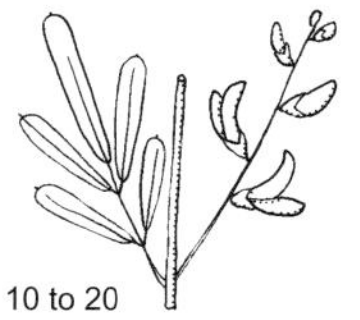	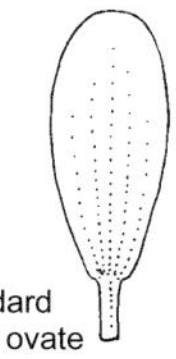	***C. reghidensis*** Endemic 598

Crotalaria

1	**Tree**		***C. socotrana*** Endemic **267**, 600
	Herb		2

2	**Leaves** simple or absent	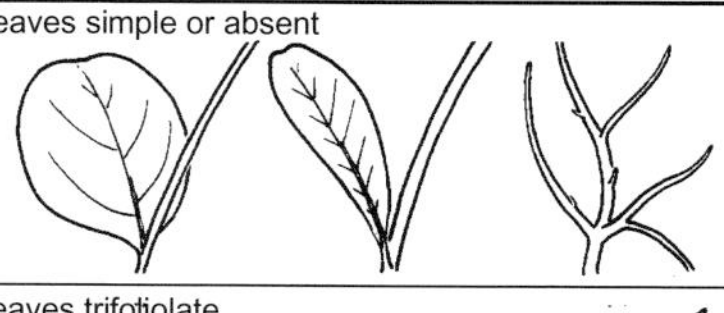	3
	Leaves trifoliolate	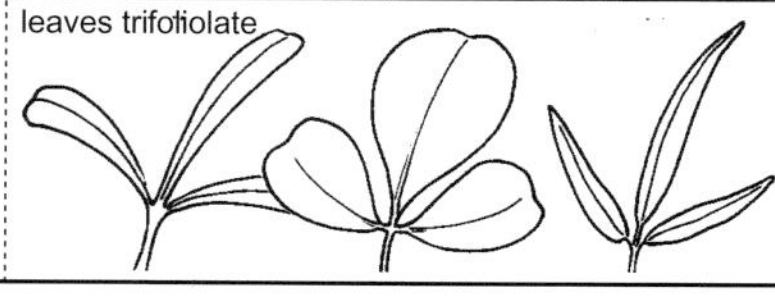	4

3	**Stems** erect; **leaves** <5cm long, narrowly oblong-obovate, often with a retuse tip	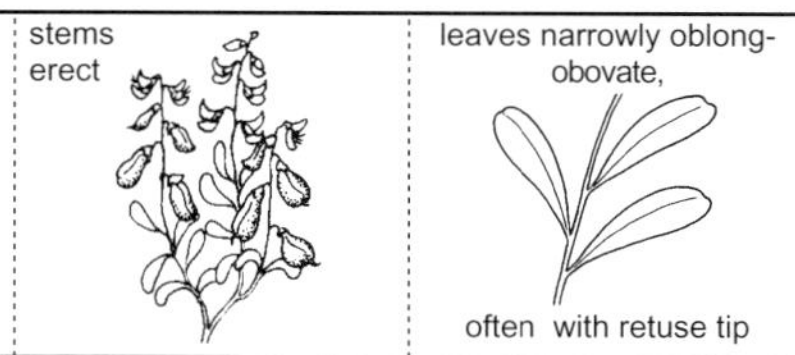	***C. retusa*** 600
	Stems prostrate; **leaves** <2cm long, linear-oblong to broadly obovate with acute, rounded or retuse tips	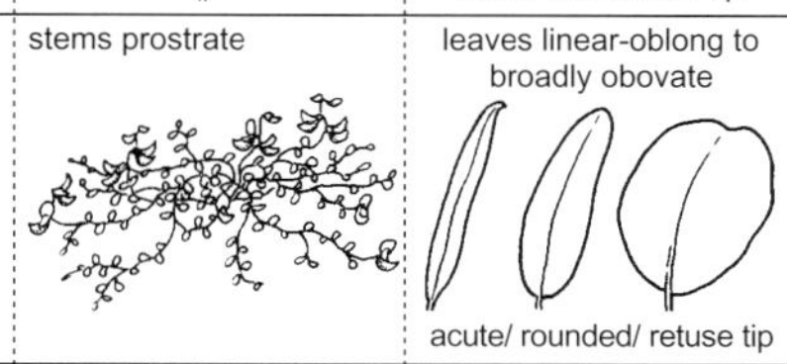	***C. strigulosa*** Endemic **267**, 600

4	**Branches** spinose (at least some); **leaflets** cuneate, emarginate	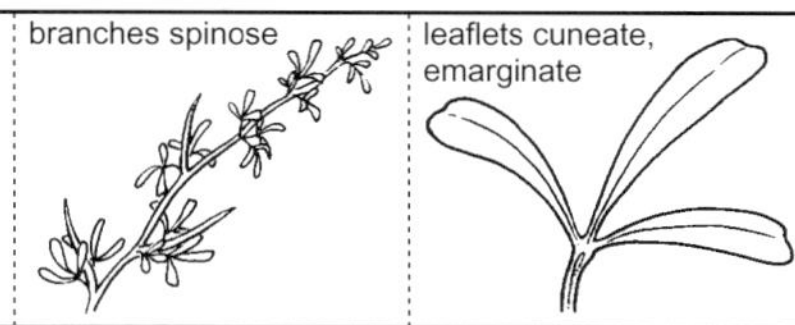	***C. spinosa*** 600
	Branches not spiny; **leaflets** linear to narrowly elliptic or obovate	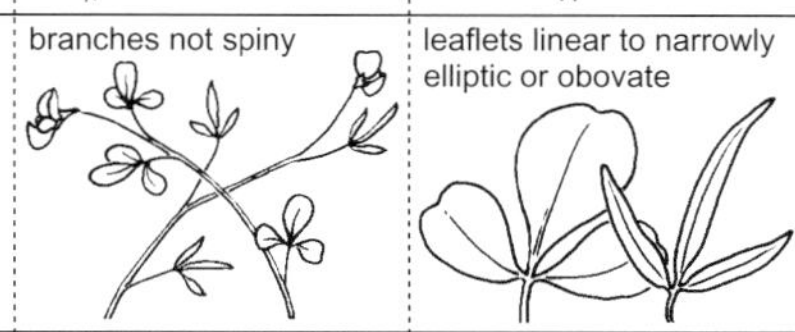	5

5	**Calyx** enlarged and papery in fruit; **fruits** strongly inflated	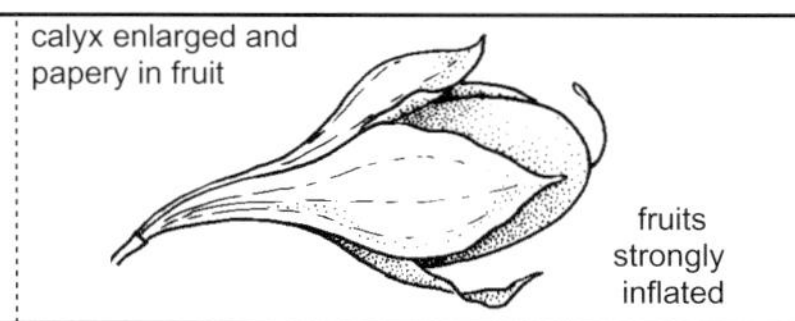	***C. pteropoda*** 599
	Calyx not enlarged and papery in fruit; **fruits** flattened or only slightly inflated	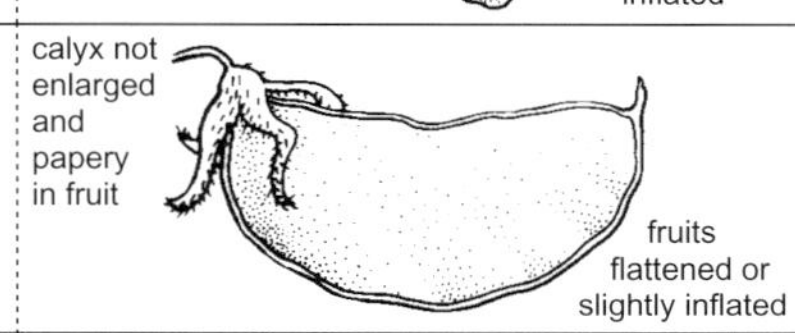	***C. leptocarpa*** 599

Indigofera

1	**Leaves** simple		2
	Leaves pinnate (leaves with more than 2 leaflets)	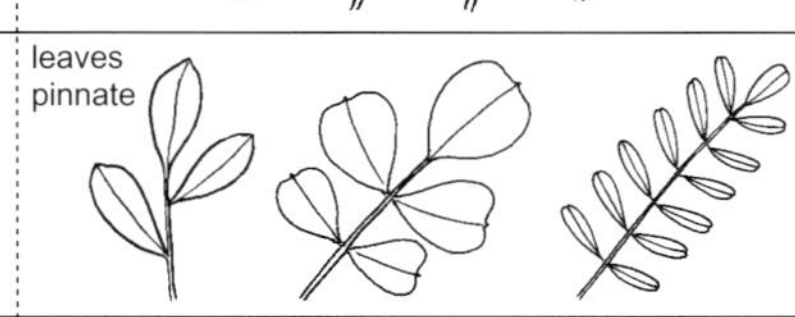	3

2	**Leaves** ovate, ± sessile and cordate at base; **fruits** somewhat torulose, 2-seeded; prostrate **annual**; **inflorescence** sessile, few-flowered	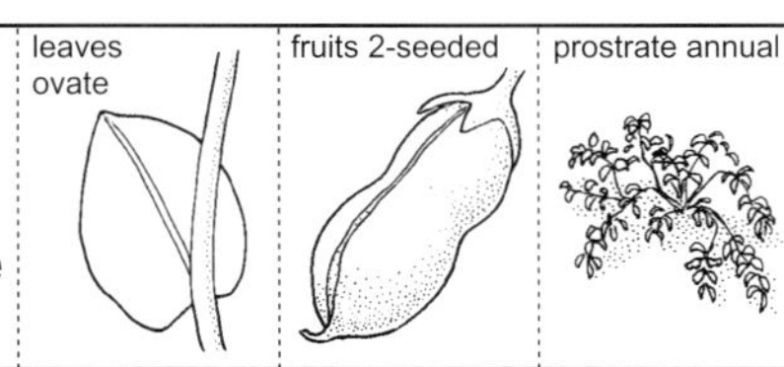	***I. cordifolia*** 604
	Leaves linear, petiole very short; **fruits** straight, c.5-seeded; erect, densely-branched **shrublet**; **inflorescence** 1or 2-flowered	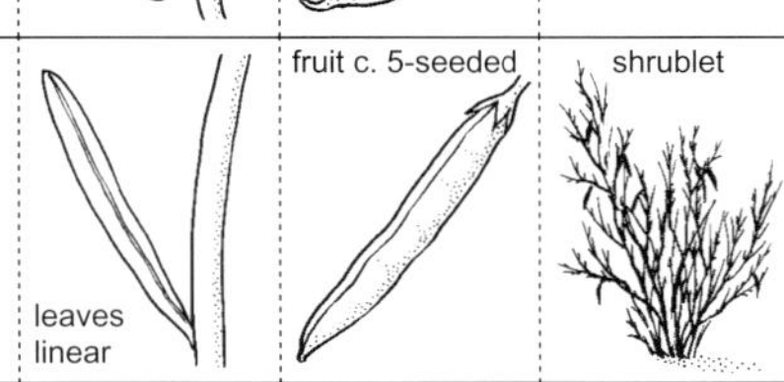	***I. pseudointricata*** **268**, 606
	Leaves oblong, petiolate; **fruits** curved, torulose, 6 to 8-seeded; erect, little-branched **shrublet**; **inflorescence** many-flowered	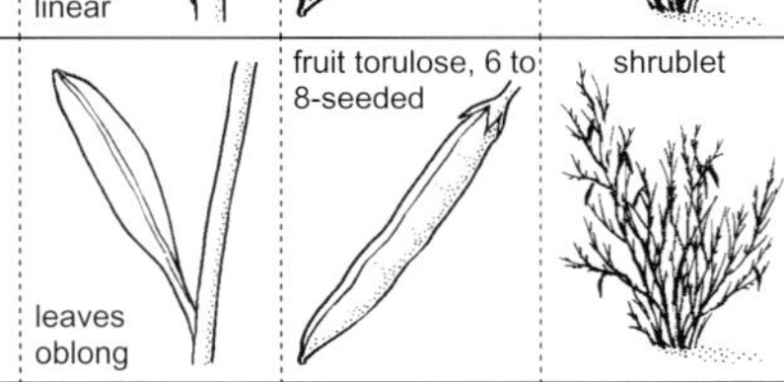	***I. oblongifolia*** 605

3	**Shrub** or small **tree** <2m tall; **corolla** c.15mm long	***I. socotrana*** Endemic **268**, 606
	Herbs or **shrublets** <1m tall; **corolla** <5mm long	4

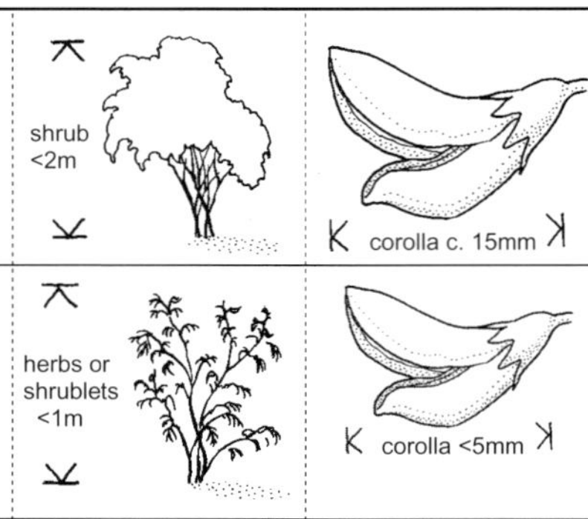

4	**Inflorescence axis** spinescent	***I. spiniflora*** 607
	Inflorescence axis not spinescent	5

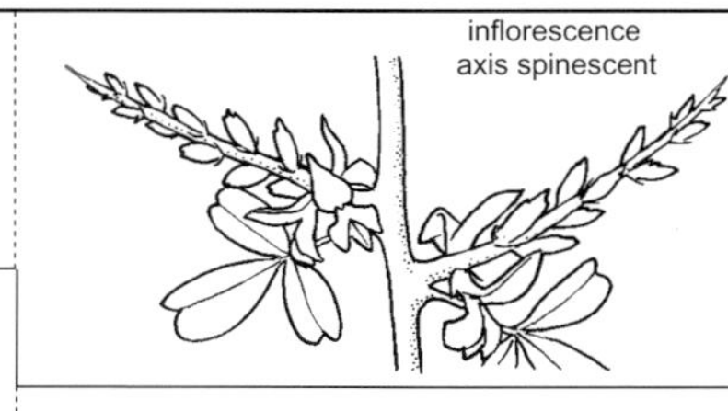

5	**Fruits** +/- circular; **leaflets** alternate	6
	Fruits longer than broad; **leaflets** opposite or alternate	7

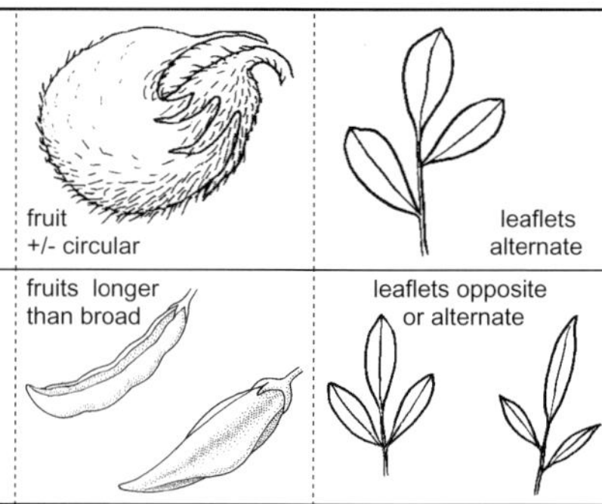

6	**Hairs** on stem and leaves crisped (slightly spreading); **leaflets** elliptic to circular; **peduncles** <5mm long	***I. nephrocarpa*** **268**, 605
	Hairs on stem and leaves appressed; **leaflets** oblong-lanceolate or linear; **peduncles** usually >5mm long	***I. nephrocarpoides*** Endemic 605

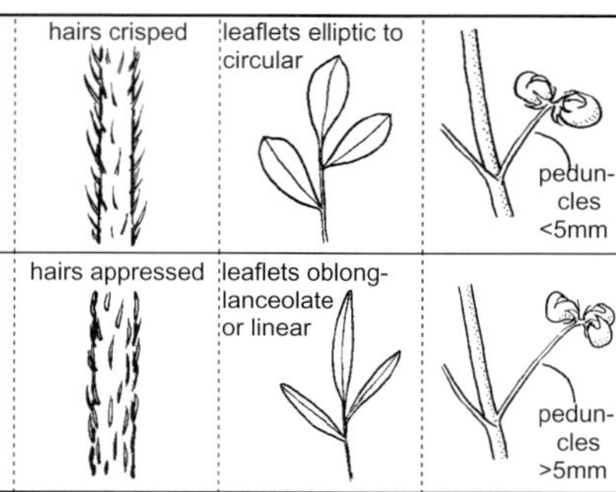

7	**Fruits** flattened, appearing jointed	***I. hochstetteri*** 604
	Fruits rounded, not jointed	8

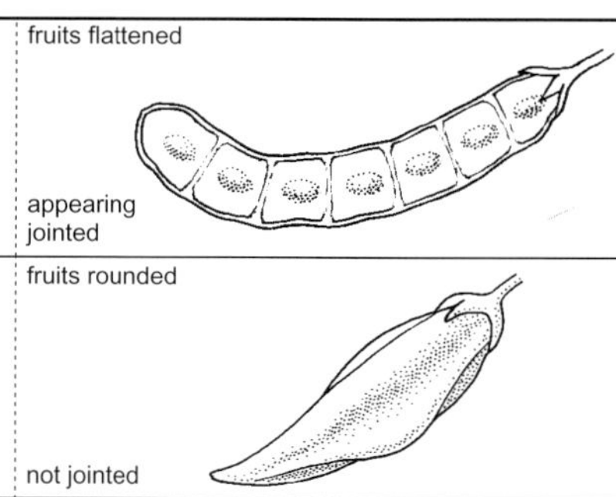

8	Annual or perennial **herbs**; **fruits** spreading	9
	Shrublets; **fruits** reflexed	10

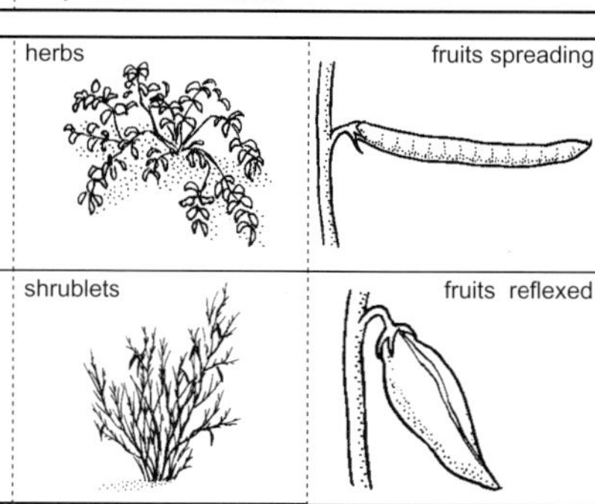

9	**Fruits** with long-stalked glandular hairs, 8 to 14-seeded; **leaves** with 9 to 11 leaflets	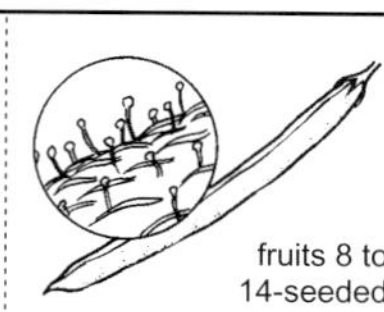	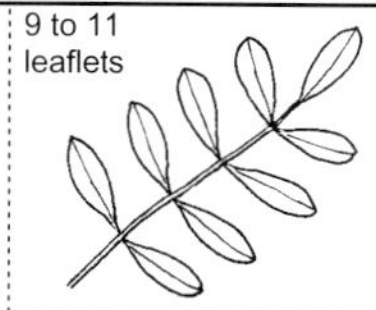	***I. colutea*** 603
	Fruits with short-stalked glandular hairs , 2 to 5-seeded; **leaves** with 5 to 7 leaflets	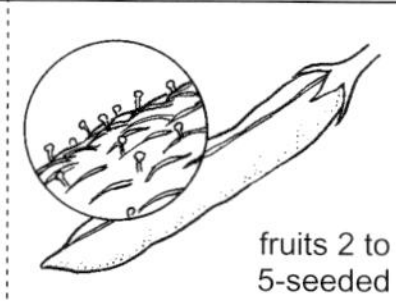	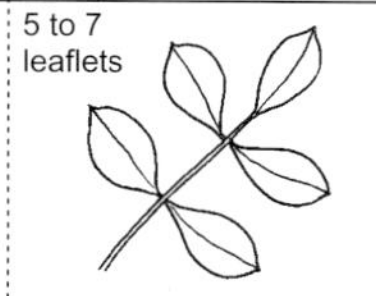	***I. argentea*** 602

10	**Fruits** deeply longitudinally grooved; **leaflets** marbled	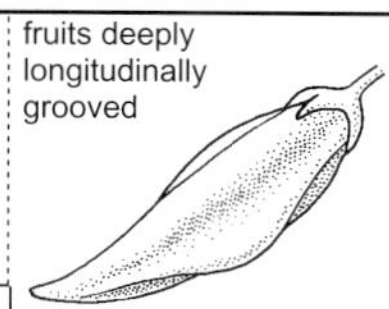	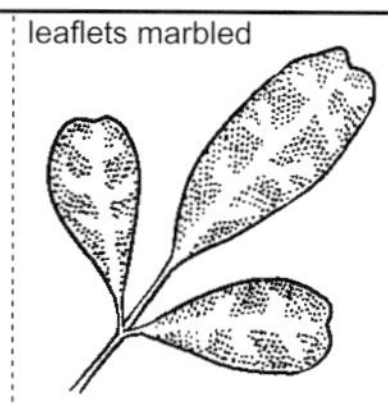	***I. marmorata*** Endemic **268**, 604
	Fruits rounded, not grooved; **leaflets** not marbled			11

11	**Leaflets** alternate, the terminal leaflet longer than the rest	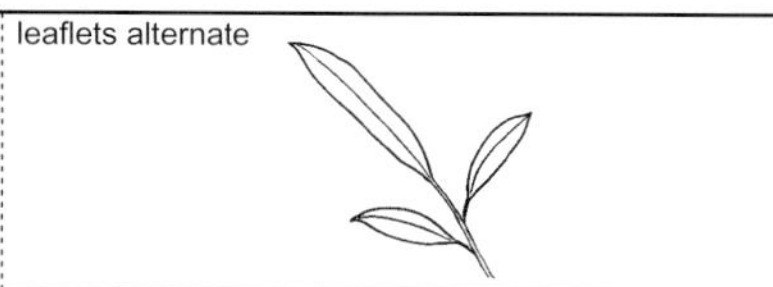	***I. oblongifolia*** 605
	Leaflets opposite, all leaflets similar	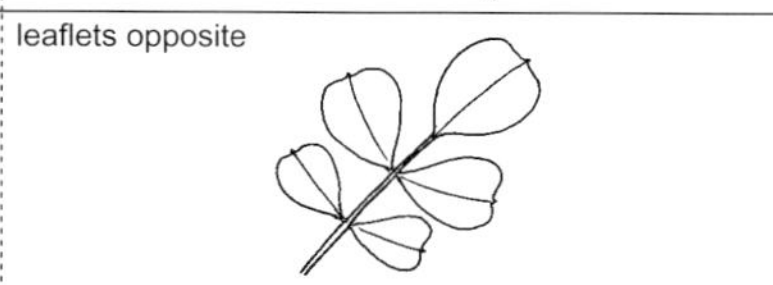	12

12	**Fruits** straight, 2 to 4-seeded; **leaves** with 3(or 5) leaflets	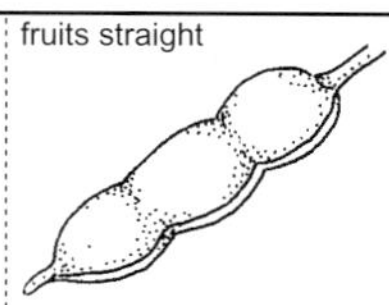	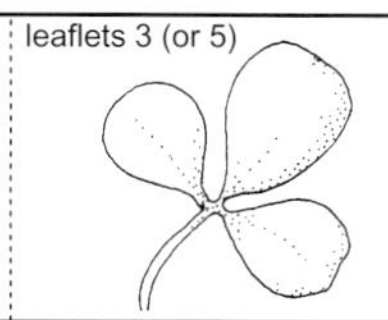	***I. nugalensis*** 605
	Fruits torulose, 2 to 4-seeded, the segments globular, sutures narrow; **leaves** with 3 to 5(or 7) leaflets	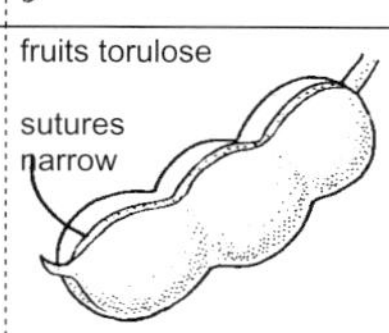	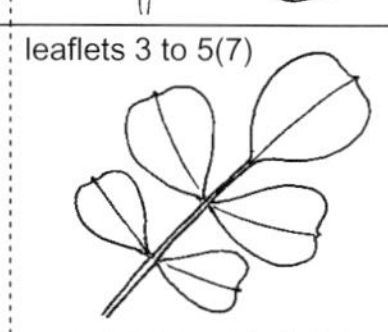	***I. articulata*** 602
	Fruits torulose, 3 to 4-seeded, the segments oblong-elliptic, sutures broad; **leaves** with 5 to 9(or 11) leaflets	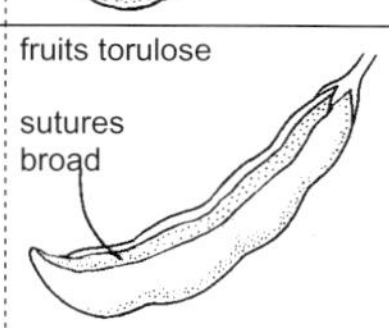	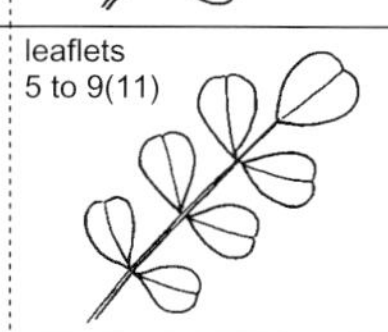	***I. coerulea*** 603
	Fruits straight, 8 to 12-seeded, sutures narrow; **leaves** with 9 to 13(or 17) leaflets	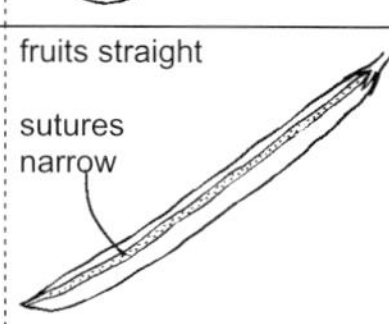	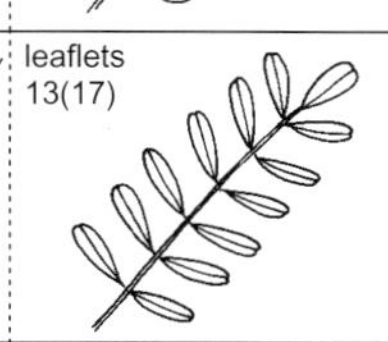	***I. tinctoria*** 607

Lotus

1	**Corolla** bright yellow, 5--12mm long, long-exserted from calyx; **pods** 1--2cm long; silvery grey or green plants (mountains)	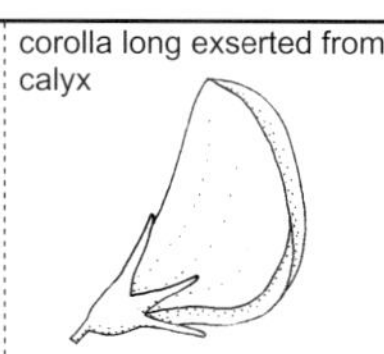	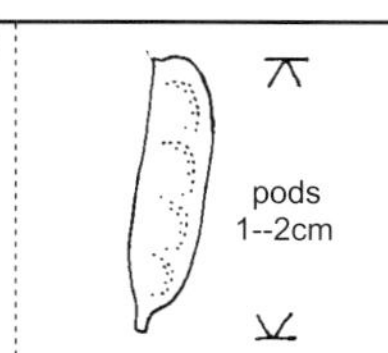	2
	Corolla pink or brownish (rarely pale yellow), 3--6mm long, included in calyx or shortly-exserted; **pods** <1cm long; glaucous (coastal plains)	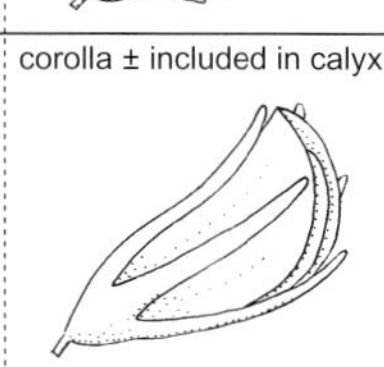	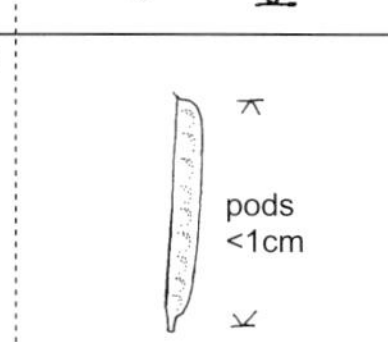	3

2	**Plant** silvery-grey, densely pubescent; **flowers** sessile; **corolla** 5--8mm long; **pods** 1--1.5cm long	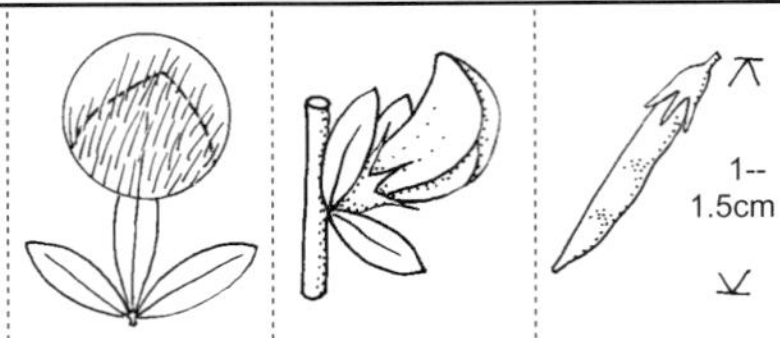 pubescent; flowers sessile	***L. mollis*** Endemic **268**, 608
	Plant green, glabrous or thinly hairy; **flowers** on 2--15mm long pedicels; **corolla** 8--12mm long; **pods** 1.5--2cm long	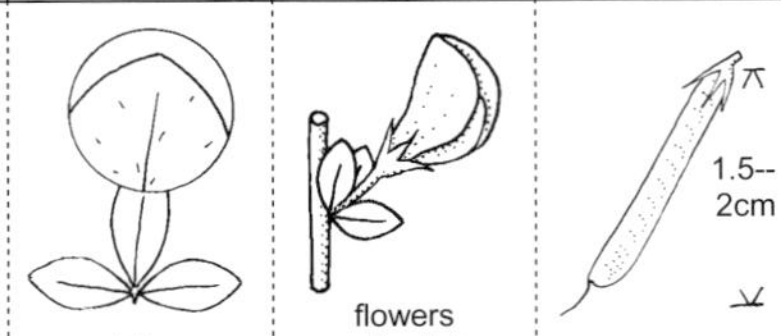 ± glabrous; flowers pedicellate	***L. ononopsis*** Endemic **268**, 608

3	Upper **leaflets** on a short stalk; **calyx** 5--8mm, densely hairy; **pod** often surrounded by persistent calyx	calyx densely hairy 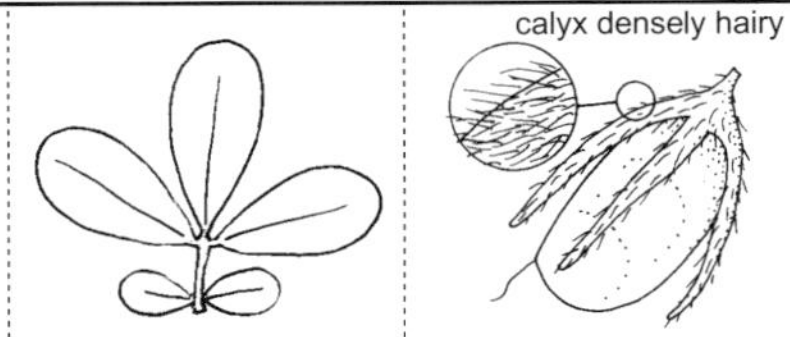 upper leaflet stalked; pod surrounded by calyx	***L. garcinii*** **268**, 608
	All **leaflets** sessile; **calyx** 1.5--3mm, sparsely hairy; **pod** exserted from persistent calyx	calyx sparsely hairy 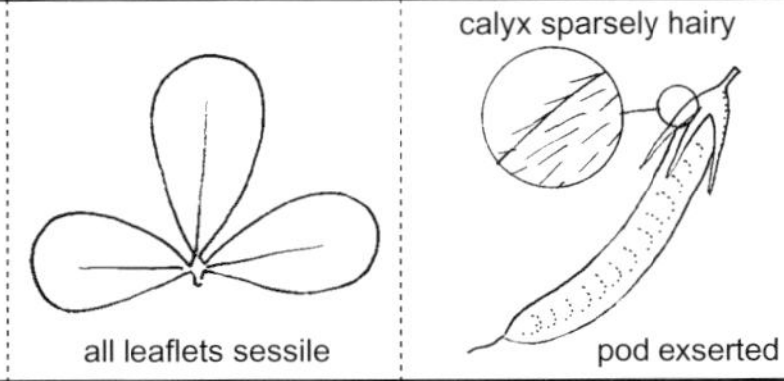 all leaflets sessile; pod exserted	***L. schimperi*** 609

Medicago

1	**Stipules** entire or with few teeth below; **plant** densely pubescent; **pod** subglobose, grooves clearly visible between the nerves on each turn of the pod	stipules entire; pod subglobose 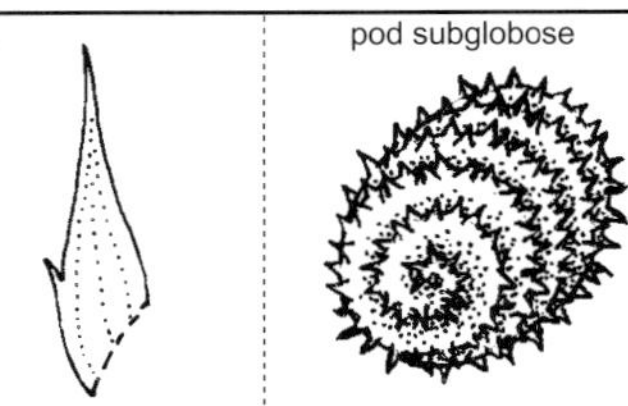	***M. minima*** 610
	Stipules laciniate; **plant** glabrous or sparsely hairy; **pod** subglobose, grooves not visible between the nerves	stipules laciniate; pod subglobose 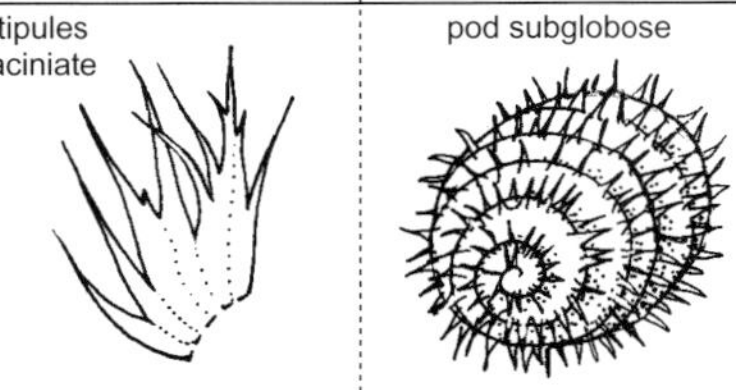	***M. laciniata*** 609
	Stipules laciniate; **plant** glabrous or sparsely hairy; **pod** flattened, grooves widely separated and clearly visible between the turns of the pod	stipules laciniate; pod flattened 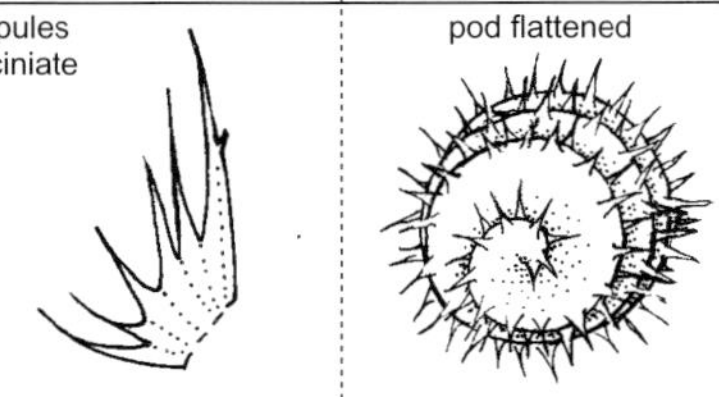	***M. polymorpha*** 610

Rhynchosia

1	**Plant** glabrescent, green; **leaflets** acute or acuminate at tip	leaflets acute or acuminate; glabrescent 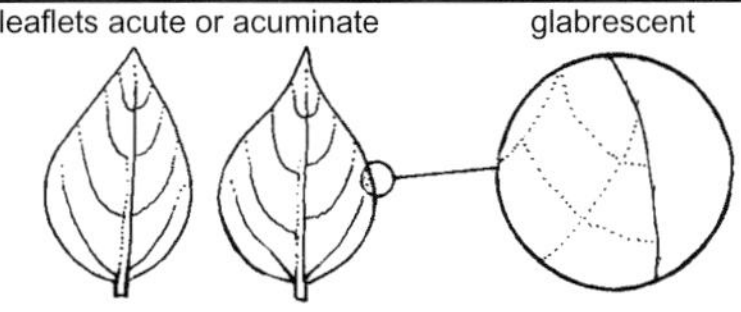	***R. minima*** 612
	Plant densely tomentose, grey-green; **leaflets** obtuse or rounded at tip	leaflets obtuse or rounded; densely tomentose 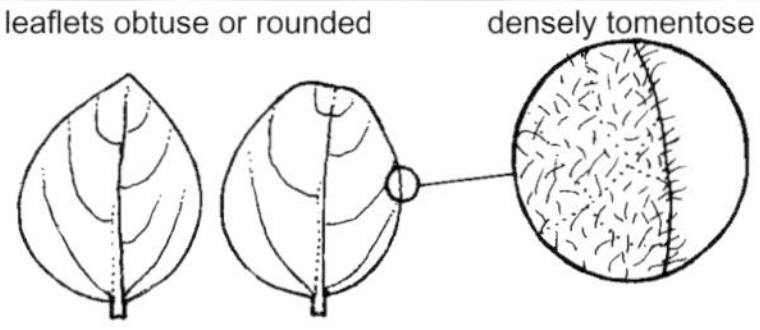	***R. pulverulenta*** 612

Tephrosia

1	**Leaves** simple; delicate annual; **fruits** <2cm long	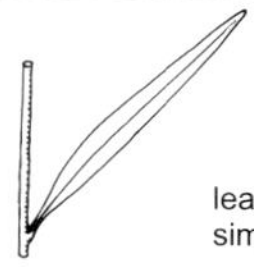	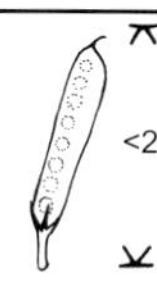	***T. strigosa*** 614
	Leaves compound; robust annual or perennial herbs; **fruits** >3cm long	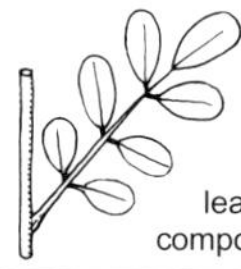	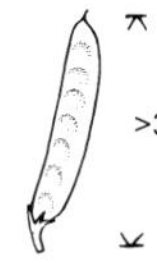	2

2	**Flowers** in leaf axils only	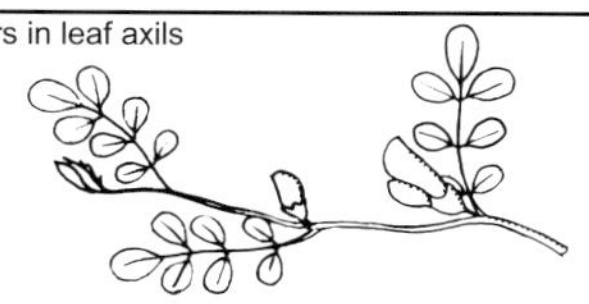	3
	Flowers in racemes		5

3	**Leaves** with 3 leaflets; **flowers** dark purple, solitary in leaf axils	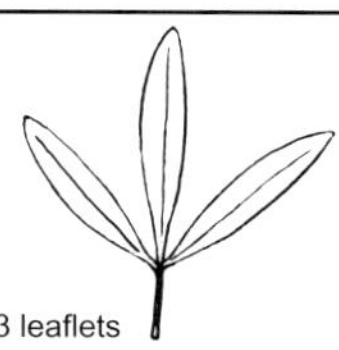		***T. odorata*** Endemic **269**, 614
	Leaves usually with >7 leaflets; **flowers** pink or pale purple, 2 to 7 in leaf axils		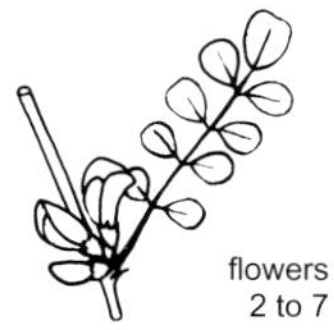	4

4	**Style** ribbon-shaped, not twisted, 1.5--2.5mm long	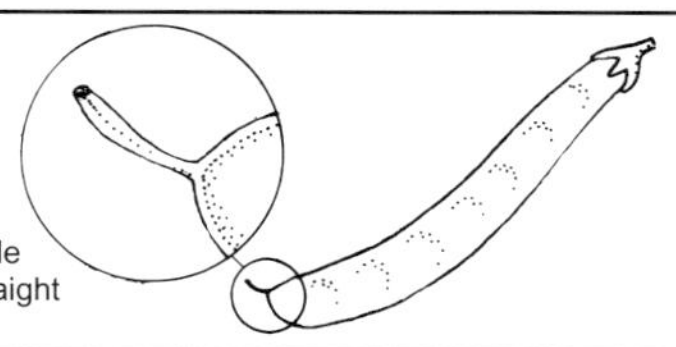	***T. subtriflora*** 615
	Style tapered, twisted, 2.5--3.5mm long	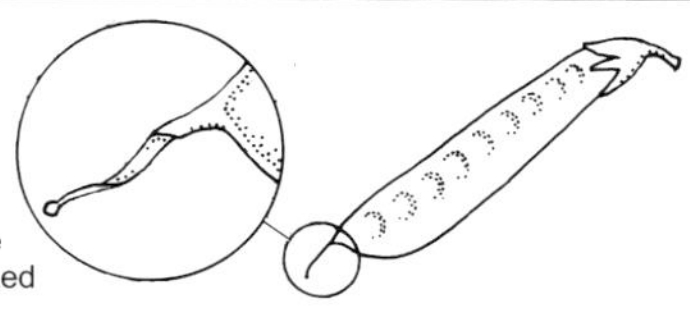	***T. uniflora*** 615

5	**Calyx** covered with dark brown hairs	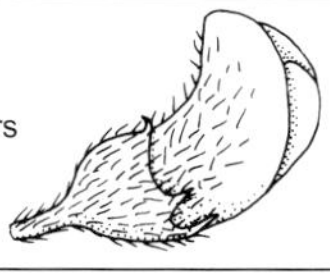	***T. socotrana*** Endemic 614
	Calyx covered with white or yellowish hairs	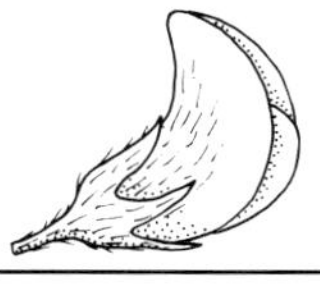	6

6	**Leaflets** 11 to 17	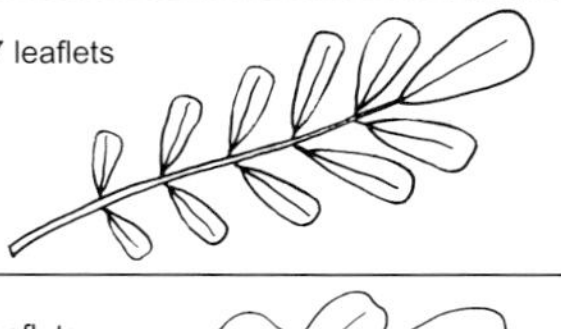	***T. purpurea*** 614
	Leaflets 5 to 9	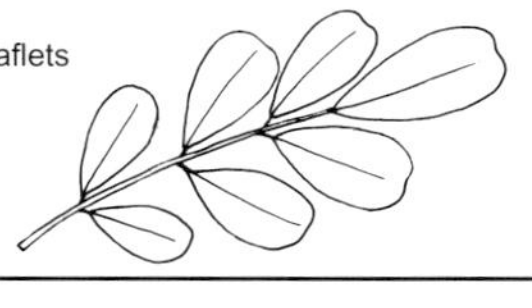	***T. apollinea*** 613

LENTIBULARIACEAE

1	Delicate epiphytic **herb** <5cm; **leaves** with bladder-like traps; **flowers** 2-lipped with a long spur	delicate epiphyte 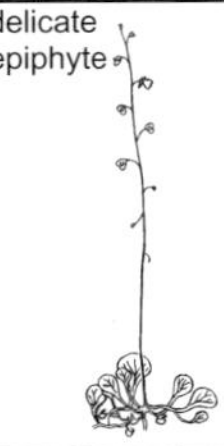	 leaves with bladder-like traps	 flowers with a long spur	***Utricularia striatula*** 619

LINACEAE

1	Erect, annual **herb**; **leaves** alternate, linear, oblanceolate; **flowers** yellow, c.5mm across; **petals** 5, free at base; **fruit** a globose capsule	erect herb  leaves alternate, linear/oblanceolate	flowers yellow 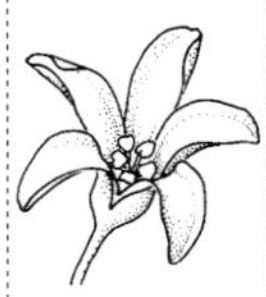 petals 5, free at base	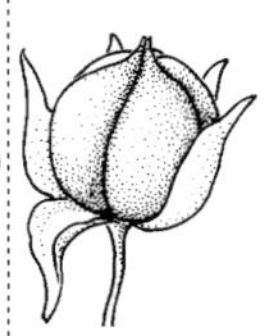 fruit a globose capsule	***Linum trigynum*** 619

LYTHRACEAE

1	**Leaves** alternate; **flowers** solitary, lilac-pink, c. 4 mm across; **calyx** cylindrical, 5 to 12-toothed	leaves alternate, flowers solitary 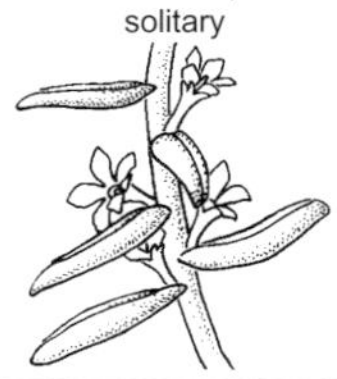	flowers lilac-pink 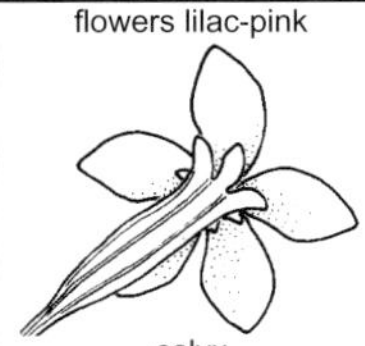 calyx cylindrical, 5 to 12-toothed	***Lythrum hyssopifolium*** 620
	Leaves opposite; **flowers** clustered, pink or greenish; **calyx** bell-shaped	leaves opposite, flowers clustered 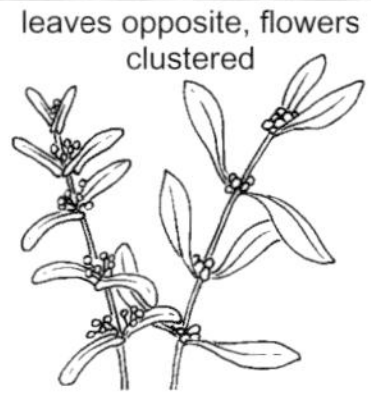	flowers pink or greenish  calyx bell-shaped	***Ammannia*** (2 spp.)

Ammannia

1	Prostrate annual **herb;** **flowers** clustered	prostrate herb 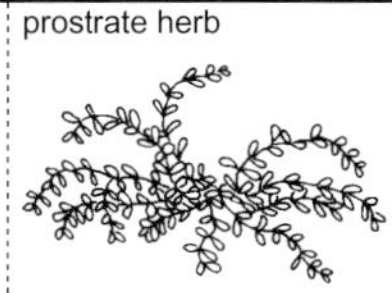	flowers clustered 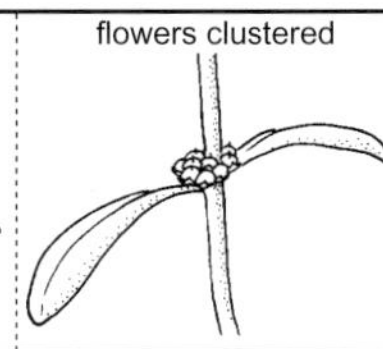	***A. baccifera*** 620
	Erect annual **herb; flowers** in short cymes	erect herb 	flowers in short cymes 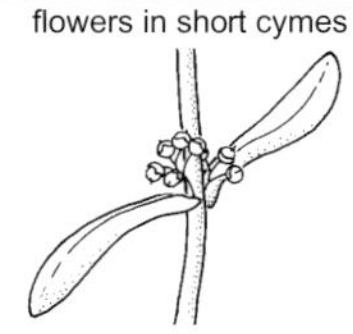	***A. multiflora*** 620

MALPIGHIACEAE

1	Small **tree**; **leaves** alternate; **flowers** bright yellow; **petals** fringed and clawed; **stamens** 10; **fruit** winged	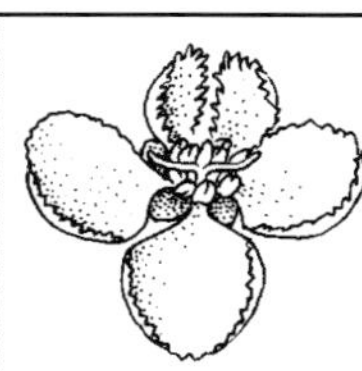 flowers with clawed petals and 10 stamens	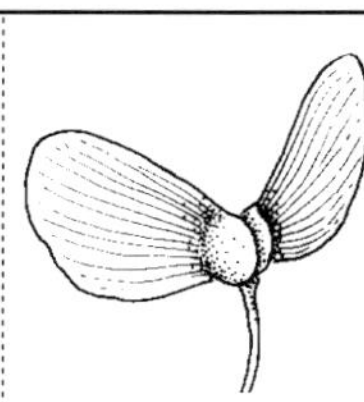 fruit winged	***Acridocarpus socotranus*** Endemic **269**, 621

MALVACEAE

1	**Fruit** a capsule, splitting to shed seeds	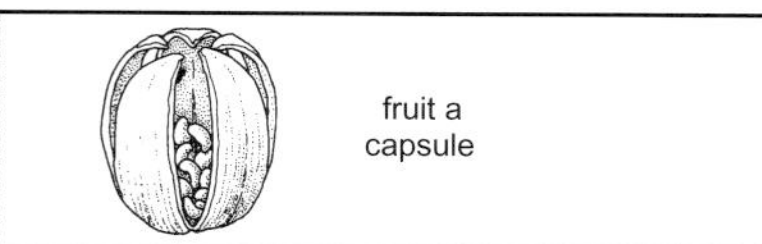	2
	Fruit a schizocarp, usually splitting into separate seed units (mericarps)	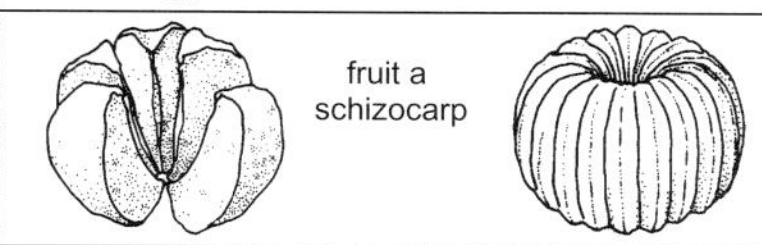	4
2	**Epicalyx bracts** absent or > 5 in number, smaller than the calyx	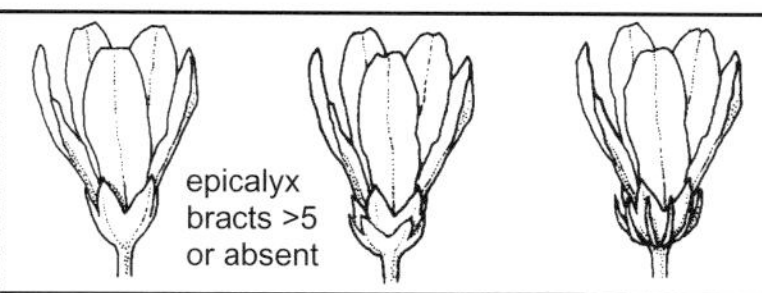	***Hibiscus*** (15 spp.)
	Epicalyx bracts 3, much larger than the calyx	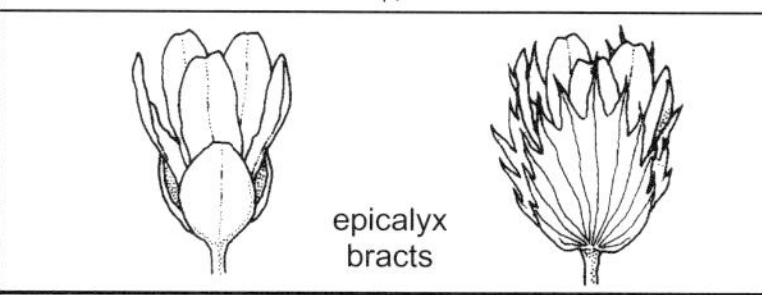	3
3	Hirsute **undershrub**; **epicalyx bracts** deeply toothed; **corolla** yellow with maroon centre; **seeds** covered in woolly hairs (cotton)	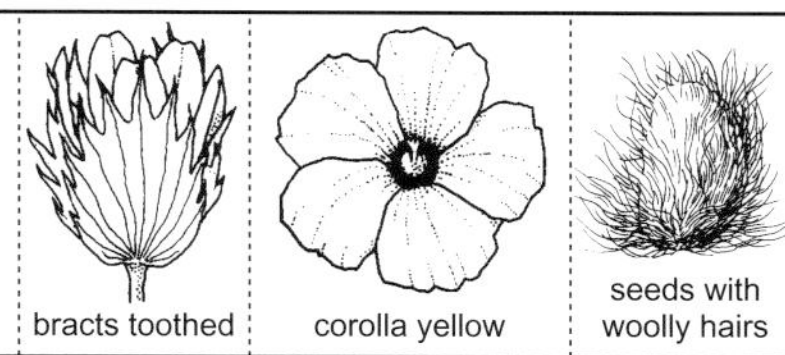	***Gossypium barbadense*** 623
	Softly tomentose **undershrub**; **epicalyx bracts** entire; **corolla** violet or maroon; **seeds** hirsute	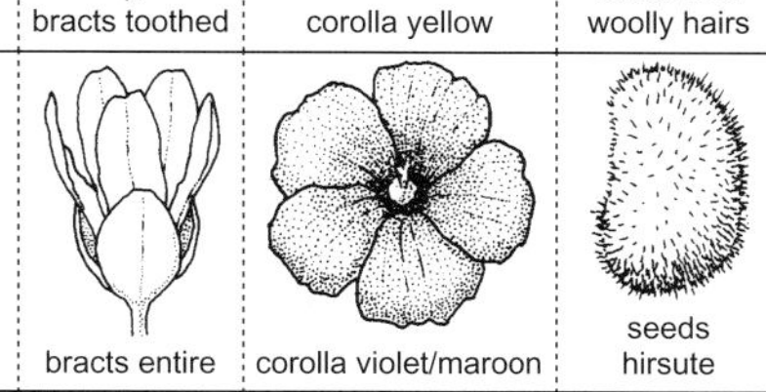	***Senra incana*** 631
4	**Epicalyx bracts** present; **corolla** mauve or pink	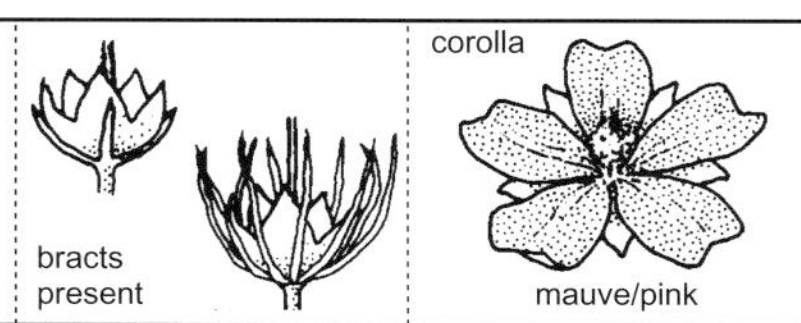	5
	Epicalyx bracts absent; **corolla** cream, yellow or orange	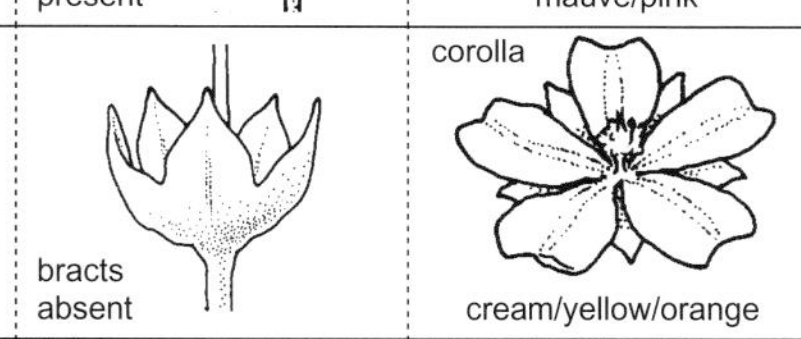	6
5	**Epicalyx bracts** 3; usually prostrate, weedy annual **herb**; **leaves** with palmate venation, more or less orbicular in outline; **corolla** mauve	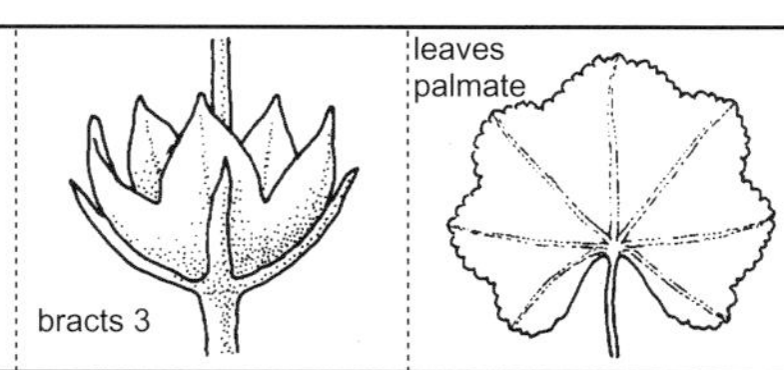	***Malva parviflora*** 630
	Epicalyx bracts 6 to 12; erect perennial **herb**; **leaves** with pinnate venation, ovate to elliptic in outline; **corolla** pink	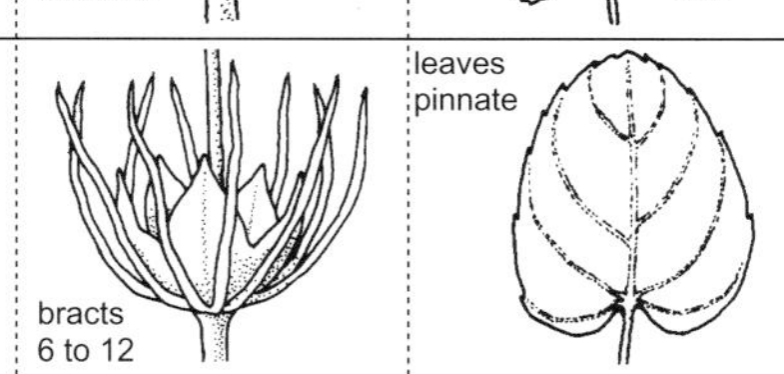	***Pavonia arabica*** 631
6	**Mericarps** 2 or more-seeded, **flowering** in the evening	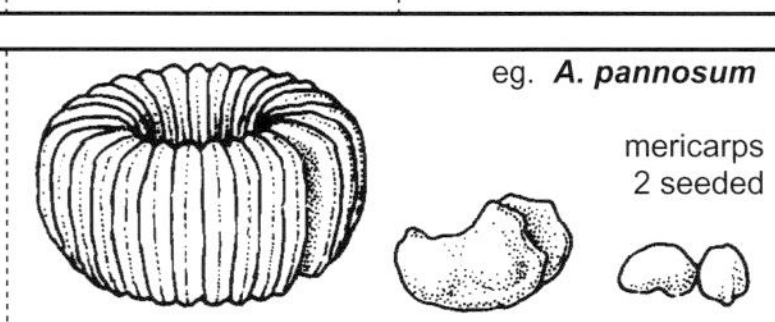	***Abutilon*** (2 spp.)
	Mericarps 1-seeded; **flowering** in the morning	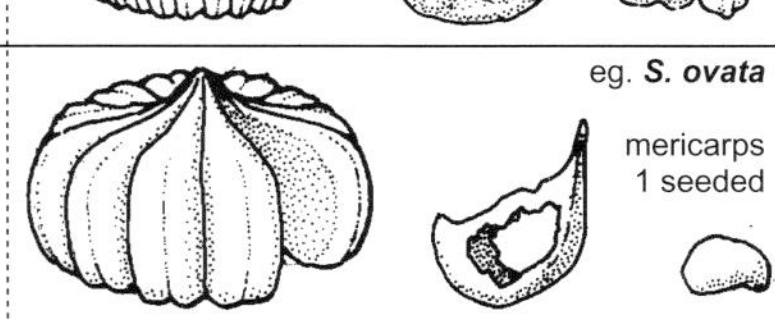	***Sida*** (4 spp.)

Abutilon

1	**Stems** covered with short, dense stellate hairs; **mericarps** 10 to 12, acute or shortly awned at the apex			***A. fruticosum*** 622
	Stems stellately tomentose and with longer, simple hairs, often glandular; **mericarps** 25 to 35, rounded at the apex		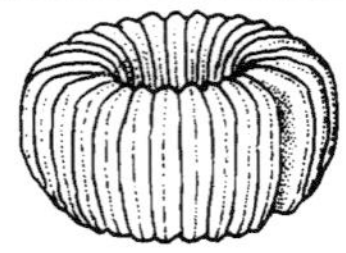	***A. pannosum*** 622

Hibiscus

1	**Epicalyx** absent; erect annual **herbs**			2
	Epicalyx present; **trees**, **shrubs,** perennial **herbs** or prostrate annual **herbs**			3
2	**Calyx** longer than fruit; **corolla** yellow; **stem** pubescent, long hairs absent		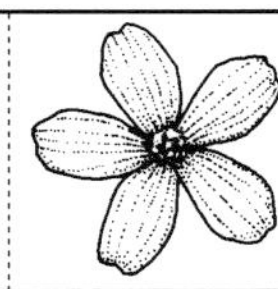	***H. sidaeformis*** 629
	Calyx shorter than fruit; **corolla** white; **stem** pubescent, with long weak hairs	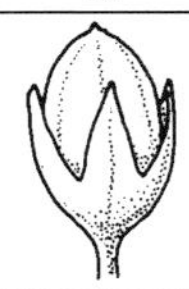		***H. lobatus*** 626
3	**Leaves** palmately lobed or divided; **stem** with simple hairs	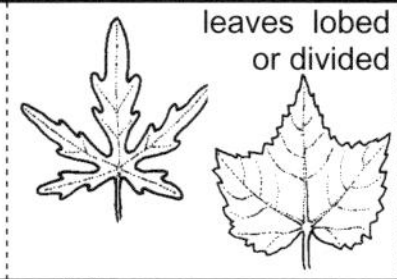	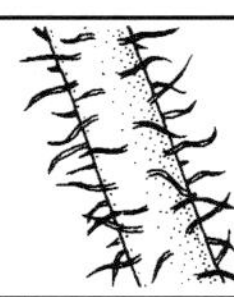	4
	Leaves not lobed or divided; **stem** with stellate hairs	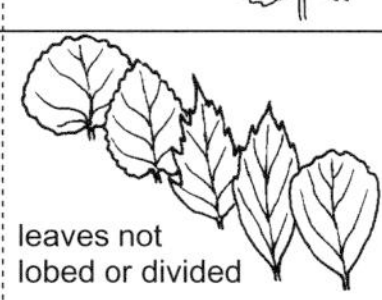		5
4	Prostrate annual **herb**; **leaves** deeply lobed or divided; **capsule** apiculate; **petals** c.1cm long	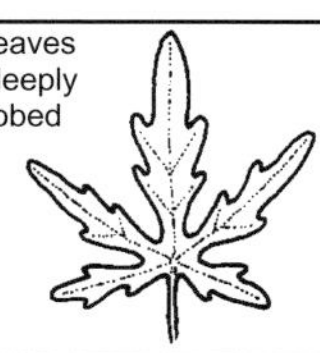		***H. palmatus*** 627
	Erect or ascending perennial **herb** or **shrublet**; **leaves** shallowly lobed; **capsule** winged; **petals** >3cm long	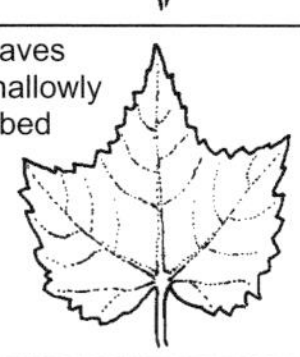	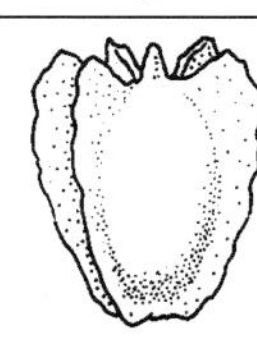	***H. vitifolius*** 630
5	Woody-based **herb** or **subshrub**; **corolla** deep red, small, <3cm long	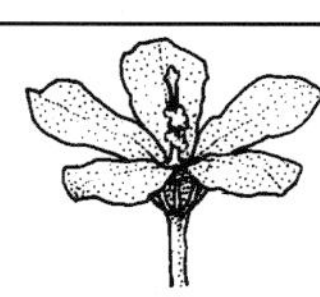		6
	Shrub or tree; **corolla** yellow, often with a red centre, large, >4cm long	 		7

	Character	Illustration	Result
6	**Leaves** obovate, tip rounded; **calyx** 6--12mm long; **stipules** spinescent; midvein of **fruit** valves green	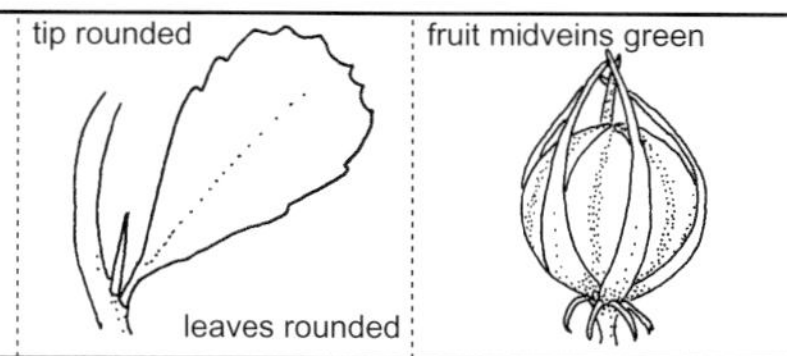	***H. scindicus*** 628
	Leaves ovate to elliptic, tip acute; **calyx** 3--4mm long; **stipules** not spinescent; midveins of **fruit** not green	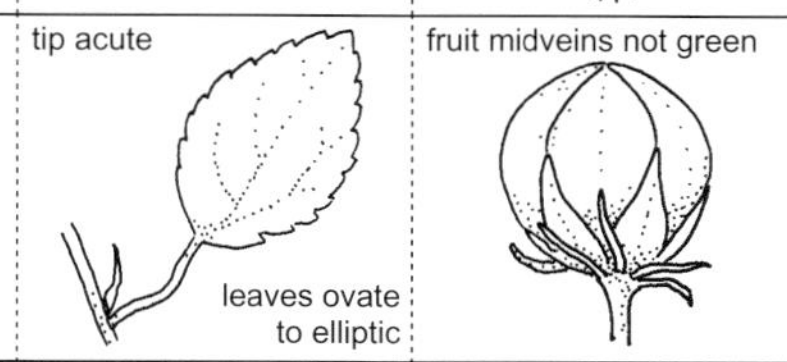	***H. micranthus*** 626
7	**Epicalyx** fused to halfway (plant of cliff faces)	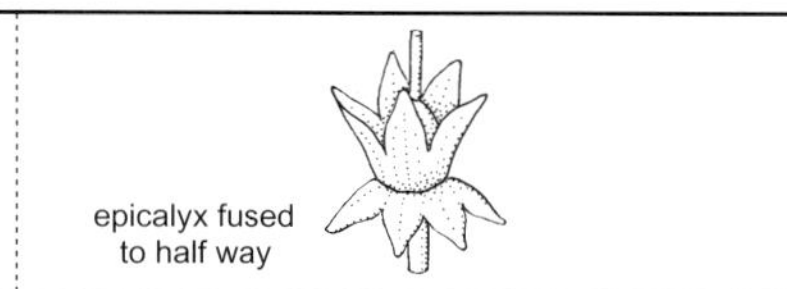	***H. socotranus*** Endemic **270**, 629
	Epicalyx free or fused only at base	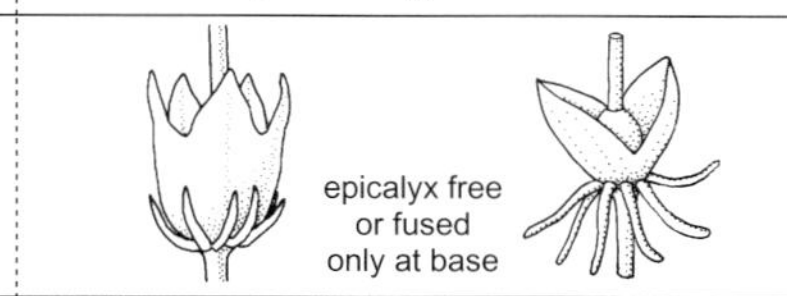	8
8	**Corolla** forming a narrow tube below and spreading above; **flowering calyx** shortly cylindrical with triangular teeth	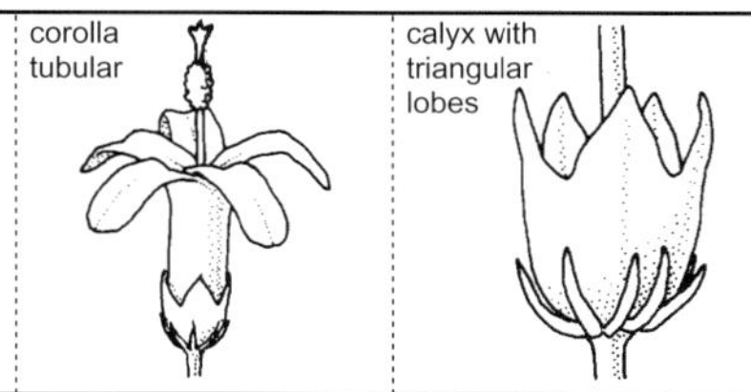	***H. stenanthus*** Endemic **270**, 629
	Corolla funnel-shaped; **flowering calyx** cup-shaped, 5-toothed or entire, sometimes splitting unequally	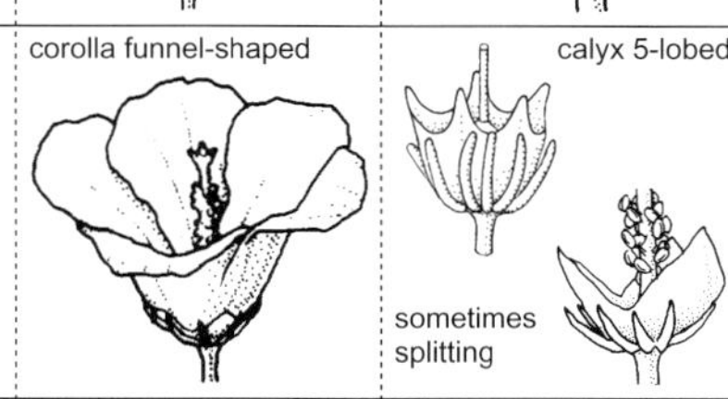	9
9	**Calyx** 5-lobed, not splitting into 2 segments	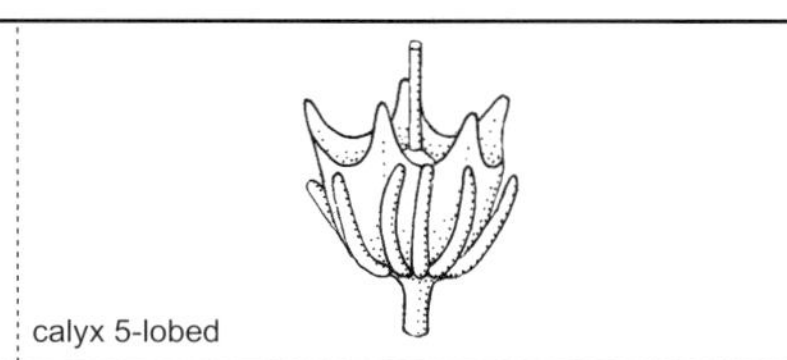	10
	Calyx splitting into 2 segments, the segments shortly toothed or entire	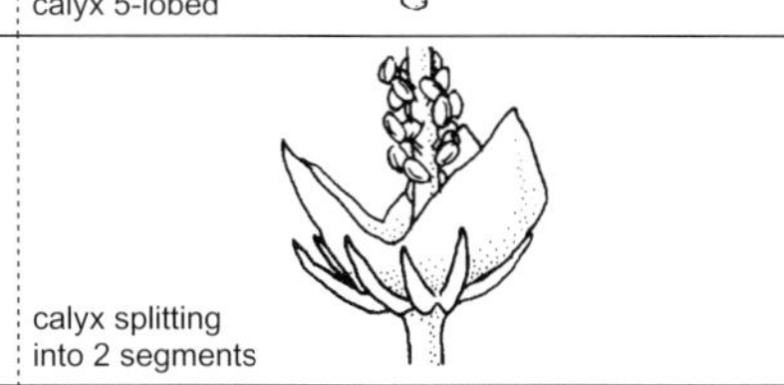	11
10	**Epicalyx lobes** linear, c.1mm across; **calyx lobes** broadly triangular; pedicels, epicalyx and calyx with small stellate **hairs** only	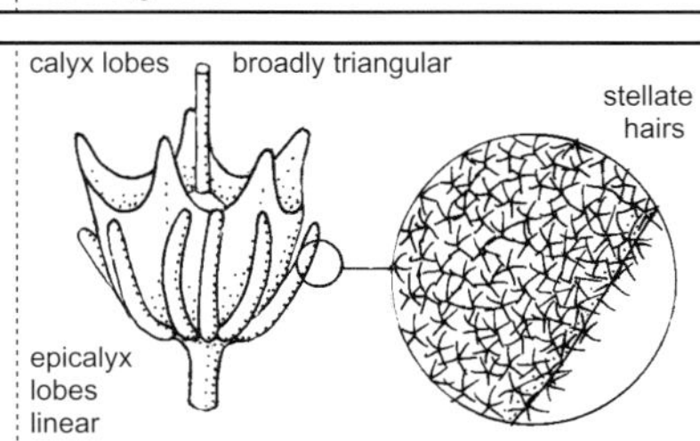	***H. dioscorides*** Endemic 624
	Epicalyx lobes filiform, c.0.5mm across; **calyx lobes** triangular ovate with acuminate tips; pedicels, epicalyx and calyx fringed with long simple **hairs**	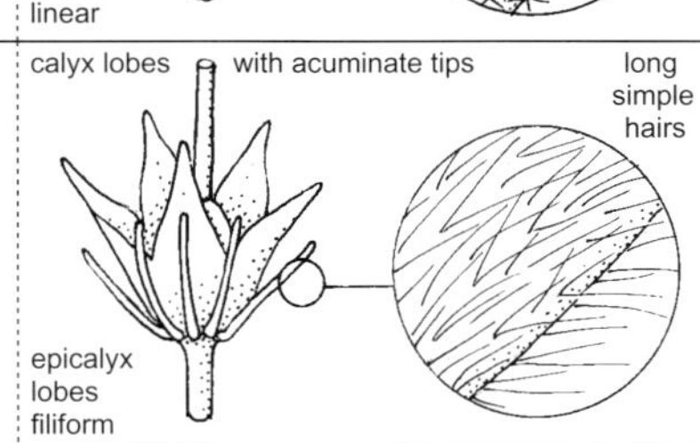	***H. noli-tangere*** Endemic **270**, 627

11	**Epicalyx lobes** reflexed, oblanceolate		12
	Epicalyx lobes erect, linear		13

12	**Epicalyx** and **calyx** glabrescent; **calyx** entire		***H. quattenensis*** Endemic **270**, 628
	Epicalyx and **calyx** densely stellate-hairy; **calyx** toothed	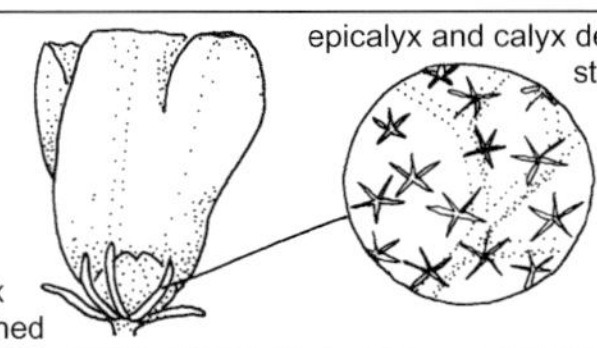	***H. diriffan*** Endemic **270**, 625

13	**Leaves** tomentose (more thinly so above) with only stellate hairs; **calyx** teeth obscure	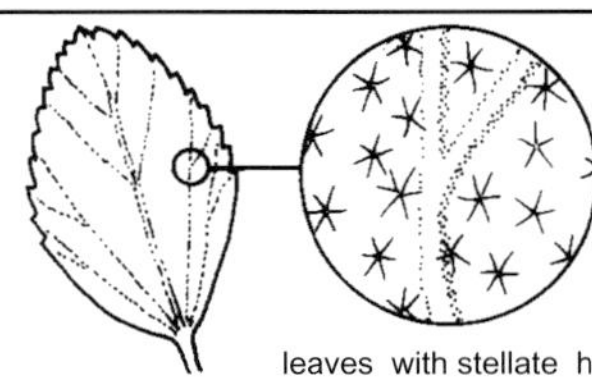	***H. malacophyllus*** Endemic 626
	Leaves tomentose with both stellate and larger (2--3 mm), detachable, 3 or 4-branched hairs; **calyx** shortly 5-toothed or with triangular lobes	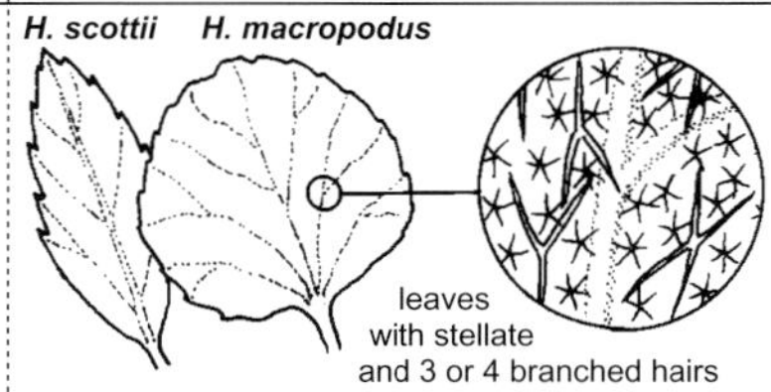	14

14	**Outer surface of petals** covered with yellowish hairs; **corolla** yellow		***H. macropodus*** Endemic 626
	Outer surface of petals covered with rust-coloured hairs; **corolla** orange-yellow	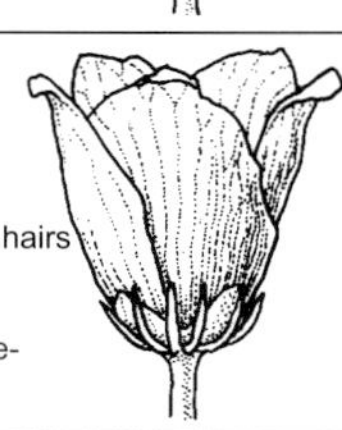	***H. scottii*** Endemic **270**, 628

Sida

1	**Mericarps** 5; **leaves** broadly ovate, cordate at the base	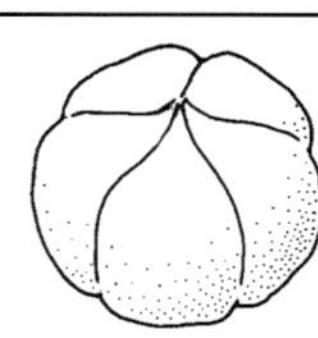	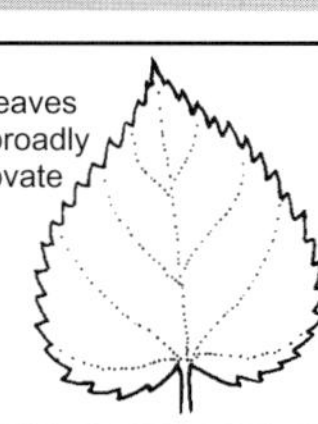	***S. cordata*** 632
	Mericarps 7 to 10; **leaves** ovate to rhombate, rounded or at most shallowly cordate at the base	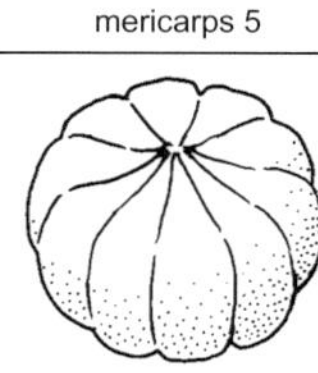	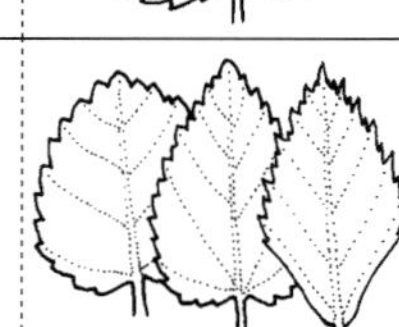	2

2

Stems with short, stellate hairs and long, simple hairs; **mericarps** with awns >2mm long

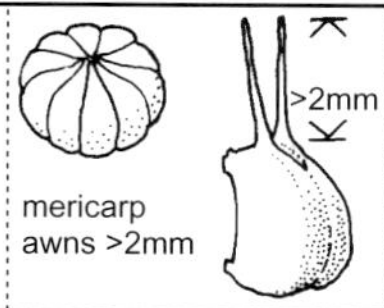

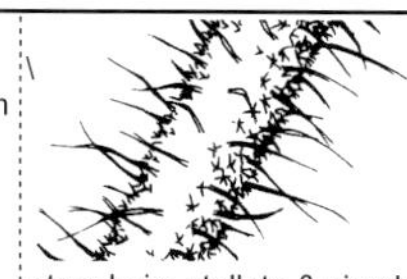

S. cordifolia 632

Stems with short, stellate hairs only; **mericarps** without awns or with awns <2mm long

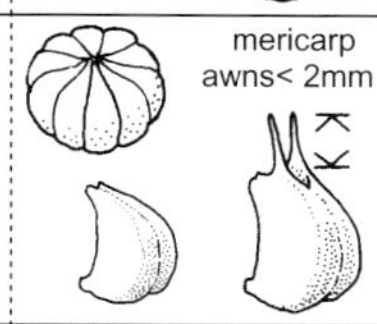

3

3

Mericarps without awns; **leaves** ovate

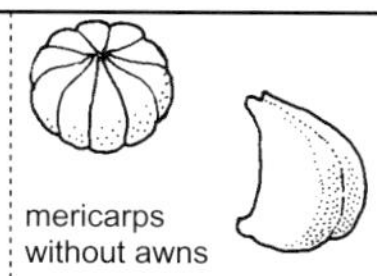

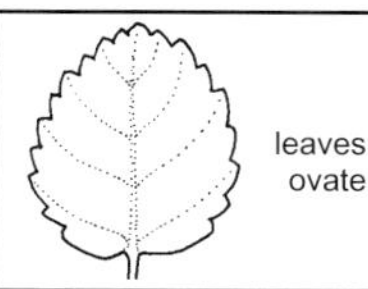

S. ovata 632

Mericarps awned; **leaves** oblong-lanceolate

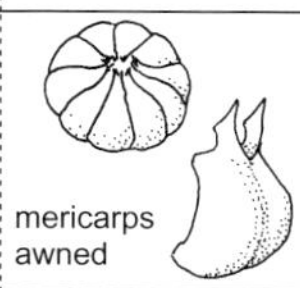

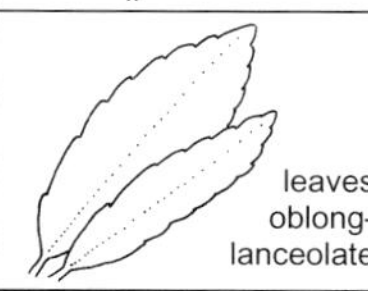

S. acuta 632

MELIACEAE

1

Small **tree**; **leaves** entire, clustered on spurs; **flowers** cream with linear petals; **stamens** 10, united into a tube; **fruit** a leathery capsule

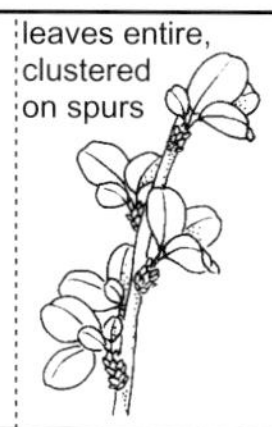

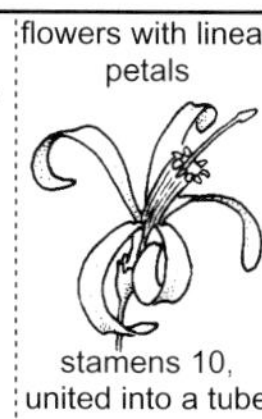

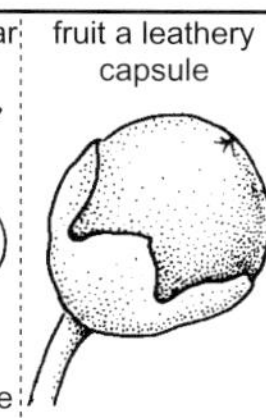

Turraea socotrana Endemic **271**, 633

MENISPERMACEAE

1

Shrub bearing spine-tipped flattened cladodes; **flowers** cream, in sessile clusters either on the cladodes or on the stem below the cladodes; **leaves** soon deciduous

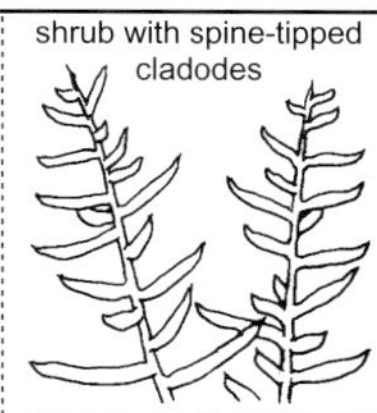

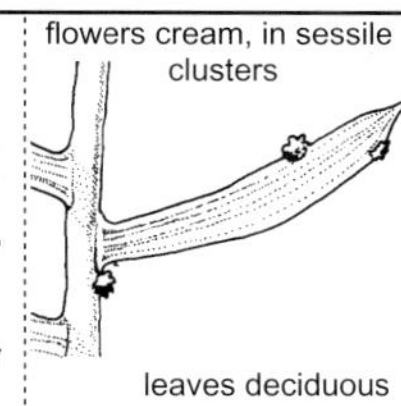

Cocculus balfourii **271**, 633

Twining or trailing **liane**; **flowers** greenish yellow, in pedunculate clusters either axillary or on leafless branches; **leaves** persistent

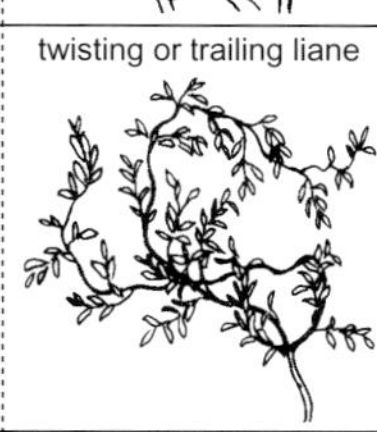

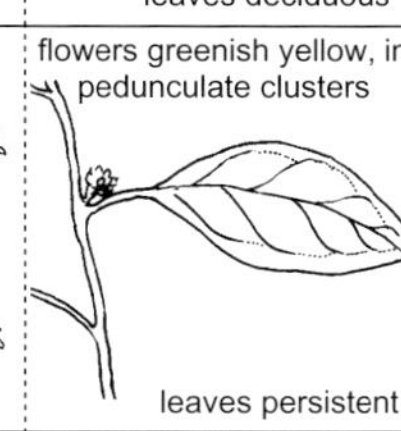

Cocculus pendulus 634

MORACEAE

1

Herbs or **shrubs** either with a swollen, succulent stem or lacking an aerial stem but with an underground tuber; **flowers** on a flattened receptacle surrounded by ray-like appendages

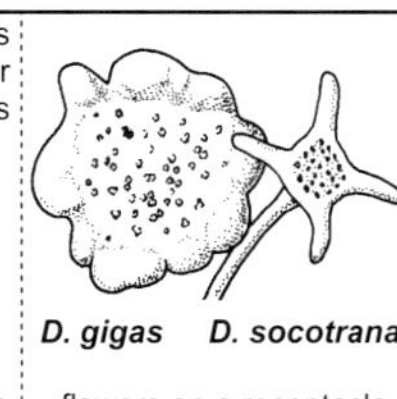

Dorstenia (2 spp.)

Woody **trees** or **shrubs**, sometimes epiphytic; **flowers** borne inside a globose or obovate hollow receptacle (the fig)

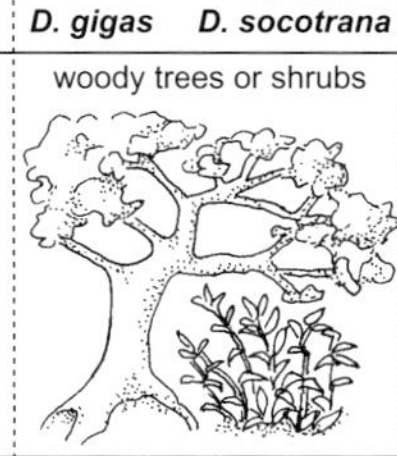

Ficus (2 spp.)

Dorstenia

	Description	Illustrations	Taxon
1	Small **herb** with underground tubers and lacking aerial stems; **leaves** cordate to reniform	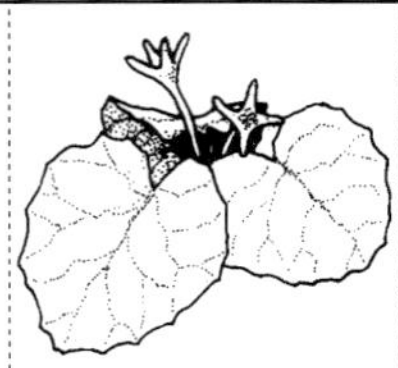 small herb 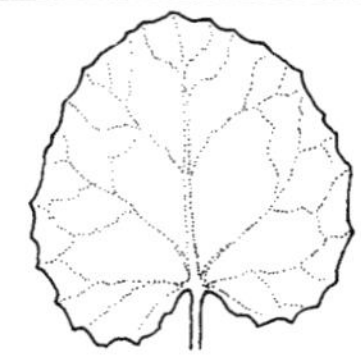 leaves cordate to reniform	***D. socotrana*** Endemic **271**, 635
	Shrub to 2.5m with bottle-shaped stems; **leaves** narrowly obovate	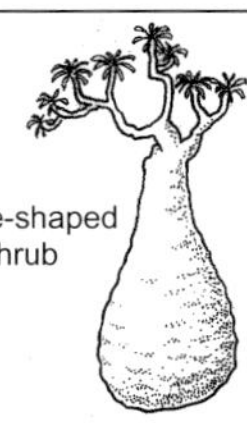 bottle-shaped shrub 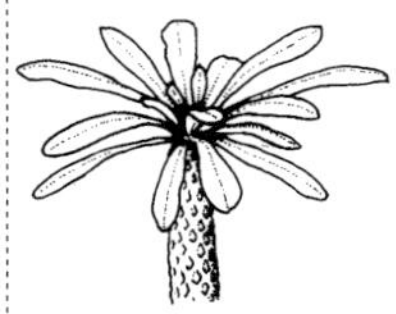 leaves narrowly obovate	***D. gigas*** Endemic **271**, 635

Ficus

	Description	Illustrations	Taxon
1	**Leaves** lanceolate, glabrous; **figs** 7--10mm across, glabrous to puberulous, white to purplish when ripe	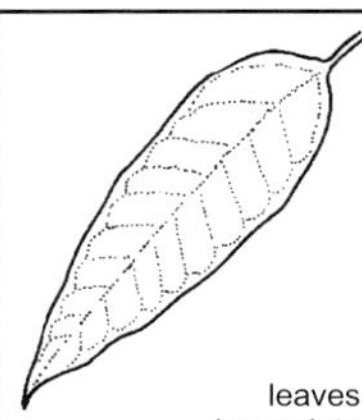 leaves lanceolate 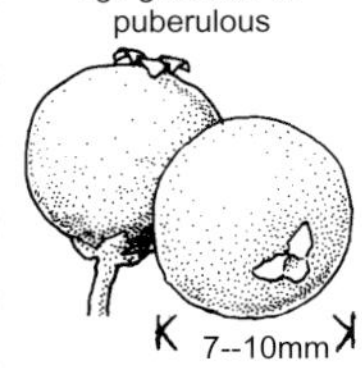 figs glabrous to puberulous	***F. cordata*** **271**, 636
	Leaves cordate, glabrous to glabrescent above, glabrescent to velutinous beneath; **figs** 10–15 (–20)mm across, tomentose to glabrescent, green with paler spots when ripe	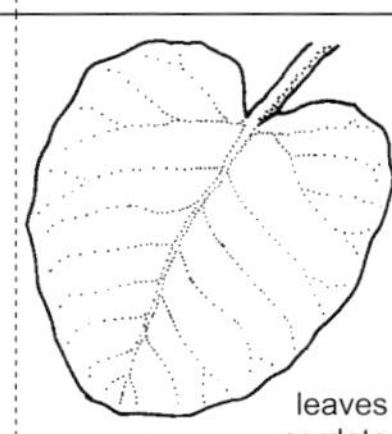 leaves cordate 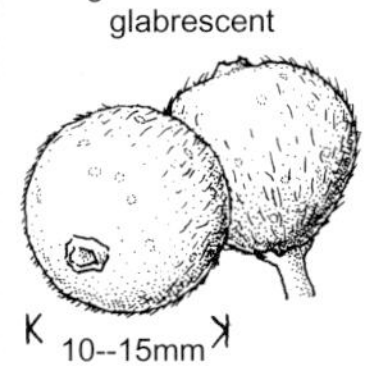 figs tomentose to glabrescent	***F. vasta*** **271**, 637

MYRSINACEAE

	Description	Illustrations	Taxon
1	Evergreen **shrub**; **leaves** alternate, serrulate; **flowers** reddish, in axillary clusters, c. 3mm; **fruit** globose, dark purple, c. 5mm across	leaves alternate, serrulate 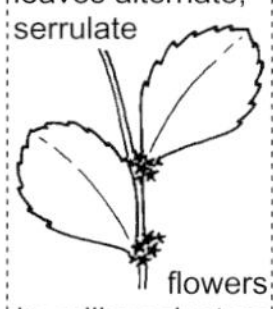 flowers in axillary clusters; flowers 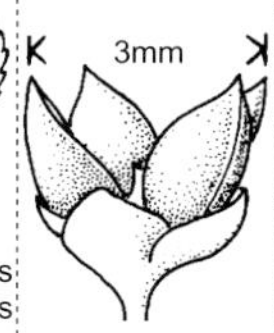 fruit globose 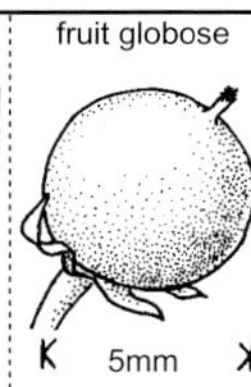	***Myrsine africana*** 639

NYCTAGINACEAE

	Description	Illustrations	Taxon
1	**Fruits** 5-ribbed, glabrous or glandular-hairy, <4mm long; prostrate annual or perennial **herb** with a tuberous root	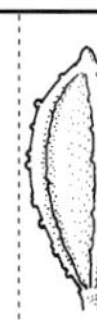 fruits 5-ribbed 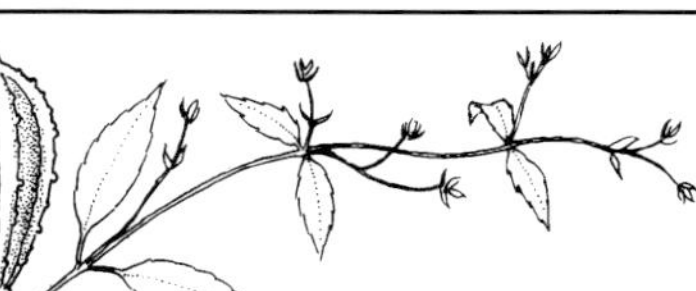 prostrate herb with tuberous root	***Boerhavia diffusa*** 639
	Fruits obscurely 10-ribbed, with conspicuous sessile or stalked, wart-like glands, >4mm long; erect or ascending perennial **herb**	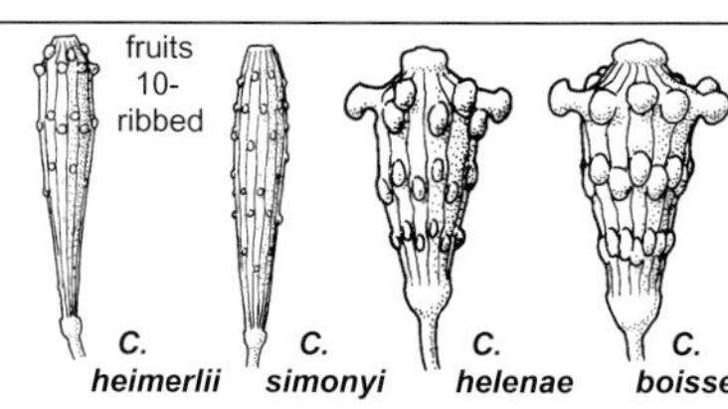 fruits 10-ribbed	***Commicarpus*** (4 spp.)

Commicarpus

1	**Stems**, at least in the region of the inflorescence, conspicuously glandular-pilose and very sticky	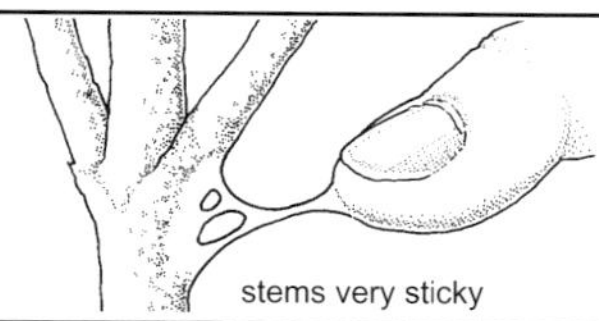		***C. heimerlii*** Endemic **272**, 641
	Stems glabrous, scabridulous or clothed with a short indumentum of crispate hairs, not sticky	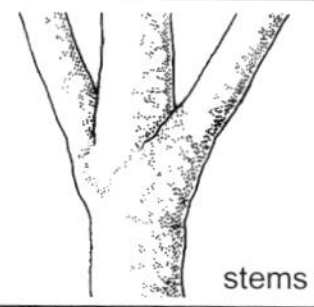		2
2	**Inflorescence** umbellate, forming a leafless mass of slender peduncles and pedicels; **perianths** narrowly funnel-shaped, with a distinct basal tube; **fruits** fusiform, inconspicuously glandular	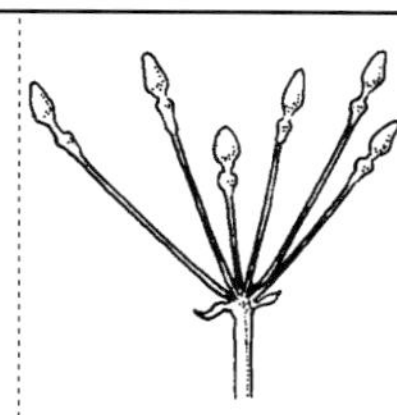	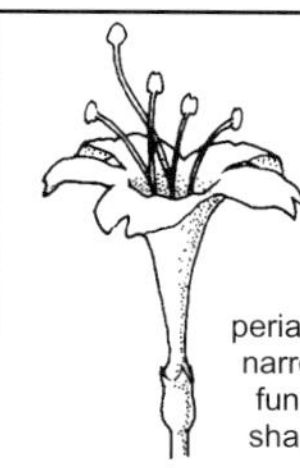	***C. simonyi*** Endemic **272**, 641
	Inflorescence normally of 1 or 2(or 3) whorls of flowers; **perianths** widely funnel-shaped, without a distinct basal tube; **fruits** clavate, prominently glandular	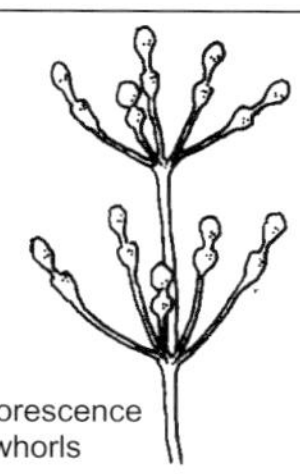	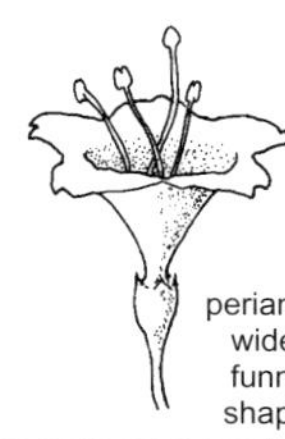	3
3	**Flowers** sessile or very shortly pedicellate; **pedicels** usually <4mm long (those in the lowest whorl sometimes longer); **perianths** <2.5mm long; **stamens** usually 2	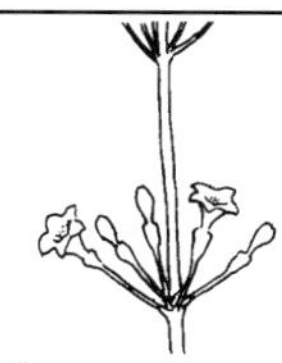	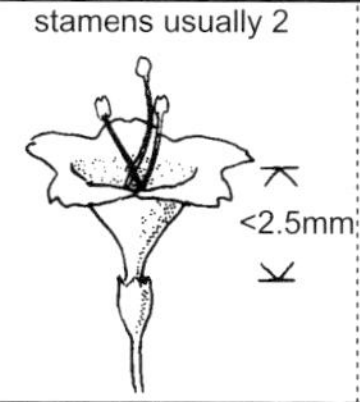	***C. helenae*** 641
	Flowers distinctly pedicellate; **pedicels** often >1cm long; **perianths** 3--5mm long; **stamens** 3(or 4)	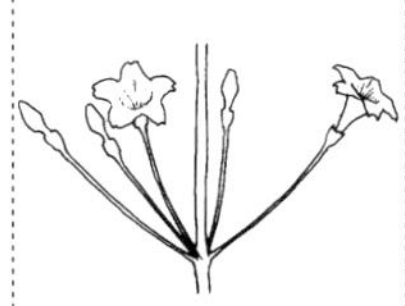	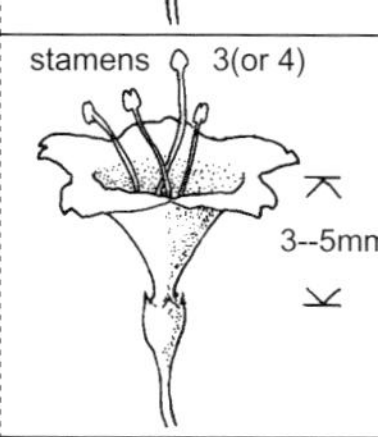	***C. boissieri*** 640

OLEACEAE

1	**Climber; leaves** opposite, trifoliolate; **flowers** white or yellow, fragrant, 5 to 9-lobed with 2 stamens; **fruit** a berry, sometimes 2-lobed		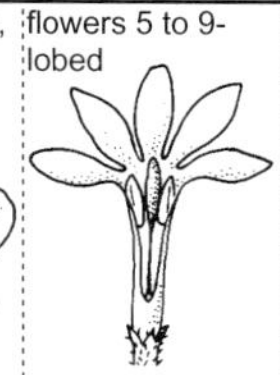	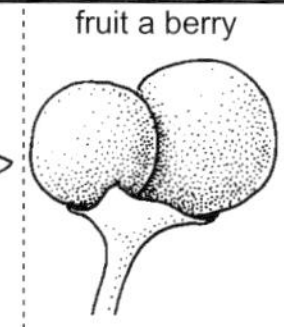	***Jasminum fluminense*** Endemic subsp. 642

ONAGRACEAE

1	Creeping **herb**, often near water; **leaves** opposite, with winged **petioles**; **flowers** in leaf axils, small, greenish	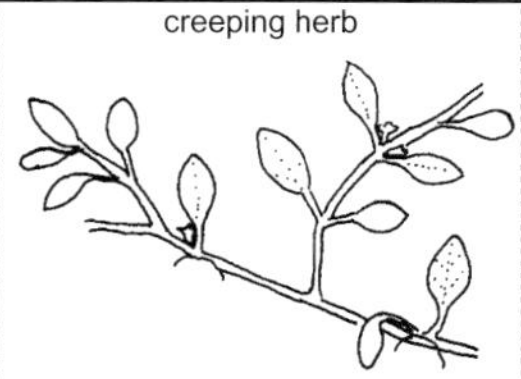	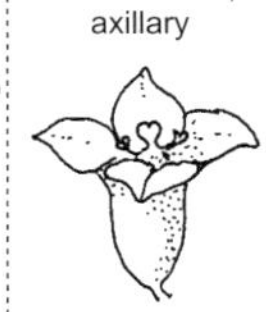	***Ludwigia palustris*** 643

OROBANCHACEAE

1 Robust **plants**; **flowers** large, <5cm long; **corolla** ± equally 5-lobed

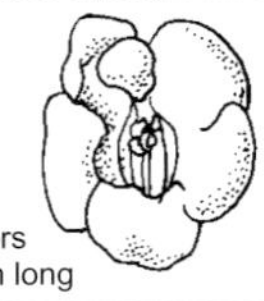

Cistanche phelypaea **272**, 643

Slender **plants**; **flowers** small, <2cm long; **corolla** 2-lipped

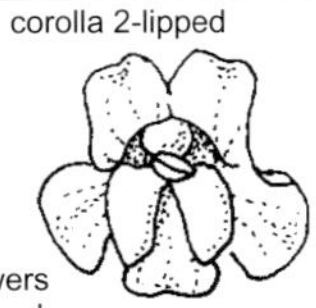

Orobanche (3 spp.)

Orobanche

1 **Flowers** in lax spikes; **bracteoles** present; **corolla tube** not inflated or glossy

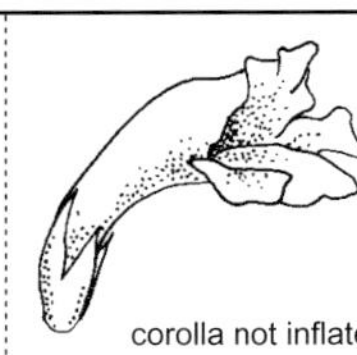

O. ramosa 644

Flowers in lax spikes; **bracteoles** absent; **corolla tube** not inflated or glossy

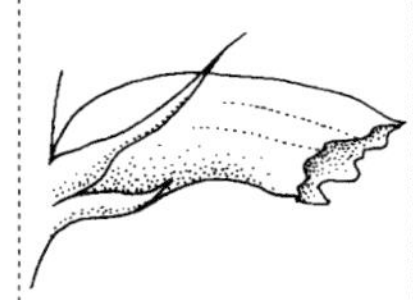

O. minor 644

Flowers in dense spikes; **bracteoles** absent; **corolla tube** inflated, glossy (especially when dry)

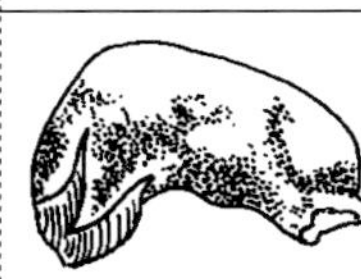

O. cernua 644

OXALIDACEAE

1 Prostrate **herb**; **leaves** alternate, 3-foliolate; **flowers** 5--10 mm across; **petals** 5, yellow; **fruit** a 5-celled capsule, the **seeds** explosively dehisced

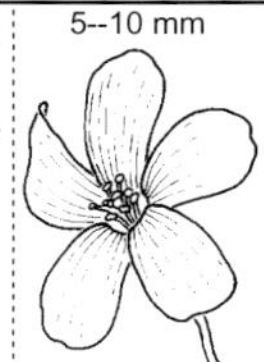

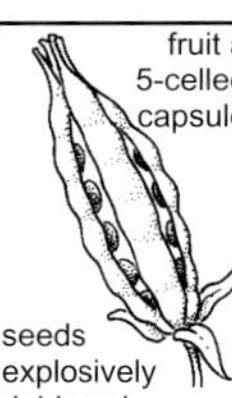

Oxalis corniculata 644

PAPAVERACEAE

1 Prickly **herb**; **leaves** spinose; **flowers** yellow, large; **fruit** a spiny capsule

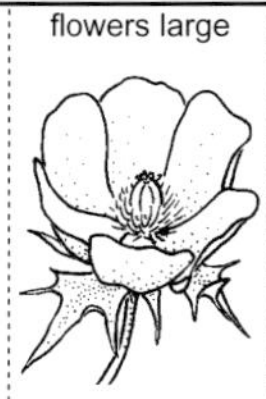

Argemone mexicana **272**, 645

PEDALIACEAE

1 Erect succulent **herb**; **leaves** opposite; **corolla** 2-lipped, yellow; **fruit** bearing spines

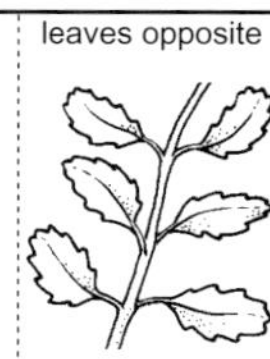

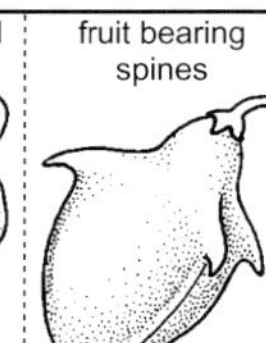

Pedalium murex 646

PIPERACEAE

1 Prostrate **herb**, often rooting at the nodes; **leaves** glabrous or glabrescent, mostly whorled, < 12mm long

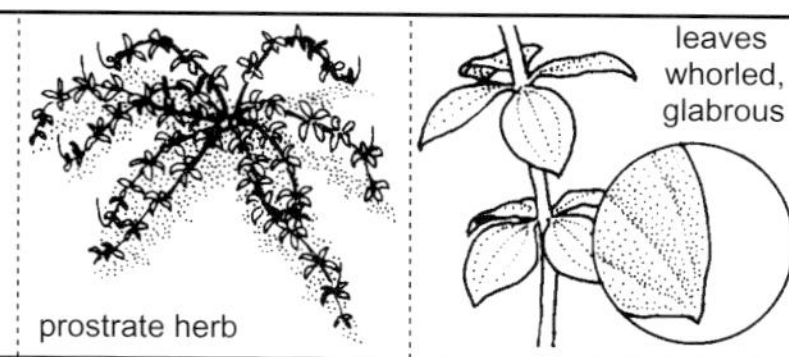

Peperomia tetraphylla 647

Erect or ascending **herb**; **leaves** sparsely pubescent, mostly opposite, > 15mm long

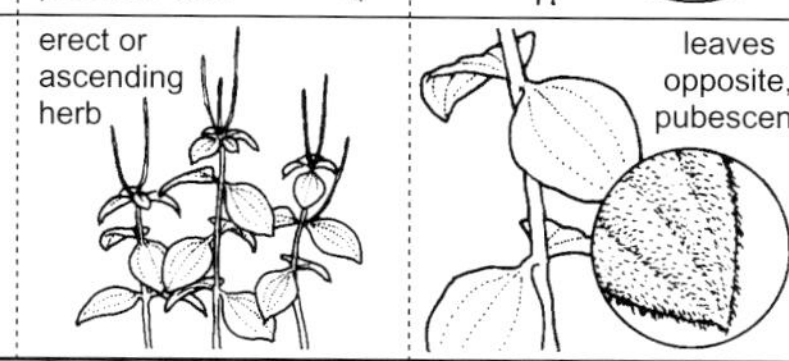

Peperomia blanda **272**, 646

PITTOSPORACEAE

1 **Shrub** or **tree**; **leaves** alternate; **flowers** bisexual, 5-merous; **fruit** a capsule

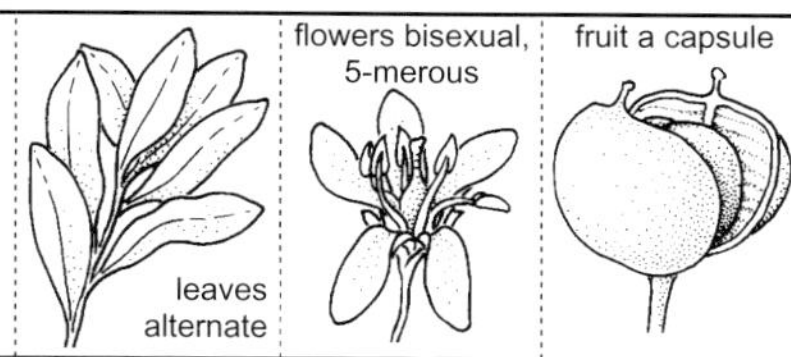

Pittosporum viridiflorum 647

PLANTAGINACEAE

1 Short **stem** present; **leaves** linear-lanceolate, <15mm wide, thinly hairy

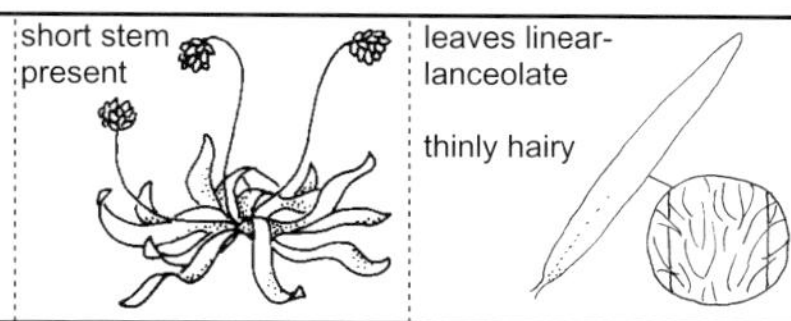

Plantago amplexicaulis 648

Stem absent; **leaves** linear, <2mm wide, densely grey-hairy

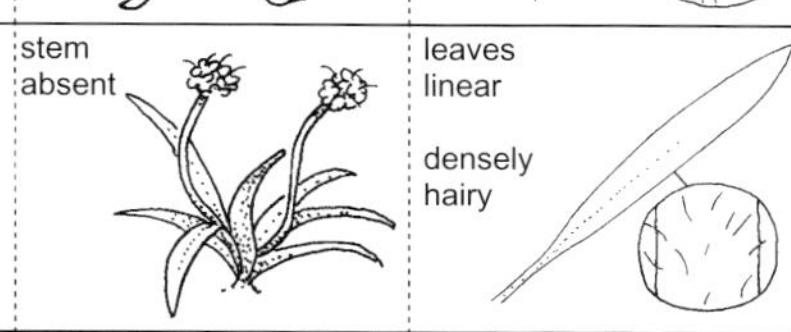

Plantago ovata 648

PLUMBAGINACEAE

1 **Shrubs** or small **trees** with open spreading or pendulous branches; **leaves** distant; **calyx** deeply 5-lobed; **corolla** yellow, funnel-shaped; **styles** 1

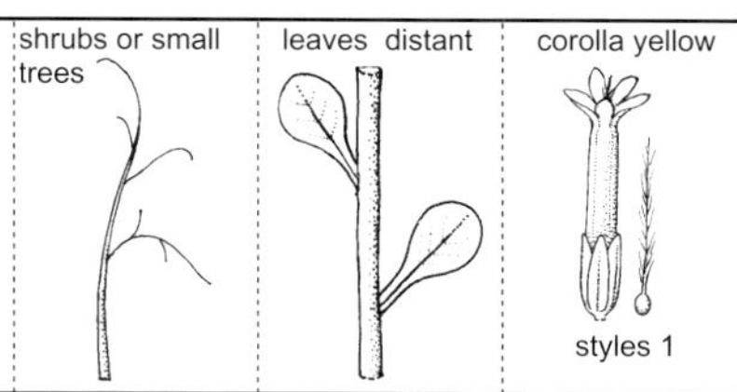

Dyerophytum (2 spp.)

Subshrubs, tufted or with a branched woody stock; **leaves** congested; **calyx** funnel-shaped, scarious; **corolla** pink, deeply 5-lobed; **styles** 5

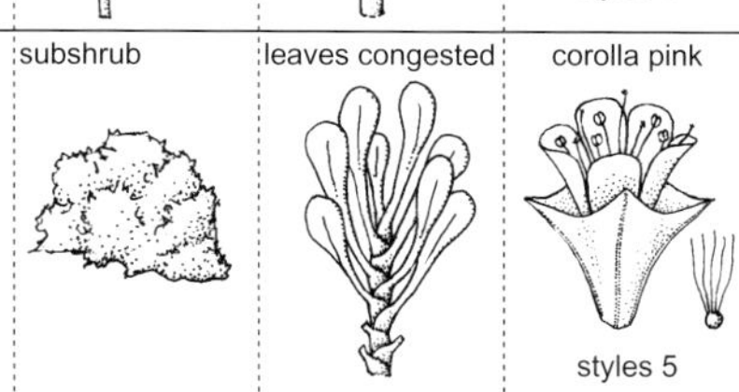

Limonium (2 spp.)

Dyerophytum

1 **Leaves** petiolate, lamina attenuate at base; **calyx** c. 10mm long

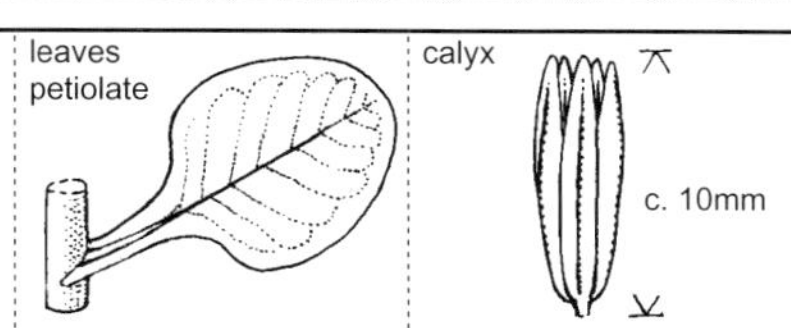

D. pendulum Endemic 649

Leaves sessile, +/- amplexicaul; **calyx** c. 5mm long

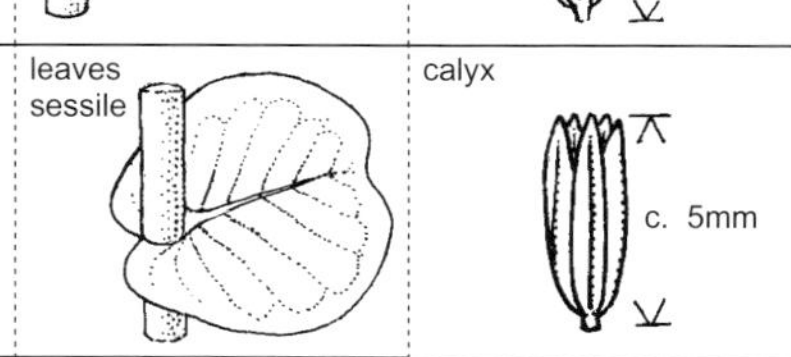

D. socotranum Endemic **272**, 649

Limonium

1	**Leaves** spathulate, flattened	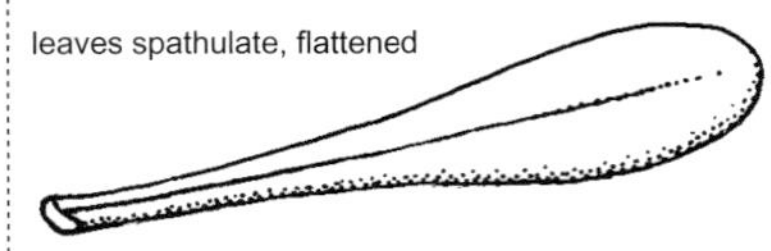	***L. paulayanum*** Endemic **272**, 650
	Leaves linear-cylindrical	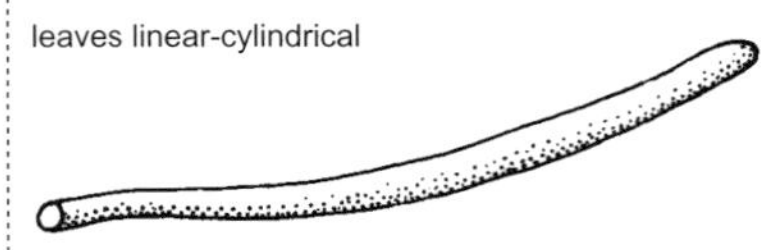	***L. sokotranum*** Endemic **272**, 650

POLYGALACEAE

1	**Flowers** in axillary racemes; **stems** glabrous; **capsule** hairy, narrowly winged; **seeds** with 3 short basal appendages	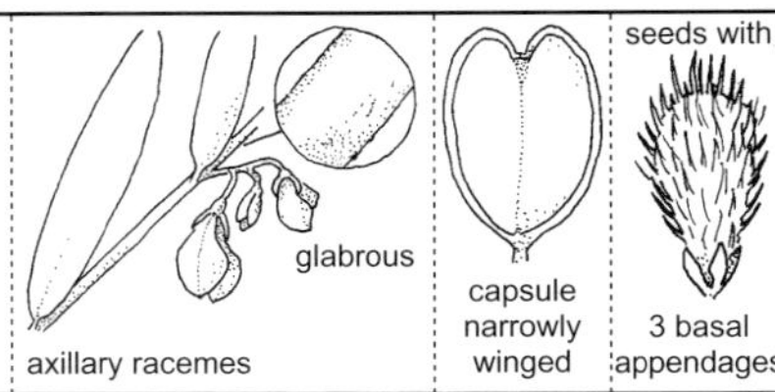	***Polygala erioptera*** 651
	Flowers in terminal racemes; **stems** with short crisped hairs; **capsule** glabrous, not winged; **seeds** with 2 short appendages	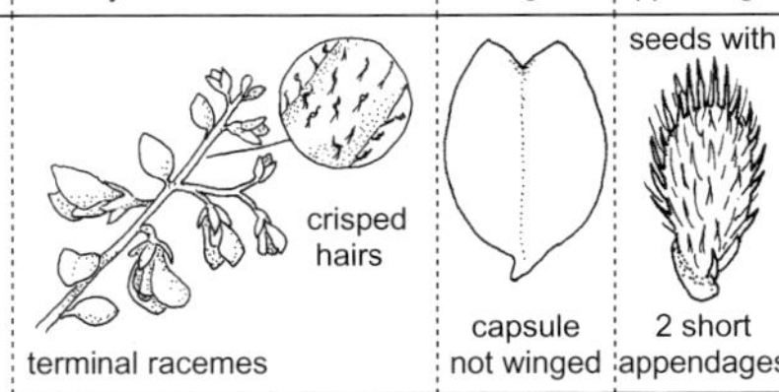	***Polygala kuriense*** Endemic **273**, 651
	Flowers in terminal racemes; **stems** glabrous; **capsule** glabrous, winged; **seeds** with a single long appendage	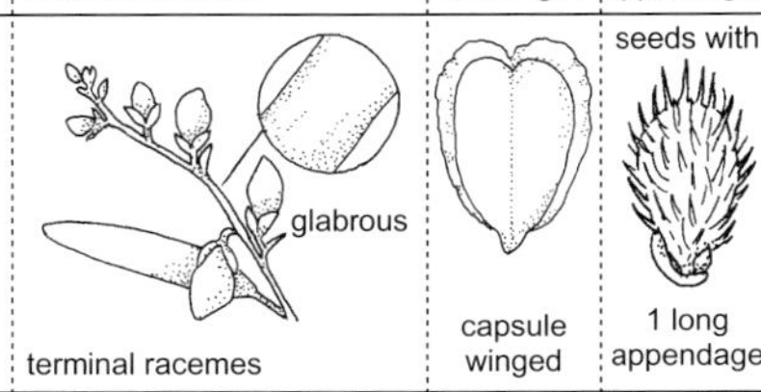	***Polygala abyssinica*** 651

POLYGONACEAE

1	**Shrubs**; **flowers** in axillary fascicles; **leaves** minute and soon deciduous; **ochreae** short, inconspicuous; **fruit** covered with stiff setae	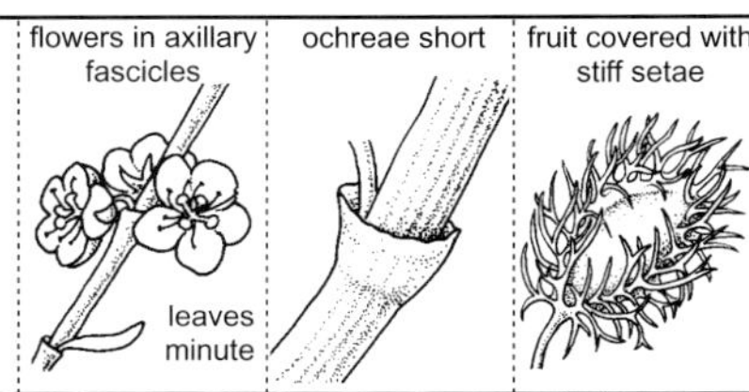	***Calligonum comosum*** 652
	Perennial **herbs**; **flowers** in terminal spikes; **leaves** lanceolate; **ochreae** conspicuous, sometimes with bristles; **fruit** a smooth nutlet	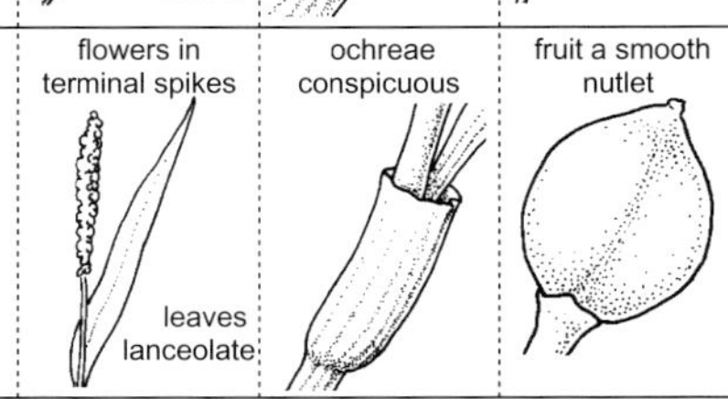	***Persicaria*** **(2 spp.)**
	Annual **herbs**; **flowers** in axillary clusters; **leaves** ovate; **ochreae** membranous; **fruit** pitted, with 3 terminal spines	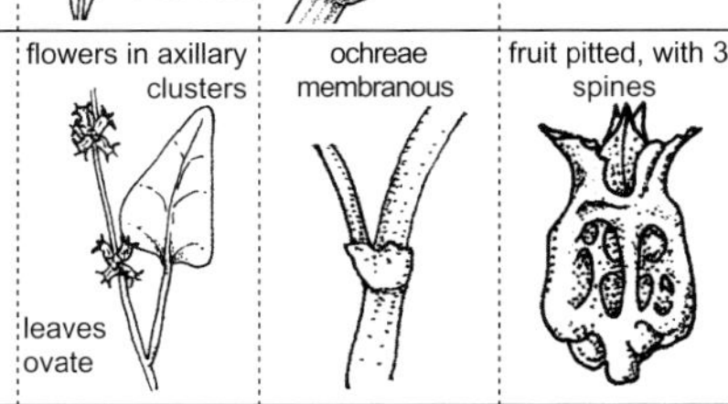	***Emex spinosus*** 652
	Annual **herbs**; **flowers** in axillary clusters; **leaves** ovate to lanceolate; **ochreae** fringed with fine setae; **fruit** with 3 spines arising from near middle	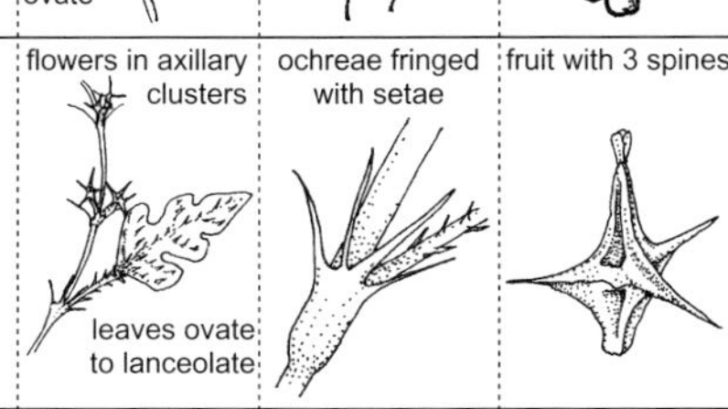	***Oxygonum sinuatum*** 653

Persicaria

1 **Ochreae** with a terminal fringe of fine bristles up to 1.5cm; **leaves** hairy beneath

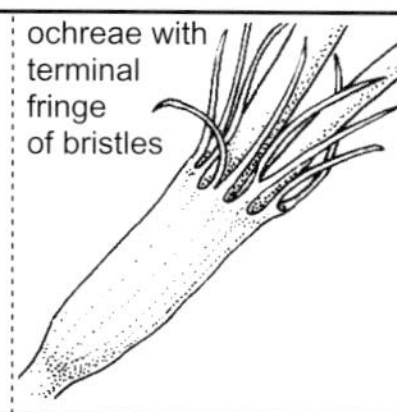

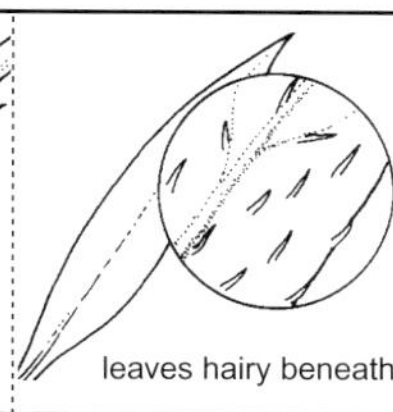

P. barbata 653

Ochreae truncate, without bristles; **leaves** glabrous

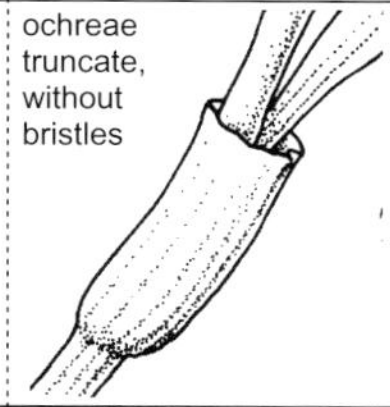

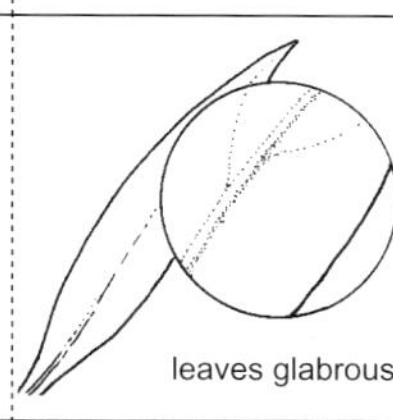

P. glabra 653

PORTULACACEAE

1 Erect **herb; fruit** dehiscing by valves; **flowers** red, pink or purple

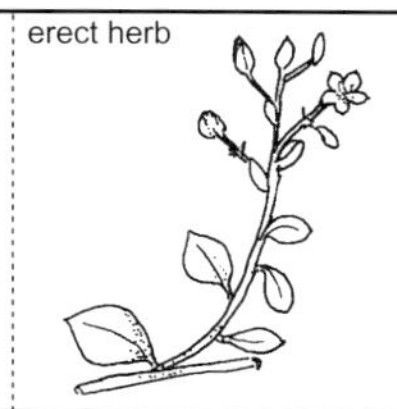

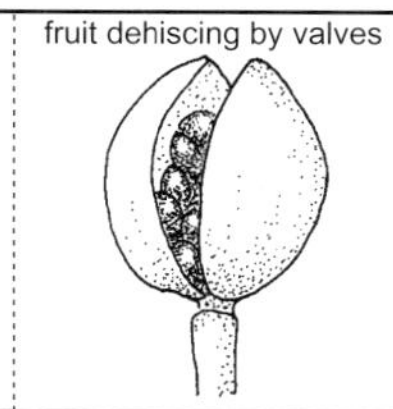

Talinum portulacifolium **273**, 656

Prostrate or decumbent **herb**; **fruit** dehiscing by a lid; **flowers** yellow

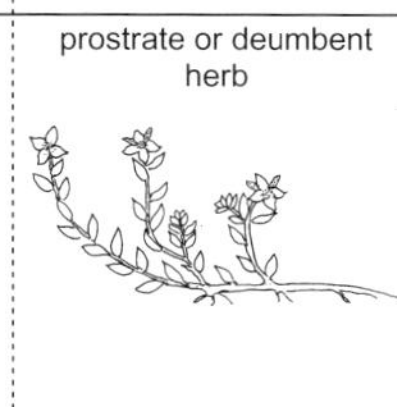

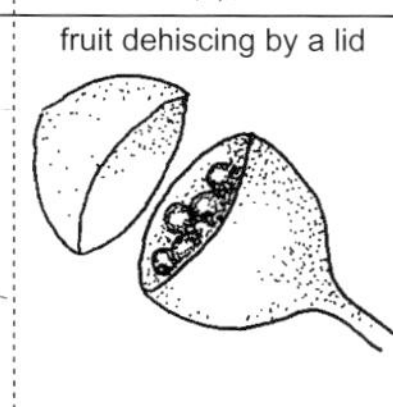

Portulaca (5 spp.)

Portulaca

1 **Leaves** globose; **flowers** stalked; **petals** 6

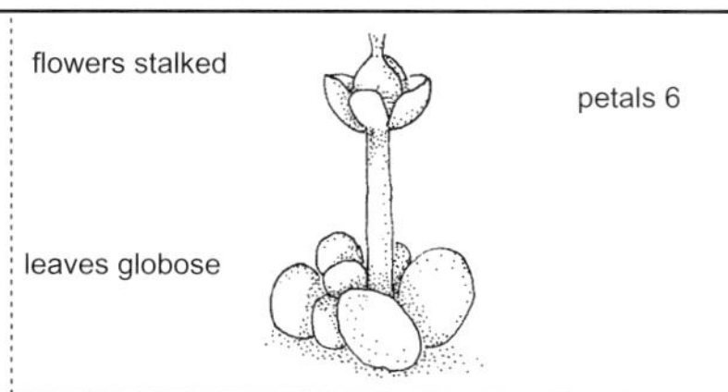

P. sedifolia Endemic **273**, 656

Leaves flattened or cylindrical; **flowers** sessile; **petals** 4 or 5

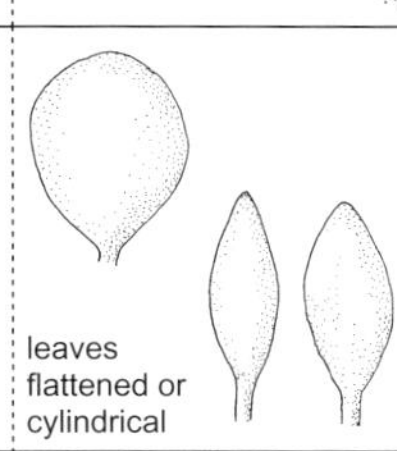

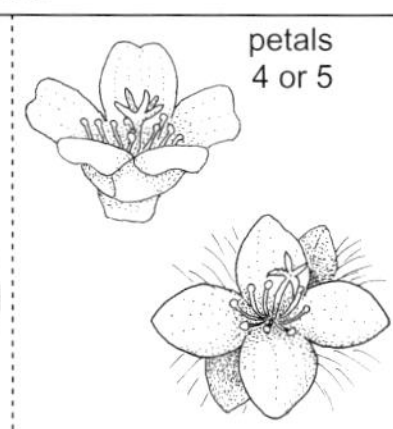

2

2 **Petals** 5; **leaves** alternate; **stipular hairs** axillary, soon falling

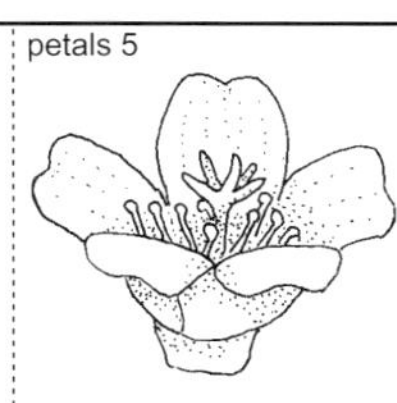

P. oleracea 655

Petals 4; **leaves** opposite; **stipular hairs** surrounding nodes, conspicuous

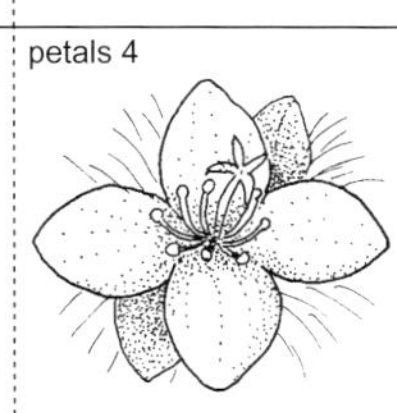

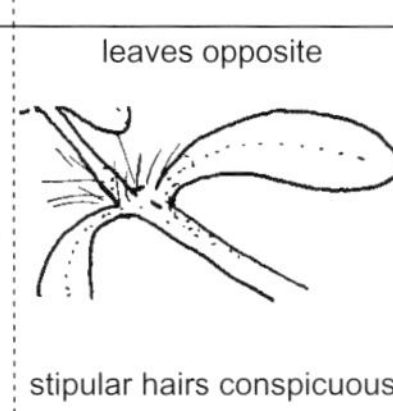

4

3

Stems ascending from a stout tuberous rootstock; **leaves** narrowly lanceolate, <25mm long

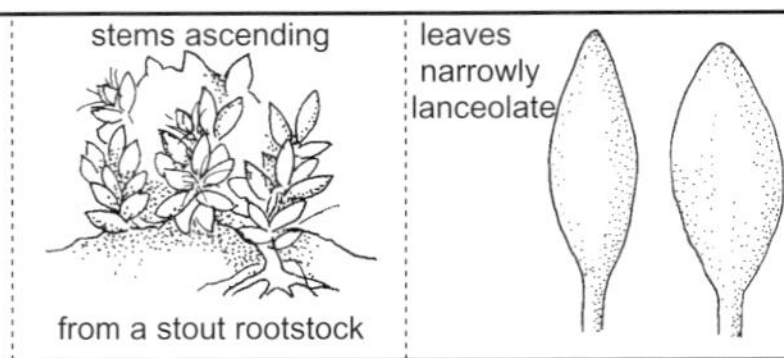

P. kuriensis
Endemic
273, 654

Stems mat-forming from narrow succulent roots; **leaves** obovate, <10mm long

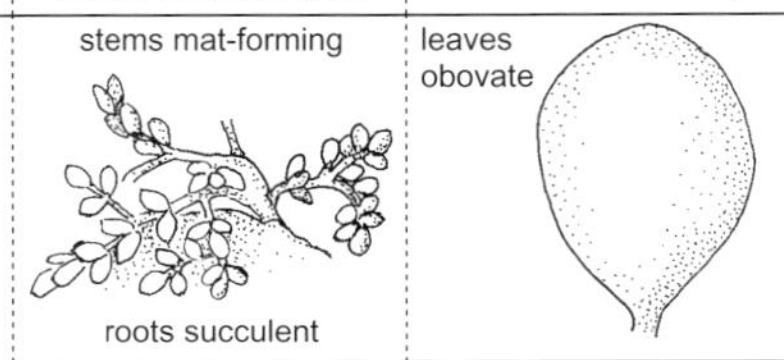

P. samhaensis
Endemic
273, 655

Stems decumbent from a narrow tuberous rootstock; **leaves** elliptic to oblong-obovate, <12mm long

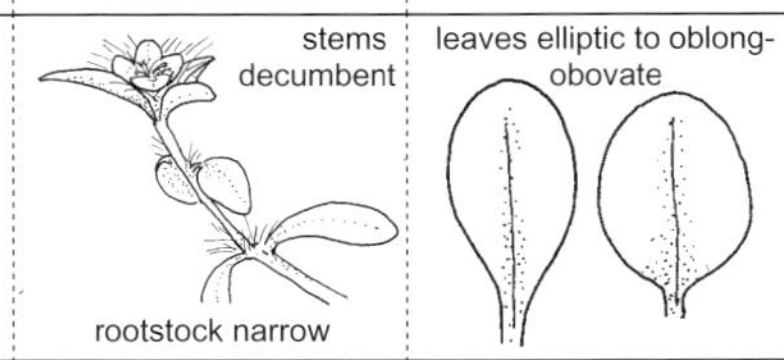

P. quadrifida
273, 655

PRIMULACEAE

1

Annual **herb**; **leaves** opposite, sessile; **flowers** blue, c.10mm across; **fruit** a capsule opening by a lid

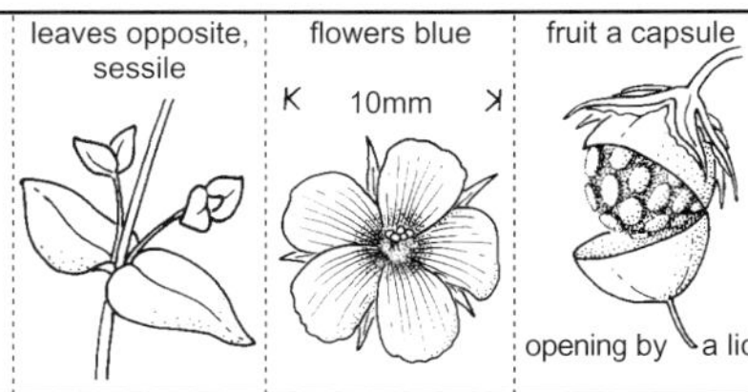

Anagallis arvensis
657

PUNICACEAE

1

Shrub or small **tree**; **leaves** opposite; **flowers** bright red, thick-textured, with many stamens and tiny petals; **fruit** globose, ripening brownish red

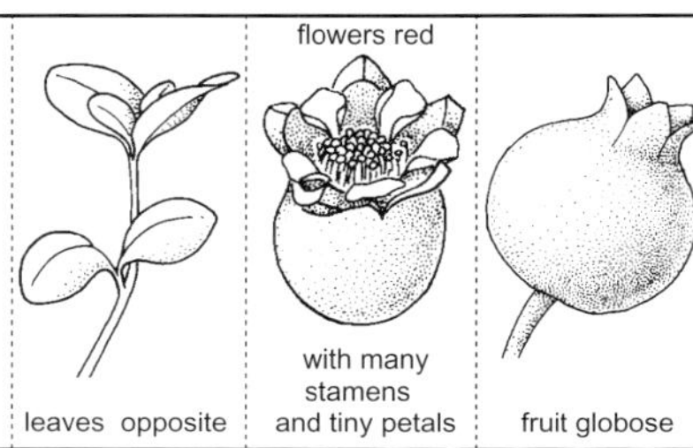

Punica protopunica
Endemic
273, 657

RESEDACEAE

1

Usually leafless tangled **shrub**; **leaves** (when present) linear; **ovary** sessile, of 3 fused carpels; **fruit** a berry

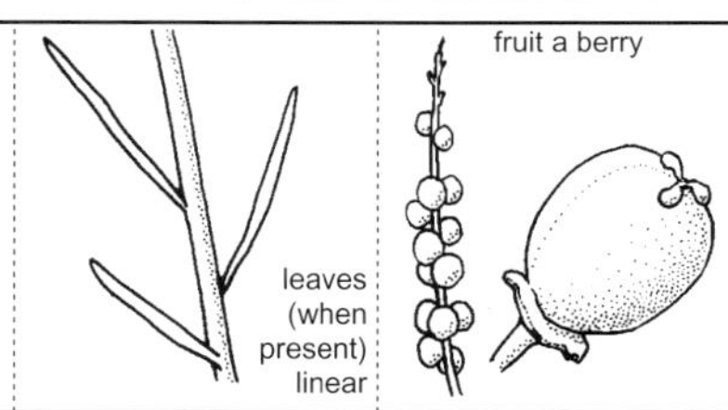

Ochradenus socotranus
Endemic
274, 659

Herb; **leaves** narrowly obovate to linear-obovate; **ovary** distinctly stipitate, of 5 or 6 carpels, fused at base and free above; **fruit** of divergent carpels

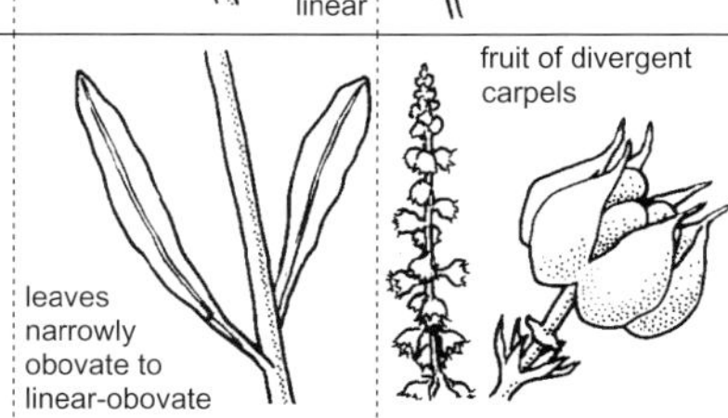

Caylusea hexagyna
658

Little-branched **shrub**; **leaves** simple to trifoliolate, ovate to circular; **ovary** stipitate, of 3 or 4(or 5) fused carpels; **fruit** a capsule, usually open at apex

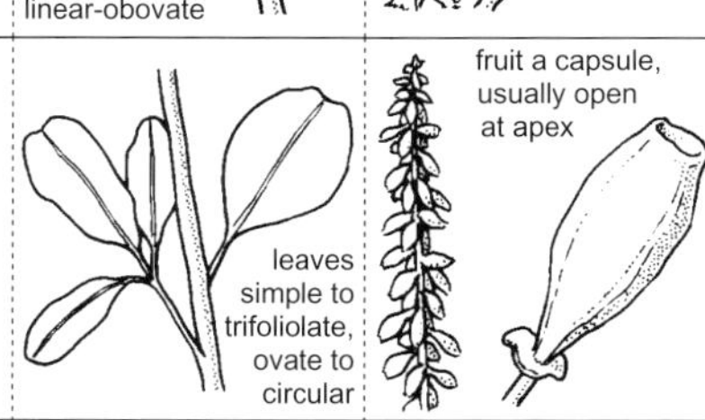

Reseda viridis
Endemic
274, 659

RHAMNACEAE

1 **Shrub** or large **tree**; **branches** spiny, without tendrils; **leaves** 3 to 5-nerved from base; **flowers** in cymes; **fruit** a yellow or reddish berry, 1--1.5cm diam. — ***Ziziphus spina-christi*** **274**, 660

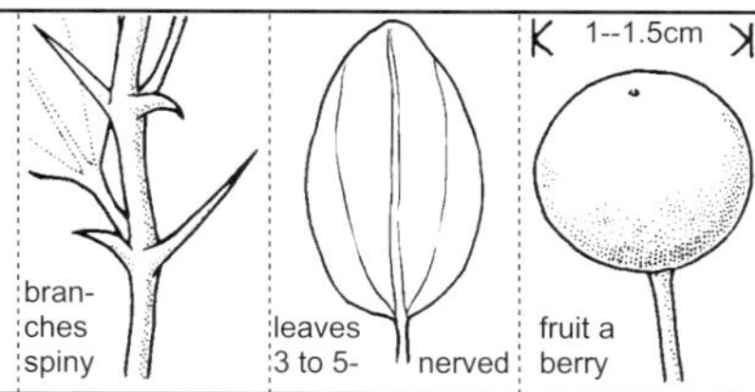

Woody **climber**; **branches** unarmed, bearing tendrils; **leaves** pinnately nerved; **flowers** in umbels; **fruit** splitting into 3, c.5mm diam. — ***Helinus integrifolius*** 660

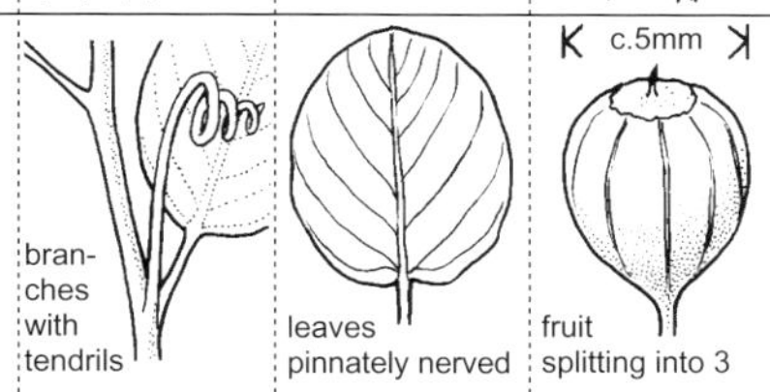

RUBIACEAE

1 **Leaves** in whorls of 4 or more, without a stipular sheath between the petioles; **flowers** inconspicuous — 2

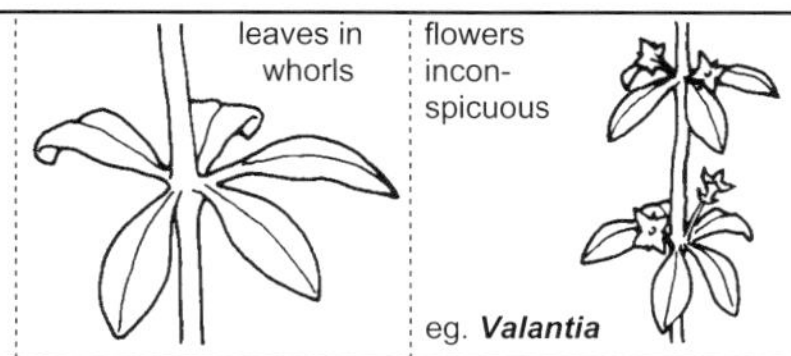

Leaves opposite (sometimes clustered at nodes but never whorled), with a stipular sheath; **flowers** conspicuous — 3

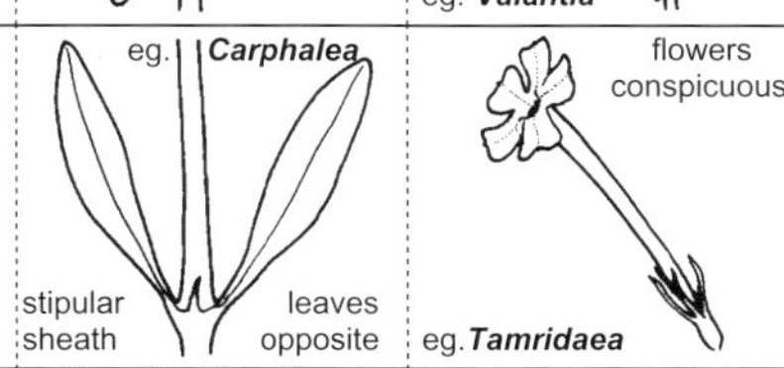

2 **Fruit** ± sessile, hidden by spiny appendages (the recurved pedicels); **leaves** oblong or ovate — ***Valantia hispida*** 671

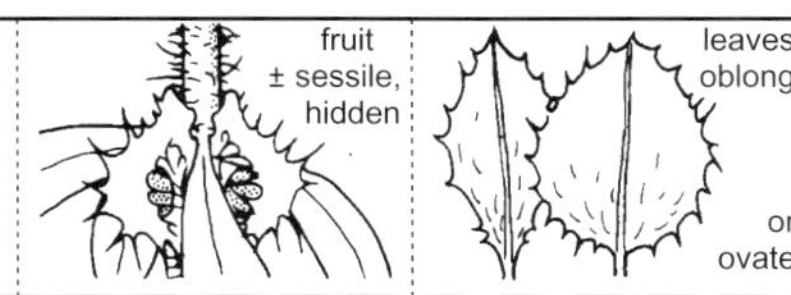

Fruit distinctly stalked, not concealed by spiny appendages; **leaves** linear to narrowly oblanceolate — ***Galium*** (2 spp.)

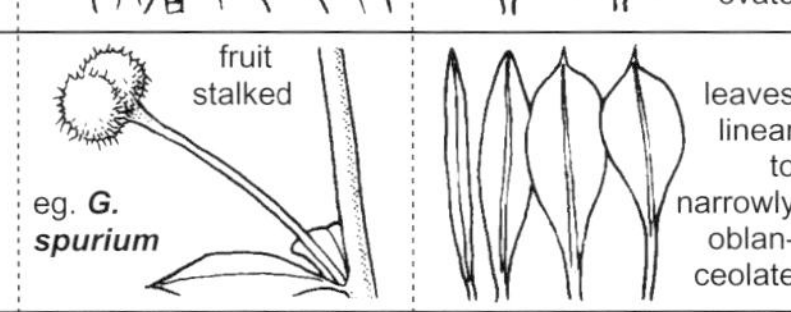

3 **Shrubs** or small **trees**; **leaves** elliptic to obovate, more than 1cm across — 4

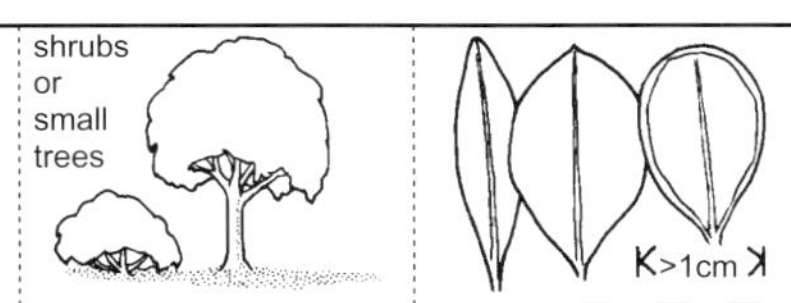

Herbs, sometimes woody-based, if subshrubs then **leaves** less than 0.5cm across — 7

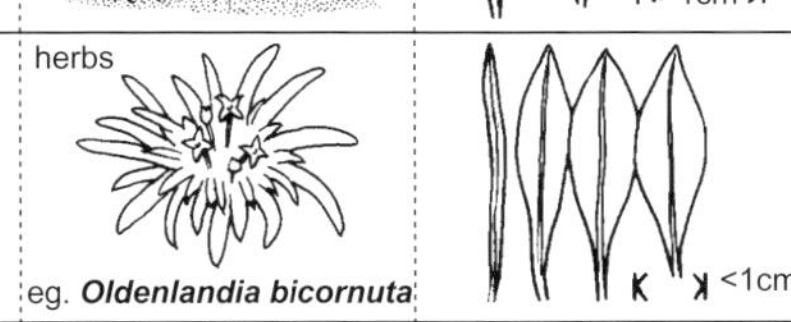

4 **Calyx** with a leaflike wing or 5 unequal, leaflike lobes; **corolla tube** longer than corolla lobes; **fruit** dry, dehiscent or indehiscent — 5

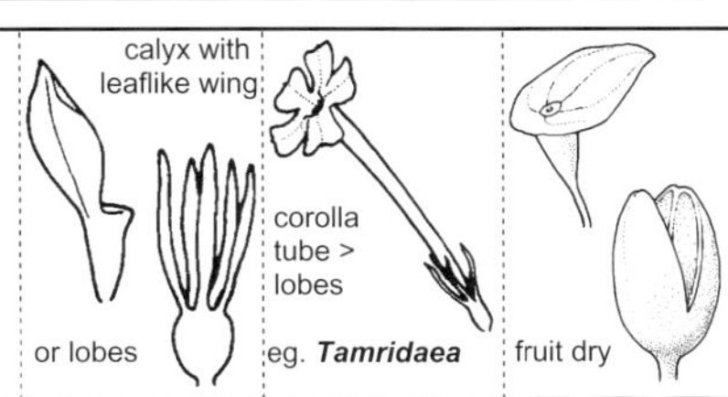

Calyx with 5 triangular lobes; **corolla tube** shorter than or equalling corolla lobes; **fruit** spherical, fleshy — 6

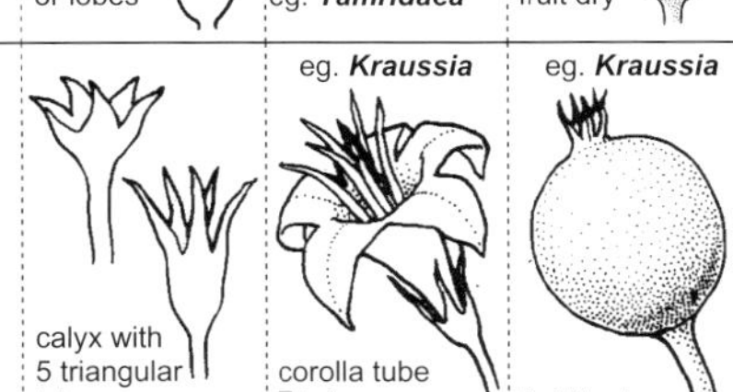

5

Calyx with a leaflike wing; **petals** acute; **fruit** indehiscent, hidden by the expanded calyx

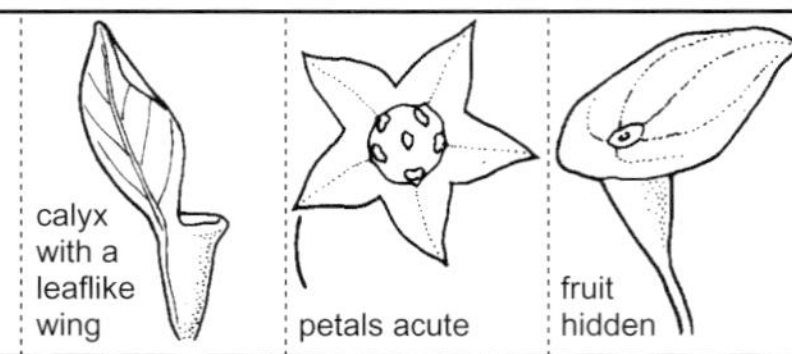

Carphalea obovata
Endemic
274, 663

Calyx with 5 unequal, leaflike lobes; **petals** emarginate; **fruit** dehiscent, not hidden

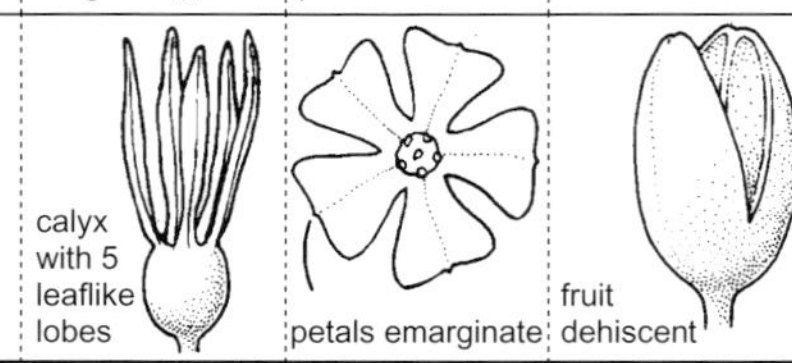

Tamridaea capsulifera
Endemic
Endemic genus
276, 670

6

Corolla tube c.2.5mm long; **stigma** capitate; **leaves** obovate with inrolled margins, bearing sunken cavities (domatia) beneath

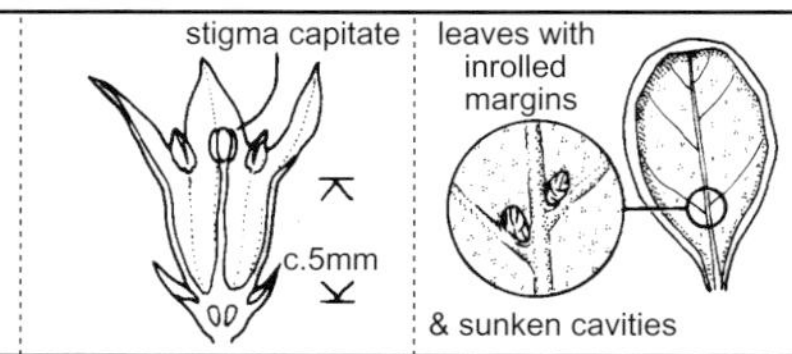

Pyrostria socotrana
Endemic
670

Corolla tube c.5mm long; **stigma** linear; **leaves** oblong-elliptic, margins not inrolled, not bearing cavities beneath

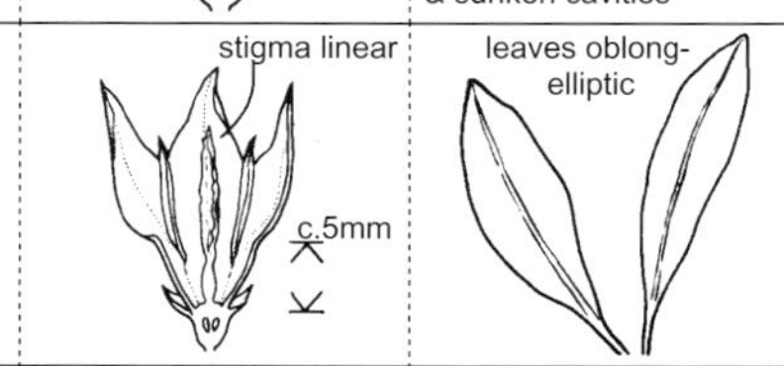

Kraussia socotrana
Endemic
275, 667

7

Subshrub; **corolla lobes** densely hairy; **fruiting calyx** strongly ribbed

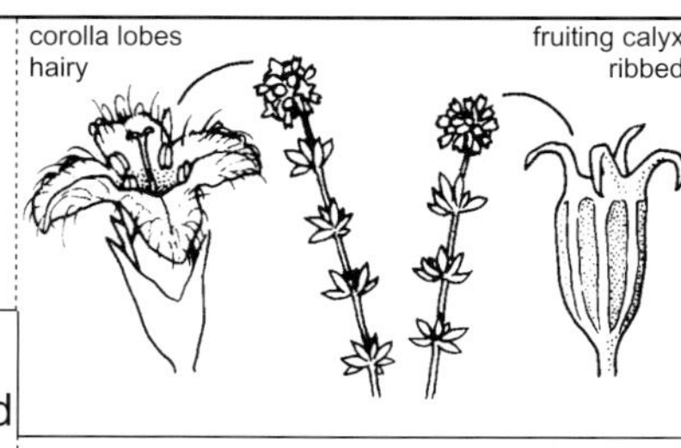

Placopoda virgata
Endemic
Endemic genus
276, 669

Herb; **corolla lobes** glabrous or thinly hairy; **fruiting calyx** unribbed

8

8

Flowers 5-merous; **plant** often foetid (foul-smelling)

Gaillonia
(4 spp.)

Flowers 4-merous; **plant** not foetid

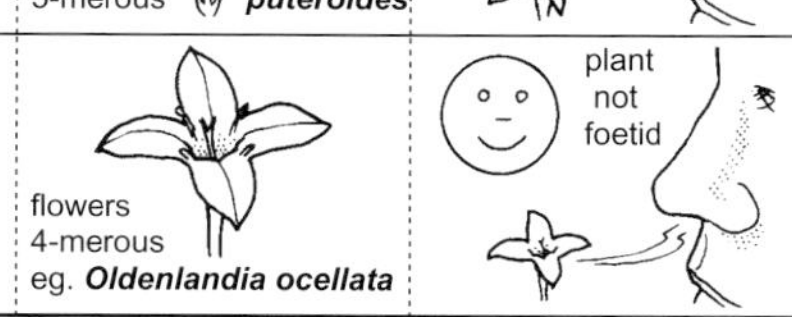

9

9

Corolla brownish yellow with linear lobes; **stamens** and **stigma** hidden within the corolla tube

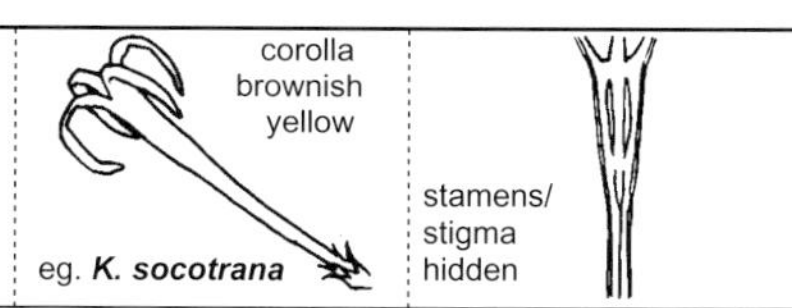

Kohautia
(2 spp.)

Corolla white or pale blue with oblong or ovate lobes; **stamens** and **stigma** visible at the mouth of the corolla tube

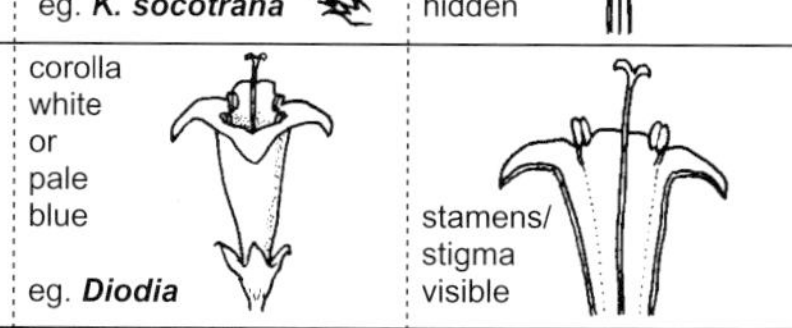

10

10

Branched **herb** with sessile, axillary flowers; **calyx** hairy; **capsule** 2-seeded, the seeds ribbed

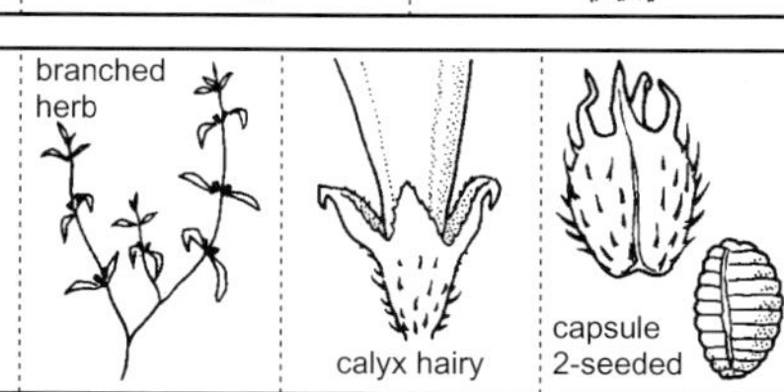

Diodia aulacosperma
664

Cushion- or mat-forming **herbs**, or if loosely branched then flowers pedicellate; **calyx** glabrous; **capsule** many-seeded, the seeds smooth

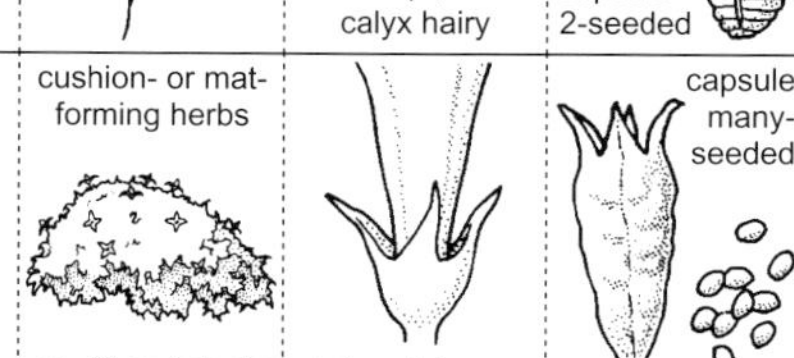

Oldenlandia
(7 spp.)

Gaillonia

1	**Corolla lobes** oblong, reflexed; **anthers** and **stigma** long-exserted; **plant** few-leaved, internodes longer than leaves; **leaves** linear to linear-elliptic or oblanceolate, flattened	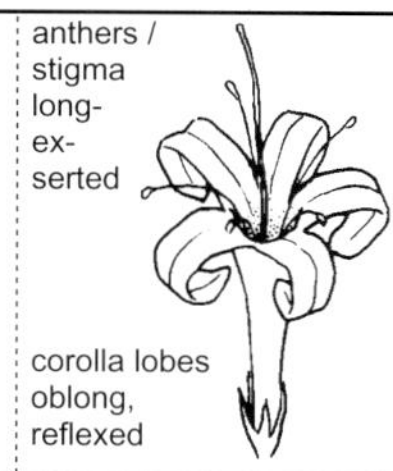	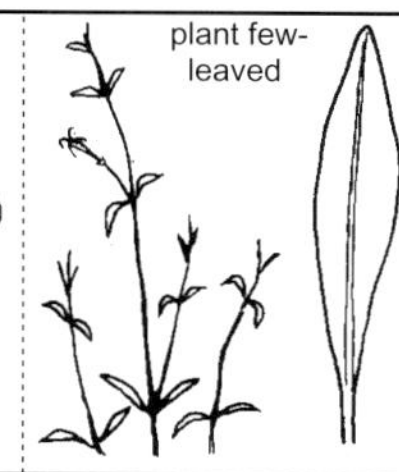	***G. puberula*** Endemic **274**, 664
	Corolla lobes narrowly lanceolate; **anthers** and **stigma** shortly exserted; **plant** few-leaved, internodes longer than leaves; **leaves** linear to linear-elliptic or oblanceolate, flattened	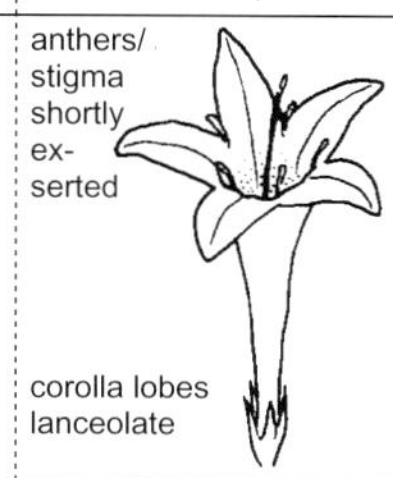	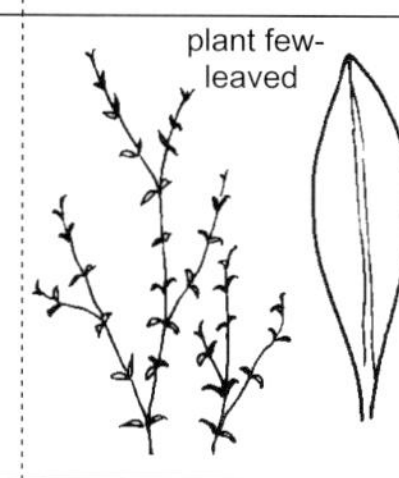	***G. thymoides*** Endemic **275**, 665
	Corolla lobes narrowly lanceolate; **anthers** and **stigma** shortly exserted; **plant** densely-leaved, internodes shorter than leaves; **leaves** oblanceolate, flattened	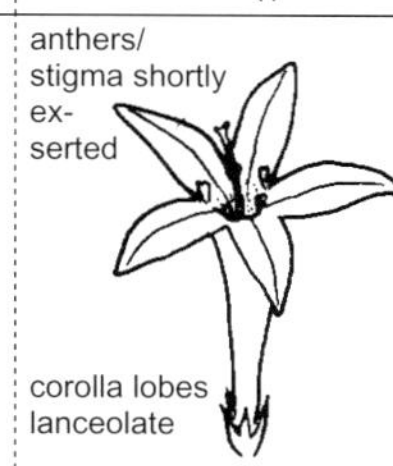	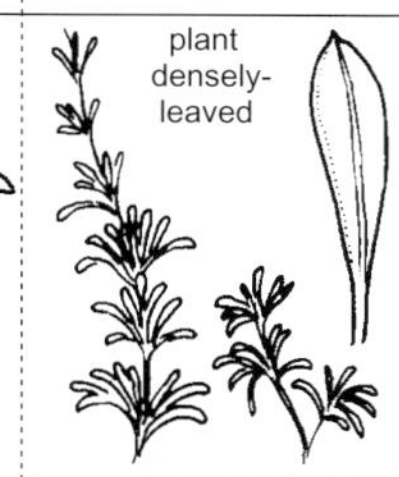	***G. putorioides*** Endemic **274**, 665
	Corolla lobes narrowly lanceolate; **anthers** and **stigma** shortly exserted; **plant** densely-leaved, internodes shorter than leaves; **leaves** linear, cylindrical	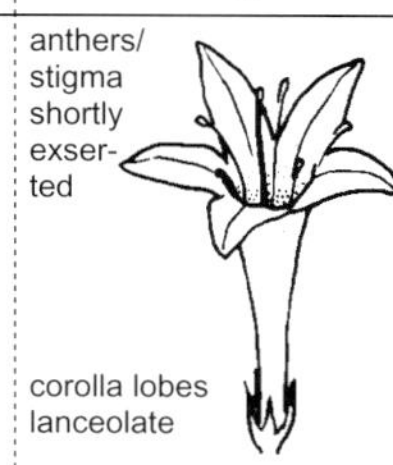	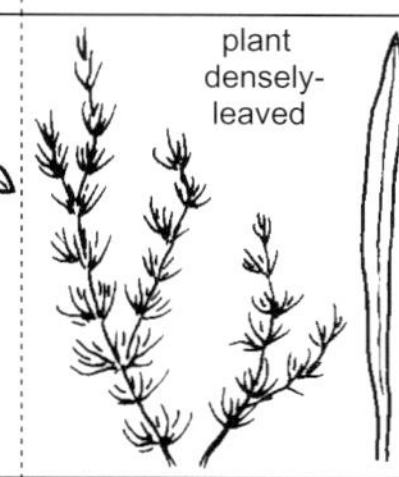	***G. tinctoria*** Endemic **275**, 665

Galium

1	Scrambling, "clinging" **herb**; **stems** with retrorse hairs; **leaves** mostly 1--3cm long, with a bristled tip; **fruit lobes** 3--4mm across	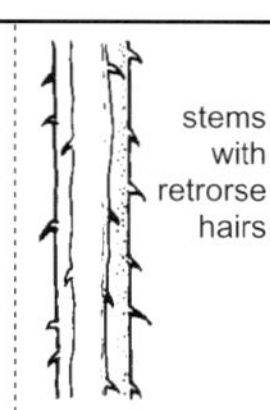	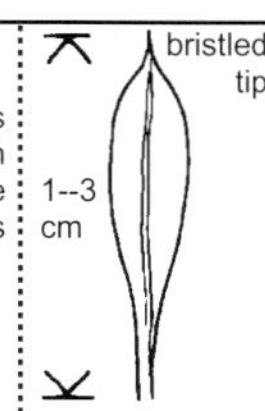	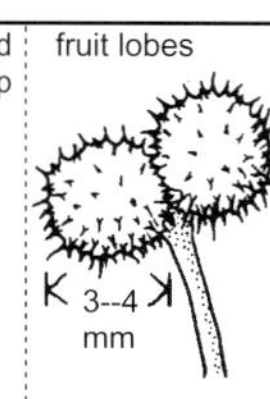	***G. spurium*** 666
	Erect **herb**; **stems** glabrous; **leaves** up to 1cm long, with an acute or obtuse tip; **fruit lobes** c.1mm across	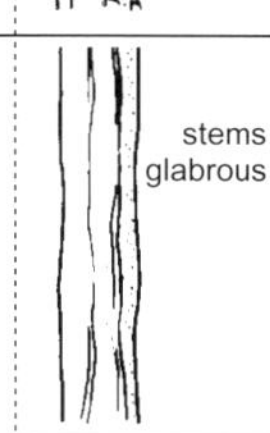	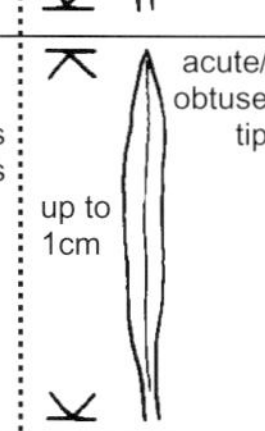	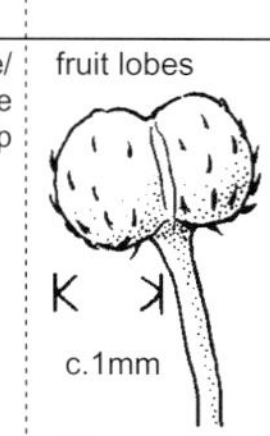	***G. setaceum*** 665

Kohautia

1	Ascending **perennial**; **leaves** linear to narrowly oblong; **flowers** pedicellate	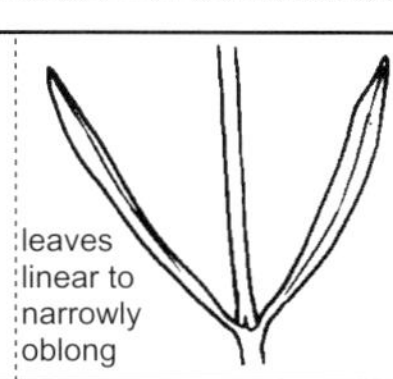	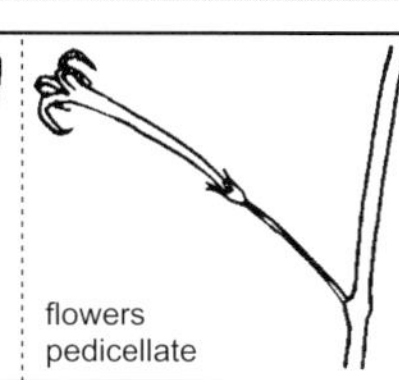	***K. socotrana*** Endemic **275**, 666
	Erect **annual**; **leaves** lanceolate; **flowers** sessile	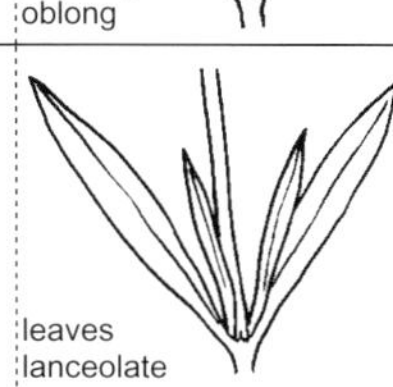	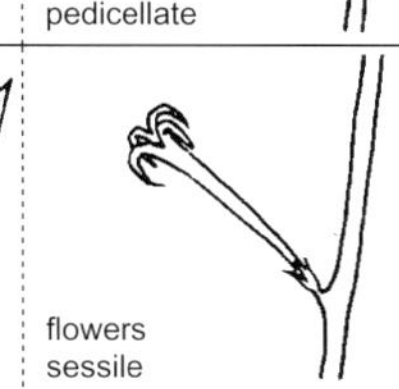	***K. subverticellata*** 666

Oldenlandia

1	Open-branched **herbs**			2
	Cushion- or mat-forming **herbs**	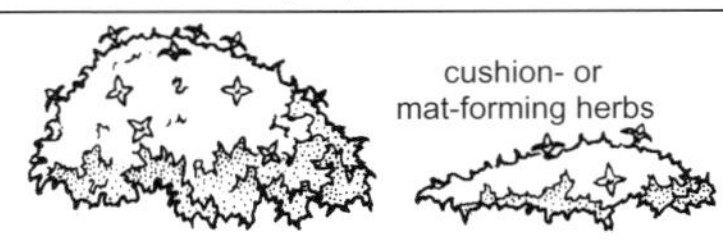		3
2	**Herb** to 60cm; **flowers** solitary or paired in the leaf axils, pedicels shorter than leaves; **fruiting calyx** round	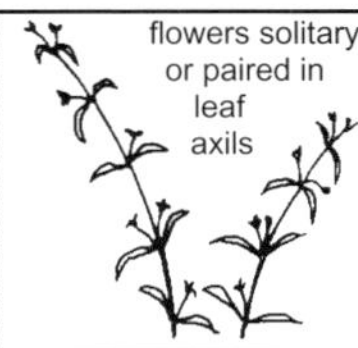	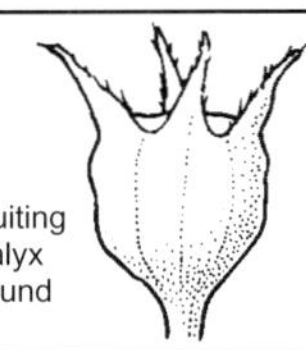	***O. herbacea*** 668
	Herb to 20cm; **flowers** in lax, terminal inflorescences, pedicels longer than leaves; **fruiting calyx** obtriangular			***O. balfourii*** Endemic **275**, 668
	Herb to 30cm; **flowers** solitary, paired or clustered in leaf axils; **fruiting calyx** round		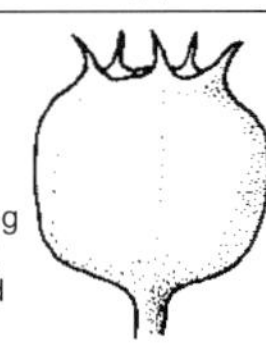	***O. corymbosa*** 668
3	**Annual**; **flowers** crowded at centre; **fruits** obtriangular, emarginate		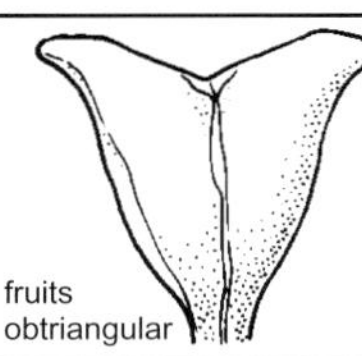	***O. bicornuta*** Endemic **275**, 668
	Perennial; **flowers** solitary, scattered; **fruits** round			4
4	**Leaves** spathulate, rounded with clear eye at tip (often succulent); **stipular sheath** not fringed; **calyx lobes** hyaline with a green dot	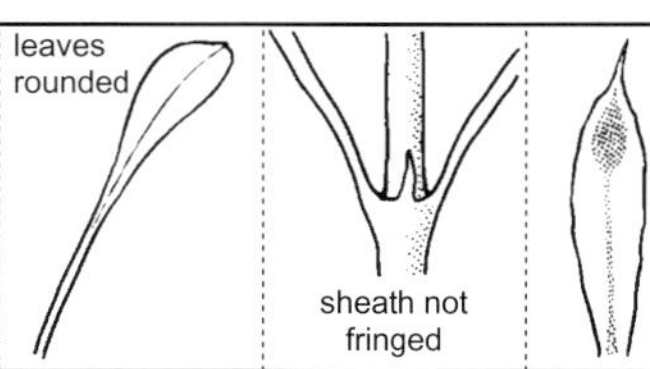		***O. ocellata*** Endemic **276**, 669
	Leaves linear to linear-oblong, acute; **stipular sheath** fringed; **calyx lobes** hyaline with a green midrib	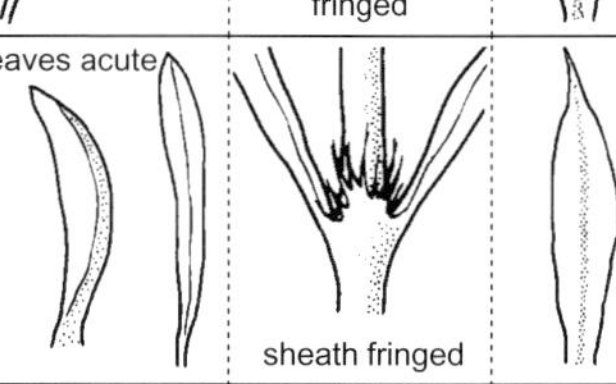		5
5	**Leaves** c.0.5cm long, the upper part angled			***O. aretioides*** Endemic 667
	Leaves 1--5cm long, the upper part straight			***O. pulvinata*** Endemic **276**, 669

RUTACEAE

	Description	Illustrations	Taxon
1	**Herb**; **leaves** pinnate, without spiny stipules; **fruit** a 4-lobed capsule; **petals** 4, yellow, fringed	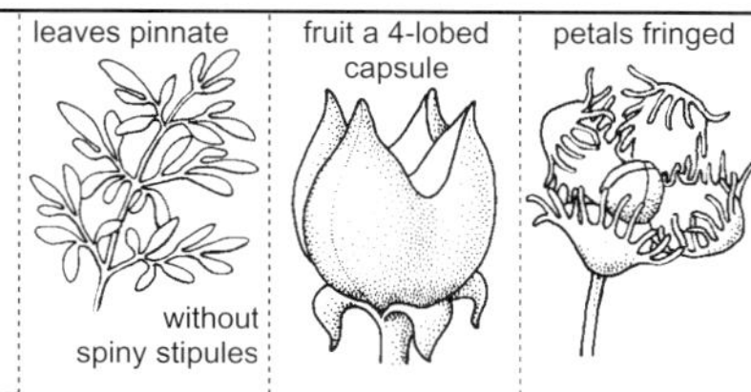	***Ruta chalapensis*** 672
	Tree; **leaves** simple, with a winged petiole and spiny stipules; **fruit** large, globose, orange (an orange); **petals** 5, white, entire	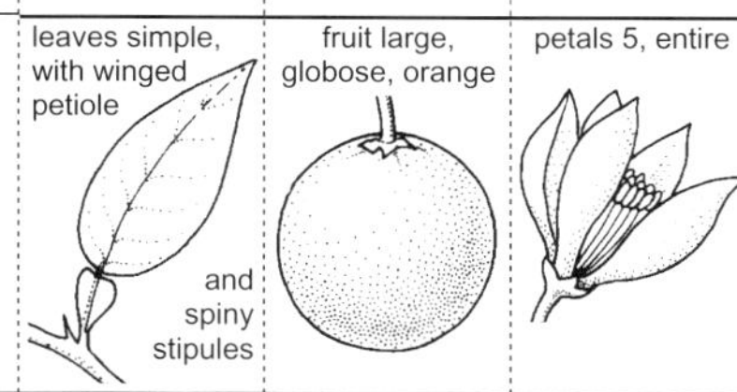	***Citrus aurantium*** 671
	Herb or **shrublet**; **leaves** simple, sessile, without spiny stipules; **fruit** a 2-lobed capsule; **petals** 4, pinkish or purplish, entire	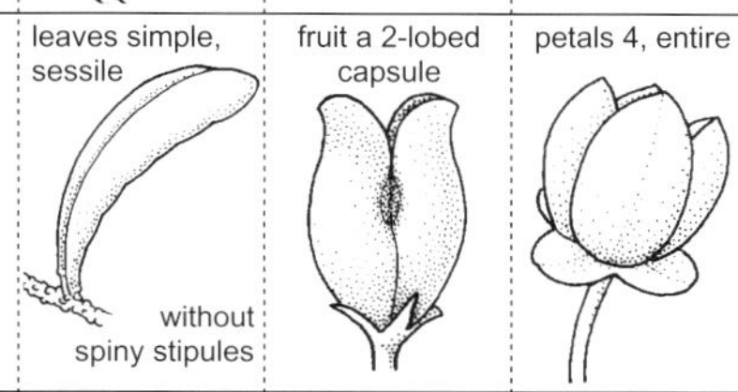	***Thamnosma*** (2 spp.)

Thamnosma

	Description	Illustrations	Taxon
1	**Shrublet**; **leaves** narrowly oblanceolate with strongly inrolled margins, congested at the ends of branches; **seeds** spiny	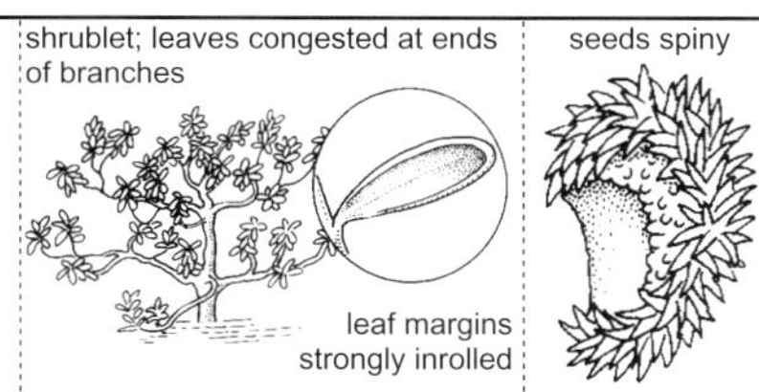	***T. socotrana*** Endemic **276**, 673
	Herb; **leaves** linear with weakly inrolled margins, laxly arranged along the stems; **seeds** minutely tuberculate	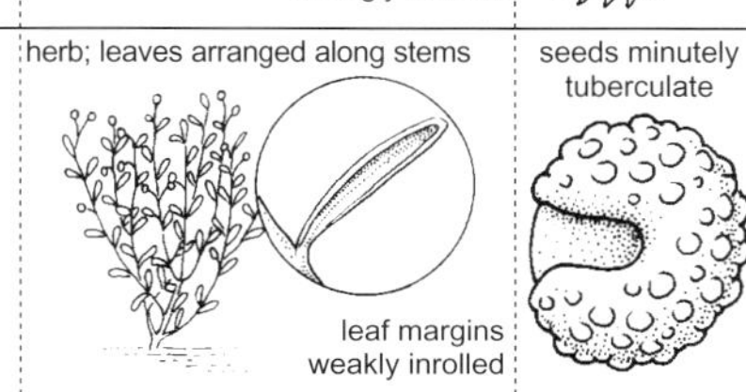	***T. hirschii*** 673

SALVADORACEAE

	Description	Illustrations	Taxon
1	Evergreen **shrub** forming tangled clumps; **leaves** bright green, leathery; **flowers** small, in panicles; **fruit** a globose red drupe	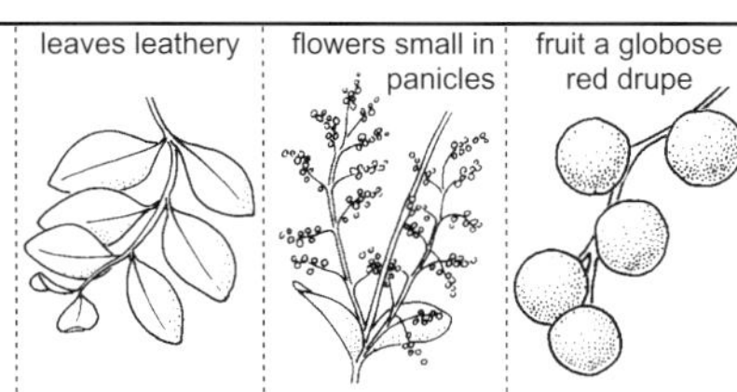	***Salvadora persica*** 674

SANTALACEAE

	Description	Illustrations	Taxon
1	**Shrub** or small **tree**, often with pendant branches; **leaves** alternate, simple, entire; **perianth** 3-lobed; **fruit** globose, bright red	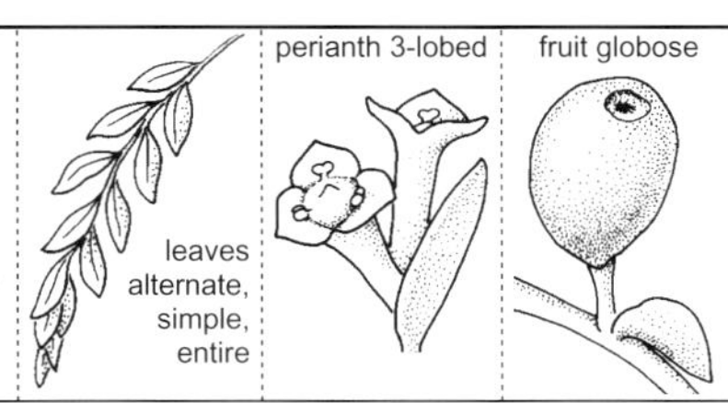	***Osyris quadripartita*** 675

SAPINDACEAE

1	**Leaves** simple, sticky, glossy; **petals** absent; **fruit** with 2 to 4 wings	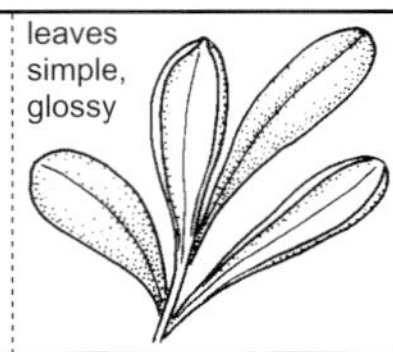	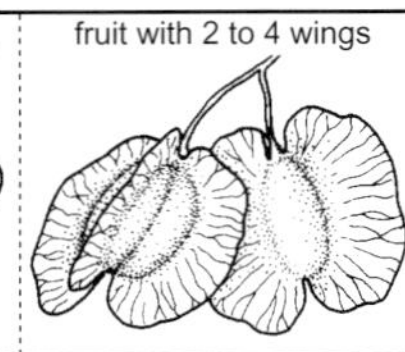	***Dodonaea viscosa*** 677
	Leaves 3-foliolate, not sticky; **petals** present; **fruit** a drupe	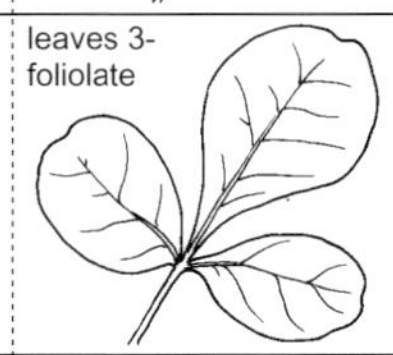	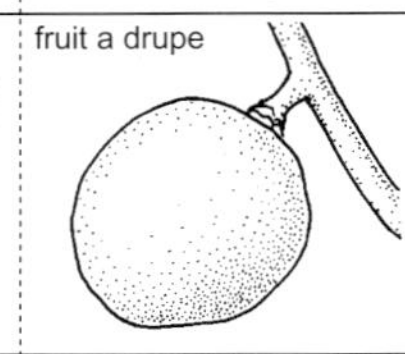	***Allophylus rubifolius*** **276**, 676

SAPOTACEAE

1	**Leaves** glabrous, green; **flowers** densely clustered in axils of leaves; **stamens** exserted from corolla	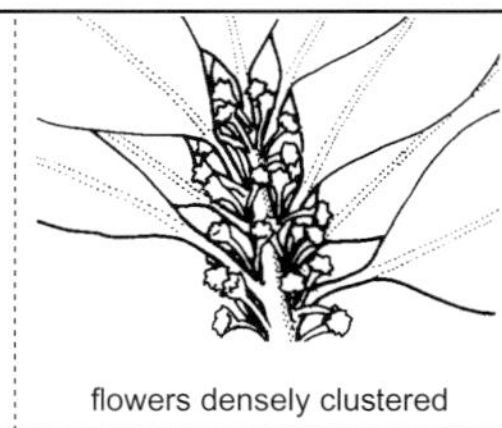	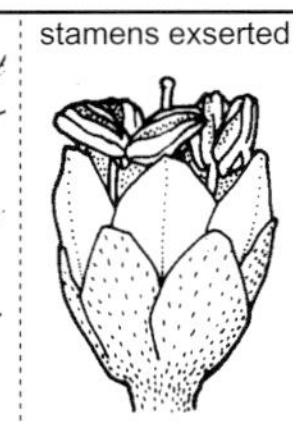	***Sideroxylon fimbriatum*** Endemic 678
	Leaves glabrous and green above, hairy and reddish beneath; **flowers** scattered in axils of leaves; **stamens** included within corolla	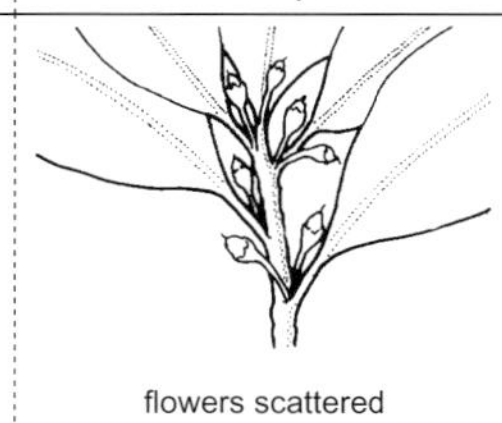	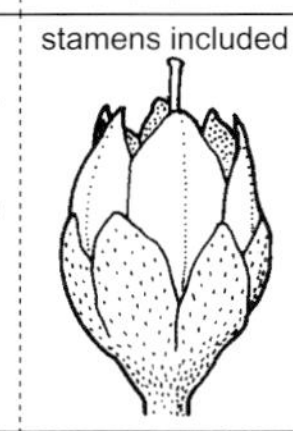	***Spinuluma discolor*** Endemic 678

SCROPHULARIACEAE

1	**Leaves** all in a basal rosette, no stem leaves present; **corolla** 2-lipped, bearing 2 yellow protuberances on the underside of the upper lip	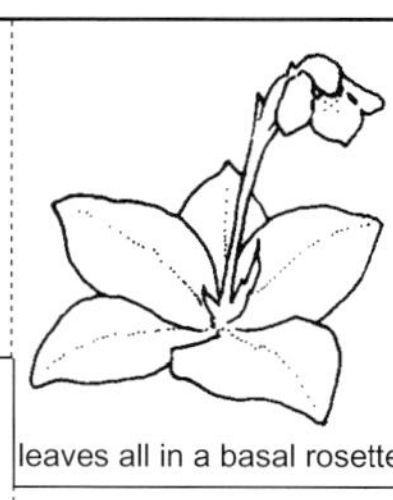	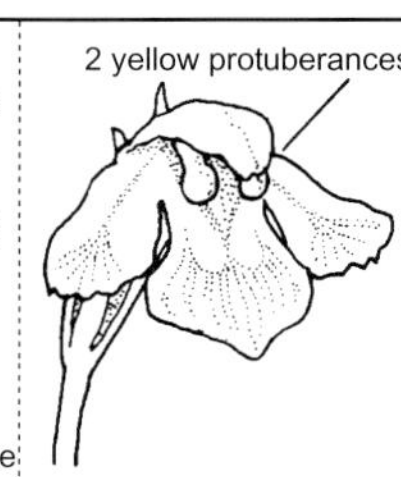	***Craterostigma pumilum*** **277**, 681
	Stem **leaves** present; **corollas** various			2

2	**Plant** of wet places; **stems** rooting at the nodes; **corolla** pink, subequally 5-lobed; **pedicels** lengthening and becoming recurved in fruit		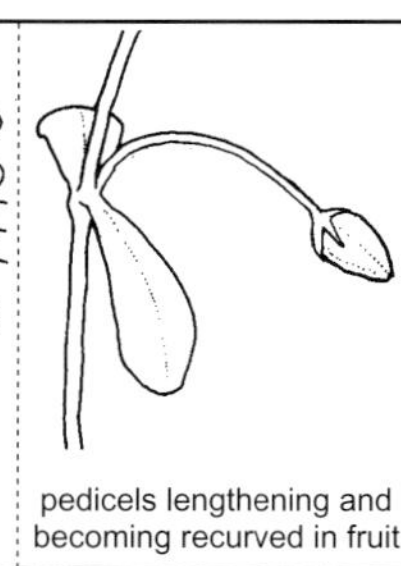	***Bacopa monnieri*** 680
	Plant of dry places; **stems** not rooting at the nodes; **corolla** and fruiting **pedicels** various			3
3	**Corolla** 2-lipped, the mouth of the corolla tube closed by a swollen protuberance on the lower lip	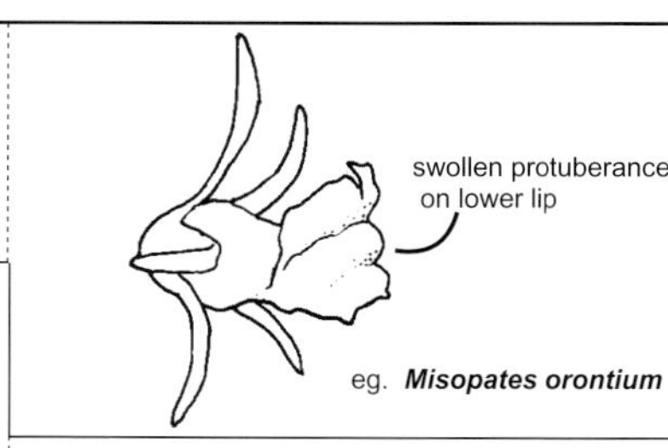		4
	Corolla 2-lipped or not, with no swollen protuberance			5

4

Corolla yellow, spurred at the base; **capsule** with equal loculi, opened by 2 apical pores; **leaves** linear to suborbicular, sometimes hastate

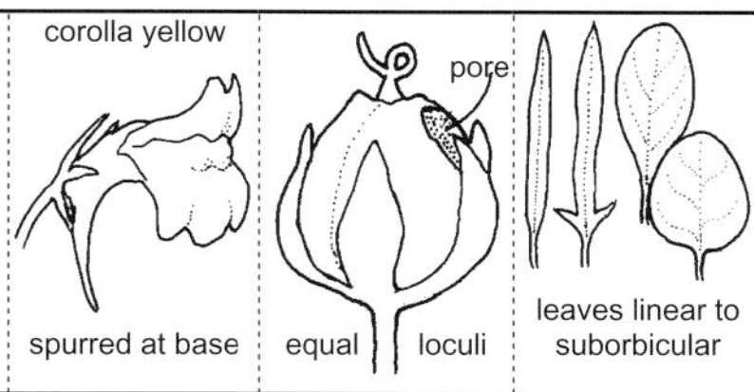

Nanorrhinum (2 spp.)

Corolla pink or cream, tube pouched at the base; **capsule** with unequal loculi, the upper loculus indehiscent, the lower opening by 2 apical pores; **leaves** linear

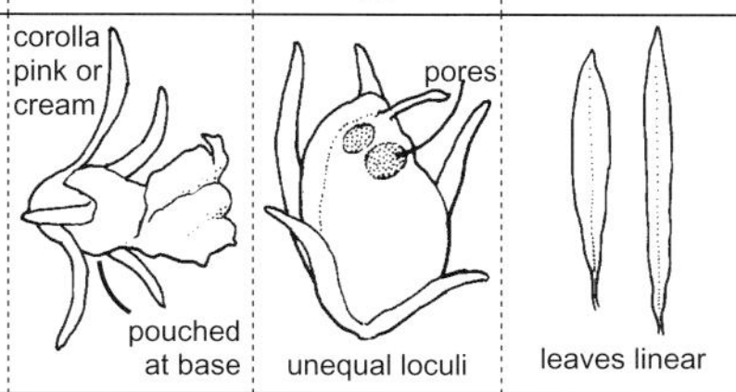

Misopates orontium 682

Corolla whitish or pinkish, not spurred or pouched; **capsule** with unequal loculi, the upper loculus indehiscent, the lower rupturing irregularly at the base; **leaves** linear

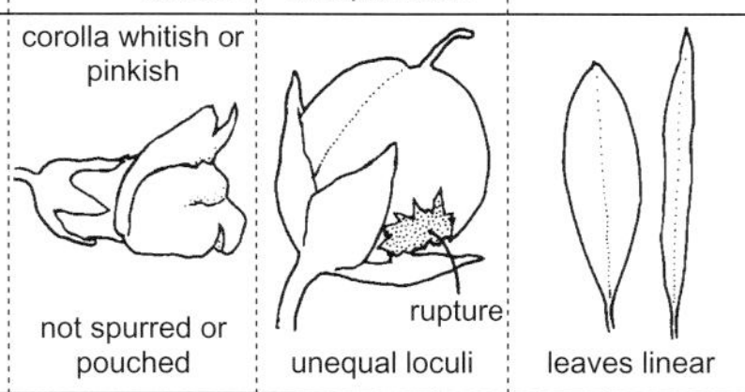

Schweinfurthia pedicellata 683

Corolla yellow, often with a red upper lip, not spurred or pouched; **capsule** with equal loculi, splitting lengthways into 2 valves; **leaves** ovate, crenate-serrate or entire

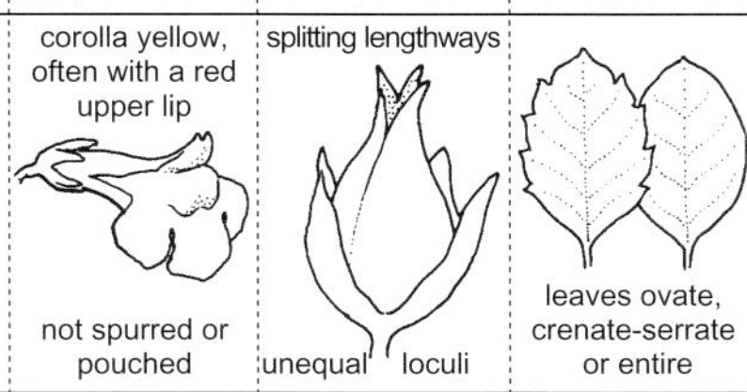

Lindenbergia sokotrana Endemic 682

5

Shrubs or **subshrubs**; **stems** woody

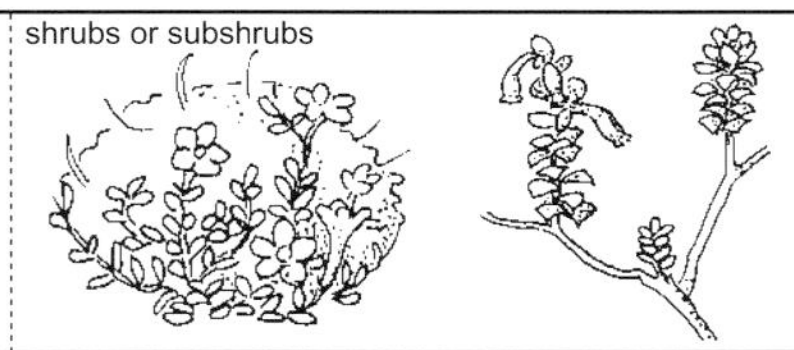

6

Herbs; **stems** herbaceous

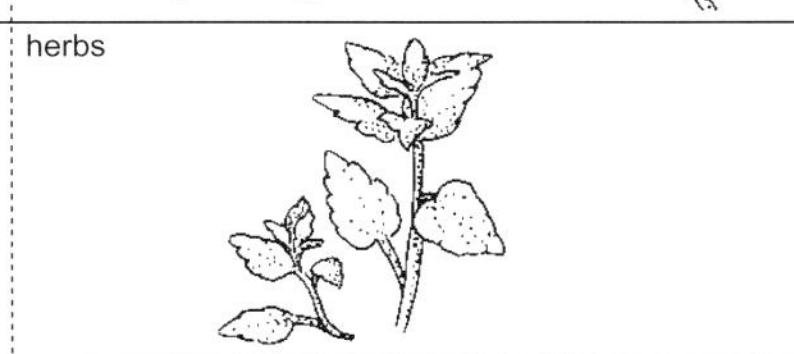

7

6

Shrub; **corolla** red with short recurved lobes and a long, rather dilated tube; **stamens** 4; **leaves** oblong-elliptic, scabrid with recurved prickles

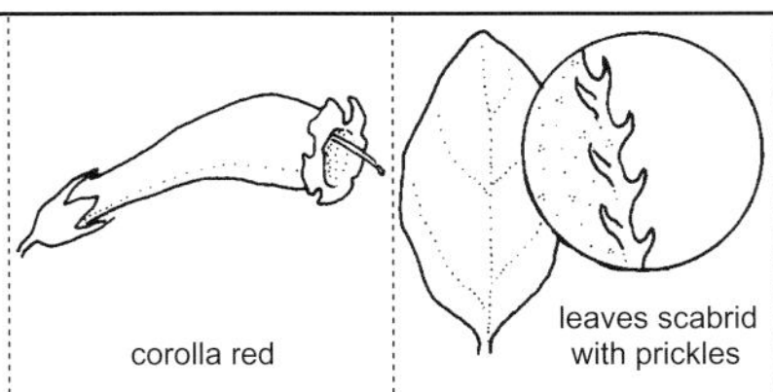

Graderia fruticosa Endemic **277**, 681

Spiny **subshrub**; **corolla** pink or mauve with spreading lobes and a long cylindrical tube; **stamens** 2; **leaves** linear or oblanceolate, with appressed silvery hairs

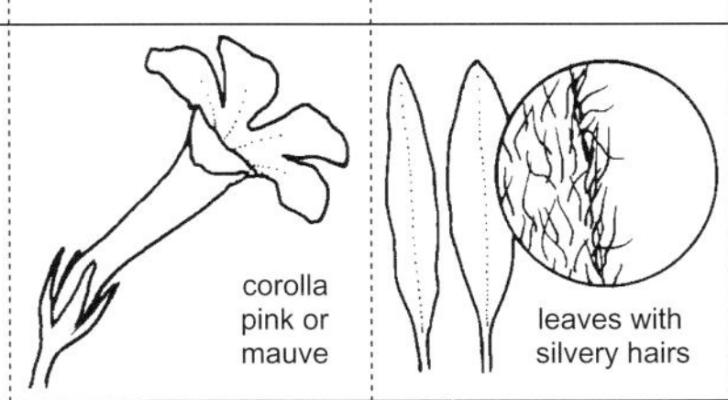

Campylanthus spinosus **277**, 680

Low **subshrub**; **corolla** white or pink with spreading lobes and short tube; **stamens** 4; **leaves** ovate, subsucculent; **calyx** becoming woody in fruit, the bracteoles adherent to its base and forming a pair of horn-like projections

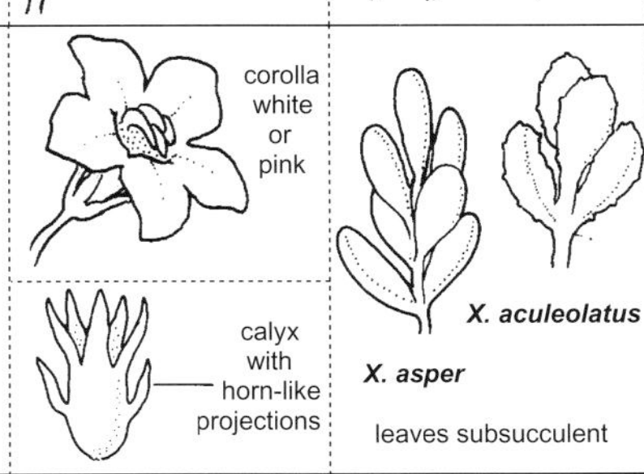

Xylocalyx (2 spp.)

7	Parasitic or semi-parasitic **herbs**; **flowers** sessile in a terminal spike; **corolla tube** long, with a bend near the middle; **leaves** linear or scale-like	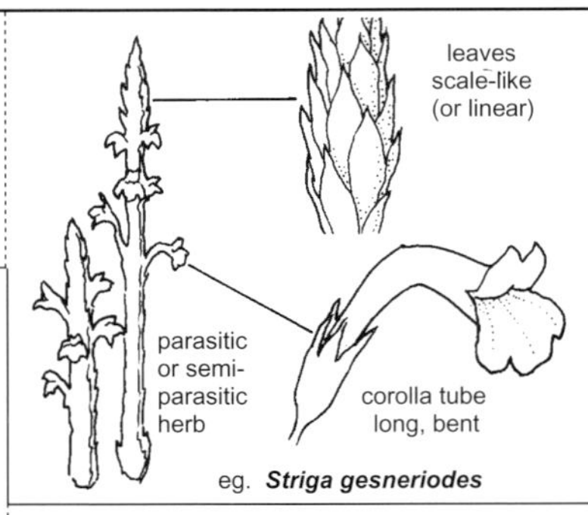 eg. ***Striga gesneriodes***	***Striga*** (3 spp.)
	Free-living **herbs**; **flowers** pedicellate; **corolla tube** short or if longer than the lobes then without a bend; **leaves** never scale-like		8

8	Low, tufted **herb**; **flowers** and **fruits** sessile; **leaves** linear; **corolla** regularly 5-lobed	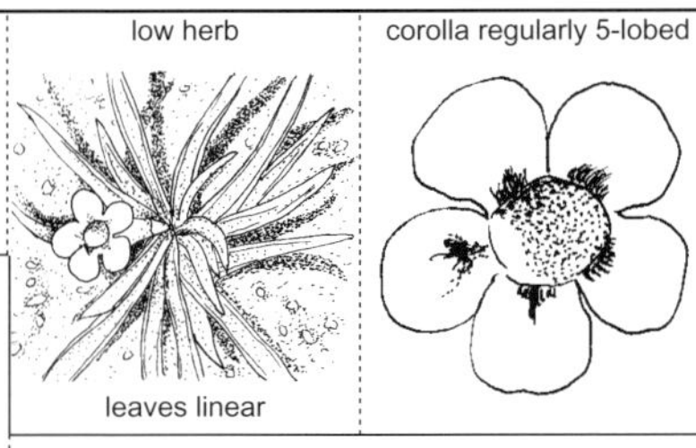	***Aptosimum pumilum*** 679
	Herbs various; **flowers** and **fruits** not sessile; **leaves** not linear; **corolla** various		9

9	**Stems** prostrate or decumbent, not viscid; **corolla** pale blue, 4-lobed with a very short tube; **stamens** 2	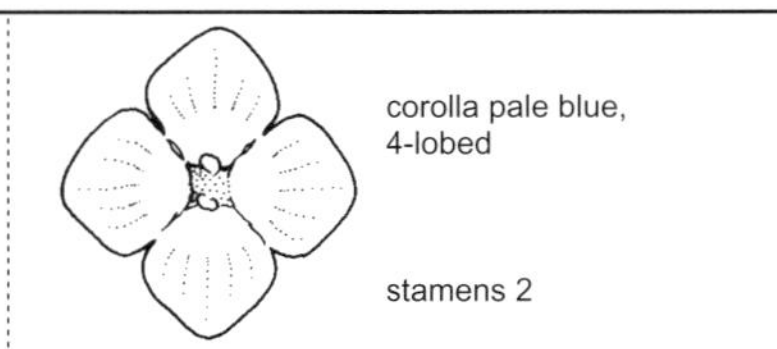	***Veronica polita*** 685
	Stems erect, whole plant sticky; **corolla** 5-lobed; **stamens** 2 or 4	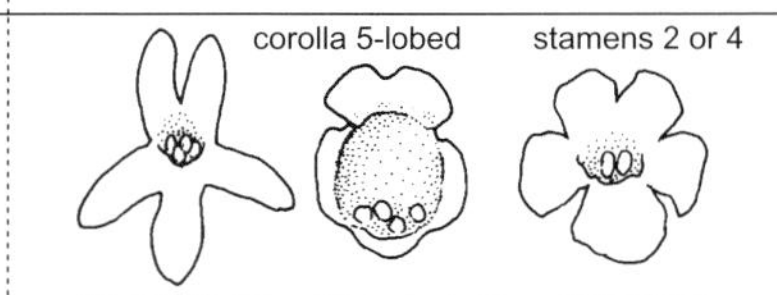 ***Camptoloma*** ***Scrophularia*** ***Anticharis***	10

10	**Leaves** alternate, cordate, crenate-dentate; **corolla** 2-lipped with a cylindrical tube; **stamens** 4	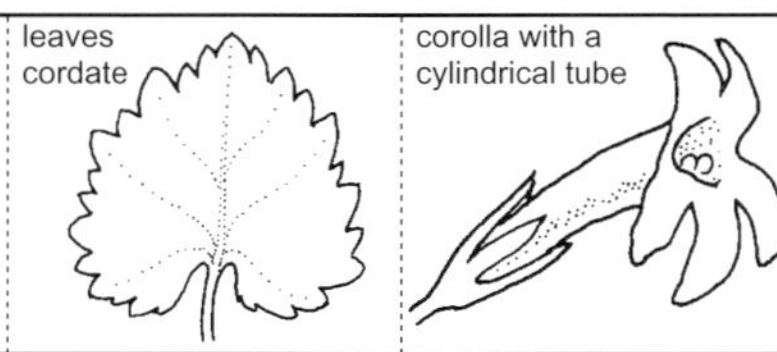	***Camptoloma lyperiiflorum*** **277**, 680
	Leaves opposite, ovate, serrate; **corolla** 2-lipped with a globular tube; **stamens** 4	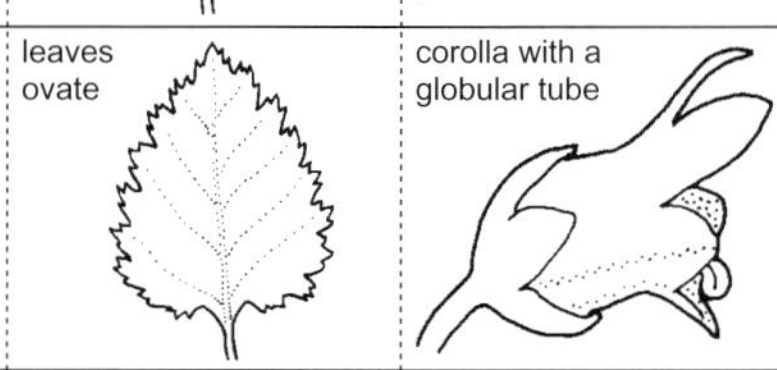	***Scrophularia arguta*** **277**, 684
	Leaves opposite, linear to elliptic, entire; **corolla** very weakly 2-lipped with a somewhat dilated tube; **stamens** 2	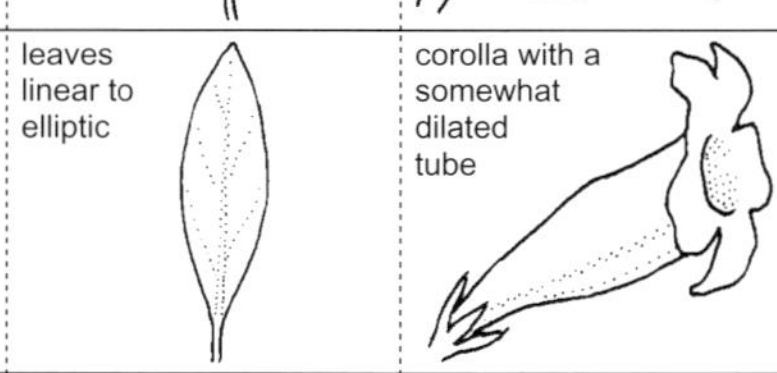	***Anticharis arabica*** 679

Nanorrhinum

1	Lower and upper **leaves** similar, both ovate or suborbicular; prostrate or ascending **herb**	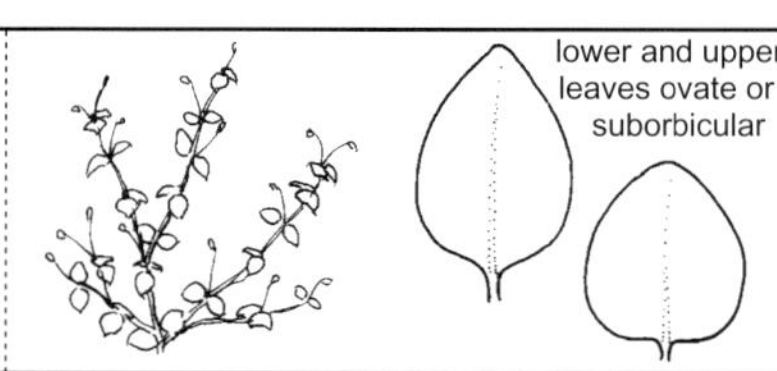	***N. kuriensis*** Endemic **277**, 683
	Lower **leaves** obovate, becoming linear above, often hastate; erect, delicately-branched **herb**	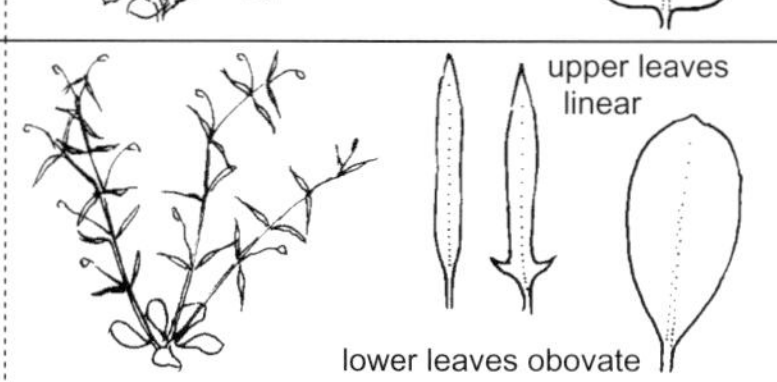	***N. hastatum*** 683

Striga

1 **Calyx** 10-ribbed; **corolla** white, yellow or scarlet — ***S. asiatica*** 684

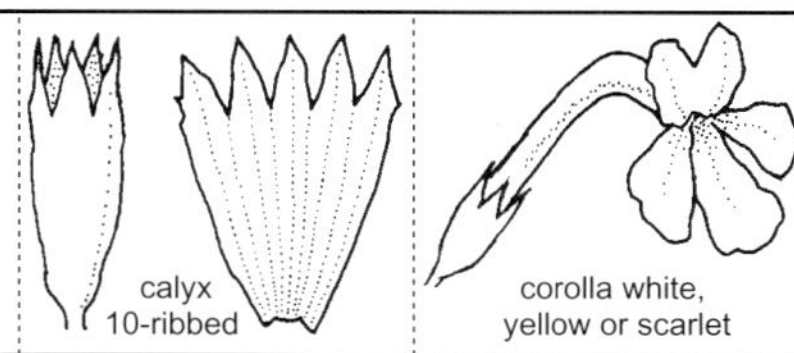

Calyx 5-ribbed; **corolla** white or purple — 2

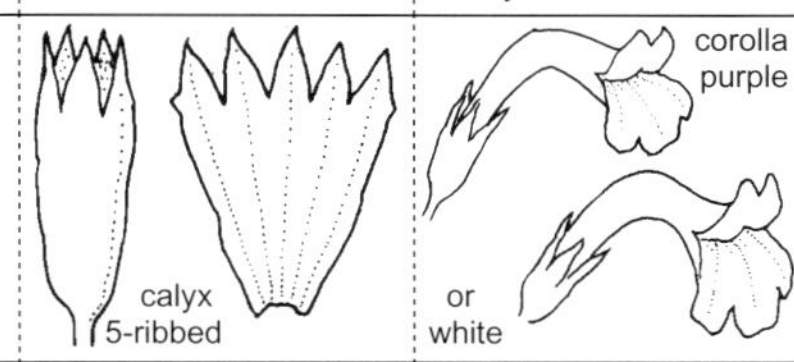

2 **Stem** pink or purple; **corolla** purple; **leaves** scale-like, appressed to the stem — ***S. gesnerioides*** 685

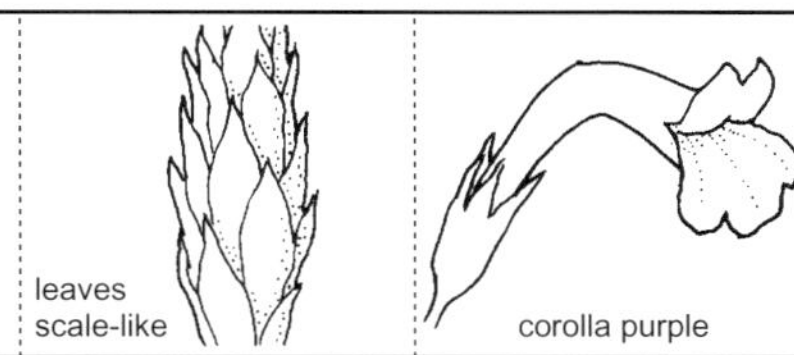

Stem green; **corolla** white; **leaves** linear — ***S. densiflora*** 684

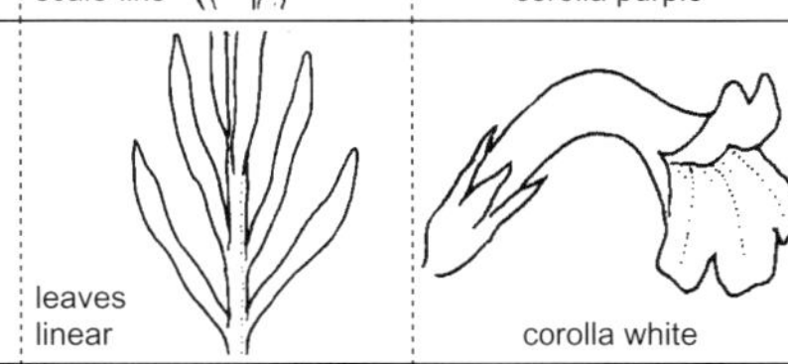

Xylocalyx

1 **Leaves** and **young branches** glabrous or covered with tubercles; **fruits** sessile — ***X. asper*** Endemic **277**, 686

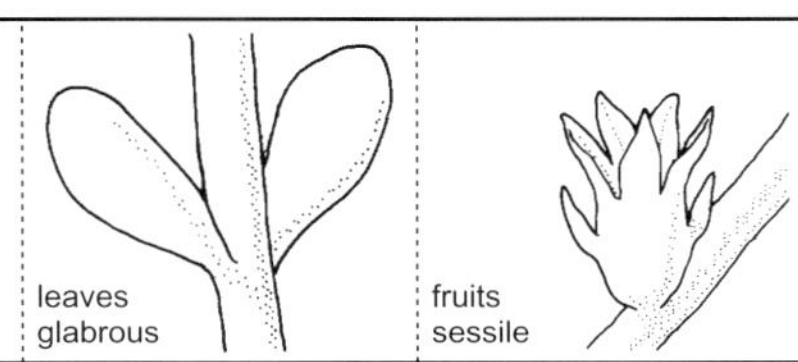

Leaves and **young branches** covered with short recurved prickles; **fruits** shortly stalked — ***X. aculeolatus*** Endemic **277**, 685

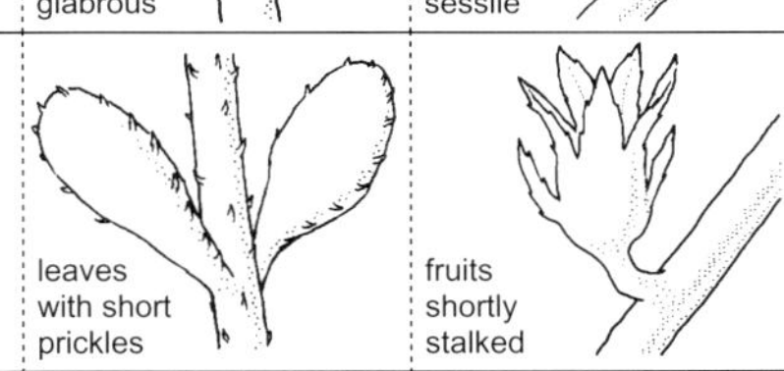

SOLANACEAE

1 **Corolla** tubular or funnel-shaped — 2

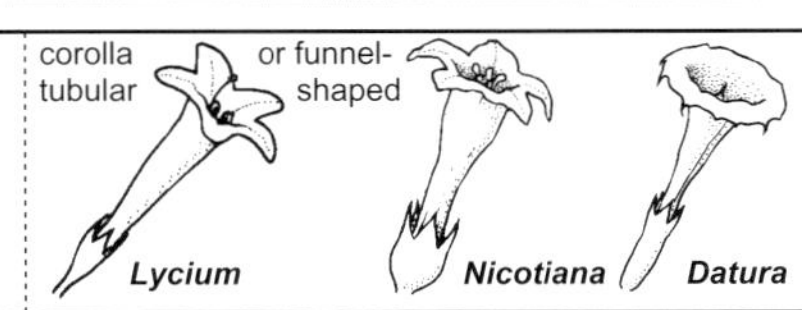

Corolla star-shaped (rotate) or cup-shaped (campanulate) — 4

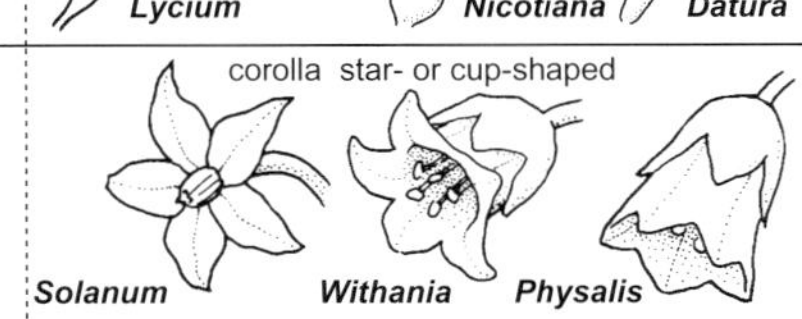

2 **Shrub** with spiny branches; **leaves** obovate, mostly fascicled, 0.5--2cm long; **corolla** c. 1cm long; **fruit** a red berry — ***Lycium sokotranum*** Endemic **278**, 687

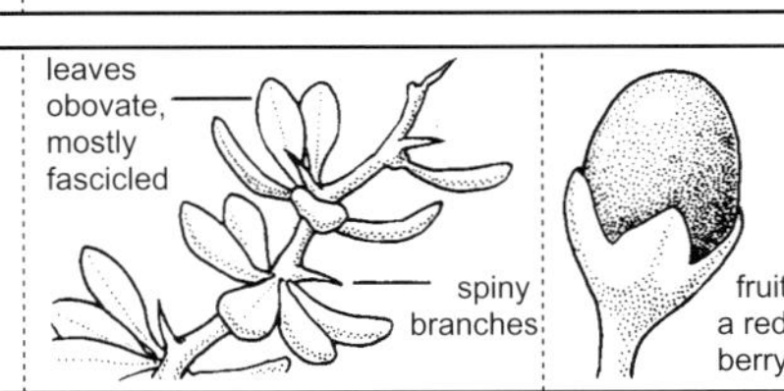

Robust **herb** with unarmed branches; **leaves** ovate, never fascicled, 5--15cm long; **corolla** 3--15cm long; **fruit** a greenish capsule — 3

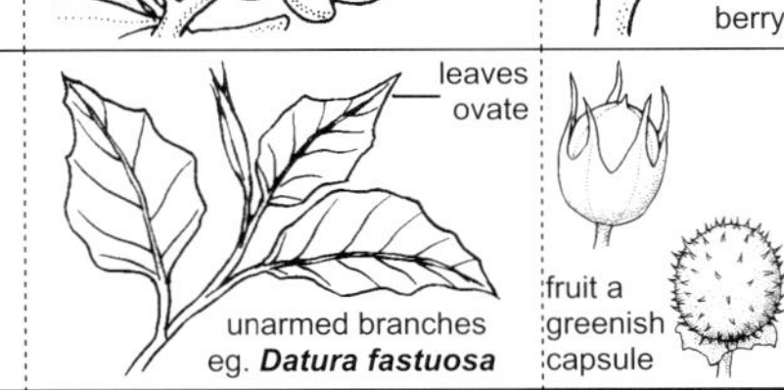

3

Corolla tubular, 4--5cm long; **flowers** in terminal panicles; **capsule** smooth; **leaves** entire

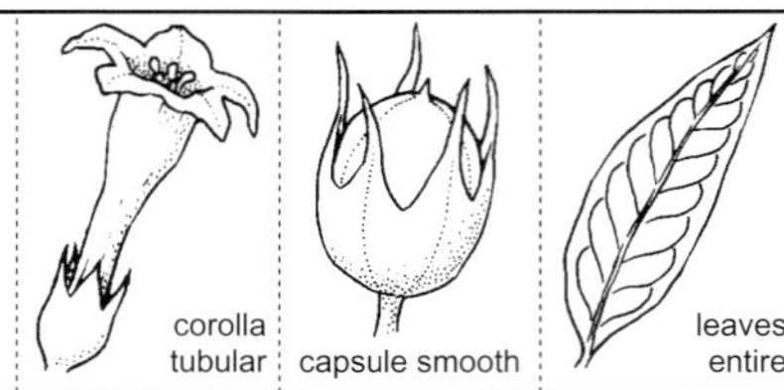

Nicotiana tabacum 688

Corolla funnel-shaped, 12--20cm long; **flowers** solitary in leaf axils; **capsule** spiny; **leaves** sinuate

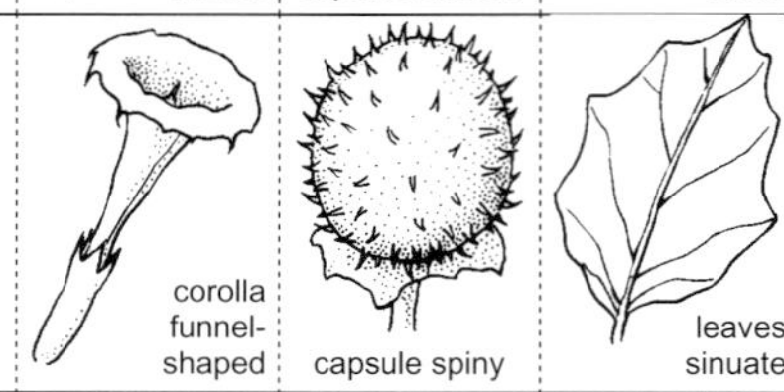

Datura fastuosa 686

4

Corolla rotate; **anthers** joined into a central cone, dehiscing by apical slots; **berry** visible, not enclosed by calyx

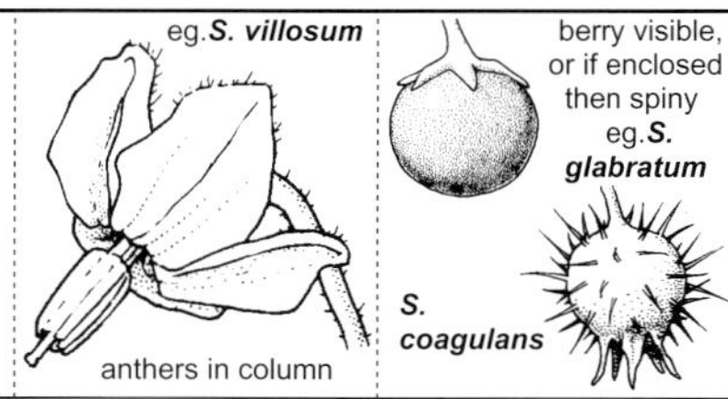

Solanum (3 spp.)

Corolla campanulate; **anthers** inserted at base of corolla, dehiscing longitudinally; **berry** partly or completely enclosed by a papery calyx

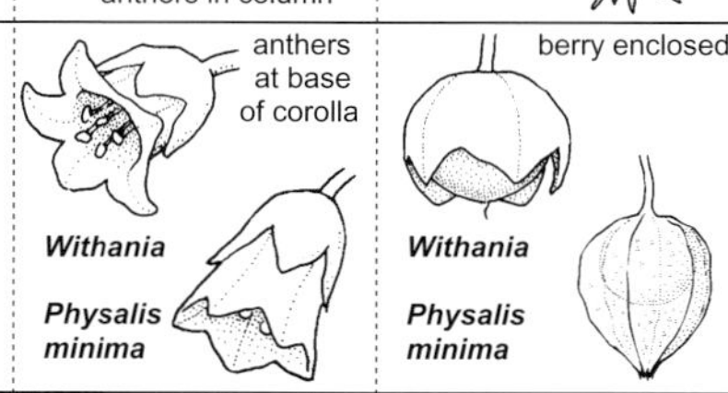

5

5

Shrublet; **flowers** clustered; **stems** stellate-tomentose; **leaves** entire

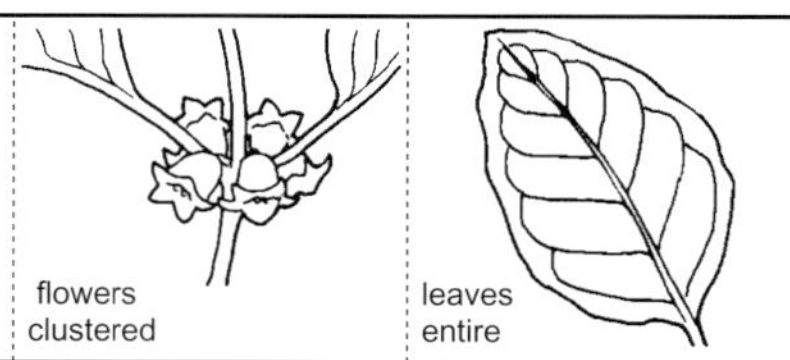

Withania (3 spp.)

Slender **herb**; **flowers** solitary; **stems** thinly hairy with unbranched hairs; **leaves** slightly dentate

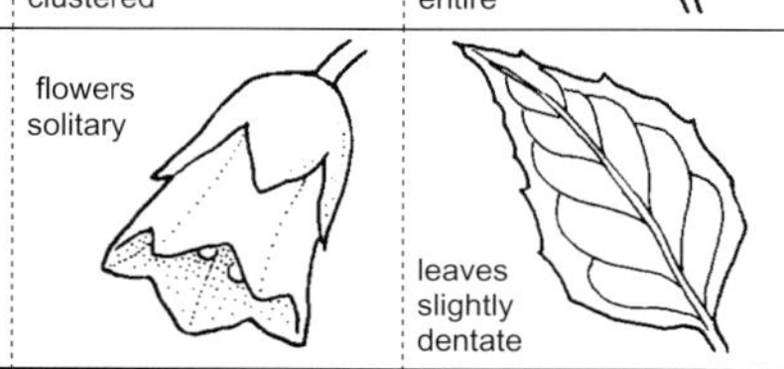

6

6

Calyx lobes ± fused; **corolla** pale yellow or white

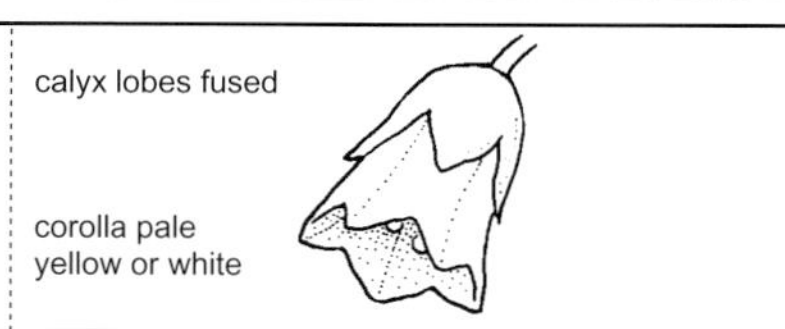

Physalis minima 689

Calyx lobes free; **corolla** purple or mauve

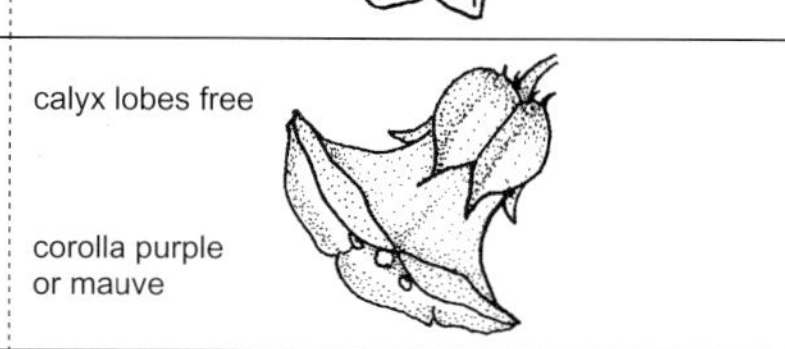

Nicandra physaloides 688

Solanum

1

Herb with non-woody stems; **stems** unarmed (not spiny); **flowers** white

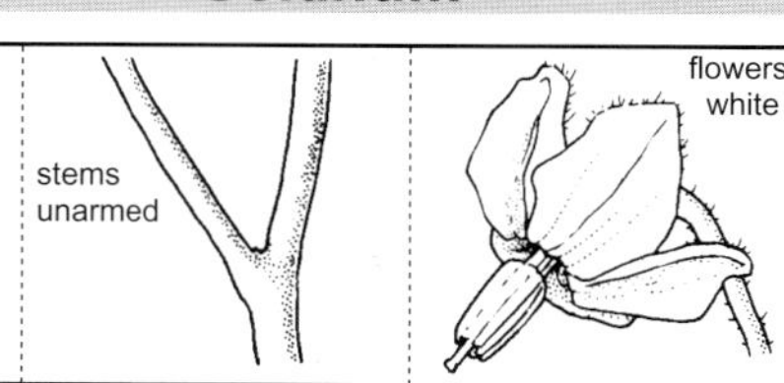

S. villosum 691

Shrubs or robust, woody-stemmed **herbs**; **stems** spiny; **flowers** blue or violet

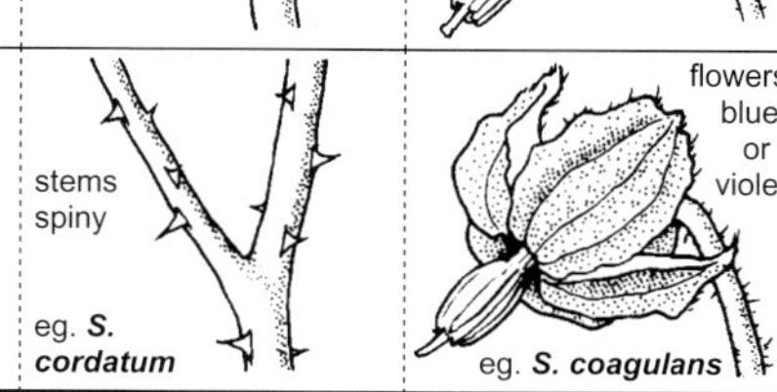

2

2

Leaves broadly ovate to round, subentire, less than 3cm long, not spiny; **fruit** c.0.7cm diam., orange-red

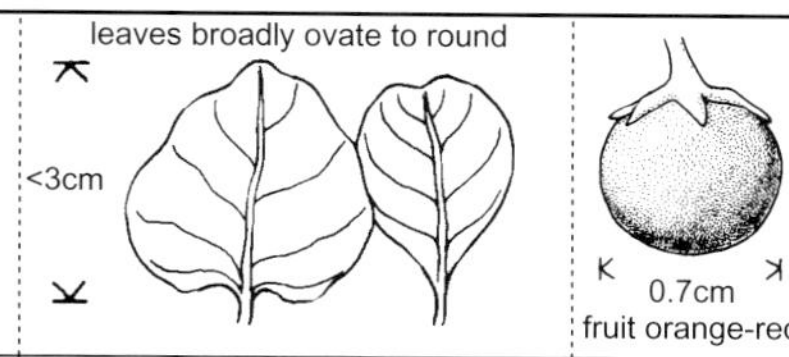

S. cordatum 690

Leaves lanceolate or ovate to oblong, sinuate, (2--)5--15cm long, the midrib spiny; **fruit** 3--4cm diam., yellow

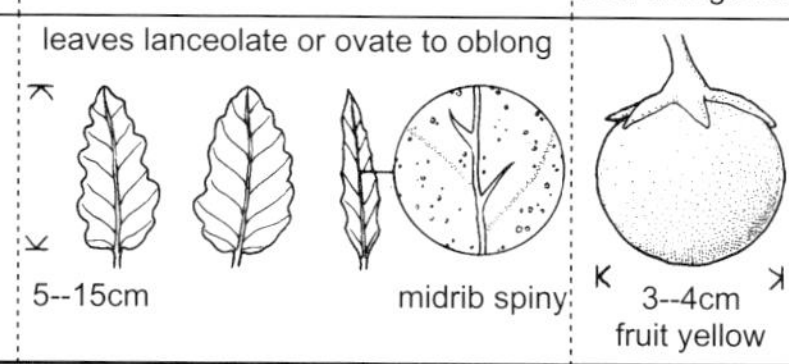

S. incanum **278**, 690

Withania

1

Fruiting calyx enlarged and papery, completely enclosing the fruit

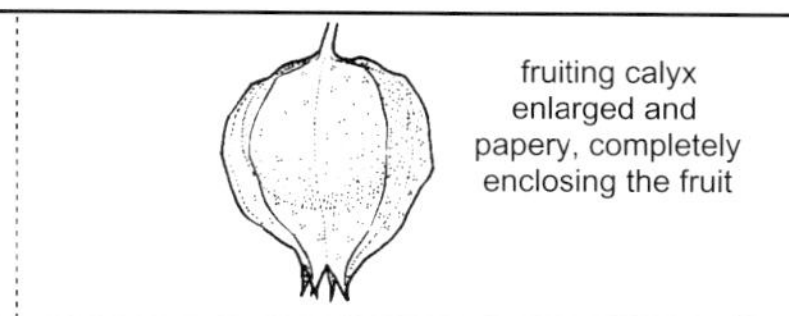

W. somnifera 692

Fruiting calyx enlarged, partly enclosing the fruit

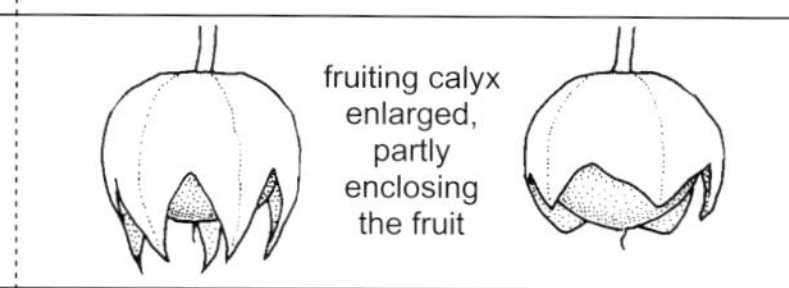

2

2

Lobes of **fruiting calyx** narrowly triangular

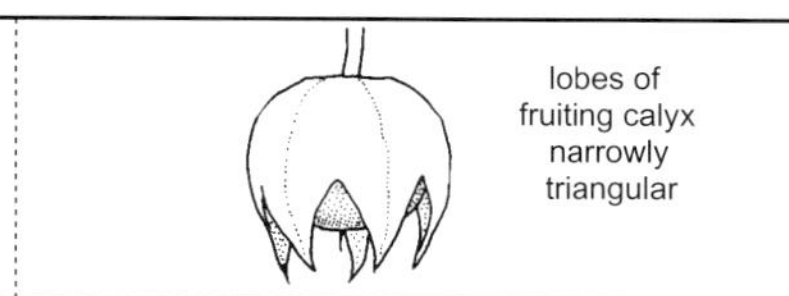

W. riebeckii Endemic 692

Lobes of **fruiting calyx** broadly triangular

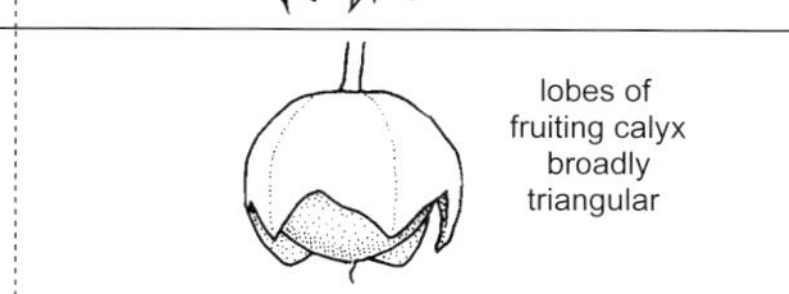

W. adunensis Endemic **278**, 691

STERCULIACEAE

1

Large **tree; leaves** 3 to 7-lobed, >10cm long; **fruit** 3 to 5-lobed

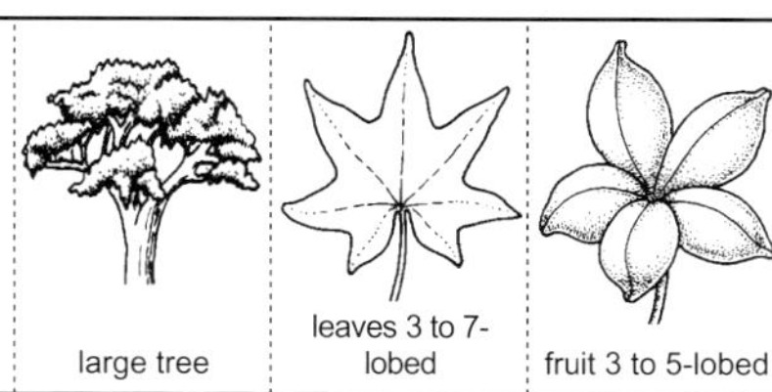

Sterculia africana Endemic var. **278**, 694

Woody-based **herb; leaves** simple, <10cm long; **fruit** unlobed

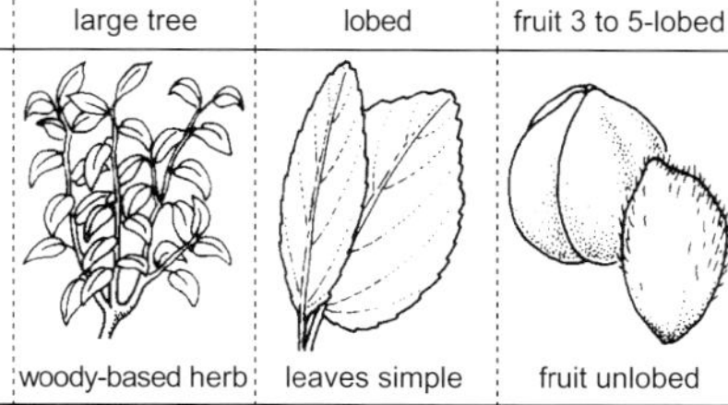

2

2

Flowers 1 or 2 in the leaf axils; **fruit** a 5-valved capsule, hidden by an **epicalyx** of 3, enlarged membranous bracts

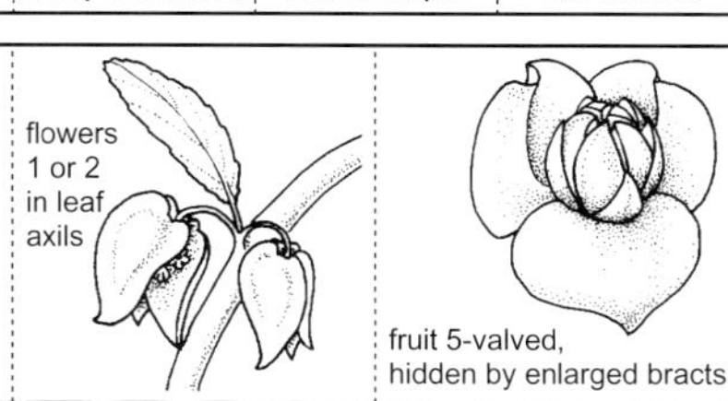

Melhania (2 spp.)

Flowers in dense axillary clusters; **fruit** 1-locular, surrounded by the persistent petals, **epicalyx** absent

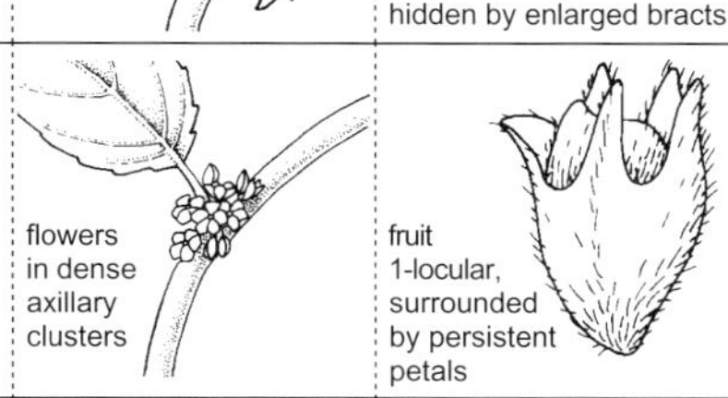

Waltheria indica 695

Melhania

1	**Epicalyx bracts** enlarging in fruit; **fruit** hidden	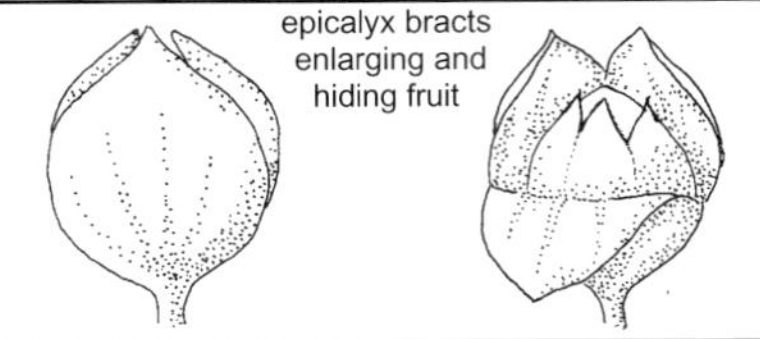 epicalyx bracts enlarging and hiding fruit	***M. muricata*** 693
	Epicalyx bracts not enlarging in fruit; **fruit** clearly visible	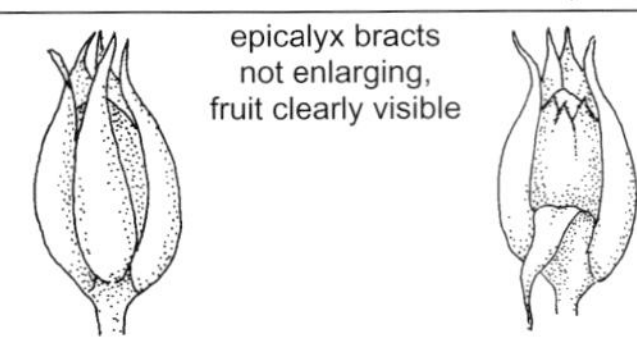 epicalyx bracts not enlarging, fruit clearly visible	***M. milleri*** Endemic 693

TAMARICACEAE

1	**Scale leaves** with non-sheathing bases, overlapping; **smaller branches** not appearing jointed	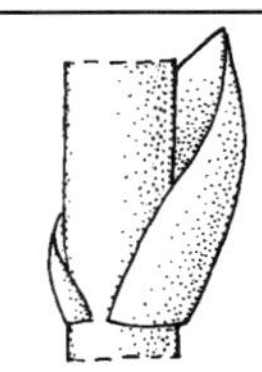 scale leaves not sheathing	 leaves overlapping	***Tamarix nilotica*** 697
	Scale leaves with sheathing bases, not overlapping; **smaller branches** appearing jointed	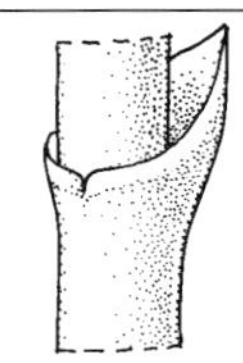 scale leaves sheathing	 branches jointed	***Tamarix aphylla*** 696

THYMELAEACEAE

1	Evergreen **shrub**; **leaves** alternate, entire; **flowers** bright yellow, in a head surrounded by overlapping bracts	 leaves alternate, entire	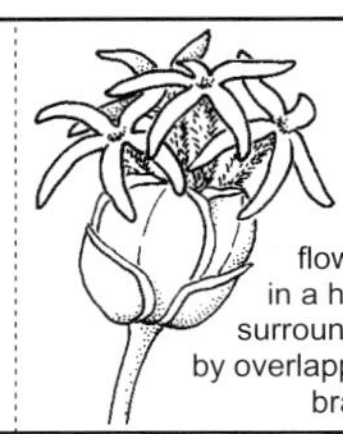 flowers in a head surrounded by overlapping bracts	***Gnidia socotrana*** Endemic **278**, 697

TILIACEAE

1	**Herbs**; **fruit** an elongate or oblong-ovoid capsule, without bristles	herbs  eg. ***C. erodioides***	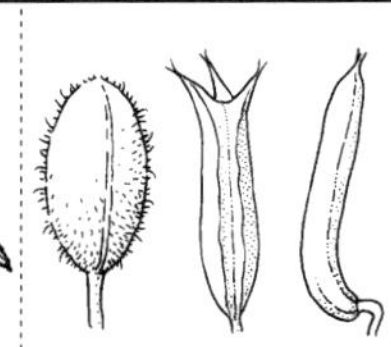 fruit a capsule	***Corchorus*** (3 spp.)
	Herbs; **fruit** an ovoid capsule, densely covered with bristles	herbs  fruit a capsule, densely covered with bristles		***Triumfetta* sp. A** Endemic 701
	Trees or **shrubs**; **fruit** an entire or 2 to 4-lobed drupe	 trees or shrubs	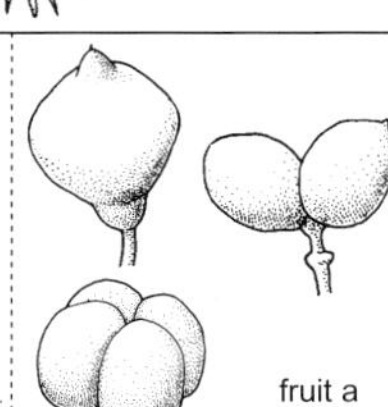 fruit a drupe	***Grewia*** (5 spp.)

Corchorus

	Description	Illustration	Result
1	**Leaves** oblong-ovate, simple to pinnate, serrate to biserrate; **fruit** oblong-ovoid, rounded at apex, 2-locular; prostrate annual or perennial **herb**	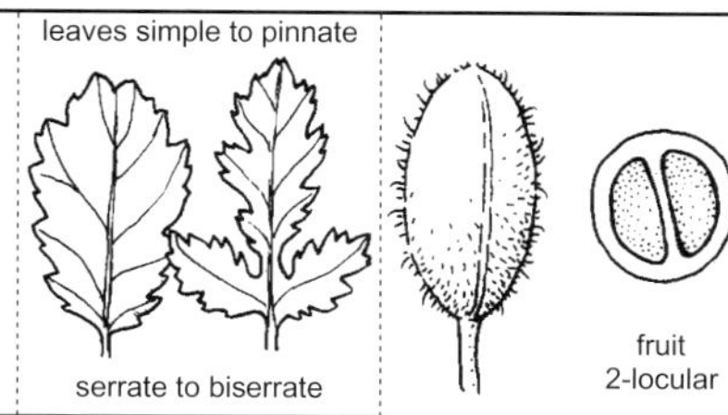	***C. erodioides*** Endemic **278**, 699
	Leaves ovate, crenate-serrate, some with linear appendages at base; **fruit** oblong-cylindrical, 3-winged, with 3 bifid teeth at apex, 3-locular; ascending annual **herb**	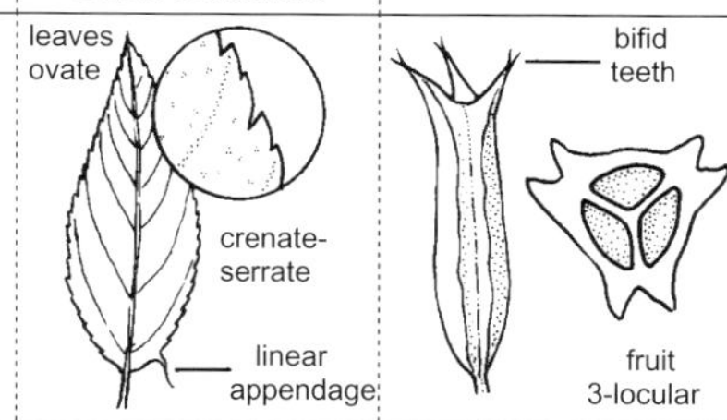	***C. aestuans*** 698
	Leaves oblong-elliptic, crenate-serrate; **fruit** linear-cylindrical, often curved, minutely 4-toothed at apex, 4-locular; prostrate, often woody-based, perennial **herb**	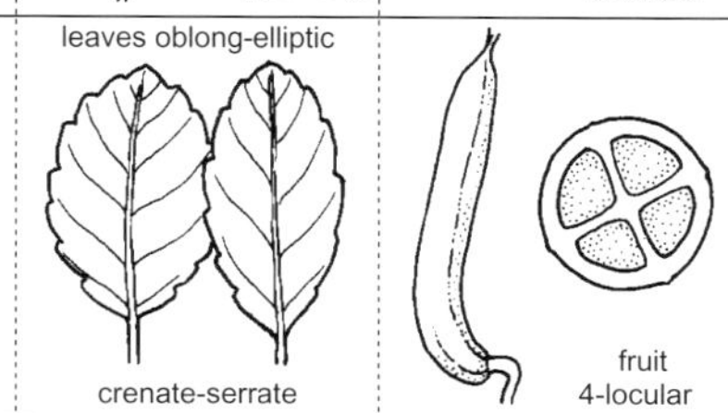	***C. depressus*** 698

Grewia

	Description	Illustration	Result
1	**Leaves** velvety-hairy at least beneath, obscuring leaf surface; **fruit** spherical, slightly 2-lobed or obovoid	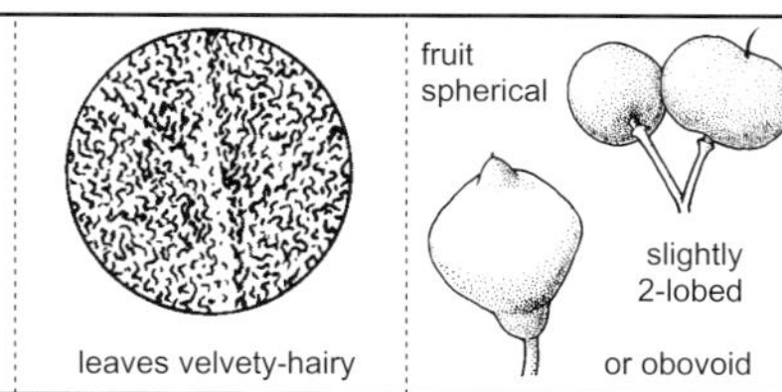	2
	Leaves glabrous or thinly hairy, not obscuring leaf surface; **fruit** distinctly 2 to 4-lobed	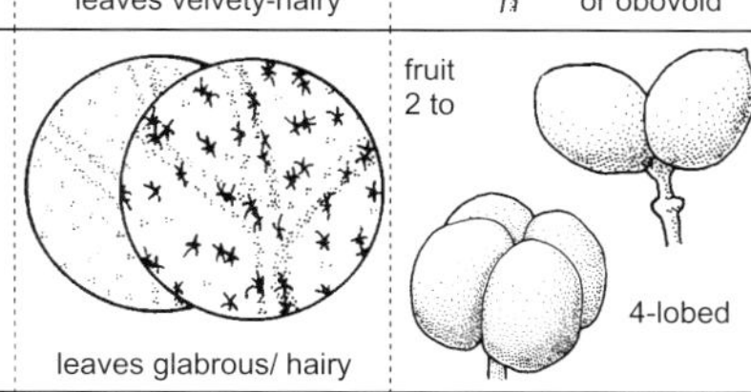	3

	Description	Illustration	Result
2	**Leaves** ovate, yellowish green, 3--5cm long, obscurely crenate or entire, shortly and softly velvety; **fruit** spherical or slightly 2-lobed	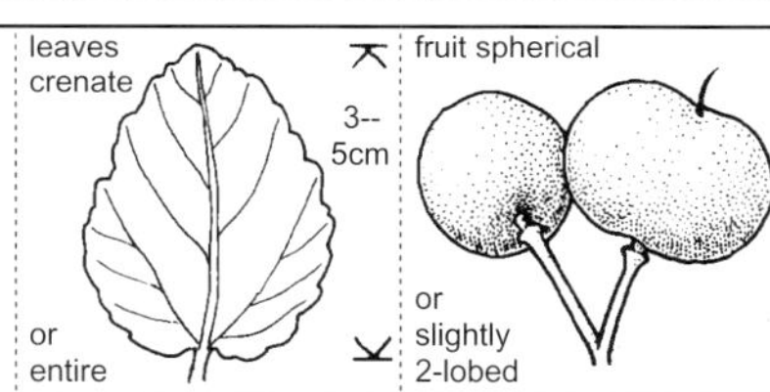	***G. milleri*** Endemic 701
	Leaves oblong-lanceolate, green above, lighter green beneath, 5--10cm long, serrate, glabrous above, softly velvety beneath; **fruit** obovoid	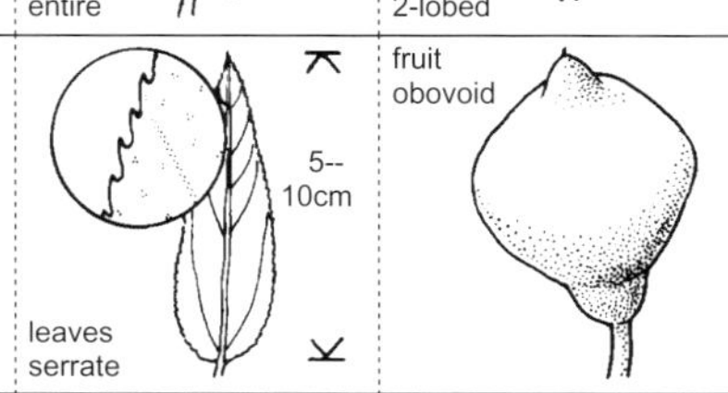	***G. turbinata*** Endemic 701

	Description	Illustration	Result
3	**Trees**; **leaves** ovate or elliptic, 2--15cm long with obtuse or acute tips	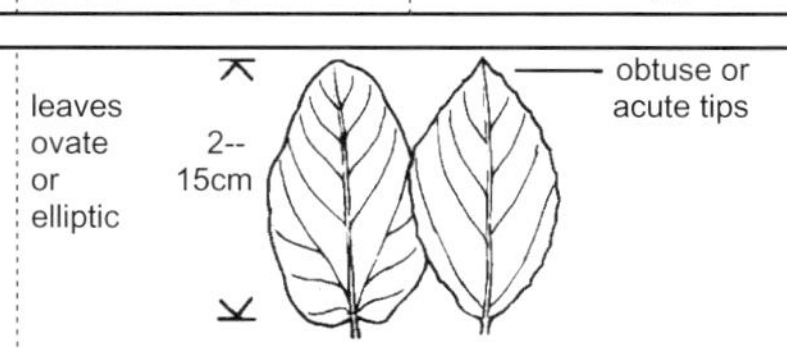	4
	Low **shrubs**; **leaves** round to obovate, 0.5--2cm long with rounded tips	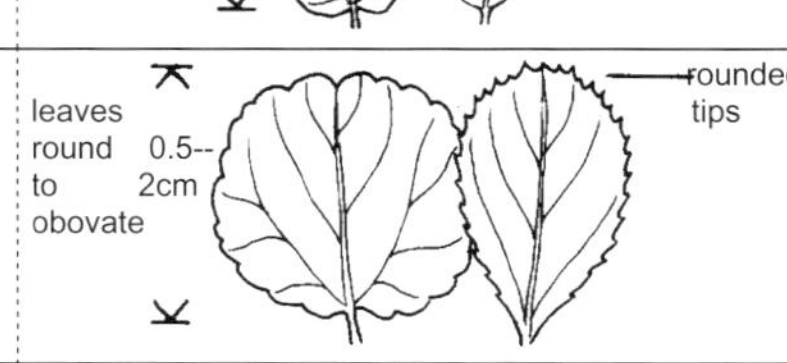	***G. erythraea*** 700

4

Leaves 4--15cm long with obtuse tips; **fruit** c.2cm across

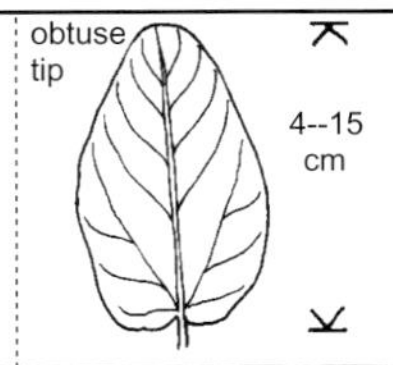

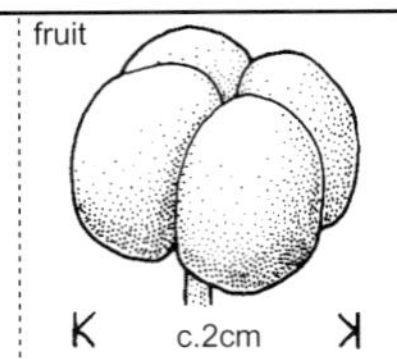

G. bilocularis
Endemic
699

Leaves 2--6cm long with acute tips; **fruit** c.1cm across

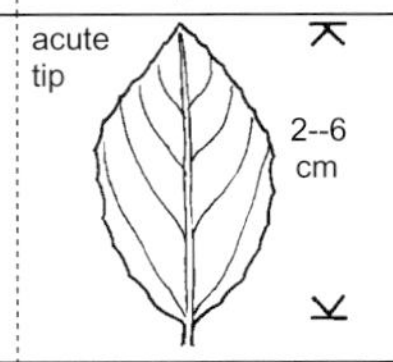

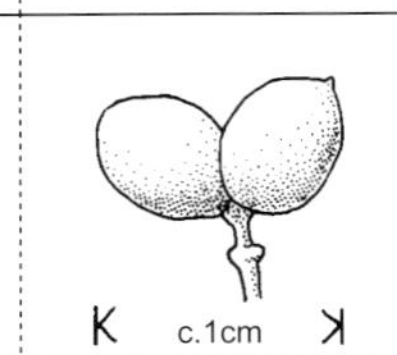

G. mansouriana
278, 701

UMBELLIFERAE

1

Prostrate **herb** with creeping stems; **leaves** subcircular with a cordate base

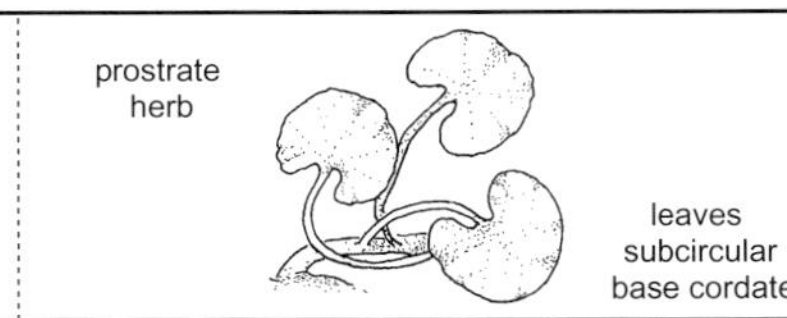

Centella asiatica
702

Erect **herbs** or **shrubs**; **leaves** various

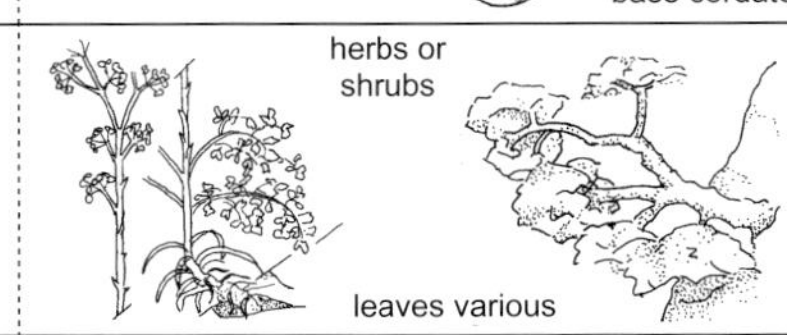

2

2

Leaves simple, subcircular; **shrublet** with woody stems

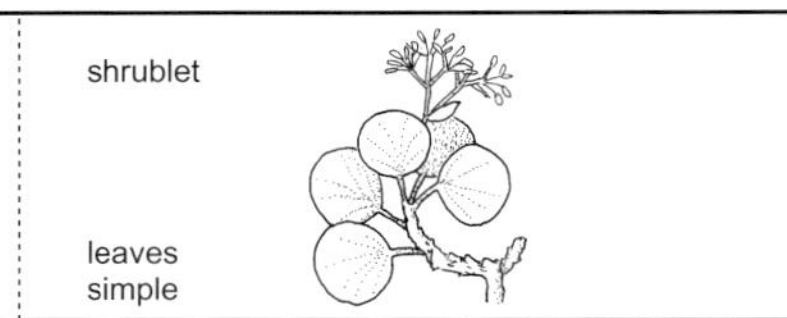

Nirarathamnos asarifolius
Endemic
Endemic genus
279, 703

Leaves divided; erect **herbs** or **shrubs**

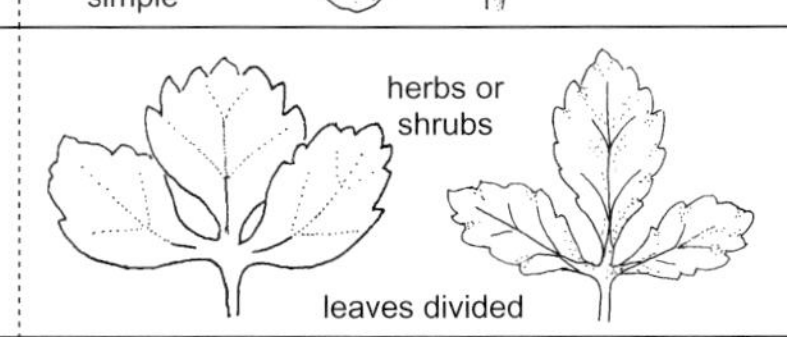

3

3

Leaves once-pinnate; **umbels** opposite leaves; **fruits** glabrous, unwinged, ridged

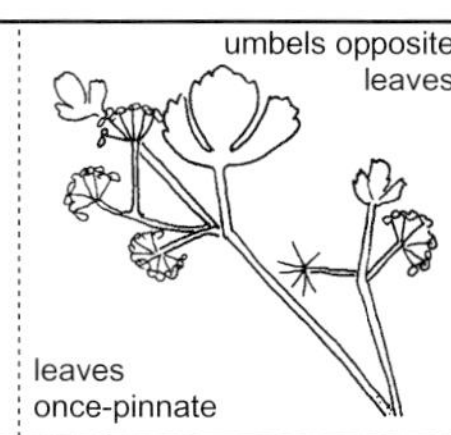

Apium graveolens
702

Leaves (2 or) 3-pinnate; **umbels** terminal; **fruits** glabrous or hairy, winged or unwinged

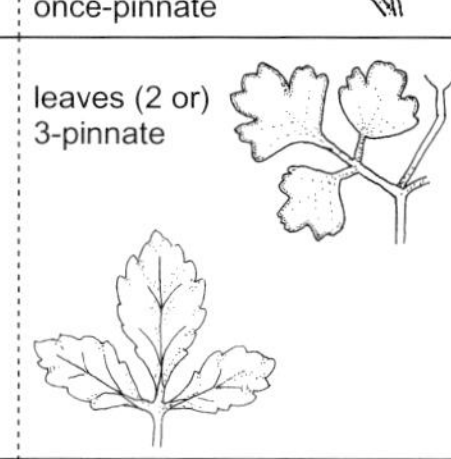

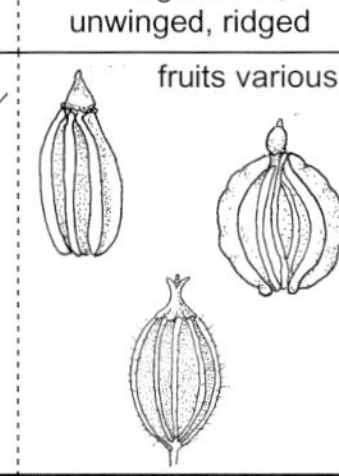

4

4

Small annual **herb**, to 25cm; **leaves** with narrow segments

Trachyspermum pimpinelloides
279, 705

Perennial **herbs** or **shrublets**, over 30cm; **leaves** with broad segments

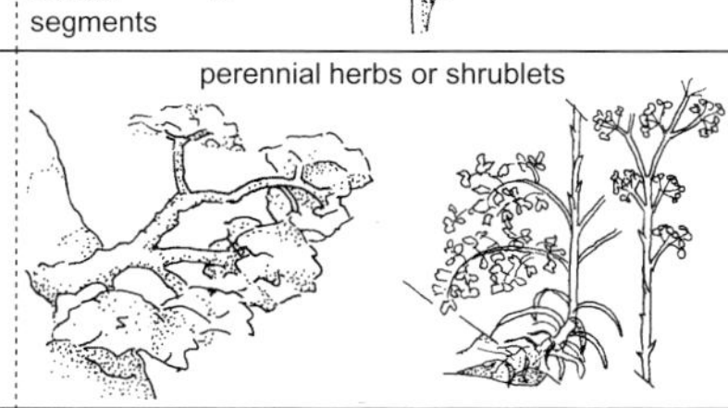

5

5	Erect perennial **herb**; **fruits** rounded-elliptic, cordate at base, compressed, ribs extended into broad wings; **leaf segments** acute	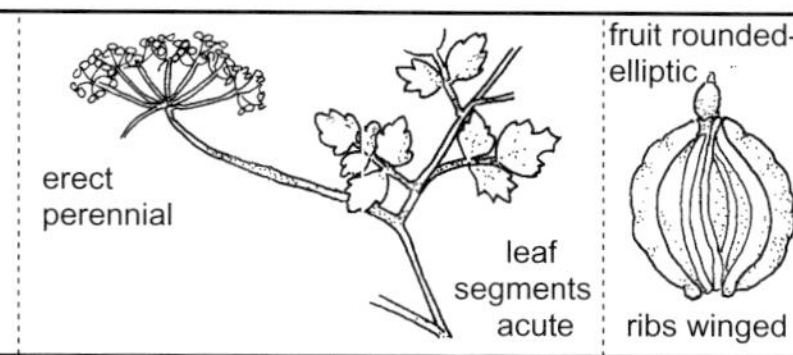	***Rughidia cordata*** Endemic Endemic genus **279**, 703
	Cliff-dwelling **shrub** with gnarled woody stems; **fruits** narrowly ovate, ribs prominently to slightly winged; **leaf segments** acute	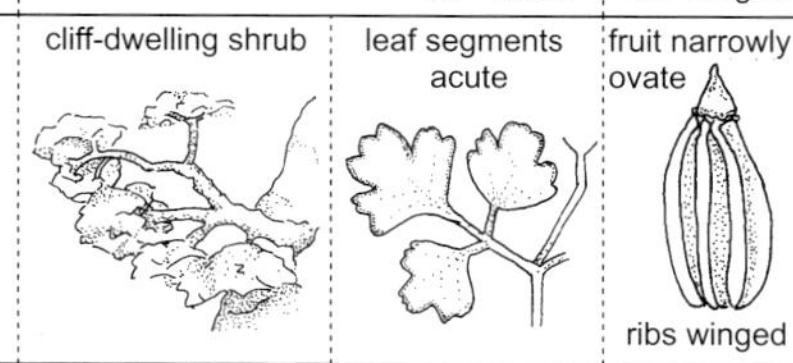	***Rughidia milleri*** Endemic Endemic genus **279**, 704
	Erect **herb** with a short woody stem; **fruits** oblong-oblanceolate, unwinged but prominently ribbed, papillose-hairy; **leaf segments** obtuse	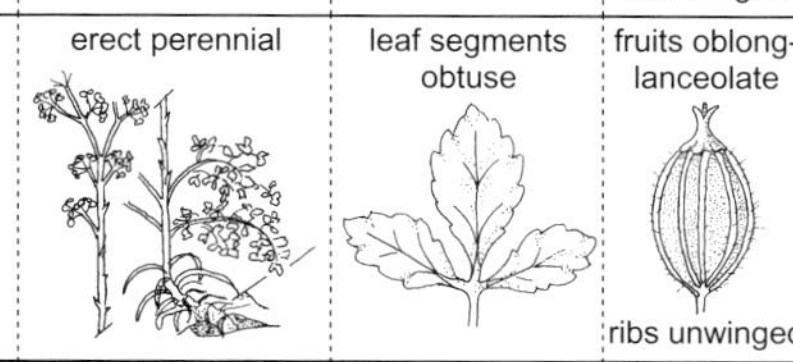	***Oreofraga morrisiana*** Endemic Endemic genus **279**, 703

URTICACEAE

1	**Leaves** dentate	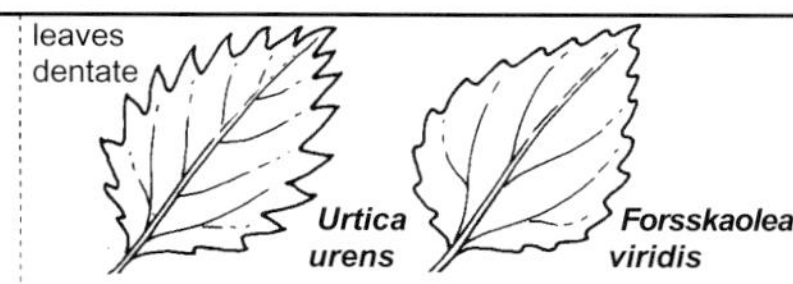	2
	Leaves entire	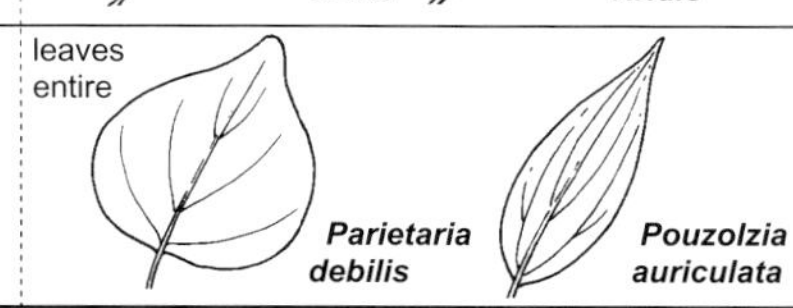	3

2	**Leaves** alternate; **leaves** and **stems** clothed with clinging, hooked hairs; **flowers** arranged in axillary receptacles	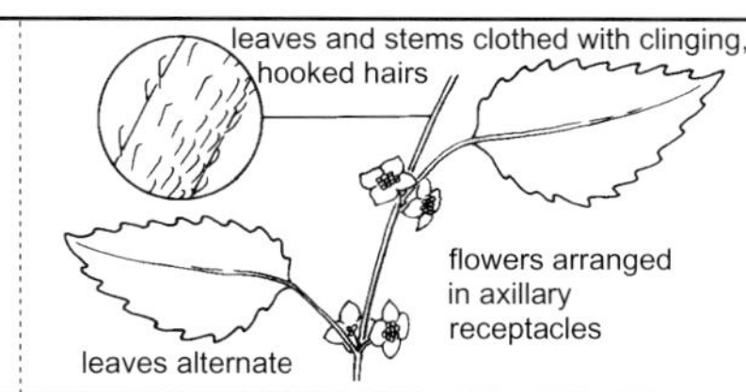	***Forsskaolea viridis*** 705
	Leaves opposite; **leaves** and **stems** clothed with stinging hairs; **flowers** arranged in axillary clusters	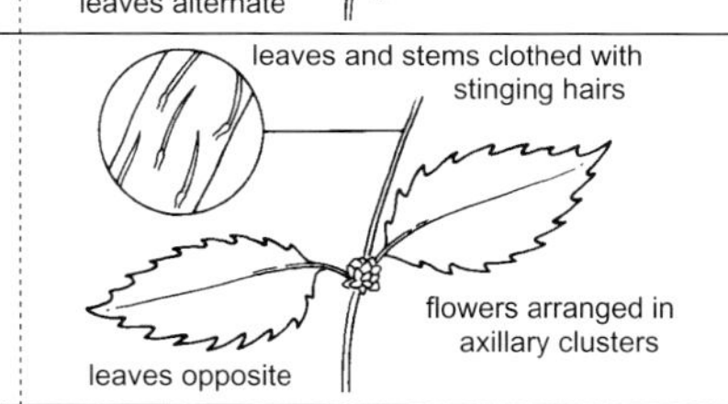	***Urtica urens*** 707
	Leaves opposite, upper 2 pairs appearing whorled; **leaves** and **stems** glabrous; **flowers** in a dense cross-shaped terminal panicle	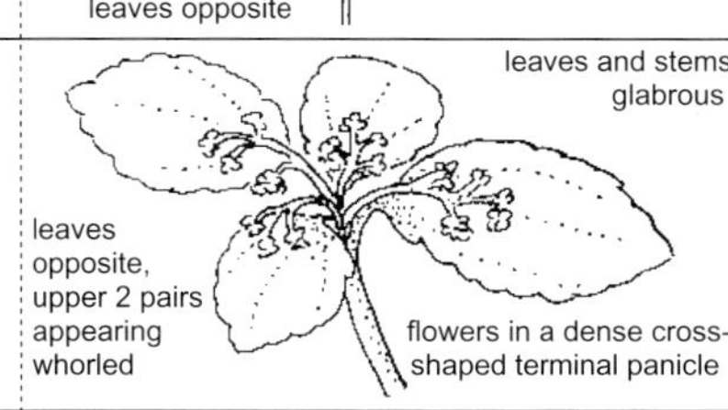	***Pilea tetraphylla*** 706

3	Delicate, prostrate, weak-stemmed **herb**; **leaves** broadly ovate with a blunt tip; **stipules** absent	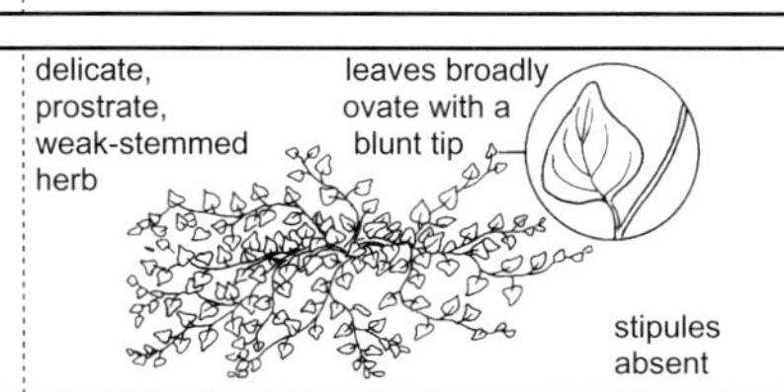	***Parietaria debilis*** 706
	Erect, robustly-stemmed **herb**; **leaves** ovate, acute; **stipules** present	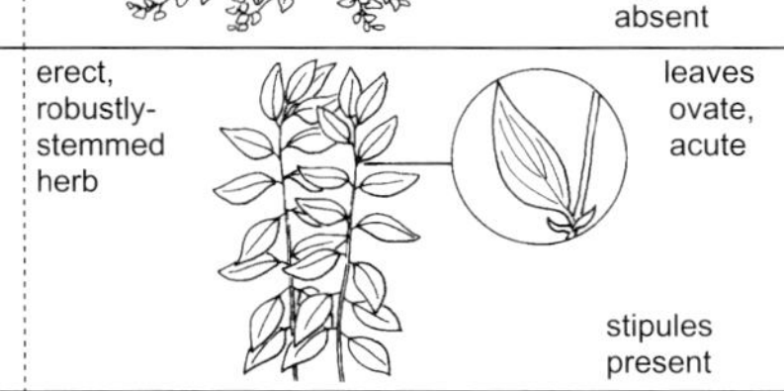	***Pouzolzia auriculata*** 707

VALERIANACEAE

1	Delicate annual **herb**, with lax paired branches ; **fruiting calyx** with 5 or 6 hooked lobes	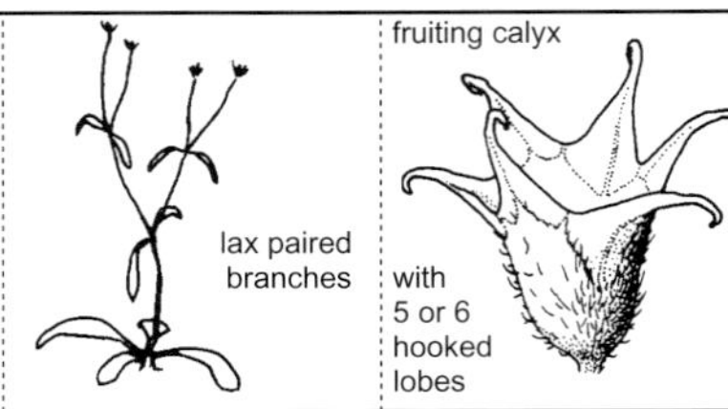	***Valerianella affinis*** Endemic 708

VERBENACEAE

1	**Trees** or **shrubs**		2
	Herbs	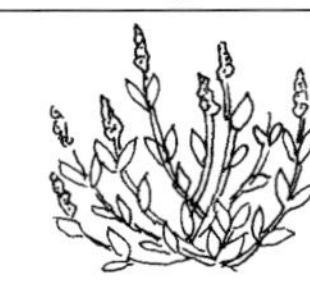	3
2	**Leaves** entire; **corolla** with spreading lobes and a short tube; **anthers** long exserted; **flowers** in few-flowered cymes	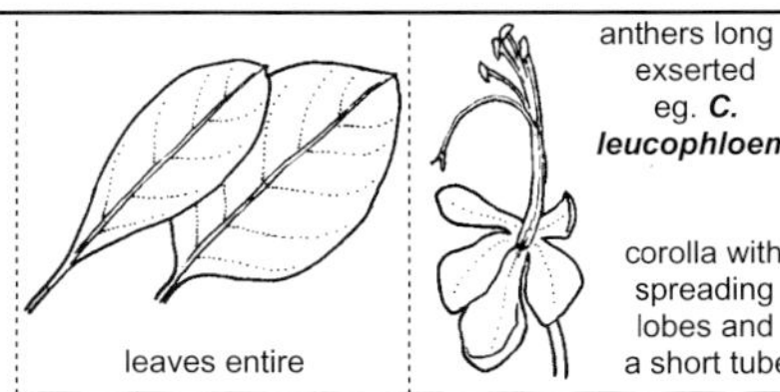	***Clerodendrum*** (2 spp.)
	Leaves toothed; **corolla** with short lobes and a long tube; **anthers** hidden; **flowers** in many-flowered racemes	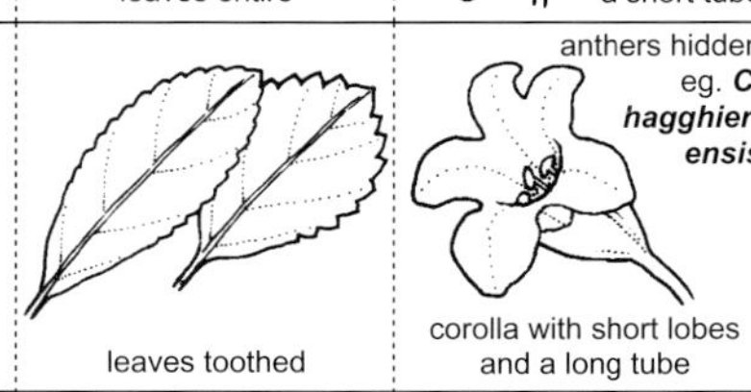	***Coelocarpum*** (2 spp)
3	Prostrate **herb** rooted in mud by water; **flowers** pink, in short, congested spikes; **fruits** smooth	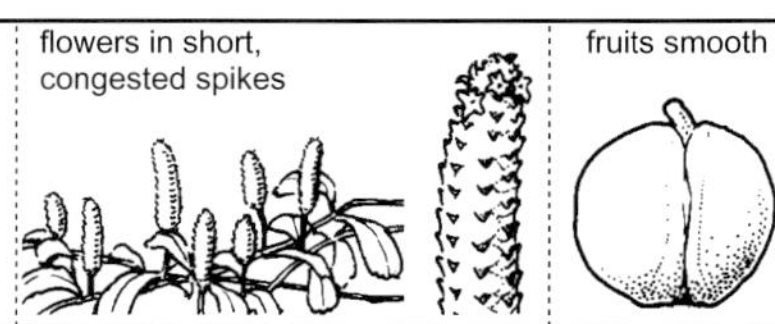	***Phyla nodiflora*** 710
	Ascending **herb,** not rooted in mud; **flowers** white or pink, in long, open spikes; **fruits** spiny		***Priva socotrana*** Endemic **280**, 711

Clerodendrum

1	**Leaves** broadly ovate, more than 4cm across, densely tomentose beneath, more thinly so above	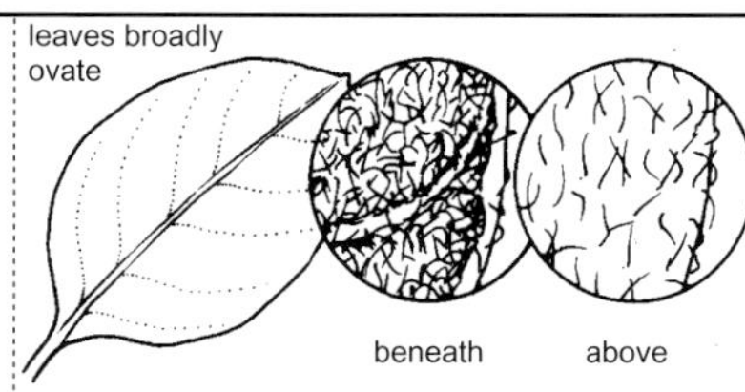	***C. galeatum*** Endemic **280**, 708
	Leaves obovate or oblong-elliptic, less than 3.5cm across, thinly tomentose beneath, glabrescent above	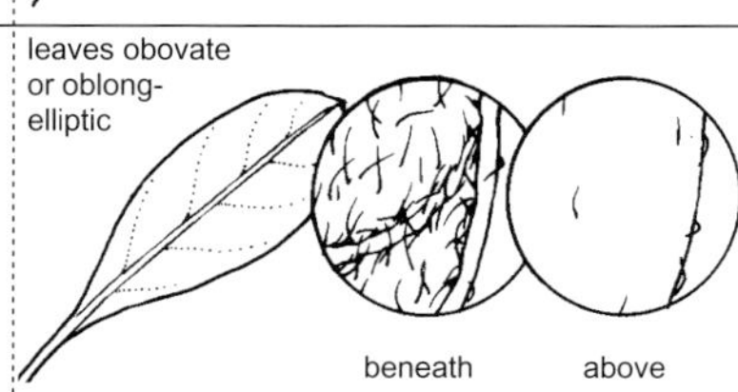	***C. leucophloeum*** Endemic **280**, 709

Coelocarpum

1 Low **shrub** to 40cm tall; **leaves** 2--4cm long; **flowers** white, in few-flowered racemes; **bracts** usually shorter than the pedicels — ***C. socotranum*** Endemic 710

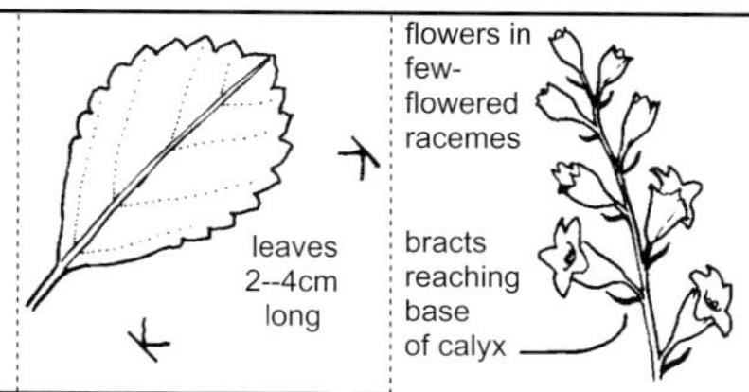

1 **Shrub** or small **tree** to 3m tall; **leaves** 3--10cm long; **flowers** mauve, in densely-flowered racemes; **bracts** usually longer than the pedicels — ***C. haggierensis*** Endemic **280**, 709

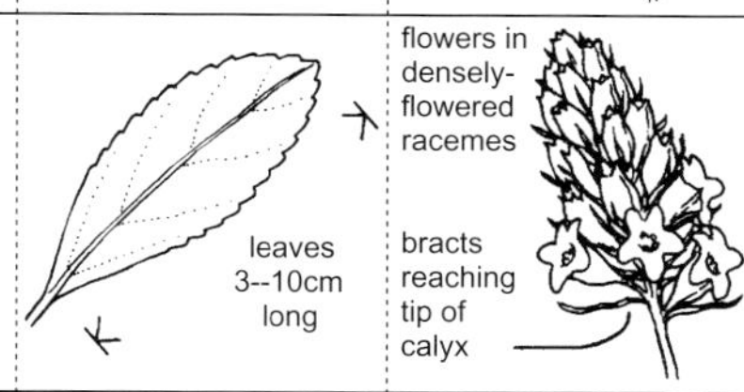

VIOLACEAE

1 **Leaves** elliptic to oblanceolate; **petals** subequal — ***Viola cinerea*** 712

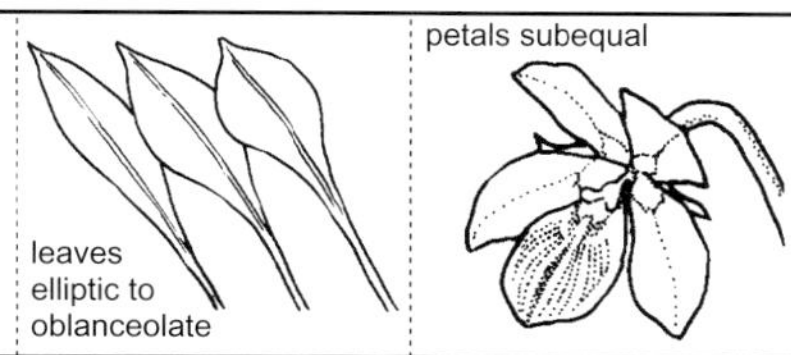

1 **Leaves** linear to linear-oblong; **lower petal** much larger than upper petals — ***Hybanthus enneaspermus*** 711

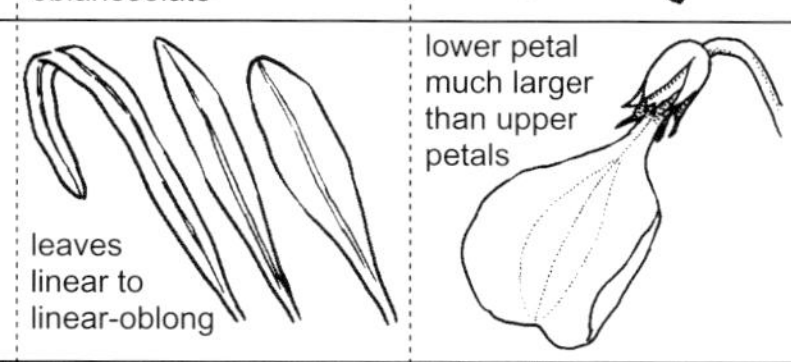

VISCACEAE

1 Parasitic **shrub**; **flowers** minute, globose, in dense axillary clusters — ***Korthalsella taenioides*** **280**, 712

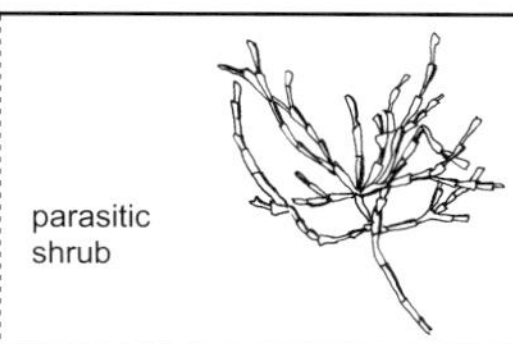

VITACEAE

1 Climbing **vine; leaves** persistent; **stems** green, rounded — ***Cissus paniculata*** Endemic **280**, 714

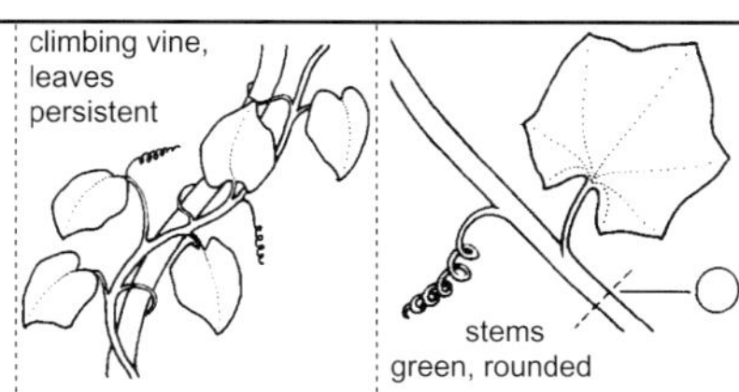

1 Trailing **shrub; leaves** soon dropping; **stems** grey-green, flattened, elliptic in section — ***Cissus subaphylla*** Endemic **280**, 715

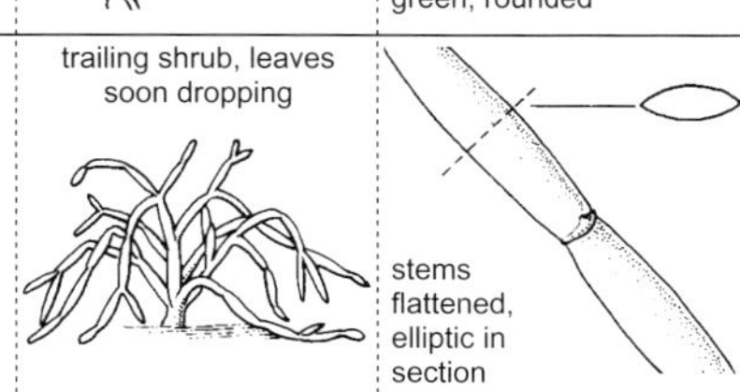

1 Trailing **shrub; leaves** soon dropping; **stems** grey-green, flattened, rectangular in section (becoming oval when mature),the angles winged — ***Cissus hamaderohensis*** Endemic **280**, 713

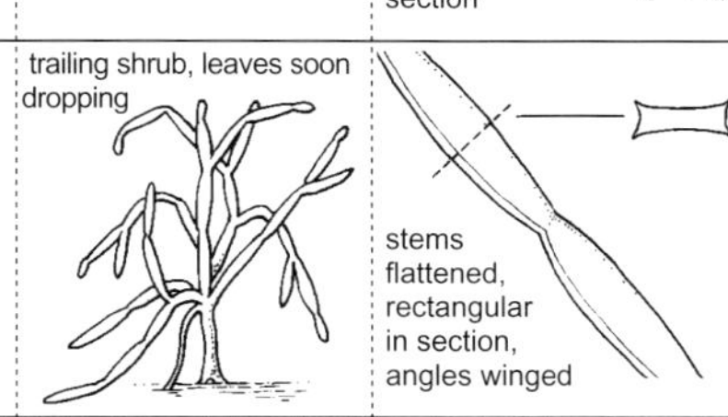

ZYGOPHYLLACEAE

1	**Leaves** pinnate, densely hairy; **fruits** spinescent, splitting into 5, indehiscent segments	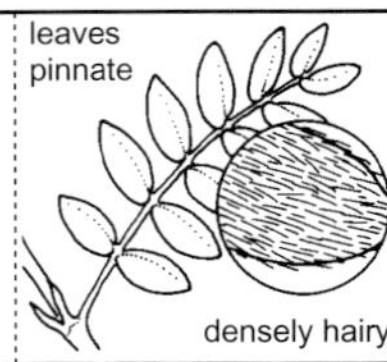	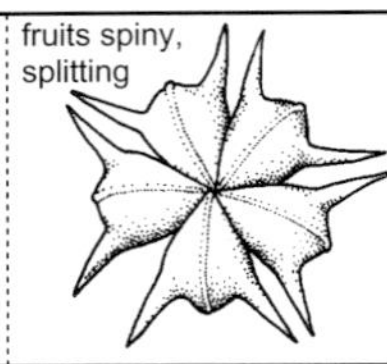	***Tribulus terrestris*** 716
	Leaves 2-foliate or simple, glabrous or glandular, rarely with a few hairs; **fruits** a non-spinescent, dehiscent capsule	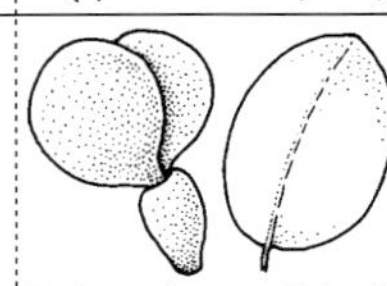	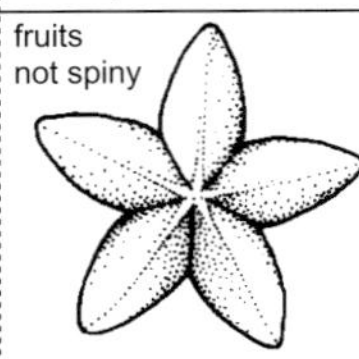	2
2	**Stipules** spinescent or not; **leaves** flattened, simple; **petals** pink; **fruits** not fleshy, smooth when dried	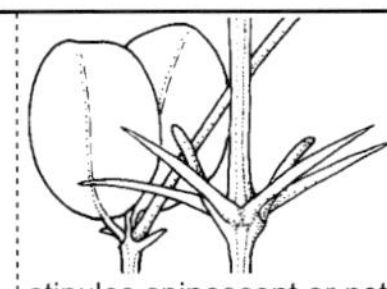	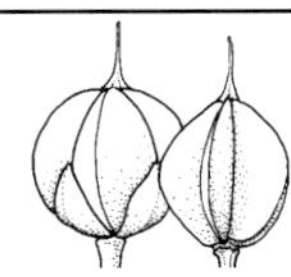	***Fagonia*** (2 spp.)
	Stipules not spinescent; **leaves** cylindrical or obovoid, if flattened then 2-foliate; **petals** white or yellow; **fruits** fleshy, wrinkled when dried	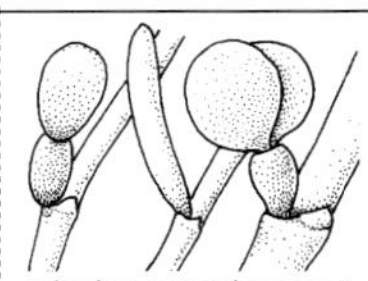	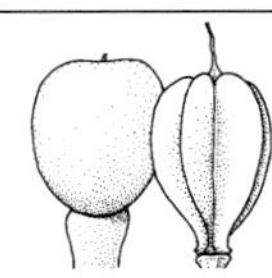	***Zygophyllum*** (3 spp.)

Fagonia

1	**Leaves** linear to narrowly elliptic; **stipules** 5--15mm long, spinescent	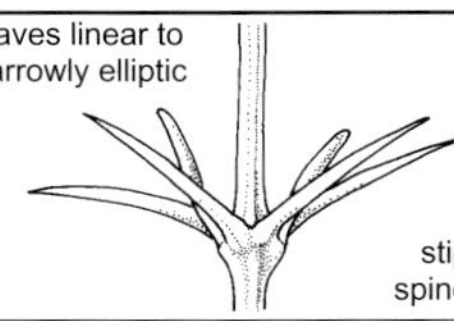		***F. paulayana*** 716
	Leaves ovate to suborbicular; **stipules** <2mm long, spinescent or not	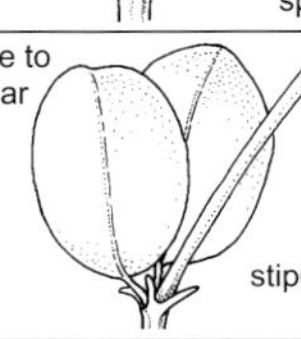		***F. luntii*** 715

Zygophyllum

1	Bright green, annual or perennial **herb**; **leaves** sessile, simple, cylindrical; **petals** yellow; **fruits** 2--2.5mm long, shallowly 5-lobe	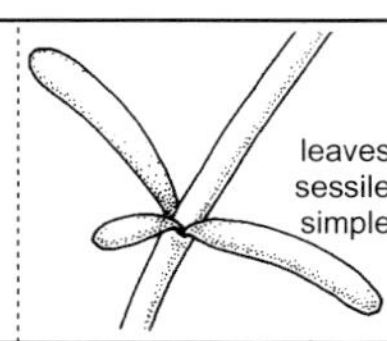		***Z. simplex*** 718
	Grey-green **shrublet**; **leaves** petiolate, 2-foliolate with obovoid-cylindrical leaflets; **petals** white; **fruits** c.7mm long, deeply 5-lobed		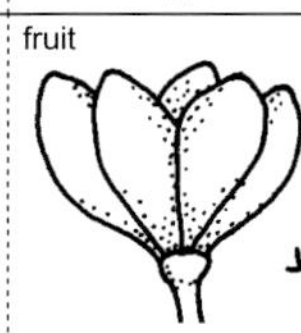	***Z. album*** 717
	Glaucous **shrublet**; **leaves** petiolate, 2-foliate with obovate, flattened leaflets; **petals** white; **fruits** 4--10mm long, shallowly 5-lobed	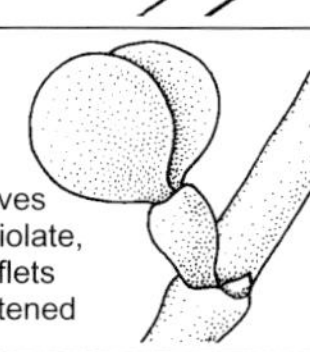		***Z. decumbens*** 717
	Green **shrublet**; **leaves** petiolate, simple or 2-foliolate, leaflets obovoid; **petals** white; **fruits** 3--6mm long, shallowly 5-lobed	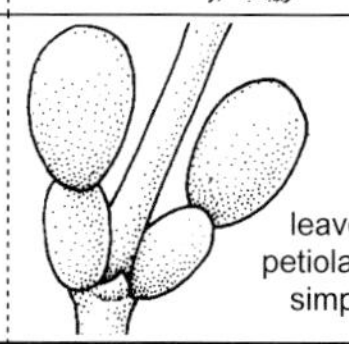	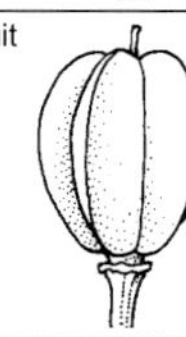	***Z. qatarense*** 717

PTERIDOPHYTES

ACTINIOPTERIDACEAE: *Actiniopteris semiflabellata*

ADIANTACEAE: *Adiantum balfourii*

ADIANTACEAE: *Negripteris sciona*

ASPLENIACEAE: *Asplenium schweinfurthii*

DENNSTAEDTIACEAE: *Pteridium aquilinum* (DA)

OLEANDRACEAE: *Nephrolepis undulata* (RA)

PTERIDACEAE: *Pteris vittata*

MONOCOTYLEDONS

ALOACEAE: ***Aloe perryi*** (AM) (DA)

ALOACEAE: ***Aloe squarrosa*** (DW)

ALOACEAE: ***Aloe jawiyon***

AMARYLLIDACEAE:
Crinum balfourii (DW)

ANTHERICACEAE:
Chlorophytum filifolium

ANTHERICACEAE:
Chlorophytum graptophyllum (DW)

ANTHERICACEAE:
***Chlorophytum* sp. A** (DW)

ASPARAGACEAE: *Asparagus africanus*

COLCHICACEAE: *Iphigenea socotrana*

DIOSCORIACEAE: *Dioscorea lanata*

DRACAENACEAE: Draceana cinnabari

GRAMINEAE: *Andropogon bentii*

HYACINTHACEAE: *Dipcadi balfourii*

HYACINTHACEAE: *Dipcadi biflorum*

HYACINTHACEAE: *Dipcadi guichardii* (DW)

HYACINTHACEAE: *Dipcadi kuriensis* (DW)

HYACINTHACEAE: *Ledebouria grandifolia*

HYACINTHACEAE: *L. insularis* (DW)

HYACINTHACEAE: *Ledebouria* aff. *revoluta* (NO)

HYPOXIDACEAE: *Hypoxis angustifolia*

IRIDACEAE: *Babiana socotrana*

IRIDACEAE: *Romulea fischeri*

JUNCACEAE: *Juncus socotranus*

ORCHIDACEAE: *Angraecum dives*

ORCHIDACEAE: *Eulophia petersii* (DH)

ORCHIDACEAE: *Habenaria socotrana*

ORCHIDACEAE: *Holothrix socotrana*

PALMAE: *Phoenix dactylifera* (left) and *Borassus aethiopum* (right)

DICOTYLEDONS

ACANTHACEAE: *Angkalanthus oligophylla*

ACANTHACEAE: *Barleria aculeata*

ACANTHACEAE: *Barleria popovii*

ACANTHACEAE: *Anisotes diversifolius*

ACANTHACEAE: *Ballochia amoena*

ACANTHACEAE: *Barleria tetracantha*

ACANTHACEAE: *Blepharis kuriensis*

ACANTHACEAE: *Blepharis spiculifolia*

ACANTHACEAE: *Chorisochora striata*

ACANTHACEAE: *Dicliptera effusa*

ACANTHACEAE: *Hypoestes pubescens*

ACANTHACEAE: *Justicia alexandri*

ACANTHACEAE: *Justicia rigida*

ACANTHACEAE: *Justicia takhinensis*

ACANTHACEAE: *Neuracanthus aculeatus*

ACANTHACEAE: *Rhinacanthus scoparius*

ACANTHACEAE: *Ruellia carnea*

ACANTHACEAE: *Ruellia dioscorides* (DA)

ACANTHACEAE: *Ruellia insignis*

ACANTHACEAE: *Ruellia kuriensis*

ACANTHACEAE: *Ruellia paulayana*

ACANTHACEAE: *Trichocalyx obovatus*

AMARANTHACEAE: *Aerva microphylla*

AMARANTHACEAE: *Psilotricum aphyllum*

AIZOACEAE: *Aiozoon canariense*

AIZOACEAE: *Corbichonia decumbens*

AMARANTHACEAE: *Aerva revoluta*

ANACARDIACEAE: *Lannea transulta*

ANACARDIACEAE: *Rhus thyrsiflora*

ANACARDIACEAE: *Rhus thyrsiflora*

APOCYNACEAE: *Adenium obesum* subsp. *sokotranum* (DA)

ASCLEPIADACEAE: *Calotropis procera*

ASCLEPIADACEAE: *Caralluma socotrana*

ASCLEPIADACEAE: *Ceropegia affinis*

ASCLEPIADACEAE: *Ceropegia tihamana*

ASCLEPIADACEAE: *Secamone cuneifolia*

ASCLEPIADACEAE: *Cryptolepis intricata*

ASCLEPIADACEAE: *Cryptolepis macrophylla*

ASCLEPIADACEAE: *Cryptolepis socotrana*

ASCLEPIADACEAE: *Duvaliandra dioscoridis*

ASCLEPIADACEAE:
Echidnopsis bentii (DW)

ASCLEPIADACEAE:
Echidnopsis inconspicua (DW)

ASCLEPIADACEAE:
Echidnopsis insularis (DW)

ASCLEPIADACEAE: *Echidnopsis milleri*

ASCLEPIADACEAE: *Echidnopsis socotrana*

ASCLEPIADACEAE: *Edithcolea grandis*

ASCLEPIADACEAE: *Periploca visciformis*

ASCLEPIADACEAE: *Secamone socotrana*

ASCLEPIADACEAE: *Socotrella dolichocnema* (DW)

ASCLEPIADACEAE: *Vincetoxicum linifolium*

AVICENNIACEAE: *Avicennia marina*

BEGONIACEAE: *Begonia socotrana*

BEGONIACEAE: *Begonia samhaensis*

BORAGINACEAE: *Cordia obtusa*

BORAGINACEAE: *Cystostemon socotranus*

BORAGINACEAE: *Echichilon pulvinata* (DA)

BORAGINACEAE: *Heliotropium aegyptiacum*

BORAGINACEAE: *Heliotropium balfourii*

BORAGINACEAE: *Heliotropium dentatum*

BORAGINACEAE: *Heliotropium nigricans*

BORAGINACEAE: *Heliotropium socotranum* (yellow and white forms)

BORAGINACEAE: *Trichodesma laxiflorum*

BORAGINACEAE: *Trichodesma microcalyx*

BORAGINACEAE: *Trichodesma scottii* (flowers)

BORAGINACEAE: *Wellstedia socotrana*

BORAGINACEAE: *Trichodesma scottii*

BURSERACEAE: *Boswellia ameero*

Boswellia ameero (flowers)

BURSERACEAE: *Boswellia bullata*

BURSERACEAE: *Boswellia dioscorides*

B. dioscorides (left);
B. bullata (right)

Boswellia bullata (trunk and flowers)

BURSERACEAE: *Boswellia elongata*

BURSERACEAE: *Boswellia* sp. A

BURSERACEAE: *Boswellia nana*

BURSERACEAE: *Boswellia nana* x *socotrana*

B. popoviana (form 1)

B. popoviana (form 3)

BURSERACEAE: *Boswellia popoviana* (form 2) (DA)

BURSERACEAE: *Boswellia socotrana*

BURSERACEAE: *Commiphora ornifolia* (DA)

B. *socotrana* (flowers)

C. ornifolia (fruits)

BURSERACEAE: *Commiphora kua*

C. planifrons

BURSERACEAE: *Commiphora planifrons*

BURSERACEAE: *Commiphora parvifolia*

C. parvifolia

BUXACEAE: ***Buxanthus pedicellatus***

CAMPANULACEAE: ***Campanula balfourii***

CAPPARACEAE: ***Cadaba insularis*** (DA) (AM)

M. angolensis **(flower)**

CAPPARACEAE: ***Capparis decidua***

CAPPARACEAE: ***Capparis cartilaginea*** (DA)

CAPPARACEAE: ***Maerua angolensis*** (DA)

M. angolensis (DA)

CARYOPHYLLACEAE: *Gymnocarpos bracteatus*

CARYOPHYLLACEAE: *Gymnocarpos kuriensis*

CARYOPHYLLACEAE: *Haya obovata*

CARYOPHYLLACEAE: *Polycarpaea caespitosa*

CARYOPHYLLACEAE: *Polycarpaea kuriensis*

CARYOPHYLLACEAE: *Sphaerocoma hookeri*

CHENOPODIACEAE: *Arthrocnemum macrostachyum*

CHENOPODIACEAE: *Lagenantha cycloptera*

COMPOSITAE: *Ageratum conyzoides*

COMPOSITAE: *Blainvillea gayana* (RA)

COMPOSITAE: *Dicoma cana*

COMPOSITAE: *Euryops arabicus*

COMPOSITAE: *Helichrysum aciculare*

COMPOSITAE: *Distephanus qazmi*

COMPOSITAE: *Helichrysum arachnoides* (RA)

COMPOSITAE: *Helichrysum balfourii*

COMPOSITAE: *Helichrysum gracilipes*

COMPOSITAE: *Helichrysum nimmoanum*

COMPOSITAE: *Helichrysum paulayanum*

COMPOSITAE: *Helichrysum rosulatum*

COMPOSITAE: *Helichrysum* sp. A

COMPOSITAE: *Helichrysum* sp. B

COMPOSITAE: *Helichrysum* sp. D (RA)

COMPOSITAE: *Helichrysum sphaerocephalum*

COMPOSITAE: *Kleinia scottii*

COMPOSITAE: *Launaea crepoides*

COMPOSITAE: *Launaea socotrana*

COMPOSITAE: *Tripteris vaillantii*

COMPOSITAE: *Pluchea obovata* (flower)

COMPOSITAE: *Pluchea obovata*

COMPOSITAE:
Prenanthes amabilis (DA)

COMPOSITAE:
Pulicaria elegans

COMPOSITAE: *Pulicaria diversifolia*

COMPOSITAE: *Pulicaria aromatica*

COMPOSITAE: *Pulicaria lanata*

COMPOSITAE: *Pulicaria dioscorides*

COMPOSITAE: *Pulicaria vieraeoides*

COMPOSITAE: *Pulicaria stephanocarpa*

COMPOSITAE: *Vernonia cockburniana*

CONVOLVULACEAE: *Convolvulus grantii*

CONVOLVULACEAE: *Convolvulus hildebrandtii*

CONVOLVULACEAE: *Convolvulus sarmentosus*

CONVOLVULACEAE: *Metaporana obtusa*

CONVOLVULACEAE: *Seddera semhaensis*

CONVOLVULACEAE: *Seddera spinosa*

CRASSULACEAE: *Crassula schimperi*

CRASSULACEAE: *Kalanchoe rotundifolia*

CRASSULACEAE: *Kalanchoe farinacea*

CRASSULACEAE: *Kalanchoe robusta* (DW)

CRUCIFERAE: *Diceratella incana*

CRUCIFERAE: *Erucastrum rostratum*

CRUCIFERAE: *Farsetia socotrana*

CRUCIFERAE: *Farsetia inconspicua*

CRUCIFERAE: *Lachnocapsa spathulata*

CUCURBITACEAE: *Cucumis prophetarum* subsp. *prophetarum*

CUCURBITACEAE: *Dendrosicyos socotrana*

CUCURBITACEAE: *Dendrosicyos socotrana* (DA)

CUCURBITACEAE: *Eureiandra balfourii*

DIRACHMACEAE: *Dirachma socotrana* (DA)

EBENACEAE: *Euclea divinorum*

EUPHORBIACEAE: *Acalypha indica*

EUPHORBIACEAE: *Cephalocroton socotranus*

EUPHORBIACEAE: *Croton sarcocarpus*

EUPHORBIACEAE: *Croton socotranus* (RH)

EUPHORBIACEAE: *Croton sulcifructus*

EUPHORBIACEAE: *Euphorbia* aff. *arbuscula*

EUPHORBIACEAE: *Euphorbia abdelkuri* (DA)

EUPHORBIACEAE: *Euphorbia arbuscula* (DA)

Euphorbia abdelkuri

EUPHORBIACEAE: *Euphorbia balsamifera*

EUPHORBIACEAE: *Euphorbia hadramautica*

EUPHORBIACEAE: ***Euphorbia hamaderohensis*** (DA)

EUPHORBIACEAE: ***Euphorbia kischenensis***

EUPHORBIACEAE: ***Euphorbia kuriensis***

EUPHORBIACEAE: ***Euphorbia obcordata***

EUPHORBIACEAE: ***Euphorbia schimperi***

EUPHORBIACEAE: ***Euphorbia schweinfurthii*** **(form A)**

Euphorbia schweinfurthii (form D) (DA)

EUPHORBIACEAE: *Euphorbia spiralis* (DA)

EUPHORBIACEAE: *Euphorbia socotrana*

EUPHORBIACEAE: *Jatropha unicostata*

EUPHORBIACEAE: *Phyllanthus rotundifolius*

Jatropha unicostata (flowers and fruit)

EUPHORBIACEAE: *Tragia balfouriana*

FLACOURTIACEAE: *Dovyalis abyssinica*

GENTIANACEAE: *Exacum affine* (DA)

GENTIANACEAE: *Exacum affine* (DA)

GENTIANACEAE: *Exacum caeruleum*

GENTIANACEAE: *Exacum caeruleum*

GERANIACEAE: *Geranium biuncinnatum*

GERANIACEAE: *Pelargonium insularis* (DW)

GLOBULARIACEAE: *Poskea socotrana*

GOODENIACEAE: *Scaevola socotraensis*

GUTTIFERAE: *Hypericum balfourii*

GUTTIFERAE: *Hypericum scopulorum* (DA)

Hypericum balfourii

Hypericum tortuosum

GUTTIFERAE: *Hypericum socotranum* subsp. *smithii*

GUTTIFERAE: *Hypericum tortuosum* (DA)

LABIATAE: *Endostemon tenuiflorus*

LABIATAE: *Lavandula nimmoi* (DA)

LABIATAE: *Leucas flagellifolia*

LABIATAE: *Leucas hagghierensis*

LABIATAE: *Leucas kischenensis*

LABIATAE: *Leucas spiculifolia*

LABIATAE: *Leucas virgata*

LABIATAE: *Micromeria remota*

LABIATAE: *Orthosiphon ferrugineus*

LABIATAE: *Teucrium balfourii*

LABIATAE: *Teucrium sokotranum*

LEGUMINOSAE: *Acacia edgeworthii*

LEGUMINOSAE: *Acacia pennivenia*

Acacia sarcophylla

LEGUMINOSAE: *Acacia pennivenia*

LEGUMINOSAE: *Acacia sarcophylla*

LEGUMINOSAE: *Chapmannia sericea*

LEGUMINOSAE: *Crotalaria socotrana*

LEGUMINOSAE: *Crotalaria strigulosa*

LEGUMINOSAE: *Dichrostachys dehiscens*

LEGUMINOSAE: *Indigofera marmorata*

LEGUMINOSAE: *Indigofera nephrocarpa*

LEGUMINOSAE: *Indigofera pseudointricata*

LEGUMINOSAE: *Indigofera socotrana*

LEGUMINOSAE: *Lotus garcinii*

LEGUMINOSAE: *Lotus mollis*

LEGUMINOSAE: *Lotus ononopsis*

LEGUMINOSAE: *Senna holosericea*

LEGUMINOSAE: *Senna socotrana*

LEGUMINOSAE: *Taverniera sericophylla*

LEGUMINOSAE: *Zygocarpum coeruleum*

LEGUMINOSAE: *Paracalyx balfourii*

LEGUMINOSAE: *Tephrosia odorata*

MALPIGHIACEAE: *Acridocarpus socotranus*

MALVACEAE: ***Hibiscus diriffan*** (DA)

MALVACEAE: ***Hibiscus scottii***

MALVACEAE: ***Hibiscus socotranus***

MALVACEAE: ***Hibiscus noli-tangere***

MALVACEAE: ***Hibiscus quattenensis***

MALVACEAE: ***Hibiscus stenanthus***

MELIACEAE: *Turraea socotrana*

MENISPERMACEAE: *Cocculus balfourii*

MORACEAE: *Dorstenia gigas*

MORACEAE: *Dorstenia gigas* (flowers and leaves)

MORACEAE: *Dorstenia socotrana*

MORACEAE: *Ficus cordata*

MORACEAE: *Ficus vasta* (DA)

NYCTAGINACEAE: *Commicarpus heimerlii*

NYCTAGINACEAE: *Commicarpus simonyi*

OROBANCHACEAE: *Cistanche phelypaea*

PAPAVERACEAE: *Argemone mexicana*

PIPERACEAE: *Peperomia blanda*

PLUMBAGINACEAE: *Dyerophytum socotranum*

PLUMBAGINACEAE: *Limonium paulayanum*

PLUMBAGINACEAE: *Limonium sokotranum*

POLYGALACEAE: *Polygala kuriense*

PORTULACACEAE: *Portulaca kuriensis*

PORTULACACEAE:
Portulaca quadrifida

PORTULACACEAE:
Portulaca samhaensis

PORTULACACEAE:
Portulaca sedifolia

PORTULACACEAE: *Talinum portulacifolium*

PUNICACEAE: *Punica protopunica* (flowers and fruit)

PUNICACEAE: *Punica protopunica* (DA)

RESEDACEAE: *Ochradenus socotranus*

RESEDACEAE: *Reseda viridis*

RHAMNACEAE: *Zizyphus spina-christi*

RUBIACEAE: *Carphalea obovata*

RUBIACEAE: *Gaillonia puberula*

RUBIACEAE: *Gaillonia putorioides*

RUBIACEAE: *Gaillonia thymoides*

RUBIACEAE: *Gaillonia tinctoria*

RUBIACEAE: *Kohautia socotrana*

RUBIACEAE: *Oldenlandia balfourii* (white form)

RUBIACEAE: *Kraussia socotrana*

RUBIACEAE: *Oldenlandia balfourii* (yellow form)

RUBIACEAE: *Oldenlandia bicornuta*

RUBIACEAE: *Oldenlandia ocellata*

RUBIACEAE: *Oldenlandia pulvinata* (DA)

RUBIACEAE: *Placopoda virgata* (flower)

RUBIACEAE: *Placopoda virgata*

RUBIACEAE: *Tamridaea capsulifera*

RUTACEAE: *Thamnosma socotrana*

SAPINDACEAE: *Allophylus rubifolius*

SCROPHULARIACEAE: *Camptoloma lyperiflorum*

SCROPHULARIACEAE: *Campylanthus spinosus*

SCROPHULARIACEAE: *Craterostigma pumilum*

SCROPHULARIACEAE: *Graderia fruticosa*

SCROPHULARIACEAE: *Nanorrhinum kuriensis*

SCROPHULARIACEAE: *Scrophularia arguta*

SCROPHULARIACEAE: *Xylocalyx aculeolatus*

SCROPHULARIACEAE: *Xylocalyx asper*

SOLANACEAE: *Lycium sokotranum*

SOLANACEAE: *Solanum incanum*

SOLANACEAE: *Withania adunensis*

THYMELAEACEAE: *Gnidia socotrana* (flowers)

STERCULIACEAE: *Sterculia africana* (DA)

Gnidia socotrana

Sterculia africana (fruit) (DA)

TILIACEAE: *Corchorus erodioides*

TILIACEAE: *Grewia mansouriana*

UMBELLIFERAE: ***Nirarathamnos asarifolius*** (DA) (AM)

UMBELLIFERAE: ***Oreofraga morrisiana*** (MM)

UMBELLIFERAE: ***Rughidia milleri***

UMBELLIFERAE: ***Rughidia cordata***

UMBELLIFERAE: ***Trachyspermum pimpinelloides*** (DA)

VERBENACEAE: *Clerodendrum leucophloeum*

VERBENACEAE: *Clerodendrum galeatum*

VERBENACEAE: *Coelocarpum haggierensis*

VERBENACEAE: *Priva socotrana*

VISCACEAE: *Korthalsella taenioides*

VITACEAE: *Cissus hamaderohensis*

VITACEAE: *Cissus paniculata*

VITACEAE: *Cissus subaphylla*

PHOTO CREDITS

All photographs taken in the Archipelago between 1989 and 2004, by Anthony Miller unless indicated:

DA – Diccon Alexander
RA – Ruth Atkinson
DH – Dylan Hannon
RH – Roger Hyam
MM – Miranda Morris
NO – Neil Oakham
DW – Debbie White (plants cultivated at the RBGE)
MT – Mats Thulin

ENCYCLOPAEDIA

SOQOTRAN ALOE (ALOE PERRYI) DICCON ALEXANDER

PTERIDOPHYTES (THE FERNS)

The ferns are flowerless plants with two distinct free-living generations: a conspicuous, herb-like sporophyte generation (reproducing by spores) alternating with the inconspicuous thalloid, gametophyte generation. The underground stems (rhizomes) are usually clothed with scales. The leaves (the fronds) are usually circinate in bud (except in *Ophioglossum*) and comprise a stalk (the stipe) and a lamina with a central nerve (the rhachis). The lamina is either simple or divided into pinnae and pinnules. The sporangia are usually borne on the lower surface of the fronds in clusters (the sori) or borne in specialised organs (as in *Marsilea* and *Ophioglossum*). The sori are either naked or covered with specialised scale-like organs (the indusia) or by the reflexed leaf margin.

Botanical notes. No family descriptions are given for ferns.

Synoptical arrangement of the families of ferns

Actiniopteridaceae
  Actiniopteris
Adiantaceae
  Adiantum
  Doryopteris
  Negripteris
  Onychium
  Pellaea
Aspleniaceae
  Asplenium
  Ceterach
Dennstaedtiaceae
  Pteridium
Dryopteridaceae
  Hypodematium
Marsileaceae
  Marsilea
Oleandraceae
  Nephrolepis
Ophioglossaceae
  Ophioglossum
Parkeriaceae
  Ceratopteris
Polypodiaceae
  Loxogramme
  Pleopeltis
Pteridaceae
  Pteris
Thelypteridaceae
  Christella

ACTINIOPTERIDACEAE

Actiniopteris Link

Fronds tufted, dichotomously branched with linear segments; sterile segments toothed at the apex, fertile entire. Sori submarginal, linear-elongate, covered by reflexed margin of frond. 5 species. Old World tropics.

Actiniopteris semiflabellata Pic.Serm.

Actiniopteris dichotoma sensu Balf.f.

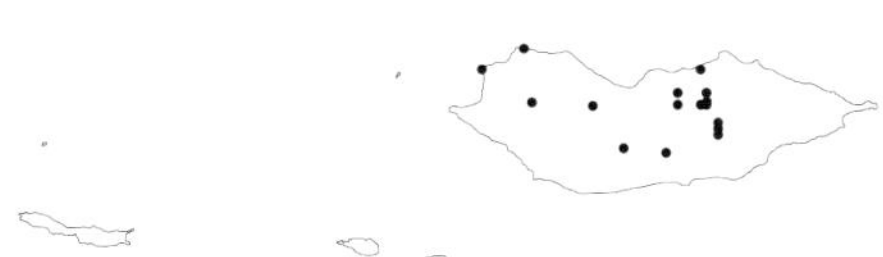

Fronds 5–30cm, green when fresh, silvery-grey when dry; lamina wedge-shaped, 1–7cm long, with 4 to 18 segments, twisted sideways when dry; segments acute and laterally toothed at apex.

Ecology. Common in shady cracks in cliffs and boulders on both limestone and granite. 50–1050m.

Distribution. NE tropical Africa, Arabia, Madagascar and Mauritius.

Botanical notes. One of the most drought-tolerant ferns.

ETHNOGRAPHIC NOTES

Name(s)

– Soqotri: *mhāṭḥˈʌṭən*, *mḥāṭˈoh* 'limp, flaccid'; *māṣel* 'dried out'; *gəfeˀhen* 'shade-loving'; *rˈʌḥmʌm*, *rˈʌxmʌm* (western Soqotra) 'grey-coloured'; *di ḥaṭˈarṭəhʌr*, *di khoṭˈorṭəhor* (western Soqotra).

– Soqotri simplified: *mhathatin*, *mhatoh*; *mazel*; *gife'hen*; *rahmam*, *rakhmam*; *di hatartihar*, *di khotortihor*.

Food (livestock). Not normally eaten by livestock, but the drying fronds are nibbled by small stock when they are very hungry in the dry season.

Fuel. Dried ferns were often used as tinder when starting a fire with firesticks.

Medicinal / veterinary. The ferns are gathered and used as a dressing for wet sores, ulcers, abscesses, wounds and lacerations of all kinds in people and livestock. They are either scorched at the fire and then ground to a powder and applied in that form; or water is added to the powder, boiled, cooled and then applied as a lotion. The same lotion can also be used when it has dried out with no further preparation. For a purulent and discharging wound, the wound is first cleaned with water and salt and then treated with a fern preparation made by gathering them when sun-dried but still green, and then crumbling them and storing them for future use. Alternatively the ferns are gathered when fresh and green and put in a dry clay pot to 'sweat' out any moisture. They are then taken out, dried in the sun, and crushed to a powder.

Notes. Islanders regard this as the 'male' version of the 'female' *Adiantium balfourii*.

Comparative uses. In Yemen it is used to treat burns. *A. australis* is said to have antibiotic properties and is used as an antiseptic and styptic. It is also used as an anti-fertility drug in some places. (Ambasta 1982; Fleurentin & Pelt/Fleurentin *et al.* [1982, 1983a,b,c]; Jain & De Fillips 1991)

ADIANTACEAE

ETHNOGRAPHIC NOTES†

Food (livestock). Ferns of the limestone plateaus can be toxic, and in particular the Adiantaceae and the Aspleniaceae. Goats are rarely tempted to eat ferns if there is good grazing. However, in the long dry season, especially in the limestone plateaus, hungry goats resort to ferns, which are tempting when green. If they eat a lot they initially seem unaffected, but often go on to develop the condition called *ᶜʌṣʌbhən'*. This is immediately obvious in some animals, which appear dizzy and disorientated, call continuously in a strange, high tone, and start to tremble, stagger and fall over. In other animals the condition does not become apparent until they are moved on. After walking a little distance the sick goat quickly tires, becomes unsteady on its feet and soon starts to tremble and stagger. The only treatment is to move the affected animals right out of the area where the ferns grow. However, this must not be done until the sun is well up, or until they have full bellies, otherwise the condition worsens. When being herded to a new area, they must be driven along slowly, with frequent rest stops. They only recover when all traces of the fern fodder have left their system, after about a week in the new area. Usually there are no long-lasting effects.

†Also apply to Aspleniaceae.

Adiantum L.

Fronds tufted or spaced, 1 to 4-pinnate; stipe dark brown, reddish brown or black, polished. Sori borne on the inner surface of the reflexed leaf margins. 200 species. Cosmopolitan.

Adiantum balfourii Baker

Fronds tufted, pinnate, narrowly oblong, 8–35 x 2–3.5cm; pinnae subopposite, fan-shaped to subcircular, 1–2 x 1–2.75cm, the margins often with a hardened black or brown edge, rounded to cuneate at the base, glabrous; petioles <1.5mm; stipe black or dark reddish-brown, glabrous.

Ecology. Widespread and common in a wide range of vegetation types and from almost sea-level to all but the highest peaks. 50–1100m.

Distribution. Somalia, Ethiopia and Djibouti.

Botanical notes. The commonest fern in the Archipelago.

ETHNOGRAPHIC NOTES

Name(s)

– Soqotri: *ṣɔ̄wṣuh* 'dots'; *mḥāṭoh* 'limp, flaccid'; *di ᶜalimləhum* 'marked'; *ᶜʌṣʌbhən* 'toxic'; *di ḥəwnhitin* 'lichen-like'; *di ḥəwritin* 'dark-coloured'; *di ḳīfuf, di ḳyūḳʌf* 'open, disingenuous'; *di ḥʌṣʌhʌn* 'large centipede'.

– Soqotri simplified: *zowzuh*; *mhatoh*; *di 'alimlihum*; *'azabhin*; *di huwnhitin*; *di huwritin*; *di qiyfuf, di qyuwqaf*; *di hazahan*.

Food (livestock). Small stock eat these when hungry (but see Adiantaceae).

Fuel. Dried ferns are often used as tinder with firesticks.

Adiantum capillus-veneris L.

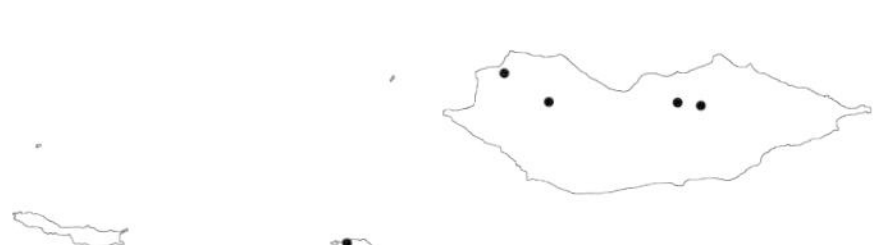

Fronds shortly spaced, arching, 1 or 2(or 3)-pinnate, narrowly ovate to ovate-triangular, 10–40 x 3–12cm; pinnules alternate, fan or wedge-shaped, 7–20(–25) x 4–18mm, shallowly to deeply lobed, glabrous; sterile pinnules crenate-dentate, the veins ending in teeth; stipe blackish-brown or reddish-brown, glabrous.

Ecology. On damp, shady cliffs, usually by permanent water or seepages. 100–1000m.

Distribution. Warm temperate and tropical regions.

ETHNOGRAPHIC NOTES

Name(s)

– Soqotri: *mḥāṭḥiṭən* 'limp, flaccid'; *di ḥəwnhitin* 'lichen-like'; *di ḥəwritin* 'dark-coloured'; *ᶜeźerəh* (Samha).

– Soqotri simplified: *mhat'hitin*; *di huwnitin*; *di huwritin*; *'edherih*.

Food (livestock). Small stock and cattle eat these when hungry (but see Adiantaceae).

Fuel. Dried ferns are used as tinder with firesticks.

Comparative uses. In Iraq this plant is used as an insect repellent. The leaves are used in the treatment of dysentery and intestinal colic, to expel gases, and as a general tonic and carminative. In India the plant is seen as a pectoral demulcent, expectorant and tonic. It is boiled in wine and given in cases of hard tumours of the spleen, liver and other viscera. The fern also contains an essential oil. (Al Rawi & Chakravarty 1964; Ambasta 1982; Jain & De Fillips 1991; Kunkel 1984)

Adiantum poirettii **Wikstr.**
Adiantum aethiopicum sensu Balf.f. non L. p.p.

Similar to *A. capillus-veneris* but a more delicate plant with 3(or 4)-pinnate fronds; sterile pinnules shallowly lobed and crenate with the veins ending in sinuses.

Ecology. No habitat details available. Balfour (1888) recorded it from hills [the Haggeher mountains] south of Tamarida [Hadibo].
Distribution. Tropical and S Africa, India, C and S America.
Botanical notes. Not seen since it was first recorded in the 19th century. Similar to and easily confused with the preceding species. It should be looked for in damp, shady places amongst the highest peaks.

Doryopteris J.Sm.

Fronds tufted, 2 to 4-pinnate, glabrous. Sori marginal, continuous or interrupted; indusia narrow, membranous, formed from reflexed leaf margins. 35 species. Tropical and warm regions.

Doryopteris concolor **(Langsd. & Fisch.) Kuhn**
Cheilanthes concolor (Langsd. & Fisch.) R.M.Tryon & A.F.Tryon
Doryopteris kirkii (Hook.) Alston
Pellaea concolor (Langsd. & Fisch.) Baker

Fronds tufted, pentagonal, 2 to 4-pinnatifid, <20 x 18cm, glabrous, lowest pinnae triangular, upper pinnae oblong-ovate; stipe reddish black, clothed with scales.

Ecology. Found only amongst the high peaks of the Haggeher mountains; in shady, humid places in dense thickets and on shady rock faces. 1100–1250m.
Distribution. Throughout the tropics.

ETHNOGRAPHIC NOTES
Name(s)
– Soqotri: *ṣeberbəher* 'sour, astringent'.
– Soqotri simplified: *ziberbiher*.
Food (livestock). Eaten by goats and cattle where it persists through the dry season in woodland, but it is not regarded as having any nutritional value.

Negripteris Pic.Serm.

Fronds tufted, 1 to 3-pinnate. Sori continuous around the margin, covered by the reflexed margin of the frond. 1 species. Tropical NE Africa and Arabia.

Negripteris sciona **Pic.Serm.**
Cheilanthes farinosa sensu Balf.f. non (Forssk.) Kaulf.
Cheilanthes farinosa sensu O.Schwartz p.p. non (Forssk.) Kaulf.

Fronds green above, powdery-white (at least when young) beneath, 5–20 x 3–7cm, narrowly ovate or ovate to pentagonal; pinnae triangular at base of frond, oblong above. Sori black or dark brown.
Ecology. Cliffs and wadi sides on the coastal foothills and limestone plateaus. 100–650m.
Distribution. Arabia and tropical NE Africa.
Botanical notes. Readily distinguished by the conspicuously contrasting dark brown or black sporangia on the powdery-white undersurface of the frond.

ETHNOGRAPHIC NOTES
Name(s)
– Soqotri: nil or *mḥāṭhʌṭən* 'limp, flaccid'; *diḳereś* 'sweepings', 'dangerous'; *ruᶜud di tīyo* 'grazing of cliffs'; *di ḥəwnhitin* 'lichen-like'; *di ḥəwritin* 'dark-coloured'; *ḥʌlḥʌləhīn* 'brown-grey'; *ṣɔ̄wṣuh* 'dots'; *raḥmam, rʌxmʌm* (western Soqotra) 'grey-coloured'.
– Soqotri simplified: *mhat'hatin*; *diqeresh*; *ru'ud di tiyo*; *di huwnitin*; *di huwritin*; *halhalihiyn*; *zowzuh*; *rahmam, rakhmam*.
Food (livestock). Small stock and cattle eat these when hungry (but see Adiantaceae).
Fuel. Dried ferns are used as tinder with firesticks.

Onychium Kaulf.

Fronds irregularly 3 or 4-pinnate, fertile and sterile fronds different. Sori marginal, covered by the reflexed leaf margins. 6 species. Old World tropics.

Onychium divaricatum (Poir.) Alston

Allosorus melanolepis Decne.
Onychium melanolepis (Decne.) Kunze

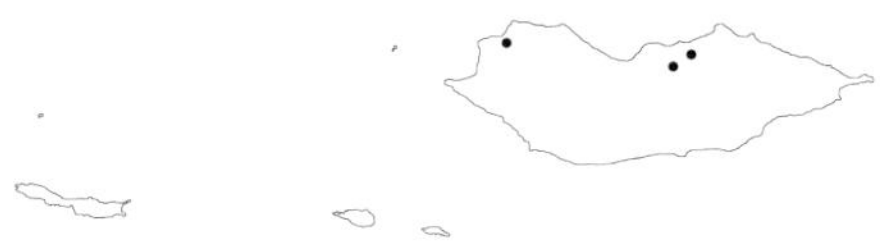

A small fern with finely dissected fronds. Fronds tufted, ovate, 7–20 x 5–13cm, glabrous; sterile segments ovate or wedge-shaped, 4–7 x 3–5mm, divided into 2 to 5 narrowly elliptic lobes; fertile segments narrowly oblong, 1.5–15 x 1–1.25mm, entire, acute.

Ecology. Cliffs and rock faces on the limestone plateaus. 450–650m.

Distribution. NE Africa, Arabia and Iran to SE and E Asia.

Botanical notes. One of the most drought-tolerant ferns. At first glance not obviously a fern; somewhat resembles a species of the flowering plant genus *Fumaria*.

ETHNOGRAPHIC NOTES

Name(s)

– Soqotri: nil or *ra̍ḥmam*, *rʌ̍xmʌm* (western Soqotra) 'grey-coloured'; *di tīyo* 'grazing of cliffs'; *di ḥə̍wni̍tin* 'lichen-like'; *di ḥə̍writin* 'dark-coloured'.

– Soqotri simplified: *rahmam*, *rakhmam*; *di tiyo*; *di huwnitin*; *di huwritin*.

Food (livestock). Goats eat this at times of hunger, if they can reach it (but see Adiantaceae).

Fuel. Dried ferns are used as tinder with firesticks.

Comparative uses. In India a decoction of the fronds of *O. siliculosum* is used to treat dysentery. (Ambasta 1982; Jain & De Fillips 1991)

Pellaea Link

Fronds tufted, 1 to 4-pinnate, clothed with scales or glabrous. Sori marginal, continuous, covered by the reflexed margin of the frond. 80 species. Tropical and warm regions.

Pellaea involuta (Sw.) Baker

Cheilanthes involuta (Sw.) Schelpe & Anthony
Pellaea viridis sensu Balf.f. non (Forssk.) Prantl p.p.

Similar to *P. viridis* but rhachis clothed with entire, not minutely serrulate, scales.

Ecology. In shady, humid places in semi-deciduous submontane woodland. 600m.

Distribution. S Africa, E tropical Africa, Madagascar and Arabia.

Botanical notes. A rare or possibly overlooked fern, closely related to and easily confused with *P. viridis*, of which it is sometimes considered to be merely a variety.

On Soqotra it is known only from a single specimen (*Schweinfurth* 596) collected on the northern slopes of the Haggeher mountains at Kishin in Wadi Di Negehen at 600m. The Schweinfurth collection consists of several plants all but one of which have been determined as *P. viridis* (a relatively common fern of the Haggeher mountains). The remaining plant, because of the presence of abundant scales on the fronds, has been determined as *P. involuta*. Scattered scales are sometimes found on the fronds of *P. viridis* and it seems likely that the plant determined as *P. involuta* is merely an extreme form of the more common *P. viridis*.

ETHNOGRAPHIC NOTES

Not seen, no information recorded.

Pellaea viridis (Forssk.) Prantl

Cheilanthes viridis Balf.f.

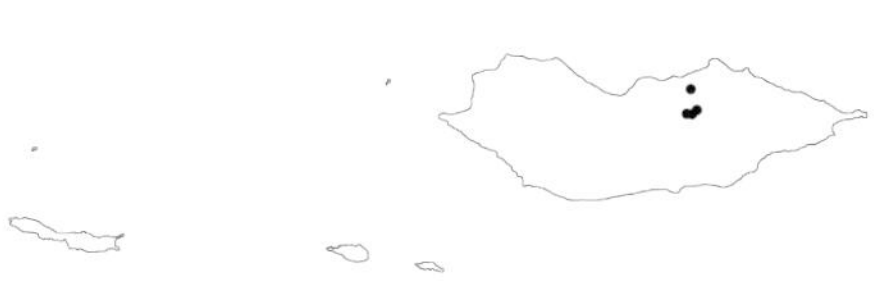

Rhizome scales minutely serrulate, pale brown and concolorous or with a dark central stripe. Fronds 2-pinnate, lanceolate to narrowly ovate, 10–40 x 5–10cm; stipe reddish black, glabrous or clothed with scales below.

Ecology. A locally common fern of semi-deciduous submontane woodlands in the Haggeher mountains; typically in deep shade in the relative humidity of dense thickets of *Rhus thyrsiflora*. 550–1100m.

Distribution. S Africa, E tropical Africa, Madagascar and Arabia.

Botanical notes. See comments under *P. involuta*.

ETHNOGRAPHIC NOTES

Name(s)

– Soqotri: *mḥāṭo̍h* 'limp, flaccid'; *di ḥə̍wni̍tin* 'lichen-like'; *di ḥə̍wri̍tin* 'dark-coloured'.

– Soqotri simplified: *mhatoh*; *di huwnitin*; *di huwritin*.

Food (livestock). Small stock and cattle eat these when hungry, towards the end of the dry season (but see Adiantaceae).

Fuel. Dried ferns are used as tinder with firesticks.

Comparative uses. In India the fronds of *P. calomelanos* are smoked for asthma and to treat a cold in the head and chest. The pounded leaves are taken as an anthelmintic. In S Africa a decoction of the rhizome of *P. calomelanus* is given for boils, cankers of the mouth and other oral infections. (Ambasta 1982; Jain & De Fillips 1991; Lewis & Elwin-Lewis 1977)

ASPLENIACEAE

ETHNOGRAPHIC NOTES
See Adiantaceae.

Asplenium L.

Fronds 2-pinnatifid. Sori elongate, borne along the veins; indusia narrow, opening inwards towards the midrib, membranous or absent or obsolete. 1 endemic. 650 species. Cosmopolitan.

Asplenium aethiopicum (Burm.f.) Bech.
Asplenium praemorsum Sw.

Fronds densely tufted, dark green, somewhat leathery, narrowly ovate, 10–30 x 5–10cm, pinnate with 2 or 3-pinnatifid pinnae, thinly covered with hair-like scales; pinnae narrowly ovate to trapeziform-acuminate, with wedge-shaped to narrowly oblong segments, serrate towards the tips. Stipe short, dark brown becoming green above, with mixed scales and hairs.
Ecology. In shade on cliffs, amongst boulders and in dense thicket in montane woodland. 1000–1400m.
Distribution. Tropical and S Africa.

ETHNOGRAPHIC NOTES
Name(s)
– Soqotri: *gɛˈsibiher, gisiˈbirəh, ginˈsibiher.*
– Soqotri simplified: *gesibiher, gisibirih, ginsibiher.*
Food (livestock). Small stock and cattle eat these when hungry (but see Aspleniaceae).
Fuel. Dried ferns are used as a tinder when starting a fire with firesticks.
Notes. Most ferns of the Central Highlands area are called *ginˈsibiher*, both small and large varieties.
Comparative uses. The fronds of *A. affine* are eaten, as are the very young fronds of *A. antiquum*, and the roots and young shoots of *A. bulbiferum*. The fronds of *A. trichomanes* are given as an anthelmintic and to treat chest complaints. *A. adiantoides* is used to treat an enlarged spleen, jaundice, malaria and incontinence of urine. A decoction of the fronds of *A. macrophyllum* is a powerful diuretic. The ash of the burned plants of *A. amboinense* and of the burned fronds of *A. nidus* is used as a salt substitute. (Al Rawi & Chakravarty 1964; Ambasta 1982; Jain & De Fillips 1991; Kunkel 1984)

Asplenium schweinfurthii Baker

Fronds tufted, dark green, leathery, 8–20 x 2–8cm, pinnate; pinnae oblong, acute or obtuse, with serrate margins, truncate at the base, with scattered hair-like scales below. Stipe green, covered with brown scales.
Ecology. Common in shade in dense thickets in montane woodland and shrubland in the Haggeher mountains; also on Abd al Kuri. (300–)600–1450m (the low altitude is from Abd al Kuri).
Distribution. Endemic.
Conservation status. Vulnerable (VU B2 a b). This fern has a fragmented distribution. It has not been seen on Abd al Kuri since the end of the 19th century when it was recorded by Forbes, suggesting a decline in the quality of habitat and a decline in its extent of occurrence.

Botanical notes. The only endemic fern in the Soqotra Archipelago.

ETHNOGRAPHIC NOTES
Name(s)
– Soqotri: *gɛˈsibiher, gisiˈbirəh, ginˈsibiher; ṣəˈberbəher* 'sour, astringent'; *ḥʌlˈḥʌləhīn* 'grey-brown'.
– Soqotri simplified: *gesibiher, gisibirih, ginsibiher; ziberbiher; halhalihiyn.*
Food (livestock). Retains its green colour into the dry season and is eaten by all livestock when there is little alternative green fodder. Not regarded as having any nutritional value.
Fuel. Dried ferns are used as tinder with firesticks.
Bedding, stuffing. Gathered and used as bedding, or, before lying down to sleep, to provide padding beneath matting made from woven date-palm fibre or leather.

Asplenium trichomanes L.

Fronds tufted, dark green, wiry, persistent, 1-pinnate, linear-lanceolate, 5–13 x 0.8–2cm; pinnae oblong, crenate-dentate, unequal and broadly cuneate at the base, subsessile, glabrous. Stipe reddish-brown to almost black, glabrous, glossy. Sori almost completely covering the surface of the pinnae at maturity.

Ecology. No habitat details available. Apparently on both limestone and granite.
Distribution. Widely distributed in temperate regions and on tropical mountains throughout the world.
Botanical notes. Recorded twice in the 19th century: by Balfour 'On the Haghier hills south of Tamarida [Hadiboh]' and by Forbes on 'Gebal Raggit [Reyged] limestone outlier of Haghier', southwest of Hadibu. This is one of several ferns not seen for over 100 years on the islands.

Ceterach Willd.

Fronds pinnatifid, densely clothed with scales beneath. Sori elongate, borne along the veins; indusia obsolete.

Botanical notes. Often included in *Asplenium*.

Ceterach phillipsianum Kümmerle

Asplenium phillipsianum (Kümmerle) Bir, Fraser-Jenkins & Lovis
Gymnogramma cordata sensu Balf.f. non (Thunb.) Schltdl.

Fronds coriaceous, tufted, deeply pinnatifid, narrowly elliptic, 10 x 2cm, densely clothed with scales beneath and glabrous above; pinnae alternate, narrowly oblong, rounded, with weakly crenate margins and auriculate or weakly decurrent at the base. Stipe densely clothed with scales.
Ecology. In shrubland on granite. 800m.
Distribution. Tropical and S Africa.
Botanical notes. Soqotran plants differ from typical, African, material in having weakly crenate pinnae rather than distinctly crenate or pinnatifid pinnae.

ETHNOGRAPHIC NOTES
Not seen, no information recorded.

DENNSTAEDTIACEAE

Pteridium Gled. ex Scop.

Fronds spaced, 3 or 4-pinnate, hairy. Sori linear, continuous along the leaf margins, covered by the reflexed leaf margin and an inner indusium. 1 species. Cosmopolitan.

Pteridium aquilinum (L.) Kuhn

Rhizome long-creeping. Fronds <2m, closely spaced. Lamina broadly triangular to broadly ovate, glabrescent above, tomentose beneath. Pinnae ovate-triangular to oblong, acute; pinnules oblong, acute, 3 or 4-pinnatifid. Indusia and reflexed leaf margins membranous and ciliate. Stipe 20(–40)cm, straw-coloured, hairy and swollen at the base, glabrous above.
Ecology. Restricted to grassland and woodland clearings on gentle granite slopes amongst the highest pinnacles of the western Haggeher mountains. It is abundant and sometimes dominant in the montane mosaic of grassland, dwarf *Hypericum* shrubland and evergreen woodland of *Rhus thyrsiflora* and *Dracaena cinnabari*. 1450–1550m.
Distribution. Cosmopolitan.
Botanical notes. *P. aquilinum* is characteristic and sometimes dominant in the western Haggeher mountains, the highest peaks on Soqotra, but does not occur around the lower peaks of the central and eastern Haggeher. It has an interesting worldwide distribution being almost cosmopolitan and frequently abundant in the areas where it occurs. However, in southern Arabia and NE Africa it is rare; in the Arabian Peninsula it is known only from a single patch (covering about an acre) of grassland at 3000m on Jebel Melhan in Yemen. It is also found on leached soils amongst the granite peaks of other Indian Ocean islands such as Mauritius.

ETHNOGRAPHIC NOTES
Name(s)
– Soqotri: *gεsˈibiher, gisibˈirəh, ginsˈibiher*.
– Soqotri simplified: *gesibiher, gisibirih, ginsibiher*.
Food (livestock). In bad years in the Central Highlands *P. aquilinum* is gathered by herders to feed to goats and cattle towards the end of the dry season. It is more readily eaten by hungry cattle than by goats. It is not regarded as having any nutritional value, but can keep at least some of the animals going until the winter rains bring up new graze and browse.
Fuel. Dried *P. aquilinum* makes a good tinder, and bunches of it are sometimes stored at home for this purpose.
Bedding, stuffing. It is cut for bedding, or is laid beneath palm-fibre matting or skins and hides in the sleeping or sitting area as padding.
Comparative uses. The plant contains thiaminase which destroys thiamine and results in vitamin B1 deficiency. It can cause serious poisoning, particularly in horses and cattle. Extracts have been shown to possess carcinogenic and mutagenic properties. In sufficient concentrations they are acutely toxic to a wide range of domestic animals. The toxicity is passed through the milk of cows, and is a potential health hazard where cattle eat this plant. The rhizomes are boiled or roasted and eaten in times of scarcity. They can also be ground into flour to make bread: the starch in the flour is bitter but this can be removed by washing. The rhizomes are also mixed with malt to brew beer, and are fed to livestock, especially pigs. The rhizomes contain a bitter saponin which is toxic to fish. The tender fronds are eaten as a vegetable and are added to soups. The green fronds are used as a fodder, and the dried ones as packing material. (Ambasta 1982; Jain & De Fillips 1991; Kunkel 1984; Lewis & Elwin-Lewis 1977)

DRYOPTERIDACEAE

Hypodematium Kunze

Fronds tufted, 2 to 4-pinnatifid, pilose. Stipe densely clothed with scales at the base. Sori subcircular, in 2 rows on the pinnules; indusia reniform, hairy. 3 species. Old World tropics.

Hypodematium crenatum (Forssk.) Kuhn

Nephrodium crenatum (Forssk.) Baker

Fronds tufted, softly herbaceous, ovate-triangular, very finely 2 to 4 times pinnately divided, 12–30 x 15–25cm, hairy; lowest pinnae triangular-ovate, 1 to 3-pinnate; upper pinnae oblong-acute, pinnate; ultimate segments oblong, with crenate-serrate margins. Stipe straw-coloured, with a dense tuft of scales at the base.

Ecology. Shady crevices on cliffs. 620–800m.

Distribution. Tropical and S Africa, Arabia and eastwards to Japan.

Botanical notes. Readily distinguished from all other ferns on Soqotra by the dense tuft of hair-like scales at the base of the stipes (stalks of the fronds). Not seen since originally collected in the 19th century.

MARSILEACEAE

Marsilea L.

Aquatic or semi-aquatic ferns with slender, creeping rhizomes. Fronds with 4 leaflets arranged in a terminal cluster. Sporangia contained in inconspicuous sporocarps (specialised pinnae), borne on short pedicels. 65 species. Widespread in temperate and tropical regions.

Marsilea coromandeliana Willd.

Leaflets fan-shaped, 2–10 x 2–12mm, glabrous or occasionally hairy, with pellucid streaks between veins. Sporocarps 2.5–3 x 1.5–2mm, broadly elliptic to oblong-elliptic, with 2 distinct teeth and ribbed lateral walls.

Ecology. Rooted in mud at the margins of ponds and streams. 30–500m.

Distribution. Tropical and S Africa, Madagascar and India.

Botanical notes. Plants growing in water are normally sterile with large, entire leaflets but those growing in mud at the margins of ponds and streams tend to be fertile with smaller, often toothed, leaflets.

ETHNOGRAPHIC NOTES

Name(s)

– Soqotri: nil or *di rihóytin* 'of the water'.
– Soqotri simplified: *di rihoytin*.

Food (livestock). Sheep are said to nibble it, but goats usually avoid plants growing in or around water.

Comparative uses. The leaves and sporocarps of *Marsilea* species are eaten. (Ambasta 1982; Jain & De Fillips 1991; Kunkel 1984)

OLEANDRACEAE

Nephrolepis Schott

Rhizomes often forming tubers. Fronds pinnate. Sori circular, in a row either side of the midrib; indusia reniform. 30 species. Tropical and warm regions.

Nephrolepis undulata (Afzel ex Sw.) J.Sm.

Nephrolepis cordifolia sensu Balf.f. non C.Presl

Fronds 1-pinnate, narrowly elliptic, <30–70cm, herbaceous, glabrous; pinnae narrowly oblong, acute, weakly crenate, cordate at the base, sessile. Stipe brown, with sparse brown scales.

Ecology. Shady rock crevices in montane thicket. 850–1350m.

Distribution. Tropical Africa, Arabia and Madagascar.

ETHNOGRAPHIC NOTES

Name(s)

– Soqotri: *səbérəh* (pl. *sébəher*) 'in the undergrowth'.
– Soqotri simplified: *siberih* (pl. *sebiher*).

Food (human). These ferns are commonly found under large-crowned trees in the wetter, higher parts of the island. Each plant produces three or more underground tubers, which are about the size of the upper thumb joint. They are dug up and baked slowly in hot ash and then peeled and eaten. They are never boiled in water. They have to be well cooked or they can cause indigestion. They have a delicious flavour and are filling, and used to be an important food in the Central Highlands during the long, lean summer months. This and *Dioscorea lanata* were considered to be the best-tasting and most filling of the bulbs and tubers of the island, and in hungrier times were actively harvested.

Food (livestock). In shady areas these plants remain green well into the dry season and are eaten by all livestock, attracted by their greenery when there is little alternative fresh forage. Cattle tend to eat the fronds more than goats. They are not regarded as having any particular nutritional value.

Notes. A large vocabulary has grown up to describe the various sizes, shapes, colours, textures and stages of development of the tubers (also see *Dioscorea lanata*).

Comparative uses. The young shoots and rhizomes of the related *N. biserrata* and *N. auriculata* are eaten, as are the underground parts of *N. cordifolia* and the young fronds of *N. hirsutula*. A decoction made from the fresh fronds of *N. cordifolia* is given to treat a cough. (Ambasta 1982; Jain & De Fillips 1991; Kunkel 1984)

OPHIOGLOSSACEAE

Ophioglossum L.

Herbs with leaf-like, sterile fronds with a fertile segment inserted at the base. Fertile segment, without a lamina, with 2 rows of sporangia arranged in a spike. 40 species. Cosmopolitan.

Ophioglossum polyphyllum A.Braun

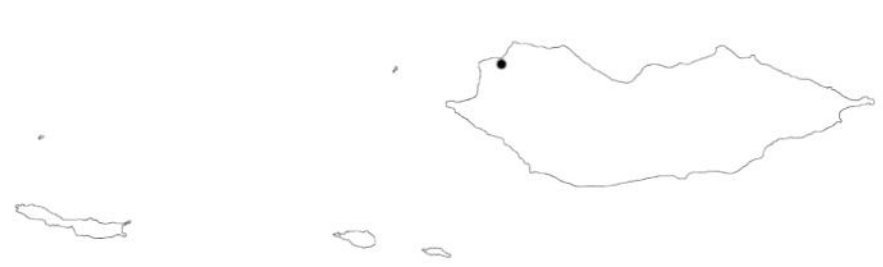

Sterile fronds 1 or 2, surrounded at the base by fibrous remains of leaf bases, narrowly ovate to elliptic, 2–4 x 1–2cm, with acute or obtuse apex and cuneate base; petioles 1–5cm. Fertile segments 3–8cm.

Ecology. Characteristic of shallow depressions on the limestone plateaus where water accumulates; appears soon after rain. c.600m.

Distribution. Tropical and S Africa to Arabia, Iran, Afghanistan and N India.

ETHNOGRAPHIC NOTES†

Name(s)

– Soqotri: nil or *di źaˤˈarihun* 'biting'; *raˈḥmam, rˈʌxmʌm* (western Soqotra) 'grey-coloured'; (rare) *baˤatət əl-xheyr*.

– Soqotri simplified: *di dha'arihun*; *rahmam, rakhmam*; *ba'atit il-kheyr*.

Food (human). The bulbs are not eaten.

Food (livestock). The foliage of *O. polyphyllum* does not seem to be palatable to livestock, and *O. reticulatum* usually lies beyond their reach.

Other. Children sometimes play with the small, dried bulbs.

Comparative uses. The ribbon-like fronds of *O. pendulum* are shredded in coconut oil and applied to the head to encourage the hair to grow. The fronds of *O. reticulatum* are eaten, and *O. ovatum* is also edible. *O. vulgatum* is regarded as a styptic, detergent, antiseptic and vulnerary; a decoction of the fronds is used for angina and a warm decoction of the rhizomes applied to boils. (Ambasta 1982; Jain & De Fillips 1991; Kunkel 1984)

†Also apply to *O. reticulatum*.

Ophioglossum reticulatum L.

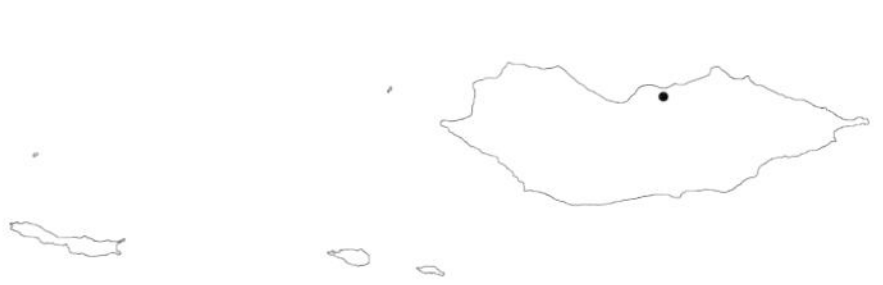

Similar to *O. polyphyllum* but the sterile lamina broadly ovate, 3–6 x 1.5–3cm, with a cordate or truncate base.

Ecology. In shade on granite cliffs. c.1000m.

Distribution. Pantropical.

ETHNOGRAPHIC NOTES

Name(s)

See *O. polyphyllum* (also uses).

PARKERIACEAE

Ceratopteris Brongn.

Aquatic ferns. Fronds tufted, fertile and sterile fronds very different. Sori arranged along the lower surface of the fronds and protected by the reflexed margins of the pinnules. 4 species. Tropical and warm regions.

Ceratopteris cornuta (P.Beauv.) Lepr.

Ceratopteris thalictroides Brongn.

Fronds tufted, light green, succulent, rather brittle, <60cm, glabrous. Sterile fronds lamina ovate or ovate-triangular, pinnate to bipinnate-pinnatifid, lobes triangular or ovate; fertile fronds lamina ovate, (2 or)3 or 4-pinnate, lobes linear.

Ecology. Rare in water holes. c.50m.
Distribution. Tropical Africa, Arabia and S Asia to N Australia.
Botanical notes. Known only from the Hadiboh plain. The only aquatic fern on Soqotra.

ETHNOGRAPHIC NOTES
Comparative uses. The fronds of *C. siliquosa* are eaten as a vegetable and are also used as a poultice for skin complaints. The fronds of *C. pteridiodes* and *C. thalictroides* are also edible. (Ambasta 1982; Jain & De Fillips 1991; Kunkel 1984)

POLYPODIACEAE

Loxogramme (Blume) C.Presl

Fronds simple, entire. Sori elongate. Indusia absent, without scales. 35 species. Old World tropics and Mexico.

Loxogramme abyssinica (Swartz) C.Presl

Fronds rather leathery, thickened, shortly spaced, glabrous, narrowly elliptic, 10–12(–35) x 0.5–1.5(–2.8)cm, entire, narrowing gradually into the short stipe below. Sori linear-oblong.
Ecology. On moist, shaded boulders in dense montane thicket. 1400m.
Distribution. SW Arabia, S and tropical Africa, Madagascar and the Mascarenes.

Botanical notes. Rare, collected only once but probably overlooked. Found in dense thicket amongst the highest pinnacles of the Haggeher mountains. Similar to *Pleopeltis macrocarpa*, which can be distinguished by its rounded not elongate sori and the presence of dark-centred peltate scales on the underside of the fronds.

ETHNOGRAPHIC NOTES
Name(s)
– Soqotri: *gɛsˈibiher*, *gisibˈirəh*, *ginsˈibiher*.
– Soqotri simplified: *gesibiher*, *gisibirih*, *ginsibiher*.
Food (livestock). Feral and wild goats are said to eat this species when there is little other forage available.
Notes. In the Central Highlands both small and large ferns are given the same name, even though clearly very different in appearance as well as in size.

Pleopeltis Humb. & Bonpl. ex Willd.

Rhizomes creeping, clothed in scales. Fronds simple, entire. Sori round, with dark-centred peltate scales. Indusia absent. 40 species. Tropics.

Pleopeltis macrocarpa (Bory ex Willd.) Kaulf.

Fronds rather leathery, thickened, shortly spaced, narrowly elliptic, entire, 10–12 x 0.5–1.4cm, glabrous above, with dark-centred peltate scales beneath. Stipe glabrous or with scattered scales.
Ecology. No habitat details available.
Distribution. S Africa, tropical Africa, Arabia, Madagascar, India and America.

Botanical notes. Found only once on Soqotra, in the 19th century; not seen recently. Should be looked for on damp cliffs or possibly as an epiphyte on trees in dense thickets in the higher mountains.

ETHNOGRAPHIC NOTES
Comparative uses. A decoction of *P. lanceolata* is used for colds and a sore throat, and to soothe an itch. The tender shoots are eaten by stock under scarcity conditions, and are said to be rich in nutrients. The young fronds of *P. nigrescens* are edible. (Ambasta 1982; Goodin & Northington 1979; Jain & De Fillips 1991; Kunkel 1984)

PTERIDACEAE

Pteris L.

Fronds 1 to 4-pinnatifid, glabrous, fertile and sterile fronds sometimes different. Sori marginal, usually continuous except for sterile apex, covered by reflexed membranous leaf margins. 250 species. Cosmopolitan.

***Pteris cretica* L.**

Fronds tufted, light green, ovate-triangular, 12–20 x 12–25cm, 1-pinnate above and with lowest pinnae 2-lobed, glabrous; pinnae and lobes of lowest pinna linear-elliptic, 6–13(–20) x 0.5–1.5(–2.5)cm, attenuate, sterile pinnae broader than fertile and minutely serrulate. Stipe long, 20–40(–60)cm, straw-coloured, glabrous except for a few scales at the base.

Ecology. Densely vegetated, moist and shaded north-facing gulley below the highest peaks of the western Haggeher mountains. 450m.
Distribution. Widespread in the Old World tropics and subtropics.
Botanical notes. Collected only once and apparently very rare on Soqotra. There are many heavily vegetated gullies and cliffs in the western Haggeher which are still botanically unexplored but which are likely to hold more populations of this fern and are also likely to harbour further new novelties for the islands.

***Pteris quadriaurita* Retz.**
Pteris biaurita sensu auct. non L.

Fronds tufted, papery, oblong-ovate, c.50 x 25cm, 3-pinnatifid, glabrous but with minute spines on upper surface; upper pinnae narrowly oblong, acute, pinnatifid; lowest pinnae with basal pinnules pinnatifid and resembling upper pinnae; ultimate segments narrowly oblong, 1–2.5 x 0.3–0.4cm. Stipe straw-coloured, glabrous except for a few scales at the base.
Ecology. In shade in evergreen montane thicket dominated by *Rhus thyrsiflora*, *Cephalocroton socotranus* and *Dracaena cinnabari*. 1050–1250m.
Distribution. Tropical Africa and Arabia.
Botanical notes. Part of a difficult species complex in need of revision. Soqotran material is referable to subsp. *catoptera* (Kunze) Schelpe. Distinguished from all other ferns on Soqotra by the presence of minute spines on the upper surface of the fronds.

ETHNOGRAPHIC NOTES†
Name(s)
– Soqotri: *gɛsˈibiher*, *gisibˈirəh*, *ginsˈibiher*; *ˈiˀišhel* (more commonly given to *Juncus* species); *ḥʌlḥˈʌləhīn* 'grey-brown'.
– Soqotri simplified: *gesibiher*, *gisibirih*, *ginsibiher*; *i'ish'hel*; *halhalihiyn*.
Food (livestock). The fronds are gathered and fed to goats and cattle in a drought or towards the end of the dry season in a bad year. Cattle find this more palatable than other livestock. It is said that feral and wild goats graze them in a prolonged dry season.
Fuel. Dried fronds make a good tinder, and bunches of it are sometimes stored at home for this purpose.
Bedding, stuffing. The fronds are cut for bedding, or are used as padding beneath palm-fibre matting or skins and hides in the sleeping or sitting area.
Notes. In the Central Highlands both small and large ferns are given the same name, even though clearly very different in appearance as well as in size.
Comparative uses. The young fronds of some species are eaten, or (as those of *P. ensiformis*) used as a flavouring. A decoction of the fresh fronds of *P. ensiformis* is given for dysentery and the juice of the rhizomes is applied to glandular swellings of the neck. A decoction of the rhizomes and fronds of *P. multifida* is given for dysentery, and is also said to be a good vermifuge. The toasted fronds and a paste of the crushed rhizomes are applied to cutaneous affections. (Ambasta 1982; Jain & De Fillips 1991; Kunkel 1984)

†Also apply to *P. vittata*.

***Pteris vittata* L.**
Pteris longifolia sensu Balf.f. non L.
Pteris longifolia sensu O.Schwartz non L.

Fronds tufted, leathery, narrowly elliptic to elliptic-oblong, 12–80(–100) x 4–25(–40)cm, 1-pinnate, glabrous; pinnae simple, 10 to 40 pairs, oblong to linear-oblong, 1–15 x 0.5–1.8cm. Stipe straw-coloured, with pale brown scales below.

Ecology. In shade, often by water, in semi-deciduous submontane woodland (with *Buxanthus pedicellatus*, *Boswellia* species and *Commiphora* species) and montane evergreen woodland (with *Rhus thyrsiflora, Buxanthus pedicellatus* and *Dracaena cinnabari*). 550–1150m.
Distribution. SW Arabia, tropical and S Africa.

ETHNOGRAPHIC NOTES
Name(s)
See *P. vittata* (also uses).

THELYPTERIDACEAE

Christella H.Lév.

Fronds 2-pinnatifid. Sori circular, borne midway between the veins and the margin. Indusia reniform, hairy. 51 species. Old World tropics.

Christella dentata (Forssk.) Brownsey & Jermy
Nephrodium molle sensu Balf.f. non Desv.
Thelypteris dentatus (Forssk.) E.P.St.John

Fronds tufted, herbaceous, <80cm, pinnate with pinnatifid pinnae, elliptic to narrowly elliptic, 15–60 x 6–18cm, acuminate, thinly hairy beneath; pinnae narrowly oblong, gradually decreasing in size and more widely spaced below, acuminate, deeply pinnatifid into oblong rounded lobes.
Ecology. In shade in montane semi-deciduous woodland on the Haggeher mountains; often near water. 500–1200m.
Distribution. S Europe, and widespread in Africa.

ETHNOGRAPHIC NOTES
Name(s)
– Soqotri: *gɛsˈibiher, gisibˈirəh, ginsˈibiher.*
– Soqotri simplified: *gesibiher, gisibirih, ginsibiher.*
Food (livestock). Not normally eaten by livestock, but the drying fronds are nibbled by small stock and cattle when they are very hungry in the dry season.
Notes. In the Central Highlands both small and large ferns are given the same name, even though clearly very different in appearance as well as in size.

GYMNOSPERM

EPHEDRACEAE

Dioecious shrubs. Leaves opposite or whorled, fused towards base, linear or reduced to a membranous sheath. Flowers in small cones; male flowers each subtended by a bract and consisting of two united scales surrounding a column bearing 3 to 9 anthers; female flowers subtended by 2 to 4 pairs of free or fused bracts, bracts swollen and fleshy in fruit. 1 genus. 40 species. N Hemisphere and S America.

Ephedra L.

See description of family.

Ephedra foliata **Boiss. ex C.A.Mey.**

Climbing or prostrate shrub to 1(–4)m; stems smooth or minutely hispidulous. Leaves <1.5cm long; margins of leaf sheaths and bracts ciliate. Male cones at tips of slender branches; flowers with 3 or 4 sessile anthers. Ripe female cones fleshy, translucent white.

Ecology. Rare in succulent shrubland on limestone cliffs. 50–350m.

Distribution. N Africa, C and SW Asia to India.

Botanical notes. Found only once, on seaward-facing cliffs in the southwest of the island.

ETHNOGRAPHIC NOTES

Name(s)

– Soqotri: *fəṭˈʌḥ*, *fetʌx* (western Soqotra) ‘split in two’ (more commonly given to *Periploca visiciformis*).

– Soqotri simplified: *fitah*, *fitakh*.

Food (livestock). Uncommon on Soqotra. The foliage and stems are occasionally nibbled by goats and sheep in a drought.

Comparative uses. The fruits of this and other *Ephedra* species are edible. (Kunkel 1984)

MONOCOTYLEDONS

ALOACEAE

Perennials with succulent leaves, often more or less stemless. Leaves with spiny margins, spirally arranged, usually in rosettes. Inflorescence of simple or branched racemes. Flowers bisexual. Perianth tubular, 6-lobed in 2 whorls, more or less actinomorphic. Stamens 6, free. Ovary superior, 3-celled, placentation axile; style simple. Fruit a many-seeded capsule. 7 genera. 400 species.

Botanical notes. Included in the Asphodelaceae by Dahlgren *et al.* (1985), but retained here.

Aloe L.

See description of family. 3 endemics. c.330 species. Tropical Arabia and drier regions of Africa and Madagascar.

Botanical notes. Most species have very limited distributions.

Aloe jawiyon **Christie, Hannon & Oakham** (see Appendix 2, p. 723)

Succulent perennial, suckering to form small colonies. Stem decumbent or erect if supported. Leaves in rosettes, spreading, 20–30cm long, recurved at tip, olive-green to bronze, margin toothed. Inflorescence 10–20cm long, unbranched. Perianth cylindrical, c.20 x 3mm, medium orange with green tips.
Ecology. Dry areas, prefers flat to gentle slopes on limestone pavement. 500–800m.
Distribution. Endemic.
Conservation status. Near Threatened (NT). Appears to be under no immediate threat. The population structure and plants are generally healthy. They are occasionally harvested but the sap is inferior to that of *A. perryi*. However, the distribution is rather fragmented and if harvesting increased it could rapidly become under threat.

Botanical notes. *A. jawiyon* is similar to *A. perryi*, with which it sometimes grows. However, it differs in a number of points. Its leaves are yellowish green, matte and finely scabrous, with the leaf tips long-acuminate and the marginal teeth smaller and more closely spaced. The bracts are as long as or longer than the pedicels (whereas those of *A. perryi* are minute and much shorter than the pedicels). The inflorescence is unbranched, with the lowest flowers held immediately above the leaves, while those of *A. perryi* are frequently branched, with the lowest flowers held well above the foliage. The perianth is spreading, medium orange with green tips and slight distal swelling; in *A. perryi* it is declinate, reddish orange to yellowish with a conspicuous distal swelling. Where they grow together *A. perryi* typically flowers before *A. jawiyon* which flowers in mid-winter (December–January).

ETHNOGRAPHIC NOTES
Name(s)
See *A. perryi* (also uses).

Aloe perryi **Baker**
Aloe forbesii Balf.f.

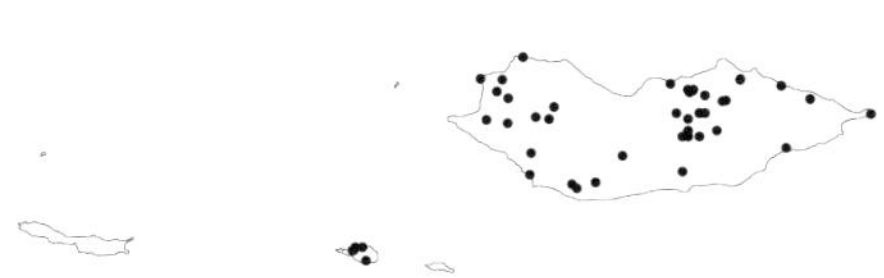

Succulent perennial, often suckering to form colonies. Stem usually decumbent, occasionally partly upright, <1m. Leaves in a basal rosette, spreading, 20–45cm long, weakly recurved at the tip, dark green to almost glaucous, becoming pink to bronze in dry areas, the margin toothed. Inflorescence <1.5m tall, frequently branched. Perianth cylindrical, c.25 x 5mm, red with green or yellow tips, gradually turning yellow.
Ecology. Widely distributed and sometimes abundant on Soqotra and Samha; mainly in dry areas; on flat or gentle slopes, primarily on limestone pavement but occasionally on sandy plains or granite mountains. Sea-level–900m.
Distribution. Endemic.
Conservation status. Near Threatened (NT). Widely distributed and in places abundant. Bitter aloes – the juice of *A. perryi* – has important pharmaceutical and medicinal properties. At present it is under no immediate threat but some populations are potentially vulnerable to pests, climate change and overharvesting.
Botanical notes. Delicate, sometimes apparently annual, plants with narrow stems found at high altitudes in the Haggeher mountains have been recognised as *A. forbesii*. However, there are many intermediates with typical *A. perryi* and so we have not found it possible to give them any formal status here. Flowers in summer and winter rains and comes into new leaf after the least rain.

ETHNOGRAPHIC NOTES†
Name(s)
– Soqotri: aloes in general *ṭayf*; *A. squarrosa* *ṭayf ᶜábdəher* 'spotted aloe'; *A. jawiyon* *žaᶜawiyɔ̄n*; aloes juice *ṭayf*; aloe leaf *eṣəlīˀoh* (pl. *ə́ṣlaˀ*).
– Soqotri simplified: aloes in general *tayf*; *A. squarrosa* - *tayf 'abdiher*; *A. jawiyon* - *ja'awiyon*; aloes juice *tayf*; aloe leaf *eziliy'oh* (pl. *izla'*).
Food (livestock). Both sheep and goats like to eat the buds, flowers and flowering spikes. In the drier and less well-vegetated areas of the island, the dead leaves are grazed by both sheep and goats towards the end of the dry season. Sheep in particular will nibble at the dead

outer leaves and root-bark of the living plants, sometimes damaging them in the process. The harvested leaves, dried out and drained of their bitter juice, are much liked by hungry small stock in the dry season. Ideally, they should be left for a full day to dry out before being fed to livestock, and make even better fodder if left for a week. A good and careful harvester gathers up the drained leaves and puts them out of reach of goats and sheep in a rock crevice or cave, and tells the local herders he has done so. Traditionally the dried harvested leaves belong to the owners of the land, not to the women who come in from outside to harvest the plants. The local herders go around after the harvest and collect the dead leaves for their animals. However, this is a feed of a bad year only, or of a prolonged dry season. On a diet of the dried leaves the animals initially lose weight, but then recover and look well. However, some sheep that eat a lot of this, to the exclusion of other forage, fall sick, sometimes fatally so.

In periods of severe drought goats and sheep also eat the tips of the fresh, green leaves. If starving, they can gnaw the plants right down to the ground, eating the root system as well. If a hungry goat is newly pregnant and eats a lot of live material, it sometimes aborts the foetus. Goatherds normally herd their newly pregnant goats away from areas where the plants grow, or hand feed them on alternative forage until their pregnancy is well advanced. However, the breeding of goats and sheep is usually managed to ensure that they are not pregnant during the severe dry season that follows a summer without rain. Sheep in early pregnancy seem less prone to abortion from eating the green plant. In general, it is animals which are not accustomed to this as a forage, or which gorge on the plants in the absence of other forage, that are prone to abortion. Native goats and those that eat aloe leaves as part of a mixed diet are less likely to be affected. In the wetter, better-vegetated areas of the island, aloe plants are rarely, if ever, eaten by livestock, except the odd dead leaf towards the end of the dry season. In the areas where the juice is not harvested and the plants therefore are of no economic importance, such as the interior valleys and Central Highlands, many herders regard them as useful dry season forage. Animals feeding on them are less thirsty, and more likely to survive the dry months in reasonable condition. In such areas animals normally seek out the tips of the live leaves, especially the tender growing tips at the heart of the plant, and prefer these to the dead leaves. They also dig up and expose the root system and crunch away at it. This is in sharp contrast to the areas in which the plants play a major role in the local economy. Here, any goat which shows a tendency to eat the living plants is driven off, and if it persists, is slaughtered in case it teaches its followers to do the same. On Samha sheep eat the dry plants and leaves and their herders collect these as fodder for them.

Bees / honey. Honeybees are attracted to the flowers. The honey made from this plant is 'hot' and highly valued for its medicinal properties. The best honey comes from aloes brought into flower by the summer rains.

Flavouring / food additive. Before the arrival of sugar, the flowers were sucked for their sweet nectar. After the summer rains especially, it is possible to run the hand along the flowering spike and come away with fingers sticky with nectar. Although the leaves, dried or fresh, are very bitter, it does not affect the taste and richness of the milk of an animal feeding on them as part of a mixed diet. If feeding heavily on them, however, the milk develops a slightly bitter edge and aftertaste. Similarly the taste of the viscera, and the water in which the carcass is cooked, becomes slightly bitter.

Fuel. In areas where the plants are plentiful, dried leaves and pieces of dried root-bark are gathered and stored for use as tinder. If there has been no rain in the preceding months, the inner core of the root-stem of a dead plant can be used as a firestick.

Medicinal / veterinary. In comparison to the widespread use of aloes on the mainland and elsewhere, the medicinal use of aloes in the islands is surprisingly limited. New uses have begun to appear over the last 20 years or so, as the islands increasingly become open to outside influence. For instance, there are a large number of Soqotran émigrés in the Gulf; increasing numbers of mainland Yemenis spend time on the islands; and many more Soqotrans have visited the mainland.

Traditionally, on Soqotra, the principal medicinal use of aloes juice was fresh for eye problems, and dried for stomach problems. Most island households keep some dried aloes juice in the home medicine chest. Some other traditional uses are described below.

Many illnesses are treated by purging and aloes juice is commonly used for this. Bodily health is viewed as dependent on an internal balance based on a properly functioning digestive system: it is through digestion that the body acquires its necessary nourishment. However, impurities inevitably accumulate in the digestive system, and these have to be eliminated by purging. Fresh aloes juice is more gentle in its effects than dried, and is therefore preferred, but not everyone has access to the fresh material. Some islanders regularly take a little powdered or fresh aloes juice in water first thing in the morning in order to keep the digestive system clean and functioning properly. Others use the juice only when constipated, taking a larger dose on an empty stomach. For children, a little fresh juice is inserted into a date, or is mixed with some sugar. A similar treatment is given to constipated livestock, or to animals suffering from stomach pain or colic. All sorts of stomach problems, stomach ache and colic, are treated with the dried or fresh juice. A distended abdomen of an unknown underlying cause is treated with aloes juice, as is a swollen liver or spleen. An animal with a painful abdomen and tenderness over the liver or spleen area is given aloes juice, some also being smeared around the anus. A sick animal displaying symptoms of rigor and unsteadiness on the feet has some aloes juice inserted into the nostrils and anus.

The fresh juice is painted over the skin to soothe pruritus. Dry eczema and scabies are treated in the same way. A small piece of dried juice can be used for this, rubbed back and forth in water until it softens to a paste. This is then applied to the affected area of skin just like the fresh juice. Drops of juice are put into an itchy eye.

Aloes juice is used to treat mastitis and inflammation generally. For mild mastitis in livestock, the hot and tender udder and teats are smeared with the fresh juice, if this is available, or with softened dried material. Women treat mastitis in the same way. The swellings of mumps and tender, raised glands are treated with an application of the juice, and it is dripped into a painful or infected ear to soothe the pain and reduce inflammation. Sore, red eyes, or if they are inflamed but painless, are treated by dripping in some fresh juice, though this treatment leaves an unpleasant taste in the mouth. Conjunctivitis is treated either by applying drops of fresh juice to the eye, or by painting the juice around the eyes, or both at once. An attempt is made to improve poor sight by mixing a powder of dried aloes juice with powdered antimony and applying the mixture to the rims of the eyes. Animals with eyesight problems are treated with an application of the fresh juice in the eyes. In humans and livestock, an eye that has suffered some trauma, or that waters constantly, is treated with repeated applications of the fresh juice.

Fresh juice is used to anoint cuts, sores and contusions to 'cool' the pain. An infected and discharging wound is treated with fresh or dried aloes juice. Lacerations, abscesses and sores in livestock are treated by washing them with a strong salt and water solution, and then applying a layer of aloes juice to deter flies. A pointed sore or boil is painted with aloes juice to bring the pus to the surface. A large, fresh laceration is treated by washing it out with salt and water, and then mixing fresh aloes juice with ground lichen or powdered, soft rock (probably gypsum), and packing the wound with this. The aloes juice keeps flies away as well as speeding the healing process. In livestock, the infected sores of ticks and of the biting flies of the rainy season are treated with an application of aloes juice. The fresh juice is painted thickly over a sore back, knee or other joint to soothe the pain. As the coating dries, it pulls the skin taut and sets hard like a sticking plaster. Dried aloes juice can also be used, after softening it by rubbing a small piece in water.

Aloes are much used in the treatment of fractures. A leaf is carefully split along its length and fitted around the fractured limb. A broken finger is treated with the liquid medicament made from the inner bark of *Commiphora* species and then bound with aloe root-bark that has been beaten out until flat and wide enough to make a bandage. This is then wound around the broken finger and held in place with a cloth bandage. It is inspected after three or four days, and if badly swollen is re-strapped less tightly. If there is no initial swelling at all, the strapping is not firm enough and it is re-strapped. On Samha, it is more common to cut four strips from the stem of a dead plant, place these along the fractured finger and then strap them in position with further strips of stem-bark or with cloth. Some people on Soqotra prefer to treat a simple fracture by first binding it with cloth and then strapping it with root-bark. Further support is given by taking four lengths of date palm rib, and binding these around the fractured limb. The wooden supports are loosened daily for about an hour at a time, the patient being encouraged to flex his muscles 'to get the blood moving'. As far as possible the fractured limb is immobilised and supported in a sling. Once the bone has healed and the strapping is removed, the limb is warmed with poultices and massaged with heated cloths. Many islanders say it is best to apply plenty of fresh aloes juice to the site of the fracture before strapping it to lessen the pain and inflammation. To treat a fractured leg in small stock or a young calf, strips of the root- and stem-bark are used in a similar way. After beating it flat, it is used to strap the fractured limb: tightly on day one and then progressively loosened until the broken bone has knitted. If it is a compound fracture and the bone has broken through the skin, the animal is held down, the bone manipulated back to the correct position, and then the limb is strapped as above. The strapping must be loosened immediately if the limb becomes puffy and swollen, but must not be removed entirely until the animal is able to bear weight. It is shut in a small pen and kept as still as possible, and is hand fed and watered until the fracture heals. Once the bandage is removed, the leg is bathed in cool water and supported with a cloth bandage until the animal has fully recovered. Sometimes the aloe bark strapping is applied over a layer of latex from *Ficus* or *Euphorbia* plants, often mixed with sheep's wool or goat hair to make a layer of padding, which then sets hard, like plaster of Paris. Or fresh aloes juice is mixed to a paste with fine red earth and moulded around the break; this sets hard and supports the limb. It is also said to help it heal more quickly and to prevent infection. A fractured hock is treated in the same way. A fractured thighbone is regarded as virtually untreatable, in humans and livestock alike. A bad sprain is supported with thick strips of stem-bark. The whole stem is used: either cut along one side, opened out and beaten flat, or slit up one side and opened just enough to fit around the sprained limb. More recently a new treatment has arrived from the mainland. If the damaged flesh around a fracture is discoloured and threatening to become gangrenous, a whole fresh aloes leaf is held at the fire until it begins to singe and bubble. It is then carefully split in half along its length and strapped over the damaged area. The dressing is replaced every day until the condition improves.

Another recent treatment is the use of aloes juice (and other bitter plant extracts) to treat malaria. The diagnosis of this illness and its treatment with bitter chloroquine is recent on the island. Many islanders substitute the bitter juice, taking it both prophylactically and to treat the illness. Some eat small pieces of fresh leaf instead of taking the juice. Someone who has inadvertently swallowed the water creature called *dilf* (unidentified), which is believed to grow to a great size in the stomach and make its host extremely ill, takes a small glass of liquid aloes, thinned with a little water. This must be drunk all at once, first thing in the morning and on an empty stomach. Severe diarrhoea follows. This treatment is said to work because at this time of day the creature is 'sitting in the stomach, with its mouth open', waiting for breakfast, and the heavy dose of aloes juice it gets instead kills it outright. Its corpse is said to be visible in the liquid faeces. Various insect larvae which are also inadvertently ingested in water can cause colicky pains or a swollen liver. The treatment is the same: taking liquid aloes juice to flush them through the digestive system. The more careful islanders filter suspect water through fine cloth before drinking it. Intestinal worms, either suspected or seen in the faeces, are likewise treated by taking a dose of aloes juice. Botfly attack to the mouth or nostrils is treated by taking fresh or dried aloes juice as soon as possible after the attack. This usually causes the sufferer to vomit, and thus, it is hoped, to expel the dangerous eggs and larvae. If the juice cannot be taken immediately, it is still worth taking it later, as the diarrhoea that follows may flush the system clear. Aloes juice is very useful as a deterrent for blowflies and other insect pests. Blowflies in particular are always attracted to any festering wound or sore, especially in livestock, and this can be prevented by cleaning the skin and then applying a thick layer of aloes juice. They are an especial nuisance at the beginning of the rains, when many small stock develop scour from feeding hungrily on the new rains grazing. Once the monsoon winds drop, blowflies arrive in great numbers, and their favourite target is sheep. They settle beneath the animal's tail and lay their eggs. A boil-like sore develops and later the larvae hatch out, and begin to burrow their way in. If this passes unnoticed, the animal usually dies. In the rainy season, shepherds must always be on the lookout for the fly, and, if they see the characteristic sores around the anus, they apply a dressing of *Euphorbia arbuscula* latex, fresh aloes juice, or aloes juice mixed with fine wood ash. This stops any eggs from hatching out, kills the larvae, and also discourages further blowflies from landing to lay their eggs.

An antiseptic dressing, called *naqf*, similar to the one made from the bark of *Commiphora* species, can be made from the dead, dried leaves. They are scorched at the fire until quite blackened, and then ground down to a powder. This dressing is used especially to treat burns, both accidental ones and those resulting from cauterisation. It is also applied to a pointed sore or boil to bring the pus to the surface. An abscess or wound that has become maggoty is first washed with a strong salt and water solution, and then treated with this *naqf* preparation. For young children who cannot stand the pain of the *naqf* treatment, the wound is usually washed out with fresh aloes juice and then covered with another layer of thicker aloes juice. This is done every day until the wound heals over. Generally speaking, a surface or fresh wound is best treated with aloes juice, and an older, deeper or infected one with the *naqf* powder. Multiple sores which have become infected are especially common in children with scabies, or who have scratched at head lice, or at the bites of fleas and other biting insects. These are treated by roasting a fresh aloe leaf until it is quite black and dry. This is then crushed to a powder and applied to the sores.

Commercial value (past, present). The aloes of Soqotra have long been famous worldwide, and continue to be harvested today. Traditionally, the most famous areas for harvesting aloes are all in western Soqotra. The aloes which produce the best-quality juice are also to be found in similar high, dry, limestone plateau areas of eastern Soqotra, but, for a variety of reasons, harvesting on a commercial scale has long since stopped here. As is common in other aloe-growing countries, the juice of the plants of hotter, drier areas is considered to be superior to that of cooler, wetter areas. Early visitors to the island reported that the best aloes were traded in Qalansiyah, the western port, though small amounts also appeared in the market towns of Qadheb, Di Hamdh and Hadiboh further east. In western Soqotra the areas that produce the best-quality aloes are Hager d-il Ba'a near Ras Isfer; the lowland Meyhah area, and especially Bit Hagi; the interior plain area of Zahar; and the high western Dimidi plateau area, extending as far as Rikbo di Risaham. Aloes of harvestable quality grow most plentifully in the Qatariyah area, running from Tiyde'eh to Ras Isfer; the low-lying Meyhah watercourse area; and in the Riyyed and Zahar interior plains. They are less plentiful but still of good quality in the dips and hollows where water gathers in the Qatn area of Ma'alah, and in the area called Rowkeb di Risaham, which is near Ras Isfer and Ras Biyodh, and includes Rowkeb and Hager as far as Da'anho, where the Qatariyah area begins.

Types of aloe. Islanders who harvest aloes commercially differentiate between four main types: (1) *ṭayf ṣáʿalhil* (*tayf za'alhil*) 'broad- or large-leaved aloes', also called *ṭayf ṣáʿabeb* (*tayf za'abeb*) 'pale-coloured aloe', whose juice dries to a very dark colour. This type is found principally in the regions east of the great watercourse system of Terr Ditrur, 'anywhere cattle are raised'. These aloes have large leaves which are bright green and very fleshy, and their juice is of little or no commercial value; (2) *ṭayf ʿábdəher* (*tayf 'abdiher*) 'spotted aloes'. These are not harvested because they grow in inaccessible places such as rock faces and cliff ledges; (3) *žəʿáwiyon* (*ji'awiyon*), whose juice dries to an orange-yellow. When dried it weighs less than the other type of aloe juice and its colour does not darken with age. This kind of aloe is found mainly in Qatariyah, in the area running along the edge of the escarpment, overlooking the southern plain. It is also found in Ras Biydoh, Jebel Mtoloh, and in the Qatn area of Ma'alah where the large, communal rainwater catchments are located. Its juice sells for a slightly lower price than that of the main commercial aloe. The plant is generally smaller and lighter-coloured than the *tayf* type below and its juice foams and bubbles when poured out, so has to be stored in a separate container. If it is mixed with juice of the *tayf* type, it causes it to behave in the same way, thereby making it considerably less valuable. The skin used to store the *ji'awiyon* juice has to have one of its four legs left untied until the bubbling has stopped; (4) *ṭayf* (*tayf*), whose juice dries to a dark orange-red colour, redder than the orange of the dried *ji'awiyon* juice. The best dried *tayf* is described as 'glittering' and 'like gold', unlike the dull, dark colour of the dried juice of aloes of the wetter areas. This type of aloe grows mainly in the high limestone plateaus of the east and west, and in the western dry interior plains. Commercially speaking, it is the most important type of aloe, and its juice sells for the highest price. The plant is commonly a dark orange colour. It is usually darker than the *ji'awiyon* type and is rarely green like the aloes of wetter regions. Islanders say it grows in association with *Ballochia*, *Trichocalyx*, *Anisotes* species and *Buxanthus pedicellatus*.

The aloe harvest. This description of the aloe harvest relates to how it is traditionally practised in western Soqotra. Oral history relates that in eastern Soqotra aloes were never harvested commercially on the scale they were in the west. Some harvesting was carried out in the Feling area (overlooking the fishing settlement of Matyaf and the southern plain), where the best-quality *tayf* aloes grow, and also in the interior plains of Mishilhi. In central Soqotra, it is said that aloes were harvested in years when the demand and price were high, though the aloes juice of this region was always regarded as inferior, and never fetched as good a price as that of the dry regions. Islanders say that this is because the Central Highlands and valleys behind are too cold, and aloes need burning heat for most of the year if they are to produce the best-quality juice. Now that there are signs that the aloes juice market may be reviving, women in both eastern and central Soqotra have started to harvest limited quantities of aloes juice, storing it at home until the price is right.

Traditionally, aloes juice is harvested by women. A hollow is scooped around the stem of the aloe plant or to one side of it. This is lined with leather to make a basin. The outer leaves are cut or snapped across at the base, always making sure that three or four leaves are left to protect the growing heart of the plant. The broken leaves are then laid at an angle around the edge of the basin so that the juice drains down into it. The harvester stays close to the plants to keep an eye on the juice, ready to transfer the oozing leaves to another basin if the first fills. Once the leaves are drained dry, the juice is transferred to a leather skin, and the harvester moves on to another area. Between five and eight *ratl* of juice (a *ratl* on Soqotra is around 500g) can be gathered by a hard-working woman in a day. The plants are best harvested in the summer months when the weather is dry and the sun very hot. The work continues until midday, when the harvesters rest. It comes to a stop if there is any wind. Some people start harvesting the plants as soon as the cool weather and cloud of winter have finally disappeared. This is so that the harvest can be finished before the time of the summer rains, or before the arrival of cloud and morning mists of the monsoon. If summer rains fall, the harvest has to stop. After rain, the plants become turgid, the quality of the juice deteriorates and it will not readily dry. In some years the harvest can recommence at the end of the monsoon, and continue through the short, still period of great heat between the cessation of the monsoon winds and the rising of the winter ones. Although short, this period is often preferred by islanders for whom the harvest is particularly important commercially, as the rains are more reliable in winter than in summer, and consequently the harvested plants have more chance to recover. Aloes are never harvested in a drought year since, if no rain falls after the harvest, the plants are seriously weakened and many will die.

Aloes can legitimately be harvested only by those with traditional rights to the land where they grow. Others who wish to join in the harvest have to seek permission from the landowners. Those in areas where the aloes are particularly numerous are usually happy to allow incomers to harvest it for them in exchange for half the yield. The harvester is given either a specific area of land to harvest, or a fixed quantity of juice to gather: five skins, ten skins and so on. Incoming women and girls commonly work beside local women, so that the latter can oversee their work and ensure that the plants are harvested properly and sustainably. The groups of women bring their own food, and settle in suitable caves or other temporary accommodation in the aloe areas. They are visited regularly by other family members, who re-supply them with food and collect the skins of aloes juice and take them back to the more permanent home settlements to be sorted and stored.

In the areas where the plants are harvested commercially, the harvest never takes place more than once a year, and sometimes, in a run of years with poor rains, only every second year. The most careful islanders harvest each individual plant only on alternate years. If harvested more frequently, a plant's growth is affected, and it is likely to produce fewer flowering spikes and fewer or no suckers. If both summer and winter rains fail (not a rare occurrence in the dry western end of Soqotra) then three or even four years are left between harvestings. The local aloe experts inspect the plants and decide which areas can and which cannot withstand harvesting that particular season.

Storage and care of the drying aloes juice. Once the small leather skins of liquid aloes juice (*ṭayf ṭūfef*, *tayf tuwfef*) are brought in from the harvesting area, the juice is transferred to larger goatskins for storage and a note is made of the exact quantity collected by each harvester. The skins of juice are then laid on a bed of ash, with another ash layer on top, for 24 hours. The next day they are put out in the sun during the hours of daylight, and re-buried in ash again once the sun goes down. This ash treatment is done principally to toughen the leather of the skin. If not treated in this way, the aloes juice 'burns' the leather and causes the skin to split. Also, if the bags of juice are left out for too long, the juice darkens in colour. This is then seen as 'spoiled', since dark-coloured aloes juice is not wanted and, even if the demand is so high that even this is being purchased, it sells for a much lower price. Nowadays, fine, dry, red earth is mostly used instead of ash. After this treatment, the skins are stored in the driest caves of the hottest, low-lying areas. The skins have to be turned weekly to redistribute the thickening juice, and to ensure that the contents dry out evenly. They must also be checked at regular intervals to see if any have become distended with air, in which case one of the legs must be untied to let it out. The juice begins to congeal, becoming *ṭayf ˈlibʌḳ* (*tayf libaq*) 'thickened aloes juice'. Finally it dries out completely, becoming as hard as a stone, and is called *ṭayf ḳeˈśaᶜ* (*tayf qesha'*) 'dried aloes juice'. Depending on the weather and ambient temperature, this process takes around a month. Once dried, it needs no further care beyond protection from damp. The bitterness of the juice protects the skins from attack by insect pests and rodents. Nowadays the aloes juice is more likely to be stored in the large Dano milk-powder tins, which hold about six or seven *ratl*.

The aloes juice trade. Aloes juice is sold by weight. Some traders purchased the aloes juice in its liquid state, and had it processed for them down in the coastal trading settlements. Once dried it was then stored there, ready for the arrival of the trading ships. Less scrupulous harvesters or aloes-juice processors would sometimes cheat to make the skins heavier and consequently get a better price. They could add a little water to the bags to be sold as liquid juice; or they could collect inner bark from one of the larger-leaved *Boswellia* species, shrub *Hibiscus*, *Ficus vasta* or *Commiphora kua*, and grind it to a paste, which they then added to the skins. Or they could instead pulverise aloe leaves with one of the barks and mix this in with the aloes juice. A purchaser would realise that he had been deceived only when the aloes juice failed to dry out properly. However, this was not a trick to play too often, as local traders would thereafter boycott those producers, just as their overseas purchaser would refuse to trade with them on their next visit to the island. There was little else that could be done to dishonestly bulk out the juice other than adding fine sand or earth to the liquid to make it weigh more once it had dried out.

In the poorer western areas aloes juice and sheep's wool cloth were the only source of income, apart from ghee. The people of the west traditionally sold their aloes juice to the traders closest to them, in Qalansiyah, or sometimes to traders in Di Hamdh or Qadheb. It was more common, however, for the local traders to work out agreements with the people of an aloe-harvesting area whereby they would send up camels laden with food – mainly cereals and dates – to a selected individual. He or she would then trade these for aloes juice with the harvesters and their families. The same individual was responsible for measuring out the juice, keeping a record of who had collected what, and storing the harvested juice in the approved fashion until such time as the trader sent up his camels to fetch it down to the coast. The camel trains of food would arrive in early summer and again, once the date harvest was over, at the end of the monsoon. Depending on the demand, and on that year's weather in the region, some would send up another camel train halfway through the winter months. The local traders normally bought juice in liquid or semi-liquid state, and measured it out and stored it in containers ready for sale once it had reached the coast. Those in the interior who could, would try to hold on to the harvested juice and process it themselves so that they could sell it directly to the overseas traders when they arrived. This gave them the best price, since liquid aloes juice always sells for less than dried. However, many people in the interior were too urgently in need of the food offered by the local traders to be able to manage this. Others were anyway only too pleased to be saved the trip down to the coastal markets to purchase it for themselves since they owned no baggage animals of their own and usually had to carry their food supplies back home on their shoulders.

Within living memory there has always been a market for aloes juice, and anyone prepared to harvest it could be sure of finding a buyer. As with other Soqotran commodities, the recent 20th century market for aloes juice developed (or rather, in the case of aloes juice, revived) with the rise of Aden as an important trading port under the British. This was a time when many more trade boats began to visit the island, from the Arabian mainland as well as from the more traditional East African–Oman–Arabian Gulf trade. The Sultan and a few local merchants purchased sea-going dhows of their own. At this time islanders remember that a leather bag, from the skin of a two-month kid, filled with dried aloes juice sold for four or five rupees or 30 *qurz*-measures of grain. (A *qurz* is roughly equivalent to the modern *ratl*, around 500g, and was considered to be sufficient for the main meal for an average-sized family of four to six adult-equivalents.) A similar-sized bag now sells for around 1500 to 2000 Yemeni riyals, and a group such as the Da'anho people, who have a lot of high quality aloes growing in their area, sold 400 leather bags in one year to a local trader. He then sold it on to traders in Sana'a.

Pest control. The fresh juice is smeared onto the emerging horns of young kids to stop rodents chewing them. It is also rubbed over leather skins of food, especially cereals, to deter rodents, mites and other pests. The large basket made from date palm fibre and used to store finger millet was also treated with aloes juice with the ash of certain woods to keep insects and other pests away. Leaves are crushed and the paste smeared around the neck of young kids and lambs to stop wild and feral cats attacking them.

Other. Small stock eating the leaves can go longer without water. The aloes juice is smeared on the nipples to wean young infants, and sometimes on the teats of house-stock to wean their young. The juice is also painted on the thumb or fingers to stop children sucking them. Fresh aloes juice makes an effective glue, which can be used to glue together handwritten sections of the Holy Qur'an to make a booklet.

In areas where these plants occur naturally, many islanders plant, or transplant, some to the slopes close to their home settlement for medicinal use. It used to be customary to plant aloes on a grave, but this practice is now frowned on. In the areas where aloes are important commercially, islanders have experimented with breaking up colonies and transplanting the young plants, either closer to settlements for ease of harvest, or to populate areas bare of the plant. They also sometimes remove young plants from wilder places and transplant them to a more accessible area. When removing a young plant, care is taken to ensure that it has a well-developed root system and is not still dependent on the 'mother' plant. In general, young plants are best transplanted when the ground is damp. Some women have been successful with collecting and sowing aloe seed at the onset of the winter rains. Seeds of Soqotran aloes have also been offered for sale on the Yemen mainland.

Notes. Recently the aloes in some areas have become infected with scale insects, visible as tiny white spots, which spread from the core of the plant out along the leaves. The plant finally yellows and dies. Islanders call this *rimidihʌn* ('ash-like'). If the harvester's hands become stained with the very bitter aloe juice they rub them clean with ash.

Al Hamdani (died AD 945) wrote of Soqotra that it grew Soqotran aloes and many palms, and produced ambergris and dragon's blood. In the late 1800s, Balfour (1888) reported that aloes juice was collected year round. He was told about three kinds. The first, thin, watery and amber-coloured, went in this liquid state to Muscat, Zanzibar and Arabia. It was the cheapest, costing 3 Maria Theresa dollars for a 30lb skin. The second kind was sent once it had started to thicken, about a month after the first, to Zanzibar and Arabia. It cost 5 Maria Theresa dollars for a 30lb skin. The third kind, fully hardened and dried out, a fortnight to a month after the second kind, was sold everywhere. It was the most expensive, at 7 Maria Theresa dollars for a 30lb skin. Balfour said that the best quality came from the area above the western market town of Qalansiyah. Bent (1897) writes of Soqotran aloes: 'now a very little of it is collected, and everybody takes what he likes from the nearest source, whereas, I believe, in former times, when aloes were an object of commerce here, the plantations were strictly divided off by walls, and the owners jealously looked after their property'.

Comparative uses. The official drug aloes species listed are 'Barbados aloes' from *A. vera* ex L. var. *officinalis*; 'Socotrine aloes' from *A. perryi*; 'Zanzibar aloes' also from *A. perryi*; 'Cape aloes' from *A. ferox* and its hybrids (also 'Natal aloes' probably from *A. candelabrum*); 'Mocha aloes' from *A. succotrina* and 'Jaffarabad aloes' from *A. vera*. The long history of use of aloes as a purgative is too well known to bear repeating here (those interested in learning more should consult Grindlay & Reynolds 1986). It is enough to note that medicinally, Soqotra aloes have for centuries been considered the best of all aloes. The plant contains barbaloin, a pale yellow crystalline glucoside soluble in water. It is more of an irritant than cascara, senna or rhubarb. Its action is largely limited to the colon and it does not clean the whole alimentary canal, but is useful in chronic constipation. Nowadays, the use of both crude aloes and aloin as a purgative is discouraged due to their irritant action, though *A. perryi* is reported to be milder and less irritating than other drug aloes. It produces pronounced pelvic congestion and is used for uterine disorders, generally in combination with iron and carminatives.

As a food, the fruit of various aloes species are eaten, their flowers are sucked for their sweet nectar, or, like the flowers of *A. boylei*, are cooked as a vegetable. The leaves of *A. abyssinica* are soaked in water and salt and pickled, those of *A. ferox* soaked and then cooked, the leaves, flower stalks and seeds of *A. barbadensis* are eaten in times of famine, and the very juicy leaves of *A. littoralis* are eaten both raw and cooked. Medicinally, on the mainland, the juice of the leaves of *A. perryi* is used as a stomachic, tonic and purgative, and is useful in dyspepsia, jaundice, piles and amenorrhoea. In Yemen the juice is also taken to regulate menstruation and to induce abortion. The many endemic aloes of Yemen are all used medicinally in some way or other, principally as a laxative, antiseptic and anti-inflammatory, and in the treatment of wounds and fractures. The juice of various aloes is widely used in the treatment of eye diseases, diseases of the skin and to treat malaria. It is smeared on the nipples to wean children and is used to stop bleeding in circumcision. The plant is burned and the ash applied to wounds of all sorts in livestock and man to keep flies away. The plants are also used in funeral ceremonies, planted by sons of the dead. Some examples of additional uses of aloes in surrounding areas are given below.

The leaves of *A. barbadensis* yield a fibre, and a dye is prepared from the plant. The dried juice is cathartic and the pulp is used to reduce or prevent menstruation. The leaf juice is also used to treat vitiligo. In East Africa *A. secundiflora* is used to make beer: the root is peeled, cleaned, de-poisoned, and boiled with molasses of maize beer. Lung disease in goats is treated by boiling branches in water, mixing the result with soda salt and feeding it to the goat. The juice is used in blessing, the plant is sown at the entrance to the place of circumcision, and a single leaf is placed under the sleeping mat of a newly married couple. In Oman the fresh leaves of *A. vera* are rubbed over the body to treat a fever, and are ground with oil and vinegar and applied to the forehead for a headache. A leaf extract of *A. tomentosa* is applied to the skin for rashes and sunburn, and a *kohl* applicator is thrust into the leaf and then applied to swollen or red eyes to soothe irritation or pain. In Yemen the wax and leaf gel of the same plant is applied for two weeks to a broken bone: it is believed that after this the bones heal more easily and can be reset. In the Philippines the juice is used to treat alopecia and is applied to swellings and wounds. In Dhofar, the juice of *A. dhufarensis* is: used fresh on the forehead to treat a headache and stuffy nose; rubbed on the chest to treat coughs and colds (often with turmeric and salt); rubbed to relieve pain in limbs and joints; used with powdered indigo to treat skin rashes, burns and on wounds as a disinfectant; applied around the eyes to relieve itching and pain, or to sharpen sight; dripped directly into an infected or damaged eye; poured over an infected wound; inserted into the nostrils to stop a nosebleed; painted over the chest to relieve coughs and chest pain; applied to burns and after cauterisation as an antiseptic and to relieve the pain; applied to the skin cosmetically to lighten its colour and to protect against cold and disease; sniffed up the nose to relieve the itchiness of allergy and hay fever; painted over the damaged areas after a stroke. Dried, it is powdered and: taken as a purge and a vermifuge; applied around a painful and discharging ear; taken in water to relieve stomach pain, to treat generalised oedema and to treat retention of urine; mixed with powdered indigo and used to anoint a newborn baby. (Abdulfatih 1987; Al-Dubaie 1993/Al-Dubaie & Al Khulaidi; Ambasta 1982; Dalziel 1948, 1956; Davis *et al.* 1983; Fleurentin & Pelt/Fleurentin *et al.* [1982, 1983a,b,c]; Ghazanfar 1994; Ghazanfar & Al-Sabahi 1993; Heine *et al.* 1988; Jain & De Fillips 1991; Kunkel 1984; Miller & Morris 1988; Oliver-Bever 1986)

†Also apply to *A. jawiyon*, *A. squarrosa*.

***Aloe squarrosa* Baker**

Succulent perennial with short prostrate stems forming small colonies, usually on cliff faces. Leaves spaced along the stem, <15cm long, strongly recurved, green with white speckles, the margin toothed. Inflorescence up to 20cm tall. Perianth cylindrical, c.25 x 5mm, pale dusty red, the perianth lobes with pale margins and becoming green at the tips.

Ecology. Restricted to cliffs on the limestone plateau at the western end of Soqotra; also rarely found away from cliffs in the protection of shrubs such as *Buxanthus pedicellatus*. 300–530m.

Distribution. Endemic.

Conservation status. Vulnerable (VU B2 a bii). Known from a few scattered sites on the western limestone plateau. In the Red Data Book (Lucas & Synge 1978) it is classed as Endangered with the following comment: 'in 1967 it was seen to be still surviving but only in critically low numbers and severely threatened by large herds of goats'. It is certainly still true that the numbers are low but populations are typically found on inaccessible cliffs and it is unlikely to be eaten except in times of severe drought. However, even small-scale collecting of this attractive succulent could threaten the survival of subpopulations. Furthermore, there is a 19th century record (Forbes 1903) of this species from Jebel Bitzobur [Bit Gobihir] on the Garieh [Qariyah] plain in the northeast of the island, an area in which it has not recently been seen, suggesting that there has been a decline in its area of occupancy.

ETHNOGRAPHIC NOTES

Name(s)

See *A. perryi* (also uses).

AMARYLLIDACEAE

Perennial herbs with bulbs. Leaves in a basal rosette, entire, strap-shaped. Flowers showy, bisexual, slightly zygomorphic, in umbel-like inflorescences. Tepals 6, in 2 whorls, fused into a tube below. Stamens 6, free. Ovary inferior, 3-celled, placentation axile; style slender. Fruit a several-seeded capsule. 60 genera. 800 species. Widely distributed, especially tropical.

Crinum L.

See description of family. 1 endemic. c.100 species. Pantropical.

***Crinum balfourii* Baker**

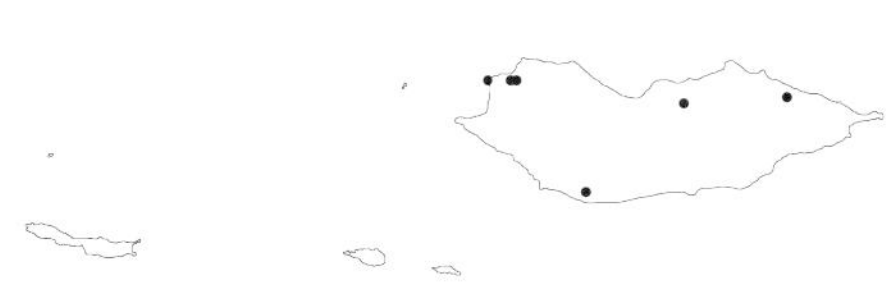

Robust perennial herb, often forming large colonies. Bulbs <15cm diameter, with a terminal neck <30cm. Leaves green, strap-shaped, <1.5m x 10cm. Flowering stems <1m; umbels 10 to 20-flowered. Perianth tube (including ovary) c.6.5cm, green; lobes somewhat reflexed, white, narrowly oblong-obovate, c.50 x 10–12mm. Filaments white. Style pink.

Ecology. Frequent on shady cliffs in semi-deciduous woodland and woody-herb communities, particularly on the limestone plateau. 500–650m.

Distribution. Endemic.
Conservation status. Least Concern (LC). Present in several vegetation types and under no present or perceived threat.
Botanical notes. One of the most dramatic plants on Soqotra, with umbels of large white flowers that are strongly and pleasantly scented. It is the largest and most robust bulbous plant on Soqotra, often forming large colonies, particularly on shady cliff ledges where it is immediately recognised by the bright green leaves which can grow up to 1.5m in length. The large bulb has a distinct stem-like 'neck' formed from the leaf bases and surrounded by the remains of old leaves.

ETHNOGRAPHIC NOTES
Name(s)
– Soqotri: *fitaḥeh*, *fiṭaḥer* 'new shoots, new growth'; *fəṭāwə* (pl. *fiṭāwitin*); *di aᶜarib* 'of the raven', *di sɔ̄ᶜīdo* 'of the vulture', names indicating the plant's unpalatability (similar names are given to the Soqotra hyacinth, *Ledebouria grandifolia*).
– Soqotri simplified: *fitaheh*, *fitaher*; *fitawi* (pl. *fitawitin*); *di a'arib*, *di sow'iydo*.
Food (human). Soqotrans describe two large bulbs, one edible and the other not. It seems most probable that *C. balfourii* is the inedible one, which suggests that there is another large-bulbed plant still to be found and identified on the island.
Food (livestock). Neither foliage nor bulb are eaten by livestock.
Notes. The edible bulb is said to grow in the high limestone areas. Its multi-layered bulb is said to resemble a large onion. Unlike *C. balfourii*, its foliage is not only palatable to livestock, but is even considered to be a nourishing feed, although bulb and foliage initially cause diarrhoea in both humans and livestock. It is not much liked by people for this reason, and also because, when eaten hot, it causes severe flatulence. It is cooked in water with added salt, if available, and a layer of *i'iri* grass or its own broad, strap-like leaves above and below, with a stone placed on top to hold the bulbs below the water. Alternatively it can be roasted in hot ash. Either way it must be peeled before eating. Some people find it more digestible if they mash the cooked bulbs to make porridge.
Comparative uses. Many *Crinum* species are regarded as poisonous to livestock, and the dried seeds of some are used in fish poisons. Other species have bulbs that are rich in starch and eaten as a famine food. The bulbs and leaves of many species are used medicinally. (Ambasta 1982; Dalziel 1948, 1956; Heine *et al.* 1988; Jain & De Fillips 1991; Kunkel 1984; Lewis & Elwin-Lewis 1977)

ANTHERICACEAE

Rhizomatous, perennial herbs, roots often tuberous. Leaves in basal rosette, usually linear. Flowers bisexual, actinomorphic, cup-shaped; pedicels jointed. Inflorescence raceme-like with 1 to 4 flowers at each node, sometimes appearing before the leaves. Tepals 6, in 2 whorls, free, all similar. Stamens 6, free. Ovary superior, 3-celled, placentation axile; style slender. Fruit a many-seeded, 3-angled capsule. Seeds black, glossy, flattened. 30 genera. 600 species. Subcosmopolitan.

Chlorophytum Ker Gawl.

See description of family. 2 endemics. c.300 species. Old World tropics, especially Africa and India.

Chlorophytum filifolium Nordal & Thulin

Herb with a short rhizome; roots with terminal cylindrical tubers. Leaves basal, linear, 30–150 x 0.2–2mm, pale green, folded, margin minutely denticulate, often slightly curved; basal sheath white. Inflorescence appearing with leaves, prostrate, lax, c.6–10cm long, unbranched, with 1 or 2 flowers per node. Tepals 2–2.5mm, white with green midrib.
Ecology. *Croton socotranus* shrubland in coastal and inland plains. 10–350m.
Distribution. Somalia.
Botanical notes. An ephemeral which appears in a flush with other annuals after rain. Soqotran plants differ from plants from Somalia by the presence of tubers on the roots.

ETHNOGRAPHIC NOTES†
Name(s)
– Soqotri: *di ḥagubgub* 'tiny palms'; *ḥɔ̄wgəb* 'young palm'; *ḳʌźbəhʌn* (general term for perennial Gramineae and Cyperaceae of damp ground).
– Soqotri simplified: *di hagubgub*; *howgib*; *qadhbihan*.
Food (human). Respond very quickly to rain, so their tender, green foliage and tiny tuber offers a welcome food after the barrenness of the long dry season. When still young and sweet the small tubers and greenery are eaten but they are too small to be filling and are not an important food.
Food (livestock). Small stock graze the foliage which in the drier areas is seen as a useful forage.
Comparative uses. The tubers and leaves of many other species are edible, and are eaten both raw and cooked. (Jain & De Fillips 1991; Kunkel 1984; Miller & Morris 1988)

†Uses also apply to *C. graptophyllum*, *C.* sp. *A*.

Chlorophytum graptophyllum **(Baker) A.G.Mill.**
Anthericum graptophyllum Baker

Herb with a short rhizome, covered with the fibrous remains of leaf bases; roots fleshy, without tubers. Leaves basal, narrowly oblanceolate, 10–15 x 0.8–1.2cm, somewhat glaucous when young, becoming pale green with a distinct network of dark green broken stripes across the width of the leaf, margin with conspicuous hairs, sometimes undulate or undulate-crisped. Inflorescence appearing before leaves, erect, c.35cm, simple or slightly branched, with 2 or 3(or 4) flowers per node. Tepals 6–9 x 1.5–3mm, white with the midrib greenish towards the tip. Flowers appear in morning, shut by midday.
Ecology. Semi-deciduous woodland (with *Buxanthus pedicellatus*, *Allophylus rubifolius*, *Dracaena cinnabari*, etc.). 600–700m.
Distribution. Endemic.
Conservation status. Least Concern (LC). Common in several vegetation types and under no present or perceived threat.

Botanical notes. Readily distinguished from all other narrow-leaved monocots on the island by its leaves which are pale green with a distinct network of dark green broken stripes across their width and with conspicuous hairs along their margin. Two distinct forms of *C. graptophyllum* occur on Soqotra. The first, more widespread, form (e.g. Muqadrihon) has generally broader leaves (usually >1cm broad) with straight or slightly undulate margins. The second form is restricted to the central and eastern slopes of the Haggeher (e.g. Kishin in Wadi Di Negehen) and has narrower leaves (usually <1cm broad) and distinctly undulate-crisped margins. The description here is based largely on the first form.

ETHNOGRAPHIC NOTES
Name(s)
– Soqotri: *ḥárgog* 'crouched'; *ḥárgod* 'unkempt, unwashed', 'hairy, fibrous' (more commonly given to *Hypoxis angustifolia*); 'like' *ɛ́rgílo* (more commonly given to *Ceropegia* species); *ḥárʾhi* (more commonly given to edible orchids).
– Soqotri simplified: *hargog*; *hargod*; 'like' *ergilo*; *har'hi.*
Uses. See *C. filifolium.*

Chlorophytum **sp. *A***

Herb to 15cm; rhizome thick, c.3 x 1cm, erect, capped by fibrous remains of old leaf bases; roots fleshy, without tubers. Leaves basal, lanceolate, 3–13 x 1.5–2.2cm, green, margin minutely denticulate. Inflorescence appearing with leaves, erect, unbranched, rather densely flowered, with 1 or 2 flowers at each node. Tepals white, 8–9 x c.3mm.
Ecology. Known only from the summit ridge of the limestone plateau of Samha. This ridge attracts low cloud and drizzle (principally from the NE monsoon) and, with the cliffs below, provides a unique habitat on Samha. 650m.
Distribution. Endemic.
Conservation status. Endangered (EN B2 a biii). The small niche occupied by *C.* sp. *A* amounts to an area of probably less than 2km^2. We strongly suspect that this marginal habitat is threatened by the trend of lower rainfall in the region and would be particularly vulnerable to the effects of global warming. The situation is particularly critical for *C.* sp. *A* which, like *Ledebouria insularis*, occupies a niche at the highest point on Samha and so, with increasing aridification, would have no possibility of mitigating the effects of gradual drying by migrating upwards.
Botanical notes. This distinctive new species has not been seen flowering on Soqotra. The description is based on a rhizome collected on Samha and cultivated at the Royal Botanic Garden Edinburgh. However, the description should be treated with caution: cultivated material can sometimes be rather anomalous. For instance, Nordal & Thulin (1993) note that plants of *C. petraeum* which flowered in the first year of cultivation had dimensions well within that found in nature, but when the same plant flowered the following year it had much larger leaves, a taller inflorescence and somewhat larger flowers.

ETHNOGRAPHIC NOTES
Name(s)
– Soqotri: *ɛrgílo ḥarhīʾo* '*Ceropegia*-like edible orchid'; *kəndīyə* '(edible) root stem, base' (Samha).
– Soqotri simplified: *ergilo harhiy'o*; *kindiyih.*
Uses. See *C. filifolium.*

ARACEAE

Herbs with tubers. Sometimes with modified erect stems bearing clusters of spiny bulbils. Leaves large, broad, peltate-cordate, long-petiolate. Flowers minute, unisexual or bisexual, arranged on a fleshy spike (spadix) enclosed in a bract-like spathe; spathe closed in the lower part and open above. Perianths small. Fruit a many-seeded berry. 108 genera. 2500 species. Mostly tropical and subtropical.

Colocasia

Bulbil-bearing shoots absent. Spathe weakly constricted between the tubular lower part and the oblong to lanceolate blade. 6 species. Tropical Asia.

Colocasia esculenta (L.) Schott

Robust terrestrial herb with a starch-rich tuber. Leaves several, ovate, <45cm long, with cordate-peltate base and acute tip; petioles <1m. Spadix 20–30cm, spathe shorter than spadix, the blade orange or yellow.

Ecology. No habitat details available.

Distribution. Widely cultivated in the tropics.

Botanical notes. Formerly cultivated in the Haggeher mountains and now rarely naturalised? The presence of this species on Soqotra is based on a photograph of a sterile plant taken in the Haggeher.

Remusatia Schott

See description of family. 4 species. Tropical Africa, Himalayas to Taiwan.

Remusatia vivipara Schott

Herbs, usually epiphytic. Leaves 1 or 2; blade broadly ovate, 15–30 x 10–20cm, base peltate-cordate, tip shortly acuminate; petioles 30–40cm. Spadix 30–45cm; spathe 70–100cm, blade yellow. Sterile bulbils brown, on erect shoots, covered in hooked prickles.

Ecology. No habitat details are available but it would be expected to be found in damp, shady places; possibly on cliffs or even as an epiphyte. 900m.

Distribution. Tropical Asia, Africa and Madagascar.

Botanical notes. Collected only once on Soqotra, by Forbes in 1903 at 'Adho Dhimellus. In swampy places in the valley north and south of our camp'. *R. vivipara* is a very distinctive plant and is unlikely to be overlooked. Furthermore, Adho Dhimellus [Adho di Melhoh] is a classic collecting locality visited by most expeditions visiting the island. It seems very likely that this plant is now extinct on Soqotra. On the Arabian mainland *Remusatia* is known only from the monsoon woodlands of Dhofar in southern Oman. Here it never flowers but reproduces through sterile bulbils which are borne on erect specialised shoots. In Dhofar it is frequently epiphytic.

The single collection of *R. vivipara* was apparently made in the same area as the site for *Colocasia esculenta* discussed above. In the vegetative state the two plants are very similar and could perhaps be confused; it is therefore possible that the record of *Remusatia* could in fact be a misidentification for *C. esculenta*. However, as no specimen of *R. vivipara* can be traced, the status of this plant on Soqotra remains a mystery.

ASPARAGACEAE

Scrambling shrubs or subshrubs. Leaves reduced and scale-like or forming spines; leaf-like cladodes present, solitary or fascicled. Flowers in axillary clusters, actinomorphic, bisexual. Tepals 6, free, in 2 whorls. Stamens 6, fused to the perianth. Ovary superior, 3-celled, with axile placentation. Fruit a globose, 1-seeded berry. 1 genus. 50 to 60 species. Old World.

Asparagus L.

See description of family. 2 endemics.

Asparagus africanus Lam.

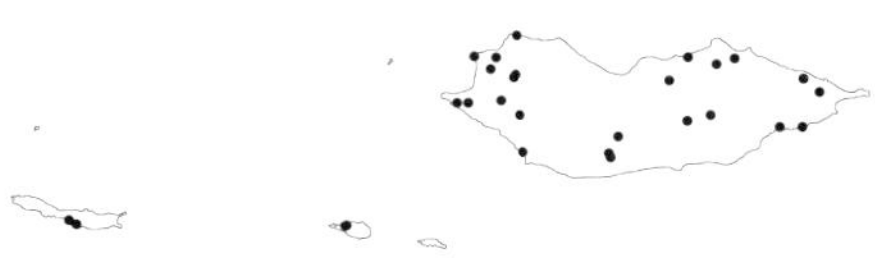

Low, intricately branched spiny shrub to 50cm. Branches often zigzagging, pale brown; spines 1–3mm. Cladodes clustered, stiff, linear, 5–15(–25)mm long. Flowers axillary, solitary or in 2 or 3-flowered fascicles. Pedicels 2–3mm, articulated below the middle. Tepals cream, 2–3mm. Fruit globose, 2–4mm diameter, red when ripe; seeds black.

Ecology. Soqotra: widespread and sometimes abundant on the limestone hills and plateau, apparently absent from the granitic Haggeher mountains; in dwarf shrubland, succulent shrubland and semi-deciduous woodland, sea-level–800m. Samha: frequent in dense shrubland and dwarf succulent shrubland, 300–500m. Abd al Kuri: common in dwarf shrubland and succulent shrubland, 30–500m.

Distribution. Widespread from India to E and S Africa; var. *microcarpus* is endemic to the Soqotra Archipelago.

Conservation status. Least Concern (LC). Common in several vegetation types and under no present or perceived threat.

Botanical notes. Soqotran plants are all referable to var. *microcarpus* Balf.f. There is some variation: on Abd al Kuri it is typically a low, rather compact, shrub whereas on Samha and Soqotra it is generally a taller, more open shrub.

ETHNOGRAPHIC NOTES†

Name(s)

– Soqotri: *məsḥábeb*, *məsḥ̄ɔwbeb* 'spreading'; *məśḥ̄ɔwməhen* 'prickly'; *məśḥ̄ɔwgən*, *məśḥ̄ɔwgeg* 'scratchy'; *məśḥérmhim di gírbʌg* 'wild cat's *Caralluma*'; *məśḥ̄ɔwbeb* 'hard, unyielding' (Samha); *móshiṭ* 'broom' (Abd al Kuri).

– Soqotri simplified: *mis'habeb*, *mis'howbeb*; *mish'howmihen*; *mish'howgin*, *mish'howgeg*; *mish'hermhim di girbag*; *mish'howbeb*; *mos'hit*.

Food (livestock). Goats and donkeys are very fond of *Asparagus*. Donkeys are not generally liked in the interior and one of the reasons is that they are often in direct competition with goats for favoured fodders. After rains the tender new *Asparagus* shoots are also much liked by sheep, and plants in the high mountains are eaten by cattle. *Asparagus* is considered to be a valuable fodder for small stock, especially since it persists through the dry season. However, if they have little else to eat, grazing too heavily on this plant can cause shivering attacks and general malaise. It is common in the northern plains and foothills and the goats of the region rely heavily on it in the dry season, but have to compete for it with the many donkeys and camels. Herders say that goats are especially fond of it because they 'long for something to crunch'. They bite off chunks of the stem and can then be seen chewing these between their back teeth with their heads on one side. They search out the plant in leaf and when dry and leafless. When there is dewfall the plant becomes more succulent and tender, but once the sun is well up, its stiffness returns.

Medicinal / veterinary. A stem is snapped off and the reddish sap that collects at the broken end is dabbed on a cut, sore or other wound to speed its recovery. This is seen as being as effective as *Jatropha unicostata* sap. The plant has most sap in the winter months. To treat fissures in the plantar skin, the whole plant is crushed and the exuding liquid heated over the fire and rubbed into the fissures.

Other. The stems are bound together to make a broom for sweeping out living quarters and mucking out livestock pens and byres. A similar brush is made for brushing away the chaff when grinding and processing cereals. A stem of the plant is peeled and whittled to a point and then used to hollow out a length of *Euphorbia arbuscula* wood to make the stem for a hubble-bubble pipe or *narghile*. This used to be widely smoked in the more settled coastal areas.

Comparative uses. The young shoots are edible, and the leaves are chewed to relieve breathing problems. The wiry stems are used to make traps and snares for small creatures, and twisted into cordage used for tying up animals. The leaves are crushed with milk and applied locally to treat guinea-worm, and as an ingredient in an ointment used by women to stimulate hair growth. The thickened roots are regarded as a diuretic and are taken as a remedy for syphilis. The whole plant is used to treat haematuria. Other related species are also used medicinally and the roots eaten. In Uganda the So eat the white root of *A. flagellaris* and twist the bark of the root to a thread used to sew containers. Girls make beads from the stems and, more esoterically, a stem is burned and then pointed at a rain cloud to stop the rain falling. The Rendille of Kenya also make thread from *A. buchanii*, a plant which also symbolises age, status and importance. Strings made from it are worn around the head by married women and other groups. (Abdulfatih 1987; Dalziel 1948, 1956; Ghazanfar 1994; Goodin & Northington 1979; Hedberg *et al.* [1982–1986]; Heine *et al.* 1988; Kunkel 1984)

†Also apply to *A.* sp. *A.*

Asparagus* sp. *A

Similar to *A. africanus*, but a weakly spiny climbing shrub with stems to 3m. Cladodes linear, flattened, 15–20mm, rather thinner and more flexible than in *A. africanus*. Flowers and fruits not known.
Ecology. Dense *Dracaena cinnabari* woodland in the western Haggeher near Skand (the highest mountain on Soqotra). 1400m.
Distribution. Endemic.
Conservation status. Data Deficient (DD). Poorly known; inadequate evidence is available to make an assessment.
Botanical notes. This weakly climbing shrub with its long herbaceous stems is known only from a single sterile gathering (*Miller* 17136) from the western Haggeher near Skand. *A. africanus* is not found in the Haggeher mountains, and is a low, spiny shrub.

ETHNOGRAPHIC NOTES
Name(s)
See *A. africanus* (also uses).

ASPHODELACEAE

Rhizomatous herbs, rhizomes yellow when cut. Leaves in a basal rosette, linear, rounded, hollow. Flowers bisexual, actinomorphic, in simple racemes, 1 per bract, on leafless stems; pedicels distinctly jointed. Tepals 6, free. Stamens 6, free. Ovary superior, 3-celled, placentation axile; style slender. Fruit a many-seeded capsule. c.17 genera. Old World, especially S Africa.

Asphodelus L.

See description of family. c.18 species in the Mediterranean and east to Himalayas and south to Somalia and Soqotra.

***Asphodelus fistulosus* L.**
Asphodelus fistulosus var. *tenuifolius* Cav.
Asphodelus tenuifolius Baker

Annual herb. Leaves tufted, linear, 15–40cm long, subsucculent. Flowers c.1cm across; tepals spreading, white or pink with a brownish-purple midvein. Fruit globose, wrinkled, on short erect pedicels.
Ecology. In degraded grassland, along tracksides and in clearings in drought-deciduous woodland, particularly in the east of the island (Homhil and Momi) where it is sometimes abundant. 150–750m.
Distribution. Mediterranean, Eritrea, Somalia and Arabia.
Botanical notes. This plant is on the increase in certain areas, particularly in the east, of Soqotra. It is interesting that it was recorded by Forbes from Abd al Kuri at the end of the 19th century but has not been seen there recently. It appears after the first rains in winter but does not come up after the summer rains. It is seasonally infested with large caterpillars.

ETHNOGRAPHIC NOTES
Name(s)
– Soqotri: *gizhiritin* 'little milk-skins'; *ḳīribrub* 'little goat skins'; *ṣifʌṭ d-oġərib* '*Dioscorea* of the ravens' (which are said to eat them) (western Soqotra).
– Soqotri simplified: *gizhiritin*; *qiyribrub*; *zifat d-ogherib*.
Food (livestock). In its immature stage, this is classed as one of the valuable annual plants. It needs plenty of rain and so does not usually come up in the lighter summer rains. Although all livestock, especially of the drier areas, like it when it is young and green, it usually gives them diarrhoea or worse. Thirsty animals especially are attracted to the young plants. At hungry times of the year, goats, sheep and cattle graze the mature, grey-coloured plants only when other fodder is exhausted. Goats in the areas where this plant is common usually eat only the flowering stalk and the leaves when they are drying out. Camels and donkeys graze the plants green and dried. It is a common plant of the flat plains of eastern Soqotra and is regarded by some as a sign of overgrazing. Nevertheless, it is an important drought fodder in this area, and the herders of the eastern Haggeher (where this plant is rare), when transhuming to Momi after the summer and winter rains, regard this plant as a valuable feed for their goats and say it increases their milk yield and makes them fat. In western Soqotra it is less common, and sheep and goats rarely graze it.
Other. Children pick the hollow mature leaves to play with. They rub them to soften them and to loosen the outer skin, and then they inflate them. They either stamp on them to make a loud bang, or use them as a popgun, or release them to float through the air. They make a rattle by holding a bunch, thicker end uppermost, and by shaking them from side to side, or tie them together in long strings.
Notes. Eaten green and in quantity, the plants can cause severe scour or *'uqabah* in goats unaccustomed to it. It is considered to be one of the most dangerous of the *'uqabah*-causing plants. There is a mnemonic on this for herders: *ʌl itə di gizhiritin ᶜan šaᶜāḳʌb lə śāḳer*: 'Don't let her [the goat] feed on asphodel or when the dry season comes, she will develop *'uqabah* '. Goats browsing heavily on this plant also develop the condition called *ḥaniḥəno*, whose signs are swellings under the chin. There is no successful treatment for this often fatal disease other than trying to feed the animal up on the best fodder.
Comparative uses. In India the little bulbs are eaten at times of hunger, and the young leaves and stems of this and *A. tenuifolius* are eaten as a potherb. The whole plant is boiled and

applied topically for ulcers, as a diuretic and as a laxative. The seeds are taken as a diuretic, and are ground and applied externally to ulcers and inflamed skin. In Oman the plant is crushed and applied to ulcers, or boiled in water and drunk as a diuretic and as a laxative. Some bedouin use it to curdle milk to make *iqt*, dried balls of curdled milk. In northern Israel the juice squeezed from the tuber-roots of *A. microcarpus* is applied to the affected area to treat skin eczema, blisters and cuts, and elsewhere in the Middle East a cataplasm of the leaves of *A. microcarpus* is used as an anti-rheumatic. Rubbing the body with roasted tubers and drinking a decoction of leaves and scapes is seen as a good remedy for withering and paralysis. Dried tubers are cooked in oil which is then dripped into the ear lying opposite to an aching tooth to relieve the pain, or dripped into the ear to treat an infected and discharging ear. The powdered and dried tubercles are used as eye powder against leucoma of the cornea and are applied locally to abscesses. The fruit too are used to treat earache and toothache. (Boulos 1983; Dalziel 1948, 1956; Goodin & Northington 1979)

COLCHICACEAE

Perennial herbs with corms. Leaves scattered along the stem, linear, sheathing at base. Flowers 1 to 3 in leaf axils. Tepals 6, in 2 whorls, free, similar. Stamens 6, free. Ovary superior, 3-celled, with 3 styles. Fruit a capsule. 17 genera. 170 species. Old World, especially S Africa.

Iphigenia Kunth

See description of family. 1 endemic. 9 species. Old World tropics.

Iphigenia socotrana Thulin

Stems slightly zigzagging. Leaves linear, grass-like, <80 x 4mm. Tepals purplish brown, c.1cm diameter. Fruit 3-lobed, c.5 x 4mm, pendent on a long reflexed pedicel.
Ecology. Occasional in succulent shrubland on the northern limestone escarpments, usually in shady places amongst boulders or in cracks in rocks. 50–150m.
Distribution. Endemic.
Conservation status. Least Concern (LC). Common in several vegetation types and under no present or perceived threat.

Botanical notes. A rather inconspicuous plant with tufted grass-like leaves and hanging fruits.

ETHNOGRAPHIC NOTES
Name(s)
– Soqotri: *di bəṭáḥ(ḥ)ʌn* 'prostrate', 'tiny Iridaceae-like plant'.
– Soqotri simplified: *di bitahan*.
Food (human). The tiny corms are not generally eaten.
Food (livestock). The aerial parts provide incidental grazing for small stock.
Comparative uses. The corms of *I. indica* are used in colic, and its flowers yield a red dye. (Ambasta 1982; Dalziel 1948, 1956; Jain & De Fillips 1991)

COMMELINACEAE

Herbs with fibrous roots, somewhat succulent. Leaves basal or cauline, alternate, with a ciliate sheathing base and an entire blade. Flowers bisexual, actinomorphic or zygomorphic, enclosed in spathes. Sepals 3, free or fused. Petals 3, free or fused. Stamens 6 or 3 plus 2 or 3 staminodes; filaments glabrous or densely hairy. Ovary superior, 2 or 3-celled; style simple. Fruit a capsule. 40 genera. 650 species. Tropical to subtemperate.

Commelina L.

Flowers zygomorphic. Lower pair of sepals fused. Petals free, upper pair larger and clawed, lower petal much smaller. Stamens 3, dissimilar; staminodes 3; anthers cross-shaped, sterile; filaments glabrous. Capsule 2 or 3-celled, cells 1 to 5-seeded. c.170 species. Cosmopolitan.

ETHNOGRAPHIC NOTES
Comparative uses. Elsewhere the mucilaginous and starchy rhizomes are cooked and eaten, and in times of scarcity the leaves are eaten raw. The herb is used as an emollient, laxative and refrigerant, and is given in the treatment of leprosy. Many species are eaten both raw and cooked. The leaves of *C. communis* and *C. nudiflora* are eaten as a vegetable, and the seeds of *C. nudiflora* in scarcity. The bruised plant is applied as a poultice for sores as well as to boils, itches and burns. Among livestock herders in Kenya, *C. africana*, *C. latifolia* and *C. benghalensis* are regarded as a valuable fodder for all livestock, making them fat and lessening their need for water. The crushed leaves are applied to the wound of circumcision and the juice squeezed into the eyes to soothe them. *Commelina* species are a common weed of cultivated ground and a pest in groundnut fields, readily sprouting from any fragments left in the soil after hoeing. As well as being a good fodder for animals and fowls, in Sudan they are sold as camel food and regarded as a tonic for horses. However, *C. nudiflora* is said to be harmful to sheep, causing foaming at the mouth and even death if eaten in excess. This is thought to be due to the mucilage in the young inflorescence. (Ambasta 1982; Heine *et al.* 1988; Jain & De Fillips 1991; Kunkel 1984)

Commelina albescens Hassk.

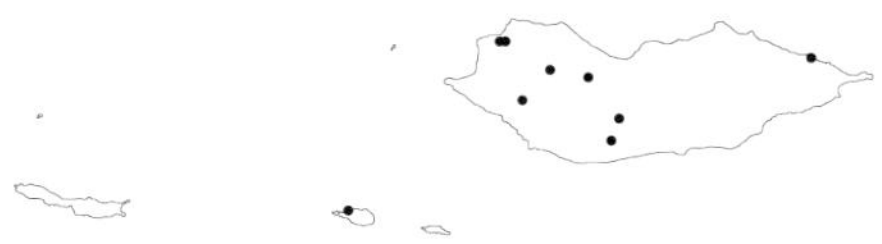

Perennial herb with prostrate or ascending stems to 30cm. Leaves glaucous or whitish; blade linear, <10 x 1cm, margins straight; leaf sheath puberulous at apex. Spathe with strongly recurved margins. Flowers mauve or pale blue. Capsule with smooth ventral cells and spinulose dorsal cell.

Ecology. Widespread in rocky places on the coastal plains and limestone plateau usually growing in the protection of shrubs or between rocks. 50–650m.

Distribution. Tropical NE Africa and Arabia.

Botanical notes. The most drought-resistant species of *Commelina* on Soqotra, immediately recognised by its linear-oblong, whitish leaves.

ETHNOGRAPHIC NOTES

Name(s)

– Soqotri: *ḳəˈwḳɔ̄r, gɔ̄ˈgor, ḳɔ̄wˈḳoh, ḳɔ̄wḳʌʾ* 'pointed'; *əˈṣənnoh, ʾəṣəˈninoh; aˈṣinnən* 'small-meshed' (Samha) (also given to *Ipomoea* species).

– Soqotri simplified: *qewqor, gowgor, qowqoh, qowqa'; ezinnoh, ezininoh; azinnin.*

Food (human). The leaves, stems and starchy rhizomes are eaten when fresh and green, especially in areas poor in other edible plants. In hungry times the fibrous roots are dug up and eaten.

Food (livestock). In the high mountain areas this is one of the most important green graze for goats and cattle. Generally associated with the upland grazings, these plants respond very quickly to rain and because of this are an important fodder until the grasses develop. They need good rains to grow well, so are more widespread in the winter than after the lighter summer rains. Along with fresh grass, they are considered to be one of the most nutritious fodders for cattle, but if the cattle eat a lot in the early rains before the grass has come up, they can develop severe scouring. Once the plants are fully mature, they are considered to improve lactation in all milk animals and to be a good, fattening feed for all livestock. Goats and sheep native to the regions where the plants occur do well on it, but incoming goats, and even sometimes sheep unaccustomed to the feed develop scour which can become serious, and even prove fatal in animals weakened by hunger after the long dry season. All goats can eat it safely once the plants begin to dry out. The plants are keenly grazed by lambs, kids and calves, and herders often take their young animals to graze in areas where the plants grow. The green aerial parts die back quickly, but the stems and roots remain and are keenly sought by small stock. Sheep and goats also dig them up to get at the underground parts when other forage is exhausted.

Flavouring / food additive. The plants give a sweet and rich flavour to the milk of animals feeding on it. This milk is much liked.

Medicinal / veterinary. Green plants are gathered and given to sick and recuperating livestock. Cows withholding their milk are tempted with handfuls of the plant to encourage them to relax and allow themselves to be milked. Cows losing weight towards the end of their pregnancy are brought this fodder to fatten them up. If this is noticed only in the last few weeks of the pregnancy, every attempt is made to rescue the calf but the mother is usually slaughtered. After giving birth, a cow which fails to pass the placenta is brought handfuls of her favourite foods, such as these plants, and encouraged to eat as much as possible. It is believed that a full belly encourages the cow to bear down and so excel the placenta.

Cleansing. Fresh, succulent stems and leaves are crushed and used for cleansing.

Cosmetic. The plant is chewed to freshen the breath.

Other. Animals feeding on these plants can go longer without water. To repair pottery, the leaves and stems of *Commelina* or *Cyanotis* are crushed with a shard of pottery and ground to a paste. This is applied to the inside of the pot, over and around the hole or crack. Then a heavily pregnant cow or goat is milked into the hand, and the sticky secretion smeared on top of the paste and around the whole area. The pot is then put beside the fire for the mixture to set and harden. Alternatively, if the hole is quite large, the paste is smeared around a suitably sized small pebble or piece of clay which is then inserted and the inner and outer surfaces around the hole are given a coating of paste. The pot is then put beside the fire for the paste to set hard.

Notes. Can cause *'uqabah* in goats unaccustomed to it, and in goats in early pregnancy can bring on abortion. The plants are a common weed of cultivated plots, especially finger millet, and have to be removed.

Commelina benghalensis L.

Annual herb with erect or ascending stems to 30cm. Leaves green; blade ovate to lanceolate, <8 x 3cm, the margin straight; leaf sheath with long, reddish hairs at apex. Spathe with more or less straight margins. Flowers blue. Capsule smooth.

Ecology. Drought-deciduous woodland. 500–900m.

Distribution. Old World Tropics, introduced elsewhere.

Botanical notes. This species has not been recorded on the Soqotra Archipelago since the 19th century. However, it can be difficult to distinguish *C. benghalensis* from *C. forskalaei* and it is possible that records of these two species have been confused. On the Yemen mainland *C. benghalensis* occurs in areas of higher rainfall than *C. forskalaei* and it is interesting that the old records of *C. benghalensis* (with the exception of a record from Abd al Kuri!) from the Soqotra Archipelago are from areas of reasonable rainfall (above Homhil and in the Haggeher mountains near Adho di Melhoh). This species should be searched for in wetter areas in the mountains.

Commelina forskalaei Vahl

Perennial herb with ascending stems to 50cm, sometimes rooting at nodes. Leaves green; blade narrowly oblong-elliptic, <6 x 1cm, margin often undulate; leaf sheath with short white hairs at apex. Spathe with more or less straight margins. Flowers blue, sometimes cleistogamous flowers present on underground stems. Capsule crenate and tuberculate.
Ecology. Widespread amongst rocks in drought-deciduous woodland and dwarf shrubland. 50–650m.
Distribution. Africa, Madagascar, Arabia and India.

Botanical notes. Distinguished from *C. albescens* by its bright green oblong-elliptic leaves. See also notes under *C. benghalensis*.

ETHNOGRAPHIC NOTES
Name(s)
– Soqotri: *ḳəwḳōr*, *gōgór*; *ḳōwḳóh*, *ḳōwḳʌʾ* 'pointed'; *əṣə́nno*, *əṣənínnoh*, *ɛṣínnən* 'small-meshed' (Samha) (also given to *Ipomaea* species); *məzḥéyməhim* 'crunchy'.
– Soqotri simplified: *qewqor*, *gowgor*, *qowqoh*, *qowqa'*; *izinnoh*, *izininnoh*, *ezinnin*; *mizheymihim*.
Notes. Uses as *C. albescens*, but these *Commelina* species with their bright green leaves and blue flowers are believed to be much more likely to cause *'uqabah* in goats, and the illness to be more severe.

Cyanotis D.Don

Flowers actinomorphic. Sepals free. Petals united into a tube below, all similar. Stamens 6, all similar, filaments densely hairy. Capsule 3-celled, the cells 1 or 2-seeded. c.50 species. Old World tropics.

Cyanotis cristata Roem. & Schult.

Annual herb with prostrate stems rooting at nodes. Leaves narrowly oblong to narrowly lanceolate, <4 x 1cm, glabrous apart from the ciliate margins, sheath c.5mm. Flowers several, enclosed in a lanceolate spathe. Corolla c.1cm diameter, purple. Filaments densely covered with blue hairs.
Ecology. Local on granite pavement in clearings in open shrubland and semi-evergreen montane woodland. 700–900m.
Distribution. Asia. Apparently introduced to Soqotra.

ETHNOGRAPHIC NOTES
Name(s)
– Soqotri: *əṣəníno ḳōwḳʌʾ*; *reḥmímhim* 'grey-coloured'.
– Soqotri simplified: *izinino qowqa'*; *rehmimhim*.
Food (human). The leaves, stems and starchy rhizomes are eaten when fresh and green, and in hungry times the roots are dug up and eaten.
Food (livestock). Not much liked by kids and lambs, but other livestock are very keen on it, and it is one of the best plants for improving lactation in cows. On Abd al Kuri it is a very highly valued fodder for sheep and goats.
Flavouring / food additive. It gives a sweet and rich flavour to the milk of animals feeding on it. This milk is much liked.
Medicinal / veterinary. Used in the same way as the *Commelina* species of the Haggeher mountains.
Cleansing. The fresh and succulent stems and leaves are used for cleansing.
Cosmetic. Chewed to freshen the breath.
Other. Animals feeding on this can go longer without water. In repairing pottery, *Cyanotis* is used in the same way as *Commelina*.
Notes. It can cause *'uqabah* in goats unaccustomed to it and can bring on abortion in goats in early pregnancy.
Comparative uses. The seeds and shoots of many species are eaten raw. *Cyanotis* species are regarded as good forage plants. However, *C. lanata* is said to cause a sort of eczema to the muzzle of horses browsing on it. In India *C. tuberosa* is given as a febrifuge. The root of *C. caespitosa* is used in a magical potion when hunting the cane-rat so that the hunters can come on it unawares. (Ambasta 1982; Dalziel 1948, 1956; Jain & De Fillips 1991; Kunkel 1984)

CYMODOCEACEAE

Dioecious marine herbs with creeping rhizomes. Leaves simple, flat or terete, sessile. Flowers solitary or forming cymose inflorescences; perianths absent. Male flowers consisting of 1 stamen; female flowers consisting of 2 free carpels. Fruit a 1-seeded nut. 5 genera. 16 species. Tropical and subtropical marine.

Botanical notes. Some of these marine flowering plants have not been recorded but are expected to occur in Soqotra's coastal waters.

Cymodocea König.

See description of family. 4 species. Coasts of W Adrica and Canary Is. to Mediterranean, Indo-Pacific.

Cymodocea rotundata Ehrenb. & Hempr. ex Aschers

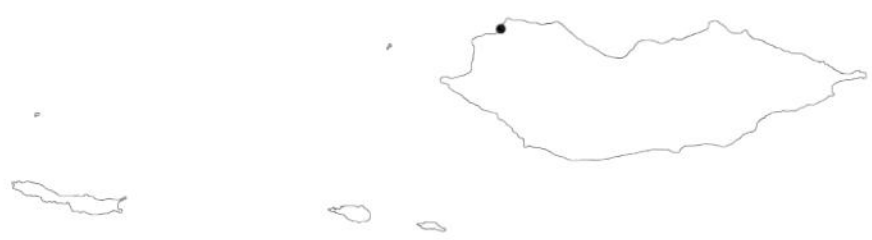

Rhizome with an erect leafy shoot and 1 or 2 roots at each node. Leaves linear, flat, 10–30 x 0.2–0.4mm, tip rounded or emarginate, margin entire or slightly toothed when young; leaf sheaths closed, 2–4cm long.

Ecology. In shallow water on sand and mudflats below the low tide level; on Soqotra recorded only from a lagoon near Qalansiyah.
Distribution. Tropical coasts of Red Sea and Indian and Pacific Oceans.

ETHNOGRAPHIC NOTES
Not seen, no information recorded.

Cymodocea serrulata (R.Br.) Aschers & Magnus

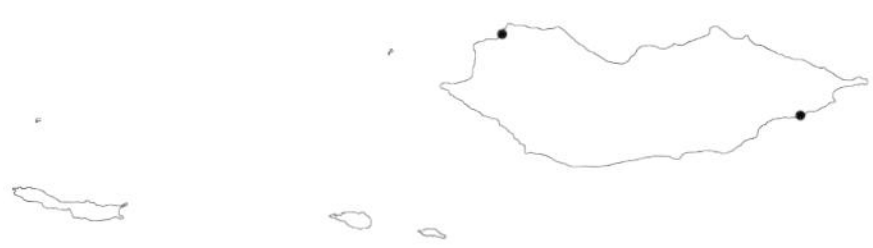

Rhizome with an erect leafy shoot and 1 or 2 roots at each node. Leaves linear, flat, 15–30 x 0.4–1cm, tip obtuse, margin with conspicuous teeth; leaf sheaths open, 1.5–3.5cm long.

Ecology. In shallow water on sand and mudflats below the low tide level; on Soqotra recorded only from a lagoon near Qalansiyah and in the southeast of the island.
Distribution. Tropical coasts of Red Sea and Indian and Pacific Oceans.

ETHNOGRAPHIC NOTES
Not seen, no information recorded.

Halodule Endl.

See description of family. 6 species. Shallow tropical seas.

Halodule uninervis (Forssk.) Aschers

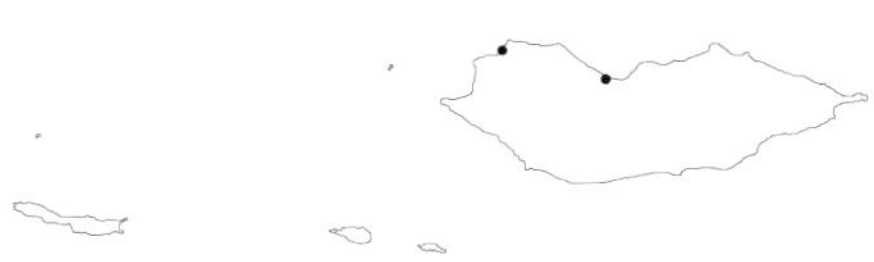

Rhizome with an erect leafy shoot and 1 to 6 roots at each node. Leaves linear, flat, 10–15 x 0.1–0.3cm, tip with 2 short lateral teeth and a rounded median tooth; leaf sheath with 2 short appendages (auricles) at top.

Ecology. In shallow water on sand and mudflats below the low tide level; on Soqotra recorded only from a lagoon near Qalansiyah and Khawr Girmah.
Distribution. Tropical coasts of Red Sea and Indian and Pacific Oceans.

ETHNOGRAPHIC NOTES
Not seen, no information recorded.

Syringodium Kütz.

See description of family. 2 species. Indian and W Pacific Oceans, Caribbean.

Syringodium isoetifolium (Aschers) Dandy

Rhizome bearing an erect leafy shoot, 1 to 3 roots and a scale leaf at each node. Leaves cylindrical, 6–30cm x 1–3mm. Flowers in a conspicuous, branched inflorescence.

Ecology. Sandy mud, often forming extensive beds.
Distribution. Tropical coasts of Red Sea and Indian and Pacific Oceans.

ETHNOGRAPHIC NOTES
Not seen, no information recorded.

Thalassodendron Hartog

See description of family. 1 species. Red Sea, Indian Ocean, Australia.

Thalassodendron ciliatum (Forssk.) den Hartog

Rhizome bearing an erect branch at every 4th node, the branches bearing conspicuous annular scars and with an apical tuft of leaves. Leaves strap-shaped, 10–15 x 0.8–1cm, the tips rounded or obtuse and the margin serrate near the tip, the leaf blade and sheath shed together.

Ecology. Habitat details not known for Soqotra. In Somalia it is reported to form extensive beds in the sublittoral zone.
Distribution. Tropical coasts of the Red Sea and Indian and Pacific Oceans.

ETHNOGRAPHIC NOTES
Not seen, no information recorded.

CYPERACEAE

Annual or perennial grass-like herbs. Stems usually solid, 3-angled. Leaves usually in a basal tuft, consisting of a closed sheath and linear blade. Inflorescence paniculate or umbel-like, open or congested, often subtended by leafy bracts. Flowers unisexual or bisexual, each subtended by a scale-like glume and arranged in 1 to 200-flowered spikelets. Perianth of 3 to 6 hairs, bristles or scales, often absent. Stamens 1 to 3; ovary superior, of 2 or 3 carpels, the style with 2 or 3 branches. Fruit a trigonous or biconvex nutlet. 90 genera. 4000 species. Cosmopolitan, especially temperate regions.

Botanical notes. Many species are rare and undercollected on Soqotra. Further exploration will undoubtedly increase the numbers of species recorded from the islands. Two specimens (*Miller* 16077, *Miller* 17134) of a distinctive species of *Carex* were collected in the western Haggeher (in dense thickets of *Rhus thyrsiflora*, 1200–1350m). Unfortunately, they cannot now be traced and represent the only record of this genus from Soqotra.

This account was prepared by Liesbeth Vandershaeve.

Bulbostylis Kunth

Herbs. Leaves flat or filiform, near base of stem only. Leaf sheaths with long hairs at mouth; ligules absent. Inflorescence a congested head or reduced panicle of spikelets. Spikelets with spirally arranged glumes. Flowers bisexual. Perianth absent. Stamens 1 or 3. Styles 3-branched, swollen at base, the style base sometimes persisting as a small projection on the nutlet. c.150 species. Pantropical.

ETHNOGRAPHIC NOTES
Food (livestock). Grazed by all livestock, though goats usually avoid them when growing in damp ground. Camels and donkeys graze the plants year round, cattle usually only in the dry season. They are considered to be especially important to small stock, and are seen as nourishing when young and green.

Bulbostylis barbata (Rottb.) C.B.Clarke

Slender, delicate, tufted annual to 15cm. Leaves <6cm, filiform. Inflorescence a head of 1 to 4 sessile spikelets. Spikelets lanceolate, <4 x 1.5mm, brownish. Stamen 1. Nutlet obovoid, triangular, <0.8 x 0.6mm, light brown, smooth; style base persisting as a small projection.
Ecology. Damp sandy soil by wadis running through date gardens. 20–50m.
Distribution. Widely distributed in warmer parts of the Old World.
Botanical notes. An inconspicuous plant, recorded only from date gardens on the Hadiboh plain but probably overlooked and under-recorded.

ETHNOGRAPHIC NOTES†
Name(s)
– Soqotri: *ḳórźʌʾ*; *ḳʌ́źəbə* (general term for Gramineae and Cyperaceae of damp ground).
– Soqotri simplified: *qordha'*; *qadhibih*.

†Also apply to *B. hispidula*.

Bulbostylis hispidula **(Vahl) R.Haines**
Fimbristylis hispidula (Vahl) Kunth

Slender, tufted annual to 11cm. Leaves <5cm, flat or caniculate, densely hairy. Inflorescence a head of 1 to 3 sessile spikelets and sometimes 1 or 2 stalked spikelets. Spikelets <6 x 3mm, reddish brown. Stamens 3. Nutlet obovate, triangular, c.1 x 1mm, light reddish brown, transversely wrinkled; style base deciduous.
Ecology. Habitat details not available.
Distribution. Tropical Africa, Malaysia and tropical America.

ETHNOGRAPHIC NOTES
Name(s)
See *B. barbata*.

Cladium P.Browne

Herbs. Stems hollow. Leaves large, coarse; ligule absent. Inflorescence a large panicle. Spikelets with spirally arranged glumes. Flowers bisexual. Perianth segments absent. Stamens 2. Style 3-branched. Nutlet triangular. 1/2 species. Cosmopolitan.

Cladium mariscus **R.Br.**

Large, stoloniferous perennial, forming dense tufts to 2m; stems rounded-triangular, glabrous. Leaves <150cm long, flat or V-shaped, margins and midrib with spiny teeth. Spikelets ovate, <5 x 2mm, sessile or subsessile, arranged in clusters. Glumes linear, <3.5mm, pale to darker brown. Nutlet irregularly wrinkled.
Ecology. Common in damp soil along the large wadis flowing from the Haggeher mountains; under-recorded on the distribution map. 450–750m.
Distribution. Cosmopolitan.

ETHNOGRAPHIC NOTES
Name(s)
– Soqotri: *ʿārəf, ʿūrəf* 'scooping water'; sometimes also *iʾišhel* (more commonly given to *Juncus* species); (rare) *ḳʌźʌbhən*.
– Soqotri simplified: *'aref, 'uwrif; i'ish'hel; qadhabhin.*
Food (human). The seeds were ground and eaten as flour in times of famine.
Food (livestock). Goats eat the new growth, flowers and fruit when they can reach it, and cattle like the fruiting heads. Camels and donkeys are said not to be keen on the plant. The greenery persists throughout the year and both cattle and goats eat the large, coarse leaves in the dry season. In drought, herders cut the leafy stems and take them to feed their hungry animals. It is not thought to be a nutritious feed, but a useful maintenance one.
Other. When other forage is in short supply, herders sometimes fire the plants to encourage new, tender and more palatable growth. This is also done when the growth has become so dense that the flow of water is impeded. Individual clumps are also burned if they are believed to be sheltering harmful vermin. If a herder comes across a goat giving birth unattended, he commonly burns some of its dried foliage beside her to keep away the predatory wild cats until the kid has been safely delivered and is able to walk.
Comparative uses. Elsewhere the plant is used as raw material for the manufacture of paper pulp. (Ambasta 1982; Jain & De Fillips 1991)

Cyperus L.

Mariscus Vahl

Herbs; tufted or with creeping rhizomes or stolons, sometimes bulbous. Leaves grass-like, sometimes scale-like, arranged in 3s; ligules absent. Inflorescence of spikes or digitate clusters. Glumes 2-ranked. Flowers bisexual, perianth absent. Style branches usually 3 (or 2). Stamens 3 (or 1). Nutlets triangular or biconvex. 600 species. Cosmopolitan.

Botanical notes. *Pycreus* and *Kyllinga* are sometimes considered as subgenera of *Cyperus*.

ETHNOGRAPHIC NOTES
Comparative uses. Many *Cyperus* are valued as fodder plants, or are used for weaving into mats. The larger sedges are used for thatching and for making brushes. Some have swollen root systems, bulbils or tubers which are eaten raw or roasted, and some produce grains which are eaten in times of famine. The best-known edible *Cyperus* is probably *C. esculentus*, the tiger nut, which is a popular food and is also regarded as having aphrodisiac effects. *Cyperus* rhizomes are considered to be tonic, stimulant and astringent, and are given as an anthelmintic, diuretic and febrifuge. In East Africa the roots and seeds of *C. digitatus* are used as necklace beads, and elders chew the root or seeds and spit the liquid over a sick person to heal him. The tubers of *C. articulatus* are fragrant and used for perfuming clothing, as are the roots and rhizomes of *C. brevifolius. C. exaltatus* is suitable for paper-making, and in Bahrain, its tubers are powdered and used as an insecticide. The tubers of *C. articulatus* are strung into necklaces and waist-girdles and worn to keep insects away. The stems are used for making mattresses, mats and sacks. The dried and pulverised rhizomes are used as a fumigant, usually with added scented resins, to perfume clothing and sweeten the air of rooms in the rainy season. The dried, fragrant tubers of *C. maculatus* are sold as 'perfume'. (Abbas *et al.* 1992; Ambasta 1982; Bebawi & Neugebohrn 1991; Heine *et al.* 1988; Jain & De Fillips 1991; Kunkel 1984; Lewis & Elwin-Lewis 1977)

Cyperus amabilis Vahl

Slender annual to 15cm, glabrous. Leaves inrolled, scabrid on margins; leaf sheaths membranous, striate. Inflorescence an open panicle with a sessile cluster of spikelets and several stalked groups of digitately arranged spikelets, sometimes reduced to a cluster of sessile spikelets; bracts leaf-like, longer ones overtop inflorescence. Spikelets linear, <17mm, golden to reddish brown; glumes linear-lanceolate, 1–1.5mm, margin golden to chestnut brown, midrib green, shortly mucronate. Stamen 1. Nutlet <0.8 x 0.6mm, obovoid, triangular, brownish, minutely papillose.

Ecology. Damp soil by wadis. 50–c.800m.

Distribution. Tropical Asia, Africa and America.

ETHNOGRAPHIC NOTES†

Name(s)

– Soqotri: *ḥámžig* 'mean; milkless' (general name for small Cyperaceae growing in damp ground or beside water); *ḳórźʌʾ*; (*ḳāźub*) *di mɔ̄źi* '(*ḳāźub*) of damp ground'; (when very young) *ḳə́ṣəho* 'erect'; (tussock-forming Cyperaceae) *saʿadifo* (pl. *saʿaduf*) 'clump'; (larger species) *iʾišhel* (more commonly given to *Juncus* species); *ḳeṣádə* 'sharp, cutting', 'oversalted' (more commonly given to large, coarse Gramineae); *ḳʌ́lḳəhʌl* 'swaying in the breeze' (a common name for mostly halophytic species); *ʾišɛ́lo* (Abd al Kuri); (rare) *di ḳerṭómoh* 'needs to be chewed' (from the inflorescence).

– Soqotri simplified: *hamjig*; *qordha'*; (*qadhub*) *di mowdhi*; *qiziho*; *sa'adifo* (pl. *sa'aduf*); *i'ish'hel*; *qezadih*; *qalqihal*; *ishelo*; *di qertomo*.

Food (livestock). Grazed by all livestock, though goats usually avoid them if they are growing in damp ground. Camels and donkeys graze them year round, cattle usually only in the dry season. They are considered to be especially important to small stock, and are seen as nourishing feed when young and green.

Notes. These are all annual species, generally considered to provide a richer feed than perennial Cyperaceae.

†Also apply to *C. compressus*, *C. rubicundus*.

Cyperus bulbosus Schrank

Perennial with a basal bulb to 20cm, glabrous. Leaves flat; leaf sheaths brownish. Inflorescence a spike of 6 to 10 sessile spikelets, sometimes also with 1 or 2 stalked spikelets; bracts leaf-like, <5cm, margins slightly scabrid. Spikelets linear-lanceolate, <15mm, dark reddish brown; glumes lanceolate, 3–4mm, dark reddish brown to almost black, with a paler midrib and 6 to 8 lateral veins. Nutlet c.1.2 x 0.5mm, triangular, greyish, papillose.

Ecology. Grassland. 550–1100m.

Distribution. Widespread in warmer parts of the Old World.

Botanical notes. One of the first plants to come up after rain. Typically found in hollows where water collects.

ETHNOGRAPHIC NOTES†

Name(s)

– Soqotri: *di ḥɔ̄wgəb, di ḥə́wgəb* 'little palm'; (smaller specimens) *di ḥəgúbgub* 'tiny palms'; *lə́ʾəh, lóʾoh* 'little pearl'; *saʿīdə* (pl. *sáʿyid*); (when very young) *ḳə́ṣəho* 'erect' (general term for first shoots of herbs and grasses).

– Soqotri simplified: *di howgib, di hewgib*; *di higubgub*; *li'ih, lo'oh*; *sa'iydih* (pl. *sa'yud*); *qiziho*.

Food (human). The small bulbils and tuberous rhizomes are eaten, usually raw. They have a sharp, ginger-like flavour. If plentiful they can also be boiled or baked.

Food (livestock). Respond very quickly to rain and the new shoots are much liked by goats and cattle. They are regarded as a valuable and nutritious feed for all livestock until they mature and begin to die back.

Comparative uses. *C. rotundus* is a very well-known medicinal plant, though it is also seen as a troublesome weed on sandy soils in tropical countries. It is used as an anti-bacterial and anti-inflammatory. The essential oil from the tubers shows oestrogenic activity and has anti-spasmodic action on the uterus. It is also used to repel insects and in perfumes. In Iraq the tubers are considered to be diaphoretic and astringent, and are given in disorders of the stomach and irritation of the bowels. In Arabia the dried or fresh tubers are powdered as insecticide, and are eaten to regulate menstruation and as an anthelmintic. The seeds are heated and used with oil as eardrops. An ointment made from the crushed tubers is applied to bee stings and bites, and rubbed on the breasts as a galactagogue. In Yemen the powdered rhizomes are used to perfume the clothes and hair. In Qatar the plant is used as an abortifacient and to treat uterine haemorrhage. The tubers are used as an analgesic, anti-spasmodic, anti-tussive and carminative. It is also used as an emmenagogue, sedative, stimulant and tonic, and is given to treat malaria, headaches and bronchitis. In India the dried tuberous roots yield an essential oil and are an aromatic used in perfumes. They were used for embalming corpses by the Scythians. In Africa the rhizome is used as a cough medicine, the pulp of the roots is given for oedema and rheumatism, and an infusion of the plant and roots is given to cattle suffering from certain types of poisoning. The roasted tubers of *C. bulbosus* are said to have a taste like potatoes, and are baked into bread or cooked into puddings. The rhizomes are crushed and applied locally to treat a variety of skin problems. (Al Rawi & Chakravarty 1964; Ambasta 1982; Batanouny 1981; Bebawi & Neugebohrn 1991; Dalziel 1948, 1956; Ghazanfar 1994; Jain & De Fillips 1991; Kunkel 1984; Oliver-Bever 1986; Rizk & El-Ghazaly 1995)

†Also apply to *C. rotundus*, *C. squarrosus*.

Cyperus compressus L.

Robust annual to 22cm, glabrous. Leaves flat, slightly scabrid on margins and midrib; leaf sheaths light brown to purplish with reddish spots on wings. Inflorescence an open panicle with a sessile cluster of spikelets and 3 to 5 stalked groups of spikelets, sometimes reduced to 1 cluster of sessile spikelets; bracts leaf-like. Spikelets linear-lanceolate, strongly compressed, <20 x 4mm, greyish to golden brown, 7 to 20-flowered; glumes ovate-elliptic, 3.5–5mm, grey to brownish with a green midrib, mucronate. Nutlet <1.8 x 1.3mm, obovoid, triangular, dark brown, smooth.

Ecology. No habitat details available; on nearby African mainland found in seasonally wet habitats.

Distribution. Pantropical.

Botanical notes. Recognised by its annual habit, compressed spikelets and large smooth nutlets. Recorded from the Haggeher near Hadiboh by Balfour in the 19th century but not seen anywhere on the island since.

ETHNOGRAPHIC NOTES

Name(s)

See *C. amabilis* (also uses).

Cyperus conglomeratus Rottb.

Cyperus amblyostachyus Fenzl
Cyperus conglomeratus var. *effusus* (Rottb.) Kük.
Cyperus conglomeratus var. *socotranus* Balf.f.
Cyperus curvulus Boeck.
Cyperus proteinolepis Boeck.

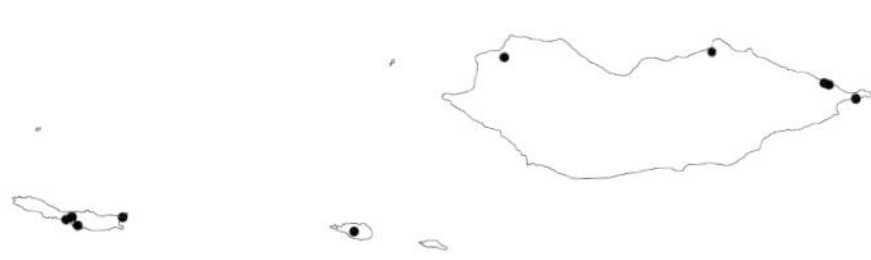

Robust tussock-forming perennial to 20cm with woody rhizome; roots covered with dense grey hairs. Leaves often wiry, channelled, strongly curving; leaf sheaths reddish brown. Inflorescence a terminal head of 3 to 12 crowded spikelets, sometimes subtended by 1 or 2 pedunculate heads of spikelets; bracts leaf-like, longest usually overtopping inflorescence. Spikelets lanceolate, strongly compressed, straw-coloured, 3 to 10-flowered. Glumes oblong, 2.5–4mm, greyish white to straw-coloured, margin hyaline, many-nerved, midrib greenish, shortly mucronate. Nutlet c.1.5 x 0.8–1mm, obovate, obscurely triangular, blackish, minutely papillose.

Ecology. Locally common in woody-herb communities on the coastal plains, particularly on sandy soil and wind-blown sand overlying limestone. Sea-level–300m.

Distribution. Mediterranean region, N and tropical Africa and SW Asia to India.

Botanical notes. A very variable species. Common on Abd al Kuri and Samha and an important component of the vegetation near the coast, particularly on sand.

ETHNOGRAPHIC NOTES†

Name(s)

– Soqotri: *ḥámžig* 'mean; milkless' (general name for small Cyperaceae growing in damp ground or beside water); *ḳórźʌʾ*; (*ḳāźub*) *di mɔ̄źi* '(*ḳāźub*) of damp ground'; (when very young) *ḳə́ṣəho* 'erect'; (tussock-forming Cyperaceae) *saʿadífo* (pl. *saʿaduf*) 'clump'; (larger species) *iʾišhel* (more commonly given to *Juncus* species); *ḳeṣádə* 'sharp, cutting', 'oversalted' (more commonly given to large, coarse Gramineae); *ḳʌ́lḳəhʌl* 'swaying in the breeze' (a common name for mostly halophytic species); *ʾišέlo* (Abd al Kuri); (rare) *di ḳerṭómoh* 'needs to be chewed' (from the inflorescence).

– Soqotri simplified: *hamjig*; *qordha'*; (*qadhub*) *di mowdhi*; *qiziho*; *sa'adifo* (pl. *sa'aduf*); *i'ish'hel*; *qezadih*; *qalqihal*; *ishelo*; *di qertomo*.

Food (livestock). Grazed by all livestock, though goats usually avoid them if they are growing in damp ground. Camels and donkeys graze them year round, cattle usually only in the dry season. They are considered to be especially important to small stock, and are seen as nourishing feed when young and green. On the outer islands, the widespread *C. conglomeratus* is a key plains fodder for livestock. *C. stramineo-ferrugineus* on Abd al Kuri is also a very important forage.

Notes. These are all perennial species. While not generally considered as nutritious as annual Cyperaceae when mature, their persistence through the dry season makes them an important dry season maintenance feed.

Comparative uses. An infusion of the seeds of *C. cyperoides* is used for toothache in the Philippines. (Lewis & Elwin-Lewis 1977)

†Also apply to *C. cyperoides*, *C. dilatatus*, *C. stramineo-ferrugineus*.

Cyperus cyperoides (L.) Kuntze

Slender perennial to 35cm, stems swollen at base. Leaves flat, slightly scabrid on margins; leaf sheaths membranous, reddish brown or brown. Inflorescence umbel-like, of 1 to several, sessile or stalked spikes; spikes <12 x 6mm, with numerous spikelets; bracts leaf-like. Spikelets linear-lanceolate, <5 x 1mm, brownish to somewhat golden; glumes linear-lanceolate, c.1.5–3mm, brownish, midrib greenish, sometimes very shortly mucronate. Nutlet <3 x 0.8mm, elliptic, triangular, brown, minutely papillose.

Ecology. Grassland in montane thicket. 600–900m.

Distribution. Pantropical.

Botanical notes. This is a very variable species, with seven subspecies described from E Africa alone. On Soqotra it has been collected only twice and it has not been possible to assign these collections to subspecies. In one of the collections the inflorescence is reduced to a solitary sessile spike and thus appears rather atypical. No mature nutlets were seen from Soqotra and the description was completed from African material.

ETHNOGRAPHIC NOTES

Name(s)

See *C. conglomeratus* (also uses).

Cyperus difformis L.

Annual to 30cm. Leaves flat, scabrid on margins and midrib; leaf sheaths brownish to purple, the basal leaves without leaf blades. Inflorescence a solitary sessile head and 5 to 10 stalked heads of digitately arranged spikelets, heads spherical, 5–10mm diameter; bracts leaf-like, overtopping the inflorescence. Spikelets <20-flowered, linear, <4mm, brown, axis zigzagging; glumes obovate, 0.6–0.8mm, reddish brown, margin pale, midrib greenish. Nutlet <0.8 x 0.4mm, obovoid, triangular, yellowish, minutely papillose.

Ecology. No habitat details available.

Distribution. Tropical and warm temperate regions of the Old World.

Botanical notes. Recorded from the Haggeher near Hadiboh by Balfour in the 19th century but not seen anywhere on the island since.

Cyperus dilatatus Schumach. & Thonn.

Robust stoloniferous perennial to 20cm. Leaves flat, scabrid on margins and midrib; leaf sheaths light brown. Inflorescence of a sessile spike and 4 or 5 stalked spikes, spikes with 5 to 10 spreading spikelets; bracts leaf-like. Spikelets 14 to 18-flowered, linear-lanceolate, <15 x 3mm, compressed, straw-coloured to reddish brown. Glumes ovate, 3–4mm, reddish brown, margin hyaline, midrib green, 3 to 5-nerved. Nutlet <2 x 0.5mm, oblong, light brown, minutely papillose.

Ecology. Wet places. 30–550m.

Distribution. Tropical Africa.

Botanical notes. Difficult to separate from *C. rotundus* but differs by the reddish spots on the leaf sheaths, the spreading spikelets and the glumes with broad, colourless margins.

ETHNOGRAPHIC NOTES

Name(s)

See *C. conglomeratus* (also uses).

Cyperus laevigatus L.

Rhizomatous perennial to 30cm, stems ridged, covered with short scales below. Leaf blades absent or very short, if present inrolled and resembling stem; sheaths dark brown to somewhat purplish. Inflorescence a crowded head of 2 to 12 spikelets, sometimes a solitary spikelet; bracts stem-like, erect, continuing direction of stem. Spikelets linear-lanceolate, often curved, 7 x 2.5mm, flattened, pale yellowish grey. Glumes ovate, 2–2.5mm, keeled near tip, keel greenish yellow with reddish brown dots, shortly mucronate. Nutlets <1.2 x 1mm, obovoid, flat on one side, rounded on the other, greyish brown, slightly papillose.

Ecology. In mud around brackish inlets. Sea-level.

Distribution. Widespread in warm and tropical regions.

Botanical notes. Recorded from near Qalansiyah in the 19th century but not seen since; possibly overlooked.

Cyperus longus L.

Stoloniferous perennial to 65cm. Leaves few, <30cm x 7mm; leaf sheaths dark reddish brown. Inflorescence an open panicle with a sessile spike and 4 to 8 stalked spikes, stalks 3–10cm; bracts leaf-like, erect, <40cm, scabrid on margins and midrib. Spikelets 4 to 6-flowered, linear-lanceolate, <13 x 1.5mm, reddish brown; glumes narrowly ovate, 3–3.2mm, reddish brown with a broad hyaline margin and greenish midrib, shortly mucronate. Nutlets <1.5 x c.0.5mm, elliptic, triangular, light brown, almost smooth.

Ecology. No habitat details available.

Distribution. Tropical Africa, Mediterranean region, C and SW Asia to India.

Botanical notes. Collected in the Haggeher mountains in the 19th century but not seen since; possibly overlooked.

Cyperus rotundus L.

Stoloniferous perennial to 35cm, stems black, base somewhat swollen. Leaves flat, scabrid on margins and midrib; leaf sheaths brownish; wings of sheaths with reddish spots. Inflorescence an open panicle with a sessile spike and 2 to 5 stalked spikes; bracts leaf-like, erect or spreading, <6cm, sometimes overtopping inflorescence. Spikelets 10 to 30-flowered, linear-lanceolate, <20 x 2mm, reddish brown; glumes linear-lanceolate, 2.5–3mm long, reddish brown with pale margins and a greenish midrib. Nutlet <1.3 x c.0.5mm, elliptic, triangular, light brown, minutely papillose.

Ecology. In wet places and damp grassland. 50–500m.

Distribution. Warmer regions.

Botanical notes. One of the first plants to come up after the rains. Typically grows in depressions where water collects.

ETHNOGRAPHIC NOTES

Name(s)

See *C. bulbosus* (also uses).

Cyperus rubicundus Vahl

Low tufted annual with fibrous roots to 15cm. Leaves flat, glabrous; leaf sheaths <10mm wide, grey to purplish. Inflorescence a terminal head of crowded spikelets; bracts leaf-like, <5cm. Spikelets <30-flowered, linear-lanceolate, <15 x 4mm, strongly compressed, reddish brown; glumes ovate, 2.5–3mm, reddish brown, 5 to 8 paler nerves on either side of midrib, margins whitish, tip recurved and shortly mucronate. Nutlet <1.1 x 0.7mm, obovoid, sharply triangular, brownish grey, minutely papillose.

Ecology. Drought-deciduous woodland, often in shallow pockets of soil in rock. 350–750m.

Distribution. Africa, Arabia, SW Asia and N Australia.

Botanical notes. One of the most distinctive species of Cyperaceae on Soqotra; readily distinguished by the strongly compressed, reddish brown spikelets.

ETHNOGRAPHIC NOTES

Name(s)

See *C. amabilis* (also uses).

Cyperus squarrosus L.

Cyperus aristatus Rottb.
Mariscus squarrosus (L.) C.B.Clarke

Low tufted annual with fibrous roots to 12cm. Leaves flat, glabrous; leaf sheaths membranous, purplish to reddish brown. Inflorescence of 1 to 3 sessile and 1 to 4 stalked spikes; bracts leaf-like, at least some overtopping inflorescence. Spikes <1.5 x 1cm, oblong, brownish, consisting of numerous crowded spikelets. Spikelets linear-oblong, <5 x 3mm, flattened, yellowish to reddish brown; glumes linear-lanceolate, 1.4–1.7mm, yellowish to reddish brown, midrib green, 3 or 4 reddish nerves on either side, tip recurved and mucronate. Nutlets <1 x 6mm, obovoid, triangular, greyish-brown, minutely papillose.

Ecology. Semi-deciduous woodland, typically in pockets of soil in rock. c.600m.

Distribution. Warmer regions.

Botanical notes. One of the first plants to come up after the rains. Typically grows in depressions where water collects. The dried leaves smell of curry powder when crushed.

ETHNOGRAPHIC NOTES

Name(s)

See *C. bulbosus* (also uses).

Cyperus stramineo-ferrugineus Kük.

Mariscus stramineo-ferrugineus (Kük.) Napper

Tussock-forming perennial with crowded stems and leaves to 16cm. Leaves filiform, flat or caniculate, glabrous, often curved at tip; upper sheaths membranous, lower thicker, brown to blackish. Inflorescence a solitary head of crowded spikelets, 8–12mm in diameter, straw-coloured to reddish brown; bracts leaf-like, erect or spreading. Spikelets 3 or 4-flowered, linear-lanceolate, <4 x 1.5mm, reddish brown; glumes ovate, 2.5–3mm, strongly concave, reddish brown, pale hyaline margin, nerves 4 to 6, midrib greenish, shortly mucronate. Nutlets <1.2 x 0.6mm, obovoid, triangular, golden brown with dark brown angles and short acute tip.

Ecology. Sheltered places on exposed limestone ridges. 550–600m.

Distribution. Africa and S Asia.

Botanical notes. A tufted perennial with hair-like leaves. The base of the plant is surrounded by the fibrous remains of old leaf bases.

ETHNOGRAPHIC NOTES

Name(s)

See *C. conglomeratus* (also uses).

Cyperus tenuiculmis **Boeck.**

Perennial to 30cm. Leaves flat, scabrid on margins and ribs; sheaths light brown, winged. Inflorescence of a sessile spike and 2 to 6 stalked spikes, stalks 2–7cm long; bracts leaf-like, erect or spreading, <6cm. Spikes 4–5cm long with spreading spikelets. Spikelets linear-lanceolate, <5cm; glumes ovate, 2.5–3mm, reddish brown, 6-nerved, margins whitish, midrib green, shortly mucronate. Nutlets <1.5 x 0.5mm, lanceolate, triangular, blackish, papillose.
Ecology. No habitat details available.
Distribution. Tropical Africa.
Botanical notes. This record is based on a single inadequate specimen collected near Qalansiyah in the 19th century but the species has not been seen since.

Eleocharis R.Br.

Heleocharis T.Lestib.

Small herbs. Leaf blades absent; ligule absent. Inflorescence a single spikelet. Perianth of 6 or 7 bristles, glabrous or with toothed margins. Stamens 2. Styles with 2 branches. Nutlet biconvex, style base persistent as a small triangular appendage. 180 species. Warm and temperate regions.

ETHNOGRAPHIC NOTES

Comparative uses. *E. dulcis* produces the well-known Chinese water chestnut, which is rich in starch. Other *Eleocharis* species are used for making matting, though this is said to not be very durable. (Ambasta 1982; Jain & De Fillips 1991)

Eleocharis caduca **(Del.) Schultes**

Annual, sometimes stoloniferous, to 10cm. Spikelet <10-flowered, ovoid, <4 x 2mm, light brown; glumes ovate, 1.5–2mm, reddish brown, midrib green, hyaline margins. Perianth bristles with numerous recurved, spine-like teeth. Nutlet c.1 x <0.7mm.
Ecology. In wet places, typically in silt by wadis and pools. 450–900m.
Distribution. Pantropical.
Botanical notes. Known only from Wadi Di Negehen but probably overlooked elsewhere.

ETHNOGRAPHIC NOTES†

Name(s)
– Soqotri: *ḳāźəb* 'tiny'; *ʾiʾišhel* (more commonly given to *Juncus* species).
– Soqotri simplified: *qadheb*; *i'ish'hel.*
Food (livestock). Grazed by all livestock, though goats usually avoid them if they are growing in damp ground. Camels and donkeys graze them year round, cattle usually only in the dry season. They are considered to be especially important to small stock, and are seen as nourishing feed when young and green.

†Also apply to *E. geniculata*.

Eleocharis geniculata **(L.) Roem. & Schult.**
Eleocharis capitata R.Br.

Tufted annual to 35cm. Spikelet many-flowered, globose, <3.5mm diameter; glumes ovate, c.2mm, pale with reddish spots below, reddish brown above, midrib obscure. Perianth bristles glabrous. Nutlets c.0.8 x 0.6mm.
Ecology. In wet places, typically in silt by wadis and pools. Sea-level–50m.
Distribution. Warmer regions.
Botanical notes. All records from Soqotra and Samha are from near sea-level, but on the adjoining Arabian mainland it is a plant of the mountains. It is certainly undercollected on Soqotra and it is probable that further fieldwork will show it to be more widespread than the map suggests.

ETHNOGRAPHIC NOTES

Name(s)
See *E. caduca* (also uses).

Fimbristylis Vahl

Tussock-forming herbs. Leaves mainly basal; ligule a rim of dense hairs. Inflorescence a panicle of spikelets. Glumes spirally arranged. Flowers bisexual. Perianth absent. Stamens 1 to 3. Style 2 or 3-branched. Nutlets triangular or biconvex. 100 species. Warm and tropical regions.

ETHNOGRAPHIC NOTES

Comparative uses. Like other sedges, many *Fimbristylis* species were used for weaving, and especially to make matting. One species, *F. exilis*, has a fragrant rhizome, and the root of another, *F. junciformis*, is given to treat dysentery. (Ambasta 1982; Dalziel 1948, 1956; Jain & De Fillips 1991)

Fimbristylis complanata (Retz) Link

Densely tufted perennial to 15cm, stem base covered with stiff scales. Inflorescence a panicle of clusters of spikelets, each cluster with 1 sessile and 2 or 3 stalked spikelets; bracts slightly scabrid. Spikelets <3 x 1.5mm, dark brown; glumes ovate, c.2mm, dark brown, margin paler, midrib greenish, tip with a recurved mucro. Stamens 3. Styles 3-branched. Nutlets <1 x c.0.5mm, triangular.

Ecology. No habitat details available.

Distribution. Pantropical.

ETHNOGRAPHIC NOTES†

Name(s)

– Soqotri: *ḳórźʌʾ*; *ḳʌ́źəbə*; (tussock-forming Cyperaceae) *saʿadífo* (pl. *sáʿaduf*) 'clump'.

– Soqotri simplified: *qordha'*; *qadhibih*; *sa'adifo* (pl. *sa'aduf*).

Food (livestock). Grazed by all livestock, though goats usually avoid them if they are growing in damp ground. Camels and donkeys graze the plants year round, cattle usually only in the dry season. They are considered to be especially important to small stock, and are seen as nourishing feed when young and green.

†Also apply to *F. cymosa*, *F. ferruginea*.

Fimbristylis cymosa R.Br.

Densely tufted, glaucous perennial to 30cm, stems rigid. Leaves obtuse at tip. Inflorescence a panicle of clusters of sessile spikelets, branches <1.5cm; bracts scabrid on margins. Spikelets ovate, <6 x 1.5mm, terete, greyish brown to light brown; glumes ovate, c.1.5mm, pale brown, margins broad, whitish, midrib green. Stamens 2. Style 2-branched. Nutlets <1 x 0.8mm, obovoid, biconvex.

Ecology. Locally common in wet places, typically in silt by wadis and pools. 30–600m.

Distribution. Pantropical.

Botanical notes. Distinguished by its biconvex nutlets and obtuse tips to the leaves.

ETHNOGRAPHIC NOTES

Name(s)

See *F. complanata* (also uses).

Fimbristylis dichotoma (L.) Vahl

Fimbristylis annua (All.) Roem. & Schult.
Fimbristylis diphylla Vahl

Tufted perennial with a short woody rhizome to 25cm, stems compressed. Leaves flat or caniculate, margin with short spine-like teeth, tip rounded with a short spine-like tooth. Inflorescence an open panicle of clusters of spikelets, each cluster with 1 sessile and 2 to 4 stalked spikelets; bracts leaf-like, slightly scabrid. Spikelets ovate, <7 x 2.5mm, cylindrical; glumes broadly ovate, 2–3mm, reddish brown, margin paler, midrib greenish, tip shortly mucronate, glabrous and shiny. Stamens 2. Stigmas 2. Nutlets <1.2 x c.1mm, biconvex, longitudinally grooved with transverse ribs over the grooves.

Ecology. No habitat details available, probably in wet grassland by water.

Distribution. Warmer regions.

Botanical notes. Collected in the Haggeher mountains in the 19th century but not seen since; possibly overlooked.

Fimbristylis ferruginea **Vahl**

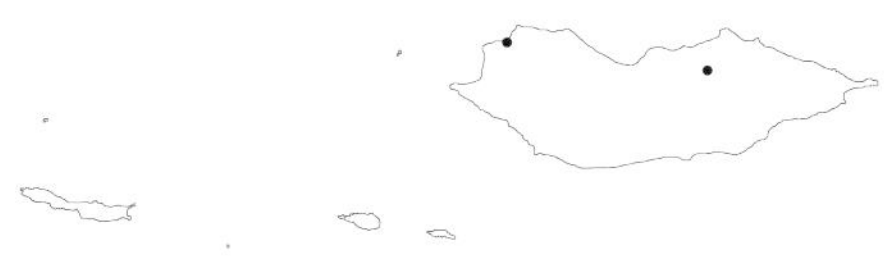

Robust perennial with short creeping woody rhizome to 30cm, stems compressed, sometimes with scattered spine-like teeth. Leaves flat or caniculate, margin and ribs with short spine-like teeth. Inflorescence of 1 sessile and 4 to 10 stalked spikelets; bracts leaf-like. Spikelets ovate, <8 x 4mm, terete; glumes broadly ovate, 3–4mm, reddish, midrib green, tip shortly mucronate, densely covered with short, whitish hairs in upper half. Stamens 3. Style 2-branched. Nutlets <1.9 x 1.4mm, obovoid, biconvex, almost smooth.

Ecology. Local in wet places, typically in silt by wadis and pools. 30–600m.

Distribution. Tropical Africa, SW Asia to Malaysia and tropical Australia.

Botanical notes. All Soqotran material is referable to subsp. *sieberiana* (Kunth.) K.Lye.

ETHNOGRAPHIC NOTES

Name(s)

See *F. complanata* (also uses).

Fuirena Rottb.

Annual herbs. Leaves grass-like. Inflorescence a cluster of spikelets. Glumes spirally arranged. Perianth of 6 segments, outer 3 short, inner 3 stalked. Stamens 2 or 3. Style 3-branched. Nutlets triangular. 30 species. Warmer regions throughout the world.

Fuirena ciliaris **(L.) Roxb.**

Fuirena glomerata Lam.

Delicate annual to 25cm, hairy, leafy throughout. Leaves flat, reduced to a basal sheath towards base of stem. Inflorescence of a cluster of spikelets in upper leaf sheaths; bracts hairy. Spikelets ovoid, <7 x 4mm, greenish, densely hairy. Glumes ovate, <4mm, dark brown, awn 1mm, strongly hairy. Inner perianth segments with square 3-nerved lamina. Stamens 3. Stigmas 3. Nutlets <1.1 x c.0.5mm, whitish, smooth, narrowed into a beak.

Ecology. Wet ground by pools or wadis. 250–450m.

Distribution. Widespread in the Old World tropics.

ETHNOGRAPHIC NOTES

Name(s)

– Soqotri: *ḳòrźʌˀ.*

– Soqotri simplified: *qordha'.*

Food (livestock). Grazed by all livestock, though goats usually avoid them if they are growing in damp ground. Camels and donkeys graze the plants year round, cattle usually only in the dry season. They are considered to be especially important to small stock, and are seen as nourishing feed when young and green.

Comparative uses. In India *F. umbellata* is used as green manure, and is burned for ashes to make salt. (Ambasta 1982; Dalziel 1948, 1956; Jain & De Fillips 1991)

Kyllinga Rottb.

Perennial herbs with a creeping rhizome or stolons; stems triangular or ridged. Leaves grass-like, arranged in 3s; ligules absent. Spikelets in globose heads, 1 or 2-flowered. Glumes 2-ranked. Flowers bisexual, perianth absent. Style branches 2. Stamens 1 or 2. Nutlets flattened. 50 species. Widespread in tropics.

Botanical notes. Sometimes treated as a subgenus of *Cyperus.*

Kyllinga brevifolia **Rottb.**

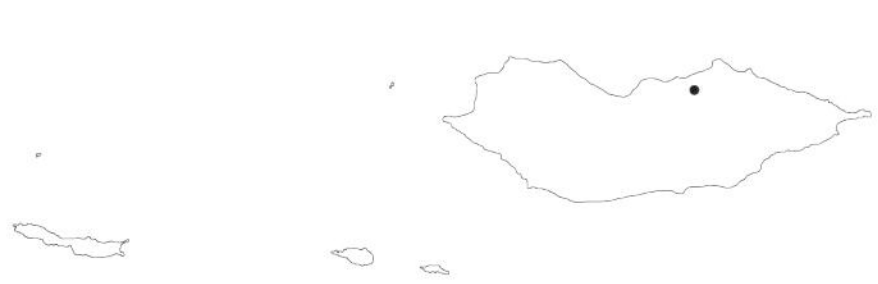

Slender perennial herb to 25cm, triangular. Leaves scabrid, at least on margins; basal leaves without blade. Inflorescence a globose head, 4–6mm diameter, usually 1 spike, yellowish to greenish; bracts usually 3, one erect and two spreading. Spike with numerous spikelets. Spikelets 2–3 x c.1mm, 1-flowered; glumes usually 2, <3mm, with a spinulose keel. Stamen 1. Nutlet <1.2 x 0.8mm, obovoid, yellowish.

Ecology. Rare, in gravel by wadis. 500–600m.

Distribution. Widespread in tropical and warm temperate regions.

Botanical notes. Only recently recorded, from Wadi Di Negehen, but possibly overlooked elsewhere.

ETHNOGRAPHIC NOTES†

Name(s)

– Soqotri: *ḳə̀źʌź* (*ḳə̀ḏoḏ*) 'low growing', 'grazed down'; *ḳòrźʌˀ.*

– Soqotri simplified: *qidhadh, qidhodh; qordha'.*

Food (livestock). Grazed by all livestock, though goats usually avoid them if they are growing in damp ground. Camels and donkeys graze the plants year round, cattle usually only in the dry season. They are considered to be especially important to small stock, and are seen as nourishing feed when young and green.

Comparative uses. In India the roots and rhizomes of this plant are used as an aromatic and the leaves used to treat diarrhoea. In East Africa the roots of various *Kyllinga* species are eaten raw for stomach pains, and the seeds are worn on string as a charm to protect against snakebite and other evil. The tiny black seeds and tubers of *K. alba* are eaten by children. The tuberous rhizome of *K. erecta* and other *Kyllinga* species is fragrant and often used as a fumigant, and the juice may be added to flavour cooked food or medicines. (Ambasta 1982; Dalziel 1948, 1956; Heine *et al.* 1988; Jain & De Fillips 1991)

†Also apply to *K. microstyla*.

Kyllinga microstyla **C.B.Clarke**
Cyperus microstylis (C.B.Clarke) Mattf. & Kük.

Slender, delicate, tufted perennial to 12cm, base surrounded by fibrous remains of old leaf bases. Leaves filiform, slightly scabrid. Inflorescence a yellowish head of usually 3 spikes with numerous spikelets, central spike 2.5–3.5 x 1.5–2mm; inflorescence bracts usually 3, leaf-like, much exceeding spikes, <2cm. Spikelets elliptic, c.1 x 0.5mm, 2-flowered; glumes c.0.8mm. Stamens 2. Stigmas 2. Nutlets c.0.9 x 0.5mm, elliptic, yellowish, minutely papillose.
Ecology. Wet ground by wadis. 50–1100m.
Distribution. Somalia, Kenya.

ETHNOGRAPHIC NOTES
Name(s)
See *K. brevifolia* (also uses).

Pycreus P.Beauv.

Annual herbs; stems triangular. Leaves grass-like, arranged in 3s; ligules absent. Inflorescence a panicle, 20 to 30-flowered. Glumes 2-ranked. Flowers bisexual, perianth absent. Style branches 2. Stamens 1 or 2. Nutlets flattened. 80 species. Mainly tropical and subtropical regions.

Botanical notes. Sometimes treated as a subgenus of *Cyperus*.

Pycreus pumilus **(L.) Nees**
Cyperus nitens Retz.
Cyperus pumilus L.

Small, tufted annual to 9cm. Leaves flat, scabrid on margins; leaf sheaths grey to purplish with reddish nerves, wings with reddish spots. Inflorescence of 1 sessile and 1 or 2 stalked spikes; bracts 2 or 3, leafy, erect. Spikelets 2–10mm x 1–2mm, linear-lanceolate. Glumes ovate, <1.2mm, greyish to reddish brown with a green 3-nerved keel, distinctly mucronate; glumes imbricate when young, but widely spaced later. Nutlet <0.6mm x 0.4mm, obovoid, biconvex, minutely tuberculate, apex apiculate.

Ecology. By running water in date gardens. 50–200m.
Distribution. Widespread in the tropics.
Botanical notes. Only recorded from date gardens near Hadiboh, but possibly overlooked elsewhere.

ETHNOGRAPHIC NOTES
Name(s)
– Soqotri: *ḳórźʌ̓*.
– Soqotri simplified: *qordha'*.
Food (livestock). Grazed by all livestock, though goats usually avoid them if they are growing in damp ground. Camels and donkeys graze the plants year round, cattle usually only in the dry season. They are considered to be especially important to small stock, and are seen as nourishing feed when young and green.

DIOSCOREACEAE

Twining herb with a large underground tuber, dioecious. Leaves alternate, ovate-cordate. Flowers unisexual, in axillary spikes. Perianth 6-lobed, in 2 whorls. Male flowers with 6 stamens, rudimentary ovary present. Female flowers with inferior ovary, ovary 3-celled, hairy, with axile placentation; styles 3, staminodes present. Fruit a 3-winged capsule. 8 genera. 630 species. Mostly tropical and warm regions.

Dioscorea L.

See description of family. 1 endemic. c.600 species. Tropical and warm regions.

Dioscorea lanata **Balf.f.**

Twining or scrambling perennial herb, with a large tuber; stems flexuous, to 3m. Leaves cordate, 3–20 x 2–10cm, 7 to 9-nerved, hairy. Flowers in axillary spikes, male flowers in groups of 2 or 3, female solitary. Perianth cream, c.1.5mm long. Capsule winged, c.1.5cm long.
Ecology. Semi-deciduous woodland in mist-affected areas of Soqotra. 250–1100m.
Distribution. Endemic.
Conservation status. Least Concern (LC). Common in several vegetation types and under no present or perceived threat.

Botanical notes. Leaves appear in winter but die back rapidly at the onset of summer. Readily distinguished by its large, heart-shaped leaves, long trailing stems and 3-angled fruits.

ETHNOGRAPHIC NOTES

Name(s)

– Soqotri: (tuber) *ṣīfəṭ* (pl. *ṣīfʌṭ*); (with aerial parts) *məzūwkən* 'twining, clinging'.

– Soqotri simplified: (tuber) *ziyfit* (pl. *ziyfat*); (with aerial parts) *mizuwkin*.

Food (human). The most important wild food of Soqotra. Indeed, after dates, for those with little access to cereals it was the most important non-livestock food in former times. It is particularly valuable because the tubers are at their best in the hungry summer months. Also, the tubers develop whether there have been good winter rains or not, so are available in most years. They were so highly valued that they were usually considered to 'belong' to those in whose clan or tribal area they grew. Someone digging them up outside his own territory, without having first sought permission, risked getting into trouble. The tubers can grow to a great size, are extremely filling, and their taste is much liked. The greenery, which dies back very quickly, is not edible.

The plants are found growing between the roots of shrubs and trees, and on steep slopes growing beside rocks. Although present year round, the tubers are at their best and biggest in mid to late summer. Towards the end of the monsoon they begin to shrink and dry out, and by the beginning of the winter rains have become shrivelled and tasteless. The tubers can be eaten raw but their juice irritates if it touches the skin. Usually they are cooked. They are either baked in hot ash, or cooked in a very small amount of water in a tightly covered pot, so that they are steamed rather than boiled. They are always peeled before eating. If there are only a few, they are generally baked. Islanders say a person can survive on a diet of nothing but these, but they can be quite indigestible and cause terrible flatulence. Those found growing in deep pockets of soil in limestone are particularly tender and flavourful, and are called 'brains' – a favourite part of the carcass – because these are more easily digested.

Herders would mark the plants as they moved around their area with the animals, and then, in the summer months, would return to dig up the tubers, sometimes staying away for two days or more until they had filled their skin bag. The tubers then had to be cooked, as they do not keep for more than four or five days once they have been dug up. There are many favourite recipes for cooking the tubers, but a common way is to put the unpeeled tubers into a large clay pot, and sprinkle them with a little salt, if available. There is no need to add more than a very little water to cover the base of the pot as the tubers exude a watery liquid themselves. The tubers are then covered with a thick layer of greenery, such as bunches of *rihiśt* grass, foliage from the shrubs *Rhus thyrsiflora*, *Sterculia africana* or *Ficus vasta*, or leafy stems of *Commelina* species, *Ipomoea* species, and, especially in the Haggeher mountains, *Pulicaria elegans*. It is common to place a layer of greenery in the bottom of the pot, underneath the tubers, as well. The contents are then weighed down with a stone, and the pot put to simmer – the longer the cooking, the more digestible the result. The tubers are then peeled and eaten, either hot or cold. They do not keep well once cooked. The hot tubers are sometimes mashed with ghee, fresh butter or soured milk (and now sometimes with sugar) as a treat, but the tubers are more usually eaten cold.

Food (livestock). The aerial parts are liked by both cattle and goats. The plant is usually more accessible to goats, for whom it is an important and nutritious fodder in the rugged areas where it grows. Sheep also eat the greenery if they come across it.

Flavouring / food additive. Before the arrival of sugar, the tuber was regarded as the best and most filling substitute for sugar.

Medicinal / veterinary. The tubers are given to anorexic or pregnant women suffering from morning sickness. A cow losing weight towards the end of her pregnancy is hand fed on the greenery to build up her strength.

Commercial value (past, present). The tubers used to be sold in the coastal markets.

Other. People eating this food can go for longer without water, as can animals grazing the foliage.

Notes. The importance of the tubers in the Soqotran culture is reflected in the extensive vocabulary used to describe their differing size, shape and maturity, and where and how they grow.

Comparative uses. The tubers of many other members of the yam family are edible. All need lengthy cooking, but some are also potentially poisonous and need special preparation to make them safe for eating. The tubers of many species are rich in saponins and are used for washing wool, silk, and the hair. Roasted tubers are used as a substitute for, or adulterant of, coffee. The juice of yam tubers generally has a numbing effect on the skin, and medicinally, the rind and fresh sap of certain varieties is a rubefacient and is used as a vesicatory. The scraped pulp of various yams is applied to burns, yaws and other sores, and is also used as a fish poison and to kill lice. The tubers of some species in Uganda are boiled, soaked in water, and the liquid fed to cattle to treat symptoms of blood in the stools. A tea made from the roots of *D. villosa* is taken to relieve the pain of childbirth. (Ambasta 1982; Dalziel 1948, 1956; Davis *et al.* 1983; Jain & De Fillips 1991; Kunkel 1984; Lewis & Elwin-Lewis 1977)

DRACAENACEAE

Evergreen trees with umbrella-shaped crowns. Leaves in dense rosettes, linear with an expanded base. Inflorescence a panicle. Flowers bisexual, actinomorphic; tepals 6, united at base. Stamens 6, attached to the perianth tube. Ovary superior, 3-celled; style simple. Fruit a berry. 2 genera. c.130 species. Macaronesia and tropical Africa, Asia to N Australia.

Dracaena Vand. ex L.

See description of family. 1 endemic. 60 species. Old World tropics.

Botanical notes. There are relatively few species in the genus with distinct trunks and these have a remarkably disjunct geographical distribution: *D. draco* and *D. tamaranae* in the Canary Islands, Cape Verde Islands, Madeira and Morocco, and *D. cinnabari*, *D. ombet* and *D. serrulata* in NE tropical Africa, SW Arabia and Soqotra. This pattern of distribution suggests the great antiquity of the genus and indicates that it is a relict of the ancient flora which surrounded the Mediterranean basin during the Miocene. For biogeographic notes see Marrero *et al.* (1998).

Dracaena cinnabari Balf.f.

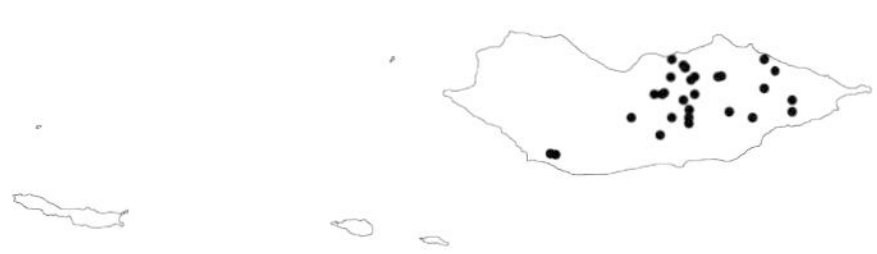

Large, single-trunked tree to 10m, repeatedly branched above to form a dense crown. Leaves densely tufted, dark green, sword-shaped, rigid, 30–60 x 2.5–4cm. Flowers in large terminal panicles. Perianths cream, c.6mm across. Fruit globose, 1–1.5cm diameter, ripening red to black.

Ecology. Common and often abundant on the granite of the Haggeher mountains and adjacent limestone plateaus (for instance at Diksam, Reyged, Rewged and Firmihin) where it is frequently dominant in evergreen and semi-deciduous woodland. There are also important outliers at Homhil and Hamaderoh and the Igalis plateaus (the raised plateau areas in between the major wadis running south from the Haggeher in the centre and east of the island). Over much of the eastern and southern limestone plateaus it is less common, occurring as small relict populations or as isolated trees. It is completely absent from the western end of the island and sparsely distributed or even absent from large areas of the eastern Haggeher. This pattern of distribution (with the exception of the eastern Haggeher and high areas in the western limestone) closely matches areas receiving frequent low cloud, rain and drizzle during the monsoon. 500–1550m.

Distribution. Endemic.

Conservation status. Vulnerable (VU B2 a b). This flagship species is fairly widespread but has a fragmented distribution with different subpopulations showing varying degrees of vigour: in the granite Haggeher mountains (particularly in the centre and western parts) *D. cinnabari* is abundant at higher altitudes, with populations showing good age structure and active regeneration (particularly in sheltered gullies and in the dense woodland amongst the granite pinnacles); on the limestone plateaus adjacent to the Haggeher (e.g. Rokeb di Firmihin) the woodland has a good age structure but shows no sign of regeneration; on certain outlying sites on the eastern limestone plateau (e.g. Hamaderoh) there are no signs of regeneration except in places inaccessible to livestock – these woodlands appear in good health but consist almost entirely of mature trees; finally, over large areas the tree is absent except for small relict populations (Iksha) or as isolated individuals (Qatariyah).

The current distribution of *D. cinnabari* suggests that the species was formerly widely distributed over large parts of Soqotra but has declined in recent times. This seems unlikely to be a result of changes in grazing practices. Local informants and the historic record suggest that there has been no change in this over the last hundred years and in any case the leaves and fruit cause scour if fed in more than small, carefully controlled amounts. The market for the resin has declined over recent years and the use of the growing leaflets for cordage has almost died out with the introduction of nylon rope. Thus it appears that *Dracaena* on the whole is currently at little risk from overexploitation and it seems that grazing is a factor preventing regeneration only when populations are already under stress because of increasing aridity. Regeneration is problematic in relict populations that lack a dense understorey of shrubs, which provide protection for young plants in healthy woodland. Herders who feed their cattle on *Dracaena* fruit report observing seeds sprouting in areas where there appears to be no active regeneration but these young seedlings do not survive the subsequent dry season.

Islanders have observed that in areas where *Dracaena* is failing to regenerate younger trees are not branching out to form the typical umbrella-shaped canopy, but are producing taller trunks instead. The exception to this is the high Haggeher mountains, where what are seen as 'normal' young trees are still common. Islanders offer various possible explanations for this. The trees do best in the areas affected by what they call *ʿilihil di ḥorf*, that is the mists, low cloud and, in particular, the constant drizzle of the monsoon. Indeed, they regard the trees as constituting a natural indicator of this microclimate. Oral history reports that the extent of this area has decreased over past generations, and this is one reason for the trees failing to flourish. Oral history also reports that in earlier times the crowns were frequently completely covered in flowering spikes, whereas now it is more common for only part, or one side, of the crown to have flowering spikes, and these are less productive. Another contributing factor is believed to be the increased feeding of livestock with the flowering spikes and fruit. This is due to increased livestock numbers, the high cost of supplementary cereal feeds, and the pressure of the years of drought in the 1970s and 1980s. This sometimes resulted in the complete stripping of the fruiting stalks from trees. Herders say that if too few fructescences are left on a tree year after year, the tree begins to die back.

The best *Dracaena* woodland occurs in Rokeb di Firmihin and the northern end of the Igalis plateaus which run from behind the Haggeher mountains down to the southern escarpment. After the Haggeher mountains themselves, these are the areas most affected by the monsoon clouds and mists. Islanders also say that the best trees occur in areas where the terrain consists of solid rock pavement with extensive cracks, down which water and soil flows after rains, providing moisture and nourishment for the roots of the trees. It is not difficult to produce seedlings from seed: cattle herders even in the drier areas have observed that when they feed the ripe fruit to their cows many seeds sprout later in the rains and cloud of the winter season. However, as soon as the cloud lifts and the sun comes out, they are burned off, or, if the seed has sprouted in areas of herbage, they are grazed down along with the grasses and herbs.

It seems likely that a major cause of the decline in extent and quality of *Dracaena* woodland is the very gradual drying of the Archipelago. No direct palaeoclimatic data are available from Soqotra, but there is strong circumstantial evidence that the islands have been drying over the last few hundred years. Local expertise describes *Dracaena* as occurring only in those high plateau regions affected by the mists and drizzle of the monsoon season (a period when the rest of the Archipelago suffers strong and desiccating winds). Oral history and tradition reports that not only has the extent of this drizzle/mist coverage diminished over the years, but also its duration and continuity: what used to be a predictable four to five months of monsoon cloud and drizzle followed by the rains of winter in the area (resulting in the total evacuation of both people and livestock from the region down to the lower hills) has become instead patchy and discontinuous.

Other potential threats to the long-term survival of *D. cinnabari* are the overexploitation of dragon's blood, a resin obtained from the bark of the tree. At present there is little demand for the resin but an increase in demand could lead to rapid overcollecting. The almost complete eradication of the closely related *D. draco* on the Canary Islands has been linked to overexploitation of the trees for dragon's blood in the Middle Ages (Lucas & Synge 1978). Another threat is from the new use of the trunks to make beehives. There was recently a case when over 20 trees were felled to make beehives for export to the mainland. This incident was generally condemned on the island but illustrates how a breakdown in traditional practices poses a very real threat to species on the islands.

Botanical notes. *Dracaena cinnabari* – the dragon's blood tree – is perhaps the most famous and distinctive plant on Soqotra. It is a major component of the landscape and is important not only as a flagship species (a rallying point for conservation awareness) but also as an indicator species (a species which flags changes in biotic and abiotic conditions) and as an umbrella species (a species which if protected will bring other species under that protection). It comes into flower in *ḳe̍yʌṭ* (i.e. February) and takes five months from flower to fully ripe fruit, in this resembling the date palm. Indeed, many islanders call *Dracaena* the date palm of the Central Highlands.

ETHNOGRAPHIC NOTES

Name(s)

– Soqotri: *aᶜarhīyib* (pl. *aᶜarhīyub*); fresh resin *emṣo̍lo*; processed inner bark and resin *ˀīˀdəḥa̍ˀ*; leaves *ḥa̍nžəher* (*di aᶜarhīyib*) 'knife', (*di*) *ša̍ᶜaf*, *śkəˀa̍mo*, *eṣəlīᶜo*; dead, dried leaves *ša̍ᶜaf*; fruit *šəne̍bher*; chunks of dried bark *se̍ned*; leaves in the growing heart (for cordage) *śkəˀa̍mo d-il ᶜa̍yn*.

– Soqotri simplified: *a'arhiyib* (pl. *a'arhiyub*); fresh resin *emzoloh*; processed inner bark and resin *'iy'diha'*; leaves *hanjiher* (*di a'arhiyib*), (*di*) *sha'af*, *shki'amo*, *ezili'oh*; dead, dried leaflets *sha'af*; fruit *shinebher*; chunks of dried bark *sened*; leaves in the growing heart (for cordage) *shki'omo d-il 'ayn*.

Food (human). Within living memory people have not eaten the fruit. However, there is a tradition in the hungrier areas of the island that people once gathered the red berries, boiled them in water and ate them at times of hunger.

Food (livestock)

Leaves. With the exception of camels, all livestock, including donkeys, eat fallen, dried leaves in hunger, and herders gather these as feed for their sheep, goats and cattle in a prolonged dry season. In a bad drought, goats have also been observed chewing the leaves of young plants if they can reach them. At such times, herders sometimes gather green leaves, crack them open and chop them roughly to feed to their goats. However, eating green leaves causes severe scour in goats, so this is a desperate measure, and goats normally avoid eating them. Dried leaves do them no harm, and, while they are regarded as having little nutritional value, they are seen as a useful subsistence feed. However, if goats eat a lot of them over a long period, they are more likely to develop scour once the rains fall. In a very bad year herders occasionally cut down a *Dracaena* tree and let the leaves die back as feed for their animals. Cattle owners feed their cows on the inner fibre of the leaves as a severe drought feed; this is less harmful than it is to goats. Cattle have not been seen eating the green leaves of their own accord.

Flowers. All livestock eat the flowering spikes except donkeys. Rarely, due to adverse weather conditions, the flowers fall before they set fruit. This is disastrous. In the bad drought year of 1999 few *Dracaena* of the Central Highlands came into flower at all. In areas where this is an important livestock feed, herders ask anxiously whether the trees has set fruit, just as they do with date palms.

Fruit. The fruit are one of the most important dry season feeds in the areas where these trees grow. However, care has to be taken when using these, since too many can make the animals sick. The skill lies in getting the balance right and judging exactly how many fruit can safely be fed to each individual animal. Other factors have to be taken into consideration, such as what other grazing and browse is available, or can be gathered and brought to the animals; how much water is available and at what distance; the condition of the individual animal, and how distant the winter rains are.

The trees come into flower in early summer, around February. This means that they are fruiting through much of the dry season, just when other fodder is scarce – hence their importance as a feed at this time. From flower to fully ripe fruit takes five months (just like date palms in most of the island). These trees are commonly called the 'date palm of the high mountains' and much of the specialist terminology used in describing the two trees is the same. As in date palms, there are considered to be different varieties of *Dracaena* that produce different kinds of fruit and foliage. The flowers give way to small green fruit (*fi̍ˀhu*, *fo'ho*), which then grow and begin to darken. If there are summer rains, these then swell

and darken further, when they are called *ḥīləh* (*hiylih*). These are the best fodder for livestock. Early in the monsoon months, the fruit begin to change to red, when they are called *ḳārīni* (*qariyni*). Finally, they ripen fully to a dark red, when they are called *bišərəh* (*bishirih*) (like ripe dates). They then drop to the ground. Once the fruit begin to redden, they are dangerous for most livestock.
– *fo'ho* fruit: can safely be eaten by goats, and are the best fruit for cattle. However, they are very small and it is generally best to leave plenty on the trees to become *hiyleh*.
– *hiyleh* fruit: with the flowers are best for goats as they are the least harmful to them. Cattle can eat enough of these to fill a head shawl, but no more.
– *qariyni* fruit: if goats eat too many of these they fall ill, and the meat, fat, viscera and bone marrow of the carcass is also stained red. In extreme cases they become inedible. Cattle can eat enough of these to fill a head shawl, but no more.
– *bishereh* fruit: as above for goats. Cattle can be given a double handful of the fruit only, or they fall ill.

For all livestock the safest fruit are green, followed by the black ones. The danger is greater if there is little other forage available for the animals. In general, in the Central Highlands where the trees are most common, the fruit are regarded as good for cows, but a risky feed for goats. They are seen as relatively safe in the summer months and early monsoon, but becoming progressively more dangerous as the monsoon progresses and the fruit redden and ripen. The danger lies in the seeds – it is the fruit with mature seeds that cause the problem. Herders say that these hard seeds collect in the digestive tract and cause blockages and other stomach problems. This is why the greatest use of the feed is in the summer months and early monsoon, when the fruit are still immature and there is still dry grass, herbage and other browse available for their animals. In the plateau areas, if the summer rains have failed, less use is made of the feed, because other forage will soon be exhausted.
– *Haggeher Draceana.* Trees growing in the Haggeher mountains proper (i.e. only the relatively small area of granite pinnacles, enveloped in cloud for much of the year) are not considered to produce toxic fruit. The berries of these trees are regarded as 'watery' and safe, compared with those of the limestone plateau below, which are 'dry' and dangerous. In periods of drought, goats and cattle can be herded up to the Haggeher to feed on the 'watery' fruit of these trees quite safely. They are completely safe for all livestock, regardless of how much they eat. This is explained as being because their seeds never become dry and hard, but remain small and soft. The Haggeher trees produce little or no *emzoloh* resin, and their bark is no good for processing into *'iy'diha'*, but they fruit reliably. Their fruit tend to be large and fat, growing from much shorter, upright stalks than those of the trees below, which are considerably longer and more lax. The leaves of the Haggeher trees are short and dense, while those of the limestone plateau are longer and grow less closely together. The islanders say that the trees of the Haggeher differ because they grow in constant moisture, in the 'sweat' of the cloud and mists which envelop the cliff faces where the trees grow.
– *Harvesting Dracaena fruit.* Herders climb the trees to gather the fruit at the *fo'ho* or *hiylih* stage. The fruit are stored in sacks or leather bags and will keep like this for one to two weeks. The riper *qariyni* and *bishirih* fruit cannot be stored in the same way, and keep for two or three days only. Herders go up from the lowland grazings to collect a few bags of the fruit from the trees of the high plateaus, and will give their animals a handful every three or four days. However, in areas which are heavily dependent on small stock rather than cattle, and which have only a few trees, herders sometimes climb the trees and remove all the young fruit to prevent them ripening and falling to the ground and making their animals sick. This lopping of the unripe fructescences was much more widely practised in the past, partly because herders were almost entirely dependent on their stock for their survival, and partly, they say, because in earlier times, unlike today, the trees produced such large numbers of fructescences that the fallen fruit became a real problem. (Indeed, today it is noticeable that the trees that produce the greatest numbers of fructescences are the very tall trees, which are the hardest to harvest.) Herders usually harvested the fresh *emzoloh* resin that emerges from the leaf bases in the crown of the tree at the same time as removing the fruiting stalks. They have observed that if all the fructescences are removed from a tree it produces less the next year and appears 'weaker'. However, if left untouched for that year, it usually recovers.

The trees are seen as being the property of those to whom the land belongs. When someone from outside the area wants to harvest some fruit for his animals, he normally asks permission and is given a tree, or a number of trees, to harvest. He is given trees distant from settlements or from goat or cattle pens whose inhabitants have prior rights to any nearby trees. If the harvester is seen to damage the leaf rosettes by removing leaves to make it easier to harvest the flowering or fruiting sprays, he is cautioned.
– *Goats.* Goats have an absolute passion for the ripe, red fruit. The main danger comes from them eating the ripe fruit when they have dropped to the ground. If they eat a lot, they can die. Goats in good condition lose it dramatically on turning to a diet of *Dracaena* fruit. However, if they are very thin, and are then fed on *Dracaena* fruit, even the red ones, they initially put on weight and look well. However, if they are fed on the fruit continuously for a month or so, and have little else to eat, they lose their appetite and sicken. In areas where there are few *Dracaena* trees, and goats are not accustomed to this feed, the fruit are seen as even more dangerous. Goats are herded away from the trees when the fruit are ripe and are fed the fruit a handful at a time. For instance, even in a prolonged dry season, a mature goat in poor condition will be given only five or six berries at a time, in the same way she might be fed individual dates. Even a healthier goat will be given no more than a palmful, some 10 to 15 berries.
– *Sheep.* Sheep seem able to manage the fruit better than goats. However, if they eat a lot of the fallen, ripe fruit when they are very hungry and have little other forage, they too fall ill, sometimes fatally. Both goats and sheep love the fallen fruit, and the problem of small stock falling ill from gorging on these has increased over recent years, as herders spend much less time actively managing their animals.
– *Cattle.* A key dry season feed for them wherever these trees grow, particularly for cows in milk. If adequate dry forage is available as roughage, the fruit makes a cow fat and healthy, and increases its milk, especially a young cow. If fed carefully, when the rangeland is not exhausted, the fruit can keep a cow in milk until the winter rains fall. If fed the fruit with inadequate roughage, there

is a delay before a cow starts to give good milk once the rains fall. An adult cow can safely manage a circular mat full of the green or black berries, or two to four sprays of fruit, at one feed, her yearling calf half that amount. If there is no standing hay left, or other herbage, this amount must be cut down. For example, in the drought of 1999 when the summer rains failed, a man on the east coast fed his two cows first on green and then on red berries throughout the summer months. When the winter rains finally fell, both the cows died. Many cattle herders say that it is unwise to allow cattle to become too accustomed to this feed because it is so laborious to gather enough fruit to keep them going. Nevertheless, some herders still use the fruit as a supplementary dry season feed, as others now use barley or wheat, and consider it to be just as good, and it costs nothing.

– *Signs of illnesses resulting from overeating ripe Dracaena fruit.* The whites of the eyes become stained orange or red; the animal develops an unsteady gait, staggers and has fits of shivering; some are unable to pass faeces; pregnant goats are likely to abort; in cattle the liver is often swollen. The animals grow progressively thinner and small stock sometimes die.

– *Avoiding action.* At times when there is no standing hay or other graze or browse, herders of the Central Highlands generally remove their goats and cattle from the *Dracaena* woodland and drive them up to the Haggeher. Camels and sheep can usually be left below. In eastern Soqotra, herders usually move their livestock out of the *Dracaena* areas and keep them away until such time as the fallen berries have dried out. The last red fruit remaining on the trees generally fall with the first rains of winter.

Bees / honey. An important forage for honeybees, especially in the summer when *Dracaena* are in flower. The honey is very 'white' and is highly thought of. If there is absolutely no trace of bitterness in honey, then it is likely to be from *Dracaena* flowers. Bees also nest in a hollow *Dracaena* tree. If this is noticed, the tree is left alone for a year and then three holes are bored above the entrance hole to the colony in order to remove some of the combs which are stacked in layers inside. More recently, under the influence of beekeepers from the Yemeni mainland, *Dracaena* trunks have been cut and sold as beehives to the mainland.

Flavouring / food additive. Milk of livestock that eat a lot of the fruit has a distinctive taste, especially if the animal has eaten too many and become unwell as a result. The taste and colour of the carcass, especially the viscera, is also altered, though the brain remains normal. The viscera in particular are stained a rather disconcerting bright red, and in the stomachs lumps of imperfectly digested matter, consisting mostly of seeds, can be found. In extreme cases the meat has to be washed again and again in several changes of water to make it palatable. The bones usually have to be discarded as the marrow becomes inedible. This problem is particularly common in wild or feral goats, or domesticated sheep and goats that have strayed from the main herd and have been eating the fallen berries. In cattle the marrow becomes a bright yellow-orange and has a bitter taste. The water the meat is cooked in is turned the same colour and is also bitter, and sometimes undrinkable.

Fuel. The dried fibrous heartwood of a dead tree makes excellent tinder, and is gathered and stored in the home for this purpose. The dead wood makes little smoke and is suitable for cooking. The inner and outer barks are one of the preferred woods for firing pottery, usually with added cattle dung, as it colours the pots a good clear red if they are completely covered in the material. The thick bark of the tree burns with a very bright flame, and in the areas where these trees are common, some is usually kept at home for this. It also burns slowly, and pieces of resin-containing bark make a very good torch. It does not go out even in a high wind, and there are many stories of people travelling great distances at night with only these to light their way. The fruiting stalks can be used as fire-sticks, most successfully when both the vertical and the horizontal sticks are of the same material.

Timber. In areas where these trees are common the dead leaves are gathered and used like date palm leaves to thatch temporary or seasonal housing, livestock byres and storehouses. The leaves are tied down with *Dracaena* cordage, or held down by rocks or branches. If the weather looks uncertain, the thatching of leaves is usually covered with a layer of earth. *Dracaena* thatch is said to be very warm, and if well made, relatively waterproof. It usually has to be replaced every year however. The larger branches of dead trees are cut in half longitudinally, the fibrous material inside removed, and the curved shell used to make guttering placed along the edge of the roof, with a gap for the area of roof above the main entrance door. A well-developed tree can be made into a temporary, seasonal milking yard. The herder lights the smudge in the shade of the tree and herds his animals there in the morning. After a time, they become accustomed to this, and come of their own accord as soon as they see or smell the smoke of the smudge.

Fibre / cordage. The leaves yield a fibre that is worked into cordage of all thicknesses. They should be gathered from an infertile tree, preferably one with long leaves. Only a single leaf at a time should be plucked from the heart of each rosette, or the tree's growth will be adversely affected. Ideally the leaves should be left to ret down in water for a week or more until the hard, outer casing has cracked open. This is then peeled off and the fibre inside is removed. It is then washed, dried, and twisted to make a string and then plied into whatever thickness is required. Cordage can also be made from the leaves without prior soaking in water. For this, the outer casing is beaten to crack it open, the interior fibre is extracted, spread out to dry, and then beaten to soften it. It is then usually wound around a stick to straighten and order the fibres, and is then worked. The resulting cordage is not thought to be as strong and long-lasting as that made from soaked leaves. *Dracaena* cordage is amongst the best on the island, equal to that made from the fibrous material around the leaf base of date palms. It is used for the heaviest work: to secure baggage in pack animals, for camel and donkey harnesses, for restraining camel and donkey stallions in rut, for raising and lowering well buckets, and for manoeuvring lintel or sill stones or other heavy objects. The ropes are used to lower people over a cliff edge to collect honey, water and fodder, or to reach the ledges where the nestlings of edible seabirds lie. It is used for climbing into the crowns of trees to gather foliage, or to climb into date palms to pollinate them and to work on the dates. However, it is said to be of little use for sea work as it cannot take prolonged exposure to wet. *Dracaena* fibre was also used to make the long and narrow nets that used to be used to trap feral donkeys and wild and feral goats. The strong but fine twine was rolled into balls and when enough of these had been made, they were knotted in the same way as a fishing net. The finished net was supple and flexible, and of contrasting dark and reddish-white colours, making it practically invisible when strung between trees.

Wood. The *dəbɔ̄bəh* (*dibobih*) drums, played mainly by the Soqotrans of African origin in the coastal areas, were made from a hollowed-out section of a *Dracaena* trunk. The wood was left to dry out and then the fibrous core was removed, either leaving both ends open, or one closed, depending on the type of drum being made. Cured sheepskin was then stretched to cover one end, if the drum was to be played sitting down, or both ends if it was to be played standing up. Then a sounding hole was bored in the middle of the drum. These drums were popular at all celebrations, but there is little demand for them now. The wood was also used to make the measure used by ghee traders and merchants. It was lined with a layer of melted *emzoloh* resin to make it watertight, as *Dracaena* wood is rather porous. The traders used this to measure out the ghee when transferring it from the clay or skin containers of the livestock owners to their imported metal containers. When full, these were sealed with lead and exported.

Medicinal / veterinary. It is nearly always the fresh resin lumps, *emzoloh*, which are used medicinally. A piece is crushed to a powder, a little water added, and rubbed over swellings or areas of inflammation. The same treatment is used for sores and rashes, spots and boils. Acne is treated with dried, powdered *emzoloh* and the powder, with or without added water, is applied to the skin to relieve itching. To treat multiple sores all over the body, usually in children with insect or flea bites that have turned septic through repeated scratching, the scabs are removed, the skin washed with a salt and water solution, and then the powder applied. Sores all over the head, again common in children from scratching at head lice, are treated by cutting the hair or shaving it off, washing the scalp with a salt and water solution, and then applying a paste of crushed *emzoloh* and water.

If a broken bone has pierced through the skin in a compound fracture, the wound is packed with dry, very finely ground *emzoloh* before being strapped. In cases of burns, the layer of burned skin is removed, any blisters lanced and the dead skin cut away, and then a paste of powdered *emzoloh* and water is spread over the burned area. The same treatment is used on burns caused by medicinal cauterisation.

Lumps of fresh, still soft resin are chewed to promote oral health, as a general tonic, to stimulate the appetite and as a remedy for a variety of stomach complaints.

The resin is much used in eye complaints, in livestock as well as in people. The finely powdered resin is mixed with water and used as eyedrops to treat sore or inflamed eyes, or as a dressing after an eye operation. It is mixed with powdered antimony and applied dry, or along the rims of the eyes, like antimony, to soothe itching and to treat conjunctivitis. The same mixture is used to promote good eyesight or to treat poor vision. Some believe the powdered *emzoloh* is more effective if added to goats' milk or breast milk first and then used as eyedrops. Infectious conjunctivitis, a particular problem in the monsoon months when the air is filled with windblown dust, is treated with ground powder added to water and used to bathe the eyes at regular intervals. Small abscesses in the rim or inner corner of the eye, and corneal opacity, are treated with eyedrops made from finely powdered *emzoloh* with added water applied daily until the condition improves. Corneal opacity with reddening of the eye, itch and tears, and inflammation of the eyes with corneal thickening and opacification, are treated in the same way, sometimes with added powdered antimony.

Skin tags around the eye, especially in livestock, are surgically excised and the skin then treated with powdered resin, usually mixed with fresh milk. In camels, corneal ulcers are treated by applying powdered resin mixed with the thick milk of a goat at the end of lactation. Cattle and goats with infected and inflamed eyes are treated with an application of dry powdered resin mixed with powdered salt. Small stock suffering from loss of sight, or poor sight, are kept in the byre out of the light, hand fed on rich fodder, and treated with eyedrops of the ground resin mixed with hot milk. This treatment is carried out on alternate days until the condition has improved. Poor sight in livestock in general is treated with an application of dry powdered resin.

The powdered resin is also taken to treat stomach pains. For sharp or colicky pains, the resin must be taken on an empty stomach and nothing else taken until the treatment has had time to work.

The resin is also used as a haemostatic. The powdered resin is added to hot milk and drunk for post-partum haemorrhage or for any other abnormal, heavy bleeding, internal or external. The powdered resin can also be added to hot water or hot ghee. For post-partum haemorrhage the patient should lie flat on her back while taking the treatment, which should be given at least four times. The powdered resin is added to water and smeared over fresh lacerations to stop the bleeding.

The resin appears to have oxytocic properties and is given, powdered and added to hot milk or water, to a woman with a retained placenta to drink. The resin must never be taken internally by a woman in early pregnancy – up to the fourth month – as it can cause abortion. The fruit are used to treat sickly livestock. Goats and cattle which are failing to thrive, are sickly or are recovering from an illness are hand fed a few of the fruit, as are cows losing weight towards the end of their pregnancy to help them to regain their strength.

Bedding, stuffing. Dead leaves are used as bedding material, the best being those that have been dead longest.

Cleansing. If a tree has large roots, the bark stripped from these is aromatic when burned. Some of the thick resin-bearing bark also makes a pleasant smell when burned. The prepared *'iy'diha'* bark can be chewed like a chewing gum to sweeten the breath and strengthen the gums.

Cosmetic. Both the fresh resin and the prepared *'iy'diha'* bark can be chewed to stain the gums, lips and tongue red. Women grind the hardened lumps of fresh resin to a powder and apply it to the face. They leave it on for one or two hours and then wash it off. This makes the skin soft and smooth. The resin is also used in this way to disguise moles and freckles, or the scars of acne. To look their best, women also mix the ground resin with powdered turmeric and a little water, and spread this paste evenly over the face, neck and bosom and leave it on for quarter to half an hour. Then, using a razor or sharp knife, they shave off the coating, removing at the same time all the hair, even the finest, from the face, and also the hair around the hairline, especially at the temples, to shape the hairline to produce the desired broad forehead. They then wash off any paste left and, if they have enough turmeric, apply a thin dusting of this to the face and bosom. Children use fresh resin to stain their fingernails or dot their fingers and palms with patterns like henna. More recently it has been discovered that by spraying a lump of the resin with perfume-spray the surface of the resin dissolves and can then be applied to the lips as a lipstick, or to the nails as a nail varnish. This is mostly used by girls dressing up.

Commercial value (past, present). *Fresh emzoloh resin.* The pure droplets of soft resin, called *śa̍ḥaz di emṣo̍lo*, are gathered from a flowering or fruiting tree in early summer (approximately January to March), and in particular from a 'female' tree. The best droplets are gathered by climbing right into the crown of the tree, and picking off the small lumps that ooze from the base of the flowering/fruiting stalks, along the leaves growing along these stalks, and around the base of the leaves. Those that have already dried out and hardened are knocked off with a stone or knife. The droplets sometimes fall by themselves to the ground and can be picked up from there. Droplets can also be found beading the upper edge of the old scars on the trunks left by previous harvestings. Resin also oozes out when harvesting the bark. These oozings become soft and sticky in the heat of the sun, and then congeal into droplets which are collected and sold with the good grade of *emzoloh*.

Dried fresh resin. This is generally called *emṣo̍lo* (*emzoloh*). There are different grades: (1) *emṣo̍lo ga̍ˁalhil* 'round balls of *emzoloh*' or *fuṣūṣ* 'essence of *emzoloh*': these are dried droplets of pure resin – they are best for medicinal use, and are sold separately for the highest price; (2) *śaḥs di emṣo̍lo* 'slices of *emzoloh*-wood': these are pieces of resinous bark with droplets of *emzoloh* attached; (3) *ḥo̍yhi di emṣo̍lo* 'dust of *emzoloh*': these are fragments of broken-up droplets. The second highest price is given for a mixture of pure *emzoloh* and *śaḥs di emṣo̍lo*. Next in price is a mixture of *śaḥs di emṣo̍lo* with *ḥo̍yhi di emṣo̍lo*. The latter, and *śaḥs di emṣo̍lo*, are usually mixed together and sold separately at a lower price. The lowest price of all is given for *emṣo̍lo di īˀdəḥa̍ˁ*, made while preparing the cooked bark. As it cooks down and thickens, small pieces are removed from the pot and rolled into small balls. These are of least use medicinally.

Cooked, processed resin (ˀīdaˀ or ˀīˀdəḥa̍ˀ). The processed resin is not used locally, and there is no tradition of it ever having been so. Again, the best bark for making *ˀīdaˀ* comes from 'female' trees. That made from the 'male' tree is said to be a very dark colour and the cut bark on the trunk turns very black.

Harvesting the bark. The tools needed are simply a broad metal blade, or a piece of flint with a sharp edge, or a sharp-edged stone and a leather container for collecting the pieces of bark. The best knife for harvesting the bark is a short blade with a rounded tip called *mi̍ṣəhil* (*mizihil*). The leather container is made by cutting a small goat or sheepskin in half and tying off the legs to make two 'collecting buckets'. The bucket is dampened and then placed in a hole dug to the size required. The leather is smoothed round the sides of the hole until it sits snugly, and then it is filled with warm, dry gravel and left to dry out. As it dries it hardens, and takes on the shape of the hole. The gravel is then tipped out and the 'collecting bucket' is ready for use. This, like the harvesting knife, is called *mizihil*. The harvester holds the bucket in his left hand, chopping pieces of bark straight into it with his right. When it is full, its contents are tipped out into another larger skin container and the work continues.

Only specialist harvesters make the very first cuts into a virgin tree. Others harvest the trees with scars from previous harvestings. These are clearly visible, and, depending on the weather and age of the tree, appear either as areas of ridged, black and smooth wood quite different from the rest of the trunk, or as areas of bark oozing with small red droplets of resin.

The harvesters cut around the blackened bark of the earlier cuttings with the knife, like removing a scab from a sore. The same tree can be harvested every year, the same scar tissue being cut and removed each year, since the resin-bearing, red, inner bark regrows. However, it is said to be better to rest the tree for a year or more between harvestings. If a tree is left undisturbed, the area around the scar becomes swollen and full of resin (the resin in the bark is given the same name as the fresh resin droplets, *emzoloh*). The resin is understood as blood, flowing along the 'arteries' and 'veins' of the tree, just as it does in a person or animal. The aim in harvesting is to tap into one of these. The red inner bark alone carries the 'blood', so only this is removed. The pale inner bark, called *śʌbḥ* (*shabh*), does not contain resin, so there is no point in removing it, especially since this would harm the tree. Unscrupulous workers do cut the paler inner bark to increase their yield, because once crushed, mixed with the resin-bearing bark, and cooked, it is stained red by the other bark and becomes virtually indistinguishable from it.

The tree can be harvested at any season of the year since there is no season when the resin is of better quality or more copious. The only time they cannot be harvested is immediately after rain. An interval of two weeks or longer must be left for the trees to dry out before going to harvest the bark.

The first cuts made into previously unharvested or virgin trees are made by known experts who can recognise, after carefully inspecting the tree, where the 'arteries' of resin are likely to be. They then remove small pieces of bark at intervals along the 'artery', and the tree is then left alone for a year. When these cut areas have turned black, or have begun to ooze small droplets of resin, the blackened parts can be removed. Small amounts only are harvested in the first few years until it is clear that the tree is able to sustain the 'bleeding' and can bear heavier cutting.

Those who control the areas where the trees grow prefer all trees to be harvested at the same time. This is so that they can oversee the workers who come to join the harvest from outside, and make sure that they do not remove too much bark or the wrong kind of bark. Incomers are given a specific area, or a number of trees, to harvest. Usually they stay with a local family and work alongside local people so they can be supervised. If it is done carefully a single tree can produce as much as 45 *ratl* (about 500g on Soqotra) and still recover to be harvested again.

Processing the bark. At the end of the day the containers of bark are taken back to the settlement and sorted out. On one heap go the droplets of pure resin picked off the bark. While not as good as the droplets removed from the crown of the tree, this is still high quality resin. On another go the small fragments of resin, resin dust, bark debris and so on which are sold as low-grade *emzoloh*. On a third go the pieces of red inner bark, picked free of all droplets or fragments of resin, and any pieces of pale bark which might have been removed from the tree by accident. The pieces of red inner bark are left to dry out for a day or two and are then pounded on a rock with a smooth rounded stone until they have been transformed into a rough paste. This is put into old clay pots on a gentle fire (the stain cannot be removed, and generally the same pots are used year after year), and allowed to heat up slowly, being stirred continuously. If the paste is very dry, a little water can be added, but the results are better with none. The mixture must be stirred until it forms a bright red and soft mass. If the bark was not well ground, it might have to be strained at this point. The smooth, cooked paste is then tipped out onto an area of flat rock, and spread out with a knife or piece of wood to make a single thin layer. As it becomes cool enough to handle, the hot paste is cut up roughly into squares. A piece at a time is taken up and moulded into flat, oblong cakes like bread (and given the same name, *eṣhīro*). This must be done very gently, and while it is still warm, so that it will not stick to the fingers. The cakes are then laid out to dry. The square shape is the easiest to stack, and to pack into sacks for sale, though of course the pieces can be moulded into any shape.

The *ʾīdaʾ* is either made from the pure red inner bark alone, as described above, or additional *Dracaena* material is added. This might include red chunks of bark found lying on the ground, which are ground and added to the pot, or the small fragments of resin, resin dust, bark debris and so on which were separated out earlier, which are similarly ground and added to the pot. When harvesting the tree some harvesters poke the powdery red substance out of cracks and crevices in the tree, or hit the bark with a rock or heavy knife wherever there is a hole or crack, and collect the powder which comes out, this too is added to the pot.

Some prefer to collect all these various fragments together and process them to add to the lower-grade *emzoloh*. This is done in various ways. They can be cooked in a separate pot in the same way as the bark, spread out over a flat rock, and then, when cooled but still warm enough to manipulate, rolled into a thin sausage and cut across to make small pieces; or the cooked paste is spread out and allowed to cool and harden and is then broken into fragments with a stone or knife; or the bits and pieces are spread in the sun to melt and are then squeezed hard to make small lumps; or they are chewed to soften them and are then squeezed to make suitably sized fragments to add to the lower-grade *emzoloh*.

Generally, it is women and older people who harvest the resinous bark and younger men who climb the trees to gather the resin droplets. Access to the bark is often facilitated by the construction of a low wall beside the trees, or a rough stairway of rocks built against the trunk.

Marketing Dracaena products. The trade demand used to be for the processed bark rather than for the pure droplets. However, the demand for the processed bark is much reduced, while that for the pure droplets remains high, on the Arabian mainland and in India, as well as locally. Both were sold to visiting traders when they arrived in their sailing boats after the monsoon. However, when the merchant families of Soqotra began to own their own trading ships, they bought up the produce themselves and travelled with it to the mainland markets. By this time, the greater demand was for the pure resin. When Sultan Ali Abdullah Salih became the owner of a dhow, he and the Sultans that followed were keen to corner the *Dracaena* market, and became actively involved in the harvest of the *Dracaena* woodland behind Hadiboh in particular. They paid rent to the owners of the land and employed local people to harvest the *Dracaena* on their behalf in exchange for food. With the arrival of the Socialist Government in 1968 this came to an end. In other areas the processed bark and the resin were stored in dry caves until news came of the arrival of a trading boat asking for either. Alternatively, a local merchant, or the local representative of an overseas merchant, would send camels laden with cereals into the *Dracaena* areas and would trade this for the resin.

In areas where the trees are numerous and the trade in resin important, the harvest is still closely organised, the trees being harvested together and rested together. The trees are rested, sometimes for two years or more, since it is recognised that overfrequent harvesting weakens them. Although the period of rest is primarily to give the trees time to fully recover, it is also to some extent a ploy on the part of those who own the trees to keep the price of the resin high. In such areas there is little speculative harvesting carried out by individuals, as is the case in the areas where there are few trees. However, in recent years, with the increasing breakdown of traditional controls, there is much less organised harvesting of the trees. Anyone who can manage to do so gathers the resin droplets and sells them in the coastal markets. Some community leaders are trying to insist that traders take a certain amount of the processed bark as well as the pure resin, arguing that otherwise the older and weaker members of the community, who are unable to climb the trees to harvest the resin, lose out completely. Indeed, when the owners of the trees manage to organise themselves, the returns can be good: for instance, in 2000 an elderly woman made YR15,000 from processed bark, and a young man YR80,000 for the pure resin. Attempts are being made in the *Dracaena* areas to establish an agreement that no individual collect the resin droplets outside the agreed harvest season, so that they can all can sell at an agreed price together. Under this system, a trader would make an order in advance for so many pounds of resin and processed bark, secure in the knowledge that this would be delivered, at an agreed price, rather than, as now happens, having to buy up small amounts at different prices throughout the year until he has the amount required. However, in many areas it still happens that the droplets are gathered throughout the year, by those in whose areas the trees grow, as well as illicitly by outsiders. The collector can be sure of finding a buyer and, in these times of economic uncertainty, cash in the hand now is understandably preferred to a possible future payout as part of a co-operative. This continual and unregulated harvesting of the trees, combined with the increased removal of the fruit as a dry season fodder, may well be additional factors in the steady decline of the *Dracaena* woodland, even if the primary cause is climatic.

Pest control. In areas where the trees are plentiful the dead wood is used on livestock smudges. The pieces of blackened bark which are removed during the harvest are especially effective as an insecticide against the many species of biting fly that arrive after the winter rains.

Other. Sheep and cattle that eat the juicy fruit can go longer without water.

Children especially often plant seeds in the ground. They need a lot of water to start sprouting but after that they grow well until they come to a stop at around knee height. They remain as a single rosette, and never branch out or grow taller. The soil under a mature *Dracaena* can be collected and used to enrich soil in a garden, or used to plant out a young date palm.

Well-grown trees with spreading crowns give a dense shade, and are favoured meeting places for people. Herders often choose to site their living quarters close to such a tree and during daylight hours most household business is carried out in its shade. At times of celebration many communal meals are held under this tree, and livestock too cluster around the trunk to escape the midday heat. As a result there is often good grazing to be found under and around the trees.

The resin is used as a dye or paint. It is used principally on pottery to decorate pots which never go on the fire: water pots, serving dishes, storage pots, incense burners, the pots for serving coffee and so on. A lump of the resin is wrapped in a piece of cloth and used to draw patterns on the red-hot pots, held firm with a stick hooked across the mouth, as soon as they have been removed from firing. Small clay animals and tiny coffee cups are also decorated in this way. Some prefer to heat the lump of resin to melting point and use this quickly to make a series of dots or lines before it cools and hardens. In the more settled communities of the coast the resin has been used to decorate whitewashed stone houses, mainly around the doors and windows, though sometimes the inner whitewashed walls were also decorated. The resin, either lumps of the fresh resin or the processed bark, has to be heated to melting point. This 'paint' is applied by hand, or with a pad of cloth or of the fibrous material that grows around the leaf base of the date palm. Leather is also dyed with powdered resin, with or without the red powder made by grinding up a 'red stone', found mainly in the Central Highlands. The powder is rubbed by hand directly onto the leather to be dyed. It is used in particular on the leather of the sheath of the knife-set which was the most important personal possession of the Soqotran, or that of the intricately plaited leather belts men used to wear. The leather is then oiled. The hands are stained dark red as a result and the stain takes some time to wear off. Tradition says that the powdered resin also used to be used to dye pale or white wool, but exactly how this was done is not clearly remembered.

The resin is used to repair cracked or holed pottery. A cracked pot, common especially in the pottery made from the clay of the foothills of the northern plains, is firstly tied tightly around the neck with something not too flammable, such as a leather thong. It is then heated beside the fire, placing the cracked side nearest the heat. When it is hot, a lump of the resin is heated until it begins to drip. It is then wrapped up in a scrap of cloth and rubbed up and down along the crack, inside and out, dabbing blobs of the resin where the crack is widest. The tie usually has to be left around the neck, as the resin has a tendency to soften each time the pot is put on the fire. It is best used to repair a pot that does not go directly on the fire for any length of time, such as food storage pots, or the pots used to store ghee. If the pot is holed, a patch can be made from a lump of the fresh resin. The pot is heated at the fire, holed side towards the heat, and then a suitably sized lump of the resin, either fresh or the processed bark, is pressed into the hole and it is turned to the heat again. It melts and stops up the hole. Alternatively a shard of pottery is shaped to fit the hole, placed in it, and then tiny fragments of resin are wedged around the edges. A burning stick is brought up to the hole and the hot smoke blown over the resin until it begins to bubble and melt. It is then quickly smeared over any gaps, and the pot removed from the fire for the resin to cool and set hard. The resin can also be used as glue. It is stronger than the latex of *Ficus cordata*, but not as good as the latex of some *Euphorbia* species. However, it is much more difficult than these to handle, as once melted it hardens very quickly.

The crushed berries can be used to clarify ghee in the absence of cereals. The sediment of buttery fruit that settles on the bottom is eaten, and is a great treat.

The bright red resin is also used as a marking ink to mark property. Date palms outside the owner's tribal or clan territory, or growing in a communal date garden, can be marked on the trunk with the owner's brand mark (the same as that of his livestock). Skins of food left in someone else's safekeeping or in a communal store down on the coast could be marked in the same way.

Starlings and ravens both frequent *Dracaena* trees. Both love the ripe fruit, and starlings commonly make their nests in them. People used to climb up into the crown of the trees to raid the nests of their plump nestlings as food. Starlings in particular are famously fond of the ripe fruit, and their droppings are bright red and full of the seeds at that time. Herders have noticed that if the starlings are eating a lot of the red fruit and then drink water, they die as a result. Small stock that inadvertently eat a lot of the starling droppings when grazing under a *Dracaena* tree fall ill and some die as a result.

If a herder passes a goat giving birth on its own, out in the open and unattended, and has no time to sit with her until she had safely given birth, he or she lights a fire of the dead wood beside her. This is done to keep away the feral cats which may attack and kill a newborn kid or even go for the mother in her weakened state.

Notes. Many islanders distinguish between the 'male' tree (called *aᶜarhīyib ḳarṭat* 'with short, curled leaves', *miʾšer* 'breeding male', *meḳereš* 'sterile' or *ṭāhʌḳ* 'smooth, soft'), which produces no *emzoloh* resin, has short leaves and makes few fruit, and the 'female' tree, which is rich in resin, has longer leaves and flowers and fruits profusely. Traditionally it is said that if there is a very heavy crop of fruit then a year of drought will follow. Herders of the Haggeher say that if the *Dracaena* in the peaks of the Haggeher come into flower at the beginning of summer, then summer rains are likely to fall. In the star-calendar of Soqotra, the tree is said to be the symbol of the two stars, 'young Sud' and 'old Sud', occurring one after the other first in October and then again in May. Al Hamdani (d. 945) writes about Soqotra that it grows Soqotran aloes and many palms, and produces ambergris and dragon's blood, and Serjeant reports that the dragon's blood from Soqotra is mentioned by a Hijaz poet as far back as the seventh century. Balfour reports that the resin is made at two seasons: in the summer heat from March onwards and again after the winter rains. He notes that the large tears of *emzoloh* sold for 10 dollars for 2.5 pounds, the cooked 'dust' resin was 1 dollar for 4 pounds, and the cooked resin lumps were 1 dollar for 5 pounds. Bent reports about the 'collection of dragon's blood' that the drops are knocked off into bags. Unbroken drops are most valued, and called *edah amsello*. Then the nice, clean, broken bits are picked out, and called *edah dakkah*; the refuse, with bits of dirt, bark and leaves stuck in it, *edah*. This is made up into cakes with a little resin and sold very cheap. Visitors from the Ministry of Agriculture in 1967 noted that the price of the cooked resin then was three to five shillings per *ratl* (c.500g on Soqotra) and on another visit in 1988 that the gum was collected in the summer, June to September.

Comparative uses. Some *Dracaena* species are edible: the young leaves of *D. angustifolius* and *D. elliptica* are cooked and eaten; the core of immature leaves forming young shoots of *D. mannii* is cooked and eaten like asparagus; the tuber of *D. ellenbeckiana* is eaten; the flowers of *D. nitens* are edible and reportedly the fruit of *D. draco* are eaten. The resin of *D. cinnabari* is used in the traditional medicine of many countries. It is also used as a colouring agent in varnishes, toothpastes, tinctures and plaster, and as a red dye. It is sold in herbal shops as a drug called *dam al-khuwayn*, principally used as a haemostatic. Other *Dracaena* species are used medicinally: *D. mannii* contains saponins, and is used as a local anti-fungal and anti-protozoan; the red resin of *D. scizantha* is used as an astringent and to stop haemorrhage; the juice of *D. fragrans*, with palm oil, is used as a liniment which is rubbed on the body of a febrile patient; the tubers and roots of *D. ellenbeckiana* are used for headache, to treat mental problems in children, for snakebite and to eliminate a tapeworm; in Dhofar the resin which oozes from the trunks and main branches of *D. serrulata* is used to treat haemorrhage and skin infections, as a healing agent, and, with added water, is applied to the skin or rubbed on legs and feet to relieve pain; the leaves of *D. laxissima* are used medicinally, though the sap is said to blind people; the root-bark of *D. mannii* is anti-bacterial. Many *Dracaena* species yield fibre from which cordage can be made, and others, such as *D. arbora* and *D. fragrans*, are planted as a boundary and fencing plant. Soap is made from the leaves of *D. mannii*, and its stems yield a light yellow dye. Hunters in Africa believe that a few leaves of *D. humilis* put into their gun will make them a good shot. (Ambasta 1982; Dalziel 1948, 1956; Goodin & Northington 1979; Heine *et al.* 1988; Jain & De Fillips 1991; Kunkel 1984; Lewis & Elwin-Lewis 1977; Miller & Morris 1988; Oliver-Bever 1986; Watt & Breyer-Bandwijk 1962)

GRAMINEAE

Annual or perennial herbs; forming a tuft or rosette and sometimes spreading laterally by underground rhizomes or surface stolons. Stems cylindrical, herbaceous, usually hollow. Leaves arising at nodes, alternate, 2-rowed, comprising a sheath (encircling the stem) and a blade which is usually linear and flat; at the junction of the blade and sheath is a hairy or membranous rim – the ligule. The inflorescence consists of spikelets arranged in spikes, racemes or panicles. The spikelet consists of 2 rows of scales arranged along an axis (the rachilla). The lowest pair of scales are the glumes, the remainder – the lemmas – each form part of a floret. The floret consists of 3(to 6) stamens, (1 to)2(or 3) feathery stigmas surmounting the superior ovary, and (0 to)2(or 3) tiny scales (lodicules); it is enclosed by the lemma on the outside and a membranous scale (the palea) on the inside. The lemmas and glumes may bear one or more bristles (the awns). The fruit grain is hard and achene-like. 660 genera. 10,000 species. Cosmopolitan.

Botanical notes. The largest family on Soqotra. The basic pattern of spikelet structure described above is consistent throughout the family. However, it is much modified by reduction, suppression and elaboration of parts. All characters used in the key can be seen with a hand lens, and grasses are not as difficult to identify as is popularly supposed. Nevertheless, the grasses are still poorly known on Soqotra and the distribution maps provided here are probably the most incomplete in this book. A number of species are included on the basis of a single collection; this does not necessarily imply that they are rare – very often the Ethnographic notes give a better indication of the distribution of the grasses on Soqotra. A few recent collections from the island have no location or habitat details attached. These were all certainly made on the islands and as they include some new records for the Archipelago they have been included. Descriptions are not given for genera with only one species.

This account was prepared by Hannah Atkins and Roger Hyam.

ETHNOGRAPHIC NOTES

Grass uses and management: introduction. The pasturelands of Soqotra are made up of both perennial and annual grasses. Generally perennials produce more leaf material than annuals, which favour seed production, and after sufficient rainfall can mature and set seed within six weeks. Soils with perennial grass cover retain more moisture than soils with annual grass cover. Annual grasses grow very well in a good year but very poorly in a bad one whereas perennials are much less susceptible to rainfall variation. Plant composition shifts in response to season of use, intensity of use and type of livestock. For instance, grazing by sheep tends to encourage grass dominance, whereas cattle grazing encourages shrubs.

Annual grasses: last for a single season, bear seed and then die back with the onset of the dry season (or even earlier). This cycle is clearly defined in temperate regions or in areas with well-defined dry seasons but in tropical areas with a poorly defined dry season annuals may survive longer than one season. Annuals, unlike perennials, do not have remnants of old leaves and stem bases left from the previous season and can normally be easily pulled out from the soil. They often constitute the main grazing in arid and semi-arid areas where there is not enough moisture in the soils during the dry season to support perennial grasses, even in a semi-dormant state. Annuals can also occur in appreciable quantities in bush or light woodland in less dry areas. They are also found in overgrazed pastures and are widespread as weeds of cultivation.

Perennial grasses: last for several seasons. They normally have a more complex structure than annuals, and are usually tufted or creeping. They spread by repeated tillering (branching from the base of the stem) or by means of rhizomes (horizontal stems that are either underground or prostrate on the soil surface) or stolons (stems creeping horizontally on the ground surface and rooting at the nodes – usually above ground).

Tufted perennials: usually develop two types of stems: either fertile, seed-producing shoots with elongated internodes and distant leaves, or sterile shoots with very short internodes and crowded leaves – these often form the main bulk of grazeable herbage. The shoots can remain sterile for years until they finally elongate and flower. Grass tufts usually live for four to eight years.

Creeping perennials: send out horizontal rhizomes or stolons. They form more even stands than tufted grasses.

Rhizomatous grasses: can grow between tufted perennials or form almost pure stands. They usually occur in moist areas or on sands but not in dry areas with hard and dry soil, since their rhizomes cannot penetrate the crust. They are common in swampy areas but not in seasonally waterlogged soil.

Stoloniferous grasses: occur predominantly in the tropics and warm countries. In a number of grasses the first internode or two of young plants raised from seed arch slightly and only the next internodes grow horizontally.

Stoloniferous, and sometimes rhizomatous, grasses are often pioneer species able to quickly occupy the bare ground of denuded pastoral or arable land when it is left fallow. They can represent a certain phase in plant succession, often as a second stage after the dominance of annual weeds in fallows, and are in their turn eventually replaced by tufted grasses or bush.

Classification of grasses on Soqotra. There are some 130 species of grass on Soqotra. The local names differ slightly from region to region, but they are identified principally according to:

- rooting system, i.e. perennial or annual;
- overall shape, i.e. clump-forming, spreading, low or tall growing;
- production or otherwise of inflorescence and fructescence, and the shape of these;
- where they grow, i.e. in the upper or lower grazings, and in association with which other plants.

It is recognised that the same grass grows differently at different altitudes and in different local environments, so they grow taller in deeper soil, more sheltered places, or in the cooler upland areas. It is also recognised that different varieties respond to different rain: some respond to the lighter rains of summer, others only to the heavier rains of winter and yet others to the mists and fine drizzle of the monsoon season. In an average year the main flowering season of the 'light rain grasses' is from the end of July onwards, and of the 'heavy rain grasses' from mid-November onwards.

Some grasses are not differentiated by name in some areas, especially on Abd al Kuri, but are given a more general term like *ruˁud* (*ru'ud*) (grazing plant) or *ḳəṣəho* (*qiziho*) (a general term for the immature green leaves of grasses and herbs). If only the stump or dried clump remains, a term such as *saˁadifo* (*sa'adifo*) or *kənṣ̌əfo* (*qinchifo*) is used, or if dying back or standing hay, a term such as *śaˁar* (*sha'ar*) (see below). Grasses that were not given a specific name on the outer islands were: *Brachiaria arida*, *Eragrostis cilianensis*, *E. ciliaris*, *Lepturus calcareus*, *Panicum rigidum* and *Sporobolus minutus.* And on Soqotra: *Acrachne racemosa*, *Brachiaria arida*, *Eragrostis barrelieri*, *E. lepida*, *Hackelochloa granularis*, *Lepturus calcareus*, *Panicum atrosanguineum* and *Tragus berteronianus.*

Within a region, different areas are known for their different grasses. For example, in the Momi region of eastern Soqotra the following grasses are known to occur:

– *di ḳerṭomo* (*di qertomo*), *di ṣəffo* (*di ziffo*), *ḳeṣādə* (*qazadih*): Tiyda' (wadi grazings).

– *mərṣ̌if* (*mirjif*), *di ʾɔ̄žihir* (*di 'owjihir*), *di ḳerṭomo* (*di qertomo*), *ʾibihir* (*'ibihir*): Shibereh (upland plateau grazings).

– *ʾowśiḳ* (*'owshiq*), *mʌsḥiṭ* (*mos'hit*), *ḳʌṭəhān* (*qatihan*): Mshilhi (lowland plains grazings).

– *di ḳerṭomo* (*di qertomo*), *di ʾɔ̄jihir* (*di 'owjihir*): Hantehiyoh (lowland grazings).

– *mərṣ̌if* (*mirjif*), *di ʾɔ̄jihir* (*di 'owjihir*): Gemgemo (headland, cliff top grazings).

– *di ḳerṭomo* (*di qertomo*), *di ʾɔ̄žihir* (*di 'owjihir*), *ɛ̄rśido* (*ershiydo*): Riydi and Digdegeh (corridors of flat plains grazing).

– *ɛ̄rśido* (*ershiydo*), *di ḳerṭomo* (*di qertomo*): Tahk (areas of gravel plain).

Or in the Shu'ub area in the extreme west of Soqotra:

– *di ʾɔ̄jihir* (*di 'owjihir*): lowland grazings, present in quantity; upland grazings, little.

– *di ḳerṭomo* (*di qertomo*): lowland grazings, absent; upland grazings, present in quantity.

– *ḳorźʌ* (*qordha*): lowland grazings, absent; upland grazings, present.

– *ḳeṣadə* (*qazadih*): lowland grazings, absent; upland grazings, absent.

– *mərž̃əf* (*mirjif*): lowland grazings, present; upland grazings, present in quantity.

– *mʌsḥiṭ* (*mos'hit*): lowland grazings, present; upland grazings, absent.

– *ɛ̄rśido*(*ershiydo*): lowland grazings, present; upland grazings, absent.

– *ṣift* (*zift*): lowland grazings, absent; upland grazings, little.

Soqotran taxonomy of grasses

(A) The key grass groups

ibihir **group**

Representative species	*Acrachne racemosa*, *Digitaria ciliaris*
Life cycle	Annual
Distribution	Not widespread; found in rangeland rather than around settlements and livestock quarters; mainly in upland grazings and in interior plains; less common in lowland grazings; not found on coastal plains or along riverbeds.
Description	Summary: upright, slim, weak. Size: short but grow taller in favourable conditions; similar in appearance to the *iʾiri* grasses, but smaller. Rooting system: shallow-rooted, deeper rooted in good soils, root system similar to that of *Commelina* species Leaves: very leafy; leaves soft and bright green, similar to *Commelina* species. Inflorescence and fruiting head: inflorescence insignificant, these grasses are quickly grazed and it is rarely seen. Response to rain: respond quite quickly to rains, in higher areas after the light summer rains as well as the main winter rains; in Central Highlands they respond to the monsoon mist and drizzle. Persistence: die back quickly after winter, except in the Central Highlands where they are more persistent. *ˁāṣəl* / non-*ˁāṣəl*: in the second rank of *ˁāṣəl* grasses, along with *məržif*, *ɛrśīṭo* and *ḳʌṭəhan.* Other: they can smell rather unpleasant; they are described as being as rich for livestock as ghee is for humans.
Livestock	One of the richest fodders on the island; very succulent, very nutritious grazing; excellent forage for sheep and goats; much liked by cattle and particularly good for young stock; become less nutritious as they dry out. However, if small stock, especially sheep, graze them heavily at the beginning of the winter rains, when they are weakened by a severe dry season, they can develop scour and sicken.

Palatibility of roots	The roots are considered particularly delicious. Sheep and goat dig them up when other grazing is scarce. This damages them and they do not grow well the next rains.
Hay	Gathered for hay, especially in the Central Highlands; considered to make a superior hay with an appetizing smell that keeps its colour well.

ˈiʾiri / ʾīyri group

Representa-tive species	*Brachiaria* and *Setaria* species
Life cycle	Annual
Distribution	Widespread grasses commonly found growing beside walls, settlements, byres, livestock pens and handling areas generally; also grow in date palm gardens and other cultivated areas; after the winter rains, transhuming herders find these grasses grown tall and thick all around the living quarters and livestock pens on their return.
Description	Summary: tall with single-plume. Clinging fruiting head. Size: tall grasses; with the *di ʿɔ̄ẓihir* type, the tallest. Rooting system: shallow-rooted with a thick, single root system similar to that of finger millet. Leaves: leafy grasses with broad, bright green leaves rather like those of finger millet, branching off alternately from the stalk. Inflorescence and fruiting head: fruiting head adheres to clothing and the coats of animals, hence these grasses are sometimes called *ləsˈḳano*, 'sticky, clinging'. Response to rain: they need good rains to develop well. Persistence: those growing in the shelter of walling or large rocks persist well into the dry season, but lose their colour and become rather tough, stemmy and unpalatable. *ʿāṣəl* / non-*ʿāṣəl*: they are not *ʿāṣəl* grasses. Other: when immature these grasses make lush *midyˈɛmə*, growing up to the knees of cattle.
Livestock	Important for livestock in the drier areas of the island, especially cattle and goats; less valued in the Central Highlands where alternatives are abundant. Well liked when immature by all livestock and considered one of the best feeds for calves. Once mature, eaten only by hungry livestock when alternative forage is scarce, though donkeys graze them at all stages of growth. Mature plants are regarded as valueless nutrition, but since they survive well into the dry season they are seen as a good subsistence feed for small stock.
Palatibility of roots	Cattle dig them up for their roots.
Hay	Make good hay which keeps for two years and sometimes longer if dried and stored carefully. When gathering for hay the grasses should be twisted off low down on the stalk and not pulled up by the roots. The best hay is made by gathering while immature, and before seed is set. Best dried in shade or partial shade, and should be turned regularly before being bagged and stored. The bristly fruiting heads can be a real nuisance when the plants grow around doorways or alongside much-used tracks, and herders often remove plants from such places before they mature, and lay them along the top of walling or on rocks out of the reach of livestock to dry out. Indeed, plants growing around living quarters are often yanked up by the roots to discourage regrowth.
Notes	Can successfully be sown from seed. Fruiting heads can be used to clarify butter. Handfuls of leaves used to encourage livestock to let down their milk. Bunches placed over *Dioscorea* and other edible bulbs and rhizomes before cooking them in water; sometimes a layer of grass also placed at the bottom of the clay pot underneath the bulbs; this is said to improve the flavour of the food. Long-haired goats, especially in highland areas, are trimmed of their hair in the thigh area and in the underbelly around the udder, because the fruiting heads cause the hair to become matted, and can also cause sores which attract blowflies.

di ʿɔ̄ẓihir group

Representa-tive species	*Heteropogon contortus*; *Cenchrus*, *Chrysopogon*, *Dichanthium* and *Themeda* species
Life cycle	Perennial
Distribution	Mainly found in the colder, upland grazings. In lower pastureland they mainly grow beside walling, or near dung or ash. Herders transhuming back from lowland grazings to the upper pastures find these grasses grown thick and tall everywhere; they grow very thickly, and cattle are said to be able to 'lick their legs to quench their thirst after walking through them'; grow more luxuriantly in wetter areas.
Description	Summary: grasses with needle-like seeds and long, twisted awns, strongly jointed stems, and a tough, wiry texture. One of the three most important grasses that 'never die'. Size: tall grasses; with the *ˈiʾiri* type of grasses above, the tallest. Rooting system: weak, short roots. Leaves: longer leaves than *məˈržif*; they are narrow, and softly hairy, growing alternately on the stem. Inflorescence and fruiting head: generally have 'multi-fingered' inflorescences; produce spiked seeds – because of this, often called *mədˈəwhīn* at the mature stage, a name given to other plants with spiny, damaging fruits, such as *Bidens*; spiked seeds and bristly awns catch on clothing and animal coats, and can damage the eyes and mouths of small stock particularly; do not produce a visible 'grain', and once the seeds fall the grasses remain as a rather stalky hay. Response to rain: respond quickly to rain but need good rain to develop fully, or the mists and steady drizzle of the Central Highland grazing. Persistence: leafage dies back quite quickly once the clouds disappear and the sun becomes strong; but they are one of the last grasses to die back, and the stalks of the standing hay are palatable. *ʿāṣəl* / non-*ʿāṣəl*: these are *ʿāṣəl* grasses. Other: one of the group of grasses which make 'honey', particularly so in the lowland grazings. They produce a

	sticky 'nectar' early in the morning in early summer when the flowers have dried out and the fruit are beginning to form. This only occurs if there is no dew, which melts the 'honey' away.
Livestock	These, with the *mə́rźif* group, are the most important fodder grasses of the island, and are called 'the backbone of the rangeland'. Apart from camels all livestock readily graze them, even when the leafage has died back. Small stock prefer them at the *di ʿḗḳereh* stage; cattle eat the fresh, green leaves after the rains and, as the rangeland dries out, move on to the fruiting head, and then finally the stems of the standing hay. Herders in areas where these grasses have been grazed out will transhume into areas where they are still growing.
Palatibility of roots	Not dug up.
Hay	Gathered for hay in the drier areas of the island, but must be cut before they set seed as once happens they are useless for hay.
Notes	Immature plants generally called *di ɔ̄źihir ḥaher*, 'dark-coloured *di ɔ̄źihir*'. Can successfully be sown from seed. Very attractive to honeybees when they are in flower and when they are making 'honey'. Decreasing in areas that have suffered low rainfall over the past years. In many areas of the island the inflorescence is called *di ḳerṭómo*. This name is sometimes used for the grasses.

di ḳerṭómoh group

(Description below is for the *di ḳerṭomo* grasses rather than for those grasses which are called *di ḳerṭómo* when in flower/ fruit)

Representa-tive species	Name given both to certain annuals (*Brachiaria* species, *Chloris pycnothrix*, *Dactyloctenium aristatum*, *Digitaria* species, *Eragrostis ciliaris*, *Hackelochloa granularis*, *Panicum atrosanguineum*, *Setaria* species) and to the inflo-rescence and fruiting head of some perennials (*Bothriochloa insculpta*, *Cenchrus* species, *Chrysopogon* species, *Lepturus* species, *Sporobolus diander*, *S. spicatus*, *Themeda triandra*)
Life cycle	Annual (and inflorescence and fruiting head of certain perennials)
Distribution	Mainly in the hotter, lowland grazings or in the highest dry pastures; uncommon in the intermediate rangeland. Commonly grow in clumps and can be quite tall. Especially common near walling, rocks and around settlements.
Description	Summary: one of the three most important grasses of the drier areas of the island (with *di ʾɔ̄źihir* and *mə́rźif*). Pro-duce one stalk with a single plume head. Especially found in plains areas. They like heat. Size: medium-sized grasses. Rooting system: single root system; grow in a single clump and do not spread. Leaves: narrow rather than broad. Inflorescence and fruiting head: a single flowering and fruiting stalk with a plume; seeds generally visible as tiny 'grains'. Response to rain: in upland pastures generally come up after the *di ʾɔ̄źihir* group of grasses; need a lot of rain to get going well. Persistence: retain their foliage and green colour longer than most other annuals. *ʿāṣəl / non-ʿāṣel*: among the most *ʿāṣəl* grasses (with the *ḳʌ́źəbə*, *di ṣə́ffo* and *mə́sḥiṭ* groups). Plants growing in hot, low-lying areas, 'out of the sight and sound of the sea and sheltered from the sea breezes', are the most *ʿāṣəl* of all. Other: one of the group of grasses of the hotter lowland areas which can make 'honey'. On mornings with no dew after good summer rains when the grasses are growing well and thickly, they produce a sweet, sticky substance like 'nectar'.
Livestock	Considered a good and rich fodder for cattle and sheep especially, and a superior feed for young stock. Promote lactation. Fully developed plants considered good for livestock as are immature grasses (*ḳə́ṣəho* and *əḳṣ́ibo*). In the outer islands they are considered good graze for young weaned kids and lambs when drying out. Key fodder for sheep in northern and southern plains. Herders in areas where these grasses have been grazed out will trans-hume into areas where they are still growing.
Palatibility of roots	Not dug up.
Hay	Some is gathered and made into hay n the drier areas of the island, but it is not as good as that made from the *iʾiri* grasses.
Notes	Can successfully be sown from seed, best collected and sown at the onset of the winter rains. Seeds of this and *ḳeṣ́adə*-type grasses have the most potential as edible grains. There is a tradition that the fruit of some species were formerly collected and ground to make a coarse flour in times of starvation. Fruiting heads can be used to clarify butter, and the strained remnants are edible. Recognised as not having an *ʿḗḳereh* stage 'because they have a single root, grow in a single clump, and do not spread'. Name *di ḳerṭómo* given to the fruiting heads of grasses of the *di ʿɔ̄źihir* group in some areas, a name sometimes extended to the grasses themselves. Other forms of the name occur: *di ḳiriṭ́amo* (pl. *di ḳerṭhaminítin*) grasses are a diminutive form of the *di ḳerṭómo* ones; *di ḳérṭəhem* is the plural form of *di ḳerṭómo*.

ḳúrźaʿ group

Representa-tive species	*Paspalum* and *Sporobolus* species
Life cycle	Perennial
Distribution	Usually in saline or alkaline ground, often around lagoons and other brackish permanent water.
Description	Summary: tough grasses that like salty ground and spread out across the ground to make a carpet. Size: usually short grasses. Rooting system: a well-established and tough rooting system. Leaves: tough often prickly leaves. Inflorescence and fruiting head: generally small and insignificant, the seeds often invisible.

	Response to rain: need heavy or persistent rain to spread; in the dry season the size of the patches remains the same. Persistence: never die right back. *ʿāṣəl* / non-*ʿāṣəl*: not *ʿāṣəl* grasses.
Livestock	Important maintenance for sheep in particular, but not considered to be particularly nutritious. More valuable as a subsistence feed for hungry livestock during the long dry months and in a drought, when even cattle will eat them. Donkeys and sheep graze them year round. Grazed heavily in coastal areas by livestock accustomed to it.
Palatability of roots	Hungry sheep and goats occasionally dig them up to get at the roots.
Hay	Never made into hay.
Notes	In some areas there is an overlap between this group and the *ḳʌźəbə* group. The name *ḳurźaʾ* is also the name given to various Cyperaceae, and on Abd al Kuri the name is widespread for grasses and Cyperaceae with names on Soqotra.

ḳeṣādə group

Representative species	*Hyparrhenia hirta*, *Pennisetum* species, *Tetrapogon villosus*, *Urochondra setulosa*
Life cycle	Perennial
Distribution	Commonly grow in places which are called 'salty', either because they face into the monsoon winds, or because they grow beside brackish water (one of the meanings of the name is 'oversalted'). Found in rocky places, often along dry riverbeds and in stony foothill areas and in cracks and crevices in the mountains.
Description	Summary: always tall, clump-forming, very stiff and hard'; 'recognisable by roughness, wiriness to the touch. Size: among the tallest grasses of the islands. Rooting system: deep, multi-branched and large rooting system, but it is rather weak. Leaves: many, long, thin leaves which are tough when green, becoming more brittle as they dry out; leaves generally grow alternately from the stalk. Inflorescence and fruiting head: generally produce a single plume with visible seeds. Response to rain: take the longest of all grasses to grow green after rain. Persistence: die back slowly and last longer into the dry season than most grasses; dormant clumps remain throughout the year. *ʿāṣəl* / non-*ʿāṣəl*: not *ʿāṣəl* grasses.
Livestock	An important subsistence, dry season fodder in the drier, more poorly vegetated regions. Found in and around date palm gardens on the southern coast where they are an important fodder for cattle. Donkeys and camels graze them at most stages of growth but large stock prefer the leafage as it dies back and becomes less hard, often ignoring the tougher green foliage. In wetter areas of the Central Highlands it is readily grazed by goats when there are few fresh alternatives, but cattle here graze it only when very hungry. However, *Andropogon bentii* is sometimes cut as a dry season fodder for them, or they are driven to areas where they grow. Goats elsewhere only nibble a little of the new growth when hungry, or eat the inflorescence. In a prolonged dry season In a prolonged dry season excess salt is washed off the cut grasses in areas exposed to the sea to make them more palatable to livestock.
Palatability of roots	Not dug up.
Hay	Harvested to make a rather tough and inferior hay in years when they produce a lot of green leafage and other grasses are scarce. This can last for two years or more. They must be cut with a knife as they are too sharp and painful to gather by hand. In some areas, they are washed first to remove excess salt.
Notes	Livestock eating a lot without other forage are prone to develop scour from the excess salt. Wild and feral goats are renowned for grazing these tough grasses. They are used for roofing in coastal areas where they are abundant, such as parts of the southern plain. The seeds (with those of the *di ḳerṭomo* group) have the most potential as edible grains. The plumed inflorescence is referred to as *di ḳerṭomo* in some areas and this name is sometimes used for the whole plant. Shorter-growing species can be called *ḳʌźəbə* in some areas.

ḳʌṭəhān group

(Description below is for the *ḳʌṭəhān* grasses rather than for those grasses called *ḳʌṭəhān* when in flower/fruit)

Representative species	*Eragrostis* species
Life cycle	Mostly delicate annuals (*Chloris virgata*, *Digitaria velutina*, *Echinochloa colona*, *Eragrostis barrelieri*, *E. lepida*, *E. cilianensis*, *E. minor*, *Melanocenchris abyssinica*, *M. jacquemontii*, *Tetrapogon tenellus*). The name is also given to the slender inflorescence of certain, usually low-growing, perennials (*Apluda mutica*, *Brachiaria arida*, *Desmostachya bipinnata*, *Enneapogon desvauxii*, *E. persicus*, *Lepturus pulchellus*, *Paspalidium geminatum*, *Sporobolus ioclados*, *S. ruspolianus*)
Distribution	Prefer hard or rocky ground and are particularly associated with *Croton socotranus* shrubland. They are common in the northern plains, like the *di ṣəffo* group of grasses. However, unlike this group, they also grow in the higher, colder regions, and are common in the dry limestone plateaus. Also found growing along cliff ledges and in rocky areas of the wetter highlands.
Description	Summary: all small, fine and delicate grasses (name means 'slender'). Size: among the shorter grasses. Rooting system: a weak root system like *di ʿɔ̄jihir*. Leaves: small and very narrow like those of *di ṣəffo*. Inflorescence and fruiting head: inflorescence and fruiting head are insignificant. Response to rain: quite slow to respond to rain. Persistence: dry out very quickly, but persist longer than grasses of the *di ṣəffo* and *ʾibihir* groups. *ʿāṣəl* / non-*ʿāṣəl*: in the second rank of *ʿāṣəl* grasses, along with *məržif*, *ɛršīṭo* and *ʾibihir* grasses.

Livestock	Important for all livestock. Larger species are important for cattle in the Central Highlands and shorter species in drier areas are commonly a key fodder for small stock. Regarded as very nutritious when green but their value quickly drops once they dry out.
Hay	Larger plants in the Central Highlands can be cut and made into hay but they are not as good for this as other upland grasses.

ḳʌ́źəbə, ḳāźəb (centre and west), mʌ́ḳźʌb (east) group

Representative species	*Panicum* and *Sporobolus* species
Life cycle	Perennial
Distribution	Commonly found spreading across flat plains areas. They can cover great stretches of ground in some areas, such as the Seyhun area in the southern plain, or the area around Dihamdh and Qadheb on the northern plains. Prefer light, sandy and 'red' soils in areas free of rock but sometimes found in damp ground and in saline soils. There are two kinds: one much tougher and more 'stringy', the other softer and more like grasses of the *ɛrśīṭo* group.
Description	<u>Summary</u>: tough, creeping 'stringy' grasses. Vital grazing in drier areas because sweet and edible all the year round. Produce a flat area of permanent grazing which rarely dries out. <u>Size</u>: always short grasses. <u>Rooting system</u>: spreading, with sweet and succulent roots and stems, even sweeter than *ɛrśīṭo* grasses. <u>Leaves</u>: short and often quite sharp. Curl back on themselves after heavy grazing. <u>Inflorescence and fruiting head</u>: usually insignificant. <u>Response to rain</u>: heavy or persistent rain needed for growth of patches. <u>Persistence</u>: persist right through the dry season and never quite die back. <u>*ʿāṣəl* / non-*ʿāṣəl*</u>: considered to be among the most *ʿāṣəl* grasses. Normally the most *ʿāṣəl* grasses are considered to be those that grow in hot, low-lying areas, 'out of the sight and sound of the sea and sheltered from the sea breezes' but this is not the case for the *ḳʌ́źəbə* group, which are considered to be *ʿāṣəl* wherever they grow. <u>Other</u>: lowland grasses that can make 'honey' in the great heat of the short summer season. They produce sticky spots with an oily appearance, which make the grasses appear pale and glistening white. Goats grazing heavily on the grasses at this time grow thin and sometimes develop scour.
Livestock	An extremely important fodder everywhere, but especially in the dry season and in time of drought. In some of the drier areas, like parts of the southern plains, they are the staple survival diet for goats and sheep. Even in years when no rains fall, the animals can survive as long as these grasses remain. They keep animals in milk even in the dry months, and help maintain weight and are extremely popular with donkeys. However, it has been noticed that if there is severe drought over a period of years, the patches can become heavily overgrazed if there is little alternative forage. If the roots are also dug up the plants can be destroyed and do not revive, even after rain. Goatherds accuse sheep of destroying precious areas of this grazing if left on it for too long, but shepherds say that digging up the plants does not destroy the grasses and that they revive after the next rains.
Palatibility of roots	Dug up by sheep in particular.
Hay	Not used to make hay.
Notes	Stems and parts of the rooting system are chewed for their sweetness. Soil conservation appears to have been practiced in many areas of Soqotra, most noticeably perhaps in Momi and the Qa'arah end of the southern plain. This is said to have been particularly for this group of grasses. The measures usually consist of curved lines of rocks and stones placed at an angle to capture moisture, soil and seed behind them so semi-circular patches of grassland can develop. This group of grasses is said to be the source of the name of the Qadheb area on the northern plains, a rather saline region, much of which used to be covered with *ḳʌ́źəbə* grasses. Various Cyperaceae are also called *ḳʌ́źəbə*. There is some overlap between the *ḳúrźaʾ* and *ḳāźəb* group in some areas. In the Central Highlands such plants found growing on damp or waterlogged ground are called *ḳʌ́źəbə di mowźi* ('*ḳʌ́źəbə* of damp areas'). *mʌ́ḳźʌb* is sometimes used as a diminutive form of *ḳʌ́źəbə* to describe very short grasses of this group.

mə́ržif group

Representative species	*Digitaria velutina*, *Echinochloa colona*, *Eragrostis* and *Lepturus* species
Life cycle	Perennial (but also the name for the immature growth-stage of some annuals)
Distribution	Hard and stony ground and generally grow in pale soils. More common in upland grazings, but also occur lower down. Two kinds are distinguished: those of the cooler, better-watered areas, called *mə́ržif*, and those of the drier areas, in both upland and lowland pastures. These are called *mə́ržif ḥáher* ('dark-coloured *mə́ržif*') and are said to produce no or insignificant flowers or fruit.
Description	<u>Summary</u>: one of the three most important grazing grasses. <u>Size</u>: short grasses, growing taller in the Central Highlands. <u>Rooting system</u>: each plant has a single root, but plants grow together in spreading clumps. Even if grazed right down, the plants continue to survive underground and after rain the roots spread below the surface of the soil to form new plants. <u>Leaves</u>: short, narrow and pointed, bright, glossy green, rather curled. Plants grow tight to the ground, spreading in patches across it. <u>Inflorescence and fruiting head</u>: those of the drier areas produce no or very insignificant inflorescence or fruiting head. However, after good winter rains they produce tiny seeds which are then blown by monsoon winds and lie protected by rocks and crevices ready to produce new plants the next rains. They do not usually seed after the lighter summer rains. <u>Response to rain</u>: produce some greenery soon after the first rains, but need good rains to develop fully or to spread.

	Persistence: slow to die back, and persist through the dry season in the shelter of rocks and in cracks in the rock pavement. *ʿāṣəl* / non-*ʿāṣəl*: in the second rank of *ʿāṣəl* grasses, along with *ʾibihir, ɛrśīṭo* and *ḳʌṭehan.* Other: the most renowned for making 'honey'. This is most abundant at the beginning of the short summer season, especially in the upland grazings.
Livestock	With the *di ʾɔ̄źihir* group of grasses, these are 'the backbone of the rangeland'. The *mərźif* grasses are always the first choice for goats all over Soqotra, except in the Central Highlands. Generally too short for cattle, except for the better-grown varieties of the Central Highlands. Small stock, particularly their young, prefer the grasses at the *ḳəṣəho* stage. When in flower and fruit, goats eat these first, and then the leaves. These grasses lose much of their nutritional value as they die back, but continue to be useful as a maintenance feed where other grazing is in short supply. Grasses of the upland, rocky areas improve lactation and give the milk a distinctive and sweet smell and flavour. The carcass of livestock that grazed heavily on these has particularly delicious meat and fat. Herders in areas where these grasses have been grazed out will transhume into areas where they are still growing.
Palatability of roots	Not dug up.
Hay	In years when they grow well, cut to make a hay which keeps well and retains its colour.
Notes	Shoots are eaten and are very sweet. Not sown from seed since they are too small to collect and their supply uncertain. In the Central Highlands it is common to use the name *mərźif* for any perennial grass with this appearance and distinctive colour and that grows in cracks in the rock.

mʌshiṭ (east and centre), *zidirə, zeydiro* (west) group

Representa-tive species	*Dactyloctenium hackelii, D. robecchii, Panicum rigidum, P. turgidum*
Life cycle	Perennial
Distribution	Generally prefer flat ground and soft, sandy soil in lowland areas. Spread to carpet an area and then send up flowering stalks; common in coastal and interior plains areas, less often found in the high plateau areas.
Description	Summary: creeping and spreading tough grasses with stiff but sweet leaves and stems. Size: medium-sized grasses. Rooting system: creeping grasses with spreading stems or stolons. Leaves: spiky green leaves not as tough as the stems. Both sweet to taste and become quite prickly as they dry. Inflorescence and fruiting head: insignificant. Response to rain: need heavy or persistent rain to spread and in the dry season the patches remains the same size. During or just after rains the plants become particularly tough and prickly. Persistence: persist well into the dry season – goats in particular always find something edible on them – but they tend to die back before the grasses of the *ḳʌźəbə* and *ḳəṣadə* groups. *ʿāṣəl* / non-*ʿāṣəl*: among the most *ʿāṣəl* grasses (with grasses of the *ḳʌźəbə, di ṣəffo* and *di ḳerṭomo* groups). Plants growing in hot, low-lying areas, 'out of the sight and sound of the sea and sheltered from the sea breezes', are the most *ʿāṣel.*
Livestock	Grazed both green and dried. Larger plants are broken up by herders for their animals. In the drier areas they provide vital dry season grazing, second only to the grasses of the *ḳʌźəbə* and *ɛrśīṭo* groups. Stems sweet and much liked by small stock. Cattle of drier areas also chew away at the plants in the dry season. They are a key grazing resource, especially for goats in the low-lying interior plains of western Soqotra and in coastal plains. Generally offer more than just subsistence grazing, helping hungry animals to maintain weight and keeping them in milk. They can be destroyed by overgrazing and there are areas where these plants have been wiped out.
Palatability of roots	Creeping stems and/or stolons dug up by sheep and goats.
Hay	Not used to make hay.
Notes	Name means 'broom', a name also given to grasses with a large, plume flowering head and either the whole or part of the plant is commonly used to sweep out sheep and goat pens and byres. Parts of the dried plants are gathered as kindling and firelighters. Where they are common, dead plants are gathered and laid under leather mats to make a sleeping area or bolster. Used for roofing where they are common. The stems of the *Dactyloctenium* species especially are used to teach fishermen's children to weave miniature *qerqor* fish traps (the real ones are usually made from tamarisk, *Salvadora persica* or *Croton socotranus*). Renowned for causing mouth sores in goats which graze on them in the early morning or soon after rain, when the grasses are especially tough and prickly. The sores can interrupt the goat's grazing and cause it to lose condition and it has then to be moved out of the area of *zidire* grazing. They share their name with *Indigofera* species, which are widespread in the same areas, and cause the same mouth problems. Grasses of the *zidire / mʌshiṭ* group and *ḳʌźəbə* group are never found growing together. Rarely, the name *mʌshiṭ* becomes '*bʌshiṭ*'. There is overlap in some areas between *mʌshiṭ* and *ḳʌźebə* grasses; on the southern plain there is some overlap between these grasses and the *ḳeṣadə* group. In other areas *Dactyloctenium* species are called *ʾibəṭ*, and on Abd al Kuri *ʾɔ̄śiyə.* In some areas the *moshiṭ* and *zidire* groups are recognised as being different. The *moshiṭ* group tend to grow in the high, dry limestone plateaus and the *zidire* group in the low-lying interior plains areas. However, it is not clear whether this is because the *moshiṭ* group have a large, plume inflorescence and are given the 'broom' name for that reason.

ɛ̄rśīṭo group

Representa-tive species	*Cynodon dactylon, Enneapogon* and *Tricholaena* species
Life cycle	Perennial

Distribution	Found in both high and low pastures. Common at the edge of wadi floors and smooth-floored gulleys and around rocks and in cracks in the rock pavement. Rarely found in the low-lying plains areas or coastal plains, but appear in enclosed, cultivated areas, such as date palm gardens. Grow best in dark-coloured and gravelly soils.
Description	Summary: short, spreading grasses with sweet stems and roots. Size: among the shortest grasses. Rooting system: form mats and are frequently deeply rooted. Creeping grasses and usually stoloniferous. Leaves: short, pointed leaves, rather similar to those of the grasses of the *mə̍rz̃if* group, but softer. Inflorescence and fruiting head: a short 'multi-fingered' inflorescence and seeds that drop very quickly. Response to rain: slow to respond to rain and need heavy or persistent rain to spread. Persistence: retain their green colour well into the dry season. *ᶜāṣəl* / non-*ᶜāṣəl*: in the second rank of *ᶜāṣəl* grasses, along with grasses of the *mə̍rz̃if*, *ᵓi̍bihir* and *ḳa̍ṭəhan* groups.
Livestock	For small stock in general, the most important grasses are the perennial, spreading grasses such as the *ḳʌ̍źəbə* and *ɛrśīṭo* grasses. and they are described as 'the backbone' of their forage. They are especially liked by sheep, for whom they are the key grazing resource in some areas. The young of small stock especially like them at the *ḳə̍ṣəho* stage and this is seen as being one of the best forages for them. Usually not important for cattle as they are too short, though some varieties grow taller in the Central Highlands where they are seen as being not as good for cattle and goats as grasses of the *mə̍rz̃if* group. Goats readily graze them, as do donkeys. Together with the *ḳʌ̍źəbə* group, these are the sweetest of grasses and the meat of stock grazing heavily on them is described as being most delicious and 'tasting of ghee'.
Palatability of roots	Renowned for being dug up by goats and sheep during a drought or prolonged dry season.
Hay	They grow taller in the Central Highlands and are cut for hay if plentiful here, especially if other, better hay-grasses are in short supply. Otherwise too short to be useful for hay.
Notes	The stems, roots and young leaves are eaten by people, though are not as sweet as those of less common *ḳʌ̍źəbə* grasses. In times of great hunger, bunches are pulled up from the ground, cleaned and eaten. Common weed of finger-millet gardens and cultivated areas generally. Not successfully sown from seed as it is too small and usually falls too quickly. Roots and underground stems are used to clarify butter when making ghee; the strainings are a delicious food. Various Cyperaceae are also given this name.

***di ṣə̍ffo*, *ṣift* group**

Representa-tive species	*Aristida*, *Lolium* and *Melinis* species
Life cycle	Annual
Distribution	Particularly associated with *Croton socotranus* shrubland everywhere on the island and common in the foothills area of the northern plain. They also occur also in date palm gardens and areas of cultivation. Like the *ḳʌ̍ṭəhān* group, they prefer smooth, hard and gravelly ground, rather than rocky ground. Mainly found in the lowland grazings, rarely in the colder upland pastures.
Description	Summary: the weakest grasses and among the smallest. They 'like heat'. Size: among the shortest grasses of the island. Rooting system: weak and shallow rooting system. Grow in small clumps. Leaves: short, fine and narrow. Inflorescence and fruiting head: plants send up a single, slender and short flowering stalk. Inflorescence and fruiting head insignificant, but when the grass dies back, fruit are sharp and spiky. Response to rain: respond quickly to rain, and develop even in the lighter summer rains. Persistence: die back more quickly than other grasses. *ᶜāṣəl* / non-*ᶜāṣəl*: amongst the most *ᶜāṣəl* grasses of the island. The plants grow in the hottest, low-lying areas 'out of the sight and sound of the sea and sheltered from the sea breezes'. Other: one of the group of grasses of the hotter lowland areas which can make 'honey'. If there have been good summer rains and the grasses are growing well and thickly, they produce a sweet, sticky substance like 'nectar' on the mornings without dew.
Livestock	Livestock eat them readily when still green, except for camels and many cattle, which are unenthusiastic. Not considered to be especially nutritious in areas where there are alternative grasses, but they are important for small stock. Once they begin to die back they are considered to be little more than maintenance fodder. When the fruiting head has fallen, goats accustomed to it readily eat the stalks. While the dried fruiting head is still present, the spiked seeds and awns are troublesome for livestock, as they tend to collect in a ball at one side of the mouth and this plug has to be removed manually. The seeds also collect on the 'beards' of goats and finally hamper the goat's feeding, and have to be removed or the matted clumps of hair cut. The seeds also pierce the skin and draw blood around the eyes and lips, tongue and gums.
Palatability of roots	Not dug up.
Hay	Never used to make hay.
Notes	The close association of these grasses with *Croton socotranus* shrubland has given rise to the term *ɛrsi̍mo* to describe the terrain where the two occur together. The seeds lodge in the coats of sheep and goats passing through the drying grasses and can make small sores that attract blowflies. Can successfully be sown from seed. Decreasing in lowland areas that have suffered low, or no, rainfall over the past years.

***ẓ̌ɔ̄tīni* group**

Representa-tive species	*Cymbopogon jwarancusa*
Life cycle	Perennial

Distribution	Prefer light sandy soils or dry and crumbly rock, and are particularly common in coastal areas where date palms grow, and in shallow, dry riverbeds. Less commonly found in the Central Highlands.
Description	<u>Summary</u>: grasses which smell of lemon. <u>Size</u>: of medium size. <u>Rooting system</u>: a strong and deep rooting system. <u>Leaves</u>: long, slender and rather tough. <u>Response to rain</u>: respond quickly to rain. <u>Persistence</u>: persist well into the dry months. <u>*ʿāṣəl* / non-*ʿāṣəl*</u>: not *ʿāṣəl* grasses.
Livestock	Cattle of the Central Highlands are generally unenthusiastic although other cattle graze them. In drier lowland areas they are considered a reasonably nutritious fodder for cattle. They are an important graze for goats of the Central Highlands when they come down to the lowland grazings during the cold and wet months. Donkeys and camels readily eat the grasses, as do sheep of the coastal and lowland regions. Once they begin to die back, they are considerably less nutritious but are still regarded as a useful maintenance fodder. The milk of livestock grazing heavily on them develops a distinctly sharp fragrance and flavour which is not liked by all herders. The taste of the buttermilk is also affected, most consider adversely (one meaning of the name is 'to begin to go off, start to smell bad').
Hay	Rarely gathered for hay.
Notes	On the decrease, particularly along the foothills of the northern plain, though they survive in the enclosed date palm plantations. Wild and feral goats are renowned for grazing on them. Dead pieces are sometimes placed on the livestock smudge to attract cows to the milking yard. The living plant placed on a smudge is said to discourage ticks, fleas and the biting flies in the monsoon months. Honeybees are attracted to the nectar in the flowers. When there is a shortage of salt or other flavouring agent, pieces of the green plant that are put on the fire give a pleasant smell and savour to whatever is cooking, and can even be added to the cooking water to replace salt in an emergency.

(B) Localised grass groups

di ʿare̍ffen group

(Name not recognised everywhere, being mainly a term of the cattle-rearing Central Highlands and valleys behind)

Representative species	*Brachiaria* species, *Digitaria ciliaris*
Life cycle	Annual
Distribution	Higher, cooler and wetter areas of the Central Highlands. They are also common in the valleys running south from the Haggeher mountains. Occasionally found in interior plains areas.
Description	<u>Summary</u>: not widespread. Look like a smaller type of finger millet. <u>Size</u>: medium-sized grasses. <u>Rooting system</u>: grow from a single root that produces a single stem. <u>Leaves</u>: pale green, very glossy and strap-shaped, rather similar to some Cyperaceae. <u>Inflorescence and fruiting head</u>: growth habit similar to that of a small finger millet, producing a thick single stem branching into multiple fruiting heads. <u>Response to rain</u>: need good rain to develop. <u>Persistence</u>: persist well into the dry season in higher, cooler areas. <u>*ʿāṣəl* / non-*ʿāṣəl*</u>: not *ʿāṣəl* grasses.
Livestock	Relished by cattle especially. Goats readily graze them but they are considered inferior to *ʾi̍bihir* grasses for them.
Palatibility of roots	Not dug up.
Hay	Rarely gathered to make hay.

ḥu̍bə̍chihin, *ḥa̍bchiher* group

(Name not recognised everywhere, being mainly a term of the cattle-rearing Central Highlands and valleys behind)

Representative species	*Arthraxon* species, *Lolium persicum*
Life cycle	Annual
Distribution	Grasses of the higher, cooler and wetter areas of the Central Highlands, *Arthraxon micans* often dominates the upper grasslands. Also found in the interior valleys running south from the Haggeher mountains.
Description	<u>Summary</u>: not widespread and generally restricted to the Central Highlands. <u>Size</u>: short grasses. <u>Rooting system</u>: shallow-rooted. <u>Leaves</u>: leaves plentiful, narrow and tender. <u>Inflorescence and fruiting head</u>: a 'multi-fingered' inflorescence with no visible 'grain'. <u>Response to rain</u>: need good rain to develop. <u>Persistence</u>: persist well into the dry season in dried-out clumps. <u>*ʿāṣəl* / non-*ʿāṣəl*</u>: not *ʿāṣəl* grasses.
Livestock	All livestock readily eat them, but they are particularly important for cattle.
Palatibility of roots	Not dug up.
Hay	Not gathered to make hay.

ḥərə̀wṭʌṭ group

(Name not recognised everywhere, being mainly a term of the cattle-rearing Central Highlands and valleys behind)

Representative species	*Panicum rigidum*
Life cycle	Perennial
Distribution	Found mainly in the wetter, high plateau areas, such as the Reyged plateau.
Description	Summary: stoloniferous and spreading like *ḳʌ́źəbə*'. The name means 'leaves that can be stripped off'. Size: tall grasses. Rooting system: spreading roots. Leaves: narrow, long leaves. Inflorescence and fruiting head: very small flowers on an open head. Response to rain: clumps re-sprout after even light rain. Persistence: never quite die back. *ʿāṣəl* / non-*ʿāṣəl*: not *ʿāṣəl* grasses.
Livestock	Particularly liked by cattle, though sheep and goats also graze them.
Palatibility of roots	Sometimes dug up by sheep and goats.
Hay	When plentiful, gathered to make hay.

rìhiśt, *rìhilt* group

(Name not recognised everywhere, being mainly a term of the cattle-rearing Central Highlands and valleys behind)

Representative species	*Arthraxon* and *Oplismenus* species
Life cycle	Annual (with the exception of the delicate, trailing perennial *Oplismenus compositus*)
Distribution	Grasses of the wetter highland areas. Commonly found beside walling and slabs of rock in the highest reaches of the mountains and in deep ravines. Also grow in the interior valleys south of the Central Highlands in areas kept free of livestock. They favour 'black' soil and are often found growing under trees. Commonly spread to form patches – one of their names, *məsḥàybibo*, means 'spreading'.
Description	Summary: soft cattle-grasses of the Central Highland areas. Size: medium-sized grasses. Rooting system: generally quite shallow-rooted. Leaves: very leafy; leaves are soft and lush. Inflorescence and fruiting head: inflorescence and fruiting head of the summer months very red-brown, and goats walking through patches of the grass look as if they have been dyed with henna. Response to rain: respond quite quickly to rain. Some come up after the lighter summer rains, but do better in the winter. Persistence: die back once the summer heat gets going unless growing in the shelter of woodland or rock. *ʿāṣəl* / non-*ʿāṣəl*: not *ʿāṣəl* grasses. Other: when immature these grasses make lush *midyɛ̀mə*, growing up to the knees of cattle. They become covered in 'honey', like grasses of the *mə̀ržif* group, in some years at the end of the winter and once the sun starts to warm up.
Livestock	Not considered nutritious when green but once they begin to die back they are seen as very fattening for cows and also as promoting lactation. At milking bunches of the tempting green foliage are offered to cows to encourage milk let-down. Cows eating a lot of the half-dried plants give characteristically sweet milk.
Palatibility of roots	Not dug up.
Hay	Poor hay and rarely gathered for this.
Notes	Dried grasses gathered and laid under skin mats as bedding.

Development of grasses. Herders broadly describe the development of grasses according to the number of days after rain:

Days after rain	*Stage of development*
3	Tiny green dots become visible
6	Small, fine shoots appear above ground
9	Shoots are developed and can provide grazing for calves and goats
20 days or less	The ground is flushed green
20 days and more	Grasses are now 'strong', nutritious, filling, and very sweet and juicy
45	Flowers emerge and some fruit begin to set: this is the very best grazing of all for mature livestock and 'tastes of honey'
90	Grasses start to become 'weaker' and less nutritious; from this stage on they deteriorate in value, unless further rain falls

Weather and grasses. Grasses on Soqotra are said to be very sensitive to wind. For instance, during the non-monsoon months, if the wind blows from the south, this is regarded as being beneficial to the grassland. It revives it and encourages grasses to put out new shoots. On the other hand, if the wind blows from the east during the winter months, this is harmful. This wind, called 'the burning wind', is hot and strong; it sears the grass and this can be catastrophic if it blows soon after the first winter rains, as it can cause the new shoots brought up by the rain to wither and die. If there is dewfall at the same time, the searing effect is exacerbated.

General grassland terminology. The following terms are used in Soqotri grass taxonomy.

– *əḳṣìbo* (eqsibo): first shoots of grasses and herbs as they emerge from the ground after rain. The exact type of grass or herb cannot yet be determined. This grazing is regarded as the best possible for sheep in particular, as well as for weaned lambs and kids.

– *ḳə̀ṣəho* (qiziho): immature, fresh green leaves of all

grasses and herbs. It is not always possible to determine the exact species of herb or grass at this stage except by the expert herder. This grazing is regarded as the best possible one for all livestock. It is fattening, improves milk yield and lessens the need for water.

– *di ᵓeḳereh* (*di 'eqereh*): this term is more commonly used in the grass-rich Central Highlands rather than in the plateau or lowland areas. It describes thickly growing, bright green, immature grasses which have not yet grown to their full height, nor come into flower. These grasses are commonly described as low growing and spreading, with leaves folded over inside one another and a very sweet, white heart. Thus the *di ḳerṭomo* (*di qertomo*) grass is recognised as not having a *di ᵓeḳereh* (*di 'eqereh*) stage because it grows in isolated clumps, has a single root, and does not spread. Together with the *midyɛməh* (*midyamih*) grasses this grazing is regarded as providing optimal grazing for cattle of all ages, as well as being excellent forage for young and sickly stock. Such grasses make excellent hay, and are gathered for this purpose if plentiful. They are more common after the heavier winter rains and if they occur after exceptional summer rains they are quickly burned off by the sun. They are one of the group of grasses that make 'honey' (see below) when the wild bees can be seen buzzing around the lush pasture. To some extent these grasses, growing in swathes across the rangeland in the Central Highlands, take the place of the grasses of the *ɛ̄rśido* (*ershiydo*) group of the lowland grazings.

– *midyɛməh* (*midyamih*): this term is more commonly used in the grass-rich Central Highlands rather than in the plateau or lowland areas. It describes areas of tall and thick grass in full, mature leaf, but which have not yet flowered. Such grasses are tall (taller than *di ᵓeḳereh* (*di 'eqereh*) above: 'up to a cow's knees'), luxuriant, and a rich, bright green. They occur particularly after prolonged or good rains. At other times they can be found growing along walling, around patches of dung or ash, at the base of slabs of rock, and in other sheltered places of deeper or richer soil. With the *di ᵓeḳereh* (*di 'eqereh*) grasses, *midyɛmeh* (*midyamih*) provides the best possible grazing for cattle. The grasses are also gathered as feed for young or sickly stock. They make excellent hay, and are gathered for this purpose when plentiful.

– *śʌṭəher* (*shatiher*): this term is more commonly used in the grass-rich Central Highlands rather than in the plateau or lowland areas. It describes an area of pure (often cleared) grassland. The same term also refers to the flesh of a carcass after all the bones and fat have been removed.

– *śaᶜar* (*sha'ar*): dried grasses and hay, either cut or standing.

– *saᶜadifoh* (*sa'adifo*): clumps of grass, normally those that have been eaten back or are in a dormant state.

– *ḳənchəfo* (*qinchifo*): stumps of grasses remaining after the grass has been chewed down to the roots, making the exact type of grass normally unidentifiable except by the expert herder.

– *ḳālihim* (*qalihim*): the grass inflorescence.

– *ṣʌfrer* (*zafrer*): individual flowers.

– *səbūlə* (*sibulih*): the ripe inflorescence when the seeds are clearly visible and look like miniature grain. It is also the general term for the fruiting head of finger millet, maize and other cereals.

– *dɛ̄ri* (*deyri*): the seed of any plant.

– *mədəwᵓḥin* (*midewhin*): the troublesome spiked seeds or awns of certain grasses and herbs.

– *biśaᵓ* or *fiśaᵓ* (*bisha'* or *fisha'*): the tillers, stolons or rhizomes produced by many perennial grasses.

– *bisum* (*bisum*): new, secondary grass growth which emerges, if there is further rain, from the stumps of grazed-down grasses. This particularly occurs when there are enough light summer rains to revive the perennial grasses. The mixture of standing hay plus this new *bisum* (*bisum*) growth is regarded as excellent fodder for all livestock, and one of the best for sickly livestock.

– *kor inābut* (*kor inabut*): this means 'so that it might revive, renew its growth'. This resting of the rangeland is the reason for transhumance between upper and lower grazings. The term *nābut* (*nabut*) also refers to the artificial pollination of the date palm.

Grass terminology and livestock. Grazing is often described in terms of the livestock it is suitable for.

– *di ᶜifif* (*di 'ifif*): the first new green shoots 'for kids and lambs'.

– *ḳəṣəho di sūwred* (*qiziho di suwred*): short, immature green leaves 'for an older kid or lamb'.

– *śebaᶜ di irəhān* (*sheba' di irihan*): mature green leaves of the shorter grasses 'to fill mature goats'.

– *rəḥūwḥa di folihi* (*rihuwha di folihi*): immature leaves of the longer grasses 'for a calf to get its tongue around them'.

– *ḳʌbź di iləhe* (*qabdh di ilihe*): fully grown, mature leaves of the longer grasses 'for a cow to tear off bunches'.

Grazing is also described in terms of livestock management. Thus, as the grasslands develop, a herder will describe the state of the rangeland to an enquirer successively as: *bəṭəl mḳanhim* (*bitil mqanhim*), when the first new greenery (herbs and shrubs as well as grasses) has come up it is 'enough to bring to an end the need to lop foliage'. This is followed by *ḳɛ̄rhe ṭaᶜamo ḥīdəbə* (*qerhe ta'amo hiydibi*) 'just enough for the local livestock to have a taster'; then *śiboᶜo diš ḥīdəbə* (*shibo'o dish hiydibi*) 'enough for the local livestock to be able to eat their fill'; then *ḳəṣəho di irəhān: ṭūwᶜan diš* (*qiziho di irihan: tuw'an dish*) 'feed for mature goats: enough for people with small stock to transhume to'; then *ḳəṣəho di folihi wə mənḳoᶜ: ṭūwᶜan diš* (*qiziho di folihi we minqo': tuw'an dish*) 'feed for weaned calves: enough for people with young cattle to transhume to'; and finally, *ḳəṣəho di iləhe: ṭūwᶜan diš* (*qiziho di ilihe: tuw'an dish*) 'feed for cattle: enough for people with pregnant and milking cows to transhume to'. And if the grasslands are especially lush, they might be described as *ḳālaᶜ bi chādihir wə ᶜa itɛ̄bur* (*qala' bi chadihir wi 'a iteybur*) 'enough pasture for a clay pot to be thrown to the ground and not break'.

Soqotri descriptions of the varying conditions of the rangeland. The vocabulary used on Soqotra to describe the various states of the rangeland is extensive. However, the basic terms, as the rangeland progressively dries out after the rains, are:

– *ḥigwəhe* (*higwihe*): grassland starts to dry out, and the bright green fades.

– *yəḳərīmīm* (*yiqiriymiym*): grasses begin to die back and leaves curl back.

– *yaᶜfirir* (*ya'firir*): grasses turn red-brown – occurs after dewfall when grasses are burned by the sun.

– *ḥalḥal* or *źerbib* (*halhal* or *dherbib*): grasses fade.

– *ḳeśaᶜ* or *śaᶜar* (*qesha'* or *sha'ar*): grasses dry out and turn to standing hay.

– *śəmɛ̄raḥ* (*shimeyrah*): leaves fall or have been eaten and only dried stalks remain.

– *ḳənchəfo* (*qinchifo*): standing hay has been eaten right down and only the chewed stumps remain.

– *māṣəl* (*mazil*): grazed out rangeland – grasses have been consumed, herbs have withered, and shrubs look desiccated.

– *ḥɔ̄sir* or *ḳənchəfo ḥəwro* (*howsir* or *qinchifo hewro*): exhausted rangeland – grasslands grazed bare and even the stumps dug up. At this stage it is normally necessary to transhume.

– *eṣlaḥ* or *məṣlaḥ* (*ezlah* or *mizlah*): overgrazed and degraded rangeland: worse than *ḥɔ̄sir* (*howsir*) rangeland above, with topsoil blowing away; the most degraded state for rangeland.

It is recognised that areas of overgrazing can occur around water sources when these are under heavy pressure. But herders say that the underground roots of the *mərẑif* (*mirjif*), *di ʾɔ̄ẑihir* (*di 'owjihir*) and *ɛ̄rśido* (*ershiydo*) groups remain viable even when the greenery above ground has been eaten right back, and that these grasses will resprout after the next rains. However, it is also recognised that sheep, with their propensity for digging up the roots of perennial grasses, can be a problem, and it is for this reason that sheep are never allowed to linger around a water source, but must be herded off as soon as they have watered.

Food (livestock). Grasses are regarded as the best possible feed for all livestock, regardless of species, age or condition. Obviously, some grasses are seen as better than others, and for most livestock, most grasses are improved by the addition of browse to the diet, but green grass is always cited by herders as the best fodder to increase milk yield and to help animals gain weight. Unlike other plants, there are no 'harmful' grasses: all are beneficial, and, above all, no grasses cause *'uqabah*. Island wide, the grasses which are seen as most valuable for all livestock are the *di ʾɔ̄ẑihir* (*di 'owjihir*) and *mərẑif* (*mirjif*) grasses, and these are described as 'the backbone of the rangeland'.

Cattle and sheep are the most reliant on grasses; they eat much the same grasses except that cattle rarely graze on the *di ṣiffoh* (*di ziffo*) grasses, and sheep do not come into the Central Highlands to feed on the grasses particular to that region. For both sheep and goats the most important grasses are the spreading perennials such as the *ḳʌźəbə* (*qadhibih*) and *ɛ̄rśido* (*ershiydo*) grasses: these are called 'the backbone of small stock grazing'. They are very sweet (especially the *ḳʌźəbə* (*qadhibih*) grasses), rarely die right back and thus provide some sort of feed all year round. The third most important grasses for them are the *mośḥiṭ* (*mos'hit*) and *iʾiri* (*i'iri*), both of which survive long into the dry season.

The most nutritious grasses are: those described as being like 'a young unmarried girl', that is, 'not yet fully grown, tender but with plenty of flavour'; for small stock, grasses at the *ḳəṣəho* (*qiziho*) or *eḳṣībo* (*eqziybo*) stages; for cattle, grasses at the *midyɛməh* (*midyamih*) or *di ʾeḳereh* (*di 'eqereh*) stages.

The least nutritious grasses are: those of the *ḳeṣadəh* (*qezadih*) group; any still standing at the end of the monsoon season, by which time they have become tough and salty; all dried grasses.

Grazing and browse for goats. All herders consider the ideal feed for goats to be a mix of grasses and browse. Those who graze their goats in the cooler upland areas believe that they have the best of both worlds: plenty of grass as a result of the cool and cloud, but also plenty of woody plants so that their animals always have both graze and browse available, the ideal diet. Herders say that a diet of mainly grasses produces a lot of froth on milk, whereas one of mainly browse very little. The water in which a goat whose main diet has been browse is cooked develops a very rich flavour; the meat is paler than that of a grass-eating goat, and is thought to be more delicious. When goats are grazing heavily on the grasses of the lowland areas the taste of the milk is affected; in particular it develops a citrus tang from eating the *Cymbopogon* grasses of these areas. Goats that graze the grasses of a purely *'aazel* area are considered by many to produce the best milk and meat.

Using grass to encourage milk let-down. Both fresh grass and hay are used to tempt a reluctant cow to let down her milk. This is especially done for a cow whose male calf has been slaughtered or whose calf has died. The most tempting grasses vary from area to area, but generally any grass that is soft, green and sweet smelling is used, along with the green foliage of Commelinaceae and Convolvulaceae species. Any hay which retains some green colour and smells sweetly can be used. A bunch of grass is placed across the three cooking stones, or on the smudge in the milking yard once the fire has burned down low. The hot grass gives off a smell that is said to be irresistible to cows, and they are encouraged to let down their milk while they are feeding on the warmed grass.

Feeding grass to cows losing weight towards the end of pregnancy. There is a condition recognised in cows towards the end of their pregnancy. It causes the animal to lose weight and become listless. To combat this, the cowherd searches out green grass or Commelinaceae or Convolvulaceae foliage and feeds this to the cow to encourage her to put on weight. If there is no fresh greenery available, dry grasses or herbs which still retain some green colour and smell sweetly are given instead.

Digging up the rooting system of grasses. In a prolonged dry season or period of drought, both goats and sheep dig up perennial grasses to get at their succulent roots. This damage is usually repaired by the next rains but, if too many roots are destroyed and the rains are delayed or fail, whole areas of these valuable grasses can be completely destroyed. This is especially true of the *ḳʌźəbəh* (*qadhibih*) and *mośḥiṭ* (*mos'hit*) groups. These tend to be very localised in their distribution, and, since these grasses are commonly the key dry season grazing resource of such areas, this can seriously damage the grazing potential of the area. Nevertheless, it is recognised by herders that this is a useful survival technique. Indeed, when selecting an animal to slaughter for their own use, they will not choose one which is skilled in finding and digging up grasses, since such an animal has a better chance than others of surviving drought. Conversely, when selecting an animal to slaughter for a guest during a period of drought, this is exactly the animal they will choose, since it will be more likely to produce a carcass with some meat and fat on it.

Flavouring / food additives

Winter. A sweet, sticky substance appears on the grasses in the high grazings. This is called *bɛ̄si* (*besi*), *rub* (*rub*) 'sweet liquid' or *'asil* (*'asil*) 'honey', and the rangeland at such a time is described as *feg* (*feg*) or *infigo ḥādəb* (*infigo hadib*), and offers the best possible grazing for all livestock. It only occurs in years when the rangeland is healthy and in peak condition, towards the end of the winter in the short period called *məźərib-hibitən* (*midhiribhibitin*) just before the onset of the short summer season. At this time the flowers of grasses of the *di ʾɔ̄z̃ihir* (*di 'owjihir*) and *mə̍rz̃if* (*mirjif*) types are drying out. Soqotrans say that the legs of livestock and people walking through the grassland at this time become covered with sweet and sticky stains. This phenomenon is most common in areas of hard but open ground, with few trees or shrubs. It lasts for around a month until the grasses begin to dry out as the summer begins to heat up. It is rare in lowland grazings or in upland areas when it rains as this dissolves the sticky liquid. With the *mə̍rz̃if* (*mirjif*) group of grasses, this can also occur earlier in the winter season, when the grasses are in bud but before they come into flower.

Summer. In the lowland plains during the greatest heat of summer, a similar phenomenon known as *māṭiḳ* (*matiq*) 'sweetness' occurs, when glistening, sticky spots, like oil, appear over the grasses of the *ḳʌ̍źəbə* (*qadhibih*) group. The same phenomenon sometimes occurs in hot upland and lowland plains areas if there have been good summer rains and the grasses of the *di ḳerṭo̍mo* (*di qertomo*), *di ḳʌ̍ṭəhān* (*di qatihan*) and *di ṣə̍ffo* (*di ziffo*) groups are growing strongly and thickly, and are fully mature.

Food (human). Many different parts of grasses are nibbled by people as well as by animals. Some grasses are also used to clarify butter when making the all-important ghee, and the strained remnants are eaten.

Bedding. Some dried grasses are laid under skins or woven palm-fibre mats as bedding or padding.

Tinder and kindling. Pieces of the tougher, clump-forming grasses are often gathered and stored for use as tinder and kindling.

Grass processing

Washing unpalatable grasses. In areas facing the sea, the coarser, tougher grasses are generally unpalatable from mid-monsoon onwards due to excess salt; very hungry livestock that graze them can develop severe scour. When other forage is scarce, bunches of these grasses are sometimes gathered, washed in sweet water, and spread out to dry. Sheep and cattle can be fed the grasses while damp, but goats will not tolerate damp fodder and the washed grasses must be completely dried before being fed to them.

Collection of standing hay and haymaking. There is a widespread island saying: 'start preparing for the dry season from the moment the rangeland greens over'. In areas where grasses are abundant, they are gathered from places that are either inaccessible to, or unfrequented by, livestock. This is practised especially by cattle herders, both those who live in the wetter and better-vegetated Central Highlands, and those of the high limestone plateaus. Indeed, cattle herders who live around the Central Highlands commonly have agreements whereby they are permitted to go into the highlands to gather standing hay once their own grazing is exhausted. In periods of severe drought herders scale isolated peaks to gather fodder for their animals, or are lowered on ropes from cliff tops to gather the grasses and herbs of the cliff ledges and those that grow around the escarpment springs. The aim is to accumulate, by the end of the winter season, a store of hay which can be used as supplementary feed should the summer rains fail or the dry season be particularly prolonged.

The best hay is made either from grass cut when it is already beginning to die back but still has the taste and smell of sweet green grass, or from grass cut while still growing and which has not been grazed. The harvester is careful to twist, cut or snap the grass well above the root system so as not to prohibit regrowth. The grasses are then spread out to dry, though not in full sun as this causes the green colour to fade too quickly. Once completely dried, the grass is stored. It is either placed on a bed of clean stones in a waterproof cave or other well-protected and dry place, or laid on a cliff ledge, or poked into a hole in a boulder or cliff face out of the reach of livestock. Sometimes it is tucked into the rafters of the byre or homestead, in an area well away from the smoke of the smudge or cooking fire. If carefully stored and not allowed to get damp, it can last for up to two years, though the very best 'green' hay has a shorter life. A wide variety of mainly broad-leaved herbs are also gathered and dried, such as various *Commelina, Pouzolzia auriculata*, *Ipomoea* species, *Parietaria* species, *Boerhavia diffusa*, *Zaleya pentandra* and *Canavalia* species. Herders with finger-millet enclosures used to dry the stalks and foliage, and grasses and herbs were also harvested from enclosed date palm gardens, tobacco fields and kitchen gardens. The dried foliage of sweet potatoes is also an important dry season cattle feed.

In grass-rich areas in normal years there are few restrictions about where grasses or standing hay can be harvested. Although there is a general agreement not to collect grasses from the areas of open and communal rangeland, there are plenty of suitable grasses to be found growing in the shelter of walling, beside large boulders or piles of rock, around animal and human quarters and so on. In bad years when the rains are poor or fail altogether, it is usually agreed that each herder should gather grasses or hay only from his own area.

In grass-poor areas green grasses are rarely gathered for hay, and restrictions are more commonly applied: a herder may harvest standing hay only from within his own area, moving outside this only with prior agreement. In such areas, where managing livestock through the long, dry season is always a struggle, many herders are against harvesting standing hay altogether, or at least insist that this be done only when it is clear that there will be enough rain to repair the loss.

Hay is given mainly to milk animals to keep them in milk. If there is enough, it is given to pregnant animals that are losing condition and to sickly animals. In a severe drought the precious hay is fed to any animal that seems likely to survive until the rains.

Firing standing hay. In the rich grassland areas of the Central Highlands, where it was felt that large amounts of standing hay discouraged the growth of tender, new grass, the firing of standing hay (but never scrub) just before the summer rains was occasionally practised. Only grassland in privately controlled areas was fired, never that of the communal grazing areas, unless all those with user rights agreed that this should be done. It was mostly carried out when, after very good winter rains, there was a lot of standing hay remaining in the summer months, and in years when the cattle population was low. The decision to fire the hay was not made until it was certain that the summer rains had begun to fall in the herder's summer grazing areas. The person in charge of firing the hay was careful to select a time of day when the wind was in the right quarter, and always remained with the fire until it had completely burned itself out. He generally had a leafy branch of a tree such as *Croton sarcocarpus* or *C. sulcifructus* ready to beat out the flames should they begin to get out of control. If the summer rains failed or were inadequate, the burned area did not revive until the arrival of monsoon drizzle and mists. This practice was more common formerly 'because there was so much more grass and fewer animals'.

Soil conservation and sowing grass seed. There are signs of former soil conservation in many areas on the island. This is noticeable for instance in the Momi uplands and in the Qa'arah area of the southern plain, but traces of the curved lines of rocks and the semi-circular patches of grassland which have developed inside are visible to the practised eye in other areas. Stones are placed in curved lines, of varying lengths and complexity, at an angle to flood stream-flow where they gather wind- and water-borne soil and seed, and conserve moisture. In the Qa'arah area, for instance, patches of grasses of the *ḳʌźabə* (*qadhibih*) group are clearly visible from the escarpment above, extending from the foot of the escarpment to about a third of the way across the plain. Islanders say these soil conservation techniques were practised by their ancestors but are rarely done now. However, it is noticeable that in some of the more arid areas, herders are once again experimenting with this: in many of the flat, plains corridors in the drier areas of both eastern and western Soqotra, herders are making stone semi-circles to trap wind-blown seeds, plant debris, soil and water, or are scraping out a hollow behind a stone windbreak to encourage soil and seed to collect. Other herders are experimenting with gathering grass seed (especially from grasses of the *ʿiʾiri* (*'i'iri*), *ʾibihir* (*'ibihir*), *di ḳerṭomo* (*di qertomo*) and *di ʾōžihir* (*di 'owjihir*) groups) and sowing it at the onset of the rains. They then build little bunds of earth and stones to hold the seed and prevent it from being washed away.

Grasses and soils. Islanders recognise that different soils produce different kinds of grass, generally attributing this to their differing ability to retain water as well as to their depth.

– *ḳerḳəher* (*qerqiher*): fine, reddish soils; water remains visible on surface for one or two hours and retains water well; one of the best soils for most grasses.

– *taḥk* (*tahk*): pebbly, gravelly soils; rainwater disappears quickly (more quickly than that falling on *digdēgə* (*digdegih*) – see below); produces little grass; the best for cultivating date palms.

– *ḳʌz* (*qaz*): pale, marl or chalky soils; retains water well, as well as clay soils, but also dries out quickly and sets hard; produces little grass; much used for roofing and for plastering internal walls.

– *ḳʌbho* (*qabho*): coarse conglomerate soil with pebbles of mixed colours; rain drains through quickly but is retained well below the surface; produces good grazing; looks hard but breaks up quite easily; good for cultivating date palms.

– *šīmi libīni* and *šīmi ḳerḳəher* (*shiymi libiyni* and *shiymi qerqiher*): pale sandy soils; water percolates through even more quickly than in a *digdēgə* (*digdegih*) – see below); produces little grass.

– *ṭuʾḥur* (*tu'hur*): clay soils; holds water for one to two weeks, but then dries up and sets hard; produces little grass; unsuitable for date palm cultivation.

– *ṣʌḳerə* (*zaqerih*): powdery grey-coloured soils; holds water well and retains it below the surface better than other soils; one of the best soils for most grasses. This kind of soil is also found around trees such as *Ficus cordata*, *Sterculia africana*, *Dracaena cinnabari*, *Boswellia* species and *Commiphora* species, and around smaller clump-forming plants such as *Cissus* species; not very good for date palms; good for growing finger millet and tobacco.

– *maḥḳāḳə* (*mahqaqih*): white, chalky soils; holds water well; looks white; very hard, but is friable; good for cultivating date palms (also a type of lichen).

In the rich grasslands of the Central Highlands, the association of certain types of soil with particular grasses is marked.

– *ḥoyhi ḥahər* (*hoyhi hahir*): 'black soil'; retains water well; fertile; good for growing everything; produces herbs of the *ʿažbib* (*'ajbib*) type, and *rihiśt* (*rihisht*) and *ērśido* (*ershiydo*) grasses; these are the best soils for cultivation – they are 'sticky', form lumps and, in general, produce plants with dark-coloured mature foliage; similar dark, fine and powdery soil can also be dug up from beneath trees such as *Ficus cordata* subsp. *salicifolia*.

– *ṭuʾḥur* (*tu'hur*): clay soils; produces *məržif* (*mirjif*) grasses after good rains.

– *ṭuʾḥur ʿamʾham* (*tu'hur 'am'ham*): pale-brown clay soils; produces *məržif* (*mirjif*) grasses after good rains; this type of clay is also used for pottery.

– *ṣʌḳerə* (*saqerih*) or *rʌmələ ḥəwro* (*ramili hewro*): dark-coloured, powdery and sandy soils; crumbly and black and commonly found in cattle-handling areas; produces herbs of the *ʿažbib* (*'ajbib*) and *ʿažərib* (*'ajirib*) types; can be used for roofing but becomes very hard.

– *ḳʌṣ* (*qaz*): fine, pale marly soils; even slower than clay soils to produce grass and will only do so if there is persistent rain or repeated showers; after heavy rain storms it can produce *məržif* (*mirjif*) grasses; these are the best soils to use for roofing.

– *ḳerḳəher* (*qerqiher*): fine reddish soils; grows anything and everything; greens over quickly after rain; the best soils for grass production; free draining, so not good as roofing material.

– *taḥk*: gravelly and pebbly soils; produces *ērśido* (*ershiydo*) and *di ḳerṭomo* (*di qertomo*) grasses.

The best soils for grasses are considered to be: darker soils (*ḥoyhi ḥahər*, *hoyhi hahir*); fine, reddish soils (*ḳerḳəher*, *qerqiher*); the area immediately around livestock and human quarters (*giśiś*, *gishish*), due to the high level of human and livestock waste; and the coarse conglomerate soils (*ḳʌbəho*, *qabiho*) found over much of the northern plains.

Grasses and terrain. Different types of terrain are recognised as producing different herbs and grasses. For instance, low-lying areas are characterised as follows:

– *digdēgə* (*digdegih*): flat, corridor or narrow plain; mostly *ḳerḳəher* (*qerqiher*) soils; responds quickly to rains but water is not retained well, tending to pass over without sinking in; produces plenty of grass and herbs.

– *mʌṭerə* (*materih*): flat-bottomed gully or long shallow depression; mostly *ḳerḳəher* (*qerqiher*) soils; responds quickly to rains; produces the most grass and also rich in herbs; sometimes dominated by a single plant species.

– *rīdi* (*riydi*): extensive shallow basin or broad plain; mostly *ḳerḳəher* (*qerqiher*) soils; responds quickly to rains; produces the most grass but also rich in herbs; better than a *digdēgə* (*digdegih*) for grasses as its soil tends to retain moisture better.

– *ṣʌləhil* (*zalihil*): small streambed, run-off or ravine; commonly *ḳerḳəher* (*qerqiher*) soils; responds quickly to rains; produces much grass; rich in herbs.

– *nōgər* (*nowgir*): sink-hole or pit; mostly *ḳerḳəher* (*qerqiher*) soils; responds quickly to rains; produces grass and herbs.

– *giśiś* (*gishish*): area around living quarters of people and livestock; rich, well-fertilised soil; produces rich greenery of all kinds.

– *gˈɛḥi* (*gehi*): wadi, ravine or valley; rain retained in odd pools; produces very little grass, more herbs.
– *bˈustān* (*bustan*): area enclosed for cultivation, usually a date palm garden, but also kitchen garden, tobacco or finger-millet plot; water remains in date palm pits for a few hours; produces little grass, more herbs. In general enclosed areas produce grasses according to their underlying soil type.

Acrachne Wight & Arn. ex Chiov.

3 species. Old World tropics.

Botanical notes. *Eleusine*, *Acrachne* and *Dactyloctenium* are closely related genera. *Acrachne* is intermediate between the other two in having an abortive spikelet at the 'raceme' tip (Phillips 1995).

Acrachne racemosa Wight & Arn. ex Chiov.

Eleusine verticillata Roxb.

Delicate annual to 80cm. Leaves flat, narrowly elliptic-ovate. Ligule a fringed membrane. Inflorescence of 1 to 5 finger-like, one-sided 'racemes', <11cm long, clustered at tip of stem, often with one below. Spikelets many-flowered, sessile, 5–13mm long, glabrous, whitish-green. Glumes subequal, shorter than lemmas, upper with a short awn. Lemmas with a short awn.

Ecology. Succulent shrubland. 50–150m.
Distribution. Old World tropics.

ETHNOGRAPHIC NOTES
Name(s)
– Soqotri: *ʾˈibihir*.
– Soqotri simplified: *'ibihir*.
Food (livestock). A nutritious fodder grass of thin soils or rocky ground. Its flat, soft leaves are grazed especially by cattle, though all livestock readily graze it.
Comparative uses. This is considered to be a good fodder grass. (Ambasta 1982; Bogdan 1977; Jain & De Fillips 1991; SEPASET)

Aeluropus Trin.

3 species. Mediterranean to India.

Aeluropus lagopodioides (L.) Trin. ex Thwaites

Aeluropus lagopoides (L.) Trin. ex Thwaites

Variable perennial, sometimes densely tufted, sometimes with long prostrate stems to 15(–30)cm long. Leaves bluish-green, distichous, pungent, flat or inrolled, rigid or soft, glabrous or hairy. Ligule a ring of hairs. Inflorescence a dense, contracted panicle of 1 to several one-sided 'racemes', 5–15mm long. Spikelets many-flowered, 2.5–4.5mm long, awnless. Glumes unequal, hairy. Lemmas 9 to 11-nerved.
Ecology. Coastal plains; brackish sand and mud by inlets. Sea-level.
Distribution. Mediterranean to India.
Botanical notes. Very variable vegetatively especially in the length, stiffness and arrangement of the leaf blades. It is extremely salt tolerant, sometimes forming almost pure swards where little else will grow (Phillips 1995).

ETHNOGRAPHIC NOTES
Name(s)
– Soqotri: *ḳˈurźʌʾ* 'windpipe', 'beside water'; *ḳāźəb*, *ḳˈʌźəbə*, *ḳāźəb di mɔ̄źi* '*ḳāźəb* of damp areas' (names given to a variety of mainly perennial Gramineae and Cyperaceae of damp ground); *diḥˈalś*, *diḥaź* (Samha).
– Soqotri simplified: *qurdha'*; *qadhib*, *qadhibih*, *qadhib di mowdhi*; *dihalsh*, *dihadh*.
Food (livestock). When growing in water or damp ground it is avoided by goats. It is regarded as a good fodder grass which is grazed by all livestock, though it is especially favoured by goats and sheep.
Notes. A period spent grazing halophytic rangeland at least once a year is considered essential for the health of small stock and camels especially. Such feed initially brings on diarrhoea which is seen as being beneficial and 'cleansing impurities from the blood'. When recovered from this, the animals begin to put on weight. Grazing heavily on saline forage can affect the taste of the milk.
Comparative uses. This is regarded as a good fodder grass, often sward-forming in sandy, salty and gypseous soils, salt marshes, mangrove swamps, waterholes, cultivated areas and wasteland. (Bogdan 1977; Goodin & Northington 1985; SEPASET)

Andropogon L.

Heterochloa Desv.
Heterochloa dura Boiss.

1 endemic. c.100 species. Tropical and warm regions.

Botanical notes. One of the dominant savanna genera, but some species have adapted to tropical highlands (Phillips 1995).

Andropogon bentii Stapf

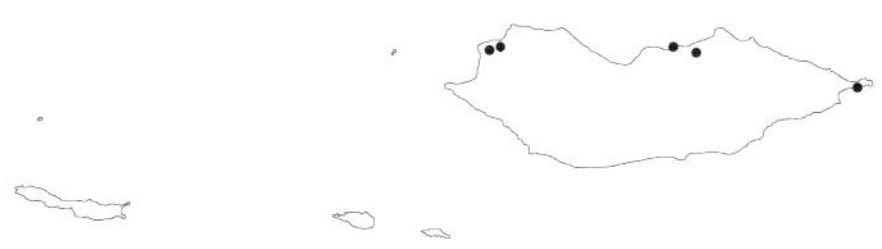

Shrubby, much-branched perennial to 40cm. Leaves flat, linear, glabrous. Ligule membranous. Inflorescence solitary or paired silky-hairy 'racemes' 1.5–2.5cm long. Spikelets 2-flowered, paired, one sessile and one pedicelled, each subtended by fine, white hairs. Sessile spikelet 5–5.5mm long, fertile, awned (awn fine, 7–10mm long, bent in the middle, glabrous). Glumes subequal. Pedicelled spikelet sterile and unawned, 4–5mm long, glabrous.
Ecology. Locally common in succulent shrubland and dwarf shrubland on limestone escarpments and plateaus; dominant in a small area of grassland on Ma'alah. 15–600m.

Distribution. Endemic.
Conservation status. Data Deficient (DD). Inadequate evidence is available to make an assessment.

ETHNOGRAPHIC NOTES
Name(s)
– Soqotri: *māsḥʌt*, *mˈʌshʌt* 'sweeping brush'; *ḳeṣˈadə* 'sharp, cutting', 'oversalted'.
– Soqotri simplified: *mas'hat*; *qezadih*.
Food (livestock). Tough grass which commonly grows on gravelly plains areas. In good years it is of little significance as a fodder grass, though camels and donkeys graze it year round, and cattle and small stock eat the new, tender growth. However, it is an important dry season fodder for goats.
Other. As its name suggests, whole plants are used as brushes to sweep out livestock and human living quarters.

Apluda L.

1 species. Mauritius to Taiwan and New Caledonia.

Botanical notes. This genus probably comprises only one highly variable polymorphic species, which is very common throughout SE Asia, India and Sri Lanka. A number of different species and varieties have been identified in the past but all have eventually been sunk as intermediates. *Apluda* is a relative of *Ischaemum* distinguished by its highly modified pedicels (Clayton & Renvoize 1986).

Apluda mutica L.

Apluda aristata L.

Perennial, stems erect or ascending to 1(–3)m. Leaves flat, linear, dark green. Ligule membranous. Inflorescence an open panicle comprised of a number of 'racemes', each enclosed in a boat-shaped bract. 'Racemes' of three 2-flowered spikelets, one sessile and two on short, flat pedicels; sessile spikelet, 2–6mm long, enclosed within a triangular box made up of the two pedicels, the upper lemma usually with awn 4–12mm long; pedicelled spikelets broadly lanceolate, the larger 2–5mm long, the smaller 1–2mm long.
Ecology. Semi-deciduous woodland. Sea-level–750m.

Distribution. Arabian Peninsula, tropical Asia to Australia.

ETHNOGRAPHIC NOTES
Name(s)
– Soqotri: *ḳˈʌṭəhʌn* 'slender', from the inflorescence; *rˈɛmhen* 'longed-for'.
– Soqotri simplified: *qatihan*; *remhen*.
Food (livestock). This perennial grass with its distinctive inflorescence grows in both tall and short grassland, as well as in wet woodlands. It is well liked by all livestock and is regarded as a good fodder.
Comparative uses. The root is rubbed on wounds made by poisonous snakes and is said to be effective. The stem and leaves are pounded and used as a remedy for fungoid feet and are also applied to fissures in the feet. In India the stalks are used for making hats and the young grass is eaten by buffaloes. (Ambasta 1982; Jain & De Fillips 1991; Perry & Metzger 1980)

Aristida L.

Tufted annuals or perennials. Leaves linear, mostly basal, flat or inrolled. Ligule a line of hairs. Inflorescence an open panicle. Spikelets 1-flowered. Glumes long, scabrid. Lemma convolute, glabrous with a conspicuous awn. Awns glabrous, distinctly trifid. 1 endemic. c.250 species. Tropics and subtropics.

Botanical notes. A large genus divided according to its awn characters. It is suggested that this classification may be at least partly artificial (Clayton & Renvoize 1986).

Aristida adscensionis L.

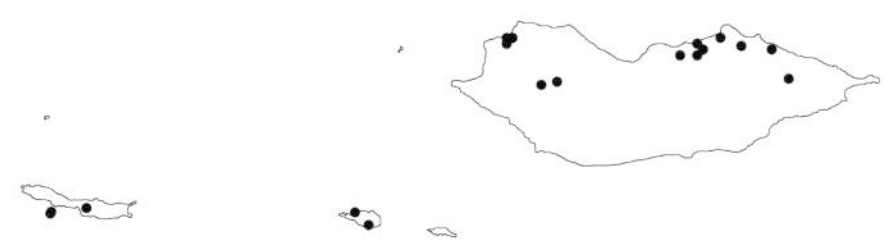

Tufted annual or short-lived perennial, stems ascending to 75cm (usually much less). Leaves flat or inrolled. Inflorescence a linear panicle <25cm long, primary branches erect or ascending. Spikelets with unequal linear-lanceolate glumes, lower shorter than upper, both shorter than lemma. Lemma 5–13(–17)mm long, scabrous, passing directly into the awn without column or articulation. Awn branches equal or laterals <1/3 shorter, central awn not reflexed.

Ecology. Common on coastal plains and drier places in the foothills; *Croton socotranus* shrubland, succulent shrubland and drought-deciduous woodland. Sea-level–600m.

Distribution. Widespread in the tropics.

Botanical notes. A very variable species often divided into several varieties or even species. It has been separated into *A. caerulescens* in the past but differs from this species only in its perennial habit. This difference is not supported by secondary characters and is thought to be caused by environmental factors.

ETHNOGRAPHIC NOTES†

Name(s)

– Soqotri: (*di*) *ṣə̍ffoh*, (*di*) *ṣə̍fət*; *ṣift* (western Soqotra); *nizá̍fə* (Abd al Kuri).

– Soqotri simplified: (*di*) *ziffoh*, (*di*) *zifit*; *zift*; *nizafih*.

Food (livestock). Often called 'common needle grass' due to their very sharp callus. After rain, the young plants are readily eaten by all livestock, but become almost unpalatable in the late-flowering and seeding stages, when animals avoid eating them because of the sharp seeds and stiff awns. Small stock graze the stalks once the seeds have fallen if more palatable grasses are not available. They produce little leaf and are low in digestible crude protein, though they are more valuable before flowering or at the early flowering stage. They are common in dry grassland, in bush or on rocky ground, and can be indicators of overgrazing and pastoral digression. They increase under heavy grazing or in the years that follow droughts, and can invade pastures composed of more valuable perennial grasses, where the latter are weakened by drought or mismanagement. They are also often pioneer grasses of bare soils.

Other. The sharp seeds can penetrate a sheep's wool to pierce the skin. This causes considerable discomfort to the animals which lose appetite and weight as a result. The seed hairs can cause great loss especially in sheep when they form masses in the stomach which cause gastro-intestinal obstruction. The skins of sheep and goats can be damaged by the sharp seeds, making them unsuitable in leatherwork as containers for liquids.

Comparative uses. *A. adscensionis* plants are used as a rather inferior thatching grass, for matting, and the wiry stems to make sieves. They are used to stuff camel saddles and the dried plants are used as a firelighter. The fine straw of *A. funiculata* is used to plait hats and wristlets. The whole plant of *A. adscensionis* and its tender shoots are considered to be palatable in India, and the seeds of the related *A. pungens* are a famine food in North Africa.

A. adscensionis (described as tufted; can be propagated by seed; survives on low and average rainfall, in both warm and cold temperatures, in both shallow or deep soil; survives in sand or loam; grows on spoils and steep slopes; grows on disturbed and undisturbed ground; drought resistant; can grow on alkaline soils) has been used for erosion control. (Bogdan 1977; Dalziel 1948, 1956; Goodin & Northington 1985; Heine *et al.* 1988; le Houerou 1980; Watt & Breyer-Bandwijk 1962)

†Also apply to *A. anaclasta*, *A. funiculata*.

Aristida anaclasta Cope

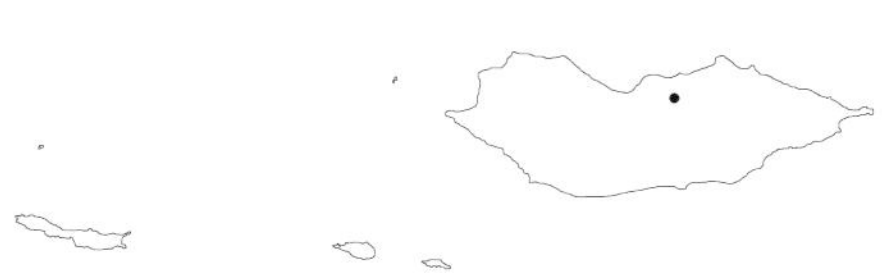

Tufted perennial; stems erect to 35cm. Leaves wiry, inrolled. Inflorescence an ovate panicle 5–7cm long, sparingly branched and bearing few spikelets. Spikelets with lanceolate glumes, lower shorter than upper, both shorter than lemma. Lemma scabrous, 9–10mm long, passing directly into the awn without either column or articulation. Awn not or slightly twisted. Awn branches unequal with the lateral branches shorter, erect or weakly divergent. Central awn reflexed.

Ecology. Rewged plateau. 300m.

Distribution. Endemic.

Conservation status. Data Deficient (DD). Inadequate evidence is available to make an assessment.

Botanical notes. Known only from the type. Characterised by its reflexed central awn.

ETHNOGRAPHIC NOTES

Name(s)

See *A. adscensionis* (also uses).

Aristida funiculata **Trin. & Rupr.**
Aristida murina Cav.

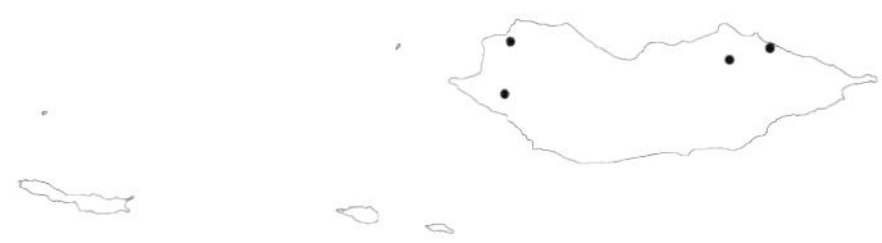

Slender, densely tufted annual; stems to 30cm. Leaves flat. Inflorescence a linear panicle <20cm long. Spikelets with linear glumes, lower longer than upper, both longer than lemma. Lemma 4–7.5mm long. Awn column articulated at the lemma tip, twisted. Awn branches subequal, 4.5–8.5cm long.
Ecology. Drought-deciduous woodland and open shrubland. 50–300m.
Distribution. Tropical Africa to India and Pakistan.

ETHNOGRAPHIC NOTES
Name(s)
See *A. adscensionis* (also uses).

Arthraxon P.Beauv.

Slender annuals with short, ovate-lanceolate leaves. Ligule a short fringed membrane. Inflorescence of slender finger-like 'racemes' clustered at the tip of the stem. Spikelets 2-flowered, paired with one sessile and one pedicelled, the latter sometimes suppressed, pale or purplish. Glumes equal. Lemmas acute with a conspicuous bent awn. 7 species. Old World tropics; introduced in America.

Botanical notes. A rather isolated genus with no obvious close relatives. Its ovate-lanceolate 'herbaceous' leaves are very distinctive. The genus is very homogeneous vegetatively and the spikelet characters often seem to vary independently of each other, leading to ill-defined specific boundaries. There are consequently large differences in opinion as to how many species should be recognised and how to delimit them (Phillips 1995).

Arthraxon lancifolius **(Trin.) Hochst.**
Arthraxon molle Benth.

Delicate, trailing annual; stems slender, to 30cm long. Inflorescence of 3 to 9 'racemes' 1–2cm long. Spikelets paired. Sessile spikelet linear, 2–3.5mm long. Pedicelled spikelet lanceolate, 1.5–2mm long, rarely absent or present only at the tips of the 'racemes'.
Ecology. Grassland clearings in montane woodland. 650–1200m.
Distribution. Tropical Africa to E Asia.
Botanical notes. A more delicate plant than *A. micans*, often with smaller, thinner leaves and smaller, more closely overlapping spikelets.

ETHNOGRAPHIC NOTES†
Name(s)
– Soqotri: *ḥʌ̍bə̍žihin*, *ḥʌ̍bə̍chihin* 'pubes'; *məśḥēbəbo* 'spreading'; *re̍hiśt*, *rehilt* 'chapped, cracked skin'.
– Soqotri simplified: *habijihin*, *habichihin*; *mish'heybibo*; *rehisht*, *rehilt*.
Food (livestock). These slender annuals produce a lot of leaf and are well liked by all livestock. *A. lancifolius* grows mainly in grasslands and on shaded slopes, *A. micans* in the upper grasslands and wet woodland of the central mountains. *A. micans* is very common, often dominant, in the grasslands of the Central Highlands, where it is regarded as especially valuable for cattle.

†Also apply to *A. micans*.

Arthraxon micans **(Nees) Hochst.**

Slender annual; stems trailing to 50(–100)cm long. Inflorescence of 2 to 12 'racemes' 2–4cm long. Spikelets appearing single as a result of the extreme reduction of the pedicelled spikelet. Sessile spikelet lanceolate (2–)3–5.5mm long. Pedicelled spikelet absent, pedicel reduced to an inconspicuous stump.
Ecology. Locally common, sometimes dominant in grassland clearings in montane woodland. 1000–1100m.
Distribution. Tropical Africa to E Asia.

ETHNOGRAPHIC NOTES
Name(s)
See *A. lancifolius* (also uses).

Bothriochloa Kuntze

Amphilophis Nash

35 species. Warm regions.

Botanical notes. Very similar to *Dichanthium* from which it is distinguished by having distinctly grooved pedicels and rhachis internodes.

Bothriochloa insculpta (Hochst. ex A.Rich.) A.Camus

Amphilophis insculpta (Hochst. ex A.Rich.) Stapf
Andropogon insculptus Hochst. ex A.Rich.
Andropogon pertusus (L.) Willd.
Andropogon pertusus var. *insculptus* (Hochst. ex A.Rich.) Hack.
Bothriochloa pertusa Kuntze

Tufted perennial; stems to 2m, stem nodes hairy. Leaves linear, flat, hairy. Ligule a short, fringed membrane. Inflorescence of 3 to 12(to 20) finger-like 'racemes', 2–3 (–8)cm long, clustered at tip of stem or along a short axis (seldom >3cm). Spikelets 2-flowered, paired, one sessile and one pedicelled. Glumes equal, shiny, purplish, usually pitted, awned. Awn 15–25mm long, glabrous, bent at the middle and spirally twisted. Rhachis and pedicels conspicuously silky-hairy.
Ecology. Coastal and inland dwarf shrubland and semi-deciduous woodland. Sea-level–600m.
Distribution. Tropical and S Africa, Arabian Peninsula.
Botanical notes. A widespread apomictic species, varying considerably in both habit and inflorescence characters. *B. pertusa* (L.) Camus is a very similar Asiatic species, barely distinguishable from *B. insculpta*. The two, however, are reported not to interbreed and are also geographically separated (Phillips 1995).

ETHNOGRAPHIC NOTES
Name(s)
– Soqotri: *di ˀɔ̄žhir*, *ˀēžəhər*; *ˀājˈeyr* (western Soqotra) 'spiky', 'painful'; with inflorescence *di ḳerṭˈomo* 'needs chewing'; (rare) *ḳeṣˈadə* 'sharp, cutting', 'oversalted'.
– Soqotri simplified: *di 'owjihir*; *'eyjihir*; *'ajeyr*; *di qertomo*; *qezadih*.
Food (livestock). A common grass of the wetter limestone plateaus which retains its palatability well into the dry season. Small stock graze it more enthusiastically than cattle, though all livestock readily graze new growth. It forms spreading tufts and also produces stolons, which develop freely when the grass is grazed or when isolated tufts grow on bare ground. It is common particularly on plains areas with heavy soils, and can be a pioneer grass invading abandoned arable land. It can withstand heavy grazing and its proportion in natural grasslands tends to increase with the increase in grazing density.
Notes. The panicles are strongly aromatic but the leaves are only slightly scented.
Comparative uses. In Kenya it has been tried with moderate to good success for reseeding denuded pastoral land in moderately dry areas with an annual rainfall of 700–800mm – it was established without any soil cultivation and responded to a tilled seedbed or to covering with cut branches as a protective measure against early grazing. It tends to be stemmy, but under management it can form a low, leafy sward that is well grazed by cattle. In India it is much valued as fodder grass for both grazing and stacking. The related *A. odorata* smells like ginger and yields a volatile oil. (Ambasta 1982; Bogdan 1977; Jain & De Fillips 1991)

Brachiaria (Trin.) Griseb.

Annuals or perennials. Leaves linear or lanceolate, flat. Ligule a line of hairs. Inflorescence of 'racemes' along a central axis, the 'racemes' sometimes with short secondary branchlets. Spikelets 2-flowered, single or paired, ovate, awnless. Glumes unequal, the lower very small, the upper more or less the same length as the lower lemma. 90 species. Old World tropics.

Botanical notes. 'A variable genus for which no satisfactory infrageneric classification is available' (Clayton & Renvoize 1986). It is reported to intergrade at various points into the following genera: *Panicum*, *Urochloa*, *Eriochloa* and *Acroceras*.

Brachiaria arida (Metz) Stapf

Tufted rhizomatous perennial to 50cm. Leaves linear. Inflorescence of 4 or 5 'racemes', 2.5–4cm long on an axis 8–14cm long. Spikelets usually pedicelled, 4–4.5mm long, glabrous, green or bluish green. Upper lemma rugose. Lower glume c.1/3 spikelet length.
Ecology. Semi-deciduous woodland and open shrubland on limestone plateaus. 550–950m.
Distribution. Somalia.

ETHNOGRAPHIC NOTES
Name(s)
– Soqotri: *ɛ̄rśˈido*; with inflorescence *ḳˈʌṭəhān* 'slender'.
– Soqotri simplified: *ershiydo*; *qatihan*.
Food (human). The new shoots, rhizomes and juicy green stems are eaten by people.
Food (livestock). Rhizomatous perennial well liked by all livestock. It is considered to be a very nutritious grass, especially for small stock, promoting lactation and, when in fruit, helping livestock to put on weight. It is especially important as a dry season fodder in the less well-vegetated areas of the island.

Brachiaria deflexa **(Schumach.) C.E.Hubb. ex Robyns**
Panicum nudiglume Hochst.
Panicum nudiglume var. *major* Hochst.

Tufted annual to 70cm, often weak and ascending. Leaves linear. Inflorescence of 6 to 15 'racemes', 2–8 (–10)cm long on an axis 6–15cm long. Spikelets usually pedicelled, scattered along the primary branches and on short secondary branchlets, 2.5–3.5mm long, glabrous or shortly hairy, green or bluish green. Upper lemma rugose. Lower glume 1/3 to 1/2 spikelet length.
Ecology. Drought-deciduous woodland. 300–550m.
Distribution. Tropical and S Africa, Arabian Peninsula, India and Pakistan.

ETHNOGRAPHIC NOTES†
Name(s)
– Soqotri: *ʾˈiʾiri, ʾīri; di ʿaˈreffen* 'like a small *Cladium*'; *sədˈʌḥan* 'prostrate'; with inflorescence *di ḳerṭˈomo* 'needs chewing'.
– Soqotri simplified: *i'iri, 'iyri; di 'areffen; sidahan; di qer-tomo.*
Food (human). There is no tradition of gathering the grains as food on Soqotra.
Food (livestock). Leafy grasses much liked by all livestock. They persist well into the dry season and are considered to be very nutritious grasses, especially in fruit. They are a particularly important forage in the drier areas of the island. Common in dry grassland, scrubland and on stony, rocky areas.
Comparative uses. These grasses, like the related *B. distichophylla*, *B. fulva*, *B. reptens* and *B. stigmatista*, yield edible wild grain. *B. ramosa* is cultivated in some areas for its grain which is considered superior to that of *Panicum sumatrense*. As a flour it is usually mixed with that of *Eleusine coracana*. Its straw is much liked by cattle. (Ambasta 1982; Dalziel 1948, 1956; Jain & De Fillips 1991; Kunkel 1984)

†Also apply to *B. lata*, *B. ovalis*, *B. ramosa*.

Brachiaria eruciformis **(Sm.) Griseb.**
Panicum eruciforme Sibth.

Loosely tufted annual to 30cm. Leaves narrowly lanceolate, glabrous or shortly hairy. Inflorescence of 4 to 10 (1–3cm long) narrow 'racemes' on an axis 3–6cm long. Spikelets borne singly in 2 rows, 1.8–3mm long, hairy. Upper lemma smooth and shiny, readily deciduous. Lower glume tiny, 1/5 to 1/4 spikelet length.
Ecology. No details of habitat or location available.
Distribution. Mediterranean to S Africa and India and Pakistan.
Botanical notes. The presence of this species is based on an inadequate specimen and should be treated with caution.

Brachiaria lata **(Schumach.) C.E.Hubb.**

Tufted annual to 70cm. Leaves lanceolate, glabrous or sparsely hairy. Inflorescence of one-sided, spike-like 'racemes', slightly bristly, 2–5cm long on an axis 5–15cm long. Spikelets on short pedicels, densely packed, 2.5–4mm long, usually glabrous, pale green. Upper lemma rugose. Lower glume c.1/3 spikelet length.
Ecology. *Dracaena* woodland. c.600m.
Distribution. Tropical and S Africa, Arabian Peninsula, tropical Asia.

ETHNOGRAPHIC NOTES
Name(s)
See *B. deflexa* (also uses).

Brachiaria ovalis **Stapf**
Brachiaria glauca Stapf

Tufted annual to 80cm. Leaves narrowly lanceolate, velvety hairy. Inflorescence of 4 to 8 (2–6cm long) 'racemes' on an axis 5–18cm long. Spikelets 2.8–4mm long, cartilaginous, smooth and glabrous. Upper lemma rugulose. Lower glume 1/4 to 1/2 spikelet length.
Ecology. Succulent shrubland. 100m.
Distribution. Tropical Africa, Arabian Peninsula, India and Pakistan.
Botanical notes. Possibly under-recorded.

ETHNOGRAPHIC NOTES
Name(s)
See *B. deflexa* (also uses).

***Brachiaria ramosa* (L.) Stapf**
Panicum nudiglume Hochst.
Panicum nudiglume var. *major* Hochst.
Panicum petiveri Trin.

Tufted annual to 50cm. Leaves narrowly lanceolate, velvety hairy. Inflorescence of 5 to 10 (3–6cm long) 'racemes' on an axis 6–13cm long. Spikelets 2.5–3.5mm long, glabrous or hairy. Upper lemma rugose. Lower glume 1/3 to 1/2 spikelet length.
Ecology. *Croton socotranus* shrubland and dry woodland. 50–750m.
Distribution. Tropical and S Africa, Arabian Peninsula, tropical Asia.

ETHNOGRAPHIC NOTES
Name(s)
See *B. deflexa* (also uses).

Cenchrus L.

Annuals or perennials. Leaves flat, glabrous or shortly hairy. Ligule a line of hairs. Inflorescence a single 'raceme'. Spikelets 2-flowered, sessile, with an involucre of fused purplish or white bristles below the spikelet. Spikelets falling complete with subtending bristles. Glumes unequal. Lower glume <1/2 spikelet length, sometimes absent. Upper glume a little shorter than spikelet. 22 species. Tropics and warm temperate regions.

Botanical notes. The genus can be recognised by its involucre of bristles subtending the spikelets. It is, however, very similar to *Pennisetum*, the main difference being that the bristles are fused in *Cenchrus* and not in *Pennisetum*. In some species of *Cenchrus*, most noticeably *C. ciliaris*, the bristles are only fused very shortly at the base of the involucre. In addition, the bristles in *Cenchrus* are flattened and grooved, a feature not seen in *Pennisetum*.

***Cenchrus ciliaris* L.**
Pennisetum ciliare (L.) Link

Tufted perennial; stems ascending or erect to 150cm. Inflorescence usually 2–5(–14)cm long. Spikelets 2–5.5mm long. Involucral bristles soft, united only at the base, the inner slender and flexuous, one much longer than the rest, the outer scabrid and shorter.
Ecology. Widespread and abundant in a variety of dry habitats. Sea-level–600m.
Distribution. Africa to India and Pakistan. Introduced elsewhere in the Old World.
Botanical notes. One of the commonest grasses on Soqotra; certainly under-recorded on the distribution map.

ETHNOGRAPHIC NOTES
Name(s)
– Soqotri: *di ᵓə̄žhir*; *ᵓēžəhər*, *ᵓājeyr* (western Soqotra) 'spiky', 'painful'; with inflorescence *di ḳerṭomo* 'needs chewing'.
– Soqotri simplified: *di 'owjihir*; *'eyjihir*, *'ajeyr*; *di qertomo*.
Food (human). There is no tradition of gathering the grains as food on Soqotra.
Food (livestock). This tuft-forming perennial, also called 'the common African foxtail', is one of the most important rangeland plants. It grows in grasslands, bush and in rocky places. It generally avoids heavy clays or soils deficient in calcium, and occurs in many distinct forms most of which are of high grazing value. It often produces short, stout rhizomes, and is a highly drought-resistant grass that can grow under an annual rainfall from 270–300mm upwards in areas with two rainy seasons, and from 400mm in areas with a single wet season. Its drought resistance can be partly explained by the very deep root system. It thrives at high air temperatures, the optimum temperature for photosynthesis being about 35°C. It responds well to close grazing, which gives better yields and better herbage quality. Green shoots often sprout during the dry season from old, apparently dried out stems, and these shoots contribute to the volume of dry season grazing. During the early growth stage the crude protein content is high, but this then decreases as the season advances. It is an excellent fodder which is rich in carbohydrates and proteins. It is well grazed by all kinds of stock and retains its palatability at the later stages of growth better than the majority of other tropical grasses. It is even grazed when quite dried out.
Comparative uses. It is an excellent grazing perennial, but species with hard and retrorsely scabrid involucres are undesirable in pastures because the involucres penetrate into sheep's wool creating a problem for wool producers. The only serious diseases of *C. ciliaris* are smut and ergot which attack the spikelets, especially in rainy seasons. It is used for erosion control. It has been successfully introduced into Australia and America where it is valued mainly for its remarkable drought resistance, as well as its good herbage quality and palatability. It is easily established and thus much used for reseeding denuded pastoral land or for improving worn-out pastures. In parts of India it becomes dominant and invades overgrazed pastures. However, it is resistant to invasion under heavily stocked conditions. It can be grown as a fodder with very little irrigation, and can be cut and fed green or as silage. It is sometimes used for roof-thatching. In India the seeds of this and *C. biflorus* are gathered and used for making porridge and bread. (Bogdan 1977; Dalziel 1948, 1956; Goodin & Northington 1985; Heine *et al.* 1988; le Houerou 1980; Kunkel 1984)

Cenchrus pennisetiformis **Hochst. & Steud.**

Similar to *C. ciliaris* but an annual or short-lived perennial and inner involucral bristles fused to 2.5mm.
Ecology. Locally common and sometimes abundant on *Croton* shrubland and succulent shrubland. Sea-level–750m.
Distribution. Tropical Africa to India and Pakistan.
Botanical notes. The boundary between *C. ciliaris* and *C. pennisetiformis* is very indistinct and there are many intermediates. *C. ciliaris* is typically a perennial with the involucral bristles united only into a disc at the base, whereas *C. pennisetiformis* is more usually annual with a cupular involucre (Cope unpubl.).

ETHNOGRAPHIC NOTES
Name(s)
– Soqotri: *di ʾɔ̄žhir*; *ʾēžəher*, *ʾāje̍yr* (western Soqotra) 'spiky', 'painful'; with inflorescence *di ḳerṭo̍moh* 'needs chewing'; *ḳīriṭa̍mo* 'little *ḳerṭo̍moh*'.
– Soqotri simplified: *di 'owjihir*, *'eyjihir*, *'ajeyr*, *di qertomo*; *qiyritamo*.
Food (livestock). A valuable fodder grass, well liked by all livestock. Its nutritive value approaches that of *C. ciliaris*. It grows commonly in sandy, stony or rocky places and remains green for a considerable part of the dry season.

Cenchrus setigerus **Vahl**
Cenchrus schimperi Hochst. & Steud.

Clump-forming perennial; stems to1m. Inflorescence dull green or dark bluish-purple, 5–9(–12)cm long. Spikelets 3–5mm long. Involucral bristles rigid, the inner erect, conspicuously flattened, united for 1/4 to 1/3 of their length, the outer absent or very few, not flattened, much shorter than the inner.
Ecology. Locally common on the northern coastal plains and limestone foothills. 0–500m.
Distribution. Tropical Africa to India and Pakistan. Introduced elsewhere in the tropics.

ETHNOGRAPHIC NOTES
Name(s)
– Soqotri: *di ʾɔ̄žhir*; *ʾēžəher*, *ʾāje̍yr* (western Soqotra) 'spiky', 'painful'; with inflorescence *di ḳerṭo̍moh* 'needs chewing'.
– Soqotri simplified: *di 'owjihir*, *'eyjihir*, *'ajeyr*, *di qertomo*.
Food (human). There is no tradition of gathering the grains as food on Soqotra.
Food (livestock). A valuable fodder grass, well liked by all livestock. It grows commonly in dry grassland and bush. Though it does not produce rhizomes or stolons, it is as drought resistant and heat tolerant as *C. ciliaris*. It has even been reported to grow under an annual rainfall of 200mm.
Comparative uses. It is an excellent grazing perennial, but species with hard and retrorsely scabrid involucres are undesirable in pastures because the involucres penetrate into sheep's wool creating a problem for wool producers.The seeds are a useful wild grain and are used in India for making porridge. It has been introduced into cultivation in dry areas such as North Australia and Queensland, where it has been successful both in sown pastures and in improving worn-out or denuded natural pastures. The seedlings are more tolerant of soil salinity than *C. ciliaris*. It has the advantage over *C. ciliaris* that its seeds are not fluffy and are easy to plant and to harvest. (Bogdan 1977; Dalziel 1948, 1956; Goodin & Northington 1985; Kunkel 1984)

Chloris Sw.

Annuals or perennials; stems erect. Leaves linear-triangular, flat or folded, mainly aggregated at the base. Ligule a fringed membrane. Inflorescence of 4 to 15 one-sided finger-like 'racemes' at the tip of the stem, sometimes partly enclosed in the sheath of the uppermost leaf. Spikelets 2 or 3-flowered, sessile, crowded, awned. Glumes shorter than the florets, unequal with the upper longer, lanceolate. Lemmas awned. 55 species. Tropical and warm regions.

Botanical notes. Habitat variable, but mainly in short grassland on poorer soils and in disturbed places.

Chloris pycnothrix **Trin.**

Stoloniferous, mat-forming annual. Leaves flat or folded, obtuse at apex. Inflorescence of 2 to 7 'racemes' <6cm long. Spikelets 2-flowered. Lemmas glabrous, the upper lemma very narrow. Awns 8–18mm long, purple.
Ecology. No details of habitat or location available.
Distribution. Tropical Africa, S America.

ETHNOGRAPHIC NOTES
Name(s)
– Soqotri: *di ʾɔ̄žhir*; *ʾēžəher*, *ʾāje̍yr* (western Soqotra) 'spiky', 'painful'; with flowering spike *di ḳerṭo̍moh* 'needs chewing'; *ḳīriṭa̍mo* 'little *ḳerṭo̍moh*'.
– Soqotri simplified: *di 'owjihir*, *'eyjihir*, *'ajeyr*, *di qertomo*; *qiyritamo*.
Food (livestock). This is regarded as a valuable fodder grass in the drier areas of Soqotra. It is well liked by all livestock.
Comparative uses. The related *C. gayana* is an important cultivated pasture species. (Bogdan 1977)

Chloris virgata **Sw.**
Chloris barbata sensu Balf.f. non Sw.

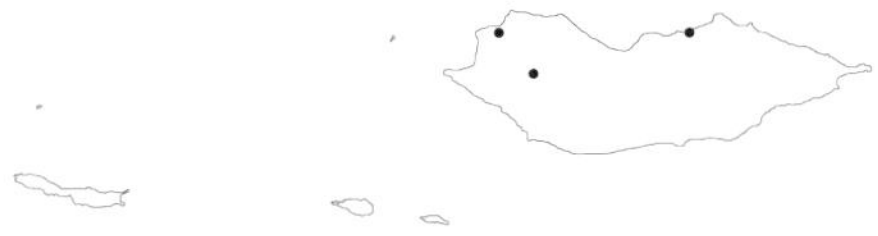

Annual to 70cm. Leaves flat or folded, linear, acute or tapering at apex. Inflorescence of 4 to 9 'racemes' <6cm long. Spikelets 2(or 3)-flowered. Lemmas conspicuously hairy, awned. Awns 2.5–8.5mm long, pale or with a slight purplish tinge.
Ecology. Open habitats on coastal plain and foothills. 30–500m.
Distribution. Tropics.
Botanical notes. A widespread and variable species.

ETHNOGRAPHIC NOTES
Name(s)
– Soqotri: *ḳʌ́ṭəhān* 'slender', from the inflorescence.
– Soqotri simplified: *qatihan*.
Food (livestock). A soft and palatable grass which is highly regarded as a fodder grass on Soqotra, where it provides valuable seasonal grazing. In years of good rainfall it can be abundant. It grows commonly in dry grassland, bush, on rocky ground, in abandoned cultivations and as a weed of arable land in dry areas.
Comparative uses. In India this is regarded as a nutritious fodder, where it occasionally constitutes the only vegetation in saline tracts of the desert region. It has been used with some success for reseeding denuded pastoral land at medium altitudes. A decoction is used for baths to treat rheumatism. It is said to be suitable for erosion control. (Ambasta 1982; Bogdan 1977; Jain & De Fillips 1991)

Chrysopogon Trin.

Tufted perennials. Leaves linear. Ligule membranous or hairy. Inflorescence an open panicle of whorled filiform, flexuous branches. Spikelets in groups of 3 at the tips of the branches, one sessile and two pedicelled. Spikelets 2-flowered, the lower floret reduced to a lemma. Upper glume and upper lemma of all spikelets awned, but the awn of the upper lemma of the sessile spikelet bent in the middle and much longer than the others. c.26 species. Old World tropics and temperate regions.

Botanical notes. A well-defined genus but the species are variable and intergrading, and the taxonomy of the genus remains to be finalised (Phillips 1995). The species found on Soqotra are part of the same intergrading complex and their taxonomy is not clearly understood.

Chrysopogon aucheri **Stapf**
Chrysopogon gryllus sensu Balf.f. non Trin.

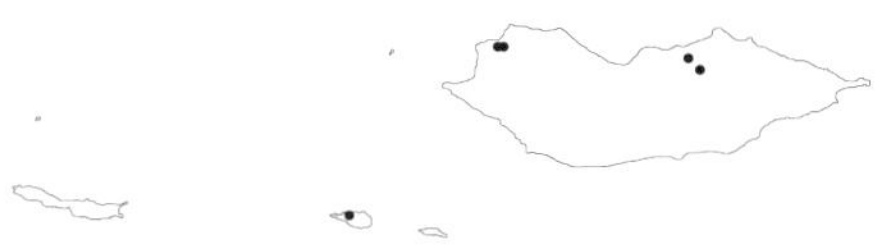

Tufted perennial to 60(–100)cm. Leaves velvety with tubercle-based hairs on the margins. Inflorescence 5–10cm long. Spikelets c.6mm long, white or purplish, usually conspicuously hairy at the top. Awns hairy, the awn of the upper lemma of the sessile spikelet 2–3cm long, brown, bent in the middle and with a twisted column, the other awns straight, 5–10mm long.
Ecology. Widespread in rocky places in shrubland and woodland. 200–600m.
Distribution. Arabian Peninsula to India and Pakistan.

ETHNOGRAPHIC NOTES†
Name(s)
– Soqotri: *di ʾɔ̄žhir*; *ʾēžəher*, *ʾājéyr* (western Soqotra) 'spiky', 'painful'; with inflorescence *di ḳerṭómoh* 'needs chewing'; *ḳīriṭámo* 'little *ḳerṭómoh*'.
– Soqotri simplified: *di 'owjihir*; *'eyjihir*, *'ajeyr*; *di qertomo*; *qiyritamo*.
Food (livestock). All the *Chrysopogon* species are regarded as nutritious and valuable fodder grasses in the islands. They are enthusiastically grazed by all livestock, green and as standing hay. They are said to be especially liked by cattle. *C. aucheri* grows commonly in dry grassland, bush and rocky places and forms broad cushions growing in scattered clumps.
Comparative uses. These grasses are recommended for reseeding denuded pastureland. The grains of the related *C. montanus* are gathered as a famine food. The culms of the related *C. aciculatus* are used in the brush industry and woven into cigarette cases. (Ambasta 1982; Heine *et al.* 1988; Jain & De Fillips 1991; Kunkel 1984)

†Also apply to *C. plumulosus*, *C. serrulatus*.

Chrysopogon plumulosus **Hochst.**

Similar to *C. aucheri*, differing only in the shorter awns of each spikelet plumose rather than hairy.
Ecology. Rolling limestone plateau. c.700m.
Distribution. Tropical Africa, Arabian Peninsula.
Botanical notes. No Soqotran material of this species has been examined and its presence on the island has not been confirmed.

ETHNOGRAPHIC NOTES
Name(s)
See *C. aucheri* (also uses).

Chrysopogon serrulatus Trin.

Tufted perennial to 150cm. Leaves glabrous or sparsely hairy. Inflorescence 3–15cm long. Spikelets 4–8mm long, white or purplish. Awns glabrous or sparsely hairy, awn of the upper lemma of the sessile spikelet 6–12mm long, other awns up to 7mm long.

Ecology. Locally common on the western limestone plateaus; in open low shrubland. 400–800m.
Distribution. Tropical and S Africa, Arabian Peninsula, tropical Asia.

ETHNOGRAPHIC NOTES
Name(s)
See *C. aucheri* (also uses).

Cymbopogon Spreng.

56 species. Old World tropics and warm regions.

Botanical notes. Similar to *Hyparrhenia* and *Andropogon*. It differs from the former in the glabrous column of the awn and from the latter in its compound inflorescence with the racemes about as long as the spatheole. It differs from both in its racemes which are deflexed at maturity and in the essential oils which are always present in the leaves. The aromatic flavour of the leaves when chewed is a useful distinguishing character for the genus. *Cymbopogon* is divided into several species complexes, within which the individual species are both variable and intergrading, and often based on ill-defined characters. There is as yet no general consensus as to the most appropriate level within this variation for the definition of species (Phillips 1995).

Cymbopogon jwarancusa (Jones) Schult.

Densely tufted perennial to 60cm. Leaves linear, wiry, usually aromatic. Ligule membranous. Inflorescence an open panicle, usually short but <40cm long, of short, paired 'racemes' 1–2.5cm long. Each 'raceme' pair (comprising one sessile 'raceme' and one pedicelled 'raceme') partially enclosed by a green or reddish bract (spatheole). Spikelets 4.5–6mm long, paired with one sessile spikelet and one pedicelled spikelet. Sessile spikelet 2-flowered; upper floret fertile and awned, awn 7–10mm long, bent, with a glabrous, twisted column. Pedicelled spikelet sterile, awnless. Pedicel and rhachis internodes densely hairy.
Ecology. Stony desertic plains and foothills. Sea-level–250m.
Distribution. Arabian Peninsula to India and Pakistan.
Botanical notes. All Soqotran material is referable to *C. jwarancusa* subsp. *olivieri* (Boiss.) Soenarko. *C. schoenanthus* (L.) Spreng. subsp. *proximus* has been recorded but all specimens examined previously placed in this subspecies have now been determined as *C. jwarancusa*.

ETHNOGRAPHIC NOTES
Name(s)
– Soqotri: *źotīni* 'odoriferous'; *rəḥáymímhin* 'grey-coloured'; (rare) *minhēyək*.
– Soqotri simplified: *dhowtiynih*; *riheymimhin*; *minheyik*.

Food (human). The stems are not used in cooking as lemon grass on Soqotra.
Food (livestock). *C. jwarancusa* is commonly called 'camel grass', and camels and donkeys on Soqotra readily browse it. All livestock in areas where it grows readily graze it, and herders regard it as useful and nutritious fodder. However, livestock unaccustomed to this feed only eat the younger parts of the plant when other grasses are not available, and their herders regard it as inferior fodder. It is more highly valued when the annual grasses have died back.
Flavouring / food additive. The milk of animals grazing heavily on it develops a distinctive flavour and perfume.
Comparative uses. *C. schoenanthus* is sometimes cultivated for its oil, which is distilled from the roots and leaves. The oil is considered suitable for perfuming soaps, but it contains carene which is a skin irritant and there is a wide variation in oil content. It is widely used medicinally: as a liniment to treat rheumatism, coughs, cholera, dyspepsia and gout. The roots are also considered of great value in treating a number of fevers, including malaria, and the internal part of the rhizome is eaten as an aphrodisiac. An infusion of the inflorescence is given as a febrifuge and of the leaves as a diuretic, stomachic and to lower blood pressure. The grass is also used as an emmenagogue and to treat gonorrhoea. Dried tufts of the plant are burned and the fumes inhaled to treat flu and some neurotic diseases, or the dried plants are soaked in water for a week and a cold infusion drunk. The Xhosa perfume their huts by keeping one or two plants in them. The plant is often used in areas where it grows to flavour tea, giving it an aromatic smell and lemony flavour. The fresh leaves are also used in curries and as a tea substitute, and the stems are used by nomadic people to thatch their roofs. (Dalziel 1956; Davis *et al.* 1983; Goodin & Northington 1985; Kunkel 1984; Perry & Metzger 1980; SEPASET)

Cynodon Rich.

8 species. Tropical and warm regions.

Botanical notes. A complex genus of morphologically similar and genetically introgressing species at two ploidy levels. It is related to *Chloris*, with which it will hybridise (Clayton & Renvoize 1986).

Cynodon dactylon (L.) Pers.

Stoloniferous perennial spreading mainly by rhizomes; stems erect to 40cm. Leaves flat or slightly inrolled. Ligule a fringed membrane. Inflorescence of 4 to 6 slender, finger-like, one-sided 'racemes', 1.5–5cm long. Spikelets 1-flowered, 2–2.5mm long, in 2 rows. Glumes subequal, shorter than the floret. Lemma 2–3mm long, nearly glabrous to silky hairy on the margins.
Ecology. Locally common in disturbed ground and as a weed in permanently wet places. 400–700m.
Distribution. Pantropical weed.
Botanical notes. A common lawn grass in the tropics; known as bermuda grass or star grass.

ETHNOGRAPHIC NOTES
Name(s)
– Soqotri: *ε̄rśido*.
– Soqotri simplified: *ershiydo*.
Food (human). The new shoots and juicy green stems are eaten by people. The leaves are rich in vitamin C.
Food (livestock). This sward-forming perennial, commonly also called 'couch grass', is one of the most important rangeland grasses on Soqotra. It has both underground rhizomes and surface stolons, and is a very variable grass. The small forms with their numerous small leaves are of low grazing value because of their numerous wiry stems; the larger forms are also of low palatability due to their thick stems and harsh leaves; the best forms are the grasses of medium size which produce plentiful leafy herbage of a high grazing value. It occurs in natural grassland and in bush, as well as in soil loosened by cultivation or by digging animals, and can become the dominant grass in overgrazed areas. It can withstand considerable grazing pressure; indeed its proportion in the herbage and the ground cover increases under reasonably heavy grazing. It is drought resistant so is especially valuable in the long dry season. All livestock are very keen on it.
Comparative uses. It is grown as a lawn grass, an effective sand binder, and as a grazing grass. However, it is often a serious weed of annual and perennial crops, as the deep-rooted rhizomes make it difficult to eradicate. It is considered suitable for erosion control. Under certain environmental conditions it can produce high concentrations of hydrocyanic acid which have been said to cause poisoning in cattle.

In Ethiopia the whole plant is given as an anthelmintic, in India to treat diarrhoea, and in Nigeria as an anti-convulsant in children. The whole plant is bruised and mixed with sodium bicarbonate and other substances and taken for heartburn; the bruised plant is applied as a styptic to wounds. A decoction of the plant mixed with fermented rice is taken after serious blood loss from wounds or parturition; this is also used to treat swellings, as a diuretic, for anasarca, and for menorrhagia. It is said to be useful in biliousness, thirst, vomiting, burning sensation, a bad taste in the mouth, hallucinations, epileptic fits, fatigue, leprosy, scabies and other skin diseases, dysentery and fevers.

The rhizomes are diuretic, aperient and emollient and are used in genito-urinary troubles. An infusion of the rhizomes and leaves is used to treat piles, dropsy, syphilis, and as an astringent for wounds. A decoction of the rhizomes is given for renal and urinary troubles, and as a depurative, emmenagogue, refreshing agent, sudorific, emollient and disinfectant vulnerary. A similar decoction is also used to treat a cough, for the suppression of urine and vesical calculus, and to purify the blood.

The roots are used to make a cooling drink; a decoction of the root is given for indigestion and as a blood purifier; it is also used on sores and swellings.

The sprouts are chewed and spat onto a wound, or the expressed juice is used as a locally applied astringent to stop bleeding.

Among the So of Uganda this grass is a traditional symbol of blessing and is carried by the bride at her wedding to receive a blessing. Its epiphyte is placed on the door to the cattle enclosure for the same reasons. (Abbas *et al.* 1992; Al Rawi & Chakravarty 1964; Ambasta 1982; Batanouny 1981; Bebawi & Neugebohrn 1991; Bogdan 1977; Dalziel 1948, 1956; Ghazanfar 1994; Ghazanfar & Al-Sabahi 1993; Goodin & Northington 1985; Heine *et al.* 1988; Jain & De Fillips 1991; Kunkel 1984; Lewis & Elwin-Lewis 1977; Perry & Metzger 1980; Rizk & El-Ghazaly 1995; Watt & Breyer-Bandwijk 1962)

Dactyloctenium Willd.

Annuals or perennials, sometimes stoloniferous. Leaves flat, linear. Ligule membranous. Inflorescence of thick one-sided finger-like 'racemes' of spiky appearance, tip of 'raceme' bare of spikelets. Spikelets in 2 rows, many-flowered. Upper glume with a short awn-like point. Lemmas acute or pointed, awnless. 1 endemic. 13 species. Old World tropics and temperate regions.

Botanical notes. Distinguished from all similar genera by the 'racemes' which are bare of spikelets at the tip. One specimen collected on Abd al Kuri (*Ogilvie-Grant-Forbes* 25) has been identified as 'probably' *Dactyloctenium scindicum*. This species has distinctive curved 'racemes' and spreads by stolons forming extensive mats.

Dactyloctenium aegyptium (L.) Willd.

Eleusine aegyptia Pers.

Robust annual spreading to form small patches; stems decumbent at first, sometimes ascending to 70cm, rooting at the nodes. Leaves flat, hairy or glabrous, often with purple markings. Inflorescence of 2 to 9 spreading or ascending 'racemes', 1–5cm long. Spikelets 3 or 4-flowered, 3.5–4.5mm long, glabrous, whitish-green. Glumes subequal. Grain transversely rugose.

Ecology. Stony ground and disturbed places. 450–550m.

Distribution. Old World tropics and temperate regions.

ETHNOGRAPHIC NOTES

Name(s)

– Soqotri: *ʾiʾiri, ʾīri; ʾibihir*; with inflorescence *di ḳerṭomo* 'needs chewing'.

– Soqotri simplified: *'i'iri, 'iyri; 'ibihir; di qertomo*.

Food (human). There is no tradition of gathering the grains as food on Soqotra.

Food (livestock). Considered to be one of the best annual grasses and is well liked by all livestock. It is an excellent fodder grass of high nutritive value and provides good seasonal grazing in the drier areas of the island. However, it dies back quickly at the onset of the dry season. With the increase of grazing pressure the plants acquire a more spreading habit and produce more seeds; the seeds are, however, smaller than those formed on taller plants grown under light grazing. It grows in dry grasslands, rocky ground, beside tracks, in areas of abandoned cultivation, and especially along sea shores as it can tolerate a considerable degree of salinity. It is widespread in areas of overgrazing.

Comparative uses. The seeds are collected by nomads as food and ground up to make a porridge, though the bruised seed is also said to be used as fish poison in parts of Africa. The dry plant or a decoction of the plant is given for dysentery and haemoptysis. A decoction is also used as an emetic and to treat coughs in children. The grains are taken by women who suffer stomach pains after childbirth, and the seeds are renowned in Africa for treating renal pains. The stems are applied to ulcers. It is sometimes sold in markets as fodder and in some areas is kept as hay. It is a weed of arable land. (Ambasta 1982; Batanouny 1981; Dalziel 1948, 1956; Goodin & Northington 1985; Heine *et al.* 1988; Jain & De Fillips 1991; Kunkel 1984; Rizk & El-Ghazaly 1995; Watt & Breyer-Bandwijk 1962)

Dactyloctenium aristatum Link

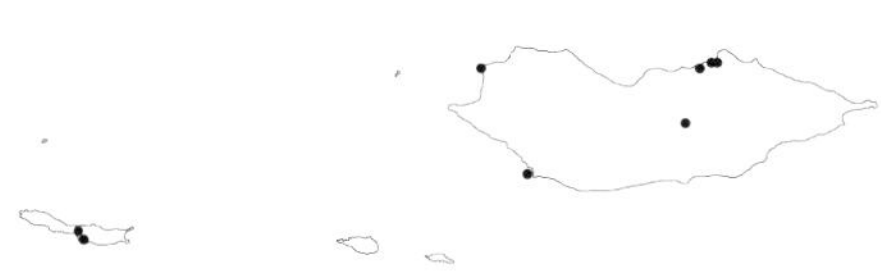

Sprawling, tufted annual; stems ascending to 40cm, often rooting at the nodes. Leaves flat, usually conspicuously hairy. Inflorescence of 4 to 8 'racemes' 1–2cm long in a dense, often subglobose head. Spikelets 3 to 5-flowered, 4–5mm long, glabrous, whitish-green. Lemmas conspicuously pointed giving inflorescence a spiky appearance. Grain granular.

Ecology. Common in gravel and sand on coastal plains; in dwarf shrubland and low succulent shrubland, rare at higher altitudes. Sea-level–650m.

Distribution. Tropical Africa to India and Pakistan.

ETHNOGRAPHIC NOTES

Name(s)

– Soqotri: *ɛ̄rśido; ḳurźaʿ* 'windpipe', 'beside water' (name given to many mainly perennial Gramineae and Cyperaceae of damp ground); with inflorescence *di ḳereṭhanitin* 'little *di qertomo*'; *nizafə* (Abd al Kuri).

– Soqotri simplified: *ershiydo; qurdha'; di qeret'hanitin; nizafih*.

Food (human). The new shoots and juicy green stems are eaten by people.

Food (livestock). This grass is readily eaten by camels and donkeys, but is most important for sheep and goats who are particularly fond of the young shoots or greener parts during the drier months when there is little other grazing available. It is regarded as an excellent fodder grass when green.

Comparative uses. The grains are used as an emergency food but can cause stomach upsets. (Bogdan 1977)

Dactyloctenium hackelii Wagner & Vierh.

Stoloniferous cushion-forming perennial with short woody stems arising from smooth woody stolons. Leaves short, distichous, rigid and pungent. Inflorescence of 4 or 5 short 'racemes' forming a compact head. Spikelets 3-flowered, c.6mm long. Glumes subequal. Grain not seen.

Ecology. Sandy coastal plains. Sea-level–20m.

Distribution. Endemic.

Conservation status. Data Deficient (DD). Poorly known; inadequate evidence is available to make an assessment.

ETHNOGRAPHIC NOTES

Name(s)

– Soqotri: *zˈidirə*, *zēydiroh* 'fat-forming'; *mˈosḥiṭ* 'sweeping brush'; *ʾɔ̄śīyə* (Abd al Kuri).

– Soqotri simplified: *zidirih*, *zeydiroh*; *mos'hit*; *'owshiyih*.

Food (livestock). Readily eaten by camels and donkeys in the coastal areas of Soqotra and Abd al Kuri. Other livestock eat young shoots or greener parts when there is little other grazing available. For the goats of the drier and more marginal coastal areas which are accustomed to the grass, it is an important subsistence fodder. However, it is regarded as having little nutritional value.

Other. As its name suggests, whole plants are used as brushes to sweep out livestock and human living quarters. Children of fishermen on Soqotra and Abd al Kuri are taught to weave miniature *qerqor* fish traps with the stems of this grass. Pieces of the dry plant are gathered for use as kindling and firelighters, and dead plants are used in roofing.

Dactyloctenium robecchii **(Chiov.) Chiov.**

Tough, mat-forming perennial with scaly stolons, stems to 30cm. Leaves short, distichous, rigid and pungent. Inflorescence of 2 to 5 'racemes' 1–2cm long forming a compact head. Spikelets 3–4mm long, glabrous. Glumes subequal. Grain granular.

Ecology. Stony plains. c.250m.

Distribution. Tropical Africa, Arabian Peninsula.

ETHNOGRAPHIC NOTES

Name(s)

– Soqotri: *zˈidirə*, *zēydiroh* 'fat-forming'; *mˈosḥiṭ* 'sweeping brush'; *ʾˈibəṭ*; *ʾɔ̄śīyə* (Abd al Kuri).

– Soqotri simplified: *zidirih*, *zeydiroh*; *mos'hit*; *'ibit*; *'owshiyeh*.

Food (human). The new shoots and juicy green stems are eaten by people. They are surprisingly sweet.

Food (livestock). It is a very important fodder grass in the drier areas of the island, indeed the most important grass of all in some marginal regions. It is drought resistant and goats especially seem able to find some edible material even at the height of a drought when the plants appear quite dead. It is readily grazed by camels and donkeys. In better-vegetated areas of the island, other livestock eat young shoots or nibble at the greener parts of the plant when there is little other grazing available.

Other. As its name suggests, whole plants are used as brushes to sweep out livestock and human living quarters. Children of fishermen on Soqotra are taught to weave miniature *qerqor* fish traps with the stems of this grass. Dead plants are used in roofing.

Desmostachya (Stapf) Stapf

Botanical notes. Closely related to *Eragrostis*.

Desmostachya bipinnata **(L.) Stapf**

Eragrostis cynosuroides Roem. & Schult.

Coarse, tufted perennial to 1.5m, spreading by stolons. Leaves inrolled. Ligule a dense ring of hairs. Inflorescence narrow, linear, comprised of many short, crowded 'racemes', 1–4cm long, arranged in clusters along an axis 25–40(–60)cm long. Spikelets many-flowered, in 2 rows along one side of the 'raceme', 3–10mm long, glabrous, pale, awnless. Glumes unequal.

Ecology. No detailed locations available but recorded from Qeyson and Qariyah (see also Ethnographic notes).

Distribution. Old World tropics.

ETHNOGRAPHIC NOTES

Name(s)

– Soqotri: *śebeṭˈano* 'never quite dies back'; 'type of *ʾˈiʾiri*, *ʾīri*'; 'type of *ḳˈurźaʾ*' 'windpipe', 'beside water' (name given to many mainly perennial Gramineae and Cyperaceae of damp ground); with inflorescence *ḳʌṭəhan*.

– Soqotri simplified: 'type of *i'iri*, *'iyri*'; *shebetano*; *qatihan*.

Food (human). The new shoots and juicy green stolons are eaten by people.

Food (livestock). This tussocky perennial grows on sandy banks, waterlogged ground and in brackish water. It is rather tough when mature, but is grazed by hungry livestock when young and green, unless it is growing in water, when it is generally avoided by goats. It is an important dry season, subsistence fodder grass.

Comparative uses. In India the younger shoots are eaten only by buffalo. Otherwise it is used as a fodder only in the absence of other grasses, when it is given mixed with grain. The culms are considered to be diuretic, and are used in dysentery and to treat menorrhagia. It is also used for coarse ropes and for thatching. It is used in the Middle East as a useful soil and sand binder. (Ambasta 1982; Bebawi & Neugebohrn 1991; Goodin & Northington 1985; Jain & De Fillips 1991; Rawi & Chakravarty 1964)

Dichanthium Willemet

Eremopogon Stapf

Tufted annuals or perennials; stem nodes hairy. Leaves usually flat, occasionally slightly inrolled, glabrous or hairy, sometimes aromatic. Ligule membranous. Inflorescence a solitary 'raceme' or a cluster of several finger-like 'racemes' at the tip of the stem. Rhachis internodes and pedicels slender, solid and conspicuously hairy. Spikelets paired, one sessile, one pedicelled. Upper lemma of sessile spikelet with a glabrous or slightly hairy awn. Pedicelled spikelet awnless. 16 to 20 species. Old World tropics.

Botanical notes. Very similar to *Bothriochloa* from which it can be distinguished by the pedicels and rhachis internodes which are solid i.e. not grooved. As in *Bothriochloa*, the spikelets are in pairs in slender, villous 'racemes', the glumes are subequal and the glabrous awn is bent in the middle. The genus is also very similar to *Andropogon*. In *Andropogon*, however, the awn arises at the tip of the lemma from between two hyaline teeth. These teeth are not present in *Dichanthium*. Pitted glumes are also very rare in *Dichanthium*.

Dichanthium annulatum (Forssk.) Stapf

Tufted perennial with numerous erect stems to 1m. Leaves flat, glabrous or hairy. Inflorescence a cluster of 2 to 9(to 15) finger-like 'racemes', purplish, densely hairy, 2–6cm long. Spikelets 3mm long. Lower glume of sessile spikelet pilose, purple, rounded at the apex, not pitted. Upper lemma with an awn 8–25mm long.

Ecology. Grassland clearings in montane woodland; also a weed of date gardens at lower altitudes. (50–) 700–1100m.

Distribution. Tropical Africa to E Asia.

ETHNOGRAPHIC NOTES

Name(s)

– Soqotri: *di ʾɔ̄žihir* 'spiky', 'painful'; *ḳeṣādə* 'sharp, cutting', 'oversalted'.

– Soqotri simplified: *di owjihir*; *qezadih*.

Food (livestock). This perennial is grazed by all livestock and is regarded as a useful and nutritious grass when young. Cattle are very keen on the young plants or plants in flower, and at such times the grass is regarded as an excellent fodder. The larger forms are often called *ḳezādə*, becoming tough when mature like all grasses of this name. At this stage, they are less liked by livestock, though camels and donkeys continue to graze them. It grows along dried riverbeds and among date palms as well as in open glades in woodland.

Other. Pieces of dry plant are gathered for use as kindling and firelighters.

Comparative uses. In India it is regarded as a valuable fodder grass for cattle. It is cultivated, mainly on an experimental scale, from seed or from tuft splits. It is recommended for young stock and breeding ewes. The culms are said to contain a manna-like substance. In the Sudan it is regarded as a serious weed of fruit gardens. It tolerates salinity but not acidity. (Ambasta 1982; Bebawi & Neugebohrn 1991; Bogdan 1977; Goodin & Northington 1985; Jain & De Fillips 1991; Kunkel 1984)

Dichanthium foveolatum (Delile) Roberty

Dichanthium foveolatus (Delile) Roberty

Tufted perennial; stems slender, wiry to 50cm. Leaves flat or slightly inrolled, often becoming reddish-brown when old, densely papillose. Inflorescence a solitary 'raceme', pale brown, densely hairy, 2–4cm long. Spikelets 4mm long. Lower glume of sessile spikelet ± glabrous, with an acute, orange tip and usually with a single pit. Upper lemma with an awn 12–18mm long.

Ecology. Granite gravel on north-facing slopes. c.250m.

Distribution. Tropical Africa to India and Pakistan.

Botanical notes. Usually with a pitted glume, cf. *Bothriochloa insculpta*. Recorded only from Abd al Kuri.

ETHNOGRAPHIC NOTES

Name(s)

– Soqotri: *lātaʾ*; type of *nizáfə* (Abd al Kuri).

– Soqotri simplified: *lata'*; type of *nizafih*.

Food (livestock). Grazed by all livestock when young and regarded as a useful and nutritious grass at this stage. The larger forms are often called *ḳezādə*, becoming tough when mature like all grasses of this name, and are then less liked by livestock, though the donkeys of Abd al Kuri continue to graze such plants, and goats gnaw at the stems. It is considered a valuable subsistence fodder in the dry season, especially for goats.

Other. Pieces of the dry plant are gathered for use as kindling and firelighters.

Digitaria Haller

Annuals. Leaves flat, linear or lanceolate, glabrous or hairy. Ligule membranous. Inflorescence of long, slender, finger-like 'racemes' arranged either at the tip of the stem with one or two below or in small whorls along a short axis or arranged singly along a long central axis. Spikelets 2-flowered, sessile or shortly pedicellate, solitary or in pairs, usually elliptic, c.2mm long, awnless. Glumes very unequal, lower sometimes absent, upper usually hairy. c.230 species. Old World tropics and temperate regions.

Botanical notes. There is a high degree of variability within and between species in the genus which has led to many superfluous species being described in the past. Complexes of intergrading species are commonplace. Whilst the core characters of the species in such complexes may be easily recognised, it is not unusual for a specimen to be atypical in one or more respects due to introgression from other species in the complex (Phillips 1995). Two of the species on Soqotra, *D. velutina* and *D. ciliaris*, are part of the same intergrading species complex and are consequently not always easy to separate.

Digitaria ciliaris (Retz.) Koeler

Annual to 100cm. Leaves linear, glabrous or sparsely hairy. Inflorescence of 3 to 6 finger-like 'racemes' 5–15 (–22)cm long either at the tip of the stem or in whorls along a short central axis. Spikelets solitary or in pairs, closely overlapping, 2.5–3.5mm long, hairy, whitish-green. Lower glume absent or a tiny scale. Upper glume 2/3 to 3/4 spikelet length.
Ecology. Semi-deciduous woodland on Soqotra; also in sheltered gulleys on summit ridge of Jebel Saleh on Abd al Kuri. 450–1000m.
Distribution. Tropics.

ETHNOGRAPHIC NOTES
Name(s)
– Soqotri: *di rˈehiśt, di rˈehilt* 'chapped, cracked skin'; *di ʿarˈeffen* 'like a small *Cladium*'; *ʾˈibihir*; *ḳˈerəhān* 'horned', from the inflorescence; with inflorescence *di ḳerṭˈomo* 'needs chewing'.
– Soqotri simplified: *di rihilt, di rihisht*; *di 'areffen*; *'ibihir*; *qerihan*; *di qertomo*.
Food (livestock). One of the most important rangeland grasses of the islands. It is considered to be excellent fodder, for small stock in particular, and said to improve lactation. It is the preferred feed for all young stock. However, it dies back quickly at the approach of the dry season. It occurs in heavily grazed grassland, in cultivated areas and in areas of sandy soil beside water.
Comparative uses. It is regarded as a good fodder, especially for goats and sheep, in India and Africa. (Ambasta 1982; Heine *et al.* 1988; Jain & De Fillips 1991)

Digitaria rivae (Stapf) Chiov.

Tufted perennial, 30–120cm. Leaves linear. Inflorescence narrow, 6 to 30 'racemes' 2–10cm long, irregularly arranged along an axis 5–25cm long. Spikelets paired, elliptic, 1.5–2.5mm long, white silky-hairy with hairs slightly exceeding the spikelet. Lower glume absent or a tiny scale. Upper glume 3/4 to almost as long as spikelet, hairy.
Ecology. No details of habitat or location available. In Somalia it is reported as growing in grassland on sandy or alluvial soil.
Distribution. Tropical Africa, Arabia.

ETHNOGRAPHIC NOTES
Not seen, no information recorded.

Digitaria velutina (Forssk.) P.Beauv.

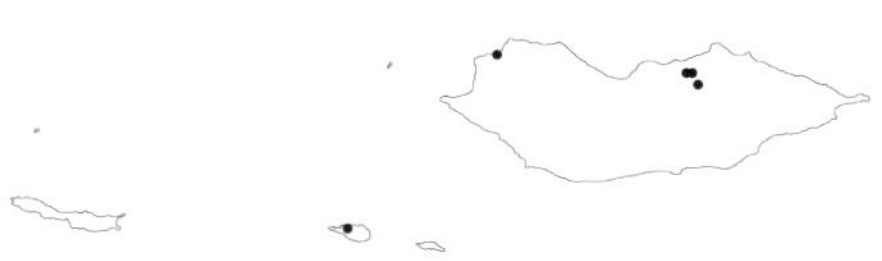

Annual to 80cm. Leaves broadly linear to lanceolate, hairy. Inflorescence of 8 to 15 'racemes' 3–8(–13)cm long arranged along a long central axis, <6cm long (shorter than the longest 'raceme'). Spikelets in pairs, loosely imbricate (overlapping by approximately 1/2 their length), 1.5–2.5mm long, glabrous or sparsely hairy, whitish-green. Lower glume absent or a tiny scale. Upper glume 2/3 or more spikelet length.
Ecology. Drought-deciduous woodland and succulent shrubland on Soqotra; in shady north-facing cliffs on Samha. 200–550m.
Distribution. Tropical Africa, Arabia.

ETHNOGRAPHIC NOTES
Name(s)
– Soqotri: *di rˈehiśt, di rˈehilt* 'chapped, cracked skin'; *di ʿarˈeffen* 'like a small *Cladium*'; *ʾˈibihir*; 'type of *mˈuržif*'; *ḳˈʌṭəhān* 'slender', from the inflorescence; with inflorescence *di ḳerṭˈomoh* (Samha).
– Soqotri simplified: *di rihilt, di rihisht*; *di 'areffen*; *'ibihir*; 'type of *murjif*'; *qatihan*; *di qertomoh*.
Food (human). There is no tradition of gathering the grains as food on Soqotra.
Food (livestock). An important rangeland grass, considered to be excellent fodder and said to improve lactation. The foliage is especially liked by cattle, but is grazed enthusiastically by all livestock. The creeping stems are relished by small and young stock.
Comparative uses. The seeds are eaten at times of scarcity. This and other *Digitaria* species are often sold in markets as fodder grasses for horses and cattle and are sometimes cut for hay. It is a common weed of arable land. (Dalziel 1948, 1956; Heine *et al.* 1988)

Echinochloa P.Beauv.

35 species. Warm regions.

Botanical notes. A difficult genus of intergrading forms. There is much uncertainty as to how this complex should be divided into species, and no reliable key to distinguish them (Clayton & Renvoize 1986).

Echinochloa colona (L.) Link

Panicum colonum L.

Tufted annual to 1m, often much less. Leaves linear, flat, ± glabrous. Ligule absent. Inflorescence of 4 to 12 one-sided, sessile 'racemes', 1–3cm long, arranged along an axis 3–12cm long. Spikelets 2-flowered, 2–3mm long, in 4 rows, greenish-purple. Glumes and lemma shortly hairy.

Ecology. Usually in damp ground near seepages and wadis. 30–1100m.

Distribution. Tropics and subtropics.

ETHNOGRAPHIC NOTES

Name(s)

– Soqotri: 'type of *ʾiʾiri, ʾīri*'; *mərẑif*; with inflorescence *ḳʌṭəhān* 'slender'.

– Soqotri simplified: 'type of *i'iri, 'iyri*'; *merjif*; *qatihan*.

Food (human). There is no tradition of gathering the grains as food on Soqotra.

Food (livestock). This small grass of wetter pastures is a nutritious grass well liked by all livestock, especially cattle. It is a valuable fodder grass for all livestock on Soqotra and is said to improve lactation. It is considered to be especially nutritious when in seed. It can be made into hay which lasts well.

Other. The seeds of the mature plants become lodged in clothing and in the coats of livestock.

Comparative uses. The grains are eaten in times of scarcity, and this grass is said to have been formerly cultivated in Egypt as a cereal – grains were found in the intestines of mummies dating back 6000 years. It is much valued as quick-growing fodder grass, said to fatten cattle and help their healthy growth. It is also reported to increase the milk output of buffaloes. It is a common weed of arable land. Hay made from it can be sacked for 5–6 years.

It is said to be an incomplete fodder for horses: if too much is fed, it causes diseased joints and excessive action of the kidneys. The related *E. haplocada* is said to be poisonous when small, but eaten when mature by all stock. The stems of the related *E. pyramidalis* are sweet and are eaten raw. (Ambasta 1982; Dalziel 1948, 1956; Goodin & Northington 1985; Jain & De Fillips 1991; Kunkel 1984; Perry & Metzger 1980)

Eleusine Gaertn.

Botanical notes. Active speciation in E Africa has made this a taxonomically difficult genus. It is close to *Acrachne* and *Dactyloctenium* (Clayton & Renvoize 1986).

Eleusine coracana (L.) Gaertn.

Eleusine indica sensu Balf.f.
Eleusine indica sensu Balf.f. subsp. *coracana*

Tufted annual to 1m. Leaves folded or flat. Inflorescence of 'racemes' 3–7cm long, 3–7mm wide, usually with a single 'raceme' below the main cluster. Spikelets 3 to 9-flowered, elliptic, glabrous, whitish-green.

Ecology. Once widely cultivated now rarely so – under-recorded on the distribution map.

Distribution. Cultivated in the Old World tropics and subtropics.

Botanical notes. Widely grown as a cereal (finger millet) in Africa, India and China, especially in upland areas. *E. coracana* subsp. *coracana* is considered to have been domesticated from *E. africana* Kenn.-O'Byrne (= *E. coracana* subsp. *africana* (Kenn.-O'Byrne) S.M.Phillips). It is a robust annual with thick stems and large leaves. The 'racemes' are thicker and the spikelets are plumper and rounder than in *E. africana*. The spikelets do not disarticulate between florets and the grain is brown, almost spherical and exposed when ripe. In other respects it resembles *E. africana*. In E Africa *E. africana* occurs as a weed in crops of *E. coracana*, with which it hybridises giving rise to a range of intermediates.

ETHNOGRAPHIC NOTES

Name(s)

– Soqotri: *bómbə* (pl. *bənɔ̄bihin*).

– Soqotri simplified: *bombih* (pl. *binowbihin*).

Food (human). This was the only cereal traditionally cultivated on Soqotra. With maize, it was also imported, mainly from East Africa. It was sold by the *qaḥf* measure, a standard measure used throughout Soqotra which was based on the merchant's measure made from a coconut shell. The islanders made a clay or wood version of exactly the same size. Even the grain for family household use was carefully measured out by the *qaḥf*-full to keep track of their consumption. The grain was usually ground coarsely and cooked in water to make a porridge. A household usually had two grindstones, one for maize and one for finger millet. Women also took their grain to grind it on communal grindstones which consisted of large rounded stones ('*migᶜélilo*', *mig'eliloh*) on an area of flat bedrock. These are quite common in the high limestone plateaus. The whole grains were also put on a clay dish on the fire and stirred around until they popped, like popcorn ('*migílᶜimo*', *migil'imoh*). This was commonly done if there was a shortage of water, or as a treat for children. The greenery of the young plants ('*cháᶜihum il sε̄bi*', *cha'ihum ilseybi*) was eaten, raw, for its sweetness and juiciness.

Food (livestock). Eleusine was cultivated primarily as a cereal, but the foliage and stems, green and dried, were also an important livestock feed. After the harvest, livestock were allowed in to the enclosures to feed on the stems of the harvested plants and dung the ground; or the plants were cut off at the base, the animals only being allowed in to graze the stumps, while the stems were dried elsewhere and stored for later use as fodder. Bundles of the harvested plants were laid out to dry in the sun for three days to a week. They were then wrapped up loosely in a sheep's wool rug and kept in a stone-built hay store or dry cave. As long as the hay was never allowed to become damp, it could last for up to four years or more and still be useful as a dry season fodder. Some people preferred to leave the plants to dry out in the enclosure before cutting and storing them. For two to three months after being cut, the foliage and stems were considered to be nutritious feed, for cattle especially. After this they were considered to have little nutritional value and were used only as a dry season feed – 'to give the animals something in their stomachs'.

Other. Finger millet was cultivated everywhere it was remotely possible to do so – above and along the edges of wadi beds, around springs and rainwater catchments, and on rock terraces: anywhere where there was an adequate supply of water within carrying distance. It was particularly cultivated around the wells and watercourses of the northern plains around Hadiboh and in the eastern Haggeher mountains. No cultivation of cereals was practised in the western half of Soqotra or along the southern plains.

In the Noged plain, only limited date palm cultivation was possible around the few brackish wells at the eastern end (at Heyf, Abataro, between Handaq and Abataro and in Di Azerhoh). The cultivation of finger millet was principally the job of women. From sowing of the seed to harvest took around three months. There were two principal seasons for sowing seed: at the very end of the monsoon/beginning of the winter rains, or at the very end of summer/beginning of the summer rains.

The cultivation of finger millet

Access to water for millet plots was always more important than access to suitable soil: poor soils could always be improved. Small seedbeds were prepared in which the soil was worked to a fine tilth, as well as larger plots to which the finger-millet seedlings were transplanted. Seed was collected from the previous year's crop, and stored separately in a carefully covered clay pot on a bed of ash, covered with more ash. The seed was inspected regularly for insect damage. The best fertilizer for finger millet was considered to be livestock manure, the best being aged sheep dung. This was carried to the plot and laid out in heaps which were then fired and the ash mixed in with the soil. This was done every year before sowing the seed. Once the seed was sown, the plot was watered regularly and kept scrupulously free of weeds. Once the seedlings were well established, after about three weeks, they were transplanted to the main plot. The need for frequent watering progressively lessened until once a day was sufficient, then every second day until the fruiting head ('*səbūlə*', *saboulih*) began to form. The main labour now was weeding rather than watering. Once the fruiting head formed, the plot was usually treated with another good load of manure, and sticks or pieces of date-rib were planted beside each plant and the fruiting head tied to it to support it and prevent it drooping as the grains ripened. The plot was then given another very thorough watering and then left without water for the week to ten days it took for the grains to ripen. It was customary at this time for members of the cultivator's family to move to live beside the plots to protect their crop from birds, rodents and human thieves.

Harvest

At harvest, the fruiting heads were cut off at the base and laid out on the ground for an hour or two in the sun. Once dried out, some of the best fruiting heads were put aside as seed, and the rest were put into sacks or cloth bags and left in the shade for about a week. At this stage gleaners were allowed to come in and gather any remaining grain. Some cultivators did not allow their animals in to feed on the stalks at this stage, but left the plants in the ground in the hopes that it might rain and the plants produce new shoots ('*fśaᶜ*', *fsha'*) and further small fruiting heads.

Threshing

The young and strong men of the neighbourhood did the threshing, and groups would often go round at harvest time doing the work in return for a share of the grain. The threshing was done by trampling the finger millet to loosen the grain from the head. Sometimes this was done indoors, the men supporting themselves on a length of rope looped over a roof beam, and twisting and turning and stamping on the grain below, surrounded by a circle of men and women who continually scooped the scattered heads back under the whirling feet. Sometimes a group of men shared the task, sitting in a circle around the pile of fruiting heads, one at a time getting up to trample the corn while the others scooped the flying fruiting heads from the edge of the pile back to the middle. It was done either on leather mats, or on an area of hard ground or flat rock swept clean, or on a threshing floor made by spreading a layer of damp clay-like soil over the ground, smoothing it as flat as possible and then leaving it to harden. The grains and bits of husk and other debris were then spread out on cloths or palm fibre mats in the sun and left to dry for two days.

Winnowing

This was done by women after having removed bits of stalk, gravel, husk and other debris from the piles of grain. They then went to an area where there was a good breeze and stood facing into it, pouring scoops of grain through a channel made by folding a circular, palm fibre mat in half, onto another larger, oblong mat below. After winnowing, the grain was left in the sun for at least a day to dry out completely, the women stirring it around and turning it at intervals. The fresh grain is pale cream in colour but grows steadily redder with age.

Storing the grain

The grain was stored in various ways:

(i) In clay pots with a lid of leather tied on while still damp so that on drying it shrunk to make a tight fit. Over this was placed a layer of dry earth or another clay pot to stop rodents nibbling away at the leather lid (aloes juice was never used on the porous clay for storing grain in case it tainted the grain). At times when the risk of theft was high, the pots were sometimes buried deep in the ground and the site carefully disguised by scattering plant debris and stones here and there to make it appear undisturbed.

(ii) In bags of leather, after treating the leather with the *tabtab* material. They were first soaked in water to soften the leather, and then packed with grain, ramming it down hard with a stick, while the bag was still damp. It was then left to dry, and when stiff as a board was smeared with fresh aloes juice or a thick layer of ash to deter rodents and other pests. Those who cultivated large quantities of grain, as in some areas of the northern plain and the eastern Haggeher, stored their grain in a large container called *'mósfi'* (*mosfi*), woven from date palm fibre. This was a large oblong basket with a matching lid that completely covered the top and sides. A platform of stones was laid out, either in a corner of the homestead or in a reliably dry and draught-proof cave, and the *mosfi* carried (it took four men to carry a full one) and laid on top. The whole area was then covered with a layer of ash and then a layer of dry earth.

Nowadays, many of the former millet fields have been given over to the cultivation of dates or, on the northern plain, to the cultivation of vegetables and fruits, and in eastern Haggeher, to the cultivation of sweet potatoes.

Comparative uses. A common name for finger millet in Africa is *wimbi*, from which it is probable that the Soqotran *bombə* is derived. Finger millet is occasionally found at roadsides as an escape, and as a weed of arable land. It is widely cultivated as a cereal in Africa and India, both as a staple food and for malt. In Nigeria it is grown also to make beer (which is regarded as a food) and to make a sweet black gruel to mix with molasses. In many parts of Central and Eastern Africa it is still cultivated. In India the straw is often used as low-quality fodder, and in northern, subtropical Punjab it is grown for green fodder or hay and then the herbage is said to be more nutritious.

Medicinally, in India the grains are seen as being tonic and astringent, and are considered to be useful in biliousness and are specially recommended as a wholesome food for diabetics. In parts of Africa the grains, with *Plumbago zeylanica*, are used as an internal remedy for leprosy, and the leaf juice is given to women in childbirth. In Sumatra the plant is used as a vermifuge; in Indo-China as a diaphoretic and remedy for diseases of the liver; and in Vietnam as a diuretic. In Goa the flowers are used as a remedy for chest conditions. Elsewhere in Asia the plant is boiled and eaten by people with stomach troubles and to treat fevers. The whole plant of the related *Eleusine indica*, especially the roots, is said to have sudorific and anti-febrile qualities. It is given to treat malaria, and the leaf juice is drunk to promote discharge of the afterbirth and to make a medicine to treat worms. A decoction of the roots is taken as an astringent for the bowels. Mixed with *Entada phaseoloides*, the plant is used to make a wash for the hair to treat dandruff and prevent hair loss. The plant is also burned and smeared over the body of a patient suffering from a relapse of any illness, the patient also inhaling the fumes. (Abbas *et al.* 1992; Ambasta 1982; Bogdan 1977; Dalziel 1948, 1956; Jain & De Fillips 1991; Kunkel 1984; Perry & Metzger 1980)

Enneapogon Desv. ex P.Beauv.

Tufted perennials with wiry stems. Ligule a fringe of hairs. Leaves usually inrolled, occasionally flat, shortly hairy. Inflorescence a contracted panicle at the tip of the stem. Spikelets 3 or 4-flowered, 1 fertile floret. Glumes slightly unequal, longer than the lemmas. Lemmas hairy, 9-nerved, with 9 awns. Uppermost floret reduced to a small tuft of awns. 28 to 30 species. Tropics and subtropics.

Botanical notes. A very uniform genus whose species resemble one another (Clayton & Renvoize 1986). Technical characters are needed to separate them.

Enneapogon desvauxii P.Beauv.

Enneapogon brachystachyus (Jaub. & Spach) C.E.Hubb.
Enneapogon elegans (Nees ex Steud.) Stapf
Pappophorum aucheri sensu Balf.f. non Jaub. & Spach
Pappophorum elegans Nees

Tufted perennial, or sometimes annual, to 18(–40)cm. Inflorescence a contracted panicle 2–8cm long, greyish-white. Spikelets 3-flowered, third floret reduced to a small tuft of awns, 0.5–2mm long.

Ecology. *Croton socotranus* shrubland; 50–200m.

Distribution. Africa, Arabia, India and Pakistan.

ETHNOGRAPHIC NOTES

Name(s)

– Soqotri: *ɛ̄rśído*; with inflorescence *ḳʌ́ṭəhān* 'slender'.

– Soqotri simplified: *ershiydo*; *qatihan*.

Food (human). The new shoots and green stems are nibbled by people, but are not very sweet.

Food (livestock). These perennial grasses of rocky and stony ground are grazed by all livestock, particularly small stock, but are not considered to be particularly nutritious. Their value to stock lies in their persistence throughout the long months of dry season, and at this time they are a very important fodder grass all over the island.

Comparative uses. Elsewhere in the Middle East this is regarded as a useful fodder and pasture plant, but of poor grazing value. (Goodin & Northington 1985)

Enneapogon persicus Boiss.

Enneapogon schimperianus (Hochst. ex A.Rich.) Renvoize

Tufted perennial, 15–45(–60)cm. Leaves often diverging at a wide angle from the stem. Inflorescence a contracted panicle <18cm long, usually grey. Spikelets 3 or 4-flowered. Third lemma often well developed (but sterile), 3–5(–9)mm long, including awns, sometimes accompanied by a vestigial fourth floret.

Ecology. Desertic plain; doubtfully recorded from a single gathering.

Distribution. Tropical Africa to India and Pakistan.

Botanical notes. *E. schimperianus* is considered to be a synonym of *E. persicus* (Phillips 1995). This species is recorded from the island but no specimens have so far been seen. It is possible that the single record (*BCS* 110) is the result of an incorrect identification as *E. persicus*.

Eragrostis Wolf

Annuals or perennials. Leaves linear, flat or inrolled. Leaf sheaths glabrous or hairy. Ligule a fringe of hairs. Inflorescence an open or contracted panicle. Spikelets many-flowered, breaking up either from the bottom upwards, the top downwards or between the florets. Glumes usually subequal, shorter than the spikelet. Lemmas usually glabrous, awnless. Paleas equalling lemmas, sometimes persistent, glabrous or conspicuously fringed. c.350 species. Tropics and subtropics.

Botanical notes. There is no modern overall account of the genus, but its taxonomy appears to be related to the mode of spikelet disarticulation which is remarkably diverse. This character is, however, probably subject to some parallel evolution (Clayton & Renvoize 1986).

Eragrostis aspera (Jacq.) Nees

Tufted annual to 1.1m. Leaves flat, sheaths hairy. Inflorescence a large, diffuse, open panicle 10–50cm long, with long, delicate branches. Axils of primary branches with short tufts of hair. Spikelets 5 to 20-flowered, linear, 3–8mm long, often purplish, breaking up from the top downwards. Glumes subequal.
Ecology. Drought-deciduous woodland. 550m.
Distribution. Tropical and S Africa, Arabia, India and Pakistan.
Botanical notes. Known on Soqotra only from a single, rather anomalous specimen.

Eragrostis barrelieri Daveau

Annual to 60cm. Leaves flat or inrolled, sheaths glabrous. Inflorescence an open or ± contracted panicle, 3–15cm long, with short, often glandular, branches and pedicels. Spikelets 8 to 25-flowered, linear, 6–15mm long, 1.5–2mm wide, purple or grey, breaking up from the bottom upwards, the palea persistent after the lemma has fallen. Glumes slightly unequal, the lower smaller than the upper.
Ecology. Grassland on limestone plateaus. 550–550m.
Distribution. Mediterranean, tropical Africa, Arabia.

ETHNOGRAPHIC NOTES†
Name(s)
– Soqotri: *mə̍rẑif*; (rare) *mi̍śraᶜ*; with inflorescence 'type of *ḳʌ̍ṭəhān*' 'slender'.
– Soqotri simplified: *merjif*; *mishra'*; 'type of *qatihan*'.
Food (livestock). This grass of dry grassland and stony ground is a common weed of arable land. It is grazed by livestock in the absence of other grasses, but is not regarded as a particularly nutritious feed.

†Also apply to *E. lepida*.

Eragrostis cilianensis (All.) Vignolo ex Janch.

Tufted annual to 75cm. Leaves flat or inrolled, glabrous or hairy, sometimes with small wart-like glands at the edges, sheaths usually hairy above. Inflorescence an open panicle, 5–15cm long, with relatively short, stiff branches, occasionally very contracted to a dense head, the pedicels often glandular. Spikelets 8 to 30-flowered, oblong to linear, 4.5–15mm long, 2.5–4mm wide, greyish, breaking up from the bottom upwards, the palea persistent after the lemma has fallen. Glumes subequal.
Ecology. Dwarf shrubland. 30–650m.
Distribution. Old World tropics and temperate regions.

ETHNOGRAPHIC NOTES
Name(s)
– Soqotri: nil or with inflorescence *ḳʌ̍ṭəhān* 'slender'; *mʌ̍shiṭ* 'sweeping brush', from the large plumed panicle.
– Soqotri simplified: *qetimamhion*; *qatihan*; *mas'hit*.
Food (human). There is no tradition of gathering the grains as food on Soqotra.
Food (livestock). Rather stemmy, but not hard, and well liked by all livestock and regarded as a useful and fairly nutritious grass. It grows in open ground in scrub, bush, in old cultivation, in overgrazed areas and as a weed of arable land. Islanders say it varies greatly in size according to how much rain it receives. When green it is seen as promoting lactation. Cattle are very keen on it and search it out when it is in seed. It is a very useful dry season fodder, especially for small stock, and is highly valued in the outer islands.
Comparative uses. A number of *Eragrostis* annuals are common but not very serious weeds of arable land. Many (such as *E. curvula*, *E. lehmannia*, *E. superba*, *E. tef*) are regarded as suitable for erosion control. Where *E. cilianensis* proliferates it can be a sign of overgrazed pastures. In Africa this is often a dominant or subdominant grass in areas with about 250mm of annual rainfall where perennial grasses cannot survive. It is a pioneer grass which colonises denuded land in higher rainfall areas. The seeds are abundantly produced and recommended for reseeding denuded pastoral land in arid, annual-grass zones. It is a common weed of arable land. The grains are eaten in times of hunger, and the straw used in parts of Africa for matting and thatching. (Ambasta 1982; Bogdan 1977; Dalziel 1948, 1956; Jain & De Fillips 1991; Kunkel 1984)

Eragrostis ciliaris **(L.) R.Br.**

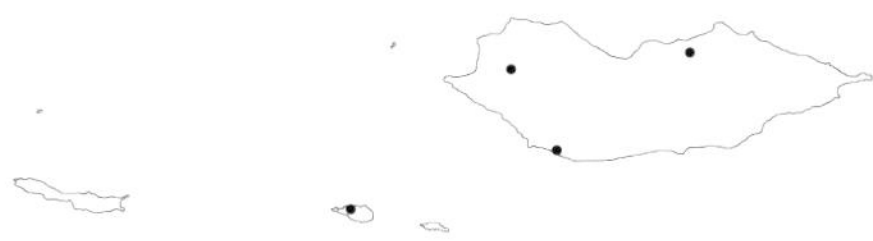

Small tufted annual to 50cm. Leaves flat, leaf sheaths hairy. Inflorescence a contracted panicle (either spicate or subglobose), 1–12cm long, with short primary branches and densely packed spikelets. Spikelets 6 to 12-flowered, ovate, 1.5–4mm long, greenish-purple with a conspicuously fringed palea, breaking up from the top downwards. Glumes subequal.
Ecology. Succulent shrubland and drought-deciduous woodland. Sea-level–600m.
Distribution. Tropics.

ETHNOGRAPHIC NOTES
Name(s)
– Soqotri: 'type of small *ˈiʾiri*' (Soqotra, Samha); *ḳeṭəmˈʌmhion* 'producing butterfat'; with inflorescence *di ḳerṭˈomo* 'needs chewing'.
– Soqotri simplified: 'type of *i'iri*'; *qetimamhion*; *di qertomo*.
Food (human). There is no tradition of gathering the grains as food on Soqotra.
Food (livestock). This is regarded as a valuable and nutritious fodder after rain, especially in the drier areas of the island and in Samha. It is seen as promoting lactation (as its name 'producing butterfat' suggests). It is palatable to all livestock, especially small stock, but cattle also readily graze it. It produces a lot of seed, and is regarded as especially nutritious at this time.
Comparative uses. It is a common weed of arable land and occurs widely in eroded soil and in overgrazed areas. The grains are collected in times of scarcity. In parts of Africa this and other *Eragrostis* species are some of the most valued fodders, and they are cut and sold and sometimes stacked for hay. (Ambasta 1982; Dalziel 1948, 1956; Goodin & Northington 1985; Jain & De Fillips 1991; Kunkel 1984)

Eragrostis lepida **(A.Rich.) Hochst. ex Steud.**

Tufted annual to 40cm. Leaves flat, leaf sheaths sparsely hairy. Inflorescence an open, delicate panicle, 3–10(–15)cm long. Spikelets 4 to 10-flowered, elliptic, 2–3.2mm long, greenish-purple with a conspicuously fringed palea, breaking up between the florets. Glumes unequal, the lower slightly shorter than the upper.
Ecology. Drought-deciduous woodland. 550m.
Distribution. NE Africa, Arabia.

ETHNOGRAPHIC NOTES
Name(s)
See *E. barrelieri* (also uses).

Eragrostis minor **Host**

Tufted annual to 60cm. Leaves flat or, occasionally, inrolled, usually glabrous with wart-like glands at the edges. Inflorescence a small open panicle, 3–20cm long, with short, usually glandular, pedicels. Spikelets 8 to 16-flowered, narrowly oblong, 0.5–1cm long, 1.5–2.2(–2.5)cm wide, greyish-green, breaking up from the bottom upwards, the palea persistent after the lemma has fallen. Glumes subequal.
Ecology. No details of habitat or location available.
Distribution. Old World subtropics and warm temperate regions.

ETHNOGRAPHIC NOTES
Not seen, no information recorded.

Eragrostis papposa **(Roem. & Schult.) Steud.**

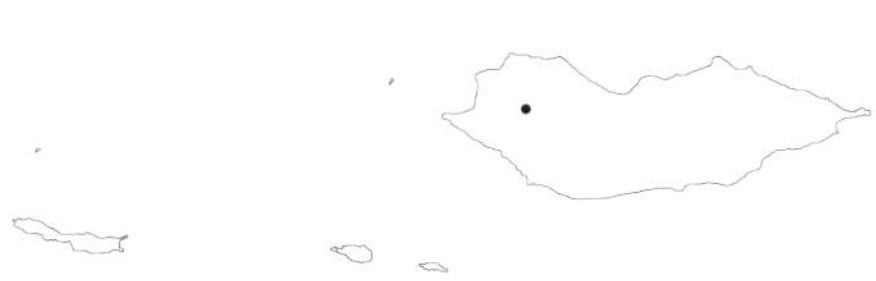

Tufted perennial to 75cm. Leaves inrolled, stiff, forming a basal cushion, leaf sheaths glabrous. Inflorescence an open or compact panicle, 5–22cm long. The axils of the primary branches glabrous. Spikelets 5 to 18-flowered, linear, 3–8.5mm long, grey-green, breaking up from the bottom upwards, the palea persistent after the lemma has fallen. Glumes small, markedly unequal, the lower shorter than the upper.
Ecology. Shrubland dominated by *Croton socotranus* and *Jatropha unicostata* on limestone plateau. 500m.
Distribution. Mediterranean, NE Africa, Arabia, India and Pakistan.
Botanical notes. Known only from one rather inadequate specimen (*Miller* 14215).

ETHNOGRAPHIC NOTES
Name(s)
– Soqotri: *mˈurẑif*; *ḳˈorźʌʾ* (more commonly given to various mainly perennial Gramineae and Cyperaceae of damp ground); *śebətˈano* 'never dies right back'.
– Soqotri simplified: *murjif*; *qordha'*; *shebitano*.
Food (livestock). Commonly occurs in dry and stony soils or on rocky ground, especially in the limestone plateaus. It is regarded as having little nutritional value, its main value lying in its persistence through the dry season, especially in damp areas.

Eragrostis viscosa **(Retz.) Trin.**

Sticky, tufted annual to 40cm. Leaves flat, sticky along the midnerve below, leaf sheaths hairy, sticky on the keel and margins. Inflorescence an open panicle, 5–15cm long. Sticky, yellow glandular patches on the axis and branches. Spikelets 5 to 15-flowered, oblong, <4.5mm long, green or purplish, with a fringed palea, breaking up between the florets. Glumes subequal, glandular.
Ecology. No details of habitat or location available.
Distribution. Tropical Africa to tropical Asia.
Botanical notes. Only a single specimen seen (*Hyam* 127) (see Botanical notes under family entry).

ETHNOGRAPHIC NOTES
Name(s)
– Soqotri: *mu̍ržif; ḳeṭəmʌ̍mhion* 'producing butterfat'.
– Soqotri simplified: *murjif; qetimamhion.*
Food (livestock). This short annual with viscous stems and leaf sheaths is eaten by all livestock and is regarded as a superior fodder (as its name, 'producing butterfat', suggests). It is well liked by cattle, and is seen as an excellent forage for all young stock. It is found mainly in the uplands and on wooded slopes.

Gastridium P.Beauv.

2 species. Canary Is., Mediterranean to Iran.

Gastridium phleoides **(Nees & Meyen) C.E.Hubb.**

Annual to 60cm. Leaves flat, linear. Ligule membranous. Inflorescence a narrow, pale green, contracted panicle 2–8cm long. Spikelets 1-flowered, overlapping, linear, 5–6mm long, glabrous. Glumes lanceolate, unequal, much longer than floret. Lemma shortly hairy, awned (awn 4–7mm long).
Ecology. Low submontane shrubland (with *Hypericum scopulorum, Euryops arabicus*). 1100m.
Distribution. Mediterranean, NE Africa, Arabia, SW Asia.

ETHNOGRAPHIC NOTES
Name(s)
– Soqotri: *re̍hišt, re̍hilt* 'chapped, cracked skin'.
– Soqotri simplified: *rehisht, rehilt.*
Food (livestock). This grass of the Central Highlands is well liked by livestock, by cattle especially, for whom it is regarded as an excellent and nutritious feed, putting on weight and improving lactation.

Hackelochloa Kuntze

2 species. Tropical regions.

Hackelochloa granularis **Kuntze**

Annual to 1m. Leaves coarse, linear-lanceolate, hairy. Ligule a short, hairy membrane. Inflorescence a single 'raceme' 5–15mm long, these gathered into a false panicle. Spikelets 2-flowered, paired, one sessile, one pedicelled. Sessile spikelet 1–1.3mm long. Lower glume pitted and tuberculate. Upper glume crested at the tip. Pedicelled spikelet 1.5–2.5mm long, narrowly winged.
Ecology. Disturbed ground in woodland and along wadis. 50–300m.
Distribution. Tropics.
Botanical notes. A very distinctive species with much-modified globose, tuberculate spikelets.

ETHNOGRAPHIC NOTES
Name(s)
– Soqotri: nil or with inflorescence *di ḳerṭo̍mo* 'needs chewing'.
– Soqotri simplified: *di qertomo.*
Food (livestock). This is grazed by all livestock after rain. When still young it is seen as a good feed, but as it ages is not regarded as a particularly valuable or nutritious grass.
Comparative uses. Commonly a weed of arable land. Medicinally, it is used in the treatment of enlarged spleen and liver. (Ambasta 1982; Bogdan 1977; Jain & De Fillips 1991)

Halopyrum Stapf

1 species. Indian Ocean coasts.

Botanical notes. A genus of uncertain affinity, perhaps distantly related to *Leptochloa* (Clayton & Renvoize 1986).

Halopyrum mucronatum (L.) Stapf

Robust stoloniferous perennial to 1m. Stems stiff and woody. Leaves stiff, linear, inrolled, wiry at the tip. Ligule a fringe of hairs. Inflorescence of short 'racemes' appressed to an elongated axis <26cm long, appearing as a single terminal 'raceme'. Spikelets 8 to 25-flowered, pedicellate, large, 1–3cm long, glabrous, straw-coloured, unawned.
Ecology. Locally abundant on coastal sand dunes. Sea-level.
Distribution. Indian Ocean coasts.

ETHNOGRAPHIC NOTES
Name(s)
– Soqotri: 'type of *ḳeṣádə*' 'sharp, cutting', 'oversalted'.
– Soqotri simplified: 'type of *qizadih*'.
Food (livestock). This tough and tussocky perennial with its woody stems is little liked by livestock, though hungry animals graze new and tender growth. Camels and donkeys graze it when there is little else to eat.
Other. Pieces of the dry plant are gathered for use as kindling and firelighters.

Heteropogon Pers.

6 species. Africa and S Europe.

Heteropogon contortus (L.) P.Beauv. ex Roem. & Schult.
Heteropogon hirtus Pers.

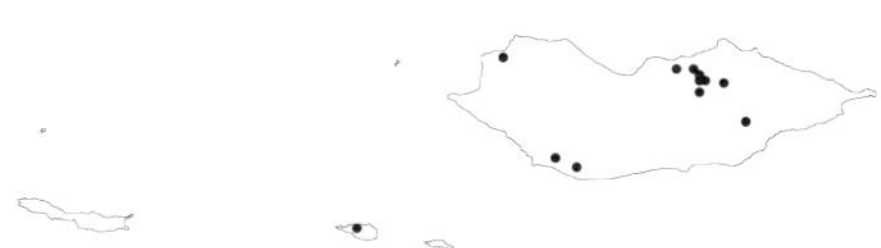

Tufted perennial to 1m. Leaves linear, flat, glabrous or hairy, leaf sheaths conspicuously flattened. Ligule a short, finely hairy membrane. Inflorescence a single, terminal 'raceme' 4–7(–10)cm long, sometimes partially enclosed by the uppermost leaf sheath. Spikelets 2-flowered, paired, one sessile and one pedicelled. Sessile spikelets 5.5–10mm long, fertile, dark brown, hairy, awned (awn hairy, brown, 5–10cm long, bent at the middle with a long twisted column). Tips of successive awns often twisted tightly together. Pedicelled spikelets 5–15mm long, male or sterile, larger than the sessile spikelets, awnless, glabrous or hairy.
Ecology. Abundant in grassland and woodland clearings in high rainfall areas; on Samha on humid north-facing cliffs. 300–1100m.
Distribution. Tropics and warm temperate regions.
Botanical notes. A characteristic grass of woodland and grassland in areas which attract cloud and rain. The stout, brown intertwining awns are very characteristic of this species. The entangled awns often cause the sessile spikelets to spring out together in a bunch. These awns readily catch in clothing and the coats of livestock and can be extremely irritating. The species is extremely polymorphic in habit and hairiness of the spikelets. It is known to be apomictic and to possess several different chromosome numbers (Phillips 1995).

ETHNOGRAPHIC NOTES
Name(s)
– Soqotri: *di ʾɔ̄žhir*; *ʾēžəhər*, *ʾājéyr* (western Soqotra) 'spiky', 'painful'; with inflorescence *di ḳerṭómo* 'needs chewing'.
– Soqotri simplified: *di 'owjihir*; *'eyjihir*, *'ajeyr*; *di qertomo*.
Food (livestock). This grass, commonly known as 'spear grass', is one of the most important and widespread rangeland grasses of the island. It dominates the plateau areas around the Haggeher mountains, the Diksam area, for instance. It is well liked by larger livestock when green and tender, and is a key cattle graze. When immature it is nutritious, but it becomes unpalatable at the seeding stage, when it becomes a nuisance to livestock and people alike. This is due to its sharp seeds and long, twisted awns with their hard pointed base.
Comparative uses. The seed, enclosed in a hard-pointed and retrorsely barbed scale, penetrates the hides of animals, causing irritation and damage. Both seeds and awns can cause real injury when the grass is dense, to sheep especially, as they readily become entangled in the wool and the sharp callus irritates and even penetrates the skin. This can result in economic loss of mutton, wool and hides. In Australia, where it covers some 30 million hectares, it is of considerable economic importance. It is also highly resistant to grass fires (due to the twisting movement of seeds caused by the movement of the awn in response to changes in air humidity, the seed is buried in the soil, thus contributing to the fire tolerance of the grass; moreover, high soil temperature stimulates germination). It thrives under light grazing, whereas heavy grazing can suppress or eliminate it (there is, however, conflicting evidence about the effect of heavy grazing). Sheep consume only moderate amounts of the herbage and digest it poorly. It is used for silage and hay, sometimes also for thatching and woven into matting. The cellulose content of the grass is sufficiently high to warrant being used for paper manufacture. The roots are considered to be stimulant and diuretic. The roots, shoot and flowers are used as a demulcent and anti-febrile. With *Tribulus terrestris* it is used to treat rheumatism in the hands. (Ambasta 1982; Bogdan 1977; Dalziel 1948, 1956; Jain & De Fillips 1991; Perry & Metzger 1980; Watt & Breyer-Bandwijk 1962)

Hyparrhenia E.Fourn.

c.55 species. Africa, Madagascar, tropical America and Asia.

Botanical notes. Distinguished from all other similar genera by the awn which is hairy in the lower part (Wood 1997). The genus is notorious for introgression among its species, many of which are difficult to distinguish (Clayton & Renvoize 1986).

Hyparrhenia hirta (L.) Stapf

Andropogon hirtus L.

Tufted perennial to 1m. Leaves flat, bluish-green, spread along the length of the stem. Ligule a short membrane. Inflorescence of paired finger-like 'racemes', 2–3cm long, at the tip of the stem, enclosed by a large bract (spatheole), 3–8cm long, green or dark red when old, sometimes hairy at the top. Spikelets in pairs (one sessile and one pedicelled). Sessile spikelet fertile, 4–6.5mm long, 2-flowered and awned (awn 1.5–2.5cm long, hairy, brown). Pedicelled spikelet sterile and unawned, 5–7mm long. Glumes unequal.

Ecology. Rocky ground. 300–650m.
Distribution. Mediterranean to S Africa, Arabia, India and Pakistan.
Botanical notes. Unaccountably not seen recently on the island – possibly overlooked and under-recorded. On the mainland it is one of the most abundant and drought-tolerant grasses.

ETHNOGRAPHIC NOTES

Comparative uses. Like other *Hyparrhenia* species, it is used for thatch and matting, but only after the sharp spikelets have fallen. It was tried under cultivation in Australia, but without much success. It has been considered suitable for erosion control. *H. rifa* has been introduced into cultivation and is of considerable importance in Latin America, especially in Brazil. (Bogdan 1977; Dalziel 1948, 1956)

Imperata Cirillo

8 species. Tropical and warm regions.

Imperata cylindrica (L.) Rauesch.

Imperata arundinacea Cirillo

Coarse, densely tufted perennial to 120cm, spreading by rhizomes to form dense clumps. Ligule scarious. Leaves mostly basal, stiff and wiry, flat or inrolled. Inflorescence a dense, hairy, white panicle, 10–20cm long. Spikelets 2-flowered, in pairs, 3–6mm long, enveloped in long, white hairs. Glumes subequal, hairy, longer than the lemmas. Lemmas hairy.
Ecology. No habitat details available.
Distribution. Old World tropics and temperate regions, S America.
Botanical notes. Reported in the 19th century (Forbes 1903) but not seen since. A pernicious weed in the Old World tropics.

ETHNOGRAPHIC NOTES

Comparative uses. It is a very persistent and vigorous grass weed. Its hard, almost spine-like top spikes the heels and injures the hoofs of animals. In India it is used in mixture with long-fibred paper pulp. It is considered suitable for land reclamation and is a good soil binder. It is used for ropes, mats and cowry bags, baskets and plates. The rhizomes and young inflorescences are eaten. The ash of the plant is used as a salt substitute. The flossy inflorescences are used to stuff cushions. Medicinally, the most important part is the rhizome. It can be collected at any time: the best is white, very sweet and cooling. It is considered to be emollient, haemostatic and anti-febrile. It is used to make medicine for chest colds in children, for hiccups and as an indigestion remedy, and is boiled as an antipyretic and taken as a drink to relieve painful joints. It is used to treat flu, haematuria, internal haemorrhage, as a haemostatic and diuretic, to treat oedema, jaundice, coughing and kidney disease. The ash of the rhizome is given in tea to treat nosebleed, and used alone to treat rheumatism. It is also put into the bath of a child with itch. The flowers or sprouts are thirst-relieving and digestive, and are also prescribed for haemorrhage and wounds and to treat the bites of poisonous insects. (Abbas *et al.* 1992; Ambasta 1982; Bogdan 1977; Dalziel 1948, 1956; Ghazanfar 1994; Ghazanfar & Al-Sabahi 1993; Goodin & Northington 1985; Jain & De Fillips 1991; Kunkel 1984; Perry & Metzger 1980; Watt & Breyer-Bandwijk 1962)

Ischaemum L.

1 endemic. 60 species. Tropical and warm regions.

Botanical notes. There is no adequate treatment of the genus available (Clayton & Renvoize 1986).

Ischaemum sp. *A*

Leaves linear. Ligule membranous. Inflorescence of paired 'racemes', sometimes locked back to back and appearing as a single 'raceme'. Rhachis internodes and pedicels stout.

Ecology. No details of habitat or location available.
Distribution. Endemic.
Conservation status. Data Deficient (DD). Inadequate evidence is available to make an assessment.
Botanical notes. Collected only once (*Hyam* 17); the specimen is not adequate for a full description.

ETHNOGRAPHIC NOTES
Comparative uses. The grains of the related *I. rugosum* are gathered as famine food. (Kunkel 1984)

Leptothrium Kunth

2 species. Caribbean and Senegal to Pakistan.

Botanical notes. The spikelets or spikelet pairs represent reduced racemes in which the pedicels have united to form a single, flattened peduncle bearing the spikelets on its tip (Phillips 1995).

Leptothrium senegalense (Kunth) Clayton

Tufted annual or short-lived perennial to 75cm. Ligule a ring of hairs. Leaves flat, narrow, bluish-green. Inflorescence a single, terminal 'raceme' 2–15cm long. Spikelets 1-flowered, solitary or in pairs on short, flattened, green peduncles, dark purple, awnless. Lower glume flat and recurved, with hooked hairs at the edges, extending beyond the rest of the spikelet. Upper glume densely covered with spiny tubercles and tightly enclosing the flower.

Ecology. Wadi on limestone. 100m.
Distribution. Tropical Africa, Arabia, SW Asia.
Botanical notes. Rare, possibly overlooked. The dark purple spikelets are highly modified and very characteristic, with hooked margins and a flat recurved tail.

ETHNOGRAPHIC NOTES
Comparative uses. It has been successfully used for reseeding denuded areas, for example in central Kenya – it seeds abundantly and the seed is easy to collect. The grains of *L. senegalense* are gathered as food in parts of Africa. (Bogdan 1977; Kunkel 1984)

Lepturus R.Br.

Ischnurus Balf.f.

Perennials, often stoloniferous. Leaves flat. Ligule a fringed membrane. Inflorescence a single, terminal 'raceme'. Spikelets 1-flowered, embedded in the rhachis. Lower glume minute or absent (although well developed in the terminal spikelet). Upper glume leathery, pressed up against the side of the rhachis covering the sunken floret. 3 endemics. 8 species. Indian and W Pacific Oceans.

Botanical notes. The precise affinity of the genus is contentious. Its nearest relatives are thought to be *Oropetium* in the Eragrostideae and *Lepturopetium* in the Cynodonteae. For the time being it has been placed in its own monotypic tribe, Lepturеae (Clayton & Renvoize 1986).

Lepturus calcareus Cope

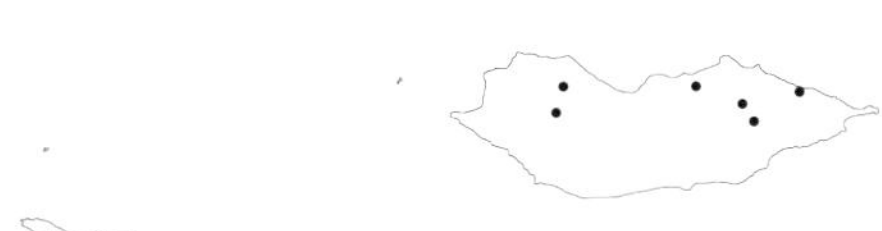

Loosely tufted perennial to 25cm. Inflorescence a single 'raceme' 1.2–2.5cm long. Spikelets upper glume broadly ovate, 1.6–1.8mm long, flat on the back, its margins flush with those of the rhachis joint.
Ecology. Succulent shrubland, drought-deciduous woodland and along wadis. 100–600m.

Distribution. Endemic.
Conservation status. Data Deficient (DD). Poorly known; inadequate evidence is available to make an assessment.

ETHNOGRAPHIC NOTES†
Name(s)
– Soqotri: *múržif*; *di ʾɔ̄žhir*; *ʾēžəhər*, *ʾājéyr* (western Soqotra) 'spiky', 'painful'; with inflorescence *di ḳerṭómo* 'needs chewing'; *L. tenuis* *ʿówśiḳ*; *L. tenuis* 'type of *ḳeṣádə*' 'sharp, cutting', 'oversalted' (Abd al Kuri); *saʿadífo* 'clump with persisting rootstock' (Samha).
– Soqotri simplified: *murjif*; *di 'owjihir*, *'eyjihir*, *'ajeyr*; *di qertomo*; *'owshiq*; 'type of *qezadih*'; *sa'adifo*.

Food (livestock). These grasses are liked by all livestock when green, but are not considered to have a high nutritional value. Their main value lies in their persistence through the dry season when other fodder is scarce. In the drier areas of Soqotra and on the outer islands, *L. tenuis* is a key rangeland grass, and on Abd al Kuri it is the dominant grass in areas below the top of the summit. On Samha the sheep and goats of the island graze *L. tenuis* heavily and it is regarded as an important feed for them.

Comparative uses. Many *Lepturus* species are considered to make useful sand binders. (Bogdan 1977)

†Also apply to *L. tenuis*.

Lepturus pulchellus **(Balf.f.) Cope**

Ischnurus pulchellus Balf.f.

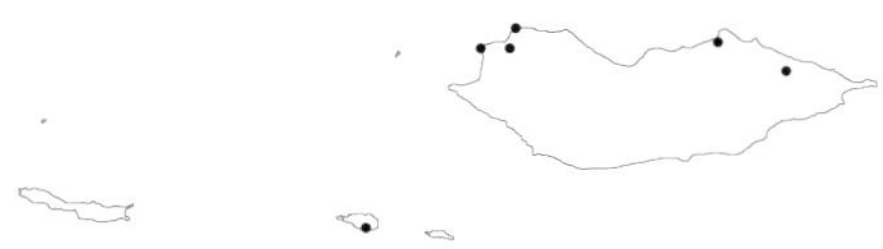

Densely tufted perennial to 15(–40)cm, with keeled, fan-shaped lower leaf sheaths. 'Raceme' 1.5–2.5cm long. Otherwise as above but with the upper glume concave on the back and the spikelets 'embraced by a triangular wing on each margin of the rhachis joint'.

Ecology. Coastal dwarf shrubland on Soqotra; desertic foothills on Samha. 10–200m.

Distribution. Endemic.

Conservation status. Data Deficient (DD). Poorly known; inadequate evidence is available to make an assessment.

ETHNOGRAPHIC NOTES

Name(s)

– Soqotri: *iʾiri*, *ʿiyri*; 'type of *ʾibihir*'; *riśiṭəno* 'small *ɛrchīṭo* grass'; beside sea 'type of small *ḳeṣādə*' 'sharp, cutting', 'oversalted'; with inflorescence *ḳʌṭəhan* 'slender'.
– Soqotri simplified: *i'iri*, *'iyri*; 'type of *ibihir*'; *rishitino*; 'type of small *qezadih*'; *qatihan*.

Food (livestock). In the drier coastal areas this is an important livestock fodder. It is regarded as a good fodder for all livestock when young and tender, but as it matures it becomes tough and wiry (hence its inclusion among the *ḳezədə* type of grasses) and less palatable to livestock. However, like all perennial grasses, its value lies in its persistence through the dry season. It is more highly valued on the island of Samha.

Lepturus tenuis **Balf.f.**

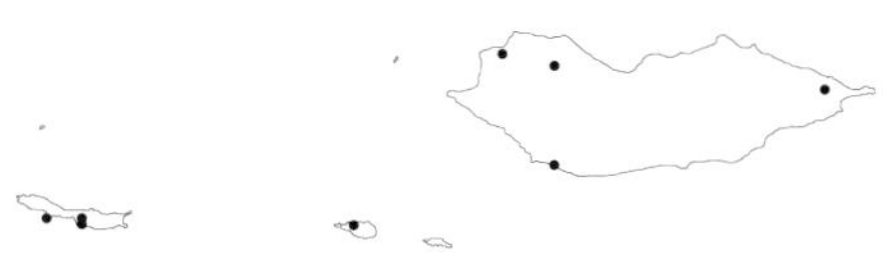

Loosely tufted perennial to 30cm. 'Raceme' 4–6cm long. Otherwise as above but with the upper glume lanceolate, 5–5.5mm long, convex, 'scarcely embraced by a narrow wing on each margin of the rhachis joint'.

Ecology. Dwarf shrubland on limestone. 200–700m.

Distribution. Endemic.

Conservation status. Data Deficient (DD). Poorly known; inadequate evidence is available to make an assessment.

Botanical notes. A single specimen (*Miller* 19075) collected on Samha probably represents a new species. Unfortunately, the material is inadequate for a formal description. It resembles *L. tenuis* but is a stouter plant with differences in the shape of the spikelet. It was found in low shrubland on north-facing limestone cliffs at 400–600m.

ETHNOGRAPHIC NOTES

Name(s)

See *L. calcareus* (also uses).

Lolium L.

Annuals or perennials. Leaves linear, flat or inrolled. Ligule membranous. Inflorescence a single, terminal 'raceme'. Spikelets 3 to 15-flowered, arranged alternately. Lower glume absent except on the terminal spikelet and there similar to the upper. Upper glume large. Lemma awned or awnless. 8 species. Temperate Eurasia, N Africa.

Botanical notes. All the species are more or less interfertile and are consequently difficult to separate as they intergrade morphologically (Clayton & Renvoize 1986).

Lolium persicum **Boiss.**

Tufted annual to 60cm. Inflorescence a 'raceme' 5–12 (–21)cm long. Spikelets 4 to 11-flowered, <22mm long. Glume as long as or usually shorter than the spikelet. Lemma always awned (awn <18mm long).

Ecology. Montane woodland and shrubland on granite. 500–1200m.

Distribution. Arabia, temperate Asia.

ETHNOGRAPHIC NOTES

Name(s)

– Soqotri: (*di*) *ṣəffo*; *ṣift*, *di ṣəfət*, *ṣāfhan* (western Soqotra); *məntɛyoh* 'of cliffs'; *ḥʌbəžihin*, *ḥʌbəchihin* 'pubes'.
– Soqotri simplified: *di ziffoh*; *zift*, *di zifit*, *zafhan*; *minteyoh*; *habijihin*, *habichihin*.

Food (livestock). This annual grass is regarded as a valuable feed when young and green, for small stock as well as for cattle. Its nutritional value drops sharply as it dries out.

Comparative uses. *Lolium* species have the highest nutritive value of temperate grasses. Most are large-seeded with excellent establishment characteristics. They generally demand better moisture and fertility conditions than other temperate species. Some species are widely cultivated as pasture grasses.

This species of ryegrass was in ancient times known as a poisonous plant and may have been the plant referred to in the Bible as 'tares'. The seed is said to be the poisonous part, producing dizziness, headache, somnolence and general intoxication in man, dog, sheep and horse, but not in the hog, cow or in poultry. The plant is said not to be toxic before it comes into seed. In Morocco a decoction of the plant is given to treat haemorrhage and incontinence of urine. The seed is used as a poultice in skin diseases and internally in the menopause. Some *Lolium* species, such as *L. perenne* and *L. multiflorum*, are regarded as valuable fodder grasses for grazing and hay, and are widely cultivated as pasture grasses. The grains of *L. perenne*, which contain gluten similar to wheat gluten, are ground to make bread in times of want. (Ambasta 1982; Bogdan 1977; Jain & De Fillips 1991; Kunkel 1984; Watt & Breyer-Bandwijk 1962)

Lolium temulentum **L.**

Annual to 120cm; stems solitary or tufted. Inflorescence a 'raceme' 5–40cm long. Spikelets 2 to 10-flowered, <30mm long. Glume toughened and as long as or longer than the spikelet. Lemma turgid at maturity, with or without awns (awn <23mm long).

Ecology. Semi-deciduous submontane woodland. 900m.

Distribution. Europe, N Africa, Arabian Peninsula, temperate Asia.

Botanical notes. The presence of this species is based on two records made in the 19th century; these specimens could not be traced.

ETHNOGRAPHIC NOTES

Comparative uses. It is a common weed of cultivation. Its seeds are frequently poisonous, due to a fungus: cases of poisoning in man and animal due to consumption of fungus-infected grains in admixture with wheat and other cereals have been reported. The toxic principle is considered to be a liquid narcotic alkaloid temuline produced by the fungus *Endoconidium temulentum*. A decoction of the entire plant is prescribed for haemorrhage and incontinence of urine. The mature grains are used to prepare a tincture in homeopathy to treat neuralgia, rheumatism and arthritis, nosebleeds, intestinal cramps and trembling limbs. Although some say the grains are toxic, they are said to be used for pinole in California. (Ambasta 1982; Bogdan 1977; Goodin & Northington 1985; Jain & De Fillips 1991; Kunkel 1984)

Melanocenchris Nees

3 species. N and tropical Africa, Arabia to India and Pakistan.

Botanical notes. Apparently has no Old World relatives and is allied to the otherwise New World Boutelouinae.

Melanocenchris jacquemontii **Jaub. & Spach**

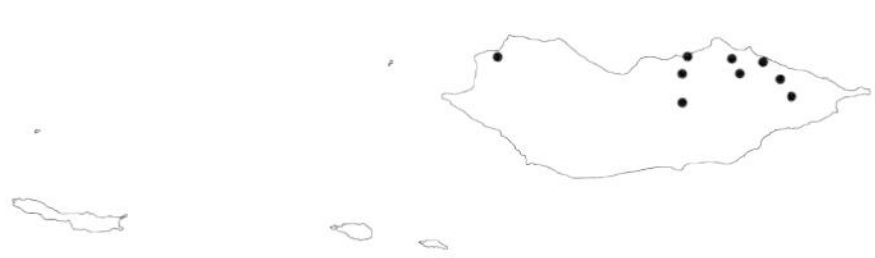

Annual. Leaves linear, hairy. Ligule a fringe of hairs. Inflorescence axis a single 'raceme', 1.5–3.5cm long, bearing 5 to 10 'spikelet clusters', the clusters <8mm long. Each 'spikelet cluster' composed of 5 to 8 spikelets, the lowest 2 fertile, the rest sterile and progressively smaller. Spikelets 2 or 3-flowered. Glumes white or purplish hairy, awned.

Ecology. Open shrubland and clearings in deciduous woodland; mainly on limestone. 50–850m.

Distribution. India and Pakistan.

ETHNOGRAPHIC NOTES

Name(s)

– Soqotri: (*di*) *ṣə̍ffo*, *ṣift*, *di ṣə̍fət*, *ṣāfhan* (western Soqotra); in wetter areas before flowering *muržif*; with inflorescence *ḳʌ̍ṭəhān* 'slender'.

– Soqotri simplified: (*di*) *ziffoh*; *zift*, *di zifit*, *zafhan*; *murjif*; *qatihan*.

Food (livestock). Grazed by all livestock, but mainly when there is a shortage of alternative grasses. It is more palatable to small stock, though when immature it is also eaten by cattle. However, it is not regarded as a particularly nutritious fodder. Small stock graze the stalks once the seeds have fallen if more palatable grasses are not available. It is found in sandy, stony or barren ground.

Melinis P.Beauv.

Rhynchelytrum Nees

11 species. Tropical and S Africa.

Botanical notes. The genera *Melinis* and *Rhynchelytrum* have traditionally been regarded as distinct, the number of nerves in the upper glume being the diagnostic character separating them. However, recent research has shown that the character is artificial and that they are best included in a single genus (Phillips 1995).

Melinis repens (Willd.) Zizka

Melinis repens subsp. *grandiflora* Hochst.
Rhynchelytrum repens (Willd.) C.E.Hubb.
Rhynchelytrum microstachyum Balf.f.
Rhynchelytrum microstachyum var. *albicomum* Balf.f.
Rhynchelytrum villosum (Parl.) Chiov.

Tufted annual or short-lived perennial. Leaves linear, flat. Ligule a line of hairs. Inflorescence an open, fluffy panicle 8–20cm long. Spikelets 2-flowered, gaping, often silky-hairy, 2–12mm long. Glumes unequal. Lower glume small, separated from the upper glume by a distinct gap. Upper glume as long as the spikelet, swollen in the lower half, 5-nerved, conspicuously pointed or with a short, straight awn. Upper lemma smooth.

Ecology. See Ethnographic notes.
Distribution. Tropical Africa to India and Pakistan.

ETHNOGRAPHIC NOTES
Name(s)
– Soqotri: 'type of (*di*) *ṣə̍ffoh*, (*di*) *ṣə̍fət*'; *ṣift* (western Soqotra).
– Soqotri simplified: 'type of (*di*) *ziffoh*, (*di*) *zifit*'; *zift*.
Food (livestock). These grasses grow mainly in the sandy, stony and thinner soils of the low-lying areas in Soqotra. They are well liked by livestock when green, but die back quickly with the approach of the dry season. When green they are considered to be a good fodder.
Comparative uses. Related species are commonly weeds of arable land and abandoned cultivation. *M. repens* is sometimes grown for its ornamental pink feathery panicles. It is grown too for hay or pasture in Brazil and Florida, and is considered to be a good fodder grass. (Bogdan 1977)

Microchloa R.Br.

4 species. Africa.

Botanical notes. The distinctive curling rhachis links this genus to *Harpochloa* and *Ctenium* (Clayton & Renvoize 1986).

Microchloa kunthii Desv.

Tufted perennial to 30cm, the base surrounded by the fibrous remains of old leaf sheaths. Leaves wiry. Ligule a short, hairy membrane. Inflorescence a solitary, slender, one-sided curved 'raceme', 2–20cm long. Spikelets 1-flowered, sessile, 2.5–5mm long, green or with a tinge of purple. Glumes subequal, as long as the spikelet, glabrous. Lemma enclosed by the glumes, awnless.
Ecology. Amongst rock debris and gravel on exposed ridge. 950m.
Distribution. Tropics.

ETHNOGRAPHIC NOTES
Name(s)
– Soqotri: *ḳeṣādə* 'sharp, cutting', 'oversalted'.
– Soqotri simplified: *qezadih*.
Food (livestock). Like other *ḳəṣ́ádə*-type grasses, this is usually more palatable to camels and donkeys. Other livestock eat new shoots and green leafage, and gnaw at the woody stems in hunger. It is not regarded as a valuable fodder grass, but is useful in a prolonged dry season.
Other. Pieces of the dry plant are gathered for use as kindling and firelighters.

Oplismenus P.Beauv.

Trailing annuals and perennials. Leaves broad, lanceolate to ovate. Ligule a hairy membrane. Inflorescence of 'racemes' along a central axis. Spikelets 2-flowered, mostly in pairs. Glumes subequal, 1/2 to 3/4 spikelet length, both terminating in an awn. Lower lemma the same length as spikelet. 5 species. Tropics.

Botanical notes. A genus of very closely related species. The sticky secretion on the awn is an unusual means of fruit dispersal in the grasses. *O. burmannii* is regarded as being a distinct species with the remaining entities forming part of a single pantropical polymorphic complex with some geographical separation but with many intermediates (Phillips 1995).

Oplismenus burmannii **(Retz.) P.Beauv.**
Oplismenus burmanni (L.) P.Beauv.

Slender annual, stems trailing, 10–60cm long. Leaves lanceolate to narrowly ovate. Inflorescence of 4 to 6 'racemes' spaced along a central axis 5–11cm long. Rhachis and pedicels silky hairy (hairs <4mm long). Spikelets 2.5–3.5mm long, shortly hairy with longer hairs on the margins, green. Lower glume awned, awn 4–13mm long, upper glume awned, awn <10mm long, both awns slightly roughened. Lower lemma with a band of hairs across the middle and with a short awn <2mm long.
Ecology. In shade in drought-deciduous woodland. 250–450m.
Distribution. Tropical Africa, Asia and America.

ETHNOGRAPHIC NOTES
Name(s)
– Soqotri: *məsḥeybibo* 'spreading'; *rəhiśt*, *rəhilt* 'chapped, cracked skin'.
– Soqotri simplified: *misheybibo*; *rihisht*, *rihilt*.
Food (livestock). Grazed by all livestock. The broad and soft leaves are a favourite feed of the cattle of the Central Highlands, and it is regarded as a very good and nutritious fodder for them, especially the younger plants. In years when it is plentiful it is gathered to make a superior and fattening hay. It retains its green colour well when dried and is liked by all animals.
Comparative uses. The sticky awns and inflorescence stick to clothing and to the coats of animals. In India this is a grass much relished by cattle, and it also yields good hay. Elsewhere the sticky inflorescence is used to catch rats. A decoction of the plant is taken by pregnant women and the root is taken as an antidote. (Ambasta 1982; Dalziel 1948, 1956; Jain & De Fillips 1991; Perry & Metzger 1980)

Oplismenus compositus **(L.) P.Beauv.**

Perennial, stems trailing, 15–150cm long. Leaves narrowly lanceolate to narrowly ovate. Inflorescence of 5 to 10 'racemes' widely spaced along a central axis 10–20cm long. Rhachis and pedicels glabrous or shortly hairy. Spikelets 2.5–4mm long, shortly hairy, green or purple. Lower glume awned, awn up to 3mm long, upper glume awned, awn <10mm long, both awns smooth. Lower lemma without a band of hairs and with a very short awn <1mm long.
Ecology. In shade in dense thicket. 600–900m.
Distribution. Tropics.

ETHNOGRAPHIC NOTES
Name(s)
– Soqotri: *məsḥeybibo* 'spreading'; *rəhiśt*, *rəhilt* 'chapped, cracked skin'.
– Soqotri simplified: *misheybibo*; *rihisht*, *rihilt*.
Food (livestock). When green, regarded as being of equal value to *O. burmannii* above, but its persistence through the dry season makes this grass even more valuable.

Panicum L.

Annuals or perennials. Leaves linear to lanceolate, flat or inrolled. Ligule a short, hairy membrane. Inflorescence an open panicle. Spikelets 2-flowered, pedicelled or sessile, awnless. Glumes unequal. Lower lemma similar in length to the upper glume, both more or less equal to the spikelet. Upper lemma leathery. 2 endemics. c.470 species. Tropics and warm temperate regions.

Botanical notes. A large genus with fairly uniform spikelets but with much variation in habit. Although there are few formal characters to separate *Panicum* from related genera, it is almost always recognisable by the lax, open panicle with well-developed secondary branches (Wood 1997).

Panicum atrosanguineum **Hochst. ex A.Rich.**

Tufted annual to 40cm. Leaves lanceolate, hairy, sheaths hairy (tubercle-based hairs). Inflorescence loose and open, 6–20cm long, branches not whorled. Spikelets all pedicellate, 1.5–2mm long, dark purple or greenish. Lower glume 1/3 to 2/3 spikelet length. Upper lemma shiny, brown or blackish.
Ecology. Semi-deciduous woodland. 500–850m.
Distribution. Tropical Africa, Arabia, India and Pakistan.

ETHNOGRAPHIC NOTES
Name(s)
– Soqotri: with inflorescence *di ḳerṭomo* 'needs chewing'.
– Soqotri simplified: *di qertomoh*.
Food (human). There is no tradition of gathering the grains as food on Soqotra.
Food (livestock). This small annual of dry, short grassland and rocky ground is palatable to all livestock, including cattle, and provides particularly useful forage in the drier areas and poorer pastures of the island. It is not considered a very nutritious grass once it dies back.
Comparative uses. It is often a weed of arable land. The grains are collected for food in times of hunger. (Ambasta 1982; Bogdan 1977; Jain & De Fillips 1991; Kunkel 1984)

Panicum maximum Jacq.

Tufted, rhizomatous perennial to 1.5(–2)m. Leaves linear to narrowly lanceolate, glabrous or hairy, sheaths fringed at the mouth and at the nodes. Inflorescence a loose, open panicle, 12–45cm long, the lower branches whorled. Spikelets mostly pedicelled, 3–4mm long, usually glabrous, sometimes shortly hairy, purple. Lower glume 1/4 to 1/3 spikelet length. Upper lemma pale, shiny, slightly rugose.
Ecology. No habitat details available.
Distribution. Tropical and S Africa, Arabia.
Botanical notes. Collected once in the 19th century; not seen since.

ETHNOGRAPHIC NOTES
Comparative uses. This grass is one of the most important grazing and fodder grasses, and it can be used extensively with high stocking rates. It is often a pioneer grass covering land cleared from woodland, and grows in grassland with scattered trees where it tends to grow in light shade, under the trees. It is cultivated to a moderate extent in Africa and has been introduced elsewhere: in India, for example, it has become one of the important fodder grasses, though it can become dominant and invade overgrazed pastures. It can be grazed, or cut for silage or hay. Generally it is an excellent fodder and pasture species, and as a cultivated grass it is much valued for its high productivity, palatable herbage and good persistence. It is rich in carotene and also contains vitamins C and B1 and tocopherol. It is often cut and sold in markets. It is liked by livestock, and compares favourably with maize fodders, though under poor soils and growing conditions it can produce hydrocyanic acid and may be poisonous to cattle in the dry season (and in South Africa is suspected of causing *Dikoor* disease in sheep). It is said to be less toxic as the rainy season progresses.

The pithy stems are used with leaves for thatching, the inflorescences for making brooms, and the seeds are cooked as a food in times of hunger. In Uganda the plant is used in tribal healing and rain-making ceremonies, and is tied around the head for relief of a headache. (Ambasta 1982; Bogdan 1977; Dalziel 1948, 1956; Jain & De Fillips 1991; Kunkel 1984; Lewis & Elwin-Lewis 1977; Watt & Breyer-Bandwijk 1962)

Panicum rigidum Balf.f.

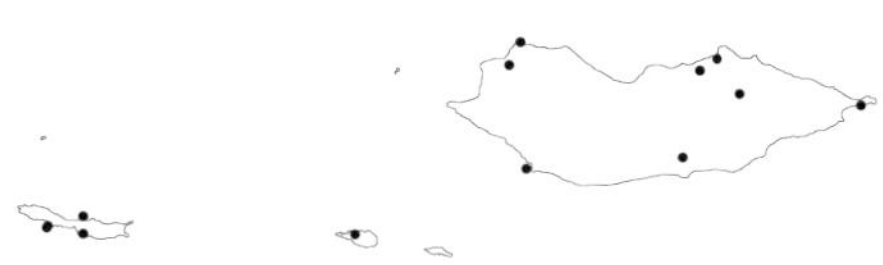

Tufted perennial, stems creeping, wiry, to 60cm long arising from a persistent knotty base. Leaves rigid, inrolled, distichous. Inflorescence ovate, 1.5–4cm long, branches stiff, not whorled, bare of spikelets towards the base. Spikelets elliptic, 1.5–1.8mm long, glabrous. Lower glume 1/3 to 1/2 spikelet length. Upper lemma pale, smooth, shiny.
Ecology. Dwarf shrubland and woody-herb communities on desertic plains and rocky slopes. Sea-level–600m.
Distribution. Endemic.
Conservation status. Least Concern (LC). Common in several vegetation types and under no present or perceived threat.

ETHNOGRAPHIC NOTES
Name(s)
– Soqotri: *ḳˈʌźəbə* (name given to various mainly perennial Gramineae and Cyperaceae, often of damp ground); *mˈʌḳźəb*; *zˈidirə, zēydiroh* 'fat-forming'; *mˈoshịṭ* 'broom'; *śˈibiṭə* 'never dies right back'; *ḥərˈəwṭʌṭ* 'with leaves that can be stripped off'; (rare) 'type of *ɛrśˈido*'; *lāṭaʾ*, 'type of *ḳˈʌrźʌʾ*' or *nizˈafə* (Abd al Kuri); *saʿadˈifo* 'clump with persistent rootstock' (Samha).
– Soqotri simplified: *qadhibih*; *maqdhab*; *zidirih*, *zeydiroh*; *mos'hit*; *shibitih*; *hirewtat*; 'type of *ershiydo*'; *lata'*; 'type of *qardha*' or *nizafih*; *sa'adifo*.
Food (human). The green stems are chewed raw.
Food (livestock). This perennial grass is one of the key rangeland plants of the drier, low-lying areas of the Soqotra Archipelago. Small stock especially depend heavily on this to survive the long dry season. The whole plant is eaten and when young is especially nutritious. As it matures the leaves remain palatable, but the stems are generally left uneaten. Herders say that it continues to have savour even when apparently dried out. In some areas of the Archipelago it is the chief food source for small stock for months at a time.
Other. As its name suggests, whole plants are used as brushes to sweep out livestock and human living quarters. Dead plants are used in roofing and as kindling and firelighters.

Panicum socotranum Cope

Tufted perennial, stems wiry, to 30cm, arising from a knotty base. Leaves linear, flat or inrolled, somewhat pungent. Inflorescence ovate, 4.5–5.5cm long, branches stiff, lowermost whorled, bare of spikelets towards the base. Spikelets elliptic, <2mm long, glabrous. Lower glume 1/2 spikelet length. Upper lemma pale, smooth, shiny.
Ecology. Wet flushes. 90m.
Distribution. Endemic.
Conservation status. Vulnerable (VU D2). Rare, with a very limited area of occupancy.

ETHNOGRAPHIC NOTES
Name(s)
– Soqotri: *ḳˈʌźəbə* (given to various mainly perennial Gramineae and Cyperaceae, often of damp ground); *mˈʌḳźəb*; (rare) 'type of *ɛrśˈido*'.
– Soqotri simplified: *qadhibih*; *maqdhab*; 'type of *ershiydo*'.
Food (human). The green stems are chewed raw.
Food (livestock). As *P. rigidum* above, this is a key grazing resource in the areas where it grows.
Notes. A period spent grazing halophytic rangeland at least once a year is considered essential for the health of small stock and camels especially. Such feed initially brings on diarrhoea which is seen as being beneficial and 'cleansing impurities from the blood'. When recovered from this, the animals begin to put on weight. Grazing heavily on saline forage can affect the taste of the milk.

Panicum turgidum **Forssk.**

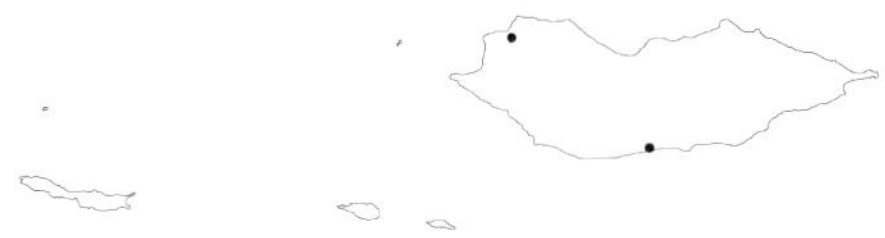

Branched perennial, stems woody, numerous, 1(–2)m. Leaves linear, flat or inrolled, stiff and pungent, glabrous. Inflorescence a small panicle, 2.5–15cm long, with few branches or spikelets, no branches whorled. Spikelets shortly pedicelled, c.4mm long, glabrous, bluish-green. Lower glume nearly as long as the spikelet. Upper lemma smooth, pale, shiny.
Ecology. Rare on coastal plains. c.20m
Distribution. Tropical Africa to India and Pakistan.
Botanical notes. In the 19th century recorded as abundant near Qalansiyah; recently only seen on the Noged plain.

ETHNOGRAPHIC NOTES
Name(s)
– Soqotri: *ḳʌ̍źebə* (given to various mainly perennial Gramineae and Cyperaceae, often of damp ground); *mʌ̍sḥiṭ* 'broom'.
– Soqotri simplified: *qadhibih*; *mas'hit*.
Food (human). There is no tradition of gathering the grains as food on Soqotra. The spongy green stems are chewed raw.
Food (livestock). One of the most important fodder grasses in the areas where it grows. When green it is liked by all livestock, but as it dries out, mainly camels and donkeys eat it. In areas with little alternative forage, goats also graze it in the dry season. It can survive severe drought, mostly in a dormant state, and recovers quickly even when little rain has fallen.
Other. As its name suggests, whole plants are used as brushes to sweep out livestock and human living quarters. Dead plants are used in roofing, and as kindling and firelighters.
Comparative uses. It commonly forms round, bushy clumps a metre or more high, which often break off at the base and roll before the wind. It is an important forage, especially for camels, in North African and South Western Asian deserts and semi-deserts, possibly due to lack of softer alternatives. The grains are gathered and eaten in times of scarcity. In some areas it is gathered and stored against hungry times – it is also found in the nests of harvesting ants. The grass is used as thatch by nomads and the straw is woven into matting. The Tuareg weave the stiff straw (as the weft) with thin threads of leather (as the warp) into mats which can only be rolled one way. It is a useful sand binder. Medicinally, the plant is used as a vulnerary agent and for removing white spots of the eye. (Ambasta 1982; Batanouny 1981; Bogdan 1977; Dalziel 1948, 1956; Goodin & Northington 1985; Jain & De Fillips 1991; Kunkel 1984; Rizk & El-Ghazaly 1995)

Paspalidium Stapf

40 species. Tropical.

Paspalidium geminatum **(Forssk.) Stapf**
Panicum paspaloides Pers.

Stoloniferous perennial to 60cm. Stems sometimes rooting from lower nodes. Leaves flat, linear, glabrous. Ligule a line of hairs. Inflorescence of 'racemes' 0.5–4cm long, arranged along a central axis, 5–30cm long, rhachis winged. Spikelets 2-flowered, pressed into hollows in the rhachis, 1.5–3mm long, glabrous. Glumes unequal. Lower glume 1/4 to 1/3 spikelet length. Upper glume 2/3 to 4/5 spikelet length. Upper lemma granulose.
Ecology. Grows in mud by permanent water.
Distribution. Old World tropics.

ETHNOGRAPHIC NOTES
Name(s)
– Soqotri: *ḳāźəb*, *ḳāźub di mɔ̄źi* '*ḳāźub* of damp ground'; with inflorescence *ḳʌ̍ṭəhān* 'slender'.
– Soqotri simplified: *qadhib*, *qadhub di mowdhi*; *qatihan*.
Food (human). The spongy stolons and stems are eaten when young and tender.
Food (livestock). This creeping grass grows commonly in damp and often brackish ground, where it can form pure colonies. It is also found growing beside streambeds where it can be seen floating in the water. Despite its tough leaves, it is well liked by livestock, except by goats, which abhor damp grazing. It is considered to be good and nutritious grazing.
Comparative uses. It prefers alkaline ground, and often grows in shallow water and in seasonally flooded grasslands. In some areas it can block smaller streams. The grains of the related *P. flavidum* are gathered and eaten in times of hunger. (Ambasta 1982; Bogdan 1977; Dalziel 1948, 1956; Heine *et al.* 1988; Jain & De Fillips 1991)

Paspalum L.

Perennials. Leaves flat, linear to narrowly lanceolate. Ligule membranous and truncate. Inflorescence of 2 or more one-sided 'racemes' arranged like fingers at the tip of the stem or along a central axis. Spikelets 2-flowered, arranged in 2 rows, awnless. Lower glume absent or very small. Upper glume membranous. Lower floret barren, without a palea, the lemma resembling the upper glume. c.330 species. Tropics.

Botanical notes. Similar to *Brachiaria*, differing in the absence of the lower glume and the shape of the upper lemma which is always acute in *Paspalum*.

***Paspalum scrobiculatum* L.**

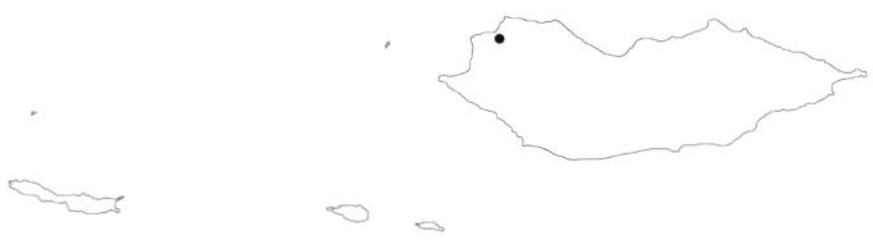

Tufted perennial to 150cm, rooting from the lower nodes. Leaves linear, light green becoming bronze when older. Inflorescence of 1 to 5(to 10) (usually >2) 'racemes', 3–6cm long, at the tip of the stem or along a central axis <8cm long. Spikelets 1.5–3mm long, rounded, glabrous. Upper lemma striate, brown at maturity.
Ecology. In mud by permanent water. Sea-level.
Distribution. Old World tropics.

ETHNOGRAPHIC NOTES
Name(s)
– Soqotri: *ḳúrźʌ'* 'windpipe', 'beside water'.
– Soqotri simplified: *qurdha'*.
Food (human). There is no tradition of gathering the grains as food on Soqotra.
Food (livestock). Eaten with enthusiasm at all stages of growth, although such damp forage is generally avoided by goats. Commonly found on damp ground or growing beside water, it is considered to be a good and nutritious grass for livestock. It is readily eaten by goats when dry.
Comparative uses. This is one of the wild grasses with edible grain and it is often collected in times of scarcity and ground on stones like millet. However, it is recognised as causing digestive disturbances if a lot is eaten. In India it is grown as minor crop, the grain being cooked like rice and used for making a type of bread. Only grain matured for six months is used for this as newly gathered grain is poisonous. It is readily grazed by cattle, especially buffaloes. Medicinally, a decoction of the root and rhizome have been used during and following childbirth, and the juice of the stem used to clear corneal opacity. Most *Paspalum* grasses are good grazing grasses, the better known being *P. dilatatum* and *P. notatum*, which have been introduced into a number of countries and are often grown on a farm scale. The grains of the related *P. ciliatum* and *P. commersonii* are used as a cereal. (Ambasta 1982; Bogdan 1977; Dalziel 1948, 1956; Jain & De Fillips 1991; Kunkel 1984; Lewis & Elwin-Lewis 1977)

***Paspalum vaginatum* Sw.**
Paspalum distichum sensu Balf.f. non L.

Stoloniferous perennial to 60cm, sometimes forming mats. Leaves linear. Inflorescence of 2 to 5 (usually paired) 'racemes', 1–2cm long, at the tip of the stem or on a very short axis. Spikelets 3–4mm long, elliptic, glabrous. Upper lemma smooth, pale at maturity.
Ecology. In mud (usually brackish) by permanent water. Sea-level–150m.
Distribution. Tropics and subtropics.
Botanical notes. Under-recorded on the distribution map.

ETHNOGRAPHIC NOTES
Name(s)
– Soqotri: *ḳúrźʌ'* 'windpipe', 'beside water'.
– Soqotri simplified: *qurdha'*.
Food (human). The green stems are chewed raw.
Food (livestock). It is enthusiastically eaten by sheep when immature and when other forage is scarce, but is generally avoided by goats when wet. It is commonly found on damp ground or growing beside water, mostly on alkaline or saline soils, where it can form dense, often pure stands. It can withstand a considerable degree of grazing and trampling. It has very tough leaves and stems but is still highly palatable to grazing animals. It is grazed by donkeys and camels in all stages of growth. However, it is generally not considered to be a nutritious grass.
Notes. A period spent grazing halophytic rangeland at least once a year is considered essential for the health of small stock and camels especially. Such feed initially brings on diarrhoea which is seen as being beneficial and 'cleansing impurities from the blood'. When recovered from this, the animals begin to put on weight. Grazing heavily on saline forage can affect the taste of the milk.
Comparative uses. It is salt tolerant and can withstand water-logging and flooding. It acts as an efficient sand binder. In coastal and tidal swamps it can form almost a turf in places with its long, creeping, rooting and branching stems. It has been introduced to Australia where it is popular and mainly used as a green cover for saline seepages. (Ambasta 1982; Bogdan 1977; Dalziel 1948, 1956; Jain & De Fillips 1991)

Pennisetum Rich.

Annuals or perennials. Leaves linear to lanceolate, inrolled or flat, rigid. Ligule a line of hairs. Inflorescence a single terminal 'raceme'. Spikelets 2-flowered, glabrous, solitary or in clusters of <4, the spikelet clusters surrounded by an involucre of 1 to many slender, sometimes plumose, bristles, free to the base. Glumes unequal. 80 species. Tropics.

Botanical notes. A variable genus with no satisfactory modern revision available (Clayton & Renvoize 1986).

***Pennisetum orientale* Rich.**

Very similar to *P. setaceum* but differing in its looser, subshrubby habit and flat or loosely folded, linear to linear-lanceolate leaves without a conspicuously thickened midrib.
Ecology. Rocky places in a variety of vegetation types. 200–1000m.
Distribution. N Africa to India and Pakistan.

ETHNOGRAPHIC NOTES†
Name(s)
– Soqotri: *ḳeṣádə* 'sharp, cutting', 'oversalted'; with inflorescence *di ḳérṭomoh*.
– Soqotri simplified: *qezadih*; *di qertomoh*.

Food (livestock). Like other grasses of the *ḳeṣádə* type, only the more tender new growth is palatable to livestock, though camels and donkeys manage it in its mature or dormant state when it has grown tough. In this state it is not regarded as a nutritious fodder, but the young shoots are seen as having more value. Goats can be seen chewing at it in a prolonged dry season when other forage is scarce. These grasses are commonly found in wadi beds.
Other. Parts of the dead plants are gathered for use as kindling and firelighters.

Comparative uses. A number of *Pennisetum* grasses provide excellent fodder and grazing grasses, such as elephant grass (*P. purpureum*), pearl millet (*P. americanum*) and their hybrids, Kikuyu grass (*P. clandestinum*) and several others. However, there are also fibrous, tough and barely palatable species (such as *P. schimperi*) which are widespread but undesirable in pastures. In the desert areas of India *P. orientale* plants are collected for fodder, and are relished by horses and donkeys. This grass is also a good soil binder. (Ambasta 1982; Bogdan 1977; Goodin & Northington 1985; Jain & De Fillips 1991)

†Also apply to *P. setaceum*.

***Pennisetum setaceum* (Forssk.) Chiov.**
Pennisetum ruppellii Steud.

Densely tufted perennial to 130cm. Leaves <3mm wide, rigid, inrolled with a thickened midrib on the upper surface. Inflorescence a single, linear, sometimes purplish 'raceme' 4–20(–30)cm long. Spikelets 6–8mm long, bristles <40mm long, some loosely plumose. Lower glume vestigial or <2mm long. Upper glume 1/5 to 2/3 spikelet length.
Ecology. Rocky places. 550m.
Distribution. N and NE tropical Africa, Arabia, SW Asia.

ETHNOGRAPHIC NOTES
Name(s)
See *P. orientale* (also uses).

Polypogon Desf.

18 species. Warm temperate regions.

Botanical notes. A heterogeneous genus without clear subdivisions (Clayton & Renvoize 1986).

***Polypogon monspeliensis* (L.) Desf.**

Tufted annual to 80cm. Leaves flat, scaberulous. Ligule membranous. Inflorescence a contracted, bristly panicle, densely cylindrical or lobed (on bigger plants). Spikelets 1-flowered, 2–3mm long, green, falling entire. Glumes subequal, as long as spikelet, scaberulous, hairy at the edges, awned (awns 4–7mm long). Lemma about half as long as the glume, awnless or awned (awn fragile, 1.5–2mm long).
Ecology. No details of habitat or location available.
Distribution. Old World temperate and tropical mountains.

Botanical notes. Known only from a single, recent record (*Hyam* 121) (see Botanical notes under family entry).

ETHNOGRAPHIC NOTES
Name(s)
– Soqotri: *ḳúrźaʿ* 'windpipe', 'beside water' (given to various mainly perennial Gramineae and Cyperaceae, often of damp ground).
– Soqotri simplified: *qurdha'*.
Food (livestock). This annual grass is often found growing in wadi beds, where it is readily grazed by all livestock when green, though they are less enthusiastic as it dies back. It it is not considered to be an important grass nutritionally.

Schmidtia Steud. ex J.A.Schmidt

2 species. Africa, Cape Verde Is., Pakistan.

***Schmidtia pappophoroides* Steud.**

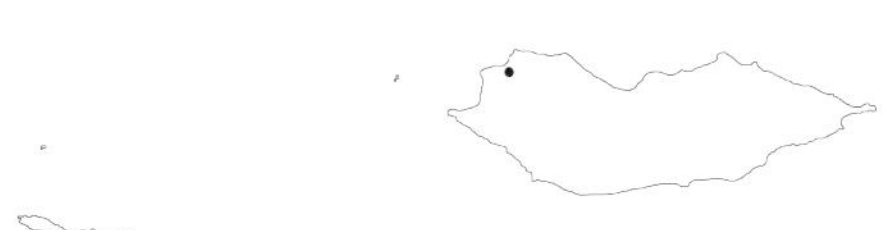

Rhizomatous perennial, often also with surface stolons, to 90cm. Leaves flat, bluish green and with gland-tipped hairs. Ligule a fringe of hairs. Inflorescence an open panicle. Spikelets 4 to 9-flowered. Glumes with 7 to 11 nerves. Lemmas hairy below the middle, 6-lobed, with 5 roughened awns 4–8mm long.
Ecology. No habitat details available.
Distribution. Tropical Africa, Pakistan.
Botanical notes. Recorded in the 19th century from near Qalansiyah but not seen since.

Setaria P.Beauv.

Annuals. Leaves flat or folded. Ligule a line of hairs. Inflorescence a single terminal 'raceme'. Spikelets 2-flowered, solitary or in clusters, surrounded by an involucre of bristles which persist after the spikelet has fallen, glabrous, whitish-green, awnless. Glumes unequal, the lower often very small. Lower lemma membranous. Upper lemma leathery, convex. Upper glume as long as or shorter than spikelet. c.100 species. Tropics and warm temperate regions.

Botanical notes. A heterogeneous genus the most characteristic feature of which are the bristles subtending the spikelets.

Setaria intermedia Roem. & Schult.

Ascending annual to 150cm. Leaves linear, with a fairly dense covering of tubercle-based hairs along the main nerves. Inflorescence contracted, narrowly lanceolate, 3–12cm long. Spikelets c.2mm long, subtended by 5 or more pale bristles.Upper glume 1/2 to 2/3 spikelet length, shorter than upper lemma. Lower glume 1/3 to 1/2 spikelet length. Upper lemma rugose, pale to golden.
Ecology. In gravel by wadis and rocky places. 150–300m.
Distribution. Tropical Africa, Arabia, India and Pakistan.

ETHNOGRAPHIC NOTES
Name(s)
– Soqotri: *ˈiʾiri, ʾˈiyri; śebˈiṭə* 'never dies right back'; with inflorescence *di ḳerṭˈomo* 'needs chewing'; in seed *mədˈəwʾḥin* (also given to other nuisance or adhesive plants with spiked seeds, such as *Bidens*).
– Soqotri simplified: *'i'iri, 'iyri; shebitih; di qertomoh; midewhin.*
Food (livestock). Grazed by all livestock and regarded as a valuable and nutritious fodder, especially for cattle. It persists through much of the dry season, and even in a drought persists in sheltered places beneath trees or beside rocks. It can be gathered and dried to make hay.
Other. *Setaria* grasses are common around dwellings and on old cultivated land. The bristles of the mature plant stick to clothing and the coats of livestock. The plants growing around livestock and human quarters are often cut down before they mature to lessen the nuisance, and are then laid out to dry and stored as hay.

Setaria pumila (Poir.) Roem. & Schult.
Setaria glauca P.Beauv.
Setaria pallidifusca (Schumach.) Stapf & C.E.Hubb.

Slender, tufted annual to 130cm. Leaves linear, more or less glabrous. Inflorescence contracted, cylindrical, 1–10cm long. Spikelets <3.5mm long, subtended by 5 to 10 brownish yellow bristles. Upper glume shorter than upper lemma. Upper lemma rugose. Lower glume 1/3 to 1/2 spikelet length.
Ecology. Montane woodland in the Haggeher mountains. 900–1050m.
Distribution. Old World tropics and warm temperate regions.
Botanical notes. A widespread and variable weed.

ETHNOGRAPHIC NOTES
Name(s)
– Soqotri: *ˈiʾiri, ʾˈiyri; śebˈiṭə* 'never dies right back'; with inflorescence *di ḳerṭˈomo* 'needs chewing'; in seed *mədˈəwʾḥin* (also given to other nuisance or adhesive plants with spiked seeds, such as *Bidens*).
– Soqotri simplified: *'i'iri, 'iyri; shebitih; di qertomoh; midewhin.*
Food (human). There is no tradition of gathering the grains as food on Soqotra.
Food (livestock). This grass is widespread in the Central Highlands, and is grazed by all livestock, which relish it when green and tender. It lasts well into the dry season and can be gathered for hay. It is considered to be a very nutritious grass, especially for cattle.
Other. As *S. intermedia.*
Comparative uses. Some African herdsmen regard it as harmful to cattle. It is often a weed of arable land. The grains are gathered and eaten, boiled or ground into flour to make a sort of bread. They are also used for making alcohol, and sometimes as an adulterant of anise (*Pimpinella anisum*). The tender shoots are occasionally eaten as a vegetable, and medicinally it is seen as being emollient and diuretic. It is considered to be a good lawn grass. The grain of the related *S. italica* is also gathered as food, and is also a popular medicine for alleviating pains after childbirth, while a preparation of the whole plant is used as a sedative to the gravid uterus. One species of *Setaria*, foxtail millet, is cultivated. (Ambasta 1982; Bogdan 1977; Dalziel 1948, 1956; Jain & De Fillips 1991)

Setaria verticillata (L.) P.Beauv.

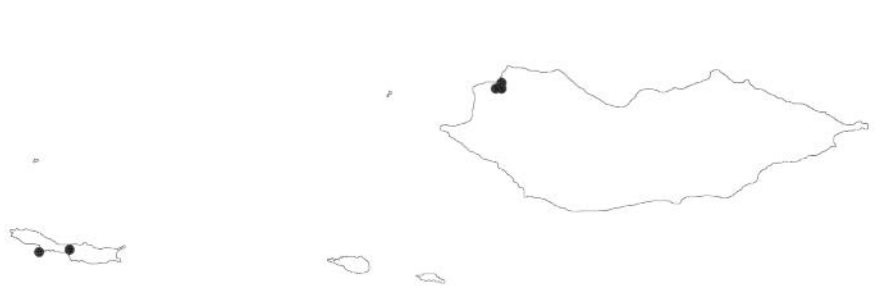

Weakly tufted annual to 1m. Leaves linear, usually sparsely hairy. Inflorescence contracted, linear to untidily lobed, 2–6cm long. Spikelets c.2mm long, whitish-green, subtended by 1 to 5 often purplish, retrorsely barbed bristles. Upper glume as long as upper lemma. Upper lemma finely rugose. Lower glume 1/3 to 1/2 spikelet length.
Ecology. Rocky slopes. 50–500m.
Distribution. Tropics and warm temperate regions.

ETHNOGRAPHIC NOTES

Name(s)

– Soqotri: *ˈiˀiri, ˀˈiyri; di raˈḥmam* 'grey-coloured'; *śebiˈṭə* 'never dies right back'; with inflorescence *di ḳerṭˈomo* 'needs chewing'; in seed *mədˈəwˀḥin* (also given to other nuisance or adhesive plants with spiked seeds, such as *Bidens*); type of *nizaˈf̣ə, ḳuˈrźaˁ* 'windpipe', 'beside water' (Abd al Kuri).

– Soqotri simplified: *'i'iri, 'iyri; di rahmam; shebitih; di qertomo; midewhin*; type of *nizafih, qurdha'*.

Food (human). There is no tradition of gathering the grains as food on Soqotra.

Food (livestock). This grass is liked by all livestock when young and before it comes into flower. It is less liked once it matures as the stem becomes rather tough and the bristles spiny. It is regarded as a good and nutritious fodder for all livestock, but especially cattle. It is less persistent through the dry season than the other *Setaria*.

Other. The bristles of this species are the most tenacious of the *Setaria* on Soqotra, and stick hard to clothing and the coats of livestock. This can be troublesome in the case of sheep and make the wool difficult to spin. Like *S. intermedia*, the plants are commonly cut down around the living quarters of both livestock and people, and along the edges of much-used tracks.

Comparative uses. It is often a weed of arable land. In parts of Africa this is cultivated for its grain which is used to prepare an alcoholic drink. The grains are also eaten by desert tribes. The ash of the plant is used as a cooking salt in Uganda. The mature, rigid bristles are used to cover granaries as a protection against rats or the spiky inflorescence is placed in the open mouth of the grain store. The bristles are also stuck together and used as strainers for milk. In veterinary medicine, the burrs are pressed into wounds and withdrawn to remove maggots. They are also used to fasten turbans. The straw is used to weave hats. (Ambasta 1982; Bogdan 1977; Dalziel 1948, 1956; Heine *et al.* 1988; Jain & De Fillips 1991; Watt & Breyer-Bandwijk 1962)

Sporobolus R.Br.

Annuals or perennials. Leaves flat or inrolled. Ligule a ring of hairs. Inflorescence an open or contracted panicle. Spikelets 1-flowered, glabrous, without awns. Glumes unequal, the upper usually the longer, both shorter than the lemma, awnless. Lemma usually 1-nerved, shiny. 60 species. Tropics and subtropics.

Botanical notes. The genus can be roughly partitioned according to whether the inflorescence branches are whorled or not, but formal subdivisions are reported to be impractical. The species intergrade so much that the species limits are often difficult to define. The genus is distinguished from *Eragrostis* by its 1-flowered spikelets, but the boundary between the genera is blurred by a number of intermediates (Clayton & Renvoize 1986).

Sporobolus diander P.Beauv.

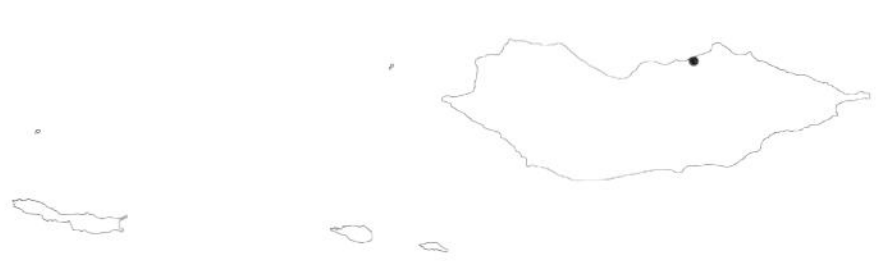

Tufted perennial to 75cm. Leaves inrolled, rarely flat. Inflorescence open, <20cm long, primary branches erect or spreading, bare of spikelets at the base. Spikelets 1.2–1.6mm long. Lower glume 1/4 to 1/3 spikelet length. Upper glume about 1/2 spikelet length.

Ecology. No habitat details available.

Distribution. Arabia, tropical Asia to Australia.

Botanical notes. Collected near Hadiboh in the 19th century but not seen since.

Sporobolus ioclados Nees

Sporobolus iocladus Nees

Tufted stoloniferous perennial to 80cm. Leaves flat or inrolled. Inflorescence an open panicle, ovate, 3–20cm long, the primary branches spreading, whorled, bare of spikelets towards the base. Spikelets 1.5–2.2mm long, grey-green or pale whitish-green. Lower glume 1/4 spikelet length. Upper glume 2/3 to as long as the spikelet.

Ecology. On silty soil, sometimes brackish. 50m.

Distribution. Tropical Africa to India and Pakistan.

ETHNOGRAPHIC NOTES†

Name(s)

– Soqotri: *ḳˈʌźəbə, ḳˈʌrźʌˀ* 'windpipe', 'beside water', *ḳerźˈaˀihun* 'little *ḳˈʌrźʌ*' (given to various mainly perennial Gramineae and Cyperaceae, often of damp ground); *śˈebəreḳ*; with inflorescence *ḳˈʌṭəhan* 'slender'.

– Soqotri simplified: *qadhibih, qardha', qerdha'ihun; shebireq; qatihan*.

Food (human). The green stems and stolons are eaten raw.

Food (livestock). These are palatable to all livestock, but are especially important for sheep on Soqotra. *S. ruspolianus* is widespread on the high limestone plateaus, the traditional sheepwalks of the island. This smaller species is more palatable to sheep than the larger, more tufted species. Both are regarded as good and nutritious fodder grasses, and *S. ioclados*, in particular, is said to promote lactation. Both are resistant to drought, and persist through the dry season. In a prolonged dry season, small stock dig up the drying plants to get at the more juicy roots, and can destroy the plants in this way. *S. ioclados* is important for goats and sheep in the coastal areas throughout the long, dry months when there is little alternative forage.

Comparative uses. *S. ioclados* can form dense, slowly spreading colonies in grassland and along lake shores, usually on saline or alkaline soil. It is greatly appreciated in arid areas for its effect on milk yield. The dried stems are occasionally used for roof thatching. (Bogdan 1977; Dalziel 1948, 1956; Heine *et al.* 1988; Watt & Breyer-Bandwijk 1962)

†Also apply to *S. ruspolianus*.

Sporobolus minutus Link

Tiny tufted annual. Leaves inrolled or flat. Inflorescence an open lanceolate panicle 2–3.5cm long, the lowermost branches whorled, upper branches thinly scattered with glandular dots, spikelets attached along the whole length of the primary branches. Spikelets 1.1–1.3mm long, purplish. Lower glume 1/2 to 3/4 spikelet length. Upper glume 3/4 spikelet length.

Ecology. Succulent shrubland. 350–400m.
Distribution. Tropical Africa, Arabia.

ETHNOGRAPHIC NOTES
Name(s)
– Soqotri: *(di) ṣəffo; ṣift, di ṣəfət, ṣāfhan* (western Soqotra); 'type of *ḳōnhitin*' 'little threads' (Abd al Kuri).
– Soqotri simplified: (*di*) *ziffoh*; *zift*, *di zifit*, *zafhan*; 'type of *qownhitin*'.
Food (livestock). This is well liked by sheep and goats on Abd al Kuri but it is not regarded as a particularly nutritious feed, and it quickly dies back.

Sporobolus ruspolianus Chiov.

Loosely tufted stoloniferous perennial to 40cm. Leaves flat. Inflorescence an open or contracted panicle, ovate or narrowly ovate in shape, 2.5–6.5cm long, primary branches short and stiff, bare of spikelets at the base, the spikelets gathered into clumps at the tips. Spikelets 1.4–2.3mm long. Glumes conspicuously scabrid, lower glume 1/2 spikelet length, upper glume 2/3 to 3/4 spikelet length.
Ecology. Coastal and inland sandy and gravel plains. Sea-level–200m.
Distribution. NE Africa, Arabia.

ETHNOGRAPHIC NOTES
Name(s)
See *S. ioclados* (also uses).

Sporobolus spicatus (Vahl) Kunth

Tufted perennial to 70cm, spreading by means of looping stolons. Leaves flat or inrolled, stiff with pungent spiny tips. Inflorescence a dense, cylindrical, spicate panicle <20cm long and 4mm wide, the primary branches closely appressed to the axis. Spikelets 1.5–2.5mm long, whitish. Lower glume 1/4 spikelet length. Upper glume 2/3 spikelet length.
Ecology. Silty, often salty, soils in dwarf shrubland on coastal plains; also between rocks on limestone ridges (Abd al Kuri) and in *Dracaena* woodland (Diksam plateau). Sea-level–600m.
Distribution. Africa to India and Pakistan.

ETHNOGRAPHIC NOTES
Name(s)
– Soqotri: *ḳāźəb, ḳʌźəbə; ḳʌrźʌˀ* 'windpipe', 'beside water' (given to various mainly perennial Gramineae and Cyperaceae, often of damp ground); with inflorescence *di ḳerṭomo* 'needs chewing'; *nizafə, ḥēmʌr* (Abd al Kuri).
– Soqotri simplified: *qadheb*, *qadhibih*; *qardha'*; *di qertomo*; *nizafih*, *heymar*.
Food (livestock). Rather fibrous and tough, but the new shoots and green parts are grazed by all livestock, especially in the drier months when the annual grasses have died back. It is not regarded as very nutritious grass, but is more highly valued on the drier, outer islands. Camels and donkeys find it palatable even when drying out. It can form extensive, often pure colonies.
Comparative uses. *S. spicatus* is said to cause diarrhoea in livestock and formerly stock were taken to feed on this as a treatment for tripanosomiasis. (Ambasta 1982; Dalziel 1948, 1956; Heine *et al.* 1988; Jain & De Fillips 1991)

Stipagrostis Nees

50 species. Old World deserts and semi-deserts.

Botanical notes. Sometimes included in *Aristida*, but has plumose awns and a preference for drier habitats (Clayton & Renvoize 1986).

Stipagrostis sokotrana (Vierh.) De Winter

Aristida sokotrana Vierh.

Densely tufted perennial to 30cm. Leaves inrolled, bluish-green. Ligule a line of hairs. Inflorescence a single terminal 'raceme'. Spikelets 2-flowered, all similar. Glumes 5 to 7-nerved, unequal, upper glume considerably longer than the lower. Lemma inrolled with a trifid awn. Awn with two short and one long (central) branch, the top of the central branch plumose.
Ecology. Sandy and gravel soil on coastal plains in low woody-herb communities. Sea-level–20m.
Distribution. Arabia.

ETHNOGRAPHIC NOTES

Name(s)

– Soqotri: *ḳeṣādə* 'sharp, cutting', 'oversalted'; 'like (*di*) *ṣi̍ffo*, *ṣift*'; *lātaʾ* (Abd al Kuri); *nizāfə* (Abd al Kuri); *ʾišɛ̍lo* (more commonly given to various Cyperaceae) (Abd al Kuri).

– Soqotri simplified: *qezadih*; 'like (*di*) *ziffoh*, *zift*'; *lata'*; *nizafih*; *'ishelo*.

Food (livestock). This is an important forage for goats on Abd al Kuri, though sheep and donkeys also graze it. It is regarded as a good maintenance feed of limited nutritional value. Its main value lies in its tolerance to drought, as rainfall on the island is extremely erratic.

Other. Dead plants are used in roofing and are gathered for use as kindling and firelighters.

Comparative uses. The grains of some African varieties are used as famine food. (Kunkel 1984)

Tetrapogon Desf.

Tufted annuals or perennials. Leaves usually flat. Ligule a fringed membrane. Inflorescence of single or paired 'racemes' at the tip of the stem. Spikelets 4–9-flowered. Glumes similar, sometimes purplish, glabrous. Lemmas awned. 5 or 6 species. Mediterranean to S Africa, Arabia, India and Pakistan.

Tetrapogon tenellus (K.D.Koenig ex Roxb.) Chiov.

Loosely tufted annual to 50cm. Inflorescence of 1 or 2 'racemes' 3–6cm long. Spikelets 5 or 6-flowered, the three lower florets fertile and with a reddish-purple awn, the upper two or three sterile and awnless. Upper glume reddish-purple, not mucronate. Lemmas whitish, glabrous or with short hairs on the nerves, awned.

Ecology. *Croton socotranus* shrubland. 50m.

Distribution. Tropical Africa, Arabia, India and Pakistan.

ETHNOGRAPHIC NOTES

Name(s)

– Soqotri: with inflorescence *ḳʌ̍ṭəhān* 'slender'.

– Soqotri simplified: *qatihan*.

Food (livestock). This grass is much liked by all livestock and is valued as an excellent and nutritious fodder in the drier areas of the island, where it grows on rocky ground and dry grassland.

Other. Dead plants are used in roofing and are gathered for use as kindling and firelighters.

Comparative uses. It is a common weed of sorghum fields. It can be used for roof thatching. (Heine *et al.* 1988)

Tetrapogon villosus Desf.

Densely tufted perennial to 60cm, basal leafy shoots flattened. Inflorescence of 2 'racemes' stuck together back to back, appearing solitary, 3–8.5cm long. Spikelets 4 to 6-flowered, each with a white or purple awn. Upper glume often purplish, mucronate. Lemmas whitish, conspicuously silky-hairy, awned.

Ecology. Limestone escarpment. 450–750m.

Distribution. Macaronesia, N and NE Africa, Arabia.

ETHNOGRAPHIC NOTES

Name(s)

– Soqotri: *ḳeṣa̍də* 'sharp, cutting', 'oversalted'.

– Soqotri simplified: *qezadih*.

Food (livestock). This tough grass provides only sparse grazing. Most livestock eat the new shoots and green leafage, but camels and donkeys also graze it when mature. All livestock chew away at the stumps in periods of prolonged drought.

Other. Parts of dead plants are gathered for use as kindling and firelighters.

Themeda Forssk.

Anthistiria L.f.

Annuals or perennials. Leaves linear. Ligule a hairy membrane. Inflorescence an open, 'leafy' panicle. Basic unit of inflorescence a single, short 'raceme' embraced by a boat-shaped spatheole. 'Racemes' then grouped together to form an open panicle. Each 'raceme' comprised of three different types of spikelet: two pairs of homogamous spikelets at the base and paired pedicelled and sessile spikelets above. Spikelets 2-flowered, upper lemma of sessile spikelet passing directly into an awn (awn bent at knee, column hairy). Pedicelled spikelets larger than the sessile, awnless. 18 species. Old World tropics and subtropics.

Botanical notes. Related to *Heteropogon*, although the 'raceme' is not always easy to understand as they are highly modified and often compacted into dense clusters (Clayton & Renvoize 1986).

Themeda quadrivalvis **(L.) Kuntze**
Anthistiria ciliata L.

Annual to 90cm, often supported at the base by stilt roots. Inflorescence with involucral spikelets 5–8mm long.

Ecology. No habitat details available.
Distribution. Arabia, India.
Botanical notes. Collected on Abd al Kuri in the 19th century; not seen since.

Themeda triandra **Forssk.**

Tufted perennial sometimes to 2m, leaves mostly confined to a basal cushion around slender, flowering stems. Inflorescence with involucral spikelets 6–14mm long.
Ecology. Grassland on granite in Haggeher mountains. 750–1000m.
Distribution. Old World tropics and subtropics.
Botanical notes. A widespread polymorphic species, and a common or sometimes dominant constituent of grassland in Africa, especially in the east and south where it is a favoured pasture grass. It is often dominant in the S African veld ('Rooigrass').

ETHNOGRAPHIC NOTES
Name(s)
– Soqotri: *di ʾɔ̄žhir*; *ʾēžəhər*, *ʾāˈjeyr* (western Soqotra) 'spiky', 'painful'; with inflorescence *di ḳerˈṭomo* 'needs chewing'.
– Soqotri simplified: *di 'owjihir*; *'eyjihir*, *'ajeyr*; *di qertomo*.
Food (human). There is no tradition of gathering the grains as food on Soqotra.
Food (livestock). 'Red oat grass' is regarded as a valuable and nutritious grass for all livestock, especially when young. It is particularly liked by cattle. Before it flowers and sets seed it is seen as promoting lactation and helping animals to put on weight. It makes reasonable hay if cut when still immature.
Comparative uses. Burning seems to encourage the spread of the grass. Wilted plants contain hydrocyanic acid and may poison livestock. It is regarded as a valuable fodder in South and East Africa. The grains are used as food, and the leaves are also eaten. The culms are used in thatching and paper manufacture. The awns in the mature spikelets of some *Themeda* species are formidable and can cause mouth abscesses. (Ambasta 1982; Bogdan 1977; Dalziel 1948, 1956; Jain & De Fillips 1991; Kunkel 1984)

Tragus Haller

7 species. Old World tropics.

Tragus berteronianus **Schult.**

Loosely tufted annual. Leaves flat with distinctive hairy margins. Ligule a ring of hairs. Inflorescence a single terminal 'raceme' 2–5cm long. Spikelets 1-flowered, paired, awnless. Glumes unequal, the lower reduced to a transparent scale or absent, the upper densely covered in hooked spines. Lemma membranous.
Ecology. Widespread but never common. 100–250m.
Distribution. America, Africa, Arabia to India and Pakistan, China.

Botanical notes. One of the most distinctive grasses; characterised by the spikelets which are densely covered in hooked spines.

ETHNOGRAPHIC NOTES
Name(s)
– Soqotri: *mˈʌsḥiṭ* 'broom'; *śˈebiṭə* 'never dies right back'; *śˈebereḳ*; 'type of *ɛ̄rˈśido*'.
– Soqotri simplified: *mas'hit*; *shebitih*; *shebereq*; 'type of *ershiydo*'.
Food (livestock). This grass is especially liked by small stock after rain and before it matures and the prickly inflorescence develops. In the drier areas of the island it is regarded as a good feed for small stock in particular, promoting lactation and helping them gain weight.

Tricholaena Schrad. ex Schult. & Schult.f.

Tufted perennials. Leaves linear, flat or inrolled. Ligule a fringe of hairs. Inflorescence an open panicle. Spikelets 2-flowered, on delicate pedicels, silky-hairy, awnless. Glumes unequal, lower glume minute or occasionally absent. Upper glume and lower lemma similar, as long as spikelet. 1 endemic. 4 species. Mediterranean and Africa to India and Pakistan.

Tricholaena teneriffae (L.f.) Link

Panicum teneriffae R.Br.
Tricholaena setacea C.E.Hubb.

Densely tufted perennial. Inflorescence 4–12cm long. Spikelets c.3mm long, silky-hairy with hairs much longer than the spikelet.
Ecology. Semi-deciduous woodland. 450m.
Distribution. Macaronesia to India and Pakistan.

ETHNOGRAPHIC NOTES
Name(s)
– Soqotri: *ɛ̄rśído.*
– Soqotri simplified: *ershiydo.*
Food (human). The sweet, green stems are eaten.
Food (livestock). This grass is grazed by all livestock but is not believed to be of significant nutritional value, particularly in areas rich in alternative forage. In the drier areas of the island, where it sometimes spreads to make a thick carpet, it is a valued fodder for cattle and especially for goats in the dry season.

Tricholaena vestita (Balf.f.) Stapf & C.E.Hubb.

Eriochloa vestita Balf.f.

Low growing densely short hairy perennial to 15cm, the stems and branches intermingling to form a carpet close to the ground. Inflorescence 2–3cm long. Spikelets c.2mm long, shortly hairy, hairs scarcely exceeding the tip of the spikelet.

Ecology. On the limestone plateau. c.500m.
Distribution. Endemic.
Conservation status. Data Deficient (DD). Although not collected for over 100 years the grasses are poorly known on Soqotra and it is too early to suggest that this species is extinct. The location given by Balfour (see below) is imprecise – for comments on Balfour's collecting localities see notes under *Pluchea glutinosa*.
Botanical notes. Collected by Balfour on the limestone plains southwest of Galonsir [Qalansiyah] in the 19th century but not seen since.

Trichoneura Andersson

7 species. Tropical Africa, Africa, S USA, Peru, Galápagos.

Trichoneura mollis (Kunth) Ekman

Slender, tufted annual. Leaves linear, flat. Ligule membranous. Inflorescence loose, of 10 to 40 'racemes', 1.5–5cm long. Spikelets 5 to 7-flowered, 6–8mm long, green or with a purple tinge, overlapping by about half their length. Glumes scabrid. Lemmas hairy along the nerves; awn <2mm long.
Ecology. No details of habitat or location available.
Distribution. NE Africa, Arabia.

ETHNOGRAPHIC NOTES
Not seen, no information recorded.

Urochondra C.E.Hubb.

NE tropical Africa to Pakistan.

Urochondra setulosa (Trin.) C.E.Hubb.

Robust, tussock-forming perennial of coastal sand dunes. Leaves tough, glaucous, sharp-tipped. Inflorescence cylindrical, 4–13cm long. Spikelets 2.5–4mm long.
Ecology. Coastal sand dunes. Sea-level.

Distribution. Tropical Africa to India and Pakistan.
Botanical notes. Similar in appearance to *Sporobolus spicatus*, another plant of coastal sand dunes. However, *S. spicatus* is less robust and has a strongly stoloniferous habit.

ETHNOGRAPHIC NOTES
Name(s)
– Soqotri: *ḳeṣádə* 'sharp, cutting', 'oversalted' (Soqotra, Samha, Abd al Kuri); *lātaʾ* (Abd al Kuri).
– Soqotri simplified: *qezadih*; *lata'*.

Food (livestock). Like other *ḳeṣádə*-type grasses, this is palatable to most livestock only when young. The new shoots and young green leave are grazed by all livestock, but only camels and donkeys graze the more mature plant. In a prolonged dry season goats gnaw the juicier stems.
Other. Pieces of the dry plant are gathered for use as kindling and firelighters.

HYACINTHACEAE

Perennial herbs with bulbs. Leaves all basal, linear or broad. Flowers bisexual, actinomorphic, in spikes or racemes. Tepals 6, in 2 whorls, free or fused into a tube below. Stamens 6, free. Ovary superior, 3-celled; style simple. Fruit a many-seeded capsule; seeds black and subglobose or flattened. 40 genera. 900 species. Temperate and tropical regions of Africa, Mediterranean and SW Asia.

Dipcadi Medik.

Leaves linear, in basal whorl. Flowers dull greenish brown or white, 2 to several in one-sided racemes, appearing at same time as leaves. Bracts not spurred. Tepals fused into a tube below, free part of inner tepals erect, outer tepals spreading and recurved, sometimes with appendaged tips. Capsule 3-lobed; seeds black, flattened. 3 endemics. c.30 species. Mediterranean, Africa and SW Asia.

ETHNOGRAPHIC NOTES
Comparative uses. Many species have edible bulbs and leaves. In Qatar, *D. erythreum* is used as an expectorant, and in Bahrain the leaves are taken as a laxative and used as an ointment for wounds. In India the bulbs of the same plant are used as a substitute and adulterant for Indian Squill (*Urginea indica*) which resembles digitalis in action, and is used as an expectorant. (Abbas *et al.* 1992; Ambasta 1982; Batanouny 1981; Jain & De Fillips 1991; Kunkel 1984; Miller & Morris 1988; Rizk & El-Ghazaly 1995)

Dipcadi balfourii Baker

Robust plant to 45cm. Leaves dark green, 3 or 4, straight, linear, 20–35 x 0.3–0.5cm, channelled. Flowers c.30, on a long (60–75cm) dark reddish-green peduncle. Perianth cylindrical, c.16 x 2mm, yellowish green, tinged red; lobes recurved: lobes of inner tepals 5 x 2mm; lobes of outer tepals 9 x 2mm. Fruit not seen.
Ecology. Succulent shrubland; on rocky limestone slopes. 50–500m.
Distribution. Endemic.
Conservation status. Vulnerable (VU D2). Rare, with a very limited area of occupancy.

Botanical notes. This bulb was originally collected by Balfour in the 19th century. He comments that it flowered at Kew but that he had no record of its collection and therefore could say nothing regarding its distribution on the island.

ETHNOGRAPHIC NOTES
Name(s)
– Soqotri: *ḥídḥud* 'sea-urchin'; *di bəṭáḥ(ḥ)ʌn* 'prostrate'.
– Soqotri simplified: *hidhud*; *do bitahan*.
Food (human). The small bulbs are edible and are either eaten raw, or, if plentiful, are cooked in water or baked in the embers.
Food (livestock). All small stock eat the foliage.

Dipcadi biflorum Ghazanfar

Low herb to 5cm. Leaves grey-green, 2 or 3, coiled, linear, 3–6 x 0.3–0.5cm. Flowers 2 or 3, on a short, c.3–4cm peduncle. Perianth tubular, c.8mm long, yellowish green; lobes equal, erect, c.6 x 2mm. Fruits 3-lobed, globose, c.1 x 1cm.
Ecology. Sandy coastal plains. Sea-level.
Distribution. SE Arabia (Oman and United Arab Emirates).

Botanical notes. An uncommon plant known only from fruiting material. The Soqotran plants closely resemble *D. biflorum* from Oman and the United Arab Emirates but have rather smaller bulbs (c.5mm diameter versus 1.8–2cm diameter), leaves (3–6 x 0.3–0.5cm versus 10–16 x 0.7–1cm) and fruits (10 x 10mm versus 14–15 x 9–10mm). Flower measurements are based on plants from Oman. The Soqotran plants are therefore tentatively placed in *D. biflorum*. Further material is required (Ghazanfar 1996).

ETHNOGRAPHIC NOTES
Not seen, no information recorded.

Dipcadi guichardii **Radcl.-Sm.**

Delicate plant to 20cm. Leaves 2 to 4, straight, filiform, 5–15cm long, channelled. Flowers 10 to 20 on an erect, 10–20cm long peduncle. Perianth tubular, 9–11 x 1–1.5mm; lobes erect, subequal, 3.5–6 x 1–1.5mm, pale reddish brown. Fruits deeply 3-lobed, 3 x 4mm.
Ecology. Coastal plains in *Croton socotranus* shrubland and in clearings in drought-deciduous woodland. 10–550m.
Distribution. Endemic.
Conservation status. Least Concern (LC). Common in several vegetation types and under no present or perceived threat.
Botanical notes. Sometimes abundant in the flush of ephemerals that come up after rain on the northern coastal plain.

ETHNOGRAPHIC NOTES
Name(s)
– Soqotri: *ḥīrɛ̄rə d-ˈʌġərib* 'little bulb of the raven' (indicates the plant's unpalatability); *di bəṭˈaḥ(ḥ)ʌn* 'prostrate'.
– Soqotri simplified: *hiyreyrih d-aghirib*; *di bitahan*.
Food (human). The small bulb is not generally eaten.
Food (livestock). All small stock eat the foliage.

Dipcadi kuriensis **A.G.Mill.** (see Appendix 2, p. 723)
Dipcadi Medik. sp. *A*

Similar to *D. balfourii* but leaves broader, pale green (not reddish); flowers 15 to 20, on a pale green peduncle; perianth white, tube rather broader and tepals more strongly recurved. Fruit deeply 3-lobed, 8 x 9mm, broadly obovoid, on erect pedicels.
Ecology. In crevices on exposed limestone ridge and rocky slopes. 450–550m.
Distribution. Endemic to Abd al Kuri.
Conservation status. Vulnerable (VU D2). Rare, with a very limited area of occupancy.
Botanical notes. Apparently closely related to *D. balfourii* but found only on the summit ridge and upper slopes of Jebel Saleh on Abd al Kuri. It has white, not yellowish-green, flowers and lacks the reddish tinge typical of *D. balfourii*.

ETHNOGRAPHIC NOTES
Name(s)
– Soqotri: *ḥʌmbəsīs* 'to make the special call to bring the breeding male to female goats in oestrous'.
– Soqotri simplified: *hambisiys*.
Food (human). In Abd al Kuri these small bulbs offered important seasonal nourishment and were actively harvested.
Food (livestock). The foliage is considered to be good and nutritious rains grazing for sheep and goats.

Drimia Jacq. ex Willd.

Leaves in a basal whorl. Flowers white, in lax racemes, appearing before or with the leaves. Bracts spurred. Tepals free or fused, spreading. Capsule 3-lobed. Seeds flattened. 1 endemic. c.120 species. S Europe, Africa, Asia.

Drimia porphyrostachys **(Baker) A.G.Mill.**
Urginea porphyrostachys Baker

Leaves c.5, spirally arranged, narrowly ovate, 6–12 x 1–2.5cm, glabrous. Flowering spike 15–35cm, the inflorescence c.50-flowered, pyramidal in bud, elongating in flower; bracts 2–5 x c.1.5mm, shortly spurred at base and shortly stalked thus appearing scale-like. Flowers c.10mm across; tepals white with a purple-pink midrib, ovate to elliptic, 4–5.5 x 1.5–3mm.
Ecology. In montane thicket and semi-deciduous woodland on granite in the Haggeher mountains. 600–1400m.
Distribution. Endemic.
Conservation status. Data Deficient (DD). Poorly known; inadequate evidence is available to make an assessment.

ETHNOGRAPHIC NOTES
Name(s)
– Soqotri: *mən ṭˈɛyoh* 'of cliffs'.
– Soqotri simplified: *min teyoh*.
Food (human). The bulb is not generally eaten.
Food (livestock). Generally grows in places inaccessible to all but the most agile goats. It is thought that the wild and feral goats of the high Haggeher areas might eat the foliage, but it is not known whether domesticated goats eat it. Related species are said to be poisonous to livestock.
Comparative uses. *Drimia* species are much used medicinally and as rodenticides. The bruised bulbs of some species are applied locally to bruises, or are scorched, dried and mixed with the ash of various plants and butter or oil, and used as a liniment for rheumatic knees and painful joints. (Ambasta 1982; Dafni *et al*. 1984; Dalziel 1948, 1956; Jain & De Fillips 1991; Miller & Morris 1988)

Ledebouria Roth

Leaves ovate, paired or in a basal whorl, sometimes with purple spots. Flowers pink, in lax or dense racemes, appearing at the same time as the leaves. Bracts not spurred. Tepals spreading, fused into a short cup below. Ovary 6-lobed, shortly stalked. Seeds subglobose. 2 endemics. c.30 species. Africa and India.

Ledebouria aff. *revoluta* (L.f.) Jessop

Soqotri material: bulb with 1 or 2 leaves. Leaves narrowly ovate, c.3–5 x 2–3cm, spotted purplish green. Descriptive notes from the Flora of Somalia (vol. 4, p. 55, 1995): bulbs c.5cm diameter; leaves 1 to several, lanceolate or narrowly elliptic, <15 x 3cm, often spotted purplish green, glabrous; inflorescence many-flowered; tepals 3–7mm, greenish, sometimes tinged purplish; ovary deeply 3-lobed, 6-angled, sometimes with horn-like appendages.
Ecology. Semi-deciduous woodland. 690m.
Distribution. Tropical and S Africa, India, Sri Lanka.
Botanical notes. *Scilla hyacinthina* (Roth) J.F.Macbride (= *L. revoluta*) is recorded from Soqotra on an old checklist kept at the Royal Botanic Garden Edinburgh. However, the origin of this record could not be traced and so it was discounted. However, a photograph of several plants, each with a solitary spotted leaf and which seem to be a good match for *L. revoluta*, was taken on the Muqadrihon Pass by Dylan Hannon, in spring 2001. This photograph was used as the basis of the first part of the description. However, until material of this plant can be examined the record of this species on the island must be treated with caution.

ETHNOGRAPHIC NOTES
Name(s)
See *L. grandifolia* (also uses).

Ledebouria grandifolia (Balf.f.) A.G.Mill.
Haemanthus grandifolius Balf.f.

Bulb c.5 x 4cm. Leaves basal, always paired, appressed to ground, not blotched, ovate to broadly ovate, 13–23 x 8–15cm, the base rounded to cordate. Inflorescence a lax 30 to 100-flowered raceme. Tepals pinkish-purple, 6–7 x 1.5–2.25mm.
Ecology. Locally common on both limestone and granite in the centre and northeast of Soqotra; growing in shady places in relatively high rainfall areas; particularly common in the dense drought-deciduous woodlands in the valleys draining to the north and northwest of the Haggeher mountains and in the east of the island on the northern, seaward-facing cliffs and broken limestone ridges (characteristically in solution hollows and pockets of soil on the limestone pavement). 50–700m.
Distribution. Endemic.
Conservation status. Least Concern (LC). Common in several vegetation types and under no present or perceived threat.
Botanical notes. Originally described as *Haemanthus grandifolius* (Amaryllidaceae) by Balfour in the 19th century on the basis of the large paired leaves which are a familiar sight in the northeast of the island. However, it was not until 1993 that it was first found in flower and its true affinities with the Hyacinthaceae recognised. *L grandifolia* flowers at the beginning of the winter rains (October to November), and if conditions remain humid, the leaves persist, increasing in size, until well into the following year (January to February) (Miller & Alexander 1996).

ETHNOGRAPHIC NOTES†
Name(s)
– Soqotri: *di fəˈṭaḥer, di fəˈṭaḥə, di fəṭāwə* (pl. *fəṭāwitin*) 'new shoots, new growth'.
– Soqotri simplified: *di fitaher, di fitahih, di fitawih* (pl. *fitawitin*).
Food (human). The bulbs are edible, but not much liked. They have to be cooked with care or they become slimy and sticky, and have to be peeled before eating. They are better baked in hot ash, but even so the texture is unpleasant and they cause flatulence. Eaten in desperation only, and not regarded as a good or useful food.
Food (livestock). The foliage is not palatable to livestock.

†Also apply to *L.* aff. *revoluta*, *L. insularis*.

Ledebouria insularis A.G.Mill. (see Appendix 2, p. 723)

Bulb c.1.5–2cm diameter. Leaves 2 to 4, ascending, ovate, 3–5 x 2–3.5cm, narrowed into a petiole-like base, not blotched. Inflorescence a lax c.15 to 25-flowered raceme; peduncle 4–10cm. Tepals pale pinkish purple, c.4.5–5 x 2mm.
Ecology. Rare, in tall herb communities on shady ledges of north-facing cliffs on Samha. 700m.
Distribution. Endemic.
Conservation status. Endangered (EN B2 a biii). This pretty, blue-flowered bulb is known only from the summit ridge on the limestone plateau of Samha where it occupies an area of probably less than 2km^2. The ridge catches low cloud and, with the north-facing cliffs below, provides a unique moist habitat on the otherwise desertic island of Samha. We strongly suspect that this marginal habitat is threatened by the trend of lower rainfall in the

region and would be particularly vulnerable to the effects of global warming. The situation is particularly critical for *L. insularis* which occupies a niche at the highest point on Samha and so, with increasing aridification, has no possibility of mitigating the effects of gradual drying by migrating upwards in altitude. The foliage of this plant is not particularly palatable to livestock; however, although the numbers of feral goats and sheep are at present small, these animals are known to graze on the limestone plateau, and as their numbers increase, this could pose a threat to the plant in the future.

Botanical notes. Similar to *L. grandifolia* but generally smaller in stature and with smaller flowers. However, it is most clearly distinguished from that species with its 2–4 smaller leaves which are narrowed, not rounded, at the base and ascending, not paired and appressed to the ground.

ETHNOGRAPHIC NOTES
Name(s)
See *L. grandifolia* (also uses).

HYDROCHARITACEAE

Dioecious or monoecious herbs of fresh or salty water. Leaves simple, sessile or petiolate. Flowers enclosed in a simple or split spathe: male 1 to many, female solitary. Calyx of 3 petaloid sepals. Petals membranous or 0. Stamens 3 to 12. Ovary inferior, 1-celled; placentation parietal. Fruit globose or ovoid. Seeds few to many. 16 genera. 90 species. Cosmopolitan.

Botanical notes. There are several marine species, which are expected to occur but have not been recorded from Soqotra. They are included in the key but no descriptions are given.

Lagarosiphon Harv.

See description of family. 9 species. Tropical Africa, Madagascar.

Lagarosiphon roxburghii **Benth.**
Nechamandra alterniflora Thwaites

Submerged aquatic of brackish or fresh water. Rhizome filiform. Leaves alternate, becoming clustered, linear-lanceolate, 15–60 x 2–6mm, often twisted; leaf sheaths open. Flowers at first hidden and then rising to the surface on filiform stalks.

Ecology. No details of habitat or location available.
Distribution. Old World tropics.
Botanical notes. Balfour (1888) notes that neither he nor Schweinfurth collected this species on Soqotra. However, Schweinfurth comments that there was a specimen from Soqotra in the Berlin herbarium collected by Boivin in 1847. It has not been seen since. A native of the Indian subcontinent.

HYPOXIDACEAE

Herbs with corms. Leaves grass-like, in a basal rosette, linear, hairy. Flowers bisexual, actinomorphic, in racemes. Tepals 4 to 6, free or united below, hairy. Stamens 4 to 6. Ovary inferior, 2 or 3-celled; style simple. Fruit a capsule, surmounted by the remains of the perianth; seeds globose, black. c.6 genera. c.150 species. Mostly S Hemisphere, especially S Africa.

Hypoxis L.

See description of family. 80 species. Warm and tropical regions, especially the S Hemisphere.

Hypoxis angustifolia **Lam.**

Perennial herb with hairy, grass-like leaves; corms clothed in fibrous leaf remains. Leaves linear, 6–15 x 0.3–0.5cm. Inflorescence 2 or 3-flowered. Tepals 4(to 6), bright yellow, c.10mm long.

Ecology. Locally common in shrubland and woody-herb communities on wetter parts of the limestone plateau. 550–650m.
Distribution. Tropical and S Africa.
Botanical notes. Across the species' wide range in tropical Africa, flowers of *H. angustifolia* typically have 6 tepals. However, plants on Soqotra almost always have 4 tepals; only very rarely are flowers with 6 tepals seen. *H. angustifolia* is readily recognised by its 4-merous, bright yellow flower and its grass-like leaves which are fringed with long, weak hairs.

ETHNOGRAPHIC NOTES

Name(s)

– Soqotri: *ḥargyod, ḥarged* (*śaᶜerher*) 'unkempt, unwashed', 'hairy, fibrous'.

– Soqotri simplified: *hargyod, harged* (*sha'erher*).

Food (human). Formerly, the corms were an important food source. After peeling, they were eaten raw or cooked, either baked in hot ash, or boiled in water. The best ash for cooking them is said to be that of dung. They were not liked as much as Iridaceae corms, but were generally considered to be as good as those of *Habenaria* and *Holothrix*.

Food (livestock). The foliage is eaten by all livestock and is considered a valuable and nourishing rains feed, especially for cattle and goats. Apparently, goats and sheep do not dig up the plants to get at the corms.

Comparative uses. Many *Hypoxis* species have edible corms. (Kunkel 1984)

IRIDACEAE

Herbs with corms. Leaves basal and cauline. Flowers bisexual, actinomorphic, solitary or in spikes, each flower enclosed by a pair of bracts. Tepals 6, in 2 whorls, fused below. Stamens 3. Ovary inferior, 3-celled, placentation axile; style 3-branched. Fruit a capsule. 92 genera. 1800 species. Cosmopolitan, especially diverse in S Africa, E Mediterranean, C and S America.

Babiana Ker Gawl. ex Sims

Corms globose, with a fibrous tunic. Leaves deciduous, lanceolate, 2-ranked. Flowers blue, actinomorphic, arising from between the overlapping leaf bases. Tepals spreading, nearly equal, fused in a long, narrow tube below. Ovary enclosed in leaf bases. 1 endemic. 63 species. Tropical and S Africa, especially Cape.

Botanical notes. Goldblatt (pers. comm.) has recently moved *B. socotrana* into a monotypic genus, endemic to Soqotra. This has been separated from the largely southern African genus *Babiana* on account of its pollen morphology, chromosome number, corm ontogeny and seed morphology. DNA sequence studies have also confirmed that *B. socotrana* was misallied generically. A paper describing the new genus will appear in *Edinburgh Journal of Botany* in early 2004.

Babiana socotrana Balf.f.

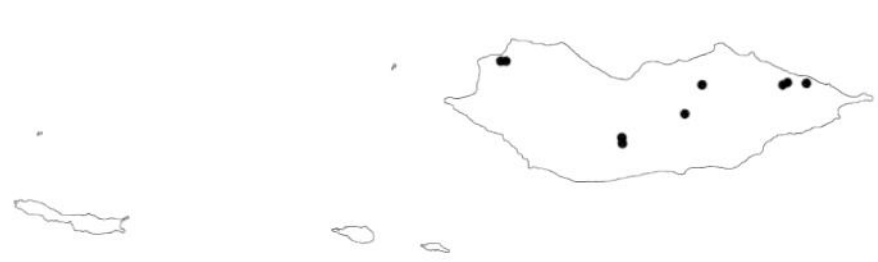

Herb to 15cm. Leaves 2-ranked, 2–20 x 0.4–1.5cm. Flowers solitary, subtended by bracts. Tepals blue, ovate to elliptic and narrowed below, 10–18 x 3–8mm; corolla tube pale blue or white, 30–50 x c.0.5mm. Fruit enclosed by the leaf bases; seeds reddish brown, ovoid, c.2.5 x 1.5mm, reticulate.

Ecology. Locally abundant in low shrubland and grassland on the limestone plateaus, in accumulations of soil in solution hollows and rock crevices on the limestone pavement. 500–900m.

Distribution. Endemic.

Conservation status. Least Concern (LC). Common in several vegetation types and under no present or perceived threat.

Botanical notes. See notes under genus.

ETHNOGRAPHIC NOTES†

Name(s)

– Soqotri: *bəṭaḥ, bəṭaᶜ, bōṭiḥ* (pl. *boṭaᶜ, bəṭaḥ, bəṭāᶜ*) 'prostrate'; (rare) *biššiі, bišᵓhi* (commonly given to plants of cliffs or with striking flowers).

– Soqotri simplified: *bitah, bita', bowtih* (pl. *bota', bitah, bita'*).

Food (human). The corms are a popular food. They are either peeled and eaten raw as and when they are found, or are gathered in large quantities and then cooked. The raw corms can be bitter, especially when mature. Their flavour is best when they are peeled and then baked in the embers or in hot ash. The best ash for cooking these is said to be from dung. When boiled in water they have a tendency to become unpleasantly slimy. Some crush the raw, peeled corms and add them to a little goat's milk and then cook them to a mush eaten hot. Others grind the cooked, peeled corms and add fresh or soured milk to the paste to make thick soup.

Food (livestock). The foliage is eaten by all livestock and is considered to be a valuable and nourishing rains fodder, especially for cattle and goats. Apparently, goats and sheep do not dig up the plant to get at the corm.

Bees / honey. Honeybees are attracted to the flowers.

Other. If no grain is available, the corms are peeled, lightly crushed and then added to boiling ghee to clarify it. This gives the ghee a delicious flavour and the butter-impregnated corms make very good eating afterwards.

Comparative uses. Many *Babiana* species have edible corms. (Kunkel 1984)

†Also apply to *Romulea fischeri*.

Romulea Maratti

Corms covered with membranous scales. Leaves linear, spirally arranged. Flowers solitary, subtended by 2 bracts. Tepals blue, yellow at base, fused into a short tube below. c.95 species. S Africa, Mediterranean and Middle East.

Romulea fischeri Pax

Romulea purpurascens var. *edulis* sensu Balf.f. non Baker

Plant to 20cm. Leaves 3 to 5, linear, 8–15cm long. Flowers solitary, blue with yellow centre. Perianth tube c.4mm; tepals ovate, c.12 x 3mm.

Ecology. In semi-deciduous woodland and shrubland mainly on granite but also recorded on limestone. 600–1200m.

Distribution. Kenya, Ethiopia, Somalia and Saudi Arabia.

Botanical notes. Locally common in the central mountains. Recorded in the 19th century from near the summit of Jebel Saleh on Abd al Kuri; not seen since. Also recorded in the 19th century from the eastern limestone plateau on Soqotra but likewise not seen since.

ETHNOGRAPHIC NOTES

Name(s)

See *Babiana socotrana* (also uses).

Comparative uses. The half-ripe fruits of some *Romulea* species are eaten, usually by children. (Kunkel 1984)

JUNCACEAE

Annual or perennial herbs. Leaves alternate or basal, sheathing at the base with a membranous ligule. Flowers in compact or diffuse cymes, bisexual, actinomorphic. Tepals 6, free, scarious. Stamens 6. Ovary superior, 1 or 3-celled; stigmas 3. Fruit a many-seeded capsule. 8 genera. 300 species.

Juncus L.

See description of family. 220 species. Widespread in temperate regions, mainly on mountains in the tropics.

Juncus bufonius L.

Slender annual to 20cm. Leaves soft, filiform. Inflorescence an open, leafy panicle occupying most of the plant. Tepals 3–6mm. Capsule oblong, blunt, 2–3mm long, included within tepals, reddish brown.

Ecology. Damp sand on exposed granite pinnacles and in boggy areas in grassland. 1200–1500m.

Distribution. Europe and mountains of tropical E Africa.

ETHNOGRAPHIC NOTES

Name(s)

– Soqotri: *ḳʌˈźəbə, ḳʌźˈʌbhen* (a widespread name for various Gramineae and Cyperaceae of damp ground).

– Soqotri simplified: *qadhibih*, *qadhabhen*.

Food (livestock). Provides incidental grazing for the goats of the high Haggeher, as well as for wild and feral goats.

Juncus rigidus Desf.

Juncus arabicus (Asch. & Busch) Adans.
Juncus maritimus var. *arabicus* Lam.

Robust perennial to 1.5m. Rhizome stout, horizontal, producing dense rows of stems. Leaves and bracts stiff, terete, spinescent, the lowest reduced to sheaths. Inflorescence a many-flowered panicle, subtended by an erect spinescent bract that continues in the direction of the stem. Tepals 2–3mm. Capsule narrowly ellipsoid, 3.5–5mm long, pale green, exserted from the tepals.

Ecology. In brackish mud in coastal inlets. Sea-level.

Distribution. Europe, Africa and SW Asia.

ETHNOGRAPHIC NOTES†

Name(s)

– Soqotri: *ˈiˀšhel, ˈiˀšel* (pl. *ˈiˀšhilˈitin, ˈeˀšəlˈitən*); *ḳʌźˈʌbhen* (commonly given to Gramineae and Cyperaceae of damp ground).

– Soqotri simplified: *i'ish'hel*, *i'ishel* (pl. *i'ish'hilitin*, *e'shilitin*); *qadhabhan*.

Food (livestock). Donkeys and camels readily eat rushes, but cows and goats only eat them when they are very hungry in the dry season. In a prolonged dry season or at times of drought, cattle herders take their animals down to feed in areas in which they grow, or cut the rushes and carry them home to their cattle.

Fibre / cordage. In some areas, such as the interior valleys behind the Central Highlands, rushes are cut and spread out in the heat of the sun until they have shrivelled and softened. They are then crushed to crack open the outer skin, which is stripped off, and the inner fibre is rolled and twisted to make cordage, which lasts for around six months. In these areas it is much used to tie up the ripening dates before harvest, as well as for making a leading rope or halter for pack animals. If prepared carefully and twisted finely it can be used as string for rough sewing after piercing holes with an awl. Stronger cordage can be made by plying two lengths of this finer string together. Cordage from rushes is regarded as being as strong as that made from *Ficus vasta* root-bark, and it is used where reliability is important, for suspending milk-skins for churning into butter for instance, or for suspending the little hammock in which babies are put to sleep. The cordage is best made in the summer months when the sun is at its hottest.

Other. Children play with the rushes. They make whips, or weave them to make tiny containers, or twist and plait them into bracelets and anklets. A sort of pipe used to be made from the largest reeds. The more sophisticated instruments were made from a metal tube with two lengths of reed inserted in one end as a mouthpiece. It was said to have been principally used by Soqotrans of Hadhrami or African origin in the coastal settlements.

Notes. Unlike *Cladium mariscus*, these rushes are not fired, as they are too valuable as a dry season fodder for cattle.

Comparative uses. Some Juncaceae are used as fodder in times of scarcity, but not all species are palatable. *J. inflexus*, for example, has poisoned cattle in Europe. Some species in Africa have a rhizome which is eaten raw and some species are suitable for fixing sand dunes, or absorbing salt from brackish soils. Many are used in making mats, baskets and chair seats. The seeds of *J. acutus* contain an oil rich in organic acids, sugars and amino acids, and its pulp has been considered for paper-making in Israel. (Ambasta 1982; Dalziel 1948, 1956; Davis *et al.* 1983; Jain & De Fillips 1991; Kunkel 1984; Lewis & Elwin-Lewis 1977)

†Also apply to *J. socotranus*.

Juncus socotranus **(Buchenau) Snogerup**
Juncus maritimus var. *socotranus* Lam.

Stout, clump-forming perennial to 2m. Rhizome inconspicuous, stems densely tufted. Leaves and bracts stiff, terete, spinescent. Inflorescence a many-flowered panicle, subtended by an erect spinescent bract that continues in the direction of the stem. Tepals 2–3mm. Capsule subglobose, 2.5–3mm long, glossy-brown.

Ecology. In marsh and along the larger watercourses draining from the Haggeher mountains. 250–950m.
Distribution. Saudi Arabia, Yemen, Oman and Iran.
Botanical notes. A large tufted plant, often forming almost impenetrable thickets along wadis.

ETHNOGRAPHIC NOTES
Name(s)
See *J. rigidus* (also uses).

NAJADACEAE

Aquatic herbs growing in fresh or marine water. Leaves submerged, opposite or whorled, with toothed margins, sheathing at base. Flowers inconspicuous, 1 to 3 in leaf axils, unisexual: male flowers consisting of a solitary stamen enclosed in a 2 to 4-lobed membranous sheath; female flowers consisting of an apparently naked ovary. Fruit flask-shaped. 1 genus. 35 species. Cosmopolitan.

Najas L.

See description of family.

Botanical notes. For a revision see Triest (1987).

Najas graminea **Delile**

Similar to *N. marina* but monoecious; leaves narrower (c.1mm broad), denticulate with 30 to 60 inconspicuous teeth on each margin; leaf sheaths with short appendages; fruits oblong-ellipsoid, 1.5–2.5 x c.0.5mm.
Ecology. Streams near Qaryih (Balfour 1888).
Distribution. Old World tropics and subtropics.
Botanical notes. Not seen recently but probably overlooked; should be looked for in slow-moving fresh water.

ETHNOGRAPHIC NOTES†
Name(s)
– Soqotri: nil or *di rihóytin* ‘of water’; *ḳúrźaʾ* (commonly given to perennial Gramineae and Cyperaceae of damp ground).
– Soqotri simplified: *di rihoytin*; *qurdha’*.
Food (livestock). Goats on Soqotra intensely dislike damp forage, but graze it when it dries out if they are very hungry. Other livestock eat it at low water if alternative forage is in short supply.
Comparative uses. *N. marina* is edible. (Kunkel 1984)

†Also apply to *N. marina*.

Najas marina **L.**

Dioecious submerged aquatic, rooting in mud; stems to 80cm, sometimes spiny. Leaves in whorls of 3, linear-lanceolate, 17–25 x 1.5–2mm, rather coarsely spiny with 5 to 15 teeth on each margin and sometimes with a few teeth on midrib; teeth sharp, 1–1.5mm, sometimes red-tipped; leaf sheaths rounded. Fruit ovoid, 2–2.5 x 1.3–1.5mm.

Ecology. Streams near Qalansiyah (Balfour 1888).
Distribution. Cosmopolitan.
Botanical notes. Not seen recently but probably overlooked; should be looked for in slow-moving fresh water. The Soqotran material is referable to subsp. *ehrenbergii* (A.Br.) Triest.

ETHNOGRAPHIC NOTES
Name(s)
See *N. graminea* (also uses).

ORCHIDACEAE

Terrestrial or epiphytic herbs: terrestrials with root tubers, rhizomes or pseudobulbs (swollen, tuber-like storage organs), epiphytes with succulent leaves. Flowers bisexual, zygomorphic in spikes, racemes or panicles. Tepals 6, in 2 whorls: outer whorl sepaloid or petaloid; inner whorl petaloid of 2 lateral tepals entire, and a median lobe (the lip) which is variously lobed and crested, sometimes produced into a spur. Stamen and style fused into a specialised column. Ovary inferior, 1-celled. Fruit a capsule. c.800 genera. 17,500 species. Cosmopolitan.

Angraecum Bory

Epiphytic herbs with short stems. Leaves leathery. Flowers in axillary spikes. Sepals and lateral petals entire; lip concave, acuminate, spurred at base. 206 species. Africa, Mascarenes, Sri Lanka.

Angraecum dives **Rolfe**
Angraecopsis tenerrima sensu Balf.f.

Perennial herb, rooted on branches. Leaves c.5–10 x 1.5–2cm, overlapping, stiff. Spikes c.10 to 15-flowered, from the axils of the leaves, appearing jointed. Flowers green, c.7mm across, with apparently 6 lobes; spur c.2.5mm.
Ecology. Rare, in dense semi-deciduous thicket; epiphytic on shady branches. 550–750m.

Distribution. Somalia, Tanzania and Kenya.
Botanical notes. The only epiphytic orchid on the islands.

ETHNOGRAPHIC NOTES
Name(s)
– Soqotri: nil or *eźərīʾyo* (*men mèssə*) 'residue or sediment (of rain)' (a general term for mosses).
– Soqotri simplified: *edheriyoh* (*men messih*).
Fuel. Lichens and mosses are gathered when dry and stored for use as tinder, and it is possible that this epiphyte is used in the same way.
Comparative uses. The leaves of *A. fragrans* are used as flavouring and as a tea. (Kunkel 1984)

Eulophia

Terrestrial herbs. Leaves stiff with toothed margins, erect, arising from swollen pseudobulbs. Flowers in large, many-flowered panicles. Sepals and lateral petals entire, recurved, green with purple-brown markings; lip 3-lobed, spurred at base. c.200 species. Tropics.

Eulophia petersii **(Rchb.f.) Rchb.f.**

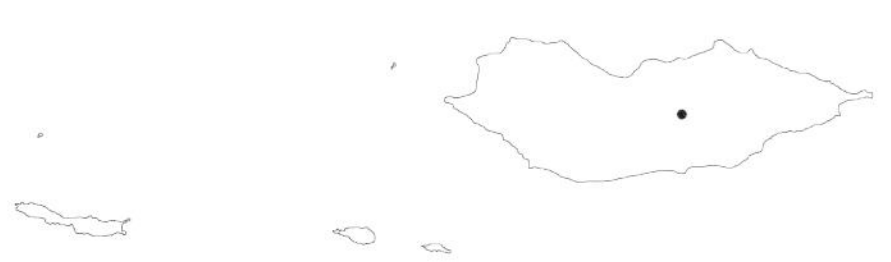

Robust herb, forming large colonies. Pseudobulbs erect, cylindrical, c.10 x 3cm, in large groups. Leaves leathery, somewhat succulent, c.20 x 2cm, margin spiny. Inflorescence not seen. Flowers not seen.
Ecology. On small cliffs and amongst rocks in open semi-deciduous woodland. c.700m.
Distribution. Africa and tropical Arabia.

Botanical notes. This record is based on sterile plants possessing only leaves and pseudobulbs. However, these are a good match for those of *E. petersii*, which is common on the adjacent African and Arabian mainland. Islanders living in the area report tall inflorescences 'swaying in the wind' – these also suggest a good match with *E. petersii*.

ETHNOGRAPHIC NOTES
Name(s)
– Soqotri: *śaʿàngəg* (*di ḳìrəhān*) 'untrimmed stem, branch (horned)' (referring to the pseudobulbs).
– Soqotri simplified: *sha'angig* (*di qirihan*).

Food (livestock). Goats eat the flowers and, when hungry, the softest growing tips. Sheep eat the fallen flowers. Otherwise this tough and spiky plant is not eaten by livestock.
Comparative uses. In Dhofar the bulbous part just above the ground is snapped off and held over the flames until it blackens and begins to ooze a liquid. This is collected and used to treat various cutaneous infections, such as scabies, ringworm, eczema, and itching weeping sores between the finger and toes, on the inside of the elbow and in the groin. The liquid is most effective and profuse when the plant is not in flower. (Miller & Morris 1988)

Habenaria Willd.

Habenaria Nimmo

Terrestrial herbs with root tubers. Leaves in a basal rosette. Flowers green, in many-flowered spikes. Sepals and lateral petals entire, the dorsal sepal hooded; lip 3-lobed, spurred at base. c.500 species. Tropics and subtropics.

Habenaria socotrana Balf.f.

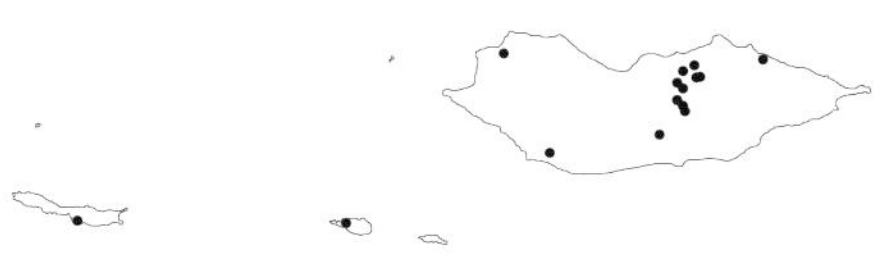

Erect herb to 30cm. Leaves 3 to 5, oblanceolate, <10 x 2cm. Spikes one-sided. Flowers green, c.8mm diameter, with 3 subequal lobes and an upper hood; spur c.10mm.
Ecology. In shade beneath woodland and on shady cliffs; common in the Haggeher and adjacent limestone plateaus; also found in high, moist areas such as Hamaderoh in the east, Ma'alah in the west and Qatariyah in the south; also on the north-facing cliffs on Samha and in rock crevices on the summit of Jebel Saleh on Abd al Kuri. 250–1400m.
Distribution. Somalia.
Botanical notes. The distribution of this species, the commonest and most widespread orchid in the Archipelago, reflects the extent of moist areas in the Archipelago, including the moist refugia on Samha, Abd al Kuri and at Qatariyah in southern Soqotra.

ETHNOGRAPHIC NOTES

Name(s)
– Soqotri: *ḥarhīʔo* (pl. *ḥárʔhi*, *ḥarhɛ̄tin*); *ḥiyrɛ̄r*, *ḥērə* (pl. *ḥīrəḥer*, *ḥirhítin*) (western Soqotra).
– Soqotri simplified: *harhiyoh* (pl. *har'hi*, *harheytin*); *hiyreyr*, *heyrih* (pl. *hiyrhiher*, *hirhitin*).
Food (human). This common plant used to be an important source of nourishment and was actively harvested, though the tubers were not liked as much as Iridaceae corms. The tubers could be peeled and eaten raw, or were cooked. They were either baked in hot ash, or cooked in milk or water. The best ash for cooking them was considered to come from dung. The cooked tubers were best when mashed, spread flat out and fresh butter dripped on top.
Food (livestock). The foliage is grazed by all livestock, but is not considered to be a particularly nutritious feed, and in areas where plenty of grass has also come up with the rains, it is generally ignored. On Samha and Abd al Kuri, however, the foliage is considered a good sheep forage. Goats and sheep have not been observed digging up the plant to get at the tuber.
Other. Sheep feeding on it can go longer without water.
Comparative uses. In Dhofar the tubers of *H. myodes*, *H. malacophylla* and *H. keayi* were all important foods in times of hunger, especially the first two. They bear the same name as the Soqotran orchid and were also eaten raw or cooked. Many other species are known to have edible tubers, and the tubers of *H. commelinifolia* are a source of *salep*. (Ambasta 1982; Jain & De Fillips 1991; Kunkel 1984; Miller & Morris 1988)

Holothrix Rich. ex Lindl.

Terrestrial herbs with small tubers. Leaves basal, paired, orbicular. Flowers greenish, in one-sided spikes. Sepals entire; lateral petals entire; lip 3-lobed, spurred at base. 1 endemic. 55 species. Tropical and S Africa, Arabia.

Holothrix socotrana Rolfe

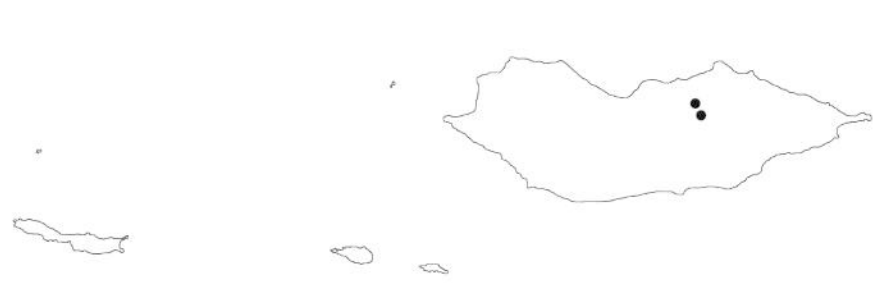

Erect herb to 20cm. Leaves 2, flat on ground, broadly ovate, <6 x 4cm diameter, pilose. Spikes 10 to 20-flowered, pilose. Flowers green, c.6mm diameter, with 5 subequal lobes from a cup-shaped base; spur c.2mm.
Ecology. Rare, in dense shade beneath thickets in semi-deciduous montane woodland in the Haggeher mountains. 800–1000m.
Distribution. Endemic.
Conservation status. Vulnerable (VU D2). Rare, with a very limited area of occupancy.

ETHNOGRAPHIC NOTES

Name(s)
– Soqotri: nil or *ḥarhīʔo* (pl. *ḥárʔhi*, *ḥarhɛ̄tin*).
– Soqotri simplified: *harhiyoh* (pl. *har'hi*, *harheytin*).
Food (human). Not common enough to be important as food, but the clumps are dug up when they are come across and the peeled tubers are eaten raw. If large numbers are found they are cooked like *Habenaria* tubers.
Food (livestock). The foliage provides incidental forage for goats.

PALMAE

Trees usually with a simple unbranched trunk, covered with old leaf bases or leaf scars. Leaves massive, forming a dense crown at top of trunk, palmately or pinnately divided; petioles spiny. Inflorescence paniculate, enclosed by spathes when young. Flowers bisexual or unisexual, 3-merous, tepals all similar or differentiated into a calyx and corolla. Stamens 3 to many. Fruit a drupe. 198 genera. 2650 species. Tropical and warm regions.

Borassus L.

Dioecious palms. Leaves palmate. Fruit massive, 1 to 3-seeded, smooth. 7 species. Old World tropics.

Borassus aethiopum Mart.

Borassus flabelliformis L.

Tree to 30m; trunk unbranched. Leaves palmate, <2m, with many folded segments; petiole edged with curved teeth. Male inflorescence <1.5m; female inflorescence <1.5cm. Fruit ovoid, <12 x 14cm.

Ecology. Planted in date gardens.

Distribution. Widespread but with a patchy distribution in the moister parts of tropical Africa.

ETHNOGRAPHIC NOTES

Name(s)

– Soqotri: *keri̍gi*.

– Soqotri simplified: *kerigi*.

Notes. These palms were brought from the mainland by the ruling Sultanic family, who are said to have planted one to mark each of their residences. There are still six left standing: five along the northern coast and one at Go'o di Salatini a little further inland. They were the private property of the family, and consequently no traditional uses are reported. Interestingly, the islanders often call an exceptionally tall 'male' *Dracaena* by the same name. Bent (1897) describes these palms: 'near Tamarida (Hadiboh) we saw just a few fan-palms, and one I remember looked very odd, as it still retained every leaf it had ever had, and looked like a yellow tower, with the green leaves at the top. All the rest were bristling, withered down to the ground. And when leaving Ferigi, we ride a little way along the river, passing a single fan-palm-tree, very tall and bare'.

Comparative uses. In parts of Senegal, several scores of these palms are planted at a child's birth and serve to some extent as a dowry. *Sap*: fresh from male or female trees makes a pleasant drink, and, ferments to palm wine. The fresh sap may be used as yeast or made into vinegar. *Trunk, stem*: the stem of the old male palm is solid, hard, and resistant to insect attack or decay and does not rot in salt water. The stems are hollowed out to make guttering and irrigation conduits, drums, beehives and bowls. *Leaves*: are used for cordage, thatching, basketwork, brooms, etc. The soft down around undeveloped leaves is given as a haemostatic. *Fruit*: the orange-yellow fruit pulp is turpentine flavoured. It can be eaten beaten up in milk, or sliced and roasted. The fruit are also taken as a medicinal food. The rind of ripe fruit yields a yellow dye used for matting. *Seeds*: immature seeds contain a sweet jelly, which is pleasant and refreshing. *Roots*: a decoction of young roots is used for respiratory complaints. The germinating radicle, forced by earthing the nuts, is edible and can be eaten cooked. This is an important famine food. The terminal bud in immature palms is tender and can be eaten raw or cooked. The young shoots are also taken as a medicinal food. (Ambasta 1982; Dalziel 1948, 1956; le Houerou 1980; Jain & De Fillips 1991; Kunkel 1984)

Phoenix L.

Dioecious palms. Leaves pinnate. Fruit small, smooth, 1-seeded. 10 to 12 species. Africa and Asia to India and Sri Lanka.

Phoenix dactylifera L.

Tree to 20m, trunk unbranched, often forming colonies. Leaves pinnate, <5m long, with numerous folded, spine-tipped segments; petioles spiny. Inflorescences large, subtended by a large bract. Fruits ovoid, 2–5cm long, fleshy.

Ecology. Cultivated in date gardens. Sea-level–500m.

Distribution. N Africa, Arabia and Middle East.

Botanical notes. The cultivated date palm.

ETHNOGRAPHIC NOTES

Name(s)

– Soqotri: *ti̍mərə* (pl. *ti̍mihər*); seedling, young palm *ḥə̍ggub*; leaflet *ḥiṣ*; frond *fɛ̄riᶜ*; petiole, leaf rib *mi̍žərid*; trunk *gidö̍ᶜo*; leaf sheath *nɔ̄si*; fruit, full-sized but still yellow *ḥa̍lḥal*; fruit, half-ripened (one end yellow, the other brown) *ḳārīni*; fruit, fully ripened on the tree *bi̍šərəh* or *məḳəmīm*; pressed dates, dates in general *ta̍mer*; fruit, dried out, malformed, stunted *li̍ʾim*, *līm*; liquidised dates, thick *mɛ̄si*; liquidised dates, thin *tətfiyeh* or *tətbīyəh di mɛ̄si*; date-pip *ᶜə̍źʌź*; palm heart *gi̍diyib*; inner core of palm trunk (eaten raw or cooked) *məki̍bihir*; de-barked palm trunk cut up and fed raw to cattle, or cooked as a human famine food *ki̍dihir*.

– Soqotri simplified: *timirirh* (pl. *timihir*); seedling, young palm *higgub*; leaflet *hiz*; frond *feyri*; petiole, leaf rib *mijirid*; trunk *gido'o*; leaf sheath *nowsi*; fruit, full-sized but still yellow *halhal*; fruit, half-ripened (one end yellow, the other brown) *qariyni*; fruit, fully ripened on the tree *bishirih* or *miqimiym*; pressed dates, dates in general *tamar*; fruit, dried out, malformed, stunted *li'im*, *liym*; liquidised dates, thick *mesi*; liquidised dates, thin *titfiyih di mesi*, *titbiyih di mesi*; date-pip *'idhadh*; palm heart *gidiyib*; inner core of palm trunk (eaten raw or cooked) *mikibihir*; de-barked palm trunk cut up and fed raw to cattle, or cooked as a human famine food *kidihir*.

Food (human). The fruit are very nutritious: though low in fat and proteins, they are rich in carbohydrates. For example, 100g of date pulp contains around 260 calories (a banana has 97, fresh figs 80 and tomatoes 22). The sugar content of the fruit is high at 60 to 70%, and they also contain 7% proteins, pectins, gums, vitamins A, B1, B2 and nicotinic acid, carotenoids, anthocyanins, flavonols and flavones.

After milk and meat, dates were the main food on the islands, more important than cereals. It was considered possible to manage without cereals, but not without dates. When transhuming or at times when livestock were unproductive, dates were often the only ingredient of the main meal. They are also a suitable food to put before a guest, offered with a small dish of ghee to dip them in.

Fresh dates. On Soqotra they are considered edible once grown to full size but still yellow. These unripe dates are bitter due to the presence of tannins in the pulp, but they are rich in vitamins, making them particularly valuable on Soqotra where fresh fruit is scarce. They are much liked once they have begun to turn brown and one end is yellow and astringent, the other brown and sweet. The most valued fresh dates are those that ripen fully on the tree. There is no tradition of making date-syrup on Soqotra, as there is on the mainland, but the best ripe dates, when pressed, exude a sweet, dark syrup.

Processed dates. The harvested crop is sorted to separate out good dates from unripe or imperfect ones. If they are not fully ripened they are spread in the sun to finish ripening, and then trodden into a solid mass and packed down hard into goatskins. These are stored in as hot and dry a place as possible, and with the pots of clarified butter represent the most valuable part of the household's food store. Another way to artificially ripen dates is to soften them by sprinkling them with salty water, and stirring until each one is thoroughly dampened. They are then left out in the sun for another few days. They taste slightly salty and eating a lot causes diarrhoea. These dates do not keep well, and cannot be stored in the goatskins with the other dates. Children especially prepare small quantities of unripe dates by putting them in a clay pot with some salty water, covering it and leaving it for a few days for the dates to soften. They are then taken out and left in the sun to dry, after which they are eaten. In the date plantations of the southern plains and the southern end of the interior valleys, the fruit do not ripen fully on the palm. They have to be slit open with a small-bladed knife, the pip extracted, and the flesh spread out to continue ripening in the sun. Once fully sun-ripened they are treated like other dates, but they do not keep as well and are eaten before the others. For celebrations, or when a lot of dates are wanted, the goatskin full of pressed dates is sliced across, like a sausage, and the slices shared out.

Liquidised dates. Pressed dates are made into a drink by repeatedly squeezing a handful in water until they disintegrate, and the water changes colour and thickens to a syrup. This is regarded as one of the best foods of the island, and one with great medicinal value. It is said to be particularly beneficial when made with last year's dates, *ta̍mer mšo̍rḳiḥ*. When made with a lot of dates and only a little water, it is regarded as a generous food to offer to guests. Made with just a few dates and a lot of water it is a good way of stretching a limited amount of the fruit, allowing everyone to at least have a taste. If the mixture is left to stand until the pips and date-fragments sink to the bottom, the lighter syrup is poured off and is considered to be the best food for a sick person.

Palm heart. The terminal or apical bud can be rather bitter or very sweet, and can be eaten both raw and cooked. Removing the terminal bud kills the palm so this is done only when cutting down an old palm or with basal shoots or suckers. In a mature palm, the outer leaves are chopped off, the fibrous material at the base of the leaf sheath is pulled off, and then the outer bark chopped away to reveal the white palm heart beneath. Women especially are said to be very fond of this. Young palms produce great quantities of suckers, which have to be removed to encourage the palm to produce more fruit. Some are planted out, others are eaten: the outer leaves are cut off to reveal the inner core of palm heart, and this is eaten raw.

Sprouted seedlings. Date pips are commonly fed to livestock towards the end of the dry season. After rain, pips lying near the surface often sprout. People and animals like to eat these. Not all varieties of palm produce good sprouts, but the palms of the southern plains and interior valleys are renowned for this.

Famine food. In times of great hunger the core of a mature date palm is eaten, though never one in fruit. One average-sized trunk is usually cut into three sections, one of which supplies food for a family of 5 to 10 people (including children) for about four days. The best part for eating is nearest the crown, especially from a male tree. If the palm looks healthy and succulent, it is cut down close to the ground, otherwise it is cut off about halfway up. The tender heart of the crown is eaten raw, and then the length of trunk is sliced across into sections, and one section is prepared and eaten at a time after laborious preparation.

Firstly the outer bark is chopped off to reveal the white core underneath. This is then cut up into small regular chunks. A large clay pot is packed with the pieces until it is full. Water, and, if available, a little salt is added and another clay pot put on top as a lid. This is simmered for between four and six hours, and then removed from the fire. When cool enough to handle, the chunks are taken out and spread out on a flat rock surface; the cooking liquid is saved. This can be drunk as a meal in place of food, and is a sweet and fortifying drink. It must be used up quickly: the next morning it is still good to drink, but by midday it will have started to turn sour and ferment. The cooked chunks, however, can be left for up to a fortnight without further processing, but after this will start to go off.

Some people chew the warm chunks just as they are, discarding the fibrous parts, whilst others crush each chunk with a rock to soften it until it can be broken up into pieces which are then squeezed hard over the pot to extract the last of the nutritious liquid. Others spread the chunks out to dry (and if they have already chewed some, to extract the sweet juice, the fibrous remains are added to the chunks) until they can be broken up and crumbled into a powder. Alternatively they can be ground to a smooth paste and spread out on a flat rock surface to dry. Once dry, the fibrous strings are extracted, and the rest is crushed to make coarse flour. The flour is stored in a clay pot, and is eaten in handfuls at mealtimes without further processing.

Wild/uncultivated dates. Like domesticated palms, wild palms also produce basal suckers that can be eaten. The unfertilised flower sometimes produces fruits, though these are very small and malformed, with a large pip and relatively little pulp. They rarely ripen and are usually inedible. If they do ripen enough to eat, they are indigestible, and can cause stomach upset.

Food (livestock). Traditionally, all livestock are excluded from date gardens once the dates start to ripen and throughout the harvest. Once the harvest is over livestock are allowed in, and most date plantations then remain open until pollination begins. Some owners prefer to keep goats out all the time, as they tend to eat the growing hearts of the young palms when they are hungry. However, in general, cattle, donkeys and sheep of the owner and his family are allowed to graze around the palms and this permission is sometimes extended to close neighbours. The suckers that grow around the base of the cultivated palms belong to the owner of the palm and may not be cut for fodder. Outside the date plantation, any green foliage of the clumps of self-sown young palms, or of suckers grown up around a dead palm, can be used by anyone as a feed for their animals.

Leaves. In a prolonged dry season both goats and cattle are fed on the leaves, both dry and green, and this is considered to be more than just a subsistence feed. The dried leaves must not be so old that they lack 'bite', however. The leaflets are split into smaller strands for younger goats, and even sheep will chew on some of this if it is cut up small enough for them. Cattle and goats both chew away at the growing tips of young palms or suckers if they can get at them. In the date-growing area of the southern plains, where alternative forage is in short supply but date palms are plentiful, palm leaves are an important part of the livestock diet.

Perianths and fruiting stalks. Both these are fed to all livestock. The green stalks available at harvest time are preferred by the animals, and are considered to be more nutritious, but people also crack them open, and dry them as a drought fodder for cattle and goats.

Pips. Pips are removed from the split dates of the southern plains and interior valleys and fed to small stock in the dry season. Sometimes they are crushed or soaked in water to make them more digestible. After rain, fallen pips sprout and these are much liked by all livestock.

Poor quality dates normally unfit for human consumption. When they are sorted after harvest, various inferior dates are put aside for livestock. Those that are slightly rotten and sour-tasting or hard, dried out or undersized, or a good size but give off a 'puff of coloured smoke' when pressed (probably fungal spores) are sorted into different heaps and then bagged and stored separately from those for human consumption.

Good quality dates. Good quality dates are rarely given to livestock, other than one or two to a favoured house-cow or goat. However, sometimes dates are given to a cow who has lost her calf, to feed her up and keep her in milk.

Liquidised dates. Favoured animals, orphaned females, sickly, pregnant or parturient animals are sometimes given a very watery mixture of liquidised dates as a nourishing feed to build up their strength. This is often fed to milk cows to increase their milk yield during the monsoon months when milk is otherwise scarce. Orphaned female young are fed the strained liquid from a leather 'feeding bottle' made from a goatskin.

Date palm core. In extreme drought, when cattle are at risk of dying from starvation, they are fed on the chopped-up core of a date palm. This is not given to goats as they are thought not to have strong enough teeth or digestive systems to cope with it, although the fibrous remains of palm heart processed for human food are fed to goats. Usually an infertile, very old, or extremely tall palm, which is difficult to cultivate, is chosen. Every trunk contains some palm heart, the quality improving closer to the crown, but the amount varies enormously from tree to tree. Only an expert can judge how good an individual tree is likely to be, and such advice is usually sought before going to the considerable expense of purchasing a date palm for cattle feed. The paler the palm heart the better for feed; a dark, red-brown core is almost inedible. The herders take the tender palm heart from the crown of the tree for themselves, and then chop one section at a time for their cattle. This is then pulped, usually out of sight of the cattle, as it is normally fed only to productive cows. If used sparingly, one palm trunk is enough for three or four cows for about a fortnight, or for 8 to 10 for around a week.

Bees / honey. The male inflorescence and the fruit when they start to ripen are very attractive to honeybees. The fruit attract many insects, especially butterflies, but they are usually only after spoiled fruit, or fruit which have dropped to the ground, and are of no danger to the crop. Honey with no trace of bitterness is often made from the date palm.

Flavouring / food additive. Feeding date products of any kind to livestock gives both meat and milk a distinctively sweet, much-liked, taste.

Fuel. Dry fronds make the best tinder and firelighter. If people need to see in the dark, they throw a handful on the fire, and they burn with a bright light which can illuminate a large area. The dried leaves are gathered and stored in the home for this purpose, or to use as a firelighter in wet, cold months. The trunk makes poor fuel.

Timber. In the areas where they were cultivated, palms provided the principal building material. For instance, within living memory, nearly all the housing along the heavily populated northern coast was made from palm material, and stone or coral-rag houses were relatively uncommon. The petiole, trunk and fronds were all used to build shelters and accommodation of all kinds. These could be constructed entirely from date palms or were combined with stone walling. Unfortunately both petiole and fronds are much liked by termites, and ideally the petiole should be put to ret down in salt water for about a week before being used.

Trunks. Trunks were used to make the ridgepole for a pitched roof, with a thatch of palm fronds. Although light and soft, this roofing timber lasts for a long time.

Petiole. Huts were built entirely from the petiole, especially beside the sea, where they were used to store fishing equipment and to provide shade for fishermen and their families. The walls were made with vertical petioles crossed by horizontal ones to make a lattice or grid. They were tied together in a square and a roofing of palm fronds strapped on.

Fronds. Fronds were used as roofing material, and were also interwoven between a double layer of petioles to make walling and enclosures.

Trunks, petioles and fronds. More permanent housing was made by erecting an oblong framework of strong poles with a palm-trunk roof tree, and then strapping on sections of walling made by carefully interweaving palm fronds between a double layer of petioles. A thatch roof of palm fronds was then tied on top. If there was no wood available for the poles, these were made by binding five or six petioles together in a bundle and then tying the bundles together, one on top of the other, to the length required. Larger houses were made by extending the framework and adding supporting pillars down the middle, with supporting beams, of palm trunks or other wood, running from the pillars to the outer wall. A circular house was similarly constructed, but had only a single central supporting pillar and a conical thatched roof. In areas with suitable building stone, the walls were made of stone, and only the roofing was palm material: a layer of petiole latticework, and then a layer of fronds, all strapped together with cordage from the fruiting stalks. Some houses had walling built in an oblong, but built up to a triangular peak at the short ends. This allowed a pitched roof, using a palm trunk as the roof tree, and then palm-frond thatching to allow rain to run off. This style of building was common in the interior valleys behind the Haggeher mountains, which received a lot of the drizzle of the monsoon months, as well as heavier winter rains.

Fibrous material growing around the leaf base. This was used as an infill for double stone walling, usually mixed with earth and pebbles.

Fibre / cordage. In areas where dates are grown, most cordage is made from date palms, finer threads and strings from the fronds, and thicker ropes from the fibrous material that grows around the leaf base.

Fronds. The leaflets are used both dry and green. Green leaflets are removed from the heart of a sucker shoot; the more mature green leaflets are no use for cordage. They are softened by beating and are then separated into fine strands and worked to make threads and string. The green material makes softer cordage, used for pack animal harness, for instance. The dried leaflets must first be soaked in water to soften them, but do not need beating. After soaking, they are separated out into fine strands and worked into cordage. The finer string is used for all sorts of miscellaneous tying jobs, as well as for stitching together the lengths of plaited fibre for basketry and matting, or sewing pack gear for baggage animals. The thicker cords are used in fishing, especially shark fishing, and for pack animal harness.

Fibrous material growing around the leaf base. This is also used for cordage. Sheets of it are pulled off and put to soak in water. When thoroughly softened, the strands are separated out and worked. It can be worked without the initial soaking, but retted down fibre is said to give better results. This makes very strong ropes, as good as or better than *Dracaena* fibre. They are used for jobs where strength is of key importance, such as tying up boats, lowering people over cliffs to hunt for honey or nestlings, or to gather drought fodder or look for water. It is also used for pack animal harness and for restraining camels and donkey stallions in rut. The net formerly used to trap feral donkeys or goats was often made from this fibre.

Fruiting stalks. These have many uses. Indeed, some islanders purposefully leave some unfertilised flowers on the date palm so that sterile fruiting stalks will be produced. They can be used with no further preparation, but it is better to soften them in water, or by beating them before working them. They are separated out into individual strands and then twisted or plaited as required. They are used for such jobs as tying up date-bearing branches or palm-rib fencing, or to strap on palm thatch and palm-frond walling.

Wood

Trunks. These are split in half and hollowed out to make guttering and water channels. They are one of the best woods for quicklime manufacture. Lashed together, they were also used to make rafts used as fishing platforms and for loading and unloading visiting ships.

Leaf midribs and petioles. These are primarily used to make fencing, bound together with the stalks of the fruiting branches. The fencing stands alone or is placed on top of a stone wall to raise it higher. Around cultivated areas this fencing is often placed to slant at an angle outwards to discourage goats from leaping over into the enclosure. They are also used to make the distinctive Soqotri wattle door which can be seen on houses, gardens, pens and byres all over the island. The line of vertical palm ribs is lashed together with fruiting stalks, and the door is held in place by hooking a loop of the binding over one of the poles that frame the doorway. The petioles are also used to make shelves, sometimes suspended above the fire to hold such things as food and animal skins that benefit from smoke. They are also used as a sort of hanging cupboard to store food out of the way of rats and other vermin, as well as hungry children, suspended from the roofing beams. A variety of traps for birds, civet cats, and fish are also made from these, and they feature in many games played by both adults and children. The dried petioles of the *Libininoh* date palm can be used as firesticks, but are most effective when both sticks are of the same material. More recently petioles have been used to make pyramidal tripods over which clothing or bedding is hung to be fumigated.

Medicinal / veterinary. Dates are the main medicinal plant of Soqotra.

Livestock. Dates or liquidised dates are fed to favoured livestock that are sick or failing to thrive. If a cow does not pass the placenta she is fed on liquidised dates. If she is too weak to drink it is poured down her throat through the leather 'feeding bottle'. This is said to 'cool' her overheated body and give her new energy. A cow beginning to lose weight in the last weeks of pregnancy is also given this feed to fatten her up.

Humans. Wherever possible, the liquidised dates remedy should be made from last year's dates, called *tamer mishorqih*. This treatment is something of an island cure-all. It is given for every sort of stomach complaint, to treat diarrhoea or vomiting, to a woman exhausted by childbirth, as a food for those recovering from a serious illness, or to someone who has been stung by a scorpion, centipede or wolf spider. It is also used to treat a distended stomach, especially if this is thought to be from constipation: the patient is given as much as he can force down. Someone with anorexia or persistent nausea is tempted with this food. A very liquid form of the drink was taken by pregnant women who often tried to limit their food intake in the hopes of restricting the growth of their baby to make birth easier. Sometimes the exact type of date was specified by the healer. The most widespread cultivar of date palm on Soqotra is the *Serefaneh* palm, and the year-old dates of this are especially prescribed for certain illnesses. However, if a woman suffers a post-partum haemorrhage, she is prescribed a drink made from the year-old dates of the superior, and much rarer, *Howqem* variety. The spines and pointed leaflet tips are used to probe wounds, prick blisters and so on, and the flexible petioles of a green frond are used to strap a bad sprain or to splint a fracture.

Bedding, stuffing. Dried fronds are used as bedding. The fronds are also woven into sleeping mats.

Cosmetic. The hair is dressed with the spine removed from the base of the frond, though for temporary use the pointed tip of a palm leaflet can be used instead. One of these spines is often kept in the woman's personal leather bag, and can be used for a variety of purposes. In the days when *kohl* antimony was a rare luxury, the date pips were roasted slowly beside the fire until dried out and blackened, and were then ground to a powder used instead of *kohl*. The pointed leaf tips are used to pick the teeth, and after a meal it is a common courtesy to pass strips of dry leaflet around to pick the teeth.

Commercial value (past, present). All parts of the date palm have commercial value: young trees, the male inflorescence, the fruit, the wood, the palm-frond ribs and the fibre, worked or unworked, are all still sold and bartered.

Pest control. The midribs of the *Libininoh* date palm make a lot of smoke and are used for livestock smudges.

Other. Goats eating the ripening dates from the ground can go longer without water.

Leaflets. These are tied in a bunch to make a brush. They are also used as a fertiliser for growing finger millet. The green leaflets are separated into fine strands used as thread to repair holed leather bags and containers. If the hole is very small, it is repaired by knotting one end of the thread, threading the other through the hole until the knot is against the hole, and then knotting the other end as close to the hole as possible and cutting the spare thread off. Cracked clay pots are repaired with this. Small holes are bored each side of the crack which is then laced up with thread made from green or dried leaflets, or from the fibrous material that grows around the leaf base. These repaired pots cannot be used for cooking, but are adequate for storing grain and even for holding liquids or for making the liquidised date drink. Although it oozes slightly the first few times, it soon becomes watertight.

Leaflets removed from the heart of a sucker are dried and plaited into long, narrow strips. These are then sewn together to make a wide variety of items. *ḥāṣir* (*hazir*) is a large oblong mat used as a floor covering. *meˈraḥt* (*meraht*) is a circular mat on which food is served. *ˈmosfi* (*mosfi*) is a very large container, which takes at least four men to carry when full. It is mostly used to store cereals, especially locally grown finger millet. The base is made like the circular food mat and then the sides are gradually built up with strips of plaited fibre. It is filled with grain and then sewn up tightly, leaving a stretch of lacing which can be undone to remove grain. *ˈdizihen* (*dizihen*) is an oblong container with a cover that exactly fits over it with a carrying handle extending from one (long) side to the other, and is used mostly by women to store their private possessions. *ˈmillug* (millug) is a container similar to the *dizihen* but smaller, and generally used to house a copy of the Holy Qur'an or sections of it. *ˈdaḥanə* (*dahanih*) is a two-handled open basket, usually used for transporting fish on the coast. *ˈgifir* (*gifir*) is a cone-shaped food cover used in the coastal villages to protect cooked food. *mṣˈʌllə* (*mizallih*) is an oblong mat used mainly in the coastal villages as a prayer mat. *ˈśiṭfəh* (*shitfih*) is a small two-handled basket. All these containers are made from plaited date fibre.

Fibrous material growing around the leaf base. This is used as stuffing and padding material, especially for pack animal harness.

Spines. These are removed from the base of the leaves and are used as an awl to pierce finer leather and material for sewing.

Fruiting and inflorescence stalks. These are also tied together in a bunch to make a coarse brush for sweeping out living quarters of livestock and people.

Date pips. In the poorer, drier areas of the island where dates are rare and seen as possessing special qualities, the pips are laid across a container of food or water, or over wrapped food, to protect the contents from the Evil Eye and other malignant forces.

Notes. Date palm cultivation is a very important part of the island economy and used to play a major role in the lives of all islanders, since those without date palms worked with and for those who owned them in exchange for dates. The 1985 agricultural survey reported 121,146 date palms, equivalent to seven palm trees per islander (the population then being given as 17,383 people). However, within living memory dates have also always been imported, formerly from Basra and more recently from the Hadhramawt and Saudi Arabia. No date palm cultivation is practised on the outer islands.

The Bents (1900) reported that 'dates form the staple food of the islanders. And out of the date-tree they get branches for their hedges, stems for their roofs; the leaf provides them with their sleeping mats, and, when beaten on stones, with fibre, with which they are exceedingly clever in making ropes. Our camel-men were always at it, and produced, with the assistance of fingers and toes, the most excellent rope at the shortest possible notice. They also make strong girdles with this fibre which the slaves, who are employed in fertilising the palm-trees, bind round their bodies and the trees so as to facilitate their ascent, and provide them with a firm seat when the point of operation is reached. They weave, too, baskets, or rather stiff sacks, in which to hang their luggage on either side of the camel'.

In the star-calendar, the date palm symbolises the star *shibilloh* which occurs around May and then again around October: the crown of the palm the first half of the star period, the trunk the latter half.

Comparative uses. A few date kernels reported from Egypt, Iran and Pakistan have been dated to the sixth and fifth millennia BC but they probably represent material collected from the wild. It is now generally agreed that wild date palms are widely dispersed near brackish rivers and springs, all along the Sahara from the Atlantic Ocean to the Persian Gulf, forming wild oases inhabited by saline or semi-saline plant communities. The earliest remains of what seem to be cultivated dates were found by Seton Lloyd in the Ubaidian horizon (c.4000 BC) at Eridu, Lower Mesopotamia; in Palestine carbonised stones were found in the classic Chalcolithic (3700 to 3500 BC) site of Teleilat Ghassul; a single date stone is available from early Bronze Age Jericho, and date palm imprints were found in the c.3000 BC Hili Oasis in the Oman Peninsula.

Some date palms live to be 200 years old, though the best fruit production starts when the palm is about 30 years old and it can continue to fruit until it is about 80 years old. One male palm can produce around 20 inflorescences with enough pollen to fertilise some 50 female palms. Female palms can be very productive, and can yield as much as 100kg of fruit per tree.

In parts of Africa, the sharp tips of the palm fronds are used as razors in surgery. In the Egyptian desert the leaf-sheath fibre is used as a filter when making coffee, and as a sponge for washing and cleaning. The pips, ground or softened in water, are used for feeding camels, goats and horses, and have been successfully substituted as a poultry feed. They can also serve as the raw material for preparing oxalic acid, and, after roasting and grinding, can be added to ground coffee to give a pleasant flavour and perfume. The sap is sweet and nutritious, though rather laxative, and it is used to prepare jaggary and sugar. It has long been used to make an alcoholic drink rich in sugars and minerals: a single date palm can be drained of up to 400 litres of sap. The ancient Sumerians drank *shikaru*, date wine, as did the ancient Egyptians, who also offered cakes made with date pulp to their gods. A plaster of date pulp is applied to improve the hair, and a paste of dates mixed with soap has been used in the Sahara to seal leaks in car radiators.

Medicinally, the leaves are seen as aphrodisiac and good for the liver. The flowers are purgative, expectorant, a liver tonic, and useful in fever and blood complaints. The terminal bud is used for intestinal haemorrhages, diarrhoea and jaundice. The pollen grains have gonadotrophic activity due to a glucoprotein, and an infusion of the pollen grains and spadix is taken as a tonic and aphrodisiac. The pollen is also an aero-allergen causing allergic rhinitis, bronchial asthma and/or hypersensitivity pneumonia. The volatile oil of the spathe has anti-bacterial and anti-fungal activity. The fruit are cooling, tonic, diuretic, laxative, fattening and aphrodisiac, and are useful in leprosy, thirst, asthma, bronchitis, fatigue, tuberculosis, abdominal complaints, fevers, vomiting, wandering of the mind and loss of consciousness. Lactating mothers are advised to eat plenty of dates to improve their milk supply. Dates are said to have a fertility regulating action, and an oestrone is reported from kernels. For instance, in northern Nigeria a mix of dates, bran and *Sterculia tomentosa* seeds are given to young heifers long before they mature to make them become prolific breeders. The date kernels are used against ulcers of the genital organs. In Oman an infusion made from male flowers is used as an aphrodisiac and general tonic. A paste made from the dates is applied to the forehead and eyes to relieve a headache, and, with added salt, is applied to bruises. The tree also yields a gum used to treat diarrhoea. (Abbas *et al.* 1992; Ambasta 1982; Barrow 1998; Batanouny 1981; Bebawi & Neugebohrn 1991; Dalziel 1948, 1956; Farooqi 1989; Fleurentin & Pelt/Fleurentin *et al.* [1982, 1983a,b,c]; Ghazanfar 1994; Goodin & Northington 1979; Goodman & Hobbs 1988; le Houerou 1980; Jain & De Fillips 1991; Kunkel 1984; Lewis & Elwin-Lewis 1977; Munier 1973; Oliver-Bever 1986; Popenoe 1924, 1973; Rizk & El-Ghazaly 1995; Watt & Breyer-Bandwijk 1962; Zohary 1982)

POTAMOGETONACEAE

Aquatic herbs growing in fresh or brackish water. Leaves submerged or floating, alternate, entire; stipules either free or fused to the leaf base to form a stipular sheath and then with a free upper part or ligule. Flowers inconspicuous, bisexual, arranged in lax or dense spikes. Tepals 4, free. Fruit a nutlet or drupe, ovoid or circular, sometimes compressed, shortly beaked. 3 genera. c.100 species. Subcosmopolitan, fresh water and brackish water.

Potamogeton L.

See description of family. c.90 species. Cosmopolitan.

Potamogeton filiformis Pers.

Stems slender, to 50cm. Leaves filiform, 30–150 x c.0.8mm, entire; stipules with tubular sheaths (10–30mm) enclosing the stem and free (5–15mm) ligules, the sheath splitting eventually. Flowers in distant clusters along a slender elongated spike; carpels with sessile stigma. Fruits clustered, ovoid, c.2.5 x 1.8mm.

Ecology. Submerged in freshwater pool. c.600m.
Distribution. Widespread in Europe, Asia and N America.
Botanical notes. Known only from a permanent pool in Qatariyah. Similar to *P. pectinatus* and *Ruppia maritima* but the latter species can be recognised by its stipular sheaths without a ligule. Distinguishing the two species of *Potamogeton* can be rather difficult. The leaves of *P. filiformis* tend to stick together when drawn out of the water whereas those of *P. pectinatus* usually remain separate.

ETHNOGRAPHIC NOTES
Not seen, no information recorded.

Potamogeton nodosus Poir.

Potamogeton fluitans sensu Balf.f. non Roth.
Potamogeton natans sensu Balf.f. non L.
Potamogeton natans subsp. *plantagineus* Ducros
Potamogeton semicoloratus A.Benn.

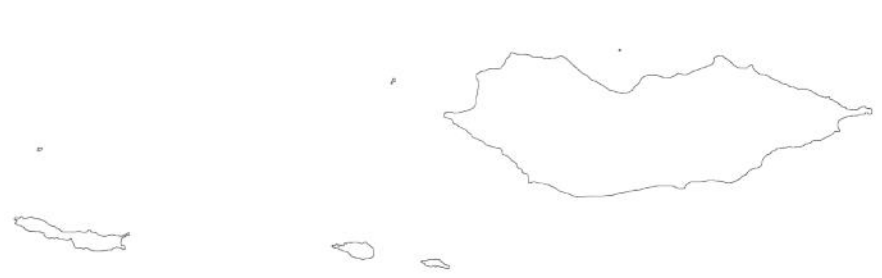

Stems robust, to 2m. Leaves petiolate, the floating and submerged different: submerged leaves elliptic, 15–25 x 2–4cm; floating leaves broadly elliptic, 7–13 x 3.5–5cm; stipules free, 4.5–12cm, enfolding the stem. Flowers in a congested spike on a robust stalk. Fruits 3–4 x 2–3mm.
Ecology. Freshwater pools. No location details available.
Distribution. Widespread in Europe, Asia and N America.
Botanical notes. Not recorded since the 19th century. Readily distinguished by its floating elliptic leaves.

Potamogeton pectinatus L.

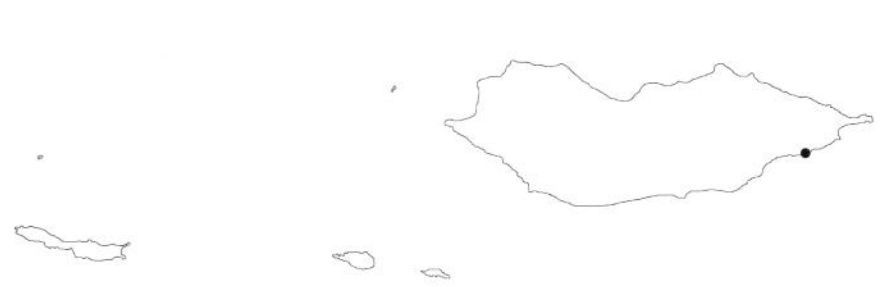

Similar to *P. filiformis* but the leaves are stiffer; the stipular sheaths are overlapping not fused (most clearly seen by cutting a transverse section through the base of the leaf) and the stigmas are borne on short (0.2mm) styles.
Ecology. Submerged in brackish lagoons. Sea-level.
Distribution. Cosmopolitan.
Botanical notes. See notes under *P. filiformis*.

ETHNOGRAPHIC NOTES
Not seen, no information recorded.

Ruppia L.

See description of family. 1 to 7 species. Temperate and subtropical regions.

***Ruppia maritima* L.**

Stems slender, to 50cm. Leaves filiform, 3–10cm long, minutely denticulate at apex; sheaths 0.5–2cm long. Fruits pyriform, 2.5 x 2.5mm, shortly beaked and on a 1–3cm stalk.

Ecology. Brackish lagoons. Sea-level.
Distribution. Temperate regions of Europe, Asia and America.
Botanical notes. Tolerant of a wide range of salinity from virtually fresh to brackish water.

ETHNOGRAPHIC NOTES
Not seen, no information recorded.

DICOTYLEDONS

ACANTHACEAE

Herbs, shrubs or small trees. Leaves opposite, stipules absent, often with cystoliths visible as streaks or bumps on the leaves, sometimes spiny. Flowers in compact inflorescences, or in few-flowered cymes or solitary, bracts present, often large and sometimes coloured. Calyx 5-lobed (sometimes 4). Corolla ± actinomorphic with 5 subequal lobes, or zygomorphic with an upper and lower lip, the lips entire or divided into <4 lobes. Stamens 2 or 4; anther cells 1 or 2, arranged side by side or one displaced above the other, sometimes with appendages. Ovary superior, 2-celled, each cell with usually 2 ovules arranged in rows. Fruits a 2 or 4-seeded capsule with longitudinal dehiscence. Seeds on hook-like jaculators. 357 genera. 4350 species. Mostly tropical, especially drier areas.

Angkalanthus Balf.f.

Ancalanthus Balf.f.

Shrubs. Leaves entire, petiolate. Flowers in spikes, bracts inconspicuous. Corolla 2-lipped, lower 3-lobed, upper 2-lobed, flared at mouth. Fertile stamens 2, held close to upper lip; anther cells 2, side by side. Fruits 2-seeded. 1 species. Endemic.

Botanical notes. For comments see Vollesen (1994).

Angkalanthus oligophylla Balf.f.

Ancalanthus paucifolius Balf.f.

Many-stemmed twiggy shrub with pendent inflorescences to 1m. Leaves ovate, 2.5–4 x 1.5–3cm, base truncate or cordate, petiole <3cm. Corolla red, fading yellow, <3cm long.

Ecology. Succulent and drought-deciduous woodland on rocky slopes along the wadis draining into the southern plains. 30m.

Distribution. Endemic.

Conservation status. Endangered (EN D). Since the 19th century only recorded with certainty from a single locality (on rocky slopes where a large wadi enters the southern plain) where fewer than 10 plants were seen. Local informants say that it is locally common in similar places along the southern escarpments but surveys in the region have failed to reveal more sites. Due to the uncertainty over its range a threatened category has been assigned.

Botanical notes. A distinctive many-stemmed shrub with attractive orange-red flowers which hang in racemes at the ends of pendulous branches. The flowers curve upwards and backwards thus regaining their original upright position despite hanging upside down. It is reported that *A. oligophylla* is common in dry ravines and at the mouths of the south-draining wadis which feed into the Noged plain. However, only two specimens are known – the original specimen collected from the mouth of Wadi Ireh by Balfour in 1880 and a second specimen collected in 1990 at apparently exactly the same spot.

ETHNOGRAPHIC NOTES

Name(s)

– Soqotri: *miśḳere̍r*, *enśḳere̍r* (pl. *miśḳero̍r*, *enśḳero̍r*) (also given to *Chorisochora* species).

– Soqotri simplified: *mish'qerer* (pl. *mish'qeror*).

Food (livestock). The leafy stems and fallen leaves are useful dry season forage, especially for goats. Camels browse the shrubs and sheep eat fallen leaves.

Bees / honey. Honeybees are attracted to the flowers.

Fuel. The dead wood makes adequate firewood.

Timber. It makes poor timber.

Anisotes Nees

Shrubs or small trees. Leaves entire, petiolate, cystoliths present. Flowers few in axillary racemes, bracts inconspicuous. Corolla 2-lipped, lower lip narrow, 3-lobed, upper lip hooded, entire. Fertile stamens 2; anther cells 2, one above the other, tails absent. Fruits 4-seeded. 1 endemic. 19 species. Tropical Africa, Madagascar.

Anisotes diversifolius Balf.f.

Anisotes diversifolius var. *brevicalyx* Balf.f.

Straggling shrub or small tree to 3m. Leaves obovate, 1.5–3 x 1–2cm. Corolla red, fading to yellow, <4cm long.

Ecology. Common in deciduous woodland on granite in the Haggeher mountains from 200m (Wadi Di Negehen) to 900m below the pinnacles; occasionally found in sheltered patches of *Buxanthus* and *Rhus* woodland in ravines and on cliffs on the seaward-facing limestone escarpments in the east of Soqotra, where it can occur down to 50m. 50–900m.

Distribution. Endemic.

Conservation status. Least Concern (LC). Common in several vegetation types and under no present or perceived threat.

Botanical notes. *A. diversifolius* is typically a many-stemmed shrub with erect, rather straggling stems. It is very similar to species of *Trichocalyx* but lacks the numerous thread-like bracts which surround the flowers in that genus. Recorded from near Qalansiyah in the west of the island in the 19th century but not seen there recently.

ETHNOGRAPHIC NOTES

Name(s)

– Soqotri: *ᶜʌləhʌl* (pl. *ᶜalaʾhal*); (western Soqotra) *ᶜʌlġʌl* 'lovely, beautiful'.

– Soqotri simplified: *'al'hal* (pl. *'ala'hal*); *'alghal*.

Food (livestock). An important and nutritious forage for all livestock. In eastern Soqotra it is found mainly in cliff areas, and is regarded as a valuable cattle feed. Leafy stems and twigs are lopped as a dry season fodder for cattle and goats, and these animals also chew at leafless twigs and stems when other forage is scarce. Fallen foliage is gathered as dry season feed for cattle in particular, but goats are also given this. Shepherds also gather foliage and plant litter for their sheep. The foliage is fattening and improves lactation in all milk animals.

Bees / honey. The fragrant flowers are important bee forage. Honeybees are said to love this plant as much as *Ballochia* species. Islanders say that honey with no trace of bitterness is likely to be from the flowers of this, *Ballochia* or *Trichocalyx* species. Bees feeding on these species, which flower after summer rains, are considered to make the very best honey of all.

Flavouring / food additive. Some herders of the Haggeher say that the carcass of cattle and goats feeding heavily on this produces slight bitterness in the cooking water. The nectar was a welcome source of sweetness (otherwise available only from dates or wild honey). In good years the flowers drip nectar if held upside down.

Fuel. The dead wood makes good firewood. Smaller stems produce a lot of smoke, but the larger ones do not, with the least smoke of all being produced by a dead stump, which is used to keep a fire in overnight. In the high mountains, the dead wood and that of *Cephalocroton socotranus* are gathered and stored for use during the monsoon and rainy weather. Both woods light readily even in damp weather. This wood is regarded as one of the best for illumination and for giving off heat.

Timber. Not highly regarded as building timber.

Wood. In the Haggeher mountains firesticks are often splinted into lengths from this. Otherwise it is little used for implements, and it is too valuable a fodder source to be used in the manufacture of charcoal or quicklime. Whippy stems are used to chastise children, and, at a time when imported metal fish-hooks were not readily available, this wood was used to fashion wooden ones.

Cleansing. Sprigs of the sweet-smelling flowers are tucked into the clothing and hair. The sniffing of sweet-smelling plants is held to protect from the 'noxious air', which can cause disease, and to help protect against infection.

Medicinal / veterinary. To treat a painful or infected ear, the fresh green leaves are crushed and the juice dropped into the ear. If the ear is discharging but there is no pain, the preferred treatment is to cauterise the area immediately in front of the ear opening and directly under the ear, over the neck gland. However, if there is associated pain, the ear and the area around it are cleaned, and the drops applied. This same procedure is followed to remove the tick called *śerb* from the ear, which, if left *in situ*, can cause considerable pain. In the Haggeher mountains, a cow losing weight towards the end of her pregnancy is fed on leafy stems to help her put on weight.

Pest control. The ash from the wood is regarded as an effective insecticide when rubbed into goatskins used to store food, especially dates.

Other. Animals feeding on this can go longer without water. Slender stems are often used as a 'brush' to incise lines and patterns on damp clay as decoration prior to firing.

Notes. Islanders distinguish between the 'male' tree (called *ᶜʌləhʌl ḳʌmses* or *kartab*) and the 'female', which has larger leaves.

Comparative uses. In Yemen the leaves of *A. trisculus* (Forssk.) (*maḏ̣īḏ̣*) are used to treat kidney stones and are prescribed as a diuretic. (Fleurentin & Pelt/Fleurentin *et al.* [1982, 1983a,b,c])

Asystasia Blume

Herbs. Leaves entire, petiolate. Flowers few, in axillary or terminal lax scorpioid cymes (one-sided racemes). Corolla 2-lipped, lower lip 3-lobed, upper 2-lobed. Fertile stamens 4. Fruits 4-seeded; seeds glabrous, rough. 70 species. Old World tropics.

Asystasia gangetica (L.) T.Anderson

Asystasia ansellioides Balf.f.
Asystasia coromandeliana Nees

Prostrate or ascending herb to 30cm. Leaves ovate, 1.5–5 x 1–3cm, sometimes notched at tip. Corolla white with lilac markings, <1cm long. Fruits <2cm long.

Ecology. Common, sometimes weedy; in succulent shrubland and drought-deciduous woodland. 200–700m.

Distribution. Pantropical weed.

ETHNOGRAPHIC NOTES

Name(s)

– Soqotri: (*di*) *ḥəwrirhin*, (*di*) *ḥəwrrihin* (pl. (*di*) *ḥəwrirhun*, (*di*) *ḥəwrrihun*) 'small and dark-coloured'; *mʌḥeriś* 'undergrowth'.

– Soqotri simplified: (*di*) *hewririhin* (pl. (*di*) *huwrirhun*, (*di*) *hewrrihon*); *maharish*.

Food (livestock). Cattle graze this enthusiastically, goats less so. However, it responds quickly to rain, and at times when there is little other greenery available, goats eat it and often develop scour as a result. In an animal weakened by hunger after a long dry season this can be fatal, and in a pregnant goat can bring on abortion. It is regarded as a nourishing feed for cattle and also increases their milk yield.

Other. Cattle grazing this plant can go longer without water.

Notes. One of the group of plants believed to cause *'uqabah*. Island herders say that if you rub the plant between the fingers it disintegrates to a sticky slime. This quality damages the gut of small stock and makes them ill.

Comparative uses. Used as a potherb – the tender leaves are a good source of thiamine. The plant juice is given to children with swellings and/or rheumatism, and also to expel worms. A decoction is taken internally by women during the latter months of pregnancy to lighten the pains of childbirth. A similar preparation is applied to the wound made by cosmetic piercing of the earlobe. In Africa the plant is considered to be a galactagogue. The leaf is used to treat the bite of a non-poisonous snake, the powdered leaf being applied to the site. The plant is used as an anti-pruritic. (Ambasta 1982; Dalziel 1948, 1956; Jain & De Filips 1991)

Ballochia Balf.f.

Shrubs or small trees. Leaves entire, subsessile, cystoliths prominent. Flowers solitary or in few-flowered, usually axillary, inflorescences. Calyx equally 5-lobed. Corolla red to orange (or yellow), 2-lipped, upper with 2 lobes, lower with 3. Fertile stamens 2, one cell per anther. Fruits 4-seeded. 3 species. Endemic to Soqotra.

Botanical notes. This genus is typical of the drought-deciduous woodlands of Soqotra where it is often abundant. The species are all distinguished by orange-red flowers with tubular corollas which have a characteristic slight bend towards the middle. They are visited by the endemic Soqotran sun-bird. The genus is in need of revision: the species are variable and poorly defined. Identifications, and hence the distribution maps, should be treated with some caution. The three species are difficult to distinguish. *B. amoena* seems to be consistently defined by its small, narrow, fasciculate leaves but the other two species are less clearly differentiated. Although the glandular calyx appears to be a useful character separating *B. rotundifolia* from *B. atro-virgata* they are otherwise so similar in overall appearance that this may represent infraspecific variation.

Ballochia amoena Balf.f.

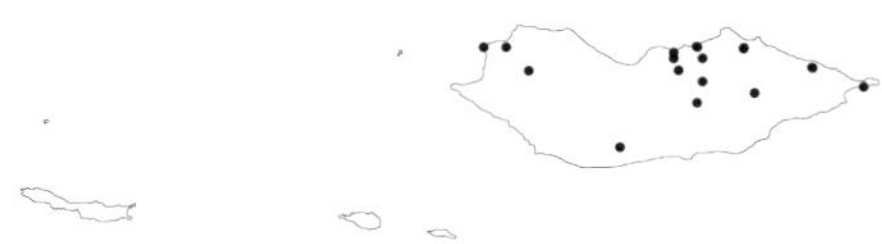

Intricately branched shrub to 2m. Leaves narrowly elliptic, <0.8 x 0.3cm, fasciculate. Calyx ± glabrous. Corolla red to orange, <3cm long.

Ecology. Widespread on the plains and foothills of the limestone escarpments; usually in succulent shrubland and drought-deciduous woodland; also occasionally in *Dracaena* woodland (e.g. Firmihin). 30–750m.

Distribution. Endemic.

Conservation status. Data Deficient (DD). Part of a taxonomically difficult group on Soqotra. Inadequate evidence is available to make an assessment.

ETHNOGRAPHIC NOTES†

Name(s)

– Soqotri: *mīṣeṣ* (from the verb 'to suck', referring to the flowers which are sucked for their nectar).

– Soqotri simplified: *miyzez*.

Food (livestock). Among the most important and nutritious forage plants, especially for small stock. They come into leaf in the summer months without rain, so in the drier areas they are a key fodder source for goats and sheep: these and the smaller *Hibiscus* bushes are some of the few shrubs which sheep can reach to browse. Sheep are said to be able to survive on a diet of the green foliage of these shrubs even when there is no grass. Cattle nibble some of the tender, new shoots, but the bushes are generally too prickly for them and the leaves too small for them to grasp. Hungry cattle of the lowland and eastern areas are more accustomed to feed on the leafless twigs than those of the better-vegetated Central Highlands. In a prolonged dry season, leafy branches are lopped for goats, and both sheep and goats eat the plant litter. All parts of the plant are regarded as a rich and fattening feed, increasing milk yield. Young and tender twigs are given to newly weaned young stock. During the dry monsoon months, with the strong winds blowing in from the sea, the bushes become rather salty, but are still browsed by hungry livestock. A certain intake of saline feed is regarded as essential for the maintenance of good health in livestock, especially small stock and camels. On Samha, this is one of the most important plants for sheep and goats.

Bees / honey. An important plant for honeybees and the honey is sweet and without a trace of bitterness. The flowers are particularly rich in nectar after summer rains, and honey at this period is considered to be the best the island produces.

Flavouring / food additive. The nectar sucked from the flowers of these plants was an important source of sweetness in the days before sugar was widely available. In good years the flowers ooze nectar when held upside down.

Fuel. The dead wood makes good firewood. It burns slowly and gives off little smoke, especially if the bark is removed first, making it good for cooking. It also gives off a good heat and is the wood of choice to keep children and young stock warm in cold weather. It burns well even in wet weather, and is much used during the rainy season. It gives a very bright light, and it is common to keep a small store in the home for this purpose.

Timber. Normally considered too valuable to be cut for timber, and is anyway usually too small to be useful for anything other than roofing infill, for which plenty of alternative woods exist. It is relatively resistant to termite attack. It is too important a fodder source to be cut for the manufacture of charcoal or quicklime.

Wood. Good, hard wood and, when the shrubs grow tall enough, it makes good sticks and herding staffs. The more slender stems are often used to make the 'bit' of the weaning device worn by young kids and lambs.

Other. The pegs that secure the skin stretched over the frame of the large tambourine (*tarānāt*) are commonly made from this wood (see Introduction).

†Also apply to *B. atro-virgata*, *B. rotundifolia*.

Ballochia atro-virgata **Balf.f.**
Ballochia puberula Vierh.

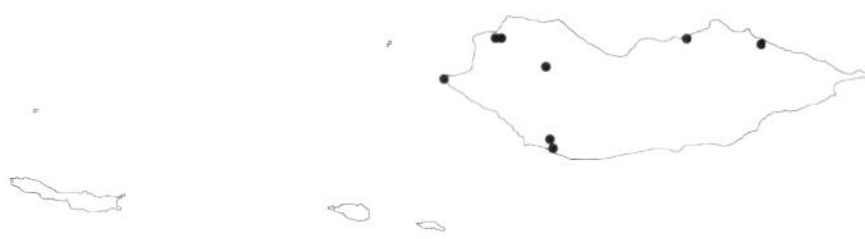

Shrub or small tree to 3m. Leaves oblong-elliptic, 1–2.5 x 0.6–1cm, not fasciculate but sometimes crowded on short branches. Calyx ± glabrous. Corolla red to orange, <3cm long.
Ecology. Succulent shrubland and drought-deciduous woodland (commonly associated with *Gnidia socotrana* and *Buxanthus pedicellatus*); dry inland plains, particularly along wadis and the foothills of the limestone plateaus of the west; also at Hamaderoh and Hoq in the east. 20–600m.
Distribution. Endemic.
Conservation status. Data Deficient (DD). Part of a taxonomically difficult group on Soqotra. Inadequate evidence is available to make an assessment.

ETHNOGRAPHIC NOTES
Name(s)
See *B. amoena* (also uses).

Ballochia rotundifolia **Balf.f.**

Small, intricately branched shrub to 0.5m. Leaves oblong-elliptic or rotund, 0.5–1.5 x 0.5–0.7cm, not fasciculate. Calyx glandular-pubescent. Corolla red to orange, <3cm long.
Ecology. Occasional in drought-deciduous woodland. 450–600m.
Distribution. Endemic.
Conservation status. Data Deficient (DD). Part of a taxonomically difficult group on Soqotra. Inadequate evidence is available to make an assessment.

ETHNOGRAPHIC NOTES
Name(s)
See *B. amoena* (also uses).

Barleria L.

Shrubs, spinose subshrubs or small trees. Leaves entire or spine-tipped, petiolate. Flowers terminal or axillary, solitary or clustered. Calyx deeply 4-lobed; outer 2 lobes longer than inner. Corolla lilac, white or orange-yellow, 2-lipped, with 1 upper lobe and 4 lower or with 5 subequal lobes. Fertile stamens 2; anther cells 2, side by side. Fruits 2 or 4-seeded. 3(4?) endemics. 250 species. Tropical.

ETHNOGRAPHIC NOTES
Comparative uses. In Kenya the whole plant of *B. acanthoides* is used to 'cleanse' a parturient woman: it is boiled in water, the water drunk, or used to wash the body, or the steam inhaled. This plant and other *Barleria* species symbolise peace, and stems are placed on the roof of the wedding house. In south and east Africa, a thorn of *B. mucronata* is used to prick around the site of snakebite, the limb then being fumigated in the smoke of a fire on which twigs of the plant and sheep's wool have been thrown. A decoction of the root and stem is also given as an emetic. *B. prionitis* has an enormous number of medicinal uses in India and elsewhere. The leaves and roots of many *Barleria* species are used for coughs and inflammation, and their flowers sucked for their nectar. (Dalziel 1948, 1956; Goodin & Northington 1985; Heine *et al.* 1988; le Houerou 1980)

Barleria aculeata **Balf.f.**

Intricate shrub to 1m. Leaves with c.3 spines and white margins, rotund or obovate, 0.5–1 x 0.5–1cm, cystoliths prominent. Flowers terminal, solitary. Corolla pale lilac, <1.5cm long, 5 subequal lobes. Fruits 4-seeded.
Ecology. Uncommon and scattered on Soqotra occurring in semi-deciduous woodland (commonly associated with *Rhus thyrsiflora* and *Buxanthus pedicellatus*) on limestone in the east, in submontane woodland (associated with *Cephalocroton socotranus*) in the Haggeher mountains and also as an outlier in the relict woodland at Qatariyah in the southwest of the island. 400–850m.
Distribution. Endemic.
Conservation status. Least Concern (LC). Found in several vegetation types and under no present or perceived threat.

Botanical notes. The distribution of this species is typical of those of plants found in areas which catch mist and low cloud. It is readily distinguished by its suborbicular, spinose leaves.

ETHNOGRAPHIC NOTES†
Name(s)
– Soqotri: *ṣáᶜyud*, *məsɔ̄ᶜidēn* 'prickly, spiny'; *śkaᶜhinítin*, *śkaˀḥinítin* 'thorny, spiny'; *məśḥūwgeg* (more commonly given to *Asparagus*); *fēlʌḳi*; (rare) *źʌ́fəraᶜ* (more commonly given to *Plectranthus* species).
– Soqotri simplified: *za'yud*, *mazow'iden*; *shka'hinitin*; *mish'howgeg*; *feylaqi*; *dhafira'*.
Food (livestock). Nutritious and fattening fodder, especially for goats, and considered to be a good feed for lactating animals. Camels and donkeys also eat them.

†Also apply to *B. argentea*, *B. tetracantha*.

Barleria argentea Balf.f.

Subshrub to 50cm with pubescent stems. Leaves narrowly elliptic or spathulate, 1–3cm long, cystoliths present. Flowers in c.3-flowered cymes, sometimes solitary, axillary. Corolla lilac-blue, 2-lipped, upper lip with 1 large lobe, lower lip of 4 lobes, <1.5cm long. Fruits 2-seeded with long beak. Flowers in c.3-flowered cymes, sometimes solitary, axillary. Corolla lilac, 2-lipped, upper lip with 1 large lobe, lower lip of 4 lobes, <1.5cm long. Fruits 2-seeded with long beak.
Ecology. No habitat details available. Balfour (1888) found it on the Kadhab [Qadhub] plain; probably in *Croton socotranus* shrubland.
Distribution. ?Endemic – also recorded from Yemen and NE Africa but these plants apparently represent another taxon.
Conservation status. Extinct (EX). *B. argentea* has not been seen since it was originally collected on the northern coastal plain of Soqotra in the middle of the 19th century (Balfour 1888). The coastal plains of Soqotra are the habitat most at risk from the developments proposed for the island. They are already showing the signs of decline in quality of habitat as people migrate from the interior to the coast. The area of plain where the plant was originally found is showing clear signs of overgrazing and, as the plant has not been seen for 100 years and the site at which it was originally found has been surveyed on a number of occasions, the plant must now be considered extinct.

ETHNOGRAPHIC NOTES
Name(s)
See *B. aculeata* (also uses).

Barleria popovii Verdc.

Small tree to 4m. Leaves ovate, 3–6 x 2–3cm, greyish tomentose. Flowers in terminal clusters, with large leafy bracts. Corolla white, <6cm long, 2-lipped, upper lip reduced, lower lip of 4 overlapping lobes. Fruits 4-seeded.
Ecology. A rare tree known from a handful of sites in sheltered submontane woodland in the central Haggeher. 600–750m.
Distribution. Endemic.
Conservation status. Endangered (EN B2 a biii). With the increasing demand for timber on the island there is pressure on all tree species. Therefore, this tree, which has an area of occupancy of less than 20km^2 and is known with certainty from only two small populations, must be considered under particular threat.
Botanical notes. Also recorded by Popov (in 1957) in a limestone gorge, known as the RAF Pass; this locality has not been traced. An attractive tree with large white flowers. It is night-flowering.

ETHNOGRAPHIC NOTES
Name(s)
– Soqotri: *di-ˈraḥmam* 'grey, ash-coloured' (also given to the dead wood of *Cephalocroton* and *Dirachma socotrana*, and often, by extension, to the trees themselves).
– Soqotri simplified: *di-rahmam.*
Food (livestock). The foliage is not palatable to livestock.
Wood. Provides a good, hard wood, but the tree is not common enough to be important as a wood, either for use as timber or for implements.
Commercial value (past, present). The dead wood is gathered and offered for sale.
Other. This is one of the plants whose dead wood, called *di ˈraḥmam*, is burned for their pleasant smell. The most widely used is the wood of *Cephalocroton socotranus*, but this species and *Dirachma socotrana* are also used. Slivers of wood are burned on a slip of pottery or thrown on the embers to give living quarters a pleasant smell and to fumigate them. Clothing is also held in the smoke of the burning wood to perfume it. It is burned especially during childbirth (but not usually at a death, when frankincense is more commonly used).
Notes. The wood is also burned to purify the surrounding area of evil spirits and influences.

Barleria tetracantha Balf.f.

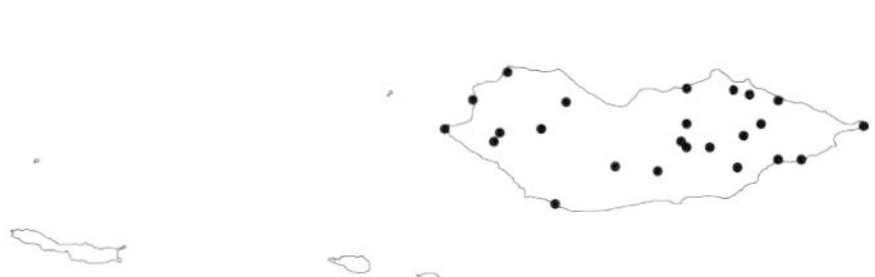

Dense subshrub with short axillary branches terminated by stellate spines. Leaves narrowly ovate or elliptic, 0.5–1 x 0.2–0.4cm, spine-tipped. Flowers solitary in leaf axils. Corolla orange-yellow, with 5 equal lobes <1.5cm. Fruits 2-seeded.
Ecology. Common in succulent shrubland and woody herb communities on the plains and limestone hills and plateaus. Sea-level–600m.
Distribution. Endemic.
Conservation status. Least Concern (LC). Found in several vegetation types and under no present or perceived threat.
Botanical notes. A low, intricately branched shrublet readily distinguished by its spine-tipped leaves and distinctive four-branched spines.

ETHNOGRAPHIC NOTES
Name(s)
See *B. aculeata* (also uses).

Blepharis Juss.

Herbs or small shrubs, often spiny. Leaves entire or lobed, often spine-tipped, sessile. Flowers axillary in short spikes, or solitary, bracts conspicuous, spiny. Corolla white or lilac, 1-lipped. Fertile stamens 4; anther cells 1, white pilose. Fruits 2-seeded. 1 endemic. 80 species. Old World tropics to S Africa and the Mediterranean.

Botanical notes. For a revision of the genus see Vollesen (2000).

ETHNOGRAPHIC NOTES

Comparative uses. In Oman the fruiting heads of *B. ciliaris* are scorched, ground and the ash applied to wounds, especially those of circumcision, as a haemostatic and disinfectant. The seeds are prescribed as an anti-inflammatory and to treat haemorrhoids. A paste made from the fresh seed-heads is used to arrest bleeding, and the dried roots are scorched and pulverised to make *kohl* for the eyes to improve vision. In Africa the spiny bracts of *B. linariifolia* are used to trace patterns on earthenware before firing, and the seeds are also collected by nomads as food. The seeds are boiled in milk and taken as a tonic. A tincture of *B. capensis* is widely used as an anthrax remedy, and a decoction is taken by mouth, as well as a leaf applied locally, to treat the bite of a snake or poisonous insect. The whole plant is pounded, left to steep in water and the liquid is used to wash the body as a remedy for measles and smallpox. Alternatively, the root is boiled and the steam inhaled. (Ambasta 1982; Dalziel 1948, 1956; Ghazanfar 1994; Goodin & Northington 1979; Heine *et al.* 1988; Jain & De Fillips 1991)

Blepharis kuriensis Vierh.

Blepharis ciliaris auct. non (L.) B.L.Burtt
Blepharis edulis Pers.
Blepharis linariifolia Vierh.

Prostrate spiny herb to 30cm. Leaves linear, 2–4 x 0.2–0.4cm, lobed, lobes with short spine tips. Flowers in short, dense spikes, bracts <3cm long, with c.5 to 7 long spines. Corolla lilac, <1.5cm long.
Ecology. Dwarf shrubland on the plains and foothills; not common on Soqotra but relatively frequent on Samha and Abd al Kuri. Sea-level–250m.
Distribution. Yemen (Mahra).
Botanical notes. A prostrate herb with spine-tipped leaves, bracts and calyces and attractive lilac flowers. It is closely related to the widespread *B. edulis* but differs with its shorter reflexed bract-spines and by its generally smaller flowers.

ETHNOGRAPHIC NOTES†

Name(s)
– Soqotri: nil or *źulḍ*; *di ʾārdəhān*, *di hārdəhān* 'a spreading area of herbage'; *məʿeyib di libihan* 'kitten excreta' and when young and green *ḥarḥērə* 'tiny acacia' (Abd al Kuri).
– Soqotri simplified: *dhaldh*; *di 'ardihan*, *di hardihan*; *mi'eyib di libihan*, *harheyrih*.
Food (livestock). Goats and sheep are very fond of these species in the drier areas of the Archipelago, especially the immature plants. Camels eat it at all stages of growth. It is not regarded as being important nutritionally.
Notes. Can cause *'uqabah* in goats unaccustomed to the feed.

†Also apply to *B. spiculifolia*.

Blepharis maderaspatensis (L.) Roth

Blepharis boerhaviifolia Juss.

Prostrate, softly hairy herb to 30cm. Leaves narrowly ovate or elliptic, 1–3.5 x 0.4–1.5cm, entire. Flowers with many bracts below a single terminal flower, bracts <1cm long, with long barbed bristles. Corolla white, <1.5cm long.
Ecology. Scattered throughout the centre and east of the island in a range of vegetation types including succulent shrubland, drought-deciduous woodland and submontane woodland. 10–700m.
Distribution. Widespread in Africa, Arabia, India and Sri Lanka.
Botanical notes. A prostrate herb with distinctive spine-fringed bracts, the fringes resembling eyelashes.

ETHNOGRAPHIC NOTES

Name(s)
– Soqotri: *di ʾārdəhān*, *di hārdəhān* 'a spreading area of herbage'; *miriśiyitin* 'of settlements'.
– Soqotri simplified: *di 'ardihan*, *di hardihan*; *mirishiyitin*.
Food (livestock). Provides incidental forage mainly for small stock, but it is not nutritionally important.
Notes. Can cause *'uqabah* in goats unaccustomed to the feed.

Blepharis spiculifolia Balf.f.

Blepharis spiculifera Juss.

Intricate shrub to 1m. Leaves fasciculate, oblanceolate, 1.5–2 x 0.3–0.4cm, spine-tipped, with 2 marginal spines. Flowers solitary, bracts spine-tipped. Corolla cream, <1.5cm long.
Ecology. Coastal plains and foothills of NE Soqotra where it is often found in *Croton socotranus* shrubland and succulent shrubland. 50–100m.
Distribution. Endemic.
Conservation status. Vulnerable (VU D2). Rare, with a very limited area of occupancy.

Botanical notes. A low, intricately branched shrub apparently endemic to NE Soqotra. It is distinguished by its narrow, spine-tipped leaves which often bear 1 or 2 marginal spines. Vollesen (2000) comments that *B. spiculifolia* is a most remarkable species which has a number of seemingly primitive traits. It is the sole member of subgenus *Ebracteata* Vollesen which is endemic to Soqotra. Vollesen suggests that *B. spiculifolia* could be a primitive forerunner to sect. *Blepharis*.

ETHNOGRAPHIC NOTES
Name(s)
See *B. kuriensis* (also uses).

Chorisochora

Woody herbs or subshrubs. Leaves entire, petiolate. Flowers in terminal racemes, bracts reddish green, conspicuous. Corolla white or pink; lobes 5, subequal. Calyx 5-lobed; lobes ± equal. Fertile stamens 2; anther cells 2, side by side. 2 endemics. 3 species. SW Africa.

Botanical notes. A genus with an interesting disjunct geographical distribution with two species on Soqotra and a third in S Africa (Transvaal) (Vollesen 1994). This pattern of distribution (cf. *Graderia*, *Wellstedia* and *Camptoloma*) offers evidence for the existence of a dry corridor during the Pleistocene connecting NE and SW Africa. It has been argued that this corridor, during dry periods in the past, would have allowed migration between these regions.

***Chorisochora minor* (Balf.f.) Vollesen**
Ecbolium striatum var. *minor* Balf.f.

Shrubby herb or subshrub to 1m. Leaves ovate or elliptic, 4–6.5 x 1.7–3.5cm. Flowers in racemes <10cm long, bracts <0.7cm long, not overlapping or covering the stem axes or calyces. Corolla white, <1cm long. Corolla 2-lipped, lower lip entire, upper 3-lobed. Fertile stamens 2, anther cells 2, one above the other, tails absent. Fruits 4-seeded.
Ecology. See *C. striata*.
Distribution. Endemic.
Conservation status. Vulnerable (VU D2). Rare, with a very limited area of occupancy.

ETHNOGRAPHIC NOTES†
Name(s)
– Soqotri: *miśḳerér*, *enśḳerér* (pl. *miśḳerór*, *enśḳerór*) (also given to *Angkalanthus oligophylla*) 'gives tainted milk'.
– Soqotri simplified: *mish'qerer*, *ensh'qerer* (pl. *mish'qeror*, *ensh'qeror*).

Food (livestock). All livestock eat the foliage and shoots when hungry or when there is little alternative forage. Leafy stems are lopped to feed cattle if the dry season is prolonged and other sources of fodder have been exhausted. Goats chew leafless twigs and gnaw at the bark in a drought. The fallen leaves are also collected as a dry season feed for all stock.
Flavouring / food additive. The milk of animals browsing on this has a disagreeable smell and an unpleasantly bitter taste, so much so that some herders will not drink it. Even the flavour of the raw buttermilk and fresh butter is affected, though this disappears once the processed milk has been cooked. The flavour of the meat of a goat that has been browsing heavily on it is impaired, even the taste of the fat is affected, and the water the meat has been cooked in is undrinkable. Indeed, the effects of the plant are so powerful that the islanders say the taste of the meat of a goat that has been browsing this plant, even for a single day, is adversely affected.
Timber. No use as timber.
Wood. Soft and little used for implements.
Comparative uses. In Oman the dried leaves of *Ecbolium viride* are powdered and applied to the skin to treat skin problems. A similar preparation is applied around the genital area as a diuretic. (Ghazanfar 1994; Ghazanfar & Al-Sabahi 1993)

†Also apply to *C. minor*.

***Chorisochora striata* (Balf.f.) Vollesen**
Ecbolium striatum Balf.f.

Vegetatively similar to *C. minor*. Distinguished by the overlapping inflorescence bracts, <2cm long, which obscure the calyces. Corolla magenta, <3.5cm long.
Ecology. Usually on sheltered cliffs in evergreen submontane thicket but also less commonly amongst shrubs in clearings in submontane woodland (Kishin). 700–1000m.
Distribution. Endemic.
Conservation status. Vulnerable (VU D2). Rare, with a very limited area of occupancy.
Botanical notes. More common than the previous species. Found mainly on granite cliffs in the Haggeher mountains but also recorded on cliffs at the headwaters of south-draining wadis (i.e. Wadi Da'arho).

ETHNOGRAPHIC NOTES
Name(s)
See *C. minor* (also uses).

Dicliptera Juss.

Herbs. Leaves entire, petiolate. Flowers solitary in leaf axils, bracts enclosing the calyx, free at base. Corolla 2-lipped, lower lip entire, upper 3-lobed. Fertile stamens 2, anther cells 2, one above the other, tails absent. Fruits 4-seeded. 1 endemic. 150 species. Tropical and warm regions.

Dicliptera effusa Balf.f.

Dicliptera ovata Balf.f.

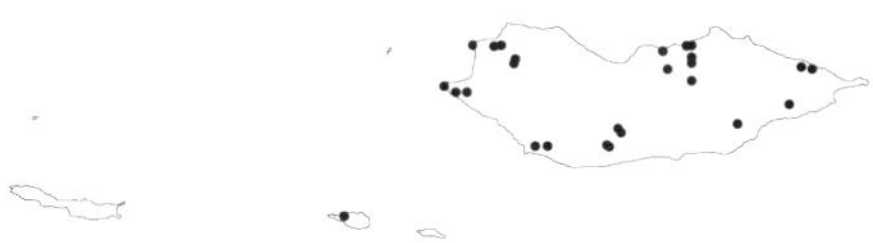

Ascending herb to 30cm. Leaves ovate, 0.7–2 x 0.5–1.5cm. Bracts spine-tipped. Corolla pink or lilac, c.1cm long. Fruits c.0.6cm long.

Ecology. Mainly on the limestone plateaus, less commonly on the granite; in a variety of vegetation types including drought-deciduous woodland and dwarf shrubland; often associated with *Buxanthus pedicellatus*, *Rhus thyrsiflora*, *Croton socotranus*, *Jatropha unicostata* and *Pulicaria stephanocarpa*. 300–600m.

Distribution. Endemic.

Conservation status. Least Concern (LC). Common in several vegetation types and under no present or perceived threat.

ETHNOGRAPHIC NOTES

Name(s)

– Soqotri: (*di*) *ḥəwrˈirhin*, (*di*) *ḥˈəwrrihin* (pl. (*di*) *ḥəwrˈirhun*, (*di*) *ḥˈəwrrihun*) 'small and dark-coloured'; *mˈʌḥeriś* 'undergrowth, understorey'; *sōf* 'woolly'.

– Soqotri simplified: (*di*) *hewririhin* (pl. (*di*) *huwrirhun*, (*di*) *hewrrihun*); *maherish*; *sowf*.

Food (livestock). Provides incidental forage for small stock but is not nutritionally important. It is considered an excellent feed for cattle when it is fresh and green.

Notes. Can cause *'uqabah* in small stock unaccustomed to the feed.

Comparative uses. The shoots and leaves of various related species are eaten. (Kunkel 1984)

Hypoestes Sol. ex R.Br.

Herbs. Leaves entire, petiolate. Flowers in short terminal spikes and axillary clusters, bracts enclosing the calyx. Corolla 2-lipped, the upper lip with 3 lobes, lower lip ± entire. Fertile stamens 2, anther cells 1. Fruits 4-seeded. 1 endemic. 40 species. Old World tropics.

Hypoestes pubescens Balf.f.

Hypoestes forskalei (Vahl) Sol. ex Roem. & Schult.

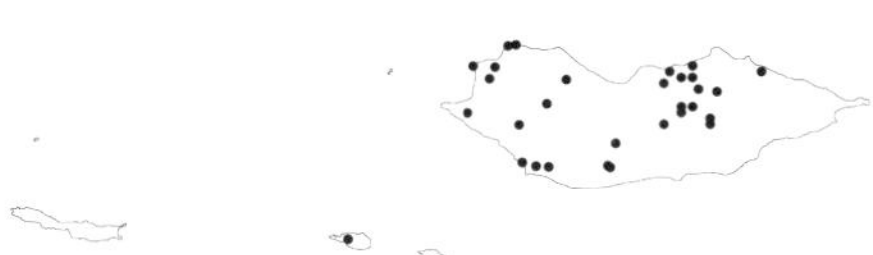

Erect, softly pubescent herb to 50cm. Leaves ovate, 1.5–5 x 1–3cm, paler beneath. Corolla white to lilac, <1.5cm long. Fruits <1cm long.

Ecology. Widespread and common in a variety of vegetation types. Sea-level–800(–1000)m.

Distribution. Endemic.

Conservation status. Least Concern (LC). Common in several vegetation types and under no present or perceived threat.

Botanical notes. *H. pubescens* has sometimes been considered to be a form of the widespread and variable *H. forskalei*. However, plants on Soqotra are consistently grey-green herbs with clear lilac flowers. For the time being I have preferred to maintain *H. pubescens* as a distinct species.

ETHNOGRAPHIC NOTES

Name(s)

– Soqotri: (*di*) *ḥəwrˈirhin*, (*di*) *ḥˈəwrrihin* (pl. (*di*) *ḥəwrˈirhun*, (*di*) *ḥˈəwrrihun*) 'small and dark-coloured'; *mˈʌḥeriś* 'undergrowth, understorey'; *di ʿˈažbub*, *di ʿˈažərib* 'quick to die back'.

– Soqotri simplified: (*di*) *hewririhin* (pl. (*di*) *huwrirhun*, (*di*) *hewrrihun*); *maherish*; *di 'ajbub*, *di 'ajirib*.

Food (livestock). Like all the *ʿˈažərib* plants that come up after good rains, this is considered to be a valuable fodder for cattle. Camels and donkeys eat it, as do goats, before other more palatable plants respond to the rain, but less enthusiastically than cattle. If they are weakened by a long dry season, and eat a lot of this sort of plant, they can develop scour, and if pregnant, abort the foetus.

Comparative uses. In nearby Dhofar this is recognised as a toxic plant that can be fatal to small stock or sickly animals. It causes shivering, head-shaking and fever. The aerial parts of various African species are cooked and eaten. *H. verticillaris* is crushed and used to rub on a cross-bow to clean it and to drive away flies; also for polishing floors. *H. purpurea* is used as a remedy for coughs and the spitting of blood, and is applied also to swellings. (Dalziel 1948, 1956; Heine *et al.* 1988; Kunkel 1984; Miller & Morris 1988)

Justicia L.

Herbs or shrubs. Leaves entire, petiolate or sessile. Flowers axillary, solitary or in few-flowered cymes, bracts inconspicuous, triangular or enclosing calyx, lanceolate. Corolla white or lilac, 2-lipped, lower lip with 3 teeth, upper ± entire. Fertile stamens 2; anther cells 2, one above the other, tails present. Fruits 4-seeded (sometimes 1-seeded). 3 endemics. 420 species. Tropics and warm temperate N America.

ETHNOGRAPHIC NOTES

Comparative uses. Various *Justicia* species are cooked and eaten, and *J. insularis* is sometimes cultivated as a potherb. In nearby Dhofar the related *J. diffusa* is toxic to goats and also on occasion to young or sickly calves, sometimes with fatal results. *Justicia* species are used in pest control. For instance *J. adhatoda* is active against fleas, flies, mosquitoes and pests of stored grain; *J. gendarussa* against clothes moths and *J. schimperiana* against body lice. In Kenya the bark of *J. cordata* is pounded and mixed with fat, dried in the sun, mixed again with fat and clay and applied to their hair by men to give it a pleasant smell. The leaves are also dried and crushed, mixed with oil and red ochre and applied to the head and/or neck as a perfume. The root is boiled and fed to infants, with or without milk, to promote good health. Many species are used medicinally, mainly to treat inflammation and problems of the skin and stomach. (Grainge & Ahmed 1988; Heine *et al.* 1988; Kunkel 1984; Lewis & Elwin-Lewis 1977; Miller & Morris 1988; Oliver-Bever 1986)

Justicia alexandri R.Atkinson (see Appendix 2, p. 723)

Shrub to 1m. Leaves ovate to elliptic, base cuneate, tip obtuse, 1.5–3 x 0.7–1.5cm, grey tomentose beneath, glossy above. Flowers solitary, bracts enclosing calyx, lanceolate. Corolla magenta, <1cm long. Fruits all elliptic, probably 4-seeded.

Ecology. A rare shrub of sheltered slopes on the highest granite pinnacles of the western Haggeher; associated with *Euryops arabicus*, *Helichrysum rosulatum*, *Hypericum tortuosum* and *Thamnosma socotrana*. c.1500m.
Distribution. Endemic.
Conservation status. Vulnerable (VU D2). Rare, with a very limited area of occupancy.
Botanical notes. See comments under the closely related *J. takhinensis*.

ETHNOGRAPHIC NOTES

Not seen, no information recorded.

Justicia heterocarpa T.Anderson

Annual, thin-stemmed herb to 20cm. Leaves ovate, 1–3 x 0.5–1.5cm. Flowers in cymes. Corolla pale lilac, <0.5cm long. Fruits of two types on the same plant, either elliptic and 4-seeded or winged, hooked and 1-seeded.
Ecology. Sporadic and appearing after rain in drought-deciduous woodland and *Croton socotranus* shrubland. 30–650m.
Distribution. Arabia, NW India, NE and tropical Africa.
Botanical notes. Unusually, two types of fruit are found on the same plant. On the upper part of the plant they are smooth and towards the base they are spiny. All material on Soqotra is referable to subsp. *heterocarpa*.

ETHNOGRAPHIC NOTES

Name(s)
– Soqotri: nil or *ᶜážərib* 'quick to die back'; *ᶜážərib śərśīroh* 'prickly *'ajirib* plant'; (rare) *ʾíbəliṭ* (more commonly given to *Amaranthus* species).
– Soqotri simplified: (*di*) *'ajirib*; *'ajirib shirishiyro*; *'ibilit*.
Food (human). The fresh greenery can be gathered and cooked like *Amaranthus*.
Food (livestock). Like all the *ᶜážərib* plants that come up after the rains, it is a very good fodder for cattle, and highly regarded by cattle owners. Camels and donkeys eat it, as do goats, but less enthusiastically than cattle. If there is little alternative fodder available and goats eat a lot of it, they can develop scour.

Justicia rigida Balf.f.

Intricately branched shrub to 50cm. Leaves fasciculate, linear to spathulate, <1cm long. Flowers solitary. Corolla white, <0.5cm long. Fruits all elliptic, 4-seeded.
Ecology. Locally common on coastal and inland plains where it is typically an undershrub in *Croton socotranus* shrubland; co-dominant with *Croton socotranus* and *Pulicaria stephanocarpa* in the dwarf shrubland of the western inner plains. 0–250m.
Distribution. Endemic.
Conservation status. Least Concern (LC). Common in several vegetation types and under no present or perceived threat.
Botanical notes. Often heavily grazed and appearing as a dark, almost black, mound beneath *Croton socotranus*.

ETHNOGRAPHIC NOTES

Name(s)
– Soqotri: (western Soqotra) *xε̄rṭʌh* 'can be stripped of foliage'; *ᶜefírə*, *ᶜífirə*, *ᶜeféro* 'red-coloured'; *gifíʾhiñ* 'stunted, bowed'; 'type of *ḥamaḥámo*' 'breeding male goat or ram to become sexually aroused' (as from eating this plant which is very good for them).

– Soqotri simplified: *kheyrtah*; *'efirih*, *'efero*; *gifi'hin*; *hamahamo*.

Food (livestock). This tough, drought-surviving plant is an important dry season and drought fodder in the drier areas of the island. In some coastal areas it is the key dry season forage. Goats, donkeys and camels all eat it, and it is seen as a very valuable maintenance feed.

Fuel. Dead plants make excellent firelighters and are often gathered and stored for this purpose or for use as tinder.

Justicia takhinensis **R.Atkinson** (see Appendix 2, p. 723)

Intricately branched shrub to 1m. Leaves obovate, base acuminate, tip rounded or notched, 0.5–1 x 0.5–1cm, glandular, cystoliths prominent, fleshy. Flowers solitary (or in few-flowered cymes?), bracts small, not enclosing calyx. Corolla magenta, <1cm long. Fruits all elliptic, 4-seeded.

Ecology. Seaward-facing limestone cliffs in shrubland dominated by *Rhus thyrsiflora* and *Ballochia amoena*. 500m.

Distribution. Endemic.

Conservation status. Endangered (EN B2 a biii). A very rare shrub known only in shrubland at a single location with a known area of occupancy of less than 20km^2. This shrubland is dependent on precipitation from mist and cloud and is thus at risk from the suspected trend of reduction in the extent and quality of these mists.

Botanical notes. Restricted to limestone cliffs at Reqadihon (above Hoq). These cliffs capture mist and cloud during the winter and are an eastern outlier for a number of species (e.g. *Coelocarpum haggierensis* and *Dirachma socotrana*) which are otherwise found only in areas of high rainfall in the Haggeher mountains. This species is closely related to *J. alexandri*.

ETHNOGRAPHIC NOTES

Name(s)

– Soqotri: *ʿefírə*, *ʿifírə*, *ʿeféro* 'red-coloured'.

– Soqotri simplified: *'efirih, 'ifirih, 'efero*.

Food (livestock). The foliage, stems and leafless twigs are liked by goats especially, but all livestock eat the foliage in the dry season. It is regarded as a good, nutritious feed.

Wood. Suitable for firesticks. It is a hard wood and useful for a wide variety of implements.

Neuracanthus Nees

Woody herbs. Leaves crenate, petiolate or sessile. Flowers axillary, in spikes (or globose clusters), bracts linear, spine-tipped, entire. Corolla purple; lobes 5, subequal. Fertile stamens 4; anther cells 2, tomentose. Fruits 4-seeded. 1 endemic. 20 species. Tropical Africa and Arabia to India.

Neuracanthus aculeatus **Balf.f.**

Erect, grey-pubescent woody herb to 30cm. Leaves linear to spathulate, 1–2cm long, strongly undulate, sessile. Flowers pinkish purple, in conspicuous elongate spikes, bracts tipped with a slender reflexed spine <1cm long.

Ecology. Locally common in *Croton socotranus* shrubland on the northeastern coastal and inland plains. Sea-level–120m.

Distribution. Endemic.

Conservation status. Endangered (EN B1 2a biii). Restricted to a small area on the northern plains. It is liked by livestock and the gradual decline in the quality of habitat due to overgrazing is likely to place this plant under some threat. The coastal plains on Soqotra are the habitat most at risk from the developments proposed for the island. They are already showing the signs of decline in quality of habitat as people move from the interior to the coast.

ETHNOGRAPHIC NOTES

Name(s)

– Soqotri: *ḥamaḥámo* 'breeding male goat or ram to become sexually aroused' (as from eating this plant which is very good for them).

– Soqotri simplified: *hamahamo*.

Food (livestock). Much liked by goats. Camels and donkeys also browse the plants, and sheep eat the litter and tender new growth. It is regarded as a very rich and nutritious feed for goats. In sheep, however, the spines can cause damage to the mouth and also occasionally to the feet.

Bees / honey. Honeybees are attracted to the flowers.

Fuel. Dead plants make good firelighters and are often gathered and stored for this purpose or for use as tinder.

Neuracanthus argyrophyllus **Chiov.**

Prostrate woody herb to 20cm. Leaves spathulate, sericeous, prominent leaf scars present. Flowers in dense spikes hidden in the leaf axils; conspicuous stiff white hairs present on fertile bracts. Corolla lilac.

Ecology. Locally common in desertic dwarf shrubland in the Zahr basin and on the granite ridge above Bandar Shu'ub. 200–500m.

Distribution. Somalia.

Botanical notes. Outside Soqotra known only from a single collection from the Hafen Peninsula in NE Somalia.

ETHNOGRAPHIC NOTES
Name(s)
– Soqotri: *l̍ibəhʌn* 'pale-coloured'; *ḳūwnu rəxa̍mmə* 'grey threads'.
– Soqotri simplified: *libihan*; *quwnu rikhammih*.
Food (livestock). Regarded as good, nutritious fodder for goats and sheep. It is an important dry season forage in the drier areas of the island.

Neuracanthus capitatus **Balf.f.**

Prostrate woody herb. Flowers in dense, globose clusters.
Ecology. No habitat details are available but Balfour (1888) found it on the Kadhab [Qadhub] plain; probably in *Croton socotranus* shrubland.
Distribution. Also recorded from Somalia.
Botanical notes. Not seen since it was originally collected on the northern coastal plain of Soqotra in the 19th century (Balfour 1888). The area of plain where the plant was found is showing clear signs of overgrazing, but it has not been extensively explored and so it is too early to suggest that it has become extinct.

Peristrophe Nees

Herbs. Leaves entire, petiolate. Flowers in lax panicles formed of pedunculate, axillary cymes, outer bracts unequal in size, one almost twice as long as the other, fused at base. Corolla 2-lipped, lower lip with 2 lobes, upper with 3. Fertile stamens 2; anther cells 2, one above the other, tails absent. Fruits 4-seeded. 15 species. Old World tropics.

Peristrophe paniculata **(Forssk.) Brummitt**
Peristrophe bicalyculata Nees

Erect, slender herb to 50cm. Leaves ovate, 2–4 x 1.2–2.5cm. Corolla pinkish mauve, <1cm long.
Ecology. A weed of date gardens, also sometimes in drought-deciduous woodland. 50–600m.
Distribution. Widespread in Africa, Arabia, India and Madagascar.

ETHNOGRAPHIC NOTES
Name(s)
– Soqotri: nil or *ᶜa̍žərib* 'quick to die back'.
– Soqotri simplified: *'ajirib*.
Food (livestock). Provides incidental grazing for livestock and like all the *ᶜa̍žərib* plants that come up after the rains, it is regarded as a very good fodder for cattle. Camels and donkeys eat it, as do goats, but less enthusiastically than cattle. If there is little alternative fodder available and goats eat a lot of it, they can develop scour.
Comparative uses. In India this is fed to horses and is also sown as a green manure. The plant yields an essential oil that shows tuberculostatic activity *in vitro*. The leaves of some *Peristrophe* species are cooked and eaten. (Ambasta 1982; Dalziel 1948, 1956; Jain & De Fillips 1991; Kunkel 1984)

Rhinacanthus Nees

Herbs. Leaves entire, petiolate. Flowers in lax, terminal scorpioid cymes (one-sided racemes), bracts lanceolate. Corolla 2-lipped, lower lip broad, 3-lobed, upper lip narrow. Fertile stamens 2; anther cells 2, one above the other, tails absent. Fruits 2-seeded. 1 endemic. 15 species. Old World tropics.

Rhinacanthus scoparius **Balf.f.**

Prostrate herb to 30cm. Leaves narrowly lanceolate, 2–3 x 0.2–0.3cm. Corolla white, c.1.2cm long. Fruits c.1cm long.
Ecology. Widespread on the coastal plains and limestone plateaus; in a number of vegetation types including *Croton socotranus* shrubland, succulent shrubland and dwarf shrubland. 20–600m.
Distribution. Endemic.
Conservation status. Least Concern (LC). Widely distributed in a variety of vegetation types but never abundant – under no present or perceived threat.

ETHNOGRAPHIC NOTES
Name(s)
– Soqotri: nil or *mʌ̍ḥeriś* 'undergrowth, understorey; *ḥamaḥa̍mo* 'breeding male goat or ram to become sexually aroused' (as from eating this plant, which is very good for them).
– Soqotri simplified: *maherish*; *hamahamo*.
Food (livestock). Seen as a good and nutritious fodder for small stock. Donkeys and camels also eat it. The small leaves make it attractive to young stock, and lambs and kids are often herded to feed in areas where it grows.
Bees / honey. Honeybees are attracted to the flowers.
Comparative uses. The roots of *R. nasuta* are boiled in milk and taken as an aphrodisiac, said to have extraordinary powers. In India the fresh roots with the leaves or seeds and added lime juice are used to treat ringworm and other skin diseases. The leaves are used in the treatment of cancers. The leaves of *R. calcaratus* are edible and are also taken as an anthelmintic. The root of this species also is taken as an aphrodisiac. (Ambasta 1982; Dalziel 1948, 1956; Jain & De Fillips 1991; Lewis & Elwin-Lewis 1977)

Ruellia L.

Herbs or small shrubs. Leaves entire, petiolate. Flowers solitary in leaf axils with inconspicuous bracts. Calyx 5-lobed. Corolla lilac or white, sometimes red; lobes 5, subequal. Fertile stamens 4. Fruits 4-seeded, terete. Seeds pubescent or with scarious margin. 4 endemics. 150 species. Tropics and temperate N America.

Ruellia carnea Balf.f.

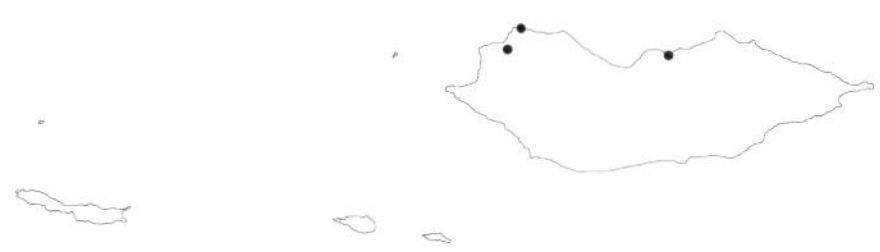

Hoary shrub to 1m. Leaves broadly ovate, 0.5–2 x 0.5–2cm, truncate or cordate, obtuse, roughly white hairy. Corolla red, <3cm long. Fruits <1cm long.

Ecology. A rather rare plant of rocky areas on the coastal plains and foothills; found in *Croton socotranus* shrubland and succulent shrubland. Sea-level–300m.

Distribution. Mainland Yemen (Mahra), Ethiopia, Somalia.

Botanical notes. Readily distinguished by its striking scarlet flowers.

ETHNOGRAPHIC NOTES

Name(s)

– Soqotri: nil or *ṣefərírihun* 'many-flowered'; *máḥeriś* 'fleet'.

– Soqotri simplified: *zefiririhun*; *maherish*.

Food (livestock). Good forage, for small stock especially.

Notes. Many *Ruellia* species have no individual name but are collectively called *ṣefərírihun* when in flower.

Comparative uses. The pounded leaves of species such as *R. repens* are applied to wounds, scalds and ulcers, and are taken for toothache and stomach ache. In Africa the root is boiled and added to milk as a food for children. Other species have edible leaves, those of *R. prostrata* being used in the production of rice beer. (Ambasta 1982; Heine *et al.* 1988; Jain & De Fillips 1991)

Ruellia dioscoridis Napper

Ruellia patula var. *minor* Jacq.

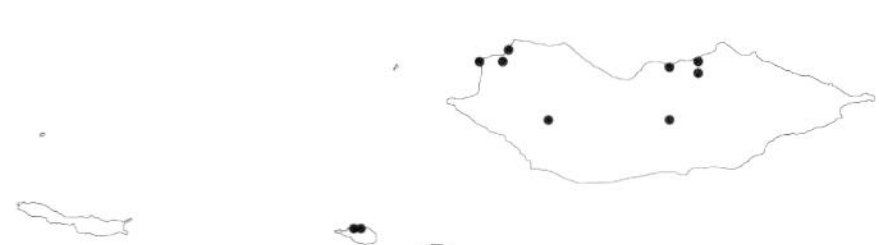

Similar in habit and leaves to *R. patula* but corolla usually white, 1.2–1.5cm long, tube narrow at mouth. Anthers level with top of corolla tube.

Ecology. Locally common on the coastal plains and limestone plateaus in succulent shrubland and dwarf shrubland. 50–650m.

Distribution. Endemic.

Conservation status. Least Concern (LC). Widely distributed in a variety of vegetation types; under no present or perceived threat.

ETHNOGRAPHIC NOTES

Name(s)

– Soqotri: nil or *ṣefərírihun* 'many-flowered'; *noi* 'of high areas'; *sɔ̄f* 'woolly'.

– Soqotri simplified: *zefiririhun*; *noy*; *sowf*.

Food (livestock). Provides incidental forage for small stock, but it is not thought to be of particular nutritional value.

Ruellia insignis Balf.f.

Shrub or small tree to 4m. Leaves broadly ovate or oblong, 1.5–4 x 0.5–3cm, cuneate, cystoliths prominent. Corolla bluish-purple, <4cm long, tube broad at mouth, lobes conspicuously overlapping. Anthers included in tube. Fruits <2cm long.

Ecology. Locally common in sheltered thicket in semi-deciduous and submontane woodland. 250–1050m.

Distribution. Endemic.

Conservation status. Least Concern (LC). Locally common and under no present or perceived threat.

Botanical notes. An attractive small tree with large bluish flowers and distinctive glandular-hairy calyces.

ETHNOGRAPHIC NOTES

Name(s)

– Soqotri: *zīzəhīn*.

– Soqotri simplified: *ziyzihiyn*.

Food (livestock). Regarded as an excellent fodder for all livestock and one of the more important fodder plants of the mountains. It responds well to the lighter summer rains, when it provides excellent forage for both cattle and goats. Small stock also like the flowers. Fallen foliage is gathered to feed to livestock, especially young animals, in the dry season.

Bees / honey. Honeybees are attracted to the flowers.

Flavouring / food additive. The flowers are sucked for their sweet nectar.

Notes. Can cause *'uqabah* in small stock unaccustomed to it. It is also said to cause 'stones' in the urine of goats browsing heavily on it.

Ruellia kuriensis **Vierh.**

Small, compact subshrub with corky branches to 20cm. Leaves somewhat fleshy, on young stems, oblong or ovate, 0.5–1.5 x 0.3–1.2cm, cuneate. Corolla pale lilac, <3cm long, tube broad at mouth. Anthers included in tube. Fruits <1cm long.
Ecology. Common on the rocky foothills of Jebel Saleh on Abd al Kuri where it is found in dwarf shrubland and succulent shrubland (with *Euphorbia balsamifera* and *E. schimperi*). 100–500m.

Distribution. Endemic to Abd al Kuri.
Conservation status. Vulnerable (VU D2). Rare, with a very limited area of occupancy.

ETHNOGRAPHIC NOTES
Name(s)
– Soqotri: nil or *ḥe̍y* 'rains grazing'; *rəte̍bib*; *rəśɛ̄riś*; *dīrbeb* '(animal) with a pale streak or flash on its coat' (Abd al Kuri).
– Soqotri simplified: *hey*; *ritebib*; *risheyrish*; *diyrbeb*.
Food (livestock). Important forage for both sheep and goats on Abd al Kuri.

Ruellia patula **Jacq.**
Ruellia patula var. *pubescence* Balf.f.
Ruellia patula var. *pubescens* Balf.f.
Ruellia sokotrana Vierh.

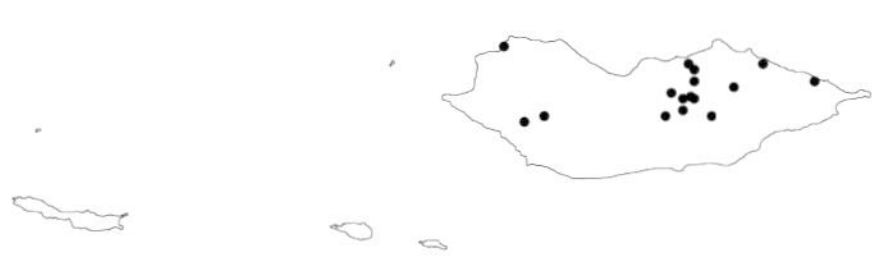

Prostrate herb, somewhat woody at base, to 30cm. Leaves on long, straight stems, oblong, 0.5–1.5 x 0.3–1cm. Corolla lilac, <2.5cm long, tube broad at mouth. Anthers included in tube. Fruits <1cm long.
Ecology. Widespread in dwarf shrubland and deciduous woodland on the limestone plateaus and submontane shrubland on the granite. 150–900m.
Distribution. Widespread in Africa, Madagascar, India and Sri Lanka.

ETHNOGRAPHIC NOTES
Name(s)
– Soqotri: *ḥərūṭ* 'can be stripped of foliage'; *ṭi̍fher* 'finger- and toenails' (more commonly given to *Tephrosia* species); *zīzəhīn*; *mʌ̍ḥeriš* 'undergrowth, understorey; *di hārdəhān*, *di ʾārdəhān* 'a spreading area of herbage'; *śe̍bəriḳ* 'glimpsed out of the corner of the eye'; *ertērit* 'forming green patches on the ground' or *di hāri̍dəhān* (Abd al Kuri).
– Soqotri simplified: *hiruwt*; *tifher*; *ziyzihiyn*; *maherish*; *di haridhan*, *di 'ardihan*; *shebiriq*; *erteyrit*.
Food (livestock). Regarded as a good and nutritious fodder for all livestock. On Abd al Kuri it is an important fodder for sheep and goats.
Bees / honey. Honeybees are attracted to the flowers.
Flavouring / food additive. The flowers are sucked for their sweet nectar.
Notes. Can cause *'uqabah* in small stock unaccustomed to it and is also said to cause 'stones' in the urine of goats browsing heavily on it.

Ruellia paulayana **Vierh.**

Prostrate woody herb to 20cm. Leaves on short, crowded stems, obovate, 0.5–1.5 x 0.5–1cm, petiolate, grey-hairy beneath. Corolla lilac, <3.5cm long, tube broadening into a wide mouth. Anthers included in tube. Fruits c.0.5cm long.
Ecology. Restricted to a small area of SE Soqotra where it is found mainly on the coastal escarpment cliffs and lip of the limestone plateau. 50–300m.

Distribution. Endemic.
Conservation status. Vulnerable (VU D2). Rare, with a very limited area of occupancy.

ETHNOGRAPHIC NOTES
Name(s)
– Soqotri: nil or *ṣefəri̍rihun* 'many-flowered'; *mʌ̍ḥeriś* 'undergrowth, understorey'.
– Soqotri simplified: *zefiririhun*; *maherish*.
Food (livestock). Provides good and nutritious forage when accessible, especially for small stock.
Bees / honey. Honeybees are attracted to the flowers.

Trichocalyx Balf.f.

Shrubs. Leaves entire, subsessile or petiolate. Flowers axillary, in dense, globose clusters, bracts and calyx lobes thread-like, glandular. Corolla 2-lipped, lower lip with 3 lobes, upper hooded. Fertile stamens 2; anther cells 2, one above the other, tails present. Fruits 4-seeded. 2 species. Endemic.

Botanical notes. Distinguished by the globose clusters of flowers partly hidden by numerous thread-like bracts and calyx lobes. Their bright orange-red flowers are visited by the endemic Soqotran sun-bird. The genus is in need of revision: the species are variable and poorly defined. Identifications, and hence the distribution maps, should be treated with some caution.

Trichocalyx obovatus Balf.f.

Very similar to, and perhaps conspecific with, *T. orbiculatus*. It can be distinguished by the narrower, definitely petiolate, obovate leaves.
Ecology. Common in succulent shrubland and deciduous woodland. 50–650m.
Distribution. Endemic.
Conservation status. Data Deficient (DD). Part of a taxonomically difficult group on Soqotra. Inadequate evidence is available to make an assessment.
Botanical notes. Plants from Wadi Ayheft and the Rewgid plateau seem to most consistently match Balfour's (1888) description of this taxa. However, no plants with purple flowers (reported by Balfour for *T. obovatus*) have been seen amongst recent collections. See also comments under the generic description.

ETHNOGRAPHIC NOTES†

Name(s)
– Soqotri: *ᶜʌ̍ləhʌl* (pl. *ᶜala̍ʾhal*); (western Soqotra) *ᶜʌ̍lġʌl* 'lovely, beautiful'; (rare) *mīṣeṣ* (from the verb 'to suck' referring to the flowers which are sucked for their nectar).
– Soqotri simplified: *'al'hal* (pl. *'ala'hal*); *'alghal*; *miyzez*.
Food (livestock). One of the key fodders on the island, regarded as a rich feed which promotes lactation. It is well liked by all livestock, particularly young stock. The smaller plants are among the few shrubs browsed by sheep, who are very fond of the small, green leaves. Camels and donkeys also eat it and camels in particular are seen as very destructive of the plants. In the drier areas of the island it is very important forage during the long, dry season. Both sheep and goats eat the fallen leaves and other plant litter.
Bees / honey. An important honeybee forage.
Flavouring / food additive. More than any other, the flowers of these plants are sucked for their sweet nectar.
Fuel. The dead smaller branches do not make a good fuel as they burn quickly and make a lot of rather pungent smoke. Larger pieces make better firewood, especially if the bark is removed first. It burns with a good bright light so some is usually kept ready in the home for this purpose. The best fuel is provided by the stump. It makes less smoke, burns more slowly and is good for cooking. A piece of this wood is often half-buried in the ash to keep a fire in overnight, though it does make the room rather smoky.
Leatherwork. The wood burns down to a fine, pale ash, which is excellent for treating skins and hides in leatherworking.
Timber. Larger branches and trunks make a good, relatively termite-resistant, timber.
Wood. Good and hard and is much used to make the important herding staff and other implements. Dried stems are used as firesticks. The dead wood makes a reasonable charcoal. At a time when imported metal fishhooks were not readily available, the heartwood was used to fashion wooden ones.
Medicinal / veterinary. The sap is applied locally to treat burns, ulcers, boils and other sores. The juice extracted from crushed leaves is dripped into a painful ear to dislodge the tick called *śerb* which, if left in place, can cause considerable pain. The same medicament is used to treat a suppurating and painful ear, after cleaning it with a salt and water solution. The leaf juice is dripped into the nostrils of goats and sheep with a runny nose. Leafy twigs are gathered and fed to a cow losing weight towards the end of her pregnancy to help build up her strength.
Pest control. The wood burns down to a fine ash, which is a good insecticide. It is rubbed on the outside of goatskins used to store food such as cereals or dates.
Other. Animals feeding on this can go longer without water. Slender stems were whittled to a point and used to decorate the damp clay of pottery with lines, dots and wavy patterns before firing.
Notes. Islanders differentiate between the 'male tree (*ᶜʌ̍ləhʌl ḳʌ̍mses* or *ka̍rtab*) and the 'female', and say that the 'male' has smaller leaves than the 'female'. The significance and uses of *Trichocalyx* are very similar to those of *Ballochia* and *Anisotes*.

†Also apply to *T. orbiculatus*.

Trichocalyx orbiculatus Balf.f.

Straggling shrub to 2m. Leaves opposite or fasciculate, orbiculate, 1–2 x 0.8–2cm, subsessile, cystoliths prominent. Corolla red, fading to orange, <3cm long. Fruits c.1.5cm long.
Distribution. Endemic.
Conservation status. Data Deficient (DD). Part of a taxonomically difficult group on Soqotra. Inadequate evidence is available to make an assessment.
Botanical notes. Perhaps more drought tolerant than the preceding species. See comments under the generic description.

ETHNOGRAPHIC NOTES

Name(s)
See *T. obovatus* (also uses).

AIZOACEAE

Annual or perennial herbs, often succulent. Leaves opposite or alternate, simple, entire; stipules absent. Flowers regular, bisexual, axillary or terminal, solitary or in cymes or fascicles. Sepals 5, free or united below. Petals absent, often replaced by petaloid staminodes. Stamens 5 to many. Staminodes often present, sometimes showy and petaloid. Ovary superior or inferior, 1 to 5-celled, cells free or fused; placentation axile, parietal, basal or apical; ovules 1 to many per cell. Fruit a capsule opening by valves or transversely, or achenes, or splitting into 1-seeded mericarps. 114 genera. 2400 species. Tropical and subtropical, especially S Africa.

Botanical notes. A broad circumscription of the Aizoaceae has been adopted here; several genera are sometimes placed in segregate families (i.e. Gisekiaceae, Molluginaceae).

Aizoon L.

Annual or perennial herbs. Leaves alternate, succulent, petiolate. Flowers solitary in the leaf axils. Calyx 5-lobed. Petals absent. Stamens many, in 5 bundles, inserted on the calyx tube. Ovary superior, of (4 or)5 many-ovulate cells; styles (4 or)5. Fruit a many-seeded capsule, dehiscing by 5 valves. 11 species. Spain, Middle East, N, E and S Africa.

Aizoon canariense L.

Prostrate annual herb to 30cm, somewhat pilose. Leaves fleshy, orbicular to obovate, 10–35 x 4–20mm, apex rounded or bluntly acuminate; petioles 3–15mm. Flowers greenish or yellowish with a yellow centre, sessile. Calyx lobes triangular, 2–3.5 x c.2mm. Stamens 12 to 25. Fruit red or pink, (4 or)5-angled, depressed centrally, valves not recurving on dehiscence, 4–6mm diameter.
Ecology. Common on the coastal and inland plains where it prefers sandy soils in *Croton socotranus* shrubland and dwarf shrubland. 0–300m.
Distribution. Tropical and N Africa, tropical Asia to Pakistan.
Botanical notes. One of the commonest small herbs on the lowland plains of Soqotra. The small, pentagonal fruits persist on the dead plants and are sometimes the only evidence of herbaceous vegetation on the sandy plains when all other herbs have been burnt off by the sun.

ETHNOGRAPHIC NOTES
Name(s)
– Soqotri: *kibiˈdino, kibiˈdinəh* 'little liver', 'small hill(ock) of earth' (other small, compact plants have the same name).
– Soqotri simplified: *kibidinoh, kibidinih*.
Food (livestock). All livestock eat this when it is young and green. In drier areas it is regarded as a good feed which improves lactation. Generally it is important dry season forage in those poorly vegetated areas where there are few alternatives. Older plants and fruiting plants are less palatable. Cattle generally avoid more mature plants, but will feed on them in lowland grazings where there are few alternatives. If they eat too many they develop sores in the gut that are visible after slaughtering. However, this does not seem to effect the overall health of the animal. Small stock feeding heavily on older plants sometimes develop blockages in the gut which can be dangerous. In the outer islands this is one of the most important plants for sheep.
Comparative uses. The leaves, young branchlets and seeds are eaten in times of hunger. An extract of the leaves is used as a fish poison in the Cape Verde Islands. (Dalziel 1948, 1956; Kunkel 1984)

Corbichonia Scop.

Orygia Forssk.

Annual or perennial herbs. Leaves alternate, petiolate. Flowers in terminal or leaf-opposed cymes. Sepals 5, free. Petaloid staminodes many. Stamens many, in 2 rows. Ovary superior, of 5 many-ovulate cells; styles 5. Fruit a many-seeded, 5-valved capsule. 2 species. SW Africa, tropical Africa to Asia.

Botanical notes. Sometimes placed in the family Molluginaceae.

Corbichonia decumbens (Forssk.) Exell

Orygia decumbens Forssk.

Stems prostrate or decumbent, to 30cm, often pink-tinged. Leaves glaucous, subsucculent, elliptic to broadly obovate, 8–30 x 4–30mm, apiculate. Flowers 5 to 10 per cyme, opening late afternoon. Sepals green with a white membranous margin, c.4mm long. Staminodes pink or magenta, many becoming longer than the sepals, delicate, soon perishing. Fruit globose, 4–5mm diameter.
Ecology. Rare, found amongst rocks on the limestone. 30m.
Distribution. Drier parts of tropical Africa, Arabia, Iran, Pakistan and India.

ETHNOGRAPHIC NOTES
Name(s)
– Soqotri: nil or *ruˈʿud* 'grazing plant'; *raʿˈadihon* 'small grazing plant'.
– Soqotri simplified: *ru'ud*; *ra'adihon*.
Food (livestock). This is one of the *ru'ud* group of small plants which come up with the rains and provide incidental green grazing until they die back and the grasses appear. Though palatable to livestock, it is not regarded as an important forage for them.

Gisekia L.

Annual herbs, marked throughout with numerous linear raphides. Leaves opposite, subsucculent. Flowers in congested axillary fascicles, the fascicles sessile or pedunculate. Sepals 5, free. Petaloid staminodes absent. Stamens 5. Ovary superior, of 5 free, 1-ovulate cells. Fruit a cluster of 5 achenes. 7 species. Old World tropics.

Botanical notes. Often placed in the Phytolaccaceae and sometimes in a family of its own (Gisekiaceae).

Gisekia pharnaceoides L.

Glabrous herb; stems prostrate, to 30cm. Leaves subsessile, linear-oblong to narrowly oblanceolate or elliptic, 5–30 x 1–8mm. Flowers greenish or reddish. Sepals c.2mm long. Achenes c.1.25mm long, papillose.
Ecology. A rare weed of cultivation on the coastal plains. 30m.
Distribution. NE Africa, Arabia, Middle East.

ETHNOGRAPHIC NOTES
Name(s)
– Soqotri: nil or *ruˈʿud* 'grazing plant'; *raʿˈadihon* 'small grazing plant'.
– Soqotri simplified: *ru'ud*; *ra'adihon*.
Food (livestock). This is one of the *ru'ud* group of small plants which come up with the rains and provide incidental green grazing until they die back. Though palatable to livestock, it is not important forage for them.
Comparative uses. This plant often has a dark red tinge due to a fungus. In India it is sometimes gathered as fodder. It is eaten as a potherb in times of scarcity and is sold as a purgative. The crushed leaves are rubbed on swellings and are given as an anthelmintic. (Ambasta 1982; Dalziel 1948, 1956; Jain & De Fillips 1991)

Glinus L.

Annual herbs. Leaves opposite or verticillate. Flowers in axillary clusters. Sepals 5, free. Petals absent. Staminodes present. Stamens 11 to 30. Ovary superior, of (3 to)5 many-ovulate cells; styles (3 to)5. Fruit a (3 to)5-valved, many-seeded capsule. c.12 species. Warm regions.

Botanical notes. Often placed in the Molluginaceae.

Glinus lotoides L.
Mollugo hirta Thunb.

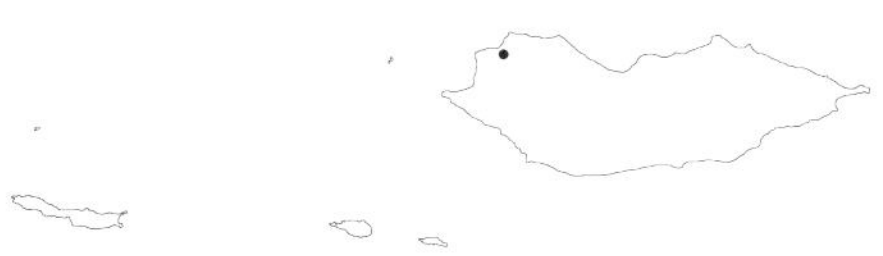

Annual herb, softly stellate-tomentose throughout; stems prostrate or decumbent, to 50cm. Leaves elliptic or broadly obovate to suborbicular, 5–30 x 3–20mm, apex obtuse or rounded, the margins entire or obscurely undulate, cuneate below; petioles 3–12mm. Flowers 2 to 10 per cluster, opening in the afternoon. Sepals greenish white, ovate, 4–8mm long. Staminodes white, c.15, linear, deeply bifid at the apex. Capsule oblong-ovoid, c.5 x 3mm diameter.
Ecology. A rare weed of cultivation on the coastal plains.
Distribution. Tropics and subtropics.

Botanical notes. Comes up after rain and disappears in the dry season.

ETHNOGRAPHIC NOTES
Name(s)
– Soqotri: nil or *ruˈʿud* 'grazing plant'; *raʿˈadihon* 'small grazing plant'.
– Soqotri simplified: *ru'ud*; *ra'adihon.*
Food (livestock). This is one of the small *ru'ud* group of plants which come up with the rains and provide incidental green grazing until they die back. Though palatable to livestock, it is not important forage for them.
Comparative uses. In India tender shoots are eaten as a potherb. The plant is also used to treat abdominal disorders. In Yemen the shoots are used to treat eczema. (Ambasta 1982; Dalziel 1948, 1956; Fleurentin & Pelt/Fleurentin *et al.* [1982, 1983a,b,c]; Heine *et al.* 1988; Jain & De Fillips 1991)

Mollugo L.

Glabrous annual herbs. Leaves opposite or verticillate or in a basal rosette. Flowers in dichasial cymes forming a leafless panicle. Sepals 5, free. Petaloid staminodes absent. Stamens 3 to 5. Ovary superior, of 3 to 5 many-ovulate cells; styles 3. Fruit a capsule opening by 3 valves. c.35 species. Tropical and warm regions.

Botanical notes. Often placed in the Molluginaceae.

Mollugo cerviana Ser.

Delicately branched annual herb with ascending stems to 12cm. Leaves sessile, glaucous; basal leaves forming a rosette, linear to spathulate, <10 x 3mm, soon withering; stem leaves linear, <15mm. Sepals greenish with white margins, 1–3mm, obtuse, persistent. Capsule subglobose, 1.5–2mm.
Ecology. Open, woody-herb communities (dominated by *Pulicaria stephanocarpa*) in the western inland desertic basin (Zahr basin). 180m.
Distribution. Widespread in the tropics and subtropics of the Old World.
Botanical notes. Rare on Soqotra.

ETHNOGRAPHIC NOTES
Name(s)
– Soqotri: nil or *rúʿud* 'grazing plant'; *raʿádihon* 'small grazing plant'; *źʌ́fḥan* (western Soqotra) *źʌ́fxan.*
– Soqotri simplified: *ru'ud*; *ra'adihon*; *dhaf'han, dhaf'khan.*
Food (livestock). This is one of the small *ru'ud* group of plants which come up with the rains, providing incidental green grazing until they die back. Though palatable to livestock, it is not important forage for them.

Mollugo nudicaulis Lam.

Delicate herb with erect or ascending stems to 25cm. Leaves in a basal rosette, obovate to oblanceolate, <40 x 10mm, tip rounded, attenuate below. Sepals white, 2–3mm long, persistent. Capsule subglobose to oblong-ovoid, 2.5–3mm diameter.
Ecology. Common and often rather weedy; in date gardens and *Croton socotranus* shrubland on the northern coastal plains. 10–200m.
Distribution. Pantropical.

ETHNOGRAPHIC NOTES
Name(s)
– Soqotri: *rúʿud* 'grazing plant'; *raʿádihon* 'small grazing plant'.
– Soqotri simplified: *ru'ud*; *ra'adihon*.
Food (livestock). This is one of the small *ru'ud* group of plants which come up with the rains, providing incidental green grazing until they die back. Though palatable to livestock, this is not an important forage for them.
Comparative uses. The leaves macerated overnight in water are taken with lime juice as a vermifuge. A decoction of the plant, taken cold, is held to be good for coughs. In India the whole plant is applied to boils to draw out the pus. (Dalziel 1948, 1956; Goodin & Northington 1979)

Tetragonia L.

Annual herbs. Leaves alternate. Flowers minute, paired in the leaf axils. Calyx 5-lobed, tubular below. Petaloid staminodes absent. Stamens 5, alternating with the calyx lobes. Ovary inferior, 1-celled; styles 3. Fruit dry, indehiscent, 1-seeded. 85 species. Tropical and warm regions, especially S Africa and N America.

Tetragonia pentandra Balf.f.

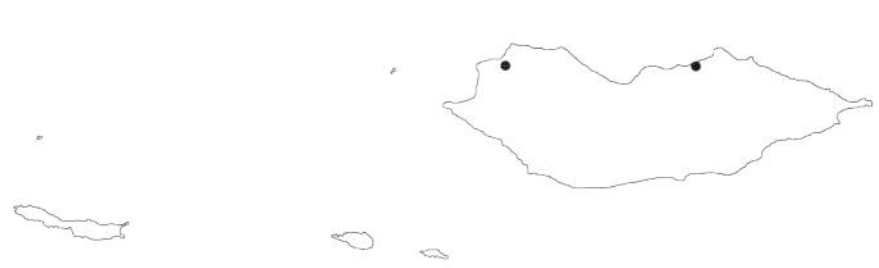

Stems prostrate to 60cm. Leaves broadly elliptic or ovate, 1–2 x 0.5–1.5cm, obtuse, contracted abruptly below; petiole 0.5–1cm. Calyx lobes triangular, c.0.5mm long. Fruit obconical, 5-angled, smooth, c.1.75mm diameter.
Ecology. No habitat details available but possibly in *Croton socotranus* shrubland on the coastal plain. 15m.
Distribution. Somalia.
Botanical notes. An inconspicuous and rather undistinguished herb discovered by Balfour near Galonsir [Qalansiyah] in 1880 and seen by Georg Popov near Hadiboh in 1957, but not seen since. Recently found in Somalia.

Zaleya Burm.f.

Annual or perennial herbs. Leaves opposite, subsucculent. Flowers subsessile in congested axillary clusters. Calyx 5-lobed, lobes with membranous margins. Stamens 5. Petaloid staminodes absent. Ovary superior, of two 2-ovulate cells; styles 2. Fruit a 4-seeded capsule, dehiscing by a 2-valved lid, the valves usually separating. 3 species. NE and E Africa, tropical Asia to Australia.

Zaleya pentandra (L.) C.Jeffrey

Trianthema pentandra L.

Prostrate annual or perennial herb to 30cm, glabrous or minutely papillose. Leaves elliptic to obovate, 10–30 x 3–20mm, obtuse; petioles 3–15mm, membranous-winged and sheathing at the base. Calyx lobes greenish or pinkish tinged with a white margin, c.2mm long. Capsule black, rectangular, 1.5–4 x 1.5–2mm.

Ecology. *Croton socotranus* shrubland and succulent shrubland, also a weed of date gardens. Sea-level–200m.

Distribution. Tropical Africa, Madagascar, Arabia and Palestine.

ETHNOGRAPHIC NOTES

Name(s)

– Soqotri: *ḥ'idis, ḥ'edist* 'secured, strengthened', '(animal) unwilling to leave the settlement to go out to graze' (also given to *Boerhavia diffusa*); (rare) *ˀ'ibəliṭ* (more commonly given to *Amaranthus* species).

– Soqotri simplified: *hidis, hedist*; *'ibelit.*

Food (human). In hunger the rooting system is dug up, cooked and eaten.

Food (livestock). Grazed mainly by small stock especially, though donkeys also like it. Hungry animals also dig up the plant to get at the roots. It is not regarded as particularly nutritious fodder.

Other. The leaves are used to repair the clay pottery of the islands. To mend a holed pot, a shard of pottery is ground down and shaped to fit the hole, the side appearing on the outer surface of the pot being broader than that of the inner. The leaves are pounded to a paste with a soft chalky stone called *maḥkāḳeh*, probably gypsum, and the paste is packed tightly around the inserted shard. The pot is then put beside the fire to harden and set the paste. If a pot is cracked, the same paste is smeared along and around the outer surface of the crack and the pot put beside the fire to set. The plant is also used in the decoration of pottery to make a dark, red-brown wash. The leaves are ground to a paste with a *Kalanchoe* plant, and a handful of the paste rubbed around the neck of the pot (usually the pot for storing ghee) when it is still hot from firing.

Notes. Used to be widespread in the northern plains and foothills, but is now much less common.

Comparative uses. In India the plant is burned to extract potash and to make a type of salt. The dried plant is pulverised and taken with millet beer for gonorrhoea. Ingested, it causes acute inflammation of the genito-urinary tract with haematuria, vomiting and bloody diarrhoea. It is capable of causing death by acute nephritis. (Dalziel 1948, 1956; Heine *et al.* 1988)

AMARANTHACEAE

Herbs or shrubs with simple, opposite or alternate leaves. Flowers unisexual or bisexual, in heads, spikes or panicles, usually subtended by a bract and 2 bracteoles sometimes also subtended by 1 or 2 sterile flowers modified into hooks or spines. Perianth of 3 to 5 free or fused tepals. Stamens (1 or 2) to 5 sometimes alternating with pseudostaminodes. Ovary superior, 1-celled with 1 to several ovules. Fruit a capsule dehiscing transversely or rupturing irregularly or a nutlet; seeds usually black or brown and shining. 71 genera. 800 species. Mainly warmer regions.

Achyranthes L.

Perennial herbs. Leaves opposite, petiolate. Flowers bisexual, in terminal and axillary spikes, erect at first but becoming deflexed and appressed to the axis in fruit; bracts and bracteoles spine-tipped. Tepals 5, free. Stamens alternating with oblong pseudostaminodes. Style filiform, stigma capitate. Fruit 1-seeded, indehiscent, falling with the bracteoles and perianth. 6 species. Tropical and subtropical Old World.

Achyranthes aspera L.

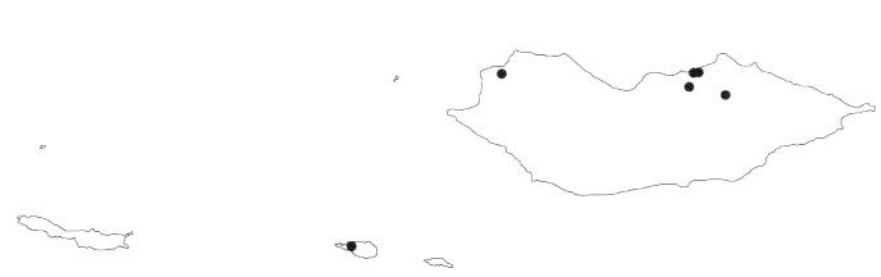

Sprawling herb to 1.5m. Leaves ovate to broadly obovate, <8 x 4.5cm, densely pubescent or glabrescent. Spikes <30cm. Tepals whitish, greenish or pink, lanceolate. Pseudostaminodes bearing a fimbriate scale. Fruit oblong-ovoid, 1–3mm diameter.

Ecology. Widespread and found in most vegetation types, sometimes rather weedy. 15–1300m.

Distribution. Tropical and warm temperate regions.

ETHNOGRAPHIC NOTES

Name(s)

– Soqotri: *mərišīˀo* (pl. *di m'iriš'itən, di mərišɛ̄tən*) 'of settlements'.

– Soqotri simplified: *mirishiy'oh* (pl. *di mirishitin, di mirisheytin*).

Food (livestock). Mainly grazed by cattle, but goats and donkeys also eat it. It is not seen as important nutritionally but, since it survives into the dry season, is a useful fodder then. It is not generally liked by herders because the spiny fruits cause sores in the teats, udder and hooves and this can lead to inflammation and cause the animal to develop a limp. If the sores on the teats or udder become infected, milk production can be affected.

Comparative uses. The young leaves are cooked as a vegetable. The ash is rich in potash and is of possible value as manure. They are used in the Middle East as a source of alkali in dyeing. Medicinally, a decoction of the herb is diuretic and is used to treat renal problems. In Oman the roots are crushed and applied to scorpion bites to reduce irritation and swelling, and in northern Sudan the root is chewed and used as a styptic on cuts. The seeds are emetic, and are also used in the treatment of leprosy. The stems are used to treat hydrophobia and snakebite as well as being used as chewsticks. An extract of the stembark is used as a contraceptive. Studies have shown the plant to possess hypotensive, cardiac depressant, vasodilating and respiratory analeptic actions. (Ambasta 1982; Bebawi & Neugebohrn 1991; Dalziel 1948, 1956; Ghazanfar 1994; Ghazanfar & Al-Sabahi 1993; Goodin & Northington 1979; Heine *et al.* 1988; Jain & De Fillips 1991; Lewis & Elwin-Lewis 1977; Oliver-Bever 1986)

Aerva Forssk.

Perennial herbs or subshrubs. Leaves alternate, rarely opposite. Flowers unisexual or bisexual, in terminal and axillary spikes which often form a terminal panicle. Tepals 5. Stamens alternating with oblong or triangular pseudostaminodes. Stigmas bilobed or capitate. Capsule 1-seeded, thin-walled, rupturing irregularly. 2 endemics. 10 species. Warm and tropical Old World.

Aerva javanica (Burm.f.) Schult.

Aerva persica (Burm.f.) Merr.

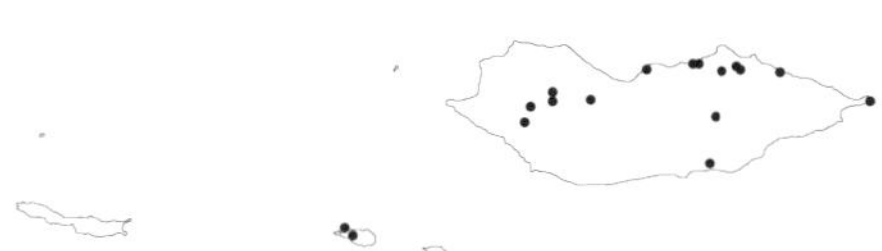

Woody-based herb to 50cm, densely whitish or yellowish tomentose throughout. Leaves linear-oblanceolate to broadly elliptic, <10 x 3cm. Flowers unisexual, in cylindrical spikes forming a leafless panicle. Female flowers with filiform stigmas. Male flowers with a rudimentary ovary.

Ecology. Common on coastal and inland plains and low hills, often in *Croton socotranus* shrubland and succulent shrubland; also a weed of date gardens. Sea-level–700m.

Distribution. Drier parts of the Old World tropics and subtropics from Africa east to Myanmar (Burma).

Botanical notes. Readily distinguished by its woolly spike-like inflorescences. A very common, variable and widespread species. Two varieties are found on Soqotra: var. *bovei* Webb differs from the commoner var. *javanica* in having narrower, linear-oblong leaves and more slender and often interrupted spikes. Male plants are apparently much rarer than female on Soqotra.

ETHNOGRAPHIC NOTES†

Name(s)

– Soqotri: *féʾ, fə́ʾ*.
– Soqotri simplified: *fe'*.

Food (livestock). Generally regarded as donkey and sheep fodder, but goats, camels and cattle also graze it when they are hungry or when the plants are young and green. In a prolonged dry season donkeys can be seen digging it up by the roots. It is not regarded as valuable nutritionally. On the outer islands it is important forage for goats and sheep, and for the few donkeys on Abd al Kuri.

Flavouring / food additive. The fat of the carcass of an animal that has eaten a lot of this becomes noticeably yellow. However, the flavour is unaffected.

Medicinal / veterinary. The roots are roasted and ground to a powder used to treat eyes which are painful and hot and/or itchy. To treat the condition known as *fiṣṣəh* in which there is corneal opacity and the eyes are sore and inflamed, the fresh root is peeled and then chewed until it froths up. The froth is then put in the sore eyes at repeated intervals until the condition clears up. The root juice is also used to treat any inflamed or painful eye. To treat sores on the scalp (a widespread problem, mainly the result of infected head lice bites), the hair is shaved off and the scalp washed. Then the roasted root is pounded with ghee and the paste rubbed all over the scalp. The same treatment is used to treat the scalps of children whose hair is not growing well, or people who have suffered acute localised hair loss. Other hair loss caused by poor nutrition, exhaustion or severe infestation of head lice is treated with the same remedy.

Bedding, stuffing. The flowering heads are used as padding and to stuff bolsters and saddle bags. It is said that only the 'female' plants have flowering heads suitable for stuffing, those of a 'male' plant being suitable only for livestock fodder.

Cosmetic. The roots are roasted at the fire and then ground with fat or ghee and smeared on the scalp. This not only acts as a dye to darken the hair, but also promotes hair growth. Women use this to make their hair black and lustrous, leaving the mixture on overnight and washing it out the next morning. In general, light-coloured hair is not considered attractive, and both men and women use this treatment to darken their hair colour. More recently, people have been experimenting with mixing the paste with imported henna or with a strong solution of tea.

Commercial value (past, present). The flowering heads are collected in sacks and offered for sale.

Comparative uses. The crushed plant of *A. javanica* and *A. tomentosa* is applied to wounds to stop bleeding. The juice extracted from the roots of these plants is used to treat eye complaints in cattle, goats and sheep. In Yemen the inflorescence is used as a hypnotic and narcotic. In Bahrain and Oman a paste of the flowers is used as a wound dressing and to stop bleeding. In Dhofar the roots are roasted and powdered as an antimony substitute. In Qatar the plant is used for toothache and a decoction is given to remove swellings. In India it is regarded as a diuretic and demulcent and is given to horses and camels as a purgative and emetic. Early Muslim authorities recommend *A. javanica* as a diuretic, an antidote to arsenic poisoning, and to disperse stones in the bladder. The roots of *A. tomentosa* are used as a chewstick and the powdered plant is applied locally to ulcers in livestock. (Abbas *et al.* 1992; Abdulfatih 1987; Ambasta 1982; Batanouny 1981; Bebawi & Neugebohrn 1991; Dalziel 1948, 1956; Fleurentin & Pelt/Fleurentin *et al.* [1982, 1983a,b,c]; Ghazanfar 1994; Goodin & Northington 1979; Jain & De Fillips 1991; Lewis & Elwin-Lewis 1977; Rizk & El-Ghazaly 1995)

†Also apply to *A. lanata*.

***Aerva lanata* (L.) Juss.**

Aerva lanata var. *robusta* Balf.f.

Woody-based herb to 60cm, lanate throughout. Leaves obovate or broadly elliptic to circular, <5 x 3cm. Flowers usually bisexual, in globular to cylindrical spikes, solitary or clustered in the leaf axils and forming a narrow leafy panicle. Style and stigma lobes short.

Ecology. Common on the plains, foothills, limestone plateau and at low altitudes on the granite foothills of the Haggeher; in succulent shrubland and drought-deciduous woodland. Sea-level–800m.

Distribution. Drier parts of the Old World tropics and subtropics from Africa east to Malaysia and the Philippines.

Botanical notes. Typically occurs at slightly higher altitudes than *A. javanica* from which it can be distinguished by its shorter flowering spikes which are arranged in the axils of upper leaves; in *A. javanica* the spikes are arranged in leafless, terminal panicles. An extremely variable species in which the infraspecific taxa recognised by various authors are impossible to distinguish on Soqotra (Townsend 1974).

ETHNOGRAPHIC NOTES

Name(s)

See *A. javanica* (also uses).

***Aerva microphylla* Moq.**

Aerva microphylla var. *humilis* Vierh.

Subshrub to 60cm, glabrous. Leaves fleshy, obovate, <15 x 7mm, lanate in axils but otherwise glabrous. Flowers bisexual, in short globular or cylindrical spikes, solitary or clustered at the ends of branches. Style short; stigma bifid.

Ecology. On cliffs and rock surfaces, often overlooking the sea; on Soqotra it is rarely found above 200m (350m above Ras Shu'ub) but on Abd al Kuri and Samha it occurs at higher altitudes (up to 650m). Sea-level–650m.

Distribution. Endemic.

Conservation status. Least Concern (LC). Common in several vegetation types and under no present or perceived threat.

Botanical notes. Dwarf plants with broader, relatively smaller leaves have been placed in var. *humilis* Vierh.; this, however, seems to be merely a grazed and stunted form.

ETHNOGRAPHIC NOTES

Name(s)

– Soqotri: *fə̍ʾ*; *di raḥmi̍miyun* 'grey-coloured'; *ḳʌ̍lḳəhʌl* 'swaying in the breeze' (general term for a variety of mainly halophytic plants); *ti̍lihin* 'remnants'; (*di*) *e̍mlʌḥ* 'saline'; (western Soqotra) *śe̍rxʌ erxāmməh* 'grey-coloured roots'; *ṣə̍ṣəb*; *tili̍lyə* 'encourages animals to follow'; *i̍rbib*, *irbi̍bo* '(animal) with a pale streak or flash on its coat' (Abd al Kuri); *tilili̍yə* (Samha).

– Soqotri simplified: *fe'*; *di rahmimiyun*; *qalqihal*; *tilihin*; (*di*) *emlah*; *sherkha erkhammih*; *zizib*; *tililyi*; *irbib*; *irbibo*, *tililiyih*.

Food (livestock). Not much liked by livestock, but grazed by donkeys and camels. Goats browse it only if very hungry. It is not regarded as useful fodder.

***Aerva revoluta* Balf.f.**

Subshrub to 1m, shortly tomentose. Leaves obovate, 1–2.5 x 0.6–1.5cm, margins revolute, glabrous or glabrescent above, densely tomentose beneath. Flowers bisexual, in short, sessile spikes forming a narrow leafless panicle. Style short; stigma bifid.

Ecology. Restricted to granite slopes in the Haggeher mountains where it is locally common in clearings in semi-deciduous and evergreen submontane woodland. (250–)500–950m.

Distribution. Endemic.

Conservation status. Least Concern (LC). Common in several vegetation types and under no present or perceived threat.

ETHNOGRAPHIC NOTES

Name(s)

– Soqotri: *fəʾ ḥagərhīyo* 'Haggeher *Aerva*'; *ḳe̍rəher* 'with a distinctive smell'; *šɛ̄də* (more commonly given to *Ruta chalapensis*); *chɛyɛ̍yaʾ* 'streaks, rays of light' (also given to the aromatic *Pluchea obovata*).

– Soqotri simplified: *fe' hagirhiyo*; *qeriher*; *sheydih; chayaya'*.

Food (livestock). Not normally liked by livestock, but eaten by cattle and goats when the dry season is severe or prolonged.

Amaranthus L.

Monoecious or dioecious annual herbs, glabrous or thinly pubescent. Leaves alternate. Flowers arranged in sessile clusters in the axils of the leaves or sometimes also in leafless terminal spikes or panicles; bracts and bracteoles scarious with a green midrib. Tepals 3 to 5, free, membranous. Stigmas 2 or 3; ovules 1. Fruit 1-seeded, dehiscing by a circumscissile lid or rupturing irregularly or indehiscent. Seeds black, shiny. c.60 species. Tropical and temperate.

ETHNOGRAPHIC NOTES

Comparative uses. The leafy parts of *Amaranthus* species are widely cooked and eaten; their protein- and starch-rich seeds are also eaten in hunger. They are also a good source of green feed for poultry. In Nepal, tender *Amaranthus* leaves are cooked and eaten to relieve diarrhoea. In Africa the plants are seen as an excellent fodder plant, and are called 'the milk plants'. The leaves are dried and used as a tobacco substitute. In Oman the leaves of *A. graecizans* are crushed and applied to scorpion stings, snakebites and itchy rashes, and it is also taken internally to treat generalised oedema, painful joints and as a remedy for swellings and inflammation, such as mastitis. Some *Amaranthus* species, however, such as *A. palmeri* and *A. retroflexus*, accumulate high concentrations of nitrates, especially when growing on heavily fertilised soil, and have been implicated in cattle and horse poisoning. (Ambasta 1982; Batanouny 1981; Ghazanfar 1994; Heine *et al.* 1988; Jain & De Fillips 1991; Kunkel 1984; Lewis & Elwin-Lewis 1977; Miller & Morris 1988; Rizk & El-Ghazaly 1995)

Amaranthus dubius Thell.

Erect herb to 50cm. Leaves ovate to ovate rhombate, <5 x 2cm, blunt or obtuse. Flowers in terminal and axillary spikes, the male mainly at the tips. Bracts and bracteoles ovate, bristle-tipped. Female tepals 5, narrowly oblong to obovate, <5mm, acute or obtuse, mucronate. Stigmas 3. Capsule ovoid, with an inflated beak, dehiscent, the lid wrinkled.

Ecology. A weed of date gardens and around habitation. Sea-level–450m.

Distribution. Tropical regions.

ETHNOGRAPHIC NOTES†

Name(s)

– Soqotri: *ʾibəliṭ* (Soqotra and Samha); nil or *ertērit* 'forming green patches on the ground' (Abd al Kuri).

– Soqotri simplified: *'ibelit*; *erteyrit.*

Food (human). The Amaranthaceae are among the most important edible plants of the Archipelago. They respond quickly to rain and rapidly colonise disturbed ground. They are common around human and livestock living quarters. The plants are gathered in bundles and taken home for cooking. They must not be carelessly uprooted, but the leafy stems cut or twisted off the stem, in order to allow the plant to regrow (which it does readily). The leafy parts are broken up and put into a clay cooking pot. Additional water is not needed but some people prefer to add a small amount. If salt is available, some is added to the pot. A stone is placed on top of the foliage to press the plants down into the pot, and embers are piled around the outside. Like spinach, the leaves cook down considerably. Some eat the leaves in this form, straight from the container, others add the greens to the main cereal dish. Men say that it is principally women and children who gather and cook this, but women say men are just as keen to eat it when it has been prepared for them. Some islanders say that not all *Amaranthus* are equally palatable, and that the *ibəlit* 'with large leaves' (called *ʾibəliṭ ṣʌʿʌlhʌl*, 'large-leaved', possibly *A. dubius*) is not eaten, even though it too is commonly found growing around human and livestock quarters. Research has shown the plants to be a good source of iron, calcium, vitamin C, folic acid and other minerals and micronutrients.

Food (livestock). Regarded as excellent fodder for cattle, though as less beneficial to sheep and goats. On the outer islands *A.* subsp. *thellunganianus* is an important fodder for goats and is also grazed by sheep.

Flavouring / food additive. The milk of animals feeding on the plants has a distinctive and sweet flavour.

Other. More recently some herders have started to scatter seeds around their settlements as fodder for household goats.

Notes. Can cause *'uqabah* in goats unaccustomed to the feed. The widespread use of this plant is reflected in the verb based on its name: *ʾibēlṭin tɛwɛ*, 'he is gathering *ʾibəliṭ* for food', and *zərf di ʾibāliṭ*, 'the clay pot for cooking *ibəlit*'.

†Also apply to *A. graecizans*, *A. sparganiocephalus*.

Amaranthus graecizans L.

Amaranthus polygamus sensu Blatter non. L.

Herb to 45cm. Leaves linear-lanceolate, ovate or obovate, <4.5 x 2.5cm. Flowers in axillary clusters, male and female mixed but male commonest in the upper clusters. Bracts and bracteoles narrowly ovate-oblong, bristle-tipped. Female tepals 3, narrowly oblong to narrowly ovate, <2mm, mucronate or shortly aristate. Stigmas 3. Capsule subglobose to ovoid, 1.5–2mm diameter, usually strongly wrinkled.

Distribution. Warm temperate and tropical Old World.

Botanical notes. Two subspecies are recorded from Soqotra.

ETHNOGRAPHIC NOTES

Name(s)

See *A. dubius* (also uses).

Subsp. *graecizans* L.
Amaranthus blitum sensu Balf.f. non L.

Ecology. Common on the northern coastal and inland plains and low hills; often in *Croton socotranus* shrubland; also a weed of date gardens and found around habitations. Sea-level–700m.

Subsp. *thellunganianus* (Nevski) Gusev

Ecology. In sandy and coral soil on the coastal plains of Samha and Abd al Kuri, particularly near habitation. Sea-level–100m.

Distribution. Tropical Africa, S Arabia and India.
Botanical notes. Distinguished from subsp. *graecizans* by its smaller and narrower leaves and the fine, hair-like tips to the tepals. Subspecies *thellunganianus* is unaccountably absent from Soqotra. This parallels its tendency in southern Arabia to prefer small islands: in Oman it is absent from the mainland but occurs on the Kuria Muria Islands.

***Amaranthus sparganiocephalus* Thell.**

Erect herb to 60cm. Leaves ovate to oblong-ovate, <4 x 3cm. Flowers in compact axillary clusters, male and female throughout. Bracts and bracteoles small, oblong-ovate, acute. Female tepals 3, elliptic-oblong to oblong-obovate, <1.25mm, apiculate. Stigmas 2. Fruiting clusters stellate; capsule somewhat woody, ovoid, conical above, <4mm diameter, longitudinally ridged.
Ecology. No habitat details available. Probably a weed of date gardens and near habitation.
Distribution. S Arabia, E and NE tropical Africa.

ETHNOGRAPHIC NOTES
Name(s)
See *A. dubius* (also uses).

Digera Forssk.

Annual herbs. Leaves alternate. Flowers bisexual, in long axillary spikes, each subtended by 2 antler-shaped sterile flowers. Tepals (4 or) 5, the two outer stiff and strongly nerved, the inner membranous. Style filiform; stigma bilobed. Fruit a nutlet, enclosed by the persistent perianth. 1 species. Old World tropics.

***Digera muricata* (L.) Mart.**
Digera arvensis Forssk.

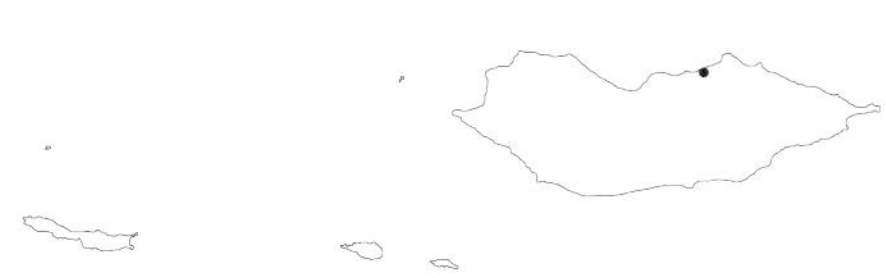

Erect herb to 45cm, glabrous or thinly pubescent. Leaves ovate to oblong-elliptic, 2–7 x 1–4cm. Flowers pink. Outer tepals ovate, 3–4mm long, 3 to 12-nerved; inner tepals slightly shorter, blunt. Fruit subglobose, c.2mm diameter.
Ecology. Uncommon on coastal and inland plains and low hills; in *Croton socotranus* shrubland and succulent shrubland; also a weed of date gardens. Sea-level–50m.
Distribution. S Arabia, tropical Africa, Madagascar, S Asia to Indonesia.
Botanical notes. This rather weedy plant is uncommon on Soqotra and represented by few collections, but amongst these three subspecific taxa have been recorded: subsp. *muricata*, subsp. *trinervis* var. *trinervis* C.C.Towns. (recorded from Soqotra but no authenticated material has been seen) and subsp. *trinervis* var. *patentipilosa* C.C.Towns. These are all similar in appearance and differ only in rather small and technical (but nevertheless easily observed) characters.

ETHNOGRAPHIC NOTES
Name(s)
– Soqotri: *di ḳe̍rəhān* 'horned (fruit)' (given to other plants with podded fruits).
– Soqotri simplified: *di qerihan.*
Food (livestock). Liked by all livestock, but especially cattle. They are regarded as a valuable and nutritious fodder.
Comparative uses. *D. muricata* is eaten as a potherb. The leaves are used to treat indigestion and the flowers and seeds are prescribed for urinary discharges. (Bebawi & Neugebohrn 1991; Goodin & Northington 1979; Heine *et al.* 1988; Kunkel 1984)

Gomphrena L.

Annual herbs. Leaves opposite. Flowers bisexual, in globose or spicate terminal heads; bracteoles keeled and strongly laterally compressed, midrib crested. Tepals 5, free. Stamens ?, filaments united into a tube; tube toothed at apex. Style long, stigmas 2. Fruit 1-seeded, indehiscent, compressed. 100 species. SE Asia, Australia, tropical New World.

Gomphrena celosioides Mart.

Prostrate herb to 30cm. Leaves oblong to oblong-elliptic, <3 x 1cm, glabrous to densely lanate. Flower head globose, becoming cylindrical in fruit, <7 x c.1.25cm, white. Bracteoles 5–6mm long, mucronate; crest small, irregularly serrulate, confined to the upper third. Fruit ovoid, c.1.5mm diameter.

Ecology. Disturbed ground by tracks and in semi-deciduous woodland (with *Buxanthus pedicellatus, Rhus thyrsiflora* and *Dracaena cinnabari*). 500–700m.

Distribution. Native of S America now widely distributed in the tropics and subtropics.

Botanical notes. An introduced weed, first recorded from Soqotra in 1989 and possibly on the increase.

ETHNOGRAPHIC NOTES

Name(s)

– Soqotri: nil or *dīrbeb* '(animal) with a white streak or flash on its coat'.

– Soqotri simplified: *diyrbeb.*

Food (livestock). Not liked by livestock.

Comparative uses. *G. globosa* is cooked and eaten as a vegetable. (Ambasta 1982; Jain & De Fillips 1991)

Psilotrichum Blume

Psilostachys Hochst.

Perennial herbs or shrubs. Leaves opposite, sometimes reduced to scales. Flowers bisexual, in lax or compact spikes, often forming open panicles. Tepals 5, free. Style slender; stigma capitate. Capsule 1-seeded, thin-walled, rupturing irregularly. 1 endemic. 15 species. Tropical Old World, especially Africa.

Psilotrichum aphyllum C.C.Towns.

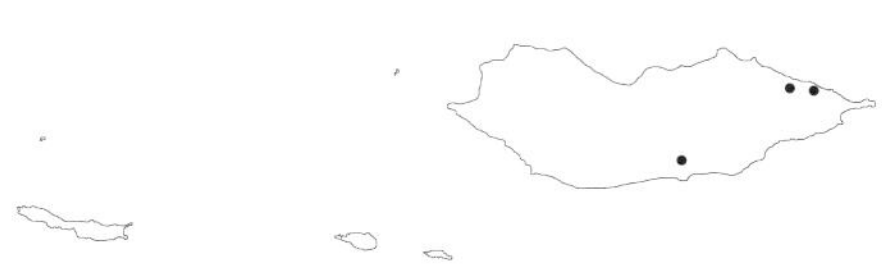

Aphyllous shrub with intricately branched stems, appressed-hairy or glabrescent. Leaves, when present, scale-like, c.4mm long. Flowers in short, compact, <6-flowered spikes.

Ecology. Limestone cliffs in drought-deciduous woodland. 15–550m.

Distribution. Endemic.

Conservation status. Endangered (EN B2 a biii). Only known from two locations on limestone cliffs in the northwest of Soqotra where it occupies an area of less than $50km^2$. It is strongly suspected that these cliffs are threatened by the long-term trend of lower rainfall in the region and would be particularly vulnerable to the effects of global warming.

Botanical notes. First recorded by Georg Popov in 1953 from cliffs, just above sea-level, on the Noged plain, but not seen there recently.

ETHNOGRAPHIC NOTES

Name(s)

– Soqotri: nil or *gēśīyo* 'of the cliffs' (*Ochradenus arabicus* of cliffs'); *ləskáno* 'sticky, clinging' (Samha).

– Soqotri simplified: *geyshiyoh*; *liskano.*

Food (livestock). Goats are said to be very keen on this if they can reach it.

Psilotrichum sericeum (Roxb.) Dalzell

Psilostachys sericea (Roxb.) Hook.f.

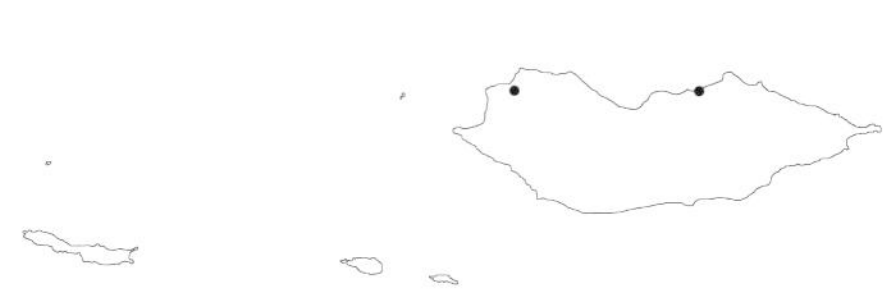

Erect or scrambling, much-branched herb to 2m, glabrescent, with tufts of silky hairs at the nodes. Leaves ovate, <4 x 2cm, silvery-sericeous beneath. Flowers in lax spikes forming open panicles, the ultimate branches zigzagging. Tepals green; outer 3-nerved; inner with broad membranous margins.

Ecology. Rare herb of stony ground on the northern coastal plain. Sea-level–50m.

Distribution. Kenya, Tanzania, Somalia and India.

ETHNOGRAPHIC NOTES

Name(s)

– Soqotri: nil or *rúʿud* 'grazing plant'; *ləskáno* 'sticky, clinging' (Samha).

– Soqotri simplified: *ru'ud*; *liskanoh.*

Food (livestock). This is one of the small, ephemeral *ru'ud* plants that come up with the rains, providing incidental green grazing until they die back. Though palatable to livestock, it is not important forage for them.

Pupalia Juss.

Herbs. Leaves opposite. Flowers in clusters along an elongated terminal spike; clusters consisting of 3 bisexual flowers surrounded by sharply hooked spines derived from sterile flowers. Tepals 5. Style slender; stigma capitate. Fruit 1-seeded, rupturing irregularly, shed as a highly adhesive burr. 4 species. Tropical Old World.

Pupalia lappacea (L.) Juss.

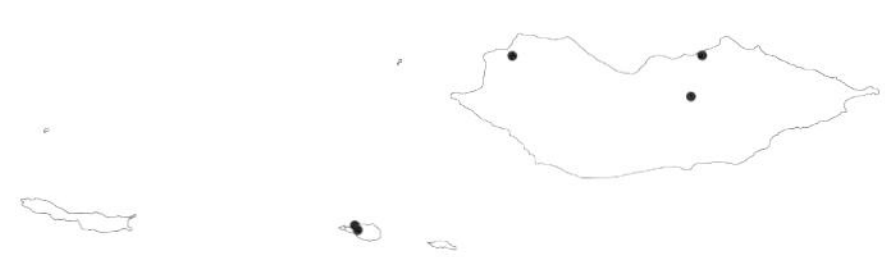

Sprawling herb to 1.5m. Leaves ovate to elliptic or circular, 1–6 x 1–3cm, sericeous or glabrescent. Tepals oblong-ovate, 4–5mm long, lanate on back. Fruiting burr 7–13mm diameter, spines <6mm.

Ecology. Often abundant on rocky slopes on coastal and inland plains and low hills; often in *Croton socotranus* shrubland and succulent shrubland; also a weed of date gardens. 10–700m.

Distribution. Tropical Old World.

Botanical notes. Readily distinguished by the dispersal units or diaspores which are burr-like and consist of a cluster of small fruits covered with sharply hooked spines (derived from sterile flowers). These diaspores are highly adhesive and readily attach themselves to livestock or clothing from which they can be extremely difficult to remove. All Soqotran material is referable to var. *velutina* (Moq.) Hook.f. characterised by its lanate tepals and sterile flowers with straw-coloured spines.

ETHNOGRAPHIC NOTES

Name(s)

– Soqotri: *ləsˈkano* 'clinging, sticky' (Soqotra and Abd al Kuri); *ḥˈʌźərʌt* 'greenery'; *dˈi məriśēˈtən, di mˈiriśˈitən* 'of settlements'; *di śkaˁanˈitin* 'prickly, spiny'.

– Soqotri simplified: *liskano*; *hadhirat*; *di mirisheytin*, *di mirishitin*; *di shka'anitin.*

Food (livestock). Eaten by all livestock and, when still green and relatively young, it is considered a nutritious feed. Mature, the plants are not liked by goatherds as the hooked spines of the burrs can cause sores on the teats and udder of livestock, which if they become inflamed can affect lactation. Lactating, thick-haired goats often have the hair around the udder shaved off to prevent this. The burrs can also become wedged in between the 'claws' of the hoof, sometimes causing lameness.

Medicinal / veterinary. The burrs are used to scarify the skin, especially in cases of ringworm. Once the affected area has been scratched until it bleeds further treatment is carried out (with an application of rock salt, extract of *Euclea divinorum* root-bark, *Cryptolepis* latex or crushed tobacco). Milk goats feeding in areas where this plant is abundant must be inspected at each milking for scratches on the teats and udder, and if found these must immediately be treated with ghee, honey (or nowadays face or hair cream), and the animal left until last to be milked.

Notes. Regarded as one of the most irritating plants on the island: the burrs catch on clothing and scratch the skin. To stop the burrs getting caught, goats with particularly long-haired thighs have much of this hair cut off if these plants are abundant. Scratches from the burrs may develop into irritating sores that can attract blowflies and become infected.

Comparative uses. In Uganda the burrs are used as lids for gourds, and in Kenya they are stuck together and used as strainers for milk. They are also pressed into wounds and withdrawn to remove maggots. In Nigeria they are an ingredient in rat poison and elsewhere they are sold for use in scarifying the skin before blood-letting. They are used as a cough medicine given in soup and applied locally for cuts, and with added oil are used as a dressing for boils and so on. The ashes of the burned plant are drunk with water for flatulence in Sudan and are applied to leprosy sores after making them bleed. (Ambasta 1982; Dalziel 1948, 1956; Heine *et al.* 1988; Jain & De Fillips 1991)

ANACARDIACEAE

Trees or shrubs. Leaves alternate, 3-foliolate or simple; stipules absent. Dioecious (sometimes bisexual flowers present on male plants). Inflorescence paniculate or spike-like. Flowers small, actinomorphic, petals 4 or 5, stamens inserted onto a ring-like disc. Ovary superior. Fruit a drupe. 60 genera. c.600 species. Mostly tropical but also Mediterranean, E Asia and America.

Lannea A.Rich.

Odina Roxb.

Trees. Stellate hairs usually present, especially on young plants. Dioecious, sometimes bisexual flowers present on male plants. Leaves alternate, simple. Flowers actinomorphic, often produced before leaves in spike-like inflorescences. Petals 4, styles 4, free. Fruit a drupe with 4 prominent style bases. 1 endemic. 40 species. Mostly tropical Africa.

Lannea transulta (Balf.f.) Radcl.-Sm.
Elaeocarpus transultus Balf.f.
Odina asplenifolia Balf.f.

Tree to 7(–15)m. Bark slightly flaky or smooth. Leaves papery, lanceolate, 6–20 x 1.5–10cm, base acute to obtuse, entire, tip acute. Flowers small. Fruit ovoid, 0.8–1.3 x 0.4–0.6cm.

Ecology. Succulent shrubland and drought-deciduous woodland, particularly on limestone escarpments. 50–650m.

Distribution. Endemic.

Conservation status. Near Threatened (NT). Whilst abundant and reasonably widespread at present, *L. transulta* is lopped as livestock fodder in dry periods. If livestock numbers increase greatly, or a succession of drought years occur, then this species will come under increasing threat.

Botanical notes. One of the largest trees on the island and a characteristic emergent in succulent shrubland on the limestone escarpments. It is deciduous, the leaves dropping in summer and then appearing at the outset of the winter rains; it flowers at the end of the summer monsoon rains.

ETHNOGRAPHIC NOTES

Name(s)

– Soqotri: *kˈənəhor* (pl. *iknīrə*) 'dangerous' (sometimes also given to *Turrea socotrana*, and, in western Soqotra, to *Euclea divinorum*); edible root of young plant *kənˈirhin* (pl. *kinˈerihon*); *bˈilbil*.

– Soqotri simplified: *kinihor* (pl. *ikniyrih*); edible root of young plant *kinirhin* (pl. *kinerihon*); *bilbil*.

Food (human). The fruits are eaten by birds and goats, but not humans. In some areas sections of root or a sapling root are chewed for liquid refreshment.

Food (livestock). An important dry season and drought fodder for goats. Both leafless twigs and branch tips are lopped. Fallen foliage is also gathered. Dried foliage is fed to cows during a prolonged dry season. When fed in adequate amounts it promotes lactation in both goats and cows. Fresh, the foliage is not regarded as a nutritious feed and livestock show little interest in it. If they can reach them goats will chew the roots and bark. In areas with more nutritious alternatives (as the interior valleys) it is little valued as a livestock feed.

Fuel. The dead wood makes adequate firewood.

Timber. A well-grown tree is sometimes referred to as *kˈənəhor ḥˈahər* ('black *Lannea*') inferring that it provides good building timber. It is relatively termite resistant after de-barking, and is seen as good for roofing. The wood of *L. transulta* from wetter areas, however, is soft and prone to cracking and splitting.

Leatherwork. Where better alternatives are absent, the inner bark and growing stems can be crushed and used to treat skins and hides to strengthen the leather.

Wood. Dead wood which is completely dry can be used as a firestick.

Medicinal / veterinary. Quicklime made from the wood is used medicinally for both livestock and people. It is ground to a fine powder and applied externally to open wounds and the surgical wounds of circumcision to staunch the bleeding. The powder is also sprinkled on weeping sores and abrasions to dry them out. In the absence of bark from one of the *Commiphora* species, the inner bark or bast can be used to make the *ˈirziz* medicament (see *Commiphora*).

Notes. The foliage can cause severe scour in goats unaccustomed it and in some cases this is fatal. Well-developed trees are appreciated for the shade they offer and are frequently used as meeting places for people, and as noon shelters for livestock. The tree features in a curse: 'May you come up against a *Lannea*!' (*itˈe kˈənəhor*!).

Comparative uses. Many species of *Lannea* have edible fruit, and one in Uganda, *L. humilis*, an edible tuberous root. Some half dozen species are also important medicinally. In nearby Dhofar, the green fruit of the closely related *L. triphylla* are well liked by pregnant women for their sourness, and pieces of the root and inner bark of young saplings were eaten as a food and to quench thirst. Species such as *L. coromandelica* in India provide good wood which is used in construction and to make furniture and implements. The wood of *L. humilis* is used in Uganda to make stools, water troughs, honey containers, milk vessels and so on. The bark of some species yields paper pulp, and that of *L. humilis* is chewed to soften it and then used to make cordage. In East Africa, the seeds of *L. alata* are eaten raw or cooked, and the white interior of the bark is added to milk and drunk for strength. (Heine *et al.* 1988; Kunkel 1984; Miller & Morris 1988)

Rhus L.

Trees or shrubs. Dioecious. Leaves 3-foliolate (or simple). Flowers very small, arranged in panicles. Petals 5, styles 3, free. Fruit a drupe, style bases absent. 2 endemics. c.200 species. Temperate and warm regions.

Rhus sp. *A*

Similar to *R. thyrsiflora* but leaves always 3-foliolate, thinly puberulous or glabrescent; leaflets ovate (to deltoid), 1–2 x 0.8–1.5cm, base obtuse, margin inrolled. Flowers in short (2–3cm) little-branched panicles. Fruit not wrinkled when dry.

Ecology. Occasional on cliffs and rocky slopes along the sides of the large wadis draining south into the Noged plain; in succulent shrubland and drought-deciduous woodland (with *Buxanthus pedicellatus* and *Euclea divinorum*). 30–600m.

Distribution. Endemic.

Conservation status. Vulnerable (VU D2). Rare, with a very limited area of occupancy.

Botanical notes. The leaves of this undescribed new species of *Rhus*, which are always trifoliate, are bright green, smaller and not as glossy as those of *R. thyrsiflora*. It comes into new leaf at the onset of rain and drops its leaves in summer. The flowers appear in winter and are in little-branched panicles which are shorter than the leaves; in *R. thyrsiflora* the flowers are in open panicles which greatly exceed the leaves. Its fruits, which are similar to those of *R. thyrsiflora*, are said to be sweeter and, unlike those of *R. thyrsiflora*, are not wrinkled when dry.

ETHNOGRAPHIC NOTES

Name(s)

– Soqotri: *ḥasəlām*, *ḥasə̍ləham* (pl. *ḥa̍səlu̍m*, *ḥaslo̍mo*) ‘thin, slim’; *di raḥmi̍mihun di tɛ̄tin* ‘sheep’s grey plant’, ‘sheep’s consolation’; (rare in Haggeher) *ʌ̍mlʌḥ* ‘salty’.
– Soqotri simplified: *hasilam*, *hasiliham* (pl. *hasilum*, *haslomo*); *di rahmimihun di tetin*; *amlah*.

Food (human). As *R. thyrsiflora*, but the fruit are said to be sweeter when fully ripe.

Food (livestock). This plant is an important feed in the drier areas where it occurs.

Notes. See *R. thyrsiflora* (also additional uses).

Rhus thyrsiflora Balf.f.

Shrub or small tree to 5m. Leaves 3-foliolate, sometimes simple but not both on the same tree, glabrous, veins characteristically pale beneath; leaflets subequal, elliptic to oblong-obovate, 3–8 x 1–4cm, base acute or attenuate, entire, tip rounded. Flowers pale yellow or cream, c.1.5–2mm diameter, arranged in axillary, many-flowered panicles. Fruit ripening red, subglobose, wrinkled when dry.

Ecology. A major component of semi-deciduous woodland on the limestone escarpments and the semi-deciduous and evergreen woodlands of the granite of the Haggeher mountains. (220–)400–1220m.

Distribution. Endemic.

Conservation status. Least Concern (LC). Common in several vegetation types and under no present or perceived threat.

Botanical notes. The commonest and dominant tree over large areas of the Haggeher mountains. Its leaves drop in summer and it comes into leaf in winter, even if there is no rain. In some areas it seems to be virtually evergreen. It flowers in the winter and comes into fruit in the summer. *R. thyrsiflora* has an interesting distribution on Soqotra. It is common in areas which attract low cloud and rain in the Haggeher and the limestone of the north-east of the island (Hoq and Hamaderoh), but is absent from the high rainfall areas on Ma’alah in the west. However, there is a small population on the remote limestone ridge at Heger above Ras Shu’ub. The limestone pinnacles along this ridge trap mist and support a small outlying relict of semi-deciduous woodland in the otherwise dry southwest of the island. It is interesting that on this ridge *R. thyrsiflora* occurs at only 220m – its lowest recorded altitude on the island. It is possible that the long isolation of this remote refuge has resulted in the telescoping of habitats. The phenomenon of telescoping is found on old oceanic islands and has the effect of compressing habitats into a smaller altitudinal range than would be expected – it thus effectively provides islands with a relatively greater range of habitats than similar areas on the mainland.

ETHNOGRAPHIC NOTES

Name(s)

– Soqotri: *zūrif* (pl. *zīrʌf*) ‘growing thin, losing weight’.
– Soqotri simplified: *zewrif* (pl. *ziyraf*).

Food (human). This widespread plant is often to be found growing beside the walls that criss-cross the plateau areas of the island. Its fruit was formerly an important component of the island diet. Both the green fruit, with their sour, lemony taste, and the sweeter red fruit were eaten. The fruit from the high, dry plateau areas is regarded as having a better flavour than that from the better-watered regions. The immature fruits are also roasted to make miniature popcorn and in this form are less astringent than the raw fruit. They are gathered by plucking the fructescences and rubbing them hard between the palms to detach the fruit from the stalks. Alternatively, they are picked and spread out on a cloth which is then folded over and rubbed to separate the fruit from the stalks. In good years the fruit can be found throughout the summer and monsoon seasons, a time when other foods are scarce.

Food (livestock). In the Haggeher mountains, where the plant is extremely common, the foliage is a key dry season fodder for both cattle and goats. Elsewhere, the leafy stems are lopped and the fallen foliage gathered and taken to the animals throughout the worst of the dry season. In a prolonged dry season cattle as well as goats tear off and gnaw at the leafless twigs, and in drought chew off the bark. Outside the dry season, camels, cattle and goats browse the young leaves when they have exhausted other more palatable forage, but at this time it is not seen as a very useful feed. Sheep eat fallen leaves throughout the year.

Flavouring / food additive. The foliage and bark can be used as a salt substitute. They are added to cold water before the food is added, and removed once the water has boiled.

Fuel. After de-barking, it makes a good fuel for cooking as it produces little smoke. If the bark is left on, it crackles and throws up sparks, so care must be taken when using it as firewood. It is a hard wood that burns slowly and a big chunk is often used to keep the fire in overnight. It is considered one of the best firewoods of the island for the intense heat and bright light it produces. It can also be used in wet weather, as it burns well even when slightly damp.

Timber. The wood is relatively termite resistant, but is not considered to be a particularly good timber. The shrubs rarely grow to a size useful for building. The branches are used in the roofing of temporary dwellings, especially the summer shelters of the high mountains.

Leatherwork. The inner bark is ground to a paste used to treat skins and hides to strengthen the leather. The skin of a dead calf is made into a tulchan and is rubbed all over with the same mixture. It is then presented to the cow, and some of her milk is sprayed over it. She is given it to sniff to encourage her to let down her milk. The paste is also rubbed onto a female calf to disguise its smell and persuade a cow whose own calf has been slaughtered to adopt her. The calf is presented to the cow and some of her milk is sprayed over it to encourage her to lick it and allow it to suckle.

The wood burns down to a fine ash which is collected and used in the preparation of skins and hides for leatherwork.

Wood. The wood is hard and is used to make a variety of implements. Where a bush grows tall enough, a slim branch is used to make the important herding stick. The dead wood also makes good charcoal.

Medicinal / veterinary. The fruit is given to someone suffering from anorexia, and as a general tonic. The sour, green fruit are commonly sought by pregnant women. Painful joints are treated by placing the sufferer in a small enclosed space, such as a small cave, at the mouth of which a fire of green *R. thyrsiflora* wood is lit. The smoke and the sweat brought on by the intense heat are both considered to be beneficial.

Cleansing. People gather the heavy dew that collects on the leaves for washing.

Pest control. The green wood and the bark are burned for livestock smudges. These give off a lot of thick smoke, which is beneficial both as an insecticide and to provide warmth during the winter months and during wet weather. In the areas affected by monsoon mists and drizzle, smudges of both dead and live wood are lit daily to warm livestock, and to protect them from the biting flies which proliferate during the rains.

Other. The wood of a well-grown shrub is considered to be one of the best for firing pottery. Good soil for cultivation can be gathered from beneath the bushes. When clearing new areas for cultivation the bushes are cut down and the brushwood heaped in piles over the stumps and set alight. The ash is considered to be a good fertiliser.

Notes. In the Central Highlands, where this plant grows profusely, it is often cut off at ground level from around the most frequented pathways so that the bushes do not tear at clothing, or, after dew or rain, soak those passing. The shrubs are also removed from milking yards, and from calf pens and paddocks. It is said that this is because grass will not grow under these shrubs, even for some time after they have been cut down. For this reason they are also often removed from enclosed grazing areas generally, and from around homesteads, byres and other areas where livestock gather. In the large livestock enclosures of the central highlands they are left untouched, as they offer shelter from wind and rain.

Comparative uses. Many *Rhus* species have edible fruit and leaves. Their leaves are also used as a tobacco substitute. *Rhus* are widely used to tan and prepare leather, and the same astringent and tannic qualities of the plant are used medicinally. *Rhus* twigs are used as toothsticks and to make mouthwashes, and the fruit can be used to curdle milk. Elsewhere in the Middle East other species of *Rhus* are used as antiseptic, astringent and haemostatic. A black dye for dyeing wool is made from the berries, and the ripe berries are taken as a febrifuge and diuretic. A paste is made from them to treat ringworm and other skin diseases. Tisanes made from the fruit of *R. coriaria* are given to treat diarrhoea, and used as a gargle to treat tooth and gum pain. The fruits are crushed, soaked and applied to swellings, or are cooked in water and added to cooked egg to treat stomach aches. In Africa *R. natalensis* is an important fodder plant for all livestock. The leaves are chewed to relieve fever, or are ground, added to water, and taken for diarrhoea. The bark is ground, soaked in water and taken for malaria. The root is boiled, with or without leaves, and drunk with milk, or used to wash the body after childbirth, as well as to treat measles. In India the galls on the leaves of *R. chinensis* are used for high quality tanning, and, medicinally, to treat dysentery, colic and diarrhoea. A vegetable wax is extracted from the fruit and used to make candles. Some species, e.g. *R. succedanea*, *R. verniciflua* and *R. wallichii*, exude a substance which is used as a varnish and in lacquerwork, to waterproof wood and to coat leather. (Ambasta 1982; Dafni *et al.* 1984; Heine *et al.* 1988; Jain & De Fillips 1991; Kunkel 1984; Lewis & Elwin-Lewis 1977; Miller & Morris 1988)

APOCYNACEAE

Shrubs or trees. White latex present in all parts. Branches sometimes spiny. Leaves opposite or alternate; stipules absent. Flowers in corymbose clusters. Sepals 5, fused at base. Corolla tubular, 5-lobed; lobes overlapping in one direction. Stamens 5; anthers connate. Ovary superior, 2-celled; ovules many. Fruit a berry or paired follicle. 215 genera. 2100 species. Worldwide, mostly in the tropics.

Botanical notes. Many species are very poisonous.

Acokanthera G.Don

Shrubs or small trees. Leaves opposite. Flowers white. Corolla tubular, lobes spreading, broadly ovate. Fruit an ellipsoid or globose berry. 5 species. Arabia, tropical E and S Africa.

ETHNOGRAPHIC NOTES

Comparative uses. The fruit are edible and are used to make jams. An infusion of the roots is used to treat syphilis. *Acokanthera* species contain cardiac glycosides, and, especially *A. longiflora* and *A. schimperi*, provide the main ingredients of arrow poisons in Africa. Except for the ripe fruit, all parts of the plant are poisonous, becoming more toxic during the dry season, and livestock die quickly after grazing it. Some species of arid areas are palatable to livestock. (Heine *et al.* 1988; le Houerou 1980; Lewis & Elwin-Lewis 1977)

Acokanthera schimperi (A.DC.) Schweinf.

Carissa schimperi DC.

Shrub or small tree to 3(–6)m. Leaves ovate-elliptic, 3–7 x 2–4cm, roughly short hairy, base acuminate, tip obtuse or acute. Flowers <1cm long. Fruit purple when ripe, 1.5–2cm diameter.

Ecology. No habitat details available.

Distribution. S Africa and S Arabia.

Botanical notes. Not seen since found by Balfour and Schweinfurth in 1880.

Adenium Roem. & Schult.

Succulent shrubs. Leaves alternate, clustered at end of branches. Flowers pink. Corolla funnel-shaped. Fruit a pair of cylindrical follicles, joined at base. 6 species. Tropical and subtropical Africa and Arabia.

Botanical notes. A single species widespread throughout the hot, dry regions of tropical Africa and Arabia. Across this broad range the flower is more or less constant, but there is a great deal of variation in vegetative features, for instance body shape, leaf shape, hairiness, etc. Extreme forms have been recognised as separate species; however, here we have followed Rowley (1999) in treating them as subspecies of one large and polymorphic species.

Adenium obesum Roem.

Adenium multiflorum Klotzsch
Adenium socotranum Vierh.

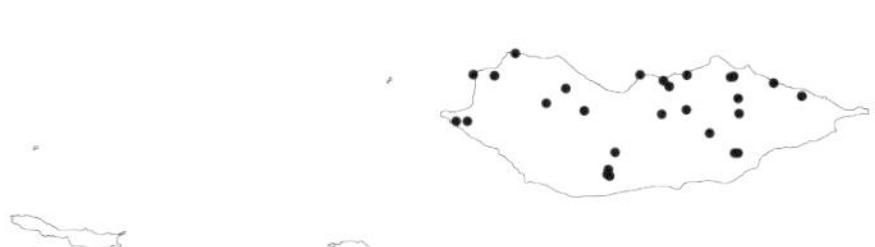

Succulent shrub to 5m; stems grotesquely swollen; bark reddish brown, smooth. Leaves clustered at branch tips, obovate to oblanceolate, <10cm long. Flowers pink, <5cm long. Fruit <10cm long.

Ecology. Widespread and often abundant in succulent shrubland on the escarpments and rocky foothills; extending into *Croton socotranus* shrubland in rocky places on the plains and into open semi-deciduous woodland at higher altitudes on the hills. Sea-level–700(–1050)m.

Distribution. Widespread in tropical Africa and Arabia; subsp. *sokotranum* is endemic to Soqotra.

Conservation status. Least Concern (LC). Widely distributed in several vegetation types where it is frequently abundant. It is highly poisonous and unpalatable to livestock.

Botanical notes. With its grotesquely swollen, almost human-like trunks and large, bright pink flowers *A. obesum* is one of the most spectacular and characteristic plants of Soqotra. *A. obesum*, the desert rose, is widely distributed in Africa and Arabia. Across this wide range five subspecies have been recognised. The Soqotran plants have been placed in subsp. *sokotranum* (Vierh.) Lavranos. This is endemic to Soqotra and is distinguished by its massive size and erect, conical stems. They can be up to 5m high and 3m across. Where they grow on the edge of escarpment their stems are covered in lichen and they appear to merge into the surrounding rocks. *Adenium* comes into leaf in winter, drops its leaves in summer and then comes into flower; it is one of the shrubs that does not need rain to flower.

ETHNOGRAPHIC NOTES

Name(s)

– Soqotri: *tərīmo* (pl. *tərε̄m*, *təˈrem*) (mainly eastern Soqotra); *ˈəsfəd* (pl. *ˈəsfud*) ‘useless, valueless’ (mainly central and western Soqotra).

– Soqotri simplified: *tiriymo* (pl. *tirem*); *isfid* (pl. *isfud*).

Food (livestock). Never grazed or browsed by any livestock. At times of great hunger, the fallen, dead leaves are eaten. In poorly vegetated areas at such times these are even gathered by herders and fed to their goats in a desperate attempt to keep them alive. Male camels occasionally eat dead leaves, but female camels are said never to do so. Desperately hungry or thirsty goats sometimes mistake this plant for the edible *Dendrosicyos socotrana* and strip off the bark and chew at the white underflesh to relieve their hunger and also to quench their thirst. They usually die as a result.

Bees / honey. The flowers are attractive to honeybees, especially after the summer rains, when most of the best honey is made. Bees also nest in the trunks of hollowed *A. obesum*. When this is noticed, the tree is left alone for a year, and then one or more holes are made above the perceived entrance to the colony and some of the honeycombs stacked up inside are removed.

Flavouring / food additive. The milk and the meat-stock of animals that have eaten dead *A. obesum* foliage is much altered in taste and immediately recognisable. It is much disliked and is only drunk if there is little other food.

Timber. Completely useless for timber or for making implements. However, large and hollowed trees, modified where necessary with stonework, make useful kid and lamb pens.

Medicinal / veterinary. The latex is applied to an infected and discharging wound. It is also applied to the site of a scorpion bite to soothe the pain. It is sometimes dripped into sore or inflamed eyes, but this is a painful treatment and usually only carried out on adults and fully grown animals. After bathing the area with a salt and water solution, sores, ulcers and wounds in livestock are treated with the latex, as are maggot-infested wounds. This is said to poison the grubs. The latex is also used to kill ticks (see Pest control). It is also used in cases of persistent pruritus: this is said to be a 'hot' treatment which cases temporary discomfort, but is widely used to relieve a long-lasting itch. It has been tried on animals suffering from mange with limited success. Mange is believed to be transmissible from livestock to humans, and consequently the same treatment is used for people with similar symptoms. For scabies and other persistent and localised irritations, a slice of the inner bark rubbed over the itchy site is often sufficient. A similar treatment is sometimes given for dry eczema and various fungal diseases of the skin. The form of dry eczema that occurs in the milking fingers of herdsmen (said especially to be the result of milking an animal of her first and 'overheated' milk) is treated with *A. obesum* latex. Some islanders use the latex in place of *Jatropha unicostata* sap on fresh cuts and grazes. To treat a serious, deep laceration, the sides of the wound are held together (sometimes also roughly stitched together with thorns of thread), and then a layer of inner bark is laid on top. This is then held in place with a cloth bandage. If there is severe bleeding, either *Jatropha unicostata* or *A. obesum* sap is poured into the wound. A pointed sore or boil is treated with an application of the latex to encourage the pus to emerge. The latex is massaged into a painful joint or aching limb to ease the pain. Mild or early signs of mastitis in livestock are treated with an application of latex over the teats and udder; more advanced cases have to be treated with cauterisation.

The leaves are crushed and rubbed over the skin to soothe the spots of measles and other forms of mild pruritus. For more persistent skin conditions, the flesh of a thick branch is chopped up small and added to cold water. It is left to soak, and then the water used as a skin wash. Thick strips of the bark and bast are also used in the treatment of fractures in small stock and calves: the broken bone is set, anointed with juice from aloes, *Commiphora ornifolia* or latex from *Euphorbia arbuscula*, and then bound firmly with the long strips of *A. obesum* bark. Over this a cloth bandage is wound. The animal is immobilised as far as possible and fed and watered by hand. The binding is checked every few days, and, if all is well, it is progressively loosened.

Pest control. Strips of bark (inner and outer together) are cut and tied around the neck of young lambs and kids to deter wild cats and ravens, which both attack young stock, commonly going for the head and neck. To further deter potential predators, sometimes the lower edge of the strip is slashed to make a fluttering fringe. Sometimes chunks of the wood are hung around the young animal's neck instead. These are usually worn until the animal is fully weaned. Ticks on livestock of all ages are treated with the latex to poison them. These are generally at their worst in the summer and monsoon months. If the tick infestation is severe enough to cause hair loss, the latex is mixed with a strong salt and water solution which is then painted over the affected skin. When this has dried out, a layer of ghee or butter is smeared on.

The burned bark produces a very acrid smoke. This is believed to have strong insecticidal qualities, so is often used for livestock smudges.

Other. These trees are seen as an indicator of forthcoming precious summer rains: if a lot come into plentiful flower at the same time in the summer, then the summer rains are said to be about to fall. Children make toy bracelets and anklets out of narrow strips of the bark. Boys and young men use the tree as a target for knife or javelin throwing contests. In the absence of *Euphorbia arbuscula*, the latex (like that of the other '*sheber*' plants, *Euphorbia schimperi* and *E. spiralis*) or crushed twigs are used to paralyse lagoon fish. However, none of these are as effective as the copious latex of *Euphorbia arbuscula*. In the absence of fine leather, the ornate *sibet* belt worn by men could be made from the underbark of a young plant. This belt, ideally, was made up of as many as 10 to 20 finely plaited strands. The pale cream colour of the *A. obesum* bast was not liked, so it was usually dyed the favoured red-brown by grinding a soft red stone to powder. This was then mixed with water and the thick paste was rubbed again and again over the fibres of the belt until the required red shade was achieved.

The plant also has more esoteric associations. It was often transplanted to stand beside the entrance to a dwelling or living area to keep away evil, or the dead wood was burned outside the living area for the same reason. Similarly, some herders planted a tree close to (but never inside) the goat pen. The tree was cut close to the ground, and the cut part planted in damp earth. Both stump and cutting normally regenerated. As part of a ritual ceremony carried out by the Soqotran shaman or exorcist, the '*mekoli*', a young tree was pulled up by the roots, chopped small and added to the water inside a goatskin. This ceremony was part of the treatment carried out for sick or bewitched livestock (the two are virtually synonymous on Soqotra).

Notes. The islanders differentiate between 'male' and 'female' trees, the 'female' being markedly fatter and shorter than the 'male'. They note that termites never attack this wood. However, this shrub, like other soft-wooded trees such as *Commiphora* species and *Sterculia africana*, is apparently vulnerable to large, white larvae, the size of a large termite queen, which eat away at the living wood from the inside. Two or three such creatures can be found inside the trunk of the dead tree when it is broken open.

Comparative uses. *A. obesum* contains several cardiac glycosides and is very toxic to livestock: the sap of stem, leaves and fruits have a steroid alkaloid which is a cardiotonic. It is used as a fish and arrow poison in eastern Africa. In nearby Dhofar the plant was regarded with caution as it was said snakes derived their venom from it, and the flowers were also said to cause the eyes to become inflamed. The bark is pounded, soaked in warm water and then used to treat dislocations, sprains, paralysis and swellings. The sap is applied to wounds and skin infections. In Kenya children use the trunk as a target, but otherwise consider the plant to be dangerous and useless. The Chamus of Kenya note that goats may eat the bark during the dry season, and that the stem is sometimes used by wild bees as a beehive. Elsewhere in Kenya a bark infusion is used to get rid of ticks and lice on camels and cattle. In West Africa it is used locally on carious teeth and is applied to chronic wounds and ulcers. In Arabia the juice extracted from the stem and crushed bark is used to heal wounds. (Abdulfatih 1987; Dalziel 1948, 1956; Fleurentin & Pelt/Fleurentin *et al.* [1982, 1983a,b,c]; Heine *et al.* 1988; Ghazanfar 1994; Lewis & Elwin-Lewis 1977; Miller & Morris 1988; Oliver-Bever 1986)

Carissa L.

Shrubs. Branches spiny. Leaves opposite. Flowers white. Corolla tubular, lobes spreading, lanceolate. Fruit a globose berry. 37 species. Africa, Asia and Australia.

Carissa spinarum L.
Carissa edulis (Forssk.) Vahl

Much-branched shrub to 1(–4)m. Spines c.1cm long. Leaves ovate to elliptic, 0.8–3 x 0.6–1.7cm. Flowers 0.7–0.9cm long. Fruit purple to black when ripe, 0.5–1cm diameter.
Ecology. Rare in montane shrubland (with *Cephalocroton socotranus*, *Croton* species and emergent *Dracaena cinnabari*) on gentle granite slopes in the Haggeher mountains (Adho di Melhoh). 900m.
Distribution. Widespread in tropical Africa, also S Arabia and to SW Africa.

Botanical notes. Collected only twice, on the saddle of Adho di Melhoh at 900m, by Popov in 1957 and Alexander in 1999. Popov described it as a 6m tree and Alexander as a 2.5m climber!

ETHNOGRAPHIC NOTES
Name(s)
– Soqotri: *di məḳtə̄nəh, di məḳtə̄naḥiyun.*
– Soqotri simplified: *di miqtownih, di miqtownahiyun.*
Food (human). The fruit are not eaten.
Food (livestock). Goats and camels browse the shrub in a drought, but cattle ignore it. It is too uncommon to have any importance for livestock fodder.
Timber. No use as timber.
Comparative uses. In India the fruits are popular and also used for pickles and chutneys. (Ambasta 1982; Jain & De Fillips 1991)

ASCLEPIADACEAE

Herbs, shrubs, small trees or climbers, often succulent, often with milky latex. Leaves opposite, entire, sometimes very reduced or absent. Stipules minute or absent. Flowers regular, bisexual. Sepals 5, ± free. Petals 5, free to fused; 5-lobed single or double corona usually present. Anthers much reduced and usually fused to the gynoecium to form a central structure – the gynostegium; the pollen either united into waxy masses – the pollinia – or (in subfamily Periplocoideae) the pollen is free and transferred by spoon-shaped translators. Ovary ± superior, of 2 almost separate carpels. Fruit a pair (sometimes only one developing) of follicles containing many silky-haired seeds. 348 genera. 2900 species. Tropical and warm regions, especially Africa.

Botanical notes. Recent research has shown that Asclepiadaceae are nested within and should be combined with Apocynaceae (Endress & Bruyns 2000). However, the traditional circumscription of the two families has been maintained here.
This account was prepared by Peter Bruyns and Anthony Miller.

Calotropis R.Br.

Small trees or shrubs, with copious white latex. Leaves opposite, sessile. Flowers cup-shaped, in lateral umbels. Pollen united into pollinia; pollinia paired. Fruit a globose, much inflated solitary follicle. 3 species. Old World tropics.

Calotropis procera (Ait.) W.T.Aiton

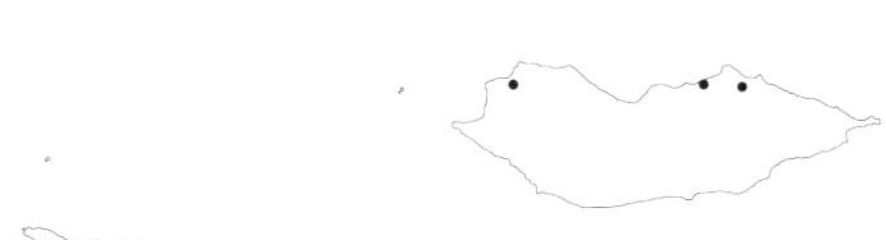

Stems to 5m, with deeply fissured bark, white tomentose throughout. Leaves leathery, ovate to obovate with a cordate base and shortly acuminate tip, 10–15cm long. Flowers white outside, pink or purple within, c.20mm across, in dense cymes. Fruits usually solitary, large, inflated, smooth, c.20cm long.
Ecology. Widespread on the plains, often near cultivation and habitation. Under-recorded on the distribution map. Sea-level–100m.
Distribution. Widespread across the dry regions of N Africa and SW Asia.
Botanical notes. A characteristic plant of waste-land near habitation. A long-established introduction on Soqotra (recorded by Balfour in 1880). Not eaten by

livestock and increasing on overexploited rangeland and thus a good indicator of overgrazing.

ETHNOGRAPHIC NOTES

Name(s)

– Soqotri: *ˈiˀiś'hur* (pl. *eˀɛ̄śir*); *gˈənigəno* (given to other species which are regarded as 'useless', especially as timber, e.g. *Ricinis communis* and *Datura fastuosa*).
– Soqotri simplified: *i'ish'hur* (pl. *e'eyshir*); *giniginoh*.

Food (livestock). Livestock eat the fallen, dried leaves in hunger, especially small stock. The green foliage and the fruit are not palatable, but the flowers are eaten by both sheep and goats.

Fuel. The dead wood is of little use as firewood, but is used if there is nothing else available. It makes a lot of smoke and has a distinctive smell, so is little used for cooking.

Timber. No use for timber.

Wood. The wood is too soft to be of any use.

Medicinal / veterinary. The latex is sometimes dripped into a painful tooth, or it is put on a scrap of cloth which is then wedged into the tooth.

Comparative uses. The latex contains caoutchouc and trypsin, and a cardiac poison. The leaves and stalks contain calotropin and calatropagenin. The related *C. gigantea* found from southern Africa to China yields one of the most virulent poisons known to man, 15 to 20 times as toxic as strychnine. *C. procera* is widely used medicinally.

Latex. It is very toxic, and has strong vesicatory (raising blisters) properties. It is given as a purgative and vermifuge. It is used to treat ringworm and eczema – one application is sufficient, the affected area turns black, and causes great local irritation. It is used on scorpion stings, venereal sores, and to extract guinea worms. It is used to treat mange and scabies in camels and goats and to treat sores in baggage camels. It is given as a rubifacient for rheumatic pains and is applied to the chest for a cough. It is used on itching skin, but never put in the eyes, since it can cause blindness. It is used in tanning and dyeing and to remore the hair of skins and hides. The Fulani use the latex to coagulate milk in cheese-making, and it is also used as a fermenting agent in making *marrisa*, a beer made from dates. In Yemen the sap is used as a local anaesthetic, and is placed on wounds to speed up the formation of pus. It is placed on a carious tooth to cause it to disintegrate, and used to extract a thorn which has gone in deep, especially in the hands and feet. In agriculture, farmers put broken-up twigs in watercourses leading to their fields so that the seeping latex destroys various kinds of insect pests.

Sap. The juice of young flowering shoots is sucked by children.

Leaves. They are used in fermenting *durra* to make *merissa*, a native beer of West Africa. Two or three leaves are placed in holes in the sand into which water may percolate to clarify the water. Leafy branches are used to deter termites. In Yemen they are used as an antiseptic for skin infections. The leaves and outer bark are scorched and then placed on a swelling. They are also used to set fractures and are applied as a hot compress to relieve the pain. A decoction of the dried leaves and new twigs is used to relieve difficult breathing. In Oman, heated leaves are placed over painful joints after smearing these with oil. The juice from warmed leaves is dropped into the nose to bring on sneezing as a remedy for a headache and catarrh. Dry leaves and pithy stems are burned and an asthmatic told to inhale the fumes, or, to treat coughs and breathing difficulties, the leaves are smoked as a tobacco. The dried leaves are also made into a snuff.

Seeds. The silky seed fibres are used for stuffing bolsters and camel saddle cushions, but the gathering of pods can cause eye damage. The seeds contain a semi-drying oil.

Bark. The inner bark yields a fibre used in cordage, and in India the bark fibre was spun into a very fine cloth. The fibre is used to make fishing nets and lines, bow strings and twine. However, this cordage quickly becomes waterlogged, does not stand rough usage and has low tenacity. An infusion of the root-bark is given as a diaphoretic, an emetic, and as a poison antidote, and was also used as a liniment. Dried root-bark is given in a soup to treat colic, as a stomachic and galactagogue, and is also useful in mild dysentery.

Roots. They are used as toothsticks, and to treat toothache and snakebite. The burned root is made up as an ointment for skin eruptions and ulcers. An infusion is used to treat coughs in camels.

Flowers. An ethanol extract of flowers has an anti-fertility action in males.

Wood. It makes a good charcoal which was formerly an essential ingredient in local gunpowders. The charcoal is powdered and made into an ointment to treat skin complaints in camels, and to polish the horns of bulls. The pith was used for gunpowder and as a tinder. The twigs are used as a stomachic, antidiarrhoetic, diuretic and to treat asthma. The stems are used as floats for fishing nets. The wood is strongly termite resistant, and, in arid areas, the stems are used to make the framework for thatched roofs and for grain stores.

The plant is also placed over doorways as a preventative against witchcraft. (Abbas *et al.* 1992; Al-Dubaie 1993/Al-Dubaie & Al Khulaidi; Ambasta 1982; Bebawi & Neugebohrn 1991; Dalziel 1948, 1956; Davis *et al.* 1983; Fleurentin & Pelt/Fleurentin *et al.* [1982, 1983a,b,c]; Ghazanfar 1994; Goodin & Northington 1979; Heine *et al.* 1988; le Houerou 1980; Jain & De Fillips 1991; Miller & Morris 1988; Oliver-Bever 1986)

Caralluma R.Br.

Boucerosia Wight & Arn.

Succulent herbs. Stems succulent, erect, 4-angled, with soft, fleshy teeth. Leaves reduced to tiny scales or absent. Flowers produced towards the apex of stems. Corolla 5-lobed, cup-shaped; corona double. Pollen united into pollinia; pollinia paired. Fruit a pair of erect, fusiform follicles. 120 species. Africa, Mediterranean to Myanmar (Burma).

Caralluma socotrana (Balf.f.) N.E.Br.

Boucerosia socotrana Balf.f.
Caralluma rosengrenii Vierh.

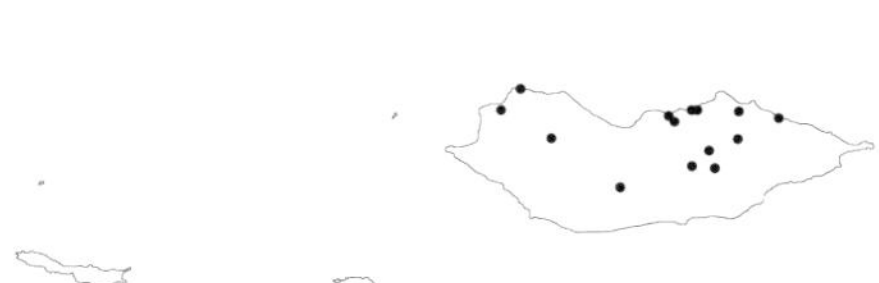

Stems erect, grey-green, to 45cm, rooting from the central stem only. Corolla 1.5–2.5cm across; pedicels 1–3mm long, bright red throughout or sometimes with yellowish tinged lobes. Fruits erect, to 10cm long.

Ecology. Widespread and often common on the coastal plains, foothills and limestone plateau, less commonly on the foothills of the granite mountains; occurring in a variety of dry vegetation types including coastal woody-herb communities, *Croton socotranus* shrubland, succulent shrubland and open drought-deciduous woodland. Sea-level–900m.

Distribution. Somalia.

Botanical notes. One of the most attractive succulent species on the island. The flower colour varies considerably; typically the flowers are deep crimson throughout but less commonly they have greenish-yellow lobes or are brownish yellow throughout. *C. rosengrenii* Vierh. was based on one of these colour forms; it is not recognised here.

ETHNOGRAPHIC NOTES
Name(s)
– Soqotri: *məśḥermihim* (pl. *məśḥerməhum*); (western Soqotra) *məśxayrem* (pl. *məśxayrʾhum*) 'eaten to still hunger'; (sometimes) *məśḥerməhum di eʾefo* '*Caralluma* for people', i.e. edible, to distinguish the plant from the inedible *məśḥerməhum di girbaḳ* '*Caralluma* for wild cats'; *miśḥāhir*.
– Soqotri simplified: *mish'hermihim* (pl. *mish'hermihum*), *mish'khayrem* (pl. *mish'khayr'hum*); *mish'hermihum di e'efo; mish'hahir*.
Food (human). The flowers, green tips and fruit are all eaten, and, in hunger, even the darker, drier stems. They are always eaten raw, and some prefer to peel the stems before eating as the outer skin is bitter. The plants respond very quickly to drizzle, rain and cloud and they were an important food of the hungry period immediately after the first rains.
Food (livestock). Goats eat the flowers, fruit and green stems after rain. Sheep are less keen on the stems, but like the fruit and flowers.
Cleansing. The succulent plant can be crushed and used to clean the hands.
Notes. People eating this suffer less from thirst. When it darkens in colour and becomes red-grey, it is sometimes called *məśḥerməhim ʿafər* ('red *C. socotrana*'), as is *Edithcolea grandis*; *məśḥerməhim* is also used to describe a 'family' of plants, which includes *C. socotrana*, *Duvaliandra dioscoridis*, *Edithcolea grandis* and all *Echidnopsis* species. In some areas, surprisingly, this 'family' can also include *Euphorbia spiralis* and *Asparagus* species. In different areas of the island, the names for each can differ, as can views on the palatability of each (except for *E. spiralis* and *Asparagus* species, which all agree are inedible for people). Thus in one area, members of the family 'covered in soft prickles' are edible, while the smooth ones are not. Or, the *Echidnopsis* species, which are 'green-coloured with reddish flowers' are edible, and the 'pale-coloured ones' are not. Or the flowers of the 'red' *məśḥerməhim* are edible for both livestock and people, and those of the 'one of the wild cats', which stinks, is not. Even more specific: the flowers of the 'one of the wild cats' are palatable to sheep and goats; *C. socotrana* which is eaten by people is not eaten by livestock; and the 'red' *C. socotrana* is likewise not eaten by livestock but people eat the flowers and fruit. Clearly, there is more research yet to be carried out into this fascinating group.
Comparative uses. In nearby Dhofar, the flowers, fruit and stems of both *C. flava* and *C. quadrangula* are eaten. In Yemen, sap from stems of *C. quadrangula* is used as an ophthalmic antiseptic, and the stems are used against smallpox. In East Africa, the sap of *C. speciosa* is applied to open wounds, and the whole plant is boiled and rubbed on the teat of a camel with a blocked teat. The woody part of the plant is soaked in warm milk and given to a sick dog to put it out of its misery. The flowers are used as a dye. (Fleurentin & Pelt/Fleurentin *et al.* [1982, 1983a,b,c]; Heine *et al.* 1988; Miller & Morris 1988)

Ceropegia L.

Herbs with twining, sometimes succulent, stems; sap clear; roots fusiform. Leaves with oblong lamina or scale-like. Corolla with lobes remaining fused at tips, the lobes strongly folded outwards down the midrib. Pollen united into pollinia; pollinia paired. Fruit a pair of slender, erect to spreading fusiform follicles. 160 species. Old World tropics.

Ceropegia affinis Vatke

Stems thin (less than 1.5mm thick), twining to 2m. Leaves narrowly oblong and usually at least 15mm long, herbaceous and persistent, with short petioles. Corolla c.20mm long; lower part of corolla tube scarcely inflated, lobes becoming nearly horizontal at apex.
Ecology. Common in dense thickets in semi-deciduous and evergreen montane woodland. 500–1550m.
Distribution. Eritrea, Ethiopia, Kenya, Somalia and Uganda.
Botanical notes. Common but overlooked. It characteristically grows in dense vegetation where its thin twining stems and inconspicuous leaves are difficult to spot.

ETHNOGRAPHIC NOTES†
Name(s)
– Soqotri: *riyēbə* (pl. *rīyəb*) 'prostrate with exhaustion'; (sometimes) *ḥalawhin*.
– Soqotri simplified: *riyebih* (pl. *riyib*); *halawhin*.
Food (human). Every part of the plant is edible: stems, tendrils, leaves, flowers, fruit and underground parts. They are eaten raw and when young have a sweet flavour, becoming pleasantly lemon-tasting as the plant matures. The plants are at their best in the early winter. They then become progressively tougher and less palatable. This species is much less common than *C. tihamana*. Another kind is called *riyɛbə di tɛ̄tin*, 'sheep's *Ceropegia*' and, as the name implies, this is less liked as a food, since it is rather astringent.
Food (livestock). All livestock eat all parts of the plants which are regarded as a good, nutritious feed.
Notes. People eating these suffer less from thirst. Some islanders do not differentiate between the two species of *Ceropegia*, and call them all *ergilo*, the name *riyēbə* being unknown.
Comparative uses. Many *Ceropegia* species are edible, the leaves and roots being eaten both cooked and raw. In nearby Dhofar, the tuber, twining stems, leaves and fruit of *C.bulbosa* are an important food of the early rains season. Elsewhere, the roots of many species are considered to be tonic and digestive, and the tubers of *C. tuberosa* are used in the treatment of diarrhoea and dysentery, as well as being eaten. (Ambasta 1982; Jain & De Fillips 1991; Kunkel 1984; Miller & Morris 1988)

†Uses also apply to *C. tihamana*.

Ceropegia tihamana **Lavranos**

Stems succulent, c.3mm thick, twining to 0.5m or less. Leaves scale-like and caducous, mostly 3–6mm long and fleshy, without petiole. Corolla c.20mm long; lower part of corolla tube noticeably inflated, lobes ascending steeply towards the tips.
Ecology. Heavily wooded slopes on steep side of a deep gorge on the Diksam plateau. 600m.
Distribution. Saudi Arabia and Yemen.
Botanical notes. The only collection of this rare but probably over-looked plant possesses only immature flowers and thus its identification as *C. tihamana* remains provisional.

ETHNOGRAPHIC NOTES
Name(s)
– Soqotri: *ʾergílo*, *ʾɛrgílə* (pl. *ʾérgəhul*, *ʾírgəhel*) 'swollen' (also given to *Anthericum* and *Corchorus* species, which also produce edible tubers).
– Soqotri simplified: *'ergilo*, *'ergilih* (pl. *'ergihul*, *'irgihel*).
Uses. See *Ceropegia affinis*.

Cryptolepis R.Br.

Diffuse shrubs, climbers or small trees; latex milky. Leaves conspicuous. Flowers white or cream, in terminal or axillary umbels. Sepals acute. Corona lobes filiform or club-shaped, straight or bent. Pollen free and transferred by spoon shape translators. Fruit a pair of woody, fusiform, ovoid or suborbicular follicles. 4 endemics. 12 species. Old World tropics.

ETHNOGRAPHIC NOTES
Comparative uses. The stems and leaves of African species are boiled and eaten in hunger. *C. buchananii* yields a fibre used for cordage and for making a kind of cloth. The root of *C. nigritana* is eaten as a famine food, and it is also boiled and taken as a remedy for griping stomach pains. The root is worn around the waist by Fulani women as a charm to attract men. The tough, stringy bark is used in making fishing nets. (Ambasta 1982; Jain & De Fillips 1991; Kunkel 1984)

Cryptolepis arbuscula **(Radcl.-Sm.) Ventner**
Mitolepis arbuscula Radcl.-Sm.

Shrub or small tree to 4m. Leaves often in axillary clusters, <3 x 1.75cm, ovate to obovate, rounded or retuse (rarely bilobed) at tip. Flowers yellow, 1–1.5cm long, solitary in leaf axils. Corolla lobes 10–12mm long. Corona lobes filiform, straight. Fruit reddish green, paired, fusiform, pointed, <4–6 x 0.5cm.
Ecology. Semi-deciduous woodland and shrubland on both the granite and limestone. 450–950m.
Distribution. Endemic.
Conservation status. Least Concern (LC). Common in several vegetation types and under no present or perceived threat.

ETHNOGRAPHIC NOTES†
Name(s)
– Soqotri: *gə́sso*, *jə́ssoh* 'small-leaved nibble' (also given to *Euphorbia schweinfurthii* which also has white latex); *mɛ̄rə* (pl. *ʾémrəhor*) (also given to *Marsdenia robusta*, *Cocculus pendulus*, *Vincetoxicum linifolium*, *Cryptolepis socotrana*, *Euphorbia obcordata* and *E. socotrana*); *nāmina*; *məʿázəhil* 'spindle'; (rare) *mīśəṭ* (more commonly given in western Soqotra to *Punica protopunica*).
– Soqotri simplified: *gissoh*, *jissoh*; *meyrih* (pl. *'emrihor*); *namina*; *mi'azihil*; *miyshit*.
Food (human). The green underbark is chewed both as a food and for its sweetness. It is quite a filling food and was important in the hungrier areas. The leaves too can be eaten and also have a sweet aftertaste. Both leaves and the bark were eaten as a famine food. The plants also sometimes exude a resin (called *śáḥaz*, a common term for frankincense, but which locally refers to any plant exudate that can be chewed), which is also eaten.
Food (livestock). Goats and sheep browse these plants keenly, and the small leaves are much liked by young weaned lambs and kids. Goats also chew the bark. In the areas where these plants grow they are a key fodder for small stock, especially goats. They are seen as being a nutritious and fattening feed, which also promotes lactation. Cattle eat the foliage of the better-grown specimens. Camels and donkeys also browse the plants.
Bees / honey. Honeybees are attracted to the flowers. If honey has absolutely no trace of bitterness it is likely to be from these plants and is a honey of superior quality.
Flavouring / food additive. The sweetness of the green underbark and leaves was much liked before the days of sugar. The milk of livestock browsing heavily on these plants is rich and sweet in flavour.
Fuel. The dead wood makes reasonable firewood. It burns slowly, and a large chunk of this wood was used to keep the fire in. It was poked deep into the pile of hot ash and embers and left there until the fire was needed again, when it was pulled out and blown on to re-ignite the flame.
Timber. The wood does not make a good timber.
Wood. Dead twigs and strips of bark are collected and kept at home for use as tinder and kindling – they light very quickly, and stay dry even in damp weather. The slimmer stems are supple and the wood soft. Thicker, straight branches are used to make the important herding staff, and slimmer stems to make the 'bit' for the weaning device worn by kids and lambs. Slender twigs are also whittled to make *kohl* applicators. As one of the names for this plant suggests, the wood is also used to make the spindle for spinning sheep's wool, and to make the 'bow' for beating the wool after sheasring to prepare it for spinning.

Medicinal / veterinary. The latex is collected by striking a trunk or branch with a rock to make it bleed, and then twisting a stick round and round in the latex. The latex is dripped directly onto the skin in cases of ringworm or fungal disease after scarifying the skin and making it bleed. Sometimes salt and ground, dried *Euclea divinorum* root is added to the latex. Raised neck glands and other glandular swellings are treated with an application of the latex. The inner bark is chewed by people feeling under the weather as a tonic, or by someone with a headache or head cold to 'clear the head'. The copious latex of better-grown specimens is smeared over the chest for chest pain or for a chronic cough to relieve the pain.
Other. Animals feeding on this can go longer without water. The 'pen' used for writing with locally produced inks was commonly made from a stem of the plant, sharpened to a point. A tiny notch was then cut in the tip to hold the ink.

†Also apply to *C. intricata*.

Cryptolepis intricata **(Balf.f.) Ventner**
Mitolepis intricata Balf.f.

Intricately branched shrub to 1m. Leaves linear-oblong to linear-obovate, 10–15 x 2–3mm. Flowers white or cream, solitary in leaf axils. Corolla lobes c.4mm long. Corona lobes filiform, straight. Fruit fusiform, pointed at tip.
Ecology. Widespread and in places abundant in shrubland and open semi-deciduous woodland, mainly on the limestone plateau, rarely on the granite. 50–750m.
Distribution. Endemic.
Conservation status. Least Concern (LC). Common in several vegetation types and under no present or perceived threat.
Botanical notes. Usually a low, rounded shrub, distinguished by the small, succulent, rather glossy leaves, small white flowers and white latex.

ETHNOGRAPHIC NOTES
Name(s)
See *C. arbuscula* (also uses).

Cryptolepis macrophylla **(Radcl.-Sm.) Ventner**
Curroria macrophylla Radcl.-Sm.

Erect shrub or small tree to 3m; stems rather zigzagging with prominent leaf scars. Leaves dark green, broadly elliptic or orbicular, <4 x 3cm, rounded or retuse at tip, obtuse at base. Flowers white or cream, in few-flowered terminal umbels. Corolla lobes c.9mm long. Corona lobes short, bent. Fruit reddish brown, fusiform, <12 x 1cm.
Ecology. On Abd al Kuri common in succulent shrubland at 200–600m; on Soqotra in shrubland and deciduous woodland on the western limestone plateau and in semi-deciduous thicket on the Haggeher mountains. 100–900m.
Distribution. Endemic.
Conservation status. Vulnerable (VU D2). Rare, with a very limited area of occupancy.
Botanical notes. *C. macrophylla* was originally described on the basis of a number of specimens from Abd al Kuri and a single specimen, the type, from Soqotra. Recently more collections have become available from Soqotra and have shown that the plants from Abd al Kuri consistently differ from those on Soqotra in having paler, broadly elliptic to broadly obovate (not circular) leaves which are rounded or obtuse (not retuse) at the tip and acute (not obtuse or rounded) at the base. Unfortunately, I have seen no flowering material from Abd al Kuri and so am reluctant to formerly recognise them at present. On Soqotra *C. macrophylla* is not common and is found in two rather different habitats: in montane thicket on the Haggeher and in drought-tolerant shrubland on the western limestone plateau. On Abd al Kuri it is relatively common in succulent shrubland on the middle and upper slopes of Jebel Saleh and is more or less the only tree on the island.

ETHNOGRAPHIC NOTES
Name(s)
– Soqotri: *mə̄rə* (pl. *ᵓémrəhor*) (also given to *Cocculus pendulus*, *Cryptolepis* species, *Vincetoxicum linifolium*, *Cryptolepis socotrana*, *Euphorbia obcordata*, *E. socotrana* and *Marsdenia robusta*); *ᶜéḳḳəhīn* (pl. *ᶜaḳāhan*), (western Soqotra) *ġéḳḳəhīn* (also given to *Cocculus pendulus* and *Secamone* species); (rare) *míləhil*.
– Soqotri simplified: *merih* (pl. *'emrihor*); *'eqqihiyn* (pl. *'aqahan*), *gheqqihiyn*; *milihil*.
Food (livestock). This provides a dry season fodder for all livestock. Goats eat the flowers and fruit, and in the wetter areas the plant is seen as a beneficial feed for goats. Herders collect leafy stems to feed to their goats in the dry season, and goats can be seen climbing up into the trees to get at the foliage. Cattle eat leaves within reach and also graze any fallen material. Camels are very fond of it. On Abd al Kuri it is an important fodder for goats; sheep do not eat the living material but occasionally eat plant litter.
Bees / honey. Honeybees are attracted to the sweet-smelling flowers.
Flavouring / food additive. Camels browsing heavily on this give a lot of rich, sweet milk. On Abd al Kuri the milk of goats browsing heavily on this develops a slightly bitter taste.
Fuel. The dead wood makes reasonable firewood, but it burns down very quickly. On Abd al Kuri, the dead wood makes good firewood, though inferior to that of *Acacia sarcophylla*. However, the trees are not common enough on the island for it to be an important source of fuel.
Timber. Not suitable as timber.
Wood. The thicker, stronger stems can be used to make the important herding staff. Slimmer and flexible stems

are used to make the 'bow' for beating sheep's wool after shearing to prepare it for spinning, and the wooden rim of the leather watering bucket. A suitable crooked stem is sometimes cut for use as a fodder hook to bring down foliage within reach. The wood makes a good charcoal. On Abd al Kuri the dead wood is used for firesticks with the dead wood of *Euphorbia balsamifera*, the former providing the upper, vertical stick and the latter the lower, horizontal one.

Medicinal / veterinary. On Abd al Kuri the charcoal made from the plant is used medicinally; it is powdered and applied to contusions, swellings and inflammation. It is also used in this way to treat the swellings of mumps and occasionally on an open wound to staunch the bleeding.

Cleansing. Sprigs of the sweet-smelling flowers are tucked into the clothing and hair. The sniffing of sweet-smelling plants is held to protect from the 'noxious air' that can cause disease, and to help protect against infection.

Notes. Animals feeding on this can go longer without water.

Cryptolepis socotrana **(Balf.f.) Ventner**
Socotranthus socotranus (Balf.f.) Kuntze

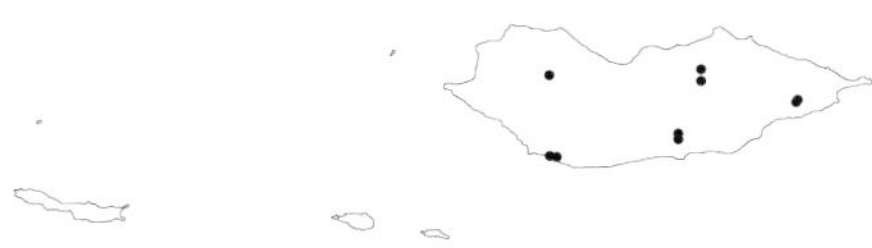

Shrub to 1m, often hanging on cliffs; much branched from base. Leaves sessile, orbicular, 3–4cm across, retuse at apex, cordate at base. Flowers white, in terminal panicles. Corolla glabrous within; corolla lobes narrowly triangular, 1–1.5cm long, twisted. Corona lobes short, club-shaped. Fruits ovoid to orbicular, c.5 x 2cm, spreading.

Ecology. Widely scattered but never common on limestone cliffs in succulent shrubland. 50–200m.

Distribution. Endemic.

Conservation status. Vulnerable (VU D2). Rare, with a very limited area of occupancy.

Botanical notes. This attractive shrub typically hangs from vertical cliff faces and is readily distinguished by its sessile, orbicular leaves and twisted corolla lobes. There are unlikely 19th century records of this species from the Haggeher mountains.

ETHNOGRAPHIC NOTES

Name(s)

– Soqotri: *dṓḳəš, dṓgəš, dāgəš* (more commonly given to *Euphorbia socotrana*); *mḗrə* (pl. *ˀemrəhor*) (also given to *Cocculus pendulus*, *Cryptolepis* species, *Vincetoxicum linifolium*, *Euphorbia obcordata*, *E. socotrana* and *Marsdenia robusta*).

– Soqotri simplified: *dowqish*, *dowgish*, *dagish*; *merih* (pl. *'emrihor*).

Food (livestock). Hungry goats and camels eat the flowers, fruit and foliage when they can reach them, but the plants commonly grow out of reach. Sheep and goats eat fallen leaves and plant litter. Some herders say that in a drought the thicker stems are pulped for goats, others say that this is not the case.

Bees / honey. Honeybees are attracted to the sweet-smelling flowers.

Flavouring / food additive. The flesh of goats that have been browsing on this plant develops an unpleasant taste, especially the viscera and cooking water.

Fuel. The dead wood makes reasonable firewood, but is little used for cooking, as the smoke is very pungent.

Timber. The wood is not suitable as a timber

Wood. The wood is hard and can be used to make implements. It is liked for making the heavy stick used by camel drivers and the important herding staff. A suitable crooked stem is sometimes cut for use as a fodder hook to bring foliage within reach.

Cleansing. Sprigs of the sweet-smelling flowers are tucked into the clothing and hair. The sniffing of sweet-smelling plants is held to protect from the 'noxious air' which can cause disease, and to help protect against infection.

Notes. The latex can be used like that of *Euphorbia socotrana*, but is considered to be much inferior to it.

Cryptolepis volubilis **(Balf.f.) Schwartz**
Curroria decidua var. *volubilis* (Balf.f.) Bullock

Climbing shrub to 3m; stems twining. Leaves often clustered, linear-elliptic to narrowly obovate or obtriangular, <4 x 1(rarely 1.5)cm across, tip rounded or retuse. Flowers white, in axillary clusters. Corolla lobes c.4mm long. Corona lobes short, bent. Fruit ovoid to subglobose, <3 x 2cm.

Ecology. Locally common in drought-deciduous woodland and less common in succulent shrubland. 30–550m.

Distribution. S Arabia.

ETHNOGRAPHIC NOTES

Name(s)

– Soqotri: *zˈiktəh*; *ᶜeḳḳəhīn*, *ᶜāgəhən* (pl. *ᶜaḳāhan*, *ᶜagāhan*), (western Soqotra) *ġeḳḳəhīn* (also given to *Cocculus pendulus* and *Secamone* species); *gˈisso* (also given to *Cryptolepis* species).

– Soqotri simplified: *ziktih*; *'eqqihiyn*, *'agihin* (pl. *'aqahan*, *'agahan*), *gheqqihiyn*; *gisso*.

Food (livestock). All livestock eat the foliage when they can reach it, and it is regarded as a nutritious feed. Goats and camels also eat the stems. Herders lop leafy stems or strip the leaves from the stems to take to their cows and goats in the dry season.

Bees / honey. Honeybees are attracted to the sweet-smelling flowers. Bees also nest in the thick foliage. If there is absolutely no trace of bitterness in honey, then it is likely to be from this plant, and to be of a very high quality.

Wood. The flexible stems are used for the wood of the 'bow' which is used to 'beat' the sheared wool to prepare it for spinning. They are also used to make the wooden rim of the skin watering bucket. A suitable crooked stem is sometimes cut for use as a fodder hook to bring foliage down within reach.

Cleansing. Sprigs of the sweet-smelling flowers are tucked into the clothing and hair. The sniffing of sweet-smelling plants is held to protect from the 'noxious air' that can cause disease, and to help protect against infection. All sweet-smelling plants are endowed with beneficial medicinal and prophylactic qualities.
Notes. Animals feeding on this can go longer without water.

Cynanchum L.

Twining herb; stems jointed, succulent, leafless, pencil-like. Flowers in sessile axillary clusters. Corolla with short tube and spreading or reflexed lobes; corona single, tubular, toothed. Pollen united into pollinia; pollinia paired. Fruit a solitary erect, fusiform follicle. 150 species. Tropical and warmer regions.

Cynanchum gerardii (Harv.) Liede

Stems green to grey-green, twining to 2m. Flowers yellow-green, c.8–12mm across; corolla lobes reflexed; corona orange, bluntly toothed. Fruits smooth, pointed.
Ecology. Widespread but not common in semi-deciduous woodland on limestone escarpments and plateau. 150–650m.
Distribution. S Africa to Arabia, Madagascar.
Botanical notes. A leafless climber with twining succulent stems and clusters of small greenish flowers.

ETHNOGRAPHIC NOTES
Name(s)
See *Vincetoxicum linifolium* (also uses).

Duvaliandra M.G.Gilbert

Leafless herb with succulent 4-angled stems and soft, fleshy teeth. Leaves absent. Flowers produced at or below middle of stems. Pollen united into pollinia; pollinia paired. Fruit a pair of erect, fusiform follicles. 1 species. Endemic.

Duvaliandra dioscoridis (Lavranos) M.G.Gilbert
Caralluma dioscoridis Lavranos

Stems prostrate, forming large colonies up to 1m across. Flowers brownish red. Corolla rotate, 35–55mm diameter, covered with club-shaped hairs. Fruits <6cm long.
Ecology. Very rare on exposed granite outcrops in the Haggeher. 900m.
Distribution. Endemic.
Conservation status. Critically Endangered (CR D). This small, *Caralluma*-like succulent is known only from a single locality where its estimated area of occupancy is extremely small – under 100m^2. The site is at 900m on a granite outcrop in the relatively wet central mountains of Soqotra. This is an unexpected place to find a succulent; however, its niche is specialised – on dry, exposed, south to southeast-facing slabs of granite. A search of likely habitats at similar altitudes in the mountains has failed to find further populations. The population is healthy and actively regenerating; nevertheless, due to the popularity of such plants with succulent enthusiasts, the low number of individuals (under 50) and the relative accessibility of the only known site, I have no hesitation in considering the plant critically endangered.

ETHNOGRAPHIC NOTES
Name(s)
– Soqotri: *di g̍iribəgh̍itin* 'of the wild cats'.
– Soqotri simplified: *di giribig'hitin.*
Food (human). Not generally eaten by people, though some herders report that the flowers and young fruit are edible.
Food (livestock). The goats of the Central Highlands, where these plants grow, are said to nibble at the new growth.

Echidnopsis Hook.f.

Succulent herbs; stems prostrate, finger-like, rooting along their length, 6 to 8-angled, covered with soft obtuse tubercles. Leaves scale-like or absent. Flower at tips of stems. Corolla tubular, cup-shaped or rotate (flattened). Pollen united into pollinia; pollinia paired. Fruit a pair of erect, fusiform follicles. 5 endemics. c.30 species. E Africa to Arabia.

Botanical notes. All species on Soqotra are very similar in vegetative characters and therefore flowers are essential for accurate identification. Unfortunately flowering specimens are rather rare in the field and the plants are all inconspicuous, easily overlooked and consequently under-recorded. Most of the species on Soqotra are only known from a few collections and it is likely that more species await discovery. For accounts of this genus see Bruyns (1988), Plowes (1993) and Lavranos (1993).

ETHNOGRAPHIC NOTES†

Name(s)

– Soqotri: *ˈtabaᶜš*, *tɔ̄baᶜš* (also the name for a kind of caterpillar), (western Soqotra) *ˈtʌbəġʌš*; *ṣəˈberbəher* 'sour, astringent'.

– Soqotri simplified: *taba'sh*, *towba'sh*, *tabighash*; *ziberbiher*.

Food (human). Islanders differentiate between *ˈtabaᶜš di tɛ̄tin* ('sheep's *Echidnopsis*') which are not eaten, and *ˈtabaᶜš* or *ˈtabaᶜš di ˈirəhān* ('goat's *Echidnopsis*') which are edible. These are not necessarily different species; the differing names possibly reflect plants at different stages of development and growing in different soils in wetter or drier areas. The edible ones are eaten raw and, when young, are quite pleasant and sweet. However, they grow very bitter with age. They are generally only eaten in the hungrier parts of the island; in the better-vegetated areas they are rarely eaten, if at all.

Food (livestock). The cattle of the high limestone plateaus eat the whole plant in hunger. Goats and sheep also eat the plants in the dry season if they are still tender and green, but only when they are hungry. They are not considered to be a nutritious feed for livestock. *E. bentii* is less palatable.

Cleansing. The succulent plant can be crushed and used to clean the hands.

Notes. Animals feeding on the plants can go longer without water. They also lessen thirst in humans.

Comparative uses. In nearby Dhofar the stem, flowers and fruits of *E. scutellata* are also eaten. (Miller & Morris 1988)

†Also apply to *Socotrella dolichocnema*.

Echidnopsis bentii N.E.Br. ex Hook.f.

Stems to 15cm long, 8–20mm thick. Corolla deep maroon, 10–13mm across, with strongly spreading, ovate-triangular lobes and a short tube, finely papillate inside. Corona c.2mm tall; inner lobes c.1.5mm long rising up in a small column; outer corona lobes c.2mm long, erect, fused into a continuous ring.

Ecology. On soil beneath dense deciduous woodland on the granite. 500–700m.

Distribution. Endemic.

Conservation status. Vulnerable (VU D2). Rare, with a very limited area of occupancy.

Botanical notes. Distinguished by the deep maroon, cup-shaped flowers. The only species on the island to grow in the relatively wet conditions of the Haggeher mountains. Originally collected by Theodore Bent in the 19th century. Unfortunately, Bent died in Aden on his return from Soqotra and it was not known whether his collection was made on Soqotra or on the mainland. The mystery was only solved when the plant was rediscovered on the southern slopes of the Haggeher in 1998 where it is relatively common over a small area, and flowers prolifically.

Echidnopsis inconspicua Bruyns

Stems to 10cm long, 5–10mm thick. Corolla tube cream or pale pink with dark maroon rings at base, 7–9mm across, finely papillate within; lobes erect to slightly spreading, broadly triangular lobes with abruptly acuminate tips; tube short. Inner corona lobes <0.5mm long, not rising up in centre of flower; outer corona lobes forming continuous ring around top of column.

Ecology. Rare, on shady cliffs in succulent shrubland on the inland slopes of the western limestone plateau. 200m.

Distribution. Endemic.

Conservation status. Vulnerable (VU D2). Rare, with a very limited area of occupancy.

Botanical notes. The smallest-flowered species on Soqotra. Only known from a single gathering.

Echidnopsis insularis Lavranos

Stems to 10cm long, 5–11mm thick. Corolla pinkish cream, tubular, somewhat swollen below and constricted at tip, 9–11mm long x 4–5mm across; corolla lobes short, triangular, 1.5–3mm long. Corona 2–2.5mm long, yellow; outer lobes fused in a 5-angled structure around the column, 10-lobed around margin; inner lobes 0.5–1mm long, broad at base becoming linear above.

Ecology. Widespread but rare on the limestone plateaus, usually in thin soil in cracks and in the shade of boulders. 200–350m.

Distribution. Endemic.

Conservation status. Vulnerable (VU D2). Rare, with a very limited area of occupancy.

Botanical notes. Related to several other species with tubular corollas from E Africa and SW Arabia. There is some variation in the shape of the corolla and the length of the corolla lobes in the few specimens seen on Soqotra.

Echidnopsis milleri **Lavranos**

Stems to 10cm long, 6–10mm thick. Corolla pale pink, 10–12mm across, smooth within, with spreading, almost flat broadly triangular lobes and a short tube. Inner corona lobes c.1mm long, rising in small column in centre of flower; free part of outer corona lobes flattened, whitish, spreading on surface of corolla, clearly divided by deep groove opposite inner lobes.

Ecology. Rare, on the western limestone plateau; in thin soil on limestone pavement in succulent shrubland. 550m.

Distribution. Endemic.

Conservation status. Vulnerable (VU D2). Rare, with a very limited area of occupancy.

Botanical notes. Only collected once; distinguished by the pink, almost flat flowers.

Echidnopsis socotrana **Lavranos**

Stems to 10cm long, 6–10mm thick. Corolla greenish yellow with reddish rings at base, cup-shaped, 3–5mm across; lobes triangular, acute, erect to spreading, 2–3mm long. Outer corona lobes forming 5 discrete pockets between swollen lower parts of inner lobes, lower than height of anthers, inner lobes consisting only of swollen lower parts and apices, broadly truncate, touching and covering only bases of anthers.

Ecology. Scattered and locally common on the limestone plateaus, typically growing in pockets of soil in the shade of rocks and bushes in dwarf succulent shrubland. 500–700m.

Distribution. Endemic.

Conservation status. Vulnerable (VU D2). Locally common but with a very limited area of occupancy.

Botanical notes. Distinguished by the yellow flowers with red centre. Possibly the commonest and most widespread species of *Echidnopsis* on the island.

Edithcolea N.E.Br.

Succulent herb. Stems roughly 5-angled, bearing tubercles each tipped with a hard conical tooth. Flowers solitary, large. Corolla with a very short tube and large spreading lobes. Pollen united into pollinia; pollinia paired. Fruit a pair of spreading slightly flattened fusiform follicles. 1 species. E and NE Africa.

Edithcolea grandis **N.E.Br.**
Edithcolea sordida N.E.Br.

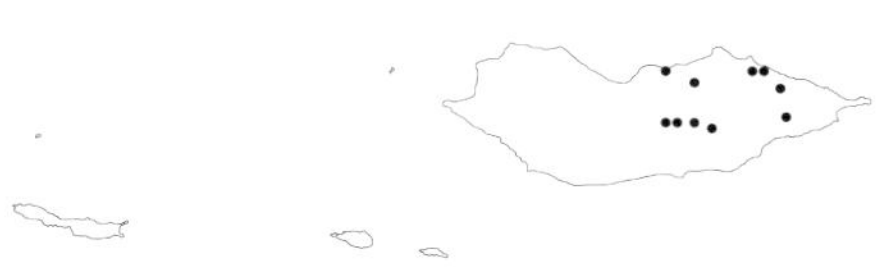

Stems branched, prostrate, to 50cm long. Flowers large, 8–12cm diameter, deep maroon and yellow; corolla lobes ovate, covered with purple hairs.

Ecology. Widespread but never common; in open woodland and succulent shrubland, mainly on limestone. 50–700m.

Distribution. Tropical E Africa and S Arabia.

Botanical notes. A spectacular succulent with very large but foetid-smelling flowers.

ETHNOGRAPHIC NOTES

Name(s)

– Soqotri: *məśḥˈermәhīm di gˈirbag, məśḥˈermәhīm di gˈirboy, di gˈiribighˈitin* 'of the wild cat'; (western Soqotra) *məśxˈayrem di ᶜˈʌġərib* 'of the raven' (plants with the name 'of the wild cat/of the raven' are always inedible, often ill-smelling or unpleasant in some other way).

– Soqotri simplified: *mish'hermihiym di girbag, mish'hermihiym di girboy, di girbig'hitin*; *mish'khayrem di 'aghirib*.

Food (human). This is not normally eaten by people, though the tender growing tips and young fruit are nibbled in hunger.

Food (livestock). Not palatable to livestock, though goats and sheep have been seen to eat the flowers.

Medicinal / veterinary. Sores on the teats or udder of livestock are treated by pounding the stems to a paste which is then smeared over the sores. The same paste is also used to treat dry eczema, especially on the hands, and deep fissures in the thickened plantar skin of the heel. In sheep, camels and goats with mange, the paste is rubbed over the bare skin where the hair has fallen out.

Cleansing. The succulent plant can be crushed and used to clean the hands.

Notes. This is also sometimes called *məśḥˈermәhīm ᶜˈafər*, 'red *Caralluma socotrana*'. Birds pull the silky fibres from the pods and use them to line their nests.

Comparative uses. In East Africa, the tips of the stems are chewed as food. The root is boiled or the whole plant added to milk and given to increase strength, especially for small children. In circumcision ceremonies the plant is mentioned for a blessing by the elders. (Heine *et al.* 1988)

Glossonema Decne.

Annual or perennial herbs. Flowers in sessile umbels. Corolla with a very short tube and 5 spreading lobes. Pollen united into pollinia; pollinia paired. Fruit a solitary, ovoidal follicle, covered in soft, fleshy spines. 4 or 5 species. Tropical Africa and SW Asia.

Glossonema revoilii Franch.

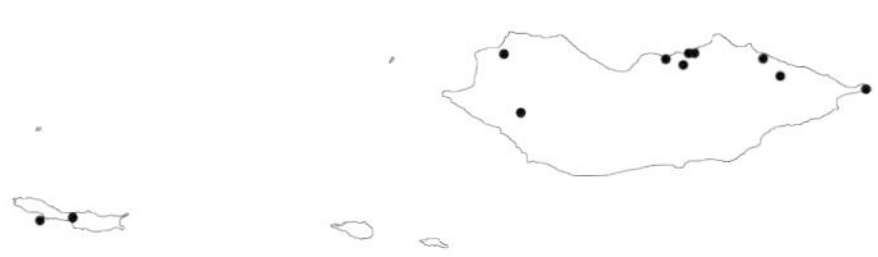

Ascending herb to 20cm. Leaves petiolate, oblong-elliptic, <3cm long, rounded at base and apex. Flowers yellowish, glabrous within. Fruits ovoid with an acute tip, <5cm long, not conspicuously inflated, with soft fleshy spines.

Ecology. Widespread in dwarf shrubland and woody-herb communities. Sea-level–600m.

Distribution. Tropical NE Africa.

Botanical notes. Distinguished by the relatively large, softly spiny fruits.

ETHNOGRAPHIC NOTES

Name(s)

– Soqotri: *fəḳʌ́li*, *fə́ḳəla* (pl. *fəḳā́lihi*) (Soqotra and Abd al Kuri), *fāllə* (Abd al Kuri); fruit *ḥárśīm di fə́ḳəla*.

– Soqotri simplified: *fiqali*, *fiqila* (pl. *fiqalihi*); *fallih*; *harshiym di fiqila*.

Food (human). The fruit are gathered and eaten, and when other food is scarce they are actively harvested. They are very sweet when young and tender and taste like fresh peas. They become fibrous and tasteless with age. They are always eaten raw, and do not keep well. They are not an important food but are much appreciated as a fresh-tasting treat.

Food (livestock). They are slow to respond to rain. All livestock eat the plants, and goats in particular like the fruit. In the drier areas and in the outer islands the plant is regarded as a valuable and nutritious fodder for sheep and goats.

Comparative uses. Many related species are also eaten. The fruit are commonly eaten by breastfeeding women to increase their milk. (Ambasta 1982; Dalziel 1948, 1956; Jain & De Fillips 1991; Miller & Morris 1988)

Marsdenia R.Br.

Shrub. Leaves petiolate, large. Flowers in dense axillary clusters. Corolla lobes spreading and slightly inwardly curved towards edges. Pollen united into pollinia; pollinia paired. Fruit a solitary follicle. 1 endemic. c.100 species. Tropical and warm regions.

Marsdenia robusta Balf.f.

Dregea arabica sensu Bullock non Decne.

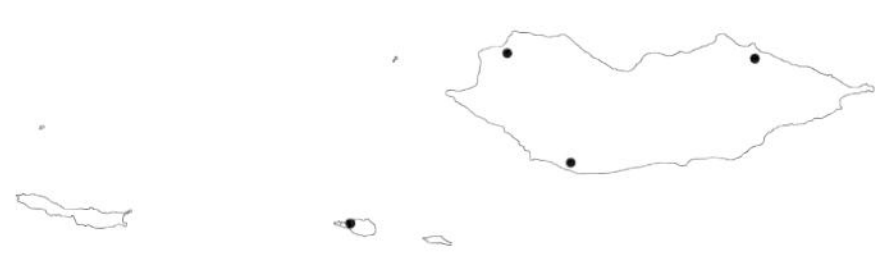

Stems erect, to 1.5m, often many-stemmed from base following coppicing. Leaves petiolate, broadly ovate to triangular, <6 x 5cm, acute or rounded at tip, cordate or truncate at base. Corolla cream, c.4mm across, corolla lobes ovate. Fruit ovoid.

Ecology. Widespread but never common, typically on cliffs in dry deciduous woodland. 50–500m.

Distribution. Endemic.

Conservation status. Vulnerable (VU D2). Rare, with a very limited area of occupancy.

Botanical notes. Readily distinguished by its petiolate, heart-shaped leaves and the axillary clusters of tiny, sweet-smelling white flowers.

ETHNOGRAPHIC NOTES

Name(s)

– Soqotri: nil or *mārə* (pl. *ʾémrəhor*) (also given to *Cocculus pendulus*, *Cryptolepis* species, *Vincetoxicum linifolium*, *Euphorbia obcordata* and *E. socotrana*); *ṣáʿalhil* (Samha) (a common name for shrubs and trees with a single trunk and large leaves, often to differentiate them from similar, smaller plants with small(er) leaves).

– Soqotri simplified: *meyrih* (pl. *'emrihor*); *za'alhil*.

Food (livestock). On Samha, this is an important fodder plant for small stock, especially sheep, as it persists through the dry season. Livestock are herded to graze in the Nuff area where many of these shrubs grow.

Bees / honey. Honeybees are attracted to the very fragrant flowers.

Fuel. On Samha the dead wood is used as firewood.

Wood. The dead wood is important on Samha for firesticks. It is one of only three suitable woods for this on the island, *Ficus cordata* and *Withania* sp. being the other two. The wood is quite hard and on Samha is used for making a variety of implements. In weaving, it is also used for the sword beater, the 'bow' for beating sheared wool to prepare it for spinning, and other wooden parts of the ground loom. The bark is used to make the fish trap. It is cut off in strips which are then split lengthways and put to soak in a pool of seawater for three or more days. Once sufficiently softened, they are woven into the various sections of the fish trap.

Cleansing. Sprigs of the sweet-smelling flowers are tucked into the clothing and hair. The sniffing of sweet-smelling plants is held to protect from 'noxious air' which can cause disease, and to help protect against infection.

Comparative uses. The leaves, shoots and fruits of some *Marsdenia* species are eaten, but others, such as *M. lucida*, are known to be poisonous to both man and animal. Species such as *M. roylei*, *M. tinctoria* and *M. tenacissima* yield a fibre used for fishing nets and lines. The roots of *M. tenacissima* constitute the drug White Turpeth, used as a purgative. The leaves of *M. tinctoria* yield a dye similar to indigo and are used to treat stomach ache. A decoction or cold infusion of the pounded leaf stems of *M. latifolia* with added lime juice is given for colic and the intestinal symptoms usually associated with worms. The leaves are rubbed on the joints of small children to help them to walk, and are used medicinally by Muslims as a purge during Ramadhan. The stems are used as chewsticks. (Ambasta 1982; Dalziel 1948, 1956; Jain & De Fillips 1991; Kunkel 1984; Lewis & Elwin-Lewis 1977)

Pergularia L.

Daemia Poir.

Herb with erect or twining stems; latex white. Leaves cordate, stalked. Flowers in stalked cymes. Corolla with a short tube and erect lobes; corona double, prominent. Pollen united into pollinia; pollinia paired. Fruit a pair of swollen follicles, covered with soft teeth. 3 to 5 species. Drier parts of Africa and Asia.

Pergularia tomentosa L.

Stems erect or twining, at least when young, to 75cm, covered with soft, grey-green hairs throughout. Leaves ovate with a cordate base. Flowers greenish with conspicuous hairs within. Fruit grey, paired, erect and pressed against each other, covered with short fleshy spines.

Ecology. Widespread and common in dry open places in *Croton socotranus* shrubland, succulent shrubland and deciduous woodland. Sea-level–700m.

Distribution. SW Asia and Africa.

Botanical notes. Readily distinguished by the grey-green, heart-shaped leaves which produce copious, and rather caustic, white latex when cut.

ETHNOGRAPHIC NOTES

Name(s)

– Soqotri: *ʾˈirəhem.*
– Soqotri simplified: *ʻirihem.*

Food (livestock). Although this plant is widely regarded as unpalatable to livestock, animals on Soqotra readily eat it. Small stock that are used to this fodder like it, especially goats. Even incoming goats eat the fallen foliage and the fruit, though sheep are less keen. All parts of the plant are regarded as a useful feed for livestock in the dry season when there is little other green to be found. Hungry donkeys and camels also eat the plant.

Leatherwork. The foliage and green stems are crushed with a little water and rubbed into the non-haired side of the freshly removed skin or hide. This is used especially on skins which are to be used for the more delicate or decorative leather items. Every last hair has to be removed for these, and using this latex produces fine but thin leather. The latex is never used on water- or milk-skins, or on the skins for storing dates, because it makes the leather too thin. Great care must be taken using this depilatory material as it acts very quickly and, if left on too long, can make the skin too fragile or can burn holes right through it. The treated skin is rolled up and left for a short time, sometimes only 10 minutes, and its progress checked repeatedly. It makes a leather 'like paper', and this is used especially on the sheepskins used to make the intricately plaited belts formerly worn by men. It was also used to make the carefully sewn satchel in which a man carried his personal possessions, and to prepare the skin, removed from the back leg of a goat, from which the sheath for the precious three-knife set of the islands was made.

Medicinal / veterinary. The latex is dripped into chronic ulcers or persistent sores to 'burn them out' or it is spread on a leaf and the leaf put over the affected area like a sticking plaster. It is used in the same way for boils, swellings and sores which have not yet burst to soften the skin and encourage the pus to emerge. The latex is used as a haemostatic on fresh lacerations and is applied to soothe the pain of a bee sting. If a thorn is proving difficult to extract, the latex is dripped onto the surrounding area to soften the skin so that it can more easily be removed. A discharging sore, especially on the foot, is treated by nicking it with something sharp until it starts to bleed, and then passing a thorn or needle around and under it, from side to side. The area is then smeared with the latex. The latex is also used to relieve itching anywhere on the body, and to treat scabies. The leaves and stems are scorched at the fire until blackened and then powdered and added to water; this lotion is then used as a wash to treat wounds and lacerations.

Comparative uses. The young tender leaves and tendrils are used as a potherb, and the flowers are also eaten. The plant is used by tanners to de-hair hides: the latex or whole pulped plant is spread on the hide for one night, or, with water and wood ash, it is made into a bath and allowed to ferment for two or three days to enable the hide to absorb the tannin readily and evenly. This and *P. daemia* are widely used medicinally.

Whole plant. It is used as an external application for rheumatism, sores, to treat a fever and for eye complaints. The plant has emmenagogue properties with a pituitrin-like action on the uterus, and is given in childbirth and to regulate menstruation. The plant is fed to turkeys to cure diarrhoea and the Fulani give it to cows to improve their milk yield. An extract of the plant is taken as a purgative and expectorant, and is used to treat haemorrhoids. An extract is taken as a purgative, vermifuge and as an abortifacient.

Leaves and stems. After circumcision a leaf-bearing shoot of the plant is warmed over the fire and the warm juice squeezed onto the wound. The leaves are used for infantile and general diarrhoea, as an expectorant and emetic, and have anthelmintic properties. They are rubbed in water which is then used as a cold infusion to bathe the head for a headache. A soap containing the leaves is used for general debility, to remove excessive oils from a baby's skin, as an analgesic and to relieve weakness and sluggishness in pregnant women. The leaves are added to a soup which is given to a parturient woman. They are crushed to make a poultice for boils, abscesses and carbuncles, and their juice, with added lime, is applied to rheumatic swellings. A decoction is used in asthma. The stems are used as a substitute for flax and fishing lines, and are soaked in warm milk which is added to the milk-skin to coagulate the milk.

Latex. This is used to treat skin diseases, especially ringworm and cases of chronic pruritus. It is sometimes also used as an adhesive.

Bark. It is fried in oil and rubbed on the skin to treat superficial dry skin lesions.

Root. To treat coughs it is chewed, or pounded and soaked in cold water, or boiled and used as an infusion. An infusion is also given to treat stomach pains and as an abortifacient in East Africa. (Ambasta 1982; Dalziel 1948, 1956; Fleurentin & Pelt/Fleurentin *et al.* [1982, 1983a,b,c]; Ghazanfar 1994; Goodman & Hobbs 1988; le Houerou 1980; Jain & De Fillips 1991; Miller & Morris 1988; Sofowora 1982)

Periploca L.

Socotora Balf.f.

Shrub; latex white. Leaves minute or absent. Flowers in few-flowered axillary cymes. Corolla tube short, the lobes spreading. Corona present. Anthers woolly; pollen free and transferred by spoon-shaped translators. Fruit a pair of narrowly cylindrical, smooth, paired, follicles diverging at 180 degrees. 10 species. Africa, SW and SE Asia.

Periploca visciformis (Vatke) K.Schum.

Socotora aphylla Balf.f.
Socotora visciformis (Vatke) K.Schum.

Densely branched shrub, the branches glabrous, stiff or flexuous. Leaves minute, triangular, soon deciduous, c.5mm long. Flowers 10–15mm across, cream or yellow, Fruits 5–7cm long.

Ecology. On cliffs in succulent shrubland and dry woodland. 50–450m.

Distribution. S Arabia, Somalia and Ethiopia.

Botanical notes. A densely branched leafless shrub with stiff or rather wiry branches. Usually found on cliffs, where it often forms large clumps up to 2m across.

ETHNOGRAPHIC NOTES

Name(s)

– Soqotri: *ʿˈarṭah* (also given to *Crotalaria strigulosa*), (western Soqotra) *xˈarṭah*; (western Soqotra) *ftˈʌx* 'split in two'; *šəfˈʌṭehʌn*; (rare) *ḳˈerʿiš* (more commonly given to *Jasminum fluminense* and *Turraea socotrana*).

– Soqotri simplified: *'artah*, *khartah*; *ftakh*; *shifetehan*; *qer'ish*.

Food (livestock). This is little eaten by livestock, though goats chew away at it if the dry season is prolonged, and sheep and goats eat the flowers, fruit and a few leaves. It is not regarded as having any nutritional value.

Flavouring / food additive. If a goat or sheep has been eating the plant a lot, the taste of the viscera, especially the stomachs, is altered and becomes unpleasant.

Fuel. Dead plants are gathered and stored at home as tinder and firelighters. The smell of the burning material is very strong and distinctive and can be recognised at a great distance.

Other. The plant is used to strengthen and darken cotton fishing equipment. This is to make it less visible in the water when fishing in rocky areas. The whole plant is crushed, added to water and left to steep. The cotton fishing equipment is lowered in and left there until the required shade has been achieved. Skeins and balls of imported cotton twist can be treated in the same way.

Notes. The plant is removed from grazing paddocks and enclosures to make room for grass and because it is not thought to be a good feed.

Comparative uses. In nearby Dhofar, the whole plant is likewise crushed and added to water to dye and toughen cotton fishing equipment. The inner bark yields a fibre, mainly used for temporary repairs to fishing and camel tackle. The buds and flowers of the closely related *P. aphylla* are eaten, and the dried stems were scorched and ground as a snuff. The bush is much used in the construction of livestock pens and the root is used by some specialists to treat snakebite. (Miller & Morris 1988)

Sarcostemma R.Br.

Low shrubs; stems erect or decumbent, cylindrical; latex white. Leaves small and quickly deciduous. Flowers in axillary clusters. Corolla lobes erect, free almost to base; corona double, joined to the stamens. Pollen united into pollinia; pollinia paired. Fruit a smooth, usually erect, fusiform solitary follicle. 1 endemic. 10 species. Tropical and warm regions.

Botanical notes. For revisions see Meve & Liede (1996) and Lavranos (1972).

Sarcostemma socotranum Lavranos

Stems green to grey-green, rigid, delicate, succulent and pencil-like. Leaves more or less absent. Corolla cream, <8mm across. Corona with conspicuously swollen inner lobes behind the anthers and small outer lobes forming a short skirt around the base of the column.

Ecology. Widespread but never common in succulent shrubland and dwarf shrubland. 200–650m.

Distribution. Endemic.

Conservation status. Data Deficient (DD). Inadequate evidence is available to make an assessment.

Botanical notes. Most plants seen on Soqotra were not in flower and so could not be identified with certainty. The map must therefore be treated with caution. A specimen from the limestone plateau on Samha is rather atypical: the flowers are small, as is typical of *S. socotranum*, but the corona is different. It has been included under *S. socotranum* for the time being.

ETHNOGRAPHIC NOTES†

Name(s)

– Soqotri: *məndūrəb* (Soqotra and Samha); *məndūrʌṣ*; *ʿˈʌlihʌṭ* (also given to *Cissus paniculata*); *di ḥˈalbeb*; *riy-ɛ̄bə* 'prostrate with exhaustion' (more commonly given to *Ceropegia* specis); *bˈilbil* 'pearl oyster'.

– Soqotri simplified: *minduwrib*; *minduwraz*; *'alihat*; *di halbeb*; *riyebih*; *bilbil*.

Food (human). The plant stems are chewed to relieve thirst and for their pleasantly astringent flavour. The flowers and fruit are also eaten and the root system can be peeled and eaten.

Food (livestock). Livestock graze the new green growth, and chew the mature stems towards the end of the dry season if this is prolonged. Cattle eat the stems if they are extremely hungry and there is little else. Donkeys and camels also eat the plant. Goats and sheep eat the flowers and fruit.

Other. People chew and suck the stems to relieve thirst.

†Also apply to *S. viminale*.

Sarcostemma viminale **R.Br.**
Sarcostemma daltoni sensu Balf. non Decne.

Similar to *S. socotranum* but a more robust plant; flowers larger, at least 10mm across; outer corona forming a low continuous ring around the base of the swollen inner lobes.
Ecology. No habitat details available.
Distribution. S Africa to Arabia, W Asia to Australia.
Botanical notes. John Lavranos, in his original description of *S. socotranum*, mentions another, larger, *Sarcostemma* from lower altitudes in the Haggeher (Wadi Di Negehen) which he thought might be *S. viminale*. There is also an old 19th century record of this species from NE Soqotra (Balfour 1888); unfortunately, the specimen on which it is based could not be traced and therefore the presence of this species on the island is still in question.

ETHNOGRAPHIC NOTES
Name(s)
See *S. socotranum* (also uses).
Comparative uses. In nearby Dhofar the stems and flowers of *S. viminale* were an important source of food in the early rains. Convulsions have been observed in cattle grazing *S. viminale*. In the Middle East the stems are eaten, and the whole plant is seen as a general cleanser and tonic. In East Africa, the root and stems are boiled and eaten for general health, warriors and elders drinking this with soup, children with milk. The stalks are chewed to relieve a cough and fever. The latex is abundant and is used as a fish poison. It is also used as a galactagogue for women and cows. (Ambasta 1982; Ghazanfar 1994; Heine *et al.* 1988; Jain & De Fillips 1991; Kunkel 1984; Lewis & Elwin-Lewis 1977; Miller & Morris 1988; Oliver-Bever 1986)

Secamone R.Br.

Shrubs or woody climbers; latex white. Leaves opposite, often appearing whorled on short side branches. Flowers in axillary clusters. Corolla tube short; corona lobes inconspicuous. Staminal column arising at base of corolla. Style head protruding beyond anthers. Pollen united into pollinia; pollinia in groups of 4. Fruit a pair of slender, smooth, fusiform, spreading follicles. 2 endemics. c.100 species. Old World tropics.

Secamone cuneifolia **Bruyns**

Shrub, small tree or woody climber to 3m with rather jointed stems. Leaves linear oblanceolate, <40 x 4mm, attenuate at base; abruptly acute at tip. Flowers pale yellow, in axillary clusters. Pedicels and calyx more or less glabrous. Corolla c.3mm, the lobes c.1.5mm long. Staminal corona lobes c.0.15mm long. Style head exserted from staminal column by 0.2mm.
Ecology. Rare in succulent shrubland and semi-deciduous woodland. 100–700m.
Distribution. Endemic.
Conservation status. Vulnerable (VU D2). Known only from three locations with a total area of occupancy of less than 20km^2. Its extreme rarity makes it vulnerable to the effects of human activities.
Botanical notes. Distinguished by the narrow, obcuneate leaves.

ETHNOGRAPHIC NOTES
Name(s)
See *S. socotrana* (also uses).

Secamone socotrana **Balf.f.**
Secamone socotrana var. *glauca* Radcl.-Sm.
Secamone socotrana var. *socotrana* Balf.f.

Shrub or woody climber to 3m. Leaves obovate, <3 x 1.5cm, rounded and sometimes mucronate at tip. Flowers cream or pale to bright yellow, in axillary clusters. Pedicels and calyx with whitish, silverish or reddish hairs. Corolla 4–5mm long; corolla lobes oblong, 2.5–3mm long. Staminal corona lobes c.1mm long. Style head exserted from staminal column by c.1mm.
Ecology. Climbing over rocks and shrubs in succulent shrubland and semi-deciduous woodland. 20–800m.
Distribution. Endemic.
Conservation status. Least Concern (LC). Common in several vegetation types and under no present or perceived threat.
Botanical notes. Plants with silvery or white versus reddish hairs on the calyx and ovate versus oblong corolla lobes have been separated off as var. *glauca*. However, recent collections have shown that *S. socotrana* is much more variable in these characters than previously thought and var. *glauca* is not recognised here.

ETHNOGRAPHIC NOTES†
Name(s)
– Soqotri: nil or *di rikəbə*, 'riding' (the tree or shrub it is growing in) (also sometimes given to *Jasminum fluminense*, which is not palatable to livestock); (sometimes) *rikəbə di ᵓikšeh* '*Secamone* of *Commiphora ornifolia*' (it is often found climbing in this tree), *ᶜeḳḳəhīn* (pl. *ᶜaḳāhan*), (western Soqotra) *ġeḳḳəhīn*.
– Soqotri simplified: *di rikibih*, *rikibih di 'iksheh*; *'eqqehiyn* (pl. *'aqahan*) (*gheqqihiyn*).
Food (livestock). All livestock love the green stems and foliage when they can reach it, but it commonly climbs right up and over the top of the host tree where it is out of reach of all but the occasional camel. Wherever possible, goats climb up into the host trees to get at the foliage. Sheep have to make do with fallen leaves and other

plant litter. Herders lop leafy stems or strip the leaves from the stems to take to their livestock in the dry season. It is regarded as a valuable feed for all animals, and as increasing milk yield.
Bees / honey. Honeybees are attracted to the fragrant flowers.
Wood. The flexible stems are used for the wood of the 'bow' used to beat the wool after shearing to prepare it for spinning. It is also used to make the wooden rim of the leather watering bucket.
Cleansing. Sprigs of the sweet-smelling flowers are tucked into the clothing and hair. The sniffing of sweet-smelling plants is held to protect from the 'noxious air' which can cause disease, and to help protect against infection.
Comparative uses. In the Philippines the fruit of *Secamone* species are heated and pounded with a little salt to a paste which is inserted into a painful and carious tooth. The copious latex of *S. myrtifolia* is applied to boils, etc., and a decoction of the leaves is widely used as a purge. (Dalziel 1948, 1956; Lewis & Elwin-Lewis 1977)

†Also apply to *S. cuneifolia*.

Socotrella Bruyns & A.G.Mill.

Stems 4-angled, with soft, fleshy teeth. Leaves absent. Flowers produced towards apex of stems. Corolla campanulate. Pollen united into pollinia; pollinia paired. Fruit unknown. 1 species. Endemic.

Botanical notes. Endemic monotypic genus – sole species *S. dolichocnema*. Differs from *Caralluma* by the absence of leaf rudiments (Bruyns & Miller 2002).

Socotrella dolichocnema Bruyns & A.G.Mill.

Stems up to 15cm long, 4–7mm thick, ascending from horizontal and often rhizomatous base, dark green. Flowers held upright on long (25–42mm) ascending pedicels. Corolla yellow with longitudinal reddish markings in tube, 18–22mm across, cup-shaped; tube shallow, 2mm deep; lobes 7–8mm long. Corona 2mm tall; outer lobes reduced to minute flaps; inner lobes 0.5mm long, erect.
Ecology. Rare on limestone pavement, in dwarf succulent shrubland on the western plateau. 650–800m.
Distribution. Endemic.
Conservation status. Vulnerable (VU D2). Rare, with a very limited area of occupancy.
Botanical notes. The specific epithet refers to the long, erect pedicels which hold the flowers facing upwards.

ETHNOGRAPHIC NOTES
See notes under *Echidnopsis*.

Vincetoxicum Wolf

Climbing plant; latex clear. Leaves opposite, often inconspicuous. Corolla rotate. Pollen united into pollinia; pollinia paired. Fruit a solitary fusiform follicle. 1 endemic. 15 species. Temperate Eurasia.

Vincetoxicum linifolium Balf.f.
Cynanchum linifolium (Balf.f.) Bullock

Densely branched twining herb, the flexuous, linear stems often forming a dense tangled mass. Leaves linear or scale-like, inconspicuous. Flowers small, reddish brown, 1.5–2mm across, in clusters on short axillary branches. Fruit fusiform, elongated into a long sharp point, c.6 x 1cm.
Ecology. Common in dry woodland and shrubland. 15–600m.
Distribution. Endemic.
Conservation status. Least Concern (LC). Common in several vegetation types and under no present or perceived threat.
Botanical notes. A leafless climber with dense, often tangled stems and minute reddish brown flowers.

ETHNOGRAPHIC NOTES†
Name(s)
– Soqotri: *ᶜeḳḳəhīn* (pl. *ᶜaḳāhan*), (western Soqotra) *ġeḳḳəhīn*; *məzūwkan* 'twining'; *mārə* (pl. *ʾemrəhor*) (also given to *Cocculus pendulus*, *Cryptolepis* species, *Euphorbia obcordata*, *E. socotrana* and *Marsdenia robusta*); (rare) *təᶜabəhor* 'plant that comes into leaf at the end of the dry season'.
– Soqotri simplified: *'eqqihiyn* (pl. *'aqahan*), *gheqqihiyn*; *mizuwkan*; *merih* (pl. *'emrihor*); *ta'abihor*.
Food (livestock). These plants normally grow out of the reach of livestock, but those that can reach the green stems or the leaves, flowers or fruit readily consume them. However, this is not considered a useful forage, except towards the end of the dry season when there is little else in the drier areas of the island.
Wood. The thicker flexible stems of *V. linifolium* are used for the 'bow' made to beat the wool after shearing to prepare it for spinning. They are also used to make the wooden rim of the leather watering bucket, the 'bit' of the weaning device worn by kids and lambs, and the upright for the spindle. A suitable crooked stem is sometimes cut for use as a fodder hook to bring foliage within reach. Thicker stems can be made into the important herding staff. The dead stems can be used as firesticks.
Medicinal / veterinary. The latex is used to treat swellings, raised neck glands especially.
Comparative uses. The leaves and tubers of many related species are eaten. The leaves of *Cynanchum arnottianum* are used as an insecticide, and are ground to a powder to kill maggots. The root of *C. vincetoxicum* is given as an emetic and diuretic. (Ambasta 1982; Heine *et al.* 1988; Jain & De Fillips 1991; Kunkel 1984; Lewis & Elwin-Lewis 1977)

†Also apply to *Cynanchum gerardii*.

AVICENNIACEAE

Trees. Young twigs 4-winged. Leaves opposite, entire; stipules absent. Flowers in dense cymes, somewhat zygomorphic. Calyx 4-lobed. Corolla 4-lobed. Stamens 4. Ovary superior, 4-celled; ovules 1 per cell, axile; style 1. Fruit a drupe. 1 genus. 14 species. Widespread on tropical coasts.

Botanical notes. Often included in the Verbenaceae.

Avicennia L.

See description of family.

***Avicennia marina* (Forssk.) Vierh.**
Avicennia officinalis L.

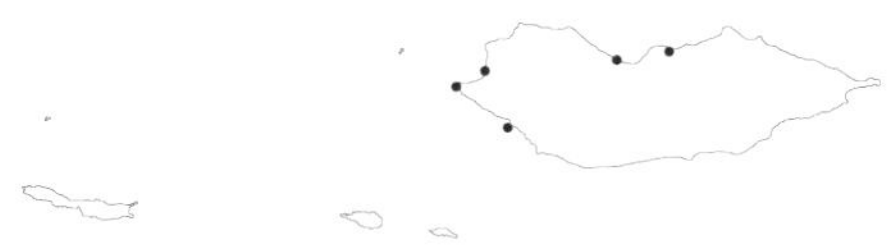

Mangrove tree growing close to the sea, to 5m, with erect breathing roots (pneumatophores). Leaves ovate to narrowly ovate, 3–8 x 1–2.5cm, base acute, entire, tip acute, greyish below. Flowers yellow or orange, c.0.5cm long. Fruit ovoid, <2cm long.

Ecology. Forms pure stands of mangrove woodland; scattered around the coasts of the north and west of the island. Sea-level.

Distribution. Widespread along the coasts of Africa, Arabia, Asia to Australia and New Caledonia.

Botanical notes. The only mangrove species on the island. The mangrove woodlands on Soqotra are apparently declining. The reasons for this are not clear; it could be due to cutting for timber or perhaps because of changes in the water table. There is evidence of recent destruction: for instance, the mangrove at Ghubbah was intact (but not particularly healthy) in 1989 but had been almost completely destroyed by cutting by 1990. There is some indication that this decline in the mangroves is not a recent phenomenon. Balfour (1888) states that mangrove grew in abundance around a creek near Qadhub, and Bent & Bent (1900) described how the lagoons at Qalansiyah and Di Hamdh were once renowned for their thick and extensive mangroves; there is no longer mangrove at any of these places. The most extensive areas of and healthiest mangrove woodland is now at found at Niyt in the west of the island.

ETHNOGRAPHIC NOTES

Name(s)

– Soqotri: *ˀˈeḳrəhəm*, *ˀˈeḳərhum* ‘longed for, addicted to’.
– Soqotri simplified: *‘eqrihim*, *‘eqirhum.*

Food (livestock). Camels along the coast browse the foliage the year round. For goats the leaves and stems are a drought fodder only. Sheep and goats eat plant litter in the dry season.

Fuel. The dead wood makes good firewood.

Timber. Provides good, termite-resistant building timber, and in the more settled areas of the northern coast (as around Di Hamdh) it was formerly heavily cut for this. Sadly, the cut plants have failed to regenerate and there is little to cut for timber now.

Wood. Hard, and was much used to modify the imported dugout canoes. It was also used to make oars, rollers, masts, and the handles of gaffs and harpoons. It was used to make a variety of implements, and the dead wood made excellent charcoal.

Notes. The mangrove appears in many stories in Soqotra where an ‘angel from God’ appears in disguise to test the charity of the wealthier, date palm-owning islanders. When the owner is reluctant to part with any of his dates, he is punished by his date palms being miraculously transformed into mangroves. The mangroves of Di Hamdh and the Shu’ub area are popularly said to be the result of such punishment.

Comparative uses. In Oman the leaves and stems are used as a camel fodder and a decoction of the bark is drunk as an astringent and aphrodisiac. In Qatar the astringent quality of the bark was used in treating smallpox. The roots were taken as an aphrodisiac, the leaves as a poultice, and the unripe seeds to accelerate suppuration of boils and abscesses. In Bahrain a decoction of the bark was used as an astringent. In tropical Africa and Asia the aromatic and bitter juice is used as an abortive. In India the wood is much used in building, the bark in tanning, and the bitter juice taken as an abortifacient. In Kenya the branches were used to make dhow ribs, the wood in lime burning and the bark to make a brown dye. In Africa the root and leaves of the related *A. germinans* are used as a vegetable salt, the leaves as an enema, the fruit are eaten in times of famine, and the bark used to treat dermatitis and in childbirth. The roots are used to treat lower intestinal complaints. The termite-proof wood of the related *A. nitida* in Africa is much used in boat- and house-building and in constructing wharves and piles. It makes a good firewood and excellent charcoal. The leaves and roots are used to make a salt, and the bark, as well as being used in tanning and as a red dye, is also used medicinally to treat itch and parasitic skin diseases. (Abbas *et al.* 1992; Batanouny 1981; Dalziel 1948, 1956; Ghazanfar 1994; le Houerou 1980; Lewis & Elwin-Lewis 1977; Rizk & El-Ghazaly 1995)

BEGONIACEAE

Herbs. Stems succulent, jointed. Leaves alternate, peltate, fleshy; stipules large. Monoecious. Flowers in bracteolate cymes. Male flowers with 4 tepals (petal-like segments), subequal or unequal; anthers many. Female flowers with (5 or)6 tepals, subequal; styles 3. Ovary 3-celled; placentation axile. Fruit a 3-winged, dehiscent capsule. 2 genera. c.900 species. Tropical and warm regions.

Begonia L.

See description of family. 2 endemics. c.900 species. Tropical and warm regions, especially New World.

Begonia samhaensis M.Hughes & A.G.Mill.

Similar to *B. socotrana* but separated by the asymmetrically ovate leaves and the unequal tepals in the male flowers; outer tepals broadly orbicular, 1.5–2.2 x 1.7–2.5cm; inner obovate elliptic, 1.4–2 x 0.8–1.4cm.

Ecology. Samha: damp and shady cracks in north-facing limestone cliffs. 600–700m.

Distribution. Endemic.

Conservation status. Endangered (EN B2 a biii). *B. samhaensis* is endemic to north-facing cliffs on the summit pinnacle and escarpments of the limestone plateau at the centre of Samha. Its area of occupation is less than 10km^2. The cliffs where it grows catch precipitation and mists (principally from the NE monsoon) and represent a unique habitat on Samha. However, it seems that this habitat is threatened by the trend of lower rainfall in the region and would be particularly vulnerable to the effects of global warming (see also *Ledebouria insularis* and *Pelargonium insularis*). The situation is particularly critical for *B. samhaensis* which occupies a niche at the highest point on Samha and so, with increasing aridification, has no possibility of mitigating the effects of gradual drying by migrating upwards in altitude (cf. the situation of *B. socotrana* on Soqotra island). A further and potentially very important problem has been a recent change in livestock management on the island. Until very recently goats and sheep were actively herded and were moved down from the limestone plateau during periods of drought. However, both the sheep (now largely feral) and the goats (actively managed only when in milk) are now left to roam freely for most of the year. Both animals are known to spend much of their time grazing on the limestone plateau and this could quickly lead to over-grazing and the eradication of plants on all but the most inaccessible ledges.

Botanical notes. Closely related to *B. socotrana*. However, unlike that species *B. samhaensis* flowers after rain and is not reliant on short daylength. See also Hughes & Miller (2002) and Hughes *et al.* (2003).

ETHNOGRAPHIC NOTES

Name(s)

See *B. socotrana* (also uses).

Begonia socotrana Hook.f.

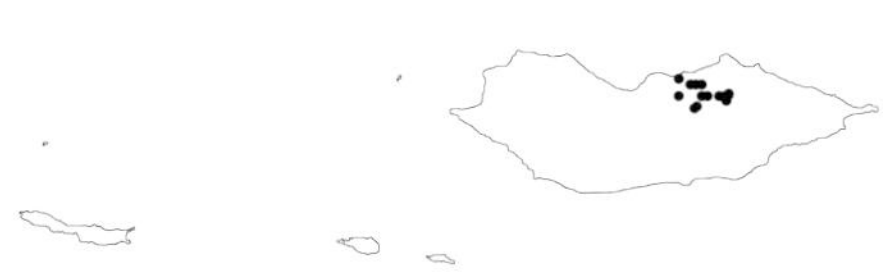

Perennial, fleshy herb to 45cm. Leaves orbicular, c.10 (–20)cm diameter; petiole <20cm long; stipules 0.9 x 0.9cm, papery when old. Male flowers pink, tepals subequal or equal, 1.8–2 x 1.3–1.9cm; anthers yellow. Female flowers pink, tepals subequal, 1.4–1.7 x 0.7–1cm; styles yellow.

Ecology. Locally common on damp and shady cracks in north-facing cliffs and occasionally on soil under dense vegetation; restricted to limestone cliffs in semi-deciduous woodland (Reyged and Rewgid plateau) and on granite cliffs and pinnacles in montane woodland and shrubland across the Haggeher mountains. 550–1300m.

Distribution. Endemic.

Conservation status. Least Concern (LC). Sparsely distributed but abundant in certain niches (north-facing cliffs and shady gullies) in two vegetation types and covering a wide altitudinal range. It has an area of occupancy of substantially less than 100km^2 but, because of its specialised niche, it is largely protected from grazing by livestock and from the destruction of its habitat. It is not liable to become threatened in the near future by the drying out of the Archipelago (discussed under *B. samhaensis*). Its relatively wide altitudinal range gives it the option of coping with gradual environmental change by migrating upwards.

Botanical notes. First collected by Balfour in 1880. He brought plants back to Britain where they were crossed with southern African species and rapidly gained horticultural importance as the hybrid parent of a host of winter-flowering begonias.

ETHNOGRAPHIC NOTES†

Name(s)

– Soqotri: *ṣʌbe̍rbiro* (pl. *ṣʌbe̍rbəher*), *ṣe̍berbir* 'sour, astringent'; *ṣəbe̍rhen* (Samha); begonias everywhere have a similar name, although larger-leaved varieties may be called *ṣəbe̍rbəher ṣa̍ᶜaləhʌl* 'large-leaved' to distinguish them from smaller-leaved plants; (rare) *źʌ̍fərə* 'plait' (and *źʌ̍fərə ḳi̍ḳəhe* 'smaller-leaved', *źʌ̍fərə ṣʌ̍ᶜəlhʌl* 'larger-leaved').

– Soqotri simplified: *zaberbiro* (pl. *zaberbiher*); *zeberbir*; (*zaberbir za'alhal*), *ziberhen*; *dhafiri* (*qiqiheh*, *za'alhal*).

Food (human). Leaves are picked and eaten for their lemony, astringent flavour, especially by those with no access to the popular tamarind fruit. Children and pregnant women are particularly fond of this food. Leaves are also chewed to relieve thirst.

Food (livestock). The plants are not normally accessible to cattle but they eat the foliage if they can. Goats and sheep nibble the plants when they find them, but tend to ignore them if there is alternative forage. In a prolonged dry season hungry livestock venture further afield in their search for feed and are more likely to graze these plants. They are not regarded as important nutritionally.

Flavouring / food additive. A handful of leaves are crushed and added to a little milk in the butter-skin to start the souring process. The soured milk is then used to ripen fresh milk before churning it to make butter.

Medicinal / veterinary. With the arrival of modern medicine and consequent new diagnosis of malaria, some islanders consider that regularly eating some leaves protects against malaria.

Other. Animals feeding on this, especially goats, can go longer without water.

Comparative uses. The leaves of many *Begonia* species are eaten both raw and cooked, and are also chewed to relieve thirst. Species such as *B. cucullata* and *B. sanguinea* are recognised diuretics. The aqueous extract of the leaves and flowers of *B. cucullata* and *B. heracleifolia* are active against gram-positive and gram-negative bacteria. A decoction of the leaves of *B. luxicans* is taken as a febrifuge, and an extract of the stalks of *B. palmata* to treat venereal diseases. (Ambasta 1982; Jain & De Fillips 1991; Kunkel 1984)

†Also apply to *B. samhaensis*.

BORAGINACEAE

Herbs, shrubs or trees. Leaves alternate or opposite, simple. Inflorescence of cymes, often spirally inrolled. Flowers bisexual or unisexual (*Cordia*), actinomorphic or rarely slightly zygomorphic. Calyx (3 to)5-lobed. Petals (4 or)5, fused, actinomorphic or zygomorphic. Stamens 5, epipetalous. Ovary superior, entire or deeply 4-lobed, 2 or 4-celled. Fruit 2 to 4 nutlets, or a drupe or a capsule. 154 genera. 2500 species. Cosmopolitan.

Arnebia Forssk.

Annual herbs. Inflorescence bracteate, spike-like. Calyx deeply divided, scarcely accrescent in fruit. Corolla funnel-shaped; throat scales absent. Anthers included or shortly exserted, subsessile. Style filiform; stigmas 2, flabellate. Nutlets 4, ovoid-pyramidal. 25 species. Mediterranean, tropical Africa, SW Asia, Himalayas.

Arnebia hispidissima (Lehm.) DC.

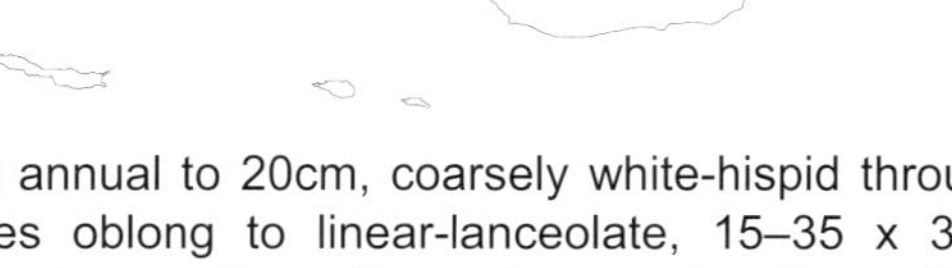

Erect annual to 20cm, coarsely white-hispid throughout. Leaves oblong to linear-lanceolate, 15–35 x 3–4mm. Bracts longer than the calyx, acute. Calyx 5–7mm, scarcely or not accrescent in fruit. Corolla yellow, 10–15mm; tube exserted from the calyx. Style simple; stigmas 2. Nutlets c.1.5mm, light orange-brown, verruculose.

Ecology. Scattered in dwarf shrubland on desertic inland and coastal plains. 15–200m.

Distribution. N and NE tropical Africa, Arabia, SW Asia to Pakistan.

ETHNOGRAPHIC NOTES

Name(s)

– Soqotri: nil or *di śkaˁinˈitin* 'prickly, spiny'; *rəḥˈammə*, (western Soqotra) *rˈʌxam* 'grey-coloured'.

– Soqotri simplified: *di shka'initin*; *rihammih, rakham.*

Food (livestock). This is one of the ephemeral *ru'ud* plants which come up with the rains, providing incidental green grazing until they die back, especially for small stock. Though palatable to livestock, it is not important forage.

Comparative uses. Toxic pyrrolizidine alkaloids, echimidine and monocrotaline, are found in this species, which make it hepatotoxic to mammals. Medicinally the plant is widely used as an anti-inflammatory. In Arabia the plant is boiled in water with added sugar and the liquid drunk to lessen fevers and as a treatment for malaria. The flowers and roots are used as a cosmetic, and the blood-red roots yield a purple-blue dye rubbed on the face as a cosmetic. In Peru the yellow latex of the stems is also used as a yellow dye. Other species also yield dyes: the roots of *A. nobilis* in Afghanistan, for instance, produce a superior red dye, Ratanjot, which is used to dye foodstuffs as well as silk and wool. (Ambasta 1982; Dalziel 1948, 1956; Ghazanfar 1994; Jain & De Fillips 1991; Mossa *et al.* 1987; Oliver-Bever 1986)

Cordia L.

Trees or shrubs. Leaves alternate or opposite, entire or toothed. Flowers in cymes, hermaphrodite or unisexual. Calyx tubular-campanulate, smooth, 3 or 4-lobed, persistent in fruit. Corolla white or cream, 4 to 6-lobed. Throat scales absent. Stamens as many as the corolla lobes. Ovary 4-locular. Style terminal, twice forked; stigmas linear or spathulate. Fruit a 1 or 2-seeded drupe, orange, ovoid, subtended by the persistent, cup-like calyx. 250 species. Tropical.

Botanical notes. Sometimes separated into the family Ehretiaceae.

ETHNOGRAPHIC NOTES

Comparative uses. Many *Cordia* species have edible fruit. In East Africa the bark of *C. sinensis* is chewed and applied to the inflamed or sore eyes of livestock, or it is soaked in water and given to cattle and small stock to drink. It is also used to relieve pains in the ribs. Here, when asking someone for livestock, it is traditional to spit on a branch and present this to the would-be donor. The roots are used as an abortifacient and the leaves to treat fevers. The bark is used for cordage and for caulking boats, and the twigs as firesticks. In Yemen the leaves are used to treat a nosebleed and as an ophthalmic antiseptic. In Dhofar, the seeds of *C. myxa* are split open and the kernel eaten to expel worms and as a remedy for diarrhoea. The leaves are boiled in water to make a tea which is drunk for stomach pains. Dried, they are used as a poultice on wounds. Goats browsing heavily on the plant in wet weather are said to develop halitosis. The leaves of *C. perrottettii* in Dhofar are used in the treatment of ophthalmic problems. The leaves of some species, such as *C. ovalis*, are so coarse that they are used as sandpaper. The bark of *C. quercifolia* exudes a gum and yields a strong fibre used for cordage. The wood is used for arrow shafts. (Dalziel 1948, 1956; Fleurentin & Pelt/Fleurentin *et al.* [1982, 1983a,b,c]; Goodin & Northington 1979; Heine *et al.* 1988; le Houerou 1980; Kunkel 1984; Lewis & Elwin-Lewis 1977; Miller & Morris 1988)

Cordia obovata Balf.f.

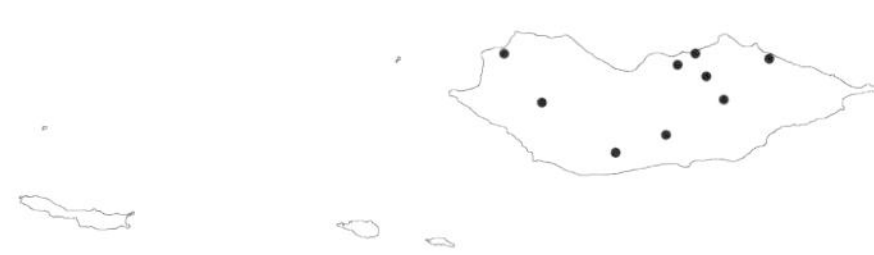

Shrub or small tree to 4m. Leaves alternate, bright green, obovate, <7.5 x 4cm, margin weakly toothed in upper half, weakly scabrid above or if not scabrid then covered with tubercules, ± glabrous beneath. Calyx 3 or 4-lobed, densely tomentose. Corolla creamy white; tube <6mm long, 4-lobed. Drupe <11 x 8mm.

Ecology. Widespread but rarely common in semi-deciduous and drought-deciduous woodland and shrubland. 100–500m.

Distribution. NE Somalia and Oman.

Botanical notes. Differs from *C. obtusa* in its leaves which are nearly always toothed near the apex and in its upper leaf surface which is scabrid with tuberculate-based hairs or, if not scabrid to the touch, then the tuberculate bases of the hairs are prominent (seen with x10 lens).

ETHNOGRAPHIC NOTES†

Name(s)

– Soqotri: *ˀəbiṭə* (pl. *ˀabuyṭ*); *šezizhīyo*; *ˀəbiṭə ṣaʿalhil* 'large-leaved *Cordia*'.

– Soqotri simplified: *'ibitih* (pl. *'abuyt*); *shezizhiyo*; *'ibituh za'alhil.*

Food (human). The ripe red fruit are eaten, especially by children. In hungry times adults also eat them, though not everyone likes the rather viscid fruit and some report that they damage the throat and the tonsils.

Food (livestock). Where other alternatives are present, the green foliage is little eaten by livestock other than camels. In a drought or if the dry season is prolonged, goats eat the leaves and also chew on the leafless twigs and gnaw at the bark. Such material is often gathered by herders in the drier areas to feed their goats at these times. Cattle eat the green leaves in the dry season, and sometimes their owners gather these as a feed for them. While not considered to be a particularly nutritious feed, it is good enough as a maintenance feed and to keep animals in milk. However, too much foliage is toxic for goats, especially at the end of the monsoon season and before the winter rains have brought up new forage. Herders say it causes hard lumps to form in the digestive system, and sometimes a blockage in the intestine which can lead to the death of the animal. Sometimes instead the animal behaves strangely for a few days, has no appetite and stumbles and staggers around, but then recovers. The fruit and fallen leaves are readily eaten by both cows and goats. However, again, if goats eat too many fruit their abdomen swells and they fall sick. This is thought to be the result once more of an internal blockage brought on by the sticky nature of the fruit. Sometimes this is fatal. Pregnant small stock that eat a lot of the fruit commonly abort the foetus, and sometimes too the adult animal dies.

Bees / honey. Honeybees are attracted to the flowers, especially after the summer rains when much of the best quality honey is made.

Flavouring / food additive. In the days before the arrival of sugar the ripe fruit (*gelḥum di ʿibiṭə*) were appreciated as a source of sweetness.

Fuel. The dead wood makes good firewood that produces little smoke and is therefore much liked for cooking. Large chunks of wood are half-buried in ash to keep fires in overnight or during an absence. A well-chosen piece can keep a fire in for two nights and a day at least. The smaller stems burn with a bright light and a store of these is often kept in the home for that purpose.

Timber. Regarded as one of the better, harder timbers of the island. The trunks and branches of the larger trees are used for the supporting pillars and main roofing beams, while the slimmer branches are used for interweaving roofing.

Leatherwork. The growing tips and young leaves are used to prepare leather from skins and hides to toughen the leather and dye it a clear red-brown colour.

Wood. Hard, and used for making a variety of implements. Slimmer, straighter branches are also made into the important herding staff. Twigs can be used as firesticks.

Pest control. In the absence of better woods, the dead wood is used for smudges lit inside the byre to keep goats and sheep warm in cold or wet weather.

Notes. The islanders distinguish between the 'female' tree, *ˀəbiṭə*, which has larger fruit and smaller leaves, and the 'male' tree, (*ˀəbiṭə*) *ṣʌʿəlhil* 'large-leaved *Cordia*' or (*ˀəbiṭə*) *ḥəwro* 'black *Cordia*', which has fewer, smaller fruit and larger leaves. In the eastern area, *C. obovata* is called *šezizhīyo* and islanders say that this is distinguished by its very fissured bark and coarse, rather tough leaves. This is regarded as a 'male' plant which does not produce fruit.

†Also apply to *C. obtusa*, *C. quercifolia*.

Cordia obtusa Balf.f.

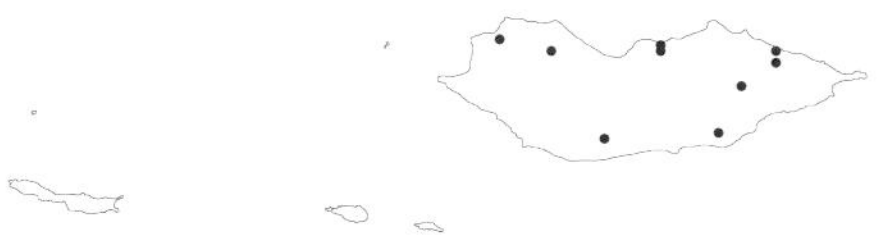

Shrub or small tree to 4m. Leaves alternate, elliptic or obovate, <6.5 x 4cm, margin entire or weakly toothed near the apex; both surfaces glabrous. Calyx 3-lobed, glabrous or softly hairy at tips. Corolla white; tube <8mm; 5 or 6-lobed. Drupe <20 x 15mm.

Ecology. Drought-deciduous and succulent shrubland, mainly on the limestone escarpments. 20–500m.
Distribution. NE Somalia.
Botanical notes. Distinguished by its smooth, glabrous leaves the margins of which are practically entire.

ETHNOGRAPHIC NOTES
Name(s)
See *C. obovata* (also uses).

Cordia quercifolia Klotzsch
Cordia gharaf (Forssk.) Asch.
Cordia rothii Roem. & Schult.

Spreading, many-stemmed shrub to 2m. Leaves opposite or subopposite, oblong-obovate, <9 x 4.5cm, rounded at tip, margin entire, prominently reticulate above and beneath. Calyx 3 or 4-lobed, densely pubescent or tomentose. Corolla cream or white, 4-lobed. Drupe <13 x 9mm.

Ecology. No habitat details available.
Distribution. N and E tropical Africa, Oman, Pakistan, India and Sri Lanka.
Botanical notes. Collected by Balfour in the 19th century but not seen since. *C. quercifolia* is distinguished from the other species on the island by its opposite or subopposite versus alternate leaves which are prominently reticulate above and beneath (the other two species are prominently reticulate beneath only).

Cystostemon Balf.f.

Herbs. Leaves alternate, entire. Inflorescence a branched panicle; flowers bracteate. Calyx 5-lobed, divided almost to the base. Corolla actinomorphic; tube cylindrical. Stamens 5; filaments with a broad, basal appendage; anthers with sterile, terminal appendages that stick together along their margins. Style gynobasic; stigma capitate. Nutlets 1 to 4, triquetrous, papillose. 1 endemic. 13 species. SW and NE Africa.

Cystostemon socotranus Balf.f.

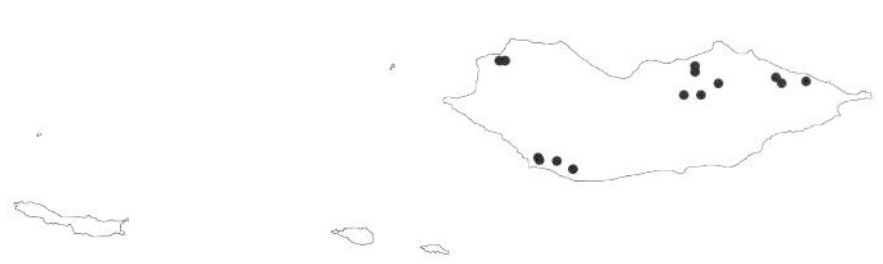

Erect perennial to 60cm, grey-green, hispid with white, tuberculate-based hairs throughout. Leaves narrowly obovate or narrowly elliptic, <12 x 2.5cm. Calyx lobes linear-oblong, 5–7mm, enlarging to c.10mm in fruit. Corolla white, c.9mm. Staminal appendage <5mm.
Ecology. Grassland and clearings in dwarf shrubland, deciduous woodland and montane woodland, also once recorded as a weed of date gardens on the Hadiboh plain at 50m. 450–1250m.
Distribution. Endemic.
Conservation status. Least Concern (LC). Common in several vegetation types and under no present or perceived threat.
Botanical notes. Readily distinguished by the azure blue flowers, marbled leaves and strigose leaves. A plant of grassland in high rainfall areas.

ETHNOGRAPHIC NOTES
Name(s)
– Soqotri: *miśḥɔ̄wgəg* 'scratchy' (commonly given to *Asparagus* and *Fagonia* species); *ḳāṣəhʌn* 'hard-packed earth'; *mḥaṭḥ'oṭen* 'limp, flaccid' (commonly given to species of fern); *marḥūməhem* 'grey-coloured'; *śkaᶜin'itin* 'prickly, spiny'.
– Soqotri simplified: *mish'howgig*; *qazihan*; *mhat'hoten*; *marhuwmihem*; *shka'initin.*
Food (livestock). Camels and donkeys readily eat this but goats only do so when alternatives are limited. Sheep generally avoid it. If a goat eats a lot the prickly hairs can damage their mouthparts. Such superficial scratches can become a real problem and cause the animal to lose all appetite.
Medicinal / veterinary. The leaves are crushed to a paste used as a dressing for sores, abscesses, ulcers and lacerations. The same paste is used to treat swellings and raised glands, especially of the neck area.

Echiochilon Desf.

Perennial herbs or subshrubs. Leaves alternate. Inflorescences racemose, bracteate. Calyx 5-lobed. Corolla funnel-shaped, actinomorphic or somewhat zygomorphic. Stamens inserted in corolla tube. Style simple or shortly bilobed. Nutlets 4, one or more often aborting, ovoid. 1 endemic. 6 species. N Africa and Arabia.

Echiochilon persicum (Burm.f.) I.M.Johnst.

Erect or cushion-forming perennial or subshrub to 35cm; foliage silvery, older branches with whitish or greyish flaking bark. Leaves narrowly oblanceolate, <1.7 x 1.2cm, densely covered with appressed tuberculate-based hairs. Flowers in terminal leafy cymes. Calyx lobes 2–4mm. Corolla white or yellowish, <7mm long. Nutlets ovoid, verrucose.

Ecology. Open semi-deciduous woodland. 500m.

Distribution. Sudan, Somalia, Arabia, Iran and Pakistan.

Botanical notes. Collected by Forbes in the 19th century at Homhil. Not seen since.

Echiochilon pulvinata A.G.Mill. & L.Urb. (see Appendix 2, p. 723)

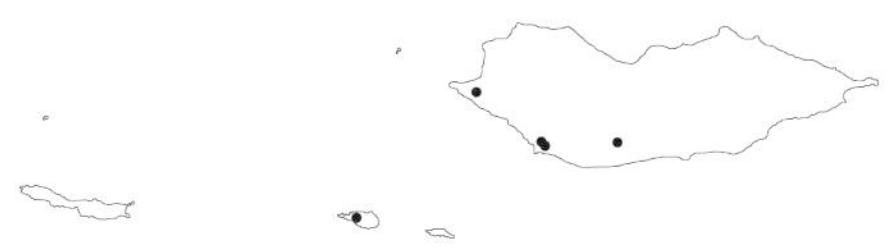

Compact, cushion-forming perennial, silvery hairy throughout. Leaves spirally arranged, densely packed, overlapping and completely obscuring the stem, oblanceolate, 2–6 x c.1mm. Flowers solitary, appearing from between the closely packed leaves. Calyx lobes linear, c.2.5mm long. Corolla white, c.4mm across; tube c.3mm. Fruit not seen.

Ecology. Locally common on Samha and SW Soqotra; in dwarf shrubland on exposed and windswept desertic limestone plateaus and cliffs. 450–700m.

Distribution. Endemic.

Conservation status. Vulnerable (VU D2). Rare, with a very limited area of occupancy.

Botanical notes. The only cushion-forming species within the genus. The flowers are almost actinomorphic and bright white.

ETHNOGRAPHIC NOTES

Name(s)

– Soqotri: nil or *ḳʌ̍ṣḳʌṣ* (pl. *ḳʌṣḳʌ̍ṣihon*) 'cream-coloured, fragmented'; *sədʌ̍ḥan* 'prostrate'.

– Soqotri simplified: *qazqaz* (pl. *qazqazihon*).

Food (livestock). Mainly grazed by small stock, especially on Samha, where it is eaten by the sheep and goats feeding on the high plateau. Where it is plentiful it is a useful maintenance feed, but it is not regarded as being of any real nutritional value.

Ehretia P.Browne

Trees or shrubs. Leaves alternate. Inflorescence paniculate. Calyx deeply 5-lobed. Corolla 5-lobed, funnel-shaped. Stamens inserted at corolla throat, exserted. Style terminal, divided into 2; stigmas capitate. Fruit a drupe. 50 species. Tropical and warm regions.

Botanical notes. *E. braunii* Vatke was recorded from the hills by Balfour in 1880. The specimen was sterile and inadequate for accurate identification. It has not been seen since.

Ehretia obtusifolia Hochst.

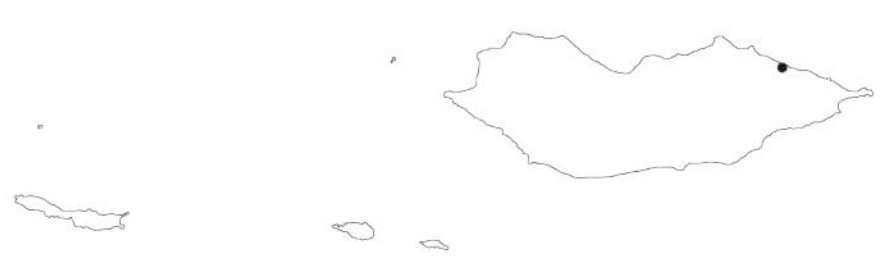

Dense shrub or small tree to 4m. Leaves obovate, <8 x 4cm, margin entire. Calyx lobes <2mm. Corolla pale pink, <7mm; lobes <5mm, obtuse. Fruit subglobose, <6mm diameter, orange or black when mature.

Ecology. Rare in drought-deciduous woodland on the seaward-facing limestone escarpments of NE Soqotra (below Homhil). 160m.

Distribution. Tropical E and S Africa, Arabia, Pakistan and NW India.

ETHNOGRAPHIC NOTES

Name(s)

– Soqotri: *eḳśəre̍r* 'terrain of sharp and cutting rocks'; *ʾəḳśəre̍h* (also given to *Reseda viridis*).

– Soqotri simplified: *'eqshirer*; *'iqshireh*.

Food (human). The fruit are not eaten on Soqotra.

Food (livestock). Goats browse this when they can get at it on the cliffs where it grows. However, it is not common enough for it to be important as a fodder, or for herders to lop leafy stems to feed their animals in the dry season.

Comparative uses. Many *Ehretia* species have edible fruit which are eaten raw and pickled. The berries of *E. cymosa* are used as a bait for birds. The inner bark of some species, such as *E. laevis*, provide a famine food. Medicinally a decoction of the fresh roots of *E. aspera*/*E. obtusifolia* are given in venereal disease; *E. thonningiana* is used as an evacuant, and to treat dysentery, fever and tetanus, and its stems are used as a tooth-stick. An infusion of the leaves of *E. cymosa* are given as a drink or used as a wash to bring down a fever and to treat convulsions in children. The juice of the leaves is given to children as a mild laxative, and the leaves and roots are used for infantile tetanus. (Ambasta 1982; Dalziel 1948, 1956; Jain & De Fillips 1991; Kunkel 1984; Lewis & Elwin-Lewis 1977)

Heliotropium L.

Herbs or subshrubs. Leaves alternate or subopposite. Inflorescence dense or lax, spike-like, often spirally inrolled. Flowers small, actinomorphic. Calyx 5-lobed. Corolla 5-lobed, white or yellowish, tubular or funnel-shaped. Ovary 4-celled with 1 ovule in each cell. Stigma conical, globose or filiform. Fruit breaking into 4 separate or 2 pairs of fused nutlets. 10 endemics. c.250 species. Tropical and temperate regions.

ETHNOGRAPHIC NOTES

Comparative uses. *Heliotropium* species in the drier areas of the Middle East and Africa are particularly grazed by camels, goats resorting to them mainly in the dry season. However, some species, such as *H. steudneri* and *H. somalense*, are said to cause diarrhoea in camels. Some species of *Heliotropium* are known to contain pyrrolizidine alkaloids, the highest content being in the roots and inflorescence, and these can affect livestock through liver toxicity and atrophy, often leading to the death of the animal. The effect is cumulative, so that losses tend to occur in older animals. Elsewhere in the Middle East plasters made from the dried and powdered leaves are used to treat infected wounds, abscesses, boils, sprains, contusions, oedema and swellings of all kinds. They are also used for anti-scabies preparations in veterinary medicine. In Yemen the leaves and flowers of *H. subulatum* are given to treat fevers. In Bahrain a poultice of the leaves or a tisane made from the leaves is used to treat snakebites, abscesses, swellings and mouth ulcers. In Oman the leaves of *H. crispum* are pounded with water to make a poultice for ulcers, snakebites and swellings. A tea made of the leaves is taken orally to treat mouth ulcers. The lightly crushed leaves of *H. fartakense* are soaked in water and a solution taken for indigestion and colic. A solution of crushed leaves with hot water is applied round painful or red eyes to soothe them and also to improve the vision. A poultice of leaves crushed with salt, turmeric and ginger is applied to swellings and sprains or dislocated joints, burns and ulcers, and a paste of the leaves and stems is used to treat ringworm and eczema, and is applied around a wound or sore. Elsewhere, *H. undulatum* is applied locally for a headache and is taken internally for gonorrhoea and as a galactagogue. In India, *H. indicum* is regarded as an emollient, vulnerary and diuretic, and is used locally to treat ulcers, sores, wounds, gum boils and skin problems. A decoction of the leaves is used in urticaria and fevers, and of the roots for coughs and fevers. The flowers are given in small doses as an emmenagogue, and in large doses as an abortifacient. The seeds are masticated as a stomachic. The leaves also yield a dye. (Abbas *et al.* 1992; Abdulfatih 1987; Ambasta 1982; Batanouny 1981; Dalziel 1948, 1956; Fleurentin & Pelt/Fleurentin *et al.* [1982, 1983a,b,c]; Ghazanfar 1994; Ghazanfar & Al-Sabahi 1993; Goodin & Northington 1979; Heine *et al.* 1988; Jain & De Fillips 1991; Miller & Morris 1988; Rizk & El-Ghazaly 1995)

Heliotropium aegyptiacum **Lehm.**
Heliotropium cimalense Vierh.

Erect annual herb to 50cm, softly grey-hairy throughout. Leaves ovate, 3–6 x 2–3cm, truncate at base, petiolate. Inflorescence dense, one-sided spikes. Calyx lobes linear-oblong, c.3mm long, densely covered with spreading hairs. Corolla white; tube 3–4mm. Stigma elongate-triangular. Nutlets 4, c.2mm long, slightly rugose on back.
Ecology. Occasional on Soqotra, Samha and Abd al Kuri in dwarf shrubland and deciduous woodland, particularly in overgrazed grassland and around habitation. (250–)600–670m.
Distribution. Arabia and NE Africa.

ETHNOGRAPHIC NOTES

Name(s)
– Soqotri: nil or *məgērəd* (given to a variety of plants used to cure leather); *eḥfílhin* 'tiny figs'; *ṣifinítin* 'plaited fibre' (Abd al Kuri).
– Soqotri simplified: *migeyrid*; *ehfilhin*; *zifinitin*.
Food (livestock). One of the ephemeral *ru'ud* plants which come up with the rains, providing incidental green grazing until they die back. Though palatable to livestock, especially small stock, this is not an important forage for them, and is not seen as having any nutritional value. It is more highly valued on the outer islands.
Leatherwork. In the absence of alternative tanning material, the whole plant is crushed to a paste used to cure skins and hides to toughen the leather.

Heliotropium "bacciferum" complex

Botanical notes. Vierhapper (1907) described a number of *Heliotropium* species from the Soqotra Archipelago, which have been maintained here because the extreme forms are distinct. However, there is considerable overlap in the distinguishing characters used by Vierhapper: for example, presence or absence of calluses or hairs on the nutlets; fruit splitting into 2 pairs of fused nutlets or 4 separate nutlets; length of stigma; and flower colour. It is interesting that the flowers of this complex are predominantly white on Abd al Kuri but yellow on the other islands of the Archipelago. The following treatment is arranged geographically by island.

---

1. Abd al Kuri

A complex of clearly related species with the extremes distinct but the limits obscured by many intermediates. *H. paulayanum* is distinguished by its roundish leaves, long petiole, ill-defined inflorescence and white flowers. *H. kuriense* is similar to *H. paulayanum* but has bigger and relatively narrower leaves, shorter petioles and clearly defined inflorescence. *H. wagneri* (which is also found on Samha) is similar to *H. paulayanum* but is distinguished by its narrowly elliptic or lanceolate leaves, poorly defined inflorescence and thinner indumentum. There is considerable variation recorded in flower and fruit characters.

"*Heliotropium kuriense* Vierh."

Similar to *H. paulayanum* but the leaves ovate, bigger and relatively narrower (<2 x 1.5cm), petiole shorter than the lamina and acute leaf base; inflorescence relatively densely flowered and clearly defined; corolla white or yellow; nutlets 4, hairy.
Ecology. Dwarf shrubland on coastal plains and rocky hill-slopes. Sea-level–520m.
Distribution. Endemic.
Conservation status. Vulnerable (VU D2). Rare, with a very limited area of occupancy.

ETHNOGRAPHIC NOTES
Name(s)
– Soqotri: *ṭʌ́bṭʌb* 'tanning plant' (Abd al Kuri).
– Soqotri simplified: *tabtab.*
Food (livestock). On Abd al Kuri and Samha this is an important and nutritious forage for goats. It is also grazed by sheep. The islanders say that the young plants are particularly nourishing and fattening.
Leatherwork. On the outer islands this, with *Limonium cylindrifolium*, is one of the most important plants used to prepare leather. The soft tips, leaves and fruit are crushed to a paste which is rubbed into the de-haired skins and hides to toughen them and colour them red-brown. It is used especially on skins that suffer the heaviest wear, such as water-skins and those used to churn butter.

"*Heliotropium paulayanum* Vierh."

Prostrate herb, sometimes rather woody, densely hirsute with short hairs and long, bulbous-based hairs. Leaves elliptic to broadly obovate or suborbicular; lamina 0.2–1 x 0.1–1cm, apex obtuse or acute, margin entire, base rounded or truncate, petiole c.1 to 3 times as long as the lamina. Inflorescence few-flowered, poorly defined. Corolla white, 3.5–6mm long. Calyx lobes equal, c.2/3 to 1/2 as long as the corolla, villous. Stigma shortly conical on a short style. Nutlets 4, glabrous or puberulous, smooth or verrucose.
Ecology. Sandy soil on coastal plains in dwarf shrubland. Sea-level–100m.
Distribution. Endemic.
Conservation status. Vulnerable (VU D2). Rare, with a very limited area of occupancy.
Botanical notes. Distinguished from all other species of *Heliotropium* in the Archipelago by its suborbicular, long-petiolate leaves.

ETHNOGRAPHIC NOTES
Name(s)
– Soqotri: *ṣifinítin* 'plaited fibre' (Abd al Kuri and Samha).
– Soqotri simplified: *zifinitin.*
Food (livestock). On Abd al Kuri and Samha this is regarded as a valuable forage for sheep and goats. On Abd al Kuri donkeys also graze it.

2. Abd al Kuri and Samha

"*Heliotropium wagneri* Vierh."

Similar to *H. paulayanum* but is a rather lax, profusely branched prostrate herb with thin indumentum; leaves bigger and narrower, narrowly ovate to narrowly elliptic (<3 x 0.8cm), acute apex and base attenuate to the petiole which is about half as long as the lamina; inflorescence few-flowered, poorly defined; corolla white or yellow; nutlets fused in 2 pairs, glabrous.
Ecology. Coastal plains and limestone slopes; dwarf shrubland. Sea-level–200m.
Distribution. Endemic.
Conservation status. Vulnerable (VU D2). Rare, with a very limited area of occupancy.
Botanical notes. *H. wagneri* is distinguished by its shortly petiolate, narrowly elliptic or lanceolate leaves and glabrous or hairy fruits breaking into two pairs of fused nutlets. A specimen from Samha (*Miller* 11447) resembles typical *H. wagneri* in its flowers and leaves but the fruits breaks into four nutlets. It is interesting that the flowers of this species are consistently white on Abd al Kuri whilst they are always yellow on Samha.

ETHNOGRAPHIC NOTES
Name(s)
– Soqotri: nil or *ṣifinítin* 'plaited fibre' (Abd al Kuri); *gələ́shen* (Samha).
– Soqotri simplified: *zifinitin*; *gilis'hen.*
Food (livestock). On Abd al Kuri and Samha this is regarded as valuable forage for sheep and goats. On Abd al Kuri donkeys also graze it.

3. Samha and Darsa

"*Heliotropium* aff. *socotranum* Vierh."

Similar to *H. socotranum* but the flowers brighter yellow and probably slightly larger; the leaves linear or linear-obovate and smaller (<0.7 x 0.2cm); inflorescence few-flowered, poorly defined.
Ecology. Coastal sandy plains, raised coral beaches on Samha and Darsa. Sea-level–200m.
Distribution. Endemic.
Conservation status. Vulnerable (VU D2). Rare, with a very limited area of occupancy.
Botanical notes. This rather distinct plant with its bright yellow flowers and small narrow leaves is clearly related to *H. socotranum* but is found only on Samha and Darsa.

ETHNOGRAPHIC NOTES
Not seen, no information recorded.

4. Soqotra

Vierhapper described four species from Soqotra: *H. riebeckii* Schweinf. & Vierh., *H. socotranum* Vierh., *H. shoabense* Vierh. and *H. derafontense*, based largely on fruiting characters and the form of the calyx, corolla and stigma. Certainly, across Soqotra there is a great deal of variation, and 'morphotypes' in different areas can be distinguished by a mosaic of characters including flower colour, leaf shape, differences in indumentum and inflorescence structure. Unfortunately, the characters used by Vierhapper for distinguishing species (especially nutlet characters) are unreliable and do not correlate with geographically distributed 'morphotypes'. For instance, Vierhapper recorded type material of *H. riebeckii*, *H. socotranum* and *H. shoabense* from Bandar Shu'ub (at the extreme west of Soqotra). I have examined his specimens and consider them all to belong to the same form.

Because of the difficulty of relating Vierhapper's species to the various 'morphotypes' across the island and because I have found it impossible to formally recognise any of these forms I have preferred to treat all the material of the complex under *H. socotranum* Vierh. However, I have not formally sunk Vierhapper's species. The technical characters used by Vierhapper are given to distinguish his species.

"*Heliotropium derafontense* Vierh."

Annual herb. Stigma short and stubby. Fruits splitting into 2 pairs of fused nutlets, hairy, warted.
Ecology. No habitat details available.
Distribution. Endemic.
Conservation status. Data Deficient (DD). Inadequate evidence is available to make an assessment.
Botanical notes. Recorded by Vierhapper from Jebel Derafonte. No material has been seen, so its status is uncertain. The type is apparently lost.

"*Heliotropium riebeckii* Schweinf. & Vierh."

Heliotropium bacciferum Balf.f. non Gurke
Heliotropium pterocarpum sensu Balf.f. non Hochst. & Steud.
Heliotropium undulatum sensu Balf.f. non Vahl

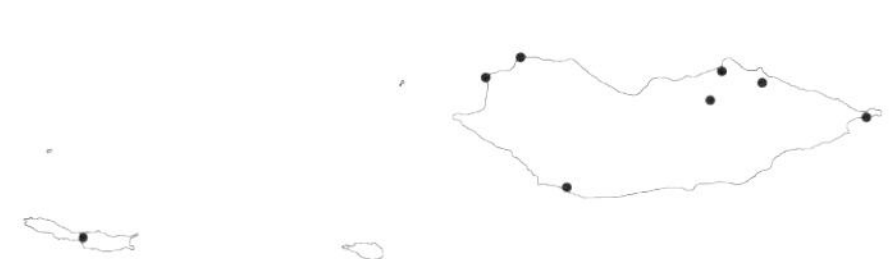

Perennial herb. Stigma long. Corolla yellow. Fruits splitting into 2 pairs of fused nutlets, glabrous, warted.
Ecology. Dwarf shrubland on northern and western coastal plains. Sea-level–50m.
Distribution. Endemic.
Conservation status. Data Deficient (DD). Part of a taxonomically difficult group on Soqotra. Inadequate evidence is available to make an assessment.

ETHNOGRAPHIC NOTES
Name(s)
– Soqotri: *d-ìbʿar* 'of camels'.
– Soqotri simplified: *d-ib'ar.*
Food (livestock). This coarse plant is browsed by camels (hence its name), and occasionally also by donkeys and hungry goats in a prolonged dry season. It is not thought to be a nutritious feed, except for camels.

"*Heliotropium shoabense* Vierh."

Heliotropium derafontense Vierh.

Perennial herb. Stigma short and stubby. Fruits splitting into 2 pairs of fused nutlets, hairy, warted.
Ecology. Dwarf shrubland. Sea-level–700m.
Distribution. Endemic.
Conservation status. Data Deficient (DD). Part of a taxonomically difficult group on Soqotra. Inadequate evidence is available to make an assessment.
Botanical notes. Vierhapper also recorded it from Abd al Kuri but I have examined his material and consider it to be referable to *H. wagneri*.

ETHNOGRAPHIC NOTES

Not seen, no information recorded.

"*Heliotropium socotranum* Vierh."

Prostrate or ascending woody-based herb with a dense indumentum of long spreading bulbous-based hairs (particularly on the calyx) and shorter, often rather matted hairs. Leaves variable from narrowly oblong to elliptic, <2 x 1cm. Inflorescence more or less congested or few-flowered and poorly defined. Flowers white or yellow. Stigma long or short. Fruits splitting into 2 pairs of fused nutlets or separate nutlets, glabrous or hairy, smooth or rugose, sometimes warted.
Ecology. Sandy and coral-derived soils on coastal plains and thin soil overlying limestone; dwarf shrubland and *Croton socotranus* shrubland. Sea-level–350m.
Distribution. Endemic.
Conservation status. Least Concern (LC). Common in several vegetation types and under no present or perceived threat.
Botanical notes. Vierhapper considered *H. socotranum* to have hairy fruits separating into four glabrous, rugose nutlets and yellow flowers. Recorded by Vierhapper from Bandar Shu'ub.

ETHNOGRAPHIC NOTES

Name(s)
– Soqotri: *ḥˈariḥer*; *šibəḥˈitin* 'with outstretched legs'; *ṣifinˈitin* 'plaited fibre' (Samha).
– Soqotri simplified: *hariher*; *shibihitin*; *zifinitin.*
Food (livestock). On Soqotra this is grazed mainly by sheep and goats in the dry season. Camels and donkeys also eat it year round. It is regarded as a maintenance feed of little nutritional value. It is more highly regarded on the outer islands, where it is considered to provide good forage for sheep and goats.

Heliotropium balfourii Gürke

Heliotropium odorum Balf.f.

Subshrub to 50cm, strigose. Leaves dark green, ovate to oblong-ovate, <6 x 3cm, distinctly petiolate, somewhat bullate. Inflorescence of paired, densely flowered, one-sided spikes, 3–10cm long, pedunculate. Calyx divided almost to the base; calyx lobes linear, c.3mm, lengthening to c.5mm in fruit. Corolla white, c.4–5mm across; tube c.5mm long. Nutlets in 2 fused pairs, 2–2.5 x 3mm, obscurely furrowed, somewhat rugose on back, glabrous.
Ecology. Widespread and sometimes common in dwarf shrubland and drought-deciduous woodland. 300–700m.
Distribution. Endemic.
Conservation status. Least Concern (LC). Common in several vegetation types and under no present or perceived threat.
Botanical notes. This shrubby species is distinguished by the yellowish green, bullate (wrinkled) leaves.

ETHNOGRAPHIC NOTES

Name(s)
– Soqotri: *ḳeṭəmˈʌmhion*, *ḳʌṭəmˈʌmhin* 'fresh butter lumps'; *ṭˈʌbṭʌb*, *mṭˈʌbihon* 'tanning plant'; *šε̄baᶜ*, *šābaḥ* 'stretched out, extended'; (western Soqotra) *ṣābaġ(ʌn)* 'dye plant'; *di raḥimˈimihon* 'grey-coloured'; *ṣaᶜləfˈitin* 'many-leaved; *mˈʌḥəriś* 'undergrowth, understorey'; *ˈəwrˈeśᵓhʌn*, *ᵓəwrˈešʌn.*
– Soqotri simplified: *qetimamhion, katimamhim*; *tabtab, mtabihon*; *shaba'*, *shabah*; *shabagh(an)*; *di rahimimihon*; *za'lifitin*; *maharish*; *'ewresh'han*, *'ewreshan.*
Food (livestock). Goats readily eat the plant which is considered to promote lactation. Sheep are less enthusiastic. Cattle of the drier plateau areas graze the plant, as do donkeys and camels everywhere. When green it is considered to be a valuable feed, especially for goats. In the dry season stems are sometimes gathered and taken to cattle and goats as a feed.
Leatherwork. This is the most widely used tanning plant on the island. The soft tips, leaves and fruit are crushed to a paste which is rubbed into skins and hides to cure and toughen them, and to stain them red-brown. It is used especially to treat the skins which suffer the most wear, such as water-skins and the skins used to churn butter. Its use for this in particular is reflected in its most common name, *ḳeṭəmˈʌmhion*, which means lumps of freshly formed butter. The use of this plant to tan leather is believed to encourage the butter to come. Sometimes the leaves of *Indigofera coerulea* are added to the tanning paste to darken the stain. This plant was used especially to dye the intricately plaited leather belts formerly worn by men, since more than other plants it was considered to result in leather of a good clear red, a colour which was much admired.
Medicinal / veterinary. The roots are peeled, sun-dried and then crushed to a paste which is applied to treat dry eczema, especially common on the hands of those who regularly milk livestock.

Cleansing. If a clay pot has developed an unpleasant smell the leaves of the plant are crushed and used to rub the inside of the pot all over to remove the smell.
Other. A paste of this plant was also used to dye the coarse, unbleached calico that was imported to the island for clothing. It was added to water and simmered over a slow fire until the water was well-coloured. The material was then lowered in and left to soak for some hours before being dried in the shade. The colours were not fast, but faded on washing and exposure to the sun. The same dye was occasionally used on pale sheep's wool. The small balls of spun wool were put in the water with the dye material and left to soak until they had changed colour. They were then hung up to dry. The paste was also used to colour and decorate clay pots, but this had to be done quickly as soon as the pots were removed from firing.

Heliotropium dentatum Balf.f.

Prostrate annual herb to 30cm, hispid; stems slender. Leaves narrowly oblong, characteristically undulate-dentate; stem leaves <3 x 0.3cm, distant; basal leaves 3–10 x 0.2–0.4cm, in a rosette. Flowers 1 to 4, at the end of branches, lax. Calyx lobes narrowly triangular, c.1.5mm. Corolla white, c.1.5mm across; tube c.3mm long. Fruit not seen.

Ecology. Drought-deciduous woodland and on wadi sides in *Croton socotranus* shrubland; seems to prefer damp sand and is a frequent weed along the wadis running through date gardens on the Hadiboh plain. 50–250m.
Distribution. Endemic.
Conservation status. Vulnerable (VU D2). Rare, with a very limited area of occupancy.
Botanical notes. Distinguished by the basal rosettes of long-petiolate, dentate-undulate leaves (at least on young plants) and long prostrate stems.

ETHNOGRAPHIC NOTES
Not seen, no information recorded.

Heliotropium nigricans Balf.f.

Shrub to 1m, almost prostrate in exposed places, all parts thinly covered with short, appressed, tuberculate-based hairs. Leaves subopposite, dark green, glossy, broadly elliptic to subcircular, 0.5–1.2 x 0.5–1cm, rounded, entire. Flowers solitary or paired at ends of branches. Calyx narrowly triangular, c.1mm long. Corolla white, 6–7mm across; tube very short. Stigmas triangular. Nutlets 4.
Ecology. Locally abundant on the southern limestone plateau in *Dracaena* woodland (Firmihin) and semi-deciduous woodland (Qatariyah). 400–600m.
Distribution. Endemic.
Conservation status. Vulnerable (VU D2). Rare, with a very limited area of occupancy.
Botanical notes. A very distinctive species distinguished by the subopposite, almost circular glossy leaves. Its affinities within the genus are not at all obvious.

ETHNOGRAPHIC NOTES
Name(s)
– Soqotri: nil or *ṣəbe̍rhon*, *ṣəbe̍rbəher* 'sour, astringent'; *maḥlu̍məhəm* 'relaxed'.
– Soqotri simplified: *ziberhon*, *ziberbiher*, *mahlumihim*.
Food (livestock). Eaten by all livestock, including camels and donkeys. Cattle normally only graze larger *Heliotropium* species when they are very hungry, and the smaller ones are eaten unintentionally when growing amongst grasses and other palatable herbs under trees and shrubs. Nevertheless, they are keen on this plant and search it out.

Heliotropium ophioglossum Stocks

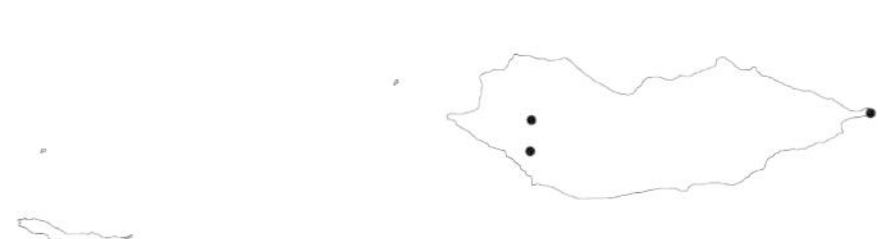

Perennial herb to 30cm; indumentum villous, with short thin hairs and longer strigose hairs. Leaves lanceolate or narrowly elliptic, 1–4 x 0.3–1cm, with acute tip, entire margin and attenuate base. Inflorescence of solitary or paired, one-sided spikes. Calyx with long, spreading, strigose hairs; lobes linear-oblong, 3–4.5mm long. Corolla white, c.4–5mm across; tube c.4mm. Stigma elongate, bifid with filiform lobes, exserted from the corolla. Nutlets 4, ovoid to subglobose, c.1.5mm long, papillose and slightly rugose.
Ecology. Dwarf shrubland (with *Croton socotranus* and *Pulicaria stephanocarpa*). 200–250m.
Distribution. Somalia, S Arabia and E Pakistan.
Botanical notes. Readily distinguished from all other species of *Heliotropium* on Soqotra by the stiff, spreading hairs which clothe the calyx and the thread-like, forked stigma lobes exserted from the corolla and which resemble a minute snake's tongue.

ETHNOGRAPHIC NOTES
Name(s)
– Soqotri: nil or *tʌ̍bṭʌb* 'tanning plant'.
– Soqotri simplified: *tabtab*.
Food (livestock). This is one of the ephemeral *ru'ud* plants which come up with the rains, providing incidental green grazing until they die back. The plant is palatable to livestock, but is not an important or valuable forage for them, and is seen as having little nutritional value.
Notes. Other uses as *H. odorum*.

Heliotropium ovalifolium Forssk.

Prostrate perennial herb to 50cm, with silvery-silky hairs throughout. Leaves obovate or elliptic, 1–5 x 0.25–2cm, notched or acute at apex. Inflorescence of short, paired one-sided spikes, the flowers two-ranked. Calyx 1.5–3mm, deeply divided into unequal lobes. Corolla white; tube 2.5–3.5mm long. Stigma conical with a basal crown. Nutlets 4, c.1mm long, densely covered with white appressed hairs.

Ecology. Rare on Soqotra, only found once recently; in mud by wadi (Diksam plateau). 670m.

Distribution. Tropical Africa, Madagascar, Egypt, Arabia and India.

ETHNOGRAPHIC NOTES

Name(s)

– Soqotri: *ḳʌlḳəhʌl* 'swaying in the breeze' (general term for a variety of mainly halophytic plants); *libinhʌn* 'pale-coloured'; *maḥluməhəm* 'relaxed'.

– Soqotri simplified: *qalqihal*; *libinhan*; *mahlumihim.*

Food (livestock). This is one of the ephemeral *ru'ud* plants which come up with the rains, providing incidental green grazing until they die back. Not normally liked by livestock unless they are very hungry. It is not considered to have any nutritional value.

Comparative uses. This plant is said to be poisonous, causing vomiting and diarrhoea. Its most common medicinal application is a local one to treat syphilitic ulcers and so on. It is taken internally in northern Nigeria where its drastic effects are thought to purge syphilis from the system. (Ambasta 1982; Dalziel 1948, 1956; Jain & De Fillips 1991)

Heliotropium rariflorum Stocks

Subshrub to 20cm, covered with appressed silvery hairs; older stems with peeling white bark. Leaves linear, 0.5–1.5 x 0.1–0.2cm. Inflorescence spike-like, the flowers subtended by bracts. Calyx 1–2mm, lobes unequal. Corolla white, campanulate, c.2.5mm diameter. Stigma globose, sometimes with a short apical projection. Nutlets 4, ovoid, 1 x 1.5mm, hairy.

Ecology. Rare on Soqotra; in dwarf shrubland on desertic plains. 200m.

Distribution. Tropical Africa, Arabia, Iran and Pakistan.

Botanical notes. Distinguished by the white peeling bark, globose stigma and small, campanulate corolla.

ETHNOGRAPHIC NOTES

Name(s)

– Soqotri: nil or *zēdirhin*; *śebəriḳ* 'glimpsed'.

– Soqotri simplified: *zeydirhin*; *shebiriq.*

Food (livestock). All livestock readily eat this, including camels and donkeys. However, it is not seen as being a nutritious feed, except for camels.

Heliotropium strigosum Willd.

Heliotropium strigosum var. *bicolor* DC. & A.DC.
Heliotropium strigosum var. *marifolium* Retz.
Heliotropium strigosum var. *scabrum* Retz.

Erect or spreading herb to 30cm, with appressed strigose hairs throughout. Leaves linear to narrowly oblong, <2.5 x 0.5cm. Inflorescence spike-like, flowers subtended by bracts. Calyx divided to base; lobes linear-lanceolate, 2–3mm. Corolla white, c.1.5mm across; tube 1.75mm long. Stigma conical with a basal disc. Nutlets 4, subglobose, c.1.25mm long, appressed hairy.

Ecology. Coastal plains; in *Croton socotranus* shrubland and succulent shrubland; also associated with human habitation. 30–50m.

Distribution. From W Africa to Australia.

Botanical notes. A very variable species with an extensive distribution.

ETHNOGRAPHIC NOTES

Name(s)

– Soqotri: nil or *ᶜažərib* (pl. *ᶜažərub*) 'quick to die back'; *maḥluməhem* 'relaxed'.

– Soqotri simplified: *'ajirib* (pl. *'ajirub*); *mahlumihim.*

Food (livestock). This weedy plant is one of the ephemeral *ru'ud* plants which come up with the rains, providing incidental green grazing until they die back. The plant is palatable to livestock, but is not much liked by them and is not seen as being of nutritional value. Where it grows in damp ground it is normally avoided by goats, which intensely dislike damp forage.

Comparative uses. In India the plant is used as a laxative and diuretic. Its juice is applied to sore eyes, and locally to treat boils, wounds and ulcers. (Ambasta 1982; Jain & De Fillips 1991)

***Heliotropium zeylanicum* Lam.**
Heliotropium subulatum (Hochst. ex A.DC.) Vatke

Erect woody-based herb to 50cm, sparsely hairy. Leaves narrowly lanceolate to elliptic, 1–10 x 0.2–2cm, acute at tip, attenuate at base. Inflorescence of 2 or 3, elongated, one-sided spikes. Calyx divided to base; lobes ovate, 1.5–3mm. Corolla white or yellowish green, tube cylindrical; lobes narrow, tapering, c.10mm long. Stigma conical. Nutlets 4, c.1.5mm long, glabrous, verrucose.
Ecology. Succulent shrubland and a weed of date gardens. 50–150m.
Distribution. Tropical Africa, Arabia and India.
Botanical notes. Readily distinguished by its long, narrow and tapering corolla lobes.

ETHNOGRAPHIC NOTES
Name(s)
– Soqotri: nil or *ʿaˈžərib* (pl. *ʿaˈžərub*) 'quick to die back'; (rare) *məgḗrəd* (also given to other tanning plants).
– Soqotri simplified: *'ajirib* (pl. *'ajirub*); *migeyrid.*
Food (livestock). Goats principally eat this, except where it grows in damp ground when it is avoided. Other livestock eat it mainly in the dry season. It is not seen as having nutritional value.
Leatherwork. In the absence of better tanning plants it can be used to prepare the skins and hides in leatherwork, to toughen them and colour them red-brown.

Trichodesma R.Br.

Herbs or large shrubs. Leaves opposite (uppermost sometimes alternate). Calyx deeply divided. Corolla campanulate. Filaments short; anthers with terminal ribbon-like appendages that are ± spirally twisted together. Style filiform; stigma capitate. Nutlets 4, with a dentate wing, sometimes absent. 3 endemics. 35 species. Tropical and warm Old World.

ETHNOGRAPHIC NOTES
Comparative uses. The leaves of *T. zeylanicum* and *T. indicum* are eaten and their seeds are a source of oil. Medicinally *T. africanum* is a diuretic and emollient. Its leaves are boiled, mixed with cereal flour, butter or oil, and natron to a paste which is given to treat diarrhoea. A hot poultice is made from the leaves and applied for febrile and inflammatory conditions. *T. indicum* is an emollient and diuretic, and is prescribed for the expulsion of a dead foetus. The roots are used in the treatment of dysentery, and are also pounded and applied to swellings of the joints. The flowers are sudorific and pectoral. The leaves of *T. zeylanicum* are emollient, demulcent and diuretic, and are used to ensure fermentation when making beer. The roots are pulverised and applied to wounds as an analgesic. The seeds yield a drying oil. In nearby Dhofar, the whole plant of *T. hildebrandtii* is pounded to a paste applied to treat cases of mastitis, in both women and livestock. (Ambasta 1982; Dalziel 1948, 1956; Jain & De Fillips 1991; Kunkel 1984; Miller & Morris 1988)

***Trichodesma laxiflorum* Balf.f.**
Trichodesma atrichum Vierh.

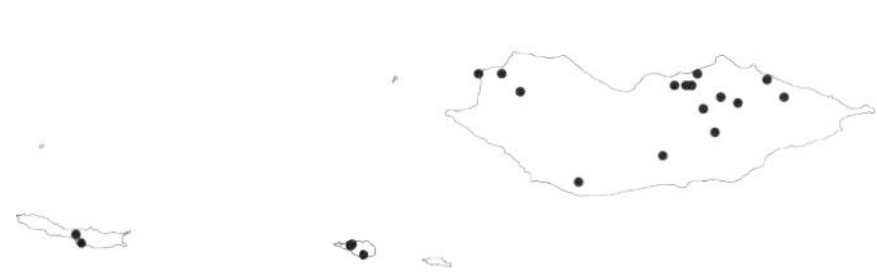

Erect herb to 80cm, with appressed or spreading setiform hairs. Leaves ovate, <8 x 4cm, both surfaces strigose. Flowers in lax, 3 to 5-flowered cymes. Calyx lobes linear-lanceolate, 3–5.5mm long. Corolla blue, <8.5mm. Nutlets with dentate wings.
Ecology. Common in drought-deciduous woodland on the limestone escarpments and in sheltered places on the limestone plateau, also in succulent shrubland and submontane woodland. 15–1220m.
Distribution. Endemic.
Conservation status. Least Concern (LC). Common in several vegetation types and under no present or perceived threat.
Botanical notes. Plants from Abd al Kuri are generally smaller in all parts and have been recognised as a separate species, *T. atrichum* Vierh. However, the differences are small and here they are considered an extreme form of this variable species.

ETHNOGRAPHIC NOTES
Name(s)
– Soqotri: nil or *ʿaˈžərib* (pl. *ʿaˈžərub*); *mḥaˈṭḥoˈṭən* 'limp, flaccid' (more commonly given to fern species); *ṣʌʿlˈʌləhon* 'broad-leaved'; *ʿoˈwźəṣ* 'overborn'; *di niˈfəziˈtin* 'full-uddered'; *handūrib* (more commonly given to *Sarcostemma* species) (Samha).
– Soqotri simplified: *'ajirib* (pl. *'ajirub*); *mhat'hotin*; *za'lalihon*; *'owdhiz*; *di nifizitin*; *handuwrib.*
Food (livestock). One of the *ʿaˈžərib* plants which flourish after the winter rains especially. Camels and donkeys like the plant, and it is readily grazed by cattle before the grasses come up. It is little eaten by goats and sheep, except in the more vegetation-poor areas, such as the outer islands, where it is a more important forage. However, even there it is not seen as having any real nutritional value.

***Trichodesma microcalyx* Balf.f.**

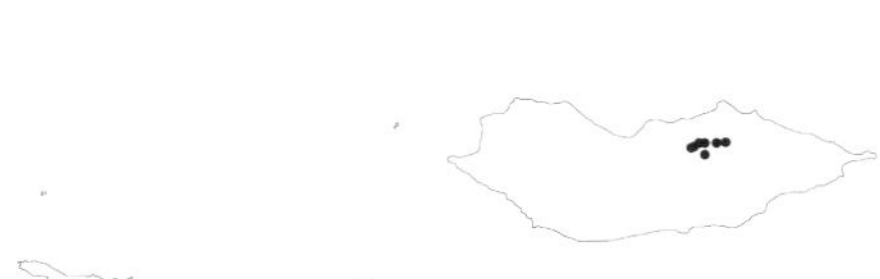

Erect herb to 90cm, with spreading unequal setiform hairs. Leaves broadly ovate or broadly elliptic, 8 x 4cm. Flowers in lax, few-flowered cymes. Calyx lobes linear-lanceolate, 4.5–6mm long. Corolla blue, <1cm. Nutlets unwinged.
Ecology. Restricted to higher parts of the Haggeher mountains in montane evergreen woodland and shrubland. 900–1500m.
Distribution. Endemic.

Conservation status. Least Concern (LC). Common in several vegetation types and under no present or perceived threat.

ETHNOGRAPHIC NOTES

Name(s)

– Soqotri: nil or *ḳə́ṣəho* (general term for new shoots of herbs and grasses coming up after rain); *dí mərišε̄tən* 'of settlements'; *mḥāṯhótən* 'limp, flaccid' (more commonly given to fern species); *məźḥɔ̄mhən di ílәhe* 'the *məźḥɔ̄mhən* of cattle'; *saḥámhəm* 'diuretic'.

– Soqotri simplified: *qiziho*; *di mirisheytin*; *mhat'hotin*; *midh'howmhin di iliheh*; *sahamhim.*

Food (livestock). All livestock eat it, but cattle are particularly fond of it and it is said to be a nourishing feed for them.

Notes. The clinging fruit often make small scratches on the udders and teats of animals, which can become infected and affect lactation.

Trichodesma scottii **Balf.f.**

Shrub to 4m. Leaves broadly ovate, <25 x 15cm. Flowers pendent, in terminal clusters. Calyx 2–4cm, deeply 5-lobed. Corolla creamy-white with yellow-brown markings at base, <5cm. Nutlets with dentate wings.

Ecology. A locally common understorey shrub in montane thicket in the central Haggeher mountains. (650–)850–1250m.

Distribution. Endemic.

Conservation status. Vulnerable (VU B2 a bii). Known from a few scattered sites in the central Haggeher mountains with a total area of occupancy of less than 20km^2. There is a 19th century record (Forbes) from Homhil on the NE limestone plateau. It has not recently been seen there, suggesting that there has been a decline in its area of occupancy.

Botanical notes. A distinctive understorey species, with large, soft leaves and big, creamy-white, funnel-shaped flowers which are arranged in pendent inflorescences that resemble the head of an upturned mop.

ETHNOGRAPHIC NOTES

Name(s)

– Soqotri: *ṣaˁlʌ́ləhon* 'broad-leaved'; *aḥtərέś* (commonly given to species with poor and soft wood).

– Soqotri simplified: *za'lalihon*; *ahtiresh.*

Food (livestock). The large leaves and dramatic flowers of this plant are generally ignored by all livestock unless the dry season is prolonged or there is a drought. At such times, if there is no alternative, hungry goats and cattle will nibble at the leaves and eat fallen foliage and other plant litter. However, in the Central Highlands where this plant grows, preferable alternatives are usually plentiful. The plant is not seen as having any nutritional value and is rarely cut by local herders for their animals.

Fuel. Does not make good firewood.

Timber. No use as timber.

Wellstedia Balf.f.

Subshrubs. Leaves alternate. Flowers small, actinomorphic, in dense axillary cymes. Calyx 4-lobed. Corolla cylindrical, 4-lobed. Ovary 2-celled. Stigma bifid. Fruit a flattened, obcordate 1-seeded capsule. 1 endemic. 5 species. Tropical NE Africa and SW Africa.

Botanical notes. This small genus has a strikingly disjunct distribution with four species in tropical NE Africa and Soqotra and one in Namibia. This pattern of distribution is usually explained by once continuous distributions along an arid corridor between the two regions during the Pleistocene or late Tertiary. It is the only genus within the Boraginaceae which has capsular fruits and is taxonomically isolated. Placed in a separate subfamily within the Boraginaceae or sometimes considered a separate family, the Wellstediaceae. For a revision see Thulin (1998).

Wellstedia socotrana **Balf.f.**

Prostrate, cushion-forming subshrub, silvery-pubescent throughout. Leaves congested in small rosettes, narrowly obovate, <1.5 x 0.5cm. Calyx c.1.5mm. Corolla c.3mm. Capsule <5mm long, asymmetrically obcordate.

Ecology. Rare on the SE limestone plateau where it is found on an exposed and sparsely vegetated desertic plain. 700m.

Distribution. Endemic.

Conservation status. Vulnerable (VU D2). Rare, with a very limited area of occupancy.

Botanical notes. A compact, cushion-forming shrublet distinguished by its silvery sericeous leaves and its congested, capsular fruits.

ETHNOGRAPHIC NOTES

Name(s)

– Soqotri: *kéṣkʌṣ, kʌ́ṣkʌṣ* 'cream-coloured'.

– Soqotri simplified: *qezqaz, qazqaz.*

Food (livestock). In the drier plateau areas of the island this is regarded as a valuable feed for sheep and goats, providing forage after the rains and persisting into the dry season. If the plants are plentiful they make a nutritious and fattening feed for small stock.

BURSERACEAE

Trees or shrubs with aromatic wood and bark. Leaves alternate, often clustered on short spur shoots, simple or pinnate; stipules absent. Flowers unisexual or bisexual, actinomorphic, solitary, fascicled or in raceme-like cymes or panicles. Calyx 4 or 5-lobed. Petals 4 or 5, free; stamens 4, 8 or 10. Ovary superior, 2 to 5-celled. Fruit a 3 to 5-valved capsule or a leathery or fleshy, 2-valved drupe. 18 genera. 540 species. Tropical, especially New World and NE Africa.

Boswellia Roxb. ex Colebr.

Trees or sometimes shrubs, rooted on the ground or on cliffs; bark often flaking, wood with milky latex and inner bark with watery, aromatic resin. Leaves alternate, simple or pinnate. Flowers in raceme-like panicles, bisexual, actinomorphic; calyx 5-lobed; petals 5, free, oblong to ovate; stamens 10. Fruits a 3 to 6-valved capsule with 1 seed per cell. 8 endemics. 24 species. Old World tropics.

Botanical notes. *Boswellia* species on Soqotra have two distinct habits: they are either cliff-rooting or ground-rooting. The cliff-dwelling species have swollen 'holdfasts' at the base of their trunks that enable them to hang on to cliffs. Intensive searching has rarely if ever found cliff-dwelling species on the ground or ground-dwelling species on cliffs. *B. nana* is the only example of a species which apparently has both habits; it is a cliff-rooting species but occasionally ground-rooting trees are found at the foot of the cliffs on which it grows. However, these individuals are suspected hybrids (but see *B.* sp. *A*). For accounts of this genus on Soqotra see Mies *et al.* (2000) and Thulin & Gifri (1998). Local tradition on Soqotra also recognised the distinction between ground-dwelling and cliff-dwelling frankincense trees (see below).

ETHNOGRAPHIC NOTES

Note on frankincense trees and gums

Frankincense trees. Island experts describe two different types of frankincense-producing *Boswellia*: one they call *ṣamaᶜano* and one *tēliyəh, tilīᵓo* (or equivalent). They clearly see these as two different plants. It is not clear to us what these represent, taxonomically speaking, but they are discussed separately to accurately reflect the views of the Soqotrans.

Further research is needed to clarify the situation regarding the Soqotri nomenclature of *Boswellia* species. However, provisionally speaking, the situation can be summarised as follows:

(i) All *Boswellia* species on Soqotra are basically called *emᶜīro* (*em'iyro*) (western Soqotra *emġīro*, *emghiyro*) (closely related to the southern Arabic term for frankincense tree (*maġareh* (*maghareh*), and to its name in the other pre-literate languages of southern Arabia (*meġerot* (*megherot*)).

(ii) All cliff-dwelling or 'constricted' *Boswellia* species are called *źifḥa* (*dhif'ha*).

(iii) *Boswellia* species that produce the best gum are called *ṣamaᶜano* (*zama'ano*) (possibly related to the Arabic *ṣamġ* (*zamgh*), 'gum'). Thus we find names such as *emᶜīro ṣamaᶜano* (*em'iyro zama'ano*) or *tiliyə ṣamaᶜano* (*tiliyih zama'ano*).

(iv) There is also a *Boswellia* species which grows commonly on coastal plains or on escarpments overlooking the sea, which is called *tīlīᵓo* etc. (*tiliy'oh*) to differentiate it from the *ṣamaᶜano Boswellia* specie which grow in the interior, and usually at higher altitudes. *tīlīᵓo* trees are more widespread in western Soqotra, *ṣamaᶜano* in eastern Soqotra, while in central Soqotra, both occur. Where they occur together, as in central Soqotra, they are definitely seen as being two very different species, producing foliage and gum of a different quality.

Soqotrans experienced in harvesting frankincense gum say that the best gums on Soqotra come from the group of plants called in general *emᶜīro* (*em'iyro*). There are three kinds of these: (*emᶜīro*) *tilīyə* which produces the best quality gum; (*emᶜīro*) *ṣamaᶜano*, which produces the second quality of gum; and finally *emᶜīro* itself, which produces the lowest quality of gum. As is the case with *B. sacra* in Dhofar, the frankincense-producing trees of wetter areas produce little or no gum, and their gum is regarded as inferior to that of the plants of the drier areas.

Gum resin. Any resin with a gum-like texture is called *śaḥaz* on Soqotra, not just that of the *Boswellia* species. Thus the resin of most of *Commiphora* species and that of *Dracaena cinnabari* is also called *śaḥaz*. The local definition of *śaḥaz* is of 'something that can be chewed', and the gum resins of all these can be chewed: for their pleasant taste, as an appetite stimulant and to cleanse the mouth and stomach.

Comparative uses. The history of the use of frankincense gum for sacred and ritual purpose is too well known and too lengthy to be rehearsed here. Instead, comparative uses for other parts of the *Boswellia* species in surrounding regions will be looked at. *Buds, flowers and seeds* of many species are eaten, as is the gum. The foliage and young stems are browsed by livestock. *Wood*: the trees are planted, or transplanted, to make a living fence. The wood is not very strong, it seasons rapidly and decays quickly in the open. However, it can be used for small poles, and is one of the only non-coniferous sources of turpentine and rosin. The wood is used for a variety of products: for example, the wood of *B. serrata* is used for making newsprint and other types of paper; for plywood, veneers, and hard boards; to make cheap furniture, boat masts, toys, sheaths for knives and so on. Ammunition boxes made from this wood are said to be better than from teak. *Bark* is widely used to dye leather. Medicinally, the bark of *B. serrata* is used as a diaphoretic, to treat rheumatism, diarrhoea, nervous disorders and skin diseases. *B. dalzielii* bark is boiled to make a wash to treat fevers and rheumatism, and is also taken internally for gastrointestinal troubles. The fresh bark is eaten to bring about vomiting after a few hours, and to relieve symptoms of giddiness and palpitations.

Oleo-gum resin is used for a variety of purposes: the volatile oil obtained from *B. serrata* is used in quick-drying varnishes, and the rosin is in high grade paints, varnishes, lacquers and printing inks. The oleo-resin-free gum is used in calico-printing and the textile industry for sizing and finishing. Medicinally the gum is expectorant, diuretic, stomachic and used to treat diarrhoea and dysentery, pulmonary affections and cutaneous troubles. In Yemen the gum of *B. sacra* is chewed by pregnant women, and is used to treat emotional and psychological problems, while elsewhere in Arabia it is used as a diuretic, purgative and, chewed or added to coffee, to improve the memory. (Ambasta 1982; Dalziel 1948, 1956; Fleurentin & Pelt/Fleurentin *et al.* [1982, 1983a,b,c]; Goodin & Northington 1979; Heine *et al.* 1988; Jain & De Fillips 1991; Kunkel 1984; Lewis & Elwin-Lewis 1977; Miller & Morris 1988)

"*Boswellia* growing in cramped conditions"

ETHNOGRAPHIC NOTES

Name(s)

– Soqotri: *źifḥa* 'squeezed, constricted; crouched down out of sight'.

– Soqotri simplified: *dhif'ha*.

Food (livestock). The bark of these shrubs is considered better as a dry season feed for livestock than that of the larger trees. Goats can tear off the bark themselves, but shepherds have to prepare it for their sheep. Islanders say that these smaller trees are particularly swift in their response to rain and recover condition quickly, unless a goat has chewed off the growing point or have effectively de-barked the shrub.

Notes. Applies to species of *Boswellia* growing in cramped conditions, especially *B. bullata*, *B. nana*, *B. popoviana* and *B.* sp. *A*. The name is applied to any form of *Boswellia* species growing on cliffs, or emerging from a crack or crevice in the rock. The islanders say that *źifḥa* is not a specific species of plant, but a particular habit and growth form. Some say that *emᶜīro źifḥa* is the 'male' of the 'female' *emᶜīro*.

Boswellia ameero Balf.f.

Tree to 5 to 8m, trunk grey with yellowish flaking bark. Leaves 10–30 x 2–15cm, pinnate, winged rhachis absent; leaflets 14 to 25, dark green above and below, 1.5–9 x 2–6cm, entire to crenate-serrate, glabrescent above, thinly tomentose beneath. Flowers in dense (8–15cm), paniculate racemes; petals red or dark pink, glabrous, oblong, <7 or 8 x 2mm, glabrous. Fruits 10(–15) x 8mm.

Ecology. Locally common in dry, semi-deciduous woodland. (120–)500–750(–1050)m.

Distribution. Endemic.

Conservation status. Vulnerable (VU B2 a b). A tree with a scattered and somewhat fragmented distribution. Some of the peripheral populations are not regenerating and considered relict in nature, strongly suggesting a decline in the extent and quality of its habitat.

Botanical notes. A hillside of trees of *B. ameero* in full flower is one of the most dramatic sights on the island. There is some variation: certain populations, for instance in the dense woodland of the Muqadrihon Pass, differ in having pale, not bright, pink flowers. No trees bearing bright pink flowers have been observed growing on cliffs; however, there is some evidence for trees with pale pink flowers on cliffs. An extreme form of these is described under *Boswellia* sp. *A* below.

ETHNOGRAPHIC NOTES†

Name(s)

– Soqotri: *emᶜīro* (pl. *ʾimʾḥer*) (western Soqotra *emġīro*); *emᶜīro ṣʌᶜəlhil* 'large-leaved *emᶜīro*'; *emᶜīro ṣamaᶜano* 'smaller-leaved *emᶜīro*' like *B. socotrana*; gum *dūr* 'blood' or *śaḥaz* 'gum'.

– Soqotri simplified: *em'iyro* (pl. *'im'her*), *emghiyro*; *em'iyro za'ilhil*; *em'iyro zama'ano*; gum *duwr*, *shahaz.*

Food (human). People nibble at the new growth or buds, and occasionally chew at small pieces of gum that can be picked off the branches in the hotter months. However, the gum of these species tends to stick to the teeth in rather an unpleasant way and *B. socotrana* is better for this purpose. All the gums are avoided by those with bad teeth as it gets stuck in cavities and can loosen rotten teeth.

Food (livestock). In many areas, the trees come into leaf at the end of the monsoon, before the winter rains, when other foliage is scarce. Elsewhere, they respond very quickly to rain. The foliage is considered to be a rich and nutritious fodder for goats, sheep, cattle and camels in the dry season. Donkeys also graze the foliage, and the smaller leaves are a good dry season fodder for weaned kids and lambs. When other forage is available, cattle eat little of the green foliage, but camels actively seek it out, and goats like the tender new growth – the mature leaves can be very coarse. Herders lop the foliage and gather up fallen leaves for their livestock, especially cattle and goats, in the dry months when other fodder is scarce, or to fatten up a breeding or lactating animal. Cattle only eat the leaves, but goats and sheep eat the leafless stems and twigs as well, and goats in particular gnaw at the bark when hungry. Like other favoured forage, leafy stems are offered to a difficult cow or goat to encourage her to let down her milk. The foliage is also gathered fresh, spread out to dry on the stone floor of a cave or other smooth rock surface, and then bundled up and stored in a dry place as a fodder reserve, where it can last for up to two years. In the less well-vegetated areas of the island, chunks of the inner and outer bark are cut from the larger branches, peeled of their papery outer bark and then cut into small pieces and fed to sheep in a prolonged dry season. The bark of these species makes their noses run, and the bark of the smaller-leaved *Commiphora* species is preferred. The foliage of *Boswellia* species is, in general, considered to be inferior to that of *Commiphora* species in terms of nutritional value. However, the comparatively large leaves of this group of *Boswellia* species make up in bulk what they lack in protein as a dry season subsistence feed. In the areas where these trees are numerous they are often the main dry season fodder.

Bees / honey. Important honeybee forage. Islanders say that bees use the gum-resin in the construction of their combs. The dead wood and bark are one of the best woods for smoking out bees when robbing them of their honey.

Fuel. The dead wood burns with a sweet smell, but is not good firewood. It burns down very quickly and its thick and strongly aromatic smoke means that it can taint food so is not used for cooking. Smaller fragments make good kindling.

Timber. The wood is of little use as timber and, unlike *Commiphora* species, even the heartwood of the dead tree is not suitable for building.

Leatherwork. The inner bark is crushed to a paste and used to prepare skins and hides to cure the leather and dye it a warm red-brown. Chunks of bark are sliced from the trunk and main branches of a well-grown tree, the outer papery bark removed, and the rest pounded with a rock to a coarse paste. This is mixed with a little water to make the thinner paste used on the skins. This material is particularly liked for tanning the leather which will be made into the water buckets used in the interior to water date palms, finger millet, livestock and tobacco. The skins that get the heaviest use, such as water- and milk-skins, and those used to store the pressed dates and churn the butter, are commonly treated with material from these plants. Skins that have grown thin and worn are treated again with the paste to prolong their life. Skins that have become smelly or mouldy are also rubbed with the paste to sweeten them.

Wood. Soft, and not suitable for making implements.

Medicinal / veterinary. The same paste used to prepare leather is occasionally used to treat sores and skin ulcers. Goats and cattle that have just given birth, are failing to thrive, or are recovering from an illness are commonly hand fed on the younger leaves of these plants.

Cleansing. The outer bark is aromatic and is burned for its pleasant smell and to fumigate living quarters. Chunks of the inner and outer bark together are sometimes left in nooks and crannies in the house for their smell to permeate the living area, and so that slivers can be cut off and put on the fire to burn. Pieces of bark with dried pieces of resin clinging to the outer surface are preferred. Small pieces of the underbark are burned on a slip of pottery and carried around living quarters to perfume them. The red underbark is burned and clothing or bedding is hung in the smoke to absorb the sweet smell. Nowadays this is most commonly achieved by draping the material in layers over a conical frame set over a clay incense burner in which fragments of the bark or resin have been lit. When returning to a settlement after an absence the living quarters are first fumigated with this to sweeten the stale air. Similarly, in wet and cold weather lumps of the gum are lit and carried around to make the living quarters smell less damp and musty. The gum oozes out spontaneously in hot weather, or when the trees are lopped for fodder. The soft oozing gum, the snapped twigs and the leaves are all aromatic, and sprigs of the leaves are sometimes tucked in the clothing for this reason. Plants that impart a perfume to the skin or clothing have an importance beyond their sweet smell alone: they are also a means to maintaining good health.

Cosmetic. The gum is chewed to clean the teeth and gums and as a remedy for bad breath. However, the gum of *B. socotrana* is preferred.

Commercial value (past, present). At times when the prices for frankincense gum were high, the gum was collected and sold, though strictly speaking only that of *B. socotrana* was considered to be of a suitable quality. Islanders report that these trees ooze much less gum in hot weather than *B. socotrana*, and that when the bark is cut, any gum that emerges does so only very slowly and very little at a time. Some people dislike the smell of *emᶜīro* gum and will not use it. When the demand for aloes juice was high, the more unscrupulous aloe harvesters or traders would crush the inner bark of these trees with that of any of the *Hibiscus* shrubs and succulent aloes leaves. They would then add the watery paste to the genuine aloes juice in the skins. The only way of knowing for sure that such adulteration had been carried out was that the product never set hard like genuine, pure aloes juice but always remained soft and semi-liquid.

Pest control. The ash of the burned wood is smeared on the exterior of the skins used to store pressed dates to deter insects. The dead wood is used for smudges in livestock pens and byres. The animals are said to like the smell and to find the smoke of the *Boswellia* species soothing as well as insecticidal. This is also a good wood to burn at milking time to calm the animals and stop them being bothered by biting flies. The more rotten the wood, the denser and more effective the smoke it gives off. Before moving into new quarters it is common to fumigate them with a fire of the dead wood to get rid of creatures believed to be harmful to man and his livestock. These are collectively called *rèʾeś*, and include some, but not all, reptiles, such as the the chameleon, skink, nocturnal *Hemidactylus*, and all snakes, as well as scorpions, wolf spiders, biting ants and the giant *Scolopendra* centipede. It is said that these cannot stand the smell of certain burning woods, so emerge from their hiding places and leave the area. On the other hand, islanders say that care must be taken when gathering dead wood, as this and the dead trees often shelter the same noxious creatures.

Other. The gum is chewed to lessen thirst. It is also burned to clear an area or dwelling of evil influences, or to prevent evil spirits from approaching.

The ash of the wood is an important fertiliser in the areas where these trees grow plentifully, especially for growing finger millet and tobacco. Small piles of the dead wood, preferably also with aged sheep dung, are laid here and there across the area to be planted, and are then fired and the cooled ash mixed with the soil before sowing the seed. The dead wood, usually with additional dried dung, is one of the best for firing pottery, especially that made from paler clays because, as the smoke turns the pottery from cream to red-brown. For this to work, the pots to be fired must be completely buried in the dead wood. The sap can be used as a glue in the absence of better alternatives. It is particularly used to repair cracked clay pottery: it is chewed to make it soft, or used when still soft from the tree, and smeared over the crack. The pot is then put beside the fire for the gum to harden and set.

The red inner bark was used to dye cotton twist or the imported coarse, undyed calico from which island clothing was made. The inner bark was ground to a paste with a little water, and this was added to a pot of boiling water and left to simmer for two to three hours until a good clear colour had been achieved. The twist or material was then lowered in and left to steep until it had reached the desired shade. It is then spread out to dry in the shade. The colour is not fast, but fades on washing and exposure to the sun. The same dye and the same method was sometimes used to colour pale or white wool. A paste of inner bark was used to dye the intricately plaited leather belts formerly worn by men. It is also used to decorate the paler clay pots as soon as they are removed from firing.

Notes. Young trees often grow in the protection of clumps of unpalatable shrubs such as *Buxanthus pedicellatus*. As they grow tall and emerge from the clump, livestock nibble the top leaves, but this does not apparently prevent the young tree from growing on. In a prolonged dry season hungry livestock feeding on the emerging foliage can develop severe scour which can last for around a fortnight until their digestive systems become accustomed to the feed. Bent (1897) writes about 'valleys entirely full of frankincense-trees with rich red leaves, like autumn tints, and clusters of blood-red flowers'.

†Also apply to *B. bullata*, *B. dioscorides*, *B. elongata*, *B. nana*, *B. popoviana*, *B.* sp. *A*.

Boswellia bullata Thulin & Gifri

Tree to 5m, rooted on cliffs; trunk pale grey, smooth (sometimes flaking). Leaves 8–15 x 3–8cm, pinnate, winged rhachis absent; leaflets 9 to 17, dark green above, white-felted beneath, 1–4 x 0.5–1.5cm, crinkly, deeply irregularly crenate-serrate, puberulent above, thickly tomentose beneath. Flowers in open raceme-like panicles; petals greenish yellow or red, oblong, 3–4mm, puberulous outside. Fruits 6-valved, globose, c.6 x 7mm.
Ecology. Rare on limestone cliffs in dry, semi-deciduous woodland. 100–300m.
Distribution. Endemic.
Conservation status. Vulnerable (VU D2). A rare tree with a fragmented distribution; known from only three locations with a very restricted total area of occupancy.
Botanical notes. A distinctive cliff-dwelling tree, characterised by its pinnate leaves and crinkly, deeply irregularly crenate-serrate leaflets. There is some variation: plants from the single population known in the east of the island (at Bit Gobihir) differ from trees from the west (at Wadi Ayeq and the cliffs above Qaysoh) in having red (not greenish yellow) petals and leaves with fewer (9 to 13), more regularly and less deeply crenate-serrate leaflets.

ETHNOGRAPHIC NOTES
Name(s)
See *B. ameero* (also uses).

Boswellia dioscorides Thulin & Gifri

Tree to 7m, rooted on cliffs; trunk brown or greenish, flaking. Leaves 15–20 x 4–8cm, pinnate, winged rhachis absent; leaflets 11 to 17, glaucous, 2–4 x 1.5–4.5cm, obscurely crenate-serrate, glabrous. Flowers in long, open panicles; petals cream or white sometimes tinged pink, ovate, 5–6 x c.3mm, glabrous. Fruits 7–11 x 4–5mm.
Ecology. Widespread and sometimes abundant on limestone cliffs in dry, semi-deciduous woodland in the south and east of the island. 200–750m.
Distribution. Endemic.
Conservation status. Vulnerable (VU D2). Found in a single vegetation type and with a limited area of occupancy.
Botanical notes. An exclusively cliff-dwelling species distinguished by its glaucous leaves and strongly peeling bark.

ETHNOGRAPHIC NOTES
Name(s)
See *B. ameero* (also uses).

Boswellia elongata Balf.f.

Tree to 8m, trunk pale brown, strongly flaking bark. Leaves 15–35 x 6–20cm, pinnate, sometimes simple, winged rhachis absent; leaflets 13 to 29, glossy green above, whitish beneath, 3–10 x 1–3cm, shallowly crenate to deeply crenate-serrate, glabrescent above, densely tomentose beneath. Flowers in long (15–30cm) panicles or racemes, petals red or cream to yellowish-green, pale greenish-pink, glabrous, oblong, 6–8mm. Fruits 10–13 x 7–11mm.
Ecology. Widely scattered in dry, semi-deciduous woodland and less commonly in succulent shrubland. 100–650m.
Distribution. Endemic.
Conservation status. Vulnerable (VU B2 a biii). *B. elongata* has a scattered and extremely fragmented distribution with many populations not regenerating and relict in nature. This pattern of distribution strongly suggests a decline in the extent and quality of its habitat.
Botanical notes. A ground-dwelling species characterised by its large, pinnate leaves which are white-felted beneath and its long, lax panicles or racemes, of typically yellowish-green flowers. Trees bearing both simple and pinnate leaves have been observed, for instance in Wadi Terubah; these are possibly the result of hybridisation with *B. popoviana* (see notes under that species).

ETHNOGRAPHIC NOTES
Name(s)
See *B. ameero* (also uses).

Boswellia nana Hepper

Shrub or small tree to 4m, sometimes prostrate, rooted on cliffs; trunk pale grey. Leaves 4–18 x 0.5–5cm, simple or lobed, margins irregularly sinuate or crenate, a few basal pinnae sometimes present, glabrous or minutely puberulous. Flowers in short (<c.5cm long), few-flowered panicles; petals pink or cream and reddish tinged, lanceolate, 4–5mm, glabrous. Fruit 3 to 5-valved, c.10 x 6mm, ripening purple.
Ecology. Rare on limestone cliffs and limestone pavement, in dry, semi-deciduous woodland in the northeast of the island. 300–550m.

Distribution. Endemic.
Conservation status. Vulnerable (VU D2). Known only from two locations, with a total area of occupancy of less than 20,000km^2.
Botanical notes. *B. nana* is recognised here as a complex of low, sometimes totally prostrate, shrubs and small trees found on limestone cliffs and areas of flat limestone pavement in a relatively restricted area of NE Soqotra. Thulin & Gifri (1998) regarded *B. nana* as a reduced form of *B. popoviana*. However, it differs from that species in its smaller, more reduced panicles of red not yellowish flowers and the glabrous not white-felted undersurface of its leaves. Certain plants from inland cliffs in NE Soqotra, included here under *B. nana*, show great variation in the degree of the division of their leaves and in their habit. I strongly suspect that these plants are the result of hybridisation with *B. socotrana*. The clearest evidence for this is a tree (*Miller* 17045) found at the base of limestone cliffs at Hamaderoh which is somewhat intermediate in appearance between *B. socotrana* and 'typical' *B. nana*. It bears both simple and pinnate leaves with the pinnate leaves having the winged rhachises which are otherwise diagnostic of *B. socotrana*. It seems also to be intermediate in its ecology: typical *B. nana* is restricted to cliffs, *B. socotrana* is ground-rooted whereas the intermediate plant was found growing amongst rocks at the base of cliffs. Populations of *B. nana* found on the seaward-facing escarpments of NE Soqotra bear only simple leaves.

ETHNOGRAPHIC NOTES
Name(s)
See *B. ameero* (also uses).

Boswellia popoviana **Hepper**

Tree to 5m, rooted on cliffs; trunk pale grey, smooth or with yellowish flaking bark. Leaves of two types: either (form 1) uniformly simple, 6–12 x 2–5cm, margins regularly crenate, or (form 2) usually lobed below, 7–16 x 3.5–10cm, margins irregularly crenate and sinuate often somewhat undulate, both forms glabrous or minutely puberulous above and densely white-felted beneath. Flowers in long (10–20cm) many-flowered panicles; petals cream and reddish tinged, ovate, 4–5mm, glabrous. Fruit 10–15 x c.5mm, 3 to 5-valved.
Ecology. Scattered on limestone cliffs in dry, semi-deciduous woodland. 20–550(–1050)m.
Distribution. Endemic.
Conservation status. Vulnerable (VU D2). Rare, with a fragmented distribution and limited area of occupancy; several populations show no sign of regeneration, suggesting a decline in the quality of its habitat.

Botanical notes. A narrower concept of this species than that of Thulin & Gifri (1998) has been adopted here, including all cliff-rooting trees bearing leaves with white-felted undersurfaces. It is found on limestone (and occasionally granite) cliffs in central-northern and central-southern Soqotra. Two distinct forms can be recognised: those with entire, rather regularly crenate leaves, found in the wadis draining to the south of the island; and those with simple or deeply lobed leaves with irregularly crenate and somewhat sinuate margins from the limestone areas to the northwest of the Haggeher mountains. Trees in Wadi Terubah bearing both entire and simple leaves are possibly the result of hybridisation with *B. elongata*. A dwarf, prostrate form of *B. popoviana* (cf. forms of *B. nana*) was recently also found on the limestone ridge at Heger (between Ras Izfer and Ras Sherubrub) in the extreme west of Soqotra above Bandar Shu'ub.

ETHNOGRAPHIC NOTES
Name(s)
See *B. ameero* (also uses).

Boswellia socotrana **Balf.f.**
Lannea aspleniifolia (Balf.f.) Engl.

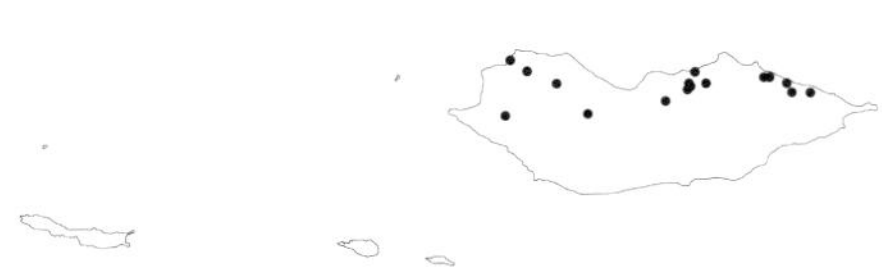

Tree to 5m; trunk grey or reddish brown, smooth or somewhat flaking. Leaves 1–12 x 0.5–2cm, pinnate, winged rhachis present; leaflets 7 to 31, pale or greyish green, 0.1–1.5 x 0.1–0.8cm, entire or lobed, glabrous or puberulent. Flowers in short (c.5cm), sparse panicles; petals cream or pale yellow, ovate, c.3mm, glabrous. Fruits subglobose, c.6 x 5mm, ripening red.
Ecology. Widely distributed in dry, semi-deciduous woodland, and less commonly in *Croton socotranus* shrubland. 50–600m.
Distribution. Endemic.
Conservation status. Vulnerable (VU D2). Rare, with a very limited area of occupancy; several populations show no sign of regeneration, suggesting a decline in the quality of its habitat.

Botanical notes. A ground-dwelling tree, readily distinguished from all other *Boswellia* species on the island by the leaves which have winged stalks and numerous, small leaflets and the inconspicuous pale yellow flowers which are borne in short, sparse panicles.

ETHNOGRAPHIC NOTES
Name(s)
– Soqotri: *ṣamaˤˈano* (pl. *ṣamaˤhənˈitən*); *tilīˀo*, *tālyˈoh*, *tʌlīˀoh*, *tilīy*, *tāliyə*, *tilīˀəh* (see notes below).
– Soqotri simplified: *zama'ano* (pl. *zama'hinitin*); *tiliy'o*, *talyoh*, *taliy'oh*, *tiliy*, *taliyeh*, *tiliy'ih*.
Food (human). People nibble at the new growth, and occasionally eat the buds and flowers. The pieces of gum that emerge in hot weather are chewed to relieve thirst and as a chewing gum to strengthen the teeth and gums. The root of a young plant can be eaten. It is very juicy so it lessens thirst as well as being a food.
Food (livestock)
– *ṣamaˤˈano*. The foliage is an important dry season feed for livestock in the areas where they grow, especially for small stock and camels, and herders lop leafy twigs as a

dry season feed for their goats. It is less sweet and therefore not considered as good as that of the *tilīʔəh* trees. The small leaves of this species are considered amongst the best feeds for young stock, especially weaned lambs and kids. Cattle do not usually feed on the leaves, nor are the leaves cut for them except in areas where none of the larger-leaved *Boswellia* species grow. Camels also browse the trees. Sheep and goats eat fallen leaves and other plant litter, and chew the leafless twigs and stems. Goats and sheep both chew the bark (inner and outer) when hungry. For sheep and young stock hepherds usually have to prepare the bark by cutting it up into small pieces and giving it a preliminary crushing.

– *tilīʔəh*. The foliage of this is considered by some to be a better dry season feed than that of *ṣamaʕano* and herders lop leafy stems for their sheep and goats. Camels also browse the trees. It is not normally fed to cattle, unless there is little alternative. The bark is apparently less liked by goats than that of *ṣamaʕano*, but goats still chew at it when very hungry.

Bees / honey. Both *ṣamaʕano* and *tilīʔəh* are an important honeybee forage, and islanders say that, as for other *Boswellia* species, the bees use the resin in the construction of their combs. The dead wood and bark of the trees are one of the best woods for smoking out bees when robbing them of their honey. It is not uncommon for bees to build hives in holes in the trunks of both.

Flavouring / food additive. The milk of goats eating a lot of *tilīʔəh* smells slightly of frankincense, as does the meat of the carcass and the water it is cooked in.

Fuel. The smoke of the burning wood is pleasantly fragrant, but neither the wood of *tilīʔəh* nor of *ṣamaʕano* makes good firewood. It burns quickly and the aromatic smoke can taint food, making these woods unsuitable for cooking.

Timber

– *ṣamaʕano*: no use as timber.

– *tilīʔəh*: acceptable timber, if not of the highest quality. It is strong enough to be used for roofing beams.

Leatherwork. The inner bark of *tilīʔəh* is crushed to a paste and used as other *Boswellia* species in leatherwork. In 1834 Wellstead (1835) described the bark of '*taleo*' being used to 'tan' skins. Skins that have developed an unpleasant smell or which have become thin and worn through use are treated again with the paste to restore their strength and sweeten them.

Wood. Both types have wood that is too soft to be useful for making implements. The dry and papery outer bark is collected and stored for use as tinder.

Medicinal / veterinary

Gum. *ṣamaʕano* and *tilīʔəh* gum is chewed to sweeten the breath and to strengthen the gums. The gum of *ṣamaʕano* is better than that of *tilīʔə*, which is rarely chewed. The saliva resulting from chewing the gum is regarded as a general tonic and soothes a disturbed stomach. When the chewing gum produces no more saliva, it is discarded. A small piece of fresh gum is inserted into a cavity to relieve the pain and to 'burn out' the diseased area of the tooth. Pieces of gum are given to people who are sickly or who have lost their appetite, and it is burned throughout childbirth to keep evil forces at bay. Frankincense gum is burned in many esoteric treatments for illnesses caused by witchcraft or the Evil Eye, for example to treat a patient suffering from the condition known as *di so'o*, whose signs are fever, general weakness, and a rash of large, flat and itchy spots all over the body, including the face. This is contracted by urinating or defaecating in an area inhabited by *jinn* or similar spirits, or by washing in water they consider to be their for their own use. It is commonly treated with a mixture of frankincense gum, donkey dung, hair from a goat or sheep's wool, and some trimmings from a goat horn. These are gathered together and then put to smoulder on hot embers and the patient is encouraged to bathe in the fumes. The gum used to be much used by the traditional healer or '*mekoli*' who was able to 'sniff out' witches and was called in whenever other remedies had failed or sorcery was suspected. On Abd al Kuri, where the gum was usually imported from Somalia or from the Arabian mainland, some droplets of frankincense are burned on a slip of pottery and given to someone with a head cold. The patient shrouds himself in a cloth and inhales the fumes.

Wood. A woman suffering from a retained placenta is treated by putting some green or dead *tilīʔəh* wood on the fire and encouraging her to crouch over it to fumigate her whole person. Some say that after this she should lie down and her stomach be massaged to encourage expulsion of the placenta.

Cleansing. *ṣamaʕano* and *tilīʔəh*: fragments of the gum, or of the bark with gum adhering to it, are put somewhere in the living quarters to give them a pleasant smell. Slices of bark or stems of the wood are put on the fire for their sweet smell. The inner bark is normally used, rather than the inner and outer together as for the larger-leaved *Boswellia* species. Sometimes a fragment of gum is put on the embers too, or the smoking gum is used to perfume and fumigate clothing and bedding. At a time when sorcery was greatly feared, people were nervous of burning the gum in their homes unless advised to do so by the traditional healer, the '*mekoli*'. Otherwise someone smelling the smoke might suspect them of involvement in sorcery. In general, it was felt best to leave the use of the gum to this traditional healer who had a special relationship with the spirit world. For this reason the gum was rarely burned, but was instead set somewhere where its smell could gently permeate the living quarters. If a clay pot, especially the butter or water pot, becomes smelly, it is filled with boiling water, left to stand for a while, and then fumigated with the smoke of the burning gum. The pot is then stood open to the sunlight for the rest of the day.

Commercial value (past, present)

ṣamaʕano and *tilīʔəh*:

Harvesting: by gathering droplets. The same verb '*imāṣilin*', meaning 'to pick at, pick off something', is used for gathering gum resin from *Boswellia* and *Commiphora* species and from *Dracaena cinnabari*, as well as for gathering various lichens. The harvesting of *Commiphora* and *Boswellia* gums usually consists of picking off any droplets that have dropped to the ground or have oozed through the bark. The plants exude resin either spontaneously at times of the greatest heat, or when the bark has been damaged in some way (as by herders lopping foliage for their livestock, goats gnawing at it or by wild and feral cats scratching the bark). Many islanders say that the best quality gum is that which results from injury to the tree, and often call such fresh gum '*dūr*' or 'blood'.

Whatever may have been the case in the past, the accepted situation now is that no-one 'owns' frankincense trees, and that anyone can collect the droplets of gum resin from the ground or from the tree itself. However, if someone is planning to gather any considerable amount, it is considered courteous to seek permission from those in whose immediate area the trees grow.

Harvesting: by cutting into the bark. Harvesting the gum by cutting into the bark is the prerogative of those who own the land on which the trees grow. Slanted cuts are made through the outer and inner bark of the larger branches and trunk, and narrow slices of the bark removed. The cuts are made where the expert cutters believe the main channels of resin – the 'arteries' of the 'blood' – to lie. The tree is then left for ten days to a fortnight, during which the exudate gathers and coagulates. This is then picked off by the harvester. In hot weather, the old scars ooze gum of their own accord, without further cutting. At this time the gum has a very strong smell and is regarded as being of a very high quality. The collected gum is not further processed or sorted out into grades or types (as happens with the gum of *Boswellia sacra* in Dhofar, for instance). It is considered to be all of the same quality.

Marketing. Even within living memory, the gum had some economic importance, and at times of greatest demand, people would come up from the coast when fishing was slack and gather it. It was sold in the coastal markets, and to the ruling and merchant families, or was used to trade with the people of the more heavily populated coastal plains. It was purchased mainly for ritual use. However, even in earlier times, the frankincense that could be purchased on occasion from visiting trade ships was usually considered to be more 'powerful' for ritual use than that from the local trees. The droplets are bigger and said to smell stronger when burned.

The greatest quantities of the Soqotran gum came from the western area, especially from the plains and foothills between Di Hamdh and Qalansiyah, where the trees grow in *Croton* shrubland. Here the gum was, and to a certain extent still is, gathered commercially for a small but steady market. In the eastern area there were not enough trees to make its harvest worthwhile commercially, and people just collected a little for their own use and to give to friends. The gum is of little importance economically now. Some is still sold in coastal markets, mainly for the few tourists, a little is used for barter, but most is used in the home, sent to émigré Soqotrans in Oman and the Gulf, or given as a gift. Its ritual use on Soqotra has much lessened, and the more esoteric applications are now generally frowned on. However, people still occasionally chew it, and it also seems likely that, under influence from the Arab mainland, its use as an incense, especially as a hospitable gesture at the close of a meal, will increase.

Pest control. *ṣamaʿano* and *tilīʾəh*: the gum or slices of resinous bark are burned when re-occupying living quarters that have been left unoccupied for a time. This is done for the same reasons as described above for large-leaved *Boswellia* species. The burning of *tilīʾəh* gum is said to be especially effective for this. Bark from both is burned in living quarters to getr rid of biting insects. The dead wood is used for smudges in livestock pens and byres, as for other *Boswellia* species.

Other. *ṣamaʿano* and *tilīʾəh*: chewing the gum relieves thirst. The gum is used to repair cracked clay pottery as described above. The gum of these species especially is used for sacred and ritual purposes. A lump is lit in the home after dusk, especially in the rains, to keep away evil. The smoke of the burning gum is used to fumigate the sick and the bewitched, usually synonymous on Soqotra, and to cleanse human and livestock living quarters, as well as to protect people and their livestock from malign influences. Date gardens and cultivated plots of finger millet and tobacco are also treated in the same way to protect them from the effects of the Evil Eye. Ritual curing of both people and their animals was done with burning frankincense. This procedure was called '*raḥaź*', 'washing' or 'cleansing'. This was normally done only on the recommendation of the traditional healer or '*mekolī*', and the gum was never set alight until the '*mekolī*' arrived and ordered this to be done. Gum was also burned when calling on God for aid. But again this was usually only as prescribed by the '*mekolī*' or by a member of one of the other groups considered to have a special relationship with the supernatural, such as the Seyyids, respected for their relationship to the family of the Prophet Muhammad. The '*mekolī*' were said to be able to magically manufacture their own frankincense: they would blow on a live coal held in their hands and transform it instantly into a lump of smouldering gum. During the circumcision ceremony, formerly carried out on adolescents, the surgeon, the *mizēdihir*, would put some gum on a slip of pottery, light it, and then place it under the 'circumcision seat', the *miskido*, where the operation would be performed. Once the smoke of the gum had risen above the stone, he would give a signal, and all present, except the young men to be circumcised, would shout *xob ʿayn*! *xob ʿayn*! to clear the area of evil, while the first of the young men approached the stone to undergo the operation. Traditionally, many in the more settled communities of the coast burned some of this wood in their living quarters every Wednesday evening, and in the mosques on Wednesday and Thursday nights, the fire being extinguished first thing on Friday morning.

It is said that long ago, marriages were solemnised by the bridegroom placing a knife across the lap of the bride and giving her family a strip of meat. The united couple then sat together under a sheep's wool blanket and were fumigated with the smoke of burning sheep dung or frankincense gum.

Those who had the expertise, and dared to do so, would treat the sick by putting a piece of the gum in a slip of pottery and lighting it. They would then carry it around the sick person or animal, making a certain sound and naming those people or spirits believed to have caused the illness. The smoke was wafted around and under the patient and God's protection called down on them. This is usually done in secret, out of fear of being accused of sorcery. Many islanders say that the gum can only properly be burned by 'ordinary people' during childbirth, and to fumigate the clothing and bedding of the childbed. The gum is burned continuously beside the woman from the beginning of the birth until the placenta is safely delivered. Some also say it may be used for the dead, being burned continually from the moment of death until the bier reaches the grave, but most islanders say that this is only acceptable among the communities of the coastal market settlements, and is not acceptable in the interior.

tilīʾəh trees produce less gum than *ṣamaʿano*, but it is of a better quality and receives a better price. *tilīʾəh* gum is also said to be more 'powerful' against sorcery than that of *emʿīro* or *ṣamaʿano* trees. However, a group of *tilīʾəh* trees growing close together are said to harbour *jinn* and other dangerous spirits, and should therefore be avoided. All *Boswellia* species are thought to shelter harmful creatures, both natural and supernatural, and care must always be taken when approaching them to remove anything from them.

On Samha frankincense is imported from Soqotra and is used especially to treat 'madness'. People who seem strange, who sleepwalk or are restless at night are treated by being fumigated in the smoke of the burning gum. They, or the spirit which has taken them over, might struggle to avoid the smoke, but must be forced to remain in it, even if this means restraining them forcibly. More recently, fumigation with frankincense smoke has been used with some success to treat women suffering from post-partum depression. On Abd al Kuri the gum is also used to perfume clothing and bedding, and is burned throughout childbirth, at weddings and beside the deathbed. It is also burned inside clay water pots to clean them and give the water a pleasant flavour.

Notes. Islanders say that the 'male' *ṣamaʿˈano* has fewer leaves and produces much less gum than the 'female'. They say that *tilīʔəh* has no 'male' or 'female'.

Boswellia* sp. *A

Tree to 3m, rooted on cliffs; trunk pale brown or greyish, strongly flaking. Leaves 18–25 x 6–12mm, pinnate, winged rhachis absent; leaflets 17 to 19, dark green, 2–6 x 2–3cm, bullate, obscurely crenate, glabrous or with scattered hairs on the veins above. Flowers in many-flowered raceme-like panicles; petals pale pink, ovate, c.7 x 3mm, glabrous. Fruits 5-valved, obovoid, c.13 x 9mm, ripening reddish.

Ecology. On limestone cliffs in semi-deciduous woodland dominated by *Dracaena cinnabari*, *Buxanthus pedicellatus* and *Croton socotranus*. 700–800m.

Distribution. Endemic.

Conservation status. Vulnerable (VU D2). Known only from a single location and occupying an area of less than 20km^2.

Botanical notes. This small tree is common on the limestone plateau of Firmihin. It is clearly related to *B. ameero*, which does not grow in the area, but differs in its cliff-rooting habit and pale pink petals. It grows side by side with *B. dioscorides* (the only place where two cliff-dwelling species have been observed growing together). It is readily distinguished from *B. dioscorides* by its green, not glaucous, leaves and in differences in the shape of the 'holdfast' which attaches the trunk to the cliff face.

ETHNOGRAPHIC NOTES

Name(s)

See *B. ameero* (also uses).

Commiphora Jacq.

Balsamodendrum Kunth

Trees or shrubs, sometimes spiny, bark smooth or peeling; wood with milky latex and inner bark with aromatic resin. Leaves alternate, simple or pinnate with up to 23 leaflets. Flowers inconspicuous, sessile or in raceme-like cymes, unisexual, actinomorphic. Calyx lobes and petals 4, linear; stamens 4 or 8. Fruit leathery or fleshy, 2-valved, with a single stone; stone surrounded by a variously shaped, fleshy pseudoaril. 4 endemics. 190 species. Africa, Madagascar, Arabia to W India, S America.

ETHNOGRAPHIC NOTES

Comparative uses. Many *Commiphora* species of neighbouring regions are palatable to cattle, camels, sheep and goats. Both *Commiphora* species and *Boswellia* species strike readily from cuttings and are an excellent material for making living hedges. The gum resin of many species is used as an astringent and as an ingredient in mouthwashes. The oleo-resin fraction also shows anti-arthritic and anti-inflammatory activity and has been found to lower serum cholesterol. The gum resin of many species is chewed as a food or to alleviate thirst and the seeds are crushed and applied to painful gums. Many gums are also used in fumigation and as incense. Other parts of the plants provide food: the tuberous roots of *C. angolensis* are eaten; the stems of *C. bruceae* chewed; the soft heartwood of *C. harveyi* eaten as a famine food; the roots of *C. mollis* are eaten raw; the peeled roots of *C. neglecta* are eaten, and the leaves of *C. pilosa* are cooked and eaten. The twigs are used as chewsticks and firesticks, and the wood is used to make a variety of containers, bowls and buckets. The barks are used in tanning and dyeing.

On the Arabian mainland the gum resin of *C. mukul* is used as incense, and as a fixative in perfumery. Medicinally it is regarded as astringent and antiseptic, and is taken as a stomachic, and for muscular rheumatism. It is taken to stimulate the appetite, and as a diaphoretic, diuretic, expectorant and uterine stimulant. The gum is burned and the smoke allowed to rise between the legs of a parturient mother to help with the expulsion of the placenta and cause the blood to flow.

The gum resin of *C. myrrha*, the source of true myrrh, is used as an amulet and as an analgesic. It is burned and the smoke inhaled to treat a fever, or drops of the resin are put in water and drunk, or a piece of resin is placed inside the nose or rubbed over the body. The resin is applied to the breast to wean infants and is also smeared on black cloth and left to harden, and then used to strap a fractured limb. Powdered, it is used as a dentifrice and to treat spongy gums. (Ambasta 1982; Fleurentin & Pelt/Fleurentin *et al.* [1982, 1983a,b,c]; Ghazanfar 1994; Heine *et al.* 1988; le Houerou 1980; Jain & De Fillips 1991; Kunkel 1984; Lewis & Elwin-Lewis 1977; Miller & Morris 1988; Oliver-Bever 1986)

Commiphora kua **(Royle) Vollesen**
Balsamodendrum mukul sensu Balf.f. non Hook.
Commiphora habessinica (O.Berg) Engl.

Low shrub to 1m with intricate, spiny branches; bark grey or brown, strongly peeling into papery strips. Leaves clustered on spur shoots or (if ungrazed) on long shoots, simple, obovate, 1–3 x 0.5–1.5cm, obscurely crenate-dentate. Flowers yellowish green and red-tinged, c.3mm long, sessile, 1 to 3 on spur shoots. Stamens 8. Fruits greenish red, ellipsoid, 5–8mm long, sessile; pseudoaril yellow with 4 linear arms.

Ecology. On Soqotra locally common on sandy soils in dwarf shrubland in the western inland plain and less common on rocky slopes in succulent shrubland; on Abd al Kuri on desertic rocky slopes. Sea-level–200m.

Distribution. Tropical E Africa south to Zambia, and in S Arabia.

Botanical notes. I have followed Thulin (Flora of Somalia, 1999) in including *C. habessinica* in this widespread and variable species. *C. kua* is a low, often prostrate, spiny shrub which is immediately distinguished by its grey or white bark which peels off in papery strips.

ETHNOGRAPHIC NOTES

Name(s)

– Soqotri: *ḥárḥer*, *ḥárḥéyr* (pl. *ḥaráḥeyr*) (also given to *Acacia sarcophylla*, *A.* aff. *tortilis*, *A. edgeworthii* and *Dichrostachys dehiscens*); *ḥarḥéyro* (pl. *ḥárḥeyr*) (Abd al Kuri); (rare) *ḥarṭānəh* (Abd al Kuri).

– Soqotri simplified: *harher*, *harheyr* (pl. *haraheyr*); *harheyro* (pl. *harheyr*); *hartanih.*

Food (human). The resin is not chewed on the islands.

Food (livestock). The tender stems and foliage are much liked by goats, sheep and camels. Small stock like the buds, flowers and fruit as well. The fruit are seen as being a nourishing and fattening food for small stock, and eating these and the tender new foliage as increasing milk yield. It is regarded as a valuable and nutritious dry season fodder for goats and sheep, especially as it never quite dies back, and there is always some nourishment to be found in the stems. Goats also gnaw the bark. On Abd al Kuri this is a key fodder for goats.

Fuel. The dead wood makes reasonable firewood which gives off a pleasant smell when burned. However, it burns quickly, and is rather smoky. On Abd al Kuri it is used especially in firing pottery, along with *Acacia* wood. The papery outer bark can be used as tinder.

Medicinal / veterinary. On Soqotra the bark or some of the resin is burned after the safe delivery of a baby. A small piece of the resin is ground to a powder which is given in warm milk or water to a toddler or young child with a sore stomach. On Abd al Kuri this is the island's most important medicinal plant. Slices of bark are scorched at the fire until thoroughly dried out and blackened and are then crushed to a powder. This is applied to fissures in the thickened plantar skin and to deep lacerations. Wet eczema, or a sore or abrasion slow to heal, is dressed with the powder to dry the skin out and help it heal. A maggoty wound is cleaned out with salt water and then packed with the powder. Multiple lacerations are treated by washing in hot water and then applying warm oil. The wounds are then dressed with the powder, with or without added salt. This treatment is followed daily until the lacerations are healing well. To treat eczema and other skin conditions which cause pruritus, a slice of the outer and inner bark together is cut from the trunk and put close to the fire to make it 'sweat'. This liquid is collected, and then applied to the area of skin after scarifying it to make it bleed.

Cleansing. Slices of the bark, or, even better, pieces of the resin are burned to fumigate living quarters as well as clothing and bedding. This is often used in preference to material from *Boswellia* species in western Soqotra where these *Commiphora* species grow. In water-poor areas, the fumigation of clothing and bedding in place of washing is important. Slices of the bark are burned to sweeten the air in living quarters which have stood unoccupied for a time.

Commercial value (past, present). The resin has recently acquired commercial significance and it is now being harvested and offered for sale to traders from the mainland. The resin is called by its Arabic name, *ṣamġ*, *zamgh*). The mainland traders say the demand is only for the *dhikr* or 'male' resin. The islanders have been told that it is in demand for its 'special' curative properties, but they are not sure how it is used. The best time to harvest it is in the hot summer months, and especially during the hot, dry winds of the monsoon when the struggling trees exude small quantities of the resin spontaneously.

Pest control. Like gum from *Boswellia* species the aromatic resin or the bark is burned around livestock and human living quarters to rid an area of potentially harmful creatures.

Other. In the areas where this plant grows it is customary to burn the bark beside the deathbed, and keep it burning until the body is removed for burial.

Comparative uses. In Yemen the resin is applied to wounds and snakebites. It is also used as an aphrodisiac, applied as a paste to the penis. With added egg yolk and honey, it is taken for digestive problems. In Oman the smoke of the burned resin is inhaled as a remedy for breathing difficulties, chest colds and swollen glands. It is applied externally as a disinfectant, and taken orally for coughs, dyspepsia, diarrhoea and as an anthelmintic. The resin is burned ceremonially at new births. A small piece is placed on an aching or caried tooth and the resin then cauterised with a heated metal rod. The resin is sometimes dissolved in water to flavour coffee or other foods. The roots and smaller stems are peeled and chewed for their sweet taste and the ripe berries are edible. (Fleurentin & Pelt/Fleurentin *et al.* [1982, 1983a,b,c]; Ghazanfar 1994; Miller & Morris 1988)

***Commiphora ornifolia* (Balf.f.) Gillett**
Lannea ornifolia (Balf.f.) Engl.
Odina ornifolia Balf.f.

Tree to 5m, branches rather tortuous; bark smooth, grey. Leaves clustered, pinnate,with 3 to 7 leaflets; leaflets ovate to broadly elliptic, 4–9 x 2–6cm, entire, glabrous or thinly hairy to velvety. Flowers in raceme-like cymes, c.5mm. Fruits fleshy, clustered on stalks 5–15cm long, bright green, grape-like, globose, c.10mm long.

Ecology. A common and widespread tree in semi-deciduous woodland, particularly on the limestone escarpments, less commonly in succulent shrubland; typically at altitudes of 100–650m but up to 1050m on the Diksam plateau.

Distribution. Endemic.

Conservation status. Near Threatened (NT). *C. ornifolia*, whilst widespread and abundant at present, is lopped as livestock fodder in dry periods. In some areas it suffers from a lack of regeneration for reasons perhaps connected with the progressive desiccation of the islands. If livestock numbers increase greatly, or a succession of drought years occur, this tree could rapidly become threatened.

Botanical notes. Radcliffe-Smith (1971) recognised two varieties: var. *glabra* (from eastern Soqotra) with glabrous leaves and var. *ornifolia* (from western Soqotra) with hairy leaves. However, there are many intermediates and they have not been maintained here.

ETHNOGRAPHIC NOTES

Name(s)

– Soqotri: *ˀiˈkšəh* (pl. *ˀikε̄šəh*); (rare) *ḳeˈrˁʌš* (in western areas); fruit *gilˁˈəmo* (pl. *geˈlḥum*); firesticks *ˀiˈźḥar di śeˈyʌṭ*: horizontal stick *źoˈhiro*; vertical stick *mošk*.

– Soqotri simplified: *'ikshih* (pl. *'ikeshih*); *qer'ash*; fruit *gil'imoh* (pl. *gelhum*); firesticks *'idh'har di sheyat*: horizontal stick *dhohiro*; vertical stick *moshk*.

Food (human). The berries are at their best in the hot, dry season. They do not keep well, and must be eaten fresh. They are acidic and well liked, especially by pregnant women and children. Chewing them also lessens thirst. The resin that oozes from cracks in the bark in hot weather is not chewed or eaten.

Food (livestock). This tree is a very important livestock feed all over the island. The large leaves are a key dry season forage for livestock, especially the cattle of drier areas, and small stock. It is seen as being a very nutritious feed, causing the animals to put on weight and increasing their milk yield. Herders cut leafy branches for their animals, and gather leaves, both dried and green, for their cattle and sheep. The fallen leaves are especially liked by goats and sheep. Goats also gnaw at the bark and chew at leafless stems. In a prolonged dry season herders gather strips of bark to feed to their goats, though they usually have to cut these into small pieces for sheep. Camels are very keen on the trees and search them out wherever they grow. However, they can strip a young tree of its foliage very quickly, and this has led to some ill-feeling towards camel-owners on the part of herders of the drier areas who are dependent on small stock.The smaller leaves are gathered and fed to young stock when there is no grass. In some areas the trees come into new leaf at the end of the monsoon, before the winter rains, when animals are at their most hungry, and at this time are an especially important fodder. All livestock eat the fruit when they can reach it.

Bees / honey. In the drier areas they are important honeybee forage, and islanders say that bees use the resin in the construction of their combs. The dead wood and bark are among the best woods for smoking out bees when robbing them of their honey.

Fuel. The dead wood burns down quickly but gives off a good light, so larger branches, or a dead stump, are burned when good illumination is important. Chunks of the wood are also stored in the home for this purpose. In areas with few alternative woods, the dead wood is collected and used for night-time burning. The thick smoke it produces means that it is rarely used for cooking, and it is also said to taint food cooked on it. If there is a severe shortage of firewood, live branches can be cut from the tree and completely de-barked. The inner wood can then be burned and makes acceptable firewood.

Timber. Not very hard, and does not make a good timber. However, the red heartwood of a 'male' tree can be used for building, after first testing it with the point of a knife to see whether it has rotted from the inside. If it seems hard, the outer bark and most of the bast is removed and the inner heartwood used. This can sometimes be improved by standing it beside the fire to harden it, or suspending it above the fire to smoke it. For the best timber, the tree must be cut at the very end of the dry season, preferably in a hot, dry area. The larger branches and the trunk of a dead tree can be split open lengthwise and the core removed to make guttering which is placed along the edge of the roof, leaving a gap above the main entrance door.

Wood. Known island-wide for its fire-making properties. The best firesticks are made from dead wood. It is most successful when both vertical and horizontal sticks are of same wood, though others can be used. The vertical stick is usually spliced into a stick of a stronger, more flexible wood, such as *Withania* or *Ficus cordata*. The wood of this *Commiphora* heats up and produces a spark much more quickly than any other on the island. The wood must be completely dry, and suitably sized sticks used to be gathered and kept at home in a dry place for this purpose.

The heartwood can be carved and was widely used for making implements. It is hard enough to be used for making locks and keys for the food storehouses. Even the wood of a dead tree can be used if it is first de-barked and then fire hardened. A short length of stem can be hollowed out with a metal awl and used to store antimony or another precious substance, such as civet paste or scented oils. A plug of the wood is fashioned to make a tight stopper. Stirrers and spoons are also carved from the wood. Two flat stems were carved to make the handles for the blacksmith's goatskin bellows. Larger branches were sliced across and fashioned into bowls of varying sizes for serving cereals or for washing before prayer. A knot in the wood can be hollowed out to make a small bowl used to feed young children or for serving ghee or honey. All utensils made of this wood were kept oiled with ghee and were stored in the shade to prevent them cracking. Strips of the bark were used to weave fish traps. The dead wood makes a good charcoal.

Medicinal / veterinary. *Commiphora* species are among the most important medicinal plants on Soqotra.

Bark. The bark and underbark are used to make the most widely used powder dressing, called *nʌḳf* (*naqf*). This is used on sores, ulcers and wounds of all sorts on both humans and livestock. Slices are cut off and scorched at the fire until quite dried and blackened. They are then reduced to a powder which is kept stored in a closed container in the home, ready for use. The *naqf* powder can also be made from the papery outer bark alone when none of the other is to hand, but this is not thought to be as effective. The thin bark is held above the flames until it blackens and curls up, and then it is pounded to a fine powder and applied to the sore area. It is also possible to make *naqf* from the bark of the small-leaved *Commiphora* species, and also from the root-bark of aloes, but neither of these is as effective as that made from *C. ornifolia*. The *naqf* dressing is applied to a wound which has become infected to draw the pus to the surface. It is also used on burns, both accidental and those resulting from medicinal cauterisation, to dry out and disinfect the burned area. The skin is washed with warm water, allowed to dry, and then the *naqf* applied, and this procedure repeated daily until the skin has healed. If the burned area becomes infected, the cleansing and application of *naqf* is continued and the area kept covered. Deep fissures of the plantar skin, especially of the heel, are treated with repeated applications of the powder, and deep lacerations are treated in the same way. To treat the surgical wounds of circumcision, the powder is used after the initial washing with the *ˈirziz* (*irziz*) liquid prepared from the small-leaved *Commiphora* species. Quantities of both medicaments are prepared in the days before the operation and kept ready at home. A malignant ulcer is treated by cauterising it all around to prevent it spreading further, washing it with the *irziz* liquid and then, once it appears cleaned of all purulent matter, applying the *naqf* powder. The same is done to treat a maggoty wound, washing it first with a strong salt and water solution. If a fractured limb becomes infected (especially common in a compound fracture) the top layer of skin can come off with the strapping when this is removed after a week or so to check on progress. The flesh and any bone visible will appear blackened. This is cleaned with the *irziz* liquid and dusted with *naqf* daily until it improves, although this is a very painful treatment. The bark is burned to treat a woman suffering from a retained placenta. She is helped to crouch over the fire and encouraged to inhale the smoke deeply and fumigates her whole body. Some say she should then lie down on her back and her stomach be massaged to help expel the placenta.

Sap and resin. The fresh sap is used to treat a variety of sores and skin ailments. An open, fresh laceration is treated with an application of the liquid sap mixed with a powder made by burning goat bone at the fire and then pulverising it. If available, powdered antimony is added instead. The sides of the wound are then brought together and a bandage bound tightly around. The fresh sap is also smeared over a boil to bring the pus to the surface, and applied to sores or blisters on the sole of the foot or hands. A layer of the sap is applied to a fractured limb before strapping it up. A mixture of resin, sea salt and sheep's wool or goats' hair is applied right around a fractured limb under the final strapping; this sets hard to support the limb, like plaster of Paris. The resin is not chewed, but can be inserted into a cavity to relieve toothache and to 'burn out' the diseased area of the tooth.

Foliage. This is considered to be the ideal feed to treat a sick camel, goat or sheep. Goats with scour are given the foliage to settle their stomachs. A parturient animal with a retained placenta is fed on the foliage to give her new strength to help expel the placenta. A cow losing weight in the last two months of pregnancy is brought fresh and fallen leaves to fatten her up.

Fruit. These are regarded as cleansing to the digestive system and to act as a general tonic. They are given to anyone recuperating from an illness, or who is sickly and listless. They are much liked by pregnant women and are given to someone with anorexia. They are also given to cool a fever and to treat nausea.

Cleansing. The resin, and the resinous bark and underbark, are burned for their pleasant smell and to fumigate clothing and bedding in the absence of anything better, such as ther dead wood of *Cephalocroton socotranus* or *Dirachma socotrana*, or the bark of *Boswellia* species.

Commercial value (past, present). At times when prices for all gums were high, the dried resin from these trees was also sold. There is some demand in the markets now for the fresh fruit, which are seen as having a medicinal value.

Pest control. The wood and the bark, both dead and green, are used for livestock smudges. Dead wood is more effective, and the older, more rotten and crumbling it is the more dense and acrid the smoke it will give off. The crumbling interior of the dead wood also produces a lot of thick insecticide smoke. This discourages biting insects, and the heat warms and relaxes the animals before milking.

Other. Animals eating the fruit need less water. Goats and sheep eating a lot of the dead, fallen foliage also need less water. The large leaves are held over the mouth and nose when going into shady or wooded areas with small stock to protect from botfly attack. The resin can be used as a glue, but is not as effective as the latex of some *Euphorbia* species.

The trees are commonly planted for their foliage, fruit, shade, and because they grow more readily and more quickly than others that the islanders have tried. Young seedlings or saplings are transplanted close to the home settlement into a hole filled with good soil, and are well fenced in. They have to be watered daily for a week or so, and thereafter at regular intervals for three or four months until they are properly established. Alternatively a large branch is cut from a well-grown tree, and set in a hole filled with good soil and well-rotted dung. It is protected with fencing and watered regularly until it has put out good roots. Some people prefer instead to plant a young sapling in the midst of a clump of one of the unpalatable plants, such as *Cissus* species or *Buxanthus pedicellatus*, to protect the young plant from livestock. It is quite common to plant longer branches as supports for the roof of the shade area, *sebeleh*, close to the homestead, or inside the compound walls. It is also possible to cut a young tree in half, transplanting the upper half to the chosen site. The bottom half regrows and sprouts new foliage. However, if this has been cut too close to the ground, any new foliage is usually quickly eaten by goats, and the tree then rarely recovers. It is always best to plant out young trees or cuttings just before the winter rains, or at the very beginning of the rains when it is cool and the sun is covered with cloud. Some islanders plant the young tree beside their rainwater catchments so that they can easily water it, and in the hopes that the tree once well grown will give them and their animals shade.

The dead wood is one of the best for firing pottery as, like the wood of other *Commiphora* species and *Boswellia* species, its smoke turns the pottery from cream to red-brown. For this to work, the pots must be completely buried in the wood.

When there is no thickened, soured milk available, the fruit are used to coagulate fresh milk and start the souring process. A handful of fruit is crushed and added to a little fresh milk in a pot. The pot is covered and left for some hours beside the fire to ferment. Once thickened and soured it is added to the skin of fresh milk to ripen it before churning it to make butter. In the absence of gum from *Boswellia* species, this resin can be used by the traditional healer or '*mekoli*'. The gum can be burned to rid an area of undesirable and harmful creatures. However, most islanders agree that *Boswellia* gum is more 'powerful' for both purposes.

A well-grown tree can be used as temporary goat quarters for milking and handling goats if there is no pen available. The owner climbs into the tree and beats it with a stick while calling to his goats, other family members rounding them up and herding them towards the tree. When they come he knocks off some leafy twigs for them to eat. After a few days of this, they come to accept it as their home and come of their own accord at milking time.

The foliage is a useful seasonal indicator in some areas. For instance, at the western end of the southern plain, when the ageing leaves change colour from green to yellow and fall to the ground at the end of the winter season, this signals that the time for moving the goats up from the plains to the limestone escarpment above has arrived.

Notes. Islanders distinguish between the 'male' trees, *ˀəkšəh ḳarṭaṭ*, 'with small and deeply wrinkled leaves', which bear fewer or no fruit, and the 'female' trees, *ˀəkšəh* or *ˀəkšəh baˁaləh*, 'married female *'ikshih*', with larger, smoother leaves and bearing many fruit. Although *Commiphora* species are loosely called 'myrrh trees' by some, none of the species on Soqotra are regarded as producing myrrh or anything which can be used as a substitute for myrrh. The myrrh used medicinally on the island has always been imported.

This species, like the other soft-wooded trees *Sterculia africana* and *Adenium obesum* and the other *Commiphora* species, is apparently vulnerable to large, white larvae – the size of a large termite queen – which eat away at the living wood from the inside. Two or three such creatures can be found inside the trunk of the dead tree when it is broken open.

The lack of regeneration of these handsome and useful trees is a cause for concern on the island. The islanders say that young trees can now only be seen in areas 'where there is no hunger', that is, in inaccessible or lightly-populated areas. Elsewhere they are eaten as soon as they appear, either by camels or by goats, who adore the tender growing shoots and new leaves of the small young trees. Some saplings survive under cover of clumps of unpalatable *Cissus* species or *Buxanthus pedicellatus*, or grow in the shelter of *Euphorbia schimperi* and *Ochradenus arabicus* in cliff areas and ravines. These last come into flower in the dry season, thus concealing the young *C. ornifolia* from view at a time when the animals are most hungry.

Commiphora parvifolia **(Balf.f.) Engl.**
Balsamodendrum parvifolium Balf.f.
Commiphora parviflora Jacq.

Tree to 4m with intricate, spreading, non-spiny branches; bark smooth. Leaves clustered on spur branches, pinnate; leaflets (3 to)7 to 11, oblong, 1–6mm long, entire, glabrous or thinly hairy. Flowers not seen. Stamens 4. Fruit unknown.

Ecology. Locally common as an emergent in *Croton* shrubland and succulent shrubland on the coastal and inland plains of W Soqotra; rare in the east. 20–350m.

Distribution. Endemic.

Conservation status. Near Threatened (NT). *C. parvifolia*, whilst widespread and abundant at present, is an important livestock fodder in dry periods. If livestock numbers increase greatly, or a succession of drought years occur, this tree could rapidly become threatened.

ETHNOGRAPHIC NOTES†

Name(s)

– Soqotri: *lėḳehem* (pl. *leḳhəmitin*, *ləḳʌhʌm*); (a rather prostrate plant) *lūwḳāmo* 'little *Commiphora*'; exudate or gum resin *gis*.

– Soqotri simplified: *leqehem* (pl. *leq'himitin*, *liqaham*); *luwqamo*; exudate or gum resin *gis*.

Food (human). The gum resin is eaten as a famine food. It is filling, and the best of all *Boswellia* and *Commiphora* species to quell both hunger and thirst. The trees of the cooler areas rarely exude resin spontaneously, except when the plant is dying, when the reddish resin emerges. Trees of hotter areas exude resin in very hot weather, or from breaks in the bark where they have been lopped. The roots are peeled and chewed as liquid refreshment and for their sweet taste.

Food (livestock). Occurring in a wide variety of habitats and in most regions of the island, these plants are one of the most important livestock fodders on the island. The foliage is a key dry season fodder for small stock, as well as for camels. For goats and camels it is considered to be a more nutritious and fattening feed than the larger-leaved *C. ornifolia*: it increases milk yield, makes animals fat and healthy, and their coats sleek. Goats like the buds, flowers and fruit and also chew the leafless stems and gnaw the bark when hungry. The best feed for older, weaned kids is leafy twigs snapped from the tree for them; younger kids are also fed the small leaves if there are no *Zygocarpum coeruleum* leaves for them. Sheep eat the fallen leaves and plant litter. Cattle eat little of the plant, but nibble the fallen leaves when hungry.

If the dry season is prolonged or in periods of drought, the bark is cut off and fed to goats and sheep. For the latter the outer bark has to be peeled off and the inner bark cut up small and slightly crushed to make it more manageable. The inner bark is also prepared to make a liquid feed for young stock. It is cut up small and crushed with a little water to a paste which is then squeezed into a container and mixed with water or boiled buttermilk. It is left to steep and then fed to young orphan calves and kids. Others prepare it by grinding the outer and inner bark together with a little water, then add the paste to a pot of water, bring it to the boil, simmer it until the liquid has turned a bright red, then strain it and leave it to cool before feeding it from a leather feeding bottle, with a little milk if they have any. Many young animals are brought through a drought on this unusual feed, though they cannot survive on it for long without other food. In the dry season milk goats are hand fed on branches to keep them in milk. As the dry season progresses and the leaves drop, the branch tips and leafless twigs are lopped to feed goats, and this is thought to be a nutritious rather than just a subsistence feed.

Bees / honey. Honeybees are attracted to the flowers.

Flavouring / food additive. The roots are peeled and chewed for their sweet taste. The milk of livestock feeding heavily on the plants has a distinctive flavour and perfume, and is rich and sweet. The taste has a faint echo of the leaves and gum resin, especially if there have been summer rains. The meat and cooking water of the carcass also smells delicious and has a characteristic and much-liked flavour.

Fuel. The dead wood makes reasonable firewood, but it is not used for cooking unless the wood is absolutely dry and the outer bark has been removed because otherwise it makes too much smoke. It burns with a bright light, and some pieces are generally kept in the home for this purpose. If there is no alternative firewood, the live wood can be used as firewood after de-barking.

Timber. Ggenerally useless as timber unless it has been cut at the very end of the dry season and left until thoroughly dried out. The dark heartwood of a dead 'male' tree, however, can be used as timber, but not for roofing as it is very susceptible to damp and prone to rot. The timber should always be tested with the point of a knife to ensure that it is not rotten inside. It can sometimes be improved by fire-hardening or smoking.

Leatherwork. The inner and outer bark of a well-grown tree are ground to a paste with a little water and used to prepare skins, to strengthen the leather and dye it the desired red-brown. This is the preferred tanning material over much of the island. It is used especially for those items which receive the greatest degree of wear and tear: the water- and milk-skins and the skin in which the butter is churned or pressed dates stored. It is also used to beautify the more personal items, such as the leather satchels in which a man carries his personal possessions, or the intricately plaited leather belt men used to wear. It is used to prepare the skin of a male kid or calf which has been slaughtered to conserve its mother's milk. This is roughly sewn into a cylindrical shape, and then stuffed with fragrant herbs. It is presented to a milker and some of her milk sprayed over it. It is then given to her to sniff at and lick to encourage her to let down her milk. The same paste is used when trying to persuade a milker whose kid or calf has been slaughtered to accept the young of another as her own. The strange young animal is rubbed with the paste to disguise its own smell and is then presented to the milker. Some of her milk is sprayed over it to encourage her to lick it and allow it to suckle.

The skin of a mature male goat or bullock, especially if it is to be used as a floor mat, is treated with the paste. The hair of these skins is not removed. The paste is rubbed long and hard into the non-haired side of the freshly slaughtered skin and then spread out on the ground to dry, pegged firmly down to stop vultures pecking at it or making off with it. The paste is re-applied whenever the skin grows too stiff, or if it becomes smelly.

Wood. It is hard enough to carve, especially the dark-coloured heartwood of a dead tree, and it is widely used for making implements. The heartwood is used for making the locks and keys for the food storehouse. Single or double edged combs for the hair are carved from it, and the narrow pointed rods used to part the hair and divide it into sections for plaiting. Little pots to store antimony or other precious substances are made by hollowing out a stem with a metal awl, and then shaping a wooden stopper to plug it. Stirrers and spoons are also made and the two flattened sticks used to work the blacksmith's goatskin bellows. Slices of trunk, or knots in the wood, could be hollowed out to make bowls. Smaller ones were made for young children to drink from, or to serve ghee or honey, or water for ritual purification before prayer. Larger ones were used as serving dishes. The wood is oiled with ghee and kept out of the sun to prevent it cracking. The dead wood makes a reasonable charcoal, and is sometimes used to make quicklime. The dried wood is not as good as that of *C. ornifolia* for firesticks, though it can be used.

Medicinal / veterinary. With *C. ornifolia*, this tree was one of the most important medicinal plants of the island.

Bark. The outer and inner bark could be prepared like that of *C. ornifolia* to make the *naqf* powder, though that of *C. ornifolia* was considered to be superior. More important is the liquid antiseptic medicament called *irziz* or *mirziz*, now described by the islanders as being 'like iodine'. This is made by cutting thick slices of bark, peeling off the outer bark, and then crushing the inner bark, and adding it to a little water with some salt, if available. The water is brought to the boil and left to simmer until judged to have acquired the requisite strength. To use it, the hot liquid is squeezed through a cloth directly onto the affected area. It has to be boiling hot to be effective, though for use as an antiseptic on old skin infections, wounds, sores and ulcers, it is used cooled as a wash. This method of using freshly cut bark is considered to make the best *irziz*, but dried bark can also be used, and strips of bark are cut, crushed and dried, and t kept in the home ready to use on lacerations and wounds of all kinds. When needed the dried bark is boiled up in a small amount of water and salt, left to steep and then strained, re-heated and used. This is the most widespread treatment for the surgical wounds of circumcision. Suitable bark is gathered ready for the day of the operation, both bark for the *naqf* powder and bark for the *irziz* liquid. Both treatments are considered essential to the successful healing of this wound. Immediately after the operation the young men would retire to the living quarters set aside for their use, where they would be treated and looked after until their wounds had healed sufficiently for them to go home. The wound was first treated with the boiling *irziz*: this makes the skin 'smoke' and turn white and is extremely painful. Once the wound had turned white and the skin had dried out, the powdered *naqf* was applied. The wound was then covered with cloth, preferably the fine black cloth that was imported for women's *qineyqi* overshirts. Thereafter cloths were soaked in the cooled *irziz* liquid and used to clean the area daily, the bandaged penis being supported on a forked twig strapped to the hips. Once the wound looked clean and had started to heal, it was treated with the *naqf* powder alone. If infection set in, the wound was treated with a strong salt and water solution, or the young man was sent to wash in the sea. The boiling *irziz* treatment is used on all dirty wounds, and to treat a malignant ulcer after the skin around the ulcer has been cauterised to stop the malignancy spreading. Once this dries out and starts to heal, the powdered *naqf* is applied. A maggoty or fly-blown wound is treated with hot *irziz* and the dry area smeared with ghee. This is done two or three times a day until the wound begins to heal. A fresh, deep laceration is treated by pulling the sides of the wound together and sticking them in place with the latex of certain *Euphorbia* species or that of *Ficus cordata*. However, if the wound becomes infected and the surrounding area swollen, red and sore, this seal is broken and the wound treated with boiling *irziz* until all the purulent matter has been 'burned out'. Then powdered *naqf* is applied. To treat a fractured finger, the whole area is bathed in cooled *irziz*, and the finger bandaged, initially with a layer of fine cloth, and then with a strapping of aloes root-bark, and finally with another layer of cloth holding it all together. A broken toe is treated similarly, except a layer of *Ficus vasta* latex is usually painted on the toe first and allowed to set firm before washing it with cool *irziz*. If a fractured limb becomes infected (not uncommon in a compound fracture) the top layer of skin sometimes comes away with the bandage and strapping once this is removed after the first week to check on progress. Often the ends of the broken bone are visible and the bone and the skin are blackened and ill-smelling. The area is treated daily with boiling hot *irziz* and powdered *naqf* until it begins to heal. The *naqf* made from the bark of these trees is used just like that made from *C. ornifolia* bark.

Wood ash. Quicklime made from the dead wood is also used medicinally. It is ground to a fine powder and used to dry out wounds, abrasions and weeping sores on people and their animals.

Gum resin. Those with strong jaws chew the gum resin to soothe a painful or upset stomach, and to strengthen the jaws. Some chew and swallow lumps of the resin to bring on retching and vomiting to treat botfly attack to the mouth or nostrils.

Foliage. Leafy stems are brought to sickly or parturient goats and camels, or to animals recovering from illness, to restore their strength.

Cleansing. A green twig is used as a toothstick, especially after eating meat, to freshen the mouth and sweeten the breath. In the absence of other fragrant gums, the gum resin is burned on a slip of pottery to fumigate the living quarters of both livestock and people. It is also used to fumigate clothing and bedding.

Cosmetic. A twig is used as a chewstick to whiten the teeth and strengthen the gums. The swallowed saliva is also believed to be good for the digestive system.

Pest control. The dead wood is used for livestock smudges, especially for cattle in cold and wet weather in the Central Highlands. The smoke produced by the bark in particular is thick and acrid and keeps biting insects away from the cows. They are also said to enjoy the fragrance of the smoke and find it soothing and relaxing. This has the effect of encouraging them to let down their milk. In wet weather smudges of the wood are lit for sheep and goats, inside their night-time quarters and at milking time. If a herder passes a goat giving birth on its own, out in the open and unattended, and has no time to sit with her until she had safely given birth, he or she lights a fire of the dead wood beside the goat. This is done to keep away the feral cats which may attack and kill a newborn kid or even go for the mother in her weakened state.

Other. Chewing on twigs lessens thirst for humans and livestock. Eating peeled sections of root also quenches thirst for people. The underbark is ground between two rocks and added to a little water, with salt if available, and boiled hard to produce a thick red liquid which is used as a red ink to write with or to mark property. The ink is used with a pen made from a slim peeled twig with the writing end notched to hold the ink. The red ink was much liked by Qur'an writers for writing the names of God or His Prophet, or for embellishing and decorating the main body of the script which was written in black ink (mainly made from soot and lamp black). The red ink is also used to mark property with the owner's name or brand mark (similar to that of his livestock). Traditional healers used this ink to write charms on scraps of cloth or paper, which were then wrapped in leather and worn around the neck, waist or upper arm by the patient. An ink can also be made by hitting the trunk or larger branches with a stone and returning the next day to collect the yellowish lumps of gum resin that emerge from the damaged bark. This can be used as a marking ink just as it is, or can be further processed with inner bark and water. The resin is put in a clay pot with an equal amount of water or a little more. Soot black and a little rock salt is added. Then two or three strips of the black *qineyqi* cotton imported to make a woman's overshirt were added, and the contents thoroughly stirred together. The pot is then covered and put aside for about a week. Then it is inspected: if very thick a little water is added. It is then stirred briskly and is then ready for use. This process produces a dark red ink which is not entirely waterproof, but more so than the resin or bark mixtures described above.

Like the dead wood of other *Boswellia* and *Commiphora* species, this is one of the best woods for firing pottery as its smoke turns the pottery from cream to red. For this to work, the pots must be completely buried in the wood.

The underbark was also used to make a dye for the imported coarse, undyed, calico cloth. It was crushed and then ground to a paste with a little water. This was boiled for some hours until a good, clear red colour had been achieved. Then the cotton material was lowered in and left to soak until it turned the required shade. It was then stretched out to dry, preferably in the shade. This dye was not a permanent one, but faded badly on washing and on too much exposure to direct sunlight. The same dye used to be used for the cotton twist imported for making fishing equipment before nylon equipment was imported. Thread spun from local cotton was dyed to make a coloured stripe in the sheep's wool cloth and woven cotton shoulder wraps. Pale or white sheeps wool was also dyed: the small balls of spun thread were put soaked and were then taken out, unwound and hung up to dry.

These trees are easily transplanted. A small tree is cut in half and the upper half planted in a prepared hole, protected with fencing. As with *C. ornifolia* above, this is watered at regular intervals until it produces roots and sprouts new leaves or flowers. It is generally planted beside settlements and is often more successful than the large-leaved *Commiphora* species. In the absence of other and better material, the sticky gum produced by striking the main branches or trunk with a stone can be used to lime birds or as a glue.

The gum resin can be burned as a fumigant, like frankincense.

Notes. Islanders differentiate between the larger-leaved 'female' tree, *le̍ḳehem ṣa̍ᶜalhil*, and the smaller-leaved 'male' tree, *le̍ḳehem ka̍ᶜasis*. Some say that *C. planifrons* is 'male' and tends to grow at higher altitudes, and *C. parvifolia* is 'female' and tends to grow in lowland areas. In general, the 'female' is considered to be best for fodder and medicinal use, and the 'male' to have better, harder wood. They say that *C. socotrana* can be either, and tends to grow at lower altitudes and on coastal foothills. Thus the islanders of the western highlands say that they have mainly 'male' trees, with only a few 'females' on the seaward slopes. Like the larger-leaved *Boswellia* species, these *Commiphora* species rarely exude gum spontanteously, but rather when damaged or dying.

These trees, like the other soft-wooded trees *Adenium obesum*, *Sterculia africana* and *C. ornifolia*, are apparently vulnerable to large, white larvae that eat the living wood from the inside. Two or three such creatures can be found inside the trunk of the dead tree when it is broken open.

The fondness of camels for this tree and their ability to strip a tree bare of foliage at one sitting is one of the reasons behind the dislike felt for this animal in the poorly vegetated areas of western Soqotra. In such areas the small-leaved *Commiphora* species are a vital fodder resource.

Comparative uses. It is interesting to compare the uses of this key Soqotran plant with the closely related *C. africana*, a similarly important plant of parts of the African mainland.

Bark. It is burned for its perfume and fragrant smoke, and used in fumigation. It is ground, soaked in water and used to dye hides red. Medicinally the bark is used to treat scorpion stings, snakebites, malaria and inflamed eyes. It is also chewed with natron and put on scorpion bites.

Roots. Roots are dug up by children and chewed to extract liquid from them. The pith of the roots is a famine food and the root itself is chewed as food. It is also used to clean the teeth. Medicinally it is used as a vermifuge and for neck and stomach pains. It is boiled and given to children to promote good health. A maceration of root-bark is given orally to treat rheumatic diseases.

Leaves. The leaves and stems are liked by all livestock, especially goats and camels. Medicinally the leaves are used to treat stomach problems and are given as a sedative.

Gum. It is eaten, and is also used for curing leather, as a perfume, cosmetic, termite repellent and as a glue for fixing axe-heads. Medicinally it is boiled, and the face held over the steaming pot to treat inflammation of the eyes. It is also used to treat spasms and fevers, and is made into plasters.

Sap. The root sap and stem of young plant is chewed as a food.

Twigs. These are used as chewsticks, are burned for their fragrant smoke and used as firesticks.

Seeds. Seeds are used for rosaries. Medicinally they are taken as a vermifuge.

Tree. These are planted or transplanted to make fencing.

Fruit. Medicinally, fruit are chewed and the juice spat into the patient's mouth or eyes to treat blisters or ulcers in a child's mouth, and various eye diseases. Care must be taken not to swallow the saliva.

Wood. Is used as a firewood and to make stools, milking vessels, beehives and gourds. (Heine *et al.* 1988; le Houerou 1980; Lewis & Elwin-Lewis 1977; Oliver-Bever 1986)

†Also apply to *C. planifrons*, *C. socotrana*.

Commiphora planifrons **(Balf.f.) Engl.**
Balsamodendrum planifrons Schweinf.
Commiphora planifrons Schweinf.

Umbrella-shaped tree to 4m with horizontal branches; bark smooth. Leaves clustered on short spur branches, pinnate; leaflets 9 to 23, overlapping, often appearing globular, 1–1.5mm long, concave with strongly revolute margins, thinly or densely hairy. Flowers deep red, c.3mm long, sessile, solitary on spur branches, appearing before or with the leaves. Stamens 4. Fruits ellipsoid, c.5mm long, apiculate, glabrous.

Ecology. Common in semi-deciduous and *Dracaena* woodland in the central mountains (from 700 to 1050m); locally common as an emergent in shrubland (with *Croton socotranus*, *Jatropha unicostata* and *Buxanthus pedicellatus*) on the western plateau (from 500 to 600m); rare (at c.700m) in the relict woodland at Qatariyah in the southwest of the island. 500–1050m.

Distribution. Endemic.

Conservation status. Near Threatened (NT). *C. planifrons*, whilst locally abundant at present, is an important livestock fodder in dry periods. If livestock numbers increase greatly, or a succession of drought years occur, this tree could rapidly become threatened.

Botanical notes. Locals say this is male *C. socotrana* (see *Miller* 14226).

ETHNOGRAPHIC NOTES

Name(s)

See *C. parvifolia* (also uses).

Commiphora socotrana **(Balf.f.) Engl.**
Balsamodendrum socotranum Balf.f.
Commiphora socotrana Balf.f.

Tree to 5m with intricate, spreading branches, branches spinescent; bark white or black, smooth, not peeling. Leaves clustered on short spur branches, simple or pinnate with 3 leaflets; leaflets 3–20mm long, obovate or oblong, entire or somewhat crenate-dentate or shallowly lobed, glabrous or thinly hairy. Flowers red, c.3mm long, sessile, 1 to 3 on spur shoots. Stamens 4. Fruits ellipsoid, c.7mm long, apiculate, glabrous.

Ecology. Locally common on the northern and NW inland and coastal plains; in *Croton socotranus* shrubland, open succulent shrubland and drought-deciduous woodland. 500–700m.

Distribution. Endemic.

Conservation status. Near Threatened (NT). *C. socotrana*, whilst widespread and abundant at present, is an important livestock fodder in dry periods. If livestock numbers increase greatly, or a succession of drought years occur, this tree could rapidly become threatened.

Botanical notes. Locals say this is female *C. planifrons* (see *Miller* 14226).

ETHNOGRAPHIC NOTES

Name(s)

See *C. parvifolia* (also uses).

BUXACEAE

Trees or shrubs. Leaves leathery, opposite, simple, entire; stipules absent. Flowers small, unisexual, actinomorphic, arranged in axillary cymes with a single female flower surrounded by 2 (or more) male flowers. Male flowers with 4 tepals; stamens 4. Female flowers with 4 to 6 tepals; staminodes absent. Ovary superior, 3-celled; ovules 2 per cell on axile placentas; styles 3. Fruit a capsule splitting into 3, each segment 2-horned. 5 genera. c.60 species. Nearly cosmopolitan.

Buxanthus Tiegh.

See description of family. 30 species. Europe, temperate Asia, Africa, W Indies, C America.

Buxanthus pedicellatus **Tiegh.**
Buxus hildebrandtii Baill.

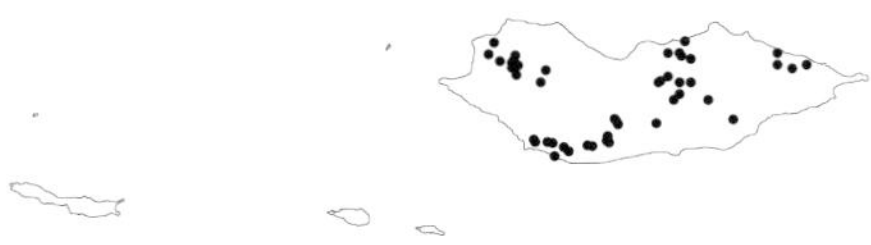

Evergreen tree or shrub to 3m. Leaves obovate to narrowly obovate, 1.5–6 x 0.5–1.5cm, stiffly erect. Flowers yellow. Fruit globose, c.1cm diameter.
Ecology. Widespread and often abundant on the mountains, on both limestone and granite. Typically above about 500m but rarely down to almost sea-level on the coastal (Noged recorded at 10m) and inland (Zahr recorded at 200m) plains. An important component of semi-deciduous shrubland and often the dominant shrub over large areas, typically with *Dracaena cinnabari*, *Gnidia socotrana*, *Trichocalyx* species, *Jatropha unicostata*, etc. Also sometimes found in succulent shrubland and in dwarf shrubland, particularly in places where the semi-deciduous woodland occurs at lower altitudes in sheltered valleys; for instance it occurs at 50m in Wadi Ayheft. (10–)100–1050m.
Distribution. Ethiopia, Somalia, Djibouti.
Botanical notes. Evergreen shrub which flowers in summer with the small fruits often persisting on the plants throughout the year. The leathery leaves often turn yellow or reddish in the dry season. One of the most characteristic shrubs on Soqotra. Outside Soqotra known only from tropical NW Africa where it occurs in *Acacia–Commiphora–Combretum* woodland typically above 1000m.

ETHNOGRAPHIC NOTES
Name(s)
– Soqotri: *mìṭəhān* (pl. *māṭin*).
– Soqotri simplified: *mitihan* (pl. *matin*).
Food (livestock). No animals eat any part of this bush. In times of great hunger, sheep and goats of the drier areas have been seen to nibble at fallen dead leaves.
Bees / honey. The flowers are richly perfumed and the bush in flower is very attractive to honeybees. Along with plants such as *Euphorbia arbuscula* and *E. schimperi*, this is considered to be one of the '*sheber*' plants and its honey is likewise pungent and considered 'hot'. *B. pedicellatus* honey is usually viscid and dark-coloured. Although not particularly highly thought of as a food, it is regarded as one of the best of the island's medicinal honeys.
Fuel. The dead wood is hard and burns slowly, and a smouldering log is often buried in ash to keep fires over long periods. Its slow-burning qualities are also used when long, slow cooking is required, as when boiling down buttermilk to make cheese, or simmering it steadily to make the thickened buttermilk which is fed to calves. The burning wood gives out a good heat and makes little smoke, so is popular for cooking and for heating enclosed spaces. In cold or wet weather the dead wood is the one commonly chosen to put on the fire to warm children and sick people as well as livestock. It burns brightly so is used for lighting, and burning brands are used as torches. It catches and burns well even in damp weather, so islanders generally keep a store in their houses.
Timber. The wood is hard, and amongst the most termite resistant of the island. Consequently it does not have to be de-barked before use. The larger shrubs of the wetter areas provide good timber, which is used for beams, king posts and rafters. If not allowed to get wet, such timber lasts down the generations. All timber is best cut in hot, dry weather; if cut in damp weather, it is attractive to wood-boring insects. Branches of even the smaller bushes can be used in roofing and for constructing enclosures, wall partitions and screens in cave settlements. In place of the usual layer of pebbles, some people lay a layer of the slimmer branches as a final layer over the roofing before adding the final thick coating of earth. Leafy branches are also used to build livestock pens, with the wood end facing outwards, and the unpalatable leafy crowns facing inwards.

Temporary dry season accommodation was often built using *B. pedicellatus* material and indeed these huts were called '*mʌyòṭihīn*', '*B. pedicellatus* huts'. A circle of low walling was built, and then poles of any available timber were used to make a framework over which leafy branches were tied, layer on layer, leafiest parts facing outwards. If summer rains threatened a further layer of *Dracaena* leaves was strapped to the roofing. Sometimes similar huts were made entirely from *B. pedicellatus* wood: three rows of four upright poles were erected, and these were then tied together horizontally and vertically to make a sturdy frame. Then thickly leaved branches of *B. pedicellatus* were tied in layers over the framework. These leafy stems produce very dense shade and such huts were renowned for remaining cool even in the greatest heat. If the framework was made from good timber poles, well dug in and tied, such a hut could last for many years. However, the leafy branches have to be replaced every year.
Leatherwork. The wood burns down to a fine ash considered one of the best for toughening skins and hides in leatherwork. It is used especially to prepare the skins in which the milk is churned, water is fetched, and dates are stored. All of these need to be very strong.
Wood. The wood makes good stakes for fencing. Firesticks are often made from it. Items such as locks, keys

and combs are not normally made from the wood as it is seen as being too hard to work. However, the sword beater used in weaving is best worked from this wood. Stakes cut from the plant are hammered at intervals into the trunk of *Sterculia africana* trees to permit access to the crown of the tree to enable herders to lop branches as dry season fodder for cattle. Dead twigs are gathered and stored for use as tinder and kindling. The wood makes good charcoal for the blacksmith. The great strength of the wood is exploited to lever and move the heaviest weights, such as slabs of rock for foundation stones, sills and lintels, or to remove a large rock encountered when trying to dig a well. Strips of the outer and inner bark are cut and used to weave fish traps.

Medicinal / veterinary. The 'hot' and pungent honey made from its flowers is considered to be powerful medicinally, and is used to treat most persistent illness not responding to other remedies. It is used especially to treat long-standing chest complaints.

Bedding, stuffing. Leafy branches are considered to make the best mattress for sleeping on, usually placed under the haired skin of a mature male goat or bullock.

Pest control. Fine ash is smeared over the goatskins in which dates are stored in order to deter insects and bugs. Burned green, the wood and foliage make great quantities of acrid smoke and burn slowly, making them ideal for the insecticide livestock smudges. *B. pedicellatus* is the most widely used material for lighting smudges in the mornings (and sometimes also evenings) of the winter months when biting flies are most prevalent and irritant to livestock. The most effective smudge is made by laying green, leafy branches over a fire of dead wood mixed with added cattle dung. To reduce flea infestation in the home, bedding and matting is taken outside and left in the sun while small fires of the green wood are lit here and there and left to smoulder for some hours, any gaps in the walling or roofing being blocked to prevent the smoke escaping. Empty livestock pens and byres are also fumigated to purify them and reduce insect numbers. Fires of green wood are lit inside the byre to treat small stock suffering from coughs and colds, or which have been attacked by botfly in order to loosen the nasal mucus. Otherwise *Buxanthus* wood is considered to produce too acrid a smoke for enclosed spaces and is used only in the open air. Fires of this wood are lit inside a newly built house, and kept burning for as long as possible in order to impregnate the timbers with the smoke. This is said to protect the wood from insects, and especially from termites. This is repeated after rains if the timbers have become at all damp.

Other. Bushes are not removed from the large *zeribah* enclosures of the Haggeher as they provide deep cover from wind and rain for all animals. In the hot and dry limestone plateaus, the shrubs give shade to sheep and goats, and, when the monsoon winds howl across the treeless tableland, they give shelter and protection to animal and herder alike. Branches are held over a fire to loosen the leaves, and the remaining twigs are used as a brush to muck out livestock quarters. Leafy branches are commonly used to cover dishes of food to keep off flies. In water-scarce areas, it was common to conceal what little water there was from others by covering the pools or crevices with flat rocks and then with leafy *B. pedicellatus* branches. This deterred people from coming too close as it gave it the appearance of a burial site (branches are used to cover the corpse of a sheep or goat to disguise the smell and discourage vultures).

Comparative uses. Trees and shrubs of the boxwood family contain the bitter-tasting alkaloid buxine. Both leaves and twigs contain a volatile oil whose ingestion causes vomiting and diarrhoea. Ingestion of large amounts can lead to death from respiratory failure. Nevertheless, the leaves and bark of some species are used medicinally. The close-grained wood of many boxwoods is used in carpentry and wood carving. (Ambasta 1982; le Houerou 1980; Jain & De Fillips 1991; Lewis & Elwin-Lewis 1977)

CAMPANULACEAE

Herbs. Leaves alternate, simple, entire; stipules absent. Flowers solitary or in lax panicles, actinomorphic, 5-merous. Calyx tubular, united with ovary. Corolla tubular with 5 lobes. Stamens 5. Ovary ± inferior; 2 to 5-celled. Fruit a capsule, dehiscing by apical valves or basal pores. 87 genera. 1950 species. Cosmopolitan.

Campanula L.

Herbs. Flowers solitary. Calyx 5-lobed with reflexed appendages between lobes. Corolla campanulate, 5-lobed. Fruit dehiscing by basal pores. 1 endemic. c.300 species. N temperate, especially Mediterranean.

Campanula balfourii Wagner & Vierh.

Campanula dichotoma sensu Balf.f. non L. p.p.

Delicate, erect herb to 15cm, with spreading hairs. Leaves broadly oblong-elliptic to orbicular, c.0.5–1 x 0.3–0.7cm. Flowers lilac-blue, <0.8cm long, ± sessile.

Ecology. Clearings in semi-deciduous woodland. 500–750m.

Distribution. Endemic.

Conservation status. Least Concern (LC). Common in several vegetation types and under no present or perceived threat.

Botanical notes. This delicate annual with its small, pretty, bell-shaped flowers appears soon after rain but rapidly dies back to nothing in the dry season.

ETHNOGRAPHIC NOTES

Name(s)

– Soqotri: nil or *ra̍ʿadihon* 'little grazing plant'.

– Soqotri simplified: *ra'adihon.*

Food (livestock). This is one of the small *ru'ud* group of plants which come up with the rains, providing incidental grazing until they die back. Though palatable to livestock, especially small stock, this is not important forage.

Wahlenbergia Schrad. ex Roth

Herbs. Flowers in lax panicles, long-pedicellate. Corolla ± deeply 5-lobed. Fruit dehiscing by apical valves. c.150 species. Subcosmopolitan.

ETHNOGRAPHIC NOTES

Comparative uses. The leaves of *Wahlenbergia* species are cooked and eaten. *W. marginata* is crushed and used to treat skin troubles and to strengthen loose teeth, while its roots are used to treat pulmonary infections. In Kenya *W. abyssinica* and *W. virgata* are much liked by livestock. (Ambasta 1982; Heine *et al.* 1988; Jain & De Fillips 1991; Kunkel 1984)

Wahlenbergia lobelioides DC.

Erect herb. Leaves linear to lanceolate, 2–8 x 0.4–2cm. Flowers <0.7cm long, pedicels <6cm long.

Ecology. No habitat details available.
Distribution. Africa and Mediterranean.
Botanical notes. Not seen on Soqotra for over 100 years. However, a tiny, easily overlooked ephemeral. The Soqotran plant is referable to subsp. *nutabunda*.

CAPPARACEAE

Herbs, shrubs or trees. Leaves alternate, simple or digitately 2 to 7-foliolate, usually entire; stipules present. Flowers hermaphrodite, actinomorphic or zygomorphic, axillary or terminal and either solitary, clustered or in racemes. Sepals 4, free or fused. Petals 4 or absent, free. Stamens few to many, often borne on a stalk (the androphore). Ovary sessile or borne on a stalk (the gynophore), 1 to several-celled with few to many ovules. Fruit a berry, nutlet or capsule. 45 genera. 680 species. Tropical and subtropical, some in arid temperate regions.

Cadaba Forssk.

Trees or shrubs. Leaves alternate or clustered. Flowers in terminal corymbs or racemes or rarely solitary, zygomorphic. Sepals 4, free. Petals 4, clawed, or absent. Nectarial appendages tubular, upper part sometimes petaloid. Stamens 4 or 5; androphore present. Gynophore present (combined with androphore to form the androgynophore). Fruits cylindrical, often somewhat torulose, tardily dehiscent. 1 endemic. 30 species. Tropical Old World, especially Africa.

Cadaba farinosa Forssk.

Shrub to 2m. Leaves farinose, grey-green, narrowly elliptic or oblong to obovate or more or less orbicular, <2.5 x 1.2cm, rounded at the tip. Flowers solitary or in few-flowered racemes at the tips of branches. Sepals <10 x 4mm. Petals 4, cream to greenish yellow, 10–15mm long, clawed. Stamens 4 or 5. Androgynophore 5–8mm long. Nectarial appendage c.5mm long, tubular, shortly toothed at the tip. Fruits <5.5 x 0.5cm, farinose.
Ecology. Coastal plains and rocky slopes in dwarf shrubland. Sea-level–200m.
Distribution. Tropical Africa, Egypt, S Arabia, Pakistan and India.
Botanical notes. In the Soqotra Archipelago found only on Abd al Kuri.

ETHNOGRAPHIC NOTES

Name(s)
– Soqotri: *jɛ̄lis*; *ḥarṭánəh* (Abd al Kuri).
– Soqotri simplified: *jelis*; *hartanih*.
Food (livestock). On Abd al Kuri this is one of the key fodders for goats and sheep. The bark, leaves, flowers, fruit and young twigs are all eaten, and are regarded as particularly nutritious and fattening. Donkeys are less enthusiastic about the plant.
Fuel. The dead wood makes a good fuel. It gives out a good heat and is much liked for cooking.
Wood. Used to make a variety of implements. It also makes good charcoal.
Comparative uses. The fodder value of *Cadaba* species is generally high. *C. farinosa* is a good browse for all livestock, though horses and donkeys do not usually eat it. The wood is used to make arrows and toothbrushes, and provide thin poles used in roofing, walling and fencing. The dried leaves are used for flavouring and the macerated flowers as a sweetener. The root is ground to a powder and used to treat eye diseases in cattle, as well as coughs, rheumatism and anthrax. The wood ash is used to treat general body pains, the bark to make cordage, the young shoots to treat stomach ache, and the leaves for coughs and skin complaints. In Oman an infusion of the leaves is used for colic and stomach ache, and the leaf juice is used as eyedrops to treat conjunctivitis. Elsewhere the leaves and twigs are pounded with grain, and then spread out to dry. This is sold in markets for use as a dry brown pudding or cake. In Sudan pounded leaves are a common addition to couscous. In Senegal the pulverised leaves are mixed with rust and taken as an iron tonic or added to coffee. In Kenya young men sleep on the branches during the circumcision ceremony. In India the roots and leaves are used as a vermifuge and deobstruent, a decoction is given in uterine obstructions, and the pulverised leaves are used as a poultice on sores. (Ambasta 1982; Bebawi & Neugebohrn 1991; Dalziel 1948, 1956; Ghazanfar 1994; Heine *et al.* 1988; le Houerou 1980; Jain & De Fillips 1991; Kunkel 1984; Miller & Morris 1988)

Cadaba insularis **A.G.Mill.** (see Appendix 2, p. 723)

Small tree or shrub to 3m, bark fissured; stems thinly glandular-hairy. Leaves subsucculent, oblong-elliptic to suborbicular, <2 x 1.5cm, rounded or retuse, the base rounded or slightly cordate, densely glandular-hairy beneath, glabrous or sparsely glandular above. Flowers in few-flowered racemes. Sepals c.3.5mm long. Petals 4, white, 8mm long, clawed. Stamens 4. Androgynophore c.8mm long. Nectarial appendage white, narrowly funnel-shaped, 3mm long. Fruit ovoid, densely covered with stalked glandular hairs.

Ecology. Rare in open dwarf shrubland (Zahr) and degraded succulent shrubland (Kilissan). 100–350m.

Distribution. Endemic.

Conservation status. Critically Endangered (CR A2/B2 a biii). This small tree is known only from two, widely separated locations: at one a survey revealed about 10 scattered individuals and at the other (admittedly poorly explored) location only a solitary tree was seen. At the first site, local informants say that considerable numbers of trees were lopped to provide fodder during recent droughts. The tree is not considered to provide particularly good browse (cf. *Dendrosicyos socotrana*), and it is a reflection of the severity of the droughts in the east of the island that it has been so drastically reduced in numbers. No regeneration has been observed at either location, which strongly suggests a decline in the quality of the habitat.

Cadaba insularis is representative of a group of plants on Soqotra which are threatened by overexploitation. This is not so much because of changes in livestock management but because of recent increased demand for timber, and drought. The recent building boom has been caused by herding families moving out of their traditional cave settlements and starting to build houses. This has meant that good fodder trees are more often felled for timber than previously, putting increased pressure on less suitable fodder species such as *C. insularis*. The recent severe drought also greatly stressed the rangeland, at a time when livestock numbers were high because of the previous few years of good rains. When the summer rains failed all over the Archipelago in 1999 (a rare occurrence) large numbers of animals had to be hand fed. The plants of the rangeland, especially browse species, were put under great pressure and people searched far and wide for greenery to feed their starving animals (even lowering themselves down cliff faces on ropes).

Botanical notes. Gnarled tree with fleshy dark green leaves and fissured bark. Flowers and fruits at the end of the winter and beginning of summer. In habit and leaf shape *C. insularis* is similar to *C. rotundifolia*; however, it differs in its floral characters: the stamens are attached near the middle of the gynophore (at the base in *C. rotundifolia*) and it has four petals. The flowers of *C. insularis* are similar to those of *C. longifolia* but the fruits are densely covered in stalked glandular hairs whereas in *C. longifolia* they are covered with sessile glands.

ETHNOGRAPHIC NOTES

Name(s)

– Soqotri: *ṣɔ̄mhir*, *ṣɔ̄mheyr* 'singing, song' (also given to *Capparis decidua*), *ṣɔ̄mhil*.

– Soqotri simplified: *zowmhir*, *zowmheyr*; *zowmhil*.

Food (livestock). Goats are not enthusiastic about the foliage, but will nibble at the twig tips, leaves or fruit if very hungry.

Timber. Not common enough to be an important timber, but it was used for building in the areas where it grew and was regarded as a useful timber. However, it rarely grows to a large enough size to provide main beams or rafters.

Cleansing. Bark is sweet smelling and produces a fragrant smoke.

Cadaba longifolia **DC.**

Shrub to 1.5m. Leaves bright green, linear-elliptic to narrowly ovate, <7 x 1.2cm. Flowers in terminal corymbs. Sepals <15mm long. Petals 4, bright yellow, maturing red, 10–15mm long, clawed. Stamens 4. Androgynophore 10–15mm long. Nectarial appendage c.10mm long, the tip broadened and slightly lobed. Fruits <2 x 0.6cm, densely glandular.

Ecology. No habitat details available.

Distribution. NE Africa and SW Arabia.

Botanical notes. Not seen since it was collected by Balfour in the 19th century. A distinctive shrub with bright green oblong leaves and glandular fruits – not easily overlooked and apparently extinct on Soqotra.

Cadaba rotundifolia **Forssk.**

Shrub to 2m. Leaves dark green, broadly ovate to orbicular, <5 x 4.5cm, retuse or rounded at the tip, glabrescent. Flowers in terminal corymbs. Sepals c.7 x 4mm. Petals absent. Stamens 4 or 5, attached at the base of the androgynophore. Nectarial appendage yellow, c.2cm long, tubular, petaloid at the tip. Fruits <5 x 0.7cm, non-glandular.

Ecology. No habitat details available; Balfour recorded it from near Galonsir [Qalansiyah].

Distribution. NE Africa and SW Arabia.

Botanical notes. Not seen since it was collected by Balfour in the 19th century. A distinctive shrub with dark green round leaves and non-glandular fruits – not easily overlooked and apparently extinct on Soqotra (cf. *C. longifolia*).

Capparis L.

Shrubs or small trees. Leaves simple, entire, often leathery, with a pair of recurved stipular thorns. Flowers solitary or clustered in the leaf axils. Sepals 4, free, unequal with the upper sepal larger and strongly hooded. Petals 4, free. Stamens 8 to numerous; androphore absent. Gynophore present. Fruit a globose to obovoid berry. 250 species. Tropical and warm temperate regions.

Capparis cartilaginea Decne.

Capparis galeata Fresen.
Capparis spinosa sensu Balf.f. non L. p.p.
Capparis spinosa sensu Forssk. non L.

Sprawling evergreen shrub to 2m. Leaves leathery, elliptic to ovate or orbicular, <5 x 5cm, with a small recurved spine inserted below the tip. Flowers solitary, on stout pedicels. Sepals 1–4cm long. Petals white becoming pink or purple, unequal, ovate to orbicular, 1–3cm long, the upper pair appearing fused. Stamens many, c.5cm long. Gynophore c.3.5cm. Fruit bright red, obovoid, 2–5 x 2–3cm.

Ecology. On cliffs, large boulders and rocky slopes in succulent shrubland. 30–350m.

Distribution. E and SW Africa, Arabia east to India.

Botanical notes. This sprawling evergreen shrub is widespread on cliffs over the dry foothills of Soqotra – it is also recorded from Samha. The dramatic white or pink-tinged flowers appear without rain and are amongst the most beautiful flowers on Soqotra.

ETHNOGRAPHIC NOTES

Name(s)

– Soqotri: *ləṣā́fə, lóṣəf* 'smooth'; fruit *gəlʿómo* (pl. *gílʿum, gílʾḥum*).

– Soqotri simplified: *lizafih, lozif*; fruit *gil'omoh* (pl. *gil'um, gil'hum*).

Food (human). The ripe fruit are occasionally eaten in the hungrier areas of the Archipelago at times when other food is scarce. Studies have shown the fruit to be very high in vitamin C, as well as containing significant amounts of protein and carbohydrate.

Food (livestock). Persistent through the dry months, and consequently an important dry season and drought fodder for goats and camels. Both living and dead parts are eaten. In times of drought herders take their animals to feed in the areas where this plant grows. It is seen as improving milk yield in milk camels feeding on it, and to put on weight in goats. In areas where it grows plentifully, as in parts of the southern plain, it is a key fodder plant for all livestock. The fruit are much liked by goats.

Bees / honey. Honeybees are attracted to the showy flowers.

Medicinal / veterinary. The leaves are crushed to a paste, dried to a powder and then sniffed up the nose to treat a head cold. Alternatively, the water in which fresh leaves have been cooked is sniffed up the nose. The skins of the fruit are roasted at the fire, and their powder used to treat all sorts of sores and wounds. The leaves are boiled, and the strained liquid drunk to relieve shortness of breath. Someone with an itchy nose, a sign that a cold is coming on, crushes the green leaves and sniffs up the juice to ward it off. Locally applied, leaf juice heated over the fire is used to relieve itching. The leaves are scorched, ground to a powder and added to water; this lotion is then used to wash the body to treat fever, aches and pains and relieve general malaise. Fallen, dead leaves are ground up, added to water and used as a wash to reduce a fever. Goats attacked by botfly are treated by squeezing leaf juice into their nostrils. Sheep and goats with runny noses are treated in the same way. Scabies and other irritations of the skin are treated by grinding leaves to a paste which is applied locally. The illness *di so'o* (whose main signs are itching and skin rashes all over the body, very itchy eyes, fever and sweating) is treated either by crushing the leaves to a paste and applying this to the rash, or by grinding the leaves and adding water, leaving the mixture overnight and then using the lotion as a wash in the morning. The treatment is more effective if the patient also inhales the fumes of the burning stems. Where the illness is believed to be of supernatural origin, and to have been caused by the patient washing or relieving himself in an area occupied by *jinn* or other inhabitant of the spirit world, seven pebbles are fetched from that same place and hung in a bag over the fire to 'imprison' the illness, as well using as the remedies described above. The pebbles have then to be returned to their place of origin.

Cleansing. The succulent leaves are crushed and used for washing and cleaning.

Other. Animals feeding on this can go longer without water.

Notes. A period spent grazing halophytic rangeland at least once a year is considered essential for the health of small stock and camels especially. Such feed initially causes diarrhoea which is seen as being beneficial and 'cleansing impurities from the blood'. A very healthy plant is referred to as *ləṣā́fə ṣáʿalhil* (broad-leaved *C. cartilaginea*) and a straggly one as *ləṣā́fə ṭífəh* (lanky *C. cartilaginea*).

Comparative uses. In nearby Dhofar, the leaves were steeped in water and used as a wash by parturient women; the leaves and growing tips were crushed and heated in water and the filtrate used externally on bruises, to relieve inflammation, oedema, mastitis, and to soothe the swelling and pain of snakebite; the leaves and stems were pounded to a paste applied to swollen and painful joints, sprains and to treat paralysis; the same paste was used to treat tick and lice infested livestock; infected tick and lice bites were treated by applying the papery skin removed from a leaf, and the juice of a leaf was dripped onto a deeply embedded tick to make it lose its hold; dead leaves were scorched, powdered and packed into the site of a malignant ulcer; a paste of leaves crushed with a little water was applied to the head for a headache; the leaves were put on the fire and the fumes inhaled by someone suffering from pruritus or allergy. In northern Oman, juice squeezed from the leaves was used to treat inflammation and, with added vinegar, as eardrops to treat loss of hearing. In Yemen the leaves are used against snakebite, and in Kenya they are ground, soaked in water and used for eye diseases in people and livestock. The stems sometimes exude a sweet, sticky and edible substance. The fruits are eaten raw in some areas and the buds pickled. In the Egyptian desert the fruits are used as a remedy for fever and headache, and the dried fruits are the main ingredient of a spicy nutritious drink which can be kept in a water-skin for one to two months. The fruits are also cooked with salt and oil and the result applied as a poultice to treat rheumatism. (Fleurentin & Pelt/Fleurentin *et al.* [1982, 1983a,b,c]; Heine *et al.* 1988; Kunkel 1984; Miller & Morris 1988)

Capparis decidua **(Forssk.) Edgew.**
Capparis aphylla Roth

Shrub or small tree to 3m with whip-like, leafless stems. Leaves soon dropping, linear, <2.5 x 0.3cm, sparsely pubescent. Flowers solitary or in clusters. Sepals 0.5–1cm long. Petals pale orange or red; inner pair narrowly ovate, <12 x c.2mm; outer pair ovate, <12 x 8mm. Stamens 8 to 20, c.1.5cm long. Gynophore 1–1.5cm long. Fruit yellow to scarlet, globose, 5–15mm diameter, pubescent.

Ecology. Rare in *Croton socotranus* shrubland and succulent shrubland. 150–350m.

Distribution. Arabia, tropical NE and N Africa, Middle East to NW India.

Botanical notes. Leafless shrub with long whippy stems frequently from a gnarled base. It is the only leafless shrub on Soqotra that has paired spiny stipules.

ETHNOGRAPHIC NOTES

Name(s)

– Soqotri: *ṣɔ̄mhir*, *ṣɔ̄mheyr* 'singing, song' (also given to *Cadaba insularis*); *ləṣāfə ṣáʿalhil* 'large-leaved *Capparis*'.

– Soqotri simplified: *zowmhir*, *zowmheyr*; *lizafih za'ilhil.*

Food (livestock). This plant is little liked by sheep and goats, even the fallen leaves, though the distinctive fruit are eaten by goats if they can reach them. At times of drought goats will gnaw at the twigs and eat any dried foliage, and sheep nibble the fallen dead leaves and other plant litter. Camels browse the tree.

Fuel. Not liked as fuel.

Timber. Little used as timber.

Comparative uses. Several *Cadaba* species are considered to be toxic for livestock (*C. tomentosa*, *C. corymbosa*, *C. fascicularis*). *C. decidua* has been demonstrated to have anti-bacterial, anti-inflammatory and anti-fungicidal properties and the fruit contains vitamin C, proteins and carbohydrates. These are astringent and seen as useful in cardiac troubles and biliousness, and in India the flower buds are also eaten and the fruit pickled. In Arabia the fruits are taken as an anthelmintic, to relieve constipation, and to treat asthma and coughs. They are also prescribed to treat hysteria and other psychological problems. The tender leaves and branches are used as a plaster for boils and swellings, and they are chewed to relieve toothache. The bark is described as acrid, laxative, diaphoretic, alexeteric, anthelmintic and useful against coughs, asthma and inflammations. The very bitter roots appear in the Indian and Farsi pharmacopoeia. The wood is hard and termite resistant, and is used for the knees of boats, the axles of cartwheels and tool handles. Both root and bark are pungent and bitter, and extracts are given in cases of intermittent fever and rheumatism. In Africa the wood ash is used as a vegetable salt and the root or leaves are used to treat venereal diseases, jaundice and an enlarged spleen. In the Egyptian desert the ripe fruits are eaten, but only after removing the seeds, which can bring on vomiting. The wood is used here to make parts of the camel saddle, and in Sudan, the larger stems are hollowed out to make water-pipes. In Oman the leaves and young stems are crushed in water, heated and the strained solution applied to bruises, snakebites and swellings. It is smeared on the forehead for headaches and over the skin to relieve itching. The leaves are crushed in water, boiled and the strained solution used to wash the parturient mother as an antiseptic. Juice extracted from the leaves is applied to swellings and, with added vinegar, as eardrops to treat deafness. A paste of leaves and stems is applied to swollen and painful joints, and to relieve muscle pain and strengthen paralysed limbs. (Ambasta 1982; Baumer 1983; Dalziel 1948, 1956; Goodin & Northington 1979; Goodman & Hobbs 1988; le Houerou 1980; Jain & De Fillips 1991; Mossa *et al.* 1987; Oliver-Bever 1986)

Cleome L.

Gynandropsis DC.

Herbs or subshrubs. Leaves alternate, simple or digitately compound. Flowers in racemes, often zygomorphic. Sepals 4, free or fused at the base. Petals 4, equal or unequal, sessile or clawed. Stamens 4 to 20, free; androphore usually absent (present in *C. gynandra*). Staminodes present or absent. Gynophore usually absent (present in *C. gynandra*). Fruit a broadly elliptic to linear capsule, dehiscing by 2 valves. 1 endemic. 150 species. Mostly tropical.

ETHNOGRAPHIC NOTES

Comparative uses. The leaves and seeds of some species are cooked and eaten. Medicinally, *Cleome* species are used as a vermifuge and to treat skin problems. In the Egyptian desert an infusion of *C. droserifolia* is used for skin eruptions and as a cough suppressant. The plant is also used as a remedy against the larvae of the botfly, and the leaves are rubbed in the moustache and hair for their pleasant smell. The foliage is thought to repel ticks; camels have been observed making for it and rubbing their heads in the greenery to rid themselves of ticks. (Dalziel 1948, 1956; Ghazanfar & Al-Sabahi 1993; Goodman & Hobbs 1988; Kunkel 1984; Miller & Morris 1988)

Cleome austroarabica **D.F.Chamb. & Lamond**

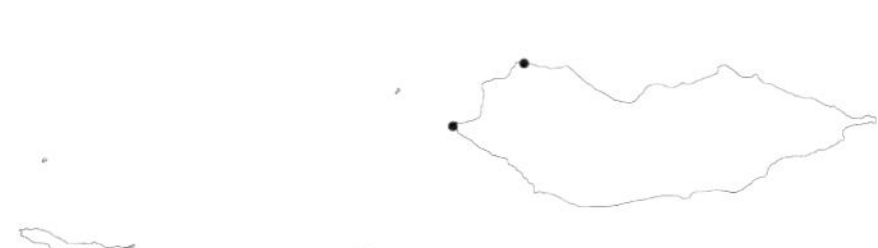

Dwarf shrub to 50cm, densely stipitate-glandular to glandular-villous. Leaves simple, broadly ovate to orbicular, <0.8cm long. Inflorescence lax, few-flowered; bracts leaf-like. Petals yellow with a central reddish-brown stripe, dimorphic, 5–8mm long. Stamens 4. Fruit broadly elliptic, <18mm long. Seeds minutely papillate when mature.

Ecology. Wadi beds in *Croton socotranus* shrubland. Sea-level–50m.

Distribution. Oman, Yemen, United Arab Emirates.

Botanical notes. This densely glandular dwarf shrub is restricted to plains near the sea in the west of Soqotra.

ETHNOGRAPHIC NOTES

Name(s)

– Soqotri: *tɔ̄mhor* 'little dates' (Abd al Kuri) (given to a variety of delicate and often sweet-smelling small plants).

– Soqotri simplified: *tomhor.*

Food (livestock). Regarded as a valuable fodder for small stock in the drier areas of the Archipelago.

Flavouring / food additive. The milk of animals eating a lot of it is said to have a characteristically sweet flavour.
Cleansing. The fragrant leaves are crushed between the hands and rubbed over the skin and clothing to impart a pleasant smell. The whole plant is used to clean the hands to remove unpleasant odours. In some areas, sprigs are tucked into the clothing and hair. The sniffing of sweet-smelling plants is held to protect from the 'noxious air' that can cause disease, and to help protect against infection. All sweet-smelling plants are endowed to some extent with beneficial medicinal and prophylactic qualities.

Cleome brachycarpa **Vahl ex DC.**
Cleome brachycarpa var. *filicaulis* Schweinf.
Cleome diversifolia Hochst. & Steud. nom. nud.

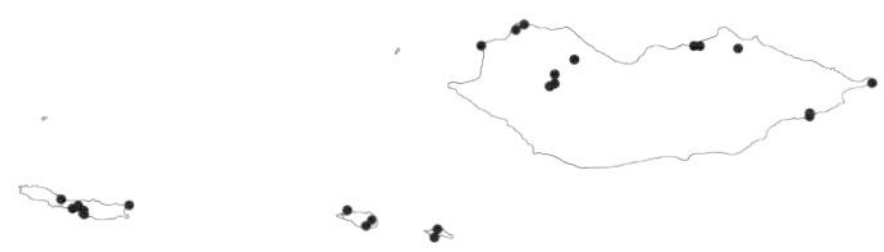

Aromatic herb to 35cm, shortly stipitate-glandular. Leaves 3(to 5)-foliolate; leaflets elliptic, <1.5cm long, sparsely stipitate-glandular. Inflorescence lax, few-flowered; bracts leaf-like, 1 to 3-foliolate. Petals yellow, sometimes with reddish markings, rarely whitish-yellow, 3–8mm. Stamens 6. Fruit elliptic, <12mm long. Seeds finely reticulate, glabrous.
Ecology. Desertic dwarf shrubland and *Croton socotranus* shrubland. Sea-level–200(–400)m.
Distribution. N and tropical Africa, Arabia east to Afghanistan and NW India.
Botanical notes. A widespread and often common perennial herb which is distinguished by the short, oblong, long-styled fruits.

ETHNOGRAPHIC NOTES
Name(s)
– Soqotri: *tàmərhen*; *təmèrhen* (Samha); *tɔ̄mhor* 'little dates' (given to a variety of delicate and often sweet-smelling small plants) (Abd al Kuri).
– Soqotri simplified: *tamirhen*; *timerhen*; *tomhor.*
Food (livestock). In the drier areas of Soqotra and in Abd al Kuri and Samha this is regarded as a valuable fodder for small stock.
Flavouring / food additive. The milk of animals eating a lot of this plant is said to have a characteristically sweet flavour.
Bedding, stuffing. The leaves and small plants are strewn among bedding to perfume it. This is done particularly for sick people to hasten their recovery.
Cleansing. The fragrant leaves are crushed between the hands and rubbed over the skin and clothing to impart a pleasant smell. The whole plant is used to clean the hands to remove unpleasant odours. In some areas, sprigs of the plant are tucked into the clothing and hair.
Comparative uses. In Qatar it is used to treat scabies, rheumatism, inflammations and leucoderma. In Oman the juice from either this crushed plant or from *C. rupicola* is painted around sore or inflamed eyes, and the aromatic leaves of this and *C. brachycarpa* are rubbed on the hands and clothing for their fresh, lemony smell. (Ghazanfar & Al-Sabahi 1993; Goodman & Hobbs 1988; Miller & Morris 1988)

Cleome gynandra **L.**
Cleome pentaphylla L.
Gynandropsis pentaphylla (L.) DC.

Faintly aromatic herb to 100cm, villous and stipitate-glandular. Leaves 3 to 5-foliolate; leaflets obovate, <70mm. Inflorescence many-flowered; bracts small, 3-foliolate below, simple above. Petals white or yellow to mauve-pink, 2–6mm long. Stamens 6, borne on an 8–20mm long androphore. Fruit borne on a 5–12mm long gynophore (arising from the androphore), linear, <80 (–115)mm. Seeds coarsely reticulate, glabrous.
Ecology. Weed of date gardens. c.50m.
Distribution. A pantropical weed.
Botanical notes. Distinguished by the palmate leaves and the long stalk (androphore) between the calyx and the fruit.

ETHNOGRAPHIC NOTES
Name(s)
– Soqotri: nil or *fègəhun*, *fèjəhen*; *fìžihin* 'sprouting, throwing up new shoots' (given to a variety of delicate, sweet-smelling plants).
– Soqotri simplified: *fegihun*, *fejihen*; *fijihin.*
Food (livestock). Browsed by goats if the date gardens where it grows are accessible to them. It is not seen as being important nutritionally.

Cleome scaposa **DC.**
Cleome papillosa Steud. nom. nud.

Herb to 20cm, sparsely villous, glandular-hairy above. Leaves simple, broadly ovate to orbicular, <2cm long, strigillose with glandular-hairy margins. Inflorescence lax, few-flowered; bracts absent above. Petals pale yellow, 3–4mm long. Stamens 6. Fruit linear, <35mm long. Seeds minutely granulate, glabrous.
Ecology. *Croton socotranus* shrubland on coastal plains. 50–150m.
Distribution. N and tropical Africa, Arabia, Pakistan.

ETHNOGRAPHIC NOTES
Name(s)
– Soqotri: nil or *fègəhun*, *fèjəhen*; *fìžihin.*
– Soqotri simplified: *fegihun*, *fejihen*; *fijihin.*
Food (livestock). In the poorly vegetated areas of Soqotra this is regarded as a useful forage plant for small stock.

Cleome socotrana **Balf.f.**

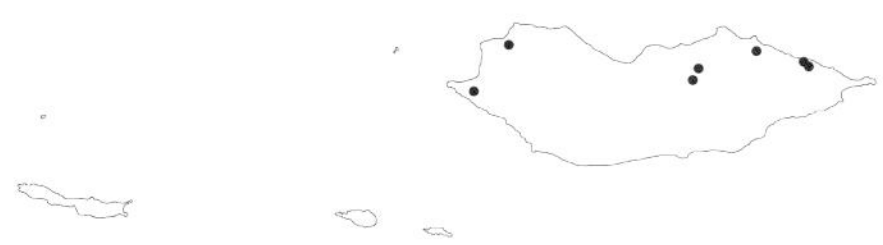

Woody-based herb to 70cm, shortly stipitate-glandular, occasionally glabrous. Leaves 3-foliolate; leaflets broadly elliptic to obovate, <20mm long. Inflorescence dense, many-flowered, lax in fruit; bracts simple, broadly obovate to orbicular. Petals white to yellow, with red or brown markings, 2–5mm long. Stamens 6. Fruit narrowly elliptic to narrowly oblong, <45mm long. Seeds densely hairy when mature.

Ecology. The ecology of this species is not clear. It is found in two very different habitats: at low altitudes in *Croton socotranus* shrubland by the coast and in montane shrubland amongst the granite pinnacles of the Haggeher. Sea-level–1100m.

Distribution. Endemic.

Conservation status. Least Concern (LC). Common in several vegetation types and under no present or perceived threat.

Botanical notes. Said to be on the increase in some areas (for instance at Halle in NE Soqotra).

ETHNOGRAPHIC NOTES

Name(s)

– Soqotri: nil or *təme̍rhen* 'little dates' (given to a variety of delicate and often sweet-smelling small plants) (Samha); *ginsi̍biher* (more commonly given to various species of fern).

– Soqotri simplified: *timerhen*; *ginsibiher*.

Food (livestock). On Samha this provides incidental forage for goats and sheep. On Soqotra it is not generally liked by livestock, though is nibbled by goats and sheep when they are hungry and little alternative forage is available. It is not regarded as important forage.

Cleome tenella **L.f.**

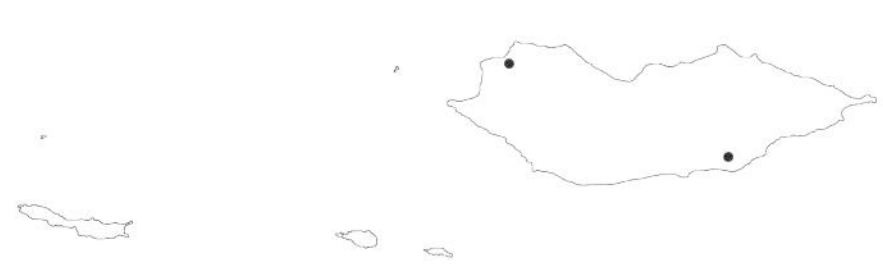

Delicate herb to 30cm, glabrous. Leaves 3-foliolate; leaflets filiform, <20mm long. Flowers in lax 1 to 5-flowered inflorescences; bracts 1 to 3-foliolate, filiform. Petals, colour not known, c.2.5mm long. Stamens 6. Fruit linear, <35mm, glabrous. Seeds reticulate, glabrous.

Ecology. An apparently rare weed of date gardens. c.100m.

Distribution. Tropical Africa and India.

Botanical notes. Not seen recently on Soqotra.

Cleome viscosa **L.**

Foetid herb to 70cm, densely glandular-villous and stipitate-glandular. Leaves 3(to 5)-foliolate; leaflets ovate-elliptic, <20mm long. Inflorescence few-flowered; bracts leaf-like. Petals yellow (drying white), 4–10mm long. Stamens 10 to 20. Fruit linear, <80mm, prominently nerved. Seeds transversely ridged, glabrous.

Ecology. Succulent shrubland and a weed of date gardens. 50–200m.

Distribution. Pantropical weed.

ETHNOGRAPHIC NOTES

Name(s)

– Soqotri: nil or *di ḳe̍rəhān* 'horned', referring to the fruit.

– Soqotri simplified: *di qerihan*.

Food (livestock). Not liked by livestock, though goats will chew the green stalks in hunger.

Maerua Forssk.

Trees. Leaves alternate, simple. Flowers in axillary and terminal racemes, actinomorphic. Hypanthium cylindrical or campanulate. Sepals 3 or 4, free. Petals absent. Stamens numerous, free, borne on an androphore. Gynophore present. Stigma capitate. Fruit cylindrical, indehiscent. 1 endemic. 50 species. Tropical and S Africa, Arabia, India.

Maerua angolensis **DC.**

Maerua socotrana (Schweinf. ex Balf.f.) Gilg

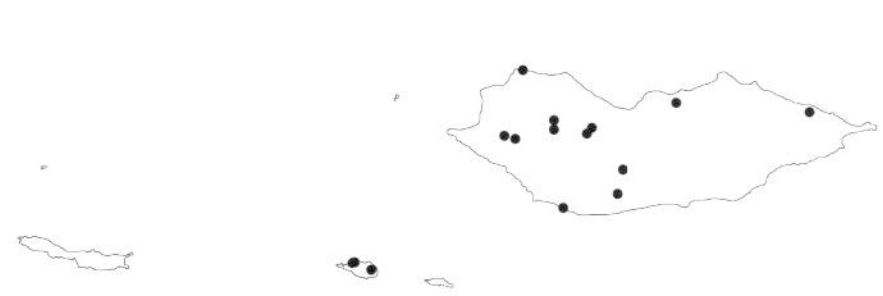

Small tree to 6m, with a single trunk and graceful pendulous branches. Leaves narrowly obovate to ovate or subcircular, 2–4 x 1–3cm, glabrous. Racemes few-flowered. Sepals cream, greenish or pale yellow, 5–10mm. Hypanthium 2–4mm long. Fruits cylindrical, torulose, 40–50 x 3–5mm, pale green when mature.

Ecology. On rocky slopes, sometimes on cliffs in succulent shrubland and drought-deciduous woodland. Sea-level–250(–500)m.

Distribution. Var. *socotrana* is endemic to Soqotra.

Conservation status. Near Threatened (NT). *M. angolensis* var. *socotrana*, whilst abundant at present, is likely to suffer from indiscriminate cutting for timber as a consequence of the development of the island and the almost inevitable breakdown of the traditional practices which at present protect this and certain other tree species. There is minimal regeneration of *M. angolensis* var. *socotrana* in the more inaccessible areas of the island, and, in the absence of alternative building woods, this straight-trunked and hardwood species is likely to come under increasing threat for building purposes.

Botanical notes. One of the most distinctive small trees on the island, readily distinguished, even when not in leaf, by its gracefully weeping branches. Two varieties are recognised within *M. angolensis* subsp. *socotrana*: var. *socotrana* which is endemic to Soqotra, and var. *africana* which is endemic to Ethiopia and Somalia and differs from var. *socotrana* in its longer sepals and hypanthium. For a discussion of the *M. angolensis* complex in NE Africa and Arabia see Kers (1993).

ETHNOGRAPHIC NOTES

Name(s)

– Soqotri: *ˀˈišḥʌb*, *ˀˈišḥib* (pl. *ˀˈišḥob*, *iˀihāḥəb*) 'outstretched, extended'; fruit *ḳaḥf di ˀˈišḥʌb*.

– Soqotri simplified: *'ish'hab, 'ish'hib* (pl. *'ish'hob, i'ihahib*); fruit *qahf di ish'hab.*

Food (human). The fruit are liked by birds but are not eaten by people.

Food (livestock). The foliage persists through the dry months, making it an important dry season fodder. The dead, fallen leaves are eaten by goats and sheep, as are the flowers, young shoots and fruit. Herders lop green branches as a drought feed for their animals, and gather dead, fallen foliage for their cattle and goats. They also gather the small, fresh leaves to feed weaned kids in the dry season. Native goats tolerate the fresh foliage, and give more milk when feeding on it, but incoming goats eat it only when hungry. The foliage is bitter, and goats unaccustomed to it can be seen pulling a face at the taste as they eat it. If goats eat too much of the green foliage to the exclusion of other fodders they lose weight and often become afflicted with temporary paralysis of the hind quarters accompanied by shivering and fits of staggering. Also, goats eating a lot of it tend to lose their appetite for other plants, so herders are not keen to encourage them to spend too much time browsing the tree. In general, small stock rarely eat a lot of the green foliage at a time as it is very 'drying' to the mouth, and makes the animal very thirsty. Sheep tend to do less well on this fodder than goats. Cattle in areas where there is little alternative browse will eat it a little. Donkeys also eat it and it is much liked by camels and is regarded as improving lactation in milk camels. This has led to some friction between camel and goat owners in areas where there are few alternative browse plants for camels. On the outer islands it is a very important dry season feed for sheep and goats.

Bees / honey. Honeybees are attracted to the flowers.

Fuel. The dead wood makes good firewood, but generally only smaller branches and twigs are used, the larger branches being more valued as timber.

Timber. The slim upright trunk provides one of the island's best building timbers. It is particularly prized in areas where other woods are scarce, such as western Soqotra and the outer islands. The trunk provides supporting pillars and the main roofing beams, and smaller cross beams and roofing infill are made from the branches. The wood is attractive to termites and other wood-boring insects so it is usually de-barked before use. Like all timber, the trees must be cut down in hot, dry weather. If kept dry, it can be used again and again by successive generations, but if it is allowed to become wet it quickly rots. In times when houses were rare, the wood from this tree was considered to be the best for modifying the former cave dwellings of the drier areas. For example, on Abd al Kuri the cave living quarters were extended outwards using this timber to support a roofing made from material of *Cissus* species and *Ochradenus arabicus* covered with a top layer of earth.

Wood. Good for implements and used to make the important herding staff. Strips of bark were used to weave fish traps. On the outer islands where trunked trees are otherwise rare, the wood was very important and was used to make most of the necessary everyday utensils. It also provided the stick for the quern, and, on Samha, was used in the manufacture of the small drum and the larger tambourine drum. It makes good charcoal.

Medicinal / veterinary. The green foliage is collected and fed to sickly goats to restore their strength. The leaves are scorched, powdered and made into a paste with water. This is rubbed over the body to treat fever, aches and general malaise. An itch is treated by applying a paste of the crushed leaves. Like *Capparis cartilaginea*, the leaves are used to treat the illness known as *di so'o*: the leaves are ground, added to water and left to steep overnight. The next morning the liquid is used as a wash for the whole body. Some prefer to mix *Capparis cartilaginea* leaves with the leaves of this tree to make the lotion, although this smells unpleasant.

Other. Attempts by herders to grow this valuable tree by transplanting young saplings have so far proven unsuccessful, though many have tried it. In parts of the southern plains livestock movements were determined by stages of growth of this tree and *Commiphora ornifolia.* After the winter rains, when their leaves begin to dry out, lose colour and fall to the ground, this was the signal for bringing the goats and sheep down from the escarpment to the plain.

Notes. Tradition says that if this tree (along with *Sterculia africana*, *Commiphora ornifolia*, *Grewia* species and *Acacia pennivenia*) comes into new leaf at the end of the monsoon, then the winter rains will soon follow. Formerly in western Soqotra it was not permitted to cut this tree without good reason and without first obtaining permission from other users of the area. It is virtually the only tree in this extensive area which grows tall and straight enough to provide good timber, and if cutting were widespread the trees would have been quickly exhausted. Because of this ruling, the *M. angolensis* trees of the region were able to grow to maturity, and develop into graceful, weeping trees such as are rarely seen elsewhere in the islands.

Comparative uses. The fodder value of many *Maerua* species is high. The fruit of *M. angolensis* are reported to be poisonous, but the leaves can be cooked and eaten. The fruit are used in northern Nigeria as a love charm, and, mixed with an antimony substitute, are rubbed over the eyelids. The wood is fine-grained and takes a fine polish; although brittle, it is used to make tools. The crushed leaves are applied as a poultice to relieve pain. The closely related *M. crassifolia* is highly regarded as a fodder plant for goats and camels, less so for cattle and sheep. Its wood is used to make tool handles, bowls, water troughs, mortars and fencing. It is used as firewood, and burning sticks are used to fumigate food containers to give them a good smell. In West Africa twigs are used as toothsticks. The Borana treat someone with a swollen belly or leg by fumigating the patient in the smoke of twigs laid over charcoal in a hollow. The leaves are boiled in water and the filtrate taken orally for abdominal colic and constipation. Powdered leaves are made into a paste with powdered henna (*Lawsonia inermis*) and applied to the site of a fracture to relieve the pain. (Baumer 1983; Bebawi & Neugebohrn 1991; Ghazanfar 1994; Heine *et al.* 1988; Kunkel 1984; Lewis & Elwin-Lewis 1977)

CARYOPHYLLACEAE

Herbs or subshrubs. Leaves opposite, sometimes fasciculate or appearing whorled, entire; stipules scarious or absent. Flowers actinomorphic, bisexual, usually 5-merous, usually subtended by bracts, typically in dichasial cymes, sometimes clustered or umbellate, occasionally solitary and terminal. Sepals free or united. Petals free or rarely absent, entire or variously dissected, often clawed. Stamens 5 to 10. Ovary superior, 1-celled or 2 or 3-celled below; styles free or connate; placentation free-central; ovules 2 to many. Fruit a few to many-seeded capsule or indehiscent and 1-seeded. 2070 genera. 89 species. Cosmopolitan, especially temperate and warm N Hemisphere.

Arenaria L.

Herbs. Leaves opposite; stipules absent. Flowers few, in lax terminal cymes. Sepals 5, free. Petals 5, white, entire or slightly emarginate. Stamens 10. Styles 3, free. Fruit a many-seeded capsule, dehiscing by 6 teeth. Seeds reniform. 150 species. N temperate, especially mountainous and arctic regions.

Arenaria foliacea Turrill

Arenaria serpyllifolia sensu Balf.f. non L. p.p.

Annual herbs to 15cm. Leaves ovate, <0.8cm long, sparsely hairy, hairs sometimes gland-tipped. Cymes lax, 2 to 6-flowered; pedicels <10mm; bracts well developed, leaf-like. Sepals 3–5mm long. Petals narrowly lanceolate, 2–4mm long. Seeds black.

Ecology. Grassland clearings in montane woodland at high altitudes in the Haggeher. c.1130m.

Distribution. Saudi Arabia, Yemen, Oman, Tanzania, Somalia.

Botanical notes. This species belongs to a complex of very closely allied taxa. It is not clear whether all the records for this complex from the Arabian Peninsula are referable to *A. foliacea* but it seems likely that they are.

ETHNOGRAPHIC NOTES

Name(s)

– Soqotri: nil or *raˈʿadihon* ‘small grazing plant’.

– Soqotri simplified: *ra’adihon.*

Food (livestock). This is one of the small *ru’ud* plants which come up with the rains, providing incidental green grazing until they die back. Though palatable to livestock, this is not important forage.

Cometes L.

Perennial herbs or subshrubs. Leaves opposite; stipules inconspicuous. Inflorescences of terminal and axillary 3(to 5)-flowered pedunculate heads, subtended by feather-like bracts. Sepals 5, free. Petals 5. Stamens 5. Style solitary, 3-lobed at the apex. Fruit 1-seeded, indehiscent. 2 species. NE Africa, Arabia, NW India.

Cometes abyssinica (R.Br.) Wall.

Cometes abyssinica subsp. *suffruticosa* Wagner & Vierh.

Low, woody-based herb or subshrub to 35cm. Leaves linear to lanceolate, rarely ovate or obovate, <3 x 0.5mm. Bracts lengthening to 15mm in fruit, needle-like segments spreading. Sepals <5mm, papillose. Petals white, 2/3 as long as the sepals. Fruits enclosed by the cream to reddish-brown, conspicuous, feather-like, prickly bracts.

Ecology. Locally common in dwarf shrubland on drier parts of the eastern limestone plateau (Wadi Faleng) and in rocky clearings in succulent shrubland. 300–450m.

Distribution. Arabian Peninsula, Somalia, Ethiopia, Sudan.

Botanical notes. Abundant near Wadi Faleng in the southwest of the island, rare elsewhere. In fruit *C. abyssinica* is quite unmistakeable: the entire plant becomes a mass of fruiting bodies, each consisting of a reddish brown ball of feather-like, needle-tipped bracts surrounding the small fruits.

ETHNOGRAPHIC NOTES

Name(s)

– Soqotri: *ḳˈʌlḳəhʌl* ‘swaying in the breeze’ (general term for a variety of mainly halophytic plants).

– Soqotri simplified: *qalqihal.*

Food (livestock). Much liked by camels and donkeys. Goats also browse it, but less enthusiastically. It is not regarded as a useful feed for small stock.

Other. Animals feeding on this can go longer without water.

Notes. A period spent grazing halophytic rangeland at least once a year is considered essential for the health of small stock especially. Such feed initially brings on diarrhoea which is seen as being beneficial and ‘cleansing impurities from the blood’. Once the animal recovers from this, it starts to put on weight.

Gymnocarpos Forssk.

Subshrubs. Leaves opposite, often fascicled on younger branches, fleshy, mucronate; stipules interpetiolar, scarious. Inflorescence terminal or subterminal, shortly pedunculate; subtended by scarious bracts. Sepals 5, free, with a membranous margin. Petals absent. Stamens 5, alternating with 5 staminodes. Style long, slender; stigma 3-lobed. Fruit 1-seeded, indehiscent, rupturing irregularly in the middle or near the base. 2 endemics. 4 species. Macaronesia and C Asia.

Botanical notes. See Petrusson & Thulin (1996).

Gymnocarpos bracteatus (Balf.f.) Thulin

Lochia bracteata Balf.f.

Subshrub to 35cm. Leaves linear-oblanceolate, terete, <1.3 x 0.2cm, olive-green. Inflorescence dense; bracts broadly ovate to suborbicular, <6 x 5mm, brown or streaked with brown throughout or with white margin in upper part only, the lowermost ones sometimes with a well-developed lamina. Seed ellipsoid, dark brown.

Ecology. Restricted to the western Haggeher mountains where it is found in low shrubland on rocky granite slopes. 450–900m.

Distribution. Endemic.

Conservation status. Vulnerable (VU D2). Rare, with a very limited area of occupancy.

Botanical notes. Distinguished by its large, entirely brown bracts and the petiolate, linear-oblanceolate, olive-green leaves. Previously *G. bracteatus* was used in a broad sense to include all plants found on Soqotra and Abd al Kuri and in Oman. However, here *G. bracteatus* is used strictly for plants from Soqotra. *G. bracteatus* is probably more closely related to species (*G. argenteus* and *G. dhofarensis*) from the Arabian mainland than to *G. kuriensis*.

ETHNOGRAPHIC NOTES

Name(s)

– Soqotri: *ṣʌ̍bəher* 'sour, astringent'; *ṣɔ̄hor* (more commonly given to *Ocimum forskolei*).

– Soqotri simplified: *zabiher*, *zowhor*.

Food (livestock). Not generally palatable to livestock, though the flowers are eaten by small stock. Goats eat it in periods of drought if there is little alternative.

Gymnocarpos kuriensis (Radcl.-Sm.) Thulin

Lochia bracteata subsp. *abdulkuriana* Chaudhri

Lochia kuriensis Radcl.-Sm.

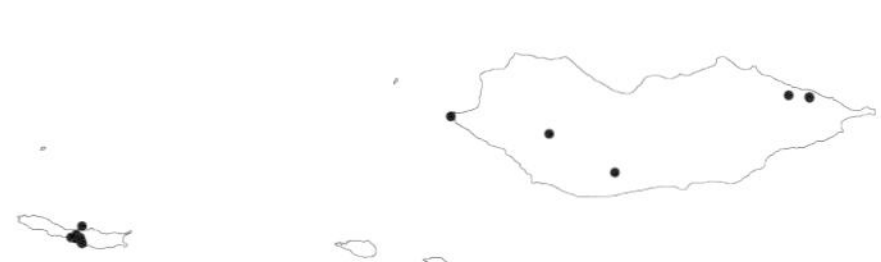

Subshrub to 30cm. Leaves narrowly obovoid to subglobose, <1.2 x 0.5cm, grey-green. Inflorescence and bracts similar to those of *G. bracteatus* but the bracts with a narrow to broad white margin all round. Seed ovoid, brown.

Ecology. Abd al Kuri: dwarf shrubland on low sand dunes, gravel plains and broken rocky slopes; 30–500m. Soqotra: exposed and windswept places on the limestone plateau in dwarf shrubland and drought-deciduous woodland; 500–700m.

Distribution. Endemic.

Conservation status. Vulnerable (VU D2). Rare, with a very limited area of occupancy.

Botanical notes. Most closely related to *G. rotundifolius* from the coastal desert of central Oman. Distinguished from *G. bracteatus* by its grey-green, obovoid to subglobose leaves.

ETHNOGRAPHIC NOTES

Name(s)

– Soqotri: *dīrbeb* '(animal) with a pale streak or flash on its coat') (more commonly given to *Pulicaria stephanocarpa*) (Samha); *ḳʌ̍lḳəhʌl* 'swaying in the breeze' (general term for a variety of mainly halophytic plants); *ḥāṭem* (Abd al Kuri) (general term for a variety of halophytic plants on this island).

– Soqotri simplified: *diyrbeb*; *qalqihal*; *hatem*.

Food (livestock). Much liked by donkeys. Sheep and goats also eat it and in dry areas, such as Abd al Kuri and around Qalansiyah, it is regarded as important forage for goats.

Fuel. On Abd al Kuri in the absence of alternative fuel, dried plant material is used as fuel, especially for cooking.

Gypsophila L.

Herbs. Leaves opposite; stipules absent. Inflorescence a large, compound cyme. Bracts small. Calyx 5-toothed, hyaline between the nerves. Petals 5. Stamens 10. Styles 2. Fruit a many-seeded capsule, dehiscing by 4 valves. Seeds auriculate with a prominent radicle, tuberculate or ridged. 125 species. Mostly temperate Eurasia.

Gypsophila montana Balf.f.

Gypsophila montana var. *viscida* Balf.f.
Gypsophila somalensis Franch.

Robust, erect woody-based herb to 90cm, glabrous or pubescent to extremely viscid-hairy. Leaves spathulate, 0.9–3 x 0.4–1cm. Calyx narrowly obconical, 2–3.5mm long. Petals white sometimes tinged pink, <5.5mm long. Capsule oblong-elliptic. Seeds c.0.8 x 0.9mm.

Ecology. Rocky slopes and cliffs in the Haggeher. c.1000m.

Distribution. Yemen, Oman, Somalia.

Botanical notes. Balfour (1888) described three varieties (var. *montana*, var. *viscida* and var. *diffusa*) based on indumentum and inflorescence form. However, because of numerous intermediates they have not been recognised here. Totally glandular-pubescent plants have been placed in subsp. *somalensis* (Franch.) Gilbert.

ETHNOGRAPHIC NOTES

Name(s)

– Soqotri: *di maḥˈaḥon* (pl. *il maḥˈaḥon*) 'fatty' (more commonly given to *Pulicaria lanata* and *P. vieraeoides*).

– Soqotri simplified: *di mahahon* (pl. *il mahahon*).

Food (livestock). Cattle readily graze on this when they can reach it. Goats browse it and it is seen as being a nutritious feed for them.

Comparative uses. Many *Gypsophila* have root systems extremely rich in saponisides, and this species contains up to 16% saponins. Elsewhere in the Middle East *G. rokejeka* is used in the preparation of the popular sweetmeat halva. (Davis *et al.* 1983)

Haya Balf.f.

Perennial herbs. Leaves in whorls; stipules scarious. Inflorescence dense axillary clusters of contracted 1 to 3-flowered spikes. Flowers subtended by 5 or more scarious bracts. Sepals 5. Petals 5. Stamens 5, alternating with minute staminodes. Ovary trigonous, with a solitary style and capitate stigma. Fruit a 1-seeded capsule, ovoid, dehiscing towards the base into 3 valves. Seeds ellipsoid. 1 species. Endemic.

Botanical notes. There is a striking resemblance between this genus and *Polycarpaea*, especially *P. hayoides*, though the latter technically differs in its several-seeded capsules that dehisce from the apex.

Haya obovata Balf.f.

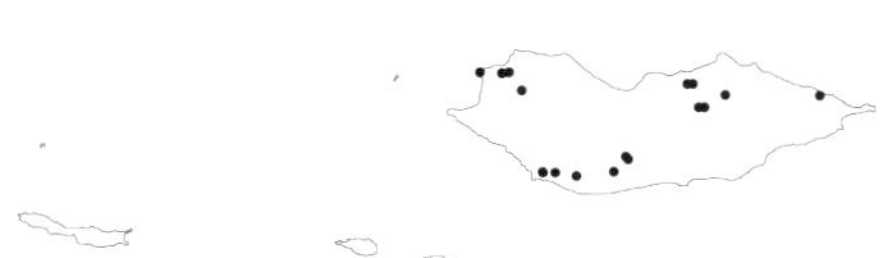

Prostrate herb to 20cm. Leaves sessile, 3 to 5 per whorl, obovate, 0.5–2 x 0.5–1cm, abruptly acuminate at tip. Bracts reddish brown, scarious, ovate-lanceolate, c.4mm, cuspidate. Sepals oblong, white, c.1mm.

Ecology. Ravines and hillsides, on limestone and granite, typically in drought-deciduous shrubland. 200–1300m.

Distribution. Endemic.

Conservation status. Least Concern (LC). Common in several vegetation types and under no present or perceived threat.

Botanical notes. A rather insignificant prostrate herb. The small flowers are almost completely hidden within the globose clusters of brownish scarious bracts. *H. obovata* is superficially similar to *Polycarpaea hayoides* but differs in its one-seeded capsules and sessile, not stalked, clusters of flowers.

ETHNOGRAPHIC NOTES

Name(s)

– Soqotri: *ləskˈano* 'clinging, sticky'; *śˈerśer* (also given to other plants with prickly or clinging fruit); *miśḥɔ̄wji* 'scratchy'; *nˈoi*, *nˈoyə* (given to a variety of plants with fragrant and/or delicate flowers); *ḳūwnuh* 'little thread'.

– Soqotri simplified: *liskanoh*; *shersher*; *mish'howji*; *noy*; *quwnuh*.

Food (livestock). Mainly eaten by goats. Sheep are less enthusiastic. In the drier areas where the plant grows, it is regarded as a good fodder for goats.

Bees / honey. Honeybees are attracted to the flowers.

Polycarpaea Lam.

Herbs, sometimes woody-based, or subshrubs. Leaves opposite, usually appearing whorled; stipules scarious. Inflorescence of terminal or axillary, often spicate, cymes. Bracts scarious. Sepals 5, free, unequal, three inner longer than two outer, not keeled, margins scarious. Petals 5. Stamens 5. Style solitary, 3-lobed at the apex. Fruit a many-seeded capsule, dehiscing by 3 valves. 7 endemics. 50 species. Tropical and warm regions, especially Old World.

ETHNOGRAPHIC NOTES

Comparative uses. In India the leaves and flowers of *P. corymbosa* are used as a snakebite antidote, and its leaves are used in jaundice and as a poultice for boils and inflammatory swellings. In Oman the whole plant of *P. repens* is crushed and applied to the site of snakebite, and the ash of crushed leaves is used to treat sarcoptic mange in camels. Many uses of *P. corymbosa* and *P. linearifolia* are superstitious, being used in charms to ensure success in love and war, and the juice drunk to combat fatigue while on the march. (Ambasta 1982; Dalziel 1948, 1956; Ghazanfar 1994; Jain & De Fillips 1991; Lewis & Elwin-Lewis 1977)

***Polycarpaea balfourii* Briq.**
Polycarpaea divaricata Balf.f.

Erect annual herb to 30cm, 2 or 3x branched above, glabrous. Basal leaves spathulate, <3.2 x 1cm; cauline leaves whorled, linear to narrowly obovate, <3.2cm long, mucronate. Flowers in dense heads made up of 2 to 4 spikes. Bracts white or brown. Sepals c.3.5(–4)mm long, midrib narrow, white suffused orange-brown. Petals linear-oblong, c.1–1.2mm, apex bidentate (or tridentate). Capsule narrowly ovoid.
Ecology. Gravelly plains, rocky slopes and wadi sides in *Croton socotranus* shrubland and succulent shrubland. 15–600m.
Distribution. Endemic.
Conservation status. Least Concern (LC). Common in several vegetation types and under no present or perceived threat.
Botanical notes. A delicate annual herb with linear stem leaves.

ETHNOGRAPHIC NOTES
Name(s)
– Soqotri: nil or *riʿud* 'grazing plant'; *ṣeferher* 'many-flowered'; *ṣəfṣifo* 'ripple, flutter'.
– Soqotri simplified: *ru'ud; zeferher, zifzifo.*
Food (livestock). This is one of the ephemeral *ru'ud* plants which come up with the rains, providing incidental green grazing until they die back. Though palatable to livestock, this is not important forage.

***Polycarpaea caespitosa* Balf.f.**

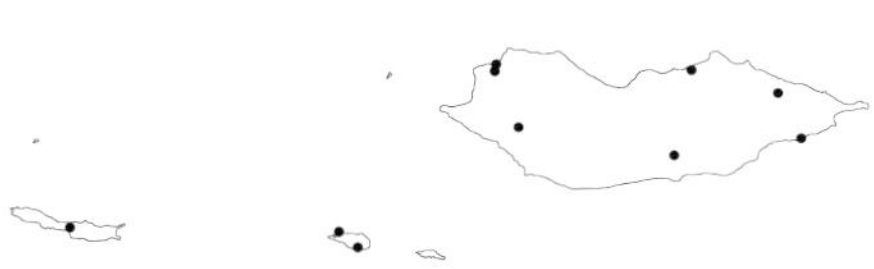

Woody-based, erect, densely tufted herb to 25cm, glabrous. Leaves tufted at base, whorled above, linear, <2 (–3)cm. Flowers in dense heads made up of 2 to 4 spikes. Bracts brown. Sepals c.3mm long, with a narrow midrib. Petals roughly ≤ length of sepals, entire or irregularly toothed at apex, margin minutely fimbriate towards the base. Capsules ovoid.
Ecology. On large boulders, rocky wadi sides and cliffs on limestone in dwarf shrubland, succulent shrubland and semi-deciduous woodland. Sea-level–500m.
Distribution. Endemic.
Conservation status. Least Concern (LC). Common in several vegetation types and under no present or perceived threat.
Botanical notes. A rather brittle perennial herb with the numerous stems bearing whorls of linear leaves and arising from a woody stock.

ETHNOGRAPHIC NOTES
Name(s)
– Soqotri: nil or *ṣeferher* 'many-flowered'; *ṣəfṣifo* 'ripple, flutter'; *šəbiyaḥ, di ṣēbḥitin* 'with outstretched legs, relaxed'.
– Soqotri simplified: *zeferher, zifzifo; shibiyah, di shebhitin.*
Food (livestock). Grazed by small stock. On Samha it is seen as a valuable sheep fodder.
Fuel. Dried plants are gathered and stored for use as kindling.

***Polycarpaea corymbosa* (L.) Lam.**

Erect annual herb to 30cm, sparsely lanate. Leaves linear, <2.5cm long. Flowers in dense corymbose heads. Bracts silvery-white. Sepals 2–3mm long, completely scarious, silvery-white or pink or pale brown. Petals as long as sepals, jagged at apex. Style obscurely 3-lobed. Capsule elliptic.
Ecology. No habitat details available; probably a weed of date gardens.
Distribution. Pantropical.
Botanical notes. Recorded by Balfour in the 19th century on the plain near Kadhab [Qadhub]; not seen since and probably now extinct on Soqotra. The only hairy species of *Polycarpaea* recorded on Soqotra – all the others are glabrous.

Polycarpaea hassalensis D.F.Chamb.

Erect herb with woody rootstock to 5cm, glabrous. Basal leaves densely tufted, narrowly spathulate, 1–1.7 x 0.3–0.4cm; cauline leaves linear-spathulate, <0.7cm long. Flowers in dense heads made up of 2 to 4 spikes; peduncle 1.2–2cm. Bracts scarious with a narrow brown midrib. Sepals c.2.5mm long, reddish brown with a green midrib and narrow hyaline margin. Petals ≤ length of sepals, margin fimbriate towards the base, entire above. Capsule broadly ovoid.

Ecology. Rocky slopes in desertic dwarf shrubland on Abd al Kuri. Sea-level–150m.
Distribution. Endemic.
Conservation status. Vulnerable (VU D2). Rare, with a very limited area of occupancy.
Botanical notes. Differs from other *Polycarpaea* species on Abd al Kuri by its narrowly spathulate, grey-green leaves (cf. *P. caespitosa* and *P. kuriensis*).

ETHNOGRAPHIC NOTES
Name(s)
– Soqotri: nil.
Food (livestock). On Abd al Kuri this is regarded as useful forage for both sheep and goats.

Polycarpaea hayoides D.F.Chamb.

Prostrate annual herb to 10cm, glabrous. Leaves whorled, spathulate, lamina 0.5–1.5 x <0.6cm; petiole <4cm. Flowers in dense heads made up of 2 to 4 spikes. Bracts reddish-brown, with a scarious margin. Outer sepals c.1.5mm long; inner sepals c.3mm long. Petals minute, c.1/2 length of capsule. Capsule ovoid, 3-seeded, dehiscing from apex.
Ecology. Rocky wadi sides on limestone and in date gardens in *Croton socotranus* shrubland and succulent shrubland. 50–450m.
Distribution. Endemic.
Conservation status. Least Concern (LC). Common in several vegetation types and under no present or perceived threat.
Botanical notes. Superficially resembling *Haya obovata* but differing in its three-seeded capsule that opens from the apex and the stalked clusters of flowers.

ETHNOGRAPHIC NOTES
Name(s)
– Soqotri: nil or *ruʿud* 'grazing plant'.
– Soqotri simplified: *ru'ud*.
Food (livestock). This is one of the ephemeral *ru'ud* group of plants which come up with the rains, providing incidental green grazing until they die back. Though palatable to livestock, this is not important forage.

Polycarpaea kuriensis Wagner

Woody-based prostrate herb, forming dense mats, to 15cm, glabrous. Leaves tufted below, whorled above, narrowly spathulate, 1.5–4 x 0.2–0.6cm. Flowers in few-flowered heads; peduncles swollen below the heads. Sepals 3–4mm long, with a narrow midrib. Petals 1/2 to ≤ length of sepals, margin entire. Capsule ovoid.
Ecology. Rocky slopes in dwarf shrubland (Soqotra and Samha) and low succulent shrubland (Abd al Kuri). 250–600m.
Distribution. Endemic.
Conservation status. Vulnerable (VU D2). Rare, with a very limited area of occupancy.
Botanical notes. Distinguished from all the other species of *Polycarpaea* in the Archipelago by its peduncles which are characteristically swollen beneath the flowering heads. In exposed places or when it has been intensively grazed it forms dense cushions, the short stems arising from a woody, plate-like basal rootstock.

ETHNOGRAPHIC NOTES
Name(s)
– Soqotri: nil or *ṣəberhen*, *ṣabirhen*, *ṣabirbiher* 'sour, astringent'; *ṣefənitən* 'plaited'; *gəfiʾhin* 'prostrate, bowed'; *di śfaʾhanitin* 'spreading out'; *səgenhen* 'curved over' (Samha).
– Soqotri simplified: *ziberhen*, *zabirhen*, *zebirbiher*; *zefinitin*; *gifi'hin*; *di shfa'hanitin*; *sigenhen*.
Food (livestock). On the outer islands this is regarded as a valuable fodder for goats and sheep.

Polycarpaea paulayana Wagner

Annual herb to 45cm, glabrous. Leaves whorled, broadly spathulate, <6 x 1.5cm; petiole <4cm. Flowers in dense heads made up of 2 to 4 spikes. Bracts scarious with a narrow brown midrib, about 1/2 length of the sepals. Sepals longer than capsules, apex rounded. Petals ± equalling capsule length.
Ecology. Succulent shrubland (Soqotra) and low shrubland dominated by *Atriplex griffithii* (Samha). 50–500m.
Distribution. Endemic.
Conservation status. Vulnerable (VU D2). Rare, with a very limited area of occupancy.

Botanical notes. A delicately stemmed prostrate annual with narrowly spathulate stem leaves.

ETHNOGRAPHIC NOTES

Name(s)

– Soqotri: nil or *ḥʌmż* 'sour, lemony'.
– Soqotri simplified: *hamdh.*

Food (livestock). Both sheep and goats on Samha like this plant, and it is seen as especially good for lactating females.

Polycarpaea spicata **Wight ex Arn.**

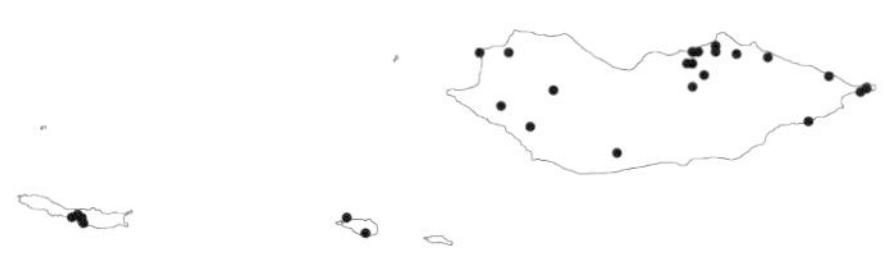

Erect annual herb to 15cm, 2 or 3x branched above, glabrous. Basal leaves in a rosette; cauline leaves whorled, narrowly linear or narrowly obovate to spathulate, <3 x 0.7cm, glaucous. Flowers in dense heads made up of 2 to 4 spikes. Bracts white or brown. Sepals c.3mm long; midrib on outer sepals broad or narrow, greenish brown. Petals oblong, c.1.3mm, apex bidentate. Capsule narrowly ovoid.
Ecology. Dwarf desertic shrubland, *Croton socotranus* shrubland, succulent shrubland and rarely in semi-deciduous woodland. Sea-level–550m.
Distribution. Endemic.

Conservation status. Least Concern (LC). Common in several vegetation types and under no present or perceived threat.
Botanical notes. All plants on Soqotra are referable to var. *capillaris* Balf.f. Var. *capillaris* differs from var. *spicata* in several significant respects and should probably be given independent specific status.

ETHNOGRAPHIC NOTES

Name(s)

– Soqotri: nil or *ḳʌ́lḳəhʌl* 'swaying in the breeze' (general term for a variety of mainly halophytic plants); *məṣírbihān* 'of the winter rains'; *ṭaháḳhen* 'smooth' (Samha).
– Soqotri simplified: *qalqihal*; *mizirbihan*; *tahaq'hen.*

Food (livestock). Grazed by small stock. It is not seen as being important nutritionally.

Silene L.

Herbs. Leaves opposite; stipules absent. Inflorescence a lax terminal cyme. Calyx tubular, 5-toothed. Petals 5, deeply bilobed, corona lobes present. Stamens 10. Styles 3(to 5), free. Fruit a many-seeded capsule, usually dehiscing by 6 teeth. Petals, stamens and ovaries borne on a stalk (carpophore) that usually elongates in fruit. 500 species. N temperate regions.

Silene apetala **Willd.**

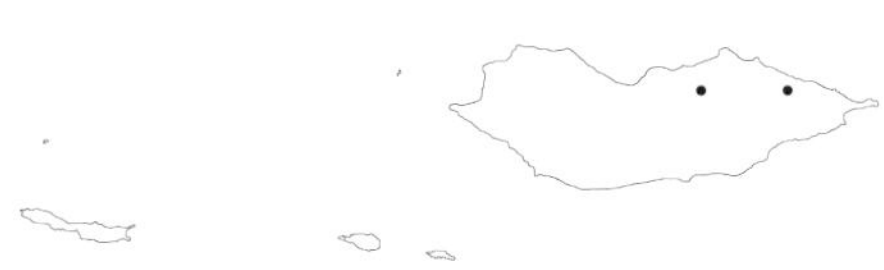

Erect annual herb to 60cm, minutely pubescent. Leaves elliptic to oblong, <4cm long, basal petiolate, cauline sessile, clasping at base. Flowers in 3 to 15(or 25)-flowered dichasial cymes. Calyx 10-ribbed, 8–10mm long, cylindrical in flower, inflating in fruit; teeth 2–3mm. Petals white, greenish or pink, 6–8mm long. Capsule 4–6.5mm diameter; carpophore 1–3mm long.
Ecology. Rare in dwarf evergreen shrubland in the Haggeher and semi-deciduous woodland on the eastern limestone plateau. 500–1100m.
Distribution. Mediterranean and SW Asia.

ETHNOGRAPHIC NOTES

Name(s)

– Soqotri: *lə́bənhən* 'white, pale'.
– Soqotri simplified: *libinhin.*

Food (livestock). Provides incidental grazing for livestock, but is not seen as important forage.

Comparative uses. Young plants of other *Silene* species are eaten as potherbs. In India the roots and leaves of *S. griffithii* are mixed with natural, impure carbonate of soda and used as a soap substitute: many species contain saponins. Medicinally, some species are used mainly to treat eye conditions and as an emollient. (Ambasta 1982; Jain & De Fillips 1991; Kunkel 1984)

Spergula L.

Annual herbs. Leaves whorled, linear; stipules scarious, free. Flowers in lax, terminal, dichasial cymes. Sepals 5, free. Petals 5, white, entire. Stamens usually 10. Styles 3, free. Fruit a many-seeded capsule, splitting into 3 valves. Seeds winged. 5 species. Temperate regions, especially Europe (4 species).

Spergula fallax (Lowe) Krause

Prostrate herb to 40cm, glabrous. Leaves linear, <3.5cm long; stipules scarious, broadly ovate, c.2mm long. Inflorescence 5 to 10-flowered; pedicels 5–17mm long. Bracts scarious, 1–1.5mm long. Sepals lanceolate, 4–5mm long, with a scarious margin. Petals c.4mm long. Capsule subglobose, 4–5mm diameter.
Ecology. Grassland clearlings in woodland, often near habitation. 550–700m.
Distribution. Canary Is., N Africa and SW Asia to India.

ETHNOGRAPHIC NOTES
Name(s)
– Soqotri: nil or (rare) *ʾibihiṭ* (more commonly given to *Orobanche* species).
– Soqotri simplified: *'ibihit.*
Food (livestock). Provides incidental grazing for all livestock, but is not seen as important forage.
Comparative uses. In Arabia *S. marginata* is used to make Syrian 'snakeroot'. In Europe and South Africa *S. arvensis* is cultivated as a fodder plant, and is recommended as a green manure and soil conservator. Its seeds yield a fatty oil and are used in bread in times of famine. Young plants are also eaten. (Ambasta 1982; Jain & De Fillips 1991; Kunkel 1984; Mossa *et al.* 1987)

Sphaerocoma T.Anderson

Small shrubs. Leaves usually fascicled, fleshy; stipules very small. Inflorescences long-pedunculate, subtended by scarious bracts. Sepals 5, free, unequal, with a broad scarious and fimbriate margin and a conspicuous deflexed apical awn. Petals 5. Stamens 5. Style solitary, obscurely 2-lobed at the apex. Fruit 1-seeded, indehiscent. 2 species. NE Sudan, S Arabia, Iran.

Sphaerocoma hookeri T.Anderson

Densely branched shrub to 60cm, glabrous or rarely with a few hairs on the young growth; older wood grey with persistent leaf bases at the nodes. Leaves linear, <2cm long. Inflorescence c.10mm across with numerous sterile flowers, enlarging and becoming spherical and spinescent in fruit; peduncles 1.2–3cm. Sepals 2.5–3mm long, the deflexed apical awns increasing to 6.5mm in fruit. Petals white. Fruit globose, shorter than the sepals.
Ecology. Dwarf shrubland and low woody-based herb communities in Samha. 50–690m.
Distribution. Somalia, Sudan, Egypt and mainland Yemen.
Botanical notes. A distinctive dwarf shrub with linear, fleshy leaves and the flowers arranged in stalked heads composed of many scarious awns each tipped with a deflexed awn.

ETHNOGRAPHIC NOTES
Name(s)
– Soqotri: *záhwə.*
– Soqotri simplified: *zahwih.*
Food (livestock). On Samha this is eaten by both sheep and goats, but is not seen as important forage.

CELASTRACEAE

Armed trees. Leaves simple, alternate; stipules present. Flowers in cymes, small, usually functionally unisexual, the male flowers with pistillodes, the female with staminodes. Sepals 5. Petals 5. Disc present. Stamens 5, attached to base of disc. Ovary superior, partially enclosed by disc, 3-celled, 2 ovules per cell; placentation axile; style 1, very short; stigmatic lobes 2 or 3. Fruit a 3-lobed capsule, dehiscing longitudinally. Seeds glossy, with a fleshy aril. 94 genera. 1300 species. Tropics and temperate areas.

Maytenus

See description of family. 1 endemic. 225 species. Tropical and warmer regions.

Maytenus sp. *A*

Tree to 3m. Branches light grey, with sparse spines, 0.5–1cm long. Leaves clustered on short shoots, obovate to orbicular, 0.8–2 x 0.5–1.5cm, glabrous, base acuminate, tip obtuse or retuse; venation prominent on both surfaces; margin entire; petiole short, c.1.5mm; stipules caducous. Flowers fasciculate; pedicel 0.8–1.2cm, articulated. Capsule obconic, 3-angled, red, glabrous to sparsely hairy, c.0.8 x 1cm. Seeds 6, brown, c.6mm long; aril fleshy, attached to basal part of seed.

Ecology. Rare in relict semi-deciduous woodland on limestone cliffs. c.580m.

Distribution. Endemic.

Conservation status. Vulnerable (VU D2). Known only from a single gathering made in the relict woodland at Qatariyah where it was found on sheltered cliffs which capture cloud and drizzle during the monsoon. Little is known of this rare tree – it seems under no immediate threat but, due to its extreme rarity and limited distribution, it has been placed in a threatened category.

Botanical notes. The fasciculate flowers in the inflorescence are unusual in *Maytenus* in this area although the same structure is also found in *M. undata*, a species found in Yemen, Ethiopia and Somalia and widespread in tropical and southern Africa. However, the small leaves with entire margins arranged on short spur shoots and the presence of spines distinguishes the taxon on Soqotra from *M. undata*. *M. arbutifolius* has rather similar size and shape of leaves but they are usually serrate and the inflorescence is clearly branched. In addition *M. arbutifolius* is a much more spinose plant.

ETHNOGRAPHIC NOTES

Name(s)

– Soqotri: *məḳtɔ̄nəh* (also given to *Carissa spinarum*).
– Soqotri simplified: *miqtownih*.

Food (human). The fruit are not eaten by people.

Food (livestock). Not generally palatable to livestock, though islanders report that camels are not deterred by the vicious thorns and occasionally browse it.

Notes. Rare and little known and no uses have been recorded for it.

Comparative uses. In Dhofar, *M. dhofarensis* and *M. senegalensis* are rarely eaten by livestock other than camels and the fruit are toxic to goats. The spiny branches are used in constructing livestock pens and enclosures. Many *Maytenus* species are important in traditional medicine, and one species of Celestraceae is the well-known *Catha edulis*, or *qat*, whose fresh leaves are chewed for their mildly narcotic effect. (Miller & Morris 1988)

CHENOPODIACEAE

Herbs or shrubs, rarely small trees; stems often jointed and/or succulent. Leaves alternate, rarely opposite, simple, often succulent, sometimes reduced and scale-like; stipules absent. Flowers small, often green, usually in axillary cymes, or in panicles, usually regular. Perianth of (1 to)5 free tepals, sometimes connate at the base or absent. Stamens (1 to)5. Ovary superior, 1-celled, ovules 1. Fruit usually a nutlet, often enclosed by the persistent perianth or bracteoles; fruiting perianth often winged. 120 genera. 1300 species. Cosmopolitan, especially desert and semi-desert regions.

Anabasis L.

Herbs, subshrubs or shrubs; stems articulate, often succulent. Leaves opposite, fleshy, connate and amplexicaul, often reduced to scales. Flowers bisexual and female, subtended by 2 bracteoles, in axillary spikes; perianth segments 5, free, 3 or 5 of them developing a transverse wing on the back in fruit. Stamens 5; staminodes 5, shorter than and alternating with the stamens; style short; stigmas 2. Fruit a utricle included within the winged perianth; seeds vertical. 30 species. Mediterranean, Middle East, C Asia.

Anabasis ehrendbergii Schweinf. ex Boiss.

Spreading or prostrate shrub to 75cm, often rooting at the nodes. Young leaves cylindrical, <0.5 x 0.1cm; mature leaves overlapping, obovoid, <0.3 x 0.2cm. Flower spikes 1.5–3cm long; bracteoles fleshy; perianth segments 2.5mm long, scarious; anthers exserted, bright crimson; stigmas purple.

Ecology. Open dwarf shrubland on rocky slopes on desertic coastal plains on Abd al Kuri. Sea-level–30m.

Distribution. Arabian Peninsula, Sudan, Ethiopia and Somalia.

Botanical notes. Readily distinguished by the overlapping and opposite scale-like leaves.

ETHNOGRAPHIC NOTES

Name(s)

– Soqotri: nil or *ḳʌ́lḳəhʌl* (pl. *ḳəlʌ́ḳəhʌl*) 'swaying in the breeze' (general term for a variety of mainly halophytic plants).

– Soqotri simplified: *qalqihal* (pl. *qilaqihal*).

Food (livestock). Provides grazing for small stock and donkeys on Abd al Kuri where it is regarded as useful dry season forage.

Flavouring / food additive. Saline forage can affect the taste of the milk of animals feeding heavily on it.

Other. Small stock feeding heavily on it need more frequent watering.

Notes. A period spent grazing halophytic rangeland at least once a year is considered essential for the health of small stock and camels especially. Such feed initially brings on diarrhoea, which is seen as being beneficial and 'cleansing impurities from the blood'. When recovered from this, the animals begin to put on weight.

Comparative uses. *A. setifera* exudes a kind of edible manna. (Kunkel 1984)

Arthrocnemum Moq.

Subshrubs. Leaves inconspicuous, fused in opposite pairs forming a sheath around the stem which appears segmented. Flowers sessile, in 3-flowered clusters immersed in spike-like terminal segments. Stamens 2. Fruiting perianth spongy. 10 species. Cosmopolitan on coasts.

Arthrocnemum macrostachyum (Moric.) Moris & Delponte

Arthrocnemum glaucum Ung.-Sternb.

Succulent, erect subshrub; segments of sterile parts of stems cylindrical, <8mm long. Terminal spikes 2–6cm. Flowers slightly exceeding the subtending bracts. Seeds 1mm long, black, glossy, papillose.

Ecology. An uncommon but probably under-recorded salt marsh plant. On Soqotra, locally common in the succulent shrubland on the seaward side of the mangrove woodland at Nit on Bandar Shu'ub; also recorded from dwarf shrubland on desertic plains on Abd al Kuri. Sea-level–10m.

Distribution. Coasts of N Africa, E Mediterranean and SW Asia to the Indus Delta.

ETHNOGRAPHIC NOTES

Name(s)

– Soqotri: nil or *kúlḳəhʌl* (pl. *ḳəlʌ́ḳəhʌl*) 'swaying in the breeze' (general term for a variety of mainly halophytic plants).

– Soqotri simplified: *qalqihal* (pl. *qilaqihal*).

Food (livestock). Provides a useful and important forage in the drier regions of Soqotra.

Flavouring / food additive. Saline forage can affect the taste of the milk of animals feeding heavily on it.

Cleansing. Succulent plants like this are crushed and used to wash the skin or wipe things clean, especially in fishing communities.

Other. Animals feeding on this can go longer without water.

Notes. A period spent grazing halophytic rangeland at least once a year is considered essential for the health of small stock and camels especially. Such feed initially brings on diarrhoea, which is seen as being beneficial and 'cleansing impurities from the blood'. When recovered from this, the animals begin to put on weight.

Comparative uses. *A. fruticosum* is eaten in salads. The fleshy parts of *A. glaucum* are eaten, as are the young shoots of *A. indicum* when pickled. The seeds of *A. subterminale* are ground and eaten. (Kunkel 1984)

Atriplex L.

Herbs or shrubs. Leaves alternate, flattened, mealy. Flowers in terminal, leafless panicles. Male flowers with 5 tepals and 5 stamens. Female flowers without tepals but with 2 bracteoles which enlarge and conceal the fruit. c.100 species. Temperate and warm regions, salt tolerant and often found on coasts.

ETHNOGRAPHIC NOTES

Comparative uses. All *Atriplex* species are palatable to livestock, some having a high fodder value – some 10 species are cultivated as fodder crops. The salt content of leaves increases with the aridity of habitat, making it less palatable. Grazing heavily on these plants increases thirst, and an excess of salt reduces the rate of food assimilation and may provoke diarrhoea, loss of weight and fainting. The seeds of many related species are eaten and ground to flour, and their leaves are eaten cooked and raw. In Bahrain and Oman the seeds of *A. halimus* are boiled in water and taken as an emetic. In India the leaves of *A. hortensis* are cooked like spinach, and eaten in salads and soups. Its leaves and seeds are rich in proteins. The seed flour helps in vitamin A deficiency, but must be used with caution due to the existence of certain toxic substances. (Abbas *et al.* 1992; Ambasta 1982; Baumer 1983; Ghazanfar 1994; Goodin & Northington 1979; le Houerou 1980; Jain & De Fillips 1991; Kunkel 1984)

Atriplex farinosa Forssk.

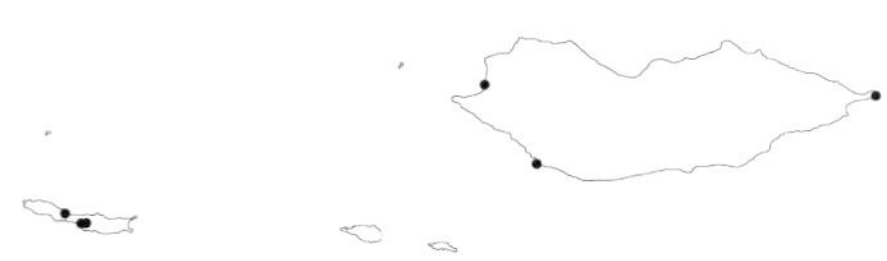

Shrub or woody-based herb to 1m, densely mealy on all parts, whitish-grey. Leaves whitish grey, petiolate, ovate-elliptic, 1.5–4 x 0.8–2.2cm, obtuse at tip, auriculate or cordate at the base, entire. Fruiting bracteoles variable in size and shape, obconical to broadly elliptic, 2–4 x 2–3mm, entire, connate at base.
Ecology. Coastal succulent shrubland often on sand dunes. Sea-level–40m.
Distribution. Coasts of NE Africa, Egypt, W and S Arabia and Jordan.
Botanical notes. All Soqotran material is referable to subsp. *farinosa*.

ETHNOGRAPHIC NOTES
Name(s)
– Soqotri: nil or *ḥātəm* 'coastal area bare of other vegetation' (Abd al Kuri); *ʿifirəh* 'having an unpleasant smell', 'red' (plants growing near the sea are commonly red-brown) (Samha).
– Soqotri simplified: *hatim*; *'ifirih.*

Food (livestock). On Abd al Kuri, and in other poorly vegetated areas, this is a key fodder for sheep and goats. On Abd al Kuri, it is also grazed by donkeys.
Flavouring / food additive. Saline forage can affect the taste of the milk of animals feeding heavily on it.
Fuel. Dried stems are gathered and used as firewood.
Cordage. On Abd al Kuri the outer and inner bark are used to make cordage.
Wood. Where other woods are in short supply, as on Abd al Kuri, the plant is used to make charcoal. Small dried pieces are also collected and stored for use as tinder.
Medicinal / veterinary. On Abd al Kuri its charcoal is ground to a fine powder and used topically to treat swelling and inflammation.
Pest control. Produces a dense and acrid smoke that is an effective insecticide.
Other. Small stock feeding heavily on it need more frequent watering.
Notes. A period spent grazing halophytic rangeland at least once a year is considered essential for the health of small stock and camels especially. Such feed initially brings on diarrhoea which is seen as being beneficial and 'cleansing impurities from the blood'. When recovered from this, the animals begin to put on weight.

Atriplex griffithii Moq.

Shrub to 1m. Leaves greenish grey, petiolate, broadly ovate to almost orbicular, 0.4–2 x 0.4–1.6cm, retuse or rounded at the tip, rounded to cuneate at the base, entire or undulate. Fruiting bracteoles broadly triangular-ovate to cordate, 3.5–5 x 3.5–5mm, reticulate on both sides, entire or with 2 small basal teeth.
Ecology. Local on Soqotra and Abd al Kuri but common on Darsa and abundant on Samha where it dominates in the shrubland at the foot of the limestone escarpment. Sea-level–50m on Soqotra, Darsa and Abd al Kuri but up to 500m on Samha.
Distribution. Coasts of Somalia, S Arabia and Pakistan.
Botanical notes. On Samha this plant is seen to having steadily increased. It forms a dense, almost impenetrable, belt below the limestone escarpment. All Soqotran material is referable to subsp. *stocksii* (Boiss.) Boulos.

ETHNOGRAPHIC NOTES
Name(s)
– Soqotri: *ʿifirə* 'having an unpleasant smell', 'red' (plants growing near the sea are commonly red-brown) (Samha, Abd al Kuri and Soqotra).
– Soqotri simplified: *'ifirih.*

Food (livestock). In the drier coastal areas this is a key fodder for livestock. On the Noged plain, camels and cattle also eat the plant and it is important dry season forage for them. In periods of drought herders bring their animals into the areas where these plants grow. It is seen as improving milk yield in camels and as a nutritious and fattening fodder in general, even in the dry season. On the outer islands it is an important fodder for goats and sheep, especially in the dry season. It stays green longer than other plants on the outer islands. When green, it is particularly liked by sheep.
Flavouring / food additive. Saline forage can affect the taste of the milk of animals feeding heavily on it.
Fuel. On the outer islands dead material is used as firewood. It produces a bright flame and not too much smoke. Stores of it are often kept in the home for lighting.

Cordage. The outer and inner bark are used to make cordage on the outer islands.
Wood. Where other woods are in short supply, as on Abd al Kuri, the plant is used to make charcoal. Small dried pieces of the plant are also collected and stored for use as tinder. It lights easily and is thus especially important in wet weather.
Other. On Samha where most islanders liked to smoke or chew tobacco when they could, the ash of this plant was used to stretch the small amount of tobacco they were able to import from Soqotra. The ash also made the tobacco much more pungent.
Notes. A period spent grazing halophytic rangeland at least once a year is considered essential for the health of small stock and camels especially. Such feed initially brings on diarrhoea, which is seen as being beneficial and 'cleansing impurities from the blood'. When recovered from this, the animals begin to put on weight.

Chenopodium L.

Herbs. Leaves alternate, flattened, often mealy (covered in white powder). Flowers in clusters arranged in leafy panicles, bisexual or female, sessile. Tepals 5, green, united, persistent but not much altered or enlarged in fruit. Stamens 5. Stigmas 2 (rarely 3 to 5), simple (rarely 2-lobed). Fruit depressed-globular, with a thin membranous pericarp. 150 species. Mostly temperate.

Chenopodium murale L.

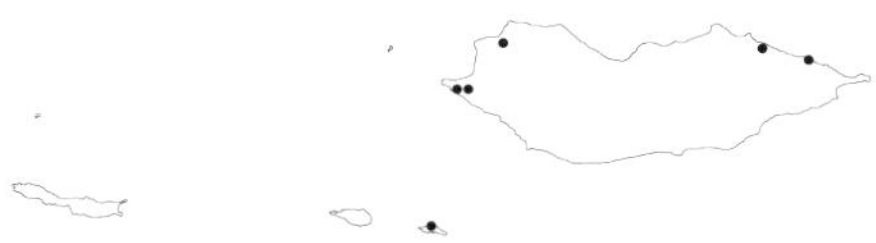

Erect annual herb to 60cm, juvenile parts mealy, otherwise glabrous. Leaves dark green, rather fleshy, ovate-rhombic or deltoid, less often lanceolate, 1.5–10 x 1–7cm, coarsely toothed. Tepals 5, fused half way, keeled on back. Seeds horizontal, black, glossy, strongly keeled, 1.2–1.5mm diameter.
Ecology. A common and widespread weed found near caves and habitation. Under-recorded on the distribution map. Sea-level–600m.
Distribution. Cosmopolitan.

ETHNOGRAPHIC NOTES
Name(s)
– Soqotri: *ʿa̍žərib* 'quick to die back'; *(di) ri̍žbeb* 'stalky'.
– Soqotri simplified: *'ajirib*; *(di) rijbeb*.
Food (human). In times of hunger, this can be gathered from around the settlement and cooked and eaten like *Amaranthus*.
Food (livestock). Like all plants of the *ʿa̍žərib/ri̍žbeb* group it is readily eaten by all livestock when it comes up after the rains and before the grasses appear. At this time it is considered to be nutritious feed.
Notes. Can cause *'uqabah* in goats unaccustomed to the feed.
Comparative uses. The leaves of many species are eaten, both cooked and raw. The fodder value of the leaf is high but it also has laxative properties. *C. ambrosioides* yields an essential oil used as a powerful anthelmintic against many internal parasites, among them tapeworms, roundworms, hookworms and intestinal amoebae. By 1980 this plant had already appeared 50 times or more in the NAPRALERT computer database of medicinal plants at the Department of Pharmacognosy and Pharmacology, University of Illinois. In Iraq the related *C. album* is used as a diuretic, aperient, anthelmintic, hepatic and mild sedative. In Qatar an extract of the same plant is used as dusting powder to allay irritation of external genitalia in children. The powdered herb is mixed with food to suppress the oestrous cycle, and the plant is also considered to be useful in treating biliousness, abdominal pains, eye diseases, throat troubles, piles and diseases of the blood, heart and spleen. It is seen to possess aphrodisiac and tonic qualities and is also used as an enema for intestinal ulceration. This plant has been reported to possess contraceptive properties. (Ambasta 1982; Batanouny 1981; Bebawi & Neugebohrn 1991; Goodin & Northington 1979; Jain & De Fillips 1991; Kunkel 1984; Rizk & El-Ghazaly 1995)

Cornulaca Delile

Shrubs. Leaves alternate, sessile, amplexicaul, often decurrent, with tufts of white woolly hairs in the axils. Flowers bisexual, axillary, surrounded by dense tufts of white hair and subtended by a leaf-like bract and 2 bracteoles; tepals 5, hyaline and free at the base, becoming indurated and fused in fruit; stamens 2. 7 species. Egypt to C Asia.

Cornulaca ehrenbergii Asch.

Sprawling shrub to 0.35 x 2m. Leaves leathery, broadly triangular-ovate, spine-tipped, base 4–6mm long, clasping and completely concealing young stems, free part 2–3mm long. Flowers solitary; tepals linear-oblong, 2–2.5mm long.
Ecology. An uncommon but probably under-recorded salt marsh plant. On Soqotra locally common in succulent shrubland on coastal sand dunes by the mangrove woodland at Nit on Bandar Shu'ub. Sea-level–10m.
Distribution. Egypt, Sudan, Ethiopia, Somalia and west coast of Arabia.

ETHNOGRAPHIC NOTES
Name(s)
– Soqotri: nil or *ḳulḳəhʌl* (pl. *ḳəlʌ̍ḳəhʌl*) 'swaying in the breeze' (general term for a variety of mainly halophytic plants).
– Soqotri simplified: *qalqihal* (pl. *qilaqihal*).
Food (livestock). Provides incidental grazing for small stock and camels in the better-vegetated areas, but is useful forage in drier regions.
Flavouring / food additive. Saline forage can affect the taste of the milk of animals feeding heavily on it.
Notes. A period spent grazing halophytic rangeland at least once a year is considered essential for the health of small stock and camels especially. Such feed initially brings on diarrhoea, which is seen as being beneficial and 'cleansing impurities from the blood'. When recovered from this, the animals begin to put on weight.

Lagenantha Chiov.

Similar to *Sevada* Moq., but the mature perianth develops a prominent transverse wing around the top of the tube, wing entire or divided into lobes with radial veins. 1 species.

Lagenantha cycloptera (Stapf) M.G.Gilbert & Friis

Choriptera semhahensis (Vierh.) Botsch.
Salsola cycloptera Stapf
Salsola semhahensis Vierh.

Low glabrous shrub to 40cm; stems whitish, older stems with cracking brownish bark. Leaves subglobose to narrowly cylindrical, <2 x 0.3cm, usually curved. Flowers forming short lateral spikes. Fruiting perianth enlarged, whitish to straw-yellow, wings 5–7mm diameter; fruit with a distinct thickened rim; seeds horizontal.

Ecology. Common on Abd al Kuri; in dwarf shrubland on the coastal plains and also in the tangled shrubbery which often colonises stabilised coastal sand dunes. Also known from Darsa but not so far recorded from Soqotra or Samha. Sea-level–30m.

Distribution. Somalia.

Botanical notes. The flowers are inconspicuous but in fruit the five perianth lobes develop into a conspicuous membranous wing which is initially straw-coloured but turns bright pink.

ETHNOGRAPHIC NOTES

Name(s)

– Soqotri: *ḥaṭāmhìn* 'coastal area bare of other vegetation' (Abd al Kuri).
– Soqotri simplified: *hatemhin.*

Food (livestock). On Abd al Kuri this is a key fodder for sheep and goats. It is also grazed by donkeys.

Flavouring / food additive. Saline forage can affect the taste of the milk of animals feeding heavily on it.

Fuel. Dried material is gathered and used as tinder and firewood, especially for cooking.

Salsola L.

Subshrubs. Leaves small, often succulent, alternate, sessile, entire, the base clasping or not. Flowers bisexual or less often unisexual, solitary in the leaf axils, each subtended by 2 bracts. Tepals 5, persistent and concealing the fruit, developing horizontal, subequal wings. 150 species. Cosmopolitan, coasts and salty places.

Salsola spinescens Moq.

Salsola forskalii Schweinf.

Rigid shrub to 1.5m, older stems and branches strongly spinescent; young shoots bud-like, minute, formed of overlapping leaves. Leaves ovate, 1.5–2 x 1.5mm, rounded at the tip, scarious on the margins, densely covered with adpressed medifixed hairs when young. Fruiting perianth 3.5–5mm diameter (including wings), wings pinkish white.

Ecology. Locally common in dwarf shrubland and in succulent shrubland on rocky wadi sides on the coastal plains. 20–250m.

Distribution. Sudan, Djibouti, Ethiopia and W and S Arabia.

Botanical notes. Unlike most other shrubby Chenopodiaceae on Soqotra not associated with coastal or brackish soils.

ETHNOGRAPHIC NOTES

Name(s)

– Soqotri: *ḥʌ̀ṭəhīm*, *ḥìṭəhʌm* 'coastal area bare of other vegetation'; *xērṭʌh* (western Soqotra) (also given to *Justicia rigida*).
– Soqotri simplified: *hatihiym*, *hitiham*; *kheyrtah.*

Food (livestock). Important fodder for small stock in drier, poorly vegetated areas, and, in the absence of grass, is especially important for sheep. It is a good browse for camels accustomed to it, and even the cattle of the coastal plains eat the plant. It is said to improve milk yield in camels, and, when green, to be a nutritious and fattening feed.

Flavouring / food additive. Saline forage can affect the taste of the milk of animals feeding heavily on it.

Fuel. Dried material is gathered as tinder and kindling.

Other. Whole plants are used in the construction of the roofing of living quarters built from date palm material. A large bunch is tied to the pointed tip of the conical roof and the tips of the palm leaf thatch are then fed into it to hold them in place.

Notes. A period spent grazing halophytic rangeland at least once a year is considered essential for the health of small stock and camels especially. Such feed initially brings on diarrhoea, which is seen as being beneficial and 'cleansing impurities from the blood'. When recovered from this, the animals begin to put on weight.

Comparative uses. The young leaves of several species are eaten as a famine food. In India *S. baryosma* is cut as a camel fodder, and the ash is used to treat itch. A type of manna is also obtained from the leaves. *S. kali* is used in glass manufacture and, medicinally, is used as an anthelmintic, emmenagogue, diuretic and cathartic. This plant has also been grown to reduce soil salinity. In Qatar and Bahrain the flowers of *S. baryosma* are used as a diuretic and anti-inflammatory. In Dhofar *S. imbricata* was burned and the ash used as a snuff. Many species of *Salsola* and *Suaeda* are rich in soda, and have been used since earliest times in the manufacture of both glass and soap. A mixture of sodium and potassium carbonate is obtained by slow oxidation of the plants burned to ashes in a pit. (Abbas *et al.* 1992; Ambasta 1982; Batanouny 1981; Jain & De Fillips 1991; Kunkel 1984; Miller & Morris 1988; Rizk & El-Ghazaly 1995)

Seidlitzia Bunge ex Boiss.

Shrubs. Leaves opposite, cylindrical, succulent. Flowers bisexual, in few-flowered axillary clusters; floral bracts 2; tepals 5, shortly fused at the base; stamens 5, exserted. Fruiting perianth persistent and concealing the fruit, developing unequal wings. 8 species. Macaronesia, Mediterranean to C Asia and Iran.

Seidlitzia rosmarinus Bunge ex Boiss.

Shrub to 1m, with opposite branches; stems much branched, the younger white, glossy. Leaves cylindrical, <3.5cm long, thickening towards the tip and tapering towards the clasping and decurrent base, often curved, with a tuft of white hairs in the axils. Fruit 1–1.2cm diameter (including wings).

Ecology. No habitat details available.
Distribution. Deserts from Egypt and SW Asia east to Afghanistan and Tadzhikistan.
Botanical notes. Known only from a single specimen from Abd al Kuri collected in the 1960s. Rather surprisingly not seen since – should be searched for, probably on the coastal plains.

ETHNOGRAPHIC NOTES

Not seen, no information recorded.

Sevada Moq.

Shrubs. Leaves alternate or opposite, often in axillary clusters, succulent, terete, with a short, hardened base. Flowers bisexual and female, in dense axillary and terminal spikes or panicles; bracteoles small, scarious. Perianth segments 5, fused into a short tube, the lobes incurved; stamens 5, with exserted anthers, alternating with small staminodal scales; ovary free; style tapered; stigmas 2 or 3, short, recurved. Fruit ovoid with acute beak; seeds vertical. 1 species.

Sevada schimperi Moq.

Densely branching shrub to 40cm, glabrous or minutely papillose; stems woody, older stems with fissured bark. Leaves linear, <1.5cm long, yellowish brown, with an axillary tuft of short white hairs. Flowers in clusters of 3 to 10; perianth segments subglobose, c.1mm; anthers yellow. Fruiting perianth c.2mm, tube longitudinally ribbed, fruit ovoid, with an acute beak.
Ecology. An uncommon but possibly under-recorded plant on Soqotra, locally common in succulent shrubland on coastal sand dunes by the mangrove woodland at Nit on Bandar Shu'ub; on Samha found in open dwarf shrubland dominated by *Atriplex griffithii* on desertic foothills. Sea-level–100m.
Distribution. Saudi Arabia, Yemen, Oman, Egypt and E tropical Africa.

ETHNOGRAPHIC NOTES
Name(s)
– Soqotri: nil or *ḳʌ́lḳəhʌl* (pl. *ḳəlʌ́ḳəhʌl*) 'swaying in the breeze' (general term for a variety of mainly halophytic plants).
– Soqotri simplified: *qalqihal* (pl. *qilaqihal*).
Food (livestock). Provides incidental grazing for small stock and camels in the better-vegetated areas, but is a useful forage in drier regions.
Flavouring / food additive. Saline forage can affect the taste of the milk of animals feeding heavily on it.
Other. Animals feeding on this can go longer without water.
Notes. A period spent grazing halophytic rangeland at least once a year is considered essential for the health of small stock and camels especially. Such feed initially brings on diarrhoea, which is seen as being beneficial and 'cleansing impurities from the blood'. When recovered from this, the animals begin to put on weight.

Suaeda Forssk. ex Scop.

Shrubs or rarely small trees. Leaves alternate, succulent, linear to subglobose, entire. Flowers bisexual or unisexual, solitary or in axillary or terminal clusters, sometimes forming panicles; bracts 2 or 3, scarious, persistent. Stamens 5. Tepals 5, persistent, partly concealing the fruit. 110 species. Cosmopolitan, coasts and salty places.

Suaeda monoica Forssk. ex J.F.Gmel.

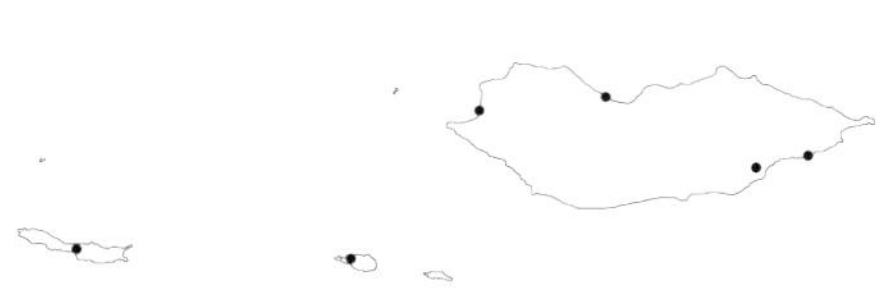

Small, erect tree or shrub to 4m; stems much branched, glabrous. Leaves sessile or shortly petiolate, flattened on both sides, linear to linear-oblong, 1–4 x 0.1–0.4cm, obtuse to subacute. Flowers unisexual, greenish, with a large (c.2.75mm diameter) male flower surrounded by several smaller female ones.
Ecology. Locally abundant and sometimes dominant in succulent shrubland on coastal sand dunes (for instance by the mangrove woodlands at Nit and Ghubbah); on Samha found in shrubland dominated by *Atriplex griffithii*. Under-recorded on the distribution map. Sea-level–500m.

Distribution. Cape Verde Is., Chad, E Africa from Mozambique north to Sudan, Egypt and Arabia.

ETHNOGRAPHIC NOTES

Name(s)

– Soqotri: *ḳʌ̍lḳəhʌl* (pl. *ḳəlʌ̍ḳəhʌl*) 'swaying in the breeze' (general term for a variety of mainly halophytic plants); *ḥāṭəm* (Abd al Kuri); *ḥʌ̍ṭəhʌm* 'coastal area bare of other vegetation' (Samha).

– Soqotri simplified: *qalqihal* (pl. *qilaqihal*); *hatim*; *hatiham.*

Food (livestock). Provides incidental grazing for small stock in the better-vegetated areas, but is a good forage in drier regions, and especially on the outer islands where it is particularly valuable for sheep. On Soqotra it is an important camel fodder.

Flavouring / food additive. Saline forage can affect the taste of the milk of animals feeding heavily on it.

Fuel. Dead material is gathered as kindling. In the absence of alternative material, on the outer islands dried material is used as fuel, especially for cooking.

Cleansing. Succulent plants such as this are crushed and used to wash the skin or wipe things clean, especially in fishing communities.

Other. Animals feeding on this can go longer without water.

Notes. A period spent grazing halophytic rangeland at least once a year is considered essential for the health of small stock and camels especially. Such feed initially brings on diarrhoea, which is seen as being beneficial and 'cleansing impurities from the blood'. When recovered from this, the animals begin to put on weight.

Comparative uses. The leaves and seeds of many species are eaten. In India this plant is considered to be a good fodder for camels, sheep and goats and the plant is also grown to reclaim saline soils. In Yemen the stems are mixed with henna to obtain a black dye, and in Africa the roots are used to treat sore throats. The woolly excrescences on the tips of the branches of *S. fruticosa* are used as an oily medium to cure sores on camel backs, and an aqueous infusion is given as an emetic. The plant is also recommended for treating waterlogged and saline soils. In Oman the stems and leaves of *S. aegyptiaca* are boiled in water and used as a gargle for gum and tooth infections. They are also burned and used as a snuff to treat giddiness, hysteria and nausea, as well as to calm the nervous system and improve poor vision. (Ambasta 1982; Fleurentin & Pelt/Fleurentin *et al.* [1982, 1983a,b,c]; Ghazanfar 1994; le Houerou 1980; Jain & De Fillips 1991; Kunkel 1984; Miller & Morris 1988)

Suaeda vermiculata **Forssk. ex J.F.Gmel.**

Suaeda fruticosa Forssk. ex J.F.Gmel.

Suaeda paulayana Vierh.

Shrub to 1.5m; stems much branched, lower usually sprawling, upper spreading, young branches minutely puberulent becoming glabrous. Leaves sessile or shortly petiolate, obovate-oblong to subglobose, terete or curved and flattened on one side only, <2(–3) x 0.4cm, usually drying black. Flowers bisexual, similar in size, usually in shortly pedunculate axillary 2 to 5-flowered clusters, sometimes solitary.

Ecology. Locally abundant in coastal salt marsh. Under-recorded on the distribution map. Sea-level.

Distribution. N Africa and SW Asia east to India and south to Kenya.

ETHNOGRAPHIC NOTES

Name(s)

– Soqotri: *ḳʌ̍lḳəhʌl* (pl. *ḳəlʌ̍ḳəhʌl*) 'swaying in the breeze' (general term for a variety of mainly halophytic plants); *ḥāṭe̍mhen* 'little *ḥāṭəm*' (Abd al Kuri); *ḥʌ̍ṭəhim* 'coastal area bare of other vegetation' (Samha).

– Soqotri simplified: *qalqihal* (pl. *qilaqihal*); *hatemhen*; *hatihim.*

Food (livestock). Provides incidental grazing for small stock in the better-vegetated areas, but is a useful forage in drier regions, and especially on the outer islands where it is particularly valuable for sheep. On Soqotra it is an important camel fodder.

Flavouring / food additive. Saline forage can affect the taste of the milk of animals feeding heavily on it.

Fuel. Dead material is gathered as kindling. In the absence of alternative material, as on the outer islands, dried material is gathered and used as fuel, especially for cooking

Cleansing. Succulent plants such as this are crushed and used to wash the skin or wipe things clean, especially in fishing communities.

Other. Animals feeding on this can go longer without water.

Notes. A period spent grazing halophytic rangeland at least once a year is considered essential for the health of small stock and camels especially. Such feed initially brings on diarrhoea, which is seen as being beneficial and 'cleansing impurities from the blood'. When recovered from this, the animals begin to put on weight.

Comparative uses. In Bahrain and Oman the green parts are burned and the smoke inhaled to relieve breathing difficulties and asthma. The plant contains alkaloids, coumarins and sterols. (Abbas *et al.* 1992; Ghazanfar 1994; Goodin & Northington 1979)

COMPOSITAE

Herbs, shrubs or trees, sometimes succulent or spinose, sometimes with latex. Leaves opposite or alternate, simple, sometimes pinnate; stipules absent. Inflorescence made up of clusters of flowers aggregated into heads (capitula) which sit on a pad of expanded tissue or receptacle, naked or sometimes with scales or hairs. Receptacle surrounded by 1 to several rows of bracts or phyllaries; green or scarious and papery. Flowers reduced, unisexual or bisexual; tubular flowers with 5 small, equal lobes, ray flowers with an expanded, often 3-lobed ligule; tubular flowers occur in centre of capitulum surrounded by ray flowers, e.g. *Bidens* species, or other species have only tubular flowers, e.g. *Pulicaria stephanocarpa*, or only ray flowers, e.g. *Launaea* species. Calyx a pappus of hairs or scales (sometimes absent) which is persistent on the fruit and helps in dispersal. Ovary inferior, ovules 1. Fruit a 1-seeded achene. 1314 genera. 21,000 species. Cosmopolitan; particularly diverse in drier regions.

Acanthospermum Schrank

Herbs. Leaves opposite, entire. Capitula axillary; florets yellow. Fruit spiny, formed from the enlarged bracts of the outer (female) florets, enclosing the achene. 6 species. Tropical Americas.

Acanthospermum hispidum DC.

Erect or sprawling, coarsely hairy, woody herb. Leaves subsessile, elliptic, 1.5–3cm long, minutely gland-dotted. Fruit 4 or 5 per capitulum, with hooked spines, the 2 apical spines much the longest.

Ecology. *Croton socotranus* shrubland. 50m.

Distribution. Native of the Neotropics, now a widespread weed.

ETHNOGRAPHIC NOTES

Name(s)

– Soqotri: *ʿáẑərib śirśíroh* 'spiny plant quick to die back'; *sɔ̄f* 'woolly'.

– Soqotri simplified: *'ajirib shirshiroh*; *sowf.*

Food (livestock). This is one of the *ʿáẑərib* group of plants which come up quickly after the winter rains. They are grazed by cattle before the grasses come up, but are less liked by goats and sheep. This plant is seen as good grazing for cows, but of little nutritional value for small stock. The spiny fruit can cause damage to their mouths, and such minor scratches, if they become infected, can cause an animal to lose its appetite. Camels and donkeys also eat the plant.

Comparative uses. The 'star burr' can become a serious nuisance in countries to which it has been accidentally introduced. In India it has long been used in dermatological medicaments and to treat leprosy. The essential oil of the herb has shown anti-bacterial and anti-fungal activity. (Ambasta 1982; Dalziel 1948, 1956; Jain & De Fillips 1991)

Achyrocline (Less.) DC.

Similar to *Helichrysum* but separated because it has more female than hermaphrodite flowers in the capitulum. c.20 species. Tropical America and Africa.

Achyrocline luzuloides Vatke

Helichrysum glumaceum DC.
Helichrysum luzuloides (Vatke) Lanza

Slender, erect, woody-based herb to 30cm. Leaves widely spaced or somewhat crowded, white hairy beneath, linear-elliptic, 1–3.5 x 0.1–0.5cm, entire. Capitula in subglobose clusters, 1–2cm diameter; phyllaries white above, acute; flowers white.

Ecology. Grassland and clearings in semi-deciduous woodland. 300–1100m.

Distribution. Widespread in Africa and S Arabia.

ETHNOGRAPHIC NOTES†

Name(s)

– Soqotri: nil or *líbinhan* 'white, pale'; *raḥámhʌm* 'grey-coloured'.

– Soqotri simplified: *libinhan*; *rahamham.*

Food (livestock). Grazed by all livestock when hungry, but is not regarded as nutritionally important.

Comparative uses. In Ethiopia the roots of *A. schimperi* are used to expel intestinal parasites. (Dalziel 1948, 1956)

†Also apply to *A. schimperi*.

***Achyrocline schimperi* Schultz**
Helichrysum schimperi (Sch.Bip.) Moeser

Erect woody-based herb to 60cm. Leaves widely spaced, white hairy beneath, ovate, 1–4.5 x 0.5–1.5cm, entire or crenate. Capitula in umbellate clusters, 2.5–6cm diameter; phyllaries pale straw-coloured, translucent, acute; flowers white.
Ecology. In the Haggeher mountains. c.900–1000m.
Distribution. Widespread in Africa and S Arabia.
Botanical notes. Not seen recently on Soqotra.

ETHNOGRAPHIC NOTES
Name(s)
See *A. luzuloides* (also uses).

Ageratum L.

Herbs. Leaves opposite, simple, glandular. Capitula clustered; phyllaries in 2 rows, green with scarious margin; tubular flowers only, white (rarely mauve). Achenes columnar, 4 or 5-angled. Pappus of 5 scales. 44 species. C and S America, Caribbean.

***Ageratum conyzoides* L.**

Erect annual herb to 50cm. Pappus ± length of achene.
Ecology. A locally abundant weed of clearings and tracksides. 100–900m.
Distribution. An introduced weed from America – now cosmopolitan.

ETHNOGRAPHIC NOTES
Name(s)
– Soqotri: *ᶜaˈžərib* (pl. *ᶜaˈžərub)* 'quick to die back'; *di ᶜaˈšbib* (pl. *di ᶜašbub*, *di ᶜašˈebeb*) 'wilting'; *di-rihoˈytin* 'water-liking'; *ṣefirˈirihon di gɛˈḥi* 'many-flowered plant of the riverbed' (more commonly given to *Exacum affine*).
– Soqotri simplified: *'ajirib* (pl. *'ajirub*); *di 'ashbib* (pl. *di 'ashbub*, *di 'ashebeb*); *di rihoytin*; *zefiririhon di gehi.*
Food (livestock). This is one of the *ᶜaˈžərib* group of plants which come up quickly after the winter rains. They are grazed by cattle before the grasses come up, but are less liked by goats and sheep. This plant is seen as good for cows, increasing their milk yield, but as of little nutritional value for small stock. Camels and donkeys also eat it.
Notes. Goats which eat a lot of this immediately after the rains often develop scour. It can also cause *'uqabah* in small stock unaccustomed to the feed.
Comparative uses. *Ageratum* species are often toxic, containing cyanogenic glycosides. *A. conyzoides* contains hydrocyanic acid, coumarin and an alkaloid and is said to be poisonous to rabbits and cows. It has a distinctive smell, and is called the 'billy-goat weed' in Australia. It has been shown to contain an anti-bacterial and to have anti-convulsant properties.
Leaves. The leaves are an important remedy for craw-craw. They are also applied to chronic ulcers and used intra-vaginally for uterine troubles. Crushed in water they are given as an emetic. A decoction is used as skin lotion for ulcers and is taken internally for fever. The leaf juice is used as an eye medicine and in Nepal as nasal drops to treat catarrh. In West Africa the leaves are used to heal wounds, especially burns. The leaf juice is squeezed into a wound which is then covered by a bruised leaf: untreated leaves of the plant have proved significantly superior to vaseline gauze as a wound dressing. In Nigeria the chest of a patient with pneumonia is rubbed with leaves. Women lie on a bed of leaves during delivery in Uganda. Elsewhere the leaves are used as a herb tea for overheating, flu, coughs, pneumonia, cystitis, dysmenorrhoea and menorrhagia and as an abortifacient.
Root. The root is used to treat diabetes, coughs, cystitis, flu, abdominal problems, digestion, colic, ague, a prolapsed uterus, syphilis and ulcers. (Batanouny 1981; Dalziel 1948, 1956; Heine *et al.* 1988; Lewis & Elwin-Lewis 1977; Oliver-Bever 1986; Rizk & El-Ghazaly 1995; Sofowora 1982)

Bidens L.

Herbs. Leaves opposite, pinnately lobed. Capitula solitary; phyllaries in 2 rows, outer herbaceous, inner with scarious margin; tubular flowers yellow; ray flowers white or yellow (sometimes absent). Pappus of few, retrorsely barbed bristles. 233 species. Cosmopolitan, especially Mexico.

Botanical notes. The two species here are naturalised weeds (from N America). It is not certain whether both of these closely related species are present; both have been recorded but only plants with white ray flowers have been observed by the authors.

***Bidens biternata* (Lour.) Merr. & Sherff**
Bidens abyssinica Sch.Bip.

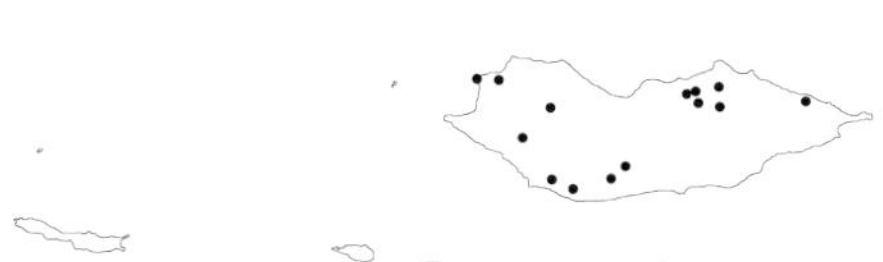

Erect annual to 50cm. Ray flowers yellow. Outer phyllaries longer than inner. Outer achenes shorter than inner.
Ecology. Locally abundant on disturbed ground and particularly near settlements and livestock pens, etc. 150–1300m.
Distribution. Pantropical weed.
Botanical notes. The commonest *Bidens* on Soqotra. One of the first plants to come up after rain.

ETHNOGRAPHIC NOTES†
Name(s)
– Soqotri: *mədə̄whīn* (pl. *mədə̄wihon*) (given to many

plants, including grasses, with spiky or bristly seed parts); *ᶜaẑərib* 'quick to die back'; *di ᶜalimlihon, di ᶜalimləhum* 'striped, marked' (the leaves have three parallel 'stripes' along their length).
– Soqotri simplified: *midewhiyn* (pl. *midewihon*); *'ajirib*; *di 'alimlihon, di 'alimlihum.*

Food (livestock). Grazed by all livestock when immature, and before other better herbs and grasses respond to the rains, and is considered to be a valuable feed. As it matures, however, it becomes less palatable and is regarded as having little nutritional value. The rod-like seeds of both species stick in clothing and livestock hair. Cattle, camels and donkeys graze it more than sheep and goats. Small stock, in particular, avoid the plant in its mature state as the bristly seeds hurt their eyes and mouth, and cause their coat to become matted. Shepherds are not keen on their sheep frequenting areas where these plants are common as their wool becomes very tangled and difficult to wash and shear. The seed also causes scratches which can become infected and cause a serious problem, especially in the udder area. The teats can become inflamed and lactation is adversely affected. Infected scratches in the mouth can also lead to loss of appetite. Interestingly, the seeds of plants from wetter areas are said to be much more vicious than those from drier areas.

Bees / honey. In the Haggeher mountains the plant is considered to be a good bee plant.

Other. The paddocks and pens of young livestock are often cleared of this damaging plant. Milking yards are also cleared of it once it sets seed.

Notes. When the plants begin to set seed many herders trim the long hair from the flanks and udder area of small stock to lessen damage to the skin. It was traditional for herders of the wetter mountain areas to pull up the seeding plants and pile them into a heap and burn them, especially those growing luxuriantly in the rich earth of the pens and small paddocks. This is done both to protect the animals and also to encourage the growth of grasses which are smothered by these plants.

Comparative uses. The leaves are cooked and eaten like spinach and the tips of young plants are made into a tea. The leaves of many other related species are also edible. The leaves of *B. pilosa* have a strong UV-mediated toxicity to bacteria and *Candida*. They are also toxic to insects and larvae, and in India they are fed to horses with intestinal parasites. The plant is regarded as a tonic and stimulant, and is used in leprosy and other skin troubles, and to treat fistulae and tumours. A decoction is used for prickly heat and as a diuretic and febrifuge. An infusion of the leaves is given as a sudorific, and the leaf juice is used for eye and ear troubles. The flowers are used in diarrhoea and the seeds as an anthelmintic. An infusion of the roots is given for colic. In East Africa, the twigs are boiled in water and the water drunk to treat malaria. The Borana tie the plant to doors and gates to deter evil in times of war, disease or when the rains fail. (Ambasta 1982; Dalziel 1948, 1956; Heine *et al.* 1988; Jain & De Fillips 1991; Kunkel 1984; Oliver-Bever 1986)

†Also apply to *B. pilosa*.

***Bidens pilosa* L.**

Erect annual to 50cm. Ray flowers white or absent. Outer phyllaries not longer than inner. Achenes uniform size.

Ecology. Rare in grassland clearings in montane semi-deciduous woodland. 900–1350m.

Distribution. Pantropical weed.

Botanical notes. Much rarer than the preceding species

ETHNOGRAPHIC NOTES

Name(s)

See *B. biternata* (also uses).

Blainvillea Cass.

Herbs. Leaves opposite, simple. Capitula clustered; phyllaries in 1 row, herbaceous; tubular flowers only. Pappus of 2 to 4 short bristles. Achenes columnar, 3 or 4-angled. 10 species. Pantropical.

***Blainvillea gayana* Cass.**

Blainvillea dichotoma (Murray) Luetzelb.

Blainvillea rhomboidea Cass.

Erect annual to 40cm. Leaves ovate, 2–10 x 1–5cm, dentate. Flowers yellow.

Ecology. No habitat details available. 100–500m.

Distribution. Pantropical weed.

Botanical notes. Not recorded recently on Soqotra.

ETHNOGRAPHIC NOTES

Not seen, no information recorded.

Conyza Less.

Herbs. Leaves simple, alternate. Capitula loosely clustered, conical; phyllaries in 2 rows, green with scarious margins; flowers yellow, tubular. Pappus of hairs. c.60 species. Tropical and subtropical regions.

Conyza hochstetteri Sch.Bip.

Erect, perennial herb to 30cm. Leaves narrowly elliptic, 2–5 x 0.5–1.5cm, dentate, auriculate at base. Capitula 0.5–1cm diameter. Flowers yellow. Pappus creamy yellow.

Ecology. Rare on Soqotra; in semi-deciduous montane woodland. c.1050m.

Distribution. Yemen, Saudi Arabia.

Botanical notes. Not recorded recently on Soqotra.

ETHNOGRAPHIC NOTES

Comparative uses. In Oman the leaves of *C. incana* are heated and applied locally to relieve muscle and joint pain, and the fumes of burning branches are used to repel insects. Some related species are toxic to small stock, especially sheep, but are also used medicinally, especially to treat internal complaints. (Abdulfatih 1987; Batanouny 1981; Ghazanfar 1994; Lewis & Elwin-Lewis 1977; Rizk & El-Ghazaly 1995)

Cyanthillium Blume

Botanical notes. For description see *Vernonia*.

Cyanthillium cinereum (L.) H.Rob.

Vernonia cinerea Less.

Annual herb to 60cm. Leaves ovate, truncate, shortly petiolate, 2–4 x 0.8–1.5cm, entire, sessile. Capitula few, in loose clusters, c.0.5cm diameter; phyllaries all acute. Pappus white.

Ecology. No details of location or habitat available. Rare or possibly under-recorded on Soqotra; a widespread weed in surrounding regions.

Distribution. Pantropical weed.

Botanical notes. See Robinson (1999).

ETHNOGRAPHIC NOTES

Not seen, no information recorded.

Dichrocephala L'Hér. ex DC.

Herbs. Leaves alternate, simple. Capitula solitary, globose; flowers yellow, tubular. Pappus absent. 4 species. Tropical and S Africa, Madagascar and SW Asia.

Dichrocephala chrysanthemifolia (Blume) DC.

Dichrocephala chrysanthemifolia var. *chrysanthemifolia* (Blume) DC.

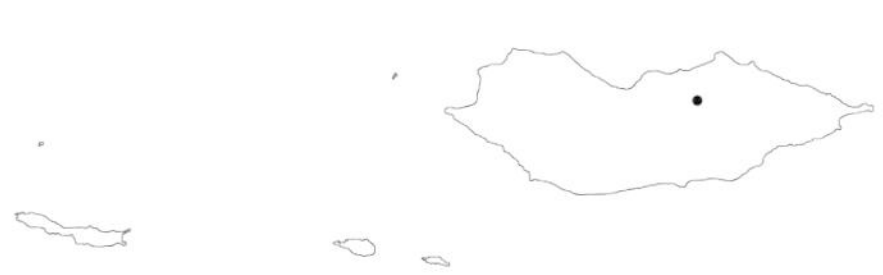

Erect, pubescent annual to 50cm. Leaves elliptic or obovate, 2–5 x 1–2cm, coarsely dentate, auriculate at base. Capitula 0.5–1cm diameter.

Ecology. No habitat details available but probably in semi-deciduous montane woodland. c.1200m.

Distribution. Yemen, Saudi Arabia.

Botanical notes. Rare or possibly under-recorded on Soqotra; widespread in surrounding regions.

ETHNOGRAPHIC NOTES

Comparative uses. The roots of *D. integrifolia* are eaten both raw and cooked. In Kenya *D. chrysanthemifolia* is poisonous to sheep, producing symptoms similar to those of ergot poisoning. (Kunkel 1984; Lewis & Elwin-Lewis 1977)

Dicoma Cass.

Woody, spinescent herbs. Leaves alternate, simple, tomentose. Capitula sessile, axillary; phyllaries long spine-tipped, straw-coloured; flowers tubular. Achenes covered with dense silky hairs. 1 endemic. c.65 species. Mainly S and tropical Africa and Madagascar with some species in NE Africa and Asia.

Dicoma cana Balf.f.

Clump-forming herb to 15cm. Leaves linear, <1cm long, entire. Flowers red. Pappus plumose.

Ecology. Locally abundant in open dwarf shrubland on the drier parts of the limestone plateaus. 300–700m.

Distribution. Endemic.

Conservation status. Least Concern (LC). Common in several vegetation types and under no present or perceived threat.

Botanical notes. The only densely spiny clump-forming perennial on Soqotra. Both leaves and phyllaries (the bracts around the flowers) are stiff and very sharp.

ETHNOGRAPHIC NOTES

Name(s)

– Soqotri: *ḥʌ̍mḥīm, ḥamaḥə̍mo* (pl. *ḥʌ̍mḥo̍m, ḥama̍ḥam*) 'breeding male goat or ram to become sexually aroused' (eating young plants is very good for them) (also given to *Campylanthus spinosus*, *Justicia rigida* and *Neuracanthus aculeatus*); *di rʌ̍xmʌm* 'grey-coloured' (western Soqotra); *di śkaʿani̍tin* 'spiny, thorny'.

– Soqotri simplified: *hamhiym*, *hamahamo* (pl. *hamhom*, *hamaham*); *di rakhmam*; *di shka'anitin.*

Food (livestock). Goats like the new shoots and younger plants. The hard spines of mature plants can damage their mouths and cause sores. This can make it difficult for the goat to eat and as a result it loses weight, so herders are not keen on their goats eating it at this mature stage. Camels and donkeys particularly like it and sheep nibble at the tender new growth. However, if camels eat too much they develop a weakness in the back legs which can affect them permanently. Islanders report that in some areas it has been wiped out by a period of severe drought.

Fuel. Dried stems are gathered and stored for use as tinder and kindling.

Dicoma tomentosa Cass.

Erect, branched herb to 30cm. Leaves oblanceolate, <5cm long, serrate. Flowers pale creamy yellow. Pappus papery.

Ecology. An uncommon weed. 50–100m.

Distribution. Yemen and Saudi Arabia.

ETHNOGRAPHIC NOTES

Name(s)

– Soqotri: nil or *ʿa̍žərib śerśīroh* 'spiny plant quick to die back'.

– Soqotri simplified: *'ajirib shershiroh.*

Food (livestock). Provides incidental grazing for goats but is not important nutritionally. Camels and donkeys also eat the plant and sheep nibble the leaves.

Comparative uses. In India the root is used as a febrifuge and in parts of Africa it is applied locally to putrescent wounds. The root is also used as a toothstick. The leaves of other *Dicoma* species are eaten. (Ambasta 1982; Dalziel 1948, 1956; Goodin & Northington 1979; Jain & De Fillips 1991; Kunkel 1984)

Distephanus

Herbs. Leaves in rosettes, simple, pinnately veined. Capitula solitary; phyllaries dark green with a narrow scarious margin, narrowly ovate, acute; flowers yellow, tubular. Pappus 1 row, plumose. Achenes with dense appressed hairs. 1 endemic. c.40 species. Mainly in Madagascar, also tropical Africa and SE Asia.

Distephanus qazmi N.Kilian & A.G.Mill.

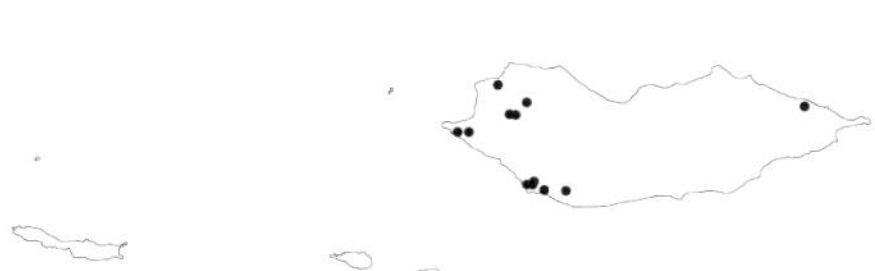

Tiny, rosette-forming herb to 3cm with thick, woody taproot. Leaves narrowly oblanceolate, ≤2.8 x 0.6cm, shallowly lobed, fleshy. Each rosette with 1 or few 1–2cm long peduncles. Capitula 1–1.3cm diameter.

Ecology. Locally common in open dwarf shrubland and cushion vegetation in exposed and windswept places on the limestone plateaus of W and NE Soqotra where it is restricted to thin soils and to crevices in limestone pavement. 330–660m.

Distribution. Endemic.

Conservation status. Least Concern (LC). Common in several vegetation types and under no present or perceived threat.

Botanical notes. This insignificant but remarkable endemic is a member of *Distephanus*, an Old World genus centred on Madagascar. It possesses several attributes unique in the genus including a low, rosette-forming habit, fleshy leaves with distinctive venation and pappose hairs arranged in a single row (Kilian & Miller 2000). Several Soqotran endemics are more or less restricted to the specialised niche occupied by *D. qazmi* including

Melhania milleri, *Lotus mollis*, *Farsetia inconspicua*, *Echiochilon pulvinata* and *Launaea crepoides*.

ETHNOGRAPHIC NOTES
Name(s)
– Soqotri: nil or *gērdə*, *gˈired* 'small-leaved'.
– Soqotri simplified: *geyrdih*, *gired.*

Food (livestock). Provides incidental grazing for small stock. It is considered a good feed for weaned lambs and kids.

Eclipta L.

Herbs. Leaves opposite, simple. Capitula few in loose clusters; phyllaries in 2 rows, herbaceous; flowers white, ray and tubular. Pappus absent. Achenes columnar. 4 species. Pantropical.

Eclipta prostrata (L.) L.
Eclipta alba (L.) Hassk.

Sprawling annual herb to 30cm. Leaves narrowly elliptic to ovate, 2–3 x 0.5–1cm, roughly hairy.
Ecology. Damp sandy and gravelly soils by permanent water, locally common in suitable habitats and typically associated with *Phyla nodiflora*. 10–700m.
Distribution. Widespread weed.
Botanical notes. A tiny white-flowered daisy characteristic of damp soils by permanent water. Under-recorded on the distribution map.

ETHNOGRAPHIC NOTES
Name(s)
– Soqotri: nil or *ruˈʿud* 'grazing plant'; *di rihˈoytin* 'of the water'.
– Soqotri simplified: *ru'ud*; *di rihoytin.*

Food (livestock). Grazed after rains, before the grasses come up, principally by cattle. Goats generally avoid it near water as they intensely dislike damp grazing. However, they eat the plant as it dries out and other livestock also eat it when they are hungry.
Notes. If a hungry goat eats much of this it can develop scour.
Comparative uses. *E. prostrata* is widely used medicinally as a tonic and deobstruent. It is used in the treatment of hepatic and spleen enlargements and in skin troubles. It is the source of a black stain, which enters into preparations for darkening the hair, and in Sudan, for tattooing and making beauty spots. The root is emetic and purgative, and is also applied to wounds in livestock. The young shoots are eaten as a vegetable. The leaves have strong UV-mediated toxicity to bacteria and *Candida* and are also toxic to insects and larvae. (Ambasta 1982; Batanouny 1981; Dalziel 1948, 1956; Jain & De Fillips 1991; Lewis & Elwin-Lewis 1977; Oliver-Bever 1986; Rizk & El-Ghazaly 1995)

Euryops (Cass.) Cass.

Shrubs. Leaves alternate, linear, terminating in 2 or 3 narrow lobes. Diamond-shaped leaf scars prominent. Capitula solitary on naked peduncles; flowers yellow, ray and tubular. Pappus many fine bristles, deciduous. 97 species. Africa and Arabia.

Euryops arabicus Steud.
Euryops pinnifolius A.Rich.
Euryops socotranus Balf.f.

Shrub to 1m. Leaves crowded, 5–6 x 0.3–0.5cm. Capitula 0.8–1cm diameter.
Ecology. Abundant and sometimes dominant in semi-deciduous submontane shrubland in the Haggeher mountains. 500–1350m.
Distribution. Yemen, Oman, Saudi Arabia.
Botanical notes. Plants from Soqotra consistently differ from plants on the nearby African and Arabian mainlands in having two or three-lobed (not entire) leaves; in all other ways they are identical. Because of this difference plants from Soqotra have sometimes been recognised as a separate (endemic) species, *E. socotranus* Balf.f. However, rarely plants are found on the mainland which also possess lobed leaves, and so, failing any other differences, we prefer to treat this as one, widespread species. *E. arabicus* is considered to be on the increase in the eastern Haggeher and Diksam. Because it is unpalatable to livestock and grass does not tend to grow in areas where it spreads, this could be interpreted as an indication of overgrazing in these areas. The situation warrants further study and should be monitored.

ETHNOGRAPHIC NOTES
Name(s)
– Soqotri: *bəṭˈiʾhin* 'slow, idle', 'prostrate'; (rare) *fəṭˈiʾhin.*
– Soqotri simplified: *biti'hin*; *fiti'hin.*
Food (livestock). Generally unpalatable to livestock, though in a prolonged dry season they have been observed to eat the flowers. In a severe drought herders cut it and as it dries out very hungry goats eat it.
Fuel. In the areas where it grows the dead wood is regarded as being good firewood, burning with a bright light and producing little smoke. It has a very distinctive smell and islanders say that they can always tell someone from the Central Highlands by the smell of this smoke on their person. Because of this strong smell it is little used for cooking.
Wood. The de-barked dry stems make excellent firelighters and good kindling, and these are collected in the dry summer months and stored in the home for use during the damp monsoon months and rains of the winter.

Pest control. The stems are burned in the Haggeher in livestock smudges – elsewhere the strong smell of the smoke is generally disliked.
Other. In the Central Highlands young children make dolls with the stems as a framework over which they arrange scraps of clothing. They also use the leaves to make a fringe for the female doll. Islanders have observed that the leaves put in water turn it a dark red-brown like tea leaves. However, there is no tradition of its having been used on the island as a dye.

Notes. Herders generally root it up from grazing paddocks and livestock pens since it prevents grasses from growing.

Comparative uses. In Saudi Arabia the heated leaf and stem sap of *E. arabicus* is used to heal wounds in the feet. In Yemen this is used to protect against the Evil Eye. (Fleurentin & Pelt/Fleurentin *et al.* [1982, 1983a,b,c]; Mossa *et al.* 1987)

Helichrysum Mill.

Herbs or shrubs. Leaves alternate, simple. Capitula terminal, solitary or clustered; phyllaries variable but always scarious, petaloid or glossy, never green, sometimes translucent; flowers white, rarely pink or lilac, tubular, outer sometimes zygomorphic. Pappus 1 row of bristles. 12 endemics. c.600 species. Widespread in Africa (especially S), Asia and Europe.

Botanical notes. A further plant of which only inadequate specimens are known may also be referable to *Helichrysum*. It is a mat-forming perennial with silvery leaves and is locally common at 500–550m on north-facing cliffs and broken rocky slopes on Samha, seeming to prefer areas that attract drizzle and low cloud. Only very old capitula have been seen and more material is required before its affinities can be assessed

---

Helichrysum aciculare Balf.f.

Cushion-forming woody herb to 15cm. Leaves in dense rosettes, dark green above, white hairy beneath, needle-shaped, 1–2.5 x 0.1cm, revolute. Capitula in dense, conical clusters, 0.5–1cm diameter, on non-leafy peduncles, 2–7cm; phyllaries golden-brown in neatly overlapping rows, obtuse, margins translucent; flowers white.
Ecology. A common cushion plant of rock faces on the granite pinnacles of the Haggeher mountains. 600–1300m.
Distribution. Endemic.
Conservation status. Least Concern (LC). Common in several vegetation types and under no present or perceived threat.
Botanical notes. Readily recognised by its rigid, sharp-tipped, linear leaves and the flowering heads held erect on long stalks. It is one of the characteristic, and most common, cushion plants of the Haggeher where it frequently grows with *Hypericum scopulorum* and *Helichrysum rosulatum*, the only other cushion-forming species of *Helichrysum* on Soqotra. There is a puzzling 19th century record of this species from the northeast of Soqotra (Homhil) in Forbes (1903). This refers to fruits collected by the expedition. However, no material can be traced and this is a highly unlikely area for this species to occur. Nevertheless, it has been included on the distribution map.

ETHNOGRAPHIC NOTES
Name(s)
– Soqotri: *dǭḳəhīn* (commonly given to certain lichens and to the mosses of the high Haggeher); *di ni'śʌz* 'solitary'; *śkaᶜini'tin* 'prickly, thorny'.
– Soqotri simplified: *dowqihiyn*; *di nishaz*; *shka'initin.*
Food (livestock). Said to be eaten by feral and wild goats when the dry season is prolonged and there are few alternatives.
Notes. Although this plant shares its name *di nishaz* with *H. rosulatum*, it does not have the same sweet smell.

Helichrysum arachnoides Balf.f.

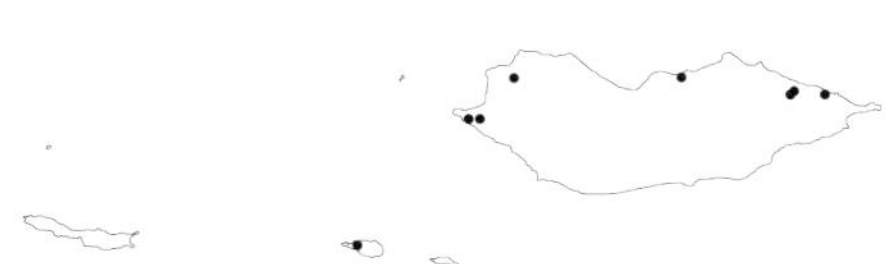

Erect or somewhat sprawling delicate herb to 10cm sparsely clothed in long, woolly hairs. Leaves evenly spaced, grey-green, broadly spathulate, 0.3–1.5 x 0.2–0.7cm, entire. Capitula single or in loose conical clusters, 0.5–1cm diameter; outer phyllaries straw-coloured, obtuse, inner phyllaries white, petaloid; flowers white.
Ecology. Typically growing in shade in grassy clearings and at the foot of cliffs in semi-deciduous woodland. 200–750m.
Distribution. Endemic.
Conservation status. Least Concern (LC). Common in several vegetation types and under no present or perceived threat.
Botanical notes. The only annual species of *Helichrysum* on Soqotra clothed in long woolly hairs.

ETHNOGRAPHIC NOTES
Name(s)
– Soqotri: nil (Soqotra and Samha); *ləse'kek* 'adhesive'.
– Soqotri simplified: *lisekek.*
Food (livestock). Provides incidental grazing for small stock, but is not important on Soqotra. On Samha it is considered to be useful forage for sheep and goats.

Helichrysum balfourii **Vierh.**
Helichrysum gracilipes var. *stoloniferum* Balf.f.

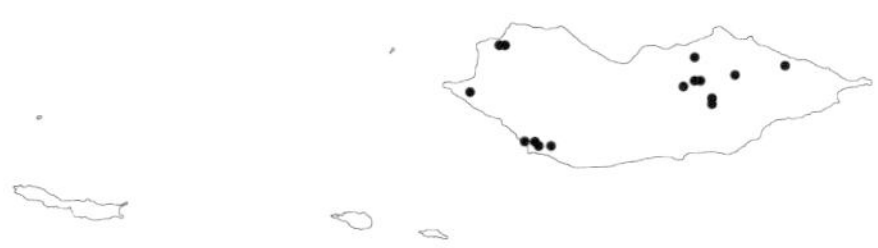

Rosette-forming herb to 15cm. Leaves grey-green, densely white hairy beneath, broadly spathulate, 1–3 x 0.4–0.7cm, entire. Capitula solitary, terminal, conical, 0.5–1cm diameter, on wiry, curving, non-leafy peduncles; phyllaries spathulate, dark below, straw-coloured above, translucent; flowers white, outer weakly zygomorphic.
Ecology. In grassy clearings and amongst rocks in semi-deciduous woodland on both the granite and limestone. 370–1067m.
Distribution. Endemic.
Conservation status. Least Concern (LC). Common in several vegetation types and under no present or perceived threat.
Botanical notes. One of the group of *Helichrysum* species on Soqotra with zygomorphic outer florets (see notes under *H. gracilipes*). Distinguished from all other helichrysums on Soqotra by its wiry flowering stalks which characteristically recurve with age.

ETHNOGRAPHIC NOTES
Name(s)
– Soqotri: nil or *kəbədino* 'little liver', 'small hillock'; *libinhan* 'white, pale'; *mərḥōmhəm* 'grey-coloured'; *di ṣefiririhun* 'many-flowered'.
– Soqotri simplified: *kibidino*; *libinhan*; *mirhowmhim*; *di zefiririhun.*
Food (livestock). Provides incidental grazing for small stock, but is not important forage on Soqotra.

Helichrysum gracilipes **Oliv. & Hiern**
Helichrysum gracilipes var. *gracilipes* Oliv. & Hiern
Helichrysum gracilipes Oliv. & Hiern forma *genuinum*

Sprawling, delicate, wiry herb to 20cm. Basal leaves forming a loose rosette, grey-green, white hairy beneath, broadly spathulate, 1–5 x 0.5–2.5cm, upper leaves elliptic, entire. Capitula solitary, cylindrical, <0.5cm diameter; phyllaries straw-coloured, acute, glossy, translucent; flowers white or pink, outer strongly zygomorphic.
Ecology. Widespread in a variety of vegetation types. 30–600m.
Distribution. Somalia.
Botanical notes. Amongst the species of *Helichrysum* found on Soqotra, capitula with zygomorphic outer florets are found only in *H. gracilipes*, *H. balfourianum* and *H. paulayanum.* This unusual feature is rare elsewhere in this large and diverse genus. The zygomorphic florets are most clearly developed in *H. gracilipes* which is also distinguished by the few florets in the capitula.

ETHNOGRAPHIC NOTES
Name(s)
– Soqotri: *libininhin* 'white, pale'; *raḥamhʌm* 'grey-coloured'; *di ṣefiririhun*, *di ṣeferher* 'many-flowered'; *ḥʌlḥʌlihin* 'pale-coloured'; *mhāṭhiṭən* 'limp, flaccid' (more commonly given to fern species); *ḥʌmź* 'astringent, sour'.
– Soqotri simplified: *libininhin*; *rahamham*; *di zefiririhun*, *di zeferher*; *halhalihin*; *mhat'hitin*; *hamdh.*
Food (livestock). Provides incidental grazing for small stock, but is not important on Soqotra.

Helichrysum nimmoanum **Oliv. & Hiern**

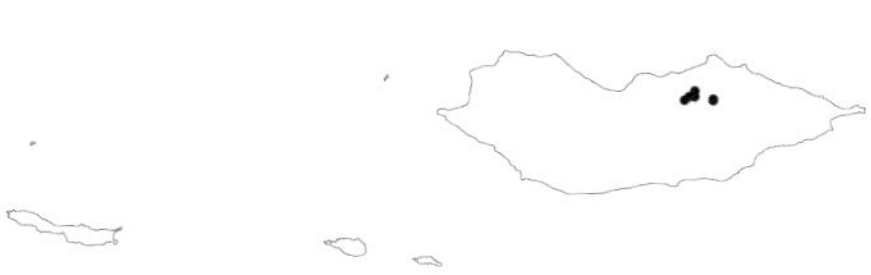

Shrub to 100cm. Leaves evenly spaced, pale green, white hairy beneath, elliptic, 1–2.5 x 0.5–1.2cm, entire. Capitula in loose clusters, 1– 4cm diameter; phyllaries straw-coloured, acute, glossy, translucent; flowers white (pinkish on some dried specimens).
Ecology. Common in the Haggeher mountains in semi-deciduous submontane woodland; particularly abundant, on sheltered ridges and in the dense thicket around the base of the granite pinnacles, in the eastern Haggeher. 900–110m.
Distribution. Endemic.
Conservation status. Vulnerable (VU D2). Rare, with a very limited area of occupancy.
Botanical notes. Similar to *H. suffruticosum* but a larger shrub with smaller, elliptic (not spathulate) leaves with entire margins. Reported to be increasing in the eastern Haggeher.

ETHNOGRAPHIC NOTES†
Name(s)
– Soqotri: *libinhan* 'white, pale'.
– Soqotri simplified: *libinhan.*
Food (livestock). This and *H. suffruticosum* are regarded as a valuable feed for all livestock, especially goats. It is one of the most important dry season fodders for the cattle of the Central Highlands. It remains in leaf when other shrubs and trees have lost their leaves, and herders gather armfuls to take to their animals at this time. Even in a drought, cows can be kept in milk if they are fed enough of this. Cattle and goats also chew the leafless twigs of accessible shrubs. Camels browse the shrub, and wild and feral goats of the Central Highlands are said to depend on it.
Fuel. Makes poor firewood.
Timber. The shrubs are little used for building, not only because they make inferior timber, but also because it is too valuable as dry season fodder.

†Uses also apply to *H. suffruticosum.*

Helichrysum paulayanum **Vierh.**

Helichrysum gracilipes var. *lanatum* Balf.f.

Similar to *H. gracilipes* but somewhat less robust; capitula ≥0.5cm diameter; phyllaries with a dark midrib; outer flowers weakly zygomorphic.
Ecology. Widespread in a variety of vegetation types. 350–1500m.
Distribution. Endemic.
Conservation status. Least Concern (LC). Common in several vegetation types and under no present or perceived threat.

ETHNOGRAPHIC NOTES
Name(s)
– Soqotri: *ḥʌlḥʌlihin* 'pale-coloured'.
– Soqotri simplified: *halhalihin*.
Food (livestock). This provides incidental grazing for small stock, but is not important on Soqotra.

Helichrysum rosulatum **Oliv. & Hiern**

Cushion-forming woody herb. Leaves in dense rosettes, white hairy beneath, narrowly elliptic, 1–2.5 x 0.2–0.4cm, entire. Capitula in dense, globose clusters, c.1.5cm diameter, on robust, leafy peduncles, 1–2cm; phyllaries pale straw-coloured, translucent; flowers white.
Ecology. A common cushion plant of rock faces on the granite pinnacles of the Haggeher mountains. 750–1300m.
Distribution. Endemic.
Conservation status. Least Concern (LC). Common in several vegetation types and under no present or perceived threat.
Botanical notes. One of the characteristic cushion plants of the Haggeher, frequently growing with *Hypericum scopulorum* and *Helichrysum aciculare*, the only other cushion-forming species of *Helichrysum* on Soqotra. *H. rosulatum* is readily recognised by its soft, narrowly elliptic leaves and the more or less sessile flowering heads. It is less common than *H. aciculare* and seems more ecologically demanding, being less widespread and not growing at such low altitudes as that species.

ETHNOGRAPHIC NOTES
Name(s)
– Soqotri: *di nīśʌz* 'solitary'.
– Soqotri simplified: *di nishaz*.
Food (livestock). Hungry goats and sheep eat the leaves if other forage is scarce, but it is not considered a useful feed for them. Cattle only graze it if they are really hungry.
Cleansing. The fragrant leaves are scattered amongst bedding and stored clothing to perfume it. They are also added to the stuffing of bolsters and pillows. Some people tuck a sprig into their clothing and pinch it as they walk along for its sweet smell. The sniffing of sweet-smelling plants is held to protect from the 'noxious air' that can cause disease, and to help protect against infection. All sweet-smelling plants are endowed with beneficial medicinal and prophylactic qualities.
Commercial value (past, present). People add leaves of *H. rosulatum* to gatherings of the grey *di ḥānə* lichen for sale overseas. It is said to be used in fumigatory mixtures on the mainland but is not used in this way on the island.
Other. The leaves are gathered, sun-dried and ground with dried *di ḥānə* lichen to make a red dye mainly used for special leather items, in particular the intricately plaited men's belt.

Helichrysum **sp. *A***

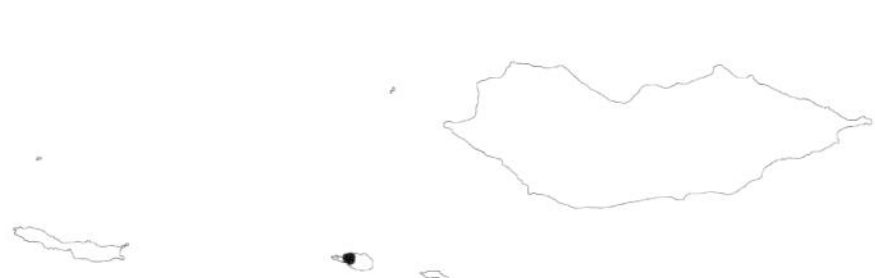

Erect (or somewhat sprawling), woody-based, aromatic herb to 30cm. Leaves evenly spaced, whitish hairy, persistent, fleshy, narrowly spathulate, 0.6–1.5 x 0.2–0.3cm, nerves radiating from base, revolute. Capitula in loose elongate clusters; phyllaries petaloid, white, broadly ovate, obtuse; flowers white.
Ecology. Cliffs and in gorges on the summit plateau on Samha; prefers areas that attract low cloud and drizzle. 500–550m.
Distribution. Endemic.
Conservation status. Vulnerable (VU D2). Rare, with a very limited area of occupancy.
Botanical notes. Woody-based herb with fleshy leaves and white flowers.

ETHNOGRAPHIC NOTES
Name(s)
– Soqotri: *tāmərhen* 'little dates' (given to many small and fragrant plants); *ləbənhən* 'white, pale'; *ṣaᶜələffən* 'many-leaved'; *ṣefiririhon* 'many-flowered'; *zəmōṭə*.
– Soqotri simplified: *tamirhen*; *libinhin*; *za'iliffin*; *zefiririhon*; *zimowtih*.
Food (livestock). Provides incidental forage for small stock especially but not considered to be nutritionally important.

Helichrysum sp. *B*

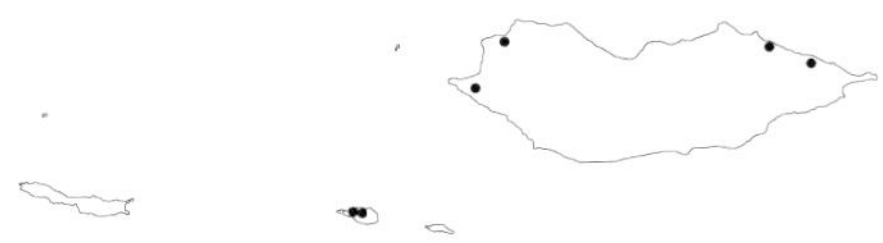

Aromatic shrub. Leaves loosely clustered, greyish-green, white hairy beneath, spathulate, 2.5–3.5 x 0.5–1cm, nerves radiating from base, entire. Capitula in dense umbellate clusters, 2–3cm diameter; phyllaries straw-coloured, acute, glossy, translucent; flowers pink or lilac.

Ecology. Hanging on limestone cliffs in shrubland. 500–800m.

Distribution. Endemic.

Conservation status. Vulnerable (VU D2). Rare, with a very limited area of occupancy.

Botanical notes. Aromatic shrub with greyish-green leaves and pink or lilac flowers. Similar to *H. suffruticosum* but differing in the narrower, more densely hairy leaves with nerves radiating from the base and entire margin.

ETHNOGRAPHIC NOTES

Name(s)

– Soqotri: *ləbə̍nhən* 'white, pale'; *tāmərhen* 'little dates' (given to many small and fragrant plants); *ṣefiri̍rihọ̍n* 'many-flowered'; *ṣaʿəlˈəffən* 'many-leaved'; *ẓʌfəri̍ʿhin* '*Plectranthus*-like'.

– Soqotri simplified: *libinhan*; *tamerhen*; *zefiririhon*; *za'iliffin*; *dhafiri'hin*.

Food (livestock). Liked especially by small stock. On Samha it is regarded as nutritious forage for sheep and goats.

Bees / honey. Honeybees are attracted to the flowers.

Helichrysum sp. *C*

Hanging shrub to 100 x 100cm. Leaves evenly spaced, whitish (densely woolly), fleshy, obovate, 1–2 x 0.5–1cm, entire. Capitula in dense, subglobose clusters, 2–2.5cm diameter; phyllaries acute, outer whitish with short, woolly hairs, inner straw-coloured, glossy; flowers pink.

Ecology. Hanging on limestone cliffs. 50–150m.

Distribution. Endemic.

Conservation status. Endangered (EN B2 a biii). Known only from a few scattered populations growing near freshwater springs and seepages on the southern limestone escarpment. There are very few of these springs and therefore the total number of plants is likely to be very low. This habitat is extremely vulnerable to water development and thus a designation of endangered seems appropriate (cf. *Exacum socotranum*, *Erythroxylum socotranum* and *Scaevola socotraensis*).

Botanical notes. Shrub with fleshy, whitish green leaves and pink flowers. Found only on vertical rock faces around springs and seepages.

ETHNOGRAPHIC NOTES

Name(s)

– Soqotri: nil or *ṣefiri̍rihon* 'many-flowered'; *li̍binhān* 'white, pale'.

– Soqotri simplified: *zefiririhon*; *libinhan.*

Food (livestock). Normally grows out of reach of livestock, but goats browse it when they can reach it. It is not considered a valuable feed for livestock.

Helichrysum sp. *D*

Sprawling herb to 30cm. Leaves evenly spaced, persistent, yellow-green, pale beneath, broadly spathulate, 1.5–3.5 x 1–1.5cm, 3-nerved, revolute, with long woolly hairs. Capitula in loose, elongate clusters; phyllaries pale green, acute with ragged, scarious margins.

Ecology. Restricted to cliffs on a windswept ridge at the western end of Soqotra. 380–530m.

Distribution. Endemic.

Conservation status. Endangered (EN B2 a biii). *Helichrysum* sp. *D* is restricted to a small refugium with a total estimated area of occupancy of less than 10km^2. At present the rapid development of the island poses no threat to this unique habitat: the area is inhospitable, inaccessible by road and there is a lack of potable water. However, its restricted distribution and the very limited extent of the niche which it occupies means that it is vulnerable to any change. It is strongly suspected that this marginal habitat is threatened by the long-term trend of lower rainfall in the region and would be particularly vulnerable to the effects of global warming.

Botanical notes. Similar to *Helichrysum* sp. *A* but distinguished by the yellow-green, non-fleshy, broadly spathulate leaves. Known only from two collections, both lacking mature flowers, from the isolated ridge at Heger (between Ras Izfer and Ras Shu'ub) in the extreme west of the island. This ridge probably captures mists from the monsoon (cf. *Nesocrambe* and *Pulicaria dioscorides*).

ETHNOGRAPHIC NOTES

Name(s)

– Soqotri: *tāmərhen* 'little dates' (given to many small and fragrant plants); *ləbə̍nhən* 'white, pale'; *ṣaʿəlˈəffən* 'many-leaved'; *ṣefiri̍rihọ̍n* 'many-flowered'; *zəmɔ̄ṭə*.

– Soqotri simplified: *tamirhen*; *libinhin*; *za'iliffin*; *zefiririhon*; *zimowteh*.

Food (livestock). Provides incidental forage for small stock but is not considered to be nutritionally important.

Helichrysum sphaerocephalum **Balf.f.**
Helichrysum sphaerocephalum var. *sarmentosum* Balf.f.

Sprawling to erect woody herb to 30cm. Leaves evenly spaced, white hairy beneath, elliptic to obovate, 1.5–5 x 0.5–1.5cm, entire. Capitula in dense, globose clusters, 1.5–3cm diameter; outer phyllaries straw-coloured, obtuse, inner white, petaloid; flowers white.
Ecology. In the Haggeher mountains locally common in semi-deciduous montane woodland and shrubland (850–1300m). In the western limestone plateau on rocky limestone slopes (650–800m) in low shrubland (with *Pulicaria stephanocarpa* and *Buxanthus pedicellatus*). 650–1300m.
Distribution. Endemic.
Conservation status. Least Concern (LC). Common in several vegetation types and under no present or perceived threat.

ETHNOGRAPHIC NOTES
Name(s)
– Soqotri: *məlbɔ̄nhan* 'white, pale'; *di śəmraḥyun* 'stalky'; *ʾəmlaḥ* 'salty'; (rare) *cheyɛyaʾ* (more commonly given to sweet-smelling plants such as *Micromeria* and *Teucrium socotrana*).
– Soqotri simplified: *milbuwnhan*; *di shimrahyun*; *'imlah*; *cheyaya'*.
Food (livestock). Eaten by most livestock when they can reach it, but is not seen as being nutritionally important.

Helichrysum suffruticosum **Balf.f.**

Aromatic subshrub. Leaves evenly spaced, pale beneath, spathulate-ovate, 1.5–5 x 0.7–2cm, pinnately nerved, crenate, softly hairy. Capitula in dense, terminal, umbellate clusters, 3–5cm diameter; phyllaries straw-coloured, acute, translucent; flowers white.
Ecology. Rare in montane semi-deciduous woodland. 1350m.
Distribution. Endemic.
Conservation status. Vulnerable (VU D2). Rare, with a very limited area of occupancy.
Botanical notes. Last seen by Alan Radcliffe-Smith in 1967.

ETHNOGRAPHIC NOTES
Name(s)
– Soqotri: *libinhan* 'white, pale'.
– Soqotri simplified: *libinhan.*
Notes. Uses as *H. nimmoanum.*

Kleinia Mill.

Succulent-stemmed shrubs. Leaves alternate, reduced, caducous. Capitula solitary, cylindrical; phyllaries in 1 row, linear, acuminate; flowers pink, tubular. Pappus white, many long bristles. 1 endemic. c.40 species. Mainly tropical Africa, also Middle East and India.

Kleinia scottii **(Balf.f.) P.Halliday**
Senecio scottii Balf.f.

Capitula 2–2.5cm long; flowers tubular.
Ecology. Occasional in semi-deciduous woodland on both the limestone and granite. 750–800m.
Distribution. Endemic.
Conservation status. Vulnerable (VU D2). Rare, with a very limited area of occupancy.
Botanical notes. *Kleinia* is a genus of succulent shrubs related to the widespread genus *Senecio*. The rounded, pencil-like stems bear distinctive markings radiating downwards from the leaf scars. This is the only succulent Compositae on Soqotra.

ETHNOGRAPHIC NOTES
Name(s)
– Soqotri: nil or 'kind of *ʿašbak*' (*Euphorbia schimperi*); *ʿašbak dimidi* '*Euphorbia schimperi* of the high limestone plateaus'.
– Soqotri simplified: *'ashbak* (*dimidi*).
Food (livestock). Largely unpalatable to livestock, though goats have been seen to eat it in a prolonged dry season. However, it is not seen as having any nutritional value.
Comparative uses. Other species are also considered to be unpalatable to livestock. In Dhofar the stems of *K. odora* are steeped in water overnight and the liquid rubbed on a painful back, and over paralysed or stiff limbs. In Yemen it is planted as an amulet on graves, and the sap of the plant is used as a diuretic. In Dhofar the juice extracted from the stems of *K. saginata* is applied to wounds as haemostatic and around red and swollen eyes. (Fleurentin & Pelt/Fleurentin *et al.* [1982, 1983a,b,c]; Heine *et al.* 1988; Kunkel 1984; Miller & Morris 1988)

Launaea Cass.

Heterachaena Fresen.

Herbs or spinescent subshrubs. Leaves alternate, simple or pinnate. Capitula cylindrical, solitary or in paniculate inflorescences; phyllaries in several rows, the inner linear-lanceolate. Flowers yellow, all 5-lobed, all ray. Pappus dimorphic, consisting of downy and setose bristles or simple with setose bristles only. Achenes ribbed, notched at base, glabrous or papillose, sometimes beaked. 4 endemics. 54 species. Mostly Africa and SW Asia.

ETHNOGRAPHIC NOTES

Comparative uses. Many wild *Launaea* species are used as salad vegetables and potherbs, and some, such as *L. taraxacifolia*, are cultivated as such. This is also fed to cows as a galactagogue and, with added natron, to sheep and goats to produce multiple births. Many *Launaea* species are also used medicinally. Some species, such as the wild lettuce, *L. scariola*, are toxic to cattle when the plants are young and eaten in large quantities. The related species, *L. quercina* and *L. virosa*, are cultivated in France as substitutes for opium: the extracted lactucarium is reputed to be a mild sedative and is used in cough mixtures instead of opium. (Ambasta 1982; Dalziel 1948, 1956; Jain & De Fillips 1991; Kunkel 1984; Lewis & Elwin-Lewis 1977)

Launaea crassifolia **(Balf.f.) C.Jeffrey**

Lactuca crassifolia Balf.f.
Lactuca kossmatii Vierh.
Lactuca paulayana Vierh.
Lactuca salehesis Vierh.

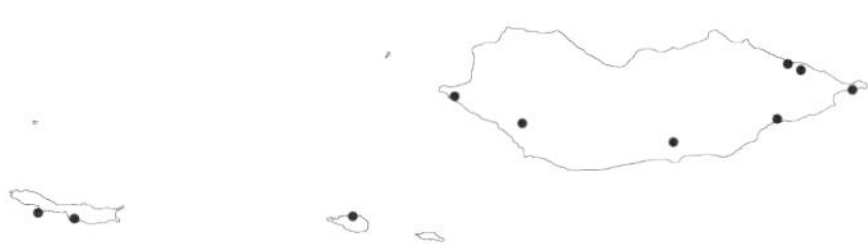

Erect branched herb to 30cm, often with tuberous taproot. Basal leaves narrowly obovate, irregularly lobed, lobes sometimes white-tipped, 2–8 x 0.6–3cm, coarsely dentate. Capitula in often frequently branched inflorescence; narrowly cylindrical becoming swollen and somewhat hardened at the base in fruit, 1–1.5cm long; phyllaries with reflexed tips. Achenes with short beak.
Ecology. Succulent shrubland, dwarf shrubland and open woody-herb communities. 10–500(–760)m.
Distribution. Somalia, Yemen and Oman.

ETHNOGRAPHIC NOTES

Name(s)
– Soqotri: *ḥədəho*; *kididhin* (more commonly given to *Euphorbia hadraumatica*); *dūwrīr* 'bleeds a little'; *ḥʌmź* 'astringent, sour'; *šiberhen* 'with pungent sap' (Samha); 'type of *gəlʿəmo*' (*L. hafunensis*) 'edible fruit' (Abd al Kuri); *lərġəbə* (Abd al Kuri).
– Soqotri simplified: *hidiho*; *kididhin*; *duwriyr*; *hamdh*; *shiberhen*; *gil'imo*; *lirgiabih*.
Food (human). Produces a swollen base and root system, which is gathered, peeled and eaten raw. In the hungrier areas of the Archipelago this was of some importance as a food after the rains. The tender, young leaves were also nibbled raw.
Food (livestock). In the drier areas of the Archipelago this is regarded as extremely valuable, fattening and nutritious forage for all livestock, but especially sheep and goats. On the outer islands it is a key livestock fodder, in particular for sheep since it persists even through the dry season.
Medicinal / veterinary. The green leaves are boiled in water and the filtrate drunk to treat a painful stomach. The leaves are ground to a paste and applied to sores in humans and livestock to draw pus.

Launaea crepoides **Balf.f.**

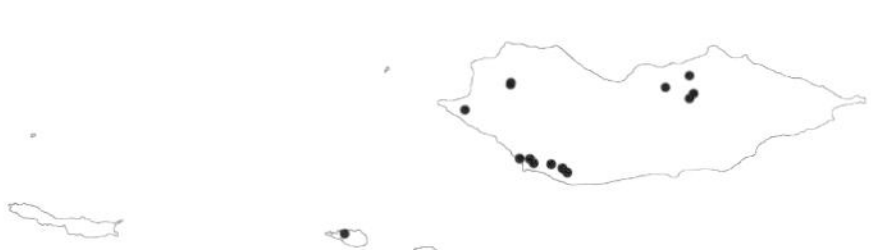

Herb to 15cm with small woody taproot. Basal leaves in a dense rosette, spathulate, 1.5–4 x 0.1–0.2cm, sinuate. Capitula in little-branched inflorescence; cylindrical, 0.5–1cm long.
Ecology. Dwarf shrubland, open woodland and semi-deciduous woodland. Restricted to mist-affected regions on Soqotra including the Haggeher mountains, Ma'alah and the areas of relict woodland on the SW plateaus. It is also found in the refugium of relatively mesic conditions on the north-facing limestone cliffs on Samha (cf. *Begonia samhaensis*). 400–800m.
Distribution. Endemic.
Conservation status. Vulnerable (VU D2). Rare, with a very limited area of occupancy.

ETHNOGRAPHIC NOTES

Name(s)
– Soqotri: nil (Soqotra, Samha and Abd al Kuri); *ḥʌmź* 'astringent, sour'; (rare) *məsʿɔ̄wbeb* 'pale'.
– Soqotri simplified: *hamdh*; *mis'owbeb*.
Food (human). Produces a swollen base and root system, which is gathered, peeled and eaten raw. In the hungrier areas of the Archipelago this was of some importance as a food after the rains. The tender, young leaves were also nibbled, but were not seen as an important food.
Food (livestock). Provides incidental grazing for small stock, and particularly for weaned young lambs and kids. It can be a valuable forage in the drier areas of the Archipelago.
Medicinal / veterinary. The leaves can be used in the same away as *L. crassifolia*.

***Launaea hafunensis* Chiov.**

Spinescent subshrub to 50cm. Leaves inserted on branches, narrowly obovate, pinnately lobed, 3–10 x 1–4cm, entire or coarsely dentate. Capitula in widely branched, spinescent inflorescence; cylindrical becoming swollen and somewhat hardened at the base in fruit, 1–1.5cm long; phyllaries with reflexed tips. Outer achenes long papillose.
Ecology. Locally common on Abd al Kuri and Samha in desertic dwarf shrubland and woody-herb communities on the coastal plains and rocky mountain slopes. Sea-level–300m.
Distribution. NE tropical Africa, S Yemen and Oman.

ETHNOGRAPHIC NOTES
Name(s)
– Soqotri: *šibérhen*, 'with bitter exudate'; *ḥʌṭʌ́mhən* 'coastal area bare of other vegetation'; (Samha); *gəlˁə́mo* 'edible fruit' (Abd al Kuri).
– Soqotri simplified: *shiberhen, hatamhin*; *gil'imo.*
Food (human). Produces a swollen base and root system, which is gathered, peeled and eaten raw. In the outer islands this was of some importance as a food after the rains, and was actively harvested. The tender, young leaves were also nibbled raw.
Food (livestock). In the outer islands this is regarded as extremely valuable, fattening and nutritious forage for sheep and goats. For sheep it is a key dry season fodder, since the greenery persists even through the dry season.

***Launaea massauensis* (Fresen.) Kuntze**
Heterachaena massauensis Fresen.
Launaea kuriensis Vierh.

Erect herb to 40cm. Leaves in a loose rosette, oblanceolate, irregularly pinnately lobed, the terminal lobe sometimes enlarged, 2.5–9 x 0.5–2.5cm, finely dentate. Capitula in frequently branched inflorescence; narrowly cylindrical, c.1cm long. Outer achenes black.
Ecology. Widespread in a variety of vegetation types on Soqotra and Abd al Kuri. 50–1100m.
Distribution. Sudan, Ethiopia, Eritrea, southern Arabia and Iran.

Botanical notes. See comments under *L. socotrana*.

ETHNOGRAPHIC NOTES
Name(s)
– Soqotri: *ṣʌfərírihən* 'many-flowered'; *ḥʌmź* 'astringent'; *rížəbeb.*
– Soqotri simplified: *zafiririhan*; *hamdh*; *rijibeb*.
Food (livestock). This is one of the *rížəbeb* group of plants which flourish after the winter rains. All livestock eat the aerial parts of the plant before the grasses come up. In less well-vegetated parts of the islands it is regarded as useful forage for goats and especially for sheep, since the greenery persists through the dry season. Weaned lambs and kids are fond of the young leaves.
Medicinal / veterinary. The leaves can be used in the same away as *L. crassifolia*.

***Launaea rhynchocarpa* (Balf.f.) B.A.Mies**
Lactuca rhynchocarpa Balf.f.

Erect herb to 10(rarely 30)cm. Leaves in a basal rosette, oblanceolate, irregularly pinnately lobed (rarely simple, sinuate), lobes sometimes white-tipped, 1.5–4(–10) x 0.1–0.8(–1.3)cm, entire. Capitula in little-branched synflorescence; narrowly cylindrical, c.1cm long. Achenes with long beak.
Ecology. Succulent shrubland and drought-deciduous woodland. Sea-level–100m.
Distribution. Endemic.

Conservation status. Least Concern (LC). Common in several vegetation types and under no present or perceived threat.

ETHNOGRAPHIC NOTES
Name(s)
– Soqotri: nil or *ˀírsib* (pl. *ˀirsíbihon*) (also given to *Portulaca* species, *Corchorus erodioides*, *Eclipta alba*); *ḥʌmź* 'astringent, sour'.
– Soqotri simplified: *'irsib* (pl. *'irsibihon*); *hamdh.*
Food (human). The tender, young leaves have a pleasant flavour and are nibbled raw.
Food (livestock). Provides incidental grazing for small stock, and particularly weaned young lambs and kids. It can be valuable forage in the drier areas of the Archipelago as some greenery persists through the dry season.

***Launaea socotrana* N.Kilian**

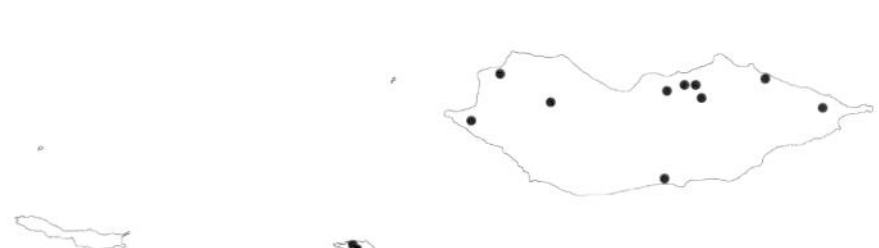

Similar to *L. massauensis* but outer achenes are brown.

Ecology. Widespread in a variety of vegetation types on Soqotra and Samha. Sea-level–800m.
Distribution. Endemic.
Conservation status. Least Concern (LC). Common in several vegetation types and under no present or perceived threat.

Botanical notes. Following Kilian (1997) we have retained *L. socotrana* as distinct from *L. massauensis* but they are very difficult to separate using the key characters. Life span (annual in *L. massauensis* and perennial in *L. socotrana*) is a difficult character to determine since both taxa have a thin taproot and no obviously well-developed storage organ. Outer achene colour is similarly not a clear defining character and specimens identified as *L. socotrana* (e.g. *Miller et al.* 8287) have blackish outer achenes.

ETHNOGRAPHIC NOTES

Name(s)

– Soqotri: *rˈižəbeb*; *ḥarˈiṭrəṭ* (more commonly given to *Leucas virgata* and *Teucrium balfourii*); *səbbˈaḥ* (Samha); *ṣˈubəhur* 'sour'; *ṣefˈerrihun*, *ṣefirirˈihˈon*, *ṣˈeferrhen* 'many-flowered'; *ḥʌmź* 'astringent'.

– Soqotri simplified: *rijibeb*; *haritrit*; *sibbah*; *zubihur*; *zefiririhon*, *zeferrhen*; *hamdh*.

Food (human). The tender, young leaves are nibbled raw.

Food (livestock). This is one of the *rˈižəbeb* group of plants which flourish after the winter rains. All livestock eat the aerial parts of the plant before the grasses come up. In the less well-vegetated parts of the islands it is regarded as a useful forage for goats and especially for sheep, since the greenery persists through the dry season. Weaned lambs and kids are fond of the young leaves.

Medicinal / veterinary. The leaves can be used in the same away as *L. crassifolia*.

Launaea* sp. *A

Herb to 30cm with thick woody taproot, sometimes branched. Leaves glaucous, in basal rosette, narrowly obovate, 2.5–6 x 0.5–1.5cm, entire; stem leaves auriculate at base. Capitula in branched inflorescence; cylindrical, 1–1.5cm long.

Ecology. Rare, collected only once, from a vertical rock face on granite pinnacles in the western Haggeher; growing with *Helichrysum rosulatum*, *Nirarathamnos asarifolius* and *Hypericum tortuosum*. c.1550m.

Distribution. Endemic.

Conservation status. Data Deficient (DD). Poorly known; inadequate evidence is available to make an assessment.

Botanical notes. Kilian (1997) includes this specimen in *L. crepoides*. However, in facies it is very distinct, differing in its generally larger size, habit, leaf shape and colour. It also has a different geographical distribution and ecological habitat. Further collections are needed before the correct status of this interesting collection can be assessed.

Orbivestus

Botanical notes. For description see *Vernonia*.

***Orbivestus cinerascens* (Sch.Bip.) H.Rob.**

Vernonia cinerascens Sch.Bip.

Vernonia spathulata Hochst.

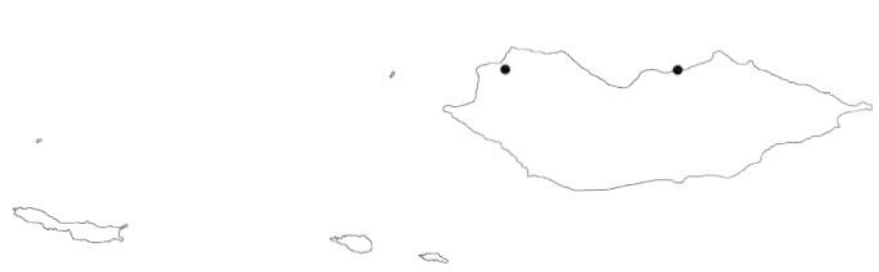

Much-branched subshrub to 60cm. Leaves narrowly obovate, 2–4 x 0.5–1cm, entire (sometimes lobed at base), glandular, sessile. Capitula few, in clusters, c.0.8cm diameter; phyllaries dark-tipped, the outer obtuse. Pappus white.

Ecology. Apparently rare but possibly overlooked. In succulent shrubland. 50m.

Distribution. Widespread weed.

Botanical notes. On the Arabian mainland this is a frequent species in dry *Acacia* woodland.

ETHNOGRAPHIC NOTES

Name(s)

– Soqotri: *ḥfˈoˀ* (pl. *ḥfˈot*); *ḥfˈe* (western Soqotra).

– Soqotri simplified: *hfo'* (pl. *hfot*); *hfe'*.

Notes. Uses as *Vernonia*.

Pluchea Cass.

Glandular subshrubs. Leaves alternate, simple. Capitula campanulate; phyllaries in many rows; flowers purple, ray and tubular. Pappus 1 row of bristles. Achenes hairy. 2 endemics (1 extinct). c.80 species. Asia, N and S America, Africa, Australia.

Pluchea glutinosa Balf.f.

Erect subshrub. Leaves lanceolate to oblanceolate, 3–12 x 1–3cm, entire to dentate, acute. Capitula densely clustered, 0.6–0.8cm long, sessile or pedunculate; phyllaries glandular.

Ecology. No habitat details available. c.800m.

Distribution. Endemic.

Conservation status. Extinct (EX). Not seen since it was collected on Soqotra by all the principal collectors in the 19th century: Balfour (near Qalansiyah and in the Haggeher), Schweinfurth (Haggeher) and Forbes *et al.* (Haggeher). Balfour in his *Botany of Socotra* (1888) comments that *P. glutinosa* is common and distributed in the hill regions of Soqotra. However, Balfour's comments on the distribution of plants should be treated with caution. A study of his diary and field books (held in the archives at the Royal Botanic Garden Edinburgh) strongly suggests that he must have made these notes from memory on his return from Soqotra. Nevertheless, it is surprising that this distinctive plant has not been seen again during the last 100 years at either of the areas (Haggeher and near Qalansiyah) where it was originally recorded – both are botanically well-known areas which have been extensively surveyed. This species must now be considered extinct.

Botanical notes. A subshrub with erect stems and lilac flowers in dense terminal panicles. See notes under *Pulicaria aromatica*.

Pluchea obovata Balf.f.

Prostrate subshrub. Leaves obovate, 1–4 x 0.4–1.5cm, entire or upper dentate, obtuse or acute. Capitula solitary or loosely clustered, 0.7–1.2cm long, pedunculate; phyllaries densely glandular, tomentose.

Ecology. Locally common on cliffs in the higher parts of the Haggeher mountains. 700–1100m.

Distribution. Endemic.

Conservation status. Vulnerable (VU D2). Rare, with a very limited area of occupancy.

Botanical notes. A distinctive purple-flowered composite often forming large mats hanging on cliffs and steep slopes.

ETHNOGRAPHIC NOTES

Name(s)

– Soqotri: (*di*) *chayāʾyáʾ*; *fígəhun*, *fižíʾhin* 'sprouting' (also given to other sweet-smelling species); *ʿaṣábṣʌb* 'branch tips'; *di chīreb* (more commonly given to *Hypericum scopularium*).

– Soqotri simplified: (*di*) *chaya'ya'*; *figihun, fiji'hin*; *'azabzab*; *di chiyreb.*

Food (livestock). Regarded as a valuable fodder for all livestock, but in particular goats. It is fattening and promotes lactation.

Bees / honey. Honeybees are attracted to the flowers.

Leatherwork. A paste of the plant is used to strengthen and cure skins and hides, in the Central Highlands especially.

Medicinal / veterinary. The skin of a dead calf or of a male calf that has been killed to conserve its mother's milk is made into a tulchan which is stuffed with sprigs of this plant. Some of her milk is sprayed over it and it is then presented to the cow to sniff to encourage her to let down her milk. When trying to persuade a cow whose calf has been slaughtered to accept the female calf of another as her own, the calf is rubbed with a paste of the crushed plant to disguise its own smell and presented to the cow. Some of her milk is sprayed over it to encourage her to lick it and allow it to suckle.

Bedding, stuffing. Sprigs of the fragrant plant are tucked amongst bedding or added to the stuffing of bolsters and pillows. In areas where the plant is plentiful it is gathered and strewn to make a sweet-smelling mattress.

Cleansing. Sweet-smelling sprigs are tucked into the clothing and pinched for their sweet smell, or rubbed between the hands and then over the skin and clothing to give it a sweet smell. All sweet-smelling plants are endowed with beneficial medicinal and prophylactic qualities.

Comparative uses. Related species are used in teas either as flavouring or as a tea substitute. Leaves of species such as *P. indica* and *P. odora* are eaten. Anti-inflammatory and anti-ulcer actions of *P. indica* have been reported. In Dhofar the whole plant of *P. arabica* is pounded and boiled in water and the filtrate applied to boils and swellings. The fresh leaf juice is used as eardrops to treat infected ears and applied locally to skin sores. The plant is also used to treat swellings in livestock. (Heine *et al.* 1988; Kunkel 1984; Miller & Morris 1988)

Prenanthes L.

Herbs. Leaves alternate, simple. Capitula in large panicles; phyllaries in 2 rows, outer reduced; flowers pink or purple, all ray. Pappus of many bristles. Achenes 3 or 4-ribbed. 1 endemic. c.30 species. Mainly Asia and N America, also in C and S Europe, Canary Is. and tropical Africa.

Prenanthes amabilis Balf.f.

Tap-rooted, rosette-forming herb to 30cm. Leaves lyrate, 4–8 x 2–6cm.

Ecology. On limestone in semi-deciduous woodland (with *Buxanthus pedicellatus* and *Gnidia socotrana*). 500–550m.

Distribution. Endemic.

Conservation status. Endangered (EN B2 a biii). Known only from the limestone escarpment above Qalansiyah where it occupies an area of less than 10km^2. It is strongly suspected that this marginal habitat is threatened by the long-term trend of lower rainfall in the region and would be particularly vulnerable to the effects of global warming.

Botanical notes. Originally described by Balfour from rocks southwest of Galonsir [Qalansiyah]. Both recent collections were made at or near the type locality. Forbes (1903) observed the plant below the Adho Dhimellus [Adho di Melhoh] pass in the Haggeher mountains. It has not been seen there recently and this record should be treated with caution. A distinctive low herb with pretty lilac flowers and succulent leaves. See notes under *Pulicaria aromatica*.

ETHNOGRAPHIC NOTES

Name(s)

– Soqotri: nil or *sine̍mhen* (more commonly given to the sweet-smelling *Clerodendrum* and, in the western area, *Coelocarpum* species); *reḥe̍mimhin* 'grey-coloured'.

– Soqotri simplified: *sinemhen*; *rehemimhin.*

Food (livestock). An uncommon plant. Herders say that it provides incidental grazing for small stock but is not important forage.

Comparative uses. *P. serpentaria* is used as a snakebite antidote and in homeopathic medicine in eastern North America. (Lewis & Elwin-Lewis 1977)

Psiadia Jacq.

Sticky shrubs. Leaves alternate, simple, entire. Phyllaries straw-coloured, in 2 or 3 rows. Flowers yellow, tubular. Pappus 1 row of bristles. 1 endemic (extinct). c.60 species. Madagascar, Mascarene Is., tropical and S Africa, also tropical Asia.

Psiadia schweinfurthii Balf.f.

No description; see Botanical notes.

Ecology. No habitat details available. c.700m.

Distribution. Endemic.

Conservation status. Extinct (EX). Collected by Schweinfurth at Kishin in the northern Haggeher mountains in the 19th century but not seen since. It is surprising that this distinctive plant has not been seen again during the last 100 years. Kishin is on the main path up to the Haggeher mountains from Hadiboh and is a botanically well-known area which has been extensively surveyed. This species must now be considered extinct.

Botanical notes. Balfour (1888) noted that his new species was similar to *P. arabica*, a widespread species on the Arabian mainland, but differed in the absence of sticky glandular hairs. No specimen has been traced and the identity and status of this species remains a mystery.

Pulicaria Gaertn.

Herbs or clump-forming subshrubs. Leaves alternate, simple, glandular. Capitula campanulate with many rows of ovate phyllaries, terminal, solitary or in paniculate inflorescences. Flowers yellow, heterogamous with 3 or 4-lobed outer, ray flowers and 5-lobed inner, tubular flowers or all tubular. Achenes smooth, glabrous or sparsely hairy. Pappus in 2 rows: the inner of long bristles, the outer a ring of connate scales (difficult to see in *P. aromatica*). 7 endemics. 50 to 60 species. Eurasia and Africa.

Pulicaria aromatica (Balf.f.) King-Jones & N.Kilian

Pluchea aromatica Balf.f.

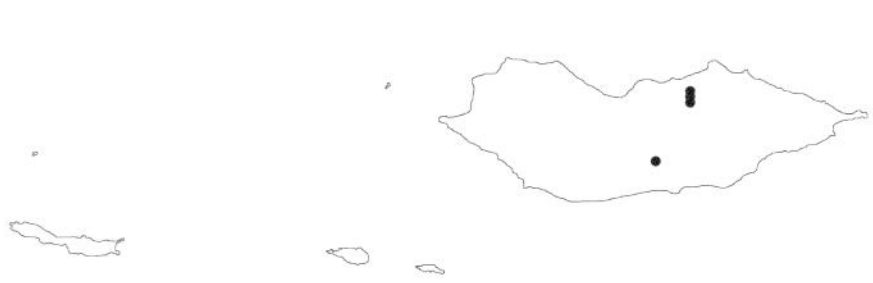

Aromatic subshrub with erect, woody branches. Leaves clustered towards branch tips, narrowly oblong to obovate, 3–7 x 1–2cm, entire, densely glandular. Capitula terminal, solitary, 1–1.5cm diameter; peduncles 1.5–7cm; ray flowers absent. Pappus outer scales absent (much reduced, appearing as a 'rim' to the achene).

Ecology. On cliffs in sheltered limestone ravine in deciduous woodland. 550m.

Distribution. Endemic.

Conservation status. Endangered (EN B2 a biii). Known only from a ravine in the south of Soqotra where it occupies an area of less than 50km^2. It is strongly suspected that this area is threatened by the long-term trend of lower rainfall in the region and would be particularly vulnerable to the effects of global warming.

Botanical notes. This is one of several plants (e.g. *Prenanthes amabilis*, *Pluchea glutinosa*, *Sida cordifolia*) collected in the 19th century from near Kishin in the northern Haggeher mountains but not seen there since. However, *P. aromatica* was recently found on cliffs on an isolated ravine in southern Soqotra and perhaps some of these other plants should be searched for in some of the well-wooded valleys draining south from the Haggeher. It is not clear what the relationships of *P. aromatica* are. It is possibly related to *P. vieraeoides* on the basis of achene characters, habit and inflorescence.

ETHNOGRAPHIC NOTES

Name(s)

– Soqotri: *fˈegəhun*, *fˈižihin* 'sprouting' (given to other sweet-smelling species such as *Thamnosma* and *Cleome*).

– Soqotri simplified: *fegihun*, *fijihin*.

Food (livestock). All livestock eat this when hungry. It is not regarded as having nutritional significance, but is a useful dry-season maintenance feed.

Bees / honey. Honeybees are attracted to the flowers.

Medicinal / veterinary. The leaves are crushed to a paste and used on abscesses, boils and infected sores to draw pus.

Bedding, stuffing. The sweet-smelling leaves are sometimes tucked amongst bedding or added to the stuffing of bolsters and pillows. This is especially done for sick people as their sweet smell is believed to hasten recovery.

Cleansing. The leaves are crushed between the fingers and rubbed over the skin or clothing to give it a sweet smell. The sniffing of sweet-smelling plants is held to protect from the 'noxious air' which can cause disease, and to help protect against infection. All sweet-smelling plants are endowed with beneficial medicinal and prophylactic qualities.

Comparative uses. The related *P. orientalis* is used in Yemen to treat digestive disorders, as an anti-pyretic, anti-malarial, appetite stimulant and analgesic. In Yemen and Oman, a decoction of the leaves and flowers of *P. jaubertii* is also used to stimulate the digestion and as a tonic. With other ingredients, its dried leaves are given to treat epilepsy, to regulate the menses and as a vermifuge. An infusion of the seeds is taken as a lactogogue and the fresh leaves are used in Yemen to spice fenugreek (*Hilba*) soup. The leaves of *P. crispa* are used as a diuretic in Bahrain, and elsewhere are applied locally to swellings and bruises and rubbed on the temples for a headache. The plant is sold in markets as a medicinal plant and a decoction of the plant is taken for a fever. *P. undulata* is used as a tonic, antispasmodic, hypoglycaemic and as a substitute for tea. (Batanouny 1981; Dalziel 1956; Fleurentin & Pelt/Fleurentin *et al.* [1982, 1983a,b,c]; Ghazanfar 1994; Rizk & El-Ghazaly 1995)

Pulicaria dioscorides R.Atkinson (see Appendix 2, p. 723)

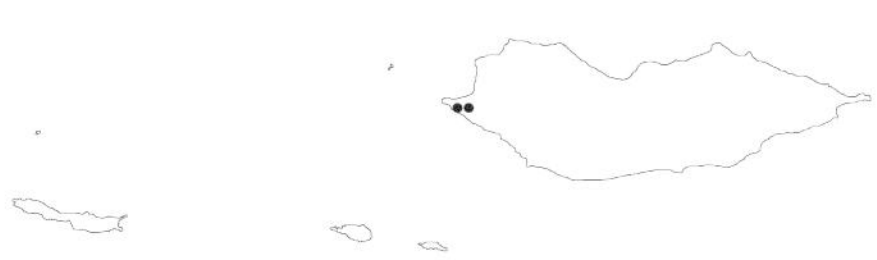

Cushion-forming woody herb to 15cm. Branches prostrate, leaf scars persistent, forming overlapping scales, bark somewhat corky at base. Leaves tightly clustered at branch tips, narrowly spathulate, leaf base clasping stem, 0.3–1.2 x 0.1–0.4cm, entire, covered with dense, long, silvery hairs. Capitula terminal, solitary, 0.8–1cm diameter; peduncles 0.5–1cm; ray flowers present.

Ecology. Restricted to exposed rock faces on a windswept ridge at the western end of Soqotra. 380–530m.

Distribution. Endemic.

Conservation status. Endangered (EN B2 a biii). *Pulicaria dioscorides* is restricted to a small refugium with a total estimated area of occupancy of less than 10km^2. At present the rapid development of the island poses no threat to this unique habitat: the area is inhospitable, inaccessible by road and there is a lack of potable water. However, its restricted distribution and the very limited extent of the niche which it occupies means that it is vulnerable to any change. It is strongly suspected that this marginal habitat is threatened by the long-term trend of lower rainfall in the region and would be particularly vulnerable to the effects of global warming (Miller *et al.* in press).

Botanical notes. Known only from the ridge at Heger (between Ras Izfer and Ras Shu'ub) in the extreme west of Soqotra. It is an attractive cushion-forming species clearly distinguished by the tightly clustered silver-grey leaves and bright yellow flowers. It is apparently related to *P. pulvinata* E.Gamal-Eldin, a mat-forming perennial, endemic to the coastal desert of central Oman. The Soqotran plant differs in its more robust habit and the presence of ray florets.

ETHNOGRAPHIC NOTES

Name(s)

– Soqotri: *di ḳˈʌṣḳʌṣ* 'cream-coloured'.

– Soqotri simplified: *di qazqaz.*

Food (livestock). Regarded as good goat browse in the high, dry areas of Soqotra where it grows. It never completely dies back and can provide some forage throughout the dry season. Sheep nibble new growth or fallen leaves and other plant litter.

Pulicaria diversifolia Balf.f.

Pulicaria shoabensis Vierh.

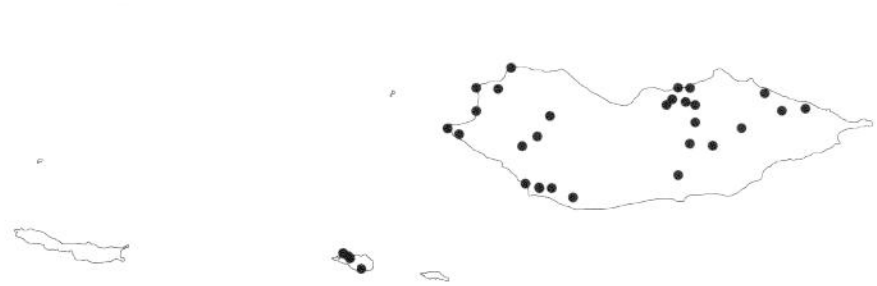

Erect non-woody herb to 40cm. Leaves forming a loose 'basal rosette' and evenly spaced up stem; spathulate to obovate, 4–9 x 1–3cm, irregularly dentate, glandular hairy. Capitula in loose inflorescences, 2–2.5cm diameter; peduncles 3–13cm; ray flowers present.

Ecology. Common in a variety of vegetation types, particularly in rocky places, at low and middle altitudes on the limestone escarpments and plateaus of Soqotra; much rarer on the granite; also found on Samha. 10–750m.

Distribution. Endemic.

Conservation status. Least Concern (LC). Common in several vegetation types and under no present or perceived threat.
Botanical notes. Probably the commonest yellow-flowered herbaceous composite on Soqotra. Very variable in habit and leaf shape.

ETHNOGRAPHIC NOTES
Name(s)
– Soqotri: *di ᶜalimləhum*, (*ḳōwḳʌˀ*) *di ᶜalumhum* 'marked'; *kibidino* 'little liver, small hill(ock)'; *mədḥōməhəm*; *ḥīreyr*; *ḥʌźḳeḳ di 'ilhitin* 'underbrush for cattle' (also given to *Parietaria debilis*); *ḥawririhim*; *tenəbeko*, 'small tobacco plant'; *di ˀʌlḳʌṣ*; *ᶜažərib* 'quick to die back'; *reḥeb* 'broad-leaved'; *ṣeferrihon* 'many-flowered'; *śkaᶜinitin* 'prickly'.
– Soqotri simplified: *di 'alimlihum*, (*qowqa'*) *di 'alumhum*; *kibidino*; *mid'howmihim*; *hiyreyr*; *hadhqiq di 'ilhitin*; *hawririhim*; *tenibeko*; *di 'alqaz*; *'ajirib*; *reheb*; *zeferrihun*; *shka'initin*.
Food (livestock). Like many of the *ᶜažərib* group, this is greedily eaten by livestock, especially cattle and goats, when it flourishes after the rains and until grasses, etc. become established. It is not seen as being valuable nutritionally, but is a useful maintenance feed.
Bees / honey. Honeybees are attracted to the flowers.
Medicinal / veterinary. The leaves are crushed to a paste and used on abscesses, boils and infected sores to draw pus.
Notes. Goats that eat a lot of this immediately after the rains often develop scour. It can cause severe *'uqabah* in small stock unaccustomed to the feed.

Pulicaria elegans **E.Gamal-Eldin**

Prostrate woody herb to 30cm. Leaves and leaf scars evenly spaced; oblong to obovate, 0.8–1.3 x 0.3–0.4cm, dentate, densely glandular. Capitula terminal, 2 or 3 per inflorescence, 1–2cm diameter; peduncles 1–5cm; ray flowers present.
Ecology. Locally common in rocky places in semi-deciduous montane shrubland and woodland. 700–1000m.
Distribution. Endemic.
Conservation status. Least Concern (LC). Common in several vegetation types and under no present or perceived threat.

ETHNOGRAPHIC NOTES
Name(s)
– Soqotri: *mədūwnəhin* 'like *źʌfəraᶜ*' 'like *Plectranthus*', from its smell.
– Soqotri simplified: *miduwnihin*; 'like *dhafira*".
Food (livestock). Livestock eat this only when hungry. It is not regarded as having any nutritional significance.
Cleansing. The fragrant leaves are crushed between the fingers and rubbed over the skin or clothing to give it a sweet smell. All sweet-smelling plants are endowed with beneficial medicinal and prophylactic qualities.
Notes. Some herders use the leaves as an insecticide in place of *Plectranthus*.

Pulicaria lanata **E.Gamal-Eldin**

Clump-forming shrub or woody herb. Branches erect with prominent leaf scars at base. Leaves evenly spaced on young growth, sessile, clasping stem at base; spathulate, 2.5–4 x 0.7–1cm, irregularly dentate, pale green with woolly hairs, dense beneath. Capitula terminal, solitary, 1.5–3cm diameter; peduncles 2–9cm; ray flowers present.
Ecology. Locally common in shrubland around the highest granite pinnacles of the Haggeher mountains. 1100–1500m.
Distribution. Endemic.
Conservation status. Least Concern (LC). Common and under no present or perceived threat.
Botanical notes. Readily distinguished by the woolly indumentum on the leaves and stems.

ETHNOGRAPHIC NOTES†
Name(s)
– Soqotri: *di maḥaḥōn*, *moḥoḥʌˀ* (pl. *il məḥaḥōn*) 'fatty'; 'kind of *źʌfəraᶜ*' (*Plectranthus*) from its smell.
– Soqotri simplified: *di mahahon*, *mohoha'* (pl. *il mihahon*); 'kind of dhafira'.
Food (human). This is the most popular plant of the Central Highlands for cooking with *Dioscorea lanata*. Layers of leafy stems are placed above and below the tubers before cooking. This is said to impart a particularly delicious flavour to the cooked tubers. Once tea and sugar became known on the island, some people tried using dried leaves of the plant either to flavour tea or as a substitute for it.
Food (livestock). Goats browse the plant when it has dried out but rarely eat it green. Cattle on the whole avoid it, but if very hungry in a prolonged dry season will graze it. Sheep also avoid it, though occasionally they eat the plant litter. It is regarded as having little nutritional value, but can be a useful dry season maintenance feed.
Bees / honey. Honeybees are attracted to the flowers.
Flavouring / food additive. When salt was not available, handfuls of this were dipped into the water used for cooking meat as a salt substitute. This was especially successful during or just after the monsoon season when it had absorbed salt from the winds.
Cleansing. The fragrant leaves are crushed between the fingers and rubbed over the skin or clothing to give it a sweet smell. All sweet-smelling plants are endowed with beneficial medicinal and prophylactic qualities.
Medicinal / veterinary. The leaves are crushed to a paste and used on abscesses, boils and infected sores to draw pus.
Bedding, stuffing. Sprigs of the plant are tucked in between stored clothing or scattered amongst bedding to give it a pleasant smell.
Pest control. Some herders use the leaves of this as an insecticide instead of *Plectranthus*.

†Uses also apply to *P. vieraeoides*.

Pulicaria stephanocarpa Balf.f.

Clump-forming shrub to 50cm. Branches erect, woody with prominent leaf scars. Leaves evenly spaced on young growth, spathulate, 0.5–2 x 0.1–0.5cm, entire or coarsely dentate, densely glandular, sometimes ± succulent. Capitula terminal, solitary, 0.5–1cm diameter; peduncles 0.5–4cm; ray flowers absent.
Ecology. An abundant, often dominant, plant of dwarf shrubland over the plains and limestone plateaus of western Soqotra; much less common in eastern Soqotra; also found on Samha and Abd al Kuri. Sea-level–800m.
Distribution. Endemic.
Conservation status. Least Concern (LC). Common in several vegetation types and under no present or perceived threat.
Botanical notes. This rather neat, rounded shrub is characteristic of large areas of western Soqotra. It is readily distinguished by the strong but pleasant aroma of the crushed leaves.

ETHNOGRAPHIC NOTES
Name(s)
– Soqotri: *dīrbeb* (Soqotra, Abd al Kuri and Samha); *gīrbeb* '(animal) with a pale streak or flash on its coat'; (rare) *(d-) irbóbo* (Soqotra, Abd al Kuri and Samha); (rare) *səgénhen* 'curved over' (Samha).
– Soqotri simplified: *diyrbeb*; *giyrbeb*; *(d-) irbobo*; *sigenhen.*
Food (livestock). In the drier areas where this plant is widespread it is the key dry season grazing for goats. Camels and donkeys eat it at all times of the year, and even cattle graze it when very hungry. Sheep are less enthusiastic, but eat the plant litter when there are few alternatives. It survives the worst droughts and is therefore one of the most important fodder plants at such times. On the outer islands the sheep eat it with more enthusiasm and it is regarded as an important feed for them.
Flavouring / food additive. The milk of livestock eating a lot of it has a distinctive taste and smell, very similar to that of the plant itself. The water used for cooking the meat of an animal that has eaten a lot of this plant also develops a strong smell and taste, the viscera being particularly affected. Both meat and milk are said to have a rather 'hot' flavour, disliked by some.
Medicinal / veterinary. The leaves are crushed to a paste used on abscesses, boils and infected sores to draw the pus.
Cleansing. When there is no water to hand, the leaves are crushed and used to cleanse the skin.
Other. The plant cannot withstand prolonged wet conditions; after heavy rains waterlogged plants die back. The plant is seen as retarding and even preventing the growth of grasses, and in the low-lying areas where it is on the increase, people have noticed that the grasses are suffering.
Notes. There is a saying on Soqotra: 'The goats have the cool winds of the monsoon (so are not seriously bothered by thirst) and plenty of *diyrbeb* to eat (so are not seriously bothered by hunger)'. This is said to have been the reply of a herder who was accused of giving insufficient attention to his goats during the monsoon. The phrase is used nowadays to imply that things are not always as they seem.

Pulicaria vieraeoides Balf.f.

Woody herb or shrub. Branches erect with prominent, closely spaced leaf scars at base. Leaves clustered at branch tips; oblong-obovate, 2–3.5 x 0.5–2cm, dentate, dense, short hairs beneath. Capitula terminal, solitary, 2.5–3cm diameter; peduncles 2–4cm; ray flowers present.
Ecology. Rather local amongst rocks in low shrubland (associated with *Hypericum* species, *Buxanthus pedicellatus* and *Cephalocroton socotranus*). 700–1550m.
Distribution. Endemic.
Conservation status. Vulnerable (VU D2). Rare, with a very limited area of occupancy.

ETHNOGRAPHIC NOTES
Name(s)
– Soqotri: *di maḥaḥōn ləṣāfə* 'broad-leaved *P. lanata*'.
– Soqotri simplified: *di mahahon lizafih.*
Notes. Uses as *P. lanata.*

Reichardia Roth

Rosette-forming herb to 30cm. Leaves oblanceolate, pinnately lobed or simple, 5–13 x 1–3cm. Achenes small (c.2mm long). 8 species. Mediterranean, S Europe, N Africa, Turkey, Middle East, Canary Is.

Reichardia tingitana Roth

Rosette-forming herb to 30cm. Leaves oblanceolate, pinnately lobed or simple, 5–13 x 1–3cm. Achenes small (c.2mm long).
Ecology. An occasional plant of disturbed places in drought-deciduous woodland. 300–500m.
Distribution. Arabia.
Botanical notes. From other similar plants on Soqotra (e.g. *Launaea* and *Sonchus*) readily distinguished by the phyllaries which have white margins and a cordate base.

ETHNOGRAPHIC NOTES
Name(s)
– Soqotri: *ʿážərib* 'quick to die back'; *ḥʌmź* 'astringent, sour'.

– Soqotri simplified: *'ajirib*; *hamdh*.

Food (livestock). This is one of the *ᶜažərib* group of plants which come up after the winter rains. They are grazed heavily and well liked by cattle before the grasses come up. They are little eaten by goats and sheep and are not seen as having much nutritional value.

Comparative uses. Other species, such as *R. picroides*, have edible roots and leaves. In Yemen the leaves of this plant are taken to treat disorders of the liver. In Oman the dried powdered leaves are mixed with water and taken orally for colic, and the ground leaves are taken in a little oil to relieve constipation. An extract of the fresh leaves, mixed with leaves of *Euphorbia helioscopia*, myrrh and aloes juice, are used as eyedrops. (Ambasta 1982; Fleurentin & Pelt/Fleurentin *et al.* [1982, 1983a,b,c]; Ghazanfar 1994; Jain & De Fillips 1991; Kunkel 1984)

Sigesbeckia L.

Herbs. Leaves opposite, simple, glandular. Capitula axillary; phyllaries in 2 rows, outer long, spreading, spathulate, conspicuously glandular, inner phyllaries ± enclosing the outer achenes; flowers yellow, ray and disc. Achenes obconical. Pappus absent. 3 species. Pantropical.

Sigesbeckia orientalis L.

Erect annual herb to 50cm. Leaves lanceolate or elliptic, 5–10 x 2.5–5cm, dentate.

Ecology. Semi-deciduous woodland in the Haggeher and northern limestone plateaus. 500–1100m.

Distribution. Pantropical weed.

Botanical notes. Readily recognised by the spathulate, conspicuously glandular-hairy phyllaries.

ETHNOGRAPHIC NOTES

Name(s)

– Soqotri: nil or *ᶜažərib* 'quick to die back'; *ḥəwririhin* 'dark-coloured'; *məltuwsək* 'sticky, adhesive' (from the glands on the fruit); *miśkaᶜihen* 'prickly'; (rare) *ᶜāṣʌbṣʌb* 'branch tips'.

– Soqotri simplified: *'ajirib*; *hewririhin*; *miltuwsik*; *mishka'ihen*; *'azabzab*.

Food (livestock). This is one of the *ᶜažərib* group of plants which come up after the winter rains. They are grazed heavily and well liked by cattle before the grasses come up. They are little eaten by goats and sheep and are not seen as having any nutritional value.

Notes. Goats that eat a lot of this immediately after the rains commonly develop scour. It can cause severe *'uqabah* in small stock unaccustomed to the feed.

Comparative uses. The leaves of *S. pubescens* are eaten as a pot herb. Medicinally, it is regarded as diaphoretic, cardiotonic and anti-scorbutic. It is used to treat a gangrenous ulcer and skin sores, and a tincture of the plant with added glycerine is used externally to treat ringworm and other parasitic infections. It is also given for rheumatism and renal colic. The roots contain an essential oil. (Ambasta 1982; Jain & De Fillips 1991; Kunkel 1984; Lewis & Elwin-Lewis 1977)

Sonchus L.

Herbs. Leaves alternate, simple. Capitula in panicles, peduncles glandular; phyllaries in 2 or 3 rows, outer reduced; flowers yellow, all ray. Pappus dimorphic, of deciduous bristles and ± persistent tufts of fine hairs. Achenes flattened, ribbed, entire at base. c.60 species. Europe, Asia, Africa, N America, Australia.

Sonchus oleraceus L.

Erect, little-branched herb with glabrous hollow stems to 1m. Leaves broadly lanceolate, deeply pinnately lobed, the base clasping the stem with large auricles, 10–20 x 6–12cm.

Ecology. Overgrazed ravine on limestone plateau. c.650m.

Distribution. Cosmopolitan weed.

Botanical notes. Reported to be common in the 19th century (Balfour 1888) but now apparently rather rare; however, probably under-recorded. From other similar plants on Soqotra (e.g. *Launaea* and *Reichardia*) readily distinguished by the dimorphic pappus of deciduous bristles and softer cottony hairs.

ETHNOGRAPHIC NOTES

Name(s)

– Soqotri: *rižəbeb* 'stalky'; *fərśʌ*.

– Soqotri simplified: *rijibeb*; *firsha*.

Food (livestock). This is one of the *ᶜažərib* group of plants which come up after the winter rains. They are grazed heavily and well liked by cattle before the grasses come up. They are little eaten by goats and sheep and are seen as having little nutritional value.

Comparative uses. The leaves and young shoots of this and other related species are eaten in salads or used as potherbs. Bundles of the young shoots and leaves are sold in markets as a food. The roots are also eaten. The leaves and stem contain 23 to 30mg per 100g of vitamin C, and alkaloids, coumarins and saponins have also been detected. In the Middle East an infusion of the whole plant is taken as a tonic, diuretic, coolant and laxative. The plant juice is used for cleansing and healing ulcers and as eyedrops. Extracts are also used for liver troubles, jaundice and as a blood purifier. The roots are used as a vermicide and the latex of the stem is used to get rid of warts. The brownish gum obtained from evaporating the plant juice is said to be a powerful cathartic and has been used to treat the opium habit. (Abbas *et al.* 1992; Batanouny 1981; Bebawi & Neugebohrn 1991; Ghazanfar 1994; Goodin & Northington 1979; Kunkel 1984; Rizk & El-Ghazaly 1995)

Tripteris Less.

Herbs. Leaves alternate, simple. Capitula solitary, on long peduncles; phyllaries in 1 row, green with a scarious margin; flowers yellow, ray and tubular present. Pappus absent. Achenes with conspicuous papery wings. c.20 species. S and tropical Africa north to Egypt, Arabia and Jordan.

Tripteris vaillantii **Decne.**
Osteospermum vaillantii (Decne.) Norl.
Tripteris lordii Oliv. & Hiern
Tripteris lordii var. *racemosa* Balf.f.
Tripteris racemosa (Balf.f.) Vierh.

Branched herb to 60cm. Leaves oblong, 2–4 x 0.5–1.5cm, coarsely dentate. Capitula c.1cm diameter.
Ecology. Widespread on coastal plains and limestone plateaus in *Croton* shrubland and dwarf shrubland. Probably under-recorded on the distribution map. 10–650m.
Distribution. Arabia.

Botanical notes. Rather unremarkable in flower but readily recognised by the conspicuous papery wings on the fruits.

ETHNOGRAPHIC NOTES
Name(s)
– Soqotri: nil or *di ḳəwrɛ̄ni* 'with two little horns' (from the winged fruit).
– Soqotri simplified: *di qewreyni.*
Food (livestock). Provides incidental fodder for livestock when they graze the grasses amongst which it grows. It is regarded as having little nutritional value.
Comparative uses. In Yemen the leaves are given to treat disorders of the liver. The seed oil of various *Osteospermum* species contains acids of possible use to industry, like castor oil, in the manufacture of plastics, paints, etc. (Davis *et al.* 1983; Fleurentin & Pelt/Fleurentin *et al.* [1982, 1983a,b,c])

Vernonia sensu lato

Herbs, shrubs or small trees. Leaves alternate, simple. Capitula campanulate with c.4 rows of greenish phyllaries; flowers tubular, purple. Achenes ribbed. Pappus 2 rows of bristles, inner long and outer short. 2 endemics. 2 endemics. c.500 species. Old World tropical and warm regions to N America.

Botanical notes. *Vernonia* was recently split into a number of segregate genera (Robinson 1999) which are largely distinguished by technical characters, especially pollen and hair types. Two of these genera, *Cyanthillium* and *Orbivestus*, occur on Soqotra, and since *Vernonia* sensu stricto is now regarded as restricted to N America, it is likely that the two endemic species remaining in *Vernonia* will soon be given new generic combinations. All four Soqotran species, which previously were placed in *Vernonia*, are kept together in the key to facilitate identification.

Vernonia cockburniana **Balf.f.**

Shrub or small tree to 3m. Branches not corky. Leaves in fascicles on short spurs evenly spaced along slender branches, narrowly obovate or obovate, 0.6–2 x 0.2–1.2cm, apetiolate. Capitula 1 to few in loose clusters, 0.5–1cm diameter; phyllaries green with scarious tip, outer phyllaries broadly ovate, obtuse. Pappus straw-coloured. Achenes sparsely glandular.
Ecology. Widespread and in places abundant in drought-deciduous woodland and succulent shrubland on both limestone and granite on the escarpments and plateaus; also in patches of woodland by wadis on the inland plains. 50–1100m.
Distribution. Endemic.
Conservation status. Least Concern (LC). Common in several vegetation types and under no present or perceived threat.

ETHNOGRAPHIC NOTES†
Name(s)
– Soqotri: *ḥfóˀ* (pl. *ḥfót*); *ḥfé* (western Soqotra).
– Soqotri simplified: *hfo'* (pl. *hfot*); *hfe'.*
Food (livestock). One of the most important fodders for goats on Soqotra, though all livestock eat it, with the exception of donkeys. Goats eat the green and dried foliage, the leafless stems and even the bark in the dry season. Leafy twigs are lopped by herders for young weaned stock and for cattle in the dry season. Sheep like and can reach the foliage of the smaller shrubs, and also eat fallen leaves and other plant litter. The larger-leaved species of the better-vegetated areas are seen as being a nutritious feed, the smaller-leaved shrubs of the drier areas less so. In the drier areas the shrubs are seen more as a valuable maintenance feed for livestock, since they provide browse of some sort throughout the dry season when other forage is scarce. For the cattle of the high plateau areas it is one of the most important dry season feeds.
Bees / honey. An important forage for honeybees.
Fuel. The dead wood makes adequate firewood.
Timber. The trunks and bigger branches are used as timber for roofing beams. The smaller branches provide timber for interweaving in roofing, or for constructing pens and walling. However, in general, the shrubs are not cut as timber unless there is no alternative, since they are too important as livestock fodder.
Wood. The wood, of the smaller species in particular, is not hard, and is little used for implements. Instead it is used for items where flexibility and softness are desirable, such as the mouthpiece of the weaning device. Strips of the outer and inner bark are cut, split and put to soak in seawater for three days or so, and are then woven to make fish traps. The straight, slim branches of

the larger species of the wetter areas are sometimes used to make the important herding staff. The pegs that secure the skin stretched over the frame of the large tambourine drum (*tarānāt*) are commonly made from this wood.
Comparative uses. The fodder value of the foliage of *Vernonia* species is high. The leaves of related species are cooked and eaten. In Oman the fresh leaves of *V. cinerea* are rubbed on scorpion bites to relieve pain and inflammation; a decoction of the plant is used to reduce fevers; an extract of the leaves is applied for piles; the bitter roots are used as an anthelmintic and as a diuretic; the seeds are eaten as an anthelmintic, and a paste made from them is applied to skin to treat eczema, ringworm and other skin problems. In India the same plant is used as a tonic, stomachic, astringent, anthelmintic, diuretic and in treating consumption. An infusion of *V. cinerea* with quinine is considered effective for malaria. In Saudi Arabia the leaves of this and of *V. roxburghii* are applied locally for snake and scorpion bite. In Ethiopia *V. cineranescens* is planted as a hedge and as a shade plant for growing coffee. In Yemen a paste of the fresh leaves of *V. leopoldii* is used to treat skin disorders. Elsewhere, the juice is given in incontinence of urine and to treat a cough and relieve colic. A decoction of the roots is given to treat diarrhoea and stomach ache. The flowers are used in fevers, rheumatism and conjunctivitis. The seeds are considered effective against threadworms and roundworms and are also given for cough, flatulence, intestinal colic, dysuria, psoriasis and other skin diseases. (Ambasta 1982; Abdulfatih 1987; Bebawi & Neugebohrn 1991; Dalziel 1948, 1956; Ghazanfar 1994; Heine *et al.* 1988; le Houerou 1980; Jain & De Fillips 1991; Kunkel 1984; Oliver-Bever 1986)

†Also apply to *V. unicata*, and uses to *Orbivestus cinerascens*.

Vernonia unicata **C.Jeffrey**

Similar to *V. cockburniana* but differs in: branches corky; leaves white hairy beneath, not fasciculate, obovate, shortly petiolate, stems winged below the petiole; phyllaries acute; achenes glabrous.
Ecology. Rare in semi-deciduous woodland on granite in the Haggeher mountains. 450–1400m.
Distribution. Endemic.
Conservation status. Vulnerable (VU D2). Rare, with a very limited area of occupancy.
Botanical notes. Much rarer than the preceding species. Far less ecologically tolerant, and restricted to the eastern Haggeher where it is found as isolated individuals.

ETHNOGRAPHIC NOTES
Name(s)
See *V. cockburniana* (also uses).

Volutaria Cass.

Herbs. Leaves alternate, simple. Capitula solitary; phyllaries in 3 or 4 rows with scarious margins and black tips; flowers purple, tubular. Pappus of oblanceolate scales. Achenes obovoid, glabrous. 16 species. Mediterranean, SW Asia, tropical E Africa.

Volutaria lippii **Cass.**
Volutarella lippii Cass.

Little-branched herb to 30cm. Leaves irregularly pinnately lobed, 5–12 x 2–5cm.
Ecology. Occasional in drought-deciduous woodland on the limestone escarpment (often with *Buxanthus pedicellatus*). 350–400m.
Distribution. N and NE Africa, Canary Is., SW Europe.

ETHNOGRAPHIC NOTES
Name(s)
– Soqotri: nil or *ḥʌmź* 'astringent, sour'.
– Soqotri simplified: *hamdh.*
Food (livestock). The young plants provide incidental but useful grazing for all livestock, especially small stock. It is not regarded as being nutritionally important.
Comparative uses. In India *V. ramosa* is used as an aperient, febrifuge and tonic, and is given to relieve a cough. (Ambasta 1982; Jain & De Fillips 1991)

CONVOLVULACEAE

Herbs, shrubs or small trees. Leaves alternate or rarely opposite, simple or occasionally palmately or pinnately divided; stipules absent. Flowers cymose or solitary, actinomorphic, bisexual. Sepals 5, free or fused at the base. Petals 5, fused. Stamens 5, epipetalous. Ovary superior, 2 to 5-celled with 2 basal ovules per cell; styles 1 or 2, terminal or rarely gynobasic; stigmas 1, 2 or 4. Fruit usually a dry, loculicidal or irregularly dehiscent capsule, rarely indehiscent and berry- or nut-like; seeds usually 1 to 4. 58 genera. 1650 species. Cosmopolitan, especially warm regions.

Convolvulus L.

Herbs, shrubs or small trees. Leaves sessile or petiolate. Flowers 1 to several in axillary clusters. Corolla infundibular or campanulate, lobed or unlobed. Stamens ± unequal. Style 1; stigmas 2, linear-cylindrical or filiform. Fruit an unlobed capsule. 3 endemics. c.250 species. Cosmopolitan, especially temperate.

Convolvulus arabicus Hochst.

Convolvulus glomeratus Choisy

Prostrate herb to 2m diameter, appressed-pubescent, sometimes with long spreading hairs. Leaves shortly petiolate, oblong-triangular, triangular-ovate or narrowly oblong-lanceolate, 10–55 x 5–32mm, base truncate or shortly cordate. Flowers 3 to 8, in pedunculate clusters. Corolla white or pink, 6–10mm diameter.

Ecology. Dwarf shrubland (with *Pulicaria stephanocarpa* and *Croton socotranus*) and drought-deciduous shrubland on the inland and coastal plains; also in the relict *Boswellia elongata* woodland on the Homhil plateau. 100–450m.

Distribution. N and NE tropical Africa, Arabia, Iran, Afghanistan, Pakistan and NW India.

ETHNOGRAPHIC NOTES

Name(s)

– Soqotri: *ḳʌ̍ʾ* (pl. *ḳɔ̄wnˈitin*, *ḳɔ̄wnhˈitin*), *ḳʌ̍ʾ di tɛ̄tin* 'threads', 'threads for sheep' (Soqotra and Abd al Kuri); *ʿˈowźiḳ*; *ləbˈənno* 'white, pale'; (rare) *ḥariˈśimo* (more commonly given to smaller *Grewia* species).

– Soqotri simplified: *qa'* (pl. *qownitin*, *qown'hitin*), *qa' di tetin*; *'owdhiq*; *libinno*; *harishimo*.

Food (livestock). Considered a valuable and fattening graze for small stock and as being especially good for weaned lambs and kids. Cattle also graze it.

Comparative uses. Elsewhere it is used as a purgative, as are other *Convolvulus* species. (Ambasta 1982; Batanouny 1981; Jain & De Fillips 1991; Rizk & El-Ghazaly 1995)

Convolvulus grantii Balf.f.

Prostrate annual herb to 30cm, appressed-puberulent and with sparse longer, spreading hairs. Basal leaves in a rosette, spathulate, to 3cm with undulate margin; stem leaves narrowly ovate or ovate-oblong. Flowers solitary. Sepals ovate, c.4mm. Corolla white or pale pink, c.6mm diameter, shallowly lobed, hairy throughout.

Ecology. Common on Abd al Kuri in dwarf shrubland and open woody-herb communities on the coastal plain, rocky slopes and in limestone ravines on the mountains (Jebel Saleh). Sea-level–480m.

Distribution. Endemic.

Conservation status. Least Concern (LC). Common in several vegetation types and under no present or perceived threat.

Botanical notes. The only species of *Convolvulus* recorded from Abd al Kuri, where it is a widespread and common endemic. It is distinguished by its spathulate leaves which have characteristically undulate margins.

ETHNOGRAPHIC NOTES

Name(s)

– Soqotri: *ḳɔ̄wnhˈitin* 'threads' (Abd al Kuri).

– Soqotri simplified: *qown'hitin*.

Food (livestock). Key fodder for sheep and goats on Abd al Kuri, where it is seen as being fattening and promoting lactation. It is considered to be particularly valuable for young stock.

Medicinal / veterinary. Sheep and goats which are failing to thrive, or which have fallen ill, are fed on the fresh green parts, after removing any dead or discoloured material.

Convolvulus hildebrandtii Vatke

Convolvulus filipes Balf.f.

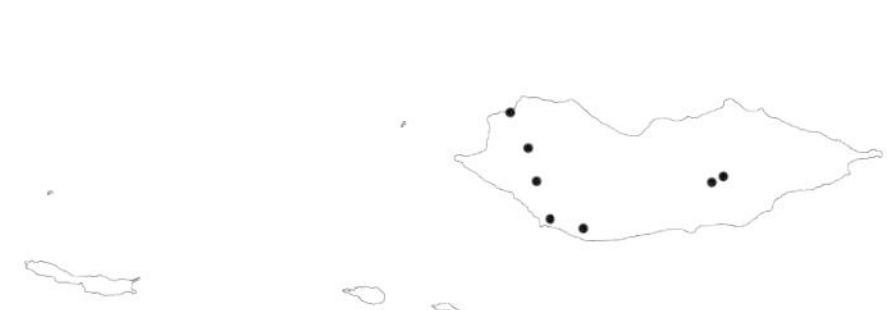

Woody-based herb with numerous slender stems, sericeous. Leaves linear, <2.5 x 0.3cm, basal rosette of narrowly ovate leaves sometimes present. Flowers 1 or 2 in pedunculate clusters, forming a long raceme. Sepals ovate, 2.5–3mm. Corolla white or off-white, 7–9.5mm diameter, unlobed, hairy throughout.

Ecology. *Croton socotranus* shrubland, succulent shrubland, dwarf shrubland (with *Pulicaria stephanocarpa*) and drought-deciduous woodland (with *Buxanthus pedicellatus* and *Cryptolepis intricata*) on the coastal plains and limestone plateau. 20–650m.

Distribution. Endemic.

Conservation status. Least Concern (LC). Common in several vegetation types and under no present or perceived threat.

Botanical notes. There is some variation within this species on Soqotra: plants from the Hadiboh plain have hairy styles whereas plants from elsewhere on the island have glabrous styles. There is also a variable correlation between this character and filament type, glabrous-styled plants having terete filaments and hairy-styled plants having filaments that are flattened towards the base. No other characters can be found that separate the Hadiboh plants as a distinct taxon. A specimen closely resembling *C. hildebrandtii* has recently been seen from S Yemen. It has flattened filament bases and glabrous styles and ovary, thus differing from both the two forms found on Soqotra in the combination of these characters.

ETHNOGRAPHIC NOTES

Name(s)

– Soqotri: *nɔ̄i*, *nˈoyi* (commonly given to sweet-smelling and/or delicate small plants); *šˈʌbiyʌḥ* 'prostrate'; *tˈaʿabəhur* 'coming into leaf at the end of the dry season'.

– Soqotri simplified: *noy*; *shabiyah*; *ta'abihur*.

Food (livestock). Important and nutritious feed for small stock in drier areas of the island, especially when green. Elsewhere it is unimportant, except for lambs and kids, though in a prolonged dry season it is more valued.

***Convolvulus rhyniospermus* Choisy**

Prostrate annual herb to 30cm, appressed-pubescent. Leaves linear to obovate, 1.5–4.5 x 0.3–0.6cm. Flowers 3 to 6, in subsessile clusters. Sepals green, lanceolate. Corolla white, pale pink or pale blue, c.5mm diameter, deeply lobed.
Ecology. Rare, but probably under-recorded, in *Croton socotranus* shrubland and succulent shrubland on the coastal plains and limestone escarpment. 50–100m.
Distribution. Egypt, Sudan, Ethiopia, Somalia and SW Arabia.

ETHNOGRAPHIC NOTES
Name(s)
– Soqotri: *nɔ̄i, no̍yi* (commonly given to sweet-smelling and/or delicate small plants).
– Soqotri simplified: *noy.*
Food (livestock). Provides incidental grazing for small stock when green. It is considered a valuable feed for young animals.

***Convolvulus sarmentosus* Balf.f.**

Prostrate, woody-based herb to 45cm, with grey or white silky appressed hairs. Leaves linear, <2.5 x 0.6cm. Flowers solitary, forming a raceme. Sepals ovate, c.3 x 1mm. Corolla white, 8.5–10.5mm diameter, unlobed, hairy throughout. Style white, stigmas red.
Ecology. Widespread in a variety of vegetation types including dwarf shrubland, drought-deciduous woodland and submontane semi-deciduous thicket; on both granite and limestone, usually in clearings or in exposed places. 50–1050m.
Distribution. Endemic.
Conservation status. Least Concern (LC). Common in several vegetation types and under no present or perceived threat.
Botanical notes. Young and old plants look very different, due to the much longer stems and subglabrous, not densely sericeous, leaves of mature specimens.

ETHNOGRAPHIC NOTES
Name(s)
– Soqotri: *ḥalɛ̄bo̍h, ḥalbe̍boh* 'milk-making'; (*di*) *ḳʌ̍ṣḳʌṣ* 'cream-coloured'; *ḳʌ̍ʾ* (pl. *ḳɔ̄wni̍tin, ḳɔ̄wnhi̍tin*) 'threads' (Soqotra and Abd al Kuri); *ta̍ʿabəhur* 'coming into leaf at the end of the dry season'; *mʌ̍ḥeriś* 'undergrowth'; *sədʌ̍ḥan* 'prostrate'; (rare) *məri̍gənoh.*
– Soqotri simplified: *haleyboh, halbeboh*; *qazqaz*; *qa'* (pl. *qownitin, qown'hitin*); *ta'abihur*; *maharish*; *sidahan*; *miriginoh.*
Food (livestock). Important and nutritious feed for small stock in the drier areas of the island, less so elsewhere, except for young lambs and kids. Cattle also eat it.
Medicinal / veterinary. Sickly sheep and goats failing to thrive are fed on the fresh green parts after removing any dead or discoloured material.

***Convolvulus siculus* L.**

Prostrate or ascending annual herb to 40cm, appressed-pubescent. Leaves ovate to oblong-ovate, 1.5–6 x 0.5–2.5cm. Flowers 1 to 3 in pedunculate axillary clusters. Sepals obovate, 6–7 x 2–2.5mm. Corolla blue, pink or lilac, 7–8mm diameter, deeply lobed, lobes hairy at tips.
Ecology. No habitat details available.
Distribution. Tropical E and NE Africa and Arabia.
Botanical notes. Collected in the 19th century but not seen recently. Probably overlooked and under-recorded. The Soqotran material is referable to subsp. *agrestis* (Hochst. ex Schweinf.) Verdc. which differs from subsp. *siculus* in its pale blue or pink, not bright blue, flowers and its pedicellate, not sessile, flowers. It has sometimes been confused in the field with *Evolvulus alsinoides* (which has two styles, each bifid), probably because of the deeply lobed, bluish corollas.

Cressa L.

Halophytic herbs. Leaves entire. Flowers subsessile in bracteate clusters. Sepals (4 or)5. Corolla campanulate, small. Stamens and styles exserted from corolla. Ovary 2-celled; ovules 4. Styles 2, separate to base; stigmas relatively large, globose. Capsule ovoid, 2 to 4-valved; seeds 1 (rarely 2). 5 species. Tropical and warm regions.

Cressa cretica L.

Prostrate, greyish-green herb to 40cm. Leaves ovate, 0.3–0.6 x 0.1–0.4cm, pubescent. Flowers initially with very strong, sweet scent. Corolla white or cream at first, later pinkish, 5–6mm diameter; lobes pubescent towards apex. Capsule 3–4mm, thinly hairy.
Ecology. Brackish soils on the coastal plains. Sea-level.
Distribution. Around the coasts of the Mediterranean region, N and tropical Africa, Arabia, SW and S Asia.
Botanical notes. This prostrate, small-leaved herb of brackish places is widespread in suitable areas (around mangrove woodland and coastal inlets) on the adjoining African and Arabian mainlands. It is apparently rare on Soqotra but is probably under-recorded.

ETHNOGRAPHIC NOTES
Name(s)
– Soqotri: nil or *ʾātel* (more commonly given to *Tamarix* species).
– Soqotri simplified: *'atel.*
Food (livestock). Rarely grazed by goats as it tends to grow on damp ground, which they avoid. Sheep and camels graze it, but it is not considered especially nutritious.
Bees / honey. Honeybees are attracted to the flowers.
Comparative uses. Elsewhere in the Middle East, the whole plant is boiled and taken as a tonic, aphrodisiac, expectorant and to stimulate the appetite. It is also given as a vermifuge, to treat constipation and as a purge in jaundice. In India the plant is used as an antiseptic, and the leaves, roots and seeds are taken as a tonic. The seeds are also ground into flour and this and the plants themselves are eaten in hunger. (Abbas *et al.* 1992; Al Rawi & Chakravarty 1964; Ambasta 1982; Batanouny 1981; Ghazanfar 1994; Goodin & Northington 1979; Jain & De Fillips 1991; Kunkel 1984; Lewis & Elwin-Lewis 1977; Rizk & El-Ghazaly 1995)

Dichondra J.R.Forst. & G.Forst.

Prostrate, mat-forming herbs. Leaves alternate, cordate. Flowers solitary, axillary, inconspicuous. Calyx deeply 5-lobed. Corolla campanulate. Stamens 5. Ovary deeply 2-lobed, each lobe with 2 ovules. Styles 2, gynobasic; stigmas capitate. Fruit a 2-lobed capsule; lobes subglobose, normally 1-seeded. 9 species. Tropical and warm regions.

Dichondra repens J.R.Forst. & G.Forst.

Creeping herb to 60cm, rooting at the nodes. Leaves <3.5 x 4cm, palmately veined from base. Calyx 2–3.5mm in flower, 4–6mm in fruit. Corolla white or greenish-white. Capsule 1.5–2mm diameter, 2-lobed, lobes subglobose.
Ecology. Growing in turf near water in wet montane grassland and woodland in the Haggeher mountains. 1000–1300m.
Distribution. Old and New World tropics and subtropics, north to Himalayas and China.
Botanical notes. An easily overlooked creeping herb with heart-shaped leaves and tiny, insignificant flowers. It typically grows through damp grass near water and superficially resembles *Centella asiatica* (Umbelliferae) which grows in similar places. A polymorphic species needing worldwide revision. It is split into several infraspecific taxa but only var. *repens* occurs on Soqotra.

ETHNOGRAPHIC NOTES
Name(s)
– Soqotri: nil or 'type of *ḥīlibə*' 'sour like buttermilk'.
– Soqotri simplified: 'type of *hiylibih*'.
Food (livestock). Goats avoid it in damp ground, but eat it when it grows under trees and shrubs. It is also grazed by sheep and cattle but is seen as having little nutritional importance.

Evolvulus L.

Herbs. Leaves alternate, entire. Flowers 1 or 2, in axillary clusters. Corolla broadly infundibular. Ovary 2-celled, ovules 4. Styles 2, filiform, each one bifid at apex; stigmas clavate to filiform. Fruit a globose, 4-valved capsule; seeds 4. 98 species. Mostly New World.

Evolvulus alsinoides L.

Decumbent herb to 60cm, with long, spreading, silky hairs. Leaves elliptic, <2.6 x 1cm. Sepals ovate-lanceolate, 2.5–3.5 x 0.5mm. Corolla bright blue, occasionally lilac or white, 5–7.5mm diameter; middle bands paler and sparsely villous beneath. Stamens white. Capsule 3–4mm.
Ecology. Occasional in clearings and amongst rocks in woodland in the northern Haggeher. 550–650m.
Distribution. Pantropical.
Botanical notes. Superficially similar to and easily confused with *Convolvulus siculus*.

ETHNOGRAPHIC NOTES

Name(s)

– Soqotri: *di ṣefirˈirihon* 'many-flowered'; *ṣəwˈeyni* 'closely knotted, woven'.

– Soqotri simplified: *di zefiririhon*; *ziweyni.*

Food (livestock). Grazed by all livestock but is not seen as having nutritional importance. It is considered to be a good feed for young stock.

Comparative uses. In India it is taken as a tonic and febrifuge. It is given as a vermifuge and applied with oil to promote growth of the hair. Elsewhere women burn the plant to fumigate the living quarters during the puerperium, and use a warm infusion as a wash during the 40 days purification. The plant is worn as an amulet to procure love and favour, and, with other herbs, it is an ingredient of a charm to exorcise evil spirit causing diseases. The Fulani add it to milk to bring success. (Ambasta 1982; Dalziel 1948, 1956; Jain & De Fillips 1991)

Ipomoea L.

Herbs or shrubs, sometimes twining. Leaves entire or palmately or pinnately lobed. Flowers axillary, solitary or in 2 to many-flowered clusters. Corolla tubular or funnel-shaped, not conspicuously lobed. Stamens 5, unequal. Ovary 2 to 4-locular, with 4 (rarely 6 to 10) ovules. Style filiform; stigma biglobular. Capsule with 3 to 10 valves; seeds 4. 500 species. Tropical and warm temperate regions.

Ipomoea cairica (L.) Sweet

Climbing perennial to 3m. Leaves broadly ovate to orbicular, deeply 3 or 5-lobed, <15 x 13cm; lobes elliptic. Sepals ovate, 5–6.5mm. Corolla mauve or purple with whitish base, funnel-shaped, 35–50mm diameter. Capsule subglobose, 1–1.2cm; seeds dark grey to orange-brown, densely short tomentose.

Ecology. Apparently rare but probably under-recorded; in semi-deciduous shrubland dominated by *Buxanthus pedicellatus* on the eastern limestone plateau. 300m.

Distribution. Native of America; widely naturalised in Africa and Asia.

ETHNOGRAPHIC NOTES

Name(s)

– Soqotri: *eṣənīno* (*di beʿˈeyr*) 'male camel's *Ipomoea*'; *eṣˈinnoh* 'close-meshed'.

– Soqotri simplified: *ezinino* (*di be'eyr*); *ezinnoh*.

Food (livestock). Like all plants of this name, it is one of the best fodders for cattle, especially for milk cows. It makes them fat and increases their milk. All parts of the plant are eaten: foliage, stems, flowers and fruit. Small stock dig the plant up by the roots and eat these when they are hungry. Goats and camels eat the plant and grow sleek and fat on it. It is gathered to feed to calves being weaned and is offered to a bereaved cow whose male calf has been slaughtered, in order to encourage her to let down her milk.

Flavouring / food additive. Milk animals feeding on a lot of it give particularly sweet and rich milk.

Medicinal / veterinary. The aerial parts are gathered and fed to sick animals or to those failing to thrive.

Other. To repair pottery, the fresh green leaves and stems of *Ipomoea* are crushed and then ground with a shard of pottery. The paste is applied to the inside of the pot, smeared over and around the hole or crack. Then a heavily pregnant cow or goat is milked into the hand and the sticky secretion smeared over the paste and surrounding area. The pot is then put beside the fire for the mixture to harden. Alternatively, if the hole is quite large, the paste is smeared round a suitably sized small pebble or piece of clay inserted into the hole from the outside. The paste is smeared around the area, inside and out, and the pot is put beside the fire for the paste to set hard.

Notes. One of the worst plants for causing *'uqabah* in small stock unaccustomed to the feed. The foliage of the sweet potato, *I. batatas*, which is cultivated on Soqotra, is also fed to sickly livestock and some herders in the interior have begun to grow the plant for its nutritious foliage. Tests have shown the fodder value of the leaf to be extremely high.

Comparative uses. In nearby Dhofar this plant was also collected to feed to livestock, especially to milk animals to increase their yield. The tuberous roots and stems, though bitter, are used as food in Hawaii. The roots and leaves are slightly cyanogenic. The seeds are used as a purgative. Many other *Ipomoea* species have medicinal uses, especially as purgatives. The leaves of the related *I. repens* are said to cause diarrhoea in horses, and camels eating the plant by accident are said to go mad and die. (Ambasta 1982; Bebawi & Neugebohrn 1991; Dalziel 1948, 1956; Jain & De Fillips 1991; Kunkel 1984; Miller & Morris 1988)

Ipomoea kotschyana Hochst. ex Choisy

Ipomoea laciniata Balf.f.

Prostrate perennial to 60cm, hispidulous. Leaves oblong-ovate, <3 x 2.5cm, pinnatifid to pinnatisect with 3 to 13 lobes. Flowers opening in evening. Corolla pink with darker centre, c.11mm long, narrowly tubular. Capsule broadly globose, with persistent styles, 5–7mm diameter; seeds pale brown with dark spots and patches of pubescence.

Ecology. Apparently rare but possibly overlooked; in drought-deciduous woodland on the limestone escarpment in the east (Hoq) and a 19th century record from near Qalansiyah. c.300m.

Distribution. E and NE Africa.

Botanical notes. The type of *I. laciniata* Balf.f. represents a very stunted and exposed form of *I. kotschyana*.

ETHNOGRAPHIC NOTES

Not seen, no information recorded.

Ipomoea nil **(L.) Roth**
Ipomoea scabra Forssk.

Vigorous climbing annual to 3m. Leaves broadly triangular, often 3-lobed, <15 x 15cm, base cordate, lobes abruptly acuminate. Sepals ovate, 4–6 x c.4mm, gradually attenuate to a 12–25mm awn. Corolla blue fading pink, 30–50mm diameter. Capsule 8–10mm; seeds blackish, glabrous.
Ecology. Apparently rare but possibly overlooked; in drought-deciduous woodland. c.550m.
Distribution. Native of tropical America, now widely naturalised in the Old World tropics.

ETHNOGRAPHIC NOTES
Name(s)
– Soqotri: *eṣenˈino*, *eṣˈinnoh* 'close-meshed'.
– Soqotri simplified: *ezenino*, *ezinnoh*.
Uses. See *I. cairica*.
Notes. Among the worst plants for causing *'uqabah* in small stock unaccustomed to the feed.
Comparative uses. The leaves and shoots are eaten in hunger. In nearby Dhofar the seeds of *I. nil* were ground with water to make a shampoo which was used to kill head lice as well as to wash the dead. The same paste was applied around the eyes to improve vision. The dried seeds were crushed, mixed with oil, and rubbed on the chest to relieve chest pain and alleviate breathing difficulties. Ground seeds were added to milk and used as a poultice to treat mastitis, or to water and salt and applied around the eyes to improve weak sight. Dried whole fruit were boiled in water and applied to painful and inflamed eyes. The seeds contain lysergic acid which has hallucinogenic properties. In Pakistan the plant is said to control aphids. (Ambasta 1982; Jain & De Fillips 1991; Kunkel 1984; Lewis & Elwin-Lewis 1977; Miller & Morris 1988; Mossa *et al*. 1987)

Ipomoea obscura **Ker Gawl.**

Trailing or climbing annual to 2m. Leaves cordate, <12 x 8cm, apex bluntly acuminate. Sepals ovate, 4–6mm long. Corolla cream, yellow, or orange, with dark red to blackish centre, funnel-shaped, 13–26mm. Capsule globose with persistent style base, 8–12mm diameter; seeds black, appressed-pubescent.
Ecology. Low shrubland and drought-deciduous woodland on both the granite and limestone. 450–1100m.
Distribution. Tropical and S Africa, Madagascar, Arabia and tropical Asia to China and Australia.

ETHNOGRAPHIC NOTES
Name(s)
– Soqotri: *eṣenˈino*, *eṣˈinnoh*.
– Soqotri simplified: *ezenino*, *ezinnoh*.
Uses. See *I. cairica*.
Other. See *I. cairica*.
Notes. One of the worst plants for causing *'uqabah* in small stock unaccustomed to the feed.
Comparative uses. In Dhofar the seeds of *I. obscura* were also ground and mixed with water to make a foaming shampoo, used to kill head lice or to wash with. Elsewhere the aromatic and mucilaginous leaves are cooked and eaten. They are also roasted, powdered, boiled in ghee and applied in aphthous affections. (Ambasta 1982; Jain & De Fillips 1991; Kunkel 1984; Miller & Morris 1988)

Ipomoea pes-caprae **(L.) Roth**
Ipomoea biloba Forssk.

Vigorous perennial with prostrate stems to 5m, somewhat succulent. Leaves broadly obovate, 2–6 x 2–6cm, deeply bilobed. Flowers open in morning. Sepals ovate, <11mm. Corolla pink or mauve, funnel-shaped, 40–55mm diameter.

Ecology. Beaches. Sea-level.
Distribution. Around the coasts of the Indian Ocean.
Botanical notes. Not seen recently on Soqotra but recorded from the beach at Qadhub in 1957 and from near Hadiboh in the 19th century. A conspicuous plant which is unlikely to have been overlooked in recent surveys of the island: it therefore seems likely that it is now extinct on Soqotra. Soqotran material is referable to subsp. *pes-caprae*.

Ipomoea sinensis **(Desr.) Choisy**

Trailing annual herb to 1m. Leaves cordate, <6 x 6cm, apex acute or abruptly acuminate. Sepals truncate at base, fringed with stiff hairs. Corolla pink or lilac with pale mid-petal bands, tubular below and spreading above, 7–15 x 12–20mm. Capsule globose, 6–10mm; seeds densely velutinous.
Ecology. Widespread in succulent shrubland and semi-deciduous montane woodland. Sea-level–650m.
Distribution. Tropical Africa and Arabia.
Botanical notes. Material on Soqotra is all referable to var. *blepharosepala* (A.Rich.) Meese.

ETHNOGRAPHIC NOTES

Name(s)

– Soqotri: *eṣenˈino, eṣˈinnoh; šˈɛnə.*

– Soqotri simplified: *ezenino, ezinnoh*; *shenih*.

Food (livestock). Loved by all livestock, in particular small stock and young animals.

Flavouring / food additive. See *I. cairica.*

Other. See *I. cairica.*

Notes. One of the worst plants for causing *'uqabah* in small stock unaccustomed to the feed.

Comparative uses. In East Africa the leaves are cooked and eaten as a vegetable. (Heine *et al.* 1988)

Metaporana N.E.Br.

Shrubs. Inflorescence a panicle. Sepals free, persistent, scarcely enlarging in fruit. Corolla campanulate with spreading, emarginate lobes. Stamens inserted ± at base of corolla tube. Ovary 2-celled, ovules 4. Styles 2, separate to base, filiform; stigmas capitate. Capsule ovoid, indehiscent; 1-seeded. 1 endemic. 4 species. E Africa and Madagascar.

Metaporana obtusa (Balf.f.) Staples

Porana obtusa Balf.f.

Evergreen shrub to 3m. Leaves ovate to oblong-elliptic, 2–8 x 1–5cm, base truncate, apex obtuse or retuse. Sepals 5–7mm. Corolla white, c.5mm, hairy and gland-dotted outside. Capsule 4–5mm, black, persistent calyx lobes spreading, brown.

Ecology. Dominant in the drought-deciduous woodland which follows the wadis in the Zahr basin (such as Wadi Meyhah); also rare in the western part of the Noged escarpment. 100–500m.

Distribution. Endemic.

Conservation status. Vulnerable (VU D2). Rare, with a very limited area of occupancy.

Botanical notes. This is the first shrub in the west to come into leaf after rain and last to die back with the onset of dry conditions.

ETHNOGRAPHIC NOTES

Name(s)

– Soqotri: *ṣˈaᶜalhil*, (western Soqotra) *ṣˈaġalhil* (pl. *ṣˈaᶜalhʌl, ṣˈaġalhʌl*) 'lovely', 'large-leaved', '(animal with) large and floppy ears'.

– Soqotri simplified: *za'alhil*; *zaghalhil* (pl. *za'alhal, zaghalhal*).

Food (livestock). The foliage is the most important fodder plant of western Soqotra. The large leaves remain green throughout the year, and are therefore a very important dry season feed. It is not seen as being particularly valuable nutritionally, but is a vital maintenance feed. Herders do not have to lop it for their goats as the shrubs grow low enough for the goats to be able to feed themselves. In the monsoon months the foliage can become unpleasantly salty. At this time the goats prefer to browse the branch tips, and eat the leaves when they have been brought down to the ground by the strong winds. One or two weeks after they have fallen they become less salty and are quite palatable. Sheep eat very little of the shrub, although when hungry they nibble at the plant litter.

Bees / honey. An important plant for honeybees in western Soqotra.

Flavouring / food additive. The milk of goats browsing heavily on it has a much-liked characteristically sweet flavour.

Fuel. The dead wood makes very good firewood, which burns with a bright light and gives off a good heat. Stores of this wood are commonly kept in the home for this reason. It makes little smoke, so is also good for cooking. The smell of the burning wood is much liked in the areas where the shrub grows and the smell of the smoke is readily recognised on people from these areas. Larger pieces of firewood burn very slowly and this wood was much used during the summer and monsoon months. At this time, livestock theft was widespread, and this wood remained alight throughout the night, suggesting to thieves the presence of a wakeful night-time guard. The inner wood of dead branches makes good tinder and was collected and kept at home for this purpose.

Timber. Too precious as livestock fodder to be cut and used as timber.

Wood. Not suitable for making implements, and indeed the shrub is so strongly protected that cutting branches, or even twigs, is frowned on.

Medicinal / veterinary. The green leaves and stems are gathered and fed to sickly animals, or given to parturient goats to help them regain their strength and make plenty of milk. The wood is often burned after childbirth for its pleasant smell and to fumigate the mother and her baby.

Cleansing. The dead wood is burned for its sweet smell.

Pest control. Green wood gives off a dense insecticide smoke, but the removal of green twigs for this purpose is generally frowned on, as the foliage is such an important livestock fodder.

Other. Animals feeding on this can go longer without water.

Comparative uses. The leaves and peduncles of related species are eaten in India. (Kunkel 1984)

Seddera Hochst.

Subshrubs or woody-based herbs, sometimes spinescent. Leaves small, often silky hairy. Flowers in racemes, dense clusters or solitary in axils. Corolla campanulate or funnel-shaped. Filaments often winged at base. Ovary globose or ovoid. Style bifid to base or appearing simple with a bifid tip; stigmas globose, sometimes pyriform. Capsule 4-valved; seeds few. 4 endemics. 15 species. Arabia, Africa and Madagascar.

Botanical notes. Includes an endemic section, *Socotroseddera* Hallier f., which contains three species: *S. fastigiata*, *S. semhahensis* and *S. spinosa* – each restricted to its own island and possibly representing an example of speciation through adaptive radiation.

Seddera fastigiata **(Balf.f.) Verdc.**
Breweria fastigiata Balf.f.
Convolvulus fastigiata (Balf.f.) Verdc.
Convolvulus socotrana Verdc.

Spiny shrub to 1m, grey or silvery sericeous. Leaves sessile, linear-lanceolate, <0.6 x 0.1cm, cuneate at base. Flowers solitary, subsessile. Corolla white, funnel-shaped, 5.5–6mm long, hairy on outside on exserted portion. Style c.4mm, appearing simple but separable into 2. Capsule included in calyx, c.3mm long.
Ecology. Sandy and pebbly soils on the desertic coastal plains, in dwarf shrubland and succulent shrubland. Sea-level–100m.
Distribution. Endemic.
Conservation status. Vulnerable (VU D2). Rare, with a very limited area of occupancy.

ETHNOGRAPHIC NOTES†
Name(s)
– Soqotri: (western Soqotra) *zi̍dire*, *zēdiro* 'fat-forming'; *li̍bəhʌn* 'white, pale'; *sə̍dʌk* (more commonly given to *Lavandula*); *kə̍mbəher* 'something small which has fallen into something else, as a fly into milk'.
– Soqotri simplified: *zidiri*, *zeydiro*; *libihan*; *sidak*; *kimbiher.*
Food (livestock). Regarded as an excellent and nutritious fodder for goats, especially in the drier areas of the island. When green it makes them fat and increases milk yield. Sheep also graze it and do well on it. It persists through the dry season and is also important at that time, for small stock in particular.
Flavouring / food additive. The milk of animals eating a lot of it is characteristically sweet and much liked.
Bees / honey. Honeybees are attracted to the flowers.

†Also apply to *S. glomerata*.

Seddera glomerata **(Balf.f.) Schwartz**
Bonamia glomerata (Balf.f.) Hallier f.
Breweria glomerata Balf.f.
Convolvulus glomerata Balf.f.

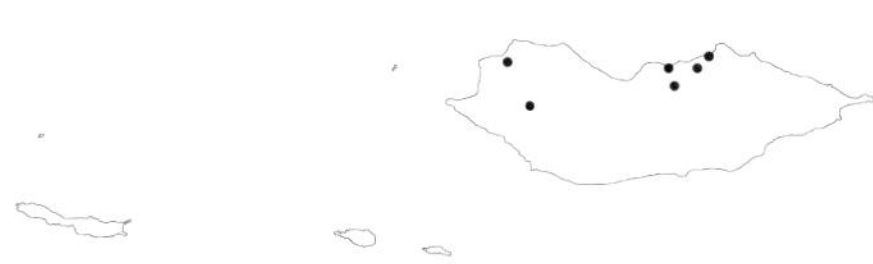

Woody-based herb to 30cm. Leaves ± sessile, elliptic, <1.5 x 1cm, silvery-hairy. Flowers in dense, terminal heads 6–20mm diameter, each flower subtended by a plumose bracteole. Calyx lobes c.6mm long. Corolla white, funnel-shaped, c.5mm long. Style bifid to base. Capsule broadly ovoid, densely sericeous.
Ecology. Dwarf shrubland and succulent shrubland in desertic plains and limestone escarpments. 50–600m.
Distribution. SW Arabia, Ethiopia and Somalia.
Botanical notes. Readily distinguished by the globose, feathery flowering head.

ETHNOGRAPHIC NOTES
Name(s)
See *S. fastigiata* (also uses).

Seddera latifolia **Hochst. & Steud.**
Breweria latifolia Benth. & Hook.f.

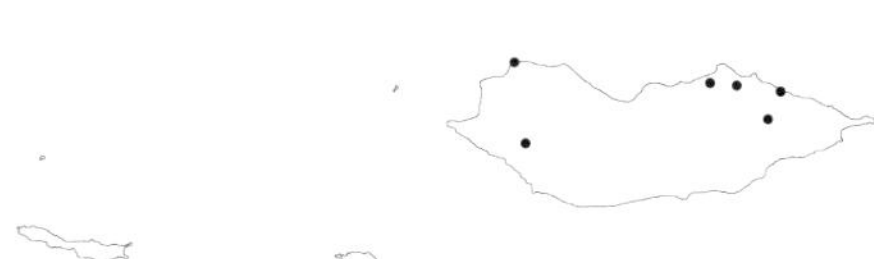

Subshrub to 50cm. Leaves sessile or subsessile, suborbicular or broadly elliptic, <1.2 x 0.6cm. Flowers subsessile in spike-like racemes. Corolla white, c.3mm diameter; lobes spreading. Stamens just exserted from corolla tube. Styles bifid to base; stigmas relatively large, globose-peltate. Capsule ovoid, c.4.5 x 3mm, loosely appressed-sericeous in upper third.
Ecology. *Croton* shrubland, succulent shrubland and dwarf shrubland. 5–250m.
Distribution. NE Africa and Arabia east to NW India.

ETHNOGRAPHIC NOTES
Name(s)
– Soqotri: *kə̍mbəher* 'something small which has fallen into something else, as a fly in milk'; *li̍bəhan* 'white, pale'; (western Soqotra) *śe̍rxʌ* 'root'; (western Soqotra) *ze̍ydiro*, *zidi̍rə* 'fat-forming' (given to various prickly or spiny, low, grey-coloured shrubs, as *Indigofera marmorata*, *I. pseudointricata*, *Campylanthus spinosus*; also to tough perennial grasses such as *Dactyloctenium* or *Panicum rigidum*).
– Soqotri simplified: *kimbiher*; *libihan*; *sherkha*; *zeydiroh*, *zidirih*.
Food (livestock). Regarded as an excellent and nutritious fodder for goats, especially in the drier areas of the island. When green it makes them fat and increases their milk yield. Camels also browse this heavily when they find it. Sheep eat the leaves when they are green.
Flavouring / food additive. Goats in particular, but also camels, browsing heavily on these plants give very sweet milk that is much liked.
Bees / honey. Honeybees are attracted to the flowers.
Fuel. The dead and dried plants are gathered and stored to use as tinder.

Seddera pedunculata (Balf.f.) Verdc.

Breweria pedunculata Balf.f.

Twiggy subshrub to 60cm. Leaves subsessile, oblong, <2.2 x 2mm, densely sericeous. Flowers subsessile in spike-like racemes. Corolla white, campanulate, c.4.5mm diameter. Styles bifid to base. Capsule broadly conical, shortly pubescent at tips.

Ecology. No habitat details available.
Distribution. Endemic.
Conservation status. Data Deficient (DD). Inadequate evidence is available to make an assessment.
Botanical notes. Known only from a poor specimen collected by Balfour on the 'plains near Galonsir' [Qalansiyah]. It is possibly synonymous with *S. hallieri* Engl. & Pilg. from Somalia but more material is needed to assess the status of *S. pedunculata*.

Seddera semhahensis R.R.Mill

Spiny shrub to 50cm. Leaves shortly petiolate, narrowly lanceolate, <0.8 x 0.2cm, base obtuse, densely grey-sericeous. Flowers solitary, subsessile. Sepals spreading at tip. Corolla white, funnel-shaped, 7–8.5mm long, hairy to base on outside. Style c.5mm long, simple, bilobed at apex; stigmas pyriform. Capsule not seen.
Ecology. Open dwarf shrubland with *Atriplex griffithii*. 100–200m.
Distribution. Endemic.
Conservation status. Vulnerable (VU D2). Rare, with a very limited area of occupancy.
Botanical notes. Endemic to Samha. A much greyer plant than either *S. fastigiata* (Soqotra) or *S. spinosa* (Abd al Kuri) and distinguished from them by the corolla tube hairy to the base, not merely on the part that is exserted from the calyx. The flowers open in the morning.

ETHNOGRAPHIC NOTES

Name(s)
– Soqotri: *mərˈigə, mərˈəgo* 'pulls out teeth' (more commonly given to *Campylanthus spinosus*).
– Soqotri simplified: *mirigih, mirigo.*
Food (livestock). On Samha this is one of the most important plants for small stock, especially sheep. It persists into the dry season and is thus an important dry season forage.
Flavouring / food additive. Animals feeding heavily on the plants give very sweet milk which is much liked.
Fuel. The dead and dried plants are gathered and stored to use as tinder and kindling.

Seddera spinosa (Vierh.) Verdc.

Bonamia spinosa Vierh.
Convolvulus kossmatii Vierh.
Convolvulus spinosa Vierh.

Spiny shrub to 60cm; branches rigid, dark green. Leaves sessile, linear-lanceolate, <1.2 x 0.2cm, sericeous or pubescent. Flowers solitary, subsessile. Corolla white, funnel-shaped, c.13mm long, hairy on outside on exserted portion. Style c.8.5mm, appearing simple but separable into 2. Capsule not seen.
Ecology. Common on the coastal plains in open dwarf shrubland and woody-herb communities. Sea-level–70m.
Distribution. Endemic.
Conservation status. Vulnerable (VU D2). Rare, with a very limited area of occupancy.
Botanical notes. Endemic to Abd al Kuri. Similar to *S. semhahensis* but leaves and stems bright green not grey-green. The flowers open in the morning.

ETHNOGRAPHIC NOTES

Name(s)
– Soqotri: *zēdirəh, zˈidirə, zˈeydiro* 'fat-forming' (Abd al Kuri) (given to various prickly or spiny, low, grey-coloured shrubs, such as *Indigofera marmorata*, *I. pseudointricata*, *I. nugalensis*, *Campylanthus spinosus*; also to tough perennial grasses such as *Dactyloctenium* or *Panicum rigidum*).
– Soqotri simplified: *zeydirih, zidirih, zeydiroh.*
Food (livestock). One of the most important and nutritious forage plants for goats on Abd al Kuri. Sheep also eat the fresh growth. It makes the animals fat and increases their milk yield.
Flavouring / food additive. Animals feeding heavily on this give very sweet milk which is much liked.
Fuel. The thicker stems are used as firewood on Abd al Kuri. They make little smoke and this makes them the preferred wood for cooking. The dead slimmer stems are gathered and stored for use as tinder. This material lights very readily and is used as kindling and to light fires. When firing pottery, a torch of this plant is used to set light to the wood piled around and over clay pots. It is used in the same way to set light to the pit in which quicklime is made.
Notes. It is said to be on the increase on Abd al Kuri.

CRASSULACEAE

Herbs or small shrubs, often succulent. Leaves opposite or alternate, usually simple and entire; stipules absent. Flowers bisexual, regular, axillary or in terminal cymes or spikes, 4 to 10-merous. Sepals and petals free or united. Stamens equal to or twice as many as petals, free or united to petals. Ovary superior; cells equal in number to sepals, free or slightly fused at base, tapering above into style; ovules few to many. Fruit free follicles. 33 genera. 1500 species. Almost cosmopolitan; especially diverse in S Africa.

Crassula L.

Tillaea L.

Herbs, often fleshy. Leaves opposite, small, united at base to form a sheath around stem. Flowers 3 to 5-merous, in 1 to 4-flowered axillary clusters. Sepals shortly united at base. Petals very shortly united at base. Stamens 3 to 5, attached to base of corolla. Follicles 3 to 5, free or slightly united at base; 2-seeded. c.200 species. More or less cosmopolitan, especially tropical and S Africa.

Crassula alata (Viv.) Berger

Glabrous ephemeral herb; stems to 5cm. Leaves lanceolate, <0.4 x 0.2cm, shortly aristate. Flowers 5-merous, 2 per leaf axil; pedicels <2mm. Calyx lobes lanceolate, acuminate, aristate. Petals white or greenish, narrowly ovate, <0.8mm long, acuminate. Seeds ridged.

Ecology. Rare but possibly overlooked in montane grassland. c.1200m.

Distribution. (subsp. *pharnaceoides*) Cameroun, tropical NE and E Africa, SW Arabia.

Botanical notes. All Soqotran material is referable to subsp. *pharnaceoides* (Fisch. & Mey.) Wickens & Bywater. This differs from subsp. *alata* (from the drier parts of the eastern Mediterranean, NE Africa and the Middle East to Iran and NW India) in its 5-merous (with five petals and sepals) flowers not 3 or 4-merous flowers.

ETHNOGRAPHIC NOTES

Comparative uses. The leaves and roots of related species are edible. (Kunkel 1984)

Crassula schimperi Fisch. & C.A.Mey.

Glabrous perennial herb; stems sprawling, to 15cm, often rooting at nodes, sometimes woody-based. Leaves succulent, linear to lanceolate, <0.8 x 0.2cm. Flowers (4 or)5-merous, 1 to 4 per leaf axil, subsessile or shortly pedicellate. Calyx lobes triangular-ovate, acute to acuminate. Petals whitish, greenish or pinkish, ovate, <1.5mm long, acute to acuminate. Seeds smooth.

Ecology. Common amongst rocks in montane semi-deciduous shrubland in the eastern Haggeher mountains. 700–900m.

Distribution. Uganda, Kenya, Tanzania, Sudan and Ethiopia.

Botanical notes. All Soqotran material is referable to subsp. *phyturus*, which differs from subsp. *schimperi* in its thick, rigid, discolorous leaves and the unspurred leaf sheaths.

ETHNOGRAPHIC NOTES

Not seen, no information recorded.

Kalanchoe Adans.

Succulent herbs or subshrubs, glabrous. Leaves opposite, simple, entire. Flowers 4-merous, in terminal panicles. Calyx lobes ± free to the base. Corolla tubular with spreading lobes. Stamens 8, in 2 whorls of 4, epipetalous. Ovary of 4 free cells; ovules numerous. 2 endemics. 125 species. Old World tropics, especially Africa and Madagascar.

ETHNOGRAPHIC NOTES

Comparative uses. Several *Kalanchoe* species are eaten as a vegetable, and the sour leaves of *K. crenata* are chewed raw. The heated leaves of various species are rubbed over the body to treat stiff joints and rheumatism and the roots used to procure an abortion. The African *K. crenata* is a widely used medicinal plant. The juice squeezed from a warmed leaf is dripped into the ear for earache; a warmed leaf is laid on the forehead for a headache; leaves are boiled or macerated in water and used as a sedative for asthma and palpitations; the leaf juice, mixed with fat or oil, is rubbed on abscesses, swellings, ulcers and burns, and over the bodies of young children with fevers. (Dalziel 1948, 1956; Kunkel 1984; Sofowora 1982; Watt & Breyer-Bandwijk 1962)

Kalanchoe farinacea Balf.f.

Herb or subshrub to 30cm. Leaves whitish green, erect, imbricate, obovate to orbicular, <5.5 x 3.5cm, subsessile, petiole not broadened at base. Flowers erect. Calyx lobes triangular, <2mm. Corolla bright red; tube 1–1.5cm; lobes ovate-oblong, apiculate. Upper anthers slightly exserted, lower anthers held at mouth of corolla tube.

Ecology. Amongst rocks in succulent shrubland and drought-deciduous shrubland. 100–800m.

Distribution. Endemic.

Conservation status. Least Concern (LC). Common in several vegetation types and under no present or perceived threat.

Botanical notes. Distinguished from the other species of *Kalanchoe* on Soqotra by the white or whitish green stems and leaves which become scaly and peeling when dry.

ETHNOGRAPHIC NOTES

Name(s)

– Soqotri: *bə̍gələhān* (or *bə̍gələhān ḳi̍ḳəhe* 'small-leaved *Kalanchoe*', to differentiate it from *K. robusta*).

– Soqotri simplified: *bigilihan* (or *bigilihan qiyqihe*).

Food (livestock). Not palatable to livestock, though goats graze it in drought when there is no other forage available. Sheep do not eat it, but it is said that camels and donkeys do so when really hungry.

Leatherwork. The whole plant is crushed and rubbed into a skin or hide after removing the hair. This makes the leather strong and supple as well as dyeing it a clear red-brown. Some people use only the root and say that this produces a darker colour.

Medicinal / veterinary. The plant is crushed and the juice squeezed over festering sores, ulcers and inflamed skin in both people and their animals. In the latter it is said to be particularly effective in treating sores on the udder.

Cleansing. The plant is beaten with a stone until the juice flows and then used as a cleansing agent. The pulp is used to wash the hands to remove dirt, sticky resin or latex and unpleasant smells. This cleanser is one of the most effective ones used by those harvesting aloes for removing the extremely bitter aloes juice from their hands. It is also used when lopping leaves from *Euphorbia arbuscula* as drought fodder for goats to remove the caustic and sticky latex from their skin.

Other. The whole plant is held over the flames until charred and hot. It is then crushed and the juice used to decorate clay pots while they are still red-hot from firing. Sometimes only the leaf juice is used. This spits and sizzles on the hot clay surface, and dyes the clay a dark red-brown. Rather than being carefully 'painted' on, the juice is usually just squeezed over and around the mouth of the hot pot and allowed to run down the sides. Sometimes *Boerhavia diffusa* or *Zaleya pentandra* leaves are crushed with the *Kalanchoe* to make a more viscid mixture. A lump of this is then held in the hand and rubbed around the neck of the pot. The plant was also used to make a dye or an ink. The whole plant was put in a clay pot on the fire and, once hot and bubbling, the expressed juice could either be used as a red-brown ink to mark property, or as a dye for specialist leather items such as the intricately plaited leather belt formerly worn by men. As a game, children rub the leaves gently between the hands to soften them and loosen the outer skin. The leaf is then inflated and popped with a loud bang.

Kalanchoe robusta Balf.f.

Kalanchoe abrupta Balf.f.

Subshrub to 50cm. Leaves glaucous, obovate, <6.5 x 3.5cm, subsessile or shortly petiolate, petiole slightly broadened at base. Flowers horizontal. Calyx lobes triangular, <4mm. Corolla orange-scarlet, slightly zygomorphic; tube 2–3cm; lobes ovate-oblong, rounded. Anthers exserted.

Ecology. Rocky slopes and amongst limestone boulders in dwarf shrubland and low succulent shrubland. 300–750m.

Distribution. Endemic.

Conservation status. Vulnerable (VU D2). Rare, with a very limited area of occupancy.

Botanical notes. A very distinctive species readily distinguished from all other species in the region by its horizontally held, slightly zygomorphic flowers in which the anthers are exserted from the corolla. Not as freely flowering as *K. farinacea*.

ETHNOGRAPHIC NOTES
Name(s)
– Soqotri: *bˈəgələhān* (or *bˈəgələhān ṣˈáʿalhʌl* 'large-leaved *Kalanchoe*', to differentiate it from *K. farinacea*).
– Soqotri simplified: *bigilihan* (or *bigilihan za'alhal*).

Food (livestock). Browsed by hungry goats in the dry season, though it is not palatable to sheep.
Leatherwork. Less effective than *K. farinacea* for working leather.
Medicinal / veterinary. As *K. farinacea*, though it is said to be less effective.

***Kalanchoe rotundifolia* Haw.**

Erect herb to 50cm. Leaves glaucous, ovate to broadly obovate, <5 x 2.5cm, shortly petiolate, petiole not broadened at base. Flowers erect. Calyx lobes subulate or narrowly triangular, <2.5mm. Corolla scarlet or red; tube <1cm; lobes linear, acute. Anthers included in corolla tube.
Ecology. Common amongst rocks in montane semi-deciduous shrubland in the Haggeher; also in the relict woodland at Qatariyah. 500–1300m.
Distribution. Tanzania, Mozambique, Zimbabwe, Botswana, Namibia and S Africa.

Botanical notes. The Soqotran populations of *K. rotundifolia* are a distant outlier of the species, the nearest station outside Soqotra being in Tanzania. There are 19th century records from Homhil in eastern Soqotra and from Abd al Kuri but it has not been seen at either place recently.

ETHNOGRAPHIC NOTES
Name(s)
– Soqotri: *bˈəgələhān*; *di ʿˈʌźəhən*.
– Soqotri simplified: *bigilihan*; *di 'adhihin*.
Food (livestock). Not normally palatable to livestock, though goats graze it in drought when no other forage is available.
Notes. Other uses as *K. farinacea*.

Umbilicus DC.

Glabrous, fleshy herbs with tuberous or rhizomatous rootstocks. Basal leaves long-petiolate, circular, peltate. Inflorescence a terminal, many-flowered raceme. Flowers 5-merous. Sepals small. Corolla tubular or urceolate with short lobes. Stamens (5 to)10, inserted on corolla. Carpels 5. Follicles many-seeded. 18 species. Europe, Mediterranean east to Iran, African and Arabian mountains.

***Umbilicus horizontalis* (Guss.) DC.**

Herb to 30cm. Leaves circular, 3–9cm diameter, margin crenate; petioles <10cm. Flowers horizontal or pendulous. Corolla pale green, sometimes reddish tinged, tubular or urceolate, <7 x 4.5mm; lobes ovate, shortly acuminate.

Ecology. Rare, only collected once, in Wadi Ayheft, in damp and shady rock crevices. 1200m.
Distribution. Mediterranean region and SW Asia.

ETHNOGRAPHIC NOTES
Name(s)
– Soqotri: nil.
Food (livestock). One of the *ru'ud* group of small plants which come up amongst other greenery with the rains, providing incidental grazing until they die back. It is not considered to be nutritionally valuable.

CRUCIFERAE

Herbs or subshrubs. Leaves alternate, entire or variously divided; stipules absent. Flowers bisexual, characteristically with 4 free sepals, 4 free petals and 6 stamens (2 short, 4 long). Ovary superior, 2-celled; placentation parietal. Fruit dry, either a siliqua (elongate) or a silicula (short), dehiscing by 2 valves, sometimes with an indehiscent upper segment; 1 to 2 to many-seeded. 390 genera. 3000 species. Mainly temperate.

Capsella Medik.

Herbs with simple, forked or stellate hairs. Leaves entire to pinnatipartite. Petals white. Fruit a many-seeded, compressed obtriangular silicula with a narrow septum. 5 species. Temperate.

Capsella bursa-pastoris L.

Erect, annual herb to 50cm. Basal leaves in a rosette, narrowly obovate, <8cm long, attenuate at base into a short petiole; stem leaves smaller, dentate to entire, clasping stem at base with acute auricles. Petals <2.5mm long. Silicula <10 x 6mm.

Ecology. Rare in grassland and between rocks in submontane semi-deciduous woodland. 1000–1200m.

Distribution. A cosmopolitan weed.

Botanical notes. In Britain known as shepherd's purse because of the distinctive shape of the fruit.

ETHNOGRAPHIC NOTES

Name(s)

– Soqotri: nil or *ruʿud* 'grazing plant'.

– Soqotri simplified: *ru'ud.*

Food (livestock). One of the small, ephemeral *ru'ud* group of plants which come up with the rains, providing incidental green grazing until they die back. Though palatable to livestock, this is not important forage for them.

Comparative uses. The leaves are used in salads and as a potherb, and the young leaves are pickled. This plant is used in Chinese medicine for ailments of the eyes and dysentery. It is used as a diuretic, febrifuge and haemostatic. The seeds yield a fatty oil. In the Middle East it is used as an astringent and is considered effective against uterine haemorrhage. A homeopathic tincture is used for nasal and internal bleeding and to treat cystitis and stones in the urinary tract. An infusion of the dried plants is given as a remedy for menorrhagia and dysmenorrhoea. (Ambasta 1982; Goodin & Northington 1979; Jain & De Fillips 1991; Kunkel 1984)

Diceratella Boiss.

Herbs, densely covered with branched and stellate hairs. Petals lilac. Fruit a tardily dehiscent siliqua, with the valves prolonged into 2 horns at the tip; seeds in 1 row with transverse septa separating them. 7 species. NE Africa and Iran.

Diceratella incana Balf.f.

Diceratella sinuata (Franch.) Oliv.

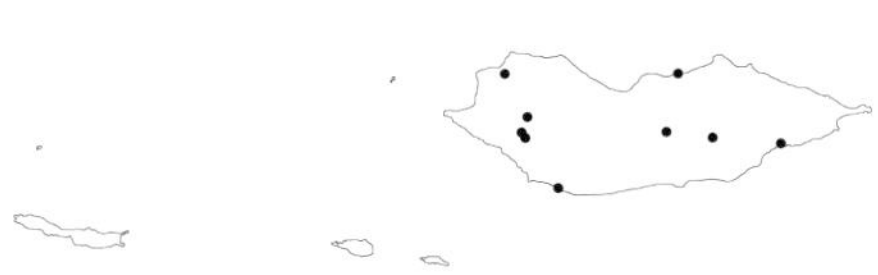

Stems erect to 60cm. Leaves ovate, <3 x 1.25cm, entire to sinuate-dentate, attenuate below into a short petiole. Petals <25mm. Siliqua linear, <17(–26) x 2.5mm, horns c.1mm.

Ecology. Sandy soils on coastal plains and rocky limestone slopes; in *Croton socotranus* shrubland, succulent shrubland and dwarf shrubland. 50–350m.

Distribution. Somalia, Ethiopia and Kenya.

Botanical notes. Readily distinguished by the pretty lilac flowers and the two horns on the tip of the fruit.

ETHNOGRAPHIC NOTES

Name(s)

– Soqotri: nil or *ʿīṭef bāṭaḥ*; *di ḳəwrēni* 'two-horned'; *məlbōwnhin* 'white, pale'; *di bəgəhun.*

– Soqotri simplified: *'iytef batah*; *di qewreyni*; *milbownhin*; *di bigihun.*

Food (livestock). Provides incidental drought fodder only.

Bees / honey. Honeybees are attracted to it, especially during the monsoon season when little else is flowering.

Comparative uses. In East Africa the bark of *D. siniata* is used for cordage and to clean out containers. (Heine *et al.* 1988)

Erucastrum C.Presl.

Herbs, glabrous or clothed with simple hairs. Leaves lyrate-pinnatifid. Petals white. Fruit a linear siliqua with an upper, indehiscent, 0 or 1-seeded segment and a lower dehiscent, 3 to 9-seeded segment, valves 3-nerved. 1 endemic. 18 species. Macaronesia, Mediterranean, C and S Europe.

Erucastrum rostratum (Balf.f.) Gomez-Campo

Brassica rostrata Balf.f.
Hirschfeldia rostrata (Balf.f.) O.E.Schulz
Hirschfeldia rostrata var. *rostrata* (Balf.f.) O.E.Schulz

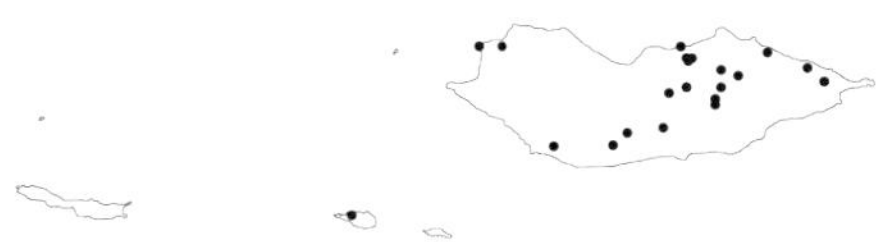

Delicate, annual herb to 50cm, glabrous or thinly to densely hirsute. Leaves lyrate-pinnatifid to pinnatifid, uppermost sometimes simple, <10 x 4cm, entire to sinuate-dentate or dentate. Petals <3.5mm. Siliqua <20 x c.1mm, spreading, glabrous; upper segment <8mm, attenuate to an acute tip.

Ecology. Common, often in damp, shady places in the mist-affected areas of the island; frequently in disturbed places near habitation; also found in the refugium of relatively mesic conditions on the north-facing limestone cliffs on Samha (cf. *Begonia samhaensis*). 100–750m.

Distribution. Endemic.

Conservation status. Least Concern (LC). Common in several vegetation types and under no present or perceived threat.

Botanical notes. Balfour (1888) recognised var. *hirsuta*, based on a single specimen, with densely hirsute indumentum and strongly dissected, dentate-serrate leaves. This probably represents an extreme form of this variable species and has not been formally recognised here.

ETHNOGRAPHIC NOTES

Name(s)

– Soqotri: nil or *ʾáž̌ərib* quick to die back'; *di ʾážbəb* 'quick wilting'; *rížəbəb* stalky'; *sibbáḥ* 'watery' (Samha).

– Soqotri simplified: *'ajirib*, *di 'ajbib*; *rijibib*; *sibbah*.

Food (livestock). This is one of the *ʾážərib* group of plants which flourish after the winter rains. It is grazed by small stock when immature, before the grasses come up. It is little liked by cattle, but camels and donkeys eat it. It is seen as having limited nutritional value.

Notes. It can cause *'uqabah* in small stock unaccustomed to this feed.

Comparative uses. Related species are eaten, and some are cultivated as a fodder or vegetable. (Ambasta 1982; Jain & De Fillips 1991; Kunkel 1984)

Farsetia Turra

Herbs or subshrubs, densely covered with parallel medifixed hairs. Leaves entire, linear to broadly obovate. Petal colour variable from white or pink to yellowish brown. Fruit a dehiscent siliqua, many-seeded; seeds broadly winged. 2 endemics. 20 species. NW Africa to NW India.

Botanical notes. For a revision see Jonsell (1986).

ETHNOGRAPHIC NOTES

Comparative uses. The tender twigs of *F. hamiltonii* and *F. jacquemontii* are eaten raw as a tonic or are used for chutney. Medicinally, *Fagonia* species are considered refrigerant, and are used to treat rheumatism. (Ambasta 1982; Dalziel 1948, 1956; le Houerou 1980; Jain & De Fillips 1991; Kunkel 1984)

Farsetia inconspicua A.G.Mill. (see Appendix 2, p. 723)

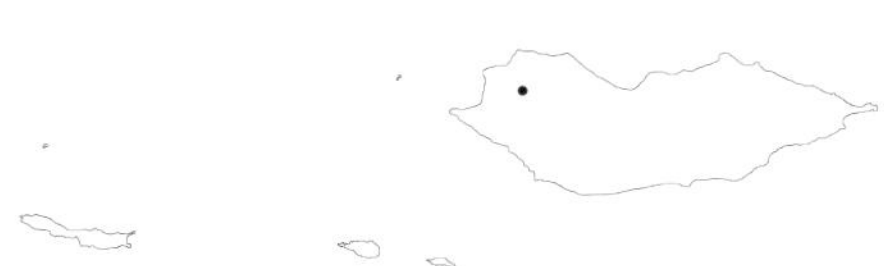

Prostrate or ascending herb to 10cm, sometimes weakly woody at base. Leaves linear-elliptic to narrowly oblanceolate, <2 x 0.5cm. Sepals c.5mm. Petals bright yellow. Siliqua narrowly oblong, <18 x 5mm; style c.0.5mm; stigma decurrent.

Ecology. Rare in shallow soil in crevices on limestone pavement in sparse dwarf shrubland. c.650m.

Distribution. Endemic.

Conservation status. Vulnerable (VU D2). Rare, with a very limited area of occupancy.

Botanical notes. This distinctive endemic has no clear allies. It is similar to *F. socotrana* in fruit shape but differs in flower colour and leaf shape and texture. It is similar to *F. stylosa* in floral characters but differs in fruit shape, style length and the shape of the stigma. It is restricted to patches of thin soil overlying the limestone pavement in exposed places on the rolling plains of the western plateau where it grows with *Distephanus qazmi*, *Melhania milleri* and *Lotus mollis*, endemics which are also associated with this particular habitat.

ETHNOGRAPHIC NOTES

Name(s)

– Soqotri: *rəḥámmə*; *rxʌ́mxʌm* 'grey-coloured' (western Soqotra)

– Soqotri simplified: *rihammih; rikhamkham.*

Food (livestock). Provides incidental grazing for livestock, but is not important nutritionally.

Farsetia longisiliqua Decne.

Subshrub to 100cm. Leaves linear, <4 x 0.5cm. Sepals <10mm. Petals pink or pale yellow fading to purple. Siliqua narrowly oblong, <50 x 5mm; style <3mm; stigma decurrent.
Ecology. Locally common in dwarf shrubland on desertic plains. Sea-level–200m.
Distribution. Borders of the Red Sea from Egypt to Somalia.

ETHNOGRAPHIC NOTES†
Name(s)
– Soqotri: *ṣɔ̄wəher* (more commonly given to *Ocimum forskolei*); *ḳʌ́żəbə* (more commonly given to perennial Cyperaceae and Gramineae of damp ground); *ḳʌ́żəbeb* 'dried out'.
– Soqotri simplified: *zowiher*; *qadhibih*; *qadhieb*.
Food (livestock). Provides useful dry season fodder for small stock and is important in the less-vegetated areas of the island for this reason. Camels and donkeys eat the plants all year round.

†Also apply to *F. socotrana*.

Farsetia socotrana B.L.Burtt
Farsetia prostrata Balf.f.

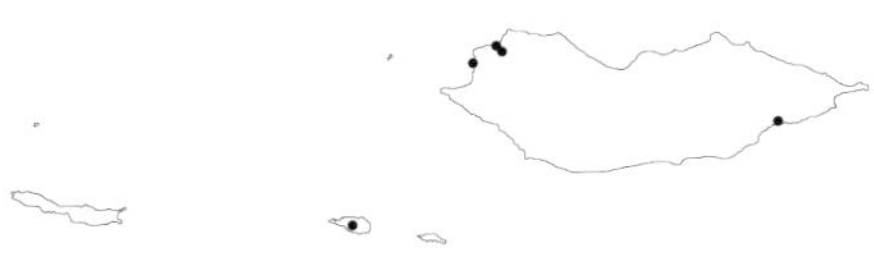

Subshrub to 30cm. Leaves grey-green, subsucculent, broadly obovate, <3.5 x 2cm, obtuse or truncate, attenuate below into a winged petiole. Sepals <6.5mm. Petals white or pale violet. Siliqua <28 x 6mm; style c.0.5mm; stigma decurrent.
Ecology. Locally common in dwarf shrubland on the coastal plains of Soqotra and Samha. Sea-level–50m.
Distribution. Endemic.
Conservation status. Vulnerable (VU B2 a biv). *F. socotrana* is known only from three widely separated locations with a total area of occupancy of less than 500km^2. Searches have failed to find it at its type locality, suggesting a decline in its extent of occurrence. The coastal plains on Soqotra are the habitat most at risk from the recent developments proposed for the island. They are already showing the signs of decline in quality of habitat as people move from the interior to the coast. The site at which it is most abundant was until recently inaccessible by road. However, there are now development plans, which would severely threaten this delicate coastal habitat and place *F. socotrana* under the threat of extinction.
Botanical notes. Sometimes grows with *Lachnocapsa spathulata*, from which it can sometimes be difficult to separate in flower; however, the flowers of *F. socotrana* are white or pale lilac and those of *Lachnocapsa* are pale yellow.

ETHNOGRAPHIC NOTES
Name(s)
See *F. longisiliqua* (also uses).

Farsetia stylosa R.Br.

Herb, sometimes woody-based, to 60cm. Leaves linear to linear-elliptic, <8.5cm. Sepals <4.5mm. Petals white to mauve or yellowish to brownish. Siliqua narrowly oblong, <30 x 4mm; style <3.5mm; stigma capitate.
Ecology. No habitat details available.
Distribution. Widespread in a belt from Mauritania to NW India.

Botanical notes. A widespread species varying greatly in petal colour and siliqua shape across its range. On Soqotra known only from a single specimen collected by Balfour in the 19th century. Balfour (1888) notes that the specimen is very fragmented and comments 'it is marked as collected near Galonsir [Qalansiyah] but I have no recollection of gathering them [it]'. He does not assign this specimen to a species but Jonsell (1986) in his revision of the genus places it in *F. stylosa*. No further collections have been made and the status of this species on Soqotra is uncertain.

Hemicrambe Webb

Fabrisinapis C.C.Towns.

Shrubs, glabrous or with simple hairs. Leaves simple, crenate-dentate. Petals white or rose. Fruit a narrowly elliptic siliqua, with a sterile, 2-valved lower segment and an upper, beak-like, 1 or 2-seeded segment. 1 endemic. 2 species. Morocco and Soqotra.

Botanical notes. A genus with an interesting disjunct distribution with one species in Morocco and the other on Soqotra. Related to *Nesocrambe* and like that genus a relict of the old Miocene flora which once bordered the Mediterranean.

Hemicrambe fruticosa **(C.C.Towns.) Gómez-Campo**
Fabrisinapis fruticosus C.C.Towns.
Hemicrambe townsendii Gómez-Campo

Cliff-dwelling shrub to 1m. Leaves ovate, <3 x 1.5cm, obtuse, irregularly crenate-dentate, cuneate below. Petals c.7mm. Siliqua <18 x 4mm, compressed, longitudinally veined; the lower sterile segment much shorter than the upper segment.
Ecology. Rare, restricted to vertical rock faces on the granite pinnacles of the Haggeher mountains. 900–1000m.

Distribution. Endemic.
Conservation status. Vulnerable (VU D2). Rare, with a very limited area of occupancy.
Botanical notes. A rare plant of vertical rock faces on the granite pinnacles. Readily distinguished by its white flowers and small, pendent fruits.

ETHNOGRAPHIC NOTES
Name(s)
– Soqotri: nil or *ṣəfṣˈifo* 'rippling'; *ḥˈaher* 'dark-coloured'.
– Soqotri simplified: *zifzifo*; *haher.*
Food (livestock). Not normally accessible to livestock, but considered to be a good feed for goats if they can reach it.

Lachnocapsa Balf.f.

Subshrubs, white-tomentose with appressed stellate and branched hairs. Leaves fleshy, simple, entire. Petals pale yellow. Fruit a compressed silicula, with a narrow septum, bilocular; 1 seed per loculus. 1 species. Endemic.

Lachnocapsa spathulata **Balf.f.**

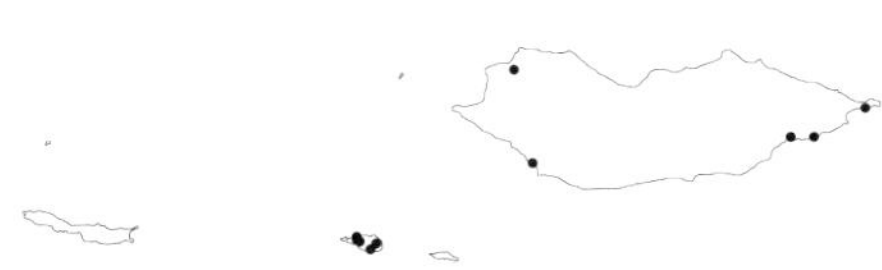

Cushion-forming subshrub to 30cm high and diameter. Leaves broadly ovate to orbicular or broadly obovate, <2.5 x 1.5cm, obtuse or rounded, truncate or cuneate below. Petals <13mm. Silicula <12 x 7mm, densely white-tomentose, closely resembling the leaves.
Ecology. Locally common in dwarf shrubland on desertic coastal plains and parts of the limestone plateau near the sea. Sea-level–700m.
Distribution. Endemic.
Conservation status. Vulnerable (VU D2). Rare, with a very limited area of occupancy.
Botanical notes. This cushion-forming subshrub has attractive pale yellow flowers and densely white-hairy fruits which can be difficult to distinguish from the fleshy leaves. It has a rather scattered distribution on Soqotra but can be locally abundant (for instance at Faleng). It was recorded from near Qalansiyah by Balfour in the 19th century but has not seen there recently (cf. *Taverniera sericophylla* and *Farsetia socotrana*). It is relatively common on Samha, particularly on the summit plateau.

ETHNOGRAPHIC NOTES
Name(s)
– Soqotri: nil or *ṣəfṣˈifo* 'rippling'; *źesˈibə* 'heaped stones' (more commonly given to *Limonium* species); (western Soqotra) *rxˈʌmxʌm* 'grey-coloured'; *ḳˈʌlḳəhʌl* 'swaying in the breeze' (general term for a variety of mainly halophytic plants); *źˈʌḥyīm* 'chewing noisily on something hard'; *źˈʌfʿan* 'crouching' (also on Samha).
– Soqotri simplified: *zifzifo*; *dhezibih*; *rkhamkham*; *qalqihal*; *dhahyiym*; *dhaf'an.*
Food (livestock). Much liked by camels. Goats browse it in the dry season and sheep nibble at the plant litter, but it is not considered to be a useful or nutritious fodder for them. On Samha it is an important forage for sheep and goats.
Other. Animals feeding on this can go longer without water.

Nesocrambe A.G.Mill.

Herbs, glabrous. Leaves simple to lyrate-pinnatisect. Petals yellow. Fruit a 2-membered siliqua, without obvious venation, both segments indehiscent, 1-seeded, septum absent, seeds embedded in pith-like tissue. 1 species. Endemic.

Botanical notes. Related to *Hemicrambe* and like that genus a relict of the ancient Miocene flora which once extended around the coasts of the Mediterranean Sea. It has survived only in a small, moist refugium on Soqotra (Miller *et al.* 2002).

Nesocrambe socotrana A.G.Mill.

Herb, somewhat woody below, stems to 20cm, often hidden in cracks in rock faces. Leaves congested on short spur shoots at tips of stems, appearing as basal rosettes above hidden stems, obovate to oblanceolate, simple with a few rounded teeth or lyrate-pinnatipartite to lyrate-pinnatisect, <3 x 2cm. Petals <4.5mm. Siliquae cylindrical, <9 x 2mm.
Ecology. Rare; on a windswept ridge at the western end of Soqotra; in shade in relatively humid cracks and gullies on limestone cliffs and pinnacles. 400–550m.
Distribution. Endemic.
Conservation status. Endangered (EN B2 a biii). *N. socotrana* is restricted to a small refugium with a total estimated area of occupancy of less than 10km^2. At present the rapid development of the island poses no threat to this unique habitat: the area is inhospitable, inaccessible by road and there is a lack of potable water. However, its restricted distribution and the very limited extent of the niche which it occupies means that it is vulnerable to any change. It is strongly suspected that this marginal habitat is threatened by the long-term trend of lower rainfall in the region and would be particularly vulnerable to the effects of global warming (Miller *et al.* in press).

Botanical notes. *Nesocrambe* is vegetatively similar to *Hemicrambe*, a genus with a remarkably disjunct distribution, with one species, *H. fruticosa*, on Soqotra and the other, *H. fruticulosa*, in Morocco. It is fascinating that *Nesocrambe*, with its yellow flowers and dissected leaves, is perhaps more comparable to the Moroccan *H. fruticulosa*. However, the fruits of both species of *Hemicrambe* are quite different from those of *Nesocrambe*: the upper segment is flattened and contains one to four seeds whilst the lower segment is sterile, but has vestigial dehiscent valves. It is astonishing that this remarkable plant has apparently survived only in a small refugium at the extreme west end of Soqotra. The extreme west of Soqotra is generally dry (as witnessed by the sparse vegetation); however, *Nesocrambe* occurs on limestone pinnacles and cliffs on the highest points of the ridge at Heger (between Ras Izfer and Ras Shu'ub, above Sherubrub). This ridge traps cloud and limited precipitation from the SW (summer) and NE (winter) monsoons and creates a small area of relatively humid conditions in this otherwise desertic region.

ETHNOGRAPHIC NOTES
Name(s)
– Soqotri: nil or *ṣəfṣ́ifo* 'rippling'.
– Soqotri simplified: *zifzifo.*
Food (livestock). Not normally accessible to livestock, but considered to be a good feed for goats if they can reach it.

Sisymbrium L.

Herbs, with simple hairs or glabrescent. Leaves pinnately lobed or divided. Petals yellow. Fruit a many-seeded dehiscent siliqua; valves 1 to 3-nerved; seeds uniseriate. 80 species. Widespread.

Sisymbrium erysimoides Desf.

Erect annual herb to 50cm, sparsely pubescent to glabrescent. Lower leaves <15 x 5cm, lyrate-pinnatipartite to lyrate-pinnatisect with 1 to 4 pairs of lateral lobes and a larger ovate terminal lobe; upper leaves similar to lower. Petals <2.5mm. Fruits erect-ascending, straight, narrowly tapering from base to tip, <40 x c.1mm, glabrous or sparsely hairy.
Ecology. Rather rare, maybe overlooked, in disturbed ground and near habitation in semi-deciduous woodland. 450–1000m.
Distribution. Mediterranean region, N Africa and SW Asia.

ETHNOGRAPHIC NOTES
Name(s)
– Soqotri: nil or *ʾážərəb* quick to die back'.
– Soqotri simplified: *'ajirib.*
Food (livestock). One of the *ʾážərib* group of plants which flourish after the winter rains. It is grazed by cattle before the grasses come up, but is little eaten by goats and sheep. It is not seen as having any nutritional value.
Notes. It can cause *'uqabah* in small stock unaccustomed to this feed.
Comparative uses. In Arabia, the leaves of the closely related *S. irio* 'London rocket' are eaten raw or cooked, and are rich in vitamin C. The seeds are boiled in water and drunk to lessen fevers and taken as an expectorant. They are also used to treat asthma, and to make stimulating poultices. The seeds yield a semi-drying oil, used for soap-making and lubrication. (Abbas *et al.* 1992; Ambasta 1982; Batanouny 1981; Ghazanfar 1994; Goodin & Northington 1979; Jain & De Fillips 1991; Kunkel 1984; Mossa *et al.* 1987; Rizk & El-Ghazaly 1995)

CUCURBITACEAE

Trailing herbs or bottle-trunked trees. Tendrils usually present. Leaves alternate, simple; stipules absent. Monoecious or dioecious. Flowers yellow. Sepals 5, free. Petals 5, united at base. Sepals and petals borne on top of a cup-like or tubular hypanthium. Male flowers solitary or in few-flowered fascicles, usually with 3 stamens, in which case 2 of the anthers are 2-thecous and the other is 1-thecous, if 5 stamens are present they are all 2-thecous. Female flowers solitary; style 1; ovary inferior, 1-celled, ovules many. Fruit a large, often fleshy, indehiscent berry. Seeds many. 121 genera. 735 species. Mostly tropical and warm regions.

Citrullus Schrad. ex Eckl. & Zeyh.

Herbs. Tendrils simple or bifid. Leaves deeply pinnately lobed. Monoecious. Flowers yellow, solitary; hypanthium cup-shaped; male with 3 stamens; female with smooth ovary. Fruits green or yellowish, smooth, firm-walled, fleshy, indehiscent. Seeds smooth. 3 species. Mostly tropical and S Africa.

Citrullus colocynthis Schrad.

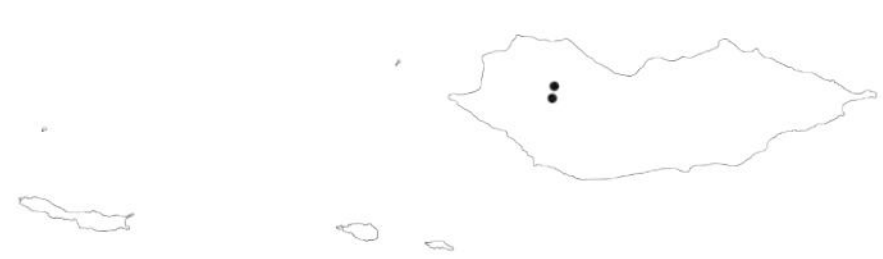

Perennial herb trailing to 3m or more, rootstock tuberous. Stems roughly hairy; tendrils simple. Leaves ovate, stiff, 3–10 x 3–6cm. Flowers <1.5cm long. Fruit green, longitudinally mottled with darker green, ripening clear yellow, globose, <10cm diameter, flesh bitter.

Ecology. Probably under-recorded but apparently restricted to inland desertic plains. 100–200m.

Distribution. Ethiopia, Somalia, semi-desert regions of N Africa, tropical Asia and Australia.

Botanical notes. The fruit dries slowly to form a yellow, spherical ball, which detaches from the plant and is said to roll across the ground in strong winds, eventually breaking open and distributing its seeds.

ETHNOGRAPHIC NOTES

Name(s)

– Soqotri: *di-aḥśā̄wə* (all small cucurbits are given this name, which refers to the fruit).

– Soqotri simplified: *di ah'shawih.*

Food (livestock). Goats are very fond of the green, unripe fruit.

Medicinal / veterinary. The poisonous nature of the fruit and its powerful purgative action is not generally recognised on Soqotra.

Other. Goats feeding on the green fruit can go longer without water. When it is ripe and yellow, children play with the fruit as a ball.

Notes. Some islanders give the same name to *Cissus paniculata.* The growth habit of this plant is rather similar to a cucurbit; it is often to be found falling down the hillside in the wetter areas, crawling over huge boulders and growing between piles of rock. However, no part of this plant is eaten either by livestock or by people.

Comparative uses. The plant is regarded as being useful as a sand binder, and it is very drought resistant. The leaves and stems are said to be palatable to all livestock, but the fruits only to donkeys and camels. Alkaloids, pectin and saponins are present in the plant; the active compound is colocynthin which is a strong purgative. The dried pulp of the unripe but full-grown fruit, freed from the rind, constitutes the commercial drug colocynth, and is officially recognised in the pharmacopoeias. It is a drastic hydragogue cathartic producing large, watery discharges. In large doses it causes violent griping, prostration and sometimes bloody discharge. It can produce severe abdominal pain, vomiting, diarrhoea, delirium and bloodstained watery stools, and is not given to nursing mothers as it can appear in the breast milk. There is, however, only a limited demand for colocynth nowadays, as it has been superseded by less drastic and toxic purgatives.

Fruit. The fruits are eaten pickled or in preserves. In Yemen part of a fruit is worn in amulets to protect against the Evil Eye, and a whole fruit is placed in a new house to protect it from evil. Ripe fruit were charred, ground and used to prepare gunpowder, tinder and fuses, and in Egypt the green fruit were crushed on a piece of flannel fabric which absorbed the juice and, dried, is a tinder. The fruit are said to keep moths at bay and pieces of dried fruit are pulverised, wrapped in paper and put amongst clothing for this purpose. A bitter black extract prepared from the rind of the fruit was smeared on water-skins and bags to keep camels away from them. Medicinally, on the Arabian mainland, the fruit pulp is cooked until it dissolves and is then applied to the foot to treat diabetes, or the cut fruit is sniffed. The dried fruit are ground and half a teaspoon taken daily to treat diabetes; alternatively the pulp is cooked and spread over the legs. A small piece of pulp is taken as a purge. The fruit are crushed and mixed with oil to massage painful joints. A steam bath is made from the fruit pulp to treat partial paralysis and spasms. To treat rheumatism, halves of green fruit are placed on each heel, the treatment being regarded as complete when the patient tastes bitterness in the mouth; alternatively two fruit are cut into pieces, fried in oil for 10 minutes and then placed as a poultice on skin; or cut halves of fruit are heated on the fire and then applied to the heel of the foot to 'suck out' the pain. The fruit are also used as an abortifacient. In Bahrain either the fruit or the root is boiled and taken as a hydragogue, a cathartic, a diuretic, and to treat jaundice, and the fruit pulp is used to treat gonorrhoea. The sap of unripe green fruit is used to treat scorpion stings, and dry fruits are used as an effective insecticide, especially for mites and weevils, and to treat dog, insect and snake bites. Elsewhere the fruit are used as a purgative, anti-pyretic and anthelmintic. They are used to treat tumours, ascites, ulcers, asthma, bronchitis, jaundice, tuberculosis glands of the neck, anaemias, amenorrhoea and elephantiasis.

Seeds. They contain a fixed oil which is used in making soap. They are a source of protein and, after very long cooking, are eaten as starvation food in Sudan and among the Tuareg, while in India the powdered seeds are used for making a special bread known locally as *sagra*. The seeds are placed amongst clothing to repel insects and especially moths. A dark-coloured, tar-like extract from the seeds is used to treat sarcoptic mange in camels. The seeds are crushed in water and used as a shampoo to darken hair colour.

Medicinally, in Yemen the seeds are used as a purgative, but are added to other foods to reduce the strong action; they are added to honey and *ḥilba* (*Trigonella foenum-graecum*) and taken to treat severe constipation. Elsewhere they are taken as an anthelmintic, to promote abortion and to treat gonorrhoea. A single seed swallowed daily for 21 days without chewing is taken as a remedy for diabetes.

Root. Medicinally, an extract of the root is used to treat ascites, jaundice, urinary disease and rheumatism. Roots are crushed and added to goats' milk as a purgative and to treat colic. In Qatar a preparation made from the root is regarded as useful in inflammation of the breasts, amenorrhoea, jaundice, rheumatism and joint pains, and is also used externally to treat opthalmia and uterine pains. A decoction of the root with added garlic is used externally to treat snakebite.
Foliage and stems. A poultice is made from the crushed leaves with added garlic and used to treat bites and stings. The powdered leaves are added to water and taken as a laxative. In Qatar a preparation of the leaves is taken as a diuretic and is used to treat jaundice and asthma. The hot sap of the plant is used to treat certain skin diseases in camels. (Abbas *et al.* 1992; Al Rawi & Chakravarty 1964; Ambasta 1982; Batanouny 1981; Bebawi & Neugebohrn 1991; Dafni *et al.* 1984; Dalziel 1948, 1956; Davis *et al.* 1983; Fleurentin & Pelt/Fleurentin *et al.* [1982, 1983a,b,c]; Ghazanfar 1994; Goodman & Hobbs 1988; Jain & De Fillips 1991; Kunkel 1984; Miller & Morris 1988; Mossa *et al.* 1987; Rizk & El-Ghazaly 1995)

Citrullus lanatus **(Thunb.) Matsum. & Nakai**

Annual usually trailing herb. Stems with soft spreading hairs; tendrils bifid. Leaves ovate or narrowly ovate, 5–20 x 5–15cm. Flowers <2cm long. Fruit green and yellow striped, ripening yellow, globose or ellipsoid, <60 x 30cm, flesh watery, not bitter.

Ecology. Rarely naturalised by tracksides on Soqotra.
Distribution. Native of SW Africa, now widely cultivated and naturalised.
Botanical notes. Widely cultivated for its edible fruits (water melon).

ETHNOGRAPHIC NOTES
Not seen, no information recorded.

Cucumis L.

Herbs. Stems roughly hairy; tendrils simple. Leaves palmately lobed. Monoecious. Flowers yellow, solitary or in small fascicles; hypanthium cup-shaped; stamens 3; ovary densely covered in soft spines. Fruit subglobose to ellipsoid, firmly fleshy, indehiscent, shortly spiny. Seeds smooth. c.30 species. Mostly tropical and southern Africa.

Cucumis dipsaceus **Ehrenb.**

Trailing herb to 2m. Stems covered in harsh, stiff, spreading hairs. Leaves broadly ovate to kidney-shaped, 3-lobed, <10 x 10cm. Fruit pale yellow on ripening, ellipsoid, 4 x 7cm, densely soft spiny.
Ecology. No habitat details available.
Distribution. Widespread in NE Africa and Arabia.
Botanical notes. Recorded at Tamarida [Hadiboh] in the 19th century but not seen since.

Cucumis ficifolius **A.Rich.**
Cucumis ficifolius var. *echinophorus* Naudin

Trailing herb to 1m. Stems roughly hairy. Leaves ovate, 3 to 5-lobed, <7 x 7cm. Fruit dark green with pale green stripes, ripening yellow, pustular, each pustule topped by a broad, blunt bristle.

Ecology. No details of habitat or location available.
Distribution. Widespread in tropical Africa.
Botanical notes. Not seen since Balfour reported it in the 19th century. Possibly a misidentification for *C. prophetarum*. The record was referable to var. *echinophorus* Naud.

Cucumis prophetarum **L.**
Perennial herb trailing to 2.5m. Stems roughly hairy. Leaves broadly ovate or round, 2–6 x 2–5cm. Flowers <0.6cm long.
Botanical notes. Two subspecies are recorded from Soqotra.

ETHNOGRAPHIC NOTES
Name(s)
– Soqotri: *di-aḥśāwəh* (Soqotra and Abd al Kuri) (all small cucurbits are given this name, which refers to the fruit).
– Soqotri simplified: *di ah'shawih.*

Food (human). People occasionally eat the softly prickly fruit, but usually only when fully ripe and yellow.
Food (livestock). The green and drying foliage is browsed by all livestock, but especially sheep and goats. This is not regarded as good, nutritious grazing, but as a useful 'filler'. The fruit are liked by all livestock, especially thirsty goats.
Comparative uses. The plant has been suggested as a possible oilseed for arid lands.
Fruit. A bitter resinous substance, myriocarpin, is reported to be present in the fruit pulp which produces nausea and is said to be toxic to animals. The fruit are eaten boiled or pickled, and chil-

dren empty them out and use them as whistles and toy milk containers. Medicinally, the liquid inside the yellow, softly prickly fruit is drunk as an emetic to treat stomach troubles, and is applied to sores and wounds caused by ticks and other external parasites in livestock. The fruit pulp is widely used as a purgative, anti-pyretic, vermifuge and emetic. A fresh fruit with the end cut off is put over a sore and inflamed finger as a dressing. To treat symptoms resembling those of rabies, the fruit are immersed in water, and the water drunk as an emetic and laxative. *Seeds.* In parts of Africa the seeds are cleaned with ash and added to vegetables. (Ambasta 1982; Batanouny 1981; Dalziel 1948, 1956; Davis *et al.* 1983; Fleurentin & Pelt/Fleurentin *et al.* [1982, 1983a,b,c]; Heine *et al.* 1988; Jain & De Fillips 1991; Kunkel 1984; Rizk & El-Ghazaly 1995)

Subsp. *dissectum* (Naudin) C.Jeffrey

Leaves grey-green, deeply lobed. Fruit pale green with dark green stripes, ripening clear yellow, ellipsoid to subglobose, 2–4cm diameter, spines <0.4cm.

Distribution. Widespread in N Africa and tropical Asia to India.

Subsp. *prophetarum* L.

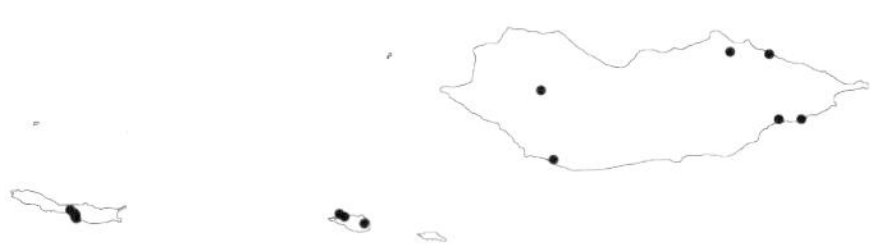

Leaves green, shallowly lobed or unlobed. Fruit pale green with dark green or purple-tinged stripes, ripening dull yellow, ellipsoid, 1–2.5cm diameter, spines 0.4–0.6cm.

Distribution. Widespread in N Africa, Yemen and Saudi Arabia.

Dendrosicyos Balf.f.

Trees. Tendrils absent. Leaves palmately nerved, not or shallowly lobed. Monoecious. Flowers solitary or in small fascicles, yellow; hypanthium tubular; male with 3 stamens. Fruit ellipsoid, smooth. 1 species. Endemic.

Botanical notes. The only tree in the cucumber family and was described by Balfour (1888) as 'a very remarkable genus'. Its squat, soft white trunk surmounted by a tuft of pendulous branches is unmistakeable.

***Dendrosicyos socotrana* Balf.f.**

Bottle-shaped tree to 6m, trunk <2m diameter, branches pendulous. Leaves broadly ovate, round, or kidney-shaped, 2–10 x 3–10cm, roughly hairy, margin spine-toothed, brittle. Flowers <3.5cm long. Fruit green ripening brick-red, 5 x 3cm.

Ecology. Widespread in dry areas on Soqotra; locally common in *Croton socotranus* shrubland on the plains and locally abundant in succulent shrubland on the limestone; less common on igneous rocks; rare on Samha. Sea-level–500m.

Distribution. Endemic.

Conservation status. Vulnerable (VU B1 2a c). At present, the main threats to this flagship species are its use as a drought fodder. In the main, with its swollen, water-storing stem it is well adapted to withstand drought conditions and should therefore be better able than many species to tolerate any drying out of the Archipelago due to climate change. However, in times of severe drought trees are cut down, pulped and fed to livestock. In some areas (Kilissan) this has resulted in its almost total eradication. Change of habitat is also an important factor in preventing regeneration. Typically, on the plains, seedlings germinate and grow, protected from livestock, in the cover of dense vegetation such as provided by the spiny shrub *Lycium sokotranum* or by colonies of the succulent shrub *Cissus subaphylla*. So where there is no *Cissus* or *Lycium* there is little, if any, regeneration. Thus, along the southern region of Soqotra and on Samha the recovery of populations of *Dendrosicyos* after drought is dependent on the presence of colonies of *Cissus*. *D. socotrana* is considered vulnerable to extinction due to the decline in quality of its habitat (due to drought) in certain areas, and the level of exploitation of populations already stressed by drought.

Botanical notes. Perhaps the best example of the swollen-trunked bottle trees for which Soqotra is famous. It is widely distributed in several vegetation types but has a rather fragmented distribution; over large areas there are only isolated trees or small, relict populations, whilst in other areas it is relatively abundant. There are a few trees on the island of Samha, but none on Darsa or Abd al Kuri.

ETHNOGRAPHIC NOTES

Name(s)

– Soqotri: *ḳˈʌmḥīn, ḳˈʌmḥīm* (pl. *ḳʌmˈʌḥin* or *ḳʌmˈʌḥim*) 'chewing greedily'.

– Soqotri simplified: *qamhiyn, qamhiym* (pl. *qamahin, qamahim*).

Food (human). The small fruit are not eaten.

Food (livestock). Goats in hunger graze the leaves, but do not otherwise eat them as they are tough and said to 'pull out the teeth'. In the dry season, small stock eat the fallen, dry leaves, but this is not regarded as a nutritious feed. Goats chew away at a young plant or one whose growth has been distorted or impeded by the rock it has grown beside. Such plants are called *ḳamaḥáno* or *ḳínimo* 'little *Dendrosicyos*'. Goats and sheep eat the fallen flowers, and any fruit within reach. Cattle on the whole are not interested in any part of the plant, except the fallen leaves and flowers when they are hungry. Goats chew the bark in a drought, sometimes killing the tree. It has been known for a goat to chew its way right through the bark and into the core of a good-sized tree, and then become trapped when the hollowed-out trunk toppled over. Thirsty goats also chew the bark to quench their thirst. In a prolonged dry season or at times of drought, goatherds go to the trees and lop off branch tips and slim stems which they then pulp to feed their goats. If the drought persists, the herders progressively cut away at the tree until finally only the trunk is left, and this too is pulped. Goats fed on this diet put on weight and have a better chance of surviving the drought. Because of this, this feed is generally reserved for goats in milk and breeding females. It is not a nutritious feed but is adequate as a maintenance feed. In areas where these plants grow they are a key dry season fodder reserve, and as such are a protected species. Only men lop and pulp the plant.

Flavouring / food additive. The carcass of goats feeding on *Dendrosicyos* is usually considered unfit for human consumption, as the meat develops an unpleasant taste. The viscera especially become surprisingly bitter and have to be discarded. Even the water in which the meat has been cooked, usually an important part of the meal, becomes undrinkable. The milk of an animal eating a lot of this plant is also affected, developing an unpleasantly bitter edge.

Wood. The wood is of no use for any purpose, including firewood. It is not attractive to termites.

Medicinal / veterinary. The wood is crushed to extract a liquid which is used medicinally, an amount sufficient to cover a fingernail for a child, a closed fistful for an adult. The extract is added to water or milk, or sometimes to butterfat or other oil, and is given to treat urinary retention, cystitis, symptoms of diabetes, problems with the liver and burns.

Other. Goats feeding on the bark and pulped wood can go longer without water. Islanders say that if they can ensure their goats are fed adequately during the dry season on this plant, *Euphorbia arbuscula*, *E. schimperi* and *Dorstenia*, they can last without being taken to water until the arrival of the winter rains.

Notes. Only goats used to this as fodder will eat it. Thus goats which spend their time in the Central Highlands, where this plant is not to be found, will not eat either the foliage or the pulped core. Seedlings and young plants are usually eaten up by goats; consequently surviving saplings are usually seen emerging from protective clumps of *Cissus* species, *Buxanthus pedicellatus* or *Ficus cordata*, which are unpalatable to livestock.

Eureiandra Hook.f.

Herbs. Tendrils simple. Leaves palmately lobed. Dioecious. Flowers yellow, solitary or in small fascicles; hypanthium tubular; male with 5 stamens, all 2-thecous. Fruits ellipsoid, smooth, beaked. 1 endemic. 8 species. Tropical Africa (2 species in Somalia).

Eureiandra balfourii Cogn.

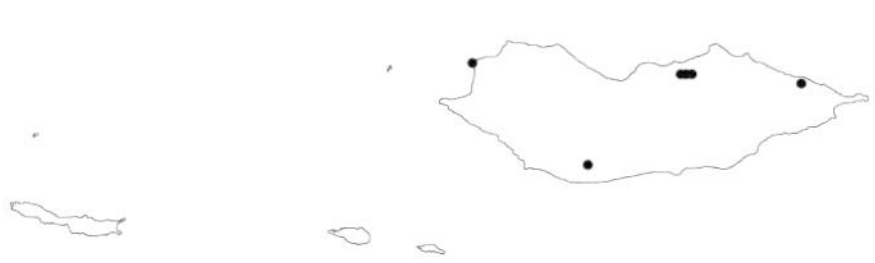

Trailing herb to 2m with a large, underground tuber. Leaves <15 x 15cm. Flowers c.3cm diameter. Fruits orange-red when mature, c.0.5 x 2.5cm.

Ecology. Widespread and locally common in drought-deciduous woodland. 250–600m.

Distribution. Endemic.

Conservation status. Vulnerable (VU D2). Rare, with fragmented distribution and a very limited area of occupancy.

ETHNOGRAPHIC NOTES

Name(s)

– Soqotri: *di-aḥśāwəh* (all small cucurbits are given this name, which refers to the fruit).

– Soqotri simplified: *di ah'shawih.*

Food (human). The fruit are not eaten by people.

Food (livestock). All livestock eat the foliage, and small stock, especially, eat every part of the plant. The aerial parts are often gathered by herders to feed their animals in the dry season. The fruit are eaten by small stock and by donkeys. The root develops during the summer months, often swelling to an amazingly large size and turning a reddish colour as it matures. In the areas where the plants grow, the roots are a key dry season fodder, for goats in particular. Although in some areas of soft ground goats can uncover much of the root for themselves, in general the herders have to search out the plants for their animals, as the swollen part of the root is often deep underground. The aerial parts start to grow in the summer months, often after the summer rains. However, at this time the root is too small to be useful, so herders generally mark the place where they saw the plant growing so that they can return later on to dig up the root when it has grown to its full size. The root is cut up small before being fed to the goats. It is regarded as an extremely nutritious feed for goats, making them fat and increasing their milk yield. The taste of the milk and meat of a goat feeding heavily on *Eureiandra* is unaffected.

Momordica L.

Herbs. Stems smooth; tendrils simple. Leaves palmately lobed. Monoecious. Flowers yellow, male flowers in pedunculate clusters; female solitary. Hypanthium cup-shaped. Fruits warty. Seeds with sculpted surface. 45 species. Old World tropics.

***Momordica balsamina* L.**

Trailing herb. Leaves sharply dentate or lobed, <9 x 12cm. Fruit bright orange, ellipsoid, with rows of small tubercles but otherwise smooth, c.5 x 3cm.

Ecology. No habitat details available.
Distribution. Drier parts of tropical and S Africa, tropical Asia and Australia.
Botanical notes. Not seen since the 19th century. A doubtful record.

***Momordica charantia* L.**
Momordica charantia var. *abbreviata* L.

Trailing herb. Leaves broadly ovate or round, lobes rounded, margin sinuate. Fruit with rows of conspicuous fleshy tubercles and scattered small tubercles.

Ecology. No habitat details available.
Distribution. Widespread in Africa and Asia.
Botanical notes. Recorded by Balfour in the 19th century but not seen anywhere on the island since.

Zehneria Endl.

Herbs. Tendrils simple. Leaves palmately lobed or unlobed. Dioecious. Flowers yellow; male 3 to many in sometimes pedunculate clusters, stamens 3; female in sessile clusters. Fruits green sometimes ripening red, clustered, globose to ellipsoid, smooth. c.35 species. Old World tropics.

***Zehneria scabra* (L.f.) Sond.**
Melothria punctata Cogn.

Trailing herb to 10m. Leaves <15 x 15cm. Flowers <1cm long x 0.5cm diameter.

Ecology. No habitat details available.
Distribution. Tropical and S Africa and tropical Asia.
Botanical notes. Not seen on Soqotra since the 19th century. A doubtful record.

CUSCUTACEAE

Parasitic twining herbs devoid of chlorophyll; stems slender, white or reddish, attached to the host by means of haustoria. Leaves absent or reduced to minute scales. Flowers in compact clusters. Sepals 4 or 5. Corolla fused, 4 or 5-merous, white or pinkish-white, subglobose or campanulate. Stamens 4 or 5, small infrastaminal scales attached to the corolla tube beneath each filament. Ovary 2-locular with 2 ovules per loculus; styles 1 or 2; stigmas capitate, globose or linear. Fruit a globose or ovoid capsule, indehiscent or circumscissile (opening by a lid). 1 genus. 145 species. Cosmopolitan.

Cuscuta L.

See description of family.

Cuscuta chinensis Lam.

Similar to *C. planiflora* but stems yellow; corolla whitish; infrastaminal appendages large, deeply fringed; stigmas globose.
Ecology. Parasitic on herbs in shrubland. 50m.
Distribution. NE Africa and Arabia to E Asia.

ETHNOGRAPHIC NOTES†
Name(s)
– Soqotri: nil or *məzúwken* 'twines around something else'.
– Soqotri simplified: *mizuwken.*
Food (livestock). These are among the ephemeral *ru'ud* group of plants which come up with the rains, providing incidental green grazing until they die back. They are grazed along with the plants they parasitise, but are thought to have little nutritional value.
Notes. Recognised as being parasitic on Soqotra.
Comparative uses. In Qatar the whole plant is used as an expectorant, carminative, tonic, anthelmintic, purgative, diaphoretic, to purify the blood and to lessen inflammation. It is seen as being useful in jaundice, in treating muscle and joint pain, lumbago, headache, diseases of the spleen and vomiting. The stem is used to treat bilious disorders. The seeds are regarded as carminative, depurative, tonic, diaphoretic, sedative and diuretic, and are given as an emmenagogue and for opthalmia, fevers and hiccups. In India the plant is taken as a purgative, to combat flatulence and for liver complaints. It is used externally for itch and the seeds are taken as a demulcent and tonic as well as being used in purgative preparations. In Africa the plant is used as a love charm, and on the Gold Coast young men use it to test the affection of a girl: if the plant placed on a tree survives and spreads he is hopeful, if not, he is regarded as rejected. In Iraq the stems are used to treat obesity and an infusion of seeds is given for catarrh. The seeds of related species are parched and ground to flour. (Al Rawi & Chakravarty 1964; Batanouny 1981; Dalziel 1948, 1956; Jain & De Fillips 1991; Kunkel 1984; Rizk & El-Ghazaly 1995)

†Also apply to *C. planiflora*.

Cuscuta planiflora Ten.
Cuscuta planiflora var. *globosa* Balf.f.

Stems reddish, to 0.5mm diameter. Flowers in compact clusters, 2–8.5mm across. Calyx campanulate, lobed; lobes broadly triangular. Corolla white or pale pink; lobes membranous, acute or obtuse. Infrastaminal appendages oblong-triangular, entire, shortly fringed at apex. Styles 2, much shorter than ovary; stigmas linear-oblong.
Ecology. Parasitic on herbs in desert grassland and semi-deciduous woodland. 50–750m.
Distribution. Worldwide.

ETHNOGRAPHIC NOTES
Name(s)
See *C. chinensis* (also uses).

DIRACHMACEAE

Trees. Leaves alternate, simple, margin crenate; stipules present. Flowers solitary, axillary, bisexual, regular, 8-merous. Sepals 8, united at base. Petals 8, free. Stamens 8. Ovary superior, 8-celled; ovules 1 per cell. Fruit a capsule splitting into 8 segments. 1 genus. 2 species. Somalia.

Dirachma Schweinf. ex Balf.f.

See description of family. 1 endemic.

Botanical notes. The only other member of this family, *D. somalensis* D.A.Link, has recently been described from Somalia (Link 1991).

Dirachma socotrana Schweinf.

Somewhat spindly tree to 5m. Leaves elliptic, 0.5–2.5 x 0.3–2cm, margin coarsely crenate; stipules linear. Flowers white, c.1.5cm diameter.

Ecology. Locally common in dense semi-deciduous woodland. 280–1000m.

Distribution. Endemic.

Conservation status. Vulnerable (VU B2 a biii). The Red Data Book (Lucas & Synge 1978) considered *D. socotrana* to be endangered. However, recent fieldwork has shown it to be fairly common in at least one location. Three subpopulations are now known with an area of occupancy of around 50km^2. In the principal locations (in the western and central Haggeher mountains) trees are relatively abundant and actively regenerating. However, at the other localities the tree is rare and possibly declining. The foliage of this species is not browsed by livestock, though goats have been known to gnaw the bark in periods of drought. The wood of the dead tree is collected for the aromatic smoke it produces when burned. Although small quantities are offered for sale (as are the dead woods of other species on the island which produce aromatic smoke) this is not considered to pose a problem at present. However, should the demand for this scented wood increase it is possible that live trees might be cut for sale and this would, of course, seriously threaten the survival of the tree. See Guarino *et al.* (1991).

Botanical notes. In the past this tree has been considered rare, even on the point of extinction. This is almost certainly because most expeditions have visited the island in January to February when the tree is not in flower; it is then rather non-descript and easily overlooked. However, it becomes far more conspicuous in summer (March to April) when it is covered in white flowers. It is particularly abundant in the dense woodland in the upper reaches of Wadi Ayheft. It is less common in other areas, for instance on the higher seaward-facing cliffs at Reqadihon (above Hoq). These cliffs capture low cloud and drizzle during the winter and are a refugium for some species (e.g. *Coelocarpum haggierensis*) which are otherwise found only in areas of high rainfall in the Haggeher mountains.

ETHNOGRAPHIC NOTES

Name(s)

– Soqotri: *ṭīfiṭ* (also given to *Placopoda virgata*, and, in western Soqotra, to *Coelocarpum socotranum*); dead wood *di ra̍ḥmam* 'grey'.

– Soqotri simplified: *tiyfit*; *di rahmam.*

Food (livestock). Camels and goats eat the leafy stems in drought. Goats also chew at the bark when very hungry.

Fuel. The dead wood is not normally used as firewood since it is valued for its fragrance. It burns with little smoke.

Timber. The wood makes a reasonable timber, but is little used for this. However, if there is no alternative timber available, the larger branches can be used as rafters or thatching spars.

Cleansing. The dead wood of this plant belongs to the group called *di ra̍ḥmam* and is burned for its pleasant smell. The most widely available of these is the wood of *Cephalocroton socotranus*, but *Barleria popovia* and *Dirachma socotrana* wood are also used. Not surprisingly, opinions vary as to which produces the most fragrant smoke. *Dirachma socotrana* wood makes little smoke until it has nearly burned down, but then gives off a strong perfume. Many keep a chunk of the wood in the home and cut slivers off and put them on the fire when visitors are coming to give the house a nice smell. These fragrant woods are commonly burned during childbirth, and clothing is often held in the smoke of the burning wood to perfume it.

Commercial (past, present). The dead wood is offered for sale in the coastal markets.

Other. The smoke from the burning wood is said to neutralise the powers of evil spirits. Traditionally, people of the more settled coastal communities would burn one of these fragrant woods in their living quarters every Wednesday evening. Such woods were also burned in the mosques on Wednesday and Thursday nights, the fire being extinguished first thing on Friday morning.

EBENACEAE

Shrubs or trees with a single leading shoot and flattened foliage branches. Bark black, gritty. Leaves ± opposite (in the Soqotran species), entire; stipules absent. Dioecious. Flowers in short axillary cymes (female flowers sometimes solitary). Calyx 4 or 5-lobed. Corolla 4 or 5-lobed, campanulate. Stamens 10 to 20. Staminodes absent in female flowers. Ovary superior, 2 or 4-celled; styles 2. Ovules 1(or 2) per cell, pendulous. Fruit a berry, calyx persistent. 2 genera. 500 species. Tropical, especially Old World.

Euclea L.

See description of family. 14 species. Tropical and S Africa.

Euclea divinorum **Hiern**
Euclea balfourii Hiern
Euclea kellau Hochst.
Euclea laurina Hiern

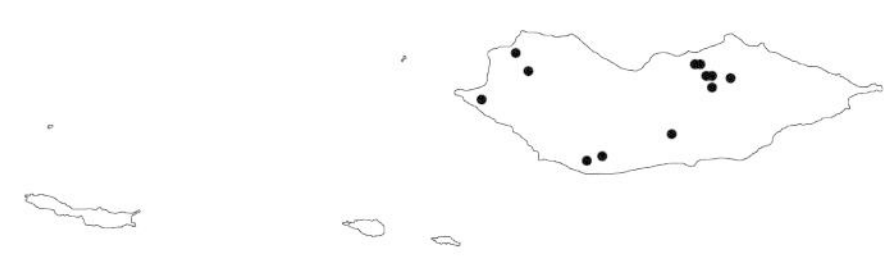

Shrub or small tree to 4m. Young branches and lower leaves with reddish, peltate scales. Leaves simple, ovate-elliptic to obovate, 1–8 x 0.7–3.5cm, tip rounded, leathery. Flowers white. Male flowers c.0.4cm diameter. Fruit globose, hard, ripening red, c.0.5cm diameter.

Ecology. Semi-deciduous woodland in mist-affected areas. Locally abundant in the woodland of the Haggeher mountains and adjoining limestone plateau at Reyged (550–1250m); in the relict woodlands at Qatariyah (c.490m), Reqadihon (600m) and on the ridge at Heger, above Bandar Shu'ub (c.500m); also rare in gullies and cliffs on Ma'alah (c.800m). 450–1250m.

Distribution. Widespread in Africa.

Botanical notes. This evergreen tree is characteristic of the semi-evergreen woodland that develops in mist-affected areas on Soqotra and its distribution more or less plots the occurrence of these. Flowers and fruits in the summer (February to March).

ETHNOGRAPHIC NOTES

Name(s)

– Soqotri: *kə̍lə, kɛ̄llɛ, ki̍liyə* (pl. *ki̍ləhe*) 'sudden gust of cold wind, sometimes with a shower of rain'; *śe̍rəmə̍no* 'sudden rain shower', 'tough, hard'; *kə̍nəhor* (in some western areas, but more commonly given to *Lannea transulta*); *ḥasəlām* (in some eastern areas, but more commonly given to *Rhus* sp. *A*); pieces of root *dūrib*.

– Soqotri simplified: *kilih, keyleh, kiliyih* (pl. *kilihe*); *sherimino*; *kinihor*; *hasilam*; pieces of root *duwrib*.

Food (livestock). This evergreen is an important dry season and drought fodder for goats in the drier areas. It is also eaten by cattle but they are less able to cope with its bitterness, and only eat it when very hungry. Donkeys occasionally eat the leaves and it is considered a good dry season fodder for camels. Herders lop leafy stems to take back to their animals, and gather the dried and fallen leaves to feed to their goats. In the better-watered and more richly vegetated areas the plant is of little importance to livestock, goats occasionally nibbling young leaves in a prolonged dry season.

Bees / honey. Honeybees are attracted to the fragrant flowers.

Flavouring / food additive. If the roots have been attacked by a certain grub (unidentified), the milk of goats eating the foliage develops a bitter flavour.

Fuel. The dead wood makes adequate firewood, but produces a lot of smoke and sparks, so is little used for cooking.

Timber. Makes poor timber and is attacked by termites unless it is de-barked before use.

Wood. Straight branches are used to make the important herding staff. The wood is also used for making a variety of implements.

Leatherwork. The inner bark is crushed and used to treat skins and hides to cure the leather and stain it red-brown. This is considered to be one of the best plants for this purpose. It is used especially to treat the leather of the floor covering called *ni̍bəʿeh* which is made from the entire, haired skin of a mature male goat. The same mixture is rubbed over the tulchan made from the skin of a dead calf and presented to the mother cow to sniff to encourage her to let down her milk. The mixture is also smeared over the back of a living calf to disguise its smell before presenting it to a strange cow, and to encourage her to accept the calf as her own and allow it to suckle.

Medicinal / veterinary. The astringent properties of the roots are exploited medicinally, especially in the treatment of skin complaints. A piece of root is either chewed, and the saliva used topically, or it is peeled, and then ground to a paste with a little water, sometimes with added salt and/or nicotine tar scraped from the bowl of a pipe. This treatment turns the skin black. To treat fungal skin complaints, such as ringworm, the affected area is first excoriated until it bleeds. Some people believe it is more effective to nick the healthy skin right around the affected area before applying the medicine, as this is said to prevent the disease from spreading any further. When struck repeatedly the living wood and bark oozes a resinous substance applied locally to treat sores, fresh wounds and abscesses. This is a painful treatment as it feels as if the skin is being burned. A piece of root is ground to a paste with salt, and inserted into a tooth cavity to alleviate toothache. The saliva produced when chewing the roots is believed to have curative and stimulant qualities, and to purify the blood.

Cleansing. Pieces of root are chewed before going to sleep to prevent halitosis and 'morning mouth'.

Cosmetic. Pieces of root are used as a chewstick to stain the gums, tongue and lips red. This is also done to cleanse the mouth.

Commercial value (past, present). The slim side-roots are dug up, cut into short lengths, and offered for sale in the coastal markets. The demand for these has increased over the last few years.

Pest control. When returning to living quarters after a long absence, a pile of the green wood is placed inside and set alight. The abundant, acrid smoke that results is said to disperse any of the undesirable creatures collectively called *reˈʾeś* which are believed to be harmful to man. This can be loosely translated as 'vermin', and includes some (but not all) reptiles, such as skinks, the nocturnal *Hemidactylus*, the chameleon and all snakes, as well as scorpions, wolf spiders, biting ants and the giant *Scolopendra* centipede. This is a done mainly in the Central Highlands where the trees are common.

Other. Some of the best soil for cultivation is found under these bushes, and is dug up and taken to the enclosed plots. When clearing new areas for cultivation the bushes are cut down, chopped up and heaped in piles over the stumps. They are then burned and the ash mixed with the soil. It is said to be an excellent fertiliser.

Notes. In eastern and western Soqotra this plant is uncommon, often growing in inaccessible cliff areas.

Comparative uses. In Africa, where the bush commonly grows on anthills, the root is boiled and added to a soup given to a young mother for post-partum problems. The root is also taken as a laxative and to expel worms. A decoction of the roots, with *Croton megalocarpus* roots, is given to treat chest pains, pneumonia or internal body swellings. An infusion of the bark is taken to stimulate the appetite. The fruit of many *Euclea* species are edible and their roots and stems widely used as chewsticks and to treat toothache. In nearby Dhofar the roots of *E. schimperi* are used cosmetically as in Soqotra, and the leaves are eaten to treat colic, diarrhoea, indigestion and mild mastitis. The inner bark is cut in long strips and used to bind a fracture to act as an antiseptic and to soothe the inflammation. It is also used to make a dye for cotton cloth. The fruit are eaten both unripe and when fully ripened. The green wood is burned as an insecticide, and is used to make implements. (Heine *et al.* 1988; Lewis & Elwin-Lewis 1977; Miller & Morris 1988)

ERYTHROXYLACEAE

Shrubs. Leaves alternate, simple, entire; stipules intrapetiolar, united. Flowers axillary, solitary or in few-flowered clusters, bisexual, actinomorphic, small. Sepals 5, united at base. Petals 5, shortly clawed, lobed nectary appendage present, inserted above base of petal. Stamens 10, united at the base to form a cup. Ovary superior, 3 or 4-celled; ovules 1 or 2 per cell; styles 3. Fruit a drupe. 4 genera. 260 species. Tropical, mostly New World.

Erythroxylon P.Browne

See description of family. 1 endemic. c.250 species. Mostly tropical Americas.

Erythroxylon socotranum Thulin

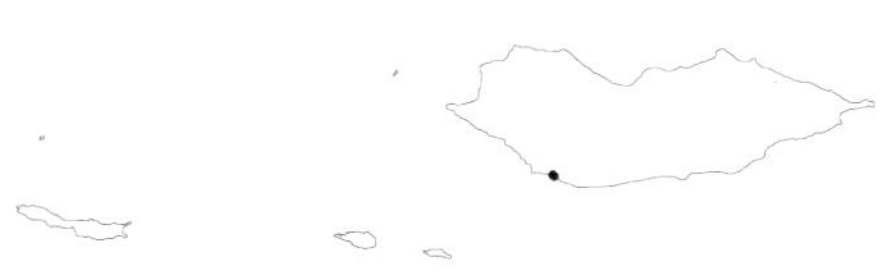

Tortuously branched small shrub to 0.5m. Stems densely covered by raised leaf scars and remains of stipules. Leaves obovate, 1–3 x 0.4–1.5cm, upper surface covered with an intricate network of raised veins. Flowers cream, c.3mm long. Fruit ellipsoid, c.5mm long.

Ecology. Very rare, on rocks and vertical cliff faces, near seepages on the limestone escarpment overlooking the Noged plain. c.150m.

Distribution. Endemic. First member of the genus to be found in SW Asia.

Conservation status. Endangered (EN B2 a biii). Known only from a few scattered populations growing near freshwater springs and seepages on the southern limestone escarpment. There are only a few of these springs and therefore the total number of plants is likely to be very low. This habitat is extremely vulnerable to water development and thus a designation of endangered seems appropriate (cf. *Helichrysum* sp. C, *Exacum socotranum* and *Scaevola socotraensis*).

Botanical notes. Probably most closely related to *E. platyclados* Boj., a species from Madagascar and the coasts of E Africa.

ETHNOGRAPHIC NOTES

Name(s)

– Soqotri: nil or *di rihˈoytin* 'of the water'.
– Soqotri simplified: *di rihoytin.*

Food (livestock). Normally grows out of the reach of livestock, but goats will eat it if they can reach it.

Comparative uses. Some related species have edible fruit. In India the leaf juice of *E. acuminatum* is taken as an anthelmintic, and in Brazil the bark of *E. citrifolium* is used to heal wounds. *E. coca* is the source of cocaine. (Ambasta 1982; Jain & De Fillips 1991; Kunkel 1984; Lewis & Elwin-Lewis 1977)

EUPHORBIACEAE

Herbs, shrubs or trees, sometimes succulent, mostly monoecious, sometimes spiny, often with latex and branched, stellate or scale-like hairs. Leaves opposite or alternate, simple or lobed, sometimes reduced or apparently absent; stipules present or absent. Flowers unisexual, 3 to 6-merous, sometimes much reduced and arranged in a specialised inflorescence, or cyathia (*Euphorbia*); petals usually absent; male flowers with 1 to numerous stamens with free or connate filaments; ovary usually 3-celled with free or united styles and solitary or paired ovules. Fruit a capsule, rarely a drupe; seeds often with a conspicuous caruncle. 321 genera. c.8000 species. More or less cosmopolitan, especially warm and tropical regions.

Acalypha L.

Monoecious herbs with unbranched hairs. Leaves alternate. Inflorescence axillary and terminal racemes, male flowers above, female flowers below. Male flowers minute, in clusters; calyx 4-lobed; petals absent; stamens 8. Female flowers in 1 to 3-flowered clusters in axils of cup-like leafy bracts; calyx 3 to 5-lobed; petals absent; ovary 3-locular; styles free, laciniate. Fruit a deeply 3-lobed capsule. 430 species. Tropical and warm regions.

Acalypha indica L.

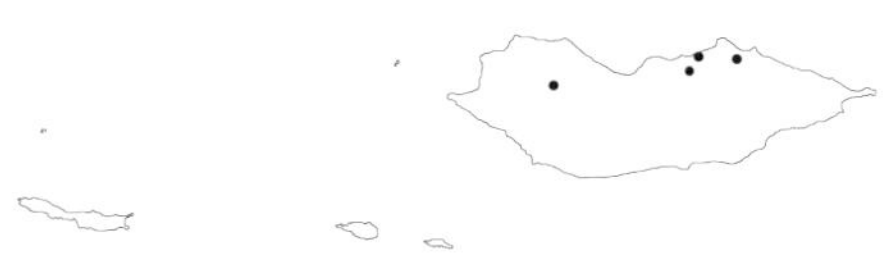

Erect herb to 50cm, hairy throughout. Leaves long-petiolate, ovate, <9 x 5cm, tip acute, margin serrate. Racemes <10cm long. Female bracts crenate-dentate, <12 x 14mm. Female flowers c.1mm diameter. Capsule 1.5 x 2mm, pubescent.
Ecology. Rocky and disturbed ground. 50–600m.
Distribution. Dry Old World tropics.

ETHNOGRAPHIC NOTES
Name(s)
– Soqotri: nil or *ʿaˈžbib di ˈriho* 'quick-wilting plant of water', *ˈrižəbeb* 'stalky'; *ʿˈʌḳelʌṣ* 'tightly curled, peppercorn hair' (also given to *Eclipta prostrata*, *Ageratum conyzoides* and *Solanum* species, all of which can also be cooked and eaten); *di ḥīdub*.
– Soqotri simplified: *'ajbib di riho*, *rijibeb*; *'aqelaz*; *di hiydub*.
Food (human). The green stems and leaves can be cooked and eaten like *Amaranthus* species.
Food (livestock). This is one of the *ʾˈažərib* group of plants which flourish after the winter rains. It is grazed especially by cattle before the grasses come up, but is little eaten by goats and sheep. It is seen as having little nutritional value.
Notes. It can cause *'uqabah* in small stock unaccustomed to this feed.
Comparative uses. More than 10 species of *Acalypha* are eaten. The plant contains a cyanogenic glucoside, poisonous in significant amounts. It also contains the alkaloid acelyophine, and extractions of the plant have expectorant and emetic properties similar to ipecac. In nearby Dhofar this plant used to be gathered as food and sold in the local markets. It was eaten both raw and cooked. Medicinally, a decoction is given as a laxative, but this may cause gastro-intestinal irritation. With added lime, the juice is used to treat a variety of cutaneous disorders. A decoction of the whole plant is used for earache and the powdered dried leaves given to children to expel worms. The leaves are also made into a poultice used to treat bed sores. In homeopathy it is used against severe cough associated with bleeding from the lungs. In Arabia, a leaf infusion of the closely related *A. fruticosa* is applied externally to treat swollen and bleeding teats of livestock and to treat bee stings, and in Yemen the leaves are used to treat eczema. (Abdulfatih 1987; Ambasta 1982; Dalziel 1948, 1956; Fleurentin & Pelt/Fleurentin *et al.* [1982, 1983a,b,c]; Hedberg *et al.* [1982–1986]; Heine *et al.* 1988; Jain & De Fillips 1991; Kokwaro 1976; Kunkel 1984; Lewis & Elwin-Lewis 1977; Miller & Morris 1988)

Andrachne L.

Herbs or subshrubs with unbranched hairs, monoecious. Leaves alternate, simple, entire; stipules laciniate. Flowers axillary, solitary or clustered; sepals 5, free, pale green with scarious margins; petals 5, small. Male flowers with 5 stamens. Female flowers with 5-locular ovary; styles 5, bifid. Fruit 3-lobed; seeds pustulate on back. 1 endemic. 25 species. Warm regions.

Andrachne schweinfurthii (Balf.f.) Radcl.-Sm.

Prostrate subshrub. Leaves grey-green, often in clusters, oblanceolate or obovate, <11 x 2mm, somewhat fleshy, papillose; stipules suborbicular, reddish-brown. Flowers solitary, greenish. Male flowers with oblong sepals, 1.5mm; petals spathulate, 1.5mm; pedicels <3mm. Female flowers with elliptic sepals, 2.5–3mm; petals minute, <0.5mm; pedicels <6mm. Capsule c.4.5mm diameter.
Ecology. Amongst rocks (both limestone and granite) in semi-deciduous woodland near the Muqadrihon Pass. 450–500m.
Distribution. Endemic.
Conservation status. Vulnerable (VU D2). Rare, with a very limited area of occupancy (possibly under 10km^2).
Botanical notes. Two varieties have been recognised on Soqotra: var. *schweinfurthii* and var. *papillosa* which is distinguished by its narrower, more densely papillose leaves. According to Radcliffe-Smith (1989), *A. schweinfurthii* includes the Somali endemic *Andrachne somalensis* Pax, but Gilbert in the Flora of Somalia (1993) maintains it as a distinct species. *A. schweinfurthii* is therefore considered to be endemic to Soqotra.

ETHNOGRAPHIC NOTES†

Name(s)

– Soqotri: nil or ‘kind of *ʿaˈžərib* ‘quick to die back’.

– Soqotri simplified: *‘ajirib.*

Food (livestock). In general these plants are not palatable to livestock. The fallen leaves provide occasional dry season forage for livestock, but the living plants are generally avoided by livestock unless they are very hungry.

†Also apply to *A. telephioides*.

Andrachne telephioides L.

Prostrate herb, sometimes woody at base. Leaves suborbicular, <6 x 6mm; stipules suborbicular, centre brownish. Flowers pale green. Male flowers clustered; sepals elliptic, c.1.5mm; petals lanceolate, c.1mm; pedicels <1.5mm. Female flowers: sepals elliptic, <3mm; petals minute, <0.5mm; pedicels 5–6mm. Capsule c.4.5mm diameter.

Ecology. Only recorded from Abd al Kuri; in rock crevices in desertic dwarf shrubland. Sea-level–150m.

Distribution. Mediterranean, Somalia and SW Asia.

ETHNOGRAPHIC NOTES

Name(s)

See *A. schweinfurthii* (also uses).

Comparative uses. *A. cordifolia* and *A. telephioides* have insecticidal properties and are used as such. Medicinally, pieces of the crushed root of *A. aspera* are applied to the eyelid to treat inflamed or swollen eyes, and *A. telephioides* has been shown to possess anti-microbial activity against *Staphylococcus aureus*, *Bacillus subtilis*, *Pseudomonas aeruginosa* and *Proteus vulgaris*. (Ambasta 1982; Batanouny 1981; Ghazanfar 1994; Jain & De Fillips 1991; Rizk & El-Ghazaly 1995)

Cephalocroton Hochst.

Monoecious shrubs with stellate hairs. Leaves alternate. Inflorescence terminal with flowers in dense, globose heads, male above and female at base. Male flowers with 3-lobed calyx and 7 to 10 stamens, petals absent. Female flowers with 6-lobed calyx, lobes entire, petals absent; ovary 3-locular; styles much divided. Fruit a 3-lobed capsule; seeds without caruncles. 1 endemic. 6 species. Tropical Africa to Sri Lanka.

Cephalocroton socotranus Balf.f.

Cephalocrotonopsis socotranus (Balf.f.) Pax

Shrub to 3m. Leaves obovate to elliptic, 1–3 x 1–2cm, tips rounded, margin entire, glabrescent above, densely stellate-tomentose beneath. Flower heads c.1cm diameter, peduncles 1.5–3cm; flowers c.1.5mm diameter; anthers yellow; stigmas yellowish green. Fruit finely stellate-tomentose, 5–8mm diameter.

Ecology. Common, sometimes dominant, in submontane semi-deciduous woodland and shrubland in the Haggeher mountains, typically around the base of granite pinnacles; often associated with *Rhus thyrsiflorus*, *Allophylus rubifolius* and *Dracaena cinnabari*. 500–1050m.

Distribution. Endemic.

Conservation status. Vulnerable (VU D2). Rare, with a very limited area of occupancy.

Botanical notes. Also doubtfully recorded from Hamaderoh in the 19th century. Comes into leaf, then flowers and fruits in the winter and drops its leaves in the summer.

ETHNOGRAPHIC NOTES

Name(s)

– Soqotri: *ṭān* (pl. *ṭˈuhun*); dead wood burned for its fragrant smoke *di rˈaḥmam* ’grey’.

– Soqotri simplified: *tan* (pl. *tuhun*); dead wood burned for its fragrant smoke *di rahmam.*

Food (livestock). The foliage is browsed by all livestock and is regarded as a very valuable and nutritious feed. It promotes lactation in camels, and makes goats and cattle fat and healthy. Sheep eat the plant litter, and cattle and goats eat the flowers and fruit. However, herders have observed that if an animal is thin and in poor condition, and then starts to feed heavily on this plant, it initially loses weight and becomes even weaker, and alternative forage has to be found for it. The leaves persist well into the dry season, and leafy stems are lopped by herders to feed their cattle and goats in the dry season. They also gather fallen foliage for their animals once the green leaves die back. In a prolonged dry season, goats chew at leafless twigs and at the bark.

Bees / honey. The sweet-smelling flowers are important for honeybees. This is regarded as the best honeybee forage of the Central Highlands.

Flavouring / food additive. The water in which the meat of an animal feeding on this is cooked is very white, rich and distinctively sweet-tasting. Indeed, the stock from the carcass of goats of the Haggeher mountains is famous all over the island for its deliciousness. The carcass of an animal (described above) which was thin and in poor condition and then ate heavily of this fodder is immediately recognisable: there is little fat on the meat, and the bone marrow is very watery and does not set firm with cooking. Nevertheless, the water the meat is cooked in is still rich and sweet, and makes a good layer of fat.

Fuel. The dead wood makes good firewood. It burns slowly and gives out a good heat, so is much used in the colder months to warm the living quarters of livestock and their owners. The dead wood burns very brightly, and burning brands are used as a torch. If the bark is removed, it gives off little smoke until on the point of going out, so is much liked for cooking. However, if the bark is left on, the smoke is acrid, and causes headaches and sore eyes when burned in an enclosed area. Twigs of the tree make a very good firelighter and are also a good kindling. They light easily even in damp weather, and in the Central Highlands the dried twigs are gathered during the dry season and stored in the home, ready for the cloudy and drizzly months of the monsoon and for the following winter. Even the green wood burns well and, in an emergency, can be used instead of dead wood.

Timber. Does not make very good timber, though the larger branches of well-developed trees are sometimes used for secondary roofing beams. It is relatively termite resistant, and does not have to be de-barked before use. The wood is more commonly used to make temporary summer shelters, forked branches being erected as a framework for brushwood walling and roofing.

Wood. Rather soft, and therefore not suitable for making implements for which durability is important. The slimmer, straighter branches are sometimes used to make the important herding staff if other, better woods are not available. It does not make very good charcoal.

Medicinal / veterinary. The fragrant dead wood is commonly burned during childbirth to protect both mother and baby from evil and to perfume the air.

Cleansing. The dead wood of this tree is the most common source of the aromatic woods collectively called *di ráḥmam*. Other woods in this group come from the dead wood of *Barleria popoviana*, *Dirachma socotrana* and *Clerodendrum* species. *D. socotrana* is generally agreed to have the sweetest smell (though not all islanders agree), but it is not widely available and most people use *C. socotranus* instead. This also produces more smoke than *D. socotrana* wood. The wood is used to fumigate living quarters, and small fragments of the wood are burned in a pottery incense burner to perfume bedding. Clothing is also held in the smoke or suspended above a burner to give it a sweet smell. Many islanders like to have at least one roof beam or rafter made from this wood, so that they can cut slivers off it to burn in the embers to perfume the air. The flowers are very fragrant, and flowering sprigs are tucked into clothing. The sniffing of sweet-smelling plants is held to protect from the 'noxious air' that can cause disease, and to be a safeguard against infection. All sweet-smelling plants are endowed with beneficial medicinal and prophylactic qualities. Sprigs are also sometimes laid amongst bedding to perfume it, particularly in bedding for a sick person, as the sweet smell is believed to relieve their discomfort and hasten their recovery.

Commercial value (past, present). The dead wood is gathered and offered for sale.

Pest control. Slivers of the wood are burned to rid living quarters of creatures believed to be harmful to man. These are collectively called *réʾeś*, and include some (but not all) reptiles, such as skinks, the nocturnal *Hemidactylus*, the chameleon and all snakes; also scorpions, wolf spiders, biting ants and the giant *Scolopendra* centipede

Other. The smoke of the burning wood is believed to keep evil at bay. In the more settled coastal communities along the coast, villagers used to burn the wood in their homes on Wednesday nights in particular, and in the mosques on Wednesday and Thursday nights, the fires being extinguished early on Friday morning.

Chrozophora Neck. ex Juss.

Monoecious herbs with stellate hairs or peltate scales. Leaves alternate. Flowers in leaf-opposed clusters, female flowers at base and male at tip. Sepals 5. Petals 5. Male flowers with 5 to 10 stamens, filaments fused into a column, ovary absent. Female flowers with 3-locular ovary; styles fused at base, bifid. Fruit a 3-lobed capsule. 12 species. Africa and Mediterranean to C and SE Asia.

Chrozophora oblongifolia (Delile) Juss.

Chrozophora obliqua (Vahl) Juss.
Chrozophora obliqua var. *frutescens* sensu Balf.f. non Schweinf.
Chrozophora tinctoria (L.) Raf.

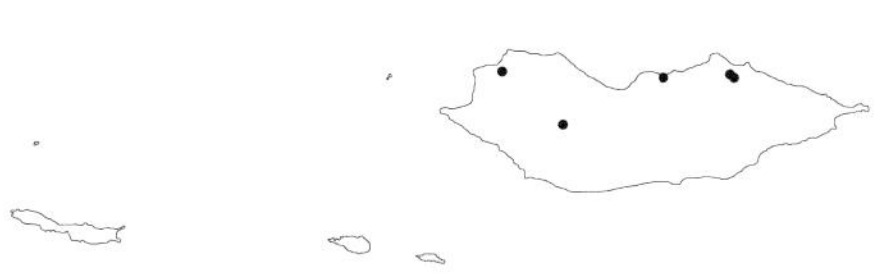

Erect herb, stellate-hairy throughout. Leaves grey-green, narrowly ovate, 4–8 x 1.5–5cm, tip acute, margin dentate. Male flowers in dense clusters, c.4mm diameter. Capsule subglobse, c.9mm diameter, covered in peltate scales.

Ecology. Disturbed ground and near habitation on the coastal plains. Sea-level–60m.

Distribution. Tropical NE Africa and SW Asia.

ETHNOGRAPHIC NOTES

Name(s)

– Soqotri: nil or *di ḥálbeb* 'quick milker' (pl. *di ḥalbíbihon*).

– Soqotri simplified: *di halbeb* (pl. *di halbibihon*).

Food (livestock). When young, this provides incidental grazing for all livestock. The mature plant is normally avoided. It is seen as having little nutritional value.

Comparative uses. *C. tinctoria* is given as a cathartic and emetic and is used to treat pimples. It produces a bluish-purple dye used for colouring liqueurs, wine, linen and cheese. (Al Rawi & Chakravarty 1964; Ambasta 1982; Jain & De Fillips 1991)

Croton L.

Monoecious or dioecious shrubs or trees, latex absent, clothed with stellate hairs or scales. Leaves alternate, entire, often discolorous and densely covered with stellate scales beneath, pinnately nerved. Flowers unisexual, in racemes, these often umbel-like. Flowers usually with 5 sepals and 5 petals: male flowers with many stamens; female flowers with a 3-celled ovary, each cell with 1 ovule; styles >2x branched. Fruits dehiscent; seeds smooth, carunculate. 3 endemics. c.800 species. Pantropical.

ETHNOGRAPHIC NOTES

Comparative uses. *Croton* species are important worldwide for their medicinal qualities. Many contain croton oil, mainly in the seeds but also in the leaves and stems, which is a powerful purgative. Ingestion of even small amounts can cause death in man and domestic animals.

It is interesting to compare the traditional uses of the larger Soqotra *Croton* with the important *Croton* of the African mainland, *C. macrostachys*, which is much used medicinally, unlike the crotons of Soqotra.

Seeds. The ground seeds show high molluscicidal activity, and are also used as a purgative and against guinea-worm.

Bark. It is regarded as cathartic, purgative and effective as a vermifuge.

Root. This is used as a vermifuge.

Leaves. Leaves are used to treat jaundice and pruritus. The leaf juice is given as a vermifuge, and the leaves are ground, soaked in cold water and drunk for stomach troubles. To treat a cough or a fever, the ground leaves are soaked in cold water for a day, and three spoonfuls drunk as a laxative and emetic.

In nearby Dhofar, *C. confertus* is also an important plant in local medicine. New shoots are dipped in butter and sucked to cure a cough, purify the blood and as a tonic, or they are made into a tea, drunk as a tonic. The leaves are pounded into paste used as a poultice to treat pain and swellings. They are lightly crushed in water and a solution taken orally for constipation and stomach ache – only in small quantities, since if taken in excess it can cause diarrhoea. They are steeped in water and the filtrate taken as an oxytocic in prolonged labour or as a remedy for a retained placenta. The growing tips of the plant are used as a salt substitute, but only a handful at a time or the food is made bitter. (Ambasta 1982; Bogdan 1977; Hedberg *et al.* [1982–1986]; Heine *et al.* 1988; Jain & De Fillips 1991; Lewis & Elwin-Lewis 1977; Miller & Morris 1988; Oliver-Bever 1986; Watt & Breyer-Bandwijk 1962)

Croton sarcocarpus Balf.f.

Dioecious tree to 5m. Leaves large, ovate, <10 x 6cm, tip acute, base cordate, strongly discolorous, green above and silvery-lepidote beneath. Flowers cream, in umbel-like groups. Fruits subglobose, 10–12 x 7–10mm, densely covered with coarse stalked stellate hairs.

Ecology. Frequent in submontane semi-deciduous woodland on granite; rare in succulent shrubland. (100–)500–1100m.

Distribution. Endemic.

Conservation status. Vulnerable (VU D2). Rare, with a very limited area of occupancy.

Botanical notes. Readily distinguished from the other species of *Croton* on Soqotra by its large, cordate leaves and larger, more coarsely and densely stellate-hairy fruits.

ETHNOGRAPHIC NOTES

Name(s)

– Soqotri: *mítrer* (pl. *mítrur*, *mitérhər*); *mítrer ṣáʿalhil* 'large-leaved *Croton*' to differentiate it from *C. socotranus*.

– Soqotri simplified: *mitrer* (pl. *mitrur*, *miterhir*); *mitrer za'alhil*.

Notes. Uses as *C. sulcifructus*.

Croton socotranus Balf.f.

Croton pachyclados Radcl.-Sm.

Croton socotranus var. *pachyclados* (Radcl.-Sm.) Radcl.-Sm.

Dioecious shrub or small tree to 3m, usually erect (prostrate in very exposed places); branches thick and rigid or thin and gracefully pendent; bark grey. Leaves very variable in size and shape, narrowly -ovate to orbicular, <4 x 2.5cm, long-petiolate and arranged along branches or, in grazed and exposed forms, shortly petiolate leaves clustered on short spur shoots, thinly or rarely densely hairy, not strongly discolorous. Flowers cream or white, in umbel-like racemes. Fruits subglobose, 5–6 x 5–6mm, with stalked stellate hairs.

Ecology. Widespread and in many places dominant on Soqotra; in shrubland on the coastal plains; succulent shrubland, submontane semi-deciduous thicket and mosaic communities on the limestone plateaus; abundant below 700m over most parts of the island. Sea-level–900m.

Distribution. Endemic.

Conservation status. Least Concern (LC). Common in several vegetation types and under no present or perceived threat.

Botanical notes. One of the commonest and most widespread shrubs or small trees of Soqotra; surprisingly absent from the outer islands. A variable species with some very distinctive forms: on exposed limestone plateaus of western Soqotra it is often a low or completely prostrate shrub with small leaves but in sheltered woodlands (e.g. in Wadi Ayheft) they are small trees with relatively large leaves and graceful branches. Another, very distinctive, form, described as *C. pachyclados*, has small, sessile leaves and thick horizontal branches, giving it a cedar-like appearance. Yet another form, found at higher altitudes, has white bark, but is otherwise morphologically indistinguishable from typical *C. socotranus*. This replaces more typical *C. socotranus* at higher altitudes (for instance on the Diksam plateau) and is considered useless by local people as its wood is soft and cannot be used for building. All these forms seem to be variants of this very common and widespread species. Small-leaved specimens mentioned by Balfour (1888) under *C. sarco-*

carpus are treated here as *C. socotranus*. *Croton socotranus* responds quickly to rain: the summer rains (April to May), the rains of the SW monsoon (June to September), and the winter rains (October to January) of the NE monsoon. Thus, according to where the rains fall in a particular year, the shrubs can be seen in leaf and flower in different areas of the island at different times of the year.

ETHNOGRAPHIC NOTES

Name(s)

– Soqotri: *mˈitrer* (pl. *mˈitrur*); *mˈitrer ḳīḳəhe* 'small-leaved' to differentiate it from *C. sarcocarpus*.
– Soqotri simplified: *mitrer* (pl. *mitrur*); *mitrer qiyqiheh.*

Food (human). The young leaves are eaten as starvation food. They are also eaten after they have been used to clarify fresh butter to make ghee.

Food (livestock). In the areas where these grow, they are the key fodder for the local goats, which readily eat the green leaves, flowers and fruit. Goats of such areas will browse these in preference to other foods, but will only browse the other species of *Croton* in hunger. The young leaves are bitter, but grow less so with age. The flowers in particular are much liked, as they are less bitter than the young leaves. The leafless twigs are eaten in the dry season. Herders regard all parts of this plant as a valuable and nutritious feed, making animals fat and increasing milk yield. Cows eat very little of the plant and donkeys are equally unenthusiastic. Sheep eat flowers and mature leaves within reach, and graze the plant litter. The areas of lowland grazing where this plant is common are regarded as the best kind of rangeland for rearing goats, especially as it is usually free of the plants which cause the *'uqabah* disease. Indeed, goatherds from outside such areas frequently seek permission to transhume into them when their own rangeland is exhausted, or at least to lop the leafy stems to feed their small stock. The plant responds very rapidly to rain, even the lightest drizzle causing it to put out new leaf and come into flower. Some foliage persists well into the dry season, and the stems and twigs remain palatable, making it one of the most important dry season forages of the island. The fallen, dead leaves and other plant litter are also a good subsistence feed for goats and sheep in the dry months.

The plants respond to rains in different ways. If it rains early in the summer months, stock cannot browse it for the first few days, as the foliage is very bitter. If it rains later on in the summer months, the foliage is even less palatable for stock. However, if there is a shower during the monsoon months, livestock can feed on the plants straight away, as the monsoon salt will have been 'washed off'. The best time for rain is when the trees have very few leaves. Any rain causes the plants to burst into flower, which is regarded as the very best feed for small stock. If it rains when the tree is already well in leaf, it produces few flowers.

Islanders have observed that goats are very choosy about this plant. They prefer the prostrate varieties of the limestone plateaus, and the stunted shrubs that emerge from cracks in the limestone pavement in the foothills areas. Then they like the plants that grow in hard-packed earth overlying a layer of a certain hard, white rock. Such plants never grow tall, and in flower their smell and taste is particularly sweet.

Bees / honey. The flowers are one of the most important forages for honeybees. After the least rain, they are humming with bees. Some of the best honey on the island comes from the areas of *C. socotranus* woodland. If honey has absolutely no trace of bitterness it is likely to be from this tree.

Flavouring / food additive. If forming a major part of the diet, the milk of the animal develops a recognisable and slightly bitter edge. In the carcass, the flavour of the viscera, and sometimes also the cooking water, is likewise slightly affected. However, this in no way makes either the milk or the meat inedible. In the monsoon months the lowland trees absorb a lot of salt from the winds blowing across the island. Islanders say that any grasses and herbs growing beneath the trees are little grazed at this time because they are too salty. Indeed, if people have no salt during these months they can use twigs from the plant instead, swishing a handful around in the cooking water to salt it. Even when the first rains of winter have fallen, the herbs below remain salty, as the rainwater dripping down on them from the leaves of the shrub above is full of salt.

Fuel. It is important as a fuel mainly because the trees are so widespread and grow so quickly. After rain, the shoots of new plants emerge from the ground everywhere around the adult trees, 'like grass', the islanders say, and grow upwards with remarkable rapidity. It is one of the few plants of the island to be purposefully dug up and dried for use as fuel. Normally, apart from a few unpalatable or otherwise useless plants, only dead wood is gathered for firewood. The wood of this *Croton* makes excellent firewood, which is said to be one of the slowest-burning woods of the island. It gives off a good heat and burns with a bright light. A good chunk of the wood is the wood of choice for night-time illumination, and burning brands are used as torches. However, if the bark is left on, it makes a lot of rather acrid smoke, the wood of the smaller, prostrate varieties especially. The de-barked lower trunk, and the stumps with the roots, make much less smoke, and are much liked for cooking. Islanders say that the smoke of this plant, like that of *Ballochia*, *Trichocalyx* and *Carphalea*, is a 'good' and beneficial smoke. Twigs and smaller branches are gathered and stored as firelighters and kindling. The wood of this plant is considered to be the best for keeping a fire in overnight or during a long absence. A chunk of the burning wood is poked well into a pile of hot ash and left there until the fire is wanted again. To keep a fire alive over even longer periods, a burning piece of the wood is poked into a pile of dry and aged sheep dung. A glowing ember is put on a slip of pottery and placed at the edge of the pile, to keep it alight from the outside as well. The heap of dung then keeps smouldering for three or four days, and then, when a fire is wanted again, the *Croton* stick is pulled out of the dung and used to kindle a fire. In earlier times when livestock theft was prevalent, this bright and slow-burning wood was often lit inside the goat pen or byre compound to give any potential thief the impression that the night guard was awake and sitting by the fire, guarding his animals.

Timber. One of the key roofing timbers of the island. As a general rule, the shrubs growing in the hotter and drier areas of the island are considered to make the best timber, the wood of *Croton* growing in the wetter areas a rather inferior one. The slimmer trunks are used especially for the ceiling, and the trunks of larger trees as secondary roofing and cross beams. The bark is attractive to termites, so is removed before use. People collect the material by walking through *Croton* shrubland, pulling plants up by the roots, and then stacking these in a pile. When enough have been collected, the pile is loaded onto a donkey or camel, or is carried on the back to the homestead. Here the plants are prepared for use. They are topped and tailed, trimmed to size and then the bark stripped off. This is either removed by slicing it off with a knife, or the plants are put to soak in water until the bark can be peeled off. The bark is dried and stored as kindling. Temporary or seasonal accommodation is commonly made using these plants. A low circular wall is built, leaving a gap for the doorway. A central pillar of a strong wood is erected in the middle. Then 10 or so poles of *Croton* or other wood are laid like spokes from the top of the central pillar down to the top of the wall, in a cone shape. Then, starting at the peak of the cone and working back down to the walls, leafy stems of *Croton* are woven in between and over the spokes until there are no gaps left. These huts are used mainly in the summer months when many islanders move uphill to escape the burning heat of the lowland plains, or when herders and their animals have transhumed in pursuit of the grazing brought up by summer rains.

Leatherwork. The wood burns down to fine and white ash, which is much used in the preparation of skins and hides in leatherwork. This is the preferred ash to use on skins which are to be used for the most demanding tasks, such as fetching water, collecting milk, storing pressed dates, or churning milk to make butter.

Wood. It is not a hard wood so is little used for making implements. The wood of the smaller, prostrate crotons of the hotter, drier upland areas is harder, but is usually too small in size to be useful for utensils. The trunks are used as a lever to move heavy objects, such as lintel or sill stones, or rocks come across when digging a well. Slimmer trunks are used for posts and poles, but are said not to be as durable as *Chapmannia* and *Ormacarpum* wood. In areas without date palms, the slimmer trunks or larger, straight branches of this shrub replace the leaf ribs so widely used in construction in date-growing areas (see *Phoenix dactylifera*). They are tied together to make the gate or wicket door for houses, pens and cultivated plots that is such a characteristic feature of island architecture. They are also used in fencing and enclosing land. For instance, kitchen gardens and precious saplings which have been sown or transplanted are commonly protected with fencing made from this wood. A stick is cut from a tree and one end sharpened to a point to make a digging stick for extracting clay, or for cleaning out the accumulated silt from the bottom of the rainwater catchment before the winter rains. Short lengths of wood are sharpened to make stakes to hammer into the trunk of a *Sterculia africana* tree to enable a herder to gain access to the crown to cut dry season fodder for his cattle. The dead wood can be used as a firestick, but only if no better wood is available, as it takes a long time to get hot, making it exhausting work. The wood of the slimmer side branches, or the inner and outer bark, is much used to make the woven sections and the bracing sticks of the fish trap, as well as its framework. The branches are cut off and put into water to ret down for several days, and then the bark is stripped off and the individual sections of the trap woven. People from the coastal fishing settlements used to go inland to collect the material for this. Indeed some islanders say that this plant is increasing in the coastal areas precisely because these fish traps are no longer made. The slimmer stems are used to make the 'bit' mouthpiece of the weaning device worn by lambs and kids. Slim and straight stems are used to make the important herding staff. The wood makes a good charcoal used for trade with the island blacksmiths, or visiting boats. The wood used to be in great demand too for making quicklime.

Commercial value (past, present). Bundles of the debarked wood are sold both internally and in the coastal markets for house construction and fencing.

Pest control. Both live and dead material is used for livestock smudges, though the dead wood is more effective. It is not de-barked for this, since it is the bark which makes the most insecticidal smoke. The pungent smoke discourages biting insects, and the heat warms and relaxes the animals before they are milked. It is used for sheep, goats and cattle mainly in the winter months when insects are at their worst. Smudges are also lit at other times of the year when it is cold or damp to warm the animals, especially just before milking. In the high mountain areas they are also lit in the cloudy and drizzling monsoon months. The fine, white ash is considered to be one of the best ashes to smear over skin food containers, especially skins of pressed dates, to deter insect pests.

Other. The wood is good for firing pottery. Leafy branches, weighed down with a rock, are used to cover food to keep off flies and other pests. Twigs are whittled flat to make the throwing sticks for the game *ṭabbānāt*. A larger and a smaller stick are cut and shaped for the game *məṣrērə*, in which the larger stick is used to flick the smaller one, placed across a hollow in the ground, as high in the sky as possible; the other player, or team, has to try and hit it back. In the absence of cereal grains, the small leaves are used to clarify fresh butter to make ghee.

Notes. *Croton* shrubland arreas are commonly dominated by the *di ṣˈəffo* group of grasses (*Aristida* and *Lolium* species). This type of association is recognised locally and is called *ersimo.* Islanders differentiate between 'male' and 'female' shrubs, the 'male' having smaller flowers and fruit than the 'female'. The importance of this plant to the islanders is reflected in the extensive and detailed lexicon describing the plant: every size, shape or growth form has its own specific term. A person who is a reliable source of help or refuge in times of trouble is called by the name of the plant. Islanders say that towards the end of the monsoon the amount of salt in the air increases, and that seeing a salty rim around the trunks of the trees is a sign that the rains of winter will soon fall. Another sign of impending rain is if spiders are seen to be making a lot of webs in the branches. The shrub is affected by two sorts of grub (unidentified). One is a tiny caterpillar which becomes obvious well into the dry season. The islanders call it *di ḳɔ̄wḥaf* 'eats anything and everything' because it chews its way through any leaves left on the tree. The other arrives halfway through the winter season, and produces what look like circular burn marks on the leaves. The affected leaves change colour, redden and drop to the ground. After about a month, new ones grow in and the shrub apparently recovers.

Croton sulcifructus **Balf.f.**
Croton elaeagnoides Balf.f.

Monoecious shrub or small tree to 6m. Leaves ovate, <6 x 4cm, strongly discolorous, green above and silvery-lepidote beneath, tip rounded or acute, base rounded or cordate. Flowers cream or white, in elongated or congested racemes. Fruits subglobose, 5–10 x 5–7mm, with sessile stellate hairs.
Ecology. Common, sometimes dominant (eastern Haggeher), in montane semi-deciduous woodland on granite; locally common in drought-deciduous woodland on limestone (Reyged and Rewgid). 350–950m.
Distribution. Endemic.
Conservation status. Vulnerable (VU D2). Rare, with a very limited area of occupancy.
Botanical notes. A rather variable species. It can be a graceful tree with cordate leaves (growing on the limestone plateaus at Rewgid and Reyged) or an untidy shrub with cordate or oblong-ovate or oblong-elliptic leaves and more congested racemes (on granite, mainly at higher altitudes in the Haggeher mountains). *C. elaeagnoides*, described by Balfour from material without fruit or male flowers, seems to be indistinguishable from *C. sulcifructus*.

ETHNOGRAPHIC NOTES†
Name(s)
– Soqotri: *ˈmitrer* (pl. *ˈmitrur*); *ˈmitrer ṣˈaᵓalhil* (pl. *ˈmitrur ṣaᶜˈlilihon*) 'large-leaved *Croton*', to differentiatie it from *C. socotranus*; *ˈmitrer ḥˈagərhi* 'Haggeher *Croton*'.
– Soqotri simplified: *mitrer* (pl. *mitrur*); *mitrer za'alhil* (pl. *mitrur za'lilihon*); *mitrer hagirhi*.
Food (livestock). Goats of the high mountains accustomed to this forage browse the rather bitter foliage, both green and drying, and it is regarded as an excellent fodder for them, especially in the dry season. They also readily eat the buds and fragrant flowers, and in the dry season their owners cut leafy stems and gather fallen leaves as a feed for them. However, goats which are not accustomed to it generally avoid the green foliage, eating only dried, fallen leaves and other plant litter. Although the leaves persist through much of the dry season, it is not often lopped by goatherds from the surrounding areas. Not only are their goats not used to the foliage, but they also say it taints the milk and flesh of their animals more than is acceptable. Hungry cattle eat a few of the green leaves, but find them too bitter and prefer the dried, fallen ones. These are gathered for them by their owners in a prolonged dry season. In a drought the inner bark is crushed and fed to any livestock which will accept such a feed. Camels readily browse the trees.
Bees / honey. Honeybees are attracted to the sweet-smelling flowers.
Flavouring / food additive. The meat, cooking water and milk of animals eating a lot of this are affected in the same way as that of animals feeding heavily on *C. socotranus*, but even more so. Some islanders find the stock and viscera too bitter to eat.
Fuel. The dead wood makes reasonable firewood. It does not make too much smoke if it is peeled first, so is used for cooking. People in the Central Highlands gather the wood in the dry months and store it against the damp monsoon months when the mountains are enveloped in cloud and mists. Dry twigs are also collected to use as kindling or tinder. The wood burns with a bright flame, so is burned a lot at night, while flaming brands can be used as a torch.
Timber. Does not make very good timber, but large branches, if cut in the heat of summer and properly dried, are used for secondary roofing beams, such as cross beams. In the absence of better timber, the trunk of a good specimen can be used for the main supporting pillar. As the better woods of Soqotra are increasingly felled for timber, the wood of this tree is growing in importance. The main advantage of the larger *Croton* species is that their wood is less prone than others, such as the smaller *Croton* species, *Maerua angolensis*, *Flueggea* species and *Buxanthus pedicellatus*, to rot. Like *C. socotranus* the bark is attractive to termites, so the wood is debarked before use.
Wood. Makes good charcoal. Otherwise it is rather soft, and not suitable for implements, though the slimmer, straighter branches are sometimes used to make the important herding staff.
Pest control. The wood is used like that of *C. socotranus* for livestock smudges.

†Uses also apply to *C. sarcocarpus*.

Euphorbia L.

Herbs, shrubs and trees, often succulent, sometimes with paired spines, milky latex present. Leaves entire, sometimes soon falling. Flowers very reduced and aggregated into a specialised inflorescence (cyathium) made up of 1 female flower (reduced to a stalked ovary), surrounded by >1 male flowers (each reduced to 1 stamen); cyathium, surrounded by a whorl of 4 or 5 glands alternating with lobes (fringed or lobed), the whole resembling a flower. Fruit a 3-seeded capsule. 11 endemics. 1600 species. Cosmopolitan, especially warm regions.

Botanical notes. A very large and variable genus. On the Soqotra Archipelago species vary from minute herbs, to succulent shrubs to large trees.

Euphorbia abdelkuri Balf.f.

Large succulent, shrub, densely branched with many erect cylindrical stems forming dense clumps to 4m tall x <5m wide; stems leafless and unarmed, 5–8cm diameter, grey-green, distinctly wrinkled and bearing distinctive 'zigzag' markings; latex yellow-green, very caustic. Cyathia small, green, in clusters towards tips of stems.

Ecology. Succulent shrubland (associated with *Euphorbia balsamifera* and *E. schimperi*) on rocky slopes; only on Abd al Kuri (Jebel Saleh). 200–350m.

Distribution. Endemic.

Conservation status. Endangered (EN B2 a biii). This remarkable succulent is endemic to the island of Abd al Kuri where c.250 mature individuals are found at a single location covering less than 10km^2. The latex of *E. abdelkuri* is toxic and the plant is therefore not eaten by livestock, so it is under no immediately perceivable threat except perhaps from collection by succulent enthusiasts. It is not found in similar habitats at other locations on the island, perhaps suggesting that there has been a decline in its extent of occurrence. That it was once more abundant is confirmed by Forbes (1903) who comments: 'This *Euphorbia* grows all up the side of Gebal Saleh [where it still grows] from about the upper two-thirds (500–1500ft), sending up its leafless stems like a forest of green candles'. It was classed as endangered in the Red Data Book (Lucas & Synge 1978). The population of *E. abdelkuri* on Abd al Kuri lies within an area designated as a Nature Sanctuary within the zoning plan for the Soqotra Archipelago.

Botanical notes. This remarkable shrub is endemic to Abd al Kuri and has been described as the most remarkable species of *Euphorbia* (Lavranos 1971). It is locally common on Jebel Saleh where it grows in sandy soils in open semi-desert shrubland and is not reported from other places on the island.

ETHNOGRAPHIC NOTES

Name(s)

– Soqotri: *fərḥʌ́ź, fərḥʌ́ḏ*.
– Soqotri simplified: *firhadh*.

Food (livestock). Not palatable to sheep and goats, nor is it grazed by donkeys.

Notes. The caustic latex has no traditional uses.

Euphorbia arbuscula Balf.f.

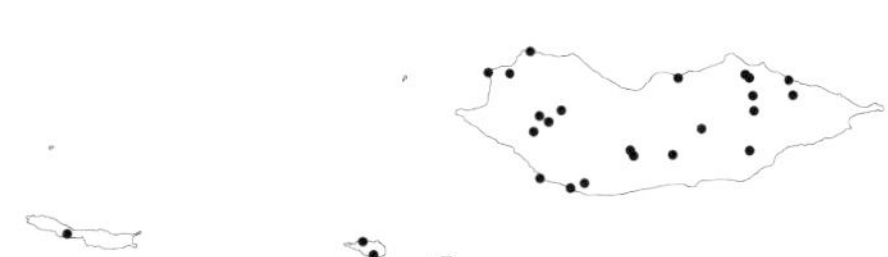

Small tree to 5m, sometimes a low shrub (var. *montana*), trunk rough. Crown flattened or rounded and profusely branched above, with leafless, pencil-like stems. Cyathia sessile, in small clusters at stem tips. Fruits brown, erect, often profuse.

Ecology. Locally common on Soqotra in *Croton socotranus* shrubland, succulent shrubland and occasional in drought-deciduous woodland; a common emergent and sometimes forming small groves on the desertic plains; there are also small populations on Samha and Abd al Kuri. Sea-level–700m.

Distribution. Endemic.

Conservation status. Near Threatened (NT). Widespread and abundant but a crucial dry season fodder. If livestock numbers increase greatly, or a succession of drought years occur, it could rapidly become threatened in some areas. The small populations on Abd al Kuri and Samha are particularly vulnerable.

Botanical notes. A common emergent over most of the island, sometimes planted, and forming small groves in places on the plains. It flowers in the hot, dry season and survives severe drought. On Abd al Kuri there is a small but flourishing colony. Here goats do not browse the young plants and trees are found at all stages of growth. The small population on Samha, where the plant is browsed, shows no signs of regeneration. On Soqotra low and windswept forms occurring at higher altitudes have been called var. *montana*.

ETHNOGRAPHIC NOTES

Name(s)

– Soqotri: *ʾímtəhe* (pl. *ʾemɛ̄ti*) (Soqotra and Samha); *ʾimṭéy* (Abd al Kuri); leaf *eṣəlīʿo* (pl. *éṣlaʿ*); bunch of leaves *ʿíšḳə* (pl. *ʿášiyuḳ*, like date palms).
– Soqotri simplified: *'imtihe* (pl. *'emeyti*); *'imtey*; leaf *eziliy'o* (pl. *ezla'*); bunch of leaves *'ishqih* (pl. *'ashiyuq*).

Food (livestock). This foliage is the key dry season fodder for goats on Soqotra. However, goats have to be accustomed to it from youth, or they are unwilling to eat it. Indeed some herders say that a goat that has never eaten this before, and then eats it out of desperate hunger, can die as a result. The very tips of the branches are cut off and then these leaves are chopped up small for the goats. Older herders complain that some herders nowadays are careless in the way they treat the trees by chopping off whole branches, or beating the branches with a heavy club to knock off bunches of leaves which they then leave strewn on the ground for the goats to eat as and when they can. This is seen as being damaging to the tree and not offering the best fodder to the animals. The leaves are eaten by goats when they are fresh and green, and later when they become grey and start to dry out. However, goats fed on a lot of this must be kept from water as if they drink a lot after eating *E. arbuscula* they develop severe diarrhoea and can even die. Plump and healthy goats initially get diarrhoea from the new diet and lose weight. However, they soon recover and begin to put on weight again, and their herders say that this diet acts like a strong purge, ridding the digestive system of all impurities. A goat on a diet of this becomes fat and healthy, and gives more milk. If a dry goat eats this fodder continuously for a fortnight or so, it becomes so strong and fat that it comes into season and calls for the male. It is such a reliable fattener, that if a herder sees that a goat on this diet continues to lose weight, develops chronic diarrhoea and looks increasingly unwell, this is a sure sign that it is suffering from *'uqabah*. If it continues on this feed, it can die as a result. The same is said to be true of goats suffering from mange. Camels browse the tree in the dry season, but mainly only the male pack camels, female camels being said to rarely eat either this or *E. schimperi*.

The rather stunted and darker-coloured trees that occur in the higher and drier areas, called *ʾˈeggəhur*, 'exhausted', are rarely cut to feed goats. Their leaves are said to 'have no flavour' and do not make a goat put on weight as the plumper, paler lowland trees do.

The latex is very caustic, and a man going to lop fodder wears old clothes or covers himself with sacking. The fodder is rarely carried to the goats. Rather, the herder climbs up into the crown of the tree and taps it with the handle of his knife, and makes the special, penetrating call to his goats which indicates that there is food. As soon as they hear him, they come running, and stand beneath the tree waiting for him to cut leaves down to them. In general, goats prefer the freshly fallen leaves of the adult trees to material from younger trees. If a goat eats a lot of this feed over a long period the hair around its muzzle is sometimes affected and begins to drop out, leaving bare patches. Some also get sore and encrusted eyes from the splashing of the latex. However, this damage disappears when the rains finally fall and the goats start to eat the fresh rains grazing. A more serious and incurable problem is the black spot or growth that appears under the tongue of some goats eating a lot of this fodder. If a goat on a diet of this fails to thrive, one of the first things its owner does is to inspect its mouth. If he finds a black area under the tongue the animal is generally slaughtered there and then as the condition will worsen and is incurable.

Interestingly, the goats of Abd al Kuri, which originate from the Arabian mainland opposite, are not accustomed to this feed, and the *E. arbuscula* seedlings and young plants remain unbrowsed. Here each adult tree is surrounded by a nursery of seedlings and saplings at different stages of development, a sight rarely seen on Soqotra, where in the dry season young plants are eaten as soon as they appear. On Soqotra, young plants sown by the islanders have to be protected with fencing. On Samha the trees are used as on Soqotra.

Bees / honey. Honeybees cluster around the flowers, but the honey made from this plant is rather bitter and 'hot' or peppery, like that made from *Buxanthus pedicellatus* and *E. schimperi*. It is regarded as a very useful medicinal honey, especially for the chest problems so common on Soqotra.

Flavouring / food additive. The milk of animals eating a lot of this plant is 'hot' to the taste, and is disliked by some. Although the carcass of goats feeding on it has plenty of fat, the viscera – especially the liver and the stomachs – and the cooking water usually have to be discarded as they are impossibly bitter. Even the skin of the slaughtered animal is said to retain the smell and taste of the latex, and is not used to store food or water.

Fuel. The wood makes terrible firewood. It burns very quickly and gives off clouds of dense and acrid smoke which taints food and irritates the eyes and nose. This and the wood of *Jatropha unicostata* and *Ficus cordata* are well known as being the worst firewoods of the islands. Its smoke, with that of *Ficus* wood, is said to be the most irritating and painful.

Timber. No use as timber. It is very resistant to termites.

Wood. Soft, and not suitable for making anything. Nor can it be used as a firestick. However, the dead wood is added to others to make quicklime.

Medicinal / veterinary. The latex is more caustic than any other on Soqotra, and burns badly if it comes into contact with the skin. As a result, it is little used for skin complaints. The exception to this is the dry eczema commonly found on the hands of those who milk a lot. Dry, cracked and itching skin is treated in the same way when other treatments have failed. The latex is spread on a large leaf and stuck over the ribs of a sheep or goat with a bad cough or breathing difficulties. It is rubbed all over a goat with a high fever, or as a final resort, over the body of a goat which seems moribund. Mixed with fine earth it is applied to bare patches in the coat of a goat or sheep with mange. A tear in the teat or udder is repaired with a dot of the latex. This seal dries up and then later drops off of its own accord. A discharging wound or one that is crawling with maggots is treated by smearing the area with fresh dung and then with the latex, and finally pasting a scrap of cloth on top. Sores in livestock are smeared with a thin layer of latex to keep off flies. It is also applied around and under the tails of sheep to deter blowflies. Only this and the latex of *E. schimperi* have the powerful odour which deters the flies that are particularly prevalent in the still period when the winds drop at the end of the monsoon and before the winds of winter start. In the early winter when many stock develop scour on the fresh rains grazing the flies are again a great nuisance and animals must be regularly inspected. If the shepherd notices the eggs or larvae of the blowfly on his animals, these are scraped off and the area painted with the latex. This prevents the flies from landing and laying their eggs. The illness in goats called *bˈikəməh*, whose signs are stiffness with sore and swollen joints, is treated by smearing the latex on and around each joint and along the ridge of the backbone. In the absence of *Ficus vasta* latex, which is less painful, a sprain is treated by an application of the latex to the swollen area, and then covering this with a layer of fine earth to make it less sticky.

Someone with a bad head cold either sniffs some latex or burns a dead leaflet on the fire and inhales the fumes. A sick person not responding to other treatment is encouraged to crouch over a fire with a cloth over his head and inhale the fumes of burning leaves. The illness called *di so'o* (whose signs are fever, general weakness, and a rash of large and hard spots which break out all over the body, including the face), is treated in the same way, though in this case more leaves are used and the whole body is bathed in smoke. A persistent fever is treated by covering the patient with a thick wool blanket and then lighting a small fire beside him, putting on some dry leaves, and encouraging him to breathe in the smoke. A fresh and deep laceration is treated by pulling the sides of the wound together and sealing them in position with the latex, like a sticking plaster. Raised and tender glands are treated with an application of the latex and fine earth, if initial treatment with *Jatropha unicostata* sap has not been successful. The very caustic latex is sometimes used instead of a red-hot iron for cauterisation, for instance, to treat mastitis in livestock, especially in goats. For both animals and people, in a simple fracture, a solid plaster is made for the broken limb by using the latex mixed with sheep's wool or goats' hair to a paste. Once the bone has been manipulated back to its original position, the limb is painted with the latex. A thick layer of wool or hair is stuck all round, and then a final coating of the latex is applied, or strips of aloe root-bark or *Adenium obesum* bark, beaten flat, are used to strap the limb. Some add salt to the latex. The limb must be immobilised, and in the case of livestock, the animal kept in a tiny pen and hand fed and watered. A similar plaster of latex and wool or hair is applied to the torso of someone with a painful chest, accompanied by wheezing and a persistent cough. Very rarely, a drop of the latex from a green leaf is used to treat chronic inflammation of the eye, corneal opacity, or a discharging eye that does not respond to other treatment. This is intensely painful, but those who have actually done this swear by it. A slightly milder remedy is to bake a leaflet at the edge of the fire, then crush it with a little water. A few drops of the filtrate are dripped into the affected eye.

Pest control. The latex and the ash of the burned wood are very effective insecticides. The burning wood has a very distinctive and strong smell much liked by both animals and their owners. The dead leaves also give off a very sweet smell and these are used with or instead of the wood. Both latex and ash deter rodents, which are a great pest on Soqotra, and are used especially on the goatskins in which the precious pressed dates are stored. The latex is smeared on the emerging horns of kids to stop rodents gnawing at them. It is also applied around the neck and anus area of lambs and kids to deter wild cats. Ravens also attack young lambs and kids and they can be destroyed by removing the eye from a carcass and putting some of the latex inside it and then leaving this beside some meat. The ravens always prefer the eye and are quick to eat it, and are poisoned as a result. The dead wood is much used for livestock smudges for sheep and goats. Although this wood does not give off great heat, it produces plentiful smoke, and is the wood of choice at times when biting flies and other insect pests are at their worst. The smoke is also regarded as being to some extent curative, and is used when animals are ill with coughs and head colds, or when they are suffering from the effects of botfly attack. The ash that remains is carefully gathered up and saved. The smudge is kept burning long after the animals have been milked and have been taken off to graze, and new material is added to thoroughly fumigate the pens and yards. Similar fires are also lit inside the byres to ensure that the livestock quarters are thoroughly fumigated. For this reason, many herders light a smudge of this material in the night quarters both in the early morning before milking, and then again in the evening. On Samha the dead wood of this tree was considered to be the best wood for all livestock smudges.

Other. Goats feeding on this can go longer without water. Herders say that if goats are fed on a diet of this or *E. schimperi* throughout the dry season they can go without water until the winter rains. Indeed, if goats drink their fill after eating this plant, they can become very ill with severe scour, which is sometimes fatal. Many herders say that goats eating a lot of this plant should be actively kept away from water. The latex is the most widely used and effective for stupefying the fish of the smaller lagoons along the coast. This is done in the hungry summer months when the dates have not yet ripened and livestock are largely unproductive. The ash of the leaves is considered to be a good fertiliser. When preparing seedbeds for tobacco or finger millet, little piles of the leaves are heaped here and there over the area and then set alight. The cool ash is mixed in with the soil, which is then ready for sowing. The latex was used to caulk the dugout canoes. It was poured into a large pot and sheep's wool or sacking fibres were added. This mixture was used to plug any gaps and cracks in the wood. The same latex/wool or hair mixture was used in making the horn handles for the Soqotri triple knife-set carried by every adult man. The latex of other plants was sometimes used, but that of *E. arbuscula* was considered to be the best. The goat's horn was boiled until it was possible to extract the core. It was then straightened by repeated heating and manipulation between slabs of stone until it was made straight and flat. The pointed tip was cut off and the horn then packed with a mixture of goats' hair or sheep's wool mixed with the latex. The tang of the knife was then heated at the fire – but not allowed to become so hot that it melted the horn – and then tapped into the horn handle until it had gone in as far as the base of the blade. It was then set aside to cool and set hard, making a firm handle. A strong handle for the larger knife blades used for chopping wood and working date palms was made by soaking strips of cloth in the latex and winding these around the tang. The final layer was rubbed with earth to stop it being sticky and then it was squeezed firmly in the hand to shape the handle to fit exactly the grasp of its owner.

The latex was also used as an adhesive, considered to be the strongest available on the island. It was used, for instance, to stick handwritten pages of the Qur'an together along one edge to make a booklet. To make a good adhesive, the latex had to be collected in the full midday heat and then left for at least two days before using it. The poisonous qualities of the latex were exploited when clearing out the small rainwater catchments and rock pools in preparation for forthcoming winter rains. When there was only very little water left, full of wriggling larvae and other undesirable insects, a cupful of the latex was stirred in. It was given time to act, and then the water was emptied out and the base and sides of the catchment scraped clean of sediment and slime. If the rains fall too soon after this treatment, the taste of the water is affected and becomes slightly bitter. This is apparently harmless to those who drink it, though some islanders say that donkeys are susceptible to water with this or *Adenium obesum* latex in it, and can die as a result. Other livestock apparently develop diarrhoea but quickly recover. On Samha, the latex is used to lime birds, but on Soqotra that of *E. socotrana* and *Ficus cordata* is preferred. The tree also has supernatural qualities. Evil spirits cannot come close to it, and a sprig of leaves can be carried for protection. It is also quite safe to go for a nap under one of the trees, as no *jinn* or other evil spirit will dare approach it. Some islanders plant a tree close to the entrance to their living quarters or at the doorway of pens and byres for this reason. Containers of food and water are protected by placing a leaf across the mouth of the pot. Leaves are burned around the homestead to keep evil at bay, or are hung around the doorway to the living quarters of people and animals when returning after a long absence. After the animals have been shut up for the night, some herders hang a leaf over the doorway to guard them through the night. If a mother has lost previous babies, she will hang leaves beside and above her baby in its hammock. Illnesses believed to be the result of sorcery or the Evil Eye are often treated by placing leaves on the embers and getting the patient to sit over the fire with a cloth over his head to inhale the fumes while others recite words to dispel the evil influences. This plant, like *Adenium obesum* and frankincense trees, was used by traditional healers in ritual cures for possession, or to lift the curse of witches. This was done for people and their animals, and also for other important possessions, such as date gardens or cultivated plots.

The tree is easily propagated from a stem and many of the groves seen around the island today are the result of such cultivation. Apart from the single tree planted for protective purposes, they were mainly planted as a dry season fodder for goats. A stem is cleanly cut off with a knife and put in a hole to which a little water has been added. It is best sown in 'red earth' and must not be watered or it dies, islanders saying that its own latex is water and fertiliser enough. After that it needs no further care other than fencing, if it is a hungry time of year and the goats have little else to eat. Or stems can be planted inside a clump of *Cissus* or *Buxanthus pedicellatus*, or other unpalatable material, to protect it from goats until it is properly established. Some herders were extremely energetic in their planting: within living memory some 50 to 100 trees were planted along various dry riverbeds. Gardens of the plant are still cultivated in the drier, vegetation-poor areas of the island. Many older men still have the habit, on seeing a dead tree, of removing some of its stems and pushing them into the ground around it to grow as a replacement. Or if they find a tree whose trunk is dead, they break up the crown, and, after planting some close to the parent tree, take the rest of the stems away to plant in their home area. It is always considered best to plant at the end of the monsoon so that the young plants will have the cool of the winter season in which to establish themselves.

Notes. Birds called *di ḥɔ̄wṭāri* 'striped', eat the fruit, making a sound like a lot of pop-guns going off. The latex is very painful if it gets in the eyes when lopping branches. The eyes should be immediately treated with fresh goats' milk, or water in which a few dates or some ripe tamarinds have been squeezed. Other islanders swear by dripping in the juice of a crushed snail of the type commonly found in the same areas as the trees. The latex is also damaging to the genitalia, causing the testicles to become very swollen and tender. This condition resolves over time, if the patient stops working the trees. To avoid this, men usually wear a length of cloth between their legs and protect their private parts with cloth padding. Any latex that falls on the skin must immediately be washed off. Initially as it comes into contact with the skin it feels cool, but soon begins to burn and become really painful. Any hair on the skin is burned off and the affected skin is left red and shiny, just like after a burn.

The plant features in oral history. Thus, in the Meyhah area of western Soqotra, one particular year of severe drought, when there had been no summer or winter rains in their area for two years, is recorded as 'The year of the *Euphorbia arbuscula* trees'. This is a reminder of the terrible months the men of the area spent walking all the way to the Heybaq area to lop the trees there, carrying it all the way back, week after week, to feed to their goats, too weakened by hunger to transhume. Bent (1897) reports 'tree-euphorbias ... the branches of which the natives throw into the lagoons so that the fish may be killed, and the poisonous milky juice of which they rub on the bottoms of their canoes to prevent leakage'.

Comparative uses. The latex of various *Euphorbia* species is carcinogenic, and most have a latex which is a powerful skin irritant, causing severe inflammation, particularly to the eyes. Not surprisingly, the consumption of various parts of *Euphorbia* species has caused illness and sometimes death in both animals and humans. Nevertheless, the latex of various species is used medicinally. On the Arabian mainland it is commonly used externally to treat warts and other cutaneous affections, or to treat snakebite, tumours of the eye, and as an irritant in rheumatism. It is also used for killing maggots in suppurating wounds and as an arrow and fish poison. Internally it is used as an emetic, a drastic purgative and an abortifacient. (Ambasta 1982; Dalziel 1948, 1956; Hedberg *et al.* [1982–1986]; Heine *et al.* 1988; Jain & De Fillips 1991; Lewis & Elwin-Lewis 1977)

Euphorbia balsamifera **Aiton**

Euphorbia balsamifera subsp. *adenensis* (Deflers) P.R.O.Bally

Dioecious shrub, forming rounded hummocks to 1m; dichotomously branched throughout; stems grey, semi-succulent. Leaves in rosettes at the ends of the branches, linear-lanceolate to narrowly obovate, <3cm long. Cyathia sessile, solitary, surrounded by a rosette of leaf-like bracts. Fruit sessile.

Ecology. Common, in places dominant, in the succulent shrubland on the foothills of Jebel Saleh on Abd al Kuri. 100–400m.

Distribution. Macaronesia and NW Africa (var. *balsamifera*); NE tropical Africa and SW Arabia (var. *adenensis*).

Botanical notes. A species with a remarkably disjunct geographical distribution, with var. *balsamifera* from Macaronesia and NW Africa and var. *adenensis* from NE tropical Africa, SW Arabia and Soqotra. In Arabia it typically grows in areas affected by mists. In the Soqotra Archipelago there are confirmed records only from Abd al Kuri. However, there are records of a similar plant from Soqotra – these have not been seen and therefore could not be identified with certainty. It seems unlikely that this distinctive shrub has been overlooked on Soqotra and therefore the identity of the Soqotri plants remains a mystery.

ETHNOGRAPHIC NOTES

Name(s)

– Soqotri: *dɔ̄ḳəš*, *dɔ̄gəš*, *dāgəš* 'mark, scar of a blow' (Soqotra and Abd al Kuri) (given to other latex-bearing *Euphorbia* species and Asclepiadaceae on Soqotra, e.g. *E. socotrana* and *Cryptolepis socotrana*).

– Soqotri simplified: *dowqesh*, *dowgesh*, *dagesh*.

Food (livestock). The greenery persists right through the dry season and on Abd al Kuri goats eat the green foliage at this time if there is little alternative forage. The islanders here believe it to be a nutritious fodder, but if there other forage is available, their livestock ignore the plant. Goats on Soqotra are said to eat the flowers, the fruit and dried, fallen foliage, but otherwise avoid it. Sheep eat a little of the fallen foliage when hungry.

Fuel. No good as firewood.

Timber. Soft and no good as timber.

Wood. On Abd al Kuri the dead wood is used as a firestick, providing the lower, horizontal stick. The vertical, upper one is usually made from the dead wood of *Cryptolepis*.

Other. The latex is sometimes used to lime birds on Soqotra, but the plant is rare here, and the latex of either *E. socotrana* or *Ficus cordata* is the one usually used.

Notes. On Abd al Kuri, although the viscosity of the latex is appreciated, it is not used either to lime birds (this is not practised on the island) or as a glue.

Comparative uses. Commonly planted for hedging. The latex was investigated as a source of rubber in the 1940s but was found to be rather inferior. The young shoots are cooked and eaten and the latex used as a chewing gum. Medicinally, the bark is used as a purge, and to treat leprosy and venereal disease. The root is anthelmintic and the powdered root in milk or boiled water is drunk as a fortifying tonic. The leaves are used to treat ringworm, and young and pregnant women wash themselves in a leaf decoction and apply fresh leaves locally as a tonic. The latex is given as an anti-toxin, and is used for snake and insect bites, sores and toothache. It is given to cattle to promote fertility and as a galactagogue, and is applied locally to soothe bony pains and to extract the guinea-worm. The latex is also used in an arrow poison, as a bird lime and to make flypaper. The aerial part of the plant with animal dung and the fat of a black goat is used to treat madness. In Yemen the latex is used to treat verrucas. In Dhofar the leaves, buds and fruit are eaten, and the latex is used as a powerful adhesive. It is also cooked with a little water to make a chewing gum and depilatory. The latex is put on the thumb to stop a child sucking it, and, mixed with livestock hair, is smeared on the teats of livestock to encourage weaning of their young. A cloth soaked in the latex is used as a plaster for fractures, and, painted over leather, is used to make tool handles. (Dalziel 1948, 1956; Davis *et al.* 1983; Fleurentin & Pelt/Fleurentin *et al.* [1982, 1983a,b,c]; le Houerou 1980; Kunkel 1984; Miller & Morris 1988; Watt & Breyer-Bandwijk 1962)

Euphorbia granulata **Forssk.**

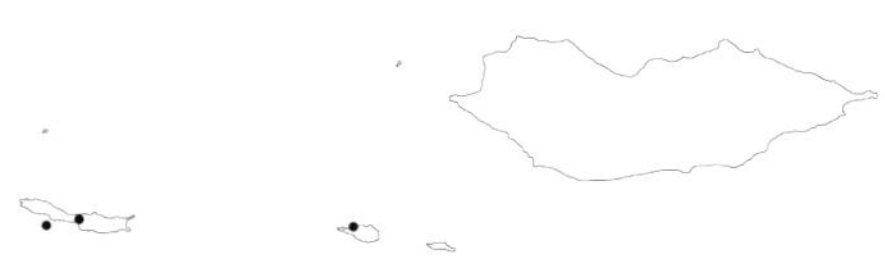

Annual herb with prostrate stems to 20cm, thinly to densely hairy. Leaves opposite, whitish green, often tinged red, ovate to oblong-obovate, <6 x 4mm, tip rounded, margins entire, base asymmetric; stipules linear, glabrous. Cyathia grouped on leafy branches, c.1mm diameter; glands minute, red; appendages pink, obscurely lobed.

Ecology. Sand and gravel soils on coastal plains and rocky slopes of the foothills; on Samha and Abd al Kuri. Sea-level–100m.

Distribution. NE Africa and Arabian Peninsula east to Pakistan.

Botanical notes. Recorded from Abd al Kuri and Samha but not from the other islands. The few plants that have been seen range from densely hairy (var. *granulata*) to sparsely hairy (var. *glabrata* (Gay) Boiss. in DC.).

ETHNOGRAPHIC NOTES

Name(s)

– Soqotri: nil or *raḥ́ammǝ* (Abd al Kuri and Samha).

– Soqotri simplified: *rahammih*.

Food (livestock). Goats and sheep graze it after the rains but it is not regarded as nutritionally important.

Euphorbia hadramautica **Baker**

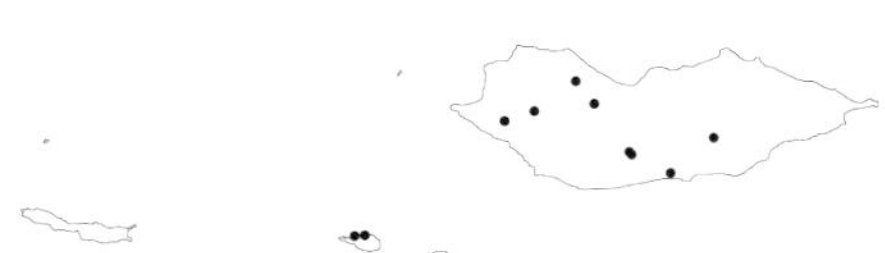

Small erect, succulent herb with a swollen, usually unbranched stem to 10cm tall and 1–2cm diameter, surmounted by a tuft of leaves and bearing distinctive leaf scars beneath. Leaves narrowly oblong, <9 x 1cm, often with undulate margins, pubescent. Cyathia clustered at the tips of the stem. Fruit shortly stalked, erect, brown, pubescent.

Ecology. Locally common on Soqotra and Samha; in desertic dwarf shrubland and low woody-herb communities. Sea-level–700m.

Distribution. NE tropical Africa and S Arabia (Dhofar and SE Yemen).

ETHNOGRAPHIC NOTES

Name(s)

– Soqotri: *kiḍ́idhin* 'small humps, hillocks'.

– Soqotri simplified: *kididhin*.

Food (livestock). Goats and sheep eat the foliage in the dry season. On Samha it is an important plant for sheep especially, as the foliage persists through the dry season. It is thought to have little nutritional value.

Medicinal / veterinary. The root is dug up and peeled. It is crushed, added to water, allowed to steep, and then the filtrate drunk to treat problems of the liver and spleen. A milder medicament, taken to treat a variety of stomach problems, is made by boiling up the green leaves and drinking the filtrate. The stronger treatment is given for malaria and jaundice (diagnosed by yellowing of the eyes and skin). It is also taken for griping pains and colic, and to get rid of intestinal worms. It has been used to procure an abortion. Very little is taken as it produces a dramatic bowel evacuation, and those who have taken this treatment generally remove themselves from the settlement as the onset of the diarrhoea can be very sudden. A small amount is taken to 'clean out the impurities from the digestive system' as a general tonic, as well as to treat constipation. Some people cannot take this treatment as it causes them to vomit violently. If too much is inadvertently taken, the patient is advised to eat lumps of ground cereal mixed with buttermilk, or to drink a lot of

soured milk to soothe the stomach. The roots are also crushed to a paste and used to treat various skin problems, mainly dry eczema and pruritus.
Notes. If sheep or goats eat a lot of this they develop diarrhoea.
Comparative uses. In Dhofar the latex taken from the part of the plant below ground is taken as a purgative and that from the part above ground as an emetic. The latex is taken as an abortifacient and is rubbed on the skin to treat a variety of skin complaints. Cloth dipped in the latex is wound around fracture of the fingers and toes, and the latex is also used as an adhesive. The seeds are eaten to treat flatulence, colic and indigestion. (Miller & Morris 1988)

Euphorbia hajhirensis **Radcl.-Sm.**

Perennial herb or subshrub to 40cm; stems prostrate, somewhat woody, glabrous or puberulent. Leaves opposite, green above, white beneath, suborbicular, 5–13 x 5–12mm, margins entire, base rounded or somewhat cordate; stipules short, puberulent. Cyathia solitary, c.2mm diameter; glands brown, appendages greenish yellow, toothed with roundish lobes.
Ecology. Rare in dwarf semi-deciduous shrubland in the Haggeher mountains. 850–1100m.
Distribution. Endemic.
Conservation status. Vulnerable (VU D2). Rare, with a very limited area of occupancy.
Botanical notes. See notes under *E. leptoclada.*

ETHNOGRAPHIC NOTES
Name(s)
– Soqotri: *ḥasəlām* 'thin, slim' (also given to *E. leptoclada*, *Euclea* species and *Hypericum balfourii*); *mə-sḥábeb* 'spreading'.
– Soqotri simplified: *hasilam*; *mishabeb.*
Food (livestock). Browsed by goats in the dry season and is thought to be useful subsistence fodder for them at this time, but with little nutritional value. Other livestock show little interest.

Euphorbia hamaderoensis **A.G.Mill.** (see Appendix 2, p. 723)

Low subsucculent shrub to 30cm; branches brownish-grey, covered with leaf scars, glabrous. Leaves glaucous, clustered at ends of branches, broadly obovate to obtriangular, 10–15 x 8–12mm, tip retuse to deeply notched, margin entire, base obtuse to cuneate, glabrous; stipules minute, glandular, red. Inflorescence 1 terminal cyathium; bracts deciduous, broadly obovate, <2.5 x 2.5mm. Cyathia 4–5mm diameter excluding glands, glabrous; glands 5, green, spreading, transversely oblong-elliptic, 1.2–1.4 x c.3mm, appendages absent, glabrous. Ovary sessile, obscurely 6-ribbed, glabrous. Capsule and seeds unknown.
Ecology. Known only from the exposed windswept summit ridge of the Hamaderoh escarpment, the highest point in the east of the island, where it is found growing in sheltered cracks between limestone blocks. 570–680m.
Distribution. Endemic.
Conservation status. Endangered (EN B2 a biii). A rare plant known from a single location with an estimated area of occupancy of less than 10km^2. There is evidence of a decline in the duration and extent of the mists affecting this location, as indicated by the lack of regeneration in the *Dracaena* woodland in the area. This apparent decline in the quality of habitat warrants the designation of an endangered category.
Botanical notes. *E. hamaderohensis* resembles a stunted form of *E. socotrana*, and is close to that species in both vegetative and floral characters: vegetatively it has the same distinctive minute, gland-like stipules, deciduous cyathial bracts and similarly shaped leaves; likewise it is very close in floral characters with its sessile ovary, minutely crenulate cyathial glands and identical floral bracts. However, it differs from *E. socotrana* in several important respects: it is a shrublet (not a tree) with smaller leaves and smaller, solitary cyathia bearing five (not six) cyathial glands.

ETHNOGRAPHIC NOTES
Name(s)
– Soqotri: nil or *gə́sso, jə́ssoh* 'of Hamaderoh', 'small-leaved', 'can be nibbled' (more commonly given to *Cryptolepis* species).
– Soqotri simplified: *gisso, jissoh.*
Food (livestock). Browsed by goats, including feral and wild ones, in the dry season especially. It is regarded as a useful forage plant at this time as it never dies right back. However, it is not considered to be important nutritionally.

Euphorbia inaequilatera **Sond.**

Annual herb with prostrate stems to 10cm, sparsely hairy. Leaves opposite, whitish green, tinged red, ovate to oblong-elliptic, <5 x 3mm, tip rounded, margin serrulate (at least around tip), base asymmetric; stipules linear, often divided, glabrous. Cyathia grouped on leafy branches, c.1mm diameter; glands pinkish, minute; appendages obscure.
Ecology. Sandy soil in dwarf shrubland on desertic coastal plain; only on Abd al Kuri. Sea-level–20m.
Distribution. Ethiopia, Somalia, Kenya and Tanzania.
Botanical notes. Distinguished from all other small, prostrate euphorbias by the minutely serrulate leaves. Known only from a single specimen, from Abd al Kuri, which is somewhat atypical in having very reduced

appendages. It seems referable to var. *inaequilatera* (thinly hairy plants).

ETHNOGRAPHIC NOTES

Name(s)

– Soqotri: nil.

Food (livestock). On Abd al Kuri this provides incidental grazing for small stock after rains, but it is thought to have little nutritional value.

Euphorbia indica **Lam.**

Erect or ascending annual herb, sparsely pilose throughout. Leaves opposite, ovate, <3 x 1.5cm, obscurely toothed; stipules triangular, toothed. Cyathia clustered in capitate cymes <1.5cm diameter, subtended by leaf-like bracts. Cyathia c.1mm diameter; glands minute, green; appendages white.

Ecology. No habitat details available; a common weed of the tropics.

Distribution. Widespread weed in tropical Africa and SW Asia.

Botanical notes. Collected in the 19th century but not seen since. Possibly overlooked. Should be looked for as a weed of cultivation.

Euphorbia kischenensis **Vierh.**

Euphorbia chamaesyce sensu Balf.f. non L. p.p.

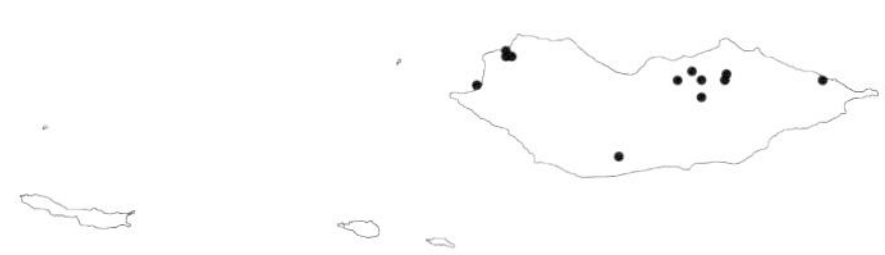

Prostrate herb to 20cm, thinly or densely hairy with short or more rarely long hairs. Leaves opposite, pale or whitish green, usually green above, white beneath, orbicular or kidney-shaped, with larger-leaved forms <9 x 7mm and smaller-leaved forms <4 x 4mm, margin entire, base rounded or cordate; stipules small, puberulous. Cyathia c.2mm diameter; glands reddish brown; appendages white, toothed, the teeth acute.

Ecology. Found in a variety of habitats from dwarf shrubland on desertic coastal plains to montane semi-deciduous woodland in high rainfall areas in the Haggeher mountains. Sea-level–1100m.

Distribution. Endemic.

Conservation status. Least Concern (LC). Common in several vegetation types and under no present or perceived threat.

Botanical notes. *E. kischenensis* is most readily distinguished from other small-leaved, prostrate euphorbias by its strongly toothed cyathial lobes, which, although small, are usually very obvious. However, within *E. kischenensis* there is a great deal of variation in degree of hairiness, hair type and habit. Typically it is a perennial plant of the mountains with kidney-shaped discolorous leaves (approaching *E. leptoclada* and *E. hajhirensis*) whilst some specimens are apparently annual; these have smaller, concolorous leaves, and come from dry areas. Some plants from the mountains approach *E. leptoclada* but their leaves are discolorous, their stems puberulent and their cyathial appendages sharply and distinctly acutely toothed. However, because of the number of intermediates it has been impossible to formally recognise even the most extreme forms. The treatment of this species is very provisional and further investigation should clarify its relationship with *E. leptoclada* and possibly lead to the formal recognition of further taxa.

ETHNOGRAPHIC NOTES

Name(s)

– Soqotri: *ni̍mənuh*, *nəmino̍w* 'changing colour'; (western Soqotra) *rəxʌ̍mhen*, *rəxa̍mmə*, *rʌ̍xmʌm* 'grey-coloured'; *ḳere̍mihon* 'longed for'; *ḳʌ̍ di mərḥūmhəm* 'grey-coloured thread'; *tili̍lyə* (more commonly given to small *Trachyspermum* species); *di ḥɔ̄wlənhi̍tin*; *gi̍sso*, 'small-leaved', 'can be nibbled' (more commonly given to *Cryptolepis* species).

– Soqotri simplified: *niminuh*, *niminow*; *rikhamhen*, *rakhammih*, *rakhmam*; *qeremihon*; *qa' di mirhuwmhim*; *tililyih*; *di howlinhitin*; *gisso.*

Food (livestock). Grazed by small stock especially, and is considered to be a rich, nutritious feed for them, making them fat and improving their milk yield.

Euphorbia kuriensis **Vierh.**

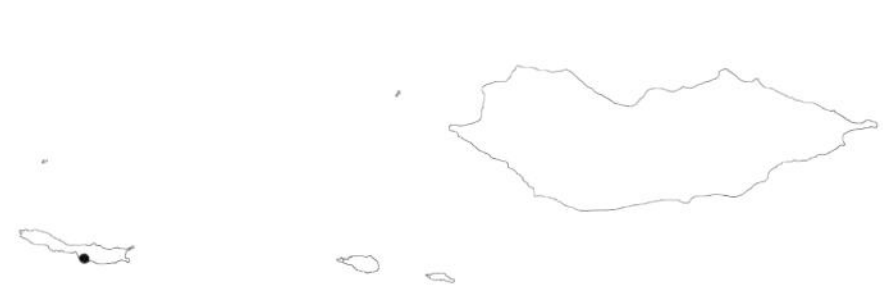

Perennial herb with a woody taproot and prostrate stems to 15cm, glabrous. Leaves opposite, green, elliptic to obovate, <12 x 5mm, apex rounded or retuse, margin entire, base obtuse or rounded, symmetric; stipules triangular, densely hairy. Cyathia c.1.5mm diameter; glands green; appendages white, obscurely toothed with rounded teeth.

Ecology. Dwarf succulent shrubland on Abd al Kuri. 200–450m.

Distribution. Endemic to Abd al Kuri.

Conservation status. Vulnerable (VU D2). Rare, with a very limited area of occupancy.

ETHNOGRAPHIC NOTES

Name(s)

– Soqotri: nil or *źāfḥan* 'crouching'; *rəśɛ̄riś*, *ˀertērit* 'forming green patches on the ground' (Abd al Kuri).

– Soqotri simplified: *dhafhan*; *risheyrish*, *erteyrit.*

Food (livestock). Grazed by small stock, and is regarded as useful dry season forage.

Euphorbia leptoclada Balf.f.

Similar to *E. hajhirensis* but a generally less robust plant with villous stems, smaller leaves (4–8 x 3–6mm) and white (not green) cyathial appendages.
Ecology. Semi-deciduous shrubland in the Haggeher mountains. 560m.
Distribution. Endemic.
Conservation status. Data Deficient (DD). Poorly known; inadequate evidence is available to make an assessment.
Botanical notes. There is some doubt as to the status of this species. It is allied to *E. hajhirensis* and somewhat intermediate between this species and extreme (montane) forms of *E. kischenensis*. Balfour (1888) notes that his description of this species is based on an imperfect specimen of *Euphorbia* found in Schweinfurth's collection, mixed with *Phyllanthus filipes*. It has not been found since it was originally collected in the 19th century.

Euphorbia obcordata Balf.f.

Slender tree or shrub to 2m, glabrous throughout. Leaves bluish-green, clustered at ends of branches, obovate-obcuneate, <6 x 3cm, tip truncate, margin entire, base attenuate. Inflorescence open, dichotomously branched, pendent, cyathia in groups of 1 to 3, subtended by transversely obcordate bracts; bracts <1.5 x 3cm. Cyathia 5mm diameter; glands yellowish brown, without appendages, transversely elliptic, 1.5–2mm diameter.
Ecology. Rare, on limestone cliffs in semi-deciduous woodland. 450–650m.
Distribution. Endemic.
Conservation status. Vulnerable (VU D2). Rare, with a very limited area of occupancy.
Botanical notes. A distinctive small tree typically found growing from vertical limestone rock faces.

ETHNOGRAPHIC NOTES
Name(s)
– Soqotri: *mārə* (also given to other asclepiad species).
– Soqotri simplified: *merih.*
Food (livestock). Not normally palatable to livestock, though it is grazed by goats in a prolonged dry season and the fallen leaves are eaten by small stock. It is thought to have minimal nutritional value.

Euphorbia schimperi Presley
Euphorbia oblanceolata Balf.f.

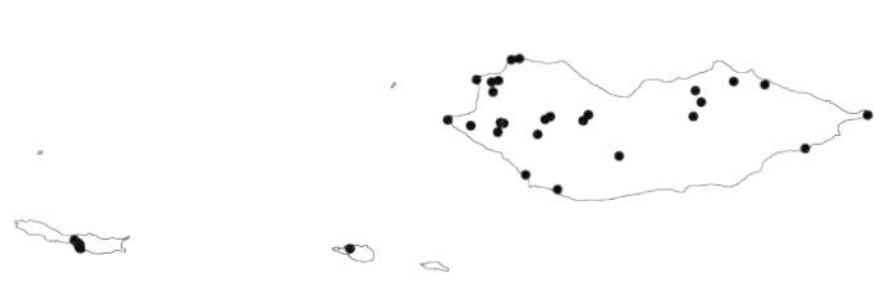

Small, often leafless, shrub, to 1m high and 2m across, branched from the base; stems glabrous, cylindrical, pencil-like, 5–8mm across. Leaves grouped at the end of the branches, narrowly oblanceolate, acute, <5 x 1cm, soon falling. Cyathia in umbels at ends of the branches; cyathial glands 4, yellowish green. Fruits shortly stalked, erect, reddish brown.
Ecology. Widespread and sometimes abundant in succulent shrubland. Sea-level–700m.
Distribution. Widespread in S Arabia and tropical and NE Africa.
Botanical notes. A variable species and part of a difficult species complex in Arabia and NE Africa; its relations with the African *E. nubica* need clarifying. *Euphorbia oblanceolata*, described by Balfour from the Haggeher, is included here. In some areas (Momi, Rokeb di Firmihin) it has died back due to droughts. It is reported to have been increasing steadily on Abd al Kuri.

ETHNOGRAPHIC NOTES
Name(s)
– Soqotri: *ᶜášbak* (also the 'female' plant *ᶜášbiko* and the 'male' plant *mέrəṭ, méyreṭ*).
– Soqotri simplified: *'ashbak* (*'ashbiko, meyrit*).
Food (livestock). In the drier areas of the island this is a key dry season fodder for the sheep and goats which are accustomed to it. Once the grasses and annual herbs have died back, goats seek out the plant. If they find enough of it, they can survive on this alone until the rains fall. In a prolonged dry season, goats also chew the bark of the thicker stems. In a drought, goats and sheep consume the dying plant, dead stems, rootstock and all. Camels also graze it, but mainly male pack animals; female camels are said rarely to eat this or *E. arbuscula*. Although mainly a dry season fodder, herders consider it to be a good, nutritious feed, as good as *E. arbuscula*, and certainly much more than just a subsistence one. Goats eating a lot of it grow fat and healthy, and give more milk. Importantly, for the water-poor regions, animals eating a lot of the plants need less water. However, if they are suffering from the *'uqabah* disease, this diet makes the condition worse. The caustic latex can cause the hair around the muzzle of animals browsing heavily on the plants to drop out. Their eyes can become dirty and encrusted, even inflamed. Once the rains fall, however, they recover. On Abd al Kuri the plant is mainly eaten in the green state, when it is regarded as very valuable fodder, indeed one of the best of the island. Sheep here also eat the dry and dead stems.
Bees / honey. Honeybees are attracted to the flowers, but the honey that results is considered to be inferior: it has a bitter edge, like that from *E. arbuscula* and *Buxanthus pedicellatus*, and is little liked as a food. It is, however, believed to be useful medicinally.
Flavouring / food additive. The meat and the viscera of livestock eating a lot of the plants develops a bitter fla-

vour, as does the water in which the meat is cooked. This is sometimes so badly affected that it has to be discarded. The milk of livestock feeding heavily on the plant is also affected, becoming slightly peppery and 'hot'.

Fuel. The dead wood burns with a very bright light, and dead plants are often collected and kept at home for this purpose. It gives off masses of dense and very acrid smoke, which is very painful to the eyes, so it is not used for cooking or indoor fires. The roots of the dead plant make much less smoke and, in less well-vegetated areas where other fuels are scarce, are considered to make a reasonable fuel. Dead stems make a good tinder and firelighter, and on the outer islands in particular, are often collected and stored for that purpose. On the outer islands, especially on Abd al Kuri where the plant is common and even increasing, the roots of the dead plant are regarded as providing a very good fuel. The dead wood of the stumpy and thickened *'meyrit' ''ashbak*-type' of plant is renowned for catching fire readily and flaming up 'like petrol'.

Wood. The wood is useless for anything other than fuel.

Medicinal / veterinary. The latex is used to treat mange. It is painted over the areas where the hair has fallen out and the skin become altered and roughened. On humans, though, it is rarely applied directly to the skin, as it is very caustic. It is smeared around and under the tails of sheep to deter blowflies. Only this and the latex of *E. arbuscula* are any use for this: that of *E. socotrana* and *Ficus cordata* does not have the powerful odour which keeps the flies away. A shepherd noticing the eggs or larvae of the blowfly on his sheep scrapes them off and then paints the latex all over the area. The illness in goats called *bíkəmə*, with signs of general stiffness and sore and swollen joints, is treated by anointing each joint and the ridge of the backbone with the latex.

Cleansing. The stems and roots of a plant of the *mérəṭ* or *ˁášbiko ''ashbak*-type' plants, if they have been dead for about three years, have a very strong and delicious smell, which hangs in the air long after they have burned out. In areas where there are few alternative fragrant fumigants, these plants are sometimes gathered and kept at home to be burned for their aromatic smell.

Pest control. The dense and pungent smoke the burning plants give off makes this a useful fuel for livestock smudges. The plants are also burned in human and livestock living quarters to fumigate them and to reduce the numbers of fleas. A little green wood added to the dead wood makes an even more powerful insecticide smoke.

Other. This plant is known island-wide for lessening the need for water of animals feeding on it. This makes it especially important in the water-poor areas and in the outer islands. On Soqotra, rainwater catchments are being built in many areas which were formerly waterless; as a consequence goats are less often seriously thirsty, and herders complain that their goats are no longer so keen on this nutritious plant. To a certain extent, stems of this plant can be used like those of *E. arbuscula* to keep evil at bay and to protect against the Evil Eye. Pieces of the stem, drained of the latex, are put across containers of food and water to protect it. The same is done for cooked food while waiting for everyone to assemble for the meal. The latex can be used to stupefy fish in the coastal lagoons, though that of *E. arbuscula* is easier to collect and also considered to be more effective. If the caustic latex gets in the eyes, this is treated by washing them with water in which either some ripe dates or tamarind fruit have been crushed.

Notes. Of the other *''ashbak*-type' plants, the one called *mérəṭ*, *méyreṭ* is stumpy and thickened. It grows mainly in the southern part of the island, and is said by some islanders to be the 'male' of the 'female' *ˁášbak* plant. It has the same properties as this plant and also makes the milk and meat of livestock grazing it slightly bitter. However, some herders report that one of the *''ashbak*-type' plants is not palatable to livestock. These euphorbias require further investigation.

This plant responds very quickly to rain, but can be totally wiped out in a prolonged drought. Many areas of Soqotra where these plants formerly grew are now bare of them.

Comparative uses. In nearby Dhofar this plant is regarded as extremely toxic to camels, which are herded away from areas where it grows. Its latex is used as an emergency adhesive and to stiffen strips of cloth to be used to bind a fractured limb. The other similar euphorbia, *E.* aff. *schimperi*, on the other hand, is eaten by camels and goats, and is regarded as a valuable fodder, increasing milk yield. The latex of the related *E. schimperiana* is used in Yemen to allay thirst. It is also used to loosen the skin to make the removal of plant spines and splinters lodged in the skin possible. The diluted latex, as well as root and leaf extracts, are used as a laxative and vermifuge. (Abdulfatih 1987; Fleurentin & Pelt/Fleurentin *et al.* [1982, 1983a,b,c]; Miller & Morris 1988)

Euphorbia schweinfurthii Balf.f.

Intricately branched subshrub to 30cm, often forming rounded cushions. Leaves opposite or clustered on short spur shoots, glaucous-grey, very variable from orbicular to linear-oblong, tip rounded, retuse or notched, margin entire, base rounded or truncate, symmetric, always densely and minutely papillose beneath (x10 lens required); stipules triangular, reddish-brown. Cyathia 2–3mm diameter; glands yellowish brown; appendages white, entire or obscurely toothed or undulate.

Distribution. Endemic.

Conservation status. Data Deficient (DD). Part of a taxonomically difficult group on Soqotra. Inadequate evidence is available to make an assessment.

Botanical notes. A broad view of *E. schweinfurthii* has been taken here. It is distinguished by the combination of shrubby habit, glaucous leaves and more or less entire cyathial appendages. However, it is extremely variable and a number of very distinct forms are found but with numerous intermediates. This treatment remains provisional and further investigation may lead to the formal recognition of more taxa within this complex.

ETHNOGRAPHIC NOTES

Name(s)

– Soqotri: *məndūrəb* (more commonly given to *Sarcostemma*); *śeriśéro* (more commonly given to *Tribulus terrestris* and *Medicago* species); *gə́sso*, *jə́ssoh* 'small-leaved', 'can be nibbled'; *ḳerémihon* 'longed for'; (western Soqotra) *rʌ́xmʌm*, *rəxʌ́mmə* 'grey-coloured'.

– Soqotri simplified: *minduwrib*; *sherishero*; *gissoh*, *jissoh*; *qeremihon*; *rakhmam*, *rakhammih*.

Food (livestock). The foliage and stems are grazed by sheep and goats in the dry season. The plants retain some green colour through the dry months, and are regarded as useful forage at this time.

Form *A* (typical)

Leaves linear-oblong, rounded or retuse at apex; petaloid appendages entire; stems puberulous.
Ecology. Montane semi-deciduous shrubland on granite; *Dracaena cinnabari* woodland on steep limestone slopes and on boulders in wadi beds; Soqotra (Haggeher and E Soqotra). 200–1000m.

Form *B*

Leaves obovate to oblong-obovate, rounded or retuse at apex; petaloid appendages entire; stems glabrous.
Ecology. Coastal succulent dwarf shrubland; Soqotra (extreme east end of the island). Sea-level–10m.

Form *C*

Leaves narrowly obovate to oblong-linear, notched at apex; petaloid appendages undulate; stems puberulous.
Ecology. Open shrubland on cliffs and rocky slopes; Samha. 200m.

Form *D*

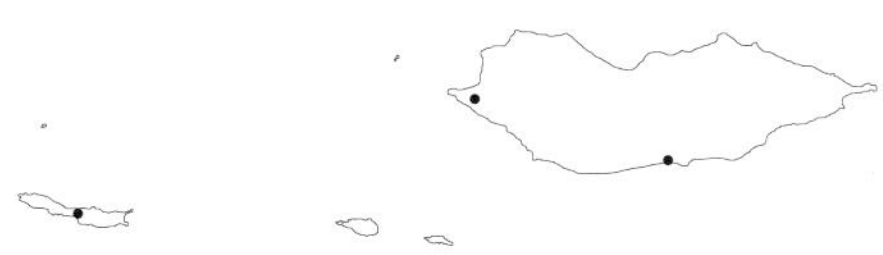

Leaves orbicular to broadly elliptic, rounded or retuse at apex; petaloid appendages undulate; stems puberulous.
Ecology. Dwarf shrubland on desertic foothills and coastal plains; Abd al Kuri and SW Soqotra. 100–200m.

Euphorbia socotrana **Balf.f.**

Glabrous tree to 6m with pale brown bark and branches bearing obvious leaf scars. Leaves glaucous, clustered at the ends of the branches, obovate, tip rounded to retuse, <7 x 5cm, shortly petiolate, glabrous. Inflorescence a solitary terminal cyathia with 3 to 6 lateral cyathia. Cyathia c.10mm diameter; marginal glands 6, transversely elliptic-oblong, without appendages. Capsule sessile, subglobose, c.10cm diameter; seeds ovoid.
Ecology. Locally common in succulent shrubland and dry, semi-deciduous woodland on the limestone plateaus; also found but less frequent in montane semi-deciduous woodland on granite in the Haggeher mountains. 450–900m.
Distribution. Endemic.
Conservation status. Vulnerable (VU D2). Rare, with a very limited area of occupancy.
Botanical notes. A tree, or rarely a shrub, which is readily distinguished by its pale brown or almost white bark and the glaucous leaves which are clustered at the tips of branches and persist on the tree well into the dry season. It comes into new leaf at the end of the dry season, before the winter rains. Reported to be on the increase in the western Haggeher.

ETHNOGRAPHIC NOTES
Name(s)
– Soqotri: *dɔ̄ḳəš*, *dāḳəš* 'mark, scar of a blow' (also given to *E. balsamifera* and *Cryptolepis socotrana*); (western Soqotra) *dāġəš*; (rare) *mərə*; edible seedling *sibɛ̄rə*.
– Soqotri simplified: *dowqish*, *daqish*; *daghish*; *merih*; edible seedling *sibeyrih*.
Food (human). In the dry, sparsely vegetated areas of Soqotra, a seedling or small sapling is dug up and eaten, by people and by small stock. It has quite a sweet taste and if a well-grown plant, makes a filling food.
Food (livestock). This belongs to an important group of plants that come into leaf at the end of the dry season before the winter rains. At this time, very hungry goats of the drier areas will eat the leaves, though they prefer fallen ones, gathered for them by their owners. Sheep also eat the plant litter. Goats in desperate hunger chew the bark, and this is sometimes chopped up for them as a drought fodder. In the better-vegetated areas the plant is generally avoided by all livestock.
Flavouring / food additive. The viscera of a goat which has been eating this becomes bitter and ill-tasting, as does the water in which the meat has been cooked. In extreme cases this has to be discarded. The milk of animals feeding on it is also altered, but is usually drinkable. The disagreeable taste disappears once the milk has been processed.
Fuel. The dead wood is no use as firewood, but makes good tinder and is also used as a firelighter. The wood is used to fire pottery in areas where it is plentiful, and is also used for making quicklime.
Timber. Useless as a timber.

Wood. Islanders remember that, before the days of widespread availability of dugout canoes, trunks and large branches of this used to be lashed together to construct makeshift rafts. These were used as fishing platforms and also for loading and unloading visiting boats.
Medicinal / veterinary. The latex is used as a sort of sticking plaster to strap fractures. A layer is pasted around the break, and then strips of *Jatropha unicostata* bark or aloe stem, beaten flat into long, thin strips, are strapped around as a bandage. The dressing is progressively loosened as the fracture heals. A cow is usually left for up to a month with this plaster on, a calf, sheep or goat for less if the fracture was a simple one. The injured animal is immobilised as much as possible and fed and watered by hand. Some apply the latex as a plaster, then cover this with cloth, and then a layer of sheep's wool or goats' hair mixed with more latex, then another cloth, and then strap it up as above. This dressing sets hard, like plaster of Paris. A fresh, deep laceration is treated by pulling the sides together and then sealing them in position with the latex. The quicklime made with the wood of the tree is considered to have medicinal properties. It was powdered very finely and used to treat the surgical wounds of circumcision, weeping sores, and all sorts of lacerations and abrasions to dry them out and speed up the healing process. This was used on livestock as well as on people. Someone with a painful chest, breathlessness and wheezing is treated by smearing the torso with a thick layer of latex. Fine earth is scattered on top to make it less sticky. The latex sets hard to make a firm coating and the support given by the 'cast' is said to relieve the pain. Some lay the latex on a length of cloth and use this to bandage the chest instead.
Pest control. The bark makes a lot of pungent smoke, so is useful for livestock smudges. Cattle herders use it especially in the summer months, even if summer rains fall. Goat and sheep owners use it in the open pens rather than inside the byres, where the acrid smoke is damaging and painful to the eyes. However, some people strongly dislike the smell of the smoke, and will not use it.
Other. The latex is collected by hitting the trunk or larger branches with a rock, and then collecting the oozing sap, either on a piece of cloth or on a stick which is wound round and round until full. It is mainly used to lime birds. The latex of plants growing in the cooler and wetter areas is not as good for this as that of plants growing in the hotter, drier areas. It is also used as a glue to mend clay pots, and to stick together handwritten pages of the Qur'an to bind them to make a booklet. To make a strong adhesive, the latex must be collected from a tree in a hot, dry area and in the full midday heat. It is then left to dry out and congeal a little before use. It was used to caulk the dugout canoes and other wooden vessels. It was collected into a clay pot and then mixed with sheep's wool or sacking fibres. This mixture was then used to plug any gaps or cracks in the boats. Congealed lumps of latex are used as a chewing gum to strengthen the teeth and gums, and to lessen thirst. The latex is smeared over the neck and anus area of young lambs and kids to protect them from predators such as ravens and wild cats. The same latex/wool or hair mixture is used in making the horn handles for the Soqotri triple knife-set carried by adult men. The latex of other plants was sometimes used, but that of *E. arbuscula* was considered to be the best. The goat's horn is boiled until the core can be extracted. The hollow horn is then straighened by repeated heating and manipulation between slabs of stone until it is made straight and flat. The pointed tip is cut off and the horn then packed with a mixture of goats' hair or sheep's wool mixed with the latex. The tang of the knife is then heated at the fire – but not allowed to become so hot that it melts the horn – and then tapped into the horn handle until it has gone in as far as the base of the blade. It is then set aside to cool and set hard, making a firm handle. A strong handle for the larger knife blades used for chopping wood and working date palms is made by soaking strips of cloth in latex and then winding these around the tang. The final layer is rubbed with earth to stop it being sticky and then it is squeezed firmly in the hand to shape the handle exactly to fit the grasp of its owner. The large leaves are held over the mouth and nose when going into shady or wooded areas in the cool months to protect against botfly attack.

Euphorbia spiralis Balf.f.

Euphorbia septemsulcata Vierh.

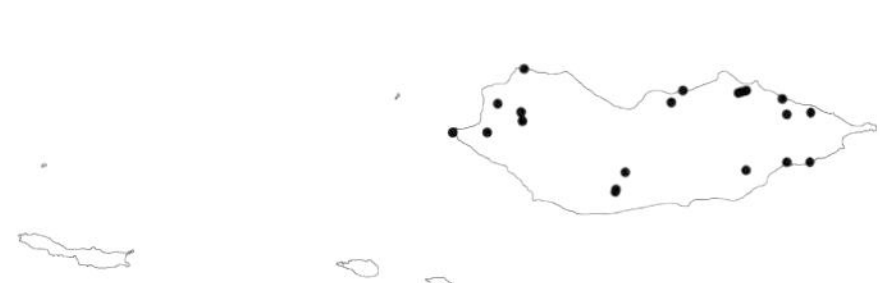

Spiny, leafless, cactus-like, succulent shrub to 50cm (rarely <1m), often densely branched; stems cylindrical, longitudinally ridged, 3–6cm diameter, 10 to 12-angled, bearing paired spines, glabrous throughout. Leaves soon deciduous. Cyathia solitary, sessile; cyathial glands green.
Ecology. Common in succulent shrubland over most of Soqotra; also in the sparse and windswept vegetation on exposed parts of the limestone plateau. Sea-level–700m.
Distribution. Endemic.
Conservation status. Least Concern (LC). Common in several vegetation types and under no present or perceived threat.
Botanical notes. This cactus-like plant is the only spiny succulent on Soqotra. It is a fairly common plant of dry open areas over much of Soqotra. The specific epithet *spiralis* coined by Balfour refers to the spiralling of the stem in specimens he observed; however, stems can be either straight or spiralling and, in fact, spiralling of the stem seems rather rare. A second species, *E. septumsulcata*, was separated by the greater number of stem angles. However, this seems to fall within the variation found in *E. spiralis* and so is not recognised here. *E. spiralis* is said to be on the increase in certain areas, for instance at Kilissan, an area badly affected by a series of severe droughts.

ETHNOGRAPHIC NOTES

Name(s)
– Soqotri: *ḳə̍ššo, ḳʌ̍šʾho; ṣo̍ʾhor, ṣo̍hor* (pl. *ṣʌ̍ʾrihin*).
– Soqotri simplified: *qisho, qash'ho; zo'hor, zohor* (pl. *za'rihin*).
Food (livestock). The living plant is never eaten by livestock, though goats who live in the dry areas where the plants are plentiful occasionally nibble at the dry stems and fallen leaves in a prolonged dry season.
Fuel. Dead stems are gathered and used as tinder.

Medicinal / veterinary. The latex is very caustic and 'burns' the skin and can be used instead of a heated iron for cauterisation. Drops are dotted here and there on swellings, especially glandular ones. This raises blisters in the skin, which in turn is said to relieve the pain and promote healing. The same is done to soften areas of skin whch have become hard and sore through repeated scratching. Dry eczema, which is a common problem on the hands of men and women who do a lot of milking, is treated by dripping the latex over the affected area. When other remedies for ringworm and fungal skin diseases have not succeeded, drops of the latex are used as a last resort.

Pest control. A stem is snapped off and the latex dripped around the house and food store to keep rodents away. The drops must never be allowed to fall directly on the skins used to store food, as they will damage the leather, but are dotted around the area where the skins lie. Grain is soaked in the latex and left in a slip of pottery as poison bait for rats and mice. Ravens are also killed by putting some latex in the eye of a recently killed sheep or goat and then leaving it out as bait. Ravens particularly go for the eyes and intestines of lambs and kids, and often take this bait and die as a result.

Notes. Cleared from grazing paddocks and enclosures as the spines and latex are seen as harmful to livestock.

Flueggea Willd.

Dioecious trees or shrubs, glabrous or with unbranched hairs. Leaves simple, entire. Flowers in axillary clusters. Sepals 5. Petals absent. Male flowers with 5 free stamens. Female flowers with 3-locular ovary. Fruit fleshy, indehiscent. 14 species. Tropics and subtropics.

Flueggea leucopyrus Willd.

Securinega leucopyrus (Willd.) Müll.Arg.

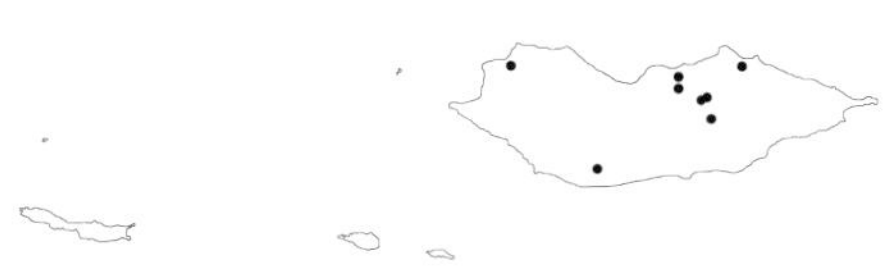

Similar to *F. virosa* but more densely branched, bark often white, side branches short, sometimes spinescent. Leaves smaller, <1.5(–2)cm long, obovate with truncate or emarginate tips.

Ecology. Semi-deciduous woodland. 100–900m.

Distribution. Tropical NE Africa, S Arabia east to India and Sri Lanka.

Botanical notes. The presence of two species of *Flueggea* on Soqotra is in some doubt. Gilbert recently suggested in the Flora of Somalia (1993) that *Flueggea leucopyrus* occurred on Soqotra. Certainly some material (for instance *Miller* 10345 from the Rewgid plateau) is perhaps referable to this species. These plants have smaller leaves with truncate or emarginate tips. However, there does not seem to be a clear ecological separation between *F. leucopyrus* and *F. virosa* on Soqotra. For instance, *Miller* 17158, with its acuminate leaves up to 3cm long, is a good match for *F. virosa* but was also found on the Rewgid plateau at a similar altitude and in a similar habitat to plants matching *F. leucopyrus*. Plants matching *F. leucopyrus* are apparently more common than *F. virosa* on Soqotra. Further investigation is needed to confirm the presence of both species on Soqotra. It is possible that the plants here tentatively placed in *F. leucopyrus* in fact represent forms of *F. virosa* which have been browsed or are plants from exposed sites.

ETHNOGRAPHIC NOTES†

Name(s)

– Soqotri: *ʿudf, ʿotf; ʿúft, ḥoft* (pl. *ʿífit, ʿífid*); *ḥasəlínž.*

– Soqotri simplified: *'udf, 'otf; 'uft, hoft* (pl. *'ifit, 'ifid*); *hasilinj.*

Food (human). The fruit are not eaten on Soqotra.

Food (livestock). All livestock graze the foliage in the dry season, especially goats and camels. Goats like the flowers, and sheep eat the plant litter and any leaves within reach. However, these plants often grow in places inaccessible to sheep. Herders lop leafy stems and collect fallen foliage to feed their goats in the dry season. Some cattle will also take this feed when they are hungry. It is considered to be a useful subsistence feed in the dry months.

Bees / honey. Honeybees are attracted to the flowers.

Fuel. The dead wood makes reasonable firewood.

Timber. Not a very good timber but is used for roofing infill. It is not strong enough for supporting beams.

Wood. Strips are cut from the bark, or slim stems are stripped of their leaves and tied in bundles. These are put to soak in water for three or four days and then the outer bark is removed and the inner bark or wood is split and used to weave the sections of the fish trap. This is thought to make a better trap than date leaf ribs, *Croton* or *Allophylus* material, as it stands up better to immersion in salt water and does not rot. Slim and straight stems are used to make the important herding staff.

Comparative uses. The leaves and fruit are eaten in hunger. The leaf juice, or a paste of the leaves with tobacco, is used to kill worms in sores. The stem-bark contains tannins and is used in fish poisons. (Ambasta 1982; Dalziel 1948, 1956; Goodin & Northington 1979; Heine *et al.* 1988; Jain & De Fillips 1991; Kunkel 1984; Lewis & Elwin-Lewis 1977; Miller & Morris 1988; Watt & Breyer-Bandwijk 1962)

†Also apply to *F. virosa*.

***Flueggea virosa* (Roxb. ex Willd.) Voigt**
Securinega virosa (Willd.) Pax & Hoffm.

Small deciduous tree to 4m with rough, brown bark. Leaves elliptic to obovate, <3.5(–6) x 1.5(–3)cm, tips rounded to shortly acuminate, glabrous. Flowers cream or white; male flowers c.2.5mm diameter, densely clustered on pedicels <8mm; female flowers loosely clustered, c.1.5mm diameter, pedicel <5mm. Fruit fleshy, white, globose, 4–5mm diameter.
Ecology. Semi-deciduous woodland. 600–650m.
Distribution. Tropical Africa and Asia east to Australia.
Botanical notes. See notes under *F. leucopyrus*.

ETHNOGRAPHIC NOTES
Name(s)
See *F. leucopyrus* (also uses).
Comparative uses. *F. virosa* is said to have anti-depressive and hallucinogenic properties.
Fruits. Fruits are eaten, both on the tree and when ripe and fallen to the ground, and they also yield a red dye. They are pounded with the twigs and soaked in cold or warm water for use as a wash for scabies and itching.
Bark. This is used for tanning and for dyeing matting black. It is also used as a fish poison, and to treat diarrhoea and pneumonia.
Roots. Roots are regarded as analgesic and aphrodisiac. A decoction is given to treat syphilis, dysmenorrhoea and oedema, and, with the root of *Terminalia* species, is drunk for dysentery. A cold water extract is drunk for stomach ache and an infusion for malaria. The roots and fruit are chewed to treat snakebite. The juice of the root and a decoction of leaves with added lemon juice is taken as a purgative. In Senegal a decoction of the roots is given for disorders of the liver, gallbladder, kidneys, bladder and genitals. In India it is used as a cure for diabetes mellitus.
Leaves. The leaf juice is used to kill worms in sores, sometimes with added tobacco juice, and is also drunk to treat epilepsy and mental illness. The pounded leaves are used as an insect repellent, and a decoction of the leaves is given as a laxative and anti-pyretic.
Gum exudate. The mucilaginous gum which exudes from the stems is used as an adhesive.
Wood. Wood is used for making agricultural implements and such things as walking sticks, chair legs, tent pegs, wicker baskets and fishing stakes. The wood ash is used for cleaning milk containers. The branches are used for weaving, and charcoal from the wood is powdered and used as a cicatrizant. The twigs are used as toothbrushes. In East Africa the branches are draped around the door and the patient inside during illness.
(Ambasta 1982; Dalziel 1948, 1956; Goodin & Northington 1979; Heine *et al.* 1988; Jain & De Fillips 1991; Kunkel 1984; Lewis & Elwin-Lewis 1977; Miller & Morris 1988; Watt & Breyer-Bandwijk 1962)

Jatropha L.

Monoecious shrubs, glabrous and with clear latex. Leaves alternate, simple. Inflorescence a panicle in the axil of the upper leaves, male above, female at base. Sepals and petals 5, larger in female flowers. Male flowers with 8 stamens. Female flowers with a 3-celled ovary; styles 3, free, bifid. Fruit a capsule; seeds caruncluate. 1 endemic. c.175 species. Dry tropics.

***Jatropha unicostata* Balf.f.**

Shrub or small tree to 3m. Leaves clustered at the tips of the branches, obovate to lanceolate, 8–15 x 3–7cm, tip acuminate, margin entire. Panicle erect, flowers 5–10mm diameter, male pale yellow, female yellow-green. Fruits globose, 8–20mm diameter, 6-lobed.
Ecology. Widespread and dominant in succulent shrubland over most of Soqotra; common in semi-deciduous woodland on the limestone; rare on the granite of the Haggeher. Sea-level–700m; up to 1050m on Diksam plateau.
Distribution. Endemic.
Conservation status. Least Concern (LC). Common in several vegetation types and under no present or perceived threat.
Botanical notes. Probably the commonest and most widespread shrub of lower and middle altitudes on Soqotra. It is abundant in rocky areas on the plains and over large areas of the limestone escarpments and plateaus. Like the other common shrub, *Croton socotranus*, it is surprisingly absent from the outer islands. However, unlike *Croton socotranus*, over its range it shows little variation. The most distinct variation is a prostrate, gnarled form found on exposed parts of the SW limestone plateau. On the Noged plain it is considered to be on the increase. Responds quickly to rain: the summer rains (April to May), the rains of the SW monsoon (June to September), and the winter rains (October to January) of the NE monsoon. Thus, according to where the rains fall in a particular year, the shrubs can be seen in leaf and flower in different areas of the island at different times of the year.

ETHNOGRAPHIC NOTES
Name(s)
– Soqotri: *sˈebərā̆* (pl. *səbɔ̄rihi*, *səbˈerhin*).
– Soqotri simplified: *sebira* (pl. *siborihi*, *siberhin*).
Food (human). The fruit are not eaten on Soqotra.
Food (livestock). No livestock eat any part of the living plant. In a severe drought, goats in the driest areas of Soqotra are hungry enough to eat the dead leaves and fallen fruit. Islanders have noticed that the partially digested stomach contents of a goat slaughtered after eating this is stained bright blood red, and the meat in close proximity to the stomach is tinged the same colour.
Bees / honey. The dead wood is regarded as one of the best for smoking out bees when stealing their honey.
Fuel. The dead wood is not good firewood, indeed it is regarded as the worst on the island. It burns badly, frequently goes out, and produces dense clouds of acrid

smoke. This badly taints food, so the wood is not used for cooking. However, it is not regarded as a 'bad' or harmful smoke, like that of *Ficus cordata* or *Euphorbia arbuscula*. The heartwood of a dead branch or tree crumbles to a fine powder, which makes excellent tinder for use with firesticks. The dead and dry wood around the heartwood is sometimes used as kindling, but it takes a lot of blowing on, or fanning, to get it going. However, *J. unicostata* wood has one important use: it is renowned for its slow burning and this quality is exploited to keep a flame going and fire alive when transhuming to a place empty of other people. A branch of the dead wood is put into the heart of a fire and kept there until well alight. Then it is carried, along with several spare 'tapers' of *J. unicostata* wood to be lit from the first if it threatens to burn out. Thus the flame is passed from 'taper' to 'taper' until the new quarters are reached, and a new fire kindled. The dead wood is also the wood of choice when firing pottery, especially pottery made from the paler-coloured clays. Usually with additional dried dung, the dead wood, like that of Burseraceae, produces smoke that changes the colour of the pottery from cream to the more desirable red-brown. For best results, the pots to be fired must be completely buried in the dead wood.

Timber. The timber is generally useless, but it can be used to build temporary human and livestock quarters in areas where other woods are scarce. Branches are cut and then, leaving a gap for the doorway, are stacked thickly in a circle, with the cut ends facing inward, to make a livestock pen. For people, a circular stone wall is built, with a gap for the doorway. Then the walls are raised and the roof is constructed with *J. unicostata* wood. Longer poles are placed across the top of the stone and brushwood walls to make a roofing support, and then the gaps are filled in with smaller branches and twigs. These constructions are not durable, and have to be re-built each year. This is partly because the wood is so soft, but also because it is very much liked by termites. The plants are so common in the drier areas that there is never a lack of wood for building such temporary constructions.

Fibre / cordage. Strips of the bark are pulled off, the outer bark removed, and the inner bark separated into fibres which are twisted together to make string of varying thicknesses for temporary use.

Leatherwork. In the vegetation-poor, drier areas of the island, where these trees are common, the inner and outer bark of a well-grown tree is crushed to a paste and used to prepare skins and hides to toughen them and colour them a deep red-brown. The bark from the main trunk produces the best colour. This material is used for skins that have the heaviest use: milk- and water-skins, and the skins used to churn butter.

Wood. The dead wood can be used as firesticks, but only the heartwood and only if it is completely dry and at least a year old. In areas with few alternative woods, the live wood is used to make stirrers, spoons and combs. They are oiled with ghee or oil and are not usually very durable. A slim branch is used to make a pipe, *ḳēreb*, for smoking the local tobacco. The branch is trimmed to the required length, and then the bark is twisted around hard to separate it from the core. The core is pulled out, leaving a hollow tube of inner and outer bark. The core is cut to make a plug shaped like a wine cork, and two or four notched channels are cut along it for the smoke to pass along. Sometimes the plug is made from clay. It is inserted into the mouthpiece of the tube, and the pipe is then ready for use. It is bent in a curve to be smoked, the lower end being placed in the smoker's mouth. Some crumbled tobacco leaf is poked down the tip over the plug and a live ember is wedged on top. The smoker normally only takes a puff or two, before passing it round to others. The same method of twisting the inner and outer bark from the core is used to make a series of tubes which are then fitted together to make a length of piping used to extract the very last water from a deep rock pool. One end of the piping is lowered into the remaining water and the other is sucked hard and the water siphoned off into a container. There is a tradition that musical pipes were once made by shepherds and goatherds in the same way, with finger holes cut in the upper surface and a notched plug fitted into the mouthpiece. Strips of bark are cut off and used to tie a front leg of a kid to a back leg to slow it down, keep it close to its mother and stop it dashing off and getting lost, or attacked by a wild cat or raven. Strips of the soft inner bark are often used to make temporary ties or hobbles.

Medicinal / veterinary. The islanders differentiate between two kinds of sap, and these are obtained and used differently. A colourless sap is extracted from the growing tips of the branches, while a darker, thicker sap is obtained by striking the trunk or principal branches with a rock until it 'bleeds'. The red sap is used as a haemostatic, painted over cuts and scratches. It is most effective if rubbed back and forth over the skin until it makes a thick soapy froth. It is used for fresh cuts and lacerations in humans and livestock alike. The coating dries out to form a waterproof coating which protects the wound from dirt and flies. The surgical wound of circumcision is treated with the sap immediately after the operation. It is also applied to soothe inflammation and swellings, especially of burns. It is useful too for ulcers and weeping sores, sometimes being mixed with *Adenium obesum* sap. If a sore or inflamed eye has not responded to milder remedies, the fresh sap is dripped into the eye. This is a very painful treatment, but seems to meet with some success. The colourless sap taken from the growing tips must be used for this. In livestock, infectious conjunctivitis can be a real problem, especially in sheep, and herders keep a close watch on their animals to detect any early signs. If they see any sign of the illness, they drip the sap into the affected eye(s) three or four times a day. If not treated in time, a badly infected animal can lose its sight altogether. In goats, an apparent lack of vision but with no outward signs in the eyes is also treated with the sap, dripped into the eyes on alternate days until the condition improves. The signs of this are a change of colour in the eye(s) which a become white, the animal becomes increasingly unable to see. Then the eyes become inflamed and the animal loses its sight altogether. It is vital to catch the condition in its early stages, before the eyes become inflamed. The animal is shut in the pen in the dark, and the sap is dripped in daily, the animal never leaving the pen, and being hand fed and watered. Once the eyes lose their whiteness the animal begins to recover.

Pain and inflammation from a scorpion sting is soothed by rubbing the area hard and applying the sap to it. The inflammation and heat of mastitis is treated with repeated applications of the sap. To draw pus to the surface and assist its drainage, fresh sap is dripped over the wound or boil. Fresh sap is dripped into a discharging and painful ear after cleaning it to remove accumulated pus. It is also painted on the skin surrounding the ear.

Raised and tender glands, especially around the neck, are treated by collecting the sap in the cupped palm and applying it thickly over the sore area. Sores on the teats and udder are treated with the sap mixed with fine earth. To treat a rash of spots, such as measles (or, in the past, smallpox), each spot or pustule is anointed with the sap. The dry exudate is chewed and swallowed as an emetic, especially as a remedy for botfly attack to the mouth and nose. Alternatively a piece of the bark, including the underbark, is ground finely and added to a small amount of water. It is left to steep for some time and then the bark is squeezed hard and the liquid drunk. This normally results in severe stomach pains, retching and vomiting, and for a few days afterwards the voice is reduced to a hoarse whisper. The patient continues to feel nauseated and cannot take any food. Despite these horrific after-effects, this treatment is given and is seen as being effective in some cases. If the botfly goes for the eyes, the treatment is drops of the sap directly into the affected eye. Someone with a wheeze and painful chest paints his chest with a layer of the sap. If a wound in an adult becomes dirty and maggot-infested, this is treated by collecting a cupped palmful of the sap and pouring it into the wound. In children and for those who cannot stand the pain of this, fresh aloes juice is used instead. A condition in small stock that causes white, foul-smelling patches in the hoof is treated by scraping away at the pale areas with a knife until the hoof bleeds and then pouring sap over. The multiple sores that occur mainly in the head and legs of young children, usually from infected insect or head lice bites, are treated with repeated applications of the sap. A stubbed and bleeding toe is treated with a dressing of the sap. Pruritus, scabies and the itchy skin condition that the islanders believe is caught from livestock with mange is treated by painting the affected skin with the sap. The same treatment is given to the sore and itchy skin of dry eczema, which is especially common in the hands of those who milk a lot. An extensive and open laceration is treated by cleaning it with salt and water salt and water and then powdered clay or chalk. Fresh aloes juice mixed with a powder made from grinding a soft, chalky stone is poured in and the clean wound is bandaged with strips of the inner bark of a *J. unicostata*. If the laceration is less deep or extensive, sap is poured in and then the sides are brought together, held in position, and more sap poured on. The sap sets hard and holds the torn skin shut. The area is then bandaged with cloth, but loosely to allow the wound 'to breathe'.

To treat someone suffering from sleeplessness, usually said to be the result of overstimulation and staying up too late talking and singing, especially if the eyes begin to feel sore and dry, a fire of *J. unicostata* wood is lit in a small enclosed space. The fire is then damped down with water and the patient shut in with the clouds of smoke. Once his eyes start to water badly he is let out. That night it is said he will sleep well.

The inner bark is supple and can be used like a bandage or plaster. For example, deep fissure(s) in the thickened plantar skin, especially of the heel, are treated by inserting a head louse into the fissure and then binding the foot with a strip of underbark. The cylinder obtained by twisting the bark away from the inner heartwood is also used to treat livestock who are suffering from the condition known as *ʿārə*. This is believed to be the result of some form of internal blockage. The animal writhes with colicky pain, loses its appetite, and develops swollen and protruding eyes. The treatment involves inserting a slim cylinder of bark into the anus of the animal and then blowing steadily through it until the animal's belly becomes as tight as a drum. Once the animal begins to fart noisily and repeatedly, it is likely to recover.

The strips of underbark are used especially in treating fractures. When fresh, the inner bark is soft and flexible, but it sets firm like a plaster cast as it dries out. It is cut from the larger branches or trunk in long, narrow strips, the outer bark removed, nd then beaten to soften them. For simple fractures, a strip is laid each side of the fracture, sometimes also above and below it, and then bound in place with a soft string made from sheep's wool. The limb is kept raised as much as possible. The strapping is removed once the pain has lessened and the bone appears to have knitted, and then the limb is massaged with ghee or oil to soften the skin and 'get the blood moving'. The same treatment is given in livestock fractures, though an animal must be discouraged from moving about by shutting it in a small pen and bringing it fodder and water there. For fractures of the small bones of sheep and goats, a firm binding of cloth soaked in the latex of various *Euphorbia* species is usually sufficient, but fractures of the larger bones are usually additionally strapped with *J. unicostata* bark. Compound fractures are much more complicated and usually an expert is called in. The bone first has to be manipulated until it is brought back into alignment. Then any torn skin is pulled together and sealed with an application of sap. Then a thick layer of the sap is applied all over and around the area. After this it is treated and bound as for a simple fracture, except the bandage is not loosened for the first week.

Cleansing. Hands and other areas of skin which are badly stained – as with aloes juice when harvesting the plants – can be cleaned by heating stems of the tree, crushing them, and then rubbing the hands with the warm juice or paste. Some crush the flowering tips instead. Twigs are peeled and chewed to clean the mouth, strengthen the gums and sweeten the breath. When water is scarce this is done after eating meat to remove the meaty smell of the breath, or as part of the ritual purification before going to pray. Stains on the teeth are removed by rubbing them hard with peeled twigs. Juicy, tender twig tips and new shoots are ground and added to water used to wash smelly clothes. Similarly, clay pots smelly with use are scrubbed down with a paste of ground twigs, and are then washed out with water to which more ground material has been added. This is especially important if the pots are to be used to feed livestock, or to milk them, as livestock are said to be upset if pots they eat from or that are milked into smell of old meat, and become reluctant to feed or to let down their milk. This would only happen in temporary quarters, or when on the move, as normally the pots for milking each type of animal are kept separate, and are used for nothing else.

Pest control. The dead wood of the tree, called *mˈʌṭəhe*, is one of the best materials for insecticide smudges. It is used especially in the winter months to deter the many kinds of biting insects and midges which arrive in swarms after the winter rains. These are at their worst from dusk onwards, and leave once the sun gets hot mid-morning. The smudges are lit first thing in the morning in the pens and milking yards, and are kept alight until the animals have been gathered in and milked. Many livestock are so bothered by biting flies at this season that as soon as the sun rises they start looking for the smoke. As soon as they see (or smell) it, they start moving towards the fire, and gather around it bathing themselves in the smoke. If a herder has a lot of animals, he has to light enough smudges for them all to have access to the smoke at this time of the year. Many cattle herders make a smudge for their cattle every day all year round, even in the dry months, and for this they usually use this wood because it is so common. When the botfly is prevalent after the winter and summer rains, this is the best wood to burn to destroy the larvae busy working their way though the nasal passages and sinuses of the sheep and goats. People remove larger leaves from the shrub and hold them over their mouth and nose when passing through shady or woodland areas with small stock in order to frustrate the attacks of the botfly. These damaging flies are known to favour this sort of territory and to target small stock especially. The wood is also burned in living quarters to get rid of mosquitoes and other insects, and to keep flea numbers down. Small fires are lit particularly around the sleeping areas to fumigate them, and the bedding is also hung up in the smoke. Fires are lit inside animal sleeping quarters for the same reason. This wood is said to be particularly discouraging to the ticks and lice which plague young stock especially. This fumigation is normally done when the occupants, animal or human, are absent, as the smoke is very pungent and hurts the eyes and nose. The same is done in the area immediately around the homestead (or within the compound walls if they exist) to keep undesirable vermin away from the living quarters. These creatures, collectively called *rè̓eś*, are believed to be harmful to both man and beast. This is a wide category, and includes some (but not all) reptiles, such as skinks, the nocturnal *Hemidactylus*, the chameleon and all snakes, as well as scorpions, wolf spiders, biting ants and the giant *Scolopendra* centipede.

Other. In the areas where it is uncommon, some people transplant one or two close to their homestead for medicinal use. The dried leaves are used as a tobacco substitute, or sometimes to stretch out a small amount of tobacco. The ash of the burned wood makes a good fertiliser for growing finger millet. Children make toy finger- and toe-rings from cylinders of bark twisted away from the inner core and then cut across to make rings. The ripe berries are used in a variety of games. Strips of bark are removed from the trunk, the outer bark removed, and the inner bark wrapped round a pebble and used as a slingshot. A narrower strip of inner bark is also used with a forked twig of some harder wood to make a catapult. A strip of bark is used to tie a foreleg to a back leg of a kid whose mother had a tendency to run wild, as this hampers its movement and thus keeps the mother close by. The headband formerly worn by men to keep their long hair away from the face, although usually made of plaited leather, could also be made from plaited strips of the inner bark. The leaves are rolled into a tube and used as whistles which make a very piercing sound. A leaf is also held across the lips, parallel to the ground, and blown across to make a loud buzzing sound. This is used as a signal to goats resting and chewing the cud after watering to get up and get moving, or to keep them moving when herding them from one place to another. The same signal is never used for sheep and cattle. A paste made from crushed twigs, or sap from thicker branches, or the trunk is used to decorate pottery. The twig tips are crushed (with a little added water if they have little sap) and then held in the hand and squeezed over the red-hot pots as they come from firing, or the sap is collected in a bark or clay cup and poured over the hot pots. The same dye was also used to decorate and colour the intricately plaited leather belts formerly worn by men. The cotton twist that used to be imported to make fishing equipment was also used to weave coloured strands into the sheep's wool rugs, and this could also be dyed with *J. unicostata* bark. Strips of bark were cut from the trunk of a well-grown shrub, and these were ground to a paste and added to a clay pot of water. This was left to steep before the cotton twist was introduced. The cotton was left to soak until the required shade was achieved. The same dye was used on white or pale sheep's wool. The small balls of spun wool were put in the water with the dye material and left to soak until they changed colour. They were then hung up to dry. The darker sap of the trunk and larger branches was used as ink, with a pen made from a slim peeled twig with the writing end notched to hold the ink. Only the main branches produce red sap; that from smaller branches and growing tips, though more copious, is much paler. The tips of the branches can be cut in the midday heat, and then put upside down in an old clay or wooden pot and left for the sap to ooze out. This darkens in time to make a good red ink. The sap takes a long time to dry out, so a good amount is collected at any one time. The ink is stored in a small stoppered clay pot, or the hollowed out leg-bone of a sheep or goat, or in the testicle or udder section of a small skin. It produces a red-brown ink, which was much liked by Qur'an writers for writing the names of God or His Prophet, or for embellishing and decorating the main body of the script which was written in black ink (mainly prepared from soot and lamp black). The *J. unicostata* ink was also used to mark property with the owner's name or mark (similar to the one used to brand his livestock). Skins of dates or cereal left by herders of the interior for safekeeping in the dry, stone-built storehouses of their friends along the coast were often marked with this. The ink was less prone to rubbing off than marks made with charcoal. Traditional healers commonly used this ink to write charms on scraps of cloth or paper wrapped in leather and worn around the neck, waist or upper arm of their patients. Someone who owned one or two date palms in a garden outside his own tribal or clan area, especially in one of the larger communal date plantations, would cut his mark in the trunk, or paint it on with the *J. unicostata* ink.

Notes. The islanders regard all *J. unicostata* as 'female', and all as capable of producing fruit. They are careful when walking through *J. unicostata* shrubland as the sap from carelessly broken stems badly stains clothing, as well as the skin.

Comparative uses. Many *Jatropha* species are used medicinally, principally as a laxative, vermifuge or emetic. The bark is commonly used in tanning and dyeing, and the seeds, or a fixed oil obtained from then, as an illuminant. Many have a translucent sap which is used to stop bleeding, and as an antiseptic. Some are poisonous and are planted as a hedge to keep snakes away, for instance, or burned to deter mosquitoes and flies. Some species have edible roots. In African ethnobotany, *J. curcas* is most commonly cited. It is interesting to compare some of its uses with those of *J. unicostata* on Soqotra: *Bark*: contains tannins and wax, and a dark blue dye use for cloth, fishing nets and lines. *Latex*: foams up, and the lather, mixed with salt, is used to clean the teeth. It is also used as a marking ink. *Seeds*: the seed oil is a well-known purgative, and the source of a semi-drying oil used to make soap, an illuminant, a textile lubricant, and, after chlorination, a plasticiser. It is also used medicinally. The detoxified oilcake has been used as a protein nutrient source in the production of antibiotics. The oil is also used for skin troubles and rheumatism. In nearby Dhofar, *J. dhofarica* plays the same important role that *J. unicostata* does on Soqotra. (Abdulfatih 1987; Al-Dubaie 1993/Al-Dubaie & Al Khulaidi; Ambasta 1982; Davis *et al.* 1983; Fleurentin & Pelt/Fleurentin *et al.* [1982, 1983a,b,c]; Heine *et al.* 1988; Jain & De Fillips 1991; Lewis & Elwin-Lewis 1977; Miller & Morris 1988; Oliver-Bever 1986; Watt & Breyer-Bandwijk 1962)

Meineckia Baill.

Subshrubs with unbranched hairs. Leaves alternate, entire; stipules dentate. Flowers in axillary clusters, pedicels long, articulated in female flowers. Sepals 5. Petals absent. Male flowers with 5 stamens, filaments united into a short column, vestigial ovary absent. Fruit a capsule; seeds pitted on sides, rugose on back. 1 endemic. 19 species. Tropical America, SW and NE tropical Africa, S Arabia, Madagascar, Assam, S India and Sri Lanka.

Meineckia filipes (Balf.f.) G.L.Webster

Phyllanthus filipes Balf.f.

Low subshrub, stems often prostrate, to 30cm. Leaves ovate to elliptic, 5–13 x 3–8mm, acute or obtuse; stipules triangular, <1.3mm long, scarious with irregularly dentate margin, brownish red. Male flowers c.1.5mm diameter, sepals obovate or suborbicular, pedicels <4mm; female flowers c.2mm diameter, sepals ovate, pedicels <20mm. Capsule 6-lobed, subglobose, 3–4mm diameter.

Ecology. Rare in succulent shrubland and drought-deciduous woodland on the limestone of northern Soqotra. 50–700m.

Distribution. Endemic.

Conservation status. Data Deficient (DD). Poorly known; inadequate evidence is available to make an assessment.

ETHNOGRAPHIC NOTES

Name(s)

– Soqotri: *mʌ́hariś* 'undergrowth, understorey'.

– Soqotri simplified: *maharish*.

Food (livestock). Eaten by livestock when it is green and when there is little else, but not with any enthusiasm. It is seen as having little nutritional value.

Phyllanthus L.

Herbs, glabrous or with unbranched hairs. Leaves spirally arranged, entire; stipules scarious, entire; stipule-like scale leaves sometimes present. Flowers axillary, minute. Sepals 5 or 6. Petals absent. Male flowers with 2 to 6 stamens; vestigial ovary absent. Female flowers usually larger; ovary 3-locular; styles 3, bifid. Fruit a dry capsule; seeds minutely tuberculate on back. 650 to 700 species. Pantropical.

ETHNOGRAPHIC NOTES

Comparative uses. The fruits of some species are edible. An infusion of the leaves is taken for a headache, and the seeds are laxative, diuretic and carminative, and contain a fatty oil. Many *Phyllanthus* species are used medicinally. (Ambasta 1982; Bebawi & Neugebohrn 1991; Dalziel 1948, 1956; Hedberg *et al.* [1982–1986]; Heine *et al.* 1988; le Houerou 1980; Jain & De Fillips 1991; Kunkel 1984; Oliver-Bever 1986; Watt & Breyer-Bandwijk 1962)

Phyllanthus maderaspatensis L.

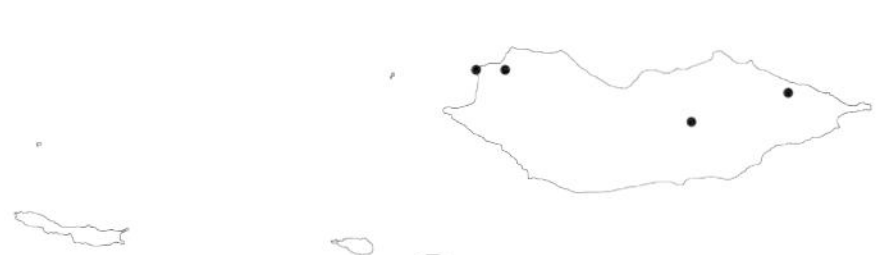

Woody-based glabrous herb. Leaves spirally arranged, linear-lanceolate, <2 x 0.6cm, tips acuminate; stipule-like scale leaves absent; stipules ovate, 1–2mm. Flowers 1–2mm diameter; female flowers in lowest clusters, upper clusters with both male and female; sepals orbicular, green with a narrow scarious margin. Capsule <2 x 3mm, smooth.

Ecology. Succulent shrubland and drought-deciduous woodland. 10–700m.

Distribution. Throughout the Old World tropics and subtropics.

ETHNOGRAPHIC NOTES

Name(s)

– Soqotri: nil or *mʌ́hariś* 'undergrowth, understorey'; *ḥīlɛ́bə* 'milk-making'; *sədíḥin* 'prostrate'.

– Soqotri simplified: *maharish*; *hiylebih*; *sidihin*.

Food (livestock). Grazed by cattle and goats and is considered to be a good and nutritious feed.

Phyllanthus rotundifolius Willd.

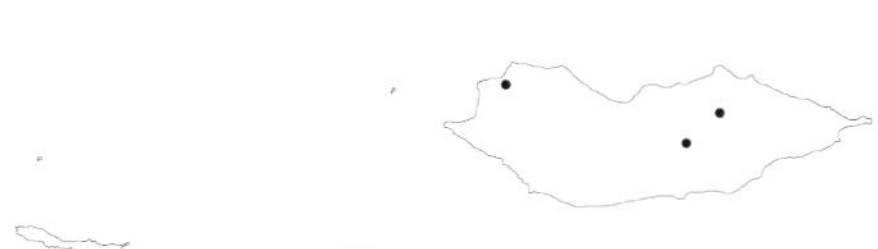

Erect annual herb to 20cm, glabrous or minutely papillose. Leaves obovate to orbicular, <12 x 8mm, rounded; 1 to several, scarious, often pink-tinged, stipule-like scale leaves and stipules in axils of leaves and side shoots. Flowers subsessile, c.1mm diameter, axillary; sepals ovate to obovate, with a green central stripe and broad scarious margin; male flowers smaller, stamens 3. Capsule 1 x 2mm, smooth.

Ecology. Amongst rocks in semi-deciduous woodland and desertic shrubland. 400–1000m.

Distribution. Tropical N and E Africa east to India.

ETHNOGRAPHIC NOTES
Name(s)
– Soqotri: nil or *eźḥēro* (Abd al Kuri).
– Soqotri simplified: *edh'heyro.*

Food (livestock). Regarded as valuable forage for both sheep and goats in the outer islands. It is less highly valued on Soqotra.

Ricinus L.

Herbs. Leaves alternate, palmately lobed, peltate. Inflorescence raceme-like, leaf-opposed, male flowers below, female above. Male flowers: calyx 3 to 5-lobed; petals absent; stamens many. Female flowers: sepals 5; petals absent; ovary 3-celled, styles free, bifid, plumose. Fruit a spiny 3-lobed capsule; seeds caruncalate. 1 species.

Ricinus communis L.

Robust, glabrous herb to 1.5m. Leaves deeply 7 to 9-lobed, <50cm diameter, lobes with glandular-serrate margins and acuminate tips. Racemes 10–30cm long. Flowers in large, erect, terminal panicles. Fruit softly spiny, 10–18 x 10–15mm.
Ecology. Near habitation on the plains, in date gardens, waste ground and roadsides. Sea-level–100m.
Distribution. Pantropical.
Botanical notes. Presumably an ancient introduction. Under-recorded on the distribution map.

ETHNOGRAPHIC NOTES
Name(s)
– Soqotri: *ginε̄gino*, *gənigəno* 'nodding', *aḥtərэś* (both names often given to plants whose wood is useless as timber); *di gilʿamihun* 'round-fruited'.
– Soqotri simplified: *gineygino*, *ginigino*; *ahtiresh*; *di gil'amihun.*
Food (livestock). The living tree is never eaten by livestock but the dead, fallen leaves and other plant litter are eaten by all livestock in times of hunger. In a drought goats have even been seen to chew the green stems. If a fat and healthy goat eats the fruit it gets persistent, severe diarrhoea, which can sometimes be fatal. Because of this, some herders cut the trees down when they find them and fire the stumps.
Wood. Described by the islanders as being 'like froth', i.e. quite useless. It can, however, be used for firesticks, if no better material is available.
Medicinal / veterinary. The sap is sometimes used on a sore or boil to bring the pus to the surface. It is also put into a rotten and painful tooth to 'kill' the 'worm' that lives inside and has caused the cavity by eating its way through the soft inside of the tooth.
Notes. In the drier areas of Soqotra plants can now be seen within compound walls where, because they grow so readily, they are planted for the pleasure of seeing something green. The islanders report that after very heavy rains seedlings spring up in areas where they were not known before. Unfortunately, few islanders seem aware of the toxicity of the seeds.
Comparative uses. Castor oil obtained from the plant has long been famous as a laxative. The ancient Egyptians, who thought that all disease came from food, believed that three purges a month were necessary, and they took castor with beer as their standard laxative. The oil has been shown to possess anti-bacterial and fungistatic properties. It is also a major industrial product, being used in lubricants, cosmetics, brake fluids, in the production of crumb rubbers and speciality soaps, and as a raw material for the manufacture of synthetic resins and fibres. Hydrogenated castor oil is useful in polishes, candles, insulating materials, plastic moulding compositions, heat-sealing adhesives and for preserving fruits and vegetables. The seeds also yield a semi-drying oil which is used widely in the manufacture of paints and varnishes. The stems are used as a source of cellulose for the manufacture of cardboard and newsprint, and are made into flutes in East Africa. The plant is sown to anchor sand dunes in Morocco and the leaves are used to feed silkworms in India. It is widely used medicinally in traditional medicine, and not only as a purgative or vermifuge. Some examples of its uses are:
Leaves. A poultice of green leaves is applied to cuts and wounds to prevent further complications. In Oman the leaves are burned and the smoke inhaled to cure bad breath. An extract of crushed leaves is used as eyedrops to relieve the pain of inflamed or swollen eyes. A decoction of the leaves, and fumigation of the lower stomach with dried leaves roasted on the fire, is recommended for bilharzia. A decoction is also taken as an emmenagogue. The leaves are applied directly on the forehead for a headache, and are rubbed over the body as a febrifuge, or over the chest for pneumonia. Boiled leaves are applied locally as a lotion, and cooked leaves are taken to promote lactation, the leaves applied to breasts or drunk as a decoction.
Fruit. The cinders of roasted fruits are powdered and placed on purulent wounds.
Roots. A decoction of the roots is used in lumbago and to expel the afterbirth. The root is chewed raw for wounds, and, with added corn stalk, to relieve toothache. Alternatively an extract of the roots is used as a mouthwash to relieve the pain. In northern Oman the roots and fruit are mixed with other ingredients and cooked as a remedy for madness.
Flowers. In Dhofar the honey made from the flowers is said to be intoxicating.
Seeds. The seeds are peeled, roasted, boiled in water and crushed to an oily paste used for a variety of skin conditions. In northern Oman the oil is locally applied for protection against the Evil Eye and *jinn*, as well as to relieve aching and stiff joints. The peeled seeds are taken as an emetic and they are also taken for some mental and nervous illnesses. However, the greatest care must be taken with the seeds: as few as six seeds are a lethal dose for a child, 10 for an adult, and there is no known antidote. The seeds are also said to destroy mosquitoes.
(Abdulfatih 1987; Ambasta 1982; Davis *et al.* 1983; Fleurentin & Pelt/Fleurentin *et al.* [1982, 1983a,b,c]; Ghazanfar 1994; Goodin & Northington 1979; Jain & De Fillips 1991; Kunkel 1984; Lewis & Elwin-Lewis 1977; Miller & Morris 1988; Oliver-Bever 1986; Watt & Breyer-Bandwijk 1962)

Tragia L.

Herbs with stinging hairs. Leaves alternate, simple. Dioecious, flowers in axillary racemes. Male flowers: calyx 3-lobed; petals absent; stamens 3. Female flowers: calyx 5 or 6-lobed, palmatifid, with prominent stinging hairs; petals absent. Ovary 3-celled. Seeds globose, caruncle absent. 1 endemic. c.125 species. Pantropical.

Tragia balfouriana J.B.Gillett

Tragia dioica Balf.f.

Scrambling or twining herb to 3m. Leaves ovate to oblong-ovate, 6–12 x 5–8cm, cordate, sharply dentate, with tufts of stinging hairs between teeth. Racemes 5–20cm long, flowers distant; male flowers c.2.5mm diameter, sepals ovate; female flowers c.5mm diameter, lobes ± linear, 1–2mm, lengthening to >15mm in fruit, covered with stinging hairs. Fruit 3-lobed, c.1cm diameter, surrounded by enlarged calyx.

Ecology. Locally common in drought-deciduous woodland on both limestone and granite. 100–650m.

Distribution. Endemic.

Conservation status. Least Concern (LC). Common in several vegetation types and under no present or perceived threat.

Botanical notes. The only plant with a serious sting on Soqotra.

ETHNOGRAPHIC NOTES

Name(s)

– Soqotri: *ṣafāḳa* 'sharp slap'.

– Soqotri simplified: *zafaqah*.

Food (livestock). Although some goats will not eat this plant, others become accustomed to it, and are recognised as such with the description *mənṣifīḳo*, '*Tragia*-eater'. These goats grow very fat and sleek on the diet and give plenty of rich milk. Camels accustomed to it like it very much, especially when it is still young and tender. Herders consider it to be one of the most fattening and nutritious fodders for livestock if they can be persuaded to eat it, and they gather leafy stems to feed to their goats and camels. They collect it by cutting the plant close to the ground and then winding the stems on a stick to protect themselves from the stinging hairs. Some goats develop sores and cracks in and around the mouth from eating a lot of this, especially in the dry season when it is rather tough and fibrous. Sheep are less keen on it and cattle will only eat it if there is little alternative forage available. They prefer *Scrophularia arguta*, called '*zafaqah* for cows'.

Flavouring / food additive. The meat of a goat that eats a lot is a very bright 'red' colour, and the fat has a rather yellow tinge. However, the taste of both remains unaffected.

Wood. The fibrous stems are used to make the 'bit' for the mouthpiece of the weaning device worn by kids and lambs. They are also used for the rim of the leather watering bucket.

Medicinal / veterinary. To treat 'pain in the liver', a large leaf is placed over the painful part. If the pain is severe and the abdomen distended, the abdomen is rubbed all over with a whole plant. This treatment is also given for a distended abdomen not responding to other treatment, even where the liver is not thought to be involved. This causes the patient to sweat and the skin to come up in weals. Some say the shock of the pain causes his the heart to 'leap aside', thereby 'making more room for the liver', and thus relieving the pain and distension.

Pest control. Food, and anything precious, is wrapped around with leafy stems of the plant and suspended from the ceiling to deter rodents and other pests, including hungry children.

Other. Goats that eat a lot need watering less frequently. Children tease one another with the stinging stems.

Notes. This is seen as being a browse plant, not a grazing one. If it touches the skin it can bring up dramatic weals, which are treated by rubbing them with powdered termite earth.

Comparative uses. The leaves of some species, such as *T. plukenetii* in East Africa, are cooked as a vegetable. Some *Tragia* species cause contact dermatitis. In Africa *T. benthami* is used as an oxytocic, and many parts of the plant are used in magical cures. Sometimes the plant is used as a poison, and sometimes to cause madness. (Ambasta 1982; Heine *et al.* 1988; Jain & De Fillips 1991; Watt & Breyer-Bandwijk 1962)

FLACOURTIACEAE

Shrubs or trees. Leaves alternate, small translucent spots present; stipules absent. Dioecious. Flowers actinomorphic, 1 to 3 (male) or solitary (female) in leaf axils, ring-like disc present. Calyx lobes (4 or)5(or 6). Petals absent. Stamens many. Styles 4 or 5, free. Ovary superior. Fruit a berry. 88 genera. 875 species. Pantropical.

Dovyalis E.Mey. ex Arn.

Aberia Hochst.

See description of family. 15 species. Africa, 1 species in Sri Lanka.

Dovyalis abyssinica (Rich.) Warb.

Aberia abyssinica Clos

Large shrub or small tree to 3m. Leaves ovate, 5–9 x 1.5–5cm, glossy green above, entire, base rounded, tip characteristically 'folded'. Flowers greenish white, 0.5–0.8cm diameter.

Ecology. Locally common in drought-deciduous woodland on limestone. 200–1000m.

Distribution. NE Africa.

Botanical notes. This small tree, which is typically found in small groves, is distinguished by its glossy, bright green leaves which have a characteristically folded tip. It has a widespread, but rather sporadic, distribution.

ETHNOGRAPHIC NOTES

Name(s)

– Soqotri: *ingélḥeš* (pl. *ingélḥéšion, ingeléḥeš*) 'stony-hearted, lacking in sympathy, compassion' (also given to *Datura fastuosa* and *Ricinis communis*).

– Soqotri simplified: *ingel'hesh* (pl. *ingelheshion, ingelehesh*).

Food (human). The fruit are eaten in poorer and hungrier parts of the island, especially by children. The unripe green ones are very sour, but they grow sweeter as they ripen and redden. The ripe fruit do not keep.

Food (livestock). Goats, donkeys and camels only eat the leafy stems when very hungry. The foliage is not palatable to cattle. It persists through the dry season, when other greenery is rare, and this makes it more attractive to livestock. However, it is not lopped by herders to feed to their goats, as it is not considered to be a useful feed. At other times of the year, goats and sheep only eat the dried and fallen leaves. Goats in particular love the ripe pods, which start to fall from the tree at the end of summer. However, if a goat unused to this feed eats too many, it can develop severe scour and lose condition drastically. Sheep also eat the ripe fruit. Herders generally try to keep their animals away from areas where the trees grow when the pods are ripe.

Bees / honey. The flowers are good forage for honeybees, especially after summer rains.

Fuel. The dead wood makes adequate firewood, though its smoke is particularly irritating to the eyes and chest. It is never used for cooking as the smoke affects the taste of food.

Timber. Makes indifferent building timber and it has to be de-barked before use to deter termites. When there is no better timber available, the trunks and branches of good-sized shrubs are used for roofing beams and rafters, and the smaller branches for roofing infill. Branches are also used to make the temporary huts used as summer quarters.

Wood. Can be used as a firestick.

Other. Indicates the presence of water underground, especially if there is also *Ficus cordata* nearby. When clearing new areas for cultivation this tree is removed.

Notes. Islanders differentiate between 'male' trees (*ingélḥeš mí'šər*) and 'female'. The 'male' is said to have no 'flowers' but to make a lot of fruit.

Comparative uses. The fruit of many species are eaten raw and cooked. In Kenya the bark is used for cordage and to clean the teeth. (Ambasta 1982; Heine *et al.* 1988; Jain & De Fillips 1991)

FRANKENIACEAE

Salt-tolerant herbs. Leaves opposite, simple; stipules absent. Flowers solitary, axillary, bisexual, actinomorphic. Calyx tubular, 4 or 5-lobed. Petals equal in number to calyx lobes, each with a long, narrow claw in the calyx tube, expanded and spreading above. Stamens 6. Ovary superior, 1-celled; ovules many on parietal placentas. Fruit a capsule. 4 genera. c.90 species. Temperate and tropical regions, in saline environments.

Frankenia L.

See description of family. c.70 species. Temperate and tropical regions.

Frankenia pulverulenta L.

Prostrate annual herb. Leaves elliptic, 2–6mm long, clustered, often covered with salt crystals. Flowers pink, c.4mm diameter. Fruit ovoid, seeds c.5mm long.

Ecology. On sandy, brackish soils, on the northern coastal plains. Sea-level.

Distribution. Widespread in Africa, Europe, Asia and Australia.

Botanical notes. Rare, but possibly overlooked, on Soqotra. Known only from the coastal plain near Qadub.

ETHNOGRAPHIC NOTES

Name(s)

– Soqotri: nil.

Food (livestock). Provides incidental grazing for coastal livestock.

Flavouring / food additive. Saline forage can affect the taste of the milk of milking animals feeding heavily on it.

Notes. A period spent grazing halophytic rangeland at least once a year is considered essential for the health of small stock and camels especially. Such feed initially brings on diarrhoea, which is seen as being beneficial and 'cleansing impurities from the blood'. When recovered from this, the animals begin to put on weight.

Comparative uses. In Qatar this mucilaginous plant is seen as having diuretic qualities and is used in empyema. In parts of Africa and Arabia the leaves of *F. revoluta* are added to butter as a condiment, while the leaves of *F. portulacaefolia* are made into a stimulating tea. Salt is obtained from burned plants of *F. berteroana.* (Ambasta 1982; Batanouny 1981; Jain & De Fillips 1991; Lewis & Elwin-Lewis 1977; Rizk & El-Ghazaly 1995)

GENTIANACEAE

Herbs. Leaves opposite, entire, usually sessile; stipules absent. Flowers in terminal cymes, actinomorphic. Calyx 5-lobed. Corolla 5-lobed. Stamens 5, inserted on corolla, alternate with lobes. Ovary superior, 2-celled; stigma 2-lobed; ovules many on parietal placentas. Fruit a capsule. 74 genera. 1200 species. Cosmopolitan, but especially diverse in temperate regions.

Centaurium Hill

Erythraea L.

Herbs. Leaves with indistinct venation, sessile. Flowers in terminal cymes. Calyx lobes unwinged. Corolla tubular, tube longer than lobes. Fruit a cylindrical capsule. c.30 species. Mostly N temperate.

Centaurium tenuiflorum Fritsch

Erythraea centaurium Pers.

Small, erect herb to 15cm. Leaves ovate to elliptic, <1 x 0.5cm. Flowers pink, c.1cm long.

Ecology. In montane grassland (Tinirih, W Haggeher). 1200m.

Distribution. NE Africa, S Europe, Mediterranean.

Botanical notes. Rare, known only from a single gathering; rather insignificant and probably overlooked in other areas.

ETHNOGRAPHIC NOTES

Name(s)

– Soqotri: nil or 'kind of *nūmḥel'*' 'winnowings, chaff' (*Leucas* species).

– Soqotri simplified: 'kind of *nuwmhel'*'.

Food (livestock). This is one of the small *ru'ud* group of plants which come up with the rains, providing incidental grazing until they die back. Though palatable to livestock, this is not an important forage for them.

Comparative uses. Several species are widely used medicinally, especially to treat stomach complaints, fevers and to make a bitter-tasting tonic. Some species, such as as *C. beyrichii* and *C. calycosum*, are believed to be toxic to sheep. (Ambasta 1982; Dalziel 1948, 1956; Jain & De Fillips 1991)

Exacum L.

Herbs, usually beside streams or in seepages. Leaves with 3 to 5 parallel veins, sessile or petiolate. Flowers in lax, terminal cymes. Calyx lobes winged. Corolla rotate, tube shorter than lobes. Fruit an elliptic or globose capsule. 3 endemics. c.65 species. Africa, Madagascar (where it is most diverse), tropical Asia to N Australia.

Botanical notes. About 65 species with a disjunct geographical distribution around the Indian Ocean basin with centres of diversity in Madagascar (extending to tropical Afica); Soqotra and S Arabia; Sri Lanka and S India (extending to the Himalayas and N Australia). This pattern of distribution was suggested as a typical example of a disjunct distribution following the breakup of the ancient supercontinent of Gondwana from c.165 to c.60 million years ago (Klackenberg 1985). However, recent molecular evidence suggests that the divergence between the Madagascar, Soqotra and India–Sri Lanka centres occurred c.21 to 17.5 million years ago and is a result of long-distance dispersal.

ETHNOGRAPHIC NOTES

Comparative uses. Some species are used to treat stomach complaints; *E. lawii* and *E. quinquenervium* are both used as a purgative. (Ambasta 1982; Dalziel 1948, 1956; Jain & De Fillips 1991)

Exacum affine Balf.f. ex Regel

Exacum gracilipes Balf.f.

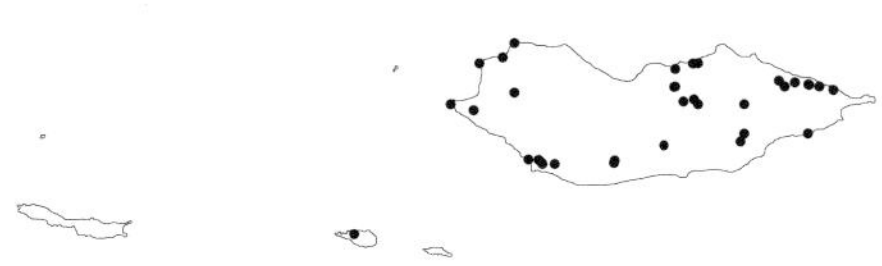

Delicate to robust, erect, leafy herb to 30cm. Leaves thin, papery when dry, upper leaves ovate to elliptic, 1.5–8 x 1–5cm, petiole long, narrow. Flowers violet, 1–1.5cm diameter; pedicels slender. Calyx lobes caudate, tip >0.2cm long.

Ecology. Widely distributed, typically in turf by water or in rock crevices and hollows where water accumulates; on both granite and limestone. Sea-level–1050m.

Distribution. Endemic.

Conservation status. Least Concern (LC). Common in several vegetation types and under no present or perceived threat.

Botanical notes. A rather variable species which in this treatment includes *E. gracilipes*, an endemic described by Balfour. The type of *E. gracilipes* is a much more delicate plant than typical plants of *E. affine*. However, the difference in habit of these two specimens is comparable to that found within most populations of *E. affine* studied during recent visits to Soqotra and it is difficult to find other characters to justify maintaining the two species. *E. socotranum* is also rather similar to *E. affine* but the characters defining it are consistent within populations. Plants from Dhofar, the southern region of Oman, recorded as *E. affine* and *E. gracilipes* have recently been shown to represent two separate species endemic to Oman (Thulin 2001). *Exacum affine* is now considered to be endemic to Soqotra and Samha. *E. affine* is widely cultivated as a house plant in Europe under various names including the Arabian, Persian or Socotran Violet. All seed in the horticultural trade originated from collections made by Schweinfurth during his expedition to Soqotra in 1881.

ETHNOGRAPHIC NOTES

Name(s)

– Soqotri: *di ḥoyhi* 'of the ground'; *di rihoytin* 'of water'; *di raʿadihun* 'tiny grazing plant'; *ṣifərirrihon* (*di geḥi*) 'many-flowered riverbed plant'.

– Soqotri simplified: *di heyhi*; *di rihoytin*; *di ra'adihun*; *zifiririhon di gehi*.

Food (livestock). Not normally grazed by livestock, but eaten in drought, mainly by small stock.

Notes. It can cause scour in small stock unaccustomed to this feed.

Exacum caeruleum Balf.f.

Exacum forbesii Balf.f.

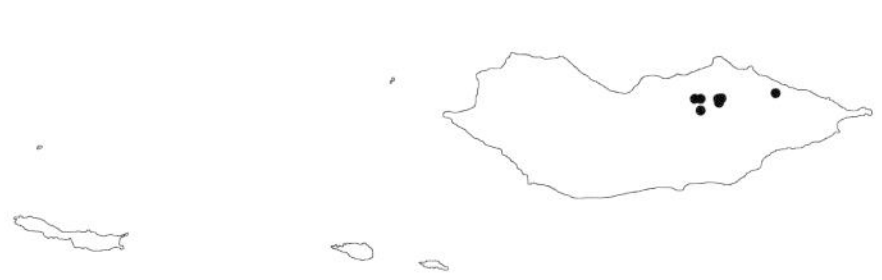

Similar to *E. affine* but separated by the ovate, thick and leathery, sessile leaves, 1.2–5 x 1–3cm, which sometimes clasp the stem.

Ecology. Dense shrubland in the higher parts of the Haggeher mountains. 900–1250m.

Distribution. Endemic.

Conservation status. Vulnerable (VU D2). Rare, with a very limited area of occupancy.

Botanical notes. Less common than the preceding species. The attractive foliage and strongly scented flowers give this plant great horticultural potential. Recorded in the 19th century from above Homhil in the northeast of Soqotra. It has not been seen there since and seems an unlikely record.

ETHNOGRAPHIC NOTES

Name(s)

– Soqotri: *ḳerəźahiyun* (diminutive form of *ḳurźaʾ*, name given to perennial Gramineae and Cyperaceae of damp areas); *di rihoytin* 'of water'; *šədak* 'to strip, as side twigs from a branch'; (rare) *źʌfəraʾ* (referring to the smell, more commonly given to *Plectranthus* species).

– Soqotri simplified: *qeridhahiyun*; *di rihoytin*; *shidak; dhafira'*.

Food (livestock). Persists into the dry season, and cattle and sheep graze the greenery when they are hungry at this time. Goats are less enthusiastic as they strongly dislike damp forage, and will nibble it only when it is dry.

Notes. It can cause scour in small stock unaccustomed to this feed.

Exacum socotranum **Balf.f.**

Similar to *E. affine* but separated by the upper leaves which are elliptic to spathulate, with a rather short, broad petiole, stout pedicel and the mucronate calyx lobes, tip <0.1cm long.
Ecology. Hanging near seepages on vertical cliffs on the southern limestone escarpment. c.130m.
Distribution. Endemic.
Conservation status. Endangered (EN B2 a biii). Known only from a few scattered populations growing near freshwater springs and seepages on the southern limestone escarpment. There are only a few of these springs and therefore the total number of plants is likely to be very low. This habitat is extremely vulnerable to water development and thus a designation of endangered seems appropriate (cf. *Helichrysum* sp. *C*, *Erythroxylum socotranum* and *Scaevola socotraensis*).
Botanical notes. *E. socotranum* is rather similar to *E. affine* but the characters defining it are consistent within populations.

ETHNOGRAPHIC NOTES
Name(s)
– Soqotri: nil or *di rihóytin* 'of water'.
– Soqotri simplified: *di rihoytin.*
Food (livestock). Normally inaccessible to livestock.

Sebaea

Herbs. Leaves sessile. Flowers in lax, terminal cymes. Calyx divided almost to base, unwinged, translucent with green veins. Corolla campanulate, included in calyx. c.60 species. Africa to India, Australia and New Zealand.

Sebaea microphylla **Knobl.**

Delicate, erect herb to 10cm. Leaves elliptic, <0.5 x 0.2cm. Flowers pale yellow, c.0.3cm long; corolla lobes <1mm long.
Ecology. In damp silty soil by wadi in date garden. c.50m.
Distribution. *S. microphylla* occurs in Yemen, NE Africa, NW Himalayas and N India.
Botanical notes. Known only from a single collection made by a wadi in a date garden on the Hadiboh plain.

ETHNOGRAPHIC NOTES
Name(s)
– Soqotri: *rúʿud* 'grazing plant', *ráʿadihon* 'tiny grazing plant'.
– Soqotri simplified: *ru'ud, ra'adihon.*
Food (livestock). This is one of the small *ru'ud* group of plants which come up with the rains, providing incidental grazing until they die back. Though palatable to livestock, this is not important forage for them.

GERANIACEAE

Annual or perennial herbs. Leaves alternate or opposite, palmately or pinnately lobed or dissected; stipules present. Inflorescence cymose or umbellate or flowers solitary. Flowers bisexual, actinomorphic or somewhat zygomorphic. Sepals free or united at the base. Petals 5, free. Stamens 10, all fertile or with some reduced to staminodes. Ovary superior, 5-locular. Fruit dehiscent, consisting of five 1-seeded mericarps (segments); mericarps united to a beak-like style. 14 genera. 730 species. Mostly temperate.

Erodium L'Hér. ex Aiton

Herbs. Leaves pinnately dissected. Flowers actinomorphic, in umbels. Stamens 5, alternating with 5 staminodes. Fruit long-beaked, splitting at maturity into 5 awned mericarps; awns spirally twisted. c.60 species. Europe to C Asia.

Erodium cicutarium **(L.) L'Hér. ex Aiton**

Prostrate annual herb. Leaves pinnate, the leaflets deeply pinnatifid with narrow, acute or laciniate lobes. Umbels 3 to 5-flowered. Petals pink, 8–10mm long. Beak of fruit 25–30mm long; mericarps c.4mm.
Ecology. Open shrubland on granite. c.700m.
Distribution. Europe, N and E Africa to SW Asia.
Botanical notes. Only seen once on Soqotra; probably an introduced weed.

ETHNOGRAPHIC NOTES
Name(s)
– Soqotri: nil or *rúʿud* 'grazing plant', *ráʿadihon* 'tiny grazing plant'.
– Soqotri simplified: *ru'ud, ra'adihon.*
Food (livestock). This is one of the small *ru'ud* group of plants which come up with the rains, providing incidental grazing until they die back. Though palatable to livestock, this is not important forage for them.
Comparative uses. The leaves are eaten as a salad and used as a potherb. (Kunkel 1984)

Geranium L.

Herbs. Leaves palmately lobed. Flowers actinomorphic, 1 to 3 on peduncles. Stamens 10. Fruit long-beaked, outer part of beak separating longitudinally and coiling upwards at maturity leaving mericarps often attached to the tip of the beak by an elastic awn. 300 species. Mainly temperate regions.

ETHNOGRAPHIC NOTES

Comparative uses. Many species are used medicinally, especially as an astringent, and the roots are used in tanning leather. The tuberous roots of *G. dissectum* and *G. tuberosum* are a famine food and the leaves of *G. incanum* are a tea substitute. The whole plant of *G. macrorrhizum* is used as an aphrodisiac in Eastern Europe. The American cranesbill, *G. maculatum*, shows experimental hypoglycaemic activity and a root infusion is taken for sore gums and pyorrhea. The plant is boiled and the resulting astringent tea is taken for dysentery. In Arabia the dried plant of *G. trilophum* is ground and locally applied to relieve backache. (Abdulfatih 1987; Lewis & Elwin-Lewis 1977)

Geranium biuncinatum Kokwaro

Similar to *G. mascatense* but differs in the mericarps 3–5mm long, minutely ridged and with a revolute toothed margin and 2 (c.2mm), hooked, apical horns.
Ecology. Semi-deciduous woodland. 250–1220m.
Distribution. NE tropical Africa and S Arabia.
Botanical notes. Material from Soqotra has smaller fruits than mainland material and possibly represents a new subspecies.

ETHNOGRAPHIC NOTES†

Name(s)
– Soqotri: *tə̍fer* 'goat/sheep droppings'; *ṣəfāṭyan*.
– Soqotri simplified: *tifer; zafatyan.*
Food (livestock). One of the small *ru'ud* plants which come up with the rains, providing incidental grazing until they die back. Though palatable to livestock, these plants are not important forage for them.
Notes. Can cause *'uqabah* in small stock unaccustomed to it, and small stock grazing heavily on the new plants after rain can develop scour. If in the first few weeks of pregnancy, they may well abort the foetus.

†Also apply to *G. trilophum.*

Geranium mascatense Boiss.

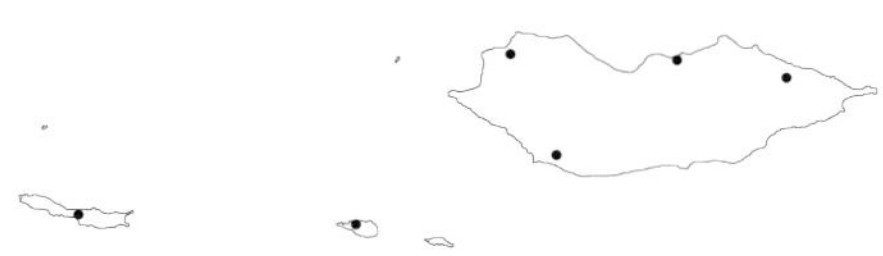

Annual herb with prostrate or ascending stems, glandular-pubescent throughout. Leaves suborbicular, 3–5cm across, deeply 5-lobed; lobes also toothed and divided. Flowers mauve with a dark centre. Petals 6–8mm long. Mericarps 2–3.5mm long, deeply transversely rugose, the ridges with short longitudinal crests.
Ecology. Semi-deciduous woodland. 400–750m.
Distribution. NE tropical Africa, S Arabia to Iran and Pakistan.

ETHNOGRAPHIC NOTES

Name(s)
– Soqotri: *tə̍fer* 'goat/sheep droppings'; *ḥʌ̍ṭəhīn* (Samha); *tá̍mərhen* 'little dates' (name given to many small plants with delicate and/or fragrant flowers); *ṣəfāṭyan*.
– Soqotri simplified: *tifer*; *hatihiyn*; *tamirhen; zafatyan.*
Food (livestock). Provides incidental grazing for small stock when young and green after rain, but is not nutritionally important.
Comparative uses. Flavonoid compounds, kaempferol, quercatin, myricetin, rutin, naringin and formononetin have been isolated from this plant, some of which have been recorded to possess anti-bacterial activity. (Ghazanfar 1994)

Geranium trilophum Boiss.

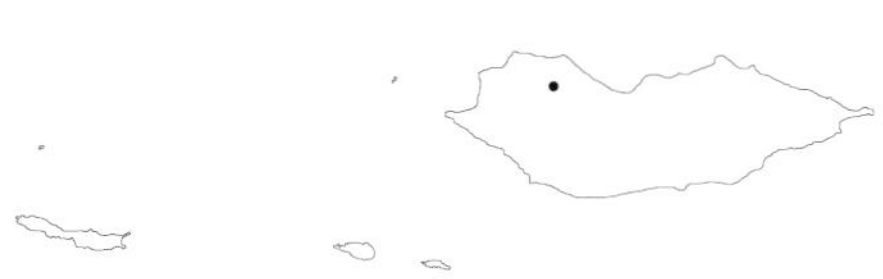

Similar to *G. mascatense* but differs in the mericarps 4–6mm long, minutely ridged and with a revolute, toothed margin.

Ecology. Semi-deciduous woodland. 600m.
Distribution. NE tropical Africa, S Arabia to Iran.

ETHNOGRAPHIC NOTES

Name(s)
See *G. biuncinatum* (also uses).

Pelargonium

Robust, perennial herbs with palmately lobed leaves. Flowers in umbels. Stamens 7 to 10 (not all fertile). Upper sepal produced into a long spur which is fused to the pedicel. Petals unequal. Fruit beaked, splitting from the base into 5 awned mericarps, the awns spirally twisted. 1 endemic. 280 species. Mainly S Africa.

Pelargonium insularis **Gibby & A.G.Mill.** (see Appendix 2, p. 723)

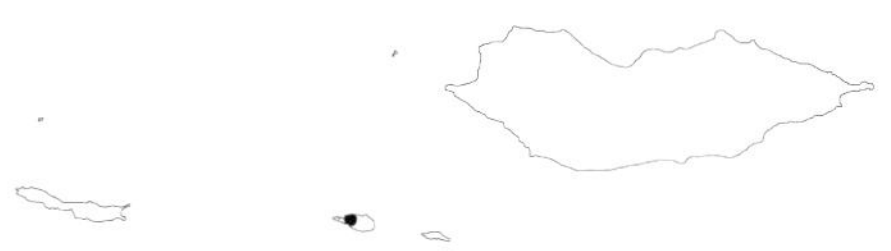

Robust perennial herb from a stout stock, glandular-pubescent throughout. Leaves subcircular in outline, up to 10 x 10cm, 5(to 7)-lobed, the margins rather irregularly crenate-serrate. Umbels 4 or 5-flowered. Sepals oblong, 6–10 x 1.5–3mm, spur apparently extending the full length of the pedicel. Petals pale pink, subequal, ovate, 20–24 x 13–15mm. Stamens 7. Fruit 5-lobed with a c.30mm beak; mericarps with 30mm awns.

Ecology. Samha; on north-facing limestone cliffs. c.650m.

Distribution. Endemic.

Conservation status. Critically Endangered (CR B2 a biii/D). This species of *Pelargonium* was discovered in 1999 on the north-facing limestone escarpment on the island of Samha. It represents the first record of the genus from the Soqotra Archipelago. At the time of its discovery only a single plant was found and an extensive search failed to find more. The new species was not seen on two previous expeditions which visited the cliffs in 1992 and 1994, and an expedition the year after its discovery, in January 2000, also failed to locate it. The cliffs on which it occurs are frequently enveloped in low cloud and provide a relatively moist refugium, occupying less than 5km^2, on an otherwise desertic island. Local informants know of the plant and comment that it is found along the cliffs and, when fresh, provides incidental grazing for livestock. Until very recently goats and sheep were actively herded on Samha, being moved down from the limestone plateau during periods of drought, thus protecting the vegetation. However, both the sheep (now largely feral) and the goats (actively managed only when in milk) are now left to roam freely for most of the year, and both types of livestock are known to spend much of their time grazing the limestone plateau. It is likely that this will lead to overgrazing and could possibly lead to the eradication of palatable plants on all but the most inaccessible ledges. This may account for apparent scarcity of the plant and the lack of recent sightings.

The IUCN listing is based on: the small population size (possibly under 50 individuals); the limited extent of occurrence and area of occupancy (under 5km^2); the restriction to a single location – where it occupies a specific microclimate; the continuing decline in the quality of this habitat based on (1) the trend of lower rainfall in the region and the effects of global warming and (2) recent changes in livestock management on the island. The limestone escarpment cliffs have been designated a Nature Sanctuary within the zoning plan for the Soqotra Archipelago.

Botanical notes. Most closely related to *P. alchemilloides* (a plant of SW Arabia and tropical NE Africa) from which it differs in its pink (not white) flowers and the development of a pronounced woody stock.

ETHNOGRAPHIC NOTES

Name(s)

– Soqotri: nil or *riˈʿud* ‘grazing plant’.

– Soqotri simplified: *ru’ud*.

Food (livestock). When young and green after rain this provides incidental grazing for small stock, but is not nutritionally important.

Comparative uses. In Yemen the leaves and stalks of *P. zonale* are eaten. The leaves and fleshy roots of other species are also edible. (Kunkel 1984)

GLOBULARIACEAE

Herbs or subshrubs. Leaves alternate, entire; stipules absent. Flowers in dense spikes, ± zygomorphic. Calyx tubular, 5-lobed. Corolla 2-lipped, upper lip 2-lobed, lower lip 3-lobed. Stamens 4. Ovary superior, 1-celled with a single pendulous ovule. Fruit a 1-seeded achene, persistent within the calyx. 10 genera. 250 species. Europe, W Asia, Africa.

Poskea Vatke

Cockburnia Balf.f.

See description of family. 2 species. Somalia.

Poskea socotrana **(Balf.f.) G.Taylor**

Cockburnia socotrana Balf.f.

Woody-based herb or small shrub, greyish tomentose. Leaves broadly to narrowly obovate, 1–2 x 0.4–1.2cm, margin entire, tip often notched. Flowers white, in a dense spike, c.1–2cm long. Corolla c.0.2cm long.

Ecology. On cliffs in succulent shrubland and semi-deciduous woodland. 100–950m.

Distribution. Recently found in Yemen (Mahra).

ETHNOGRAPHIC NOTES

Name(s)

– Soqotri: nil or *di ʾˈitəhi, di tīyə* ‘cliff-growing’; *bˈišši* (commonly given to cliff-growing plants or those with

striking flowers); *di ṣefəririhon* 'many-flowered'; *ṣaˁlifitin di ḥōsir* 'leaves for time of scarcity'; *ḥasəlinž*, *ḥasəling* 'the two poles used to strap the load onto the baggage camel'; *libinhin* (*di didəhīn*) 'white, pale plant of the central limestone plateau'.
– Soqotri simplified: *di 'itihi, di tiyih*; *bishi*; *di zefiririhon*; *za'lifitin di howsir*; *hasilinj*, *hasilin'g*; *libinhin* (*di didihiyn*).

Food (livestock). Not normally accessible to livestock, but goats eat it if they can reach it.
Bees / honey. Honeybees are attracted to the flowers.
Leatherwork. In the absence of alternative material, the crushed inner bark and new stems can be used to cure leather.

GOODENIACEAE

Shrubs. Leaves alternate, entire; stipules absent. Flowers solitary, axillary, zygomorphic. Calyx 5-lobed. Corolla tubular, 5-lobed, split along upper side almost to base. Stamens 5. Ovary inferior, 2-celled; 1 style; placentation axile. Fruit a drupe. 16 genera. 430 species. Tropical, mostly Australian.

Scaevola L.

See description of family. 1 endemic. 130 species. Tropical, especially Australia.

Scaevola socotraensis St.John

Pendulous shrub with branches to 1m. Leaves tending to be clustered on end of branches, elliptic to obovate, c.4 x 1.5cm, slightly fleshy. Corolla white, c.1.8cm long.
Ecology. Very rare, on rocks and vertical cliff faces, near seepages on the limestone escarpment overlooking the Noged plain. c.150m.
Distribution. Endemic.
Conservation status. Endangered (EN B2 a biii). Known only from a few scattered populations growing near freshwater springs and seepages on the southern limestone escarpment. There are only a few of these springs and therefore the total number of plants is likely to be very low. This habitat is extremely vulnerable to water development and thus a designation of endangered seems appropriate (cf. *Helichrysum* sp. *C*, *Erythroxylum socotranum* and *Exacum socotranum*).

ETHNOGRAPHIC NOTES
Name(s)
– Soqotri: nil or *di rihoytin* 'of water'.
– Soqotri simplified: *di rihoytin.*
Food (livestock). Not normally accessible to livestock, but goats eat it if they can reach it.
Comparative uses. In India the leaves of *S. frutescens* are eaten as a vegetable and smoked as tobacco. Medicinally, they are given for indigestion and made into a poultice to treat tumours and swollen legs. The juice of the fruits is applied externally for eye troubles and taken internally to induce menstruation. A decoction of the roots is used in dysentery. The wood from the basal part of the shrub is hard and withstands salt water, and is used to make nails and pegs for boats. The pith is used for sun-hats, floats and for making a kind of rice paper. (Ambasta 1982; Jain & De Fillips 1991)

GUTTIFERAE

Herbs, shrubs or trees. Leaves opposite, simple, often glandular with translucent dots and streaks and sometimes with black gland dots; stipules absent. Flowers bisexual, actinomorphic, solitary or in terminal cymes or panicles. Sepals 5, free. Petals 5, free, contorted in bud. Stamens numerous, in 3 or 5 bundles. Ovary superior, 1 to 5-celled; ovules numerous; styles 3 or 5, free or fused. Fruit a many-seeded, 3 or 5-valved capsule. 47 genera. 1350 species. Tropical and N temperate regions.

Hypericum L.

See description of family. 5 endemics. c.370 species. Mostly temperate.

Botanical notes. Robson (1987) suggests that the Soqotran endemics provide a key to an understanding of the intersectional relationships of a large part of *Hypericum*: *H. balfourii* provides a link between the E African *H. revolutum* subsp. *keniense* and the E Asian sect. *Ascyreia*, and *H. socotranum* links *H. revolutum* subsp. *keniense* with three wholly or mainly Mediterranean sections (sections *Adenotrias*, *Heterophylla* and *Triadenioides*).

ETHNOGRAPHIC NOTES
Comparative uses. Many *Hypericum* species are used medicinally, and *H. perforatum* (St John's wort) appears in many remedies in the West. (Ambasta 1982; Jain & De Fillips 1991; Lewis & Elwin-Lewis 1977)

Hypericum balfourii N.Robson

Hypericum mysorense sensu Balf.f. non Heyne

Glabrous shrub or small tree to 3m. Leaves sessile, narrowly elliptic to narrowly lanceolate, 10–45 x 4–12mm; base cuneate to subauriculate; venation parallel (with 3 to 7 main veins) near base, becoming pinnate towards tip. Flowers in 1 to 3(to 10)-flowered terminal cymes. Sepals 5–9mm long. Petals golden yellow, 20–35mm long, deciduous. Stamen bundles 5. Styles 5, united below. Capsule 5-valved, ovoid, 13–15 x 7–9mm.

Ecology. Low montane shrubland and dwarf *Hypericum–Cephalocroton* shrubland; mainly on granite in the Haggeher, but also less commonly on limestone (Reyged); typically above 600m, lower altitudes are atypical. (120–)600–1100m.

Distribution. Endemic.

Conservation status. Vulnerable (VU D2). Rare, with a very limited area of occupancy.

Botanical notes. Closely related to *H. mysorense*, a species from S India and Sri Lanka.

ETHNOGRAPHIC NOTES

Name(s)

– Soqotri: *bi̍šši* (commonly given to cliff-growing plants or those with striking flowers); *ḥasəlɛ̍m* 'slim'; *ʿaṣʌ̍bṣʌb* 'branch tips'; (rare) *ḥari̍ṭriṭ* 'physically weak' (more commonly given to *Leucas virgata*).

– Soqotri simplified: *bishi*; *hasilem*; *'azabzab*; *haritrit.*

Food (livestock). All livestock eat it when they are hungry and there is little alternative fodder, but they are not enthusiastic, and it is not seen as nutritionally useful. If the dry season is prolonged, herders cut green and leafy branches to tempt cattle and goats to eat.

Bees / honey. Honeybees are attracted to the flowers.

Fuel. The dead wood makes reasonable firewood.

Timber. The trunks and larger branches are used for cross beams and roofing generally. It is not as durable as other timbers, so is not considered to be among the best woods for building. However, it is quite resistant to termites. The trees tend to grow in areas where other better timbers are to be found, so locally it is not an important timber.

Wood. Mainly used for temporary purposes, such as making barriers for pens and byres, or covering a waterhole. The longer, straighter branches can be used to make the important herding staff.

Hypericum fieriense N.Robson

Low shrub to 1m with stems and leaves densely to sparsely pubescent. Leaves petiolate, triangular-ovate to oblong-ovate, 10–17 x 7–12mm; base broadly cuneate to truncate; petiole 4–6mm; venation pinnate, obscure between main lateral veins. Flowers in subumbellate 3 to 5-flowered terminal cymes. Sepals 4–5mm long. Petals yellow. Stamen bundles 3. Styles 3, free. Capsule 3-valved, pyramidal-ovoid, 6–7.5 x 4.5–5mm.

Ecology. Low *Hypericum–Cephalocroton* shrubland with emergent *Dracaena cinnabari*. 1350–1500m.

Distribution. Endemic.

Conservation status. Vulnerable (VU D2). Rare, with a very limited area of occupancy.

Botanical notes. The rarest of the shrubby *Hypericum* species on Soqotra; known from only two locations. Most likely to be confused with *H. scopulorum* or *H. tortuosum* but readily distinguished by the pubescent stems and leaves.

ETHNOGRAPHIC NOTES

Name(s)

– Soqotri: nil or 'like *chīreb*' (*H. scopulorum*) 'sloppy, semi-liquid'.

– Soqotri simplified: 'like *chiyreb*'.

Food (livestock). Said to provide incidental grazing for the wild and feral goats of the high mountains, but is not grazed by domesticated livestock.

Hypericum scopulorum Balf.f.

Erect, glabrous shrub to 1m. Leaves sessile, obovate to oblong-elliptic, 10–24 x 4–11mm; base subauriculate; venation pinnate, obscure between main lateral veins. Flowers solitary in axils of uppermost leaves. Sepals 5–7.5mm long. Petals golden-yellow to golden-orange, 5–11mm long, persistent. Stamen bundles 3. Styles 3, free. Capsule 3-valved, subglobose, c.5 x 4mm, valves vittate.

Ecology. Common, sometimes dominant, on granite in dwarf *Hypericum–Cephalocroton* shrubland on the higher slopes of the Haggeher mountains; co-dominant with *Helichrysum rosulatum* on pinnacles in the eastern Haggeher; less commonly on limestone cliffs. (550–)800–1350m.

Distribution. Endemic.

Conservation status. Least Concern (LC). Common in several vegetation types and under no present or perceived threat.

Botanical notes. Reported to be on the increase in the eastern Haggeher.

ETHNOGRAPHIC NOTES

Name(s)

– Soqotri: *chīrəb*, *di ḳīyrəb* 'sloppy, semi-liquid'; (rare) *śeru̍brub* 'ticks'.

– Soqotri simplified: *chiyrib*, *di qiyrib*; *sherubrub.*

Food (livestock). Not normally grazed by livestock, but they will nibble at the leafless stems and eat the fallen leaves in drought.

Bees / honey. Honeybees are attracted to the fragrant flowers.

Fuel. The dead wood makes reasonable firewood, but burns very quickly. It gives out a good, bright light and it is common to keep a small store of this for this purpose.
Leatherwork. The leafy stems are crushed and used to treat skins and hides to strengthen the leather and stain it red-brown. It is considered to be one of the best plants for this purpose. It is used especially to treat the leather of the floor covering called *ni̍bəʿeh*, which is made from the entire, haired skin of a mature male goat.
Wood. Dead plants are gathered and stored in the home for use as firelighters and kindling. This material lights well even in wet weather, but it is best with the bark removed.
Other. Some of the best soil for cultivation is gathered from under and around these plants.
Notes. Usually removed from grazing paddocks and enclosures in the Haggeher mountains because no grass will grow beneath them.

***Hypericum socotranum* Good**

Glabrous shrub or small tree to 4.5m. Leaves sessile, elliptic to obovate or subcircular, 12–30 x 5–22mm; base cuneate; venation mainly pinnate with 1 or 2 pairs of basal lateral veins extending around the margin towards the tip. Flowers in 1 to 5-flowered terminal cymes. Sepals 4.5–6.5mm long. Petals golden yellow, 15–40mm long, deciduous. Stamen bundles 5. Styles 5, united below. Capsule 5-valved, ovoid, 12–13 x 6–7mm.
Distribution. Endemic.
Botanical notes. Two subspecies are recorded on Soqotra.

ETHNOGRAPHIC NOTES
Name(s)
– Soqotri: *ḥase̍ləhan, ḥasəlān, ḥasəle̍yn* 'thin, slim'.
– Soqotri simplified: *haselihan, hasilan, hasileyn.*
Food (livestock). Goats and cattle only eat the foliage when they are very hungry and there is little else.
Fuel. The dead wood makes acceptable firewood.
Wood. The longer, straighter branches can be made into the important herding staff.

Subsp. *socotranum* Good
Hypericum lanceolatum sensu Balf.f. non Lam.

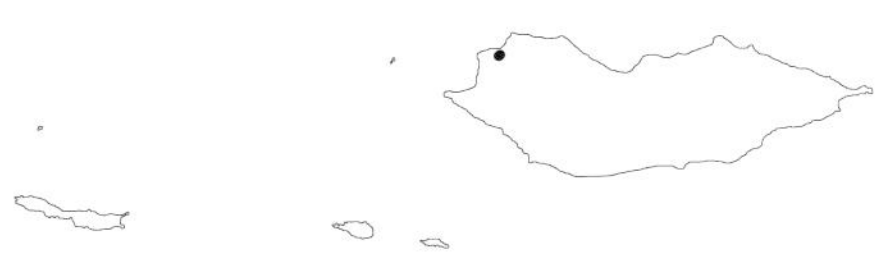

Leaves elliptic to oblanceolate, 5–10mm wide, acute to obtuse. Flowers solitary; pedicels 1–6mm.
Ecology. Restricted to drought-deciduous woodland on the limestone escarpments of NW Soqotra. 200–300m.
Conservation status. Endangered (EN B2 a biii). Known only from the north-facing limestone escarpment near Qalansiyah at the western end of Soqotra where it occupies an area of less than 10km^2. It is strongly suspected that this marginal habitat is threatened by the long-term trend of lower rainfall in the region and would be particularly vulnerable to the effects of global warming.
Botanical notes. There are several local endemics and plants of note on the escarpment above Qaysoh near Qalansiyah including *Boswellia bullata*, *Dorstenia gigas*, *Prenanthes amabilis*, *Aloe squarrosa*, *Gaillonia thymoides* and this subspecies.

Subsp. *smithii* N.Robson

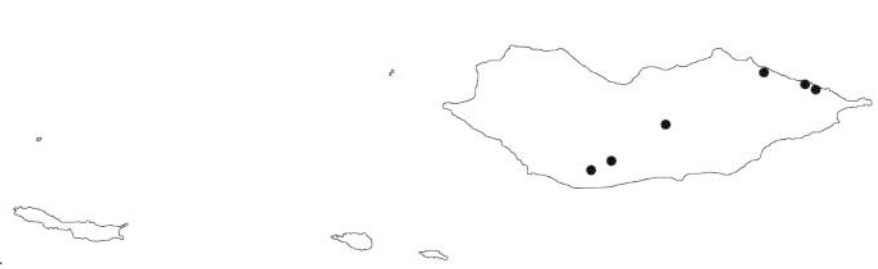

Leaves broadly elliptic to subcircular, 8–22mm wide, apiculate to rounded. Flowers in 1 to 5-flowered clusters; pedicels 5–25mm.
Ecology. Drought-deciduous woodland on limestone in the central and eastern parts of Soqotra. 400–650m.
Conservation status. Least Concern (LC). Common in several vegetation types and under no present or perceived threat.
Botanical notes. The closest relative of *H. socotranum* subsp. *smithii* is *H. dogonbadanicum* Assadi from the *Quercus brandtii* forests of the Zagros mountains in SW Iran. Robson (1967) suggests that *H. dogonbadanicum* is a relative of a basically African group of species which migrated northward, into regions which now have a Mediterranean-type climate, during the Tertiary. Subsequent drying out led to a disruption of this once continuous distribution. It is a rare example of an Iranian–Soqotran disjunction.

***Hypericum tortuosum* Balf.f.**

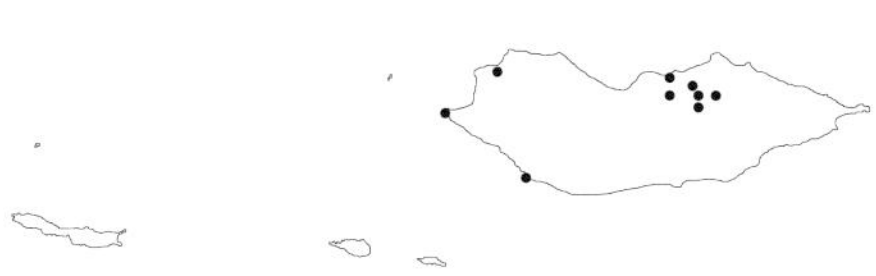

Glabrous shrub to 50cm. Leaves shortly petiolate, becoming sessile above, obovate to oblong-elliptic, 8–25 x 4–20mm; base cuneate, sometimes auriculate; venation pinnate, obscure between the main lateral veins. Flowers in subumbellate, many-flowered, terminal cymes. Sepals 3.5–6mm long. Petals yellow, 5–7.5mm long, persistent. Stamen bundles 3. Styles 3, free. Capsule 3-valved, subglobose, c.4.5 x 4mm, valves verrucose.
Ecology. Shrubland and semi-deciduous woodland on both granite and limestone, often on cliffs. (30–)600–1000m.
Distribution. Endemic.
Conservation status. Least Concern (LC). Common in several vegetation types and under no present or perceived threat.

ETHNOGRAPHIC NOTES
Name(s)
– Soqotri: nil or 'like *chiyrib*' (*H. scopulorum*) 'sloppy, semi-liquid'; *bíšši* (commonly given to cliff-growing plants or those with striking flowers).
– Soqotri simplified: 'like *chiyrib*'; *bishi.*
Food (livestock). Said to provide incidental grazing for the wild and feral goats of the high mountains, but is rarely grazed by domesticated livestock.

LABIATAE

Herbs or shrubs. Stems often 4-angled. Leaves opposite, simple (sometimes bipinnately lobed); stipules absent. Flowers usually in whorls along the stem, sometimes solitary or in subcapitate clusters. Calyx tubular, actinomorphic or zygomorphic, sometimes 2-lipped; lobes 5(to 10). Corolla tubular, 2-lipped, sometimes ± actinomorphic or upper lip absent. Stamens 4. Ovary superior, 4-celled; 1 ovule per cell. Style rises from the base of the ovary between the nutlets; stigma entire or 2-lobed. Fruits <4 single-seeded nutlets. 221 genera. 5600 species. Cosmopolitan, especially drier areas.

Botanical notes. Several species are characteristically aromatic.

Endostemon N.E.Br.

Herbs. Leaves simple. Flowers pedicellate, in terminal racemes, bracts inconspicuous. Calyx zygomorphic; lobes 5, upper expanded, reflexed, laterals short, lower 2 long, spinose; characteristic tuft of white hairs in throat. Corolla subequally 5-lobed. Stamens included in tube. Style equally 2-lobed. 17 species. Widespread in tropical Africa, 1 species in India.

Endostemon tenuiflorus (Benth.) Ashby
Orthosiphon tenuiflorus Benth.

Erect herb to 20cm. Leaves narrowly elliptic to oblanceolate, 1–2 x 0.2–0.5cm, softly grey hairy, sparsely dentate or crenate. Mature calyx c.0.5cm long. Corolla white to pink, c.1cm long.
Ecology. Locally common on dry rocky hills. 450–600m.
Distribution. S Arabia, NE Africa, S Africa, Madagascar.
Botanical notes. Superficially similar to *Orthosiphon pallidus* but distinguished by the tuft of white hairs in the throat of the calyx.

ETHNOGRAPHIC NOTES
Name(s)
– Soqotri: nil or *támərhen* 'little dates', *nói* (both widespread names for small, delicate and often fragrant plants); *luʾlúʾuh* 'pearls'; *di ḥōwlhinítin*; *di ʿášbib* 'quick to wilt'.
– Soqotri simplified: *tamirhen*; *noy*; *lu'lu'uh*; *di howlhinitin*; *di 'ashbib*.
Food (livestock). Provides incidental forage for livestock, especially for small stock. It is regarded as a good fodder if plentiful.
Bees / honey. Honeybees are attracted to the sweet-smelling flowers.
Cleansing. Flowering sprigs are tucked into the clothing and hair. The sniffing of sweet-smelling plants is held to protect from the 'noxious air' which can cause disease, and to help protect against infection. All sweet-smelling plants are endowed with beneficial medicinal and prophylactic qualities.
Notes. Goats unaccustomed to this forage can develop scour if they eat a lot of young, green plants just after the rains.
Comparative uses. The fodder value of the foliage of Labiatae is high. *E. rubicundus* has tubers which are eaten and which yield starch. Medicinally, a decoction of *E. glabratus* is used in diarrhoea and piles, and the leaves as a febrifuge. The leaves are pounded and applied to cuts and wounds. The leaves of *E. spiralis* contain tannin and an essential oil, and an infusion of the plant is given in kidney and bladder diseases, and to treat gout and rheumatism. (Ambasta 1982; Bebawi & Neugebohrn 1991; Dalziel 1948, 1956; Jain & De Fillips 1991)

Lavandula L.

Herbs. Leaves bipinnately lobed. Flowers ± sessile, in terminal spikes, bracts inconspicuous. Calyx actinomorphic; lobes 5, equal. Corolla subequally 5-lobed. Aromatic. 1 endemic. 30 species. Widespread in S Europe, SW Asia, Arabia and N Africa.

Lavandula nimmoi Benth.
Lavandula subnuda sensu Balf.f. non Benth.

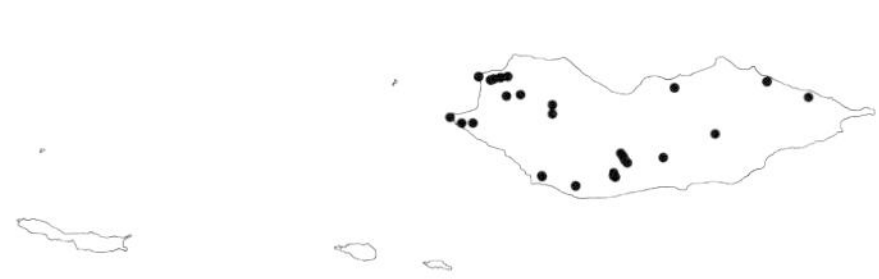

Erect, spindly herb to 1m. Leaves ovate to elliptic, 1–5 x 0.5–3cm. Mature calyx c.0.5cm long. Corolla lilac or blue, c.1cm long.
Ecology. Common and widespread on limestone in a variety of vegetation types. 50–800m.
Distribution. Endemic.
Conservation status. Least Concern (LC). Common in several vegetation types and under no present or perceived threat.

ETHNOGRAPHIC NOTES
Name(s)
– Soqotri: *sídʌk* (sometimes given to other very fragrant herbs).

– Soqotri simplified: *sidak*.

Food (human). People nibble the sweet-smelling flowers.

Food (livestock). Goats and sheep especially graze it, and in particular like the flowers. It is considered to be more nutritious when still green and before it dries out and becomes reddish in colour. However, even in this mature state it is seen as a good maintenance, dry season feed. Cattle also graze the plant, especially when young, and camels and donkeys eat it, both green and dried out.

Bees / honey. Honeybees are attracted to the flowers.

Wood. Dead stems are gathered and stored at home for use as tinder and kindling. When it burns it gives off a very pleasant smell.

Leatherwork. The green plant, especially when in flower, is widely used to treat skins and hides. It is one of the best plants for toughening leather and is therefore in demand for the skins used for the heaviest work of carrying water or churning butter. The soft stems are often added to many other plant mixtures used in leatherwork. It makes them softer and more manageable, but also gives the finished leather a lovely smell. A paste of the plant is also used to re-treat a dried-out and stiff, or ill-smelling, skin.

Bedding, stuffing. The plants are gathered and scattered over the sleeping area, especially during the monsoon months. Sprigs are laid among bedding to perfume it, and are added to the stuffing for bolsters and pillows. It is particularly used in bedding for the sick, as its sweet smell is believed to relieve their discomfort and hasten their recovery.

Cleansing. Flowering stems are tucked amongst the clothing and pinched to release the sweet smell. The sniffing of sweet-smelling plants is held to protect from 'noxious air' that can cause disease, and to help protect against infection. Sprigs are also rubbed between the hands and then over the skin, hair and clothing to perfume it. This is done especially when hunting the wild goat to disguise the human odour.

Other. The skin of a dead calf, or of a male killed to conserve its mother's milk, is made into a tulchan stuffed with these plants. After being sprayed with some of her milk, it is presented to a cow to sniff, to encourage her to let down her milk. A paste of crushed plants is also smeared over the back of a living calf to disguise its smell before presenting it to a strange cow. Some of her milk is sprayed over the calf, which is then led around for her to sniff, to encourage her to accept it as her own and allow it to suckle. The paste is also used to dye leather for finer, more personal, items a dark red-brown colour. It was much used in making the intricately plaited belts formerly worn by men. The paste is also used as a dye to decorate clay pottery, but this must be done when the pots are still red-hot from firing.

Notes. Wild and feral goats are especially known for browsing on *Lavandula*.

Comparative uses. Many *Lavandula* species yield an essential oil which is used chiefly in perfumery, though it is also said to have carminative and stimulant properties, and is used for colic, headaches, chest complaints and biliousness. In the Middle East an infusion of the inflorescences of *L. dentata* is taken first thing in the morning and before going to bed to treat urinary retention and remove kidney or ureter stones. In Saudi Arabia the fresh leaves are tied to the forehead to relieve a headache, and to joints to relieve rheumatic pain. The steam of the boiled plant is inhaled as a cold remedy. In Yemen *L. pubescens* and *L. stoechas* are used as a diuretic, and as a remedy for broncho-pulmonary infection. In Dhofar the leaves of *L. dhofarensis* are added to milk and drunk to relieve stomach aches. The plant is also used as a vermifuge, and the smoke of burning leaves is inhaled to improve the memory. A decoction of the leaves is taken orally for kidney problems. (Abdulfatih 1987; Ambasta 1982; Fleurentin & Pelt/Fleurentin *et al.* [1982, 1983a,b,c]; Goodin & Northington 1979; Jain & De Fillips 1991; Miller & Morris 1988)

Leucas R.Br.

Lasiocorys Benth.

Herbs or shrubs. Leaves simple. Flowers ± sessile, solitary or clustered in the axils of leaf-like bracts or in dense whorls with linear or minute bracts. Calyx 5 or 10-lobed, actinomorphic or oblique due to the lengthening of one side. Corolla white, 2-lipped, upper lip with characteristic fringe of white hairs. Style unequally 2-lobed. 6 endemics. 7 endemics. c.150 species. Africa, Arabia, tropical Asia.

Botanical notes. All species are characterised by similar flowers which have a fringe of white hairs on the upper lip of the white corolla and contrasting orange anthers. For a revision of this genus see Sebald (1980).

ETHNOGRAPHIC NOTES

Comparative uses. Many *Leucas* species are used as a pot-herb, and their tender shoots as a vegetable. Medicinally, the leaves of *L. aspera* and *L. zeylanica* are used externally to treat skin problems and as a snakebite antidote. The fragrant flowers of many species are made into syrups drunk to soothe a cough or a cold. The leaves of some species, such as *L. zeylanica*, yield an essential oil. (Ambasta 1982; Jain & De Fillips 1991; Kunkel 1984; Lewis & Elwin-Lewis 1977)

Leucas flagellifolia (Balf.f.) Guerke

Lasiocorys flagellifera Balf.f.

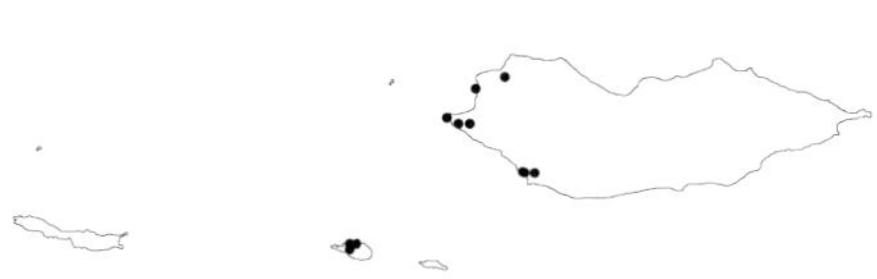

Prostrate or mat-forming perennial herb to 30cm. Leaves obovate, c.0.5 x 0.2cm, distal margin with c.7 spines, densely white hairy. Flowers solitary in the axils of leaf-like bracts. Calyx 5-lobed, c.0.5cm long. Corolla <1cm long.

Ecology. Dwarf shrubland and open woody-herb communities on desertic coastal plains and slopes, mainly on limestone. Sea-level–700m.

Distribution. Endemic.

Conservation status. Vulnerable (VU D2). Rare, with a very limited area of occupancy.

Botanical notes. Restricted to coastal plains and desertic foothills in the southwest of Soqotra and on Samha.

ETHNOGRAPHIC NOTES

Name(s)

– Soqotri: nil or *ḥarišimhin* 'unproductive'; *ləskano* 'sticky, clinging' (Samha); *śkaʿhinitin* 'spiny, prickly'.

– Soqotri simplified: *harishimhin*; *liskano*; *shka'hinitin.*

Food (livestock). Provides incidental grazing for all livestock, and is considered to be a good forage when young and green. As it dries out it is of little nutritional value.

Bees / honey. Honeybees are attracted to the flowers.

Leucas hagghierensis **Gifri & Cortés-Burns** (see Appendix 2, p. 723)

Weak shrub to 3m. Leaves ovate, 2–6 x 1–3cm, crenate. Flowers solitary or few, in the axils of leaf-like bracts, spreading. Calyx 5-lobed, <1.5cm. Corolla c.3cm long, softly hairy.

Ecology. Dense submontane semi-deciduous thicket around granite pinnacles. 1050–1500m.

Distribution. Endemic.

Conservation status. Vulnerable (VU D2). Rare, with a very limited area of occupancy.

Botanical notes. The largest *Leucas* on Soqotra. Rather rare in dense thicket amongst the highest slopes of the central and western Haggeher.

ETHNOGRAPHIC NOTES

Name(s)

– Soqotri: *di feylʌḳ*; *ṣaʿalilihun* 'large-leaved'; *zīzəhīn* (more commonly given to *Ruellia patula* and *R. insignis*).

– Soqotri simplified: *di feylaq*; *za'alilihun*; *ziyzihiyn.*

Food (human). People suck the flowers for the nectar.

Food (livestock). The foliage is not palatable to livestock.

Bees / honey. Very attractive to honeybees.

Fuel. The dead wood is only used for firewood if no better alternative is available.

Wood. Poor and little used, either for making implements, or as a timber.

Leucas kischenensis **(Radcl.-Sm.) Sebald**

Leucas lanata sensu Balf.f. non Benth.

Leucas thymoides Baker

Leucas thymoides subsp. *kischenensis* Radcl.-Sm.

Shrub to 1m. Leaves elliptic to orbicular, >3cm long, crenate, densely grey hairy beneath. Flowers few, in the axils of leaf-like bracts, spreading. Calyx 10-lobed, c.0.6cm long. Corolla c.1cm long.

Ecology. Open semi-deciduous shrubland on granite. 650–1000m.

Distribution. Endemic.

Conservation status. Least Concern (LC). Common in several vegetation types and under no present or perceived threat.

ETHNOGRAPHIC NOTES

Name(s)

– Soqotri: *məgērəd* (more commonly given to the sweet-smelling *Coelocarpum socotranum*, which it resembles, and *Micromeria* species); *nōwḥel* 'chaff, winnowings'; *nōmhʌl* 'little ants' (from the tiny fruit) (more commonly given to *Leucas virgata*).

– Soqotri simplified: *migeyrid*; *nowhel*, *nowmhal.*

Food (livestock). Livestock eat this only in a prolonged dry season when very hungry and when there is little other forage available.

Bees / honey. Honeybees are attracted to the flowers.

Fuel. The dead wood is not highly regarded as firewood.

Wood. Poor and little used, either for making implements, or as a timber.

Leucas martinicensis **(Jacq.) R.Br.**

Erect unbranched annual herb to 40cm. Leaves ovate to elliptic, 2–3 x 1–1.5cm, crenate. Flowers axillary in dense, globose clusters; bracts linear, c.0.5cm long. Calyx 10-lobed, c.0.8cm long, bent, the upper lobes > 2x as long as others, setose, spreading. Corolla c.1cm long.

Ecology. Locally common in disturbed ground and grassland. 200–600m.

Distribution. Pantropical.

ETHNOGRAPHIC NOTES

Name(s)

– Soqotri: *ʿafolhi*; *ʿažərib* 'quick to die back'; (rare) *kirbilitin* 'tufts if hair at the tail tip'.

– Soqotri simplified: *'afolhi*; *'ajirib*; *kirbilitin.*

Food (livestock). This is one of the *ʿažərib* group of plants which come up after the winter rains. It is readily grazed by cows before the grasses come up, but is little eaten by goats and sheep, although they like the buds and flowers and also the fruit. Sheep eat the plant litter and dried-out plants only, and camels eat the growing tips. It is not seen as having any nutritional value. In a drought goats, and sometimes also sheep, dig up the roots and eat them.

Notes. It is said to be on the increase in heavily grazed coastal areas in eastern Soqotra.

Comparative uses. The leaves are eaten for their minty flavour. The herb is insecticidal and is burned to expel mosquitoes. An infusion of the plant is given for gastro-intestinal disorders, and as a wash or steam for fevers. It is drunk hot for colds. The plant is also considered to be a charm for good luck or riches: the seeds are rubbed with butter between the palms of the hands while the person asks good luck to come. (Ambasta 1982; Dalziel 1948, 1956; Jain & De Fillips 1991)

Leucas neuflizeana **Courbai**

Erect unbranched annual herb to 40cm. Leaves narrowly obovate to spathulate, 1–3 x 0.3–1cm, crenate. Flowers axillary in many-flowered whorls; bracts minute. Calyx 10-lobed, c.0.6cm long, mouth oblique, lower 3 lobes drawn forward. Corolla <1cm long.
Ecology. Widespread on the plains and foothills, usually in disturbed places; also a weed of date gardens. 50–700m.
Distribution. Tropical Africa, Yemen, Oman.

ETHNOGRAPHIC NOTES
Name(s)
– Soqotri: nil or *ʿážərib* 'quick to die back'; *ʿážbəb* 'quick wilting'.
– Soqotri simplified: *'ajirib*; *'ajbib*.
Food (livestock). This is one of the *ʿážərib* group of plants which come up after the winter rains. It is grazed by all livestock when still green, and, before the grasses come up, is especially liked by cattle. It is seen as having little nutritional value.
Notes. It can cause *'uqabah* in small stock unaccustomed to this feed. It is said to be on the increase in heavily grazed areas.

Leucas penduliflora **Gifri & Cortés-Burns** (see Appendix 2, p. 723)

Erect shrub to 50cm. Leaves elliptic to orbicular, <1.5cm long, entire; petiole c.3/4 length of leaf. Flowers in the axils of leaf-like bracts, few, pendulous. Calyx 5-lobed, <1.5cm long. Corolla <3cm long.
Ecology. Deciduous shrubland on rocky granite slopes. c.550m.
Distribution. Endemic.
Conservation status. Vulnerable (VU D2). Rare, with a very limited area of occupancy.
Botanical notes. Only collected once, above the Muqadrihon Pass. Readily distinguished by its pendulous flowers.

ETHNOGRAPHIC NOTES
Name(s)
– Soqotri: *nɔ̄wḥel ḥégirhi* '*L. virgata*-type of the Haggeher mountains'.
– Soqotri simplified: *nowhel hegirhi.*
Food (livestock). Regarded as superior to *L. virgata* as forage, being fattening for both goats and cattle.
Notes. Uses as *L. virgata*.

Leucas samhaensis **Cortés-Burns & A.G.Mill.** (see Appendix 2, p. 723)

Erect shrublet to 50cm. Leaves orbicular to kidney-shaped, 0.5–1 x 0.5–1cm, entire, velvety below. Flowers subterminal in 3 or 4-flowered clusters. Calyx 10-lobed, c.0.5cm long, densely hairy. Corolla c.1cm long.
Ecology. Locally abundant in low shrubland at the base of the north-facing limestone escarpment on Samha. 400–500m.
Distribution. Endemic.
Conservation status. Vulnerable (VU D2). Rare, with a very limited area of occupancy.

ETHNOGRAPHIC NOTES
Name(s)
– Soqotri: *ṣaʿlə́ffən* 'many-leaved'; *kəbédino* 'little liver', 'small hill(ock), mound' (Samha).
– Soqotri simplified: *za'liffin*; *kibedino.*
Food (livestock). Provides grazing for small stock but is not seen as being nutritionally important. It is useful dry season subsistence feed.

Leucas spiculifera **(Balf.f.) Guerke**
Lasiocorys spiculifolia Balf.f.
Leucas spicata R.Br.

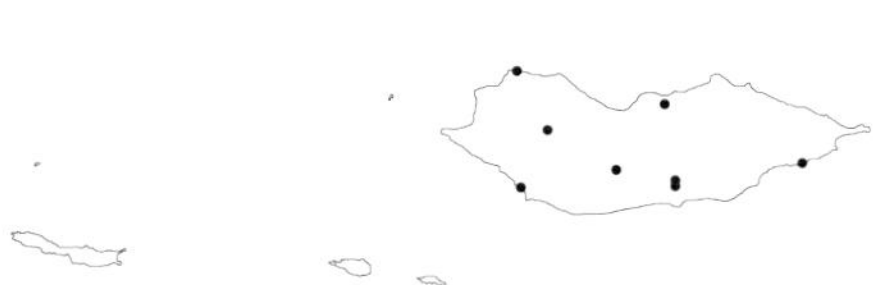

Woody herb, often hanging on cliffs, to 1m. Leaves narrowly elliptic, c.1 x 0.1–0.5cm, with spinose tip and 2 lateral spines, sometimes entire. Flowers solitary or few in the axils of leaf-like bracts. Calyx 5-lobed, c.0.5cm long, lobes spinose. Corolla c.1cm long.
Ecology. On cliffs and large boulders along wadis and on the escarpments. Sea-level–350m.
Distribution. Endemic.
Conservation status. Least Concern (LC). Common in several vegetation types and under no present or perceived threat.
Botanical notes. More or less restricted to inaccessible cliffs.

ETHNOGRAPHIC NOTES
Name(s)
– Soqotri: *ḥəwrírihim* 'dark-coloured'; *śkaʾḥinítin* 'spiny, prickly'; *śkáʿhinítin di məźɔ̄wźʌʾ* 'spiny plant of damp places'; *źeyṭ* 'like *Ziziphus*'; *sɔ̄wgihīn* 'curved'; *chayéyεʿ* (more commonly given to sweet-smelling *Micromeria* species).
– Soqotri simplified: *hewririhim*; *shka'hinitin* (*di midhowdha'*); *dheyt*; *sowgihiyn*; *dhayaya'.*

Food (livestock). Goats eat it if they can reach it and when they are hungry. The spines can damage their mouths and, if the scratches become infected, can cause them to go off their food and lose condition.
Bees / honey. Honeybees are attracted to the flowers.

Leucas urticaefolia **R.Br.**

Erect annual herb with little-branched stems to 50cm. Leaves ovate to elliptic, <6cm long, crenate. Flowers many, in dense axillary whorls; bracts c.0.8cm long, linear-lanceolate, 3-veined. Calyx 10-lobed, c.1cm long, mouth oblique, lower 3 lobes drawn forward.
Ecology. No habitat details available; on adjacent mainland it is a field weed or typically in wadi beds.
Distribution. Tropical Africa, Yemen, Oman, India.
Botanical notes. Recorded from Hadiboh and Qalansiyah in the 19th century but not seen since.

Leucas virgata **Balf.f.**

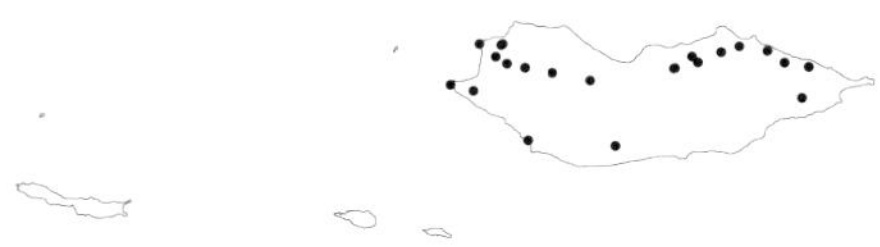

Spindly, much-branched shrub to 1m. Leaves elliptic to orbicular, 0.5–1 x 0.3–0.7cm, entire, densely short grey hairy below. Flowers solitary or few, in the axils of leaf-like bracts, spreading. Calyx 10-lobed, c.0.5cm long. Corolla 1–1.5cm long, softly hairy.
Ecology. Widespread and abundant in open shrubland and drought-deciduous woodland over much of the limestone plateaus. 100–700m.
Distribution. Endemic.
Conservation status. Least Concern (LC). Common in several vegetation types and under no present or perceived threat.
Botanical notes. One of the commonest shrubs on the island.

ETHNOGRAPHIC NOTES†
Name(s)
– Soqotri: *ḥariˈṭrət* 'weak'; *nɔ̄wḥel* 'chaff, winnowings'; *nɔ̄mhel* 'little ants' (from the tiny fruit); *śfeˈṭəhʌn*; *śeˈṭəher* 'tears at clothing'; *śeˈṭəhin* 'steep-sided gully'; *nˈoi*, *tˈamərhen* (both widespread names for sweet-smelling herbs); (rare) *di ḥɔ̄wləˈnitin*.
– Soqotri simplified: *haritrit*; *nowhel*, *nowmhel*; *shfet'han*; *shetiher*; *shetihin*; *noy*, *tamirhen*; *di howlinitin*.
Food (human). When tea reached the island, some islanders began to use the dried leaves as a tea substitute or to stretch limited tea supplies. They dye the hot water red-brown just like tea. More recently the leaves have been used to flavour tea, and the islanders say that this plant is the *za'tar* tea of the island (this is a common herb tea of the mainland which includes wild marjoram). Some take the leaves in boiling water as a tisane. The best flavour is obtained from leaves picked green and partially sun-dried. Branches are gathered and hung up to dry, and then leaves picked off and used or crushed to a powder which is kept in a watertight container. Many islanders agree that browning the sugar before adding it to the kettle of tea further improves the colour and flavour.
Food (livestock). The green aerial parts are eaten by cattle and goats, but not with any great enthusiasm, and only if there is little other forage. Animals eat more in the dry season, once the plant has begun to dry out and when they are hungrier. Sheep eat it both green and dry, and in the areas where it is widespread it is regarded as important forage for them.
Bees / honey. An important plant for honeybees.
Fuel. Dead stems are gathered and used as tinder and kindling. They give off a pleasant smell when burned.
Leatherwork. The green plant is sometimes ground and added to the paste used to strengthen skins and hides in leatherwork. This softens the mixture and makes it easier to work, as well as giving the finished leather a pleasant smell. A paste of the crushed plant is also used to re-treat a dried-out and stiff, or ill-smelling, skin.
Medicinal / veterinary. Sprigs are crushed and sniffed by someone with heartburn or indigestion. The leaves are crushed and sniffed long and hard by someone who fears falling ill, or when approaching a sick person or animal. This is also done to blot out an unpleasant smell or 'noxious air' which is seen as contributing to disease, for instance when passing by a dead animal. More recently, some people have begun to drink a tisane of the leaves to treat a variety of stomach problems, and the plant in general is nowadays regarded as something of a cure-all.
Bedding, stuffing. Sprigs are gathered and scattered over the sleeping area, especially during the monsoon months. They are laid amongst bedding to perfume it and are added to the stuffing for bolsters and pillows. It is particularly used in bedding for the sick as its sweet smell is believed to relieve their discomfort and hasten their recovery.
Cleansing. Flowering stems are tucked amongst the clothing and pinched to release the sweet smell. The sniffing of sweet-smelling plants is held to protect from infection and disease. Sprigs are also rubbed between the hands and then over the skin, hair and clothing to perfume it.
Commercial value (past, present). The islanders now gather it as a gift, or to sell in the Gulf to émigré Soqotrans and friends on the mainland. It was these people who originally suggested to them the medicinal qualities of the plant and its usefulness as a drink. On their return visits to the island they often ask for it to be made into a tisane for them.
Other. The skin of a dead calf, or of a male killed to conserve its mother's milk, is made into a tulchan stuffed with this plant. It is then taken to the bereaved cow, some of her milk is sprayed over it, and then it is presented to her to sniff. This is to encourage her to let down her milk. A paste of crushed plants is also smeared over the back of a living calf to disguise its smell before presenting it to a strange cow. Some of her milk is sprayed over the calf, which is then led around for her to sniff to encourage her to accept it as her own and allow it to suckle.

†Uses also apply to *L. penduliflora*.

Micromeria Benth.

Woody herbs or subshrubs. Leaves simple. Flowers pedicellate, clustered along the stem, bracts leaf-like, caducous. Calyx 5-lobed, actinomorphic, lobes spreading. Corolla 2-lipped, upper lip glabrous. Style entire at tip. 1 endemic. 70 species. Canary Is. and Mediterranean to the Himalayas and SW China.

Micromeria remota Balf.f.

Micromeria microphylla Benth.
Micromeria microphylla var. *imbricata* Balf.f.
Micromeria microphylla var. *remota* Balf.f.
Satureja remota (Balf.f.) Vierh.

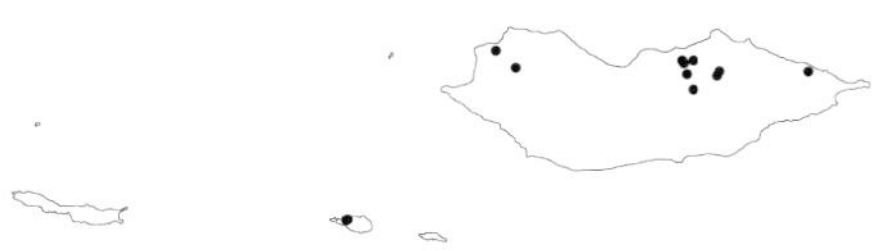

Prostrate woody herb or subshrub to 40cm. Leaves ovate, <0.5 x 0.3cm, entire. Calyx c.0.5cm long. Corolla pink, c.1cm long.

Ecology. Open rocky slopes in montane woodland on granite; also in high areas on limestone in the east and west of the island; on Samha on north-facing limestone escarpment. 400–1550m.

Distribution. Endemic.

Conservation status. Least Concern (LC). Common in several vegetation types and under no present or perceived threat.

ETHNOGRAPHIC NOTES

Name(s)

– Soqotri: *chayˈɛyɛᶜ*, *sˈədʌk* (*di jˈəfun*) 'lavender-like plant for the tulchan'; 'kind of *sˈədʌk*'; *ṣˈeyhimˈɛro*; *tˈomhor*, *tˈamˈerhen* 'little dates' (a widespread name for sweet-smelling plants); *ləṭēnˈuh* 'unmoving'; *məstˈəwrəher*; *gˈɛləs* (Samha).

– Soqotri simplified: *chayaya'*, *sidak* (*di jifun*); 'kind of *sidak*'; *zeyhimero*; *tomhor*, *tamerhen*; *liteynuh*; *mistewriher*; *gelis*.

Food (livestock). In the dry season, after the grasses have died back, this is regarded as useful forage feed, for cattle especially. It is also browsed by goats, which chew at the leafless twigs. All livestock eat the plant litter when hungry.

Bees / honey. Valuable forage for honeybees.

Fuel. Dead stems are gathered and stored to use as tinder and kindling. They give off a pleasant smell when burned.

Leatherwork. The sweet-smelling leaves and flowering twigs are sometimes ground and added to the plant paste used to strengthen skins and hides in leatherwork. This softens the mixture and makes it easier to work, as well as giving the finished leather a pleasant smell. A paste of the crushed plant is also used to re-treat a dried-out and stiff, or ill-smelling, skin.

Bedding, stuffing. Stems are strewn over the floor of the sleeping area to perfume it. The aromatic foliage is also added to the stuffing of bolsters and pillows. It is particularly used in bedding for the sick as its sweet smell is believed to relieve their discomfort and hasten their recovery.

Cleansing. Flowering stems are tucked amongst the clothing and pinched to release their sweet smell. The sniffing of sweet-smelling plants is held to protect against infection and disease. All sweet-smelling plants are endowed with beneficial medicinal and prophylactic qualities. Sprigs are also rubbed between the hands and then over the skin, hair and clothing to perfume it. This is done especially to disguise the smell of hunters setting off to track down wild goats.

Other. The skin of a dead calf, or of a male killed to conserve its mother's milk, is made into a tulchan stuffed with this plant. It is then taken to the bereaved cow, some of her milk is sprayed over it, and then it is presented to her to sniff to encourage her to let down her milk. A paste of crushed plants is also smeared over the back of a living calf to disguise its smell before presenting it to a strange cow. Some of her milk is sprayed over the calf, which is then led around for her to sniff to encourage her to accept it as her own and allow it to suckle.

Comparative uses. The leaves of many *Micromeria* species are unpalatable to livestock, but are often used as flavourings. In India *M. biflora*, Indian wild thyme, is used to treat worm-infested wounds in cattle. In Israel a tisane of the leaves of *M. fruticosa* is taken as a general tonic for indigestion, coughs and colds. It is also given for heart disorders and high blood pressure. A poultice is made with the leaf extract and applied locally to wounds. *Satureja* (*Micromeria*) *thymbra* is used in much the same way, but a compress of the leaves is also applied to swellings and dropsy, and a tisane with added sugar taken as a laxative. In East Africa a decoction of the leaves of *S. punctata* is drunk against swellings on the cheeks and face, boiled leaves being used to mop the swollen area. *S. hortensis* (summer savory) and *S. montana* (winter savory) both produce Oil of Savory. This possesses anti-bacterial and anti-fungal activity, and their seeds yield a drying oil as good as linseed oil. (Ambasta 1982; Dafni *et al.* 1984; Hedberg *et al.* [1982–1986]; Jain & De Fillips 1991; Kunkel 1984; Lewis & Elwin-Lewis 1977)

Ocimum L.

Herbs. Leaves simple. Flowers shortly pedicellate, in axillary whorls, bracts inconspicuous. Calyx 5-lobed, zygomorphic, upper lobes broad, rounded, forming the upper lip of the calyx, lateral and lower lobes bristle-like. Corolla 2-lipped; tube straight. Stamens exserted, spreading beyond upper lip. Style 2-lobed. Strongly aromatic. c.150 species. Worldwide, tropical and temperate, especially rich in Africa.

Ocimum forskolei Benth.

Ocimum canum Sims
Ocimum hadiense Forssk.
Ocimum menthaefolium Hochst.

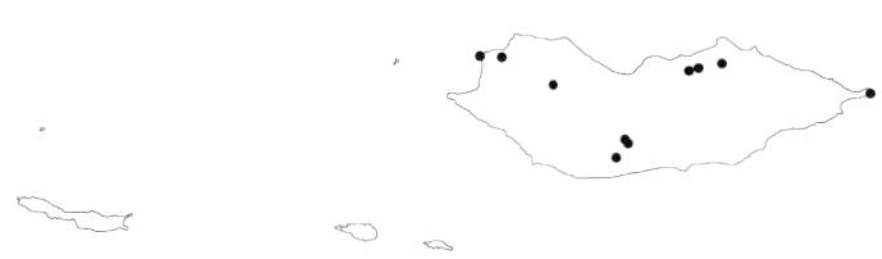

Bushy, woody herb to 50cm. Leaves ovate, 1–3 x 0.7–1.5cm, sparsely dentate. Calyx c.0.5cm long. Corolla pink or cream, c.1cm long.

Ecology. Open woodland, wadi beds, tracksides, date gardens. 200–700m.

Distribution. Yemen, Oman, Saudi Arabia.

ETHNOGRAPHIC NOTES

Name(s)

– Soqotri: *ṣɔ̄wʾhər*, *ṣɔ̄wbəher* 'odoriferous'; *širemhən* 'small plant, herb'; *luʾluʾuh* 'pearls'.

– Soqotri simplified: *zow'hir*, *zowbiher*, *shiremhin*; *lu'lu'uh*.

Food (livestock). Important forage of the lowland grazings. All livestock eat it and it is regarded as a good feed, especially when young and green. Goats and cattle also chew the leafless stems in the dry season. In cows it is seen as promoting lactation, less so in milk goats.

Bees / honey. Honeybees are attracted to the flowers.

Flavouring / food additive. Not used in cooking in any way, though, since the arrival of tea on the island, some people have begun to add the leaves, dried or green, to tea to flavour it and give it a pleasant smell. Some make a tisane from the leaves alone, with added sugar. The milk of animals eating a lot of it develops a distinctive and delicious flavour.

Fuel. Dead stems of larger plants are gathered and used as kindling or tinder. They burn with a pleasant smell.

Leatherwork. The sweet-smelling leaves and flowering twigs are occasionally added to the plant mixture used to treat skins and hides in leatherwork. This softens it and makes it easier to work, as well as giving the finished leather a pleasant smell. The same mixture is also used to re-treat a dried-out and stiff, or ill-smelling, skin.

Medicinal / veterinary. The leaves are crushed and the juice dripped into the ear to relieve earache. If the ear is discharging, it is cleaned first with a salt and water solution before putting in the drops. The leaves are crushed and sniffed long and hard by someone who fears falling ill, or when approaching a sick person or animal. This is also done to blot out an unpleasant smell or 'noxious air' seen to contribute to disease, for instance when passing by a dead animal. More recently, some have begun to drink a tisane made from the leaves to treat various stomach problems.

Bedding, stuffing. Flowering stems are strewn over the floor of the sleeping area to perfume it. The aromatic foliage is also added to the stuffing of bolsters and pillows. It is particularly used in the bedding of the sick as its sweet smell is believed to relieve their discomfort and hasten their recovery.

Cleansing. Sprigs are tucked into the hair or waistband and pinched for their lovely smell. The sniffing of sweet-smelling plants is held to protect against infection and disease. Leafy stems are also rubbed between the hands and then over the skin, hair and clothing to perfume it. This is done especially to disguise the smell of hunters before they set off to track down wild goats. The leaves are rubbed over the hands if they have become tainted with something unpleasant-smelling.

Other. The skin of a dead calf, or of a male killed to conserve its mother's milk, is made into a tulchan stuffed with this plant. It is then taken to the bereaved cow, some of her milk is sprayed over it, and then it is presented to her to sniff to encourage her to let down her milk. A paste of crushed plants is also smeared over the back of a living calf to disguise its smell before presenting it to a strange cow. Some of her milk is sprayed over the calf, which is then led around for her to sniff to encourage her to accept it as her own and allow it to suckle. In the more settled villages of the island, this plant is often cultivated for its pleasant smell and to flavour tea. In the interior it is sometimes planted within the compound walls or beside rainwater catchments for the same reason.

Comparative uses. In Arabic this plant, and other *Ocimum* species, is called *ar-rīḥān*, a name that is mentioned twice in the Qur'an (Sura LV, v. 12 and Sura LVI, v. 89). Throughout the Middle East *Ocimum* species are cultivated or grown as house plants. The leaves are added to tea or drunk as a tisane to relieve an upset stomach, to soothe a fever and for colds. Basil is considered to be stimulant, diuretic and carminative. The juice of leaves is sniffed up to bring on a sneeze to clear the brain, and as a remedy for mild nervous or hysterical disorders. The mucilaginous seeds are highly astringent and are used to treat diarrhoea. They are also useful in urinary disorders such as gonorrhoea and cystitis. The roots are used as a balm. In Oman the ground leaves of *O. basilicum* with *Prosopis cineraria* pods and lemon juice are used as eyedrops to treat cataracts. The seeds crushed in water are taken as a drink for colds and, when boiled in water, to treat diarrhoea. The leaf juice or crushed leaves are placed on wounds to stop bleeding. The plant is widely used as a deodorant, and women mix it to a paste with oil and use it cosmetically. It is also considered to be an aphrodisiac, and men commonly wear some when visiting their women. In Yemen, a twig of basil is wrapped in the shroud before wrapping up the body, and a paste made from the crushed leaves is applied to the hair to make it smell sweet and stop it turning grey. The dried leaves are smoked as a tobacco. In Dhofar the crushed leaves are smelled to ease a headache and dizziness; they are also crushed and placed in the nose for a cold, or in the ear to treat an ear infection. The juice from young leaves is applied around inflamed or sore eyes, and is given as an antipruritic, especially for itchy insect bites. The oil extracted from *O. viride* is used to prevent mosquito bites. In Africa, the volatile oil of *O. gratissimum* has been shown to possess both antimicrobial and anthelmintic properties. Sweet basil, *O. basilicum*, is used as an insect repellent, and the juice of the leaves is used as a nasal douche and for ringworm. In East Africa circumcised girls sleep on a layer of *O. suave* to perfume their bodies, and the plant is also used to treat scorpion stings. (Abdulfatih 1987; Ambasta 1982; Fleurentin & Pelt/Fleurentin *et al.* [1982, 1983a,b,c]; Ghazanfar 1994; Hedberg *et al.* [1982–1986]; Heine *et al.* 1988; Jain & De Fillips 1991; Miller & Morris 1988; Oliver-Bever 1986; Watt & Breyer-Bandwijk 1962)

Orthosiphon Benth.

Herbs or shrubs. Leaves simple. Flowers pedicellate, solitary or in whorls on short axillary or terminal axes, bracts inconspicuous. Calyx 5-lobed, zygomorphic, upper lobes broad, rounded, lateral and lower lobes bristle-like. Corolla 2-lipped. Stamens exserted from corolla tube but shorter than upper corolla lip. Style entire at tip. 1 endemic. c.40 species. Old World tropics.

Orthosiphon ferrugineus Balf.f.

Shrub to 1m. Leaves broadly ovate, 1–2 x 0.5–1.5cm, base truncate, crenate, long petiolate. Flowers solitary or in short racemes, axillary. Calyx weakly 2-lipped, c.0.6cm long. Corolla pink, c.1cm long, tube much longer than calyx.

Ecology. Deciduous shrubland on rocky granite slopes. 550–850m.

Distribution. Endemic.

Conservation status. Vulnerable (VU D2). Rare, with a very limited area of occupancy.

Botanical notes. This rather insignificant shrub is locally common around the Muqadrihon Pass but scarce elsewhere.

ETHNOGRAPHIC NOTES

Name(s)

– Soqotri: nil or *šire̍mhən* 'small plant, herb'; *di ḥɔ̄wlənhi̍tin* (common name for Labiates in the Central Highlands).

– Soqotri simplified: *shiremhen*; *di howlinhitin.*

Food (livestock). Goats and cattle eat the foliage and young stems of the green plant when there is little alternative. Goats and sheep eat the litter in the dry season. It is not regarded as being very valuable nutritionally.

Bees / honey. Honeybees are attracted to the flowers.

Flavouring / food additive. The milk of goats eating a lot becomes rather watery and thin, and also develops a rather unpleasant taste which is not liked.

Orthosiphon pallidus Royle

Woody-based herb to 30cm. Flowers in whorls on short terminal axes. Calyx strongly 2-lipped, upper lobe expanded and forming the upper lip of the calyx (similar to *Ocimum*), c.0.5cm long. Corolla white, c.0.5cm long, tube not much longer than calyx.

Ecology. Rocky limestone slopes. 200–600m.

Distribution. Yemen, Oman, Saudi Arabia.

Botanical notes. Superficially similar to *Endostemon tenuiflorus* but distinguished by the lack of a tuft of white hairs in the throat of the calyx.

ETHNOGRAPHIC NOTES

Name(s)

– Soqotri: nil or *luˀlu̍ˀuh* 'pearls'; *di ḥɔ̄wlənhi̍tin*; *di ᶜa̍žbib* 'quick wilting'; 'kind of *nɔ̄wḥelˀ* (*Leucas virgata*); *ta̍merhen*, *n̍oi* (both widespread names for small, delicate and often fragrant plants).

– Soqotri simplified: *lu'lu'uh*; *di howlinhitin*; *di 'ajbib*; 'kind of *nowhel'*; *tamerhen, noy.*

Food (livestock). This is one of the *ˀa̍žərib* group of plants which flourish after the winter rains. It is readily grazed by cattle before the grasses come up, but is little eaten by goats and sheep, though they eat the fallen leaves in the dry season, and goats chew the leafless twigs in a drought. It is not seen as having any nutritional value.

Bees / honey. Honeybees are attracted to the flowers.

Plectranthus L'Hér.

Subshrubs or herbs. Leaves simple. Flowers pedicellate, whorled, on elongate axes, bracts inconspicuous. Calyx 5-lobed, weakly zygomorphic, upper lobe rounded, lateral and lower lobes triangular. Corolla 2-lipped, tube bent down at mouth of calyx, lower lobe boat-shaped, ± enclosing the stamens. Aromatic. 1 endemic. c.300 species. Tropical and warm Old World.

Plectranthus socotranus Radcl.-Sm.

Erect, fleshy shrublet or bushy herb to 40cm. Leaves ovate or elliptic, 1.5–2.5 x 0.75–1.5cm, crenate. Corolla lilac to purple, 0.8–1.5cm long.

Ecology. Open rocky slopes and cliffs in semi-deciduous woodland. 450–1100m.

Distribution. Endemic.

Conservation status. Least Concern (LC). Common in several vegetation types and under no present or perceived threat.

Botanical notes. The only succulent-leaved labiate on Soqotra. The leaves when crushed are strongly aromatic.

ETHNOGRAPHIC NOTES

Name(s)

– Soqotri: *źʌ̍fərʌᶜ* (pl. *źʌfɛ̍raᶜ*, *źʌfəriᶜhi̍tin*); *mṭʌ̍mhir* 'odoriferous'.

– Soqotri simplified: *dhafira'* (pl. *dhifeyra'*, *dhafiri'hitin*); *mtamhir.*

Food (livestock). Remains green throughout the dry season, but is little liked by livestock. Goats and cattle only eat a little of the drier material when they are very hungry. Camels and donkeys sometimes eat it, and sheep eat the dried and fallen leaves.
Bees / honey. Honeybees are attracted to the flowers.
Medicinal / veterinary. Sprigs are crushed and sniffed deeply by someone suffering from heartburn or indigestion.
Bedding, stuffing. Sprigs are occasionally scattered amongst bedding or added to stuffing of bolsters and pillows, but this is less common than with other aromatic labiates, as the smell is less liked.
Cleansing. The leaves are crushed and rubbed over the hands and clothing to disguise an unpleasant smell. The succulent leaves are also used to wash with.
Pest control. The leaves are used to discourage fleas by rubbing crushed leaves over household equipment such as bedding, sheep's wool rugs, floor mats and so on. This lessens the numbers of fleas but does not get rid of them. Sprigs are also placed amongst bedding and clothing to deter fleas and other vermin. A paste of the leaves is rubbed over the coats of goats, kids and young calves to deter mites, fleas and lice.
Other. If animals feed on this they can go longer without water.

Notes. The islanders distinguish between two varieties: one they call *źʌfərʌʿ ḳiḳəḥe*, 'small-leaved', and the other *źʌfərʌʿ ṣaʿalhʌl*, 'large-leaved'. Both have the same very strong smell, reminiscent of 'Vick' ointment. Thirsty and hungry goats that eat more than a little, in the absence of other green forage, can develop scour.
Comparative uses. The tubers of some species are used as a vegetable. In Oman the milk of livestock browsing on *P. barbatus* is sweet and rich with a distinctive odour. In Yemen the leaves of *P. cylindraceus* are used to treat digestive disorders. In Saudi Arabia the leaf extract of *P. tenuiflorus* is used to treat ear infections. The leaves of *P. asirensis* are applied to wounds as an antiseptic, and, dried and powdered, are used as a baby powder to treat nappy rash and itching. In East Africa, a decoction of the roots of *P. barbatus* is used against stomach pains, including delayed delivery in labour, and to treat infertility in women. The leaves are pounded and drunk against rheumatism, or added to water and the filtrate drunk for a cough and tonsillitis. In India *P. rugosa* is used to repel fleas, and in Simla the best honey is believed to come from areas where these plants are plentiful. (Abdulfatih 1987; Ambasta 1982; Fleurentin & Pelt/Fleurentin *et al.* [1982, 1983a,b,c]; Hedberg *et al.* [1982–1986]; Heine *et al.* 1988; Jain & De Fillips 1991; Kunkel 1984; Miller & Morris 1988)

Teucrium L.

Woody herbs. Leaves simple. Flowers in subcapitate, terminal clusters, sessile, bracts inconspicuous. Calyx 5-lobed, actinomorphic, lobes equal. Corolla white, upper lip absent. 2 endemics. c.100 species. Worldwide, especially Mediterranean.

Teucrium balfourii Vierh.

Teucrium prostratum Balf.f.

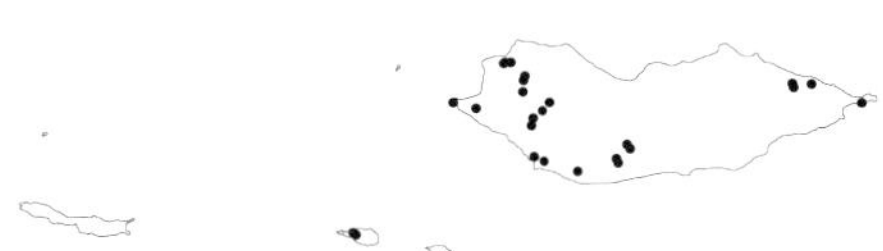

Mat-forming woody herb to 30cm. Upper stems densely white hairy. Leaves linear, c.0.5cm long, margin inrolled, completely obscuring lower surface. Flowers in clusters <1.5cm diameter. Calyx c.0.4cm long. Corolla c.0.5cm long.
Ecology. Widespread and locally common on open rocky slopes in a variety of vegetation types. 100–800m.
Distribution. Endemic.
Conservation status. Least Concern (LC). Common in several vegetation types and under no present or perceived threat.
Botanical notes. This strongly aromatic plant often forms extensive and dense mats over rocks. See comments under *T. sokotranum*.

ETHNOGRAPHIC NOTES

Name(s)
– Soqotri: *gīrbeb, dīrbeb* (pl. *gīrbub, dīrbub*) (also given to *Pulicaria stephanocarpa*); *di kibiher* 'little ball of wool'; *di ḳʌṣkʌṣ* 'cream-coloured'; *riḥemimhin* 'grey-coloured'; *mēlḥo* 'salty'; *cheyēyaʾ* (more commonly given to aromatic *Micromeria* species); *šēdə* (more commonly given to *Ruta chalapensis*); *səgenhen* 'bowed' (Samha); *ḥariṭrət* 'weak' (more commonly given to *Leucas virgata*).
– Soqotri simplified: *giyrbeb, diyrbeb* (pl. *giyrbub, diyrbub*); *di kibiher*; *di qazqaz*; *rihemimhin*; *meylho*; *cheyaya'*; *sheydih*; *sigenhen*; *haritrit.*

Food (livestock). Important fodder for small stock especially. It is common in the drier areas of the island where it is one of the key fodders. It makes goats and sheep fat and sleek, and, when green and in flower, promotes lactation. Cattle also eat it but are less enthusiastic than sheep and goats.
Bees / honey. Important honeybee forage.
Leatherwork. The sweet-smelling leaves and flowering twigs are occasionally added to the usual leatherwork plant mixtures to soften and perfume them. The same mixture is also used to re-treat a dried-out and stiff, or ill-smelling, skin.
Pest control. After shearing, women put sprigs among their wool to discourage insects.
Comparative uses. Many related species are used as flavouring and to make a tisane. In Oman the leaves of *T. mascatense* are commonly sold in herbal shops. They are boiled in water or rose-water for a long time and the solution cooled and taken as a drink for stomach ache and colic. The dried leaves are burned and the smoke placed under bed covers and inhaled to treat a fever. The leaves are boiled in water with leaves of *Rhazya stricta*, *Fagonia indica*, myrrh, and sea salt and black salt (potassium chloride), and the warm tea taken for abdominal colic, fever and diabetes. The water in which leaves have been boiled is drunk to reduce blood flow during menstruation. In Oman too, the fresh or dried leaves of *T. polium* are boiled in water and taken as a drink or dripped onto bites and abscesses. The leaves are also used to treat malarial fever and jaundice, and dried, are smoked as a treatment for rheumatism. In Bahrain the leaves of the same plant are boiled and drunk for infections of the large intestine and for malaria, and the whole plant is used as an analgesic, antibiotic, purgative and anti-pyretic. In Israel a tisane made from the leaves is given as a bitter drink to adults to treat stomach aches; children are given a little of the drink, or are fumigated in a steam bath with the leaves added. The

crushed leaves are applied to open wounds to aid healing. A tisane from the leaves is given to treat diabetes, and a steam bath and/or a tisane of the leaves is used to treat colds and fever. In Egypt it is used as an appetiser and expectorant; a hot infusion of the tender parts is taken for stomach troubles and the plant is used to treat colds, smallpox and itch. An infusion is given as a vermifuge, stimulant, tonic, vulnerary and to treat female sterility. (Abbas *et al.* 1992; Ambasta 1982; Batanouny 1981; Dafni *et al.* 1984; Ghazanfar 1994; Jain & De Fillips 1991; Kunkel 1984; Rizk & El-Ghazaly 1995)

Teucrium sokotranum **Vierh.**
Teucrium petiolare Balf.f.
Teucrium petiolare var. *pubescens* Balf.f.

Straggling or hanging woody herb to 40cm. Upper stems sparsely grey hairy. Leaves elliptic, 0.8–1.5 x 0.5–1cm, crenate, pale beneath. Flowers in clusters <2.5cm diameter. Calyx c.0.7cm long. Corolla <1cm long.
Ecology. Locally common on open rocky slopes in the Haggeher; also scattered on limestone in the northeast of Soqotra. 330–1500m.
Distribution. Endemic.
Conservation status. Least Concern (LC). Common in several vegetation types and under no present or perceived threat.

Botanical notes. Some collections from the limestone in the northeast of Soqotra approach *T. balfourii*: they have similar whitish stems and the smaller flowers and leaves of *T. balfourii*. The status of these plants deserves further investigation. It was recorded from the western limestone plateau in the 19th century but has not been seen there recently.

ETHNOGRAPHIC NOTES
Name(s)
– Soqotri: *lǝbǝhān*, *lǝbǝnhǝn* 'white, pale'; *di ḳʌṣkʌṣ* 'cream-coloured'; *nifǝzitin* 'full-uddered'; *ṣǝfiririhʌn* 'many-flowered'; *ṣʌfṣʌf* 'rippling'.
– Soqotri simplified: *libihan*, *libinhan*; *di qazqaz*; *nifizitin*; *zifiririhan*; *zafzaf.*
Food (livestock). Eaten by all livestock when green, especially small stock, but only if there is little alternative. It is not important for livestock and is seen as having limited nutritional value.
Bees / honey. Honeybees are attracted to the flowers.

LEGUMINOSAE

Herbs, shrubs or trees, sometimes climbers. Leaves alternate, usually pinnate or 3-foliolate, sometimes simple or digitately 3-foliolate; stipules usually present. Sepals 5, united. Petals 5, free or partially united. Fertile stamens (3 to)10 or many. Ovary superior, consisting of a single cell with 1 to several ovules inserted on the adaxial (lower) suture. Fruit a pod dehiscing along the upper suture or indehiscent and breaking into segments. c.650 genera. 17,500 species. Cosmopolitan.

Botanical notes. The family is treated here in three subfamilies. The subfamilies are clearly defined on Soqotra and provide a useful way of splitting up this large family.

Leguminosae Juss. subfam. Caesalpinioideae

Unarmed herbs, shrubs or trees. Leaves paripinnate. Flowers solitary or in racemes, partially zygomorphic. Petals ± similar. Stamens 10 (or fewer). 150 genera. 2500 species. Pantropical.

Botanical notes. The species on Soqotra are apparently parasitic on a wide range of species.

Chamaecrista Moench

Herbs. Leaves with scale-like gland between each pair of leaflets. Flowers in racemes; pedicels with bracteoles. Fertile stamens 5. Pods linear, oblong, flat, elastically dehiscent with twisting valves. c.250 species. Pantropical, especially diverse in the Neotropics.

Chamaecrista absus **(L.) Irwin & Barneby**
Cassia absus (L.) Irwin & Barneby

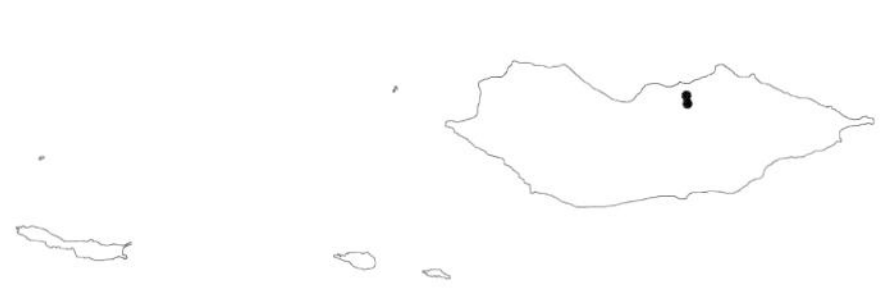

Erect or spreading herb to 1m. Leaves 4-foliolate; leaflets elliptic to obovate, 0.8–2 x 0.5–1.5cm. Petals yellow, c.0.5cm long. Pods with spreading hairs, 4–6 x 0.6–0.8cm. Seeds brown to black, glossy, c.0.5 x 0.4cm.
Ecology. Clearings in drought-deciduous woodland. 150–550m.

Distribution. Widespread in the Old World tropics.

ETHNOGRAPHIC NOTES
Name(s)
– Soqotri: *ḥabbat al ʿayn* 'the seed for the eye'.
– Soqotri simplified: *habbat al 'ayn.*
Food (livestock). Provides incidental grazing for livestock, but is not important forage.
Medicinal / veterinary. As its name implies, this plant is used principally to treat eye problems, particularly the condition known as *di ʿeyḥʌntin* in which the eyes are inflamed and painful. This problem is especially widespread during the monsoon months when the air is filled

with wind-blown dirt and dust. The seeds are removed from the small pods, and cracked between the teeth to enable removal of the black, outer skin. The kernel is then ground to a powder and put in the affected eye(s). The same medicament is used after carrying out any operation on the eye to soothe the pain and prevent infection. Some add hot water and use the solution as eyedrops, but most agree that it is more effective as a dry powder. This treatment is also given to livestock with sore eyes.

Senna Mill.

Herbs or shrubs. Leaves sometimes with conspicuous glands. Flowers in racemes; pedicels without bracteoles. Petals yellow. Fertile stamens 10. Pods indehiscent or dehiscing with difficulty. 1 endemic. c.240 species. Pantropical, especially diverse in the Neotropics.

Senna holosericea (Fresen.) Greuter

Cassia holosericea Fresen.

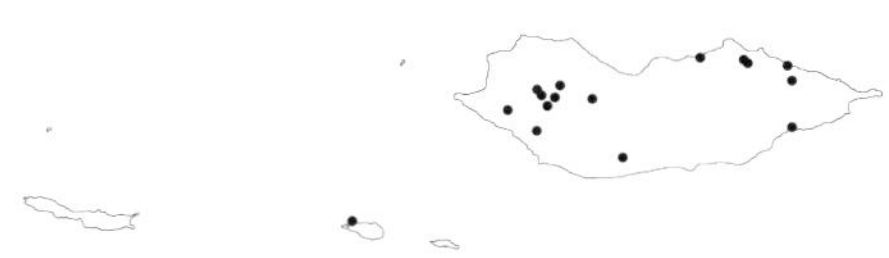

Spreading or prostrate herb or small shrub. Branches densely hairy. Leaves 8 to 16-foliolate; stipules triangular, sometimes subulate, c.0.5cm long; leaflets oblong-elliptic, 1–2.5 x 0.5–1cm, densely hairy. Petals 0.4–0.8cm long. Pods oblong, flattened, c.3 x 1.5cm, crest absent. Seeds c.0.7 x 0.4cm.

Ecology. Widespread and sometimes abundant in *Croton socotranus* shrubland and succulent shrubland. Sea-level–500m.

Distribution. Sudan to Somalia, Arabia to Pakistan (Sind).

Botanical notes. Reported to be increasing on the Noged plain.

ETHNOGRAPHIC NOTES†

Name(s)

– Soqotri: *ferˈiro, fɛ̄rrə; ṭˈifher* 'finger-, toenail'; *ḥəwēr, ḥəwɔ̄r* 'dark-coloured' (Samha); *sˈenneh mˈəkkī* (from Arabic) 'Mecca senna'.

– Soqotri simplified: *feriro, feyrrih; tifher; hiweyr, hiwor; senneh mikkiy.*

Food (livestock). Sheep and goats from drier areas can become accustomed to it as a dry season feed, eating the leaves, buds and flowers, both fresh and fallen to the ground. The pods are rarely eaten fresh, but once they have dried out, they are eaten by small stock, especially the large pods of *S. socotrana*. Hungry cows occasionally chew on the stems and eat the leaves, and camels and donkeys also eat the plants when there is little else. While regarded as being marginally better forage than *Tephrosia* plants, which are very bitter, these plants can have a laxative effect if a lot are eaten. This weakens the animal and causes it to lose weight, which is dangerous when so many dry months still lie ahead. Even so it is said that goats used to this as forage are well able to survive the dry season on this diet if other dry season forage is also available. Incoming livestock, and livestock in areas where the plants are uncommon, generally avoid them.

Bees / honey. Seen as good honeybee forage in the drier areas, especially in the summer months.

Flavouring / food additive. The carcass of a goat eating a lot of this plant makes little fat, and the water the meat is cooked in is noticeably darker than usual. However, the taste of the meat is not affected. The milk of goats and sheep eating any quantity produces a lot of froth.

Medicinal / veterinary. The leaves and stems are crushed to a paste and applied around swellings, boils and abscesses to draw pus. It is also smeared on the udder in cases of mastitis. For spots and rashes, measles for example, the root is roasted, ground to a paste and applied to the affected areas. The dried seeds are gathered when black and fully ripe and stored in the home. When needed, they are ground with a little water and smeared over the chest to treat chest pain and a chronic cough. The paste adheres to the skin and dries to make a hard coating like sticking plaster. To treat problems of the liver and the yellow eyes of jaundice, the pods with their seeds are added to water and brought to the boil, four or five times. The liquid is left to cool and the filtrate then drunk. This is also used as a laxative to clean out the stomach and relieve stomach pain or bloating. The leaves are gathered, dried, and then ground to a powder, which is added to water (with a tamarind pod and sugar if available) and boiled. The cooled water is then drunk on an empty stomach. If the results are too drastic the patient is 'cooled' by washing in cool water and is given a drink made by squeezing ripe tamarind fruits in water with added sugar or honey to soothe the stomach. Many islanders recommend taking this drink regularly, ideally fortnightly, as a general tonic to cleanse for the whole digestive system. Sometimes the fruit are added to the leaves, which is said to make the purging action more gentle, and sometimes the seeds alone are used. Some believe the plant should only be treated with appropriate ceremony if it is to be used medicinally: it should be washed seven times and passed between the fingers while pronouncing the *shihādah* ('There is no God but Allah'). This is done, it is said, in case the plant had been growing in polluted ground, such as where a corpse is buried.

Other. *Senna* plants are commonly dug up and removed from date palm plantations as they are said to attract a grub (unidentified) which, when young, lives on the roots. As it grows, it transfers its attention to the roots of the young date palms, and over time can destroy them.

Notes. There is a certain overlap between *Senna* and *Tephrosia* plants. Islanders say that the smaller plants with small fruit and leaves are the 'female' and the larger plants with larger fruit and leaves are the 'male' ones.

Comparative uses. Many *Senna* species are toxic, though not fatal, to sheep and other livestock. Senna is widely valued for its cathartic properties, but human deaths have also been reported from overdoses of herbal remedies containing senna material. It is especially useful in habitual constipation, as it increases the peristaltic movements of the colon. The pods have the same therapeutic action as the leaves, but cause less griping. The young leaves of *S. holosericea* are eaten in soups, and the seeds are also edible, but have a laxative action. In Dhofar the leaves are gathered and boiled in water, and the cooled filtrate drunk for a variety of stomach problems. The leaves are also crushed and used as a cleaning material. The seeds of this and other *Senna* species are also used as a coffee substitute. In Yemen the leaves are given to treat nervous problems, along with other ingredients. (Dalziel 1948, 1956; Fleurentin & Pelt/Fleurentin *et al.* [1982, 1983a,b,c]; Kunkel 1984; Lewis & Elwin-Lewis 1977; Miller & Morris 1988)

†Also apply to *S. socotrana*.

Senna italica **Mill.**
Cassia italica (Mill.) Lam. ex Steud.
Cassia obovata Collad.

Erect to prostrate herb or small shrub. Leaves 8 to 12-foliolate; stipules triangular, <1cm long; leaflets obovate-elliptic, 1–4 x 0.5–2cm; glands absent. Sepals usually blackish or brownish. Petals 0.8–2cm long. Pods oblong, flattened, 3–6 x 1.3–2cm, with crest running along middle. Seeds similar to *S. holosericea*.
Ecology. No habitat details available.
Distribution. Widespread in Africa, India and Pakistan.
Botanical notes. Recorded from Hadiboh in the 19th century; not seen recently.

ETHNOGRAPHIC NOTES
Comparative uses. In Yemen the leaves are ground to a paste with added salt, and placed on wounds or fractures to bring out trapped blood. The shoots are also used to soothe coughs. In Dhofar the seeds are eaten, and in northern Oman the powdered leaves and seeds are mixed with thyme, lemon and sugar, boiled in water, and the filtrate taken orally once a week to treat chronic constipation. In Saudi Arabia the leaves are boiled in water with added lemon and rose water (no sugar) and drunk to treat stomach cramps and constipation. A decoction of the leaves and seeds is taken as a laxative and vermifuge. Elsewhere in the Middle East the crushed seeds are used in ophthalmic diseases. In East Africa the plant is boiled and drunk with milk or mutton soup for malaria and various diseases of the blood. (Abbas *et al.* 1992; Abdulfatih 1987; Al-Dubaie 1993/Al-Dubaie & Al Khulaidi; Batanouny 1981; Fleurentin & Pelt/Fleurentin *et al.* [1982, 1983a,b,c]; Ghazanfar 1994; Goodin & Northington 1979; Heine *et al.* 1988; Miller & Morris 1988; Rizk & El-Ghazaly 1995)

Senna obtusifolia **(L.) Irwin & Barneby**
Cassia tora sensu Balf.f. non L.

Erect herb or shrub to 2.5m. Leaves 6-foliolate; stipules linear; leaflets obovate, obtuse, 1.5–6 x 1–4cm; conspicuous, cylindrical glands present between some pairs of leaflets. Flowers 1 or 2 in leaf axils. Pods linear, ± round in section, 13–23 x c.0.5cm, dehiscent. Seeds many, brown, c.0.5 x 0.3cm.
Ecology. Disturbed ground near habitation.
Distribution. Pantropical weed.
Botanical notes. Not seen recently; possibly under-recorded.

ETHNOGRAPHIC NOTES
Comparative uses. The leaves are eaten in South America. (Kunkel 1984)

Senna socotrana **(Serrato) Lock**
Cassia socotrana Serrato
Cassia sophera sensu Balf.f. non L.
Senna sophera sensu Balf.f. non (L.) Roxb.

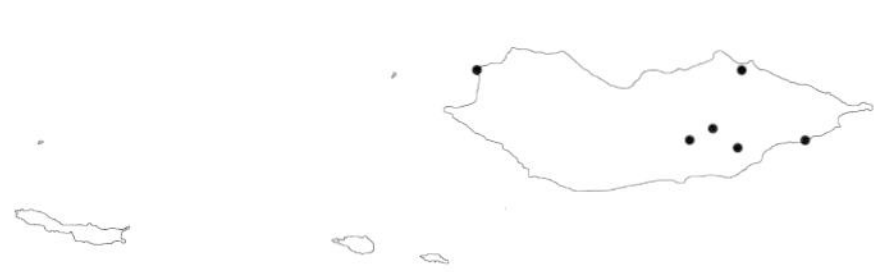

Erect herb or shrub. Leaves c.10-foliolate; stipules caducous; leaflets lanceolate, 2–3 x 0.5–1cm, acute; conspicuous glands present, positioned near base of petiole. Pods linear, c.6cm long, ± round in section.
Ecology. Disturbed ground by tracks and wadi beds. Sea-level–500m.
Distribution. Endemic.
Conservation status. Least Concern (LC). Common in several vegetation types and under no present or perceived threat.

ETHNOGRAPHIC NOTES
Name(s)
– Soqotri: *di ḳerəhān* 'horned' (from the podded fruits); *ṭifher* 'finger-, toenail'; *ḥahər* 'dark-coloured'; *senneh məkkī* (from Arabic) 'Mecca senna'.
- Soqotri simplified: *di qerihan*; *tifher*; *hahir*; *senneh mikkiy*.
Food (livestock). Hungry camels browse the plants if there is little alternative. The dried leaves and pods are eaten by sheep. Towards the end of the dry season goats eat the flowers and fallen leaves, and finally chew at the dried stalks. If the plant is widespread in an area, local goats can become accustomed to it as a dry season feed. However, it is very bitter and incoming goats are reluctant to eat it.
Bees / honey. Seen as a good honeybee forage in drier areas in the summer months.
Medicinal / veterinary. When there is no *S. holosericea* the foliage is sometimes used instead to move the bowels, but is not seen as being as effective.

Tamarindus L.

Trees. Flowers in racemes, young buds enclosed by caducous bracteoles. Sepals 4. Petals reddish-brown, 3 large and 2 minute, setiform. Fertile stamens 3, filaments united. Pods oblong, thick and pulpy, edible. 1 species.

Tamarindus indica L.

Large, evergreen tree to 25m. Leaves 20 to 40-foliolate; leaflets narrowly oblong, <c.3 x 1cm. Large petals 1–1.5cm long. Pods 7–14 x 2–3cm. Seeds chestnut brown, 1–2 x 1–1.2cm.

Ecology. Widespread, usually in or near wadis. 100–600m.

Distribution. Widely planted in tropical and subtropical regions.

Botanical notes. One of the largest trees on Soqotra.

ETHNOGRAPHIC NOTES

Name(s)

– Soqotri: *ṣu̍bəhur* (pl. *eṣbɛ̄rə*); fruit *ḥe̍bi̍lə* (pl. *ḥo̍bəhel*).
– Soqotri simplified: *zubihur* (pl. *ezbeyrih*); fruit *hebilih* (pl. *hobihel*).

Food (human). The fresh, green leaves and flowers are chewed for their refreshing astringent flavour. The immature fruit are eaten raw and are much liked by everyone, but especially by pregnant women, who can develop a craving for them. The mature fruit are ripened in the sun and then stored in skins or a covered clay pot. They are a popular ingredient in cooking as well as having many medicinal uses, and most families try to keep some in the home. They are used for flavouring sauces and to make a lemony drink. This is made by crushing immature or ripe fruit in water and leaving them to steep, or by cooking them in water for a long time and then leaving it to cool. The fruit were also dried and packed in small skins and taken to the coastal markets to sell. When they have ripened to a red colour and become too hard to crack with the teeth, the seeds are also made into flour They are taken out of the pods and baked in hot ash until the outer skin splits open. The inner kernel is then removed and ground to flour, which is kept in a cloth or leather bag. It is cooked with water or buttermilk to make a sort of porridge, and has a taste reminiscent of finger-millet flour. However, it does not keep well, so only small amounts are made at a time. When ingredients for a sauce for the basic cereal dish are scarce, a pod or two are ground and added to water to make a flavouring or relish. In the more settled coastal villages the ripe pods are traditionally used to make a rich sauce for meat and fish: the ripe, sticky fruit are squeezed in water and left to stand for a while before removing the seeds. The meat or fish is fried in clarified butter with spices, especially turmeric if available, and then the tamarind water is added and the pot left to simmer. It is tasted for salt and then served.

Food (livestock). Small stock eat the dried, fallen foliage, buds and flowers, and the tree is browsed by camels. The green foliage is harmful to livestock unaccustomed to it, as it gives them severe diarrhoea, and can even cause pregnant livestock to abort. Goats attracted by the greenery in the dry summer months have been observed eating a few leaves, but make a face at its bitterness, and turn away. Herders say that if a hungry goat is allowed to eat the fresh, green foliage in any quantity it will lose its appetite altogether and sicken. In a prolonged dry season green, leafy stems are cut for goats, sheep and cattle, but are only fed in very small amounts. It makes a better feed if allowed to dry out first, or is mixed with other dried plant material to make it more appetising. Goats and cattle also chew the leafless stems, and goats nibble the fallen fruit, while cows are quite keen on them. The fallen leaves are gathered and fed to all livestock towards the end of the dry season when other forage is very scarce. In general, the foliage is considered to have little nutritional value, but in the better-watered valleys where the trees are common, and the livestock are used to it, it can be quite an important dry season feed.

Bees / honey. Important honeybee forage. If honey pours easily through the fingers, and has absolutely no trace of bitterness, then it is likely to be from *Tamarindus* or *Ziziphus* flowers, and is of the highest quality.

Flavouring / food additive. The flowers are sucked for their nectar.

Fuel. The dead wood does not make good firewood, as it burns down too quickly.

Timber. Makes a good timber, highly resistant to termites. However, it is rarely cut for this since the living tree is too valuable a resource.

Wood. The dead wood is very hard. Indeed, islanders say that 'once the wood is dead the knife makes no impression'. The dark, purple-brown heartwood is used for a variety of implements where hard wood is necessary, such as wooden locks and keys, combs, the sword beater and shed stick for weaving, spoons, stirrers, small bowls, grain and ghee measures and antimony containers. Tamarind wood is also used to make the tambourine drum (*tarānāt*) (see Introduction). The dead wood makes good charcoal.

Medicinal / veterinary. The sick are given the raw pods, or a drink made from them, as a tonic and restorative. Someone with a high fever is wrapped up in wool blankets and given water in which ripe pods have been soaked. This drink causes the patient to sweat and gives him mild diarrhoea, both of which are said to 'remove all impurities from the body'. If someone feels unwell, but has no specific symptoms, they take the same treatment for the same reason. If a more powerful purgative has been taken (such as aloes juice, *Euphorbia hadramautica* or *Gnidea*) and the patient is suffering from griping pains in the stomach, he is given a drink of warm water in which some ripe pods have been soaked, preferably with added sugar, to soothe the stomach. Colicky pain is treated in the same way. Constipation is treated by taking the same drink, or by eating pods baked in hot ash and then peeled. Someone suffering from a bloated or swollen stomach is treated by taking a drink of warm water in which ripe pods have been squeezed and then left to soak. Tamarind pods are also used to tempt patients with anorexia or the nausea of pregnancy.

Cosmetic. The seeds are baked in hot ash until quite black and dry. They are then crushed to a powder used in place of antimony or *kohl*. Although a little kohl used to be imported, it was too expensive for most islanders.

Commercial value (past, present). The fruit were dried, packed in small skins and taken to the coastal markets for sale. Another method of preparing the fruit for sale was to gather them when green, and cook them in water slowly until quite soft. They were then spread out in the sun to dry, after which the outer skin was peeled off. This dried pulp was also put in skin containers and either sold in the markets or given as gifts. The pods were also exported to the outer islands, mainly for medicinal use. The ripe pods are still offered for sale in the coastal markets.

Other. A pod is chewed to reduce thirst. Unripe pods are used to sour milk: one or two are crushed and added to water with a little salt, and then simmered steadily until the water has evaporated. The remaining pulp is added to a little milk in the butter-skin to start the souring process; this soured milk is then used to ripen fresh milk before churning it to make butter. Alternatively two or three pulverised pods are added to a little milk in a clay pot which is then covered and left in a warm place for some hours. If prepared in the morning, by mid-afternoon the milk will have soured and thickened. Some islanders prefer to place one or two crushed pods in the milk-skin with only a small amount of fresh milk; they then inflate the skin and hang it in a warm place near the fire for the milk to thicken and sour. On the outer islands, where the pods were rare, as they had to be imported, only part of a single pod was crushed and put with a little milk in a small clay pot. The mixture was stirred briskly, then the pot covered and left for up to two days for the milk to sour and thicken.

Young *Tamarindus* saplings are dug up and transplanted close to a settlement, and, more recently, islanders have been experimenting with sowing the seeds. The seedlings are then planted out in the compound area, near a rainwater catchment or in a date palm enclosure. The best soil for cultivating the seedlings is that dug up from under a mature tamarind tree. This rich, black soil is also collected and taken to cultivated areas to enrich kitchen gardens, or used to plant out young date palms. Mature trees offer thick shade, and are a popular meeting place. Milking yards for cattle are often established under a tamarind tree by lighting the smudge to one side of it and herding the cows there every day to be milked until they learn to come of their own accord as soon as they see or smell the smoke.

Notes. Some islanders say that the 'male' tree (*ṣ'ubəhur mi'šer*) produces little or no fruit, but others say that all trees are 'female' and all produce fruit. The trees are generally considered to be 'owned' by people in the locality, who have prior rights to the fruit. Others can harvest fruit only with the express permission of the 'owners'. In the dry season, if there is little else in their stomachs, the fresh green foliage can cause severe scour in hungry and malnourished animals and this can last for up to two weeks until their digestive system adjusts. It can also cause *'uqabah* in small stock unaccustomed to it. Islanders have observed that the fruit are sometimes attacked by caterpillars, one kind eating the pulp and the other the seeds. In the star-calendar of the island, ziziphus and the tamarind are said to symbolise the star *turrīyəh* (the Pleiades), because they come into new leaf at this time even if there is no rain.

Comparative uses. A tamarind tree can reach an age of 200 years, and remain fertile throughout its life. They are slow to grow, and do not produce fruit before the age of 10 or 12 years. The average annual fruit production of a single tree can be as much as 150–200kg. The sugar content of the ripe fruit is 30–40%, but they also contain organic acids (citric, ascorbic, acetic, tartaric), vitamins and minerals, particularly calcium. In India the trees are planted in rows in forests as a fire-break as almost nothing can grow under its dense permanent shade.

Seeds. The seed coat is high in tannin. They yield a fatty oil like peanut oil which is used in paints and varnishes, and as an illuminant. The kernels are used as food in times of scarcity, alone or mixed with cereal flour. The seeds can be roasted, then soaked in cold water and eaten the next day. Kernel powder is used as a sizing material in the textile industry, and its polysaccharides make it an excellent substitute for fruit pectins. It is also used as an adhesive in bookbinding, and in making cardboard and plywood. Medicinally, a thick paste of ground seeds is used as a cast to protect and heal broken bones. In Yemen the seeds are powdered and made into a paste applied to boils and infected sores. The seeds can be used to turn fruit juices into jellies, and they are also eaten fried or roasted.

Fruit pulp. This is used as an auxiliary in dyeing and tanning, and for polishing and cleaning metal-ware. The pulp was used by sailors instead of lime or lemon juice for its anti-scorbutic properties. It is sucked to relieve thirst on a long journey, and a drink made from the fruits is believed to prevent sunstroke and fevers. The boiled pulp accelerates the healing of open sores. The pulp juice is given to treat stomach problems and colds, and is regarded as refrigerant, carminative and laxative. An infusion is given for biliousness and febrile conditions. For smallpox the fruits are boiled in water and drunk. To stop post-partum bleeding the fruit is soaked in cold water, stirrred and drunk – if there are no fruit, pounded leaves can be used instead. In Oman the pods are soaked in water and a solution drunk as a blood cleanser and general tonic: taken first thing in the morning this acts as a mild laxative and vermifuge. In Yemen the flesh of ripe fruits is used to relieve nausea, haemorrhoids and as a febrifuge.

Bark. Bark is used to dye cloth dark yellow or dark brown, and the young stems and bark yield a fibre. Medicinally, a decoction of the bark is given as an astringent in diarrhoea and is used as an eye lotion. It is regarded as a tonic and febrifuge, and is used to treat stock with gastro-enteritis.

Leaves and flowers. The young leaves, flowers and very young pods are slightly bitter and are used to dress rice, fish or meat soups, curries and stews. Both leaves and flowers are used in dyeing. Medicinally, the leaves are boiled, and the lotion used for washing wounds and sores. The fresh leaves are also crushed to make a poultice for inflammatory swellings.

Wood. It is a good strong wood, and termite resistant. It is used for utensils and in building. It is recommended as a substitute for teak and sal to make beams, rafters, purlins and trusses. It is excellent for turning and carving, and for making tool handles and furniture. It is a good fuel and makes excellent charcoal customarily used for gunpowder or for burning in gas generators during World War II. The ashes of the wood are used in tanning goatskins. Twigs are used as chewsticks.

Root. This is used to treat venereal diseases. (Abdulfatih 1987; Ambasta 1982; Baumer 1983; Dalziel 1948, 1956; Fleurentin & Pelt/Fleurentin *et al.* [1982, 1983a,b,c]; Heine *et al.* 1988; le Houerou 1980; Jain & De Fillips 1991; Kunkel 1984; Miller & Morris 1988)

Leguminosae Juss. subfam. Mimosoideae

Trees or shrubs, often spiny. Leaves bipinnate. Flowers clustered in dense spherical heads or spikes, actinomorphic, very small. Petals free or fused below, inconspicuous. Stamens 10 or many, exserted and often conferring a brush-like appearance to the inflorescence. 50 genera. 2800 species. Pantropical, especially dry regions.

Acacia Mill.

Trees or shrubs. Conspicuous, paired, stipular spines usually present. Flowers bisexual, in small, dense, spherical heads. Stamens many. Fruit a dehiscent or indehiscent pod. 2 endemics. 1200 species. Pantropical, especially diverse in Australia.

ETHNOGRAPHIC NOTES

Comparative uses. The gums from various *Acacia* species have emollient and bechic properties. They also have an emulsifying power of which much use is still made in pharmaceutics and in the manufacture of confectionery, glues, finishes and so on. All *Acacia* barks are rich in tannins, and have good wood. Over 80 *Acacia* species provide food through their gum, seeds, pods or bark. *A. tortilis* is an important fodder tree wherever it occurs. During drought, the pods are eaten after removing the seeds, and the gum too is edible. The inner bark is exceptionally strong and is used for cordage. The bark, leaves and pods are all used for tanning leather: they are dried and then wrapped in cloth and submerged in water. The tannins leach out within a couple of days, and the skin is then immersed and dried numerous times. When the dried hide remains flexible the tanning process is regarded as complete. Another method is to shave the bark and then boil and cool it. This solution is poured over the skin or hide, which has been staked out, hair-side underneath. This is done several times over the course of a few days and then the skin is left to dry. The wood is much used in building and to make implements. Medicinally, the gum is used to treat broncho-pulmonary infections and as an analgesic. (Fleurentin & Pelt/Fleurentin *et al.* [1982, 1983a,b,c]; Goodman & Hobbs 1988; Heine *et al.* 1988; Kunkel 1984)

Acacia edgeworthii T.Anderson

Acacia balfourii Woodrow
Acacia socotrana Balf.f.

Spreading shrub to 1.5m. Bark smooth, reddish with prominent lenticels. Spines straight, 1–6cm long. Leaves 2–5cm long, 4 to 12 pairs of pinnae; leaflets linear-oblong, c.1mm long. Flowers white. Pods thick, woody, straight, dehiscent.

Ecology. Coastal sand dunes and gravel hills. Sea-level–20m.

Distribution. Yemen, Oman, Somalia, NE Kenya, Ethiopia.

Botanical notes. Restricted to areas by the sea, along the central north coast of Soqotra. It is reported that within living memory this tree extended along the coast from Howlef east to Ras di Hamri. It was not cut but when the old trees died there were no young ones to replace them.

ETHNOGRAPHIC NOTES

Name(s)

– Soqotri: *ḥarḥéro* (pl. *ḥárḥeyr*) (Abd al Kuri) (also given to *A.* aff. *tortilis*, *A. sarcophylla* and *Dichrostachys dehiscens*).

– Soqotri simplified: *harhero* (pl. *harheyr*).

Food (livestock). One of the richest fodders for goats on Abd al Kuri. They eat the foliage, twigs, buds and flowers, and in particular like the pods, which are regarded as an extremely nutritious feed. Sheep also eat any foliage within reach, and eat the fallen litter. On Soqotra it is also browsed by camels. All parts of the plant are considered to be a valuable feed for all livestock.

Bees / honey. Honeybees are attracted to the flowers.

Fuel. The dead wood makes excellent firewood, giving off great heat and little smoke. It is liked for cooking and for warming and lighting enclosed places. It also makes very good, slow-burning charcoal.

Timber. Good, hardwood timber, but the shrub is usually too small, too uncommon, and too precious as a livestock feed for it to be cut for this.

Wood. Hard and close-grained, and suitable for making a variety of implements, especially if it is fire-hardened first.

Acacia nilotica (L.) Willd. ex Delile

Large tree to 8m. Bark fissured, blackish. Spines straight, <8cm long. Leaves 5–8cm long, 2 to 4 pairs of pinnae; leaflets linear-oblong, 0.3 x 0.1cm. Flowers yellow. Pods woody, straight or slightly curved, sometimes constricted between the seeds, indehiscent.

Ecology. A few trees cultivated in Hadiboh.

Distribution. Yemen, Oman, Saudi Arabia, India, widespread in Africa.

ETHNOGRAPHIC NOTES

Name(s)

– Soqotri: *ṭalḥ* (from Arabic).

– Soqotri simplified: *talh.*

Food (livestock). Goats in the Hadiboh area, where a few of these trees have been planted, are not enthusiastic about the foliage, but are said to eat the pods occasionally when very hungry.

Bees / honey. Honeybees are attracted to the flowers.

Comparative uses. The young pods are eaten as a vegetable, and the seeds roasted as a food or used as flavouring. The bark is ground to make flour in times of famine. The gum is used in confectionery. Medicinally, the gum is fried in oil and given to women after childbirth. It is also taken for general debility and as an aphrodisiac. In Oman the resin is mixed with egg white, and applied to the eyes to treat cataracts. For diabetes the seeds are cleaned, crushed lightly, left in water overnight, and half a glass of the liquid drunk the next morning. Alternatively the seeds are put in fresh milk and two glasses are drunk daily. The leaves are eaten to relieve diarrhoea. They have an anti-protozoal action, are molluscicidal and are also used as an anthelmintic. In Dhofar the leaves are pounded to a paste used on boils and swellings. The pods are burned and the smoke inhaled for colds. (Ghazanfar 1994; Kunkel 1984; Miller & Morris 1988; Oliver-Bever 1986)

Acacia pennivenia Schweinf.

Entada species? Balf.f.

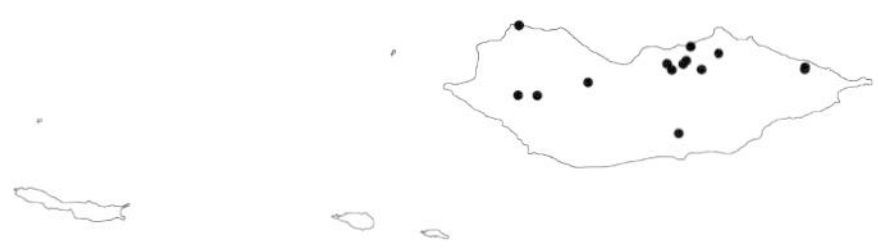

Small tree to 4m. Bark rough, brown. Spines very small or absent. Leaves c.10cm long, 2 or 3 pairs of pinnae; leaflets linear-oblong, c.1 x 0.3cm. Flowers cream or pale yellow. Pods not seen.

Ecology. Widespread in drought-deciduous woodland. 50–650m.

Distribution. Endemic.

Conservation status. Near Threatened (NT). *A. pennivenia*, whilst abundant at present, is lopped as livestock fodder in dry periods. If livestock numbers increase greatly, or a succession of drought years occur, then this species could rapidly become threatened.

Botanical notes. Balfour in his *Botany of Socotra* (1888) has a record of an *Entada* species. No specimen can be traced and the identity of his plant is a mystery. He notes this as a 'A beautiful and graceful tree of which material is too fragmentary to permit identification, [which] is provisionally referred to this genus'. He goes on to say that it has some resemblance to *A. pennivenia* Schweinf. and that the inhabitants give it the same name (*tom'hor*). No species of *Entada* has been recorded from the island and it seems likely that Balfour's plant was in fact *A. pennivenia*.

ETHNOGRAPHIC NOTES

Name(s)

– Soqotri: *tˈomʾhor* (pl. *tˈimirhin*) 'little dates' (also given to *Dichrostachys dehiscens*).

– Soqotri simplified: *tom'hor* (pl. *timirhin*).

Food (livestock). Cows and goats eat the foliage in the dry months. Herders gather leafy branches to feed to weaned kids, for whom it is considered a valuable feed. Camels in general avoid the tree. The lopped leafy stems and the fallen leaves are fed to cattle and goats towards the end of the dry season, when it is regarded as a useful maintenance feed. The new leaves that appear at the very end of the monsoon are much liked by goats, and are seen as being a nutritious fodder. In general, it is more highly valued in the drier areas of the island than in the Central Highlands, where there are many alternative fodders.

Bees / honey. Honeybees are attracted to the flowers.

Flavouring / food additive. The milk of goats eating a lot develops a distinctive flavour, not liked by some. The water in which the carcass is cooked also has a recognisable taste, and the flavour of the viscera is affected.

Fuel. The dead wood makes poor firewood that burns badly and gives off a lot of unpleasant smoke.

Timber. Poor timber, which is very attractive to termites. Sometimes the quality can be improved by cutting the tree at the hottest time of day in the hottest and driest months and then fire-hardening it. Even so, it has a tendency to rot and crumble, and can never be used for roofing beams.

Wood. Poor and little used.

Medicinal / veterinary. Leafy stems are gathered to feed to a sickly camel, but only in small quantities, as eating too much can make it worse. A cow losing weight towards the end of her pregnancy is fed on leafy stems to build up her strength. The leaves are ground to a paste with a little water and this is smeared around the breasts of a woman with mastitis, and she is advised to stop breastfeeding for two days. Sometimes the same treatment is given to a cow or goat.

Notes. Islanders say that the 'female' trees (*tˈomʾhor ṣˈaʿəlhʌl*) have larger leaves and a denser crown than the 'male' (*tˈomʾhor mˈiʾšer*). Tradition says that if this tree and the large *Grewia* species, *Commiphora ornifolia*, *Maerua angolensis* and *Sterculia africana* trees all come into new leaf at the end of the monsoon, then the winter rains will soon follow. If a goat suffering from *'uqabah* is allowed to feed on this, it develops severe diarrhoea which can prove fatal.

Acacia sarcophylla Chiov.

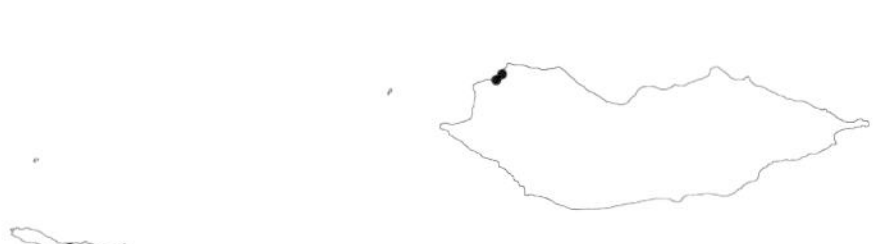

Spreading shrub or small tree to 3m. Bark corky, whitish. Spines straight, <4cm long. Leaves 2–5cm long, 1 or 2 pairs of pinnae; leaflets broadly elliptic, fleshy, c.0.5 x 0.2cm. Flowers cream. Pods straight, constricted between the seeds, dehiscent, softly hairy. The plant has a characteristic smell of garlic.

Ecology. Coastal sand dunes. Sea-level–50m.

Distribution. Somalia.

Botanical notes. Common on the central plains of Abd al Kuri. On Soqotra rare on the coast near Qeysoh. There is also a record from the 19th century from the beach near Mori. On Abd al Kuri it is seen as having steadily increased over the years. Boulos (1995) sinks *A. sarcophylla* under *A. oerfota* var. *brevifolia*. However, the taxon on Soqotra has small leaves with fleshy leaflets and straight, long spines and is quite distinct from the mainland collections he places under both *A. oerfota* var. *brevifolia* and *A. oerfota* var. *oerfota* which have non-fleshy, smaller leaflets, and shorter, usually curved spines.

ETHNOGRAPHIC NOTES

Name(s)

– Soqotri: *ḥˈarḥer*, *ḥˈarḥeyr* (pl. *ḥarˈaḥeyr*) (Soqotra and Abd al Kuri) (also given to *A.* aff. *tortilis*, *A. edgeworthii* and *Dichrostachys dehiscens*).

– Soqotri simplified: *harher*, *harheyr* (pl. *haraher*).

Food (livestock). Considered to be superlative browse

for goats on Abd al Kuri and on Soqotra. On Abd al Kuri it is considered to be equal to *A. edgeworthii* in value, and is one of the richest and best fodders for goats on the island. On Soqotra it used to be widespread along the northern coast, especially in the Hallah di Setmo area, a region well known for goat rearing. Leafy twigs are lopped for goats, and the fallen leaves and pods are also gathered as a very nutritious dry season feed. Sheep eat the fallen leaves and pods, and camels also browse the tree. It is generally considered to be a very valuable fodder tree.

Bees / honey. Honeybees are attracted to the flowers.

Fuel. The dead wood makes excellent firewood that burns slowly and gives off little smoke. It also makes excellent, slow-burning charcoal. On Abd al Kuri it is an important fuel, used especially for cooking. It is regarded as the best firewood of the island: it burns very brightly and also gives out an intense heat. It was also much used here for firing pottery.

Timber. The wood is hard and makes a good timber. However, the plants are too valuable as a livestock forage to be much cut for building.

Cordage. On Abd al Kuri the root-bark was used for all sorts of cordage. It was often used to bind the woven sections of the fish trap together.

Wood. Rarely, the slimmer and straighter branches are cut to make the important herding staff, but in general any cutting of the wood of this valuable fodder tree is frowned on. The stump and roots of a dead tree were used for making quicklime on Abd al Kuri, layered with coral rag. This was used to treat cotton lines and nets to be used when fishing in 'white' or sandy areas (*Limonium sokotranum* was used to dye the lines and nets a red-brown for fishing in 'red' or rocky areas).

Medicinal / veterinary. On Abd al Kuri the charcoal made from the wood is regarded as having medicinal value. It is ground to a powder and used on contusions, swellings, ulcers, abscesses and all sorts of inflammation.

Notes. Islanders say that formerly, within living memory, both *A. edgeworthii* and this species extended in a strip right along the northern coast from Tsirihin to Ra's Damar. It is not known what caused the trees to stop regenerating and die back.

Acacia aff. *tortilis* (Forssk.) Hayne

Small tree to 5m. Bark purplish. Spines of two sorts on the same plant: short, <0.5cm long, recurved; and long, <5cm long, straight, white. Leaves 1–3cm long, 2 to 10 pairs of pinnae; leaflets small, c.0.1cm long, sparsely short-hairy. Flowers cream or pale yellow. Pods contorted, sometimes into loose spirals, slowly dehiscent or indehiscent.

Ecology. Rare on inland gravel plain. c.100m.

Distribution. Yemen, Oman, Saudi Arabia, Somalia, Ethiopia, Israel.

Botanical notes. A description is given for *A. tortilis*, a common plant on the Arabian mainland but doubtfully represented here by one sterile collection, from the Zahr basin (*Miller et al.* 8710). This plant has similar leaves to *A. tortilis* but lacks the characteristic dimorphic spines found on mainland Arabia, instead having uniform, ± straight, spines 0.5–1.5cm long.

ETHNOGRAPHIC NOTES

Name(s)

– Soqotri: *ḥa̍rḥeyr* (pl. *ḥara̍ḥeyr*) (also given to *A. edgeworthii*, *A. sarcophylla* and *Dichrostachys dehiscens*).

– Soqotri simplified: *harheyr* (pl. *haraheyr*).

Food (livestock). Not common, but provides very valuable grazing for goats where it occurs. Goats eat the foliage, twigs, buds and flowers, but especially love the pods, which are regarded as an extremely nutritious food. Sheep eat the fallen pods and plant litter, and camels also browse the trees.

Bees / honey. Honeybees are attracted to the flowers.

Fuel. The dead wood makes excellent firewood. It burns slowly and gives off little smoke, so is excellent for cooking.

Timber. Hard and would be suitable for timber but the rare plants are too valuable as a livestock forage to be cut for timber.

Wood. Rarely, the slimmer and straighter branches are cut to make the important herding staff, but in general any cutting of the wood of this valuable fodder tree is frowned on.

Acacia sp. *A*

Low, apparently heavily grazed, intricately branched shrub with short (<0.4cm) spines; globose galls present, c.0.5cm diameter. Leaves with c.2 or 3 pairs of pinnae; leaflets linear-oblong, c.0.1cm long, glabrous with a thickened midrib.

Ecology. Semi-deciduous woodland on limestone escarpment. 200–350m.

Distribution. Endemic.

Conservation status. Data Deficient (DD). Poorly known; inadequate evidence is available to make an assessment.

Botanical notes. This taxon is represented by a single, sterile, collection (*Miller et al.* 8629) from Dihof in eastern Soqotra. It cannot be assigned with confidence to any of the species known to occur on the island. It bears some resemblance to *Miller et al.* 8710, *A.* aff. *tortilis* (see above), but it differs in habit, the shorter spines and glabrous leaflets. It has been seen on a number of other occasions but never with flower or fruit.

ETHNOGRAPHIC NOTES

Name(s)

– Soqotri: *ḥa̍rḥeyr* (also given to *A.* aff. *tortilis*, *A. edgeworthii*, *A. sarcophylla* and *Dichrostachys dehiscens*); *śərəmə̍no* 'strong, tough, hard' (also sometimes given to *Euclea divinorum*); (rare) *to̍mᵊhor* 'little dates'.

– Soqotri simplified: *harheyr*, *sheremeno*; *tomhor.*

Notes. Uses as *A. edgeworthii* and *A. sarcophylla*.

Dichrostachys (A.DC.) Wight & Arn.

Trees. Spines formed from short lateral twigs (not paired). Flowers in spikes, the lower flowers sterile, pink, the upper fertile, yellow, smaller. Stamens 10. Pods dehiscent, papery or leathery. 1 endemic. 9 species. Old World tropics.

Dichrostachys dehiscens Balf.f.

Small tree to 5m. Bark greyish. Spines <2cm long. Leaves 2.5–3.5cm long, with 3 to 5 pairs of pinnae; leaflets linear-oblong, c.0.2 x 0.1cm. Flower spikes c.1.5–4cm long. Pod 3–4 x 0.5–1cm, straight or slightly curved.
Ecology. Drought-deciduous woodland. 250–600m.
Distribution. Endemic.
Conservation status. Vulnerable (VU D2). Rare, with a very limited area of occupancy.
Botanical notes. Often heavily grazed.

ETHNOGRAPHIC NOTES

Name(s)
– Soqotri: *ḥárḥeyr* (also given to *A.* aff. *tortilis*, *A. edgeworthii* and *A. sarcophylla*); *tómhor* 'little dates'; *śərəmə́no* 'strong, tough, hard' (also sometimes given to *Euclea divinorum*).
– Soqotri simplified: *harheyr*; *tom'hor*; *sheremeno.*
Food (livestock). Considered one of the best browse plants for goats and camels in the drier areas of the island, especially in the valleys leading down to the southern Noged plain. It is fattening and increases milk yield in all livestock. Goats eat the foliage, shoots, twigs, buds and flowers. They also love the pods, which are seen as being especially nutritious. Herders cut leafy twigs to feed to young lambs and recently weaned kids. It is also an important dry season fodder in such areas, and herders cut leafy stems and collect fallen pods and foliage for their animals. In areas where other alternatives are scarce it is also lopped for cattle towards the end of the dry season. The new growth put out at the very end of the monsoon is regarded as a particularly valuable feed for goats, which are generally in poor condition by this time.
Bees / honey. Important honeybee forage in the drier areas of the island.
Fuel. The dead wood makes good firewood that burns slowly, with a bright light and giving off little smoke.
Timber. Inferior timber and, in the drier areas of the island, too valuable as a livestock forage to be cut for timber. However, islanders say that heartwood from the drier, hotter areas of the island is hard and makes adequate timber. However, the tree must be a 'male' one, since its wood is always more close-grained and harder than that of the 'female', and as a precaution the wood must be tested with a knife first. Sometimes the timber is improved by fire-hardening: the cut wood is placed beside or above the fire to dry every last bit of moisture out before use.
Wood. The heartwood is close-grained and hard and, rarely, straighter branches are cut to make the important herding staff. However, in general any cutting of wood from this valuable fodder tree is frowned on.
Medicinal / veterinary. Sickly and parturient livestock are fed on the leafy stems to build up their strength.
Cleansing. The sweet-smelling flowers are sometimes tucked amongst clothing to perfume it, or are worn somewhere on the person for their pleasant smell. Plants that perfume the skin and clothing have an importance beyond that of their sweet smell and are seen as actively promoting good health.
Notes. Islanders say that the 'female' trees (*tóm'hor ṣáᶜalhʌl*) have larger leaves and a denser crown than the 'male' (*tóm'hor mí'šer*). Tradition says that if this tree and the large *Grewia* species, *Commiphora ornifolia*, *Maerua angolensis* and *Sterculia africana* trees all come into new leaf at the end of the monsoon, then the winter rains will soon follow. If a goat suffering from *'uqabah* is allowed to feed on this, it develops severe diarrhoea, which can be fatal.
Comparative uses. The fruits and leaves of some species are edible. *D. cinerea*, in particular the pods, provides important fodder for all livestock. The resin from the root of a young plant is chewed, and the seeds are used in soups. The wood makes good firewood, and is used for cog-wheels, walking sticks, tent-pegs and other implements. The bark yields a fibre, and twigs are used as a toothbrush. Medicinally, the flowers, fruit, bark and branches are used to treat rheumatism, and by women during pregnancy as a safeguard against miscarriage. The tender shoots are bruised and used in ophthalmia. The bark is used to treat snakebite, and as an anthelmintic and expectorant. The roots are used for snakebite and scorpion stings, and also as a purgative, expectorant, diuretic, and to treat venereal diseases and carious teeth. The leaves are used for scorpion bites, indigestion and diarrhoea. *D. glomerata* is an efficient insecticide and also molluscicidal. (Ambasta 1982; Dalziel 1948, 1956; Goodin & Northington 1979; Heine *et al.* 1988; le Houerou 1980; Jain & De Fillips 1991; Kunkel 1984; Lewis & Elwin-Lewis 1977; Oliver-Bever 1986)

Prosopis L.

Small trees or shrubs. Paired stipular spines usually present. Flowers in spike-like racemes, all fertile. Stamens 10. Pod indehiscent, woody or leathery with fleshy partitions between the seeds. c.40 species. Mostly New World.

Prosopis juliflora DC.

Straggling, spiny small tree or shrub to 5m. Spines 0.5–5cm long. Leaves with 1 to 3 pairs of pinnae, each with 10 to 20 pairs of leaflets; leaflets linear to linear-oblong, 0.5–2.5 x 0.2–0.5cm, glabrous. Flowers cream, small, c.3mm; spikes c.5–10cm long. Pod 10–30 x 1–2cm, straight or slightly curved, compressed, pale yellow when ripe.

Ecology. Near settlements on coastal plain. Sea-level–50m.

Distribution. Native of the New World.

Botanical notes. Widely used as a hedging plant and for stabilising sand in the Middle East because of its drought tolerance. On mainland Yemen it has become a severe invasive pest in some areas. On Soqotra it has been seen at a few localities around the airport and is now the subject of a programme of eradication.

ETHNOGRAPHIC NOTES

Name(s)

– Soqotri: nil or *ġāf, ḳāf* (from Arabic).

– Soqotri simplified: *ghaf, qaf.*

Notes. A recent introduction with no traditional uses.

Comparative uses. The seeds propagate vigorously after passing through the digestive tracts of animals, and it also produces suckers from stock and roots, making it an important pioneer for arid zones where it can stabilise sand dunes and provide firewood and forage. The leaves are evergreen and are liked by sheep, and donkeys and goats eat the foliage at the end of the dry season. Some varieties with low tannin content are also browsed by cattle. The mature pods produce a farinaceous substance relished by small stock. However, they must be ground before being given to stock otherwise they pass intact through the digestive tract. The pods are eaten raw or ground into a flour by people. The flour keeps for long periods of time and can be consumed either fresh or cooked. The pods are also a source of sugar and are used to make refreshing drinks. The nutritive value of the seeds is close to that of barley grain. The trunk exudes an edible gum. The wood makes reasonably good timber and good charcoal. (Baumer 1983; Kunkel 1984)

Leguminosae Juss. subfam. Papilionoideae

Herbs or shrubs, rarely trees or climbers. Unarmed (rarely armed). Leaves simple, 3-foliolate or imparipinnate, rarely paripinnate with only 2 leaflets. Flowers in racemes or solitary, zygomorphic. Petals arranged into the characteristic pea flower – consisting of 1 outermost standard petal, 2 lateral wing petals and 2 innermost petals joined at the lower edge to form a keel. Stamens 10, variously united. 440 genera. 12,000 species. Cosmopolitan.

Alysicarpus Desv.

Herbs. Leaves simple; stipules scarious, sheathing. Flowers in terminal racemes. Calyx scarious, appearing 4-lobed. Corolla red. Pods transversely segmented; segments indehiscent. 25 species. Old World tropics.

Botanical notes. The scarious stipules and simple leaves of *Alysicarpus* give them a superficial resemblance to species of *Polygala.*

Alysicarpus monilifer (L.) DC.

Alysicarpus glumaceus A.G.Mill. non (Vahl) DC.

Similar to *A. vaginalis* but differs in its appressed-pubescent stems and pods, which are constricted between the segments.

Ecology. Grassland clearings in semi-deciduous woodland on granite. c.650m.

Distribution. Sudan, Ethiopia, Somalia, Pakistan and India.

ETHNOGRAPHIC NOTES†

Name(s)

– Soqotri: nil or *məsḥɔ̄wbəb* 'spreading'.

– Soqotri simplified: *mis'howbib.*

Food (livestock). Liked by all livestock and is regarded as a very nutritious feed. When it comes up after rains it is quickly grazed down, especially by cattle and goats.

Notes. Young goats and goats unaccustomed to it can develop severe scour if they eat a lot in the early rains.

Comparative uses. The fodder value of *Alysicarpus* species is high, and some are cultivated as a pasture or fodder crop. Some species have been recommended for the improvement of grasslands, since they regenerate easily from the seeds that fall to the ground. *A. vaginalis* is cut after fruiting and stored as hay. It is also used as a soil improver. However, it is said to cause mucous diarrhoea if given in excess in a young or immature state during rains. The seeds of this and *A. rugosus* are eaten as emergency food. (Ambasta 1982; Bebawi & Neugebohrn 1991; Bogdan 1977; Dalziel 1948, 1956; Jain & De Fillips 1991; Kunkel 1984)

†Also apply to *A. vaginalis.*

Alysicarpus vaginalis DC.

Prostrate annual herb. Stems glabrescent. Leaves narrowly ovate or obovate, <6 x 2cm. Calyx linear, c.0.4cm long, not exceeding the first segment of the pod. Corolla red or mauve, c.0.5cm long. Pods linear, 1.5–2cm long, not constricted between the 4 to 7 segments, slightly wrinkled.
Ecology. Semi-deciduous woodland. 450–900m.
Distribution. Widespread in the Old World tropics.

ETHNOGRAPHIC NOTES
Name(s)
See *A. monilifer* (also uses).

Astragalus L.

Herbs. Leaves pinnate; leaflets entire. Flowers in axillary racemes. Calyx 5-lobed, 2-lipped. Corolla white. Pod curved, longitudinally divided by a deep groove. 2500 species. Temperate regions, especially diverse in W and C Asia.

Astragalus hauarensis Boiss.

Erect, softly hairy herb. Leaves 5 to 9-foliolate; leaflets ovate to elliptic, 0.5–1.5 x 0.3–0.6cm. Pods 1–4cm long.
Ecology. Sandy coastal plain; only recorded from Abd al Kuri. 20m.
Distribution. Somalia.

ETHNOGRAPHIC NOTES
Name(s)
– Soqotri: nil or *ruʿud* 'grazing plant' (Abd al Kuri).
– Soqotri simplified: *ru'ud.*
Food (livestock). Provides incidental grazing for the sheep and goats of Abd al Kuri, but is not common enough to be important forage.

Canavalia DC.

Climbers. Stems twining. Leaves 3-foliolate; leaflets entire. Flowers in axillary racemes. Calyx 5-lobed, 2-lipped. Corolla mauve. Pod narrowly oblong, dehiscent. 50 species. Tropical, mainly New World.

Canavalia africana Dunn
Canavalia ensiformis Balf.f. non (L.) DC.
Canavalia virosa Balf.f. non (Roxb.) Wight & Arn.

Robust, climbing perennial to 5m. Leaflets ovate, 5–15 x 3–12cm. Calyx c.12mm long. Corolla 3–3.5cm. Pod 10–20 x c.3cm. Seeds 1.5–2cm.
Ecology. Date gardens along wadis. c.500m.
Distribution. Tropical Africa, Yemen and India.
Botanical notes. Robust climber blanketing shrubs and small trees in date gardens; found along the main wadis draining south from the Haggeher.

ETHNOGRAPHIC NOTES
Name(s)
– Soqotri: *ʿərʿero* 'ripple, eddy'.
– Soqotri simplified: *'ir'ero.*
Food (human). The seeds are cooked and eaten in times of hunger, but they are considered a rather shameful and 'improper' food so are usually eaten in private.
Food (livestock). It remains green throughout the monsoon months, and so is an important dry season fodder for cattle especially, though goats also eat it. Cattle eat the pods, which are seen as a good and nutritious feed. Sheep are less enthusiastic, but eat fallen leaves and pods when very hungry. The foliage alone is not considered to have great nutritional value, but when mixed with dry grasses of this season is seen as being a valuable feed, maintaining weight and lactation. Where the plant is in plentiful supply, at the end of the winter, the foliage is gathered along with grasses and other herbs and dried to make hay.
Medicinal / veterinary. The seeds are crushed and eaten, or are ground and added to water, left to steep, and the filtrate drunk as a laxative. This remedy has to be taken on an empty stomach. The foliage is gathered and fed to a cow with a retained placenta or an animal with colic. It is also fed to a cow to give her new strength if labour is prolonged and the animal visibly tiring.
Comparative uses. *Canavalia*, originating from South and Central America, has been grown from pre-historic times for its edible seeds, and is now cultivated throughout the tropics for its seeds, as a cover or green manure crop, or for fodder. It retains its leaves while fruiting and its deeply penetrating and widely branching roots loosen the deeper layers of soil. The seeds can be toxic when eaten by humans or fed to cattle in large quantities, and the young shoots of some varieties are said to be indigestible and even poisonous. The very young pods of the related *C. ensiformis* are eaten as snap beans, and the young shoots are also eaten. The seeds are roasted as a coffee substitute, and the seed of the bean is a source of urease, an enzyme used as a specific reagent for urea in biological chemistry. The pods are a popular rattle and at one time in Angola the seeds were a trivial form of currency. The green pods and beans of *C. gladiata* are also extensively used as vegetable, and both of these varieties are cultivated for their edible seed and as a green manure and fodder. (Ambasta 1982; Bogdan 1977; Jain & De Fillips 1991; Kunkel 1984; Oliver-Bever 1986)

Chapmannia Torr. & A.Gray

Arthrocarpum Balf.f.

Shrubs or small trees. Leaves imparipinnate; leaflets entire, minutely reddish-reticulate beneath. Flowers yellow, sessile in spikes. Calyx 2 to 5-lobed, 2-lipped; upper lip 3 or 4-lobed, lower entire. Corolla yellow. Stamens 10, five long alternating with five short, filaments united into a tube below. Pods transversely segmented, flat; segments 1-seeded, indehiscent. 4 endemics. 7 species.

Botanical notes. A genus with a remarkably disjunct geographical distribution: 2 species in the New World (one in Florida and one in C America and Venezuela) and 5 species in the Old World (one in Somalia and four on Soqotra). This form of distribution with a disjunction between the Horn of Africa and the New World can be explained by a land bridge between N America and Eurasia during the early Tertiary. The species all share a distinctive reddish reticulation on the undersurface of the dried leaves (Thulin 1999b).

Chapmannia gracilis (Balf.f.) Thulin

Arthrocarpum gracile Balf.f.

Shrub or small tree to 5(–9)m. Bark grey, deeply fissured. Leaves 3 to 7-foliolate; leaflets oblong-obovate, 1–3 x 0.6–1.2cm, glabrous above, moderately to densely covered in appressed hairs beneath. Spike 1 or 2-flowered. Calyx c.0.8–1cm, upper lip 3-lobed. Corolla c.1cm long. Pod <2.5 x 1cm; 1 to 4 segments with appressed hairs.

Ecology. Widespread and common in semi-deciduous woodland on both the limestone and granite. 200–800m.

Distribution. Endemic.

Conservation status. Near Threatened (NT). Whilst abundant at present, *C. gracilis* is an important fodder, the trees are lopped as livestock fodder in dry periods and the wood makes one of the best timbers on the island. If livestock numbers increase greatly, or a succession of drought years occur, then this species could rapidly become threatened.

Botanical notes. The most variable species (particularly as regards leaf shape) of *Chapmannia* on the island. Readily distinguished from all other species by its one or two-flowered inflorescences.

ETHNOGRAPHIC NOTES†

Name(s)

– Soqotri: *ṭʌləhon* (*C. reghidensisis* also called *ṭʌləhon ʿarabi*); (in some areas) *ḥaymʾhər*, *ḥimihər* (pl. *ḥomihir*, *ḥəmʌhʌr*) (western Soqotra) *khaymə her*, etc.; *hamirmiro*.
– Soqotri simplified: *talihon* (*talihon ‘arabi*); *haym’hir*, *himihir* (pl. *homihir*, *himahar*); (smaller shrubs) *hamirmiro.*

Food (human). The tender new leaves are nibbled as a fresh-tasting treat.

Food (livestock). This is one of the most important fodders for livestock on Soqotra, like *Zygocarpum coeruleum*. Animals that eat a lot become sleek and fat, give plenty of rich milk, and make a good-tasting and rich carcass well laced with fat. It is especially important for all livestock during the drier months. At this time both cattle and goats also chew at the leafless twigs. As the leaves are good for milk production, leafy stems are often cut and fed to milk animals failing to thrive. The leaves are also considered the best feed for young weaned kids, and, along with *Ziziphus*, to be an optimal feed for camels. Sheep eat any leaves within reach, and also eat fallen foliage and other plant litter. This fodder is so liked by livestock that leafy stems are used as a lure or bribe to control recalcitrant animals – to encourage them into a pen or enclosure, or to let down their milk, for instance. In the the dry season the trees and shrubs are lopped for pregnant and lactating animals, and then, if the dry season is a hard one and prolonged, the twigs and dried foliage are gathered for all favoured animals.

Flavouring / food additive. The milk of livestock eating a lot is particularly rich and sweet, and is renowned for producing much butter.

Bees / honey. Important bee forage, one of the best after *Ziziphus* and *Tamarindus*. The honey is said to be one of the best of the island.

Fuel. The dead wood makes one of the best firewoods of the island. It lights easily, burns slowly and makes little smoke, especially if the bark is removed first, making it ideal for cooking, and warming enclosed spaces. It gives off a good heat and is much used for warming the house, and young or sickly livestock in bad weather. A store is kept in the home for use in wet or cold weather, especially in the high mountain areas, which become enveloped in drizzle and cloud during the monsoon and winter months. The wood burns very brightly, making it ideal for night fires, and burning sticks are used as a torch. A chunk of *ḥamirmiro* wood is used to keep fires in, by tucking a big chunk of the wood into a heap of hot ash and embers. When the fire is needed again, the herder pulls it out, taps it firmly on the fire-stones, and then blows on it to re-ignite the flames. When these shrubs and trees were very common, their wood was gathered, and sold or bartered as firewood in the market towns of the coast.

Timber. The wood is hard and is considered to make one of the best timbers of the island, equivalent in its hardness to that of *Buxanthus pedicellatus*. It is very termite resistant, but, as with all timber, it is best cut in hot, dry weather in order not to attract wood-boring insects. The trunks and larger branches are used for the main roofing beams and supporting pillars. Timber from this species, like that of *Buxanthus pedicellatus*, *Ziziphus* and *Maerua angolensis*, lasts several generations and can be re-used again and again, as long as the wood is never allowed to get wet, and is regularly fumigated with smoke.

Leatherwork. The dead wood burns down to a very fine ash which is very good for preparing skins and hides.

Wood. Close-grained and hard, and much used for making implements. It is considered to be as good as *Buxanthus pedicellatus* wood for making a variety of the more important implements and utensils which need hard wood, such as wooden locks and keys, grain and ghee measures, pack saddles, combs, the sword beater and shed stick for weaving, and small bowls. The slimmer, straight branches are used to make the important herding staff, and the wood is widely used for poles and straining posts in fencing, and in the construction of pens, byres and temporary accommodation. It also makes the highest grade of charcoal, and this was highly valued by the island blacksmiths. That of *ḥímiher* was preferred to that made from *ḥamirmíro*, as its heat was said to be easier to control, and its coals quicker to douse when necessary.

Medicinal / veterinary. The fodder of choice for sick animals, especially small stock and camels. It is also fed to an animal with a retained placenta, or to one in poor condition or which has lost weight. A cow losing weight towards the end of her pregnancy is brought leafy branches to build up her strength. Like other much-liked fodders, leafy stems are offered to a milk animal reluctant to let down her milk; as she starts to chew the leaves she is likely to relax and allow herself to be milked. To treat someone suffering from painful joints, a fire of this wood is lit at the mouth of a small cave or other enclosed space and the patient sat inside. It is kept burning until the patient is covered in sweat, which is said to improve the condition.

Pest control. The dead wood is used for livestock smudges in cold weather, either outside in the pen or milking yard, or, if the weather is wet, inside the byres. It is mainly used to warm and soothe the animals, rather than to deter the many biting insects that flourish after the rains.

Other. When there are no cereals available, the leaves can be used to clarify butter when making ghee, though in this case the residue is not usually eaten. It is the wood of choice for firing pottery. In the days before metal fish-hooks were readily available, fish-hooks of all sizes were whittled from the strong heartwood. Before going into wooded areas in the winter months, a place and time when the botfly was most prevalent, the face of a child was rubbed with powdered charcoal made from the wood. It was believed that by painting the children's faces in this way they were made invisible to the botfly.

Notes. There is considerable overlap between the traditional uses of these species and *Zygocarpum coeruleum*.

†Also apply to *C. reghidensis*, *C. sericea* and *C. tinireana*.

Chapmannia reghidensis **Thulin & McKean**

Similar to *C. gracilis* but distinguished by the narrow, linear-oblong to narrowly oblanceolate leaflets <4 x 0.4cm and 10 to 20-flowered spikes.

Ecology. Locally common in semi-deciduous woodland in the Muqadrihon Pass and Reyged plateau. 450–550m.

Distribution. Endemic.

Conservation status. Endangered (EN B2 a biii). A tree with a very restricted distribution, known only from woodland at one location where it occupies an area of considerably less than 500km^2. Exploitation of this woodland is predicted to increase in the next few years, leading to a decrease in the quality of its habitat.

Botanical notes. Distinguished from all other species of *Chapmannia* on Soqotra by its narrow leaflets. It could, perhaps, be confused with narrow-leaved forms *C. gracilis* (with which it sometimes grows) but can be separated from that species by its many-flowered inflorescence and the general absence of indumentum on its stem, leaves and calyx.

ETHNOGRAPHIC NOTES

Name(s)

See *C. gracilis* (also uses).

Chapmannia sericea **Thulin & McKean**

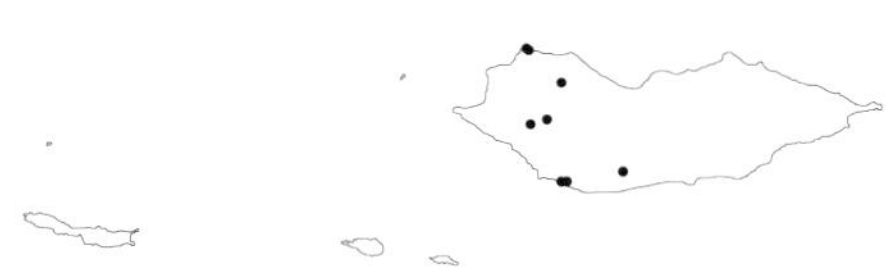

Shrub or small tree to 4m. Bark fissured. Leaves 5 to 11-foliolate; leaflets narrowly oblanceolate to linear-oblong, 0.5–1.5 x 0.1–0.3cm, silvery green, conspicuously silky hairy. Spike 10 to 20-flowered. Calyx 0.5–0.8cm long, upper lip 4-lobed, silky hairy. Corolla 1–1.2cm long. Pod <2 x 0.7cm; 1 to 5 segments, silky hairy.

Ecology. Widespread and locally common on rocky limestone slopes and along wadis in the west of Soqotra. 5–700m.

Distribution. Endemic.

Conservation status. Near Threatened (NT). Whilst widespread and under no immediate threat at present, *C. sericea* is an important fodder; the trees are lopped as livestock fodder in dry periods and the wood makes one of the best timbers on the island. If livestock numbers increase greatly, or a succession of drought years occur, then this tree could rapidly become threatened.

Botanical notes. There is considerable variation in the number of leaflets, sometimes even on the same branch, and there is also regional variation in the conspicuous silky indumentum. Plants from NW Soqotra have tubercle-based hairs on the axis of the inflorescence whilst these are absent on plants from SW Soqotra.

ETHNOGRAPHIC NOTES

Name(s)

See *C. gracilis* (also uses).

Chapmannia tinireana **Thulin**

Small tree to 3m. Leaves (3 to)5-foliolate; leaflets elliptic, 0.7–2 x 0.4–1.1cm, glabrous above with appressed hairs beneath. Spikes <10-flowered. Calyx c.0.8cm long, upper lip 4-lobed, densely covered in tubercle-based hairs. Corolla c.1cm long. Pod <2 x 0.8cm; 1 or 2 segments with appressed hairs.

Ecology. Rare, known only from a few trees, in semi-deciduous woodland (with *Punica protopunica* and *Croton* species) on granite slopes in the western Haggeher. 1100m.

Distribution. Endemic.

Conservation status. Endangered (EN B2 a biii). A tree with a very restricted distribution, known only from woodland at one location where it occupies an area of considerably less than 500km^2. Exploitation of this woodland is predicted to increase in the next few years, leading to a decrease in the quality of its habitat.

Botanical notes. Easily recognised by its three to five elliptic leaflets which are glabrous above and by the c.10-flowered inflorescences and viscid hairs on the bracts and calyces.

ETHNOGRAPHIC NOTES

Name(s)

See *C. gracilis* (also uses).

Crotalaria L.

Priotropis Wight & Arn.

Herbs, sometimes trees. Leaves simple or digitately 3-foliolate, all the leaflets arising from the same point; leaflets entire. Flowers yellow, in terminal or axillary racemes, keel strongly hooked. Pods inflated. 2 endemics. c.600 species. Mostly Africa and Madagascar.

ETHNOGRAPHIC NOTES

Comparative uses. *Crotalaria* (rattlebox) are mostly avoided by grazing or browsing animals, apparently because of their bitter or otherwise unpleasant taste, and the presence of toxic glycosides. Some species, such as *C. incana*, *C. retusa*, *C. spectabilis* and *C. verrucosa*, contain the toxic alkaloid monocrotaline, and are toxic to a wide range of livestock. There are no outstanding fodder or pasture plants amongst the 500 or so species, though some are used as cover crops or for green manure. The leaves and seeds of some species are edible; the leaves are often cooked with potash. The seeds of *C. pallida* are used as a coffee substitute. Some *Crotalaria* provide fibre for cordage and in making canvas, and others are used medicinally. (Bogdan 1977; Dalziel 1948, 1956; Heine *et al.* 1988; Kunkel 1984; Lewis & Elwin-Lewis 1977)

Crotalaria leptocarpa **Balf.f.**

Crotalaria dubia Balf.f.

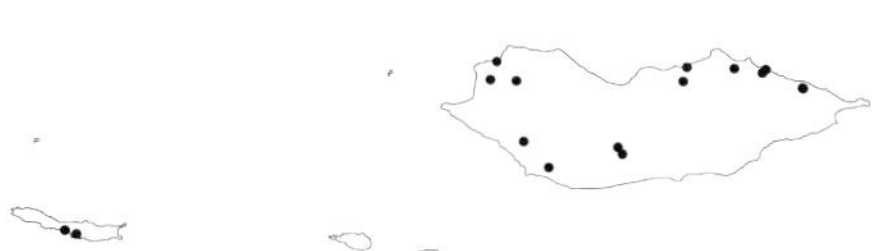

Prostrate, slender, rather wiry herb. Leaves sparse, 3-foliolate, 0.5–3cm long; leaflets linear to narrowly elliptic (sometimes obovate), 0.1–1.5 x 0.1–0.5cm. Flowers 1 to few, axillary. Pods flattened or only slightly inflated, 1–1.5cm long.

Ecology. Widespread and common on the plains and foothills. Sea-level–800m.

Distribution. Ethiopia, Somalia, S Arabia.

ETHNOGRAPHIC NOTES

Name(s)

– Soqotri: nil or *taʿabəhor* 'persisting or coming into new leaf during the monsoon'; *ṣefīrīrihin* 'many-flowered'; *śimērah̩* 'stalky'.

– Soqotri simplified: *ta'abihor*; *zefiririhin*; *shimeyrah.*

Food (livestock). Liked by small stock and regarded as a good and nourishing feed for them, making them fat and healthy. Cattle also browse it, but it is less important forage for them. The plant is more highly valued on Abd al Kuri.

Bees / honey. Honeybees are attracted to the flowers.

Crotalaria pteropoda **Balf.f.**

Prostrate, white-hairy herb. Leaves 3-foliolate, 1.5–2.5cm long; leaflets obovate, 0.7–1 x 0.4–0.8cm, middle one enlarged. Flowers 1 to few, axillary, pedicel conspicuously winged. Calyx enlarged and papery in fruit. Pods strongly inflated, c.1.5cm long.

Ecology. No habitat details available.

Distribution. Somalia, Oman (Dhofar).

Botanical notes. Recorded by Balfour in the 19th century on the hills southwest of Galonsir [Qalansiyah] but not seen since. However, recently found in Oman (Dhofar).

Crotalaria retusa **L.**

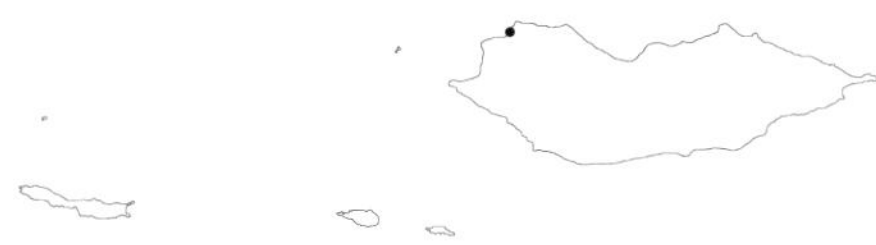

Erect herb. Leaves simple, <5cm long, narrowly oblong-obovate, often with a retuse tip.

Ecology. No habitat details available.
Distribution. Widespread in Africa, tropical Asia, Australia and the Neotropics.
Botanical notes. Recorded by Balfour in the 19th century on the plains near Galonsir [Qalansiyah] but not seen since. A showy species with bright yellow flowers, not easily overlooked. Now possibly extinct on Soqotra.

Crotalaria socotrana **(Balf.f.) Thulin**
Priotropis socotrana Balf.f.

Tree to 5m. Leaves 3-foliolate, 3–7cm long; leaflets oblong-elliptic, 2–4 x 0.7–2cm. Flowers c.10 to 20, in short axillary racemes. Pods slightly inflated, hard, c.1.5cm long.
Ecology. Semi-deciduous woodland. 550–650m.
Distribution. Endemic.
Conservation status. Vulnerable (VU D2). Rare, with a very limited area of occupancy.
Botanical notes. The only tree species within the genus on Soqotra.

ETHNOGRAPHIC NOTES
Name(s)
– Soqotri: *miśḳerer ṣʌˤʌlhil*, *enśḳerer* 'like a large-leaved *Angkalanthus* or *Chorisochora*').
– Soqotri simplified: *mish'qerer za'alhil*, *enshqerer.*
Food (livestock). Goats browse it and it is considered to be quite a good fodder. Cattle sometimes browse the foliage, and sheep and goats chew at the bark in hunger. However, the tree is too uncommon to be important forage for livestock.
Flavouring / food additive. Like other plants with the same name, *Chorisochora minor* and *Chorisochora striata*, the milk of animals eating a lot of this develops a distinctive and unpleasant flavour. These plants are said to be the only ones on the island that make milk practically undrinkable. The taste is still perceptible in the processed milk products, such as buttermilk and fresh butter, but it is considerably less in products processed by cooking. The meat and fat of an animal that has been eating a lot is equally tainted, as are the viscera and the water in which the meat has been cooked, and some people will not touch any of it. Indeed, islanders say that even if a goat has eaten the plant the day before it is slaughtered much of its carcass is quite inedible.

Crotalaria spinosa **Hochst.**

Erect, spiny herb to 50cm. Leaves 3-foliolate, leaflets narrowly wedge-shaped, 0.5–1 x 0.1–0.2cm, notched at apex. Flowers few, on spinose axes. Pods strongly inflated, c.1cm long.
Ecology. Clearings in semi-deciduous montane woodland. 900–1000m.
Distribution. Tropical Africa, Arabia.

ETHNOGRAPHIC NOTES
Name(s)
– Soqotri: nil or *ṣʌfərir's ihon* 'many-flowered'; *śkaˤinitin* 'prickly, spiny'; *məsɔ̄ˤiden* 'spiny'; *sigirhen* 'dome-shaped'.
– Soqotri simplified: *zafiririhon*; *shka'initin*; *misow'iden*; *sigirhen.*
Food (livestock). Much liked by all livestock, donkeys and camels included, but is not regarded as a valuable feed, except for camels.
Notes. It can cause *'uqabah* in small stock unaccustomed to this feed.
Comparative uses. In Yemen the new shoots are used as a urinary antiseptic. In East Africa, a twig is placed in the milk container to ensure plentiful butter. (Fleurentin & Pelt/Fleurentin *et al.* [1982, 1983a,b,c]; Heine *et al.* 1988; le Houerou 1980)

Crotalaria strigulosa **Balf.f.**
Crotalaria abdelkuriensis Vierh.
Crotalaria kuriensis Vierh.
Crotalaria schweinfurthii Vierh.

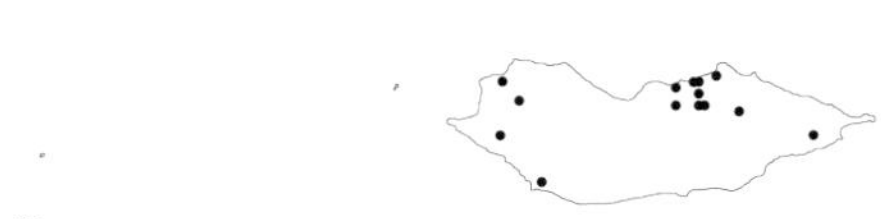

Prostrate herb. Leaves simple, 0.4–2.5cm long, linear-oblong to broadly obovate. Flowers c.5 or 6 in open racemes. Calyx teeth long and papery in fruit. Pod strongly inflated, 0.5–1cm long.

Ecology. Widespread in a variety of vegetation types. 10–1100m.
Distribution. Endemic.
Conservation status. Least Concern (LC). Common in several vegetation types and under no present or perceived threat.
Botanical notes. As recognised here, this is a very variable species found in a variety of habitats. Vierhapper described two further species, *C. schweinfurthii* and *C. kuriensis*, based mainly on leaf characteristics but they are not maintained here because of numerous intermediates. Plants with broader leaves tend to be found in the mountains (*C. schweinfurthii*) and on the outer islands

(*C. kuriensis*) whilst plants with more or less linear leaves are found on the coastal plains.

ETHNOGRAPHIC NOTES

Name(s)

– Soqotri: *ṭʌlḳʌt* (also given to *Polygala* species); *ḳɔ̄wnuh*, *ḳʌʾ di tɛ̄tin* 'little thread', 'thread for sheep'; *məsḥɔ̄wbəb* 'spreading'; *gilʿamihin* 'with rounded fruit'; *sigirhen* 'dome-shaped'; *śimeraḥ*, *śimərahiyun* "stalky'; *noi* 'of mountain areas'; *məriśīʾo* 'undergrowth, understorey'; (rare) *ʾarṭʌh* (more commonly given to *Periploca*).

– Soqotri simplified: *talqat*; *qownuh*, *qa' di tetin*; *mis'howbib*; *gil'amihin*; *sigirhen*; *shimerah*, *shimirahiyun*; *noy*; *mirishiy'oh*; *'artah*.

Food (livestock). Much liked by livestock, especially cattle. It makes them fat and increases their milk yield. It is regarded as a good fodder, but it dies back with the grasses and so is not available in the long dry season. It is highly valued on the outer islands and in the drier areas of Soqotra.

Flavouring / food additive. The milk of a goat that has eaten a lot is very rich and produces a lot of froth.

Bees / honey. Honeybees are attracted to the flowers.

Cullen Medik.

Herbs, conspicuously black-glandular warty. Leaves simple, toothed. Flowers in axillary racemes. Pod small, 1-seeded, indehiscent, glabrous. c.35 species. S Asia to Australia, drier parts of Africa.

Botanical notes. Sometimes treated as a segregate of *Psoralea* L.

Cullen corylifolia (L.) Medik.

Psoralea corylifolia L.

Bushy herb to 1m. Leaves ovate to subcircular, <9 x 7.5cm. Flowers yellowish or purplish, racemes dense with <c.20 flowers. Pod <0.5cm long.

Ecology. No habitat details available.

Distribution. Somalia, Djibouti, S Arabia and east to China.

Botanical notes. Recorded by Balfour about dwellings in the 19th century; not seen recently.

ETHNOGRAPHIC NOTES

Comparative uses. The seeds are eaten, and contain a fixed oil and an essential oil. They are made into a seed cake rich in nitrogen and minerals, and suitable for green manure as well as feed. In Dhofar the seeds are crushed with milk or butter and applied to treat mastitis and painful swellings. Thinned down with water, the paste is applied over the body and forehead to cool a fever, or rubbed over the chest to relieve breathing problems. When crushed, the seeds give out a pleasant perfume and are rubbed over the limbs as a deodorant. In Yemen the seeds are crushed with *Thymus vulgaris* to create a paste applied for skin disorders. They are also used as an aphrodisiac. Elsewhere the fruits are regarded as laxative, diuretic and diaphoretic, and, used internally and externally as a paste, are especially recommended for leucoderma, leprosy, psoriasis and inflammatory diseases of the skin. Pieces of the root are inserted into a painful tooth cavity. (Ambasta 1982; Fleurentin & Pelt/Fleurentin *et al.* [1982, 1983a,b,c]; Ghazanfar 1994; Jain & De Fillips 1991; Kunkel 1984; Lewis & Elwin-Lewis 1977; Miller & Morris 1988)

Desmodium Desv.

Herbs. Leaves 3-foliolate; leaflets entire; stipules scarious. Flowers few in leaf axils, long pedicellate. Pods jointed. c.300 species. Warm regions, especially E Asia, Brazil and Mexico.

Desmodium triflorum (L.) DC.

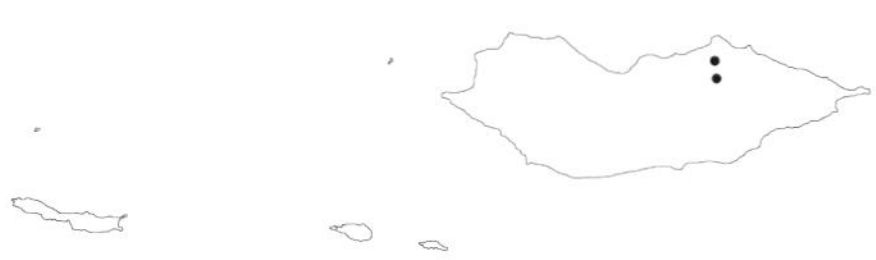

Sprawling, wiry herb. Stems and young leaves with soft, spreading hairs. Leaves 2–3cm long including petiole; leaflets broadly obovate; stipules c.0.3cm long, persistent. Corolla blue. Pods 0.7–1.5cm long, breaking into 2 to 5 segments.

Ecology. Occasional in woodland. 100–950m.

Distribution. Widespread in the tropics.

ETHNOGRAPHIC NOTES

Name(s)

– Soqotri: nil or *ru'ud* 'grazing plant', *ra'adihon* 'small grazing plant'; *ṣʌfəririhon* 'many-flowered'; *mḥaṭḥiṭin* 'limp, flaccid' (more commonly given to species of fern).

– Soqotri simplified: *ru'ud*, *ra'adihon*; *zafirifihon*; *mhat'hitin*.

Food (livestock). Liked by all livestock, especially cattle. It is regarded as a nutritious and fattening feed which also increases milk yield.

Comparative uses. Many *Desmodium* varieties are used for fodder or for green manure. *D. intortum* and *D. unicinatum* have become important pasture legumes in Australia and Africa. A number of species contain appreciable amounts of tannins, which may reduce their palatability, and some species are regarded as pasture weeds. *D. triflorum* may be regarded as replacing *Trifolium* in the tropics and is recommended as a ground cover, but its matted root system prevents soil aeration. It is good for green manure and provides a good fodder. It is easily propagated but the seed is difficult to collect in quantity, and it dies out in the dry season. It is used for lawns and pastures. Medicinally, the leaves of *D. triflorum* are used to treat dysentery, diarrhoea and convulsions, and are also given to promote lactation. (Ambasta 1982; Bogdan 1977; Dalziel 1948, 1956; Jain & De Fillips 1991)

Erythrina L.

Trees. Branches armed with curved prickles. Leaves 3-foliolate; leaflets entire. Flowers large, in many-flowered terminal or axillary racemes. Calyx with brownish, stellate hairs. Corolla large. Pods indehiscent. 108 species. Tropical.

Erythrina melanacantha Taub. ex Harms

Erythrina melanacantha subsp. *somala* (Chiov.) Gillett

Rather bushy tree to 20m. Bark fibrous. Leaflets rhombic, broadly elliptic or suborbicular, <3.5 x 3.5cm, covered with fine, dense, stellate hairs. Corolla red, 4–7.5cm long. Pod oblong, curved, 10–20cm long. Seeds bright yellow.

Ecology. No habitat details available.

Distribution. Somalia, Ethiopia.

Botanical notes. Not seen since the 19th century. It is surprising that this showy species with its large red flowers has not been seen since, and the record must be treated with some caution.

Indigofera L.

Herbs or subshrubs. Hairs medifixed, parallel. Leaves simple, 3-foliolate or pinnate with more than 3 leaflets; leaflets opposite, sometimes alternate. Inflorescence an axillary raceme. Flowers usually bright red, sometimes pink; petals hairy. Anthers apiculate. Pods rounded in section (sometimes flattened), apiculate. 3 endemics. c.700 species. Tropical and subtropical regions.

ETHNOGRAPHIC NOTES

Comparative uses. A number of *Indigofera* species are grazed or browsed, but some are toxic or suspected of being toxic to animals, especially to horses. No *Indigofera* species are of great importance as a source of fodder, though *I. spicata* and *I. hirsuta* are cultivated, mostly in mixtures with fodder or pasture grasses. (Ambasta 1982; Bogdan 1977; Jain & De Fillips 1991)

Indigofera argentea Burm.f.

Indigofera burmanni Boiss.

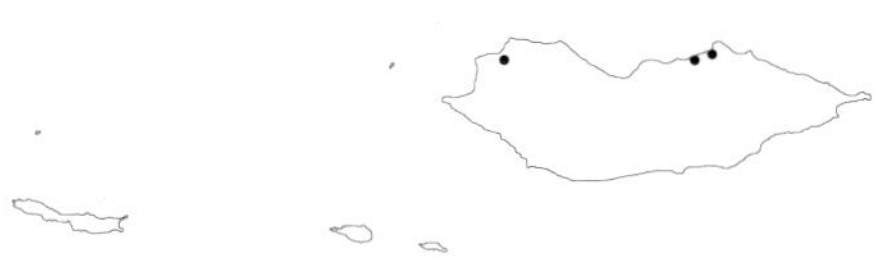

Sprawling, grey-hairy herb to 15cm. Leaves 5 to 7-foliolate; leaflets obovate, c.0.3cm long. Flowers ± solitary. Pods spreading, c.1cm long, with short-stalked glandular hairs. Seeds 2 to 5.

Ecology. Sandy soils overlying limestone on the coastal plains. 10m.

Distribution. Widespread in drier parts of Africa and SW Asia.

Botanical notes. In the 19th century recorded from Hadiboh and Qalansiyah; recently seen only in the *Acacia edgeworthii* woodland at Ras Hawlaf.

ETHNOGRAPHIC NOTES

Name(s)

– Soqotri: nil or *rúʼud* ‘grazing plant’, *ráʼadihon* ‘small grazing plant’.

– Soqotri simplified: *ru'ud, ra'adihon.*

Food (livestock). Provides incidental grazing for hungry livestock, but is not regarded as having any particular nutritional value.

Comparative uses. The roots and leaves of *I. argentea* are given as a bitter tonic, and the seeds as an anthelmintic. The seeds are also used as food in times of scarcity. (Ambasta 1982; Bogdan 1977; Jain & De Fillips 1991)

Indigofera articulata Gouan

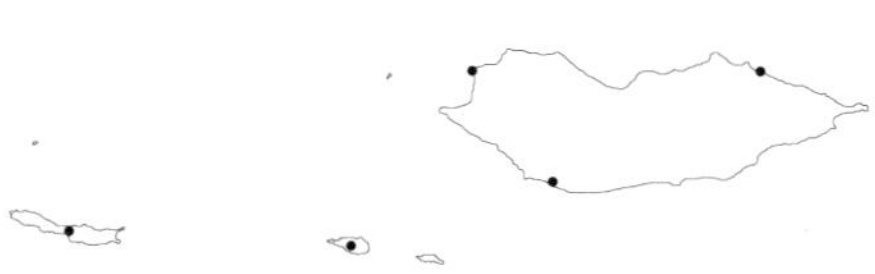

Spreading, softly white-hairy subshrub to 50cm. Leaves 3 to 5-foliolate; leaflets obovate, 0.5–2cm long. Flowers in short racemes. Pod reflexed, constricted between seeds, 0.7–1.2cm long; segments globular, sutures narrow. Seeds 2 to 4.

Ecology. Widespread and sometimes abundant on the coastal plains. Sea-level–50m.

Distribution. Drier parts of tropical and subtropical Africa and Asia.

ETHNOGRAPHIC NOTES

Name(s)

– Soqotri: *dáərfə* ‘itch, irritating’ (Soqotra and Abd al Kuri) (also given to *I. pseudointricata*); *śébərʌḳ* ‘glimpsed’.

– Soqotri simplified: *dirfih*; *shebiraq.*

Food (livestock). Provides dry season grazing for small stock in particular, though camels and donkeys also eat it. It is not important nutritionally, though is more highly valued as a fodder on the outer islands. It is said to become very tough when dry and to ‘pull out the teeth’ of animals tugging at twigs.

Comparative uses. On the Arabian mainland the roots are used as toothbrushes to whiten the teeth, and to soothe painful gums. In Oman they are chewed to relieve toothache. The leaves yield black and blue dyes. The roots and leaves are given as a bitter tonic and the seeds are used as a vermifuge as well as being eaten in times of scarcity. (Abdulfatih 1987; Ambasta 1982; Batanouny 1981; Ghazanfar 1994; Jain & De Fillips 1991; Kunkel 1984; Rizk & El-Ghazaly 1995)

Indigofera coerulea Roxb.

Erect, sparsely hairy subshrub to 50cm. Leaves 5 to 9(to 11)-foliolate; leaflets obovate, 1–2 x 0.5–1.5cm. Flowers many, in dense racemes. Pod reflexed, somewhat constricted between the seeds, 0.5–1.5cm long; segments oblong-elliptic, sutures broad. Seeds 3 to 4.

Botanical notes. Two subspecies of this widespread and variable species are found on Soqotra.

ETHNOGRAPHIC NOTES

Name(s)

– Soqotri: *ḥəwīr* (occasionally with 'l' *ḥəwīl*); *ḥáhər* 'dark-coloured'; *ḥə́wrīr* (pl. *ḥōwrírihin*) 'dark-coloured' (Samha and Soqotra).

– Soqotri simplified: *hiwiyr* (*hiwiyl*); *hahir*; *hewriyr* (pl. *howririhin*).

Food (livestock). Little liked by livestock, though hungry sheep and goats eat it in the drier areas of the island and in the outer islands.

Leatherwork. The plant is added to plant pastes used to treat skins and hides to cure the leather and stain it red-brown. It deepens the colour and makes the leather softer and easier to work.

Medicinal / veterinary. The leaves are crushed to a paste applied around, but not over, a dirty wound, abscess or sore.

Cosmetic. The leaves and stems are chewed to darken the gums, tongue and lips, preferably after chewing a piece of *Euclea divinorum* root. The latter dyes the mouthparts red, and the material from *Indigofera* species deepens and enhances the colour. Some people grind the plant to a paste and smear this over the gums. This is done to strengthen the teeth and gums as well as for cosmetic reasons. It is used mainly by women in this way, though men freely make use of the *Euclea divinorum* chewstick. The leafy stems are crushed with the lichens used as a henna substitute (or even to the henna powder imported from the Arabian mainland) to deepen the colour of the final stain. Children play with the paste and pretend it is henna, practising decorating their nails, hands and feet with it.

Comparative uses. In Dhofar both these plants are important as dye plants for colouring cotton cloth, date palm leaflets and leather. The roasted and ground roots are also used as red-brown dye, mainly for leather, and in the absence of indigo, as a cosmetic, after mixing in oil or ghee. The roots are also used medicinally as an anti-inflammatory, to treat snakebite, the swellings of mumps, and swollen and painful joints. An extract of the leaves and root together is taken as an oxytocic, and a paste of the leaves is applied to the skin to soothe burns and animal bites. (Miller & Morris 1988)

Var. *coerulea* Roxb.

Leaflets more or less glabrous on upper surface.

Ecology. Coastal plains and foothills; *Croton socotranus* shrubland and succulent shrubland. Sea-level–200m.

Distribution. Drier parts of tropical and subtropical Africa and Asia.

Var. *occidentalis* Roxb.

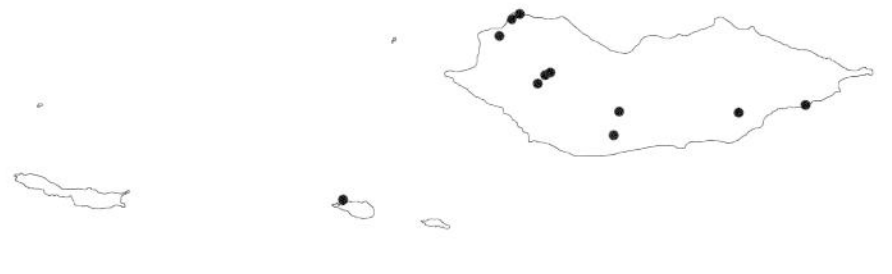

Differs from var. *coerulea* by its leaflets which bear parallel hairs on both surfaces.

Ecology. Dry plains and foothills; *Croton socotranus* shrubland and succulent shrubland. Sea-level–550m.

Distribution. Drier parts of tropical and subtropical Africa and Asia.

Indigofera colutea (Burm.f.) Merr.

Indigofera viscosa Lam.

Sprawling herb to 30cm. Leaves 9 to 11-foliolate; leaflets linear-oblong, <1.5 x 0.5cm. Flowers in short racemes. Pods spreading, c.1.5cm long, with long-stalked glandular hairs. Seeds 8 to 14.

Ecology. Drought-deciduous woodland. 90m.

Distribution. Pantropical.

ETHNOGRAPHIC NOTES

Name(s)

– Soqotri: nil or *šēdə* (pl. *šədɛ́ʔ*) 'calming', 'shared' (more commonly given to *Ruta chalapensis* and *Aerva revoluta* in the Central Highlands).

– Soqotri simplified: *sheydih* (pl. *shideh*).

Food (livestock). Provides incidental grazing for hungry livestock, but is not regarded as important or nutritious forage.

Indigofera cordifolia Roth

Prostrate, whitish herb to 15cm. Leaves simple, ovate, 0.5–1 x 0.3–0.5cm, cordate or obtuse, covered with long shaggy hairs. Flowers in small, ± sessile clusters in the leaf axils. Pods spreading, somewhat constricted between seeds, 0.3–0.5cm long. Seeds (1 or)2.

Ecology. Desertic dwarf shrubland and clearings in *Croton socotranus* shrubland. 10–250m.

Distribution. Widespread in Old World tropics.

Botanical notes. Readily distinguished by its simple, cordate-based leaves.

ETHNOGRAPHIC NOTES

Name(s)

– Soqotri: *məsḥ̄ɔ̄ẁbeb* 'spreading' (also given to *Alysicarpus monilifer*, *A. vaginalis* and *Crotalaria strigulosa*); *ṣɔ̄wbəher* 'sour, astringent'; nil (Abd al Kuri).

– Soqotri simplified: *mis'howbeb*; *zowbiher.*

Food (livestock). Liked by all livestock and regarded as nutritious feed, making animals fat and increasing milk yield. It is especially valued in the drier areas of Soqotra and on Abd al Kuri.

Comparative uses. The white seeds are used like poppy seeds in making bread, but are harmful if eaten on their own. The young shoots and powdered bark are eaten during periods of extreme food scarcity. (Ambasta 1982; Bogdan 1977; Goodin & Northington 1979; Jain & De Fillips 1991)

Indigofera hochstetteri Baker

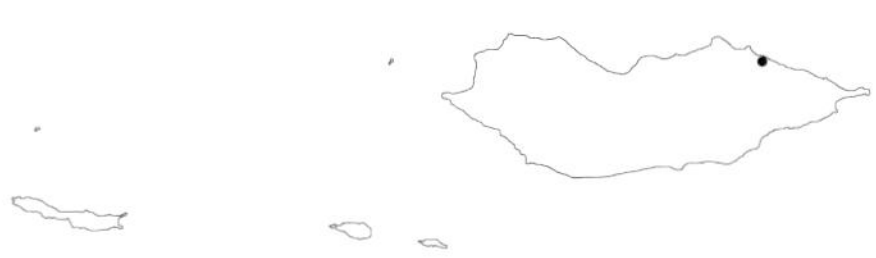

Sprawling, greyish-hairy herb to 30cm. Leaves 3 to 7-foliolate; leaflets elliptic, 0.5–1.5 x 0.3–0.6cm. Flowers clustered in short racemes. Pods reflexed, flattened, characteristically appearing jointed, 1–1.5cm long. Seeds 5 to 8.

Ecology. Rare in *Croton socotranus* shrubland. 50m.

Distribution. NE Africa west to Mali, E Africa, E Zaire, Arabia to Pakistan and India.

Botanical notes. Readily distinguished by the flattened fruits.

ETHNOGRAPHIC NOTES

Name(s)

– Soqotri: *ṭ'ifher* 'finger- and toenail' (more commonly given to *Tephrosia*, *Senna* and *Ruellia* species); *di bət'aḥan* 'prostrate'.

– Soqotri simplified: *tifher*; *di bitahun.*

Food (livestock). Grazed by livestock mainly in the dry season, but not with great enthusiasm or when alternative forage is available. It is not regarded as being a valuable fodder.

Notes. There is a certain overlap between this, *Senna* and *Tephrosia* species. Islanders regard the smaller *tifher* plants with small fruit and leaves as 'female', and the larger plants with larger fruit and leaves as 'male'.

Comparative uses. Said to be toxic to livestock. Herders in East Africa report that although livestock like it, when they feed on it during the rainy season they sometimes die for unknown reasons. (Bebawi & Neugebohrn 1991; Heine *et al.* 1988)

Indigofera marmorata Balf.f.

Erect shrub with many parallel ascending branches to 1(–2)m. Leaves fasciculate, 3(to 5)-foliolate, c.0.5cm long; leaflets marbled. Flowers pink, solitary or few. Pods reflexed, 0.5–1cm long, deeply longitudinally grooved. Seeds 3 to 5.

Ecology. Locally common in open shrubland and woodland clearings on granite on the northern slopes of the Haggeher mountains. 400–600m.

Distribution. Endemic.

Conservation status. Least Concern (LC). Common in several vegetation types and under no present or perceived threat.

Botanical notes. Readily identified by the marbled pattern formed by hairs on the upper leaf surface and its characteristic trigonous and beaked fruits.

ETHNOGRAPHIC NOTES

Name(s)

– Soqotri: *zēydiro* 'fat-forming'; *zēdiro di 'irəhān* 'goats *zēdiro*'; *līyəf.*

– Soqotri simplified: *zeydiro*; *zeydiro di 'irihan*; *liyif.*

Food (livestock). Grazed by all livestock when hungry. Goats in particular eat the new growth, and the mature plant when there is little alternative. They also chew the leafless twigs. It is not seen as being a nutritious feed but as an adequate maintenance one.

Fuel. Dead plants are gathered and used as tinder and kindling, and bunches of it are often kept in the home for this purpose. It catches light very quickly and is often used as a firelighter. It burns with a very bright flame, and thicker stems are used as a torch.

Bedding, stuffing. The dead plants are sometimes pulled up and laid stalk end uppermost to make a mattress over which wool rugs or animal skins are laid.

Pest control. Dry plants are piled over the floor of badly flea-infested livestock and human quarters after blocking air outlets. The plants are then set alight and left to burn right down. This brings flea numbers down, though does not get rid of them altogether. This is most commonly done in the smaller pens which house young stock, as fleas are particularly harmful to the health of young animals.

Indigofera nephrocarpa Balf.f.

Compact or somewhat sprawling, silvery herb to 15cm. Hairs on stem and leaves slightly spreading. Leaves 3(to 5)-foliolate; leaflets alternate, elliptic to circular, c.0.1cm long. Flowers solitary. Peduncle <0.5cm long. Pods ± circular, c.0.3cm diameter. Seed 1.

Ecology. Widespread and common in dry, open areas on the plains and limestone plateaus. Sea-level–800m.

Distribution. Somalia, Oman and Pakistan.

Botanical notes. Distinguished by the inconspicuous, comma-shaped fruits. It tends to form a dense cushion with tiny leaves (c.0.3cm long), sending out short-lived sprawling shoots with larger leaves. See also comments under *I. nephrocarpoides*.

ETHNOGRAPHIC NOTES

Name(s)

– Soqotri: nil or *ḳūwnuh* 'little threads'; *noi*; *ruʿud* 'grazing plant'; *məsḥabeb*, *məsḥɔ̄wbeb* 'spreading' *ḥʌmź*, *ḥɔ̄wmʌź*, *ḥəmʌźəhen* 'sour, astringent'; *rətɛ̄rit* 'forming green patches on the ground' (Abd al Kuri); *rəḥamma* 'grey-coloured' (Samha).

– Soqotri simplified: *qownuh*; *noy*; *ru'ud*; *mis'habeb*, *mis'howbeb*; *hamdh*, *howmadh*, *himadhihen*; *riteyrit*; *ri-hammih*.

Food (livestock). Regarded as a valuable and nutritious fodder for small stock, especially in the drier areas of the island. The stems and some foliage persist though the dry months, when they are important forage. On Samha *I. nephrocarpa* is one of the most important plants for stock, especially sheep, and on Abd al Kuri it and *I. nephrocarpoides* are highly valued.

Indigofera nephrocarpoides J.B.Gillett

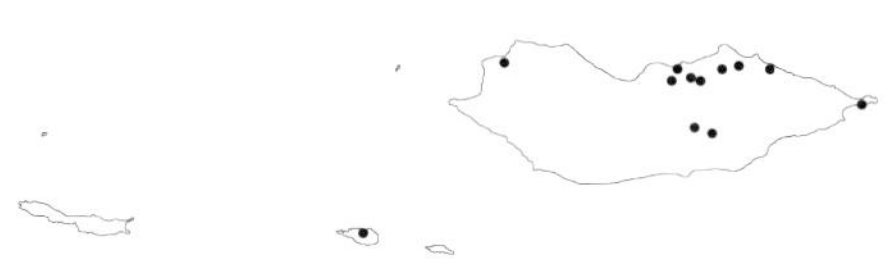

Similar to *I. nephrocarpa* but differs in its appressed hairs on stems and leaves; leaflets oblong-lanceolate or linear; and peduncle usually >0.5cm long.

Ecology. Clearings in drought-deciduous woodland. 200–850m.

Distribution. Endemic.

Conservation status. Least Concern (LC). Common in several vegetation types and under no present or perceived threat.

Botanical notes. Closely related to *I. nephrocarpa* from which it can be distinguished by the appressed (not crisped) indumentum on the stems and leaflets; narrower leaflets; longer peduncles; and discoid, flattened (not subglobose) fruits. However, there is a considerable degree of overlap in these characters with some rather intermediate specimens. Further study might result in these two species being better treated as varieties. Generally found at higher altitudes than *I. nephrocarpa* and (unlike that species) apparently never on the coastal plains.

ETHNOGRAPHIC NOTES

Name(s)

See *I. nephrocarpa* (also uses).

Indigofera nugalensis Thulin

Procumbent, silvery herb to 10cm. Leaves 3(to 5)-foliolate; leaflets obovate, <0.5cm long; petiole expanded, c.1.5cm long. Flowers few in short racemes. Pods reflexed, straight, c.1cm long. Seeds 2 to 4.

Ecology. Dwarf shrubland on foothills. c.50m.

Distribution. Somalia.

Botanical notes. Similar to *I. spiniflora* but differing in fruit and seed characters (see notes under that species). Reported to be steadily increasing on Abd al Kuri.

ETHNOGRAPHIC NOTES

Name(s)

– Soqotri: *zidirə* 'fat-forming' (Abd al Kuri); *riśɛ̄riś*, *rətɛ̄rit* 'forming green patches on the ground' (Abd al Kuri).

– Soqotri simplified: *zidirih*; *risheyrish*, *riteyrit*.

Food (livestock). Key fodder plant for goats on Abd al Kuri. It is seen as nourishing and fattening, and increases milk. Sheep also graze it.

Fuel. The dead plants are an important fuel on Abd al Kuri, especially for cooking. It is also gathered and stored as kindling. It lights easily and is often used as a fire-lighter, especially for lighting the brushwood piled over clay pots in firing, and in the manufacture of quicklime. It burns with a very bright flame, and the thicker stems are used as a torch.

Indigofera oblongifolia Forssk.

Indigofera paucifolia Delile

Erect, little-branched, silvery shrub to 1m. Leaves simple or 3 to 5-foliolate; leaflets alternate, oblong, 0.5–2 x 0.3–0.5cm, terminal leaflet much larger than others. Flowers many, in crowded racemes. Pods curved, deeply constricted between seeds, 1.5–2.5cm long; segments globose. Seeds 6 to 8.

Ecology. Open shrubland on coastal plains with *I. articulata* and *I. pseudointricata*. c.10m.

Distribution. Widespread in drier parts of tropical and subtropical Africa and Asia.

Botanical notes. On Soqotra strangely restricted in its

distribution, being found only in a relatively small area of the eastern Noged plain where it dominates a coastal community. Balfour (1988) records it from 'about Galonsir [Qalansiyah] and elsewhere'. It is apparently extinct in the west of the island now.

ETHNOGRAPHIC NOTES

Name(s)

– Soqotri: *ḥayriro* 'blackened'; *heḳeh* 'tough, difficult to break'.

– Soqotri simplified: *hayriro*; *heqeh*.

Food (livestock). Browsed by camels year round, and goats and donkeys too when very hungry, though in general the plant is too salty and bitter to be very appealing to small stock. The plant retains its colour throughout the dry season, and towards the end of this period it is more tempting to other animals. However, the stems are very tough and damage the mouths of goats in particular, to the extent of loosening their teeth and occasionally dislodging them. Herders have observed that eating a lot also changes the colour of the teeth of a goat. The sore mouths which result from browsing this can make goats lose their appetite and so lose condition. In a prolonged dry season sheep eat the plant litter and any half-dry leaves within reach. If it forms a large part of a camel's diet the animal loses condition and grows thin, but if mixed with browse of other plants it remains fit and well. Herders of areas where it is plentiful are not keen on it and in some places have tried to eradicate it, so far with little success.

Comparative uses. Regarded as a good camel fodder. Medicinally, the root is boiled in milk and taken as a purgative. Extracts of the plant are given for enlargement of the spleen and liver, and to improve the appetite. Elsewhere in Arabia a decoction of the leaf is used as a shampoo to remove dandruff and keep the hair soft and healthy. In Dhofar the stems are used to weave fish traps and the root, crushed with or without the leaves, is applied to snakebites, insect stings and swellings as an analgesic and anti-inflammatory. (Abdulfatih 1987; Dalziel 1948, 1956; Heine *et al.* 1988; le Houerou 1980; Miller & Morris 1988)

Indigofera pseudointricata **J.B.Gillett**

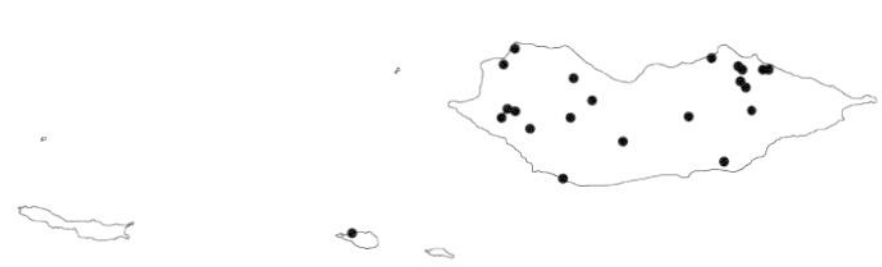

Erect, densely branched, silvery subshrub to 50cm. Leaves simple, linear, 0.2–0.7cm long, revolute, petiole very short. Flowers ± solitary. Pods straight, c.1.5cm long. Seeds c.5.

Ecology. Coastal and inland plains; *Croton socotranus* shrubland and desertic shrubland; widespread and sometimes abundant. Sea-level–550m.

Distribution. Somalia.

Botanical notes. A common low shrub on the coastal and inland plains. Characterised by its intricately branched woody branches and grey-green unifoliolate leaves.

ETHNOGRAPHIC NOTES

Name(s)

– Soqotri: *zēydiro*, *zidirə* 'fat-forming'; *zēydiro di ḥəmār* 'donkey's *zēydiro*'; *zēydiro di bəᶜeyr* 'baggage camel's *zēydiro*'; *śebərʌḳ* 'glimpsed'; *dirfə* 'itch, irritation' (Soqotra and Samha); *dirbeḳ* 'on heat'.

– Soqotri simplified: *zeydiroh*, *zidirih*; *zeydiroh di himar*; *zeydiroh di bi'eyr*; *shebiraq*; *dirfih*; *dirbeq*.

Food (livestock). One of the key fodders for goats and sheep in the drier areas of the island. It is regarded as a nutritious and fattening feed, which also increases milk production. Weaned kids and lambs are taken to areas where this grows, as it is seen to be an easily digestible fodder. Camels and donkeys eat the plant the year round, as the various names suggest. Cattle, however, rarely graze it. It is an important dry-season feed, al though when the plant is very dry it can damage the mouths of small stock eating a lot of it, making it hard for them to continue feeding and thus causing them to lose condition. In the southern plains and other dry and vegetation-poor areas, the larger plants are gathered by herders and chopped up for their goats as the dry season progresses and other fodders become increasingly scarce. It is not seen as being a nutritious feed but as a very valuable maintenance one. On the outer islands it is highly appreciated as livestock forage.

Bees / honey. Honeybees are attracted to the flowers.

Fuel. The dead plants are an important fuel on Abd al Kuri and Samha, especially for cooking as it makes little smoke. It lights easily and is often used as a firelighter, and is often stored as kindling. It burns with a very bright flame, and the thicker stems are used as a torch.

Bedding, stuffing. The dead plants are sometimes pulled up and laid stalk end uppermost to make a mattress over which wool rugs or animal skins are laid.

Pest control. Dry plants are piled over the floor of badly flea-infested livestock and human quarters after blocking air outlets. The plants are then set alight and left to burn right down. This brings flea numbers down, though does not get rid of them altogether. This is most commonly done in the smaller pens which house young stock, as fleas are particularly harmful to the health of young animals.

Indigofera socotrana **Vierh.**

Indigofera gerardiana Graham

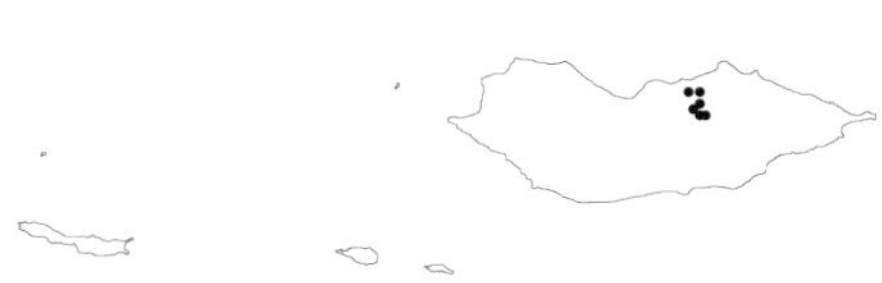

Shrub or small tree to 2m. Leaves 7 to 11-foliolate; leaflets oblong-elliptic, 1–2 x 0.5–1cm. Flowers pink, c.1cm long, many, in crowded racemes. Pods straight, flattened, 2–3cm long. Seeds 5 to 8.

Ecology. Evergreen and semi-deciduous montane woodland. 500–1350m.

Distribution. Endemic.

Conservation status. Least Concern (LC). Common in several vegetation types and under no present or perceived threat.

Botanical notes. Distinct from other species on the island in both its habit and habitat. It is the only species that grows as a tree rather than as a herb or shrublet and is the only species restricted to the relatively moist submontane vegetation of higher altitudes in the Haggeher mountains. This suggests that *I. socotrana* has a different evolutionary history to most other Soqotran *Indigofera*, which diversified in response to the advance of arid conditions throughout NE Africa during the Pliocene (Schrire 1995).

ETHNOGRAPHIC NOTES
Name(s)
– Soqotri: *ḥāhər* 'dark-coloured'; 'kind of *ṭífher*'; (rare) *ḥáyríro* 'blackened' (more commonly given to *I. oblongifolia*).
– Soqotri simplified: *hahir*, 'kind of *tifher*'; *hayriro.*
Food (livestock). Goats native to the areas where the plants grow show little interest in this evergreen plant. Even in the dry season they rarely eat the foliage, as there are many other alternative dry season fodders in these areas. However, hungry goats transhuming into the area from outside are attracted to the greenery, and even sheep eat the fallen leaves. However it is seen as maintenance feed only, not as having any nutritional value.
Notes. The *tifher* plants with small leaves and fruit are regarded by islanders as the 'female' of those *tifher* plants with larger leaves and fruit.

Indigofera spiniflora Boiss.

Silvery subshrub to 30cm. Leaves 3 to 5-foliolate; leaflets oblong-elliptic or linear, 0.3–2 x 0.1–0.4cm. Flowers in a lax raceme; inflorescence axis spinescent, persistent. Pods straight, reflexed, 1–1.5cm long, somewhat constricted between the seeds. Seeds 4 to 6.

Ecology. Open shrubland on limestone plateaus; usually above 200m on Soqotra, the lower altitudes refer to collections from Samha. Sea-level–550m.
Distribution. Arabia, Ethiopia, Somalia, NE Kenya.

ETHNOGRAPHIC NOTES
Name(s)
See *I. nephrocarpa* (also uses).

Indigofera tinctoria L.
Indigofera tinctoria L. var. *tinctoria*

Erect, woody herb to 50cm. Leaves 9 to 13-foliolate; leaflets oblong-elliptic, 0.7–2 x 0.3–1cm. Flowers many, in crowded racemes. Pods straight, reflexed, 2.5–3cm long. Seeds 8 to 12.
Ecology. Northern foothills of the Haggeher, in succulent shrubland. 50–300m.
Distribution. Widely distributed in dry regions of the tropics and subtropics.

ETHNOGRAPHIC NOTES
Name(s)
– Soqotri: *ṭífher* 'finger- and toenails' (more commonly given to *Tephrosia*, *Senna*, *I. hochstetteri* and *Ruellia* species).
– Soqotri simplified: *tifher.*
Food (livestock). In areas where the plants are plentiful, livestock accustomed to this forage eat some green material and plant litter when very hungry. Goats like to nibble the flowers in the dry season. It is generally not palatable to livestock who have transhumed in from other areas.
Flavouring / food additive. The taste of the meat, especially the viscera, and the cooking water of an animal that has eaten a lot can be unpleasantly bitter, sometimes so much so that it has to be discarded.
Notes. The *tifher* plants with small leaves and fruit are regarded by islanders as the 'female' of those *tifher* plants with larger leaves and fruit.
Comparative uses. In India this plant was formerly grown for indigo. It has now largely been replaced by *I. sumatrana*, *I. suffruticosa* and *I. erecta*, but is still cultivated for the indigo dyeing in tropical West Africa. It is sometimes grown for green manure. Medicinally, an extract is used in epilepsy and other nervous disorders, and it is an ingredient in an ointment used for sores, old ulcers and piles. A decoction of the leaves is given in blennorrhagia. The roots are used to treat urinary complaints and hepatitis. (Ambasta 1982; Dalziel 1948, 1956; Jain & De Fillips 1991; Kunkel 1984)

Lotus L.

Herbs. Leaves 3 to 5-foliolate; leaflets entire, the lower 2 stipule-like; stipules minute. Flowers solitary or in few-flowered umbels, axillary. Calyx 5-lobed. Corolla yellow or pink. Stamens 10, nine filaments fused, one free. Pods dehiscent, cylindrical, straight. 2 endemics. c.100 species. N temperate regions.

ETHNOGRAPHIC NOTES
Comparative uses. The fodder value of the leaf is high. *L. corniculatus* (bird's-foot trefoil) is widely cultivated in temperate areas, and sometimes also at high altitudes in the tropics and subtropics. It is a good soil binder, free from pests and diseases, and may be used as pasture or fed as silage or hay. However, some species, such as *L. arabicus*, are toxic to horses, sheep and goats. The immature plant contains prussic acid, and death is often sudden, often within half an hour of eating the plant. When it has matured and set seed, it is a good fodder. The young pods of some species, such as *L. corniculatus* and *L. edulis*, are edible, and the pods of *L. gebelia* can be eaten like string beans. (Ambasta 1982; Bebawi & Neugebohrn 1991; Bogdan 1977; Dalziel 1948, 1956; Jain & De Fillips 1991; Kunkel 1984; Lewis & Elwin-Lewis 1977)

Lotus garcinii DC.

Lotus stocksii Boiss.

Prostrate herb, glaucous. Leaves digitately 3 to 5-foliolate; leaflets 0.3–0.8 x 0.2–0.5cm, obovate. Flowers solitary, sessile. Calyx 0.5–0.8cm. Corolla pale pink, 0.4–0.6cm, included within, or shortly excluded from, calyx. Pod 0.4–0.8 x 0.2cm, surrounded by the persistent calyx.
Ecology. Sandy and gravel coastal plains. Sea-level–10m.
Distribution. Egypt, Somalia, Arabia, Iran, Pakistan.
Botanical notes. Readily distinguished by the calyx which persists around the fruits.

ETHNOGRAPHIC NOTES
Name(s)
– Soqotri: nil or *raʿádihon* 'little grazing plant'.
– Soqotri simplified: *ra'adihon.*
Food (livestock). On Abd al Kuri this is highly valued as a nutritious feed for sheep and goats. If plentiful, it is fattening and increases milk yield.

Lotus mollis Balf.f.

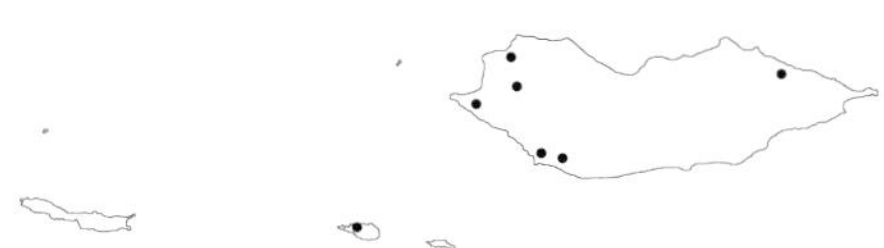

Prostrate perennial herb, covered in dense, appressed, silvery-grey hairs. Leaves 3-foliolate; leaflets narrowly elliptic to narrowly obovate, c.0.5 x 0.2cm. Flowers solitary, sessile. Calyx c.0.4–0.5cm long. Corolla bright yellow, <0.8cm long, long-exserted from calyx. Pod 1–1.5 x 0.2cm.
Ecology. In thin soil/gravel overlying limestone pavement on exposed limestone plateaus and cliffs. 400–750m.
Distribution. Endemic.
Conservation status. Vulnerable (VU D2). Rare, with a very limited area of occupancy.

ETHNOGRAPHIC NOTES
Name(s)
– Soqotri: nil or (Soqotra and Samha) *raʿádihon* 'little grazing plant'; *kə́ṣkʌṣ* 'cream-coloured'.
– Soqotri simplified: *ra'adihon*; *qizqaz.*
Food (livestock). Valued as nutritious forage for sheep and goats on Samha. On Soqotra it provides incidental forage, but is considered to be of limited nutritional value.

Lotus ononopsis Balf.f.

Prostrate perennial herb, bright green, glabrous or thinly hairy. Leaves 3-foliolate; leaflets narrowly elliptic to ovate or circular, 0.3–1.2 x 0.1–0.5cm. Flowers solitary, shortly pedicellate. Calyx 0.3–0.6cm. Corolla bright yellow, 0.8–1.2cm long, long-exserted from calyx. Pod 1.5–2 x c.0.2cm.
Ecology. Clearings in semi-deciduous woodland on the limestone in NE Soqotra and grassland on granite in the Haggeher mountains. 500–1500m.
Distribution. Endemic.
Conservation status. Least Concern (LC). Common in several vegetation types and under no present or perceived threat.
Botanical notes. Very variable in leaf shape and size.

ETHNOGRAPHIC NOTES
Name(s)
– Soqotri: *kə́ṣkʌṣ* 'cream-coloured'; *ríhiśt* (more commonly given to various Gramineae of the Central Highlands); *źerʌ́ṭhinítin* 'little farts' (from the sound made by treading on the fruiting plant).
– Soqotri simplified: *qizqaz*; *rihisht*; *dherat'hinitin.*
Food (livestock). Regarded as very rich and fattening feed for cattle and goats. It promotes lactation and is considered to be nutritious at all stages of development. The pods are seen as particularly valuable. Plants are sometimes cut and carried to livestock which need to have their strength built up, or to tempt a cow reluctant to let down her milk.
Flavouring / food additive. The milk of animals grazing on it heavily is recognisably sweet and rich, and makes plentiful and superior butter.

Lotus schimperi **Steud.**

Lotus arabicus var. *trigonelloides* sensu Balf.f. non Webb & Berthel.
Lotus glinoides L.

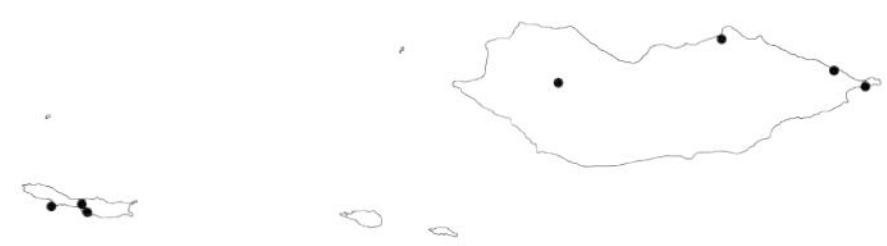

Erect or prostrate annual herb, glaucous. Leaves 3 to 5-foliolate; leaflets obovate, 0.5–1 x 0.2–0.5cm. Flowers 1 to 3(to 5), sessile or subsessile. Calyx 0.2–0.3cm. Corolla yellow, pink or brownish, <0.4cm, shortly exserted from calyx. Pod 0.5–1 x c.0.2cm.

Ecology. Desertic vegetation on dry coastal and inland plains on Soqotra (sea-level–120m); exposed mountain ridge on Abd al Kuri (c.500m). Sea-level–500m.

Distribution. Egypt, Sudan, Ethiopia, Arabia, Iran, Pakistan.

ETHNOGRAPHIC NOTES

Name(s)

– Soqotri: nil or *raᶜadihon* 'little grazing plant'; 'kind of *ḳʌlḳəhʌl*'; *kubuś* 'of the ram' (Abd al Kuri).
– Soqotri simplified: *ra'adihon*; 'kind of *qalqihal*'; *kubush* (Abd al Kuri).

Food (livestock). Grazed by livestock when they are brought down to the coastal plains after rain, but it is not regarded as a nutritious forage, and there are usually plenty of more valuable forage plants available at such times. On Abd al Kuri it is more highly valued, and is grazed by sheep and goats after rains.

Medicago L.

Herbs. Leaves 3-foliolate; leaflets toothed; stipules entire or laciniate. Flowers in axillary racemes. Calyx of 5 subequal lobes. Corolla yellow. Anthers 10; filaments 9 fused, 1 free. Pod spiny, indehiscent, coiled. c.50 species. Mediterranean, SW Asia, NE and S Africa.

ETHNOGRAPHIC NOTES

Comparative uses. Lucerne or alfalfa, *M. sativa*, is of primary importance for agriculture. It is cultivated for pasturage, hay, lucerne meal and silage, and is highly valued as a fodder, though excessive feeding on it may cause bloating. It is also useful for bee pasturage, and its seeds, as well as containing a yellow dye, also yield a drying oil suitable for paints and varnishes. *M. laciniata* is said to be a palatable fodder which, when abundant, can contribute to the quality and the bulk of herbage. However, it can be a nuisance in sheep pastures because of the spiny pods adhering to the wool from which they are difficult and costly to separate. *M. minima* is also of some importance in pasturelands, but is suspected of causing photosensitisation. (Ambasta 1982; Bogdan 1977; Jain & De Fillips 1991; Kunkel 1984)

Medicago laciniata **All.**

Spreading annual herb, glabrous or sparsely hairy. Leaflets obovate-cuneate, c.0.5–1cm long; stipules laciniate. Racemes 1 or 2-flowered. Corolla <0.5cm long. Pods subglobose, c.0.5cm diameter, in a spiral of c.½ turns; grooves not visible between the nerves on each turn; spines hooked, usually <0.3cm long.

Ecology. Rare in grassland. 750m.

Distribution. Africa and SW Asia.

Botanical notes. Soqotran material is referable to var. *brachyacantha* Boiss. which has stipules divided to below the middle.

ETHNOGRAPHIC NOTES†

Name(s)

– Soqotri: *śəriśiroh* 'little *śerśer*' (*Tribulus terrestris*); 'like *śerśer*' (because of its spiny fruit); *ḥʌmź* 'sour, astringent'; *di ᵓaridəhān, di hārdəhān* 'area of spreading herbage'.
– Soqotri simplified: *sherishiroh*; 'like *shersher*'; *hamdh*; *di aridihan, di hardihan*.

Food (livestock). Regarded as valuable forage plants for livestock; they are equated with the better grasses. They are seen as being fattening, and as enriching and increasing milk in lactating animals. In small stock the spiny pods can lodge between the claws of the hoof, causing sores that can become infected and even lead to lameness. The pods also become entangled in the coats of both sheep and goats, matting the hair and scratching the skin. In sheep this is troublesome when washing and shearing. Donkeys also eat the plants.

Bees / honey. Honeybees are attracted to the flowers.

Notes. These plants can cause *'uqabah* in goats unaccustomed to them. Longer-haired goats and sheep in areas where these plants are common often have the hair of the thigh and udder area trimmed to avoid the scratches caused by the prickly fruit lodging in their coats. The scratches frequently develop into sores and attract flies, when they can become a problem.

†Also apply to *M. minima*, *M. polymorpha*.

Medicago minima Lam.

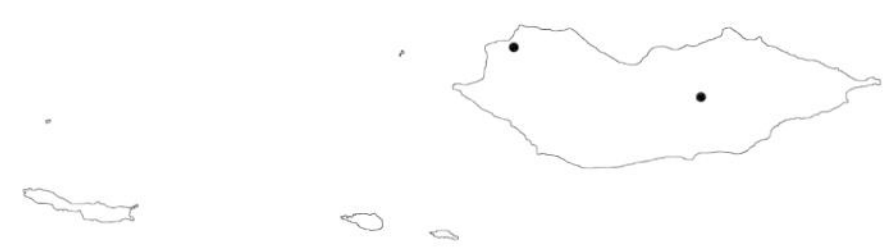

Spreading annual herb, softly hairy throughout. Leaflets obovate, c.0.5–1cm long; stipules entire or slightly toothed. Racemes 1 to 8-flowered. Corolla <0.5cm long. Pod subglobose, 0.3–0.5cm diameter, in a spiral of 3 to 5 turns; grooves clearly visible between the nerves on each turn; spines hooked, length variable.

Ecology. Grassland on granite and limestone. 650–700m.
Distribution. Somalia, Ethiopia, N Africa, Europe, S Asia.

ETHNOGRAPHIC NOTES
Name(s)
See *M. laciniata* (also uses).

Medicago polymorpha L.

Medicago denticulata Willd.
Medicago hispida Gaertn.
Medicago hispida var. *denticulata* (Willd.) Engl.

Spreading annual herb, glabrous or sparsely hairy. Leaflets broadly obovate-cuneate, 0.5–2cm long; stipules laciniate. Racemes 1 to 8-flowered. Corolla <0.5cm long. Pod flattened, 0.4–1cm diameter, in a spiral of c.3 turns; grooves widely separated and clearly visible between the nerves on each turn; spines usually hooked, length variable.
Ecology. Grassland on granite in the Haggeher mountains. 700–1200m.
Distribution. Europe, E and N Africa and Asia.

ETHNOGRAPHIC NOTES
Name(s)
See *M. laciniata* (also uses).

Melilotus Mill.

Herbs. Leaves 3-foliolate; leaflets toothed. Flowers yellow, in spike-like racemes. Pods very small, ovoid, wrinkled. 20 species. Temperate and subtropical Eurasia, N Africa, Ethiopia.

Melilotus indica (L.) All.

Melilotus parviflora Desf.

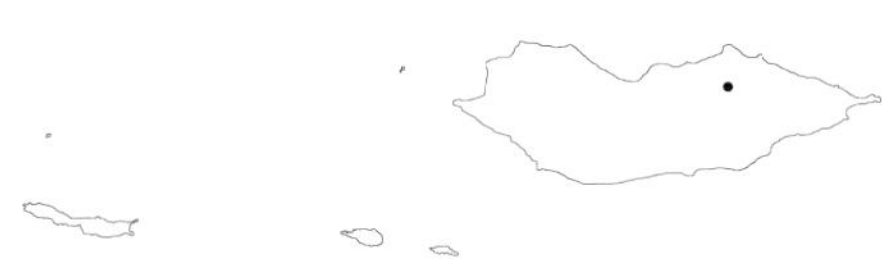

Prostrate herb. Leaflets narrowly obovate to obovate, 1–2 x 0.2–1cm. Flowers c.0.3cm long. Pods c.1.5mm diameter.
Ecology. Grassland on granite in the Haggeher mountains. c.950m.
Distribution. Middle East to India.

ETHNOGRAPHIC NOTES
Name(s)
– Soqotri: *ruˈʿud* 'grazing plant'.
– Soqotri simplified: *ru'ud.*
Food (livestock). Regarded as rich and valuable feed for all livestock, increasing weight and promoting lactation. Its value is equated with that of the better grasses.
Bees / honey. Honeybees are attracted to the flowers.

Comparative uses. This common weed of arable land is also grown on a small scale for fodder in India, pure or mixed with annual grasses. It contains coumarin and the hay or silage, if poorly prepared, can cause internal bleeding. The plant is also used as green manure and to improve saline soils. When fed alone, it may cause lethargy, tympanites and paralysis. The leaves of related species are used in soups and salads, and to flavour butter and cheese. Medicinally, *Melilotus* contain an antibiotic active against broad-spectrum bacteria. The plant is also used as a discutient and emollient and is also considered to be narcotic. The seeds are used in bowel complaints and infantile diarrhoea, and for diseases of the genital organs in both sexes. They are useful too against colds and moths. On the Arabian mainland both *M. indica* and *M. alba* are used as a stimulant and emollient and as a poultice or plaster on swellings. A decoction is given for urinary bladder pains and as a diuretic, and an infusion of the flowering branches as an anti-spasmodic. (Ambasta 1982; Batanouny 1981; Bogdan 1977; Ghazanfar 1994; Goodin & Northington 1979; Jain & De Fillips 1991; Kunkel 1984; Lewis & Elwin-Lewis 1977; Mossa *et al.* 1987; Rizk & El-Ghazaly 1995)

Microcharis Benth.

Similar to *Indigofera* but differentiated by the glabrous petals and pedicel with 90 degree bend. 33 species. Africa, Arabia and Madagascar.

Botanical notes. Close to *Indigofera* but differing by its glabrous petals, prolonged rostrum at the apex of the keel, discoid stigma, columnar style and fruits (often) held at right angles to the pedicels.

Microcharis disjuncta (J.B.Gillett) Schrire

Indigofera disjuncta var. *fallax* J.B.Gillett
Indigofera disjuncta J.B.Gillett
Microcharis disjuncta var. *fallax* (J.B.Gillett) Schrire

Prostrate, grey-hairy, annual herb to 15cm. Lower leaves simple, upper 3-foliolate; leaves/leaflets ovate, 0.5–1.5 x 0.5–1cm. Flowers red, few. Pods c.1.5cm long. Seeds c.5.
Ecology. Sandy and gravel coastal plains. Sea-level–20m.
Distribution. (var. *fallax*) Middle East, Oman.
Botanical notes. Material from Soqotra is all referable to var. *fallax* (J.B.Gillett) Schrire which differs from var. *disjuncta* in having three-foliolate as well as one-foliolate leaves.

ETHNOGRAPHIC NOTES
Name(s)
– Soqotri: *səd'ʌḥan* 'prostrate'; *kibid'inəh* 'little liver', 'small hill(ock) of earth'; nil (Samha); *ru'ʿud* 'grazing plant' (Abd al Kuri).
– Soqotri simplified: *sidahan*; *kibidinih*; *ru'ud.*
Food (livestock). Provides incidental grazing for hungry livestock, but is not regarded as important or nutritious forage. On the outer islands it is a more highly valued fodder for sheep and goats, as it often persists well into the dry season.

Paracalyx Ali

Cylista Aiton

Shrubs. Branches twining or trailing. Leaves 3-foliolate. Flowers yellow with red veins, in short racemes. Calyx scarious, strongly accrescent, lower lobe boat-shaped. Pods short, enclosed within expanded calyx lobes. 1 endemic. 6 species. Old World tropics.

Paracalyx balfourii (Vierh.) Ali

Cylista balfourii Vierh.
Cylista scariosa Roxb.
Cylista schweinfurthii Wagner & Vierh.
Paracalyx schweinfurthii (Wagner & Vierh.) Ali

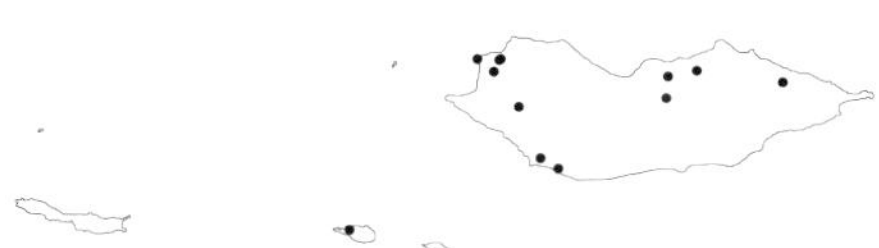

Vigorous shrub with silvery hairs. Leaflets broadly wedge-shaped or obovate, 2–5 x 1.5–4cm. Pods c.1.5cm long.
Ecology. Locally common over the limestone plateaus of Soqotra; rare in relict woodland at the base of the limestone escarpment on Samha. 500–700m.
Distribution. Endemic.
Conservation status. Least Concern (LC). Common in several vegetation types and under no present or perceived threat.

ETHNOGRAPHIC NOTES
Name(s)
– Soqotri: *s'idəd, s'idad* 'strong, forceful' (Soqotra and Samha); (rare) *ʿotf* (more commonly given to *Flueggea* species).
– Soqotri simplified: *sidid, sidad*; *'otf.*
Food (livestock). Key fodder for livestock on Soqotra and Samha, although not considered as valuable a feed as *Chapmannia* species or *Zygocarpum coeruleum*. The green leaves are enthusiastically browsed by cattle, camels and goats, especially when the grasses are drying out. Sheep eat both green and fallen leaves. It makes animals fat and sleek, and increases their milk yield. In the dry season the drying leaves and leafless stems are eaten by all livestock. Herders search for leafy stems to cut for their animals, and gather fallen foliage. In the dry months the best foliage is given to pregnant animals to keep them in good condition, and to milkers to keep them in milk. Like species of *Chapmannia* and *Zygocarpum coeruleum*, it is so liked by livestock that leafy stems are used by herders as a lure or bribe to manage recalcitrant animals – to induce them to go into an enclosure, or to let down their milk, for instance. These shrubs often come into new leaf towards the end of the monsoon when there is little other forage, making it doubly valuable for livestock which are usually in poor condition by this time.
Bees / honey. Surprisingly, honeybees do not seem to be attracted to the flowers.
Fuel. The dead wood makes only average firewood. It is rather smoky and burns too quickly to be useful. Smaller pieces are often gathered as kindling.
Timber. Makes poor timber and is little used in building.
Wood. Soft and has a tendency to split, so it is of little use for making implements. In the absence of better wood, straight branches can be used to make the herding staff, but these are rarely very durable, even if fire-hardened.
Medicinal / veterinary. The green foliage is almost as effective as *Zygocarpum coeruleum* and *Chapmannia* species for sick animals.
Other. The skin of a dead calf, or of a male slaughtered to conserve its mother's milk, is made into a tulchan stuffed with this plant material. It is then presented to the

cow to sniff and some of her milk is sprayed over it to encourage her to let down her milk. When trying to persuade a cow whose male calf has been slaughtered to accept the female calf of another as her own, the calf is rubbed with a paste of the crushed plant to disguise its own smell and presented to the cow. Some of her milk is sprayed over it to encourage her to lick it and allow it to suckle.

Comparative uses. *P. cytisoides* is used as a green manure for paddy. (Ambasta 1982; Jain & De Fillips 1991)

Rhynchosia Lour.

Herbs, sometimes woody-based. Stems prostrate or climbing. Leaves 3-foliolate; leaflets entire, gland-dotted beneath. Flowers in axillary racemes. Calyx 5-lobed; lobes unequal. Corolla yellow. Anthers 10; filaments 9 fused, 1 free. Pods circular or shortly oblong, flattened. Seeds 1 or 2. c.200 species. Pantropical.

Rhynchosia minima (L.) DC.

Climbing or prostrate herb, glabrescent green. Leaves ovate to rhomboid, 1–4 x 1–3cm, acute or acuminate. Inflorescence lax. Calyx c.0.5cm. Corolla 0.5–0.6cm. Pod oblong, 1–1.7 x 0.4–0.5cm, glabrescent. Seeds 2.

Ecology. Limestone plateau. 300–650m.

Distribution. Tropical Africa, Arabia east to China and the Neotropics.

Botanical notes. Soqotran material is referable to var. *minima*.

ETHNOGRAPHIC NOTES†

Name(s)

– Soqotri: *ləska̍no̍h* 'sticky, clinging'; *məzə̍wkʌn* 'twining, clinging'; *di rīheś, di ra̍hśe, di re̍hiś* (more commonly given to various Gramineae of the Central Highlands); *ṭefə̍ri̍rihin* 'finger- and toenails', 'buds about to break open'; *məṣḥe̍ybibo* 'spreading'.

– Soqotri simplified: *liskanoh*; *mizewkan*; *di riyhesh, di rah'sheh, di rehish*; *tefiririhin*; *mis'heybibo*.

Food (human). The seeds are eaten raw, and are at their best when ripe and brown. This is a casual snack for children especially, not an important food. The aerial parts are occasionally strewn over the main cereal dish as a relish.

Food (livestock). Considered very nutritious and valuable feed for all livestock, cattle especially being very fond of it. Animals eat every part of it, stems, leaves and fruit. If there is plenty, animals put on weight and give more milk.

Comparative uses. Some species are grown as a green manure and as a fodder for cattle and horses, and others have edible leaves and roots. *R. minima* is said to be toxic to fish, and its seeds to be poisonous. Medicinally, the leaves are used as an abortifacient and to treat skin diseases. The seeds of *R. longeracemosa* and *R. pyramidalis* are also said to be toxic, and in Mexico are thought to be hallucinogenic. In Tanzania a decoction of the roots of *R. albissima* is used to treat impotence and given as an aphrodisiac. (Ambasta 1982; Batanouny 1981; Dalziel 1948, 1956; Goodin & Northington 1979; Jain & De Fillips 1991; Kunkel 1984; Lewis & Elwin-Lewis 1977; Rizk & El-Ghazaly 1995)

†Also apply to *R. pulverulenta*.

Rhynchosia pulverulenta Stocks

Rhynchosia memnonia Balf.f. non DC.

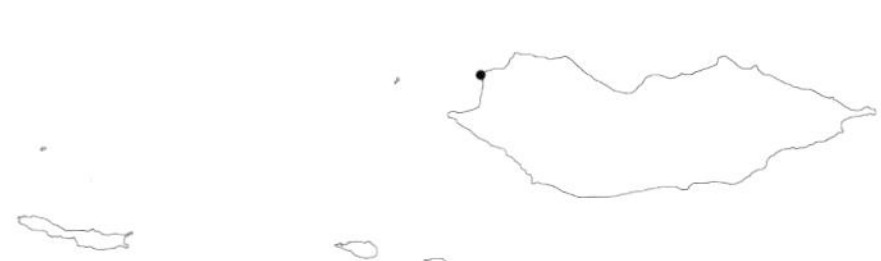

Similar to *R. minima* but differs in its grey-green, densely tomentose leaves and stems, rounded or obtuse leaflets and pubescent pods.

Ecology. Succulent shrubland on limestone escarpment. c.50m.

Distribution. NE tropical Africa, Arabia to India.

ETHNOGRAPHIC NOTES

Name(s)

See *R. minima* (also uses).

Taverniera DC.

Subshrubs. Leaves 3-foliolate; leaflets entire. Flowers solitary or few, axillary. Calyx expanded and papery in fruit. Pods jointed, spiny, hidden by the calyx. 1 endemic. 10 species. W India, Middle East, NE Africa.

Taverniera sericophylla Balf.f.

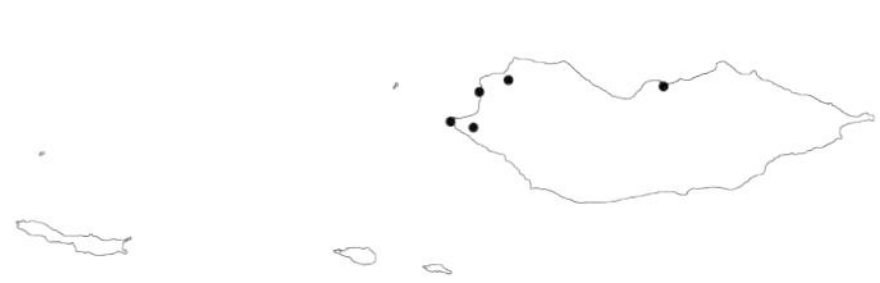

Prostrate subshrub to 0.4m. Leaflets obovate, c.0.7 x 0.4cm, semi-succulent. Flowers pink with darker lines. Pods c.1cm long.

Ecology. Locally common in dwarf shrubland on the coastal plain and foothills of western Soqotra (with *Pulicaria stephanocarpa*, *Limonium* species and *Farsetia socotrana*). Sea-level–300m.

Distribution. Endemic.

Conservation status. Vulnerable (VU D2). In the Red Data Book (Lucas & Synge 1978) *T. sericophylla* is categorised as extinct. However, in 1998 it was rediscovered on an isolated coastal plain some 15km from the type locality, where it occupies a narrow belt parallel with the beach. More recently it has been found on the dry foothills of the adjoining mountains. It now has an estimated area of occupancy of considerably less than 20km^2. The failure to find it at its original site suggests that there has been a decline in its area of occupancy. The new coastal site was until recently inaccessible by road but there is now a vehicle track and there are plans for developing the coastal plain, which would severely threaten this delicate coastal habitat. However, the recent discovery of the plant on the dry foothills much reduces the perceived threat to this species. The plant is, however, highly palatable to livestock and therefore any increase in grazing in the area could lead to it becoming critically endangered in a short time period. It has recently also been reported, but not confirmed, from Erisil in the northeast of the island and at the western end of Qa'arah on the southern plain.

ETHNOGRAPHIC NOTES

Name(s)

– Soqotri: nil or *ʿamihaṭ*.

– Soqotri simplified: *'amihat.*

Food (livestock). In the coastal areas where it grows it is regarded as an excellent fodder for goats and sheep. Camels also eat it, and it can be cut and fed to cattle.

Bees / honey. Honeybees are attracted to the flowers.

Comparative uses. The sweet roots of *T. cuneifolia* are used as a substitute for liquorice. The leaves are crushed to make a poultice for sloughing wounds, and the flowers are eaten. (Ambasta 1982; Jain & De Fillips 1991)

Tephrosia Pers.

Herbs or shrubs. Hairs simple. Leaves simple or pinnate; leaflets silky hairy, lateral nerves closely parallel. Pods narrowly oblong, flattened, often explosively dehiscent, the valves then becoming twisted. 2 endemics. c.400 species. Seasonal tropics, especially Africa.

ETHNOGRAPHIC NOTES

Comparative uses. Most species are unpalatable to livestock, but *T. ansellii*, *T. bracteolata* and *T. linearis* are said to be eaten by stock. Some are very poisonous. For instance, *T. vogelii* is insecticidal and piscicidal, especially the seeds. A decoction of its leaves, bark and green pods are an abortifacient, and are used to kill parasites of dogs and goats as well as for various skin troubles. The pounded leaves are used to stupefy fish. In Oman the leaves of *T. apollinea* are ground, put in the nostrils and inhaled to relieve nasal congestion. They are boiled in water and used as eardrops, and dried and powdered and applied to wounds, or heated to a paste applied to fractures and swollen joints to relieve pain. The dried and powdered stem-bark is added to water and poured into the ear to extract camel ticks. Elsewhere, the plant is used to adulterate senna. *T. purpurea* (wild indigo) and *T. uniflora* are grown as a green manure and cover crop. In India *T. purpurea* and *T. tinctoria* are used to extract a dark blue dye similar to indigo. On the mainland a decoction of the seeds is given for the retention of urine and other problems of the urinary tract. The plant is used as a tonic, laxative, diuretic and deobstruent to treat bronchitis, bilious febrile attacks, boils, pimples and bleeding piles.

Root. This is added to milk to flavour it. A decoction of the roots is given in dyspepsia, diarrhoea, rheumatism, asthma and urinary disorders. The roots are given with black pepper for colic, and a liniment prepared from them is used in elephantiasis. The pulverised roots are smoked to relieve asthma and a cough. They are also used as an insecticide.

Pods. A decoction of the pods is used as a vermifuge and to stop vomiting.

Seeds. The seeds yield an oil said to be good for scabies, itch, eczema and other skin conditions. They are also used as a coffee substitute and as an insecticide. (Abdulfatih 1987; Ambasta 1982; Bebawi & Neugebohrn 1991; Dalziel 1948, 1956; Ghazanfar 1994; Jain & De Fillips 1991; Kunkel 1984; Oliver-Bever 1986)

Tephrosia apollinea DC.

Tephrosia apollinea var. *brevistipulata* Vierh.
Tephrosia purpurea sensu Balf.f. non (L.) Pers.

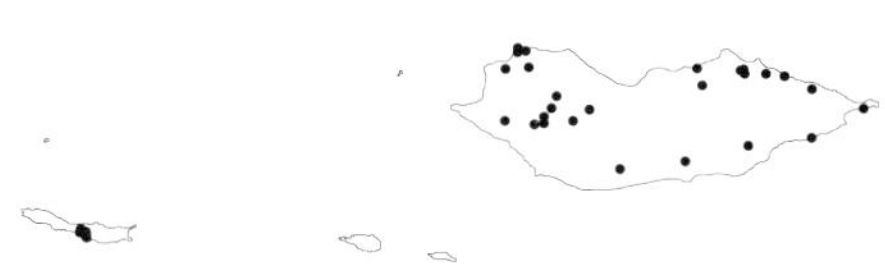

Separated from *T. purpurea* on the basis of its usually fewer leaflets (5 to 9) and more erect habit.

Ecology. Often abundant on the plains and foothills, particularly in overgrazed areas. Sea-level–400m.

Distribution. Arabia, NE Africa, Egypt.

Botanical notes. Very widespread and often abundant on Soqotra. It is particularly common in degraded *Croton socotranus* shrubland on the coastal plains, and is probably an indicator of overgrazing. Reported to be on the increase in many areas on the island. Thulin (1997a) suggests that this is best regarded as a form of *T. purpurea*, but the combination has not yet been made.

ETHNOGRAPHIC NOTES†

Name(s)

– Soqotri: *ṭifher* 'finger-, toenail'.

– Soqotri simplified: *tifher.*

Food (livestock). Goats that are accustomed to it eat the dry parts when hungry, and sheep eat dead and fallen material. Both readily eat the flowers and, as the dry season progresses, the fallen pods. Cattle nibble the fallen leaves when there is no alternative feed, and camels browse the green plant when very hungry. Otherwise this plant is avoided by livestock: the green leaves and pods are very bitter, and animals which are not accustomed to it can become ill if they feed on it.

Bees / honey. Important honeybee forage in the drier areas, especially in the summer months.

Flavouring / food additive. The milk of animals eating a lot develops rather a bitter taste. The meat of the carcass, especially the viscera, and the water in which the meat is cooked also become unpleasantly bitter.

Fuel. The dead stalks are gathered and stored for use as tinder and kindling.

Other. *Tephrosia* species cannot stand too much wet, and die back if waterlogged. Where they come up in the small grazing paddocks and enclosures they are pulled up and destroyed, as they prevent the grasses and more delicate herbs from growing. *Tephrosia* species are on the increase in most areas of the island and are becoming a serious nuisance, especially in coastal areas such as Hallah in eastern Soqotra.

Notes. Islanders regard the smaller *tifher* plants with small fruit and leaves as 'female', and the larger plants with larger fruit and leaves as 'male'. There is often a lack of distinction between *Senna* and *Tephrosia* species.

†Also apply to *T. purpurea*, *T. socotrana*.

Tephrosia odorata Balf.f.

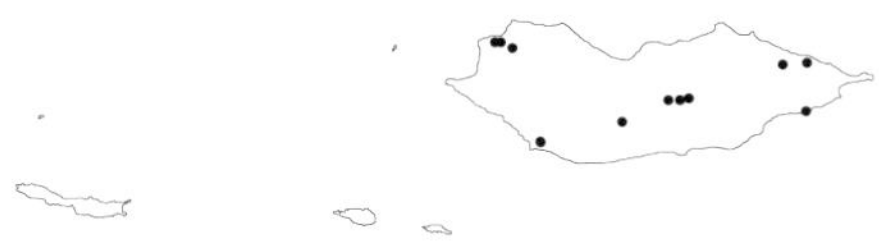

Small prostrate herb, sometimes with large woody rootstock. Leaves 3-foliolate; leaflets narrowly oblong to linear, 0.4–2cm long. Flowers purple with some white, solitary in leaf axils, c.1.2cm long.
Ecology. Widespread in grassland on the limestone plateaus. 300–700m.
Distribution. Endemic.
Conservation status. Least Concern (LC). Common in several vegetation types and under no present or perceived threat.
Botanical notes. A most distinctive species with its relatively large purple flowers.

ETHNOGRAPHIC NOTES
Name(s)
– Soqotri: nil or *di hibāləno* 'little tamarind fruit', 'spots of a rash'; *śəmraḥ* 'stalky'; *noi* (given to various small and sweet-smelling plants); *śebərʌḳ* 'glimpsed'; *ḥaṭarṭəhʌr* 'striped' (more commonly given to fern species); *maḥśɔ̄mhin* 'sticks in the throat'; *ṭifher* 'finger-, toenail'.
– Soqotri simplified: *di hibalino*; *shimrah*; *noy*; *shebiraq*; *hatartihar*; *mah'showmhin*; *tifher.*
Food (livestock). See *T. apollinea*, but it is said to be even less palatable.
Bees / honey. Honeybees are attracted to the flowers.
Cleansing. Flower sprigs are sometimes tucked into the clothing or amongst bedding for their pleasant smell. The sniffing of sweet-smelling plants is held to protect from the 'noxious air' that can cause disease, and against infection. All sweet-smelling plants are endowed with beneficial medicinal and prophylactic qualities.
Notes. Can cause *'uqabah* in small stock unaccustomed to it.

Tephrosia purpurea sensu Balf.f. non (L.) Pers.

Erect to prostrate herb to 80cm, often silvery hairy. Leaves 11 to 17-foliolate; leaflets narrowly oblong to linear, 1–3cm long. Flowers reddish purple, in lax, axillary racemes, c.0.8cm long. Pods 3–4cm long.
Ecology. No details of habitat or location available.
Distribution. Africa, Arabia, Pakistan to Indonesia.
Botanical notes. Thulin (1997a) regards this as part of the widespread and variable species *T. apollinea* – see notes under that species. All records are included under the map of *T. apollinea.*

ETHNOGRAPHIC NOTES
Name(s)
See *T. apollinea* (also uses).

Tephrosia socotrana Thulin

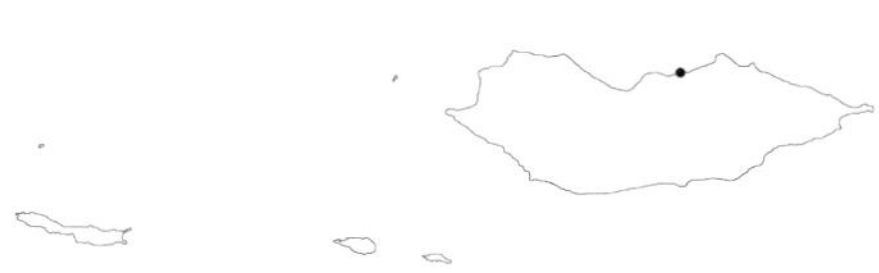

Slender shrub to 2m, often densely white hairy. Leaves 7 to 11-foliolate; leaflets narrowly obovate, 1.4–3.5 x 0.5–1cm. Flowers purplish-red, in dense terminal racemes. Calyx covered with dense brown hairs.
Ecology. On cliffs on the coastal limestone escarpment. 300m.
Distribution. Endemic.
Conservation status. Vulnerable (VU D2). Rare, with a very limited area of occupancy.
Botanical notes. Seen only on the limestone cliffs at Ras Heybaq.

ETHNOGRAPHIC NOTES
Name(s)
See *T. apollinea* (also uses).

Tephrosia strigosa (Dalzell) Santapau & Maheshwari

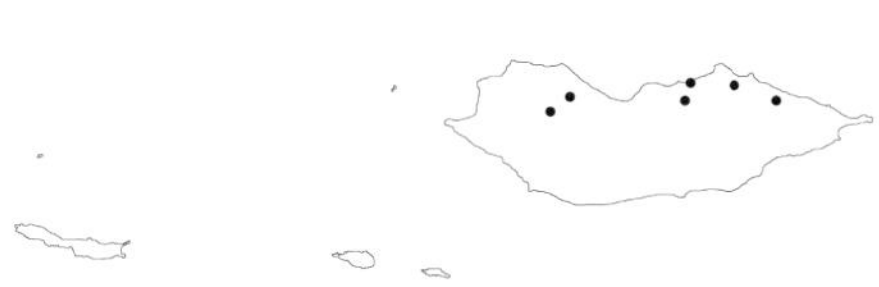

Erect or somewhat spreading, delicate herb. Leaves simple, narrowly elliptic, 1.5–4 x 0.1–0.3cm. Flowers bluish, solitary in leaf axils. Pods 1.5–2cm long.
Ecology. *Croton socotranus* shrubland and clearings in deciduous woodland. 10–250m.
Distribution. Oman, United Arab Emirates, Pakistan, India.

ETHNOGRAPHIC NOTES
Name(s)
– Soqotri: *ṭifher* 'finger-, toenail'; *di bətaḥan*, *bəṭaḥiyun* 'prostrate'; *di həbālino* 'little tamarind fruit', 'spots of a rash'.
– Soqotri simplified: *tifher*; *di bitahun*, *bitahiyun*; *di hibalino.*
Food (livestock). Weaned lambs and kids nibble this when there is little else available but it is otherwise avoided by livestock.
Notes. Can cause *'uqabah* in small stock unaccustomed to it.

Tephrosia subtriflora **Hochst.**

Spreading herb. Leaves 5 to 9-foliolate; leaflets elliptic, 1–3 x 0.3–1cm. Flowers pinkish, several in leaf axils. Styles straight, ribbon-shaped, 1.5–2.5mm long. Pods c.3cm long, somewhat curved.

Ecology. Dwarf shrubland on the coastal plain. Sea-level.

Distribution. NE Africa, Arabia, Madagascar, Pakistan, India, Myanmar (Burma).

ETHNOGRAPHIC NOTES

Name(s)

– Soqotri: *ṭifher* 'finger-, toenail'.

– Soqotri simplified: *tifher.*

Food (livestock). Apparently more palatable to small stock than the other *Tephrosia* species, and they are said to search it out in the dry season. However, it is considered to have little nutritional value.

Tephrosia uniflora **Pers.**

Tephrosia anthylloides Hochst.

Prostrate or ascending slender herb, often silvery hairy. Leaves 5 to 17-foliolate; leaflets elliptic to obovate, 0.4–1 x 0.1–0.6cm. Flowers yellowish or purple, 1 or 2 in leaf axils. Styles twisted, tapering, 2.5–3.5mm long. Pods 1–5cm long.

Ecology. Dwarf shrubland on the coastal plains. Sea-level–100m.

Distribution. Drier parts of Africa, Arabia, Pakistan and India.

ETHNOGRAPHIC NOTES

Name(s)

– Soqotri: *ṭifher* 'finger-, toenail'; *ḥewīr* 'dark-coloured' (western Soqotra); *ḳerəhān* 'horns' (from the podded fruit); *ṭifher* or *ṭīfiṭ* (Abd al Kuri).

– Soqotri simplified: *tifher*; *hewiyr*; *qerihan*; *tiyfit.*

Food (livestock). When dry, this is quite palatable to small stock in the more poorly vegetated regions of Soqotra and on the outer islands. However, it is considered to have little nutritional value.

Teramnus P.Browne

Herbs. Leaves 3-foliolate; leaflets entire. Flowers pinkish white, solitary in leaf axils. Stamens 10, five fertile anthers alternating with five sterile. Pods slightly curved, with an upturned beak. 8 species. Tropics.

Teramnus repens **(Taub.) Baker f.**

Teramnus labialis sensu Balf.f. non Spreng.

Teramnus labialis var. *mollis* sensu Balf.f. non Spreng.

Teramnus repens subsp. *gracilis* (Chiov.) Verdc.

Scrambling slender herb with spreading hairs. Leaflets ovate to subcircular, 1–2.5 x 0.7–2cm. Pods 2.5–4cm long.

Ecology. Drought-deciduous woodland. 50–450m.

Distribution. Arabia and NE Africa.

Botanical notes. All material on Soqotra is referable to subsp. *gracilis* (Chiov.) Verdc.

ETHNOGRAPHIC NOTES

Name(s)

– Soqotri: *ḥabbat al ʿayn* 'seed for the eye' (also given to *Chamaechrista absus*); *eritrit* 'forming green patches on the ground'; *di ḥʌmźəhʌn* 'sour, astringent'; *ʿafolhi ḳʌṭən* 'slim *ʿafolhi*' (*Leucas martinicensis*).

– Soqotri simplified: *habbat al 'ayn*; *eritrit*; *di hamdhihan*; *'afolhi qatin.*

Food (livestock). Very much liked by all livestock, especially cattle, and is seen as being a good feed for them, on a par with grass. Animals feeding on it become fat, and increase their milk yield.

Medicinal / veterinary. As its name implies, this plant is used principally to treat eye problems, and particularly the condition known as *di ʿeyḥʌntin*, in which the eyes are inflamed and painful. This problem is especially widespread during the monsoon months when the air is filled with wind-blown dirt and dust. The seeds are removed from the small pods, and cracked between the teeth to remove their black, outer skin. The kernels are then ground to a powder and put in the affected eye. The same medicament is used after carrying out any operation on the eyes to soothe the pain and prevent infection. Some add hot water and use the solution as eyedrops, but most say that it is more effective as a dry powder. This treatment is also given to livestock with sore eyes.

Notes. It can cause severe scour if a lot is eaten at the beginning of the rains.

Comparative uses. Many *Teramnus* species provide excellent grazing, and cultivation of some as fodder or pasture plants has been tried. *T. labialis* is grown for grazing, as a green manure and as a cover crop. Its young leaves are eaten and the plant is also used to treat rheumatism, tuberculosis, nervous affections and catarrh; the fruits are taken as an astringent, stomachic and febrifuge. (Ambasta 1982; Bogdan 1977; Jain & De Fillips 1991; Kunkel 1984)

Trigonella L.

Herbs. Leaves 3-foliolate; leaflets toothed. Flowers yellow, in short axillary racemes. Pods flattened, strongly curved. 1 endemic. 80 species. Mediterranean, S Africa, Australia.

Botanical notes. Pungent smell (fenugreek).

Trigonella falcata Balf.f.

Small, rather delicate, prostrate herb. Leaflets obovate, 0.4–1.2 x 0.4–1.2cm. Flowers c.0.5cm long. Pods c.1.5cm long.

Ecology. Semi-deciduous woodland on limestone escarpment. 650m.

Distribution. Endemic.

Conservation status. Data Deficient (DD). Inadequate evidence is available to make an assessment.

Botanical notes. Apparently a rather rare endemic; possibly overlooked.

ETHNOGRAPHIC NOTES

Name(s)

– Soqotri: *ru'ud* 'grazing plant'; *ḳə̀ṣəho* (general term used to describe new shoots of herbs and grasses coming up after rain).

– Soqotri simplified: *ru'ud*; *qiziho*.

Food (livestock). Liked by all livestock, especially small stock, and regarded as useful feed for them.

Comparative uses. Many species are grown as potherbs. They are also used as condiments and, in Arabia and Africa, as flavouring agents for cheese and butter. Medicinally they are used chiefly in the treatment of diarrhoea. *T. foenum-graecum* is fenugreek, widely used in cooking. (Ambasta 1982; Jain & De Fillips 1991; Kunkel 1984)

Vigna Savi

Herbs. Leaves 3-foliolate, entire. Flowers purple, in short terminal or axillary racemes. Anthers 10; filaments 9 fused, 1 free. Pods flattened. 150 species. Mostly Old World tropics.

Vigna membranacea A.Rich.

Vigna luteola Balf.f. non (Jacq.) Benth.

Scrambling vine with spreading hairs. Leaflets ovate, sometimes hastate, 2–6 x 1–4cm. Flowers purple, c.2cm long. Pods c.5cm long.

Ecology. No habitat details available. 150–450m.

Distribution. NE Africa and Arabia.

Botanical notes. Recorded from the Haggeher in the 19th century but not seen since. Soqotran material is referable to subsp. *caesia* (Chiov.) Verdc.

ETHNOGRAPHIC NOTES

Comparative uses. Many of these wild vetches provide good grazing for livestock, and others are grown for food (such as the adzuki bean, black and green gram). Some also have edible roots. The cowpea, *V. unguiculata*, is an important food crop, which is also grown for fodder and as a green manure. (Ambasta 1982; Bogdan 1977; Jain & De Fillips 1991)

Zornia J.F.Gmel.

Herbs. Leaves 2-foliolate; leaflets entire, glandular-punctate. Inflorescence a lax, terminal spike; flowers enclosed in paired bracts; bracts slightly spurred at base. Filaments united into a tube. Pod transversely segmented; segments indehiscent, covered with bristles. c.75 species. Tropical.

Zornia glochidiata DC.

Zornia diphylla Pers.

Erect annual herb to 40cm. Leaflets lanceolate, 1–2 x 0.2–0.8cm. Bracts lanceolate, 0.8–1cm. Calyx colourless, c.0.3cm long. Corolla yellow, c.0.5cm long, much shorter than the bracts. Pod 1–2cm long, constricted between the seeds, segments round; bristles c.0.2cm long, stiff and minutely retrorsely hispid.

Ecology. Grassland clearings in semi-deciduous woodland; on granite in Haggeher mountains. c.650m.

Distribution. Tropical Africa and Madagascar.

ETHNOGRAPHIC NOTES

Name(s)

– Soqotri: nil or *raʿàdihon* 'tiny grazing plant'.

– Soqotri simplified: *ra'adihon.*

Food (livestock). Of no importance to livestock, but it is eaten along with the grasses amongst which it grows. It is considered to be as nutritionally valuable as grass.

Comparative uses. *Z. diphylla* is grown as a cover crop and a green manure. It is a nutritious fodder for cattle. It is pounded up and eaten with couscous and, mixed with butter or oil, is used to rub on the body for feverish chills, etc. The plant is also used to treat dysentery and the root is given to children as a soporific. Some *Zornia* species, e.g. *Z. diphylla*, form a slight lather with water and can be used as soap. (Ambasta 1982; Dalziel 1948, 1956; Jain & De Fillips 1991)

Zygocarpum Thulin & Lavin

Shrubs. Leaves c.5-foliolate. Flowers blue or purple, usually in pairs, long pedicellate. Pods jointed, breaking transversely into 1-seeded segments. 1 endemic. 6 species. SW Arabia and Somalia.

Botanical notes. For notes on phylogeny and biogeography see Thulin & Lavin (2001).

Zygocarpum coeruleum (Balf.f.) Thulin & Lavin

Ormocarpum caeruleum Balf.f.

Shrub or small tree with long, slender branches. Leaves c.1cm long; leaflets obovate-orbicular, with conspicuous blackish midrib. Pod segments c.1.5cm long.
Ecology. Common in semi-deciduous woodland and shrubland, mainly on the limestone plateaus but also occurs on granite in the Haggeher mountains. (50–)200–800m.
Distribution. Endemic.
Conservation status. Least Concern (LC). Common in several vegetation types and under no present or perceived threat.
Botanical notes. Recognised by the violet-blue flowers. Usually a slender tree or shrub but can be prostrate in windswept places on the limestone plateau. However, it is easily recognised, even when not in flower, by the distinctive purplish black line along the midrib on the undersurface of the leaflets.

ETHNOGRAPHIC NOTES

Name(s)
– Soqotri: *ḥˈaymᵓher*, *ḥˈimihər* (pl. *ḥˈomihir*, *ḥəmˈʌhʌr*) (western Soqotra) *khˈayməher*, etc.; smaller specimens *ḥamirmˈiro* 'little *ḥˈimihər*' (pl. *ḥamˈirmirhinˈitin*); *ṭˈʌləhon*.
– Soqotri simplified: *haym'her*, *himiher* (pl. *homihir*, *himahar*); *khaymiher*; smaller specimens *hamirmiro* (pl. *hamirmirhinitin*); *talihon*.
Food (human). Tender new leaves are eaten as a fresh-tasting nibble.
Food (livestock). One of the key fodders for livestock on Soqotra, especially for goats, though it also provides dry season fodder for cows. It promotes lactation and is often cut and fed to milk animals for that reason. It is the best feed for young weaned kids, making them fat and sleek. With *Ziziphus*, it is considered to be the optimal feed for camels. Sheep are fed the small leaves in the dry season, or if they are sickly or have just given birth. Indeed, it is so liked by livestock that leafy stems are used as a lure or bribe to manage recalcitrant animals. As well as being eaten fresh, it is lopped in the dry season for pregnant and lactating animals, and then, if the dry season is prolonged, for all favoured animals. Both cattle and goats chew at the leafless twigs at this time. For goats it is the best of all dry season fodders, and the second best for cattle after *Sterculia africana*.
Bees / honey. Important honeybee forage.
Flavouring / food additive. The milk of livestock eating a lot is characteristically rich and sweet, and makes a lot of excellent butter.
Fuel. The dead wood is one of the best firewoods of the island. It lights easily, burns slowly and makes little smoke, especially if the bark is removed first, making it ideal for cooking and warming enclosed spaces. It gives off a good heat and is much used for warming the home and young or sickly livestock in cold or wet weather. A store is kept in the home for this use, especially in the high mountain areas that are enveloped in mists and cloud for much of the monsoon and winter. The wood burns with a very bright flame, making it ideal for night-time fires and illumination generally. Burning sticks are used as torches. In addition, the wood gives off a very pleasant smell as it burns.
Timber. The wood is hard and is considered to provide one of the best timbers on the island. It is not attractive to termites but, like all timber, it must be cut in hot, dry weather. Wood from wetter areas makes better timber if de-barked and dried fully before use. The timber is considered equal in hardness to that of *Buxanthus pedicellatus*. The trunks and larger branches are used for the main roofing beams and supporting pillars. Timber from this and *B. pedicellatus*, like that of *Ziziphus* and *Maerua angolensis*, lasts through several generations and can be re-used again and again, as long as it is never allowed to get wet and is regularly fumigated with smoke. A big chunk of *ḥamirmˈiro* wood is used to keep fires in by tucking it deep into a heap of hot ash and then leaving it until the fire is needed again. Then the herder removes it, taps it firmly on the fire-stones, and then blows to rekindle the flame.
Leatherwork. The dead wood burns down to a very fine ash which is very good for preparing skins and hides.
Wood. The wood is considered to be as hard as that of *Buxanthus pedicellatus*. It is close-grained and used for implements and utensils where hardness of wood is of prime importance, such as wooden locks and keys, ghee and grain measures, pack saddles, combs, the sword beater and shed stick for weaving, and small bowls. The slimmer, straight branches are used to make the important herding staff, and the wood is widely used for poles and straining posts in fencing, and in the construction of pens, byres and temporary accommodation. It also makes the highest grade of charcoal, and this was highly valued by the island blacksmiths. That of *ḥˈimiher* was preferred to that made from *ḥamirmˈiro*, as its heat was said to be easier to control, and the coals quicker to douse when necessary.
Medicinal / veterinary. The fodder of choice for sick animals, especially sheep, goats and camels. It is also fed to an animal with a retained placenta, or one in poor condition or which has lost weight. A cow losing weight towards the end of her pregnancy is brought leafy stems to build up her strength. Like other favourite fodders, leafy stems are offered to a milk animal reluctant to let down her milk; once she starts to chew the leaves she is likely to relax and allow herself to be milked. To treat someone suffering from painful joints, a fire of this wood is lit at the mouth of a small cave or other enclosed space and the patient sat inside. It is kept burning until the patient is covered in sweat, which is said to improve the condition.
Cleansing. The wood and especially the dried roots give off a very aromatic smoke when burned, and pieces of the root are kept at the home for this purpose.

Pest control. The dead wood is used for livestock smudges, either outside in the pen and milking yard, or, if the weather is bad, inside the byre. It both warms and soothes the animals, and deters the many biting insects that flourish after the rains.

Other. It is the wood of choice for firing pottery. In the days before metal hooks were readily available, fish-hooks of all sizes were whittled from the strong heart-wood. Before going into wooded areas in the winter months, a time when the botfly was most prevalent, the face of a child was rubbed with powdered charcoal made from the wood. The botfly always haunts such shady, still places, and it was believed that by painting the children's faces they were made invisible. When there are no cereals available, the leaves can be used to clarify butter when making ghee, though the residue is not usually eaten.

Notes. The islanders distinguish between *ḥimiher* and *ḥamirmiro*: the latter is not only smaller in stature, but also has smaller leaves and is inferior to the *ḥimiher* in every way, even as firewood, though its charcoal is preferred by blacksmiths. It is possible that *ḥimiher* refers to *Chapmannia* species and *ḥamirmiro* to *Zygocarpum coeruleum* shrubs, but further research is needed to establish this. It is further complicated by the fact that in different parts of the island the same species can be called both *ṭáləhon* and *ḥimiher*.

Comparative uses. This species was formerly included in *Ormocarpum* and these notes refer to species in that genus. In Dhofar, *O. dhofarense* is one of the most important fodder plants for livestock, and the flowers are sucked for their nectar. The leaves are also used medicinally. They are crushed to a paste rubbed into the scalp to kill head lice. Other *Ormocarpum* species, such as *O. trachycarpum* in Africa, are also highly prized as fodder trees. Its twigs are used as a toothbrush and the roots are boiled in water and drunk for stomach problems. (Ambasta 1982; Heine *et al.* 1988; le Houerou 1980; Jain & De Fillips 1991; Miller & Morris 1988)

LENTIBULARIACEAE

Herbs. Leaves alternate, bearing bladder-like traps; traps with a complex arrangement of glands at the mouth (view with hand lens). Flowers in lax racemes on a slender terminal peduncle, bisexual, zygomorphic. Calyx 2-lobed. Corolla 2-lipped, the upper lip with 2 teeth, the lower with 3 lobes and a spur. Stamens 2, enclosed by corolla. Ovary superior, 1-celled; ovules many on a large, free-central placenta; style 1. Fruit a capsule. 3 genera. c.250 species. Cosmopolitan.

Utricularia L.

See description of family. 214 species. Cosmopolitan.

Utricularia striatula Sm.

Tiny, mat-forming, epiphytic, stoloniferous herb to 5cm. Leaves basal, often hidden amongst moss, c.0.5 x 0.2cm. Corolla white, 0.4–1cm long (much smaller in cleistogamous flowers). Capsule globose, keeled, c.2.5mm long.

Ecology. Epiphytic, in shade, in semi-deciduous montane woodland. 1200m.

Distribution. W Africa, Ethiopia, China, Arabia (Dhofar), India, SE Asia. Apparently rare in E Africa. On the nearby Arabian mainland known only from the monsoon woodlands of Dhofar in southern Oman.

Botanical notes. Minute herb growing with moss on branches. Ephemeral, growing only during the wet season.

ETHNOGRAPHIC NOTES

Name(s)

– Soqotri: *eźərīʾo* (*men mèssə*) 'residue or sediment (from the rain)' (a general term for mosses).

– Soqotri simplified: *edheriy'oh* (*men messih*).

Fuel. It is gathered inadvertently with dried mosses and lichens which are collected and stored for use as tinder in the wet season.

LINACEAE

Herbs. Leaves alternate, simple, entire; stipules absent. Flowers in terminal lax, bracteate cymes, actinomorphic. Sepals 5. Petals 5. Stamens 5, filaments expanded and united at base. Ovary superior, 5-celled; ovules 2 per cell on axile placentas. Styles 5. Fruit a 5-part capsule. 15 genera. c.300 species. Cosmopolitan.

Linum L.

See description of family. c.200 species. Temperate and subtropical; especially diverse in Mediterranean region.

Linum trigynum L.

Linum gallicum L.

Erect, delicate, annual herb to 10cm. Leaves linear to narrowly lanceolate, 0.5–1.2 x c.0.1cm, entire, tip acute. Flowers yellow, c.0.5cm across. Fruit globose, c.2cm diameter, surrounded by persistent calyx with sepals <0.3cm.

Ecology. Semi-deciduous woodland on limestone plateaus. 600–650m.

Distribution. Mediterranean and Middle East.

ETHNOGRAPHIC NOTES

Name(s)

– Soqotri: *ḳə̀ṣəho* (general term for the tender new shoots of herbs and grasses); *lə̀bənhən* 'white, pale'.

– Soqotri simplified: *qiziho*; *libinhan.*

Food (livestock). This is one of the *ru'ud* group of plants which come up with the rains, providing incidental green grazing until they die back. Though palatable to livestock, this is not important forage for them on Soqotra.

Comparative uses. Some *Linum* species are toxic and contain cyanogenic glycosides: both *L. rigidum* and *L. neomexicanum* have been implicated in livestock poisoning, and *L. catharticum* of Europe is also a hazard to livestock. *L. usitatissimum* is an important plant, as its name suggests. The stalks are the source of flax, while the seeds yield a quick-drying oil, linseed oil, and are made into a cake as a protein supplement for livestock. In Yemen the seeds, called *mūma*, are used to treat kidney stones. It is used as a laxative and anti-inflammatory. (Ghazanfar 1994; Kunkel 1984; Lewis & Elwin-Lewis 1977)

LYTHRACEAE

Herbs, often in or near water. Leaves opposite or alternate, simple; stipules absent. Flowers actinomorphic. Sepals and petals inserted on a cup-like hypanthium; sepals triangular sometimes with subulate appendages between them; petals pink or white, delicate, often with claw, crumpled in bud. Ovary superior, 2 to 4-celled. Fruit a capsule. 26 genera. 580 species. Tropical and some temperate.

Ammannia L.

Annual herbs, glabrous; stems 4-sided. Leaves opposite. Flowers in axillary cymes, or in cymes arranged on short axillary branches. Hypanthium bell-shaped. Sepals 4. Petals absent. Stamens 4. c.20 species. Wet habitats throughout the tropics and subtropics.

Ammannia baccifera L.

Prostrate herb to 15cm, rooted in mud by springs and wadis. Leaves oblanceolate to narrowly oblong-elliptic, 1–3 x 0.1–0.7cm, attenuate at base. Flowers white or pinkish, 0.1–0.2cm diameter, densely clustered in axils. Fruits ripening red, subglobose, c.0.2cm diameter.
Ecology. In mud by permanent water. 50–200m.
Distribution. Widespread in the Old World.

ETHNOGRAPHIC NOTES†
Name(s)
– Soqotri: *mṣirbihin di rīho* 'winter-rains plant of water'; *ḳerəżaʾihun* 'little *ḳurżaʾ*' (given to various Gramineae and Cyperaceae of damp ground).
– Soqotri simplified: *mizirbihin di riyho*; *qeridha'ihun*.

Food (livestock). Provides incidental grazing for cattle and sheep. Goats on Soqotra intensely dislike damp forage, but graze the plants once they have dried out. They are considered to be a useful fodder.
Comparative uses. The fodder value of *A. baccifera* is high. In Sudan it is grazed by livestock but the leaves raise blisters on the animal's skin. In Iraq this property is exploited and the leaves are used in skin diseases as a rubifacient. Taken internally they are acrid and vesicant. In India the leaves are bruised and used to treat ringworm and other parasitic skin conditions, and the plant is reported to have anti-typhoid and anti-tubercular properties. Elsewhere the whole plant of *A. crassicaulis* is crushed and placed in drinking troughs as a purge for cattle. In Iraq *Lythrum salicaria* is regarded as an astringent and tonic, and is used in the treatment of haemorrhoids, diarrhoea and dysentery. *L. anceps* and *L. salicaria* are eaten as potherbs. (Al Rawi & Chakravarty 1964; Bebawi & Neugebohrn 1991; Dalziel 1948, 1956; Jain & De Fillips 1991; Kunkel 1984)

†Also apply to *Lythrum hyssopifolium*.

Ammannia multiflora Roxb.

Similar to *A. baccifera* but distinguished by erect stems, cordate leaf bases clasping the stem, and flowers in cymes on short axillary branches.

Ecology. No habitat details available, but probably growing in mud by wadis.
Distribution. Widespread in the Old World.
Botanical notes. Recorded by Balfour (1888) from Qalansiyah about dwellings in the 19th century; not seen recently.

Lythrum L.

Herbs. Leaves alternate. Flowers solitary in leaf axils. Hypanthium cylindrical. Sepals c.5, alternating with short, subulate appendages. Stamens 4 to 6. Capsule cylindrical. c.30 species. More or less cosmopolitan.

Lythrum hyssopifolium L.

Prostrate annual to 30cm. Leaves narrowly oblong, 1–1.5 x c.0.2cm. Flowers lilac or pink, c.0.4cm diameter, solitary in leaf axils.

Ecology. In mud by permanent water. 50m.
Distribution. Cosmopolitan weed.

ETHNOGRAPHIC NOTES
Name(s)
See *Ammannia baccifera* (also uses).

MALPIGHIACEAE

Trees with appressed, medifixed hairs. Leaves alternate, simple; stipules absent. Flowers in racemes, actinomorphic. Sepals 5, united at base. Petals 5, clawed. Stamens 10. Ovary superior, 3-celled, styles 2. Fruit a winged samara. c.60 genera. 1200 species. Pantropical, especially New World.

Acridocarpus Guill. & Perr.

See description of family. 1 endemic. 30 species. Tropical Africa with 1 species in Madagascar and 1 in New Caledonia.

Acridocarpus socotranus Oliv.

Acridocarpus orientalis sensu Balf.f.

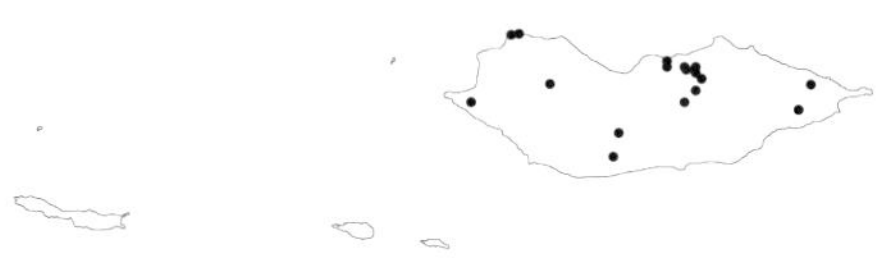

Small tree to 5m. Leaves elliptic to obovate, 3–8 x 1.5–4.5cm, leathery, veins prominent beneath, entire, tip rounded. Flowers bright yellow, 1–1.5cm diameter, in terminal racemes, peduncle and pedicels reddish hairy. Fruit 1 or 2-winged; wings 1.5–2 x 1cm.

Ecology. Drought-deciduous woodland, succulent shrubland and thickets along wadis. 20–650m.

Distribution. Endemic.

Conservation status. Least Concern (LC). Common in several vegetation types and under no present or perceived threat.

Botanical notes. Readily distinguished by its panicles of bright yellow flowers, winged fruits and leathery, evergreen leaves.

ETHNOGRAPHIC NOTES

Name(s)

– Soqotri: *kərə̍llo* (pl. *kəre̍l*); *ahtəre̍ś*, *aḥte̍ś* (central and western Soqotra); *məstə̍wrəher* (western Soqotra when cliff-dwelling).

– Soqotri simplified: *kirilloh*; *ahteresh*, *ahtesh*; *mistewriher*.

Food (livestock). The leaves remain green through most of the summer months, making it important dry season forage all over the island. In drier areas the foliage is considered to be quite good for goats, but they will not browse it unless there is nothing preferable to eat. In the better-vegetated areas the plant is less liked and goats usually only eat young foliage, buds and flowers. Hungry sheep eat leaves within reach, but normally prefer the fallen, dead leaves. If the dry season is severe, herders lop leafy stems for their animals, and gather fallen leaves to feed to young kids and lambs. However, herders treat it with great caution. If a lot is eaten by livestock, especially in the summer months when other forage is scarce, and its greenness makes it particularly attractive to the hungry animals, it can be harmful. Once the monsoon winds blow, and the whole plant becomes impregnated with salt, it becomes even more dangerous. In goats it can cause partial paralysis of the hindquarters. Even cattle, who only eat the fallen foliage if the dry season is prolonged or during drought, start to stumble and stagger if they eat a lot at this time. This is especially true of livestock that are not accustomed to it. Although most animals make a full recovery, sickly animals, or those already weakened by hunger, can be seriously affected, and sometimes die. Camels and donkeys readily eat the plant year round, but for camels the foliage is particularly toxic. The toxic effects show themselves quickly: the camel starts to totter and stumble, and is soon hardly able to drag itself along. Sometimes the condition can be improved by taking the animal down to the coast to drink brackish water for some time. Some cameleers also cauterise the animal on the backbone, above the tail, and burn a cross on the back of each thigh. Although the camel may recover from the worst effects, it will always suffer from residual weakness in the hindquarters. If it is a baggage animal it will no longer be able to bear heavy loads. It can be given lighter loads to carry on level ground, and can manage gentle slopes going uphill, but is usually unable to go downhill without stumbling. For this reason it is no longer safe to carry people, and many camel owners feel that it is better to slaughter such a disabled animal. The toxic effects of *Acridocarpus* are made even worse if the camel has also eaten *Allophylus* foliage. It becomes seriously ill and usually dies. Some camel owners say that *Acridocarpus* foliage is less toxic after good rains, when the salt has been 'washed off'. Many say that its toxicity in the dry months lies not so much in the plant itself and the raised summer salinity, but in a multi-coloured caterpillar or larva called *bəda̍ʿaš* (unidentified), which is to be found wrapped up inside the leaves of the bush at this season. Whatever the reason, cameleers who depend on their pack animals for their livelihood generally round up their camels and tie them up, or hobble them closely to keep them from the bushes. Alternatively they herd their animals away from areas where *Acridocarpus* grows and watch them closely so that they do not stray. The camels are usually restrained in this way from the end of the short summer season, right through the monsoon months, only being released when the winter rains fall.

Bees / honey. Important honeybee forage, especially in the summer months.

Fuel. The dead wood makes good firewood. It burns slowly with little smoke, and gives out a good heat, so is much used for cooking. It is liked for warming children and the sick in the living quarters, and for lighting fires in livestock byres and pens in cold weather.

Timber. The well-grown trees of the Central Highlands are often referred to as *ši̍rəhem ḥa̍hər*, 'black tree', referring to its desirability as timber. The trunk and branches are used for cross beams, supporting pillars and roof trees, the slimmer branches for roofing infill and for making a sturdy framework for walling and partitions. The better-grown specimens are a key timber of the drier areas of the island, but some of the smaller shrubs are soft-wooded and make a poor timber, lacking durability and with a tendency to crumble. The name in western Soqotra for such plants, *aḥte̍ś*, refers to this. The wood is attractive to termites, so is usually de-barked before use.

Wood. The important herding staff is made from slim branches of the stronger wooded trees. Dead, small stems can be used as firesticks.

Pest control. The green wood is good for livestock smudges, as it produces a lot of dense smoke.
Notes. Islanders differentiate between the 'male' trees with smaller leaves, and the larger-leaved 'female' tree (*kərˈəllo bˈaʿalə*), which is more common in the interior. The shrubs are usually removed from calf paddocks and smaller enclosures in the Haggeher mountains. Due to its reported effects on livestock, many mainland Yemenis jokingly refer to this plant as the Soqotran *qat*.
Other. The earth from around and under the plants is gathered and used for cultivation.

Comparative uses. In Oman, headaches, paralysis and muscle or tendon pain are treated by rubbing the seed oil and leaf extract of *A. orientalis* over the affected area. The oil, called '*ḥal al qafaṣ*', is made by soaking the dried seeds in water to remove the outer covering, and then grinding them to a coarse powder. Salt is added and the mixture is then kneaded with a little water to extract the oil. Cosmetically it is applied to the face and body to soften the skin. However, if it is drunk it can cause diarrhoea and vomiting. In Dhofar, the inner bark and new shoots are used as a tanning agent and are ground together and applied to the teat and udder to treat mastitis. The crushed leaves are added to water, and the filtrate drunk as a remedy for internal swellings, or applied locally to external swellings. (Dalziel 1948, 1956; Ghazanfar & Al-Sabahi 1993; Miller & Morris 1988)

MALVACEAE

Herbs, shrubs or small trees, often with an indumentum of stellate hairs. Leaves simple or palmately lobed; stipules usually present. Flowers actinomorphic, bisexual, usually solitary in leaf axils. Epicalyx sometimes present. Sepals 5, usually fused below. Petals free. Stamens many; filaments united into a column which is often fused to the base of the corolla. Ovary superior, 5 to many-celled; style with branches equalling or 2x length of carpels. Fruit a capsule or schizocarp. 116 genera. 1550 species. Cosmopolitan, especially tropical.

Abutilon Mill.

Perennial herbs or subshrubs. Leaves simple, rarely lobed. Flowers axillary, solitary or in panicles, opening in the evening. Epicalyx bracts absent. Calyx 5-lobed. Petals yellow, sometimes with a purple base. Fruit a schizocarp; mericarps 2 or more seeded. c.100 species. Tropics and subtropics.

Abutilon fruticosum Guill. & Perr.

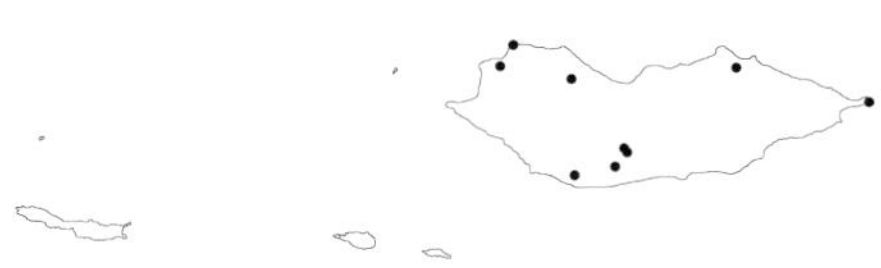

Perennial herb or subshrub covered with a grey, velvety indumentum of stellate hairs. Leaves broadly ovate, 3–5(–11) x 2–4(–8)cm, base cordate, margin denticulate or entire, tip acute or rounded. Flowers yellow, c.1.5cm across. Fruit cylindrical, (5–)7–8(–9)mm long; mericarps (5 to)10 to 12, acute or shortly awned at the apex.
Ecology. Widespread in stony places on the coastal plains, coastal foothills and in date gardens; characteristically in *Croton* shrubland and succulent shrubland. 10–500m.
Distribution. Tropical Africa and Arabia to India.
Botanical notes. *Miller* 19113 (a rather depauperate specimen) is anomalous in having fruits with only five mericarps. However, in all other respects it agrees with typical material.

ETHNOGRAPHIC NOTES
Name(s)
– Soqotri: nil or *mərˈʌnnəhʌ*; *ṣaʿlifˈitin* 'many-leaved'; *ṣefˈerher* 'many-flowered'; *šəźˈarhən* 'green'.
– Soqotri simplified: *miranniha*; *za'lifitin*; *zeferher*; *shidharhin*.
Food (human). In the hungrier, more poorly vegetated areas of the island in the days before sugar, the flowers were gathered and sucked for their nectar.
Food (livestock). The velvety leaves are browsed by goats, which also like the flowers and fruit. Sheep eat plant litter. It is not common enough to be useful forage and is regarded as having little nutritional value.
Medicinal / veterinary. The leaves and soft growing tips are crushed, and the paste applied to sores and cuts on both humans and livestock.
Comparative uses. In East Africa the wood is used for firesticks, and the dry stems as a firelighter. (Ambasta 1982; Dalziel 1948, 1956; Heine *et al.* 1988; Jain & De Fillips 1991)

Abutilon pannosum Webb

Abutilon graveolens sensu Balf.f.
Abutilon muticum G.Don

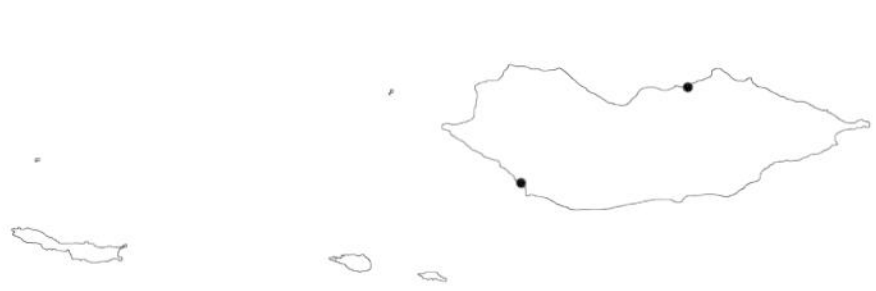

Perennial herb or subshrub, shortly stellate-tomentose. Leaves broadly ovate to orbicular, sometimes 2 or 3-lobed, 3–10(–17) x 3–8(–13)cm, base cordate, margin dentate, tip rounded to acuminate. Flowers yellow with a purple centre, c.3cm diameter. Fruit depressed-globular, 10–15mm long; mericarps 25 to 35, rounded at apex.

Ecology. Uncommon on the coastal plains in *Croton socotranus* shrubland and on the limestone plateau. c.350m.
Distribution. Tropical N Africa and the Middle East.

ETHNOGRAPHIC NOTES
Name(s)
– Soqotri: nil or *mərˈʌnnəhʌ*; *ṣaʿlifˈitin* 'many-leaved'; *ṣefˈerher* 'many-flowered'; *gˈumuṣ*, *gīmiṣ* (also given to *Malva parviflora*); *šəźˈarhən* 'green'.
– Soqotri simplified: *miranniha*; *za'lifitin*; *zeferher*; *gumuz*, *giymiz*; *shidharhin*.
Food (human). The fruit, seeds and leaves are edible. The green seeds are sweet but grow tougher and taste-

less with age. If enough can be gathered, they make a filling meal. In the hungrier, more poorly vegetated areas of the island in the days before sugar, the flowers were gathered and sucked for their nectar.
Food (livestock). All livestock eat this in the dry season, but it is not regarded as a particularly valuable feed.
Medicinal / veterinary. The leaves and soft growing tips are crushed, and the paste applied to sores and cuts on both humans and livestock.
Comparative uses. The seeds are eaten in hunger. The flowers and seeds of other *Abutilon* are also eaten, the flowers are eaten both raw and cooked. The dried fruits of *A. theophrastis* are ground into flour used for bread and soups. (Kunkel 1984)

Gossypium L.

Shrubs or small trees. Leaves deeply 3 to 5-lobed. Flowers solitary or clustered in leaf axils. Epicalyx bracts 3, large, heart-shaped, deeply toothed at apex, cordate at base. Calyx subentire or obscurely 5-lobed. Ovary 3 to 5-celled. Fruit a capsule; seeds with long white hairs (cotton). c.50 species. New and Old World tropics and tropics.

Botanical notes. New World species are widely cultivated and escaped in the Old World.

Gossypium barbadense L.

Shrub to 3m. Leaves <25cm across. Epicalyx lobes <3.5cm long, each with 2 to 6 teeth; teeth lanceolate, at least 3x as long as wide. Petals yellow, c.5cm long. Capsule ellipsoid, pitted with conspicuous glands.
Ecology. Near habitation. Sea-level–50m.
Distribution. A S American species, widely cultivated throughout the tropics.

ETHNOGRAPHIC NOTES

Name(s)
– Soqotri: *ˈiʾiśīrə* (pl. *eʾε̄śir*, *ʾε̄śər*) ‘fresh and soured milk mixed together’; cotton fibre *ēʾəˈśor*, *‘ɔ̄ˈśor*; fruit capsule *giˈridi*.
– Soqotri simplified: *i’ishiyri* (pl. *e’eyshir*, *‘eyshir*); cotton fibre *‘ey’shor*, *‘oshor*; fruit capsule *giridih*.
Food (livestock). These cultivated shrubs are usually protected with fencing or walling, as goats and cattle love the foliage, buds and flowers. Goats also chew at the bark when hungry.
Wood. The dead wood can be used as a firestick.
Other. Seeds are planted in protected seedbeds and then transplanted. They used to be quite widely cultivated in areas where water was available: beside main riverbeds, in house compounds and in farms around the main towns of Hadiboh and Qalansiyah. The odd shrub was also planted in the better-watered parts of eastern Soqotra, in the interior valleys behind the Haggeher, and in the larger date gardens. The shrubs began to be neglected when imported cloth became cheaper and more widely available. Then the protective walling was breached and cattle and goats were able to get in and eat the shrubs. Some plants are still cultivated in settlements in more remote areas. They are carefully tended and the cotton tufts are collected and picked clean. The cotton wool is most commonly used in laying out the dead. Small tufts are also used to make little pads, which are soaked in perfume oil and tucked in the outer ear or into clothing. To make thread, the fluffy balls are put in the sun to dry, and then the black seeds laboriously extracted. The cotton wool is then either beaten with the bow, like sheep's wool, or the fibres teased out between the fingers. In earlier times the spun thread used to be dyed and used to weave the coloured stripes of the finer wool rugs, especially those worn as clothing by women. Before the arrival of nylon, string made from cotton wool was also used to make fishing lines, and finer fibres were twisted to make sewing thread. When cotton twist began to be imported, this was commonly used instead to make fishing equipment, including nets, in the main fishing settlements and in areas close to the coastal markets.

A very fine and strong cloth called *ḥāf* used to be woven entirely from locally grown cotton, using the same ground loom as that used to weave the sheep's wool cloth. The *ḥāf* was made from five woven strips stitched together, and had a carefully made fringe at each end. Lengthways along the middle of each strip was a single red or green stripe, dyed usually with imported dyes, but sometimes paler and less watertight colours were achieved with local plant dyes. Unlike the sheep's wool cloth, the cotton cloth was woven with equally sized threads for both weft and warp, and was ‘beaten’ hard as it was woven to produce a close, tight weave. This cloth took a long time to make which made it very expensive. It was worn as a shoulder wrap by men dressed up to look their best. Like the best tightly woven woollen cloth, the fibres were said to expand in the wet, making the cloth water repellent. The more careful weavers treated the cloth to make it especially water repellent. The seeds picked out of the fibre were put to one side, and when enough had been gathered, they were ground to a fine paste which was then rubbed into the completed cloth always going in the same direction. This made the fibres lie flat and gave the cloth a waxy or polished appearance. The cloth was then put in the sun to dry out, and after being tapped all over to knock off the debris, was ready for use.
Notes. The islanders differentiate between the ‘Arab’ cotton bush (*ʾˈiʾiśīrə ˤˈarabī*) which is ‘male’ and has smaller leaves, and the ‘Soqotri’ cotton bush (*ʾˈiʾiśīrə soˈḳotəri*), which is ‘female’ and has larger leaves.
Comparative uses. Cotton shrubs provide fibre, cottonseed, cottonseed oil, cottonseed cake and cottonseed flour. The leaves of some species are eaten, and the seed can be used as a source of cooking oil. Cotton seeds are rich in B-complex vitamins and are used as a feed, but the cottonseed meal – the residue left after the oils have been pressed out – can be poisonous to cattle, sheep and pigs due to presence of sesquiterpene phenol and gossypol. The seeds of the related *G. herbaceum* are used as a coffee substitute. In Dhofar the fibre from *G. stocksii* was used much as cotton wool is today, and threads spun from it were used for wicks, and to repair clothing and fishing equipment. In northern Oman, the cotton plant appears in remedies for toothache and earache, and a decoction of the plant was mixed with ghee and taken to treat shortness of breath. It was also one of the ingredients of a remedy for treating fits. The plants are widely used medicinally. The plant is a

demulcent, laxative, galactagogue and expectorant. A decoction is given to treat intermittent fevers, and a cold infusion of the leaves with lime juice to treat dysentery. The leaves and crushed seed kernels are applied to sores or as a poultice to bruises and swellings. The lint is used as a dressing for wounds. The root-bark is oxytocic and emmenagogic. The boiled root tea of various *Gossypium* species is taken to ease labour. A decoction of the roots is taken as a contraceptive, and the root is also taken to procure an abortion. The seed oil has an anti-fertility action in males, and drops of the oil are used to treat earache. (Abdulfatih 1987; Ambasta 1982; Jain & De Fillips 1991; Kunkel 1984; Lewis & Elwin-Lewis 1977; Miller & Morris 1988; Oliver-Bever 1986)

Hibiscus L.

Herbs, subshrubs, shrubs or small trees, with an indumentum of stellate hairs, the hairs often extreme irritants. Leaves entire to deeply palmately lobed, margins entire or toothed. Flowers sometimes large and showy, solitary in leaf axils. Epicalyx lobes absent (in annual species) or 8 to 15, linear or lanceolate, free or fused (*H. socotranus*). Calyx 5-lobed. Petals yellow (sometimes with a dark red base), red or white. Ovary 5-celled. Fruit a capsule; seeds glabrous or hairy. 9 endemics. c.300 species. Warm temperate to tropical regions.

Botanical notes. WARNING. Extreme caution should be exercised when near the tree species: they have extremely irritating hairs which are readily shed and can cause discomfort for some considerable time. The hairs on the leaves are less irritating when the leaves are new and green but are increasingly so in dry weather.

***Hibiscus dioscorides* A.G.Mill.** (see Appendix 2, p. 723)

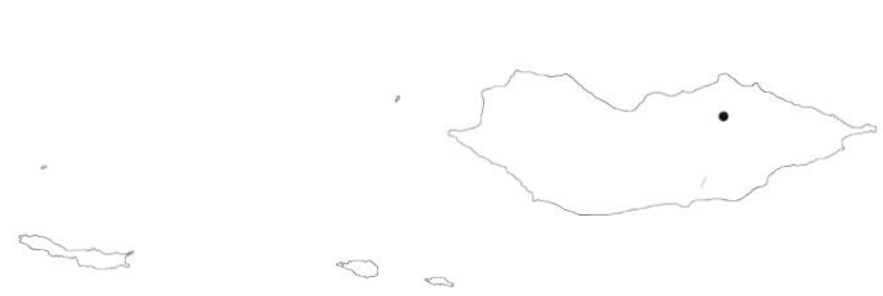

Shrub to 2m. Leaves ovate to broadly ovate, sometimes obscurely 3-lobed, 1.5–2.5 x 1–2cm, base broadly cuneate, margin serrate, tip rounded, thinly to densely appressed hairy with small (c.1mm across) stellate hairs above and mainly large (c.2mm across) 3-armed hairs mixed with a few stellate hairs beneath. Epicalyx lobes 10 or 11, free, linear, 4–5 x c.1mm, erect. Calyx cup-shaped, 5-toothed, 7–9mm long; teeth triangular, c.4 x 4–5mm, with shortly acuminate tips. Epicalyx and calyx densely stellate-hairy. Corolla funnel-shaped, bright yellow with a purplish brown base, c.3.5cm long, outer surface covered with reddish-brown 3-armed hairs.

Ecology. On granite cliffs amongst granite boulders, in shrubland. 950m.

Distribution. Endemic.

Conservation status. Data Deficient (DD). Part of a taxonomically difficult group on Soqotra. Inadequate evidence is available to make an assessment.

Botanical notes. Distinguished from the other woody *Hibiscus* species on Soqotra by the combination of five-lobed, cup-shaped calyx and filiform epicalyx lobes. It is perhaps closest to *H. noli-tangere* but differs in the indumentum of the calyx and epicalyx, the linear not filiform epicalyx lobes and the broadly triangular calyx lobes. Known only from the type gathering on Jebel Jaaf in the eastern Haggeher. This is the only local endemic recorded from the eastern Haggeher although there is evidence of floristic differences between the eastern, central and western Haggeher.

ETHNOGRAPHIC NOTES†

Name(s)

– Soqotri: *dirˈʌffān, dirˈeffen,* (western Soqotra) *dˈərfən* (pl. *di rˈʌfˈʌntə, dirəfˈontən*).

– Soqotri simplified: *diraffan, direffen, dirfin* (pl. *dirafantih, dirifonten*).

Food (livestock). The plants survive drought well and for this reason are one of the key dry season fodders for all livestock. The new, fresh growth is one of the best foods for young stock, calves included. The plants commonly come into new leaf at the end of the monsoon before the rains, at a time when most other graze and browse is exhausted. At this time it is particularly valuable for livestock. The foliage, buds, flowers and younger stems are all eaten, and animals eat fallen leaves and other plant litter as well. It is regarded as a nutritious fodder, making animals fat and increasing their milk. In hot and dry weather, the mature foliage is covered with irritant hairs, but when the leaves are still young and bright green, the hairs are fewer and much less irritating (*H. malacophyllus* seems to be an exception to this). The foliage is lopped and carried to cattle and goats, sometimes also to sheep. Only leaves are gathered for cattle, but goats can manage leafless twigs and stems, and also chew at the bark when hungry. The bark in particular is said to be very sweet, and certainly goats are very keen on it. These plants are one of the key browse plants for camels. However, if a pack camel has been feeding on these bushes, because of the very fine, penetrating hairs it cannot be ridden bareback but has to be covered with a rug first. In the dry season camels are herded to areas where the shrubs are plentiful to browse. In the better-vegetated areas, goats are less enthusiastic, but it is one of the most important browse plants for the goats of the high plateaus and other drier areas, just as it is a key dry season fodder for the cattle of the same areas. The smaller species are important for sheep, since, like *Ballochia* and *Trichocalyx* shrubs, the foliage grows within their reach. When gathering it for fodder, herders cover the face and other exposed skin with a cloth, and when they have finished, rub their hands with fine earth to get rid of the worst of the irritant hairs. If goats or cattle eat too much of the foliage in the dry season at a time when there is little else to eat, and especially on the high limestone plateaus or the plains foothills, they can fall ill with the condition known as *ʿāṣub*. The signs of this are weakness and trembling in the legs, especially the hind legs, and the animal begins to stagger and stumble. The condition does not show itself until the animal is herded any distance, when it is quickly seen that it tires quickly, is unsteady on its feet, and has to be given frequent rests or be allowed to lie down. Occasionally this happens to camels too. The animals generally recover the following day, and there are no long-term effects. Eating some varieties of fern can bring on the same problem in hungry small stock, as does the chewing of charcoal, or ingesting bird droppings, especially those of ravens or vultures, when grazing around and under trees.

Bees / honey. The flowers are an important bee forage, and make good honey.
Flavouring / food additive. The fat of a carcass of an animal that has eaten a lot of these plants is noticeably yellow. However, the taste of the fat, and of the rest of the carcass, is not affected.
Fuel. The dead wood makes reasonable firewood, and smaller twigs can be used as wet season tinder if other better tinders are not available.
Timber. While not among the best timbers of the island, the larger trees make a useful timber. Some shrubs are too small to be useful for other than roofing infill, while others are large enough to provide roofing beams. However, the wood readily rots if allowed to become damp and some people are reluctant to use it for roofing for this reason. The shrubs that grow in the wetter areas are commonly de-barked before use to deter termites. The wood of the shrubs of the drier and hotter areas is harder than that of wetter, cooler areas, and, like all timber, the trees should be cut in the hottest months at the hottest time of day.
Fibre / cordage. Strips of bark are removed, the outer bark peeled off and the inner bark separated out into fine strands, twisted together to make cordage of differing thickness. The strands are fine and strong enough for even quite fine threads to be worked. *Hibiscus* cordage is stronger than that made from *Ficus vasta* and is even strong enough to be used for camel gear. The finer threads are used for sewing, and for making and repairing equipment. The leaves, crushed to a paste with *Portulaca* and *Corchorus* leaves, can also be twisted to make threads, but these are not very durable and are mainly for temporary use.
Wood. The wood is hard enough to be useful for implements. Slim, straight branches are cut to make the important herding staff and other sticks. The dead wood makes reasonable charcoal.
Medicinal / veterinary. Someone whose hair is falling out or who wants to make their hair grow thick and long gathers green leaves and crushes them to a paste which is then rubbed into the scalp. Goats and cattle failing to thrive or which are recovering from an illness are hand fed on the foliage to restore their strength. The same feed is given to a cow losing weight towards the end of her pregnancy.
Cleansing. The leaves are crushed to a paste with *Portulaca* and *Corchorus* plants. A little is added to the paste to make a froth used to cleanse the scalp and wash the hair.
Other. Animals feeding on this can go longer without water. The strong, inner, white bark is used with a pebble to make a sling shot, or is stretched between the prongs of a forked twig like a rubber band to make a catapult. When the demand for aloes juice was high, the more unscrupulous aloe harvesters or traders of the interior would collect the inner bark of these shrubs and that of *Boswellia* species and succulent aloe leaves. They would then crush these together and add the pulp to the genuine aloes juice in the skins. This deception was possible because the skins of aloes juice were purchased directly from the interior, no further processing being carried out in the market towns. The only way of knowing for sure that such adulteration had been carried out was that the product never set hard like genuine, pure aloes juice, but always remained slightly liquid.
Notes. The fine hairs cause great pain and are potentially damaging if they get into the eyes. If hairs do get into the eyes, the treatment is to wash them with a solution of *Dracaena* resin or *Curcuma longa* ground to a fine powder and added to water. If this is not successful, a local healer with a steady hand is approached; he removes the hairs with a scrap of material rolled up and brushed swiftly across the surface of the eyeball. The three-armed irritant hairs that make this plant such a nuisance have led to these shrubs becoming a symbol of anything intensely irritating. Thus a person who is a trouble-maker is referred to as *di raffan*, and a person who has become fed up with constant pestering from someone asking him for something will make it clear that he has absolutely no intention of giving it by telling him angrily to 'Go and eat *di raffan*!'.

†Also apply to *H. diriffan*, *H. macropodus*, *H. malacophyllus*, *H. noli-tangere*, *H. quattenensis*, *H. scottii*, *H. stenanthus*.

Hibiscus diriffan **A.G.Mill.** (see Appendix 2, p. 723)

Small tree to 5m. Leaves elliptic to triangular-ovate, 2–4 x 2–3.5cm, base broadly cuneate to cordate, margin irregularly serrate, sometimes obscurely 3-lobed, tip rounded, thinly tomentose with stellate hairs above and densely hairy with stellate hairs and large (2–3mm across) 3-armed hairs beneath. Epicalyx lobes 9 to 13, apparently fused at base, oblanceolate, 6–12 x 1.5–3mm, with a distinct midvein and green 'lamina', erect at first becoming reflexed. Calyx 11–12mm, splitting to below the middle into 2 ovate lobes; lobes entire or with obscure teeth. Pedicels, epicalyx and calyx densely appressed stellate-hairy. Corolla yellow with a purplish brown base, funnel-shaped, 3.5–5cm long.

Ecology. Distributed on the southern limestone plateaus from Diksam to Wadi Irih and extending onto the granite of the Haggeher mountains. 20–1300m.
Distribution. Endemic.
Conservation status. Least Concern (LC). Common in several vegetation types and under no present or perceived threat.
Botanical notes. Similar to *H. quattenensis* but differs in the densely stellate-hairy, not glabrescent, pedicels, epicalyx and calyx; also the leaves are narrower with a distinctly cordate base and crenate margins, and have a tendency to become lobed.

ETHNOGRAPHIC NOTES

Name(s)

See *H. dioscorides* (also uses).

Hibiscus lobatus **Kuntze**
Hibiscus solandra L'Hér.

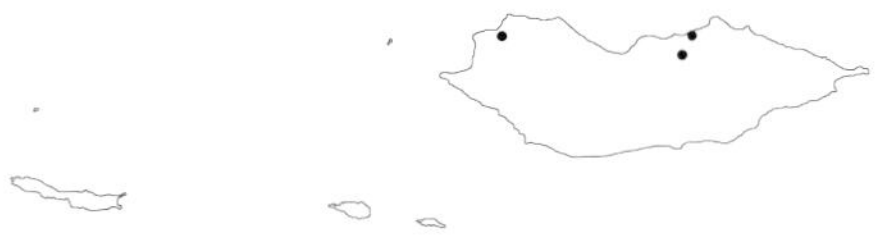

Erect, annual herb to 1m, pubescent throughout and with long weak hairs on the stems. Leaves variable, ovate to orbicular below to linear or deeply 3 to 5-lobed above, margins crenate. Epicalyx absent. Calyx lobes lanceolate, 6–10mm. Petals white, c.1cm. Fruit ovoid, obtuse, c.1cm, exceeding calyx; seeds glabrous.
Ecology. Amongst grass in drought-deciduous woodland. c.450m.
Distribution. Tropical Africa and Asia.

ETHNOGRAPHIC NOTES
Not seen, no information recorded.

Hibiscus macropodus **Wagner & Vierh.**

Shrub to 1.5m. Leaves ovate or rounded, 1–2 x 1.5–2.2cm, sometimes obscurely 3-lobed, base rounded or cordate, margin dentate or crenate-dentate, tip rounded or obtuse, thinly tomentose with stellate hairs above and densely hairy with stellate hairs and large (1–1.5mm across) 3-armed hairs beneath. Epicalyx lobes 8 to 10, linear, 6 x 1mm, free, densely stellate-hairy. Calyx c.12mm, splitting into 2 ovate shortly toothed lobes, densely stellate-hairy. Corolla yellow, obovate, c.4cm long.
Ecology. No habitat details available.
Distribution. Endemic.
Conservation status. Data Deficient (DD). Inadequate evidence is available to make an assessment.
Botanical notes. Vierhapper and Wagner (Vierhapper 1907) based their description of this species on two collections from Bandar Shu'ub and from Wadi Meige on the northern slopes of Djebal Rahmen. These specimens have not been seen by the authors and a photograph (Vierhapper 1907, t. 7, 2) is not particularly helpful. According to the description this species is characterised by the calyx splitting into two segments, linear epicalyx lobes and long pedicels. Several specimens (*Miller* 11207, *Gwynne* 323 and *Popov* 276) match this description in leaf and epicalyx shape and indumentum but lack the very long pedicels described by Vierhapper and Wagner (and shown in the photograph). Lengthening of the pedicel in fruit is recorded in *H. quattenensis*, *H. scottii* and *H. dioscorides*. The identify of this plant remains a mystery. The geographical distribution of the cited specimens also casts doubt on the integrity of this species.

ETHNOGRAPHIC NOTES
Name(s)
See *H. dioscorides* (also uses).

Hibiscus malacophyllus **Balf.f.**

Shrub or small tree to 4m. Leaves rhombic to subcircular, 1.5–3 x 1–2cm, base rounded to cuneate, margin serrate, tip acute to rounded, densely tomentose with stellate hairs above and beneath. Epicalyx lobes c.9, free, linear, c.8 x 0.75mm, erect. Calyx splitting into 2 ovate lobes; lobes with broadly triangular teeth. Pedicels, epicalyx and calyx densely stellate-hairy. Corolla yellow with a dark red base, funnel-shaped, c.3cm long.
Ecology. Drought-deciduous woodland and succulent shrubland on limestone. 100–600m.
Distribution. Endemic.
Conservation status. Vulnerable (VU D2). Rare, with a very limited area of occupancy.
Botanical notes. Distinguished by its dense indumentum of stellate hairs and the complete absence of the large three to five-armed irritant hairs which are an unpleasant characteristic of the other shrubby or arborescent species of *Hibiscus* on Soqotra. It seems on the basis of recent collections that *H. malacophyllus* is a plant with a rather narrow distribution in the north of the island from the Muqadrihon Pass to Ras Heybaq.

ETHNOGRAPHIC NOTES
Name(s)
See *H. dioscorides* (also uses).

Hibiscus micranthus **L.f.**

Erect woody-based herb to 50cm; stems hispid. Leaves narrowly ovate to elliptic, up to 10–20 x 5–12mm, base rounded, margin serrate, sometimes shallowly 3-lobed, tip acute. Flowers solitary in leaf axils. Epicalyx lobes 7 or 8, linear, 1.5–3mm long. Calyx 3–4mm long. Petals pink or red, c.5mm. Capsule globose, 5–10mm across, puberulous; seeds with woolly hairs.
Ecology. Clearings in semi-deciduous woodland. 600m.
Distribution. Widespread from tropical Africa to India.
Botanical notes. In Somalia and Ethiopia the flowers are recorded as red or white.

ETHNOGRAPHIC NOTES†
Name(s)
– Soqotri: *di ʾārdəhʌn, di hārdəhʌn* 'a spreading area of herbage' (a name given to other plants such as *Blepharis* species, *Pavonia arabica*, *Sida* species, *Melhania muri-*

cata and *Ruellia patula*, all of which are prickly or have stellate hairs and are suspected of causing *'uqabah* in small stock); *ləsēyʌk* 'sticky'; *śɔ̄whiḋitin* 'damaging to the mouth'.

– Soqotri simplified: *di 'ardihan, di hardihan*; *liseyaq*; *showhiditin*.

Food (livestock). All livestock eat this in the dry season. Cattle especially graze it when other grasses and herbs have died back as it retains its green colour. It is not thought to be useful nutritionally. The name *śɔ̄whiḋitin* refers to the fact that this tough plant hurts the mouths of cattle who have to yank it hard when grazing.

Other. All the plants of this *di 'ardihan* type with 'sticky' leaves can be used to repair holed pottery, even that used to hold liquid. A pebble of the size of the hole, or a shard of pottery ground down to fit, is tapped gently to plug the hole. Then the leaves are ground to a paste with either cooked colostrum or meconium of a newborn animal. This is smeared around the hole, inside and outside, sometimes with added ground feldspar. The pot is then put to heat beside a gentle fire until the paste sets hard.

Notes. All plants with this name cause *'uqabah* in small stock, especially goats, which are unaccustomed to this feed. Island herders say that if you rub such plants between the fingers it disintegrates to a sticky slime, and it is this quality which they believe damages the gut of goats and makes them ill.

Comparative uses. The bark is used to make containers and to make toy arrows. The fresh plant is crushed in water and the juice applied to snakebite in which a small incision has been made. The leaf juice is also a renal remedy. A decoction of the root is used to treat bronchitis and is drunk for snakebite and of fresh roots against convulsive fevers in children. Many *Hibiscus* species yield fibre which is used for cordage and for fishing lines, and many have edible leaves and shoots which can be eaten either raw or cooked. *H. rosa-sinensis* and *H. sabdariffa* especially are widely used medicinally. (Ambasta 1982; Batanouny 1981; Dalziel 1956; Heine *et al.* 1988; Jain & De Fillips 1991; Kunkel 1984; Lewis & Elwin-Lewis 1977; Oliver-Bever 1986; Rizk & El-Ghazaly 1995)

†Uses also apply to *H. palmatus*.

***Hibiscus noli-tangere* A.G.Mill.** (see Appendix 2, p. 723)

Shrub or small tree to 3m. Leaves orbicular to broadly ovate-triangular, sometimes obscurely 3-lobed, 10–25 x 10–30mm, tip rounded, margin serrate, base shallowly cordate or truncate, thinly to densely appressed hairy with small (c.1mm) stellate hairs above and mainly large (2–3mm) trifurcate hairs mixed with a few stellate hairs beneath. Epicalyx lobes 9 to 11, free, filiform, 5–6mm long, erect. Pedicels, epicalyx and calyx with simple, stellate and 2 or 3-armed hairs. Calyx 5-lobed, 7–12mm long; lobes triangular-ovate, with shortly acuminate tip, 7–9 x c.5mm. Corolla funnel-shaped, bright yellow with a dark red base, 3–4.5cm long.

Ecology. Drought-deciduous woodland on the coastal limestone escarpment. 100–200m.

Distribution. Endemic.

Conservation status. Endangered (EN B2 a biii). A very rare tree known only from a single location with a known area of occupancy of less than 20km^2. The woodlands of the coastal escarpments of Soqotra are particularly susceptible to the human activity leading to a rapid decline in the quality of habitat, hence the designation of an endangered category.

Botanical notes. Distinguished from the other woody *Hibiscus* species on Soqotra by its combination of five-toothed calyx, filiform epicalyx segments and the distinctive indumentum on its epicalyx and calyx: the surface of these are clothed with stellate hairs but along their margins the indumentum is apparently composed of simple hairs; these are, on closer examination, a mixture of simple and few-armed stellate hairs with long (c.1mm) arms. Known only from Hoq woodlands on the coastal limestone escarpment below Homhil in the northeast of Soqotra. These woodlands are floristically rich and perhaps the best example on Soqotra of the woodland found on the seaward-facing slopes on the limestone escarpments.

ETHNOGRAPHIC NOTES

Name(s)

See *H. dioscorides* (also uses).

***Hibiscus palmatus* Forssk.**

Hibiscus intermedius A.Rich.

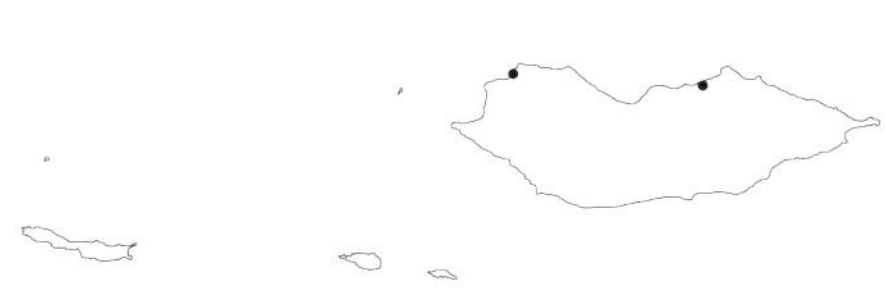

Prostrate herb to 1m, sparsely hispid with simple hairs on stems. Leaves deeply palmately lobed; lobes usually narrow, coarsely dentate. Epicalyx lobes c.8 to 10, narrowly lanceolate, c.12 x 1mm. Calyx lobes lanceolate, connate near base, c.8 x 1.5mm in flower, increasing to 15 x 5mm in fruit, scarious but with distinctive green veins; petals cream to pale yellow, c.1.5cm. Fruit globose, obtuse, c.7mm diameter, with a c.3mm apical extension, hispid; seeds with red-brown hairs.

Ecology. Sparse dwarf shrubland on sandy, desertic plain. 200m.

Distribution. Tropical Africa and SW Asia.

Botanical notes. Soqotran material is all referable to var. *aristaevalvis* Hochr. which has cordate leaves which are orbicular in outline and lobed almost to the base. Its most distinctive feature is the terminally awned fruit.

ETHNOGRAPHIC NOTES

Name(s)

– Soqotri: *di ʾārdəhʌn, di hārdəhʌn* 'a spreading area of herbage' (given to a variety of prickly plants and those with stellate hairs, which are suspected of causing *'uqabah* in small stock); *ləsēyʌk* 'sticky'.

– Soqotri simplified: *di 'ardihan, di hardihan*; *liseyak*.

Notes. Uses as *H. micranthus*.

Hibiscus quattenensis A.G.Mill. & Thulin (see Appendix 2, p. 723)

Shrub or small tree to 3m. Leaves broadly ovate or suborbicular, 1–4.5 x 1–4cm, base broadly cuneate to truncate, margin entire, tip rounded to obscurely crenate, glabrescent above with sparse, large (2–4mm) 3-armed hairs beneath. Epicalyx lobes 9 or 10, linear-oblanceolate, 5–9 x 1–4mm, with a distinct midvein and lamina-like margin, becoming strongly reflexed. Calyx cup-shaped, 10–15mm long, splitting to below the middle into 2, broadly ovate, entire lobes, glabrescent. Corolla funnel-shaped, bright yellow with a purplish brown base, 4–5cm long.

Ecology. Widespread on the coastal plains and dry limestone foothills and plateaus of SW Soqotra; mainly in *Croton socotranus* shrubland and succulent shrubland. Sea-level–600m.

Distribution. Endemic.

Conservation status. Least Concern (LC). Common in several vegetation types and under no present or perceived threat.

Botanical notes. Readily distinguished from the other woody *Hibiscus* species on Soqotra by its glabrescent pedicels, epicalyx and calyx. It is closest to *H. diriffan* with its oblanceolate, reflexed epicalyx lobes.

ETHNOGRAPHIC NOTES

Name(s)

See *H. dioscorides* (also uses).

Hibiscus scindicus Stocks

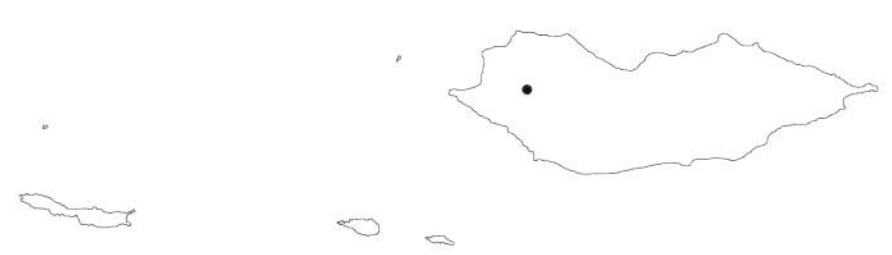

Subshrub to 20cm. Leaves obovate, 0.5–1.5 x 0.3–0.8cm, base cuneate, margin serrate towards tip, sometimes lobed, tips rounded; stipules 2–3mm, spinescent. Epicalyx lobes c.8, filiform, 2–4mm. Calyx lobes lanceolate, c.6mm lengthening to c.12mm in fruit. Petals red, 6mm. Fruit subglobose, c.8mm diameter, midvein of valves green; seeds densely covered with cottony hairs.

Ecology. Rare in open dwarf shrubland on desertic plain. c.200m.

Distribution. Somalia, Oman and Pakistan.

Botanical notes. Known only from a single, recent collection made in the Zahr basin. Several plants which are widespread in deserts in surrounding areas are, on Soqotra, known only from the Zahr basin. The single specimen from Soqotra has bright red flowers whereas flower colour outside Soqotra is recorded as white or cream; however, in all other respects the Soqotran plant resembles plants from other areas. More collections are necessary to assess the significance of this difference in flower colour.

ETHNOGRAPHIC NOTES

Not seen, no information recorded.

Hibiscus scottii Balf.f.

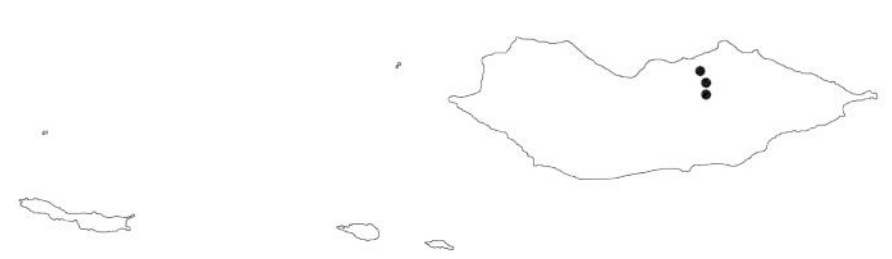

Small tree to 3m. Leaves lanceolate or rhombic, 1–4 x 1–2.5cm, base rounded to cuneate, margin serrate, tip acute to rounded, thinly tomentose with stellate hairs above and densely hairy with stellate hairs and large (2–3mm) 3-armed hairs beneath. Epicalyx lobes 8 to 15, free, finger-like, 5–12 x c.1mm, erect. Calyx splitting into 2 ovate lobes; lobes with triangular teeth, c.3 x 3mm. Pedicels, epicalyx and calyx densely appressed stellate-hairy. Corolla golden- or orange-yellow with a purplish brown base, with rust-coloured 3-armed hairs on outer surface, funnel-shaped, c.5cm long.

Ecology. Restricted to dense semi-deciduous woodland on the granitic slopes of the Haggeher mountains. 550–800m.

Distribution. Endemic.

Conservation status. Vulnerable (VU D2). Rare, with a very limited area of occupancy.

Botanical notes. Distinguished by its orange-yellow corollas which are clothed with rust-coloured, three-armed hairs on their outer surface. It is also characterised by the calyx which splits into two, toothed segments and the ovate or rhombate leaves. The plants included here match the plant illustrated in Balfour's *Botany of Socotra* (1888, t. 5a). Balfour in his treatment apparently included all plants with funnel-shaped corollas and three-armed hairs, a somewhat broader definition than the one used here.

ETHNOGRAPHIC NOTES

Name(s)

See *H. dioscorides* (also uses).

Hibiscus sidaeformis **Baill.**
Hibiscus ternatus Mast.

Erect annual herb to 30cm, pubescent throughout. Leaves very variable; lower usually orbicular, upper usually 3-lobed. Epicalyx absent. Calyx lobes lanceolate, c.1cm. Petals yellow, c.1cm. Fruit globose, obtuse, c.5mm, much shorter than the calyx; seeds glabrous.
Ecology. Coming up after rain in drought-deciduous woodland. 50–650m.
Distribution. Tropical Africa, Arabia and Madagascar.

ETHNOGRAPHIC NOTES
Name(s)
– Soqotri: *ʾˈažərib* 'quick to die back'; *ləsēyʌk, lūwsʌk* 'sticky'.
– Soqotri simplified: *'ajirib*; *liseyak*, *luwsak.*
Food (livestock). This is one of the *ʾˈažərib* group of plants which flourish after the winter rains. It is readily grazed by cattle before the grasses come up, but is little eaten by goats and sheep. All livestock like the large leaves in the dry season and cattle especially graze these when other grasses and herbs have died back. It is not thought to be valuable nutritionally.
Notes. It can cause *'uqabah* in small stock unaccustomed to this feed.

Hibiscus socotranus **G.Ll.Lucas**

Low shrub. Leaves suborbicular to transversely ovate, <22 x 22mm, base rounded to shallowly cordate, crenate or repand-dentate margin, tip rounded, densely stellate-hairy above and beneath but with larger hairs beneath. Epicalyx 7-lobed, fused in lower half, lobes triangular, stellate-hairy. Calyx 5-lobed; lobes ovate, c.14mm. Corolla yellow, <3cm long, outer surface with large stellate hairs.
Ecology. Cliffs on limestone escarpment. 400–600m.
Distribution. Endemic.
Conservation status. Endangered (EN B2 a biii). *H. socotranus* was originally discovered by John Lavranos and Alan Radcliffe-Smith, in 1967 on the north-facing limestone escarpment above Qaysoh near Qalansiyah at the western end of Soqotra. However, after being searched for on a number of occasions, it was only in 2004 that it was seen again when a single plant was found growing on cliffs close to the route of a new road being constructed up the escarpment. At present this plant is under no direct threat from the road but the situation needs careful monitoring. It must be considered endangered due to its rarity, small area of occupation and the trend of lower rainfall in the region.
Botanical notes. Distinguished from all other woody species of *Hibiscus* on Soqotra by its fused epicalyx. The relationship between *H. socotranus* and the other Soqotran endemic *Hibiscus* species is not clear. The degree of fusion of the epicalyx (to about one third) is unusual within *Hibiscus* and unique amongst the species on Soqotra; the only other species with a similarly fused epicalyx is *H. erlangeri* from Somalia which, on account of its epicalyx, has sometimes been separated into the monotypic genus *Symphochlamys*. *Hibiscus erlangeri* differs from *H. socotranus* in its calyx, which splits into two, more or less entire lobes; in this it resembles some of the other endemic Soqotran species (e.g. *H. scottii*, *H. malacophyllus*, *H. macropodus*, *H. diriffan* and *H. quattenensis*). There are several other local endemics and plants of note associated with *H. socotranus* on this escarpment: these include *Boswellia bullata*, *Dorstenia gigas*, *Prenanthes amabilis*, *Aloe squarrosa*, *Gaillonia thymoides* and *Hypericum socotranum* subsp. *socotranum.*

ETHNOGRAPHIC NOTES
Not seen, no information recorded.

Hibiscus stenanthus **Balf.f.**

Prostrate or low shrub to 1.5m. Leaves orbicular to broadly ovate-triangular, sometimes obscurely 3-lobed, 10–15 x 10–20mm, base shallowly cordate or truncate, margin obscurely serrate, tip rounded, thinly to densely appressed hairy, with small (0.5–1mm across) stellate hairs above and large (c.2mm across) 3-armed hairs and a few small stellate hairs beneath. Epicalyx lobes 7 to 9, filiform, <6mm long, erect. Calyx cup-shaped, 5-lobed, c.10mm long; lobes broadly triangular, with shortly acuminate tips, 2–3 x 3–5mm. Epicalyx, calyx and pedicels densely hairy with large 3-armed hairs and scattered stellate hairs. Corolla yellow, tubular with spreading lobes c.3cm long. Anthers confined to top quarter of staminal column. Stigmas red.
Ecology. Common, sometimes abundant, in the north-east of Soqotra in rocky areas on the limestone plateau. 330–600m.
Distribution. Endemic.
Conservation status. Least Concern (LC). Common in several vegetation types and under no present or perceived threat.
Botanical notes. Readily distinguished by its tubular corolla, without a red base but with red stigmas, its cylindrical calyx, and anthers restricted to the top quarter of the staminal column. Its short filiform epicalyx lobes are also distinctive. On Soqotra similar epicalyx lobes are found only in *H. noli-tangere*, the geographically closest species. Usually a low shrub, *H. stenanthus* never grows very tall and, on more windswept sites (for instance above Ras Momi), it is virtually prostrate.

ETHNOGRAPHIC NOTES
Name(s)
See *H. dioscorides* (also uses).

Hibiscus vitifolius L.

Robust herb or subshrub to 1.5m; stems glandular with simple hairs. Leaves shallowly 3 to 5-lobed, ± orbicular, <20cm long, base cordate, pilose and with shorter stellate hairs; lobes ovate, acute, margins serrate-dentate. Flowers often in terminal racemes. Epicalyx lobes 6 to 10, linear, c.10mm. Calyx lobes ovate-acuminate, fused to middle, c.1.5 x 0.5cm, widening in fruit, stellate tomentose with glandular hairs. Petals yellow with a purple base, 3.5–5.5cm. Fruit subglobose, winged, c.12mm long, with 1–5mm awns at tip, becoming scarious; seeds glabrous.
Ecology. Clearings in semi-deciduous woodland and along wadis. 400–750m.
Distribution. Tropical and subtropical regions of the world.
Botanical notes. Distinguished from all other species on Soqotra by its winged fruits.

ETHNOGRAPHIC NOTES
Name(s)
– Soqotri: nil or (rare) *bɔ̄wnə̍no* 'tiny *Sterculia africana*'.
– Soqotri simplified: *bownino.*
Food (livestock). The leaves, buds and flowers are eaten by small stock and cattle towards the end of the dry season, and goats chew the stems in a drought. Donkeys and camels also eat it but it is not considered to be nutritionally valuable. It is a relatively uncommon plant.
Comparative uses. The stems yield a fibre for cordage, twines, ropes and fishing lines, and the plant also possesses anti-inflammatory properties. (Ambasta 1982; Jain & De Fillips 1991; Oliver-Bever 1986)

Malva L.

Herbs. Leaves shallowly palmately lobed. Flowers in axillary clusters. Epicalyx bracts 3, linear. Petals pink. Fruit a schizocarp; mericarps 10, 1-seeded. c.30 species. Europe, temperate Asia.

Malva parviflora L.

Weedy annual herb. Stems decumbent <50cm, glabrous or pubescent. Leaves suborbicular, shallowly 5 to 7-lobed, <c.7cm across, margins crenate. Epicalyx bracts c.3mm. Petals c.4mm, pale pink, scarcely longer than the calyx. Mericarps strongly ribbed with sharp edge.
Ecology. A widespread weed found near caves and habitation. Under-recorded on the distribution map. Sea-level–700m.
Distribution. Cosmopolitan weed.

ETHNOGRAPHIC NOTES
Name(s)
– Soqotri: nil or *gīmiṣ*, *gu̍muṣ*; *šəźa̍rhen* 'green'.
– Soqotri simplified: *giymiz*, *gumuz*; *shidharhen.*
Food (human). The fruit are edible and often eaten by children. The leafy stems are sometimes gathered and cooked like *Amaranthus*.
Food (livestock). These plants spring up all around settlements after rain. They are readily eaten by all livestock and are considered to be a useful feed, especially when other herbs and grasses are in short supply.
Comparative uses. The leaves of this and many other *Malva* species are eaten raw and cooked. In the Middle East the plant is often cultivated and sold in vegetable markets. The fruit are edible and the young shoots are eaten as a famine food. The plant is a rich source of calcium, iron and carotene. It is reported to be poisonous to bovines and horses. Medicinally, an infusion of the leaves is taken as a nerve tonic, and a decoction is used as a taeniacide and to treat profuse menstruation. The leaves are given as a douche for uterine troubles and applied as a hot poultice for wounds and swellings. The seeds yield a fatty oil and are demulcent, being taken for coughs, ulcers and to treat fevers. An infusion of the seeds and leaves is applied to soothe the skin. The root is chewed or rubbed into the gums for pyorrhea. An infusion of the flowering and fruiting branches is used as a gargle for its astringent properties, and is prescribed for people on diets and for gastro-intestinal ailments. (Abbas *et al.* 1992; Ambasta 1982; Batanouny 1981; Dafni *et al.* 1984; Ghazanfar 1994; Goodin & Northington 1979; Jain & De Fillips 1991; Kunkel 1984; Rizk & El-Ghazaly 1995)

Pavonia Cav.

Herbs, stellate-hairy. Leaves simple or lobed. Flowers solitary or clustered in leaf axils. Epicalyx of 5 to 25 free lobes. Calyx 5-lobed. Petals pink. Fruit a schizocarp; mericarps 5, 1-seeded, sometimes winged. c.250 species. Warmer regions.

Botanical notes. *P. odorata* was recorded by Bent from Soqotra. However, many of his collections were almost certainly made on the Arabian mainland. There are no confirmed sightings from Soqotra.

Pavonia arabica Hochst. & Steud.

Erect perennial herb to 50cm, sometimes woody-based. Leaves oblong-ovate, 2–6 x 1–3cm, base rounded or cordate, margin entire, tip rounded or notched, stellate hairy. Epicalyx lobes 10 to 15, linear, 5–15mm, stellate hairy and with long simple hairs. Sepals shorter than epicalyx. Petals pink, slightly exceeding epicalyx. Fruit subglose, 3–7mm diameter; mericarps sometimes villous, unwinged or slightly winged.
Ecology. Rare in semi-deciduous woodland dominated by *Dracaena cinnabari*; under-recorded. c.650m.
Distribution. East and NE tropical Africa, and Arabia west to NE India.

ETHNOGRAPHIC NOTES
Name(s)
– Soqotri: *di ʾārdəhʌn, di hārdəhʌn* 'a spreading area of herbage' (given to a variety of prickly plants or those with stellate hairs, which are suspected of causing *'uqabah* in small stock); *həbālino* (rare *həbādino*) 'tiny spots of a rash', 'tiny tamarind fruit'.
– Soqotri simplified: *di 'ardihan, di hardihan*; *hibalino* (*hibadino*).
Food (livestock). This plant responds to heavy rains, after which it seeds and spreads everywhere. Small stock eat it, but it is regarded as a dangerous plant for goats if eaten in any quantity. It can cause severe, sometimes fatal, scouring in hungry animals in poor condition. It can also cause goats in early pregnancy to abort. It often retains its green colour at a time when other herbs and grasses have yellowed and died back, and at this time it is tempting to all livestock. Goats rarely eat the immature plants, as when they are at this stage there is plenty of alternative forage. However, when other forage has died back, they graze the mature plants and it is then that they fall ill. Cattle are keen on the plant, and it is considered a good feed for them. Sheep and donkeys also graze it with little problem. Herders who are dependent on goats for their livelihood uproot and destroy the plants whenever they come across them. However, after heavy rains the plants become too plentiful and widespread to eradicate in this way.
Other. These plants are removed from the grazing paddocks and pens because when they grow well they smother other more useful herbs and grasses. All the *di 'ardihan* type with 'sticky' leaves can be used to repair holed pottery, even those used to hold liquid. A pebble the size of the hole, or a shard of pottery ground down to fit, is gently tapped in to plug the hole. Then the leaves are ground to a paste with either cooked colostrum, or meconium from a newborn animal. This is smeared around the hole, inside and outside, sometimes with added ground feldspar. The pot is then put to heat beside a gentle fire until the paste sets hard.
Notes. This is one of the group of plants believed to cause *'uqabah*. Island herders say that if you rub such a plant between the fingers and it disintegrates to a sticky slime it is this quality that damages the gut of goats. Some islanders say that it causes scour because it is commonly covered in tiny caterpillars.
Comparative uses. Some *Pavonia* species, such as *P. odorata*, are cultivated for their edible leaves and fruits, and for their fragrant flowers. The bark of others yields a fibre, which can be made into string and fishing tackle. They are often grazed by cattle: indeed when a wounded old cow gives blood instead of milk, Borana healers throw a root of *P. zeylanica* into the fire to cure the cow, and roots of other species are added to fresh milk to assist the rapid production of butter. (Ambasta 1982; Dalziel 1948, 1956; Heine *et al.* 1988; Jain & De Fillips 1991; Kunkel 1984)

Senra Cav.

Herbs or shrubs, softly hairy throughout with simple hairs. Leaves shallowly 3-lobed, cordate at base. Flowers solitary or clustered in leaf axils. Epicalyx lobes 3, large, heart-shaped, entire, becoming scarious in fruit. Calyx 5-lobed. Ovary 5-celled. Fruit a 5-winged capsule; seeds hairy. 3 species. E Africa, Arabia and India.

Senra incana Cav.

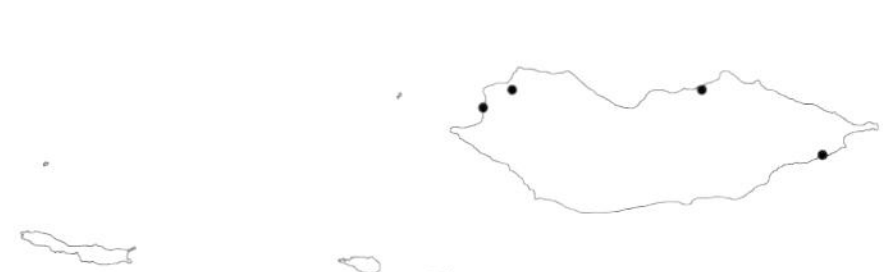

Herb or weak shrub to 1.5m. Leaves suborbicular, 2–4 x 2–5cm, 3-lobed, entire or irregularly serrate. Epicalyx lobes 1–3 x 1–2.5cm, cordate at base. Corolla dark purple or yellow with purple base; petals 2–3cm. Capsule subglobose, c.5mm across, hidden by scarious epicalyx.
Ecology. No recent habitat details available but in the 19th century found near villages on the coastal plains.
Distribution. NE Africa, Arabia and India.
Botanical notes. Not seen on Soqotra recently, but in the past it was apparently relatively common. Vierhapper (1907) recorded it at several locations and Balfour (1888) notes that it was common around Qalansiyah and other villages. It was most recently seen in 1967 on the northern coastal plain near Mori.

Sida L.

Herbs. Leaves simple. Epicalyx bracts absent. Flowers usually solitary and axillary, usually opening in the morning. Petals yellow. Fruit a schizocarp, mericarps 1-seeded. 150 species. Tropics and subtropics.

Sida acuta Burm.f.

Erect perennial herb. Stems <50cm, thinly puberulous. Leaves oblong-lanceolate, <6 x 2cm, margin dentate. Mericarps 6 to 8, with short awns, reticulate on back.

Ecology. Locally abundant in grassland in the eastern Haggeher. c.1000m.

Distribution. Widespread in the Old World tropics.

ETHNOGRAPHIC NOTES

Name(s)

– Soqotri: *di ᵓārdəhʌn, di hārdəhʌn* 'a spreading area of herbage' (given to a variety of prickly plants or those with stellate hairs, which are suspected of causing *'uqabah* in small stock); (rare) *di ṣefəririhun* 'many-flowered'.

– Soqotri simplified: *di 'ardihan, di hardihen; di zefiririhun.*

Food (livestock). See *Pavonia arabica.*

Other. See *Pavonia arabica.*

Comparative uses. Several *Sida* species have edible leaves and seeds. The bark of many yields fibres useful for cordage. Medicinally, *S. acuta* is used as an abortifacient and its leaves have anti-bacterial properties. (Ambasta 1982; Dalziel 1948, 1956; Goodin & Northington 1979; Jain & De Fillips 1991; Kunkel 1984; Oliver-Bever 1986)

Sida cordata (Burm.f.) Borss.Waalk.

Sida humilis Willd.

Slender prostrate or ascending herb. Stems with short, stellate and long, simple hairs. Leaves broadly ovate, <5 x 5cm, base cordate, margin crenate to serrate. Mericarps 5, without awns.

Ecology. Haggeher mountains. c.600m.

Distribution. Pantropical.

Botanical notes. Recorded from the Haggeher in the 19th century but not seen since.

Sida cordifolia L.

Similar to *S. cordata* but an undershrub with ovate to oblong leaves, base shallowly cordate or rounded; mericarps c.10, with awns 3–5mm long.

Ecology. Haggeher mountains. c.500m.

Distribution. Pantropical.

Botanical notes. Recorded from the Haggeher in the 19th century but not seen since.

ETHNOGRAPHIC NOTES

Comparative uses. *S. cordifolia* can be a troublesome weed. It is also a source of fibre which compares favourably with jute. The stems are used as brooms and the bark fibre to make girths for horses. The leaves are a fattening food for horses, and have anti-bacterial properties. Ephedrine has been found in this plant and it is listed in the Indian Pharmaceutical Code for the relief of hay fever and asthma. The plant is regarded as a demulcent and a febrifuge, and is used to treat dysentery. The roots are astringent, diuretic and tonic. Their infusion is given in urinary troubles, cystitis, strangury and haematuria. With asafoetida and rock salt they are used in the treatment of hemiplegia, sciatica and facial paralysis. Powdered roots with milk are given in leucorrhoea and frequent micturition. The root-bark is crushed with sesame oil and milk, and is used to treat facial paralysis and sciatica. (Ambasta 1982; Dalziel 1948, 1956; Goodin & Northington 1979; Jain & De Fillips 1991; Kunkel 1984; Oliver-Bever 1986)

Sida ovata Forssk.

Sida grewioides Guill. & Perr.
Sida rhombifolia sensu Balf.f. non L.
Sida subrotunda Hochst.
Sida urens L.

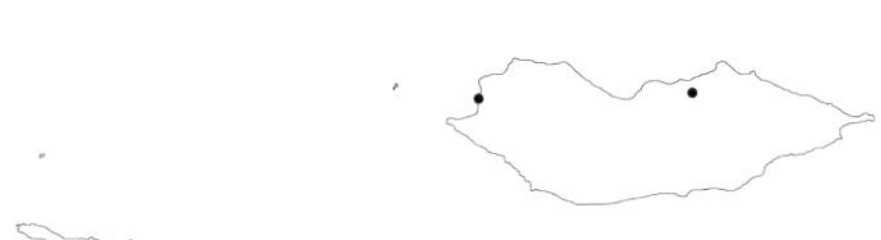

Similar to *S. cordata* but a subshrub, densely covered with short stellate hairs only; leaves ovate, base rounded; mericarps 7 to 10, without awns.

Ecology. On the limestone plateaus.

Distribution. Drier parts of tropical Africa, Arabia and eastwards to NW India.

Botanical notes. Recorded from the Haggeher in the 19th century but not seen since.

MELIACEAE

Trees or shrubs. Leaves alternate, simple, entire; stipules absent. Flowers in panicles, actinomorphic. Calyx 6-merous, fused. Corolla 4 or 5-merous, petals free. Stamens c.8 to 10, filaments united into a conspicuous staminal tube. Ovary superior, c.5-celled, style 1. Fruit a leathery capsule dehiscing longitudinally by 5 (or 10) valves. 51 genera. 575 species. Mostly tropical.

Turraea L.

See description of family. 1 endemic. 60 species. Tropical and S Africa, Madagascar, tropical Asia to Australia.

Turraea socotrana Styles & F.White

Shrub to 2m. Shoots densely covered with persistent leaf bases. Leaves grey-green, clustered on spur shoots, ovate to obovate, 2–5 x 1.5–3.5cm. Flowers clustered on spur shoots, sometimes appearing before leaves. Calyx cup-shaped, shallowly toothed. Petals pale yellow, narrowly oblanceolate, 3–4 x 0.3–0.5cm. Staminal tube 2.5–3cm long. Style long exserted with prominent ovoid stigma. Fruit pale red or orange, a globose capsule, c.0.8cm diameter.

Ecology. A local tree of cliffs in ravines on limestone plateaus in the west of Soqotra; less commonly on granite in the Haggeher. 500–700m.

Distribution. Endemic.

Conservation status. Vulnerable (VU D2). Rare, with a very limited area of occupancy.

Botanical notes. The attractive sweet-smelling cream flowers emerge before the leaves.

ETHNOGRAPHIC NOTES

Name(s)

– Soqotri: *məstə̍wrəher*; *ḳe̍rʿʌš* 'mean, ungenerous' (also given to the sweet-smelling *Jasminum fluminense*); *bə̍ššə* (commonly given to cliff-growing plants or those with striking flowers); *kə̍nəhor* (more commonly given to *Lannea transulta*); (*ši̍rəhām*) *ḥa̍hər* 'black tree' (referring to its good timber).

– Soqotri simplified: *mistewriher*; *qer'ash*; *bishih*; *kinihor*; (*shiriham*) *hahir.*

Food (livestock). The green foliage is not palatable to livestock, though very hungry goats eat the fallen, dead leaves.

Bees / honey. Honeybees are attracted to the perfumed flowers.

Fuel. The dead wood makes a good, slow-burning firewood, but it is not common enough for this to be an important source of fuel.

Timber. Makes good timber, and the trunks and larger branches are used for roof beams and supporting pillars. It is durable, and termite resistant, so it does not have to be de-barked before use. However, it is not common enough to be an important timber.

Wood. Close-grained and hard, and good for making a variety of implements. The straighter, slender branches make a good herding staff.

Notes. The fact that this very striking tree is uncommon across the Archipelago is reflected in the wide variety of non-specific names that have been given it.

Comparative uses. Species such as *T. nilotica* have edible fruit, those of *T. virens* are pickled and their juice used as a dye. The leaves and bark of some species are used as a purgative. Many species are known for their good hard wood. (Ambasta 1982; Dalziel 1948, 1956; Jain & De Fillips 1991; Kunkel 1984)

MENISPERMACEAE

Dioecious lianes or shrubs. Leaves alternate, simple, sometimes soon falling, then bearing cladodes (spiny leaf-like appendages). Stipules absent. Flowers unisexual, actinomorphic, in clusters; sepals free, 6 to 15, in 2, 4 or 5 whorls; petals free, 6, in 2 whorls; male flowers with 6 to 9 stamens, free or united into a central column; female flowers with 3 to 6, free carpels. Fruits fleshy, the single seed surrounded by a bony horseshoe-shaped endocarp. 78 genera. 520 species. Mostly tropical and warm regions.

Cocculus DC.

See description of family. 11 species. Tropical and warm regions.

Cocculus balfourii Schweinf. ex Balf.f.

Shrub to 1m, bearing flattened, spine-tipped cladodes, 3–9 x 0.1–1cm. Leaves soon deciduous, ovate to obovate, 1–2cm. Flowers cream, in sessile clusters on cladodes or on stem below. Sepals c.1.5mm. Petals acute, c.1.25mm. Fruits c.4mm, bright red.

Ecology. Common and widespread on granite in the Haggeher. 700–1000m.

Distribution. Oman (Dhofar) and SE Yemen.

Botanical notes. A very spiny shrub, often forming almost impenetrable patches in the Haggeher mountains, particularly in the grassland and low shrubland around the western peaks. Forman (1980) described *C. balfourii* as the most remarkable species in the Menispermaceae: with few exceptions the Menispermaceae are woody climbers, while *C. balfourii* is a dense shrub bearing sharply pointed cladodes (flattened spine-like appendages which perform the photosynthetic function of leaves). The leaves, which form below the cladodes, soon fall so the plants are usually more or less leafless.

ETHNOGRAPHIC NOTES

Name(s)

– Soqotri: *di ḥanižirihun, di ʿanižirihun* 'knife-blades'; *śʌmḳʌs* 'hooks, catches onto things', 'kindling'.

– Soqotri simplified: *di hanijirihun, di 'anijirihun*; *shamqas.*

Food (livestock). Cattle, goats and even sheep, in drier areas, eat the tender, new growing tips. The berries, when green and ripening to red, are sometimes eaten by goats. Camels and donkeys also feed on the plant when other forage is scarce.

Fuel. The larger pieces of dead wood make reasonable firewood and the smaller stems are used as kindling. Broken fragments make a very good tinder or firelighter, and are collected and stored at home for this.

Wood. The spiny branches are used to make barriers to livestock pens and doorways, and to fence off enclosures.

Notes. In the highest Haggeher this and other prickly plants grow closely together presenting an impenetrable wall of thicket, called *ṭamhum* or *ṭʌmhur*. The plants are not removed from the large livestock enclosures of the Central Highlands because they provide shelter and limited forage, but they are removed from the small grazing and calf paddocks where their fierce spines can cause problems.

Cocculus pendulus **(J.R.Forst.) Diels**

Cocculus laeba DC.

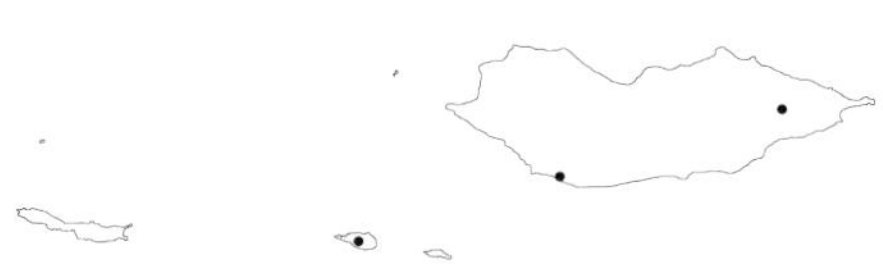

Twining or trailing liane to 10m; young branches whitish pubescent; trunk white. Leaves ovate to oblong-ovate or deltoid, sometimes 1 or 2-lobed at base, 0.5–5 x 0.3–2.5cm. Flowers greenish yellow; clusters axillary or on leafless branches. Sepals 0.7–2mm. Petals emarginate, 1–2mm. Fruits 4–6mm long, dark red.

Ecology. Uncommon and rather scattered across Soqotra and Samha. Sea-level–300m.

Distribution. N Africa, tropical E Africa south to Kenya, Arabia, east to S Iran, Pakistan and India.

Botanical notes. The characteristic trailing stems of this rather uncommon plant can reach up to 10m and clothe both trees and cliffs.

ETHNOGRAPHIC NOTES

Name(s)

– Soqotri: nil or *rikəbə* 'riding high'; *tilihīn* 'remnants'; 'kind of *ʿeḳḳəhin*' 'twisting and twining around something else'.

– Soqotri simplified: *rikibih*; *tilihiyn*; 'kind of *'eqqihin*'.

Food (livestock). Camels and goats like the tender stems, foliage and flowers, when they can reach it. Goats eat the litter when hungry in the dry season. On Samha it is considered a good feed for both sheep and goats.

Fuel. The dead wood makes reasonable firewood.

Wood. The flexible stems are used to make the 'bow' for beating or carding shorn wool before spinning it. They are also used to make the rim of the leather watering bucket and spinner's bag and for the sticks wedged across the watering bucket to keep it open.

Other uses. Children play on the hanging stems, sometimes using them as a swing.

Comparative uses. The flowers and fruit are eaten. The red fruit and juice of the plant are sometimes fermented to make an alcoholic drink. The leaves are also edible. The plant contains alkaloids reported to possess hypotensive and anti-cancer activities, and is used mainly as a purgative and to treat skin problems, and in India and Pakistan to treat malarial and other fevers. An infusion of the plant is used to help remove thorns from the feet. Roots are given as a febrifuge, vermifuge and tonic and are used as a fish poison. They are also taken as an emetic and used to treat hepatitis, jaundice, yellow fever, leprosy and syphilis. Leaves are given as a febrifuge, vermifuge and emetic. They are also given to treat biliousness. Stems are given for intermittent fever. (Ambasta 1982; Batanouny 1981; Dalziel 1948, 1956; Goodin & Northington 1979; le Houerou 1980; Jain & De Fillips 1991; Kunkel 1984; Lewis & Elwin-Lewis 1977; Miller & Morris 1988; Oliver-Bever 1986; Rizk & El-Ghazaly 1995)

MORACEAE

Monoecious or dioecious trees, shrubs or herbs; sap milky. Leaves alternate or subopposite, simple or lobed; stipules present. Flowers minute, arranged on disc-shaped receptacles or inside hollow receptacles (figs), unisexual, actinomorphic; perianth segments 2 to 4 or absent; male flowers with 1 to 4 stamens; female flowers with an inferior or superior 1-celled ovary, stigmas 1 or 2. Fruit an achene or drupe, enclosed or immersed in the fleshy receptacle. 48 genera. 1200 species. Tropical and warm regions.

Dorstenia L.

Monoecious herbs or shrubs; either with swollen succulent stems or stemless with an underground tuber. Leaves alternate, crowded at ends of branches or rising directly from tuber, entire or toothed. Receptacle fringed by linear or rounded appendages. Fruit a minutely tuberculate achene, explosively expelled. 2 endemics. 170 species. Tropical.

Botanical notes. For a revision see Friis (1983).

Dorstenia gigas **Schweinf. ex Balf.f.**

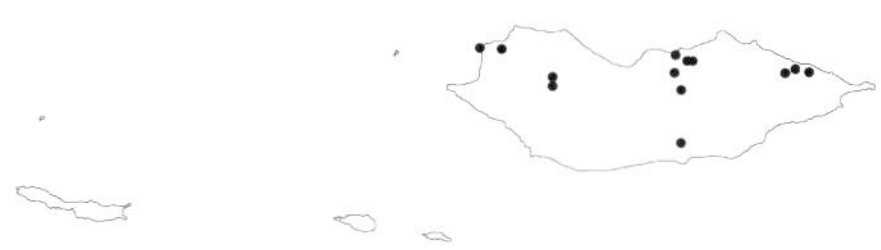

Shrub to 2.5m; stems swollen, <1.5m diameter. Leaves dark green, bullate, narrowly obovate, 3–15 x 1–3cm, acute or obtuse, obscurely crenate, attenuate into a short petiole. Receptacle circular, yellowish green, 5–25mm diameter; appendages rounded, 1.5–3 x 1–3mm.

Ecology. Widespread and locally common on cliffs and steep slopes on limestone and on the granite pinnacles of the Haggeher. 30–1100m.

Distribution. Endemic.

Conservation status. Near Threatened (NT). Although widespread and in places abundant this species has a rather fragmented distribution and there is the suggestion that in some locations it is in decline. The possibility of overharvesting by either succulent enthusiasts or during times of drought could lead to some populations qualifying for a threatened category in the near future.

Botanical notes. One of the characteristic pachycaul or bottle trees of Soqotra. When not in flower it can be confused with *Adenium obesum*; however, the characteristically bullate leaves with their revolute margins readily distinguish the two. It is the largest species in the genus and can reach up to 2.5m in height (rarely more than 1.5m) and the stems can be up to 1m in diameter. It is almost invariably found on inaccessible cliffs and can sometimes be abundant, occupying almost every suitable niche.

ETHNOGRAPHIC NOTES

Name(s)

– Soqotri: *kˈartab* (pl. *kˈʌrətˈibihin*, *kərˈitib*) 'dried out, withered; stunted'.

– Soqotri simplified: *kartab* (pl. *karitibihin*, *kiritib*).

Food (livestock). Not normally accessible to livestock, but goats sometimes find them and nibble the leaves, though they do not chew the succulent stem unless very hungry. In a drought or prolonged dry season the whole plant is gathered by herders and pulped for their goats. Their cliff habit makes them difficult to collect but they can sometimes be knocked down by a thrown stone or be reached by herders lowering themselves on ropes down the cliff faces. The strongest ropes are used for this, usually made from *Dracaena* fibre or from the fibrous, sacking-like leaf sheath of the date palm. The swollen stems are not seen as a nutritionally valuable feed, but as useful subsistence. They are not considered to be as good as *Euphorbia arbuscula* as a drought feed, but are regarded as being at least as good as pulped *Dendrosicyos*. The plants are not normally fed to goats in the central highlands, an area rich in alternative forage, even in a drought, and goats here are not used to it and reluctant to eat it.

Flavouring / food additive. The meat of goats eating a lot turns the cooking water a creamy colour, and makes it very sweet and delicious. However, the meat itself is tainted and sometimes even inedible. The taste of the milk of goats eating a lot of the plant is recognisably altered, developing a distinctive bitter flavour.

Medicinal / veterinary. The bark is used to strap a fracture, especially in small stock. It has also been successfully used on young calves, but they have to be immobilised and hand fed and watered until the fracture is healed. The strapping is applied after the fracture has been manipulated and set, and the surrounding areas dressed with one of the preparations made from bark from *Commiphora* species, aloes juice or *Euphorbia arbuscula* latex. The bark strapping is then bound round with a cloth bandage. The fractured limb is inspected at regular intervals and the strapping progressively loosened. To treat mange in small stock the swollen stem is de-barked and the white core chopped up finely. This is added to cold water, stirred for a time, and then left to steep for some hours. It is then used as a wash for the affected animal.

Other. Animals feeding on this can go longer without water. Islanders say that goats fed on a steady diet of this, *Euphorbia arbuscula*, *E. schimperi* and *Dendrosicyos* can go without water right through the dry season.

Notes. Islanders have noticed that termites are not attracted to the plant.

Comparative uses. In Oman the seeds of *D. foetida* are eaten to treat flatulence and indigestion. In Dhofar the underground parts are a well-liked food, eaten both raw and cooked, though the more mature plants are rather hot and peppery. They are at their best after rain. The abundant milky juice of *D. gourmaensis* is acrid and irritating to the eye. The dried rhizomes of *D. contrajerva* are ground and used to flavour tobacco. (Dalziel 1948, 1956; Ghazanfar 1994; Miller & Morris 1988)

Dorstenia socotrana **A.G.Mill.**

Perennial herb, stemless with an underground tuber; tuber c.1cm diameter. Leaves cordate to reniform, entire, 8–25 x 8–30mm, rounded, entire or irregularly denticulate, cordate at the base. Receptacles circular, 3–5mm diameter (excluding appendages); appendages linear, 1–3mm, evenly distributed.

Ecology. Restricted to seaward-facing escarpments in NE Soqotra where it grows on shady limestone cliffs and boulders in drought-deciduous woodland. 20–300m.

Distribution. Endemic.

Conservation status. Vulnerable (VU D2). Rare, with a very limited area of occupancy. Common in a fairly narrow altitudinal band on seaward-facing limestone escarpment cliffs in NE Soqotra. Its total area of occupancy is probably under 10km^2. It is under no immediate or perceived threat but its restriction to a small, specialised niche makes it vulnerable to environmental change. It is also of potential interest to succulent enthusiasts and could be threatened by overcollection.

Botanical notes. In habit very different from *D. gigas*, but in flower its affinities with that species are more obvious.

ETHNOGRAPHIC NOTES

Name(s)

– Soqotri: nil or *bɔ̄wnˈəno* 'tiny *Sterculia africana*'.

– Soqotri simplified: *bownino.*

Food (human). The small tuber is edible and is eaten raw when it develops after rains. However, the plant is not common enough for it to be an important food.

Food (livestock). The aerial parts provide incidental grazing for goats if they can reach it, but it is not nutritionally important.

Ficus L.

Monoecious or dioecious trees or shrubs. Leaves alternate, entire; stipules free or partly fused. Flowers borne inside a hollow receptacle (fig); figs with a small apical opening (ostiole) fringed by small bracts. Flowers of three kinds: male, female and gall. Perianths 2 to 6-lobed; male flowers with 1 to 3 stamens; female flowers with a filiform style; gall flowers with a reduced style and stigma. Achenes enclosed within fig. 800 species. Tropical and warm regions, especially SE Asia.

Ficus cordata Thunb.

Ficus cordata subsp. *salicifolia* (Vahl) C.C.Berg
Ficus salicifolia Vahl

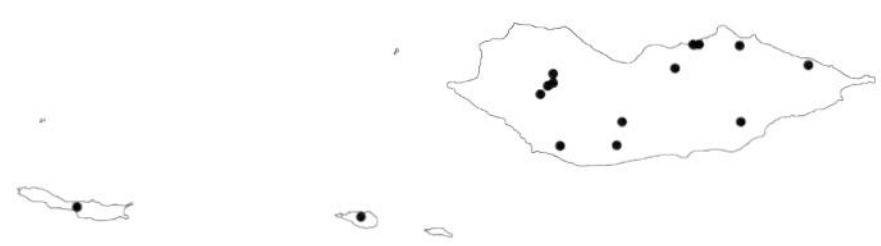

Shrub or small tree to 8m. Leaves evergreen, lanceolate, 2–20 x 1–3cm, acute to acuminate, entire, rounded or slightly cordate at the base, glabrous; petioles 1–5cm. Figs 1 to 3 in the leaf axils, globose, 7–10mm diameter, glabrous to puberulous, white to purplish when ripe, sessile or pedunculate.

Ecology. Widespread and common along wadis and on cliffs, usually near water, at lower altitudes in the drier areas of the island. 50–650m.

Distribution. Throughout Africa and Arabia.

Botanical notes. Soqotran material is all referable to subsp. *salicifolia* (Vahl) C.C.Berg.

ETHNOGRAPHIC NOTES

Name(s)

– Soqotri: *ʾˈitʾib* (pl. *ʾˈitʾub*); (western Soqotra) *ʾˈiṯib*.
– Soqotri simplified: *'it'ib* (pl. *'it'ub*); *'ithib*.

Food (human). People in the hungrier areas of the Archipelago, especially children, compete with birds for the ripe fruit.

Food (livestock). The green foliage is not palatable to livestock, though islanders say that in a drought goats have even been seen to nibble at the green leaves, in particular the tender, new growth, and donkeys are said to occasionally eat the leaves. The dried and fallen foliage is eaten in a prolonged dry season, and herders gather the fallen leaves to feed their cattle, sheep and goats in a drought. Sheep and goats also eat the fallen, dried fruit. However, if goats eat too many they fall ill, and if they eat a lot of the fresh fruit are said sometimes to become purblind. The fruit of trees at higher altitudes is less toxic and goats can eat these safely. Small stock on the poorly vegetated island of Samha are more used to eating the foliage and fruit, but still only do so when very hungry.

Bees / honey. The dead wood is one of the best for smoking out bees when stealing their honey.

Flavouring / food additive. Herders on Samha say that the milk of stock feeding on the dead or living material becomes slightly bitter.

Fuel. The dead wood makes good firewood, and in areas where there is a shortage of firewood, it is collected and a store kept in the home. Unlike other trees and shrubs of the islands, it is sometimes cut and dried for use as firewood, as it quickly regrows, and is of little importance as a livestock food. The firewood is little used for cooking as it gives off clouds of dense and acrid smoke. It is a preferred wood for firing pottery.

Timber. Not very good timber but is used in areas where there are few alternatives. It is not attractive to termites and is not normally de-barked before use. The wood of well-watered trees makes poor timber, and only the wood of a tree from a dry area, cut at the hottest time of the year, is of any use.

Fibre / cordage. Cordage is made from the bark, except on Abd al Kuri, where the islanders agreed that the limited number of trees were only to be used for firesticks (see below). On Samha, where the tree is more common and there are few other fibre-producing trees, cordage from the bark was much in demand. It was used to make nets and fishing lines before imported fishing equipment became available, and was twisted into string and then plied to produce ropes. On Soqotra this cordage is regarded as rather inferior and lacking in durability, and other fibre-producing plants are preferred.

Leatherwork. The inner bark can be crushed and used to treat skins and hides to strengthen the leather and stain it dark red-brown. The best bark is from the base of the trunk. Sappy bark is no good for leatherwork, so that of the upper branches and stems is not used.

Wood. The dead wood is considered second only to *Commiphora ornifolia* as an effective firestick. It was often used as the horizontal base, with wood from another source spliced into a handle of a third wood being used for the vertical upper stick. This was the only tree that reliably produced fire on Abd al Kuri, and the wood of the few trees on the island was protected for this use alone. On Samha they also had *Withania* and *Marsdenia* to make fire, but *Ficus* was always preferred as it produces a spark more quickly than the others. The wood is little used for implements, but the rollers for hauling out dugout canoes were commonly made from branches of this tree. Usually three rollers were made for each canoe. When the canoes were dragged right up the beach at the beginning of the monsoon to an area sheltered from the monsoon wind and waves, they were turned upside down and placed on the rollers, and left there until the sea grew calm again. The wood was much used in charcoal and quicklime manufacture.

Medicinal / veterinary. The quicklime is used medicinally. It is ground to a fine powder and used as a dressing, mainly for lacerations, ulcers and the surgical wounds of circumcision. It is also rubbed over weeping sores and abrasions to dry them out. The latex is used medicinally. It is smeared over a pointed sore or boil to bring the pus to the surface. For an extensive laceration, it is painted over both sides of the tear and these are then held together until they stick firmly. The latex is smeared over a cloth used to strap the chest of someone with a painful cough or other chest pain. Sometimes the latex is applied directly to the skin of the chest, especially to treat a bad wheeze. In the absence of *Jatropha unico-*

stata bark, strips of the inner bark of this are used to strap and splint a fracture. Broad strips are beaten to soften them and are placed each side of the fracture, and other slimmer strips are bound around.

Pest control. Both the green and dead wood are used in insecticide smudges as they produce great quantities of acrid smoke.

Other. These trees generally grow near water and are regarded by the islanders as a good indicator of the presence of water, especially if there are also *Dovyalis* trees present. The latex makes the best bird lime. The thinner branches contain most latex, and are struck with a stone or the blunt edge of a knife to cause the latex to run. This is collected by twisting a stick round and round beneath the cut until it can hold no more. It is regarded as a better bird lime than *Euphorbia socotrana* latex because the latter melts and runs if the sun is strong, and then the trapped bird can pull itself free. The *Ficus* latex remains adhesive even in great heat. The latex is also used as a glue, to make a knife handle for example. If a goat horn is not available for the smaller knives, or if making a handle for the larger, broader blades used to chop wood and lop branches for fodder, strips of cloth are soaked in the latex and wound round and round the tang. This handle is then squeezed hard to compress the layers and to shape it to fit the hand. Once it has dried out, it is rubbed with fine earth to stop it being sticky. The latex of *Euphorbia socotrana* is considered to be superior for this. The latex is also used to stick together handwritten pages of the Qur'an to make a booklet.

The black soil collected from under these trees is one of the best soil improvers for cultivated plots, and it is also mixed with dung and used to enrich the soil of kitchen gardens and young date palms. The rotting leaves are gathered for the same purpose, and, when mixed with sheep dung, are regarded as one of the richest fertilisers. Dead wood is gathered and piled with dung on seedbeds, now usually for kitchen gardens but formerly for finger millet and tobacco plots. They are burned and the ash is mixed with the soil to enrich it. The new fresh growth is scorched and crushed, and smoked as a substitute for tobacco or to stretch a limited amount. A paste made from the ground inner bark is used as a red dye for the more special and personal leather items, such as the intricately plaited belt formerly worn by men. A more liquid version of the paste is used as a red dye to decorate clay pots, but this has to be applied quickly while the pots are still hot from firing.

Notes. The islanders differentiate between the 'male' tree that bears no fruit and grows low to the ground and the fruit-bearing 'female' with a single trunk and spreading branches. They also note that nothing grows in the shade of these trees.

Comparative uses. In Yemen the sap is used as an ophthalmic antiseptic. In Dhofar slivers of the inner bark are pounded to a paste which is used in the preparation of leather, and is rubbed into a smelly milk-skin to sweeten it. The paste is also used to treat festering skin sores. The leaf sap is applied to warts, and the leaves are crushed and eaten with lemon to cure indigestion. Ground with *Calatropis procera* flowers, they are applied to bruised fingers and toes. It is reported that figs that fall into riverbed pools are chewed by fish. The inner bark yields a strong fibre. In Saudi Arabia the latex of the related *F. palmata* is added to milk to curdle it and make yoghurt. The proteolytic enzyme ficin digests living *Ascaris* and *Trichuris* worms and occurs in *Ficus* latex. This is well tolerated and non-toxic to man when taken internally, and is widely used as a vermifuge by people in South America and Panama. (Fleurentin & Pelt/Fleurentin *et al.* [1982, 1983a,b,c]; Ghazanfar 1994; Lewis & Elwin-Lewis 1977; Miller & Morris 1988)

***Ficus vasta* Forssk.**
Ficus socotrana Balf.f.

Large tree to 20m, starting as epiphyte or on rocks, often with aerial roots. Leaves deciduous, broadly ovate to suborbicular, 7–30 x 6–20cm, rounded or shortly acuminate, entire, cordate at the base, glabrescent above, glabrescent or velutinous beneath; petioles 4–10cm. Figs 1 or 2 in leaf axils, subglobose, 1–1.5cm diameter, tomentose to glabrescent, green with paler spots when ripe; sessile or pedunculate.

Ecology. On cliffs, in ravines and on boulders in drought-deciduous woodland, mainly on limestone. 200–650m.

Distribution. E and NE tropical Africa; S and SW Arabia.

Botanical notes. One of the largest trees on Soqotra. Starting off on cliffs or rarely as an epiphyte on branches or trunks of large trees. It produces aerial roots which thicken and ultimately form a trunk.

ETHNOGRAPHIC NOTES

Name(s)

– Soqotri: *ṭəḳ* (pl. *əṭwāḳ*, *əṭwˈʌḳ*) 'softened'; green immature fruit *ˈibgəher*, *ḳˈəṭəhar*; half-ripe fruit *bəgɔ̄lhi*; fully ripe fruit *ḥfˈol*; aerial roots *ʿāṭəl*.
– Soqotri simplified: *tiq* (pl. *itwaq*); green immature fruit *ibgihir*, *qitiher*; half-ripe fruit *bigowlhi*; fully ripe fruit *hfol*; aerial roots *'atil*.

Food (human). The fruiting season is not long and the fruit rarely ripen fully before the short summer season. Immature fruit are usually preferred to riper ones. They are normally eaten fresh, as and when they are found, but in the hungrier areas, or in the better-vegetated areas when a fig tree has produced a good crop of ripe fruits, they are harvested. The sun-dried fruit are then stored in goatskins. They are considered best eaten with a little ghee, like dates. In the better-vegetated parts of the island the fruit are not an important food, and only children really search them out. Birds are extremely fond of the fruit and humans have to compete with them for both ripe and immature fruit.

Food (livestock). The large leaves remain green throughout the dry season and are an important dry season fodder for goats, especially in areas where there are few alternatives. They browse the foliage, and their owners also lop leafy stems for them. Only the stems with green leaves are cut; those with yellowing leaves are avoided, as goats will not normally eat the older leaves. In the drier regions the foliage is sometimes also gathered to feed cattle, but, unlike goats, they prefer the older, withered leaves. Fallen leaves and other plant litter are eaten by livestock, and are also gathered by herders. The fallen fruit are eaten by all livestock. However, islanders have observed that, unusually, no part of the tree is palatable to camels, neither fresh nor dried leaves. Apart from the fruits, this tree is not seen as being of nutritional importance for livestock.

Bees / honey. Attractive to honeybees in the drier areas.
Flavouring / food additive. The figs were a much-appreciated source of sweetness before the arrival of sugar.
Fuel. The dead wood makes very poor firewood, which produces clouds of acrid smoke.
Timber. Not a good timber, but it is used if alternatives are not available. Like *F. cordata*, the wood of well-watered trees makes poor timber, and only the wood of a tree of a very dry area cut at the hottest time of the year is of any use. The bark is attractive to termites, so must be removed before use. The larger branches can be used as roofing beams, but it is rarely used for the main supporting pillars or cross beams.
Fibre / cordage. The aerial roots are used as ropes, but do not last well. Cordage for short-term use is made from the inner bark. This too is not very durable and has a tendency to break as it dries out. The inner bark from the roots (*ᶜẹ́ṣiyo di ṭiḳ*) makes the best cordage. A cut is made high in the bark of the root and another lower down. The bark between the cuts is then peeled off and the inner bark removed. This is then separated into individual strands used as cordage just as they are, without twisting or plying. The strands can be further teased out to make quite fine string used for such things as tying on the weaning harness or the contraceptive 'apron'. However, this cordage is not considered to be as strong as that from *Hibiscus*.
Leatherwork. The inner and outer barks are ground to a paste used to strengthen leather and stain it red-brown. This paste is also used to prepare the skin of a male kid or calf slaughtered to preserve its mother's milk. The skin is made into a tulchan and stuffed with fragrant herbs. This is presented to a milk cow or goat to sniff at, after spraying with some of her milk, to encourage her to let down her milk. The mixture is also smeared over the back of a living calf to disguise its smell before presenting it to a strange cow, to encourage her to accept the calf as her own and allow it to suckle.
Wood. The dead wood is used as a firestick, but is not as effective as that of *F. cordata*. Otherwise the wood is little used.
Medicinal / veterinary. The latex is heated and applied while still hot to sores, ulcers and lacerations to speed the healing process, or to boils and infected swellings to bring the pus to the surface. If a badly infected wound is not responding to treatment, it is cleaned with salt and water and then latex applied. The surrounding area is cauterised to make a circle round the wound 'to stop the poison spreading'. The latex is spread over the chest of someone with a painful cough or sore chest, or sometimes spread over a cloth which is then bound around the chest. A layer of fine earth is spread over the upper surface to make it less sticky. This is regarded as the most effective latex for this purpose. Layers of the latex are spread over a damaged or sprained limb to make a sort of sticking plaster. It is also painted over the site of a fracture once the bone has been set, in small stock and calves as well as in humans. The animal must be immobilised as much as possible and hand fed and watered until the bone has mended. Sometimes a piece of cloth, or the inner bark of a medicinal plant such as *Jatropha unicostata* or aloe, is spread with a thick layer of latex and then used to strap the fractured limb. If a finger or toe has been damaged, it is painted with the latex before being dressed with the medicament made from bark of *Commiphora* species and then bound in a cloth bandage. For major fractures, bone-setters often apply a thick layer of latex, then a layer of sheep's wool or goats' hair, and finally strap the limb.
Commercial value (past, present). The fruit were not sold formerly, but more recently have occasionally appeared in the markets for sale.
Pest control. The burned wood makes no sparks and produces masses of thick, acrid smoke, so it is ideal for insecticide livestock smudges. The best for this is the crumbly core of dead wood, or green wood. For cattle, this wood is used for the smudges of the summer months only, other woods being preferred in the cold, rainy months. The smoke is too pungent for use in living quarters.
Other. These trees generally grow near water and are regarded as a good indicator of the presence of water. The inner bark is chewed to extract a sweet, thirst-quenching liquid. If clearing an area for cultivation, these trees are usually removed since islanders have observed that nothing grows beneath them. The inner bark of the aerial roots is used with a pebble as a sling-shot, or to make the 'rubber band' of a catapult made from a forked twig of a hard wood. The aerial roots are used as swings by children. The large leaves are held over the mouth and nose when entering wooded areas to protect against botfly attack. A large leaf is folded and held to the mouth by goatherds to increase the volume of the sound made by blowing hard though pursed lips at goats to get, or keep, them moving. Hollowed-out stems are used as a pipe to smoke locally grown tobacco. The unripe fruit are used as balls and the smaller ones in games instead of pebbles.
Notes. The islanders say that 'male' trees (*ṭiḳ mí'šər*) produce fewer fruit than 'female' ones. There is a widespread saying on Soqotra: '*šəmɛ̄daᶜ bə ṭiḳ*!', the equivalent of don't count your chickens until they're hatched! The story goes that these words were said by a man after he found a fig tree covered in fruit, and killed a goat so he could use its skin to store the figs he would harvest. When he had finished slaughtering and skinning the goat, he looked up at the tree and saw that starlings had stripped it of its fruit.
Comparative uses. Figs are everywhere an important food, and the inner bark of various species is considered to be suitable for paper manufacture. In Dhofar the inner bark of *F. vasta* is chewed both as food and as liquid refreshment. The bark is also important as a source of a good, strong fibre for making cordage. Slices of inner bark are used to strap a fractured limb, which hardens as it dries, making a rigid casing, like a plaster. The ash of the burned bark is used as a snuff. The sticky latex of immature fruits, and the resin that oozes from the tree in great heat, are used as a depilatory and glue. The resin is also used medicinally as a haemostatic, and the latex is boiled in water, strained and poured into a large or deep wound to dry it out and prevent infection. The aerial roots are used to make the curved hoops of a cradle. In East Africa the fruit are an important food, and the latex is used as glue and to fill cracks and holes. It is also chewed by children as a chewing gum. In Yemen a mixture of figs, dates, raisins and honey is given against depression or nervous tension. In Saudi Arabia the latex of the closely related *F. carica* is used to extract thorns, and drops are added to milk to curdle it. In Oman the latex of *F. carica* is applied to warts to burn them out, and to blisters and boils to draw the pus. It is also ingested as an emetic and vermifuge. The leaves are used as a poultice on swollen or bruised fingers and toes, and the crushed, dried leaves are mixed with honey and applied for leprosy, and to lighten freckles and other skin discoloration. The figs are eaten to help remove kidney stones, and, after boiling in water, are taken as a cough mixture. (Abdulfatih 1987; Ambasta 1982; Fleurentin & Pelt/Fleurentin *et al.* [1982, 1983a,b,c]; Ghazanfar 1994; Heine *et al.* 1988; Jain & De Fillips 1991)

MYRSINACEAE

Shrubs. Leaves alternate, with faint gland dots, serrulate; stipules absent. Subdioecious, flowers on male plants with no ovaries but stamens present on female plants. Flowers in axillary clusters, small, actinomorphic. Calyx 5-lobed. Corolla campanulate, 5-lobed, lobes free almost to base. Ovary superior, 1-celled; style 1; ovules in 1 row on free-central placenta. Fruit a 1-seeded berry. 37 genera. 1250 species. Tropical and subtropical, Old World.

Myrsine L.

See description of family. 5 species. Africa and tropical Asia.

Myrsine africana L.

Evergreen shrub to 2m. Leaves narrowly elliptic to suborbicular, 1–3 x 0.4–1.5cm, serrulate. Flowers reddish, c.0.3cm diameter. Fruit globose, dark purple, <0.5cm diameter.
Ecology. In the Haggeher.
Distribution. S and E Africa, SW Asia to China.

Botanical notes. Collected in the 19th century but not seen since.

ETHNOGRAPHIC NOTES
Comparative uses. In India the fruit are eaten and are taken as an anthelmintic, especially against tapeworm. The fruits are also taken as a laxative, and are one of the constituents of an ointment used to treat ringworm and other skin complaints. Elsewhere the seeds are an adulterant of pepper. It contains myrsine-saponin, which is under investigation for use as a drug in cancer chemotherapy. (Jain & De Fillips 1991; Kunkel 1984; Lewis & Elwin-Lewis 1977)

NYCTAGINACEAE

Herbs. Leaves opposite, simple, exstipulate. Flowers in terminal or axillary cymes, sometimes whorled or umbellate; actinomorphic, bisexual. Perianths 5-lobed, tubular, petaloid, lower part persistent around ovary. Stamens 1 to 10. Ovary superior, 1-celled, 1-ovulate; style simple. Fruit (the anthocarp) a 1-seeded achene enclosed by persistent part of perianth. 34 genera. 350 species. Tropical and warm regions, especially New World.

Boerhavia L.

Herbs, sometimes woody-based. Leaves entire. Flowers bisexual in terminal or axillary umbellate heads or open panicles. Perianths funnel-shaped or campanulate. Stamens 1 to 3, shortly exserted. Fruit 5-ribbed, glabrous or glandular-hairy. 20 species. Tropical regions.

Boerhavia diffusa L.
Boerhavia repens sensu Balf.f. non L. p.p.

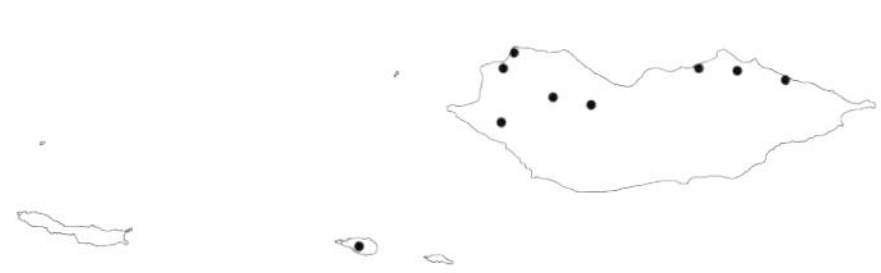

Herb with prostrate stems to 1m long, glabrous, sparsely pubescent or sparsely to densely glandular. Leaves ovate to subcircular, 10–50 x 5–45mm, often paler beneath. Flowers in long-stalked 3 to 12-flowered clusters either axillary or in leafless panicles. Perianths usually pink, rarely magenta, 1.25–1.5mm long. Fruit 3–4 x 1.25–1.5mm, puberulous and usually glandular-viscid.
Ecology. Widespread on the plains and foothills. Sea-level–700m.
Distribution. Pantropical.

ETHNOGRAPHIC NOTES
Name(s)
– Soqotri: *ḥidəho* (pl. *ḥidhoytin*) 'strong'; *ḥidis*, *ḥedist* 'to make secure, strengthen', 'animal unwilling to leave the settlement to go out to graze' (also given to *Zaleya pentandra*).
– Soqotri simplified: *hidiho* (pl. *hid'hoytin*); *hidis*, *hedist*.

Food (human). This used to be an important food of the hungry summer months for people of the coastal plains, and in drier areas where it is common, as around Qalansiyah and Shu'ub. The islanders look for the plants in flat, open and usually low-lying areas. The foliage develops after the winter rains, and the roots are swollen and ready to eat by the time the foliage dies back in the summer. Sometimes the tip of the swollen root is visible above ground, and is easily dug up; at other times a deep hole has to be dug. Since the outer skin is very bitter, this is removed, and the root then either eaten raw or cooked. It is baked in embers or hot ash, or is boiled in water. It is best when mashed and eaten with soured milk or ghee. Some of the more mature roots have a rather unpleasant taste, and generally they are regarded as a filling rather than nutritious food. The taste is improved by cooking the roots in water with a layer of *iʾiri* or *di ʾɔ̄žihir* grass, or *Amaranthus* foliage, placed above and below.
Food (livestock). The mature plant is not much liked except by sheep, though all livestock eat it when hungry and little alterative forage is available. Younger plants from better-watered areas are considered good feed, and to promote lactation in sheep and cows. In a drought, goats dig them up to get at the roots. On Samha they are

grazed by sheep in the plateau area, and are regarded as nutritious feed.

Flavouring / food additive. When fully grown, and especially if there are summer rains, this plant secretes a sweet and sticky substance (*bēsi*), which tastes like nectar. This substance also occurs in certain grasses, and before sugar arrived on the islands was much liked, by children in particular. It is also said to impart a sweet flavour to the milk of animals grazing heavily on the foliage.

Other. Like *Zaleya pentandra*, the leaves are used to repair the local clay pottery. If a pot is holed, a shard is ground down and shaped to fit the hole, the side appearing on the outer surface of the pot being broader than that of the inner. Then leaves of the plant are pounded to a paste with a soft chalky stone called *maḥkāḳeh* (possibly gypsum) and the paste is packed tightly around the inserted shard. The pot is then put beside the fire for the paste to set and harden. If a pot is cracked, the same paste is smeared along and around the crack on the outer surface, and the pot put beside the fire to dry. The plant is also used to decorate pottery. The leaves are ground to a thick paste with stems of *Kalanchoe*. This is rubbed around the neck of the pot (commonly the *ḳɔ̄məh*, the pot for storing clarified butter) while it is still hot from firing. It leaves a dark red-brown stain. At the end of winter the foliage is gathered along with grasses to make hay. The grass and herbs are laid out to dry, preferably in shade to retain its green colour. When completely dried the hay is stored in a cave or similar weatherproof place as dry season fodder for pregnant and lactating stock.

Notes. It used to be widespread in the northern plains and foothills, but is now much less common.

Comparative uses. In India the roots are taken as an expectorant, a laxative, and a powerful diuretic. An extract is used in asthma. The active constituent is the alkaloid punarnavine, and this plant is considered to be the source of the Ayurvedic drug Punarnava. The leaf is used as a potherb in soup and is fed to sheep and cattle. The seeds are added to cereals and other foods. In Nigeria the roots are added to cakes. An infusion of the whole plant is given as a mild laxative and febrifuge for children. It is also given for convulsions and to procure regular menstruation. A decoction of the roots is given for jaundice. Both roots and leaves are regarded as being emetic, and in larger doses are used in asthma. Externally the thick roots, after softening by boiling, are applied locally as a poultice for ulcers, to draw abscesses and to encourage extraction of the guinea-worm. In the Middle East the plant is regarded as purgative, emetic and antisyphilitic. (Ambasta 1982; Dalziel 1948, 1956; Goodin & Northington 1979; Jain & De Fillips 1991)

Commicarpus Standl.

Herbs, sometimes woody-based. Leaves entire or sinuously lobed. Flowers bisexual, in umbels or whorls forming an open panicle. Perianths funnel-shaped, tubular below. Stamens 2 to 6, exserted. Stigma capitate, exserted. Fruit 10-ribbed, studded with conspicuous, sessile or stalked wart-like glands. 2 endemics. c.27 species. Tropics.

Botanical notes. For a revision see Meikle (1978).

Commicarpus boissieri (Heimerl) Cufod.

Boerhavia boissieri Heimerl

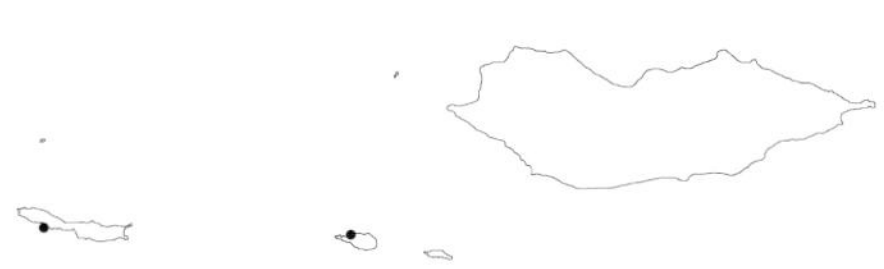

Herb to 1.5m, glabrous. Leaves ovate to subcircular, <7cm, entire or sinuate. Flowers distinctly pedicellate, in whorls; pedicels 6–15mm. Perianths pink, purple or magenta, 3–5mm long, widely funnel-shaped, with a very short, inconspicuous basal tube, externally glabrous. Stamens usually 3(or 4). Fruits clavate, with prominent warty glands at the tip, and scattered prominent glands below.

Ecology. Coastal plains and desertic foothills of Samha and Abd al Kuri. Sea-level–100m.

Distribution. Oman, Pakistan and India.

ETHNOGRAPHIC NOTES†

Name(s)

– Soqotri: *ᶜáyṭift*, *ᶜīṭift* 'folded, wrapped' (Soqotra, Samha and Abd al Kuri); *ᶜayṭáfʾʌn* 'small *ᶜáyṭift*'; *ḥédist* 'to make secure, strengthen', 'animal unwilling to leave the settlement to go out to graze' (more commonly given to *Boerhavia diffusa* and *Zaleya pentandra*).

– Soqotri simplified: *'aytift*, *'iytift*; *'aytaf'en.*

Food (human). *C. boissieri*, as suggested by one of its names, *ḥédist*, resembles *Boerhavia diffusa* and *Zaleya pentandra* in that its finger-like root can be eaten raw or cooked in the same way as these two plants.

Food (livestock). Eaten by all livestock, but it is not regarded as good fodder. Donkeys and camels enjoy it, and in drier areas it is regarded as useful for goats. Elsewhere goats only eat it in hunger. On Abd al Kuri and Samha, *Commicarpus* are important fodder for goats and sheep, but on Soqotra sheep rarely eat it.

Fuel. The dead stems are gathered and stored in a dry cave for use as kindling during the rains.

Other. Animals feeding on it can go longer without water.

Notes. It can cause *'uqabah* in goats unaccustomed to it. It is seen as being on the increase across the islands.

Comparative uses. In Dhofar the new shoots of *C. boissieri* were eaten, as a food and to allay thirst. In times of hunger the roots were dug up and roasted. Other species, such as *C. plumbagineus*, are also edible. In Kenya the milk of animals feeding heavily on *Commicarpus* species is reported as bitter. In Africa the twigs of *C. plumbagineus* are pounded and used as a poultice for sores and stings. Plant parts of *C. helenae* are mixed with water and poured into a cow's nose to treat various complaints. (Heine *et al.* 1988; Kunkel 1984; Miller & Morris 1988)

†Also apply to *C. heimerlii*, *C. helenae*, *C. simonyi*.

Commicarpus heimerlii **(Vierh.) Meikle**
Boerhavia heimerlii Vierh.
Boerhavia scandens sensu Balf.f. non L. p.p.

Herb to 1.5m, glabrous towards the base, viscid above. Leaves fleshy, ovate to subcircular, <5cm, entire. Flowers distinctly pedicellate, in umbels, seldom in whorls or irregular; pedicels 5–7mm. Perianths pink or purple, 6–7.5mm long, narrowly funnel-shaped with a well-developed basal tube, externally glabrous. Stamens 3 to 5. Fruit clavate, prominently glandular at the tip, inconspicuously glandular below.

Ecology. Rocky areas on limestone. Sea-level–700m.
Distribution. Endemic.
Conservation status. Least Concern (LC). Common in several vegetation types and under no present or perceived threat.
Botanical notes. Covered in glandular hairs which make it very sticky to the touch.

ETHNOGRAPHIC NOTES
Name(s)
See *C. boissieri* (also uses).

Commicarpus helenae **(Roem. & Schult.) Meikle**

Herb to 1.5m, glabrous or sparsely scabrid. Leaves ovate-triangular, <5(–9)cm, entire or sinuously lobed. Flowers sessile or shortly pedicellate, in narrow whorls; pedicels <4mm. Perianths pale pink or purple, 1.5–2.5mm long, widely funnel-shaped, with a very short, inconspicuous basal tube, externally glabrous. Stamens usually 2. Fruits clavate, with prominent, long-stalked glands at the tip, inconspicuously glandular below.

Ecology. Amongst rocks in desertic foothills and coastal plains. Sea-level–100m.
Distribution. Tropical Africa, Egypt, Arabia and Middle East.

ETHNOGRAPHIC NOTES
Name(s)
See *C. boissieri* (also uses).

Commicarpus simonyi **(Heimerl & Vierh.) Meikle**
Boerhavia plumbaginea var. *socotrana* Heimerl
Boerhavia scandens sensu Balf.f. non L. p.p.
Boerhavia simonyi Heimerl & Vierh.

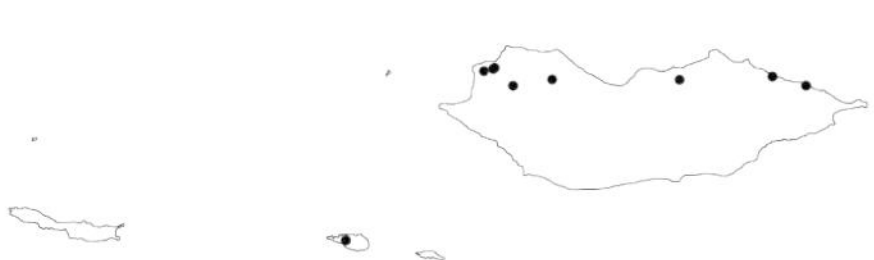

Herb to 1.5m, glabrous or sparsely scabrid. Leaves ovate-triangular to subcircular, <5cm, entire. Flowers distinctly pedicellate, in umbels or rarely in whorls; pedicels <2cm. Perianths mauve, pink or purple, 7–7.5mm long, narrowly infundibuliform with a well-developed basal tube. Stamens 3(to 5). Fruit narrowly spindle-shaped, inconspicuously glandular.

Ecology. Rocky slopes in drought-deciduous woodland on limestone and granite. 100–800m.
Distribution. Endemic.
Conservation status. Least Concern (LC). Common in several vegetation types and under no present or perceived threat.

ETHNOGRAPHIC NOTES
Name(s)
See *C. boissieri* (also uses).

OLEACEAE

Shrubs. Leaves opposite, trifoliolate (sometimes simple); stipules absent. Flowers solitary or in few-flowered cymes, actinomorphic. Calyx campanulate, 4-lobed. Corolla tubular, 5 to 9-lobed. Stamens 2. Ovary superior, 2-celled; ovules 2 per cell. Fruit a berry, often deeply bilobed. 24 genera. 900 species. Subcosmopolitan, especially Asia.

Jasminum L.

Jasminium Dumort.

See description of family. 1 endemic. 450 species. Old World tropics, some temperate.

Jasminum fluminense Vell.

Jasminum fluminense subsp. *socotranum* P.S.Green
Jasminum rotundifolium Balf.f.

Climbing or scrambling shrub to 2m. Leaflets broadly ovate to elliptic, 3–4.5 x 1.5–3.5cm, tips rounded. Flowers white or yellow, c.2cm long, fragrant. Fruit c.0.7cm diameter.

Ecology. Moist woodland on both granite and limestone. 600–1100m.

Distribution. Subsp. *socotranum* is endemic.

Conservation status. Least Concern (LC). Common in several vegetation types and under no present or perceived threat.

Botanical notes. All Soqotran material is referable to the endemic subsp. *socotranum* P.S.Green.

ETHNOGRAPHIC NOTES

Name(s)

– Soqotri: *ḳerʿʌš*, *ḳerʿiš* (pl. *ḳerʿišo*) 'mean, ungenerous' (also given to *Turraea socotrana*); *niśazə* 'solitary' (also given to species of *Helichrysum*); *rikəbə* 'riding high' (also given to *Cocculus pendulus*, and *Cryptolepis* and *Secamone* species); *ḥahər* (given to various hard-wooded species); (rare) *ṣamaʿano* 'odoriferous' (more commonly given to *Boswellia* species).

– Soqotri simplified: *qer'ash*, *qer'ish* (pl. *qer'isho*); *nishazih*; *rikibih*; *hahir*; *zama'ano.*

Food (livestock). The foliage is not palatable to livestock, though the litter is nibbled by goats when there is little else to eat. Camels and cattle avoid it.

Bees / honey. Honeybees love the fragrant flowers.

Timber. The larger stems make good timber, and some are even large enough to use as roof beams. However, the plant is often rather straggly and is not widespread enough to be an important timber on Soqotra.

Wood. The close-grained and hard wood is used to make implements in areas where it grows.

Cleansing. The flowers are very fragrant and sprigs of flowers are put amongst clothing or tucked into the turban or waistcloth for their sweet perfume. The sniffing of sweet-smelling plants is held to protect from the 'noxious air' that can cause disease, and to help protect against infection. All sweet-smelling plants are endowed with beneficial medicinal and prophylactic qualities.

Comparative uses. The flowers of various species are widely used to flavour tea, and the leaves are used as a condiment. Jasmines are generally unpalatable to livestock, though *J. fruticans* in Africa is reported as occasionally being eaten by camels and goats. In Dhofar the leaves of *J. grandiflorum* are pounded to a paste used as a soothing dressing for burns, or packed into a deep wound, or applied to an infected and discharging wound. The growing tips and stems are crushed and rubbed into the leather of an ill-smelling milk- or butter-skin to sweeten it. In Kenya the fruit or root parts of jasmine species are ground and added to water and then used to wash calves to kill lice. In Yemen the flowers of various jasmines are used as a febrifuge. In Oman the dried leaves of *J. grandiflorum* are boiled in water and drunk to relieve digestive problems. In Qatar *J. sambac* is said to be an abortifacient, and its root to be a useful emmenagogue. A preparation from the root is also given as a sedative and is taken fresh to treat insomnia, and to relieve a headache and the pain of a fracture. Its flowers yield a yellow pigment used as a substitute for saffron. (Ambasta 1982; Batanouny 1981; Fleurentin & Pelt/Fleurentin *et al.* [1982, 1983a,b,c]; Ghazanfar 1994; Heine *et al.* 1988; le Houerou 1980; Jain & De Fillips 1991; Kunkel 1984; Miller & Morris 1988; Oliver-Bever 1986; Rizk & El-Ghazaly 1995)

ONAGRACEAE

Herbs, usually in or near water. Leaves opposite, simple, entire; stipules absent. Flowers solitary in leaf axils, actinomorphic. Sepals 4. Petals absent. Stamens 4. Ovary inferior; ovules parietal; style 1. Fruit a capsule. 24 genera. 650 species. Cosmopolitan; especially diverse in N America.

Ludwigia L.

See description of family. 75 species. Cosmopolitan, especially N America.

Ludwigia palustris Elliot

Glabrous herb, creeping and rooting at the nodes. Stems and petioles winged. Leaves broadly elliptic or ovate, 0.8–4 x 0.5–2.5cm. Flowers green, 0.2–0.3cm diameter. Fruit c.0.4 x 0.2cm.

Ecology. In mud by permanent water. 50m.

Distribution. N America, Europe to Caucasus, N Africa, S Africa. Naturalised in Hawaii and New Zealand.

Botanical notes. Recorded as common in the 19th century by Balfour (1888); rare or possibly overlooked recently.

ETHNOGRAPHIC NOTES

Name(s)

– Soqotri: nil or *di rihóytin* 'of water'.

– Soqotri simplified: *di rihoytin.*

Food (livestock). Sheep are said to nibble it, but goats usually avoid plants growing in or around water.

Comparative uses. The stem peel of *L. caparosa* is used for a tea; the tips of the stems of *L. repens* are eaten as a vegetable, and the leaves of *L. abyssinica* are edible. Medicinally, *L. perennis* is boiled in oil and applied to the body to bring down a fever. The leaves of *L. prostrata* are used for toothache and to soothe aching muscles. (Ambasta 1982; Jain & De Fillips 1991; Kunkel 1984)

OROBANCHACEAE

Parasitic herbs, lacking chlorophyll and growing on the roots of other plants. Stems erect, usually unbranched. Leaves alternate, scale-like. Flowers in terminal spikes. Calyx 5-lobed. Corolla 5-lobed, zygomorphic. Stamens 4. Ovary superior, 1-celled; ovules numerous, on 4 parietal placentas. Fruit a many-seeded capsule. 17 genera. 230 species. N Hemisphere, temperate and subtropical.

Botanical notes. The species on Soqotra are apparently parasitic on a wide range of species.

Cistanche Hoffmanns. & Link

Robust herbs. Stems stout, unbranched, swollen below ground. Flowers in dense spikes, large, subtended by 1 large bract and 2 smaller bracteoles. Calyx with obtuse lobes. Corolla ± equally 5-lobed. 16 species. Mediterranean, N Africa to China.

Cistanche phelypaea (L.) Cout.

Cistanche lutea Hoffmanns. & Link
Cistanche tubulosa Wight

Stout herb to 50cm. Flowers yellow, <5cm long.

Ecology. Often parasitic on coastal salt bush plants. Comes up on dunes after rains. Sea-level–1300m.

Distribution. N Africa, Arabia and SW Europe.

ETHNOGRAPHIC NOTES

Name(s)

– Soqotri: nil or *ʾíbihʌṭ*, *ʾíbihiṭ* (Soqotra); *behíṯ* (Abd al Kuri).

– Soqotri simplified: *'ibihat*, *'ibihit*; *behith.*

Food (livestock). Not grazed by livestock.

Comparative uses. In some areas of northern Africa the young stem, swollen at the base, is considered a delicacy when cooked. In the Middle East it is regarded as having diuretic properties, and the thick lower part of the plant is dried and powdered and used to treat diarrhoea. The plant is also used as an aphrodisiac and tonic to treat spermatorrhea impotence. Dried and powdered, the stem is mixed with camel's milk and applied as a cataplasm for contusions. In Yemen the sap of the *Cistanche* species called '*ṭāj*' is used to treat verrucas. (Dalziel 1948, 1956; Fleurentin & Pelt/Fleurentin *et al.* [1982, 1983a,b,c]; Goodin & Northington 1979)

Orobanche L.

Erect herbs. Stems slender, sometimes branched. Flowers in lax or dense spikes, subtended by 1 large bract and sometimes by 2 smaller bracteoles. Calyx lobes acute. Corolla 2-lipped. 150 species. Mostly temperate.

Orobanche cernua Loefl.

Erect herb to 30cm. Stems unbranched. Flowers in dense spikes, pink or pale lilac, 1.2–2cm long. Corolla tube white, inflated, glossy on dry specimens. Bracteoles absent.
Ecology. Semi-deciduous woodland on the granite of the Haggeher mountains. 450–1050m.
Distribution. Africa, Arabia, Middle East, C Asia, India and Australia.

ETHNOGRAPHIC NOTES†
Name(s)
– Soqotri: nil or (*məśḥerəmhim*) *di aʿarib* '(*Caralluma*) of the ravens'; *ʾibihiṭ*; (rare) *məśḥerəmhim ʿafər* 'red *Caralluma*'.
– Soqotri simplified: (*mish'heremhim*) *di a'arib*; *'ibihit*; *mish'hermihim 'afir.*
Food (livestock). Not grazed by livestock.

†Also apply to *O. minor*.

Orobanche minor Sm.
Orobanche abyssinica Ach.

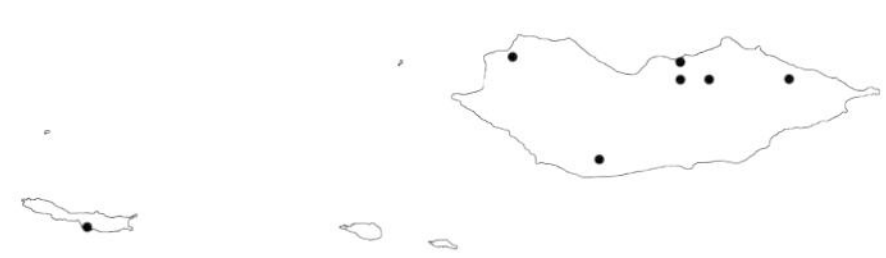

Erect herb to 60cm. Stems unbranched. Flowers in lax spikes, mauve or yellowish, 1–1.8cm long. Bracteoles absent.
Ecology. Semi-deciduous woodland on the limestone plateau and escarpments. 300–750m.
Distribution. Africa, Arabia, S Europe, Middle East, N America.

ETHNOGRAPHIC NOTES
Name(s)
See *O. cernua* (also uses).
Comparative uses. The Borana in Africa pound the root of *O. minor*, add sheep fat, and drink this as a soup for venereal disease. If a cow refuses to suckle, a root is cut into pieces and inserted in the cow's vagina, and within an hour or so she will start suckling. In Sudan livestock are said to graze this plant. (Ambasta 1982; Bebawi & Neugebohrn 1991; Heine *et al.* 1988; Jain & De Fillips 1991)

Orobanche ramosa L.

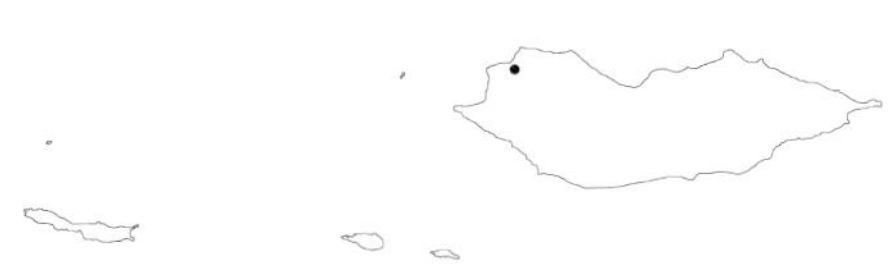

Slender herb to 20cm. Stems often branched. Flowers in lax spikes, dirty pink, c.1.2cm long. Bracteoles present.
Ecology. 'On *Lactuca* [*Launaea*] near Galonsir [Qalansiyah]' (Balfour 1888).
Distribution. Africa, Arabia, Middle East, C Asia, N America.
Botanical notes. Collected in the 19th century but not seen since.

OXALIDACEAE

Herbs. Leaves alternate, 3-foliolate; stipules absent. Flowers in axillary few-flowered umbels, actinomorphic. Petals 5, free. Stamens 10. Ovary superior. Fruit a 5-celled capsule, explosively dehiscent. 7 genera. c.900 species. More or less cosmopolitan.

Oxalis L.

See description of family. c.500 species. More or less cosmopolitan, especially S America and Cape region of S Africa.

Oxalis corniculata L.

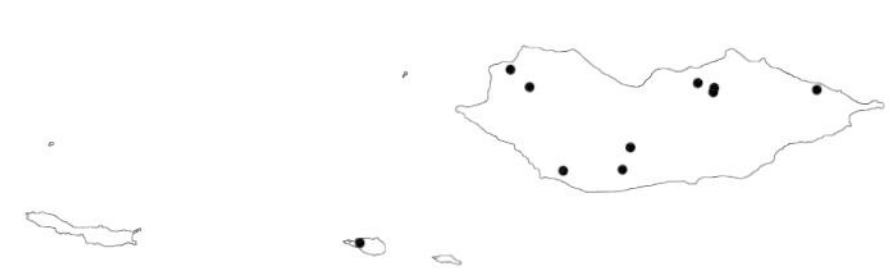

Prostrate, mat-forming herb to 20cm. Leaflets obtriangular, 0.3–1.5 x 0.3–1.5cm, 2-lobed, rounded. Flowers yellow, c.0.7cm diameter. Fruit cylindrical, 1–2 x 0.3–0.4cm.
Ecology. Widespread and common in open rocky ground. 100–800m.
Distribution. Cosmopolitan.
Botanical notes. One of the commonest and most widespread herbs on Soqotra.

ETHNOGRAPHIC NOTES
Name(s)
– Soqotri: nil or *ṣebərher*, *ṣōbəher* 'tart, astringent'; *ḥʌmźʌʾ*, *ḥemʌźəhen* 'sour, astringent'.

– Soqotri simplified: *zebirher*, *zowbiher*; *hamdha'*, *hemadhihen*.

Food (human). People pick the leafy stems and nibble them raw for their tart, lemony flavour.

Food (livestock). Grazed by small stock and regarded as good for their general health.

Medicinal / veterinary. Especially liked by pregnant women, and is recommended as an appetite stimulant for them and as a restorative for those recovering from illness.

Comparative uses. The leaves are a good source of vitamin C and carotene, but are harmful if eaten to excess. In India the leaves are eaten raw as a salad or cooked as a vegetable, and the seeds are eaten in times of hunger. Medicinally, the fresh juice is given in dyspepsia, piles, anaemia and tympanitis, and is also given to children with rickets. An infusion of the leaves is used to cure opacity of the cornea, and the leaf juice is given to counteract *Datura* poisoning and as an antidote to arsenic and mercuric poisoning. It is also applied to soothe bites and stings. In Yemen this plant, called *ḥumaiḍah*, is used to treat vertigo. In Qatar it is used as a diuretic, emmenagogue and anthelmintic, and is used to treat bruises, diarrhoea, enteritis, fever, flu, snakebite, sprains, swellings, and urinary infections. *Oxalis* species are used for green manuring, and are grown on slopes to check soil erosion. Some species are implicated in sheep poisoning; e.g. eating *O. pes-caprae* leads to death of sheep after around seven or eight weeks. (Ambasta 1982; Dalziel 1948, 1956; Jain & De Fillips 1991)

PAPAVERACEAE

Herbs, armed throughout with sharp prickles; with white or yellow latex. Leaves alternate, often lobed or deeply dissected; stipules absent. Flowers actinomorphic, bisexual, solitary. Sepals 2(or 3), free, spine-tipped. Petals 4 to 6, free. Stamens numerous. Ovary superior, 1-celled with parietal placentation. Fruit a many-seeded capsule, splitting by 3 to 6 valves at the top. 23 genera. 210 species. Mostly N temperate.

Argemone L.

See description of family. 28 species. N and S America.

Argemone mexicana L.

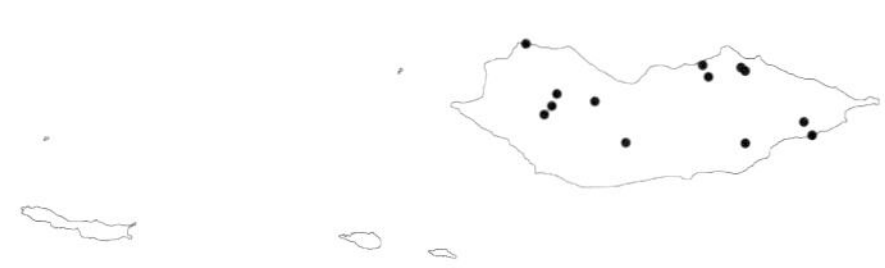

Annual herb to 80cm. Leaves somewhat fleshy, glaucous with conspicuous white veins, pinnatifid, <12 x 5cm, margins and veins beneath armed with sharp prickles. Flowers sessile. Petals golden yellow, obovate, 2–3.5cm long. Capsule oblong, 3–4cm long armed with prickles.

Ecology. Disturbed ground, particularly on the plains and limestone plateaus. Sea-level–200m.

Distribution. Native of the New World, now a pantropical weed.

ETHNOGRAPHIC NOTES

Name(s)

– Soqotri: *mərʌ́nnəha*; (rare) *śē'ənítin* 'rough'.

– Soqotri simplified: *miranniha*; *shey'initin*.

Food (livestock). In poorly vegetated areas livestock eat young plants and new shoots after rain, but they are not highly regarded as a feed. The mature plants are less palatable, but are sometimes eaten in the dry season when there is little else available. Livestock from the better-vegetated regions ignore it.

Medicinal / veterinary. Latex expressed from the crushed plant is used as an antiseptic. The seeds are crushed and mixed with water, or fresh or soured milk, to make a paste smeared over inflamed skin, swellings and boils. They are also ground with a little water and applied to swollen glands and the throat area to soothe a sore throat. The juice extracted from the ground seeds is dripped into a painful or infected ear. The same extract is rubbed over the udder and teats of a milk animal, or the inflamed breast of a woman, to treat mastitis. Powdered seeds are applied to the eyes like *kohl* to treat inflamed, discharging or sore eyes.

Comparative uses. This plant is widely used medicinally: by 1980 it had already appeared more than 50 times in the NAPRALERT computer database of medicinal plants at the Department of Pharmacognosy and Pharmacology, University of Illinois. It is said to grow wherever cattle have pastured. The seeds have a cannabis-like effect, and in many countries the herb, juice and flowers are reputed to be narcotic. The Hausas say that honey made from the plant drives men mad.

Seeds. Yield from 22 to 36% of a nauseous, bitter, non-edible oil considered a remedy for many skin diseases. The semi-drying oil is used as a lubricant in India, to make soap, and in paint manufacture in America. In small amounts the seed oil is cathartic, and in larger doses causes violent purging and vomiting. It is sometimes used as a substitute for, or adulterant of, mustard or linseed oils, but the toxic nature of the oil makes it unsafe even for non-edible purposes in view of possible misuse. The seeds are also considered to be laxative, emetic, expectorant and demulcent.

Sap. The yellow sap has slightly corrosive properties and is applied to warts, chancres, and suchlike. It is also used externally in scabies, dropsy, jaundice, skin complaints and ophthalmia. In Nigeria both the seed oil and the yellow sap are widely used as a preventative of white ants.

Roots. In Africa, an infusion of the roots is used as a wash by drummers to sustain their vigour for all night play. A grub reputedly found in the root is taken by acrobats or dancers to stimulate energy. The roots are used by women for spools on which to wind thread, and women are commonly given an infusion of the plant to drink at the onset of childbirth. (Bebawi & Neugebohrn 1991; Dalziel 1948, 1956; Davis *et al.* 1983; Ghazanfar 1994; Lewis & Elwin-Lewis 1977; Oliver-Bever 1986)

PEDALIACEAE

Herbs. Often glandular, becoming slimy when wet. Leaves opposite (uppermost sometimes alternate), margin dentate; stipules absent. Flowers solitary in axils, yellow, zygomorphic. Sepals 5, small. Corolla yellow, tubular, 5-lobed, zygomorphic. Stamens 4, 1 small staminode often present. Ovary superior, 2-celled; style 1, stigmas 2; ovules 2 per cell, axile. Fruit a 4-angled capsule with 4 spines. 18 genera. 95 species. Tropical and warm regions, especially coasts and deserts.

Pedalium L.

See description of family. 1 species.

Pedalium murex L.

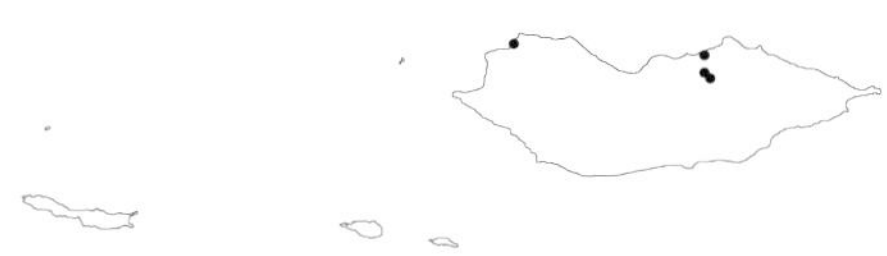

Erect succulent, glandular herb to 40(–70)cm. Leaves oblong-elliptic, <5 x 3.5cm, irregularly and coarsely dentate. Corolla tube c.2cm long. Fruit 1–2cm long.

Ecology. Weed of date gardens and cultivated areas. Sea-level–300m.

Distribution. Tropical and NE Africa, Madagascar, Arabia, India, Sri Lanka.

ETHNOGRAPHIC NOTES

Name(s)

– Soqotri: nil or *tˈibbˈeyhon* 'drooling'.

– Soqotri simplified: *tibbeyhon.*

Food (livestock). Not palatable to livestock.

Comparative uses. In India a viscid mucilage is made by swirling the stems, leaves, fruit or seeds in water or milk. This drink is then used medicinally to treat dysuria, gonorrhoea and other urino-genital diseases. It is also taken as a diuretic and tonic. The fruits are taken as an aphrodisiac and a decoction is given for incontinence of urine, nocturnal emissions, spermatorrhoea and impotence. In some areas the leaves are eaten as a vegetable. (Ambasta 1982; Jain & De Fillips 1991; Kunkel 1984; Lewis & Elwin-Lewis 1977)

PIPERACEAE

Succulent herbs. Leaves simple, alternate, opposite or whorled, entire, palmately veined at the base; stipules absent. Flowers bisexual, in terminal and leaf-opposed spikes, minute, borne in the axils of peltate bracts. Perianth absent; stamens 2; ovary superior, 1-celled, with 1 ovule per cell; stigma sessile and brush-like. Fruit a drupe. 14 genera. 1940 species. Tropical.

Peperomia Ruiz & Pav.

See description of family. 1000 species. Tropical and warm regions, especially America.

Peperomia blanda (Jacq.) Kunth

Peperomia arabica Decne.
Peperomia blanda var. *leptostachya* (Hook. & Arn.) Dull.
Peperomia goudotii sensu Balf.f. non Miq.

Erect or ascending herb to 30cm, finely tomentose. Leaves mainly opposite, a few in whorls of 3 above, obovate, 1.5–5 x 1–3cm, with a shortly acuminate or rounded tip and rounded or narrowly cuneate base, sparsely pubescent; petiole 0.5–1.5cm. Spikes solitary or in groups of 2 to 5, in the axils of the upper leaves and terminal, 30–70 x c.1.5mm, glabrous. Drupe cylindrical, c.1mm.

Ecology. In shade amongst large boulders in low shrubland. 600–1100m.

Distribution. New World tropics, tropical Africa, Yemen and Sri Lanka.

Botanical notes. All Arabian material belongs to var. *leptostachya* (Hook. & Arn.) Dull.

ETHNOGRAPHIC NOTES†

Name(s)

– Soqotri: nil or *ḥonḍˈol* 'useless, lazy, impotent'; (*di*) *ḥə-wˈririhin* 'black, dark-coloured'; *mˈʌhariś* 'undergrowth, understorey'.

– Soqotri simplified: *hondhol*; (*di*) *hewririhin*; *maharish*.

Food (livestock). Not normally palatable to livestock, but in extreme hunger they eat it.

Notes. One of the group of plants believed to cause *'uqabah*. Island herders believe that the sticky slime that results when you rub these plants between the fingers damages the gut of goats and makes them ill.

Comparative uses. Many species have edible leaves. *P. pellucida* is grown in gardens as a vegetable. It is also a common ingredient in infusions to treat convulsions in adults and children, and is made into a poultice for sores and boils. In Kenya the root of *P. abyssinica* is soaked in water and the filtrate drunk to treat malaria and gynaecological problems of women who have not yet given birth. (Dalziel 1948, 1956; Heine *et al.* 1988; Kunkel 1984)

†Also apply to *P. tetraphylla*.

Peperomia tetraphylla **Hook. & Arn.**
Peperomia reflexa (L.f.) A.Dietr.
Peperomia reflexa (L.f.) A.Dietr. var. *parviflora*
Peperomia reflexa var. *parvifolia* C.DC.

Prostrate herb, often rooting at the nodes. Leaves opposite or mostly in whorls of 3 or 4, obovate or broadly elliptic to suborbicular, 7–12 x 5–8mm, with a rounded tip and rounded or broadly cuneate base, glabrous or glabrescent; petiole 1–3mm. Spikes solitary, terminal, 10–20(–30) x c.1.5mm, pubescent. Drupes ellipsoid, c.1mm.
Ecology. In shade between large boulders in low shrubland. 900–1500m.
Distribution. Pantropical.

ETHNOGRAPHIC NOTES
Name(s)
See *P. blanda* (also uses).

PITTOSPORACEAE

Evergreen trees or shrubs. Leaves alternate, simple, entire; stipules absent. Flowers actinomorphic, in terminal or axillary panicles. Sepals 5, free or fused at base. Petals 5, free. Stamens 5, alternating with petals. Ovary superior, unilocular, with 2 parietal placentas; style simple; stigma truncate. Fruit a loculicidally dehiscing capsule, valves woody. 9 genera. 240 species. Tropical and warm Old World, especially Australia.

Pittosporum Banks ex Sol.

See description of family. 200 species. Tropical Old World, Africa to Pacific.

Botanical notes. For a revision see Friis (1987).

Pittosporum viridiflorum **Sims**

Shrub or tree to 5m. Leaves crowded towards ends of branches, narrowly oblanceolate, 8–15 x 1.5–4.5cm, tips acuminate, prominently veined beneath; petioles <2cm. Flowers fragrant in many-flowered panicles. Sepals narrowly ovate, <1.5 x 0.8mm. Petals white, cream or yellowish green, narrowly oblong, 4–7mm. Capsule globose, 5–10mm diameter; capsule valves spreading or reflexed, <2mm thick, 4-seeded.
Ecology. Rare in dense thicket in the western Haggeher mountains. 1450–1500m.
Distribution. S and tropical Africa, SW Arabia and S India.
Botanical notes. Also reported from Abd al Kuri. However, not seen there recently and an unlikely record. A polymorphic and widespread species in tropical Africa. Arabian plants differ somewhat from African and were placed in subsp. *arabicum* by Cufodontis (1952). Friis (1987) considered Arabian and Soqotran material best treated as an informal '*arabicum*' grouping within the species.

ETHNOGRAPHIC NOTES
Name(s)
– Soqotri: *di ʿāṣʌbṣʌb* 'branch tips' (also given to *Pluchea obovata* and *Hypericum balfourii*); *ḥáhər* 'black, dark-coloured' (referring to its excellence as a timber).
– Soqotri simplified: *di 'azabzab*; *hahir.*

Food (livestock). Cattle and goats eat the foliage in hunger. The greenery persists into the dry season and leafy branches are lopped for goats at this time. Herders also gather the fallen leaves to feed their cattle and goats towards the end of the dry season. Camels browse the trees year round.
Bees / honey. The fragrant flowers are very attractive to honeybees.
Fuel. The dead wood makes adequate firewood.
Timber. The wood is hard and makes good timber. It is termite resistant and does not have to be de-barked before use. The tree rarely grows large enough to provide the larger beams and supporting pillars but is used for rafters and roofing infill.
Wood. Close-grained and good for making a variety of implements, including the important herding staff.
Comparative uses. Some related species have edible seeds, but they are often bitter and indigestible. In Africa goats and sheep eat the foliage but the fruit are extremely toxic. A string is made from the bark and used to fasten beehives. Medicinally, the bark is crushed, boiled and the filtrate drunk cold to stimulate the appetite, and taken as an emetic and laxative to clean out the system. In India the bark extract of *P. dasycaulon* shows anti-bacterial and anti-fungal activity, and the leaves and fruit of *P. ferrugineum* are used as a fish poison. The wood of other *Pittosporum* species is used for rafters, as fuel, and in carving. (Ambasta 1982; Heine *et al.* 1988; Jain & De Fillips 1991; Kunkel 1984)

PLANTAGINACEAE

Herbs. Leaves alternate, in a loose rosette, sparsely dentate; stipules absent. Flowers small, in a compact spike borne on a long, terminal peduncle, actinomorphic. Wind pollinated. Calyx 4-lobed. Corolla 4-lobed, papery. Stamens 4. Ovary superior, 2-celled; placentation axile. Fruit a capsule, dehiscing transversely. 3 genera. 255 species. Cosmopolitan.

Plantago L.

See description of family. 250 species. Cosmopolitan.

Plantago amplexicaulis Cav.

Loose rosette-forming herb to 20cm, usually with a short stem. Leaves and stem thinly hairy with spreading hairs, linear-lanceolate, 2–10 x 1–1.5cm, sparsely dentate. Flower spike 0.5–2cm long, peduncle 1–15cm long. Fruit c.0.6 x 0.3cm.
Ecology. Widely distributed on the plains and limestone plateau. 150–650m.
Distribution. Throughout Arabia.

ETHNOGRAPHIC NOTES†
Name(s)
– Soqotri: nil or *ṣéfərher* 'flowers'; *ḳewrēni* 'two horned' (from the fruit).
– Soqotri simplified: *zefirher*, *qewreyni*.
Food (livestock). Eaten by all livestock, but not regarded as a nutritious feed.
Notes. Can cause severe *'uqabah* in goats unaccustomed to it.
Comparative uses. Many species have edible leaves. In Yemen the seeds of *P. major*, called *bizr*, are taken as an antipyretic. In Arabia the seeds of *P. major* are warmed over a fire, crushed, and swallowed with water as a general tonic. They are also boiled in water and given to treat diarrhoea, dysentery and fevers. The fresh leaves are crushed and applied as a poultice for abscesses. The whole plant of *P. coronopus* is boiled and taken as a laxative, and is crushed to a paste to treat wounds. In Yemen *Plantago* seeds are soaked and used as a poultice for boils and ulcers, and as a cosmetic for the hair, and in Bahrain and Oman as a laxative and wound poultice. In Qatar a paste is applied to the eyes for ophthalmia and the plant is also used to treat pulmonary infections and as an antidote to snake venom. The cooling property of the seeds is used in treating biliousness, dysentery, leprosy, inflammation, coughs and chronic diarrhoea. (Abbas *et al.* 1992; Al-Dubaie 1993/Al-Dubaie & Al Khulaidi; Batanouny 1981; Davis *et al.* 1983; Fleurentin & Pelt/Fleurentin *et al.* [1982, 1983a,b,c]; Ghazanfar 1994; Ghazanfar & Al-Sabahi 1993; Kunkel 1984; Rizk & El-Ghazaly 1995)

†Also apply to *P. ovata*.

Plantago ovata Forssk.

Similar to *P. amplexicaulis* but stem absent; leaves and stems densely grey-hairy; leaves linear, up to 2mm across.
Ecology. Grassland clearings in Haggeher mountains. 950m.
Distribution. N Africa, Mediterranean region and SW Asia.

ETHNOGRAPHIC NOTES
Name(s)
See *P. amplexicaulis* (also uses).
Comparative uses. The dried seeds of *P. ovata*, known as ispaghula or isubgol, are important commercially and it is cultivated in India and Pakistan. In 1976–77, some 4770 tonnes of the husk were exported from India for use as a stabiliser in ice cream, as a thickening agent, in cosmetics and printing, and after treatment with a caustic soda solution as an agar-agar substitute. The seeds have a mucilaginous husk, which is the active principle for the main use of the plant as a gentle laxative and emollient. (Abbas *et al.* 1992; Al-Dubaie 1993/Al-Dubaie & Al Khulaidi; Batanouny 1981; Davis *et al.* 1983; Fleurentin & Pelt/Fleurentin *et al.* [1982, 1983a,b,c]; Ghazanfar 1994; Ghazanfar & Al-Sabahi 1993; Kunkel 1984; Rizk & El-Ghazaly 1995)

PLUMBAGINACEAE

Subshrubs, shrubs or small trees. Leaves alternate, simple, entire; stipules absent. Inflorescence paniculate. Flowers hermaphrodite, actinomorphic, subtended by bracts. Calyx persistent, deeply or shallowly 5-lobed. Corolla 5-lobed, either tubular below or divided almost to base. Stamens 5, attached to and opposite corolla lobes. Ovary superior with a single ovule; styles 1 or 5. Fruit a 1-seeded capsule surrounded by the persistent calyx. 22 genera. 440 species. Cosmopolitan, especially maritime.

Dyerophytum Kuntze

Vogelia Lam.

Shrubs or small trees with open, spreading or pendulous branches. Leaves evergreen, petiolate or sessile, distant, glabrous. Inflorescence a lax, terminal panicle composed of pendulous spikes. Calyx deeply 5-lobed. Corolla funnel-shaped, 5-lobed. Style 1. 2 endemics. 4 species. Disjunct distribution: 2 species in Soqotra, 1 from India to Arabia and 1 in SW Africa.

Dyerophytum pendulum **(Balf.f.) Kuntze**
Vogelia pendula Balf.f.

Slender tree to 3.5m; branches pendulous. Leaves pale glaucous-green, often mealy white below, 15–50 x 5–20mm, broadly elliptic to obovate and gradually or abruptly narrowing into an attenuate petiole, tip obtuse or apiculate. Calyx green with red stripes, 10mm long. Corolla pale yellow, 15–20mm long.
Ecology. Deciduous woodland; usually on cliffs. 600–1000m.
Distribution. Endemic.
Conservation status. Vulnerable (VU D2). Rare, with a very limited area of occupancy.

ETHNOGRAPHIC NOTES†
Name(s)
– Soqotri: *ˈəmlaḥ* 'salty'; *ˤˈeṭʌf*, *ˤˈayṭift*, *ˤayṭˈʌffʌn* 'small form of *ˤˈayṭift*' 'folded, wrapped'; *ṣˈaˤalhil* 'large-leaved' (more commonly given to the more widespread *Commicarpus*); (rare) *mɛ̄rə*.
– Soqotri simplified: *imlah*; *'etaf*, *'aytift*, *'aytaffan*; *za'alhil*; *meyrih*.
Food (livestock). Camels like this forage when they can reach it, but goats and cattle only when they are very hungry towards the end of the dry season. Sheep and donkeys eat both living and dead parts.
Wood. The thicker stems are used to make the important herding staff.
Notes. Can cause *'uqabah* and small sores (*ḳˈʌmṭiṭ*) in the intestines in stock unaccustomed to it.
Comparative uses. In nearby Dhofar the new shoots of *D. indicum* used to be eaten, and in times of salt shortage the more mature, salty shoots are used as a salt substitute in cooking. The stems are dried, crushed and smoked as a tobacco to ease breathing difficulties. The stems are also used to weave bird traps. (Miller & Morris 1988)

†Also apply to *D. socotranum*.

Dyerophytum socotranum **J.R.Edm.**
Dyerophytum indicum var. *socotrana* (Wight) Kuntze
Vogelia indica var. *socotrana* Balf.f.

Small tree or shrub to 3m; branches pendulous. Leaves pale glaucous-green, 10–50 x 10–50mm, circular to obovate, sessile, lower leaves amplexicaul, tip retuse. Calyx reddish brown, c.5mm. Corolla yellow, c.15mm.

Ecology. Limestone cliffs, boulders and wadi sides. 30–650m.
Distribution. Endemic.
Conservation status. Vulnerable (VU D2). Rare, with a very limited area of occupancy.

ETHNOGRAPHIC NOTES
Name(s)
See *D. pendulum* (also uses).

Limonium Mill.

Statice L.

Tufted or much-branched subshrubs. Leaves evergreen, succulent, in crowded rosettes. Inflorescence a leafless panicle made up of congested, 1-sided groups of 2-flowered bracteate spikelets. Calyx funnel-shaped, white-scarious. Corolla deeply 5-lobed. Styles 5. 2 endemics. c.150 species. Cosmopolitan.

Botanical notes. *L. paulayanum* and *L. sokotranum* show a great deal of ecotypic variation in leaf shape, habit, inflorescence size, calyx length and calyx hairiness. They both grow in saline conditions along the coast and, less commonly, inland on cliffs. Inland plants tend to be more prostrate and tufted, with smaller, less robust leaves and shorter inflorescences. In some areas there are more or less mixed populations of the two species but in other places they are rather more zoned, with *L. sokotranum* near the sea and *L. paulayanum* inland of it. Forms with intermediate leaves (i.e. narrowly spathulate) are sometimes found in both mixed and 'zoned' populations. Within the genus there is evidence of introgressive hybridisation in some species and this may account for intermediate forms here. *L. paulayanum* is closely related to, and perhaps conspecific with, *L. axillare*, a variable species widely distributed in coastal regions from Somalia to Pakistan. *L. paulayanum* differs in its hairy calyx. *L. kossmatii*, described from Abd al Kuri, is treated as an extreme, inland, form of *L. paulayanum*. It may, on further study, deserve subspecific recognition.

ETHNOGRAPHIC NOTES
Comparative uses. In nearby Dhofar the roots and stems of *L. axillare* were ground and mixed with water and applied as an astringent. The roots and stalks were used to dye fishing equipment and cordage, and the ash of the dead plant was mixed with tobacco and taken as a snuff. Other species are also astringent and often used to treat diarrhoea and dysentery. The leaves of some species, such as *L. beaumieranum*, *L. ornatum* and *L. tetragonum*, are eaten as a vegetable. (Ghazanfar 1994; Kunkel 1984; Lewis & Elwin-Lewis 1977; Miller & Morris 1988)

Limonium paulayanum **(Vierh.) J.R.Edm.**
Limonium axillare Balf.f. non Kuntze
Limonium kossmatii (Wagner & Vierh.) Verdc.
Statice axillaris Balf.f. non Forssk.
Statice kossmatii Wagner & Vierh.
Statice paulayana Vierh.

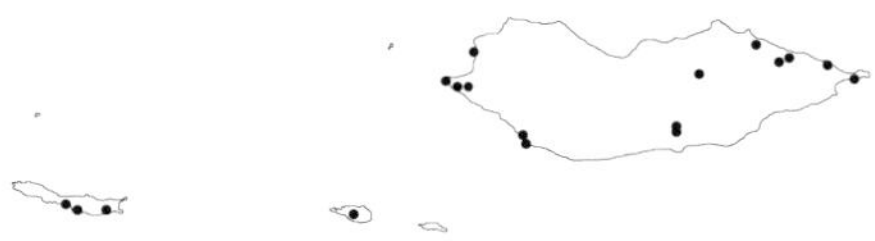

Erect or sometimes more or less prostrate subshrub, rarely with a distinct trunk, to 1.5m. Leaves grey-green, spathulate (sometimes narrowly spathulate in forms intermediate with *L. sokotranum*), flattened, 15–50 x (3–)5–10mm, glabrous. Bracts reddish. Calyx 3–5mm long. Corolla pink.
Ecology. Coastal dwarf shrubland and inland cliffs. Sea-level–550m.
Distribution. Endemic.
Conservation status. Least Concern (LC). Common in several vegetation types and under no present or perceived threat.

ETHNOGRAPHIC NOTES
Name(s)
– Soqotri: *źoṣˈibə*, *leṣˈibə* 'pile of stones' (plants growing beside the sea are usually given this name), *teṣˈibə* (Abd al Kuri); *məśḥɔ̄wrɛ̄ro* 'tough, enduring'; *miśḥɔ̄wtə* 'dry, tough, hard'; (western Soqotra) *məźxɔ̄mhəm* 'crunchy'.
– Soqotri simplified: *dhozibih, lezibih, tezibih*; *mish'howreyro*; *mish'howtih*; *midh'khowmhim.*
Food (livestock). In the drier areas the plant is browsed by camels and goats; sheep mostly eat the flowering stems and fallen leaves. It is not regarded as nutritious feed, though a useful one for camels. On the outer islands it is more important as a fodder for sheep and goats.
Notes. Other uses as *L. sokotranum.*

Limonium sokotranum **(Vierh.) Radcl.-Sm.**
Limonium cylindrifolium Balf.f. non (Forssk.) Verdc.
Statice cylindrifolia Balf.f. non Forssk.
Statice socotrana Vierh.

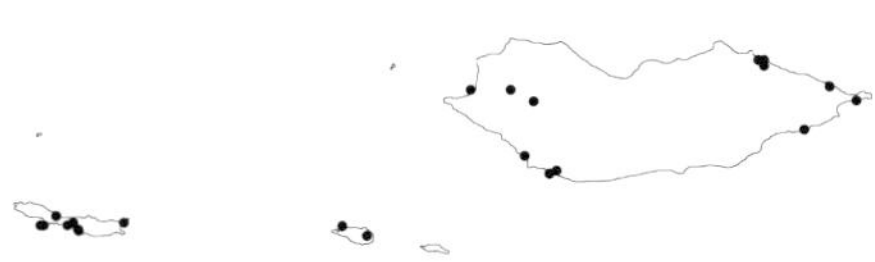

Similar to *L. paulayanum* but differs in its linear-cylindrical not spathulate leaves.
Ecology. Coastal dwarf shrubland and inland cliffs. Sea-level–550m.
Distribution. Endemic.
Conservation status. Least Concern (LC). Common in several vegetation types and under no present or perceived threat.

ETHNOGRAPHIC NOTES
Name(s)
– Soqotri: *źoṣˈibə*, *ləṣˈibə* 'pile of stones' (plants growing beside the sea are usually given this name) (Soqotra, Samha and Abd al Kuri); *təṣˈibə* (Abd al Kuri); *ḳˈisʌs* 'ridge'.
– Soqotri simplified: *dhozibih*, *lizibih*; *tizibih*; *qisas.*
Food (livestock). Important dry season forage in the drier coastal areas of the islands. Browsed by goats, sheep and camels, and donkeys also eat it. On the mainland it is regarded as a useful forage for camels in particular. On the outer islands it a valued feed for goats and sheep.
Bees / honey. Honeybees are attracted to the flowers.
Flavouring / food additive. Such saline forage can affect the taste of the milk of animals feeding heavily on it.
Fuel. On the outer islands the dead plants are used as fuel when there is nothing better to hand. They are the preferred fuel for firing pottery, both here and in coastal areas of Soqotra where alternative fuels are in short supply.
Leatherwork. On the outer islands and in coastal areas of Soqotra this is the chief plant used to strengthen skins and hides in leatherwork, and to dye them a dark red-brown. On Soqotra it is especially used on sheepskins. The living plants are ground with a little water to a paste smeared over both sides of the skin from which the hair has already been removed.
Other. The roots are used to dye cotton fishing equipment. They are lightly crushed and handfuls are then rubbed along the length of cotton fishing lines and other nets. Alternatively, the roots are ground more finely, and put in water to steep until the colour has leached out. Then the cotton material is lowered in and weighed down with a rock. It is left overnight, or longer, until it has acquired the desired shade. Some use the bark instead, but follow the same procedure. Fishing equipment dyed like this was used to fish in 'red', or rocky, areas. Nets used for fishing in 'white', or sandy, areas were dyed with quicklime. A container large enough to take the nets and lines was filled with powdered quicklime and water and stirred vigorously to make a thick sludge. The fishing equipment was lowered in, stirred around and then taken out and spread in the sun to bleach. *Limonium* plants were also used, on the outer islands especially, to colour the imported white, undyed calico cloth from which most clothing was made. The plants were ground to a paste and added to water, and then the cloth put in and left to soak until it had turned a reddish-brown colour. On Samha this dye paste was commonly added to a pool of seawater, as it was believed the salt helped to fix the dye. The sheep's wool ropes of the fishing nets were dyed in the same way. The same paste was used to decorate pottery. When the pots began to turn red in firing, they were lifted out with a stick, and quickly painted around the rim with the paste.
Notes. A period spent once or twice a year grazing halophytic rangeland is considered essential, especially for the health of small stock and camels. Such feed initially brings on diarrhoea, which is seen as being beneficial and as 'cleansing impurities from the blood'. When recovered from this, the animals begin to put on weight. The islanders say that if you walk among such plants at dawn, they cover the legs with a layer of salt.

POLYGALACEAE

Low shrubs or herbs. Leaves alternate, simple, entire; without stipules. Flowers bilaterally symmetrical, bisexual, in terminal or axillary racemes. Sepals 5, the 2 lateral (the wings) enlarged and petaloid. Petals 3 or 5, free or fused, the lowest forming a keel; the laterals obscure or absent; the 2 upper as long as the keel. Stamens 8, with filaments united into a tube around the ovary and slit down one side. Ovary 2-loculate with a single ovule in each loculus. Fruit a flattened, winged or unwinged capsule; seeds silky, with an aril bearing 1 to 3 papery appendages. 18 genera. 950 species. Cosmopolitan.

Polygala L.

See description of family. 1 endemic. 500 species. Cosmopolitan.

Polygala abyssinica Fresen.

Erect perennial herb; stems glabrous, woody below. Leaves linear to narrowly elliptic, 10–40 x 2–3mm. Flowers in terminal, more or less leafless racemes <20cm long. Wings whiter or mauve, with darker veins. Capsule glabrous, winged; seeds with a single long appendage.
Ecology. Uncommon, in semi-deciduous woodland. 500m.
Distribution. Tropical E Africa and Arabia.

ETHNOGRAPHIC NOTES†
Name(s)
– Soqotri: nil or *ṣáʿafə* 'fish scale'; *ṭáʌlḳʌṭ*; *kibíʾihin* 'small piece of animal dung'; *di ḳáʌṣḳʌṣ* 'cream-coloured' (given to many small plants of the high plateau areas).
– Soqotri simplified: *za'afih*; *talqat*; *kibi'hin*; *di qazqaz*.

Food (livestock). Grazed by goats and sheep and when they are green they are considered to be a good feed and to increase milk yield. The perennial *Polygala* are more highly valued as forage on the outer islands.
Comparative uses. *Polygala* species are widely believed to increase the milk of animals feeding on them, and their common name is 'milkwort'. Many species are used medicinally for the relief of coughs and asthma, and the leaves and seeds of some are edible. *Polygala* species contain saponins and methyl salicylate; the former is now known to have anti-inflammatory properties, and methyl salicylate is similar to oil of wintergreen and is readily absorbed through the skin. In India the roots of *P. abyssinica* are taken as an expectorant, and the tender leaves of *P. chinensis* are eaten in times of hunger. An infusion of the leaves of the latter is given to treat asthma, chronic bronchitis and catarrhal affections, and the roots are used to treat fevers and dizziness. (Ambasta 1982; Jain & De Fillips 1991)

†Also apply to *P. erioptera*.

Polygala erioptera DC.
Polygala paulayana Vierh.

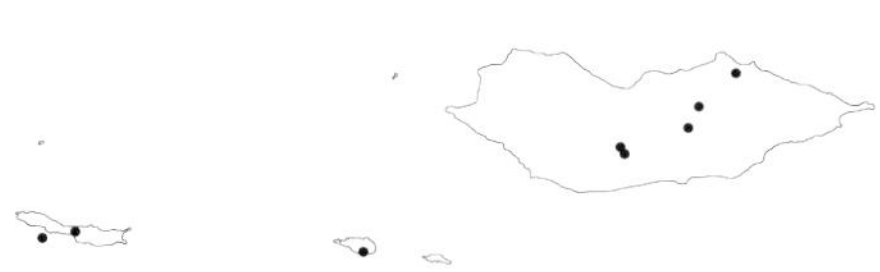

Prostrate or erect annual or perennial herb; stems glabrous. Leaves linear, 10–30mm long. Flowers in short axillary racemes. Wings white or pink with greenish veins. Capsule hairy, narrowly winged; seeds with 3 short basal appendages.

Ecology. Widespread in *Croton socotranus* shrubland, succulent shrubland, desertic plains and open, drought-deciduous woodland. 50–700m.
Distribution. Tropical Africa and Arabia to India.
Botanical notes. I consider *P. paulayana* Vierh., described from Samha, as merely a small form of this extremely variable species.

ETHNOGRAPHIC NOTES
Name(s)
See *P. abyssinica* (also uses).

Polygala kuriense A.G.Mill. (see Appendix 2, p. 723)

Prostrate, woody-based perennial to 10cm; stems somewhat flattened, with short crisped hairs. Leaves narrowly oblong-elliptic, <8mm long. Flowers in terminal, more or less leafless racemes <2cm long. Wings white with green veins. Capsule glabrescent, not winged; seeds with 2 short appendages.
Ecology. Sparse dwarf shrubland on dry rocky slopes. 30–50m.

Distribution. Endemic to Abd al Kuri.
Conservation status. Vulnerable (VU D2). Rare, with a very limited area of occupancy.
Botanical notes. An inconspicuous plant, growing between rocks and easily overlooked. It is readily distinguished by its rather flattened, prostrate stems.

ETHNOGRAPHIC NOTES
Name(s)
– Soqotri: nil or *ḳɔ̄níitin* 'small threads'; *gilʿəmhiníitin* 'tiny fruit'.
– Soqotri simplified: *qownitin*; *gil'imhinitin.*
Food (livestock). An important rains fodder for small stock in Abd al Kuri, and considered to be a rich feed, increasing milk yield.

POLYGONACEAE

Herbs or shrubs, sometimes aquatic. Leaves alternate, simple or lobed, sometimes absent; stipules connate forming an often membranous sheath (the ochrea). Flowers bisexual or unisexual, actinomorphic, in racemes, panicles or clusters. Perianth with 3 to 6 segments, free or fused, often accrescent and enclosing the fruit, sometimes winged or toothed or bearing spines. Stamens 6 to 15. Ovary superior, 1-celled with a solitary, basal ovule; styles 2 to 4. Fruit a trigonous or biconvex nutlet, sometimes beset with bristles, sometimes hidden by the accrescent perianth lobes. 51 genera. 1150 species. Cosmopolitan, especially N temperate.

Calligonum L.

Desert shrubs. Leaves minute, soon falling. Ochreae short. Flowers bisexual in axillary fascicles. Perianth segments 5, free, pale pink or whitish green with a darker midvein. Stamens c.15. Styles 4. Fruit hardened, winged or densely covered with stiff, simple or branched setae. 80 species. Mostly Mediterranean.

Calligonum comosum L'Hér.

Calligonum polygonoides subsp. *comosum* (L'Hér.) Soskov

Shrub to 120cm; branches rigid, jointed, swollen at nodes, whitish-grey; young branches green, thin, flexuous, often fascicled. Leaves subulate, 1–5mm, soon falling. Perianth lobes c.3mm; pedicels equalling or longer than the flowers. Fruits red or pale yellowish green, ellipsoid, 7–8 x 5–6mm; setae branched, 3–5mm, arising from 4 pairs of longitudinal wings.

Ecology. No habitat details available.

Distribution. Deserts from Egypt east to Pakistan.

Botanical notes. A single sterile specimen from Abd al Kuri is probably referable to this species.

ETHNOGRAPHIC NOTES

Comparative uses. In Arabia the dried leaves and stems are used to soothe toothache. The twigs are used as firewood because they make very little smoke. Powdered or slim twigs are added to milk as a tonic or flavouring. A leaf poultice is used to relieve rheumatic pain and swellings. In India a decoction of the root mixed with catechu is used for sore gums and as a gargle, and the flowers and young buds are eaten with buttermilk or salt in times of scarcity. Dried branches are used to form screens that, after wetting with water, help ward off desiccating hot winds. Dried whole plants are used as fencing. (Abdulfatih 1987; Ambasta 1982; Goodin & Northington 1979; Jain & De Fillips 1991; Mossa *et al.* 1987)

Emex Neck. ex Campd.

Monoecious herbs, glabrous throughout. Ochreae membranous. Leaves alternate, simple. Flowers green or greenish red in axillary clusters. Male flowers with 3 to 6 perianth segments, fused at base; stamens 4 to 6. Female flowers with 6 perianth segments, tubular below, accrescent and hardening in fruit, outer 3 becoming spinescent; styles 3. Nutlets trigonous, enclosed within pitted, spinescent perianth. 2 species. Mediterranean, S Africa, Australia.

Emex spinosus (L.) Campd.

Rather fleshy annual herb to 30cm. Leaves oblong-ovate to ovate or triangular, 2.5–8 x 1.5–4.5cm, rounded, truncate or cordate below; petioles 2–10cm. Male flowers pedicellate, c.1.5mm long. Female flowers sessile, 5–6mm long. Fruiting perianth 4–5 x 3–4mm (including spines), pitted and with 3 terminal spines; spines recurved, 1–3mm.

Ecology. A weed of cultivation and near habitation. 500m.

Distribution. Mediterranean east to Pakistan: introduced and weedy elsewhere.

ETHNOGRAPHIC NOTES

Name(s)

– Soqotri: nil or (when young) *ri̍žəbeb*; (when mature) *ʿa̍žərib* 'quick to die back'.

– Soqotri simplified: *rijibeb*; *'ajirib*.

Food (livestock). As with other plants of the *ʿa̍žərib* group, these plants provide useful grazing after the winter rains while young and green and before the grasses come up. Cattle especially like them, though goats are less enthusiastic.

Notes. If goats that are unaccustomed to it eat a lot they can develop scour.

Comparative uses. Elsewhere the roots are eaten raw or cooked, and the leaves treated as spinach. (Kunkel 1984)

Oxygonum Burch. ex Campd.

Herbs. Ochreae tubular, fringed with fine setae. Leaves alternate, sinuously lobed (in Arabia). Flowers pink or white, in axillary spike-like racemes. Male flowers with 4 or 5 perianth segments, fused at base. Hermaphrodite flower with 5 perianth segments, tubular; stamens 8; styles 3. Fruiting perianth enclosing the nutlet, fusiform with 3 spines arising from near the middle. 30 species. Tropical and S Africa.

Oxygonum sinuatum (Hochst. & Steud.) Dammer

Glabrous or pubescent annual herb; stems to 30cm. Leaves ovate to lanceolate, 2–5 x 1–3cm, with 1 to 3 pairs of acute or rounded lobes; petiole 1–2cm. Inflorescence <25cm, flowers distant. Flowers 2–3mm. Fruiting perianth 5–6mm long and broad (including spines); spines spreading, 1–2mm long.

Ecology. A weed of cultivation and near habitation. 150m.
Distribution. Arabia, tropical and S Africa.

ETHNOGRAPHIC NOTES
Name(s)
– Soqotri: (when young) *rížəbeb*; (when mature) *ʿážərib* 'quick to die back'.
– Soqotri simplified: *rijebeb; ʻajirib.*
Food (livestock). See *Emex spinosus.*
Comparative uses. Elsewhere, the leaves are eaten with fatty meals. (Kunkel 1984)

Persicaria (L.) Mill.

Annual or perennial herbs, sometimes aquatic. Ochreae cylindrical, truncate or with a terminal fringe of bristles. Leaves oblong-ovate or lanceolate, entire. Flowers in globose heads or spikes. Perianth 4 or 5-lobed. Stamens 5 to 8, dilated at base. Styles 2 or 3. Nuts trigonous or biconvex. c.150 species. Subcosmopolitan.

Botanical notes. Sometimes included in *Polygonum.*

Persicaria barbata (L.) Hara
Polygonum barbatum L.

Similar to *P. glabra* but the ochreae brown, <2cm, appressed-setulose, fringed with fine bristles <1.5cm.

Ecology. No details of habitat or location available.
Distribution. Tropical Africa, Asia and Australia.
Botanical notes. Recorded in the 19th century but not seen since.

Persicaria glabra (Willd.) M.Gómez
Polygonum glabrum Willd.

Robust, erect herb to 1m, glabrous. Ochreae reddish brown, <3cm, truncate, not fringed. Leaves lanceolate, <25 x 5cm, acuminate, cuneate below, minutely yellow-glandular. Racemes 2 to 4 at ends of branches, densely flowered, 5–10cm; peduncles and racemes glabrous or very sparsely glandular. Perianth pink, c.3.5mm. Nut biconvex or trigonous without dimpled faces, c.2.5mm, shining.

Ecology. By permanent water in date gardens. c.50m.
Distribution. Tropical Africa, Asia and America.

ETHNOGRAPHIC NOTES
Name(s)
– Soqotri: nil or *di rihóytin* 'of water'.
– Soqotri simplified: *di rihoytin.*
Food (livestock). Nibbled by livestock, but rarely by goats, who intensely dislike damp fodder. They graze it only when it has dried out and there is little alternative forage.
Comparative uses. Elsewhere the plant is used as a potherb, and an infusion of the leaves is taken to relieve the pain of colic. (Bebawi & Neugebohrn 1991; Kunkel 1984)

PORTULACACEAE

Annual or perennial herbs, often succulent. Leaves opposite or alternate, simple, entire. Stipules absent or represented by a tuft of hairs, otherwise plants glabrous. Flowers bisexual, actinomorphic, solitary or in clusters at the ends of branches or in terminal racemes or panicles. Sepals 2, free or united below. Petals 4 to 6, free or united below. Stamens 4 to numerous, free or attached to the base of the petals. Ovary inferior or semi-inferior, 1-celled; ovules many; placentation free-central or basal; style simple with 2 to 9 branches. Fruit a capsule, dehiscing by valves or circumscissile, 1 to many-seeded. 21 genera. 400 species. Cosmopolitan.

Portulaca L.

Annual or perennial herbs. Leaves alternate or opposite, fleshy, subsessile or shortly petiolate. Stipules a tuft of hairs. Flowers sessile or stalked, solitary or in 2 to 10-flowered clusters. Petals 4 to 6, free or united at the base. Stamens 7 to 20. Ovary semi-inferior; style simple, 3 to 6-branched. Capsule dehiscing by a lid. 3 endemics. 40 species. Tropical and warm regions.

Botanical notes. In all species the flowers are short-lived, opening in the late morning and closed by late afternoon. For a review see Gilbert & Phillips (2000).

Portulaca kuriensis M.G.Gilbert

Tufted perennial with a tuberous taproot; stems to 5cm; not rooting at nodes; reddish tinged. Leaves opposite, flattened-terete, narrowly lanceolate, 8–25 x 2–5mm, tip acute or obtuse at tip; stipular hairs white, 2–6mm long. Flowers solitary, c.5mm across, surrounded by an involucre of 4 leaf-like bracts. Sepals 2–4 x 1.5mm. Petals 4, clear yellow within, paler on back, ovate, c.4 x 2.5mm, rounded. Stamens 6 to 8. Stigmas 4. Epidermal cells of seeds with sinuate margins.

Ecology. Dwarf shrubland on rocky slopes. 50–500m.

Distribution. Endemic.

Conservation status. Vulnerable (VU D2). Rare, with a very limited area of occupancy.

Botanical notes. The thickened taproot readily distinguishes *P. kuriensis* from other species of *Portulaca* in the Archipelago. It is probably most closely reated to *P. dhofarensis* M.G.Gilbert from Oman (Dhofar) but differs in seed characters.

ETHNOGRAPHIC NOTES†

Name(s)

– Soqotri: *məḥələbboh*, *maḥləbiboh*, *maḥleybəbo*; *ḥaliblibo* (pl. *ḥaliblib*), *ḥalūbɛboh*, *ḥalɛboh* 'like buttermilk' (all referring to the sour flavour of the plants); *məṣūrib* (pl. *mṣirbihon*) 'of the winter rains' (given to various types of edible greenery that come up with rain); *ʾirsib* (also given to *Corchorus* species); *mərsūbəhān* (Samha, and swollen underground part *kəndīyə*); *mirsem* (Abd al Kuri).

– Soqotri simplified: *mihilibboh*, *mahlibiboh*, *mahleybibo*; *haliblibo* (pl. *haliblib*), *haluwbeyboh*, *haleyboh*; *mizewrib* (pl. *mizirbihon*); *'irsib*; *mirsewbihan* (*kindiyih*); *mirsem.*

Food (human). In hungrier areas of Soqotra and in the outer islands, all these plants played an important role in the human diet. They are eaten to quench thirst, and are also much liked for their sour, lemony flavour. They respond very quickly to the lightest shower of rain, and are ready to eat before other greenery has come up. Many plants persist in pockets throughout the dry season, and although not so luscious, they are still a popular nibble. When the plants are plentiful they are collected and taken home to cook. They are either put in a pot with minimal water and cooked until wilted, or left longer until reduced to mush. If salt is available, some is added to the pot. In the more settled coastal villages they were also added to sauces. In times of hunger, the whole plant is dug up and both underground and overground parts eaten raw. Some *Portulaca* species of Soqotra and Samha make small underground tubers, and these too are eaten. On Abd al Kuri, the swollen root system of *P. kuriensis* was an important food and was actively harvested when food supplies were low.

Food (livestock). Eaten by small stock, especially sheep, and they are seen as a valuable grazing resource in the dry months. They persist through the summer months, and although they dry out and become rather tough are still a useful forage. Cattle eat them very little. However, if a goat eats too much, it gets diarrhoea, loses condition, and sometimes also suffers from shivering attacks.

Flavouring / food additive. The milk of livestock eating the plants has a sweet and delicious flavour.

Medicinal / veterinary. A succulent plant is crushed to a paste and applied to the udder of a milk animal with mastitis to 'cool' the inflammation.

Cleansing. More succulent plants are crushed and used in washing and cleansing. Sometimes the viscid juice expressed from crushed plants is used as a hair lotion after dressing the hair in plaits. This makes the hair smooth and shiny. If a clay pot has developed an unpleasant smell, the leaves are crushed and used to rub the inside of the pot to remove the smell.

Other. Animals feeding on these species can go longer without water.

Notes. Can cause *'uqabah* in small stock unaccustomed to them. In general, the plants called *ḥaleblib* (and variants) are bright green in colour, whereas those called *mṣirbəhʌn* (and variants) are more dull and grey-coloured.

Comparative uses. *P. oleracea* is rich in calcium, iron and potassium salt and contains malic and citric acids as well as traces of free oxalic acid. It can be used as green manure but prolonged use may increase salinity in the soil. It is given as a fodder to sheep, cattle and pigs but with caution as it can cause oxalic acid poisoning. In the Middle East it is often cultivated, or collected from wild and sold in markets.

Whole plant. The whole plant is edible: root, flower, seeds, leaves and stem. In Dhofar it used to be gathered in bunches and sold in the coastal markets. In Oman it is cut into small pieces and eaten with salt to treat constipation. Elsewhere, it is used as a potherb, in soups and salads, the fleshy stems are pickled, and the roasted seeds are also eaten. The herb is diuretic due to the presence of potassium salts, and it is used to treat scurvy and diseases of the liver, spleen, kidney and bladder. It is applied to sore nipples and given for mouth ulcers, burns and scalds, and for dysuria and haematuria. It is used locally on swellings, bruises and whitlows, and is crushed with native natron and oil, and applied as a poultice for abscesses and boils to bring them to a head.

Leaves, stems and seeds. In Oman the leaves and seeds are applied locally to soothe the skin, and the plant is eaten to cleanse the blood. In Bahrain it is given for spitting blood, and the seeds are prescribed to treat dyspepsia and opacities of the cornea. The seeds are ground for mush and bread, and are taken as a vermifuge. They are also fed to a baby failing to thrive, and given as a charm to bring luck. The leaves are macerated in water and taken regularly as a heart tonic and diuretic.

Juice. This is dripped into the ear for earache.

P. quadrifida is sometimes called the 'ten o'clock' plant because the flowers open at about that time. In India it is used as a vegetable but is not thought to be very wholesome. Medicinally it is used in asthma, coughs, urinary discharge, inflammation and ulcers. A poultice of the crushed plants is applied in haemorrhoids and erysipelas. In Oman the plant is ground with salt and applied to swellings, or over wounds to draw the pus. (Other uses as *P. oleracea*.) (Abbas *et al.* 1992; Ambasta 1982; Batanouny 1981; Bebawi & Neugebohrn 1991; Dalziel 1948, 1956; Ghazanfar 1994; Goodin & Northington 1979; Heine *et al.* 1988; Jain & De Fillips 1991; Kunkel 1984; Lewis & Elwin-Lewis 1977; Miller & Morris 1988; Rizk & El-Ghazaly 1995)

†Also apply to *P. oleracea*, *P. quadrifida*, *P. samhaensis*.

***Portulaca oleracea* L.**

Annual herb; stems prostrate or spreading, up to 30cm. Leaves alternate, obovate or spathulate, 10–30 x 2–10mm, rounded at the tip, flattened; stipular hairs few, soon deciduous. Flowers solitary or in <10-flowered clusters. Sepals 2–4mm. Petals 5, yellow, obovate, 4–8mm. Seeds minutely tuberculate with star-shaped tubercles.

Ecology. Desertic plains and a weed of cultivation. Sea-level–50m.
Distribution. A widespread weed of the tropics and subtropics.
Botanical notes. *P. oleracea* is a polyploid complex within which nine subspecies have been described. The subspecies are apparently autogamous and are recognised mainly by their seed-coat sculpturing.

ETHNOGRAPHIC NOTES
Name(s)
See *P. kuriensis* (also uses).

***Portulaca quadrifida* L.**

Mat-forming perennial with a tuberous taproot; stems to 15cm, rooting at nodes, often reddish tinged. Leaves opposite, flattened, elliptic to oblong-obovate, <12 x 4–5mm, acute; stipular hairs white, 2–5mm long. Flowers solitary, surrounded by an involucre of 4 leaf-like bracts. Sepals c.2 x 2mm. Petals 4, clear yellow within, paler on back, ovate, c.4 x 2.5–3mm. Stamens 8 or 9. Epidermal cells of seeds with rounded margins.

Ecology. Widespread and often abundant in a variety of dry habitats across Soqotra. Sea-level–900m.
Distribution. A widespread weed of the tropics and subtropics.
Botanical notes. Material from Soqotra is unusual within the species in possessing a tuberous main root but in other ways is indistinguishable from *P. quadrifida* which is widely distributed, particularly in Africa, where it is highly variable (Phillips 2000).

ETHNOGRAPHIC NOTES
Name(s)
See *P. kuriensis* (also uses).

***Portulaca samhaensis* A.G.Mill.** (see Appendix 2, p. 723)

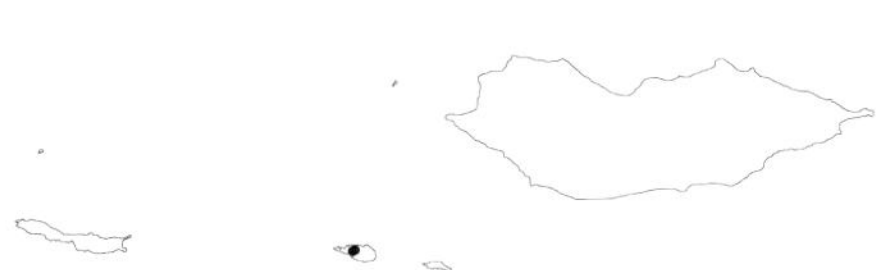

Mat-forming perennial with narrow fleshy roots; stems to 15cm, rooting at nodes, reddish tinged. Leaves opposite, flattened, obovate, 4–10 x 3–5mm, rounded or obtuse at tip; stipular hairs white, 1–2mm long. Flowers solitary, c.11mm across, surrounded by an involucre of 4 leaf-like bracts. Sepals c.3 x 2mm. Petals 4, clear yellow within, paler on back, obovate, c.5 x 3mm. Stamens c.7.
Ecology. Open woody-herb community and dwarf shrubland on rolling limestone plateau. 650–700m.
Distribution. Endemic.
Conservation status. Endangered (EN B2 a biii). *P. samhaensis* is restricted to the summit plateau of Samha where it has a total estimated area of occupancy of less than 10km^2. It is predicted that there will be a decrease in the quality of the habitat on the plateau due to the recent change in livestock management on the island. Until very recently goats and sheep were actively herded and moved down from the limestone plateau during periods of drought. However, both the sheep (now largely feral) and the goats (actively managed only when in milk) are now left to roam freely for most of the year. Both animals are known to spend much of their time grazing on the limestone plateau and this could quickly lead to overgrazing and the eradication of plants on all but the most inaccessible ledges.
Botanical notes. Clearly closely related to *P. quadrifida* and *P. kuriensis* but differing in its smaller stature and the prostrate stems which trail across the ground giving a necklace-like appearance.

ETHNOGRAPHIC NOTES
Name(s)
See *P. kuriensis* (also uses).

Portulaca sedifolia **A.G.Mill.** (see Appendix 2, p. 723)

Minute, tufted succulent, with narrow fibrous root; stems to 2cm. Leaves opposite, densely clustered, broadly obovoid to globose, flattened adaxially, 3–6mm long x 2–4mm across, rounded; stipular hairs 1–2mm long, inconspicuous. Flowers solitary, on 10–15mm pedicels, surrounded by 4 globose bracts. Sepals c.4 x 3.5mm. Petals 6, yellow, c.5 x 2–2.5mm. Stamens c.20.

Ecology. Only known from the ridge above Jebel Shu'ub where it occurs in small depressions on limestone slabs on an exposed ridge. 400–500m.

Distribution. Endemic.

Conservation status. Endangered (EN B2 a biii). *P. sedifolia* is restricted to a limestone ridge at the extreme western end of Soqotra with a total estimated area of occupancy of less than 10km^2. At present the rapid development of the island poses no threat to this unique habitat: the area is inhospitable, inaccessible by road and there is a lack of potable water. However, its restricted distribution and the very limited extent of the niche which it occupies means that it is vulnerable to any change. It is strongly suspected that this marginal habitat is threatened by the long-term trend of lower rainfall in the region and would be particularly vulnerable to the effects of global warming.

Botanical notes. This distinctive species is readily distinguished by its diminutive stature, densely clustered globose leaves, and the involucre of 'calyx-like' bracts surrounding the solitary flowers which are held erect on long pedicels. The presence of stipular hairs, the articulated stems and the involucre of four bracts surrounding the flower clearly place *P. sedifolia* in subgenus *Portulaca* sect. *Neossia*. However, within this section its affinities are not clear. Flower measurements are based on buds; no fully open flowers have been seen.

ETHNOGRAPHIC NOTES

Not seen, no information recorded.

Talinum Adans.

Perennial herbs. Leaves alternate, succulent. Stipules absent. Flowers stalked in terminal racemose panicles. Petals 5. Stamens c.25. Ovary superior; style filiform, 3-branched. Capsule dehiscing by 3 valves. 50 species. Warm regions, especially N America.

Talinum portulacifolium **(Forssk.) Asch. & Schweinf.**
Talinum crassifolium sensu auct. non Willd.
Talinum cuneifolium Willd.

Stems erect, up to 80cm, from a thickened rootstock. Leaves subsessile, narrowly to broadly obovate, 2–6(–9) x 1–3cm, with rounded or obtuse tip and cuneate base. Sepals ovate, 4–6mm. Petals crimson, pink or purple, obovate, 9–12mm long. Capsule globose, 5–7mm diameter.

Distribution. Africa, Arabia and India.

Ecology. Rocky places on the limestone plateaus. 300–700m.

ETHNOGRAPHIC NOTES

Name(s)

– Soqotri: *ṣ̄ɔ̄bār* 'sour, astringent'; *maḥələbi̍nuh*, *maḥlēbboh* 'like buttermilk' (referring to the sour flavour of the plant); (rare) *zi̍rʾe̍h* 'cowpeas'.

– Soqotri simplified: *zowbar*; *mahilibinuh*, *mahleybboh*; *zir'eh.*

Food (human). Relished for its acid, lemony flavour. The foliage and the stems are eaten raw, and are also strewn on cooked cereals as a relish.

Food (livestock). All livestock like it, especially goats. It is seen as being a valuable and rich fodder.

Other. Animals feeding on it can go longer without water.

Comparative uses. In Dhofar the leaves and stems were gathered and eaten, both raw and cooked. The plant is widely used as a vegetable, and in Kenya the root too is boiled and eaten. Livestock who eat it too freely are said to develop a swollen stomach. In India the plant is used as an aphrodisiac. Children with stomach ache are given the boiled root mixed with milk. (Ambasta 1982; Dalziel 1948, 1956; Heine *et al.* 1988; Jain & De Fillips 1991; Kunkel 1984; Miller & Morris 1988)

PRIMULACEAE

Herbs. Leaves opposite, entire; stipules absent. Flowers solitary, axillary, actinomorphic. Calyx campanulate, 5-lobed. Corolla rotate, 5-lobed. Stamens 5, inserted opposite corolla lobes. Ovary superior, 1-celled; ovules inserted on a free-central placenta. Fruit a capsule, dehiscing around its equator. 22 genera. 800 species. More or less cosmopolitan, especially N Hemisphere.

Anagallis L.

See description of family. c.20 species. Mostly Europe.

Anagallis arvensis L.

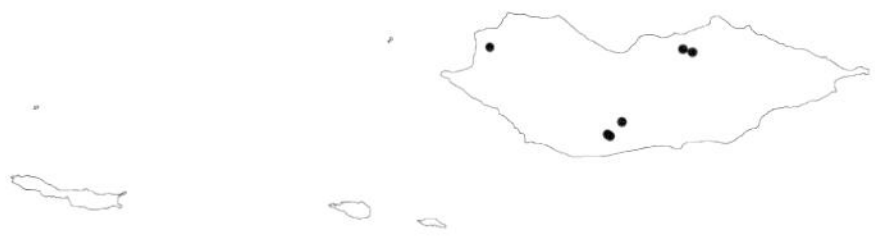

Creeping annual herb, rooting at nodes. Leaves opposite, ovate, <2 x 1cm, sessile. Flowers blue, c.1cm diameter. Fruit a globose capsule, c.0.5cm diameter, opening by a lid.

Ecology. Clearings in woodland. 50–700m.

Distribution. Europe, widely naturalised.

Botanical notes. All Soqotran material is referable to var. *caerulea* Lam.

ETHNOGRAPHIC NOTES

Name(s)

– Soqotri: *di ʿalˈimləhum* 'little tips', 'striped, marked'; *ləskˈano* 'sticky, clinging' (Samha); *ṣefərˈirihun* 'many-flowered'.

– Soqotri simplified: *di 'alimlihum*; *liskano*; *zefiririhun.*

Food (livestock). Provides incidental forage for livestock, but it is not important nutritionally.

Comparative uses. *A. arvensis* has been implicated in poisoning of sheep and calves. It has been shown to have anti-fungal and anti-viral properties, attributed to the presence of saponins. The ingestion of large amounts can cause tremors. Research has shown it has anti-poliomyelitis virus action and acts on *herpes simplex*. The whole plant has anti-fertility action in males. In Oman the crushed plant is used to treat skin rash and ulcers, and to treat snakebite and epilepsy. The leaves and seeds are put in cloth, then dipped in water, and sucked to decrease thirst on hot days. In India the plant is used as an expectorant, diaphoretic and vulnerary, and is given to treat dropsy, leprosy, hydrophobia, mania and other cerebral affections. It is also used to intoxicate fish and expel leeches from the nostrils of cattle (an acetyl saponin isolated shows marked taenicidal activity). In Bahrain the whole plant is used topically on discharges and to treat rheumatism. Elsewhere the leaves are eaten in salads and as a potherb. (Ambasta 1982; Batanouny 1981; Jain & De Fillips 1991; Lewis & Elwin-Lewis 1977; Mossa *et al.* 1987; Oliver-Bever 1986; Rizk & El-Ghazaly 1995)

PUNICACEAE

Trees or shrubs. Leaves opposite, simple, entire; stipules absent. Flowers solitary in the leaf axils, actinomorphic. Calyx 5 to 8-lobed. Petals 5 to 8. Stamens many. Ovary inferior, many-celled; style 1; ovules many, on parietal placentas in upper cells and axile placentas in lower cells. Fruit a hard-skinned pomegranate. 1 genus. 2 species. SE Europe, Middle East to Himalayas.

Punica L.

See description of family. 1 endemic.

Botanical notes. *Punica* is represented by two species: *P. protopunica*, which is endemic to Soqotra, and *P. granatum*, which grows from the Himalayas to the Balkans and is cultivated, particularly in S Europe, for its edible fruit, the pomegranate.

Punica protopunica Balf.f.

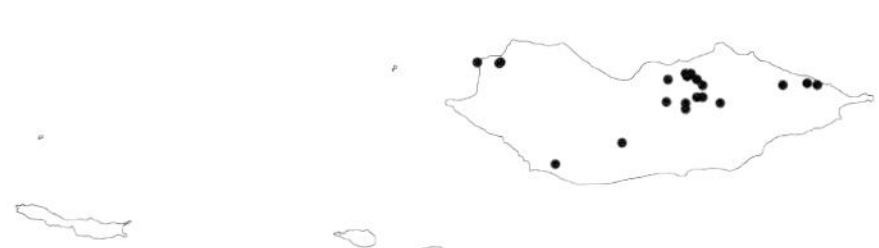

Shrub or small tree to 5m. Leaves leathery, 3–6 x 2–4cm, ovate to elliptic. Flowers bright red, c.2cm long, thick-textured, waxy; petals tiny. Fruit globose, 2–3cm diameter, ripening brownish red.

Ecology. Locally common in moist woodland on both the limestone and granite. 300–1200m.

Distribution. Endemic.

Conservation status. Vulnerable (VU D2). In the Red Data Book (Lucas & Synge 1978) *P. protopunica* is considered endangered. However, recent surveys have shown it to be widespread on the island and fairly common in places. Its total area of occupancy is probably around 100km^2. It has a fragmented distribution with different subpopulations showing varying degrees of vigour. In some upland, plateau regions it becomes prostrate, in other areas it grows to a small tree. In some areas it is common and actively regenerates. However, over large areas the tree is absent except for small relict populations with no apparent regeneration. This indicates a decline in quality of habitat – the reasons for this are uncertain. The foliage of this species is not browsed by livestock, nor is the tree lopped as drought fodder. The wood is of no importance, either as firewood or in construction, nor are the small, bitter fruit appealing as human or livestock food. However, local expertise suggests that the population of this species (unlike *Dracaena* or *Cadaba*, q.v.) is stable.

Botanical notes. Flowers and fruits from December and January through to the summer.

ETHNOGRAPHIC NOTES

Name(s)

– Soqotri: (shrub and fruit) *rihīni* (pl. *rˈɛhin*), (rare) *dihīni*; (western Soqotra) shrub *mīśəṭ* and fruit *raxānə*.

– Soqotri simplified: *rihiyni* (pl. *rehin*), *dihiyni*; *miyshit*; *rakhanih.*

Food (human). The fruit were not an important food on Soqotra. The pithy material lying under the skin and around the seeds of larger, riper fruit is the only part eaten, the skin and the seeds being discarded. The seeds are said to be bitter and caustic and to cause sores on the tongue and small mouth ulcers. In times of hunger the skin of an unripe fruit is eaten as well as the pith, but again, not the seeds. The skin of the unripe fruit is extremely sour.

Food (livestock). In the well-vegetated areas of the island the foliage is little grazed, except in a prolonged dry

season, when the fallen, dead leaves are eaten. In drier areas, goats eat the leaves and especially the new shoots of the smaller and more prostrate shrubs, while sheep eat dry plant litter. It is not palatable to cattle. Goats are fond of the immature fruit, and small stock also like the flowers, fresh or dried.
Bees / honey. Good forage for honeybees.
Flavouring / food additive. The nectar was a welcome source of sweetness (otherwise available only from dates or wild honey) in the days before sugar. In good years the islanders recall that if held upside down the flowers would ooze nectar.
Fuel. The dead wood makes good firewood, making little smoke and giving out good heat.
Timber. Makes good timber and is termite resistant, but it is usually too small to be useful for anything other than roofing infill. Branches are used to make protective or concealing covers for waterholes and rock cisterns, and is considered to be as durable and strong as the wood of *Buxanthus pedicellatus* for this purpose.
Wood. Hard and close-grained, and used to make smaller implements. The straight branches of larger specimens are used to make the important herding staff. Pieces of dead wood make good charcoal.
Medicinal / veterinary. The skins of the fruit are crushed, cooked, and the paste applied to treat sores and wounds. Alternatively they are dried and crushed to a powder which is applied topically to skin sores in the same way. Occasionally people eat the seeds as a purge to relieve stomach ache. More recently people have started using the skins to treat stomach complaints and to expel worms, a treatment apparently learned from the mainland.
Commercial value (past, present). In the 1970s and 1980s the dried fruit were apparently offered for sale in Aden, it is supposed for medicinal purposes.
Other. The skin of a pink-coloured fruit, rather than a fully ripe, red one, is crushed in water and added to a small amount of fresh milk to sour it. The soured milk is then used to ripen fresh milk before churning it to make butter.
Notes. Islanders say that the 'male' tree (*rihīni miʾšər*) bears no fruit. Bent (1897) writes about the wild pomegranate with its lovely red flowers and small yellow fruits, the flannelly coating of which only is eaten, instead of the seeds, as is the case with the cultivated one.
Comparative uses. The pomegranate belongs to the earliest ensemble of cultivated fruit trees in the Old World. As well as supplying fresh fruit they were also considered to be a symbol of fertility. The oldest illustration of the pomegranate dates from the reign of Thutmoses III (1504–1450 BC) in Karnak. Carbonised pips and fragments of pomegranate peel were obtained from early Bronze Age Jericho and Arad. They were identified among the vegetable remains of the 12th Dynasty (1970–1800 BC) Drah Abuel Naga in Egypt, and pomegranate rinds were found during the excavation of Neolithic Gezer.
Fruit, rind and bark. Both appear in pharmacology as taeniafuges. In Yemen the peel of *P. granatum*, *rummān*, is used in wound healing, as an antiseptic and to treat diarrhoea. The crushed rind is also used as a haemostatic. In Oman the dried rind is burned, pounded and applied to ulcers. The bark of the root can be poisonous if taken in excess. In North Africa the bark was once much used for tanning leather and the crushed rinds are used as a poultice on burns. Whole fruit are ground and eaten to improve a 'weak' stomach.
Flowers. In Yemen the flowers are dried, soaked in water for six days and then cooked with fish to reduce the fishy smell. In Oman the dried rind is mixed with *Thymus vulgaris* and flour, baked into bread, and eaten for diarrhoea and as an anthelmintic. Elsewhere, an infusion of the flowers is given as a vermifuge. The flowers stain the saliva and were formerly used as a red dye, and the flower buds were used in bronchitis.
Juice of fruit. In Oman the juice of the fruit is placed in the sun and then used as eyedrops to improve the vision. It is drunk fresh for jaundice, and with added vinegar is applied to skin rashes. Mixed with sugar it is drunk to improve the health of the liver.
Leaves. Young leaves are edible when boiled. (Dalziel 1948, 1956; Fleurentin & Pelt/Fleurentin *et al.* [1982, 1983a,b,c]; Ghazanfar 1994; Kunkel 1984; Zohary 1982)

RESEDACEAE

Herbs or shrubs, sometimes dioecious. Leaves alternate, entire or variously dissected; stipules = glands. Inflorescence of spike-like racemes. Flowers zygomorphic, subtended by a bract. Sepals 2 to 8, free or fused at base. Petals 0 to 8, free, unequal, usually divided, with a basal appendage. Disc present. Stamens 10 to numerous. Carpels 2 to 7, superior, ± free or fused and forming a 1-locular ovary, ovules numerous, placentation parietal or basal, stigma sessile. Fruit a berry or capsule, sometimes gaping at apex; seeds reniform. 6 genera. 75 species. N Hemisphere, especially Old World.

Caylusea A.St.-Hil.

Herbs. Leaves simple, entire. Flowers bisexual. Sepals 5, fused at base. Petals 5. Disc present. Stamens 10 to 15. Ovary stipitate, of 5 or 6 carpels, fused only at base. Fruit of divergent or erect gaping carpels, seeds 1(or 2) per carpel. 3 species. Cape Verde Is., N and E Africa to India.

Botanical notes. See Taylor (1959).

Caylusea hexagyna **(Forssk.) M.L.Green**
Caylusea canescens A.St.-Hil.

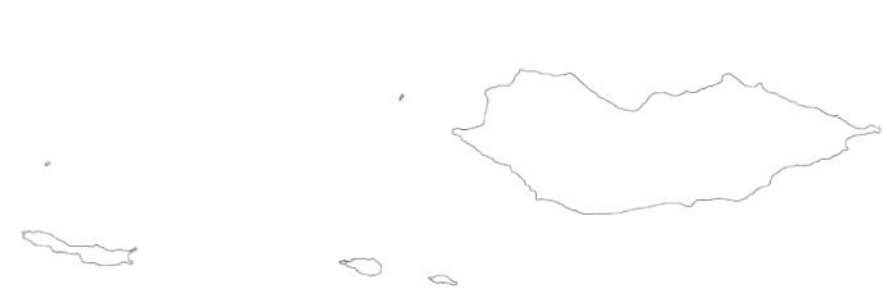

Erect or decumbent herb to 60cm, white-grey hairy or glabrous. Leaves narrowly obovate, 10–50 x 2–10mm, acute or obtuse, margin flat or wavy. Raceme <20cm. Sepals ovate, 1.2–2mm. Petals white, the upper 5 to 9-lobed, 2–3mm. Carpels strigose, 2–3mm, on 2–3mm long stipes.
Ecology. No details of habitat or location available.
Distribution. N Africa south to Ethiopia, Middle East, Arabia.
Botanical notes. Originally collected by Theodore Bent in the 19th century. Some of Bent's collections were made on the mainland and, as this species has not been found on Soqotra, it is possible that this collection was made on the mainland. It remains a very dubious record.

Ochradenus Delile

Dioecious shrubs. Leaves linear, entire, sessile, sometimes fascicled. Sepals 5 to 8. Petals absent or linear. Disc excentrically surrounding ovary, fleshy. Stamens 10 to many. Ovary sessile, of 3 fused carpels. Fruit a berry; seeds many. 1 endemic. 7 species. Middle East.

Botanical notes. See Miller (1984).

***Ochradenus socotranus* A.G.Mill.** (see Appendix 2, p. 723)
Ochradenus baccatus sensu Balf.f. non Delile

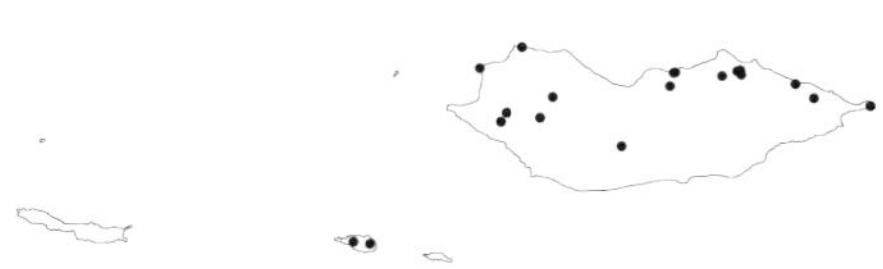

Glabrous shrub to 1.5m. Leaves linear, 1–4cm. Racemes lax, 8–20cm; pedicels 1–2mm. Flowers yellow. Sepals oblong-ovate, 1–1.5mm. Petals absent or rarely present. Disc reflexed. Stamens 10 to 18. Fruit baccate, globose, 3–6mm broad, white when mature. Seeds 1.2–1.8 x 1–1.5mm, dull, smooth, brown to black.
Ecology. Dry rocky slopes. 10–700m.
Distribution. Endemic.
Conservation status. Least concert (LC). Common in several vegetation types and under no present or perceived threat.
Botanical notes. Most closely related to *O. arabicus* Chaudhary, Hillc. & A.G.Mill. and *O. baccatus* Delile from the mainland of Yemen. It is similar to *O. arabicus* in being clearly dioecious but differs in having a fleshy fruit and dull, rather than glossy, seeds. The dull, smooth seeds also distinguish it from *O. baccatus*, which has rugose seeds and apparently hermaphrodite flowers.

ETHNOGRAPHIC NOTES
Name(s)
– Soqotri: *gerśīʾoh*, *gēśīʾoh* (pl. *gɛrśiʾ*, *gēśoy*) 'burp'; *gɛrśīyə* (Samha).
– Soqotri simplified: *gershiy'oh*, *geshiy'oh* (pl. *gershi'*, *geshoy*); *gershiyih.*
Food (livestock). In vegetation-poor areas this is an important dry season goat fodder as it retains its green colour throughout the dry season. It increases lactation in milk camels and goats and makes them put on weight. Sheep eat a little in the dry season. Camels and donkeys browse it heavily; indeed, camels are said to be able to survive on a diet of this alone. When flowering and fruiting it is much more palatable, and at this time even cattle eat the tender new shoots. In the wetter and better-vegetated areas it is only browsed in a prolonged dry season. In the outer islands it is regarded as good fodder for both sheep and goats.
Flavouring / food additive. In the dry season, the meat of the carcass of animals browsing heavily on it can develop a slightly bitter flavour.
Fuel. The dried wood is excellent for lighting fires, and in areas where it is common, most households keep a store for lighting fires in wet weather. If a bright light is needed, handfuls of this material are thrown on the fire. It is rarely used for cooking, however, as its smoke can taint the food.
Timber. On the outer islands especially, the plant was used as a roofing material under a top coating of earth.
Pest control. The root produces a lot of very pungent smoke, which is a good insecticide. To lessen the numbers of fleas in livestock and human quarters, the ventilation holes are blocked up, and then dead plants are collected and piled up inside and set alight. The fires are left until all the smoke has disappeared. This procedure is repeated until the living quarters have been made habitable again. This is done especially in the small byres and pens where young stock are housed, as fleas present a greater health hazard to them.
Other. In areas most affected by the blasts of the monsoon winds, stems are packed around the ripening fructescences of the date palm to protect the fruit. Goats feeding on it can go longer without water.
Comparative uses. In Dhofar the ash of burned *O. arabicus* is added to snuff to make it more pungent, and the dried stems are used as a firelighter. Elsewhere in the Middle East the fruit and leaves of *O. baccatus* are taken for stomach pain, the twigs, leaves and flowers are applied to wounds and sores to kill maggots, and the fruit are eaten. (Abbas *et al.* 1992; Batanouny 1981; Ghazanfar 1994; Goodin & Northington 1979; le Houerou 1980; Miller & Morris 1988; Rizk & El-Ghazaly 1995)

Reseda L.

Subshrubs. Leaves entire or lobed. Flowers, bisexual, 4 to 6(to 9)-merous. Petals laciniate. Disc present. Stamens 10 to many. Ovary shortly stipitate, of 3 or 4(or 5) carpels. Fruit a capsule, gaping at apex; seeds many. 1 endemic. 55 species. Europe, Mediterranean to C Asia.

***Reseda viridis* Balf.f.**

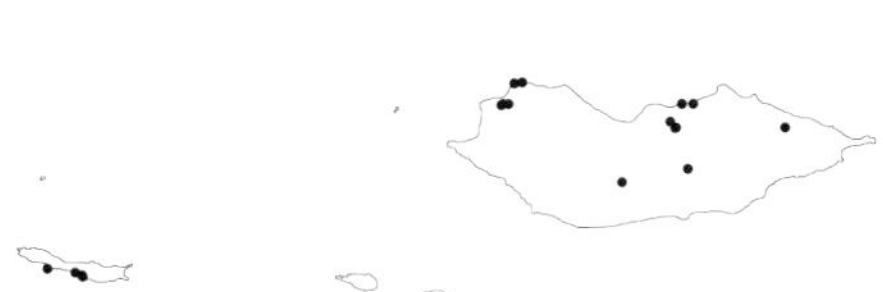

Subshrub to 1m, glabrous. Leaves entire or rarely 3-lobed, ovate to circular, 1.5–6 x 1–3cm; petioles 1–3cm. Flowers greenish white, 6-merous; upper petals 2–3mm, 7 to 9 lobes. Stamens 12 to 15. Capsule ellipsoid to obovoid, 8–15 x 3–4mm, 3-angled, 3-toothed at apex.
Ecology. Dry rocky slopes and cliffs. Sea-level–700m.
Distribution. Endemic.
Conservation status. Least Concern (LC). Common in several vegetation types and under no present or perceived threat.

ETHNOGRAPHIC NOTES
Name(s)
– Soqotri: nil or *ḥahər* 'dark-coloured'; *di ṣifərirrihan* 'many-flowered'; *ekśəreh* 'impenetrable undergrowth, difficult terrain of sharp rocks'; (rare) *ḥaliblǝbi* 'sour like buttermilk'.
– Soqotri simplified: *hahir*; *di zifiririhan*; *ek'shirih*; *haliblibi.*

Food (livestock). Rarely accessible to livestock, but the foliage is said to be eaten by goats if they are very hungry.
Notes. Said by some to be one of the *ʿūd el kəsīr* plants (see Introduction).
Comparative uses. The young shoots and leaves of some species are edible. In India *R. luteola* yields a dye used for colouring silk and wool. The flowers and fresh roots of *R. odorata* yield an essential oil, and the seeds a fatty oil. Some species are used medicinally, mainly in the treatment of stomach aches and as an anthelmintic. (Ambasta 1982; Goodin & Northington 1979; Jain & De Fillips 1991; Kunkel 1984)

RHAMNACEAE

Shrubs, trees or climbers. Leaves alternate, simple, entire; stipular spines or tendrils present. Flowers small, in axillary cymes, actinomorphic. Sepals 5. Petals 5. Stamens 5. Ovary superior, 2 or 3-celled; style 1, minute; ovules 1 per cell, placentation basal. Fruit a single-seeded drupe or dry and splitting into 3 parts. 53 genera. 875 species. Cosmopolitan, especially tropical and warm regions.

Helinus E.Mey. ex Endl.

Climbers. Stipular tendrils present. Ovary 3-celled. 5 species. Tropical and S Africa, Madagascar, NW India.

Helinus integrifolius Lam.

Woody climber. Branches with tendrils. Leaves ovate, c.1.5–3 x 1.5–2.5cm, pinnately veined. Flowers in umbels. Fruit splitting into 3 parts, c.0.5cm diameter.
Ecology. Rare on cliffs in sheltered limestone ravine in deciduous woodland. 550m.
Distribution. Tropical Africa.

ETHNOGRAPHIC NOTES
Not seen, no information recorded.

Ziziphus Mill.

Shrubs or trees. Stipular spines present. Ovary 2-celled. Fruit a single-seeded drupe. 86 species. Tropical and warm regions.

Ziziphus spina-christi (L.) Willd.
Ziziphus jububa sensu Balf.f. non Lam.

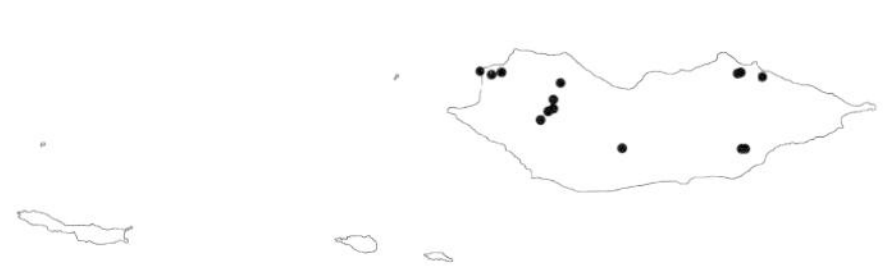

Shrub or large tree to 10m. Branches spiny. Leaves ovate to elliptic, 2–5 x 1.2–3cm, 3 to 5-nerved from base. Flowers in cymes. Fruit a yellow or reddish berry, 1–1.5cm diameter.
Ecology. Widespread and common on the plains, along wadis and in woodland on the foothills. 60–500m.
Distribution. Africa, Middle East to NW India.
Botanical notes. A large tree with a rounded crown, three to five-nerved leaves, and zigzag branchlets bearing paired spines.

ETHNOGRAPHIC NOTES
Name(s)
– Soqotri: tree *źʌʾd, źod, źoʾod maḥarito* (pl. *iźʾid*); young tree or bush (that produces no fruit) *ʾɔ̄ḳeśhən* (pl. *ʾɔ̄ḳeśyun*); fruit *žirema* (pl. *žirihem*); dried leaves *məgefef*.
– Soqotri simplified: tree *dha'd, dhod, dho'od maharito* (pl. *idh'id*); young tree or bush *'owqesh'hin* (pl. *'owqeshyun*); fruit *jiremih* (pl. *jirihem*); dried leaves *migefef*.
Food (human). The fruit are one of the most important plant foods on the island, especially as they are at their best in the summer months when other food is often scarce. They are particularly important in the drier areas of the island, where there are few other plant foods. They are easily gathered from smaller shrubs, and taller trees are harvested by throwing stones to knock down the fruit. The fruit ripen from green to yellow. They then develop brown patches and become softer and sweeter. The round pips are cracked open to get at the kernel, which is delicious and very nutritious. In hungry times the whole fruit is ground between two rocks to a paste eaten in handfuls without further preparation. When the fruit are plentiful, they are harvested and then spread out to dry. They are then ground, pip and all, and the meal left to dry in the sun on a flat rock or palm-fibre mat. When completely dry (it easily turns mouldy if at all damp) it is stored in a covered clay pot or a skin bag. Some islanders only grind the pip to make this flour, and eat the flesh. At mealtimes, a little water or milk is added to the flour to make a sort of porridge. Occasionally a group of well-developed trees growing together in a particular place, usually a watercourse, are regarded as being the property of those living nearby, but otherwise anyone can gather the fruit from any tree at any time without seeking permission first.
Food (livestock). One of the best fodders for goats and camels. For goats it is most important in the dry season, but camels browse it all year round. It is regarded as the best possible feed for camels, helping them to put on weight and increasing milk yield. In areas where these trees are common, such as the interior valleys, the foliage is the key dry season fodder for goats. Sheep eat

any leaves within reach, and are fed leaves in the dry season, and and dried plant litter at other times of the year. The foliage makes sheep and goats fat and sleek, and makes them give more milk. Herders often herd their pregnant and milk goats to browse on the trees, or bring them lopped leafy stems to eat. However, if goats are unaccustomed to it and eat a lot at times when other forage is scarce, they sometimes develop scour and temporarily lose weight. However, after a fortnight or so, their stomachs adjust and they begin to put on weight. Goats suffering from *ʿuqabah* are not allowed to eat the foliage, or they develop very severe diarrhoea, which is sometimes fatal. They are not allowed to eat too many of the fruit either, as this too is said to damage their digestive system. Cattle are put off from browsing the trees by the thorns, and only eat leaves from the ground, or fresh leaves stripped from the spiny branches for them by their owners. If they eat plenty of these, they too put on weight and give more milk. Fresh leaves are gathered to feed to young lambs and kids who have recently been weaned. The small straggly or prostrate *ʾɔ̄ḳéśhən* bushes also provide fodder for goats, sheep and camels, but make no fruit. Herders say that animals that browse these smaller plants for any length of time are prone to developing sores in and around the mouth.

Ziziphus trees come into leaf in the hot, dry months and respond well to pruning, quickly pushing out new leaves and tender new stems. Indeed, islanders in areas where the trees are common say that for the best results they must be managed actively by giving them a severe pruning at the very end of the monsoon every year, or at least every second year. The prunings are gathered and fed to goats, and the fallen leaves to kids. However, in the drier areas of the island most herders feel that the foliage and fruit are too important to permit 'wasteful' pruning in this way. Here a branch is lopped only if it has grown so tall that the foliage is out of reach. Even then, rather than removing the branch, wherever possible another suitable branch is bent and weighed down with a rock so that the goats can climb up into the tree and get at the foliage. Wild goats are famous on Soqotra for their ability to climb into these trees, and especially for their skill in leaping up to reach the foliage of the tallest branches. Because of this, in the remoter areas of the island where wild goats live, it is common for herders to hang clothing or some material over branches of the tree in such a way that it is very obvious. This is said to discourage the wild goats from approaching the tree by suggesting that there are humans close by.

Bees / honey. As on mainland Yemen, this tree is famous for its honey, especially in the summer months. If good summer rains bring the trees into heavy flower, they become thick with bees, and the resultant honey is the best and most expensive on the island. If honey is very liquid, pouring easily through the fingers, and without the least trace of bitterness, it is said to be probably from ziziphus flowers.

Flavouring / food additive. In the drier areas of the island the flowers were much liked for their sweetness. Islanders say that meat of animals that have eaten a lot of the foliage is particularly sweet and richly marbled with fat.

Fuel. The dead wood makes excellent firewood. It burns with little smoke and so is much liked for cooking and heating the home. The wood burns very brightly and pieces are often kept at home for this purpose. It burns with an intense heat so is used for warming people and livestock in cold and wet weather. Dry twigs are often also stored at home for use as firelighters and kindling. A thick chunk of the wood, tucked into the embers and covered with ash, is good for keeping a fire in.

Timber. This tree provides the best timber on Soqotra, and it is very much in demand for building. It is termite resistant, but is attacked by other insects, and because it is so valuable it is usually de-barked before use. The trees, especially if they are well managed, produce straight trunks and stout, well-shaped branches used for the main roofing beams and supporting pillars. In areas where the trees are few and far between they are rarely cut as timber, since they are too valuable for food and fodder. Where they are more common, or if an individual tree is known to be a poor fruiter or to produce little foliage, branches can be cut as timber. However, the main trunk is normally only cut with the full agreement of the local community and if genuine need can be demonstrated. A well-matured piece of timber from this tree, cut at the hottest time of the year and carefully de-barked, can last for four generations, or even longer if it is never allowed to get wet and is thoroughly impregnated with smoke to deter insects.

Leatherwork. The wood burns down to a fine ash, which is considered one of the best for processing skins and hides.

Wood. Close-grained and hard, it was much used for making implements, including locks and keys. The wood was also used to make the measuring bowl used by ghee traders and merchants when transferring the clarified butter from the clay or skin containers of the herders to the imported metal containers which, when filled, were sealed with lead and exported. The wood makes a good firestick, and small pieces of heartwood were whittled to make fish-hooks in the days before imported fishing equipment became widely available. The straighter slim branches were much sought after for making the important herding staff. The wood was used to make the frame for the Soqotri tambourine drum (*tarānāt*; see Introduction). The dead wood was formerly used to make charcoal for the island blacksmiths, and it was also regarded as one of the best woods for making quicklime and for firing pottery. However, with the sharp increase in demand for the tree as building timber (which has to be from green wood), the trees have become much less common, and are rarely available for such alternative uses.

Medicinal / veterinary. Dried or fresh leaves are ground and put in water. They are left to soak and the filtrate is drunk as a treatment for constipation and to relieve stomach pains generally. If childbirth is very prolonged, a similar mixture is drunk to hasten the process. A woman suffering from a retained placenta is also given this to drink, though an alternative treatment is to make a fire of ziziphus wood and encourage the patient to stand or crouch over it with her legs apart to fumigate her lower body with the smoke. The leaves and stems are crushed to a paste and smeared over the forehead and scalp of someone with a severe and persistent headache. It is painted over the belly of a woman suffering from post-partum pain, or over raised and tender neck glands. The paste is also applied to sores on the skin or to a scalp covered in sores caused by scratching at head lice. It was such a widely used medicament that most home 'medicine chests' contained ziziphus leaves (*migéfef d źʌʾd, di źɔ̄wʾed*), stored either whole or ground to a powder. In livestock, a normally productive animal that has grown thin and listless is given leafy stems to restore it to good health. The soft growing tips are cut and given to an animal recovering from an illness or a difficult birth. A cow that is losing weight towards the end of her pregnancy is brought ziziphus leaves to give her strength for the forthcoming birth.

Quicklime made with this wood is also used medicinally, on both humans and their livestock. It is ground to a fine powder and used as a dressing for the surgical wounds of circumcision and to treat deep lacerations. It is also spread over weeping sores and abrasions to dry them out and speed the healing process. To treat mumps, a piece of charcoal and some ziziphus bark is set alight on a slip of clay pottery and the fumes are inhaled by the patient. To get rid of warts, a fine thorn from a bushy ziziphus is used to probe around and under the wart to make it bleed. It is then left for a week or so when the scab is yanked away to bring away the wart, root and all.

Cleansing. Dried leaves, sometimes ground to a powder, are kept in the home for washing and cleansing. The powder or crushed fresh leaves are added to water and left to steep. This mixture is then used to wash the hair and to cleanse the scalp. It is also used to remove nits and head lice, and to wash the body of the dead before wrapping them in the cotton shroud. Women cleanse themselves with a similar mixture after menstruation and parturition.

Cosmetic. The hair is dressed with a paste of powdered, dried ziziphus leaves or of crushed fresh leaves after arranging the hair to make three fat plaits, one each side of the head and one down the nape of the neck. This is done especially when a woman has woven 'false plaits' of sheep's wool into her own hair to make the plaits fatter and longer. This dressing makes the sheep's wool lie flat and blend with the hair, and gives a shine to the whole head.

Commercial value (past, present). The fruit and dried leaves were sometimes used as barter. More recently the fruit have been offered for sale in the markets, and there is a brisk trade in the timber.

Other. Animals feeding on the foliage can go longer without water. Many islanders say that soil that can grow ziziphus is also suitable for cultivation, and collect it for this purpose. Young seedlings have been transplanted close to the homestead for their fruit, wood and foliage, but this is not often successful. The saplings do not always take and even those that do survive are weak and produce little foliage. Attempts have also been made by herders to grow plants from seed, but this too has rarely been successful. Large trees are sometimes used as a milking area by cattle herders who are away from their home byres and milking yards. A cattle smudge is lit beside the tree for a few days, and the cows are herded to stand in the smoke before milking. After a time they come of their own accord as soon as they see and smell the smoke. This sometimes works with goats as well.

Notes. The islanders say that the 'male' tree (*źóʾod miʾšər*) is excessively thorny and less fruitful than the 'female' tree (*źóʾod maḥariṭo*), which has fewer thorns and gives most fruit. Islanders who are expert in the management of ziziphus say that:

(1) A straggly bush (*ʾɔ̄ḳeśhen*) can be trained and shaped to become a useful tree by selecting one or more strong-growing stems and removing all the others. Any side branches growing from the selected stems are also removed. The young plant is then fenced around with rocks and the prunings to protect it from livestock. It is carefully observed to see which stem is growing most strongly, and when this is apparent, any others are removed. The fencing is maintained until the trunk has grown beyond the height of the fence, at which time the tree is considered to be able to manage on its own. This training of young plants is most easily done within a walled enclosure, in a date palm garden, for instance.

(2) Mature ziziphus trees should be cut back hard at the end of every monsoon, unless both summer and winter rain have failed, in which case it is left for another year. All the branches are lopped until only the main trunk with the stumps of the main branches remains. Once the trunk has been cleared of side branches, the forester climbs up into the tree to lop off any branches remaining in the crown. After about a month new growth appears, which is very tender and succulent. Trees treated in this way grow strongly and have long, productive lives, providing plenty of foliage and fruit. Trees of a comparable age which have not been actively managed like this tend to look rather 'dry' and leafless and produce markedly less fruit.

(3) To train a ziziphus to provide good timber the lower side branches of a young sapling are all removed to leave the trunk bare and smooth. One or several stakes are then cut and strapped to the tree to make it grow straight. As new growth appears on the trunk it is removed. This trimming and pruning is continued until a good trunk has been produced, which takes around seven years. However, the wood is still too soft and young to make good timber, and the tree is left growing until it is hard enough. This is tested by piercing it with a knife tip.

In the drought year of 1999, herders from one valley in the interior, who are knowledgeable about ziziphus management, fed their animals entirely on lopped foliage twice a day for a full month. At the end of this period the trees looked very bare, yet they had all fully recovered by the next year. The islanders say that in their experience trees are not damaged by severe pruning; they are more likely to be damaged by a hungry camel yanking at a branch so hard that it tears off, splitting the trunk. This sort of damage can be fatal.

There is a poem known all over Soqotra about the 'liking of ziziphus for the knife'. The ziziphus tree says: 'You do not bring me good news/by telling me that heavy rain clouds are gathering;/But by bringing me the curved blade/to trim me of my shoots and stems'.

Disputes over the cutting of ziziphus trees have given rise to many well known stories on Soqotra, of how people resisted various rulers of the day who were keen to cut down trees for their various building projects. One famous, or infamous, ziziphus grew in Wadi Lahaz, east of Hadiboh: hands which had been chopped from persistent livestock thieves were hung from this tree as a warning to passers by. This tree has now been cut down. ziziphus and tamarind trees feature in local star lore: they are said to symbolise the 'star' *turriyih* or *qadahum*, which comes at the end of June and again in mid-November (each 'star' occurs twice yearly). The trees come into leaf at both these times even if there is no rain.

Comparative uses. In many parts of the world *Z. spina-christi* is cultivated for its edible fruits, or planted as a living hedge, or shelterbelt. The plant develops an extremely deep taproot and spreading laterals, making it useful for stabilising sand dunes and other unstable soils. The termite-proof wood is an important timber; it is also used for furniture, in turnery and to make handles for implements, for example. The seeds yield a fatty oil and

the bark contains tannin. The leaves are sometimes cooked and eaten in soups and gruels. In Sudan a drink is made from the pulp of fruit after fermenting and drying it, and this is said to be very rich in vitamin C. Baumer notes that for propagation, firstly the pulp of the fruit must be softened in water to extract the stone, then this must be cracked with a hammer to collect the seed. He reports that germination is fairly good: seedlings grown in plastic bags are ready for planting out after 18–24 weeks.
Leaves. In Saudi Arabia a decoction is used to treat dandruff and keep hair healthy, and an infusion is used to wash a corpse. In Yemen they are used as an antiseptic and anti-pruritic. The dried leaves are crushed and used as a baby powder. The green leaves are crushed with lemon juice and the paste applied to the face and the hair to make it soft. Fresh leaves are chewed to relieve pain in the gums, or are smeared over the forehead for a headache. They are cooked and the paste applied to fractures before binding. They are boiled with water and the filtrate given to women to ease prolonged labour. In Oman they are boiled in water and the mixture applied to soothe skin. A paste of crushed leaves is applied to treat skin sores and painful and inflamed joints. Elsewhere in the Middle East an infusion of the leaves is taken as an astringent, to expel worms and as a remedy for diarrhoea. A poultice of fresh leaves is given for abscesses, and is put on swollen eyes before going to bed.
Fruit. In Yemen the fruit are considered to be a general tonic and cleanser of the whole system. They are taken daily with the crushed seed kernels to relieve chest pain and respiratory problems. Elsewhere in the Middle East the fruits are taken to treat fever, and an infusion of the fruits is taken for measles.
Kernels. In Oman the kernels are used to treat blisters and bruises and, after crushing, are eaten daily for chest pain. *Bark.* In East Africa the bark and roots are pounded and boiled in water, and the red filtrate is drunk warm as an emetic and laxative.
Ash. In the Middle East the ash of the wood is mixed with vinegar and applied locally to treat snakebite. (Abbas *et al.* 1992; Ambasta 1982; Baumer 1983; Fleurentin & Pelt/Fleurentin *et al.* [1982, 1983a,b,c]; Goodin & Northington 1979; Heine *et al.* 1988; le Houerou 1980; Jain & De Fillips 1991; Miller & Morris 1988; Mossa *et al.* 1987)

RUBIACEAE

Herbs, shrubs or trees. Leaves opposite or whorled, simple, entire; stipules present forming a sheath around the nodes or leaf-like (*Galium*). Flowers 4 or 5-merous, actinomorphic, bisexual. Calyx fused, lobed or entire with a 1-sided wing. Corolla fused, usually tubular. Stamens fused to the corolla. Ovary inferior with axile placentation and 2 to many ovules. Fruit dry or fleshy, dehiscent or indehiscent, often enclosed by the enlarged calyx. 637 genera. 10,700 species. Cosmopolitan, especially tropical and warm regions.

Carphalea Juss.

Dirichletia Klotzsch

Shrubs or small trees. Leaves opposite with a stipular sheath. Calyx an entire rim with a leaf-like wing. Corolla 5-lobed, the lobes acute and shorter than the corolla tube. Fruit dry, indehiscent, hidden by the expanded calyx. 1 endemic. 14 species. Tropical Africa, Madagascar.

Carphalea obovata (Balf.f.) Verdc.

Carphalea glaucescens (Hiern) Vatke
Carphalea obovata var. *albescens* (Balf.f.) Verdc.
Dirichletia glaucescens Hiern
Dirichletia lanceolata Balf.f.
Dirichletia obovata Balf.f.
Dirichletia venulosa Balf.f.

Shrub or tree to 3m. Leaves narrowly lanceolate to broadly obovate, 2.5–6.5 x 0.5–3cm. Corolla pink, lilac or blue; corolla lobes acute; corolla tube 2–2.5cm.
Ecology. Locally common in woodland on the mountains and along wadis. 100–1100m.
Distribution. Endemic.
Conservation status. Least Concern (LC). Common in several vegetation types and under no present or perceived threat.
Botanical notes. A common and widespread shrub with small pink or lilac flowers. It is readily distinguished by the asymmetric wing-like calyx which gradually expands and hides the developing fruit. Balfour (1888) originally described three species (under *Dirichletia*) from the island. These were later reduced to a single species, *Carphalea obovata*, by Verdcourt (1973).

ETHNOGRAPHIC NOTES
Name(s)
– Soqotri: *śīḥəṭ* (pl. *śḥe̍t*, *śīḥʌ̍t*) 'dry, tough, enduring'; small or prostrate specimen *məśəwḥāṭo*, *śīhiṭe̍no*, *śeṭe̍no*.
– Soqotri simplified: *shiyhit* (pl. *sh'het*, *shiyhat*); small or prostrate specimen *mishewhato*, *shiyhiteno*, *sheteno*.
Food (livestock). A key fodder for small stock on Soqotra, especially goats. Sheep mostly eat the foliage when it is young and green. Donkeys also like it, and it is an important dry season feed for camels. Cattle only eat the leaves when there is little else. All livestock eat the fallen leaves, and goats chew the leafless stems. It is considered to be a very valuable fodder for goats and to increase milk yield. In the dry season herders lop foliage and leafless twigs for their goats, and gather plant litter for their animals. Leafy stems are taken back to feed young stock.
Bees / honey. Honeybees are attracted to the flowers and the honey is regarded as one of the best on Soqotra.
Fuel. The dead wood makes good firewood, especially when de-barked. Its smoke is not troublesome so the wood is good for cooking. When the bark is removed it makes a good firelighter. It burns with a bright light and some dried wood is usually kept in the home for this purpose. Burning sticks are used as torches.
Timber. The wood is hard and is used as timber, especially in roofing. The trunks are used as cross beams; larger branches from the rare, well-grown trees are used as rafters, and the smaller branches for roofing infill. It is

relatively termite resistant but it is usually safer to remove the bark before use. It is important to cut the timber when it is hot and dry; if it is cut in cold or wet weather it quickly rots.
Leatherwork. The dead wood burns down to a good, fine ash which is good for processing skins and hides.
Wood. Close-grained and hard, and good for making a variety of implements, including hair combs. The straighter branches are used to make the important herding staff, as well as handles for utensils. It is good for fencing, but the wood for this must be cut in hot, dry weather. Stakes can be sharpened and used for digging, or to clear out water sources. Sticks are cut and hammered into the trunk of a *Sterculia africana* tree so that the herder can climb into the crown to lop foliage and branches as a dry season feed for his cattle. It makes a good charcoal.
Medicinal / veterinary. Sickly animals are brought leafy branches to restore them to good health. Small, forked twigs are wedged around the base of a painful tooth and then tapped with a stone to dislodge the tooth.
Pest control. The dead wood is used to warm goats and keep flies off during the cold winter weather. If it is wet, the fires are lit inside the byre, and if the skies are clear, in the open pen.
Notes. Islanders say that the 'female' trees (*śīḥəṭ báʿalə*) have larger leaves and bear many more flowers than the 'male'. This tree is considered by islanders to be the 'sister' of *Vernonia* species.

Diodia L.

Herbs. Leaves opposite with stipular sheaths. Flowers sessile, axillary. Corolla 4-lobed. Fruit a 2-seeded capsule; seeds ribbed. 30 to 50 species. Tropical Africa and Madagascar, USA south to Argentina and Uruguay.

Diodia aulacosperma K.Schum.

Spermacoce hispida sensu Balf.f. non L.

Annual herb to 40cm; stems branched, erect, decumbent or prostrate. Leaves ovate-elliptic, 1–4.5 x 0.3–2cm. Calyx hairy, 1.5–2mm long. Corolla pink, lilac, blue or white; corolla tube 2–4.5mm. Stamens and stigma visible at the mouth of the corolla tube. Fruit 2.5–4mm long.
Ecology. Drought-deciduous woodland and in sheltered ravines on the southern limestone plateau. 350–650m.
Distribution. Kenya, Tanzania.
Botanical notes. All Soqotran material is referable to var. *angustata* Verdc.

ETHNOGRAPHIC NOTES
Name(s)
– Soqotri: *məltúwsək* 'sticky, clinging'; *ḳāwrɛ́ni* 'two horned' (referring to the fruit); *ʿážbib* 'wilting'; *ʿážərib* 'quick to die back'.
– Soqotri simplified: *miltewsik*; *qewreyni*; *'ajbib*, *'ajrib.*
Food (livestock). This is one of the *ʿážərib* group of plants which flourish after the winter rains. It is grazed by cattle before the grasses come up, but is little eaten by goats and sheep. It is not seen as having any nutritional value, and is not an important forage on Soqotra.
Comparative uses. *D. maritima* is used to make medicine for infants, and the juice of *D. scandens* is warmed and dropped into the ear for earache. It is also used to prepare a cough medicine for children and in prescriptions for venereal disease. (Dalziel 1948, 1956)

Gaillonia A.Rich. ex DC.

Neogaillonia Lincz.

Herbs or subshrubs; often foetid. Leaves opposite. Flowers small, tubular. Calyx and corolla 5-lobed. Fruit a nutlet. 4 endemics. 30 species. NE Africa, Arabia and SW Africa.

Botanical notes. For a revision see Thulin (1997b).

Gaillonia puberula Balf.f.

Neogaillonia puberula (Balf.f.) Lincz.

Subshrub to 90cm. Leaves linear to linear-elliptic or oblanceolate, 1.25 x 0.5cm, internodes longer than leaves. Flowers in axillary or terminal 3-flowered cymes. Calyx lobes subequal. Corolla white, tube 4–5mm long, lobes reflexed, 4–5mm long. Anthers and stigmas long exserted, 3 stamens much longer than the other 2. Fruit 0.5cm long.
Ecology. Common in rocky places on the limestone escarpments. 30–650m.
Distribution. Endemic.
Conservation status. Least Concern (LC). Common in several vegetation types and under no present or perceived threat.

ETHNOGRAPHIC NOTES†
Name(s)
– Soqotri: *źeriṭhanítin* 'farts'; *ləṭēnuh* 'unmoving'.
– Soqotri simplified: *dherit'hanitin*; *liteynuh.*
Food (livestock). Avoided by most livestock, but they nibble at the stalks in severe drought if there is little else. It has been observed that if eaten by goats they develop halitosis and their belches are also foul-smelling.

†Also apply to *G. putorioides*, *G. thymoides*.

Gaillonia putorioides **(A.R.Sm.) Petruss. & Thulin**
Jaubertia putoroides Radcl.-Sm.
Neogaillonia putoroides (Radcl.-Sm.) Lincz.

Subshrub to 50cm. Leaves oblanceolate-spathulate, 1.2–1.7 x 0.3–0.4cm. Flowers solitary, axillary. Calyx 3mm long, lobes lanceolate, unequal. Corolla white, tube 9–10mm long, lobes narrowly lanceolate, 2mm long. Anther and stigmas shortly exserted.

Ecology. Local in low evergreen shrubland on granite in the central and eastern Haggeher. 900–1550m.
Distribution. Endemic.
Conservation status. Vulnerable (VU D2). Rare, with a very limited area of occupancy.
Botanical notes. Very bad-smelling!

ETHNOGRAPHIC NOTES
Name(s)
See *G. puberula* (also uses).

Gaillonia thymoides **Balf.f.**
Neogaillonia thymoides (Balf.f.) Lincz.

Subshrub to 60cm. Leaves linear to linear-elliptic or oblanceolate, 1.25–1.7 x 0.3cm, internodes longer than leaves. Flowers solitary, axillary. Calyx lobes unequal. Corolla tube 0.6cm long. Fruit densely hairy.
Ecology. Apparently restricted to semi-deciduous woodland on the limestone escarpment and plateau above Qalansiyah. 350–700m.
Distribution. Endemic.
Conservation status. Endangered (EN B2 a biii). Known only from the north-facing limestone escarpment near Qalansiyah at the western end of Soqotra where it occupies an area of less than 10km^2. It is strongly suspected that this marginal habitat is threatened by the long-term trend of lower rainfall in the region and would be particularly vulnerable to the effects of global warming.

ETHNOGRAPHIC NOTES
Name(s)
See *G. puberula* (also uses).

Gaillonia tinctoria **Balf.f.**
Gaillonia tinctoria var. *glabra* Radcl.-Sm.
Neogaillonia tinctoria (Balf.f.) Lincz.

Subshrub to 45cm. Leaves linear, cylindrical, 1.25–1.7 x 0.1–0.3cm, internodes shorter than the leaves. Flowers solitary, axillary. Calyx lobes unequal. Corolla pale yellow or white, tube 0.5cm long, lobes narrowly lanceolate, 0.2cm long. Anthers and stigma shortly exserted.
Ecology. Rocky places in semi-deciduous woodland on both limestone and granite. 400–1000m.
Distribution. Endemic.
Conservation status. Least Concern (LC). Common in several vegetation types and under no present or perceived threat.

ETHNOGRAPHIC NOTES
Name(s)
– Soqotri: nil or *ḥāhər* ‘dark-coloured’; *ləṭēnuh* ’unmoving’; *di ṣīnem* ‘odoriferous’; (rare) *di bəhɔ̄mə*.
– Soqotri simplified: *hahir*; *liteynuh*; *di ziynem*; *di bihomih.*
Food (livestock). Avoided by most livestock, but they nibble at the stalks in severe drought if there is little else. It has been observed that if eaten by goats they develop halitosis and their belches are also foul-smelling.

Galium L.

Annual herbs, some with stiff, retrorse hairs so that the plants cling to clothing. Leaves whorled. Flowers in small, axillary or terminal cymes. Calyx minute. Corolla small, 4-lobed, with a short tube. Fruit of two 1-seeded mericarps joined together. 300 to 400 species. Temperate and high altitude tropics.

Galium setaceum **Lam.**

Slender annual to 35cm, stems glabrous. Leaves <1 x 0.5–0.9cm, tip acute or obtuse. Flowers in lax, 3 to 9-flowered cymes. Corolla purplish or yellowish-green, 0.5–0.8mm long. Fruit lobes 0.5–0.8mm long, hairy or glabrous.
Ecology. Rare in deciduous shrubland on limestone plateaus. 500m.
Distribution. Canary Is., Mediterranean, N Africa, SW Asia.
Botanical notes. A very variable species. Only subsp. *decaisnei* (Lam.) Boiss. is recorded from Soqotra.

ETHNOGRAPHIC NOTES†
Name(s)
– Soqotri: nil or *ra'adihon* 'small grazing plant'; *liskˈano* 'sticky, clinging'; *məźḥūmhem*, *məźḥˈeymhim* 'crunchy'.
– Soqotri simplified: *ra'adihon*; *liskano*; *midh'hewmhim*, *midh'heymhim.*
Food (livestock). These are small, ephemeral *ru'ud*-type plants which come up with the rains, providing incidental green grazing until they die back. Though palatable to livestock, they are not an important forage for them.

Comparative uses. Young plants have been cooked as a famine food. The seeds of many *Galium* species, including *G. spurium*, are used as a coffee substitute or adulterant. Its roots yield a purple dye, and an infusion is given as an aperient, diuretic, refrigerant and anti-scorbutic. In Iraq the tops of *G. verum* yield a yellow dye which is used to colour butter and cheese, and the roots a red dye used for dyeing wool. The straw is used for mattresses. A decoction of the plant or its juice is used in epilepsy, hysteria and cutaneous troubles. (Al Rawi & Chakravarty 1964; Ambasta 1982; Jain & De Fillips 1991; Kunkel 1984; Lewis & Elwin-Lewis 1977)

†Also apply to *G. spurium*.

***Galium spurium* L.**
Galium aparine sensu auct. non L.
Galium spurium subsp. *africanum* Verdc.

Scrambling, 'clinging' herb to 1.8m, stems with retrorse hairs. Leaves linear to narrowly oblanceolate, 1–3 x 0.1–0.3cm, tip bristled. Flowers in 1 to 9-flowered cymes. Corolla greenish-white, white, green or yellowish, tube 0.3–0.4mm long, lobes triangular, 1mm long. Fruit distinctly stalked, fruit lobes 3–4mm diameter, hairy.

Ecology. Locally common in montane woodland and shrubland in the Haggeher mountains; less commonly on the adjacent limestone (Muqadrihon). 600–1300m.
Distribution. Tropical and S Africa.
Botanical notes. All material from Soqotra is referable to subsp. *africanum* Verdc.

ETHNOGRAPHIC NOTES
Name(s)
See *G. setaceum* (also uses).

Kohautia Cham. & Schltdl.

Herbs. Leaves opposite, linear. Flowers axillary. Calyx 4-lobed, small. Corolla 4-lobed with a cylindrical tube. Fruit a capsule crowned by the 4 persistent calyx lobes. 1 endemic. 31 species. Africa, Arabia, India, Pakistan, Iran, Madagascar, Cape Verde Is.

***Kohautia socotrana* Bremek.**
Oldenlandia schimperi T.Anderson

Herb to 40cm. Leaves linear to narrowly oblong, 3 x 0.1cm. Corolla brownish-yellow, tube 1cm long, lobes linear, 5mm long. Stamens and stigma hidden within the corolla tube. Fruit round, 2.5mm diameter.
Ecology. Widespread and recorded from a variety of habitats including exposed and windswept places on the coastal plains and limestone plateaus, deciduous woodland along wadis and semi-deciduous woodland on the limestone plateaus. Sea-level–600m.
Distribution. Endemic.

Conservation status. Data Deficient (DD). Inadequate evidence is available to make an assessment.

ETHNOGRAPHIC NOTES†
Name(s)
– Soqotri: *kibdˈinuh* 'little liver', 'small hill(ock)'; *ṣˈʌfṣʌf* 'ripples'; *śebəṭˈano* 'well-built'; 'kind of *ˀˈibiṭ*' (more commonly given to spiky, tough grasses such as *Dactyloctenium robechii*).
– Soqotri simplified: *kibdinuh*; *zafzaf*; *shibitano*; 'kind of *'ibit'*.
Food (livestock). Provides incidental forage for small stock but is not important for them. Hungry sheep dig the plants up to eat the roots.

†Also apply to *K. subverticellata*.

***Kohautia subverticellata* (K.Schum.) Mantell**

Annual to 60cm. Leaves lanceolate, 0.2–0.5 x 0.2–0.8cm wide. Flowers axillary, in pairs. Corolla white, yellowish-green or brownish-yellow, tube 3.6–5.8mm long, lobes linear. Fruit subglobose c.2.5 x 3mm.

Ecology. Dry drought-deciduous woodland and succulent shrubland. 100–150m.
Distribution. NE Africa, S Africa, Indian subcontinent.
Botanical notes. Material from Soqotra is all referable to subsp. *eritreensis* (Bremek.) Mantell ex Puff.

ETHNOGRAPHIC NOTES
Name(s)
See *K. socotrana* (also uses).

Kraussia Harv.

Shrubs or small trees. Leaves opposite with a stipular sheath. Calyx with 5 triangular lobes. Corolla 5-lobed, the lobes acute, shorter than the corolla tube. Stigma linear. Fruit spherical, fleshy. 1 endemic. 4 species. S and E Africa.

Kraussia socotrana Bridson

Shrub to 1m, typically growing on cliffs. Leaves oblong-elliptic, 2–6 x 1–2cm, the margins not inrolled. Calyx 5-lobed; lobes triangular, c.1mm long; tube 1.8mm. Corolla white or pink; tube c.5mm long, shorter than or equalling lobes. Fruit spherical, 6mm across.

Ecology. Rare on cliffs along wadis draining south into the Noged plain; in succulent shrubland. 30–100m.

Distribution. Endemic.

Conservation status. Vulnerable (VU D2). Rare, with a very limited area of occupancy.

Botanical notes. *K. socotrana* is quite distinct from the African species of *Kraussia* in its non-winged pollen receptacle, pubescent corolla throat and narrower corolla tube. In overall facies *K. socotrana* shows greatest resemblance to the most geographically remote species, *K. floribunda* Harv. from southern Africa (Bridson 1995).

ETHNOGRAPHIC NOTES

Name(s)

– Soqotri: nil or *zīrəf*, *zūrif* (more commonly given to *Rhus* species); *ˈtaˀabəˈhur* 'coming into leaf at the end of the dry season'; *aḥˈsilihin* 'puffy, waterlogged'.
– Soqotri simplified: *ziyrif*, *zewrif*; *ta'abihur*; *ahsilihin.*

Food (human). There is no tradition of eating the fruit on Soqotra.

Food (livestock). Rarely accessible to livestock, but they eat the foliage if they can reach it. Little is known of its nutritional importance, but it is not gathered by herders for their animals.

Bees / honey. Honeybees are attracted to the fragrant flowers.

Wood. The wood is hard, and slim, straight branches are cut to make the important herding staff.

Comparative uses. The fruits of *K. floribunda* and *K. schlechteri* are eaten. (Ambasta 1982; Jain & De Fillips 1991)

Oldenlandia L.

Erect or cushion-forming herbs. Leaves opposite. Flowers solitary or paired in the leaf axils or in lax, terminal inflorescences. Calyx and corolla 4-lobed. Fruit a capsule. 5 endemics. Up to 300 species. Tropics.

Botanical notes. Five endemic species on Soqotra: *O. balfourii*, sole member of the endemic subgenus *Alsinastrum* Brem., and *O. aretioides*, *O. bicornuta*, *O. ocellata* and *O. pulvinata*, all members of the endemic subgenus *Platyrrhyrnchus* Brem.

ETHNOGRAPHIC NOTES

Comparative uses. *O. corymbosa* is said to cause diarrhoea in goats and sheep, but is cited as a fodder for horses and cattle. In India a decoction is taken for remittent fever, such as malaria and undulant fever, and the juice is applied to the palms and soles to relieve the heat of a fever. The plant is also used to treat nervous depression, and its root is used as an aphrodisiac. In Zaire a decoction of the stems and leaves of *O. affinis* is drunk by women towards the end of labour to provoke strong uterine contractions and shorten delivery time. Two active principles have been isolated: serotonin and a utero-active polypeptide. The root of *O. umbellata* (Indian madder) yields a red dye which is used on calico, wool and silk. The leaves of many species are eaten, and many are also used medicinally. (Dalziel 1948, 1956; Lewis & Elwin-Lewis 1977; Oliver-Bever 1986)

Oldenlandia aretioides Vierh.

Cushion-forming herb to 5cm. Leaves linear-oblong, 5 x 1.2mm, upper part of leaf angled (patent). Flowers solitary. Calyx tube 0.4mm long, lobes 1.6mm long. Corolla not seen.

Ecology. No habitat details available; probably similar to *O. pulvinata*.

Distribution. Endemic.

Conservation status. Data Deficient (DD). Inadequate evidence is available to make an assessment.

Botanical notes. Similar to *O. pulvinata* from which it differs in its smaller leaves and calyx. It possibly represents an extreme form of that species. Not seen since it was first collected in the 19th century at opposite ends of the island near Qalansiyah and near Wadi Faleng.

***Oldenlandia balfourii* Bremek.**
Hedyotis stellarioides Balf.f.

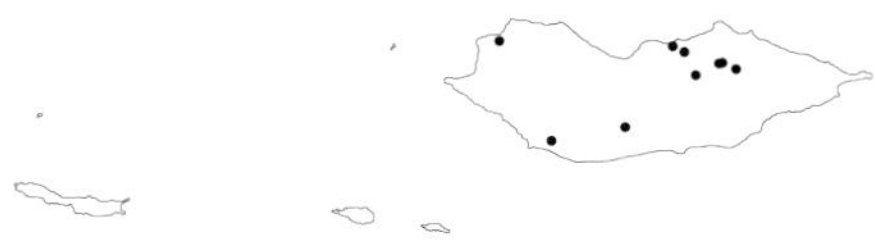

Open branched herb to 20cm. Leaves ovate or ovate-lanceolate, 5–12 x 3.5–8mm wide. Flowers in lax, terminal inflorescences, pedicels longer than leaves. Calyx tube 0.5mm long, obtriangular in fruit. Corolla yellow, tube 3mm long.
Ecology. Semi-deciduous woodland. 350–1370m.
Distribution. Endemic.
Conservation status. Least Concern (LC). Common in several vegetation types and under no present or perceived threat.
Botanical notes. This species occupies an isolated position within the genus and has no close allies. It was placed in the monotypic subgenus *Alsinastrum* Brem. by Bremekamp (1952).

ETHNOGRAPHIC NOTES†
Name(s)
– Soqotri: nil or *tʌ́mhur*, *tímihor* (pl. *támərihon*, *tamə́ritin*) 'little dates' (given to many small plants with delicate and/or fragrant flowers); *kibdínuh* 'little liver', 'small hill(ock)'; *ḳə̄wrɛ́ni* 'two horned' (from the fruit).
– Soqotri simplified: *tamhur*, *timihor* (pl. *tamirihon*, *tamiritin*); *kibdinuh*; *kewreyni*.
Food (livestock). Regarded as excellent forage for sheep and goats. When it is plentiful, animals grow fat on it and give more milk.
Flavouring / food additive. Milk from livestock feeding heavily on the plants has a distinctive sweet taste.

†Also apply to *O. bicornuta*, *O. corymbosa* var. *caespitosa*.

***Oldenlandia bicornuta* (Balf.f.) Bremek.**
Hedyotis bicornuta Balf.f.

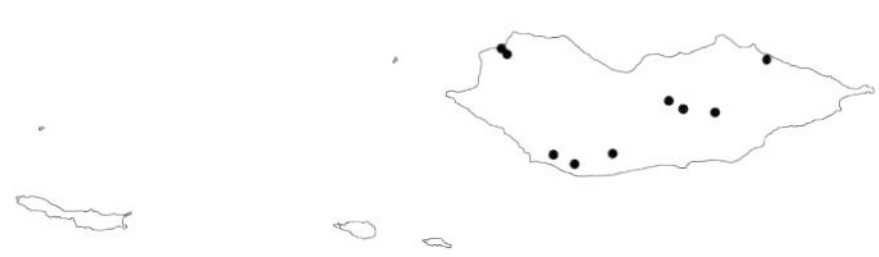

Cushion-forming, rosulate herb. Leaves linear, 15–50 x 1–3mm. Flowers crowded at the centre of the plant. Calyx tube 0.6mm long, lobes ovate-lanceolate. Corolla white, tube 13–14mm long. Fruit obtriangular, 5 x 3mm.
Ecology. Amongst rocks on limestone plateaus. 450–1000m.
Distribution. Endemic.
Conservation status. Least Concern (LC). Common in several vegetation types and under no present or perceived threat.
Botanical notes. Readily distinguished by its rosette-forming annual habit and fruit with a pair of triangular terminal teeth.

ETHNOGRAPHIC NOTES
Name(s)
See *O. balfourii* (also uses).

***Oldenlandia corymbosa* var. *caespitosa* (Benth.) Verdc.**
Oldenlandia caespitosa Hiern
Oldenlandia caespitosa var. *subpedunculata* (Kuntze) Bremek.

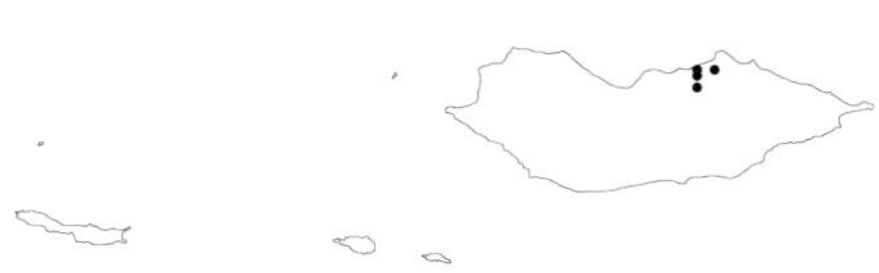

Open branched herb to 30cm. Leaves narrowly elliptic, 7–25 x 1.5–4mm. Flowers usually solitary, occasionally in pairs, at the nodes. Calyx tube 0.7–1mm long, lobes 0.5–1.8mm long. Corolla white or tinged blue, pink or purple, tube 0.6–1mm long, lobes 0.5–1.2mm long. Fruit a capsule to 2.2mm long.
Ecology. Date gardens, by water on coastal plain and deciduous woodland. 50–400m.
Distribution. Africa, Madagascar, Arabia.
Botanical notes. Easily confused with *O. balfourii* and perhaps under-recorded.

ETHNOGRAPHIC NOTES
Name(s)
See *O. balfourii* (also uses).

***Oldenlandia herbacea* (L.) Roxb.**
Oldenlandia heynei Oliv.

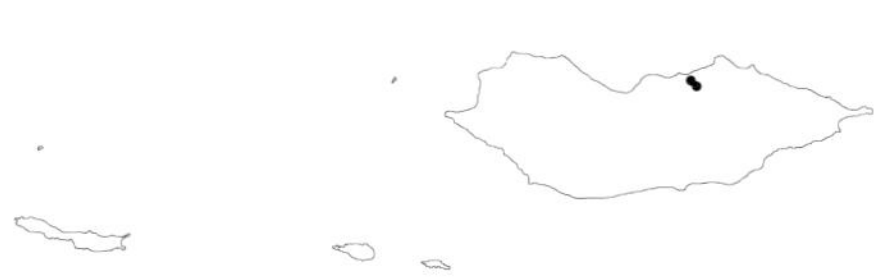

Open branched herb to 60cm. Leaves linear to linear-oblanceolate, 6–55 x 1–3.5mm. Flowers solitary or paired at the nodes. Pedicels 0.8–3cm long, as long as or longer than the leaves. Calyx tube 0.5–1mm long. Corolla white, lilac or mauve (tube sometimes green), tube 0.2–1.1cm long, lobes ovate. Fruit 2–2.5mm long.
Ecology. Semi-deciduous woodland in the Haggeher mountains. c.600m.
Distribution. Tropics and subtropics of Africa and Asia.
Botanical notes. Not seen since it was collected in the 19th century.

Oldenlandia ocellata **Bremek.**

Similar to *O. pulvinata* but the leaves are obovoid, succulent and have an 'eye' of clear tissue at the tip.
Ecology. Shady rocky slopes and ledges; particularly common in cracks on the limestone pavement on Jebel Saleh. 30–500m.
Distribution. Endemic to Abd al Kuri.
Conservation status. Vulnerable (VU D2). Rare, with a very limited area of occupancy.
Botanical notes. Readily distinguished from the other cushion-forming species by the curious clear 'eye' at the tip of the succulent leaves and calyces. This window of clear tissue possibly acts as a lense to focus the rays of the sun on photosynthetic cells around the sides of the semi-globose leaf tips. This is comparable to the leaves of the living stones of the genus *Lithops* (Aizoaceae) from southern Africa.

ETHNOGRAPHIC NOTES
Name(s)
– Soqotri: nil or *ˈtomhor* 'little dates' (Abd al Kuri) (given to many small plants with delicate and/or fragrant flowers).
– Soqotri simplified: *tomhor.*
Food (livestock). Liked by goats and sheep, but not common or accessible enough to be important nutritionally.

Oldenlandia pulvinata **(Balf.f.) Bremek.**
Hedyotis pulvinata Balf.f.
Oldenlandia pulvinata forma *congesta* Vierh.

Cushion-forming herb to 5cm. Leaves linear to linear-oblong, 9–20 x 0.5–1mm. Flowers solitary. Calyx tube 0.8mm long, lobes 3–4.2mm long, hyaline with a green midrib. Corolla mauve, tube 6.5–13.5mm long, lobes 3.5mm long. Capsule glabrous.
Ecology. On soil and rocks on the plains and limestone plateaus. Sea-level–800m.
Distribution. Endemic.
Conservation status. Least Concern (LC). Common in several vegetation types and under no present or perceived threat.
Botanical notes. The commonest cushion-forming plant on the islands. Very variable.

ETHNOGRAPHIC NOTES
Name(s)
– Soqotri: *di ḳeˈṣḳʌṣ* 'cream-coloured'; *kuˈbəhur* 'little ball of wool'; *sədaˈḥan* 'crouched, prostrate'; *mʌˈḥeriś* 'undergrowth, understorey'; *ˀammēri* 'hard working'; *ḥʌlhˈʌləhin* 'dun-coloured'; *nāyim*; *dīrbeb* '(animal) with a pale streak or flash on its coat'; *di həbālino* 'with tiny tamarind-like fruit, spots of a rash''; *rəhaˈmmə* 'grey-coloured' (Samha).
– Soqotri simplified: *di qezqaz*; *kubihur*; *sidahan*; *maharish*; *'ammeyri*; *halhalihin*; *nayim*; *diyrbeb*; *di hibalino*; *rihammih.*
Food (livestock). Palatable and available year round, and consequently one of the most important fodders for small stock in the drier areas of the Archipelago. Even cattle graze it when grasses and other herbs are exhausted. The whole plant is sweet smelling, even when dry. It is said to make small stock fat and to increase their milk. On Samha this is one of the most important plants for small stock, especially sheep.
Flavouring / food additive. Milk from livestock eating a lot has a distinctive sweet taste.
Bees / honey. Honeybees are attracted to the flowers.

Placopoda Balf.f.

Shrubs. Leaves clustered, with stipular sheaths. Flowers 4-merous, in terminal cymes. Calyx 4-lobed. Corolla lobes densely hairy. Fruiting calyx strongly ribbed. 1 species. Endemic.

Botanical notes. Possibly closely related to *Carphalea* (Balfour 1888). One collection (*Miller* 11458 from Samha) has larger fruit than typical and may represent a new species.

Placopoda virgata **Balf.f.**
Placopoda virgata var. *nana* Balf.f.

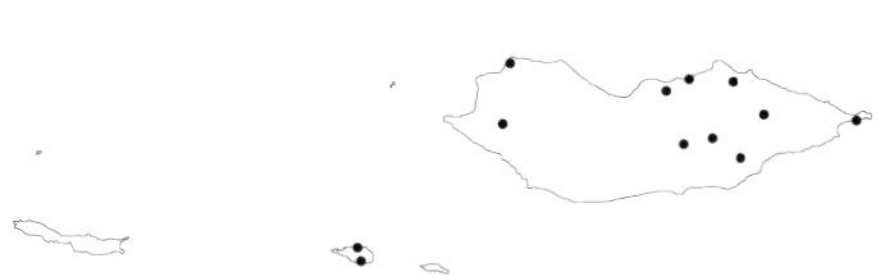

Shrub or subshrub to 2m but sometimes almost prostrate. Leaves elliptic-oblong, <1cm long. Calyx strongly ribbed in fruit. Corolla white; tube 4mm long. Fruit ribbed, 4mm long.
Ecology. Coastal plains and dry foothills; in *Croton socotranus* shrubland and drought-deciduous woodland. 10–600m.
Distribution. Endemic.
Conservation status. Least Concern (LC). Common in several vegetation types and under no present or perceived threat.

ETHNOGRAPHIC NOTES
Name(s)
– Soqotri: *ṭīfiṭ* (pl. *ṭīfʌṭ*) 'making a certain sound to get goats to move off to pasture' (also given to *Dirachma socotrana* and occasionally to *Coelocarpum socotranum* in western Soqotra).
– Soqotri simplified: *tiyfit* (pl. *tiyfat*).
Food (livestock). In low-lying, drier areas of Soqotra this is a key fodder for small stock, and especially goats. It is regarded as a very nutritious feed on which animals grow fat and produce more milk. Young stock are commonly

herded to areas where the plants grow to feed on the small leaves. If alternative forage is available cattle only eat a little of the plant, but camels and donkeys both like it. The drying leaves and leafless stems are an important dry season forage for small stock. During the monsoon months, the strong winds blowing in from the sea make the plants rather salty, but it is still browsed by hungry livestock, and a certain intake of saline fodder is regarded as essential to maintaining good health. On Samha it is regarded as one of the island's most important fodder plants.
Bees / honey. Honeybees are attracted to it, and islanders say that its honey has absolutely no trace of bitterness.
Fuel. The dead wood makes good firewood and does not make too much smoke, so it can be used for cooking. Dried twigs make useful kindling.
Timber. The wood is relatively termite resistant, but the plant is both too valuable as livestock feed and also usually grows too small to be used as timber. Better-grown plants of wetter areas sometimes provide roofing infill.
Wood. Good and hard, and straighter trunks or the larger branches of trees of the wetter areas are used in fencing. The wood is suitable for a variety of implements and it makes good charcoal.
Other. Used to produce a yellow or yellow-orange dye. The roots and/or bark are gathered and then ground with a little water. This mixture is left to steep in a pot until the plant material has softened, when more water is added. Balls of white or pale grey sheep's wool are unwound and lowered into the water. They are weighed down with a rock to keep them below the surface, and are left until the required colour has been achieved, overnight or longer. Some women put the balls of wool into the dye pot. People have also tried this with cotton fibre from the local cotton bushes, but it seems to be less successful.
Notes. Areas where these plants grow in any number are regarded as very superior rangeland for goats, free from the *'uqabah* disease.

Pyrostria Comm. ex Juss.

Shrubs. Leaves opposite, bearing sunken cavities (domatia) beneath, with stipular sheaths. Flowers in 2 or 3-flowered axillary cymes. Calyx 5-lobed. Stigma capitate. Fruit spherical, fleshy. 1 endemic. 10 species. Tropical Africa, Indian Ocean islands.

Pyrostria socotrana (Radcl.-Sm.) Bridson

Canthium socotranum Radcl.-Sm.

Shrub to 2m. Leaves obovate or spathulate-obovate, 2.5–4 x 1.5–2cm, with inrolled margins; stipules triangular, 4mm. Calyx lobes triangular, 1.5mm long. Corolla white, 5-lobed; tube 2.5mm long; lobes lanceolate, 2.5mm long. Fruit glossy green.
Ecology. Submontane semi-deciduous woodland. 850–1000m.
Distribution. Endemic.
Conservation status. Vulnerable (VU D2). Rare, with a very limited area of occupancy.
Botanical notes. Rare, in Haggeher mountains.

ETHNOGRAPHIC NOTES
Name(s)
– Soqotri: *ḥáhər* 'dark-coloured' (referring to the hardness of the wood); (rare) *šəźárher* '(always) green'.
– Soqotri simplified: *hahir*; *shidharher.*
Food (livestock). It is not clear whether the foliage of this rare tree is palatable to livestock or not, but it is thought that the fallen leaves are eaten.
Fuel. The dead wood makes good firewood, but the tree is too rare for it to be much used as fuel.
Timber. Much admired as a building timber, though the plants rarely grow to a size to produce the main beams and supporting pillars. The tree is not common enough to be an important timber on the island.
Wood. The wood is hard and close-grained, and is good for making implements when it can be found.
Comparative uses. The wood of *C. dicoccum* is used to make agricultural implements, combs, toys, posts and rafters, and its bark is given as a febrifuge and applied externally to fractures. (Ambasta 1982; Jain & De Fillips 1991)

Tamridaea Thulin & B.Bremer

Shrubs. Leaves opposite or clustered in 3s, with stipular sheaths. Flowers in terminal cymes. Calyx of 5, unequal lobes. Corolla lobes emarginate. Fruit a dry capsule. 1 species. Endemic.

Tamridaea capsulifera (Balf.f.) Thulin & B.Bremer

Mussaenda capsulifera Balf.f.
Pseudomussaenda capsulifera (Balf.f.) Wernham

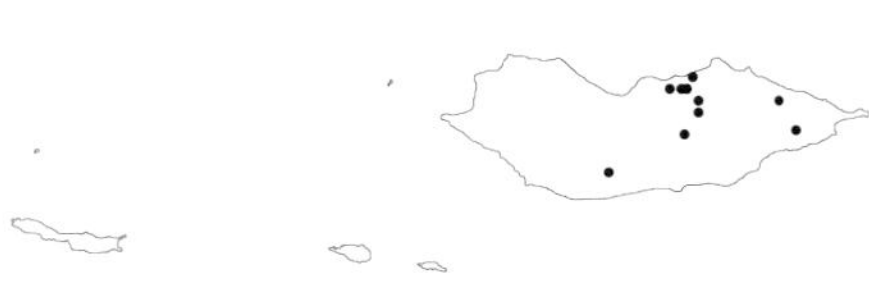

Shrub to 1m. Leaves elliptic to oblong-elliptic, <5.5 x 2.5cm. Calyx 5-lobed; tube 1.5–2mm long; lobes leaf-like, linear to spathulate, 4–8mm long. Corolla white or yellow, 5-lobed, tube 2.5–3.5cm long, lobes notched at the tip. Fruit 4–6.5mm long.
Ecology. Fairly common in the north-central and north-eastern part of the island; rocky places on both limestone and granite, often growing in rock crevices. 100–1100m.
Distribution. Endemic.
Conservation status. Least Concern (LC). Common in several vegetation types and under no present or perceived threat.
Botanical notes. The emarginate corolla lobes and sweet-smelling jasmine-like flowers are characteristic of

this small shrub. It comes into new leaf after rain, drops its leaves in summer and flowers in winter. The species was first described in *Mussaenda* and later transferred to *Pseudomussaenda* but it is now placed in a monotypic genus. The name *Tamridaea* is derived from Tamrida, an old name for the capital of Soqotra (now Hadiboh).

ETHNOGRAPHIC NOTES

Name(s)

– Soqotri: *ḥárśīm* (pl. *ḥarɔ̄śim*) 'speaking rapidly' (also given to the smaller *Grewia* species); (rare) 'kind of *ʾə́biṭə*' (more commonly given to *Cordia* species).

– Soqotri simplified: *harshim* (pl. *harowshim*); 'kind of *'ibitih*'.

Food (livestock). Grazed by all livestock when accessible, but it is not regarded as nutritionally important. It is not important as livestock forage.

Bees / honey. Honeybees are attracted to the fragrant flowers.

Fuel. The dead wood makes good firewood but its comparative rarity means it is little used for this.

Timber. A well-grown tree makes excellent timber.

Wood. Hard and close-grained and good for a variety of implements. Its straighter branches are much sought after for making the important herding staff.

Cleansing. The deliciously fragrant flowers are put amongst clothing, and a flowering sprig is tucked into the waistband or turban. The sniffing of sweet-smelling plants is held to protect from the 'noxious air' that can cause disease, and to help protect against infection. All sweet-smelling plants are endowed with beneficial medicinal and prophylactic qualities.

Notes. Some islanders say that this is one of the *ʿūd al kəsīr* plants; indeed Balfour was given this name (*'od el kseh*) for it.

Comparative uses. The leaves of *Tamridaea* species are edible or used as a substitute for tea. Fruit of *Mussaenda elegans* are edible and sweet. The leaves often have astringent qualities and are used medicinally. (Dalziel 1948, 1956; Kunkel 1984)

Valantia L.

Herbs. Leaves in whorls, without a stipular sheath between the petioles. Flowers inconspicuous, in 3-flowered axillary cymes. Calyx absent. Corolla 3 or 4-lobed. Fruits sessile, concealed by spiny appendages (the recurved pedicels and peduncles). 3 or 4 species. Canary Is., Mediterranean to Iraq and Arabia, Somalia.

Valantia hispida L.

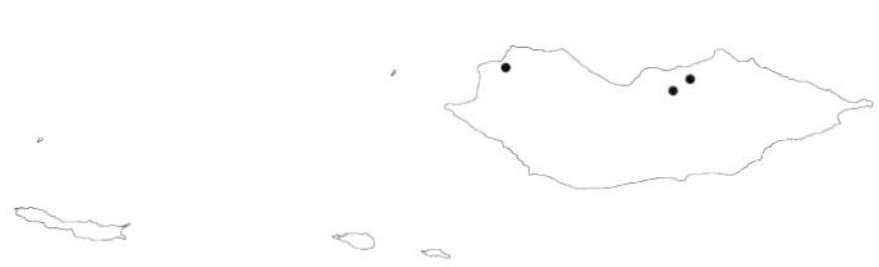

Annual herb to 25cm. Leaves narrowly obovate to oblanceolate-spathulate, 2–10 x 1–3.5mm. Corolla white, purple or yellow, cup-shaped, 1.5–2mm long.

Ecology. On rocks in semi-deciduous woodland. 450–550m.

Distribution. Canary Is., Mediterranean, Iraq, Iran.

ETHNOGRAPHIC NOTES

Name(s)

– Soqotri: nil or *mhā́ṭhíṭən* 'limp, flaccid' (more commonly given to smaller ferns).

– Soqotri simplified: *mhat'hitin.*

Food (livestock). Provides incidental grazing for hungry small stock if they can reach it, but it is not important forage for them.

RUTACEAE

Herbs, subshrubs or trees. Frequently covered in conspicuous glands making the plants strongly aromatic or foetid when crushed. Leaves simple, pinnate or bipinnate, margin entire; stipular spines sometimes present. Flowers in terminal cymes or solitary in leaf axils, actinomorphic. Calyx 4 or 5-lobed, small. Petals 4 or 5, entire or fringed. Stamens 8 to many. Thick basal disc of 8(or 10) glands. Ovary superior, deeply lobed; 4 or 5-celled; ovules many. Fruit a large fleshy berry (orange) or a 2 or 4-lobed capsule. 161 genera. 1700 species. Cosmopolitan, especially tropical.

Citrus L.

Trees. Leaves simple; petiole conspicuously winged; small stipular spines present. Flowers white, solitary in leaf axils. Calyx small, 5-lobed. Petals 5, entire. Stamens many. Fruit a large, fleshy berry. 16 species. Native to S and SE Asia but widely cultivated and naturalised in the tropics.

Citrus aurantium Willd.

Tree to 8m. Leaves glossy, ovate, 5–12 x 2.5–7cm, tip acute to acuminate; stipular spines 0.2–0.5cm long. Flowers c.1.5cm long. Fruit globose, c.4–6cm diameter, orange.

Ecology. Dense montane woodland. 850–1250m.

Distribution. Native of SE Asia, probably Vietnam.

Botanical notes. Scattered in the Haggeher; apparently naturalised; some transplanted to near settlements, and seeds sown in some places.

ETHNOGRAPHIC NOTES
Name(s)

– Soqotri: (tree and fruit) *tiˈnežə* (pl. *ˈtənuž*).
– Soqotri simplified: *tinejih* (pl. *tinuj*).

Food (human). Traditionally, the fruit were never an important food on the island. Some were eaten, or taken as gifts to friends and patrons in the coastal settlements. Herders from the Central Highlands area where the trees grow would occasionally squeeze the juice and give it to friends to use in cooking, or to make a drink for Ramadhan. Nevertheless, herders from the interior remember that many fruit were left to rot on the ground. Nowadays they are much more in use. The juice is squeezed over dishes of grain, fish and meat, and the leaves are added to glasses of tea. A refreshing drink, popular during Ramadhan, is made from the juice with lots of added sugar.

Food (livestock). Hungry camels and goats occasionally browse the young leaves, but the foliage is not palatable to cattle. All livestock eat the fallen leaves if there is little other forage available. Goats and cattle occasionally eat the fallen fruit, and cattle have been observed to wrinkle their noses at the sour taste. Most livestock nibble at the peel.

Bees / honey. Honeybees are attracted to the sweet-smelling flowers.

Flavouring / food additive. The juice is used to flavour sauces.

Fuel. The dead wood makes reasonable firewood, but not for cooking, as it gives off a lot of acrid smoke.

Timber. Since the trees grow only in the Haggeher mountains, where there are plenty of alternative timber trees, it is rarely cut for building.

Wood. Hard and close-grained, and used to make a variety of implements. Straight branches are cut to make the important herding staff.

Medicinal / veterinary. The fruit are given to pregnant women for morning sickness, and to others suffering from nausea. The juice is also given to sickly people as an appetite stimulant. It is squeezed into hot water and drunk as a tonic and restorative. Since the arrival of modern medicine, and the diagnosis and treatment of malaria with bitter chloroquine, some islanders believe that eating the bitter fruit protects against malaria.

Commercial value (past, present). The fruit were not sold in earlier times; more recently they have started to appear in the coastal markets. Young saplings are also occasionally offered for sale.

Other. The juice is squeezed into a pot of milk and left overnight to thicken and sour it. The soured milk is then used to ripen fresh milk before churning it to make butter. Young saplings are commonly transplanted to grow beside settlements or inside compound walls. Successful attempts have also been made to grow the tree from seed, by sowing the seed in earth enriched with dried dung, fencing it securely and watering it regularly. The good, black soil is gathered from under these trees to enrich seedbeds in kitchen gardens. Children play with the fruit as balls.

Notes. Islanders say that the 'male' trees (*tiˈnežə ˈmiʾšər*) produce few or no fruit. Bent (1897) refers to the fact that 'wild oranges, too, are found on Mount Haghier, of a very rich yellow when ripe, but bitter as gall to eat'.

Comparative uses. The flowers, leaves and fruits yield volatile oils used in perfumery. A perfumed oil is also obtained from fresh rinds. The fruit are a rich source of provitamin A and B1 and the peel contains citroflavonoids which have anti-inflammatory, anti-histamine and diuretic actions. Medicinally, in Africa the powdered roots and chopped fruit are mixed with leaves and taken against polio. A cold-water extract of root is taken for an upset stomach, and the root is also used in the treatment of patients believed to be possessed by the Devil. Elsewhere, leaf juice is squeezed into a sore ear, and pieces of fruit are roasted and applied to ulcers, yaws and so on, the dressing changed daily until the sore is healed. In nearby Dhofar, the leaves of the closely related *C. aurantifolia* are added to tea to give a lemony flavour and energising tonic, and the juice of the fruit is squeezed into a painful and discharging ear. Heartburn is treated by drinking hot water to which fresh juice has been added. In northern Oman, where limes have long been cultivated and were an important export in earlier times, the fruit were widely used medicinally. The peel, mixed with myrrh, salt and *ḥilṭīt* (*Ferula* sp.) is taken on an empty stomach first thing in the morning over a period of time as a remedy for pains of the liver and lungs, and to treat tuberculosis. Lime juice is squeezed over the head of a feverish child, and a paste made by grinding a dried lime with salt and water is heated and applied to the site of a deeply embedded thorn or splinter to soften the skin and aid its extraction. (Ambasta 1982; Dalziel 1948, 1956; Jain & De Fillips 1991; Miller & Morris 1988)

Ruta L.

Herbs. Leaves pinnate or bipinnate. Flowers yellow, in terminal cymes, 4-merous (terminal flower 5-merous). Petals fringed. Stamens 8. Fruit a 4-lobed capsule. 7 species. Macaronesia, Mediterranean, SW Asia, Arabia.

Ruta chalapensis L.
Ruta graveolens sensu auct. non L.
Ruta graveolens var. *angustifolia* sensu auct. non Hook.f.

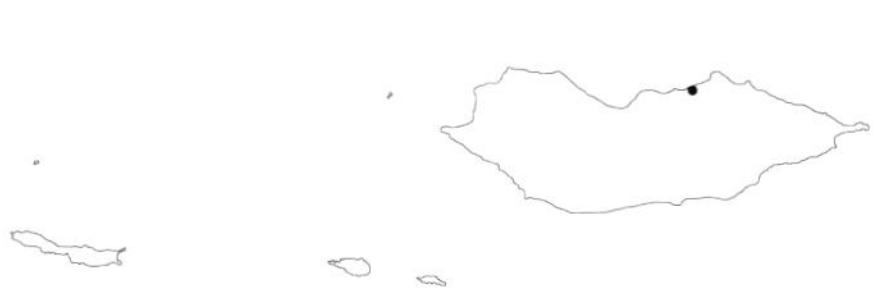

Robust herb. Leaves (including petiole) 4–12cm long; leaflets narrowly elliptic to obovate, 0.5–1.5cm long, conspicuously gland-dotted. Flowers c.1.5cm diameter. Fruit c.0.5cm diameter.

Ecology. Grows in the Haggeher. Rarely cultivated near Hadiboh.

Distribution. Macaronesia, Mediterranean, SW Asia, Arabia.

ETHNOGRAPHIC NOTES
Name(s)

– Soqotri: *šɛ̄də* (pl. *šəˈdɛʾ*) 'calming', 'shared'.
– Soqotri simplified: *sheydih* (pl. *shideh*).

Food (livestock). Not greatly liked by livestock, but hungry goats and cattle will nibble at it if they can reach it. It is not important nutritionally.

Bees / honey. Honeybees are attracted to the flowers.

Cleansing. A sprig is tucked into the clothing or head cloth for its sweet smell. Sprigs are rubbed between the hands to remove or disguise unpleasant smells.

Other. The fruit are threaded onto a string to make a toy necklace, bracelet or anklet.

Comparative uses. Widely used medicinally. By June 1980 it had already appeared 50 or more times in the NAPRALERT computer database of medicinal plants at the Department of Pharmacognosy and Pharmacology, University of Illinois.

Leaves. In India they are used in salads and stews, and as a good substitute for *R. graveolens* (garden rue). In Ethiopia a hydro-alcoholic extract of the leaf is taken orally as an oxytocic and anti-fertility remedy. In southwest Arabia the pulverised leaves are applied externally for snakebite, and used to treat headache, earache, wounds, swellings, rheumatism and certain mental problems. In Yemen the leaves, called *shadhāb*, are chewed with sugarcane for stomach ache, and the fresh leaves are considered to be an aphrodisiac. In northern Israel a decoction of the leaves with olive oil is applied to the forehead to treat a headache or fever by inducing sweating. They are used to make a steam bath to treat coughs and general body aches, and are made into a tisane drunk for stomach aches. A decoction of the leaves is applied as a compress to strained eyes, and fleeces are rinsed in a water extract. In Oman the leaves are used as an anti-rheumatic and against abdominal colic. The leaves and seeds are boiled in oil used as a rub for rheumatism.
Plants. The shoots are used against headache in Yemen, or to procure an abortion and to regulate menstruation. The plant is believed to protect against the Evil Eye and is often carried when visiting a pregnant woman or attending wedding ceremonies. An infusion of the entire plant is taken for colds, intestinal troubles and earache. A hot infusion of the plant is cooled and used as eardrops for earache, and the whole plant is boiled in milk to treat nervousness. Dried, the plant is taken as a snuff for nasal disease, and an infusion is given as nosedrops to treat vomiting and fevers in children and babies. In Oman the whole plant is rubbed on painful joints and other parts to relieve pain. Elsewhere in the Middle East the fresh plant is seen as a scorpion repellent due to its strong smell.
Roots. In northern Israel the crushed roots or a decoction of the leaves with olive oil are applied as a compress to treat partial paralysis. (Ambasta 1982; Bogdan 1977; Boulos 1983; Ghazanfar 1994; Ghazanfar & Al-Sabahi 1993; Goodin & Northington 1979; Jain & De Fillips 1991; Lewis & Elwin-Lewis 1977)

Thamnosma Torr. & Frém.

Herbs or subshrubs. Leaves simple, sessile. Flowers pinkish or purple, solitary in leaf axils. Petals 4, entire. Stamens 8. Fruit a 2-lobed capsule. Very aromatic. 1 endemic. 6 species. N America, S Africa, Yemen, Oman, Somalia.

Botanical notes. A genus with a remarkably disjunct geographical distribution between three distant arid regions: S Africa; tropical NW African, SW Arabia and Soqotra; S USA and N Mexico. Both species on Soqotra are strongly aromatic (Thulin 1999a).

Thamnosma hirschii Schweinf.

Erect herb to 25cm. Leaves laxly arranged along stem, linear, 1–2cm long, upper surface gland-dotted but not conspicuously indented, margins weakly inrolled. Fruit c.0.7cm long. Seeds minutely tuberculate.
Ecology. Locally common in desertic coastal and inland plains; typically in degraded *Croton socotranus* shrubland. 10–200m.
Distribution. Oman, Yemen and Somalia.
Botanical notes. Superficially most similar to the N American species *T. texana*.

ETHNOGRAPHIC NOTES
Name(s)
– Soqotri: *tibərerihon* 'walling'; *tamərhen* 'little dates' (given to a variety of delicate-flowered and sweet-smelling plants); *miśerʌḳ* 'comb, shapely'.
– Soqotri simplified: *tibirerihon*; *tamirhen*; *mishiraq*.
Food (livestock). An important fodder in drier areas of the island and much liked by small stock. It persists through the dry season, so it is important forage at this time. It fattens animals that eat it and increases their milk yield. Cattle also eat it.
Bees / honey. Honeybees are attracted to the flowers.
Cleansing. Much appreciated for its fragrance. Like all sweet-smelling plants, it is endowed with beneficial medicinal and prophylactic qualities. Plants are worn tucked into the hair or waistband, and sprigs are gathered and put among clothing and bedding to perfume it. It is particularly used in bedding for sick people, as its sweet smell is believed to relieve their discomfort and hasten their recovery.

Thamnosma socotrana Balf.f.

Compact woody subshrub to 50cm. Leaves congested at ends of branches, narrowly oblanceolate, 1–2.5cm long, conspicuously indented on upper surface, margins strongly inrolled. Fruit c.1cm long. Seeds spiny.
Ecology. Locally common in low shrubland on granite in the Haggeher mountains. 700–1400m.
Distribution. Endemic.
Conservation status. Vulnerable (VU D2). Rare, with a very limited area of occupancy.
Botanical notes. Distinguished by the aromatic, evergreen leaves with strongly inrolled margins.

ETHNOGRAPHIC NOTES
Name(s)
– Soqotri: *šε̄də* (pl. *šədεˀ*) 'calming, 'shared'; *fegəhun*, *fižihin* 'sprouting, throwing up new shoots' (commonly given to a variety of small, delicate and sweet-smelling plants); *ṣūwəher* (more commonly given to *Ocimum forskolei*).
– Soqotri simplified: *sheydih*; *fegihun*, *fijihin*; *zowiher*.
Food (livestock). Livestock rarely graze it except in a drought.
Bees / honey. Honeybees are attracted to the flowers.
Flavouring / food additive. In small stock that have eaten a lot of this the fat of the carcass is markedly yellow, but the taste is not affected.
Bedding, stuffing. Sprigs are laid among bedding to perfume it, and dried stems are added to the stuffing for bolsters and pillows. It is tucked amongst bedding and

stored clothing to perfume it. It is particularly used in bedding for sick people as its sweet smell is believed to relieve their discomfort and hasten their recovery.
Cleansing. Sprigs are tucked amongst clothing or into the head cloth to lend it a sweet smell.
Other. Hunters used to rub it over the skin to disguise their smell when tracking down wild goats. Sprigs are also rubbed into the coat of a young calf to disguise its own smell and persuade a cow to accept it as her calf. The skin of a dead calf, or of a male slaughtered to conserve its mother's milk, is made into a tulchan stuffed with sprigs. It is then presented to the cow to sniff and some of her milk is sprayed over it to encourage her to let down her milk.

SALVADORACEAE

Shrubs. Leaves opposite. Flowers in lax panicles, actinomorphic, small. Calyx campanulate, 4-lobed. Corolla campanulate, 4-lobed, lobes joined at base. Stamens 4. Ovary superior, 1-celled; style 1; ovules 1, erect. Fruit a globose drupe. 3 genera. 11 species. Old World, especially warm, dry regions.

Salvadora L.

See description of family. 5 species. Africa and tropical Asia.

Salvadora persica L.

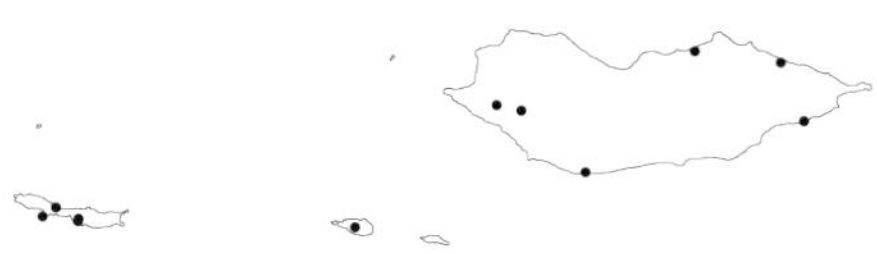

Evergreen shrub forming tangled clumps. Leaves bright green, leathery, ovate to lanceolate, c.3–6 x 0.7–3.5cm, entire. Flowers small, c.0.1cm diameter, in lax panicles. Fruit red, <0.5cm diameter.
Ecology. Dry coastal plains and foothills; generally below 100m on Soqotra but up to 500m on Abd al Kuri. Sea-level–500m.
Distribution. Africa and S Asia.
Botanical notes. Evergreen and often highly visible as the only green plant on the coastal plains in the dry season. It fruits and flowers all year round. On Abd al Kuri it is seen as having steadily increased. There is an interesting belt of dense thickets of *Salvadora* just inland from the coast along the central Noged plain.

ETHNOGRAPHIC NOTES
Name(s)
– Soqotri: *ʾírəhek* (pl. *ʾérʾəheyk*).
– Soqotri simplified: *'irihek* (pl. *er'iheyk*).
Food (human). In the poorer areas of the Archipelago the berries are eaten raw. The fully ripe, darker-coloured ones are preferred.
Food (livestock). Goats and camels in particular browse the foliage, and goats also like the buds, flowers and fruit. Donkeys and sheep also eat the leaves, and the few cattle of the southern plains have become accustomed to it as a dry season fodder. The foliage persists through the dry season and all livestock eat the fallen leaves in the dry season in areas where other forage is scarce, making it a key dry season fodder in the coastal areas where it grows. Herders also come down to the coast and cut leafy stems for their animals at this time. On Abd al Kuri it is a key fodder for goats and sheep, and it is generally seen as a nutritious and fattening feed.
Flavouring / food additive. The flavour of the milk of animals eating a lot of it has a distinctive taste due to the high salt content of the leaves. Before the days of sugar, the ripe fruit were an important source of sweetness for the inhabitants of the outer islands especially, since they have neither date palms nor active honeybees.
Fuel. The dead wood makes acceptable firewood, but it is little used for cooking or in the home as it produces a lot of rather pungent smoke. It is an important fuel on Abd al Kuri where there are few alternatives.
Wood. On Abd al Kuri the dead wood is the principal material used for charcoal. Here too, the outer and inner bark was used to weave the fish traps from the woody stems, which also provided the rigid framework.
Medicinal / veterinary. The leaves are used to treat the condition called *di so'o*, a common condition whose signs are fever, general weakness, and a rash of large, hardy and extremely itchy spots which break out all over the body, including the face. They are ground, added to water and left to steep overnight. In the morning the filtrate is used to wash the patient's body. On Abd al Kuri its charcoal is ground to a powder and used to treat contusions, swellings and inflammation.
Cleansing. As elsewhere in the Middle East, the roots and twigs are used to make toothsticks to whiten and strengthen the teeth, massage the gums and combat halitosis. A finger-length piece of root is cut and a ring of bark at one end removed to expose the inner wood. This is then frayed out to produce a stubby 'brush'.
Commercial value (past, present). Suitably sized pieces of root and twigs are sold as toothsticks.
Pest control. The dead wood is used for insecticidal livestock smudges in coastal areas, as it produces a lot of pungent smoke.
Other. Animals feeding on this can go longer without water.
Notes. A period spent grazing halophytic rangeland at least once a year is considered essential for the health of small stock and camels especially. Such feed initially brings on diarrhoea, seen as beneficial and as 'cleansing impurities from the blood'. When recovered from this, the animals begin to put on weight. The islanders say that if you walk among these plants at dawn your legs become covered in salt.
Comparative uses. The foliage has a high water content (60–70%) and is rich in minerals. The bushes coppice well for sticks and fuel, and are planted to make shelterbelts. The fruit are chewed fresh but are preferred dried, and are stored for later use by nomadic tribes. The shoots and leaves are eaten as a salad. Both fruit and leaves have a pungent and slightly spicy flavour. The seed fat is used for candles and in the manufacture of chocolates, and the seed oil, although inedible, is used in soaps and as a substitute for coconut oil. In the Egyptian desert

an extract from the seeds is used to make a bright green oil which is used as a varnish for finishing wooden tools. A vegetable salt is prepared from the ash. The wood is white, easy to work and takes a good polish. It is prone to attack by various beetle larvae and fungi, and makes a poor fuel. It is used for house poles, bed legs and fencing. The fruits are used in the preparation of a fermented drink. The plant is used to make toothpaste (sold as *Sarakan* toothpaste in Britain): the root-bark contains 27% ash, large amounts of chlorine, and trimethylamine, an alkaloid and a resin. In addition, silica, sulphur and vitamin C have also been found. The high chloride content helps to remove tartar and other stains from the teeth, while the silica by its mechanical actions is helpful in tooth whitening, and the resin forms a coat over the enamel to protect against tooth decay. The presence of trimethylamine possible has a stimulatory effect on the gingiva, and the vitamin C helps to heal spongy and bleeding gums.
Roots. Roots are boiled and drunk with soup as a laxative, and with water as an emetic. In Yemen the roots are used against inflammation and bleeding of the gums.
Bark. Fresh root-bark is used as a vesicant and as an ingredient of snuff. A decoction is taken as a tonic and emmenagogue. It is ground, soaked in cold water and drunk to clean out the stomach and to relieve back pain, or ground to a paste with water which is rubbed on the head for serious febrile illness. The stem-bark is used to treat gastric troubles and as an ascarifuge.
Leaves. A decoction of the green leaves is used to treat asthma and coughs, and a poultice used in piles and tumours. They are cooked and the liquid poured into an animal's mouth to treat diseases affecting the neck and spine. In Yemen the leaf juice is taken internally as a poison antidote. An infusion of the leaves is used to treat urine retention and bilharzia. They are ground and mixed with millet and milk to make a laxative drink, and in Dhofar pulverised fresh leaves are applied to swellings. The shoots are rubbed against the gums to strengthen them and to relieve toothache. The dried leaves are taken in small doses with water for flatulent dyspepsia. Powdered leaves and stem are applied to scorpion stings, and in Dhofar a dressing of powdered, dried leaves is applied to blisters and malignant ulcers. Elsewhere in the Middle East the powdered leaves are mixed with millet flour and honey and made into small balls taken every morning for 40 days as an anti-syphilitic.
Wood. In Dhofar the ash of the dead wood is boiled to a paste with water and oil and applied cold to the skin of livestock with mange, usually camels.
Fruits. These are seen as carminative and diuretic, and are used in biliousness and rheumatism. They are taken to settle the stomach and stimulate the appetite, and are chewed to bring down a fever and promote general health. In the Egyptian desert a syrup is made from ripe fruits and used to grease skins to make them waterproof. An infusion is taken by women to aid conception, and in Yemen the dried fruit are taken as a laxative, and, cooked with cloves, sugar and ginger, are eaten daily to regulate menstruation.
Roots with leaves. These are cooked and drunk for a retained placenta and to soothe coughs.
Seeds. These are seen as purgative, diuretic and tonic, and also yield a fatty oil which is applied to rheumatic swellings. (Baumer 1983; Dalziel 1948, 1956; Davis *et al.* 1983; Fleurentin & Pelt/Fleurentin *et al.* [1982, 1983a,b,c]; Goodin & Northington 1979; le Houerou 1980; Kunkel 1984; Miller & Morris 1988)

SANTALACEAE

Dioecious trees or shrubs. Leaves alternate, simple, entire; stipules absent. Flowers small, greenish, actinomorphic, unisexual, in umbellate cymes or terminal panicles. Perianth 1-whorled, 3(or 4)-lobed. Male flowers with 3(or 4) stamens attached to the perianth tube, opposite the lobes. Female flowers: ovary inferior, 1-locular; ovules 1 to 3; style simple; stigma 3(or 4)-lobed. Fruit a drupe. 36 genera. 500 species. Subcosmopolitan.

Osyris L.

See description of family. 6 or 7 species. Mediterranean and Africa to India.

Osyris quadripartita **Decne.**

Osyris arborea Wall.
Osyris lanceolata Hochst. ex Steud.
Osyris pendula Balf.f.

Glabrous shrub or small tree to 5m, branches sometimes pendent, ridged. Leaves elliptic to elliptic-oblong, 2–7 x 0.5–4cm, acute or mucronate, attenuate below, subsessile, coriaceous. Female flowers solitary or in 2 or 3-flowered cymes; male flowers in 5 to 13-flowered cymes. Perianths yellowish green; lobes triangular-ovate, 1–1.5mm long. Fruits globose, 5–10mm diameter, becoming bright red.
Ecology. Montane thicket in the Haggeher mountains; typically on dry cliffs. 700–750m.
Distribution. Tropical Africa, Arabia and S Asia.

ETHNOGRAPHIC NOTES
Name(s)
– Soqotri: (*di*) *ˈəmlʌḥ* 'salty'.
– Soqotri simplified: (*di*) *emlah.*
Food (livestock). Goats, camels and sheep eat the leaves and stems when they can reach it, though it often grows on cliffs out of their reach. It is not regarded as a nutritious feed, though herders lop it and also gather fallen material to feed their livestock in the dry season.
Fuel. The dead wood makes adequate firewood, and the smaller twigs are gathered as kindling.
Wood. Flexible stems are used to make the 'bit' of the weaning harness worn by young kids and lambs.
Comparative uses. Some species (e.g. *O. alba*) have edible fruits. Species that grow to a greater size provide good timber, and the leaves of many are used in tanning. The seed oil in Santalaceae is interesting in that it provides one of the few known sources of acetylenic fatty acids, whose reactivity may have industrial potential. (Davis *et al.* 1983; Kunkel 1984)

SAPINDACEAE

Trees or shrubs. Leaves alternate, simple or trifoliolate. Flowers unisexual or bisexual. Calyx 4 or 5-lobed, free or partially fused. Petals absent or 4 or 5, free. Stamens 5 to 20. Ovary superior, 1 or 8-celled; ovules 1 or 2 per cell. Fruit a capsule or drupe. 144 genera. 1325 species. Tropical and warm regions.

Allophylus L.

Small tree. Leaves 3-foliolate. Flowers bisexual. Sepals and petals 4, free. Stamens 8. Ovary 2 or 3-celled. Fruit a drupe. c.100 species. Tropical and subtropical regions.

Botanical notes. A taxonomically difficult genus. The flowers and fruit are uniform in all species whereas the leaves are very variable. Leenhouts (1967) took an extreme view, 'lumping' all species into a single superspecies with the 'species' considered as regional 'races'. This is not widely accepted.

Allophylus rubifolius (A.Rich.) Engl.

Small deciduous tree to 3m. Leaves 3-foliolate; leaflets elliptic to obovate, 2–8cm long, rounded or notched at apex, margin irregularly and shallowly crenate or dentate, hairy or glabrescent beneath. Flowers in clusters in raceme-like inflorescences. Sepals c.1mm long. Petals white or cream, c.1mm. Fruit subglobose, 6–8mm diameter, hairy.
Ecology. Widespread and often abundant in semi-deciduous woodland in the higher mountains. 450–1100m.
Distribution. Tropical Africa.
Botanical notes. Soqotran material is referable to var. *rhusiphyllus*, which is also found in NE Somalia. A very variable species.

ETHNOGRAPHIC NOTES
Name(s)
– Soqotri: *zi̍rkīn* (pl. *zərȯ̍kihīn*).
– Soqotri simplified: *zirkiyn* (pl. *zirokihiyn*).
Food (human). The slightly sticky and rather tasteless fruit are edible but are not much liked, and are only eaten by people when they are hungry and in areas where there are few alternative plant foods. Women, especially if pregnant, are said to like them, preferring ripe fruit.
Food (livestock). The foliage persists well into the dry season and so it is an important dry season fodder for livestock, especially goats, in the drier areas of the island. Cattle also browse the shrubs, and leafy branches are lopped for them in the drier cattle-rearing regions. Sheep eat any green leaves they can reach, and also the dead, fallen foliage. It is regarded as a valuable feed for camels. All livestock eat the flowers and fruit. In the dry season herders gather the fallen foliage to feed their animals. The foliage is considered to be superior nutritionally to that of *Rhus* species, another key dry season fodder, though both are seen as maintenance feeds only, especially for cattle. It is always safer when eaten along with other fodder. If cattle eat a lot of the green foliage to the exclusion of other fodders they develop the condition known as *ʿa̍ṣʌb*, a form of temporary weakness in the hindquarters, accompanied by staggering and general malaise. Goats unaccustomed to it can be affected in the same way. In the hotter, low-lying areas where there is little alternative browse, pregnant cows eating a lot can abort their foetus. If camels browse heavily on this at the same time as on *Acridocarpus socotranus*, the toxic effects of the latter are compounded and the animal becomes very ill, often fatally so.
Bees / honey. The fragrant flowers make it one of the best forage plants for honeybees, and it makes good honey, especially in the summer months. This is the case even in the lower foothills areas, which are otherwise poor in bee-forage plants.
Fuel. The dead wood makes good firewood, but the bark is usually removed when cooking in enclosed spaces as it makes a lot of bitter smoke, which can taint food and also cause great discomfort to those around the fire. With the bark is removed, the wood makes little smoke and burns with a good, bright flame.
Timber. A useful wood for building, smaller branches being used for roofing infill, and larger ones for cross beams and rafters. It is quite termite resistant and can be used without removing the bark first.
Leatherwork. The wood burns down to a fine, very pale ash good for processing skins and hides.
Wood. Close-grained and suitable for making smaller items. However, it is not always successful for the larger items, such as the important herding staff, because the dry wood has a tendency to split. Strips of inner and outer bark are retted down and then used to weave fish traps.
Cleansing. Green leaves are crushed and mixed with water to make a paste which is used to cleanse the skin and wash the hair and scalp.
Pest control. The dead wood with the bark left on is used for livestock smudges lit for goats in bad weather. These are lit inside the byre if it is wet, or, if the sky is clear, out of doors in the open pen. The dense and acrid smoke discourages biting insects, and the heat warms and relaxes the animals before they are milked.
Other. The shrubs are usually removed from areas enclosed for cultivation.
Comparative uses. In Dhofar the leaves are ground and applied to boils to draw the pus. In Tanzania a decoction of the root is drunk to treat swellings on the ribs or chest, and a paste prepared from the ground roots is smeared on incisions made in swellings. Elsewhere a medicament is prepared from the roots to ease childbirth. The liquid extracted from boiled roots is given to small children to cure diarrhoea, or is used as a mouthwash to relieve toothache. The wood of *A. africanus* is used in Africa to make arrows and walking sticks, and is so liked by bees that beehives are placed in or close to the trees. The root is used to treat coughs, colds and menstruation, and both roots and twigs are used as chewsticks. It is considered to be diuretic, and is used to treat gastritis, hookworm and venereal diseases. The leaves are taken as a galactagogue, and are crushed in the hand and inhaled to soothe a headache. A decoction of the leaves is given to treat colic and fever in children, both internally and as a lotion. The fruits are eaten as a remedy for tapeworm. (Ambasta 1982; Dalziel 1948, 1956; Heine *et al.* 1988; le Houerou 1980; Jain & De Fillips 1991; Lewis & Elwin-Lewis 1977; Miller & Morris 1988; Oliver-Bever 1986)

Dodonaea Mill.

Evergreen shrub or small tree. Leaves simple, glandular and shiny, the margin entire. Flowers unisexual or bisexual. Sepals 3 or 4. Petals absent. Stamens 7 or 9. Ovary 2-lobed. Fruit a 2 or 3-winged capsule. c.50 species. Tropical and warm regions, especially Australia.

Dodonaea viscosa Jacq.

Dodonaea angustifolia L.f.

Glabrous shrub to 3m; bark black, rough. Leaves narrowly elliptic to narrowly oblanceolate, 3–6 x 1–3cm, tip acute, base attenuate, characteristically shiny because of glandular secretions. Flowers greenish-yellow, in small terminal panicles. Sepals ovate, c.3mm. Ovary with conspicuous style. Fruit round in outline, notched at base and apex, 2(or 3)-winged; wings papery.

Ecology. Widespread but never abundant, in dry woodland. 600–1050m.

Distribution. Old and New World tropics.

Botanical notes. All Soqotran material is referable to var. *angustifolia*. Var. *viscosa* is a plant of coastal plains with a wide distribution in the tropics. The subspecies are sometimes given specific status but Verdcourt in the Flora of tropical East Africa (1998) prefers to treat them as varieties, pointing out that plants of *D. angustifolia* from Jamaica and E Africa would have to be more closely related to plants from India and Australia than to coastal populations (*D. viscosa*) in the same territory.

ETHNOGRAPHIC NOTES

Name(s)

– Soqotri: nil or *śezʌz*; *śʌ̄ṣ*, *śoṣ*, *šoʾoṣ*; (rare) *biššī*.

– Soqotri simplified: *shezaz*; *shaz*, *shoz*, *sho'oz*; *bishi.*

Food (livestock). Cattle and goats eat the foliage only if very hungry, but it is not seen as having any nutritional value. They are more enthusiastic about the flowers and the fruit. During periods of drought goats have been observed to chew at the bark and leafless stems.

Fuel. The dead wood makes adequate firewood, but the plant is not common enough on Soqotra to make it a popular fuel.

Timber. Not common enough to be important as timber, but it has been used in building, and is considered to provide a good and durable timber. It is also resistant to termites.

Other. Islanders have observed that no grass or herbs grow beneath these shrubs, and where they occur in an area of potentially good grazing they dig them up and discard them in order to encourage the grass to grow. If the shrub is too well established for this, they have found that removing the layer of fallen leaves beneath and scraping the soil makes limited growth of herbs and grasses possible.

Comparative uses. The seeds are edible and the fruits have been used as a substitute for hops and yeast. Water and ethanol extracts have been shown to be lethal to earthworms. The shrub is planted as a sand binder and to reclaim marshy land. In Africa the wood is used in building and fencing and to make charcoal. The heartwood is hard and close-grained and has the added advantage of being termite-proof. It is made into tool handles and walking sticks and is used for engraving and turnery. The gummy and resinous leafy twigs can be used as torches. In Iraq the bark is used as a fish poison, and in southwest Arabia the leaves are used in tanning leather. In Dhofar the wood is piled in heaps in cultivated areas and burned as a fertiliser. The wood ash is also added to snuff to make it more pungent. Medicinally the leaves are chewed in Peru as a stimulant and to adulterate coca leaves. Elsewhere they are chewed to relieve toothache, and in Ethiopia a preparation made from the leaves is used to get rid of intestinal parasites and to treat snakebite and skin lesions. In Qatar the plant is a popular panacea with anaesthetic properties, the leaves being used as a febrifuge, sudorific and laxative. They are given to heal wounds without scarring, and for rheumatism and gout, sore throats and haemorrhoids. The leaves are used too in wounds and bone fractures in cattle. In India a decoction of leaves is given to women to check excess menstruation. In Iraq the bark is used to make astringent baths and fomentations. (Al Rawi & Chakravarty 1964; Batanouny 1981; Dalziel 1948, 1956; Goodin & Northington 1979; Heine *et al.* 1988; le Houerou 1980; Kunkel 1984; Lewis & Elwin-Lewis 1977; Miller & Morris 1988; Rizk & El-Ghazaly 1995)

SAPOTACEAE

Trees. Bifid hairs often present. Leaves alternate, entire; stipules absent. Flowers small, solitary or in fascicles in leaf axils, actinomorphic. Calyx 5-lobed. Corolla 5-lobed. Stamens 5, alternating with 5 staminodes. Ovary superior, 5-celled; style 1; ovules 1 per cell, axile. Fruit a berry. 107 genera. 1000 species. Mostly tropical.

Sideroxylon L.

Spiniluma (Baill.) Aubrév.

Trees. Leaves on main branches only. Flowers clustered in axils of leaves. Stamens exserted from corolla; staminodes petaloid. Fruit globose, style persistent. 1 endemic. 100 species. Old World tropics.

Sideroxylon fimbriatum Balf.f.

Tree to 8m. Leaves elliptic, 3–8 x 1.5–4cm, base acute, tip acute to obtuse, reticulate veins ± conspicuous, glabrous or sparsely pubescent on both surfaces. Flowers c.3mm long. Fruit c.0.8cm diameter.

Ecology. Uncommon, in dense montane woodland in the Haggeher mountains. 550–1300m.

Distribution. Endemic.

Conservation status. Vulnerable (VU D2). Rare, with a very limited area of occupancy.

ETHNOGRAPHIC NOTES

Name(s)

– Soqotri: *ṣāmaᶜ*, *ṣɔ̄maᶜ* (pl. *ṣamáᶜo*) 'with no, or vestigial, ears'; *ḥáhər* 'dark-coloured' (referring to its excellent timber); (rare) *bíšši*, *bíšᵊhi*; (rare) *ríkəbə* 'riding high' (also given to *Jasminum fluminense*, *Cocculus pendulus*, and *Cryptolepis* and *Secamone* species).

– Soqotri simplified: *zama'*, *zowma'* (pl. *zama'o*); *hahir*; *bishi*; *bish'hi*; *rikibih.*

Food (livestock). Goats browse the leaves, but they are not considered important or nutritious forage. The greenery persists into the dry season, and herders lop leafy twigs for them in the dry season, and gather the fallen, dried leaves. Cows only eat the foliage when very hungry. Camels browse the trees and the foliage is believed to increase their milk yield.

Bees / honey. The rather insignificant flowers are liked by honeybees, and it is one of the more important bee plants of the Central Highlands.

Fuel. The dead wood makes a good fuel which produces little smoke, making it good for cooking and heating enclosed spaces.

Timber. Highly valued as timber, especially as it is relatively termite resistant and does not have to be debarked before use. However, it is not as highly thought of as *Spiniluma discolor*. The larger branches are used for the main roofing beams, rafters and supporting pillars.

Wood. Close-grained and hard, and used to make a variety of implements.

Medicinal / veterinary. In areas where goats are accustomed to the feed, fresh green leaves are gathered and fed to a sickly or parturient goat to restore it to good health.

Comparative uses. The fruit of many species are edible. The latex of *S. glabrescens* in Malaysia is used as chewing gum, and that of *S. attenuatum* and *S. kaernbachianum* used as temporary filling for tooth cavities. The wood of *S. inerme* is used in building boats and bridges. (Ambasta 1982; le Houerou 1980; Jain & De Fillips 1991; Kunkel 1984; Lewis & Elwin-Lewis 1977)

Spiniluma (Baill.) Aubrév.

Trees. Leaves often clustered on short branchlets. Flowers solitary in axils of leaves. Stamens included in corolla; staminodes narrowly triangular. Fruit elliptic. 1 endemic. 2 species. NE Africa and Arabia.

Spiniluma discolor (Radcl.-Sm.) Friis

Sideroxylon discolor Radcl.-Sm.

Tree to 5(–10)m. Leaves obovate, 2.5–4.5 x 1.5–2.5cm, base acuminate, tip rounded, reticulate veins conspicuous, ± glabrous above, reddish and hairy beneath. Flowers c.3mm long. Fruit c.1cm long.

Ecology. Uncommon, in dense montane woodland in the Haggeher mountains. 600–1000m.

Distribution. Endemic.

Conservation status. Vulnerable (VU D2). Rare, with a very limited area of occupancy.

ETHNOGRAPHIC NOTES

Name(s)

– Soqotri: *ḳāśɔ̄bə*, *ḳeśyɔ̄bə* (pl. *ḳāśóyb*) 'hard, stony'; *ḥáhər* 'dark-coloured' (referring to its excellent timber); (rare) *ṣɔ̄maᶜ ḳérṭʌṭ* '*Sideroxylon* with small, crinkled leaves'.

– Soqotri simplified: *qashobih*, *qeshyobih* (pl. *kashoyb*); *hahir*; *zoma' qertat.*

Food (livestock). Hungry goats eat a little of the foliage, but it is not considered to be a significant or useful forage. As with other palatable evergreen plants, leafy stems are lopped by herders for their goats in the dry season. Cows eat a few of the fallen, dried leaves when very hungry, but otherwise ignore it. Camels browse the trees and the foliage is believed to increase their milk yield.

Bees / honey. The rather insignificant flowers are liked by honeybees, and this is one of the more important bee plants of the Haggeher mountains.

Fuel. The dead wood makes a good fuel which produces little smoke and is good for cooking and heating enclosed spaces.
Timber. Considered one of the best timbers of Soqotra. It is hard work to cut, and is said to 'ring like iron when struck'. It is termite-proof, so does not have to be debarked before use. The larger branches are used for king posts, cross beams, rafters and supporting pillars. So many trees have been cut for timber in recent years that their number has fallen sharply.
Wood. Close-grained and hard, it is used to make a variety of implements.

SCROPHULARIACEAE

Herbs or shrubs. Leaves opposite or alternate; without stipules. Flowers bisexual, in racemes, sometimes axillary. Corolla fused into a short or long tube below, 5-lobed, regular or 2-lipped, sometimes spurred. Stamens 2 or 4, joined to the petals. Ovary superior, 2-locular, with axile placentation. Fruit a capsule containing numerous seeds. 222 genera. 4500 species. Cosmopolitan, especially temperate regions.

Anticharis Endl.

Herbs. Leaves alternate, entire. Flowers long-pedicellate, axillary, forming terminal leafy racemes. Corolla 5-lobed, funnel-shaped, very weakly 2-lipped with a somewhat dilated tube; stamens 2. c.10 species. Africa to Malaysia.

Anticharis arabica Endl.

Anticharis glandulosa sensu auct. non Asch.

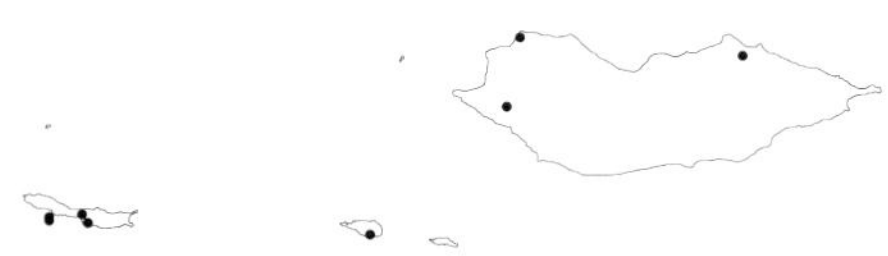

Erect, branched annual to 10cm, sticky, glandular-hairy throughout. Leaves linear to elliptic, c.2cm long. Corolla white, pink or lilac; corolla tube 6mm long.
Ecology. Desertic coastal plains. Sea-level–150m.
Distribution. NE Africa and Arabia.
Botanical notes. The flowers are white on Abd al Kuri and Samha but generally pink on Soqotra.

ETHNOGRAPHIC NOTES
Name(s)
– Soqotri: nil or *ráʿadihon* 'small grazing plant'.
– Soqotri simplified: *ra'adihon.*
Food (livestock). This is one of the small, ephemeral *ru'ud* plants which come up with the rains, providing incidental green grazing until they die back. Though palatable to livestock, it is not an important forage for them. It is more highly valued on the outer islands.
Comparative uses. In Yemen the fruit of *A. linearis* and *A. khazma* are used as an ophthalmic antiseptic. (Fleurentin & Pelt/Fleurentin *et al.* [1982, 1983a,b,c])

Aptosimum Burch. ex Benth.

Herbs. Leaves alternate, appearing whorled. Flowers axillary, sessile, regular, funnel-shaped with spreading lobes. c.40 species. Tropical and S Africa.

Aptosimum pumilum Benth.

Densely tufted herb to 10cm, sometimes ± stemless. Leaves linear, <5cm, conspicuously fringed at the base. Flowers crowded in the leaf axils at the base of the plant, nearly concealed by the leaves. Calyx lobes linear, fringed. Corolla lilac or pale blue. Fruit c.0.5mm long.
Ecology. Coastal plain, on sand in *Salvadora* woodland. 10m.
Distribution. Drier parts of tropical Africa and Arabia.

ETHNOGRAPHIC NOTES
Name(s)
– Soqotri: nil or *chemhəlínitin.*
– Soqotri simplified: *chemhilinitin.*
Food (livestock). See *Anticharis arabica.*

Bacopa Aubl.

Herbs. Leaves opposite. Flowers long-pedicellate, solitary in the leaf axils. Corolla cup-shaped, lobes subequal, weakly 2-lipped. 60 species. Tropics and subtropics.

Bacopa monnieri (L.) Pennell

Herpestis monnieria (L.) Humb., Bonpl. & Kunth

Glabrous herb; stems creeping, rooting at the nodes. Leaves oblanceolate, slightly fleshy, <1cm long. Corolla white, pink or pale blue, 6–8mm long. Pedicel lengthening and becoming recurved in fruit.

Ecology. In mud by permanent water. Sea-level–500m.

Distribution. Widely distributed in warmer parts of the world.

Botanical notes. Probably under-recorded on the distribution map.

ETHNOGRAPHIC NOTES

Name(s)

– Soqotri: nil or *maḥlūbəb* 'like buttermilk'; *məźḥūmhen*; 'crunchy'; *ḳorźʌʾ* (general term for perennial Gramineae and Cyperaceae growing in damp ground).

– Soqotri simplified: *mahluwbib*; *midh'huwmhen*; *qordha.*

Food (livestock). See *Anticharis arabica.*

Comparative uses. It is cooked and eaten in soups and in India is said to improve the intellect. Medicinally, it is used to treat epilepsy, insanity and other nervous conditions, and the leaves are used as a diuretic and aperient. (Ambasta 1982; Jain & De Fillips 1991)

Camptoloma Benth.

Densely glandular perennial herb. Leaves alternate, petiolate. Flowers in terminal racemes. Calyx divided to the base. Corolla 2-lipped. 3 species. Canary Is., Somalia, Arabia, Namibia.

Botanical notes. A genus with a remarkably disjunct geographical distribution with one species confined to the Canary Islands, another to the western deserts of Angola and Namibia and the third to the coastal fringes of Somalia, S Arabia and Soqotra. The species are remarkably similar in appearance with their round leaves and the long, glandular hairs on all vegetative parts.

Camptoloma lyperiiflorum (Vatke) Hilliard

Camptoloma villosa Balf.f.
Chaenostoma liperiflorum (Vatke) Wettst.
Chaenostoma oxypetalum Wagner & Vierh.

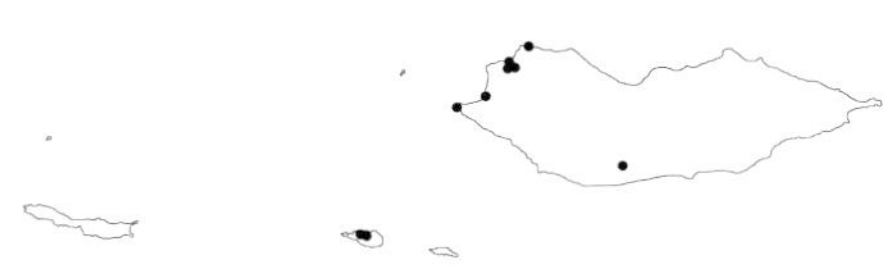

Ascending herb, all parts densely glandular, wet to the touch. Leaves orbicular, <3cm long, rounded with crenate-serrate margins and cordate base. Calyx lobes 0.5cm long. Corolla white, 0.8–1.5cm long.

Ecology. Typically found in the shade of overhanging rocks on seaward-facing cliffs. Sea-level–650m.

Distribution. Somalia, Yemen (Mahra) and Oman (Kuria Muria Islands).

Botanical notes. Often rather etiolated in appearance when growing in deep shade. Immediately recognisable by its indumentum of long, glandular hairs which make the plant feel wet, almost soapy, to the touch. Balfour (1888) records it from cliffs above 900m in the Haggeher mountains, but this seems unlikely and it is probable that Balfour has given the wrong locality.

ETHNOGRAPHIC NOTES

Name(s)

– Soqotri: nil or *mḥɔ̄wgib* '(of) enclosed areas'.

– Soqotri simplified: *mihowgib.*

Food (livestock). Not normally palatable to livestock, but is occasionally eaten by very hungry goats.

Campylanthus Roth

Subshrubs. Leaves alternate. Calyx lobes 5, free to base. Corolla tubular with double bend just above exsertion from calyx. Stamens 2. Seeds winged. 15 species. Canary Is., NE Africa, Arabia, Pakistan.

Botanical notes. A genus with a markedly disjunct geographical distribution between NE tropical Africa and Arabia and the Canary and Cape Verde Islands. With a few exceptions, all species have very restricted distributions.

Campylanthus spinosus Balf.f.

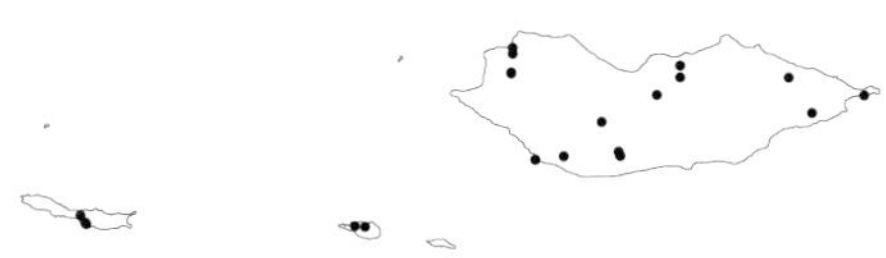

Spiny, densely branched subshrub to 60cm. Leaves linear or oblanceolate, silver-hairy, <0.5cm long. Flowers 2 or 3 at the end of the branches, below the spine. Corolla <1cm, white, pink or mauve, with spreading lobes. Fruit 3–5mm long.

Ecology. Widespread in dwarf shrubland; typically in exposed places on the coastal plains and limestone plateaus. Sea-level–700m.

Distribution. Somalia.

Botanical notes. Variable in leaf size and shape and

flower size. The variation in leaf size led to the description of *C. spinosus* var. *kuriensis* Smith from a single collection. Since then intermediates have been found, and this variety has not been retained at any rank. However, it is interesting that plants from Abd al Kuri have white flowers whereas those from Soqotra and Samha have pink or mauve flowers.

ETHNOGRAPHIC NOTES
Name(s)
– Soqotri: *ḥámḥīm* '(breeding male goat) to become sexually aroused, as when fattened on a diet of this)'; *ḥʌ́mḥīm di aḥmīri* 'hamhiym for donkeys'; *mírigo* 'gap-toothed' (referring to its toughness); *ṣʌ́frer* 'many-flowered'; *zēydiro* 'fat-forming' (given to various prickly or spiny, low, grey-coloured shrubs, such as *Indigofera marmorata* and *I. pseudointricata*; also to tough perennial grasses such as *Dactyloctenium* or *Panicum rigidum*); *zímgilo* 'swarm of bees'; nil (Abd al Kuri); *mərégə* (Samha).
– Soqotri simplified: *hamhiym*; *hamhiym di ahmiyri*; *mirigo*; *zafrer*; *zeydiroh*; *zimgilo*; *miregih*.
Food (livestock). Highly regarded fodder for all livestock in the drier areas of the island. It is much liked by both camels and donkeys; even the cattle of the drier areas graze it. It is seen as being particularly good for goats, and sheep eat the more tender shoots. It is a good dry season fodder as it persists through the dry months. However, if either camels or goats eat a lot of it, to the exclusion of other fodder, they develop a weakness in the hindquarters, sometimes even temporary paralysis, and become tottery and unable to walk any distance until the effect has worn off. They recover after a few days if they are moved to another area with different forage. On the outer islands it is one of the key fodders for sheep and goats.
Fuel. Dead dry plants are collected and stored for use as tinder and kindling. If a bright light is wanted, handfuls are put on the fire.

Craterostigma Hochst.

Rosette herbs. Leaves in a basal rosette. Flowers in a terminal raceme, arising from the centre of the basal rosette of leaves. Corolla strongly 2-lipped, the upper small, the lower broad and spreading. 10 to 15 species. Tropical and S Africa, Madagascar, Arabia, India.

Botanical notes. This genus is in need of revision (Wood 1997).

Craterostigma pumilum Hochst.

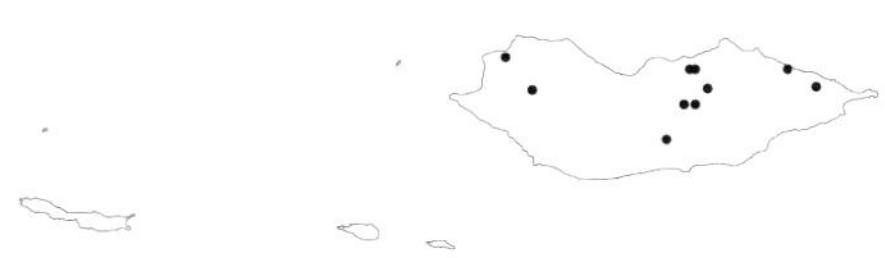

Perennial herb to 60cm. Leaves ovate to elliptic, 1–6cm long. Calyx tubular, c.6mm long. Corolla blue and white bearing two yellow protuberances on the underside of the upper lip. Fruit broadly ovoid.
Ecology. Widespread and locally common in woodland clearings; comes up after rain in shallow depressions. 350–1100m.
Distribution. S and E Africa.

ETHNOGRAPHIC NOTES
Name(s)
– Soqotri: nil or *di ṣʌfəríríhon* 'many-flowered'; *di ḥāmímḥim*; 'type of *ḥárˀhi*' (an orchid).
– Soqotri simplified: *di zafiririhon*; *di hamimhim*; 'type of *har'hī*'.
Food (livestock). The aerial parts provide incidental forage for livestock, but it is not important forage.

Graderia Benth.

Shrubs. Leaves opposite. Flowers solitary, axillary. Corolla with short recurved lobes and a long, rather dilated tube. Anthers 4. 1 endemic. 4 species. S Africa.

Botanical notes. A genus with an interesting disjunct geographical distribution: 2 species in S Africa, 1 in Tanzania, and 1 on Soqotra. This pattern of distribution (cf. *Chorisochora*, *Wellstedia* and *Camptoloma*) offers evidence for the existence of a dry corridor during the Pleistocene connecting NE and SW Africa. It has been argued that this corridor, during dry periods in the past, would have allowed migration between these regions.

Graderia fruticosa Balf.f.

Shrub to 2m, covered with short, hard, hooked hairs. Leaves brittle, drying black, oblong-elliptic, 1.8–3 x 1–1.3cm, scabrid with recurved prickles. Calyx cup-shaped, 0.8–1cm long. Corolla bright red, 3–4cm long.
Ecology. Rather uncommon in shrubland on rocky slopes and ridges in the high Haggeher mountains. 700–1300m.
Distribution. Endemic.
Conservation status. Vulnerable (VU D2). Rare, with a very limited area of occupancy.

ETHNOGRAPHIC NOTES
Name(s)
– Soqotri: nil or *širémhən* 'small plant, herb'; *di feríˀirihun* 'with branching stems, of butterflies'; *bíšši, bišˀhi* (commonly given to cliff-growing plants or those with striking flowers).
– Soqotri simplified: *shiremhin*; *di feri'irihun*; *bishi, bish'hi*.
Food (livestock). Not normally accessible to livestock, but goats will eat the foliage if they can reach it. It is not regarded as nutritious feed.
Bees / honey. Honeybees are attracted to the flowers.

Lindenbergia Lehm.

Annual or perennial herbs. Leaves opposite. Flowers in leafy terminal racemes. Corolla 2-lipped; the mouth of the corolla tube closed by a swollen, boss-like protuberance on the lower lip. Stamens 4. 1 endemic. 12 species. NE Africa, Arabia, Asia.

Lindenbergia sokotrana Vierh.

Lindenbergia kuriensis Vierh.
Lindenbergia paulayana Vierh.
Lindenbergia sinaica sensu Balf.f. non Benth.

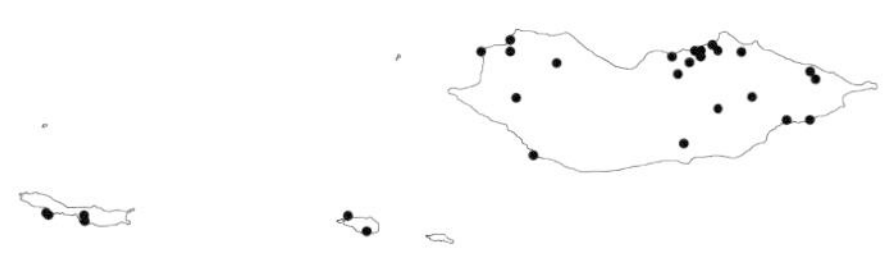

Woody-based perennial herb to 30cm; pubescent throughout. Leaves elliptic to ovate, 2–20 x 2–15mm, crenate-serrate or entire. Corolla yellow, with a red upper lip, 7–11mm long. Fruit 4–6mm long.
Ecology. Widespread in dry places on the foothills and coastal plains. Sea-level–750m.
Distribution. Endemic.
Conservation status. Least Concern (LC). Common in several vegetation types and under no present or perceived threat.

Botanical notes. Usually an annual, but under favourable conditions it can live for several years. The type of *L. paulayana* is one of these perennial forms and the type of *L. kuriensis* is an annual form from Abd al Kuri. *L. sokotrana* is closely related to the widespread species *L. indica*.

ETHNOGRAPHIC NOTES
Name(s)
– Soqotri: nil or *rúʿud* 'grazing plant'; 'like *gúmuṣ*' 'like *Malva parviflora*'.
– Soqotri simplified: *ru'ud*; 'like *gumuz*'.
Food (livestock). Provides incidental forage for livestock, but is regarded as having little nutritional value, except when fresh and green. On the outer islands, it is more highly regarded.

Misopates Raf.

Herbs. Leave alternate. Flowers subsessile in terminal racemes or solitary in leaf axils. Corolla 2-lipped, the mouth of the corolla tube closed by a swollen protuberance on lower lip, tube pouched at base. Capsule with unequal loculi, upper locule indehiscent, lower opening by 2 apical pores. 7 species. Mediterranean, Cape Verde Is., Ethiopia and NW India.

Botanical notes. The genus lacks vegetative variation and is often treated as a single species. There is, however, considerable variation in seed morphology.

Misopates orontium (L.) Raf.

Antirrhinum orontium L.

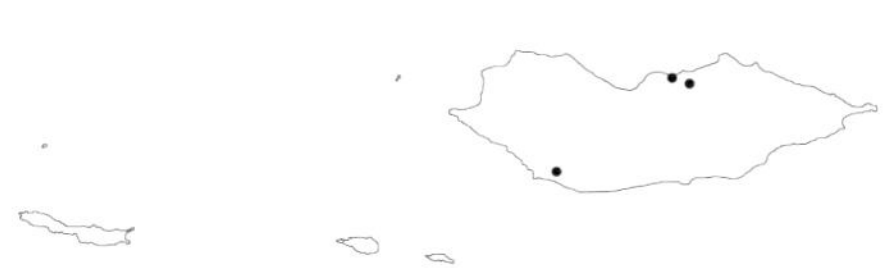

Annual herbs to 30cm; stems erect, little branched, glabrous to sparsely hairy. Leaves linear, 3–4cm long. Corolla pink or cream, 7–9mm long. Capsule 7–8mm long.
Ecology. Grassland and clearings in semi-deciduous woodland. c.600m.

Distribution. Widespread weed, widely naturalised in temperate areas.

ETHNOGRAPHIC NOTES
Name(s)
– Soqotri: nil or *rúʿud* 'grazing plant'.
– Soqotri simplified: *ru'ud.*
Food (livestock). This is one of the small and ephemeral *ru'ud* plants which come up with the rains, providing incidental green grazing until they die back. Though palatable to livestock, it is not important forage for them.

Nanorrhinum Betsche

Herbs. Leave alternate. Flowers subsessile, solitary in leaf axils. Corolla 2-lipped, the lower lip 3-lobed, the mouth of the corolla tube closed by a swollen protuberance, spurred. Capsule subglobose, the loculi equal, dehiscing by a lid. 1 endemic. 10 species. N tropical Africa and Arabia.

Botanical notes. *Nanorrhinum* is a recent generic split from *Kickxia* based on operculate (cap-like) dehiscence of the capsule; this split is supported by molecular studies. See Ghebrehiwet (2000).

Nanorrhinum hastatum (Benth.) Ghebrehewit

Kickxia hastata (R.Br. ex Benth.) Dandy
Linaria hastata R.Br.

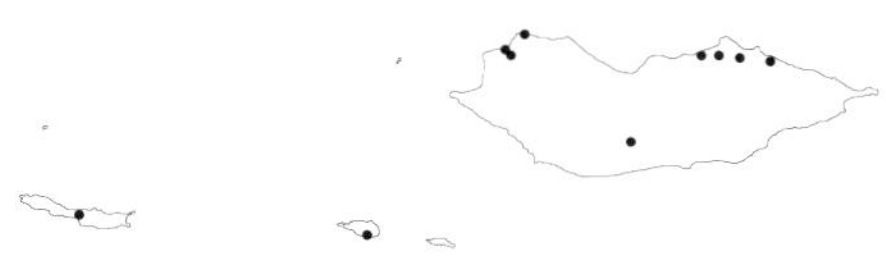

Erect annual to 50cm, glabrous except for the flowers and base of the stem. Basal leaves lanceolate to suborbicular, <12mm long, upper leaves linear, <14mm long, often hastate. Corolla 4.5–7mm long, spur 3.5–4.5mm long.
Ecology. Locally common in succulent shrubland and dwarf shrubland. Sea-level–650m.
Distribution. NE Africa, Arabia.
Botanical notes. A very variable species which it may be possible to subdivide into a number of infraspecific taxa (Sutton 1988).

ETHNOGRAPHIC NOTES
Name(s)
– Soqotri: *ṣeˈfərher, di ṣefərˈirihon* 'many-flowered'; *mˈʌhəriš* 'undergrowth, understorey'; *di ʾārdəhʌn, di hārdəhʌn* 'a spreading area of herbage' (given to other plants that are prickly or have stellate hairs and cause *'uqabah* in small stock).
– Soqotri simplified: *zefirher, di zefiririhon*; *maharish*; *di aridihan, di hardihan.*
Food (livestock). Liked by small stock when young and green, but not regarded as important forage.
Notes. Can cause *'uqabah* in goats unaccustomed to it.
Comparative uses. Many *Kickxia* are not palatable to livestock, and some, such as *K. incana*, are actually toxic. In Dhofar *K. qaraica* is much liked by goats and is often called 'little dates for she-goats' because of this. (Ambasta 1982; Dalziel 1948, 1956; Jain & De Fillips 1991; Miller & Morris 1988)

Nanorrhinum kuriensis (Radcl.-Sm.) Ghebrehiwet

Kickxia kuriensis Radcl.-Sm.

Annual herb to 15cm, glandular hairy. Upper and lower leaves similar, all ovate to suborbicular, sometimes hastate, <1cm long. Corolla 6–7mm long, yellowish-white, spur 2.5–3mm long, sometimes white. Fruit 2.5–4mm.
Ecology. Locally common dry rocky slopes and sandy plains. Sea-level–500m.
Distribution. Endemic.
Conservation status. Vulnerable (VU D2). Rare, with a very limited area of occupancy.

Botanical notes. Common on the foothills of Jebel Saleh on Abd al Kuri. One specimen (*Miller et al.* 14177) is quite different from the above description in habit, being tufted and much smaller than the other collections. There do not appear to be any other significant differences. This collection was growing on an exposed sandy plain whereas typically this species is found growing in the shade of boulders. Also found near the main settlement on Samha.

ETHNOGRAPHIC NOTES
Name(s)
– Soqotri: nil or *ḳōnhˈitin* 'little threads'; *tˈomhor* 'little dates' (Abd al Kuri) (also given to other small and delicate plants with attractive flowers).
– Soqotri simplified: *qownitin*; *tomhor.*
Food (livestock). On the outer islands regarded as good and nutritious fodder for sheep and goats.

Schweinfurthia A.Braun

Herb. Leaves alternate. Flowers long-pedicellate, solitary in leaf axils. Corolla 2-lipped; the mouth of the corolla tube closed by a swollen, boss-like protuberance on the lower lip. Capsule with unequal loculi, the upper loculus indehiscent, the lower rupturing irregularly at the base. 6 species. NE Africa to India.

Schweinfurthia pedicellata Benth. & Hook.

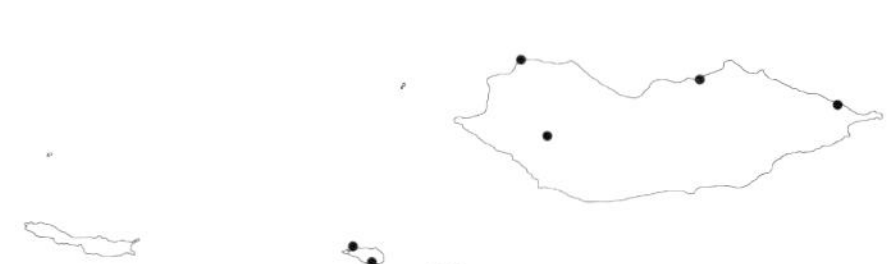

Glabrous annual herb; stems erect to 40cm. Leaves linear-oblong, 2–3cm long. Corolla white or pink, 6–9mm long. Capsule 4–6mm long.

Ecology. Comes up after rain on coastal and inland plains. Sea-level–100m.
Distribution. NE Africa, SW Asia (outliers in Tanzania and the Comoros).
Botanical notes. Sometimes treated as a synonym of *S. pterosperma*. The two taxa can be distinguished, however, by differences in habit and seed and fruit characters (Sutton 1988).

ETHNOGRAPHIC NOTES
Name(s)
– Soqotri: nil or *gilgilhítin* 'trembling in the breeze' (Samha); 'type of *ʿafólhi*' (*Leucas martinicensis*).
– Soqotri simplified: *gilgilhitin*; 'type of *'afolhi*'.

Food (livestock). Provides incidental forage for livestock, but is not regarded as a particularly useful forage. On Samha, however, it is regarded as a good fodder for both goats and sheep.

Scrophularia L.

Herbs with square stems. Leaves opposite. Flowers in axillary cymes. Corolla 5-lobed, 2-lipped with a short, swollen, almost globular tube. Stamens 4. 200 species. Widespread in N temperate regions but extending to tropical America.

Scrophularia arguta Aiton

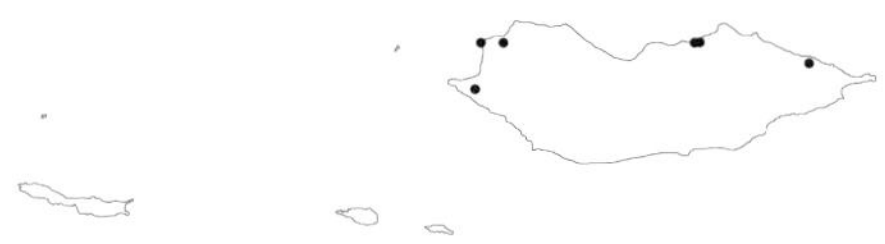

Erect annual herb to 50cm, whole plant viscid. Leaves ovate to elliptic, 4–5cm long, the margin serrate; corolla reddish brown, 6mm long. Capsule ovoid, beaked.
Ecology. Moist, shady places on foothills, often near habitation. 100–550m.
Distribution. Tropical NE Africa and Arabia.

ETHNOGRAPHIC NOTES
Name(s)
– Soqotri: *ṣafā́ḳa di ʾíləhe* '*Tragia* or *Urtica urens* for cattle'; *ʿážbib* 'wilting'.
– Soqotri simplified: *zafaqah di 'ilihe*; *'ajbib*.

Food (livestock). Like others in the *ʿážbib/ʿážərib* group, it flourishes after winter rains. It is liked by cattle before the grasses come up, but is little eaten by goats and sheep. It is seen as a good fodder for cattle until such time as the grasses fully develop.
Flavouring / food additive. Cows eating a lot give very sweet milk.
Wood. The stems are used to make the 'bit' for the weaning muzzle worn by kids and lambs.
Notes. Can cause scour in animals that eat a lot immediately after the rains, before the grasses have come up. It can also cause *'uqabah* in goats unaccustomed to it.
Comparative uses. Some species, such as *S. dentate*, are used as goat fodder, but others, *S. aquatica* for instance, contain a cardiac glycoside which causes violent purging in animals. (Ambasta 1982; Bebawi & Neugebohrn 1991; Dalziel 1948, 1956; Jain & De Fillips 1991)

Striga Lour.

Erect parasitic or semi-parasitic herbs. Leaves opposite or alternate, linear or scale-like. Flowers in a terminal spike. Corolla tube long with a bend near the middle, 2-lipped. Stamens 4. Calyx with a cylindrical tube and 5 spine-like lobes. Fruit a capsule. 40 species. Old World tropics and subtropics.

Botanical notes. Semi-parasitic on grasses, cereals and a range of other hosts, causing great economic loss in certain areas.

ETHNOGRAPHIC NOTES
Comparative uses. In Sudan *S. asiatica* is reported as a good fodder. *S. gesnerioides* is used in the treatment of diabetes. Some of the darker, purple-flowered species are chewed in Dhofar and Kenya to make the teeth bright, and dye the lips, gums and tongue indigo. Indeed, *S. senegalensis* is added to indigo dye vats to deepen the colour. In Oman the flowers of *S. hermontheca* are dried, powdered and applied to an infected or irritated eye. (Abdulfatih 1987; Ambasta 1982; Bebawi & Neugebohrn 1991; Dalziel 1948, 1956; Ghazanfar 1994; Jain & De Fillips 1991; Miller & Morris 1988)

Striga asiatica Kuntze
Striga hirsuta sensu Balf.f. non Benth.

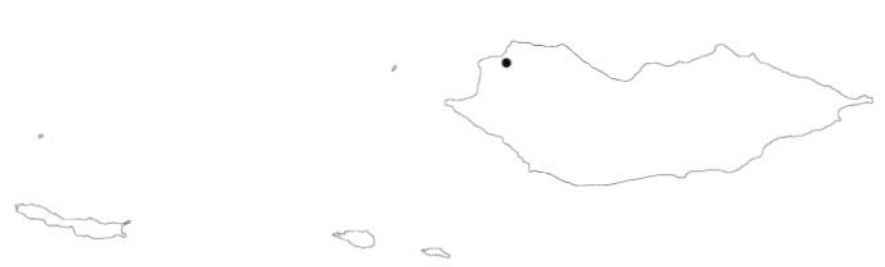

Slender, semi-parasitic herb to 20cm, bristly hairy. Leaves mostly alternate, linear, <1.5 x 0.1cm. Flowers opposite or alternate in a terminal spike. Corolla bright red or white, tube c.1cm long, bent just below the lobes. Calyx with 10 or more ribs, 6–8mm long. Capsule c.7mm long.
Ecology. 'On the plains about Galonsir [Qalansiyah]' (Balfour 1888).
Distribution. Tropical and S Africa, Arabia to SE Asia and China. Introduced in N America.
Botanical notes. Not seen since the 19th century. Semi-parasitic on wild grasses and some crops (e.g. maize).

Striga densiflora Benth.

Erect, green, semi-parasitic herb to 12cm, almost glabrous or bristly hairy. Leaves linear, <4cm long. Flowers in slender spikes with long bracts and linear bracteoles. Corolla white, c.0.8cm long. Calyx 5-ribbed, c.0.5cm long.
Ecology. Succulent shrubland. 300m.
Distribution. Indian subcontinent.

ETHNOGRAPHIC NOTES†
Name(s)
– Soqotri: nil or *məśḥərməhim ʿafər* 'red *Caralluma*-type'; (western Soqotra) *źərġanitin* 'slimy sediment, silt'; (rare) *di fiṭaḥe* 'new shoots, new growth' (more commonly given to *Crinum balfourii* and *Ledebouria* species).
– Soqotri simplified: *dhirghanitin*; *mish'hermihim 'afir*; *di fitahe*.
Food (livestock). Not palatable to livestock.

†Names also apply to *S. gesnerioides*.

***Striga gesnerioides* (Willd.) Vatke**
Striga orobanchioides Benth.

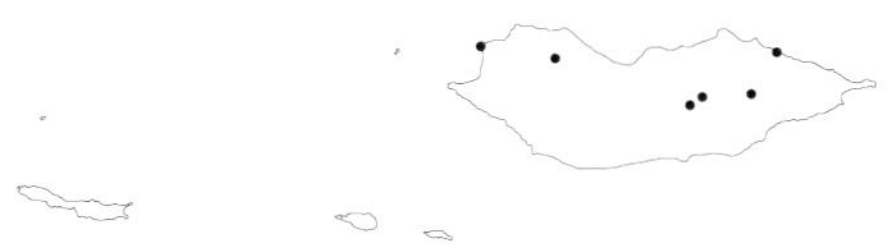

Parasitic herb to 30cm, glabrous or hairy, all parts dark reddish purple. Leaves scale-like, appressed to the stem, <1cm long. Flowers opposite or alternate, 2 or 3 per node, subtended by a scale-like bract and 2 bracteoles. Corolla pink or white, tube c.8mm long. Stamens 4. Calyx 5-ribbed, c.5mm long. Fruit c.4mm long.

Ecology. Widespread but always in small numbers in *Croton socotranus* shrubland, succulent shrubland and drought-deciduous woodland. Sea-level–750m.
Distribution. Tropical Africa and Asia.

ETHNOGRAPHIC NOTES
Name(s)
See *S. densiflora*.
Food (human). The swollen base of the plant, sometimes called *fiṭaḥə*, can be roasted and eaten in time of famine. However, it causes indigestion and flatulence, and is not popular.
Food (livestock). Not palatable to livestock.

Veronica L.

Herbs. Leaves opposite. Flowers pedicellate, solitary in the leaf axils. Corolla 4-lobed with a very short tube; stamens 2. Capsule laterally compressed. c.180 species. Temperate areas and some on tropical mountains. Several subcosmopolitan weeds.

***Veronica polita* Fr.**

Annual herb; stems branched from the base, prostrate or decumbent, hairy. Leaves ovate, 1–2cm long, with serrate margins. Corolla pale or deep blue, 4–8m across.
Ecology. Semi-deciduous woodland on limestone. 600m.
Distribution. Temperate Eurasia and N Africa. Probably also native in SW Asia, but now an almost cosmopolitan weed.

ETHNOGRAPHIC NOTES
Name(s)
– Soqotri: nil or (*di*) *ṣʌfəririhon* 'many-flowered'; *saḥamhəm* 'diuretic'.
– Soqotri simplified: *di zafiririhon*; *sahamhim*.
Food (livestock). Much liked by cattle and said to make them fat and increase their milk yield. Goats also eat it, but less enthusiastically than cattle.
Comparative uses. The leaves and shoots of many species are eaten in salads and soups. Many are regarded as anti-scorbutic and are prescribed for scurvy. They are also given as an appetite stimulant and diuretic. *V. anagallis* is used by women to dress their hair, and the leaves of *V. perottetti* are used as chewing tobacco in the absence of real tobacco. (Ambasta 1982; Dalziel 1948, 1956; Jain & De Fillips 1991; Kunkel 1984)

Xylocalyx Balf.f.

Low, woody subshrubs. Leaves opposite, ovate. Flowers axillary, subsessile or shortly pedicellate. Corolla white or pink with 5 subequal, spreading lobes. Stamens 4. Calyx cup-shaped, becoming woody in fruit, bracteoles forming a pair of horn-like projections at the base. Calyx becoming skeletonised with age. 2 endemics. 5 species. Somalia.

Botanical notes. Distinguished by the calyx which, with age, hardens and completely encloses the capsule. The resulting woody structure, consisting of the enlarged calyx and capsule, persists on the plant and is a characteristic feature of the genus. In the dry state the fruit valves are closed, preventing seed dispersal; however, when moistened, the valves move outwards and allow dispersal. Movement in response to wetting is an adaptation often found in plants growing in areas with low and erratic rainfall. It ensures that the seeds remain protected on the plant until there is sufficient rain for germination.

***Xylocalyx aculeolatus* S.Carter**

Greenish-black subshrub to 25cm, leaves and young branches covered with short, recurved prickles. Leaves <1 x 0.35cm long. Flowers shortly pedicellate. Corolla white or mauve, tube 6–7mm long. Fruiting calyx <1cm long.
Ecology. Dwarf shrubland on limestone plateaus. 200–600m.
Distribution. Endemic.
Conservation status. Least Concern (LC). Common in several vegetation types and under no present or perceived threat.

ETHNOGRAPHIC NOTES

Name(s)

– Soqotri: nil or *ḳəṭəmʌmhiyon* 'little lumps of fresh butter'.
– Soqotri simplified: *qitimamhiyon.*

Food (livestock). Liked by all livestock, especially goats, and is regarded as valuable fodder for small stock and to increase their milk yield.

Flavouring / food additive. The milk of animals eating a lot is said to be especially rich, hence the name 'little lumps of butter'.

Xylocalyx asper **Balf.f.**

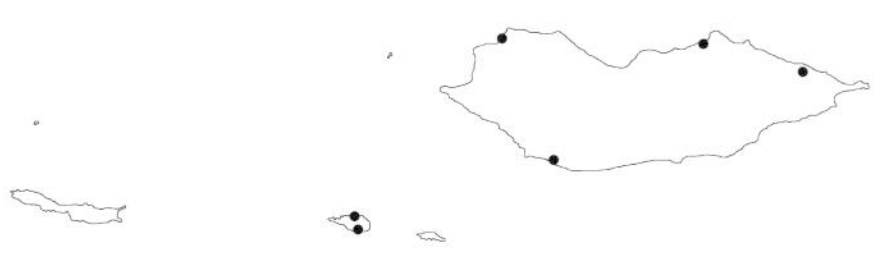

Procumbent subshrub to 10cm. Stem and leaves light green. Leaves broadly ovate, c.4mm long. Flowers subsessile, pedicel length variable. Corolla blue or white, tube 6–7mm long, scarcely exserted from the calyx. Calyx enlarging to 9mm in fruit.

Ecology. Coastal plains and desertic foothills, less commonly on limestone plateaus (Ras Geridihon). Sea-level–600m.

Distribution. Endemic.

Conservation status. Least Concern (LC). Common in several vegetation types and under no present or perceived threat.

Botanical notes. Apparently generally found at lower altitudes than the preceding species.

ETHNOGRAPHIC NOTES

Name(s)

– Soqotri: nil or *ḳəṭəmʌmhiyon* 'little lumps of fresh butter'; *handūrib* (Samha); *di kubihur* 'little balls of wool'.
– Soqotri simplified: *qitimamhiyon*; *handuwrib*; *di kubihur.*

Food (livestock). See *X. aculeolatus* (also flavouring).

SOLANACEAE

Herbs or shrubs, often armed. Leaves alternate, simple; stipules absent. Flowers solitary or clustered, sometimes in short racemes. Calyx of 5 partly fused sepals, often persistent and enlarged around fruit. Corolla actinomorphic, cup-shaped, campanulate or rotate; lobes 5. Stamens 5; anthers often touching, dehiscing longitudinally (or by apical pores in *Solanum*). Ovary superior, 2-locular (3 to 5-locular in *Nicandra*); style 1; ovules numerous on axile placentas. Fruit a berry or capsule. 90 genera. 2600 species. More or less cosmopolitan, especially S America.

Datura L.

Herbs. Flowers solitary in leaf axils, very big. Calyx tubular. Corolla long-tubular broadening to a funnel at the mouth, fleshy; lobes small, subulate. Fruit a greenish, globose, spiny capsule. 8 species. Native of S N America but now widely naturalised.

Datura fastuosa **L.**
Datura fastuosa var. *alba* Clarke

Robust, unarmed annual herb to 2m. Leaves ovate, 7–15 x 3–10cm, sinuate or with few, large teeth. Calyx 6–8cm long. Corolla white, 12–20cm long. Fruit c.3cm diameter.

Ecology. Rocky wadi sides, disturbed ground and in enclosures. 30–800m.

Distribution. Naturalised in Yemen, Oman and Saudi Arabia.

ETHNOGRAPHIC NOTES

Name(s)

– Soqotri: *mərʌnnəhʌ*; *gənigəno* (both commonly given to plants whose wood is useless); *ṣʌᶜəlhʌl* 'large-leaved'; *mgelḥeš* (more commonly given to *Dovyalis abyssinica* also with toxic fruit); (rare) *źeyṭ*.
– Soqotri simplified: *miranniha*; *ginigino*; *za'ilhal*; *mgel-hesh*; *dheyt.*

Food (livestock). If the leaves are drying out or on the ground they are sometimes nibbled by goats and sheep when there is little other forage available towards the end of the dry season. The fruit are never eaten, and normally it is avoided by livestock.

Other. Children play with the large flowers, inflating and tying them to make a tiny milk-skin.

Notes. No tradition of medicinal use on Soqotra.

Comparative uses. *Datura* species are of interest as a possible source of the alkaloid scopolamine used as a pre-anaesthetic in surgery and obstetrics. Scopolamine is a cerebral depressant, useful in agitated or maniacal conditions; it is also used to produce amnesia and partial analgesia in labour. It is very effective for preventing motion sickness at sea or in air travel. The pulverised seeds of *Datura* species are used in fermented drinks, or an infusion of the leaves and twigs are taken for intoxication and hallucinations. They were used in the preparation of 'witches' salves' to give them the sensation of flying. *D. fastuosa* and *D. metel* are abundant sources of scopolamine which is an important anti-spasmodic and is used to treat some secondary glaucomas, as an aid in refraction and for other diagnostic purposes. On the Arabian mainland the whole plant of *D. fastuosa* is boiled and taken for asthma, and as a sedative, which is very toxic in high doses. In Dhofar the dried flowers and leaves are smoked to relieve breathing difficulties, and in small doses the seeds are taken as an analgesic. In northern Oman the roots are placed around a dwelling, or worn as a protection against *jinn* and the Evil Eye. In Yemen the seeds of the ripe fruit are burned and the smoke inhaled to treat epilepsy. In Africa the seeds have been used criminally to stupefy with intent to rob. They are powdered and taken with butter to treat impotence, or are applied to the genitalia to promote sexual vigour. The crushed plant is placed on boils or applied to relieve the pain of stings. The fruits are ground and added to flour as an effective poisoned bait for rats, and a watery extract of the root is sprinkled on the floor of huts to keep away snakes. (Abbas *et al.* 1992; Dalziel 1948, 1956; Davis *et al.* 1983; Fleurentin & Pelt/Fleurentin *et al.* [1982, 1983a,b,c]; Lewis & Elwin-Lewis 1977; Miller & Morris 1988; Oliver-Bever 1986)

Lycium L.

Armed shrubs. Leaves fasciculate. Calyx tubular. Corolla tubular; lobes spreading. Fruit a globose berry. 1 endemic. c.100 species. Widespread in warm temperate regions.

Lycium sokotranum **Wagner & Vierh.**

Lycium europaeum sensu auct. non L.
Lycium socotrana L.

Much-branched shrub to 2m. Branches with short spines, <1cm long. Leaves obovate, 0.5–2 x 0.1–0.4cm. Calyx 0.2–0.3cm long. Corolla yellow, c.1cm long. Fruit red, 0.5–1cm diameter.

Ecology. Widespread and often abundant on coastal plains and limestone plateaus. Sea-level–800m.

Distribution. Endemic.

Conservation status. Least Concern (LC). Common in several vegetation types and under no present or perceived threat.

Botanical notes. On Abd al Kuri *L. sokotranum* is abundant on the central plains. The plants here are much more robust and with longer spines than those on Soqotra. It is an important 'nursery' plant, protecting seedlings and young plants of species such as *Dendrosicyos socotranus* from livestock. It is similar to *L. depressum* and *L. arabicum* from the Arabian mainland.

ETHNOGRAPHIC NOTES

Name(s)

– Soqotri: *su̍ʾhur*, *su̍hur* (pl. *se̍ʾrihan*, *se̍rhin*, *su̍ʾhərīn*) 'barrier'; fruit *gi̍ləḥum di su̍ʾhur*; *nābət* or *miṣa̍ʿ di su̍ʾhur* (Abd al Kuri); *di se̍rʾhen* (Samha).

– Soqotri simplified: *su'hur*, *suhur* (pl. *se'rihan*, *serhin*, *su'hiriyn*); fruit *di su'hur*, *nabit* or *miza' di su'hur*, *di ser'hen*.

Food (human). In the hungrier areas of the Archipelago the ripe, red fruit are eaten. They are sweet and filling and very plentiful on some bushes. They are mostly eaten raw, but if a lot are gathered they are also cooked in milk or buttermilk, and crushed to make a sort of sweet porridge.

Food (livestock). Goats, sheep, camels and donkeys all eat the foliage and new shoots. Camels and goats are most enthusiastic, but sheep and donkeys are more discouraged by the spines and only eat it when there is little else. Milk yield is increased in camels and goats that eat a lot of green leaves. Cattle from drier areas also eat the young stems and green leaves, especially larger leaves from taller shrubs. Apart from the new shoots after rain, it is not considered to be a particularly nutritious feed, but provides some sort of fodder in the form of leaves, twigs or fruit for much of the year, so is a useful maintenance feed when other forage is scarce. Small stock, especially goats, are very keen on the ripe fruit. In some of the driest areas of Soqotra, where these plants are very common, and on the outer islands, it is key forage for livestock, especially goats. Young stock in particular nibble small leaves and search out flowers and fruit. Browsing the shrubs heavily in dry weather can cause damage to an animal's mouth. At this time the spines are very hard, and can scratch the skin inside and around the mouth. If the scratches become infected, the animal can find it difficult to feed, and consequently can lose weight and fall ill. Goats eating a lot can also develop the condition known as *ḥaniḥə̍no*, in which hard and lumpy swellings develop around the lower jaw and under the chin. This can be fatal.

Bees / honey. The shrubs respond to the lightest rain and quickly come into flower. In the drier areas, it is one of the key forage plants for honeybees, especially if there are summer rains. Honey with absolutely no trace of bitterness is made from its flowers.

Flavouring / food additive. The fruit are high in vitamin C and before the days of sugar were much liked for their sweetness.

Fuel. The dead wood makes excellent firewood. It burns slowly, giving out a good heat with little smoke so is popular for cooking. It is much used in the cold months to warm the home and livestock. It burns with a very bright light, and a flaming branch is used as a torch. Dead twigs are gathered and stored as kindling and tinder. On the outer islands where wood is scarce it is especially important as firewood.

Timber. The most widely used plant for all sorts of enclosure fencing, partition building and pen construction. Indeed, the verb for 'enclosing' is from the same root as its name. The wood is not attractive to termites and does not have to be de-barked before use. Large cave settlements are divided into different areas with partitions made from *L. sokotranum* branches laid on top of a stone foundation. Branches are often placed along the top of the stone walls of cultivated areas to deter goats. The doorways to yards, pens and byres are closed when milking or handling livestock by placing the thorny branches across them, and livestock are held penned in caves or sections of caves by walling made from *L. sokotranum* brushwood. Sheep pens are often constructed entirely from this material. A transhuming family will close up their cave or hut living quarters with the brushwood to signal their absence (but continued possession of the accommodation), and to keep their belongings safe from roaming livestock.

Wood. Close-grained and hard, and that from larger shrubs is used to make all sorts of implements. The slimmer, straighter branches are used to make the important herding staff. The wood is also suitable for use as a firestick, and the dead wood makes superior charcoal.

Medicinal / veterinary. The ripe berries are recommended for a variety of stomach ailments. The spines are used to sew up lacerations made by a wild cat attacking a young lamb or kid. For this, a spiny branch is held over the fire until it glows red and the tips of the thorns have changed colour. The thorns are then broken off and used to pin the sides of the tear shut to keep the wound covered and encourage healing. If a tooth is causing a lot of trouble, it is extracted by winding a strip of the strong root-bark first around the tooth and then around a stick, and giving the stick a sharp tug. If this is not effective, the gum around the tooth is scraped down to expose the root, and then a twig of the wood is placed against the exposed base of the tooth, and the other then tapped sharply with a stone.

Pest control. The green wood makes a lot of dense, dark smoke, so is sometimes used for livestock smudges.

Other. Branches and twigs are used to keep evil forces and the Evil Eye at bay. For instance, when returning to an area after an absence, branches are placed across the entrance to the goat and sheep pens before going inside or letting the animals in. Some people do the same for their own living quarters. Branches are also hung on the entrance to pens and byres once the animals have been safely shut up for the night. Twigs are placed across food and water containers and women hang a branch of the shrub from the ceiling of the place where they go to give birth. A sprig is tied to the mouth of the butter-skin to encourage the butter lumps to form. Every time a new butter-skin is made, a lucky sprig is tied to the string that joins the legs of the goatskin. Cheese made from boiled buttermilk is spread as a thick paste over a smooth rocky surface to dry and is then surrounded with twigs to keep creatures, both natural and supernatural, away.
Notes. The plains around Qadheb were traditionally called *di sˈiʔirhin*, or 'area of *L. sokotranum* shrubs'. Many stories are told of date palms being magically transformed into *L. sokotranum* bushes by the wrath of God: usually an angel, disguised as a poor and desperate man (or woman), asks the owner of a date palm garden for a handful of dates from the branch to stave off starvation. The request is rudely refused, the owner saying that the trees are not palms, but *L. sokotranum* bushes. But his bluff is called when the lie comes true.

Islanders differentiate between two sorts of *L. sokotranum*. One is very common in the drier areas, and is a low shrub that rarely bears fruit, or if it does they are very small ones, often fail to ripen and are generally inedible. It produces foliage readily after rain, but does not flower or fruit unless the rains are heavy. Some islanders call these bushes *sˈuʔhur di źˈaʔad* ('the ziziphus-like *L. sokotranum*'). This is not the case on Abd al Kuri where the shrubs generally fruit well. The other *L. sokotranum* grows to a tall shrub or small tree, with a proper trunk, and is mainly found close to water or in the wetter areas of the island. Its leaves are generally larger than that of the other type, and it bears plentiful fruit that ripen bright red and are sweet. It is often called *sˈuʔhur di bīrhˈitin* ('the *L. sokotranum* of child-bearing women'). It is given this name partly because it is used by women during childbirth (as mentioned above), but partly also in recognition of the fact that it is women who are generally in charge of sheep, and this shrub features largely in their management.
Comparative uses. Many *Lycium* species have edible berries and leaves that can be used as a tea substitute. The foliage of some, such as *L. halimifolium* and *L. barbarum*, is poisonous to livestock. The roots of *L. europaeum* (= *L. sokotranum* sensu Balf.) are soaked in water, and then either cooked or left to stand, and used as a body wash for tick fever and polio. In East Africa the leaves are used as a vegetable. In Yemen the leaves of the closely related *L. shawii* are taken against kidney stones, and in Saudi Arabia used to treat jaundice. In Dhofar the fruit are eaten to soothe stomach pains, and in northern Oman the stems are boiled as a laxative, tonic and diuretic. The berries and growing tips are pounded together and an extraction taken to treat colic, or applied around the eyes to improve the vision. (Abbas *et al.* 1992; Ambasta 1982; Heine *et al.* 1988; Jain & De Fillips 1991; Kunkel 1984; Lewis & Elwin-Lewis 1977; Miller & Morris 1988)

Nicandra Adans.

Herbs. Flowers solitary in leaf axils. Calyx papery, inflated in fruit. Corolla campanulate. Anthers dehiscing longitudinally. Fruit a berry, enclosed by calyx. 1 species. Peru.

Nicandra physaloides Gaertn.

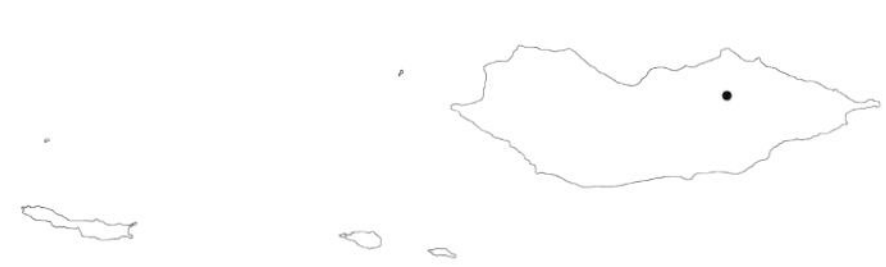

Robust, unarmed herb to 1m. Leaves ovate, 10–20 x 5–15cm, coarsely and irregularly dentate. Corolla mauve, c.1.5cm long. Fruit green, maturing orange, c.1.5cm diameter.
Ecology. A rare weed found near animal enclosures in the Haggeher mountains. 1000m.
Distribution. Native of Peru, widely naturalised in Europe, Asia and the Americas.

ETHNOGRAPHIC NOTES
Name(s)
– Soqotri: nil or *mərˈʌnnəhʌ* (commonly given to plants with useless wood).
– Soqotri simplified: *miranniha.*
Food (human). The fruit are not eaten on the island.
Food (livestock). Not a common plant and the islanders are not sure whether it is palatable to livestock or not.
Comparative uses. The berries are edible and the young leaves are boiled and eaten. (Kunkel 1984)

Nicotiana L.

Unarmed herbs. Flowers in terminal panicles. Corolla tubular. Fruit a smooth capsule, partially enclosed by calyx. 67 species. Americas, S Pacific, Australia and SW Africa.

Nicotiana tabacum L.

Robust herb with unarmed branches to 1m. Leaves ovate, 10–30cm long, entire, sessile. Corolla pinkish, 4–5cm long. Fruit greenish.
Ecology. Formerly widely cultivated and now with a scattered distribution across Soqotra; however, no detailed locations available.
Distribution. Native of tropical Americas.
Botanical notes. Formerly widely cultivated, less so now, occasionally naturalised.

ETHNOGRAPHIC NOTES

Name(s)

– Soqotri: *tə́mbák, təmbɛ́ko.*

– Soqotri simplified: *timbak, timbeko.*

Food (livestock). Not palatable to livestock.

Medicinal / veterinary. This cultivated plant is one of the most widely used medicinal plants of the island, for both livestock and their owners. Tobacco is particularly used to treat the after-effects of botfly attack. The best cure is to quickly chew some and swallow the bitter juice, or to smoke a pipe, inhaling the smoke deeply. If this cannot be done immediately after the attack, then some very strong tobacco is smoked, and inhaled so deeply that the patient becomes nauseated. If he vomits, this is all to the good, and means the cure is likely to succeed. If the chest is affected, the patient is given a pipe to smoke and again, must inhale deeply and smoke the whole pipe. If the eye is affected, a dried tobacco leaf is ground to a powder and added to water and used as eyedrops. If the nose is affected, the same tobacco and water mixture is sniffed up the nose, or some tobacco leaf is scorched, powdered and taken as a snuff. Similar treatments are given to goats and sheep, the main botfly targets. Either the leaf juice, or powdered tobacco, is added to a little water and then dropped into the affected eye(s) or into the nostrils, or it is given as a drench.

Tobacco is used as a remedy for other ailments too. The dried or fresh leaves are crushed with a little water and used as eardrops for a suppurating and/or painful ear. Fungal skin infections are treated by scarifying the area and then rubbing in the powdered scrapings from a pipe, sometimes mixed with some salt and powdered *Euclea divinorum* root. Someone with a bad head cold or with breathing difficulties smokes tobacco to relieve the symptoms. A sty in the eye is pierced with the pointed stick of a tobacco leaf and then powdered tobacco applied to the wound. Many young men used to smoke tobacco during circumcision, both as a sign of their courage and perhaps also to distract them from the pain.

Commercial value (past, present). Tobacco was formerly cultivated wherever possible and was taken to the coastal markets for sale. Now growing it is rather frowned on.

Pest control. The smoke is an effective insecticide. Cultivators spray their plants with water in which tobacco has been soaked to get rid of insect pests.

Other. On Soqotra tobacco was cultivated as follows. The seed, collected the previous year, is sown in seedbeds with soil enriched by wood ash, aged sheep dung and, wherever possible, bat dung. The beds are kept well watered, and side shoots of the young plants are removed. The small plants are then transplanted to the main growing area. Some farmers miss out the seedbed and plant seed directly into the main growing area. Once established, apart from weeding and removing the flowering stalks as they appear, the plants are left alone until they begin to change colour. They are then harvested by snipping the broad leaves close to the stalk. Secondary growth often grows from the harvested stalks, which is also harvested. The leaves are tied by the stalk in bundles of four to six leaves, and are then dried, hung upside down from the stalk, in the fresh air. Once completely dry, the bundles are bound round and round from top to bottom, and are stored in sacks ready for sale.

Notes. Tobacco used to be much liked by both men and elderly women. It was smoked, taken as a snuff and chewed, the plug being stored between the lower gums and bottom lip between chewings. Many islanders could afford only tiny amounts of the real thing, and would stretch it with dried and crushed tobacco stalks or *Ficus cordata* leaves, or by adding wood ash to the powdered tobacco.

Comparative uses. Tobacco contains the toxic alkaloid nicotine. This causes vomiting and diarrhoea, a slow pulse, collapse and respiratory failure. Deaths have been known to occur from eating the leaves as a boiled vegetable, and leaf dressings, applied to open wounds under a bandage, to cause nausea and vomiting with prostration and a feeble pulse. Livestock have also died after eating the plant. It is used medicinally as a sedative, anti-spasmodic and vermifuge. It is given for skin troubles, gastro-intestinal disorders and skin problems.

Seeds. These yield a semi-drying oil used after refining for edible purposes and in paints and varnishes. The seed cake is used as animal feed since the seeds are free from the toxin nicotine.

Leaves. The pulverised leaf is put in a carious tooth. They are heated with oil and rubbed on the chest in respiratory complaints. A wash or ointment of tobacco leaves is used to protect livestock from tsetse flies. The ash of the leaves is applied to soothe bites and stings.

Juice. This is applied to the scalp for ringworm, but this has sometimes been known to have fatal results.

Tobacco. Tobacco powder and extracts are used in agricultural insecticides and to eradicate lice and ticks in animals. Snuff is wrapped in a lump of foodstuff as fishing bait or to stupefy fish. (Ambasta 1982; Dalziel 1948, 1956; Jain & De Fillips 1991; Lewis & Elwin-Lewis 1977)

Physalis L.

Herbs. Flowers solitary in leaf axils. Calyx papery, becoming enlarged around fruit. Corolla campanulate. Fruit a berry, completely enclosed by calyx. 80 species. Cosmopolitan, especially New World.

Physalis minima L.

Slender herb to 1m. Stems with sparse, unbranched hairs. Leaves ovate, <6cm long, sparsely dentate. Corolla pale yellow. Fruiting calyx <2cm long.

Ecology. Weed, known only from the coastal plain near Hadiboh. c.50m.

Distribution. Pantropical.

Botanical notes. Readily distinguished by the papery, lantern-like calyx which completely surrounds the fruit.

ETHNOGRAPHIC NOTES

Comparative uses. The leaves and unripe fruit of *Physalis* species are suspected of poisoning sheep. The Cape gooseberry (*P. peruviana*) has edible fruit and leaves and the fruits are also made into a preserve. The plant is used as a tonic, diuretic and purgative. An infusion of *P. angulata* is drunk by women during childbirth. The crushed fruit are taken in milk to treat sterility, and a lotion made from them is used in infantile ophthalmia. The juice is dropped into the eyes of a fainting child to restore it. (Ambasta 1982; Dalziel 1948, 1956; Jain & De Fillips 1991; Kunkel 1984)

Solanum L.

Herbs or shrubs. Flowers solitary, clustered or in short racemes. Calyx shallowly campanulate. Corolla star-shaped (rotate). Anthers joined into a cone, dehiscing by apical pores. Fruit a berry, not enclosed by calyx. c.1400 species. More or less cosmopolitan.

Solanum cordatum **Forssk.**

Solanum gracilipes Decne.

Spiny subshrub to 30cm. Spines straight or slightly curved, <0.5cm long. Leaves broadly ovate to round, <3cm long, subentire. Flowers solitary. Corolla violet. Fruit orange-red, c.0.7cm diameter.

Ecology. Uncommon on limestone plateaus and the foothills of the Haggeher. 300–700m.

Distribution. Ethiopia, Somalia, Arabia, NW India and Pakistan.

ETHNOGRAPHIC NOTES†

Name(s)

– Soqotri: nil or *ʿʌ̍ḳəlʌṣ* 'tightly curled, peppercorn hair' (also given to *Eclipta prostrata*, *Ageratum conyzoides* and *Acalypha indica* which can also be cooked and eaten); *fε̄rəher*.

– Soqotri simplified: *'aqilaz*; *feyriher.*

Food (human). The red fruit can be eaten, but are rather bitter. Some people cook and eat the leaves in hunger, sometimes mixing them with *Amaranthus* leaves.

Food (livestock). Regarded as quite a nutritious livestock fodder. Cattle and goats graze it, and sheep eat dried material from the ground. Camels and donkeys also graze it.

Flavouring / food additive. Milk of livestock feeding on it has a very sweet taste.

Other. The leaves, sometimes with added fruit, are used to repair the local clay pottery. They are ground with some milk and dry clay to make a thick paste, which is kneaded and smeared over the crack, both inside and out. The pot is then put beside the fire for the paste to harden and set. If the pot is holed and the hole quite large, a small pebble, or a shard of pottery ground down to fit the hole, is inserted first and then paste is smeared around it.

Notes. Can cause *'uqabah* in small stock unaccustomed to it.

Comparative uses. Many species are used medicinally. In Dhofar the ripe black berries of *S. nigrum* were crushed, mixed with oil or ghee, and used cosmetically to stain the skin dark blue. (Ambasta 1982; Jain & De Fillips 1991; Miller & Morris 1988)

†Also apply to *S. villosum*.

Solanum incanum **L.**

Solanum anguivi sensu Balf.f. non L.
Solanum indicum sensu Balf.f. non Nees

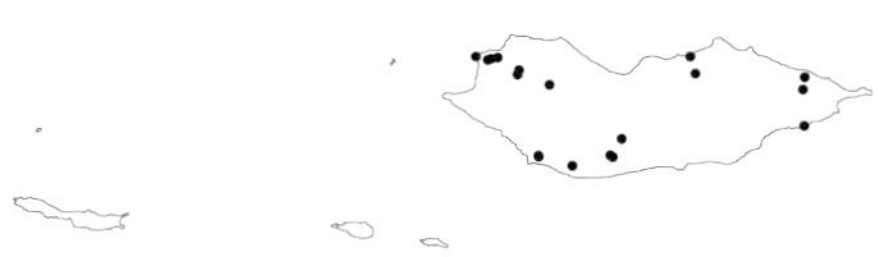

Spiny shrub to 2m. Stems densely stellate hairy. Spines broad-based and slightly hooked. Leaves broadly lanceolate to ovate, 5–15cm long, sinuate, midrib spiny, densely stellate hairy. Flowers solitary or in small racemes. Corolla blue or violet, c.1.5cm diameter. Fruit yellow, 3–4cm diameter.

Ecology. A widespread and conspicuous plant of disturbed places, grassland, wadi sides and clearings in woodland. Sea-level–800m.

Distribution. Africa, tropical Asia and the Middle East.

ETHNOGRAPHIC NOTES

Name(s)

– Soqotri: *ḥa̍ržīm*, *ḥa̍ržəhīm* (pl. *ḥa̍rgəhum*) 'greedy'; fruit *ḥolḳo̍moh* (pl. *ḥa̍lḳum*).

– Soqotri simplified: *harjiym*, *harjihiym* (pl. *hargihum*); fruit *holqomoh* (pl. *halqum*).

Food (livestock). In drier areas the younger leaves are eaten by goats, who seem undeterred by the spiny prickles, and in the dry season the plant is regarded as quite nutritious feed if they are accustomed to it. However, the spines of mature and dry plants can damage the mouths of goats unaccustomed to it. The skin in and around the mouth, and even in the gullet, is scratched as the food is swallowed. The sore mouth makes it difficult for the goat to feed, and consequently it can lose weight and fall ill. Sheep also eat the fallen leaves, but if they eat the green ones in hunger they are very prone to damage their mouths. Donkeys and camels graze the leaves in the dry season, and the cattle of the drier areas also eat younger leaves when there is little alternative. When very hungry, goats chew the green stems as well as the leaves. They also eat the green fruit when they are thirsty. In the better-vegetated areas all parts of the plant are generally ignored by livestock. Even goats here rarely eat the leaves, but still seek out the fruit in very hot weather. The spines can become lodged in the cloven hoof and cause sores which can become infected and cause lameness.

Medicinal / veterinary. The seeds of a ripe fruit are inserted into the cavity of a sore tooth. They are also scorched, then ground and the powder put into the painful tooth. The fruit are dried and ground as a dressing for sores, abscesses and ulcers. This is seen as a very 'hot' dressing, and as good at 'drawing the poison' out of the skin. Alternatively the soft inside of a ripe, yellow fruit is pounded to a paste and used in the same way. Both unripe green fruit and ripe yellow ones are crushed and the juice rubbed over skin affected by dry eczema or by scabies. This makes the skin very hot and painful for about two days, but then the condition improves. The same mixture is applied to relieve itching.

Other. Children play with the dried fruit as a ball. Goats feeding on the fruit can go longer without water.

Notes. The spiny plants are a nuisance around living quarters and they are commonly cut away from cave entrances, doorways and busy pathways.

Comparative uses. *S. incanum* contains solasodine, which is commercially important in the preparation of certain hormonal steroids. In Dhofar the fruits are crushed and used on skins and hides to remove hair.

Roots. A decoction is used against vomiting and diarrhoea, to treat gonorrhoea, abdominal pains, fever, liver troubles, carbuncles, coughs, colic, sore throat and indigestion. They can be used for toothache by scrubbing the affected tooth with a piece of the root. The root is regarded as poisonous and, with other plants, is used in the preparation of arrow poison. It is also used to treat snakebite. Boiled with sour milk and grain porridge it is given for syphilis, and the fruit juice, sometimes with pounded leaves, is rubbed on the hands for an infective eruption supposed to be syphilitic. The root and leaves are crushed together with natron, and applied as a rub for rheumatic and swollen joints.

Fruit. These are used externally to treat suppurating sores around the nails. A small hole is made in a fruit and a sore finger inserted to draw the pus of ulcers or weeping wounds. A fruit is broken open and applied to a cut or wound, and crushed fruits are used to treat epilepsy. Fruit are given to children as an emetic but with care. The fruit juice is applied to a finger with a whitlow to draw it out, and is also given for ringworm. The juice is squeezed into the nostrils of sheep to cure coughs, and is applied as a remedy for dandruff, snakebite and earache. It is used to coagulate milk to make cheese. A poultice or infusion of fruit is effective in the removal of benign external tumours. A piece of fruit is inserted into the nose of cattle for some cattle sicknesses by the Fulani, and the fruit are used as a purge for livestock. The dried fruit is put on the fire and then eaten for flatulence.

Seeds and flowers. Seeds are used externally to treat inflammation, earache and headache. Flowers with fruit are used for constipation and diarrhoea. The seeds of the fruit are burned and the smoke inhaled to ease toothache.

Leaves. A decoction of the leaves is used against vomiting and diarrhoea. A young leaf is rubbed hard on the site of a snakebite. A leaf infusion is given for earache, and a decoction against gonorrhoea. The leaf juice is given as an expectorant and to treat laryngitis. A cold-water extraction is used for eye diseases, headaches, sore teeth and caries. (Ambasta 1982; Bebawi & Neugebohrn 1991; Dalziel 1948, 1956; Hedberg *et al.* [1982–1986]; Jain & De Fillips 1991; Miller & Morris 1988)

Solanum villosum Willd.

Solanum luteum L.
Solanum nigrum sensu Balf.f. non L.
Solanum villosum subsp. *miniatum* (Bernh. ex Willd.) Edmonds

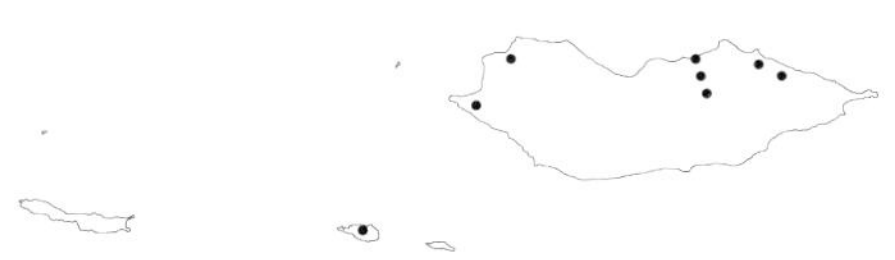

Unarmed erect or sprawling herb to 1m. Leaves ovate or elliptic, <10cm long, entire or sinuate. Flowers in few-flowered clusters. Corolla white or violet. Fruit orange-red, <1cm diameter.

Ecology. A common and widespread weed found near caves and habitation. Under-recorded on the distribution map. 300–1150m.

Distribution. E Africa, Arabia, Mediterranean, Middle East, Iran to India.

Botanical notes. All material on Soqotra is referable to subsp. *miniatum* (Bernh. ex Willd.) Edmonds.

ETHNOGRAPHIC NOTES

Name(s)

See *S. cordatum* (also uses).

Withania Pauquy

Unarmed shrubs. Stems densely covered with stellate hairs. Leaves entire. Flowers clustered in leaf axils. Calyx cup-shaped, somewhat enlarged and papery in fruit. Corolla campanulate. Fruit a berry, partially enclosed by calyx. 2 endemics. 10 species. Africa, Europe, Arabia, subtropical Asia to India.

Withania adunensis Vierh.

Shrub to 1m. Leaves ovate, 4–7 x 2.5–4cm. Corolla greenish, c.0.5cm diameter. Fruiting calyx not completely enclosing fruit; lobes broadly triangular. Fruit greenish, 0.5–1cm diameter.

Ecology. Widespread in succulent shrubland and drought-deciduous woodland. Sea-level–650m.

Distribution. Endemic.

Conservation status. Least Concern (LC). Common in several vegetation types and under no present or perceived threat.

Botanical notes. Close to and perhaps not distinct from *W. riebeckii.*

ETHNOGRAPHIC NOTES†

Name(s)

– Soqotri: *ᶜúbab, ᶜíbəb, ᶜɔ̄wbeb* (Soqotra and Samha).
– Soqotri simplified: *'ubab, 'ibib, 'owbeb.*

Food (livestock). In drier areas of the island and along the coast, sheep eat the dry and fallen leaves when they are very hungry, and sometimes dig up the plants to eat the underground parts. Goats are less enthusiastic, but sometimes eat the fallen leaves and plant litter. Towards the end of the dry season, cattle and goats eat the green leaves of plants growing in the higher, wetter areas, though this is not regarded as a nutritious feed. Nevertheless, cattle herders sometimes gather the leafy stems for their cows if the dry season is prolonged. In some areas, such as in the interior valleys, the leaves can become an important dry season feed. Birds like the ripe red berries, and they are also eaten by goats.

Fuel. In some of the wetter areas, plants grow tall with woody stems, and these are used as firewood. They burn with a strong and very pleasant smell, but the smoke is acrid and painful to the eyes. It is little liked for cooking, and some islanders say it taints the food. Dried stems are often collected and stored as firelighters, as it lights very quickly and easily. Larger stems are gathered and stored in the home as kindling.

Wood. Dried stems are often used as firesticks but they need a lot of work. This plant is renowned for making a very particular noise as it is rubbed, 'like someone chuckling'. To make fire, either both firesticks are *Withania* wood or one is *Commiphora ornifolia* (the best wood for making fire in the Archipelago).

Medicinal / veterinary. The leaves are ground to a paste with water and used as drops for a painful or discharging ear, or to kill a tick or mite that has crawled inside the ear. Great care must be taken when doing this: if too much is used or if the drops penetrate too far inside the ear it is very painful and 'burns'. Crushed leaves are rubbed on itching skin to soothe it, or the patient crouches in the smoke of a fire made from dead *Withania* wood. Measles spots and other rashes are treated in the same way. A paste of ground leaves is rubbed on painful limbs or around stiff and sore joints. Swollen ankles are treated with a poultice of hot ground leaves, and other swellings and raised glands are given the same treatment. Ground leaves mixed with water are painted over an abscess or boil to encourage it to burst and the pus to discharge. To treat a rash, crushed leaves are added to water and this is used as a wash over the whole body. The leaves are mixed with material from *Cissus* species to treat chronic ulcers and saddle sores in pack camels. To treat mastitis in milk animals, especially cows, the udder is washed with hot water and then a paste of crushed leaves is smeared over the whole udder area.

Pest control. The wood, both green and dried, is excellent for livestock smudges, producing plenty of pungent smoke.

Other. The leaves are crushed and mixed with a little water to make a paste used to decorate clay pottery when it is red-hot after firing. It dyes the clay a dark orange-yellow. Sometimes, instead, a handful of leaves is held in the hand and squeezed directly onto the hot clay. The dead leaves are added to tobacco to make it go further, and are sometimes smoked on their own when no tobacco is available.

Comparative uses. *W. somnifera* is listed as a sedative in the pharmacopoeia of India and is widely used as a narcotic and anti-epileptic. The tender shoots are used as a vegetable and as fodder for goats, though they are suspected of poisoning stock.

Leaves. Withaferin A shows antibiotic, anti-inflammatory and anti-tumour activity, and its use has been patented. In Dhofar the fruit and leaves are crushed and left to macerate in water, the filtrate then being drunk as an oxytocic by a woman in prolonged labour or who has a retained placenta. The leaves are also put into the udder cover of a camel with mastitis, and are crushed to a paste and rubbed over the udder and teats of small stock with mastitis. A leaf poultice is used to treat swellings and heal broken bones. They are given as a febrifuge and applied to painful swellings, sore eyes and lesions. The whole plant is soaked in water for a day, and the filtrate drunk for malaria. The leaves and roots are used to kill lice. In Yemen the leaves are crushed with garlic to make a poultice for burns, snakebites and wounds. The crushed leaves are also applied to soothe the burns of cauterisation, and the leaf juice is used to treat nausea and rheumatism.

Seeds. These are used in Sudan to coagulate milk. This is said to be even more effective than the Indian plant, *W. coagulans*, which is used to curdle milk in place of rennet when making cheese.

Roots. These are used to treat hiccups, coughs, dropsy, rheumatism and female disorders. They are used as a sedative in cases of senile debility, and are applied externally in scabies, ulcers and inflammatory conditions. The root is boiled and the water used to wash the skin in measles and skin infestations. It is peeled and boiled in a soup drunk for pains in the joint, and is boiled as an abortifacient and sedative. Dried powdered roots are taken in small doses by sterile women. In Yemen the roots are used as a dental analgesic. In Oman the dried roots are used to treat burns and sunburned skin.

Stems. These are used as firewood and in making fencing. They are burned to fumigate containers. In Yemen water in which the plant has been boiled is put on wounds and sores in which a thorn is embedded, and then the sore is rubbed hard until it bleeds or reddens to cause the pus to discharge and bring the thorn with it. The plant is also considered to be an aphrodisiac.

Fruit and seeds. The fruit are sometimes used as soap in northern Nigeria. They are given as a diuretic and are said to be mildly laxative. They are, however, considered to be poisonous, and to be emetic, diuretic and anaesthetic. In Oman the berries and leaves are crushed in water and the liquid given to women in painful or prolonged labour. (Abbas *et al.* 1992; Abdulfatih 1987; Al-Dubaie 1993/Al-Dubaie & Al Khulaidi; Ambasta 1982; Bebawi & Neugebohrn 1991; Dalziel 1948, 1956; Davis *et al.* 1983; Fleurentin & Pelt/Fleurentin *et al.* [1982, 1983a,b,c]; Ghazanfar 1994; Goodin & Northington 1979; Heine *et al.* 1988; le Houerou 1980; Jain & De Fillips 1991; Miller & Morris 1988)

†Also apply to *W. riebeckii*, *W. somnifera*.

Withania riebeckii Schweinf.

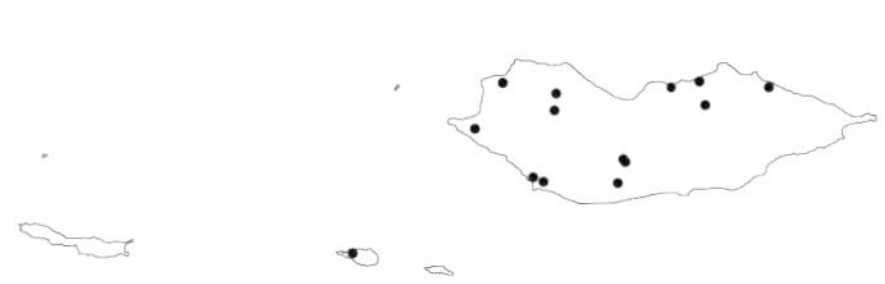

Similar to *W. adunensis* but lobes of fruiting calyx more narrowly triangular.

Ecology. Widespread in succulent shrubland and drought-deciduous woodland. Sea-level–800m.

Distribution. Endemic.

Conservation status. Least Concern (LC). Common in several vegetation types and under no present or perceived threat.

ETHNOGRAPHIC NOTES

Name(s)

See *W. adunensis* (also uses).

Withania somnifera (L.) Dunal

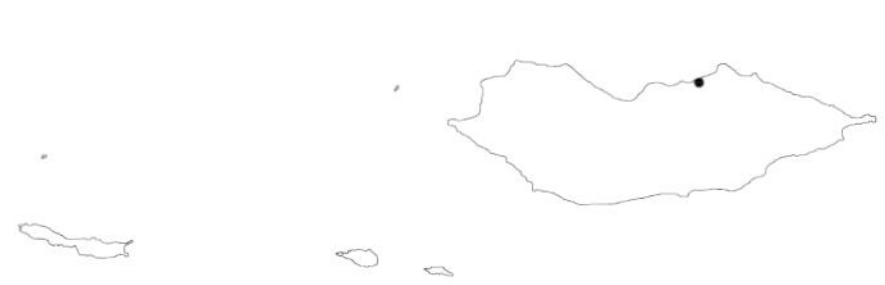

Similar to *W. adunensis* but fruiting calyx enlarged and papery, completely enclosing fruit.

Ecology. Probably a weed of cultivation.

Distribution. Africa, S Europe, Arabia, Middle East, subtropical Asia to India.

Botanical notes. Found near Hadiboh in the 19th century; not seen since.

ETHNOGRAPHIC NOTES

Name(s)

See *W. adunensis* (also uses).

STERCULIACEAE

Trees or herbs, usually with stellate hairs. Leaves alternate, simple or palmately lobed; stipules present. Flowers bisexual or unisexual, actinomorphic. Epicalyx sometimes present. Sepals 5, lanceolate, free or fused at base. Petals 5, free or absent. Stamens 2 to many, in 2 whorls, free or fused into bundles. Ovary superior, 1 to 5-celled, with axile placentation. Fruit a capsule or with 3 to 5 follicles. 72 genera. 1500 species. Tropical and warm regions.

Melhania Forssk.

Herbs or subshrub, densely stellate hairy. Leaves simple. Flowers yellow or orange, bisexual. Epicalyx segments 3, linear or cordate, persistent. Petals yellow. Stamens 5, alternating with staminodes, all united at base. Ovary 5-celled; stigmas 5. Fruit a 5-valved capsule, valves usually 2-seeded. 1 endemic. 60 species. Old World tropics.

Melhania milleri Abedin

Annual herb to 5cm. Leaves lanceolate, 0.8–2 x 0.7–1cm, serrate, green above, densely stellate and grey-green beneath; stipules deciduous. Flowers solitary or paired. Epicalyx segments linear, 3–4mm. Sepals 6–10mm long. Capsule subglobose, 4–5mm diameter; seeds smooth.

Ecology. *Jatropha–Croton* shrubland and semi-deciduous shrubland. 100–600m.

Distribution. Endemic.

Conservation status. Data Deficient (DD). Poorly known; inadequate evidence is available to make an assessment.

ETHNOGRAPHIC NOTES†

Name(s)

– Soqotri: *di ʾārdəhān*, *di hārdihān* 'a spreading area of herbage' (also given to other prickly plants or those with stellate hairs that can cause *'uqabah* in small stock); *ḥamˈʌźəhʌn* 'sour, astringent'; *ṣifṣˈifəh* 'rippling'; *gˈumuṣ* (more commonly given to *Malva parviflora*); *di gilʿˈamihun* 'with tiny fruit'; 'type of *ḳˈəzḳəz*' 'cream-coloured'.

– Soqotri simplified: *di ardehan*, *di hardihan*; *hamadhihan*; *zifzifih*; *gumuz*; *di gil'amihun*; 'type of *qizqiz*'.

Food (human). The green fruit can be eaten, but are not much liked.

Food (livestock). Cattle like the plants once they have grown to a reasonable size. Both *Melhania* species are regarded as a good and nutritious feed for cattle, but when *M. muricata* is mature it is said to damage the mouths of cattle because it becomes so tough. Palatable to small stock, but the stellate hairs can be harmful to their digestive system.

Other. All plants of the *di 'ardihan* type with 'sticky' leaves can be used to repair holed pottery, even pots used to hold liquid. *Melhania* leaves are best for this when the plant is still young. A pebble the size of the hole, or a shard of pottery ground down to fit, is tapped gently to plug the hole. Then the leaves are ground to a paste with either cooked colostrum or meconium of a newborn animal. This is smeared around the hole, inside and outside, sometimes with added ground feldspar. The pot is then put to heat beside a gentle fire until the paste sets hard. This method is suitable for repairing all clay pots and containers.

Notes. Can cause *'uqabah* in sheep and goats unaccustomed to it. *M. milleri* can cause scour in small stock when they feed on it heavily after the rains and before the grasses have come up.

†Also apply to *M. muricata*.

Melhania muricata Balf.f.

Perennial herb to 15cm. Leaves oblong, 2–5 x 0.5–1cm, entire to serrate, green, stellate hairs above, white felted hairs beneath. Flowers solitary. Epicalyx segments large, cordate, hiding the flowers, c.1.5cm across. Sepals 6–8mm long. Capsule oblong, 5–7 x 3–5mm, enclosed by epicalyx segments; seeds muricate.

Ecology. Dwarf shrubland on limestone plateaus. c.500m.

Distribution. Somalia, Yemen and Oman.

ETHNOGRAPHIC NOTES

Name(s)

See *M. milleri* (also uses).

Sterculia L.

Trees. Leaves palmate. Flowers unisexual or bisexual, in panicles. Epicalyx absent. Calyx cup-shaped, 4 or 5-lobed, petaloid. Petals absent. Stamens 10 to 15, fused into a column in male flowers and around base of ovary in female flowers. Ovary of 3 to 5 fused carpels, becoming free in fruit. Fruit of 3 to 5 spreading follicles, woody, bright yellow when dehisced. 1 endemic variety. 200 species. Tropical.

Sterculia africana (Lour.) Fiori

Large deciduous tree with massive trunk to 15m, with smooth pinkish bark. Leaves crowded at ends of branches, 3 to 5-lobed, cordate at base, 7–15cm across, lobes acute, velvety beneath, scabrous above. Calyx yellow with pinkish streaks, 1–2cm across. Follicles ovoid.

Ecology. Succulent shrubland on the limestone escarpments and wadi sides; usually below 700m but rarely extending to over 1000m on the Diksam plateau. 100–1050m.

Distribution. Endemic variety.

Conservation status. Near Threatened (NT). *S. africana*, whilst widespread and abundant at present, is an important livestock fodder in dry periods. If livestock numbers increase greatly, or a succession of drought years occur, it could become threatened in some areas.

Botanical notes. The massive trunks of this tree are one of the characteristic sights of the limestone escarpments. It is usually pollarded which adds to its gnarled appearance. On Soqotra all material is referable to the endemic var. *socotrana* (K.Schum.) Fiori.

ETHNOGRAPHIC NOTES

Name(s)

– Soqotri: *bɔ̄hīn* (pl. *bˈehen*, *ˀˈibəhen*); fruit *ḳˈʌlˀʌš di bɔ̄hīn*, *ḳˈaḥyif di bɔ̄hīn*; seeds *śkīm d-il bɔ̄hīn*; pulped wood *kˈidihir*.

– Soqotri simplified: *bowhiyn* (pl. *behen*, *'ibihen*); fruit *qal'ash di bowhiyn*, *qahyif di bowhiyn*; seeds *shkiym di-l bowhiyn*; pulped wood *kidihir*.

Food (human). The seeds were an important food where the tree occurs. When immature and still pale in colour, they are baked in hot ash. Ripe, darkened seeds are eaten raw and are also ground like grain and used as flour. They are very filling: 10 seeds are said to adequately replace a meal, and a single tree in good condition can produce up to two sacks of fruit. The fruit are ready for harvesting from late summer through the monsoon season. The swollen base of a young seedling can also be eaten. It is said to be a filling food and also to quench thirst. However, as these trees are important as a source of both human and livestock food, this is discouraged.

Food (livestock). The foliage is one of the most important fodders of the island, especially for cattle. It increases lactation, and makes the goats and cows that feed on it fat and healthy. The fruit are also eaten by cattle and are considered to be a rich and fattening feed. The leaves and fruit casings are gathered and fed to weaning calves, along with softer grasses and *Ipomoea* and *Commelina* greenery. Camels also browse the leaves, and owners of milk camels gather foliage for them in times of hunger. The foliage persists into the dry season, making it a key dry season fodder. Towards the end of a prolonged dry season, herders cut leafy stems and gather dried, fallen leaves for their animals. Leafy branches keep for two or three days before drying out, and the slimmer branches for up to a week. Once the foliage is exhausted, the leafless twigs and stems are cut, and, when these are exhausted, the smaller branches. As the more accessible parts are lopped, the herders have to climb the trees to reach more distant branches. They do this with ropes of date-palm fibre, or by hammering a series of wooden pegs into the trunk to make footholds. If the drought persists, they progress to the larger branches, until finally, if no other fodder is available, they cut down the tree itself. The woody parts are pulped for cattle. The thin and papery outer layer is removed, and then the outer bark is cut away until the red heartwood is reached. The inner bark is roughly chopped and fed to the animals, and finally the wood itself is pulped for them. The pulped trunk is rather bitter, not sweet like that of the date palm, which is also chopped up to feed cattle in a drought. Some cows take time to accept this feed and are only driven to it by hunger. Goats are unable to tolerate it, but chew at the leafless twigs, fruit casings and inner bark. Gnawing marks are occasionally visible on young saplings, but the goats do not chew right through to the heartwood. The bark of the living adult tree is too tough for them, but they have been observed chewing at the bast if the outer bark is damaged, and in some cases this has damaged the trees.

Bees / honey. Honeybees are attracted to the flowers. They also nest in any hollow parts of the tree trunk. When this is noticed, the tree is left alone for a year, or until it is judged that they have had time to fill some combs with honey. Then two or three holes are made above the entrance hole, and some of the combs stacked up inside are removed.

Flavouring / food additive. If goats eat a lot the smell of their milk is altered, and sometimes also the taste. Cows surviving mainly on the pulped wood give milk with a strong and unpleasant flavour, and the meat and fat of the carcass are similarly tainted. Some people find the taste of the milk so unpleasant that they will not drink it, likewise the water the meat has been cooked in – normally an important part of a meal. Even the buttermilk and fresh butter have an unpleasant taste when eaten in the raw state, but the odd taste disappears once they have been cooked and further processed.

Fuel. Useless as firewood.

Wood. Soft and no use for making implements. However, within living memory, the trunks and larger branches were used to make rafts (*rēmuš*) used as a fishing platform and to transport goods to and from visiting boats.

Medicinal / veterinary. Bunches of the large leaves are gathered and fed to parturient goats and cows, to sickly and recuperating stock or to animals failing to thrive. After removing the seeds, the fruit are crushed to a paste, water added and the mixture stirred briskly. The resultant frothy, mucilaginous liquid is used to wash the scalp and to kill nits and head lice. The same preparation is used by women to clean themselves after menstruation and childbirth. The inner bark or fruit casings are crushed to a paste and rubbed over an aching head.

Cosmetic. Women crush the fruit to make a mucilaginous gum to smear over their fringe, temple hair and plaits to make the hairs lie down flat and give the hair a shine.

Other. Animals feeding on the lopped branches can go longer without water. The unripe seeds are very hard and are used in games. The tree is sometimes successfully transplanted as a seedling or sapling to beside a water catchment, settlement or inside the settlement compound. This tree has always been highly valued for its foliage, fruit, bark and heartwood, and has long been protected by the islanders.

Notes. In the Central Highlands goats are unused to it and reluctant to eat it. Livestock, especially goats, that are unaccustomed to it commonly develop scour, and this can last for up to a fortnight until their digestive system becomes adjusted. This tree, like the other soft-wooded trees (*Commiphora* species and *Adenium obesum*), is vulnerable to large, white larvae, the size of the termite queen, that eat away at the living wood from the inside. Two or three such creatures can be found inside the trunk of the dead tree. The islanders differentiate between 'male' trees (*bɔ̄hīn ḳārəṭ*) which produce less fruit and come into leaf only after rain, and 'female' trees which come into leaf in the summer months and bear many fruit. The wood of the 'male' is much harder than that of the 'female'. Tradition says that if this tree and the large *Grewia* species, *Commiphora ornifolia*, *Maerua angolensis* and *Acacia pennivenia* trees all come into new leaf at the end of the monsoon, then the winter rains will soon follow.

Comparative uses. In Dhofar resin exuding from the bark is worked into a lather with water and used as soap, shampoo and disinfectant. It is also rubbed into the scalp to get rid of head lice. In Kenya the seeds are cooked and eaten and the wood used to make utensils. Other species also provide edible fruit and seeds, and bark fibre from which cordage is made. Many exude a gum resembling tragacanth, that of *S. tragacantha* being used by African blacksmiths in smelting iron and to mend broken earthenware and calabashes. *S. urens* yields a gum called Indian tragacanth, glue is made from the proteins of the seeds, and its pulverised bark is given to women to facilitate delivery. The mucilage, produced by steeping its leaves and tender shoots in water, is given to cattle with pleuro-pneumonia. Its leaves are nutritious, rich in vitamin C and widely used as fodder, and its tender roots are cooked and eaten. (Ambasta 1982; Dalziel 1948, 1956; Heine *et al.* 1988; le Houerou 1980; Jain & De Fillips 1991; Miller & Morris 1988)

Waltheria L.

Herbs or subshrubs, with stellate and simple hairs. Leaves simple. Flowers bisexual. Calyx 5-lobed. Epicalyx segments narrow. Petals yellow. Stamens 5, fused at base, staminodes absent. Ovary 1-celled. Fruit a 2-valved capsule. 67 species. Tropical Americas, Africa, Madagascar, Malay Peninsula, Taiwan.

Waltheria indica L.

Perennial herb or subshrub to 60cm. Leaves narrowly elliptic to lanceolate, 2.5–7 x 1–2cm, tip acute, serrate, stellate-hairy beneath. Flowers 5–8mm across, in dense axillary fascicles. Calyx lobes c.2mm. Petals longer than sepals. Capsule obovoid, c.3 x 2mm.

Ecology. Clearings in montane semi-deciduous woodland. 900m.

Distribution. Pantropical.

ETHNOGRAPHIC NOTES

Name(s)

– Soqotri: 'type of *ṣafāḳa*' (*Tragia balfouriana*); 'type of *ṣafáḳa di ʾíləhe*' '*Scrophularia arguta* or *Urtica urens* for cattle'.

– Soqotri simplified: 'type of *zafaqah*'; 'type of *zafaqah di 'ilihe*'.

Food (livestock). Not generally liked by livestock and is not grazed except in great hunger.

Comparative uses. In India the plant is used as a febrifuge, purgative and emollient. The roots are chewed or a decoction drunk to control internal haemorrhage and to induce fecundity in women. The leaves and flowers are mucilaginous and are used on sores, the expressed juice of the leaves being used for eye diseases and syphilis. The flowers and root-bark are used against thrush. In Tanzania a paste of fresh leaves, mixed with sodium bicarbonate, is applied to an infected, swollen finger. The crushed leaves are applied to wounds to facilitate healing, and the leaves are boiled and the vapour inhaled to treat convulsions in children. The roots are used as an astringent, to treat eye ache, and as a restorative during the labours of harvesting. A decoction of the root is given to children to strengthen their resistance to fever, for syphilis, in a cough medicine and sometimes as an anti-epileptic. Powdered roots are used externally on wounds, and a preparation of the roots is taken internally for fever, dysentery and epilepsy. (Ambasta 1982; le Houerou 1980; Jain & De Fillips 1991; Oliver-Bever 1986)

TAMARICACEAE

Shrubs or small trees with slender branches. Leaves simple, scale-like with salt-secreting glands. Flowers actinomorphic, bisexual, white or pink in spike-like racemes. Sepals 5, free. Petals 5, free. Stamens 5. Ovary superior, 1-celled; styles 3 or 4. Fruit a capsule; seeds many, with a tuft of long hairs at apex. 5 genera. 78 species. Africa, Mediterranean to C Asia.

Tamarix L.

See description of family. 54 species. Africa, Mediterranean to C Asia.

Tamarix aphylla (L.) Karst.

Shrub to 2m (elsewhere a tree <10m); young branches reddish brown. Scale leaves grey-green, c.1mm, not overlapping, encircling stem at base, the stem thus appearing jointed. Racemes 3–6 x 0.3–0.6cm long. Flowers white (also pink elsewhere). Sepals c.1mm. Petals c.3mm. Capsule c.4mm.

Ecology. Rare on gravel soils on coastal plain. c.30m.

Distribution. N Africa south to Kenya and Somalia; Arabia east to Pakistan.

Botanical notes. The description is based on a single, somewhat atypical specimen – the only record of this widespread species so far found on the island. At this site (the coastal plain at Bandar Shu'ub) it grows with *T. nilotica*, where it seems to show a preference for the gravel plains while *T. nilotica* is restricted to the wadis.

ETHNOGRAPHIC NOTES†

Name(s)

– Soqotri: *ˈiʾiṭʌl, ˈaʾaṭʌl, ʾāṭʌl* 'clothed in rags, of impoverished appearance'; *ḳˈʌlḳəhʌl* 'swaying in the breeze' (general term for a variety of mainly halophytic plants).

– Soqotri simplified: *i'ital, a'atel, 'atel; qalqihal.*

Food (livestock). Not generally palatable to livestock on Soqotra except at times of severe drought, though camels will browse it when hungry. On Abd al Kuri and Samha, an important fodder for goats, and sheep eat the plant litter.

Bees / honey. The flowers are attractive to honeybees.

Flavouring / food additive. The salty leaves can be used as a salt substitute.

Timber. Makes good timber. In coastal areas the wood is used for rafters and cross beams. In date-growing areas of the southern plains, and around Qalansiyah especially, it is used to make the framework onto which palm fronds or leaf ribs are tied to build living quarters. The wood is resistant to termites and does not have to be debarked before use.

Fuel. The dead wood makes superior firewood which gives off a good heat. It is especially important as fuel on the outer islands where other firewood is scarce.

Cordage. In the outer islands especially, cordage is made from either inner bark or root-bark.

Wood. Slim stems are used to make the framework for fish traps. On the outer islands, thin strips of the outer and inner bark are used to weave fish traps, as well as to make the frame for them. The individually woven sections of the trap are bound together with cordage made from the bark. The wood is good, and is used to make various implements. In areas short of alternative woods, it was used in quicklime manufacture.

Comparative uses. Most *Tamarix* species are relatively unpalatable to livestock. In Saudi Arabia tamarisk are planted on farms as a soil stabiliser and windbreak. It is propagated from cuttings and is very wasteful of water. *T. aphylla* is commonly sown by bedouin to provide shade. They also plant it for the soft branches for their flocks to eat, although, since it excretes salt through special glands in its leaves, it is a very salty fodder. *T. aphylla* is also the source of an edible manna. In India the wood is used in the manufacture of ploughs and wheels and the bark and galls are used in tanning and as a mordant in dyeing. The bark is bitter and astringent and is used to treat eczema and other skin conditions. It is also boiled in vinegar and used against head lice. In Yemen and Dhofar the dark resin is eaten and salt from the leaves is used in times of salt shortage. In Dhofar the dried leaves are boiled in water and a solution given to a woman in difficult or prolonged labour. The dried leaves are powdered and used as a dressing for sores and wounds. When there are no tea leaves, the inner bark is dried, powdered and made into a tea which stains the mouth of anyone who drinks a lot of it. In northern Oman dried twigs are scorched at the fire, powdered and used to dress burns. *T. nilotica* also has a relatively large water requirement. It has extensive taproots to supply it with water but this is usually to the detriment of other nearby plants. Indeed it is considered a pest in areas where irrigation is practised as it colonises irrigation channels and streambeds. Its wood makes a good fuel, and is used for fence posts, wood pulp and to make furniture. It apparently contains a principle which combats liver damage. The bark is used to treat eczema and other skin disorders, and the insect galls as an astringent. In Yemen an infusion of the fresh leaves is used to treat diarrhoea. (Abdulfatih 1987; Ambasta 1982; Davis *et al.* 1983; Fleurentin & Pelt/Fleurentin *et al.* [1982, 1983a,b,c]; Goodin & Northington 1979; le Houerou 1980; Jain & De Fillips 1991; Kunkel 1984; Miller & Morris 1988; Zohary 1982)

†Also apply to *T. nilotica*.

Tamarix nilotica **(Ehrenb.) Bunge**
Tamarix arabica Bunge
Tamarix arborea sensu auct. non (Sieber ex Ehrenb.) Bunge
Tamarix gallica sensu auct. non L.
Tamarix mannifera sensu auct. non (Ehrenb.) Bunge
Tamarix mascatensis Bunge
Tamarix socotrana Vierh.

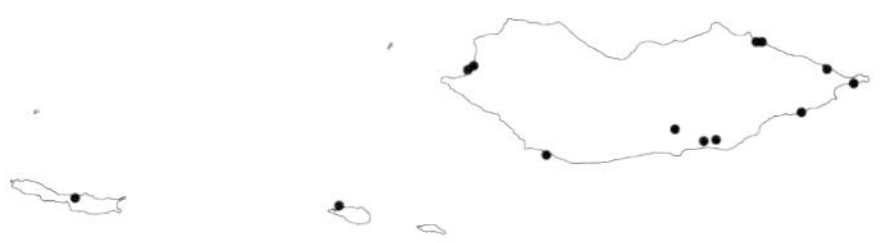

Shrub or small tree to 5m. Scale leaves grey-green, c.1–3mm, overlapping, half-clasping stem at base. Racemes 3–10 x 0.5–1cm. Flowers pink or white. Sepals c.1mm. Petals c.3mm. Capsule 2–3mm.
Ecology. Common and widespread by brackish water on the coastal plains; often forming dense thickets; occasionally growing by seepages on cliffs. Sea-level–250m.
Distribution. NE Africa, Middle East, Arabia.
Botanical notes. The genus is in need of a modern revision. *T. mascatensis* and *T. arabica*, which are separated on rather ill-defined characters, have been treated here as forms of the widespread *T. nilotica*. Records of the following species from Soqotra are all treated under this species: *T. arabica* Bunge, *T. arborea* (Sieb. ex Ehrenb.) Bunge, *T. gallica* L., *T. mannifera* (Ehrenb.) Bunge, *T. mascatensis* Bunge, *T. socotrana* Vierh.

ETHNOGRAPHIC NOTES
Name(s)
See *T. aphylla* (also uses).

THYMELAEACEAE

Shrubs. Leaves alternate, entire; stipules absent. Flowers in small terminal heads surrounded by broad, overlapping bracts, actinomorphic. Calyx with a long tube and 5 (sometimes 4) spreading lobes, densely hairy. Petals absent. Stamens 10 (or 8), sessile on calyx tube. Ovary superior, 1-celled; ovules 1 per cell, pendulous. Fruit small, dry, enclosed within persistent base of calyx. 50 genera. 720 species. Cosmopolitan, especially Australia and tropical Africa.

Gnidia L.

Lasiosiphon Fresen.

See description of family. 1 endemic. 140 species. Tropical and S Africa, Madagascar to W India and Sri Lanka.

Gnidia socotrana **(Balf.f.) Gilg**
Lasiosiphon socotranus Balf.f.

Evergreen shrub to 3m. Leaves narrowly obovate, 2–3 x 0.4–0.8cm, clustered on ends of branches. Flowers bright yellow, c.0.7cm diameter.
Ecology. Abundant in semi-deciduous woodland and shrubland on the western plateau and on the Haggeher and adjacent limestone plateaus; dominant in woodland in the northern wadis of the Haggeher mountains; also found on the summit of the coastal escarpment in the east of the island. 300–800m.
Distribution. Endemic.
Conservation status. Least Concern (LC). Common in several vegetation types and under no present or perceived threat.

ETHNOGRAPHIC NOTES
Name(s)
– Soqotri: *ḳˈiṭəhər*, *ḳēṭəher* 'very dark', 'leading astray' (western and central Soqotra); *ləˀˈəffən*, *leˀˈɛfən*, *ləˀɛ̄yef* (pl. *ləˀˈefiyun*) 'undependable, work shy'.
– Soqotri simplified: *qiytihir*, *qeytiher; li'iffin*, *le'efin*, *li'eyef* (pl. *li'efiyun*).
Food (livestock). The fresh greenery is not eaten by livestock, though they will eat fallen, dried leaves in hunger if there is nothing else available. Goats native to the area where it grows are better able to tolerate it than incoming goats.
Fuel. Much used as firewood, though not for cooking as its smoke is thick and acrid. The heartwood burns slowly with an extremely bright flame and 'sings' as it burns, so it is much used for illumination, and households generally keep some to hand for this purpose.
Wood. The stems of larger plants provide hard wood, which can be used to make small implements. Small pieces of dead wood are gathered and stored for use as tinder and kindling.
Medicinal / veterinary. The leaves are gathered and dried in the sun, before being stored in the family medicine chest. They are steeped in water or ground and added to water, and the filtrate drunk to treat constipation and as a vermifuge. Green leaves are also boiled in water and the filtrate taken in the same way. This medicament is taken for all sorts of stomach complaints with the aim of purging the stomach of the impurities believed to be at the root of the problem. It must always be taken on an empty stomach, preferably first thing in the morning. Taken in too great quantities or at too great a strength it is a powerful purgative and can be dangerous, causing abortion in women in early pregnancy, for instance. Nor is the treatment equally tolerated by everyone. In some it causes nausea and severe vomiting as well as diarrhoea. Some people advocate a regular mild dose to keep the digestive system in working order: a little taken before going to bed and again on rising. Since the arrival of modern medicine, with the diagnosis of malaria and its

treatment with bitter chloroquine, some islanders now take it as a malarial prophylactic, as well as to treat malaria. If the latter, the dose is followed some three hours later by drinking thickened or cooked soured milk. If someone has ingested the water larvae called *dilf*, and develops a painful and distended stomach, he is given a drink of powdered dry leaves in water to expel them. The illness called *horf*, whose main sign is a swollen liver and painful abdomen, is treated in the same way. After taking a strong dose, the patient is advised to remove himself from human habitation as the diarrhoea can be sudden and uncontrollable.

Cleansing. The tender green leaves are crushed with a little water, or on their own, to make a smooth paste used to wash the face and body. This is also an effective shampoo and disinfectant. Watered down, it is used as a wash to treat an aching and exhausted body or general malaise.

Pest control. The dead wood is used to make smudges, especially for sheep, during cold weather to deter biting flies and other insect pests. In areas affected by monsoon drizzle and mists, the smudges are lit every day until the weather clears. It is also used for smudges for goats in the winter months to get rid of botfly larvae and keep the fly at bay.

Other. In times when livestock theft was a problem, and herdsmen had to spend the night with their livestock to protect them, the wood was commonly used to keep the fire burning brightly throughout the dark hours to persuade a prospective thief that full watch was being kept. At times of dewfall, people go to these plants as well as to bushes of *Rhus thyrsiflora* to wash themselves in the dew that collects plentifully on them.

Comparative uses. Coumarins and the toxic mezerein are present in *Gnidia* species. In Oman a decoction of fresh leaves of *G. somalensis* is used as an emetic and purgative to 'clean out' the stomach, leaves taken from the upper part of the plant being considered to be the most effective. In Somalia it is used in the treatment of tuberculosis. The young leaves and flowering tops of *G. kraussiana* in Africa are very poisonous, and ingestion of only small amounts is sufficient to kill. The plants have the greatest toxicity when in flower. Extracts are used in West Africa as arrow and fish poisons, and also in criminal activities. (Ghazanfar 1994; Lewis & Elwin-Lewis 1977)

TILIACEAE

Herbs, shrubs or trees usually stellately hairy. Leaves alternate, stipulate. Flowers bisexual, actinomorphic, axillary or in leaf-opposed fascicles. Sepals 4 or 5, free. Petals 4 or 5, free. Stamens 10 to many. Ovary superior, 2 to 4-celled; placentation axile. Fruit a capsule or lobed or globose drupe. 48 genera. 725 species. Subcosmopolitan.

Corchorus L.

Herbs with simple or stellate hairs. Flowers solitary or in leaf-opposed fascicles. Sepals 4 or 5, free. Petals 4 or 5, yellow. Stamens c.10 or many. Ovary 2 to 4-celled. Fruit a many-seeded elongate capsule. 1 endemic. c.75 species. Tropical and subtropical regions.

Corchorus aestuans L.

Corchorus acutangulus Lam.

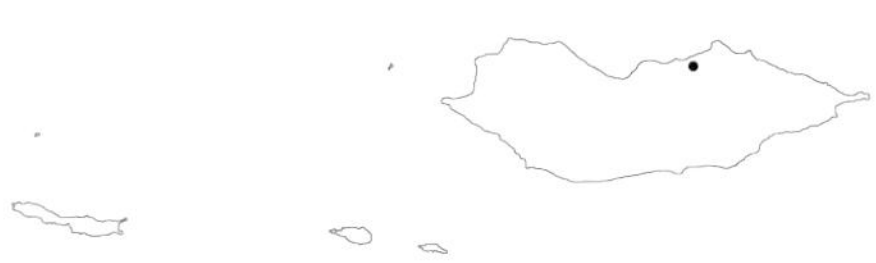

Ascending herb to 50cm. Leaves ovate, 2–12 x 1–5cm, tip acute, base rounded sometimes with basal setae, margin crenate-serrate. Flowers in 2 or 3-flowered fascicles, c.1cm across. Sepals and petals 5. Stamens many. Capsule cylindrical, 1.5–4 x 0.4–0.6cm, 3-locular, 3-winged, with 3 bifid teeth at tip.

Ecology. Weed of date gardens on the Hadiboh plain. 50m.

Distribution. Pantropical.

ETHNOGRAPHIC NOTES

Comparative uses. The leaves are used as a potherb and in salads, and in some places are cultivated as such. The plant yields a good fibre. In India the seeds are given as a stomachic. The whole plant, including the root, is used in the form of an injection for urethral discharges. (Ambasta 1982; Dalziel 1948, 1956; Jain & De Fillips 1991; Kunkel 1984)

Corchorus depressus (L.) Stocks

Corchorus antichorus Raeusch.

Prostrate, woody-based perennial herb to 20cm. Leaves oblong-elliptic, 0.5–2 x 0.2–1cm, margin crenate-serrate. Flowers in 1 or 2-flowered fascicles, 6–8mm across. Sepals and petals 4. Stamens c.10. Capsule linear-cylindrical, straight or curved, 0.8–2 x 0.2cm, 4-locular, minutely 4-toothed at apex.

Ecology. *Croton socotranus* shrubland on the coastal plains and dry foothills. 50–200m.

Distribution. N and E tropical Africa, SW Asia to India.

ETHNOGRAPHIC NOTES†

Name(s)

– Soqotri: *ˀírsib* (pl. *ˀərsíbihon*) (also given to *Portulaca* species); *mṣírbihon* 'plants of the winter rains' (a group of edible plants, including Portulacaceae); *kəndīyə* 'base of plant' (Samha).

– Soqotri simplified: *'irsib* (pl. *'irsibihon*); *mizirbihon; kindiyih.*

Food (human). These small plants are among the first to respond to rains and one of the last to die back. The

leaves are edible, as are the tiny swollen roots. Both are eaten raw. They are also gathered and strewn on the main cereal dish as a relish.
Food (livestock). The small leaves are a particular favourite with kids and lambs, but all livestock eat them, especially when grasses have died back. It is considered to be a beneficial fodder for livestock in the more poorly vegetated regions.
Fibre / cordage. It is ground with bark, with or without leaves, from shrubby *Hibiscus* (*H. macropodus*, *H. malacophylla*, *H. scottii*, *H. stenanthus* and *H. quattenensis*) to make a fibrous material that is twisted to make cordage of varying thicknesses.
Cleansing. The leaves are pulverised and used to wash the body, by women to clean themselves after menstruation and childbirth, and in laying out the dead. The plants are crushed to a paste with a little water until the mixture froths up, and this is then used as a shampoo and disinfectant. Sometimes the plants are ground with leaves of shrubby *Hibiscus* and used to wash the hair and cleanse the scalp. If a clay pot has become smelly, it is commonly cleaned out by rubbing it with a mixture of *Corchorus*, sometimes with added material from *Heliotropium odorum*. The head is dressed with a paste of the crushed plants after weaving 'false plaits' of black sheep's wool into the hair to make three fat plaits (one each side of the head and one falling down the nape of the neck). It acts as a lacquer to hold the hair in place and make it shine.
Other. Animals feeding on it can go longer without water, especially sheep. It is used to repair the clay pottery of the islands. If a pot is holed, a shard of a broken pot is ground down and shaped to fit the hole, the side appearing on the outer surface of the pot being broader than that of the inner. Then leaves of the plant are pounded to a paste with a soft chalky stone called *maḥkāḳəh*, possibly gypsum, and this is then packed tightly around the inserted shard, and the pot put beside the fire to harden and set the paste. If a pot is cracked, the paste is smeared along and around the crack on the outer surface and the pot put beside the fire to set the paste.
Notes. Small stock feeding heavily on it just after the rains can develop scour if the preceding dry season was prolonged or harsh. A common weed of cultivated plots, especially finger millet, and must be removed.
Comparative uses. In India the leaves of *C. depressus* are used as an emollient and the seeds as a tonic. In Yemen the leaves are used to treat urinary infections. In Qatar it is regarded as cooling, relieving pain when applied locally, and assisting the removal of tumours. The mucilage is used in treating gonorrhoea. The annual species, *C. olitorius*, is a popular summer potherb in Egypt, Jordan and Syria. (Ambasta 1982; Batanouny 1981; Fleurentin & Pelt/Fleurentin *et al.* [1982, 1983a,b,c]; Goodin & Northington 1979; Jain & De Fillips 1991; Rizk & El-Ghazaly 1995)

†Also apply to *C. erodioides*.

Corchorus erodioides **Balf.f.**
Corchorus erodioides var. *bicrenatus* Vierh.
Corchorus erodioides var. *pinnatus* Vierh.

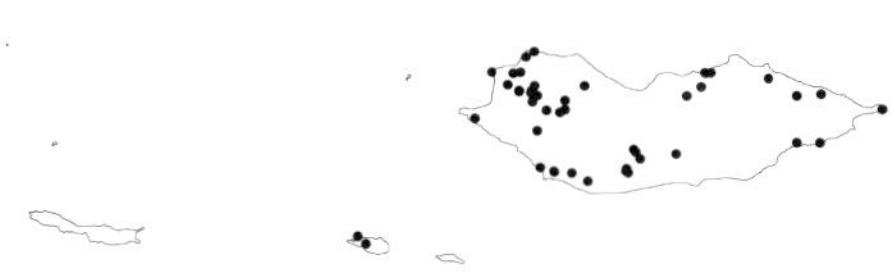

Prostrate herb to 15cm. Leaves in rosettes, oblong-ovate, 4–9 x 0.6–2cm, simple to pinnate or bipinnate, serrate to biserrate. Flowers solitary, c.1cm across. Sepals and petals 4. Stamens many. Fruits oblong-ovoid, 5–8 x 1–2mm, rounded at apex, 2-locular.

Ecology. Common and widespread on the coastal plains and limestone plateaus; absent from the granite of the Haggeher mountains. Sea-level–800m.
Distribution. Endemic.
Conservation status. Least Concern (LC). Common in several vegetation types and under no present or perceived threat.

ETHNOGRAPHIC NOTES
Name(s)
See *C. depressus* (also uses).

Grewia L.

Shrubs or trees. Flowers axillary or leaf-opposed. Sepals 5. Petals 5, yellow or white. Stamens many. Ovary 2 to 4-celled. Fruit a globose or a 2 to 4-lobed drupe with 1 to 4 nutlets; nutlets 1 or 2-seeded. 3 endemics. 150 species. Warm regions of the Old World.

Grewia bilocularis **Balf.f.**

Shrub or small tree to 4m. Leaves ovate, 4–20 x 5–8cm, tip acute, margin crenate-serrate, glabrous on both surfaces; petioles 1–2.5cm. Inflorescence leaf-opposed, 1 or 2-flowered. Fruits 2 to 4-lobed, c.2cm across, blackish red at maturity, glabrous, shining.
Ecology. Rare in semi-deciduous woodland. 60–800m.
Distribution. Endemic.
Conservation status. Vulnerable (VU D2). Rare, with a very limited area of occupancy.
Botanical notes. Flower colour not known.

ETHNOGRAPHIC NOTES†
Name(s)
– Soqotri: *ᶜərəṭ/d*, *ᶜārəyṭ/d*, (western Soqotra) *ġared* (pl. *aᶜaridə*, *ᵓaġaridə*); larger trees of the Haggeher mountains are sometimes called *ᶜərəṭ ḥahər* 'dark-coloured' (referring to their excellent timber); smaller trees of the drier areas are sometimes called *ḥariśɛmo* (more commonly given to *G. erythraea* and *G. tenax*).
– Soqotri simplified: *'irit/d*, *'areyt/d*, *ghared* (pl. *a'aridih*, *'agharidih*); *hahir*, *harishemo*.
Food (human). The fruit are eaten by birds and livestock, but not by people, although children sometimes nibble them.
Food (livestock). The fresh and fallen leaves are eaten by all livestock. The foliage is considered to provide one of the best feeds for goats and cattle, and leaves are

gathered and fed to pregnant and lactating females. Camels and donkeys also browse the foliage, and donkeys eat fallen leaves. The leaves persist into the dry season and they are an important dry season fodder for cattle and goats at this time. Leafy branches and fallen leaves are gathered by herders to feed their animals. They are seen as a nutritious and fattening feed which also increases milk yield.

Bees / honey. Good bee forage, especially trees and shrubs from higher mountain areas.

Fuel. The dead wood and bark make good firewood that burns with little smoke and is suitable for cooking. The smaller pieces burn with a clear light, so this is a preferred fuel when illumination is required. The wood burns slowly so is useful for processing milk. Large chunks are good for keeping a fire in overnight or during a long absence; alight and buried in ash it is said to continue smouldering for up to two successive nights. To relight the fire, it is removed from the ash and blown on to rekindle the flame.

Timber. One of the best timbers of the island. It is used for the main beams, supporting pillars and roof trees. The bark is attractive to termites so the wood is usually de-barked before use.

Wood. Hard and close grained. It is used to make locks and keys, weaving equipment, grain and ghee measures, camel saddles, and a variety of other implements, including combs. These are either single-sided or have teeth carved on both sides, those on one side being closer together than those on the other. The straight, slim branches are used to make the important herding staff. The wood makes an excellent charcoal.

Notes. Islanders differentiate between the 'male' tree (*ᶜərəṭ ḳiṭəher*), which bears little or no fruit, and the 'female' (*ᶜərəṭ ṣʌᶜalhʌl*) which has larger leaves and bears plentiful fruit. Tradition says that if this tree and *Sterculia africana*, *Commiphora ornifolia*, *Maerua angolensis* and *Acacia pennivenia* all come into new leaf at the end of the monsoon, then the winter rains will soon follow.

Comparative uses. Many species yield fibre for ropes and string, and provide good, strong wood for implements. In Africa and elsewhere the berries and seeds of *G. bilocularis* and other *Grewia* species are eaten, and the wood is considered to provide the best wood for sticks, clubs and bows. In East Africa the leaves are sometimes used like soap and to clean garments. The bark, which contains tannins, is boiled and the decoction drunk for miscarriage. The bark of the closely related *G. villosa* is pounded and soaked in water and given to a cow to drink for a retained placenta. In Dhofar the bark of this and *G. bicolor* is crushed and used to wash the body, clean the scalp and hair, and get rid of head lice. The edible fruits of some *Grewia* species, e.g. *G. bicolor*, are made into a fermented drink. The roots of this species are cooked in a soup given when close to giving birth. An infusion of the bark of *G. mollis* is used in beating mud floors, and is mixed with mud for the walls of huts to give a smooth surface. The bark is mucilaginous and is cooked in soup, or dried and pulverised to mix with bean meal and made into cakes. It is also used for cordage. The leaves and bark are used as chewsticks in West Africa, and, medicinally, as an evacuant and to treat fever and wounds. The leaves of many species are mucilaginous and are used to cleanse the hair of vermin. (Ambasta 1982; Dalziel 1948, 1956; Heine *et al.* 1988; Jain & De Fillips 1991; Lewis & Elwin-Lewis 1977; Miller & Morris 1988)

†Also apply to *G. mansouriana*, *G. milleri*, *G. turbinata*.

Grewia erythraea Schweinf.

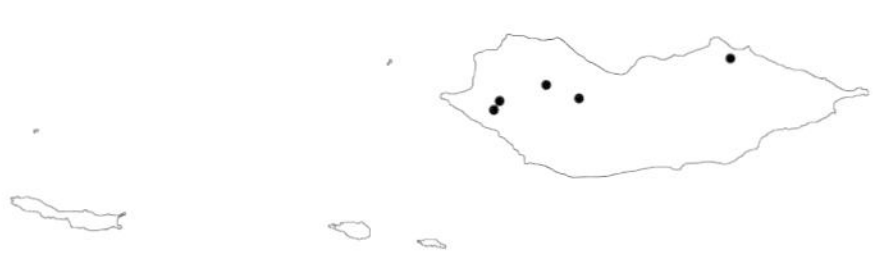

Shrub to 1m. Leaves obovate, 1–2 x 0.5–1cm long, tip rounded, margin serrate, glabrous or thinly stellate-hairy beneath. Flowers solitary or paired, 1.5–2cm across. Petals white. Fruits 2 to 4-lobed, orange turning purple, pubescent.

Ecology. Widespread but never abundant in dry shrubland on the northern and inland plains. 100–200m.

Distribution. Africa and SW Asia.

ETHNOGRAPHIC NOTES

Name(s)

– Soqotri: *ḥarśəhīm* (pl. *ḥarśum*, *ḥarɔ̄śim*); larger, better-grown specimens are sometimes called *ᶜərəṭ* like the other larger *Grewia* species; smaller plants are *hariśɛmo* 'little *ḥarśəhīm*'; even smaller plants are called *di ḥereśiyun* 'bachelor, unmarried' (because married men carry a proper herding staff, and this small shrub is too small to make one).

– Soqotri simplified: *harshihiym* (pl. *harshum*, *harowshim*); *ierit/d*; *harishemo*; *di hereshiyun.*

Food (human). The fruit are edible and sweet when fully ripe and red and were a popular food before the arrival of sugar. Birds also like the fruit.

Food (livestock). Important fodder in the drier areas of the island. It is grazed by cattle, camels, donkeys and small stock and considered a good and nutritious feed for small stock and immature animals. If the dry season is prolonged, the leafless twigs are lopped and fed to livestock.

Bees / honey. Honeybees are attracted to the flowers.

Flavouring / food additive. Before the days of sugar, the ripe fruit were an important source of sweetness for the islanders.

Fibre / cordage. String can be made from the bark, but it is not of a very high quality, and is not durable, so is only used for temporary purposes.

Fuel. The dead wood is good as firewood and is gathered for kindling, but sometimes gives off rather an unpleasant smell.

Wood. Good and can be used to carve small implements. It makes good charcoal

Comparative uses. In Dhofar the fruit are well liked, both when green and unripe, and when red and fully ripe. Elsewhere a preparation is made from the bark of *G. tenax* and given as a vermifuge. The wood and stems are used to treat coughs and pain in the sides. The bark is used to make cordage, and produces a mucilaginous gum used by women against hair vermin. (Baumer 1983; Dalziel 1948, 1956; Goodin & Northington 1979; Heine *et al.* 1988; le Houerou 1980; Miller & Morris 1988)

Grewia mansouriana **Abedin**

Small tree to 2m. Leaves ovate to elliptic, 2–6 x 1–3cm, tip acute, margin obscurely and irregularly crenate to dentate, glabrous; petioles 1–3cm long. Flowers pink, 2 to 4, in leaf-opposed racemes. Fruits 2 to 4-lobed, red at maturity, glabrous; lobes 3–5mm across.

Ecology. Scattered and never common in semi-deciduous woodland, particularly on sheltered cliffs and in gullies. 300–600m.
Distribution. S Arabia.

ETHNOGRAPHIC NOTES
Name(s)
See *G. bilocularis* (also uses).

Grewia milleri **Abedin**

Small tree to 5m. Leaves yellowish green, ovate, 3–6 x 1.5–4cm, tip obtuse, margin obscurely crenate or entire, shortly and softly velvety with stellate hairs on both surfaces; petioles 5–10mm. Inflorescence leaf-opposed. Petals yellow, 6–8mm long. Fruits spherical or obovoid, 5–10mm across, yellow to orange, glabrous, unlobed or rarely shallowly 2-lobed.
Ecology. Dry ravines and wadi sides. 10–150m.
Distribution. Endemic.
Conservation status. Vulnerable (VU D2). Rare, with a very limited area of occupancy.

ETHNOGRAPHIC NOTES
Name(s)
See *G. bilocularis* (also uses).

Grewia turbinata **Balf.f.**

Large shrub or tree to 4m. Leaves oblong-lanceolate, 5–12 x 2–4cm, tip acute or acuminate, margin serrate, glabrous above, softly velvety beneath with stellate hairs; petioles 1–2cm. Inflorescence leaf-opposed, 3-flowered. Fruits spherical or obovoid, 1–1.5cm across, green, glabrous, shining.

Ecology. Widespread but never common in semi-deciduous woodland on the Haggeher mountains and adjacent limestone plateau. 200–650m.
Distribution. Endemic.
Conservation status. Vulnerable (VU D2). Rare, with a very limited area of occupancy.
Botanical notes. Local people report that the flowers are yellow.

ETHNOGRAPHIC NOTES
Name(s)
See *G. bilocularis* (also uses).

Triumfetta

Herbs. Leaves alternate, simple or 3-lobed. Flowers yellow, in spike-like racemes. Fruit a globose or ovoid capsule covered with hooked prickles. 1 endemic? 100 species. Tropical.

Triumfetta **sp. *A***

Erect annual to 25cm; stems densely hairy with short stellate hairs and longer simple hairs. Leaves orbicular in outline, <3 x 3cm, simple or more usually obscurely 3-lobed, acute, margin irregularly crenate-serrate, sparsely hairy with simple and stellate hairs. Flowers and fruits not seen.
Ecology. In shade in dense woodland. c.850m.
Distribution. ?Endemic.

Conservation status. Data Deficient (DD). Inadequate evidence is available to make an assessment.
Botanical notes. Known only from a single sterile collection which could not be identified to any species with certainty. This specimen was found during the summer rains in woodland on the southern slopes of the Haggeher. At this time of year these slopes catch the SW monsoon and in some years receive substantial quantities of rain and drizzle. These slopes are poorly known and have not been extensively botanically explored during this season.

ETHNOGRAPHIC NOTES
Not seen, no information recorded.

UMBELLIFERAE

Herbs or shrubs; often aromatic. Leaves alternate, simple or pinnate. Flowers in simple or compound umbels. Petals free, white or yellow. Fruits dry, consisting of 2 mericarps; often ribbed or winged. 410 genera. 3100 species.

Botanical notes. *Foeniculum vulgare* Miller, a cultivated culinary herb, was recorded in the 19th century but has not been seen recently. It is an erect, yellow-flowered, strongly aromatic herb to 1.5m, the leaves are 2 to 4-pinnate and the leaf segments linear; the fruit is oblong, glabrous, ridged.

Apium L.

Herbs. Leaves simply pinnate, serrate. Umbels leaf-opposed. Fruits glabrous. 20 species. Temperate regions except Africa.

Apium graveolens L.

Erect, biennial herb to 50cm; strongly aromatic and glabrous throughout. Stems deeply grooved. Leaves dark green, pinnate, the ultimate segments triangular, usually 3-lobed. Flowers white. Fruits ovoid, ridged.

Ecology. A rare plant found in marsh by a small stream.
Distribution. Cosmopolitan.
Botanical notes. The whole plant smells strongly of celery.

ETHNOGRAPHIC NOTES

Not seen, no information recorded.

Centella L.

Stoloniferous herbs with entire leaves. Umbels compact. Fruits flattened, with 5, obscure ridges, reddish. 20 species. Mostly S Africa.

Centella asiatica (L.) Urb.

Hydrocotyle asiatica L.

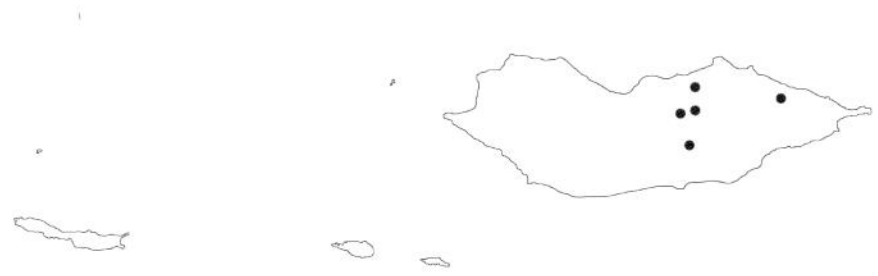

Prostrate herb with creeping stems. Leaves orbicular, 1.5–2cm across, with a cordate base and crenate margin. Umbels 1 to 5-flowered, on short (up to 1cm) axillary peduncles. Fruits reddish.
Ecology. In gravel by water and in wet grassland, particularly by seepages on the granite of the Haggeher mountains. 200–1000m.
Distribution. Pantropical.
Botanical notes. Resembles *Dichondra repens* (Convolvulaceae), also a plant of damp grassland, which can be distinguished by its 2-lobed capsule.

ETHNOGRAPHIC NOTES
Name(s)
– Soqotri: *gyāmis* 'creeping'; *ḥamžig* 'milkless, ungenerous' (more commonly given to smaller Cyperaceae); *ḳorźʌʾ*, *ḳerźaʾihʌn* (more commonly given to perennial Gramineae and Cyperaceae of damp ground).
– Soqotri simplified: *gyamis*; *hamjig*; *qordha'*, *qerdha'ihan.*
Food (livestock). Comes up quickly after the lightest rains and hungry livestock eat it at this time before better forage comes up. It is avoided by goats when growing in damp ground. It is not important forage on Soqotra.
Notes. Can cause *'uqabah* in small stock unaccustomed to it.
Comparative uses. Eaten in salads and soups and also widely used medicinally. It is considered to be diuretic and tonic, and is used in skin conditions and leprosy: a glycoside, asiaticoside, shown to be active in the treatment of leprosy has been isolated. Large doses are said to have a narcotic effect. The leaves and stems are anti-bacterial and the whole plant has an anti-fertility action in males. It is also reported to improve the mental ability and behaviour of mentally retarded children. (Kunkel 1984; Lewis & Elwin-Lewis 1977; Oliver-Bever 1986)

Nirarathamnos Balf.f.

Shrublet. Leaves simple, subcircular, parallel-veined. Umbels compound with 5 to 15 rays; umbellules 7 to 16-flowered; bracts and bracteoles similar, parallel-veined. Flowers white, yellowish or greenish. Fruit with prominent 5-ribbed mericarps; styles shorter than the conical stylopodium. 1 species. Endemic.

Botanical notes. Simple leaves are unusual within the family; because of this Balfour suggested its affinities were with *Bupleurum*, the other simple-leaved genus in the subfamily Apioideae. However, fruit anatomy has recently shown this genus to be distinct from other known genera of Umbelliferae and its closest affinities are not known.

Nirarathamnos asarifolius **Balf.f.**

Cliff-dwelling shrublet, to 50cm; stems contorted, densely covered with leaf scars. Leaves leathery, bright green, glossy, subcircular, 1–3.5cm long. Bracts and bracteoles 2–8mm long. Flowers c.1mm across. Fruit ovoid-elliptic,1.5–2.5 x 4–6mm; mericarps 5-ribbed, prominent, tuberculate; styles 0.5–1mm long.
Ecology. Locally common in the Haggeher mountains; in fissures of vertical cliff faces on granite pinnacles. 1000–1550m.
Distribution. Endemic.
Conservation status. Vulnerable (VU D2). Rare, with a very limited area of occupancy.
Botanical notes. Readily distinguished by its glossy simple leaves with their distinctive parallel venation.

ETHNOGRAPHIC NOTES
Name(s)
– Soqotri: nil or *ṙikəbə ṣʌ̇ʿalhʌl* 'large-leaved and riding high'; (rare) *dəbəhɔ̄mə*; (rare) *di tīyə* 'of cliffs'.
– Soqotri simplified: *rikibih za'alhal*; *di tiyih*; *dibihowmih*.
Food (livestock). Not normally accessible to livestock, but they are said to graze the foliage if they can reach it.

Oreofraga M.F.Watson & E.L.Barclay

Perennial herb with a short woody stem. Leaves 2 to 3-pinnate or 3-palmate. Umbels compound, pedunculate. Petals yellow. Fruits papillose-hairy, mericarps prominently ribbed; styles short. 1 species. Endemic.

Botanical notes. No obvious affinities in the family.

Oreofraga morrisiana **M.F.Watson & E.L.Barclay**

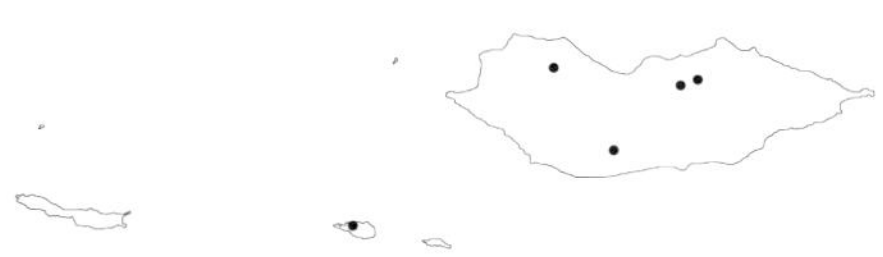

Erect perennial, 1–2.5m tall, with a short woody stem; stem solid, smooth, sparsely hairy, up to 50cm. Leaves waxy, glossy green, glabrous except for a few hairs on the veins beneath; leaflets 1–1.5 x 1–1.5mm, serrate with obtuse teeth. Umbels with 7 to 26 rays, white-hairy; bracts 4 to 14, oblong-lanceolate, 2–9mm, hairy. Umbellules 10 to 20-flowered; bracteoles 4 to 8, ovate, 1.5–3mm, hairy. Petals 0.5–1mm. Fruit oblong-oblanceolate, 2 x 3–4mm, ribs papillose-hairy; styles to 0.5mm.
Ecology. Rare, on cliffs on both the limestone and granite. 500–1100m.
Distribution. Endemic.
Conservation status. Vulnerable (VU D2). Rare, with a very limited area of occupancy.
Botanical notes. Relatively common in the zone on the limestone escarpment on Samha which captures low cloud and drizzle at certain times of year. On Soqotra it is restricted to cliffs on the limestone escarpment near Qalansiyah and is rare on the granite pinnacles of the Haggeher.

ETHNOGRAPHIC NOTES
Name(s)
– Soqotri: nil or *ṙikəbə ṣʌ̇ʿalhʌl* 'large-leaved and riding high'; *šinhīyə* (Samha).
– Soqotri simplified: *shinhiyih*; *rikibih za'alhal*.
Food (livestock). The foliage is not palatable to livestock.
Medicinal / veterinary. On Samha the root is dug up, baked in the embers, peeled, and then ground to a paste used as a wound dressing. Alternatively, the root is roasted slowly until quite blackened and dried out. It is then ground to a powder which is dried and stored in the household medicine chest. The powder is used to treat weeping sores, abscesses, ulcers and lacerations of all kinds. Both paste and powder are used to treat the surgical wound of circumcision. If this becomes infected, the skin is washed in seawater and the powder dressing re-applied daily until the wound has healed over.

Rughidia M.F.Watson & E.L.Barclay

Perennial herb or shrublet; stems hollow or solid, glabrous. Leaves 2 to 3-pinnate. Umbels compound, pedunculate, glabrous; rays 11 to 36; umbellules 9 to 16-flowered; bracteoles ovate to lanceolate. Petals white or cream. Fruits glabrous, the mericarps prominently ribbed, narrowly or broadly winged; styles short. 2 species. Endemic.

Rughidia cordata **(Balf.f.) M.F.Watson & E.L.Barclay**
Peucedanum cordatum Balf.f.

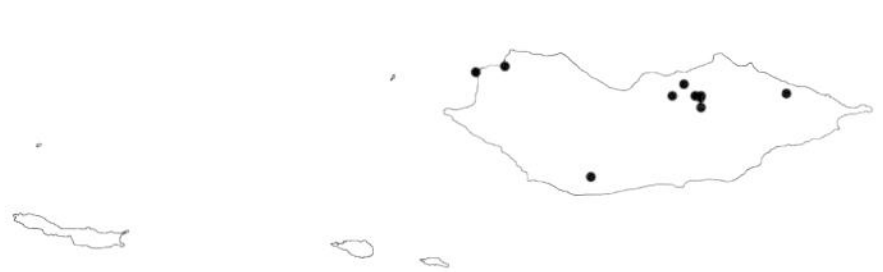

Erect perennial to 30–50cm; stems terete, smooth, hollow. Leaves shiny green, somewhat leathery; ultimate segments 1.5–4.5 x 1.8–4cm, margin coarsely serrate, entire near base; teeth acute. Umbels with 5 to 33 rays; bracts 0 to 4, oblong-lanceolate, often absent, 3–18mm. Umbellules 9 to 12-flowered; bracteoles 4 to 9, 1–7mm, oblong-lanceolate, entire. Petals white or cream, c.1mm. Fruit rounded-elliptic, cordate at base, compressed; mericarps with prominent ribs, the lateral ribs extending into broad wings, 2mm broad; styles 0.5mm.
Ecology. Locally common in shade in moist woodland. 500–1200m.
Distribution. Endemic.

Conservation status. Least Concern (LC). Common in several vegetation types and under no present or perceived threat.
Botanical notes. Balfour placed this species in the genus *Peucedanum* but commented that within the genus it was unusual on account of its peculiarly dissected leaves and cordate-based fruit. Recently molecular studies have shown it to be unrelated to any other species in *Peucedanum* but most closely related to the newly described *R. milleri*.

ETHNOGRAPHIC NOTES

Name(s)
– Soqotri: *ˈsegəhur*, *ˈsejəhər* 'stacked, with curly foliage'; (rare) *ḥarhīʾo* (more commonly given to edible orchids); edible part *śeˈriḥo di segəhur* 'root of *Rughidia*'.
– Soqotri simplified: *segihur*, *sejihir*; *harhiy'o*; edible part *sheriho di segihur.*
Food (human). This was a crucial food in the central highland area where no date palms grow and the cultivation of finger millet was limited. The roots can grow to an impressive length where the soil is deep enough. In more shallow earth, the root is shorter and tends to be more bulbous, and in some parts of the island it makes no useful root at all. It often grows in ledges on cliffs and people were lowered down on ropes to gather the roots. It is a popular food with a sweet flavour and smell. When freshly dug, it smells very like fresh carrots. In the central highlands, two sorts of root are distinguished, one with a 'white' stem and root, the other with a 'red' stem and root. The latter is regarded as a more mature version of the former, which is preferred for its tenderness. The root is eaten raw, baked in hot ash, or boiled in water. In hungrier years it was also prepared to make flour. The roots were peeled and then ground on a rock to make a coarse paste. If the root was very mature and had developed a hard core, this was removed before preparing the flour. The paste was then spread out in a layer to dry in the sun. It was stirred and turned until absolutely dry – it goes mouldy otherwise. Once dry, it was ground to a fine flour and any fibres removed. The dry flour was then put in a clay pot or goatskin and stored in a watertight cave. The flour was eaten plain or was mixed with fresh milk or buttermilk to make porridge. It was also added to water and cooked to produce a thick paste, which was best when eaten with a little ghee. If properly prepared and dried, islanders say it keeps very well and is little attacked by pests and mites.
Food (livestock). The foliage is not palatable to livestock in the high central mountains. However, herders of the limestone plateau areas below say it is eaten by goats, sheep and cattle, with no ill effects.
Commercial value (past, present). Half a sack of the flour used to be bartered for three *kīl* measures of grain.
Other. In the absence of grain to clarify butter the roots are crushed and dried, and used instead. After the clarified butter has been poured off, the butter-impregnated remains make a delicious treat.
Comparative uses. The roots of *Peucedanum palustre* are used as a ginger substitute. The leaves of *P. araliaceum* are used as a fumigant to scent clothing, and the bark of young shoots comes off in the form of a tube with which children play. Medicinally, the seeds and leaves of *P. aucheri* are taken for indigestion in the Middle East; the leaves of *P. fraxinifolium*, which smell like fennel, are taken as a vermifuge, boiled to make a lotion, and an infusion given as a diuretic and depurative. *P. ostruthium* is rich in coumarin derivates and is taken as a tonic to improve well-being. In Africa, *P. araliaceum* (syn. *P. fraxinifolium*) is an important medicinal plant. The roots are an ingredient in a preparation to treat serious abdominal pains and constipation, gonorrhoea and bilharzias. A decoction is given for malaria and as an analgesic in rheumatism (only in small amounts as it can cause dizziness). A root is chewed for a sore throat. It is used for snakebite and an infusion is drunk to treat allergic swellings. Stem-bark is burned and the ash rubbed into scarification to treat rheumatism. A decoction of bark is drunk with milk to treat dysentery and excessive flatulence. It is a remedy for asthma, and is pounded with leaves and added to water to make a bath for someone with swellings due to allergy. *P. araliaceum* is believed to cause abortion in goats. A cold-water extract of leaves and twigs is fed to goats and calves to treat sudden onset of breathing difficulties. (Dalziel 1948, 1956; Goodin & Northington 1979; Hedberg *et al.* [1982–1986]; Kunkel 1984; Lewis & Elwin-Lewis 1977)

Rughidia milleri **M.F.Watson & E.L.Barclay**

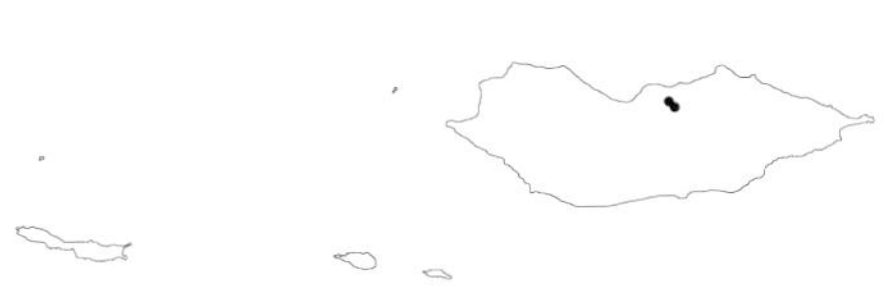

Cliff-dwelling shrublet to 50cm; stems contorted, densely covered with leaf scars. Leaves glossy green, glabrous but with a scabrous surface, somewhat fleshy; ultimate segments lobed, 8–16 x 9–16mm, serrate towards apex, attenuate at base; teeth acute. Umbels with 11 to 36 rays; bracts ovate, 1–10mm. Umbellules dense, 9 to 16-flowered; bracteoles 2 to 8, lanceolate, 1–4mm. Petals cream, c.1mm. Fruit narrowly ovate, 1.5–3 x 4–5mm; mericarp ribs prominent to slightly winged; styles 0.5mm.

Ecology. Locally common on the Rewgid plateau; on vertical rock faces on limestone cliffs. 600–750m.
Distribution. Endemic.
Conservation status. Vulnerable (VU D2). Rare, with a very limited area of occupancy.

ETHNOGRAPHIC NOTES

Name(s)
– Soqotri: nil or *ḳˈʌlʿiš* 'uneven, unstable' (in western Soqotra the name for *Sterculia africana* fruit); (rare) *bəṭˈiʾhin* 'slow', 'idle'; *ṣˈʌʿalhʌl* 'broad-leaved'.
– Soqotri simplified: *qal'ish*; *biti'hin*; *za'alhal.*
Food (livestock). Not normally accessible to livestock and they have not been observed feeding on it.
Other. Goats have been seen drinking from the cup-like leaves in dry weather.

Trachyspermum Link

Annual herb. Leaves pinnate with narrow segments. Umbels compound. Flowers white or cream. Fruits glabrous, papillose or hairy. 20 species. Tropical and NE Africa, C Asia, India and W China.

Botanical notes. For a revision of this genus on Soqotra see Barclay & Watson (1998).

***Trachyspermum pimpinelloides* (Balf.f.) H.Wolff**
Carum calcicolum Balf.f.
Carum kuriense Vierh.
Carum kuriense var. *trichocarpum* (Vierh.) C.C.Towns.
Carum pimpinelloides Balf.f.
Carum trichocarpum Vierh.
Trachyspermum calcicola Link
Trachyspermum trichocarpum (Vierh.) H.Wolff

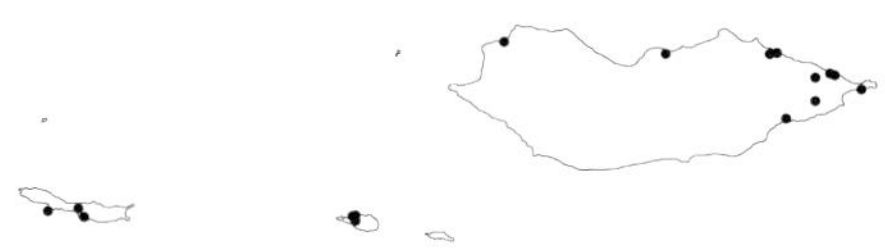

Erect annual to 25cm; stems glabrous or hairy. Leaves 1 or 2-ternate with 2 or 3 pairs of pinnae, ultimate segments narrowly obovate to linear. Umbels compound. Flowers white or cream. Fruit ovoid to rounded, 1–1.5 x 1mm, glabrous, papillose or hairy with hooked hairs.
Ecology. On sandy coastal plains and dry rocky slopes on limestone. Sea-level–500m.
Distribution. ?Somalia.
Botanical notes. Very variable in fruit indumentum; also possibly occurs in Somalia.

ETHNOGRAPHIC NOTES
Name(s)
– Soqotri: *tiˈlilyə, tilɛ̄liyə, tiˈlili, tililhˈiyo* (Soqotra and Samha); *ṣəbˈerbəher* 'astringent'; *ḥʌmź* 'sour'; *ḳˈʌṭemhem* 'butter-producing'; *di ḳīfˈiˀo, di ḳiyīfuf; sˈigərhen* 'dome-shaped'; *tˈomhor* (given to a variety of small and often fragrant herbs) (Abd al Kuri); *zəmɔ̄ṭə* (Samha).
– Soqotri simplified: *tililyih, tileyliyih, tilili, tililhiyoh; ziberbiher; hamdh; qatemhem; di qiyfi'oh, di qiyfiyfuf, sigirhen; tomhor; zimowtih.*
Food (livestock). This rarely comes up after the light summer rains in drier areas, but flourishes after the heavier winter rains. It is a very important fodder for sheep and goats, especially in the drier areas, and animals will travel great distances to feed on it. It is said to be one of the best feeds for putting on weight and improving lactation. Cattle also graze it, but it is not an important feed for them.
Flavouring / food additive. The carcass of a sheep or goat which has eaten a lot of it makes a particularly sweet and rich stock which, when cooled, makes a thick layer of fat. The meat too has a very sweet and distinctive flavour and perfume, which some say resembles that of frankincense. The milk of animals grazing heavily on it is particularly delicious, again with a faint but distinctive smell of frankincense.
Bedding, stuffing. Used to perfume bedding and put inside pillows or tucked amongst clothing. It is particularly used in bedding for sick people as its sweet smell is believed to relieve their discomfort and hasten their recovery. Plants that impart a perfume to the skin and clothing have an importance beyond that of their smell alone; they are also a means of promoting and maintaining good health.
Cleansing. Occasionally crushed and rubbed on the skin and clothing for its sweet smell.
Notes. Goats from the wetter Central Highlands who are unaccustomed to this feed develop mild staggers if they eat a lot, but soon recover. In the outer islands, the seeds of an umbel also called *zəmɔ̄ṭəh*, possibly caraway, were imported from East Africa and drunk as a tisane for various stomach complaints.
Comparative uses. The fruits of *Carum carvi* (caraway) is widely used as a spice and flavouring agent. The starchy roots of *C. carvi* and of *C. bulbocastanum* (black caraway) are eaten as a vegetable. Both yield an essential oil, Oil of Caraway or Oleum Carvi, used for flavouring and as a carminative. Other species are also eaten raw and used for flavouring. (Kunkel 1984; Sofowora 1982)

URTICACEAE

Monoecious or dioecious herbs, sometimes with stinging hairs. Leaves alternate or opposite, entire or variously dentate, simple or lobed; stipules present or absent. Flowers small and inconspicuous, usually unisexual, actinomorphic, arranged in axillary clusters or panicles or borne on receptacles. Perianth brownish or greenish, of 2 to 5 free or fused tepals; male flowers with 1 to 5 stamens; female flowers with a superior, 1-celled ovary; ovules 1; style 1 with filiform or brush-like stigma. Fruit an achene, free or enclosed in dry or fleshy persistent perianth. 52 genera. 1050 species. Tropical and few temperate.

Forsskaolea L.

Monoecious herbs, clothed with hooked hairs. Leaves alternate, toothed; stipules free. Flowers arranged on subsessile axillary receptacles; receptacles densely woolly within, surrounded by an involucrum of 2 to 8 bracts; male flowers with 3-lobed perianths and a solitary stamen, forming an outer ring; female flowers in centre of receptacle, without a perianth. Fruits enclosed in an involucrum of bracts. 6 species. Canary Is., SE Spain, Africa, Arabia, India.

***Forsskaolea viridis* Ehrenb.**

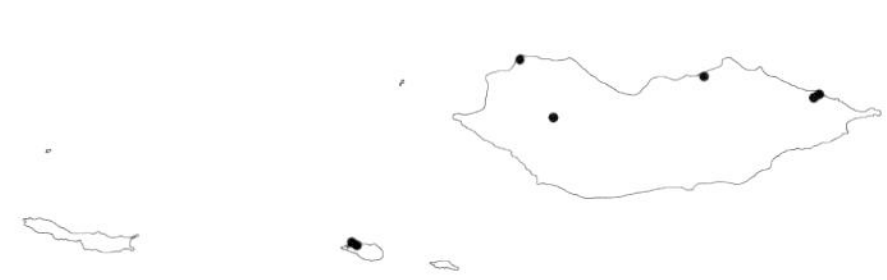

Erect annual herb to 50cm, thinly to densely covered in hooked hairs. Leaves ovate to rhombate, 1–9 x 1–4cm, acute, margins crenate-serrate, base truncate or cuneate. Involucral bracts broadly ovate to broadly obovate, 3–8 x 4–5mm, acute.
Ecology. Coastal plains and foothills. 10–300m.
Distribution. SW Africa, tropical E and NE Africa and SW Arabia.

ETHNOGRAPHIC NOTES
Name(s)
– Soqotri: *məltˈəwsek, ləskˈano* 'sticky, clinging'; *ḥɔ̄kənˈitin* (Samha).

– Soqotri simplified: *miltewsek*, *liskano*; *howkinitin*.
Food (livestock). Eaten by all livestock when green. However, it is regarded as having little nutritional value.

Comparative uses. The leaves are eaten as a potherb and the bark yields a good fibre. (Ambasta 1982; Jain & De Fillips 1991; Kunkel 1984)

Parietaria L.

Monoecious herbs. Leaves alternate, entire; stipules absent. Flowers green, bracteate, in 1 to several-flowered axillary clusters, unisexual or bisexual; perianths 3 or 4-lobed. Achenes ovoid, enclosed in persistent perianth. 20 species. Subcosmopolitan.

Botanical notes. A specimen (*Miller* 19064) collected at 650m in a shady limestone gulley on Samha is incomplete but perhaps referable to *P. alsiniifolia*, a widespread species in N Africa and SW Asia.

Parietaria debilis G.Forst.

Australina capensis sensu Balf.f. non Wedd.

Prostrate herb with weak stems to 50cm. Leaves broadly ovate to circular, <2 x 1.5cm, acute or rounded, often constricted towards tip, base rounded or subcordate, thinly hairy; petiole 5–15mm. Axillary clusters with mainly female and bisexual flowers; perianths c.1mm, ovoid in female flowers and cup-shaped in bisexual flowers. Achenes ovoid, c.1 x 0.6mm.
Ecology. In shade in montane woodland and thicket on granite in the high Haggeher mountains. 900–1450m.
Distribution. Tropical Africa, Madagascar, SW Arabia, tropical Asia, Australia, New Zealand and S America.
Botanical notes. A delicate plant which is only in evidence during the season of clouds and drizzle (cf. *Triumfetta* sp. and *Utricularia striatula*).

ETHNOGRAPHIC NOTES†
Name(s)
– Soqotri: *ṭáʾiyiḳ* 'liquid dung' (referring to the effect this has on livestock feeding on it); *mə́mbaᶜ* 'making land fertile that has not previously been cultivated'; *ḥʌ́źḳiḳ d-ilhítin* 'undergrowth plant for cows' (also given to *Pulicaria diversifolia*); *minhēyək*; *ríẑbeb* 'stalked'.
– Soqotri simplified: *ta'iyiq*; *mimba'*; *hadhqiq d-ilhitin*; *minehyik*; *rijibeb*.
Food (livestock). Grows quickly after the onset of the rains, and all livestock, except for sheep, readily graze it, and continue to do so even when the grasses have come up. It is especially liked by cattle and camels, and is regarded as an extremely nutritious feed. The whole plant is gathered and dried as hay and stored in weatherproof caves as a feed for the worst months of the dry season. It is mainly given to pregnant animals or those in milk, especially cows. In the Haggeher mountains this and *Pouzolzia auriculata* are the key non-grass fodders for cattle. Indeed, their value is equated with that of the best grasses.
Medicinal / veterinary. Like other rich and favoured fodders, it is gathered and offered to cows who are reluctant to let down their milk. This encourages them to relax and give their milk. It is also fed to sickly or parturient cattle to build up their strength.
Notes. If it is eaten by malnourished livestock in quantity at the onset of the rains, before grasses are available, it can cause scour. It can also cause severe *'uqabah* in goats unaccustomed to it.
Comparative uses. In Israel the leaves of *P. diffusa* are macerated and applied to external wounds, and in the Bermudas *P. floridana* is grown for tea. (Dafni *et al.* 1984; Kunkel 1984)

†Also apply to *Pouzolzia auriculata* (except Comparative uses).

Pilea Lindl.

Dioecious herbs. Leaves opposite, toothed; stipules fused. Flowers in dense axillary panicles. Male flowers with 2 to 4, free tepals and 2 to 4 stamens; female flowers 3 or 4-lobed, median lobe larger than lateral lobes. Achenes ovoid, compressed. c.250 species. Tropical and warm regions.

Botanical notes. For a revision see Friis (1989).

Pilea tetraphylla (Steud.) Blume

Erect herb to 10cm, glabrous. Leaves ovate, <3 x 2cm, acute or obtuse, crenate-serrate, broadly cuneate at base, upper 2 pairs appearing whorled. Flowers in a dense cross-shaped terminal panicle of mainly female flowers; male flowers in panicles in lower leaf axils. Achenes c.0.75mm.
Ecology. Carpetting shady rocks in semi-deciduous woodland. 850m.
Distribution. Tropical Africa, Madagascar and SW Oman (Dhofar).
Botanical notes. Rare, found only once during the summer rains in woodland on the southern slopes of the Haggeher. At this time of year these slopes catch the SW monsoon and in some years receive substantial quantities of low cloud and drizzle. These slopes are poorly known and have not been extensively botanically explored during this season.

ETHNOGRAPHIC NOTES
Name(s)
– Soqotri: nil or *eźərīʾyo* (*men méssə*) 'residue or sediment (from the rain)' (a general term for mosses).
– Soqotri simplified: *edheriy'oh* (*men messih*).

Food (livestock). Goats on Soqotra intensely dislike damp grazing or browse, but when hungry gnaw at the dried-out plants.

Comparative uses. Many related species have edible leaves and stems. (Kunkel 1984)

Pouzolzia Gaudich.

Monoecious herbs. Leaves alternate or rarely opposite, entire; stipules free. Flowers in axillary, bisexual clusters. Male perianth 5-lobed, stamens 5. Female perianth tubular, toothed at apex. Achenes enclosed in persistent perianth. 50 species. Tropical.

Pouzolzia auriculata Wight

Herb to 75cm. Leaves narrowly to broadly ovate, 2.5–10 x 0.75–4cm, acute or acuminate, base cuneate to rounded, pubescent; petiole 2.5–4cm. Flower clusters c.5mm across.

Ecology. In rock crevices and between boulders, in moist shade in the Haggeher mountains. 450–950m.

Distribution. India and Sri Lanka.

Botanical notes. One of the plants which come up with the cloud and mists of the monsoon season.

ETHNOGRAPHIC NOTES

Name(s)

See *Parietaria debilis* (also uses).

Comparative uses. The leaves and roots of many related species are edible, both raw and cooked. The bark of species such as *P. viminea* yields a fibre used for fishing nets and cordage, and *P. wightii* is used as a soap substitute. Medicinally, the leaves of *P. zeylanica* are believed to expel worms and are used as a vulnerary and cicatrizant for gangrenous ulcers. The juice, or a decoction of the leaves, is given to promote lactation, and a leaf poultice is applied to sores and boils. (Ambasta 1982; Jain & De Fillips 1991; Kunkel 1984)

Urtica L.

Monoecious herbs with stinging hairs. Leaves opposite, toothed; stipules free. Flowers in axillary spike-like panicles; perianth of 4, free tepals, inner pair longer than outer. Achenes smooth, enclosed by inner pair of tepals. 30 species. Subcosmopolitan, especially N temperate.

Urtica urens L.

Erect, monoecious annual herb to 50cm. Leaves ovate, 1–5 x 0.5–3.5cm, acute or obtuse, coarsely dentate; petioles 1–2cm. Panicles bisexual, spike-like, c.1cm long, clustered in leaf axils.

Ecology. In evergreen submontane woodland and shrubland. 1100–1200m.

Distribution. Temperate regions and tropical mountains.

ETHNOGRAPHIC NOTES

Name(s)

– Soqotri: nil or 'one of the *ṣafāḳa* group' (*Tragia balfouriana*).

– Soqotri simplified: 'one of the *zafaqa*'.

Food (livestock). Rarely eaten by sheep, goats or cattle, unless they are very hungry and there is no alternative forage. However, donkeys and camels find it palatable.

Comparative uses. Many *Urtica* species are used as a potherb and have nutritious seeds; their stems also provide strong fibres. Medicinally, in Saudi Arabia a leaf poultice of *U. urens* is used to treat muscular pain and rheumatism. In Qatar the plant is used to treat diarrhoea, haemorrhage and kidney ailments. The plant is also regarded as an aphrodisiac, and is used as an irritant, and as an ingredient in a fish poison. An infusion of the whole plant is given as a galactagogue and anti-haemorrhagic, and a preparation of the bark is taken for catarrh and bladder pains. The leaves are given as a diuretic. In northern Israel the powdered seeds of this and *U. pilulifera* are used to treat external wounds, and fresh leaves are applied to external inflammations. In homeopathy an extract of the fresh leaves is used for eczema, dysmenorrhoea, metrorrhagia and nosebleeds. Lotions are made from it to promote hair growth and lessen dandruff, and it is used externally to soothe wounds and ulcers. Fresh plants are used as a painful but effective rub for rheumatism. (Abdulfatih 1987; Ambasta 1982; Batanouny 1981; Dafni *et al.* 1984; Goodin & Northington 1979; Jain & De Fillips 1991; Kunkel 1984; Rizk & El-Ghazaly 1995)

VALERIANACEAE

Herbs. Leaves opposite, simple, margin undulate or slightly toothed; stipules absent. Flowers in lax cymes, somewhat zygomorphic. Calyx 5 or 6-lobed, reduced in flower but expanded in fruit with hooked lobes. Corolla tubular with 5 lobes. Stamens 3, attached near top of corolla tube. Ovary inferior, 1-celled; ovules 1 per cell, pendulous; style 1. Fruit dry, indehiscent, with 6 prominent hooks; seeds 1. 17 genera. 400 species. Cosmopolitan, especially N temperate and Andes.

Valerianella Mill.

See description of family. 1 endemic (extinct). c.50 species. N temperate to N Africa and SW Asia.

Valerianella affinis Balf.f.

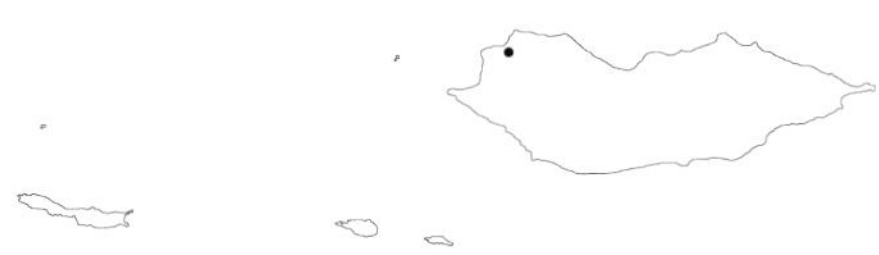

Delicate annual herb to 15cm, with lax dichotomous branches. Leaves linear-elliptic to obovate, 0.8–2 x 0.3–0.5cm.

Ecology. 'Not uncommon annual on dry hill slopes near Galonsir [Qalansiyah]' (Balfour 1888).

Distribution. Endemic.

Conservation status. Extinct (EX). Known only from a single collection made by Balfour on Ma'alah in the 19th century. Although the plant is insignificant and easily overlooked, the site where this mysterious plant was originally found has been surveyed many times over the last 100 years and consequently it must now be considered extinct.

Botanical notes. The only other species of this predominantly Mediterranean and N African genus recorded from the region is *V. microcarpa* Lois from Kenya and Ethiopia. *V. microcarpa* is widespread in the Mediterranean region but it is not clear whether it is native to or introduced into tropical NE Africa. The status of the Soqotran *V. affinis* needs further investigation. For the time being it remains an example of a rare Mediterranean–Soqotran geographical disjunction and has interesting biogeographical implications.

VERBENACEAE

Trees, shrubs or herbs. Leaves opposite, toothed or entire; stipules absent. Flowers in cymes, racemes or spikes, somewhat zygomorphic. Calyx 2, 4 or 5-lobed. Corolla 4 or 5-lobed. Stamens 4. Ovary superior, 4-celled; ovules 1 per cell, placentation axile; style 1. Fruit a drupe or 2 nutlets. 91 genera. 1900 species. Tropical with some temperate.

Clerodendrum L.

Trees or shrubs. Leaves entire. Flowers cream or greenish, in few-flowered cymes. Calyx 5-lobed. Corolla with 5 spreading lobes and a short tube. Anthers long-exserted from calyx tube. Fruit a globose drupe. 2 endemics. 400 species. Tropical and warm, especially Old World.

Clerodendrum galeatum Balf.f.

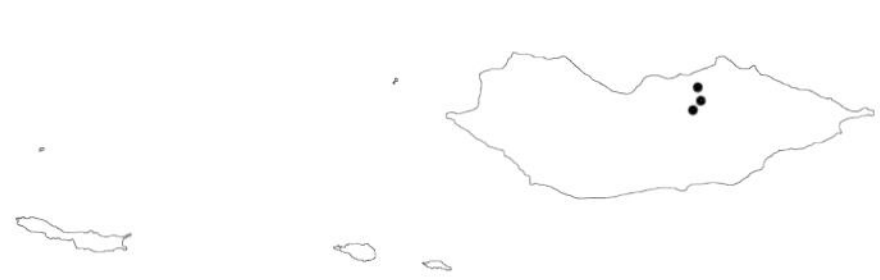

Shrub or small tree. All parts ± tomentose. Leaves broadly ovate or elliptic, 3–8 x 2–4cm, densely tomentose above, more thinly beneath.

Ecology. Dense submontane semi-deciduous thicket. 1050–1200m.

Distribution. Endemic.

Conservation status. Vulnerable (VU D2). Rare, with a very limited area of occupancy.

Botanical notes. Less common than the next species; found only in the Haggeher mountains.

ETHNOGRAPHIC NOTES†

Name(s)

– Soqotri: *sənémhən* (pl. *sənémhiyun*) 'ill-smelling' (also given to *Coelocarpum* species in western Soqotra; both these species have a very sweet smell so the name is surprising); (rare) *ṭīfiṭ* (western Soqotra) (more commonly given to *Placopoda virgata* and *Dirachma socotrana*).

– Soqotri simplified: *sinemhin* (pl. *sinemhiyun*); *tiyfit.*

Food (livestock). One of the most important dry season fodders for cattle and goats. For cattle of the drier upland areas it is the key dry season feed, though when other forage is available after rains, they eat little of the foliage. In the dry season, herders lop leafy branches and collect plant litter for their animals. In a prolonged dry season both cattle and goats tear off and chew the leafless twigs. Camels are very keen on it and browse it all year round, and sheep eat fallen leaves. It is regarded as a very good fodder for livestock, making the animals fat and increasing their milk yield.

Bees / honey. Important forage for honeybees.

Flavouring / food additive. Animals eating a lot of the foliage give very sweet milk. However, if goats eat a lot in the dry season their meat can develop an unpleasant flavour, and the water the carcass is cooked in usually has to be discarded.

Fuel. The dead wood makes good firewood and burns with little smoke, making it good for cooking and heating enclosed areas. The wood lights well even in wet weather and supplies of kindling are often stored in the home for this purpose.

Timber. Good, durable timber. The trunks and larger branches of taller specimens are used for the main roofing beams and rafters, and sometimes also for the supporting pillars. It is relatively termite resistant and does not have to be de-barked before use.

Wood. Hard and close-grained and used to make a variety of implements. The wood of this tree is considered to be the very best for making the important herding staff.

Cleansing. The dead wood, and especially the roots, give off aromatic smoke, and it is gathered and kept at home for this purpose. Some people even prefer it to *Cephalocroton socotranus* wood. Slivers are burned on a slip of pottery or thrown on the embers to give living quarters a pleasant smell and to fumigate them. Clothing is also held in the smoke of the burning wood to perfume it. The dead roots give off a very sweet smell when burned, and the flowers are perfumed. Sprigs are put among clothing and tucked into the turban or waistband. The leaves too are sweetly scented and are picked and rubbed between the hands and over the skin and clothing to give them a pleasant smell.

Other. The wood is burned to rid living quarters of creatures believed to be harmful to man. These are called *réʾeś*, and include some (but not all) reptiles, such as skinks (*gumš*), the nocturnal *Hemidactylus* (*rə́ḥḥo*) and snakes (*bikīli*). They also include scorpions (*di ḳáʿanhin*), wolf spiders (*fiṭámə*), biting ants (*díftīm*), the giant *Scolopendra* centipede (*di ḥʌ́ṣəhʌn*) and the chameleon (*məśḥéybihi*).

Notes. The reason for a name that means 'ill-smelling', when in fact all parts are appreciated for their sweet smell, is not clear. The islanders differentiate between the 'female' (*sənéməhən báʿalə*), which is more sweetly perfumed, and the 'male' (*sənéməhən ḳārəṭ*), whose wood is better for making the herding staff. There is a popular short poem known all over Soqotra about the staffs made from the wood of this tree: '*Grewia* wood breaks if you strike a blow with it,/and *Hibiscus* wood splits./Only *Clerodendrum* make a proper herding staff/that leaves its mark on whatever it strikes!'

Comparative uses. Many related species have edible leaves and flowers, but some are foul-smelling. In East Africa the root of the closely related *C. myricoides* is boiled with meat and taken as a laxative. It is also given for gonorrhoea and, with milk, for joint disease. The leaves of *C. rotundifolium* are pounded, soaked in water, and then used as a wash to get rid of fleas on livestock and bed bugs in living quarters. The Borana put the root of *C. myricoides* on the fire and inhale the fumes for a headache. In Qatar the leaves of *C. glabrum* are used as a remedy for colic, cough, fevers, intestinal parasites and convulsions in children. The plant is also used as a purgative for calves. (Ambasta 1982; Batanouny 1981; Dalziel 1948, 1956; Goodin & Northington 1979; Hedberg *et al.* [1982–1986]; Heine *et al.* 1988; Jain & De Fillips 1991; Kunkel 1984; Lewis & Elwin-Lewis 1977; Rizk & El-Ghazaly 1995)

†Also apply to *C. leucophloeum*.

Clerodendrum leucophloeum **Balf.f.**

Small tree or shrub to 3m. Leaves obovate or oblong-elliptic, 1.5–5 x 1–3.5cm, glabrescent above, thinly tomentose beneath.

Ecology. Widespread in deciduous and semi-deciduous woodland. 100–850m.

Distribution. Endemic.

Conservation status. Vulnerable (VU D2). Rare, with a very limited area of occupancy.

Botanical notes. Commoner and more widespread than the preceding species.

ETHNOGRAPHIC NOTES

Name(s)

See *C. galeatum* (also uses).

Coelocarpum Balf.f.

Shrubs or small trees. Leaves toothed. Flowers in racemes. Calyx 5-lobed. Corolla with 5 short lobes and a long tube. Stamens included in tube. Fruit an ovoid drupe. 2 endemics. 7 species (6 species in Soqotra, Somalia and Madagascar).

Coelocarpum haggierensis **A.G.Mill.** (see Appendix 2, p. 723)

Shrub or small tree to 3m. Leaves ovate to elliptic, 2–10 x 1.2–4cm. Flowers mauve, c.0.5cm long, in dense racemes; bracts 2–5mm long, usually longer than the pedicels. Fruits red, c.0.6cm long.

Ecology. Thickets in semi-deciduous submontane woodland; more rarely as a low shrub on granite slopes (above Firmihin) or on limestone cliffs (Rewgid); also found on seaward-facing limestone cliffs in the northeast of Soqotra. 700–1520m.

Distribution. Endemic.

Conservation status. Least Concern (LC). Common in several vegetation types and under no present or perceived threat.

Botanical notes. Local in dense thicket in the Haggeher mountains, but it is also less common in other areas, for instance on the higher seaward-facing cliffs at Reqadihon (above Hoq). These cliffs attract low cloud and rain during the winter and are a refugium for some species (e.g. *Dirachma*) which are otherwise found only in areas of high rainfall in the Haggeher and adjacent limestone plateaus.

ETHNOGRAPHIC NOTES

Name(s)

– Soqotri: *nifəzítin* 'full-uddered'; *di tīyə* 'of cliffs'; *ḥáhər* 'dark-coloured' (commonly given to trees that produce good timber); *ṣáʿalílihon* 'large-leaved'; *məgḗred, míjərid*

(more commonly given to *C. socotranum*, and also sometimes to *Leucas kischenensis* and *Heliotropium* species also used to cure leather).
– Soqotri simplified: *nifizitin*; *di tiyih*; *hahir*; *za'alilihon*; *migeyred*, *mijirid*.
Food (livestock). All livestock eat the foliage. Cattle like it and eat it whenever it is available. Goats usually only eat it when there is little other forage available. Sheep eat any leaves within reach, including fallen ones. Herders lop leafy branches to feed their animals during the dry season. However, it is not considered to be nutritionally important.
Fuel. Not liked as firewood.
Leatherwork. The foliage and stems are crushed and the paste used to cure leather and stain it a dark red-brown. It is popular especially for treating the smaller, softer skins used for making more personal items, such as the satchels used by men and women for carrying their private possessions. *Lavandula* foliage is often added to the paste to give it a sweet smell and to darken the colour.
Wood. Slimmer and straighter branches of the better-grown specimens are used to make the important herding staff.
Cleansing. Not as fragrant as *C. socotranum*, but sometimes used in the same way.
Other. Crushed foliage and soft stems dye the hands a dark yellow, like turmeric, but there is no tradition of using the plant as a dye.

***Coelocarpum socotranum* Balf.f.**
Coelocarpus socotranus Balf.f.

Low shrub to 40cm. Leaves 1.5–2(–4) x 1–1.5(–2)cm. Flowers white, c.0.5cm long, in few-flowered racemes; bracts 1.5–3mm long, usually shorter than the pedicels. Fruits c.0.5cm long.
Ecology. Dwarf shrubland on the eastern and western limestone plateaus. 200–650m.
Distribution. Endemic.
Conservation status. Least Concern (LC). Common in several vegetation types and under no present or perceived threat.

ETHNOGRAPHIC NOTES
Name(s)
– Soqotri: *məgēred* (also sometimes given to *Leucas kischenensis* and *Heliotropium* species also used to cure leather); *ṭʌlʿʌt əzēnət* 'beautiful/perfumed growth'; *rīḥ ədini* (from Arabic, 'perfume of the world'); (in parts of western Soqotra) *ṭīfiṭ* or *sənemhen* (see *Clerodendron galeatum*).
– Soqotri simplified: *migeyred*; *tal'at ezeynit*; *riyh idini*; *tiyfit* or *sinemhen*.
Food (livestock). All livestock eat the foliage, both green and dried, but it is not seen as being important nutritionally.
Fuel. The dead wood makes good firewood and produces a pleasantly perfumed smoke, making it popular for indoor fires.
Timber. Normally too small to be useful as timber, but larger branches and trunks are used as roofing infill.
Leatherwork. One of the more important plants of the islands for preparing skins and hides to make leather. It makes the leather supple and strong, and dyes it a clear red-brown. The roots are most effective, but stems and foliage are also used.
Wood. Not very hard but acceptable for utensils and implements for temporary use.
Cleansing. The flowers are sweet smelling and are put amongst clothing or worn tucked into the waistcloth or turban. The dead wood is often gathered and kept at home to be burned for its fragrant smoke. It is important for this purpose in the drier areas where *Dirachma socotrana*, *Clerodendrum* species and *Cephalocroton socotranus* do not grow. It is used at weddings, or by married women to fumigate their persons and their clothing. Chips of the dead wood are burned in clay burners and carried around the living quarters to give them a pleasant smell.
Other. When the green berries ripen and turn blue-black, they produce a dark blue stain, but there is no tradition of this having been used as a dye on Soqotra. Crushed leaves and stems were occasionally used to dye cotton twist or white sheep's wool an orangey-red. Spun balls of the cotton or wool fibre were put in a pot with the ground material, weighed down with a rock to keep them under the surface of the water, and left to steep until the required colour had been obtained. The colour was not fast but faded on washing and exposure to the sun.

Phyla Lour.

Herbs. Leaves toothed. Flowers pink, in short congested spikes on long, erect peduncles. Calyx 2-lobed. Corolla with short tube and 4 spreading lobes. Stamens included in tube. Fruit 2 smooth nutlets enclosed by persistent calyx. 15 species. Tropical and warm regions.

***Phyla nodiflora* (L.) Green**
Lippia nodiflora Michx.

Prostrate herb rooted in mud. Leaves obovate to oblanceolate, 1–4 x 0.3–2cm, apex toothed. Flower spikes 0.5–2cm long; peduncles 2–8cm long.
Ecology. In mud by seepages and permanent water. Sea-level–500m.
Distribution. Africa and SW Asia.
Botanical notes. One of the commonest plants of marshy areas.

ETHNOGRAPHIC NOTES
Name(s)
– Soqotri: nil or *ḳāźəb*, *ḳʌźʌbihon* (more commonly given to various small Gramineae and Cyperaceae of damp

ground); *dilḥɔ̄ź* (also given to the grass *Aeluropus lagopodioides*); *birīˀtin*.
– Soqotri simplified: *qadhib*, *qadhabihon*; *dilhowdh*; *biriy'tin*.
Food (livestock). Provides incidental forage for livestock, but is not seen as having any nutritional significance. Goats, which dislike damp forage, generally avoid it.
Comparative uses. Elsewhere the leaves are eaten. In Sudan it is regarded as a good fodder and turf plant, and it is generally valued in hot and dry countries for making lawns. Medicinally, in Qatar an infusion of the tender stalks and leaves is given to children suffering from indigestion and to women after delivery, and in Bahrain the leaves are used as a diuretic and anti-pyretic. In Iraq the whole plant is considered to be cooling and it is used to treat stoppage of the bowels and painful knee joints. A leaf poultice is given for boils, swollen cervical glands, erysipelas and indolent ulcers. The plant contains a large quantity of potassium nitrate, which might explain its diuretic action. (Abbas *et al.* 1992; Ambasta 1982; Batanouny 1981; Bebawi & Neugebohrn 1991; Dalziel 1948, 1956; Jain & De Fillips 1991; Kunkel 1984; Rizk & El-Ghazaly 1995)

Priva Adans.

Herbs. Leaves toothed. Flowers white or pink, in long, open spikes. Calyx 4-lobed. Corolla with long tube and 5 spreading lobes. Stamens included in tube. Fruit 2 spiny nutlets enclosed by persistent calyx. 1 endemic. 20 species. Tropical and warm regions.

Priva socotrana Moldenke

Priva adhaerans sensu auct. non (Forssk.) Chiov.
Priva cordifolia sensu auct. non (L.) Druce
Priva leptostachya sensu auct. non Juss.
Priva tenax Verdc.

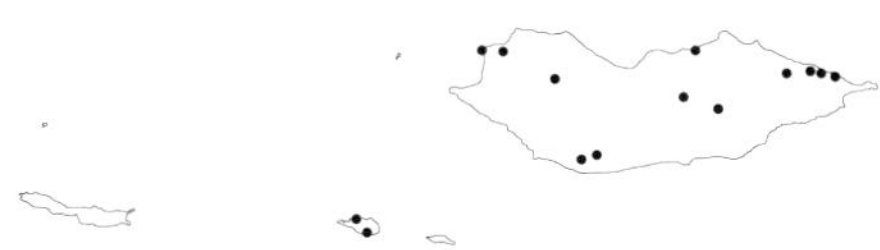

Erect herb to 50cm. Covered in hooked hairs making the plant rough to touch. Leaves narrowly ovate or ovate, 1–5 x 0.3–1.5cm, coarsely toothed. Flowers c.1cm long.
Ecology. Widespread and common on the limestone escarpments and plateaus. 50–950m.
Distribution. Endemic.
Conservation status. Least Concern (LC). Common in several vegetation types and under no present or perceived threat.

ETHNOGRAPHIC NOTES
Name(s)
– Soqotri: nil or *mˈʌhariś* 'undergrowth, understorey'; *məltˈuwsek* 'sticky, clinging'; *ləṣˈʌʿhān* 'sticky'; *ṣefirˈirihon* 'many-flowered'; *ṣɔ̄wbəher* 'sour, astringent'.
– Soqotri simplified: *maharish*; *miltuwsek*; *liza'han*; *zefiririhon*; *zowbiher*.
Food (livestock). When young and tender after rain, the plant is readily grazed by all livestock. It is seen as having limited nutritional value.
Notes. Can cause *'uqabah* in small stock unaccustomed to it.
Comparative uses. The tubers of *P. grandiflora* are eaten, and the leaves of *P. lappulacea* used as a tea. (Kunkel 1984)

VIOLACEAE

Herbs or subshrubs. Leaves simple, alternate, margins crenate or serrate; stipules present. Flowers solitary in the leaf axils, zygomorphic. Sepals 5. Petals 5. Stamens 5, forming a ring around the ovary. Ovary superior, 3-celled; ovules many, parietal; style 1. Fruit a capsule splitting into 3. 23 genera. 830 species. Cosmopolitan.

Hybanthus Jacq.

Ionidium Vent.

Herbs or subshrubs. Petals unequal, lower petal expanded. 150 species. Tropical and warm regions.

Hybanthus enneaspermus (L.) F.Muell.

Alsodeia socotrana Balf.f.
Ionidium suffruticosum Ging.

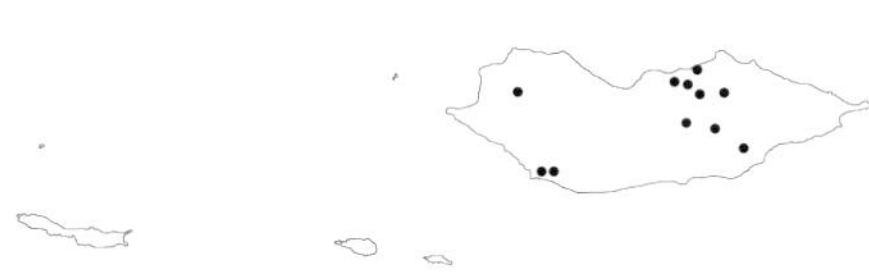

Small subshrub or creeping herb to 30cm. Leaves linear to linear-oblong, 2–3 x 0.3–0.7cm, margin serrate. Flowers blue or purple.
Ecology. Widespread and common in woodland and shrubland. 150–1400m.
Distribution. Africa, Madagascar, Arabia, India, Sri Lanka, S Asia to N Australia.

ETHNOGRAPHIC NOTES
Name(s)
– Soqotri: *mˈʌḥəriṭ* 'to strip the leaves from a stem'; *šeh* 'private parts'; *ḥəwrˈirhin* 'dark-coloured'.
– Soqotri simplified: *mahirit*; *sheh*; *hewririhin*.
Food (livestock). All livestock are very keen on it and it is usually eaten out as soon as it comes into leaf. It is considered a nutritious and fattening feed.
Notes. Can cause *'uqabah* in goats feeding heavily on it if they are not used to it.
Comparative uses. The foliage is sold in markets in Nigeria. It is added to the food of pregnant and parturient women to strengthen the child and to help the mother recover. An infusion is given to infants. In India it is regarded as a demulcent, diuretic and tonic, and the root is used for urinary infections and bowel complaints in children. (Ambasta 1982; Dalziel 1948, 1956; Jain & De Fillips 1991)

Viola L.

Herbs. Petals ± equal, lower petal shortly spurred. 400 species. Temperate regions.

Viola cinerea **Boiss.**

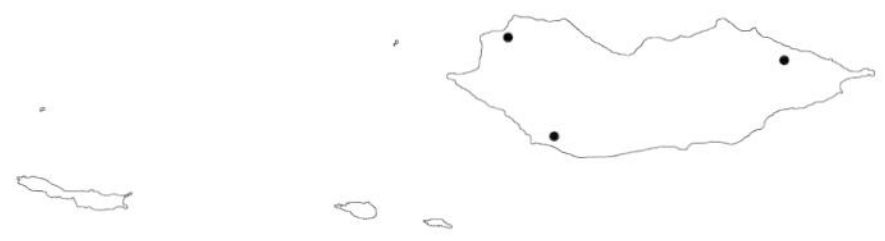

Erect annual herb to 15cm. Leaves elliptic to obovate, 1–2 x 0.5–1.5cm, margin crenate. Flowers purple.
Ecology. Amongst rocks on the limestone escarpments. c.600m.
Distribution. NE Africa, Arabia, S Iran, Afghanistan, NW India.
Botanical notes. Plants on Soqotra are referable to var. *stocksii* Boiss.

ETHNOGRAPHIC NOTES
Name(s)
– Soqotri: nil or *ráʿadihon* 'tiny grazing plant'.
– Soqotri simplified: *ra'adihon.*
Food (livestock). One of the small *ru'ud* plants which come up with the rains, providing incidental green grazing until they die back. Though palatable to livestock, it is not important forage for them.
Comparative uses. Many *Viola* species have edible leaves. In India the roots are used as an emetic, and as a substitute for, or adulterant of, ipecac. (Ambasta 1982; Jain & De Fillips 1991)

VISCACEAE

Epiphytic shrubs, hemi-parasitic on the stems of trees. Stems articulated, green. Leaves scale-like, in 2 ranks. Monoecious; flowers minute, unisexual, 3-merous. Male flowers with the anthers fused together (forming a synandrium). Female flowers with a style and a club-shaped stigma; ovary inferior, 1-celled, the ovule undifferentiated. Fruit a 1-seeded berry. 8 genera. 450 species. N tropical Africa, Madagascar, Mascarenes, Himalayas to Japan and New Zealand.

Korthalsella Tiegh.

See description of family. 45 species. N tropical Africa, Madagascar, Mascarenes, Himalayas to Japan and New Zealand.

Korthalsella taenioides **(Comm. ex DC.) Engel**

Densely branched shrub, parasitic on the branches of trees; stems up to 10cm long, the upper usually flattened. Flowers minute, globose, in dense axillary clusters. Fruits club-shaped.
Ecology. Epiphytic on branches of *Croton* species in dense montane thicket. 1250–1500m.
Distribution. NE tropical Africa and Madagascar east to Japan and Australia.
Botanical notes. The tufts of short, leafless twigs which comprise this plant are easily overlooked in the dense thicket where it is found. It is the only epiphytic parasite now known on Soqotra. However, Balfour (1888) records a species of *Loranthus* from near Qalansiyah. He notes that his specimen has twigs but not flowers and that the specimens are not sufficient for complete identification. It is surprising that this epiphytic shrub has not been seen again; species of *Loranthus* are normally evergreen and relatively easy to spot during the dry season when most other trees lose their leaves.

ETHNOGRAPHIC NOTES
Name(s)
– Soqotri: *eźərīʾo* (*men méssə*) 'residue or sediment (from the rain)' (a general term for mosses).
– Soqotri simplified: *edheriy'oh* (*men messih*).
Fuel. It can be gathered inadvertently along with dried mosses and lichens when these are collected for use as tinder in the wet season.

VITACEAE

Succulent trailing shrubs or climbing vines. Stems jointed, flattened or rounded, sometimes winged. Tendrils present or absent. Leaves deciduous or persistent, palmately veined, simple or lobed. Flowers in cymes opposite leaves or tendrils, small, actinomorphic. Calyx vinaceous, cup-shaped. Petals yellow, 4. Stamens 4, inserted on a ring-like disc. Ovary superior, 2-celled; ovules 2 per cell, erect. Style 1. Fruit a purplish berry. 13 genera. 800 species. Tropical and warm regions.

Cissus L.

See description of family. 3 endemics. c.350 species. Tropical and warm regions.

Cissus hamaderohensis **Radcl.-Sm.**
Cissus quadrangularis sensu auct. non Wall.

Trailing succulent shrub. Stems grey-green, flattened, rectangular in section, the angles winged (less marked on mature stems). Tendrils present. Leaves deciduous, ovate, c.2 x 1cm, simple or 3-lobed, sessile.

Ecology. Common and widespread on the limestone plateaus and above 300m on the limestone escarpments; in dwarf shrubland and open woodland. 300–800m.

Distribution. Endemic.

Conservation status. Least Concern (LC). Common in several vegetation types and under no present or perceived threat.

Botanical notes. Replaces *C. subaphylla* above 300m over most of the island; absent from the granite of the Haggeher.

ETHNOGRAPHIC NOTES†

Name(s)

– Soqotri: (*di*) *ʿʌ̍ṭərhe* ‘octopus’.
– Soqotri simplified: (*di*) *‘atirheh.*

Food (human). In the hungrier areas of the island, or in times of famine, the fruit were eaten when fully ripe and red. The outer skin and the pip were discarded and the juicy pulp was eaten. The islanders differentiate between two kinds of fruit of *Cissus* species, one called *ḥo̍wḳam*, after the least common but best kind of date palm, and the other *ṣərəfānə*, after the most widespread date cultivar. The rarer *ḥo̍wḳam* fruit are said to have red flesh, which is sweet, and the common *ṣərəfānə* fruit to have a much paler flesh, which is very ‘hot’ and much less liked.

Food (livestock). Very hungry goats and sheep eat the plant litter, especially the stalks. In a severe drought goats also eat the dried-out stems, and, rarely, starving goats have been seen to nibbe tender new growth. Small stock also eat the fallen fruit. However, if goats eat a lot they develop the condition known as *ʿa̍ṣʌb*, a form of temporary weakness in the hindquarters, causing them to become unsteady on their feet and tottery. They tire quickly and are generally weak and sickly looking. In a prolonged dry season, when alternative forage is exhausted, herders in the drier areas prepare the plant to make it palatable. They dig a hollow around the roots and fill this with dead plant material as tinder. This is then set alight and kept burning gently until all the outer skin (which is the harmful and indigestible part) of the stems has split, curled up and burned off to reveal the bright green underneath. The hungry animals are kept away from the plant until it has cooled. Prepared in this way, goats become quite keen on it, and if they eat enough, begin to put on weight. If done with insufficient care and fire allowed to take hold, the entire clump can be destroyed. A careful herder will light the fire to singe only one side of the plant, leaving the rooting system as intact as possible in order to ensure that the plant survives. Because it is normally not palatable to livestock, the clumps often provide shelter for young saplings of more palatable plants, and these also can be destroyed in this process of singeing. For instance, herders in one area of the southern plains remember well that, in one very bad year, they singed their way through whole swathes of *Cissus* species. They ascribe to this the lack of regeneration of the more valuable plants of the area, such as *Maerua angolensis*, *Capparis* species, *Dendrosicyos socotranus*, *Croton* species and *Commiphora* species. To avoid possible wholesale destruction of clumps, many herders recommend breaking off stems from around the outer edges of the clump, and holding them over a fire, built some distance away, until they change colour and the outer skin cracks off.

Bees / honey. Honeybees in less well-vegetated areas are attracted to the flowers.

Flavouring / food additive. On Samha, where there is no cultivation of date palms and few nectar-rich plants, the very ripe and red berries of *C. subaphylla* were valued for their sweetness. The carcass of a goat feeding on *Cissus* species makes the water it is cooked in very ‘hot’ and bitter, and is rarely drunk. Sometimes even the taste of the meat, the viscera in particular, is also affected.

Fuel. Dead stems make useful tinder, needing·only the lightest puff to get it going. Supplies are often kept in the home as firelighters. Otherwise the dead material makes poor firewood which burns down quickly and gives off a lot of bitter smoke. It can be used though for firing pottery, and, with *Jatropha unicostata* wood, is much used for this in the less well-vegetated areas of the island.

Leatherwork. The ash of the burned stems is used to prepare skins in leatherwork, particularly in areas where other wood is scarce, as on the outer islands.

Timber. In less well-vegetated areas they are sometimes trained up the walls of huts, and to cover roofing in order to strengthen and stabilise them. They are cut and used as a roofing material under a final coating of earth, especially on Abd al Kuri where timber is particularly scarce.

Wood. Dead material was much used in the manufacture of quicklime, except in coastal or treeless areas, where the large clumps give valued shelter and shade for young kids and lambs.

Medicinal / veterinary. Stems of the plant are roasted and then ground to a paste used to treat chronic ulcers, especially those that developed from pack sores in camels. The treatment is extremely painful and the camel has to be tied down while the wound is cleaned and the paste applied. The dressing is changed every day for about a fortnight and the camel is led beside the other loaded camels to keep it moving. Then a powder made from dried stems is packed into the wound. This is done every day until the wound has healed over, but even then the camel is not worked for six months or more. Some camel owners use a different method. They scorch the stems at the fire until the outer skin is blackened and can be removed. They then pound the softened interior fibre with salt (and sometimes now with soap powder) and use this to pack the wound. They then saddle the animal up and ride it hard, and at night, tie it up so that it cannot eat anything. Each day they clean the wound out with a salt and water solution and apply a new dressing and then saddle the camel and ride it hard. This continues for at least a week, during which time the animal loses weight and looks increasingly miserable. Once the ulcerated flesh is clean, and the bone of the spine clearly visible below, the area is packed with a dressing made from powdered, roasted stems. Then the camel is set free to go where it likes. If an animal has a sore between the claws of the hoof, usually from stubbing it on a stone, and the joint above has become swollen, this is treated with a preparation from the plant. A section of the stem of a young, green plant is baked beside the fire until it really hot, and then this is strapped around the leg, directly over the swollen joint. In cold weather, small stock sometimes develop the disease known as *bˈikəmə* or *bikīmhin*, which can be fatal. Its main signs are swelling and stiffness of the joints, especially in the legs, sometimes accompanied by sores that break out all over the body. It is treated by putting stems of a green plant to roast over the embers of the fire. They are left until they bubble and ooze, and then are wrapped around the swollen joints and strapped in place. A section of stem in between two 'joints' is roasted at the fire until hot and black. It is then ground, and the cooled powder applied to a laceration or weeping sore. In humans and livestock this acts as a disinfectant and dries out the sore. To bring the pus to the surface of a boil, or a swollen and infected wound, a section of stem is scorched and then crushed to make a paste smeared over the area. The same paste is used to treat dry eczema in both people and livestock, and to soothe chronic pruritus. If small stock, usually in wet weather, develop the disease known as *mīt*, where pale patches appear on the hooves (possibly a fungal disease), the pale patches are pared off with a knife and a paste made by grinding the fruit is rubbed in. This sticks fast and provides an antiseptic dressing, rather like a sticking plaster. Alternatively, the pale areas of the hoof are cauterised with a heated knife blade, and then a powder made from the roasted stems or fruit is applied as a dressing to dry it out. The sick are sometimes fumigated in the smoke of stems of the plant places on the fire.

Pest control. Dead stems give off a lot of dense and bitter smoke, and are used for insecticide smudges and to get rid of mosquitoes around the living quarters.

Other. Animals feeding on prepared stems can go longer without water. Good earth for cultivation is found beneath large clumps and is often gathered for this purpose. In date palm growing areas, the whole plant is sometimes dug up, the hole deepened, and a young date palm planted in the hole. People who want to grow *Euphorbia arbuscula* or *Commiphora ornifolia* often plant a young one in the middle of a clump of *Cissus* species; not only is the soil likely to be good, but the *Cissus* species also protect the young plant because livestock normally ignore it. In the central highlands the plant is usually removed from the calf and grazing paddocks to allow the grass to grow. A length of stem between two 'joints' is used as a lure in fishing: one end is gently beaten to fray it out to make a fringe that mimics the tail of a small fish. Dried stems are burned to ward off evil, and live or dead ones are laid across containers of food and water to protect the contents from malign forces.

Comparative uses. *Cissus* species often have edible fruits, and the leaves and stems of some are eaten raw. Some are used to bind hut roofing or are placed on the roof to protect against wind, and are also said to be termite resistant. The stems of one species in India, *C. repanda*, contain large quantities of potable water. Borana girls plant *C. quadrangularis* on their father's grave and in Guinea the stems and leaves are fed to cattle to induce milk. Medicinally, a decoction of *C. subaphylla* is taken three times a day in parts of Africa, and is also used to bathe the affected parts of a fracture. In Dhofar, the fresh leaves and stems of *C. quadrangularis* are pounded and applied to burns, wounds and saddle sores. The same mixture is applied to the udder of a parturient goat withholding her milk, or to livestock with mastitis. Elsewhere, an infusion of the plant is drunk for syphilis and, with or without added tamarind, for gonorrhoea. The boiled and sugared stems are taken as a stomachic and to treat various intestinal complaints. In Kenya an infusion of the root is used as a pesticide against termites. In East Africa the plant is used to treat tuberculosis, chest complaints and stomach diseases. The root and/or stem is burned, boiled in water, and the filtrate drunk to clean out the stomach. In West Africa the fresh leaves and stems are pounded and applied to bruises and wounds. In Senegal a decoction of the stems and leaves is used as a friction and wash in pains with fever and as a remedy for malaria. The plant is also used to treat osteoporosis. In Yemen, the leaves of *C. digitata* are used for earache and to treat vertigo. (Abdulfatih 1987; Ambasta 1982; Dalziel 1948, 1956; Fleurentin & Pelt/Fleurentin *et al.* [1982, 1983a,b,c]; Hedberg *et al.* [1982–1986]; Heine *et al.* 1988; Jain & De Fillips 1991; Kunkel 1984; Lewis & Elwin-Lewis 1977; Miller & Morris 1988; Oliver-Bever 1986)

†Also apply to *C. subaphylla*.

Cissus paniculata **(Balf.f.) Planch.**
Vitis paniculata Balf.f.

Climbing succulent vine to 2(–3)m. Stems green, rounded in section. Tendrils present. Leaves persistent, broadly ovate, <18 x 20cm, shallowly palmately lobed, margin with small teeth, base strongly auriculate.

Ecology. A common vine in drought-deciduous woodland in the wadis draining the Haggeher mountains. 200–650m.

Distribution. Endemic.

Conservation status. Least Concern (LC). Common in several vegetation types and under no present or perceived threat.

ETHNOGRAPHIC NOTES

Name(s)

– Soqotri: (*di*) *ˤˈʌṭərhe*; *ˤˈaliḥʌṭ* (also given to *Sarcostemma* species).

– Soqotri simplified: (*di*) *'aterhe*; *'alihat.*

Food (livestock). Living material is not eaten by any livestock. Fallen, dead leaves are eaten by hungry sheep and goats when there is little alternative forage.

Notes. The sap causes irritation to the skin, so it is generally avoided.

Cissus subaphylla **(Balf.f.) Planch.**
Vitis subaphylla Balf.f.

Erect, intricately branched succulent shrub. Stems grey-green, elliptic in section. Tendrils absent or non-functional. Leaves deciduous, obovate, 1 x 0.5cm, simple, petiolate.
Ecology. Widespread and often abundant on the plains and in succulent shrubland on the limestone escarpments; generally occurs below c.300m (replaced by *C. hamaderohensis* at higher altitudes) but reaches up to 750m on the dry dipslopes of the western limestone plateau; co-dominant at 100m on Samha in open shrubland. Sea-level–750m.
Distribution. Endemic.
Conservation status. Least Concern (LC). Common in several vegetation types and under no present or perceived threat.
Botanical notes. *C. subaphylla* is common in dry, low-lying areas, only rarely occurring above 300m; at higher levels it is replaced by *C. hamaderohensis*. It is reported to be on the increase in upland areas. The importance of *C. subaphylla* as a nursery plant for a variety of palatable species is well known throughout the islands. Thus its absence can affect the recovery of populations of trees such as *Dendrosicyos*, *Maerua* and *Commiphora*, etc.

ETHNOGRAPHIC NOTES
Name(s)
See *C. hamaderohensis* (also uses).
Notes. Children on Samha are cautioned against this species, as its sap causes itching if it comes into contact with the skin. The fruit are said to be poisonous.

ZYGOPHYLLACEAE

Shrubs or herbs. Branches sometimes jointed at nodes. Leaves fleshy or leathery, opposite; stipules present, sometimes spiny. Flowers solitary or paired, actinomorphic. Sepals 5. Petals 5. Stamens 10. Ovary superior, 5-celled, often winged; style 1; ovules several, axile. Fruit a capsule, sometimes splitting into 5. 27 genera. 250 species. Tropical and warm regions, especially arid or saline areas.

Fagonia L.

Herbs. Leaves simple, flattened; spiny stipules often present. Flowers pink or lilac. Fruits a dry, 5-lobed capsule, dehiscing from the base. Seeds brown. 30 species. Mediterranean, SW Asia to NW India, SW Africa, N America.

ETHNOGRAPHIC NOTES
Comparative uses. In Yemen the shoots of *F. indica* are used as an anti-paralytic, and the plant has been shown to act as a stimulant to the central nervous system. In Oman the leaves, stems and root are given for abdominal colic, dyspepsia, fever and to treat venereal diseases. The leaves and roots, soaked in water or ground and boiled in water, are taken orally for colic. The leaves are boiled in water, the foam removed, and the liquid used for bathing. A paste is made from the leaves and applied to the forehead in fevers, and the whole plant is boiled in water and the filtrate used to wash the genitals to treat venereal diseases. The plant is crushed and the juice drunk to treat malaria. An extract of the plant is taken with water for spleen and liver pain, and water in which the plant with some thyme has been boiled is given for kidney stones and to treat eye problems. In northern Oman herbalists sell leaves and stems of *F. indica* mixed with *Blepharis ciliaris* flowers as a cure for fever: a handful of the powdered material is boiled in two glasses of water with a dried lemon for 10 minutes and drunk twice a day before meals. Other *Fagonia* species are used to treat tumours and ulcers. (Bebawi & Neugebohrn 1991; Fleurentin & Pelt/Fleurentin *et al.* [1982, 1983a,b,c]; Ghazanfar 1994; Miller & Morris 1988)

Fagonia luntii **Baker**
Fagonia cretica var. *luntii* (Baker) Hadidi
Fagonia cretica var. *socotrana* (Balf.f.) Schweinf.

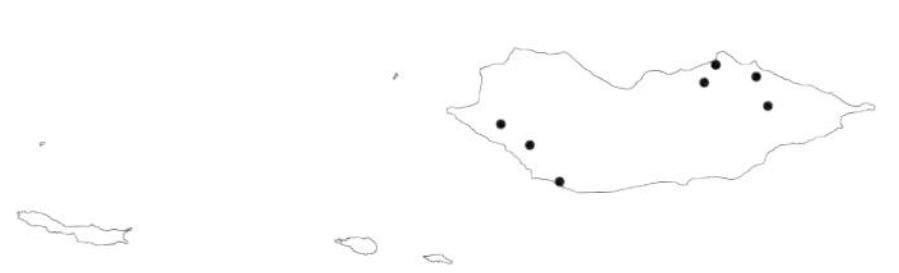

Glandular herb to 30cm, often with sand sticking to leaves. Leaves ovate to suborbicular, 0.8–2 x 0.5–2cm; stipules sometimes spiny, <0.2cm long. Flowers pink, c.1cm diameter.
Ecology. Rare in dwarf shrubland on the inland plains. 200m.
Distribution. Mainland Yemen, Oman, Somalia, Djibouti.
Botanical notes. El-Hadidi (1973) used the name *F. socotrana* and recognised the conspecificity of plants from mainland Yemen and Soqotra but he separated the mainland variety by its glandular indumentum and consequent tendency for sand to stick to the plant. This character is also found on Soqotra and it seems difficult to maintain the variety. *F. luntii* is an older name than *F. socotrana* so *F. socotrana* var. *luntii* (Baker) Hadidi is invalid.

ETHNOGRAPHIC NOTES
Name(s)
– Soqotri: *ṣifṣˈifə*, *ṣʌfṣˈifo* 'rippling'; *ḥamˈʌźəhʌn* 'sour'; *ʿˈažərib*'quick to die back'; (rare) *məsḥōwbeb* 'spreading'.
– Soqotri simplified: *zifzifih*, *zafzifo*; *hamadhihan*; *'ajirib*; *mis'howbeb*.
Food (livestock). Camels and donkeys eat this all year round and small stock also graze it. In the drier areas where it grows it is considered to be a nutritious fodder. Livestock eating a lot grow fat and healthy, but on slaughtering herders have observed that the carcass has little fat in and around the stomach. A diet high in the plants is also said to turn the blood of such animals a very bright red and to cause it to flow readily.

Fagonia paulayana **Wagner & Vierh.**
Fagonia arabica sensu Balf.f. non L.
Fagonia cretica var. *arabica* sensu Balf.f. non T.Anderson

Spiny herb to 30cm. Leaves linear to narrowly elliptic, <1.5cm long; stipules spiny, 0.5–1.5cm long. Flowers lilac, c.0.5cm diameter.
Ecology. Rare in dwarf shrubland on the inland plains. 200m.
Distribution. Arabia, Somalia, Kenya.

ETHNOGRAPHIC NOTES
Name(s)
– Soqotri: *gīrbəb* 'mangey' (Soqotra and Abd al Kuri) (more commonly given to *Pulicaria stephanocarpa*, and to *Teucrium* and *Euphorbia* species which persist through the dry season); *gifiʾhin* 'bowed'; *məṣɔ̄ʿidihīn*, *məṣɔ̄wʿədīn* 'prickly, spiny'; *məśḥermhim di girbag* 'wild cat's *Caralluma*'.
– Soqotri simplified: *giyrbeb*; *gifi'hin*; *mizow'idihin*, *mizow'idiyn*; *mish'hermhim di girbag.*
Food (livestock). Camels and donkeys eat it all year round. It provides dry season and drought grazing for small stock, but is not considered important or nutritious forage. On Abd al Kuri it is more highly regarded as a fodder.
Medicinal / veterinary. Fissures in the sole of the foot are treated with a paste made from the crushed whole plant. The exuding liquid is heated over the fire and the paste is rubbed into the fissures.

Tribulus L.

Herbs. Leaves pinnate. Flowers yellow. Fruits splitting into 5 indehiscent segments; each segment with 2 spines. 25 species. Tropical and warm regions, especially the drier parts of Africa.

Tribulus terrestris **L.**

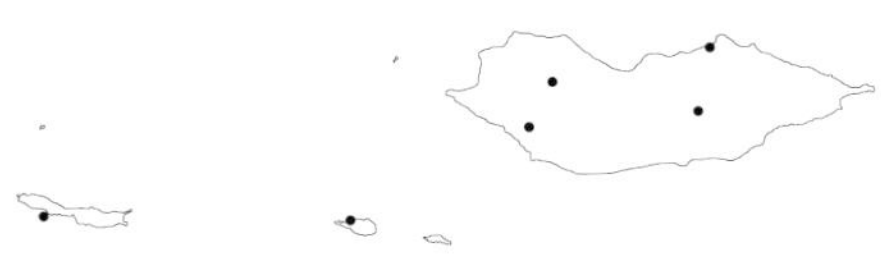

Prostrate, greyish pubescent herb to 15cm. Leaves c.12-foliolate; leaflets elliptic, 2–4mm long. Flowers c.1cm diameter. Fruits c.0.5cm diameter, segments c.3mm long.
Ecology. Widespread on the plains; commonly near habitation. Sea-level–500m.
Distribution. Pantropical.
Botanical notes. The fruits are spiny and attach themselves to livestock.

ETHNOGRAPHIC NOTES
Name(s)
– Soqotri: *śeriśero* (pl. *śerśer*).
– Soqotri simplified: *sherishero* (pl. *shersher*).
Food (livestock). Responds quickly to rain and is often present and green before the grasses have come up. At this time, all livestock, especially sheep, graze it readily. Sheep graze it at other times too, as do cattle and donkeys, but all livestock become less enthusiastic as the plant matures and sets fruit, when it becomes very tough. Cattle in particular avoid it when it is covered in the prickly fruit, but goats eat the fruit as long as they remain green. In general, it is considered to be useful forage, but one that becomes a problem once it has fruited, and the fruit dry out and become hard. They then get lodged in the hooves, and can cause inflammation and sores, often resulting in lameness. The dry fruit can also damage the mouths of animals, in particular sheep. On the outer islands this is regarded as a valuable rains forage for sheep and goats.
Notes. Longer-haired goats often have the hair trimmed around the thigh and udder areas to avoid the scratches caused by fruit lodged in their coats. The scratches frequently develop into sores and then attract flies and can become a real problem. Cattle who graze a lot of it often develop small pustules in the intestines. Otherwise they seem unaffected nor is their lactation is not disturbed.
Comparative uses. Contains a photosensitising principle (causing sunburn) which has been suspected of causing poisoning in sheep, often with associated jaundice. They develop ataxia and move in a crouching manner and this often progresses to paralysis. Some claim that the plant is toxic to livestock when eaten fresh. The poisonous principle is unknown, but the plant can accumulate nitrate and bring about sudden death in sheep and cattle who have eaten a lot of young growth. The young leaves and tender shoots have been used as a potherb, and the capsules ground to a flour used in times of scarcity – during the Sahel drought of 1980–85 many people survived on the ground seed. The leaves are rich in calcium but poor in iron, providing a cheap supplement to rice diets. Medicinally, the fruits are taken as a tonic and diuretic, and an oil prepared from them is used for rheumatism. The leaves are taken as a stomachic, and in Yemen the leaves or whole plants are given as a diuretic, anti-inflammatory and against kidney stones. The roots are taken as an aperient, and, soaked in water, then boiled and the filtrate drunk, to treat malaria. (Ambasta 1982; Bebawi & Neugebohrn 1991; Fleurentin & Pelt/Fleurentin *et al.* [1982, 1983a,b,c]; Goodin & Northington 1979; Heine *et al.* 1988; Jain & De Fillips 1991; Kunkel 1984)

Zygophyllum L.

Herbs or subshrubs. Leaves cylindrical or obovoid, if flattened then 2-foliolate; stipules not spinescent. Flowers white or yellow, 5-merous. Stamens 10. Fruits a fleshy, dehiscent capsule, sometimes conspicuously 5-lobed, wrinkled when dry. Seeds black. 80 species. Mediterranean, SW Asia, S Africa, Australia.

Zygophyllum album L.f.

Grey-green, shortly hairy subshrub to 50cm. Leaves simple or 2-foliolate; petiole expanded, cylindrical, c.0.8cm long; leaflets obovoid to cylindrical, 0.3–1cm long; large, water-filled blisters are frequent on the leaflets and petiole. Flowers white. Fruit star-shaped, conspicuously 5-lobed, c.0.7cm long.

Ecology. Coastal plains. 0–50m.
Distribution. N Africa, S Arabia, tropical E Africa.
Botanical notes. The collections from Soqotra and Abd al Kuri are referable to var. *ambylocarpum*.

Zygophyllum decumbens Del.

Zygophyllum album sensu auct. non L.

Glaucous shrublet. Leaves 2-foliolate; petiole cylindrical; leaflets obovate, flattened, <1.5 x 1cm. Flowers white. Fruits 4–10mm long.
Ecology. No habitat details available.
Distribution. NE Africa and Arabia.
Botanical notes. Not seen since recorded in the 19th century.

ETHNOGRAPHIC NOTES†

Name(s)
– Soqotri: *ḳʌlḳəhʌl* 'swaying in the breeze' (Soqotra and Abd al Kuri) (general term for a variety of mainly halophytic plants).
– Soqotri simplified: *qalqihal.*

Food (livestock). An important forage for camels. Goats and sheep only eat it in the dry season or in drought. *Z. qatarense* is an important fodder for small stock on the outer islands.
Flavouring / food additive. Saline forage can affect the taste of the milk of animals feeding heavily on it.
Other. Livestock feeding on the more succulent plants can go longer without water.
Notes. A period spent grazing halophytic rangeland at least once a year is considered essential for the health of small stock and camels especially. Such feed initially brings on diarrhoea, which is seen as being beneficial and 'cleansing impurities from the blood'. When recovered from this, the animals begin to put on weight.

†Also apply to *Z. qatarense*.

Zygophyllum qatarense Hadidi

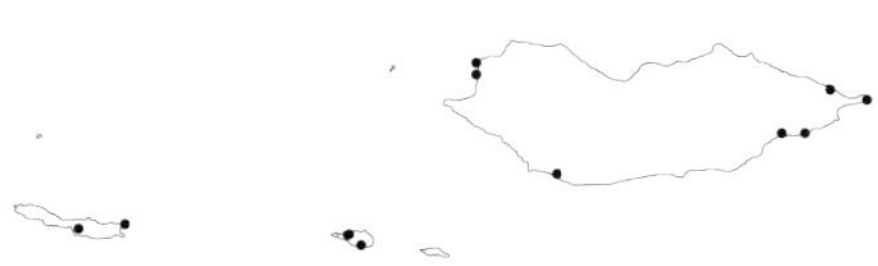

Yellowish-green, glabrous subshrub to 50cm. Leaves simple, sometimes 2-foliolate; petiole expanded, cylindrical, 0.5–1cm long; leaflets obovoid to cylindrical, 0.3–1cm long. Flowers white. Fruits globose, somewhat 5-lobed, 3–6mm long, apex truncate.
Ecology. Low succulent shrubland and coastal woody-herb communities. Sea-level–500m.
Distribution. Arabian Peninsula.
Botanical notes. This is a very variable species both on Soqotra and in mainland Arabia. The number of leaflets is variable on the same plant and leaflets can be almost globose, obovoid or cylindrical. The fruit also varies and *Miller et al.* 12787 has large (10mm long), broadly cylindrical, clavate fruit which are much bigger than the fruit of other Soqotran specimens of *Z. qatarense* examined. It is possible that this collection represents *Z. coccineum* which according to Wood (1997) is present in mainland Yemen.

ETHNOGRAPHIC NOTES

Name(s)
See *Z. decumbens* (also uses).

Zygophyllum simplex L.

Bright green annual or perennial herb to 20cm. Leaves simple, cylindrical, 0.3–1.5cm long, sessile; stipules small, reddish. Flowers yellow. Fruits globose, somewhat 5-lobed, 2–2.5mm long.

Ecology. Dry vegetation types on the plains and foothills. Sea-level–250m.

Distribution. N and E Africa and through SW Asia to W India.

ETHNOGRAPHIC NOTES

Name(s)

– Soqotri: *ḳʌ̍lḳəhʌl* 'swaying in the breeze' (general term for a variety of mainly halophytic plants); *məźḥɔ̄mhen* 'crunchy'; *fəźʌ̍ʾhan* (Samha and Abd al Kuri); *rūwḥʌź biš ʾīdi* 'for washing the hands' (Abd al Kuri).

– Soqotri simplified: *qalqihal*; *midh'howmhen; fidha'han*; *ruwhadh bish 'iydi.*

Food (livestock). Important forage for camels. Goats and sheep eat it when hungry towards the end of the dry season. On the outer islands it is regarded as important forage, and is readily eaten by all livestock.

Medicinal / veterinary. On the outer islands especially, the crushed stems and leaves are mixed with earth to a paste rubbed well into the scalp and left to dry. When the coating is scraped off and the scalp washed the nits and lice come away with it.

Cleansing. In water-poor regions, the fleshy leaves and stems are crushed and the liquid used for washing and cleansing.

Other. Livestock feeding on the more succulent plants can go longer without water.

Comparative uses. The seeds can be eaten as a food, and the plants cultivated to reduce soil salinity. Medicinally, it is said to possess anti-bacterial activity. An infusion of the leaves and seeds is applied to the eyes in ophthalmia and leucoma. The seeds are given as an anthelmintic, and the leaves used as an antibiotic and laxative. In Oman and elsewhere drops of a leaf extract are put into the eyes to treat redness and infection. (Abbas *et al.* 1992; Ambasta 1982; Batanouny 1981; Ghazanfar 1994; Jain & De Fillips 1991; Kunkel 1984; Rizk & El-Ghazaly 1995)

APPENDICES

CUCUMBER TREE (DENDROSICYOS SOCOTRANA) DICCON ALEXANDER

APPENDIX 1: NOTES ON LICHENS

The many lichens on the islands are often useful to the islanders. These have not been identified except for *Roccella tinctoria*, locally called *di ḥānə* (*di hanih*). These grow on rock and are used in dyeing. Children, and occasionally women, dye their nails and fingertips with ground lichen mixed to a paste with spittle, and use it in the same way as henna (which does not grow on the island and has to be imported). Some also paint a circle on the palm and back of the hand. The soles of the feet are rarely stained in this way, as is done with henna, though occasionally the toenails are dyed. It is rare for men to use this henna substitute, though fishermen often dye the palms of their hands with henna to toughen the skin for hauling in the lines. Only imported henna was used to dye the hair and beard, and this was normally too expensive for widespread use.

Di hanih lichens were also used to colour the unbleached calico formerly imported into the island for use as clothing and as a winding sheet. The lichen was ground to a paste with water and applied with the fingers to decorate the calico waistcloth and shawl worn by men, usually to make a single stripe running along its length. Sections of the fringe were dyed in the same way. The intricately plaited leather belts formerly worn by men were also dyed with lichen mixtures, usually by marking a series of vertical lines, or stripes, running from the top to the bottom with the paste. This group of lichens also once had a commercial value, and were gathered in sacks for sale to the mainland, where the islanders understand that it was used in fumigatory mixtures. On Samha, a lichen called *səˈdeʾ* (*sideh*) is still gathered and used by the island women to make a sweet-smelling incense for their own use. There is no doubt that all sorts of lichens were also added to the sacks and sold as *di hanih* when demand was high. Within living memory, when Aden was still flourishing as a centre of trade, one measure of cereal could be obtained for ten times its weight in lichen. Some of the *di hanih* lichens have a pleasant smell and are kept at home and rubbed between the hands and then over the skin and clothing to perfume it. *Di hanih* lichens are also used medicinally as discussed below for *hannih* lichens, and are eaten by very hungry goats.

saʿˈangəher (*sa'angiher*) 'tangled, messy (hair)' lichens, which hang from shrubs and trees, in the limestone plateau areas in particular, are rarely gathered for sale, but are collected when dry and stored in a dry place for use as tinder. This group of lichens is also known as *ˈsəwgihur* (*suwgihur*) 'tinder'. They are not eaten by livestock, but can be used when processing butter to make ghee. They are pulled up and the knobbly little roots collected. These are roasted, then peeled, cleaned, and crushed to a coarse flour which is added to the boiling butter to clarify it. It is said to make a particularly delicious smell 'that spreads for miles', and the sediment left after straining the ghee is edible.

ˈḥannə (*hannih* 'like henna') or *di ˈḳeśʾur* (*di qesh'ur* 'of rock') (western Soqotra *ˈraxmam*, *rakhmam* 'grey-coloured'), smooth, pale lichens that grow mainly on rocks like the *di hanih* ones, are used medicinally as well as in dyeing. These and the *di hanih* lichens are gathered when dry and crushed to a powder used to pack deep wounds, usually after first painting the area with aloes juice. The same treatment is used for ulcers and for sores in pack animals. Made into a paste by adding a little water, or crushed and used as a powder, the lichens are rubbed into skin sores. Children are particularly prone to these all over the body, a result of scratching flea, lice, scabies mites or other insect bites.

The very dry and white lichen of the limestone plateaus was differentiated from that of other areas with the name *mahḳāḳə* (*mahqaqih*) (also a name for chalky, pale rock), and was regarded as particularly effective for treating major wounds and lacerations. The wounds were first cleaned with powdered chalky earth, hot water and salt, and then fresh aloes juice. Then the wound was packed with the crushed *mahqaqih* lichen and bound with strips of bark sliced from a *Jatropha unicostata* shrub, each strip including a generous amount of the moist inner bark. These powdery white lichens were also much used in the treatment of fractures in both people and their livestock.

Mosses that grow in the Central Highlands are locally called *dōwḳəhīn* (*dowqihiyn*) 'little beard', and are grouped with the lichens by the islanders. Certainly, for much of the year they have a dry, grey appearance like lichen. However, they are considered to be 'cold', unlike the true lichens. The mosses are not eaten by livestock, but children peel them off the rocks and use them to make false beards! Also included in the lichen group is a red fungus which grows on aloe and *Kalanchoe* species in the wet woodland areas, and which is locally called *ʾˈidəhan di ʿˈiləhil* (*idihan di 'ilihil*) or 'ears of the monsoon rains'. This is not used in any way.

The *di ḳāṣʌl* (*di kazal*) lichen grows in the granite pinnacles, amongst the very highest peaks. It grows on plants rather than on rocks like the *di hanih* lichens and is redder than the other lichens. When the demand for *di hanih* lichen was high, the mountain people also gathered this and sold it mixed in with the *di hanih*.

APPENDIX 2: NEW SPECIES AND COMBINATIONS

NEW SPECIES

ACANTHACEAE

Justicia alexandri R.Atkinson **sp. nov.** (p. 408)
Distinguished from the other woody *Justicia* species on Soqotra by the ovate to elliptic leaves, grey tomentose beneath, glossy above.

A ceteris speciebus *Justiciae* lignosis in Socotra habitantibus foliis ovatis usque ellipticis subtus cinereotomentosis, supra nitidis differt.

Type: Soqotra, Haggeher mountains, near Skand, 12°34′N, 54°00′E, 1550m, 1 iv 1996, *A. Miller & D. Alexander* 14280 (holo. E!).

Justicia takhinensis R.Atkinson **sp. nov.** (p. 409)
Distinguished from the other woody *Justicia* species on Soqotra by the fleshy obovate leaves with rounded or notched tip and prominent cystoliths.

A ceteris speciebus *Justiciae* lignosis in Socotra habitantibus foliis carnosis obovatis apice rotundato usque emarginato et cystolithis prominentibus differt.

Type: Soqotra, above Hoq, 12°35′N, 54°21′E, 500m, *A. Miller & D. Alexander* 14305 (holo. E!).

ALOACEAE

Aloe jawiyon Christie, Hannon & Oakham **sp. nov.** (p. 297)
Similar to *A. perryi* Baker, with which it is sometimes sympatric; differs in the more yellow-green, matte and finely scabrous leaves; the long-acuminate leaf tip; the smaller and more closely spaced marginal teeth; the bracts as long as or longer than the pedicel (not minute and much shorter than the pedicel); the unbranched inflorescence, with the lowest flowers held immediately above the leaves (not frequently branched with the lowest flowers held well above the foliage); the perianth spreading, medium orange with green tips and slight distal swelling (not declinate, reddish orange to yellowish with conspicuous distal swelling).

A. perryi Baker qua interdum est sympatrica similis, sed foliis magis flavovirentibus hebetatis tenuiter scabris, apice folii longe acuminato, dentibus marginalibus minoribus magis approximatis, bracteis pedicellum aequantibus vel longioribus (haud minutis pedicello multo brevioribus), inflorescentia haud ramosa floribus infimis proxime super folia ferentibus (haud frequenter ramosa floribus infimis bene super folia portatis), perianthio patenti aurantiaco apicibus viridibus pulvino distali exiguo (non perianthio declinato rubescenti-aurantio usque flavescenti pulvino distali conspicuo) differt.

Type: Soqotra, limestone ridge at Heger [western tip of the island], 12°31′N, 53°23′E, 380–530m, 5 ii 2001, *A. Miller, R. Atkinson & A. Khulaidi* 20022 (holo. E!).

BORAGINACEAE

Echiochilon pulvinata A.G.Mill. & L.Urb. **sp. nov.** (p. 448)
Distinguished from all other species within the genus by the cushion-forming species habit and more or less actinomorphic, bright white flowers.

Ab omnibus aliis speciebus generis habitu pulviniformi et floribus candidissimis ± actinomorphis distincta.

Type: Soqotra, plateau above Noged plain, 12°22′N, 53°47′E, 500m, 6 iii 1996, *A. Miller, D. Alexander & A. Khulaidi* 14105c (holo. E!, iso. K).

CAPPARACEAE

Cadaba insularis A.G.Mill. **sp. nov.** (p. 475)
Similar to *C. rotundifolia* Forssk. but differs in the presence of 4 petals and its stamens attached near the middle of the gynophore, not near the base; differs from *C. longifolia* DC. in the broader (oblong-elliptic to suborbicular) not linear-elliptic leaves and the fruits bearing dense stalked (not sessile) glands.

C. rotundifoliae Forssk. similis sed praesentia quattuor petalorum et staminibus prope medium (haud versus basin) gynophori affixis differt; a *C. longifolia* DC. foliis latioribus (oblongo-ellipticis usque suborbicularibus, non lineari-ellipticis) et fructibus glandulos densos pedunculatos (haud sessiles) ferentibus recedit.

Type: Soqotra, Kilissan, 12°29′N, 54°19′E, 350m, 10 ii 1999, *A. Miller, D. Alexander, A.S. Sulaiman & N.M. Talib* 17062 (holo. E!, iso. K).

COMPOSITAE

Pulicaria dioscorides R.Atkinson **sp. nov.** (p. 511)
From all other species on Soqotra distinguished by its tightly clustered silver-grey leaves and bright yellow flowers; similar to *P. pulvinata* E.Gamal-Eldin, but differs in its more robust habit and the presence of ray florets.

Ab omnibus speciebus *Pulicariae* in Socotra habitantibus foliis argenteo-griseis valde fasciculatis et floribus clare luteis distincta; *P. pulvinatae* E.Gamal-Eldin similis sed habitu robustiore et florulis radiis praesentibus differt.

Type: Soqotra, limestone ridge at Heger [western tip of the island], 12°31′N, 53°23′E, 350m, 5 ii 2001, *A. Miller, R. Atkinson & A. Khulaidi* 20008 (holo. E!, iso. K).

CRUCIFERAE

Farsetia inconspicua A.G.Mill. **sp. nov.** (p. 529)
No clear allies; similar to *F. socotrana* B.L.Burtt with its narrowly oblong fruits but differing in its narrower, non-fleshy leaves and bright yellow (not white or lilac) flowers; similar to *F. stylosa* R.Br. in flower colour, septum characters, flower size and nectary shape, but differing in narrowly oblong (not linear) fruit, shorter style and decurrent (not capitate) stigma.

Affinitates proximae nullae; *F. socotranae* B.L.Burtt in fructibus anguste oblongis similis sed foliis angustioribus haud carnosis et floribus clare luteis (haud albis vel lilacinis) differt; *F. stylosae* R.Br. colore et magnitudine floris, characteribus septi, et forma nectarii fructu tamen anguste oblongo (haud lineari), stylo breviore et stigmate decurrenti (non capitato) recedit.

Type: Soqotra, Wadi Tehan, 12°35′N, 53°30′E, 650m, 6 ii 2000, *A. Miller & N.M. Talib* 19109 (holo. E!, iso. K).

EUPHORBIACEAE

Euphorbia hamaderoensis A.G.Mill. **sp. nov.** (p. 554)
Similar to *E. socotrana* Balf.f. but clearly distinguished by its low shrubby habit, smaller (up to 1.5cm long) leaves and smaller (4–5mm across) cyathia which are solitary and have 5 (not 6) cyathial glands.

E. socotranae Balf.f. similis sed habitu humiliter fruticoso, foliis minoribus (usque ad 1.5cm longis) et cyathiis minoribus (4–5mm diametro) solitariis glandulis cyathialibus 5 (haud 6) clare differt.

Type: Soqotra, Hamaderoh, 12°35′N, 54°17′E, 680m, 8 ii 1999, *A. Miller, D. Alexander, A.S. Sulaiman & N.M. Talib* 17048 (holo. E!).

GERANIACEAE

Pelargonium insularis Gibby & A.G.Mill. **sp. nov.** (p. 573)
Most closely related to *P. alchemilloides* (L.) L'Hér. (a plant of SW Arabia and tropical NE Africa) from which it differs in its pink (not white) flowers and the development of a pronounced woody stock.

P. alchemilloidis (L.) L'Hér. (planta Arabiae inter occasum solis et meridiem Africaeque tropicalis inter septentrionem et orientem) a qua floribus roseis (haud albis) et caudice lignoso valde evoluto differt.

Type: Samha, limestone escarpment, 12°09′N, 53°01′E, 400–500m, 15 ii 1999, *A. Miller, D. Alexander, A.S. Sulaiman & N.M. Talib* 17084b (holo. E!).

HYACINTHACEAE

Dipcadi kuriensis A.G.Mill. **sp. nov.** (p. 385)
Similar to *D. balfourii* Baker but leaves broader, pale green (not reddish tinged); flowers 15–20, on a pale green peduncle; perianth white, the tube rather broader, the tepals more strongly recurved. Fruit deeply 3-lobed, 8 x 9mm, broadly obovoid, on erect pedicels.

D. balfourii Baker similis sed foliis latioribus pallide viridibus (haud rubescenti tinctis); floribus 15–20 in pedunculo pallide viridi, perianthio albo tubo latiusculo tepalis magis recurvatis differt. Fructus profunde trilobatus, 8 x 9mm, late obovoideus, in pedicello erecto ferenti.

Type: Abd al Kuri, Jebel Saleh, 12°10′N, 52°15′E, 550m, 9 ii 1992, *A. Miller, M. Bazara'a, L. Guarino & J. Nyberg* 11434a (holo. E!).

Ledebouria insularis A.G.Mill. **sp. nov.** (p. 386)
Similar to *L. grandifolia* (Balf.f.) A.G.Mill. & D.Alexander but leaves 2–3, ascending, ovate with an acute base (not leaves paired, prostrate, broadly ovate with a cordate base); flowers c.4.5mm (not 6–7mm) long.

L. grandifoliae (Balf.f.) A.G.Mill. & D.Alexander similis sed foliis 2–3 ascendentibus ovatis basi acuto (haud binis prostratis late ovatis basi cordato), floribus brevioribus (c. 4.5mm, non 6–7mm) differt.

Type; Samha, summit plateau, 12°09′N, 53°03′E, 650m, 15 ii 1999, *A. Miller, D. Alexander, A.S. Sulaiman & N.M. Talib* 17084c (holo. E!).

LABIATAE

Leucas hagghierensis Gifri & Cortés-Burns **sp. nov.** (p. 579)
Differs from most species on Soqotra by its 5-toothed calyx; differs from *L. penduliflora* Gifri & Cortés-Burns, the only other woody species on Soqotra with a 5-toothed calyx, by its ovate, crenate (not entire) leaves and ascending (not pendulous) flowers.

A plurimis speciebus in Socotra habitantibus calyce quinquedentato differt; a *L. penduliflora* Gifri & Cortés-Burns (infra descripta; in Socotra alia tantum species lignosa cum calyce quinquedentato) foliis ovatis crenatis (haud integris) et floribus ascendentibus (non pendulis) recedit.

Type: Soqotra, Haggeher mountains, lower slopes of Fieri Peaks, 12°35′N, 54°02′E, 1050m, 11 ii 1996, *A. Miller, M. Bazara'a, L. Guarino & N. Kassim* 10417 (holo. E!).

Leucas penduliflora Gifri & Cortés-Burns **sp. nov.** (p. 580)
Amongst Soqotran species most similar to *L. virgata* Balf.f. in habit but differing in its 5- (not 10-)toothed calyx and pendulous (not spreading) flowers.

Inter species socotranas *L. virgatae* Balf.f. habitu maxime similis sed in calyce 5- (haud 10-)dentato et floribus pendulis (haud patentibus) differt.

Type: Soqotra, Muqadrihon Pass, 12°37′N, 54°01′E, 550m, 26 i 1990, *A. Miller, M. Bazara'a, L. Guarino & N. Kassim* 10088 (holo. E!, iso. K!).

Leucas samhaensis Cortés-Burns & A.G.Mill. **sp. nov.** (p. 580)
Differs from *L. virgata* Balf.f. in its orbicular (not elliptic to broadly elliptic) leaves and more or less zygomorphic (not actinomorphic), obscurely (not prominently) ribbed calyx.

A *L. virgata* Balf.f. foliis orbicularibus (haud ellipticis usque late ellipticis) et calyce ± zygomorpho (haud actinomorpho) obscure (non prominenter) costato differt.

Type: Samha, limestone escarpment, 12°09′N, 53°03′E, 500m, 15 ii 1999, *A. Miller, D. Alexander, A.S. Sulaiman & N.M. Talib* 17081A (holo. E!).

MALVACEAE

Hibiscus dioscorides A.G.Mill. **sp. nov.** (p. 624)
From the other woody *Hibiscus* species on Soqotra distinguished by its combination of 5-lobed, cup-shaped calyx and filiform epicalyx lobes. Differs from *H. noli-tangere* A.G.Mill. by the indumentum of the calyx and epicalyx, the linear not filiform epicalyx lobes and the broadly triangular calyx lobes.

A speciebus ceteris *Hibisci* in Socotra lignosis combinatione calycis cupuliformis quinquelobati et loborum epicalycis filiformium distinctus. A *H. noli-tangere* A.G.Mill. (species nova infra descripta) indumento calycis epicalycisque, lobis epicalycis linearibus haud filiformibus et lobis calycis late triangularibus recedit.

Type: Soqotra, eastern Haggeher mountains, Jebel Jaaf, 12°35′N, 54°06′E, 950m, 6 ii 1992, *A. Miller, M. Bazara'a, L. Guarino & J. Nyberg* 11342 (holo. E!).

Hibiscus diriffan A.G.Mill. **sp. nov.** (p. 625)
From most woody *Hibiscus* species on Soqotra differs in its reflexed, oblanceolate epicalyx lobes and calyx which splits into two entire lobes; from *H. quattenensis* A.G.Mill. & Thulin differs in the densely stellate-hairy, not glabrescent, pedicels, epicalyx and calyx.

A plurimis speciebus *Hibisci* lignosis in Socotra lobis epicalycis reflexis oblanceolatis et calyce in lobos duos integros scindenti differt; a *H. quattenensi* A.G.Mill. & Thulin (species nova infra descripta) pedicellis epicalyce calyceque omnibus dense stellatim pilosis (haud glabrescentibus) recedit.

Type: Soqotra, Noged plain, 12°23′N, 53°05′E, 30m, 6 ii 1990, *A. Miller, M. Bazara'a, L. Guarino & N. Kassim* 10290 (holo. E!, iso. K!).

Hibiscus noli-tangere A.G.Mill. **sp. nov.** (p. 627)
Distinguished from most woody *Hibiscus* species on Soqotra by its 5-toothed calyx and filiform epicalyx segments and from all Soqotra species by the distinctive indumentum of predominantly simple hairs on the epicalyx and calyx.

A plurimis speciebus *Hibisci* lignosis in Socotra calyce quinquedentato et segmentis epicalycis filiformibus differt et ab omnibus speciebus *Hibisci* socotranis indumento distincto in epicalyce calyceque e pilis plerumque simplicibus composito recedit.

Type: Soqotra, escarpment below Homhil, 12°35′N, 54°18′E, 130m, 6 iii 1996, *A. Miller, D. Alexander & N.A. Ali* 14296 (holo. E!).

Hibiscus quattenensis A.G.Mill. & Thulin **sp. nov.** (p. 628)
Similar to *H. diriffan* A.G.Mill. with its oblanceolate, reflexed epicalyx lobes but distinguished from this and all other woody *Hibiscus* species on Soqotra by its glabrescent pedicels, epicalyx and calyx.

H. diriffan A.G.Mill. (supra descripta) lobis epicalycis oblanceolatis reflexis similis sed ab hoc et omnibus aliis speciebus *Hibisci* in Socotra lignosis pedicellis epicalyce calyceque glabrescentibus differt.

Type: Soqotra, Zahr Basin, 12°29′N, 05°32′E, 120m, 2 iii 1996, *A. Miller & D. Alexander* 14052 (holo. E!).

POLYGALACEAE

Polygala kuriense A.G.Mill. **sp. nov.** (p. 651)
Similar to *P. erioptera* DC. but differs in the terminal (not axillary) racemes, glabrous (not crisped-hairy) stems, unwinged (not narrowly winged) capsule and seeds with 2 (not 3) basal appendages.

P. eriopterae DC. similis sed racemis terminalibus (haud axillaribus), caulibus glabris (haud crispato-pilosis), capsula exalata (non anguste alata) et seminum appendiculis basalibus duis (haud tribus) recedit.

Type: Abd al Kuri, Jebel Saleh, 12°10′N, 52°13′E, 30m, 19 ii 1999, *A. Miller, D. Alexander, A.S. Sulaiman & N.M. Talib* 17103 (holo. E!).

PORTULACACEAE

Portulaca samhaensis A.G.Mill. **sp. nov.** (Subgenus *Portulaca* sect. *Neossia* D.Legrand) (p. 655)
Similar to *P. kuriensis* M.G.Gilbert but mat-forming (not ascending) with narrowly tuberous roots (not with a stout rootstock) and obovate (not narrowly lanceolate) leaves, up to 10mm long.

P. kuriensi M.G.Gilbert similis sed tegetes formans (haud ascendens) radicibus anguste tuberosis (haud caudice crasso) et foliis obovatis (non anguste lanceolatis) usque ad 10mm longis differt.

Type: Samha, summit plateau, 12°09′N, 53°02′E, 690m, 16 iii 1996, *A. Miller & D. Alexander* 14204 (holo. E!, iso. K).

Portulaca sedifolia A.G.Mill. **sp. nov**. (Subgenus *Portulaca* sect. *Neossia* D.Legrand) (p. 656)
Differs from all other species in the region by its small stature, densely clustered globose leaves and the involucre of bracts subtending the solitary, long-pedicelled, flowers.

Ab omnibus aliis speciebus regionis statura pumila, foliis globosis dense fasciculatis et involucro bractearum flores solitarios longe-pedicellatos subtento differt.

Type: Soqotra, limestone ridge at Heger [western tip of the island], 12°31′N, 53°23′E, 350m, 5 ii 2001, *A. Miller, R. Atkinson & A. Khulaidi* 22012 (holo. E!, iso. K).

RESEDACEAE

Ochradenus socotranus A.G.Mill. **sp. nov.** (p. 659)
Similar to *O. arabicus* Chaudhary, Hillcoat & A.G.Mill. but differing in the baccate, not papery, fruits; differs from *O. baccatus* Delile in the smooth, not rugose, seeds and the clearly dioecious habit.

O. arabico Chaudhary, Hillcoat & A.G.Mill. similis sed fructibus baccatis haud chartaceis distincta; ab *O. baccato* Delile seminibus laevibus haud rugosis et habitu clare dioico recedit.

Type: Soqotra, Jebel Rughid, 12°36′N, 53°57′E, 620m, 29 i 1992, *A. Miller & J. Nyberg* 11190 (holo. E!).

VERBENACEAE

Coelocarpum haggierensis A.G.Mill. **sp. nov.** (p. 709)
Similar to *C. socotranum* Balf.f. but a shrub or tree up to 3m tall; leaves 3–10cm long; with mauve flowers arranged in racemes which strongly elongate in fruit.

C. socotrano Balf.f. similis sed frutex vel arbuscula usque 3m altus foliis 3–10cm longis, floribus malvinis in racemis a dispositis, racemis fructiferis valde elongatis.

Type: Republic of Yemen, Soqotra, Haggeher mountains, Fieri Peaks, 12°35′N, 54°02′E, 1050m, 11 ii 1990, *A. Miller, M. Bazara'a, L. Guarino & N. Kassim* 10418 (holo. E; iso. K, UPS).

APPENDIX 2: NEW SPECIES AND COMBINATIONS

NEW COMBINATIONS

ANTHERICACEAE

Chlorophytum graptophyllum (Baker) A.G.Mill.

HYACINTHACEAE

Drimia porphyrostachys (Baker) A.G.Mill.

UMBELLIFERAE

Rughidia cordata (Balf.f.) M.F.Watson & E.L.Barclay

REFERENCES

Abbas, J.A., El-Oqlah, Ahmed, A. & Mahasneh, A.M. (1992) Herbal plants in the traditional medicine of Bahrain. *Econ. Bot.* 46(2): 158–163 [1993].

Abdulfatih, H.A. (1987) *Medicinal Plants of Southwestern Saudi Arabia.* Al Thaghr Press, Kahmis Mushayt, Saudi Arabia.

Al-Dubaie, A.A. (1993) *Medicinal Plants in Yemen.* Yemen Update No. 32 (45–46).

Al-Dubaie, A.A. & Al Khulaidi, A.A. *Medicinal and Aromatic Plants of Yemen.* Obadi Centre for Scientific Publications, Sana'a, Yemen. (In Arabic)

Al Rawi, A. & Chakravarty, H.L. (1964) *Medicinal Plants of Iraq.* Technical Bulletin No. 15. Government Press, Baghdad.

Ambasta, S.A. (ed.) (1982) *The Useful Plants of India.* CSIR, New Delhi.

Balfour, I.B. (1888) The Botany of Socotra. *Trans. Roy. Soc. Edinb.* 31: 1–446.

Barclay, E.L. & Watson, M.F. (1998) A revision of *Carum* and *Trachyspermum* (Umbelliferae) in the Socotran Archipelago. *Kew Bull.* 53(4): 897–907.

Barrow, S. (1998) A monograph of *Phoenix* L. *Kew Bull.* 53(3).

Batanouny, K.H. (1981) *Ecology and Flora of Qatar.* University of Qatar, Qatar.

Baumer, M. (1983) *Notes on Trees and Shrubs in Arid and Semi-arid Regions.* EMASAR PHASE II, FAO, UN Environment Programme, Rome.

Bebawi, F.F. & Neugebohrn, L. (1991) *A Review of Plants of Northern Sudan with Special Reference to their Uses.* Technical Co-operation, Federal Republic of Germany. Deutsche Gesellschaft für Technische Zusammenarbeit (GTZ), Postfach 5180, D-6236 Eschborn, Federal Republic of Germany. 294pp.

Bent, J.T. (1897) The island of Socotra. *Nineteenth Century* 41: 975–992.

Bent, J.T. & Bent, M.V.A. (1900) *Southern Arabia.* Elder & Co., London.

Beydoun, Z.R. & Bichan, H.R. (1970) The geology of Socotra Island, Gulf of Aden. *Quart. J. Geol. Soc. London* 125(3): 413–436.

Bogdan, A.V. (1977) *Tropical Pasture and Fodder Plants (Grasses and Legumes).* Longman, London.

Boulos, L. (1983) *Medicinal Plants of Northern Africa.* Reference Publications Inc., Michigan.

Boulos, L. (1995) Notes on *Acacia* Mill. Studies in the Leguminosae of Arabia: 1. *Kew Bull.* 50(2): 327–337.

Bremekamp, C.E.B. (1952) *The African Species of* Oldenlandia *L. sensu Hiern & K. Schumann.* Amsterdam. 297pp.

Bridson, D.M. (1995) A new *Kraussia* (Rubiaceae: Octotropideae) from Socotra. *Kew Bull.* 50(4): 773–776.

Bruyns, P.V. (1988) A revision of *Echidnopsis* Hook.f. (Asclepiadaceae). *Bradleya* 6: 1–48.

Bruyns, P.V. & Miller, A.G. (2002) *Socotorella*, a new genus of stapeliad (Apocynaceae-Asclepiadaceae) from the Island of Socotra. *Novon* 12(3): 330–333.

Burkhill, H.M. (1956) A revision of *Useful Plants of West Tropical Africa*. Crown Agents, London.

Clayton, W.D. & Renvoize, S. (1986) *Genera Graminum: Grasses of the World.* Kew Bull. Add. Ser. 13. 389pp.

Cortés-Burns, H. (2000) *A taxonomic and biogeographic study of Socotran Indigoferae.* Unpublished MSc thesis, Edinburgh.

Cronk, Q.C.B. (1986) *Botanical Survey of Soqotra for Conservation and Plant Genetic Resources.* WWF Conservation Yearbook 1985/1986. WWF, Gland, Switzerland.

Cufodontis, G. (1952) Revision der afrikanischen Arten von *Pittosporum* (Sapindaceae). *Feddes Rep.* 55: 27–112.

Dafni, A., Yaniv, Z. & Palevitch, D. (1984) Ethnobotanical survey of medicinal plants in Northern Israel. *J. Ethnopharmacol.* 10: 295–310.

Dahlgren, R., Clifford, H.T. & Yeo, P.F. (1985) *The Families of Monocotyledons: Structure, Evolution and Taxonomy.* Springer-Verlag, Berlin.

Dalziel, J.M. (1948) *The Useful Plants of West Tropical Africa.* London.

Dalziel, J.M. (1956) *Useful Plants of West Tropical Africa.* Crown Agents for Overseas Governments, London (see also Burkhill 1956).

Davis, J.B., Kay, D.E. & Clark, V. (1983) *Plants Tolerant of Arid or Semi-arid Conditions with Non-food Constituents of Potential Use.* Tropical Products Institute, 56/62 Grays Inn Road, London WC1X 8LU.

El-Hadidi, M.N. (1973) Revision of *Fagonia* species (Zygophyllaceae) with tri- to unifoliolate and simple leaves. *Osterr. Bot. Zeitschr.* 121(5): 269–278.

Endress, M.E. & Bruyns, P.V. (2000) A revised classification of the Apocynaceae s.l. *Bot. Rev.* 66(1): 1–56.

Engler, A. & Prantl, K. (1964) *Syllabus der Pflanzenfamilien* (ed. H. Melchior).

Farooqi, M.I.H. (1989) *Plants of the Qur'an.* Sidrah Publishers, Aminabad, Lucknow 226018, India.

Fleurentin, J. & Pelt, J-M. (1982) Repertory of drugs and medicinal plants of Yemen. *J. Ethnopharmacol.* 6: 85–108.

Fleurentin, J. & Pelt, J-M. (1983a) Additional information for a repertory of drugs and medicinal plants of Yemen. *J. Ethnopharmacol.* 8: 237–243.

Fleurentin, J. & Pelt, J-M. (1983b) Cultural background of the medicinal plants of Yemen. *J. Ethnopharmacol.* 7: 183–203.

Fleurentin, J., Mazars, G. & Pelt, J-M. (1983c) Additional information on the cultural background of drugs and medicinal plants of Yemen. *J. Ethnopharmacol.* 8: 335–344.

Forbes, H.O. (ed.) (1903) *The Natural History of Socotra and Abd-el-Kuri.* Liverpool. 598pp.

Forman, L. (1980) *Cocculus balfourii* (Menispermaceae) in Oman. *Kew Bull.* 35(2): 379–381.

Friis, I. (1983) The acaulescent and succulent species of *Dorstenia* sect. *Kosaria* from NE tropical Africa and Arabia. *Nord. J. Bot.* 3: 533–538.

Friis, I. (1987) A reconsideration of the genus *Pittosporum* in Africa and Arabia. *Kew Bull.* 42: 319–335.

Friis, I. (1989) A revision of *Pilea* in Africa. *Kew Bull.* 44(4): 557–600.

Ghazanfar, S.A. (1994) *Handbook of Arabian Medicinal Plants.* CRC Press, London.

Ghazanfar, S.A. (1996) The genus *Dipcadi* (Hyacinthaceae) in the Arabian Peninsula. *Kew Bull.* 51(4): 803–807.

Ghazanfar, S.A. & Al-Sabahi (1993) Medicinal plants of Northern and Central Oman (Arabia). *Econ. Bot.* 47(1): 89–98.

Ghebrehiwet, M. (2000) Taxonomy, phylogeny and biogeography of *Kickxia* and *Nanorrhinum* (Scrophulariaceae). *Nord. J. Bot.* 20(6): 655–689.

REFERENCES

Gilbert, M.G. & Phillips, S.M. (2000) A review of the opposite-leaved species of *Portulaca* in Africa and Arabia. *Kew Bull.* 55: 769–802.

Goodin, G.A. & Northington, D.K. (eds) (1985) *Plant Resources of Arid and Semi-arid Lands: A Global Perspective.* Academic Press, Inc., Orlando, Florida and Academic Press, London.

Goodin, J.R. & Northington, D.K. (eds) (1979) *Arid Land Plant Resources.* Texas Technical University.

Goodman, S.M. & Hobbs, J.H. (1988) The ethnobotany of the Egyptian Eastern Desert: a comparison of common plant usage between two culturally distinct Bedouin groups. *J. Ethnopharmacol.* 23: 73–89.

Grainge, M. & Ahmed, S. (1988) *Handbook of Plants with Pest-control Properties.* Resource Systems Institute, Honolulu, Hawaii/John Wiley & Sons.

Griffiths, J.F. (1972) The Horn of Africa. In: *Climates of Africa: World Survey of Climatology*, Vol. 10, pp. 133–151.

Grindlay, D. & Reynolds, T. (1986) The aloe vera phenomenon: a review of the properties and modern uses of the leaf parenchyma gel. *J. Ethnopharmacol.* 16: 117–151 (and update in *J. Ethnopharmacol.* 68: 3 (1999)).

Guarino, L., Miller, A.G. & Obadi, N. (1991) *Dirachma socotrana* – back from the brink? *Oryx* 25: 229–232.

Gwynne, M.D. (1968) Soqotra. In: Conservation and vegetation in Africa south of the Sahara (eds O. & I. Hedberg). *Acta Phytogeog. Suec.* 54: 179–185.

Haines, S.B. (1845) Memoir of the South and East Coasts of Arabia. *J. Roy. Geogr. Soc.* 15.

Hall, H.T.B. (1977) *Diseases and Parasites of Livestock in the Tropics.* Intermediate Tropical Agriculture Series, Longman.

Hare, K.H. (1977) Climate and desertification. In: *Desertification: Its Causes and Consequences.* UN Secretariat, London.

Hedberg, I., Hedberg, O., Madati, P.J., Mshigeni, K.E., Mshiu, E.N. & Samuelsson, G. (1982) Inventory of plants used in traditional medicine in Tanzania. II Plants of the families Dilleniaceae – Opiliaceae. *J. Ethnopharmacol.* 6: 29–60.

Hedberg, I., Hedberg, O., Madati, P.J., Mshigeni, K.E., Mshiu, E.N. & Samuelsson, G. (1983a) Inventory of plants used in traditional medicine in Tanzania. II Plants of the families Dilleniaceae – Opiliaceae. *J. Ethnopharmacol.* 9: 105–128.

Hedberg, I., Hedberg, O., Madati, P.J., Mshigeni, K.E., Mshiu, E.N. & Samuelsson, G. (1983b) Inventory of plants used in traditional medicine in Tanzania. II Plants of the families Acanthaceae – Cucurbitaceae. *J. Ethnopharmacol.* 9: 237–260.

Hedberg, I., Hedberg, O., Madati, P.J., Mshigeni, K.E., Mshiu, E.N. & Samuelsson, G. (1986) Inventory of plants used in traditional medicine in Tanzania. II Plants of the families Papilionaceae – Vitaceae. *J. Ethnopharmacol.* 16: 23–103.

Heine, B., Heine, I. & Brenzinger, M. (1988) *Plant Concepts and Plant Use: An Ethnobotanical Survey for the Semi-arid and Arid Lands of East Africa.* Cologne Development Studies, Parts I–IV. Verlag Breitenbach Publishers.

Hemming, C.F. (1966) The vegetation of the northern region of the Somali Republic. *Proc. Linn. Soc., London* 177: 173–250.

le Houerou, H.N. (ed.) (1980) *Browse in Africa: The Current State of Knowledge* (International Symposium on Browse in Africa, Addis Ababa, 8–12 April 1980). International Livestock Centre for Africa, Addis Ababa, Ethiopia.

Hughes, M., Hollingsworth, P.M. & Miller, A.G. (2003) Population genetic structure in the endemic *Begonia* of the Socotra archipelago. *Biol. Conserv.* 113(2): 277–284.

Hughes, M. & Miller, A.G. (2002) A new endemic species of *Begonia* (Begoniaceae) from the Socotra Archipelago. *Edinb. J. Bot.* 59(2): 273–281.

IUCN (2001) *IUCN Red List Categories and Criteria: Version 3.1.* IUCN Species Survival Commission. IUCN, Gland, Switzerland and Cambridge, UK. 30pp.

Jain, S.K. & De Fillips, R.A. (1991) *Medicinal Plants of India.* 2 vols. ISBN 0 917256 39 5. Reference Publications Inc., Algonac, Michigan.

Jonsell, B. (1986) A monograph of *Farsetia* (Cruciferae). *Symb. Bot. Upsal.* 25(3): 1–107.

Kers, L.E. (1993) New taxa of *Maerua* (Capparaceae) proposed for the Flora of Ethiopia. *Novon* 3: 50–54.

Kilian, N. (1997) Revision of *Launaea* Cass. (Compositae, Lactuceae, Sonchinae). *Englera* 17: 389pp.

Kilian, N. & Miller, A.G. (2000) Studies in the Compositae of the Arabian Peninsula and Socotra – 5. *Distephanus qazmi* (Vernonieae, Gymnantheminae), a remarkable new species from the island of Socotra, Yemen. *Willdenowia* 30: 83–91.

Klackenberg, J. (1985) The genus *Exacum. Opera Bot.* 84: 5–144.

Kokwaro, J.O. (1976) *Medicinal Plants of East Africa.* East Africa Literature Bureau, Nairobi.

Kunkel, G. (1984) *Plants for Human Consumption: An Annotated Checklist of the Edible Phanerogams and Ferns.* Koenigstein.

Lavin, M., Thulin, M., Labat, J-N. & Pennington, R.T. (2000) Africa, the odd man out: molecular biogeography of Dalbergioid legumes (Fabaceae) suggests otherwise. *Syst. Bot.* 25: 449–467.

Lavranos, J.J. (1971) The island of Abd-al-Kuri and its peculiar *Euphorbia. Cact. Succ. J. (USA)* 43: 109–111.

Lavranos, J.J. (1972) A new *Sarcostemma* from Socotra. *Nation. Cact. Succ. J. (UK)* 27(2): 37–38.

Lavranos, J.J. (1993) Two new species of *Echidnopsis* (Asclepiadacae: Stapelieae) from Socotra. *Cact. Succ. J. (USA)* 65(6): 293–295.

Leenhouts, P.W. (1967) A conspectus of the genus *Allophylus* (Sapindacee). *Blumea* 15: 301–358.

Lewis, W.H. & Elwin-Lewis, M.P.F. (1977) *Medical Botany: Plants Affecting Man's Health.* Cambridge University Press, Cambridge.

Link, D.A. (1991) *Dirachma somalensis* D.A.Link sp. nov. A new species of a remarkable and highly endangered monogeneric family. *Bull. Jard. Bot. Nat. Belg.* 61: 3–13.

Lucas, G. & Synge, H. (1978) *The IUCN Plant Red Data Book.* IUCN, Gland, Switzerland.

Magallón, S., Crane, P.R. & Herendeen, P.S. (1999) Phylogenetic pattern, diversity and diversification of eudicots. *Ann. Missouri Bot. Gard.* 86: 297–372.

Marrero, A., Almeida, R.S. & Gonzalez, M.M. (1998) A new species of the wild dragon tree, *Dracaena* (Dracaenaceae), from Gran Canaria and its taxonomic and biogeographic implications. *Bot. J. Linn. Soc.* 128(3): 291–314.

Meikle, D. (1978) A key to *Commicarpus. Notes Roy. Bot. Gard. Edinb.* 36: 235–249.

Meve, U. & Liede, S. (1996) *Sarcostemma* R.Br. (Asclepiadaceae) in East Africa and Arabia. *Bot. J. Linn. Soc.* 120(1): 21–38.

Mies, B.A., Lavranos, J.J. & James, G.J. (2000) Frankincense on Soqotra Island (*Boswellia*, Burseraceae; Yemen). *Cact. Succ. J. (USA)* 72(5): 265–278.

Miller, A.G. (1984) A revision of *Ochradenus. Notes Roy. Bot. Gard. Edinb.* 41: 491–504.

Miller, A.G. & Alexander, D. (1996) *Haemanthus grandifolius* Balf.f.: an enigma resolved. *Edinb. J. Bot.* 53(1): 43–47.

Miller, A.G., Atkinson, R., al-Khulaidi, A.W. & Taleeb, N. (2002) *Nesocrambe*, a new genus of Cruciferae (Brassiceae) from Soqotra. *Willdenowia* 32: 61–67.

Miller, A.G. & Hyam, R.D. (2001) A botanical contribution towards the conservation of the Soqotra Archipelago in research and management options for protected areas. In: *Proceedings of the 1st International Symposium and Workshop on Arid Zone Environments*, pp. 89–102. ERWDA, Abu Dhabi.

Miller, A.G. & Morris, M. (1988) *Plants of Dhofar, the Southern Region of Oman: Traditional, Economic and Medicinal Uses*. Office of the Adviser for Conservation of the Environment, Diwan of Royal Court, Sultanate of Oman.

Miller, A.G., Morris, M.J. & Atkinson, R.A. (in press) Conservation on Soqotra: ravaging goats or climatic change. *Yemen. J. Sci.*

Mossa, J.S., Al Yahya, M.A. & Al Meshal, I.A. (1987) *Medicinal Plants of Saudi Arabia*, Vol. I. King Saud University Libraries, Saudi Arabia.

Munier, P. (1973) *Le Palmier Dattier.* Techniques agricoles et productions tropicales, Vol. 14. G-P Maisonneuve & Larose, Paris.

Nordal, I. & Thulin, M. (1993) Synopsis of *Anthericum* and *Chlorophytum* (Anthericaceae) in the Horn of Africa, including the description of nine new species. *Nord. J. Bot.* 13(3): 257–280.

Oliver-Bever, B. (1986) *Medicinal Plants in Tropical West Africa.* Cambridge University Press, Cambridge.

Perry, L.M. & Metzger, J. (1980) *Medicinal Plants of East and Southeast Asia: Attributed Properties and Uses*. MIT Press, Cambridge, Massachusetts and London.

Petrusson, L. & Thulin, M. (1996) Taxonomy and biogeography of *Gymnocarpos* (Caryophyllaceae). *Edinb. J. Bot.* 53(1): 1–26.

Phillips, S. (1995) *Flora of Ethiopia and Eritrea, Vol. 7: Poaceae (Graminae).* Addis Ababa, Uppsala. 420pp.

Phillips, S. (2000) Notes on *Portulaca* L. (Portulaceae) in tropical East Africa. *Kew Bull.* 55(3): 687–698.

Plowes, D.H.C. (1993) A new account of *Echidnopsis* (Asclepiadaceae: Stapeliae). *Haseltonia* 1: 65–85.

Popenoe, P. (1924) The date palm in antiquity. *Scientific Monthly* 19: 313–325.

Popenoe, P. (1973) *The Date Palm* (ed. H. Field). Field Research Projects, Coconut Grove, Miami, Florida.

Popov, G.B. (1957) The vegetation of Soqotra. *J. Linn. Soc. (Bot.)* 55: 706–720.

Rabinowitz, D. (1981) Seven forms of rarity. In: *The Biological Aspects of Rare Plant Conservation* (ed. H. Synge). John Wiley & Sons, Chichester.

Radcliffe-Smith, A. (1971) New or noteworthy species from Soqotra and Abd al Kuri. *Hook. Ic. Pl.* 7(4): t.3673–3700.

Radcliffe-Smith, A. (1989) Notes on African Euphorbiaceae: XX. *Acalypha* (ii), etc. *Kew Bull.* 44(3): 447.

Rizk, A.M. & El-Ghazaly, G.A. (1995) *Medicinal and Poisonous Plants of Qatar*. University of Qatar, Qatar.

Robinson, H. (1999) Revisions of paleotropical Vernonieae (Asteraceae). *Proc. Biol. Soc. Washington* 112(1): 229.

Robson, N.K.B. (1967) An Iranian floristic link with Socotra in *Hypericum* (Hypericaceae). *Pl. Syst. Evol.* 155: 89–92.

Robson, N.K.B. (1987) Studies in the genus *Hypericum* L. (Guttiferae): 7. Section 29. *Brathys* (part 1). *Bull. Brit. Mus. (Nat. Hist.) Bot.* 16(1): 106pp.

Rowley, G.D. (1999) *Pachypodium* and *Adenium*. Southampton. 80pp.

Schrire, B.D. (1995) Evolution in the tribe Indigoferae (Leguminosae: Papilionoideae). In: *Advances in Legume Systematics, Part 7: Phylogeny* (eds M.D. Crisp & J.J. Doyle), pp. 161–244. Royal Botanic Gardens, Kew.

Sebald, O. (1980) Die gattung *Leucas* R. Brown (Labiatae) in Afrika und auf der Arabischen Halbinsel. *Stuttgarter Beitr. Naturk., A.* 341: 1–200.

SEPASET (Survey of Economic Plants for Arid and Semi-arid Zones). Newsletters published by the Royal Botanic Gardens, Kew.

Sofowora, A. (1982) *Medicinal Plants and Traditional Medicine in Africa*. John Wiley & Sons, New York.

Sutton, D. (1988) *A Revision of the Tribe Antirrhineae*. British Museum (Natural History), London. 575pp.

Takhtajan, A. (1986) *Floristic Regions of the World*. University of California Press.

Taylor, P. (1959) The genus *Caylusea* in Africa. *Kew Bull.* 13: 283–286.

Thulin, M. (1993) *Flora of Somalia*, Vol. I. Royal Botanic Gardens, Kew.

Thulin, M. (1997a) A new species and a new record of *Tephrosia* (Leguminosae) from Socotra. *Nord. J. Bot.* 17(4): 349.

Thulin, M. (1997b) *Gaillonia* (Rubiaceae-Paederieae) in Africa and Arabia. *Nord. J. Bot.* 18(1): 31–38.

Thulin, M. (1998) A new species of *Wellstedia* (Boraginaceae) from Somalia. *Nord. J. Bot.* 18(6): 663–665.

Thulin, M. (1999a) *Thamnosma* (Rutaceae) in Africa. *Nord. J. Bot.* 19(1): 5–11.

Thulin, M. (1999b) *Chapmannia* (Leguminosae-Stylosanthinae) extended. *Nord. J. Bot.* 19(5): 597–607.

Thulin, M. (2001) *Exacum* (Gentianaceae) on the Arabian Peninsula and Socotra. *Nord. J. Bot.* 21(3): 243–247.

Thulin, M. & Gifri, A. (1998) The frankincense trees (*Boswellia* spp., Burseraceae) of Soqotra. In: *Proceedings of the First International Symposium on Soqotra Island: Present and Future* (ed. H.J. Dumont), pp. 107–113. UN Publications, New York.

Thulin, M. & Johansson, A.N.B. (1996) Taxonomy and biogeography of the anomalous genus *Wellstedia.* In: *Biodiversity of African Plants, Proceedings XIV AETFAT Congress*, pp. 73–86.

Thulin, M. & Lavin, M. (2001) Phylogeny and biogeography of the *Ormocarpum* group (Fabaceae): a new genus *Zygocarpum* from the Horn of Africa region. *Syst. Bot.* 26(2): 299–317.

Townsend, C.C. (1974) Notes on Amaranthaceae: 2. *Kew Bull.* 29(3): 461–475.

REFERENCES

Triest, L. (1987) A revision of the genus *Najas* L. (Najadaceae) in Africa and surrounding islands. *Mem. Acad. Sci. Outre-Mer, Cl. Sci. nat. med. Nouv. Ser.* 21(4): 88pp.

Verdcourt, B. (1973) A revision of the African species of *Carphalea* Juss. (= *Dirichletia* Klotzsch) (Rubiaceae). *Kew Bull.* 28: 423–428.

Vierhapper, F. (1907) Beiträge zur Kenntnis der Flora Sudarabiens und der Inseln Sokotra, Semha und 'Abd el Kuri. *Denkschr. Kaiserl. Akad. Wiss. Wien math.-nat. Kl.* 71: 321–490.

Vollesen, K. (1994) Delimitation of *Angkalanthus* (Acanthaceae: Justicieae) and the new genus *Chorisochora. Kew Bull.* 49(3): 469–479.

Vollesen, K. (2000) Blepharis *(Acanthaceae): A Taxonomic Revision.* Royal Botanic Gardens, Kew. 342pp.

Watt, J.M. & Breyer-Bandwijk, M.G. (1962) *The Medicinal and Poisonous Plants of Southern and Eastern Africa*. Livingstone, London.

Wellstead, J.R. (1835) Memoir on the Island of Soqotra. *J. Roy. Geogr. Soc.* 5: 129–229.

White, F. (1983, 1986) *The Vegetation of Africa: A descriptive memoir to accompany the UNESCO, AETFAT, UNSO vegetation map of Africa*. UNESCO, Paris.

Wood, J.R.I. (1997) *A Handbook of the Yemen Flora.* Royal Botanic Gardens, Kew. 434pp.

Zohary, M. (1982) *Plants of the Bible.* Cambridge University Press, Cambridge.

SELECT BIBLIOGRAPHY

Akerele, O., Heywood, V. & Synge, H. (eds) (1991) *Conservation of Medicinal Plants.* Cambridge University Press, Cambridge.

Al Hubaishi, A. & Muller-Hohenstein, K. (1984) *Introduction to the Vegetation of Yemen: Ecological Basis and Floristic Composition, and Human Influence.* Eschborn, Germany.

Awadh, N.A. (1999) *Myrrh and Frankincense.* Obadi Centre for Scientific Publications, Sana'a, Yemen. (In Arabic)

Ayensu, E.S. (1978) *Medicinal Plants of West Africa.* Reference Publications Inc., Michigan.

Badheeb, A.S. (1991) *Medicinal Plants of Yemen.* Maktab al-Irshad, Sana'a, Yemen. (In Arabic)

Balick, M.J. & Cox, P.A. (1996) *Plants, People and Culture: The Science of Ethnobotany.* Scientific American Library, New York.

Beyhl, F.E. (1996a) Attempts at raising the Soqotran Dragon Tree, *Dracaena cinnabari* Balf.fil., outside the Island. In: *Proceedings of the First International Symposium on Soqotra Island: Present and Future. Conservation and Sustainable Use of Biodiversity of Soqotra Archipelago.* Technical Series, Vol. 1. United Nations Publications, New York.

Beyhl, F.E. (1996b) The *emta* tree (*Euphorbia arbuscula* Balf.fil.): a succulent tree of dracoid habits from the island of Soqotra. In: *Proceedings of the First International Symposium on Soqotra Island: Present and Future. Conservation and Sustainable Use of Biodiversity of Soqotra Archipelago.* Technical Series, Vol. 1. United Nations Publications, New York.

Bogdan, A.V. (1958) *A Revised List of Kenya Grasses.* The Government Printer, Nairobi.

Bond, A.J. (1995) *Ethnobotany and Land Use in Soqotra.* Unpublished project for MSc in the Biodiversity and Taxonomy of Plants, Royal Botanic Garden Edinburgh.

Botting, D. (1958a) *Island of the Dragon's Blood.* Hodder & Stoughton, London.

Botting, D. (1958b) The Oxford University Expedition to Socotra. *Geogr. J.* 124, 2.

Boxhall, P.G. (1965) *Report of the RAF/Army Expedition to Soqotra.* Cyprus.

Brown, G.H. (1966) *Social and Economic Conditions and Possible Development of Socotra.* Unpublished report.

Chopra, I.C., Abrol, B.K. & Handa, K.L. (1983 reprint) *Medicinal Plants of the Arid Zones,* Part I (Part II Paris, R. & Dillemann, G.). UNESCO.

Cook, F.E.M. (1995) *Economic Botany: Data Collection Standard.* Royal Botanic Gardens, Kew.

Cotton, C.M. (1996) *Ethnobotany: Principles and Applications.* John Wiley & Sons, Chichester.

Cracker, L.E. & Simon, J.E. (1981) *Herbs, Spices and Medicinal Plants,* Vol. I. Onyx Press.

Crocker Jones, G. *Traditional Spinning and Weaving in the Sultanate of Oman.* Historical Association of Oman.

De Winter, B. (1971) Floristic relationships between the northern and southern arid areas in Africa. *Mitt. Bot. Staatsamul. Munch.* 10: 424–427.

Dickson, V. (1955) *The Wild Flowers of Kuwait and Bahrain.* George Allen & Unwin Ltd, London.

Doe, D.B. (1992) *Socotra: Island of Tranquillity.* Immel Publishing, London.

Edwards, D.C. & Bogdan, A.V. (1951) *Important Grassland Plants of Kenya.* Pitman & Sons, Nairobi.

Farnsworth, N.R. (ed.) (1990) *Economic and Medicinal Plant Research.* Academic Press, New York.

Groom, N. (1981) *Frankincense and Myrrh.* Longman, London and New York.

Guarino, L., Miller, A.G., Baazara, M. & Obadi, N. (1990) Soqotra: the island of bliss revisited. *Diversity* 6: 28–31.

Hannon, P.H. (1995) Succulents of Socotra. *Cact. Succ. J. (USA)* 67(5): 267–275.

Hooper, D. & Mann, H.H. (1937) *Useful Plants and Drugs of Iran and Iraq.* Botanical Series, Field Museum of Natural History, Vol. IX, No. 3. Chicago.

Humphreys, L.R. (1987) *Tropical Pastures and Fodder Crops* (2nd edition). Longman, London.

Khan, M.S. (1986) *Islamic Medicine.* Routledge and Kegan Paul, London.

Koopowitz, H. & Kaye, H. (1990) *Plant Extinction: A Global Crisis* (2nd edition). Christopher Helm, London.

Lamprey, H.F., Herlocker, D.J. & Field, C.R. (1980) *Browse in Africa: The Current State of Knowledge* (ed. H.N. le Houérou). ILCA.

Lavin, M. & Luckow, M. (1993) Origins and relationships of tropical North America in the contexts of the boreotropics hypothesis. *Amer. J. Bot.* 80: 1–14.

Miller, A.G. & Alexander, D. (1995) Soqotra's misty future. *New Scientist* 1988: 32–35.

Miller, A.G., Boulos, L. & Mill, R.R. (1994) Regional overview: South West Asia and the Middle East. In: *Centres of Plant Diversity: A Guide and Strategy for their Conservation* (eds S.D. Davis *et al.*), pp. 293–348. WWF & IUCN, Cambridge.

Miller, A.G. & Cope, T.A. (1996) *Flora of the Arabian Peninsula and Socotra.* Edinburgh University Press, Edinburgh. 586pp.

Miller, A.G. & Morris, M. (2001) *Conservation and Sustainable Use of the Biodiversity of the Soqotra Archipelago.* Final Report: Target Areas, GEF, YEM/96/G32. Royal Botanic Garden Edinburgh.

Miller, A.G. & Nyberg, J.A. (1991) Patterns of endemism in Arabia. *Fl. Veg. Mundi* 9: 263–279.

Morris, M. (1997) The harvesting of frankincense in Dhofar, Oman. In: *Profumi D'Arabia* (ed. A. Avanzini), pp. 231–251. Saggi Di Storia Antica 11, Rome.

Morris, M. (2002a) Plant names in Dhofar and the Soqotra Archipelago. *Proceedings of the Seminar for Arabian Studies* 32: 47–61.

Morris, M. (2002b) *Manual of Traditional Land Use in the Soqotra Archipelago.* GEF, YEM/96/G32. Royal Botanic Garden Edinburgh.

Morris, M. (2003) The Soqotra Archipelago: concepts of good health and everyday remedies for illness. *Proceedings of the Seminar for Arabian Studies* 33: 319–341.

Naumkin, V. (1993) *Island of the Phoenix: An Ethnographic Study of the People of Socotra.* Ithaca Press, Reading.

Neuwinger, H.D. (1996) *African Ethnobotany: Poisons and Drugs.* Chapman & Hall.

Obadi, N.A. & Al Khulaidi, A.A. (1997) *Trees and Shrubs of Yemen.* FAO/Obadi Centre for Scientific Publications, Sana'a, Yemen. (In Arabic)

ODI: Papers of the Pastoral Development Network of the Agricultural Administration Unit, Overseas Development Institute.

Osborn, D. (1968) Notes on the medicinal and other uses of plants in Egypt. *Econ. Bot.* 22: 165–177.

Owdat, M. & Laham, G. (1987) *Medicinal Plants and their Uses.* Al Ahli Press, Cairo.

Prendergast, H.D.V., Etkin, N.L., Harris, D.R. & Houghton, P.J. (eds) (1998) *Plants for Food and Medicine.* Proceedings of the Joint Conference of the Society for Economic Botany and the International Society for Ethnopharmacology, London, 1–6 July 1996.

Schopen, A. (1983) *Traditionelle Heilmittel in Jemen.* Steiner, Weisbaden.

Simeone-Senelle, M-C. (1994) *Suqutra: parfums, sucs et resins.* Saba, Revue trimestrielle, France.

Staüble, N. (1984) Etude ethnobotanique des Euphorbiacées d'Afrique de l'Ouest. *J. Ethnopharmacol.* 16: 23–103 [1986].

Strelcyn, S. (1968) *Médicine et Plantes d-Ethiopie: Les Traités Médicaux Ethiopiens.* Zaklad Orientalistyki Polskiej Akademiinank, Warsaw.

UNESCO (1960) *Medicinal Plants of Arid Zones.* Paris.

University of Aden (1982) *Socotra Island.* University of Aden Research Programme, Faculty of Education.

Verdcourt, B. (1969) The arid corridor between the north-east and south-west areas of Africa. *Palaeoecol. Afr.* 4: 140–144.

Wickens, G.E., Goodin, J.R. & Field, D.N. (eds) (1985) *Plants for Arid Lands.* Allen & Unwin, London.

Wilson, R.T. (1988) *The Camel* (2nd impression). Longman.

Also the journals *Economic Botany* (*Econ. Bot.*) and *Journal of Ethnopharmacology* (*J. Ethnopharmacol.*).

INDEX TO PLANT NAMES

Page references are distinguished as follows: Illustrated Key – *italic*; Plant Portraits – **bold**; Encyclopaedia – plain.

Aberia abyssinica see *Dovyalis abyssinica*
Abutilon ... *203*, *204*, 622
Abutilon fruticosum ... *204*, 622
Abutilon graveolens see *Abutilon pannosum*
Abutilon muticum see *Abutilon pannosum*
Abutilon pannosum ... *204*, 622
Acacia ... *191*, 591
Acacia aff. *tortilis* ... *191*, 593
Acacia balfourii see *Acacia edgeworthii*
Acacia edgeworthii ... *191*, **267**, 591
Acacia nilotica ... *191*, 591
Acacia pennivenia ... *191*, **267**, 592
Acacia sarcophylla ... *191*, **267**, 592
Acacia socotrana see *Acacia edgeworthii*
Acacia sp. *A* ... *191*, 593
Acalypha ... 542
Acalypha indica ... *177*, **260**, 542
Acanthaceae ... *75*, *77*, *87*, *127*, 400
Acanthospermum ... 495
Acanthospermum hispidum ... *159*, 495
Achyranthes ... 417
Achyranthes aspera ... *134*, 417
Achyrocline ... *161*, *163*, 495
Achyrocline luzuloides ... *164*, 495
Achyrocline schimperi ... *164*, 496
Acokanthera ... 427
Acokanthera schimperi ... *137*, 427
Acrachne ... 346
Acrachne racemosa ... *106*, *112*, 346
Acridocarpus ... 621
Acridocarpus orientalis see *Acridocarpus socotranus*
Acridocarpus socotranus ... *93*, *202*, **269**, 621
Actiniopteridaceae ... 285
Actiniopteris ... 285
Actiniopteris dichotoma see *Actiniopteris semiflabellata*
Actiniopteris semiflabellata ... *64*, **233**, 285
Adenium ... 427
Adenium multiflorum see *Adenium obesum*
Adenium obesum ... *137*, **242**, 427
Adenium socotranum see *Adenium obesum*
Adiantaceae ... 286
Adiantum ... 286
Adiantum aethiopicum see *Adiantum poirettii*
Adiantum balfourii ... *65*, **233**, 286
Adiantum capillus-veneris ... *66*, 286
Adiantum poirettii ... *66*, 287
Aeluropus ... 346
Aeluropus lagopodioides ... *106*, *107*, *111*, 346
Aeluropus lagopoides see *Aeluropus lagopodioides*
Aerva ... *85*, *135*, 418
Aerva javanica ... *135*, 418
Aerva lanata ... *135*, 419
Aerva lanata var. *robusta* see *Aerva lanata*
Aerva microphylla ... *135*, **241**, 419
Aerva microphylla var. *humilis* see *Aerva microphylla*
Aerva persica see *Aerva javanica*
Aerva revoluta ... *135*, **241**, 419
Ageratum ... 496
Ageratum conyzoides ... *159*, **253**, 496
Aizoaceae ... *76*, *84*, *133*, 414
Aizoon ... 414
Aizoon canariense ... *133*, **241**, 414
Allophylus ... 676
Allophylus rubifolius ... *220*, **276**, 676
Allosorus melanolepis see *Onychium divaricatum*
Aloaceae ... *70*, *95*, 297
Aloe ... 297
Aloe forbesii see *Aloe perryi*
Aloe jawiyon ... *95*, **234**, 297
Aloe perryi ... *95*, **234**, 297
Aloe squarrosa ... *95*, **234**, 303
Alsodeia socotrana see *Hybanthus enneaspermus*
Alysicarpus ... *192*, *195*, 595
Alysicarpus glumaceus see *Alysicarpus monilifer*
Alysicarpus monilifer ... *195*, 595
Alysicarpus vaginalis ... *195*, 596
Amaranthaceae ... *76*, *84*, *85*, *134*, 417
Amaranthus ... *135*, *136*, 420
Amaranthus blitum see *Amaranthus graecizans* subsp. *graecizans*
Amaranthus dubius ... *136*, 420
Amaranthus graecizans ... *136*, 420
Amaranthus graecizans subsp. *graecizans* ... 421
Amaranthus graecizans subsp. *thellunganianus* ... 421
Amaranthus polygamus see *Amaranthus graecizans*
Amaranthus sparganiocephalus ... *136*, 421
Amaryllidaceae ... *71*, *95*, 303
Ammannia ... *202*, 620
Ammannia baccifera ... *202*, 620
Ammannia multiflora ... *202*, 620
Amphilophis insculpta see *Bothriochloa insculpta*
Amphilophis see *Bothriochloa*
Anabasis ... 489
Anabasis ehrendbergii ... *158*, 489
Anacardiaceae ... *89*, *136*, 424
Anagallis ... 656
Anagallis arvensis ... *77*, *214*, 657
Ancalanthus paucifolius see *Angkalanthus oligophylla*
Andrachne ... *178*, *179*, 542
Andrachne schweinfurthii ... *179*, 542
Andrachne telephioides ... *179*, 543
Andropogon ... 347
Andropogon bentii ... *110*, **236**, 347
Andropogon hirtus see *Hyparrhenia hirta*
Andropogon insculptus see *Bothriochloa insculpta*
Andropogon pertusus see *Bothriochloa insculpta*
Andropogon pertusus var. *insculptus* see *Bothriochloa insculpta*
Angkalanthus ... 400
Angkalanthus oligophylla ... *128*, **238**, 400
Angraecopsis tenerrima see *Angraecum dives*
Angraecum ... 391
Angraecum dives ... *126*, **237**, 391
Anisotes ... 400
Anisotes diversifolius ... *129*, **238**, 400
Anisotes diversifolius var. *brevicalyx* see *Anisotes diversifolius*
Anthericaceae ... *72*, *96*, 304
Anthericum graptophyllum see *Chlorophytum graptophyllum*
Anthistiria ciliata see *Themeda quadrivalvis*
Anthistiria see *Themeda*
Anticharis ... 679
Anticharis arabica ... *222*, 679
Anticharis glandulosa see *Anticharis arabica*
Antirrhinum orontium see *Misopates orontium*
Apium ... 702
Apium graveolens ... *228*, 702
Apluda ... 347
Apluda aristata see *Apluda mutica*
Apluda mutica ... *109*, 347
Apocynaceae ... *88*, *137*, 427

Aptosimum ... 679
Aptosimum pumilum ... *222*, 679
Araceae ... *70*, *96*, 306
Arenaria ... 481
Arenaria foliacea ... *154*, 481
Arenaria serpyllifolia see *Arenaria foliacea*
Argemone ... 645
Argemone mexicana ... *210*, **242**, 645
Aristida ... *108*, *114*, 348
Aristida adscensionis ... *114*, 348
Aristida anaclasta ... *114*, 348
Aristida funiculata ... *114*, 349
Aristida murina see *Aristida funiculata*
Aristida sokotrana see *Stipagrostis sokotrana*
Arnebia ... 445
Arnebia hispidissima ... *144*, 445
Arthraxon ... *105*, *114*, 349
Arthraxon lancifolius ... *114*, 349
Arthraxon micans ... *114*, 349
Arthraxon molle see *Arthraxon lancifolius*
Arthrocarpum gracile see *Chapmannia gracilis*
Arthrocnemum ... 489
Arthrocnemum glaucum see *Arthrocnemum macrostachyum*
Arthrocnemum macrostachyum ... *156*, **252**, 489
Asclepiadaceae ... *77*, *88*, *137*, 429
Asparagaceae ... *71*, *96*, 307
Asparagus ... 307
Asparagus africanus ... *96*, **235**, 307
Asparagus sp. *A* ... *96*, 308
Asphodelaceae ... *72*, 308
Asphodelus ... *97*, 308
Asphodelus fistulosus ... *97*, 308
Asphodelus fistulosus var. *tenuifolius* see *Asphodelus fistulosus*
Asphodelus tenuifolius see *Asphodelus fistulosus*
Aspleniaceae ... 289
Asplenium ... 289
Asplenium aethiopicum ... *67*, 289
Asplenium phillipsianum see *Ceterach phillipsianum*
Asplenium praemorsum see *Asplenium aethiopicum*
Asplenium schweinfurthii ... *65*, **233**, 289
Asplenium trichomanes ... *65*, 289
Astragalus ... 596
Astragalus hauarensis ... *195*, 596
Asystasia ... 401
Asystasia ansellioides see *Asystasia gangetica*
Asystasia coromandeliana see *Asystasia gangetica*
Asystasia gangetica ... *127*, 401
Atriplex ... *157*, *158*, 490
Atriplex farinosa ... *158*, 490
Atriplex griffithii ... *158*, 490
Australina capensis see *Parietaria debilis*
Avicennia ... 443
Avicennia marina ... *85*, *142*, **245**, 443
Avicennia officinalis see *Avicennia marina*
Avicenniaceae ... *85*, *142*, 443

Babiana ... 388
Babiana socotrana ... *125*, **237**, 388
Bacopa ... 680
Bacopa monnieri ... *220*, 680
Ballochia ... *127*, *129*, 402
Ballochia amoena ... *129*, **238**, 402
Ballochia atro-virgata ... *129*, 403
Ballochia puberula see *Ballochia atro-virgata*
Ballochia rotundifolia ... *129*, 403
Balsamodendrum mukul see *Commiphora kua*
Balsamodendrum parvifolium see *Commiphora parvifolia*
Balsamodendrum planifrons see *Commiphora planifrons*
Balsamodendrum socotranum see *Commiphora socotrana*
Barleria ... *128*, *129*, 403
Barleria aculeata ... *130*, **238**, 403
Barleria argentea ... *130*, 404
Barleria popovii ... *130*, **238**, 404
Barleria tetracantha ... *130*, **238**, 404
Begonia ... 444
Begonia samhaensis ... *142*, **246**, 444
Begonia socotrana ... *142*, **246**, 444
Begoniaceae ... *82*, *142*, 444
Bidens ... *159*, *162*, 496
Bidens abyssinica see *Bidens biternata*
Bidens biternata ... *162*, 496
Bidens pilosa ... *162*, 497
Blainvillea ... 497
Blainvillea dichotoma see *Blainvillea gayana*
Blainvillea gayana ... *159*, **253**, 497
Blainvillea rhomboidea see *Blainvillea gayana*
Blepharis ... *127*, *130*, 405
Blepharis boerhaviifolia see *Blepharis maderaspatensis*
Blepharis ciliaris see *Blepharis kuriensis*
Blepharis edulis see *Blepharis kuriensis*
Blepharis kuriensis ... *130*, **239**, 405
Blepharis linariifolia see *Blepharis kuriensis*
Blepharis maderaspatensis ... *130*, 405
Blepharis spiculifera see *Blepharis spiculifolia*
Blepharis spiculifolia ... *130*, **239**, 405
Boerhavia ... 639
Boerhavia boissieri see *Commicarpus boissieri*
Boerhavia diffusa ... *208*, 639
Boerhavia heimerlii *Commicarpus heimerlii*
Boerhavia plumbaginea var. *socotrana* see *Commicarpus simonyi*
Boerhavia repens see *Boerhavia diffusa*
Boerhavia scandens see *Commicarpus heimerlii*
Boerhavia scandens see *Commicarpus simonyi*
Boerhavia simonyi see *Commicarpus simonyi*
Bonamia glomerata see *Seddera glomerata*
Bonamia spinosa see *Seddera spinosa*
Boraginaceae ... *81*, *95*, *143*, 445
Borassus ... 393
Borassus aethiopum ... *126*, **237**, 393
Borassus flabelliformis see *Borassus aethiopum*
Boswellia ... *89*, *148*, 457
Boswellia ameero ... *149*, **248**, 458
Boswellia bullata ... *149*, **248**, 460
Boswellia dioscorides ... *149*, **248**, 460
Boswellia elongata ... *149*, **249**, 460
Boswellia nana ... *148*, **249**, 460
Boswellia popoviana ... *148*, **249**, 461
Boswellia socotrana ... *149*, **250**, 461
Boswellia sp. *A* ... *149*, **249**, 464
Bothriochloa ... 350
Bothriochloa insculpta ... *105*, 350
Bothriochloa pertusa see *Bothriochloa insculpta*
Boucerosia socotrana see *Caralluma socotrana*
Brachiaria ... *112*, *115*, 350
Brachiaria arida ... *115*, 350
Brachiaria deflexa ... *115*, 351
Brachiaria eruciformis ... *115*, 351
Brachiaria glauca see *Brachiaria ovalis*
Brachiaria lata ... *113*, *115*, 351
Brachiaria ovalis ... *115*, 351
Brachiaria ramosa ... *115*, 352
Brassica rostrata see *Erucastrum rostratum*
Breweria fastigiata see *Seddera fastigiata*
Breweria glomerata see *Seddera glomerata*
Breweria latifolia see *Seddera latifolia*
Breweria pedunculata see *Seddera pedunculata*
Bulbostylis ... *98*, 313

Bulbostylis barbata ... *99*, 313
Bulbostylis hispidula ... *99*, 314
Burseraceae ... *89*, *92*, *148*, 457
Buxaceae ... *150*, 472
Buxanthus ... *86*, 472
Buxanthus pedicellatus ... *86*, *150*, **251**, 472
Buxus hildebrandtii see *Buxanthus pedicellatus*

Cadaba ... *151*, 474
Cadaba farinosa ... *152*, 474
Cadaba insularis ... *152*, **251**, 475
Cadaba longifolia ... *152*, 475
Cadaba rotundifolia ... *151*, 475
Calligonum ... 652
Calligonum comosum ... *212*, 652
Calligonum polygonoides subsp. *comosum* see *Calligonum comosum*
Calotropis ... 429
Calotropis procera ... *138*, **243**, 429
Campanula ... 473
Campanula balfourii ... *150*, **251**, 473
Campanula dichotoma see *Campanula balfourii*
Campanulaceae ... *80*, *150*, 473
Camptoloma ... 680
Camptoloma lyperiiflorum ... *222*, **277**, 680
Camptoloma villosa see *Camptoloma lyperiiflorum*
Campylanthus ... 680
Campylanthus spinosus ... *221*, **277**, 680
Canavalia ... 596
Canavalia africana ... *194*, 596
Canavalia ensiformis see *Canavalia africana*
Canavalia virosa see *Canavalia africana*
Canthium socotranum see *Pyrostria socotrana*
Capparaceae ... *79*, *84*, *93*, *151*, 474
Capparis ... *151*, *152*, 476
Capparis aphylla see *Capparis decidua*
Capparis cartilaginea ... *152*, **251**, 476
Capparis decidua ... *152*, 477
Capparis galeata see *Capparis cartilaginea*
Capparis spinosa see *Capparis cartilaginea*
Capsella ... 528
Capsella bursa-pastoris ... *173*, 528
Caralluma ... 430
Caralluma dioscoridis see *Duvaliandra dioscoridis*
Caralluma rosengrenii see *Caralluma socotrana*
Caralluma socotrana ... *139*, **243**, 430
Carissa ... 429
Carissa edulis see *Carissa spinarum*
Carissa schimperi see *Acokanthera schimperi*
Carissa spinarum ... *137*, 429
Carphalea ... 663
Carphalea glaucescens see *Carphalea obovata*
Carphalea obovata ... *216*, **274**, 663
Carphalea obovata var. *albescens* see *Carphalea obovata*
Carum calcicolum see *Trachyspermum pimpinelloides*
Carum kuriense see *Trachyspermum pimpinelloides*
Carum kuriense var. *trichocarpum* see *Trachyspermum pimpinelloides*
Carum pimpinelloides see *Trachyspermum pimpinelloides*
Carum trichocarpum see *Trachyspermum pimpinelloides*
Caryophyllaceae ... *78*, *153*, 481
Cassia absus see *Chamaecrista absus*
Cassia holosericea see *Senna holosericea*
Cassia italica see *Senna italica*
Cassia obovata see *Senna italica*
Cassia socotrana see *Senna socotrana*
Cassia sophera see *Senna socotrana*
Cassia tora see *Senna obtusifolia*
Caylusea ... 658
Caylusea canescens see *Caylusea hexagyna*
Caylusea hexagyna ... *214*, 658
Celastraceae ... *93*, *156*, 488
Cenchrus ... *110*, *116*, 352
Cenchrus ciliaris ... *116*, 352
Cenchrus pennisetiformis ... *116*, 353
Cenchrus schimperi see *Cenchrus setigerus*
Cenchrus setigerus ... *116*, 353
Centaurium ... 569
Centaurium tenuiflorum ... *183*, 569
Centella ... 702
Centella asiatica ... *228*, 702
Cephalocroton ... 543
Cephalocroton socotranus ... *178*, **260**, 543
Cephalocrotonopsis socotranus see *Cephalocroton socotranus*
Ceratopteris ... 293
Ceratopteris cornuta ... *66*, 293
Ceratopteris thalictroides see *Ceratopteris cornuta*
Ceropegia ... *137*, *139*, *140*, 431
Ceropegia affinis ... *140*, **243**, 431
Ceropegia tihamana ... *140*, **243**, 432
Ceterach ... 290
Ceterach phillipsianum ... *65*, 290
Chaenostoma liperiflorum see *Camptoloma lyperiiflorum*
Chaenostoma oxypetalum see *Camptoloma lyperiiflorum*
Chamaecrista ... 586
Chamaecrista absus ... *190*, 586
Chapmannia ... *195*, *196*, 597
Chapmannia gracilis ... *196*, 597
Chapmannia reghidensis ... *196*, 598
Chapmannia sericea ... *196*, **267**, 598
Chapmannia tinireana ... *196*, 599
Cheilanthes concolor see *Doryopteris concolor*
Cheilanthes farinosa see *Negripteris sciona*
Cheilanthes involuta see *Pellaea involuta*
Cheilanthes viridis see *Pellaea viridis*
Chenopodiaceae ... *84*, *85*, *91*, *156*, 489
Chenopodium ... 491
Chenopodium murale ... *157*, 491
Chloris ... *105*, *116*, 353
Chloris barbata see *Chloris virgata*
Chloris pycnothrix ... *116*, 353
Chloris virgata ... *116*, 354
Chlorophytum ... 304
Chlorophytum filifolium ... *96*, **234**, 304
Chlorophytum graptophyllum ... *96*, **234**, 305
Chlorophytum sp. *A* ... *96*, **234**, 305
Choriptera semhahensis see *Lagenantha cycloptera*
Chorisochora ... *128*, *130*, 406
Chorisochora minor ... *130*, 406
Chorisochora striata ... *130*, **239**, 406
Christella ... 295
Christella dentata ... *67*, 295
Chrozophora ... 544
Chrozophora obliqua see *Chrozophora oblongifolia*
Chrozophora obliqua var. *frutescens* see *Chrozophora oblongifolia*
Chrozophora oblongifolia ... *178*, 544
Chrozophora tinctoria see *Chrozophora oblongifolia*
Chrysopogon ... *108*, *116*, 354
Chrysopogon aucheri ... *116*, 354
Chrysopogon gryllus see *Chrysopogon aucheri*
Chrysopogon plumulosus ... *116*, 354
Chrysopogon serrulatus ... *116*, 355
Cissus ... 712
Cissus hamaderohensis ... *231*, **280**, 713
Cissus paniculata ... *231*, **280**, 714
Cissus quadrangularis see *Cissus hamaderohensis*
Cissus subaphylla ... *231*, **280**, 715
Cistanche ... 643

Cistanche lutea see *Cistanche phelypaea*
Cistanche phelypaea *210*, **272**, 643
Cistanche tubulosa see *Cistanche phelypaea*
Citrullus *175*, 533
Citrullus colocynthis *175*, 533
Citrullus lanatus *175*, 534
Citrus *93*, 671
Citrus aurantium *219*, 671
Cladium 314
Cladium mariscus *98*, 314
Cleome *151*, *152*, 477
Cleome austroarabica *153*, 477
Cleome brachycarpa *152*, 478
Cleome brachycarpa var. *filicaulis* see *Cleome brachycarpa*
Cleome diversifolia see *Cleome brachycarpa*
Cleome gynandra *153*, 478
Cleome papillosa see *Cleome scaposa*
Cleome pentaphylla see *Cleome gynandra*
Cleome scaposa *153*, 478
Cleome socotrana *153*, 479
Cleome tenella *152*, 479
Cleome viscosa *153*, 479
Clerodendrum *230*, 708
Clerodendrum galeatum *230*, **280**, 708
Clerodendrum leucophloeum *230*, **280**, 709
Cocculus 633
Cocculus balfourii *207*, **271**, 633
Cocculus laeba see *Cocculus pendulus*
Cocculus pendulus *207*, 634
Cockburnia socotrana see *Poskea socotrana*
Coelocarpum *230*, *231*, 709
Coelocarpum haggierensis *231*, **280**, 709
Coelocarpum socotranum *231*, 710
Coelocarpus socotranus see *Coelocarpum socotranum*
Colchicaceae *72*, *97*, 309
Colocasia 306
Colocasia esculenta *96*, 306
Cometes 481
Cometes abyssinica *154*, 481
Cometes abyssinica subsp. *suffruticosa* see *Cometes abyssinica*
Commelina *97*, 309
Commelina albescens *97*, 310
Commelina benghalensis *97*, 310
Commelina forskalaei *97*, 311
Commelinaceae *72*, *97*, 309
Commicarpus *208*, 640
Commicarpus boissieri *209*, 640
Commicarpus heimerlii *209*, **272**, 641
Commicarpus helenae *209*, 641
Commicarpus simonyi *209*, **272**, 641
Commiphora *89*, *92*, *149*, 464
Commiphora habessinica see *Commiphora kua*
Commiphora kua *150*, **250**, 465
Commiphora ornifolia *150*, **250**, 466
Commiphora parviflora see *Commiphora parvifolia*
Commiphora parvifolia *150*, **250**, 468
Commiphora planifrons *150*, **250**, 471
Commiphora socotrana *150*, 471
Compositae *74*, *89*, *159*, 495
Convolvulaceae *81*, *95*, *168*, 516
Convolvulus *169*, 516
Convolvulus arabicus *169*, 517
Convolvulus fastigiata see *Seddera fastigiata*
Convolvulus filipes see *Convolvulus hildebrandtii*
Convolvulus glomerata see *Seddera glomerata*
Convolvulus glomeratus see *Convolvulus arabicus*
Convolvulus grantii *170*, **257**, 517
Convolvulus hildebrandtii *169*, **257**, 517
Convolvulus kossmatii see *Seddera spinosa*
Convolvulus rhyniospermus *169*, 518
Convolvulus sarmentosus *170*, **257**, 518
Convolvulus siculus *169*, 518
Convolvulus socotrana see *Seddera fastigiata*
Convolvulus spinosa see *Seddera spinosa*
Conyza 498
Conyza hochstetteri *162*, 498
Corbichonia 414
Corbichonia decumbens *133*, **241**, 414
Corchorus *226*, *227*, 698
Corchorus acutangulus see *Corchorus aestuans*
Corchorus aestuans *227*, 698
Corchorus antichorus see *Corchorus depressus*
Corchorus depressus *227*, 698
Corchorus erodioides *227*, **278**, 699
Corchorus erodioides var. *bicrenatus* see *Corchorus erodioides*
Corchorus erodioides var. *pinnatus* see *Corchorus erodioides*
Cordia *143*, *144*, 445
Cordia gharaf see *Cordia quercifolia*
Cordia obovata *144*, 446
Cordia obtusa *144*, **246**, 447
Cordia quercifolia *144*, 447
Cordia rothii see *Cordia quercifolia*
Cornulaca 491
Cornulaca ehrenbergii *158*, 491
Crassula *172*, 525
Crassula alata *172*, 525
Crassula schimperi *172*, **258**, 525
Crassulaceae *77*, *94*, *172*, 525
Craterostigma 681
Craterostigma pumilum *220*, **277**, 681
Cressa 519
Cressa cretica *168*, 519
Crinum 303
Crinum balfourii *95*, **234**, 303
Crotalaria *192*, *194*, *196*, 599
Crotalaria abdelkuriensis see *Crotalaria strigulosa*
Crotalaria dubia see *Crotalaria leptocarpa*
Crotalaria kuriensis see *Crotalaria strigulosa*
Crotalaria leptocarpa *197*, 599
Crotalaria pteropoda *197*, 599
Crotalaria retusa *197*, 600
Crotalaria schweinfurthii see *Crotalaria strigulosa*
Crotalaria socotrana *196*, **267**, 600
Crotalaria spinosa *197*, 600
Crotalaria strigulosa *197*, **267**, 600
Croton *178*, *179*, 545
Croton elaeagnoides see *Croton sulcifructus*
Croton pachyclados see *Croton socotranus*
Croton sarcocarpus *179*, **260**, 545
Croton socotranus *179*, **260**, 545
Croton socotranus var. *pachyclados* see *Croton socotranus*
Croton sulcifructus *179*, **261**, 548
Cruciferae *84*, *173*, 528
Cryptolepis *139*, *140*, 432
Cryptolepis arbuscula *140*, 432
Cryptolepis intricata *140*, **243**, 433
Cryptolepis macrophylla *141*, **243**, 433
Cryptolepis socotrana *141*, **244**, 434
Cryptolepis volubilis *141*, 434
Cucumis *175*, 534
Cucumis dipsaceus *176*, 534
Cucumis ficifolius *175*, 534
Cucumis ficifolius var. *echinophorus* see *Cucumis ficifolius*
Cucumis prophetarum 534
Cucumis prophetarum subsp. *dissectum* *176*, 535
Cucumis prophetarum subsp. *prophetarum* *176*, **259**, 535

Cucurbitaceae *80*, *91*, *174*, 533
Cullen 601
Cullen corylifolia *192*, 601
Curroria decidua var. *volubilis* see *Cryptolepis volubilis*
Curroria macrophylla see *Cryptolepis macrophylla*
Cuscuta 538
Cuscuta chinensis *176*, 538
Cuscuta planiflora *176*, 538
Cuscuta planiflora var. *globosa* see *Cuscuta planiflora*
Cuscutaceae *73*, *176*, 538
Cyanotis 311
Cyanotis cristata *97*, 311
Cyanthillium 498
Cyanthillium cinereum *167*, 498
Cylista balfourii see *Paracalyx balfourii*
Cylista scariosa see *Paracalyx balfourii*
Cylista schweinfurthii see *Paracalyx balfourii*
Cymbopogon 355
Cymbopogon jwarancusa 355
Cymodocea 312
Cymodocea rotundata *70*, 312
Cymodocea serrulata *70*, 312
Cymodoceaceae *69*, *70*, 312
Cynanchum 435
Cynanchum gerardii *137*, 435
Cynanchum linifolium see *Vincetoxicum linifolium*
Cynodon 356
Cynodon dactylon *107*, 356
Cyperaceae *72*, *98*, 313
Cyperus *99*, 314
Cyperus amabilis *101*, 315
Cyperus amblyostachyus see *Cyperus conglomeratus*
Cyperus aristatus see *Cyperus squarrosus*
Cyperus bulbosus *100*, 315
Cyperus compressus *101*, 316
Cyperus conglomeratus *101*, 316
Cyperus conglomeratus var. *effusus* see *Cyperus conglomeratus*
Cyperus conglomeratus var. *socotranus* see *Cyperus conglomeratus*
Cyperus curvulus see *Cyperus conglomeratus*
Cyperus cyperoides *100*, 316
Cyperus difformis *101*, 317
Cyperus dilatatus *100*, 317
Cyperus laevigatus *99*, 317
Cyperus longus *100*, 317
Cyperus microstylis see *Kyllinga microstyla*
Cyperus nitens see *Pycreus pumilus*
Cyperus proteinolepis see *Cyperus conglomeratus*
Cyperus pumilus see *Pycreus pumilus*
Cyperus rotundus *100*, 318
Cyperus rubicundus *101*, 318
Cyperus squarrosus *99*, 318
Cyperus stramineo-ferrugineus *101*, 318
Cyperus tenuiculmis *100*, 319
Cystostemon 447
Cystostemon socotranus *143*, **246**, 447

Dactyloctenium *106*, *107*, *117*, 357
Dactyloctenium aegyptium *117*, 357
Dactyloctenium aristatum *117*, 357
Dactyloctenium hackelii *117*, 357
Dactyloctenium robecchii *117*, 358
Datura 686
Datura fastuosa *224*, 686
Datura fastuosa var. *alba* see *Datura fastuosa*
Dendrosicyos 535
Dendrosicyos socotrana *91*, *174*, **259**, 535
Dennstaedtiaceae 290
Desmodium 601
Desmodium triflorum *193*, 601
Desmostachya 358
Desmostachya bipinnata *112*, 358
Diceratella 528
Diceratella incana *173*, **258**, 528
Diceratella sinuata see *Diceratella incana*
Dichanthium *105*, *109*, *117*, 359
Dichanthium annulatum *117*, 359
Dichanthium foveolatum *117*, 359
Dichanthium foveolatus see *Dichanthium foveolatum*
Dichondra 519
Dichondra repens *168*, 519
Dichrocephala 498
Dichrocephala chrysanthemifolia *162*, 498
Dichrocephala chrysanthemifolia var. *chrysanthemifolia* see *Dichrocephala chrysanthemifolia*
Dichrostachys 594
Dichrostachys dehiscens *191*, **268**, 594
Dicliptera 407
Dicliptera effusa *129*, **239**, 407
Dicliptera ovata see *Dicliptera effusa*
Dicoma *160*, *163*, 499
Dicoma cana *163*, **253**, 499
Dicoma tomentosa *163*, 499
Digera 421
Digera arvensis see *Digera muricata*
Digera muricata *135*, 421
Digitaria *107*, *112*, *113*, *117*, 360
Digitaria ciliaris *118*, 360
Digitaria rivae *117*, 360
Digitaria velutina *118*, 360
Diodia 664
Diodia aulacosperma *216*, 664
Dioscorea 322
Dioscorea lanata *103*, **235**, 322
Dioscoreaceae *70*, *103*, 322
Dipcadi *123*, *124*, 384
Dipcadi balfourii *124*, **236**, 384
Dipcadi biflorum *124*, **236**, 384
Dipcadi guichardii *124*, **236**, 385
Dipcadi kuriensis *124*, **236**, 385
Dipcadi sp. *A* see *Dipcadi kuriensis*
Dirachma 539
Dirachma socotrana *87*, *176*, **260**, 539
Dirachmaceae *87*, *176*, 539
Dirichletia glaucescens see *Carphalea obovata*
Dirichletia lanceolata see *Carphalea obovata*
Dirichletia obovata see *Carphalea obovata*
Dirichletia venulosa see *Carphalea obovata*
Distephanus 499
Distephanus qazmi *162*, **253**, 499
Dodonaea 677
Dodonaea angustifolia see *Dodonaea viscosa*
Dodonaea viscosa *220*, 677
Dorstenia *207*, 634
Dorstenia gigas *208*, **271**, 635
Dorstenia socotrana *208*, **271**, 635
Doryopteris 287
Doryopteris concolor *67*, 287
Doryopteris kirkii see *Doryopteris concolor*
Dovyalis 568
Dovyalis abyssinica *91*, *183*, **263**, 568
Dracaena 324
Dracaena cinnabari *103*, **235**, 324
Dracaenaceae *70*, *103*, 324
Dregea arabica see *Marsdenia robusta*
Drimia 385
Drimia porphyrostachys *123*, 385
Dryopteridaceae 291
Duvaliandra 435
Duvaliandra dioscoridis *139*, **244**, 435
Dyerophytum *211*, 648

Dyerophytum indicum var. *socotrana* see *Dyerophytum socotranum*
Dyerophytum pendulum *211*, 649
Dyerophytum socotranum *211*, **272**, 649

Ebenaceae *86*, *177*, 540
Ecbolium striatum see *Chorisochora striata*
Ecbolium striatum var. *minor* see *Chorisochora minor*
Echidnopsis *140*, *141*, 435
Echidnopsis bentii *141*, **244**, 436
Echidnopsis inconspicua *142*, **244**, 436
Echidnopsis insularis *141*, **244**, 436
Echidnopsis milleri *141*, **244**, 437
Echidnopsis socotrana *142*, **244**, 437
Echinochloa 361
Echinochloa colona *113*, 361
Echiochilon *144*, 448
Echiochilon persicum *144*, 448
Echiochilon pulvinata *144*, **246**, 448
Eclipta 500
Eclipta alba see *Eclipta prostrata*
Eclipta prostrata *159*, 500
Edithcolea 437
Edithcolea grandis *140*, **245**, 437
Edithcolea sordida see *Edithcolea grandis*
Ehretia 448
Ehretia obtusifolia *143*, 448
Elaeocarpus transultus see *Lannea transulta*
Eleocharis *98*, *102*, 319
Eleocharis caduca *102*, 319
Eleocharis capitata see *Eleocharis geniculata*
Eleocharis geniculata *102*, 319
Eleusine 361
Eleusine aegyptia see *Dactyloctenium aegyptium*
Eleusine coracana *106*, 361
Eleusine indica see *Eleusine coracana*
Eleusine indica subsp. *coracana* see *Eleusine coracana*
Eleusine verticillata see *Acrachne racemosa*
Emex 652
Emex spinosus *212*, 652
Endostemon 577
Endostemon tenuiflorus *186*, **265**, 577
Enhalus acoroides *69*
Enneapogon *110*, *114*, *118*, 363
Enneapogon brachystachyus see *Enneapogon desvauxii*
Enneapogon desvauxii *118*, 363
Enneapogon elegans see *Enneapogon desvauxii*
Enneapogon persicus *118*, 363
Enneapogon schimperianus see *Enneapogon persicus*
Entada sp. see *Acacia pennivenia*
Ephedra 296
Ephedra foliata *63*, 296
Ephedraceae 296
Eragrostis *104*, *118*, 364
Eragrostis aspera *119*, 364
Eragrostis barrelieri *119*, 364
Eragrostis cilianensis *119*, 364
Eragrostis ciliaris *107*, *118*, 365
Eragrostis cynosuroides see *Desmostachya bipinnata*
Eragrostis lepida *118*, 365
Eragrostis minor *119*, 365
Eragrostis papposa *119*, 365
Eragrostis viscosa *118*, 366
Eremopogon see *Dichanthium*
Eriochloa vestita see *Tricholaena vestita*
Erodium 571
Erodium cicutarium *184*, 571
Erucastrum 529
Erucastrum rostratum *173*, **258**, 529
Erythraea centaurium see *Centaurium tenuiflorum*
Erythrina 602
Erythrina melanacantha *193*, 602
Erythrina melanacantha subsp. *somala* see *Erythrina melanacantha*
Erythroxylaceae *93*, *177*, 541
Erythroxylon 541
Erythroxylon socotranum *93*, *177*, 541
Euclea 540
Euclea balfourii see *Euclea divinorum*
Euclea divinorum *86*, *177*, **260**, 540
Euclea kellau see *Euclea divinorum*
Euclea laurina see *Euclea divinorum*
Eulophia 391
Eulophia petersii *126*, **237**, 391
Euphorbia *177*, *179*, 549
Euphorbia abdelkuri *180*, **261**, 549
Euphorbia arbuscula *180*, **261**, 549
Euphorbia balsamifera *181*, **261**, 552
Euphorbia balsamifera subsp. *adenensis* see *Euphorbia balsamifera*
Euphorbia chamaesyce see *Euphorbia kischenensis*
Euphorbia granulata *182*, 553
Euphorbia hadramautica *180*, **261**, 553
Euphorbia hajhirensis *181*, 554
Euphorbia hamaderoensis *181*, **262**, 554
Euphorbia inaequilatera *182*, 554
Euphorbia indica *181*, 555
Euphorbia kischenensis *182*, **262**, 555
Euphorbia kuriensis *182*, **262**, 555
Euphorbia leptoclada *182*, 556
Euphorbia obcordata *180*, **262**, 556
Euphorbia oblanceolata see *Euphorbia schimperi*
Euphorbia schimperi *180*, **262**, 556
Euphorbia schweinfurthii 557
Euphorbia schweinfurthii form *A* (typical) .. *182*, **262**, 558
Euphorbia schweinfurthii form *B* *182*, 558
Euphorbia schweinfurthii form *C* *182*, 558
Euphorbia schweinfurthii form *D* *182*, **263**, 558
Euphorbia septemsulcata see *Euphorbia spiralis*
Euphorbia socotrana *181*, **263**, 558
Euphorbia spiralis *179*, **263**, 559
Euphorbiaceae *74*, *82*, *84*, *88*, *89*, *90*, *92*, *177*, 542
Eureiandra 536
Eureiandra balfourii *175*, **260**, 536
Euryops 500
Euryops arabicus *160*, **253**, 500
Euryops pinnifolius see *Euryops arabicus*
Euryops socotranus see *Euryops arabicus*
Evolvulus 519
Evolvulus alsinoides *168*, 519
Exacum *183*, *184*, 570
Exacum affine *184*, **264**, 570
Exacum caeruleum *184*, **264**, 570
Exacum forbesii see *Exacum caeruleum*
Exacum gracilipes see *Exacum affine*
Exacum socotranum *184*, 571

Fabrisinapis fruticosus see *Hemicrambe fruticosa*
Fagonia *232*, 715
Fagonia arabica see *Fagonia paulayana*
Fagonia cretica var. *arabica* see *Fagonia paulayana*
Fagonia cretica var. *luntii* see *Fagonia luntii*
Fagonia cretica var. *socotrana* see *Fagonia luntii*
Fagonia luntii *232*, 715
Fagonia paulayana *232*, 716
Farsetia *174*, 529
Farsetia inconspicua *174*, **259**, 529
Farsetia longisiliqua *174*, 530
Farsetia prostrata see *Farsetia socotrana*
Farsetia socotrana *174*, **258**, 530
Farsetia stylosa *174*, 530
Ficus *207*, 636

Ficus cordata *208*, **271**, 636
Ficus cordata subsp. *salicifolia* see *Ficus cordata*
Ficus salicifolia see *Ficus cordata*
Ficus socotrana see *Ficus vasta*
Ficus vasta *208*, **271**, 637
Fimbristylis *99*, 320
Fimbristylis annua see *Fimbristylis dichotoma*
Fimbristylis complanata *102*, 320
Fimbristylis cymosa *102*, 320
Fimbristylis dichotoma *102*, 320
Fimbristylis diphylla see *Fimbristylis dichotoma*
Fimbristylis ferruginea *102*, 321
Fimbristylis hispidula see *Bulbostylis hispidula*
Flacourtiaceae *91*, *183*, 568
Flueggea *178*, *183*, 560
Flueggea leucopyrus *183*, 560
Flueggea virosa *183*, 561
Forsskaolea 705
Forsskaolea viridis *229*, 705
Frankenia 569
Frankenia pulverulenta *78*, *183*, 569
Frankeniaceae *78*, *183*, 569
Fuirena 321
Fuirena ciliaris *98*, 321
Fuirena glomerata see *Fuirena ciliaris*

Gaillonia *216*, 664
Gaillonia puberula *217*, **274**, 664
Gaillonia putorioides *217*, **274**, 665
Gaillonia thymoides *217*, **275**, 665
Gaillonia tinctoria *217*, **275**, 665
Gaillonia tinctoria var. *glabra* see *Gaillonia tinctoria*
Galium *215*, 665
Galium aparine see *Galium spurium*
Galium setaceum *217*, 665
Galium spurium *217*, 666
Galium spurium subsp. *africanum* see *Galium spurium*
Gastridium 366
Gastridium phleoides *108*, 366
Gentianaceae *77*, *183*, 569
Geraniaceae *74*, *78*, *83*, *184*, 571
Geranium *184*, 572
Geranium biuncinatum *184*, **264**, 572
Geranium mascatense *184*, 572
Geranium trilophum *184*, 572
Gisekia 415
Gisekia pharnaceoides *133*, 415
Glinus 415
Glinus lotoides *134*, 415
Globulariaceae *80*, *94*, *185*, 573
Glossonema 438
Glossonema revoili *138*, 438
Gnidia 697
Gnidia socotrana *226*, **278**, 697
Gomphrena 422
Gomphrena celosioides *134*, 422
Goodeniaceae *80*, *185*, 574
Gossypium 623
Gossypium barbadense *203*, 623
Graderia 681
Graderia fruticosa *221*, **277**, 681
Gramineae *72*, *103*, 332
Grewia *93*, *226*, *227*, 699
Grewia bilocularis *228*, 699
Grewia erythraea *227*, 700
Grewia mansouriana *228*, **278**, 701
Grewia milleri *227*, 701
Grewia turbinata *227*, 701
Guttiferae *86*, *185*, 574
Gymnocarpos *154*, *155*, 482
Gymnocarpos bracteatus *155*, **252**, 482
Gymnocarpos kuriensis *155*, **252**, 482
Gymnogramma cordata see *Ceterach phillipsianum*
Gynandropsis pentaphylla see *Cleome gynandra*
Gypsophila 483
Gypsophila montana *154*, 483
Gypsophila montana var. *viscida* see *Gypsophila montana*
Gypsophila somalensis Franch. see *Gypsophila montana*

Habenaria 392
Habenaria socotrana *126*, **237**, 392
Hackelochloa 366
Hackelochloa granularis *104*, *113*, 366
Haemanthus grandifolius see *Ledebouria grandifolia*
Halodule 312
Halodule uninervis *69*, 312
Halophila ovalis *69*
Halophila stipulacea *69*
Halopyrum 367
Halopyrum mucronatum *111*, 367
Haya 483
Haya obovata *154*, **252**, 483
Hedyotis bicornuta see *Oldenlandia bicornuta*
Hedyotis pulvinata see *Oldenlandia pulvinata*
Hedyotis stellarioides see *Oldenlandia balfourii*
Heleocharis see *Eleocharis*
Helichrysum *161*, *163*, 501
Helichrysum aciculare *163*, **253**, 501
Helichrysum arachnoides *164*, **253**, 501
Helichrysum balfourii *164*, **254**, 502
Helichrysum glumaceum see *Achyrocline luzuloides*
Helichrysum gracilipes *164*, **254**, 502
Helichrysum gracilipes forma *genuinum* see *Helichrysum gracilipes*
Helichrysum gracilipes var. *gracilipes* see *Helichrysum gracilipes*
Helichrysum gracilipes var. *lanatum* see *Helichrysum paulayanum*
Helichrysum gracilipes var. *stoloniferum* see *Helichrysum balfourii*
Helichrysum luzuloides see *Achyrocline luzuloides*
Helichrysum nimmoanum *165*, **254**, 502
Helichrysum paulayanum *164*, **254**, 503
Helichrysum rosulatum *163*, **254**, 503
Helichrysum schimperi see *Achyrocline schimperi*
Helichrysum sp. *A* *164*, **254**, 503
Helichrysum sp. *B* *165*, **254**, 504
Helichrysum sp. *C* *165*, 504
Helichrysum sp. *D* *164*, **254**, 504
Helichrysum sphaerocephalum *164*, **255**, 505
Helichrysum sphaerocephalum var. *sarmentosum* see *Helichrysum sphaerocephalum*
Helichrysum suffruticosum *165*, 505
Helinus 660
Helinus integrifolius *215*, 660
Heliotropium *144*, *145*, 449
Heliotropium aegyptiacum *146*, **246**, 449
Heliotropium aff. *socotranum* 451
Heliotropium bacciferum complex 449
Heliotropium bacciferum see *Heliotropium riebeckii*
Heliotropium balfourii *146*, **246**, 452
Heliotropium cimalense see *Heliotropium aegyptiacum*
Heliotropium dentatum *145*, **247**, 453
Heliotropium derafontense 451
Heliotropium derafontense see also *Heliotropium shoabense*
Heliotropium kuriense *147*, 450
Heliotropium nigricans *145*, **247**, 453
Heliotropium odorum see *Heliotropium balfourii*
Heliotropium ophioglossum *145*, 453
Heliotropium ovalifolium *146*, 454

Heliotropium paulayanum ... *147*, 450
Heliotropium pterocarpum see *Heliotropium riebeckii*
Heliotropium rariflorum ... *145*, 454
Heliotropium riebeckii ... *146*, *147*, 451
Heliotropium shoabense ... *146*, *147*, 452
Heliotropium socotranum ... *146*, *147*, **247**, 452
Heliotropium strigosum ... *146*, 454
Heliotropium strigosum var. *bicolor* see *Heliotropium strigosum*
Heliotropium strigosum var. *marifolium* see *Heliotropium strigosum*
Heliotropium strigosum var. *scabrum* see *Heliotropium strigosum*
Heliotropium subulatum see *Heliotropium zeylanicum*
Heliotropium undulatum see *Heliotropium riebeckii*
Heliotropium wagneri ... *146*, *147*, 450
Heliotropium zeylanicum ... *145*, 455
Hemicrambe ... 531
Hemicrambe fruticosa ... *173*, 531
Hemicrambe townsendii see *Hemicrambe fruticosa*
Herpestis monnieria see *Bacopa monnieri*
Heterachaena massauensis see *Launaea massauensis*
Heterochloa dura see *Andropogon*
Heterochloa see *Andropogon*
Heteropogon ... 367
Heteropogon contortus ... *109*, 367
Heteropogon hirtus see *Heteropogon contortus*
Hibiscus ... *93*, *203*, *204*, 624
Hibiscus dioscorides ... *205*, 624
Hibiscus diriffan ... *206*, **270**, 625
Hibiscus intermedius see *Hibiscus palmatus*
Hibiscus lobatus ... *204*, 626
Hibiscus macropodus ... *206*, 626
Hibiscus malacophyllus ... *206*, 626
Hibiscus micranthus ... *205*, 626
Hibiscus noli-tangere ... *205*, **270**, 627
Hibiscus palmatus ... *204*, 627
Hibiscus quattenensis ... *206*, **270**, 628
Hibiscus scindicus ... *205*, 628
Hibiscus scottii ... *206*, **270**, 628
Hibiscus sidaeformis ... *204*, 629
Hibiscus socotranus ... *205*, **270**, 629
Hibiscus solandra see *Hibiscus lobatus*
Hibiscus stenanthus ... *205*, **270**, 629
Hibiscus ternatus see *Hibiscus sidaeformis*
Hibiscus vitifolius ... *204*, 630
Hirschfeldia rostrata see *Erucastrum rostratum*
Hirschfeldia rostrata var. *rostrata* see *Erucastrum rostratum*
Holothrix ... 392
Holothrix socotrana ... *126*, **237**, 392
Hyacinthaceae ... *72*, *123*, 384
Hybanthus ... 711
Hybanthus enneaspermus ... *231*, 711
Hydrocharitaceae ... *68*, *69*, 387
Hydrocotyle asiatica see *Centella asiatica*
Hyparrhenia ... 368
Hyparrhenia hirta ... *105*, *109*, 368
Hypericum ... 574
Hypericum balfourii ... *185*, **265**, 575
Hypericum fieriense ... *185*, 575
Hypericum lanceolatum see *Hypericum socotranum* subsp. *socotranum*
Hypericum mysorense see *Hypericum balfourii*
Hypericum scopulorum ... *185*, **265**, 575
Hypericum socotranum ... 576
Hypericum socotranum subsp. *smithii* ... *186*, **265**, 576
Hypericum socotranum subsp. *socotranum* ... *186*, 576
Hypericum tortuosum ... *185*, **265**, 576
Hypodematium ... 291
Hypodematium crenatum ... *67*, 291
Hypoestes ... 407
Hypoestes forskalei see *Hypoestes pubescens*
Hypoestes pubescens ... *127*, **239**, 407
Hypoxidaceae ... *71*, *125*, 387
Hypoxis ... 387
Hypoxis angustifolia ... *125*, **237**, 387

Imperata ... 368
Imperata arundinacea see *Imperata cylindrica*
Imperata cylindrica ... *107*, *114*, 368
Indigofera ... *193*, *194*, *195*, *197*, 602
Indigofera argentea ... *199*, 602
Indigofera articulata ... *199*, 602
Indigofera burmanni see *Indigofera argentea*
Indigofera coerulea ... *199*, 603
Indigofera coerulea var. *coerulea* ... 603
Indigofera coerulea var. *occidentalis* ... 603
Indigofera colutea ... *199*, 603
Indigofera cordifolia ... *197*, 604
Indigofera disjuncta see *Microcharis disjuncta*
Indigofera disjuncta var. *fallax* see *Microcharis disjuncta*
Indigofera gerardiana see *Indigofera socotrana*
Indigofera hochstetteri ... *198*, 604
Indigofera marmorata ... *199*, **268**, 604
Indigofera nephrocarpa ... *198*, **268**, 605
Indigofera nephrocarpoides ... *198*, 605
Indigofera nugalensis ... *199*, 605
Indigofera oblongifolia ... *197*, *199*, 605
Indigofera paucifolia see *Indigofera oblongifolia*
Indigofera pseudointricata ... *197*, **268**, 606
Indigofera socotrana ... *198*, **268**, 606
Indigofera spiniflora ... *198*, 607
Indigofera tinctoria ... *199*, 607
Indigofera tinctoria var. *tinctoria* see *Indigofera tinctoria*
Indigofera viscosa see *Indigofera colutea*
Ionidium suffruticosum see *Hybanthus enneaspermus*
Iphigenia ... 309
Iphigenia socotrana ... *97*, **235**, 309
Ipomoea ... *169*, *170*, 520
Ipomoea biloba see *Ipomoea pes-caprae*
Ipomoea cairica ... *170*, 520
Ipomoea kotschyana ... *170*, 520
Ipomoea laciniata see *Ipomoea kotschyana*
Ipomoea nil ... *170*, 521
Ipomoea obscura ... *170*, 521
Ipomoea pes-caprae ... *170*, 521
Ipomoea scabra see *Ipomoea nil*
Ipomoea sinensis ... *170*, 521
Iridaceae ... *71*, *125*, 388
Ischaemum ... 369
Ischaemum sp. *A* ... *105*, *109*, 369
Ischnurus pulchellus see *Lepturus pulchellus*
Ischnurus see *Lepturus*

Jasminum ... 642
Jasminum fluminense ... *85*, *209*, 642
Jasminum fluminense subsp. *socotranum* see *Jasminum fluminense*
Jasminum rotundifolium see *Jasminum fluminense*
Jatropha ... 561
Jatropha unicostata ... *178*, **263**, 561
Jaubertia putoroides see *Gaillonia putorioides*
Juncaceae ... *72*, *125*, 389
Juncus ... 389
Juncus arabicus see *Juncus rigidus*
Juncus bufonius ... *125*, 389
Juncus maritimus var. *arabicus* see *Juncus rigidus*
Juncus maritimus var. *socotranus* see *Juncus socotranus*
Juncus rigidus ... *125*, 389
Juncus socotranus ... *125*, **237**, 390

Justicia ... *128*, *131*, 408
Justicia alexandri ... *131*, **239**, 408
Justicia heterocarpa ... *131*, 408
Justicia rigida ... *131*, **239**, 408
Justicia takhinensis ... *131*, **239**, 409

Kalanchoe ... *94*, *172*, 526
Kalanchoe abrupta see *Kalanchoe robusta*
Kalanchoe farinacea ... *172*, **258**, 526
Kalanchoe robusta ... *172*, **258**, 526
Kalanchoe rotundifolia ... *172*, **258**, 527
Kickxia hastata see *Nanorrhinum hastatum*
Kickxia kuriensis see *Nanorrhinum kuriensis*
Kleinia ... 505
Kleinia scottii ... *160*, **255**, 505
Kohautia ... *216*, *217*, 666
Kohautia socotrana ... *217*, **275**, 666
Kohautia subverticellata ... *217*, 666
Korthalsella ... 712
Korthalsella taenioides ... *85*, *231*, **280**, 712
Kraussia ... 667
Kraussia socotrana ... *216*, **275**, 667
Kyllinga ... *99*, *102*, 321
Kyllinga brevifolia ... *102*, 321
Kyllinga microstyla ... *102*, 322

Labiatae ... *75*, *87*, *186*, 577
Lachnocapsa ... 531
Lachnocapsa spathulata ... *173*, **259**, 531
Lactuca crassifolia see *Launaea crassifolia*
Lactuca kossmatii see *Launaea crassifolia*
Lactuca paulayana see *Launaea crassifolia*
Lactuca rhynchocarpa see *Launaea rhynchocarpa*
Lactuca salehesis see *Launaea crassifolia*
Lagarosiphon ... 387
Lagarosiphon roxburghii ... *68*, 387
Lagenantha ... 492
Lagenantha cycloptera ... *157*, **252**, 492
Lannea ... 424
Lannea aspleniifolia see *Boswellia socotrana*
Lannea ornifolia see *Commiphora ornifolia*
Lannea transulta ... *136*, **241**, 424
Lasiocorys flagellifera see *Leucas flagellifolia*
Lasiocorys spiculifolia see *Leucas spiculifera*
Lasiosiphon socotranus see *Gnidia socotrana*
Launaea ... *161*, *165*, 506
Launaea crassifolia ... *166*, 506
Launaea crepoides ... *165*, **255**, 506
Launaea hafunensis ... *165*, 507
Launaea kuriensis see *Launaea massauensis*
Launaea massauensis ... *166*, 507
Launaea rhynchocarpa ... *166*, 507
Launaea socotrana ... *166*, **255**, 507
Launaea sp. *A* ... *165*, 508
Lavandula ... 577
Lavandula nimmoi ... *187*, **265**, 577
Lavandula subnuda see *Lavandula nimmoi*
Ledebouria ... *123*, *124*, 386
Ledebouria aff. *revoluta* ... *124*, **236**, 386
Ledebouria grandifolia ... *124*, **236**, 386
Ledebouria insularis ... *124*, **236**, 386
Leguminosae ... *189*, 586
Leguminosae subfam. Caesalpinioideae ... *88*, *189*, *190*, 586
Leguminosae subfam. Mimosoideae .. *88*, *189*, *191*, 591
Leguminosae subfam. Papilionoideae . *79*, *88*, *189*, *192*, 595
Lentibulariaceae ... *80*, *202*, 619
Leptothrium ... 369
Leptothrium senegalense ... *111*, 369
Lepturus ... *111*, *119*, 369
Lepturus calcareus ... *119*, 369
Lepturus pulchellus ... *119*, 370
Lepturus tenuis ... *119*, 370
Leucas ... *187*, 578
Leucas flagellifolia ... *188*, **266**, 578
Leucas hagghierensis ... *188*, **266**, 579
Leucas kischenensis ... *188*, **266**, 579
Leucas lanata see *Leucas kischenensis*
Leucas martinicensis ... *187*, 579
Leucas neuflizeana ... *187*, 580
Leucas penduliflora ... *188*, 580
Leucas samhaensis ... *188*, 580
Leucas spicata see *Leucas spiculifera*
Leucas spiculifera ... *188*, **266**, 580
Leucas thymoides see *Leucas kischenensis*
Leucas thymoides subsp. *kischenensis* see *Leucas kischenensis*
Leucas urticaefolia ... *187*, 581
Leucas virgata ... *188*, **266**, 581
Limonium ... *211*, 649
Limonium axillare see *Limonium paulayanum*
Limonium cylindrifolium see *Limonium sokotranum*
Limonium kossmatii see *Limonium paulayanum*
Limonium paulayanum ... *212*, **272**, 650
Limonium sokotranum ... *212*, **272**, 650
Linaceae ... *84*, *202*, 619
Linaria hastata see *Nanorrhinum hastatum*
Lindenbergia ... 682
Lindenbergia kuriensis see *Lindenbergia sokotrana*
Lindenbergia paulayana see *Lindenbergia sokotrana*
Lindenbergia sinaica see *Lindenbergia sokotrana*
Lindenbergia sokotrana ... *221*, 682
Linum ... 619
Linum gallicum see *Linum trigynum*
Linum trigynum ... *202*, 619
Lippia nodiflora see *Phyla nodiflora*
Lochia bracteata see *Gymnocarpos bracteatus*
Lochia bracteata subsp. *abdulkuriana* see *Gymnocarpos kuriensis*
Lochia kuriensis see *Gymnocarpos kuriensis*
Lolium ... *109*, *120*, 370
Lolium persicum ... *120*, 370
Lolium temulentum ... *120*, 371
Lotus ... *195*, *199*, 607
Lotus arabicus var. *trigonelloides* see *Lotus schimperi*
Lotus garcinii ... *200*, **268**, 608
Lotus glinoides see *Lotus schimperi*
Lotus mollis ... *200*, **268**, 608
Lotus ononopsis ... *200*, **268**, 608
Lotus schimperi ... *200*, 609
Lotus stocksii see *Lotus garcinii*
Loxogramme ... 293
Loxogramme abyssinica ... *64*, 293
Ludwigia ... 643
Ludwigia palustris ... *209*, 643
Lycium ... *95*, 687
Lycium europaeum see *Lycium sokotranum*
Lycium socotrana see *Lycium sokotranum*
Lycium sokotranum ... *223*, **278**, 687
Lythraceae ... *77*, *202*, 620
Lythrum ... 620
Lythrum hyssopifolium ... *202*, 620

Maerua ... *93*, 479
Maerua angolensis ... *151*, **251**, 479
Maerua socotrana see *Maerua angolensis*
Malpighiaceae ... *93*, *202*, 621
Malva ... 630
Malva parviflora ... *203*, 630
Malvaceae ... *83*, *93*, *203*, 622
Mariscus see *Cyperus*

Mariscus squarrosus see Cyperus squarrosus
Mariscus stramineo-ferrugineus see Cyperus stramineo-ferrugineus
Marsdenia 438
Marsdenia robusta 139, 438
Marsilea 291
Marsilea coromandeliana 64, 291
Marsileaceae 291
Maytenus 488
Maytenus sp. A 93, 156, 488
Medicago 193, 200, 609
Medicago denticulata see Medicago polymorpha
Medicago hispida see Medicago polymorpha
Medicago hispida var. denticulata see Medicago polymorpha
Medicago laciniata 200, 609
Medicago minima 200, 610
Medicago polymorpha 200, 610
Meineckia 565
Meineckia filipes 178, 565
Melanocenchris 371
Melanocenchris jacquemontii 110, 371
Melhania 225, 226, 693
Melhania milleri 226, 693
Melhania muricata 226, 693
Meliaceae 94, 207, 633
Melilotus 610
Melilotus indica 193, 610
Melilotus parviflora see Melilotus indica
Melinis 372
Melinis repens 104, 372
Melinis repens subsp. grandiflora see Melinis repens
Melothria punctata see Zehneria scabra
Menispermaceae 83, 92, 207, 633
Metaporana 522
Metaporana obtusa 168, **257**, 522
Microcharis 611
Microcharis disjuncta 193, 611
Microcharis disjuncta var. fallax see Microcharis disjuncta
Microchloa 372
Microchloa kunthii 111, 372
Micromeria 582
Micromeria microphylla see Micromeria remota
Micromeria microphylla var. imbricata see Micromeria remota
Micromeria microphylla var. remota see Micromeria remota
Micromeria remota 187, **266**, 582
Misopates 682
Misopates orontium 221, 682
Mitolepis arbuscula see Cryptolepis arbuscula
Mitolepis intricata see Cryptolepis intricata
Mollugo 133, 134, 416
Mollugo cerviana 134, 416
Mollugo hirta see Glinus lotoides
Mollugo nudicaulis 134, 416
Momordica 175, 176, 537
Momordica balsamina 176, 537
Momordica charantia 176, 537
Momordica charantia var. abbreviata see Momordica charantia
Moraceae 74, 89, 207, 634
Mussaenda capsulifera see Tamridaea capsulifera
Myrsinaceae 94, 208, 639
Myrsine 639
Myrsine africana 94, 208, 639

Najadaceae 68, 390
Najas 390
Najas graminea 68, 390
Najas marina 68, 391
Nanorrhinum 221, 222, 683
Nanorrhinum hastatum 222, 683
Nanorrhinum kuriensis 222, **277**, 683
Nechamandra alterniflora see Lagarosiphon roxburghii
Negripteris 287
Negripteris sciona 67, **233**, 287
Neogaillonia puberula see Gaillonia puberula
Neogaillonia putoroides see Gaillonia putorioides
Neogaillonia thymoides see Gaillonia thymoides
Neogaillonia tinctoria see Gaillonia tinctoria
Nephrodium crenatum see Hypodematium crenatum
Nephrodium molle see Christella dentata
Nephrolepis 291
Nephrolepis cordifolia see Nephrolepis undulata
Nephrolepis undulata 65, **233**, 291
Nesocrambe 532
Nesocrambe socotrana 173, 532
Neuracanthus 127, 131, 409
Neuracanthus aculeatus 131, **240**, 409
Neuracanthus argyrophyllus 131, 409
Neuracanthus capitatus 131, 410
Nicandra 688
Nicandra physaloides 224, 688
Nicotiana 688
Nicotiana tabacum 224, 688
Nirarathamnos 702
Nirarathamnos asarifolius 228, **279**, 703
Nyctaginaceae 76, 208, 639

Ochradenus 659
Ochradenus baccatus see Ochradenus socotranus
Ochradenus socotranus 214, **274**, 659
Ocimum 583
Ocimum canum see Ocimum forskolei
Ocimum forskolei 186, 583
Ocimum hadiense see Ocimum forskolei
Ocimum menthaefolium see Ocimum forskolei
Odina asplenifolia see Lannea transulta
Odina ornifolia Commiphora ornifolia
Oldenlandia 216, 218, 667
Oldenlandia aretioides 218, 667
Oldenlandia balfourii 218, **275**, 668
Oldenlandia bicornuta 218, **275**, 668
Oldenlandia caespitosa see Oldenlandia corymbosa var. caespitosa
Oldenlandia caespitosa var. subpedunculata see Oldenlandia corymbosa var. caespitosa
Oldenlandia corymbosa var. caespitosa 218, 668
Oldenlandia herbacea 218, 668
Oldenlandia heynei see Oldenlandia herbacea
Oldenlandia ocellata 218, **276**, 669
Oldenlandia pulvinata 218, **276**, 669
Oldenlandia pulvinata forma congesta see Oldenlandia pulvinata
Oldenlandia schimperi see Kohautia socotrana
Oleaceae 81, 85, 209, 642
Oleandraceae 291
Onagraceae 76, 209, 643
Onychium 288
Onychium divaricatum 66, 288
Onychium melanolepis see Onychium divaricatum
Ophioglossaceae 292
Ophioglossum 292
Ophioglossum polyphyllum 64, 292
Ophioglossum reticulatum 64, 292
Oplismenus 106, 120, 372
Oplismenus burmanni see Oplismenus burmannii
Oplismenus burmannii 120, 373
Oplismenus compositus 120, 373
Orbivestus 508

Orbivestus cinerascens ... *167*, 508
Orchidaceae ... *71*, *126*, 391
Oreofraga ... 703
Oreofraga morrisiana ... *229*, **279**, 703
Ormocarpum caeruleum see *Zygocarpum coeruleum*
Orobanchaceae ... *73*, *210*, 643
Orobanche ... *210*, 644
Orobanche abyssinica see *Orobanche minor*
Orobanche cernua ... *210*, 644
Orobanche minor ... *210*, 644
Orobanche ramosa ... *210*, 644
Orthosiphon ... *186*, *189*, 584
Orthosiphon ferrugineus ... *189*, **266**, 584
Orthosiphon pallidus ... *189*, 584
Orthosiphon tenuiflorus see *Endostemon tenuiflorus*
Orygia decumbens see *Corbichonia decumbens*
Osteospermum vaillantii see *Tripteris vaillantii*
Osyris ... 675
Osyris arborea see *Osyris quadripartita*
Osyris lanceolata see *Osyris quadripartita*
Osyris pendula see *Osyris quadripartita*
Osyris quadripartita ... *91*, *219*, 675
Oxalidaceae ... *83*, *210*, 644
Oxalis ... 644
Oxalis corniculata ... *210*, 644
Oxygonum ... 653
Oxygonum sinuatum ... *212*, 653

Palmae ... *70*, *126*, 393
Panicum ... *104*, *120*, 373
Panicum atrosanguineum ... *120*, 373
Panicum colonum see *Echinochloa colona*
Panicum eruciforme see *Brachiaria eruciformis*
Panicum maximum ... *120*, 374
Panicum nudiglume see *Brachiaria deflexa*, *Brachiaria ramosa*
Panicum nudiglume var. *major* see *Brachiaria deflexa*, *Brachiaria ramosa*
Panicum paspaloides see *Paspalidium geminatum*
Panicum petiveri see *Brachiaria ramosa*
Panicum rigidum ... *121*, 374
Panicum socotranum ... *121*, 374
Panicum teneriffae see *Tricholaena teneriffae*
Panicum turgidum ... *120*, 375
Papaveraceae ... *83*, *210*, 645
Pappophorum aucheri see *Enneapogon desvauxii*
Pappophorum elegans see *Enneapogon desvauxii*
Paracalyx ... 611
Paracalyx balfourii ... *194*, **269**, 611
Paracalyx schweinfurthii see *Paracalyx balfourii*
Parietaria ... 706
Parietaria debilis ... *229*, 706
Parkeriaceae ... 293
Paspalidium ... 375
Paspalidium geminatum ... *113*, 375
Paspalum ... *107*, *113*, *121*, 375
Paspalum distichum see *Paspalum vaginatum*
Paspalum scrobiculatum ... *121*, 376
Paspalum vaginatum ... *121*, 376
Pavonia ... 631
Pavonia arabica ... *203*, 631
Pedaliaceae ... *75*, *210*, 646
Pedalium ... 646
Pedalium murex ... *210*, 646
Pelargonium ... 573
Pelargonium insularis ... *184*, **264**, 573
Pellaea concolor see *Doryopteris concolor*
Pellaea ... 288
Pellaea involuta ... *68*, 288
Pellaea viridis ... *68*, 288
Pellaea viridis see *Pellaea involuta*
Pennisetum ... *110*, *121*, 376
Pennisetum ciliare see *Cenchrus ciliaris*
Pennisetum orientale ... *121*, 376
Pennisetum ruppellii see *Pennisetum setaceum*
Pennisetum setaceum ... *121*, 377
Peperomia ... 646
Peperomia arabica see *Peperomia blanda*
Peperomia blanda ... *211*, **272**, 646
Peperomia blanda var. *leptostachya* see *Peperomia blanda*
Peperomia goudotii see *Peperomia blanda*
Peperomia reflexa see *Peperomia tetraphylla*
Peperomia reflexa var. *parvifolia* see *Peperomia tetraphylla*
Peperomia tetraphylla ... *211*, 647
Pergularia ... 439
Pergularia tomentosa ... *138*, 439
Periploca ... 440
Periploca visciformis ... *137*, **245**, 440
Peristrophe ... 410
Peristrophe bicalyculata see *Peristrophe paniculata*
Peristrophe paniculata ... *129*, 410
Persicaria ... *212*, 653
Persicaria barbata ... *213*, 653
Persicaria glabra ... *213*, 653
Peucedanum cordatum see *Rughidia cordata*
Phoenix ... 393
Phoenix dactylifera ... *126*, **237**, 393
Phyla ... 710
Phyla nodiflora ... *230*, 710
Phyllanthus ... *178*, *183*, 565
Phyllanthus filipes see *Meineckia filipes*
Phyllanthus maderaspatensis ... *183*, 565
Phyllanthus rotundifolius ... *183*, **263**, 565
Physalis ... 689
Physalis minima ... *224*, 689
Pilea ... 706
Pilea tetraphylla ... *229*, 706
Piperaceae ... *73*, *211*, 646
Pittosporaceae ... *93*, *211*, 647
Pittosporum ... 647
Pittosporum viridiflorum ... *93*, *211*, 647
Placopoda ... 669
Placopoda virgata ... *216*, **276**, 669
Placopoda virgata var. *nana* see *Placopoda virgata*
Plantaginaceae ... *81*, *211*, 648
Plantago ... 648
Plantago amplexicaulis ... *211*, 648
Plantago ovata ... *211*, 648
Plectranthus ... 584
Plectranthus socotranus ... *186*, 584
Pleopeltis ... 293
Pleopeltis macrocarpa ... *64*, 293
Pluchea ... *161*, *166*, 509
Pluchea aromatica see *Pulicaria aromatica*
Pluchea glutinosa ... *166*, 509
Pluchea obovata ... *166*, **255**, 509
Plumbaginaceae ... *81*, *94*, *211*, 648
Polycarpaea ... *154*, *155*, 484
Polycarpaea balfourii ... *155*, 484
Polycarpaea caespitosa ... *155*, **252**, 484
Polycarpaea corymbosa ... *155*, 484
Polycarpaea divaricata see *Polycarpaea balfourii*
Polycarpaea hassalensis ... *156*, 485
Polycarpaea hayoides ... *156*, 485
Polycarpaea kuriensis ... *156*, **252**, 485
Polycarpaea paulayana ... *156*, 485
Polycarpaea spicata ... *155*, 486
Polygala ... 651
Polygala abyssinica ... *212*, 651
Polygala erioptera ... *212*, 651

Polygala kuriense ... *212*, **273**, 651
Polygala paulayana see *Polygala erioptera*
Polygalaceae ... *79*, *212*, 651
Polygonaceae ... *84*, *212*, 652
Polygonum barbatum see *Persicaria barbata*
Polygonum glabrum see *Persicaria glabra*
Polypodiaceae ... 293
Polypogon ... 377
Polypogon monspeliensis ... *114*, 377
Porana obtusa see *Metaporana obtusa*
Portulaca ... *213*, 654
Portulaca kuriensis ... *214*, **273**, 654
Portulaca oleracea ... *213*, 655
Portulaca quadrifida ... *214*, **273**, 655
Portulaca samhaensis ... *214*, **273**, 655
Portulaca sedifolia ... *213*, **273**, 656
Portulacaceae ... *78*, *82*, *213*, 654
Poskea ... *80*, 573
Poskea socotrana ... *94*, *185*, **264**, 573
Potamogeton ... 398
Potamogeton filiformis ... *68*, 398
Potamogeton fluitans see *Potamogeton nodosus*
Potamogeton natans see *Potamogeton nodosus*
Potamogeton natans subsp. *plantagineus* see *Potamogeton nodosus*
Potamogeton nodosus ... *68*, 398
Potamogeton pectinatus ... *68*, 398
Potamogeton semicoloratus see *Potamogeton nodosus*
Potamogetonaceae ... *68*, 398
Pouzolzia ... 707
Pouzolzia auriculata ... *229*, 707
Prenanthes ... 510
Prenanthes amabilis ... *160*, **256**, 510
Primulaceae ... *77*, *214*, 656
Priotropis socotrana see *Crotalaria socotrana*
Priva ... 711
Priva adhaerans see *Priva socotrana*
Priva cordifolia see *Priva socotrana*
Priva leptostachya see *Priva socotrana*
Priva socotrana ... *230*, **280**, 711
Priva tenax see *Priva socotrana*
Prosopis ... 595
Prosopis juliflora ... *191*, 595
Pseudomussaenda capsulifera see *Tamridaea capsulifera*
Psiadia ... 510
Psiadia schweinfurthii ... *162*, 510
Psilostachys sericea see *Psilotrichum sericeum*
Psilotrichum ... *134*, *136*, 422
Psilotrichum aphyllum ... *136*, **241**, 422
Psilotrichum sericeum ... *136*, 422
Psoralea corylifolia see *Cullen corylifolia*
Pteridaceae ... 294
Pteridium ... 290
Pteridium aquilinum ... *67*, **233**, 290
Pteris ... 294
Pteris biaurita see *Pteris quadriaurita*
Pteris cretica ... *65*, 294
Pteris longifolia see *Pteris vittata*
Pteris quadriaurita ... *67*, 294
Pteris vittata ... *65*, **233**, 294
Pulicaria ... *162*, *166*, 510
Pulicaria aromatica ... *167*, **256**, 510
Pulicaria dioscorides ... *167*, **256**, 511
Pulicaria diversifolia ... *166*, **256**, 511
Pulicaria elegans ... *167*, **256**, 512
Pulicaria lanata ... *167*, **256**, 512
Pulicaria shoabensis see *Pulicaria diversifolia*
Pulicaria stephanocarpa ... *167*, **256**, 513
Pulicaria vieraeoides ... *167*, **256**, 513
Punica ... 657
Punica protopunica ... *86*, *214*, **273**, 657
Punicaceae ... *86*, *214*, 657
Pupalia ... 423
Pupalia lappacea ... *134*, 423
Pycreus ... 322
Pycreus pumilus ... *99*, 322
Pyrostria ... 670
Pyrostria socotrana ... *216*, 670

Reichardia ... 513
Reichardia tingitana ... *161*, 513
Remusatia ... 306
Remusatia vivipara ... *96*, 306
Reseda ... *93*, 659
Reseda viridis ... *214*, **274**, 659
Resedaceae ... *79*, *93*, *214*, 658
Rhamnaceae ... *90*, *215*, 660
Rhinacanthus ... 410
Rhinacanthus scoparius ... *129*, **240**, 410
Rhus ... *136*, 425
Rhus sp. *A* ... *136*, 425
Rhus thyrsiflora ... *136*, **242**, 425
Rhynchelytrum microstachyum see *Melinis repens*
Rhynchelytrum microstachyum var. *albicomum* see *Melinis repens*
Rhynchelytrum repens see *Melinis repens*
Rhynchelytrum see *Melinis*
Rhynchelytrum villosum see *Melinis repens*
Rhynchosia ... *194*, *200*, 612
Rhynchosia memnonia see *Rhynchosia pulverulenta*
Rhynchosia minima ... *200*, 612
Rhynchosia pulverulenta ... *200*, 612
Ricinus ... 566
Ricinus communis ... *177*, 566
Romulea ... 389
Romulea fischeri ... *125*, **237**, 389
Romulea purpurascens var. *edulis* see *Romulea fischeri*
Rubiaceae ... *76*, *87*, *215*, 663
Ruellia ... *127*, *131*, 411
Ruellia carnea ... *132*, **240**, 411
Ruellia dioscoridis ... *132*, **240**, 411
Ruellia insignis ... *132*, **240**, 411
Ruellia kuriensis ... *132*, **240**, 412
Ruellia patula ... *132*, 412
Ruellia patula var. *minor* see *Ruellia dioscoridis*
Ruellia patula var. *pubescence* see *Ruellia patula*
Ruellia patula var. *pubescens* see *Ruellia patula*
Ruellia paulayana ... *132*, **240**, 412
Ruellia sokotrana see *Ruellia patula*
Rughidia ... 703
Rughidia cordata ... *229*, **279**, 703
Rughidia milleri ... *229*, **279**, 704
Ruppia ... 399
Ruppia maritima ... *68*, 399
Ruta ... 672
Ruta chalapensis ... *219*, 672
Ruta graveolens see *Ruta chalapensis*
Ruta graveolens var. *angustifolia* see *Ruta chalapensis*
Rutaceae ... *84*, *92*, *93*, *219*, 671

Salsola ... 492
Salsola cycloptera see *Lagenantha cycloptera*
Salsola forskalii see *Salsola spinescens*
Salsola semhahensis see *Lagenantha cycloptera*
Salsola spinescens ... *158*, 492
Salvadora ... 674
Salvadora persica ... *87*, *219*, 674
Salvadoraceae ... *87*, *219*, 674
Santalaceae ... *91*, *219*, 675
Sapindaceae ... *89*, *91*, *220*, 676
Sapotaceae ... *94*, *220*, 678

Sarcostemma *137*, *142*, 440
Sarcostemma daltoni see *Sarcostemma viminale*
Sarcostemma socotranum *142*, 440
Sarcostemma viminale *142*, 441
Satureja remota see *Micromeria remota*
Scaevola 574
Scaevola socotraensis *185*, **264**, 574
Schmidtia 377
Schmidtia pappophoroides *106*, *108*, 377
Schweinfurthia 683
Schweinfurthia pedicellata *221*, 683
Scrophularia 684
Scrophularia arguta *222*, **277**, 684
Scrophulariaceae *73*, *75*, *80*, *81*, *87*, *220*, 679
Sebaea 571
Sebaea microphylla *183*, 571
Secamone *139*, *142*, 441
Secamone cuneifolia *142*, **243**, 441
Secamone socotrana *142*, **245**, 441
Secamone socotrana var. *glauca* see *Secamone socotrana*
Secamone socotrana var. *socotrana* see *Secamone socotrana*
Securinega leucopyrus see *Flueggea leucopyrus*
Securinega virosa see *Flueggea virosa*
Seddera *168*, *169*, *171*, 522
Seddera fastigiata *171*, 523
Seddera glomerata *171*, 523
Seddera latifolia *171*, 523
Seddera pedunculata *171*, 524
Seddera semhahensis *171*, **257**, 524
Seddera spinosa *171*, **257**, 524
Seidlitzia 493
Seidlitzia rosmarinus *157*, 493
Senecio scottii see *Kleinia scottii*
Senna *190*, 587
Senna holosericea *190*, **269**, 587
Senna italica *190*, 588
Senna obtusifolia *190*, 588
Senna socotrana *190*, **269**, 588
Senna sophera see *Senna socotrana*
Senra 631
Senra incana *203*, 631
Setaria *106*, *110*, *121*, 378
Setaria glauca see *Setaria pumila*
Setaria intermedia *121*, 378
Setaria pallidifusca see *Setaria pumila*
Setaria pumila *121*, 378
Setaria verticillata *121*, 378
Sevada 493
Sevada schimperi *157*, 493
Sida *203*, *206*, 632
Sida acuta *207*, 632
Sida cordata *206*, 632
Sida cordifolia *207*, 632
Sida grewioides see *Sida ovata*
Sida humilis see *Sida cordata*
Sida ovata *207*, 632
Sida rhombifolia see *Sida ovata*
Sida subrotunda see *Sida ovata*
Sida urens see *Sida ovata*
Sideroxylon 678
Sideroxylon discolor see *Spiniluma discolor*
Sideroxylon fimbriatum *220*, 678
Sigesbeckia 514
Sigesbeckia orientalis *159*, 514
Silene 486
Silene apetala *153*, 486
Sisymbrium 532
Sisymbrium erysimoides *174*, 532
Socotora aphylla see *Periploca visciformis*
Socotora visciformis see *Periploca visciformis*
Socotranthus socotranus see *Cryptolepis socotrana*
Socotrella 442
Socotrella dolichocnema *139*, **245**, 442
Solanaceae *81*, *95*, *223*, 686
Solanum *224*, 690
Solanum anguivi see *Solanum incanum*
Solanum cordatum *225*, 690
Solanum gracilipes see *Solanum cordatum*
Solanum incanum *225*, **278**, 690
Solanum indicum see *Solanum incanum*
Solanum luteum see *Solanum villosum*
Solanum nigrum see *Solanum villosum*
Solanum villosum *224*, 691
Solanum villosum subsp. *miniatum* see *Solanum villosum*
Sonchus 514
Sonchus oleraceus *161*, 514
Spergula 487
Spergula fallax *154*, 487
Spermacoce hispida see *Diodia aulacosperma*
Sphaerocoma 487
Sphaerocoma hookeri *154*, **252**, 487
Spiniluma 678
Spiniluma discolor *220*, 678
Sporobolus *104*, *122*, 379
Sporobolus diander *122*, 379
Sporobolus ioclados *122*, 379
Sporobolus iocladus see *Sporobolus ioclados*
Sporobolus minutus *122*, 380
Sporobolus ruspolianus *122*, 380
Sporobolus spicatus *108*, *122*, 380
Statice axillaris see *Limonium paulayanum*
Statice cylindrifolia see *Limonium sokotranum*
Statice kossmatii see *Limonium paulayanum*
Statice paulayana see *Limonium paulayanum*
Statice socotrana see *Limonium sokotranum*
Sterculia 694
Sterculia africana *225*, **278**, 694
Sterculiaceae *82*, *90*, *225*, 693
Stipagrostis 380
Stipagrostis sokotrana *108*, 380
Striga *222*, *223*, 684
Striga asiatica *223*, 684
Striga densiflora *223*, 684
Striga gesnerioides *223*, 685
Striga hirsuta see *Striga asiatica*
Striga orobanchioides see *Striga gesnerioides*
Suaeda *157*, *158*, 493
Suaeda fruticosa see *Suaeda vermiculata*
Suaeda monoica *158*, 493
Suaeda paulayana see *Suaeda vermiculata*
Suaeda vermiculata *158*, 494
Syringodium 312
Syringodium isoetifolium *69*, 312

Talinum 656
Talinum crassifolium see *Talinum portulacifolium*
Talinum cuneifolium see *Talinum portulacifolium*
Talinum portulacifolium *213*, **273**, 656
Tamaricaceae *90*, *226*, 696
Tamarindus 589
Tamarindus indica *190*, 589
Tamarix 696
Tamarix aphylla *226*, 696
Tamarix arabica see *Tamarix nilotica*
Tamarix arborea see *Tamarix nilotica*
Tamarix gallica see *Tamarix nilotica*
Tamarix mannifera see *Tamarix nilotica*
Tamarix mascatensis see *Tamarix nilotica*
Tamarix nilotica *226*, 697

Tamarix socotrana see *Tamarix nilotica*
Tamridaea ... 670
Tamridaea capsulifera ... *216*, **276**, 670
Taverniera ... 612
Taverniera sericophylla ... *193*, **269**, 612
Tephrosia ... *192*, *195*, *201*, 613
Tephrosia anthylloides see *Tephrosia uniflora*
Tephrosia apollinea ... *201*, 613
Tephrosia apollinea var. *brevistipulata* see *Tephrosia apollinea*
Tephrosia odorata ... *201*, **269**, 614
Tephrosia purpurea ... *201*, 614
Tephrosia purpurea see *Tephrosia apollinea*
Tephrosia socotrana ... *201*, 614
Tephrosia strigosa ... *201*, 614
Tephrosia subtriflora ... *201*, 615
Tephrosia uniflora ... *201*, 615
Teramnus ... 615
Teramnus labialis see *Teramnus repens*
Teramnus labialis var. *mollis* see *Teramnus repens*
Teramnus repens ... *194*, 615
Teramnus repens subsp. *gracilis* see *Teramnus repens*
Tetragonia ... 416
Tetragonia pentandra ... *133*, 416
Tetrapogon ... *105*, *110*, *122*, 381
Tetrapogon tenellus ... *122*, 381
Tetrapogon villosus ... *122*, 381
Teucrium ... *187*, *189*, 585
Teucrium balfourii ... *189*, **266**, 585
Teucrium petiolare see *Teucrium sokotranum*
Teucrium petiolare var. *pubescens* see *Teucrium sokotranum*
Teucrium prostratum see *Teucrium balfourii*
Teucrium sokotranum ... *189*, **267**, 586
Thalassia hemprichii ... *69*
Thalassodendron ... 313
Thalassodendron ciliatum ... *69*, 313
Thamnosma ... *219*, 673
Thamnosma hirschii ... *219*, 673
Thamnosma socotrana ... *219*, **276**, 673
Thelypteridaceae ... 295
Thelypteris dentatus see *Christella dentata*
Themeda ... *109*, *123*, 381
Themeda quadrivalvis ... *123*, 382
Themeda triandra ... *123*, 382
Thymelaeaceae ... *91*, *226*, 697
Tiliaceae ... *82*, *93*, *226*, 698
Trachyspermum ... 704
Trachyspermum calcicola see *Trachyspermum pimpinelloides*
Trachyspermum pimpinelloides ... *228*, **279**, 705
Trachyspermum trichocarpum see *Trachyspermum pimpinelloides*
Tragia ... 567
Tragia balfouriana ... *177*, **263**, 567
Tragia dioica see *Tragia balfouriana*
Tragus ... 382
Tragus berteronianus ... *111*, 382
Trianthema pentandra see *Zaleya pentandra*
Tribulus ... 716
Tribulus terrestris ... *232*, 716
Trichocalyx ... *128*, *132*, 412
Trichocalyx obovatus ... *132*, **241**, 413
Trichocalyx orbiculatus ... *132*, 413
Trichodesma ... *143*, *147*, 455
Trichodesma atrichum see *Trichodesma laxiflorum*
Trichodesma laxiflorum ... *147*, **247**, 455
Trichodesma microcalyx ... *147*, **247**, 455
Trichodesma scottii ... *147*, **247**, 456
Tricholaena ... *104*, *123*, 383
Tricholaena setacea see *Tricholaena teneriffae*
Tricholaena teneriffae ... *123*, 383
Tricholaena vestita ... *123*, 383
Trichoneura ... 383
Trichoneura mollis ... *112*, 383
Trigonella ... 616
Trigonella falcata ... *193*, 616
Tripteris ... 515
Tripteris lordii see *Tripteris vaillantii*
Tripteris lordii var. *racemosa* see *Tripteris vaillantii*
Tripteris racemosa see *Tripteris vaillantii*
Tripteris vaillantii ... *160*, **255**, 515
Triumfetta ... 701
Triumfetta sp. *A* ... *226*, 701
Turraea ... 633
Turraea socotrana ... *94*, *207*, **271**, 633

Umbelliferae ... *83*, *88*, *228*, 702
Umbilicus ... 527
Umbilicus horizontalis ... *172*, 527
Urginea porphyrostachys see *Drimia porphyrostachys*
Urochondra ... 383
Urochondra setulosa ... *108*, 383
Urtica ... 707
Urtica urens ... *229*, 707
Urticaceae ... *74*, *84*, *229*, 705
Utricularia ... 619
Utricularia striatula ... *80*, *202*, 619

Valantia ... 671
Valantia hispida ... *215*, 671
Valerianaceae ... *75*, *76*, *230*, 708
Valerianella ... 708
Valerianella affinis ... *76*, *230*, 708
Verbenaceae ... *75*, *77*, *87*, *230*, 708
Vernonia ... *161*, *167*, 515
Vernonia cinerascens see *Orbivestus cinerascens*
Vernonia cinerea see *Cyanthillium cinereum*
Vernonia cockburniana ... *168*, **257**, 515
Vernonia spathulata see *Orbivestus cinerascens*
Vernonia unicata ... *168*, 516
Veronica ... 685
Veronica polita ... *222*, 685
Vigna ... 616
Vigna luteola see *Vigna membranacea*
Vigna membranacea ... *194*, 616
Vincetoxicum ... 442
Vincetoxicum linifolium ... *138*, **245**, 442
Viola ... 712
Viola cinerea ... *231*, 712
Violaceae ... *79*, *231*, 711
Viscaceae ... *73*, *85*, *231*, 712
Vitaceae ... *82*, *90*, *231*, 712
Vitis paniculata see *Cissus paniculata*
Vitis subaphylla see *Cissus subaphylla*
Vogelia indica var. *socotrana* see *Dyerophytum socotranum*
Vogelia pendula see *Dyerophytum pendulum*
Volutarella lippii see *Volutaria lippii*
Volutaria ... 516
Volutaria lippii ... *161*, 516

Wahlenbergia ... 474
Wahlenbergia lobelioides ... *150*, 474
Waltheria ... 695
Waltheria indica ... *225*, 695
Wellstedia ... 456
Wellstedia socotrana ... *143*, **247**, 456
Withania ... *224*, *225*, 691
Withania adunensis ... *225*, **278**, 691
Withania riebeckii ... *225*, 692
Withania somnifera ... *225*, 692

Xylocalyx *221*, *223*, 685
Xylocalyx aculeolatus *223*, **277**, 685
Xylocalyx asper *223*, **277**, 686

Zaleya 417
Zaleya pentandra *134*, 417
Zehneria 537
Zehneria scabra *175*, 537
Ziziphus 660
Ziziphus jububa see *Ziziphus spina-christi*
Ziziphus spina-christi *215*, **274**, 660
Zornia 616
Zornia diphylla see *Zornia glochidiata*
Zornia glochidiata *192*, 616
Zygocarpum 617
Zygocarpum coeruleum *195*, **269**, 617
Zygophyllaceae *78*, *87*, *232*, 715
Zygophyllum *232*, 717
Zygophyllum album *232*, 717
Zygophyllum album see *Zygophyllum decumbens*
Zygophyllum decumbens *232*, 717
Zygophyllum qatarense *232*, 717
Zygophyllum simplex *232*, 718

INDEX TO SOQOTRI NAMES

'aatel, a'atel — *Tamarix* spp.
a'arhiyib — *Dracaena cinnabari*
'afolhi — *Leucas martinicensis, Schweinfurthia pedicellata*
'afolhi qatin — *Teramnus repens*
ahfilhin — *Heliotropium aegypticum*
ahsilihin — *Kraussia socotrana*
ahtesh — *Acridocarpus socotranus*
ahtiresh — *Acridocarpus socotranus, Ricinus communis, Trichodesma scottii*
'ajeyr — *Bothriochloa insculpta, Cenchrus* spp., *Chloris pycnothrix, Chrysopogon* spp., *Heteropogon contortus, Lepturus calcareus, Themeda triandra*
'ajirib — *Andrachne schweinfurthii, Ageratum conyzoides, Bidens* spp., *Chenopodium murale, Diodia aulacosperma, Emex spinosus, Erucastrum rostratum, Fagonia luntii, Heliotropium zeylanicum, Hibiscus sidaeformis, Justicia heterocarpa, Leucas martinicensis, L. neuflizeana, Oxygonum sinuatum, Peristrophe paniculata, Pulicaria diversifolia, Reichardia tingitana, Sigesbeckia orientalis, Sisymbrium erysimoides, Trichodesma laxiflorum*
'ajirib shershiroh — *Acanthospermum hispidum, Dicoma tomentosa, Justicia heterocarpa*
'al'hal — *Anisotes diversifolius, Trichocalyx* spp.
'alghal — *Anisotes diversifolius, Trichocalyx* spp.
'alihat — *Cissus paniculata, Sarcostemma* spp.
'amihat — *Taverniera sericophylla*
amlah — *Rhus* sp. *A*
'ammeyri — *Oldenlandia pulvinata*
'aqilaz — *Solanum cordatum*
'aref — *Cladium mariscus*
'areyt/d — *Grewia bilocularis, G. mansouriana, G. turbinata*
'artah — *Crotalaria strigulosa, Periploca visciformis*
'ashbak — *Euphorbia schimperi, Kleinia scottii*
'ashbak dimidi — *Kleinia scottii*
'ashbiko — *Euphorbia schimperi*
'atel — *Cressa cretica*
'atirhe — *Cissus hamaderohensis, C. paniculata, C. subaphylla*
'aytaf'en, 'aytaffan — *Commicarpus* spp., *Dyerophytum* spp.
'aytift — *Commicarpus* spp., *Dyerophytum* spp.
'azabhin — *Adiantum balfourii*
'azabzab — *Hypericum balfourii, Pluchea obovata, Sigesbeckia orientalis*

ba'atit il-kheyr — *Ophioglossum* spp.
behith — *Cistanche phelypaea*
bigilihan — *Kalanchoe* spp.
bigilihan qiyqihe — *Kalanchoe farinacea*
bigilihan za'alhal — *Kalanchoe robusta*
bilbil — *Sarcostemma* spp.
biriy'tin — *Phyla nodiflora*
bish'hi, bishi, bishih — *Babiana socotrana, Dodonaea viscosa, Graderia fruticosa, Hypericum balfourii, H. tortuosum, Romulea fischeri, Sideroxylon fimbriatum, Turraea socotrana*
bita', bitah — *Babiana socotrana, Romulea fischeri*
biti'hin — *Euryops arabicus, Rughidia milleri*
bombih — *Eleusine coracana*
bowhiyn — *Sterculia africana*
bownino — *Dorstenia socotrana, Hibiscus vitifolius*
bowtih — *Babiana socotrana, Romulea fischeri*

chaya'ya', chayaya' — *Aerva revoluta, Helichrysum sphaerocephalum, Micromeria remota, Pluchea obovata, Teucrium balfourii*
chemhilinitin — *Aptosimum pumilum*
chiyreb, chiyrib — *Hypericum fieriense, H. scopulorum, H. tortuosum*

dagish, daqish — *Cryptolepis socotrana, Euphorbia balsamifera, E. socotrana*
dha'd — *Ziziphus spina-christi*
dhaf'an — *Lachnocapsa spathulata*
dhaf'han, dhafhan, dhaf'khan — *Euphorbia kuriensis, Mollugo cerviana*
dhafira' — *Barleria* spp. (not *B. popovii*), *Exacum caeruleum, Plectranthus socotranus, Pulicaria elegans, P. lanata*
dhafiri'hin — *Helichrysum* sp. *B*
dhafirih (dhafirih qiqihe, dhafirih za'alhal) — *Begonia* spp.
dhahyiym — *Lachnocapsa spathulata*
dhaldh — *Blepharis kuriensis, B. spiculifolia*
dhayaya' — *Leucas spiculifera*
dherit'hanitin — *Gaillonia puberula, G. puteroides, G. thymoides, Lotus ononopsis*
dheyt — *Datura fastuosa, Leucas spiculifera*
dhezibih — *Lachnocapsa spathulata*
dhif'ha — *Boswellia* spp. (growing in cramped conditions)
dhirghanitin — *Striga* spp.
dho'od, dhod maharito — *Ziziphus spina-christi*
dhowtiynih — *Cymbopogon jwarancusa*
dhozibih — *Limonium paulayanum, L. sokotranum*
di 'adhihin — *Kalanchoe rotundifolia*
di 'ajbib — *Erucastrum rostratum*
di 'ajirib — *Justicia heterocarpa*
di 'alimlihon, di 'alimlihum — *Adiantum balfourii, Anagallis arvensis, Bidens* spp., *Pulicaria diversifolia*
di 'alqaz — *Pulicaria diversifolia*

di 'alumhum *Pulicaria diversifolia*
di 'anijirihun *Cocculus balfourii*
di 'ardihan *Blepharis kuriensis, B. maderaspatensis, B. spiculifolia, Hibiscus palmatus, Nanorrhinum hastatum, Pavonia arabica, Ruellia patula, Sida* spp.
di 'areffen *Brachiaria deflexa, B. eruciformis, B. lata, B. ovalis, B. ramosa, Digitaria ciliaris, D. velutina*
di 'ashbib *Ageratum conyzoides, Endostemon tenuiflorus*
di 'atirhe *Cissus hamaderohensis, C. paniculata, C. subaphylla*
di 'azabzab *Pittosporum viridiflorum*
di 'itihi, di tiyih *Poskea socotrana*
di 'owjihir *Bothriochloa insculpta, Cenchrus* spp., *Chloris pycnothrix, Chrysopogon* spp., *Heteropogon contortus, Lepturus calcareus, Themeda triandra*
di a'arib *Crinum balfourii, Orobanche* spp.
di ah'shawih *Citrullus colocynthis, Cucumis prophetarum, Eureiandra balfourii*
di ajbub, di ajirib *Hypoestes pubescens*
di ardihan *Medicago* spp., *Melhania* spp.
di bigihun *Diceratella incana*
di bitahan *Dipcadi balfourii, D. guichardii, Iphigenia socotrana*
di bitahun, bitahiyun *Indigofera hochstetteri, Tephrosia strigosa*
di chaya'ya' *Pluchea obovata*
di chiyreb *Pluchea obovata*
di dha'arihun *Ophioglossum* spp.
di emlah *Aerva microphylla, Osyris quadripartita*
di feri'irihun *Graderia fruticosa*
di feylaq *Leucas hagghierensis*
di fitahe *Striga* spp.
di fitaher, di fitahih, di fitawih *Ledebouria* spp.
di gil'amihun *Melhania* spp., *Ricinus communis*
di girbighitin *Duvaliandra dioscoridis, Edithcolea grandis*
di hagubgub *Chlorophytum filifolium*
di halbeb *Chrozophora oblongifolia, Sarcostemma* spp.
di hamdhihan *Teramnus repens*
di hamimhim *Craterostigma pumilum*
di hanijirihun *Cocculus balfourii*
di hardihan *Blepharis spiculifolia, B. kuriensis, B. maderaspatensis, Hibiscus palmatus, Medicago* spp., *Melhania* spp., *Nanorrhinum hastatum, Pavonia arabica, Ruellia patula, Sida* spp.
di hatartihar *Actiniopteris semiflabellata*
di hazahan *Actiniopteris semiflabellata, Adiantum balfourii*
di hewgib *Cyperus bulbosus, C. rotundus, C. squarrosus*
di hewrirhin *Asystasia gangetica, Dicliptera effusa, Hypoestes pubescens, Peperomia* spp.
di hewrrihon *Asystasia gangetica*
di hibalino *Tephrosia odorata, T. strigosa*
di higubgub *Cyperus amabilis, C. compressus, C. difformis*
di hiydub *Acalypha indica*
di howgib *Cyperus bulbosus, C. rotundus, C. squarrosus*
di howlhinitin *Endostemon tenuiflorus, Euphorbia kischenensis, Leucas virgata, Orthosiphon ferrugineus, O. pallidus*
di hoyhi *Exacum affine*
di huwnitin, di huwnhitin *Adiantum balfourii, A. capillus-veneris, Negripteris sciona, Onychium divaricatum*
di huwritin *Adiantum balfourii, A. capillus-veneris, Negripteris sciona, Onychium divaricatum, Pellaea viridis*
di irbobo *Pulicaria stephanocarpa*
di khotortihor *Actiniopteris semiflabellata*
di kibiher *Teucrium balfourii*
di kubihir *Xylocalyx asper*
di mahahon *Gypsophila montana, Pulicaria lanata*
di mahahon lizafih *Pulicaria vieraeoides*
di mirisheytin, di mirishitin *Pupalia lappacea, Trichodesma microcalyx*
di mowdhi *Cyperus amabilis, C. compressus, C. conglomeratus, C. cyperoides, C. difformis, C. dilatatus, C. laevigatus, C. longus, C. stramineo-ferrugineus*
di nifizitin *Trichodesma laxiflorum*
di nishaz *Helichrysum aciculare, H. rosulatum*
di owjihir *Dichanthium annulatum*
di qazqaz *Convolvulus sarmentosus, Polygala abyssinica, P. erioptera, Pulicaria dioscorides, Teucrium sokotranum*
di qeret'hanitin *Dactyloctenium aristatum*
di qerihan *Cleome viscosa, Digera muricata, Senna socotrana*
di qertomo, di qertomoh *Brachiaria deflexa, B. eruciformis, B. lata, B. ovalis, B. ramosa, Cenchrus ciliaris, C. pennisetiformis, C. setigerus, Chloris pycnothrix, Chrysopogon* spp., *Cyperus amabilis, C. compressus, C. conglomeratus, C. cyperoides, C. difformis, C. dilatatus, C. laevigatus, C. longus, C. stramineo-ferrugineus, Dactyloctenium aegypticum, Digitaria ciliaris, D. velutina, Eragrostis ciliaris, Hackelochloa granularis, Heteropogon contortus, Lepturus calcareus, Panicum atrosanguineum, Pennisetum orientale, P. setaceum, Setaria intermedia, S. pumila, S. verticillata, Sporobolus spicatus, Themeda triandra*
di qewreyni *Diceratella incana, Tripteris vaillantii*
di qezqaz *Oldenlandia pulvinata*
di qiyfi'oh, di qiyfiyfuf *Trachyspermum pimpinelloides*

di qiyfuf	*Adiantum balfourii*
di qiyrib	*Hypericum scopulorum*
di qyuwqaf	*Adiantum balfourii*
di ra'adihun	*Exacum affine*
di rah'sheh	*Rhynchosia* spp.
di rahimimihon	*Heliotropium balfourii*
di rahmam	*Setaria verticillata*
di rahmimihun di tetin	*Rhus* sp. *A*
di rahmimiyun	*Aerva microphylla*
di rakhmam	*Dicoma cana*
di rehish	*Rhynchosia* spp.
di rihabih	*Colocasia esculenta*
di rihilt	*Digitaria ciliaris*, *D. velutina*
di rihisht	*Digitaria ciliaris*, *D. velutina*
di rihoytin	*Ageratum conyzoides*, *Eclipta prostrata*, *Erythroxylon socotranum*, *Exacum affine*, *E. caeruleum*, *E. socotranum*, *Lagarosiphon roxburghii*, *Ludwigia palustris*, *Marsilea coromandeliana*, *Najas graminea*, *Persicaria glabra*, *Scaevola socotraensis*
di rijbeb	*Chenopodium murale*
di rikibih	*Secamone* spp.
di riyhesh	*Rhynchosia* spp.
di shebhitin	*Polycarpaea caespitosa*
di shfa'hanitin	*Polycarpaea kuriensis*
di shimrahyun	*Helichrysum sphaerocephalum*
di shka'anitin	*Arnebia hispidissima*, *Dicoma cana*, *Pupalia lappacea*
di sow'iydo	*Crinum balfourii*
di tiyih	*Coelocarpum haggierensis*, *Nirarathamnos asarifolius*
di tiyo	*Onychium divaricatum*
di zafiririhon	*Craterostigma pumilum*, *Veronica polita*
di zeferher	*Helichrysum gracilipes*
di zeferirihon	*Evolvulus alsinoides*, *Helichrysum balfourii*, *H. gracilipes*, *Nanorrhinum hastatum*, *Poskea socotrana*, *Reseda viridis*, *Sida acuta*
di ziffoh	*Aristida* spp., *Lolium persicum*, *Melinis repens*, *Stipagrostis sokotrana*
di zifit	*Aristida* spp., *Lolium persicum*, *Melinis repens*
di ziynem	*Gaillonia tinctoria*
d-ib'ar	*Heliotropium riebeckii*
dibihowmih	*Nirarathamnos asarifolius*
dihalsh	*Aeluropus lagopioides*
dihadh	*Aeluropus lagopioides*
dilhowdh	*Phyla nodiflora*
diqeresh	*Negripteris sciona*
diraffan	*Hibiscus* spp. (only woody spp.)
di-rahmam	*Barleria popovii*
dirbiq	*Indigofera pseudointricata*
dirfih	*Indigofera articulata*
dirfin	*Hibiscus* spp. (only woody spp.)
diyrbeb	*Gomphrena celosioides*, *Gymnocarpos kuriensis*, *Oldenlandia pulvinata*, *Pulicaria stephanocarpa*, *Ruellia kuriensis*, *Teucrium balfourii*
dowgish, *dowqesh*, *dowqish*	*Cryptolepis socotrana*, *Euphorbia balsamifera*, *E. socotrana*
dowqihiyn	*Helichrysum aciculare*

duwriyr	*Launaea crassifolia*
'edherih	*Adiantum capillus-veneris*
'efero, *'efirih*	*Justicia alexandri*, *J. rigida*, *J. takhinensis*
'eqqihiyn	*Cryptolepis macrophylla*, *C. volubilis*, *Secamone* spp., *Vincetoxicum linifolium*
'eqrihim	*Avicennia marina*
'ershiydo	*Brachiaria arida*, *Cynodon dactylon*, *Dactyloctenium aristatum*, *Enneapogon desvauxii*, *Tricholaena teneriffae*
'ergilo	*Chlorophytum graptophyllum*
'etaf	*Dyerophytum* spp.
'ewresh'han, *'ewreshan*	*Heliotropium balfourii*
'eyjihir	*Bothriochloa insculpta*, *Cenchrus* spp., *Chloris pycnothrix*, *Chrysopogon* spp., *Heteropogon contortus*, *Lepturus calcareus*, *Themeda triandra*
ghared	*Grewia bilocularis*, *G. mansouriana*, *G. turbinata*
gheqqihiyn	*Cryptolepis macrophylla*, *C. volubilis*, *Secamone* spp., *Vincetoxicum linifolium*
gifi'hin	*Fagonia paulayana*, *Justicia rigida*, *Polycarpaea kuriensis*
gil'amihin	*Crotalaria strigulosa*
gil'imhinitin	*Polygala kuriense*
gil'imo	*Launaea crassifolia*, *L. hafunensis*
gilgilhitin	*Schweinfurthia pedicellata*
gilis'han	*Heliotropium shoabense*
gineygino, *giniginoh*	*Calotropis procera*, *Datura fastuosa*, *Ricinus communis*
ginsibiher	*Asplenium aethiopicum*, *A. schweinfurthii*, *Cleome socotrana*, *Christella dentata*, *Loxogramme abyssinica*, *Pleopeltis macrocarpa*, *Pteridium aquilinum*, *Pteris* spp.
gired	*Distephanus qazmi*
gisibihir, *gisibirih*	*Asplenium aethiopicum*, *A. schweinfurthii*, *Christella dentata*, *Loxogramme abyssinica*, *Pleopeltis macrocarpa*, *Pteridium aquilinum*, *Pteris* spp.
gisso, *gissoh*	*Cryptolepis arbuscula*, *C. intricata*, *C. volubilis*, *Euphorbia hamaderoensis*, *E. kischenensis*, *E. schweinfurthii*
giymiz	*Abutilon pannosum*, *Malva parviflora*
giyrbeb	*Fagonia paulayana*, *Pulicaria stephanocarpa*, *Teucrium balfourii*
gizhiritin	*Asphodelus fistulosus*
gowgor	*Commelina* spp.
gumuz	*Abutilon pannosum*, *Malva parviflora*, *Lindenbergia sokotrana*, *Melhania* spp.
gyamis	*Centella asiatica*
habbat al 'ayn	*Chamaecrista absus*, *Teramnus repens*

Soqotri name	Species
habichihin, habijihin	*Arthraxon* spp., *Lolium persicum*
hadhirat	*Pupalia lappacea*
hadhqiq di 'ilhitin	*Parietaria debilis, Pouzolzia auriculata, Pulicaria diversifolia*
hahir	*Coelocarpum haggierensis, Gaillonia tinctoria, Hemicrambe townsendii, Indigofera coerulea, I. socotrana, Pittosporum viridiflorum, Pyrostria socotrana, Reseda viridis, Senna socotrana, Sideroxylon fimbriatum, Spiniluma discolor, Turraea socotrana*
halawhin	*Ceropegia tihamana*
halbeboh	*Convolvulus sarmentosus*
haleyboh	*Convolvulus sarmentosus, Portulaca* spp.
halhalihan, halhalihin, halhalihiyn	*Asplenium schweinfurthii, Helichrysum gracilipes, H. paulayanum, Negripteris sciona, Oldenlandia pulvinata, Pteris* spp.
haliblibi	*Reseda viridis*
haliblibo	*Portulaca* spp.
haluwbeyboh	*Portulaca* spp.
hamadhihan, himadhihen	*Fagonia luntii, Indigofera nephrocarpa, I. spiniflora, Melhania* spp.
hamahamo	*Dicoma cana, Justicia rigida, Neuracanthus aculeatus, Rhinacanthus scoparius*
hambisisys	*Dipcadi kuriensis*
hamdh	*Indigofera nephrocarpa, I. spiniflora, Launaea crassifolia, L. crepoides, L. massauensis, L. rhynchocarpa, L. socotrana, Medicago* spp., *Polycarpaea paulayana, Reichardia tingitana, Trachyspermum pimpinelloides, Volutaria lippii*
hamdha'	*Oxalis corniculata*
hamhiym	*Campylanthus spinosus, Dicoma cana*
hamhiym di ahmiyri	*Campylanthus spinosus*
hamirmiro (pl. *hamirmirhinitin*)	*Chapmannia* spp., *Zygocarpum coeruleum*
hamjig	*Centella asiatica, Cyperus amabilis, C. compressus, C. conglomeratus, C. cyperoides, C. difformis, C. dilatatus, C. laevigatus, C. longus, C. stramineo-ferrugineus*
handuwrib	*Xylocalyx asper*
har'hi	*Chlorophytum graptophyllum, Craterostigma pumilum*
hargid (*sha'erher*)	*Hypoxis angustifolia*
hargog	*Chlorophytum graptophyllum*
hargyod	*Hypoxis angustifolia*
harhero	*Acacia edgeworthii*
harheyr, harher	*Acacia sarcophylla, A.* aff. *tortilis, A.* sp. *A, Commiphora kua, Dichrostachys dehiscens*
harheyrih	*Blepharis kuriensis, B. spiculifolia*
harhiy'o, harhiy'oh, harhiyoh	*Habenaria socotrana, Holothrix socotrana, Rughidia cordata*
hariher	*Heliotropium socotranum*
harishimhin	*Leucas flagellifolia*
haritrit	*Hypericum balfourii, Leucas virgata, Teucrium balfourii*
harjihiym, harjiym	*Solanum incanum*
harshihiym	*Grewia erythraea, G. tenax*
harshim	*Tamridaea capsulifera*
hartanih	*Cadaba farinosa, Commiphora kua*
haselihan	*Hypericum socotranum*
hasilam, hasilem	*Euclea divinorum, Euphorbia hajhirensis, Hypericum balfourii, Rhus* sp. *A*
hasilan, hasileyn	*Hypericum socotranum*
hasiliham	*Rhus* sp. *A*
hasilin'g, hasilinj	*Flueggea* spp., *Poskea socotrana*
hatamhin	*Launaea hafunensis*
hatartihar	*Tephrosia odorata*
hatehiym, hatihim	*Salsola spinescens, Suaeda vermiculata*
hatem, hatim	*Atriplex farinosa, Gymnocarpos kuriensis, Suaeda monoica*
hatemhen, hatemhin	*Lagenantha cycloptera, Suaeda vermiculata*
hatiham	*Suaeda monoica*
hatihiyn	*Geranium mascatense*
hawririhim	*Pulicaria diversifolia*
haym'her	*Zygocarpum coeruleum*
hayriro	*Indigofera oblongifolia, I. socotrana*
hedist	*Boerhavia diffusa, Commicarpus* spp., *Zaleya pentandra*
hemadhihen	*Oxalis corniculata*
heqeh	*Indigofera oblongifolia*
hewiyr	*Tephrosia uniflora*
hewrirhin, hewririhin	*Asystasia gangetica, Dicliptera effusa, Hybanthus enneaspermus, Peperomia* spp., *Sigesbeckia orientalis*
hewririhim	*Leucas spiculifera*
hewriyr	*Indigofera coerulea*
hewrrihin, hewrrihon	*Asystasia gangetica, Dicliptera effusa*
hey	*Ruellia kuriensis*
heymar	*Sporobolus spicatus*
hfe', hfo'	*Orbivestus cinerascens, Vernonia cockburniana, V. unicata*
hibalino	*Pavonia arabica*
hidhud	*Dipcadi balfourii*
hidiho	*Boerhavia diffusa, Launaea crassifolia*
hidis	*Boerhavia diffusa, Zaleya pentandra*
himiher	*Zygocarpum coeruleum*
hirewtat	*Panicum rigidum*
hiruwt	*Ruellia patula*
hiteham	*Salsola spinescens*
hiweyr	*Senna holosericea*
hiwiyl	*Indigofera coerulea*
hiwiyr	*Indigofera coerulea*
hiwor	*Senna holosericea*
hiylebih, hiylibih	*Dichondra repens, Phyllanthus maderaspatensis*
hiyreyr	*Habenaria socotrana, Pulicaria diversifolia*
heyrih	*Habenaria socotrana*
hiyreyrih d-aghirib	*Dipcadi guichardii*
hoft	*Flueggea* spp.
hondhol	*Peperomia* spp.
howgib	*Chlorophytum filifolium*
howkinitin	*Forsskaolea viridis*
huwenitin	*Pellaea viridis*

'i'iri, 'iyri — *Brachiaria eruciformis*, *Dactyloctenium aegypticum*, *Setaria intermedia*, *S. pumila*, *S. verticillata*
'ibelit — *Amaranthus* spp.
'ibib — *Withania* spp.
'ibihat, 'ibihit — *Cistanche phelypaea*
'ibihir — *Acrachne racemosa*, *Dactyloctenium aegypticum*, *Digitaria ciliaris*, *D. velutina*
'ibilit — *Justicia heterocarpa*, *Zaleya pentandra*
'ibit — *Dactyloctenium robecchii*, *Kohautia* spp.
'ibitih — *Cordia* spp., *Tamridaea capsulifera*
'ibitih za'alhil — *Cordia* spp.
'ifirih — *Atriplex farinosa*, *A. griffithii*, *Justicia alexandri*, *J. rigida*, *J. takhinensis*
'ikshih — *Commiphora ornifolia*
'imlah — *Helichrysum sphaerocephalum*
'imtey, 'imtihe — *Euphorbia arbuscula*
'iqshireh — *Ehretia obtusifolia*
'ir'ero — *Canavalia africana*
'irihek — *Salvadora persica*
'irihem — *Pergularia tomentosa*
'irit/d — *Grewia bilocularis*, *G. mansouriana*, *G. turbinata*
'irsib — *Corchorus* spp., *Launaea rhynchocarpa*
'irsib mirsewbihan — *Portulaca* spp.
'ish'hib — *Maerua angolensis*
'ishelo — *Stipagrostis sokotrana*
'it'ib, 'ithib — *Ficus cordata*
'iyri — *Brachiaria deflexa*, *B. lata*, *B. ovalis*, *B. ramosa*, *Desmostachya bipinnata*, *Echinochloa colona*, *Eragrostis ciliaris*, *Lepturus pulchellus*
'iytef batah — *Diceratella incana*
'iytift — *Commicarpus* spp.

kirbilitin — *Leucas martinicensis*
kirilloh — *Acridocarpus socotranus*
kubihur — *Oldenlandia pulvinata*
kubush — *Lotus schimperi*

lata' — *Dichanthium foveolatum*, *Panicum rigidum*, *Stipagrostis sokotrana*, *Urochondra setulosa*
le'efin, li'iffin — *Gnidia socotrana*
leqehem — *Commiphora parvifolia*, *C. planifrons*, *C. socotrana*
lezibih — *Limonium paulayanum*
li'eyif — *Gnidia socotrana*
li'ih — *Cyperus bulbosus*, *C. rotundus*, *C. squarrosus*
libihan — *Neuracanthus argyrophyllus*, *Seddera fastigiata*, *S. glomerata*, *S. latifolia*, *Teucrium sokotranum*
libinhan, libinhen, libinhin — *Achyrocline luzuloides*, *Helichrysum balfourii*, *H. nimmoanum*, *H.* sp. *A*, *H.* sp. *B*, *H.* sp. *C*, *H. suffruticosum*, *Heliotropium ovalifolium*, *Linum trigynum*, *Silene apetala*, *Teucrium sokotranum*
libinhin (*di didihiyn*) — *Poskea socotrana*
libininhin — *Helichrysum gracilipes*
libinno — *Convolvulus arabicus*
lirghibih — *Launaea crassifolia*
lisekek — *Helichrysum arachnoides*
liseyak — *Hibiscus micranthus*, *H. palmatus*, *H. sidaeformis*
liskano, liskanoh — *Anagallis arvensis*, *Forsskaolea viridis*, *Haya obovata*, *Leucas flagellifolia*, *Psilotrichum aphyllum*, *P. sericeum*, *Pupalia lappacea*, *Rhynchosia* spp.
liteynuh — *Gaillonia puberula*, *G. puteroides*, *G. thymoides*, *G. tinctoria*, *Micromeria remota*
liyif — *Indigofera marmorata*
liza'han — *Priva socotrana*
lizafih — *Capparis cartilaginea*
lizafih za'alhil — *Capparis decidua*
lizibih — *Limonium sokotranum*
lo'oh — *Cyperus bulbosus*, *C. rotundus*, *C. squarrosus*
lozif — *Capparis cartilaginea*
lu'lu'uh — *Endostemon tenuiflorus*, *Ocimum forskolei*, *Orthosiphon pallidus*
luwsak — *Hibiscus sidaeformis*

mah'showmhin — *Tephrosia odorata*
maharish — *Asystasia gangetica*, *Convolvulus sarmentosus*, *Dicliptera effusa*, *Heliotropium balfourii*, *Hypoestes pubescens*, *Meineckia filipes*, *Nanorrhinum hastatum*, *Oldenlandia pulvinata*, *Peperomia* spp., *Phyllanthus maderaspatensis*, *Priva socotrana*, *Rhinacanthus scoparius*, *Ruellia carnea*, *R. patula*, *R. paulayana*
mahilibinuh — *Talinum portulacifolium*
mahirit — *Hybanthus enneaspermus*
mahleybboh — *Talinum portulacifolium*
mahleybibo, mahlibiboh — *Portulaca* spp.
mahlumhim — *Heliotropium nigricans*, *H. ovalifolium*, *H. strigosum*
mahluwbib — *Bacopa monnieri*
maqdhab — *Panicum rigidum*, *Panicum socotranum*
marhuwmihim — *Cystostemon socotranus*
mas'hat, mas'hit — *Andropogon bentii*, *Eragrostis cilianensis*, *Panicum turgidum*, *Tragus berteronianus*
mazel — *Actiniopteris semiflabellata*
mazow'iden — *Barleria* spp. (not *B. popovii*)
medh'howmhen — *Zygophyllum simplex*
meltuwsek, meltuwsik — *Diodia aulacosperma*, *Priva socotrana*
merih — *Euphorbia obcordata*, *E. socotrana*
merishiy'oh, mirishiy'oh — *Achyranthes aspera*, *Crotalaria strigulosa*
merjif — *Echinochloa colona*, *Eragrostis barrelieri*, *E. lepida*
meylho — *Teucrium balfourii*
meyrih — *Cryptolepis arbuscula*, *C. intricata*, *C. macrophylla*, *C. socotrana*, *Dyerophytum* spp., *Marsdenia robusta*, *Vincetoxicum linifolium*
meyrit — *Euphorbia schimperi*

Soqotri name	Species
mhat'hatin, mhat'hitin, mhat'hotin	*Actiniopteris semiflabellata, Adiantum capillus-veneris, Cystostemon socotranus, Desmodium triflorum, Helichrysum gracilipes, Negripteris sciona, Trichodesma laxiflorum, T. microcalyx*
mhatoh	*Actiniopteris semiflabellata, Adiantum balfourii, Pellaea viridis*
mi'azihil	*Cryptolepis arbuscula, C. intricata*
mi'eyib di libihan	*Blepharis kuriensis, B. spiculifolia*
mid'howmihim	*Pulicaria diversifolia*
midewhin, midewhiyn	*Bidens* spp., *Setaria intermedia, S. pumila, S. verticillata*
midh'huwmhen, midh'howmhin	*Bacopa monnieri, Trichodesma microcalyx*
midh'howmhin di ilihe	*Trichodesma microcalyx*
midh'khowmhim	*Limonium paulayanum*
miduwnihin	*Pulicaria elegans*
migeyred, migeyrid	*Coelocarpum haggierensis, C. socotranum, Heliotropium aegypticum, H. zeylanicum, Leucas kischenensis*
mihilibboh	*Portulaca* spp.
mihowgib	*Camptoloma lyperiiflorum*
mijirid	*Coelocarpum haggierensis*
milbownhin, milbuwnhan	*Diceratella incana, Helichrysum sphaerocephalum*
milihil	*Cryptolepis macrophylla*
miltewsek, miltuwsik	*Forsskaolea viridis, Sigesbeckia orientalis*
mimba'	*Parietaria debilis, Pouzolzia auriculata*
min teyoh, minteyoh	*Drimia porphyrostachys, Lolium persicum*
minduwraz	*Sarcostemma* spp.
minduwrib	*Euphorbia schweinfurthii, Sarcostemma* spp.
minheyik	*Cymbopogon jwarancusa, Pouzolzia auriculata*
miqtownih	*Maytenus* sp. *A*
miranniha	*Abutilon fruticosum, A. pannosum, Argemone mexicana, Datura fastuosa, Nicandra physaloides*
miregih, mirigeh	*Campylanthus spinosus, Seddera semhahensis*
mirhowmhim	*Helichrysum balfourii*
miriginoh	*Convolvulus sarmentosus*
mirigo	*Campylanthus spinosus, Seddera semhahensis*
mirishiyitin	*Blepharis maderaspatensis*
mirsem	*Portulaca* spp.
mis'habeb	*Asparagus* spp., *Indigofera nephrocarpa, I. spiniflora*
mis'heybibo	*Rhynchosia* spp.
mis'howbeb, mis'howbib	*Alysicarpus* spp., *Asparagus* spp., *Crotalaria strigulosa, Fagonia luntii, Indigofera cordifolia, I. nephrocarpa, I. spiniflora*
mis'owbeb	*Launaea crepoides*
mish'hahir	*Caralluma socotrana*
mish'heremhim di a'arib	*Orobanche* spp.
mish'hermhim di girbag	*Asparagus* spp., *Fagonia paulayana*
mish'hermihim	*Caralluma socotrana*
mish'hermihim 'afir	*Orobanche cernua, Striga* spp.
mish'hermihim di e'efo	*Caralluma socotrana*
mish'hermihim di girbag	*Edithcolea grandis*
mish'heybibo	*Arthraxon* spp.
mish'howbeb	*Asparagus* spp.
mish'howgeg, mish'howgig	*Asparagus* spp., *Barleria* spp. (not *B. popovii*), *Cystostemon socotranus*
mish'howgin	*Asparagus* spp.
mish'howji	*Haya obovata*
mish'howmihen	*Asparagus* spp.
mish'howreyro	*Limonium paulayanum*
mish'khayrem di 'aghirib	*Edithcolea grandis*
mish'khayrrem	*Caralluma socotrana*
mish'qerer	*Angkalanthus oligophylla, Chorisochora* spp.
mish'qerer za'alhil	*Crotalaria socotrana*
mis'habeb	*Euphorbia hajhirensis*
mis'heybibo	*Oplismenus burmannii*
mishiraq	*Thamnosma hirschii*
mishka'ihen	*Sigesbeckia orientalis*
mishra'	*Eragrostis barrelieri, E. lepida*
misow'iden	*Crotalaria spinosa*
mistewriher	*Acridocarpus socotranus, Micromeria remota, Turraea socotrana*
mitihan	*Buxanthus pedicellatus*
mitrer	*Croton sarcocarpus, C. socotranus, C. sulcifructus*
mitrer hagirhi	*Croton sulcifructus*
mitrer qiyqiheh	*Croton socotranus*
mitrer za'alhil	*Croton sarcocarpus, C. sulcifructus*
miyshet, miyshit	*Cryptolepis arbuscula, C. intricata, Punica protopunica*
miyzez	*Ballochia* spp., *Trichocalyx* spp.
miz'heymihim	*Commelina* spp.
mizewkan	*Rhynchosia* spp.
mizewrib	*Portulaca* spp.
mizirbihan, mizirbihon	*Corchorus* spp., *Polycarpaea spicata*
mizirbihin di riyho	*Ammannia* spp., *Lythrum hyssopifolium*
mizow'idihin, mizow'idiyn	*Fagonia paulayana*
mizuwkan, mizuwken	*Cuscuta chinensis, C. planiflora, Vincetoxicum linifolium*
mohoha'	*Pulicaria lanata*
mos'hit	*Asparagus* spp., *Dactyloctenium hackelii, D. robecchii, Panicum rigidum*
mtabihon	*Heliotropium balfourii*
mtamhir	*Plectranthus socotranus*
murjif	*Digitaria velutina, Eragrostis papposa, E. viscosa, Lepturus calcareus*
nayim	*Oldenlandia pulvinata*
nifizitin	*Coelocarpum haggierensis, Teucrium sokotranum*
niminow, niminuh	*Euphorbia kischenensis*
nishazih	*Jasminum fluminense*
nizafih	*Aristida* spp., *Dactyloctenium aristatum, Dichanthium foveolatum, Panicum rigidum,*

Soqotri name	Species
	Setaria verticillata, Sporobolus spicatus, Stipagrostis sokotrana
nowhel	*Leucas kischenensis, L. virgata*
nowhel hegirhi	*Leucas penduliflora*
nowmhal, nowmhel, nuwmhel	*Centaurium tenuiflorum, Leucas kischenensis, L. virgata*
noy	*Convolvulus hildebrandtii, C. rhyniospermus, Crotalaria strigulosa, Endostemon tenuiflorus, Haya obovata, Indigofera nephrocarpa, I. spiniflora, Leucas virgata, Orthosiphon pallidus, Ruellia dioscoridis, Tephrosia odorata*
'otf	*Flueggea* spp.
'owbeb	*Withania* spp.
'owdhiz	*Trichodesma laxiflorum*
owqoh	*Commelina* spp.
'owshiq	*Lepturus tenuis*
'owshiyeh	*Dactyloctenium hackelii, D. robecchii*
qa'	*Convolvulus arabicus, C. sarmentosus*
qa' di mirhuwmhim	*Euphorbia kischenensis*
qa' di tetin	*Convolvulus arabicus, Crotalaria strigulosa*
qadhabhan, qadhabhen, qadhbihan	*Chlorophytum filifolium, Juncus bufonius, J. rigidus*
qadhabihon	*Phyla nodiflora*
qadheb, qadhib	*Aeluropus lagopodioides, Eleocharis* spp., *Paspalidium geminatum, Phyla nodiflora, Sporobolus spicatus*
qadhib di mowdhi	*Aeluropus lagopodioides*
qadhibeb	*Farsetia longisiliqua, F. socotrana*
qadhibih	*Aeluropus lagopodioides, Bulbostylis* spp., *Farsetia longisiliqua, F. socotrana, Fimbristylis* spp., *Juncus bufonius, Panicum rigidum, P. socotranum, P. turgidum, Sporobolus ioclados, S. spicatus*
qaf	*Prosopis juliflora*
qal'ish	*Rughidia milleri*
qalqihal	*Aerva microphylla, Anabasis ehrendbergii, Arthrocnemum macrostachyum, Cometes abyssinica, Cornulaca ehrenbergii, Cyperus amabilis, C. compressus, C. conglomeratus, C. cyperoides, C. difformis, C. dilatatus, C. laevigatus, C. longus, C. stramineo-ferrugineus, Gymnocarpos kuriensis, Heliotropium ovalifolium, Lachnocapsa spathulata, Polycarpaea spicata, Sevada schimperi, Suaeda monoica, S. vermiculata, Tamarix* spp., *Zygophyllum decumbens, Z. qatarense, Z. simplex*
qamhiym, qamhiyn	*Dendrosicyos socotrana*
qardha'	*Panicum rigidum, Sporobolus ioclados, S. spicatus*
qash'ho	*Euphorbia spiralis*
qashobih	*Spiniluma discolor*
qatemhem	*Trachyspermum pimpinelloides*
qatihan	*Apludica mutica, Brachiaria arida, Chloris virgata, Digitaria velutina, Echinochloa colona, Enneapogon desvauxii, Eragrostis barrelieri, E. cilianensis, E. lepida, Lepturus pulchellus, Paspalidium geminatum, Sporobolus ioclados, Tetrapogon tenellus*
qazihan	*Cystostemon socotranus*
qazqaz	*Convolvulus sarmentosus, Echiochilon pulvinata, Wellstedia socotrana*
qer'ash, qer'ish	*Commiphora ornifolia, Jasminum fluminense, Periploca visciformis, Turraea socotrana*
qerdha'ihan, qerdha'ihun, qeridha'ihun, qeridhahiyun	*Ammannia* spp., *Centella asiatica, Exacum caeruleum, Lythrum hyssopifolium, Sporobolus ioclados*
qeremihon	*Euphorbia kischenensis, E. schweinfurthii*
qerihan	*Digitaria ciliaris, Tephrosia uniflora*
qeriher	*Aerva revoluta*
qeshyobih	*Spiniluma discolor*
qetimamhion	*Eragrostis ciliaris, E. viscosa, Heliotropium balfourii*
qewqor	*Commelina* spp.
qewreyni	*Diodia aulacosperma, Plantago* spp.
qeytiher	*Gnidia socotrana*
qezadih	*Andropogon bentii, Bothriochloa insculpta, Cyperus amabilis, C. compressus, C. difformis, Dichanthium annulatum, Lepturus pulchellus, Microchloa kunthii, Pennisetum orientale, P. setaceum, Stipagrostis sokotrana, Tetrapogon villosus, Urochondra setulosa*
qezqaz	*Wellstedia socotrana*
qidhadh, qidhodh	*Kyllinga* spp.
qisas	*Limonium sokotranum*
qisho	*Euphorbia spiralis*
qitimamhiyon	*Xylocalyx aculeolatus, X. asper*
qiyribrub	*Asphodelus fistulosus*
qiyritamo	*Cenchrus pennisetiformis, Chloris pycnothrix, Chrysopogon* spp.
qiytihir	*Gnidia socotrana*
qizadih	*Halopyrum mucronatum, Lepturus tenuis*
qiziho	*Cyperus amabilis, C. bulbosus, C. compressus, C. difformis, C. rotundus, C. squarrosus, Linum trigynum, Trichodesma microcalyx, Trigonella falcata*
qizqaz, qizqiz	*Lotus mollis, L. ononopsis, Melhania* spp.
qordha, qurdha'	*Aeluropus lagopodioides, Bacopa monnieri, Bulbostylis* spp., *Centella asiatica, Cyperus amabilis, C. compressus, C. conglomeratus, C. cyperoides,*

	C. difformis, *C. dilatatus*, *C. laevigatus*, *C. longus*, *C. stramineo-ferrugineus*, *Dactyloctenium aristatum*, *Desmostachya bipinnata*, *Eragrostis papposa*, *Fimbristylis* spp., *Fuirena ciliaris*, *Kyllinga* spp., *Lagarosiphon roxburghii*, *Najas graminea*, *Paspalum scrobiculatum*, *P. vaginatum*, *Polypogon monspeliensis*, *Pycreus pumilus*, *Setaria verticillata*
qown'hitin, qownitin	*Convolvulus grantii*, *Nanorrhinum kuriensis*, *Polygala kuriense*
qownuh	*Crotalaria strigulosa*, *Indigofera nephrocarpa*, *I. spiniflora*
qowqa'	*Commelina* spp.
qowqa' di 'alumhum	*Pulicaria diversifolia*
quwnu rikhammih	*Neuracanthus argyrophyllus*
quwnuh	*Haya obovata*
ra'adihon	*Anticharis arabica*, *Arenaria foliacea*, *Campanula balfourii*, *Corbichonia decumbens*, *Desmodium triflorum*, *Erodium cicutarium*, *Gisekia pharnaceoides*, *Glinus lotoides*, *Indigofera argentea*, *Lotus garcinii*, *L. mollis*, *L. schimperi*, *Mollugo cerviana*, *M. nudicaulis*, *Sebaea microphylla*, *Viola cinerea*, *Wahlenbergia lobelioides*, *Zornia glochidiata*
rahamham	*Achyrocline luzuloides*, *Helichrysum gracilipes*
rahammih	*Euphorbia granulata*
rahmam	*Actiniopteris semiflabellata*, *Negripteris sciona*, *Onychium divaricatum*, *Ophioglossum* spp.
rakham	*Arnebia hispidissima*
rakhammih	*Euphorbia schweinfurthii*
rakhmam	*Actiniopteris semiflabellata*, *Euphorbia schweinfurthii*, *Negripteris sciona*, *Onychium divaricatum*, *Ophioglossum* spp.
reheb	*Pulicaria diversifolia*
rehisht	*Arthraxon* spp., *Gastridium phleoides*
rehmimhim	*Cyanotis cristata*
remhen	*Apludica mutica*
rihammih	*Arnebia hispidissima*, *Farsetia inconspicua*, *Indigofera nephrocarpa*, *I. spiniflora*, *Oldenlandia pulvinata*
rihemimhin, riheymimhin	*Cymbopogon jwarancusa*, *Teucrium balfourii*
rihilt	*Oplismenus burmannii*, *O. compositus*
rihisht	*Lotus ononopsis*, *Oplismenus burmannii*, *O. compositus*
rihiyni	*Punica protopunica*
rijbeb, rijibib	*Acalypha indica*, *Chenopodium murale*, *Emex spinosus*, *Erucastrum rostratum*, *Launaea massauensis*, *L. socotrana*, *Oxygonum sinuatum*, *Parietaria debilis*, *Pouzolzia auriculata*, *Sonchus oleraceus*
rikhamkham	*Farsetia inconspicua*, *Lachnocapsa spathulata*
rikibih	*Cocculus pendulus*, *Jasminum fluminense*, *Sideroxylon fimbriatum*
rikibih di 'ik'sheh	*Secamone* spp.
rikibih za'alhal	*Nirarathamnos asarifolius*, *Oreofraga morrisiana*
risheyrish	*Euphorbia kuriensis*, *Indigofera nugalensis*, *Ruellia kuriensis*
rishitino qatihan	*Lepturus pulchellus*
ritebib	*Ruellia kuriensis*
riteyrit	*Indigofera nugalensis*
riteyrit rihammih	*Indigofera nephrocarpa*, *I. spiniflora*
riyebih	*Ceropegia affinis*, *Sarcostemma* spp.
riyh idini	*Coelocarpum socotranum*
ru'ud	*Astralagus hauarensis*, *Capsella bursa-pastoris*, *Corbichonia decumbens*, *Desmodium triflorum*, *Eclipta prostrata*, *Erodium cicutarium*, *Gisekia pharnaceoides*, *Glinus lotoides*, *Indigofera argentea*, *I. nephrocarpa*, *I. spiniflora*, *Lindenbergia sokotrana*, *Melilotus indica*, *Microcharis disjuncta*, *Misopates orontium*, *Mollugo cerviana*, *M. nudicaulis*, *Pelargonium insularis*, *Polycarpaea balfourii*, *P. hayoides*, *Psilotrichum sericeum*, *Sebaea microphylla*, *Trigonella falcata*
ru'ud di tiyo	*Negripteris sciona*
ruwhadh bish 'iydi	*Zygophyllum simplex*
sa'adifo	*Panicum rigidum*, *Cyperus amabilis*, *C. compressus*, *C. conglomeratus*, *C. cyperoides*, *C. difformis*, *C. dilatatus*, *C. laevigatus*, *C. longus*, *C. stramineo-ferrugineus*, *Fimbristylis* spp., *Lepturus calcareus*
sa'iydih	*Cyperus bulbosus*, *C. rotundus*, *C. squarrosus*
sahamhim	*Trichodesma microcalyx*, *Veronica polita*
sebira	*Jatropha unicostata*
segihur	*Rughidia cordata*
sejihir	*Rughidia cordata*
senneh mikkiy	*Senna holosericea*
sha'angig (di qirihan)	*Eulophia petersii*
shaba', shabah	*Heliotropium balfourii*
shabagh(an)	*Heliotropium balfourii*
shabiyah, shibiyah	*Convolvulus hildebrandtii*, *Polycarpaea caespitosa*
shamqas	*Cocculus balfourii*
shaz	*Dodonaea viscosa*
shebetano, shebitano	*Desmostachya bipinnata*, *Eragrostis papposa*
shebiraq, shebireq, shebiriq	*Heliotropium rariflorum*, *Indigofera articulata*, *I. pseudointricata*, *Ruellia patula*,

	Sporobolus iocladus, Tragus berteronianus
shebitih	*Setaria intermedia, S. pumila, S. verticillata, Tragus berteronianus*
shedih	*Indigofera colutea, Ruta chalapensis*
sheh	*Hybanthus enneaspermus*
shenih	*Ipomoea sinensis*
sheremeno, sherimino	*Acacia* sp. *A, Dichrostachys dehiscens, Euclea divinorum*
sherishero, sherishiroh	*Euphorbia schweinfurthii, Medicago* spp., *Tribulus terrestris*
sherkha	*Seddera latifolia*
sherkha erkhammeh	*Aerva microphylla*
shersher	*Haya obovata, Medicago* spp.
shetiher	*Leucas virgata*
shetihin	*Leucas virgata*
shey'initin	*Argemone mexicana*
sheydih	*Aerva revoluta, Teucrium balfourii, Thamnosma socotrana*
shezaz	*Dodonaea viscosa*
shezizhiyo	*Cordia* spp.
shfet'han	*Leucas virgata*
shiberhen	*Launaea crassifolia, L. hafunensis*
shibihitin	*Heliotropium socotranum*
shibitano	*Kohautia* spp.
shibitih	*Panicum rigidum*
shidak	*Exacum caeruleum*
shidharhen, shidharhin	*Abutilon fruticosum, A. pannosum, Malva parviflora*
shidharher	*Pyrostria socotrana*
shifatehan	*Periploca visciformis*
shimerah, shimeyrah, shimrah	*Crotalaria strigulosa, C. leptocarpa, Tephrosia odorata*
shimirahiyun	*Crotalaria strigulosa*
shinhiyih	*Oreofraga morrisiana*
shiremhen	*Orthosiphon ferrugineus*
shiremhin	*Graderia fruticosa, Ocimum forskolei*
shiriham hahir	*Turraea socotrana*
shiyhit	*Carphalea obovata*
shka'hinitin, shka'initin	*Barleria* spp. (not *B. popovii*), *Crotalaria spinosa, Cystostemon socotranus, Helichrysum aciculare, Leucas flagellifolia, L. spiculifera, Pulicaria diversifolia*
shka'hinitin (*di midhowdha'*)	*Leucas spiculifera*
sho'oz, shoz	*Dodonaea viscosa*
showhiditin	*Hibiscus micranthus*
sibbah	*Erucastrum rostratum, Launaea socotrana*
siberih	*Nephrolepis undulata*
sidahan	*Brachiaria deflexa, B. eruciformis, B. lata, B. ovalis, B. ramosa, Convolvulus sarmentosus, Echiochilon pulvinata, Microcharis disjuncta, Oldenlandia pulvinata*
sidak	*Lavandula nimmoi, Seddera fastigiata, S. glomerata*
sidak (*di jifun*)	*Micromeria remota*
sided, sidid	*Paracalyx balfourii*
sidihin	*Phyllanthus maderaspatensis*
sigenhen	*Polycarpaea kuriensis, Pulicaria stephanocarpa, Teucrium balfourii*
sigirhen	*Crotalaria spinosa, C. strigulosa, Trachyspermum pimpinelloides*
sinemhen, sinemhin	*Clerodendrum* spp., *Coelocarpum socotranum, Prenanthes amabilis*
sowf	*Acanthospermum hispidum, Dicliptera effusa, Ruellia dioscoridis*
sowgihiyn	*Leucas spiculifera*
su'hur	*Lycium sokotranum*
ta'abihor, ta'abihur	*Convolvulus hildebrandtii, C. sarmentosus, Crotalaria leptocarpa, Kraussia socotrana, Vincetoxicum linifolium*
ta'iyiq	*Parietaria debilis, Pouzolzia auriculata*
tabtab	*Heliotropium balfourii, H. kuriense, H. ophioglossum*
tahaq'hen	*Polycarpaea spicata*
tal'at ezeynet	*Coelocarpum socotranum*
talh	*Acacia nilotica*
talihon	*Chapmannia* spp., *Zygocarpum coeruleum*
taliy'oh, taliyeh, talyoh	*Boswellia socotrana*
talqat	*Crotalaria strigulosa, Polygala abyssinica, P. erioptera*
tamerhen, tamirhen	*Cleome brachycarpa, Geranium mascatense, Helichrysum* sp. *A, H.* sp. *B, Leucas virgata, Micromeria remota, Orthosiphon pallidus, Thamnosma hirschii*
tamhur	*Oldenlandia balfourii, O. bicornuta, O. corymbosa*
tamirhen 'ashbib	*Endostemon tenuiflorus*
tan	*Cephalocroton socotranus*
tayf	*Aloe perryi*
tayf 'abdiher	*Aloe squarrosa*
tenibeko	*Pulicaria diversifolia*
tibbeyhon	*Pedalium murex*
tibirerihon	*Thamnosma hirschii*
tifer	*Geranium biuncinatum, G. mascatense, G. trilophum*
tifher	*Indigofera hochstetteri, I. socotrana, I. tinctoria, Ruellia patula, Senna holosericea, S. socotrana, Tephrosia apollinea, T. odorata, T. purpurea, T. socotrana, T. strigosa, T. subtriflora, T. uniflora*
tileyliyih	*Trachyspermum pimpinelloides*
tilihin, tilihiyn	*Aerva microphylla, Cocculus pendulus*
tililhiyoh	*Trachyspermum pimpinelloides*
tilili	*Trachyspermum pimpinelloides*
tililyi, tililyih	*Aerva microphylla, Euphorbia kischenensis, Trachyspermum pimpinelloides*
tiliy, tiliy'ih, tiliy'o	*Boswellia socotrana*
timbak, timbeko	*Nicotiana tabacum*
timerhen	*Cleome brachycarpa*
timihor	*Oldenlandia balfourii, O. bicornuta, O. corymbosa*
timirih	*Phoenix dactylifera*

tinejih — *Citrus aurantium*
tiq — *Ficus vasta*
tiriymo — *Adenium obesum*
tiyfit — *Clerodendrum* spp., *Coelocarpum socotranum*, *Dirachma socotrana*, *Placopoda virgata*, *Tephrosia uniflora*
tizibih — *Limonium sokotranum*
tom'hor, tomhor — *Acacia pennivenia*, *A.* sp. *A*, *Cleome austroarabica*, *C. brachycarpa*, *Dichrostachys dehiscens*, *Micromeria remota*, *Nanorrhinum kuriensis*, *Oldenlandia ocellata*, *Trachyspermum pimpinelloides*

'ubab — *Withania* spp.
'udf, 'uft — *Flueggea* spp.
'uwrif — *Cladium mariscus*

za'afih — *Polygala abyssinica*, *P. erioptera*
za'alhal, za'alhil — *Datura fastuosa*, *Marsdenia robusta*, *Metaporana obtuse*, *Rughidia milleri*
za'alilihon, za'alilihun, za'lalihon — *Coelocarpum haggierensis*, *Leucas hagghierensis*, *Trichodesma laxiflorum*, *T. scottii*
za'iliffin, za'liffin — *Helichrysum* sp. *A*, *H.* sp. *B*, *H.* sp. *D*, *Leucas samhaensis*
za'lifitin — *Abutilon fruticosum*, *A. pannosum*, *Heliotropium balfourii*
za'lifitin di howsir — *Poskea socotrana*
za'yud — *Barleria* spp. (not *B. popovii*)
zaberbiro — *Begonia* spp.
zabiher — *Gymnocarpos bracteatus*
zabirhen — *Polycarpaea kuriensis*
zafaqa', zafaqah — *Tragia balfouriana*, *Urtica urens*, *Waltheria indica*
zafaqah di 'ilihe — *Scrophularia arguta*, *Waltheria indica*
zafatyan — *Geranium biuncinatum*, *G. mascatense*, *G. trilophum*
zafhan — *Lolium persicum*
zafirifihon — *Desmodium triflorum*
zafiririhan, zafiririhon — *Crotalaria spinosa*, *Launaea massauensis*
zafrer — *Campylanthus spinosus*
zafzaf — *Kohautia* spp., *Teucrium sokotranum*
zafzifo — *Fagonia luntii*
zaghalhil — *Metaporana obtusa*
zahwih — *Sphaerocoma hookeri*
zama' — *Sideroxylon fimbriatum*
zama'ano — *Boswellia socotrana*, *Jasminum fluminense*
zeberbir — *Begonia* spp.
zebirbiher — *Polycarpaea kuriensis*
zebirher — *Oxalis corniculata*
zeferher, zefirher — *Abutilon fruticosum*, *A. pannosum*, *Nanorrhinum hastatum*, *Plantago* spp.,*Polycarpaea balfourii*, *P. caespitosa*
zeferrhen — *Launaea socotrana*
zeferrihon — *Pulicaria diversifolia*
zefinitin — *Polycarpaea kuriensis*
zefiririhin — *Crotalaria leptocarpa*
zefiririhon, zefiririhun — *Anagallis arvensis*, *Helichrysum* sp. *A*, *H.* sp. *B*, *H.* sp. *C*, *H.* sp. *D*, *Launaea socotrana*, *Priva socotrana*, *Ruellia carnea*, *R. dioscoridis*, *R. paulayana*
zefiririhon di gehi — *Ageratum conyzoides*
zewrif — *Kraussia socotrana*, *Rhus thyrsiflora*
zeydirhin — *Heliotropium rariflorum*
zeydirih — *Seddera spinosa*
zeydiro, zeydiroh — *Dactyloctenium robecchii*, *D. hackelii*, *Indigofera marmorata*, *I. pseudointricata*, *Panicum rigidum*, *Seddera fastigiata*, *S. glomerata*, *S. latifolia*, *S. spinosa*
zeydiro di 'irihan — *Indigofera marmorata*
zeydiro di himar — *Indigofera pseudointricata*
zidirih — *Dactyloctenium hackelii*, *D. robecchii*, *Indigofera nugalensis*, *I. pseudointricata*, *Panicum rigidum*, *Seddera fastigiata*, *S. glomerata*, *S. latifolia*, *S. spinosa*
zeydiroh di bi'eyr — *Indigofera pseudointricata*
zeyhimero — *Micromeria remota*
ziberbiher — *Asplenium schweinfurthii*, *Doryopteris concolor*, *Echidnopsis* spp., *Heliotropium nigricans*, *Socotrella dolichocnema*, *Trachyspermum pimpinelloides*
ziberhen — *Begonia* spp., *Heliotropium nigricans*, *Polycarpaea kuriensis*
zifat d-oghirib — *Asphodelus fistulosus*
ziffoh — *Aristida* spp., *Melinis repens*, *Stipagrostis sokotrana*
zifinitin — *Heliotropium aegypticum*, *H. paulayanum*, *H. socotranum*, *H. wagneri*
zifiririhan — *Teucrium sokotranum*
zifiririhon di gehi — *Exacum affine*
zifit — *Aristida* spp., *Lolium persicum*, *Melinis repens*, *Stipagrostis sokotrana*
zifzifih — *Fagonia luntii*, *Melhania* spp.
zifzifo — *Hemicrambe townsendii*, *Lachnocapsa spathulata*, *Nesocrambe socotrana*, *Polycarpaea balfourii*, *P. caespitosa*
ziktih — *Cryptolepis volubilis*
zimgilo — *Campylanthus spinosus*
zimowtih — *Helichrysum* sp. *A*, *H.* sp. *D*, *Trachyspermum pimpinelloides*
zir'eh — *Talinum portulacifolium*
zirkiyn — *Allophylus rubifolius*
ziweyni — *Evolvulus alsinoides*
ziyfat — *Dioscorea lanata*
ziyrif — *Kraussia socotrana*
ziyzihiyn — *Leucas hagghierensis*, *Ruellia insignis*, *R. patula*
zizeb — *Aerva microphylla*
zo'hor — *Euphorbia spiralis*
zow'hir — *Ocimum forskolei*
zowbar — *Talinum portulacifolium*
zowbiher — *Ocimum forskolei*, *Oxalis corniculata*, *Priva socotrana*
zowhor — *Gymnocarpos bracteatus*

zowiher — *Farsetia longisiliqua*, *F. socotrana*, *Thamnosma socotrana*
zowma' — *Sideroxylon fimbriatum*
zowma' qertat — *Spiniluma discolor*
zowmheyr — *Cadaba insularis*, *Capparis decidua*
zowmhil — *Cadaba insularis*
zowmhir — *Cadaba insularis*, *Capparis decidua*
zowzuh — *Adiantum balfourii*
zubehur, *zubihur* — *Launaea socotrana*, *Tamarindus indica*

UPPER EGYPT
Great Desert of Selima
BARABRA
GREAT DESERT OF NUBIA
DONGOLA
Cubbabeesh Arabs
DESERT OF BAHIOUDA
Beni Gerar Arabs
Beni Faisara Arabs
KORDOFAN PROVINCE conquer'd by Sennaar from Darfowar
NUBIA supposed the ISLAND of MEROE
SENNAAR
ABYSSINIA
GALLA
K.m of KAFFA
K.m of GURAGUE
K.m of HADE
K.m of IFAT
K.m of FATIGAR
K.m OF BALI
K.m of DAWARO
K.m of AMHARA
K.m of BEGEMDER
SHOA
HALLANGA TAKA
DERKIN
Suakem
Massuah
MECCA
Jidda
Medina
Taïef
Hodeida
Mocha
NEDSJED EL ARED
HAULAN
Saadieh
Saadeh
Hamdieh
Here the Portuguese found King David encamped in the Year 1520